T0276915

MABBERLEY'S PLANT-BOOK

Mabberley's Plant-book is internationally accepted as an essential reference text for anyone studying, growing or writing about plants. With some 26 000 entries, this comprehensive dictionary provides information on every family and genus of seed-bearing plant (including conifers), plus ferns and clubmosses, besides economically important mosses and algae. The book combines taxonomic details and uses with English and other vernacular names found in commerce. The third edition was recognised in the American Botanical Council's annual James A. Duke Excellence in Botanical Literature Award for 2008 and the International Association for Plant Taxonomy's Engler Medal in Silver 2009. In this new edition, each entry has been updated to take into consideration the most recent literature, notably the greater understanding resulting from molecular analyses; over 1400 additional entries (including ecologically and economically important genera of seaweeds) have been included, ensuring that *Mabberley's Plant-book* continues to rank among the most practical and authoritative botanical texts available.

To the memory of
E.J.H. ('John') Corner (1906–1996)
botanical colossus

DAVID J. MABBERLEY

Wadham College, University of Oxford, United Kingdom
Universiteit Leiden and Naturalis Biodiversity Center, Leiden, Nederland
Macquarie University and National Herbarium of New South Wales, Sydney, Australia

MABBERLEY'S PLANT-BOOK

A portable dictionary of plants, their classification and uses

utilizing Kubitzki's *The families and genera of vascular plants* (1990–)
and current botanical literature;
arranged according to the principles of molecular systematics

FOURTH EDITION

completely revised, with some 1400 additional entries

CAMBRIDGE
UNIVERSITY PRESS

CAMBRIDGE
UNIVERSITY PRESS

University Printing House, Cambridge CB2 8BS, United Kingdom

One Liberty Plaza, 20th Floor, New York, NY 10006, USA

477 Williamstown Road, Port Melbourne, VIC 3207, Australia

4843/24, 2nd Floor, Ansari Road, Daryaganj, Delhi - 110002, India

79 Anson Road, #06-04/06, Singapore 079906

Cambridge University Press is part of the University of Cambridge.

It furthers the University's mission by disseminating knowledge in the pursuit of education, learning and research at the highest international levels of excellence.

www.cambridge.org
Information on this title: www.cambridge.org/9781107115026
DOI: 10.1017/9781316335581

First edition published 1987
Second edition published 1997
Reprinted with corrections 1998, 2000, 2002, 2006
Third edition published 2008
Reprinted with corrections 2009, 2014
Fourth edition published 2017

A catalogue record for this publication is available from the British Library

Library of Congress Cataloging-in-Publication data
Names: Mabberley, D. J., author.
Title: Mabberley's plant-book / David J. Mabberley.
Other titles: Plant-book
Description: Fourth edition, completely revised, with some 1400 additional entries. | New York, NY : Cambridge University Press, 2017. | Includes bibliographical references and index.
Identifiers: LCCN 2017019747 | ISBN 9781107115026 (alk. paper)
Subjects: LCSH: Plants – Nomenclature. | Plant names, Popular – Dictionaries. | English language – Dictionaries.
Classification: LCC QK11 .M29 2017 | DDC 580.3 – dc23
LC record available at https://lccn.loc.gov/2017019747

ISBN 978-1-107-11502-6 Hardback

Contents

Introduction

Botanists in five or six centuries hence will probably be able to study affinities with few or none of the doubts which perplex those of the present day. Richard Salisbury (1761–1829)

This dictionary is an attempt to present a review of, and introduction to, plants, the organisms (with their symbionts) on which we depend for our survival in terms of food, clothing, fuel, medicine and 'ecosystem services'. This fourth edition is offered at a time when it is becoming clear that the gravity of the plight of this crucial diversity of plants is now being grasped by the general public: humans have initiated and accelerated the sixth and fastest mass extinction of biodiversity on this planet.

Besides the rising tide of alarm about extinction, there are increasing public concerns about, or general interest in, organic food, herbal remedies and supplements, biofuels, 'superfoods', aromatherapy, traditional Chinese and other Asian medicine, genetically modified organisms and invasive species. Gardening and the mere presence of plants are increasingly seen as good therapy, while travel overseas is unabated and ecotourism flourishes. Cookery books, documentary films and even popular novels are full of plant names unfamiliar to international audiences.

The practical necessity to classify plants according to their uses is at the root of plant systematics, which is therefore one of the oldest of all sciences, being evident in the historical records of all major civilisations. This book reflects the work of botanists, charged with the recognising, documenting and classifying of plants – scientists who investigate the overall structure and biochemistry, physiology, ecology and geography of plants, today in the stark reality of the danger to their very subjects of research.

Despite this longevity and an enormous literature, both printed and online, the fact remains that we still cannot accurately state how many plant species there are (or, very often, even agree on what constitutes a species). Each year some 2000 new species are described, and we do not know how many undescribed species remain undiscovered in the field or lie fallow as preserved specimens in museum collections (see D.J. Mabberley, 'The problem of 'older' names', *Regnum Vegetabile* 123(1991)123–134).

There is now an urgent need to complete an inventory of the plant species of the world before this becomes an exercise in palaeobotany. It is perverse that, faced with this crucial task, cohorts of young trained botanists (and maturer ones) are, as 'plant scientists', actively deterred from this task, to concentrate on publication of papers of theoretical interest, in 'high-impact' journals, without which truly Neronian fiddling they cannot advance their careers. It is particularly lamentable that this now also obtains in the (often developing) countries which retain most of the greatest biodiversity. Nonetheless, like the guardians of human artefacts in museums, those of us, who are able, owe it to the coming generations to at least try to collect and describe what remains. We can carry on with much of today's more fashionable biology after 'Rome burns', but the basics of systematics could be just impossible in the future.

In a world dominated by the internet, it may seem somewhat strange that a traditional dictionary of this kind is still being produced to address these concerns, but it reflects a demand for such a work. In fact, it is largely the internet that has made such a book more important than ever – in three principal ways.

Firstly, there is a need for a balanced review and synthesis of the scattered and often conflicting information about the plant world, for the internet is stuffed with information, much of it uncritical, contradictory, out of date or of dubious value. The first three editions of this dictionary were attempts to provide a handy text covering the vascular plants, their botany and relationships, their uses and their common names. This edition is in the same mould, but such is the enormous advance in our knowledge in the last decade, those earlier editions are now completely superseded.

Secondly, so much authoritative information is now available on the internet, that a key, leading to definitive taxonomic revisions and reviews now online, is needed, such that all users of this book, rather than merely the cognoscenti, can readily reach those primary research sources. Where a revision of a genus has been published in recent years, or an older one is still widely quoted, I have indicated the place of publication in a very abbreviated form. It has always seemed to me that this would give the reader with a need to follow up the literature a valuable start, and now, with much of that literature on the internet (especially useful is the Biodiversity Heritage Library (http://www.biodiversitylibrary.org)), readers can much more readily reach such primary sources. It has been possible to augment those pointers in earlier editions with many hundreds more in this new edition, though it must be noted that new species may have subsequently been described, so accounting for any differences in species numbers listed here.

Thirdly, many plant people find that browsing (surfing?) a single handy volume such as this is not only educational but also enjoyable – in the home, office or lab, or in the field. This is of course part of the phenomenon that is responsible for the survival of books in general, when their demise has been so confidently, and repeatedly, predicted over the last decades.

In preparation for the successive editions of the book, I have gathered information from modern Floras, handbooks, monographs and periodicals, particularly dwelling on the literature published since 1970: indeed, as comprehensive a scan of the germane plant literature (and appropriate websites) of this period as could be made by one man has been attempted. It would be impossible to cite within the text all sources of information, but the major ones are listed under 'Acknowledgement of sources' on p. 1009. Since the publication of the third edition, it has been possible to examine and now cite very many more, including many new journals which are published only online. I fully realize that in providing more of this information to readers, I am denying my host institutions' scores in literature citation indices, a regrettable obsession which (by bean-counting bureaucrats) is the bane of civilized endeavour in science. The abbreviations used in the references are explained, as are others found in the entries, on p. 1029.

I have maintained a cross-reference file from genera to families so that the estimated size of families in terms of numbers of genera and species more exactly reflects the information actually set out in the text under the constituent generic headings. Readers who compare this edition with the first three will see that in a world threatened with major extinctions through our own actions, there has been an encouraging international taxonomic effort (notably insightful being the revelations from molecular work) but that this effort has sometimes had the apparently detrimental effect of the changing of names of plants well known in commerce: food-plants, timber, drugs, fibres and ornamentals. There can be few academic disciplines in which advances have such an unfortunate drawback: there is in consequence a real economic cost to some sectors (see J. Valleau (2004) 'Plant name changes: good science, angry growers and confused gardeners', *Proceeding of the XXVI IHC – IVth International Symposium on Taxonomy of Cultivated Plants*, eds. C.G. Davidson and P. Trehane, pp. 63–66 (although Valleau's notion that there have been 'three hundred years of relative peace in the world of taxonomy' is illusory – as an examination of any nineteenth-century horticultural encyclopaedia will demonstrate)).

Because of provisions in the *International Code of Nomenclature*, names can be protected from upset due merely to the unearthing of earlier valid ones or to other reasons, so that the layperson's cavil against the scholastic principles of the *Code* are now redundant, and thus, for example, the correct generic name for the chrysanthemum is indeed *Chrysanthemum* (although in ensuring that, it means that the annual 'chrysanthemums', being generically distinct, have the until recently unfamiliar name *Glebionis*). Nonetheless, recent taxonomic advances, largely the result of molecular analyses (see below), often reinforcing the findings from classical work (notably that of the intricate structure of seeds brought together in the late E.J.H. Corner's classic *Seeds of Dicotyledons* (1976)), have led, since the third edition, not only to clarification of the domesticated history of, for example, rice and potatoes on the one hand, but also to the changing (and sometimes changing back) of the Latin names of, for example, African violets, alkanet, angostura bitters, Buddha's palm, brazilwood, choko, cochineal cactus, coleus, derris dust, Dutch crocus, Easter cactus, elephant grass, English elm, glory of the snow, Good King Henry, guinea grass, holy basil, Japanese knotweed,

Kikuyu grass, Koster's curse, larkspur, Livingston(e) daisy, partridge-breasted aloe, pearl millet, stavesacre, summer cypress, topinambour, vegetable sheep, violet cress, voodoo lily, wasabi, Welsh poppy, white ipecanuanha and wild thyme, on the other.

It is tempting to deprecate such name changes, but when these are based on a sound advance in scientific understanding, it is perhaps unreasonable to argue for the outmoded. An example of economic importance from my own work should suffice to explain the significance: until recently the species of orange-like fruit-trees native to Australia were in endemic (*Eremocitrus*) or subendemic (*Microcitrus*) 'genera', but intensive recent analysis has shown that they fall under the genus *Citrus* so that, contrary to received wisdom, Australia has more native *Citrus* species than does any other country. The importance of grasping this information in terms of hybridizing, rootstocks and other concerns for new crops, of breeding in pest-resistance, and of germplasm conservation and so forth is obvious.

For cogent discussions of the general issues here, see C.J. Humphries 'The implication of pragmatism for systematics', *Regnum Vegetabile* 123 (*Improving the stability of names: needs and options*) (1991)313–322 and A. Minelli, 'The changing paradigms of biological systematics', *Bull. Zool. Nomenclature* 52(1995)303–309. Nonetheless, it is clear that names of many familiar plants, names still effectively stuck in the thinking of pre-Linnaean folk-taxonomy (see below) will have to be changed if such are to reflect their modern relationships, as inconvenient as this may be in the short term. These things are not academic caprice. Recent genetic and developmental studies have shown that the genetic bases for what to humans may appear to be great differences in terms of, say flower colour and shape or fruit form, features previously used to distinguish genera (particularly if it meant one was edible; the other, not), are often slight. Very often the differences reflect evolutionary pressures in pollination or dispersal ecology. It has long been recognized that a genus such as *Linum* has species with white, yellow, red or blue flowers and that within a genus such as *Lobelia*, there are transitions between dry fruits with wind-dispersed seeds to fleshy fruits with animal-dispersed ones. Because we, too, are animals that use sight and scent to distinguish plants, these differences figure large in our perceptions of the natural world. Work on the genus *Erythranthe* (as *Mimulus s.l.*) showed that a single gene change can switch a species from bird-pollinated (red) to bee-pollinated and, in *Aquilegia*, slight differences govern the switch between hummingbird and moth pollination (see D.A. Levin, 'Ecological speciation: crossing the divide' in *Systematic Botany* 29(2004)807–816). That these shifts readily take place accounts for the enormous amount of parallel evolution in plants – parallelism that has made the understanding of the inter-relationships of vascular plants so difficult.

So as to reflect the true affinities in terms of genealogy, species with different pollination syndromes and therefore floral morphology are correctly accommodated in the same genus: *Zauschneria* in *Epilobium*, *Willdampia* in *Swainsona*, *Gaura* in *Oenothera*, *Antholyza* in *Gladiolus*, *Rigidella* in *Tigridia* and so forth (see also *Erica*, *Ornithogalum*, *Sinningia*). Again, certain life forms have arisen repeatedly (e.g. rheophytes within many unrelated genera); it becomes clear then that *Rotula* represents merely a rheophytic *Ehretia*, just as Lemnaceae are free-floating Araceae. Mycotrophism in its varying dependence on fungal symbionts has arisen many times, and the transition to complete reliance is seen in the amalgamating of *Listera* in *Neottia*, for example. Carnivory has arisen in different unrelated lines. The parasitic habit has arisen repeatedly too – within Podocarpaceae in conifers but especially in angiosperms. For example, in 1810, the great British botanist Robert Brown (1773–1858) correctly referred the parasitic *Cassytha* to Lauraceae; *Cuscuta* has long been in Convolvulaceae, and now the hemiparasitic members of former Scrophulariaceae go with parasitic Orobanchaceae.

To maintain these and similar examples as separate genera or families, effectively picking holes in the (monophyletic) generic fabric and denying us the framework within which we cannot only begin to understand ecological-evolutionary shifts but also marvel at the workings of evolution itself, is to maintain the holey relic as paraphyletic. The controversy over the retention of such generic (and indeed family) concepts effectively requiring the maintenance of paraphyletic groups has largely subsided, as the fact remains that the bulk of new work, which this book attempts to mirror, has in general long since moved on (see P.F. Stevens, 'An end to all things? – plants and their names', *Australian Systematic Botany* 19(2006)115–133) and a plan for recognising only monophyletic groups in an entire flora for one country (J.W. Kadereit *et al.*, 'Which changes are needed to render all genera of the German flora monophyletic?', *Willdenowia* 46(2016)39–91) can be proposed with confidence. However, attempts at the marrying of the underlying evolutionary processes inevitably leading to transitional paraphyletic groups, on the one hand, and a biological classification system, on the other, so ably discussed by E. Hörandl and T.F. Stuessy, 'Paraphyletic groups as natural units of biological classification', *Taxon* 79(2010)1641–1653 (see also D.J. Mabberley, 'The optimistic in pursuit of the unrecognisable: a note on the origin of angiosperms', *Taxon* 33(1984)77–79), mean that the last word has not been said on this.

Today's DNA-based revolution is in reality far less shocking than what happened 200 years ago, when the sexual system of Linnaeus was overtaken by the 'natural system' of the French School on the Continent and Robert Brown in Britain, but it is a remarkable fact that the general sweep of the modern classification based on DNA analysis reflects the one laboriously developed over centuries of scrutiny of cellulose and lignin attached to sheets of herbarium paper. In other words, the features we perceive by eye – morphological features – have, on the whole, been a remarkably helpful guide not only in identifying plants but also in classifying them in a way that reflects their evolutionary relationships. As Darwin predicted, 'Our classifications will come to be, so far as they can be made, genealogies'.

However, the fact that angiosperms are remarkable for the very high frequency of hybridizations and polyploidy (perhaps 50% of species have such in their ancestry) may well confound simplistic DNA-cladistic analyses, as persuasively argued by C.A. Stace ('Plant taxonomy and biosystematics – does DNA provide all the answers?', *Taxon* 54(2005)999–1007) because such analyses can sometimes be severely at odds with cytogenetic evidence and breeding experiments. That polyploid 'races' within particular species effectively behave as biological species (see D.E. Soltis et al., 'Autopolyploidy in angiosperms: have we grossly underestimated the number of species?', *Taxon* 56(2007)13–30) suggests that there may well be more (crypto-)species in any case. Our understanding of the vascular plants is still very imperfect. And it is likely that with the imminent destruction of so much remaining 'wild' habitat and concomitant extinctions, exacerbated by global warming, we will never know – just as we will never know all that came before the extinctions of the past.

This New Edition

For the first two editions of this book, I followed the system of Cronquist, *An Integrated System of Classification of Flowering Plants* (1981) as modified by Kubitzki (see below), for it had fresh descriptions with valuable bibliographies. Indeed, it represented a landmark in angiosperm systematics. For the third edition, the published volumes (kindly presented by the author) of Klaus Kubitzki's *The Families and Genera of Vascular Plants* were the standard to which other literature was compared and is the bedrock of this edition. It is sincerely to be hoped that the last volumes of this monumental endeavour, a new 'Genera Plantarum' will be published with all speed.

As in the earlier editions, I have been conservative in the splitting of families and genera. This has generally been the philosophy at the family level in both Kubitzki's and Cronquist's work and was convincingly argued by C.G.G.J. van Steenis in his 'Doubtful virtue of splitting families' (*Bothalia* 12(1978)425–427; see also W.R. Philipson, 'The treatment of isolated genera', in *Botanical Journal of the Linnean Society* 95(1987)19–25). Where there remains disagreement, I have continued to take the broad view. The circumscriptions in the last edition, based on consensus in workshops I held in a number of countries beforehand, were largely those adopted in the third version of the Angiosperm Phylogeny Group's system ('APG III', 2009) and since (see Angiosperm Phylogeny Group, 'An update of the Angiosperm Phylogeny group classification for the orders and families of flowering plants: APG IV', *Botanical Journal of the Linnean Society* 181(2016)1–20) and the system adopted here (p. 997).

Nonetheless, some specialists working intensively on any particular family, seeing the diversity within that family, but perhaps with less of an eye to the bigger picture, have sometimes tended to advocate the splitting up of well-known families, as for example Boraginaceae, by comparison with, for example, Malvaceae. Very often the consequence of recognising major subunits in a family at the family level often leads to the recognition of a number of monogeneric families with unfamiliar names for the minor subunits. For the general public a more satisfactory solution is to recognise these subunits at the subfamily level, thereby showing their overall affinities. Nevertheless, where there is good new evidence, families shown to be polyphyletic, e.g. Molluginaceae and Phytolaccaceae, are split here; work still needs to be done in this regard, particularly in Santalales.

Also in deference to continuity where scientifically possible, I have maintained the older of the permitted alternative names for the families Compositae, Cruciferae, Gramineae, Labiatae, Leguminosae, Palmae, Umbelliferae and others, rather than the later-coined

Asteraceae, Brassicaceae, Poaceae, Lamiaceae (older name Viticaceae), Fabaceae, Arecaceae, Apiaceae found in many works and websites: these 'alternative' names have equal standing in the *Code*. Besides the continuity argument, there are others for, as Peter Valder (*The Garden Plants of China* (1999), p. 15) notes, the original names 'are familiar, have historical associations, and are descriptive'. Such are largely names of large 'natural' families (including those of fundamental human concern, the grasses and the pulses), whose circumscription was very early recognised and which concepts have been largely unchanged over hundreds of years – such names are therefore pointers to this conceptual robustness. Practising taxonomists in Compositae and Leguminosae have strongly argued for use of those names over Asteraceae and Fabaceae – and lay people are familiar with 'cruciferous vegetables' as well as 'legumes' (for example, the odious abbreviations such as bignons, bromels, etc. derived from Latin family names, would in that case logically lead to 'fabs').

More fundamentally, the type genera of all but Arecaceae (an almost exclusively tropical group in any case) of the standardized (later) names are temperate – *Aster, Brassica, Poa, Lamium, Faba* (a name long-lost in the synonymy of *Vicia*) and *Apium* are frequently not only small and herbaceous, reflecting the depauperate nature of the flora of the north rather than the magnificence of the tropics (see E.J.H. Corner in typically expansive mode, 'On thinking big', *Phytomorphology* 17(1968)24–28), but also represent an almost colonial infliction of northern pre-occupations (similarly as with Gentianaceae, Lythraceae and Violaceae – essentially tropical woody families), on tropical peoples and the rich floras they steward for us all: not surprisingly the younger names are less popular in tropical countries (see Ruth Kiew, 'Family names for plants', *Gardenwise* [*Singapore*] 25(2005)28).

Where recent work has had a much greater effect on users of plant names is where the species binomial has changed because of re-modelling of genera, usually as a result of DNA work, much as in the last edition so much change was at the family level. In this book, following current literature, an effort has been made to recognize monophyletic groups, wherever possible, and to point out paraphyly where it has been identified; some striking examples, including genera such as *Brassica*, await the braveheart to bring the logical scientific process to fruition. A glance at the text will show that many genera described in the last edition as 'heterogeneous' have been refigured to make monophyletic entities. Sometimes this has led to their being split up, as in the continued disintegration of *Centaurea, Clerodendrum, Hedyotis, Panicum, Senecio, Stipa* and *Vernonia*, for example, and now the splitting up of, among others, *Abutilon, Aloe, Caesalpinia, Calathea, Capparis, Drimys, Eria, Fagraea, Hybanthus, Mimulus, Minuartia, Pandanus, Plectranthus, Pleurothallis, Polygala, Potentilla, Quassia, Rhus* and *Salsola*. But generally much more commonly, as also foreshadowed in earlier editions, 'outlying' or 'satellite' genera have been shown to be 'nested' in larger groupings, so that many genera have become greatly enlarged with their absorption. There is continued consolidation in *Hibiscus* and *Justicia*, for example, and, though perhaps less spectacularly, in much-needed recent treatments of genera in Cruciferae and Umbelliferae. In this edition are presented from recent literature enlarged generic concepts in, for example, *Aristolochia, Banksia, Calamus, Carex, Cenchrus* (including *Pennisetum*), *Cleome, Curcuma, Cynanchum, Cynoglossum, Cyperus, Daucus, Dendrobium, Dracaena* (including *Sansevieria*), *Hydrangea, Maxillaria, Melaleuca* (including *Callistemon*), *Miconia, Myrcia, Oncidium* (including *Odontoglossum*), *Oreocharis, Paepalanthus, Primula* (including *Dodecatheon*), *Ruellia, Sonchus, Sporobolus* (including *Spartina*), *Streptocarpus* (including *Saintpaulia*), *Syzygium* and *Urochloa*, not a small number of these now similar in scope to the generic concept that Linnaeus himself had of them over 250 years ago.

Such resolution has also removed many 'anomalous' geographical distributions of earlier editions (see *Escallonia, Meconopsis, Physalis*, for example, compared with these entries in the third); others, e.g. *Cunonia, Mirabilis* and *Turbina*, clearly now merit similar attention.

Although attempting to mirror current opinion at the generic level, I have, when faced with conflicting views as in the case of families, continued to take the conservative line in maintaining larger genera. From a fieldworker's point of view this is more satisfactory in any case – the *Gestalt* of a fig is usually unmistakable, but to split the genus *Ficus* into several on the basis of characters revealed only by lenses seems academic self-indulgence. Moreover such splitting up, as (re-)proposed in *Iris* for example, has not found favour with the horticultural industry. I entirely agree, therefore, with the wise words of P.H. Davis and V.H. Heywood (*Principles of Angiosperm Taxonomy* (1963), p. 106) long ago: 'When in doubt whether to accord generic rank to a group, there is much to be said for the *subgenus* as a suitable category; it draws attention to the group in the classification and at the same time allows people to continue to use the old binomial'.

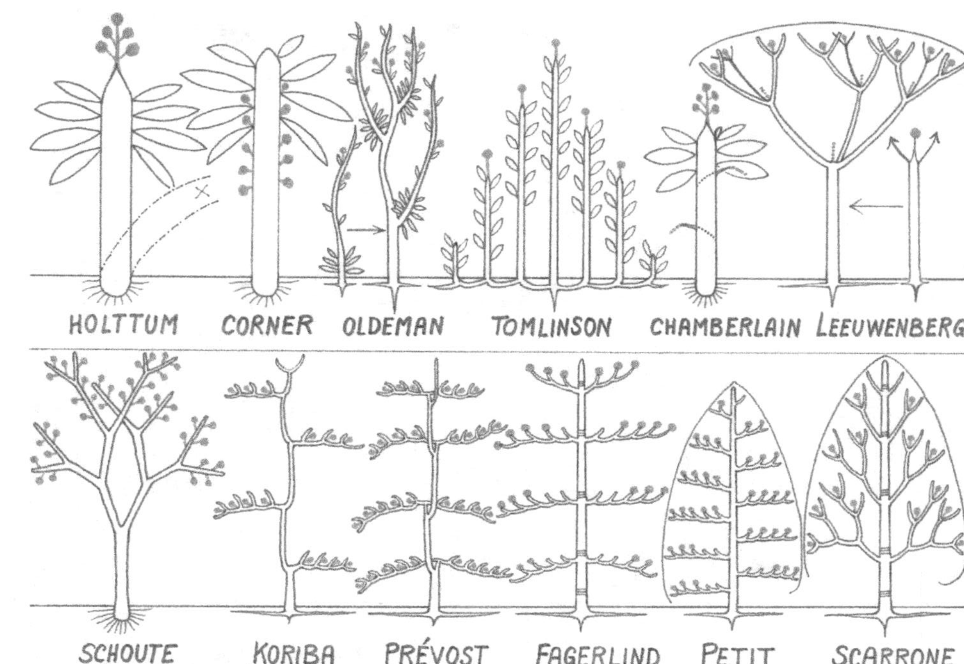

Figure 1. Plant architectural 'models', named after botanists. Reproduced with kind permission of Francis Hallé (Montpellier); previously unpublished.

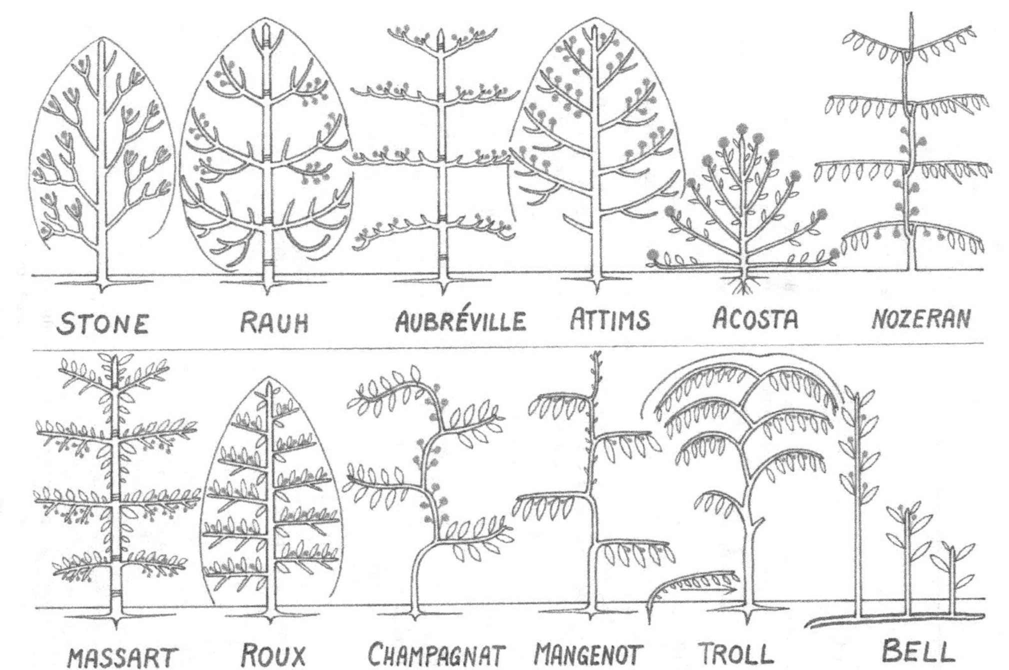

Figure 1. (*cont.*)

Nonetheless, for practical reasons, there is value in recognizing some of the formerly distinguished subdivisions within such consolidated genera and this is readily achieved informally by writing e.g. 'hebes', 'hepaticas', 'mahonias' etc. for particular clades, although some would wish to name such (and other) clades formally in an entirely new system of classification (Phylocode).

Despite all these modern circumscriptions, there are still rather few genera with large numbers (c. 500+) of species (cf. D.G. Frodin, 'History and concepts of big plant genera', *Taxon* 53(2004)753–776): *Acacia*, *Alchemilla* (mostly apomicts), *Allium*, *Anthurium*, *Ardisia*, *Aristolochia*, *Artemisia*, *Asplenium*, *Astragalus* (at 2850 likely the biggest), *Begonia*, *Berberis*, *Bulbophyllum*, *Carex* (at c. 2100 perhaps the second biggest), *Cousinia*, *Crotalaria*, *Croton*, *Cyathea*, *Cyclosorus*, *Cyperus*, *Cyrtandra*, *Dendrobium*, *Dioscorea*, *Diospyros*, *Elaphoglossum*, *Epidendrum*, *Erica*, *Eucalyptus*, *Eugenia*, *Euphorbia*, *Ficus*, *Galium*, *Habenaria*, *Helichrysum*, *Hibiscus*, *Hieracium*, *Homalomena*, *Ilex*, *Impatiens*, *Indigofera*, *Ipomoea*, *Ixora*, *Justicia*, *Lepanthes*, *Masdevallia*, *Maxillaria*, *Miconia* (at 1900 perhaps the fourth biggest), *Mimosa*, *Myrcia*, *Oxalis*, *Passiflora*, *Pedicularis*, *Peperomia*, *Phyllanthus*, *Pilea*, *Piper* (at c. 2000 perhaps the third biggest), *Pleurothallis*, *Poa*, *Psychotria*, *Quercus*, *Ranunculus*, *Rhododendron*, *Salvia*, *Schefflera* (but to be dismantled), *Selaginella*, *Senecio*, *Silene*, *Solanum*, *Stelis*, *Syzygium* and *Tillandsia*.

Nonetheless, most of these and other large genera have no modern monographs, making attempts at calculating accurately the number of plant species in the world somewhat fraught. Rarely does the modern scientific milieu permit grappling with the basic taxonomy, as opposed to the overall phylogeny, of such groups of this size, although the trend has been to have multinational teams working together on the phylogeny of such groups, such as *Astragalus*, *Begonia*, *Carex*, *Euphorbia*, *Miconia*, *Psychotria* and *Solanum* (see R.D. Stone in *Taxon* 63(2014)539–561). But few of those mentioned have received full attention with regard to the basic 'alpha' taxonomy, despite the international brouhaha about biodiversity and inventory of the planet's plant resources; one laudable exception is the recent revision of *Convolvulus* (though treated in a paraphyletic sense; J.R.I. Wood *et al.*, 'A foundation monograph of *C.*', *PhytoKeys* 51(2015)1–282). Without such critical monographic (as opposed to floristic and generalized phylogenetic) work, many of these species numbers generated mechanically from standard databases, albeit with the best of intentions, are inevitably approximations, not least because of the growing mountain of unrealised synonymy (see D.J. Mabberley, 'The problem of 'older' names', *Regnum Vegetabile* 123(1991)123–134).

Since the last edition of this book appeared, major reviews of many of the largest families including the biggest three) have been published, notably of orchids, crucifers, Boraginaceae, Compositae, Cucurbitaceae, Gesneriaceae and Caryophyllales, besides grasses, Euphorbiaceae and many other families in Kubitzki's book. A great deal of valuable information has also been brought together in Peter Stevens's admirable Angiosperm Phylogeny website (http://www.mobot.org/MOBOT/Research/APweb/welcome.html). All of these, with as much recent published literature in hard copy and online as I have been enabled to scan (see Acknowledgement of Sources, p. 1009), besides a burgeoning correspondence (see below), have been considered in the preparation of this edition. Where major overhauls of families such as Rubiaceae, for example, are still in progress, I have updated the text to show where things are moving, so that there is continuity with the last edition of this book.

As in the third edition, I have not used 'subsp.' for cultivated relations of wild plants and consistently used 'cultivar group' instead. This makes greater biological sense in leaving subspecies as ecologically or geographically morphologically distinctive subdivisions of 'wild' plants. However, in an inexorably urbanizing world, the definition of 'wild' is a moot point and, in any case, tends to support the haughty notion that the human milieu is distinct from (and superior to) the 'environment'. Indeed, the weeds in our gardens and fields are as much under evolutionary pressures imposed by cultivation as are our crops. Many of those crops, like rice, oats and rye, began as 'weeds' – at what point did they become 'cultivated'? (See D.J. Mabberley, 'Where are the Wild Things?', in *Paradisus: Hawaiian Plant Watercolors* by Geraldine King Tam (1999, '1998')). Logically many localized 'ecotypes' of 'wild' species would appear to have much in common with 'cultivars' arranged in cultivar groups.

As an aid to understanding the developmental sequences leading to the mature structures of plants (their 'architecture'), I have extended, where possible, the use of the architectural 'model' – effectively a developmental blueprint – to which such seem to conform. These models are not invariant and within species there can be variations including architectural mutants (see Hallé *et al.* 1978), but they are often very useful reference points for comparing the structures of different taxa. (See Figure 1.)

I have been enabled, through the good offices of Julian Shaw of the Royal Horticultural Society, to add for the first time the most commonly encountered orchid hybrid generic names. I have also included commercially important algae used in sushi, for example, and certain bryophytes (liverworts as well as mosses) of economic importance not included in earlier editions.

A particular effort has been made in this edition, again often with the help of Julian Shaw, but also Kanchi Gandhi (Harvard), to pin down the first publication of names of many cultivated plants often quoted elsewhere with the vague authority 'Hort[ulanorum = 'of gardeners']', i.e. in ignorance of the first use of the binomial. Increased precision in this matter allows for recognition or designation of type material leading to a scientific basis for the use of such names. In some cases, and in plant names in general, this means citing authors different from those often seen, but in general this has little effect on most users of these names.

Moreover, a particular effort has been made to ensure that generic names generally attributed to Linnaeus are ascribed to their true authors when their manuscript descriptions were published in his work, complete with his attributing authorship to them. This leads to the happy result that *Linnaea* is no longer written as if Linnaeus himself had coined the name in some form of self-commemoration (whatever the real story behind it). Where Linnaeus took up a generic name used by an earlier author he cites, this is also indicated here, demonstrating that he in fact was very much following the concepts of earlier workers, in particular Tournefort (though not with such a narrow generic concept when dealing with cultivated plants and their relatives). It shows that, insofar as his nomenclature as opposed to his standardized, handy classification system was concerned, he was not as disruptive as many commentators have intimated.

As with the last edition, I have increased the number of vernacular names (notably those non-English names used in modern English literature as well as in the trade for timbers (many tropical ones sadly now merely as rare desiderata for wood-collectors) and other products including Traditional Chinese Medicine) as more and more such names are coming into widespread use. Furthermore, a number of Latin names continue to be used in the vernacular in a sense that is different from their current technical one. Some are taxonomically or nomenclaturally outmoded, some are misidentifications, some are pre-Linnaean and some are spelling errors or even specific epithets (again, some outmoded) or abbreviations of them. Such include acidanthera, afrormosia, alyssum, amaryllis, arum, aster, aubretia, auri, azalea, bartonia, calla, canariense, cineraria, croton, dimorphotheca, dipladenia, epiphyllum, eulalia, fremontia, funkia, geranium, gloxinia, godetia, goldfussia, hortensia, ilex, kentia, lasiandra, laurustinus, leylandii, lippia, lisianthus, macrocarpa, mangium, mesembryanthemum, mespilus, mimosa, moluccana, montana, montbretia, nasturtium, pandani, poinciana, poinsettia, pyrethrum, retinospora, robusta, rochea, smilax, spinifex, statice, stephanotis, stevia, syringa, trifoliata, tuberose, utile, vallis, verbena and wellingtonia. Many are used in horticulture or for timber and the like: these words are genuinely part of current English and no amount of insistence by academic botanists will remove them. They are therefore in this volume. For readers wishing to know international equivalents of the English vernacular names, a very helpful text is Umberto Quattrocchi's *CRC World Dictionary of Plant Names* (four volumes, 1999).

In short, largely in consequence of scientific advances, there are very extensive changes in the text of *Mabberley's Plant-book*, such that the earlier editions are now obsolete. Almost every one of the original entries has had to be checked or updated in the light of the newly published work discussed above, so that this edition is a much greater advance on the third, than that was on the first two. The text of this edition is almost 10% longer than that of the third, but the use of fine paper ensures that the book still remains portable (and remains open whilst being consulted). Included are some 1400 additional entries. Of these, many refer to resurrected genera but well over 800 are newly proposed ones, though almost a third of those have not received general acceptance and so have already slipped into synonymy, pointing up how careful proposers should be before burdening the literature with yet more generic names. Indeed the full story is even more woeful in this regard in that I have not included all such new names, particularly many applied to orchids, as these are notable only for their evanescence.

In the preparation of this new edition, I am deeply indebted, once again, to Professor Kubitzki, and to many other people, notably the librarians at the Naturalis Biodiversity Center

[sic] (Leiden, especially Anja Buijsen), Linnean Society of London, Royal Botanic Gardens Kew (especially Craig Brough), Natural History Museum (London, especially Andrea Hart and Armando Mendez), Lindley Library of the Royal Horticultural Society (London, especially Thomas Pink), Bodleian Libraries (Oxford), Macquarie University (Sydney), Botanischer Garten und Botanisches Museum (Berlin), Royal Botanic Gardens Sydney (especially Miguel Garcia), and Missouri Botanical Garden (St Louis, especially Victoria McMichael). The internet and, particularly, email have made communication and checking of databases so much easier than anything imaginable when I started writing the first edition of this book (1973).

Generally through email, then, many kind users of the book have suggested amendments, additions or improvements or have generously answered my questions. The following have been especially helpful: Frédéric Achille, Frits Adema, Ihsan Al-Shehbaz, Fatima Al-Talib, Tinda van Andel, Wendy Appelquist, Pieter Baas (who also generously hosted me on many occasions when I was working in Leiden libraries), Max van Balgooy, Christine Bartram, Mark Beilstein, Paul Berry, Volker Bittrich, Thomas Borsch, Barbara Briggs, Xander van der Burgt, James Byng, Katherine Challis, Mark Chase, Martin Cheek, Maarten Christenhusz, John and Marion Clarkson, Emilio Constantino, John David, Rosemary Davies, Lars van Dijk, Stefan Dressler, Robert Easton, Robert Faden, Aljos Farjon, Mike Fay, Duane Fernandes Lima, Judith Field, Eberhard Fischer, Pádraic Flood, Stephen Forbes, Vicki Funk, Patrick Gale, Kanchi Gandhi, Rhys Gardner, Nicholas Garner, Peter Goldblatt, Rafaël Govaerts, Jason Grant, Werner Greuter, John Grimshaw, Francis Hallé (to whom I am particularly grateful for permission to reproduce the hitherto unpublished Figure 1), Raymond Harley, David Harris, Alistair Hay, Charlie Heatubun, Pat Herendeen, Simon Hiscock, Peter Hoch, Hong De Yuan, Jay Horn, Charlie Jarvis, Matthew Jebb, Mats Jertson, Barrie Juniper, Gudrun and Joachim Kadereit, Tony Kanellos, Michael Keelan, Roland Keller, Elizabeth (Toby) Kellogg, Ruth Kiew, Phillip Kodela, Rogier de Kok, Job Kuijt, Klaus Kubitzki, Walter Lack, Mark Large, Philip Le Roux, Gwilym Lewis, Memory and Walter Lewis, Chris Liffen, Russell (Rusty) Linker, Jim Luby, Laura Mabberley, Steven McKay, Karol Marhold, Serena Marner, Bruce Maslin, Brian Mathew, K.T. Mathew, Constantijn Mennes, David Middleton, H.Y. Mohan Ram, Arnaud Mouly, Josef Mullins, Gonzalo Nieto Felliner, Dean Nicolle, Kevin Noakes, Henry Noltie, Hans Nooteboom, Inger Nordal, Dick Olmstead, Andrew Orme, Alan Paton, Ching-I. Peng, Terry Pennington, Sylvia Phillips, Helen Pickering (who also kindly provided the cover photograph), A.K. Pradeep, Peter Raven, Tim Rich, Diego Salariato, Hervé Sauquet, André Schuiteman, Tanja Schuster, Julian Shaw, David Simpson, Erik Smets, Soh Wuu Kuang, Pam and Doug Soltis, Clive Stace, George Staples, Peter Stevens, Lena Struwe, Tod Stuessy, Brett Summerell, Mesfin Tadese, Paul-Robert Takàcs-Koppàndi, Nigel Taylor, John Thomson, Mats Thulin, Jimmy Turner, Peter Valder, Michiel Van Slageren, Prashant Vaze, Jan-Frits Veldkamp, Natasha de Vere, Giovanni Vitulli, Warren Wagner, Jacek Wajer, Tim Waters, James Wearn, Jun Wen, Peter Weston, Barbara Wiecek, John Wiersema, Peter Wilkie, Alastair and Karen Wilson, Michael van Winkle and Scott Zona. Particularly generous with their time and expertise were those who offered or agreed to check entries related to their specialities (any residual errors are my own): Frank Almeda, Walter Judd and Fabián Michelangeli (*Miconia*), Max Coleman (*Ulmus*), Jenny Edmonds and Sandra Knapp (*Solanum*), Pádraic Flood (*Arabidopsis*), Matthew Jebb (*Nepenthes*), Eve Lucas (*Myrcia*), Bruce Maslin (*Acacia*), Peter Michael (*Echinochloa, Erigeron, Eryngium, Gamochaeta, Glandularia, Verbena*), Alan Paton (*Coleus, Plectranthus*), and Hanneke Wilson (*Vitis*).

I would also like to acknowledge the mentally nourishing fare of classical music provided during the writing of this edition by the Australian Broadcasting Corporation, through its 24-hour service, ABC Classic FM, and thank wholeheartedly Robert Fernandez for ensuring my physical frame was kept up to the task. Above all, I thank Andrew Drummond, without whose indefatigable forbearance and understanding this new edition could never have been completed.

In conclusion, I yet again offer the lexicographer's lament, following Dr Samuel Johnson (1709–1784), author of the groundbreaking *Dictionary of the English Language* (1755), who pointed out that readers only remark on such a book when what they seek cannot be found, but would add that any suggestions for additions, improvements or emendations, backed by reference to published materials, where appropriate, will be gratefully received (and acknowledged in print – see above) in the hope that future editions of my book might more nearly meet the needs of its users.

Sydney, November 2016

How to Use This Book and Get the Most Out of It

Firstly, please read the Introduction.

Modern technical names for plants are made up of at least two parts, both italicized: an initial *generic* name (e.g. *Quercus* (oak)), and a second *specific* epithet (e.g. *rubra* (red)), as *species* (singular: *species*) are arranged in *genera* (singular: *genus*): the double name for a species is a *binomial*. Genera are arranged in *families* (e.g. Fagaceae, Compositae), and these in *orders* (e.g. Fagales, Asterales), and those in *subclasses* (e.g. Magnoliidae, Polypodiidae) and those in classes (e.g. Equisetopsida). Generally, the first published name for a genus or a species is the correct one (such names are said to have *priority*). Generic names have an initial capital letter, specific ones a lowercase one, even when commemorating people or places. Traditionally, generic names were based on Greek words, generally brought into Latin form, while species epithets were based on Latin words. However, epithets can be derived from any root whatsoever.

In zoology the generic and specific names can be the same, but such *tautonyms* are not permitted in botanical nomenclature. A third epithet is added where recognizable races, *subspecies* (subsp.) or *varieties* (var.), have been named. Such *trinomials* in botany have the linking 'subsp.' or 'var.' (but this is not done in zoology). Cultivated varieties (correctly *cultivars*) are written in roman with single quotes around them: the first letter is capitalized (e.g. *Solanum viride* 'Anthropophagorum'). In technical works, the binomial is followed by an authority, an abbreviated version of the name of the person who coined the name, for example, 'L.' (= Linnaeus). If the species is moved to a different genus, that person's name is put in parentheses and, in botany, followed by the authority that made the move, for example, '(L.) Sm.'.

This book attempts to present all currently accepted generic and family names (found for example in Greuter *et al.*'s *Names in current use for extant plant genera* (1993)), and commonly used English and other vernacular names of angiosperms and other seed plants and ferns (as well as other vascular plants with spores besides economically and ecologically significant mosses, liverworts and seaweeds), excluding wholly fossil groups. Also included are those commonly encountered synonyms found in literature published since 1970. Generic names encountered in the older literature and not found in this book should be sought first in *Index nominum genericorum* (http://botany.si.edu/ing/) and *The plant list* (http://www.theplantlist.org/), where their modern identity is often indicated.

Each generic entry includes the family to which the genus is assigned; the number of species within the genus; its distribution and other details of botanical, horticultural, agricultural, medicinal or other economic importance as well as vernacular names (English and others encountered in commerce) applied to species within the genus. As far as possible, the species are listed in alphabetical order. Where there has been a recent or recently cited monograph or review of the genus, the place of its publication is added in abbreviated form. Even with this reference, however, it is always worth referring to the family entry, where additional botanical information will be found. Each family entry includes a statistic of the number of genera followed by an oblique line and the number of species. Other information given includes the classification of the family and its sub-division, the principal genera, and the distribution, botanical details and main uses of plants within the family. English names are merely cross-referenced to the generic entries, which should be followed up as any further information on that species and its relations will be found only there.

General abbreviations and abbreviations for authors' names used in this text are given on p. 1029 and p. 1038, respectively. However, some additional explanation may be useful. 'R' refers to recent revisions, reviews, synopses or keys of the genus or family concerned, and N, E, S, W refer to compass points and not political divisions, so that, for example,

'S Afr.' = southern Africa. 'Warm' is taken to mean subtropical and/or warm temperate, whereas 'SE As.' is mainland SE Asia (Indochina), 'Mal.' (Malesia) is the area covered by *Flora malesiana* (Malay Peninsula to Bismarck Archipelago), 'N Am.' with a species number indicates the circumscription used in *Flora of North America* (i.e. Canada and United States), 'Papuasia' is New Guinea to and including the Solomon Islands, 'Macaronesia' comprises the groups of islands off the African coast in the north Atlantic and 'Eur.' (Europe) is used in the sense of *Flora europaea*.

Of unusual signs, ~ before a generic or family name indicates that the subject of the entry is sometimes included in, has recently been included in or is very close to the taxon following the sign; this can lead to the discovery of further information if the taxon has been widely misclassified or incorrectly named. Single quotes around a name in italics mean that the name has been used (widely) in the wrong sense. Brackets around a number in a floral formula indicate that the parts are united (usually by some intercalated tube and not 'fused', a term which, like 'anomalous', for rare or unusual, and 'arranged [by whom?]', referring to leaves etc. in family descriptions, has been avoided).

Technical terms have been kept as few as possible because it has not proved feasible to provide a glossary but see Henk Beentje's invaluable *The Kew plant glossary: an illustrated dictionary of plant identification terms* (2010). Here one word in particular needs some explanation, however: 'endemic' in biology means 'restricted', whereas in medicine it means 'indigenous'.

Sample Entries

For users unfamiliar with the condensed style of dictionaries such as this, the following generic, family and vernacular name entries are set out *in extenso* as examples to aid comprehension.

Adinandra Jack. Pentaphylacaceae (3; Ternstroemiaceae). c. 90 Indomal. (Borneo 32). R: JAA 28(1947)1. Roux's Model

Adinandra [see http://botany.si.edu/ing/ for more information] first described by William Jack (1795–1822). Family Pentaphylacaceae [see that entry for further details; also http://www.mobot.org/MOBOT/Research/APweb/welcome.html] (3; Ternstroemiaceae [i.e. tribe Frezierieae (which includes the formerly recognized family, Ternstroemiaceae, and other currently recognized genera including *Cleyera and Eurya*, etc. – see those entries)]. Around 90 species are native in the Indomalesian region, 32 of them in Borneo. A review of the genus is in the *Journal of the Arnold Arboretum* vol. 28 (1947), beginning on page 1 [see http://www.biodiversitylibrary.org/item/33604#page/5/mode/1up]. The trees have a branching pattern corresponding to Roux's Model (see Fig. 1)

Anisodus Link ex Spreng. (~ *Scopolia*). Solanaceae (IV 2). 4 temp. E As. R: A.T. Hunziker, *Gen. Solan.* (2001) 361. ***A. luridus*** Link ex Spreng. (*A. stramoniifolius*) – yak fodder in Himal.

Anisodus [see http://botany.si.edu/ing/ for more information] first described by Curt Polycarp Joachim Sprengel (1766–1833), who validated the name first suggested by Johann Heinrich Friedrich Link (1767–1851). Closely allied to and sometimes included in genus *Scopolia* [see that entry]. Family Solanaceae [see that entry for further details; also http://www.mobot.org/MOBOT/Research/APweb/welcome.html], subfamily Solanoideae, tribe Hyoscyameae [which also includes the genera *Atropa*, *Hyoscyamus*, *Scopolia* – see those entries]. Four species indigenous in temperate east Asia. Revision published in A.T. Hunziker, *Genera Solanacearum*, beginning on page 361. One of these is *Anisodus luridus* first described by Curt Polycarp Joachim Sprengel (1766–1833), who validated the name first suggested by Johann Heinrich Friedrich Link (1767–1851), a synonym is *Anisodus stramoniifolius*; it is used as fodder for yaks in the Himalaya

Erythroxylaceae Kunth. Magnoliidae – Malpighiales. 4/240 trop. esp. Am. Glabrous trees & shrubs oft. with alks incl. cocaine. Lvs in spirals (opp. in *Aneulophus*), simple (oft. with longit. markings), entire; stip. intrapetiolar. Fls small, reg., usu. bisex., (4)5-merous, oft. heterostylous, solit. or in axillary fascicles; K a tube with imbr. or valvate lobes, C imbr. & usu. with adaxial ± basal appendages, disk 0, A 10–12 usu. forming a tube, anthers with longit. slits, $\underline{G}$ (2 or 3 (4)) with as many locs & styles (± connate), ovule 1(2) in 1 fert. loc., axile, pend., anatr. to hemitropous, bitegmic. Fr. a 1-seeded drupe; seed with straight embryo in copious (rarely 0) starchy endosperm. n = 12

Genera: *Aneulophus, Erythroxylum, Nectaropetalum, Pinacopodium Erythroxylum* a source of narcotics etc.

Erythroxylaceae [see http://www.mobot.org/MOBOT/Research/APweb/welcome.html] described by Karl Sigismund Kunth (1788–1850). Angiosperms, order Malpighiales [see Appendix for allied families]. Four genera with 240 species indigenous in the tropics, especially tropical America. Glabrous trees and shrubs often with alkaloids including cocaine. Leaves in spirals (but opposite in species of *Aneulophus*) simple (often with longitudinal markings), entire; stipules intrapetiolar. Flowers small, regular, usually bisexual with parts in fives (rarely fours), often heterostylous, borne singly or in axillary fascicles; calyx a tube with imbricate or valvate lobes, petals imbricate and usually with adaxial, more-or-less basal appendages, disk absent, stamens 10 (rarely 12) usually forming a tube, their anthers with longitudinal slits; ovary superior, with rarely two and usually two or three (rarely four) united carpels with as many locules and more-or-less connate styles, ovules usually one (rarely two) axile, pendulous in the only fertile locule, anatropous to hemitropous, bitegmic. Fruit a one-seeded drupe; seed with a straight embryo in copious (rarely absent) starchy endosperm. Haploid chromosome number 12

Genera: *Aneulophus, Erythroxylum, Nectaropetalum, Pinacopodium* [see those entries] Species of *Erythroxylum* a source of narcotics etc.

money plant *Crassula ovata, Epipremnum pinnatum* 'Aureum'; **m. tree** *C. ovata;* **m. wort** *Lysimachia nummularia;* **Cornish m.w.** *Sibthorpia europaea*

money plant *Crassula ovata, Epipremnum pinnatum* cultivar Aureum; **money tree** *Crassula ovata;* **moneywort** *Lysimachia nummularia;* **Cornish moneywort** *Sibthorpia europaea* The entries to which these refer should then be examined. For example, under *Epipremnum*:

Epipremnum Schott. Araceae (IV 4). c. 20 SE As. to W Pac. (alleged fossils in Oligocene of N Egypt). Lianes, some medic. & cult. orn. esp. ***E. pinnatum*** (L.) Engl. (*Monstera dilacerata*, Indomal. to W Pac.) – like *M. deliciosa* with perforated lvs, many cvs esp. **'Aureum'** (poss. wild sp. – *E. aureum* (Linden & Bouché) Bunting from Society Is., 'money plant' – rarely flowering so owners of fl. pls considered 'in the money'), irreg. varieg., widely planted in trop.

Epipremnum [see http://botany.si.edu/ing/ for more information] described by Heinrich Wilhelm Schott (1794–1865). Family Araceae [see that entry for further details; also http://www.mobot.org/MOBOT/Research/APweb/welcome.html], subfamily Monsteroideae, tribe Monstereae [i.e. related to *Monstera, Scindapsus* etc.]. Approximately 20 species indigenous in south-east Asia to the western Pacific (alleged fossils known from the Oligocene of northern Egypt). Lianes, some of medicinal value and some cultivated as ornamentals especially ***Epipremnum pinnatum***, which was first described in another genus by Carl Linnaeus (von Linné, 1707–1778) and first transferred to *Epipremnum* by Heinrich Gustav Adolf Engler (1844–1930). It is indigenous in Indomalesia to the western Pacific, resembling *Monstera deliciosa* [see entry for *Monstera*], with perforated lvs; there are many cultivars grown especially **'Aureum'**, possibly a wild species from the Society Is., in which case its correct name is *Epipremnum aureum*, first described in another genus by Jean Jules Linden (1817–1898) and Peter Carl Bouché (1783–1856) and first transferred to *Epipremnum* by George Sydney Bunting (1927–2015), known as 'money plant' because it rarely flowers and owners of flowering plants are considered to be 'in the money'; it has irregularly variegated leaves and is widely planted in the tropics

Disclaimer

As a result of the evolution of defence mechanisms against predators, much of the plant world is inedible, if not poisonous, as far as humans are concerned. Most people realize this and act responsibly, but, such is our litigious world, it is sadly essential that I state here that notes on edibility and so forth in this book are merely recorded information and do not constitute recommendations. No responsibility will be taken for readers' own actions.

Bibliographic Note on the Third Edition of *Mabberley's Plant-Book*

The first printing (2008) was followed by two reprints (2009, 2014), each with additions and amendments; the 2014 reprint used finer paper, restoring the portability of the book. All were hardback and bore the same ISBN.

A

Aa Reichb.f. (~ *Altensteinia*). Orchidaceae (IV 2b). 25 Andes, Costa Rica, coastal Peru. Some protandry. See also *Myrosmodes*

Aakia J. Grande (~ *Panicum*). Gramineae (XXIII.3). 1 C Am.: ***A. tuerckheimii*** (Hack.) J. Grande. R: T 63(2014)219

Aaron's rod *Verbascum thapsus*

Aaronsohnia Warb. & Eig. Compositae (Anth.). 2 SW As., N Afr. R: BBMNHB 23(1993)156

abacá *Musa textilis*

abarco (wood) *Cariniana* spp. esp. *C. pyriformis*

Abarema Pittier (~ *Pithecellobium*). Leguminosae (II 4). Incl. *Klugiodendron* 49 trop. Am. (Amaz. c. 35). R: MNYBG 74,11(1996)41. Heterogeneous. Fls with 1 or 2 ovaries in same infl.

Abasoloa Llave = ? *Eclipta*

Abassian boxwood *Buxus sempervirens*

Abatia Ruíz & Pavón. Salicaceae (Flacourtiaceae). Incl. *Aphaerema*, 10 trop. Am. mts; in Andes above timber line. R: FN 22(1980) 48. Lvs opp. ***A. rugosa*** Ruíz & Pavón (Peru) – lvs source of black dye

Abaxianthus M.A. Clem. & D.L. Jones = *Dendrobium*

Abbo rubber *Ficus lutea*

Abdominea J.J. Sm. = *Robiquetia* (but see OB 95(1988)51)

Abdra Greene (~ *Draba*). Cruciferae. 2 N Am. R: T 61(2012)947

Abdulmajidia Whitm. = *Barringtonia*

abé *Canarium schweinfurthii*

Abebaia Baehni = *Manilkara*

abel *Canarium schweinfurthii*

abele *Populus alba*

Abelia R. Br. (~ *Linnaea*). Caprifoliaceae (Linnaeaceae). Excl. *Diabelia, Vesalea, Zabelia* 40 S & E As. Domatia. Cult. orn. esp. ***A. × grandiflora*** (André) Rehder – variable K (*A. chinensis* R. Br. (K5) × *A. uniflora* R. Br. (K usu. 2), both China)

Abeliophyllum Nakai. Oleaceae (2). 1 Korea: ***A. distichum*** Nakai – like *Forsythia* but heterostylous fls white, fr. winged like *Fontanesia*, 1 of v. few Korean endemic genera (cf. *Hanabusaya, Pentactina*), cult. orn., fewer than 20 left in wild. R: BM n.s. 15(1998)144.

Abelmoschus Medik. = *Hibiscus*

abem *Berlinia* spp.

Aberemoa Aubl. = *Guatteria*

Aberrantia Luer = *Acianthera*

Abies Mill. Pinaceae. 47 N temp. (Eur. 5) to Vietnam, C Am. R: NRBGE 46(1989)59, RV 21(1990)9. Firs. Massart's Model. Lvs borne on branches without short shoots, those on laterals twist into horiz. plane. Cones mature in 1 yr. Many valuable timbers & resins. ***A. alba*** Mill. (silver f., whitewood, S. Eur. mts) – tallest Eur. tree (to 350 yrs old) form. much grown for constr. work & telegraph poles & imp. as 'Swiss pine' for piano- & violin-making, favoured by Greeks & Romans for building fast warships, esp. for oars of triremes (loses lower branches early), but since 1900 much attacked by aphids & now replaced by *A. grandis*, source of Alsatian or Strasburg turpentine (Vosges), oil used in bath preps & med. esp. resp. when inhaled, the principal Christmas tree of the Cont.; ***A. amabilis*** (Loud.) Forbes (NW N Am.) – to 72 m tall, bole to 2.37 m diam., lives to 750 yrs, local medic.; ***A. balsamea*** (L.) Mill. (balsam f., N Am.) – pulp contains juvabione, homologue of insect juvenile hormone, & used in Am. (not Brit.) paper prods, oleoresin the source of Canada balsam used in microscop. preps, local medic. & C. pitch; ***A. bracteata*** D. Don (*A. venusta*, bristlecone f., St Lucia f., Calif.) – cones with long needle-like bracts; ***A. cephalonica*** Loud. (Gk. f., Greece) used for ships in anc. Greece; ***A. cilicica*** Antoine & Kotschy (Syria, As. Minor) – resin used by anc. Egyptians in mummification, timber for ship masts; ***A. firma*** Sieb. & Zucc. (Jap. f., C & S Japan); ***A. fraseri*** (Pursh) Poir. (Fraser f., she balsam, Alleghanies); ***A. grandis*** (D. Don) Lindl. (white or giant f., NW Am.), to 81 m tall, bole to 2.02 m diam., lives at least 500 yrs, introd. GB 1834 & > 62 m tall by 1989; ***A. lasiocarpa*** (Hook.) Nutt. (alpine f., W US); ***A. magnifica*** A. Murr. (red f., NW Am.); ***A. nebrodensis*** (Lojac.) Mattei (*A. alba* subsp. *n.*) reduced to 20 trees in N Sicily but being replanted there; ***A. nordmanniana*** (Steven) Spach (Cauc. f., Nordmann Christmas tree, E Medit.) – needles held longer than those of *Picea abies*; ***A. pindrow*** (D. Don) Royle (Himal. f.); ***A. procera*** Rehder (*A. nobilis*, noble f., NW Am.) – to 90 m tall, bole to 2.75 m diam., lives at least 500 yrs, wood oft. sold as 'larch'; ***A. sachalinensis*** (Schmidt) Masters (N Jap.) – needles mixed with those of *Picea jezoensis* to yield Jap. pine-needle oil

Abietaceae Gray = Pinaceae

Abildgaardia Vahl (~ *Fimbristylis*). Cyperaceae (II 4). 15 trop. & warm (N Am. 1), esp. Aus.

abir scented powder used in Hindu cerem., largely ground rhiz. *Hedychium spicatum*

abiu *Pouteria caimito* & other *P.* spp.

Abobra Naud. Cucurbitaceae (XV). 1 temp. S Am.: ***A. tenuifolia*** (Gillies) Cogn. – cult. orn. dioec. cli.

Abolboda Humb. Xyridaceae. 23 trop. S Am., marshy savanna. R: AMBG 79(1992)820, HPB 10(2005)138

Abolbodaceae Nakai = Xyridaceae

Aboriella Bennet (*Smithiella*) = *Pilea*

Abortopetalum Degener = *Abutilon*

aboudikro *Entandrophragma cylindricum*

Abraham, Isaac & Jacob *Symphytum officinale*

Abrahamia Randrianasolo (~ *Protorhus*). Anacardiaceae (I). 19 Madag. Large fr. & seeds

Abrodictyum C. Presl (~ *Cephalomanes*). Hymenophyllaceae. 25 trop. Pacific, Afr. & is. (7)

Abroma Jacq. (*Ambroma*). Malvaceae (Bytt.-Bytt.; Sterculiaceae). 1 trop. As. to Aus.: ***A. augustum*** (L.) L.f. (devil's cotton, Mal.) – Petit's Model, bark a source of jute-like fibre

Abromeitia Mez = *Fittingia*

Abromeitiella Mez = *Deuterocohnia*

Abronia Juss. Nyctaginaceae (6). Incl. *Tripterocalyx*, 24 SW US to N Mex. Ground r. of some spp. form. eaten by Native Americans. Cult. orn.

Abrophyllum Hook.f. Rousseaceae (Carpodetaceae; Grossulariaceae s.l.). 1–2 E Aus. R: AusSB 10(1997)861

Abrotanella Cass. Compositae (Sen.-Abr.). 18 NG, Aus., NZ, temp. S Am. R: PSE 197(1995)155. Oft. cushion-forming alpines

Abrus Adans. Leguminosae (III 17). 17 pantrop. (? introd. Am.). ***A. precatorius*** L. (coral pea, crab's eyes, Ind. liquorice, jequerity seeds, lucky or Paternoster beans, rosary pea, weedy in SE US) cont. alks, used as basis of contraceptive & abortifacient in Ind., comm. med. (coughs) Vietnam; r. yields a poor quality liquorice subs.; seeds used as beads & weights (rati, Ind. cf. *Adenanthera*) but v. poisonous esp. in contact with wounds or eyes due to abrin, a toxic glycoprotein, inhibiting protein synthesis (esp. in cancer cells) through inactivation of ribosomes, 0.5 g fatal in humans but detoxified above 65°C

absinthe *Artemisia absinthium*

Absolmsia Kuntze = *Hoya*

abura *Mitragyna stipulosa*

Abuta Aubl. Menispermaceae (II). 32 trop. S Am. R: MNYBG 22,2(1971)30. Some ed. fr., med. (dangerous, some with ecdysteroids), arrow poisons. ***A. imene*** (Mart.) Eichler bark used in kind of curare (Colombia); ***A. rufescens*** Aubl. (white Pareira r., Guianas) – med. (urinogenital tract)

Abutilon Mill. Malvaceae (Malv.-Malv.). Excl. *Callianthe*, 120 trop. & warm (48 Aus., 1 Eur – ***A. theophrasti*** Medik., imp. fibre-pl. in China (chingma, Am. or Chinese jute or hemp, Ind. mallow, Manchurian jute, velvetleaf) but major agric. weed in Am. introd. from China pre-1750 & causing losses of >$343 m per annum by 1980s). No epicalyx, some bird-poll. Fls of ***A. esculentum*** A. St-Hil. (Braz.) ed. Cult. orn., others medic. (e.g. ***A. indicum*** (L.) Sweet (trop. OW)) & fibre-pls

Abutilothamnus Ulbr. = *Bastardiopsis*

Abyssinian myrrh *Commiphora madagascariensis*

Acacallis Lindl. = *Aganisia*

Achmena H.P. Fuchs = *Erysimum*

acacia *Acacia* spp. (see also *Faidherbia, Senegalia, Vachellia*), (form. in timber trade) *Robinia pseudoacacia*, (Aus.) *Albizia* spp., weedy *Senna* spp.; **cedar a.** *Falcataria toona*; **false a.** *Robinia pseudoacacia*; **rose a.** *R. hispida*; **sweet a.** *V. farnesiana*

Acacia Mill. Leguminosae (II 3). Excl. *Acaciella, Faidherbia, Mariosousa, Senegalia, Vachellia*, 1468 trop. & warm OW esp. Aus. (c. 1053 – R: FA 11A, 11B (2000)), Hawaii 1. R: T 59(2010)16; prob. 'sister' to *Paraserianthes*. Bipinnate lvs (70 spp. in Aus.); phyllodic spp. with bipinnate lvs when seedlings (occ. persisting in adult e.g. *A. melanoxylon*); phyllode develops through intercalary growth in midrib region, s.t. decurrent on stem as in ***A. alata*** R. Br. (W Aus.); phyllotaxis of phyllodes decussate, spiral, fasciculate, whorled as in ***A. lycopodiifolia*** A. Cunn. ex Hook. (NW Aus.) & ***A. verticillata*** (L'Hérit.) Willd. (SE Aus.) or 'chaotic' as in ***A. conferta*** A. Cunn. ex Benth. (NE Aus.). Gum from phyllodes eaten by Aus. Aborigines. Many imp. prods – timber, fuel, forage, tanbark (some tannins molluscicidal, dyestuffs, gums, ayahuasca subs. (*N,N* DMT) & cult. orn. though some pestilential weeds, esp. Aus. spp. in SW Cape (S Afr.) where poll. occurs & there are no seed-pests, but also some E Aus. spp. weedy in SW Aus. ***A. acuminata*** Benth. (raspberry jam (tree), SW Aus.) – jam-scented durable timber, host for *Santalum spicatum*; ***A. ampliceps*** Maslin (salt wattle, W Aus.) – reclamation of salted soils; ***A. aneura*** F. Muell. ex Benth. (mulga, Aus.) – apomictic, dark heartwood (boomerangs etc.) & seeds (ground & eaten) used by Aborigines; ***A. aphylla*** Maslin (reindeer wattle, SW Aus.) – rare, leafless 'reindeer bush' sold at Christmas in Aus.; ***A. auriculiformis*** A. Cunn. ex Benth. (auri, trop. Aus., NG) – fuel crop tree with its hybrids with *A. mangium* imp. plantation crops (pulp, wood) in As.; ***A. baileyana*** F. Muell. (Cootamundra wattle, E Aus.) – weedy in parts of Aus., hybridizes with *A. dealbata*; ***A. bakeri*** Maiden (NE Aus., reduced through felling) – to 50 m (tallest in Aus.), seeds germ. in fr. on tree, yellowish timber ('hickory') for flooring etc.; ***A. cambagei*** R. Baker (gidgee, gidyea, S & E Aus.) & other spp. – smelly foliage in wet weather; ***A. celastrifolia*** Benth. (SW Aus.) – heads of 1–3 fls each with A to 300 (500+), G 3–7; ***A. cincinnata*** F. Muell. (NE Aus.) – imp. plantation tree China; ***A. cognata*** Domin (SE Aus.) – cult. orn. esp. 'weeping' forms; ***A. colei*** Maslin & L. Thomson (N Aus.) – cult. trop. & S Afr.; ***A. crassicarpa*** Cunn. ex Benth. (trop. As. to Pacific) – sawn wood, tried for Kraft pulp in trop. As.; ***A. dealbata*** Link (blue or silver wattle, Aus.) – florists' 'mimosa' imported to London from S France in winter, weedy S Eur., S Afr.; ***A. decurrens*** Willd. (black or green wattle, NSW) – mass-flowerings indicate hottest & driest part of 11-yr. meteorological cycle recog. by Aborigines, tanbark in S Afr.; ***A. dodonaeifolia*** (Pers.) Balbis (S Aus.) – twigs with resinous ridges; ***A. eglandulosa*** DC. (*A. cyclops*, SW Aus.) – invasive S Afr.; ***A. howittii*** F. Muell. (sticky wattle, E Victoria) – viscid cult. orn.; ***A. implexa*** Benth. (E Aus., hickory, lightwood); ***A. kempeana*** F. Muell. (Aus.) – witchetty grubs (Aboriginal food) in roots; ***A. koa*** A. Gray (Hawaii; ***A. heterophylla*** Willd. (Réunion) 18 000 km away prob. conspecific ancestor) – with 2 other Hawaiian endemics bearing extrafl. nectaries attractive to ants, though Hawaii has no native ant spp. (? phylogenetic inertia), bark a red dye-source trad. timber for sailing-canoe hulls & surfboards but also furniture, guitars etc.; ***A. leprosa*** Sieber ex DC. (SE Aus.) – '**Scarlet Blaze**' only red-flowering a., state fl. Victoria; ***A. longifolia*** (Andr.) Willd. (sally wattle, Sydney golden w., SE Aus.) – bark-strips twisted together for Aborigines' fishing-lines, seeds ed. cooked, cult. orn., lime-tolerant stock for other spp., weedy in S Eur., NZ & S Afr. fynbos where controlled by gall-forming wasp; ***A. mangium*** Willd. (mangium, E Mal., trop. Aus.) – fuelwood, woodchips for pulp, building timber, charcoal for 'activated charcoal', most planted trop. *A.* (23 m trees in 9 yrs, reaching 23 cm d.b.h.; 900 000 ha in Malaysia & Indonesia); ***A. mearnsii*** De Wild. (black wattle, SE Aus.) – imp. tanbark (tannin 60–65% of extract), charcoal, pulp for paper, rayon, widely cult., invasive S Eur., S Afr., Hawaii; ***A. melanoxylon*** R. Br. (E Aus. or Tasmanian blackwood, Aus.) – sup. timber, incl. piano soundboards, weedy S Eur & S Afr., imp. plantation tree China; ***A. omalophylla*** A. Cunn. ex Benth. (*A. homalophylla*, yarran, E Aus.) – durable timber (violet wood); ***A. papyrocarpa*** Benth. (SW Aus.) – dense wood for musical instruments esp. flute head-joints; ***A. paradoxa*** DC. (*A. armata*, kangaroo thorn, temp. Aus.) – hedging; ***A. pendula*** A. Cunn. ex G. Don (myall, E Aus.) – fine violet-scented timber, form. used for pipe-bowls; ***A. podalyriifolia*** A. Cunn. ex G. Don (NE Aus., natur. elsewhere) –

winter-flowering in Aus.; ***A. prominens*** A. Cunn. ex G. Don (golden rain wattle, NSW); ***A. pubescens*** (Vent.) R. Br. (hairy wattle, NSW) – easily grown in Eur., but endangered in wild; ***A. pycnantha*** Benth. (golden wattle, SE Aus., weedy SW Aus., S Afr.) – extrafl. nectaries attractive to poll. birds, tree imp. source of tanbark & gum (eaten by Aborigines), Aus. national emblem (1988; Wattle Day 1 Sept.); ***A. ramulosa*** W. Fitzg. (C & S Aus.) – fr. tardily dehiscent, poss. adapted to extinct megafauna; ***A. retinodes*** Schldl. (wirilda, S & SE Aus.) – cult. orn., cut-fl. in Israel; ***A. salicina*** Lindl. (cooba, Aus.); ***A. saligna*** (Labill.) H.L. Wendl. (*A. cyanophylla*, Port Jackson (willow), W Aus., weedy elsewhere in Aus., S Eur. & S Afr.) – potential wood & biofuel crop, fodder in e.g. Chile, coastal sandhills rehabilitation in Libya, pest in S Afr. fynbos where controlled by gall-forming rust fungus (*Uromycladium tepperianum*); ***A. spectabilis*** A. Cunn. ex Benth. (Mudgee wattle, arid & subarid Aus.); ***A. subporosa*** F. Muell. (river wattle, SE Aus.) – along streams; ***A. terminalis*** (Salisb.) Macbr. (sunshine wattle, SE Aus.) – winter-fl., local medic.; ***A. tetragonophylla*** F. Muell. (dead finish, S Aus.) – sharp phyllodes used to puncture & deplete warts; ***A. tumida*** F. Muell. ex Benth. (NW Aus.) – flour from roasted seeds used for damper by Aborigines; ***A. victoriae*** Benth. (gundabluey, Aus.) – emergency fodder, seeds Aboriginal food, also for coffee (wattlecino) and icecream flavouring, though ***A. notabilis*** F. Muell. (SE Aus.) better flavour

Acaciella Britton & Rose (~ *Acacia*). Leguminosae (II 4). 15 S-C US to Arg. Some herbaceous. ***A. glauca*** (L.) Rico (*Acacia g.*, Jamaica) – cult. orn., covercrop in teak plantations in Indonesia

Acaena Mutis ex L. Rosaceae (8). c. 100 S hemisph. N to Calif. (1) & Hawaii (1), many natur. elsewhere, e.g. ***A. novae-zelandiae*** Kirk ('*A. anserinifolia*', SE Aus., NZ, noxious in US) in GB, introd. early 1900s in fleeces from Aus. or NZ. R: BB 74(1910–11). Bidgee-widgee (Aus.), bidi-bidi (NZ). ***A. eupatoria*** Bitter (Urug.) – used to control fertility

açai *Euterpe edulis*

acajou *Anacardium occidentale*; *Guarea guidonia* or other 'mahoganies'

Acalypha Royen ex L. Euphorbiaceae (Acal.-Acal.-Acal.). c. 450 trop. & warm (Mal. 28 – R: Blumea 55(2010)25; Afr. 50). R: Pflanzenr. IV.147.XVI(1924)12. Orig. Afr.; colonized As. several times, Am. 3 times. Mostly monoec. or dioec. shrubs, Koriba's Model (*A. amentacea*), stigmas much branched. Some with prickly galls (cf. *Hopea ponga*). ***A. amentacea*** Roxb. 'subsp. ***wilkesiana*** (Muell. Arg.) Fosb.' (*A. tricolor*, *A. wilkesiana*, beefsteak plant, copperleaf, Jacob's coat, cultigen, Pac.) – potherb, local medic., widely cult. trop. orn. with mottled lvs incl. '*A. godseffiana* Masters' cult. orn. trop. As. esp. '**Heterophylla**' with v. narrow varieg. lvs; ***A. herzogiana*** Pax & K. Hoffm. ('*A. reptans*', Bolivia, Arg.) – cult. orn., some (red) infls with homoeotic fls with 5–8 styles in place of A; ***A. hispida*** Burm.f. (?orig. NG, Bismarcks) – cult. orn., only female known, infls coloured by branched red stigmas; ***A. indica*** L. (OW trop., widely natur.) – potherb, local medic., r. irresistible to cats; ***A. filiformis*** Poir. subsp. ***rubra*** (Muell. Arg.) Govaerts ('*A. rubrinervis*', St Helena) – 2 m treelet extinct by 1870, allied to subsp. ***f.*** (*A. reticulata*, Madag., Masc.)

Acalyphopsis Pax & K. Hoffm. = *Acalypha*

Acampe Lindl. (~ *Saccolabium*). Orchidaceae (V 16c). 8 OW trop. ***A. rigida*** (Sm.) P. Hunt (Ind.) – raindrop-mediated self-poll.

Acamptoclados Nash = *Eragrostis*

Acamptopappus (A. Gray) A. Gray (~ *Haplopappus*). Compositae (Ast.-Sol.). 2 SW N Am. deserts. R: Madroño 35(1988)247

Acanthaceae Juss. Magnoliidae-Lamiales. Incl. Avicenniaceae, Mendonciaceae, excl. Thomandersiaceae, 202/3800 largely trop., incl. open country & deserts, ext. to Medit., US, Aus. R: KB 55(2000)579. Herbs, incl. some aquatics (even mangrove – *Acanthus*, *Bravaisia*) & twiners, shrubs, few trees (incl. mangrove *Avicennia*), s.t. with unusual sec. thickening. Many typical of ground flora of trop. forests & some with gregarious flowering (*Mimulopsis*, *Strobilanthes*). Lvs simple, usu. decussate (spiral in Nelsonioideae) with cystoliths, seen as streaks in lvs (not in Nelsonioideae, Thunbergioideae nor tribe Acantheae of Acanthoideae), s.t. spiny; stip. 0. Bracts & bracteoles usu. present, oft. coloured & ± encl. fl. Fls bisex. (resupination through twisting of C evolved several times in III), usu. zygom. or 2-lobed (upper lip s.t. absent e.g. *Acanthus*), usu. with nectariferous disk below ovary. K (4–6, usu. 5), to 16-lobed, oft. sharply pointed, persistent; C (4 or 5); A 2, 4 or 5 (*Pentstemonacanthus*), epipet.; 1 or more staminodes oft. present; 1 anther lobe oft. smaller than other, connective oft. long, pollen with diverse architecture, imp. in generic delimitation. $\underline{G}$ (2), 2-loc. with axile placentation, each with 2–∞ (esp. Nelsonioideae) usu. anatr. (orthotropous in *Avicennia*)) ovules in 2 rows; style oft. with 2 stigmas. Fr. usu. 2-loc., explosive loculicidal capsule, usu. with seeds on hook-like funicular jaculators (not I, II, IV); seeds non-endospermous (exc. Nelsonioideae, where oily & ruminate), with large embryos, the testa s.t. (e.g. *Blepharis*, *Crossandra*) with hairs or scales becoming sticky on wetting (cf. *Linum*), embryo ± viviparous in *Avicennia*. X = 7–21

Classification & large genera:

I. **Nelsonioideae** (ovules ∞, cystliths & jaculators 0; R: Aliso 32(2014)18): 5 genera incl. *Elytraria*, *Staurogyne*

II. **Thunbergioideae** (incl. Mendoncioideae; lianes, ovules 4, jaculators 0, capsules or drupes): *Mendoncia*, *Thunbergia*

III. **Acanthoideae** (usu. herbs, oft. with swollen nodes; ovules 2–∞, jaculators, explosive capsules): 2 tribes (some unassigned) – 1. Acantheae (cystoliths 0 (see SB 30(2005)834); *Aphelandra*, *Blepharis*, *Crossandra*), 2. Ruellieae (cystoliths (R: IJPS 174(2013)111) being remodelled); 7 subtribes (many split from a.) – a. Ruelliinae (*Ruellia*), b. Andrographinae (*Andrographis*, *Phlogacanthus*), Erantheminae (*Eranthemum*), Hygrophilinae (*Brillantaisia*, *Hygrophila*), Mimulopsinae (*Mimulopsis*), Petaliidiinae (*Dyschoriste*, *Phaulopsis*), Strobilanthinae (*Strobilanthes*), Trichantherinae (*Sanchezia*, *Trichanthera*) [c. Justiciinae (*Asystasia*, *Dicliptera*, *Justicia*), d. Barleriinae (*Barleria*, *Lepidagathis*) to be excluded]

IV. **Avicennioideae** (mangrove trees & shrubs with swollen nodes & erect pneumatophores; ovules 4, jaculators 0): *Avicennia* (only)

Many cult. orn. esp. *Acanthus*, *Aphelandra*, *Crossandra*, *Eranthemum*, *Fittonia*, *Graptophyllum*, *Hypoestes*, *Hygrophila*, *Justicia*, *Ruellia* & *Thunbergia*

Acanthambrosia Rydb. = *Ambrosia*

Acanthanthus Y. Ito = *Echinopsis*

Acanthella Hook.f. Melastomataceae. 2 trop. S Am. R: Biollania ed. esp. 6(1997)369

Acanthephippium Blume (*Acanthophippium*). Orchidaceae (V 14). 13 trop. As. to Tonga. R: OM 8(1997)119. Jug orchids (j.-shaped fls)

Acanthocalycium Backeb. = *Echinopsis*

Acanthocalyx (DC.) Tieghem (~ *Morina*). Caprifoliaceae. 2 Himal., China. R: FOC 19(2011)649

Acanthocardamum Thell. = *Aethionema*

Acanthocarpus Lehm. Dasypogonaceae. 7 W Aus. R: FA 46(1986)92

Acanthocephalus Karelin & Kir. Compositae (Lact.-Crep.). 2 C As.

Acanthocereus (A. Berger) Britton & Rose. Cactaceae (III 1). 1 SE US to Venez.: ***A. tetragonus*** (L.) Hummelinck (*A. pentagonus.*) – ed. fr. See also *Peniocereus*, *Pseudoacanthocereus*

Acanthochiton Torrey = *Amaranthus*

Acanthochlamydaceae P.C. Kao = Velloziaceae

Acanthochlamys P.C. Kao. Velloziaceae (Acanthochlamydaceae). 1 SW China: ***A. bracteata*** P.C. Kao

Acanthocladium F. Muell. (~ *Helichrysum*). Compositae (Gnap.-Cass.). 1 SE Aus.: ***A. dockeri*** F. Muell. R: OB 104(1991)81

Acanthocladus Klotzsch ex Hassk. (~ *Polygala*). Polygalaceae (IV). 8 trop. Am. R: Novon 20(2010)318

Acanthococos Barb. Rodr. = *Acrocomia*

Acanthodesmos C. Adams & du Quesnay. Compositae (Vern.). 1 Jamaica: ***A. distichus*** C. Adams & du Quesnay

Acanthogilia A. Day & Moran. Polemoniaceae (I). 1 Baja Calif.: ***A. gloriosa*** (Brandegee) A. Day & Moran – desert shrub with persistent woody-spinose lvs & winged seeds. R: PCAS 44(1986)111

Acanthogonum Torrey = *Chorizanthe*

Acantholepis Less. = *Echinops*

Acantholimon Boiss. Plumbaginaceae (II). 165 E Med. to C As., gravelly deserts & mts (Eur. 1). R: S. Mobayen (1964) *Revision taxon. du genre A.* Cult. rock-pls

Acantholippia Griseb. = *Lippia* (but see Darw. 22(1980)511, Dominguezia 3(1982)1)

Acanthomintha (A. Gray) A. Gray. Labiatae (VII 2b). 4 Calif., 1 ext. to Mex. R: Madroño 38(1991)285

Acanthonema Hook.f. = *Streptocarpus* (but see R: BMNHN, Adans. 3(1982)416)

Acanthopale C.B. Clarke (~ *Strobilanthes*). Acanthaceae (III 2b). c. 7 trop. Afr., 5 Madag.

Acanthopanax (Decne. & Planch.) Miq. = *Eleutherococcus*

Acanthophippium Blume = *Acanthephippium*

Acanthophoenix H. Wendl. Palmae (V 14h). 3 Masc. R: Palms 50(2006)84. ***A. rubra*** (Bory) H. Wendl. – cult. orn., form. felled for palm hearts & now rare in wild

Acanthophora Merr. = *Aralia*

Acanthophyllum C. Meyer. Caryophyllaceae (III 1). Incl. *Allochrusa, Diaphanoptera, Ochotonophila, Scleranthopsis*, excl. *Czeikia*, 63 SW & C As., Siberia. Many xeroph. with prickly lvs, some cult. orn.

Acanthoprasium (Benth.) Spenn. (~ *Ballota*). Labiatae (VI). 2 Alps, Cyprus

Acanthopsis Harv. Acanthaceae (III 1). 7 S Afr.

Acanthopteron Britton = *Mimosa*

Acanthorhipsalis (K.Schum.) Britton & Rose = *Lepismium*

Acanthorrhinum Rothm. Plantaginaceae (Ant.-Ant.; Scrophulariaceae s.l.). 1 NW Afr.: ***A. ramosissimum*** (Cosson & Durieu) Rothm.

Acanthorrhiza Linden = *Cryosophila*

Acanthoscyphus Small (~ *Oxytheca*). Polygonaceae (I 1). 1 S Calif.: ***A. parishii*** (Parry) Small. R: FNA 5(2005)437

Acanthosicyos Welw. ex Hook.f. Cucurbitaceae (XIV). 1 S trop. Afr.: ***A. horridus*** Welw. ex Hook.f. (nara, narras) of sand-dunes with r. to 12 m long, paired thorns at nodes, ed. fr. & seeds. See also *Citrullus*

Acanthospermum Schrank. Compositae (Mill.-Mel.). 5–6 trop. Am., intr. OW esp. ***A. hispidum*** DC. (Leeuwenberg's Model). R: CUSNH 20(1921)383. Some med., esp. ***A. australe*** (Loefl.) Kuntze (S Am., widely natur.) – contraceptive in Uruguay

Acanthosphaera Warb. = *Naucleopsis*

Acanthostachys Klotzsch. Bromeliaceae (3). 2 E Braz., Parag., NE Arg. C papillose (unique in B.)

Acanthostelma Bidgood & Brummitt = *Crabbea* (but see KB 40(1985)855)

Acanthostyles R. King & H. Robinson (~ *Eupatorium*). Compositae (Eup.-Dis.). 1 E S Am.: ***A. buniifolius*** (Hook. & Arn.) R. King & H. Robinson. R: AusSB 24(2011)90

Acanthosyris (Eichler) Griseb. Santalaceae (Cervantesiaceae). 5 temp. S Am. R: Britt. 48(1997)578. Root-parasites; wood good. ***A. falcata*** Griseb. – ed. fr.

Acanthothamnus Brandegee. Celastraceae (I; Canotiaceae). 1 Mex.: ***A. aphyllus*** (Schldl.) Standl. (*A. viridis*)

Acanthotreculia Engl. = *Treculia*

Acanthoxanthium (DC.) Fourr. = *Xanthium*

Acanthura Lindau. Acanthaceae (III). 1 Braz.: ***A. mattogrossensis*** Lindau

Acanthus Tourn. ex L. Acanthaceae (III 1). 22 trop. & warm OW, S Eur. (3). Mostly thorny xeroph.; fl. with lower lip only; extrafl. nectaries. ***A. ilicifolius*** L. (*Dilivaria i.*, OW mangrove) – bird-poll. in Aus., local medic. (also ***A. ebracteatus*** Vahl – Mal.) esp. boils; ***A. mollis*** L. (bear's breech, Eur.) – bee-poll. cult. orn.; ***A. spinosus*** L. (oyster pl. (Aus.), E Medit.) – supposed orig. of leaf motif in Corinthian capitals

Acareosperma Gagnepain. Vitaceae. 1 Laos: ***A. spireanum*** Gagnepain

acaroid resin See *Xanthorrhoea*

Acarpha Griseb. (~ *Boopis*). Calyceraceae. Excl. *Nastanthus*, 2 temp. S Am.

Acaulimalva Krapov. Malvaceae (Malv.-Malv.). 19 Andes. R: Darw. 19(1974)9

Acca O. Berg (~ *Psidium*). Myrtaceae (II 10). 3 S Am. R: FN 45(1986) 134. ***A. sellowiana*** (O. Berg) Burret (*Feijoa s.*, feijoa, pineapple guava, S Braz., Paraguay, Urug., N Arg.) – ed. fr. for preserves

Accara Landrum (~ *Psidium*). Myrtaceae (II 10). 1 Braz.: ***A. elegans*** (DC.) Landrum. R: SB 15(1990)221

aceituno *Quassia simarouba*

Acelica Rizz. = *Justicia*

Acentra Philippi (~ *Hybanthus*). Violaceae. 1 Arg.: ***A. serrata*** Philippi

Acer Tourn. ex L. Sapindaceae (II 1; Aceraceae). Incl. *Dipteronia* (lvs pinnate; fr. winged all round) 127 N temp. & trop. mts (Eur. 14, China 101 – R: FOC 11(2008)515, W Mal. 1). R: D.M. van Gelderen *et al.* (1994) *Maples of the World*. Trees & shrubs (maples) with entire or palmately lobed lvs, occ. pinnate (sect. *Negundo*), s.t. with domatia, & fr. a pr of long-winged samaras. Monopodial (Rauh's Model) & sympodial spp., the latter able to occupy shady habitats. Some heterodichogamy & dioecy. Oligocene pollens referred here; lvs common in Miocene. Some imp. timbers & honey-pls, local fibres & medic., sources of syrup & many cult. orn. inc. street trees, oft. with bright autumn colours. Wood hard, usu. white, used for flooring, esp. squash courts & dance floors, furniture, musical instruments (e.g. violin-backs, with mahogany around spruce piano-frames), gunstocks, shoelasts etc.; figured wood known as papapsco, fiddle-back, & bird's eye maple (burrs of *A. saccharum*), used in veneers (a Victorian fad), tiger maple that having contrasting light & dark lines in grain (esp. *A. macrophyllum, A. rubrum*); bark ground for bread by E Canad. Indians; maple leaf (based on ? *A. rubrum*) on Canadian flag. ***A. campestre*** L. (field or hedge maple, Eur., inc. Br.) – can form coppicing clumps <4 m diam. in English woods, timber form. for bowls, spoons, cutlery handles, Saxon harps & tobacco pipes (Ulmer pipes), stake form. favoured for shafting hearts of vampires; ***A. macrophyllum*** Pursh (Pacific maple, W N Am.) – timber esp. flooring; ***A. negundo*** L. (box elder, N Am. to Guatemala, invasive in C & E Eur., Aus.) – pinnate lvs, cult. orn., wood soft; ***A. palmatum*** Thunb. (Jap. maple, E As.) – cult. orn. inc. many dwarf cvs; ***A. pensylvanicum*** L. (moosewood, N Am.) – cult. orn.; ***A. platanoides*** L. (Norway maple, Eur., W As., invasive GB, Macaronesia, Aus., NZ, US) – timber, cult. orn. tolerant of road salt so pop. street tree NE US; ***A. pseudoplatanus*** L. (sycamore (app. first confused with sycomore (*Ficus sycomorus*) by Shakespeare, though Gk *sykomorus* is mulberry), great or Scottish maple, form. (still in Scotland) 'plane', Eur., W As.) – poll. breeding thrips, timber for turnery, joinery, 'love-spoons', polo mallets, violin-backs, when stained grey = harewood (as in Hepplewhite furniture; orig. 'airwood', i.e. *ehren-holz*), early nectar source for bees (resultant honey green), early plantation tree in Scotland & now aggressive weed of Br. woodland (Tolpuddle Martyrs had union meetings [1830s] under one in Dorset, still there 1996), **'Corstorphinense'** (Corstorphine plane) early-shooting cv. with golden lvs; ***A. rubrum*** L. (red maple, N Am.) – populations of constant males with some labile to females, fewer constant females & even fewer labile to males though some fluctuate widely in their sexuality, timber; ***A. saccharinum*** L. (silver or white maple, N Am.) – many orn. forms; ***A. saccharum*** Marshall (rock or sugar or striped maple, N Am.) – timber & maple syrup & sugar, tapped in early spring (40 K t per annum [2006], 1/3 US prod. in Vermont mid-Feb.–end March), maple water 'the new coconut water' (2014), wind- & insect-poll., one of most spectacular trees of the fall (gold to crimson); ***A. tataricum*** L. subsp. ***ginnala*** (Maxim.) Wesm. (*A. ginnala*, E As.) – poss. source of polyphenols

Aceraceae Juss. = Sapindaceae

Aceranthus Morren & Decne = *Epimedium*

Aceras R. Br. = *Orchis*

Acerates Elliott = *Asclepias*

Aceratium DC. Elaeocarpaceae. 20 E Mal. to Solomon Is., N Aus., Vanuatu (1). ***A. oppositifolium*** DC. (Indomal.) – ed. fr.

Aceratorchis Schltr. = *Galearis* (peloric)

acerola *Malpighia emarginata*

Acetosa Mill. = *Rumex*

Acetosella (Meissner) Fourr. = *Rumex*

acha *Digitaria exilis*

achacha *Garcinia humilis*

Achaenipodium Brandegee = *Verbesina*

Achaetogeron A. Gray = *Erigeron*

Acharagma (N.P. Taylor) Glass (~ *Escobaria*). Cactaceae (III 9). 2 Mex.

Acharia Thunb. Achariaceae. 1 S Afr.: ***A. tragodes*** Thunb. R: Strel. 10(2000)45

Achariaceae Harms (Kiggelariaceae). Magnoliidae – Malpighiales. 32/170 trop., S Afr. R: KB 57(2002)171. Trees, shrubs, stemless or cli. herbs (Acharieae). Lvs in spirals or 2-ranked, simple, oft. palmately lobed. Fls solit. or in racemes, hypogynous, regular, unisexual (pls. monoecious, rarely dioec.). K 2–5, C 4- 15 oft. in 2 series, (3–5) in Acharieae, A same no. as C, attached to tube, connective broad; $\underline{G}$ (2–10), 1-loc. with 2 – several ovules on each parietal placenta, style ± deeply lobed. Fr. septicidal-capsular; seeds s.t. arillate, with straight embryo & copious endosperm

Principal genera: *Hydnocarpus, Lindackeria, Ryparosa*

Includes (arr. in 4 tribes) many genera form. in Flacourtiaceae. Many medic. esp. for leprosy – *Caloncoba, Carpotroche, Gynocardia, Hydnocarpus*; timber – *Kiggelaria*

Acharitea Benth. = *Nesogenes*

Achasma Griff. = *Etlingera*

Achatocarpaceae Heimerl (~ Phytolaccaceae). Magnoliidae – Caryophyllales. 2/c. 11 warm Am. Dioec. trees or shrubs with regular sec. thickening, s.t. spiny. Lvs in spirals, simple, entire, exstip. Fls in axillary racemes or panicles or ramifl.; K 4 or 5, persistent in fr., C 0, A 10–20, filaments s.t. connate at base, pollen diff. from other Caryophyllales, $\underline{G}$ (2), l-loc. with 1 basal, campylotropous ovule & 2 distinct styles. Fr. a berry; seed almost exarillate, with curved embryo surrounding perisperm, without true endosperm

Genera: *Achatocarpus, Phaulothamnus*

Achatocarpus Triana. Achatocarpaceae. c. 10 Mex. to Arg. ***A. gracilis*** H. Walter (Mex.) – wind- poll., dioec.

Achetaria Cham. & Schldl. Plantaginaceae (Grat. – Stem.). 6 trop. Am.

Achillea Vaill. ex L. Compositae (Anth.-Mat.). Incl. *Leucocyclus, Otanthus*, 117 Euras. (Eur. 52), Medit., few N Am. R: BBMNHB 23(1993)128. Alks. Many med. & cult. orn. (yarrow), e.g. ***A. erba-rotta*** All. subsp. ***moschata*** (Wulfen) Richardson (*A. moschata*, C Alps) – med., source of Iva liqueur & bitters; ***A. filipendulina*** Lam. (W to C As.) – yellow fls; ***A. maritima*** (L.) Ehrend. & Guo (*Otanthus m.*, cottonweed, coasts W Eur. & Nr. E) – cult. orn. with felted lvs & rayless capitula; ***A. millefolium*** L. (milfoil, N temp., natur. Aus., NZ) – med. (haemostatic properties, staunching bleeding etc.; most uses [355] of any drug pl. used by Native Americans), mosquito-repellent (stachydrine), tobacco subs. (Sweden), ed. (Ger.), herbal treatment of arthritis, with *Ephedra* sp. in Neanderthal burial head-wreaths, red-water fever treatment in stock; ***A. ptarmica*** L. (sneezeweed, sneezewort, Eur.) – form. med. & salad pl., 'double' forms = bachelor's buttons; ***A. santolina*** L. (NE Afr. to Pakistan) – Bedouin med. pl., poss. source ess. oil; ***A.* 'Taygetea'** – common yellow herb. perenn., poss. *A. millefolium* × *A. clypeolata* Sm. (SE Eur.)

Achimenes Pers. (~ *Gloxinia*). Gesneriaceae (II 5b). 23 trop. Am. Diff. spp. poll. by birds, moths, bees & butterflies – multiple orig. of floral spurs & other poll. syndromes. Cult. orn. e.g. ***A. erecta*** (Lam.) H.P. Fuchs (*A. coccinea, A. pulchella*), parent of many hybrids, & ***A. grandiflora*** (Schiede) DC. (hot water plant)

achiote *Bixa orellana*

achira *Canna indica*

Achlaena Griseb. = *Arthropogon* (but see AMBG 88(2001)368)

Achlydosa M.A.Clem. & D.L. Jones (~ *Megastylis*). Orchidaceae (IV 2f). 1 New Caled.: ***A. glandulosa*** (Schltr.) M.A. Clem. & D.L. Jones

Achlyphila Maguire & Wurd. Xyridaceae. 1 Venez.: ***A. disticha*** Maguire & Wurd. R: AMBG 79(1992)874

Achlys DC. Berberidaceae (II 2; Podophyllaceae). 3 Jap., W N Am. R: T 16 (1967) 308. Perianth aborts early; unstable arr. of organ primordia leads to irreg. floral phyllotaxis & unstable A nos. ***A. triphylla*** (Sm.) DC. (deerfoot, vanilla leaf, W N Am.) – local medic., cult. orn.

Achnatherum P. Beauv. (~ *Stipa*). Gramineae (X). c. 100 temp. (N Am. 28 – R: FNA 24(2007)114), Medit.; s.s. 21 As. R: Phytol. 74(1993)13. S. Am. spp. noxious weeds in Tasmania. ***A. brandisii*** (Mez) Z.L. Wu (*S. b.*, Himal.) – toxic (cyanogenic glycosides)

Achnophora F. Muell. Compositae (Ast.-Hin.). 1 S Aus. (Kangaroo Is.): ***A. tatei*** F. Muell.

Achnopogon Maguire, Steyerm. & Wurd. Compositae (Stifft.). 2 Venez., Guyana. Shrubs

achocha *Cyclanthera pedata*

Achoriphragma Soják = *Parrya*

Achradelpha Cook = *Pouteria*

Achradotypus Baill. = *Pycnandra*

Achras L. = *Manilkara*

Achrouteria Eyma = *Chrysophyllum*

Achuaria Gereau = *Raputia*

Achudemia Blume = *Pilea*

Achyrachaena Schauer. Compositae (Mad.-Mad.). 1 NW Am.: ***A. mollis*** Schauer – 'everlasting', pappus scales silvery. R: FNA 21(2006)258

Achyranthes L. Amaranthaceae (I 2). 8–10 OW trop. & subtrop. (Eur. 1). ***A. arborescens*** R. Br. (Norfolk Is.) – 4-merous fls like *Nototrichium* with ***A. mangarevica*** Suesseng. (Mangariva Is., Polynesia) – poss. extinct, only tree-like spp. in genus; ***A. aspera*** L. – lvs ed. (Java), ash source of salt, branches used as toothbrushes & local medic. (Arabia) as cardiac depressant & vasodilator; ***A. bidentata*** Blume (trop. OW) – medic., cult. China

Achyrobaccharis Schultz-Bip. ex Walp. = *Baccharis*

Achyrocalyx Benoist. Acanthaceae (III 1). 4 Madag.

Achyrocline (Less.) DC. Compositae (Gnap.-Gnap.). c. 35 S Am. R: OB 104(1991)148. Afr. spp. referred to *Helichrysum*. Some contraceptives (Urug.)

Achyronychia Torrey & A. Gray. Caryophyllaceae (I 2). 1 SW US, Mex.: ***A. cooperi*** Torrey & A. Gray. R: FNA 5(2005)46

Achyropappus Kunth (~ *Bahia*). Compositae (Bah.). 1-(?2) Mex.: ***A. anthemoides*** Kunth

Achyropsis (Moq.) Hook.f. Amaranthaceae (I 2). 6 trop. & S Afr. R: NS 16(1960)100
Achyroseris Schultz-Bip. = *Scorzonera*
Achyrospermum Blume. Labiatae (VI). c. 25 OW trop.
Achyrothalamus O. Hoffm. = *Erythrocephalum*
Aciachne Benth. (~ *Ortachne*). Gramineae (X). 3 high Andes (N Arg. to Costa Rica). R: NJB 7(1987)667
Acianthella D.L. Jones & M.A. Clem. = *Acianthus*
Acianthera Scheidw. (~ *Pleurothallis*). Orchidaceae (V 13c). c. 120 trop. Am. esp. Braz. R: Lindleyana 16(2001)241
Acianthopsis Szlach. = *Acianthus*
Acianthus R. Br. Orchidaceae (IV 3a). 20 NG, Aus., NZ, New Caled. (esp.). R: Allertonia 7(1995)146. Mainly fly-poll. (2 selfed), cult. orn.
Acicarpha Juss. Calyceraceae. 3 trop. Am.
Acidanthera Hochst. = *Gladiolus*
Acidocroton Griseb. Euphorbiaceae (Crot.-Crot.). 11 Carib. (Cuba 7, Hispaniola 3, Jamaica 1)
Acidonia L. Johnson & B. Briggs (~ *Persoonia*). Proteaceae (III 2). 1 SW Aus.: ***A. microcarpa*** (R. Br.) L. Johnson & B. Briggs
Acidosasa B.M. Yang (~ *Sasa*). Gramineae (IV). c. 11 S China (10, endemic), Vietnam (1). R: APS 29(1991)517. Bamboos; heterogeneous?
Acidoton Sw. Euphorbiaceae (Acal.-Pluk.-Trag.). 8 Jamaica, Hispaniola
Acilepidopsis H. Robinson (~ *Vernonia*). Compositae (Vern.-Mesa.) 1 E S Am.: ***A. echitifolia*** (DC.) H. Robinson
Acilepis D. Don (~ *Vernonia*). Compositae (Vern.-Centrap.). c. 10 S & SE As.
Acineta Lindl. Orchidaceae (V 12i). 15 trop. Am. Cult. orn. ***A. chrysantha*** (Morren) Lindl. & Paxton (C Am.) – fragrance 45% cineole, attracting male euglossine bee-poll.
Acinopetala Luer = *Masdevallia*
Acinos Mill. = *Clinopodium*
Acioa Aubl. Chrysobalanaceae (3). 6 N S Am. with ed. oily seeds. R: FOW 10(2003)1. ***A. edulis*** Prance (castanha de cutia) – imp. local fr., long known as oil seed coll. in flooded Amazon forests not described until 1975; ***A. longipendula*** (Pilg.) Sothers & Prance (*Couepia l.*, Braz.) – cv. cult. nr Manaus for ed. cotyledons & oil. For Afr. spp. see *Dactyladenia*
Acion Briggs & L. Johnson = *Chordifex*
Aciotis D. Don. Melastomataceae. Incl. *Spennera* 13 trop. Am. R: SBM 62(2002)18. Berries & capsules
Aciphylla Forst. & G. Forst. Umbelliferae (III 8). 39 NZ (37), Aus. (2). R: TRSNZ 84(1955)1. Spiny rock-pls (cult.), dioec., spines poss. deterred moas from grazing. Many hybrids, some with spp. of *Anisotome* & *Gingidia*. ***A. squarrosa*** Forst. & G. Forst. (NZ) – bayonet plant, speargrass
Acis Salisb. (~ *Leucojum*). Amaryllidaceae. 9 Medit. R: PSE 246(2004)240. Cult. orn. incl. ***A. autumnalis*** (L.) Sweet (*L. a.*, autumn snowflake)
Acisanthera P. Browne. Melastomataceae. c. 20 trop. Am.
Ackama A. Cunn. (~ *Caldcluvia*). Cunoniaceae (VI). 2 Aus., 2 NZ
ackee = akee
Ackermania Dodson & Escobar = *Benzingia* (but see Orquideologia 18(1993)211)
Acleisanthes A. Gray. Nyctaginaceae (6). Incl. *Selinocarpus*, 16 SW N Am., Somalia (1)! R (s.s.): Novon 12(2002)58
Aclisia E. Meyer ex C. Presl = *Pollia*
Acmadenia Bartling & H.L. Wendl. Rutaceae (I 4). 32 SW to E Cape. R: JSAB 48(1982)169
Acmanthera (A. Juss.) Griseb. Malpighiaceae. 7 S Am. R: CUMH 11,2(1975)41, 17(1990)39
Acmella Rich. ex Pers. (~ *Spilanthes*). Compositae (Helia.-Spil.). 30 trop. with some (now) pantrop. weeds esp. ***A. uliginosa*** (Sw.) Cass. R: SBM 8(1985). ***A. oleracea*** (L.) R.K. Jansen (*Spilanthes o.*, Braz. or Pará cress, jambú) – salad veg. though spilanthol produces tingling of tongue, local medic., & in tacacá soup to bring on a cooling sweat (Braz.), cultigen perhaps derived from *A. alba* (L'Hérit.) R.K. Jansen (C Peru); ***A. oppositifolia*** (Lam.) R.K. Jansen – ed. Chaco Indians, cult. orn.
Acmena DC. = *Syzygium*
Acmenosperma Kausel = *Syzygium*
Acmispon Raf. (~ *Lotus*). Leguminosae (III 22). Incl. *Ottleya*, *Syrmatium*, 35 N Am. (32), Chile (1)

Acmopyle Pilg. Podocarpaceae. 2 New Caled., Fiji. R: Phytol.M 7(1984)10. Mangenot's Model. Fossils of Tertiary SE Aus., Antarctic similar to ***A. sahniana*** Buchholz & N. Gray (drautabua, Fiji) – 9 small pops in mts of Viti Levu

Acmopyleaceae Melikian & A. Bobrov = Podocarpaceae

Acnistus Schott = *Iochroma*

acoccha *Cyclanthera pedata*

Acoelorrhaphe H. Wendl. Palmae (III 4b). 1 C Am.: ***A. wrightii*** (Griseb. & H. Wendl.) Becc. – cult. orn. R: J. Dransf. & al., *Genera Palmarum* (2008)272

Acokanthera G. Don. (~ *Carissa*). Apocynaceae (I h). 5 Arabia, trop. E & S Afr. R: KB 37(1982)41. Arrow poisons & drugs – cardiac glycosides, incl. ouabain with effect of digitalin. Wood extract mixed with *Euphorbia* latex & acacia gum & applied to arrow. Ouabain can be absorbed through skin; death can follow 20 mins after injection into bloodstream, effected by murderers coating prickly fr. of *Tribulus terrestris* L. with extract & leaving these in path of barefoot victims. ***A. oppositifolia*** (Lam.) Codd (*A. venenata*) – ordeal-poison (wood up to 1.1% ouabain)

acom *Dioscorea bulbifera*

Acomastylis Greene = *Geum* (but see FRB 72(1933)78)

Acomis F. Muell. Compositae (Gnap.-Ang.). 3 Aus. R: OB 104(1991)123

Aconisia J. Grande (~ *Panicum*). Gramineae (XXIII 3). 1 Panama: *A. grandis* (Hitchc. & Chase) J. Grande. R: Phytoneuron 2014–22:2

aconite *Aconitum* spp.; **winter a.** *Eranthis hyemalis*

Aconitella Spach = *Delphinium*

Aconitum Tourn. ex L. Ranunculaceae (I 4). Excl. *Gymnaconitum*, c. 100 N temp. (Eur. 7). Diterpene alks, poisonous (though ***A. bisma*** (Buch.-Ham.) Rapaics (Himal.) not, tubers med.) & drug source in med. (zhi fu zi; psychoactive alks in tubers of *A. ferox* etc.) inc. comm. cough mixtures; dried lvs with gunpowder prod. toxic gas (anc. China), insect deterrent in *Daphne* paper in anc. Tibet. Cult. orn. (R: GH 6(1945)463; aconites, monkshood). ***A. ferox*** Wall. ex Ser. (N Ind., med.); ***A. lycoctonum*** L. (badger's bane, wolfsbane, Eur., N Afr.) – cult. orn., fls purple-lilac, form. wolf poison (lambs treated with it); ***A. napellus*** L. (W & C Eur.) – source of aconitine for heart treatment, form. administered to criminals, fly control since before 1240; ***A. volubile*** Pallas ex Koelle (E As.) – twiner (in cult. '*A. volubile*' = other spp. e.g. ***A. hemsleyanum*** Pritzel (C & W China))

Aconogonon (Meissn.) Reichb. = *Koenigia*

Aconogonum Reichb. = *Koenigia*

Acopanea Steyerm. = *Bonnetia*

Acoraceae Martinov. Magnoliidae – Acorales. 1/2 OW & N Am. R: FM I,20(2012)1. Marsh or emergent aquatic rhiz. perenn. herbs with aerenchyma, aromatic; raphides 0. Lvs linr., distich.; venation parallel. Infl. a cylindric spadix-like term. spike subtended by erect leaf-like bract & borne on leaf-like peduncle with 2 sep. vasc. systems. Fls herm., bracts 0; P 3 + 3 free, persistent; A 3 + 3 with free flattened filaments & horseshoe-shaped anthers; G (2, 3), each loc. with several apical, orthotropous ovules, both integuments with trichomes. Berry with 1–5(-9) seeds with perisperm & abundant endosperm

Genus: *Acorus*

Acoridium Nees & Meyen = *Dendrochilum*

acorn single-seeded fr. of *Quercus* spp.

Acoroides Sol. ex Kite = *Xanthorrhoea*

Acorus L. Acoraceae. 2 OW & N Am. wetlands. R: FM I,20(2012)10. Rhiz. sympodial, scented. Lvs iris- or grass-like ('gladiolus' of Med. GB). ***A. calamus*** L. (calamus, sweet flag, N temp. & Ind. [?orig.] to NG) – rhiz. used med. (esp. modern Chinese herbalism; since Hippocrates (460–377 BC; also in Tutankhamun's tomb), form. for toothache, tonic & in dysentery, also spread on hall & church floors ('rushes') & in the 'oil of holy ointment' for anointing altars & sacred vessels in Old Testament ('sweet calamus' of Exodus XXX:25), hung up at night in Sumatra to keep evil spirits from children, initiation cerem. for NG men, effective insecticide (di- & tri-terpenes) & still imp. flavouring in Continental eaux-de-vie. Pls in Eur. (& temp. Ind.) all triploid (diploids in As., both in Canada, tetraploids in trop. S & temp. E As.), clonal intr. (Russia C11, Poland C13, Hungary 1517 & distrib. Clusius (Holland) 1574) reaching GB from Lyon with Gerard & natur. c. 1668; ***A. gramineus*** Aiton (E As.) comm. medic. (stomache-ache) Vietnam

Acosmium Schott. Leguminosae (III 11). Excl. *Guianodendron*, *Leptolobium*, 3 trop. Am. R: NRBGE 29(1969)349

Acosta Adans. = *Centaurea*

Acostaea Schltr. = *Specklinia* (but see MSB 15(1986)15, 24(1987)1)

Acostia Swallen. Gramineae (XXIII 3). 1 Ecuador: ***A. gracilis*** Swallen

Acourtia D. Don. Compositae (Mut.-Nass.). c. 80 S US (5) to C Am., WI. R: Phytol. 27(1973)228,38(1978)456. Herbs. ***A. microcephala*** DC. (Calif.) – cult. orn.

Acrachne Wight & Arn. ex Chiov. Gramineae (XXIX 5). 3 OW trop.

Acradenia Kipp. Rutaceae (I 1). 2 E & SE Aus. R: FA 26(2013)54. Cult. orn.

Acrandra O. Berg = *Campomanesia*

Acranthera Arn. ex Meissn. Rubiaceae (IV?). c. 40 Indomal., esp. Borneo. R: JAA 28(1947)261. Some rheophytes

Acreugenia Kausel = *Myrcianthes*

Acridocarpus Guillem. & Perr. Malpighiaceae. 29 trop. Afr. (23), Madag. (4, 1 ext. Masc. 1), Arabia & Ind. (1), New Caled. (1). lvs in spirals; stip. 0

Acriopsis Reinw. ex Blume. Orchidaceae (V 12a). 9 SE As. to Solomon Is. R: OM 1(1986)1. Epiphytes oft. assoc. with ants' nests. ***A. liliifolia*** (J. Koenig) Ormerod & Seidenf. (*A. javanica*, Sumatra to NG) – infusion used for fever, cult. orn.

Acrisione R. Nordenstam (~ *Senecio*). Compositae (Sen.-Tuss.). 2 C Chile. R: BJ 107(1985)581. Trees & shrubs

Acritochaete Pilg. Gramineae (XXIV). 1 trop. Afr. mts: ***A. volkensii*** Pilg.

Acritopappus R. King & H. Robinson (~ *Eupatorium*). Compositae (Eup.-Gyp.). 17 E Braz. R: MSBMBG 22(1987)138. Some pachycaul treelets

Acrobotrys K. Schum. & K. Krause. Rubiaceae (I 5). 1 Colombia: ***A. discolor*** K. Schum. & Krause

Acrocarpus Wight ex Arn. Leguminosae (I 4). 1 Indomal.: ***A. fraxinifolius*** Wight ex Arn. (mt. rain forests Ind., Myanmar to Java) – imp. timber (hard, brown) used esp. for tea-chests (Darjeeling), promising plantation sp. (C Am.). R: Blumea 38(1994)314

Acrocephalus Benth. = *Platostoma*. See also *Haumaniastrum*

Acroceras Stapf (~ *Panicum*). Gramineae (XXIV 2). 19 OW trop. esp. Madag., 5 Am. ***A. macrum*** Stapf (Nile grass, SE Afr.) – cult. pasture grass, highveld, Transvaal

Acrochaene Lindl. (~ *Monomeria*). Orchidaceae (V 15). 1 Himal. to Thailand: ***A. punctata*** Lindl.

Acrochaete Peter = *Setaria*

Acroclinium A. Gray = *Rhodanthe*

Acrocoelium Baill. = *Leptaulus*

Acrocomia Mart. Palmae (V 8b). Incl. *Gastrococos*, c. 3(–34) trop. Am. Monoec. prickly, fast-growing cult. orn., seeds s.t. disp. by cattle. ***A. aculeata*** (Jacq.) Mart. – human disp. to C Am. from S, in archaeological sites c. 9500 BP incl. '*A. mexicana* Karw. ex Mart.' (coyoli palm – ed. fr. (cooked), oil source, palm-sap wine & '*A. totai* Mart.' (gru-gru, mbocaya, Paraguay palm, NE Arg., Paraguay) – imp. source of palm kernel oil, locally used for soap, palmhearts; ***A. crispa*** (Kunth) Becc. (*G. c.*, Cuba) – swollen (belly) trunk stores water

Acrocoryne Turcz. (~*Metastelma*). Apocynaceae (V c). 1 Carib.: ***A. caribaea*** Turcz.

Acrodiclidium Nees & Mart. = *Licaria*

Acrodon N.E. Br. Aizoaceae (V). 5 SW Cape. R: H. E. K. Hartmann, *Ill. Handb. Succ. Pls*, Aiz. A–E(2002)23. Cult. orn.

Acroglochin Schrad. ex Schultes. Amaranthaceae (Chenopodiaceae I 1). 2 C & E As., Himal. Fr. mass prickly with axes not terminating in fls

Acrolophia Pfitzer (~ *Eulophia*). Orchidaceae (V 12b). 7 SW Cape to Zululand. R: Strel. 9(2000)156

Acronema Falc. ex Edgew. Umbelliferae (III 8). 25 Sino-himal. Embryo monocotyledonous

Acronia C. Presl = *Pleurothallis*

Acronychia Forst. & G. Forst. Rutaceae (I 2). 48 Indomal., W Pac. R: JAA 55(1974)469. Alks. Some medic. pls SE As., used for caulking boats; cult. orn. ***A. acidula*** F. Muell. (lemon aspen) – fr. extract in comm. skin cleansers; ***A. wilcoxiana*** (F. Muell.) T. Hartley (mushy-berry, SE Aus.) – fr. eaten by settlers

Acropanea Steyerm. = *Bonnetia*

Acropelta Nakai = *Polystichum*

Acrophorus C. Presl = *Dryopteris*

Acrophyllum Benth. Cunoniaceae (IV). 1 lower Blue Mts, NSW: ***A. australe*** (A. Cunn.) Hoogl. – cult. orn.

Acropogon Schltr. (~ *Sterculia*). Malvaceae (Sterc.; Sterculiaceae). 25 New Caled. R: BMNHN 4, 8(1986)357, 10(1988)93. Oft. remarkable unbranched pachycauls

Acroptilon Cass. = *Rhaponticum*

Acrorchis Dressler. Orchidaceae (V 13b). 1 C Am.: ***A. roseola*** Dressler

Acrorumohra (H. Ito) H. Ito (~ *Dryopteris*). Dryopteridaceae (I). 7 trop. As.

Acrosanthes Ecklon & Zeyher. Aizoaceae (I). 5 W Cape. R: H. E. K. Hartmann, *Ill. Handb. Succ. Pls*, Aiz. A–E(2002)25

Acrosorus Copel. Polypodiaceae (V; Grammitidaceae). 9 trop. As. to Samoa

Acrostemon Klotzsch = *Erica*

Acrostichaceae Mett. ex Frank = Pteridaceae

Acrostichum L. Pteridaceae (II). 3 pantrop. Mangrove. Marsh ferns. Upper pinnae reduced & bearing sporangia ('acrostichoid' condition achieved in unrelated fern groups). ***A. aureum*** L. (pantrop.) – found nr hot springs at 550 m in Zimbabwe (form. coast), young lvs ed. (Mal.), old ones used for thatch (Vietnam)

Acrosynanthus Urb. = *Remijia*

Acrothamnus Quinn (~ *Leucopogon*). Ericaceae (VII 7). 7 SW Pacific. R: AusSB 18(2005)451

Acrotome Benth. ex Endl. (~ *Leonotis*). Labiatae (VI). 6 trop. & S Afr.

Acrotrema Jack. Dilleniaceae (IV). 9 Indomal. esp. Sri Lanka (7). Herbs s.t. with finely divided lvs. ***A. costatum*** Jack (Myanmar to N Sumatra) used in postnatal medic. in W Mal.

Acrotriche R. Br. Ericaceae (VII 7; Epacridaceae). 15 temp. Aus. R: Telopea 1(1980)421. Cult. orn. ***A. depressa*** R. Br. (native currant) – used in chocolates in S Aus.; ***A. serrulata*** R. Br. (Tas.) – prob. poll. nocturnal mammals

Acrymia Prain. Labiatae (Cymarioideae). 1 Selangor (Malay Pen., endangered): ***A. ajugiflora*** Prain. R: KB 1908: 114

Acsmithia Hoogl. = *Spiraeanthemum*

Actaea L. Ranunculaceae (I 2). (Incl. *Cimicifuga*, *Souliea*) 27 N temp. (Eur. 3). R: T 47(1998)613. ***A. cimicifuga*** L. (*C. foetida*, bugbane, Siberia, E As.) – used to deter bugs in Siberia, med. in China, Eur. pls with this name usu. ***A. europaea*** (Schipcz.) J. Compton; ***A. racemosa*** L. (*C. r.*, black cohosh or snakeroot, E N Am.) – dried rhiz. medic. (efficacious in treatment of menopausal symptoms), cult. pls with this name usu. ***A. simplex*** (DC.) Prantl (E As.); ***A. rubra*** (Aiton) Willd. (N temp.) – nos & position of C variable as 10 primordia can develop into C or A (homoeosis); ***A. spicata*** L. (baneberry, Herb Christopher, black cohosh, Euras.) – v. poisonous, medic., Radix Christopherianae form. used against skin disease, asthma, rheumatism & esp. St Vitus Dance

Actephila Blume. Phyllanthaceae (Por.; Euphorbiaceae s.l.). 31 Indomal., China, N Aus. R: KB 63(2008)46

Actinanthella Balle. Loranthaceae (5 7). 2 SE & S trop. Afr. R: R. Polhill & D. Wiens, *Mistletoes of Afr*. (1998)99

Actinanthus Ehrenb. = *Oenanthe*

Actinea A. Juss. = *Helenium*

Actinidia Lindl. Actinidiaceae. 60 Indomal., E As. R: JAA 33(1952)1. Climbers. Monoterpenoid alk., actinidine, physiologically active in cats attracted to pls, beta-phenylethyl alc. inducing them to salivate being a widely used ingredient in scents long known to be attractive to them. Cult. E As. spp. for fr. incl. ***A. arguta*** (Sieb. & Zucc.) Miq. with glabrous small fr. (kiwiberry eaten whole) & esp. hairy-fruited ***A. deliciosa*** (A. Chev.) Liang & A.R. Ferg. ('*A. chinensis*', Chinese gooseberry, kiwi-fr. [name coined NZ as marketing 'Chinese' after 1949 problematic], yangtao) – app. bisex., pl. actually dioec. though some bisex. clones in cult., hexaploid with diploid *A. chinensis* Planch. one parent, all comm. orchards derived from NZ stocks orig. planted 1904 notably large-fruited '**Hayward**' (95%), 1 fr. more vitamin C than daily requirement, helping prevent constipation on long flights, also used in liqueurs & 'champagne' (form. for wine in China), seed oil allegedly 'collagen-boosting'. Cult. orn. E As. spp. incl. ***A. kolomikta*** (Rupr. & Maxim.) Maxim. with naturally varieg. lvs in males & ***A. polygama*** (Sieb. & Zucc.) Maxim., both attractive to cats

Actinidiaceae Engl. & Gilg. Magnoliidae – Ericales. 3/360 trop. & warm (not Afr.) & As. mts. Trees, shrubs & lianes with raphides in parenchyma; indumentum of scales & hairs. Lvs simple, in spirals, usu. serrate to dentate; stip. minute or 0. Infl. usu. cymose, axillary or on old wood, s.t. fls solit. Fls bisex. or unisexual, hypogynous; K 5, rarely 3, 4, 6–9, imbr., persistent in fr., C 5, rarely 3, 4, 6–9, imbr., A ∞ or as few as 10, oft. in

5 clusters opp. petals, anthers s.t. dehiscing by pores, G (3–30 or more) with as many locules as carpels with ∞ axile unitegmic anatr. ovules. Fr. a berry or rarely a loculicidal capsule; seeds with copious oily or proteinaceous endosperm; embryo large, usu. straight
Genera: *Actinidia, Clematoclethra, Saurauia*
Seed evidence suggesting affinity with Theaceae confirmed by DNA findings

Actiniopteridaceae Pic.-Serm. = Pteridaceae

Actiniopteris Link. Pteridaceae (III). 4 Afr. (most) to Sri Lanka. In xeric conditions, esp. ***A. radiata*** (Sw.) Link. ***A. semiflabellata*** Pic.-Serm. (E Afr. to Mauritius) – local medic. In Yemen, Socotra

Actinobole Fenzl ex Endl. Compositae (Gnap.-Ang.). 4 Aus. R: OD 104(1991)127

Actinocarya Benth. = *Microula*

Actinocephalus (Körn.) Sano = *Paepalanthus* (but see R: T 53 (2004)99, Novon 22(2012)282)

Actinocheita F. Barkley. Anacardiaceae (I). 1 Mex.: ***A. filicina*** (DC.) F. Barkley

Actinocladum McClure ex Soderstrom. Gramineae (V 2). 1 Braz. savanna: ***A. verticillatum*** (Nees) Soderstrom, regenerating from 2 ground-level nodes after fire

Actinodaphne Nees. Lauraceae (II). 100 Indomal., E As. Dioec. ***A. angustifolia*** (Blume) Nees (trop. As.) – imp. honey-source; ***A. hookeri*** Meissn. (trop. As.) – seeds with 75% oil (90% lauric acid)

Actinodium Schauer (~ *Darwinia*). Myrtaceae (II 15). 2 SW Aus. Fls resembling Compositae, pollen like Santalaceae

Actinokentia Dammer. Palmae (V 14a). 2 New Caled. R: Allertonia 3(1984)320

Actinolema Fenzl. Umbelliferae (II 1). 2 E Medit. R: Fl. Iraq 5,2(2013)134

Actinomeris Nutt. = *Verbesina*

Actinophloeus (Becc.) Becc. = *Ptychosperma*

Actinorhytis H. Wendl. & Drude. Palmae (V 14a). 1 Papuasia: ***A. calapparia*** (Blume) R. Scheffer widely cult. in SE As. & W Mal. for magic, seeds chewed like *betel* in Java, wood for pillars, floorboards & doors in NG

Actinoschoenus Benth. (~ *Fimbristylis*). Cyperaceae (II 7). 4 Madag., Sri Lanka, China

Actinoscirpus (Ohwi) R. Haines & Lye. Cyperaceae (II 2). 1 Indomal.: ***A. grossus*** (L.f.) Goetghebeur & D. Simpson (*Scirpus g.*) – used for sleeping-mats, bags etc., tubers ground for flour in Ind., locally medic.

Actinoseris (Endl.) Cabrera = *Richterago*

Actinospermum Elliott = *Balduina*

Actinostachys Wall. (~ *Schizaea*). Schizaeaceae. 16 trop. Buds at petiole-bases leading to thickets

Actinostemma Griff. Cucurbitaceae (IV). Incl. *Bolbostemma*, 3 Ind. to Jap., Russia: ***A. tenerum*** Griff. (*A. lobatum*) – cooking oil

Actinostemon Mart. ex Klotzsch (~ *Gymnanthes*). Euphorbiaceae (Euph.-Hipp.). 15 WI, S Am.

Actinostrobus Miq. = *Callitris*. (but see A. Farjon (2005) *Monogr. Cupressaceae*: 487)

Actinotaceae Konstantinova & Melikian = Umbelliferae

Actinotinus Oliv. = *Viburnum*. The sp. *A. sinensis* Oliv. was based on a trick, the infl. of a *Viburnum* having been inserted in the term. bud of an *Aesculus* (cf. *Papilionopsis*, *Stalagmitis*)

Actinotus Labill. Umbelliferae (Azor.). 18 Aus. (17 endemic), NZ. Heads congested & surrounded by bracts suggesting a Composita. ***A. helianthi*** Labill. (flannelflower, Aus.) – cult. orn., prot. in Aus.

Actites Lander = *Sonchus* (but see Telopea 1(1976)129)

Acunaeanthus Borh. & al. (~ *Mazaea*). Rubiaceae (I 5). 1 Cuba: ***A. tinifolius*** (Griseb.) Borh. R: Britt. 51(1999)226

Acunniana Orchard (~ *Wollastonia*). Compositae (Helia.-Ecl.). 1 N Aus.: ***A. procumbens*** (DC.) Orchard. R: Nuytsia 23(2013)430

Acustelma (Baill.) Venter = *Pentopetia*

Acuston Raf. (~ *Fibigia*). Cruciferae (2). 1 E Medit.: ***A. perenne*** (Mill.) Mabb. & Al-Shehbaz (*A. lunarioides*, *F. l.*)

Acyntha Medik. = *Dracaena*

Acystopteris Nakai (~ *Cystopteris*). Cystopteridaceae. 3 Indomal. to Jap.

Ada Lindl. = *Brassia*

Adam and Eve *Aplectrum hyemale*; **A.'s flannel** *Verbascum thapsus*; **A.'s needle** *Yucca* spp.

Adamanthus Szlach. = *Maxillaria*

Adamantinia Van den Berg & G. Conç. Orchidaceae (V 13b). 1 Bahia: ***A. miltonioides*** Van den Berg & G. Conç.

Adansonia L. Malvaceae (Bomb.; Bombacaceae). 8 Afr. (1), Madag. (6), NW Aus. (1). R: AMBG 82(1995)445. Trees with charac. swollen trunks; fls open at dusk, some in 30 secs. Madag. spp. seeds water disp.(?), 'Longitubae' largely hawkmoth-poll., but also visited by birds & lemurs, these the main poll. of ***A. grandidieri*** Baill. (S Madag.) 'Brevitubae'; *A. gregorii* interm. in fl. structure & poll. unknown but visited by honeyeaters & (?) marsupials. ***A. digitata*** L. (*A. kilima*, baobab, Judas's bag, monkey bread, trop. Afr.) – lives up to 1235 yrs (Namibia), the fls of equal nos of right & left-twisting convolute C, though right on average have more A, poll. by bats but also visited by insects & bushbabies; fr. distrib. by mammals esp. baboons & elephants, seeds with enhanced germ. thereafter; planted in sacred places & as groves in E Afr. coastal forest; bark used for cloth, inner bark for rope, fr. & seeds for fuel, dried fr. provides drink rich in citric & tartaric acids, fresh fr. eaten by baboons, many med. & other uses, even as rooms (prison-cells, lavatories etc. – see KB 37(1982)173). ***A. gregorii*** F. Muell. (*A. gibbosa*, gourd-tree, Aus.) – ed. seeds, trunk a water source (up to 300 litres per trunk) for birds & Aborigines, 'tubers' ed. pickled

Adarianta Knoche = *Pimpinella*

adder's meat *Stellaria holostea;* **a.'s mouth** *Malaxis;* **a.'s tongue** *Ophioglossum* spp.

Addisonia Rusby = *Helogyne*

Adelaster Lindl. ex Veitch = *Fittonia*

Adelia L. Euphorbiaceae (Acal.-Adel.). 9 trop. & warm Am. R: SB 22(2007)585

Adelinia Cohen (~ *Cynoglossum*). Boraginaceae (3.8.4?). 1 NW N Am.: ***A. grandis*** (Lehm.) Cohen. R: SB 40(2015)617

Adeliopsis Benth. = *Hypserpa*

Adelobotrys DC. Melastomataceae. 31 trop. Am. Trees (8, poss. a distinct genus) & lianes (23)

Adelocaryum Brand (*Paracaryopsis;* ~*Cynoglossum*). Boraginaceae (3.8.3?). 4 Ind., Arabia. R: EJB 67(2010)145

Adelodypsis Becc. = *Dypsis*

Adelonema Schott (~ *Homalomena*). Araceae (VII 6). 16 trop. Am. R: SB 41(2016)37. Aniseed smell when crushed

Adelonenga Hook.f. = *Hydriastele*

Adelopetalum Fitzg. = *Bulbophyllum*

Adelosa Blume = *Rotheca*

Adelostemma Hook.f. = *Cynanchum*

Adelostigma Steetz. Compositae (Inul.-Pluch.). 1 trop. Afr.: ***A. senegalensis*** Benth. – known only from lost type specimen!

Adelphacme K.L. Gibbons & al. (~ *Mitreola*). Loganiaceae. 1 W Aus.: ***A. minima*** (Conn) K.L. Gibbons & al. (*M. m.*). R: Telopea 15(2013)38

Adelphia W.R. Anderson (~ *Mascagnia*). Malpighiaceae. 4 trop. Am. R: Novon 16(2006)170

Adenacanthus Nees = *Strobilanthes*

Adenandra Willd. Rutaceae (I 4). 18 SW Cape. R: OB 32(1972). ***A. fragrans*** (Sims) Roemer & Schultes – cult. orn., aromatic lvs used as tea in Cape

Adenanthe Maguire & al. = *Tyleria*

Adenanthellum R. Nordenstam. Compositae (Anth.-Cot.). 1 S Afr.: ***A. osmitoides*** (Harv.) R. Nordenstam. R: BBMNHB 23(1993)145

Adenanthemum R. Nordenstam = *praec.*

Adenanthera Royen ex L. Leguminosae (II 1). 13 trop. As., Pacif. R: NJB 12(1992)85. Seeds hard, bright red or red & black. Timber good. ***A. pavonina*** L. (coral wood, red sandalwood, SE As. to Aus., invasive Seychelles, SE US) – street tree, red dye from heartwood used for forehead spot of Brahmins, seeds used as beads (Circassian seeds) & weights by goldsmiths (A. = goldsmith in Arabic). NB. Ganda system of Ind. based on *Abrus precatorius* (q.v.), but derived from a system with double that seed wt. & *Adenanthera* seeds are twice as heavy as *Abrus* seeds

Adenanthos Labill. Proteaceae IV 4). 33 S & W Aus. R: Brunonia 1(1978)303. Cult. orn. 'woolly bush'

Adenarake Maguire & Wurd. Ochnaceae (III). 2 Venez. R: BJ 113(1991)183

Adenaria Kunth. Lythraceae. 1 Mex. to Arg.: ***A. floribunda*** Kunth

Adenia Forssk. Passifloraceae. 97 OW trop., esp. Afr. R: MLW 71–18(1971). Many Afr. spp. with thorns or swollen stems (believed to have evolved 4 times; tubers arose c. 8 times)

some med. ***A. ellenbeckii*** Engl. (E Afr.) – fr. ed.; ***A. hondala*** (Gaertn.) de Wilde (*A. palmata*, Sri Lanka, S Ind.) – inner cambium prod. medullary steles while outer cambium continues to divide

Adenium Roemer & Schultes. Apocynaceae (IIa). 5 (s.t. treated as subspp.) trop. & subtrop. Afr., Arabia. R: MLW 80–12(1980). Thick-stemmed xerophytes with lvs in spirals, cardiac glycosides effective as fish- & other poisons & in ordeals. ***A. multiflorum*** Klotzsch (impala lily, trop. Afr.) – fish stupifier, Transvaal; ***A. obesum*** (Forssk.) Roemer & Schultes (desert rose, E Afr. to S Arabia) – arrow & ordeal poison, cult. orn. 'specially prot.' in Zimbabwe, cult. in Chinese gardens in SE As. for good luck

Adenoa Arbo. Passifloraceae (Turneraceae). 1 E Cuba: ***A. cubensis*** (Britton & P. Wilson) Arbo, on serpentine

Adenocalymma Mart. ex Meissn. Bignoniaceae (3). (Incl. *Memora*) 82 trop. Am. R: AMBG 99(2014)382. Big lianes with 4-ribbed stems due to 'anomalous' vascular structure

Adenocarpus DC. Leguminosae (III 9). 15 trop. Afr. mts, Canaries, Med. (Eur. 4). R: BSB II 41(1967)67, OBPC 5(1989)69. Alks. Cult. orn.

Adenocaulon Hook. Compositae (Mut.-Adenoc.). 5+ S As., N Am., Guatemala, Arg. R: Candollea 45(1990)514

Adenochilus Hook.f. Orchidaceae (IV 3b). 2 Aus., NZ

Adenochlaena Boivin ex Baill. = *Cephalocroton*

Adenocline Turcz. Euphorbiaceae (Crot.-Aden.-Aden.). 3 S Afr.

Adenocritonia R. King & H. Robinson (~ *Eupatorium*). Compositae (Eup.-Crit.). 3 Mex., Guatemala, Jamaica. R: Phytol. 71(1991)178

Adenodaphne S. Moore = *Litsea*

Adenoderris J. Sm. = *Polystichum*

Adenodolichos Harms. Leguminosae (III 18). 15 trop. Afr. Local veg.

Adenoglossa R. Nordenstam. Compositae (Anth.-Anth.). 1 NW Cape: ***A. decurrens*** (Hutch.) R. Nordenstam. R: Strel. 10(2000)120

Adenogramma Reichb. Molluginaceae. 11 Afr. R: JSAB 21(1995)84. Indehiscent 1-seeded fr.

Adenogrammaceae Nakai = Molluginaceae

Adenogyne Klotzsch = ? *Sebastiania*

Adenolisianthus (Progel) Gilg = *Irlbachia*

Adenolobus (Benth.) Torre & Hillc. Leguminosae (I 1). 2 SW Afr.

Adenoncos Blume. Orchidaceae (V 16c). 17 Vietnam, Thailand, Mal.

Adenoon Dalz. Compositae (Vern.-Lin.). 1 Indomal.: ***A. indicum*** Dalz.

Adenopappus Benth. = *Tagetes*

Adenopeltis Bertero ex A. Juss. Euphorbiaceae (Euph.-Hipp.). 1 Chile, Peru: ***A. serrata*** (Aiton) I. M. Johnston. R: A. Radcliffe-Smith, *Gen. Euphorb.* (2001) 369

Adenophaedra (Muell. Arg.) Muell. Arg. Euphorbiaceae (Acal.-Ber.). 3 trop. S Am.

Adenophora Fischer (~ *Campanula*). Campanulaceae (I 6). 62 temp. Euras. (Eur. 2). Differs from *Campanula* only in disk at style-base. Some ed. r. & cult. orn. ***A. liliifolia*** (L.) A. DC. (*A. communis*, C Eur. to Manchuria) – cult. r.-crop (Jap.); ***A. tetraphylla*** (Thunb.) Fischer (*A. triphylla*, sha shen, China, Jap.) – tubers imp. medic. China

Adenophorus Gaudich. (~ *Grammitis*). Polypodiaceae (V; Grammitidaceae). 9 Mal. to Pacific

Adenophyllum Pers. (~ *Dyssodia*). Compositae (Tag.-Pect.). 10 SW US to C Am. R: Sida 11(1986)372

Adenoplea Radlk. = *Buddleja*

Adenoplusia Radlk. = *Buddleja*

Adenopodia C. Presl (*Pseudoentada*, ~ *Entada*). Leguminosae (II 13). c. 7 trop. Afr. & C Am. (3). R: KB 41(1986)73

Adenoporces Small = *Tetrapterys*

Adenorachis (DC.) Nieuwl. = *Aronia*

Adenorandia Vermoesen (~ *Gardenia*). Rubiaceae (II 1). 1 trop. Afr.: ***A. kalbreyeri*** (Hiern) Robbrecht & Bridson

Adenosciadium H. Wolff. Umbelliferae (III 8). 1 SW Arabia: ***A. arabicum*** H. Wolff

Adenosma R. Br. Plantaginaceae (Grat.-Stem.; Scrophulariaceae s.l.). 15 China, Indomal., Aus. Fragrant due to ess. oil (1%), decoctions for bowel complaints, rheumatism; ***A. caerulea*** R. Br. (Indomal.) – cult. Vietnam for med.; ***A. indiana*** (Lour.) Merr. (Indomal.) – intercropped with upland rice in China

Adenostachya Bremek. = *Strobilanthes*

Adenostegia Benth. = *Cordylanthus*

Adenostemma Forst. & G. Forst. Compositae (Eup.-Aden.). 24 trop. Am., Afr., with 1 pantrop. weed, ***A. lavenia*** (L.) Kuntze (*A. viscosum*), with sticky pappus suited to animal disp., also source of a blue dye, local medic. R: MSBMBG 22(1987)58

Adenostephanus Klotzsch = *Euplassa*

Adenostoma Hook. & Arn. Rosaceae (Sorb.). 2 Calif. & nearby Mex. incl. ***A. fasciculatum*** Hook. & Arn. (greasewood) – shrub char. of chaparral, local medic. R: FNA 9(2014)392

Adenostyles Cass. Compositae (Senec.-Tuss.). 3 C & S Eur. R: Willdenowia 42(2102)58

Adenothamnus Keck. Compositae (Mad.-Mad.). 1 Baja Calif.: ***A. validus*** (Brandegee) Keck

Adesmia DC. Leguminosae (III 11). 240 S Am. R: Darw. 13(1964)9, 14(1967)463

Adhatoda Mill. = *Justicia*

Adiantaceae Newman = Pteridaceae

Adiantopsis Fée (~ *Cheilanthes*). Pteridaceae (IV). 28+ trop. Am. R: T 60(2011)1265. Lvs pinnate, pedate, palmate

Adiantum Tourn. ex L. Pteridaceae (V). c. 200 cosmop. (Eur. 1) esp. trop. Am. (maidenhair ferns), some med. & cult. orn. (C.J. Goudey (1985) *M. fs in cultivation*), many Am. spp. weedy in OW oil palm & tea plantations, all Aus. spp. prot. ***A. aethiopicum*** L. (warm OW) – med. & abortifacient; ***A. capillus-veneris*** L. (cosmop.) – med. & flavouring, in hair-tonics & syrups esp. 'Sirop de Capillaire', a cure-all, cult. orn. (many cvs); ***A. pedatum*** L. (N Am., E As.) – black petioles used in Native Am. basket-weaving; ***A. raddianum*** C. Presl (*A. cuneatum*, Am., Afr.) – fertility control in Urug., cult. orn., many cvs

Adina Salisb. Rubiaceae (I 2). Incl. *Adinauclea, Haldina, Metadina, Pertusadina*, 8 E As. R: SB 39(2014)310. ***A. eurhyncha*** (Miq.) Krüger & Löfstrand (*P. e.*, W Mal.) – fenestrated trunk

Adinandra Jack. Pentaphylacaceae (3; Ternstroemiaceae). c. 90 Indomal. (Borneo 32). R: JAA 28(1947)1. Roux's Model

Adinauclea Ridsd. = *Adina*

Adinobotrys Dunn = *Callerya*

Adinocleome Iltis & Cochrane = *Cleome* (but see Novon 23(2014)51)

Adinorchis Szlach. & al. = *Chaubardia*

Adipe Raf. = *Bifrenaria*

Adipera Raf. = *Senna*

Adiscanthus Ducke. Rutaceae (I 8). 1 Amaz.: ***A. fusciflorus*** Ducke

adlay ed. forms of *Coix lacryma-jobi*

Adlumia Raf. ex DC. Papaveraceae (Fumariaceae). 1 E N Am. (***A. fungosa*** (Aiton) Britton, Sterns & Pogg., biennial cli., cult. orn., alks), 1 Korea & Manchuria. R: OB 88(1986)19

Adolphia Meissn. Rhamnaceae (2). 1 SW N Am.: ***A. infesta*** (Kunth) Meissn. – in chaparral. R: Darw. 32(1993)188

Adonidia Becc. (~ *Veitchia*). Palmae (V 14i). 2 C Mal. to NG

Adonis Dill. ex L. Ranunculaceae (II 1). 26 temp. Euras. (Eur. 10). R: ANMW 66(1963)51, Webbia 25(1971)299. Cardiac glycosides. ***A. annua*** L. (*A. autumnalis*, S Eur., SW As., pheasant's eye) – cult. orn. with scarlet petals, each with dark basal spot, form. coll. in S Eng. for Covent Garden market, London & sold as 'red morocco'; ***A. vernalis*** L. (cult. 'doubles' with C10 & 0 nectaries) & other spp. (Eur.) med. with same effect as digitalin

Adoxa L. Viburnaceae (Adoxaceae). Incl. *Tetradoxa*, 3 N temp. (Eur. 1: ***A. moschatellina*** L., moschatel, townhall clock), S to Himal., Colorado & Illinois, seeds disp. by snails). R: BBR 7,4(1987)96

Adoxaceae E. Meyer = Viburnaceae

Adrastaea DC. = *Hibbertia*

Adriana Gaudich. Euphorbiaceae (Acal.-Ric.). 2 Aus. R. Aus SB 9(1996)757. Bitterbushes. ***A. urticoides*** (A. Cunn.) P.I. Forst. (*A. tomentosa*) – form. used like tobacco

Adromischus Lem. Crassulaceae (II). 28 S Afr. R: Bothalia 12(1978) 382, 633; J. Pilbeam & al. (1998) *A.* Succ. herbs & subshrubs allied to *Cotyledon* but infls racemose & corolla-tube slender. Cult. orn. oft. with spotted lvs

Adrorhizon Hook.f. Orchidaceae (V 16a). 1 Sri Lanka: ***A. purpurascens*** Hook.f.

adrue (Carib.) *Cyperus articulatus*

adzuki bean *Vigna angularis*

Aechmanthera Nees = *Strobilanthes*

Aechmea Ruíz & Pavón. Bromeliaceae (3). Incl. *Lamprococcus, Macrochordion, Ortgiesia, Platyaechmea, Podaechmea, Pothuava*, 260 trop. Am. R: FN 14(1979)1766. Heterogeneous. Epiphytes, scape usu. well-dev. Many cult. orn. esp. room pls. esp. ***A. fasciata*** Lindl. (*Platyaechmea f.*, urn pl., Braz.)

Aedesia O. Hoffm. Compositae (Vern.-Lin.). 3 trop. Afr. R: FFG 37(1992)362

Aegialitidaceae Lincz. = Plumbaginaceae

Aegialitis R. Br. Plumbaginaceae (II). 2 Indomal. to Aus., mangroves

Aegiceras Gaertn. Primulaceae (Myrsinaceae). 1 SE As. to Pacific mangrove – ***A. corniculatum*** (L.) Blanco – Scarrone's Model, viviparous, bark & seeds sources of effective fish-poison saponin

Aegicerataceae Blume = Primulaceae (Myrsinaceae)

Aegilemma Löve = *Triticum*

Aegilonearum Löve = *Triticum*

Aegilopodes Löve = *Triticum*

Aegilops L. = *Triticum* (but see WAUP 94–7(1994)111, 139)

Aeginetia L. Orobanchaceae (Orob.). c. 3 Indomal., E As. Parasitic on r. of monocots esp. Gramineae. ***A. indica*** L. (Indomal. to Jap.) – pest of sugar-cane etc., converting sucrose to reducing sugars, used to treat diabetes in SE As.

Aegiphila Jacq. (~ *Clerodendrum*). Labiatae (III). 102 trop. Am. R: Britt. 1(1934)245. Some dioec.

Aegle Corr. Serr. Rutaceae (II 2). 3 Indomal. Thorny decid. trees, alks. ***A. marmelos*** (L.) Corr. Serr. (bael, bel tree, beli, Bengal quince, golden apple, Ind. & Myanmar) – Hindu sacred tree (pericarp pieces also used in rosaries), good timber, fls for scenting water, bark for gum, fr. rind for ess. oil (mycotoxic), fr. pulp for drinks & in treatment of dysentery & as soap, unripe fr. a yellow dye

Aeglopsis Swingle. Rutaceae (II 2). 5 trop. Afr. Cult. orn., poss. *Citrus* stock

Aegokeras Raf. (*Olymposciadium*, ~ *Seseli*). Umbelliferae (III 8). 1 Turkey: ***A. caespitosa*** (Sm.) Raf.

Aegonychon Gray (~ *Buglossoides*, *Lithospermum*). Boraginaceae (2.2). 3 Eur. R: T 63(2014)1074

Aegopodium L. Umbelliferae (III 8). 10 temp. Euras. (Eur. 1). ***A. podagraria*** L. (ground elder, goutweed, bishop weed, Eur., natur. N Am.) – intr. GB by Romans as potherb (boiled like spinach) & med. for gout, now a pernicious weed of gardens; '**Variegatum**' (inexplicably) cult. widely US

Aegopogon Humb. & Bonpl. ex Willd. = *Muhlenbergia*

Aegopordon Boiss. = *Jurinea*

Aellenia Ulbr. = *Halothamnus*

Aeluropus Trin. Gramineae (XXIX). c. 10 Med. (Eur. 2) to N China (4) & Ind. Halophytes, good fodder & hay

Aenhenrya Gopalan. Orchidaceae (IV 2d). 1 S Ind.: ***A. rotundifolia*** (Blatter) Sath. Kumar & Rasm. R: Novon 7(1997)81

Aenictophyton A.T. Lee. Leguminosae (III 14). 2 NW Aus. R: Muelleria 29(2011)184

Aenigmatanthera W.R. Anderson (~ *Mascagnia*). Malpighiaceae. 2 S Am. R: Novon 16(2006)173

Aenigmopteris Holttum (~ *Tectaria*). Tectariaceae (Dryopteridaceae). 5 Mal. R: Blumea 30(1984)3

Aeolanthus Mart. = *seq.*

Aeollanthus Mart. ex Spreng. Labiatae (VIII 3e). (Incl. *Icomum*) 45 trop. & warm Afr. Usu. succ., oft. geoxylic, some with lvs in spirals. Explosive poll. mechanism. Some ed. & cult. ***A. myrianthus*** Bak. (*A. gamwelliae*, ninde, trop. Afr.) – distilled for oil rich in geraniol, subs. for palmarosa; ***A. suaveolens*** Mart. ex Spreng. (trop. Afr.) – sold in Braz. as local medicine (macassá) for skin & eye problems

Aeoniopsis Rech.f. = *Bukiniczia*

Aeonium Webb & Berth. Crassulaceae (I 4). (Incl. *Greenovia*) 36 Macaronesia (most), Med. to Tanzania & Arabia. R: Ho-Yih Liu (1989) *Systematics of A.* Close to *Sempervivum* but usu. caulescent; many hybrids in wild. Holttum's & Scarrone's Models. Cult. orn. succ. esp. ***A. arboreum*** (L.) Webb & Berth. (Gran Canaria, early natur. Morocco etc., NZ where an aggressive coast colonist) – sap used to harden fishermen's lines; ***A. nobile*** (Praeger) Praeger (La Palma) – rosettes to 80 cm diam.

Aequatorium R. Nordenstam (~ *Ligularia*). Compositae (Sen.-Tuss.). Excl. *Nordenstamia*, 12 Peru, Bolivia, N Arg. Trees & shrubs. R: OB 44(1978)59

Aerangis Reichb.f. (~ *Angraecum*). Orchidaceae (V 16d). Incl. *Microterangis*, c. 50 trop. & S Afr. (26, 1 ext. to Sri Lanka, R: KB 34(1979)239), Madag. Cult. orn., some with long spurs e.g. ***A. ellisii*** (B.S. Williams) Schltr. (Madag.)

Aeranthes Lindl. Orchidaceae (V 16d). c. 40 Afr. (2), Masc. Cult. orn.

Aerides Lour. Orchidaceae (V 16c). 25 trop. & E As. to NG. Cult. orn.

Aeridostachya (Hook.f.) Brieger = *Eria*

× **Aeridovanda** J. Colman. Orchidaceae. *Aerides* × *Vanda*. 230+ grexes

Aerisilvaea Radcl.-Sm. = *Lingelsheimia*

Aerva Forssk. Amaranthaceae (I 2). 10 warm & trop. OW. Some ed. ***A. javanica*** (Burm.f.) Schultes (trop. As.) – fluffy infls used to stuff pillows in Middle E, source of hair-dye Socotra, local medic.; ***A. lanata*** (L.) Juss. (trop. As.) – exported from Sri Lanka for use in kidney disease

Aesandra Pierre = *Diploknema*

Aeschrion Vell. Conc. = *Picrasma*

Aeschynanthus Jack. Gesneriaceae (III 2j). c. 160 Indomal. (China 34, Malesia c. 136). Epiphytic subshrubs or lianes with fleshy lvs, mostly bird-poll. Seeds with long hairs or emergences at chalazal end. Cult. orn.

Aeschynomene L. Leguminosae (III 11). c. 175 trop. & warm. Shrubs (pith plant) with sensitive lvs & lightweight pith, that of ***A. aspera*** L. (emergent aquatic with bulk of aerenchyma being sec. xylem, trop. As.) & ***A. indica*** L. (trop. & warm OW – seeds fatal to humans & pigs) being used in sunhats (shola or sola), head-dresses for Kerala dancers, floats etc. & charcoal of latter in fireworks & gunpowder. ***A. elaphroxylon*** (Guillemin & Perrottet) Taubert (trop. Afr.) component of sudd, wood (ambatch) used for rafts & floats; ***A. hispida*** Willd. (S N Am.) – lightest wood known (specific gravity 0.044)

Aesculus L. Sapindaceae (II 2; Hippocastanaceae). 13 SE Eur. (1), Ind. & E As. (5), N Am. (7). R: Plantsman 6 (1985)230. Buckeyes, horse-chestnuts. Orig. E As., spread to Eur. & N Am. (or poss. reverse). Scarrone's Model. *A. hippocastanum* & allies susceptible to spectacular infestations of leaf-miner (*Cameraria ohridella*) noted in Balkans since 1961 (though in herbarium specimens coll. 1879), spread to GB by 2002 with unsightly effect on lvs but not fatal. Cult. orn., some fish-poisons (US) perhaps due to aesculin, a coumarin. ***A. californica*** (Spach) Nutt. (Calif. buckeye, Calif.) – seeds form. much eaten by Native Calif., when mashed a suppository for piles; ***A. chinensis*** Bunge (China) – planted in Buddhist temples as subs. for *Shorea robusta* (under which Buddha was buried); ***A. flava*** Sol. (*A. octandra*, yellow b., E N Am.) – valuable timber, with ***A. glabra*** Willd. (S US; & *A. hippocastanum*) form. used for artificial limbs; ***A. hippocastanum*** L. (horse-chestnut, Balkans to Himal., widely cult.) – propolis from buds, nectar guides in fls change from yellow to red as fls age, some med. uses (efficacious against chronic venous insufficiency), form. fed to ailing horses by Turks (aescin a good bruise remedy, now comm.), bark yields black dye for silk & cotton, wood limited use, seeds the conkers (i.e. conquerors [or poss. from conch as snail-shells orig. used]) of children's game (oblionker) & starch source for acetone prod. WW I, parent (with *A. pavia*) of sexual hybrid, ***A. × carnea*** J. Zeyher ('**Planteriensis**' a sterile backcross with *A. hippocastanum*), & with *A. flava* Sol. of graft hybrid, ***A. +dallimorei*** Sealy, 'Baumanii' sterile ('double' fls), good street tree; ***A. indica*** (Cambess.) Hook.f. (Ind. horse-chestnut, Himal.) – horse med., molluscicidal saponins; ***A. pavia*** L. (red buckeye, red horse-chestnut, N Am.) – widely cult.

Aetanthus (Eichler) Engl. (~ *Psittacanthus*). Loranthaceae (4 4). 15 N Andes. R: PBE 131(2014)10. Fls to 25 cm long

Aetheocephalus Gagnepain = *Athroisma*

Aetheolaena Cass. = *Senecio*

Aetheolirion Forman. Commelinaceae (II 3b). 1 Thailand: ***A. stenolobium*** Forman – cli. with elongate capsules of winged seeds

Aetheopappus Cass. = *Psephellus*

Aetheorhiza Cass. = *Sonchus*

Aethiorhyncha Dressler (~ *Chondrorhyncha*). Orchidaceae (V 12j). 1 S Ecuador: ***A. andrettae*** (Jenny) Dressler. R: Lankesteria 5(2005)95

Aethephyllum N.E. Br. = *Cleretum* (but see BBP 61(1986)436)

Aethiocarpa Vollesen = *Harmsia*

Aethionema R. Br. Cruciferae (1). 45 Med. (Eur. 9). some ann. spp. incl. ***A. heterocarpum*** Gay have 2 kinds of fr.: dehiscent, many-seeded & indehiscent, 1-seeded. Many cult. orn. esp. ***A. saxatile*** (L.) R. Br.

Aethusa L. Umbelliferae (III 8). 1 Eur., W As., N Afr.: ***A. cynapium*** L. (fool's parsley) – poisonous weed (coniine), used med. with caution

Aetoxylon (Airy Shaw) Airy Shaw. Thymelaeaceae (Oct.). 1 Borneo: ***A. sympetalum*** (Steenis & Domke) Airy Shaw – a source of agarwood (see *Aquilaria*). R: FM I,4(1953)365

Aextoxicaceae Engl. & Gilg. Magnoliidae – Berberidopsidales. 1/1 C Chile. Tree, dioec. with ferrugineous, peltate indumentum. Lvs subopp. or in spirals, simple; stipules 0. Fls in axillary racemes, each encl. in bracteole in bud; K 5, rarely 4 or 6, strongly imbr. & caducous, C 5, rarely 4 or 6, imbr., A 5, rarely 4 or 6 reniform nectary glands, G 1 with bifid style & 2 anatr. bitegmic, pend. ovules, each with a massively beaked nucellus protruding beyond integuments. Fr. a dry drupe with single stone & seed with ruminate endosperm. n = 16
Genus: *Aextoxicon*

Aextoxicon Ruíz & Pavón. Aextoxicaceae. 1 S Chile & nearby Arg.: ***A. punctatum*** Ruíz & Pavón – comm. timber (olivillo) with long vessel elements & tracheids. R: BM n.s. 29(2012)182

afara or **white a.** *Terminalia superba*; **black a.** *T. ivorensis*

Affonsea A. St-Hil. = *Inga*

Afgekia = *Callerya*

afo *Poga oleosa*

Afrachneria Sprague = *Pentaschistis*

Afraegle (Swingle) Engl. Rutaceae (II 2). 4 W Afr. Thorny trees & shrubs, source of oil for food

Aframmi Norman. Umbelliferae (III 8). 2 Angola

Aframomum K. Schum. Zingiberaceae (II 4). 50 trop. Afr. Fls last 1 day or less; nectaries at ovary apex ('stylodia'). Seeds spices, cardamom subs.: ***A. angustifolium*** (Sonn.) K. Schum. (longoza, Madag. cardamom) – comm. skin products; ***A. corrorima*** (A. Br.) Jansen (korarima, NE & E Afr.) – spice for coffee & sauces; ***A. melegueta*** (Roscoe) K. Schum. (*Amomum grana-paradisi* L., Guinea grains, grains of paradise, alligator, malagueta or melegueta pepper, W Afr.) – the best, used in gin, in magic in trop. Am.

Afrardisia Mez = *Ardisia*

Afraurantium A. Chev. = *Citrus*

Africa or **African blackwood** *Dalbergia melanoxylon*; **A. bowstring hemp** *Dracaena* spp.; **A. breadfruit** *Treculia africana*; **A. canarium** *Canarium schweinfurthii*; **A. corn lily** *Ixia* spp.; **A. ebony** *Diospyros* spp.; **A. elemi** *Boswellia frereana*; **A. golden walnut** *Lovoa trichilioides*; **A. hemp** *Sparrmannia africana*; **A. kino** *Pterocarpus erinaceus*; **A. lily** *Agapanthus africanus*; **A. lotus** *Ziziphus lotus*; **A. love-grass** *Eragrostis curvula*; **A. mahogany** *Khaya* spp. esp. *K. antotheca*, *K. senegalensis*; **A. marigold** *Tagetes erecta*; **A. millet** *Cenchrus spicatus*; **A. padouk** *Pterocarpus soyauxii*; **A. peach** *Nauclea latifolia*; **A. pencil cedar** *Juniperus procera*; **A. rosewood** *Guibourtia demeusei*, *Pterocarpus erinaceus*; **A. rubber** *Landolphia* spp.; **A. sandalwood** *Excoecaria africana*; **A. satinwood** *Zanthoxylum gillettii*; **A. sheep bush** *Pentzia incana*; **A. teak** *Milicia excelsa*, *Oldfieldia africana*; **A. tulip-tree** *Spathodea campanulata*; **A. violet** *Streptocarpus ionanthus*; **A. walnut** *Lovoa trichilioides*; **A. whitewood** *Annickia* spp. esp. *A. chlorantha*; **A. yellow-wood** *Podocarpus* spp.

afrikander (S Afr.) *Gladiolus* spp.

Afroaster Goldbl. & Manning (~ *Aster*). Compositae (Ast.-Ast.). 18 SE Afr. R: MBSM 11(1973)153, Strelitzia 29(2012)792

Afrobrunnichia Hutch. & Dalziel (~ *Brunnichia*). Polygonaceae (II 2). 1–2 trop. Afr.

Afrocalathea K. Schum. (~ *Marantochloa*). Marantaceae. 1 W Afr.: ***A. rhizantha*** (K. Schum.) K. Schum.

Afrocalliandra E. Souza & Queroz (~ *Calliandra*). Leguminosae (II 4). 2 trop. Afr. R: T 62(2013)1213

Afrocarpus (Buchholz & N. Gray) Page (~ *Podocarpus*). Podocarpaceae. 3 trop. & S Afr. R: NRBGE 45(1988)383. ***A. falcata*** (Thunb.) Page – bole to 12.5 m diam. (Ethiopia); ***A. usambarensis*** (Pilg.) Page (E Afr.) – reaching 75 m, prob. Afr.'s tallest tree

Afrocarum Rauschert = *Berula*

Afrocrania (Harms) Hutch. = *Cornus*

Afrocrocus Manning & Goldbl. (~ *Syringodea*). Iridaceae (VI 5). 1 SW Afr.: ***A. unifolius*** (Goldbl.) Manning & Goldbl. – lvs 1(2). R: P. Goldblatt & J. Manning, *The Iris family* (2008)160

Afrodaphne Stapf = *Beilschmiedia*

Afrofittonia Lindau. Acanthaceae (III 2c). 1 trop. W Afr.: ***A. silvestris*** Lindau

Afroguatteria Boutique. Annonaceae (III 7). 1 trop. Afr.: ***A. bequaertii*** (de Wild.) Boutique

Afrohybanthus Flicker (~ *Hybanthus*). Violaceae. 25 OW trop. R: Phytotaxa 230(2015)43

Afroknoxia Verdc. = *Knoxia*

Afrolicania Mildbr. (~ *Licania*). Chrysobalanaceae (I). 1 trop. W Afr.: ***A. elaeosperma*** Mildbr (*L. e.*, poyok) – seeds yield a drying oil for. subs. for tung oil for paints etc. R: FOW 9(2003)175

Afroligusticum Norman. Umbelliferae (III 11). 13 S Afr., afromontane. R: T 57(2008)358

Afrolimon Lincz. (~ *Limonium*). Plumbaginaceae (II). 7 S Afr.

Afropteris Alston = *Pteris*

Afroqueta Thulin & Razafim. (~ *Piriqueta*). Passifloraceae. 1 SC & S Afr.: ***A. capensis*** (Harv.) Thulin & Razafim. R: T 61(2012)315

Afrorchis Szlach. = *Brachycorythis*

Afrorhaphidophora Engl. = *Rhaphidophora*

afrormosia *Pericopsis elata*

Afrormosia Harms = *Pericopsis*

Afrosciadium P. Winter (~ *Peucedanum*). Umbelliferae (III 11). 18 trop. & S Afr. R: T 53(2008)359. ***A. kerstenii*** (Engl.) P. Winter (*P. k.*, trop. E Afr. mts) – on Ruwenzori with woody erect stem to 2 m topped with rosette of lvs, said to be gorilla food on Virungas

Afroscirpoides García-Madrid & Muasya (~ *Scirpoides*). Cyperaceae (II 5). 1 S Afr.: ***A. dioeca*** (Kunth) García-Madrid. R: T 64(2015)698

Afrosersalisia A. Chev. = *Synsepalum*

Afrosison H. Wolff. Umbelliferae (III 8). 3 trop. Afr.

Afrosolen Goldbl. & Manning (*Psilosiphon*, ~ *Lapeirousia*). Iridaceae (III 1). 18 subSaharan Afr. R: Strelitzia 35(2015)108, Bothalia 46(2016)2024:3

Afrostyrax Perkins & Gilg. Huaceae. 2–3 trop. Afr. Bark & seeds imp. local garlicky flavouring

Afrothismia (Engl.) Schltr. Thismiaceae (Dioscoreaceae s.l.). 8 trop. Afr. R: Blumea 48(2003)478

Afrotrewia Pax & K. Hoffm. Euphorbiaceae (Acal.). 1 W trop. Afr.: ***A. kamerunica*** Pax & K. Hoffm. R: T 57(2008)140

Afrotrichloris Chiov. Gramineae (XXIX 5). 2 Somalia

Afrotrilepis (Gilly) Raynal. Cyperaceae (III 2). 2 trop. W Afr. ***A. pilosa*** (Boeck) Raynal 'arborescent'. R: BN 126(1973)331

Afrotysonia Rauschert (*Tysonia*). Boraginaceae (3.8.3). 3 S & E Afr. R: NRBGE 43(1986) 467

Afrovivella A. Berger (~ *Rosularia*). Crassulaceae (I 5a). 1 Ethiopia: ***A. semiensis*** (A. Rich) A. Berger. R: U. Eggli. *Ill. Handb. Succ. Pls*, Crass. (2005)23

Afzelia Sm. Leguminosae (I 2). c. 11 OW trop. (Mal. 2). Comm. timbers (aligna, apa) e.g. ***A. rhomboidea*** (Blanco) S. Vidal (Malacca teak, Mal.), ***A. quanzensis*** Welw. (chamfuta, pod or red mahogany, trop. Afr.) – seeds (red & black) used as beads (lucky beans)

Afzeliella Gilg = *Guyonia*

Agalinis Raf. (~ *Sopubia*). Orobanchaceae (Gerard.; Scrophulariaceae s.l.). 40 trop. & warm Am., ± parasitic on r. of other pls. ***A. purpurea*** (L.) Pennell (US) – cult. orn.

Agallis Philippi = *Tropidocarpum*

Agalmyla Blume. Gesneriaceae (III 2j). 97 Mal. R: EJB 59(2002)24, 323. Epiphytes. Cult. orn.

Aganippea Moçiño & Sessé ex DC. = *Jaegeria*

Aganisia Lindl. Orchidaceae (V 12j). 4 trop. Am. Diminutive cult. orn.

Aganonerion Pierre ex Spire (~ *Urceola*). Apocynaceae (II c). 1 SE As.: ***A. polymorphum*** Pierre ex Spire. R: Blumea 41(1996)71

Aganope Miq. (~ *Ostryocarpus*). Leguminosae (III 16). Incl *Xeroderris*, 4 trop. Afr., 2 SE As.

Aganosma (Blume) G. Don. Apocynaceae (II c). Excl. *Amphineurium*, 7 SE As. to C Mal. Coffee & tea subs., local medic.

Agapanthaceae Voigt = Amaryllidaceae

Agapanthus L'Hérit. Amaryllidaceae (Agapanthaceae). 6 S Afr. R: W. Snoeijer (2004) *A.: a revision of the genus*, p. 58. Pop. cult. orn. (esp. tub-pls,) with at least 625 cvs (incl. hybrids) of e.g. ***A. africanus*** (L.) Hoffsgg. (Afr. lily), though these less oft. grown than those of ***A. praecox*** Willd. (*A. orientalis*), 'subsp. *orientalis*' invasive in Aus.

Agapetes D. Don ex G. Don (~ *Vaccinium*). Ericaceae (VIII 5). (Excl. *Paphia*, *q.v.*, incl. most Indomal. '*Vaccinium*') c. 400 trop. As. to Mal. Everg. cult. orn. shrubs, some with lvs used in a tea (Ind.)

Agarista D. Don ex G. Don (~ *Leucothoe*). Ericaceae (VIII 2). 31 Afr. to Masc. 1 (*Agauria*), Am. R: JAA 65(1984)255

agarwood *Aquilaria malaccensis*, also *Aetoxylon sympetalum*, *Gonystylus bancanus*, *Gyrinops ledermannii*

Agastache Clayton ex Gronov. Labiatae (VII 2c). 22 mts & deserts N Am, W As. (1). R: SBM 15(1987), BZ 78,2(1993)112. Bold cult. orn., some used as flavouring (N Am.); ***A. foeniculum*** (Pursh) Kuntze (*A. anethiodora*, anise or giant hyssop) – basis of a drink & local med.

Agastachys R. Br. Proteaceae (III). 1 Tasmania: ***A. odorata*** R. Br.

Agasthiyamalaia Rajkumar & Janarth. (~ *Poeciloneuron*). Calophyllaceae. 1 W Ghats, India: ***A. pauciflora*** (Bedd.) Rajkumar & Janarth. R: JBRIT 1(2007)130

Agasyllis Spreng. Umbelliferae (III 9). 1 Cauc.: ***A. gummifera*** (L.) Dierb.

Agatea A. Gray. Violaceae (III 2a). 8 NG, W Pacific

Agathelpis Choisy = *Microdon*

Agathis Salisb. Araucariaceae. c. 13 Sumatra to NZ (1) & Fiji (New Caled. 5); kauri & dammar. R: PSE 135(1980)41. Fossils from Eocene of Patagonia. Trees usu. monoec. (Massart's Model); lvs broad, entire, leathery; cones take 2 yrs to mature. Imp. trop. timber trees (gen. purpose softwoods from masts to matches, artificial limbs, pulp & charcoal), prod. copals (white c. trad. used for caulking & burnt to drive off mosquitoes, later for varnish, linoleum, road-markings, glazing fr.-tin labels & glossy photo-prints): ***A. australis*** (D. Don) Lindl. (kauri, NZ) – to 2000 yrs old, poll. followed by fert. after 1 yr & mature seeds at 16 months, timber for hulls of Maori sailing canoes, most of pre-1906 fire San Francisco (& imp. in Sydney & Melbourne) built of it, 'fossil' copals 'mined' as subsistence living in C19 NZ & trunks preserved in bogs nr Auckland now exploited for timber; ***A. dammara*** (Lamb.) Poir. (bendang, bindang, E Ind. or Manila copal, W & C Mal.) – bark burned to deter mosquitoes; ***A. macrophylla*** (Lindl.) Masters (*A. vitiensis*, dakua, Fijian kauri, Solomon Is. to Fiji) – copal for glazing pots, shipbuilding, & torches; ***A. microstachya*** J. Bailey & C. White (bull kauri, Queensland) – to 1000 yrs old, 45m tall with bole to 6 m girth (biggest in Aus.). Saplings easily grown pot-pls, though diff. spp. scarcely distinguishable then

Agathisanthemum Klotzsch (~ *Hedyotis*). Rubiaceae (IV 1). 5–6 trop. Afr., Comoro Is.

Agathophora (Fenzl) Bunge = *Halogeton*

Agathosma Willd. Rutaceae (I 4). 150 S Afr. esp. SW Cape (143, 138 endemic). R: JSAB 16(1950)55, Strel. 9(2000)624. Heath-like shrubs with ess. oils, for which some cult. to give buchu oil (diuretic). Cult. orn. ***A. betulina*** (P. Bergius) Pill. principal source of buchu, also used in artificial blackcurrant flavourings, urinary antiseptic (volatile oil: diosphenol) also used in herbal treatment of arthritis

Agati Adans. = *Sesbania*

Agauria (DC.) Hook.f. = *Agarista*

Agavaceae Dumort. = Asparagaceae

Agave L. Asparagaceae (Agavaceae). Incl. *Manfreda* (R: CSM 30(1985)56), *Polianthes* (R: CUSNH 8(1903)8), 225 S US to W Panama, Carib. & Venez. R: U. Eggli, *Ill. Handb. Succ. Pls, Monocots* (2001)6. First ill. Eur. 1543 (***A. obscura*** Schiede ex Schldl, ?Mex.). Usu. short-stemmed pachycauls with spectacular tall, fast-growing infls, occ. herbs with thick bulb-like bases. Flowering shoots hapaxanthic (Holttum's Model), pl. reprod. by suckers or bulbils from infl. as well as seed. Fibre-pls (Tampico fibre, keratto etc.), cult. orn. incl. hybrids e.g. ***A. × taylorii*** B.S. Williams (*A. geminiflora* (Tagl.) Ker × *A. filifera* Salm-Dyck, Mex.), core the source of pulque (fermented sap) & mescal or tequila (distilled) or comm. syrup. V. imp. in Native Am. culture, where many cvs & hybrids selected: bud (heart) ed. when about to flower, pit-baked for hours, fls cooked in tortillas, soap source & for stunning fish, cloudy pulque or tequila imp. vitamin source, 'worms' or 'maguey slugs' an Aztec delicacy, still served with guacamole; lvs poss. use in paper-making. ***A. americana*** L. (century plant – through erroneous belief it flowers when 100 yrs old [actually 10–20], maguey, American aloe, E Mex., invasive Medit., S Afr.) – imp. Apache foodpl., a source of mezcal, widely cult. (even S Br.) orn. (in Eur. (Padua) since 1561), hedgepl., with varieg. cvs; ***A. bulliana*** (Bak.) Thiede & Eggli (*Prochnyanthes mexicana*, Mex.) – used medic. & in washing clothes; ***A. cantala*** Roxb. ex Salm-Dyck (cantala, Manila maguey, Bombay aloe, ? Mex.) – triploid (2n = 90), widely cult. OW (orig. introd. as hedge-pl.) for fibre softer & finer but weaker than sisal used in hard fibre twines; ***A. fourcroydes*** Lem. (henequén) – sterile, domesticated by Maya from *A. vivipara* (of which prob. best considered a cv), fibre for binder twine etc., waste for paper pulp, textiles in Yucatan, leaf-bases & peduncle ed., local medic.; ***A. funkiana*** K. Koch & Bouché (Jaumave ixtle or fibre, Mex.) – fibre for brushes; ***A. inaequidens*** K. Koch (Mex.) – fasciated infls to 1.4 m wide recorded; ***A. karatto*** Mill. – only sp. in Lesser Antilles, largely a set of cultigens; ***A. lechuguilla*** Torrey (lecheguilla, tula ixtle, Texas & N Mex.) – fibre (most Tampico f.) for brushes, ropes, textiles since 6000 BC, now as cellulose pulp, wax in soap; ***A. polianthes*** Thiede & Eggli (*Polianthes tuberosa*, tuberose, unknown wild, cult. Eur. 1601) – fls gardenia-scented waxy-white, cult. pre-Columbian Mex. ('omixochitl') & added to chocolate as flavouring, now grown in France, China, Egypt & Morocco for v. frag. fls (oft. 'double' form), ess. oil used in scent incl. by undertakers on corpses; ***A sisalana*** Perrine (sisal, named after the Mexican seaport) – sterile pentaploid (2n = 150) of unknown orig., widely cult. (c. 300K t p.a.) for fibre (Florida 1836) esp. Tanzania (industry orig. derived from 62 plantlets introd. 1893), Madag. & Braz., used in cigarette papers, banknotes, teabags, filter paper, dartboards & added to recycled paper, wax a carnauba subs. for car- & shoe-polishes in Ind., molluscicidal saponins, hecogenin a precursor in cortisone prod.; ***A. tequilana*** F. Weber (blue agave, Mex.) – a source of mescal (mezcal), tequila (true t. from T. region) prod. over 190 M l in 2016 (151 M l for export), syrup comm. in US but high in fructose; ***A. vera-cruz*** Mill. (Mex.) – comm. source of fructose, cordage; ***A. virginiana*** L. (*Manfreda v.*, SE US, Mex.) – poll. sphinx moths, snake-bite remedy; ***A. vivipara*** L. (*A. angustifolia*, Mex., C Am.) – imp. food, drink, fibre, comm. mescal, var. ***letonae*** (Trelease) P. Forster (*A. letonae*, Salvador henequen, C Am.) – fibre finer than *A. fourcroydes*, used in sacking

agba *Prioria balsamifera*

agboin *Piptadeniastrum africanum*

Agdestidaceae Nakai = Phytolaccaceae

Agdestis Sessé & Moçiño ex DC. Phytolaccaceae (Agdestidaceae). 1 Mex. to Honduras: ***A. clematidea*** Sessé & Moçiño ex DC., cli., cult. orn. R: JAA 66(1985)35. Poss. referable to Sarcobataceae

Agelaea Sol. ex Planch. Connaraceae (IV). 8 trop. Afr., As. R: AUWP 89–6(1989)136, BJBB 61(1991)72. Lianes or scramblers, some med. ***A. borneensis*** (Hook.f.) Merr. (SE As. to C Mal.) – liane, stems for ropes, e.g. supporting church-bells in Tayabas (Philippines) for 200+ yrs

Agelanthus Tieghem. (~ *Tapinanthus*). Loranthaceae (5 7). 59 Afr. & Arabia. R: R. Polhill & D. Wiens, *Mistletoes of Afr.* (1998)137

Agenium Nees. Gramineae (XXII 7). 3 Braz. to Arg. R: FR 43(1938)80

Ageomoron Raf. = *Caucalis*

Ageratella A. Gray ex S. Watson. Compositae (Eup.-Alom.). 1 Mex.: ***A. microphylla*** (Sch. Bip.) A. Gray. R: Phytol. 78(1995)204

Ageratina Spach. Compositae (Eup.-Oxy.). Incl. *Pachythamnus*, *Piptothrix*, 200+ E US (3), C & W S Am., 2 natur. OW. R: MBSMBG 22(1987)428. x = 17. ***A. adenophora*** (Spreng.) R. King & H. Robinson (Crofton weed (S Afr.), Mex. devil, Mex.) – triploid apomictic, aggressive weed in S Afr., Aus. & NZ (gall fly introd. to control it), causing acute lung disease in grazing horses & skin allergies in humans; ***A. altissima*** (L.) R. King & H. Robinson (E US) – local medic.; ***A. crassiramea*** (Robinson) R. King & H. Robinson (*Pachythamnus c.*, C Am.) – fat stems leafless in fl.; ***A. riparia*** (Regel) R. King & H. Robinson (*Eupatorium r.*, C Am.) – bad weed trop. As., controlled by *Entyloma ageratinae* (smut fungus from Jamaica) in Hawaii

Ageratinastrum Mattf. Compositae (Vern.-Erl.). 5 trop. Afr.

Ageratum L. Compositae (Eup.-Ager.). c. 40 trop. Am. R: AMBG 58(1971)6, 62(1975)901. x = 10. ***A. conyzoides*** L. (billy-goat weed, blue top (Aus.), S Am., now pantrop. weed) – Stone's Model, folk med. (e.g. diarrhoea in Burundi), fls ed. Mafia Is., Tanzania, planted China as poll. diversionary food for mites predatory on citrus; ***A. houstonianum*** Mill. (*A. mexicanum*, SE Mex., C Am.) widely cult. orn. edging pl. with blue (cvs with white or pink) fls., contains precocenes 1 & 2 (based on 2,2-dimethylcromene), which interfere with juvenile hormone activity & cause precocious metamorphosis of insects, & an oil v. toxic against *Fusarium* wilts of *Cajanus cajan*

Agiabampoa Rose ex O. Hoffm. = *Alvordia*

Agianthus Greene = *Streptanthus*

Agiorta Quinn (~ *Leucopogon*). Ericaceae (VII 7). 3 E Aus. R: AusSB 18(2005)450

Aglaia Lour. Meliaceae (II-Trich.). c. 120 Indomal. (Borneo 60), W Pacif. R: KBAS 16(1992). Heterogeneous? Some unbranched pachycauls & rheophytes. Bird-disp. spp. with fatty seeds, mammal-disp. with sweet ones. Some timbers (e.g. ***A. argentea*** Blume (Mal.) comm.) & ed. fr. (i.e. seed arils) locally eaten e.g. ***A. edulis*** (Roxb.) Wall. (Indomal.) & ***A. exstipulata*** (Griff.) Theob. (SE As., W. Mal.). ***A. cucullata*** (Roxb.) Pellegrin (Indomal.) – timber (tasua, Thailand); ***A. korthalsii*** Miq. (Himal. to Sulawesi) – cult. fr. tree Malay Pen.; ***A. odorata*** Lour. (SE As.) – cult. shrub incl. topiary work, male fls used to scent tea

& linen, also for colds & asthma, foliage for victors' chaplets in Vietnamese television quiz shows; ***A. saltatorum*** A.C. Sm. (Fiji) – introd. other is. for fls used in leis & to scent coconut oil; ***A. sexipetala*** Griff. (*A. aspera*, SE As. to NG) – ed. aril, local timber

Aglaodorum Schott. Araceae (VII 13). 1 W Mal.: ***A. griffithii*** (Schott) Schott, tidal mud-flats, viviparous

Aglaomorpha Schott. = *Drynaria*

Aglaonema Schott. Araceae (VII 13). 21 Indomal. R: SCB 1(1969)1. Local medic., cult. orn. esp. ***A. commutatum*** Schott (C Mal.), many cvs & ***A. pictum*** (Roxb.) Kunth '**Tricolor**' (Sumatra), silver-green due to separation of upper and middle leaf-layers (single gene control); ***A. modestum*** Engl. (S China, SE As.) – long cult. in E for good luck

Aglossorrhyncha Schltr. Orchidaceae (V 10b). 13 E Mal., Papuasia

Agnirictus Schwantes = *Stomatium*

Agnesia Zuloaga & Judziewicz (~ *Olyra*). Gramineae (VI 3). 1 Amaz.: ***A. lancifolia*** (Mez) Zuloaga & Judziewicz. R: Novon 3(1993)306

Agnorhiza (Jepson) W. Weber = *Wyethia* (but see Phytol. 85(1998)19)

Agonandra Miers ex Benth. Opiliaceae. 10 C & trop. S Am. R: FN 82(2000)22. Only trop. Am. genus; root-parasites. ***A. brasiliensis*** Miers ex Benth. (Braz.) – seeds an oil source, partially hydrogenated yielding a rubber subs.; ***A. racemosa*** Standl. (Mex.) – dioec., wind-poll.

Agonis (DC.) Sweet. Myrtaceae (II 14). Excl. *Paragonis*, *Taxandria*, 4 SW Aus. R: Nuytsia 16(2007)397. Cult. orn. (willow myrtles)

Agoseris Raf. Compositae (Lact.-Mic.). 11 W N Am. (9), temp. S Am. (2). Cult. orn. ***A. aurantiaca*** (Hook.) Greene form. eaten by Native Amer.

Agouticarpa C. Persson. Rubiaceae (II 1). 6 trop. Am. R: Britt. 55(2003)180. Dioec.

agretti *Salsola soda*

Agrianthus Mart. ex DC. Compositae (Eup.-Gyp.). 9 Braz. R: KB 48(1993)265

Agrimonia Tourn. ex L. Rosaceae (Ros.-Agrim.). c. 20 N temp. (Eur. 3, N Am. 7). to W Mal., C & S Afr. (1), Haiti & Braz. Agrimony. Local medic. N Am. ***A. eupatoria*** L. (common a., OW) – general prophylactic & purifier, folk med. in Eur. (liver disease), source of yellow dye; ***A. procera*** L. ('*A. repens*', '*A. odorata*', As. Minor) – cult. orn. ('scented a.')

agrimony *Agrimonia* spp.; **common a.** *A. eupatoria*; **hemp a.** *Eupatorium cannabinum*; **scented a.** *A. procera*

Agriophyllum M. Bieb. Amaranthaceae (Chenopodiaceae I 6). 6 Eur. (1) to C As. ***A. squarrosum*** (L.) Moq. (*A. gobicum*, Azerbaijan, Siberia to China) – seeds imp. food in Mongolia, forage crop; ***A. latifolium*** Fischer & Meyer (C As.) – a tumbleweed

Agriphyllum Juss. = *Berkheya*

Agrocharis Hochst. = *Daucus*

Agropyron Gaertn. Gramineae (XV). c. 15 temp. OW (Eur. 6). (For perenn. XV. with keeled glumes (*A. repens* etc.), see *Elymus*)

Agropyropsis (Battand. & Trabut) A. Camus. Gramineae (XVI 12). 1 Algeria: ***A. lolium*** (Cosson & Durieu) A. Camus

Agrostemma L. Caryophyllaceae (III 3). 2–3 Med. Corn cockle. ***A. brachylobum*** (Fenzl) Hammer, diploid, & ***A. githago*** L., tetraploid, form. troublesome weed but now scarce in GB where introd., seeds poss. poisonous due to saponins, cult. orn.

Agrostis L. Gramineae (XVI 5). 264 temp. (Eur. 25, Am. 67 [N Am. 21 – R: FNA 24(2007)633]), trop. mts. R: SBU 17,1(1960). Bent grasses, imp. in lawns & pastures. ***A. alba*** L. (red top, Euras.) – allegedly natur. NZ from discarded mattresses of Nova Scotia migrants; ***A. canina*** L. (velvet bent, brown b., Eur.) – for lawns esp. putting greens; ***A. capillaris*** L. (*A. tenuis*, common b., brown top, colonial b., Rhode Is. b., Eur., invasive Aus., NZ) – pasture & lawns esp. bowling greens, selected metal-tolerant strains in spoil-tip reclamation; ***A. gigantea*** Roth (red top, black b., Euras.) – pastures & lawns; ***A. stolonifera*** L. (creeping b., fiorin, N temp., invasive Aus.) – lawns, found in coffin of child mummified not later than 21st Dynasty, modern clones selected for glyphosate tolerance poss. 'superweeds'. See also *Bromidium*, *Lachnagrostis*, *Podagrostis*

Agrostistachys Dalz. Euphorbiaceae (Acal.-Agr.). c. 10 Ind. to NG. R: Blumea 46(2001)76. Corner's Model, ants nesting around leaf-bases. ***A. borneensis*** Becc. (*A. longifolia*, Indomal.) – lvs to 60 cm long used for thatch in Sri Lanka, latex from burnt wood to blacken teeth

Agrostocrinum F. Muell. Asphodelaceae (Phormiaceae). 2 SW Aus. R: Nuytsia 15(2004)246

Agrostophyllum Blume. Orchidaceae (V 13f). c. 100 Seychelles to W Pacif. R (sect. *Appendiculopsis* – 5): OM 8(1997)7

Agrostopoa Davidse & al. (~ *Muhlenbergia*). Gramineae (XVI 15). 3 Colombia, paramos. R: Novon 19(2009)33

aguacate *Persea americana*

aguassú *Attalea* spp. esp. *A. phalerata*

ague root *Aletris farinosa*; **a. weed** *Gentianella quinquefolia, Eupatorium cannabinum*

Aguiaria Ducke. Malvaceae (Bomb.; Bombacaceae). 1 Amaz. Braz.: ***A. excelsa*** Ducke

Ahernia Merr. Salicaceae (Flacourtiaceae). 1 Hainan, Philippines: ***A. glandulosa*** Merr.

Ahouai Mill. = *Thevetia*

ahun *Alstonia congensis*

ahuyama *Cucurbita moschata*

Ahzolia Standl. & Steyerm. = *Sicyos*

ai *Blumea balsamifera*

aibika *Hibiscus manihot*

Aichryson Webb & Berth. Crassulaceae (I 4). 14 Macaronesia, natur. Portugal. R: U. Eggli, *Ill. Handb. Succ. Pls*, Crass (2005)24. Succ. pls, some esp. ***A.* x *aizoides*** (Lam.) E. Nelson (*A.* × *domesticum*, *A. tortuosum* (Aiton) Webb & Berth. × *A. punctatum* (Link) Webb & Berth.) & '**Variegatum**' (form. common cottage window-sill pl.) cult. orn.

Aidia Lour. Rubiaceae (II 1). Incl. *Anomanthodia, Gynopachis*, excl. *Pelagodendron*, c. 48 OW trop. (Afr. 8, 3 restricted to S. Tomé. R: BMNHN 4,8(1986)259). R. Blumea 41(1996)143. Local medic. Mal.

Aidiopsis Tirv. (~ *Randia*). Rubiaceae (II 1). 2 Mal. R: BMNHN 4,8(1986)287

Aidomene Stopp = *Asclepias*

aielé *Canarium schweinfurthii*

Ailanthus Desf. Simaroubaceae. 5–10 As. (China 6) to Aus. Koriba's Model; medic. esp. dysentery (amoebicide, anthelminthic etc) – quassinoids. ***A. altissima*** (Mill.) Swingle (tree of heaven, China, natur. in N Am. (introd. 1784), C & S Eur.) – cult. orn. introd. France 1740s, used as street tree & in soil conservation, prob. fastest-growing decid. hardy tree in GB with lvs to 1.3 m long on saplings (so form. pollarded for 'subtropical bedding'), weedy in S & E Eur., Azores, N Am. (marks early settlements in Calif.), E Aus. (where causing dermatitis), allelopathic (ailanthine, a quassinoid, most effective), usu. dioec., extrafloral nectaries, males with foetid lvs & fls, form. a spiny cv. used as host ('*A. vilmoriniana*') for Chinese silkmoth, *Samia cynthia* (Drury) in prod. of Shantung silk (now from *Morus*, though tree still in old Chinese cemeteries in Calif.), honey with cat smell but delicious on standing; ***A. excelsa*** Roxb. (Ind.) – wood for matches; ***A. integrifolia*** Lam. (Indomal.) – wood for shoes, matches, pulp; ***A. malabarica*** DC. (mattipaul, Ind.) – resin used as incense in Hindu temples, lvs yield a black dye; ***A. moluccana*** DC. (Mal.) – 'ailanto', 'reaching for the sky', hence both Lat. & Engl. names

Aimara Salariato & Al-Shehbaz (~ *Menonvillea*). Cruciferae (Crem.). 1 NW Chile: ***A. rollinsii*** (Al-Shehbaz & Martic.) Salariato & Al-Shehbaz. R: T 62(2013)1230

Ainea Ravenna = *Tigridia*

Ainsliaea DC. Compositae (Perty.). 49 E As. to W Mal. R: AMBG 94(2007)97

Ainsworthia Boiss. = *Tordylium* (but see NRBGE 31(1971)109)

Aiouea Aubl. Lauraceae (I). c. 25 trop. Am. R: FN 31(1982)89, AMBG 75(1988)402. Polyphyletic?

Aiphanes Willd. Palmae (V 8b). 24 trop. Am. R: FN 70(1996)33. Prickly (? deterrent to extinct megafauna?) monoec. cult. orn. e.g. ***A. minima*** (Gaertn.) Burret (*A. erosa*, macaw palm)

Aipyanthus Steven = *Arnebia*

air fern *Sertularia argentea* L., a marine zoophyte (animal) sold in US; **a. plant** *Kalanchoe pinnata, Tillandsia ionantha*; **a. potato** *Dioscorea bulbifera*

Aira L. Gramineae (XVI 7). 8 Eur. (8) & Med. to Iran but widespread as weeds. Some cult. – 'hair grasses'

Airopsis Desv. Gramineae (XVI 5). 1 S Eur., NW Afr.: ***A. tenella*** (Cav.) Coss. & Durieu

Airosperma Lauterb. & K. Schum. Rubiaceae (?III 5). 6 NG (4), Fiji (2)

Airyantha Brummitt. Leguminosae (III 2). 1 Afr. (Guineo-Congolian): ***A. schweinfurthii*** (Taubert) Brummitt – medic., 1 NW Borneo to SW Philippines

Aisandra Airy Shaw = *Diploknema*

Aistocaulon Poelln. ex H.J. Jacobsen = *Nananthus*

Aistopetalum Schltr. Cunoniaceae (II). 2 NG. R: FM I,16(2002)96

Aitchisonia Hemsl. ex Aitch = *Plocama*

Aitonia Thunb. = *Nymania*

Aitoniaceae Harv. = Meliaceae

Aizoaceae Martinov. Magnoliidae – Caryophyllales. Excl. Molluginaceae, 125/1880 trop. & subtrop., mostly in S Afr. &, to a lesser extent, Aus. R: H. E. K. Hartmann, *Ill. Handb. Succ. Pls*, Aiz., 2 vols (2002); Willd. 45(2015)295. Succ. herbs, more rarely shrubs or subshrubs, rarely spiny (some *Ruschia*) with betalains & not anthocyanins, oft. with C_4 photosynthesis & crassulacean acid metabolism (C_3 in *Tetragonia*). Lvs simple, opp. to spiral, usu. ± succ. with centric rather than bifacial structure, rarely stipulate (e.g. *Trianthema*). Fls solit. or in small cymes, bisex., rarely pl. monoec. P (3–)5(–8), oft. basally united with A forming a tube, A (4)5 – ∞, when ∞ outer ones petaloid 1–6 whorls, the flowers resembling Compositae, nectaries commonly in a ring at inner base of A; $\underline{G}$ or $\overline{G}$ (2–5 – ∞) with as many locules, rarely 1-loc., ovules 1 – (usu.) ± ∞ in each, campylotr. to almost anatr., bitegmic. Fr. usu. loculicidal capsule (98% opening when wet with seeds expelled by 'jet action' through nozzle formed by fr. apex), oft. encl. in persistent K; seeds with large embryo curved around copious starchy or ± oily & proteinaceous perisperm, without true endosperm, s.t. arillate. x = 8, 9

Classification & principal genera:

I. **Aizooideae** (petaloid A 0, capsule loculicidal or septicidal or nut, aril 0): *Aizoon, Tetragonia*

II. **Sesuvioideae** (petaloid A 0, capsule circumscissile, seeds arillate): *Sesuvium, Trianthema*

III. [**Tetragonioideae** (Tetragoniaceae; incl. in I.)]

IV. **Mesembryanthemoideae** (petaloid A usu. united by a tube, placentation C, nectaries shell-shaped to tubular): *Mesembryanthemum* (only; here treated in a narrow sense: form. incl. spp. now referred to many segregates in V., many recognizable by aspect; R: T 56(2007)749)

V. **Ruschioideae** (petaloid A usu. free, placentation basal or parietal, nectaries crest-shaped; R [Apatesieae]: T 64(2015)517): *Argyroderma, Carpobrotus, Cephalophyllum, Conophytum, Delosperma, Dinteranthus, Drosanthemum, Faucaria, Lampranthus, Lithops*

Compared with related Cactaceae, most A. are leaf-succ. *Tetragonia* a potherb, many cult. orn. esp. spp. of *Carpobrotus, Cleretum, Conophytum, Gibbaeum, Lampranthus, Lithops, Ruschia*

Aizoanthemum Dinter ex Friedrich. Aizoaceae (I). 4 N Namibia to S Angola. R: H. E. K. Hartmann, *Ill. Handb. Succ. Pls*, Aiz. A–E (2002)28

Aizoon L. Aizoaceae (I). 13 S Spain (1), As. Minor, N, E & S Afr., introd. elsewhere. R: H. E. K. Hartmann, *Ill. Handb. Succ. Pls*, Aiz. A–E (2002)28. Valuable browse in Cape. ***A. canariense*** L. (N & S Afr. to S As.) – used as food by Tuareg

Aizopsis Grulich = *Phedimus*

Ajania Polj. (~ *Chrysanthemum*). Compositae (Anth.-Art.). 34 C & E As. R: BZ 68(1983)207. Heterogeneous. ***A. pacifica*** (Nakai) Bremer & C. Humphries (*Chrysanthemum p.*, Jap.) – yellow-flowered Dutch potpl.

Ajaniopsis C. Shih (~ *Artemisia*). Compositae (Anth.-?Art.). 1 Tibet, China: ***A. pencilliformis*** C. Shih. R: BBMNHB 23(1993)116

ajmud *Trachyspermum roxburghianum*

ajowan *Trachyspermum ammi*

ajram *Anabasis articulata*

Ajuga L. Labiatae (III). 30 OW esp. temp. (Eur. 10) but also lowland Mal. Bugle. C with no upper lip; phytoecdysteroids. ***A. chamaepitys*** (L.) Schreb. (ground pine, yellow b., Euras.) – med.; ***A. iva*** (L.) Schreb. (Medit.) – poss. antimalarial; ***A. reptans*** L. (common b., Eur.) – cult. orn. (ground-cover), many cvs incl. hybrids with *A. genevensis* L. (S Eur.) like '**Jungle Beauty**'

Ajugoides Makino (~ *Lamium*). Labiatae (V1). 1 Jap.: ***A. humilis*** (Miq.) Makino

Akania Hook.f. Akaniaceae. 1 NE Aus. (fossil from Palaeocene of Arg.): ***A. bidwillii*** (Hogg) Mabb. (turnipwood)

Akaniaceae Stapf. Magnoliidae-Brassicales. Incl. Bretschneideraceae, 2/2 SE As. (1), NE Aus. (1). Trees with alkaloids (*Akania*), mustard-oils & myrosin cells in bark & infl. (at least *Bretschneidera*). Lvs in spirals, pinnate, leaflets toothed to entire; stip. small or 0. Fls bisex. in racemes or panicles, slightly irreg. in *Bretschneidera*. K (5), imbr.; C 5, convolute or not; A 8–10, 5 outer opp. K in *Akania*; nectary-disk (*Bretschneidera*) or 0; $\underline{G}$ 3-, 3-loc. with style & 3-lobed stigma & 2 superposed ± anatr., pend., bitegmic ovules per locule. Fr. a loculicidal capsule; seeds with copious endosperm (*Akania*) smelling of bitter almonds; embryo straight

Genera: *Akania, Bretschneidera* (wood almost identical)

akeake *Dodonaea viscosa; Olearia avicenniifolia*

Akeassia Lebrun & Stork. Compositae (Ast.-Gra?). 1 trop. Afr.: ***A. grangeoides*** Lebrun & Stork. R: Cand. 48(1993)332

Akebia Decne. Lardizabalaceae. 5 (1 hybrid?) temp. E As. R: CBM n.s.29(2012)256. Monoec. twiners; female fls (usu. lower) larger; follicles fleshy, ed. (insipid). Cult. orn. esp. ***A. quinata*** (Houtt.) Decne. (fls vanilla-scented, invasive SE US) & ***A. trifoliata*** (Thunb.) Koidz. (fls scentless)

akee *Blighia sapida*

Akersia Buin. = *Borzicactus*

ako *Antiaris toxicaria*

akom *Terminalia superba*

akomu *Pycnanthus angolensis*

akoub *Gundelia tournefortii*

akra *Calotropis procera*

Akrosida Fryx. & Fuertes (~ *Bastardia*). Malvaceae (Malv.-Malv.). 2 Hispaniola, Braz. R: Britt. 59(2007)387

Akschindlium Ohashi (~*Desmodium*). Leguminosae (III 19). 1 SE As.: ***A. godefroyanum*** (Kuntze) Ohashi. R: JJB 78(2003)275

Akund fibre *Calotropis procera*

al *Morinda citrifolia*

Ala Szlach. = *Habenaria*

Aladenia Pichon = *Farquharia*

Alafia Thouars. Apocynaceae (II b). 23 trop. Afr. (15), Madag. (8). R: KB 52(1997)770. ***A. perrieri*** Jum. (Madag.) – latex used as soap

Alajja Ikonn. = *Eriophyton* (but see NS 8(1971)274)

Alamania Lex. Orchidaceae (V 13b). 1 Mex.: ***A. punicea*** Lex. R: C.L. Withner, *The Cattleyas & their relatives* V(1998)5

alan *Shorea albida*

alang-alang *Imperata cylindrica*

Alangiaceae DC. = Cornaceae

Alangium Lam. Cornaceae (Alangiaceae). 21 trop. Afr., China to E Aus. & New Caled. (1). R: BJBBuit. III,16(1939)139. Some Massart's Model. Cult. orn., timber & med. esp. ***A. salviifolium*** (L.f.) Wangerin (trop. As.) – Roux's Model, poll. passerines (& bees), fr. bird-disp., ipecacuanha subs. in Ind.; ***A. villosum*** (Blume) Wangerin (muskwood, NE Aus.) – local scented timber

Alania Endl. Boryaceae. 1 Brisbane Water & Blue Mts, SE Aus.: ***A. endlicheri*** Kunth. R: FA 45(1987)279

Alansmia M. Kessler & al. (~ *Terpsichore*). Polypodiaceae (V; Grammitidaceae). 26 trop. Am. R: Britt. 63(2011)238

Alantsilodendron Villiers (~ *Dichrostachys*). Leguminosae (II 1). 8 Madag.

Alaska pine *Tsuga heterophylla;* **A. yellow cedar** or **cypress** *Xanthocyparis nootkatensis*

Alatavia Rodionenko = *Iris*

Alatiglossum Baptista = *Gomesa*

Alatiliparis Marg. & Szlach. = *Liparis*

Alatococcus Acev.-Rodr. Sapindaceae. 1 Brazil: ***A. sigueirae*** Acev.-Rodr. R: Phytokeys 10(2012)2

Alatoseta Compton. Compositae (Gnap.). 1 S Afr.: ***A. tenuis*** Compton. R: OB 104(1991)46

albarco *Cariniana* spp.

albardine *Lygeum spartum*

Alberta E. Meyer. Rubiaceae (III 5). 1 SE Afr.: ***A. magna*** E. Meyer– cult. orn. with reddish fls, seeds need 1 yr 'after-ripening' on tree. R: T 58(2009)765. Madag. spp. = *Razafimandimbisonia*

Albertinia Spreng. Compositae (Vern.-Lychn.). 1 Braz.: ***A. brasiliensis*** Spreng. – scandent

Albertisia Becc. Menispermaceae (I). 17 trop. & warm Afr. (13), SE As. (6)

Albertisiella Pierre ex Aubrév. = *Planchonella*

Albidella Pichon = *Echinodorus*

Albizia Durazz. Leguminosae (II 4). Incl. *Balizia* c. 130 trop. (Madag. 30 (24 endemic); Am. 22 – R: MNYBG 74,1(1996)203). Heterogeneous. See also *Falcataria, Paraserianthes,*

Samanea. Trees (Troll's Model), shrubs or lianes, unarmed, arils 0. Timber ('acacia' in Aus.) & shade trees for tea, coffee etc., gums. ***A. anthelmintica*** Brongn. (trop. & S Afr.) – molluscicidal saponins; ***A. chinensis*** (Osbeck) Merr. (Ind. to Thailand & Indonesia) – tea-shade; ***A. grandibracteata*** Taubert (nongo, trop. E Afr.) – timber; ***A. julibrissin*** Durazz. (silk tree, Iran to Jap., invasive SE US) – cult. orn., long-selected (cf. *Melia azedarach*), some cvs hardy in GB, seeds 147 yrs old germ. on herbarium sheets at Natural Hist. Mus. (London) after bomb damage in World War II; ***A. lebbeck*** (L.) Benth. (EI walnut, kok(k)o, siris, trop. As., invasive Carib., S Afr.) – extrafl. nectaries on lvs attractive to (? defending) ants, cabinet wood (laurel wood) esp. in US, in railway coaches

Albizzia Benth. = *Albizia*

Albovia Schischkin = *Pimpinella*

Albraunia Speta = *Chaenorhinum* (but see D.A. Sutton, *Rev. Antirrhineae* (1988)131)

Albuca L. (~ *Ornithgalum*). Asparagaceae. 100–140 trop. & S Afr. to Ethiopia. R: T 58(2009)92. ***A. canadensis*** (L.) F.M. Leighton (S Afr. !) – cult. orn.

albuerra wood *Astronium* spp.

Alcantara Glaz. ex G. Barroso = *Heterocoma*

Alcantarea (Mez) Harms (~ *Vriesea*). Bromeliaceae (2). c. 30 E Braz. R: TSP 91(1995)11

Alcea L. Malvaceae (Malv.-Malv.). c. 50 Med. to C As. (Eur. 5). Hollyhocks, cult. orn. (to 7.39 m, UK record), esp. ***A. rosea*** L., poss. hybrid of *A. setosa* (Boiss.) Alef. (Crete & Turkey) & *A. pallida* (Willd.) Waldst. & Kit. (C & SE Eur.), or an As. sp., some med. uses, used in Neanderthal burial head-wreaths with *Ephedra* sp., hybrids with *Althaea officinalis* = × ***Alcathaea suffrutescens*** Hinsley

Alchemilla L. Rosaceae (Pot.). (Incl. *Aphanes*) c. 900 (mostly microspp.) N temp. OW trop. mts, Aus. (1), S Am. Fls inconsp., green, apetalous, with epicalyx. Subg. *A.* apomictic, A4 with sterile pollen, many high polyploids [Eur. 137]; subg. *Lachemilla* (shrubs & herbs) A2 – c. 80 C & S Am.; subg. *Aphanes* (annuals) A1 – c. 20 Eur., Ethiopia, Aus., Am.); ***A. arvensis*** (L.) Scop. (*Aphanes a.*, parsley piert, Eur., W As., natur. N Am.) – thought to be useful in bladder probs as stimulates copious secretion of lithic acid. Some cult. orn. esp. ***A. mollis*** (Buser) Rothm. (E Carpathians to Cauc.) with large lvs & consp. guttation; ***A. pinnata*** Ruíz & Pavón (*Zygalchemilla p.*, S Am.) – lvs pinnate like Potentilleae; ***A. vulgaris*** L. (s.s. = *A. acutiloba* Opiz, lady's mantle, Eur.) – medic.

Alchornea Sw. Euphorbiaceae (Alc.-Alc.). c. 42 trop. (Am. 23 – R: FN 93(2004)55) to Mongolia – ***A. davidii*** Franch.). Some adventitious embryony; some fish-disp. in Amaz.; some local medic. Mal. and Braz., effective against *Helicobacter pilori* responsible for stomach ulcers, incl. ***A. cordifolia*** (Schum. & Thonn.) Muell. Arg. (Christmas bush, trop. Afr.) – thicket-forming, foodpl. of sitatunga, source of black dye (like ***A. floribunda*** Muell. Arg. (W Afr.)), hallucinogen taken by gorillas & chimpanzees; ***A. laxiflora*** (Benth.) Pax & K. Hoffm. (trop. & S Afr.) – stem a chewing-stick in Nigeria; ***A. triplinervia*** (Spreng.) Muell. Arg. (Amaz.) – insect antifeedants; ***A. villosa*** (Benth.) Muell. Arg. (Mal.) – bark fibre used for string

Alchorneopsis Muell. Arg. Euphorbiaceae (Acal.-Car.). 1 trop. Am.: ***A. floribunda*** (Benth.) Muell. Arg.

Alcimandra Dandy = *Magnolia*

Alciope DC. = *Capelio*

alcohol fermentation of sugars (incl. honey) & starches gives ethanol, further purified by distillation. Ethanol increases levels of healthy high-density lipoprotein cholesterol & reduces blood clot risk, but liver converts it to acetaldehyde (hangovers), a mutagen. Wide range of plant sources for industrial alc. or potable (beer 3–5%, wine 10–12%), e.g. *Saccharum* for car fuel (Braz.), grapes, cereals, incl. maize, rice, barley, rye, potatoes, sugar-beet, bananas, *Agave* & many palms incl. *Arenga, Borassus, Caryota, Cocos, Elaeis, Nypa, Phoenix*, but yeast fermentation stops at 13% alc. as enzymes inhibited so spirits (37–60%; e.g. c. 55 l consumed per head per annum in London 1700s) require distillation

Aldama Llave. Compositae (Helia.-Helia.). 2 C & trop. S Am. R: SB 14(1989)580

alder *Alnus* spp.; **black, common** or **Eur. a.**, *A. glutinosa*; **a. buckthorn** *Frangula alnus*; **green a.** *A. alnobetula* subsp. *crispa*; **grey a.** *A. incana*; **hazel a.** *A. serrulata*; **Italian a.** *A. cordata*; **Japanese a.** *A. japonica*; **Oregon** or **red a.** *A. rubra*; **white a.** *A. incana*

Aldina Endl. Leguminosae (III 1). 15 trop. S Am. R: MNYBG 8(1953)103. Timber

Aldrovanda Monti ex L. Droseraceae. 1 C Eur., As. to NE Aus.: ***A. vesiculosa*** L. (waterwheel) – Tertiary relict (poss. only 50 natural pops left), rootless aquatic with whorls of 4–9 lvs, each with a trap like *Dionaea* (sister genus, so such traps evolved only once from 'fly-paper' type) for insect-capture, comprising pr of concave lobes connected by midrib each with 6–10 trigger hairs; 2 (1 in young lvs) hairs struck leads to Ca^{2+} ion influx in midrib cells perhaps activating membrane ATPase responsible for K^+ transport leading to turgor loss & lobes coming together; obligate self-poll. incl. cleistogamy. R: A. Cross (2012) *A.: the waterwheel pl.*

Aldrovandaceae Nakai = Droseraceae

ale orig., beer brewed without hops

alecost *Tanacetum balsamita*

Alectorurus Makino = *Comospermum*

Alectra Thunb. Orobanchaceae (Esc.; Scrophulariaceae s.l.). 40 trop. Afr., As. R: NBGB 15(1941)423. Some holomycotrophs. Some dyes e.g. orange-yellow from ***A. sessiliflora*** (Vahl) Kuntze. ***A. picta*** (Hiern) Hemsl. (S Afr.) – noxious weed of legumes

Alectryon Gaertn. Sapindaceae. Incl. *Heterodendrum*, 34 E Mal. (R: Blumea 33(1988)313), Aus. (13, 12 endemic), NZ, New Caled. to Hawaii. Seeds black with red arils; some timbers & cult. orn. incl. ***A. oleifolius*** (Desf.) S. Reynolds (*Heterodendrum o.*, Aus. rosewood, boonaree, boonery, Aus.) – fr. eaten fresh by Aborigines, imp. fodder tree in inland Aus., & ***A. excelsus*** Gaertn. (titoki, NZ) – tough wood for tool-handles etc.; ***A. myrmecophilus*** Leenh. (NE NG) – twigs myrmecophilous

alehoof *Glechoma hederacea*

Aleisanthia Ridl. Rubiaceae (I 5). 2 Malay Pen. R: NJB 16(1996)563

Aleisanthiopsis Tange. Rubiaceae (I 5). 2 Borneo. R: NJB 16(1996)571

Alepidea Delaroche. Umbelliferae (II 1). 40 trop. & esp. S Afr. R: BN 4(1949)257. Some medic.

Alepidocalyx Piper = *Phaseolus*

Alepidocline S.F. Blake. Compositae (Mill.-Gal.). Incl. *Cuchumatanea*, 6 Mex. to Venez. R: Phytol. 69(1990)387. ***A. sp.*** (*C. steyermarkii*, Guatemala) – montane ann. less than 1 cm tall

Alepis Tieghem (~ *Elytranthe*). Loranthaceae (3). 1 NZ: ***A. flavida*** (Hook.f.) Tieghem

Aleppo galls *Quercus pubescens* etc.; **A. pine** *Pinus halepensis*

alerce *Fitzroya cupressoides, Tetraclinis articulata*

Aletes J. Coulter & Rose. Umbelliferae (III 8). 15–20 W N Am. ***A. anisatus*** (A. Gray) W.L. Theob. & Tseng – lvs ed.

Aletris L. Nartheciaceae (Melanthiaceae s.l.). Excl. *Metanarthecium*, 20 E As., N Am. (5). Stargrass. Some cult. orn. ***A. farinosa*** L. (ague r., colic r., unicorn r., N Am.) – medic. esp. diuretic

Aleurites Forst. & G. Forst. Euphorbiaceae (Al.-Al.). (Excl. *Vernicia*) 2 Indomal., W Pacif. R: Blumea 44(1999)79. ***A. moluccanus*** (L.) Willd. (*A. trilobus*, candlenut oil tree, candleberry, balucanat, kemiri nuts, kukui nut, Otaheite walnut; SE As., widely grown (prob. from 'Hoabinhian' times (6000–1000 BC) on), even on otherwise useless land, & natur. in trop.) – source of China wood oil, candlenut oil, lumbang oil with similar uses, also in curry & comm. shampoos, & seeds used like conkers (*Aesculus hippocastanum*) in Malay Pen., soot from burnt ones form. used in tattooing in Tonga, State tree of Hawaii where wood used for canoes, r. bark for a red dye (also in ikat dyeing in Bali), seeds for illumination (oil) & condiments, in leis & Western costume jewellery

Aleuritopteris Fée = *Allosorus*

Alexa Moq. = *Castanospermum* (but see AMBG 82(1995)551, BZ 95(2010)1160)

alexanders *Smyrnium olusatrum*

Alexandra Bunge = *Suaeda*

Alexandrian senna *Senna alexandrina*

Alexeya Pakhamova = *Paraquilegia*

Alexfloydia B. Simon (~ *Panicum*). Gramineae (XXIV 4). 1 Coffs Harbour, NSW: ***A. repens*** B. Simon – endangered. R: Austrobaileya 3(1992)670

Alexgeorgea Carlq. Restionaceae. 3 SW Aus. R: AusSB 3(1990)752. Geocarpic

Alexitoxicon St-Lager = *Vincetoxicum*

alfa = halfa

alfalfa *Medicago sativa*

Alfaroa Standl. Juglandaceae. 5 Mex. to Colombia. R: AMBG 65(1978)1078

Alfaropsis Iljinskaya = *Engelhardtia*

Alfredia Cass. Compositae (Card.-Card.). 4 C–E As.

algarobilla *Balsamocarpon brevifolium*

algarroba *Ceratonia siliqua* (Eur.); *Prosopis* spp. esp. *P. chilensis, P. glandulosa* (Am.)

Algerian fibre *Chamaerops humilis*; **A. grass** *Macrochloa tenacissima*; **A. oak** *Quercus canariensis*

Algernonia Baill. Euphorbiaceae (Euph.-.Hur). Incl. *Tetraplandra*, 11 E Braz. R: Britt. 65(2013)313

Algrizea Proença & NicLugh. Myrtaceae (II 10). 1 Bahia: ***A. macrochlamys*** (DC.) Proença & NicLugh. R: SB 31(2006)320

algum or **almug** Classical E Med. wood used for harps & lutes, erroneously believed to be *Pterocarpus santalinus*

Alhagi Gagnebin. Leguminosae (III 25). 1 Med. incl. Eur. to Nepal: ***A. maurorum*** Medik. – xerophyte, withered pls blow around in dry season; branches exude sap, hardening into brownish lumps (manna), effective laxative. R: KB 57(2002)440

Alibertia A. Rich. ex DC. Rubiaceae (Ix.-Cord.). Incl. *Borojoa*, 35 trop. Am. Some ed. fr.

× **Aliceara** Iwanaga & M. Wreford. Orchidaceae. *Brassia* × *Miltonia* × *Oncidium* – 270+ grexes

Alicia W.R. Anderson (~ *Mascagnia*). Malpighiaceae. 2 S Am. R: Novon 16(2006)174

Aliciella Brand (~ *Gilia*). Polemoniaceae (III). 21 W N Am. R: Aliso 17(1998)25

Aliella Qaiser & Lack = *Phagnalon* (but see BJ 106(1986)488)

Alifana Raf. An older name for *Brachyotum* (q.v.)

Aligera Suksd. (~ *Plectritis*). Caprifoliaceae (Valerianaceae). 15 W N Am.

aligna *Afzelia* spp.

Aliniella Raynal = *Alinula*

Alinorchis Szlach. = *Habenaria*

Alinula Raynal = *Cyperus* (but see BJBB 58(1988)457)

Aliopsis Omer & Qaiser (~ *Gentianella*). Gentianaceae. 1 C As.: ***A. pygmaea*** Omer & Qaiser – 1–5cm tall. R: Willd. 21(1991)190

Alisma L. Alismataceae. 11 N temp. (Eur. 4), Aus. (2). R: OB 19(1968)98. Water plantains, esp. ***A. plantago-aquatica*** L. (Euras. to NE Aus., introd. Alaska) s.t. cult. orn., medic. & ed., though base believed poisonous

Alismataceae Vent. Magnoliidae – Alismatales. Incl. Limnocharitaceae, 14/100 cosmop. esp. N temp. Perenn. (ann. in seasonal water), usu. glabrous aquatics (some free-floating) & marsh pls with rhizomes, vessels only in r., laticifers present. Lvs in spirals, oft. dimorphic, juvenile linr. submerged & mature linr. to ovate or sagittate emergent, usu. with petiole, sheathing at base. Infls usu. thyrsoid with whorls of branches or umbel-like. Fls hermaph. to unisexual (dioec. in *Burnatia*), solit. or in cymose umbels. K 3; C 3, usu. white, ephemeral, occ. 0 in female fls of *Burnatia* & *Wiesneria*; A 3 or 6 – ∞, when 6 in prs next to petals, when ∞ in dense whorls, developing centripetally (centrifugally in *Butomopsis*, *Hydrocleys* & *Limnocharis* s.t. separated as Limnocharitaceae), anthers extrorse, pollen 9–29-porate, rarely 2(3)-porate (*Caldesia*); $\underline{G}$ 3 or 6 – ∞, each carpel (s.t. distally unclosed) with 1 (rarely more) anatr. to campylotropous bitegmic ovules. Fr. a head of achenes (or follicles), usu. dehiscent; seeds without endosperm, embryo horseshoe-shaped. x = 5–13 Chief genera: *Alisma, Echinodorus, Sagittaria.* Generic distinctions unsatisfactory esp. *Caldesia, Ranalisma & Echinodorus.* Some cult. orn. & ed. lvs

aliso *Platanus occidentalis*

alison, hoary *Berteroa incana*

Alistilus N.E. Br. Leguminosae (III 18). 3 S trop. Afr., Madag.

alizarin *Rubia tinctorum*

alkali grass (Am.) *Distichlis*, or *Puccinellia* spp.

alkaloids small organic compounds, many derived from amino-acids, & cont. heterocyclic nitrogen. Some are deterrent to animals & several are of importance in medicine: they are also involved in moving nitrogen around the plant. Generally named after the genus of the pl. where orig. found, e.g. digitalin (*Digitalis*), nicotine (*Nicotiana*), strychnine (*Strychnos*). Colchicine (from *Colchicum*) is used in plant breeding to double chromosome numbers somatically: it inhibits spindle functioning in mitosis so that chromosomes do not sep. to form 2 sep. nuclei, but remain together, forming 1

alkanet *Alkanna tinctoria, Pentaglottis sempervirens*; **bastard a.** *Buglossoides arvensis*

Alkanna Tausch. Boraginaceae (2.2). c. 40 S Eur. (17), Med. to Iran. ***A. tinctoria*** Tausch (*A. lehmannii, A. matthioli, A. tuberculata*, Med., alkanet) – r. source of red dye (form. extracted from camel urine), form. used to tint inferior port & colour thermometer fluids, chemists' shop bottles & to detect fats (now replaced by Sudan IV etc.)

Alkekengi Tourn. ex Mill. (~ *Physalis*). Solanaceae (1). 1 C & S Eur., W. As. to Jap.: ***A. officinarum*** Moench (*P. alkekengi, P. franchetii*, Chinese lantern, winter cherry) – cult. orn. 'everlasting' with vermilion inflated calyces (toxic) around ± ed. berry (calyces skeletonizing as frs redden), form. medic., f. ***monstrosa*** (Miq.) J.M.H. Shaw – peduncles sterile, pendent catkin-like shoots

Allaeophania Thwaites = *Metabolos*

Allagopappus Cass. Compositae (Inul.-Inul.). 2 Canary Is.

Allagoptera Nees. Palmae (V 8a). Incl. *Polyandrococos* 5 Braz., Bolivia, Parag., Arg. R: FN 73(1996)1. Schoute's Model. Some cult. orn., some ed. fr. ***A. caudescens*** (Mart.) Kuntze (*P. c.*, E Braz.) – unbranched infl. covered with A

Allamanda L. Apocynaceae (I 9). c. 15 trop. Am. R: RBB 9(1980)125. Champagnat's Model; seeds hairy. ***A. cathartica*** L. (S Am.) widely cult. orn. liane, dangerous purgative

Allanblackia Oliv. Guttiferae (2.). 9 trop. Afr. R: BJBB 39(1969)347. Seeds a source of oils, used as butter subs. during World War II. ***A. floribunda*** Oliv. (W Afr., kisidwe), ***A. oleifera*** Oliv. (C Afr., kagné butter), ***A. stuhlmannii*** (Engl.) Engl. (E Afr., mkani fat), bark a source of red dye

Allantoma Miers. Lecythidaceae (I). 15 NE S Am.

Allantospermum Forman. Irvingiaceae. 1 Mal., 1 Madag. Form. placed in Simaroubaceae or Ixonanthaceae

Allardia Decne (~ *Waldheimia*). Compositae (Anth.-Han.). 8 C & E As. R: BBMNHB 23(1993)98

Alleghany blackberry *Rubus allegheniensis;* **A. plum** *Prunus umbellata*

Alleizettea Dubard & Dop = *Danais*

Alleizettella Pitard. Rubiaceae (II 4). 2 SE As.

Allemanda L. = *Allamanda*

Allenanthus Standl. Rubiaceae (III 4). 3 C Am.

Allendea Llave = *Liabum*

Allenrolfea Kuntze. Amaranthaceae (Chenopodiaceae II 2). 2–3 SW US. ***A. occidentalis*** (S. Watson) Kuntze – seeds ed.

Allexis Pierre. Violaceae (III 1a). 4 trop. W Afr. Some 'litterbox pls'. ***A. cauliflora*** (Oliv.) Pierre – Corner's Model, 1-seeded fr. expodes

allgood *Blitum bonus-henricus*

allheal *Stachys palustris, Valeriana officinalis*

Alliaceae Borkh. = Amaryllidaceae

Alliaria Heister ex Fabr. Cruciferae (47). 1 Eur., temp. As. ***A. petiolata*** (M. Bieb.) Cavara & Grande (Jack-by-the-hedge, garlic mustard, hedge garlic, Eur., invasive N Am.) – exudes glucosinolates inhibiting competing pls & their mycorrhizae, form. used as subs. garlic flavouring (higher in vitamin A than spinach & in C than orange juice). R: JAA 69(1988)218

alligator apple *Annona* spp.; **a. pear** *Persea americana;* **a. pepper** *Aframomum melegueta;* **a. weed** *Alternanthera philoxeroides;* **a. wood** *Guarea guidonia*

Allionia L. Nyctaginaceae (6). 1–2 C & W USA to Chile & Arg. R: Phytol. 77(1994)46. Fr. glandular like *Pisonia*

Allioniella Rydb. = *Mirabilis*

Allittia P. Short = *Brachycome* (but see JABG 28(2014)26)

Allium Tourn. ex L. Amaryllidaceae (Alliaceae). Incl. *Caloscordum* (without odour, united P), *Milula* (spicate infl.), *Nectaroscordum* (semi-inf G), c. 880 Euras. (Eur. 117) esp. C As., Am. (N Am. 96), to Afr., Sri Lanka & Mex. (R: M. Gregory et al. (1998) *Nomenclator Alliorum;* (sect. *A.*) B. Mathew (1996) *A review of A. sect. A.*). Mostly perenn. bulbous or rhiz. herbs, with strong smell (largely aliphatic disulphides) when bruised (allinase combines with alliin to form allicin), antifungal, antibacterial (delaying growth of salmonella in meat noted by Pasteur, 1858; toxic to cats & dogs; avoided by some Hindus) some med.; hybrids almost unknown. Many cult. ed. (e.g. ***A. altaicum*** Pallas (C As.) – poss. ancestor of *A. fistulosum* & *A. sativum*, ***A. ramosum*** L. (C As.) & ***A. victorialis*** L. (Euras.) in Mongolian markets) & orn. (D. Davies (1992) *The ornamental onions*) esp. ***A. cristophii*** Trautv. (*A. bodeanum*, '*A. christophi*', Turkey, Iran); ***A. ampeloprasum*** L. (Euras., N Afr.) – Russian garlic infl. scapes ed. comm.; 3 Groups: **Ampeloprasum** = Levant garlic, large bulbs (elephant garlic) used as seasoning, **Porrum** = leek, narrow bulbs & leaf-bases eaten (15000 t prod. per annum in UK alone; leek shows in NE Engl. since 1880s), known to Sumerians, infusion used to wash windows deters flies (saponins effective against leek moth larvae), that of seeds as a kind of Anglo-Saxon toothpaste, outer layers of leeks giving a hair-bleach, national emblem of Wales, allegedly commemorating victory of Cadwalladr over Saxons (AD 640) when worn as distinguishing markers, **Kurrat** = kurrat (poss. leek of the Bible), lvs ed.; *A. ascalonicum* – name misapplied to shallot; ***A. canadense***

L. (Canada garlic) – bulbs form. consumed by Native Americans; ***A. cepa*** L. (cultigen, onion, poss. derived from *A. vavilovii* Popov & Vved., but known from Dead Sea Chalcolithic (c. 3500–3000 BC)) – flavour due to dipropyldisulphide, propanethiol), 2 groups: **Cepa** = onion, single bulbs to over 5 kg without bulbils in infl. (6 types grown at time of Pliny) – world prod. 12 M tonnes per annum (esp. Med., Jap., US), much dehydrated or powdered as well as marketed fresh ('2 lbs' a day enough to cause anaemia, though high in anti-oxidant quercetin), blanched forms in Catalonia = calçots (Calçotada Festival in Jan.), damaged tissue releasing sulphurous volatiles (hence weeping – avoided by breathing through nose only), 10 compounds interacting to inhibit platelet aggregation & thereby blood-clotting & thrombosis, infl. stalk pith used as reeds in Turkish wind instruments, **Aggregatum** = shallot ('*A. ascalonicum*' the original 'scallion', a name now referred to 'spring onions' (young Cepa) or (US) leek), multiplier or potato o., produces lat. bulbs & none in infl. (oft. sterile), widely used as pickled o.; ***A. cernuum*** Roth (lady's leek, N Am.) – ed. (strong), local medic.; ***A. chinense*** G. Don (chiao t'ou (China), rakkyō (Jap.), As.) – pickled; ***A.* x *cornutum*** Clementi ex Vis. (*A. cepa* × ?) – sterile triploid evergreen 'shallot' with infl. bulbils; ***A. fistulosum*** L. (Welsh (i.e. Ger. *welsche*, foreign) o., Jap. bunching o., cultigen app. derived from *A. altaicum*, natur. on turf roofs in Norway) – ed. lvs for salads; ***A. insubricum*** Boiss. & Reut. (N Italy) – cult. orn., pendent 1-fld scapes, ***A. moly*** L. (moly, S France, Spain) – cult. orn. yellow fls; ***A. oleraceum*** L. (field garlic, Eur.) – occ. gathered for food; ***A.* × *proliferum*** (Moench) Willd. (*A. cepa* × *A. fistulosum*, Egyptian or tree o.) – diploids & triploids in cult., prop. by bulbils in infl.; ***A. sativum*** L. (garlic, cultigen, poss. derived from *A. altaicum* but known from Dead Sea Chalcolithic, c. 3500–3000 BC; 6 types cult. at time of Pliny, world prod. (2012; 77% in China) 17+ M t) – strong flavouring (due to di-2-propenyldisulphide) for food ('black garlic' fashionable in US is caramelized), also in chewing-gum, some Finnish vodka & garlic salt for cooking, bactericidal & much used med. (e.g. toothache in Sumatra, insect bites in S Eur., supplements in Br. reducing risk of colds by half) since time before Galen, against infections, allicin effective against the antibiotic-resistant hospital 'superbug' MRSA, also effective in circulatory (lowers blood pressure) & resp. disorders, cholesterol levels & preventing tumours (colon, prostate etc.), & in hort. (since time of Pliny) where effective against foot-rot of runner beans, slug poison, also in superstition in bride's bouquet in Greece & worn round neck to ward off Trolls, vampires etc. (var. ***ophioscorodon*** (Link) Doell is true rocambole); ***A. schoenoprasum*** L. (chives, Euras.) – lvs used in cooking; ***A scorodoprasum*** L. (sandleek, Euras.) – bulbs ed.; ***A. sphaerocephalon*** L. (round-headed garlic or leek, Eur. & Med.) – bulbs ed.; ***A. tricoccum*** Aiton (ramps, E N Am.) – 'gourmet' ingredient in US (spring Ramps Festival in W Virginia); ***A. triquetrum*** L. (Eur., Medit.) – invasive in Aus.; ***A. tuberosum*** Rottler ex Spreng. (garlic chives or Chinese c., SE As.) – ed.; ***A. ursinum*** L. (ramsons, wild or wood garlic, Eur.) – medic. (corr. long-believed in Aran Is. to be effective against blood-clots – allicin lowers cholesterol), lvs cross over in development so that 'lower' surfaces are on top; ***A. vineale*** L. (crow or false garlic, Eur.) – bad weed (wheat-sized bulbils in US wheat-fields gave garlic-flavoured bread, cows in infected pastures yielding garlicky milk), molluscicidal steroid saponins, '**Hair**' – viviparous, head of plantlets with wispy green tendril-like processes

Allmania R. Br. ex Wight. Amaranthaceae (I 2). 1–2 trop. As.

Allmaniopsis Süsseng. Amaranthaceae (I 2). 1 E Kenya: ***A. fruticulosa*** Süsseng.

Alloburkillia Whitm. = *Burkilliodendron*

Allocalyx Cordemoy = *Bacopa*

Allocarya Greene = *Plagiobothrys*

Allocaryastrum Brand = *Plagiobothrys*

Allocassine N. Robson. Celastraceae (I). 1 SE trop. & S Afr.: ***A. laurifolia*** (Harv.) N. Robson. R: SAJB 64(1998)189

Allocasuarina L. Johnson (~ *Casuarina*). Casuarinaceae. 58 Aus. (esp. S). R: JABG 6(1982)73. Dioec. or monoec. Sheoaks (some), wood exported early C19 as Botany Bay oak, B. B. wood or beefwood. ***A. littoralis*** (Salisb.) L. Johnson – ash a subs. for lye for cleaning printing presses in early Vic.; ***A. luehmannii*** (R. Baker) L. Johnson (E Aus.) – bull oak; ***A. torulosa*** (Aiton) L. Johnson (NE Aus.) – form. imp. as shingles

Allocephalus Bringel & al. Compositae (Vern.-Vern.). 1 Brazil: ***A. gamolepis*** Bringel & al. R: SB 36(2011)785

Allocheilos W.T. Wang. Gesneriaceae (III 2j). 2 S China

Allochilus Gagnepain = *Goodyera*

Allochrusa Bunge ex Boiss. = *Acanthophyllum*

Alloeochaete C.E. Hubb. Gramineae (Arund.). 6 S trop. Afr. R: KB 30 (1975)570. ***A. oreogena*** Launert (Mt Mlanje) – tree-trunk-like tussocks to 1.5 m

Alloispermum Willd. (~ *Galinsoga*). Compositae (Mill.-Gal.). c. 15 Mex., C Am., N Andes. R: Phytol. 38(1978)411, 68(1990)134

Allolepis Soderstrom & Decker (~ *Distichlis*). Gramineae (XXIX 1). 1 S US & Mex.: ***A. texana*** Soderstom & Decker. R: SCB 87(1997)22

Allomaieta Gleason. Melastomataceae. Incl. *Cyphostyla*, 8 Colombia. R: IJPS 172(2011)1175

Allomarkgrafia Woodson (~ *Mesechites*). Apocynaceae (II d). 10 trop. Am. R: Britt. 49(1997)337

Allomorphia Blume (~ *Oxyspora*). Melastomataceae. 25 SE As., local medic.

Alloneuron Pilg. (~ *Loreya*). Melastomataceae. 4 Colombia, Peru. R: IJPS 172(2011)1175. See also *Wurdastom*

Allophylastrum Acev.-Rodr. (~ *Allophyllus*). Sapindaceae (IV 1). 1 N S Am.: ***A frutescens*** Acev.-Rodr. R: Phytokeys 5(2011)40

Allophyllum (Nutt.) A.D. & V. Grant. Polemoniaceae (III). 6 W US. R: Aliso 3(1955)93

Allophylus L. Sapindaceae (IV 1). 1 v. polymorphic sp., ***A. cobbe*** (L.) Forsyth f., trop. & warm, which at a local level can be subdivided into c. 175 app. distinct biological spp. R: Blumea 15(1967)301. Cult. orn., useful wood, somewhat medic., fr. used to stun fish in New Ireland

Alloplectus Mart. Gesneriaceae (II 5c). Excl. *Crantzia, Glossoloma*, 5 NW S Am., Am. R: Selbyana 25(2005)186. Cult. orn. close to *Columnea*

Allopterigeron Dunlop. Compositae (Inul.-Pluch.). 1 trop. Aus.: ***A. filifolius*** (F. Muell.) Dunlop. R: FA 37(2015)429

Allosanthus Radlk. = *Thinouia*

Alloschemone Schott (~ *Scindapsus*). Araceae (IV 4). 2 Amaz. Braz. R: Aroideana 24(2001)85

Alloschmidia H. Moore = *Basselinia*

Allosidastrum (Hochr.) Krapov. & al. (~ *Pseudabutilon*). Malvaceae (Malv.-Malv.). 4 trop. Am. R: BSBM 48(1988)23

Allosorus Bernh. (*Aleuritopteris*; ~ *Cheilanthes*). Pteridaceae. ?45 OW trop. ext to Medit. (8; R: Willd. 42(2012)284)

Allospondias (Pierre) Stapf (~ *Spondias*). Anacardiaceae (II). 2 SE As.

Allostigma W.T. Wang. Gesneriaceae (III 2j). 1 S China: ***A. guangxiensis*** W.T. Wang

Allosyncarpia S.T. Blake. Myrtaceae (II 11). 1 N Aus. (Arnhem Land): ***A. ternata*** S.T. Blake

Alloteropsis J. Presl. Gramineae (XXIV 2). 5 OW trop. R: SB 13(1988)587. ***A. semialata*** (R. Br.) Hitchc. (cockatoo grass) – seeds ed. cockatoos, diploid (2n = 18) C_3, hexaploid (2n = 64) C_4!

Allotoonia J. Morales & J.K. Williams (~ *Echites*). Apocynaceae (II e). 5 Mex., C Am. Carib. R: Sida 21(2004)135

Allotropa Torrey & A. Gray. Ericaceae (II 3; Monotropaceae). 1 W US: ***A. virgata*** Torrey & A. Gray. R: FNA 9(2009)391

Allowissadula Bates. Malvaceae (Malv.-Malv.). 9 Texas & Mex. R: GH 11(1978)329

Allowoodsonia Markgraf. Apocynaceae (II b). 1 Solomon Is.: ***A. whitmorei*** Markgraf

Alloxylon P. Weston & Crisp (~ *Oreocallis*). Proteaceae V 3). 4 NG & E Aus. (3 endemic – satin oak)

allseed *Radiola linoides*; **four-leaved a.** *Polycarpon tetraphyllum*

allspice *Pimenta dioica, Chimonanthus praecox*; **Calif. a.** *Calycanthus occidentalis*; **Carolina a.** *C. floridus*, **a. jasmine** *Gelsemium* spp.; **wild a.** *Lindera* spp.

allthorn *Koeberlinia spinosa*

Alluaudia (Drake) Drake. Didiereaceae. 6 S & SW Madag. R: FMad. 121(1963)12. Koriba's Model. 2n = 240. Stems used for building

Alluaudiopsis Humbert & Choux. Didiereaceae. 2 S & SW Madag. R: FMad. 121(1963)6. 2n = 48

Almaleea Crisp & P. Weston (~ *Pultenaea*). Leguminosae (III 13). 5 SE Aus. R: Telopea 4(1991)307

Almeidea A. St-Hil. Rutaceae (I 8). 5 NE S Am.

almeidina see *Euphorbia tirucalli, Hevea brasiliensis*

almon *Shorea almon*

almond *Prunus dulcis*; **Barbados a.** *Terminalia catappa*; **bitter a.** *P. dulcis* var. *amara*; **country a.** (Ind.) *Terminalia catappa*; **Cuddapah a.** *Buchanania lanzan*; **dog a.** *Andira inermis*; **dwarf Russia a.** *Prunus tenella*; **earth a.** *Cyperus esculentus*; **Ind. a.** *Terminalia catappa*; **Java a.**

Canarium luzonicum; **red a.** *Alphitonia* spp.; **wild a.** (S Afr.) *Brabejum stellatifolium*, *Terminalia catappa*

almondette *Buchanania lanzan*

almug See algum

Almutaster Löve & D. Löve (~ *Aster*). Compositae (Ast.- Sym.). 1 W N Am.: ***A. pauciflorus*** (Nutt.) Löve & D. Löve. R: Phytol. 77(1994)250

Alniphyllum Matsum. Styracaceae. 3 SW China, Taiwan, SE As. Cult. orn.

Alnus Mill. Betulaceae (I). 35 N temp. (Eur. 4), S to Assam, SE As. & Andes (Colombia 1). R: JAA 71(1990)18. Alders. Attims's Model. Male catkins long, females short, becoming woody cones with 5-lobed scales & minute winged nutlets. Fls before lvs in many spp. but several, notably trop. spp., fl. in autumn when lvs still present (cf. *Quercus*). Useful timber & dyes, r. with nitrogen-fixing actinomycete symbionts, twigs accumulate gold, pollen significant in hay-fever, cult. orn. Planted spp. incl.: ***A. acuminata*** Kunth (Mex. to N Arg.) – fuelwood; ***A. alnobetula*** (Ehrh.) K. Koch (*A. viridis*, Eur., E N Am.), incl. subsp. ***crispa*** (Aiton) Raus (*A. crispa*, E N Am., green a.) – N Am. local medic., red dye; ***A. cordata*** (Loisel.) Duby (Italian a., Corsica & S Italy, NW Albania); ***A. glutinosa*** (L.) Gaertn. (black, common or Eur. a., Euras., Med.) – wood for carving &, allegedly, Stradivarius violins & form. clogs (wood a poor conductor of heat) & gunpowder charcoal, for building scaffolding in Elizabethan times & (Virgil) for the first boats, the piles of Venice (durable under water), bark for tanning (reddens leather), medic. (decoction of 'cones' for gout in W Eng.) esp. gargle, many orn. cvs (incl. '**Imperialis**' with highly lacinate lvs, 1859), 20 000 killed UK 1993–99 by hybrid *Phytophthora*; ***A. incana*** (L.) Moench (white or grey a., N temp.) – in N Am. local medic., fibre, dye (orange); ***A. japonica*** (Thunb.) Steud. (Jap. a., NE As., Jap.) – charcoal for gunpowder; ***A. rubra*** Bong. (*A. oregona*, Oregon or red a., W N Am.) – said to be best for fish-smoking, canoes, local medic., red dye; ***A. serrulata*** (Aiton) Willd. (*A. rugosa*, hazel a., E N Am.) – local medic.

Aloaceae Batsch ('Aloeaceae') = Asphodelaceae

Alocasia (Schott) G. Don. Araceae (VII 25). 80–100 Indomal. (R (W, C Mal.): GBS 50(1998)236, 51(1999)3) to Aus. (13, R: Blumea 35(1991)499). Differs from *Colocasia* in having basal ovules. Fly-poll., fr. bird.-disp. Some ed. rhiz., local medic., cult. orn. with many hybrids. ***A. cucullata*** (Lour.) G. Don (cult. SE As., wild form = '*A. odora* (Lodd.) Spach', Jap. to SE As.) – unsealed G, no fr., introd. Hawaii by Chinese; ***A. macrorrhizos*** (L.) G. Don (giant taro, cunjevoi, Indomal.) – wild earwig-poll. (thermogenesis) pops in Vanuatu, cult. for ed. rhiz. & orn., lvs to 3.02 x 1.92m, with ***A. robusta*** M. Hotta (Borneo), largest undivided lvs known, sap an antidote to *Dendrocnide gigas* stings; ***A. longiloba*** Miq. (*A. picta*, *A. veitchii*, S. China to C Mal.) – housepl.

Alococarpum Riedl & Kuber (~ *Prangos*). Umbelliferae (III 5). 1 Iran: ***A. erianthum*** (DC.) Riedl & Kuber

aloe *Aloe* spp. (of Psalms = lign-aloes, q.v.); **Am. a.** *Agave americana*; **Bombay a.** *A. cantala*; **Cape a.** *Aloe ferox*; **Curaçao a.** *A. vera*; **partridge-breasted a.** *Gonialoe variegata*; **a. wood** *Aquilaria malaccensis* (= aloe (ahaloth) of Bible, 'aloexylum')

Aloe Tourn. ex L. Asphodelaceae. Incl. *Chortolirion*, excl. *Aloiampelos*, *Aloidendron*, *Aristaloe*, *Gonialoe*, *Kumara*, 400–560 (+ 32 nat. hybrids) trop. & esp. S Afr., Madag., Arabia, Canary Is. R: S. Carter & al., *Aloes* (2011). Treelets to sessile pls with term. rosettes of succ., non-fibrous (cf. *Agave*) lvs, usu. toothed or spiny at margin, & infls of yellow or red bird-poll. fls with coloured nectar, ? some pseudogamous apomixis. Cult. orn. inc. hybrids with *Gasteria* spp. (× *Gasteraloe*). Yellow juice of cut & drained lvs dried = drug (bitter) aloes, purgative (active principle an anthraquinone, barbaloin acting on colon) & used against nail-biting, esp. from *A. ferox* & *A. vera* though some with toxic hemlock alks, e.g. ***A. ballyi*** Reynolds (E Afr.); ***A. ferox*** Mill. (Cape aloe, S Afr.) – bird- & bee-poll., source of Cape aloes; ***A. humilis*** (L.) Mill. (S Afr.) – sessile cult. orn., several cvs; ***A. maculata*** All. ('*A. saponaria*', *A. latifolia*, S Afr.) – used in Afr. treatment of boils etc., form. for tanning; ***A. perryi*** Bak. (Socotra) – best & most famous 'aloes' (Socotrine or Zanzibar a.); ***A. polyphylla*** Schönland ex Pill. (Lesotho) – nat. fl., highly endangered; ***A. vera*** (L.) Burm.f. (*A. barbadensis*, Barbados or Curaçao a., SW Arabia, long natur. Medit., intr. warm Am.) – mentioned in Babylonian texts, leaf parenchyma comm. source of aloes esp. in Carib. (worth $70–80 M US), used in shampooing, constituent of many cosmetics, antibiotic barbaloin active against tubercle bacilli & long used in skin infections (penetrates 4 times faster & 7 times deeper than water) in Arabia & used to coat (esp. sliced) fruit to extend storage time, modern remedy for burns incl. sunburn & dentistry, mouthwash etc. with over 200 active ingredients (incl. all ess. amino-acids for humans & 12 vitamins)

Aloeaceae Batsch = Asphodelaceae

aloes See under *Aloe*

aloexylum *Aquilaria* spp.

Aloiampelos Klopper & G. Sm. (~ *Aloe*). Asphodelaceae. 10 S Afr. R: Phytotaxa 76(2013)10. Scramblers

Aloidendron (A. Berger) Klopper & G. Sm. (~ *Aloe*). Asphodelaceae. 6 Afr. R: Phytotaxa 76(2013)9. Trees (Leeuwenberg's Model), ***A. barberae*** (Dyer) Klopper & G. Sm. (*Aloe bainesii*, S Afr.) to 18 m

Aloinopsis Schwantes (~ *Nananthus*). Aizoaceae (V). 8 W & C S Afr. R: H. E. K. Hartmann, *Ill. Handb. Succ. Pls*, Aiz. A–E (2002)33. Cult. orn.

Aloitis Raf. = *Gentiana*

Alomia Kunth. Compositae (Eup.-Alom.). 5 Mex. R: MSBMBG 22(1987)239

Alomiella R. King & H. Robinson. Compositae (Eup.-Ayap.). 2 Braz. R: Phytol. 56(1984)256

Alona Lindley = *Nolana*

Alonsoa Ruíz & Pavón. Scrophulariaceae (Hem.). 11 trop. Am., S Afr. (2). Some with elaiophores attractive to bees. Cult. orn. pot-pls, esp. ***A. warscewiczii*** Regel (Peru)

Alopecurus L. Gramineae (XVI 15). 52 N temp. (Eur. 14), S Am. R: Turk. J. Bot. 23(1999)245. Foxtail, meadow grasses esp. ***A. pratensis*** L. (meadow f., blackgrass, Euras., natur. N Am.)

Alophia Herb. Iridaceae (VII 5). c. 5 warm Am. Lvs pleated. Cult. orn. as 'Eustylus'. 'Alophia' in cult. = *Herbertia*

Aloysia Palau. Verbenaceae (Lant.). Incl. *Acantholippia*, *Xeroaloysia*, c. 30 Am. ***A. citrodora*** Palau (*A. triphylla*, *Lippia c.*, lemon verbena, cultigen?) – cult. orn., lemon flavouring in French liqueurs & used as a fragrant sedative tea in S Am., effective in potpourri

alpam root *Thottea* spp.

Alpaminia O. Schulz = *Weberbauera*

alpenrose *Rhododendron ferrugineum*

Alphandia Baill. Euphorbiaceae (Crot.-Trig.). 3 Melanesia

Alphitonia Reisseck ex Endl. Rhamnaceae (inc. sed.). c. 15 Mal., Aus. (5, 3 endemic), W Pac. R: KB 1925: 168. Roux's Model. Some fine red timbers (red almond, red ash (Aus.)) inc. ***A. ponderosa*** Hillebrand (kauila, Hawaii) – carving & beams, form. used for javelins etc., sinks in water, lvs & bark yield a bluish dye; ***A. excelsa*** (Fenzl) Benth. (soap tree, Aus.) – lvs yield soap subs.; ***A. petriei*** Braid & C. White (NE Aus.) – partly beetle-poll., few adventitious cli. because of allelopathic bark

Alphonsea Hook.f. & Thomson. Annonaceae (IV 7). 23 China, Indomal. R: BJ 118(1996)81

alpine ash (Aus.) *Eucalyptus delegatensis*, *Fraxinus nigra*; **a. azalea** *Kalmia procumbens*; **a. campion** *Silene suecica*; **a. poppy** *Papaver alpinum s.l.*; **a. rose** *Rhododendron ferrugineum*; **a. sow-thistle** *Cicerbita alpina*

Alpinia Roxb. Zingiberaceae (II 4). (Excl. *Pleuranthodium*) c. 200 warm As. to Pac. R: EJB 49(1990)1. Heterogeneous. Cult. orn. & medic. (antiseptic cineol). Rhiz. scented like ginger. Some monoec. ***A. galanga*** (L.) Willd. (galangal, Siamese ginger, trop. As.) – source of ess. oil, rhiz. a condiment, fls ed. Java; ***A. globosa*** Horan. (China) – rhiz. medic., seeds a condiment; ***A. kwangsiensis*** T.L. Wu & Senjen Chen (China) – heterodichogamous (2 morphs, 1 shedding pollen a.m., stigmas receptive p.m., the other the reverse); ***A. officinarum*** Hance (E & SE As., cult. China) – similar, form. used in some Russian tea (Nastroika) & liqueurs; ***A. zerumbet*** (Pers.) B.L. Burtt & R.M. Sm. ('*A. speciosa*', shell ginger) with hexagonal rhiz. system uniformly exploiting environment, cult. Hawaii

Alposelinum Pim. = *Lomatocarpa*

Alrawia (Wendelbo) Persson & Wendelbo. Asparagaceae (Hyacinthaceae). 2 Iran, Iraq

Alseianthiopsis Tange (~ *Greenea*). Rubiaceae (I 5). 2 Borneo. R: NJB 16(1996)571

Alseis Schott. Rubiaceae (Cond.). 16 S Mex. to Braz. Decid. with protogynous fls before or with new lvs

Alseodaphne Nees (~ *Dehaasia*). Lauraceae (I). Incl. *Nothaphoebe* 40 Indomal. R: Candollea 28(1973)95. Light timbers

Alseuosmia Cunn. Alseuosmiaceae. 5 NZ, some hybridizing. R: NZJB 38(2000)159. Cult. orn.

Alseuosmiaceae Airy Shaw. Magnoliidae – Asterales. 5/10 W Pacif. R: Blumea 29(1984)387. Shrubs, *Pittosporum*-like in habit, s.t. epiphytic. Lvs in spirals to subopp. or in 3, 4 (or 5)-verticillate pseudowhorls, simple with hairs in axils, usu. serrate. Fls reg., bisex., oft. incl. cleistogamous ones, heavily scented; K ((4)5–7); C((4)5(–7)), valvate, oft. crenate to erose; A ((4)5)–7), attached to C or free, alt. with C-lobes, anthers introrse with longit. slits; $\overline{G}$ (2) s.t. only half inferior, 2- or 3-loc. with nectary-disk & axile placentation, ovules 2 – ∞ per locule. Fr. a 2-loc. berry (caps. in *Platyspermation*) with 1 – ∞ endospermous seeds

Genera: *Alseuosmia, Crispiloba, Periomphale, Platyspermation, Wittsteinia*

Formerly considered allied to Caprifoliaceae, which have opp. lvs & valvate corolla-lobes with free A & different pollen; the constituent genera have been placed in Ericaceae, Gesneriaceae, Rutaceae etc.

alsike *Trifolium hybridum*

Alsinaceae Bartl. = Caryophyllaceae

Alsinidendron H. Mann = *Schiedea*

Alsinula Dostál = *Stellaria*

Alsmithia H. Moore = *Heterospathe*

Alsobia Hanst. (~ *Episcia*). Gesneriaceae (II 5c). 2 C Am. R: Selbyana 5(1978)28

Alsodeiopsis Oliv. Icacinaceae. 11 trop. Afr. Troll's Model. R. of some spp. supposed aphrodisiac. ***A. zenkeri*** Engl. – rheophyte

Alsomitra (Blume) Spach. Cucurbitaceae (I). 1 Indomal.: ***A. macrocarpa*** (Blume) M. Roem. – juv. v. small thread-like lvs on forest floor, seeds winged, to 120 mm across. R: FM I,19(2010)25

Alsophila R. Br. = *Cyathea*

Alstonia R. Br. Apocynaceae (I a). Excl. *Tonduzia*, 41 China, Indomal., W Pac., Afr. (2). R: Blumea supp. 11 (1998). Pagoda trees (Leeuwenberg's, Koriba's & Prévost's Models) with whorled lvs & many alks. Some useful timbers (milkwood for carving in N Aus.) & latex, used to adulterate rubber or jelutong, some med. (***A. angustiloba*** Miq. (Thailand, W Mal.) & ***A. boonei*** de Wild. (W Afr.) imp. local antimalarials), some ed. fr. ***A. congensis*** Engl. (pattern wood, alstonia, ahun, awun, trop. Afr.) – soft white wood for boats, war-drums etc.; ***A. constricta*** F. Muell. (bitter bark, fever bark, E Aus.) – med. & tonic, bitters; ***A. scholaris*** (L.) R. Br. (dita bark, Indomal. to Aus.) – long used med. as vermifuge, allegedly antimalarial, bark decoction for accelerating childbirth in Indonesia, wood used for coffins, *wayang golek* puppets (Java) & formerly slates (writing rubbed off with *Tetracera* lvs in Malaysia), hence specific name, now for gaudy ('dancer's') masks in Sri Lanka; ***A. spatulata*** Blume (SE As. to NG) – v. light root wood (Siamese balsa; driftwood reaching Marshall Is.) used for pith helmets, floats etc., to stop up corpse orifices, med.

Alstroemeria L. Alstroemeriaceae. 79 S Am. R: PL 8(1952)41. Lvs twisted at base during development so 'upper' surface facing downwards. rhiz. local starch sources (el chuño from *A. ligtu*). Cult. orn. esp. ***A. aurea*** Graham (*A. aurantiaca*, Peruvian lily, Chile (!), invasive Aus., poss. not distinct from *A. versicolor* Ruíz & Pavón) but being replaced by 'Ligtu hybrids' (*A. ligtu* L. × *A. haemantha* Ruíz & Pavón, both Chile)

Alstroemeriaceae Dumort. (Liliaceae s.l.). Magnoliidae – Liliales. Incl. Luzuriagaceae, 4/191 Aus., NZ, C & S Am. esp. Andes. Erect or twining perenn. (1 ann.) herbs, s.t. epiphytic, with vessels & symp. rhiz., some r. tuberous. Lvs in spirals or distichous, ovate or lanceolate to linear, usu. twisted at base during development so 'upper' surface faces downwards. Infls umbel-like helicoid cymes or fls solit. Fls 3-merous, bisex., ± reg.; P 3(4) + 3(4), similar or outer shorter, all oft. dotted, with nectaries at base of 2 or all inner; A 3 + 3 with longit. introrse dehiscence (or apical pores); $\overline{G}$ ((1)3(4)), rarely 1-loc., with ∞ anatr. bitegmic ovules on axile or parietal placentae. Fr. loculicidal caps. or berry; seeds ± globose. x = 8 or 9

Genera: *Alstroemeria, Bomarea, Drymophila, Luzuriaga*

Cult. orn. & some ed.

Altamiranoa Rose = *Sedum*

Altensteinia Kunth. Orchidaceae (IV 2b). 7 Andes

Alternanthera Forssk. Amaranthaceae (II 2). c. 80 trop. & warm, esp. Am. Chaff-flower, joyweed, broad path (WI). Some ed. like spinach, others much used as trop. bedding pls for spectacular lvs. ***A. ficoidea*** (L.) P. Beauv. (Mex. to Arg.) – widely cult. esp. '**Bettzichiana**', with blotched yellow & red lvs ('Jacob's coat'); ***A. philoxeroides*** (Mart.) Griseb. (alligator weed, S Am.) – introd. N Am. as cult. orn. & as crayfish fodder now a weed (app. by veg. reprod.) rivalling water hyacinth (controlled in SE US by S Am. stem-boring moth, leaf-eating flea-beetle & bud-feeding thrips), also in trop. As., Aus. (controlled by beetle, *Agascicles hygrophila*, in NSW), NZ & natur. Italy (Pisa, 1999); ***A. pungens*** Kunth (khaki weed, S Am.) – natur. Aus., where lawn weed & harmful to stock; ***A. sessilis*** (L.) DC. (mukunawanna, trop., invasive in S Afr., Aus.) – ed. like spinach

Althaea L. Malvaceae (Malv.-Malv.). 12 Eur. (5) to NE Siberia. ***A. officinalis*** L. (marsh mallow, Eur., natur. in US) – source of orig. marshmallow, r. medic. & used in salads, fibres form. used in paper-making

Althenia Petit. Potamogetonaceae (Zannichelliaceae). 1 Medit.: ***A. filiformis*** Petit introd. S Afr. R: Lagascalia 14(1986)102

Althoffia K. Schum. = *Trichospermum*

Altingia Noronha = *Liquidambar*

Altingiaceae Lindl. = Hamamelidaceae

Altoparadisium Filg. & al. = *Arthropogon* (but see AMBG 88(2001)363)

alum root *Heuchera* spp. esp. *H. sanguinea*

aluminium plant *Pilea cadierei*

alunqua *Marsdenia australis*

Aluta Rye & Trudgeon (~ *Thryptomene*). Myrtaceae (II 15). 5 W & C Aus. R: Nuytsia 13(2000)347

alva marina *Zostera marina*

Alvaradoa Liebm. Picramniaceae. 6 trop. Am. R: FNA 9(2014)10

Alvesia Welw. Labiatae (VII 3e). 3 trop. Afr. R: KB 25(1971)407, 26(1972)564

Alvimia Calderón ex Soderstrom & Londoño (~ *Atractantha*). Gramineae (V 2). 3 Braz. Climbers with fleshy fr. R: AJB 75(1988)833

Alvimiantha Grey-Wilson. Rhamnaceae (4). 1 Braz.: ***A. tricamerata*** Grey-Wilson. R: Bradea 2(1978)287

Alvordia Brandegee. Compositae (Helia.-Helia.). 4 Calif., Mex. R: PCAS 30(1964)157

alyce clover *Alysicarpus* spp. esp. *A. vaginalis*

Alyogyne Alef. Malvaceae (Malv.-Goss.?). 4 Aus. R: Austr. Pl. 4(1966)16. Cult. orn. ***A. hakeifolia*** (Giord.) Alef. – long-lived seeds germ. in response to fire, extrafl. nectaries

Alysicarpus Necker ex Desv. Leguminosae (III 19). 25–30 OW trop., 1 introd. Am. Alyce clover, good fodder esp. ***A. vaginalis*** (L.) DC. (Indomal. natur. pantrop.) – as nutritious as alfalfa

Alyssoides Mill. Cruciferae (2). 2 S & SE Eur.(2), Turkey. Cult. orn.

Alyssopsis Boiss. Cruciferae (3). 2 Iran to C As.

alyssum *Lobularia maritima*; **hoary a.** *Alyssum alyssoides*; **sweet a.** *L. maritima*

Alyssum Tourn. ex L. Cruciferae (2). Incl. *Clypeola*, excl. *Hormathophylla*, 195 Med. to Siberia (Eur. 70, Balkans 45 with 20 endemics; Turkey 90 with 50 endemics), N Am. 1. R: JAA 45(1964)60,358 & 46(1965)183. Many accumulate nickel, 46 to a level of more than 1000 μg per g dry matter (all in sect. *Odontarrhena* & almost all E Med. & Turkey). Some cult. orn. ***A. alyssoides*** (L.) L. (hoary alyssum, Eur., natur. N Am.)

Alyxia Banks ex R. Br. Apocynaceae (I i). 106 Indomal. (69; R: Blumea 45(2000)1) to W Pac. (39; R: Blumea 47(2002)1; Madag. spp. = *Petchia*). Rauh's & Roux's Models. New Caled. spp. accumulate manganese, up to 1.15% dry wt. in 1 sp. Some medic., also cult. for fls & bark (sold as white cinnamon) to scent clothes. ***A. oliviformis*** Gaudich. (~ *A. stellata* (Forst. & Forst.f.) Roem. & Schultes; Hawaii) – stripped lvs & bark used in leis

Alzatea Ruíz & Pavón. Crypteroniaceae (Alzateaceae). 1 Costa Rica to Bolivia: ***A. verticillata*** Ruíz & Pavón

Alzateaceae S. Graham = Crypteroniaceae

amacha drink prep. from *Hydrangea macrophylla* subsp. *serrata* in Jap.

Amaioua Aubl. Rubiaceae (Ix.-Cond.). 25 trop. S Am.

Amalocalyx Pierre. Apocynaceae (II e). 3 SE As.

Amalophyllon Brandegee (~ *Phinaea*). Gesneriaceae (II 5b). 12–13 trop. Am. R: Selbyana 29(2008)160

amaltas *Cassia fistula*

Amana Honda (~ *Tulipa*, *Erythronium*). Liliaceae. 5 E As. R: APS 43(2005)269. ***A. edulis*** (Miq.) Honda – bulb ed.

Amanoa Aubl. Phyllanthaceae (Brid.-Aman.; Euphorbiaceae s.l.). 16 trop. Am. (13, R: Britt. 42(1990)260), 3 trop. Afr. & Madag. Some fish-disp. in Amazon

Amaraboya Linden ex Masters = *Blakea*

Amaracarpus Blume. Rubiaceae (IV 7). (Excl. *Dolianthus*) 22 Seychelles, Indomal. to Aus., Solomon Is. R: Blumea 49(2004)28. Local medic.

Amaracus Gled. = *Origanum*

amaranth orig. an imaginary never-fading fl., later applied to a number of OW pls incl. *Celosia argentea*; (timber, usu. **amaranthe**) *Peltogyne* spp.; **globe a.** *Gomphrena globosa*

Amaranthaceae Juss. Magnoliidae – Caryophyllales. Incl. Chenopodiaceae, 173/2125 trop. & warm, few temp. Herbs (mostly) to climbers (e.g. *Hablitzia, Stilbanthus*), oft. succ., shrubs or rarely trees (e.g. *Achyranthes* (some), *Charpentiera, Haloxylon, Nototrichium*) with sec. thickening of concentric rings of vascular bundles, prod. betalains but not anthocyanins nor tannins. Lvs in spirals or opp., simple, usu. entire, oft. with Kranz anatomy. Fls bisex. or unisexual (pls dioec., polygamous or monoec.) small, solit. or (usu.) in cymose or thyrsoid infls oft. with bristly bracts & bracteoles, these s.t. conspicuously pigmented, wind- or insect-poll., rarely unisex. P 3–5 rarely 0–2, rarely connate at base, usu. falling with fr.; A usu. same no. as P & opp. them, usu. connate at base into a tube, oft. with teeth or lobes; nectary oft. a ring within tube; G 1-loc., with 1 style & 1 basal (rarely apical), campylotropous, bitegmic ovule, rarely several (e.g. *Celosia*) on a basal short placenta. Fr. an irreg. rupturing capsule, rarely a berry, surrounded by persistent P (anthocarp) bracts & bracteoles, the whole the disseminule; seeds usu. with abundant starchy perisperm, true endosperm ± absent. x = (6-) 8–13, 17+

Classification & principal genera of former A. s.s.:

I. **Amaranthoideae** (A 4-loc., 1–many ovules)
  1. **Celosieae** (ovules ∞ to many): *Celosia*
  2. **Amarantheae** (ovule 1): *Amaranthus, Ptilotus*

II. **Gomphrenoideae** (A 2-loc., 1 ovule)
  1. **Pseudoplantageae** (fert. fls subtended by 1 or 2 spinous modified fls): *Pseudoplantago*
  2. **Gomphreneae** (fls not so): *Alternanthera, Gomphrena, Iresine, Pfaffia*

For classification & principal genera of 'Chenopodiaceae' (only solit. ovules) not yet re-arr. here, see *Plant-book* ed. 2 (excl. III 1 = Sarcobataceae) & Willd. 45(2015)333; Willd. 42(2012)13 (Chenopodioideae – 4 tribes), T 60(2011)71 (Camphorosmoideae), T. 62(2013)108 (Polycnemoideae)

Many halophytes & also weeds (e.g. *Chenopodium, Salsola*), others imp. constituents of desert floras in W US & Aus. Several types of C_4 photosynthesis & c. 17 diff. kinds of leaf anatomy (some without Kranz syndrome). Grain crops (*Amaranthus, Chenopodium*; see also *Agriophyllum, Cycloloma*), sugar (*Beta*) & veg. (*Atriplex, Beta*, Celosia) incl. spinach (*Spinacia*). Some medic (*Dysphania, Nanophytum*) & firewood sources (*Arthrophytum, Haloxylon*); some cult. orn. (*Alternanthera, Amaranthus, Celosia, Gomphrena, Iresine, Kochia*)

Amaranthus Tourn. ex L. Amaranthaceae (I 2). 75 trop. & temp. (Eur: 2 native, many NW natur.). Coarse ann. herbs – weeds, cult. orn., grain crops, some used like spinach (bhaji); 65 monoec. (i.e. subgg. *A., Albersia*; R: AMBG 101(2015)271). ***A. albus*** L. (pigweed, Mex., natur. subcosmop.) – a tumbleweed; ***A. caudatus*** L. (cat-tail, Inca wheat, love-lies-bleeding, quihuicha, derived from *A. hybridus* subsp. *quitensis* (Kunth) Costea & Carratero, S Am.) – a tumbleweed, cult. orn., protein-rich seeds (high in lysine) & lvs ed. in Andes but displaced by colonists' cereals; ***A. hybridus*** L. (incl. *A. hypochondriacus* L., prince [of Wales]'s feather, poss. distinct & derived from hybridization with *A. powellii* S. Watson (SW N Am.), SW N Am.) – ed. lvs & seeds (esp. form. Mex.), esp. '*A. cruentus* L.' (*A. paniculatus*) a cultigen with ed. grain selected in C Am. 4000 BC), cult. orn. esp. 'var. *erythrostachys* Moq.' with red spikes; ***A. palmeri*** S. Watson (W US) – herbicide-resistant in cottonfields; ***A. retroflexus*** L. (pigweed, C & E N Am., natur. subcosmop.) – weed, grains & lvs ed. N Am.; ***A. tricolor*** L. (Indomal., natur. pantrop.) – cult. orn. (Chinese spinach, Joseph's coat, tampala), potherb & quinoa subs.

× **Amarcrinum** Coutts = *Amaryllis* × *Crinum*

amarelo, pau *Hortia excelsa*

amaretto, amaretti *Prunus dulcis* (*P. armeniaca, P. persica*)

× **Amaristes** Hannibal = *Amaryllis* × *Cybistetes* (= *Ammocharis*). Poss. corr. name for '× *Amarygia*' hybrids

Amaroria A. Gray (~ *Soulamea*). Simaroubaceae. 1 Fiji: ***A. soulameoides*** A. Gray

× **Amarygia** Cif. & Giac. = *Amaryllis* × *Brunsvigia*. Cult. orn. esp. × ***A. parkeri*** (W. Watson) H. Moore (× *A. bidwillii*) allegedly *Amaryllis belladonna* × *Brunsvigia* sp., though poss. truly × *Amaristes*

Amaryllidaceae J. St-Hil. Magnoliidae – Asparagales. Incl. Agapathaceae, Alliaceae, 72/1825 trop. & warm to Eur., esp. S Afr. & Andes. Perenn. herbs with bulbs, rarely rhiz. (e.g. *Clivia, Scadoxus*); r. contractile, with vessels. Lvs usu. flat, sheathing at base, usu. fleshy, glabrous, distich. or in spirals. Infls term., umbel-like comprising condensed helicoid cymes. Fls 3-merous, bisex., usu. reg. (irreg. in e.g. *Sprekelia*); P 3 + 3, s.t. atop a tube, corona present in *Narcissus*; A 3 + 3 (3 in *Zephyranthes*, up to 18 in *Gethyllis*) inserted at

P bases, filament bases s.t. united by a staminal 'corona' (*Hymenocallis*, *Pancratium* etc.), anthers longit. dehiscent (apical pores in *Galanthus*, *Leucojum* etc.); $\overline{G}$ (3), 3-loc., each loc. with several to ∞± anatr. ovules with (0 –)2 integuments. Fr. a loculicidal capsule or berry. Seeds globose, s.t. with elaiosome, or flattened & winged

Classification & principal genera – if subfamilies recog.: **Agapanthoideae** (*Agapanthus*); **Allioideae** (3 tribes – Allieae (*Allium*), Gillesieae (*Ipheion*, *Nothoscordum*), Tulbaghieae (*Tulbaghia*)); **Amaryllidoideae** (13 tribes; *Brunsvigia*, *Crinum*, *Cyrtanthus*, *Gethyllis*, *Haemanthus*, *Hippeastrum*, *Hymenocallis*, *Lycoris*, *Narcissus*, *Stenomesson*, *Zephyranthes*)

Form. placed with fams also with inferior G now known to be parallelism. Wind-disp. of infls freq. Imp. crop pls (*Allium*) & many favoured cult. orn. (all genera above, *Acis*, *Amaryllis*, *Ammocharis*, *Beaucarnea*, *Caliphruria*, *Clivia*, *Galanthus*, *Eucharis*, *Griffinia*, *Habranthus*, *Ismene*, *Leucojum*, *Nerine*, *Proiphys*, *Scadoxus*, *Sternbergia*, *Worsleya* etc.) incl. intergeneric hybrids

amaryllis Usu. *Hippeastrum* spp. & hybrids, see also *Ammocharis*; **blue a.** *Worsleya procera*

Amaryllis L. Amaryllidaceae. 2 Namaqualand & W Cape. ***A. belladonna*** L. (belladona lily) – cult. orn. fls in autumn without lvs, signifying 'Pride' in Language of fls. '*Amaryllis*' of greenhouses etc. = hybrids of *Hippeastrum* (q.v.) & efforts to restore the generic name *Amaryllis* to them failed (cf. *Chrysanthemum*)

Amasonia L.f. (~ *Aegiphila*). Labiatae (III, Verbenaceae). 8 trop. Am. R: FR 46(1939)194. Lvs in spirals

Amatlania Lundell = *Ardisia*

amatungulu *Carissa bispinosa*

Amauria Benth. Compositae (Perit.-Perit.). 3 SW N Am. R: Madroño 21(1972)516. Pappus 0

Amauriella Rendle = *Anubias*

Amauriopsis Rydb. = *Hymenothrix* (but see Phytoneuron 2010–10(2010)1)

Amauropelta Kunze = *Thelypteris*

amazakoué *Guibourtia ehie*

Amazon lily *Eucharis grandiflora*

ambal *Phyllanthus emblica*

ambarella *Spondias cytherea*

Ambassa Steetz (~ *Vernonia*). Compositae (Vern.-Erl.). 3 E Afr.

ambatch *Aeschynomene elaphroxylon*

Ambavia Le Thomas. Annonaceae (II). 2 Madag.

Ambelania Aubl. Apocynaceae (I d). 3 trop. S Am. R: AUWP 87–1(1987)23. Disk 0

amber fossil resins from largely extinct conifers esp. in Baltic region, occ. preserving insects caught in the fresh exudation; **a. bell** *Erythronium americanum*

Amberboa (Pers.) Less. (~ *Centaurea*). Compositae (Card.-Cent.). 6 Med. (Eur. 1) to C As. ***A. moschata*** (L.) DC. (*C. moschata*, sweet sultan, E Med.) – cult. orn. ann. with yellow, white, pink or purple scented fls

ambila *Pterocarpus angolensis*

Ambilobea Thulin & al. (~ *Boswellia*). Burseraceae (V). 1 Madag.: ***A. madagascariensis*** (Capuron) Thulin & al. R: NJB 26(2008)223

Amblostoma Scheidw. = *Epidendrum*

Amblovenatum J. Roux = *Cyclosorus*

Amblyanthe Rauschert = *Dendrobium*

Amblyanthopsis Mez. Primulaceae (Myrsinaceae). 2 Himal.

Amblyanthus A. DC. Primulaceae (Myrsinaceae). 3 E Himal., NG

Amblyanthus (Schltr.) Brieger = *Dendrobium*

Amblygonocarpus Harms. Leguminosae (II 1). 1 trop. Afr. savannas: ***A. andongensis*** (Oliv.) Exell & Torre – timber (banga wanga), cult. orn. R: Strel. 10(2000)276

Amblygonum (Meissn.) Reichb. = *Polygonum*

Amblynotopsis J.F. Macbr. = *Antiphytum*

Amblynotus (A. DC.) I.M. Johnston = *Eritrichium*

Amblyocalyx Benth. = *Alstonia*

Amblyocarpum Fischer & C. Meyer (~ *Inula*). Compositae (Inul.-Inul.). 1 Caspian: ***A. inuloides*** Fischer & C. Meyer

Amblyolepis DC. (~ *Hymenoxys*). Compositae (Hele.-Tet.). 1 Texas, Mex.: ***A. setigera*** DC. R: FNA 21(2006)420

Amblyopappus Hook. & Arn. Compositae (Mad.-Bae.). 1 Calif., NW Mex., Peru, Chile: ***A. pusillus*** Hook. & Arn. R: FNA 21(2006)348

Amblyopetalum (Griseb.) Malme = *Oxypetalum*

Amblyopyrum Eig = *Triticum*

Amblysperma Benth. (~ *Trichocline*). Compositae (Mut.-Mut.). 2 SW W Aus. R: KB 69[9507](2014)2

Amblystigma Benth. = *Philibertia*

Ambongia Benoist. Acanthaceae (III 2c). 1 Madag.: ***A. perrieri*** Benoist

Amborella Baill. Amborellaceae. 1 New Caled.: ***A. trichopoda*** Baill. – wind- & insect-poll. sweetly night-scented fls, staminodes in ♀ fls mimicking fertile A, 31 mitochondrial protein genes received (horiz. gene transfer) from other spp. incl. 3 diff. moss spp.

Amborellaceae Pichon. Magnoliidae – Amborellales. 1/1 New Caled. See IJPS 161 suppl. (2000)237. Dioec. everg. shrub (Troll's Model) with pachycaul shoots, accum. aluminium; wood without vessels. Lvs in spirals, distich. later, simple, toothed, stip. 0. Fls in axillary cymes, functionally unisexual, hypogynous to somewhat perigynous; transitions from bracts to P & P to G; P 7, 8 (female) or 9–11 (male), in spiral but outer almost whorled, weakly adnate at base; A 12–21 sessile anthers in several cycles (female with 1 or 2 staminodes), outer ones basally adnate to P, dehiscing by slits, pollen without apertures; $\underline{G}$ 5 or 6, not closed at tips, ovule solit., bitegmic, pend., orthotropous. Fr. of sep. drupe-like carpels with sculptured stone derived from mesocarp; seeds with copious endosperm & minute basal embryo. 2n = 26
Genus: *Amborella*
'Sister' to all other angiosperms

Amboroa Cabrera. Compositae (Eup.-Crit.). 2 Bolivia, Peru. R: MSBMBG 22(1987)376

Amboyna wood *Pterocarpus* spp. esp. *P. indicus*

Ambrella Perrier. Orchidaceae (V 16d). 1 Madag.: ***A. longituba*** Perrier

ambrette seed *Hibiscus abelmoschus*

ambrevade *Cajanus cajan*

Ambroma L.f. = *Abroma*

Ambrosia Tourn. ex L. Compositae (Helia.-Amb.). Incl. *Franseria, Hymenoclea*, 30 cosmop. (Eur. 1) esp. Am. (US & Canada 22 – R: FNA 21(2006)10). R: JAA 45(1964)401. Ragweeds, imp. cause of hay-fever through copious pollen, molluscicides, fls used in liqueurs. Capitula > 1-fld, unisexual; fr. encl. in involucre. ***A. artemisiifolia*** L. (hogbrake, N Am.) – one of commonest hay-fever causes in US, poss. oil source; ***A. maritima*** L. (Medit.) – molluscicidal sesquiterpene lactones used on field scale, flavour in liqueurs; ***A. peruviana*** Willd. (warm Am.) – green dye in Peru; ***A. psilostachya*** DC. (N Am.) – medic. tea prep. by Indians, plant inhibits growth of nitrogen-fixing bacteria, algae & poss. some grasses through allelopathy; ***A. salsola*** (Torr. & A. Gray) Strother & Baldwin (*H. salsola*, cheeseweed) – crushed lvs cheese-smelling; ***A. tenuifolia*** Spreng. – a fertility control in Urug.

Ambrosina Bassi. Araceae (VII 22). 1 Medit. (inc. Eur.): ***A. bassii*** L. – elaiosomes. R: Aroid. 28(2005)37

Amburana Schwacke & Taubert. Leguminosae (III 1). 3 trop. S Am. R: Phytotaxa 212(2015)249. Timber (amburana, umburana) & volatile oil from seeds, which cont. coumarin

amchur *Mangifera indica*

Ameghinoa Speg. Compositae (Mut.-Nass.). 1 arid Patagonia, Arg.: ***A. patagonica*** Speg. – shrub

Amelanchier Medik. Rosaceae (16). Incl. *Malacomeles, Peraphyllum*, c. 23 N temp. (Eur. 1, As. 1) S to Mex., Guatemala. R: BSBF 122(1975)243, CJB 68(1990)2231. Many hybrids, some apomicts (apospory, pseudogamy). Fr. ed. N Am., sarvis or servis berry, serviceberry, saskatoon, eaten overripe (cf. *Mespilus germanica*) incl. hybrids esp. of ***A. alnifolia*** (Nutt.) M. Roemer (N Am.) – comm. preserves, cold remedy, other medic.; ***A. arborea*** (Michaux f.) Fern. (E N Am.) – seed germ. enhanced after passage through cedar waxwing. Cult. orn. (shad, service, Juneberry) esp. ***A. lamarckii*** F. Schroeder (*A. laevis* of gardens, E N Am., natur. SE England, apomict poss. microsp. of hybrid orig.) & ***A. ovalis*** Medik. (*A. rotundifolia, A. vulgaris*, snowy mespilus, C & S Eur. mts)

Amelichloa Arriaga & Barkw. (~ *Nassella*). Gramineae (X). 5 trop. Am. R: Sida 22(2006)146

Amellus L. Compositae (Ast.-Hom.). 12 S Afr. R: MBSM 13(1977)579. ***A. asteroides*** (L.) Druce (*A. lychnitis*) – cult. orn.

Amentiferae old name for certain catkin-bearing angiosperm families, now separated, e.g. Salicaceae to Malpighiales, Fagaceae to Fagales

Amentotaxaceae Kudô & Yamam. = Taxaceae

Amentotaxus Pilg. Taxaceae. 4 China, SE As. R: BR 64(1998)296

American ash *Fraxinus americana*; **A. barberry** *Berberis canadensis*; **A. basswood** *Tilia americana*; **A. beech** *Fagus grandifolia*; **A. cherry** *Prunus serotina*; **A. chestnut** *Castanea dentata*; **A. cowslip** *Dodecatheon* spp.; **A. ebony** *Brya ebenus*; **A. elder** *Sambucus nigra* var. *canadensis*; **A. elemi** *Bursera* spp.; **A. elm** *Ulmus americana*; **A. frogbit** *Limnobium spongia*; **A. grass** (WI) *Thysanolaena maxima*; **A. holly** *Ilex opaca*; **A. hornbeam** *Carpinus caroliniana*; **A. jute** *Abutilon theophrasti*; **A. laurel** *Kalmia* spp.; **A. lime** *Tilia americana*; **A. lotus** *Nelumbo lutea*; **A. plane** *Platanus occidentalis*; **A. red gum** *Liquidambar styraciflua*; **A. red pine** *Pinus resinosa*; **A. senna** *Senna marilandica*; **A. spikenard** *Aralia racemosa*, *Maianthemum racemosum*; **A. walnut** *Juglans nigra*; **A. waterweed** *Elodea canadensis*; **A. white pine** *P. strobus*; **A. whitewood** *Liriodendron tulipifera*; **A. wormseed** *Dysphania ambrosioides*

Americus Hanford = *Sequoiadendron*

Ameroglossum E. Fischer & al. Scrophulariaceae (Russ.). 2 Braz. R: SB 41(2016)428

Amerorchis Hultén = *Galearis* (but see FNA 26(2002)550)

Amerosedum Löve & D. Löve = *Sedum*

Amesiella Schltr. ex Garay. Orchidaceae (V 16c). 3 Philipp.

Amesiodendron Hu. Sapindaceae. 1 China to W. Mal.: ***A. chinense*** (Merr.) Hu – furn. timber. R: FM I,11(1994)465

Amethystanthus Nakai = *Isodon*

Amethystea L. Labiatae (III). 1 Turkey to Jap.: ***A. caerulea*** L. – cult. orn.

Amherstia Wall. Leguminosae (I 2). 1 Myanmar, found only twice in wild (extinct there?): ***A. nobilis*** Wall. – cult. orn. with beautiful pink fls (orchid tree, pride of Burma), lvs in limp flushes with brown spots, later becoming stiff & green (cf. *Brownea*, *Saraca*), young lvs & fls ed.

Amianthum A. Gray (~ *Zigadenus*). Melanthiaceae. 1 E N Am.: ***A. muscitoxicum*** (Walter) A. Gray – toxic to stock, Cherokee skin cure, mixed with honey or molasses a housefly insecticide. R: FNA 26(2002)89

Amicia Kunth. Leguminosae (III 11). 7 trop. Am. R: SBM 98(2015). ***A. zygomeris*** DC. (Mex.) – subpachycaul cult. orn., lvs with sleep movements & prot. by stip. in bud

amioki *Phyllanthus emblica*

Amischophacelus R. Rao & Kamm. = *Cyanotis*

Amischotolype Hassk. Commelinaceae (II 1d). 22 trop. As. (R: GBS 64(2012)54), 1–4 Afr. Some cauliflory

Amitostigma Schltr. = *Hemipilia*

amla *Phyllanthus emblica*

Ammandra Cook (~ *Phytelephas*). Palmae (IV 3). 1 Colombia, Ecuador.: ***A decasperma*** Cook – veg. ivory (cf. *Phytelephas*). R: OB 105(1991)41

Ammannia Houst. ex L. Lythraceae. Incl. *Hionanthera*, *Nesaea*, c. 75 cosmop. (Eur. 2, Afr. 43), mostly wet places. ***A. baccifera*** L. (*A. multiflora*, jerry-jerry, warm OW introd. Am.) – local skin medic, form. Aboriginal foodstuff; ***A. senegalensis*** Lam. (trop. Afr.) – aquarium pl.

Ammanthus Boiss. & Heldr. ex Boiss. = *Anthemis*

Ammi Tourn. ex L. Umbelliferae. (III 8). Excl. *Visnaga*, c. 9 Macaronesia, Med., W As. Medic. ***A. majus*** L. (bullwort, Med.) – cult. for cut-flower trade, methoxsalen used against psoriasis

Ammiaceae Barnh. = Umbelliferae

Ammiopsis Boiss. = *Daucus*

Ammobium R. Br. Compositae (Gnap.-Cass.). 3 E Aus. R: OB 104(1991)83. ***A. alatum*** R. Br. – cult. as 'everlasting' fl.

Ammobroma Torrey = *Pholisma*

× **Ammocalamagrostis** P. Fourn. Gramineae (*Ammophila* × *Calamagrostis*). × ***A. baltica*** (Schrad.) P. Fourn. useful sand-binder

Ammocharis Herb. Amaryllidaceae. Incl. *Cybistetes*, 7 trop. & S Afr. R: JLSB 52(1939)169, Bothalia 41(2011)311. Alks. Cult. orn. bulbs incl. ***A. longifolia*** (L.) M. Roem. (*C. l.*, S Afr.) – umbel falls off & is blown away scattering seeds, introd. repeatedly from early C17 as '*Amaryllis*', poss. parent of '× *Amarygia*' hybrids

Ammochloa Boiss. Gramineae (XVI 8). 3 Med. (Eur. 1) to Mid. E. R: KBAS 13(1986)109

Ammocodon Standl. = *Acleisanthes*

Ammodaucus Cosson & Durieu (~ *Thapsia*). Umbelliferae (III 2). 2 Macaronesia to trop. Afr. R: T 65(2016)585. ***A. leucotrichus*** (Cosson & Durieu) Cosson & Durieu (Afr.) – oft. sold in markets, cult. as condiment & medic.

Ammodendron Fischer ex DC. Leguminosae (III 2). 4–5 W & C As. to NW China

Ammoides Adans. Umbelliferae (III 3). 2 Med. (Eur. 1)

ammoniacum, gum *Dorema ammoniacum*

Ammophila Host. Gramineae (XVI 5). 3 E N Am., Eur. & N Afr. Hybrids with *Calamagrostis*. ***A. arenaria*** (L.) Link (marram, beach or mel grass – coastal Eur., invasive Aus., NZ, W US) – sand-binder used for thatch, baskets, chair seats, brooms, mats etc., fibres found in 'seaballs' on Irish coast (cf. *Posidonia*), texture of woven fibre poss. inspiration for char. curves & scroll-work of Celtic design

Ammopiptanthus Cheng f. Leguminosae (III 6). 1–2 C As. Antifreeze proteins

Ammopursus Small = *Liatris*

Ammoselinum Torrey & A. Gray. Umbelliferae (III 8). 2 N Am., 1 Urug. R: Phytoneuron 2012–87:11

Ammosperma Hook.f. Cruciferae (12). 2 N Afr.

Ammothamnus Bunge (~ *Sophora*). Leguminosae (III 2). 2 C As. ***A. mongolicus*** (Komarov) Cheng f. – antifreeze proteins

Amoana Leopardi & Carnevali (~ *Epidendrum*). Orchidaceae (V 13). 2 Mex. R: Phytotaxa 65(2012)29

Amoebophyllum N.E. Br. = *Phyllobolus*

Amolinia R. King & H. Robinson (~ *Eupatorium*). Compositae (Eup.-Heb.). 1 Mex., Guatemala: ***A. heydeana*** (Robinson) R. King & H. Robinson. R: MSBMBG 22(1987)401

Amomis O. Berg = *Pimenta*

Amomum Roxb. Zingiberaceae (II 4). c. 150 As. to Aus. (Borneo 25). Creeping aromatic rhiz., fls on sep. scapes. ***A. aculeatum*** Roxb. (Indomal.) used to tranquillize wild bees (*Apis dorsata*) in Andamans before honey-collecting; ***A. dallachyi*** F. Muell. (NE Aus.) – ed. rhiz.; ***A. villosum*** Lour. (SE As.) – seeds imp. med. pl. (esp. gastric) Vietnam. Seeds of several spp. used like cardamom (most imp. non-timber forest prod. in Laos): ***A. aromaticum*** Roxb. (Bengal cardamom, Ind.), ***A. compactum*** Sol. ex Maton (round c., Java), ***A. maximum*** Roxb. (Java c., Mal.) cult.

Amomyrtella Kausel. Myrtaceae (II 10). 1 N Arg.: ***A. guili*** (Speg.) Kausel

Amomyrtus (Burret) Legrand & Kausel. Myrtaceae (II 10). 2 Chile, Arg. R: Bonpl. 13(2004)23. ***A. luma*** (Molina) Legrand & Kausel (cauchao, palo madroño) – cult. orn.

Amoora Roxb. = *Aglaia*

Amoreuxia Moçiño & Sessé ex DC. Bixaceae (Cochlospermaceae). 3 SW US to C Am. R: BJ 101(1980)244. ***A. palmatifida*** Moçiño & Sessé ex DC. – ed. fr. & r.; ***A. wrightii*** A. Gray (SW US) – almost extinct

Amorimia W.R. Anderson (~ *Mascagnia*). Malpighiaceae. 10 S Am. R: Novon 16(2006)176

Amorpha L. Leguminosae (III 10). 15 N Am. R: Rhodora 77(1975)337. Wings & keel 0, standard folded around base of staminal tube. Cult. orn. ***A. canescens*** Pursh – anthelminthic. ***A. fruticosa*** L. (bastard indigo, invasive S & E Eur.) used as bedding by Native Americans

Amorphophallus Blume ex Decne. Araceae (VII 18). (Incl. *Pseudodracontium*) c. 290 OW trop. (Afr. & Madag. 35 – R: Englera 25(2003)). Unusual in rainforest pls in having well-defined ann. dormancy. Corms with solit. leaf produced after flowering. Cult. orn. curiosities for enormous lvs (petioles supported by turgor pressure incl. in aerenchyma in core) & infls or for ed. corms. Some with epiphyllous bulbils. Alks incl. coniine (!) must be removed by boiling. ***A. bulbifer*** Blume (NE Ind. to Myanmar) – apomict (2n = 39), bulbils on lvs; ***A. konjac*** K. Koch (*A. rivieri*, konjaku, China to SE As.) – starch source as flour & comm. source of mannose for diabetic diets, blood cholesterol control, etc.; ***A. paeoniifolius*** (Dennst.) Nicolson (*A. campanulatus*, elephant yam, suram, Indomal.) – ed. corms, 3rd after rice & maize as carbohydrate source in Indonesia; ***A. titanum*** (Becc.) Arc. (Sumatra) – carrion-beetle-poll. (thermogenesis to 36°C, eye-watering stench due to ? butyric aldehyde), leaf to 4.6 m tall & across, corm to 50 cm diam. & 138+ kg & spathe to 2.4 m, growing 7.5 cm a day, oft. considered largest infl. of herbaceous (cf. *Corypha*) pls but ***A. gigas*** Teijsm. & Binn. (*A. brooksii*, Sumatra) alleged to have infl. incl. peduncle to 4.36 m (& corm to 70 kg)

Amorphospermum F. Muell. (~ *Niemeyera*). Sapotaceae (IV). 2 NG, E Aus. R: T 62(2013)760

amourette *Brosimum guianense*

Ampalis Bojer ex Bureau = *Streblus*

Amparoa Schltr. = *Rhynchostele*

Ampelamus Raf. = *Cynanchum*

Ampelaster Nesom. Compositae (Ast.-Sym.). 1 E US: ***A. carolinianus*** (Walter) Nesom. R: Phytol. 77(1994)250

Ampelocalamus S.L. Chen & al. Gramineae (IV). c. 13 Himal. to China (13, 12 endemic – R: FOC 22(2006)99). Heterogeneous

Ampelocera Klotzsch. Ulmaceae. 10 Mex. to Braz. R: AMBG 76(1989)1087

Ampelocissus Planch. Vitaceae. Excl. *Nothocissus*, c. 95 trop. (Am. 4, prob. = *Vitis*). Some ed. fr. e.g. ***A. abyssinica*** (A. Rich.) Planch. (E Afr.) & local med.

Ampelodesma P. Beauv. ex Benth. = seq.

Ampelodesmos Link. Gramineae (X). 1 Med. (inc. Eur.): ***A. mauritanicus*** (Poir.) T. Durand & Schinz (dis or diss grass) – grows with esparto & used for paper, fish-nets, ropes etc. locally, fodder when young

Ampelopsis Michaux. Vitaceae. Excl. *Nekemias*, 16 temp. & subtrop. Am. & As. Cult. orn. climbers with forked tendrils but no discoid tips (cf. *Parthenocissus* which is loosely referred to as '*Ampelopsis*' in hort.), incl. ***A. brevipedunculata*** (Maxim.) Trautv. (E As.) – invasive in US. Some local medic.

Ampelopteris Kunze = *Cyclosorus*

Ampelosycios = seq.

Ampelosicyos Thouars. Cucurbitaceae (IX). 3 Madag.

Ampelothamnus Small = *Pieris*

Ampelozizyphus Ducke. Rhamnaceae (9). 2 NE S Am. R: HPB 20(2015)161 ***A. amazonicus*** Ducke (Braz.) – bark a soap subs.

Amperea A. Juss. Euphorbiaceae (Acal.-Amp.). 8 Aus. esp. SW (6 endemic). R: AusSB 5(1992)1

Amphiachyris (DC.) Nutt. Compositae (Ast.-Sol.). 2 C US, Texas. R: SB 4(1979)178

Amphianthus Torrey = *Grateola*

Amphiasma Bremek. Rubiaceae (IV 15). 5–6 trop. & SW Afr.

Amphiblemma Naud. Melastomataceae. 13 trop. W Afr. R: Adansonia 13(1973)429, 14(1974)469. Cult. orn.

Amphiblestra C. Presl = *Tectaria*

Amphibolia L. Bolus ex Herre (~ *Eberlanzia*). Aizoaceae (V). 5 S Namibia & Karoo to Cape. R: H. E. K. Hartmann, *Ill. Handb. Succ. Pls*, Aiz. A–E (2002)37

Amphibolis Agardh. Cymodoceaceae. 2 off coasts of W & S Aus. Submerged marine aquatics forming 'meadows' imp. as fish-breeding grounds & orig. of some 'seaballs' (cf. *Posidonia*); at least ***A. antarctica*** (Lab.) Asch. (to depths of 27 m) with submarine poll. (shorelines strewn with 1000s yellow-green empty anthers when male fls shed) & vivipary, fr. with 4-lobed 'comb' anchoring young seedling to substrate

Amphibologyne Brand = *Antiphytum*

Amphibromus Nees (~ *Helictotrichon*). Gramineae (XVI 1). 12 Aus. (10; R: Telopea 2(1986)715), NZ (1), S Am. (2). Cleistogamy common

Amphicarpa = seq.

Amphicarpaea Elliott ex Nutt. Leguminosae (III 18). 4–5 E As., N Am., Afr. R: SW Nat. 9(1964)207. Some have cleistogamous fls giving rise to subterr. fr. (cf. *Arachis*) esp ***A. bracteata*** (L.) Fern. (*A. monoica*, hog peanut, N Am.) – seeds ed. Native Americans

Amphicarpon Raf. = seq.

Amphicarpum Kunth. Gramineae (XXIV 2). 2 SE US sandy pinewoods. ***A. amphicarpon*** (Pursh) Nash (*A. purshii*) – self-fert. aerial (also cross-fert.) & subterr. cleistogamous florets, fr. released after culm death or fire (fire-prone pine-barren habitat)

Amphicome (R. Br.) G. Don = *Incarvillea*

Amphidasya Standl. (~ *Pauridiantha*). Rubiaceae (I 9). 13 trop. Am.

Amphidetes Fourn. = *Matelea*

Amphidoxa DC. = *Gnaphalium*

Amphigena Rolfe = *Disa*

Amphiglossa DC. Compositae (Gnap.-Rel.). (Incl. *Pterothrix*) 11 S Afr. R: Bothalia 29(1999)65

Amphilophium Kunth. Bignoniaceae (3). Incl. *Glasiova*, *Pithecoctenium*, 45 C & trop. S Am. R: AMBG 98(2014)399. Lianes, some with adhesive disks (no 'glue', cells growing into fissures etc.) at tendril tips (cf. *Parthenocissus*). Local medic. & cult. orn. (monkey comb) esp. ***A. crucigerum*** (L.) L. Lohmann (*P. crucigerum*, *P. echinatum*)

Amphimas Pierre ex Harms. Leguminosae (III 2). 3 Afr. (Guineo-Congolian). Fleshy covering of seed prob. endocarp & not aril

Amphineurion (A. DC.) Pichon (~ *Aganosma*). Apocynaceae (II c). 1 Indomal.: ***A. marginatum*** (Roxb.) Middleton. R: FM I,18(2007)129

Amphineuron Holttum = *Cyclosorus*

Amphinomia DC. = *Lotononis*

Amphiodon Huber (~ *Poecilanthe*). Leguminosae (III 4). 1 S Am.: ***A. effusus*** Huber. R: SB 39(2014)1150

Amphiolanthus Griseb. = *Microcarpaea*

Amphipappus Torrey & A. Gray. Compositae (Ast.-Sol.). 1 SW US: ***A. fremontii*** Torrey & A. Gray – spiny. R: AJB 30(1943)481

Amphipetalum Bacig. Talinaceae. 1 Paraguay, Bolivia: ***A. paraguayense*** Bacig. R: Cand. 43(1988)409

Amphiphyllum Gleason. Rapateaceae (I 1). 2 Venez.

Amphipogon R. Br. Gramineae (Arund.). 9 Aus. R: FA 44B(2005)9

Amphipterum (Copel.) Copel. = *Hymenophyllum*

Amphipterygium Schiede ex Standl. Anacardiaceae (I). 4 Mex., Peru. ***A. adstringens*** (Schldl.) Standl. (Mex.) – dioec., wind-poll., source of red dye

Amphirrhox Spreng. Violaceae (III 1d.). 1 trop. Am.: ***A. longifolia*** (A. St.-Hil.) Spreng.

Amphiscirpus Oteng-Yeboah (~ *Scirpus*). Cyperaceae (II 1). 1 W N Am., Arg., Chile: ***A. nevadensis*** (S. Watson) Oteng-Yeboah. R: NRBGE 33(1974)308

Amphiscopia Nees = *Justicia*

Amphisiphon W. Barker = *Daubenya*

Amphistemon Groeninckx. Rubiaceae (IV 15). 2 Madag. R: BJLS 163(2010)450

Amphitecna Miers (~ *Dendrosicus*). Bignoniaceae (7). 18 trop. Am. R: FN 25(1980)50. ***A. regalis*** (Linden) A. Gentry (Mex.) – lvs to 1 m × 35 cm

Amphithalea Ecklon & Zeyher. Leguminosae (III 7). Incl. *Coelidium*, 42 Cape. R: OB 80(1985), Strel. 9(2000)462

Amphitoma Gleason = *Miconia*

Amphoradenium Desv. = *Grammitis*

Amphoricarpos Vis. Compositae (Card.-Card.). 3–4 SE Eur. (1) to Cauc.

Amphorocalyx Bak. Melastomataceae. 5 Madag. Leeuwenberg's Model

Amphorogynaceae Nickrent & Der. See Santalaceae

Amphorogyne Stauffer & Hürlimann. Santalaceae (Amphorogynaceae). 2 New Caled.

Ampliglossum Campacci = *Gomesa*

Amrad gum *Vachellia nilotica*

Amsinckia Lehm. Boraginaceae (B. 3.8.4). 15 W US (2 natur. Eur.), W temp. S Am. R(N Am.): AJB 44(1957)529, JAA suppl. 1(1991)98. Distyly, alks, deeply lobed cotyledons; cult. orn.

Amsonia Walter. Apocynaceae (I b). 19 N Am., Jap. R: AMBG 15(1928)379. Bluestar. Alks. Cult. orn. ***A. hirtella*** Standl. (N Am.) – 2–5% rubber

Amydrium Schott. Araceae (IV 4). 5 SE As. to Mal. R: KB 54(1999)381. Some cult. orn. ('*Epipremnopsis*')

Amyema Tieghem. Loranthaceae (5 3). 91 SE As. (Mal. 59; R: FM I,13(1997)228) to Aus. (36, 32 endemic) & Samoa. R: Blumea 36(1992)293. ***A. miraculosa*** (Miq.) Tiegh. oft. hyperparasite (rare in L.) of ***A. miquelii*** (Miq.) Tiegh. in Aus.; ***A. quandang*** (Lindl.) Tieghem (Aus.) only on *Acacia platycarpa* F. Muell. in Aus., reliant on just 2 bird spp., mistletoe bird & spiny-cheeked honeyeater, for disp., the latter getting most of energy requirement from nectar & fr. (available all yr.) & poll.

Amygdalus Tourn. ex L. = *Prunus*

Amylotheca Tieghem. Loranthaceae (3). 5 SE As. to New Caled. (Aus. 2, 1 endemic). R: Blumea 38(1992)209

Amyrea Leandri. Euphorbiaceae (Acal.-Pyc.-Necep.). 11 Madag. R: KB 53(1998)438

Amyris P. Browne. Rutaceae (I 9). 40 trop. Am. (S 8; R: Cand. 46(1991)227). ***A. balsamifera*** L. (esp. Cuba) – timber (candlewood, lignum rhodium, WI sandalwood) also in incense & source of oil; ***A. elemifera*** L. (torchwood) – oily wood used as fuel; ***A. plumieri*** DC. (Mex.) – Yucatán elemi used in lacquers

Amyrsia Raf. = *Myrteola*

Amyxa Tieghem ex Domke. Thymelaeaceae (I). 1 Borneo: ***A. pluricornis*** (Radlk.) Domke. R: FM I,4(1953)363

Anabaena A. Juss. = *Romanoa*

Anabasis L. Amaranthaceae (Chenopodiaceae III 3). Incl. *Brachylepis*, c. 50 Medit. (Eur. 4), C As. Alks (effective agent in ***A. aphylla*** L. (W As.) used as insecticide); ***A. articulata*** (Forssk.) Moq. (ajram, Medit.) – foams, so used in washing

Anacampseros L. Anacampserotaceae (?Cactaceae; Portulacaceae). Excl. *Grahamia*, incl. *Avonia, Talinaria, Xenia* 34 S Afr., warm S Am. R: BJ 113(1992)471; G. Rowley (1995), *A., Avonia, Grahamia*: 30. Xerophytic herbs with fleshy lvs; buds prot. by bundles of hair (? stip.); ***A. lanceolata*** (Haw.) Sweet – 4 cotyledons. Cult. orn. esp. ***A. telephiastrum*** DC. (S Afr.)

Anacampserotaceae Eggli & Nyffeler (~ Portulacaceae/Cactaceae). Magnoliidae-Caryophyllales. 3/36 E & S Afr., Aus., Am. T 59(2010)232. Shrubs to pachycaul perenn. herbs. Lvs succ., usu. terete to globose, in spirals. Fls bisexual in thyrses; K 2 fleshy, dry in fr.; C 5; A 5–25, $\underline{G}$ (3) with calyptra of marcescent P & A. Capsules with loculicidal dehiscence; endocarp with basket-like valves; seeds with separate coats & straight embryo.

Genera: *Anacampseros, Grahamia, Talinopsis*

Anacampta Miers = *Tabernaemontana*

Anacamptis Rich. Orchidaceae (IV 4d). Incl. *Orchis p.p.*, 11 Eur., Med. to Iran, esp. on calcareous soil. R: H. Kretzschmar & al. (2007), *The orchid genus A.* ed. 2:47. No poll. reward (attraction by deceit): ***A. morio*** (L.) R. Bateman & al. (*Orchis m.*, green-winged orchid, Euras.); ***A. pyramidalis*** (L.) Rich. (pyramid orchid) – poll. butterflies, moths

Anacantha (Iljin) Soják = *Jurinea*

Anacaona Alain = *Penelopeia*

Anacardiaceae R. Br. Magnoliidae – Sapindales. 82/950 trop., subtrop., Med., temp. N Am. Trees, shrubs, lianes or rarely geoxylic suffrutices, with vertical resin-ducts (s.t. latex-channels) in bark, larger leaf-veins etc., the resin oft. allergenic, black or becoming so on exudation. Lvs in spirals, rarely opp. (e.g. *Bouea*) or whorled, pinnate or trifoliolate (leaflets not articulated), less oft. simple, stip. 0 or vestigial (rare). Infls thyrsoid, terminal or axillary, rarely cauliflorous or epiphyllous (females of *Campylopetalum, Dobinea*) oft. bisex. or usu. unisexual fls (oft. with functionless parts of other sex), reg., usu. hypogynous; pedicels often articulate. K & C usu. (3-)5-merous, s.t. more or fewer, valvate or imbr., K usu. connate basally, P s.t. 0, K also rarely; A 10–5 (in 1 or 2, rarely more, whorls) rarely more (to 100) or only 1 fertile, rarely connate basally, borne outside, on, or rarely in a nectary-disk, s.t. 5-lobed, or the disk a gynophore, $\underline{G}$, rarely (*Drimycarpus*) $\overline{G}$, (3–1) to (5) or rarely (12), pluriloc. or usu. 1-loc., or rarely carpels discrete (then only 1 fertile), styles distinct or not; ovule 1 in each locule, usu. anatropous, bitegmic or unitegmic. Fr. usu. a drupe with ± resinous mesocarp, s.t. a samara, berry or syncarp; seeds with oily, s.t. chlorophyllous, embryo, endosperm scanty or 0. x = 7–16, polyploidy common

Classification & chief genera:

I. **Anacardioideae** (incl. Dobineae, Rhoeae, Semecarpeae, Julianieae; usu. trees & shrubs, often causing contact dermatitis, lvs simple to multifoliolate, G 1–3): 60 genera incl. *Buchanania, Campylopetalum, Dobinea, Gluta, Mangifera Melanochyla, Ozoroa, Rhus, Schinus, Searsia, Semecarpus, Sorindeia, Toxicodendron, Trichoscypha*

II. **Spondioideae** (trees, shrubs only rarely causing dermatitis, lvs usu. pinnate, G ((1–)4 or 5(–12)); R: SB 31(2006)342): 21 genera incl. *Lannea, Spondias, Tapirira*

Toxins from resin seep out from lvs during rain, though monkeys, at least, are immune to this (Hemiptera do not feed on those with alkylcatechols). Alkali or antihistamine counteract effect of resin, but mango-eating & lacquered articles can cause a reaction in sensitive humans. Insect- to wind-poll.; animal or water disp., some wind (samaras or by enlarged K or C, or trichomes on fr.). Endocarps split regularly or not or have 'shutters, stoppers, plugs, caps or lids' (Kubitzki). Cult. orn. & fr. trees incl. sumachs (*Rhus*), cashew (*Anacardium*), pistachio (*Pistacia*), mango (*Mangifera*), *Bouea, Buchanania, Schinus, Spondias* etc., dyes (*Amphipterygium, Cotinus, Semecarpus*), timbers (*Antrocaryon, Astronium, Dracontomelon, Koordersiodendron, Lannea, Loxopterygium, Parishia, Swintonia* etc.), lacquer (*Gluta, Rhus*), tannin & medic.

Anacardium L. Anacardiaceae (I). 11 trop. Am. R: MNYBG 42(1987)1, Britt. 44(1992)331. Trees (Scarrone's Model e.g. ***A. excelsum*** (Kunth) Skeels (C Am.) – timber (espavé), dug-out canoes in Colombia) to geoxylic suffrutices with domatia; ants attracted to extrafl. nectaries a potential for defence. ***A. occidentale*** L. (cashew-nut, acajou (F.)) – n = 12–29, widely cult. in Mal. poll. by bee-flies (grasshopper parasites), G 1 giving kidney-shaped nut with hard acrid (oil irritant due to cardol & anacardic acid) pericarp around seed (promotion-nut, coffin-nail), pedicel swells into ed. pear-like body (cashew-apple, from which a drink, cajuado (Braz.) & a liqueur, fenni (Goa) prep.), pericarp yields cashew-nutshell liquid (CNSL) used in brake-linings, clutches, plastic resins etc., stem source of a gum like g. arabic (bookbinding – acajou gum) & bark an indelible ink from sap (held to be contraceptive), resin coll. euglossine bees for nest-building & by humans for tarring boats, fish-nets & for varnish, seeds (30 M t p.a. by 2009, India biggest producer) must be steamed or roasted & usu. sold salted (also c. cream in low-dairy diets), molluscicidal alkylsalicylic acids used on field scale

Anacharis Rich. = *Elodea*

Anacheilum Reichb. ex Hoffsgg. = *Prosthechea*

Anaclanthe N.E. Br. = *Babiana*

Anacolosa (Blume) Blume. Olacaceae (Aptandraceae). 22 OW trop. (Afr. 0, Madag. 2, Aus. 1). ***A. frutescens*** (Blume) Blume (*A. luzoniensis*, SE As. to C Mal.) – promising nut

Anacyclia Hoffsgg. = *Billbergia*

Anacyclus L. Compositae (Anth.-Mat.). 12 Med. (Eur. 4). R: BBMNHB 7(1979)83. Alks. Cult. orn. ***A. pyrethrum*** (L.) Lag. (pellitory) medic. (*Radix Pyrethri*), mouthwashes, liqueur flavour etc.

Anadelphia Hackel. Gramineae (XXII 7). 14 W to SE trop. Afr. R: KB 20(1966)275

Anadenanthera Speg. Leguminosae (II 1). 2 trop. Am. Cult. orn. ***A. peregrina*** (L.) Speg. (*Piptadenia p.*) seeds with alkaline ash give niopo (cohoba, yopo), a potent hallucinogenic snuff (tryptamines & β carbolines); ***A. colubrina*** (Vell. Conc.) Brenan used similarly esp. in Arg., also timber

Anadendrum Schott. Araceae (IV 2). 30 Indomal. Lvs used in curries

Anaectocalyx Triana ex Hook.f. = *Miconia*

Anagallidium Griseb. = *Swertia*

Anagallis Tourn. ex L. (~ *Lysimachia*). Primulaceae (Mysinaceae). (Incl. *Asterolinon*, excl. *Centunculus*) c. 30 Eur. (5), Afr. mts, S Am. (2), with 1 pantrop. Cult. orn.: ***A. arvensis*** L. (common or scarlet pimpernel, poor man's or shepherd's weather-glass, Eur. but widely natur.) – red fls close in dull or cold weather, form. medic. (extracts active against polio & herpes), subsp. ***caerulea*** Hartman (*A. foemina*) – blue fls; ***A. monelli*** L. (blue p., Medit.); ***A. nemorum*** (L.) Büscher & Loos (*L. n.*, yellow p., W & C Eur.) – form. medic.; ***A. tenella*** (L.) L. (bog p., Eur.)

Anaglypha DC. = *Gibbaria*

Anagyris Tourn. ex L. Leguminosae (III 6). 2 Medit. to Iran, Canaries. ***A. foetida*** L. form. used as (dangerous) purgative, toxic

Anakasia Philipson. Araliaceae. 1 W NG: ***A. simplicifolia*** Philipson

Anamaria Souza = *Stemodia* (but see BolB 19(2001)43)

Anamirta Colebr. Menispermaceae (IV). 1 Indomal.: ***A. cocculus*** (L.) Wight & Arn. (*A. paniculata*) – liane with furrowed bark, acarodomatia & poisonous fr. (fish-berry, Ind. berry, 'cocculus indicus') used as fish-poison, in beer to give bitter flavour & 'heady' char., & (dangerous) parasiticide (alks), also used (picrotoxin) in barbiturate poisoning & treating both head lice & schizophrenia

Anamomis Griseb. = *Myrcianthes*

anamú *Petiveria alliacea*

Ananas Mill. Bromeliaceae (3). 1 (?–7) trop. Am. R: FN 14(1979)2051. Hybridize with *Pseudananas* spp. ***A. comosus*** (L.) Merr. (*A. sativus*, pineapple, 'nana' of the Tupi Indians) a seedless cultigen, prob. derived from bird-poll. seeded pls in Paraguay, but selected forms widely distrib. in pre-Columbian times, terr. stem with terminal infl. becoming fleshy syncarp of 100–200 berry-like fr., bracts & axis, topped with a tuft of lvs, which can be used in propagation, first known to Europeans from Guadeloupe (Columbus in 1493), introd. St Helena 1505, Philippines 1558, now a weed in some countries, & grown under glass in GB in early 1700s, when v. fashionable & the model for much garden statuary, now imp. crop in Hawaii (esp. spineless '**Smooth Cayenne**'), Malaysia & Kenya for canning & juice & for fresh fr. export (esp. spiny '**Queen**'), bromelain comm. in treatment of swollen joints, fibre from lvs (crowa) strong & soft but diff. to extract, exported from Philippines & Taiwan to Spain for fine embroidery; ***A. lucidus*** Mill. (curagua, NE S Am.) – fibre-pl.

Anangia W. de Wilde & Duyfjes = *Zehneria*

Ananthocorus L. Underw. & Maxon (~ *Vittaria*). Pteridaceae (V). 1 trop. Am.: ***A. angustifolius*** (Sw.) Underw. & Maxon

Ananthura H. Robinson & Skvarla. Compositae (Vern.). 1 C Afr.: ***A. pteropoda*** (Oliv. & Hiern) H. Robinson & Skvarla. R: Novon 21(2011)253

Anapalina N.E. Br. = *Tritoniopsis*

Anaphalioides auctt. = *Anaphaloides*

Anaphalis DC. (~ *Helichrysum*). Compositae (Gnap.-Cass.). 110 As., N Am. (natur. Eur.), 1 Chile (Andes). Subdioec. cult. orn. 'everlastings'. ***A. margaritacea*** (L.) Benth. (pearly e., E As., N Am., natur. Eur. esp. on tips, where an early colonist) – form. medic. (whooping cough in Ireland, smoked like tobacco for coughs & headache in Suffolk), first Am. herb. pl. cult. in Eur. (C 16), natur. Wales by 1698

Anaphaloides (Benth.) Kirpiczn. Compositae (Gnap.-Cass.). 7 NZ (5), NG (2). Subdioec. subshrubs. R: NZJB 35(1997)465

Anaphyllopsis A. Hay (~ *Cyrtosperma*). Araceae (V). 3 trop. S Am. R: Aroid. 11(1988)25

Anaphyllum Schott. Araceae (V). 2 S Ind.

anardana *Punica granatum*

Anarrhinum Desf. Plantaginaceae (Ant.-Ant.; Scrophulariaceae). 6–8 Medit. to Ger. (Eur. 5–6) & Ethiopia. R: D.A. Sutton, *Rev. Antirrhineae* (1988)249. Some cult. orn.

Anarthria R. Br. Restionaceae (Anarthriaceae). 7 SW Aus.

Anarthriaceae D. Cutler & Airy Shaw = Restionaceae

Anarthrophyllum Benth. Leguminosae (III 9). 15 Andes. R: Darwiniana 18(1974)453

Anarthropteris Copel. = *Loxogramme*

Anartia Miers = *Tabernaemontana*

Anaspis Rech.f. = *Scutellaria*

Anastatica L. Cruciferae (4). 1 Morocco to S Iran: ***A. hierochuntica*** L. (rose-of-Jericho, resurrection-pl.) – ann., lvs fall as seed matures, branches fold inwards & dead pl. is blown as a ball, dispersing seeds, branches opening out in moist conditions, dead pls oft. sold as curiosities, opening & closing on wetting & drying (cf. *Selaginella lepidophylla*)

Anastrabe E. Meyer ex Benth. Stilbaceae (Scrophulariaceae s.l). 1 S Afr.: ***A. integerrima*** E. Meyer ex Benth. – elaiophores attractive to bees. R: Strel. 10(2000)514

Anastraphia D. Don (~ *Gochnatia*). Compositae (Gochn.). 31 trop. Am. R(Cuba): CN 49(2011)28,39

Anastrophea Wedd. = *Sphaerothylax*

Anathallis Barb. Rodr. (~ *Pleurothallis*). Orchidaceae (V 13c). c. 150 trop. Am. R: Lankesteriana 13 (2014)327

Anatherostipa (Kuntze) Peñailillo = *Lorenzochloa*. But see Gayana (Bot.) 53(1996)278

Anatropanthus Schltr. Apocynaceae (V a; Asclepiadaceae III 4). 1 Borneo: ***A. borneensis*** Schltr.

Anatropostylia (Plitm.) Kupicha = *Vicia*

anatto *Bixa orellana*

Anaueria Kosterm. (~ *Beilschmiedia*). Lauraceae (I). 1 Amaz.: ***A. brasiliensis*** Kosterm.

Anax Ravenna = *Stenomesson*

Anaxagorea A. St-Hil. Annonaceae (I). c. 27 trop. Am., 3 Sri Lanka to W Mal. R: BJ 105(1984)73. Troll's Model. Fr. explosive, thrown to 6 m. Seeds used like camphor for clothes, local medic. ***A. dolichocarpa*** Sprague & Sandw. (NE S Am.) – seeds for jewellery; ***A. prinoides*** (Dunal) A. DC. (Braz.) – poll. nitidulid beetles; ***A. rheophytica*** Maas & Westra (Venez. Amazonia) – only rheophyte in Amazonia

Anaxeton Gaertn. Compositae (Gnap.-Cass.). 10 SW Cape. R: OB 104(1991)92

Ancana F. Muell. = *Meiogyne*

Ancathia DC. Compositae (Card.-Card.). 1 C As., China, Mongolia: ***A. igniaria*** (Spreng.) DC.

anchan *Clitoria ternatea*

Anchietea A. St-Hil. Violaceae (III 2b). 5 trop. S Am. ***A. salutaris*** A. St-Hil. – liane with medic. r.- bark

Anchistea C. Presl = *Woodwardia*

Anchomanes Schott. Araceae (VII 12). 5 trop. Afr. Famine food

Anchonium DC. Cruciferae (5). 2 W & C As. R: Cand. 39(1984)715

anchor plant (Aus.) *Discaria pubescens*

anchote *Coccinia abyssinica*

anchovy pear *Grias cauliflora*

Anchusa L. Boraginaceae (B.2.1.1). Excl. *Lycopsis*, c. 35 Eur. (20), N & S Afr., W As. Bugloss. Fls blue or yellow. Cult. orn. esp. ***A. azurea*** Mill. (*A. italica*, Med., natur. N Am.), ***A. capensis*** Thunb. (Cape forget-me-not, S Afr.) & ***A. officinalis*** L. (Eur., W As.) – form. med. (melancholia etc.) & a veg., also rouge from r. (cf. *Alkanna*). See also *Anchusella*

Anchusella Bigazzi & al. (~ *Anchusa*). Boraginaceae (B.2.1.1). 2 C & E Medit. R: PSE 205(1997)253

Ancipitia (Luer) Luer = *Pleurothallis*

Ancistrachne S.T. Blake = *Cleistochloa*

Ancistragrostis S.T. Blake (~ *Deyeuxia*). Gramineae (XVI 5). 1 NG, Aus.: ***A. uncinioides*** S.T. Blake

Ancistranthus Lindau. Acanthaceae (III 2c). 1 Cuba: ***A. harpochiloides*** (Griseb.) Lindau

Ancistrocactus (K. Schum.) Britton & Rose = *Sclerocactus*

Ancistrocarphus A. Gray (~ *Stylocline*). Compositae (Gnap.-Gnap.). 1Calif.: ***A. filagineus*** A. Gray. R: OB 104(1991)175

Ancistrocarpus Oliv. Malvaceae (Grew.-Apeib.; Tiliaceae). 4 trop. Afr.

Ancistrocarya Maxim. (~ *Lithospermum*). Boraginaceae (B.2.2). 1 Jap., Korea: ***A. japonica*** Maxim.

Ancistrochilus Rolfe. Orchidaceae (V 14). 2 trop. Afr.

Ancistrochloa Honda = *Calamagrostis*

Ancistrocladaceae Planch. ex Walp. Magnoliidae – Caryophyllales. 1/27 OW trop. Lianes, branching sympodially, branch-tips hooked or twining, s.t. alks. Lvs in spirals, oft. in rosettes on lat. flowering shoots, simple, entire, small waxy glands present; stip. minute, ephemeral. Fls small, in axillary or terminal spikes to panicles, oft. app. dichot. cymes, articulated, regular exc. K, ± epigynous. K 5, adnate to ovary, imbr., unequal, accrescent in fr.; C 5, s.t. connate at base, convolute, ± fleshy; A 10, 5 somewhat larger than others, rarely only 5, or 15, filaments ± connate at base & adnate to bases of C, anthers opening by longit. slits; G (3), semi-inf., apex free with 3 ± free styles, 1-loc. with 1 basal hemitropous bitegmic ovule. Fr. a nut crowned with accrescent sepals, floating in water; seed with hard, starchy ruminate endosperm; embryo straight

Only genus: *Ancistrocladus*

Form. assoc. with Dipterocarpaceae because of winged fr. but now known to be close to Nepenthaceae & other carnivorous Caryophyllales

Ancistrocladus Wall. Ancistrocladaceae. 18 W Afr. & Kenya (R: KB 55(2000)873), 9 Sri Lanka to S China & Borneo. Dioncophylline (& other alks) larvicidal, molluscicidal, fungicidal, anti-HIV. ***A. korupensis*** D.W. Thomas & Gereau (W Afr.) – only source of michellamine B, a naphthyl isoquinoline alk. active against HIV; ***A. tectorius*** (Lour.) Merr. (*A extensus*, SE As., W Mal.) – r. form. used in treatment of dysentery & malaria, young lvs as flavouring (Thailand)

Ancistrophora A. Gray (~ *Verbesina*). Compositae (Helia.-Verb.). 1 Cuba: ***A. wrightii*** A. Gray

Ancistrophyllum (G. Mann & H. Wendl.) H. Wendl. = *Laccosperma*

Ancistrorhynchus Finet. Orchidaceae (V 16d). 17 trop. Afr.

Ancistrostylis Yamaz. = *Staurogyne*

Ancistrothyrsus Harms. Passifloraceae. 1–2 W trop. S Am.

Ancistrotropis (~ *Vigna*). Leguminosae (III 18). 6+ trop. Am. R: AJB 98(2011)1704

Ancrumia Harv. ex Bak. = *Solaria*

Ancylacanthus Lindau = *Ptyssiglottis*

Ancylanthos Desf. Rubiaceae (III 2). 1 trop. Afr.: ***A. rubiginosus*** Desf. – fr. ed. R: KB 51(1996)350

Ancylobothrys Pierre = *seq.*

Ancylobotrys Pierre. Apocynaceae (I c). 7 trop. Afr., 1: ***A. petersiana*** (Klotzsch) Pierre (used against venereal disease on Mafia Is.) ext. to Madag. R: WAUP 94–3(1994)4

Ancylostemon Craib = *Oreocharis*. But see NRBGE 21(1954)215

Ancylotropis Eriksen (~ *Monnina*). Polygalaceae (IV). 2 Braz. R: PSE 186(1993)48

anda-assy oil *Joannesia princeps*

Andeimalva J. Tate. Malvaceae. 4 Andes. R: Lundellia 6(2003)13

Anderbergia R. Nordenstam. Compositae (Gnap.). 6 W Cape. R: ANMW 98B, suppl. (1996)407

Andersonia R. Br. Ericaceae (VII 3; Epacridaceae). c. 35 SW Aus. R: KB 16(1962)85

Andersonglossum Cohen (~ *Cynoglossum*). Boraginaceae (B.3.8.4?). 3 N Am. R: SB 40(2015)618

Anderssoniopiper Trel. = *Piper*

Andes berry *Rubus glaucus*; **A. rose** *Bejaria* spp. esp. *B. racemosa*

Andesia Hauman = *Oxychloe*

Andinia (Luer) Luer (~ *Salpistele*). Orchidaceae (V 13c). 13 trop. Am. R: Lindleyana 16(2001)251

Andinopuntia Guiggi = *Australocylindropuntia*

Andira Lam. Leguminosae (III 11). 30 trop. Am., 1 ext. to W Afr. R: SBM 64 (2003) 36. Champagnat's Model. Malodorous vertebrate (? bat)-disp. drupes (andirá = bat in 'lingua geral' of Braz.). Timber, coffee-shade, bark (cabbage bark) medic. esp. vermifuge. ***A. inermis*** (Wright) DC. (angelin, dog almond, Am., Afr.) – in dry forest fls biennially, in rainforest only 3 times in 12 yrs, timber (bastard mahogany, kurara, kuraru, partridge-wood) & planted for shelter belts in WI (cabbage-tree), bark (cabbage b.) medic., seeds emetic & anthelminthic. See also *Vataireopsis*

andiroba *Carapa guianensis*

andoung *Aphanocalyx heitzii*

Andrachne L. Phyllanthaceae (Phyll.-Por.; Euphorbiaceae s.l.). 22 S As., Afr., Mex., Peru. R: KB 63(2008)55. Cult. orn. See also *Leptopus*

Andradea Allemão. Nyctaginaceae (1). 1 SE Braz.: ***A. floribunda*** Allemão

Andrea Mez (~ *Canistropsis*). Bromeliaceae (3). 1 C Braz.: ***A. sellowiana*** (Bak.) Mez. R: T 54(2005)66

Andreadoxa Kallunki. Rutaceae (I 8). 1 Bahia: ***A. flava*** Kallunki. R: Britt. 50(1998)59

Andreettaea Luer = *Pleurothallis*

Andresia Sleumer = *Cheilotheca*

Andriana B.-E. van Wyk (~ *Heteromorpha*). Umbelliferae (III 8). 3 Madag.

Andringitra Skema (~ *Dombeya*). Malvaceae (Domb.). 6 Madag. R: T 61(2012)623

Androcalva C.F. Wilkins & Whitlock (~ *Commersonia*). Malvaceae (Bytt.-Lasio). 33 Aus. R: AusB 24(2011)286.

Androcalymma Dwyer. Leguminosae (I 3). 1 Amaz. Braz.: ***A. glabrifolium*** Dwyer

Androcentrum Lem. = *Bravaisia*

Androcera Nutt. = *Solanum*

Androchilus Liebm. ex Hartman = *Ponthieva*

Androcorys Schltr. (~ *Herminium*). Orchidaceae (IV 4d). 10 Himal., China mts, Honshu (1)

Androcymbium Willd. = *Colchicum*

Andrographis Wall. ex Nees. Acanthaceae (III 2b). 20 trop. As. Local med., some., e.g. ***A. wightiana*** Arn. ex Nees (Ind.), endangered through over-collecting. ***A. paniculata*** (Burm.f.) Nees (creat, kariyat, Indomal.) – bitter stomachic & tonic, subs. for *Munronia pinnata*, for hypertension & diabetes (Sabah), claimed to stimulate immune system

Androlepis Brongn. ex Houllet. Bromeliaceae (3). 1 C Am.: ***A. skinneri*** Brongn. ex Houllet

Andromeda L. Ericaceae (VIII 3). 1–2 N temp. (Eur. 1). ***A. polifolia*** L. (bog rosemary, 'marsh andromeda') – cult. orn., lvs & twigs for tanning in Russia, andromedotoxin lowers blood pressure, causes breathing problems, dizziness, cramps etc.

Andromycia A. Rich. = *Asterostigma*

Andropogon L. Gramineae (XXII 7). 122 trop. & warm (Eur. 1; N Am. 13). R: HIP 37(1967)sub t. 3644. Bluestem; dominant savanna genus. Extrafl. nectaries. Local medic. N Am., e.g. ***A. floridanus*** Scribner (SE US). Thatching, erosion control & some fodder esp. ***A. gayanus*** Kunth (Gamba grass, trop. Afr.) – invasive in trop. Aus., trop. S Am.; ***A. virginicus*** L. (Am.) invasive in Aus., Hawaii

Andropterum Stapf. Gramineae (XXII 4). 1 trop. Afr.: ***A. stolzii*** (Pilg.) C. Hubb.

Andropus Brand = *Nama*

Androsace Tourn. ex L. Primulaceae. Incl. *Douglasia, Pomatosace, Vitaliana*, c. 100 N temp. (Eur. 24, China 73). R: G. Smith & D. Lowe (1997) *The genus A.* ± xerophytes, usu. heterostyled; differs from *Primula* in C-tube shorter than K & constricted at throat. Cult. orn. rock-pls, 'rock-jasmine', esp. ***A. carnea*** L. (Eur.), ***A. lanuginosa*** Wall. & ***A. sarmentosa*** Wall. (Himal.)

Androsiphon Schltr. = *Daubenya*

Androsiphonia Stapf (~ *Paropsia*). Passifloraceae. 1 Liberia.: ***A. adenostegia*** Stapf – Cook's Model

Androstachyaceae Airy Shaw = Picrodendraceae

Androstachys Prain. Picrodendraceae (Picr.-Misch.; Euphorbiaceae s.l.). (Incl. *Stachyandra*) 5 Madag., 1 ext. SE trop. Afr.: ***A. johnsonii*** Prain (mecrusse, mzimbeet) – timber allegedly termite-proof

Andostephanos Fern. Casas. = *Hieronymiella*

Androstephium Torrey (~ *Muilla*). Asparagaceae (Alliaceae s.l.). 3 W US (2), Mex. Cult. orn.

Androstoma Hook.f. (~ *Cyathodes*). Ericaceae (VII 7). 1 NZ, 1 Tasmania. R: AusSB 18(2005)450

Androstylanthus Ducke = *Helianthostylis*

Androtium Stapf. Anacardiaceae (I). 1 W Mal.: ***A. astylum*** Stapf

Androtrichum (Brongn.) Brongn. Cyperaceae (II 5). 1 N Arg.: ***A. trigynum*** (Spreng.) H. Pfeiffer

Androya Perrier. Scrophulariaceae (Myoporaceae). 1 S Madag.: ***A. decaryi*** Perrier

Andruris Schltr. = *Sciaphila*

Andryala L. Compositae (Lact.-Hier.). c. 25 Med. (Eur. 5) to Canary Is. ***A. ragusina*** L. (SW Eur.) – r. latex used as chewing-gum in C Spain

Andrzeiowskia Reichb. Cruciferae. 1 Balkans to Cauc.: ***A. cardamine*** Reichb.

Anechites Griseb. Apocynaceae (I 9). 1 trop. Am.: ***A. nerium*** (Aubl.) Urb. – alleged to improve memory. R: Britt. 35(1983)228

anegré *Pouteria* spp.

Aneilema R. Br. Commelinaceae (II). 64 warm (Afr. 60; Am. 1, also in Afr.). R: R. Faden (1975) *Biosyst. study of the genus A.*, SCB 76(1991). Some andromonoec.

Anelsonia J.F. Macbr. & Payson. Cruciferae (11). 1 W US: ***A. eurycarpa*** (A. Gray) J.F. Macbr. & Payson. R: FNA 7(2010)347

Anelytrum Hackel = *Avena*

Anemanthele Veldk. (~ *Stipa*). Gramineae (X). 1 NZ: ***A. lessoniana*** (Steud.) Veldk. (pheasant's tail grass) – cult. orn.

Anemarrhena Bunge. Asparagaceae (Anthericaceae). 1 China, Korea: ***A. asphodeloides*** Bunge – med., molluscicidal steroid saponins, cult. orn.

Anemarrhenaceae Conran & al. = Asparagaceae

Anemia Sw. Anemiaceae (Schizaeaceae s.l.). Incl. *Mohria* (R: Bothalia 25(1995)1), 115 trop. & warm esp. Am. R: ISCS 36(1962)349. Cult. orn. ***A. caffrorum*** (L.) Christenh. (*M. c.*, frankincense fern, trop. & S Afr., to Masc.) – fronds scented; ***A. tomentosa*** (Savi) Sw. (Urug.) – contraceptive

Anemiaceae Link (~ Schizaeaceae). Polypodiidae – Schizaeales. 1/115 trop. & warm esp. Am. R: T 55(2006)711. Terr. ferns with hairy creeping to suberect rhizomes. Fronds determinate; veins dichot., occ. anastomosing. Sporangia usu. on basal pr of skeletonized pinnae (or all pinnae modified & fertile); spores 128–156 per sporangium, tetrahedral. Gametophyte green, cordate. x = 38

Genus: *Anemia*

Anemocarpa Paul G. Wilson. Compositae (Gnap.-Ang.). 3 Aus. R: Nuytsia 8(1992)452

Anemoclema (Franch.) W.T. Wang = *Anemone*

Anemone Tourn. ex L. Ranunculaceae (II 2). Incl. *Anemoclema*, *Barneoudia*, *Hepatica*, *Knowltonia* (fleshy drupelets), *Oreithales*, *Pulsatilla* (persistent long plumose styles), c. 200 Euras. (Eur. 17), Sumatra, Tas. (1), NZ (1), S & E Afr. (mts), N Am. to Chile (S hemisph. 21 – R: JJB 81(2006)195). R: SB 37(2012)146. Perianth petaloid, involucre in some resembling K. Tepals white, yellow, red or blue. Toxic: glycoside, ranunculin, hydrolyses to volatile toxic irritant oil, protoanemonin. Cult. orn., some med. ('windflower' a confusion as generic name not from *anemos* (Gk. wind) but prob. Gk. from old Semitic Naaman (i.e. Adonis) whose blood held to have produced ***A. coronaria*** L. (in Mediaeval Engl. '*anemone*' name for poppies)). *A.* used as cut-fls orig. in Turkey but dev. later in Italy usu. *A. coronaria*, prob. the Biblical 'lilies of the field', (domesticated c. 400 yrs, some 180 cvs grown 1780s–1790s Pavolvsk nr St Petersburg; 'single' fls = 'de Caen') with its semidouble cvs (= 'St Brigid'), also ***A. pavonina*** Lam. (single fls = 'St Bavo') & ***A. hortensis*** L. (all Med.) & hybrid between last two: ***A.*** × ***fulgens*** Gay; Spring spp. incl. ***A. apennina*** L. (S Eur.) – K 10–15 & ***A. blanda*** Schott & Kotschy (SE Eur. to Cauc.) – K <20, fr. head nodding; hepaticas are usu. ***A. actutiloba*** (DC.) Lawson (*Hepatica a.*, N Am.) – ant-disp. seeds & ***A. hepatica*** L. (*H. nobilis*, *H. triloba*, Euras.) – supposed medic., many cvs selected Jap., esp. homoeotic mutants with A replaced by P; Jap. anemones of autumn are *A.* × ***hybrida*** Paxton (*A. vitifolia* Buch.-Ham. ex DC. (Afghanistan to W China & Myanmar) × *A. scabiosa* H. Lév. & Vaniot (*A. hupehensis* var. *japonica*, China, natur. Jap.)) cvs esp. pure white '**Honorine Jobert**'; ***A. multifida*** Poir. – Am. bipolar distrib. (cf. *Carex macloviana*, *Koenigia islandica*, *Osmorhiza berteroi*); ***A. nemorosa*** L. (wood anemone, Euras.) – indicator of anc. woodland in GB where seed rarely viable, pls spreading c. 2 m per C, cult. orn., lvs & rhiz. imp. in bank-vole diet, nectar 0 but mainly bee-poll.,? ant-disp.; ***A. pulsatilla*** L. (*Pulsatilla vulgaris*, pasque-flower [orig. passeflower], Eur.) – indicative of grassland at least a C old in GB, form. medic. esp. C Eur., used to dye Mediaeval Easter eggs green

Anemonella Spach (~ *Anemone*). Ranunculaceae (II 2). 1 E N Am.: ***A. thalictroides*** (L.) Spach –ed. tubers; cult. orn. incl. 'doubles'

Anemonidium (Spach) Löve & D. Löve = *Anemone*

Anemonopsis Sieb. & Zucc. Ranunculaceae (I 2). 1 Jap. (C Honshu): ***A. macrophylla*** Sieb. & Zucc. – cult. orn.

Anemopaegma Mart. ex Meissn. Bignoniaceae (3). 41 trop. Am. Lianes, some cult. orn. & aphrodisiac

Anemopsis Hook. & Arn. Saururaceae. 1 SW US, Mex.: ***A. californica*** (Nutt.) Hook. & Arn. – herb of wet places, infl. resembles single fl., cult. orn., aromatic stock medic. & used as beads (Apache b.). R: FOW 11(2005)6

Anepsias Schott = *Rhodospatha*

Anerincleistus Korth. Melastomataceae. 30 S China & Ind. to Philippines. R: PANSP 141(1989)29. Herbs & shrubs, some rheophytes incl. ***A. rupicola*** (Nayar) Maxw.

Anetanthus Hiern ex Benth. Gesneriaceae (II 3b). 2 C Colombia, SE Braz., Peru, Bolivia

Anetholea Peter G. Wilson (~ *Syzygium*). Myrtaceae (II 9). 1 NSW: ***A. anisata*** (Vickery) Peter G. Wilson – aniseed odour (anethole) cf. *Backhousia*, aniseed myrtle oil used in liqueurs (NSW). R: AusSB 13(2000)434

Anethum Tourn. ex L. Umbelliferae (III 10). 1 SW As.(?): ***A. graveolens*** L. (incl. *A. sowa*, dill, (E.) Ind. d., natur. Euras., N Am.) – cult. since at least 400 BC (tax payable acc. to St Matthew's Gospel), ann. herb with fennel-like flavour for fish & pickling gherkins, seeds & lvs in soup & salads, dill water used for infant flatulence

Anetium (Kunze) Splitg. = *Polytaenium*

Anettea Szlach. & Mytnik = *Gomesa* (but see PJB 51(2006)49)

Aneulophus Benth. Erythroxylaceae. 1–2 W trop. Afr. Roux's Model

Aneurolepidium Nevski = *Leymus*

Angadenia Miers. Apocynaceae (II e). 2 Florida, WI

Angasomyrtus Trudgen & Keigh. Myrtaceae (II 14.). 1 SW Aus.: ***A salina*** Trudgen & Keigh. R: Nuytsia 4(1983)435

angel, death *Justicia pectoralis*; **a. wings** *Caladium* spp.

Angelesia Korth. (~ *Licania*). Chrysobalanaceae (1). 3 SE As. R: Blumea 59(2014)103

Angelica L. (~ *Peucedanum; Ostericum* poss. dist.). Umbelliferae (III 9). c. 110 N hemisph. (Eur. 8), NZ(?). Some decaploids; some orn., medic. & ed. ***A. archangelica*** L. (Euras., natur. GB – angelica) – lvs a veg. & petioles (US 'French rhubarb'), infl. axes etc. candied esp. nr Niort, France, since C 18) for cake-decoration etc., their form the inspiration for fluted (Doric) columns of Anc. Greece, tonic & flavouring in wines, gin, etc. esp. char. taste of Benedictine & Chartreuse, though r. infusion taken 3 times a day alleged to create distaste for alcohol; ***A. atropurpurea*** L. (NE N Am.) – similar uses; ***A. sinensis*** (Oliv.) Diels (dang gui, dong quai, China) – used in menopause, tonics etc.

angelin *Andira inermis*

angelique *Dicorynia guianensis*

Angelocarpa Rupr. = *Angelica*

Angelonia Bonpl. Plantaginaceae (Ang.; Scrophulariaceae s.l.). 25 trop. Am. esp. Braz. Poll. by female oil-collecting bees (*Centris* spp.). Cult. orn. incl. cut-fls esp. ***A. biflora*** Benth. (Braz.) – natur. trop. OW

Angelphytum G. Barroso = *Dimerostemma* (but see PBSW 97(1984)96)

angel's hair *Cucurbita ficifolia*; **A.'s tears** *Narcissus triandrus* 'Albus'; **a. trumpet** *Brugmansia* spp.

Angianthus Wendl. Compositae (Gnap.-Ang.). 22 S Aus. (SW Aus. 18). R: Muelleria 5(1983)153, OB 104(1991)131. Heads compound

angico gum *Parapiptadenia rigida*

Anginon Raf. (*Rhyticarpus*). Umbelliferae (III 8). 12 Afr. R: NJB 17(1997)561. Woody

Angiopteridaceae Fée ex Bommer = Marattiaceae

Angiopteris Hoffm. Marattiaceae (Angiopteridaceae). (Incl. *Archangiopteris, Macroglossum*) c. 20 OW trop. (form 1: ***A. evecta*** (Forster f.) Hoffm., s.t. split into up to 200 microspp.). R: T 57(2008)751. Massive short stem with fronds to 3 m, sporangia distinct, annulus complicated; local medic., oil used to perfume coconut oil in Pac., stem ed. Ind. (starch) & basis of intoxicating drink

Angiospermae Lindl. = Magnoliidae

Angkalanthus Balf.f. Acanthaceae (III 2c). 1 Socotra: ***A. oligophylla*** Balf.f. – firewood. R: KB 49(1994)473

Angolaea Wedd. Podostemaceae (III). 1 R. Cuanza, Angola: ***A. fluitans*** Wedd.

Angolluma Munster = *Orbea* (but see Excelsa 16(1993)104)

Angophora Cav. (~ *Eucalyptus*). Myrtaceae (II 11). 14 E Aus. R: Telopea 2(1986)749. Diff. from *Eucalyptus* in sep. petals – gum myrtles, 'apples' (Aus.). Cult. orn. esp. ***A. costata*** (Gaertn.) Britten (Sydney red gum, E Aus.) with striking bark; ***A. woodsiana*** Bailey (NE Aus.) – poll. by 12 spp. of beetle

Angoseseli Chiov. Umbelliferae (III 3). 2 Angola

angostura *Angostura trifoliata*

Angostura Roem. & Schultes. Rutaceae (I 8). 8 trop. S Am. R: KB 53(1998)259. Fls sympetalous, C zygomorphic; local med. in Venez. where few stomach disorders. Bark of ***A. trifoliata*** (Willd.) Elias (*Galipea officinalis*) yields angostura used in pink gins etc. See also *Conchocarpus, Rauia*

Angostyles Benth. = *Angostylis*

Angostylidium (Muell. Arg.) Pax & K.Hoffm. = *Tetracarpidium*

Angostylis Benth. Euphorbiaceae (Acal.-Pluk.-Pluk.). 1 Braz.: ***A. longifolia*** Benth. R: A. Radcliffe-Smith, *Gen. Euphorb.* (2001) 244

Angraecopsis Kraenzlin. Orchidaceae (V 16d). 22 trop. Afr., Masc. Cult. orn.

Angraecum Bory. Orchidaceae (V 16d). c. 220 trop. Afr., Masc., Sri Lanka. R: KB 28(1973)496. Heterogeneous. Epiphytes (some minute e.g. ***A. humile*** Summerh. (E Afr.) 1–4 cm tall, cult. orn. ***A. arachnites*** Schltr. (Madag.) – exclusively poll. by hawk-moth (*Panogena lingens*) which also poll. other spp.; Masc. spp. (Madag. ancestors hawkmoth-poll.) poll. white-eyes, ***A. cadetii*** Bosser (Réunion) poll. crickets (unique); ***A. sesquipedale*** Thouars (E Madag.) – spur to 45 cm, poll. moth (*Xanthopan morgani praedicta*) with corr. proboscis length predicted by Darwin before discovery. See also *Jumellea*

angsana *Pterocarpus indicus*

Anguillaria R. Br. = *Wurmbea*

Anguloa Ruíz & Pavón. Orchidaceae (V 12g). 9 trop. S Am. R: Orquideologia 21(1999)161, CBM 22(2005)63. R: H.F. Oakeley (2008) *Lycaste, Ida & A.*: 328. Cult. orn. ('baby in cradle' due to lip mobile when rocked)

Angylocalyx Taubert. Leguminosae (III 2). 7 trop. Afr. ***A. oligophyllus*** (Bak.) Bak.f. (W Afr.) – Corner's Model

× **Angulocaste** DeBievre. Orchidaceae. *Anguloa* × *Lycaste* – 130 grexes

Anhaloniopsis (Buxb.) Mottram = *Matucana*

Ania Lindl. (~ *Tainia*). Orchidaceae (V 14). 11 Indomal. R: OM 6(1992)49

Aniba Aubl. Lauraceae (I). 41 Andes, Colombia to Bolivia R: FN 31(1982)18. Rauh's Model. Source of Braz. sassafras & bois-de-rose oil. ***A. coto*** (Rusby) J.F. Macbr. (Bolivia) – medic. 'coto bark'; ***A. perutilis*** Hemsl. (comino) – good timber (form. railway sleepers, now Rolls Royce dashboards); stumps to 300 yrs old used for artefacts; ***A. rosodora*** Ducke (N S Am.) –heterodichogamy, poll. stingless bees attracted by oil glands, disp. by toucans, exp. Braz. (c. 28 M t p.a.) esp. to Fr., Switzerland, distillate from steamed chips used in Chanel No 5

aniègre *Pouteria* spp.

Anigozanthos Lab. Haemodoraceae (II). Excl. *Macropidia*, 11 SW Aus. R: Aus.JB 25(1977)524, Aus.SB 4(1991)663. Cult. orn. (many hybrids) with woolly fls (kangaroo-paw), incl. ***A. humilis*** Lindl. (cat's-paw)

animé (gum) *Hymenaea coubaril*

Aningeria Aubrév. & Pellegrin = *Pouteria*

Anisacanthus Nees (*Idanthisa*). Acanthaceae (III 2c). Excl. *Thyrsacanthus*, 8 SW US, Mex. Cult. orn.

Anisachne Keng = *Calamagrostis*

Anisadenia Wall. ex Meissn. Linaceae (II). 2 Himal. to C China. R: FOC 11(2008)37

Anisantha K. Koch. See *Bromus*

Anisantherina Pennell (~ *Agalinis*). Orobanchaceae (Gerard.). 1 trop. Am.: ***A. hispidula*** (Mart.) Pennell

anise *Pimpinella anisum* (aniseed balls), (USA) *Foeniculum vulgare*; **Chinese** or **star a.** *Illicium verum*; **Japanese star a.** *I. anisatum*; **purple** or **tree a.** *I. floridanum*

aniseed myrtle *Anetholea anisata*

Aniseia Choisy. Convolvulaceae (1). 4 trop. Am., ***A. martinicensis*** (Jacq.) Choisy pantrop., ed. young shoots

Aniselytron Merr. (~*Calamagrostis*). Gramineae (XVI 15). 2 N Ind. to Jap. (China 2). R: FOC 22(2006)310. Hybrid origin?

Aniserica N.E. Br. = *Erica*

anisette *Foeniculum vulgare, Pimpinella anisum*

Anisocalyx L. Bolus = *Jacobsenia*

Anisocampium C. Presl (~ *Athyrium*). Athyriaceae. 4 S & E As. R: T 60(2011)827

Anisocapparis Cornejo & Iltis (~ *Capparis*). Capparaceae. 1 Bolivia, W Braz., Parag., N Arg.: ***A speciosa*** (Griseb.) Cornejo & Iltis. R: JBRIT 2(2008)62

Anisocarpus Nutt. (~ *Madia*). Compositae (Mad.-Mad.). Incl. *Raillardiopsis*, 2 W N Am. R: Novon 9(1999)462. App. allopolyplody led to *Argyroxiphium* alliance (Hawaii)

Anisochaeta DC. Compositae (Gnap.). 1 S Afr.: ***A. mikanioides*** DC. R: OB 104(1991)47

Anisochilus Wall. ex Benth. Labiatae (VII 3e). 16 Indomal. R: KB 64(2009)236. Oft. succ.

Anisocoma Torrey & A. Gray. Compositae (Cich.-Cich.-Micr.). 1 SW N Am.: ***A. acaulis*** Torrey & A. Gray. R: FNA 19(2006)310

Anisocycla Baill. Menispermaceae (I). 3 trop. Afr., 3 Madag.

Anisodontea C. Presl. Malvaceae (Malv.-Malv.). 19 S Afr. R: GH 10(1969)215. Cult. orn.

Anisodus Link ex Spreng. (~ *Scopolia*). Solanaceae (IV 2). 4 temp. E As. R: A.T.Hunziker, *Gen. Solan.* (2001) 361. ***A. luridus*** Link ex Spreng. (*A. stramoniifolius*) – yak-fodder in Himal.

Anisomallon Baill. = *Apodytes*

Anisomeles R. Br. Labiatae (VI). (Incl. *Epimeridi*) 26 OW trop. R: Austrobaileya 9(2015)333. Some med. incl. ***A. indica*** (L.) Kuntze (Afr. to Jap., Aus.) also eaten in sago cakes (Mal.), a good bee-pl.

Anisomeria D. Don. Phytolaccaceae (I). 2–3 Chile

Anisomeris C. Presl = *Chomelia*

Anisopappus Hook. & Arn. Compositae (Athro.). 17 trop. & S Afr., China (1). R: AJBM 54(1996)379, PSE 199(1995)168

Anisopetala (Kraenzlin) M.A. Clem. = *Dendrobium*

Anisophyllea R. Br. ex Sabine. Anisophylleaceae. 67 trop. (OW 25). R: Phytotaxa 229(2015)10. Massart's Model; serial buds. ***A. laurina*** R. Br. ex Sabine (W Afr.) – fr. ed. (monkey-apple)

Anisophylleaceae Ridl. Magnoliidae – Cucurbitales. 4/71 trop. R: Phytotaxa 229(2015)9. Trees & shrubs, oft. accumulating aluminium. Lvs in spirals, simple, usu. dimorphic, up to 6 serial buds in axil; stip. 0 (or 2–4 minute, ?colleters). Fls in axillary racemes or panicles on leafless shoots, usu. unisexual (monoecy), mostly 4-merous, epigynous. $\underline{\mathrm{K}}$ valvate, entire to dissected or 0; A usu. 8 in 2 whorls, opening by longitudinal slits; $\overline{\mathrm{G}}$ (4), rarely (3), with sep. styles; ovules 1 or 2 in each locule, pend., oft. unitegmic. Fr. indehiscent, woody to drupaceous, s.t. winged; seeds 1–4 with no endosperm. n = 7, 8
Genera: *Anisophyllea, Combretocarpus, Poga, Polygonanthus*
Form. placed nr. Rhizophoraceae, but differ in spirals of ± exstipulate lvs, distinct styles & pollen features (Cronquist) besides molecular evidence

Anisopoda Bak. Umbelliferae (III 8). 1 Madag.: ***A. bupleuroides*** Bak.

Anisopogon R. Br. Gramineae (IX). 2 SE Aus. R: FA 44A(2009)14

Anisoptera Korth. Dipterocarpaceae (Dipt.-Dipt.). 10 Bangladesh to NG; fossils from NW Ind. G semi-inf. Imp. timbers & veneers, though high silica levels blunt saws. ***A. curtisii*** Dyer ex King (W Mal.) – krabak; ***A. laevis*** Ridl. (W Mal.) etc. – mersawa; ***A. scaphulla*** (Roxb.) Kurz (SE As. to W Mal.) – kaunghmu; ***A. thurifera*** (Blanco) Blume etc. – palosapis (Philippines), resin used to line tunnels in termite nests in N Moluccas by *Chalicodoma pluto* (world's biggest bee)

Anisopus N.E. Br. Apocynaceae (V c; Asclepiadaceae). 2 trop. W Afr. R: KB 49(1994)739

Anisosciadium DC. Umbelliferae (III 1). 3 SW As. R: Fl. Iraq 5,2(2013)121

Anisosepalum Hossain. Acanthaceae (I). 3 C Afr. R: Aliso 32(2014)19

Anisosperma A. Silva Manso (~ *Fevillea*). Cucurbitaceae. 1 Braz.: ***A. passiflora*** (Vell.) Silva Manso

Anisostachya Nees = *Justicia*

Anisotes Nees. Acanthaceae (III 2c). Incl. *Macrorungia, Metarungia*, 22 trop. Afr., Madag. R: NJB 1(1981)623. Alks

Anisothrix O. Hoffm. Compositae (Gnap.). 2 SW Cape. R: OB 104(1991)46

Anisotoma Fenzl. Apocynaceae (V b; Asclepiadaceae). 2 S Afr.

Anisotome Hook.f. Umbelliferae (III 8). 15 NZ, subAntarctic is. R: UCPB 33(1961)1. Dioec. Hybrids with *Aciphylla* spp.

Anisum Hill = *Pimpinella*

anjan *Clitoria ternatea*

Ankyropetalum Fenzl. Caryophyllaceae (III 1). 4 E Med. to Iran. R: Wentia 9(1962)170

Anna Pellegrin. Gesneriaceae (III 2j). 4 S China, N Vietnam

Annaea Kolak. = *Campanula*

Annamocalamus H.N. Nguyen & al. (~ *Melocanna*). Gramineae (V 5). 1 Vietnam: ***A. kontumensis*** H.N. Nguyen & al. R: Cand. 68(2013)160

Annamocarya A. Chev. = *Carya*

annatto *Bixa orellana*

Annea Mackinder & Wieringa (~ *Hymenostegia*). Leguminosae (I 2). 2 trop. Afr. R: Phytotaxa 142(2013)1

Anneliesia Brieger & Lueckel (*Gynizodon*) = *Miltonia*

Annesijoa Pax & K. Hoffm. Euphorbiaceae (Ricinod.). 1 NG: ***A. novoguineensis*** Pax & K. Hoffm. R: Blumea 49(2004)426

Anneslea Wall. Pentaphylacaceae (2; Ternstroemiaceae). 3 Indomal., China. R: JAA 33(1952)79. G semi-inf. ***A. fragrans*** Wall. –source of beautifully marked timber, local medic.

Annesorhiza Cham. & Schldl. Umbelliferae (III 8). 12 S Afr. R: NJB 21(2001)626. ***A. macrocarpa*** Ecklon & Zeyher – form. a market r. veg.; ***A. nuda*** (Aiton). B.L. Burtt (*A. capensis*) – ed. r.

Annickia Setten & Maas (*Enantia*). Annonaceae (IV 1). 8 trop. Afr. (E Afr. 1). R: SGP 77(2007)94. Some timbers esp. ***A. chlorantha*** (Oliv.) Setten & Maas (*E. chlorantha*, Afr. whitewood, W Afr.) & dyes. Alks

Annona L. Annonaceae (III 5). Incl. *Raimondia, Rollinia, Rolliniopsis*, 137 trop. Am. (129) & Afr. (incl. *Anonidium*, 8). R: AHB 10(1931)197, 12(1934)112,190. Troll's & Roux's Models. ***A. coriacea*** Mart. (Braz.) fls heat up over 2 nights, ***A. crassiflora*** Mart. (Braz.) fls 1 night then dropped, ***A. cornifolia*** A. St- Hil. (Braz.) fls over 1–2 nights, scarab-poll.; ***A. sericea*** Dunal (NE S Am.) – thick fleshy petals never fully opening, encl. A & G, c. 7 p.m. fl. temp. rises to 6°C above ambient with odour like chloroform & ether attractive to small chrysomelid beetles & flies which enter, stigmas then fall & anthers become erect & release pollen, A then fall off 1 by 1 & petals also to release insects. Fr. a large fleshy syncarp formed by amalgamation of pistils & receptacle, some fish-disp. (e.g. ***A. hypoglauca*** Mart. – obligate) in Amazon, many ed. (custard, alligator or monkey apples). ***A. cherimola*** Mill. (cherimoya, custard apple, sharifa, Andes of Peru & Ecuador); ***A. diversifolia*** Saff. (ilama, C Am.); ***A. glabra*** L. (pond apple, trop. Am. & W Afr., weed of nat. significance in Aus. prob. introd. as r.-stocks, now disp. by cassowaries & flying foxes) – excellent epiphyte host in Everglades, seeds with 6–10 layers of buoyancy cells in endotesta, sea-disp., r. used for corks; ***A. mucosa*** Jacq. (*Rollinia m., R. deliciosa*, biribá) ed. football-sized fr., 3-winged fls, beetles force way in, deposit pollen, remain until C falls & then carry off pollen; ***A. muricata*** L. (soursop, guanábana, graviola, trop. Am.) – domatia; ***A. purpurea*** Moçiño & Sessé ex Dunal (soncoya, Mex.); ***A. reticulata*** L. (bullock's-heart, custard apple, sugar apple, trop. Am.) – inferior to *A. cherimola*; ***A. scleroderma*** Saff. (poshte, C Am.) – thick skin quality to be bred into other spp. to ease shipping problems; ***A. squamosa*** L. (sweetsop, custard apple, sita phal, trop. Am.) – some heterodichogamy, seeds with insecticides (acetogenins) inhibiting electron transport chain, local med. inc. in malaria, '**Thai seedless**' mutant has no seed outer integument (gene identified), widely grown hybrids (***A. × atemoya*** Mabb., atemoya) with *A. cherimola* raised several times

Annonaceae Juss. Magnoliidae – Magnoliales. 107/2450 trop. (incl. Am. – Candollea 49(1994)389), *Asimina* ext. N to Michigan. Trees (many Petit's & Roux's Models, s.t. buttressed), shrubs or lianes (30% of OW spp., few Am.), usu. with alks, resin canals & septate pith; twigs oft. drying blackish. Lvs simple, typically distich. & with glaucous or metallic sheen, stip. 0. Fls solit. or in var. basically cymose infls, s.t. cauliflorous or even on underground suckers (some *Duguetia*), fragrant, usu. nodding, freq. opening before all parts fully dev., poll. by insects esp. beetles, which may be trapped inside (some A. thermogenic), also thrips, flies & even bees, bisex., rarely not so, hypogynous & commonly 3-merous, usu. brittle or fleshy. P usu. 3 whorls of 3, rarely more or fewer & rarely connate at base, fleshy; A ∞, in spirals, rarely 3 or 6 cyclic or connate at base, filaments 1-veined, anthers opening by longit. slits, pollen shed in monads, tetrads or even polyads, v. varied. $\underline{G}$ ∞, rarely connate to form compound ovary (*Monodora, Isolona*); ovules 1–(usu.) ∞, anatr. or s.t. campylotr., bitegmic or rarely with a new integument intercalated between other 2. Fr. of distinct berries to dry indehiscent or explosive (*Anaxagorea*) or united into syncarp by development of receptacle; seeds bitegmic or tritegmic (e.g. *Cananga*, the third one lying between the others, a condition unique in angiosperms) oft. arillate, endosperm copious, ruminate, oily (& sometimes starchy). x = 7, 8, 9

Classification & large genera (BotJLS 169(2012)32):

I. **Anaxagoreoideae** (trees, G free, seed with no 'middle' integument nor aril) – *Anaxagorea* (only)

II. **Ambavioideae** (trees, G free, seed with 'middle' integument & s.t. aril) – *Cananga, Cleistopholis*

III. **Annonoideae** (incl. Monodoroideae; trees or lianes, G free (or not), seed with no 'middle' integument, s.t. arillate) – 7 tribes: 1. Bocageeae – *Mkilua, Porcelia;* 2. Xylopieae – *Artabotrys, Xylopia* ; 3. Duguetiae – *Duguetia;* 4. Guatterieae – *Guatteria* (only); 5. Annoneae – *Annona, Goniothalamus;* 6. Monodoreae – *Isolona, Monodora;* 7. Uvarieae – *Desmos, Fissistigma, Friesodielsia, Monanthotaxis, Uvaria*

IV. **Malmeoideae** (lvs in spirals, G free, seeds with 0 middle integument or aril) – 7 tribes: 1. Piptostigmateae – *Annickia, Sirdavidia;* 2. Malmeeae – *Oxandra, Unonopsis;* 3. Maasieae – *Maasia* (only); 4. Fenerivieae – *Fenerivia* (only); 5. Dendrokingstonieae – *Dendrokingstonia* (only); 6. Monocarpieae – *Monocarpia* (only); 7. Miliuseae – *Alphonsea, Marsypopetalum, Meiogyne, Miliusa, Mitrephora, Monoon, Polyalthia, Stelechocarpus*

Acetogenins acting on pests & tumours by depleting ATP levels through inhibition of Complex I of mitochondria & NADH oxidase of plasma membranes. Several *Annona* cult. for fr. (see also *Asimina, Porcelia*), others for spice (*Monodora, Xylopia*) & scent, e.g. *Cananga, Mkilua* (see also *Desmos, Stelechocarpus*); timbers incl. spp. of *Annickia, Cleistopholis, Duguetia, Oxandra;* cult. orn. incl. *Monoon*

Anochilus (Schltr.) Rolfe = *Pterygodium*

Anoda Cav. Malvaceae (Malv.-Malv.). 23 S US & (esp.) Mex. to Bolivia, Arg. & Chile. R: Aliso 11(1987)485. Cult. orn.

Anodendron A. DC. Apocynaceae (II c). 17 Indomal. to Jap. & Vanuatu. R: Blumea 41(1996)38. ***A. candolleanum*** Wight (Thailand to Java, Philippines.) – form. imp. fibre from this liane; ***A. paniculatum*** A. DC. (Indomal.) – fibre for fishing nets

Anodiscus Benth. Gesneriaceae (II 3). 1 Andes of Peru, Ecuador: ***A. xanthophyllus*** (Poepp. & Endl.) Mansfeld (*A. peruvianus*) – roadside weed. R: Selb. 6(1982)174

Anodopetalum A. Cunn. ex Endl. (III). Cunoniaceae. 1 SW Tasmania: ***A. biglandulosum*** (Hook.) Hook.f. (horiz. scrub) – trunk oft. bends to horiz., branches forming thickets

Anoectocalyx Benth. = *Anaectocalyx*

Anoectochilus Blume. Orchidaceae (IV 2d). c. 40 trop. As. to Hawaii. Cult. orn. (jewel orchids), 'vogue' foliage stove-pls of mid-late C19 & local medic.

Anogeissus (DC.) Wall. Combretaceae (II 2a). Excl. *Finetia*, 7 OW trop. to Arabia. R: KB 33(1979)555. Timber, dyes, gums (Ind. gum) & medic. ***A. acuminata*** (DC.) Wall. (Ind. to SE As.) – comm. timber (yon); ***A. latifolia*** (DC.) Wall. (Ind.) – gatty gum, lvs produce a black dye & used in tanning (dhawa); ***A. leiocarpa*** (DC.) Guillem. & Perr. (trop. Afr.) – termite-proof wood, chewing-sticks in Nigeria, lvs source of yellow dye, plant a vermifuge for stock

Anogramma Link. Pteridaceae (III). 5 Azores & SW Eur. (1), OW trop. (***A. ascensionis*** (Hook.) Diels of Ascension Is. extinct) to NZ, trop. Am. Subterr. perenn. prothallus in ***A. leptophylla*** (L.) Link, which has delicate ann. sporophytes

Anoiganthus Bak. = *Cyrtanthus*

anokye *Guibourtia ehie*

Anomacanthus R. Good (*Gilletiella*). Acanthaceae (II; Mendonciaceae). 1 Congo, Angola: ***A. congolanus*** (De Wild. & T. Durand) Brummitt. R: KB 45(1990)710

Anomalanthus Klotzsch = *Erica*

Anomalesia N.E. Br. = *Gladiolus*

Anomalluma Plowes = *Pseudolithos*

Anomalocalyx Ducke. Euphorbiaceae (Crot.-Ost.). 1 nr Manaos, Braz.: ***A. uleanus*** (Pax) Ducke. R: A. Radcliffe-Smith, *Gen. Euphorb.* (2001)338

Anomalostylus R. Foster = *Trimezia*

Anomanthodia Hook.f. = *Aidia*

Anomatheca Ker-Gawler = *Freesia*

Anomianthus Zoll. = *Uvaria*

Anomochloa Brongn. Gramineae (I). 1 Bahia, Braz.: ***A. marantoidea*** Brongn. R: SCB 68(1989)2. Forest grass of marantoid habit rediscovered 1976

Anomochloaceae Nakai = Gramineae
Anomoctenium Pichon = *Amphilophium*
Anomopanax Harms = *Mackinlaya*
Anomostachys (Baill.) Hurus. (? *Sarothrostachys*; ~ *Excoecaria*). Euphorbiaceae (Euph.-Hipp.). 3(?+) trop. Afr., Madag.
Anomospermum Miers. Menispermaceae (II). 6 trop. Am. R: MNYBG 22,2(1971)61
Anomostephium DC. = *Wedelia*
Anomotassa K. Schum. Apocynaceae (V c; Asclepiadaceae). 1 Ecuador: ***A. macranthus*** K. Schum.
Anona Mill. = *Annona*
Anonidium Engl. & Diels = *Annona*
Anoplocaryum Ledeb. Boraginaceae (B.3.3). 5 C As. to Siberia
Anopteris (Prantl) Diels = *Pteris*
Anopterus Lab. Escalloniaceae (Grossulariaceae s.l.). 2 SE Aus. Cult. orn. ***A. macleayanus*** F. Muell. – pachycaul with lvs to 30 cm long
Anopyxis (Pierre) Engl. Rhizophoraceae (I). 1 trop. Afr.: ***A. klaineana*** (Pierre) Engl.- Attims's Model
Anosporum Nees = *Cyperus*
Anotea (DC.) Kunth = *Hibiscus*
Anotis DC. = *Arcytophyllum*
Anotites Greene = *Silene*
Anplectrum A. Gray = *Diplectria*
Anredera Juss. Basellaceae. 12 warm Am. R: KB 62(2007)309. Lianes, cult. orn. esp. ***A. cordifolia*** (Ten.) Steenis (Madeira or mignonette vine, subtrop. S Am., invasive Aus., NZ) – local medic. (anti-inflammatory), natur. in S Eur. where cult. as veg. ('*Boussingaultia baselloides*') & Indomal. to Aus. reprod. only by aerial tubers; ***A. vesicaria*** (Lam.) Gaertn. f. (SW N Am.) – local medic. Mex.
Ansellia Lindl. Orchidaceae (V 12b). 1 trop. & S Afr.: ***A. africana*** Lindl. – cult. orn. R: KM 10(1993)32
Antarctic beech *Nothofagus antarctica*
Antegibbaeum Schwantes ex C. Weber (~ *Gibbaeum*). Aizoaceae (V). 1 W Cape: ***A. fissoides*** (Haw.) C. Weber. R: H. E. K. Hartmann, *Ill. Handb. Succ. Pls*, Aiz. A–E (2002)40
antelope bush *Purshia* spp.; **a. grass** *Echinochloa pyramidalis*; **a. horn** *Asclepias viridis*; **a. orchid** *Dendrobium canaliculatum*
Antennaria Gaertn. Compositae (Gnap.-Cass.). c. 40 temp. (Eur. 6) & arctic-alpine, warm (not Afr.). R: OB 104(1991)96. Small dioec. stoloniferous sexual diploids but also oft. derived autotetraploids in 1 sp., & sexual & apomictic allopolyploid (triploid to octoploid) herbs (pussy's toes) cult. rock gardens esp. ***A. dioica*** (L.) Gaertn. (mt. everlasting, cat's-foot, Euras., Aleutians). ***A. alpina*** (L.) Gaertn. (circumpolar) – apomictic; ***A. rosea*** (Eaton) Greene – polyploid complex (3n, 4n) with wider range than diploids & poss. involving up to 8 of them
Antenoron Raf. = *Persicaria*
Anteremanthus H. Robinson. Compositae (Vern.-Lych.). 2 Braz. R: SB 39(2014)661
Anteriorchis E. Klein & Strack = *Anacamptis*
Anthacanthus Nees = *Oplonia*
Anthaenantia P. Beauv. = *Anthenantia*
Anthaenantiopsis Mez ex Pilg. Gramineae (XXIII 3). 4 warm S Am. R: SB 18(1993)434
Antheliacanthus Ridl. = *Pseuderanthemum*
Anthemis L. Compositae (Anth.-Anth.). Excl. *Cota*, incl. *Archanthemis*, c. 155 Eur., Med. to Iran & E Afr. (1). R: BBMNB 23(1993)132. Aromatic, medic., some cult. orn. ***A. cotula*** L. (mayweed, Euras., cosmop. weed) – source of insecticide & alleged mouse-repellent, taints cows' milk & blistered harvester's hands; ***A. pseudocotula*** Boiss. (E Med., N Afr.) – in floral collars in Tutankhamun's tomb
Anthenantia P. Beauv. (*Anthaenantia*). Gramineae (XXIII 2). 5 SE US pine barrens, trop. Am. R: Sida 21(2004)294
Anthephora Schreb. Gramineae (XXIV 1). 12 trop. Afr., Arabia with ***A. hermaphrodita*** (L.) Kuntze ext. to Am. R: SB 13(1988)589
Anthericaceae J. Agardh = Asparagaceae
Anthericopsis Engl. (~ *Murdannia*). Commelinaceae (II 2). 1 trop. E Afr.: ***A. sepalosa*** (C.B. Clarke) Engl. R: FTEA Commel.(2012)49

Anthericum L. Asparagaceae (Anthericaceae). 65 Afr. (S Afr. 3), Eur. (3), W As. Afr. spp. with r.-tubers = *Chlorophytum;* Am. spp. = *Echeandia;* poss. all Afr. spp. to be excl. ***A. liliago*** L. (St Bernard's lily, S Eur.) – cult. orn. border pl.

Antherolophus Gagnepain = *Aspidistra*

Antheropeas Rydb. = *Eriophyllum*

Antheroporum Gagnepain. Leguminosae (III 16). 1 (or 4) China

Antherostele Bremek. Rubiaceae (I 9). 4 Philippines. R: JAA 21(1940)26

Antherostylis C. Gardner = *Velleia*

Antherothamnus N.E. Br. Scrophulariaceae (Scroph.). 1 C & S Afr.: ***A. pearsonii*** N.E. Br. R: O.M. Hilliard, *Manuleae* (1994)75

Antherotoma (Naud.) Hook.f. Melastomataceae. 2 trop. Afr., Madag.

Anthobembix Perkins = *Steganthera*

Anthobolus R. Br. Santalaceae (? Opiliaceae). 3 Aus. R: MBMZ 213(1959)101. Root-parasites; ovary sup.; fr. on coloured fleshy receptacle

Anthobryum Philippi = *Frankenia*

Anthocarapa Pierre. Meliaceae (II-Trich.). 1 Aus., NG to Rotuma: ***A. nitidula*** (Benth.) Mabb. R: Blumea 31(1985)132

anthocephalus *Neolamarckia cadamba*

Anthocephalus A. Rich. = *Breonia*. See also *Neolamarckia*

Anthocercis Lab. Solanaceae (III 1). 10 Aus. R: Telopea 2(1981)174. Some cult. orn.

Anthochlamys Fenzl. Amaranthaceae (Chenopodiaceae I 6). 3 SW & C As.

Anthochloa Nees & Meyen = *Poa*

Anthochortus Nees. Restionaceae. 7 SW & S Cape. R: Bothalia 15(1985)484

Anthocleista Afzel. ex R. Br. Gentianaceae. c. 50 trop. Afr., Madag., Masc. Trees (Leeuwenberg's & Scarrone's Models) with big leaves (cabbage-trees) e.g. ***A. vogelii*** Planch. (W Afr.) – sapling lvs to 2.5 m long. ***A. grandiflora*** Gilg (S Afr.) – local malaria cure, antibacterial compounds found

Anthoclitandra (Pierre) Pichon = *Landolphia*

Anthodiscus G. Meyer. Caryocaraceae. 9 trop Am. R: OB 92(1987)179. Fish-poisons

Anthodon Ruíz & Pavón. Celastraceae (III). 2 trop. Am.

Anthogonium Wall. ex Lindl. Orchidaceae (V 10a). 1 E Himal. to SE As.: ***A. gracile*** Wall. ex Lindl. – cult. orn. R: Fl. Bhutan 3,3(2002)279

Antholyza L. = *Gladiolus*

Anthonotha P. Beauv. (~ *Macrolobium*). Leguminosae (I 2). 17 trop. Afr. forests. R: PEE 143(2010)71. Timber, dyes, local medic. & ed. (seeds). ***A. macrophylla*** P. Beauv. (W Afr.) – same growth form as *Scaphopetalum amoenum* (*q.v.*)

Anthophyta = Magnoliidae

Anthopteropsis A.C. Sm. Ericaceae (VIII 5). 1 C Panamá: ***A. insignis*** A.C. Sm.

Anthopterus Hook. Ericaceae (VIII 5). 11 Andes. R: Britt. 48(1997)606

Anthorrhiza Huxley & Jebb. Rubiaceae (IV 7). 9 NG. R: Blumea 36(1991)21. Ant-pls

Anthosachne Steud. = *Elymus*

Anthosiphon Schltr. = *Maxillaria*

Anthospermopsis (K. Schum.) Kirkbride (~ *Staelia*). Rubiaceae (IV 15). 1 Braz.: ***A. catechosperma*** (K. Schum.) Kirkbride. R: Britt. 49(1997)373

Anthospermum L. Rubiaceae (IV 13). 39 Afr., Madag.

Anthostema A. Juss. Euphorbiaceae (Euph.-Anth.). 2 trop. W Afr., 1 Madag. Nozeran's Model. Fls in cyathium like *Euphorbia* but have P

Anthotium R. Br. Goodeniaceae. 3 SW Aus. R: Nuytsia 7(1989)49

Anthotroche Endl. Solanaceae (III 1). 4 Aus. R: FA 29(1982)30

Anthoxanthum L. Gramineae (XVI 4). (Incl. *Hierochloe*) c. 50 Euras. (Eur. 13), Arctic, trop. mts, N (5) & C Am. Strong scent (coumarin – anticoagulant used [as Coumadin] in surgery to prevent clotting, but metabolized by aspergillus to dicoumarol, inducing vitamin K deficiency etc. in stock so mouldy hay must be discarded). ***A. horsfieldii*** (Kunth) Mez (*H. h.*, Java) – good fodder, form. for stock of princes in Java; ***A. odoratum*** L. (sweet vernal grass, spring grass, Euras., widely natur.) – widespread tetraploid prob. hybrid of diploid N & S races, fodder (little food value); ***A. nitens*** (Weber) Schouten & Veldk. (*H. odorata*, holy grass, N temp.) – similar scent, strewn on church floors (introd. Scotland by Prussian monks for this; cf. *Myrrhis*), burned as incense in New Mex. & used to scent clothes, local medic., basketry, cerem. etc. in N Am.

Anthriscus Pers. Umbelliferae (III 2). c. 20 Euras. to Afr. mts. R: PBS 13(1997)19. ***A. cerefolium*** (L.) Hoffm. (chervil) – cult. since time of Pliny, lvs aniseed-flavoured for seasoning (a soup in Netherlands) & salad, in *fines-herbes* (with parsley & chives) for omelettes etc.; ***A. sylvestris*** (L.) Hoffm. (cow-parsley, Queen Anne's lace, range of genus) – common roadside pl., still persisting superstition (Engl.) that bringing fls into house leads to death of Mother

Anthurium Schott. Araceae (III 2). c. 1500 [c. 900 named] trop. Am. (e.g. Panamá 152) in 18 sections. R(C Am.): AMBG 70(1984)211, SBM 14(1987). Terr. & epiphytic, in 19 sects (Aroid. 6(1983)85) incl. sect. *Pachyneurium* (114; R: AMBG 78(1991)539) usu. 'bird's nest' habit, s.t. ant-inhabited e.g. ***A. gracile*** (Rudge) Lindl. Fls bisex. with P, the spadix s.t. in brightly coloured spathe (black in 1 Colombian sp.; usu. dull & small, here long-lived & poss. attractive to disp. birds), some with scent coll. male euglossine bees. Fr. a berry hanging from spadix by 2 threads from P when ripe, oft. brightly coloured. Some dried lvs used to perfume tobacco in S Am., many local med.; c. 30 spp. widely cult. orn. (flamingo fls) for foliage & spathes esp. ***A. scherzerianum*** Schott (Costa Rica) & ***A. andraeanum*** Linden ex André (Colombia, Ecuador) but usu. ***A. ×ferrierense*** Masters & T. Moore (*'A andraeanum'*, *A. a.* × *A. hoffmannii* Schott (Costa Rica, Panama) × *A. nymphaeifolium* K. Koch & Bouché (Colombia, Venez.)) & other hybrids, some with ***A. amnicola*** Dressler (Panamá) – scented mauve infls, form. (prob. now extinct) on boulders in streams. ***A. armeniense*** Croat (Guatemala) – lilac-scented white pollen; ***A. harrisii*** (R. Graham) G. Don (Braz.) – infls smell of red wine; ***A. pendulifolium*** N.E. Br. (trop. Am.) – lvs to 1.8 m long

Anthyllis L. Leguminosae (III 22). 22 Med., W Eur., Macaronesia, NE Afr. Some cult. orn. & fodder esp. ***A. vulneraria*** L. (kidney-vetch, lady's fingers, Eur. & Medit.) – complex sp. of some 35 sspp., host pl. for larvae of Small Blue (*Cupido minimus*)

Antiaris Leschen. Moraceae (III). 1 OW trop.: ***A. toxicaria*** Leschen. (*A. africana*, upas-tree, ipoh) – Roux's Model, timber in Afr. (ako) for canoes, plywood core, bark-cloth for wrapping rubber, latex an adulterant of *Funtumia* rubber, arrow & ordeal poison (cardiac glycosides), fabled tree allegedly poisoning surroundings & fatal to approach

Antiaropsis K. Schum. Moraceae (II). 2 NG. R: Blumea 50(2005)539. ***A. decipiens*** K. Schum. – dioec., thrips-poll., thrips ovipositing in ♂ infls (nymphs eat pollen), but also ♀ (no reward) & effect poll. (cf. *Ficus*)

Anticharis Endl. Scrophulariaceae (Apt.). 14 Afr. to Mal.

Anticheirostylis Fitzg. = *Genoplesium*

Anticlea Kunth (~ *Zigadenus*). Melanthiaceae. 11 As., N Am. to Guatemala. R: Novon 12(2002)301. Some cult. orn. (*'Stenanthium'*) esp. ***A. elegans*** (Pursh) Rydb. (*Z. glaucus*, white camas(h), N Am.)

Anticoryne Turcz. (~ *Baeckea*). Myrtaceae (II 15). 2 SW Aus.

Antidaphne Poeppig & Endl. Santalaceae (Eremolepidaceae). 7 W trop. S Am. R: SMB 18(1988)17

Antidesma Burm. ex L. Phyllanthaceae (Ant.-Ant.; Euphorbiaceae s.l). 150+ OW trop. & warm esp. As. (Afr. 10). Troll's Model; some rheophytes. High ploidy levels. Bird-disp. pink, red or black fr. Timber & fr. trees, local medic., cult. orn. ***A. bunius*** (L.) Spreng. (bignay, Chinese laurel, Indomal.) – fr. bitter to some but sweet to others, preserved, fls smell of powdered fish; ***A. ghaesembilla*** Gaertn. (Indomal. to Aus.) – fr. valued N Qld; ***A. montanum*** Blume (Mal. to Japan) introd. Afr., NE S Am., fr. used to adulterate pepper; ***A. pulvinatum*** Hillebrand (Hawaii) – domatia

Antigonon Endl. Polygonaceae (Erio-Brunn.). 3–6 trop. Am. Lianes with tendrils (infl. axes), cult. orn. esp. ***A. leptopus*** Hook. & Arn. (coral vine, corallita, Mex.) – bright pink fls, tubers ed. (nut-like flavour)

Antigramma C. Presl = *Asplenium*

Antillanorchis Garay = *Tolumnia*

Antillanthus R. Nordenstam (~ *Senecio*). Compositae (Senec.-Senec.). 17 Cuba. R: CN 44(2006)51. Shrubs

Antillia R. King & H. Robinson (~ *Eupatorium*). Compositae (Eup.-Crit.). 1 Cuba: ***A. brachychaeta*** (Robinson) R. King & H. Robinson. R: MSBMBG 22(1987)304

Antimima N.E. Br. (~ *Ruschia*). Aizoaceae (V). 96 SW Namibia & S Afr. R: H. E. K. Hartmann, *Ill. Handb. Succ. Pls*, Aiz. A–E (2002)41

Antinisa (Tul.) Hutch. = *Homalium*

Antinoria Parl. (~ *Aira*). Gramineae (XVI 5). 2 Medit. R: KBAS 13(1986)132

Antiostelma (Tsiang & P.T. Li) P.T. Li = *Micholtzia* (but see Novon 2(1992)218)

Antiotrema Hand.-Mazz. Boraginaceae (B.3.8.2). 1 SW China: ***A. dunnianum*** (Diels) Hand.-Mazz.

Antiphiona Merxm. Compositae (Inul.-Pluch.). 2 trop. & SW Afr.

Antiphytum DC. ex Meissn. Boraginaceae (B.1). 10 Mex. & trop. Am. R: CGH 68(1923) 48

Antirhea Comm. ex Juss. Rubiaceae (III 3). 36 Madag. & Mal. to Samoa (Am. spp = *Pittoniotis* or *Stenostomum* – bisex. fls). Dioec. Some timber trees

Antirrhinaceae Pers. = Plantaginaceae

Antirrhinum Tourn. ex L. Plantaginaceae (Ant.-Ant.; Scrophulariaceae s.l.). 25 Med. esp. Iberia (24), natur. temp. R: D.A. Sutton, *Rev. Antirrhineae* (1988)67. Mouth of fl. closed & nectar accessible only to bees, which force an entry. ***A. majus*** L. (snapdragon, Medit.) – cult. orn., many cvs incl. tetraploids, peloric forms, 'azalea-flowered' (with extra whorl of small C), etc. (trailing cvs poss. hybrids); ***A. cirrhigerum*** (Ficalho) Rothm. (*A. majus* subsp. *c.*, SW Iberia) a scrambler

Antistrophe A. DC. Primulaceae (Myrsinaceae). 4–5 Indomal.

Antithrixia DC. Compositae (Gnap.-Rel.). 1 Namaqualand: ***A. flavicoma*** DC. – shrub. R: OB 104(1991)65

Antizoma Miers. Menispermaceae (V). 2 arid S Afr. R: JSAB 46(1980)1

Antonella Caro = *Tridens*

Antongilia Jum. = *Dypsis*

Antonia Pohl. Loganiaceae (Strychnaceae). 1 S Am.: ***A. ovata*** Pohl

Antoniaceae Hutch. = Loganiaceae

Antonina Vved. = *Clinopodium*

Antopetitia A. Rich. Leguminosae (III 22). 1 trop. Afr. mts: ***A. abyssinica*** A. Rich. – close to *Ornithopus*

Antoschmidtia Boiss. = *Schmidtia*

Antrocaryon Pierre. Anacardiaceae (II). 2–3 trop. W Afr., 1 trop. Am. Poss. oilseeds. ***A. micraster*** A. Chev. & Guill. (Afr.) – timber for planks & furniture

Antrophora I.M. Johnston = *Lepidocordia*

Antrophyopsis (Benedict) Schuttp. (~ *Antrophyum*). Pteridaceae (V; Vittariaceae). 3 trop. Afr., Indian Ocean. R: T 65(2016)717

Antrophyum Kaulf. Pteridaceae (V; Vittariaceae). 40 trop. esp. Mal. Cult. orn.

Antunesia O. Hoffm. = *Distephanus*

añu *Tropaeolum tuberosum*

Anubias Schott. Araceae (VII 7). 8 C & W Afr. R: Aqua-Planta Sonderh. 1 (1987). Some rheophytes. Cult. orn. aquarium pls esp. ***A. afzelii*** Schott (Senegal to Mali)

Anulocaulis Standl. = *Boerhavia* (but see SB 35(2010)868)

Anura (Juz.) Tscherneva = *Arctium*

Anuragia Raizada = *Pogostemon*

Anurosperma (Hook.f.) H. Hallier = *Nepenthes*

Anvillea DC. Compositae (Inul.-Inul.). 2 N Afr., Middle East. R: NJB 2(1982)297

Anvilleina Maire = *praec.*

anyaran *Distemonanthus benthamianus*

Anychia Michaux = *Paronychia*

Anzybas M.A. Clem. & D.L. Jones = *Corybas*

aoi *Asarum caulescens*

Aoranthe Somers (~ *Porterandia*). Rubiaceae (II 1). 5 trop. Afr. R: BJBB 58(1988)47

Aorchis Vermeulen = *Galearis*

Aosa Weigend (~ *Loasa*). Loascaceae (I 1). 6 E Braz., 1 Hispaniola

Aostea Buscal. & Muschler = *Vernonia*

Aotus Sm. (~ *Pultenaea*). Leguminosae (III 13). c. 24 Aus.

apa *Afzelia* spp.

Apache beads *Anemopsis californica*

Apacheria C. Mason. Crossosomataceae. 1 Arizona: ***A. chiricahuensis*** C. Mason. R: FNA 9(2014)10

Apalanthe Planch. (~ *Elodea*). Hydrocharitaceae. 1 trop. S Am.: ***A. granatensis*** (Humb. & Bonpl.) Planch. R: AB 21(1985)157

Apalatoa Aubl. = *Crudia*

Apalochlamys Cass. (~ *Cassinia*). Compositae (Gnap.-Cass.). 1 Aus.: ***A. spectabilis*** (Lab.) Steud. R: OB 104(1991)85

Apaloxylon Drake = *Neoapaloxylon*

Apama Lam. = *Thottea*

Apargidium Torrey & A. Gray = *Microseris*

Aparisthmium Endl. (*Conceveibum*). Euphorbiaceae (Alc.-Alc.). 1 trop. S Am.: ***A. cordatum*** (A. Juss.) Baill. R: A. Radcliffe-Smith, *Gen. Euphorb.* (2001) 194

Apassalus Kobuski = *Dyschoriste*

Apatesia N.E. Br. Aizoaceae (V). 3 SW Cape. R: H. E. K. Hartmann, *Ill. Handb. Succ. Pls, Aiz. A–E* (2002)65

Apatophyllum McGillivray. Celastraceae (I). 5 Aus. R: Austrobaileya 5(2000)696

Apatostelis Garay = *Stelis*

Apatzingania Dieterle = *Echinopepon*

Apedium Chiron & al. = *Selenipedium* (peloric form!)

Apeiba Aubl. Malvaceae (Grew.-Apeib.; Tiliaceae). 7 trop. S Am. Timber trees (Troll's Model). ***A. tibourbou*** Aubl. (Braz.) – oil from seeds used in treatment of rheumatism

Apera Adans. Gramineae (XVI 15). 5 Eur. (3) to Afghanistan. Silky bents, cult. orn.

Apetahia Baill. (~ *Sclerotheca*). Campanulaceae (III 4). 7 Society Is., Marquesas, Rapa. Arborescent

Aphaenandra Miq. = *Mussaenda*

Aphaerema Miers = *Abatia*

Aphanactis Wedd. Compositae (Mill.-Gal.). 13 Mex., C Am., N Andes. R: BSAB 19(1980)35

Aphanamixis Blume. Meliaceae (II-Trich.). 3 Indomal. to Solomons. R: Blumea 31(1985)136. ***A. polystachya*** (Wall.) R. Parker (throughout range) – timber (tasua), semi-drying oil from seeds, cult. orn., s.t. with hollow ant-infested shoots

Aphanandrium Lindau = *Neriacanthus*

Aphananthe Planch. Cannabaceae (Ulmaceae II). 3 Madag. (1), Indomal. to Jap. & E Aus. (1: ***A. aspera*** (Thunb.) Planch. – fast-growing timber tree, lvs used like sandpaper), Mex. (1).

Aphandra Barfod. Palmae (IV 3). 1 Ecuador: ***A. natalia*** (Balslev & A. J. Hend.) Barfod. R: OB 105(1991)44

Aphanelytrum Hackel = *Poa*

Aphanes L. = *Alchemilla*

Aphania Blume = *Lepisanthes*

Aphanisma Nutt. ex Moq. Amaranthaceae (Chenopodiaceae I 1). 1 Calif., NW Mex.: ***A. blitoides*** Nutt. ex Moq.

Aphanocalyx Oliv. Leguminosae (I 2). (Incl. *Monopetalanthus*) 14 W & C Afr. (Guineo-Congolian forest). R: WAUP 99–3(1999)115. ***A. heitzii*** (Pellegrin) Wieringa (*M. h.*, andoung, Gabon) – timber

Aphanocarpus Steyerm. Rubiaceae (IV 7). 1 Venez.: ***A. steyermarkii*** (Standl.) Steyerm.

Aphanococcus Radlk. = *Lepisanthes*

Aphanopetalaceae Doweld (~ Cunoniaceae). Magnoliidae – Saxifragales. 1/2 S Aus. Scandent shrubs with lenticellate shoots. Lvs simple, opp., usu. serrate; stip. 0 but minute colleters at nodes. Fls reg., herm., 4-merous, solit., or in lax cymes, axillary. K spreading, accrescent, at base coalescing with base of C & A forming a tube; C minute or 0; A 8, anthers with longit. slits; G 4 semi-inf., 4-loc. with 1 apical anatropous, bitegmic ovule per loc. Fr. nut-like with 1 seed; embryo straight, endosperm fleshy.
Genus: *Aphanopetalum*

Aphanopetalum Endl. Aphanopetalaceae. 2 S Aus. Form. in Cunoniaceae

Aphanopleura Boiss. Umbelliferae (III 8). 6 C As. to Afghanistan

Aphanosperma Daniel. Acanthaceae (III 2c). 1 NW Mex.: ***A. sinaloense*** (Leonard & Gentry) Daniel. R: AJB 75(1988)545

Aphanostelma Schltr. = *Cynanchum*

Aphanostemma St-Hil. = *Ranunculus*

Aphanostephus DC. = *Erigeron*

Aphanostylis Pierre = *Landolphia*

Aphelandra R. Br. Acanthaceae (III 1). 175 trop. Am. R: SCB 18(1975). Hummingbird-poll. Cult. housepls for showy bracts & foliage, esp. ***A. squarrosa*** Nees '**Louisae**' with white veins, ***A. scabra*** (Vahl) Sm. (*A. deppeana*) – extrafl. nectaries in bracts attract ants prot. fr. (9 times unprotected ones matured) & ***A. liboniana*** Linden (Braz.) etc.

Aphelandrella Mildbr. = praec.

Aphelexis D. Don = *Edmondia*

Aphelia R. Br. Restionaceae (Centrolepidaceae). 6 Aus. R: JABG 16(1995)95

Aphloia (DC.) Benn. Aphloiaceae (Flacourtiaceae). 1 (polymorphic; -8?) trop. E Afr., Madag., Masc., Seychelles: ***A. theiformis*** (Vahl) Benn. – Troll's Model, complex juvenile lvs as in unrelated taxa in Masc., lvs used as tea in Masc.

Aphloiaceae Takht. (~ Flacourtiaceae). Magnoliidae – Crossosomatales. 1/1 E Afr. to Seychelles. R: KB 57(2002)174. Everg. glabrous tree or shrub without cyanogenic glycosides. Lvs distich., serr(ul)ate; stip. minute. Fls solit. or in fascicles or racemes, bisex. K 4 or 5 (6), valvate; C 0; A ∞ inserted nr rim of receptacle, anthers small. Disk 0. $\underline{G}$ 1-loc. with parietal placentation with 6–8 campylot. ovules in 2 rows. Fr. a berry with persistent K & A; seeds c. 6 with incurved embryo & scant endosperm
Genus: *Aphloia*

Aphoma Raf. = *Iphigenia*

Aphragmia Nees = *Ruellia*

Aphragmus Andrz. ex DC. Cruciferae (6). Incl. *Lignariella, Staintoniella,* 12 Himal. (10) to NE Siberia. R: HPB 5(2000)110

Aphyllanthaceae Burnett = Asparagaceae

Aphyllanthes Tourn. ex L. Asparagaceae (Aphyllanthaceae). 1 Portugal to Italy, N Afr.: ***A. monspeliensis*** L. – rhiz. with sec. thickening, cult. orn.

Aphyllarum S. Moore = *Caladium*

Aphyllocladus Wedd. (~ *Hyalis*). Compositae (Mut.-Onis.). 4 Andes of S Bolivia, N Chile, NW Arg. R: Darw. 9(1951)367. Shrubs

Aphyllodium (DC.) Gagnepain (*Dicerma*; ~ *Hedysarum*). Leguminosae (III 19). 7 trop. As. to Aus. R: Taiwania 42(1997)143

Aphyllon Michaux = *Orobanche*

Aphyllorchis Blume. Orchidaceae (V 1). 22 Sri Lanka to Jap. & Aus. Mycotrophs, usu. self-poll.

Apiaceae Lindl. See Umbelliferae

Apiastrum Nutt. Umbelliferae (III 5). 1 SW N Am.: ***A. angustifolium*** Nutt. R: Phytoneuron 2012–87:9, 89:1

Apinagia Tul. Podostemaceae (III). 50 trop. S Am. Incl. *Bladowia*, first described as a hepatic!

apio *Arracacia xanthorrhiza*

Apiopetalum Baill. Umbelliferae (Mack.; Araliaceae). 2 New Caled.

Apios Fabr. Leguminosae (III 18). 6 E As., N Am. (2 – R: Castanea 70(2005)89). ***A. americana*** Medik. (*A. tuberosa*, potato-bean, groundnut, N Am.) – bee-poll. (not fly- as oft. claimed) oft. triploid & seeds sterile, ed. sweet tubers boiled or roasted imp. Native Am. food, occ. cult.

apitong *Dipterocarpus* spp. timber

Apium Tourn. ex L. Umbelliferae (III 8). Excl. *Helosciadium*, 19 temp. (Eur. 1) & warm esp. S Am. Some contraceptives (S Am.). ***A. graveolens*** L. (widespread) – flavour due to 3-butylphthalide, 3-butyltetrahydrophthalide, apiole & myristicin, lvs in an Egyptian garland of 1200 BC, local medic. (seeds diuretic poss. contrib. to wt loss!) Ireland, cult.: 2 forms – 'var. *dulce* (Mill.) Poir.' (celery) with blanched petioles eaten raw or cooked (chewing & digestion alleged to consume more calories than in the celery eaten) & fr. ('seed') used in flavouring, celery salt & treatment of arthritis & rheumatism (trad. Chinese med. to reduce blood pressure; phthalides reduce stress hormone levels); 'var. *rapaceum* (Mill.) DC.' (celeriac) with ed. turnip-like r; wild forms with apiumoside (glycoside). ***A. prostratum*** Labill. ex Vent. (sea celery, s. parsley, Aus., NZ) – bushfood Aus. See also *Cyclospermum*

Aplanodes Marais. Cruciferae (16). 2 S Afr. R: Both. 9(1966)111

Aplectrum (Nutt.) Torr. Orchidaceae (V 13e). 1 N Am.: ***A. hyemale*** Torr. – terr., mycotrophic when young, leaf 1 ephemeral, tuber med. & used in cementing earthenware ('putty-root'), cult. orn. See also *Cremastra*

Apleura Philippi = *Azorella*

Apluda L. Gramineae (XXII). 1 Mauritius & Socotra to Taiwan & New Caled.: ***A. mutica*** L. R: FOC 22(2006)614

Apoballis Schott (~ *Schismatoglottis*). Araceae (VII 8). 20 trop. As.

Apocaulon R. Cowan. Rutaceae (I 8). 1 Venez.: ***A. carnosum*** R. Cowan. R: MNYBG 8(1953)119

Apochaete (C.E. Hubb.) J. Phipps = *Tristachya*

Apochiton C.E. Hubb. Gramineae (XXIX 5). 1 Tanzania: ***A. burttii*** C.E. Hubb.

Apochloa Zuloaga & Morrone (~ *Panicum*). Gramineae (XXIII 1). 15 NE S Am. R: SB 33(2008)288

Apochoris Duby = *Lysimachia*

Apoclada McClure. Gramineae (V 3). 1 Braz.: ***A. simplex*** McClure & L.B. Sm.

Apocopis Nees. Gramineae (XXII 6). 16 Indomal., China. R: KB 1952:101.Racemes usu. paired

Apocynaceae Juss. Magnoliidae – Gentianales. Incl. Asclepiadaceae, 345/4675 mostly trop. & few temp. R: BR 66(2000)31. Usu. lianes, less oft. trees (many with Leeuwenberg's Model), shrubs or herbs or succ. with ubiquitous laticifer systems with white latex (rarely yellow or red; clear in *Pachypodium*), glycosides & alks & s.t. unusual sec. thickening, internal phloem usu. present. Lvs simple, entire, opp. s.t. whorled or condensed spirals (*Pachypodium*), rarely truly spiral, s.t. much reduced, oft. with close parallel lat. veins; stip. small or 0. Fls in cymes or racemes or solit., usu. bisex., oft. showy, ± regular, usu. (4)5-merous exc. G: K (5) with imbr. or valvate lobes, C (5) usu. funnel- or salver-shaped, convolute or rarely valvate or imbr., A epipet. or (5) in a short sheath around style, alt. with C, anthers coherent or connate around style-head (in IV & V forming a columnar gynostegium, where pollen in pollinia, cf. Orchidaceae, extracted by translators – solidified secretions of anthers, style-head or both, consisting of arms (retinacula) joined at middle by a 2-parted gland (corpusculum), 1 arm attached to pollinium(a) of 1 theca of 1 anther, the other to that (those) of adjoining theca or anther), 5 nectary glands nr. ovary-base, s.t. confluent into disk, reduced or 0, $\underline{G}$ or semi-inf. (2(–8)) or 2 united by style, or more, 1- or 2-loc., with 2 – ∞ ovules in each locule, oft. pend., anatropous, unitegmic, style usu. simple with thickened head & ring of hairs below it. Fr. diverse (follicle usu. paired, berry or drupe etc.), seeds usu. flattened, oft. with crown of hairs, embryo straight, ± oily endosperm. x = 8–12+ incl. triploids

Classification & chief genera:

I. **Rauvolfioideae** (Plumerioideae; C aestivation usu. to left, A usu. free from style, anthers usu. full of pollen, seeds usu. hairless) – a. Alstonieae (*Alstonia, Aspidosperma*); b. Vinceae (*Catharanthus, Rauvolfia, Vinca*); c. Willughbeeae (*Hancornea, Landolphia, Saba*); d. Tabernaemontaneae (*Tabernaemontana, Voacanga*); e. Melodineae (*Dyera*); f. Hunterieae (*Hunteria, Picralima*); g. Plumerieae (*Allamanda, Cerbera, Plumeria, Thevetia*); h. Carisseae (*Acokanthera, Carissa* (only)); i. Alyxieae (*Alyxia*)

II. **Apocynoideae** (C aestivation usu. to right, A usu. adnate to style-head, anthers usu. empty at base, seeds usu. hairy) – a. Wrightieae (*Adenium, Nerium, Strophanthus, Wrightia*); b. Malouetieae (*Funtumia, Pachypodium*); c. Apocyneae (*Apocynum, Beaumontia*); d. Mesechiteae (*Mandevilla*); Echiteae (*Parsonsia, Prestonia*)

III. [Asclep. I.] **Periplocoideae** (pollen granular, in tetrads, transferred by spoon-shaped translator ending in a sticky disk; OW)-*Cryptostegia, Periploca, Raphionacme*

IV. [Asclep. II]. **Secamonoideae** (pollen massed in pollinia; anthers 4-loc.): genera form. incl. in V – *Secamone, Toxocarpus*

V. [Asclep. III]. **Asclepiadoideae** (pollen massed in pollinia; anthers 2-loc.; nectar on stigmas, 3 of 5 stigmatic chambers transmitting pollen tubes to 1 'ovary', the other 2 to the second): a. Marsdenieae (*Dischidia, Fockea, Hoya, Marsdenia, Telosma*); b. Ceropegieae (*Brachystelma, Caralluma, Ceropegia, Hoodia, Huernia, Orbea, Stapelia*); c. Asclepiadeae (R: T 46(1997)236); *Asclepias, Astephanus, Calotropis, Cynanchum, Gonolobus, Matelea, Oxypetalum, Pachycarpus, Vincetoxicum*)

Generic limits being resolved with many mono- or oligo-specific genera recently 'sunk'. III–V (Asclepiadaceae – journal: *Asklepios*) with elaborate insect-poll., the corpuscula attaching to the legs (guided by grooves in the column) of nectar-seeking insects; fls with disagreeable scent in many & poll. by flesh-flies, some of which lay eggs in the fls of Ceropegieae (incl. Stapelieae – up to 40 cm across); in *Ceropegia* C-tube forming a trap in which flies are temporarily imprisoned, with hairs shrivelling to allow exit afterwards (cf. *Arum*) & oft. app. v. specific scents for particular poll. *Dischidia* & some *Hoya* spp. are ant-pls

Many sources of drugs (e.g. *Hoodia*) & poisons, e.g. strophanthin (*Strophanthus*) in treatment of heart disease & as a cortisone precursor; see also *Acokanthera, Adenium, Catharanthus* (vincristine), *Cerbera, Picralima, Rauvolfia* (reserpine), *Voacanga*; many ornamentals esp. 'Stapelieae' (R: T 40(1991)381) with 6 endemic genera & c. 160 endemic spp. in Karroo-Namib (succ.), also *Allamanda, Asclepias, Catharanthus, Ceropegia, Fockea, Hoya, Mandevilla, Marsdenia* (incl. stephanotis), *Nerium, Pachypodium, Plumeria, Saba, Thevetia & Vinca*; timber from *Alstonia, Aspidosperma, Dyera, Gonioma, Ochrosia*, fibre from *Anodendron, Asclepias, Chonemorpha*; seed-hairs used as floss (e.g. *Calotropis*); rubber from *Clitandra, Cryptostegia, Funtumia, Hancornia, Landolphia, Mascarenhasia, Raphionacme, Urceola, Willughbeia*; dyes from

Tabernaemontana, Wrightia; poss. oil (*Asclepias*); some anti-sweeteners (*Marsdenia*); some bad weeds (*Cryptostegia*)

Apocynum Tourn. ex L. Apocynaceae (II c). Incl. *Trachomitum*, 9 S Russia (Eur. 3) to China, temp. Am. Seeds hairy. Cult. orn. (dogbane), form. medic. ***A. cannabinum*** L. (Ind. hemp, Choctaw r.) – bark a source of fibre for ropes, sails etc., r. emetic & cardiac stimulant; ***A. venetum*** L. (*T. v.*, E Eur., W As.) – kendyr fibre locally used for sails & nets, seeds with useful floss

Apodandra Pax & K. Hoffm. = *Plukenetia*

Apodanthaceae Takht. (~ Rafflesiaceae). Magnoliidae – Cucurbitales. 2/10 trop. Am. & warm. R: Phytokeys 36(2014)47. Chlorophyll-less r. or stem parasites, monoec. or dioec. No stems, rs or lvs, just endophytic cell-system in host parenchyma, Fls small, on host stems. P (outer 1 or 2 whorls = bracts?) 2 + 4 + 4 or 3 + 6 + 6 (rarely 4 whorls). A (c. 15), pollen-sacs in 1, 2 or 4 rings; pollen tricolpate or inapertuate. G (4[5]), 1-loc., ± inf. (pistillode in males), opp. inner P, style v. short, placentation parietal with 50–300 anatropous ovules with 2 integuments. Fr. a fleshy berry. Seeds with endosperm. n = 16
Genera: *Apodanthes, Pilostyles*. Poll. flies and bees, poss. also wasps; disp. prob. birds

Apodanthera Arn. Cucurbitaceae (XIII). 16 warm Am. ***A. undulata*** A. Gray (*A. aspera* melon-loco, SW N Am.) – minor oilseed, since Pre-Columbian

Apodanthes Poit. Apodanthaceae (Rafflesiaceae II 2). 1 trop. S Am.: ***A. caseariae*** Poit. – parasitic on *Casearia* spp. R: PhytoKeys 36(2014)49

Apodasmia Briggs & L. Johnson (~ *Leptocarpus*). Restionaceae. 4–5 SW & SE (1) Aus., NZ (1: ***A. similis*** (Edgar) Briggs & L. Johnson (oioi) – form. thatch on raupo houses), Chile (1). R: Telopea 7(1998)371. Salt-tolerant

Apodicarpum Makino. Umbelliferae (III 8). 1 E Jap.: ***A. ikenoi*** Makino

Apodiscus Hutch. Phyllanthaceae (Ant.-Mart.; Euphorbiaceae s.l.). 1 W Afr.: ***A. chevalieri*** Hutch. – monoec. R: A. Radcliffe-Smith, *Gen. Euphorb.* (2001) 63

Apodocephala Bak. Compositae (Ast.-Ast.). 9 Madag. Trees & shrubs, ***A. pauciflora*** Bak. attaining 30 m

Apodolirion Bak. Amaryllidaceae. 6 Cape to Transvaal. R: Willd. 15(1986)466. Cult. orn.

Apodostigma R. Wilczek. Celastraceae (III). 1 trop. Afr., Madag.: ***A. pallens*** (Oliv.) R. Wilczek

Apodytes E. Meyer ex Arn. Metteniusaceae (Icacinaceae s.l.). 5 OW trop. esp. Afr. e.g. ***A. dimidiata*** E. Meyer ex Arn. (white pear) – good timber) to Queensland

Apoia Merr. = *Sarcosperma*

Apollonias Nees = *Persea*

Apomuria Bremek. = *Psychotria*

Aponogeton L.f. Aponogetonaceae. c. 50 OW trop. (Aus. 12; R: Telopea 8(1998)7, 11(2006)130), S Afr. R: BB 33(137)(1985). Orig. Aus.? Aquatics grown in aquaria, some ed. tubers. ***A. distachyos*** L.f. (Cape pondweed, water hawthorn, Cape asparagus, Cape, invasive in Aus.) – cult. orn., fr. spikes ed. like spinach or pickled; ***A. madagascariensis*** (Mirbel) Bruggen (*A. fenestralis*, lace-leaf, Madag.) – (difficult) aquarium pl. with lace-like lvs due to patches of tissue between veins dying during leaf development

Aponogetonaceae Planch. Magnoliidae – Alismatales. 1/c. 50 OW trop., S Afr. Perenn., glabrous hydrophytes with short starchy rhiz. or corm & secretory canals, vessels in r. or 0. Lvs with midrib & parallel veins & pseudopetiole, floating or all submerged. Infl. emergent with spathe early lost, simple or 2–10-fld or contracted into a head. Fls small, pls usu. monoec. or dioec. P (1)2(–6) distinct, oft. persistent, when 1 with broad base & resembling a bract, or 0; A 3 + 3 (or ∞ in 3 or 4 whorls), pollen grains monosulcate, yellow; $\underline{G}$ 2–9 with short styles, each loc. with (1)2–8(–14) basal, anatropous, usu. bitegmic ovules. Fr. of distinct follicles, seeds without endosperm, embryo straight with single terminal cotyledon. x = 8
Only genus: *Aponogeton*

Apophyllum F. Muell. Capparaceae. 1 NE Aus.: ***A. anomalum*** F. Muell.

Apoplanesia C. Presl. Leguminosae (III 10). 1 dry C Am.: ***A. paniculata*** C. Presl – hard wood (cabinetwork), yellow dye from bark, cult. orn.

apopo *Lovoa trichilioides*

Apopyros Nesom = *Erigeron* (but see Phytol. 76(1994)177)

Aporocactus Lem. = *Disocactus*

Aporopsis (Schltr.) M.A. Clem. & D.L. Jones = *Dendrobium*

Aporosa Blume. Phyllanthaceae (Scep.; Euphorbiaceae). c. 75 Indomal. (Borneo 30) to Solomon Is. R: Blumea suppl. 17(2004)151. Massart's Model. Some timber, dyes, ed. fr., local medic. High Al^{3+} conc. in some spp., so useful as mordants, e.g. ***A frutescens*** Blume (Mal.) – bark (sasah) form. used in batik-making (Java)

Aporosella Chodat & Hassler = *Phyllanthus*

Aporostylis Rupp & Hatch. Orchidaceae (IV 13b). 1 NZ : ***A. bifolia*** (Hook. f.) Rupp & Hatch. R: Gen. Orch. 1(2001)89

Aporrhiza Radlk. Sapindaceae. 4–6 trop. Afr.

Aporum Blume = *Dendrobium*

Aporusa Blume = *Aporosa*

Aposeris Necker ex Cass. (~ *Hyoseris*). Compositae (Cich.-Cich.-Hyos.). 1 C Eur.: ***A. foetida*** (L.) Less.

Apostasia Blume. Orchidaceae (I). 6 trop. As. to Aus. Local medic., ***A. nuda*** R. Br. (Malay Pen.) for diarrhoea

Apostasiaceae Lindl. = Orchidaceae

Apostates Lander. Compositae (Bah.). 1 N Rapa: ***A. rapae*** (F. Brown) Lander – only 2 pls left, last seen in fl. 1921. R: AusSB 2(1989)2, T 65(2016)1076. See also closest ally, *Picradeniopsis*

Apowollastonia Orchard (~ *Wedelia*). Compositae (Helia.). 8 Mal. (1), Aus. R: Nuytsia 23(2013)406

Appendicula Blume. Orchidaceae (V 15). c. 150 trop. As. to Tonga. ***A. rupestris*** Ridl. (Malay Pen.) – rheophyte

Appendicularia DC. Melastomataceae. 1 Guianas: ***A. thymifolia*** DC.

Appendiculopsis (Schltr.) Szlach. = *Agrostophyllum*

Appertiella C. Cook & Triest. Hydrocharitaceae. 1 Madag.: ***A. hexandra*** C. Cook & Triest

apple usu. *Malus* spp. esp. *M. domestica* long cult. in Eur. Many other fr. trees etc. also called apples (*Angophora* spp. (Aus.); that of the Garden of Eden perhaps *Strychnos*!): **akee a.** *Blighia sapida*; **alligator a.** *Annona* spp.; **Argyle a.** *Eucalyptus cinerea*; **balsam a.** *Momordica balsamina*; **a. banana** *Musa* × *paradisiaca*; **a.berry, purple a.** *Billardiera macrantha*; **black** or **brush a.** *Planchonella australis*; **a. box** *E. bridgesiana*; **Chinese a.** *Malus prunifolia*; **cocky a.** *Planchonia careya*; **crab a.** *M. sylvestris*, but also applied to any natur. seedling a. & some orn. flowering spp.; **a. cucumber** *Cucumis sativus* 'Crystal Apple'; **custard a.** *Annona* spp.; **earth a.** *Smallanthus sonchifolius*; **elephant a.** *Limonia acidissima*; **emu a.** *Owenia acidula*; **golden a.** *Aegle marmelos*, *Spondias cytherea*, **(of the Hesperides)** *Cydonia oblonga*; **gopher a.** *Licania michauxii*; **a.jack** *Malus pumila*; **kangaroo a.** *Solanum aviculare*, *S. laciniatum*; **kei a.** *Dovyalis caffra*; **lady a.** *Syzygium suborbiculare*; **love a.** = tomato; **Malay a.** *Syzygium malaccense*; **mammee a.** *Mammea americana*; **a. mango** *Mangifera* × *odorata*; **may a.** *Podophyllum peltatum*; **monkey a.** *Annona* spp., *Anisophyllea laurina*, *Strychnos* spp.; **oak a.** insect galls on *Quercus* spp.; **Otaheite a.** *Spondias cytherea*; **pond a.** *Annona glabra*; **rose a.** *Syzygium jambos*, *S. malaccense*; **rose a., water** *S. aqueum*; **sand a.** *Parinari capensis*; **a. of Sodom** *Solanum linnaeanum* etc., *Calotropis procera*; **star a.** *Chrysophyllum cainito*; **sugar a.** *Annona* spp.; **thorn a.** *Datura stramonium*; **velvet a.** *Diospyros* spp.; **wood a.** *Limonia acidissima*

Appunettia R. Good = *Morinda*

Appunia Hook.f. (~ *Morinda*). Rubiaceae (IV 10). 15 trop. Am. R: Adans. 33(2011)303

Aprevalia Baill. = *Delonix*

apricot *Prunus armeniaca*; **Briançon a.** *P. brigantina*; **Japanese a.** *P. mume*; **a. plum** *P. simonii*; **San Domingo a.** *Mammea americana*; **a.vine** *Passiflora incarnata*

aprono *Mansonia altissima*

Aptandra Miers. Olacaceae (Aptandraceae). 4 trop. S Am. (3), Afr. (1). R: FN 38(1984)111. Troll's Model. ***A. tubicina*** (Poepp.) Miers (*A. spruceana*, Amaz. Braz., Peru & Bolivia) – r.-bark source of an oil

Aptandraceae Miers. See Olacaceae

Aptandropsis Ducke = *Heisteria*

Aptenia N.E. Br. = *Mesembryanthemum*

Apterantha C.H. Wright = *Lagrezia*

Apteria Nutt. Burmanniaceae. 1 S US & WI to S Am.: ***A. aphylla*** (Nutt.) Small – first named a *Lobelia*! R: FNA 26(2002)488

Apterigia (Ledeb.) Galushko = *Noccaea*

Apterokarpos Rizz. (~ *Loxopterygium*). Anacardiaceae (I). 1 NE Braz.: ***A. gardneri*** (Engl.) Rizz.

Apteropteris (Copel.) Copel. = *Sphaerocionium*

Apterosperma H.T. Chang. Theaceae. 1 China: ***A. oblatum*** H.T. Chang

Apterygia Baehni = *Sideroxylon*

Aptosimum Burchell. Scrophulariaceae (Apt.). 20 Afr.

Apuleia Mart. Leguminosae (I 3). 1NE Peru, SE Braz., Arg., variable: ***A. leiocarpa*** (J. Vogel) J.F. Macbr. (*A. praecox*) – andromonoec. (males A 3, 4 x hermaph. A2, G1), extrafl. nectaries; tough timber. R: KB 65(2010)228

Apurimacia Harms. Leguminosae (III 16). 2 drier S Am. Fish-poisons & insecticides

aquiboquil, aquibuquil *Lardizabala biternata*

Aquifoliaceae Bercht. & J. Presl. Magnoliidae – Aquifoliales. Excl. Sphenostemonaceae (= Paracryphiaceae) & Phellinaceae, 1/600+ almost cosmop. Trees, usu. small, or shrubs, rarely cli., usu. everg. Lvs almost always in spirals, oft. with resiniferous & laticiferous cells in mesophyll, stip. small or 0. Fls small, regular, hypogynous, usu. unisexual (pls dioecious), in axillary, cymose infls (usu. thyrses), s.t. reduced to fascicles or even 1 fl., 4–6 (–23)-merous: K imbr., ± connate to 0; C ± connate at base, imbr., rarely free; A usu. same no. as & alt. with C, usu. basally connate; $\underline{G}$ ((2–)4–6(–23)), style short or 0, ovule 1(2) per locule, pend., anatr. or ± campylotropous, unitegmic, funicle oft. with ventral protuberance poss. rep. suppressed 2nd ovule. Fr. a usu. bird-disp. drupe with as many pyrenes as carpels; seeds with small embryo nr. micropyle, endosperm copious, oily & proteinaceous, no starch, germ. after 1–3 yrs. x = 17, 18, 19, 20 (usu. 2n = 40, though polyploids known)

Genus: *Ilex* (incl. *Nemopanthus*). Poss. incl. Helwingiaceae, Phyllonomaceae

Aquilaria Lam. Thymelaeaceae (Thym.-Aq.; Aquilariaceae). 17 Indomal. Decaying heartwood (agarwood, gaharu [wood]) saturated with an oleoresin, the basis of incense, when distilled (oud) oil used in scent & medicine (comm. for intestinal parasites), endangered through overexploitation. ***A. malaccensis*** Lam. (*A. agallocha*, agarwood, aloewood, calambac, eaglewood, lign-aloes, Indomal.) – fibre used for rope & textiles, for writing-paper pre C14; the aloe (ahaloth) of Bible, 'aloexylum'. See also *Aetoxylon, Gyrinops, Gonostylus*

Aquilariaceae R. Br. ex DC. = Thymelaeaceae

Aquilegia Tourn. ex L. Ranunculaceae (III 1). 80 N temp. (Eur. 19). R: GH 7(1946)1. P with long spurs (diff. lengths due to diff. cell elongation assoc. with diff. pollinators) secreting nectar, coll. honeybees; A oft. >50 in whorls of 5. Alks. Cult. orn. assoc. with (Virgin Mary's) sorrow (columbines, granny-bonnets) mostly hybrids, those with hooked spurs usu. derived from ***A. vulgaris*** L. (Eur., violet-blue fl. form native GB), those with long straight spurs from ***A. chrysantha*** A. Gray & ***A. coerulea*** James (N Am.); form. medic. incl. ***A. canadensis*** L. (E N Am.) – scarlet fls visited by hummingbirds, local medic. incl. wash for poison ivy lesions, seeds used to scent tobacco & clothing (esp. bachelors') by Native Americans, & ***A. formosa*** Fischer ex DC. (W N Am.) – seeds used to discourage head lice; ***A. fragrans*** Benth. (W Himal.) – creamy fls scented

Arabian coffee *Coffea arabica;* **A. violet** *Exacum affine*

Arabidella (F. Muell.) O. Schulz. Cruciferae (37). 7 Aus. R: TRSSA 89(1965)177, Nuytsia 17(2007)457

Arabidopsis (DC.) Heynh. Cruciferae (15). 11 N temp. (Eur. 9) to trop. Afr. mts, S Afr. R: Novon 7(1997)323. ***A. thaliana*** (L.) Heynh. (thale cress, range of *A.*) – subject of genetic research (2n = 10, though at least one polyploidy event in ancestry; life-cycle complete in 1 month in lab – favoured model system in experimental botany), used in space flights to test theory of gravity perception

Arabis L. Cruciferae (7). Excl. *Boechera, Turritis*, etc. 60 N temp. (Eur. 35, N Am. 15), Med. to trop. Afr. mts. Heterogeneous as type sp. close to *Draba* & *Aubrieta* rather than other *A.* spp. Cult. orn. herbs oft. mat-forming grown in rock gardens etc. (rock or wall cresses) esp. ***A. caucasica*** Willd. ex Schldl. (*A. alpina* subsp. *caucasica, A. albida*, SE Eur. to Iran) – white fls, usu. wingless seeds (dry rocks), oft. confused with ***A. alpina*** L. (Eur. & Afr. mts, Himal., Greenland & NE N Am., discovered on Skye 1887), rarely cult. (seeds winged, moist rocks), ***A. pumila*** Jacq. (Eur. mts) – prop. by leaves

Aracamunia Carnevali & I. Ramírez. Orchidaceae (IV 2h). 1 S Venez.: ***A. liesneri*** Carnevali & I. Ramírez – glandular structures on leaf-sheaths poss. carnivorous. R: Gen. Orch. 3(2003)168

Araceae Juss. Magnoliidae – Alismatales. Incl. Lemnaceae, 119/6450 mostly trop. (Borneo 36/1000 – 95% endemic) & subtrop. with few temp. R: R. Govaerts & D.G. Frodin (2002) *World checklist & bibliography of A. (& Acoraceae)*, D. Bown, *Aroids* ed. 2(2000); BJ

113(1991)396; Willd. 21(1991)35. Scrambling shrubs or climbers with aerial r., herbs (oft. enormous) with corms or tubers, rarely true epiphytes or free-floating aquatic (*Pistia*, Lemnoideae where small to minute free-floating thalloid pls with 0-several unbranched r. & 0 xylem (tracheids in *Spirodela* r.)), usu. with bundles of raphides (H-shaped in T.S., oft. with lat. barbs) throughout shoot, oft. with laticifers, oft. cyanogenic & s.t. with alks or other toxins; vessels mostly only in r. Stems sympodial, rarely monopodial; r. mycorrhizal, without roothairs. Lvs in spirals or distich., variously parallel- or net-veined, usu. developing acropetally (like typical dicots). Infl. unbranched spadix (oft. smelly), usu. terminal & subtended by a ± prominent usu. coloured spathe, in Lemnoideae with 2 reproductive pouches in thallus, new pls, rarely infls, forming within. Fls many, small, without bracts, poll. by insects, esp. flies (rarely wind) when volatiles incl. ammonia, indole (both stimulating oviposition in carrion flies), skatole, trimethylamine etc., bisex. or unisexual (monoec. with males in upper spadix, rarely pseudo-dioec.). P of 4 or 6(8) distinct or connate tepals in 2 whorls, but reduced or 0 in unisexual fls or always 0 as in Lemnoideae; A (1 –)4, 6 or 8(–32), s.t. ± connate, anthers opening by term. pores or slits or longit. slits; G ((2)3(– 47 in *Philodendron*) usu. with axile placentation, style short or 0, ovules 1–∞ per loc., oft. anatropous, bitegmic. Fr. usu. berry, rarely dry or leathery & splitting irreg. or whole spadix forming syncarp; seeds 1 – ∞, s.t. fleshy, embryo large s.t. with small 2nd cotyledon, linr., embedded in copious oily (& s.t. also starchy) endosperm or endosperm 0. x = 5–17 +. Journal: *Aroideana*

Classification & chief genera:

I. **Gymnostachydoideae** (lvs grass-like with striate venation; spathe inconspic., fls bisex., micropyle 0; E Aus.: *Gymnostachys* (only)

II. **Orontioideae** (swamp pls; rhiz. erect, leaf-blade entire, fls bisex., endosperm little or 0; N temp.): *Lysichiton*, *Symplocarpus*

III. **Pothoideae** (climbers; lvs entire with lat. veins all reticulate or at least those of 2nd & 3rd so, spathe persistent): 1. Potheae (*Pothos*); 2. Anthurieae (*Anthurium* (only))

IV. **Monsteroideae** (climbers or short-stemmed; lvs oft. perforate or pinnatifid with varied venation, spathe usu. decid.; 12 genera in 4 tribes): 1. Spathiphylleae (*Spathiphyllum*); 2. Anadendreae (*Anadendrum* (only)); 3. Heteropsideae (*Heteropsis* (only)); 4. Monstereae (*Epipremnum*, *Monstera*, *Rhaphidophora* (all with 'holes' in lvs, cf. *Monstera*), *Scindapsus*)

V. **Lasioideae** (land (oft. cli.) or marsh pls; lvs with reticulate venation; fls bisex./unisexual, spathe rarely constricted, never decid.: *Cyrtosperma*

VI. **Calloideae** (land or marsh pl.; venation striate, fls usu. bisex.; N temp: *Calla* (only)

VII. **Aroideae** (land or marsh (some true aquatic) pls; tubers or rhizomes, lvs net-veined, spathe usu. constricted, fls usu. unisexual, A or (A); several temp., some with epiphyll. infls; 26 tribes (prob. too finely drawn): 1. Zamioculcadeae (*Zamioculcas*); 2. Stylochaetoneae (*Stylochaeton* (only)); 3. Dieffenbachieae (*Dieffenbachia*); 4. Spathicarpeae (*Spathicarpa*); 5. Philodendreae (*Philodendron*); 6. Homalomeneae (*Homalomena*); 7. Anubiadeae (*Anubias* (only)); 8. Schismatoglottideae (heterogeneous with Philonotieae (NW) often recog.; *Schismatoglottis*); 9. Crytocoryneae (*Crytocoryne*); 10. Zomicarpeae (*Zomicarpa*); 11. Caladieae (*Caladium*, *Syngonium*, *Xanthosoma*); 12. Nephthytideae (*Nephthytis*); 13. Aglaonemateae (*Aglaonema*); 14. Culcasieae (*Culcasia*); 15. Montrichardieae (*Montrichardia* (only)); 16. Zantedeschieae (*Zantedeschia* (only)); 17. Callopsideae (*Callopsis* (only)); 18. Thomsonieae (*Amorphophallus*); 19. Arophyteae (*Arophyton*); 20. Peltandreae (*Typhonodorum*); 21 Arisareae (*Arisarum* (only)); 22. Ambrosineae (*Ambosina* (only)); 23. Areae (*Arum*, *Dracunculus*, *Typhonium*); 24. Arisaemateae (*Arisaema*); 25. Colcasieae (*Alocasia*, *Colocasia*); 26. Pistieae (*Pistia* (only))

VIII. **Lemnoideae** (Lemnaceae; thalloid free-floating pls; app. not close to free-floating *Pistia* (VII 26) but a parallel neotenous line (between II & III on pollen evidence); R: IBM 34(1965)1, Blumea 18(1970)355): *Lemna*, *Spirodela*

Many genera mono- or oligospecific with nrly half spp. in *Anthurium* & *Philodendron*. Relic *Protarum* in Seychelles but exc. VIII fam. absent from New Caled., NZ & Hawaii (& those poss. introd.). Incl. oldest monocot fossils known early Cretaceous from NE Braz. – *Spixiarum*). Distinction between fl. & infl. unclear. Poll. oft. involves traps & lures (incl. thermogenesis due to uncoupling of proteins & mediation of an alternative oxidase for beetles & flies, interior of spathe of some trop. spp. concentrating light by reflection & refraction, besides 'window-panes' (cf. *Aristolochia*), while foul smells imitate allelochemicals that elicit feeding or oviposition behaviour in carrion & dung flies). Of enormous economic importance as foodpls in trop.: *Alocasia, Amorphophallus, Colocasia, Cyrtosperma, Xanthosoma* – cocoyams etc., *Monstera* with ed. fr., *Calla & Arum* minor starch sources. Lemnoideae imp. food for wildfowl & fish, some grown on dairy waste-water & fed. to cattle (*Spirodela*); *Wolffia* used to test herbicide levels in water. Some invasive esp. aquatic *Lemna, Pistia.* Many familiar housepls incl. spp. of *Aglaonema, Anthurium, Caladium, Cryptocoryne* (aquaria), *Dieffenbachia* (poisonous), *Monstera, Philodendron* (poisonous), *Spathiphyllum, Typhonium, Zamioculcas* & other cult. orn. esp. *Epipremnum, Homalomena, Schismatoglottis, Scindapsus, Zantedeschia*

Arachis L. Leguminosae (III 11). 69 S Am. R: Bonpl. 8(1994)17, 16 suppl. (2007). Non-protein amino-acids in lvs. ***A. hypogaea*** L. (peanut, groundnut, monkeynut, earthnut) – ann., fls bending down after fert., pedicel forcing young fr. underground to ripen, tetraploid ('*A. monticola*' in wild) derived from diploids *A. duranensis* Krapov. & W. Gregory (Arg., etc.) & *A. ipaensis* Krapov. & W. Gregory (Bolivia etc.), widely cult. (42% in China) for seeds (38 M t p.a.) crushed for c. 50% mono-unsaturated oil for cooking, margarine, soap & cosmetics etc. (used in first diesel engines, 1892; yields twice as much as soy), the cake good for animal feed, raw or roasted for humans (though peanut allergy (e.g. 3% children in Aus., 3.3 M people US) exacerbated by proteins altered by roasting), processed to peanut butter (most in US with aflatoxin!), satay rojak (Mal.), shells used for insulation (USDA Agric. Mon. 19(1954)1) & poss. paper, alleged remains in China 3300–2800 BCE. Some spp. grown as forage & per. ground-cover esp. Braz. ***A. glabrata*** Benth. & ***A. pintoi*** Krapov. & W. Gregory

Arachniodes Blume (~ *Dryopteris*). Dryopteridaceae (I). c. 140 trop. & warm esp. As. & Am. (Afr. 1, Madeira 1). Some cult. orn.

Arachnis Blume. Orchidaceae (V 16c). Incl. *Armodorum, Esmeralda,* 14 Himal., SE As. to Solomon Is. Cult. orn. (scorpion orchids; some with moble lips set in motion by light breeze) esp. ***A* x *maingayi*** (Hook.f.) Schltr. '**Maggie Oei**' (*A. flos-aeris* (L.) Reichb.f. × *A. hookeriana* Reichb.f. (Malay Pen.), natural & artificial hybrid) – most imp. cut-fl. export of Singapore

Arachnitis Philippi. Corsiaceae. 2 Chile & Falklands to Bolivia. R: Willd. 25(1996)323. ***A. uniflora*** Philippi predicted to be poll. by fungus- gnats

Arachnocalyx = *Erica*

Arachnorchis D.L. Jones & M.A. Clem. = *Caladenia*

Arachnothryx Planch. = *Rondeletia* (But see ABAH 28(1982)68, 33(1987)301, 35(1989)309)

Araeoandra Lefor = *Viviania*

Araeococcus Brongn. Bromeliaceae (3). 6 trop. Am. R: FN 14(1979)1505

Arafoe Pim. & Lavrova = *Ligusticum* (but see BZ 74(1989)102)

Aragoa Kunth. Plantaginaceae (Dig.-Arag.; Scrophulariaceae s.l.). 19 paramos of Colombian & Venez. Andes. R: AJBM 51(1993)76. Shrubby, coniferoid

Araiostegia Copel. = *Davallia*

arak flavoured with *Pimpinella anisum*

Aralia Tourn. ex L. Araliaceae (2). Incl. *Megalopanax, Pentapanax,* 71 N Am., E As., Mal. R: Cathaya 13–14(2002)31, 15–16(2004)29, CUSNH 57(2011). ***A. merrillii*** Shang (*A. scandens,* Mal.) – a prickly cli. Cult. orn., medic. & some ed. ***A. chinensis*** L. (E As.) – young lvs a veg.; ***A. cordata*** Thunb. (E As.) – similar ('udo'); ***A. hispida*** Vent. (bristly sarsaparilla, E N Am.); ***A. nudicaulis*** L. (wild s., N Am.) – local. medic.; ***A. racemosa*** L. (American spikenard, E N Am.) – rhiz. & r. medic.; ***A. spinosa*** L. (Hercules' club, C & E US) – bark medic.

Araliaceae Juss. Magnoliidae – Apiales. 38/1625 trop. esp. Indomal., Am., few temp. (esp. S.). R: D.G. Frodin & R. Govaerts (2003) *World checklist & bibliography of A.* Trees (usu. pachycaul, s.t. unbranched & hapaxanthic as *Harmsiopanax*), shrubs, lianes, woody epiphytes or herbs (esp. Hydrocotyloideae), s.t. armed, usu. with secretory canals & multilacunar nodes. Lvs in spirals, rarely opp. or whorls, ± stip., usu. palmately or pinnately compound or lobed, s.t. to 2nd or 3rd degree, rarely simple (as *Meryta*), rounded or peltate (Hydrocotyloideae). Infls term., rarely lat., usu. umbels or heads in panicles etc., rarely solit. Fls usu. bisex., epigynous & 5-merous: K commonly small teeth to 0; C (3 –)5 (– 12), rarely connate at base or forming calyptra, valvate or imbr.; A usu. 1 or 2 x C or ∞, anthers with longit. slits; $\overline{G}$ (2 – 5(– ∞)), rarely $\underline{G}$, locules 1 (e.g. *Arthrophyllum*) – ∞, with as many styles, s.t. ± connate, swollen basally & confluent with epigynous nectary-disk, ovules 1(2) per loc., pend., anatropous, unitegmic. Fr. a drupe with as many pyrenes as G or a berry, rarely a schizocarp with persistent carpophore like Umbelliferae (e.g. *Harmsiopanax*); seeds with small embryo & oily endosperm. x = 11, 12 (usu.), or more

Classification & chief genera (partly after Harms):
Aralioideae (usu. woody, lvs usu. pinnately or palmately compound)
1. **Scheffiereae** (C valvate): *Fatsia, Hedera, Schefflera, Tetrapanax*
2. **Aralieae** (C ± imbr., sessile with broad base): *Aralia, Panax*
3. **Mackinlayeae** (C valvate, shortly clawed) = Umbelliferae – Mack.]

Hydrocotyloideae (form. Umbelliferae – Hydrocotyleae p.p.): herbaceous, lvs rounded to peltate – *Hydrocotyle, Neosciadium, Trachymene* (only)
Fossils referred to Araliaceae known from Upper Cretaceous. Close to Umbelliferae, included by some authors, *Harmsiopanax* & *Astrotricha* having characters of both fams. Some medic. & drug pls (e.g. *Aralia, Eleutherococcus, Panax*), paper (*Tetrapanax*), timber (*Eleutherococcus*), but mostly orn. – *Aralia, Eleutherococcus, Fatsia, Hedera, Polyscias, Schefflera*

Aralidiaceae Philipson & Stone = Torricelliaceae

Aralidium Miq. Torricelliaceae (Aralidiaceae, Cornaceae s.l.). 1 W Mal.: ***A. pinnatifidum*** (Jungh. & De Vriese) Miq. – dioec., alks, local medic., lvs used to repel insects from rice-fields, wood for flooring etc.

Araliopsis Engl. = *Vepris*

Araliorhamnus Perrier = *Berchemia*

aramina fibre *Hibiscus americanus*

× **Aranda** Holltum & Derx. Orchidaceae. *Arachnis* × *Vanda* – 460 grexes

Arapatiella Rizz. & A. Mattos. Leguminosae (I 4). 2 SE Braz.

arar *Tetraclinis articulata*

araroba *Andira araroba*

Araracuara Fernández-Alonso. Rhamnaceae (inc. sed.). 1 Colombian Amaz.: ***A. vetusta*** Fernández-Alonso. R: AJBM 65(2008)343

Aratitiyopea Steyerm. & L. Berry. Xyridaceae. 1 Venez.: ***A. lopezii*** (L.B. Sm.) Steyerm. & L. Berry. R: AMBG 79(1992)877

araucaria, oil of *Neocallitropsis pancheri*

Araucaria Juss. Araucariaceae. 18 SW Pac. (New Caled. 13), S Braz. to Chile (2). R: Phytol.M 7(1984)11. Massart's & Rauh's Models. S Am. spp. with broad lvs – ***A. araucana*** (Molina) K. Koch (mapuche, monkey-puzzle, Chile pine, Chile, where it is most imp. conifer) now prot., imp. timber, resin medic., a favourite architectural tree of the Victorians, living to (?) 2000 yrs with lvs lasting 10 – 15 yrs (later marcescent & prickly), coning after c. 100 yrs, ed. seeds (Chile nuts) app. disp. by parakeets, fastigiate form grown UK; ***A. angustifolia*** (Bertol.) Kuntze (Paraná pine, Braz. p., S Braz. & nearby Arg.) – imp. timber, seeds ed. Broad lvs also in ***A. bidwillii*** Hook. (bunya-bunya pine, Queensland) – hypocotyl develops into parenchymatous tuber, seedlings surviving for several yrs until a break in canopy appears, seeds ed., trees (one of few heritable things in Aboriginal culture) the focus for feasts every few yrs, As from 'miles around' invited to join, reaching 50 m in Coimbra, Portugal. Narrow lvs in ***A. hunsteinii*** K. Schum. (*A. klinkii*, klinki pine, NG) – tallest trop. tree (88.9 m)*, wood imp. esp. for plywood when peeled 'green'. Needle lvs in ***A. cunninghamii*** Mudie (hoop pine, Moreton Bay pine, E Aus.) – timber, ***A. columnaris*** (Forst. f.) Hook. (Cook pine, New Caled.) – to 60 m, often confused in cult. with next, & ***A. heterophylla*** (Salisb.) Franco (Norfolk Is. pine, Norfolk Is.) – planted on Ascension Is. for sailing-ship masts, pop. conservatory plant when a seedling, lat. branches may be rooted but never produce a leader, continuing plagiotropic growth indefinitely

Araucariaceae Henkel & W. Hochst. Pinidae. 3/32 S hemisph. (exc. Afr.) to SE As. Monoec. to dioec. everg. trees with broad to needle lvs, s.t. pungent. Male cones cylindrical, the pollen without air-bladders, females usu. large, ± globose, taking 2–3 yrs to mature & disintegrating when seeds mature. Ligule in *Araucaria*, ± adnate to carpel, absent in others. Ovule 1, free (immersed in ligule in *Araucaria*). Cotyledons 2, occ. deeply 2-cleft. x = 13
Genera: *Agathis, Araucaria, Wollemia*
Differing from Pinaceae in lvs & ovule 1; fossils back to Triassic in both hemispheres. Imp. timber trees

Araujia Brot. Apocynaceae (Vc; Asclepiadaceae (III 1)). Incl. *Morrenia*, c. 10 S Am. ***A. odorata*** (Hook & Arn.) Fontella & Goyder (*M. o.*) – pestilential climber on citrus in Florida, fr. ed.; ***A. sericifera*** Brot. (cruel plant, moth vine, S Braz., invasive in S Eur., S Afr., Aus,

* Very recently overtaken by *Shorea faguetiana* Heim (Dipterocarpaceae) at 94.1 m in Borneo

NZ) – cult. orn. cli. with fls that hold proboscides of night-flying moths until daytime, fibre used in textiles

arazá *Eugenia stipitata*

Arbelaezaster Cuatrec. (~ *Senecio*). Compositae (Sen.-Sen.) 1 Colombia: ***A. ellsworthii*** (Cuatrec.) Cuatrec. R: Cand. 15(1986)1

Arberella Soderstrom & Calderón (~ *Olyra*). Gramineae (VI 3). 7 trop. Am.

Arboa Thulin & Razafim. (~ *Erblichia*). Passifloraceae. 4 Madag. R: T 61(2012)317

arboloco *Montanoa quadrangularis*

arbor-vitae *Thuja* spp.; **American a.** *T. occidentalis*; **Chinese a.** *Platycladus orientalis*; **giant a.** *T. plicata*; **Japanese a.** *T. standishii*; **western a.** *T. plicata*

Arbulocarpus Tenn. = *Spermacoce*

Arbutus Tourn. ex L. Ericaceae (III). 10 W N Am. (3), trop. Am. (5), W Eur. (2) to Med. Heterogeneous (W N Am. spp. distinct genus)? Small trees & shrubs with red flaking bark & berries, cult. orn. ***A. menziesii*** Pursh (madroña, madrone, madroño, W N Am.) – veg. buds act as extrafl. nectaries, tanbark & timber, local medic.; ***A. unedo*** L. (strawberry tree, S Eur. & Ireland) – part of Lusitanian element in Br. flora, Scarrone's Model, female gametophyte ready for fert. at poll. but first zygotic division 5 months, tanbark, fr. ed., preserved & flavour in liqueurs (esp. in Portugal), fls while previous yr's fr. ripening

Arcangelina Kuntze = *Tripogon*

Arcangelisia Becc. Menispermaceae (III). 2 SE As. to Mal. R: KB 32(1978)333. ***A. flava*** (L.) Merr. – germicidal dye from wood, r. infusion medic. & abortifacient

Arceuthobium M. Bieb. Santalaceae (Viscaceae). 42 N to C Am., WI, Medit. (Eur. 2, usu. on *Juniperus*), NE trop. Afr. (1), Sinohimal. to W Mal. R: USDA Agr. Handbk 40(1972), 709(1996). Dwarf mistletoes (? always) on gymnosperms. Fr. explosive, seeds ejected to 16 m at 27 m/sec. surrounded in viscin, which, when hydrated by rain, allows seed to slide down needle-leaves to branch where germinating (so spp. with drooping needles not infected. ***A. minutissimum*** Hook.f. (Himal.) – 2–5 mm long (excl. haustoria), poss. smallest dicot.?

Arceuthos Antoine & Kotschy = *Juniperus*

Archakebia C.Y. Wu = *Akebia* (but see Cathaya 8–9(1997)49)

Archangel redwood or **A. yellow deal** *Pinus sylvestris*; **yellow a.** *Lamium galeobdolon*

Archangelica Wolf = *Angelica*

Archangiopteris Christ & Giesenh. = *Angiopteris*

Archanthemis Lo Presti & Oberprieler (~ *Anthemis*). Compositae (Anth.-Anth.). 4 Caucasus. Shrublets. R: T 59(2010)1454

Archboldia E. Beer & H.J. Lam = *Clerodendrum*

Archboldiodendron Kobuski. Pentaphylacaceae (3; Ternstroemiaceae). 1 NG mts: ***A. calosericeum*** Kobuski – Attims' Model. R: Brunonia 3(1980)47

Archeria Hook.f. Ericaceae (VII 2; Epacridaceae). 7 Aus. (5), NZ (2)

Archiatriplex G.L. Chu. Amaranthaceae (Atrip.; Chenopodiaceae I 3). 1 Sichuan: ***A. nanpinensis*** G.L. Chu. R: JAA 68(1987)461

Archibaccharis Heering. Compositae (Ast.-Bac.). 32 Mex. to C Panamá. R: Phytologia 32(1975)81. Functionally dioec.

Archiboehmeria C.J. Chen (~ *Boehmeria*). Urticaceae (III). 1 S China to SE As.: ***A. atrata*** (Gagnep.) C.J. Chen

Archiclematis (Tamura) Tamura = *Clematis*

Archidendron F. Muell. Leguminosae (II 4). 94 Indomal. (esp. Borneo). R: OB 76(1984). G 1 – several, dark blue to black seeds hanging by funicles from red inner wall of carpel in many spp. Timbers. ***A. jiringa*** (Jack) I. Nielsen (*Pithecellobium lobatum*, jengkol, ngapi, Myanmar to Borneo) – smelly seed ed. cooked

Archidendropsis I. Nielsen (~ *Albizia*). Leguminosae (II 4). 14 SW Pac. R: BMNHN 4,5(1983)335

Archigrammitis Parris (~ *Grammitis*). Polypodiaceae (Gram.). 6 Mal., W Pacific. R: FG 19 (2013)133

Archihyoscyamus Lu = *Hyoscyamus*

Archileptopus P.T. Li = *Leptopus*

Archineottia S.C. Chen = *Holopogon*

Archiphysalis Kuang = *Withania*

Archirhodomyrtus (Niedenzu) Burret (~ *Rhodomyrtus*). Myrtaceae (II 10). 5 New Caled., Aus. (1)

Archiserratula L. Martins (~ *Serratula*). Compositae (Card.-Cent). 1 Yunnan: ***A. forrestii*** (Iljin) L. Martins

Archivea Christenson & Jenny. Orchidaceae (V 10). 1 Braz.: ***A. kewensis*** Christenson & Jenny. R: Orchids 65(1996)457

Archontophoenix H.A. Wendl. & Drude. Palmae (V 14a). 6 E Aus. R: FA 39(2011)188. Cult. orn.: ***A. alexandrae*** (F. Muell.) H.A. Wendl. & Drude (Alexandra palm) & ***A. cunninghamiana*** (H.A. Wendl.) H.A. Wendl. & Drude (bangalow (palm), Illawara palm, piccabeen) – prot. in Aus., invasive in Braz. forests

Archytaea Mart. Bonnetiaceae (Guttiferae II). 2 NE S Am. Aubréville's Model

Arcoa Urb. Leguminosae (I 4). 1 Hispaniola: ***A. gonavensis*** Urb.

Arctagrostis Griseb. Gramineae (XVI 15). 2 Arctic inc. Eur., marshy tundra. R: FNA 24(2007)676

Arctanthemum (Tzvelev) Tzvelev. (~ *Chrysanthemum*). Compositae (Anth.-Art.). (Excl. *Hulteniella*) 3 Arctic to Japan. R: FN 19(2006)535. ***A. arcticum*** (L.) Tzvelev – cult. orn.

Arcteranthis Greene (~ *Ranunculus*). Ranunculaceae (II 3). 1 NW N Am.: ***A. cooleyae*** Greene

Arcterica Cov. = *Pieris*

Arctium L. Compositae (Card.-Card.). Incl. *Anura, Hypacanthium, Schmalhausenia,* 40 subcosmop. (Eur.4). R: T 60(2011)546. *Cousinia* & allies poss. referable here. Burdocks, the burs being fr. heads covered with involucral bracts, which become hooked & woody after fert. & get attached to animal fur, clothing etc. (allegedly inspiration for 'Velcro' fastenings) & promote disp. ***A. lappa*** L. cult. for ed. r. (gobō) in Jap. etc., young lvs ed. as salad in Scandinavia, Canada & Jap., & in comm. dandelion & burdock tea; dried r. medic. esp. skin complaints (also ***A. minus*** (Hill) Bernh.), form. for gonorrhoea, now herbal treatment of arthritis, gout, eczema & anorexia nervosa, 'Burry Man', man covered in burs (Queensferry, Edinburgh since 1687) giving & receiving gifts etc.

Arctogentia Löve = *Gentianella*

Arctogeron DC. Compositae (Ast.-Ast.). 1 C As. to China & Mongolia: ***A. gramineum*** (L.) DC.

Arctomecon Torrey & Frémont. Papaveraceae (IV). 3 Mojave Desert, SW US. R: Rhodora 95(1993)204. C s.t. persistent. Cult. orn.

Arctophila (Rupr.) Anderss. = *Dupontia*

Arctopoa (Griseb.) Probat. (~ *Poa*). Gramineae (XVI 15). 5 As., N Am.

Arctopus L. Umbelliferae (II?). 3 S Afr. R: Strel. 9(2000)275. Dioec. ***A. echinatus*** L. – r. medic. R: AMBG 95(2008)477

Arctostaphylos Adans. Ericaceae (III). Excl. *Arctous, Comarostaphylis, Xylococcus,* 66 W N Am. (62 – R: FNA 9(2009)406), circumpolar (incl. Eur.) 1. R: JEMSS 56(1940)1. Fire-adapted (pyrophytes or with persistent seedbanks; seed germ. enhanced by immersion in H_2SO_4) mostly prostrate shrubs (palo blanco), cult. orn., fr. ed. – ground as meal or in drinks. ***A. pungens*** Kunth (manzanita, W N Am.) – to 3 m, orn. wood; ***A. uva-ursi*** (L.) Spreng. (bearberry, uva-ursi, circumpolar) – lvs used in tanning 'Russian leather', or N Am. smoking mixtures (kinnikinni(c)k) & in tea in Russia, as urinary antiseptic in UK since C13, imp. local medic. N Am.

Arctotheca Wendl. Compositae (Arct.-Arct.). 5 S Afr., 2 natur. Aus., 1 in S Eur. (***A. calendula*** (L.) Levyns – Capeweed). R: FRB 11(1922)48

Arctotis L. Compositae (Arct.-Arct.). 50 S Afr. to Angola. Some poll. monkey beetles. Cult. orn. (incl. *Venidium*) esp. ***A. × hybrida*** Hort. ('*A. stoechadifolia*') – parentage unclear

Arctottonia Trel. = *Piper*

Arctous (A. Gray) Niedenzu (~ *Arctostaphylos*). Ericaceae (III). 3–4 N hemisph. ***A. alpina*** (L.) Niedenzu (black bearberry, circumpolar)

Arcuatopterus Sheh & Shan (~ *Ferula*). Umbelliferae (III 10). 6 Himal. SW China. R: FR 111(2000)556

Arcyna Wiklund = *Cynara* (but see Willd. 33(2003)63)

Arcyosperma O. Schulz. Cruciferae (7). 1 Himal.: ***A. primulifolium*** (Thomson) O. Schulz

Arcypteris Underw. = *Pleocnemia*

Arcytophyllum Willd. ex Schultes & Schultes f. Rubiaceae (IV 15). 15 trop. Am. mts. R: MNYBG 60(1990). Some weedy in OW ('*Anotis*')

Ardisia Sw. Primulaceae (Myrsinaceae). Incl. *Afrardisia, Gentlea, Synardisia, Tetrardisia,* c. 500 trop. (esp. C Mal. & C Am. (c. 120)) & warm excl. Afr., rare Aus. R: PNASP 141(1989)268, (subg. *Auriculardisia*) AMBG 90(2003)187. Massart's, Rauh's & Bell's Models, some 'litterbox' pls, some with 'throwaway branches' with swollen 'axils' at junction with axis. Some med. & ed. fls or fr. (marlberries), cult. orn. esp. ***A. crenata*** Sims (oft. confused with ***A. crispa*** (Thunb.) A. DC., which does not have crisped lvs) of NE Ind. to Jap., undergrowth treelet with crimped lvs (poss. proteinaceous deposits or bacterial nodules responsible) & red fr. retained for several seasons, invasive in S Afr., Masc., SE US; ***A. elliptica*** Thunb. (As.) – invasive Masc., Hawaii, SE US; ***A. silvestris*** Pitard (Vietnam) – comm. med. (gastric)

Ardisiandra Hook.f. Primulaceae (Myrsinaceae). 3 trop. Afr. mts

Areca L. Palmae (V 14b). 42 Indomal. R: FR 33(1933)217. Cult. orn. Alks. ***A. catechu*** L. (betelnut, pinang, areca nut, quticha, supari), cultigen (? tetraploid orig. in Sulawesi; closest ally: ***A. concinna*** Thwaites (Sri Lanka)) throughout trop. As. (prob. cult. by 'Hoabhinians' 8000–3000 BP): seed (medic., source of red dye), cut into slices & usu. chewed (by 200–400 M people) in a wad (in India paan, pan; NG buai) of betel pepper (*Piper betle*) lvs with lime, causes (oral cancer &) saliva to turn red & promotes salivation (& spitting so spreading TB) & wellbeing (arecaine a mild narcotic), dulling appetite & medic., prob. formerly used in some Eur. tooth-powders & poss. orig. of myth of anthropophagous W invaders in NG folklore, bracts used as curd containers in Sri Lanka, brown dye from wood; ***A. rheophytica*** Dransf. (Sabah) – only on streambanks on ultramafics; ***A. vestiaria*** Giseke (C Mal.) – epidermis of young lvs for textiles

Arecaceae Bercht. & J. Presl. See Palmae

Arecastrum (Drude) Becc. = *Syagrus*

Arechavaletaia Speg. = *Azara*

Aregelia Kuntze = *Nidularium*, Mez = *Neoregelia*

Areldia Luer = *Specklinia*

Aremonia Necker ex Nestler. Rosaceae (8). 1 SE Eur.: ***A. agrimonioides*** (L.) DC. – after flowering, hypanthium becomes elaiosome & both it & fr. disp. by ants

Arenaria L. Caryophyllaceae (II 1). Excl. *Odontostemma, Solitaria*, c. 95 N temp. (Eur. 54, N Am. 9), Medit., Andes. Sandworts. R: JLSB 33(1898)326. Cult. orn. & weeds. ***A. bryophylla*** Fern. (Himal.) – pl. at highest alt. (6180m) on Mt Everest. See also *Eremogone*

Arenga Lab. Palmae (III 6). c. 20 trop. As. (Mal. 15) to N Aus. Cult. orn., juicy fr. irritant. ***A. microcarpa*** Becc. (Mal.) – sago; ***A. pinnata*** (Wurmb) Merr. (*A. saccharifera*, sugar-palm, gomuti, ejow, Mal.) – widely cult., male spadices tapped for syrupy sap evaporated to produce palm sugar (jaggery), palm wine or toddy, (when distilled) arrack, sago from trunk, good fibre from leaf-sheaths, hapaxanthic (Holttum's Model), infls appearing in descending order; ***A. tremula*** (Blanco) Becc. (Philipp.) – bud ed., narcotic

Arenifera Herre (~ *Psammophora*). Aizoaceae (V). 1 N Cape: ***A. pillansii*** (L. Bolus) Herre. R: T 65(2016)258

arere *Triplochiton scleroxylon*

Arethusa Gronov. ex L. Orchidaceae (V 10). 1 E N Am.: ***A. bulbosa*** L. (laughingjackass) – cult. orn. R: FNA 26(2002)596

Aretiastrum (DC.) Spach = *Valeriana*

Arfeuillea Pierre ex Radlk. Sapindaceae (III 1). 1 Laos, Thailand: ***A. arborescens*** Pierre ex Radlk.

argan (oil) *Argania spinosa*

Argania Roemer & Schultes. Sapotaceae (III). 1 Morocco, Algeria, introd. Libya, natur. S Spain: ***A. spinosa*** (L.) Skeels (argan) – seed oil (rich in vitamin E, 80% unsaturated fatty acids) like olive oil for cooking (seeds excreted or dropped by cli. goats split open for it), Berber folk medic., now in 'anti-ageing' facials & aftershave, fr. ed. cattle, timber good, gum valuable

Argantoniella López & R. Morales = *Satureja* (but see Fl. Ib. 12(2010)421)

Argemone Tourn. ex L. Papaveraceae (IV). c. 23 N & S Am., WI, Hawaii. R: MTBC 21(1958)1, Britt. 13(1961)91. Alks. Cult. orn. annuals & 1 shrub, some medic. ***A. mexicana*** L. (Mex. poppy, prickly p., C Am., now pantrop. weed) – seeds a minor oil source though as contaminants in grain thought to cause glaucoma in Ind., latex a yellow dye in Peru

Argentina Hill = *Potentilla* (but likely to be segregated)

Argentipallium Paul G. Wilson (~ *Ozothamnus*). Compositae (Gnap.-Ang.). 6 temp. Aus. R: Nuytsia 8(1992)455

Argeta N.E. Br. = *Gibbaeum*

Argillochloa W. Weber = *Festuca*

Argocoffeopsis Lebrun. Rubiaceae (II 3). 9 trop. Afr. R: BJBB 51(1981)361

Argomuellera Pax. Euphorbiaceae (Acal.-Pyc.-Pyc.). 11 trop. Afr. (6), Madag. (6). R: BSRBB 91(1959)274

Argophyllaceae Takht. Magnoliidae – Asterales. 2/c. 21 SW Pacific. Small trees or shrubs; indumentum of T-shaped trichomes. Lvs simple in spirals (or 3- or 4-leaved fascicles on brachyblasts); guard-cells raised above rest of epidermis; stip. 0. Fls (4– or) 5(–8)-merous. K valvate, basally connate; C basally connate or free, with adaxial ligule; A opp. K; G 1–3(–6)loc., semisuperior with 1 – many anatr. ovules per loc. Fr. a loculicidal capsule (*Argophyllum*) or drupe (*Corokia*); seeds obovate or linear-elongate with minute or elongate embryo in fleshy endosperm.
Genera: *Argophyllum, Corokia*
Form. referred to Cornaceae, Grossulariaceae

Argophyllum Forst. & Forst. f. Argophyllaceae (Grossulariaceae s.l.). c. 15 trop. Aus. (5), New Caled. (10)

Argopogon Mimeur = *Ischaemum*

Argostemma Wall. Rubiaceae (IV 4). c. 100 OW trop. (Afr. 2; Borneo 28, R: AMBG 76(1989)7). Some Mal. spp. with black lvs due to complete absorption of light by compact palisade cells full of chloroplasts, others with single functional leaf (cf. *Monophyllaea*)

Argostemmella Ridl. = *Argostemma*

Argusia Boehmer = *Heliotropium*

Argyle apple *Eucalyptus cinerea*

Argylia D. Don. Bignoniaceae (1). 12 S Peru, Chile, Arg. Herbs or subshrubs

Argyll's tea-tree, Duke of *Lycium barbarum*

Argyranthemum Webb ex Schultz-Bip. Compositae (Anth.-Gleb.). 24 Macaronesia. R: BBMNH 5(1976)147. Originating from Medit. Cult. orn. (marguerites) incl. 'doubles' & hybrids, s.t. grown esp. as standards. ***A. frutescens*** (L.) Schultz-Bip. (Paris daisy, Gran Canaria) – aggressive colonist in NZ. ***A sundingii*** Borgen (Tenerife) – 1 of 8 known diploid hybrid spp., but arising at least twice from *A. broussonetii* (Pers.) Humphries (montane) & *A. frutescens* (coastal)

Argyreia Lour. (~ *Ipomoea*). Convolvulaceae (9). Excl. *Rivea*, 130+ Indomal. to Aus. (1). Some cult. orn. esp. ***A. nervosa*** (Burm.f.) Bojer (*A. speciosa*, elephant cli., Ind.) – med. in Ind., high in LSA sold as psychodelic in Netherlands, dried fr. with accrescent K sold as 'wood rose' or 'Hawaiian baby wood rose' (smaller than *Merremia* spp.) in Hawaii

Argyrochosma (J. Sm.) Windham (~ *Cheilanthes*). Pteridaceae (IV). 16 trop. & warm Am. (esp. Mex.)

Argyrocytisus (Maire) Raynaud (~ *Adenocarpus*). Leguminosae (III 9). 1 Morocco: ***A. battandieri*** (Maire) Raynaud – cult. orn. with pineapple-scented fls

Argyrodendron F. Muell. (~ *Heritiera*). Malvaceae (Sterc.; Sterculiaceae). 7 Aus. Good timber (booyong, tulip oak)

Argyroderma N.E. Br. Aizoaceae (V). 10 S Afr. R: H. E. K. Hartmann, *Ill. Handb. Succ. Pls*, Aiz. A–E (2002)70. Cult. orn. succ.

Argyroglottis Turcz. (~ *Helichrysum*). Compositae (Gnap.-Gnap.). 1 N Aus.: ***A. turbinata*** Turcz. R: OB 104(1991)80

Argyrolobium Ecklon & Zeyher. Leguminosae (III 9). c. 80 S Afr. to trop. Afr. highlands, Madag., Med. (Eur. 2) to Ind. Alks. Mostly xerophytes, some with cleistogamous fls

Argyronerium Pitard = *Epigynum*

Argyrophanes Schldl. = *Chrysocephalum*

Argyrotegium J. Ward & Breitw. (~ *Euchiton*). Compositae (Gnaph.). 4 SE Aus., NZ. R: NZJB 41(2003)608

Argyrovernonia MacLeish = *Chresta*

Argyroxiphium DC. Compositae (Mad.-Mad.). 5 Maui & Hawaii. R: Allertonia 4,1(1985)50. Woody hapaxanthic monocaulous to polyxanthic branched pachycauls (silverswords), some high alt. ***A. sandwicense*** DC. – Holttum's Model, fls after 15–50 yrs, by 1995 reduced to 500 individuals; ***A. virescens*** Hillebr. – extinct (last seen 1945)

Argythamnia P. Browne. Euphorbiaceae (Acal.-Chro.-Dit.). Excl. *Ditaxis* 18 trop. Am. (WI 10). R (subg.A.): GH 10(1966)1

arhar *Cajanus cajan*

Aria (Pers.) Host = *Sorbus*

Ariadne Urb. = *Mazaea*

Ariaria Cuervo = *Bauhinia*

Arida (R. Hartman) Morgan & R. Hartman = *Leucosyris* (but see Sida 20(2003)1410,1422)

aridan *Tetrapleura tetraptera*

Aridaria N.E. Br. = *Mesembryanthemum*

Aridarum Ridl. (~ *Schismatoglottis*). Araceae (VII 8). (Incl. *Heteroaridarum*) 12 Borneo. R: Telopea 9(2000)183, Willd. 42(2012)264. Some rheophytes incl. ***A. borneensis*** (Hotta) Bogner & A. Hay (*H. b.*)

Arikuryroba Barb. Rodr. = *Syagrus*

Arillastrum Pancher ex Baill. (~ *Stereocaryum*). Myrtaceae (II 11). 1 New Caled.: ***A. gummiferum*** (Brong. & Gris) Baill.

Ariocarpus Scheidw. Cactaceae (III 9). 6 S Texas, Mex. R: AJB 50(1963)724, 51(1964)144. Cult. orn. slow-growing, mimicking stones. ***A. kotschoubeyanus*** (Lem.) K. Schum. (Mex.) – withdraws into seasonally flooded soil & can be completely buried between growing seasons

Ariopsis Nimmo. Araceae (VII 25). 2 Ind. to Myanmar

Ariosorbus Koidz. = *Sorbus*

Aripuana Struwe & al. (~ *Macrocarpaea*). Gentianaceae. 1 Amaz. Braz.: ***A. cullmaniorum*** Struwe & al. R: HPB 2(1997)236

Arisaema Mart. Araceae (VII 24). 210 E Afr. & Arabia (10, R: KB 41(1986)261), trop. & E As. (Jap. 42), W N Am. R: G. & L. Gusman (2002) *The genus A.* Some with long spadix appendages dispersing odours attractive to poll., some poss. poll. by snails, ***A. utile*** Hook.f. ex Schott (*A. verrucosum* var. *utile*, Himal.) by fungus-gnats. Cult. orn. (cobra-lily, dragon-arum, snail-fl.; R: Plantsman 3(1982)193), some medic. & with ed. corms, esp. ***A. flavum*** (Forssk.) Schott (Yemen to W China) in Arabia & ***A. triphyllum*** (L.) Schott (Jack-in-the-pulpit, Ind. turnip, E N Am.) – smallest infertile, larger reproduce as males, largest as females, imp. local med.

Arisarum Mill. Araceae (VII 21). 3 Med. (Eur. 2). R: KM 7(1990)16. Cult. orn. ***A. proboscideum*** (L.) Savi (mouseplant, Italy, SW Spain) with long tail-like process on spathe, poll. by fungus-gnats (in spring – few fungi available!) like *Asarum* spp. (q.v.); ***A. vulgare*** Targ.-Tozz. (friar's cowl, Med.) – elaiosomes, corms a famine food

Arischrada Pobed. = *Salvia*

Aristaloe Boatwr. & Manning (~ *Aloe*). Asphodelaceae. 1 S Afr mts: ***A. aristata*** (Haw.) Boatwr. & Manning. R: SB 39(2014)69

Aristavena F. Albers & Butzin = *Deschampsia*

Aristea Aiton (?*Gemmingia*). Iridaceae (IV). 58 trop. & S Afr., Madag. (8). R: Lunds Univ. Årsskr. NFAvd2,36(1940)10. Morning flowering, some in S Afr. poll. by monkey beetles. Cult. orn.

Aristeguietia R. King & H. Robinson. Compositae (Eup.-Crit.). 21 Andes. R: MSBMBG 22(1987)343

Aristeyera H. Moore = *Asterogyne*

Aristida L. Gramineae (Aristid.). c. 300 warm (Eur. 1, China 10, Aus. 58 [52 endemic], N Am. 29). R: MRL 58, A & B (1929 – 33). Usually indicate soil disturbance or overgrazing. ***A. contorta*** F. Muell. (Aus.) – fr. round legs can immobilize sheep; ***A. funiculata*** Trin. & Rupr. (N Afr. to Pakistan) – florets in balls to 25 cm diam. disp. along ground by wind in Sudan; ***A. junciformis*** Trin. & Rupr. (S Afr.) – broom grass; ***A. pungens*** Desf. (N Afr.) – famine grain

Aristocapsa Reveal & Hardman (~ *Centrostegia*). Polygonaceae (I 1). 1 Calif.: ***A. insignis*** (Curran) Reveal & Hardman. R: Phytol. 66(1989)84

Aristogeitonia Prain. Picrodendraceae (Pic.-Misch.; Euphorbiaceae). 7 trop. Afr. (Angola 1, E Afr. 2), Madag. (4). R: KB 43(1988)627

Aristolochia Tourn. ex L. Aristolochiaceae (II). Incl. *Einomeia, Endotheca, Euglypha, Holostylis, Howardia, Isotrema*, excl. *Pararistolochia*, c. 500 trop. & warm OW (Eur. 19 native; E & S As. 68, R: APS 27(1989)321; Afr. 11, R: BotJLS 151(2006)220). R: SB 40(2015)671. P (K) a fly-trap attracting Diptera by smell to poll. after sliding down a 'slip zone', their phototropism attracting them to windows at tube-base in some, and unable to escape until it expands later, when A have matured & deposited pollen on them; in other spp. insects detained until downward-pointing hairs wither & in yet others insects attracted by prod. of heat (cf. *Arum*), though some can be regularly self-poll.; ***A. inflata*** Kunth & ***A. maxima*** Jacq. (trop. Am.) – no trap, poll. insects ovipositing in decomposing fls, larvae living in fallen ones. Poisonous alks or aristolochic acid. Many with large fleshy funicle attractive to seed-dispersers. Many medic. (Gk. *aristos* = best, *lochia* = childbirth, from the curved fl. of *A. clematitis* recalling human foetus in right position prior to birth, Doctrine of Signatures indicating that the plant would ease parturition (the pls actually efficacious abortifacients!), some cult. orn. ***A. albida*** Duch. (*A. petersiana*, E Afr.) – toxic to stock, arrow poison; ***A. clematitis*** L. (birthwort, Eur.) – abortifacient ± natur. GB; ***A. fimbriata*** Cham. (*Howardia f.*, Urug.) – fertility control; ***A. gorgona*** M. Blanco (Costa Rica) – fls to 31 × 30 cm, one of largest Am. fls; ***A. glaucescens*** Kunth (*H. g.*, S Am.) – source of yellow pareira (pereira) used as diuretic etc.; ***A. grandiflora*** Sw. (*H. g.*, WI) – perianth to 20 cm across with 'tail' to 60 cm; ***A. macrophylla*** Lam. ('*A. durior*', *A. sipho*, Dutchman's pipe, E N Am.) – rapid cli.; ***A. serpentaria*** L. (*Endotheca s.*, Virginia snakeroot, E N Am.) – rhiz. suggestive of snakes (Doctrine of Signatures) & actually efficacious in snakebite treatment!

Aristolochiaceae Juss. Magnoliidae – Piperales. Incl. Asaraceae, Hydnoraceae, Lactoridaceae, 8/660 trop. & warm esp. Am. Aromatic lianes, scramblers, shrubs or rhiz. herbs, s.t. r.-parasites (Hydnoroideae), with alks or aristolochic acid. Lvs in spirals, simple, oft. cordate, s.t. lobed, marcescent, venation palmate (0 in *Hydnora* – only angiosperms without even scales), true stip. 0. Fls solit. or in terminal or lat. racemes or cymes, bisex., reg. or irreg., oft. smelling of rotting meat. P (prob. K) usu. (3) with S-shaped tube usu. & with 0–3 lobes or reg. & 3-lobed, oft. petaloid; C 0 or small, but 3 in *Saruma*; nectaries (patches of secretory hairs in C-tube) oft. present; A (5)6–c. 40 (fewer in some *Aristolochia*) in 1 or 2 (to 4 in *Thottea*) whorls free or ± united with style to form gynostemium, anthers extrorse (rarely introrse); G ((3)4–6), almost distinct in *Saruma* to inferior, (1)4–6-loc., s.t. with incomplete partitions; ovules ∞ per loc., unitegmic (H.) or bitegmic, usu. anatr. Fr. usu. septicidal capsule, s.t. with fleshy endocarp, many-seeded, rarely follicular (*Saruma*) or indehiscent & 1-seeded, dehiscing basipetally (*Thottea*) or acropetally (most *Aristolochia*); seeds (to 90k in H.) with minute embryo & copious oily (s.t. also starchy) endosperm. x = 4–7, 12, 13, 16 poss. orig. 7

Classification & genera:

I. [1.] **Asaroideae** (herbs, not twining, fls not constricted between P & G; s.t. treated as distinct fam.): *Saruma, Asarum*

[2.] *Lactoris* (shrub)

[3.] **Hydnoroideae** (root-parasites): *Hydnora, Prosopanche* (only)

II. [4.] **Aristolochioideae** (woody or herbaceous, oft. cli.; fls constricted between P & G): *Aristolochia, Pararistolochia, Thottea*

Early Cretaceous fossils (*Hexagyne* Coiffard & al.) from Braz. prob. referrable here. Generic limits of *Aristolochia* debated. Some thermogenesis

Cult. orn. (*Aristolochia, Asarum*), some medic. (*Aristolochia, Thottea*)

Aristopsis Guerra = *Aristida*

Aristotelia L'Hérit. Elaeocarpaceae. 5 E Aus., NZ, Peru to Chile. R: KB 40(1985)491. Cult. orn. esp. ***A. chilensis*** (Molina) Stuntz (*A. macqui*, maqui, Chile, invasive on Juan Fernandez – small ed. fr. preserved & used to colour wine) & ***A. serrata*** (Forst. & Forst. f.) W. Oliver (*A. racemosa*, NZ) – useful wood incl. inlay-work

Arivela Raf. = *Cleome*

Arizona poppy *Kallstroemia grandiflora*

Arjona Comm. ex Cav. Santalaceae (Schoepfiaceae). 12 temp. S Am. ***A. tuberosa*** Cav. with ed. tubers (macachi, Chile)

arjun *Terminalia arjuna*

Armatocereus Backeb. (~ *Lemaireocereus*). Cactaceae (III 1). 7 Colombia, Ecuador, Peru

Armeniaca Mill. = *Prunus*

Armeria Willd. Plumbaginaceae (II). c. 90 N temp. (Eur. 43), Andes to Tierra del Fuego. Cult. orn. tufted perennials. After fert. K becomes a membranous organ aiding wind-disp. Common on coasts, mts etc. Most cvs derived from ***A. maritima*** (Mill.) Willd. (range of genus (polymorphic), thrift, sea-pink) – punningly figured on C20 English threepenny bit, form. used in treatment of obesity, r. sliced & boiled in milk ('arby') used against TB in pre-1700 Orkneys

Armodorum Breda = *Arachnis*

Armoracia P. Gaertner & al. Cruciferae (16). 3 E & SE Eur. (2) to Siberia, E N Am. (1). R: JAA 69(1988)160. ***A. rusticana*** P. Gaertner & al. (horse-radish, ? cultigen (or poss. native SW C As.) – clones with irreg. meiosis perhaps of hybrid orig., seeds not set) – prop. by r.-cuttings for over 2000 yrs, r. source (with salt, oil & vinegar) of relish for beef, oysters etc., local medic. N Am., now with garlic used in hay-fever cures

Armouria Lewton = *Thespesia*

Arnaldoa Cabrera (~ *Barnadesia*). Compositae (I). 3 S Am. R: Novon 12(2002)418. Spirals of floral primordia not perfect, style without brushing hairs, pollen attaching by pollenkitt – all primitive features

Arnanthus Baehni = *Pichonia*

arnatto *Bixa orellana*

Arnebia Forssk. Boraginaceae (B.2.2). Excl. *Huynhia*, c. 30 Med. (Eur. 1), trop. Afr., Himal. Some, incl. ***A. pulchra*** (Roemer & Schultes) Edmondson (*A. echioides*, *Echioides longiflora*, prophet flower, Armenia, Cauc. to Iran), have black spots on C, fading as it matures. ***A. benthamii*** (G. Don) I.M. Johnston (Himal.) – pachycaul

Arnebiola Chiov. = *Arnebia*

Arnhemia Airy Shaw. Thymelaeaceae (Oct.). 1 Aus.: ***A. cryptantha*** Airy Shaw. R: FA 18(1990)124

Arnica L. Compositae (Mad.-Arn.). 29 N temp. & arctic (Eur. 2; US & Canada 26 – R: FNA 21(2006)366). R: Britt. 4(1943)386. Cult. orn. incl. ***A. montana*** L. (arnica, mt. tobacco, C & N Eur.) – r. & capitula medic. (tincture of a., astragalin (flavonoid) inhibiting histamine release; though in US often *Heterotheca inuloides*!) now promoted for jet-lag, ***A. fulgens*** Pursh (N Am.) more efficacious

Arnicastrum Greenman. Compositae (Tag.-Pect.). 2 Mex. R: SB 11(1986)277

Arnicratea Hallé. Celastraceae (III). 3 Indomal. R: BMNHN 4,6(1984)12. ***A. grahamii*** (Wight) Hallé – seeds ed.

Arnocrinum Endl. & Lehm. Asphodelaceae. 3 SW Aus. R: FA 45(1987)246

Arnoglossum Raf. Compositae (Senec.-Tuss.). 8 E US. R: FNA 20(2006)622. NW '*Cacalia*' spp. ***A. reniforme*** (Hook.) H. Robinson (*A. muhlenbergii*, *C. m.*) – poss. oil source

Arnoseris Gaertn. Compositae (Lact.-Hier.). 1 Eur.: ***A. minima*** (L.) Dumort. (*A. pusilla*, swine or lamb's succory)

Arnottia A. Rich. = *Cynorkis*

arogyapaccha *Trichopus zeylanicus*

Aromadendron Blume = *Magnolia*

Aronia Medik. Rosaceae (Ros.-Mal.). Excl. *Photinia*, *Stranvaesia* 2 EN Am. R: FNA 9(2014)445. Cult. orn. (chokeberries): ***A. arbutifolia*** (L.) Pers. (*A. pyrifolia*, *P. p.*, red chokeberry), ***A. melanocarpa*** (Michaux) Elliott (black c.) – several cvs, fr. high in vitamin C, eaten with meat in Tallinn & their hybrid, ***A. ×prunifolia*** (Marshall) Rehder – cult., '**Nero**', fr. for preserves & wine (Russia), juice & food-colouring

Arophyton Jum. Araceae (VII 19). 7 Madag. R: BJ 92(1972)24

Aropsis Rojas = *Spathicarpa*

Arpitium Necker ex Sweet. Older name for *Endressia*

Arpophyllum Lex. Orchidaceae (V 13b). 35 trop. Am. R: Orquidea 4(1974)16. Cult. orn.

Arquita Gagnon & al. (~ *Caesalpinia*). Leguminosae (I 4). 5 Andes. R: T 64(2015)479

Arrabidaea DC. = *Fridericia*

arracacha *Arracacia xanthorrhiza*

Arracacia Bancr. Umbelliferae (III 5). 55 trop. Am. ***A. xanthorrhiza*** Bancr. (arracacha, apio, Peruvian parsnip, N S Am.) – Andean r.-crop with strong parsnip taste, locally replacing potatoes

arrack potable spirit distilled from sugary sap of palms, esp. spp. of *Arenga*, *Borassus*, *Caryota*, *Cocos*, *Corypha*, sugar-cane or rice

arrayán *Luma apiculata*

Arrhenatherum P. Beauv. (~ *Helictotrichon*). Gramineae (XVI 2). 8 Eur. (5), Med., N & W As. R: KBAS 13(1986)124. Like *H.* but florets dimorphic, though dimorphism varying within same panicle. ***A. elatius*** (L.) J. & C. Presl (false oat, French rye-grass, Eur., invasive NZ, W US) – form. in seed mixtures for hay; s.t. (var. ***bulbosum*** (Willd.) Spenn., onion couch) with basal internodes swollen to 1.6 cm diam. – propagules fed to pigs

Arrhenechthites Mattf. (~ *Senecio*). Compositae (Senec.-Senec.). 5 mts of NG, Aus. (1). R: AMBG 43(1956)74

Arrhostoxylum Nees = *Ruellia*

Arrojadoa Britton & Rose. Cactaceae (III 3). 4 E Braz. R: Bradleya 6(1988)90, 7(1989)35. Cult. orn. Stem tubers in ***A. dinae*** Buining & Brederoo

Arrojadocharis Mattf. Compositae (Eup.-Gyp.). 2 E Braz. R: MSBMBG 22(1987)118

Arrojadoopsis Guiggi = *Arrojadoa*

arrow poisons Usu. mixtures of pl. toxins with adhesive gums & s.t. animal substances (oft. of magical significance only). In Afr. *Adenium* & *Acokanthera*, Am. *Strychnos* etc. (q.v.)

arrowgrass *Triglochin* spp.

arrowhead *Sagittaria sagittifolia* & other spp.

arrowroot *Maranta arundinacea* (**WI, St Vincent** or **Bermuda a.**). Other starches with small grains used as subs.: **Afr. a.** *Tacca leontopetaloides*, **Bombay a.** *Curcuma angustifolia*; **Braz. a.** *Ipomoea batatas*, *Manihot esculenta*; **Chinese a.** *Nelumbo nucifera*; **East Indian a.** *C. angustifolia*; **Fiji a.** *T. leontopetaloides*; **Florida a.** *Zamia pumila*; **Guyana a.** *Dioscorea alata*; **Hawaiian a.** *T. leontopetaloides*; **Indian a.** *C. angustifolia*; **Japanese a.** *Pueraria montana* var. *thomsonii*; **marble a.** *Myrosma cannifolia*; **Pará a.** *Manihot esculenta*; **Portland a.** *Arum maculatum*; **Queensland a.** *Canna indica*; **Rio a.** *M. esculenta*; **Tahiti a.** *T. leontopetaloides*

Arrowsmithia DC. (~ *Macowania*). Compositae (Gnap.). 1 E Cape: ***A. styphelioides*** DC. R: OB 104(1991)51

arrow-weed *Tessaria integrifolia*

arrow-wood *Viburnum acerifolium*, *V. dentatum*

arsesmart *Persicaria hydropiper*

Artabotrys R. Br. Annonaceae (III 2). 100+ OW trop. (Afr. 31; Aus. 1). Roux's Model. Alks. Some ed. fr. ***A. hexapetalus*** (L.f.) Bhand. (*A. odoratissimus*, *A. uncinatus*, Ind., Sri Lanka) – plagiotropic shoots grow into veg. shoots, recurved hooks (thorns) or sympodial shoots with hooks & fls, with more thorns produced in shade, cult orn., a stimulant tea made from aromatic fls

Artanacetum (Rzazade) Rzazade = *Artemisia*

Artanema D. Don. Linderniaceae (Plantaginaceae Grat.-Lind.; Scrophulariaceae s.l.). 4 trop. OW. R: Willd. 43(2013)220

Artedia L. Umbelliferae (III 3). 1 E Med.: ***A. squamata*** L. R: Fl. Iraq 5,2(2013)258

Artemisia Tourn. ex L. Compositae (Anth.-Art.). Incl. *Crossostephium* (R: BBNMHB 23(1993)120), *Neopallasia*, *Picrothamnus*, *Seriphidium* – all fls bisex., *Sphaeromeria*, 540 N temp. (Eur. 55; China 170; N Am. 50), Pac. Is., W S Am., S Afr. (1); usu. dry areas. R: BBMNHB 23(1993)117,120 (subg. *Tridentatae*: SBM 89(2009)30). Aromatic shrubs & herbs (ragweeds) charac. of Russian steppes & US 'sage-brush': vermifuges, stimulants, culinary herbs, cult. orn. Capitula small, wind-poll., causing hay-fever problems (in US 1 sq. mile emits 16 t in 2 wks (Aug.–Sept.). ***A. abrotanum*** L. (southernwood, lad's love, old man, E Turkey, long cult. S Eur.) – v. rarely flowering, medic., also tea, form. for deterring moths from clothes; ***A. absinthium*** L. (absinthe, wormwood, temp. Euras., N Afr., natur. N Am.) – a ketone (thujone – similar to THC in cannabis) medic., disinfectant, pl.-pest control since time of Pliny (AD 77) & absinthe (liqueur harmful prob. due to methanol impurities), digestive used in Pernod 1797 (recipe from a Swiss Dr. Ordinaire), until recently banned (when anise used); ***A. afra*** Jacq. ex Willd. (trop. Afr. to E Cape) – local medic.; ***A. annua*** L. (qinghao, Euras., natur. N Am.) – efficacious antimalarial (huanghuahaosu, qinghaosu) in China (no patent as developed during Cultural Revolution) & med. trials, due to artemisinin (sesquiterpene lactone in comm. drugs, in erythrocytes with Fe releasing peroxides lethal to *Plasmodium falciparum*; analogues now synth. using yeast), lvs used for burns, grafting-stock for chrysanthemums; ***A. arborescens*** L. (Medit.) – hedge-pl. in SE Aus. (natur.), NZ; ***A. argyi*** H. Lév. & Vaniot (E As.) – flavouring, medic., used in moxibustion in China & anc. Japan; ***A. chinensis*** L. (*C. c.*, *C. artemisioides*, E As.) – hairs from young lvs coll. China used like cotton wool (moxa), local medic., smoke used as insecticide; ***A. cina*** Berg ex Polj. (Levant wormseed, C As. to China) – medic., anthelminthic (santonin) known as santonica; ***A. dracunculoides*** Pursh (Russian tarragon, N Am.) – rank flavour but oft. sold in Aus. as *A. dracunculus*; ***A. dracunculus*** L. (tarragon, estragon, S & E former USSR, natur. N Am.) – lvs for flavouring (terpineol), esp. sauce tartare & t. vinegar used with fish, like ***A. frigida*** Willd. & ***A. ludoviciana*** Nutt. (N. Am.) – imp. local medic. N Am.; ***A. glacialis*** L. & ***A. mutellina*** Villars (*A. laxa*, *A. umbelliformis*, Alps) – flavouring for genépi liqueur; ***A. herba-alba*** Asso (Med. to Himal.) – wormwood of Bible; ***A. lactiflora*** Wall. ex DC. (China) – cult. orn. herb. perenn.; ***A. maritima*** L. (*S. m.*, sea wormwood, Euras.) – source of santonin (vermifuge); ***A. mexicana*** Willd. ex Spreng. (Mex.) – used by Aztecs for throat-sores, dandruff etc.; ***A. norvegica*** Fries (3 summits in N Scotland, Norway, Urals); ***A. pontica*** L. (Roman wormwood, SE Eur., natur. N Am.) – flavour of vermouth; ***A. princeps*** Pamp. (E As.) – lvs ed. esp. in rice dumplings in Jap.; ***A. tilesii*** Ledeb. (Arctic) – medic. (Inuit), properties like codeine; ***A. tridentata*** Nutt. (*S. t.*, SW US) – sage-brush dominant, imp. local medic.; ***A. vulgaris*** L. (mugwort [OE *mucg* = midge], N temp., natur. Aus.) – tundra veg. of Ice Age GB (now only anthropogenic), one of oldest herbs (terpenoids & sesquiterpene lactones) known worn on Tynwald Day in Isle of Man, lvs a condiment, improving digestion, relieving depression, restoring menstrual flow & easing child-delivery, smoked as tobacco subs. (docko) in Berks to late C19, also used in yomogi soba noodles, magic & superstition (GB until C19) & against gangrene in horses

Artemisiastrum Rydb. = *Artemisia*

Artemisiella Ghafoor. Compositae (Anth.-Art.). 1 Himal. to China: ***A. stracheyi*** (C.B.Clarke) Ghafoor. R: Cand. 47(1992)636

Artemisiopsis S. Moore. Compositae (Gnap.). 1 S trop. Afr.: ***A. villosa*** (O. Hoffm.) Schweick. R: OB 104(1991)55

Arthraerva (Kuntze) Schinz. Amaranthaceae (I 2). 1 SW Afr.: ***A. leubnitziae*** (Kuntze) Schinz. R: U. Eggli, *Ill. Handb. Succ. Pls*, Dicots (2002)6

Arthragrostis Lazarides (~ *Panicum*). Gramineae (XXV 6). 4 Aus. R: Nuytsia 5(1984)285

Arthraxon P. Beauv. Gramineae (XXII). 27 warm As. (China 12) to Aus. (2, 3 also pantrop.). R: Blumea 27(1981)255. ***A. hispidus*** (Thunb.) Makino (*A. ciliaris*, trop. & warm OW, natur. Am.) – yellow dye, medic. in China

Arthrobotrya J. Sm. = *Teratophyllum*

Arthrocarpum Balf. f. = *Chapmannia*

Arthrocereus A. Berger. Cactaceae (III 4). 4 W & SE Braz. Cult. orn.

Arthrochilus F. Muell. (~ *Spiculaea*). Orchidaceae (IV 3e). 15 Aus. (esp.), NG. R: PRSQ 86(1975)155

Arthroclianthus Baill. Leguminosae (III 19). c. 30 New Caled., 1 ext. to Vanuatu

Arthrocnemum Moq. Amaranthaceae (Chenopodiaceae II 2). Excl. *Sarcocornia*, 2 Euras. (1), N & C Am. (1). Alks. Some ed.

Arthromeris (T. Moore) J. Sm. Polypodiaceae (II). 9 N Ind. to S China & (?) Borneo

Arthrophyllum Blume = *Polyscias*

Arthrophytum Schrenk. Amaranthaceae (Chenopodiaceae III 3). 9 W & C As. ***A. arborescens*** Litv. (Turkestan) – valuable source of firewood, cult. in Sahara

Arthropodium R. Br. Asparagaceae (Anthericaceae). (Incl. *Dichopogon*) c. 16 Madag. (1), Aus. (12), NZ (2), New Caled. (1). Cult. orn. (vanilla lilies, i.e. scented thus) esp. ***A. cirrhatum*** (Forst.f.) R. Br. (rock-lily, NZ); ***A. milleflorum*** (DC.) J.F. Macbr. & ***A. minus*** R. Br. (Aus.) – ed. tubers

Arthropogon Nees. Gramineae (XXIII 1). Incl. *Achaena, Altoparadisium*, 6 Braz., WI. R: AMBG 88(2001)366

Arthropteridaceae H.M. Liu & al. = Tectariaceae

Arthropteris J. Sm. ex Hook.f. Tectariaceae. Incl. *Psammiosorus*, 12–15 OW trop. (esp. Madag. & NG) to NZ & Juan Fernandez (1). Form. in Oleandraceae. Climbers; cult. orn.

Arthrosamanea Britton & Rose = *Albizia*

Arthrosolen C. Meyer = *Gnidia*

Arthrostemma Pavón ex D. Don. Melastomataceae. 4 trop. Am. Cult. orn.

Arthrostylidium Rupr. Gramineae (V 2). 32 trop. Am. R: CUSNH 39(2000)13. Heterogeneous. Cli. & scrambling bamboos, some prickly

Arthrostylis R. Br. Cyperaceae (I 9). 1 trop. Aus.: ***A. aphylla*** R. Br. R: FR 53(1944)192

Artia Guillaumin. Apocynaceae (II e). 4 China, SE As., New Caled.

artichoke, Chinese or **Japanese** *Stachys affinis*; **common, globe** or **French a.** *Cynara cardunculus*; **Jerusalem a.** *Helianthus tuberosus*

artillery plant *Pilea microphylla*

Artocarpus Forst. & Forst. f. Moraceae (II). Incl. *Prainea*, 62 Indomal. (Mal. 32 – R: FM I, 17(2006)71) to Aus. R: JAA 40(1959)113, 298, 327, 41(1960)73, 111; Blumea 22(1975)409; SB 35(2010)777. Timber & fr. trees (Rauh's Model), thick white latex used as a bird-lime; lvs simple, lobed or pinnate; stip. usu. large, conical, covering buds & leaving scars; monoec. with fls in heads, females swelling to become (oft. huge; poss. assoc. with former megafaunal disp.) fr. with large seeds embedded in head & covered with waxy or pulpy layer (enlarged K-tube), strap-shaped parts between seeds being undeveloped female fls. ***A. altilis*** (Parkinson) Fosb. (*A. communis, A. incisus*, breadfruit, C Mal. to Melanesia (seeded forms), most Melanesian & Polynesian ones derived from '*A. camansi* Blanco' (NG, where seeds ed.), Micronesian from hybrids with ***A. mariannensis*** Trécul (Palau, Marianas), also domesticated) – greatest diversity of clones in Carolines & E Polynesia, cult. esp. Pac. for fr. (starchy, rich in Vitamin B with some A & C), cvs usu. seedless, orig. introd. from Bligh's 2nd voyage (1793); ***A. chama*** Buch.-Ham. (*A. chaplasha*, chaplash, Ind., Burma, Andamans, Nicobars) – timber; ***A. elasticus*** Reinw. ex Blume (SE As., W Mal.) – bark-cloth, lvs to 2 m in sapling stages; ***A. heterophyllus*** Lam. (jak, trop. As., ? native S Ind.) – poll. flies breed in male infls (cf. *Ficus*), fast-growing (15 m in 3 yrs), huge fr. (barrel-shaped, to 90 cm long & 40 kg) on major branches or trunk, ed. raw or cooked

(11 main cvs), seeds ed. (jak-nuts), timber excellent for furniture & fuel (wood chips form. exclusive source of dye for Buddhist robes in SE As.), local medic., hybrids with *A. integer* incl. '**Cheena**' cult.; ***A. hirsutus*** Lam. (S Ind.) – boat-building; ***A. integer*** (Thunb.) Merr. (chempedak, Mal.) – poll. gall-midges attracted by a fungus in ♂ fls, where eggs laid (larvae feeding on mycelium) & by ♀s because of similar smell, fr. ed., seeds boiled; ***A. lacucha*** Roxb. ex Buch.-Ham. (*A. lakucha*, lacoocha, Ind. to S China & NG) – ed. fr., timber good, seeds ed. fodder tree in Nepal; ***A. nitidus*** Trécul (Indomal.) – lvs used as plates in Philippines; ***A. nobilis*** Thw. (Sri Lanka) – lvs crenate, seeds ed. boiled or fried; ***A. odoratissimus*** Blanco (marang, terap, Borneo, (? natur.) Philipp.) – ed., like jak; ***A. tamaran*** Becc. (Borneo) – loincloths

Artorima Dressler & G. Pollard. Orchidaceae (V 13b). 1 Mex.: ***A. erubescens*** (Lindl.) Dressler & G. Pollard. R: C.L. Withner, *The Cattleyas & their Relatives* V(1998)13

Artrolobium Desv. = *Coronilla*

arugula *Eruca vesicaria*

Arum Tourn. ex L. Araceae (VII 23). 29 Eur. (10), Medit. R: P. Boyce (1993) *The genus A.*; Aroideana 29(2006)132 Alks. Cult. orn. incl. ***A. maculatum*** L. (cuckoo-pint, i.e. lively penis (Anglo-Saxon; poss. Shakespeare's 'long purples'), jack-in-the-pulpit, lords-&-ladies, wake-robin, Eur.) – Chamberlain's Model, self-incompatible, spadix generates high temp. & stench attractive to flies, principally owl-midges, which are imprisoned by hairs (rudimentary male fls) at mouth of spathe, until male fls below them mature & hairs wither, allowing flies dusted with pollen to pass out to another infl. with mature female fls, tubers leafless in first yr., ed. cooked & source of Portland arrowroot (also ***A. italicum*** Mill. (W Eur., natur. Arg.)), also form. used to starch linen (cypress powder) & esp. ruffs, since time of Dioscorides used in bonesetting (as in Crete today); ***A. rupicola*** Boiss. (*A. conophalloides*, E Med., W As.) – poll. blood-sucking insects

arum-lily *Zantedeschia aethiopica*; **yellow a.-l.** *Z. elliottiana*

Aruncus L. Rosaceae (Ros.-Spir.). 1 (variable) N temp. & subarctic (incl. Eur.): ***A. dioicus*** (Walter) Fern. – cult. orn. (goat's-beard), local medic. (N Am.). R: FNA 9(2014) 422

Arundina Blume. Orchidaceae (V 10a). 1 Himal. to Pac.: ***A. graminifolia*** (D. Don) Hochr., to 3 m, cult. orn., invasive Carib., Hawaii. R: OB 89(1986)16

Arundinaria Michaux. Gramineae (IV). Excl. *Bashania, Kuruna, Sarcocalamus*, 3 US. ***A. gigantea*** (Walter) Muehlenb. (giant or switch cane) – ed. young shoots, fr. eaten by Native Americans & early settlers. Many cult. orn. spp. trad. referred here now seg. in other genera, e.g. *Fargesia, Oldeania, Pseudosasa, Sasa, Thamnocalamus*

Arundinella Raddi. Gramineae (XXII 1). c. 50 warm (China 20, 8 endemic – R: FOC 22(2006)564; Mal. 7). R: KB 10(1955)377, CJB 45(1967)1047

Arundo Scheuchz. ex L. Gramineae (Arund.). 5 Medit. (Eur. 3 + *A. donax* L. natur.), Taiwan. R: T 61(2012)1222. ***A. donax*** L. (Med., invasive in S Eur., Azores, S. Afr., Aus., NZ, W US) – the 'reed shaken by the wind' of the Bible, used for 5000 yrs for pipe instruments (orig. Pan pipes), the reed for oboes, clarinets (& other double-beating reeds) & organ-pipes (Spanish cane or reed, though French deemed best); stems used by Romans for garden-fence posts, still for walking-sticks, fishing-rods, also source of cellulose in rayon-making & poss. useful for paper

Arundoclaytonia Davidse & R. Ellis. Gramineae (XIX). 1 Amazonas, Braz.: ***A. dissimilis*** Davidse & R. Ellis

arvi *Colocasia esculenta*

Arytera Blume. Sapindaceae. 28 Indomal. to E Aus. (10, 8 endemic). R: Blumea suppl. 9(1995)149. Wood for tool-handles

Asaemia (Harv.) Benth. (~ *Athanasia*). Compositae (Anth.-Urs.). 1 S Afr.: ***A. minuta*** (L.f.) Bremer (*Athanasia m.*). R: Strel. 9(2000)306

asafoetida *Ferula* spp. (q.v.)

asamela *Pericopsis elata*

asanfona *Pouteria cuspidata*

Asanthus R. King & H. Robinson = *Steviopsis*

asarabacca *Asarum europaeum*

Asaraceae Vent. = Aristolochiaceae

Asarca Lindl. = *Chloraea*

Asarina Mill. Plantaginaceae (Plant.-Ant.; Scrophulariaceae s.l.). Excl. *Lophospermum, Maurandya*, 1 SW Eur.: ***A. procumbens*** Mill. cult. orn.

Asarum L. Aristolochiaceae (I; Asaraceae). Incl. *Hexastylis*, c. 80 N temp. (Eur. 1; Jap. 30; N Am. c. 15). Myrmechory in N Am. spp. Dark brown resiny-scented fls visited by flies, fungus-gnats etc. In latter, found in 8 sects. of genus, there are fungus-like appendages & gnats lay eggs but larvae cannot eat tissue. Cult. orn. (many cvs Jap.) & medic., figured in designs of C6 Jap. buildings. Cult. spp. incl. ***A. canadense*** L. (wild ginger, E N Am.) – ginger subs., rhiz. medic., ***A. caulescens*** Maxim. (aoi, Jap.) – revered at Kamo shrine, Kyoto, & involved in imperial revival of sumo, & ***A. europaeum*** L. (asarabacca, Eur.) – form. medic. esp. after alcoholic excess (emetic), constituent of some snuffs

Ascarina Forst. & Forst. f. Chloranthaceae. 10–11 Madag. (1), Mal., Polynesia, NZ. Stone's & Scarrone's Models

Ascarinopsis Humbert & Capuron = *Ascarina*

Aschenbornia Schauer = *Calea*

Aschersoniodoxa Gilg & Muschler. Cruciferae (27). 4 Andes. R: KB 67(2012)483

Aschistanthera C. Hansen. Melastomataceae. 1 Vietnam: ***A. cristanthera*** C. Hansen. R: NJB 7(1987)653

Asciadium Griseb. Umbelliferae (I?). 1 Cuba: ***A. coronopifolium*** Griseb.

Ascidieria Seidenf. (~ *Eria*). Orchidaceae (V 15). 8 Indomal.

Ascidiogyne Cuatrec. Compositae (Eup.-Ager.). 2 Peru. R: MSBMBG 22(1987)152

Asclepiadaceae Medik. ex Borkh. = Apocynaceae

Asclepias Tourn. ex L. Apocynaceae (V c; Asclepiadaceae III 1). Incl. *Odontostelma, Trachycalymma*, excl. *Gomphocarpus*, c. 120 N (R: AMBG 41(1954)1, 261) & C Am., esp. US (some natur. OW (Eur. 1: ***A. syriaca*** L., E N Am.!)), 80 trop. (38) & S Afr. Cult. orn. (milkweed, silkweed), shoots form. cooked & local medic. N Am., rubber (milkweed r.), some weeds; cardiac glycosides but Monarch butterfly immune & becoming toxic to jays after feeding on *A.* ***A. crispa*** Bergius (S Afr.) – cardiac stimulant in Cape; ***A. curassavica*** L. (bloodflower, swallow-wort, matac, Ind. r., bastard ipecacuanha, S Am., now pantrop. weed) – cult. orn., medic.; ***A. eriocarpa*** Benth. (Calif.) – fibre for cordage, latex for chewing-gum (Indians); ***A. speciosa*** Torrey (Canada, W US) – roadside weed, 75% latex is a refinable oil, rest suitable for food & cosmetics industries; ***A. syriaca*** form. cult. for fibre (stems), bee-forage & seed-floss (made up into veg. silk), green fr. inverted & sold to tourists in Switzerland (budgerigar flowers); ***A. tuberosa*** L. (butterfly-weed, pleurisy-r., N Am.) – medic.

Asclepiodora A. Gray = *Asclepias*

Ascocentropsis Senghas & Schild. = *Vanda*

Ascocentrum Schltr. ex J.J. Sm. = *Vanda*

Ascochilopsis Carr = *Grosourdya*

Ascochilus Ridl. = *Grosourdya*

Ascoglossum Schltr. = *Renanthera*

Ascogrammitis Sundue (~ *Terpsichore*). Polypodiaceae (Gram.). 17 trop. Am. R: Britt. 62(2010)361

Ascolabium Ying = *Vanda*

Ascolepis Nees ex Steud. = *Cyperus*

Ascopholis C. Fischer (~ *Cyperus*). Cyperaceae (II 5). 1 S Ind.: ***A. gamblei*** C. Fischer

Ascotainia Ridl. = *Tainia*

Ascotheca Heine. Acanthaceae (III 2c). 1 trop. W Afr.: ***A. paucinervia*** (C.B. Clarke) Heine

Ascyrum L. = *Hypericum*

Asemanthia (Stapf) Ridl. = *Mussaenda*

Asemeia Raf. (~ *Polygala*). Polygalaceae. 25–30 S Am. to US (1)

Asemnantha Hook.f. Rubiaceae (III 4). 1 Mex.: ***A. pubescens*** Hook.f.

Asepalum Marais. Orobanchaceae (Cyclocheilaceae). 1 NE & E Afr.: ***A. eriantherum*** (Vatke) Marais – shrub. R: KB 35(1981)809

ash *Fraxinus* spp., esp. *F. excelsior* (**common, Eur., French, Polish** or **Slavonian a.**); **alpine a.** *F. nigra, Eucalyptus delegatensis* (Aus.); **Am., Can.** or **white a.** *F. americana, F. nigra, F. pennsylvanica*; **black** or **brown a.** *F. nigra*; **blue a.** *F. quadrangulata*; **Blue Mt. a.** *E. oreades*; **blueberry a.** *Elaeocarpus obovatus*; **bumpy a.** *Flindersia schottiana*; **Canary a.** *Beilschmiedia bancroftii*; **Cape a.** *Ekebergia capensis*; **claret a.** *Fraxinus angustifolia* 'Raywood'; **crow's a.** *Flindersia australis*; **flowering a.** *Fraxinus ornus*; **Japanese a.** *F. nigra* subsp. *mandshurica*; **manna a.** *F. ornus*; **mountain a.** *Sorbus aucuparia, Eucalyptus regnans* (Aus.); **Oregon a.** *F. latifolia*; **prickly a.** *Orites excelsa*; **a. pumpkin** (or **gourd**) *Benincasa hispida*; **red a.** *F. pennsylvanica, Alphitonia* spp. (Aus.); **silver a.** *Flindersia schottiana*; **silvertop a.** *E. sieberi*;

weeping a. *F. excelsior* 'Pendula'; **white a.** *Fraxinus americana, E. fraxinoides* (Aus.); **yellow a.** *Cladrastis kentukea, Emmenosperma alphitonioides, Eucalyptus luehmanniana*

Ashanti blood *Mussaenda erythrophylla;* **A. pepper** *Piper guineense*

ashok *Saraca asoca*

ashplant *Leucophyllum frutescens*

Ashtonia Airy Shaw. Phyllanthaceae (Scep.; Euphorbiaceae s.l.). 2 Malay Pen., Borneo

ashwagandha *Withania somnifera*

Asian mint *Persicaria odorata*

Asiasarum F. Maek. = *Asarum*

Asimina Adans. Annonaceae (III 5). 8 E N Am. (Eocene of London). R: Britt. 12(1960)241. Alks. Natural hybrids with *Deeringothamnus* spp. ***A. reticulata*** Shuttlew. ex Chapman (dog apple, Seminole tea, Florida) – medic. tea; ***A. triloba*** (L.) Dunal (papaw, pawpaw, E US) – Troll's Model, suckering so clones anc., fr. (largest native US sp., to c. 400 g, poss. megafaunal disp. agents 'lost') ed. (can be baked in a 'bran bread') & as nutritious as banana, buds naked but prot. by rusty indumentum, yeasty-smelling fls attract pollinators, few insect pests but zebra swallowtail butterfly (*Protographium marcellus*) utilizes plant's acetogenins (asimicin affecting mitochondrial electron transport system) against its predators (birds)

Asiphonia Griff. = *Thottea*

Askellia W. Weber (~ *Crepis*). Compositae (Cich.-Cich.(-Crep.)). 7 C As., N Am. R: Phytol. 55(1984)6

Asketanthera Woodson. Apocynaceae (II e). 4 trop. Am.

Askidiosperma Steud. Restionaceae. 12 W Cape. R: Bothalia 15(1985)431, NJB 21(2001)195

asna *Terminalia alata*

asoka *Saraca asoca*

Aspalathus L. Leguminosae (III 8). 279 S Afr. esp. SW Cape (273, 258 endemic). R: Strel. 9(2000)466. Oft. heath-like habit, cult. orn. ***A. linearis*** (Burm.f.) R. Dahlgren (*A. contaminatus*, Rooibos tea, W Cape) – tea used by Khoikhoi, free of caffeine & low in tannin, 'Rotbuschtee' imported to Ger.

Asparagaceae Juss. Magnoliidae – Asparagales. Incl. Agavaceae, Anthericaceae (R:KB 51(1996)670), Aphyllanthaceae, Behniaceae, Convallariaceae, Dracaenaceae, Eriospermaceae, Herreriaceae, Hyacinthaceae (R: EJB 80(2004)533), Laxmanniaceae, Lomandraceae (R: KB 51(1996)670), Ruscaceae & Themidaceae (R: T 45(1996)446) 110/2450 subcosmop. incl. arid. Sparsely branched trees (Agavaceae, Dracaenaceae), pachycaul rosette-pls, herbs (s.t. with ann. or woody aerial parts, s.t. prickly; Asparagaceae s.s.) with rhizomes, corms or bulbs, some v. poisonous, or lianes, with vessel elements, freq. with oxalate raphides. Lvs usu. flat, in spirals, rarely distich. or verticillate, s.t. prickly (Agavaceae) or with leaf-like photosynthesizing organs intermediate between lvs & stems ('cladodes', 'phylloclades') s.t. bearing fls & oft. scale lvs (Asparagaceae s.s.), or lvs reduced to scales with scapes photosynthetic (some Anthericaceae, Aphyllanthaceae, Hyacinthaceae). Fls in spikes to thyrses, umbels or solit., bisex. or unisexual; P 3 + 3 free or with basal tube, green to yellow, white, red or blue, A 3 + 3, s.t. united into a column, $\underline{G}$ to $\overline{G}$ (3), (1)3-loc. with axile placentation & 1–12 ovules per loc. Fr. a globose coloured berry or capsule; seeds oft. black, s.t. flattened, occ. winged, some with elaiosomes (?arils)

Classification & principal genera (if subfamilies recog.: **Agavoideae** (*Agave, Anthericum, Chlorophytum, Hosta, Yucca*), **Aphyllanthoideae** (Aphyllanthes), **Asparagoideae** (*Asparagus*), **Brodiaeoideae** (Dichelostemma), **Lomandroideae** (*Cordyline, Lomandra, Thysanotus*), **Nolinoideae** (*Aspidistra, Dracaena, Nolina, Ophiopogon*), **Scilloideae** (4 tribes – *Bellevalia, Drimia, Muscari, Ornithogalum, Scilla*))

Finely drawn fams incl. here 'difficult to distinguish' (Stevens), former Agavaceae, Aphyllanthaceae, Hyacinthaceae & Themidaceae comprising a basal grouping but not clearly morphologically characterizable

Imp. fibre-pls (*Agave* (also beverages), *Furcraea*) & veg. (*Asparagus, Camassia*, also *Ornithogalum, Ruscus*); soap subs. (*Agave, Beschorneria, Chlorogalum*). Many with showy fls etc. so imp. comm. cult. orn. (*Agave, Anthericum, Arthropodium, Asparagus, Aspidistra, Bellevalia, Beschorneria, Brodiaea, Camassia, Chlorophytum, Convallaria, Cordyline, Danae, Dasylirion, Dracaena, Hosta, Hyacinthoides, Hyacinthus, Liriope, Maianthemum, Muscari, Nolina, Ornithogalum, Ophiopogon, Paradisea, Polygonatum, Puschkinia, Rohdea, Ruscus, Scilla, Semele, Thysanotus, Veltheimia, Yucca*)

Asparagopsis (Kunth) Kunth = *Asparagus*

asparagus *Asparagus officinalis;* **Bath a.** *Ornithogalum pyrenaicum;* **a. bean** *Vigna unguiculata;* **Cape a.** *Aponogeton distachyos;* **a. fern** *Asparagus setaceus;* **a. lettuce** *Lactuca sativa;* **a. pea** *Lotus tetragonolobus;* **white a.** *A. albus*

Asparagus Tourn. ex L. Asparagaceae. Incl. *Protasparagus*, c. 120 OW excl. Australasia. R (S Afr. incl. *Myrsiphyllum*): Bothalia 9(1966)31, 15(1984)77, 25(1995)205. Usu. rhizomes with photosynthetic infls, s.t. with longer-lived woody, thorny stems, some with molluscicidal steroid saponins. Bird-disp. fr., so many invasive. Cult. orn. & ed. (young shoots). ***A. acutifolius*** L. (Medit.) – shoots coll. from wild in Spain; ***A. aethiopicus*** L. '**Sprengeri**' (S Afr., natur. elsewhere) – cult. orn.; ***A. albus*** L. (white a., W Med.) – ed. shoots; ***A. asparagoides*** (L.) Druce (*M. a.*, bridal creeper, trop. & S Afr., invasive Aus., NZ) – the smilax of florists; ***A. cochinchinensis*** (Lour.) Merr. (SE As.) – raw herb used in treatment of whooping cough in modern Chinese herbalism; ***A. densiflorus*** (Kunth) Jessop (S Afr.) – invasive in Aus., SE US; ***A. officinalis*** L. (garden a., Eur. to N Afr., invasive in Aus.) – McClure's Model, 2n = 20, 40, a. of commerce (early cvs incl. C18 '**Violet Dutch**'), eaten (though characteristic smell of urine due to break down of sulphurous compounds) fresh & tinned (*Chrysopogon zizanioides* added to enhance flavour), prob. native in S Eur. but cult. since anc. Greece, ground seeds said to be good subs. for coffee, & aerial parts suggested for high-quality paper-making; ***A. prostratus*** Dumort. (*A. o.* subsp. *p.*, Eur. Atlantic coasts) – wild sp. in GB, 2n = 40, ?80; ***A. setaceus*** (Kunth) Jessop (*A. plumosus*, a. fern, trop. & S Afr.) – much used ('fern') by florists in buttonholes etc. though largely replaced by *Rumohria adiantiformis*, molluscicidal steroid saponins

Aspasia Lindl. (~ *Brassia*). Orchidaceae (V 12h). 7 trop. Am. R: Britt. 26(1974)333. Cult. orn. poll. euglossine bees

Aspazoma N.E. Br. = *Mesembryanthemum*

aspen *Populus tremula*, (N Am.) *P. tremuloides*, *P. grandidentata;* **lemon a.** *Acronychia acidula*

Asperella Humb. = *Hystrix*

Asperuginoides Rauschert (*Buchingera*). Cruciferae (inc. sed.). 1 Middle E, C As.: ***A. axillaris*** (Boiss. & Hohen.) Rauschert – glochidiate spines

Asperugo Tourn. ex L. Boraginaceae (B.3.3). 1 Eur.: ***A. procumbens*** L. (madwort, so-called because its r. once a subs. for madder)

Asperula L. (~ *Galium*). Rubiaceae (IV 16). c. 200 Euras. esp. Med. (Eur. 66, Turkey 41), Aus. (16, dioec. & poss. another genus). Cult. orn. incl. ***A. cynanchica*** L. (squinancywort [squinancy = quinsy, tonsilitis], Euras. incl. GB), ***A. tinctoria*** L. (Medit.) – r. source of red dye

asphodel *Asphodelus ramosus;* **bog a.** *Narthecium ossifragum;* **giant a.** *Eremurus* spp.; **Scottish a.** *Tofieldia pusilla;* **yellow a.** *Asphodeline lutea*

Asphodelaceae Juss. Magnoliidae – Asparagales. Incl. Aloeaceae ('subfam. Alooideae' – R: SB 39(2014)66)), Hemerocallidaceae, Xanthorrheaceae, 39/1050 Eur. to C. As., Afr. esp. S Afr., As. to W. Pacific esp. Aus., Andes. Pachycaul treelets to climbers (*Geitonoplesium*) & rhiz. herbs when r. s.t. tuberous, s.t. with vessels (sec. thickening in *Xanthorrhoea*). Lvs narrow to linear, oft. succ., in spirals, s.t. marcescent. Infls spikes to panicles with usu. leafless peduncles (s.t. woody as in *Corynotheca*), app. lat. as stem sympodial. Fls usu. bisex., P 3 + 3 or (3 + 3), oft. somewhat zygomorphic, free or atop a tube, A 3 + 3 (s.t. some staminodes), s.t. with hairs, G (3), (1–)3-loc. with axile placentation & 1–∞ anatr. to orthotropous ovules per loc. Fr. a loculicidal capsule, nut or berry; seeds s.t. arillate, elongate & ovoid to angled (winged in *Eremurus*). x = 7, 11

Classification & principal genera (if subfams recog.): **Asphodeloideae** (*Aloe, Asphodelus, Eremurus, Gasteria, Haworthia, Kniphofia*); **Hemerocallidoideae** (*Hemerocallis, Phormium*); **Xanthorrhoeoideae** (*Xanthorrhoea*)

Superficially resembling 'Hyacinthaceae' (e.g. *Drimia*), i.e. Asparagaceae, & v. similar morphologically to 'Anthericaceae' (also Asparagaceae), such that field distinction oft. unclear, but lacking steroidal saponins though freq. with anthraquinones

Many cult. orn.; Aloe medic. & in cosmetics etc. Fibre (*Phormium*), dyes & gums (*Asphodelus*)

Asphodeline Reichb. Asphodelaceae. 16 Medit. (Eur. 3) to Cauc. R: Candollea 53(1998)423. Cult. orn. incl. ***A. lutea*** (L.) Reichb. (yellow asphodel) – r. ed. like potatoes & ***A. taurica*** (Pallas) Endl. (E Med.)

Asphodelus Tourn. ex L. Asphodelaceae. 16 Medit. (Iberia 13), W As. Cult. orn. incl. ***A. albus*** Mill. – source of alcohol, ***A. fistulosus*** L. – bulbs ed. Bedouin, noxious weed ('onion weed', but see also *Bulbine, Nothoscordum*) in Aus., Calif., Mex.; ***A. ramosus L.*** ('*A.*

aestivus', A. microcarpus, asphodel, Medit.) – r. a source of a yellow dye for carpets (Egypt) & a gum used by bookbinders in Turkey (tchirish), local medic. (skin complaints) Socotra

Aspicarpa Rich. Malpighiaceae. 12 warm Am.

Aspidiaceae Burnett = Tectariaceae

Aspidistra Ker-Gawler. Asparagaceae (Convallariaceae). 90 E As. Mushroom-shaped pistil forms a lid over cavity formed by 6–8-lobed P; amphipod-, not snail-poll. ***A. elatior*** Blume ('*A. lurida*', aspidistra, China) – most cult. orn. pl. in China, v. tolerant housepl. incl. varieg. forms pop. in C19 ('oft. regarded as a symbol of dull, middle-class respectability' – Oxford English Dictionary)

Aspidocarya Hook.f. & Thomson. Menispermaceae (III). 1 NE Ind. to SW China: ***A. uvifera*** Hook.f. & Thomson – fr. ed. R: KB 39(1984)101

Aspidogenia Burret = *Myrcianthes*

Aspidoglossum E. Meyer. Apocynaceae (V c; Asclepiadaceae III 1). 36 trop. & S Afr. R: F. Albers & U. Meve, *Ill. Handb. Succ. Pls*, Asclep. (2002)10

Aspidogyne Garay. Orchidaceae (IV 2d). c. 50 trop. Am. R: Bradea 2(1977)200

Aspidonepsis Nicholas & Goyder (~ *Aspidoglossum*). Apocynaceae (V c; Asclepiadaceae III 1). 5 Drakensberg, S Afr. R: Bothalia 22(1992)24

Aspidophyllum Ulbr. = *Ranunculus*

Aspidopterys A. Juss. ex Endl. Malpighiaceae. 15–20 Indomal. ***A. indica*** (Willd.) Theob. (Ind.) – common cli.

Aspidosperma Mart. & Zucc. Apocynaceae (I a). 70 trop. & S Am. Massart's Model; many alks, some anti-microbial. Good timber (peroba rosa), e.g. ***A. excelsum*** Benth. (paddlewood, locally dominant) for tool-handles etc., ***A. polyneuron*** Muell. Arg. (S & SE Braz., overexploited) & ***A. tomentosum*** Mart. & Zucc. (lemonwood, Braz.), bark (quebracha) used in tanning

Aspidostemon Rohwer & H. Richter (~ *Cryptocarya*). Lauraceae (I). 28 Madag. R: Adans. 28(2006)10

Aspidotis (Hook.) Copel. (~ *Cheilanthes*). Pteridaceae (IV). 4 SW N Am.

Aspilia Thouars = *Wedelia*

Aspiliopsis Greenman = *Podachaenium*

aspirin See *Filipendula, Salix*

Aspleniaceae Newman. Polypodiidae – Polypodiales. 1/c. 700 cosmop. R: T 61(2012)523. Non-arborescent ferns incl. climbers, epiphytes & rheophytes. Rhiz. bearing clathrate scales at apices & petiole-bases; rs blackish, wiry. Fronds usu. monomorphic, simple to multipinnate. s.t. with bulbils; veins pinnate or forking, usu. free. Sori usu. single, elongate, dorsal on the veins, indusium attached to side of vein, usu. long & narrow & with free tapering ends. x = 36, but 38, 39 in '*Hymenasplenium*'

Genus: *Asplenium*

See also Dryopteridaceae, Lomariopsidaceae form. incl. here

Aspleniopsis Mett. ex Kuhn (~ *Austrogramme*). Pteridaceae (III). 3 Mal. to Pacific

Asplenium L. Aspleniaceae. Incl. *Camptosorus, Ceterach, Ceterachopsis, Diellia, Holodictyum, Loxoscaphe, Phyllitis, Pleurosorus, Schaffneria, Sinephropteris*, but *Hymenasplenium* (diff. r. & x no.) poss. a segregate, c. 700 subcosmop. esp. trop. (Eur. 31; 1 of few genera evenly trop. distrib.: 33% As., 22% Afr., 30% Am., 10% Aus. & Pac.; many is. endemics (Madag. 40, Hawaii, Tristan da Cunha, NZ, Réunion etc.); centres (sec.) of diversity in Appalachians, C Am. mts, Andes, Himal.). Some with buds on long whip-like leaf-tips, s.t. on a naked rachis. Hybrids & polyploids common, many local medic. & cult. orn. incl. ***A. adiantum-nigrum*** L. (subcosmop.); ***A. bulbiferum*** G. Forst. (SW Pac.) – viviparous; ***A. ceterach*** L. (rusty-back, Eur.); ***A. marinum*** L. (sea spleenwort, Eur.): ***A. montanum*** L. (mt. s., E US); ***A. nidus*** L. (bird's nest fern, OW trop.) – pioneering epiphyte forming a nest of simple fronds in which leaf-detritus etc. collects & r. grow, some forms with ed. young fronds for cooking, hairwash to improve growth, local medic. & contraceptive in SE As.; ***A. rhizophyllum*** L. (*Camptosorus r.*, N Am) – 'walking fern', cult. orn. with tips of fronds taking r. & thus spreading pl., prothallus drought-resistant; ***A. ruta-muraria*** L. (wall-rue, w. spleenwort, Euras.); ***A. scolopendrium*** L. (Eur., hart's- tongue fern) – simple fronds used in treatment of colds, erysipelas & burns; ***A. trichomanes*** L. (maidenhair spleenwort, subcosmop.) – cough cure ('maidenhair') in Ireland. ***A. acrobryum*** Christ (NG) – salt source (K^+, Na^+, Cl^-)

Asplundia Harling. Cyclanthaceae. c. 100 trop. Am. R: AHB 18(1958)139

Asplundianthus R. King & H. Robinson (~ *Eupatorium*). Compositae (Eup.-Crit.). 10 N Andes. R: MSBMBG 22(1987)346

Asraoa Joseph = *Wallichia*
assacu *Hura crepitans*
assagai wood *Curtisia dentata*
assai (palm) *Euterpe edulis*
Assam indigo *Strobilanthes cusia*
assié *Entandrophragma utile*
Assoella J. Monts. = *Arenaria*
asta *Oxandra lanceolata*
Asta Klotzsch ex O. Schulz. Cruciferae (8). 1 Mex.: ***A. schaffneri*** (S. Watson) O. Schulz. R: CGH 214(1984)19
Astartea DC. Myrtaceae (II 15). 22 SW Aus. R: Nuytsia 23(2013)211
Astelia Banks & Sol. ex R. Br. Asteliaceae (Liliaceae s.l.). Incl. *Collospermum*, c. 30 Masc., NG, Aus., NZ, Polynesia to Hawaii, Chile. Dioecious; some bromeliad-like epiphytes NZ. R: KSVH III, 14(2)(1934)3. Cult. orn. inc. ***A. nervosa*** Hook.f. (bush flax, NZ) – berries ed. Maoris. Some locally useful fibres
Asteliaceae Dumort. Magnoliidae – Asparagales. 3/35 Pac., Masc. Herbs with tuberous rhiz. & saponins, oft. dioec. Lvs in spirals linr. to elliptic, usu. with white to silvery indumentum (poss. water-absorbing). Infl. of bracteate racemes or spikes; P 3 + 3 s.t. with basal tube, dull, A 3 + 3, $\underline{G}$ (3(-7)), (1)3-loc., each loc. with 4 – c. 15 anatr. ovules. Fr. a berry or a globose capsule (*Milligania*); seeds ovate, oft. angular, testa black, endosperm copious
Genera: *Astelia, Milligania, Neoastelia*
Epiphytic forms with rosettes encl. water-collecting cisterns, resembling Bromeliaceae in habit; lvs to 2 m long. App. close to Hypoxidaceae
Astelma Schltr = *Marsdenia*
Astemma Less. = *Monactis*
Astenolobium Nevski = *Astragalus*
Astephania Oliv. = *Anisopappus*
Astephanus R. Br. Apocynaceae (V c; Asclepiadaceae (III 1). 2 S Afr. R: KB 58(2003)878 (Am. spp. (Braz. 7 – R: Bradea 4(1987)377) referable to other genera)
aster (garden or **Chinese)** *Callistephus chinensis*; **Mexican a.** *Cosmos* spp.; **sea a.** *Tripolium pannonicum*
Aster Tourn. ex L. Compositae (Ast.-Ast.). Incl. *Heteropappus, Kalimeris* (R: AMBG 84(1997)773 – polyploid series (x = 9), 2–12-ploid), excl. *Afroaster, Chloracantha, Eurybia, Galatella, Symphyotrichum, Tripolium*, c. 160 Euras. (Eur. 15; Himal. (R: NRBGE 26(1964)67)), N Am. (1). R: Phytol. 77(1994)141. Form. incl. *T. pannonicum* (*A. tripolium*), which prob. received name 'Michaelmas daisy' because of late flowering, the name transferred to the familiar garden pls (now *Symphyotrichum, q.v.*) introd. after 1752, when the adoption of the Gregorian calendar caused Michaelmas Day to fall 11 days earlier to coincide with their flowering: M. daisies (starworts or frost-flowers, US) in 5 major groups, 1 derived from ***A. amellus*** L. (Euras.), e.g. ***A.* × *frikartii*** Silva Tarouca & C. Schneider (*A. a.* × *A. thomsonii* C.B. Clarke (Himal.)), the rest from N Am. *S.* spp. Cult orn. incl. ***A. alpinus*** L. (Pyrenees, Alps; subsp. ***verhappii*** Onno in N Am.) – rock-pl., ***A. hispidus*** Thunb. (*Heteropappus hispidus, H. meyerdorffii*, Jap.)
Asteraceae Martinov. See Compositae
× **Asterago** Everett = × *Solidaster*
Asteranthaceae Knuth = Lecythidaceae
Asteranthe Engl. & Diels. Annonaceae (III 6). 2 trop. E Afr. R: BN 133(1980)53
Asteranthera Hanst. Gesneriaceae (II 4b). 1 Chile: ***A. ovata*** (Cav.) Hanst. – woody cli., cult. orn.
Asteranthos Desf. Lecythidaceae (IV; Asteranthaceae). 1 N Braz.: ***A. brasiliensis*** Desf. – bush, corona of staminal orig., bark purgative
Asteridea Lindl. (~ *Athrixia*). Compositae (Gnap.-Ang.). 9 Aus. R: MBSM 16(1980)129, AusSB 13(2000)739
Asterigeron Rydb. = *Erigeron*
Asteriscium Cham. & Schldl. Umbelliferae (Azor.). 8 Chile, Arg. R: UCPB 33(1962)99
Asteriscus Tourn. ex Mill. ('*Nauplius*' auct.). Compositae (Inul.-Inul). Excl. *Ighermia, Pallenis*, 8 Macar., Med. (Eur. 2.). R: NJB 7(1987)1. Cult. orn. ***A. pygmaeus*** (DC.) Coss. & Durieu (Medit. to C As.) etc. – Leeuwenberg's Model
Asterocarpus Ecklon & Zeyher. Older name for *Pterocelastrus*
Asterochaete Nees = *Carpha*, but perhaps distinct

Asterogyne H. Wendl. ex Hook.f. Palmae (V 11). 5 trop. Am. R: Britt. 55(2003)345. Cult. orn. ***A. martiana*** (H. Wendl.) Hemsl. (C Am.) – crown acts as trap for nutrients which drip down trunk

Asterohyptis Epling. Labiatae (VII 3c). 4 Mex. to Costa Rica. R: Phytoneuron 2011–2:1. Gynodioec.

Asterolasia F. Muell. Rutaceae (I 3). 16 temp. Aus. R: FA 26(2013)416

Asterolinon Hoffsgg. & Link = *Anagallis*

Asteromoea Blume. = *Kalimeris*

Asteromyrtus Schauer (~ *Melaleuca*). Myrtaceae (II 4). 13 N Aus. R: AusSB 1(1988)373. Cult. orn.

Asteropeia Thouars. Asteropeiaceae (Theaceae s.l.). 8 Madag. R: Adansonia III,21(1999)257

Asteropeiaceae Takht. ex Reveal & Hoogl. Magnoliidae – Caryophyllales. 1/8 Madag. Everg. trees or scandent shrubs. Lvs in spirals, entire, shortly petiolate; stip. 0. Fls small in term. or axillary thyrses; pedicels s.t. with 2–8 minute caducous bracts. K5 imbr., basally united; C 5 free, imbr., decid. A9(10-)15, basally connate, folded in bud, anthers versatile; G ((2)3) with 2(-6) apical ovules per carpel. Fr. nut-like, usu. with 1 reniform seed & accrescent K forming a wing; endosperm almost 0, embryo curved with large, thin spirally-coiled cotyledons
Genus: *Asteropeia*
Form in Theaceae (Ericales) but wood anatomy v. diff., DNA placing A. here

Asterophorum Sprague = *Christiana*

Asteropsis Less. (~ *Podocoma*). Compositae (Ast.-Pod.). 1 S Braz., Urug.: ***A. megapotamica*** (Spreng.) Marchesi & al. (*A. macrocephala*). R: Britt. 61(2009)3

Asteropterus Adans. = *Leysera*

Asteropyrum J.R. Drumm. & Hutch. Ranunculaceae (III 3). 2 China

Asterosedum Grulich = *Phedimus*

Asterostemma Decne. Apocynaceae (Va; Asclepiadaceae III 4). 1 Java: ***A. repandum*** Decne

Asterostigma Fischer & C. Meyer. Araceae (VII 4). Excl. *Croatella, Incarum,* 5 Braz.

Asterothamnus Novopokr. (~ *Aster*). Compositae (Ast.-Ast.). 7 C As. to China & Mongolia

Asterotricha V.V. Botsch. = *Fibigia*

Asterotrichion Klotzsch (~ *Plagianthus*). Malvaceae (Malv.-Malv.). 1 Tasmania: ***A. discolor*** (Hook.) Melville – usu. dioec. R: KB 20(1967)512

Asthenatherum Nevski = *Centropodia*

Asthenochloa Buese. Gramineae (XXII 5). 1 C Mal.: ***A. tenera*** Buese

asthma plant, a. weed *Euphorbia hirta, Parietaria judaica*

Astianthus D. Don. Bignoniaceae (1). 1 Mex. to Nicaragua: ***A. viminalis*** (Kunth) Baill.

Astiella Jovet. Rubiaceae (IV 1). 1 Madag.: ***A. delicatula*** Jovet

Astilbe Buch.-Ham. ex D. Don. Saxifragaceae (I 7). c. 23 E As., 1 Appalachians (***A. biternata*** (Vent.) Kearney – dioec. cf. *Aruncus*). Cult. orn. inc. hybrids esp. ***A. × arendsii*** Arends – over 60 cvs. Oft. confused with *Aruncus* & *Filipendula* (Rosaceae) but those have many A & 3 to many sep. pistils

Astilboides Engl. Saxifragaceae (I 4). 1 N China, Korea: ***A. tabularis*** (Hemsl.) Engl.

Astiria Lindl. (~ *Dombeya*). Malvaceae (Bomb.). 1 Mauritius: ***A. rosea*** Lindl. – extinct by 1860s

Astoma DC. = seq.

Astomaea Reichb. Umbelliferae (III 5). 2 E Med. to C As.

Astomatopsis Korovin = *Astomaea*

Astonia S.W.L.Jacobs. Alismataceae. 1 Aus.: ***A. australiensis*** (Aston) S.W.L.Jacobs. R: Telopea 7(1997)14

Astracantha Podlech = *Astragalus*

Astraea Klotzsch (~ *Croton*). Euphorbiaceae (Crot.-Crot.). 10 trop. Am.

Astragalus Tourn. ex L. Leguminosae (III 24). Incl. *Astracantha* (subg. *Tragacantha*), excl. *Erophaca*, c. 2850 mainly N temp. (Eur. 133, Turkey c. 400, China 287; Am. esp. W N Am c. 500), C & W As. ext. to Chile, N Ind. & trop. Afr. mts. R (OW – 2350): D. Podlech & S. Zarre (2013) *Taxonomic.....A. in the OW* (3 vols). Largest seedpl. genus. Common in steppe, prairie etc., oft. ± spiny (goat's-thorn) cushion-forming shrubs, old petiole & rachis becoming woody & leaflets sharp-tipped. Some ed. but some indicators of selenium assoc. with uranium ore e.g. in Utah, & some accumulators toxic to stock, 'locoweeds' (loco = Spanish for crazy) esp. in Midwest US & Canada; source of gum tragacanth (form. 'gum dragon') in W & C As. – tapped from stems or r., hydrophilic &

colloidal properties valuable in icecream, lotions, sizing etc. & in pharmaceuticals for suspending resinous tinctures & heavy insoluble powders, medicated creams, jellies etc.; several spp. but esp. ***A. gummifer*** Lab. (*Astracantha g.*, used since Anc. Greece, & in Iran as gum sarcocolla). Some cult. orn. ***A. brachycalyx*** Fisch. (*A. adscendens*, SW As.) – host to psyllids (*Cyamophila astragalicola*) exuding 'gaz of Khunsar' used in sweets; ***A. canadensis*** L. (N Am) – ed. r.; ***A. crassicarpus*** Nutt. (*A. caryocarpus*, W N Am., buffalo bean) – ed. pods; ***A. fasciculifolius*** Boiss. (Iran) – used as face-gloss by harem women; ***A. glycyphyllos*** L. (milkvetch, fitsroot, Euras.) – fodder & herbal tea; ***A. membranaceus*** Fisch. ex Bunge (*A. mongholicus*, C & E As.) – one of most commonly used herbs (huang qi) in modern Chinese herbalism; ***A. sinicus*** L. (E As.) – green manure for paddyland, medic. See also *Erophaca*

Astranthium Nutt. Compositae (Ast.-Astr.). 11 S US & Mex. R: PMSUB 2(1965)429

Astrantia Tourn. ex L. Umbelliferae (II I). 8 C & S Eur. (5), W As. R: BJ 121(1999)507. Whole infl. 'mimics' a single flower, the florets in *A. major* L. being bisex. but only functionally male at end of season. Cult. orn. esp. ***A. major*** & ***A. maxima*** Pallas (masterworts)

Astrebla F. Muell. ex Benth. Gramineae (XXIX 5). 4 Aus. R: FA 44C(2005)452. *A.* spp. esp. ***A. pectinata*** (Lindl.) Benth. (Mitchell grass) – drought-resistant pasture grasses, imp. food for wild budgerigars, form. Aboriginal grain

Astrephia Dufr. = *Valeriana*

Astridia Dinter. Aizoaceae (V). 12 SW Afr. R: H. E. K. Hartmann, *Ill. Handb. Succ. Pls*, Aiz. A–E (2002)81. Cult. orn.

Astripomoea Meeuse (~ *Ipomoea*). Convolvulaceae (9). 12 trop. & S Afr. Cult. orn.

Astrocalyx Merr. Melastomataceae. 1 Philipp.: ***A. calycina*** (Vidal) Merr. R: Blumea 35(1990)75

Astrocaryum G. Meyer. Palmae (V 8b). 36 trop. Am. (endocarps washed up on NW Eur. beaches). Fr. scatter-hoarded by agoutis, which s.t. peel off pericarp first (? thereby removing injurious invertebrate larvae) in Amaz. Fibre, oil, cult. orn. ***A. aculeatum*** G. Meyer (NE S Am.) – fr. pulp for juice, ice cream etc., ? incl. *A. tucuma* Mart. (tucuma, Braz. etc.) – oil like *Cocos*; ***A. jauary*** Mart. (Amaz.) – fr. used to catch fish though obligately disp. by characin fish; ***A. murumuru*** Mart. (murumuru, Amazonia) – palm kernel oil (also other spp., guere); ***A. vulgare*** Mart. (trop. Am) – found in archaeological sites 11000 BP (*A.j.* & *A.m.* with ***A. chambira*** Burret to 9000 BP), oil, fibre strongest in Amazon, poss. commercially viable (at least in handicrafts, Colombia), vitamin A in concs 3 times those of carrots, incl. ***A. tucumoides*** Drude (awarra, NE S Am.) – lvs used in mats

Astrocasia Robinson & Millsp. Phyllanthaceae (Wiel.-Astro.; Euphorbiaceae s.l.). 6 trop. Am. R: SB 17(1992)311

Astrococcus Benth. Euphorbiaceae (Acal.-Pluk.-Pluk.). 1 Amaz. Braz. & Venez.: ***A. cornutus*** Benth. R: A. Radcliffe-Smith, *Gen. Euphorb.* (2001)243

Astrocodon Fed. = *Campanula*

Astrodaucus Drude (~ *Caucalis*). Umbelliferae (III 3). 2 Euras.

Astrolepis Benham (~ *Cheilanthes*). Pteridaceae (IV). 6 SW N Am.

Astroloba Uitew. (~ *Gasteria*). Asphodelaceae. Incl. *Poellnitzia*, 6 Cape, Karoo. R: SB 39(2014)69. P ± reg. ***A. rubriflora*** (L. Bolus) G.F. Sm. & Manning (*P. r.*, SW Cape) – bird-poll., cult. orn. succ.

Astroloma R. Br. (~ *Leucopogon*). Ericaceae (VII 7; Epacridaceae). 28 Aus. Heterogeneous with 6 monospecific genera to be excl. Cult. orn. incl. ***A. humifusum*** (Cav.) R. Br. – sweet edible fr.

Astronia Blume. Melastomataceae. 59 Indomal. to Pac. (not Aus.). R: Blumea 35(1990)75

Astronidium A. Gray. Melastomataceae. 67 NG, W Pac. to Society Is. R: Blumea 35(1990)115

Astronium Jacq. Anacardiaceae (I). 7+ trop. Am. R: Phytol. 16(1968)112. Like *Rhus* but K winged, enveloping fr. Good timbers (albuerra wood) esp. ***A. fraxinifolium*** Schott (Braz. etc.) – gonçalo-alves, kingwood, locustwood, tigerwood, zebrawood, & ***A. urundeuva*** (Allemão) Engl. (S Am.) – urunday, overexploited; ***A. grande*** Engl. (Braz.) – fls every 5 yrs; ***A. graveolens*** Jacq. (Mex.) – dioec., wind-poll.

Astrophytum Lem. Cactaceae (III 9). Incl. *Digitostigma*, 6 Texas, Mex. Spineless cult. orn. esp. ***A. asterias*** (Zucc.) Lem., endangered in wild through over-collecting but easily raised from seed

Astrostemma Benth. = *Hoya*

Astrothalamus C. Robinson. Urticaceae (III). 1 W & C Mal.: ***A. reticulatus*** (Wedd.) C. Robinson

Astrotricha DC. Araliaceae. c. 20 Aus. (esp. SE, 1 in Hamersley Ranges, NW Aus.)
Astrotrichilia (Harms) Penn. Meliaceae (II). 12 Madag.
Astus Trudgen & Rye. Myrtaceae (II 15). 4 SW Aus. R: Nuytsia 15(2005)502
Astydamia DC. Umbelliferae (III 10). 1 Canaries, Madeira: ***A. latifolia*** (L.f.) Baill.
Astyposanthes Herter = *Stylosanthes*
Asyneuma Griseb. & Schenk. Campanulaceae (I 9). 22–33 Medit. (Eur. 4) to Cauc., E As. (1). R: Boissiera 17(1970)25. Cult. orn.
Asyneumopsis Contandr. & al. = *Asyneuma*
Asystasia Blume. Acanthaceae (III 2c). c. 70 OW trop. ***A. gangetica*** (L.) T. Anderson (pantrop. weed) – thiamine-rich potherb, local medic., cult. orn.
Asystasiella Lindau = *Asystasia*
ata *Lygodium circinnatum*
Atadinus Raf. (*Oreoherzogia*, ~ *Rhamnus*). Rhamnaceae. 9 Euras., Medit. R: T 65(2016)926
Ataenidia Gagnepain = *Marantochloa*
Atalanthus D. Don = *Sonchus*
Atalantia Corr. Serr. Rutaceae (II 2). Incl. *Severinia* (R: W.T. Swingle, *Bot. Citrus* (1943) 274), 16–17 trop. & E As. to Mal. ***A. buxifolia*** (Poir.) Oliv. (*S. b.*, S China & Taiwan) – hedge-pl. & citrus stock; ***A. linearis*** (Blanco) (*S. l.*, Philippines) – rheophyte; ***A. monophylla*** (L.) DC. (Ind.) – fr. an oil source, medic.
Atalaya Blume. Sapindaceae. 12 Afr. to Aus. ***A. hemiglauca*** (F. Muell.) F. Muell. (Aus.) – whitewood, poss. cabinet wood
Atalopteris Maxon & C. Chr. Older name for *Ctenitis*
atamasco lily *Zephyranthes atamasca*
Atamisquea Miers ex Hook. & Arn. Capparaceae. 1 Arizona, Mex., Bolivia, Chile, Arg. semideserts: ***A. emarginata*** Miers ex Hook. & Arn.
Ataxipteris Holttum = *Ctenitis*
Ateixa Ravenna = *Sarcodraba*
Atelanthera Hook.f. & Thomson. Cruciferae (26). 1 C As. to Himal.: ***A. perpusilla*** Hook. f. & Thomson – some anthers monothecous
Ateleia (DC.) Benth. Leguminosae (III 1). 20 trop. Am. R: Webbia 17(1962)153. C 1 (standard). ***A. herbert-smithii*** Pittier (C Am.) – dioec., wind-poll.
Atemnosiphon Leandri = *Gnidia*
atemoya *Annona* × *atemoya*
Ateramnus P. Browne = *Sapium*
Athamanta L. Umbelliferae (III 8). Excl. *Bubon*, 7 Medit. (Eur. 5–6), Macar. R: EJB 58(2001)339. ***A. cretensis*** L. (Candy carrot) – flavouring for liqueurs
Athanasia L. Compositae (Anth.-Ath.). 39 S Afr. esp. Cape. R: OB 106(1991)5. Cult. orn.
athel pine, a. tree *Tamarix aphylla*
Athenaea Sendtner = *Aureliana* (but see BSAB 26(1989)91)
Atherandra Decne. Apocynaceae (III; Asclepiadaceae I). 2 SE As., W Mal.
Atherolepis Hook.f. Apocynaceae (III; Asclepiadaceae I). 3 SE As.
Atherosperma Lab. Atherospermataceae (Monimiaceae I 1). 1 SE Aus.: ***A. moschatum*** Lab. – alks, strongly scented bark used as a tea (Tasmania), v. pale wood for toys etc. R: FA 2(2007)102
Atherospermataceae R. Br. (~ Monimiaceae). 7/16 NG, Aus., New Caled., NZ, Chile. Pls not accum. aluminium. Lvs opp., serrate. Fls small, bisex. or unisexual, oft. 2-merous, in racemes; P sepaloid or petaloid; $\underline{G}$ 4-∞. Fr. a head of plumose nutlets. n = 22, 57
Classification & genera:

1. **Atherospermateae** (leaf-hairs centrifixed, inner staminodes elongating in fr.): *Atherosperma, Laureliopsis*
2. **Laurelieae** (leaf-hairs basifixed, inner staminodes not enlarging): *Daphnandra, Doryphora, Dryadodaphne, Laurelia, Nemuaron*

Fossils from Late Cretaceous/Early Miocene Antarctic; wood known from Upper Eocene of Ger. Timber & spices (*Doryphora, Laurelia*)
Atherostemon Blume = *Atherandra*
Athertonia L. Johnson & B. Briggs. Proteaceae (V 4). 1 Queensland: ***A. diversifolia*** (C.White) L. Johnson & B. Briggs – fr. ed. rats, rat-kangaroos. R: FA 16(1995) 413
Athrixia Ker-Gawler. Compositae (Inul.). 14 trop. & S Afr. R: MBSM 16(1980)46. ***A. phylicoides*** DC. (S Afr.) – twig-brooms in S Afr.; ***A. pinifolia*** N.E. Br. (Natal) – rheophyte
Athroisma DC. Compositae (Ast.-Ath.). 12 OW trop. R: BJLS 119(1995)137

Athroostachys Benth. Gramineae (V 2). 1 E Braz.: ***A. capitata*** (Hook.) Benth., clambering bamboo

Athrotaxis D. Don. Cupressaceae (Taxodiaceae). 2 Tasmania. R: FA 48(1998)570. Timber good (Tasmanian cedar) for general & cabinet work, esp. ***A. selaginoides*** D. Don (King William (Billy) pine). ***A. cupressoides*** D. Don – to 1000 yrs old, poss. useful in dendrochronology; *A.* × ***laxifolia*** Hook – hybrid between *A. s.* ♂ & *A. c.* ♀

Athyana (Griseb.) Radlk. Sapindaceae (IV 1). 1 Paraguay, Arg.: ***A. weinmanniifolia*** (Griseb.) Radlk.

Athyriaceae Alston (~ Woodsiaceae). Polypodiidae – Polypodiales. 5/c. 650 cosmop. R: T: 61(2012)526. Terr. or saxicolous, s.t. rheophytic ferns with usu. unbranched rhiz. bearing lanceolate scales & s.t. golden hairs; rs blackish, wiry. Lvs simple to 2-pinnate-pinnatifid, usu. in spirals, sparsely to somewhat scaly & occ. pubescent, occ. with bulbils; veins free or. s.t. anastomosing. Sori usu. elongate, on top of or along 1 side of vein, solit. or paired, usu. with indusium opening along lateral margin. x = 40, 41
Genera: *Anisocarpium, Athyrium, Cornopteris, Deparia, Diplazium*

Athyrium Roth. Athyriaceae (Dryopteridaceae). Excl. *Anisocarpium*, 220 cosmop. (Eur. 2), esp. E & SE As. Heterogeneous; close to *Diplazium*. Cult. orn. esp. ***A. filix-femina*** (L.) Roth (temp. N & Am., lady fern)

Athysanus Greene. Cruciferae (7). 2 W US. R: FNA 7(2010)267

Atkinsia R. Howard = *Thespesia*

Atkinsonia F. Muell. Loranthaceae (2). 1 Blue Mts, SE Aus.: ***A. ligustrina*** (Lindl.) F. Muell. – terr. shrub to 2 m parasitic on r. of diff. hosts

Atlanthemum Raynaud (~ *Helianthemum*). Cistaceae. I Medit. inc. Eur.: ***A. sanguineum*** (Lag.) Raynaud

Atlantic or **Atlas cedar** *Cedrus atlantica*; **Atlantic ivy** *Hedera hibernica*

Atocion Adans. = *Silene*

Atomostigma Kuntze = *Myrcia* (though described in Rosaceae)

Atopostema Boutique = *Monanthotaxis*

Atractantha McClure. Gramineae (V 2). 6 E Braz. R: CUSNH 39(2000)25, SB 36(2011)310. Cli. bamboos

Atractocarpa Franch. = *Puelia*

Atractocarpus Schltr. & K. Krause (~ *Randia*). Rubiaceae (II 1). Incl. *Neofranciella, Sukunia, Sulitia, Trukia* c. 40 Philippines, SW Pacific to Aus. (7)

Atractogyne Pierre. Rubiaceae (Ix.-Sherb.). 3 W Afr. Petit's Model

Atractylis L. Compositae (Card.-Carl.). 29 Euras., Medit. (Eur. 4). R: BSBF 134 Lettres Bot. 1987(2)179

Atractylodes DC. Compositae (Card.-Carl.). 5 E As. R: BSBF 134 Lettres Bot. 1987(2)179. Imp. medic. pls in China inc. ***A. lancea*** (Thunb.) DC. & ***A. macrocephala*** Koidz. (bai zhu, ingredient of local 'viagra')

Atragene L. = *Clematis*

Atraphaxis L. Polygonaceae (II 4). 25 N Afr., SE Eur. (4) to Himal., E Siberia. Cult. orn., oft. spiny steppe-pls

Atrichantha Hilliard & B.L. Burtt. Compositae (Gnap.-Rel.). 1 SW Cape: ***A. gemmifera*** (L. Bolus) Hilliard & Burtt. R: OB 104(1991)74

Atrichodendron Gagnepain = ? (not Solanaceae)

Atrichoseris A. Gray. Compositae (Cich.-Cich.(-Micr.)). 1 SW US: ***A. platyphylla*** (A. Gray) A. Gray. R: FNA 19(2006)309

Atriplex Tourn. ex L. Amaranthaceae (Atrip.; Chenopodiaceae I 3). Excl. *Halimione, Extriplex*, c. 300 temp. & warm. Many halophytes (C3 & C4): salt-tolerant forage (saltbushes, e.g. ***A. nummularia*** Lindl. (old man s., S Aus).; ***A. repanda*** Philippi (Chile) – to 40 yrs old but overexploited for sheep & goats; ***A. vesicaria*** Heward ex Benth., Aus.). Many ed. Native Americans, 1 widely cult.: ***A. hortensis*** L. ([gold] orache, As.) – lvs used like spinach, some cvs with coloured lvs (green after cooking); ***A. semibaccata*** R. Br. (Aus.) – invasive W US

Atropa L. Solanaceae (1). 3 W Eur. (1) to Himal. R: A.T. Hunziker, *Gen. Solanacearum* (2001)335. ***A. belladonna*** L. (belladonna, deadly nightshade, Eur. to Iran) – all parts poisonous (little atropine but higher in hyoscyamine) to humans (but not rabbits or pigs which can detoxify alks) though fr. said to taste nice (!), used medic. (in comm. cough mixtures etc.), assoc. with Ger. god, Odin. Atropine dilates pupils (form. used by women to make eyes brighter & larger), used as nerve-gas antidote in First Gulf War

Atropanthe Pascher (~ *Scopolia*). Solanaceae (1). 1 China: ***A. sinensis*** (Hemsl.) Pascher. R: A.T. Hunziker, *Gen. Solanacearum* (2001)364

Atroxima Stapf. Polygalaceae (III). 2 W & C Afr. R: MLW 77 – 18(1977)14. Mangenot's Model

atta (flour) *Triticum turgidum* Durum Group

Attalea Kunth. Palmae (V 8a). Incl. *Maximiliana, Orbignya, Scheelea,* 69 trop. Am. R: Fieldiana 38(1977)36, Phytol. 36(1977)89, 37(1977)219. Oft. large, slow-growing palms with valuable palm kernel oil, the most imp. source for indigenous peoples & used in Western aromatherapy & cosmetics (aguassú, babassú, babaçú, babaco) esp. ***A. phalerata*** Mart. ex Spreng. (*O.p., O. barbosiana,* coco de macaco, Braz.) – poll. beetles & s.t. wind, source of coal-like fuel, methanol from endocarp & mesocarp, plastics, animal feed, etc.; ***A. butyracea*** (L.f.) Wess. Boer (C N trop. S Am.) – 'butter'-source, pigfood; ***A. cohune*** Mart. (*O. c.,* cohune nut, C Am.) – source of oil, lvs also ed.; ***A. colenda*** (Cook) Balslev & A.J. Hend. (W Ecuador, SW Colombia) – potential source of lauric acid; ***A. cuatrecasana*** (Dugand) A.J. Hend. & al. (*O. cuatrecasana,* táparos, Colombia) – seeds ed.; ***A. funifera*** Mart. (Bahia piassava, piassaba, coquilla, Braz.) – male when young & increasingly female as it reaches canopy, lvs (bases) sources of piassava fibre for brushes, brooms etc., nuts highly polished & used for carving etc.; ***A. insignis*** (Mart.) Drude (Colombia to N Peru) – in 9000 yr-old archaeological sites; ***A. maripa*** (Aubl.) Mart. (*M. maripa, M. regia,* kokerite, NE S Am., Trinidad) – monoec., cult. orn, oilseeds & thatch; ***A. spectabilis*** Mart. (*O. s.,* curua, Braz.) – oil & thatch

attar of roses See *Rosa*

Atuna Raf. Chrysobalanaceae (4). 8 Indomal., W Pac. (Fiji 2). R: FOW 10(2003)66. Distrib. ocean currents (? & squirrels, pigs). ***A. racemosa*** Raf. subsp. ***excelsa*** (Jack) Prance, *A. excelsa,* W Pac. – local medic. (antibacterial incl. MRSA), seed oil to scent coconut oil in Fiji, hair dressing in Carolines, boat-caulking in Solomons, mashed seeds ed. with fish etc. in Moluccas (cotyledons sold for this in Philippines markets), branches on which dead lvs remain indefinitely used for thatching walls in Fiji

Atylosia Wight & Arn. = *Cajanus*

aubergine *Solanum melongena*

Aubletiana J. Murillo (~ *Conceveiba*). Euphorbiaceae (Alc.-Conc.). 2 W Afr.

Aubregrinia Heine. Sapotaceae (IV). 1 trop. W Afr.: ***A. taiensis*** (Aubrév. & Pellegrin) Heine. R: T.D. Pennington, *S.* (1991)211

Aubrevillea Pellegrin. Leguminosae (II 1). 2 Guineo-Congolian forests

Aubrieta Adans. Cruciferae (7). 12 S Eur. (6) to Iran. R: [Q]BAGS 7(1939)157, 217. Cult. orn. esp. ***A. deltoidea*** (L.) DC. (Sicily to Turkey) & hybrids of obscure parentage, the 'aubrietia' or 'aubretia' of gardens (***A. × cultorum*** Bergmans)

Aucellia Szlach. & Sitko = *Maxillaria*

Aucoumea Pierre. Burseraceae (4). 1 trop. W Afr.: ***A. klaineana*** Pierre (okoumé, Gaboon mahogany) – Rauh's Model, good timber & resin

Aucuba Thunb. Garryaceae (Cornaceae s.l.). c. 8 Himal. to Jap., esp. China. Dark P & prominent nectar-disks like *Garrya,* with which grafts successful. Leeuwenberg's Model. Cult. orn. everg. dioec. shrubs esp. ***A. japonica*** Thunb. (E As., Jap. laurel) – offered in Jap. temples, & its vulgar yellow-spotted virus-infected cv., '**Variegata**'

Aucubaceae J. Agardh = Garryaceae

Audouinia Brongn. Bruniaceae (2). Incl. *Tittmannia,* 5 SW Cape. R: T 60(2011)1145

Auerodendron Urb. Rhamnaceae (6). 7 WI. ***A. pauciflorum*** Alain (Puerto Rico) – reduced to 5 trees in wild

Augea Thunb. Zygophyllaceae (I). 1 S Afr.: ***A. capensis*** Thunb. R: PSE 240(2003)36. Endosperm 0

Augouardia Pellegrin. Leguminosae (I 2). 1 Gabon: ***A. letestui*** Pellegin

Augusta Pohl. Rubiaceae (II 1). 1 E Braz.: ***A. longifolia*** (Spreng.) Rehder

Augustea Iamonico (~ *Polycarpaea*). Caryophyllaceae (I 1). 3 S Am. R: Phytotaxa 236(2015)72

Aulacocalyx Hook.f. Rubiaceae (II 4). 9 trop. Afr. R: KB 52(1997)639

Aulacocarpus O. Berg = *Mouriri*

Aulacolepis Hackel = *Aniselytron*

Aulacophyllum Regel = *Zamia*

Aulacospermum Ledeb. Umbelliferae (III 5). Incl. *Pseudotrachydium,* 19 Euras. (Eur. 1). R: FR 111(2000)526, 528

Aulandra H.J. Lam. Sapotaceae (II). 2 Borneo. Staminal tube. R: Tree Fl. Sabah Sarawak 4(2002)206

Aulax Bergius. Proteaceae (IV 2). 3 S Afr. R: SAJB 53(1987)464. Dioec. cult. orn.

Aulojusticia Lindau = *Justicia*

Aulonemia Goudot (~ *Arthrostylidium*). Gramineae (V 2). c. 40 C & trop. S Am. R: SCB 9(1973)53, AMBG 77(1990)353. Prob. incl. *Colanthelia*. Montane forest bamboos, valuable forage for pack animals

Aulosepalum Garay. Orchidaceae (IV 2h). 7 Mex. to Costa Rica. R: HBML 28(1982)298

Aulosolena Kozo-Polj. = *Sanicula*

Aulospermum J. Coulter & Rose = *Cymopterus*

Aulostylis Schltr. = *Calanthe*

Aulotandra Gagnepain. Zingiberaceae (II 1). 1 trop. W Afr., 5 Madag.

Aunt Lucy *Ellisia nyctelea*

Auranticarpa Cayzer & al. (~ *Pittosporum*). Pittosporaceae. 6 N Aus., 1 ext. to NSW. R: AusSB 13(2000)904

Aureliana Sendt. (~ *Capsicum*). Solanaceae (IV 5a). Incl. *Athenaea*, 12 C S Am. R: BotJLS 177(2015)329

Aureolaria Raf. Orobanchaceae (Gerard.; Scrophulariaceae). 10 E US, 1 Mex. R: PANSP 80(1928)372. Hemiparasitic mostly on Fagaceae, but also Ericaceae. Sometimes cult. orn. incl. ***A. flava*** (L.) Farw. – iridoid (aucubin) sequestered by *Euphydryas phaeton* butterflies

auri *Acacia auriculiformis*

auricula *Primula auricula*, *P.* × *pubescens* (usu.)

Auricülardisia Lundell = *Ardisia*

Aurinia Desv. (~ *Alyssum*). Cruciferae (2). 10 Eur. (9) & W As. Like *Alyssum* but infls usu. axillary & fl. buds spherical. Cult. orn. esp. ***A. saxatilis*** (L.) Desv. (gold-dust, C Eur. to Turkey) – rock-pl.

Aurinocidium Romowicz & Szlach. = *Grandiphyllum*

Austinia Buril & A.R. Simões = *Daustinia*

Australian blackwood *Acacia melanoxylon*; **A. bluebell** *Wahlenbergia gracilis*, **(honey)** *Echium plantagineum*; **A. cedar** *Toona ciliata*; **A. chestnut** *Castanospermum australe*; **A. currant** *Leucopogon* spp.; **A. daisy** *Vittadinia* spp.; **A. fuchsia** *Correa* & *Epacris* spp.; **A. grass-tree** *Xanthorrhoea* spp.; **A. heath** *Epacris* spp.; **A. honeysuckle** *Banksia* spp.; **A. kino** *Eucalyptus camaldulensis*; **A. maple** *Flindersia brayleyana*; **A. mint** *Prostanthera* spp.; **A. pine** *Casuarina equisetifolia*; **A. red cedar** *Toona ciliata*; **A. rosewood** *Alectryon oleifolius*; **A. sandalwood** *Santalum* spp.; **A. walnut** *Endiandra palmerstonii*

Australina Gaudich. Urticaceae (V). 1 SE Aus. & NZ, 1 Ethiopia & Kenya (cf. *Dietes*, *Pelargonium*). R: NJB 8(1988)53

Australluma Plowes = *Caralluma*

Australopyrum (Tzvelev) Löve (~ *Agropyron*). Gramineae (XV). 5 NG E Aus. (3 endemic), NZ. R: Telopea 13(2011)40

Australorchis Brieger = *Dendrobium*

Austrian pine *Pinus nigra*

Austroamericium Hendrych = *Thesium*

Austrobaileya C. White. Austrobaileyaceae. 1 Queensland: ***A. scandens*** C. White – fls smell of rotting fish, pollen like *Clavatipollenites* (Lower Cretaceous), young lvs on fast-growing shoots reflexed & acting as grapples, absence of phloem sieve-tubes now denied. R: FA 2(2007)17

Austrobaileyaceae Croizat. Magnoliidae – Austrobaileyales. 1/1 NE Aus. Everg. liane. Lvs opp. to subopp., simple, conduplicate; stip. 0. Fls large, solit., axillary on leafy shoots, nodding, prob. fly-poll., hypogynous. P 11–23, in compact spiral, sepaloid to petaloid; A6–11, in spirals, laminar, not differentiated into filament & anther, inner reduced to 6+ staminodes, pollen monosulcate; $\underline{G}$ (4-)6–9(-14) free, ± in spirals, style 2-lobed, ovules (4–)6–8(–14) in 2 series along ventral side of each carpel, apotropous, bitegmic. Fr. berry; seeds sarcotestal, embryo v. small with abundant ruminate endosperm, germ. epigeal. 2n = 44

Only sp.: *Austrobaileya scandens*

Austrobassia Ulbr. = *Maireana*

Austrobrickellia R. King & H. Robinson (~ *Eupatorium*). Compositae (Eup.-Alom.) 3 N S Am. R: MSBMBG 22(1987)253

Austrobryonia H. Schaef. Cucurbitaceae (X). 4 W & C Aus. R: SB 33(2008)125

Austrobuxus Miq. Picrodendraceae (Cal.-Diss.; Euphorbiaceae s.l.). c. 25 W Mal. [not NG], Qld (2), New Caled., Fiji

Austrocactus Britton & Rose. Cactaceae (III 1). 3 Arg., Chile. Cult. orn.

Austrocedrus Florin & Boutelje (~ *Libocedrus*). Cupressaceae. 1 S Chile & Arg.: ***A. chilensis*** (D. Don) Pic.-Serm. & Bizzari (*L. chilensis*, Chilean cedar) – Attims's Model. R: A. Farjon, *Monogr. Cupressaceae & Sciadopitys* (2005)456

Austrocephalocereus Backeb. = *Micranthocereus*

Austrochloris Lazarides. Gramineae (XXIX 5). 1 Queensland: ***A. dichanthoides*** (Everist) Lazarides. R: FA 44B(2005)282

Austrocritonia R. King & H. Robinson (~ *Eupatorium*). Compositae (Eup.-Crit.). 4 Braz. R: MSBMBG 22(1987)349

Austrocylindropuntia Backeb. (~ *Opuntia*). Cactaceae (II). 9 N S Am. ***A. floccosa*** (Salm-Dyck) Ritter (mts of Peru, Bolivia) – cushion-forming at 4000 m

Austrocynoglossum M. Popov ex R. Mill = *Hackelia* (but see NRBGE 46(1989)43)

Austrodanthonia Linder = *Rytidosperma*

Austroderia N. Barker & Linder (~ *Cortaderia*). Gramineae (Danth.) 5 NZ. R: AMBG 97(2010)343

Austrodolichos Verdc. Leguminosae (III 18). 1 N Aus.: ***A. errabundus*** (M. Scott) Verdc.

Austrodrimys Doweld = *Tasmannia*

Austroeupatorium R. King & H. Robinson (~ *Eupatorium*). Compositae (Eup.-Eup.). 13 trop. S Am. (esp. Braz.) to Urug. R: MSBMBG 22(1987)67

Austrofestuca (Tzvelev) Alexeev = *Poa*

Austrogambeya Aubrév. & Pellegrin = *Chrysophyllum*

Austrogramme Fournier. Pteridaceae (III). 6 E Mal. to New Caled. R: FG 11(1975)61

Austroliabum H. Robinson & Brettell = *Microliabum*

Austromatthaea L.S. Sm. Monimiaceae (V 2). 1 NE Queensland: ***A. elegans*** L.S. Sm. R: FA 2(2007)70

Austromuellera C. White. Proteaceae. 2 NE Queensland. R: FA 19B(1999)173

Austromyrtus (Niedenzu) Burret. Myrtaceae (II 10). 21 E Aus. (3), New Caled. R: SBM 65(2003)19. ***A. dulcis*** (C. White) L.S. Sm. (midyim, E Aus.) – imp. Aboriginal fr. See also *Gossia, Lenwebbia*

Austropeucedanum Mathias & Constance. Umbelliferae (III 10). 1 NW Arg.: ***A. oreopansil*** (Griseb.) Mathias & Constance

Austrosteenisia Geesink (~ *Lonchocarpus*). Leguminosae (III 16). 4 N Aus., NG. R: Austrobaileya 5(1997)80

Austrostipa S.W.L. Jacobs & J. Everett (~ *Stipa*). Gramineae (X). 63 Aus, 1 ext. to NZ. Pasture grasses. ***A. aphylla*** (Rodway) S.W.L. Jacobs & J. Everett (*S. a.*, SE Tasmania) & ***A. muelleri*** (Tate) S.W.L. Jacobs & J. Everett (*S. m.*, S & SE Aus.) – rudimentary leaf-blades

Austrosynotis C. Jeffrey (~ *Senecio*). Compositae (Sen.). 1 S trop. Afr.: ***A. rectirama*** (Bak.) C. Jeffrey – cli. by basally thickened petioles. R: KB 41(1986)878

Austrotaxaceae Nakai = Taxaceae

Austrotaxus Compton. Taxaceae (Austrotaxaceae). 1 New Caled.: ***A. spicata*** Compton. R: BR 64(1998)299. Rauh's Model

Autana Philbrick. Podostemaceae. 1 Venez.: ***A. andersonii*** Philbrick. R: Novon 21(2011)476

Autonoe (Webb & Berth.) Speta = *Scilla*

Autranella A. Chev. (~ *Mimusops*). Sapotaceae (I 1). 1 W Afr.: ***A. congolensis*** (de Wild.) A. Chev.

Autrania Winkler & Barbey = *Myopordon*

autumn crocus *Colchicum autumnale*; **a. olive** *Elaeagnus umbellata*; **a. snowflake** *Acis autumnalis*

Autumnalia Pim. Umbelliferae (III 10). 2 C As. R: BZ 74(1989)1485

Auxemma Miers = *Cordia*

Auxopus Schltr. Orchidaceae (V 7). 4 trop. Afr. Mycotrophic

avaram bark *Senna auriculata*

Avellanita Philippi. Euphorbiaceae (Acal.-Acal.-Avel.). 1 C Chile: ***A. bustillosii*** Philippi. R: A. Radcliffe-Smith, *Gen. Euphorb.* (2001) 229

Avellara Blanca & Díaz de la Guardia = *Scorzonera*

Avellinia Parl. = *Trisetaria*

Avena Tourn. ex L. Gramineae (XVI 2). 24 Eur., Medit. to Ethiopia. OW. R: B. Baum (1977) *Oats*. Like *Helictotrichon* but ann. & with smooth rounded glumes but *H. macrostachyum* (Cosson & Durieu) Henrard (N Afr.) intermediate; 2 weedy, 6–7 cult. spp. ***A. sativa*** L., cult. oats (24 M t p.a.), hexaploid (maternal parent like diploid *A. wiestii* Steud.), a weed of other cereals, domesticated c. 2000 BC & carried from Medit. to NW Eur., basis of porridge, brose (uncooked; Atholl b. with whisky), oat-cakes, oatmeal soap etc.; other hexaploids incl. ***A. fatua*** L., one of world's worst weeds, & **A.** ***sterilis*** L.– awns twist & untwist in moist surroundings & s.t. used as 'flies' in fishing; ***A. strigosa*** Schreb. (W Medit.) – animal feed, straw for trad. chairbacks (Orkney), little cult.

Avenella Bluff ex Drejer (~ *Deschampsia*). Gramineae. 1 temp. & cold: ***A. flexuosa*** (L.) Drejer (*D. f.*) – bipolar distrib.

Avenochloa Holub = *Helictotrichon*

avens, mountain *Dryas octopetala*; **water a.** *Geum rivale*; **wood** or **yellow a.** *G. urbanum*

Avenula (Dumort.) Dumort. (~ *Helictotrichon*). Gramineae (XVI 7). 30 N temp. esp. Eur.

Averia Leonard = *Tetramerium*

Averrhoa L. Oxalidaceae. 2–4, (?) E Braz. or Mal., cult. pantrop. R: Reinw. 12(2008)325. Lvs irritable. Fr. trees. ***A. bilimbi*** L. (bilimbing, bilimbi, camias, cucumber tree) – Koriba's Model, rather sour fr. used for pickles, jams, jellies etc., borne on trunk & branches; ***A. carambola*** L. (carambola, caramba) – Troll's Model, ribbed, axillary fr. pickled, or sweet forms eaten raw, transverse sections being called star-fr., medic. Juice from both spp. removes stains from hands, clothes & weapons

Averrhoaceae Hutch. = Oxalidaceae

Averrhoidium Baill. Sapindaceae (III 1). 5 Mex. (1), SW Amaz. (2), SE Braz. & Paraguay (2). R: Britt. 54(2002)114

Avetra Perrier = *Trichopus*

Avicennia L. Acanthaceae (IV; Avicenniaceae). 4–7 warm, mangroves. R: P.B. Tomlinson (1986) *Bot. Mangroves*: 186. Attims's (& ?Champagnat's) Model; aerial r. projecting above mud at low tide, viviparous. Some good timber (not splittable radially, only tangentially due to concentric bands of included phloem embedded in conjunctive parenchyma), tan-bark. ***A. marina*** (Forssk.) Vierh. (grey mangrove, OW) – termite-resistant building timber, incl. for boat-building & Aus. Aboriginal shields, brown dye for batik, cotyledons washed up on beaches boiled & eaten by Aus. Aborigines

Avicenniaceae Miq. = Acanthaceae

Aviceps Lindl. = *Satyrium*

Avignon berry *Rhamnus saxatilis*

avocado (pear) *Persea americana*

avodiré *Turraeanthus africanus*

Avonia (Fenzl) Rowley = *Anacampseros*

Avonsera Speta = *Ornithogalum*

awari *Pterygota bequaertii*

awarra (palm) *Astrocaryum vulgare*

awlwort *Subularia aquatica*

awun *Alstonia congensis*

awusa nut *Plukenetia conophora*

axe-breaker *Coatesia paniculata*

Axinaea Ruíz & Pavón. Melastomataceae. 41 trop. Am. R: Scient. Dan. B,Biol. 4(2014). Close to *Meriania*

Axinandra Thwaites. Crypteroniaceae. 2–4 Sri Lanka (1), W Mal.

Axiniphyllum Benth. Compositae (Mill.-Mill). 5 Mex. R: Madroño 25(1978)46, 34(1987)164

Axonopus P. Beauv. Gramineae (XXIII 3). Incl. *Ophiochloa*, 104 trop. & warm Am. (N Am. 3), Afr. (1). R: JAA suppl. 1(1991)291. Lawn-grasses (carpet grass). ***A. affinis*** Chase (trop. Am.) – sown in trop. pastures & ***A. compressus*** (Sw.) P. Beauv. (trop. Am.) – lawn-grass widely natur.; ***A. hydrolithicus*** (Filgueiras & al.) A. López & Morrone (*O. h.*, Braz.) – on serpentine

Axyris L. Amaranthaceae (Axyr.; Chenopodiaceae I 3). 6 E Eur. (1) to Korea. R: Willd. 41(2011)81. Heterocarpy. ***A. amaranthoides*** L. (Euras.) – weedy N Am.

ayahuasca *Banisteriopsis* spp. esp. *B. caapi*

ayama *Cucurbita moschata*

ayan *Distemonanthus benthamianus*

Ayapana Spach (~ *Eupatorium*). Compositae (Eup.-Ayap.). 16 trop. Am. R: MSBMBG 22(1987)195. ***A. triplinervis*** (Vahl) R. King & H. Robinson a medicinal tea, cult. in Braz. (ayapana)

Ayapanopsis R. King & H. Robinson (~ *Eupatorium*). Compositae (Eup.-Ayap.). 17 Andes R: MSBMBG 22(1987)197

aye *Sterculia rhinopetala*

Ayenia L. (~ *Byttneria*). Malvaceae (Bytt.-Bytt.; Sterculiaceae). 70+ S US to Arg. R: Op. Lill. 4(1960)1

Ayensua L.B. Sm. Bromeliaceae (1). 1 Venez.: ***A. uaipanensis*** (Maguire) L.B. Sm.

Aylacophora Cabrera (~ *Nardophyllum*). Compositae (Ast.-Hin.). 1 W Arg.: ***A deserticola*** Cabrera. R: SCB 92(2009)17

Aylostera Speg. (~ *Rebutia*). Cactaceae. 5 NW S Am. R: PJB 43(2011)2777

Aylthonia N. Menezes = *Barbacenia*

Aynia H. Robinson. Compositae (Vern.-Lep.). 1 Peru: ***A. pseudascaricida*** H. Robinson

ayote *Cucurbita moschata*

ayous *Triplochiton scleroxylon*

Aytonia L.f. = *Nymania*

Azadehdelia Braem = *Cribbia*

Azadirachta A. Juss. Meliaceae (II-Mel.). 2 Indomal. R: FM I,12(1995)337. ***A. excelsa*** (Jack) Jacobs (Mal.) – timber, young shoots ed.; ***A indica*** A. Juss. (neem, nim, margosa, prob. native in Myanmar) – widely cult. & natur. OW trop., spread by bats, germ. enhanced by passage through baboon guts, one of world's most useful trees though fls causing allergic reactions in some: fuel crop easily grown on poor soils, shade-tree (50 000 in plains of Saudi Arabia for ann. camp of 2 M Muslim Haj pilgrims), fodder, ed. fls & lvs, timber a mahogany subs., used in soaps, toothpaste, lotions, honey from C Ind. sold US, seeds an oil source, lvs (incorporated in nests, ? reducing parasite loads, by house-sparrows in Ind.) & seeds provide a field insecticide (oil rich in azadirachtin the most potent insect antifeedant known: 10 p.p.m. lethal to many Lepidoptera), medic. (sadao [Thailand], comm.; skin diseases such as scabies; poss. antimalarial a limonoid, gedunin, efficacious), postcoital contraceptive & for arthritis treatment) etc. etc. Nat. Research Council (1992) *Neem – a tree for solving global problems*

azalea *Rhododendron* spp. & hybrids (decid.); **alpine** or **trailing a.** *Kalmia procumbens*; **kurume a.** *R.* × *obtusum*; **spider a.** *R. stenopetalum*

Azalea L. = *Rhododendron*

Azanza Alef. = *Thespesia*

Azara Ruíz & Pavón. Salicaceae (Flacourtiaceae). 10 S Am. R: BJ 98(1977)151. Lvs in spirals but 1 stipule oft. almost as large giving appearance of opp. lvs, ***A. microphylla*** Hook.f. (Chile, Arg.) with leaf, smaller 'stipule' & glandular smaller stip.-like structure interpreted as homoeotic replacement of 1 'stipule' by a 'leaf' (complete developmental continuum from 'gland' to 'leaf'). C 0, outer A without anthers. Cult. orn. Wood bitter

azarole *Crataegus azarolus*

Azilia Hedge & Lamond. Umbelliferae (III 10). 1 Iran: ***A. eryngioides*** (Pau) Hedge & Lamond

Azima Lam. Salvadoraceae. 4 S Afr. to Hainan, Philippines, Lesser Sunda Is. Alks. Axillary thorns (prob. lvs of undeveloped shoot cf. Cactaceae). Crushed branches of ***A. sarmentosa*** (Blume) Benth. (Java) foetid

azobé *Lophira alata*

Azolla Lam. Salviniaceae (Azollaceae). 5 trop. & warm (Am. 2 – R: SGP 74(2004)301); 2 natur. Eur., ***A. filiculoides*** Lam. in GB (1910 [invasive Eng. canals with clearing costing £400K p.a. by 2010, controlled by weevils; sale banned 2014], now also invasive trop. As., S Afr.; native NW but in Eur. in earlier interglacials). R: PSE 184(1993)187. Lvs sessile, 2-lobed, upper lobe aerial & photosynthetic with large mucilage-filled cavity incl. colonies of *Anabaena azollae* Strasb., unique in using fructose to fix nitrogen to ammonia (passed to host, which returns amino-acids, proteins & ribonucleotides) even in dark, the lower with dorsal side immersed & usu. achlorophyllous. Growth of cotyledon leaf ruptures megasporangial wall allowing symbiont to reach developing lvs. R. lose r.-cap & resemble submerged lvs of *Salvinia*. Sporocarps 2(4) forming lower lobe of 1st leaf of a lat. branch, the upper forming an involucre over them, each with micro- or mega-sporangia without a dehiscence mechanism; microspores held in massulae in each sporangium by hardened frothy mucilage. Massulae usu. covered with barbed hairs (glochidia); megasporangium with 1 spore, which becomes a floating female prothallus to be anchored to the massulae by the glochidia, the union sinking to the bottom for fert. ***A. pinnata*** R. Br. (trop. As.) – doubles wt. in 7 days, used to control mosquitoes by blocking water-surface, as green manure & stock feed, in rice-fields may fix 50–150 kg ha^{-1} in 1–4 months

Azollaceae Wettst. = Salviniaceae

Azorella Lam. Umbelliferae (Azor.). Excl. *Bolax* 26 Andes to temp. S Am., Falklands, Antarctic Is. R: Darw. 32(1993)172. Prob. heterogeneous. Densely tufted cushion-pls, those in S Peru held to be up to 3000 yrs old. ***A. monantha*** Clos ex Gay ('*A. caespitosa*') forms the balsam bogs of the Falklands, an extract used medic.; ***A. compacta*** Philippi (*A. yareta*) & other spp. (llareta, yareta, Andes etc.) – imp. fuel-source burning with little smoke, now endangered through excessive cutting, local medic. with poss. value in diabetes

Azorina Feer (~ *Campanula).* Campanulaceae (I 6). 1 Azores.: ***A. vidalii*** (H. Watson) Feer (*C. v.*) – subpachycaul shrub to 2 m

Aztec clover *Trifolium amabile;* **A. pine** *Pinus teocote;* **A. tobacco** *Nicotiana rustica*

Aztecaster Nesom. Compositae (Ast. – Hin.). 2 Mex. Dioec. shrubs. R: Phytol. 75(1993)55

Aztekium Bödecker (~ *Strombocactus*). Cactaceae (III 9). 3 Mex. ***A. ritteri*** (Bödecker) Bödecker – slow-growing, threatened in wild, cult. orn., seeds arillate

azuki bean *Vigna angularis*

Azukia Takah. ex Ohwi = *Vigna*

Azureocereus Akers & J.H. Johnson = *Browningia*

B

ba ji tian *Morinda officinalis*

babaco, babaçu or babassú palm *Attalea* spp. esp. *A. phalerata*

babacó *Vasconcellea × pentagona*

babai *Cyrtosperma merkusii*

Babbagia F. Muell. = *Sclerolaena*

Babcockia Boulos = *Sonchus*

Babiana Ker-Gawler ex Sims. Iridaceae (III 3). c. 93 S Zambia to Cape (Socotra sp. = *Cyanixia*). R: Strelitzia 18(2007)11. Fourteen pollination transitions incl. 2 to bird-p.: many spp. poll. long-proboscid flies, also bees, monkey-beetles; bird-poll. spp. with 'perches' in infls form. *Antholyza;* seeds black (usu. brown in Iridaceae). Cult. orn. ***B. fragrans*** (Jacq.) Steud. (*B. plicata,* S Afr., baboon r.) – corms ed. settlers, baboons & porcupines; ***B. ringens*** (L.) Ker-Gawl. (Cape) – poll. malachite sunbirds

Babingtonia Lindl. (~ *Baeckea*). Myrtaceae (II 15). Excl. *Harmogia, Kardomia, Sannantha* 11 SW Aus. R: Nuytsia 25(2015)227

baboen *Virola surinamensis*

baboon root *Babiana fragrans*

babul bark *Vachellia nilotica*

baby blue eyes *Nemophila menziesii;* **b. in cradle** *Anguloa* spp.; **b. corn** *Zea mays;* **b.'s breath** *Gypsophila paniculata;* **b. panda bamboo** *Pogonatherum puniceum*

bacaba palm *Oenocarpus distichus*

Baccaurea Lour. Phyllanthaceae (Scep.; Euphorbiaceae s.l.). 43 Indomal., W Pac. R: Blumea suppl. 12(2000)80. Several ed. fr. esp. ***B. dulcis*** (Jack) J. Voigt (Sumatra, W Java), ***B. motleyana*** (Muell. Arg.) Muell. Arg. (rambai, Mal.) & ***B. ramiflora*** Lour. (*B. sapida*, rambai, lutqua, Indomal.) oft. cult. ***B. racemosa*** (Blume) Muell. Arg. (Mal.) – wood used in ikat dyeing in Bali

Baccharidastrum Cabrera = *Baccharis*

Baccharidiopsis G. Barroso = *Baccharis*

Baccharis L. Compositae (Ast.-Bac.). c. 350 Am. (N Am. 21). R: MBSM 21(1985)1. Some monoec. or polygamo-dioec., most dioec., shrubs, many leafless with winged or cylindrical photosynthetic stems. Some med., lvs ground for green dye & cult. orn. e.g. salt-tolerant ***B. halimifolia*** L. (consumption weed, groundsel bush, tree-groundsel, E N Am. saltmarshes) – noxious weed in S Eur., NSW; ***B. salicifolia*** (Ruíz & Pavón) Pers. (*B. viminea*, warm Am.) – rheophyte, s.t. planted for erosion control; ***B. sarothroides*** A. Gray (SW N Am.) – potential oilseed; ***B. trimera*** (Less.) DC. (Urug.) – fertility control

Baccharoides Moench (~ *Vernonia*). Compositae (Vern.-Lin.). c. 30 trop. Afr., Ind. ***B. anthelmintica*** (L.) Moench (*V. a.*) – used for skin problems ('bukchie')

bachelor's buttons usu. 'double-flowered' forms of e.g. *Achillea ptarmica, Bellis* spp., *Centaurea cyanus, Ranunculus acris*

Bachmannia Pax. Capparaceae. 1 S Afr.: ***B. woodii*** (Oliv.) Gilg. R: Strel. 10(2000)204

Backhousia Hook. & Harv. Myrtaceae (II 6). Incl. *Choricarpia*, 13 Aus. ***B. bancroftii*** Bailey & F. Muell. (Johnstone River hardwood, Queensland) – hard close-grained timber; ***B. citriodora*** F. Muell. (lemon myrtle, Queensland) – source of ess. oils (citronellal insecticidal) used as lemon subs. in WW II lemonade, now flavouring in liqueurs, pasta & cheesecake, also for a tea & in soap; ***C. subargentea*** (C. White) Harrington (*C. s.*, E Aus.) – ironwood

Baclea Fourn. = ? *Gonolobus*

bacon and eggs *Daviesia* spp.

Bacopa Aubl. Plantaginaceae (Grat.-Grat.; Scrophulariaceae s.l.). 56 warm esp. Am. Mostly aquatic or paludal. R: PANSP 98(1946)83. Some cult orn. (see also *Chaenostoma*) incl. ***B. monnieri*** (L.) Wettst. (pantrop., water hyssop) in aquaria, also medic. ('brahmi' in Ind.), comm. extract claimed to improve memory & added to hair-oil in Ind.

Bactris Jacq. ex Scop. Palmae (V 8b). 77 trop. Am. R: FN 79(2000)12. Some fish-disp. in Amazon; ed. fr., oilseeds, wood (chonta) trad. source of bows, blowpipe darts, spears etc., cult. orn. ***B. gasipaes*** Kunth (*B. ciliata*, peach-palm, peach nut, pejibay(e), pejivalle, domesticated in interAndean valleys of Pacific lowlands of Colombia & introd. Amazon) – insect-poll., also wind-, widely cult. S Am. for fleshy pulp (sold in markets) & now comm. oilseed, also attendant weevils (like shrimps); ***B. major*** Jacq. (black roseau) – cult. orn.

bacu *Cariniana* spp., *Tieghemella heckelii*

Bacularia F. Muell. ex Hook.f. = *Linospadix*

bacury *Platonia insignis*

badam *Terminalia catappa*

badger's-bane *Aconitum lycoctonum*

badi *Nauclea diderrichii*

Badiera DC. (~ *Polygala*). Polygalaceae (IV). 25+ trop. Am. ***B. venenosa*** (Poir.) Hassk. (*P. v.*, Mal.) – pachycaul

Badilloa R. King & H. Robinson (~ *Eupatorium*). Compositae (Eupat.-Crit.). 10 N Andes. R: MSBMBG 22(1987)350

badinjan (WI) *Solanum melongena*

Badula Juss. Primulaceae (Myrsinaceae). 17 Madag., Masc. (3 extinct). Poss. incl. *Oncostemum*. ***B. crassa*** A. DC. (Mauritius) – pachycaul treelet

Badusa A. Gray. Rubiaceae (I 7). 3 Palawan, NG, W Pac. R: Blumea 28(1982)145. On limestone, cf. *Bikkia*

Baeckea L. Myrtaceae (II 15). 52 Aus., ***B. frutescens*** L. ext. to China – lvs in a tea, medic. & ess. oil for scents, soap etc. See also *Astartea, Babingtonia, Ericomyrtus, Euryomyrtus, Harmogia, Hysterobaeckea, Ochrosperma, Triplarina*

bael fruit *Aegle marmelos*

Baeolepis Decne. ex Moq. = *Decalepis*

Baeometra Salisb. ex Endl. Colchicaceae. 1 S Afr.: ***B. uniflora*** (Jacq.) Lewis (beetle lily) – toxic to stock (alks), natur. Aus. R: Strel. 10(2000)588

Baeothryon Ehrh. ex A. Dietr. = *Eleocharis*

Baeria Fischer & C. Meyer = *Lasthenia*

Baeriopsis J. Howell. Compositae (Mad.-Bae). 1 Baja Calif.: ***B. guadalupensis*** J. Howell. R: U. Eggli, *Ill. Handb. Succ. Pls*, Dicots (2002)20

Bafodeya Prance ex F. White (~ *Parinari*). Chrysobalanaceae (2). 1 W Afr.: ***B. benna*** (Scott-Elliot) F. White. R: FOW 9(2003)180

Bafutia C. Adams (~ *Emilia*). Compositae (Sen.-?Oth.). 1 Cameroun: ***B. tenuicaulis*** C. Adams

bagac *Dipterocarpus grandiflorus*

Bagassa Aubl. Moraceae (II). 1 NE S Am.: ***B. guianensis*** Aubl. – promising plantation tree

bagasse crushed *Saccharum* (q.v.)

bag-flower *Clerodendrum thomsoniae*

bag-pod *Sesbania vesicaria*

bagtikan *Parashorea malaanonan*

Bahama grass *Cynodon dactylon*; **B. hemp** *Agave sisalana*; **B. pitchpine** *Pinus caribaea*; **B. whitewood** *Canella winterana*

Baharuia Middleton (~ *Ichnocarpus*). Apocynaceae (IIc). 1 Sumatra, Borneo: ***B. gracilis*** Middleton. R: Blumea 40(1995)445

Bahia Lag. Compositae (Bah.). Excl. *Picradeniopsis*, 1 Chile: ***B. ambrosioides*** Lag. See also *Hymenothrix*

Bahia fibre = piassava fibre; **B. grass** *Paspalum notatum*; **B. piassava** *Attalea funifera*; **B. rosewood** *Dalbergia nigra*; **B. wood** *Paubrasilia echinata*

Bahianthus R. King & H. Robinson (~ *Eupatorium*). Compositae (Eupat.-Gyp.). 1 NE Braz.: ***B. viscosus*** (Spreng.) R. King & H. Robinson. R: MSBMBG 22(1987)113

Bahiella J. Morales (~ *Echites*). Apocynaceae (II e). 2 Bahia, Brazil. R: Sida 22(2006)342

Bahiopsis Kellogg (~ *Viguiera*). Compositae (Helia.-Helia.). 12 SW N Am. R: BJLS 140(2002)71

bai zhu *Atractyloides macrocephala*

baib grass *Eulaliopsis binata*

Baijiania A.M. Lu & J.Q. Li (~ *Siraitia*). Cucurbitaceae (VI). Incl. *Sinobaijiania*, 5 SE As. to Borneo. R: Blumea 48(2003)279, 51(2006)494

Baikiaea Benth. Leguminosae (I 2). c. 6 trop. Afr. Most extensive decid. forests on Kalahari Sand in Zambesi basin. ***B. plurijuga*** Harms (C Afr.) – valuable timber (Zambesi redwood)

Baileya Harv. & A. Gray ex Torrey. Compositae (Hele.-Tet.). 3 SW US, Mex. R: Sida 15(1993)491. ***B. multiradiata*** Harv. & A. Gray ex Torrey – cult. orn., rubbed under arms as deodorant

Baileyoxylon C. White. Achariaceae (Flacourtiaceae). 1 Queensland: ***B. lanceolatum*** C. White

Baillonella Pierre. Sapotaceae (I 1). 1 trop. W Afr.: ***B. toxisperma*** Pierre (djave) – ed. fat from seeds, good timber (moabi)

Baillonia Bocquillon. Verbenaceae (Cith.). 1 S Am.: ***B. amabilis*** Bocq. R: Darw. 5(1941)167

Baimashania Al-Shehbaz. Cruciferae (7). 2 China. R: Novon 10(2000)321

Baissea A. DC. Apocynaceae (II c). 18 trop. Afr. R: BJBB 64(1995)89

Baitaria Ruíz & Pavón = *Calandrinia*

bai-yor *Morinda citrifolia*

baitoa *Phyllostylon brasiliensis*

Bajacalia Loockerman & al. (~ *Porophyllum*). Compositae (Tag.-Pect.). 3 Baja Calif. R: SB 28(2003)204

bajri *Cenchrus spicatus*

bakain *Melia azedarach*

bakana *Scirpus atrovirens*

Bakerantha L.B. Sm. = *Hechtia*

Bakerella Tieghem (~ *Taxillus*). Loranthaceae (5 7). 16 Madag.

Bakeridesia Hochr. (~ *Abutilon*). Malvaceae (Malv.-Malv.). 20 trop. Am. R: GH 10(1973)446. See also *Callianthe*. ***B. vulcanicola*** (Standl.) D. Bates (S Costa Rica, Guatemala) – disjunct distrib.

Bakerolimon Lincz. (~ *Limonium*). Plumbaginaceae (II). 2 Peru, N Chile

Bakerophyton (Léonard) Hutch. = *Aeschynomene*

Bakoa Boyce & S.Y. Wong. Araceae. 4 Borneo. ***B. nakamotoi*** S.Y. Wong – rheophyte with dehiscent 'berries'

bakphul *Sesbania grandiflora*

baku *Tieghemella heckelii*

bakuli pods *Lagerstroemia microcarpa*

bakupari *Garcinia brasiliensis*

bakury *Platonia insignis*

Balaka Becc. Palmae (V 14i). Incl. *Solfia*, 8–9 Fiji, Samoa. R: Palms 54(2010)161

Balakata Esser (~ *Anomostachys*). Euphorbiaceae (Euph.-Hipp.). 2 Indomal. to Aus. R: Blumea 44(1999)154

Balanitaceae M. Roem. = Zygophyllaceae

Balanites Del. Zygophyllaceae (IV; Balanitaceae). 9 trop. Afr. (7, 2 ext. to Arabia), Ind. & Myanmar (2). R: KB 56(2001)12. Champagnat's Model. Oilseeds esp. drought-resistant ***B. aegyptiaca*** (L.) Del. (trop. Afr. to E Med.) for medicinal balanus oil, soap, lvs, fls & fr. (desert date) ed., bark medic. (steroid saponins toxic to cold-blooded hosts of guinea-worm & some stages of bilharzia) & fibre, contraceptives from r., wood useful (ship-building in Classical times); ***B. maughamii*** Sprague (menduro, trop. E & S Afr.) – fr. ed., oilseed; ***B. rotundifolia*** (Tieghem) Blatter (*B. orbicularis*, kullam, Somalia) – gum resin & oilseed; ***B. wilsoniana*** Dawe & Sprague (trop. Afr.) – fr. eaten by elephants in Ghana, seeds disp. in droppings

Balanopaceae Benth. (~ Chrysobalanaceae). Magnoliidae – Malpighiales. 1/9 SW Pac. R: Allertonia 2(1980)191. Dioec. everg. trees; hairs unicellular. Lvs in spirals, the proximal on each branch scale-like, the distal normal & sometimes verticillate, all simple & coriaceous with vestigial stip. Fls small, anemophilous, with 0 or vestigial perianth, males in axillary catkins – A (1 –)3 – 6(– 14), females solit., naked ovary subtended by many deltoid bracts in spirals – $\underline{G}$ (2[3]) with as many locules (s.t. not perfectly separated) & distinct bifid (s.t. forked again) styles & each with 2 nearly basal, apotropus/epitropous, bitegmic ovules.

Fr. a drupe in a persistent involucre, resembling an acorn, with 2 or 3 pyrenes; seeds with large green embryo in thin layer of endosperm. n = 20(21)

Only genus: *Balanops*

Form. placed nr Fagaceae, then Buxaceae, & considered by some to represent a 'prefloral' condition, but now thought to be extremely simplified

Balanophora Forst. & G. Forst. Balanophoraceae. 17 OW trop. to S Jap. R: DBA 28,1(1972)1. Acarpellate; seeds c. 7 μg each. Some agamospermy. Spp. in S Jap. poll. pyralid moths laying eggs in infls. Known to parasitize at least 74 spp. in 35 families, ***B. fungosa*** Forst. & G. Forst. with at least 25 host spp. Wax from tubers used in torches in Java etc., as bird-lime in Thailand; some medic. (anti-asthmatic). Foxy smell assoc. with fly-poll.; *B. fungosa* monoec. with mousey nectar poll. beetles in N Queensland, infr. slowly wearing away & seeds disp. by rain (? insects, pigs); ***B. dioica*** R. Br. ex Royle (Himal.) – also named as if fungus (*Cordyceps racemosa* Berk.)!

Balanophoraceae Rich. Magnoliidae – Santalales. 17/67 trop., subtrop. esp. upland forest, Med. Chlorophyll-less, fleshy r.-parasites with little host specificity, usu. no stomata or guard cells. Overground parts usu. fleshy club-shaped infls ('capitula'), fungus-like in appearance, pale yellow to brown, pink or purplish, bearing many fls, some of which are the smallest known; underground parts app. modified r., tuber-like (vestigial r. in *Corynaea*), amorphous, up to the size of a baby's head, rarely with scale lvs (*Lophophytum*) entirely parasite tissue (e.g. *Dactylanthus*, NZ; *Helosis*, *Lophophytum*, *Scybalium*, Am.) or part host ('corpus intermedium'), a chimerical system unknown elsewhere in green pls, prod. 'rhizomes' which grow through ground attacking new host-r. Lvs in spirals to whorled, scale-like, without stomata, or 0. Infls terminal, oft. developing inside 'tuber', rupturing its tissue which remains as a 'volva' at base (in *Chlamydophytum* maturing completely before rupture), unbranched or with branches in spirals (*Sarcophyte*). Fls unisexual (pls dioec. or monoec., when on sep. infls or together but males towards base, infl. in *Helosis* covered with startlingly geometric hexagonal scales, each of which is surrounded by 2 concentric rings of female fls, the males occupying the corners under the scales. Males diverse, P 0 or 3, 4(– 8), discrete or connate at base, valvate, A 1 or 2 (where P 0) or same as P, anthers with 2, 4 or many locules, A in *Helosis* & *Scybalium* united at base with discrete anthers, in other genera A a tube tipped with pollen-sacs; pollen not sculptured. Females v. reduced, P 0 or minute, cup-like in *Mystropetalon*; ovules, placentas & carpels oft. not easily recognizable but $\underline{G}$ ($\overline{G}$ in *Mystropetalon*) (2 or 3) or acarpellate (*Balanophora*), with 1 or 2 embryo-sacs without recognizable nucellus or integuments; some apomixis. Fr. indehiscent, s.t. surrounded by swollen 'pedicel' or perianth-tube (*Mystropetalon*) or aggregated to form fleshy multiple fr.; seed 1 with minute embryo embedded in endosperm. x = 8, 9, 12 etc.

I. **Mystropetaloideae** (ovary inf., adnate to K; style 1, ovules 3 pendulous) – *Mystropetalon* (only)
II. **Dactylanthoideae** (ovary inf.; A 1 or 2; stylidia connate into 1 style) – *Dactylanthus*, *Hachettea*
III. **Sarcophytoideae** (infl. consp. branched; stigma sessile) – *Chlamydophytum*, *Sarcophyte*
IV. **Helosidoideae** (infl. when young covered with hexagonal peltate bracts) – *Helosis*

Most genera are monospecific, the largest being *Balanophora* (17 spp.). Clearly simplified pls, it is difficult to decide whether the fam. represents a natural assemblage or comprises end-prod. of convergent evolution, though presently the first view is preferred, the fam. being derived from the less specialized parasitic Santalales, poss. nr. the ancestors of Olacaceae. Form. split into several fams, now treated as tribes, 5 of them having a storage substance resembling starch, the Balanophoroideae having a waxy reserve (balanophorin). *Cynomorium* (Med. to As.) here separated at family level (sculptured exine, ovule integument) but *Mystropetalon* v. distinctive, notably in pollen which is unique in angiosperms in being triangular, square or pentagonal when viewed end on but almost always square when viewed from the side. App. fly-poll., though spp. of *Ombrophytum*, largely subterr., prob. apomictic; juicy elaiosomes around fr. of *Mystropetalon* spp. attractive to ants which are disp. agents. Phallic infls. have suggested aphrodisiac qualities. Balanophoroideae provide waxes used in lighting

Balanops Baill. Balanopaceae. 9 N Queensland, New Caled., Vanuatu, Fiji. R: Allertonia 2(1980)207

balanus oil *Balanites aegyptiacus*

Balansaea Boiss. & Reuter = *Conopodium*
balasier *Calathea lutea*
balata *Manilkara bidentata, Ecclinusa balata*
balau *Shorea* spp. esp. *S. glauca, S. maxwelliana;* **red b.** *S. guiso, S. kunstleri*
Balaustion Hook. Myrtaceae (II 15). Excl. *Cheyniana*, 1 SW Aus.: ***B. pulcherrimum*** Hook. R: Nuytsia 19(2009)139
Balbisia Cav. Francoaceae (Ledocarpaceae; Geraniaceae s.l.). Incl. *Wendtia*, 11 S Am. Shrubby, ovary with many ovules per carpel
Balboa Planch. & Triana = *Chrysochlamys*
bald cypress *Taxodium distichum*
Baldellia Parl. = *Echinodorus*
baldmoney *Meum athamanticum*
Balduina Nutt. Compositae (Hele.-Gai). 3 SE US. R: Britt. 27(1975)355
Balfouria (H. Ohba) H. Ohba = *Ohbaea*
Balfourodendron Méllo ex Oliv. Rutaceae (I 7). 2 S Braz. R: Brit. 50(1998)357. ***B. riedelianum*** (Engl.) Engl. – valuable furniture timber (guatambu moroti, pau marfim), cult. orn.; alks
Balgooya Morat & Meijden. Polygalaceae (II). 1 New Caled.: ***B. pacifica*** Morat & Meijden. R: BMNHP IV,13(1991)B, Adans (1991)3
Balinotella Soják = *Seseli*
Baliospermum Blume. Euphorbiaceae (Cod.-Cod.). 5 Ind. to Sumbawa. Drastic purgatives esp. ***B. solanifolium*** (Burm.) Suresh (*B. montanum*)
balisier (WI) *Calathea lutea, Heliconia bihai*
Balizia Barneby & Grimes = *Albizia* (but see MNYBG 74,1(1996)34)
Ballantinia Hook.f. ex E. Shaw. Cruciferae (37). 1 Victoria, Tasmania: ***B. pumilio*** (DC.) Mabb. (*B. antipoda*)
ballart *Exocarpos* spp.
Ballochia Balf.f. Acanthaceae (III 2c). 3 Socotra
balloon flower *Platycodon grandiflorus;* **b. pea** (S Afr.) *Lessertia frutescens;* **b. vine** *Cardiospermum halicacabum*
Ballota L. (~ *Marrubium, Stachys*). Labiatae (VI). Excl. *Acanthoprasium*, c. 25 Eur. (6), Med., W As., S Afr. (1). ***B. acetabulosa*** (L.) Benth. (Med.) – fr. used as floating wick in olive oil lamps; ***B. nigra*** L. (black horehound, Euras., natur. US) – adulterant of *Marrubium vulgare*, ess. oil, form. medic. (colds, asthma)
Ballya Brenan = *Aneilema*
Ballyanthus Bruyns = *Orbea* (but see SBM 63(2002)180)
balm *Melissa officinalis;* **bastard b.** *M. melissophyllum;* **bee b.** *Monarda didyma;* **Canary b.** *Cedronella canariensis;* **b. of Gilead** *Abies balsamea, Commiphora gileadensis, Liquidambar orientalis, Populus* ×*jackii* 'Gileadensis', *P. nigra*, ?orig. *Pistacia terebinthus;* **horse b.** *Collinsonia canadensis;* **lemon, sweet** or **tea b.** *Melissa officinalis;* **Vietnamese b.** *Elsholtzia ciliata*
Balmea Martínez. Rubiaceae (I 3). 1 Mex.: ***B. stormae*** Martínez. R: AMBG 82(1995)423
balmony *Chelone glabra*
Baloghia Endl. Euphorbiaceae (Cod.-Bal.). 15 E Aus. (3), New Caled. (12, endemic), Norfolk Is. Red sap. ***B. inophylla*** (Forst.f.) P. Green (*B. lucida*, scrub bloodwood) – sap source of indelible paint
Balonga Le Thomas = *Uvaria*
Baloskion Raf. (~ *Restio*). Restionaceae. 8 E Aus. R: Telopea 8(1995)23. ***B. tetraphyllum*** (Lab.) Briggs & L. Johnson (dingo fern) – comm. cut foliage pl.
balsa *Ochroma pyramidale;* **Siamese b.** *Alstonia spatulata*
balsam orig. poss. *Boswellia sacra* or *Commiphora gileadensis; Impatiens* spp.; **b. apple** *Momordica balsamina;* **b. bog** *Azorella monantha;* **Canada b.** *Abies balsamea;* **Copaiba b.** *Copaifera officinalis;* **b. fig** *Clusia rosea;* **b. fir** *A. balsamea;* **garden b.** *I. balsamina;* **gurjun b.** *Dipterocarpus* spp.; **Indian b.** *I. glandulifera;* **Mecca b.** *Commiphora gileadensis;* **orange b.** *I. capensis;* **b. pear** *Momordica charantia;* **b. of Peru** *Myroxylon balsamum* var. *pareirae;* **b. poplar** *Populus balsamifera;* **rock b.** *Clusia, Peperomia;* **b. root** *Wyethia sagittata;* **Tolu b.** *Myroxylon balsamum* var. *balsamum;* **b. tree** *Colophospermum mopane;* **umiry b.** *Humiria* spp.
balsamic vinegar *Vitis vinifera*
Balsaminaceae A. Rich. Magnoliidae – Ericales. 2/1000 trop. OW with few temp. Herbs (rarely subshrubby), subsucculent, nearly always glabrous, stems ± translucent, s.t. with tubers or rhiz. Lvs in spirals, whorled or opp., simple, stip. 0 or a pr of petiolar glands. Fls

bisex., solit. or in cymes, zygomorphic, usu. resupinate; K 3 (*Impatiens*) or 5 (*Hydrocera*), the app. lowermost petaloid with a slender spur-nectary; C 5 distinct in *Hydrocera*, ± connate in *Impatiens*; A 5, filaments connate at least above, anthers ± connate into calyptra over pistil, pollen-sacs divided by trabeculae separating sporogenous tissue into islands; G (4 rare, 5) with axile placentation & 1 or 5 stigmas, ovules 1 per locule (*Hydrocera*) or ∞ (*Impatiens*), anatropous, bitegmic or rarely unitegmic. Fr. a berry-like drupe (*Hydrocera*) or explosively dehiscent loculicidal capsule, the valves twisting in dehiscence, the stone in the first eventually separating into 5 pyrenes; embryo straight, endosperm 0 (*Hydrocera*) or little. x = 6 – 11

Genera: *Hydrocera* (1 sp.), *Impatiens* (many cult. orn.)

Balsamita Mill. = *Tanacetum*

Balsamocarpon Clos (~ *Caesalpinia*). Leguminosae (I 4). 1 Chile: ***B. brevifolium*** Clos (*Caesalpinia b.*, algarobilla). R: PhytoKeys 71(2016)100

Balsamocitrus Stapf. Rutaceae (II 2). 2 trop. E Afr. R: W.T. Swingle, *Bot. Citrus* (1943)463

Balsamodendron Kunth = *Commiphora*

Balsamorhiza Hook. ex Nutt. = *Wyethia*

Balsas J. Jiménez & Vega. Sapindaceae (III). 1 Mex.: ***B. guerrerensis*** Cruz & Vega. R: Novon 21(2011)197

Balthasaria Verdc. Pentaphylacaceae (3; Theaceae s.l.). 1 trop. Afr.: ***B. schliebenii*** (Melchior) Verdc.

Baltic parsley *Cenolephium denudatum*; **redwood** or **yellow deal** *Pinus sylvestris*; **B. whitewood** *Picea abies*

Baltimora L. Compositae (Helia.-Ecl.). 2 trop. Am. R: Fieldiana 36(1973)31

balucanat *Aleurites moluccanus*

balustine flowers *Punica granatum*

Bambara groundnut *Vigna subterranea*

Bambekea Cogn. Cucurbitaceae (XIII). 1 trop. C & W Afr.: ***B. racemosa*** Cogn.

bamboo spp. of some dozens of genera of grasses (name also used for pls resembling them) or woody culms of many used for building, pipes, walking-sticks, flooring, furniture etc., when split for mats, blinds, baskets, fans, hats, umbrellas, brushes etc.; fibre also for paper pulp, fuel brickettes, carpet & clothing. Culms to 37 m, ed. when young in many spp. (bamboo shoots, usu. *Phyllostachys* spp.), with deposits of silica in cell walls & fibrous when mature. Some flower annually, others hapaxanthic; Tai Khai-Chih (c. AD 460) *Chu Phu* [Treatise on bamboo] – first recorded botanical monograph of any plant group; **baby panda b.** *Pogonatherum puniceum*; **berg b.** *Bergbambos tessellata*; **black b.** *Phyllostachys nigra*; **Calcutta b.** *Dendrocalamus strictus*; **Chinese water b.** *Dracaena sanderiana*; **common b.** *Bambusa vulgaris*; **fishpole b.** *P. aurea*; **heavenly b.** *Nandina domestica*; **lucky b.** *D. sanderiana*; **madake b.** *Phyllostachys reticulata*; **male b.** *B. bambos*, *Dendrocalamus strictus*; **b. palm** *Dypsis lutescens*, *Raphia* spp.; **spiny b.** *B. b.*; **Terai b.** *Melocanna baccifera*; **Tongking b.** *Pseudosasa amabilis*; **umbrella b.** *Thamnocalamus spathaceus*; **b. vines** (N Am.) *Smilax* spp.

Bambusa Schreb. Gramineae (V 4). Excl. *Guadua*, 100+ trop. & warm OW (China 80, 67 endemic – R: FOC 22(2006)9; Am. cult. 22 – R: CUSNH 39(2000)29) & As. to Aus. Many cultigens. Cult. orn., timber, pulp, bamboo shoots. ***B. arnhemica*** F. Muell. (trop. Aus.) – hapaxanthic after 40–50 yrs, culms for spear-shafts & didjeridoos in N Terr.; ***B. bambos*** (L.) Voss (*B. arundinacea*, male or spiny bamboo, Ind., SE As.) – to 37 m in Kerala (at Kew growing 91 cm in a day = 0.63 mm per minute!), hapaxanthic after 31–54 yrs with seeding over 5–6 yrs in Ind. e.g. 1868–72, 1912–16, 1958–62, 1991–95, timber, ed. shoots, sacred in Ind., concretions of silica in stems (tabashir) medic.; ***B. multiplex*** (Lour.) Schultes & Schultes f. (*B. glaucescens*, China) – widely cult. hedging bamboo in SE As. etc.; ***B. spinosa*** Roxb. (*B. blumeana*, *B. stenostachya*, SE As., Mal.) – used for prefab. houses made in Vietnam for US occupation, strips used to cut umbilical cord in Flores; ***B. vulgaris*** Schrad. ex Wendl. (common bamboo, cultigen ?orig. SE As., S China, widely grown) – hapaxanthic after at least 150 yrs (but now not setting seed so poss. going extinct), pulp, constr., ed. shoots, culms held during prayers in Sumatra to ward off spirits, '**Wamin**' an orn. cv. with swollen internodes; ***B. wrayi*** Stapf (Malay Pen.) – clumps found in territories of Orang Asli as favoured for blowpipes. See also *Dendrocalamopsis*

Bambusaceae Burnett = Gramineae

Bamiania Lincz. (~ *Cephalorhizum*). Plumbaginaceae (II). 1 Afghanistan: ***B. pachycorma*** (Rech.f.) Lincz.

Bamlera K. Schum. & Lauterb. = *Astronidium*

Bampsia Lisowski & Mielcarek. Linderniaceae (Plantaginaceae (Grat.-Lind.); Scrophulariaceae s.l.). 2 Congo (Katanga)

ban xia *Pinellia ternata*

banak *Virola koschnyi*

banana *Musa* spp., cvs & hybrids; **apple b.** *M.* × *paradisiaca;* **bush b.** (Aus.) *Marsdenia australis;* **b. passionfruit** *Passiflora mollissima;* **b. poka** *P. tarminiana*

Banara Aubl. Salicaceae (Flacourtiaceae). 33 trop. Am. R: FN 22(1980)83. Troll's Model

bancha twig tea *Camellia sinensis*

Bancroftia R. Porter = ? *Clinopodium*

bandakai (Ind.) = okra

Bandeiraea Welw. ex Benth. = *Griffonia*

baneberry *Actaea spicata*

Banfiopuntia Guiggi = *Australocylindopuntia*

banga wanga *Amblygonocarpus andogensis*

bangalay *Eucalyptus botryoides*

bangalow (palm) *Archontophoenix cunninghamiana*

banglang *Lagerstroemia speciosa*

Banisteria Houston ex L. = *Heteropterys* + *Banisteriopsis*

Banisterioides Dubard & Dop = *Sphedamnocarpus*

Banisteriopsis C. Robinson. Malpighiaceae. 66 trop. Am. esp. Braz. R: FN 30(1982). See also *Diplopterys*. Mostly lianes. Some hallucinogens in S Am. esp. ***B. caapi*** (Griseb.) Morton (wild & cult. Amazonia) – bark infusion (ayahuasca (Peru), caapi (Braz.), yagé, yajé (Colombia)) usu. with *Psychotria viridis* (q.v.), lvs & bark s.t. smoked, psychotropic effects due to β carboline alks blocking breakdown of DMT (N,N dimethyltryptamine) in *Psychotria viridis* & *D. cabrerana*. ***B. lutea*** (Griseb.) Cuatrec. (Peru) – app. eglandular fls mimicking oil-prod. M., visited by poll. bees

Banjolea S. Bowd. = *Nelsonia*

Banksia L.f. Proteaceae (V 2). Incl. *Dryandra*, 171 Aus. (171, 170 endemic, 150 SW Aus.), NG (1: ***B. dentata*** L.f.). R: FA 17B(1999)175, 251. Oliocene-Miocene fossils in NZ. Aus. honeysuckles, nectar used as food by Aborigines. Leeuwenberg's & Fagerlind's Models; everg. with hard woody follicles encl. in woody infr. (recorded from mid-Eocene) derived from bracts & bracteoles; seeds winged. Some poll. by honey possums; some, e.g. ***B. ornata*** F. Muell., have fire-dependent (the floral remnants acting as tinder) follicle dehiscence, stress between sclereids of different types in inner & outer layers allayed by a resin which is destroyed by fire – in other spp. the resin is chemically different & the fr. opens without fire. Some timbers, tanbarks & cult. orn. ***B. aemula*** R. Br. (wallum, E Aus.) – typical of sandy coastal heathlands; ***B. ericifolia*** L.f. (NSW) – weedy S Afr.; ***B. grandis*** Willd. (SW Aus.) – nectar eaten by Aborigines

banyan *Ficus benghalensis*

baobab *Adansonia digitata*

Baolia H.W. Kung & G.L. Chu (~ *Chenopodium*). Amaranthaceae (Chenopodiaceae I 2). 1 China: ***B. bracteata*** H.W. Kung & G.L. Chu

Baphia Afzel. ex Lodd. Leguminosae (III 2). 48 trop. & S Afr. (1), Madag. (2). R: BZ 96(2011)920. Troll's Model. ***B. nitida*** Afzel. ex Lodd. (cam-wood, trop. Afr.) – form. source of red dye, wood turning red from white on exposure to air, timber used for walking-sticks, violin bows etc., chewing-stick in Nigeria

Baphiastrum Harms = *Leucomphalos*

Baphicacanthus Bremek. = *Strobilanthes*

Baphiopsis Benth. ex Bak. Leguminosae (III 1). 1 trop. Afr.: ***B. parviflora*** Benth. ex Bak.

Baptisia Vent. Leguminosae (III 6). 17 E US. R: AMBG 27(1940)119. Alks. False indigo. ***B. tinctoria*** (L.) R. Br. etc. form. used as dye-pls. Cult. orn., some med. e.g. ***B. alba*** (L.) R. Br. (*B. lactea, B. leucantha*), *B. tinctoria*

Baptistonia Barb. Rodr. = *Gomesa*

Baptorhachis W. Clayton & Renvoize = *Axonopus*

Barathranthus Danser = *seq.*

Baratranthus (Korth.) Miq. Loranthaceae (5 3). 3 Sri Lanka, W Mal.

barb grass *Hainardia cylindrica*

Barbacenia Vand. Velloziaceae. 104 S Am. R: SCB 30(1976)4

Barbaceniopsis L.B. Sm. Velloziaceae. 3 Andes. R: SCB 30(1976)37

barbadine *Passiflora quadrangularis*

Barbados almond *Terminalia catappa*; **B. cedar** *Juniperus bermudiana*; **B. cherry** *Malpighia emarginata*; **B. gooseberry** *Pereskia aculeata*; **B. lily** *Hippeastrum puniceum*; **B. mastic** *Sideroxylon foetidissimum*; **B. pride** *Caesalpinia pulcherrima*; **B. snowdrop** *Habranthus tubispathus*

Barbamine Khokr. = *Barbarea*

Barbara's buttons *Marshallia* spp.

Barbarea Tabern. ex R. Br. Cruciferae (16). 22 N temp. (Eur. 10). Weeds (wintercress), esp. ***B. vulgaris*** R. Br. (yellow rocket) – noxious in US ('**Variegata**' breeding true), & some ed. esp. ***B. verna*** (Mill.) Asch. (*B. praecox*, land cress, American, Normandy or upland c., W Med., Macaronesia, natur. elsewhere) – salad

barbary fig *Opuntia ficus-indica*; **b. nut** *Moraea sisyrinchium*

barbasco (S Am.) fish-poisoning pls, esp. *Lonchocarpus* spp.; **b. root** *Dioscorea composita*

barbatimão *Stryphnodendron adstringens*

Barberetta Harv. Haemodoraceae (I). 1 S Afr.: ***B. aurea*** Harv. R: Strel. 10(2000)609

barberry *Berberis* spp.; **Alleghany** or **American b.** *B. canadensis*; **common** or **European b.** *B. vulgaris*

Barberton daisy *Gerbera jamesonii*

Barbeuia Thouars. Barbeuiaceae (Phytolaccaceae s.l.). 1 Madag.: ***B. madagascariensis*** Steud. – large liane, blackens on drying

Barbeuiaceae Nakai (~ Phytolaccaceae). Magnoliidae – Caryophyllalles. 1/1 Madag. Liane (specimens dry black). P5, A ∞. $\underline{G}$ (2) with 1 ovule per carpel. Fr. a loculicidal capsule. Seeds 1 or 2, arillate

Genus: *Barbeuia*

Barbeya Schweinf. Barbeyaceae. 1 NE Afr., Arabia: ***B. oleoides*** Schweinf.

Barbeyaceae Rendle. Magnoliidae – Rosales. 1/1 NE Afr., Arabia. Small dioec. tree. Lvs opp., simple, with dense abaxial indumentum, stip. 0. Fls small, wind-poll., reg. in short axillary cymes without bracts or bracteoles: P 3 or 4, slightly connate at base; A 6–9(–12); $\underline{G}$ (1(–3)), ± connate, each carpel with 1 loc. & 1 pend., anatr. unitegmic ovule. Fr. a nut with accrescent P; embryo straight, endosperm 0

Genus: *Barbeya*

More-or-less distinct carpels & a primitive phloem type here corr. suggested affinity with Urticales (= Rosales)

Barbieria DC. (~ *Clitoria*). Leguminosae (III 18). 1 trop. Am.: ***B. pinnata*** (Pers.) Baill.

Barbosa Becc. = *Syagrus*

Barbosella Schltr. Orchidaceae (V 13c). 19 trop. Am.

Barbrodria Luer (~ *Masdevallia*). Orchidaceae (V 13). 1 Braz.: ***B. miersii*** (Lindl.) Luer – cult. orn.

Barcella (Trail) Drude (~ *Elaeis*). Palmae (V 8c). 1 Braz.: ***B. odora*** (Trail) Drude. R: Principes 30(1986)74

Barcelona nut *Corylus avellana*

Barclaya Wall. (*Hydrostemma*). Nymphaeaceae (Barclayaceae). 3 Indomal. R: FPM 5(2015)209. Aquatics, ***B. rotundifolia*** Hotta (W Mal.) terr. Seeds with hooked hairs adhering to coats of wild pigs in Mal. Some cult. aquaria, e.g. ***B. longifolia*** Wall. (orchid lily, Andamans to W Mal.)

Barclayaceae Li = Nymphaeaceae

Barcoo grass *Iseilema* spp.

Bardotia E. Fischer & al. Orobanchaceae. 1 N Madag.: ***B. ankaranensis*** E. Fischer & al. R: Phytotaxa 46(2012)29

bareet grass *Leersia hexandra*

barilla ash with high levels of sodium carbonate used in soap- & glass-making, derived from pls, esp. spp. of *Halogeton, Salsola, Suaeda*

Barjonia Decne. Apocynaceae (V c; Asclepiadaceae III 4). 7 Braz. R: Rodriguesia 31,51(1979)7, SB 38(2013)766

bark, French *Pinus pinaster*

Barker bush *Persoonia longifolia*

Barkeria Knowles & Westc. (~ *Epidendrum*). Orchidaceae (V 13b). 17 C Am. R: AOSB 42(1973)620. Cult. orn.

Barkerwebbia Becc. = *Heterospathe*

Barkleyanthus H. Robinson & Brettell (~ *Senecio*). Compositae (Sen.- Tuss.). 1 SW US to C Am.: ***B. salicifolius*** (Kunth) H. Robinson & Brettell. R: FNA 20(2006)614

Barklya F. Muell. (~ *Bauhinia*). Leguminosae (I 1). 1 Queensland: ***B. syringifolia*** F. Muell. R: FA 12(1998)166

Barleria Plum. ex L. Acanthaceae (III 2d). c. 100 OW trop., 1 trop. Am. R: KB 52(1997)551. Many xerophytes with bracteolar thorns. Seeds with hairs which swell when wetted. Some med. & cult. orn. incl. ***B. cristata*** L. (Philippine violet, Ind. [?], natur. trop.) – hedgepl. e.g. Christmas Is. (Ind. Ocean) incl. varieg. cvs; lvs, r. etc. chewed against toothache, esp. ***B. lupulina*** Lindl. (Mauritius)

Barleriola Oersted. Acanthaceae (III 2d). 6 WI

barley *Hordeum* spp. esp. *H. vulgare*; **meadow b.** *H. secalinum*

Barlia Parl. Orchidaceae (IV 2). 2 Canary Is., Med. (Eur. 1). R: Gen. Orchid. 2(2001)257

barna tree *Crateva religiosa*

Barnadesia Mutis. Compositae (Barn.). 19 S Am. esp. trop. Andes. R: AMBG 86(1999)70. Trees & shrubs, often thorny

Barnardia Lindl. (~ *Scilla*). Asparagaceae. 2 As.

Barnardiella Goldbl. = *Moraea*

Barnebya Anderson & Gates. Malpighiaceae. 2 Braz. R: Britt. 33(1981)275

Barnebydendron Kirkbride (*Phyllocarpus*). Leguminosae (I 2). 1 trop. Am.: ***B. riedelii*** (Tul.) Kirkbride. R: KB 63(2008)144

Barnebyella Podlech = *Astragalus*

Barneoudia Gay = *Anemone*

Barnettia Santisuk = *Santisukia*

Barnhartia Gleason. Polygalaceae (II). 1 trop. S Am.: ***B. floribunda*** Gleason – liane to tops of trees

barnyard grass *Echinochloa crus-galli*; **b. millet** [, **Jap.**] *E. esculenta*, [, **Ind.**] *E. frumentacea*

Barombia Schltr. = *Aerangis*

Barombiella Szlach. = *Plectrelminthus*

barometz *Cibotium barometz*

Barongia Peter G. Wilson & Hyland. Myrtaceae (II 5). 1 N Queensland: ***B. lophandra*** Peter G. Wilson & Hyland. R: Telopea 3(1988)257

Baronia Bak. (~ *Rhus*). Anacardiaceae (I). 1–3 Madag.

Baroniella Costantin & Gallaud. Apocynaceae (III; Asclepiadaceae I). 8 Madag. R: Cand. 52(1997)392, 57(2002)67

baros camphor *Dryobalanops aromatica*

Barosma Willd. = *Agathosma*

barrel, brown *Eucalyptus fastigata*

barrenwort *Epimedium alpinum*

Barreria L. = ? (Bruniaceae)

Barringtonia Forst. & G. Forst. Lecythidaceae (II). Incl. *Abdulmajidia*, 69 E Afr. (1), Madag. (2), trop. As. (Mal. 60) – R: FM I 21(2013)10 & Pac. R: Allertonia 12(2012). Corner's, Leeuwenberg's, Koriba's, Scarrone's Models. Seeds with saponins used as fish-poison like *Derris*. ***B. asiatica*** (L.) Kurz (*B. speciosa*, Madag. to Pacific) – fr. washed up in Ireland (1985), Cornwall (2012), early colonist of Krakatao, cult. orn. street-tree in trop.; ***B. corneri*** Kiew & Wong (Malay Pen.) – litter-trapping pachycaul with adventitious r. growing into debris; ***B. edulis*** Seem. (Fiji, spread in cult. to NG), ***B. novae-hiberniae*** Lauterb. (Moluccas to Vanuatu) & ***B. procera*** (Miers) Knuth (NG to Aus., Vanuatu) – ed. seeds (roasted)

Barringtoniaceae DC. ex F. Rudolphi = Lecythidaceae

Barroetea A. Gray = *Brickellia*

Barrosoa R. King & H. Robinson (~ *Eupatorium*). Compositae (Eupat.-Gyp.). 9 trop. S Am. R: MSBMBG 22(1987)92

Barteria Hook.f. Passifloraceae. 4 trop. Afr. R: Adansonia III, 21(1999)310. Roux's Model. Some myrmecophytes (***B. solida*** Breteler (W C Afr.) not): ***B. nigritana*** Hook.f. (Nigeria to Zaire) inhabited by small ants, ***B. fistulosa*** Mast. (*B. n.* subsp. *f.*, Nigeria to Tanzania) by big ants. Pls. without ants in Nigeria grow less well, ants deterring insect (& ? larger animal) grazing in myrmecophilous ones

Barthea Hook.f. Melastomataceae. 1 China, Taiwan: ***B. barthei*** (Benth.) Krasser

Barthlottia E. Fischer. Scrophulariaceae (Man.). 1 Madag.: ***B. madagascariensis*** E. Fischer

Bartholina R. Br. (~ *Holothrix*). Orchidaceae (IV 4d). 2 Cape. R: Strel. 9(2000)157

Bartholomaea Standl. & Steyerm. Salicaceae (Flacourtiaceae). 2 C Am.

Bartlettia A. Gray. Compositae (Bah.). 1 SW N Am.: ***B. scaposa*** A. Gray. R: SWN 8(1963) 117

Bartlettina R. King & H. Robinson (~ *Eupatorium*). Compositae (Eupat.-Heb.). 37 trop. Am. R: MSBMBG 22(1987)403. ***B. sordida*** (Less.) R. King & H. Robinson (*E. megalophyllum*, Mex.) – cult. orn., invasive Indonesia

Bartonia Muhlenb. ex Willd. Gentianaceae. 3E N Am. R: SB 34(2009)167. Mycotrophs with scale lvs & little chlorophyll

Bartonia Pursh = *Mentzelia*

Bartschella Britton & Rose = *Mammillaria*

bartsia, alpine *Bartsia alpina*; **red b.** *Odontites vernus*; **yellow b.** *Bellardia viscosa*

Bartsia L. Orobanchaceae (Rhin.; Scrophulariaceae s.l.). Excl. *Bellardia, Hedbergia*, 1 montane Eur., NE N Am.: ***B. alpina*** L.

Bartsiella Bolliger = *Odontites* (but see Willd. 26(1996)76)

barus camphor *Dryobalanops aromatica*

barwood *Pterocarpus erinaceus, P. soyauxii*

Barylucuma Ducke = *Pouteria*

Basanacantha Hook.f. = *Randia*

Basananthe Peyr. (*Tryphostemma*). Passifloraceae. c. 30 trop. & S Afr. R: Blumea 21(1974)327

Basedowia E. Pritzel. Compositae (Gnap.-Cass.). 1 C Aus.: ***B. tenerrima*** (F. Muell. & Tate) J. Black. R: OB 104(1991)85

Basella L. Basellaceae. 5 Madag. (3), E Afr. (1) & 1 pantrop. (***B. alba*** L. (*B. rubra*), Malabar spinach or nightshade, Ceylon or Ind. spinach) cult. potherb poss. not native in Am. or As., cli. with cleistogamous fls & fr. encl. in P which becomes fleshy. R: KB 62(2007) 304

Basellaceae Raf. Magnoliidae – Caryophyllales. 4/19 trop. & warm esp. Am. (?introd. As.). Perenn. glabrous rhiz. herbs with fleshy, mucilaginous, ann. cli. shoots & s.t. (*Ullucus*) tubers, with sec. cambia but no anthocyanins (betalains present). Lvs opp. to in spirals, simple, entire, oft. succ.; stip. 0. Infls. term. or axillary spikes to panicles of small, regular, bisex. (s.t. functionally unisexual) fls; bracteoles 2, K (also considered bracteoles by some) 2, oft. coloured, s.t. adnate to base of petals; C 5 (s.t. considered K), imbr. with a basal tube or 0, persistent in fr.; A (4)5(–9) opp. C, adnate at base to C, anthers opening by term. slits or pores or longit. slits; annular nectary around base of A; $\underline{G}$ (3) becoming 1-locular with 1 basal, bitegmic ovule. Fr. a thin-walled nutlet encl. in P; seeds with perisperm or 0, endosperm 0. x = 11, 12

Genera: *Anredera, Basella, Tournonia, Ullucus*

Basella & *Ullucus* are foodpls, *Anredera* some cult. orn.

Baseonema Schltr. & Rendle. Apocynaceae (III; Asclepiadaceae I). 1 trop. E Afr.: ***B. gregorii*** Schltr. & Rendle. R: FTEA Apoc.(2012)161

Bashania Keng f. & Yi (~ *Arundinaria*). Gramineae (IV). 2 China. Heterogeneous. ***B. faberi*** (Rendle) Yi (*A. f.*, SW China) – imp. panda food

Basigyne J.J. Sm. = *Dendrochilum*

basil *Ocimum basilicum* (**lemon b., sweet b.**); **bush** or **Greek b.** *O. basilicum* 'Minimum'; **hoary b.** *O. canum*; **holy b.** *O. tenuiflorum*; **Italian b.** *O. basilicum* 'Fino Verde Compatto'; **lemon b.** *O.* × *africanum*; **lime b.** *O. americanum*; **musk b.** *Basilicum polystachyon*; **Peruvian b.** *O. campechianum*; **sweet b.** *O.* × *africanum*; **Thai b.** *O. tenuiflorum*; **b. thyme** *Satureja acinos*

Basilicum Moench. Labiatae (VII 3d). 1 OW trop. to E Aus. ***B. polystachyon*** (L.) Moench (musk basil) – medic. tea

Basiloxylon K. Schum. = *Pterygota*

Basiphyllaea Schltr. Orchidaceae (V 13a). 7 WI, Florida (1). R: HPB 5(2001)487

Basisperma C. White. Myrtaceae (II 5). 1 NG: ***B. lanceolata*** C. White

Basistelma Bartlett = *Cynanchum*

Basistemon Turcz. Scrophulariaceae (Hemim.). 8 trop. Am. R: SB 10(1985)125. Shrubs, elaiophores attractive to bees

Baskervilla Lindl. Orchidaceae (IV 2b). 10 Andes, S Braz.

basket flower *Ismene narcissiflora*

basralocus *Dicorynia guianensis*

bass fibrous young bark esp. from *Tilia* spp.; **b. wood** *T. americana* etc.

bass-broom fibre = piassava

Bassecoia B.L. Burtt (~ *Pterocephalus*). Caprifoliaceae (Dipsacaceae). 3 Himal., W China, Thailand. R: BJLS 132(2000)67 (as *Pterocephalodes*)

Basselinia Vieill. Palmae (V 14c). Incl. *Alloschmidia*, 12 New Caled. R: Allertonia 3(1984)355, KB 63(2008)65

Bassia All. Amaranthaceae (Camph.; Chenopodiaceae I 4). Incl. *Kirilowia, Kochia, Panderia,* excl. *Chenolea, Spirobassia,* 10 warm. ***B. scoparia*** (L.) A.J. Scott (*K. s.*, summer cypress, S Eur to Jap.) – cult. for broom-making (Jap.), orn. foliage pl. (bee-poll., fls inconspicuous) for formal bedding, f. ***trichophylla*** (Schmeiss) Welsh (burning bush) most often cult., lvs red in autumn), widely natur., weedy in Aus., where form. used to rehabilitate salinized land

bassine fibre *Borassus flabellifer*

bassorin *Orchis* spp. (& other orchid) starch

Bassovia Aubl. = *Solanum*

basswood *Tilia americana*

bastard acacia *Robinia pseudoacacia;* **b. alkanet** *Buglossoides arvensis;* **b. balm** *Melittis melissophyllum;* **b. box** *Polygaloides chamaebuxus;* **b. burr** *Xanthium spinosum;* **b. cedar** *Chukrasia tabularis, Guazuma ulmifolia, Soymida febrifuga;* **b. cinnamon** *Cinnamomum aromaticum;* **b. grass** (NZ) *Carex* (*Uncinia*) spp.; **b. indigo** *Amorpha fruticosa;* **b. mahogany** *Eucalyptus botryoides;* **b. rosewood** *Synoum glandulosum;* **b. wild rubber** *Funtumia africana*

Bastardia Kunth. Malvaceae (Malv.-Malv.). 3 trop. Am.

Bastardiastrum (Rose) D. Bates. Malvaceae (Malv.-Malv.). 8 Mex. R: SBM 25(1988) 118

Bastardiopsis (K. Schum.) Hassler. Malvaceae (Malv.-Malv.). 10 trop. S Am.

Basutica E. Phillips = *Gnidia*

bat flower, b. lily, b. plant *Tacca integrifolia*

bataan, b. mahogany *Shorea polysperma*

Bataceae Mart. ex Perleb. Magnoliidae – Brassicales. 1/2 Pac. & Am. coasts. Maritime shrubs, (sub)glabrous. Lvs opp., simple, narrow, succ.; stip. minute. Fls unisexual (pls dioec. or monoec.), small, in strobiloid spikes, males initially encl. in sac-like organ poss. rep. K or pr of bracteoles, splitting down 1 side or into 4; P (? staminodes) 4, A 4 alt. with P, anthers opening by longitudinal slits. Female fls: P 0, $\underline{G}$ 2, 4- loc., with 1 anatr. bitegmic ovule in each & 2 stigmas. Fr. a sea-disp. drupe with 4 pyrenes; seeds without endosperm, embryo almost straight. 2n = 18

Only genus: *Batis*

Pollen referred here known from the Maestrichtian of Calif.

batai wood *Paraserianthes falcataria*

Batania Hatusima = *Pycnarrhena*

batako plum *Flacourtia inermis*

Bataprine Nieuw. = *Galium*

Batemannia Lindl. Orchidaceae (V 12j). 5 trop. S Am. Cult. orn.

Baterium Miers = *Haematocarpum*

Batesanthus N.E. Br. = *Cryptolepis*

Batesia Spruce ex Benth. Leguminosae (I 4). 1 Amazonia: ***B. floribunda*** Spruce ex Benth. – wood for furniture etc.

Batesimalva Fryx. Malvaceae (Malv.-Malv.). 4 NE Mex (3), Venez. (1)

Bath asparagus *Ornithogalum pyrenaicum*

Bathiaea Drake = *Brandzeia*

Bathiorchis Bosser & Cribb = *Goodyera*

Bathiorhamnus Capuron. Rhamnaceae (11). 7 Madag. R: Adansonia 30(2008)153

Bathysa C. Presl. Rubiaceae (Cond.). Excl. *Schizocalyx*, 4 Amaz.

Batidaceae auctt. = Bataceae

Batidaea (Dumort.) Greene = *Rubus*

Batiki blue grass *Ischaemum indicum*

batiputa *Ouratea parviflora*

Batis P. Browne. Bataceae. 2: ***B. maritima*** L. (saltwort, beachwort, Hawaii (introd.), SW US, WI, Atlantic S Am., dioec.), ***B. argillicola*** P. Royen (NG, Queensland, monoec.) – occ. used in salads, ash form. used in glass- & soap-making

Batocarpus Karsten. Moraceae (II). 3 trop. Am. R: BMNRB 37(1968)1

Batodendron Nutt. = *Vaccinium*

Batopedina Verdc. Rubiaceae (IV 1). 3 W & S trop. Afr.

Batopilasia Nesom & Noyes (~ *Erigeron*). Compositae (Ast.-Bolt.). 1 Mex.: ***B. byei*** (Sundberg & Nesom) Nesom & Noyes. R: Sida 19(2000)81

Batrachium (DC.) Gray = *Ranunculus*

Battandiera Maire = *Ornithogalum*

Baudouinia Baill. Leguminosae (I 3). 6 Madag. ***B. rouxevillei*** Perrier – endangered, yellow sapwood & dark brown heartwood used for carving canes, lamp-standards etc., form. for chieftains' sceptres

Bauera Banks ex Andrews. Cunoniaceae (II). 4 temp. E Aus. Cult. orn.

Baueraceae Lindl. = Cunoniaceae

Bauerella Borzi = *Sarcomelicope*

Baueropsis Hutch. = *Cullen*

Bauhinia Plum. ex L. Leguminosae (I 1). c. 150 s.s., pantrop. esp. S Am. (75; (s.l.): Mal. 69 – R: FM I,12(1996)442). Trees (Troll's Model) or oft. lianes with flattened stems, e.g. ***B. scandens*** L. var. ***anguina*** (Roxb.) Ohashi (snake cli., Ind.) curving in alt. directions at each node. Some dioec., some heterostylous (otherwise unknown in fam.), reduction series of A, 10, 5, 3 to 1 and to C1; some poll. by bats, others birds or insects; some with explosive fr. (seeds of ***B. purpurea*** L. (Indomal.) ejaculated to 15m), others, e.g. ***B. binata*** Blanco (SE As. to trop. Aus.) of coastal forest, with floating fr. Extrafl. nectaries. Many cult. orn. (butterfly tree, camel's foot, kachnar): ***B.* × *blakeana*** Dunn (yeung chi ging, *B. variegata* × *B. purpurea*) – floral emblem of Hong Kong (single clone – '**Sir Henry Blake**'; ***B. variegata*** L. (orchid tree, Ind. & China, like *B. purpurea* invasive S Afr.) – lvs, fls & pods ed., bark medic. & used in tanning, sacred to Buddhists; others used for cordage & food locally (e.g. fl. buds of *B. purpurea* in Sikkim), also gum, dyes & timber e.g. ***B. divaricata*** L. (*B. porrecta*, mountain ebony) in Jamaica. See also *Barklya, Gigasiphon, Lysiphyllum, Phanera, Piliostigma Schnella, Tylosema*

Baukea Vatke = *Rhynchosia*

Baumannia K. Schum. = *Knoxia*

Baumea Gaudich. = *Machaerina*

Baumia Engl. & Gilg. Orobanchaceae (Buch.; Scrophulariaceae). 1 Angola: ***B. angolensis*** Engl. & Gilg

Baumiella H. Wolff = *Berula*

bauno *Mangifera caesia*

bawchan seed *Cullen corylifolia*

Baxteria R. Br. Dasypogonaceae (Xanthorrhoeaceae s.l.). 1 SW Aus.: ***B. australis*** R. Br. R: FA 46(1986)141

Baxteriaceae Takht. = Dasypogonaceae

bay berry *Morella californica*; **b. laurel** *Laurus nobilis*; **b. rum** *Pimenta racemosa*; **b.wood** *Swietenia macrophylla*

Bayabusua de Wilde & de Wilde-Duyfies. Cucurbitaceae (Zan.-Zan.) 1 Malay Penin.: ***B. clarkei*** (King) de Wilde & de Wilde-Duyfies. R: Sandakania 13(1999)1

Baynesia Bruyns. Apocynaceae (Asclepiadaceae). 1 NW Namibia: ***B. lophophora*** Bruyns. R: Novon 10(2000)354

bayonet grass *Bolboschoenus maritimus*; **b. plant** *Aciphylla squarrosa*

Bdallophytum Eichler. Cytinaceae (Rafflesiaceae s.l.). 3 C Am. R: ABM 87(2009)9. Parasitic on Burseraceae

bdellium resin from *Commiphora* spp.

beach heath *Hudsonia ericoides*

Beadlea Small = *Cyclopogon*

beadplant *Nertera granadensis*

beak rush *Rhynchospora* spp.

Bealia Scribner = *Muhlenbergia*

bean most widely cult. are *Phaseolus* spp.; **adzuki b.** *Vigna angularis*; **asparagus b.** *V. unguiculata* subsp. *sesquipedalis*; **baked b.s** *P. vulgaris*; **Bengal b.** *Mucuna pruriens* var. *utilis*; **black b.** *Castanospermum australe, Glycine max, Kennedia nigricans*; **black-eye(d) b.** *V. unguiculata*; **Boer b.** *Schotia* spp.; **bog b.** *Menyanthes trifoliata*; **borlotti b.** *P. vulgaris*; **broad b.** *Vicia faba*; **buffalo b.** *Astragalus crassicarpus, Mucuna* spp.; **Burma b.** *P. lunatus*; **butter b.** *P. lunatus*; **Calabar b.** *Physostigma venenosum*; **cannellini b.** *Phaseolus vulgaris*; **Cherokee b.** *Erythrina herbacea*; **cluster b.** *Cyamopsis tetragonoloba*; **Congo b.** *Cajanus cajan*; **crab-eye** or **cranberry b.** *P. vulgaris*; **b. curd** *Glycine max*, see also *Vigna*; **duffin b.** *P. lunatus*; **dwarf b.** *P. vulgaris*; **field b.** *Vicia faba*; **flageolet b.** *P. vulgaris*; **Florida velvet b.** *Mucuna pruriens* var. *utilis*; **foulia b.** *V. faba*; **French b.** *P. vulgaris*; **garbanzo b.** *Cicer arietinum*; **Goa b.** *Psophocarpus tetragonolobus*; **Great Northern b.** *Phaseolus vulgaris*; **ground b.** *Macrotyloma geocarpum*; **haricot b.** *P. vulgaris*; **horse b.** *Vicia faba*; **hyacinth b.** *Lablab purpureus*; **icecream b.** *Inga edulis*; **Indian b.** *Catalpa bignonioides*; **jack b.** *Canavalia ensiformis*; **jumbie b.** orn. seeds esp. of *Leucaena leucocephala* etc.; **kidney b.** *Phaseolus vulgaris*; **Lima b.** *P. lunatus*; **locust**

b. *Ceratonia siliqua;* **long b.** *Vigna unguiculata;* **lucky b.** *Castanospermum australe, Entada gigas,* spp. of *Abrus, Afzelia, Erythrina, Thevetia;* **Madag. b.** *P. lunatus;* **maloga b.** *Vigna lanceolata;* **marama** or **morama b.** *Tylosema esculentum;* **Mary's b.** *Merremia discoidesperma;* **mat** or **moth b.** *V. aconitifolia;* **mescal b.** *Dermatophyllum secundiflorum;* **mung b.** *V. radiata;* **navy b.** *P. vulgaris;* **New Guinea b.** *Lagenaria siceraria;* **nicker b.** *Guilandina bonduc;* **Nuñas b.** *P. vulgaris;* **ordeal b.** *Physostigma venenosum;* **paternoster b.** *Abrus precatorius;* **pinto b.** *Phaseolus vulgaris,* **potato b.** *Apios americana;* **Queensland b.** *Entada rheedei* & other *E.* spp.; **phasey b.** *Macroptilium lathyroides;* **Rangoon b.** *P. lunatus;* **red b.** *P. vulgaris, Dysoxylum mollissimum;* **rice b.** *Vigna umbellata;* **rosecoco bean** *P. vulgaris;* **sabre b.** *Canavalia ensiformis;* **scarlet runner b.** *P. coccineus;* **sea b.** *Entada gigas, Mucuna* spp.; **snake b.** *Vigna unguiculata* subsp. *sesquipedalis;* **snap b.** *P. vulgaris;* **soya b.** *Glycine max;* **b. sprouts** *V. radiata;* **string b.** *P. vulgaris;* **tepary b.** *P. acutifolius;* **tic b.** *Vicia faba;* **Tonka b.** *Dipteryx odorata;* **turtle b.** *P. vulgaris;* **velvet b.** *Mucuna pruriens* var. *utilis;* **walnut b.** *Endiandra palmerstonii;* **white b.** *P. vulgaris;* **winged b.** *Psophocarpus tetragonolobus;* **yam b.** *Pachyrhizus erosus;* **wax b.** *Phaseolus vulgaris;* **yard-long b.** *Vigna unguiculata* subsp. *sesquipedalis*

bear *Hordeum vulgare*

bearberry *Arctostaphylos uva-ursi;* **black b.** *Arctous alpina*

beargrass *Yucca* spp., *Xerophyllum tenax*

bear's breech *Acanthus mollis;* **b.'s ear** *Primula auricula;* **b.'s foot** *Helleborus foetidus*

beard flower *Pogonia* spp.; **b. grass** *Polypogon monspeliensis;* **b. tongue** *Penstemon* spp.

Beatsonia Roxb. = *Frankenia*

Beaucarnea Lem. (~ *Nolina*). Asparagaceae. Incl. *Calibanus,* 12 SW N & C Am. R: T 63(2014)1206. Leeuwenberg's Model; swollen bole-base, lvs linear, s.t. dioec. ***B. hookeri*** (Lem.) Bak. (*C. caespitosus,* Mex.) – large tuber & few grassy lvs; ***B. recurvata*** Lem. (*N. b.,* bottle or pony-tail palm, elephant-foot tree, Mex.) – (ugly) housepl.

Beaufortia R. Br. = *Melaleuca*

Beaumontia Wall. Apocynaceae (IIc). 9 Ind. & China to Bali. R: AUWP 86–5(1986)3. Lvs smoked like tobacco in NG. Cult. orn. esp. ***B. grandiflora*** (Roxb.) Wall. (Himal. to Vietnam) – fr. to 30 cm long with 200–300 seeds 2.5 cm long with 7 cm plumes

Beauprea Brongn. & Gris. Proteaceae (IV). 13 New Caled. R: FNC 2(1968)20

Beaupreopsis Virot. Proteaceae (IV). 1 New Caled.: ***B. paniculata*** (Brongn. & Gris) Virot

Beautempsia Gaudich. (~ *Capparis*). Capparaceae. 1 Ecuador, Peru: ***B. avicenniifolia*** (Kunth) Gaudich. R: JBRIT 3(2009)685

beauty berry *Callicarpa americana;* **b. bush** *Kolkwitzia amabilis*

Beauverdia Herter (~ *Tristagma*). Alliaceae. 4 S. Braz., Arg., Urug. R: SB 39(2014)769

beaver poison *Cicuta maculata;* **b. wood** *Celtis occidentalis*

Bebbia Greene. Compositae (Mill.-Dysc.). 2 SW N Am. R: Madroño 24(1977)112. Xerophytes

Beccarianthus Cogn. (~ *Astronidium*). Melastomataceae. 22+ Borneo to NG

Beccariella Pierre = *Pleioluma*

Beccarinda Kuntze. Gesneriaceae (III 2d). 8 NE Ind. to Vietnam & Sumatra (1)

Beccariodendron Warb. = *Gonothalamus*

Beccariophoenix Jum. & Perrier. Palmae (V 8a). 3 Madag. R: Palms 51(2007)75, 58(2014)61. ***B. madagascariensis*** Jum. & Perrier – almost extinct (c. 40 left in wild, but cult. orn.), tapped for sap, leaf fibre for 'Manarano' hats

Becheria Ridl. = mixture of *Ixora* & *Psychotria* spp.

Bechium DC. (~ *Vernonia*). Compositae (Vern.-Centrop.) c. 2 Madag.

Becium Lindl. = *Ocimum*

Beckeropsis Figari & De Not. = *Cenchrus*

Beckmannia Host. Gramineae (XVI 15). 2 N temp. (Eur. 2). R: FOC 22(2006)364. Fodder; grains of ***B. eruciformis*** (L.) Host ed. Jap.

Beckwithia Jepson (~ *Ranunculus*). Ranunculaceae (II 3). 4 N temp. & boreal. ***B. glacialis*** (L.) Löve & D. Löve (*R. g.,* Arctic, Eur. mts, Sierra Nevada) – calcifuge cult. orn. holding alt. record for Scandinavia (2370 m) & Alps (4275 m)

Beclardia A. Rich. Orchidaceae (V 16d). 2 Mascarenes

Becquerelia Brongn. Cyperaceae (III 4). 8 trop. Am. R: MNYBG 17(1967)25

Bedfordia DC. (~ *Brachyglottis*). Compositae (Sen.-Tuss.). 3 SE Aus. (Tas. 2). R: Muelleria 19(2004)82. Trees & shrubs

bedstraw *Galium* spp.; **hedge b.** *G. album;* **lady's b.** *G. verum*

beech *Fagus* spp.; **African b.** *Faurea saligna;* **American b.** *Fagus grandifolia;* **Antarctic b.** *Nothofagus antarctica;* **black b.** *N. solandri;* **blue b.** *Carpinus caroliniana;* **brown b.**

Cryptocarya glaucescens; **Cape b.** *Myrsine melanophloeos*; **common** or **Eur. b.** *Fagus sylvatica*; **copper b.** *F. sylvatica* Cuprea Group; **Dawyck b.** *F. sylvatica* 'Dawyck'; **b. fern** *Phegopteris connectilis*; **Indian b.** *Millettia pinnata*; **Japanese b.** *Fagus crenata*; **myrtle b.** *N. cunninghamii*; **oriental b.** *F. sylvatica* subsp. *orientalis*; **purple b.** *F. s.* Atropunicea Group; **red b.** *Nothofagus fusca, Flindersia* spp.; **roble b.** *N. obliqua*; **silver b.** *N. menziesii*; **southern b.** *N.* spp.; **Southland b.** *N. menziesii*; **Tasmanian b.** *N. cunninghamii*; **Turkish b.** *Fagus sylvatica* subsp. *orientalis*; **white b.** *Elaeocarpus kirktonii, Gmelina* spp. esp. *G. leichhardtii*

beedi (Ind.) cheap cigarette rolled in a leaf, not paper

beef apple *Manilkara zapota*; **b. plant** *Iresine herbstii*; **b. steak plant** *Acalypha amentacea* subsp. *wilkesiana*; **b. suet tree** *Shepherdia argentea*; **b.wood** *Allocasuarina* spp., *Casuarina equisetifolia, Grevillea striata, Stenocarpus salignus*

beer *Hordeum vulgare*, (**b., white**) *Triticum aestivum*

Beesia Balf.f. & W.W. Sm. Ranunculaceae (I 2). 2 N Myanmar, W & SW China

beet *Beta vulgaris*: cvs include beetroot, spinach beet & sugar beet

beetle lily *Baeometra uniflora*; **b.weed** *Galax urceolata*

Befaria Mutis = *Bejaria*

beggar tick *Bidens* spp. esp. *B. pilosa*; **b.weed** *Desmodium* spp.

Begonia Tourn. ex L. Begoniaceae. 1800+ trop. & warm esp. Am. (Afr. 111; R [sect. *Tetraphila* (30; As. 1: ***B. afromigrata*** J. de Wilde (Laos, Thailand) – long distance disp.?)]: WAUP 2001–2(2002)13); As. c. 750 (Malay Pen. 52 + 2 natur. – R: R. Kiew (2005) *Bs of Peninsular Malaysia*; Borneo 111 but poss. 600), Aus. 0, in 63 sections. R: WAUP 98–2(1998)63, CUSNH 43(2002); M.C. Tebbitt (2005) *Bs: cultivation*.... Monoec. herbs (some reduced, with just 1 functional leaf – cf. *Monophyllaea*) to pachycaul shrubs with fibrous or tuberous r. or rhiz. & upright, cli. or 0 stems, many epiphytic; lvs usu. consp. asymmetric (elephant-ear) some with velvety surface appearing diff. colours at diff. light angles, e.g. ***B. thaipingensis*** King (Malay Pen.) blue-grey to gold, or with silvery spots due to raised air-filled cells, some bipinnate (e.g. ***B. sinofloribunda*** Dorr (SE China); some cauliflory (***B. cauliflora*** M. Sands (Borneo)), epiphyllous infls, branches & lvs in different spp., some prop. by leaf-cuttings; some with hooked or fleshy fr. (evolved in diff. sections, at least twice in Afr.) or seeds (animal-disp.) or seeds of some with air-filled cells acting as balloons. ***B. ferruginea*** L.f. (trop. Am.) – hummingbirds visit ♀ fls for nectar & also nectarless ♂ mimics. 2n = 14–156. Over 10 000 hybrids & cvs recorded (bedding-pl. trade in US 2004–5 worth $56M p.a.), the genus divided hort. into fibrous-rooted, tuberous & rhiz. Most cult. are fibrous-rooted ***B.* Semperflorens-cultorum Group** (*B. cucullata* Willd. (*B. c.* var. *hookeri, B. semperflorens*, SE Braz., NE Arg.) × *B. schmidtiana* Regel (Braz.) late C19 & later with other spp.), tuberous ***B.* Tuberhybrida Group** (*B* × *tuberhybrida*; cvs derived from crossing a number of Andean spp., 1860s onwards; J. Haegeman (1979) *Tuberous Bs.*) & rhiz. ***B. rex*** Putzeys (Assam, though pls. grown under this name are generally hybrids involving other spp. as well = ***B.* Rex-cultorum Group**). Other cult. spp. incl. ***B. grandis*** Dryander (*B. evansiana*, E As.) – long cult. China (Jap. by 1641), hardy in Br., ***B. masoniana*** Irmscher (iron cross, S China, ?hybrid), ***B.* × *phyllomaniaca*** Mart. (*B. incarnata* Link & Otto (Mex.) × *B. manicata* Brongn. ex Cels (Mex.) raised before 1853) – lvs on stems & lvs; ***B. socotrana*** Hook.f. (Socotra) – bulbous, winter-flowering habit now bred into fibrous-rooted & tuberous hybrids like ***B.* Hiemalis Group** (*B.* × *hiemalis* Fotsch., *B. socotrana* × *B.* Tuberhybrida Group) esp. 'Rieger' bs, grown comm. on huge scale as potpls) & ***B.* × *cheimantha*** Everett ('Christmas' b., *B. dregei* Otto & Dietr. (S Afr.) × *B. socotrana*). Some medic., beverage & pigfood in China, some Mex. spp. with ed. petioles, ***B. muricata*** Blume (*B. tuberosa*, Mal.) with ed. lvs. Journal: *The Begonian*

Begoniaceae C. Agardh. Magnoliidae – Cucurbitales. 2/1800+ trop. & warm. R: SCB 60 (1986). Succ. herbs or shrubs, s.t. cli., usu. monoec. oft. with crassulacean acid metabolism, accumulating free organic acids (e.g. malic & oxalic) in cells. Lvs in spirals, s.t. distich., usu. asymmetric & simple, s.t. palmately lobed or compound; stip. oft. large, free. Infls usu. axillary, cymose; fls oft. irreg. P all petaloid, 2 sets of 5 (the outer larger) but more oft. fewer with 2 unlike valvate sets of 2 in males & a single imbr. set of 5 in females (10 in *Hillebrandia*), rarely connate at base; A 4 to ± ∞, originating centripetally, s.t. arr. on 1 side of fl., anthers (yellow like stigmas) opening by longit. slits or term. pores, the connective oft. elongated so that pollen-sacs well-separated; $\overline{G}$ (usu. 2 or 3, up to 6) with axile (parietal in *B.* sect. *Coelocentrum*) placentas or these not quite meeting, ovary v. oft. with (1–)3(–6) prominent wings, styles distinct or connate at base, bifid; stigma resembling A (?deceit poll.), ovules ∞, anatropous, bitegmic. Fr. usu. loculicidal capsule, rarely berry; seeds ∞, small with tiny straight embryo & almost 0 endosperm. x = 10–21+

Genera: *Begonia, Hillebrandia*

Cult. orn. (*Begonia,* q.v.)

Begoniella Oliv. = *Begonia*

Beguea Capuron. Sapindaceae. 4–5 Madag.

Behaimia Griseb. (~ *Lonchocarpus*). Leguminosae (III 16). 1 Cuba: ***B. cubensis*** Griseb. – good timber

Behnia Didr. Asparagaceae (Behniaceae, Philesiaceae). 1 S & SE Afr.: ***B. reticulata*** (Thunb.) Didr. R: FZ 13(2008)30

Behniaceae Conran & al. = Asparagaceae

Behria Greene = *Bessera*

Behuria Cham. Melastomataceae. 15 S Braz.

bei mu *Fritillaria cirrhosa*

Beilschmiedia Nees. Lauraceae (I). c. 200 trop. (Madag. 9; Am. 28 – R: AMBG 86(1999)664) to Aus., NZ, C Chile. Aubréville's Model. Some good timbers (bolly gums, Aus.) incl. ***B. bancroftii*** (Bailey) C. White (Canary ash, yellow walnut, Queensland), ***B. tarairi*** (Cunn.) Kirk (taraire, NZ) & ***B. tawa*** (Cunn.) Kirk (tawa, NZ); ***B. kweo*** (Mildbr.) Robyns (mkweo, Tanzania) – form. exported to Ger. for luxury panelling etc. but supplies exhausted c. 1945; ***B. roxburghiana*** Nees (Himal.) – wood for house-building & tea-chests

Beirnaertia Louis ex Troupin. Menispermaceae (I). 1 trop. Afr.: ***B. cabindensis*** (Exell & Mendonça) Troupin

Beiselia Forman. Burseraceae (I). 1 Mex.: ***B. mexicana*** Forman – G 10–12, each with 2 superposed ovules & fr. with 10–12-flanged columella

Bejaranoa R. King & H. Robinson (~ *Eupatorium*). Compositae (Eupat.-Gyp.). 2 trop. Am. R: MSBMBG 22(1987)99

Bejaria Mutis. Ericaceae (V 1). 15 trop. & warm Am. R: FN 66(1995)69. ***B. racemosa*** Vent. (? = *B. paniculata* Cels ex Dum. Cours., SE US) & other spp. (Andes rose) like rhododendrons in veg.

Bejaudia Gagnepain = *Myrialepis*

bela *Jasminum sambac*

Belairia A. Rich. = *Pictetia*, though perhaps distinct (6 Cuba)

Belamcanda Adans. = *Iris*

Belandra S.F. Blake = *Prestonia*

Belangera Cambess. = *Geissois*

Belemia Pires. Nyctaginaceae (4). 1 Braz.: ***B. fucsioides*** Pires

Belencita Karsten. Capparaceae. 1 Colombia, Venez.: ***B. nemorosa*** (Jacq.) Dugand

beli *Aegle marmelos, Jasminum sambac*

belian *Eusideroxylon zwageri*

Belicea Lundell = *Morinda*

Beliceodendron Lundell = *Lecointea*

bell pepper *Capsicum* spp.; **Qualup b.** *Pimelea physodes*; **b. tree** *Halesia* spp.; **b.wort** *Uvularia* spp.

bella umbra *Phytolacca dioica*

belladonna *Atropa belladonna*; **b. lily** *Amaryllis belladonna*

Bellardia All. (~ *Bartsia*). Orobanchaceae. Incl. *Parentucellia*, 3 Medit., 45 Andes. ***B. latifolia*** (L.) Cuatrec. (*P. l.*, Medit.) – introd. China; ***B. viscosa*** (L.) Fisch. & C. Meyer (*P. v.*, yellow bartsia) – ann. weed in e.g. US

Bellardiochloa Chiov. (~ *Poa*). Gramineae (XVI 15). 4 SE Eur., Middle E.

Bellendena R. Br. Proteaceae (I). 1 Tasmania: ***B. montana*** R. Br. R: FA 16(1995)125, 472

Bellevalia Lapeyr. (~ *Muscari*). Asparagaceae (Hyacinthaceae, Liliaceae s.l.). 45 Med. (Eur. 10) to Iran & N Afghanistan. R: PJBJS 1(1940)42, 131, 336. Cult. orn. incl. ***B. romana*** (L.) Sweet (Roman hyacinth, Med.)

bellflower *Campanula* spp.; **b., Chilean** *Lapageria rosea*; **b., climbing** *Gloriosa modesta*; **b., ivy-leaved** *Wahlenbergia hederacea*

Bellida Ewart. Compositae (Gnap.-Ang.). 1 SW Aus.: ***B. graminea*** Ewart. R: Nuytsia 8(1992)367

Bellidastrum Scop. (~ *Aster*). Compositae (Ast.-Ast.). 1 C & S Eur. mts: ***B. michelii*** Cass.

Belliolum Tieghem = *Zygogynum*

Bellis Tourn. ex L. Compositae (Ast.-Bell.). 8 Eur. (7), Medit. Daisies (daisy = day's eye because closing at night), some medic. Cult. orn. esp. forms of ***B. sylvestris*** Cirillo (SE

Eur.) and hybrids with ***B. perennis*** L. (widespread lawn weed, 'innocence' in 'Language of Fls', medic. (coughs etc.)); '**Prolifera**' with sec. heads in axils of involucral bracts (hen-&-chickens d.), 'double' forms with all ligulate florets (bachelor's buttons) etc.

Bellium L. Compositae (Ast.-Bell.). 4 Medit. Eur. (3). Some cult. orn.

Belloa Remy (~ *Lucilia*). Compositae (Gnap.-Gnap.). 9 Andes of Venez. to C Chile. R: BJLS 106(1991)189

Bellonia Plum. ex L. Gesneriaceae (II 5a). 2 WI. Woody; axillary thorns (? infls.) produced in dry localities, 0 in wet

bells of Ireland *Moluccella laevis*

Bellucia Necker ex Raf. (~ *Loreya*). Melastomataceae. 8 trop. Am. R: MNYBG 50(1989)5, RACCE 20(1996)244. Trees & shrubs (Leeuwenberg's Model); fls 5–8-merous, G 10–14(15)-loc. Fr. ed., ***B. pentamera*** Naud. (*B. axinanthera*) grown in Mal. as fr. & orn. tree

Bellynkxia Muell. Arg. = *Morinda*

Belmontia E. Meyer = *Sebaea*

Beloglottis Schltr. (~ *Spiranthes*). Orchidaceae (IV 2h). 7 SE US to Braz. R: HBML 28(1982)302

Belonanthus Graebner = *Valeriana*

Belonophora Hook.f. Rubiaceae (II 3). 5 trop. Afr. esp. W. R: KB 55(2000)71

Beloperone Nees = *Justicia*

Belostemma Wall. ex Wight = *Vincetoxicum*

Belosynapsis Hassk. (~ *Cyanotis*). Commelinaceae (II 1c). 4 Madag., Ind. to NG. ***B. vivipara*** (Dalz.) C. Fischer (India) – 1 of few C. epiphytes

Belotia A. Rich = *Trichospermum*

Beltrania Miranda = *Enriquebeltrania*

beltia hyacinthe *Bletilla striata*

bel-tree *Aegle marmelos*

Belvisia Mirb. = *Lepisorus* (but see Blumea 37(1993)511)

Bemangidia L. Gaut. Sapotaceae (Tseb.). 1 SW Madag.: ***B. lowryi*** L. Gaut. R: T 62(2013)979

Bemarivea Choux = *Tinopsis*

Bembicia Oliv. Salicaceae (Flacourtiaceae). 4–5 Madag. $\overline{G}$ infl. strobiloid

Bembicidium Rydb. = *Poitea*

Bembiciopsis Perr. = *Camellia*

Ben, oil of *Moringa oleifera*

Bencomia Webb & Berth. = *Poterium* (but see R: BMac. 6(1980)71)

bendang *Agathis dammara*

Benedictella Maire = *Lotus*

Benedictine flavour due to *Angelica archangelica*

Beneditaea Tol. = *Ottelia*

Benevidesia Saldanha & Cogn. Melastomataceae. 2 SE Braz.

Bengal bean *Mucuna pruriens* var. *utilis*; **B. cardamom** *Amomum aromaticum*; **B. kino** *Butea monosperma*; **B. quince** *Aegle marmelos*

benge *Guibourtia arnoldiana*

Benguellia G. Taylor. Labiatae (VII 3d). 1 Angola: ***B. lanceolata*** (Guerke) G. Taylor. R: JB 69, suppl.(1931)156

beni seed *Polygala butyracea*; **black b. s.** *Hyptis spicigera*

Benin mahogany *Khaya* spp.; **B. walnut** *Lovoa trichilioides*; **B. wood** *K. grandifoliola*

Benincasa Savi. Cucurbitaceae (XIV). Incl. *Praecitrullus*, 2 trop. As. ***B. hispida*** (Thunb.) Cogn. (*B. pruriens*, wax or white gourd, ash pumpkin, petha) – cultigen growing up to 2.3 cm in 3 hrs, fr. coated with wax (vessel for scented coconut oil in Polynesia before Eur. contact), boiled as veg. with curry, candied or pickled, local medic. R: FM I,19(2010)39

Benitoa Keck (~ *Lessingia*). Compositae (Ast.-Mac.). 1 Calif.: ***B. occidentalis*** (H.M. Hall) Keck. R: SBM 20(2000)20

Benjamin, gum = benzoin; **stinking B.** *Trillium erectum*; **B. tree** *Ficus benjamina*

Benjaminia Mart. ex Benj. Plantaginaceae (Grat.-Grat.; Scrophulariaceae s.l.). 1 Venez., Braz.: ***B. reflexa*** (Benth.) D'Arcy – aquatic ann.

Benkara Adans. (~ *Randia*). Rubiaceae (II 1). Incl. *Fagerlindia*, 19 Indomal. R: Reinw. 12(2008)398. ***B. fasciculata*** (Roxb.) Rids. – fr. for sunstroke treatment

Bennettiodendron Merr. Salicaceae (Flacourtiaceae). 2–3 Indomal., China

Benoicanthus Heine & A. Raynal = *Ruellia*

Benoistia Perrier & Leandri. Euphorbiaceae (Al.-Neo.). 3 Madag.

Bensoniella Morton (~ *Mitella*). Saxifragaceae. 1 W US: ***B. oregona*** (Abrams & Bacig.) Morton

Benstonea Callm. & Buerki (~ *Pandanus*). Pandanaceae. c. 60 India to Fiji. Usu. acaulescent, often epiphytes (*P.* sect. *Acrostigma*). R: Cand. 67(2012)328

bent grass *Agrostis* spp.; **black b.** *A. gigantea*; **brown b.** *A. canina*; **common** or **colonial b.** *A. capillaris*; **silky b.** *Cynosurus* spp.; **velvet b.** *A. canina*

Benthamantha Alef. = *Coursetia*

Benthamia A. Rich. Orchidaceae (IV 4d). 29 Madag., Mascarenes. R: FMad. 49,1(1939)16

Benthamidia Spach = *Cornus*

Benthamiella Speg. Solanaceae (Benth.). 12 S Patagonia. Cushion pls in open steppe. R: BN 133(1980)67

Benthamina Tieghem (~ *Amyema*). Loranthaceae (5 3). 1 E Aus.: ***B. alyxifolia*** (Benth.) Tieghem

Benthamistella Kuntze = *Buchnera*

Bentia Rolfe = *Justicia*

Bentinckia A. Berry. Palmae (V – tribe?). 2 Ind., Nicobar Is.

Bentinckiopsis Becc. = *Clinostigma*

Bentleya E. Bennett (~ *Billardiera*). Pittosporaceae. 2 SW Aus. R: BJLS 103(1990)309. Rhizomatous shrubs

benzoin (gum Benjamin) resinous balsams derived from *Styrax* spp.

Benzingia Dodson. Orchidaceae (V 12j). Incl. *Ackermania*, 9 trop. S Am.

Benzonia Schum. Rubiaceae (?). 1 W Afr.: ***B. corymbosa*** Schum.

Bequaertia R. Wilczek (~ *Campylostemon*). Celastraceae (III). 1 trop. Afr.: ***B. mucronata*** (Exell) R. Wilczek. R: BJBB 26(1956)399

Bequaertiodendron De Wild. = *Englerophytum*

ber *Ziziphus mauritiana*

Berardia Villars. Compositae (Card.-Card.). 1 W Alps: ***B. subacaulis*** Villars. R: ANMW 99,B(1996)329

Berberidaceae Juss. Magnoliidae – Ranunculales. 14/600 N temp. to trop. mts. Trees, shrubs (some pachycaul &/or spiny at nodes) or perenn. herbs, usu. glabrous, oft. with alks & tissues coloured yellow with berberine (an isoquinoline); vascular bundles oft. ± scattered, woody spp. oft. with broad medullary rays. Lvs in spirals (opp. in *Podophyllum*), pinnate, ternate or simple (unifoliolate with articulation at base of leaflet in some *Berberis* spp.); stip. small or 0, though petiole oft. flared nr base. Fls in usu. term. racemes, cymes or solit., bisex., reg., (2, *Epimedium*)3(4)-merous, with cortical vasc. system. P (aborts in *Achlys*) usu. of 6 or 7(–9) series, typically outer 2 (? sepals but oft. considered bracts) sepaloid, oft. caducous, next 2 (? nectary-less C but oft. considered petaloid K) petaloid, inner 2 or 3 (? nectariferous C but oft. considered staminodes) usu. petaloid & nectariferous (not in *Diphylleia* nor *Podophyllum*); A (4–)6(–18) usu. same no. as nectariferous C, but s.t. 2 x & usu. opp. C, anthers usu. opening by 2 valves that lift up from base (longit. slits in *Nandina* & *Podophyllum*); $\underline{G}$ app. 1 but oft. interpreted as derived from 3, ovules anatr. or hemitropous, bitegmic, commonly ∞ on a thickened marginal placenta, or 2 or (*Achlys*) 1, basal. Fr. usu. a berry, seldom dry, dehiscent or not; seeds oft. arillate, embryo small, endosperm abundant with oils, protein & s.t. hemicellulose. x = 6, 7, 8, 10

Classification:

I. **Nandinoideae** (shrubs, A with longit. dehiscence, seeds endotegmic): *Nandina*

II. **Berberidoideae** (shrubby or herbaceous, A with valvate (longit.) dehiscence, seeds exotestal):

1. **Leonticeae** (ovules 1–4; R: CJB 67(1989)2310): *Caulophyllum, Gymnospermium, Leontice*
2. **Berberideae** (ovules several): *Berberis, Bongardia, Diphylleia, Dyosma, Epimedium, Jeffersonia, Podophyllum, Ranzania, Vancouveria*

Many cult. orn. esp. spp. of *Berberis* (some alt. hosts of stem-rusts of cereals), *Epimedium*; *Podophyllum* medic.

Berberidopsidaceae Takht. Magnoliidae – Berberidopsidales. 2/3 Aus., Chile. R: KB 57(2002)174. Everg. woody scramblers with cyanogenic glycosides. Lvs in spirals with palmate sec. venation; stip. 0. Fls bisex., solit. axillary or in term. racemes. P (9-)12(-15) in spirals with distinct K & C or transitional members. Nectary-disk extrastaminal, lobed, or 0. A 6 – ∞, in whorls or irreg., arising from sinuses in disk if present, anthers dehiscing by slits, pollen tricolpate. $\underline{G}$ (3, 5) with parietal placentation & 2-∞ ovules per carpel. Fr. a berry topped by persistent style base

Genera: *Berberidopsis, Streptothamnus*
Form. in 'dustbin' Flacourtiaceae (Malpighiales)

Berberidopsis Hook.f. Berberidopsidaceae. 2 Chile (1), E Aus. (1). R: Blumea 30(1984)21. ***B. corallina*** Hook.f. (Chile, extinct in wild?) cult. orn.

Berberis Tourn. ex L. Berberidaceae (II 2). (Incl. *Mahonia*, × *Mahoberberis*) 500+ Euras. (Eur. 2, China c. 200), N Afr., trop. Afr. mts, Am. R: JLSBot. 57(1961)1. Shrubs (barberries), usu. spiny, with alks & yellow wood (berberine) form. used in eye disease; fr. of many ed., wood a dyestuff; many cult. orn. (some with Corner's Model) incl. hybrids esp. as hedges & ground-cover. Simple lvs of leptocaul *Berberis s.s.* derived from leaflets of pachycaul ancestors with thorny pinnate lvs (*Mahonia*); lvs of long shoots tripartite spines (transitions to 'true' lvs oft. present), short leafy & flowering shoots in their axils. A sensitive, springing upwards on contact by insect & showering side of its head with pollen. Several spp. are alt. hosts of stem-rusts (first proposed by Banks (1805) but not accepted until de Bary (1865–6)) of wheat, oats, barley & rye & attempts at their eradication in US etc. have been made, though not a certain or indispensable phase in life-cycle of *Puccinia graminis*, though the more dangerous *P. striiformis* (yellow rust) will also grow on *Berberis*, while in Russia removal not advocated as *B.* spp. imp. honey-pls. ***B. aquifolium*** Pursh (*M. a.*, Oregon or Rocky Mt. grape, NW N Am.) – berries cooked by Native Americans, local medic., planted as game-cover in Eur. (invasive in C & E Eur.); ***B. aristata*** DC. (chitra, Nepal) – bitter tonic for fevers; ***B. canadensis*** Mill. (American or Alleghany barberry, E N Am.) – ed. fr., alt. host for rusts; ***B. glaucocarpa*** Stapf (trop. As.) – invasive NZ; ***B. × hortensis*** Mabb. (*M. × media*, *B. japonica* (Thunb.) Spreng. (*M. j.*, China, long cult. Jap.) × *B. oiwakensis* (Hayata) Laferr. (*M. lomariifolia*, Myanmar, W China)) – accidental cross in 1950s Ireland, commonly planted sweetly scented, winter-flowering pachycaul, several cvs. esp **'Charity'**; ***B. leschenaultii*** Wall. ex Wight & Arn. (*M. l.*, Ind., E As.) – natur. Aus.; ***B. repens*** Lindl. (*M. r.*, holly grape, W N Am.) – cult. orn., locally medic., fr. used in jellies, drinks, etc.; ***B. × stenophylla*** Lindl. (*B. darwinii* Hook. (temp. S Am., introd. Aus., NZ where a pest) × *B. empetrifolia* Lam. (Chile, Arg.)) – common hedgepl.; ***B. swaseyi*** Buckley ex M.J. Young (*M. s.*, Texas) – promising fr.-bush; ***B. thunbergii*** DC. (As.) – cult. orn., invasive C Eur., US; ***B. vulgaris*** L. (common or Eur. barberry, Eur.) – allegedly introd. Eur. by Arabs, leading to rust problems in cereals, natur. GB, form. much planted N Am. for hedges, fine wood for turning, toothpicks, a dyestuff for silk, cotton, wool & leather & staining wood, form. a hair-dye, fr. (laxative) preserved esp. a seedless form in a French jam, bark etc. form. medic.

Berchemia Necker ex DC. Rhamnaceae (6). c. 20 E Afr. to E As., New Caled. (1), W N Am. (1). Cult. orn. twiners. ***B. discolor*** (Klotzsch) Hemsl. (trop. & S Afr., Madag.) – fr. ed. Uganda.

Berchemiella Nakai. Rhamnaceae (6). 2 China, Jap. R: BBR 8,4(1988)119

berdi *Typha domingensis*

bere (= **bear**) *Hordeum vulgare*

Berendtia A. Gray = *Hemichaena*

Berendtiella Wettst. & Harms = *Hemichaena*

Berenice Tul. Campanulaceae (I). 1 Réunion: ***B. arguta*** Tul.

berg bamboo *Bergbambos tessellata*; **b. lily** *Ornithogalum candicans*

bergamot *Monarda* spp. esp. *M. didyma*; **b. orange** *Citrus × limon*

Bergbambos Stapleton (~ *Thamnocalamus*). Gramineae (IV). 1 S Afr. mts: ***B. tessellata*** (Nees) Stapleton. R: Phytokeys 25(2013)99

Bergenia Moench. Saxifragaceae (I 4). 10 temp. & subtrop. E As. R: APS 26(1988)20. Cult. orn. (elephant-ear, Siberian saxifrage) esp. as ground-cover; some a source of tannin & medic., many hybrids (high degree of self-incompatibility), most commonly cult. ***B.*** × *schmidtii* (Regel) Silva Tarouca (*B.* ***crassifolia*** (L.) Fritsch (Siberia, Mongolia)) ×***B. ciliata*** (Haw.) Sternb. (*B. ligulata*, W Pakistan to SW Nepal) – rhizomes a tea-leaf subs. in Kashmir. ***B. emeiensis*** C.Y. Wu ex J.T. Pan (China) – used to treat brain disorders

Bergera Koenig (~ *Murraya*). Rutaceae (II 1). 5 S & SE As. to New Caled. ***B. koenigii*** L. (*M. koenigii*, curry leaf, Ind., Sri Lanka, invasive Venez.) – lvs *always* used in Ind. curries, carbazole alks with antifungal activity, oil used in soap industry

Bergeranthus Schwantes. Aizoaceae (V). 12 E Cape. R: H. E. K. Hartmann, *Ill. Handb. Succ. Pls*, Aiz. A–E (2002)84. Cult. orn.

Bergerocactus Britton & Rose. Cactaceae (III 1). 1 Calif. & Baja Calif.: ***B. emoryi*** (Engelm.) Britton & Rose – fr. extrudes pulp & seeds; hybrids with *Pachycereus* & *Myrtillocactus* spp. in wild. R: CSM 10(1965)51

Bergeronia M. Micheli = *Muellera*

Berghesia Nees. Rubiaceae (?). 1 Mex.: ***B. coccinea*** Nees

Bergia L. Elatinaceae. 24 warm (Aus. 10 – R: JABG 11(1989)80)

Berginia Harv. = *Holographis*

Berhautia Balle. Loranthaceae (5 7). 1 Senegal, Gambia: ***B. senegalensis*** Balle. R: R. Polhill & D. Wiens, *Mistletoes of Afr.* (1998)118

Beringia R. Price & al. = *Transberingia*

Berkheya Ehrh. Compositae (Arct.-Gort.). 75 trop. & S Afr. R: MBSM 3(1959)104. Heterogeneous. ***B. coddii*** Roessler (S Afr.) – nickel accumulator used to revegetate contaminated soils

Berkheyopsis O. Hoffm. = *Hirpicium*

Berlandiera DC. Compositae (Helia.-Eng.). 4 S US, Mex. R: Britt. 19(1967)285. Cult. orn. esp. ***B. lyrata*** Benth. – chocolate-scented

Berlinia Sol. ex Hook.f. Leguminosae (I 2). 21 trop. Afr. R: EJB 63(2006)162; SBM 91(2011)22. Some timbers (abem, ekpogoi)

Berlinianche (Harms) Vatt. = *Pilostyles*

Bermuda arrowroot *Maranta arundinacea;* **B. buttercup** *Oxalis pes-caprae;* **B. cedar** *Juniperus bermudiana;* **B. grass** *Cynodon dactylon*

Bernardia Houston ex Mill. Euphorbiaceae (Acal.-Ber.). 50+ trop. Am. esp. Braz. (SW US 1). Dioec., wind-poll.

Bernardinia Planch. = *Rourea*

Berneuxia Decne. Diapensiaceae. 1 Himal.: ***B. thibetica*** Decne. R: Rhodora 45(1943)335

Berniera Baill. = *Beilschmiedia*

Bernoullia Oliv. Malvaceae (Malv.-Bomb.; Bombacaceae). 3 Mex. to Colombia

Berrisfordia L. Bolus = *Conophytum*

Berroa Beauverd. Compositae (Gnap.-Gnap.). 1 subtrop. S Am.: ***B. gnaphalioides*** (Less.) Beauverd. R: BJLS 106(1991)191

berry, Bay *Morella californica;* **buffalo b.** *Shepherdia argentea;* **Christmas b.** *Schinus terebinthifolius;* **Panama b.** *Muntingia calabura;* **partridge b.** *Gaultheria procumbens;* **phenomenal b.** see *Rubus;* **purple apple b.** *Billardiera longiflora;* **yellow b.** *Rhamnus saxatilis*

Berrya Roxb. Malvaceae (Brown.; Tiliaceae s.l.). 5 Indomal. ***B. cordifolia*** (Willd.) Burret (*B. ammonilla*, Trincomali wood, hamilla, S Ind., Sri Lanka) – valuable red timber e.g. for coconut arrack vats

Bersama Fres. Francoaceae (Melianthaceae). c. 8 trop. & S Afr. Chamberlain's & Rauh's Models

Bersamaceae Doweld = Francoaceae

bertam *Eugeissona tristis*

Berteroa DC. Cruciferae (2). 5 OW N temp. (Eur. 5). R: JAA 68 (1987)207. ***B. incana*** (L.) DC. (hoary Alison) – widely natur.

Berteroella O. Schulz = *Stevenia*

Bertholletia Bonpl. Lecythidaceae (I). 1 trop. S Am.: ***B. excelsa*** Bonpl. (Braz. nut, Pará nut). R: FN 21,2(1990)114. Tree to 50 m (Roux's Model). Ligule (see fam.) pressed down in A & anthers available therefore only to big bees (*Xylocopa* spp. & female euglossines, visitors to orchids for oils); self-sterile, fr. once trees c. 10 yrs old takes 14 months to mature, forming large woody capsule; operculum falls inwards & seeds with hard woody testa & oily endosperm (Braz. nuts of commerce; 100 K people employed) gnawed out by agoutis, which scatter-hoard them. Most (76 K t per annum) coll. from wild trees (logging now banned), the fr. being split with an axe; oil for foodstuffs (source of selenium, an antioxidant) & soap, hair conditioner & cosmetics from seeds (60% oil, 17% protein) in commerce

Bertiera Aubl. Rubiaceae (IX-Bert.). c. 52 trop. Am. & Afr. (41; Madag. 3). Corner's, Petit's & Roux's Models

Bertolonia Raddi. Melastomataceae. 17 SE Braz., Venez. (1). R: AJBRJ 30(1990)97. Cult. orn. foliage pls esp. ***B. maculata*** DC. (SE Braz.) & its hybrids with *Gravesia guttata* Triana (Madag.) & *Sonerila* spp. ***B. mosenii*** Cogn. (SE Braz.) – water-disp., a raindrop forcing seeds to 'corners' of fr. where expelled

Bertya Planch. Euphorbiaceae (Ricinoc.-Bert.). 28 Aus. R: Austrobaileya 6(2002)190. Resins

Berula Koch. Umbelliferae (III 8). Incl. *Afrocarum*, 5 N temp. (Eur. 1), E & S Afr. ***B. bracteata*** (Roxb.) Spalik & Downie (*Sium b.*, jellico [i.e. angelica], St Helena) – form. coll for sale & eaten like celery; ***B. erecta*** (Huds.) Colv. – aquatic, hybridizes with *Helosciadium nodiflorum* in UK, used by Apaches for rheumatism, fatal to cattle in NSW

Beruniella Zak. & Nabiev = *Heliotropium*

Berylsimpsonia B.L. Turner (~ *Proustia*). Compositae (Mut.-Nass.). 2 Greater Antilles. R: Phytol. 74(1993)351. Spiny shrubs

Berzelia Brongn. Bruniaceae (3). 16 Cape. R: T 60(2011)1147. Cult. orn., shoots sold by florists as 'Cape greens'

besan (flour) *Cicer arietinum*

Beschorneria Kunth. Asparagaceae (Agavaceae). 7 C & S Mex. mts. R: Plantsman 10(1989)194. Lvs a soap subs., fls ed., cult. orn. esp. ***B. yuccoides*** K. Koch as in S France

Besleria Plum. ex L. Gesneriaceae (II 3a). 200 warm Am. esp. Andes of Ecuador & Colombia. Trees, shrubs & perenn. herbs

Bessera Schultes f. Asparagaceae. Incl. *Behria*, 3 Calif., Mex. Cult. orn.

Besseya Rydb. = *Veronica*

Beta Tourn. ex L. Amaranthaceae (Chenopodiaceae I 1). Excl. *Patellifolia*, 7 Eur. (6), Med. R(sect. *B.*): WAUP 93–1(1993)31. 'Seeds' actually coherent frs enveloped in woody calyces. Selected forms of ***B. vulgaris*** L. 'subsp. *maritima* (L.) Arc.' (wild sea-beet) cult. since time of Assyrians, 4 cv. groups (some allegedly result of hybridisation between other groups): **Garden Beet Group** (beetroot) – swollen hypocotyl, boiled veg. used in e.g. borshch (soup), medic. Anc. Rome, high concentrations of red betalains (anti-oxidants, used in herbal treatment of cancer, though 14% UK pop. cannot metabolize red betanin leading to beeturia (red urine), natural dyes incl. for food), vitamin C, tyrosine (immune system), iron, folic acid; **Leaf Beet Group** ('subsp. *cicla*', rhubarb chard, spinach beet, Swiss c., 'silver b.' (Aus.)) – r. swollen; **Sugar Beet Group** (sugarbeet) – 30% world's sugar (272 M t p.a.); **Fodder Beet Group** (mangel-wurzel, mangold) – swollen hypocotyl, cattle-feed (UK record: 24.72 kg); all are biennial with sugar reserves in 'r.', those of sugarbeet up to 20% by wt. (first factory in Silesia, 1801), proposed as biofuel, MSG prep. from fermented molasses

bété *Mansonia altissima*

betel (nut) *Areca catechu*

Bethencourtia Choisy (*Canariothamnus*, ~ *Senecio*). Compositae (Sen.-Sen.). 3 Canary Is. R: CN 44(2006)26

Betonica Tourn. ex L. (~ *Stachys*). Labiatae (VI). 15 Euras.

betony *Stachys officinalis*

betsa-betsa fermented cane-sugar, s.t. flavoured with *Nuxia congesta*, the national drink of Malagasy Republic

Betula Tourn. ex L. Betulaceae (I). 35 N hemisph. (Eur. 4, birches). R: JAA 71(1990)32; K. Ashburner & al. (2013) *The genus B.* Decid., monoec. trees & shrubs allied to *Alnus* but catkins shattering when ripe. Timber for furniture, plywood, cotton-reels, brush- & broom-backs, skis etc., sap for sweetening, fermenting & shampoo, twigs for brooms & 'birching' schoolboys, birch bud oil form. giving char. smell to Russian leather; resin poss. medic. disinfectant now used in Finnish tooth-cleaners. Betulinic acid in bark triggers cell death in melanomas but not other cell cultures. Birch bark 1800 yrs old (Afghanistan) bears oldest known Buddhist MSS. ***B. alleghaniensis*** Britton ('*B. lutea*', yellow b., E N Am.) – good timber; ***B. lenta*** L. (American black or cherry b., E N Am.) – timber (form. imp. UK esp. for Windsor chair seats); tepee pegs for Blackfoot, distilled bark gives oil high in methyl salicylate used medic. (wintergreen), 'birch beer' made from fermenting sap, now a major flavouring in r. beer; ***B. maximowicziana*** Regel (Jap. b., Jap.); ***B. michauxii*** Spach (NE Canada) – creeping shrub; ***B. nana*** L. (dwarf b., circumpolar); ***B. neoalaskana*** Sarg. (*B. resinifera*, Russian Far East, Alaska, Yukon) – triterpenoid papyriferic acid protects against depredation by snowshoe hares; ***B. nigra*** L. (red b., E N Am.); ***B. papyrifera*** Marshall (canoe b., paper b., N. Am.) – timber for turning, shoe-lasts, pegs & pulp, bark impervious to water & form. used for canoes, baskets, cups & wigwam covers by Native Americans; ***B. pendula*** Roth (only native tree in Iceland) & ***B. pubescens*** Ehrh. (common or Eur. birches (brown b. = *B. pu.*), poss. only subspecifically distinct, many hybrids = B. × ***aurata*** Borkh.) – reach N limit of tree-growth & form. v. important for Scottish Highlanders (constr., carts, furniture, fences, fuel, fish-smoking (still used for haddock), tanbark), wood still used for furniture esp. in Scandinavia, forms with pretty markings (cause obscure) known as Karelian b. or masur, bark a famine food or eaten with sturgeon eggs in E As. as well as used for shoes, clothes etc., medic. (eczema in Ireland) & imported to anc. Egypt to decorate boxes, tar used for preserving leather & wood, lvs fermented to 'birch wine' & a tonic 'wine', & with alum give a green dye, with chalk a yellow one, oil medic., sprigs used to decorate churches at Whitsun in GB, forms of *B. pendula* (silver b.) grown for ornament esp. weeping '**Youngii**', shrubby variants in wild, e.g. '*B. oycoviensis* Besser' (SE Poland) perhaps of hybrid orig., while ***B. pubescens*** f. ***columnaris*** Ulvinen (Finland) is fastigiate; ***B. populifolia*** Marshall (NE N Am.) – used as a lead indicator in Wisconsin; ***B. schmidtii*** Regel (E As.) – wood too dense to float in water; ***B. utilis*** D. Don (Himal.) – bark form. used as writing & packing material, subsp. ***jacquemontii*** (Spach) Ashburner & McAll. – cult. orn. with whitest bark

Betulaceae Gray. Magnoliidae – Fagales. Incl. Corylaceae, 6/140 N temp., trop. mts. R: R. Govaerts & D. G. Frodin, *World checklist... Fagales* (1998)11, 83. Decid. trees or shrubs, monoec. (or dioec.), anemophilous, typically with ectotrophic mycorrhizae in r. Lvs usu. in spirals, simple, usu. conduplicate & toothed; stip. decid. Male fls in ± elongate pend. catkins, females in pend. or erect, short, oft. woody ones with dichasia subtended by bracts. K (0)1–6, scale-like; C 0; A same no. & opp. K or app. <18 (poss. a congested 3-flowered cymule), free or united at base, the pollen-sacs ± distinct (not in *Alnus*); $\underline{G}$ (2(3)) with ± distinct styles, 2(3)-loc. below, 1-loc. above, ovules axile, pend. from nr. summit of partition, 1 or 2 each locule, anatropous, unitegmic or bitegmic (*Carpinus*); fert. chalazogamous after delayed growth of pollen-tube. Fr. a nut or 2-winged samara, in Coryloideae subtended or almost encl. in 2 or 3 leafy bracts; seed usu. 1 with thin fleshy endosperm or 0, embryo with oily thickened cotyledons. x = 8, 11, 14

Classification & genera:

I. **Betuloideae** (vessel-elements not spirally thickened, nut samaroid): *Alnus, Betula*

II. **Coryloideae** (vessel-elements spirally thickened, nut large, scarcely flattened): *Carpinus, Corylus, Ostrya, Ostryopsis*

These subfams have been treated as sep. fams but serological study shows their unity (Cronquist). Pollen attrib. to Betulaceae known from Upper Cretaceous, wood attrib. to *Alnus* & *Carpinus* from Eocene. Up to 4 months between poll. & fert. Timber, nuts & cult. orn.

Bewsia Goossens (~ *Leptochloa*). Gramineae (XXIX 3). 1 C & S Afr.: ***B. biflora*** (Hack.) Goossens. R: Strel. 10(2000)684

Beyeria Miq. Euphorbiaceae (Ricinoc.-Bert.). 24 Aus. Turpentine bushes. R: Austrobaileya 7(2008)581

Beyrichia Cham. & Schldl. = *Achetaria*

bezetta rubra dye from *Chrozophora tinctoria*

bhaji *Amaranthus* spp.

bhang *Cannabis sativa* 'subsp. *indica*'

Bharbur grass *Eulaliopsis binata*

Bhesa Buch.-Ham. ex Arn. Centroplacaceae. 6 Indomal. R: Blumea suppl.4(1958)150. Form. in Celastraceae. Arils ed.

Bhidea Stapf ex Bor (~ *Andropogon*). Gramineae (XXII 7). 3 Ind.

Bhutanthera Renz. Orchidaceae (IV 4d). 5 E Himal. to SW China. R: EJB 58(2001)99

Bia Klotzsch (~ *Tragia*). Euphorbiaceae (Acal.-Pluk.-Trag.). 5 trop. Am. R: NJB 31(2013)595

Biancaea Tod. (~ *Caesalpinia*). Leguminosae (I 4). 6 trop. As. R: PhytoKeys 71(2016)69. ***B. decapetala*** (Roth) Degener (*C. d.*, *C. sepiaria*, Mysore thorn, Ind. to Jap.) – spiny scrambler, used for hedging, noted as hallucinogenic in old Chinese herbals, invasive S Afr., NZ; ***B. sappan*** (L.) Tod. (*C. s.*, Braz. wood, sappanwood, Indomal.) – heartwood (prob. first b. w., 'brazilium' (C12)), form. source red dye, cult. orn.

Biarum Schott. Araceae (VII 23). 21 Med. (Eur. 8). R: Aroideana 29(2006)2, BM 25(2008)2. ***B. tenuifolium*** (L.) Schott (S Eur.) – infls smellable at 20 m

bibolo *Lovoa trichilioides*

bicacaro *Canarina canariensis*

bicuhyba fat from seeds of *Bicuiba oleifera*

Bicuiba de Wilde (~ *Virola*). Myristicaceae. 1 Braz.: ***B. oleifera*** (Schott) de Wilde (*V. bicuhyba*, *V. oleifera*, bicuhyba) – seed-oil for candles etc.

Bidaria (Endl.) Decne = *Marsdenia*

Bidens Tourn. ex L. Compositae (Cor.). Incl. *Megalodonta* (R: AB 21(1985)99), 340 cosmop., esp. Mex. (Eur. 3 + natur. spp., N Am. 25, Afr. 63 (R: KB 48(1993)437), Hawaii 19 all interfertile). R: PFMB 16(1937)1. Heterogeneous, allied to *Cosmos*, but incl. Afr. spp. of *Coreopsis* (striate, grooved caryopses unlike true *Cor.*). Shrubs, lianes, perenn. & ann. herbs; pappus of few mostly retrorsely barbed awns involved in fr. disp. (even by migrating salamanders in Ontario). ***B. beckii*** Torrey ex Spreng. (*Megalodonta b.*, N Am.) – aquatic with entire aerial lvs & dissected submerged ones, v. sensitive to pollution; ***B. cosmoides*** (A. Gray) Sherff (Hawaii) – bird-poll. pendent capitula. Some cult. orn. e.g. ***B. aurea***

(Aiton) Sherff (*B. ferulifolia*, S US to C Am.) esp. '**Golden Eye**' in hanging-baskets, but many weeds (bur-marigold, cuckold, pitchfork, sticktight, tickseed, beggar tick), esp. ***B. pilosa*** L. (blackjack (Am.), cobbler's pegs (Aus.), trop., invasive Taiwan, S Afr., Aus.) – 3 or 4 generations a yr, 1 pl. can produce up to 6000 fr. a yr, r. exudates allelopathic to lettuce, beans, maize & sorghum; some medic. locally e.g. *B. pilosa* against diarrhoea in Burundi

bidgee-widgee *Acaena* spp.

bidi (Ind.) cheap cigarette rolled in a leaf, not paper

bidi-bidi *Acaena* spp.

Bidoupia Aver. & al. Orchidaceae (Goodyer.). 1 Vietnam: ***B. phongii*** Aver. & al. R: Phytotaxa 266(2016)289

Bidwillia Herb. = ? *Trachyandra*

Biebersteinia Stephan. Biebersteiniaceae (Geraniaceae s.l.). 5 Greece (? extinct) to C As.

Biebersteiniaceae Schnizl. Magnoliidae – Sapindales. 1/5 Greece to C As. Perenn. herbs s.t. with tubers, glandular – hairy, s.t. foetid. Lvs pinnatisect or imparipinnate with lobed to compound leaflets, in spirals; petiolar stip. Infl. an erect spike or panicle; fls reg., bisex., 5-merous. K free, imbr.; C free, imbr., oft. clawed, alt. with 5 fleshy nectary glands; A 10; G 5 on short gynophore with 1 pend. ovule per loc., styles discrete, apically connate, stigma capitate. Fr. a schizocarp with persistent columella & ± accrescent K, nutlets 5 crustaceous; seeds large, rugulose with slightly curved embryo (cotyledons leafy) & scant endosperm. n = 5

Genus: *Biebersteinia*

Form. in Geraniaceae s.l. but ethereal oils & 1 ovule per carpel, distinctive flavones & methyl esters like Rutaceae place B. in Sapindales, confirmed by DNA analysis

Bienertia Bunge ex Boiss. Amaranthaceae (Chenopodiaceae III 2). 3 Eur. to W C As. R: PB 146(2012)554. 2 incl. ***B. cycloptera*** Bunge ex Boiss. C 4 physiology but not Kranz anatomy

Biermannia King & Pantl. Orchidaceae (V 16c). 11 trop. As to Bali

Bifora Hoffm. Umbelliferae (III 4). 3 Med. (Eur. 2) to C As.

Bifrenaria Lindl. Orchidaceae (V 12g). 21 trop. Am. R: Britt. 56(2004)318. Cult. orn.

big *Hordeum vulgare*; **b. tree** *Sequoiadendron giganteum*

bigarade *Citrus* × *aurantium* Sour Orange Group

Bigelowia DC. Compositae (Ast.-Sol.). 2 SE US. R: Sida 3(1970) 451

bignay *Anitdesma bunias*

Bignonia Tourn. ex L. Bignoniaceae (3). Incl. *Clytostoma*, *Cydista*, *Phryganocydia*, *Saritaea*, 29 trop. Am. to SE US (1). R: AMBG 98(2014)415. Some toxic to stock; many cult. orn. lianes incl. ***B. capreolata*** L. – 4-armed 'anomalous' xylem & 4 conspicuous bark ribs (phloem), ***B. corymbosa*** (Vent.) L. Lohmann (*P. c.*) – fragrant pink or purple fls, ***B. magnifica*** Bull (*S. m.*, Colombia, Ecuador) – large purple to rose fls, scent coll. male euglossine bees; ***B. diversifolia*** Kunth (*Cydista d.*, Mex. to C Am.) – mass-flowering in Costa Rica; ***B. nocturna*** (Barb. Rodr.) L. Lohmann (*Tanaecium n.*, Braz.) – hallucinogen, smelling of almonds

Bignoniaceae Juss. Magnoliidae – Lamiales. 82/860 mainly trop. esp. S Am. Trees, lianes, shrubs, rarely herbs, lianes usu. with unusual vascular structure. Lvs opp., s.t. whorled, rarely in spirals, pinnate to 3-compound, less oft. simple or palmately compound, terminal leaflet s.t. a tendril; stip. 0. Fls bisex., in thyrses, racemes or solit., usu. conspic. K (5), s.t. bilobed, or unlobed, rarely with a calyptra (*Lundia* spp.); C (5), oft. 2-lipped, rarely ± reg., imbr. or rarely valvate; A alt. with C but attached to tube, 4, in 2 prs, the fifth (adaxial) staminodal or 0, rarely all 5 fertile (*Oroxylum*) or 2 fertile & 3 staminodal (*Catalpa*); pollen v. diverse in structure; annular or cupular nectary-disk usu. around $\underline{G}$ (2) with 2-lobed stigma, 2-loc. with 2 axile placentas per loc., or 1-loc. with 2 or 4 ± intruded parietal placentas, or (*Tourrettia*) 4-locular with ovules uniseriate in each locule; ovules ∞ anatr. or hemitropous, unitegmic. Fr. a bivalved capsule, v. oft. with a replum, rarely fleshy & indehiscent; seeds usu. flat, winged in capsules, endosperm usu. 0. x = 20

Classification & chief genera:

1. **Tecomeae** (fr. dehiscent perpendicular to septum, placentation axile): *Campsis*, *Catalpa*, *Incarvillea*, *Jacaranda*, *Markhamia*, *Pandorea*, *Paratecoma*, *Radermachera*, *Stereospermum*, *Tabebuia*, *Tecoma*
2. **Oroxyleae** (fr. dehiscent parallel to septum, placentation axile, As.): *Oroxylum*
3. **Bignonieae** (fr. dehiscent, lianes, Am.): *Adenocalymma*, *Arrabidaea*, *Dolichandra*, *Lundia*, *Pyrostegia*
4. **Eccremocarpeae** (fr. dehiscent, placentation parietal, wiry climbers, Andes): *Eccremocarpus* (only)

5. **Tourrettieae** (fr. dehiscent, placentation axile, capsule spiny, Am.): *Tourrettia* (only)
6. **Coleeae** (fr. indehiscent, lvs usu. pinnate, usu. bee-poll., Afr. & Madag.): *Colea, Kigelia, Phyllarthron*
7. **Crescentieae** (fr. indehiscent, bat-poll., lvs palmate &/or spiral): *Crescentia, Parmentiera*

[Schlegelieae (fr. indehiscent, bird- or insect-poll., lvs simple, opp.): here referred to own fam., Schlegeliaceae] *Paulownia* (endosperm 0, distinctive floral anatomy, embryo & seed morphology) is also best placed in own fam., Paulowniaceae

Some valuable timbers (e.g. *Cybistax, Handroanthus, Markhamia, Pajanelia, Paratecoma, Roseodendron, Stereospermum*) & many cult. orn. street-trees (*Catalpa, Jacaranda, Spathodea, Tabebuia, Tecoma*), lianes (*Amphilophium, Bignonia, Campsis, Dolichandra, Eccremocarpus, Pandorea, Podranea*), potpls (*Radermachera*) & hardy herb. perenn. *Incarvillea*

Bijlia N.E. Br. Aizoaceae (V). 2 N & W Cape. R: H. E. K. Hartmann, *Ill. Handb. Succ. Pls*, Aiz. A– E (2002)87

Bikinia Wieringa. Leguminosae (I 2). 10 W C Afr. R: WAUP 99–3(1999)187. Heterogeneous? Timbers, barks for canoes

Bikkia Reinw. Rubiaceae (I 7). Excl. *Thiollierea*, 3–10 W Pac. On coral (cf. *Badusa*)

Bilacunaria Pim. & Tikhom. = *Cachrys*

bilberry *Vaccinium myrtillus*

Bilderdykia Dumort. = *Fallopia*

Bilegnum Brand = *Cynoglossum*

bilimbi *Averrhoa bilimbi*

bilinga *Nauclea diderrichii*

bill, parrot's *Clianthus puniceus*

Billardiera Sm. Pittosporaceae. Incl. *Pronaya, Sollya*, excl. *Marianthus, Rhytidosporum*, 23 Aus. R: AusSB 17(2004)89. Strongly protandous with flower-colour changes; some with blue fls (bluebells), others with pendent yellow-orange fls. Cult. orn. (apple berries) esp. ***B. macrantha*** Hook.f. (*'B. longiflora'*, blueberry, purple apple berry, SE Aus.) & ***B. heterophylla*** (Lindl.) Cayzer & Crisp (*S. h.*, SW Aus., weedy SE Aus., natur. Portugal)

Billbergia Thunb. Bromeliaceae (3). 62 trop. Am., esp. E Braz. R: FN 14(1979)1975. Many hybrids. Cult. orn.

Billburttia Magee & Wyk (~ *Peucedanum*). Umbelliferae (Apieae). 2 Madag. R: PSE 283(2009)241

Billia Peyr. Sapindaceae (II; Hippocastanaceae). 2 S Mex. to trop. S Am.

billian *Eusideroxylon zwageri*

Billieturnera Fryx. Malvaceae (Malv.-Malv.). 1 S Texas & NE Mex.: ***B. helleri*** (Rose) Fryx. R: SBM 25(1988)131

billion dollar grass *Echinochloa esculenta*

Billolivia Middleton. Gesneriaceae. 7 Vietnam. R: Phytotaxa 161,4(2014)255, GBS 66(2014)193

Billy buttons *Craspedia uniflora*

Billya Cass. = *Petalacte*

billy-goat plum *Terminalia ferdinandiana*; **b. weed** *Ageratum conyzoides*

bilsted *Liquidambar styraciflua*

bilum (NG) bags made from *Gnetum* spp., *Pangium edule* etc.

bimble box *Eucalyptus populnea* subsp. *bimbil*

bindang *Agathis dammara*

bindi *Hibiscus esculentus*

bindweed *Convolvulus*, spp.; **black b.** *Fallopia convolvulus*; **common** or **field b.** *C. arvensis*; **great** or **hedge b.** *C. sepium*; **sea b.** *C. soldanella*

bindy-eye, bindyi *Soliva sessilis*

bine *Humulus lupulus*

Binotia Rolfe = *Gomesa*

bintangor *Calophyllum* spp.

binuang *Octomeles sumatrana*

Biolettia Greene = *Trichocoronis*

Biondia Schltr. = *Vincetoxicum*

Biophytum DC. Oxalidaceae. 50 trop. Lvs pinnate, s.t. sensitive, e.g. ***B. sensitivum*** (L.) DC. (trop. As.) – local medic.; seeds arillate. ***B. umbraculum*** Welw. (*B. petersianum*, OW trop.) – medic., e.g. for venereal disease in E Afr.

Bipinnula Comm. ex Juss. Orchidaceae (IV 2a). 11 temp. S Am.

Bipontia S.F. Blake = *Soaresia*

birch *Betula* spp; **American black b.** *B. lenta;* **brown b.** *B. pubescens;* **canoe b.** *B. papyrifera;* **common** or **European b.** *B. pendula, B. pubescens;* **dwarf b.** *B. nana;* **Japanese b.** *B. maximowicziana;* **Karelian b.wood** *B. pendula, B. pubescens;* **paper b.** *B. papyrifera;* **red b.** *B. nigra;* **silver b.** *B. pendula;* **WI b.** *Bursera simaruba;* **yellow b.** *Betula alleghaniensis*

bird cactus *Euphorbia tithymaloides;* **b. cherry** *Prunus padus;* **b. pepper** *Capsicum annuum* var. *glabriusculum;* **b.'s eye** *Veronica* spp.; **b.'s eye maple** *Acer saccharum;* **b.'s eye primrose** *Primula farinosa;* **b.'s foot** *Ornithopus perpusillus:* **b.'s foot trefoil** *Lotus corniculatus;* **b.'s nest fern** *Asplenium nidus;* **b.'s nest orchid** *Neottia nidus-avis;* **yellow b.'s nest** *Hypopitys monotropa;* **b.'s tongue** *Ornithoglossum* spp.

bird-of-paradise *Strelitzia reginae*

biribá *Annona mucosa*

birthroot *Trillium erectum*

birthwort *Aristolochia* spp. esp. *A. clematitis*

Bisboeckelera Kuntze. Cyperaceae (III 4). 4 S Am. R: MNYBG 17(1967)35

Bischofia Blume. Phyllanthaceae (Bisch.; Euphorbiaceae s.l.). 1 Indomal.: ***B. javanica*** Blume – invasive SE US, spread by Polynesians (dye for tapa cloth), bark medic. & dyestuff for rattan baskets; 1 C & SE China: ***B. polycarpa*** (A. Léveillé) Airy Shaw. R: FOC 11(2008)217. Timber trees, fr. for wine & spirits, seeds for oil

Bischofiaceae (Muell. Arg.) Airy Shaw = Phyllanthaceae

Biscutella L. Cruciferae (9). 45 S & mid Eur. to Medit. R: Britt. 38(1986)86. ***B. laevigata*** L. (SE Eur., widely natur., buckler mustard) – one of most polymorphic spp. in Eur. flora, cult. orn.

Biserrula L. (~ *Astragalus*). Leguminosae (III 24). 1 Med. to E Afr.: ***B. pelecinus*** L.

Bisglaziovia Cogn. Melastomataceae. 1 Braz.: ***B. behurioides*** Cogn.

Bisgoeppertia Kuntze. Gentianaceae. 3 Cuba, Hispaniola. R: Willd. 38(2008)179

Bishopalea H. Robinson = *Heterocoma*

Bishopanthus H. Robinson. Compositae (Liab.). 1 Peru: ***B. soliceps*** H. Robinson – known from 1 specimen

Bishopiella R. King & H. Robinson. Compositae (Eupat.-Gyp.). 1 E Braz.: ***B. elegans*** R. King & H. Robinson. R: MSBMBG 22(1987)124

bishop's cap *Mitella* spp.; **b.weed** *Aegopodium podagraria*

Bishovia R. King & H. Robinson. Compositae (Eupat.-Crit.). 2 Arg., Bolivia. R: MSBMBG 22(1987)324

Bismarckia Hildebr. & H. Wendl. Palmae (III 8a). 1 W Madag. savannas: ***B. nobilis*** Hildebr. & H. Wendl. (*Medemia n.*) – cult. orn. (?hort. cliché), lvs for thatch & basketry etc. & paper, trunk source of a sago. R: J. Dransfield & al., *Gen. Palm.* (2008)309

Bisquamaria Pichon = *Laxoplumeria*

bisselon *Khaya senegalensis*

Bistella Adans. = *Vahlia*

bistort *Bistorta officinalis*

Bistorta (L.) Scop. (~ *Polygonum, Persicaria*). Polygonaceae (II 1). c. 50 temp. ***B. affinis*** (D. Don) Greene (*Persicaria a., Polygonum a.*, Himal.) – cult. orn. ground-cover; ***B. officinalis*** Delarbre (*Polygonum bistorta, Persicaria b.*, bistort, Euras.) – dried rhiz. medic., ingredient of 'Easter Ledges Pudding' in Lake Dist., UK, at Easter poss. though legend of its aiding conception & retention of child; ***B. vivipara*** (L.) Delarbre (*Polygonum v., Persicaria v.*, N temp.) – bulbils in place of fls at infl. base, more at alt. (incomplete floral induction?)

Biswarea Cogn. = *Herpetospermum* (but see FOC 19(2011)33)

bitou bush *Chrysanthemoides monolifera*

bitter almond *Prunus dulcis* var. *amara;* **b. aloes** *Aloe* spp.; **b. apple** *Citrullus colocynthis;* **b. bark** *Alstonia constricta;* **b.berry** *Solanum aethiopicum;* **b.bush** *Adriana* spp.; **b.cress** *Cardamine* spp.; **b.cress, hairy** *C. hirsuta;* **b.leaf** *Gymnanthemum amygdalinum;* **b. nut** *Carya cordiformis;* **b. orange** *Citrus × aurantium;* **b. root** *Lewisia rediviva;* **b. sweet** *Solanum dulcamara;* **b.s., oriental** *Celastrus orbiculatus;* **b.weed** *Helenium amarum*

Bituminaria Heister ex Fabr. (~ *Psoralea*). Leguminosae (III 20). 3 Medit. ***B. bituminosa*** (L.) Stirton – tar-smelling weed, var. ***albomarginata*** Stirton (tedera) – forage legume tried in Aus. as drought-tolerant

Biventraria Small = *Asclepias*

Bivinia Jaub. ex Tul. (~ *Calantica*). Salicaceae (Flacourtiaceae). 1 E trop. Afr., Madag.: ***B. jalbertii*** Tul. – borer-proof wood, prot. Zimbabwe

Bivonaea DC. Cruciferae (10). 1 W Medit.: ***B. lutea*** (Biv.) DC.

Bixa L. Bixaceae. 5 trop. Am.: ***B. orellana*** L. – pre-Columbian cultigen grown as living fence & tried for land rehabilitation for forestry in Amazonia, but esp. for orange colouring (bixin, a carotenoid) obtained from testa, the original Amerindian bodypaint (hence 'Redskins', 'Red Indians', the ground paste (achiote) imp. from Mex. to US)), also effective insect-repellent (anatto, annatto, arnatto, roucou, urucu), form. imp. dyestuff rich in Vitamin A, now replaced by Congo red for fabric but still used in food (E160b; 10 K t seeds world prod., 7 T exported esp. Kenya, Peru) esp. cheese, butter, margarine & chocolate (consumption doubled since 1950, synthetics banned) as it is almost tasteless, & soaps & other comm. (insecticidal) skin prods. esp. for inflammation & 'golden staph'. Extrafl. nectaries: seed prod. doubled in presence of attracted ants warding off predators; ***B. platycarpa*** Ruíz & Pavón – timber for houses, pulp, rayon

Bixaceae Kunth. Magnoliidae – Malvales. 4/21 trop. Trees to rhiz. herbs with red or orange juice in secretory cells. Lvs palmate, palmately lobed to simple, in spirals; petiole oft. with complex vascular anatomy; stip. present, in *Bixa* prot. term. buds. Fls bisex., ± regular, in term. racemes to panicles; K 5 imbr., decid., C5 imbr. or convolute, A ∞, centrifugal, oft. assoc. with 5(–10) trunk bundles, anthers with slits or pores; nectary-disk intrastaminal or A on it; $\underline{G}$ (2–5) with ± deeply intruded partitions meeting basally & apically, the placentation thus partly axile, partly parietal exc. *Bixa* where wholly parietal, style 1, ovules anatropous, bitegmic. Fr. a loculicidal capsule; seeds glabrous or woolly, embryo embedded in oily & proteinaceous or (*Bixa*) starchy endosperm. x = 6–8

Genera: *Amoreuxia, Bixa, Cochlospermum, Diegodendron*

Diegodendron (wood anatomy like Sphaerosepalaceae) form. in Ochnaceae. *Cochlospermum* & *Amoreuxia* s.t. separated off as Cochlospermaceae, characterised by palmately lobed or palmate (not simple) lvs, fr. with 3–5 (not 2) valves & oily (not starchy) endosperm, but they are all more closely allied to one another than to other taxa Ornamentals, dyestuffs, kapok etc.

Bizonula Pellegrin. Sapindaceae. 1 Gabon: ***B. letestui*** Pellegrin

Blabeia Baehni = *Planchonella*

Blaberopus A. DC. = *Alstonia*

Blachia Baill. Euphorbiaceae (Cod.-Cod.). 11 Indomal.

black apple *Planchonella australis;* **b. bean** *Castanospermum australe, Glycine max, Kennedia nigricans;* **b. bearberry** *Arctous alpina;* **b. beech** *Nothofagus solandri;* **b.berry** *Rubus fruticosus* etc.; **b.b., Himalayan** *R. armeniacus;* **b. bindweed** *Fallopia convolvulus;* **b. boy** *Xanthorrhoea* spp.; **b. brush** *Coleogyne ramosissima;* **b. bryony** *Dioscorea communis;* **b.bush** *Maireana pyramidata;* **b.butt** *Eucalyptus pilularis;* **b. cap** *Rubus occidentalis;* **b. chokeberry** *Aronia melanocarpa;* **b. cohosh** *Actaea racemosa, A. spicata;* **b.currant** *Ribes nigrum;* **b. dammar** *Canarium* spp.; **b.-eye (d) bean or pea** *Vigna unguiculata;* **b.-eyed Susan** *Thunbergia alata, Rudbeckia* spp.; **b. fonio** *Digitaria iburua;* **b. garlic** *Allium sativum;* **b. gin** *Kingia australis;* **b.grass** *Alopecurus pratensis;* **b. gum** *Nyssa sylvatica;* **b. haw** *Viburnum prunifolium;* **b. henna** *Indigofera tinctoria;* **b. iris** *Iris nigricans;* **b. jack** *Bidens pilosa;* **b. Jessie** *Pithecellobium unguis-cati;* **b. laurel** *Gordonia lasianthus;* **b. locust** *Robinia pseudoacacia;* **b. mulberry** *Morus nigra;* **b. mustard** *Brassica nigra;* **b. nightshade** *Solanum nigrum;* **b. oak** *Quercus emoryi, Q. velutina;* **b. palm** *Normanbya normanbyi;* **b. pepper** *Piper nigrum;* **b. peppermint** *Eucalyptus amygdalina;* **b. raspberry** *Rubus occidentalis;* **b.root** *Veronicastrum virginicum;* **b. roseau** *Bactris major;* **b. rosewood** *Dalbergia latifolia;* **b. sapote** *Diospyros nigra;* **B. Sea walnut** *Juglans regia;* **b. sloe** *Prunus umbellata;* **b. snakeroot** *Actaea racemosa;* **b. stinkwood** *Ocotea bullata;* **b.thorn** *Prunus spinosa;* **b. walnut** *Juglans nigra;* **b. wattle** *Acacia mearnsii, Callicoma serratifolia;* **b. widow** *Geranium phaeum;* **b.wood, Afr.** *Dalbergia melanoxylon;* **b.wood, Aus.** or **Tasmanian** *Acacia melanoxylon;* **b.wood, Bombay** or **Ind.** *D. latifolia*

Blackallia C. Gardner. Rhamnaceae (5). 1 SW Aus.: ***B. nudiflora*** (F. Muell.) Rye & Kellermann (*B. biloba*). R: Nuytsia 16(2007)301

Blackiella Aellen = *Atriplex*

Blackstonia Hudson. Gentianaceae. 4 Eur. (4), Medit. ***B. perfoliata*** (L.) Hudson (yellow-wort) in GB

bladder campion *Silene vulgaris;* **b. fern** *Cystopteris fragilis;* **b.nut** *Staphylea pinnata;* **b. pod** *Lesquerella* spp.; **b. seed** *Levisticum* spp., *Physospermum* spp.; **b. senna** *Colutea arborescens;* **b.wort** *Utricularia* spp.

blady grass *Imperata cylindrica*

blaeberry *Vaccinium myrtillus*

Blaeria L. = *Erica*

Blainvillea Cass. Compositae (Helia.-Ecl.). 4 Am., ***B. acmella*** (L.) Philipson now pantrop. weeds. R: Austrobaileya 8(2012)655

Blakea P. Browne. Melastomataceae. Incl. *Topobea*, 150 trop. Am. Some hemiepiphytic spp. in cloud forest poll. by several rodent spp.; domatia. Fr. ed. Tanbark from ***B. trinervis*** Pavón ex D. Don (Guianas)

Blakeanthus R. King & H. Robinson (~ *Ageratum*). Compositae (Eup.-Ager.). 1 Guatemala, Honduras: ***B. cordatus*** (S.F. Blake) R. King & H. Robinson. R: MSBMBG 22(1987)149

Blakeochloa Veldk. = *Plinthanthesis*

Blakiella Cuatrec. Compositae (Ast.-Hin.). 1 Venez., Colombia: ***B. bartsiifolia*** (S.F. Blake) Cuatrec.

Blanchetia DC. Compositae (Vern.-Lynchn.). 1 NE Braz.: ***B. heterotricha*** DC. – raises sweat in humans

Blanchetiastrum Hassler = *Hibiscus* (*Pavonia*)

Blanchetiodendron Barneby & Grimes (~ *Albizia*). Leguminosae (II 4). 1 E Braz.: ***B. blanchetii*** (Benth.) Barneby & Grimes. R: MNYBG 74,1(1996)127

Blancoa Lindl. (~ *Conostylis*). Haemodoraceae (II). 1 SW Aus.: ***B. canescens*** Lindl. – bird-poll., self-incompatible. R: FA 45(1987)110

Blandfordia Sm. Blandfordiaceae. 4 E Aus. R: FA 45((1987)175. Trad. Aboriginal foodpls, now prot. in wild; cult. orn. (Christmas bells); fls found in gut of first emu shot in Aus. (1788)

Blandfordiaceae Dahlgren & Clifford. Magnoliidae – Asparagales. 1/4 E Aus. Herb. perennials with short rhizome, distich. linr. lvs & term. racemes of pend. bisex. reg. nodding fls. P3 + 3 with long tube, red to yellow, marcescent; A 3 + 3 inserted within tube; $\underline{G}$ (3), stipitate, 3-loc., each loc. with 40–50 anatr. ovules on axile placenta. Fr. a septicidal caps.; seeds with short brown hairs, endosperm copious with 0 starch, embryo linr.

Only genus: *Blandfordia*

App. closely allied to Boryaceae

Blandibractea Wernham = *Simira*

Blandowia Willd. = *Apinagia*

blanket flower *Gaillardia pulchella* & other *G.* spp.; **b. leaf** *Verbascum thapsus*

Blastania Kotschy & Peyr. = *Ctenolepis*

Blastemanthus Planch. Ochnaceae (III). 3 NE S Am. K 5 (+ 5 (+ 5)). R: BJ 113(1991)178

Blastocaulon Ruhl. = *Paepalanthus*

Blastus Lour. Melastomataceae. 12 Assam to W Mal. R: BMNHN 4, 4 Adans. (1982)43

blazing star *Mentzelia laevicaulis, Chamaelirium luteum, Liatris* spp.

Bleasdalea F. Muell. ex Domin (*Turrillia*, ~ *Grevillea*). Proteaceae (V 4). 5 W Pac., NE Aus. R: AJB 62(1975)138

Blechnaceae Newman. Polypodiidae – Polypodiales. 8/180 cosmop. R: T 55(2006)716. Usu. terr., s.t. small tree-ferns, or cli., oft. with stolons, scaly (not clathrate-) at apex. Fronds usu. pinnate to pinnatifid. Sori continuous or not, on veins parallel to leaflet midrib, usu. with indusia opening towards midrib (acrostichoid in *Brainea, Stenochlaena*). Gametophyte cordate, green. x = 27, 28, 31–37

Classification & genera:

Blechnoideae (scales not peltate, petiole vasc. bundles forming U in cross-sect.): *Blechnum, Brainea, Sadleria, Salpichlaena, Telmatoblechnum, Woodwardia*

Stenochlaenoideae (scales peltate, petiole vasc. bundles in 2 circles): *Stenochlaena*

Blechnum L. Blechnaceae (Blechnoideae). Incl. *Doodia, Pteridoblechum, Steenisioblechnum, excl. Telmatoblechnum*, c. 150 subcosmop. esp. S hemisph. (Eur. 1: ***B. spicant*** (L.) Roth, hard fern). Fronds uniform or dimorphic, when the fertile ones v. reduced; young fronds of many spp. ed., medic. Cult. orn. esp. ***B. gibbum*** (Lab.) Mett. (W Pac.), a tree-fern. ***B. minus*** (R. Br.) Ettingh. (Aus.) – ecdysteroids; ***B. francii*** Ros. (*B. obtusatum* var. *f.*, New Caled.) – rheophyte, poss. not a *B.* (Chambers)

Blechum P. Browne = *Ruellia*

bleeding heart *Dicentra spectabilis, Clerodendrum thomsoniae, Homalanthus populifolius*

bleedwood tree *Pterocarpus angolensis*

Bleekeria Hassk. = *Ochrosia*

Bleekrodea Blume (~ *Streblus*). Moraceae (I). 3 Madag., Malay Pen., Borneo. R: PKNAW C 91(1988)359

Blennodia R. Br. Cruciferae (37). 2 Aus. R: TRSSA 89(1965)168

Blennosperma Less. Compositae (Sen.-Tuss.). 2 Calif., 1 Chile. R: Britt. 16(1964)289. ***B. nanum*** (Hook.) Blake (Calif.) – seeds for flour

Blennospora A. Gray (~ *Calocephalus*). Compositae (Gnap.-Ang.). 2 S Aus. R: Muelleria 6(1987)354

Blepharandra Griseb. Malpighiaceae. 6 trop. Am.

Blepharidachne Hackel. Gramineae (XXIX 1). 4 W US & Arg. R: Britt. 31(1979)446

Blepharidium Standl. Rubiaceae (I 3). 1–2 C Am.

Blephariglottis Raf. = *Platanthera*

Blepharipappus Hook. (~ *Layia*). Compositae (Mad.-Mad.). 1 W US: ***B. scaber*** Hook. R: FNA 21(2006)259

Blepharis Juss. Acanthaceae (III 1). 129 OW trop. to S Afr. & Med. R: K. Vollesen (2000) *B.*, KB 57(2002)451. Seed-hairs swell when wetted. Some seeds eaten in Afr. & some used in anthrax treatment

Blepharispermum Wight ex DC. Compositae (Athro.). 15 Afr., Arabia to Sri Lanka. R: PSE 182(1992)177. Trees & shrubs, some timbers e.g. ***B. hirtum*** Oliv. – planking in town-houses in Dhofar, Arabia, where endemic

Blepharistemma Wall. ex Benth. Rhizophoraceae (I). 1 SW Ind.: ***B. serratum*** (Dennst.) Suresh

Blepharitheca Pichon = *Cuspidaria*

Blepharizonia (A. Gray) Greene. Compositae (Mad.-Mad.). 2 Calif. R: FNA 21(2006)289

Blepharocalyx O. Berg. Myrtaceae (II 10). 4 warm S Am.

Blepharocarya F. Muell. Anacardiaceae (I, Blepharocaryaceae). 2 NE Aus. R: BotJLS 95(1987)61

Blepharocaryaceae Airy Shaw = Anacardiaceae (I)

Blepharochilum M.A. Clem. & D.L. Jones = *Bulbophyllum*

Blepharodon Decne. Apocynaceae (V c; Asclepiadaceae III 1). c. 45 C & S Am., 1 N Am.

Blepharoneuron Nash = *Muhlenbergia*

Blephilia Raf. Labiatae (VII 2b). 3 E N Am. R: Rhodora 94(1992)1. Cult. orn.

blessed thistle *Centaurea benedicta*

Bletia Ruíz & Pavón. Orchidaceae (V 13a). 33 trop. Am. (Florida 2). Terr. cult. orn. Some med. esp. ***B. purpurea*** (Lam.) DC. (widespread inc. in salt-spray zone of Costa Rica) in WI, where used against poisoning from fish. See also *Bletilla*

Bletilla Reichb.f. Orchidaceae (V 10b). 5 temp. E As. Cult. orn. esp. ***B. striata*** (Thunb.) Reichb.f. – r. extract medic. ('beltia hyacinthe')

Blighia König. Sapindaceae. 3 trop. Afr. R: BJBB 21(1951)151. ***B. sapida*** König (W Afr.), the akee, grown for edible arils but the unripe fr., seeds & the raphe between aril & rest of seed poisonous with hypoglycin A, a non-protein amino-acid replacing lysine & causing sickness, coma & death in humans & other animals

Blighiopsis Veken. Sapindaceae. 2 trop. Afr. R: KB 68(2013)352

blimbing = bilimbi

blind grass *Stypandra glauca.*; **b.-your-eyes** *Excoecaria agallocha*

blinks *Montia fontana*

Blinkworthia Choisy (~ *Ipomoea*). Convolvulaceae (9). 2 Myanmar, trop. China

blite *Blitum bonus-henricus*

Blitum L. (~ *Chenopodium*). Amaranthaceae. Incl. *Monolepis*, c. 10 N temp., Aus (1). ***B. bonus-henricus*** (L.) Reichb. (*C. b.*, Good King Henry, allgood, blite, Eur., natur. N Am.) – lvs like spinach, young shoots like asparagus; ***B. nuttallianum*** Schultes (N Am.) – ed.

Blomia Miranda. Sapindaceae. 1 Mex.: ***B. prisca*** (Standl.) Lundell

blood berry *Rivina humilis*; **b. currant** *Ribes sanguineum*; **b. flower** *Scadoxus multiflorus*, *Asclepias curassavica*, *Melaleuca cruenta*; **b. leaf** *Iresine* spp.; **b. orange** *Citrus* × *aurantium* Sweet Orange Group cvs; **b. plum** *Haematostaphis barteri*; **b. root** *Sanguinaria canadensis*, *Haemodorum coccineum*; **b.wood** *Corymbia* spp. esp. *C. gummifera*, *Pterocarpus* spp.; **b. wood, scrub** *Baloghia inophylla*; **b.wort** Haemodoraceae

Bloomeria Kellogg. Asparagaceae (Alliaceae s.l.). 2 SW N Am. R: Madroño 12(1953)19. Cult. orn.

Blossfeldia Werderm. (~ *Parodia*). Cactaceae (III 5). 1 Andes of N Arg., Bolivia: ***B. liliputana*** Werderm. – a few mm diam., unique succ. resurrection pl. (poikilohydric like some mosses) with almost no stomata & hairy, arillate ant-disp. seeds unique in fam.

Blotia Leandri = *Wielandia*

Blotiella Tryon. Dennstaedtiaceae. c. 20 trop. Am. (1), Afr. to Masc. One of few fern genera strongly centred in Afr.

blue agave *Agave tequilana*; **b. amaryllis** *Worsleya procera*; **b. ash** *Fraxinus quadrangulata*; **b. beard** *Salvia viridis*; **b.bell** (English) *Hyacinthoides non-scripta*, (Scottish) *Campanula* *rotundifolia*, (Am.) *Mertensia* spp., (Aus.) *Billardiera* spp., (honey) *Echium plantagineum*, *Wahlenbergia* spp., (NZ) *Wahlenbergia* spp., (S Afr.) *Wahlenbergia* spp., *Gladiolus* spp.; **Spanish b.bell** *Hyacinthoides hispanica*; **b.bell creeper** *B. heterophylla*; **b.berry** *Vaccinium* spp., *B. macrantha* (Aus.), *Dianella nigra* (NZ); **b.b. ash** *Elaeocarpus obovatus*; **b.b., rabbit-eye** *V. ashei*; **b. bottle** *Centaurea* spp.; **b. boys** *Pycnostachys urticifolia*; **b. buttons** *Succisa pratensis*; **b. cardinal flower** *Lobelia siphilitica*; **b. cohosh** *Caulophyllum thalictroides*; **b. corn** *Zea mays* cv.; **b. couch** *Digitaria didactyla*; **b. curls** *Trichostema* spp. esp. *T. lanatum*; **b. devil** *Eryngium ovinum*, *E. pinnatifidum*; **b.-eyed grass** *Sisyrinchium* spp.; **b.-eyed Mary** *Omphalodes verna*; **b. flag** *Iris versicolor*; **b. gem** *Craterostigma plantagineum*; **b. grama** *Bouteloua gracilis*; **b.grass** *Poa* spp., *Festuca* spp.; **b. gum** *Eucalyptus globulus*; **b. gum, Sydney** *E. saligna*; **b. haze** *Pseudoselago spuria*; **b. jacket** *Tradescantia ohiensis*; **b. lace flower** *Trachymene coerulea*; **b. mahoe** *Hibiscus elatus*; **b. moor grass** *Sesleria caerulea*; **b. pine** *Pinus wallichiana*; **b. poppy** *Meconopsis* spp.; **b. sailor** *Cichorium intybus*; **b. star** *Amsonia* spp.; **b. stem** *Andropogon* spp., *Schizachyrium scoparium*; **b. toadflax** *Linaria canadensis*; **b. top** (Aus.) *Ageratum conyzoides*; **b. vine** *Clitoria ternatea*

bluet *Vaccinium* spp.; **mountain b.** *Centaurea montana*

bluets *Houstonia caerulea*

Blumea DC. Compositae (Inul.-Inul.). c. 100 OW trop., S Afr. R: Blumea 10(1960)176. ? Heterogeneous. ***B. balsamifera*** (L.) DC. source of ai or ngai camphor (As.) used in food, medic. (post-parturition aid in Sabah) etc.; other spp. imp. medic. locally. See also *Pseudoconyza*

Blumenbachia Schrad. Loasaceae (I 1). 12 S Am. R: Sendtnera 4(1997)208. Stinging hairs

Blumeodendron (Muell. Arg.) Kurz. Euphorbiaceae (Acal.-Acal.-Blum.). 5 Andamans, Mal.

Blumeopsis Gagnepain = *Blumea*

Blumeorchis Szlach. = *Cleisostoma*

blush, maiden's *Sloanea australis*; **b.wood** *Hylandia dockrillii*

Blutaparon Raf. (~ *Philoxerus*). Amaranthaceae (II 2). 4 Ryukyu Is., Am., W Afr. R: T 31(1982)113

Blysmocarex Ivanova = *Carex*

Blysmopsis Oteng-Yeboah (~ *Blysmus*). Cyperaceae (II 6). 1 circumboreal: ***B. rufa*** (Hudson) Oteng-Yeboah. R: NRBGE 33(1974)309

Blysmus Panzer ex Schultes (~ *Scirpus*). Cyperaceae (II 6). 4 temp. Euras. (Eur. 2)

Blyttia Arn. = *Vincetoxicum*

Blyxa Noronha ex Thouars. Hydrocharitaceae. 10 OW trop. R: AB 15(1983)1

boarwood *Symphonia globulifera*

boat lily *Tradescantia spathacea*

Bobartia L. Iridaceae (VII 2). 17 S Afr. esp. Cape. R: OB 37(1974). Lvs sword-like to centric

Bobea Gaudich. (~ *Timonius*). Rubiaceae (III 3). 4 Hawaii. R: Man. Fl. Pl. Hawaii 2(1990)1114. Yellow wood form. for canoes, paddles etc. Domatia

Bobgunnia Kirkbride & Wiersema (~ *Swartzia*). Leguminosae (III 1). 2 trop. & S Afr. R: Britt. 49(1997)1. Timber, insecticide. ***B. madagascariensis*** (Desv.) Kirkbride & Wiersema (*S. m.*, not Madag.!) – molluscicidal saponins (triterpenes), more effective than any drug for fungal skin infections

Bobrovia A.J. Khokr. = *Trifolium*

Bocagea A. St-Hil. Annonaceae (III 1). 3 E Braz. R: Britt. 47(1995)291. Only 6 colls; extinct?

Bocageopsis R. Fries. Annonaceae (IV 2). 4 trop. Am. ***B. multiflora*** (Mart.) R. Fries (Amaz.) – poll. exclusively by thrips

Bocconia Tourn. ex L. Papaveraceae (I). 10 warm Am. Apetalous, seeds arillate, alks. R: KB 1920: 275. Allied to *Macleaya*. ***B. frutescens*** L. (tree-celandine, trop. Am.) – pachycaul (Corner's Model), cult., natur. Java, latex used in treatment of warts (cf. *Chelidonium*) & to dye feathers

Bocoa Aubl. (~ *Swartzia*). Leguminosae (III 1). 3 trop. S Am. Timbers, local medic.

Bocquillonia Baill. Euphorbiaceae (Alc.-Alc.). 14 New Caled. R: FNC 14(1987)114

Boea Comm. ex Lam. Gesneriaceae (III 2i). Excl. *Dorcoceras*, 10 E Mal. to trop. Aus. & Solomon Is. R: T 65(2016)285. Holttum's, Corner's, Rauh's & Chamberlain's Models. Cult. orn. esp. ***B. hygroscopica*** F. Muell. (Queensland), a resurrection pl. retaining its chlorophyll when dry

Boeberastrum (A. Gray) Rydb. (~ *Dyssodia*). Compositae (Tag.-Pect.). 2 Baja Calif. R: Sida 11(1986)373

Boeberoides (DC.) Strother (~ *Dyssodia*). Compositae (Tag.-Pect). 1 Mex.: ***B. grandiflora*** (DC.) Strother. R: Phytol. M 10(1996)5

Boechera A. Löve & D. Löve (~ *Arabis*; x = 7, *A. s.s.* has x = 8). Cruciferae (11). c. 70 sexual diploid spp. W N Am. hybridizing to give (facultative) apomictic diploids, backcrossing giving c. 40 obligate apomict triploids, involving aneuploidy with non-recombining B chromosomes). ***B. holboellii*** (Hornem.) A. Löve & D. Löve – pseudogamy; ***B. pulchra*** (S. Watson) W. Weber (*A. p.*, Calif.) – sexual sp., affected by crucifer rust (*Puccinia consimilis*) to develop pseudofls of bright yellow lvs secreting nectar; ***B. stricta*** (Graham) Al-Shehbaz (*A. drummondii*, N Am.) – local medic., esp. kidneys & back

Boeckeleria T. Durand = *Tetraria*

Boehmeria Jacq. Urticaceae (III). 47 trop. & N subtrop. (OW 33 – R: Blumea 58(2013)90; New World 14 – R: OB 129(1996)15). Heterogeneous; monoec. or dioec. trees to herbs without stinging hairs (false nettles). ***B. nivea*** (L.) Gaudich. (ramie, China grass, trop. As.) cult. for longest, toughest & most silky of all known veg. fibres (c. 200 K t. p.a.), used in rope, Toyota Prius door insulation, etc., also Canton or Chinese linen & gas-mantles, but stems difficult to decorticate. Cult. in Med. & Calif. etc. (whitish lower leaf-surface); 'var. ***tenacissima*** (Roxb.) Miq.' (rhea, greenish) cult. in Mal. etc. Other spp., e.g. ***B. splitgerbera*** Koidz. ('*B. biloba*', raseita-so, Jap.), also tried for fibre & orn.

Boehmeriopsis Komarov = *Fatoua*

Boeica T. Anderson ex C.B. Clarke. Gesneriaceae (III 2d). 13 Himal., China, SE As. Heterogeneous

Boeicopsis Li = *Boeica*

Boelckea Rossow. Plantaginaceae (Grat.-Grat.). 1 Bolivia: ***B. beckii*** Rossow. R: Parodiana 7(1992)18

Boenninghausenia Reichb. ex Meisn. Rutaceae (I 10). 1(–3) Assam to C Jap., Mal mts: ***B. albiflora*** (Hook.) Meisn. – cult. orn. R: FOC 11(2008)73

Boerhavia Vaill. ex L. Nyctaginaceae (6). (Incl. *Anulocaulis*, *Commicarpus* & *Cyphomeris*) c. 50 warm (***B. chinensis*** (L.) Rottb. (trop. OW) & allies with 10-ribbed (as opposed to 5-) fr. s.t. seg. as *Commicarpus*). Weeds, local med. & veg. ***B. diffusa*** L. (OW) – antihelminthic in Ind., eaten by wild boar (*Sus scrofa*); ***B. repens*** L. (OW) – seed to seed in 10 days

Boerlagea Cogn. Melastomataceae. 1 Borneo: ***B. grandiflora*** Cogn.

Boerlagella Cogn. = ? *Pouteria*

Boerlagellaceae H.J. Lam. See Sapotaceae

Boerlagiodendron Harms = *Osmoxylon*

Boesenbergia Kuntze. Zingiberaceae (II 1). c. 80 Indomal. ***B. pandurata*** (Ridl.) Schltr – imp. in green curry paste in Thai cooking; ***B. rotunda*** (L.) Mansf. (Chinese key, gra-chai) – cult. for medic. & flavouring from rhiz. & r. esp. Thai soup

bog asphodel *Narthecium ossifragum*; **b. bean** *Menyanthes trifoliata*; **b. myrtle** *Myrica gale*; **b. orchid** *Hammarbya paludosa*; **b. pimpernel** *Anagallis tenella*; **b. rosemary** *Andromeda polifolia*; **b. rush, black** *Schoenus nigricans*; **b. spruce** *Picea mariana*; **b. violet** *Pinguicula vulgaris*

boga medalo(a) *Tephrosia candida*

Bogan flea *Calotis hispida*

Bognera Mayo & Nicolson. Araceae (VI1 3). 2 Braz.

Bogoria J.J. Sm. Orchidaceae (V 16c). 4 Java, NG

Bogotá tea *Symplocos theiformis*

Bohemian cherry *Cornus mas*

Boholia Merr. Rubiaceae (tribe?). 1 C Mal.: ***B. nematostylis*** Merr.

bois de rose oil from *Aniba* spp.; **b. de Panama** *Quillaja saponaria*, *Saponaria officinalis*; **b. fidèle** *Citharexylum* spp., *Vitex* sp.

Boisduvalia Spach = *Epilobium*

Boissiera Hochst. ex Ledeb. = *Bromus*

Bojeria DC. = *Inula*

bok choy *Brassica rapa* Chinensis Group

Bokhara clover *Melilotus albus*; **B. plum** *Prunus bokhariensis*

Bokkeveldia D. Mueller-Doblies & U. Mueller-Doblies = *Strumaria*

Bolandia Cron (~ *Cineraria*). Compositae (Sen.-Sen.). 2 S Afr. R: Novon 16(2006)224

Bolandra A. Gray. Saxifragaceae (I 6). 2 NW US. R: BJLS 90(1985)57

Bolanosa A. Gray. Compositae (Vern.-Leib.). 1 S Mex.: ***B. coulteri*** A. Gray

Bolanthus (Ser.). Reichb. Caryophyllaceae (III 1). c. 15 (Eur. 6, 4 Greece eastwards) to Israel. R: Wentia 9(1962)157

Bolax Comm. ex Juss. (~ *Azorella*). Umbelliferae (Azor.). 4–5 temp. S Am.

Bolbidium Lindl. ex Brieger = *Dendrobium*

Bolbidium (Lindl.) Lindl. = *Maxillaria*

Bolbitidaceae (Pichi-Serm.) Ching = Dryopteridaceae

Bolbitis Schott. Dryopteridaceae (II; Bolbitidaceae). c. 60 trop. & warm esp. SE As. R: LBS 2(1977)123. ***B. portoricensis*** (Spreng.) Hennipman (Am.) & ***B. quoyana*** (Gaudich.) Ching (As.) – frond-tips rooting & forming new pls

Bolboschoenus (Asch.) Palla (~ *Scirpus*). Cyperaceae (II 2). 14 cosmop. (Aus. 4–5, N Am. 5). R: NSL 36(2004)84. ***B. medianus*** (Cook) Soják & ***B. planiculmis*** (Schmidt) Egorova (*B. caldwellii*, Aus.) – tubers ed. Aborigines; ***B. maritimus*** (L.) Palla (*S. paludosus*, bayonet grass, N temp. to Pacific & S Am.) – rhiz. form. eaten

Bolbostemma Franquet = *Actinostemma* (but see FOC 19(2011)17)

boldo *Peumus boldus*

Boldoa Cav. ex Lag. (~ *Salpianthus*). Nyctaginaceae (2). 2–3 C Am.

Boleum Desv. = *Vella*

Bolivian coriander *Porophyllum ruderale*; **B. rubber** *Sapium glandulosum*

Bolivicactus Doweld = *Parodia*

Bollea Reichb.f. = *Pescatorea*

Bollwiller pear × *Sorbopyrus auricularis*

Bolocephalus Hand.-Mazz. = *Dolomiaea*

Bolophyta Nutt. (~ *Parthenium*). Compositae (Helia.-Amb.) 3 W US. R: Phytol. 41 (1979) 486

Boltonia L'Hérit. Compositae (Ast.-Bolt.). 5 C & E N Am. SB 12(1997)138. Cult. orn. esp. ***B. asteroides*** (L.) L'Hérit. (N Am.); some As. spp. used as leaf veg. in Jap.

Bolusafra Kuntze. Leguminosae (III 18). 1 S Afr.: ***B. bituminosa*** (L.) Kuntze. R: Strel. 10(2000)277

Bolusanthus Harms. Leguminosae (III 9). 1 S C Afr: ***B. speciosus*** (Bolus) Harms – cult. orn. tree with hard ant- & borer-proof timber. R: Strel. 10(2000)278

Bolusia Benth. Leguminosae (III 8). 5 Afr. S of equator

Bolusiella Schltr. Orchidaceae (V 16d). 6 trop. Afr.

bolwarra *Eupomatia* spp.

Bomarea Mirb. Alstroemeriaceae (Liliaceae s.l.). c. 107 Mex. to trop. Am. Many climbers, cult. orn., some with ed. r. tubers e.g. ***B. edulis*** (Tussac) Herb. (salsilla, trop. Am.)

Bombacaceae Kunth = Malvaceae (Bomb.)

Bombacopsis Pittier = *Pachira*

Bombax L. Malvaceae (Bomb.; Bombacaceae). Excl. *Rhodognaphalon*, 8 OW trop. R: BJBB 33(1963)84, 253. Trees (Aubréville's Model), s.t. spiny. Cult. orn., soft timber, ovary hairs a source of kapok. ***B. brevicuspe*** Sprague (*R. b.*, W Afr.) – in Ghana elephants strip off bark to 10 m; ***B. buonopozense*** P. Beauv. (W & C Afr.) – comm. timber, chessmen carved from spines; ***B. ceiba*** L. (*B. malabaricum*, (Ind.) silk cotton tree, simul, trop. As.) – inferior kapok (hairs used by *Dolichoderus bispinosus* ants for nest-building), timber for matches, canoes etc., medic. (incl. constituent of supposed aphrodisiacs), red ed. fls, nectar ed. Aus. Aborigines

Bombay aloe *Agave cantala*; **B. ebony** *Diospyros montana*; **B. hemp** *Crotalaria juncea*; **B. mastic** *Pistacia atlantica*; **B. mix** see *Trachyspermum ammi*

bombway, white *Terminalia procera*

Bombycidendron Zoll. & Moritzi = *Hibiscus*

Bombycilaena (DC.) Smoljan. (~ *Micropus*). Compositae (Gnap.-Gnap.). 2 Euras. R: Cand. 69(2014)56

Bommeria Fourn. Pteridaceae (IV). 5 SW N & C Am. R: JAA 60(1979)445. Xeric & seasonal mts. Cult. orn.

bonace *Daphnopsis tinifolia*

Bonafousia A. DC. = *Tabernaemontana*

Bonamia Thouars. Convolvulaceae (4). 50+ trop. R: Phytol. 17(1968)121

Bonania A. Rich. Euphorbiaceae (Euph.-Hipp.). 8–10 WI esp. Cuba, ?Venez.

Bonannia Guss. Umbelliferae (III 10). 1 S Italy, Greece: ***B. graeca*** (L.) Halácsy

Bonatea Willd. (~ *Habenaria*). Orchidaceae (IV 4d). 13 trop. & S Afr., Arabia. Cult. orn. esp. ***B. speciosa*** (L.f.) Willd. (S Afr.)

Bonatia Schltr. & K. Krause = *Tarenna*

bonavist *Lablab purpureus*

bonduc nut *Guilandina bonduc*

Bonellia Colla (~ *Jacquinia*). Primulaceae (Theophrastaceae). 31 trop. Am. R: Novon 14(2004)116. Alt. lvs cf. pseudoverticillate *J.*

boneseed *Chrysanthemoides monilifera*

boneset *Eupatorium perfoliatum*

Bonetiella Rzed. Anacardiaceae (I). 1 Mex.: ***B. anomala*** (I.M. Johnston) Rzed.

Bongardia C. Meyer. Berberidaceae (II 2). 1 E Med. (Greece) to Afghanistan: ***B. chrysogonum*** (L.) Spach – lvs & rhiz. ed.

bongossi *Lophira alata*

Bonia Bal. (~ *Bambusa*). Gramineae (V 4). 5 Vietnam & S China. R: KB 51(1996)567

Boninia Planch. = *Melicope*

Boninofatsia Nakai = *Fatsia*

Boniodendron Gagnepain. Sapindaceae. 1–2 China to SE As.

Bonnaya Link & Otto (~ *Lindernia*). Linderniaceae. 12 IndoPacif. R: Willd. 43(2013)220

Bonnayodes Blatter & Hallberg = *Limnophila*

Bonnetia Mart. Bonnetiaceae (Guttiferae s.l.). 29 trop. Am. R: JAA 29(1948)393. Some pachycaul candelabriform shrubs

Bonnetiaceae Beauvis. ex Nakai (~ Guttiferae, Theaceae). Magnoliidae – Malpighiales. 3/35 Mal., trop. Am. Everg., glabrous trees & shrubs. Lvs in spirals, minutely serrate, petioles short; buds long pointed. Fls in cymes. C contorted. A ∞. G (3–5), styles 1 – several. Fr. a septicidal capsule; seeds ∞, small. n = c. 150
Genera: *Archytaea, Bonnetia, Ploiarium*
Some timber (*Ploiarium*)

Bonniera Cordemoy = *Angraecum*

Bonnierella R. Viguier = *Polyscias*

Bonplandia Cav. Polemoniaceae (II). 1 Mex.: ***B. geminiflora*** Cav. – fls ± zygomorphic. R: Aliso 19(2000)61

bonsamdua *Distemonanthus benthamianus*

Bontia Plum. ex L. Scrophulariaceae (Myoporaceae). 1 WI, trop. S Am.: ***B. daphnoides*** L. – Rauh's Model, cult. orn. R: R.J. Chinnock, *Eremophila & allied genera* (2007)162

Bonyunia Schomb. ex Progel. Loganiaceae (Strychnaceae). 10 S Am. R: AMBG 96(2009)548

boobialla, boobyalla *Myoporum* spp.

boojum tree *Fouquieria columnaris*

boonaree, boonery *Alectryon oleifolius*

Boophane Herb. = *seq.*

Boophone Herb. Amaryllidaceae. Excl. *Crossogyne*, 6 subSaharan Afr. R: Strel. 9(2000)54. Alks, poisonous. ***B. disticha*** (L.f.) Herb. – onion-scented infrs blown around, disp. seeds, haemanthine medic. but also affects eyes

Boopis Juss. Calyceraceae. Incl. *Moschopsis*, c. 21 Andes, Arg., S Braz., Falklands

Boosia Speta = *Drimia*

booyong *Argyrodendron* spp.

boppel nut, red *Hicksbeachia pinnatifolia*

bopple nut = macadamia nut

Boquila Decne. Lardizabalaceae. 1 Chile, Arg.: ***B. trifoliolata*** (DC.) Decne – mimics lvs & spininess of support pl., dioec., occ. monoec. R: BM n.s. 29(2012)255,277

bora *Vigna unguiculata*

borage *Borago officinalis*

Boraginaceae Juss. Magnoliidae – Boraginales. Incl. Codonaceae, Coldeniaceae, Cordiaceae, Ehretiaceae, Heliotropiaceae, Hoplestigmataceae, Hydrophyllaceae, Lennoaceae, Namaceae, Wellstediaceae, 121/2600 subcosmop. R: T 65(2016)506. Trees, shrubs (s.t. pachycaul), freq. herbs (rarely achlorophyllous r.-parasites (Lennooideae)), rarely lianes, usu. with char. unicellular bristly hairs (with basal cystolith or similar), oft. with alks & red alkannin (in r.). Lvs in spirals (rarely with some opp.), simple, usu. entire, oft. with cystoliths &/or oxalate crystals, stip. 0. Fls usu. in term. cymes (scorpioid, some appearing helicoid), rarely solit. & axillary, usu. bisex. (gynodioecy in e.g. *Echium*, dioecy in *Cordia* etc.), ± regular, s.t. heterostylous; bracteoles 0; K (4)5(–8), free to connate, imbr., rarely valvate; C ((4)5(6)), oft. salveriform, lobes imbr. or convolute, rarely valvate, the tube with hairy appendices between lobes in Boraginoideae; A (4)5(6) alt. with C, attached to tube, anthers opening by longit. slits; annular nectary-disk around G or 0; $\underline{G}$ (2) but (4, 5) in NG spp. of *Trigonotis*, & (5–16) in Lennooideae, half-inf. in some *Nama* & *Wigandia*, with twice as many compartments, each with 1 ovule, rarely 1 carpel suppressed or 1-loc. (Hydrophylloideae); in Cordioideae G entire with terminal twice bifid style ripening to drupe usu. with 4-loc. stone; in Ehretioideae G entire to 4-lobed ripening to drupe with 2 2-seeded or 4 1-seeded stones or separating into 4 segments; in Heliotropioideae G entire or 4-lobed ripening into 2 (1)2-seeded or 4 1-seeded nutlets; in Boraginoideae G deeply 4-lobed with simple or bifid style ripening into (1–) 4 nutlets or 2-loc. ripening into 1- or 2-seeded capsule; in Lennooideae fr. a fleshy capsule tardily dehiscing irregularly circumscissile; ovules anatr. to hemitropous with 1 integument. Embryo with 2 cotyledons (deeply bifid in *Amsinckia*), straight; endosperm copious, oily to 0. x = 4–13+

Classification & principal genera (the subfams. s.t. recog. as fams (see Introduction); for most fr. differences see above):

Cordioideae (Ehretiaceae; mainly trees, style term., 4-branched; 2/350 mainly trop.): *Cordia, Varronia* (*Coldenia* (Coldeniaceae; 1/1) & *Hoplestigma* (Hoplestigmataceae; 1/2) to be excl. as monogeneric subfams)

Ehretioideae (Ehretiaceae; mainly trees, style term., 2-branched; 7/160): *Bourreria, Ehretia, Halgania, Tiquilia*

Heliotropioideae (Heliotropiaceae; commonly herbs, style term., undivided; 4/450): *Euploca, Heliotropium, Myriopus*

Hydrophylloideae (Hydrophyllaceae; pachycaul trees & shrubs, herbs, G usu. 1-loc., style term.; 12/250 subcosmop. esp. W N Am.): *Nemophila, Phacelia* (*Codon* to be excl. as **Codonoideae** (Codonaceae; 1/2)); *Eriodictyon, Nama, Turricula, Wigandia* to be excl. as **Namoideae** (Namaceae; 4/71)

Boraginoideae (usu. herbs, style gynobasic; *Wellstedia* to be excl. as **Wellstedioideae** (Wellstediaceae; 1/6) – woody herbs, fls 4-merous, fr. a loculicidal capsule; 90/1650 in 3 major groups – R: T 65(2016)535 q.v. for details):

1. **Echiochileae** (fls reg. or not, faucal scales 0, style-base flat or pyramidal): *Echiocaulon*
2. **Boragineae** (incl. Echieae, Lithospermeae; fls reg. or not, faucal scales, style-base flat or slightly convex) – 2 main subgroupings [1 subdivided]: 1. (Borag.) *Anchusa, Borago, Pulmonaria, Symphytum*; 2. (Lith.) *Alkanna, Arnebia, Cerinthe, Echium, Lithospermum, Onosma*
3. **Cynoglosseae** (incl. Eritrichieae; fls reg., faucal scales, style-base ± conical) 8 subgroupings [some subdivided] incl. 1. (Trich.) *Trichodesma*; 4. (Omphal.) *Myosotidium, Omphalodes*; 5. (Roch.) *Eritrichium, Lappula*; 7. (Myosot.) *Myosotis, Trigonotis*; 8. (Cynog.) *Cynoglossum, Cryptantha*

Lennooideae (Lennoaceae; chlorophyll-less r.-parasites; warm Am. deserts): *Lennoa, Pholisma* (only)

Diversification of primarily woody groups poss. in mid-Cretaceous. Timber & fr. trees (*Cordia, Ehretia*), dyeplants (*Alkanna, Asperugo, Cordia, Lithospermum, Onosma*, etc.), oil incl. medic. (*Borago*), potherbs (*Pholisma, Symphytum* etc.), firewood (*Echiochilon*), bee-fodder (*Echium, Phacelia*) & cult. orn. (*Echium, Glandora, Heliotropium, Hesperochiron, Lithospermum, Myosotis, Nemophila, Omphalodes, Phacelia, Pulmonaria, Romanzoffia, Symphytum, Wigandia* etc.)

Borago Tourn. ex L. Boraginaceae (B.2.1.1). 5 Med., Eur., As. R: Buletinul Fac. Stiinte Cernauti 2(1928)438. Cult. orn. esp. ***B. officinalis*** L. (borage, Eur. & Med.) – bee-fodder, field-crop for oil (starflower o.) sold like evening primrose o. (seeds high in gamma-linoleic acid, unusual fatty-acid intermediate in biosynthesis of prostaglandins, so used for menstrual pains, but now withdrawn in Eur.), fls in salads, flavouring for claret cup (replaced by cucumber), form. potherb

Borassodendron Becc. (~ *Borassus*). Palmae (I 8b). 2 Malay Pen., Borneo. R: Reinwardtia 8 (1972) 351. Rhino- and elephant-disp. seeds

Borassus L. Palmae (I 8b). 6 OW (trop. Afr. 3, Madag. 2, Indomal. 1). R: KB 62(2007)568, Palms 53(2009)41. Dioec., lvs palmate. ***B. aethiopum*** Mart. (trop. & S Afr.) – fls after 30–40 yrs (prot. Zimbabwe), fr. ed. Uganda, seeds found in elephant dung; ***B. flabellifer*** L. (*B. sundaicus*, lontar, Palmyra palm, Ind. to Myanmar), cult., timber for rafters etc. (resists salt water), lvs for thatch & *olas* or writing-paper (perhaps the oldest form; a stylus used for writing), leaf-base fibre for brushes etc. (bassine or Palmyra fibre), split lvs used for weaving mats, baskets etc., fr. eaten roasted & infl. tapped for toddy from which sugar (jaggery), arrack, vinegar etc. prep. – 120 000 litres of toddy produced in lifetime of 1 palm; seeds surrounded by fibres worked as small orn. heads for sale in Ind., seedlings ed. & yield odiyal flour for breakfast food etc. when ground (according to an old Tamil song there are 801 uses in all)

Borderea Miégev. = *Dioscorea*

Borealluma Plowes = *Caralluma*
Boreava Jaub. & Spach = *Isatis*
borecole *Brassica oleracea* Acephala Group
Borinda Stapleton = *Fargesia* (but see EJB 51(1994)284)
Borismene Barneby. Menispermaceae (III). 1 trop. S Am.: ***B. japurensis*** (Mart.) Barneby
Borissa Raf. = *Lysimachia*
boriti poles *Rhizophora mucronata*
Borkonstia Ignatov = *Aster*
borlotti beans *Phaseolus vulgaris*
Borneacanthus Bremek. Acanthaceae (II 2d). 6 Borneo
Borneo camphor *Dryobalanops aromatica*; **B. ironwood** *Eusideroxylon zwageri*; **B. mahogany** *Calophyllum inophyllum*; **B. redwood** *Shorea* spp.; **B. rubber** *Willughbeia coriacea, W. firma*; **B. tallow** *Shorea palembanica*; **B. teak** *Intsia* spp.
Borneodendron Airy Shaw. Euphorbiaceae (Ricinoc.-Cocc.). 1 NE Borneo: ***B. aenigmaticum*** Airy Shaw – lvs & fls in whorls of 3, sap red (v. unusual in E.). R: HIP 5,7(1967)t. 3633
Borneosicyos W. de Wilde. Cucurbitaceae (XIV). 1 Sabah: ***B. simplex*** W. de Wilde. R: FMB 14(2007)35
Bornmuellera Hausskn. Cruciferae (2). 5 Balkans (3), As. Minor. Nickel accumulators
Bornmuellerantha Rothm. = *Odontites* (but see Novon 20(2010)265)
Borodinia Busch (~ *Boechera*). Cruciferae (11). 1 E Siberia (***B macrophylla*** (Turcz.) O. Schulz (*B. tilingii*)), 7 E N Am. R: SB 38(2013)203
Borodiniopsis D. German & al. Cruciferae (7). 1 China: ***B. alaschanica*** (Maxim.) D. German & al. R: T 61(2012)966
Borojoa Cuatrec. = *Alibertia*
Boronella Baill. = *Boronia*
Boronia Sm. Rutaceae (I 3). Incl. *Boronella*, 148 Aus. (R: FA 26(2013)124), 6 New Caled. R (sect. *Valvatae* – 63): Muelleria 12(1999)3. Prot. in Aus.; cult. orn. esp. ***B. floribunda*** Sieber ex Reichb. as cut-fl. in Aus., ***B. megastigma*** Nees ex Bartling in Eur. for scent, this & other spp. grown comm. for that in Victoria. ***B. anemonifolia*** A. Cunn. (SE Aus.) – handling can lead to lightheadedness
Borrachinea Lavy = *Borago*
Borreria G. Meyer = *Spermacoce*
Borrichia Adans. Compositae (Helia.-Eng.). 2 SE US, WI. R: AMBG 65(1978)681
Borsczowia Bunge = *Suaeda*
borshch *Beta vulgaris*
Borthwickia W.W. Sm. Resedaceae (Borthwickiaceae). 1 Myanmar, SE Yunnan: ***B. trifoliata*** W.W. Sm. – lvs opp., C 6
Borthwickiaceae J.X. Su & al. = Resedaceae
Borya Labill. Boryaceae. 11 W Aus., Victoria, Queensland. ***B. nitida*** Labill. – lvs revive after desiccation when orange; ***B. mirabilis*** D. Churchill (Victoria) – self-incompatible, ± entirely veg. reprod.
Boryaceae M. Chase & al. (~ Asparagaceae). Magnoliidae – Asparagales. 2/12 Aus. R: KB 51(1996)676. Xeromorphic (some resurrection) pls with rhizomes, lignified r. & aerial stems. Lvs in spirals. Fls in spikes or racemes. P 3 + 3, marcescent, usu. with short tube, A 6 usu. adnate to tube, $(\underline{G}3)$ with many anatr. ovules per carpel. Fr. a loculicidal caps.; seeds with phytomelan & copious endosperm. n = 11, 14
Genera: *Alania, Borya*
Borzicactella F. Ritter = seq.
Borzicactus Riccob. (~ *Cleistocactus*). Cactaceae. 17 Ecuador, Peru
Boschia Korth. (~ *Durio*). Malvaceae (Hel.; Bombacaceae). c. 6 Myanmar, Mal. esp. Borneo
Boschniakia C. Meyer. Orobanchaceae (Orob.). 4 N & Arctic Russia, As. to Jap., NW N Am. R: ABY 9(1987)296. ***B. rossica*** (Cham. & Schldl.) B. Fedtsch. (N Euras., NW N Am.) – on *Alnus*, attractive to cats (cf. *Nepeta*)
Boscia Lam. Capparaceae. c. 30 trop. & S Afr., Arabia (20), Madag. Some ed. fr., medic. & coffee subs. (seeds, r.)
Bosea L. Amaranthaceae (I 2). 3 Canary Is., Cyprus (not Turkey!), Ind. Dioec. shrubs
Bosistoa F. Muell. ex Benth. Rutaceae (I 1). 4 E Aus. rain forests. R: JAA 58(1977) 416
Bosleria Nelson = *Solanum*
Bosqueia Thouars ex Baill. = *Trilepisium*

Bosqueiopsis De Wild. & T. Durand. Moraceae (IV). 1 trop. Afr.: ***B. gilletii*** De Wild. & T. Durand

Bossera Leandri = *Alchornea*

Bossiaea Vent. Leguminosae (III 14). Excl. *Platylobium*, 78 temp. (esp. SW) Aus. Several spp. with flattened green stems & minute scale lvs, seedlings with more typical ones; yellow bee-poll. fls to red bird-poll. Some cult. orn.

Boston fern *Nephrolepis exaltata* 'Bostoniensis'; **B. ivy** *Parthenocissus tricuspidata*

Bostrychanthera Benth. = *Chelonopsis*

Boswellia Roxb. ex Colebr. Burseraceae (III). c. 20 dry trop. Afr. (esp. NE; Socotra 8 endemics) & As. Fragrant resins used in incense & aromatherapy (incl. for jet-lag), esp. frankincense (gum olibanum) derived from ***B. sacra*** Flueckiger (Somalia, S Arabia) in classical times (in Roman times c. 3 k t p.a. (2001 – 1 k t) as valuable as gold & poss. the orig. balsam brought from Somalia (Punt) by the Queen of Sheba), ***B. papyrifera*** (Del.) Hochst. (NE Nigeria to Ethiopia) & esp. ***B. socotrana*** Balf.f. (Socotra) in antiquity & recent comm., still in Coptic churches, largely ***B. carteri*** Birdw. (? = *B. sacra*) & ***B. frereana*** Birdw. (Afr. elemi, Somalia) today. ***B. serrata*** Roxb. ex Colebr. (consp. on Ind. dry hills) – Ayurvedic med., timber for charcoal, tea-chests etc.

Botany Bay oak, B. B. wood *Allocasuarina* spp.

Botelua Lag. = *Bouteloua*

Bothriochilus Lem. = *Coelia*

Bothriochloa Kuntze (~ *Dichanthium*). Gramineae (XX 7). 37 warm Am. (N Am. 9). R: Darw. 38(2000)134. Cult. for fodder

Bothriocline Oliv. ex Benth. Compositae (Vern.-Erl.). c. 30 trop. Afr. (E Afr. 29; R: KB 43(1988)257), Madag.

Bothriospermum Bunge. Boraginaceae (B.3.8.2). 5 trop. & NE As.

Bothriospora Hook.f. Rubiaceae (Cond.). 1 trop. Am.: ***B. corymbosa*** (Benth.) Hook.f. (*Euosmia c.*) – parasiticidal, wood toxic so meat skewered & roasted on it fatal

Bothrocaryum (Koehne) Pojark. = *Cornus*

bo-tree *Ficus religiosa*

Botryarrhena Ducke. Rubiaceae (Ixor.-Cord.). 2 Peru, Braz., Venez.

Botrychiaceae Horan. = Ophioglossaceae

Botrychium Sw. Ophioglossaceae. c. 25 temp., polar, trop. mts. R: MTBC 19,2(1938)22. (Eur. 7 incl. ***B. lunaria*** (L.) Sw., moonwort). Habit of *Ophioglossum* but sterile lvs also lobed. ***B. ternatum*** (Thunb.) Sw. – lvs said to be a veg. in Jap.

Botryoloranthus (Engl. & K. Krause) Balle = *Oedina*

Botryomeryta R. Viguier = *Meryta*

Botryophora Hook.f. Euphorbiaceae (Acal.-Acal.-Blum.). 1 SE As., W Mal.: ***B. geniculata*** (Miq.) Airy Shaw. R: A. Radcliffe-Smith, *Gen. Euphorb.* (2001)174

Botrypus Michaux = *Botrychium*

Botryostege Stapf = *Elliottia*

Botschantzevia Nabiev. Cruciferae (7). 1 C As.: ***B. karatavica*** (Lipsch. & Pavlov) Nabiev

Bottegoa Chiov. Rutaceae (III; Ptaeroxylaceae). 1 Somalia, SE Ethiopia, N Kenya: ***B. insignis*** Chiov.

Bottionea Colla = *Trichopetalum*

bottle gourd *Lagenaria siceraria*; **b. palm** *Beaucarnea recurvata*; **b.-tree** *Brachychiton* spp.

bottlebrush *Melaleuca* (*Callistemon*) spp.; **b. grass** *Hystrix* spp.

Boucerosia Wight & Arn. = *Caralluma*

Bouchardatia Baill. Rutaceae (I 1). 1 E Aus.: ***B. neurococca*** (F. Muell.) Baill. (union nut, E Aus.) – fine timber, poss. box subs. R: FA 26(2013)52

Bouchea Cham. (~ *Chascanum*). Verbenaceae (Dur.). 9 trop. Am., 1 Ethiopia. R: FR 48(1940)17, 49(1940)91

Bouchetia Dunal (~ *Salpiglossis*). Solanaceae (Petun.). 3 S US to Braz.

Bouea Meissn. Anacardiaceae (I). 3 SE As., Mal. Lvs. opp. Fr. ed. esp. ***B. macrophylla*** Griff. (gandaria, plum mango, Mal.) – sambal

Bougainvillea Comm. ex Juss. Nyctaginaceae (4). c. 16 C & trop. S Am. Poll. by hummingbirds, fls arising from persistent coloured bracts in groups of 3 resembling a flower. Cult. orn. usu. spiny lianes (Champagnat's Model) to 25 m, esp. cvs of ***B. × buttiana*** Holttum & Standl. (*B. glabra* Choisy (Braz.) × *B. peruviana* Bonpl. (Colombia to Peru), resynthesized). 'Double' forms have each fl. replaced by a short shoot with variable nos of bracts. ***B. arborea*** Glaz. (Braz.) – tree

Bougueria Decne = *Plantago*

Bouletia M.A.Clem. & D.L. Jones. See *Dendrobium*

bouncing Bet *Saponaria officinalis*

bouquet, bridal *Poranopsis paniculata*

Bourasaha Thouars. Older name for *Burasaia*

bourbon *Zea mays*

Bourdaria A. Chev. = *Cincinnobotrys*

Bournea Oliv. = *Oreocharis*

Bourreria P. Browne. Boraginaceae (Ehr.; Ehretiaceae). Incl. *Hilsenbergia*, c. 50 trop. Am. to Masc. R: JAA suppl. 1(1991)59. Trees. ***B. huanita*** (Lex.) Hemsl. (huanita, Mex.) – scented fls used in tobacco, drinks etc.

Bousigonia Pierre. Apocynaceae (I c). 2 SE As.

Boussingaultia Kunth = *Anredera*

Bouteloua Lag. Gramineae (XXIX 1). Incl. *Buchloe*, *Chondrosum*, 58 Canada to Arg. (N Am. 19) esp. Mex. R: AMBG 66(1979)348, Aliso 18(1999)61. Pasture & cult. orn. grasses (C4) imp. in natural grasslands esp. ***B. curtipendula*** (Michaux) Torrey (side-oats grama, C N Am.), ***B. dactyloides*** (Nutt.) Columbus (buffalo grass) – monoec. or dioec. creeping grass of W prairies, good fodder, s.t. a lawn-grass & for erosion control, trad. for First Am. grasshouses in Oklahoma, & ***B. gracilis*** (Kunth) Griffiths (SW N Am., blue grama)

Boutiquea Le Thomas = *Neostenanthera*

Boutonia DC. Acanthaceae (III 2d). 1 Madag.: ***B. cuspidata*** DC.

Bouvardia Salisb. Rubiaceae (I 1/IV 24). 20 trop. Am. R: AMBG 55(1968)1. Some heterostylous. Cult. orn. fragrant cut-fls esp. ***B. humboldtii*** Hend. & Andr. Hend. (incl. double forms C19) &, taller, ***B. longiflora*** (Cav.) Kunth

Bouzetia Montr. = ?*Suriana*

Bovonia Chiov. = *Aeollanthus*

Bowdichia Kunth. Leguminosae (III 9). 2 trop. S Am. R: BZ 96(2011)1550. Strong timber (sucupira), e.g. ***B. nitida*** Spruce ex Benth. – wheel hubs & rims

Bowenia Hook. ex Hook.f. Zamiaceae (Stangeriaceae, Boweniaceae). 2 NE Aus. R: FA 46(1998)6361. Lvs bipinnate, sold for decoration; underground stem with blue-green algal symbionts (*Anabaena*) ed. Aborigines, poll. *Miltotranes* weevils

Boweniaceae Stevenson = Zamiaceae

bower plant *Pandorea jasminoides*

Bowiea Harv. ex Hook.f. Asparagaceae (Hyacinthaceae). 1 S & E Afr.: ***B. volubilis*** Harv. ex Hook.f. – xerophyte with 1 or 2 evanescent lvs, bulb at ground-level & ann. green, soon leafless, cli. 'stems' (infls) to 6 m & green fls, cult. orn., bulb toxic (cardiac glycosides) & locally magico-med. R: EJB 60(2004)554

Bowkeria Harv. Stilbaceae (Scrophulariaceae s.l.). 5 S Afr. Cult. trees & shrubs with floral elaiophores attractive to bees. R: NRBGE 29(1969)7

Bowles's mint *Mentha* × *spicata* 'Alopecuroides'

Bowlesia Ruíz & Pavón. Umbelliferae (Azor.). 15 S Am. R: UCPB 38(1965)1

Bowman's root *Gillenia trifoliata*

Bowringia Champ. ex Benth. = *Leucomphalos*

bowstring hemp *Dracaena* spp.

bow-wood *Maclura pomifera*

box or **boxwood** *Buxus sempervirens*, (Aus.) *Eucalyptus* spp.; **Abassian b.** *B. sempervirens*; **apple b.** *E. bridgesiana*; **bastard b.** *Polygaloides chamaebuxus*; **bimble b.** *E. populnea* subsp. *bimbil*; **Brisbane** or **brush b.** *Lophostemon confertus*; **Cape b.** *B. macowanii*, *Gonioma kamassi*; **Colombian b.** *Casearia praecox*; **E. London b.** *B. macowanii*; **b. elder** *Acer negundo*; **Florida b.** *Schaefferia frutescens*; **grey b.** *E. moluccana*; **Japanese b.** *Buxus microphylla*; **Maracaibo, Venezuelan** or **WI b.** *Casearia praecox*; **San Domingo b.** *Phyllostylon brasiliensis*; **b.thorn** *Lycium* spp.; **yellow b.** *E. melliodora*

boxty *Solanum tuberosum*

Boyania Wurdack. Melastomataceae. 1 Guianas: ***B. ayangannae*** Wurdack

Boykinia Nutt. Saxifragaceae (I 6). Incl. *Conomitella*, excl. *Telesonix*, 8 E As., N Am. (7 – R: FNA 9(2009)105, 125), woodland. R: BJLS 90(1985)35. Cult. orn.

boysenberry *Rubus* 'Boysenberry'.

Braasiella Braem & al. = *Tolumnia*

Brabejum L. Proteaceae (V 4). 1 SW Cape: ***B. stellatifolium*** L. – seeds form. eaten roasted (wild almonds, w. chestnuts), coffee subs., supposed toxic. R: Strel. 9(2000) 575

bracaatinga *Mimosa scabrella*

Brachanthemum DC. (~ *Chrysanthemum*). Compositae (Anth.-Art.). 10 C As. to China. R: BBMNHB 23(1993)113

Brachiaria (Trin.) Griseb. = *Urochloa*

Brachionidium Lindl. Orchidaceae (V 13c). c. 75 trop. Am. R: CJB 34(1956)160

Brachionostylum Mattf. (~ *Senecio*). Compositae (Sen.-Sen.). 1 NG: ***B. pullei*** Mattf. – app. dioec., monopodial pachycaul

Brachistus Miers. Solanaceae (IV 5c). 3 C Am. R: AMBG 68(1981)226

Brachtia Reichb.f. = *Brassia*. But see Orquideologia 9(1974)5

Brachyachenium Bak. = *Dicoma*

Brachyachne (Benth.) Stapf = *Cynodon*; see also *Micrachne*

Brachyactis Ledeb. = *Symphyotrichum*

Brachyandra Philippi = *Helogyne*

Brachyapium (Baill.) Maire = *Stoibrax*

Brachybotrys Maxim. ex Oliv. Boraginaceae (B.3.7). 1 Manchuria, E Siberia: ***B. paridiformis*** Maxim. ex Oliv.

Brachycarpaea DC. = *Heliophila*

Brachycaulos Dixit & Panigr. ? Saxifragaceae. 1 Sikkim: ***B. simplicifolius*** Dixit & Panigr. – coll. once

Brachycereus Britton & Rose. Cactaceae (III 4). 1 Galápagos: ***B. nesioticus*** (Robinson) Backeb., on lava

Brachychaeta Torrey & Gray = *Solidago*

Brachychilum (R. Br.) Petersen = *Hedychium*

Brachychiton Schott & Endl. Malvaceae (Sterc.; Sterculiaceae). 31 Aus., NG (2, 1 endemic). R: AusSB 1(1988)199. Bottle-trees (stems swollen) allied to *Sterculia* though seeds adhere to inside of fr., pend. from margins in *S.*; fr. with stinging hairs; cult. orn. incl. hybrids: ***B. acerifolius*** (G. Don) Macarthur (flame-tree, NE Aus.); ***B. populneus*** (Schott & Endl.) R. Br. (kurrajong, NE Aus.) – seeds ed. (roasted & ground), bark strips twisted into Aborginal fishing-lines & ***B. rupestris*** (Lindl.) K. Schum. (Queensland) cult. for fodder in E Aus.

Brachychloa S. Phillips. Gramineae (XXIX). 2 Mozambique, Natal. Sandy sites

Brachycladium (Luer) Luer = *Lepanthes*

Brachyclados Gillies ex D. Don. Compositae (Mut.-Mut.). 3 temp. S Am. Shrubs. Medic. (asthma)

Brachycodon Fed. = *Campanula*

Brachycodonia Fed. ex Kolak. = *Campanula*

Brachycome Cass. (*Brachyscome*). Compositae (Ast.-Bra.). Excl. *Hullsia, Pembertonia, Roebuckiella*, 87 Aus. (R: JABG 28(2014)27), 3 NZ, 1 NG. R: PLSNSW 74(1949)97, 75(1950)122. Some apomicts. ***B. dichromosomatica*** C. Carter (S Aus.) – 2n = 4. Cult. orn. esp. ***B. iberidifolia*** Benth. (Swan River Daisy, W to S Aus.). For New Caled. spp. see *Pytinicarpa*

Brachycorythis Lindl. Orchidaceae (IV 4d). 36 trop. & S Afr., trop. As. R: KB 10(1955)226. Some epiphytes, usu. terr. incl. mycotrophs ('*Schwartzkopffia*')

Brachycylix (Harms) R. Cowan. Leguminosae (I 2). 1 NC Colombia: ***B. vageleri*** (Harms) R. Cowan – flagelliflory, timber

Brachycyrtis Koidz. = *Tricyrtis*

Brachyelytrum P. Beauv. Gramineae (VII). 1 E As., 2 N Am. R: JAA 69(1988)253, SB 28(2003)683. Woodland grasses (? relicts of widespread Mid-Tertiary veg.)

Brachyglottis Forst. & Forst. f. Compositae (Sen.-Tuss.). Excl. *Bedfordia*, incl. *Dolichoglottis, Haastia, Traversia, Urostemon*, c. 40 SE Aus., NZ, Chatham Is. R: OB 44(1978)25. Prob. also in S Am. & Madag. Trees, shrubs, lianes (e.g. ***B. sciadophila*** (Raoul) R. Nordenstam, NZ) & shrublets, ***B. sp.*** (*H. pulvinaris* Hook.f., NZ) forming large cushions in subalpine & alpine veg. (cf. *Raoulia*, veg. sheep), showing rapid evolutionary radiation in NZ. Cult. orn. esp. shrubby ***B. x jubar*** Sell **Dunedin Group** ('D. Hybrids'), usu. **'Sunshine'** (*B. compacta* (Kirk) R. Nordenstam (NZ) × *B. laxifolia* (J. Buch.) R. Nordenstam (NZ)), widely planted everg. in towns ('*Senecio greyi*'), some also involving *B. monroi* (Hook.f.) R. Nordenstam (NZ)

Brachyhelus (Benth.) Post & Kuntze = *Schwenckia*

Brachylaena R. Br. (~ *Tarchonanthus*). Compositae (Card.-Tarch.). 11 trop. & S Afr. (6) to Madag. (5). R: KB 55(2000)2. Trees & shrubs. ***B. huillensis*** O. Hoffm. (S & E Afr.) – endangered, strong termite-proof timber for floors, sleepers, figure-carving (muhugwe or muhuhu)

Brachylepis C. Meyer = *Anabasis*

Brachyloma Sond. Ericaceae (VII 7; Epacridaceae). 8 Aus. To be split up. Cult. orn.

Brachylophon Oliv. Malpighiaceae. 3 OW trop.

Brachymeris DC. = *Phymaspermum*

Brachynema Benth. Olacaceae. (?) 1 Amaz. Peru & Braz.: ***B. ramiflorum*** Benth. R: FN 38(1984)99

Brachyotum (DC.) Hook.f. (*Alifana*). Melastomataceae. 50 S Am., paramo & subparamo. R: MNYBG 8(1953)343. Poll. by hummingbirds & *Diglossa* flower-piercers

Brachypeza Schltr. ex Garay. Orchidaceae (V 16c). 10 Mal.

Brachypodium P. Beauv. Gramineae (XIII). 16 temp. Euras. (Eur. 7; R: Boissiera 45(1991)), Mex., C & S Am. False bromes, esp. ***B. pinnatum*** (L.) P. Beauv. (tor grass); x = 5, 7, 9. Starch grains consumed 28–30 K yrs BP

Brachypremna Gleason = *Ernestia*

Brachypterum (Wight & Arn.) Benth. = *Solori*

Brachypterys A. Juss. Malpighiaceae. 3 trop. S Am., WI

Brachyrhynchos Less. (~ *Senecio*). Compositae (Sen.-Sen.). 1 S Afr.: ***B. junceus*** Less. (*S. j.*)

Brachyscias Hart & Henwood. Umbelliferae. 1 W Aus: ***B. verecundus*** Hart & Henwood. R: AusSB 12(1999)176

Brachyscome Cass. Earlier (?mistaken) name for *Brachycome*

Brachysema R. Br. = *Gastrolobium*

Brachysiphon A. Juss. Crypteroniaceae (Penaeaceae). 5 SW Cape. R: OB 18(1968)5, NJB 15(1995)63

Brachysola Rye (~ *Chloanthes*). Labiatae (IV 1). 2 SW Aus. R: Nuytsia 13(2000)331

Brachystachyum Keng = *Semiarundinaria*

Brachystegia Benth. Leguminosae (I 2). 26 trop. Afr.: 7 in Guineo-Congolian forest, rest major components (oft. forming monospecific stands) of decid. (miombo) woodlands of S C Afr. C ± 0. Alleged widespread hybridity denied. Some timbers (okwen), barkcloth, local medic. etc.

Brachystele Schltr. Orchidaceae (IV 2h). Excl. *Mesadenus*, 15 trop. Am. esp. Braz. to Arg.

Brachystelma R. Br. (~ *Ceropegia*). Apocynaceae (V b; Asclepiadaceae III 5). (Incl. *Microstemma*) c. 115 Afr. with 1 Ind., NG & Aus.: ***B. glabriflorum*** (F. Muell.) Schltr. (*Microstemma g.*) – ed. tubers sought by animals incl. humans. R: F. Albers & U. Meve, *Ill. Handb. Succ. Pls*, Asclep. (2002)20. Many cult. orn.

Brachystemma D. Don. Caryophyllaceae (II 1). 1 Himal.: ***B. calycinum*** D. Don

Brachystephanus Nees. Acanthaceae (III 2c). Incl. *Oreacanthus*, 22 trop. Afr., Madag. R: SGP 79(2009)126. ***B. giganteus*** Champluvier (*O. mannii*, C Afr.) – hapaxanthic after 9 yrs growth; ***B. glaberrimus*** Champluvier (C Afr.) – only gynomonoec. A.; ***B. myrmecophilus*** Champluvier (C Afr.) – myrmecophyte with extrafl. nectaries & stem domatia

Brachystigma Pennell (~ *Agalinis*). Orobanchaceae (Gerard.; Scrophulariaceae s.l.). 1 SW N Am.: ***B. wrightii*** (A. Gray) Pennell

Brachythalamus Gilg = *Gyrinops*

Brachythrix Wild & Pope. Compositae (Vern.-Erl.). 6 C & E Afr. R: Kirkia 11(1978)25

Brachytome Hook.f. Rubiaceae (II 1). 5 S China, Indomal. R: Candollea 42(1987)351

Bracisepalum J.J. Sm. Orchidaceae (V 10b). 2 Sulawesi. R: Blumea 28(1983)413, OM 1(1986)19

bracken *Pteridium aquilinum*

Brackenridgea A. Gray (~ *Ochna*). Ochnaceae (II). c. 7 OW trop. Some yellow dyes

Bracteantha Anderb. & Haegi = *Xerochrysum*

Bracteanthus Ducke = *Siparuna*

Bracteocarpaceae A. Bobrov & Melikan = Podocarpaceae

Bracteocarpus A. Bobrov & Melikan = *Podocarpus*

Bracteolanthus de Wit = *Lysiphyllum*

Bradburia Torrey & A. Gray (~ *Chrysopsis*). Compositae (Ast.-Sol.). 2 SE US. R: FNA 20(2006)211

Bradea Standl. ex Brade. Rubiaceae (IV 1). 5 Braz.

Bradford pear *Pyrus calleryana* 'Bradford'

Braemia Jenny (~ *Houlletia*). Orchidaceae (V 12i). 1 Amaz.: ***B. vittata*** (Lindl.) Jenny. R: GO 5(2009)402

Bragaia Esteves & al. = *Brasilicereus*

Brahea Mart. ex Endl. Palmae (I 4b). c. 10, limestone of C Am. R: GH 6(1943)177. Cult. orn., some ed. fr. & oil

brahmi *Bacopa monnieri*

Brainea J. Sm. (~ *Blechnum*). Blechnaceae (Blechnoideae). 1 NE Ind. to W Mal.: ***B. insignis*** (Hook.) J. Sm. – dwarf fire-tolerant tree-fern with acrostichoid fronds

brake *Pteridium aquilinum*

bramble *Rubus* spp.; **b., strawberry** *R. pedatus*

Brandegea Cogn. = *Echinopepon*

Brandella R. Mill = *Microparacaryum*

Brandisia Hook.f. & Thomson. Paulowniaceae (Scrophulariaceae s.l.). 13 Myanmar, China

brandy See *Vitis vinifera*

brandy-bottle *Nuphar lutea*

Brandzeia Baill. (*Bathiaea*). Leguminosae (I 2). 1 Madag.: ***B. filicifolia*** Baill.

Brasenia Schreb. Cabombaceae. 1 trop. Am., Afr., Ind., temp. E As., Aus.: ***B. schreberi*** J. Gmelin – Bell's Model, submerged parts covered with mucilaginous jelly (ed. Jap. – junsai), A 12–18, wind-poll. (cf. *Cabomba*), cult. orn. (fossils in Eur.). R: Aqua-P. Sond. 3(1992)37

brasiletto woods of spp. of *Caesalpinia* & *Peltophorum*

Brasilia G. Barroso = *Calea*

Brasilianthus Almeda & Michelang. Melastomataceae. 1 Braz.: ***B. carajensis*** Almeda & Michelang. R: Phytotaxa 273,4(2016)272

Brasiliastrum Lam. = *Comocladia* + *Picramnia*

Brasilicereus Backeb. (~ *Cereus*). Cactaceae (III 3). 2 E Braz.

Brasilidium Campacci = *Gomesa*

Brasiliocroton P.E. Berry & Cordeiro. Euphorbiaceae (Crot.). 2 NE Braz. R: SB 39(2014)229

Brasiliopuntia (K. Schum.) Berger (~ *Opuntia*). Cactaceae (II). 2 S Am.

Brasiliorchis R. Singer & al. = *Maxillaria* (but see Novon 17(2007)94)

Brasiliparodia = *Parodia*

Brasilocalamus Nakai = *Merostachys*

Brasilocycnis Gerlach & Whitten = *Lueckelia*

Brassaia Endl. = *Schefflera*

Brassaiopsis Decne & Planch. = *Trevesia*

Brassavola R. Br. Orchidaceae (V 13b). 22 trop. Am. R: C.L. Withner, *The Cattleyas & their relatives* V(1998)31. Cult. orn. incl. ***B. nodosa*** (L.) Lindl. throughout genus range incl. salt-spray zone in C Am., white night-scented fls, first trop. orchid to fl. in Eur. (NL 1698, ex Curaçao)

Brassia R. Br. Orchidaceae (V 12h). Incl. *Ada, Brachtia, Mesospinidium*, c. 65 trop. Am. (Florida 1). R (sect. *B.*): OD 43 (1979) 164. Usu. poll. female wasps coll. pollinia on heads while stinging labellum surface as if prey (usu. spiders) on which eggs laid. Cult. orn. esp. ***B. aurantiaca*** (Lindl.) M.W. Chase (*A. a.*, Colomb. Andes) – poll. hummingbirds

Brassiantha A.C. Sm. Celastraceae (I). 1 NG: ***B. pentamera*** A.C. Sm. R: FM I,6(1962)392

Brassica Tourn. ex L. Cruciferae (12). 38 Euras. (Eur. 22), Medit. R: JAA 66(1985)288. Poss. heterogeneous; *Eruca, Raphanus* (hybrids with *B.* raised; *R.* poss. derived from *B. nigra* & *B. rapa/oleracea* hybrids) & *Sinapis* prob. belong here. Herbs, oft. pachycaul, s.t. woody, e.g. ***B. somalensis*** Hedge & A. Miller (Somalia) & *B. oleracea* (Jersey longjacks); some weedy e.g. ***B. tournefortii*** Gouan (Medit.) in Aus., W US. Glucosinolates in crushed lvs broken down by myrosinase to bitter thiocyanates, nitriles etc.; crushed seeds mixed with 'must' of old wine = 'mustum ardens' i.e. mustard – flavour due to pungent allyl isothiocyanate released by enzyme within 10 mins of adding water to m. powder & 'hot' *p*-hydroxybenzyl isothiocyanate; English mustard prep. from *B. hirta* (*Sinapis alba*) with *B. nigra* (50:50) first ground to wettable dust 1720 (by a Mrs Clements) & industrialized by Jeremiah Colman later, some WWII m. *B. hirta*; Am. m. brightened with turmeric. Veg. ('greens' in GB), selected forms grown for seed, fl., stem or r. Many long cult., some known to Theophrastus (400 BC) & 12 distinct types listed in Pliny's time; poss. ancestry of most cult. from 3 diploid spp.: *B. nigra* (2n = 16), *B. oleracea* (2n = 18) & *B. rapa* (2n = 20), *B. juncea* (polyphyletic; 2n = 36) being *B. nigra* × *B. rapa*, *B. napus* (2n = 38) being *B. oleracea* × *B. rapa*, some rape kales backcrosses with *B. o.*, & *B. carinata* (2n = 34) being *B. oleracea* × *B. nigra*, though some kales may be derived from ***B. cretica*** Lam. & other spp. in E Medit. ***B. carinata*** A. Braun (Texsel greens, NE Afr.) – more protein-rich than spinach; ***B. hirta*** Moench (white mustard, Medit.) – mustard- & oil-prod. seeds; ***B. juncea*** (L.) Czerniak. (Ind. m., kai choy, rai, S & E As., natur. Eur.) – grown for spring greens, esp. **'Crispifolia'** (var. *crispifolia*, *B. japonica*, Chinese m., Jap. greens, senposai) & seeds (Dijon mustard); ***B. napus*** L. (rape, colza, a cultigen) – **Napobrassica Group** (swede (introd. GB from Holland 1755), neep(s), rutabaga) – ed. r., **Pabularia Group** (Siberian kale) – curled bluish lvs eaten in winter, 'summer races' ('subsp. *oleifera*', oilseed rape, rape-seed; 61 M t [2009], some 30% GM) consp. yellow fields in N Eur., seed 40% mono-unsaturated oil – that (esp. '**Canola**' (i.e. Canadian oilseed, canola oil)) richer than olive oil in alphalinolenic acid & low in erucic acid & glucosinolates, used in margarine, mayonnaise, salad & cooking oils & biodiesel (by esterification to RME (rape methyl ester) resembling fossil diesel – with glycerine, used in soap & converted to propylene glycol used in plastics & antifreeze, & animal feed as byproducts) like soya, that with high levels used in lubricating jet-engines etc., source of colza oil (illuminant) & birdseed, hay-fever source (UK with 64 000 ha in 1978, 348 000 ha by 1988) diminishing racehorse performance, seedlings an ersatz subs. for slower-growing *Sinapis alba*; ***B. nigra*** (L.) Koch (black or brown m., Euras.) – prob. the 'm. seed' of the Bible, source of a pungent m., seedlings = 'mustard' of 'm. & cress'; ***B. oleracea*** L. (coastal Eur., n = 9, but poss. anc. hexaploid cult. 8000 yrs NE Eur.), **Acephala Group** (kale, borecole, collards) – ed. lvs in cool season, incl. cavalo nero (Tuscan cabbage) with blackish lvs, some cvs cult. orn. with coloured lvs & others with tall woody stems used for walking-sticks & the lvs for cow fodder (e.g. Jersey longjacks in Channel Is.), **Alboglabra Group** (Chinese kale or brocolli, gai laan, kai lan) – oriental veg., **Botrytis Group** (broccoli, cauliflower, broccoflower) – ed. infls in compact heads, indole-carbinol app. protecting against hormone-related cancers (breast, prostate), **Capitata Group** (cabbage) – large ed. term. bud, excess helps to cause goitre, Savoy c. being a cv. with puckered lvs, sauerkraut prep. from slightly fermented c., coleslaw being c. dressed with mayonnaise or vinegar etc., some cvs with red lvs often pickled (red cabbage), **Gemmifera Group** ((Brussels) sprouts, selected c. 1750) – ed. compact lat. buds, cult. orn. New York, **Gongylodes Group** (kohlrabi, knol-kohl) – ed. swollen stem, **Italica Group** (purple sprouting, s. broccoli incl. calabrese) – like Botrytis but infl. not compacted into single head, alleged to reduce incidence of bladder & colorectal cancers, sulphurophane slows cartilage-destruction in osteoarthritis, **Tronchuda Group** – Portuguese cabbage; ***B. rapa*** L. (sarson, Eur.) – imp. oilseed in Ind., pest in NZ, **Chinensis Group** (*B. chinensis*, Chinese (flowering white) cabbage, chard or mustard, pe-tsai, pak-choi, bok (buk) choy, choi (choy) sum, Shantung cabbage, tatsoi) – ed. lvs usu. boiled, **Japonica Group** – salad greens (mizuna with ragged lvs; mibuna with entire), **Pekinensis Group** (*B. pekinensis*, wong-bok, wong-baak chihli, Chinese or celery cabbage) – ed. lvs ± in head, orig. kimchi (Korea) cf. sauerkraut (*B. oleracea*), 'santo serrated' with loose heads of serrate lvs, **Perviridis Group** (*B. campestris*, komatsuma, komatsuna) – rich in vitamins, cooked like spinach, **Rapifera Group** (turnip, some 'neeps') – 1 of the oldest r.-crops (UK record: 15.975 kg) orig. 'Jack O'Lantern', **Ruvo Group** – incl. rapini; '**Tyfon**' (Holland greens) cross between Pek. & Rap. Groups, grazed by dairy cows

Brassicaceae Burnett. See Cruciferae

Brassiodendron C. Allen (~ *Endiandra*). Lauraceae (I). 6 Aus., 1 ext. to NG

Brassiophoenix Burret. Palmae (V 14i). 2 NG. R: Principes 19 (1975) 100. Monoec.

Brassiopsis Szlach & Górniak = *Brassia*

× **Brassocattleya** Rolfe. Orchidaceae. *Brassia* × *Cattleya* – 500+ grexes

× **Bratonia** Moir (× *Miltassia*). Orchidaceae. *Brassia* × *Miltonia* – 169 grexes

brauna *Melanoxylum brauna*

Braunblanquetia Eskuche. Plantaginaceae (Grat.-Grat.; Scrophulariaceae s.l.). 1 Venez. to Arg.: ***B. litoralis*** Eskuche. R: BSAB 15(1974)357

Braunsia Schwantes. Aizoaceae (V). 4–6 Cape. R: H. E. K. Hartmann, *Ill. Handb. Succ. Pls*, Aiz. A–E (2002)88. Cult. orn.

Bravaisia DC. Acanthaceae (III 2c). 3 trop. Am., 2 being mangrove trees. R: Proc. Calif. Acad. Sci. 45(1988)111

Bravoa Lex. = *Agave*

Bravocactus Doweld = *Turbinicarpus*

Braxireon Raf. = *Narcissus*

Braya Sternb. & Hoppe. Cruciferae. (26). 25 N circumpolar (N Am. 7), Alps (Eur. 3), C As., Himal. R: Pfl. 86 (IV 105) (1924)226

Brayopsis Gilg & Muschler (~ *Englerocharis*). Cruciferae (27). 8 Andes. R: JAA 71(1990)94. ***B. gamosepala*** Al-Shehbaz (Bolivia) – gamosepaly (evolved independently in 11 genera of C.)

Brayulinea Small = *Guilleminea*

Brazil or **Brazilian arrowroot** *Ipomoea batatas, Manihot esculenta*; **B. cherry** *Eugenia* spp.; **B. copal** *Hymenaea courbaril*; **B. cress** *Acmella oleracea*; **B. lacewood** *Euplassa meridionalis*; **B. nut** *Bertholletia excelsa*; **B. nutmeg** *Cryptocarya moschata*; **B. pepper** *Schinus* spp. esp. *S. terebinthifolius*; **B. pine** *Araucaria angustifolia*; **B. redwood** *Paubrasilia echinata, Brosimum rubescens*; **B. rosewood** *Dalbergia* spp.; **B. satinwood** *Euxylophora paraensis*; **B. tea** *Ilex paraguariensis*; **B. tulipwood** *Dalbergia decipularis*; **B. wood** *Biancaea sappan, Paubrasilia echinata* etc. etc., now *Haematoxylum brasiletto*

Brazoria Engelm. ex A. Gray. Labiatae (VI). 3 Texas. R: PSE 203(1996)69

Brazzeia Baill. Lecythidaceae (Scytopetalaceae). 3 trop. W Afr. R: KB 70,6(2015)48

bread-and-cheese (GB) *Crataegus monogyna*

breadfruit *Artocarpus altilis*; **Nicobar b.** *Pandanus leram*

breadnut *Brosimum alicastrum, Pediomelum esculentum*

breadroot *Pediomelum esculentum*

Bredemeyera Willd. Polygalaceae (IV). c. 15 trop. Am.

Bredia Blume. Melastomataceae. 30 E & SE As.

Breea Less. = *Cirsium*

Breitungia Löve & D. Löve = *Sedum*

Bremekampia Sreemad. = *Haplanthodes*

Bremeria Razafim. & Alejandro (~ *Mussaenda*). Rubiaceae (II 6). 24 Madag., 4 Masc. R: AJB 92(2005)555

Bremontiera DC. (~ *Indigofera*). Leguminosae (III 8). 1 Réunion: ***B. ammoxylon*** DC. – almost extinct

Brenania Keay. Rubiaceae (II 1). 2 W Afr.

Brenandendron H. Robinson (~ *Vernonia*). Compositae (Vern.-Gym.). 3 trop. Afr. R: PBSW 112(1999)244

Brenaniodendron Léonard = *Micklethwaitia*

Brenesia Schltr. = *Pleurothallis*

Brenierea Humbert. Leguminosae (I 1). 1 Madag.: ***B. insignis*** Humbert – cladodes (unique in tribe), C 0, petaloid staminodes 5

Breonadia Ridsd. Rubiaceae (I 2). 1 trop. Afr. & Madag.: ***B. salicina*** (Vahl) Hepper & J.R. Wood (*B. microcephala*) – rheophyte. R: Strel. 10(2000)483

Breonia A. Rich. Rubiaceae (I 2). (Incl. *Neobreonia*) 20 Madag. R: AMBG 89(2002)11

Bretschneidera Hemsl. Akaniaceae (Bretschneideraceae). 1 China, Taiwan, Thailand, Vietnam: ***B. sinensis*** Hemsl.

Bretschneideraceae Engl. & Gilg = Akaniaceae

Breviea Aubrév. & Pellegrin. Sapotaceae (IV). 1 W Afr.: ***B. sericea*** Aubrév. & Pellegrin. R: T.D. Pennington, *Sapotaceae* (1991)211

Brevilongium Christenson = *Otoglossum*

Brevoortia Alph. Wood = *Dichelostemma*

Brewcaria L.B. Sm., Steyerm. & H. Robinson. Bromeliaceae (1). 2 Guyana Highland. R: AMBG 73(1986)714

Brewerina A. Gray = *Eremogyne*

Brexia Noronha ex Thouars. Celastraceae (I; Brexiaceae). 1(?12) coastal lowlands E Afr., Madag., Seychelles. ***B. madagascariensis*** (Lam.) Ker with ed. fr. (E Afr.: mfukufuku)

Brexiaceae Loudon = Celastraceae

Brexiella Perrier. Celastraceae (I). c. 7 Madag. R: FMad 116(1946)58

Breynia Forst. & Forst. f. (~ *Phyllanthus*). Phyllanthaceae (Phyll.-Phyll.; Euphorbiaceae s.l.). Incl. *Sauropus*, c. 85 China to New Caled. & Aus. (7). Troll's Model; ***B. heteroblasta*** (Airy Shaw) Welzen & Pruesapan (SE As.) – rheophyte. Some poll. by sp.-specific, seed-parasitic *Epicephala* moths in Jap. ***B. androgyna*** (L.) Chakrab. & Balakr. (*S. a.*, katuk, pakwan, star gooseberry, Mal.) – used for fencing & veg. (soup alleged to encourage lactation in Sumatra), lvs used to dye food green, fr. ed.; Some medic. esp. ***B. officinalis*** Hemsl. (China, Jap.) – for asthma; cult. orn., esp. ***B. disticha*** Forst. & Forst.f. '**Roseopicta**' (*B. disticha* f. *nivosa, B. nivosa, P. cernuus*, snowbush, Pac.)

Brezia Moq. = *Suaeda*

Brianhuntleya Chesselet & al. (~ *Ruschia*). Aizoaceae (V). 1 S Afr.: ***B. intrusa*** (Kensit) Chesselet & al. R: Bothalia 33(2003)161

briar root *Erica arborea*

Brickellia Elliott. Compositae (Eup.-Alom.). Incl. *Barroetea, Phanerostylis*, c. 100 W US, Mex., C. Am. to Arg. R: MSBMBG 22(1987)221, 224. Some cult. orn., local med.

Brickelliastrum R. King & H. Robinson (~ *Steviopsis*). Compositae (Eup.-Alom.). 2 SW US, N Mex. R: Phytol. 24(1972)63

bridal bouquet *Poranopsis paniculata*; **b. creeper** *Asparagus asparagoides*; **b. veil, Tahitian** *Gibasis pellucida*; **b. wreath** *Francoa sonchifolia*

Bridelia Willd. Phyllanthaceae (Brid.-Pseud.; Euphorbiaceae s.l.). 37 OW trop. (Indomal. 15 – R: Blumea 41(1996)273; Afr. 20; Madag. & Masc. 2). *B.* spp. in Afr. food of certain silkworms giving Afr. wild or Anape silk; some timbers, tanbarks, local medic. (esp. ***B. retusa*** (L.) A. Juss., Ind. to W Mal.), incl. for diabetes

Bridgesia Bertero ex Cambess. Sapindaceae (IV 1). 1 Chile: ***B. incisifolia*** Bertero ex Cambess.

Briedelia Willd. = *Bridelia*

Briegera Senghas = *Jacquiniella*

Briggsia Craib = *Oreocharis*

Briggsiopsis K.Y. Pan (~ *Briggsia*). Gesneriaceae (III 2j). 1 S China: ***B. delavayi*** (Franch.) K.Y. Pan

Brighamia A. Gray. Campanulaceae (III 4). 2 Hawaii. R: SB 14(1989)133. Pachycaul treelets (Corner's Model), seeds like *Delissea* & prob. derived from *Lobelia* subg. *Tupa* of As. ***B. insignis*** A. Gray (olulu) – reduced (goats, rarity of poll. moth) to ? 7 in wild (2005), but prop. in cult., comm. pot-pl.

Brillantaisia P. Beauv. Acanthaceae (III 2a). 13 trop. Afr., 2 ext. to Madag. R: BBMNHB 28(1998)83. 2 posterior A perfect (only genus in A. thus)

Brimeura Salisb. (~ *Hyacinthus*). Asparagaceae (Hyacinthaceae). 3 W Medit. R: FGeo 36(2001)201. Cult. orn. esp. ***B. amethystina*** (L.) Chouard (Pyrenees)

brindleberry *Garcinia gummi-gutta*

brinjal *Solanum melongena*

Brintonia Greene (~ *Solidago*). Compositae (Ast.-Sol.). 1 SE US: ***B. discoidea*** (Elliott) Greene. R: FNA 20(2006)106

Briquetastrum Robyns & Lebrun = *Plectranthus*

Briquetia Hochr. Malvaceae (Malv.-Malv.). 5 warm Am.

Briquetina Macbr. = *Citronella*

Brisbane lily *Proiphys cunninghamii*

bristle-cone fir *Abies bracteata*; **b. pine** *Pinus longaeva*

Britoa O. Berg = *Campomanesia*

Brittenia Cogn. ex Boerl. = *Phyllagathis*

brittlebush *Encelia farinosa*

Brittonella Rusby = *Mionandra*

Briza L. Gramineae (XVI 5). Incl. *Chascolytrum*, 5 temp. Euras. (Eur. 4), S Am. Quake or quaking grass, shaking grass. Cult. orn. esp. ***B. maxima*** L. (Med., invasive Aus.), ***B. media*** L. (doddering-dillies, jiggle-joggles, Euras.)

Brizochloa V. Jir. & Chrtek = *Briza*

Brizula Hieron. = *Aphelia*

broad bean *Vicia faba*; **b. path** (WI) *Alternanthera* spp.

Brocchia Vis. (~ *Tanacetum*). Compositae (Anth.-?Mat.). 1 SW As., N Afr.: ***B. cinerea*** (Delile) Vis.

Brocchinia Schultes f. Bromeliaceae (1). 21 N S Am. R: FN 14(1974)437, AMBG 73(1986)702. Mostly terr. Some cult. orn. incl. ***B. reducta*** Bak. (Venez.) – resembles *Heliamphora* (q.v.) with bright green lvs forming cylinders, nectar-like odour unique in fam. but no nectar (unlike *Heliamphora*), poss. carnivore as trichomes on lvs can absorb leucine; some spp. have *Utricularia* pls in 'pitchers'

broccoli *Brassica oleracea* Botrytis Group; **sprouting b.** *B. oleracea* Italica Group

Brochoneura Warb. Myristicaceae. 3 E Madag. R: Adansonia 13(1973)205. Arils rudimentary

Brockmania W. Fitzg. = *Hibiscus*

Brodiaea Sm. Asparagaceae (Alliaceae s.l.). 14 W N Am. R: UCPB 60(1971). Corms ed. raw or cooked by N Am. Indians. Cult. orn. but many pls sold as *B.* are spp. of *Ipheion*, *Triteleia* or *Dichelostemma*

Brodriguesia R. Cowan. Leguminosae (I 2). 1 E Braz.: ***B. santosii*** R. Cowan

Brombya F. Muell. (~ *Melicope*). Rutaceae (I 2). 2 Queensland. R: FA 26(2013)87. Ant-disp.

brome (grass) *Bromus* (*Anisantha*, *Bromopsis*, *Ceratochloa*) spp.; **false b.** *Brachypodium* spp.; **Hungarian b.** *Bromus inermis*; **soft b.** *B. hordeaceus*

bromelain *Ananas comosus*

Bromelia Plum. ex L. Bromeliaceae (3). 49 trop. Am. R: FN 14(1979)1649. Terr. cult. orn. ***B. hieronymi*** Mez (Chaco) – textiles, young shoots ed.; ***B. pinguin*** L. (WI, S Am.) – source of pinguin fibre, fr. ed.; ***B. serra*** Griseb. (Bolivia & Braz. to Arg.) – source of caraguata or chaguar fibre used for sacks, cordage & poss. useful for paper-making, young shoots ed. See also *Deinacanthon*

Bromeliaceae Juss. Magnoliidae – Poales. 59/2975 trop. Am. (1 W trop. Afr.), few subtrop. Am. Terrestrial xeromorphic pachycauls to stemless epiphytic herbs (many with Tomlinson's Model), usu. with papain-like proteolytic enzymes; vessels with scalariform perforations. Lvs in spirals, narrow, parallel-veined, entire or spinose-serrate (water-storage tissue in mesophyll), usu. with stalked peltate water-absorbing scales, usu. concave adaxially, channelling water to centre of rosette ('tank'). Fls bisex. or functionally unisexual, reg. or slightly not, 3-merous, hypogynous to epigynous, in term. spikes, racemes or heads usu. with conspicuous coloured bracts, poll. by birds, insects, bats or rarely wind (*Navia*) or cleistogamous; K 3 green & herbaceous to ± petaloid, free or connate at base, C 3 free or connate at base, usu. brightly coloured, s.t. with paired basal nectary scales, A 3 + 3, oft. connate or adnate to P, anthers opening by longit. slits, $\underline{G}$ (3) to $\overline{G}$ (3) with term., oft. 3- fid style, ovules few to ± ∞ on axile placentas, anatr. or rarely campylotropous, bitegmic. Fr. a berry or less oft. a septicidal capsule, rarely multiple & fleshy (*Ananas*); seeds in capsules winged or plumose (due to splitting of testa starting at chalazal end but remaining attached at micropylar one), endosperm mealy, copious, the starch in compound grains. x = 8–28, oft. 25

Classification & chief genera:

Although trad. arr. as follows, *Brocchinia* (septal nectary above ovules) 'sister' to rest (septal nectary below ovules), then *Lindmania* to the other 6 groups now recog.

1. **Pitcairnioideae** ($\underline{G}$, rarely half-inf., fr. a capsule, seeds winged etc. but not with plumose crown, usu. terr.; paraphyletic): *Dyckia*, *Hechtia* (distinct group – dioec.), *Navia* (with *Cottendorfia* etc. distinct group, **Navioideae**, sister to 3. & much of 1.), *Pitcairnia* (1 in W Afr.), *Puya* (distinct group, now in 3.)
2. **Tillandsioideae** (As above but seeds with plumose crown, lvs usu. entire, pls. mostly epiphytic; generic limits unclear): *Guzmania*, *Tillandsia*, *Vriesea*
3. **Bromelioideae** ($\overline{G}$ (3), rarely only half-inf.; fr. a berry or rarely multiple fleshy, seeds without appendages, lvs usu. spiny-toothed, usu. epiphytes with r. oft. only for attachment; generic limits unclear; R: SB 40(2015)119): *Aechmea*, *Ananas*, *Billbergia*, *Bromelia*, *Canistrum*, *Cryptanthus*, *Neoregelia*, *Nidularium*, *Puya*

Bromelioideae outnumber the other 2; Tillandsioideae difficult to assign to genera without flowers; 'Pitcairnioideae' held to be most arch., xeromorphy a 'pre-adaptation' to epiphytism. Epiphytes in rainforest & even deserts (on cacti), water economy marked by absorbing trichomes, succulence, dark CO_2 fixation (arisen independently a no. of times), decid. lvs etc. Some tank-forming spp. of *Brocchinia* in *Sphagnum* bogs carnivorous pls with tanks evolved in nutritional rather than water stresses (*Catopsis* prob. also carnivorous). Tanks (up to 20 l in *Glomeropitcairnia*) oft. habitats for nos. of insects & other animals & even spp. of *Utricularia* restricted to them; certain *Hyla* frogs block tanks with head thus reducing evaporation to advantage of both – 1 sp. hibernates there in dry season

Fr. crops (*Ananas*, *Bromelia*), local fibres (*Ananas*, *Bromelia*, *Chevaliera*, *Deinacanthon*, *Neoglaziovia*, *Pseudananas*, *Puya*, *Tillandsia*) & v. many cult. orn. housepls (*Flora Neotropica* 14(1974–9), W. Rauh (1981), *Bromelien*; BBP 63(1988)403) esp. *Neoregelia*, *Nidularium*

Bromelica (Thurb.) Farw. = *Melica*

Bromheadia Lindl. Orchidaceae (V 16). c. 25 Sri Lanka (1), Burma to Queensland. R: OM 8(1997)79. Fls drop in response to temp. change

Bromidium Nees & Meyen (~ *Agrostis*). Gramineae (XVI 5). 5 S Am. R: Darw. 24(1982)194

Bromopsis (Dumort.) Fourr. = *Bromus*

Bromuniola Stapf & C. Hubb. = *Chasmanthium*

Bromus Monti ex L. Gramineae (XIV). Incl. *Anisantha*, *Bromopsis*, *Ceratochloa*, c. 160 temp. (Eur. 37; China 55, 8 endemic – R: FOC 22(2006)371; N Am. 28 native + 24 natur. – R: FNA 24(2007)193); trop. mts. R: NRBGE 30 (1970)361. Some orn., forage, weeds: bromes, easily confused with fescues but usu. with hairy tubular leafsheaths & subapical lemma awns. ***B. catharticus*** Vahl (*B. unioloides*, rescue grass, S Am.) – forage; ***B. hordeaceus*** L. (*B. mollis*, soft b., Eur., natur. Aus., N Am.); ***B. inermis*** Leysser (Hungarian b., Euras., natur. N Am., invasive Canada) – hay; ***B. interruptus*** (Hackel) Druce (S England) – extinct in wild since 1972 but basis for a sculpted corbel Oxford Univ. Museum Nat. Hist. (1910)

& re-established in Oxfordshire 2005; ***B. mango*** Desv. (Chile) – old cereal cult. Araucano Indians pre-Conquest; ***B. rigidus*** Roth (*B. diandrus* subsp. *r.*, ripgut, Euras.) – pungent callus & rough awn penetrate mouth, eyes & intestine of stock; ***B. rubens*** L. (S Eur.) – invasive Aus., W US; ***B. secalinus*** L. (cheat or chess, Euras., Medit.) – poss. domesticated seed-crop now a weed of rye; ***B. tectorum*** L. (*Anisantha t.*, downy chess, Eur., Medit.) – invasive W US; ***B. trinii*** Desv. (Chile) – facultative cleistogamy in adverse conditions

Brongniartia Kunth. Leguminosae (III 4). 63 Am. ***B. alamosana*** Rydb. a poss. oilseed

Brongniartikentia Becc. = *Clinosperma* (but see Allertonia 3(1984)337)

Brookea Benth. Plantaginaceae (Ant.-Chel.; Scrophulariaceae s.l.). 5 Borneo, Sulawesi

brooklime *Veronica beccabunga*

brookweed *Samolus valerandi*

broom *Cytisus, Genista, Spartium* etc. spp.; **b.bush** *Melaleuca uncinata*; **common** or **Eur. b.** *Cytisus scoparius*; **coral b.** *Carmichaelia crassicaulis*; **Mt Etna b.** *Genista aetnensis*; **Scotch b.** *C. scoparius*; **Spanish b.** *Spartium junceum*; **white b.** *Retama monosperma*

broom-corn *Sorghum bicolor*

broomrape *Orobanche* spp.

brose *Avena sativa*

Brosimopsis S. Moore = *Brosimum*

Brosimum Sw. Moraceae (IV). 15 trop. Am. R: FN 7(1972)161, 83(2001)231. Male fls (single A), female fls & fr. embedded in fleshy receptacle. ***B. alicastrum*** Sw. (breadnut, ramón) – lvs for fodder (drought-tolerant), seeds ed., latex potable, wood fine; ***B. guianense*** (Aubl.) Huber (amourette, leopard wood, letterwood, snakewood, Guianas) – for violin bows, turnery etc.; ***B. rubescens*** Taubert (*B. paraense*, cardinal wood, brazilwood, satiné, Braz. redwood) – best from Fr. Guiana, for furniture (esp. Fr.) etc.; ***B. utile*** (Kunth) Oken (cow-tree, milk tree) – latex potable & base for chewing-gum, barkcloth

Brossardia Boiss. = *Noccaea*

Broughtonia R. Br. Orchidaceae (V 13). 6 Greater Antilles, Bahamas. R: RJBN 17–18(1998)10. Cult. orn. like *Epidendrum*

Brousemichea Bal. = *Zoysia*

Broussa tea *Vaccinium arctostaphylos*

Broussaisia Gaud. = *Hydrangea*

Broussonetia L'Hérit. ex Vent. Moraceae (I). 8 trop. & warm As. (7), Madag. (1). Dioec.: male fls in drooping catkins with explosive A (cf. *Urtica*, rare in Moraceae), females in globose heads, fr. a syncarp (cf. *Morus*) of orange-red drupelets. ***B. kazinoki*** Sieb. (kozo, Korea) – paper in use in Jap. by AD 610; ***B. papyrifera*** (L.) Vent. (paper mulberry, E As. early taken to Polynesia, natur. in N Am.) – usu. no fls, fr. in trop., tapa or kapa cloth & paper from inner bark of tree, oft. grown as coppice, c. 100 AD in China made into paper by glueing strips together, c. 600 AD in Jap. used in paper, lanterns, umbrellas, still imp. in Laos for paper esp. for altars & cerem. umbrellas, in US sold as 'mulberry paper' (or even 'rice paper') for origami, calligraphy, paper screens & now ecologically sensitive funerary urns!

Browallia L. Solanaceae (II 2). Incl. *Streptosolen*, c. 9 SE Arizona to C S Am., Carib. A 4; fr. a capsule encl. in K. Cult. orn., widely natur. esp. ann. ***B. americana*** L., perenn. ***B. speciosa*** Hook., both with several cvs & ***B. jamesonii*** Benth. (*S. j.*, Colombia, Peru) with orange fls

brown barrel *Eucalyptus fastigata*; **b. cress** *Nasturtium* × *sterile*

Brownanthus Schwantes = *Mesembryanthemum* (but see BJ 105(1985)316, KB 61(2006)378)

Brownea Jacq. Leguminosae (I 2). 12 Costa Rica & WI to Peru. Fast-growing young shoots with pink or red lvs speckled white on flaccid stems later turn green & stiffen up (cf. *Saraca, Amherstia*). Several spp. with bright red bird-poll. heads of fls. with copious nectar; supposed contraceptives in NW Amazonia (? 'Doctrine of Signatures')

Browneopsis Huber. Leguminosae (I 2). 7 Panamá to Peru

Brownetera Rich. ex Tratt. = *Phyllocladus*

brownheart *Vouacapoua* spp.

Browningia Britton & Rose (III 7). Cactaceae. 8 Andes of Bolivia, Peru, N Chile. Cult. orn. trees & shrubs incl. ***B. hertlingiana*** (Backeb.) F. Buxb. (*B. viridis*) to 10 m

Brownleea Harv. ex Lindl. Orchidaceae (IV Iv 4a). 8 trop. & S Afr., Madag. R: JSAB 47(1981)13. Poss. mimicry between ***B. galpinii*** Bolus (S Afr.) & *Cephalaria galpiniana* Szabo

Brownlowia Roxb. Malvaceae (Brown.; Tiliaceae). 25 SE As., wetter Mal. (Borneo 17) to Solomons. Fr. oft. water-disp.

Brucea J. Mill. Simaroubaceae. c. 6 OW trop. Medic. esp. ***B. javanica*** (L.) Merr. (*B. amarissima*, Indomal.) for dysentery & worms, antimalarial quassinoids

Bruckenthalia Reichb. = *Erica*

Brugmansia Pers. (~ *Datura*). Solanaceae (IV 4). 5 S Am. esp. Andes (no truly wild spp. known). R: A. Hay & al. (2012) *Huanduj. B.* Perenn. shrubs & trees with pend. fls (*Datura* herbaceous with erect fls). Alks esp. scopolamine leading to hallucinations after violent intoxication. Cult. orn. (angel's trumpets, poss. 700 cvs [A. Hay]) esp. ***B.*** × ***candida*** Pers. (*B. aurea* × *B. versicolor* Lagerh. (Ecuador), oft. labelled *B. arborea*, with C to 50 cm); ***B. arborea*** (L.) Sweet (true, Ecuador & N Chile) – comm. source of scopolamine (medic.) now overtaken by *Duboisia*, ***B. aurea*** Lagerh. (Andes etc., many cvs, incl. '*Methysticodendron*', veg. prop. by shamans) & ***B. sanguinea*** (Ruíz & Pavón) D. Don (Colombia to Chile) used as medic. & hallucinogens by Am. Indians

Bruguiera Sav. Rhizophoraceae (III). 6 E Afr. to Pac. (fossils similar in London Eocene, cf. *Nypa*). Mangroves (Aubréville's Model); conspicuous knee-r. but no aerial r. from branches (cf. *Rhizophora*). Tanbark, black dyes, timber & charcoal, chips for pulp & rayon. ***B. gymnorhiza*** (L.) Sav. – bird-poll. in Ryukyu Is.; ***B. sexangula*** (Lour.) Poir. – young seedlings ed. Sulawesi

Bruinsmia Boerl. & Koord. Styracaceae. 1 Assam, Myanmar; 1 Mal. (not Malay Pen.)

Brunellia Ruíz & Pavón. Brunelliaceae. 62 trop. Am., esp. montane forests. R: FN 2(1970)

Brunelliaceae Engl. (~ Cunoniaceae). Magnoliidae – Oxalidales. 1/62 trop. Am. Everg. trees, oft. densely brown-tomentose, commonly dioec. or gynodioec. Lvs opp. or ternate, pinnate, trifoliolate to unifoliolate or simple with opp. leaflets entire to doubly-dentate; stip. small, decid., oft. more than 2. Fls small in cymes, ± hypogynous, reg.; K ((4)5(–8)), connate basally, valvate, persistent in fr., C 0; A in 2 whorls, each same number as K or inner with up to twice as many, inserted in notches of a nectary-disk, anthers with longit. slits; $\underline{G}$ (2 or 3) up to same number as K, each with curved to almost circinate style with linr. stigmatic surface, ovules 2 per carpel, collateral, pend., anatropous, bitegmic. Fr. a follicle, developing with stem pointing outwards or downwards, usu. densely tomentose & with long pointed trichomes; endocarp ± lignified; seeds 1 or 2 per carpel with shiny testa & corky red 'aril', attached by funicle in dehisced fr., embryo large with flattened cotyledons & abundant mealy endosperm. n = 14

Genus: *Brunellia*

Poss. best incl. in Cunoniaceae but differing in stigma, distinctive pubescence, fr. & also wood characters

Brunfelsia Plum. ex L. Solanaceae (Petun.). 46 trop. Am. R: Fieldiana Bot.n.s. 39(1998)34. Alks. Cult. orn. & medic. Fls white fading to yellow or purple fading to white. ***B. americana*** L. (lady of the night, WI) – fragrant at night; ***B. australis*** Benth. (yesterday, today & tomorrow, trop. S Am.) – fls fading from purple to white, due to changing acidity; ***B. grandiflora*** D. Don (W S Am.) – lvs & bark hallucinogens in Amazon; ***B. pauciflora*** (Cham. & Schldl.) Benth. (yesterday, today & tomorrow, Braz.) – many cvs; ***B. uniflora*** (Pohl) D. Don (*B. hopeana*, manacá, Braz. & Venez.) – used against syphilis

Brunfelsiopsis (Urb.) Kuntze = *Brunfelsia*

Brunia Lam. Bruniaceae (3). Incl. *Lonchostoma*, *Raspalia*, 37 Cape. R: T 60(2011)1149. Cult. orn.

Bruniaceae R. Br. ex DC. Magnoliidae – Bruniales. 6/81 S Afr. esp. Cape. Shrubs to small trees, usu. ericoid & with long unicellular hairs. Lvs in spirals, small, simple, oft. with blackened tip due to a local cork cambium, oft. imbr. & with centric structure; stip. 0 or vestigial. Fls sessile in spikes or heads when s.t. involucrate & resembling Compositae, rarely solit., usu. small, bisex., reg. usu. epigynous; K (4)5, connate or not, imbr.; C (4)5, imbr., rarely (*Lonchostoma*) connate at base with short tube; A same no. as C & alt. with C, anthers with longit. slits; intrastaminal disk s.t. present; G (2), rarely (3) (*Audouinia*) with as many locules but app. 1 in *Berzelia* & some *Brunia*, ovules (1)2–4(–12) per locule, pend., anatropous, unitegmic. Fr. dry, oft. with persistent K, achene-like with 1 seed or 1- or 2-seeded carpels separating & opening along ventral suture; seeds small s.t. with aril, embryo small, straight, endosperm fleshy. n = 10, 11

Chief genera: *Berzelia*, *Brunia*, *Staavia*, *Thamnea*

Strikingly similar fossils from Upper Cretaceous of Sweden. Some cut-fls

Brunnera Steven. Boraginaceae (B.2.1.1). 3 E Med. (Eur. 1) to W Siberia. ***B. macrophylla*** (Adams) I.M. Johnston (Cauc., W Siberia) – cult. orn.

Brunnichia Banks ex Gaertn. Polygonaceae (Erio.-Brunn.). Excl. *Afrobrunnichia*, 1 SE US: ***B. ovata*** (Walter) Shinners – winged pedicel aids disp.

Brunonia Sm. ex R. Br. Goodeniaceae (Brunoniaceae). 1 Aus.: ***B. australis*** Sm. ex R. Br. (pincushion) – 2n = 18, 36, 72, cult. orn. R: FA 35(1992)1

Brunoniaceae Dumort. = Goodeniaceae

Brunoniella Bremek. (~ *Ruellia*). Acanthaceae (III 2a). 6 Aus. (5), NG, New Caled. R: JABG 9(1986)95

Brunsfelsia L. = *Brunfelsia*

Brunsvigia Heister. Amaryllidaceae. 18 Afr. R: PL 6(1950)63. Much confused with *Amaryllis* etc. Alks. Several bird-poll. (red tubular fls). ***B. radulosa*** Herb. – local med., reputedly psychoactive

brush apple *Planchonella australis;* **black b.** *Coleogyne ramosissima;* **b. box** *Lophostemon confertus*

Brussels sprouts *Brassica oleracea* Gemmifera Group

Bruxanellia Dennst ex Kostel. = *Blachia*

bruyère *Erica arborea*

Brya P. Browne. Leguminosae (III 11). 4 WI esp. Cuba. Spines in place of lvs on long shoots. ***B. ebenus*** (L.) DC. – principal lumber tree of WI giving Jamaica or American ebony, coc(c)us or cocos wood, the heartwood blackening with age, used for musical instruments & form. for door-handles etc.

Bryantea Raf. = *Neolitsea*

Bryantiella J.M. Porter (~ *Gilia*). Polemoniaceae (III). 2 S Calif. R: Aliso 19(2000)70

Bryanthus Gmelin. Ericaceae (Er.-Bry.). 1 Kamchatka, Jap.: ***B. gmelinii*** D. Don

Bryaspis Duvign. Leguminosae (III 11). 2 W trop. Afr.

Brylkinia F. Schmidt. Gramineae (XI). 1 E Russia, China, Jap.: ***B. caudata*** (Munro) F. Schmidt. R: FOC 22(2006)212

Bryobium Lindl. = *Eria*

Bryocarpum Hook.f. & Thomson. Primulaceae. 1 E Himal.: ***B. himalaicum*** Hook.f. & Thomson

Bryodes Benth. = *Lindernia* (but see KB 52(1997)750)

Bryodesma Soják = *Selaginella*

Bryomorphe Harv. Compositae (Gnap.-Rel.). 1 S Afr.: ***B. aretioides*** (Turcz.) Druce (*B. lycopodioides*). R: Bothalia 41(2011)325

Bryonia L. Cucurbitaceae. c. 12 (many ill-defined) Euras., N Afr., Canary Is. R: KB 23(1969)441. ***B. alba*** L. (Euras.) marks northernmost limit of fam. (Scandinavia). Some medic. tubers & cult. orn. Some spp. dioec., some either dioec. or monoec. e.g. *B. alba* (monoec. in N Eur., natur. W N Am. where largely apomictic), dioec. in ***B. dioica*** Jacq. (*B. cretica* subsp. *dioica*, tetter-berry, white bryony, Euras., Medit.) – shoots coll. like asparagus in S Eur., drastic purgative, fatally attractive to cattle but adds gloss to horses' coats, tuber form. used as counterfeit mandrake, acrid juice of berries (12 can be fatal) blistering

bryony, black *Dioscorea communis;* **white b.** *Bryonia dioica*

Bryophyllum Salisb. = *Kalanchoe*

Bryum Dill. ex L. Bryaceae (Musci). [s.l., but being split up] c. 800 cosmop. Some veg. prop. by gemmae, incl. from rhizoid system. ***B. argenteum*** Hedw. (cosmop. weedy) – tolerant of urban pollution. See also *Ptychostomum*

buah keluak *Pangum edule*

buai *Areca catechu*

buaze *Securidaca longipedunculata*

Bubalina Raf. = *Burchellia*

Bubania Girard = *Limoniastrum*

Bubbia Tieghem = *Zygogynum*

bubble *Catha edulis*

bubinga *Copaifera salikounda, Guibourtia demeusei*

Bubon L. = *Athamanta*

Bubonium Hill = *Asteriscus*

Bucephalandra Schott (~ *Schismatoglottis*). Araceae (VII 8). 5 Borneo. R: Telopea 9(2000)195. Rheophytes. ***B. motleyana*** Schott – 1.5–8 cm tall

Bucephalophora Pau = *Rumex*

Buceragenia Greenman = *Pseuderanthemum*

Buchanania Spreng. Anacardiaceae (II). 25 Indomal., W Pac. G 4–6, 1 fert. ***B. lanzan*** Spreng. (Indomal.) – ed. oily seeds (calumpang nut, cheronjee, Cuddapah almond, almondette) used like almonds, esp. in Shrikhand sweetmeats in Ind., exported as 'almondettes'

Buchenavia Eichler. Combretaceae (II 2a). c. 20 trop. Am. R: BBMNHB 3(1963)4. ***B. tetraphylla*** (Aubl.) R. Howard (*B. capitata*, WI, N S Am.) – hard timber for floors, boats, house-constr. etc.
Buchenroedera Ecklon & Zeyher = *Lotononis*
Buchholzia Engl. Capparaceae. 2 W Afr. ***B. coriacea*** Engl. – smelly elephant-disp. fr.
Buchingera Boiss. & Hohen. = *Asperuginoides*
Buchloe Engelm. = *Bouteloua*
Buchlomimus Reeder & al. = *Bouteloua*
Buchnera L. Orobanchaceae (Buch.; Scrophulariaceae s.l). c. 100 warm (NW 16) esp. Afr. Hemiparasites. ***B. leptostachya*** Benth. (trop. Afr.) – used for earache in Tanzania
Buchnerodendron Guerke. Achariaceae (Flacourtiaceae). 2 C & E Afr. R: BJ 94(1974)289
Buchtienia Schltr. Orchidaceae (IV 2h). 4 trop. S Am. R: HBML 28(1982)304, T 64(2015)359
buchu *Agathosma* spp.
Bucida L. = *Terminalia*
Bucinellina Wiehler = *Columnea*
buck bean *Menyanthes trifoliata*; **b. berry** *Vaccinium* spp. (N Am.); **b.bush** *Salsola kali*; **b.eye** *Aesculus* spp.; **Mex. b.eye** *Ungnadia speciosa*; **yellow b. eye** *A. flava*; **b. spinifex** *Triodia longiceps*
Buckinghamia F. Muell. Proteaceae (V 3). 2 Queensland. R: Muelleria 6(1988)417
Bucklandia R. Br. ex Griff. = *Exbucklandia*
buckler fern *Dryopteris* spp.; **broad b.f.** *D. dilatata*; **narrow b. f.** *D. carthusiana*; **b. mustard** *Biscutella laevigata*
Buckleya Torr. Santalaceae (Thesiaceae). 5 S US (1), Jap., C China. Decid., dioec. parasites. R: Castanea 47(1982)17. ***B. distichophylla*** (Nutt.) Torrey (S US) – parasitic on *Tsuga* spp.
buckthorn *Rhamnus* spp.; **alder b.** *Frangula alnus*; **common** or **Eur. b.** *R. cathartica*; **Pallas's b.** *R. erythroxyloides*; **sea b.** *Hippophae rhamnoides*
Buckollia Venter & R.L. Verh. (~ *Tacazzea*). Apocynaceae (III). 2 E Afr. R: SAJB 60(1994)97
buckwheat *Fagopyrum esculentum*; **Siberian, Tartary** or **Kangra b.** *F. tataricum*
Bucquetia DC. Melastomataceae. 3 trop. S Am.
Budawangia Telford = *Epacris*
Buddha's hand *Citrus medica* 'Fingered'; **B. palm** *Saribus rotundifolius*
Buddleia auctt. = *Buddleja*
Buddleja Houst. ex L. Scrophulariaceae (Buddlejaceae). (Excl. *Chilianthus, Nicodemia*) 90 warm esp. E As. Trees, shrubs, rarely herbs; wide range of pollen morphology, 2n = 38(?)–c. 456. Cult. orn. & some local med. ***B. asiatica*** Lour. (Indomal.) – strong freesia-like scent imperceptible to some (cf. *Freesia*), fish-poison; ***B. davidii*** Franch. (China) – many cvs, introd. GB 1890s & colonising wasteground by 1930s being more attractive to butterflies etc. than any native pl., large assoc. fauna by 1980s, now invasive in Eur., N Am., Aus., NZ with more than 2000 pls per ha in infected sites; ***B. officinalis*** Maxim. (SW China) – infls coll. Yunnan for extract used in med. skin products; ***B. salviifolia*** (L.) Lam. (trop. & S Afr.) – useful timber
Buddlejaceae Wilhelm = Scrophulariaceae
budgerigar flower *Asclepias syriaca*
Buergersiochloa Pilg. Gramineae (VI 1). 1 NG: ***B. bambusoides*** Pilg.
Buesia (Morton) Copel. = *Hymenophyllum*
Buesiella C. Schweinf. = *Cyrtochilum*
buffalo bean *Astragalus crassicarpus, Mucuna* spp.; **b. berry** *Shepherdia argentea*; **b. bur** *Solanum rostratum*; **b. clover** *Trifolium reflexum*; **running b. c.** *Trifolium stoloniferum*; **b. currant** *Ribes aureum*; **b. gourd** *Cucurbita foetidissima*; **b. grass** *Boutleloua dactyloides, Panicum coloratum*, (Aus.) *Stenotaphrum secundatum*; **b. nut** *Pyrularia pubera*; **b. thorn** *Ziziphus mucronata*; **b. wood** *Burchellia bubalina*
buffel grass *Cenchrus ciliaris*
Bufonia L. Caryophyllaceae (II 1). 20 Medit. (Eur. 7), Canary Is. 4-merous. The allegation that Linnaeus deliberately altered the generic name from *Buffonia*, commemorating Buffon, as a malicious pun on *bufo* (= toad) is said to be unfounded (*J. Linn. Soc.* 2(1858)183)
Buforrestia C.B. Clarke. Commelinaceae (II 2). 1 Guianas, 2 W & C Afr.
bugbane *Actaea foetida*
bugle *Ajuga* spp.; **common** or **Eur. b.** *A. reptans*; **yellow b.** *A. chamaepitys*
bugles, red *Conostylis canescens*
bugloss *Echium plantagineum*; **viper's b.** *E. vulgare*

Buglossoides Moench (~ *Lithospermum*). Boraginaceae (B.2.2). Excl. *Aegonychon*, 2 Eur., SW As. ***B. arvensis*** (L.) I.M. Johnst. (*L. a.*, bastard alkanet) – dye-pl., cosmop. weed

bugseed *Corispermum* spp.

Buhsea Bunge = *Cleome*

Buhsia Bunge = *Cleome*

buk choy *Brassica rapa* Chinensis Group

bukchie *Baccharoides anthelmintica*

Bukiniczia Lincz. Plumbaginaceae (II). 1 Afghanistan, Pakistan: ***B. cabulica*** (Boiss.) Lincz.

Bulbine Wolf. Asphodelaceae. Incl. *Jodrellia*, c. 78 trop. & S Afr. (esp. SW Cape) diploids, Aus. (5) polyploids. Many medic. uses in S Afr.; cult. orn., weeds in Aus. ('onionweed', cf. *Asphodelus*, *Nothoscordum*). ***B. bulbosa*** (R. Br.) Haw. (E Aus.) – form. favoured ed. corms by Aboriginal people; ***B. frutescens*** (L.) Willd. (S Afr.) – shrubby cult orn.; ***B. mesembryanthemoides*** Haw. (S Afr.) – resembles *Lithops* spp. with which it lives

Bulbinella Kunth. Asphodelaceae. 22 S Afr. (17; R: SAJB 53(1987)431), NZ (6). Cult. orn. (cat-tail)

Bulbinopsis Borzi = *Bulbine*

Bulbocodium L. = *Colchicum*

Bulbophyllum Thouars. Orchidaceae (V 15). c. 1800 trop. (NG 600; Florida 1). R: OM 2(1987)1 (Afr. 72), 7(1993)1. Epiphytes with 1 or 2 lvs terminating each pseudobulb. Most fly-poll. incl. overpoweringly carrion-scented like ***B. beccarii*** Rchb.f. (Borneo) & sweetly scented e.g. clove-scented ***B. macranthum*** Lindl. (Indomal.) – the 2 lat. sepals directed upwards & meet nr. tips where flies land, holding on to outside of otherwise slippery sepals while licking surface; they work down sepals which are parted below such that flies can no longer straddle them; as they slip they clutch at the solid-looking tongue-like lip but as their wt. is transferred to it, they are suddenly flung backwards & downwards, for the lip is pivoted; two springy arms nr. tip of column embrace fly as lip returns to former position; the fly soon escapes but in so doing carries pollinia away on abdomen. In other spp. flies are thrown head first against column & receive pollinia on head while others attract flesh-flies by bad smells etc. (after Proctor & Yeo); ***B. penicillium*** C. Parish & Reichb.f. (Myanmar, SW China) – moving lip presses poll. fly towards anther (unique?). Several cult. orn. ***B. minutissimum*** F. Muell. (Aus.) – pseudobulbs hollow with stomata on inner surface; ***B. nocturnum*** J. Verm. & al. (New Britain) – each fl. opens for just 1 night, only nocturnal orchid; ***B. phalaenopsis*** J.J. Sm. (NG) – lvs to 1.7 m long; ***B. rheophyton*** J. Verm. & Tsukaya (S Borneo) – rheophyte

Bulbostylis Kunth (~ *Fimbristylis*). Cyperaceae (II 4). 100 trop. & warm (Aus. 7, N Am. 8). ***B. leucostachya*** C.B. Clarke (Venez.) – pachycaul treelet

Bulbulus Swallen = *Rehia*

bulgur *Triticum* spp.

bull bay *Magnolia grandiflora*; **b. briar** or **brier** *Smilax rotundifolia*; **b. horn thorn** *Vachellia cornigera* etc.; **b. nettle** *Cnidoscolus stimulosus*; **b. oak** *Allocasuarina luehmannii*; **b.wort** *Ammi majus*

bullace *Prunus* × *domestica* 'ssp. *institia*'

Bulleyia Schltr. Orchidaceae (V 12). 1 E Himal. to SW China: ***B. yunnanensis*** Schltr. R: Fl. Bhutan 3,3(2002)323

bullock's heart *Annona reticulata*

Bulnesia C. Gay. Zygophyllaceae (II). c. 8 S Am. R: Darw. 25(1984)299. Timber e.g. ***B. arborea*** (Jacq.) Engl. (Maracaibo lignum-vitae, verawood, Colombia, Venez.) – used like lignum-vitae & ***B. sarmientoi*** Lorentz ex Griseb. (Paraguay l.-v., palo santo, Paraguay, Braz., SE Bolivia & Arg.) – CITES-listed, wood distilled for fragrant oil of guaiac for soap as well as used like l.-v.; ***B. retama*** (Hook. & Arn.) Griseb. (Arg.) – common shrub, source of retamo wax used in shoe polish etc.

bulrush prob. orig. (as in Bible) *Cyperus papyrus*, *Schoenoplectus lacustris* or *Vossia cuspidata* but name early transferred to *Typha* spp., Rubens's picture of Christ scourged showing this & all pictures of Moses in 'the bulrushes' also; **b. millet** *Cenchrus spicatus*

bulwaddy *Macropteranthes kekwickii*

bumble tree *Capparis mitchellii*

bumbo *Daniellia* spp.

Bumelia Sw. = *Sideroxylon*

bumpy ash *Flindersia schottiana*

buna *Fagus crenata*

bunch berry *Cornus canadensis;* **b. flower** *Veratrum virginicum;* **b. grass** *Schizachyrium scoparium*

Bunchosia Rich. ex Kunth. Malpighiaceae. 55 trop. Am. Fr. ed. in many spp. – 'marmelo' (Braz.)

Bungarimba Wong (~ *Porterandia*). Rubiaceae (II 1). 4 Mal. R: Sandakania 15(2004)41

Bungea C. Meyer. Orobanchaceae (Cymb.; Scrophulariaceae s.l.). 2: SW As. (1), C As. & China (1)

bungi, bungy = ponga

bungwall *Blechnum indicum*

Bunias L. Cruciferae (13). 2 Med. (Eur. 2), As. Spirolobal cotyledons. Salad & fodder, e.g. ***B. orientalis*** L. (E Eur. & W As., adventive in C & W Eur., N Am.)

Buniella Schischkin = *Bunium*

Buniotrinia Stapf & Wettst. = *Ferula*

Bunium L. Umbelliferae (III 8). c. 40 Eur. (4) to N. Afr., (esp.) SW & C As. R: BMOIPB 93,1(1988)88. Seedlings with 1 cotyledon. ***B. bulbocastanum*** L. (Eur.) – tubers & lvs ed., form. much cult., medic.

Bunochilus D.L. Jones & M.A. Clem. = *Pterostylis*

bunya-bunya pine *Araucaria bidwillii*

Buphthalmum L. Compositae (Inul.-Inul.). 3 Eur. (2), W As. Cult. orn. (ox-eye)

Bupleurum Tourn. ex L. Umbelliferae (III 8). c. 165 Euras. (Eur. 39), N Afr., Canary Is., arctic N Am. (1), S Afr. (1). Shrubs e.g. ***B. dracaenoides*** H.C. Wang & al. (SW China) – pachycaul treelet to 2 m, & herbs with entire lvs, oft. parallel-veined (hare's ear). Some medic. China & cult. orn. incl. ***B. rotundifolium*** L. (Eur., natur. N Am.) – thorow-wax, i.e. 'throw-wax' (through-grow) named because of perfoliate lvs

bur clover *Medicago* spp.; **b. cucumber** *Sicyos angulatus;* **b. dock** *Arctium* spp.; **b. grass** *Cenchrus* spp.; **b. head** *Echinodorus* spp.; **b. marigold** *Bidens* spp.; **Noogoora b.** *Xanthium* spp.; **b. nut** *Tribulus* spp.; **b. oak** *Quercus macrocarpa;* **b. parsley** *Caucalis platycarpos;* **b. reed** *Sparganium* spp.; **b. rose** *Rosa roxburghii;* **b. weed** *Medicago* spp., *Sparganium* spp., *Xanthium* spp. esp. *X. strumarium*

Burasaia Thouars (*Bourasaha*). Menispermaceae (III). 5 Madag. ***B. madagascariensis*** DC. – source of a bitter principle used in beer-making

Burbidgea Hook.f. Zingiberaceae (II 4). 5 Borneo. R: NRBGE 42(1985)262

Burchardia R. Br. Colchicaceae (Liliaceae s.l.). 6 SW, 1 ext. to E Aus. (***B. umbellata*** R. Br. – ed. Sydney Aborigines). R: Nuytsia 15(2005)352. No colchicine!

Burchellia R. Br. Rubiaceae (?II 1). 1 S Afr.: ***B. bubalina*** (L.f.) Sims – shrub with v. hard wood (buffalo wood). R: Strel. 10(2000)483

Burckella Pierre. Sapotaceae (II). 14 E Mal. to Tonga. Ed. fr. esp. ***B. fijiensis*** (Hemsl.) A.C. Sm. & S. Darwin (Fiji) & ***B. obovata*** (Forst.f.) Pierre (E Mal. to Vanuatu)

Burdachia Mart. ex A. Juss. Malpighiaceae. 4 trop. S Am.

Burdekin plum *Pleiogynium timoriense*

burgan *Kunzea ericoides*

Burgesia F. Muell. = *Leptosema*

Burgundy pitch from *Picea abies*

Burkartia Crisci. Compositae (Mut.-Nass.). 1 Patagonia: ***B. lanigera*** (Hook. & Arn.) Crisci

Burkea Hook. Leguminosae (I 4). 1 W & S Afr.: ***B. africana*** Hook. – source of a soluble gum & tanbark. R: Strel. 19(2000)278

Burkillanthus Swingle. Rutaceae (II 2). 1 W Mal.: ***B. malaccensis*** (Ridl.) Swingle

Burkillia Ridl. = *Burkilliodendron*

Burkilliodendron Sastry. Leguminosae (III 16). 1 Malay Pen.: ***B. album*** (Ridl.) Sastry – thought extinct (only 1 coll.) but recoll. 1993. R: GBS 52(2000)7

burlap (US) = hessian

Burlemarxia Menezes & Semir (~ *Barbacenia*). Velloziaceae. 3 Braz. R: T 40(1991)413

Burma padauk *Pterocarpus macrocarpus*

Burmabambus Keng f. = *Yushania*

Burmannia L. Burmanniaceae. 57 warm (Aus. 2 autotrophic, N Am. 3). Some autotrophic in grassland, some mycotrophic in lowland rain forest

Burmanniaceae Blume (~ Dioscoreaceae). Magnoliidae – Dioscoreales. Excl. Thismiaceae (~ Dioscoreaceae), 10/95 trop., widely scattered. Small mycotrophs (few autotrophs), ann. or perenn., oft. with rhizomes or tubers & usu. without r.-hairs or chlorophyll (those with s.t. with scalariform vessels). Lvs in spirals, usu. scale-like & colourless to yellowish or reddish but s.t. green, simple. Fls terminal, 1 or in cymes (usu. bifurcate cincinnus), bisex., reg. to ± irreg., epigynous; P (3 or 6), tubular or campanulate, outer valvate, inner smaller or 0, 3 (or 6) with elongate term. appendages, A 3 on tube, sessile or subsessile, s.t. connate by anthers forming tube around style & opening by lat. or longit. slits, connective oft. expanded, $\overline{G}$ (3), 1–3-loc., s.t. basally 3-, apically 1-loc., nectary glands on G or within, ovules ∞, anatropous, bitegmic, minute. Fr. a capsule, rarely fleshy, oft. winged, variously dehiscent; seeds minute with tiny embryo with as few as 4 cells when seed shed, endosperm ± 0. n = 6, 8

Principal genera: *Burmannia, Gymnosiphon*

Thismiaceae (A 6, rarely 3; P circumscissile) separated – if this were not done B. would fall into Dioscoreaceae s.l. (with Taccaceae): mycotrophy evolved twice in the order. Part of Corsiaceae poss. referable here

Burmeistera Karsten & Triana (~ *Centropogon*). Campanulaceae (III 3). 102 trop. S Am. R: Pfllanzenr. IV,276b(1943)122, (1953)767, 276c(1968)834. Bat- & bird-poll.

Burmese lacquer-tree *Gluta usitata;* **B. rosewood** *Pterocarpus indicus*

Burnatastrum Briq. = *Plectranthus*

Burnatia M. Micheli. Alismataceae. 1 trop. Afr.: ***B. enneandra*** M. Micheli – juvenile submerged, adults s.t. floating or emergent, insect-poll., nutlets & rhiz. fragments disp. water & animals. R: FZ 12,2(2009)4

burnet *Sanguisorba* spp. esp. *S. officinalis;* **b. rose** *Rosa spinosissima;* **b. saxifrage** *Pimpinella major*

Burnettia Lindl. Orchidaceae (IV 3f). 1 SE Aus.: ***B. cuneata*** Lindl. (lizard orchid) – hapaxanthic mycotroph assoc. with *Melaleuca squarrosa*. R: Gen. Orch. 2(2001)157

burning bush *Dictamnus albus, Bassia scoparia* f. *trichophylla*

Burnsbaloghia Szlach. = *Deiregyne*

burr, bastard or **Bathurst** *Xanthium spinosum;* **b. daisy** *Calotis* spp.; **Narrawa b.** *Solanum cinereum;* **Noogoora b.** *X. occidentale*

burrawang *Macrozamia spiralis*

Burretiodendron Rehder. Malvaceae (Dom.; Tiliaceae). 6 SW China, SE As. R: JAA 71(1990)375. ***B. hsienmu*** Chun & How (xianmu, SW China & N Vietnam) – valuable timber but vulnerable

Burretiokentia Pichi-Serm. Palmae (V 14c). 5 New Caled. R: Allertonia 3(1984)393

Burrielia DC. = *Lasthenia*

burrograss *Scleropogon brevifolius*

Burroughsia Mold. (~ *Lippia*). Verbenaceae (Lant.). 2 Mex.

burroweed *Isocoma tenuisecta*

Bursaria Cav. Pittosporaceae. 7 Aus. R: AusSB 12(1999)124. Usu. spiny, seeds 1 to several per fr. Cult. orn. ***B. spinosa*** Cav. (Christmas bush [S Aus.]) – street-tree in Victoria, harvested for aesculin

Bursera Jacq. ex L. Burseraceae (IV). c. 100 trop. Am. Decid. aromatic polygamo-dioec. trees & shrubs, sources of resins (copals in C Am.), oils (linaloe, linaloa oil (Mex.), ***B. linanoe*** (La Llave) Rzed. & al. (*B. delpechiana*) etc. introd. India 1912 as 'Indian lavender tree' overtaking Mex. prod.) & medic., esp. ***B. simaruba*** (L.) Sarg. (*B. gummifera*, gumbo-limbo, WI birch, incense tree, S Florida, C Am., WI) – wood for fuel & matches (allegedly form. for merryground horses), resin (American elemi, chibou) for varnish etc., medic. (bee-stings), used by Mayas for incense; ***B. glabrifolia*** (Kunth) Engl. (Mex.) – local carving, resin used as tribute; ***B. penicillata*** (DC.) Engl. (Ind. lavender, Mex.) – much grown in Ind. for ess. oil.; ***B. schlechtendahlii*** Engl. (Mex.) – resin in canals under pressure, released when canals punctured & ejected to 1.5 m, miring insect predators

Burseraceae Kunth. Magnoliidae – Sapindales. 19/675 trop. esp. Am. & NE Afr. Trees or shrubs with prominent resin ducts in (oft. flaking) bark, resin usu. almond-scented. Lvs in spirals, rarely opp., pinnate or trifoliolate, rarely unifoliolate, leaflets usu. pulvinate with prominent fine venation when dry; stip. rare. Fls in thyrses, rarely racemes or heads, bisex. or not (trees oft. dioecious), reg., usu. hypogynous; K (3)4 or 5, usu. connate basally, C (3)4 or 5, rarely 0, both imbr., rarely valvate; A 3–5 (+3–5), filaments rarely connate, without or within annular nectary-disk, anthers with longit. slits, staminodes oft. present in female fls. $\underline{G}$ ((2)3–5(–12 (*Beiselia*)) forming plurilocular ovary with terminal style, ovules pend. on axile placenta, (1 or) 2 per locule, anatr. or hemitropous to campylotropous, bitegmic or rarely unitegmic. Fr. a drupe with 1–5 1-seeded stones or 1 stone with all seeds, rarely a 'pseudocapsule' opening to reveal 1-seeded nutlets; embryo oily with lobed or cleft cotyledons & almost 0 endosperm. x = 11, 13, 23

Classification & chief genera:

Beiselia app. sister to rest, but trad. arr.:

Protieae (drupe with 2–5 free or adhering but not fused parts): *Protium, Tetragastris*

Bursereae (drupe with endocarp completely fused, exocarp dehiscing by valves; poss. best split into *Bursera-Commiphora* & *Boswellia-Garuga* groups): *Boswellia, Bursera, Commiphora, Garuga*

Canarieae (drupe with completely fused endocarp): *Canarium, Dacryodes, Haplolobus, Santiria*

Close to Anacardiaceae; chem. similar to Rutaceae & Meliaceae but limonoids prob. absent. Rainforest as well as savanna trees: some timbers, ed. seeds & fr., incense & scents (incl. frankincense (*Boswellia*) & myrrh (*Commiphora*)), varnishes

Burseranthe Rizz. = *Trichilia*

Burtonia R. Br. = *Gompholobium*

Burttdavya Hoyle = *Nauclea*

Burttia Bak. f. & Exell. Connaraceae (I). 1 C Tanzania: ***B. prunoides*** Bak. f. & Exell – seeds for poisoning animals. R: AUWP 89–6(1989)169

Buschia Ovcz. = *Ranunculus*

Buseria T. Durand = *Coffea*

bush banana *Marsdenia australis;* **b. basil** *Ocimum minimum;* **b. flax** *Astelia nervosa, Phormium tenax;* **b. lawyer** *Rubus australis;* **b. pea** *Pultenaea* spp.; **b.rue** *Cneoridium dumosum;* **b. tomato** *Solanum centrale*

Bussea Harms. Leguminosae (I 4). 5 trop. Afr., 2 Madag.

bussu palm *Manicaria saccifera*

Bustelma Fourn. = *Oxystelma*

Bustillosia Clos = *Asteriscium*

Busy Lizzie *Impatiens walleriana*

Butania Keng f. = *Sinarundinaria*

butcher's broom *Ruscus aculeatus*

Butea Roxb. ex Willd. Leguminosae (III 18). (Excl. *Meizotropis*, q.v.) 2 Indomal. R: BBSI 29(1987)202. ***B. monosperma*** (Lam.) Kuntze (*B. frondosa*, dhak, palas, flame of the forest, bastard teak, Ind. to Myanmar) – gum (Bengal kino) astringent, seed-oil (muduga oil) vermifuge, fls (tisso flowers) give a red dye thrown in Holi Festival (Hindu), lvs stitched together as plates in restaurants in Ind., bark for cordage & sails, timber good under water also for charcoal, lac insects feed on it, used to reclaim saline land & is one of the most beautiful of all flowering trees (fls bright orange-red), sacred to Brahmins in Ind.; ***B. superba*** Roxb. ex Willd. (Ind.) – marketed as male aphrodisiac

Buteraea Nees = *Strobilanthes*

Butia (Becc.) Becc. Palmae (V 8a). c. 12 S Am. R: Principes 23(1979)65. 5 grass-like spp. in Parag.-Braz. grasslands. Ed. drupes; cult. orn. (some cold-hardy)

Butomaceae Mirb. Magnoliidae – Alismatales. 1/1 temp. Euras. Emergent aquatic perenn. monopodial glab. herb with starchy rhizome. Lvs distich., parallel-veined, linr., erect, ± triquetrous. Infl. axillary, cymose umbel with 3 bracts. Fls reg., bisex., hypogynous; P 3 + 3, outer greenish, inner pink, A 3 prs + 3, anthers with longit. slits, pollen monosulcate, $\underline{G}$ 6 distally unclosed, connate at base in ring with short styles & nectaries basally, ovules ∞, over inner surface of carpels, anatropous, bitegmic. Fr. of sep. follicles; embryo straight with single term. cotyledon & endosperm 0. n = 7–13 (small)

Only sp.: *Butomus umbellatus*. Lvs poss. homologous with petioles of Alismataceae

Butomopsis Kunth (*Tenagocharis*). Alismataceae (Limnocharitaceae). 1 trop. OW: ***B. latifolia*** (D. Don) Kunth. R: FA 39(2011)2

Butomus Tourn. ex L. Butomaceae. 1 temp. Euras. (natur. N Am.): ***B. umbellatus*** L. (flowering rush) – cult. orn., rhiz. powdered for bread in N Euras.

butter and eggs *Linaria vulgaris;* **b.bur** *Petasites* spp.; **b.cup** *Ranunculus* spp.; **Bermuda b.c.** *Oxalis pes-caprae;* **dika b.** *Irvingia gabonensis;* **b.fruit** *Diospyros blancoi;* **illipe b.** *Shorea* spp.; **b.nut** *Juglans cinerea;* **b.n. squash** *Cucurbita moschata;* **shea b.** *Vitellaria paradoxa;* **b.wort** *Pinguicula* spp., esp. *P. vulgaris*

butterfly flower *Schizanthus* spp.; **b. lily** *Hedychium coronarium;* **b. orchid** *Platanthera* spp.; **b. pea** *Centrosema* spp., *Clitoria ternatea;* **b. tree** *Bauhinia* spp.; **b. weed** *Asclepias tuberosa*

button bush *Cephalanthus occidentalis;* **b. grass** *Gymnoschoenus sphaerocephalus;* **b. mangrove** *Conocarpus erectus;* **b.weed** *Spermacoce* spp.; **b.wood** *Platanus occidentalis, C. erectus, Laguncularia racemosa;* **bs, policeman's** *Jasione montana*

Buttonia McKen ex Benth. Orobanchaceae ('Butt.'; Scrophulariaceae s.l.). 2 trop. & S Afr.

Butumia G. Taylor = *Saxicolella*

Butyrospermum Kotschy = *Vitellaria*

Buxaceae Dumort. Magnoliidae – Buxales. Incl. Didymelaceae, Haptanthaceae, Stylocerataceae, 7/115 nearly cosmop. Everg. trees, shrubs, or rarely herbs, oft. with steroid alks. Lvs opp., less oft. in spirals, simple; stip. 0. Fls oft. in heads or spikes, small., reg., unisexual (pls monoec., rarely dioec.), P not petaloid, 2 + 2 (s.t. 5, or 3 + 3, 0 in staminate fls of *Didymeles* (?1 in females) & pistillate fls of *Styloceras*), anthers with longit. slits, disk 0, $\underline{G}$ ((1-)3(4)), primary locules divided into uniovulate cells in *Pachysandra* & *Styloceras*, ovules (1) 2 per primary locule, pend., anatr. (to hemitropous), bitegmic, oft. (at least) with an obdurator. Fr. a loculicidal dehiscent capsule, less oft. a drupe; seeds black, shiny, usu. caruncuIate, embryo straight, endosperm firm, oily or 0. x = 10, 14
Genera: *Buxus, Didymeles, Haptanthus, Notobuxus, Pachysandra, Sarcococca, Styloceras*
Seed anatomy suggested relict nature of B. (Corner) form. put nr. Euphorbiaceae (though no columella in fr. like E.) though wood & A suggested affinity with Pittosporaceae. Cult. orn. & high quality wood (*Buxus*)

Buxiphyllum W.T. Wang & C.Z. Gao = *Paraboea*

Buxus Tourn. ex L. Buxaceae. Excl. *Notobuxus*, c. 90 W Eur. (2), Medit. to S Afr. (9 S of Sahara, R: KB 44(1989)296) & Madag. (9, R: Adans. III,24(2002)181), temp. E As. (c. 40), WI (c. 50; Cuba – c. 40% of spp.), C Am. Monoec., terminal female surrounded by male fls. Fr. explosive, inner layer of pericarp separating from outer. Alks: lvs & seeds strongly purgative. Cult. orn. (threatened by boxwood blight (*Cylindrocladium buxicola*, orig.?) introd. UK 1990s & now all Eur., NZ, N Am.) everg. hedges withstanding pruning, topiary etc. esp. ***B. sempervirens*** L. (common or Eur. box, Abassian boxwood, Eur. (native S England) & Medit.) – planted in anc. cemeteries in Balkans, Caucasus, oft. for hedges (that at Wymondley Priory, Herts, UK allegedly from 1541), and edging fl. beds, many cvs (incl. varieg. '**Variegata**' (oregonia)), wood hardest & heaviest of Br. trees, that used in rulers, musical instruments incl. bagpipes, croquet balls, inlay etc. since anc. Egyptians (combs, flutes, furniture, lyres) & box balls still used for drawing lots for fixtures in FA Cup, now from Turkey, first used for wood engravings by Bewick c. 1800, now largely replaced by Venez. box, *Casearia praecox*, foliage used like *Salix* as 'palm' in Mediaeval GB & Holland (palmboomje) on Palm Sunday, lvs & sawdust form. used to dye hair auburn; ***B. macowanii*** Oliv. (Cape b., E. London b.wood, E Cape, Natal); ***B. microphylla*** Sieb. & Zucc. (Jap. b., ? E As.) – cult. orn. (many cvs) in Jap. since 1450 but no wild pls found

bya-rkang *Delphinium* spp.

Byblidaceae Domin. Magnoliidae – Lamiales. Excl. Roridulaceae, 1/6 NG to Aus. Carnivorus herbs & suffrutices with long-stalked & sessile oily or mucilaginous glands, the former trapping insects, the latter posited to digest them. Lvs in spirals, sublinear, abaxially curved, veins parallel; stip. 0. Fls solit. in axils, bisex., hypogynous, zygomorphic due to A, bracteoles 0; K 5, imbr., s.t. shortly connate basally, persistent, C 5, imbr. or convolute shortly connate basally, A 5 alt. with C, attached to C tube, anthers with term. pores or pore-like slits, $\underline{G}$ (2) with simple style, ovules $\pm \infty$ on axile placentas, anatropous, unitegmic. Fr. a loculicidal capsule; seeds with small straight embryo & copious starchy endosperm. n = 8, 9
Genus: *Byblis*
Poss. incl. Linderniaceae. *Roridula* now in its own fam. again (in Ericales!) – similarities being morphological convergence

Byblis Salisb. Byblidaceae. 6 Aus., 1 ext. to NG. Fire-tolerant carnivores (Corner's Model), ***B. gigantea*** Lindl. (SW Aus.) dying back to tuber in summer (? fire necessary for germ.); capitate-glandular hairs do not move (cf. *Drosera*); ? buzz-poll., bees vibrating pollen out by flapping wings at right frequency

Byrsanthus Guillemin. Salicaceae (Flacourtiaceae). 1 W Afr.: ***B. brownii*** Guillemin

Byrsocarpus Schum. = *Rourea*

Byrsonima Rich. ex Kunth. Malpighiaceae. c. 150 trop. Am. Aubréville's & Fagerlind's Models. Some fish-disp. in Amaz., some bird-disp. (unique in its subfam. but also in *Bunchosia* & *Malpighia*). Bark of some spp. for tanning, timber useful (surette); some ed. fr. (e.g. ***B. crassifolia*** (L.) Kunth (trop. Am. savannas) for icecream etc.) & cult. orn.

Byrsophyllum Hook.f. Rubiaceae (II 1). 2 Ind., Sri Lanka

Bystropogon L'Hérit. Labiatae (VII 2b). 7 Macaronesia. R: PM 18(1984)39

Bythophyton Hook.f. Linderniaceae (Phrymaceae – Micro.; Scrophulariaceae s.l.). 1 Indomal.: ***B. indicum*** (Hook.f. & Thomson) Hook.f – submerged aquatic ann.

Byttneria Loefl. Malvaceae (Bytt.- Bytt.; Sterculiaceae). 132 pantrop. R: Bonplandia 4(1976). Shrubs & lianes (***B. morii*** L. Barnett & Dorr (Fr. Guiana) a tree), some myrmecophily
Byttneriaceae R. Br. = Malvaceae (Bytt.)
by-yu *Macrozamia riedlei*

C

caa-ehe *Stevia rebaudiana*
caapi *Banisteriopsis caapi*
Caamembeca Pastore (~ *Polygala*). Polygalaceae. 11 S Am. R: KB 17(2012)437
Caatinganthus H. Robinson (~ *Stilpnopappus*). Compositae (Vern.-Ele.). 2 N Braz.
Caballeroa Font Quer = *Saharanthus*
cabbage *Brassica oleracea* Capitata Group; **c. bark** *Andira inermis*; **black c.** *Melanodendron integrifolium*; **celery c.** *B. rapa* Pekinensis Group; **Chinese c.** *B. rapa* Chinensis Group & Pekinensis Group; **he c.** *Pladaroxylon leucadendron*; **Kerguélen c.** *Pringlea antiscorbutica*; **longjack c.** *B. somalensis*; **Lundy c.** *Coincya wrightii*; **Moluccan c.** *Pisonia grandis* 'Alba'; **palm c.** bud of several palm spp., e.g. *Euterpe oleracea*, *Roystonea regia*, *Elaeis guineensis*, *Cocos nucifera*, eaten as salad fresh or tinned, in some spp. its collection leading to death of tree; **Portuguese c.** *B. oleracea* Tronchuda Group; **c. rose** *Rosa × centifolia*; **Shantung c.** *B. rapa* Chinensis Group; **she c.** *Lachanodes arborea*; **c. tree** name applied to many pachycaul trees with massive heads of lvs, e.g. spp. of *Anthocleista & Vernonia* (W Afr.), *Cussonia* (S Afr.), *Livistona australis* (NSW), *Cordyline* (NZ), *Andira inermis* (WI); **Tuscan c.** *B. oleracea* Acephala Group
cabello de angel *Cucurbita ficifolia*
cabelluda *Myrcia tomentosa*
Cabi Ducke = *Callaeum*
cabinet cherry *Prunus serotina*
Cabobanthus H. Robinson (~ *Vernonia*). Compositae (Vern.-Centrap.). 2 E Afr.
Cabomba Aubl. Cabombaceae. 5 warm Am. R: NJB 11(1991)179. Fanwort, fishgrass. Fly-poll. (cf. *Brasenia*). Aquarium pls pop. as 'oxygenators', often invasive, e.g. ***C. aquatica*** Aubl., ***C. caroliniana*** A. Gray in Netherlands, ***C. furcata*** Schultes & Schultes f. in Malaysia
Cabombaceae Rich. ex A. Rich. (~ Nymphaeaceae). Magnoliidae – Nymphaeales. 2/6 trop. & warm temp. Aquatic herbs with rhiz. & elongate leafy stems; scattered vascular bundles & no cambium; alks 0. Lvs dimorphic, in *Brasenia* in spirals with floating elongate to peltate lamina, in *Cabomba* many or all opp. or whorled, submerged & deeply dissected; stip. 0. Fls solit., aerial & entomophilous (nectaries), bisex., hypogynous; K (2)3(4), C (2)3(4), A 3 or 6 (*Cabomba*), 18–36 (*Brasenia*) with slightly flattened filaments, staminodes 0, pollen usu. monosulcate, $\underline{G}$ (1-)3–18 with terminal (*Cabomba*) or decurrent (*Brasenia*) styles, ovules (1)2 or 3(-5), nr. dorsal suture, anatropous, bitegmic (no aril). Fr. coriaceous, oft. dehiscent follicles; seeds small with little endosperm & copious perisperm, embryo small, dicotyledonous. 2n = 80, 104
Genera: *Brasenia, Cabomba*
Fossils from Lower Cretaceous of NE Braz. Differing from Nymphaeaceae in free carpels. Vessels unlike any others in 'dicots', suggesting vessels originated several times in angiosperms
Cabralea A. Juss. Meliaceae (II-Trich.). 1 trop. Am. (variable): ***C. canjerana*** (Vell. Conc.) Mart. – large ponerine ants remove seeds from predator-prone zone to nests to remove lipid-rich aril, imp. timber like *Cedrela*, sawdust a source of a red dye, bark used against fevers
Cabreraea Bonif. (~ *Chiliophyllum*). Compositae (Ast.-Hin.) 1 W Arg.: ***C. andina*** (Cabrera) Bonif. R: SCB 92(2009)20
Cabreriella Cuatrec. Compositae (Sen.-?). 2 Colombia. R: BSAB 19(1980)15. Lvs opp.
cabreuva oil *Myrocarpus frondosus*
Cabucala Pichon = *Petchia*
cabuya fibre *Furcraea cabuya*
Cacabus Bernh. = *Exodeconus*
Cacalia Tourn. ex L. OW spp. now referred to *Adenostyles* & *Parasenecio*, NW to *Arnoglossum*
Cacaliopsis A. Gray (~ *Senecio*). Compositae (Sen.-Tuss.). 1 W US: ***C. nardosmia*** (A. Gray) A. Gray. R: FNA 20(2006)627

cacao = cocoa

Caccinia Savi. Boraginaceae (B.3.1). 6 W & C As.

cachaça *Saccharum officinarum*

cachana *Iostephane madrensis*

cachibou *Calathea lutea*

Cachrys Tourn. ex L. Umbelliferae. (III 5). Incl. *Bilacunaria*, 7–8 Eur., N Afr.

Cacosmia Kunth. Compositae (Liab.). 3 Andes. R: BN 130(1977)279

Cactaceae Juss. Magnoliidae – Caryophyllales. 126/1500 NW, esp. hot & dry, OW trop. (*Rhipsalis baccifera*). R: D. Hunt (ed., 2006) *New Cactus Lexicon*; E. F. Anderson (2001) *The Cactus Family*. Xeromorphic trees or, most commonly, stem succ., s.t. epiphytic, with crassulacean acid metabolism, accumulating organic acids & usu. prod. alks & always betalains; r. shallow, widespreading. Stem unbranched, columnar & sparsely branched or cushion-forming etc.; cuticle usu. thick, shoots oft. photosynthetic for v. many yrs, most oft. with spines, these in spiralled hairy areoles, those poss. rep. axillary buds or short shoots with lvs or bud-scales replaced by spines oft. with tufts of short barbed irritant hairs (glochids); sieve-tubes with P-type plastids (globular protein crystalloid surrounded by ring of proteinaceous filaments). Lvs in spirals, simple, succ., or usu. v. small & ephemeral or 0. Fls usu. solit. at areoles, rarely at branch-tips or in terminal cymes (*Pereskia*), oft. large & conspic., poll. by hummingbirds, bees, bats or hawkmoths, bisex. (rarely not), ± reg. P (∞), in spirals but not clearly divisible into K & C, united at base in an hypanthium, A ∞, centrifugally dev., in spirals or groups from hypanthium, anthers with longit. slits, nectary a ring within hypanthium, $\overline{G}$ (3–∞) but G weakly united in *Pereskia*, with single style & as many arms as G, ovules usu. ∞, basal in *Pereskia* (G partly partitioned) but usu. 3 or more parietal placentas in uniloc. G, ovules campylotropous to rarely anatropous, bitegmic. Fr. usu. a berry, rarely dry-dehiscent; seeds (arillate in Opuntioideae) with usu. curved embryo & no true endosperm, perisperm present or 0, starchy. x = 11

Classification & principal genera:

I. **Pereskioideae** (lvs broad, glochids 0, fls term. or in term. cymes, seeds black, exarillate, ?allied to *Talinum* (Talinaceae), paraphyletic): *Maihuenia, Pereskia* (only). Mex. & Carib. etc. *P.* is 'sister' to rest of family, & rest of *P.* & *M.* better referred to their own subfams.

II. **Opuntioideae** (lvs ± terete, decid., glochids present, seeds with pale bony aril or winged; calcium oxalate as whewellite): *Consolea, Cylindropuntia, Grusonia, Miqueliopuntia, Opuntia, Pereskiopsis, Pterocactus, Quiabentia, Tacinga, Tephrocactus, Tunilla*

III. **Cactoideae** (lvs 0 or v. small, glochids present, seeds black or brown, exarillate; calcium oxalate as weddellite): 9 tribes (though *Blossfeldia* app. 'sister' to all C.):

1. **Echinocereeae** (tree-like or shrubby, tube armed; incl. 8.): *Armatocereus, Carnegiea, Cephalocereus, Echinocereus, Harrisia*
2. **Hylocereeae** (scandent or epiphytic, tube armed): *Disocactus, Epiphyllum, Hylocereus*
3. **Cereeae** (tree-like or shrubby, tube unarmed; R: Bradleya 7(1989)13): *Cereus, Melocactus, Pilosocereus*
4. **Trichocereeae** (tree-like, tube with ∞ scales & hairy areoles): *Cleistocactus, Echinopsis, Gymnocalycium, Rebutia*
5. **Notocacteae** (like 4 but usu. small pls): *Copiapoa, Eriosyce, Neoporteria, Parodia*
6. **Rhipsalideae** (epiphytic, stems usu. segmented): *Hatiora, Rhipsalis, Schlumbergera*
7. **Browningieae** (tree-like, v. spiny, fls lat., usu. nocturnal; *Lymanbensonia* & *Calymmanthium* (terr. or epiphytes, fls not spiny) sep. as **Lymanbensonieae**)
8. Pachycereeae = 1.]
9. **Cacteae** (pls usu. dwarf, usu. with many-ribbed, unjointed stems, fls diurnal; R: S&B 11(2013)104): *Echinocactus, Ferocactus, Lophophora, Mammillaria*

The number of genera is perhaps still inflated by overfamiliarization in horticulture & overstressing of trivial features conspicuous therein. It is likely that the now reduced Portulacaceae (& Anacampserotaceae) belong here

Xeromorphic features of thick cuticle & pachycaul structure of large volume to surface ratio, as well as weak sec. growth permitting extended retention of photosynthetic stem-surfaces, allow tolerance of extended periods of water-stress & diurnal temp. fluctuation. They are *the* NW stem succ., *Euphorbia* (unrelated) having similar role in Afr. & Didiereaceae (same order) similar in Madag. & S Afr. The shallow r.-system rapidly takes up whatever rainfall there is & CAM allows absorption of CO2 at night for use in photosynthesis in day, both features 'pre-adapting' the group to epiphytism (cf. Bromeliaceae) while the spines retain air, condense dew & prot. from grazing. Although typically of hot dry areas, the C. reach Br. Columbia & Patagonia & to 4000 m in the Andes. Mainly animal-disp. esp. by birds, some by water or wind (*Pterocactus*)

Extremely commonly cult. as orn. pot-pls etc., others (esp. *Hylocereus, Opuntia*) valued for fr., medic. (*Selenicereus*), hallucinogens (*Lophophora*) or cochineal (*Opuntia*), but other *O.* spp. pestilential weeds in OW, where some used as hedges. The epiphytes incl. spp. highly hybridized (cf. orchids) in cult. – 'epiphyllums' & other imp. housepls (*Schlumbergera*), as well as *Rhipsalis*, whose OW distr. has been suspected as merely natur.

cactus Strictly a Cactacea but oft. ignorantly applied to succ. prickly pls. in other families, e.g. *Agave*; **bird c.** *Euphorbia tithymaloides*; **brittle c.** *Opuntia fragilis*; **Christmas c.** *Schlumbergera* × *buckleyi*; **cholla c.** *Opuntia* spp.; **cochineal c.** *O. cochenillifera*; **crab c.** *S. truncata*; **Easter c.** *Rhipsalidopsis gaertneri*; **night-flowering c.** *Selenicereus grandiflorus*; **old man c.** *Cephalocereus senilis*; **San Pedro c.** *Echinopsis pachanoi*

Cadaba Forssk. Capparaceae. 30 OW trop. esp. Afr. Disk a tube; androphore & gynophore present. ***C. farinosa*** Forssk. (trop. Afr. to Ind.) – ed.

Cade, oil of *Juniperus oxycedrus*

Cadelium Medik. = *Vigna*

Cadellia F. Muell. Surianaceae. 1 NE Aus.: ***C. pentastylis*** F. Muell. – stip.

Cadetia Gaudich. = *Dendrobium*

Cadia Forssk. Leguminosae (III 72). 6 Madag., 1 NE trop. Afr. to S Arabia. R: ABN 19(1970) 227. Adaxial C not always outermost

Cadiscus E. Meyer ex DC. (~ *Senecio*). Compositae (Sen.-Sen.). 1 S Afr.: ***C. aquaticus*** E. Meyer ex DC. – aquatic, spongy ± succ. stems

cadushi *Cereus repandus*

Caelebogyne J. Sm. (~*Alchornea*). Euphorbiaceae (Alc.-Alc.). 1–2 NE Aus.

Caelospermum Blume (*Coelospermum*, ~ *Morinda*). Rubiaceae (IV 10). 11 SE As. to Aus. R: Adans. 33(2011)303

Caesalpinia Plum. ex L. Leguminosae (I 4). Excl. *Arquita, Biancaea, Coulteria, Denisophytum, Erythrostemon, Guilandina, Herrerolandia, Libidibia, Mezoneuron, Paubrasilia, Poincianella, Pomaria, Tara, Ticanto*, 9 trop. R: PhytoKeys 71(2016)43. ***C. pulcherrima*** (L.) Sw. (Barbados pride, Paradise flower, peacock f., prob. native trop. As., now pantrop.) – house-sparrows in Calcutta line nests with quinine-rich leaves instead of neem (poss. during malaria outbreaks), cult. orn. (Leeuwenberg's Model), medic. esp. laxative

Caesalpiniaceae R. Br. = Leguminosae (I)

Caesaria Cambess. = *Viviania*

Caesia R. Br. Asphodelaceae (Hemerocallidaceae). 12 S Afr. (2). Madag. (1), Aus. (9) & NG

Caesulia Roxb. Compositae (Inul.-Inul). 1 NE Ind.: ***C. axillaris*** Roxb.

cafta *Catha edulis*

caigua *Cyclanthera pedata*

caihuba *Virola surinamensis*

Cailliella Jacq.-Fél. Melastomataceae. 1 W Afr.: ***C. praerupticola*** Jacq.-Fél.

Caiophora C. Presl. = *Cajophora*

Cairns satinwood *Dysoxylum pettigrewianum*

caja fruit *Spondias mombin*

Cajalbania Urb. = *Poitea*

Cajanus DC. Leguminosae (III 18). (Incl. *Atylosia*) 34 OW trop. (Aus. 15) incl. 1 cultigen, ***C. cajan*** (L.) Huth (pigeon pea, Congo bean, C. pea, arhar) app. derived from arillate sp. in S As., where it may be crossed with local spp., though it has lost its aril, with large seeds & erect habit – widely cult (4 M t [2010]) for ed. seeds (dahl, dhal, red gram, catjang, ambrevade, gungo pea) & as cover crop, fuelwood, green manure, source of lac, silkworm foodpl. etc. R: AUWP 85–4(1986)43

cajeput = cajuput

Cajophora C. Presl ('*Caiophora*'). Loasaceae (I 1). (*S. s.*) 34 Andean S Am. R: Sendtnera 4(1997)225. Usu. with stinging hairs. ***C. coronata*** (Arn.) Hook. & Arn. – rodent-poll; ***C. lateritia*** Klotzch (*Loasa aurantiaca*) – climbs anticlockwise, then clockwise, until hitting a support (Darwin). Some cult. orn. incl. ***C. x herbertii*** (Herb.) Paxton (*C. lateritia* × *C. pentlandii* (Paxton) Paxton) – only hort. hybrid in L.

cajuado Braz. drink made from cashew-apple

cajuput oil *Melaleuca cajuputi*

cakes, Pontefract *Glycyrrhiza glabra*

Cakile Mill. Cruciferae (12). 7 shores of Eur. (2), Medit., Arabia, Aus. (natur.), N Am. R: CGH 205(1974)1. Taproots form. powdered & mixed with other flour for bread by Native Americans – antiscorbutic famine-food; ***C. maritima*** Scop. – sea rocket

calaba *Calophyllum calaba*

Calabar bean *Physostigma venenosum*

calabash gourd *Lagenaria siceraria*; **c. nutmeg** *Monodora myristica*; **sweet c.** *Passiflora maliformis*; **tree c.** *Crescentia cujete*

calabaza *Cucurbita moschata*

calabrese *Brassica oleracea* Italica Group

calabura *Muntingia calabura*

Calacanthus T. Anderson ex Benth. Acanthaceae (III 2). 1 Indomal.: ***C. grandiflorus*** (Dalz.) Radlk.

Caladenia R. Br. Orchidaceae (IV 3b). Excl. *Cyanicula, Elythranthera, Ericksonella, Glossodia, Pheladenia*, 267 Aus., 1 ext. to Mal. & New Caled., NZ (6 subg. s.t. recog. as genera). R: AusSB 17(2004)179. Sexually attracted male thynnid wasp-poll. (sp.-specific, where odour mimics volatile sex hormones emitted from mandibular or abdominal glands of wingless females cli. out from parasitizing beetle larvae in soil) & other deception poll. (no nectar; e.g. ***C. rigida*** R. Rogers – poss. mimic of *Burchardia umbellata*) in Aus. Some autogamy incl. cleistogamy; nat. hybrids with *Glossodia* spp. Some cult. orn.

Caladeniastrum (Szlach.) Szlach. = *Caladenia*

Caladiopsis Engl. = *Chlorospatha*

Caladium Vent. Araceae (VII 11). 18 trop. S Am. R: Selb. 5(1981)367. Alks; some cyanogenic glycosides. Cult. orn. ('angel wings') for foliage, esp. cvs of ***C. bicolor*** (Aiton) Vent. (Braz.), sap irritant, cvs incl. '*C.* × *hortulanum* Birdsey', with almost all white lvs, 'vogue' foliage stove-pls of mid- to late C19; ***C. lindenii*** (André) Madison (*Xanthosoma l.*, Ind. kale, C Panamá to Colombia) – varieg. pot-pl. pop. in US ('var. *sylvestre* Grayum' – wild form)

Calamagrostis Adans. Gramineae (XVI 5). Excl. *Aniselytron, Deyeuxia*, c. 100 subcosmop. (Eur. 14, Am. 132 [N 25 – R: FNA 24(2007)706]). R: FR 63(1960)229. Heterogeneous. Some N Am. spp. with v. restricted distrib. Hybrids, polyploidy & apomixis common

calamansi *Citrus* × *microcarpa*

calambac *Aquilaria malaccensis*

calamander wood *Diospyros quaesita*

calamint *Clinopodium* spp.; **common c.** C. *ascendens*; **wood c.** C. *menthifolia*

Calamintha Mill. = *Clinopodium*

calamondin *Citrus* × *microcarpa*

Calamophyllum Schwantes (~ *Cylindrophyllum*). Aizoaceae (V). 3 distrib. unknown. R: H. E. K. Hartmann, *Ill. Handb. Succ. Pls*, Aiz. A–E (2002)95

Calamphoreus Chinnock (~ *Eremophila*). Scrophulariaceae (Myoporaceae). 1 SW Aus.: ***C. inflatus*** (C. Gardner) Chinnock. R: R.J. Chinnock, *Eremophila & allied genera* (2007)169

Calamovilfa (A. Gray) Hackel = *Sporobolus* (but see Castanea 31(1966)145)

Calamus L. Palmae (I 3f). Incl. *Ceratolobus, Daemonorops, Pogonotium, Retispatha*, 520 OW trop. (Afr. 1: ***C. deerratus*** G. Mann & H. Wendl.) esp. Mal. R: ARBGC 11(1908)1, app.(1913)1. Usu. spiny climbers with long internodes (some with alt. whorls of upward- & downward-pointing toothed flanges encl. ant-inhabited passages – if disturbed ants beat on stem in unison providing an alarm alerting even elephants), forming the canes of the rattan industry for furniture, basketry, mats, bridges, chimney-sweeps' brushes etc.; also resins ((Sumatran) dragon's blood) used in varnishes & form. medic. derived from fr., but some small, e.g. ***C. minutus*** Dransf. (Trengganu) only 50 cm tall. Oft. distal pinnae are backward-pointing spines, which act as grapples on surrounding veg. with which the rattan grows up, the stem reaching 170 m or may be up to 200 m, the rattans growing up again from the forest floor after the collapse of their supporting trees; rattan industry in Indonesia worth $2.7 billion by 1987 ($2.2 billion in 2015) incl. ***C. trachycoleus*** Becc. (S Borneo but planted elsewhere), in Malaysia (worth $2 billion by 1977) based on ***C. manan*** Miq. (W Mal., endangered in wild) canes grow 3 m per season & ***C. caesius*** Blume (W Mal.) up to 4 m, 6 spp. used in Ind., though production now declining. Some poss. fr. crops. ***C. erinaceus*** (Becc.) Dransf. (*C. aquatilis*, W Mal.) grows in mangrove or nearby; ***C. macrophylla*** (Becc.) W. Bak. (*D. m.*) & ***C. verticillaris*** Griff. (*D. v.*) in Malay Penin. have ants' nests absorbing through-flow incl. that from debris accum. around apex so that nutrients passed to palms; *C.* ***muelleri*** H.A. Wendl. & Drude (E Aus.) – chain-lengths used as standard measures by early Eur. surveyors; *C.* ***pygmaeus*** Becc. (Borneo) – infls r. to form new pls; *C.* ***scipionum*** Lour. (W Mal.) – source of Malacca cane for walking-sticks, umbrella-handles (ground & sold as ersatz parmesan cheese 1960s) etc.; *C.* ***zollingeri*** Becc. (Sulawesi) – imp. rattan

calamus root *Acorus calamus*

Calanda K. Schum. Rubiaceae (III 8). 1 trop. Afr.: *C.* ***rubricaulis*** K. Schum. R: BJ 110(1989)546

Calandrinia Kunth. Montiaceae (Portulacaceae s.l.). Excl. *Baitaria, Cistanthe,* 14 W N & (esp.) S Am. Cult. orn. fleshy pls. Some ed.

Calandriniopsis Franz = *Montiopsis*

Calanthe R. Br. Orchidaceae (V 14). c. 200 As. to Tahiti, trop. & S Afr. (1), Madag., C. Am. (1), WI (1). All parts of pl. turn blue when damaged. Some with fls only openable by (large) euglossine bees. Cult. orn. incl. first Western orchid hybrid (Jap. hybrids in C18), *C.* × ***dominyi*** Lindl. (*C. masuca* Lindl. (Mal.) × *C. triplicata* (Willemet) Ames (Mal. to Aus.) fl. 1856)

Calanthea (DC.) Miers (~ *Capparis*). Capparaceae. 2 trop. Am. R: HPB 13(2008)119

Calantica Jaub. ex Tul. Salicaceae (Flacourtiaceae). Excl. *Bivinia,* 10 Madag. R: Adans. 36(2014)85

Calanticaria (B.L. Robinson & Greenm.) E. Schilling & Panero (~ *Viguiera*). Compositae (Helia.-Helia.). 5 SW N Am. R: BJLS 140(2002)73

Calathea G. Meyer. Marantaceae. Excl. *Goeppertia,* c. 37 trop. Am. *C.* ***lutea*** (Aubl.) G. Meyer (balisier, cachibou, WI) – lvs used for making & lining baskets, to wrap food incl. cooking tamales, also promising source of wax (cauassú); *C.* ***utilis*** H. Kenn. (Ecuador) – thatch

Calathiana Delarbre = *Gentiana*

Calathodes Hook.f. & Thomson. Ranunculaceae (II 1). 3 Himal. to Taiwan

Calathostelma Fourn. = *Ditassa*

Calatola Standl. Icacinaceae. 7 Mex. to Ecuador. *C.* ***colombiana*** Sleumer (Colombian Amaz.) – lvs chewed to blacken lips & teeth

Calawaya Szlach. & Sitko = *Maxillaria*

Calcaratolobelia Wilbur = *Lobelia* (but see Sida 17(1997)561)

Calcareoboea C.Y. Wu ex Li = *Petrocodon* (but see ABY 4(1982)241)

Calceolaria Feuillée ex L. Calceolariaceae (Scrophulariaceae s.l.). c. 300 C Mex to Patagonia esp. Andes (trop. Am. 181 – R: FN 47(1988)1). Slipper flowers. Shrubs, lianes, herbs; elaiophores attractive to oil-bees. Some local med. but several cult. orn. esp. florists' pot-pl. (**Herbeohybrida Group**, *C.* × *herbeohybrida*, *C.* × *hybrida*), hybrids of *C. crenatiflora* Cav., *C. corymbosa* Ruíz & Pavón & *C. cana* Cav. (all Chile); some hardy. *C.* ***andina*** Benth. (Chile) – naphthoquinones effective against mites, aphids etc.; *C.* ***uniflora*** Lam. (Chile) – bird.-poll. (food-body reward for non-nectarivorous poll. birds, no nectar, nor oil as no oil-bees in S S Am.)

Calceolariaceae Olmstead (~ Gesneriaceae; Scrophulariaceae s.l.). Magnoliidae – Lamiales. 2/300 NZ, trop. (upland) & S Am. (cf. *Fuchsia*). Shrubs & herbs. Fls 4-merous in thyrses. K valvate. C with short tube & saccate abaxial lobe. A 2(3, some *Calceolaria*), thecae apically distinct; staminodes 0; $\underline{G}(2)$. Capsule both loculicidal & septicidal. n = (8)9

Genera: *Calceolaria, Jovellana*

Imp. cult. orn.

Calchas P.V. Heath = *Plectranthus*

Calciphila Liede & Meve (~ *Cynanchum*). Apocynaceae (Ascl.). 2 Somalia. R: Novon 16(2006)369

Calciphilopteris Yesilyurt & H. Schneid. Pteridaceae (IV). 4 Mal. R: Phytotaxa 7(2010)55

Calcitrapa Heister ex Fabr. = *Centaurea*

Calcitrapoides Fabr. = *Centaurea*

calçot *Allium cepa*

Caldcluvia D. Don. Cunoniaceae (VI). 1 Chile, Arg.: *C.* ***paniculata*** (Cav.) D. Don. See also *Ackama, Spiraeopsis*

Calderonella Soderstrom & H. Decker = *Zeugites*

Calderonia Standl. = *Simira*

Caldesia Parl. (~ *Alisma*). Alismataceae. 5 OW trop. (1 ext. to Eur.)

Calea L. Compositae (Helia.-Mel.). c. 125 warm Am. *C.* ***ternifolia*** Kunth (*C. zacatechichi*) – hallucinogen in Mex.; *C.* ***urticifolia*** (Mill.) DC. (C Am., weedy in OW) – medic. intoxicant

Caleana R. Br. Orchidaceae (IV 3e). Excl. *Paracaleana*, 1 E Aus.: ***C. major*** R. Br. (duck orchid) – poll. male sawflies trapped by labellum flipping over into column bowl, pinning insect upside-down. (cf. *P.*)

Calectasia R. Br. Dasypogonaceae (Calectasiaceae). 11 S Aus. (1). R: Nuytsia 13(2001)417

Calectasiaceae Schnizl. = Dasypogonaceae

Calendula L. Compositae (Cal.). c. 15 Medit. (Eur. 5) to Iran, Macaronesia. ***C. officinalis*** L. (common, pot or Scotch marigold, ruddles; orig. unclear) – cultigen (poss. hybrid between spp. with 2n = 14 & 18), cult. orn. signifying 'grief' in Language of Fls, with many cvs, incl. **'Prolifera'** (breeds true) with proliferated capitula (hen-&-chickens; also veg. shoots prod. from G bases of typical form in N Ind. in spring), also medic. (fevers etc. in Portugal), effective against chilblains & warts, used in cosmetics & to colour butter & thicken soups, petals to garnish salads

Calepina Adans. Cruciferae (14). 2 Eur., Medit. to C As. R: JAA 66(1985)338

Calia Téran & Berland. = *Dermatophyllum*

Calibanus Rose = *Beaucarnea*

Calibrachoa Cerv. = *Petunia*

× **Calicharis** Meerow. Hybrids between *Caliphruria & Eucharis* spp.

calico *Gossypium* spp.; **c. bush** *Kalmia latifolia*

Calicorema Hook.f. Amaranthaceae (I 2). 2 trop. & S Afr. Heterogeneous

Calicotome Link = *Cytisus*

California Aldasoro & al. (~ *Erodium*). Geraniaceae. 1 Calif., Baja C.: ***C. macrophylla*** (Hook. & Arn.) Aldasoro & al. R: AJBM 59(2002)213

California allspice *Calycanthus occidentalis*; **C. buckeye** *Aesculus californica*; **C. laurel** *Umbellularia californica*; **C. lilac** *Ceanothus* spp.; **C. nutmeg** *Torreya californica*; **C. pepper** *Schinus molle*; **C. poppy** *Eschscholzia californica*; **C. redwood** *Sequoia sempervirens*

Caliphruria Herb. (~ *Eucharis*). Amaryllidaceae. 4: W Colombia 3, Peru 1. R: AMBG 76(1989)212. Hybrids with *Eucharis* = × ***Calicharis***

calisaya *Cinchona calisaya*

Calispepla Vved. = *Argyrolobium*

Calla L. Araceae (VI). 1 N temp. bogs: ***C. palustris*** L. (water arum) – aquatic, P 0, fls biennially, ed. starch from rhiz. (N.B. 'Calla' of florists is *Zantedeschia*)

Callaeum Small. Malpighiaceae. 10 trop. Am. R: SB 11(1986)335

Callerya Endl. (~ *Wisteria*). Leguminosae (III 16). 19 SE & E As., Aus. R: Blumea 39(1994)1. ***C. nieuwenhuisii*** (J.J. Sm.) Schot (*M. n.*, Borneo) – cli. ant-plant

Calliandra Benth. Leguminosae (II 4). Excl. *Afrocalliandra, Viguieranthus, Zapoteca*, 142 SW US to Urug. & N Chile. R: MNYBG 74,3(1998)3, T 62(2013)1213. Pollen in polyads of 8 grains. Cult. orn. with heads of fls esp. ***C. guildingii*** Benth. (cunure, Braz., natur. S US, WI), ***C. haematocephala*** Hassk. (S Am.) – incl. as bonsai in temp. countries, etc. ***C. houstoniana*** (Mill.) Standl. (*C. calothyrsus*, C Am.) – forage, fuelcrop, coppices well in Indonesia, returning fertility to soil esp. in Java

Calliandropsis H. Hernández & Guinet (~ *Desmanthus*). Leguminosae (II 1). 1 Mex.: ***C. nervosus*** (Britton & Rose) H. Hernández & Guinet (allied Madag. spp. = *Alantsilodendron*)

Callianthe Donnell (~ *Abutilon*). Malvaceae (Malv.-Malv.). 40 trop. Am. esp. E Braz. R: SB 37(2012)718. Cult. orn. (esp. pot-pls) cvs ('*A.* × *hybridum*', flowering maples) derived from ***C. picta*** (Hook. & Arn.) Donnell, ***C. striata*** (Lindl.) Donnell, etc.

Callianthemoides Tamura. Ranunculaceae (II 3). 1 S Am.: ***C. semiverticillatus*** (Phil.) Tamura

Callianthemum C. Meyer. Ranunculaceae (II 1). 14 mts of Eur. (3) & C As. R: VKZGW 49(1899)316. Cult. orn. esp. ***C. anemonoides*** (Zahlbr.) Endl. (NE Alps)

Callicarpa L. Labiatae (inc. sed.; Verbenaceae). c. 140 trop. & subtrop. Dioec. spp. (also in C Am.), heterostylous on Bonin Is., females with non-germ. pollen perhaps poll. reward. Cult. orn. shrubs (Champagnat's to Petit's Model) esp. ***C. americana*** L. (French mulberry, beautyberry, S N Am., WI) – local medic., ***C. bodinieri*** A. Léveillé (China); others medic., fish-poisons etc.

Callicephalus C. Meyer (~ *Centaurea*). Compositae (Card.-Cent.). 1 SW to C As.: ***C. nitens*** (Willd.) C. Meyer

Callichilia Stapf. Apocynaceae (I d). 7 trop. Afr. R: MLW 78–7(1978)1

Callichlamys Miq. Bignoniaceae (3). 1 trop. Am.: ***C. latifolia*** (Rich.) K. Schum. R: AMBG 98(2014)423

Callicoma Andr. Cunoniaceae (VII). 1 E Aus.: ***C. serratifolia*** Andr. (black wattle) – P 0, orig. (1788) wattle of First Fleet white settlers (name later transferred to *Acacia* spp. with similar heads of fls), cult. orn. R: AusJB 29(1981)721

Calligonum L. Polygonaceae (II 4). 80 Med. (Eur. 1). Shrubs used as sand-binders. ***C. polygonoides*** L. (W As. to Ind.) – fls ed. on bread or cooked (Ind.)

Callilepis DC. Compositae (Gnap.). 5 S Afr. R: OB 104(1991)46

Callipeltis Steven. Rubiaceae (IV 16). 3 Spain (1) & Egypt to Baluchistan

Calliphysalis Whitson (~ *Physalis*). Solanaceae (IV 5c). 1 SE US: ***C. carpenteri*** (Riddell) Whitson

Callipteris Bory = *Diplazium*

Callirhoe Nutt. Malvaceae (Malv.-Malv.). 9 N Am. R: MNYBG 56(1990). Cult. orn., many with ed. r., medic.

Calliscirpus C. Gilmour & al. (~ *Scirpus*). Cyperaceae. 2 W US. R: KB 68(2013)98

Callisia Loefl. Commelinaceae (II 1g). (Incl. *Cuthbertia, Hadrodemas*) 20 trop. Am. R: KB 41(1986)407, JAA 70(1989)117. Cult. orn. esp. ***C. navicularis*** (Ortgies) D. Hunt (*Tradescantia n.*, Mex.) with succ. lvs

Callista Lour. See *Dendrobium*

Callistachys Vent. (~ *Oxylobium*). Leguminosae (III 13). 1 SW Aus.: ***C. lanceolata*** Vent.

Callistanthos Szlach. = *Pteroglossa*

Callistemon R. Br. = *Melaleuca*

Callistephus Cass. Compositae (Ast.-Ast.). 1 China: ***C. chinensis*** (L.) Nees – the aster or China a. of gardens (cult. China for 2000 yrs, introd. Eur. 1728) with fls from white & yellow to red, purple, & blue, the disk-florets oft. replaced by ray ones. R: SBM 20(2000)22

Callisthene Mart. Vochysiaceae (I). 14 dry S & C Braz., N Paraguay, E Bolivia. R: H.F. Martins, *O gênero C.* (1981)

Callistigma Dinter & Schwantes = *Mesembryanthemum*

Callistopteris Copel. (~ *Cephalomanes*). Hymenophyllaceae. 5 SE As., Pacific

Callistylon Pittier = *Coursetia*

Callithauma Herb. = *Stenomesson*

Callitrichaceae Link = Plantaginaceae

Callitriche L. Plantaginaceae (Callitrichaceae). c. 30 almost cosmop. (Eur. 11 with 2 N Am. spp. natur.). Starworts. Genus unique in co-occurring aerial & underwater poll. systems: exineless pollen in submerged spp., a reduction or loss evolving at least twice. Some self-poll. with tubes passing even through stem to poll. another fl. Some grown in aquaria incl. ***C. stagnalis*** Scop. (water s., Eur., Medit., Macaronesia) – introd. New York 1861, by 1998 spread c. 800 km inland from both E & W coasts of N Am. Other spp. v. sensitive to pollution & their performance can be used to predict presence of particular pollutants in S Ger.

Callitris Vent. Cupressaceae. Incl. *Actinostrobus*, 19 Aus. (17, endemic), New Caled. R: A. Farjon, *Monogr. Cupressaceae & Sciadopitys* (2005)496. Cypress pines. Lvs & cone-scales whorled; cones ripen in 1–2 yrs. Some good termite-proof timbers & resins. ***C. endlicheri*** (Parl.) Bailey (? = *C. australis* (Pers.) Sweet, *C. calcarata*, black cypress, SE Aus.) – source of Aus. sandarac used in varnishes; ***C. glaucophylla*** J. Thompson & L. Johnson ('*C. glauca*', white c. p.; warm Aus.) – timber for flooring, weatherboard, panelling; ***C. rhomboidea*** R. Br. ex Rich. (Illawara pine, Oyster Bay p., Port Jackson cypress or p., SE Aus.), etc.

Callitropsis Oerst. = *Xanthocyparis*

Callopsis Engl. Araceae (VI 17). 1 Tanzania: ***C. volkensii*** Engl. – cult. orn.

Callostylis Blume (~ *Eria*). Orchidaceae (V 15). 5 Mal.

Calluna Salisb. Ericaceae (V 5). 1 Eur., As. Minor, natur. in N Am.: ***C. vulgaris*** (L.) Hull (common, white or Scottish heather, ling) – understorey shrub (Rauh's Model) in woodland or more commonly maintained as major constituent of moors by burning & cutting in Scotland etc. for 'game' bird-shooting; poll. by bees (honey a major constituent of Drambuie) & other insects but in northerly sites wind-poll. or visited by thrips, *Thrips ericae*; baled shoots used in road-constr. in New Forest etc., also for brooms; used like hops in Scotland in 'heather ale' for 4000 yrs (form. with *Filipendula ulmaria* & (to stop fermentation) *Osmunda regalis*), medic. & form. a yellow dye for wool; cattlefeed & distillate local medic. in Spain; many cvs cult. orn., invasive NZ outcompeting native tussock grasses; white cvs 'good luck' since Victorian times, now aggressively hawked by 'gypsies' etc. in C London. R: FNA 9(2009)491

Calocarpum Pierre = *Pouteria*

Calocedrus Kurz. Cupressaceae. 3 N Myanmar, SW China, Taiwan, Vietnam, Thailand, W N Am. R: A. Farjon, *Monogr. Cupressaceae & Sciadopitys* (2005)401. ***C. decurrens*** (Torrey) Florin (incense cedar, white c., W N Am.) – timber for shingles etc.; cult. orn. esp. fastigiate form in GB unknown in wild, though trees grown in Ireland & Italy broader (reason unknown)

Calocephalus R. Br. Compositae (Gnap.-Ang.). 11 temp. Aus. R: OB 104(1991)131. Cult. orn. See also *Leucophyta*

Calochilus R. Br. Orchidaceae (IV 3i). 27 Aus., NZ, NG, New Caled. R: PLSNSW 71(1947)287. Some poll. by sexual deceit (see *Caladenia*); S Aus. spp. all self-poll. ***C. paludosus*** R. Br. (E Aus.) – tubers form. ed. Aborigines

Calochlaena (Maxon) M. Turner & R. White (~ *Culcita*). Dicksoniaceae (Cyatheaceae s.l.). 6 Philipp. & Java to Aus. R: AFJ 78(1988)86

Calochone Keay. Rubiaceae (II 1). 2 W Afr.

Calochortaceae Dumort. = Liliaceae

Calochortus Pursh. Liliaceae (Calochortaceae). c. 70 W N Am. (56) to Guatemala. R: AMBG 27(1940)371. Mariposa (or sego) lilies. Bulbous pls, some ed., local medic. cult. orn. (fls white, yellow, red to purple, bluish or brownish)

Calocrater K. Schum. Apocynaceae (Id). 1 W & C Afr.: ***C. preussii*** K. Schum. – a 'litter-bin' pl. R: Fontqueria 41(1995)11

Calodecaryia J. Leroy. Meliaceae (II-Trich.). 2 Madag.

Calodendrum Thunb. Rutaceae (I 4). 1 E Afr. to Cape: ***C. capense*** (L.f.) Thunb. (Cape chestnut) – seeds source of an oil for soap, timber useful, cult. orn. (20 yrs from seed to fls). R: Strel. 10(2000)496

Calogyne R. Br. = *Goodenia*

Calolisianthus Gilg = *Irlbachia*

Calomeria Vent. (*Humea*, ~ *Cassinia*). Compositae (Gnap.-Ang.). 1 S Aus.: ***C. amaranthoides*** Vent. (*H. elegans*) – cult. orn. pachycaul biennial treelet with marcescent lvs (incense pl., Aus.), causing dermatitis in some. R: OB 104(1991)117

Caloncoba Gilg. Achariaceae (Flacourtiaceae s.l.). 10 trop. Afr. R: BJ 94(1974)120. Rauh's Model. ***C. echinata*** (Oliv.) Gilg source of gorli oil used like chaulmoogra oil in treatment of leprosy etc.; ***C. flagelliflora*** (Mildbr.) Pellegr. (W Afr.) – basal whip-like infl. branches at ground level to 11 m long

Calonema (Lindl.) Szlach. = *Caladenia*

Calonemorchis Szlach. = *Caladenia*

Calonyction Choisy = *Ipomoea*

Calopappus Meyen (~ *Nassauvia*). Compositae (Mut.-Nass.). 1 Chilean Andes: ***C. acerosus*** Meyen. R: Cald. 15(1986)57. Shrub

Calophaca Fischer ex DC. = *Caragana*

Calophanoides (C.B. Clarke) Ridl. = *Justicia*

Calophyllaceae J. Agardh (~ Guttiferae). Magnoliidae – Malpighiales. 15/500 trop. Trees & shrubs. Leaves simple, vernation often flat, paired glands at base. Flowers 4- or 5-merous; C (0-)4, 5(-8), A not in obvious bundles, anthers often with glands, style usu. long, stigmas expanded to punctate. Fr. a berry or drupe; seeds 1–many; embryo chlorophyllous or not, cotyledons oft. v. large.

Principal genera: *Calophyllum, Kayea, Kielmeyera, Mammea*

Calophyllum L. Calophyllaceae (Guttiferae III). 186 trop. (10 NW; most Indom.). R: JAA 61(1980)117. Timbers (bintangor) & oilseeds, cult. orn. ***C. antillanum*** Britt. (trop. Am.) – cult. orn., invasive SE US; ***C. brasiliense*** Cambess. (Santa Maria, galba, jacareuba, trop. Am.) – timber; ***C. calaba*** L. (calaba, Indomal.) – ed. fr., oilseed, timber; ***C. inophyllum*** L. (Alexandrian laurel, tamanu, OW) – disp. by bats & sea, oilseed (domba oil, pinnay oil) med., illuminant, mixed with coconut oil to give Tongan Oil for massage, timber (Borneo mahogany) esp. for boats (& Gauguin wood-carvings), bark medic., pericarp a dye-source (Hawaii), cult. orn.; ***C. macrocarpum*** Hook.f. (W Mal.) – rhino- & elephant-disp. seeds; ***C. neoebudicum*** Guillaumin (damanu, W Pacific) – furniture timber; ***C. polyanthum*** Wall. ex Choisy (Ind.) – much exploited W Ghats for boat-masts & plywood; ***C. tacamahaca*** Willd. (Mauritius & Réunion) oft. confused with *C. inophyllum*

Calopogon R. Br. Orchidaceae (V 10a). 5 E N Am., WI. R: FNA 26(2002)597. Bees land on lip to try to coll. pseudopollen on hair-like organs resembling anthers, but wt. causes lip to fall, dropping bee on to column, the pollinia sticking to its back. Cult. orn.

Calopogonium Desv. Leguminosae (III 18). 5–6 NW trop., 1 (***C. mucunoides*** Desv., widely grown ground-cover crop & green manure esp. under coconuts) ext. to OW. Isoflavonoid phytoalexins

Calopsis P. Beauv. ex Desv. = *Restio* (but see Bothalia 15(1985)464)

Calopteryx A.C. Sm. = *Thibaudia*

Calopyxis Tul. = *Combretum*

Calorhabdos Benth. = *Veronicastrum*

Calorophus Lab. Restionaceae. 2 Victoria, Tasmania

Caloscordum Herb. = *Allium*

Calospatha Becc. = *Calamus*

Calostemma R. Br. Amaryllidaceae. 3 E Aus. R: JABG 22(2008)48. ***C. purpureum*** R. Br. – P purple to yellow or white, 'seeds' = bulbils (cf. *Proiphys*), alks, cult. orn.

Calostephane Benth. Compositae (Inul.-Inul.). 5 trop. Afr., Madag. R: KB 69[9516](2014)1

Calostigma Decne. = *Oxypetalum*

Calotesta Karis (~ *Metalasia*). Compositae (Gnap.-Rel.). 1 SW Cape: ***C. alba*** Karis. R: OB 104(1991)74

Calothamnus Lab. = *Melaleuca*

Calotis R. Br. Compositae (Ast.-Bra.). c. 28 SE As. (2), Aus. R: PLSNSW 77(1952)146. Burr daisies, some Aus. spp. weedy, e.g. ***C. hispidula*** (F. Muell.) F. Muell. (Bogan flea)

Calotropis R. Br. Apocynaceae (V c; Asclepiadaceae III 1). 3 trop. & warm Afr. & Ind. R: NJB 11(1991)301. Bark fibre (madar, mudar), latex (alks) like gutta-percha, seed floss used like kapok. ***C. gigantea*** (L.) Aiton f. – multiple nectaries at petiole – lamina junction, lvs medic. & ed., fibre (wara, yercum), fls used in temple garlands & as architectural motif, candied by Chinese in Java; ***C. procera*** (Aiton) Aiton f. – now pantrop. weed (apple of Sodom) indicator of overgrazing in S Arabia & invasive in Aus., medic., poss. hydrocarbon source, r. used as chewsticks in Afr., strong fibre (akund fibre, French cotton), fine wood-ash form. used in gunpowder, fls (akra) offered to Hindu god Hanuman

Calpidochlamys Diels = *Trophis*

Calpocalyx Harms. Leguminosae (II 1). 11 W Afr. R: BMNHN 4,6(1974)297. timber, salt from ash

Calpurnia E. Meyer. Leguminosae (III 7). 7 E Cape, Afromontane forests, S Ind. R: Bothalia 29(1999)6. Pod winged. Cult. orn. esp. ***C. aurea*** (Aiton) Benth. (Afr., Ind.)

Caltha L. Ranunculaceae (I 1). 12 temp. (Eur. 1). R: Blumea 21(1973)119. Volatile protoanemonin irritant to skin & mucous membranes; alks. Cult. orn. esp. ***C. palustris*** L. (kingcups, marsh marigold, mayblob, N temp.) – molluscicidal saponins, fl. buds form. pickled like capers; ***C. dionaeifolia*** Hook.f. (S S Am.) & other S hemisph. spp. ±paired lobes to lvs, grows with *Drosera & Pinguicula* spp. but does not trap food; ***C. introloba*** F. Muell. (SE Aus.) – stabilizes soil in Snowy Mts, fls even under snow

caltrops *Tribulus terrestris*; **water c.** *Trapa natans*

calumba root *Jateorhiza palmata*; **false c. r.** *Coscinium fenestratum*

Caluera Dodson & Determan. Orchidaceae (V 12h). 3 N S Am.

calumpang (nut) *Buchanania lanzan*

Calvaria Gaertn. f. = *Sideroxylon*

Calvary clover *Medicago echinus*

Calvelia Moq. = *Suaeda*

Calvoa Hook.f. Melastomataceae. 19 trop. Afr. R: BJLS 136(2001)186

Calycacanthus K. Schum. Acanthaceae (III 2c). 1 NG: ***C. magnusianus*** K. Schum.

Calycadenia DC. (~ *Hemizonia*). Compositae (Mad.-Mad.). 10 W US. R: FNA 21(2006)270

Calycanthaceae Lindl. Magnoliidae – Laurales. Incl. Idiospermaceae, 3/10 trop. Aus., China, temp. N Am. Shrubs or small trees with aromatic bark; young stems with 4 inverted vasc. bundles in cortex. Lvs opp., simple, entire, aromatic; stip. 0. Fls solit., term., bisex., perigynous, beetle-poll. P 15–40, ± petaloid, in spirals., A 5–30, in spirals, ± ribbon-like with short or 0 filaments, connective extended beyond pollen-sacs with longit. slits, 10–15 staminodes, nectariferous, G 1–35 in spirals in hypanthial cup, each 1-ovulate, distal ovule abortive, proximal anatropous, bitegmic. Fr. of ∞ achenes in enlarged fleshy & oily, proteinaceous hypanthium; seeds endotestal, poisonous, cotyledons twisted, endosperm with 0 paternal contribution. x = 11, 12

Genera: *Calycanthus, Chimonanthus, Idiospermum*

Fossils similar to extant pls incl. Lower Cretaceous ones with lobed lvs (Braz.)

Cult. orn., medic., spices & scent

Calycanthus L. Calycanthaceae. 1 China (*Sinocalycanthus*), 2 SW & E US. R: Castanea 30(1965)63. Sweetshrubs (US). Fls on leafy shoots (cf. *Chimonanthus*); toxic (calycanthine like strychnine). Cult. orn.: ***C. floridus*** L. (Carolina allspice, E US) – bark form. medic., subs. for cinnamon, crossed with ***C. chinensis*** (W. C. Cheng & S. Y. Chang) P. T. Li (*S. c.*, China) to give '**Hartlage Wine**'; ***C. occidentalis*** Hook. & Arn. (Calif. a., SW US)

Calycera Cav. Calyceraceae. 9 temp. S Am. R: SB 39(2015)1230

Calyceraceae R. Br. ex Rich. Magnoliidae–Asterales. 6/55 S S Am. Herbs with inulin. Lvs in spirals (often a basal rosette), simple, entire to pinnately lobed; stip. 0. Fls bisex., s.t. functionally unisexual, in centripetally developing involucrate heads, epigynous; K (4)5(6), s.t. spiny; C ((4)5(6)), reg. or not, lobes valvate; A same no. & alt. with lobes, attached nr summit of tube, filaments ± connate, anthers opening by longit. slits & pollen released into anther-tube where pushed out by growth of style; $\overline{G}$ (2), uniloc., stigma capitate, ovule 1, pend., anatropous, unitegmic. Fr. achene-like with apical persistent K; seed with straight embryo & oily endosperm. x = 8, 15, 18, 21. R: NJB 12(1992)63

Genera: *Acarpha, Acicarpha, Boopis, Calycera, Gamocarpha, Nastanthus*

Lvs & inulin like Campanulaceae, filament-attachment, ovary & embryological features like Dipsacales, heads & pollen incl. presentation like Compositae (C.-Barn. with similar distr.) though ovule basal there & chem. different but cladistic analysis allies Calyceraceae with Compositae & Goodeniaceae. Cronquist suggests that Calyceraceae have an 'outmoded chem. arsenal' as defence & this might explain their lack of success compared with Compositae; most spp. occur in dry open veg.

Calycobolus Willd. ex Schultes. Convolvulaceae (6). 16 trop. Afr. (13), trop. Am. (3). R: PEE 146(2013)330. ***C. heudelotii*** (Oliv.) Heine (W Afr.) – mass-flowering

Calycocarpum Nutt. ex Spach. Menispermaceae (III). 1 E N Am.: ***C. lyonii*** (Pursh) A. Gray. R: JAA 45(1964)28

Calycocorsus F.W. Schmidt = *Willemetia*

Calycogonium DC. = *Miconia*

Calycolpus Berg. Myrtaceae (II 10). c. 10 trop. Am.

Calycomis D. Don = *Acrophyllum*

Calycopeplus Planch. (~ *Euphorbia*). Euphorbiaceae (Euph.-Euph.-Neo.). 5 W Aus. R: Austrobaileya 4(1995)418

Calycophyllum DC. Rubiaceae (Cond.). 10 trop. Am. 1 K-lobe oft. leaf-like. Some timbers esp. ***C. candidissimum*** (Vahl) DC. (dagami, degame) – tool-handles & turnery, bows (lemonwood – US) & charcoal, ***C. multiflorum*** Griseb. (palo amarillo, Arg.)

Calycophysum Karsten & Triana. Cucurbitaceae (XV). 5 NW trop. S Am.

Calycopteris Lam. ex Poir. = *Getonia*

Calycorectes Berg = *Eugenia*

Calycoseris A. Gray. Compositae (Cich.-Cich.). 2 SW N Am.

Calycosia A. Gray. Rubiaceae (IV 7). 5 Polynesia

Calycosiphonia Pierre ex Robbrecht. Rubiaceae (II 3). 2 trop. Afr. Seed-coat 0

Calycotropis Turcz. = ? *Polycarpaea*

Calyculogygas Krapov. Malvaceae (Malv.-Malv.). 1 Urug.: ***C. uruguayensis*** Krapov.

Calydorea Herb. Iridaceae (VII 5). c. 16 temp. S Am. R: AMBG 78(1991)510. Lvs pleated. Heterogenous, spp. referable to *Tigridia & Nemastylis*; Florida sp. already excl. as *Salpingostylis*

Calylophus Spach = *Oenothera*

Calymmanthera Schltr. Orchidaceae (V 16c). 5 NG

Calymmanthium F. Ritter. Cactaceae (Lymanb.). 1 N Peru: ***C. substerile*** F. Ritter

Calymmatium O. Schulz. Cruciferae (3). 2 C As.

Calymmodon C. Presl. Polypodiaceae (V; Grammitidaceae). c. 30 Sri Lanka to (esp. Borneo) Aus. & Polynesia. R: PJS 34(1927)259

Calymmostachya Bremek. = *Justicia*

Calypso Salisb. Orchidaceae (V 13e). 1 circumboreal, terr.: ***C. bulbosa*** (L.) Oakes – tubers form. ed. N Am., cult. orn. R: FNA 26(2002)622

Calyptocarpus Less. Compositae (Helia.-Verb.). 2 Texas to Guatemala

Calyptochloa C.E. Hubb. = *Cleistochloa* (but see Austrobaileya 8(2012)636)

Calyptostylis Arènes = *Rhynchophora*

Calyptraemalva Krapov. (*Calyptrimalva*). Malvaceae (Malv.-Malv.). 1 Braz.: ***C. catharinensis*** Krapov.

Calyptranthera Klack. (~ *Pervillea*). Apocynaceae (IV). 2 Madag. R: Candollea 53(1998)395

Calyptranthes Sw. = *Myrcia*

Calyptrella Naudin = *Graffenrieda*
Calyptridium Nutt. (~ *Cistanthe*). Montiaceae. Incl. *Spraguea*, 14 SW US
Calyptrimalva Krapov. = *Calyptraemalva*
Calyptrion Ging. (*Corynostylis*). Violaceae (III 2b). 4 trop. Am. R: T 63(2014)1335. Some medic. esp. ***C. volubilis*** (L.B. Sm. & A. Fernández) Paula-Souza (*Corynostylis v.*, Colombia) – powerful vermifuge
Calyptrocalyx Blume. Palmae (V 14g). Incl. *Paralinospadix*, c. 27 Moluccas (1), NG. R: Blumea 46(2001)211
Calyptrocarya Nees. Cyperaceae (II 4). 8 trop. Am. R: AMBG 75(1988)860
Calyptrochilum Kraenzlin. Orchidaceae (V 16d). 2 trop. Afr.
Calyptrogenia Burret (~ *Hottea*). Myrtaceae (II 10). Excl. *Neomitranthes*, 6 Carib.
Calyptrogyne H. Wendl. Palmae (V 11). Incl. *Calyptronoma*, c. 20 C Am., WI. R: Principes 39(1995)145, SB 30(2005)70. Some bat-poll. incl. ***C. ghiesbreghtiana*** (Linden & H. Wendl.) H. Wendl. – cult. orn. ± stemless
Calyptronoma Griseb. = *Calyptrogyne*
Calyptrorchis Brieger = *Pleurothallis*
Calyptrosciadium Rech.f. & Kuber. Umbelliferae (III 5). 2 Iran, Afghanistan. R: Cand. 59(2004)100
Calyptrotheca Gilg. Didiereaceae (Portulacaceae s.l.). 2 NE trop. Afr. R: FTEA Portulac. (2002)34. Shrubs
Calystegia R. Br. = *Convolvulus*
Calythropsis C. Gardner = *Calytrix*
Calytrix Lab. Myrtaceae (II 15). c. 80 Aus. esp. SW. R: Brunonia 10(1987)1. Cult. orn. (fringe-myrtles) esp. ***C. tetragona*** Lab., heath-like
Camarea A. St-Hil. Malpighiaceae. 7 E S Am. R: Hoehnea 17,1(1990)1
Camaridium Lindl. = *Maxillaria*
Camarotea Scott-Elliot. Acanthaceae (III). 1 Madag.: ***C. romiensis*** Scott-Elliot
Camarotis Lindl. = *Micropera*
camas(h) *Camassia* spp., *Zigadenus* spp.; **common c.** *Camassia quamash*; **death c.** *Toxicoscordion* spp.; **white c.** *Anticlea elegans*
Camassia Lindl. Asparagaceae (Hyacinthaceae). 6 N Am. R: AMN 28(1942)712. Differs from *Scilla* in P-lobes 3-veined. Bulbs ed. esp. ***C. quamash*** (Pursh) Greene (*C. esculenta*), the camas(h) or quamash of Native Americans; others cult. orn. esp. ***C. scilloides*** (Raf.) Cory (E US)
camatillo *Dalbergia congestiflora*
Cambajuva Viana & al. (~ *Aulonemia*). Gramineae (V 2). 1 S Braz.: ***C. ulei*** (Hackel) Viana & al. R: SB 38(2013)98
Cambessedesia DC. Melastomataceae. 23 S Braz.
× **Cambria** comm. name for several multigeneric orchid hybrid crosses, esp. × *Vuylstekeara*
cambuca *Plinia edulis*
Camchaya Gagnepain. Compositae (Vern.-Lin.). 6 SE As. R: APG 23(1968)71
camel bush *Trichodesma zeylanicum*; **c.'s foot** *Bauhinia* spp.
Camelina Crantz. Cruciferae (15). 8 Eur. (4), Med. to C As. R: JAA 68(1987)234. ***C. sativa*** (L.) Crantz (gold-of-pleasure, false flax, Med. region) – hexaploid (?hybrid orig.) domesticated in E or SE Eur., fibre, seeds for cagebirds, seed-oil (cameline oil) like rape oil now mooted as aviation fuel, imp. oil-crop in C & E Eur. until 1940s
Camelinopsis A. Mill. Cruciferae. 2 Iran, Iraq. R: NRBGE 36(1978)30
Camellia L. Theaceae. 119 Indomal. (Mal. 1), E As. R: J.R. Sealy (1958) *A revision of the genus* C., H.T. Chang & B. Bartholomew (1984) Cs., ABY 21(1999)149. Tea, oilseeds, cult. orn. evergreens incl. hybrids, esp. ***C. japonica*** L. (Korea, Jap., Taiwan) – over 32 000 named cvs (incl. hybrids with *C. reticulata* Lindl. (Yunnan) etc.) with red, white or pink single or double (A petaloid) fls, seed-oil (tsubaki oil) a hair-oil for Jap. women & cooking, lubricating & stamp-pad oil & soap in China; ***C. kissii*** Wall. (Bhutan, Cambodia, China) – cult. for seed-oil in skin- & hair-products; ***C. oleifera*** Abel (China; & ***C. chekiangoleosa*** Hu, ***C. grijsii*** Hance [*C. yuhsienensis*]) – source of comm. tea oil for cooking etc.; ***C. sasanqua*** Thunb. (Jap.) – oilseed lower quality than *C. japonica*, cult. orn.; ***C. sinensis*** (L.) Kuntze (tea, S & E As.) – long cult. by Chinese (allegedly from 2737 BC, certainly from 350 BC in Szechuan spreading along Yangtze valley to eastern provinces), next to water now world's (& GB's: 180 000 t. per annum (165 M cups a day), i.e. more than N Am. [worth $5.55 billion in 2004] & W Eur. together) most imp. (incl. most imp. caffeine cf. *Coffea*) drink (first in UK, from Jap. to Holland 1610 – medic.!, public sale in London 1657

(Pepys's first cup 1660) but at £3.50 per lb. too dear (in 1770s most in UK smuggled – taxes high) for general drinking until C18 (when GB consumption 2 lb. per person a yr, by 1900 6 lb.), & in late C17 oft. adulterated with young lvs & shoots of *Uncaria gambir*), pls introd. to Java & Ind. c. 1835 & later to Sri Lanka (thus responsible for transmigration of Ind. Tamils thither but Chinese tea paid for in opium leading to widespread addiction in China), now comm. even in Cornwall, UK; var. ***assamica*** (J. Masters) Kitam. with larger lvs native in wetter SE As., more suited to culture in Sri Lanka & Assam; trees pruned to table-top bushes & young shoots (up to 4 flushes a yr.) nipped off (finer qualities without expanded lvs), withered, rolled & fermented (exc. green tea (less caffeine) – drunk for 5000 yrs (now 3 M t p.a.), chewed, fermented or pickled, in Myanmar (lahpet), Thailand (miang); 'oolong' being 'semi-fermented', that just from twigs = bancha twig t.), once thought to be a distinct sp.!), dried & sorted into grades, e.g. pekoe, souchong, for Darjeeling, Assam & Ceylon ts & (citrus flavoured) Earl Grey (black), gunpowder (green), s.t. compressed into bricks & form. transported by camel thus across C As. to Russia; stimulant due to alks (at least 6 incl. caffeine & theobromine; theophylline – drug from it for acute asthma attacks & kidney transplant treatment), flavour (flavonoids good antioxidants against heart disease & poss. cancers, epigallocatechin in green tea reducing skin cancer tumours) s.t. supplemented with flower-petals (*Jasminum* etc.) or bergamot (as in Earl Grey t.); seeds yield an oil & var. ***sinensis*** used as hedge-pl. in US

Camelliaceae DC. = Theaceae

Camelostalix Pfitzer = *Pholidota*

Cameraria Plum. ex L. Apocynaceae (Ig). 2 WI

Camerunia (Pichon) Boit. = *Tabernaemontana*

camias *Averrhoa bilimbi*

Camissonia Link. Onagraceae. Excl. *Camissoniopsis, Chylismia, Chylismiella, Eremothera, Eulobus, Holmgrenia, Taraxia, Tetrapteron*, 12 Pac. Am. R: CUSNH 37(1969)161

Camissoniopsis W.L. Wagner & Hoch (~ *Camissonia*). Onagraceae. 14 SW N Am. esp. Calif. Usu. ann., polyploid complex of tetraploids & hexaploids

Camoensia Welw. ex Benth. Leguminosae (III 2). 2 Gulf of Guinea. Lianes. ***C. scandens*** (Welw.) J.B. Gillett (*C. maxima*) – cult. orn. with largest leguminous fl., to 20 cm across (scented)

camomile See chamomile

Campanea Decne = *Kohleria*

Campanocalyx Valeton = *Keenania*

Campanula Tourn. ex L. Campanulaceae (I 6). Incl. *Symphyandra*, excl. *Azorina, Favratia*, c. 420 N temp. esp. Medit. (Eur. 144; Turkey 95), trop. mts. Bellflowers; many cult. orn. (H.C. Crook (1951) *Campanulas*). Heterogeneous (cf. *Lobelia*). A dehisce in bud, depositing pollen on style-hairs; as fl. opens (but some cleistogamous e.g. ***C. dimorphantha*** Schweinf. (NE Afr., S & E As.) where seeds prod. more quickly in cleistogamous fls) A wither, exc. bases prot. nectar, style presenting pollen to insects; stigmas sep. & receive external pollen, eventually curling right back & effecting self-poll. Fr. a capsule, if erect dehiscing distally, if pendent proximally, so that seeds escape only when pl. shaken, as by wind. ***C. alpestris*** All. (*C. allionii*, Alps) & ***C. carnica*** Schiede ex Mert. & Koch (*C. linifolia*, Alps) – constituents of Chartreuse; ***C. rapunculus*** L. (rampion, Euras.) – tap-r. form. much eaten but old clones with parsnip-like r. app. lost by 1820s. Some almost ineradicable weeds, e.g. ***C. rapunculoides*** L. (Euras., N Afr., natur. GB) with brittle rhizomes; many beautiful garden pls. incl.: ***C. americana*** L. (*Campanulastrum a.*, E N Am.) – poss. source of oil & rubber; ***C. carpatica*** Jacq. (Carpathians) – rock-pl.; ***C. isophylla*** Moretti (N Italy) – hanging baskets; ***C. latifolia*** L. (Euras.) – fls emetic; ***C. medium*** L. (Canterbury bell, S Eur.), **'Calycanthema'** ('cup-&-saucer') with petaloid K forming saucer round C; ***C. persicifolia*** L. (Euras.) – most common garden pl., many cvs; ***C. pyramidalis*** L. (S C Eur.) – infls to 1.5 m; ***C. robinsiae*** Small (Florida) – aquatic ann. thought extinct until 1982 when 2 populations found; ***C. rotundifolia*** L. (harebell, (Scottish) bluebell, N temp.) – polymorphic; ***C. thyrsoides*** L. (Alps, Balkans) – 1 of only 2 spp. with yellow fls; ***C. versicolor*** Andrews (SE Eur.) – lvs ed. Greece

Campanulaceae Juss. Magnoliidae – Asterales. 85/2250 cosmop. R: T.G. Lammers (2007) *World checklist & bibliog. C.* Mostly herbs but some shrubs & (usu. pachycaul) trees with a network of laticifers in phloem & oft. medullary bundles or phloem, storing polysaccharides as inulin. Lvs simple (rarely pinnate), in spirals, seldom opp. or whorled (*Ostrowskia*); stip. 0. Infls racemose to cymose (rarely epiphyllous – *Ruthiella*) or solit.; fls bisex. (rarely not, e.g. dioec. spp. of *Lobelia*), epigynous to (rarely) perigynous (*Cyananthus*). K ((3–)5(–10)), imbr. or valvate with odd lobe posterior (appearing similar in Lobelioideae through resupination of fl. with anterior one), persistent, C ((4)5(–10)), s.t. free, usu. valvate ± reg. (Campanuloideae), irreg. in Lobelioideae with 3-lobed upper lip appearing as lower (resupination), other lip 2-lobed, A (3–)5(–10) connivent (Campanuloideae) or connate (Lobelioideae), forming anther-tube in which pollen shed & through which style with collecting hairs below initially adpressed stigmas grows, alt. with C, attached to annular epigynous nectary-disk or corolla-base, anthers separating after anthesis in Campanuloideae, pollen exine usu. spinulose in Campanuloideae but reticulate in Lobelioideae, $\overline{G}$ (2–5) usu. (3) in Campanuloideae & (2) in Lobelioideae, rarely G 5 (*Cyananthus*), usu. with as many locules as G but in some Campanuloideae primary locules divided by partitions from carpellary midribs, in some Lobelioideae 1-loc. with 2 parietal placentas, ovules ∞ on axile, rarely parietal, placentas, anatropous, unitegmic. Fr. a berry or capsule, dehiscing variously; seeds small, s.t. winged, with straight embryo in oily endosperm (starchy in some *Wahlenbergia* (*Cephalostigma*)). x = 6–17

Classification & chief genera:

I. **Campanuloideae** (fl. ± reg., anthers eventually usu. free; 96% spp. OW): *Adenophora, Asyneuma, Campanula, Cyananthus, Phyteuma, Wahlenbergia*

II. **Nemacladoideae** (fl. irreg., not resupinate, not all A epipet., annuals; SW N Am.): *Nemacladus*

III. **Lobelioideae** (fl. irreg. usu., resupinate; anthers connate; most genera 'nested' in *Lobelia* s.l.): *Burmeistera, Centropogon, Cyanea, Lobelia, Siphocampylus*

IV. **Cyphocarpoideae** (fl. irreg., not resupinate, all A epipet.; poss. in V): *Cyphocarpus* (only)

V. **Cyphioideae** (fl. irreg., not resupinate, not all A epipet., tubers): *Cyphia* (only)

Some genera, e.g. *Lobelia*, show transition from fleshy to dry fr. & erect pachycaul treelets to creeping rhiz. herbs with concomitant reduction in vessel-element length. Woody spp. oft. considered 'anomalous' but occur widely scattered in the family, all Hawaiian bird-poll., wind-disp. lobeliads (126 spp. derived from 1 woody sp. introd. c. 13 Myrs ago, i.e. not secondary 'insular woodiness'; most species-rich radiation from 1 introd. in any is. system; many endangered; 25% of Hawaiian endemics already extinct)

Some medic. (*Adenophora, Lobelia, Platycodon*). Many cult. orn., some ed. esp. *Adenophora, Phyteuma*

Campanulastrum Small = *Campanula*

Campanulorchis Brieger (~ *Eria*). Orchidaceae (V 15). 5 Mal.

Campanumoea Blume = *Codonopsis*

campeachy or **campeche wood** *Haematoxylum campechianum*

Campbellia Wight = *Christisonia*

Campecarpus H. Wendl. ex Becc. = *Cyphophoenix* (but see Allertonia 3(1984)384)

Campeiostachys Drobov = *Elymus*

Campelia Rich. = *Tradescantia*

Campestigma Pierre ex Costantin. Apocynaceae (Va; Asclepiadaceae III 4). 1 SE As.: ***C. purpureum*** Pierre ex Costantin

camphor (laurel) *Cinnamomum camphora*; **baros, barus, Borneo** or **Sumatra c.** *Dryobalanops aromatica*; **E. Afr. c. wood** *Ocotea usambarensis*; **ngai c.** *Blumea balsamifera*; **c. plant** *Tanacetum balsamita*; **c. weed** *Pluchea camphorata*; **c. wood** *Tarchonanthus camphoratus*

Camphorosma L. Amaranthaceae (Chenopodiaceae I 4). 4 E Med. (Eur. 3), C As. R: T 60(2011)172. ***C. monspeliaca*** L. – shrubby, medic.

Campimia Ridl. Melastomataceae. 1 Malay Pen.: ***C. wrayi*** (King) Ridl. R: Willd. 17(1988)147

campion *Silene* spp.; **alpine c.** *S. suecica*; **bladder c.** *S. vulgaris*; **moss c.** *S. acaulis*; **red c.** *S. dioica*; **rose c.** *S. coronaria*; **sea c.** *S. uniflora*; **white c.** *S. latifolia*

Campnosperma Thwaites. Anacardiaceae (I). c. 13 trop. Am., Madag. (4+), Seychelles (1), Sri Lanka (1), SE As., Mal., W Pac. Oft. forming monospecific stands in swamps. Aubréville's Model. Timber for boxes, canoes etc. ***C. coriaceum*** (Jack) Steenis (Mal.) & ***C. brevipetiolatum*** Volkens (E Mal., Pac.) – source of parasiticidal tigasco skin oil in Papua

Campomanesia Ruíz & Pavón. Myrtaceae (II 10). c. 30 S Am. R (trop. – 25): FN 45(1986)13. Ed. fr. (Pará guava, Paraguay, Arg.) esp. ***C. guaviroba*** (DC.) Kiaerskov (guabiroba, Braz.), some fish-disp. in Amaz.

Campovassouria R. King & H. Robinson (~ *Eupatorium*). Compositae (Eup.-Dis.). 1 E S Am.: ***C. cruciata*** (Vell.) R. King & H. Robinson (*C. bupleurifolia*). R: MSBMBG 22(1987)79

Campsiandra Benth. Leguminosae (I 4). 19 trop. S Am. esp. Amazonia. Medic. ***C. laurifolia*** Benth. water-disp. (aerenchyma in seed-coat)

Campsidium Seemann. Bignoniaceae (1). 1 Chile, Arg.: ***C. valdivianum*** (Philippi) Bull – cult. orn. R: FN 25,2(1992)16

Campsis Lour. Bignoniaceae (1). 2 E As. (***C. grandiflora*** (Thunb.) K. Schum., Chinese trumpet- flower medic. since anc. times; hexose-rich nectar for perching bird-poll.), E US (***C. radicans*** (L.) Bureau, trumpet cli., creeper or vine; sucrose-rich nectar for humming-bird-poll.). Adventitious r.-climbers like *Hedera*; extrafl. nectaries, attracted ants reducing herbivore attack. Cult. orn. incl. their hybrid ***C.*** × ***tagliabuana*** (Vis.) Rehder (nectar hexose-dominant!)

Camptacra N. Burb. (~ *Vittadinia*). Compositae (Ast.-Pod.). 2 NG, trop. Aus. R: Brunonia 5(1982)11

Camptandra Ridl. Zingiberaceae (II 1). 4 W Mal.

Camptocarpus Decne. Apocynaceae (III; Asclepiadaceae I 2). (Incl. *Tanulepis*) 9 Madag., Mauritius. R: BJ 120(1998)56

Camptodium Fée = *Tectaria*

Camptolepis Radlk. Sapindaceae. 4 Madag, 1 ext to coastal E & NE Afr. R: MMNHN 19(1969)108

Camptoloma Benth. (~ *Sutera*). Scrophulariaceae (inc. sed.). 3 Canary Is., Somalia, S Yemen, Namibia. R: O.M. Hilliard, *The Manuleeae* (1994)80

Camptopus Hook.f. = *Psychotria*

Camptorrhiza Hutch. Colchicaceae (Liliaceae s.l.). 1 S Afr.: ***C. strumosa*** (Bak.) Oberm., 1 Ind. Alks

Camptosema Hook. & Arn. Leguminosae (III 18). 10 S Am.

Camptosorus Link = *Asplenium*

Camptostemon Masters. Malvaceae (Malv.-Malv.; Bombacaceae). 2 C Mal., N Aus. Mangrove. A (∞)

Camptostylus Gilg (*Cerolepis*). Achariaceae (Flacourtiaceae). 3 trop. W & C Afr. R: BJ 94(1974)283

Camptotheca Decne. Nyssaceae (Cornaceae). 1 S & SE China.: ***C. acuminata*** Decne (happy tree, xi shu) – used against colon, rectal & ovarian cancers, cult. orn.

Campuloclinium DC. (~ *Eupatorium*). Compositae (Eup.-Gyp.). 14 trop. Am. R: MSBMBG 22(1987)126. ***C. macrocephalum*** (Less.) DC. (pom-pom weed, Braz.) – weedy S. Afr.; ***C. viridiflorum*** Bartl. (Braz.) – form. cult.

Campylandra Bak. = *Rohdea*

Campylanthus Roth. Plantaginaceae (Dig.; Scrophulariaceae s.l). 15 Macaronesia (2), NE Afr. & Arabia (12), Pakistan (1). R: EJB 60(2003)151. ***C. salsoloides*** (L.f.) Roth (Canary Is.) – cult. orn.

Campylocentrum Benth. Orchidaceae (V 16d). c. 65 trop. Am. (Florida 1). Some leafless

Campyloneurum C. Presl (~ *Polypodium*). Polypodiaceae (V). 20 trop. & warm Am. esp. Andes

Campylopetalum Forman. Anacardiaceae (I, Podoaceae). 1 N Thailand: ***C. siamense*** Forman – epiphyllous female fls

Campylosiphon Benth. Burmanniaceae. 1 trop. S Am.: ***C. purpurascens*** Benth. – oft. confused with *Voyria*

Campylospermum Tieghem (~ *Ouratea*). Ochnaceae. 50 trop. OW. Some 'litter-box' pls (Corner's Model), incl. ***C. amplectens*** (Stapf) Farron (*O. a.*, W Afr.)

Campylostachys Kunth. Stilbaceae. 1 W Cape: ***C. cernua*** (L.f.) Kunth. R: Strel. 10 (2000)542

Campylostemon Welw. Celastraceae (III). 8+ trop. Afr. Links Celast. (s.s.) with form. seg. Hippocrateaceae

Campylotheca Cass. = *Bidens*

Campylotropis Bunge. Leguminosae (III 19). 37 As. R: JJB 77(2002)191,251,315. Close to *Lespedeza*

Campynema Labill. Campynemataceae (Liliaceae s.l.). 1 Tasmania: ***C. lineare*** Lab. R: BMNHN4,8(1986)129. Leaf 1. Referred at diff. times to Amaryllidaceae, Colchicaceae, Hypoxidaceae, Iridaceae, Liliaceae or Melanthiaceae. See also *Campynemanthe*

Campynemanthe Baill. Campynemataceae (Liliaceae s.l.). 3 New Caled. R: BMNHN 4,8(1986)121

Campynemataceae Dumort. Magnoliidae – Liliales. 2/4 SW Pac. Rhiz. herbs with calcium oxalate raphide bundles, linr. lvs with persistent fibrous bases, & panicles of or solit. 3-merous greenish fls. P 3 + 3, enlarging after fert.; A 3 + 3 at base of P; $\overline{G}$ or half so, 1- or 3-loc. with parietal placentae & 3 – ∞ anatr. ovules. Fr. a 6-ribbed capsule with irreg. exotestal seeds (embryo minute), dehiscing by decay
Genera: *Campynema, Campynemanthe*

camu, camu-camu *Myrciaria dubia*

Camusiella Bosser = *Setaria*

camwood *Baphia nitida, Pterocarpus soyauxii*

Canaca Guillaumin (~ *Austrobuxus*). Picrodendraceae (Cal.-Diss.; Euphorbiaceae s.l.). 7 New Caled

Canacomyrica Guillaumin. Myricaceae. 1 New Caled.: ***C. monticola*** Guillaumin – on ultramafics (Eocene, Miocene NZ)

Canacorchis Guillaumin = *Bulbophyllum*

Canada balsam *Abies balsamea;* **C. black spruce** *Picea mariana;* **C. bluegrass** *Poa compressa;* **C. garlic** *Allium canadense;* **C. hemlock** *Tsuga canadensis;* **C. pitch** *Abies balsamea*

Canadanthus Nesom. Compositae (Ast.-Sym.). 1 N Am.: ***C. modestus*** (Lindl.) Nesom. R: Phytol. 77(1994)250

Canadian wild rice *Zizania palustris*

canaigre *Rumex hymenosepalus*

Cananga (Dunal) Hook.f. & Thomson. Annonaceae (II). 2 trop. As. to Aus. R: GBS 61(2009)193. ***C. odorata*** (Lam.) Hook.f. & Thomson – cult. orn. (Roux's Model) & in Madag., Comoro Is. & Réunion, Philippines etc., fls the source of ylang-ylang or cananga oil (mature tree giving 9 kg fresh fls yielding 30 g oil per annum; world prod. 100 t by 1980), a hair-dressing & constituent of Chanel 'No. 5' (1921), Revlon's 'Charlie' etc. (oft. mixed with pimento oil), P chewed with betel in Sri Lanka

Canaria Jim. Mejías & P. Vargas (~ *Seseli*). Umbelliferae. 1 Canary Is.: ***C. tortuosa*** (Webb & Berth.) Jim. Mejías & P. Vargas. R: Phytotaxa 212(2015)73

Canariellum Engl. = *Canarium*

canariense *Tropaeolum canariense*

Canarina L. Campanulaceae (I 1). 3 Canary Is. (1), trop. E Afr. (2). R: SBT 55(1961)48. Cult. orn. with tubers; like *Campanula* but fls 6-merous & fr. a berry (ed.), lvs opp. or ternate & C yellow to red. ***C. canariensis*** (L.) Vatke (bicacaro, Canary Is.) – bird-poll. by 'insectivorous' birds (no sunbirds in Canary Is.!)

Canariothamnus R. Nordenstam = *Bethencourtia*

Canarium L. (?*Rumphia*) Burseraceae. (V) c. 120 trop. Afr. (2), Madag. (33), Indomal. R: Blumea 9 (1959)275). Timbers (kedondong), resins white turning black (black dammar) & ed. seeds (oily pili or nangai nuts) esp. SE As., usu. ***C. ovatum*** Engl. (Philippines) – used in moon cakes, also ***C. indicum*** L. (galip or ngali nut, Indomal.) – source of Solomon nut oil used in tanning lotions & other cosmetics, ***C. luzonicum*** (Blume) A. Gray (Java almond, Philippines) – source of Manila elemi for varnishes etc.), ***C. zeylanicum*** (Retz.) Blume (S As.) – resin used as fumigant Sri Lanka; & fr., e.g. ***C. album*** (Lour.) Räusch. (trop. As.), ***C. odontophyllum*** Miq. (dabai, W & C Mal.) & ***C. pimela*** C. Koenig (*C. tramdenum*, Chinese olives, SE China, SE As.), those of ***C. harveyi*** Seem. (nail nut) widely sold in Polynesian markets. ***C. euphyllum*** Kurz (dhup, Ind. white mahogany, Andamans); ***C. schweinfurthii*** Engl. (abé, abel, aielé, incense tree, trop. Afr.) – Rauh's Model, timber stained as mahogany subs., church incense in Uganda, fr. sold in markets so nat. distrib. unclear

Canary balm *Cedronella canariensis;* **c. creeper** *Tropaeolum peregrinum;* **c. grass** *Phalaris canariensis;* **c. grass, reed** *P. arundinacea;* **C. ivy** *Hedera algeriensis;* **C. pine** *Pinus canariensis;* **c. whitewood** *Liriodendron tulipifera, Magnolia* spp.; **c. wood** *Morinda citrifolia,* also applied to several S Am. timbers

Canastra Morrone & al. (~ *Arthropogon*). Gramineae (XXIII 1). 2 Braz. R: Novon 11(2001)429

Canavalia DC. Leguminosae (III 18). c. 60 trop. (Hawaii 6 endemic) esp. Am. R: Britt. 16(1964)106. Alks. Green manure, stock feed & beans esp. ***C. ensiformis*** (L.) DC. (jack bean, sword bean, sabre bean, Jamaican horse bean, trop. Am.) – young pods ed. though unripe seeds considered toxic; ***C. gladiata*** (Jacq.) DC. (sword bean, poss. derived from ***C. cathartica*** Thouars, OW – fls & seeds used in leis in Hawaii), similar but pods longer; ***C. rosea*** (Sw.) DC. (*C. maritima*) – common pantrop. beach pl.

Canbya Parry ex A. Gray. Papaveraceae (IV). 2 W N Am. C marcescent in ***C. candida*** Parry ex A. Gray (W Mojave), decid. in ***C. aurea*** S. Watson (Oregon, Nevada)

cancer-root *Conopholis americana*

Cancrinia Karelin & Kir. Compositae (Anth.-? Han.). 4 C As. to China. R: BBMNHB 23(1993)99

Cancriniella Tzvelev. Compositae (Anth.-Can.). 1 C As.: ***C. krascheninnikovii*** (Rubtzov) Tzvelev [!]. R: BBMNHB 23(1993)99

candelilla (**wax**) *Euphorbia antisyphilitica*

candied peel *Citrus medica* & other *C.* spp.

candle berry *Morella cerifera, Aleurites moluccanus;* **c. bush** *Senna alata;* **c. nut (oil)** *A. moluccanus;* **c. plant** *Curio articulata;* **c. tree** *Parmentiera cereifera;* **c. wood** *Amyris balsamifera*

Candollea Lab. (1805) = *Stylidium*

Candollea Lab. (1806) = *Hibbertia*

Candolleodendron R. Cowan. Leguminosae (III 1). 1 NE S Am.: ***C. brachystachyum*** (DC.) R. Cowan

Candy carrot *Athamanta cretensis;* **cotton c.** (US), **c. floss** (UK) spun sugar; **c. tuft** *Iberis* spp.

cane see rattan; **dumb c.** *Dieffenbachia* spp.; **giant** or **switch c.** *Arundinaria gigantea;* **Malacca c.** *Calamus scipionum* etc.; **c. palm** *Dypsis lutescens;* **rajah c.** *Eugeissona minor;* **Spanish c.** *Arundo donax;* **c. sugar** *Saccharum officinarum;* **Tongking c.** *Pseudosasa amabilis*

Canella P. Browne. Canellaceae. 1–2 S Florida, WI. Scarrone's Model. ***C. winterana*** (L.) Gaertn. – source of canella bark (peppery taste like *Drimys*) used as condiment & medic. (tonic & stimulant), a fish-poison (Puerto Rico) & for flavouring tobacco; timber (Bahama whitewood)

Canellaceae Mart. Magnoliidae – Canellales. 5/17 trop. Afr. (E) & Am. Glabrous usu. aromatic trees (or shrubs), the oil-cells with terpenes in the parenchyma. Lvs in spirals, simple, entire, oft. with pellucid dots; stip. 0. Fls bisex., reg., solit. or in terminal or axillary racemes to panicles, or solit.; K (considered bracts by some) 3, leathery, imbr., C (considered K by some) (4)5–12 in 1 or 2(–4) whorls &/or spirals (outer whorl s.t. considered K), imbr., ± connate in *Canella* & *Cinnamosma*, A (6–12 (to 35 or 40 in *Cinnamodendron*)) forming a tube with extrorse anthers without, pollen monosulcate, G (2–6), 1-loc. with thick style & 2–6-lobed stigma, placentas parietal, 2–6 with 1 or 2 rows of 2–∞ ovules each, ovules hemitropous, bitegmic. Fr. a berry with 2 or more seeds with oily endosperm (ruminate in *Cinnamosma*) & s.t. vestigial arils. 2n = 22, 26, 28

Genera: *Canella, Cinnamodendron, Cinnamosma, Pleodendron, Warburgia*

Form. thought perhaps closest to Myristicaceae, the 2 from ancestors more like modern Magnoliaceae (Cronquist), but seed structure like Winteraceae (Corner) an accurate predictor. Some condiments & medic.

Canephora Juss. (~ *Fernelia*). Rubiaceae (Octotrop.). 5 Madag.

canihua *Chenopodium pallidicaule*

canistel *Pouteria campechiana*

Canistropsis Leme (~ *Nidularium*). Bromeliaceae (3). Excl. *Andrea*, 9 trop. Am.

Canistrum C.J. Morren. Bromeliaceae (3). Excl. *Edmundoa*, 7 E Braz. R: FN 14(1979)1715. Some cult. orn. esp. ***C. fragrans*** (Linden) Mabb. (*C. lindenii*)

Canizaresia Britton = *Piscidia*

cankerberry *Coptis trifolia*

Canna L. Cannaceae. 8–10 trop. Am. Black seeds (some viable for c. 600 yrs) used as beads. Cult. orn. esp. forms of ***C. indica*** L. (Ind. shot, natur. throughout trop.) & commonly ***C.* × *generalis*** L. Bailey (*C. indica* × *C. iridiflora* Ruíz & Pavón) – fls yellow, & ***C.* × *orchioides*** L. Bailey (*C.* × *generalis* × *C. flaccida* Salisb., S US to Panamá) – fls yellow & red, usu. sterile; ed. starchy rhiz. of triploid forms of *C. indica* (*C.* × *discolor*, *C. edulis*, achira, Queensland arrowroot, tous-les-mois) – starch grains suited to infants & invalids (cf. *Maranta*)

Cannabaceae Martinov (incl. Celtidaceae; ~ Urticaceae). Magnoliidae – Rosales. 9/90 N temp. to trop. R: T 62(2013)473. Trees, lianes (*Humulus*) or herbs (*Cannabis*), dioec., rarely monoec., wind-poll. with pyridine alks.; laticifers 0. Lvs distich., opp. or in spirals (serrate, palmately lobed in *Humulus*) with glandular hairs; stip. free or connate, persistent. Infls basically cymose, females smaller, male fls – K 5, A 5 opp. K all in a spiral, female fls – K tubular (merely a ring in cultivars of *Cannabis*) around $\underline{G}$ (2), 1-loc. with 2 dry stigmas & 1 ovule, subapical, anatropous, bitegmic. Fr. a drupe or achene covered in K; seeds with curved or coiled (*Humulus*) embryo in oily endosperm. x = 8 (*Humulus*), 10 (*Cannabis*)

Genera: *Aphananthe, Cannabis, Celtis, Chaetachme, Gironniera, Humulus, Lozanella, Pteroceltis, Trema*

A fam. causing much human happiness (& misery); fibre (*Cannabis*), useful timber (& dyes) from *Aphananthe, Celtis* (also fr.), *Chaetachme, Gironniera, Trema*

Cannabidaceae See Cannabaceae

Cannabis Tourn. ex L. Cannabaceae. 1 C As.: ***C. sativa*** L. (Ind. hemp), ann. to 8(–12) m, dioec. with sex chromosomes, though sex modifiable by environmental factors; v. variable (700+ cvs) & orig. wild material cannot be certainly identified but in cult. 2 groups of cvs, the more northerly cult. ('subsp. *sativa*', since 4000 BC, esp. in N & NE China where form. only fibre available, prob. used in first paper (AD 105) there, obligatory crop in Eliz. times in GB, where illegal since 1951) for saltwater-resistant fibre (hemp used for ropes ('canvas' cognate with c.), first Levi jeans & first US flag, fibre-board, paper (e.g. cigarette-papers, Magna Carta, bibles incl. first King James B.), insulator for car doors & roofs (BMW, Mercedes), with lime in 'hemcrete' blocks good insulation, etc.) & more southerly cult. ('subsp. *indica* (Lam.) E. Small & Cronq.') principally for psychotropic drugs (marijuana, marihuana (Mex.), grass or pot (US, where allegedly the biggest cash crop worth $32 billion, also in Canada (Mary Jane) at (2003) $7 billion), dagga (S Afr.), kif (Morocco)), since 1842 (Eur.) becoming *the* drug of Vietnam War period, cannabis resin (cannabinoids esp. THC like those prod. during exercise giving a 'high' & attaching to same receptors, discovered 1995), which exudes from the glandular hairs & is used like opium (effects described 2736 BC by Chinese Emperor Shen Neng). In Ind., 3 common forms: ganja (dried unripe infrs), charas or churras (resin knocked off twigs, bark etc.) & bhang (largely mature lvs of wild pl.). Smoked ('weed') to decarboxylate cannabinoids (also achieved by cooking as 'hash cakes') with or without tobacco ('skunk') by c. 300 M people (2003) in cigarettes ('joints', the smoke with tar high in lung cancer-promoting benzapyrene) or taken as an intoxicating liquid formed from it (hashish; Arabic for 'hashishtaker' = r. of word 'assassin'), in food ('shelled' hemp seeds, rich in omega-3, in salads etc.) or drink (e.g. in comm. beers in Netherlands & ice-tea flavouring) it has a stimulating & pleasantly exciting effect, relief from multiple sclerosis ('sativex' under tongue), cerebral palsy & glaucoma (in some 28 medic. preps in USA before removed from pharmacopeia 1937; Queen Victoria took it for menstrual cramps), though addictive & in excess can cause delirium & 'moral weakness and depravity' (Uphof), doubling risk of developing psychotic illnesses such as schizophrenia, THC disrupting short-term memory by weakening connections between neurons in hippocampus & assoc. with weak creative thinking, but three cannabinoid preps now comm. for pain relief. Seeds source of hemp seed-oil used in varnishes, food, soap, lip balm & fuel in Nazi tanks etc. & used as bird seed & to attract fishes. R.C. Clarke & M.D. Merlin (2013) *C.: evolution & ethnobotany*

Cannaboides B.-E. van Wyk (~ *Heteromorpha*). Umbelliferae (III 8). 2 Madag. R: T 48(1999)740

Cannaceae Juss. Magnoliidae – Zingiberales. 1/8–10 trop. & warm Am. Glabrous herbs with starchy rhizomes. Lvs in spirals with a sheath passing into petiole, simple lamina (laterally rolled in bud) with prominent midrib & ∞ lat. veins but no ligule nor pulvinus. Infl. terminal oft. with short 2-flowered cymules axillary to principal bracts. Fls large, bisex., obliquely orientated so that no organ is clearly median, K 3, spiral, persistent in fr., C 3 (1 smaller than others) joined in a basal tube with functional A 1 (middle of inner A cycle), petaloid with pollen-sac along more nearly median edge, & at least 1 staminode with (1)2(–4) additional ones, $\overline{G}$ (3), 3-loc. with axile placentation, style petaloid, ovules ± ∞ in each loc., anatropous, bitegmic. Fr. a caps. (? s.t. indehiscent) usu. bristly; seed exarillate with straight embryo in thin starchy endosperm & copious hard, starchy perisperm. x = 9

Genus: *Canna*

Petaloid staminode s.t. called labellum but not homologous with that in Zingiberaceae (Cronquist). Pollen shed on style in bud; insects land on staminode brushing stigma with foreign pollen & then the pollen already on style

Cannaeorchis M.A. Clem. & D.L. Jones = *Dendrobium*

cannelloni *Phaseolus vulgaris*

cannibal's tomato *Solanum viride* 'Anthropophagorum'

Cannomois P. Beauv. ex Desv. Restionaceae. 7 S & SW Cape. R: Bothalia 15(1985)480. ***C. virgata*** Hochst. – at 3.5 m, tallest R.

cannonball tree *Couroupita guianensis*

canoe birch *Betula papyrifera*

canola (oil) *Brassica napus* 'Canola'

Canotia Torrey. Celastraceae (I, Canotiaceae). 2 SW US. Leafless, poss. allied to *Acanthothamnus*. R: Britt. 27(1975)119

Canotiaceae Airy Shaw = Celastraceae

Canscora Lam. Gentianaceae. 9 OW trop. R: Blumea 48(2003)5. Some medic.

Cansjera Juss. Opiliaceae. 3 Indomal. to Aus. R: Willdenowia 9(1979)43. Root-parasites

cantala fibre *Agave cantala*

cantaloupe *Cucumis melo*

Canterbury bell *Campanula medium*

Canthiopsis Seem. (~ *Tarenna*). Rubiaceae (II 2). 1 Fiji: ***C. odorata*** Seem. (*T. seemanniana*)

Canthium Lam. (*Plectronia*). Rubiaceae (III 2). c. 30 trop. OW (China 4, Ind. (incl. Sri Lanka), 2). Roux's Model. Some fine timbers & ed. fr.; some Afr. spp. with hollow ant-infested twigs. See also *Keetia, Multidentia, Psydrax, Pyrostria*

Cantinoa Harley & Pastore (~ *Hyptis*). Labiatae (VII 3c). 25 trop. Am. R: Phytotaxa 58(2012)8

Cantleya Ridl. Stemonuraceae (Icacinaceae s.l.). 1 Mal.: ***C. corniculata*** (Becc.) R. Howard – fragrant timber (sandalwood subs.), ed. fr. R: FM I,7(1972)51

Canton fibre *Musa* hybrid (*M. textilis* × ?); **C. linen** *Boehmeria nivea*

Cantua Lam. Polemoniaceae (II). 12 Andes. R: Aliso 19(2000)62. Shrubs & trees, some cult. orn.

Caobangia A.R. Sm. & X.C. Zhang. Polypodiaceae. 1 Vietnam, SW China: ***C. squamata*** A.R. Sm. & X.C. Zhang. R: Novon 12(2002)546

Capanea Decne ex Planch. = *Kohleria*

Capanemia Barb. Rodr. Orchidaceae (V 12h). 9 Braz. R: Orquideologia 7(1972)215

caparrosa *Neea theifera*

Capassa Klotzsch = *Philenoptera*

Cape aloes *Aloe ferox;* **C. ash** *Ekebergia capensis;* **C. asparagus** *Aponogeton distachyos;* **C. aster** *Felicia* spp.; **C. beech** *Myrsine melanophloeos;* **C. box** *Buxus macowanii;* **C. b.wood** *Gonioma kamassi;* **C. chestnut** *Calodendrum capense;* **C. cowslip** *Lachenalia aloides;* **C. ebony** *Euclea pseudebenus, Heywoodia lucens;* **C. figwort** or **fuchsia** *Phygelius* spp. esp. *P. capensis;* **C. gooseberry** *Physalis peruviana;* **C. greens** *Berzelia* spp.; **C. gum** *Vachellia* spp.; **C. honeysuckle** *Tecoma capensis;* **C. jasmine** *Gardenia augusta;* **C. lilac** (Aus.) *Melia azedarach;* **C. lily** *Crinum* spp.; **C. mahogany** *Trichilia emetica;* **C. pondweed** *Aponogeton distachyos;* **C. primrose** *Streptocarpus* spp.; **C. shamrock** *Oxalis* spp.; **C. spinach** *Rumex spinosus;* **C. thorn** *Ziziphus mucronata;* **C.weed** *Arctotheca calendula;* **C. willow** *Salix mucronata*

Capelio R. Nordenstam (*Alciope*). Compositae (Sen.-Tuss.) 3 Cape. R: CN 38(2002)72, 39(2003)50

Capeobolus Browning (~ *Costularia*). Cyperaceae (II 7). 1 Cape: ***C. brevicaulis*** (C.B. Clarke) Browning. R: SAJB 65(1999)218

Capeochloa Linder & N. Barker (~ *Danthonia*). Gramineae (Danth.). 3 S Cape

caper *Capparis spinosa;* **c. spurge** *Euphorbia lathyrus*

Caperonia A. St-Hil. Euphorbiaceae (Acal.-Chro.-Dit.). c. 35 trop. Am., Afr. (5), Madag. (1)

Caphexandra Iltis & Cornejo (~ *Capparis*). Capparaceae. 1 C Am.: ***C. heydeana*** (J.D. Sm.) Iltis & Cornejo. R: HPB 16(2011)65

Capillipedium Stapf (~ *Dichanthium*). Gramineae (XXII 7). 18 E Afr., trop. As. to Aus. & New Caled. (1). Cult. orn. (aromatic infls)

Capirona Spruce. Rubiaceae (Cond.). 1 NE S Am.: ***C. decorticans*** Spruce. R: AMBG 82(1995)421. K-like *Mussaenda*

Capitanopsis S. Moore. Labiatae (VII 3e). 3 Madag.

Capitanya Schweinf. ex Guerke = *Coleus*

Capitularia Valcken. = *Capitularina*

Capitularina Kern (~ *Chorizandra*). Cyperaceae (I 1). 1 Papuasia: ***C. involucrata*** (Valcken.) Kern – 5-angled stems

Capnoides Mill. (~ *Corydalis*). Papaveraceae (Fumariaceae I 1). 1 NE N Am.: ***C. sempervirens*** (L.) Borkh., natur. Norway, cult. orn. R: OB 88(1986)19

Capnophyllum Gaertn. Umbelliferae (III 10). Excl. *Krubera*, 4 S Cape Province

Capparaceae Juss. Magnoliidae – Brassicales. Excl. Cleomaceae, Koeberliniaceae, Pentadiplandraceae, Physenaceae & Setchellanthaceae, 31/450 warm, few temp. arid. Shrubs, rarely herbs or trees, prod. mustard-oil glucosides & alks, s.t. with unusual sec. thickening of concentric type, r. s.t. with endotrophic mycorrhizae. Lvs in spirals, rarely opp., simple, trifoliolate or usu. palmate; stip. 0 or small, oft. glands or spiny. Fls usu. in racemes, rarely solit. & axillary, usu. bisex., ± irreg., receptacle usu. prolonged into gynophore/androgynophore, K (3)4(–6), oft. decussate, s.t. basally connate, C (2–)4(–6), rarely 0 or connate, alt. with K, oft. clawed, A rarely (1)2 or 4 alt. with C but oft. 2 or all 4 of A primordia developing to give 6–∞ A, some staminodal, but A not tetradynamous like Cruciferae, anthers with longit. slits, nectary an extrastaminal ring or merely receptacular protrusion, $\underline{G}$ (2(–8)), 1-loc. with parietal placentas (s.t. ephemerally or never meeting, giving pluriloc. ovary), style 1, ovules (1–)∞ per placenta, campylotropous or rarely anatr. before fert., bitegmic. Fr. usu. stipitate & a berry; seeds oft. reniform with ± curved or folded oily embryo, s.t. with arils, endosperm little or 0, perisperm s.t. present. n = 7–80

Principal genera: *Boscia, Cadaba, Capparis, Maerua, Ritchiea*

Excluded genera incl. *Dipterygium* referred to ?Cleomaceae, *Borthwickia, Forchhameria, Stixis & Tirania* to Resedaceae, but *Keithia* not definitely placed. Some ed. fr. esp. *Capparis* (capers) & cult. orn.

Capparicordis Iltis & Cornejo (~ *Capparis*). Capparaceae. 3 S Am. R: Britt. 59(2007)246

Capparidaceae auctt. = Capparaceae

Capparidastrum (DC.) Hutch. (~ *Capparis*). Capparaceae. 15 trop. Am. R: HPB 13(2008)230

Capparis Tourn. ex L. Capparaceae. Excl. *Anisocapparis, Beautempsia, Caphexandra, Capparicordis, Capparidastum, Colicodendron, Cynophalla, Hispaniolanthus, Mesocapparis, Monilicarpa, Neocalyptrocalyx, Preslianthus, Quadrella, Sarcotoxicum*, c. 150 trop. & warm (Eur. 1). Shrubs, scramblers or trees (Roux's Model) from deserts to rain forests but usu. with seasonal drought, usu. with white or yellowish fls lasting 1 day, some opening only at night (e.g. ***C. lucida*** (DC.) Benth., Java to Aus.); some with stip. thorns; extrafl. nectaries. Some medic., others ed. fr. (high in vitamin C). ***C. buwaldae*** Jacobs (Borneo) – myrmecophyte; ***C. mitchellii*** (F. Muell.) Benth. (bumble tree, Aus.) – ed. fr., timber like box as is that of ***C. nobilis*** (Endl.) Benth. (E Aus.); ***C. spinosa*** L. (capers, Medit. to Pacific) – cult. for fl. buds (cf. cloves) pickled (also fr. = c. berries) as a relish, esp. in France, in sauce Tartare etc., dried lvs rennet subs. in cheese-making

Capraria L. Scrophulariaceae (Leuc.). 4 warm Am. R: Lundellia 7(2004)59. Lvs in spirals

Caprifoliaceae Juss. Magnoliidae – Dipsacales. Incl. Dipsacaceae (R: T 62(2013)124), Morinaceae (R: BBM-NHB 12(1984)1), Valerianaceae, excl. Carlemanniaceae, 39/900 N temp., Medit., As., trop. mts., S Afr. Shrubs or small trees, lianes or herbs. Lvs opp. to (rarely) whorled, simple (rarely pinnate [= pinnatisect?]); stip. 0 or small. Fls bisex., epigynous, oft. constricted below K limb, usu. in cymose or mixed infls (s.t. involucrate-capitate), s.t. with epicalyx, K ((4)5), oft. small or even bristle-like to 0, lobes usu. imbr., ± accrescent in fr., C ((3-)5), usu. reg. (s.t. 2-lipped), lobes imbr. or valvate, tube oft. nectariferous, A (1-)5 attached to tube alt. with lobes, or 4 (*Linnaea*) even when P 5-merous, anthers with longit. slits, $\overline{G}$ (2– 5(–8)), rarely semi-inf., with as many locules as G with axile placentation or partitions failing to meet in ovary apex or only 1 loc. fertile, style term. or stigma(s) subsessile, ovules 1–∞ per locule, pend., anatropous, unitegmic. Fr. a berry, drupe, dry & dehiscent (s.t. with wing or plumose K), or a cypsela; seeds usu. with straight embryo & oily, fleshy endosperm (rarely 0). x = (5-)8 or 9(–12)

Principal genera: *Abelia, Centranthus, Diervilla, Dipelta, Dipsacus, Heptacodium, Knautia, Kolkwitzia, Leycesteria, Linnaea, Lonicera, Patrinia, Scabiosa, Symphoricarpos, Triosteum, Weigela*

Carlemannia & *Silvianthus* excl. as Carlemanniaceae, *Sambucus* with *Viburnum* now in Viburnaceae; form. recog. close relationship with Dipsacaceae & Valerianaceae (notable for foetid monoterpenoid & sesquiterpenoid ethereal oil-cells detectable after 100 yrs in herbarium material!) confirmed. *Zabelia* is 'sister' to D., V., Morinaceae clade. Generic limits around *Scabiosa* still debated

Scent & medic. (*Nardostachys, Valeriana*); salads (*Valerianella*); horticulturally v. imp. as many genera incl. fine hardy flowering shrubs, herbs or lianes

Capsella Medik. Cruciferae (15). 5 temp. (incl. Eur. 3), warm: diploids & ***C. bursa-pastoris*** Medik. (shepherd's purse) – recent allopolyploid (2n = 32) from *C. orientalis* Reuter (Russia, 2n = 16) & *C. grandflora* (Fauché & Chaub.) Boiss. (C Italy, W Greece, 2n = 16), now cosmop. weed (cf. *Poa annua*), poss. second most common (after *Polygonum aviculare* – Al-Shehbaz), usu. self-poll., seed mucilage attracting & trapping mosquito larvae & nematodes with a toxin, which kills them, & proteases, the resultant amino-acids absorbed by seeds; spinach subs. high in vitamins, seeds made into flour by Native Americans (also medic.), in TCM used in eye disease & dysentery, Herba Bursae Pastoris form. used as diuretic, febrifuge etc., coll. seeds found at Catal Huyuk (5950 BC) & in stomach of Tollund Man

Capsicodendron Hoehne = *Cinnamodendron*

Capsicophysalis (Bitter) Averett & M. Martínez (~ *Chamaesaracha*). Solanaceae (IV 5d). 1 Mex. & C Am.: ***C. potosina*** (Robinson) Averett & M. Martínez. R: JBRIT 3(2009)72

Capsicum Tourn. ex L. Solanaceae (IV 6). c. 32 trop. Am. R: J. Andrews (1995) *Peppers* ed. 2: 44. Leeuwenberg's Model. Coloured nectar. Fr. a many-seeded berry with twice as much vitamin C as citrus (up to 340 mg per 100 g) but also capsaicinoids deterrent to most animals (& *Fusarium* – microbial defence), attractive to disp. birds (esp. curve-billed thrasher), now 25% (c. 7 M t p.a.) of world spice market (cf. *Piper nigrum* 17%). 'Heat' (form. measured in Scoville units – 16 M = pure capsaicin (hottest cv, '**Carolina Reaper**', 1.57 M+); police pepper spray 5.3 M – but now p.p.m.; water with 1 part c. in 11 M distinctly pungent!) due to stimulation of heat-detecting receptors, best countered by casein which breaks bonds between receptor & capsaicin (a protein), which was thrown by Mayans in battle (cf. pepper spray today) & is effective in creams to dull pain of arthritis, rheumatism & neuralgia, lowers blood pressure (at least in rats), reduces size of pancreatic & prostate cancers, etc., & form. added to beer to give 'strength' or 'bite'. 4 spp. widely cult. in trop. Am. (R: T 18(1969)277), where used for at least 6000 yrs, & 2 elsewhere: ***C. annuum*** L. var. ***annuum*** (most of the cult. peppers) with 5 main groups of cvs, some of the first 3 also grown as orn.: **Cerasiforme Group** (cherry p.) – fr. small, v. pungent, **Conoides Group** (cone p.) – fr. usu. erect, ± conical, **Fasciculatum Group** (red cone p.) – fr. erect, slender, red, clustered, v. pungent, **Grossum Group** (bell p. [US], green p., sweet p., pimento) – fr. large, thick-skinned, ± bell-shaped, with depression at base, scarcely pungent (the principal salad peppers, red, green, yellow, white or black), **Longum Group** (cayenne [prob. from Tupi word, not after C. city] p., chilli p.) – fr. usu. drooping, to 30 cm long, v. pungent, the source of chilli powder, cayenne pepper (50 000 Scoville units), paprika (esp. Medit. & in Hungarian goulash) also pickled e.g. fefferoni (Croatia); ***C. annuum*** var. ***glabriusculum*** (Dunal) D'Arcy (var. *minimum*, bird pepper) incl. the wild or spontaneous forms in Am.; ***C. chinense*** Jacq. – cvs incl. 'habaneros' such as 'Carolina Reaper'; ***C. frutescens*** L. (a name much used for forms of *C. annuum*) – the source of Tabasco (5000 Scoville units) & other hot sauces, '**Naga Jolokia**' (bhut jolokia) – tiny (1 M Scoville units) allegedly *C. frutescens* × *C. chinense*), dev. for pepper spray in Ind.; ***C. pubescens*** Ruíz & Pavón – vogue fr. for sauces etc., comm. in Mal.

Captaincookia Hallé = *Ixora*

capucin *Northia hornei*

capulin *Prunus serotina*; **c. cherry** *P. salicina*

Capurodendron Aubrév. Sapotaceae (Tseb.). 23 Madag. R: FMad. 164(1974)68

Capuronetta Markgraf = *Tabernaemontana*

Capuronia Lourteig (~ *Galpinia*). Lythraceae. 1 Madag.: ***C. benoistii*** (Leandri) P. Berry (*C. madagascariensis*). R: CRHSAS 251(1960)1033

Capuronianthus J. Leroy. Meliaceae (I). 1 N & 1 S Madag. R: Adansonia 16(1976)174

Caputia R. Nordenstam & Pelser (~ *Senecio*). Compositae (Sen.-Sen.). 5 S Afr. R: CN 50(2012)59. Succ.

Caracasia Szyszyl. Marcgraviaceae. 2 Venez. C free, A 3

Caragana Fabr. Leguminosae (III 24). Incl. *Calophaca*, *Halimodendron*, c. 100 E Eur. (4), C As. to China. R: NSL 32(2000)76, PhytoKeys 70(2016)126. Cult. orn. & imp. fuel in treeless country. ***C. arborescens*** Lam. (Siberia, C As., Mongolia, Manchuria) – windbreak, bark used for ropes, young pods ed.; ***C. halodendron*** (Pall.) Dum. Cours. (*H. h.*, Eur. to C As.) – salt-steppe shrub with persistent spine-tipped leaf rachides, cult. orn.; ***C. pygmaea*** (L.) Dum. Cours. (NW China, Siberia) – oft. procumbent, cult. orn.; ***C. sinica*** (Buchoz) Rehder (*C. chamlagu*, N China) – fls ed.; ***C. spinosa*** (L.) Hornem. (Mongolia, China) – spiny branches form. stuck on tops of walls (cf. broken glass) nr. Beijing

caraguata fibre *Bromelia serra*; *Eryngium pandanifolium*

Caraipa Aubl. Calophyllaceae (Guttiferae (I1). c. 28 trop. S Am. R: MNYBG 29(1978)97. Lvs in spirals. Timber & medic. balsam, oils etc.

Carajasia Selas & al. Rubiaceae (IV 15). 1 Braz.: ***C. cangae*** Selas & al. R: Phytotaxa 206(2015)16

Carallia Roxb. Rhizophoraceae (II). 11 Madag., Indomal., trop. Aus. Massart's Model. Some timber for furniture, flooring etc. ***C. brachiata*** (Lour.) Merr. '**Honiara**' – fastigiate cult. orn.

Caralluma R. Br. Apocynaceae (Vb; Asclepiadaceae III 5). Incl. *Frerea*, excl. *Desmidorchis*, *Monolluma*, c. 34 Medit. (Eur. 2), Macaronesia to Somalia & NE Tanzania to Myanmar. R: F. Albers & U. Meve, *Ill. Handb. Succ. Pls*, Asclep. (2002) 46. Succ. cult. orn.; stems, fls &

fr. ed. S Arabia; pregnane glycosides antitumour activity etc., already in Ayurvedic med. (chungah) so over-exploited for med. (& food) in Pakistan. ***C. adscendens*** (Roxb.) R. Br. var. ***fimbriata*** (Wall.) Gravely & Mayur. (Ind., Myanmar) allegedly weight-loss agent (cf. *Hoodia*); ***C. frerei*** G. Rowley (*F. indica*, NW Ind.) – only leafy stapeliad, endangered in wild, but prop. in cult.

caramba or **carambola** *Averrhoa carambola*

Caramuri Aubrév. & Pellegrin = *Pouteria*

Carapa Aubl. Meliaceae (I-Xyl.). (3–) 27 trop. Afr. (16), Am. (11). Extrafl. nectaries esp. petiolar. ***C. grandiflora*** Sprague (Afr.) – seeds hoarded by rodents (also ***C. procera*** DC. in Am.); ***C. guianensis*** Aubl. (trop. Am.) – fls every 5 yrs, good timber (coondi, crabwood, bastard mahogany, andiroba, tallicona), seeds fish-bait & source of oil (andiroba) for lamps, soap- & candle-making, medic. (arthritis, throat infections in Braz.), insect-repellent & form. for shrinking heads

Carapichea Aubl. (~ *Psychotria*). Rubiaceae (IV 7). 23 trop. Am. R: AMBG 99(2013)113. ***C. ipecacuanha*** (Brot.) L. Andersson (*P. i.*, ipecacuanha, Braz.) – form. much coll. Mato Grosso, cult. Mal. for dried rhiz., used medic. (alk. = emetine), esp. as expectorant & for amoebic dysentery

caraway *Carum carvi*; **c. thyme** *Thymus herba-barona*

Cardamine Tourn. ex L. Cruciferae (16). (Excl. *Cardaminopsis*, incl. *Dentaria*, *Iti*, *Loxostemon*) c. 200 temp. (Eur. 31) incl. trop. mts (Afr., NG), native on all continents save Antarctica. R: JAA 69(1988)92. Some cult. orn. & weeds (bittercress) & some watercress subs. ***C. amara*** L. (Euras.) – lowest chromatin amount per nucleus known; ***C. bulbifera*** (L.) Crantz (*Dentaria b.*, Euras.) – cult. orn., axillary bulbils; ***C. chenopodiifolia*** Pers. (S Am.) – ann., amphicarpous, the geocarpic fr. with adv. r. on pedicel; ***C. corymbosa*** Hook.f. (? NZ, natur. NW N Am.) – weed in UK since 1975, leaflets rooting to form new pls; ***C. diphylla*** (Michaux) Alph. Wood (E N Am.) – rhiz. ed., local medic.; ***C. hirsuta*** L. (hairy bittercress, N temp.) – noisome ann. weed (usu. hairless!); ***C. oligosperma*** Nutt. (~ *C. hirsuta*, shot weed, W N Am.) – tiresome weed; ***C. pratensis*** L. (cuckoo flower, lady's smock, meadow cress, spinks, N temp.) – adv. buds on basal lvs, 2n = 16, 24 etc., 73–96; ***C. trifida*** (Poir.) B.M. Jones (E Eur. to China) – underground tuberiform leaf-blades for storage

Cardaminopsis (C. Meyer) Hayek = *Arabidopsis*

cardamom *Elettaria cardamomum*; also loosely applied to *Aframomum* & *Amomum* spp. (q.v.)

Cardaria Desv. = *Lepidium*

Cardenanthus R. Foster. = *Mastigostyla*

Cardenasiodendron F. Barkley. Anacardiaceae (I). 1 Bolivia: ***C. brachypterum*** (Loes.) F. Barkley

Cardiacanthus Nees & Schauer = *Carlowrightia*

Cardiandra Sieb. & Zucc. = *Hydrangea* (but see JJB 60(1985)139, 161)

cardillo *Scolymus hispanicus*

cardinal climber *Ipomoea sloteri*; **c. flower** *Lobelia cardinalis*; **c. wood** *Brosimum rubescens*

Cardiochilos Cribb. Orchidaceae (V 16d). 1 Malawi, S Tanzania: ***C. williamsonii*** Cribb

Cardiochlamys Oliv. Convolvulaceae (2). 2 Madag.

Cardiocrinum (Endl.) Lindl. Liliaceae. 3 Himal., E As. Bulbs hapax. but reprod. by offsets, lvs cordate. Cult. orn. esp. ***C. giganteum*** (Wall.) Makino (Himal.) to 3.5 m tall

Cardiogyne Bur. = *Maclura*

Cardiomanes C. Presl = *Hymenophyllum*

Cardionema DC. Caryophyllaceae (I 2). 6 W N Am. to Chile

Cardiopetalum Schldl. Annonaceae (III 1). Excl. *Froesiodendron*, 3 trop. S Am. R: Britt. 47(1995) 259

Cardiophyllarium Choux = *Doratoxylon*

Cardiopteridaceae Blume. Magnoliidae – Aquifoliales. 5/45 trop. R [p.p.]: FM 1,7(1972)93. Evergreen trees & twining lianes, s.t. with milky latex (*Cardiopteris*); accumulating aluminium. Lvs in spirals, s.t. appearing distichous, lobed or not; stip. 0. Fls. small, bisex. or not (andromonoec., dioec.), in axillary to cauliflorous (*Pseudobotrys*) spikes or fascicles to circinnate panicles or thyrses, K ((4)5), persistent, C ((4)5), A (4)5 inserted in C-tube (exc. *Citronella*) alt. with lobes, anthers with longit. slits, disk 0, $\underline{G}$ (3), 1-loc. with 0–2 styles (in *Cardiopteris* 1 elongate persistent in fr., the other decid.) & 2 pendent anatr. to orthotr. unitegmic (ategmic in *Cardiopteris*) ovules. Fr. a drupe or 2-winged samara, ± stipitate; seed 1 with minute embryo & fleshy endosperm

Genera: *Cardiopteris*, *Citronella*, *Gonocaryum*, *Leptaulus*, *Pseudobotrys*

Form. thought nr Icacinaceae (Icacinales) to which many genera form. referred (v. similar pollen)

Cardiopteris Wall. ex Royle. Cardiopteridaceae. 3 SE As., Mal. to Solomons. Twining herbs with latex & samaras. Lvs used as veg.

Cardiopterygaceae Tieghem = Cardiopteridaceae

Cardiospermum L. Sapindaceae (IV 1). 15 trop. Am., 3 ext. to Afr., 1 pantrop. weed. Climbers with inflated balloon-like frs, cult. orn. esp. ***C. halicacabum*** L. (balloon vine, trop. Am.) – pantrop. weed cult. in Amaz. for seeds worn in small bands by men to ward off snakebite, allegedly aphrodisiac in S As.; ***C. grandiflorum*** Sw. (heartseed, trop. Am. & Afr., invasive Aus., S Afr.) – black seeds used as beads (heart shape thereon), lvs a veg.

Cardioteucris C.Y. Wu = *Rubiteucris*

cardol *Anacardium occidentale*

Cardonaea Aristeg., Maguire & Steyerm. = *Gongylolepis*

cardoon *Cynara cardunculus*

Cardopatium Juss. Compositae (Card.-Card.). Excl. *Cousiniopsis*, 1 Med. incl. Eur.: ***C. corymbosum*** (L.) Pers.

Cardosoa S. Ortiz & Paiva (~ *Anisopappus*). Compositae (Athro.). 1 Angola: ***C. athanasioides*** (Paiva & S. Ortiz) S. Ortiz & Paiva. R: AJBM 67(2010)8

Carduncellus Adans. (~ *Carthamus*). Compositae (Card.-Cent.). 27 Med. esp. W

Carduus Vaill. ex L. Compositae (Card.-Card.). c. 120 Euras. (Eur. 48), Medit., E Afr. mts. R: MBSM (1963)139, (1964)279. Differs from *Cirsium* in minutely barbellate, non-plumose pappus. ***C. acanthoides*** L. (Euras.) – invasive USA; ***C. nutans*** L. (musk thistle, Euras., natur. N Am.), 'Scotch thistle' – ed. thick pith when boiled, dried fls form. used to curdle milk; ***C. pycnocephalus*** L. (Euras., Medit.) – invasive Aus., W US

Cardwellia F. Muell. Proteaceae (V 4). 1 Queensland: ***C. sublimis*** F. Muell. (silky oak) – fine cabinet wood. R: FA 16(1995)358

Carenidium Baptista = *Gomesa*

Carex L. Cyperaceae (IV). Incl. *Kobresia* (imp. pasture pls in montane Russia), *Schoenoxiphium*, *Uncinia* (bastard or hook grass; NZ); disp. facilitated by hook from infl. axis projecting beyond utricle), c. 2100 cosmop. esp. temp. & cold (Eur. 180, China 527, Jap. 202, NZ 80, N Am. 480). R: CJB 68(1990)1405. Perennials (some short-lived) with rhiz. (s.t. vertical) usu. wet places, where oft. (co-)dominant as in Arctic tundra (***C. parvula*** O. Yano [*K. pygmaea*, Himal.] over vast areas, yak pasture in Tibet); in E N Am. up to 20 spp. in a few ha of forest), ***C. nivalis*** Boott to 5475 m in Karakorum. Wind-poll. infls (s.t. unisexual; 12 spp. dioec.) of female fl. (prob. a condensed infl. unit) reduced to naked G, encl. in an utricle, &/or male of 1–3 A in axil of a glume. Most primitive group is sect. *Vigneastra* with compound panicles of bisex. spikelets (trop. OW). Some mycorrhizal. ***C. arenaria*** L. (Euras., introd. US pre-1870) on sand-dunes with habit of *Ammophila*; ***C. capitata*** L. and ***C. macloviana*** Urv. bipolar (cf. *Anemone multifida*, *Koenigia islandica*, *Osmorhiza berteroi*), ***C. microglochin*** Wahlenb. & ***C. magellanica*** Lam. poss. only subsp. distinct polar populations; some with elaiosomes & ant-disp. ***C. brizoides*** L. (Eur.) used as veg. hair packing material, others somewhat medic. or used for hat-making etc.; ***C. kobomugi*** Ohwi (E As.) – used for sand-dune stabilization US

Careya Roxb. Lecythidaceae (II; Barringtoniaceae). 4 trop. As. Like *Planchonia* but embryo without cotyledons. ***C. arborea*** Roxb. (Ceylon or patana oak) almost only tree sp. in grassy patanas of Sri Lanka, seeds & young fr. ed., lvs used for silkworms

Carib grass *Urochloa polystachya*

Caribea Alain. Nyctaginaceae (7). 1 Cuba: ***C. litoralis*** Alain

Carica L. Caricaceae. 1 warm Am.: ***C. papaya*** L. (papaya, papaw, pawpaw, melon tree) – fast-growing (everleafing (13–15 new lvs per month) & -flowering for 8–9 months) cultigen (wild in Mex.?) with pachycaul habit (Corner's Model), palmate lvs & principal support in phloem fibres, usu. dioec. (females stable, 'males' s.t. andromonoecious; male fls open for 1 day, females for 7; male specific region on autosome – ?incipient sex chromosome) though sex changeable by damage, hormones etc., poll. by thrips in Afr., moths in Malaysia; cult. throughout trop. for ed. fr. eaten fresh & tinned (11 M t p.a.), icecream & chewing-gum flavour & for papain (an antibacterial protease, the only natur. pl. one in comm.) derived from scarifying unripe fr. & used as meat tenderizer (meat wrapped in lvs becomes tender too), to reduce cloudiness in beer, to shrinkproof wool & silk & in control of termites & in upmarket toothpastes & flatulence cures, lvs locally med. & comm. ointments for skin conditions incl. burns, seeds used to adulterate pepper, vermifuge in Mex. Cvs incl. 'strawberry p.' with red-orange mesocarp. R: Ernstia 10(2000)78.

Caricaceae Dumort. Magnoliidae – Brassicales. 6/34 trop. & warm Am., trop. Afr. *(Cylicomorpha)*. R: V.M. Badillo (1971) *Monografia de la familia* C. Trees, oft. sparsely branched & pachycaul, usu. prickly, rarely prostrate herbs *(Jarilla)*, dioec., rarely monoec., with well-dev. system of anastomosing, articulated laticifers. Lvs in spirals, oft. large, palmately veined & lobed to palmate, rarely otherwise; stip. 0 or spine-like. Fls. axillary, solit. or in cymes, rarely bisex., reg., K(5), P(5), the tube v. short in female fls, lobes convolute or valvate, A 5 or 5 + 5, attached to C-tube, distinct *(Carica)* or basally connate, anthers with longit. slits, $\underline{G}$ (5), 1-loc. with deeply intrusive parietal placentas or these meeting to give pluriloc. ovary with axile placentas, styles distinct (stigmas s.t. subpetaloid), ovules ∞, anatropous, bitegmic with ± enlarged funicle. Fr. a large berry (pepo); seeds ∞ with gelatinous sarcotesta, straight embryo & oily, proteinaceous endosperm. n = 9
Genera: *Carica, Cylicomorpha, Horovitzia, Jacaratia, Jarilla, Vasconcellea*
Fr. trees, esp. *Carica, Vasconcellea*

Carinavalva Ising. Cruciferae (37). 1 S Aus.: ***C. glauca*** Ising. R: Fl. S Aus. 1(1986)388

Cariniana Casar. Lecythidaceae (I). Excl. *Allantoma*, 7 trop. S Am. R: FN 21(1979)218. A somewhat 1-sided with small flies as poll. (see fam.). Fr. like *Lecythis* (monkey pots); valuable timber (abarco, albarco, bacu, jequitiba, jiquitiba) esp. ***C. pyriformis*** Miers (NE S Am.) – form. called Colombian mahogany & exported to Eur. See also *Allantoma*

Carionia Naudin = *Medinilla*

caripé *Licania octandra*

Carissa L. Apocynaceae (I h). 7 warm OW E to Aus. & New Caled. (1 – ***C. spinarum*** L. (*C. edulis*)). R: WAUP 01–1(2001)5. Prévost's Model. Oft. with branch thorns: grown for hedging & tart fr. (conkerberry, congaberry in Aus.). ***C. bispinosa*** (L.) Brenan (amatungulu, E & S Afr.) – hedges, fr. ed.; ***C. carandas*** L. (karamda, Indomal.) – hedges, fr. pickled, s.t. called Christ's thorn; ***C. macrocarpa*** (Ecklon) A. DC. (*C. grandiflora*, Natal plum, S Afr.) – hedge-pl., fr. large, sold in markets

Carissophyllum Pichon = *Tachiadenus*

Carlemannia Benth. Carlemanniaceae (Caprifoliaceae s.l.). 3 Indomal. mts

Carlemanniaceae Airy Shaw (~ Caprifoliaceae). Magnoliidae – Lamiales. 2/5 Indomal. Shrubs & perenn. herbs. Lvs opp., joined by a line across stem, toothed; stip. 0. Infls term. & axillary cymes. Fls 4- or 5-merous (heterostyly in *Silvianthus*). K 4 or 5 unequal, adnate to ovary; C(4,5), lobes imbr. or induplicate-valvate; A 2 inserted at middle of C-tube, with anthers connivent around style, latrorse; disk cylindrical or conical; G (2) with many anatr. bitegmic ovules. Fr a dry or fleshy capsule, 2- or 5-valved with persistent K. Seeds many, embryo small in endosperm
Genera: *Carlemannia, Silvianthus*
A 2 & G(2) with many ovules, etc. exclude C. from Caprifoliaceae

Carlephyton Jum. Araceae (VII 19). 4 NW Madag. (3 on limestone, 1 on granite). R: BJ 92(1972)10, Willd. 42(2012)216

Carlesia Dunn. Umbelliferae (III 8). 1 E China: ***C. sinensis*** Dunn

Carlina Tourn. ex L. Compositae (Card.-Carl.). (Incl. *Chamaeleon*) 29 Eur., Macaronesia, Medit., W As. R: FR 83(1972)213, Oest. Akad. Wiss. Math.-Nat. Kl., Denkshr. 127, 128 (1990, 1994). Leeuwenberg's & Scarrone's Models. ***C. acaulis*** L. (Eur., common in Alps) – typical form with sessile capitula, the bracts spreading out star-like in dry air to release cypselas, others with peduncles to 25 cm (subsp. ***caulescens*** (Lam.) Schuebler & Martens (subsp. *simplex*))

Carlowrightia A. Gray. Acanthaceae (III 2c). c. 25 SW US to Costa Rica, warm to arid. R: FN 34(1983), Britt. 40(1988)245

Carlquistia Baldwin (~ *Madia*). Compositae (Mad.-Mad.). 1 Calif.: ***C. muirii*** (A. Gray) Baldwin – exposed on granite rocks. R: Novon 9(1999)463

Carludovica Ruíz & Pavón. Cyclanthaceae. 4 trop. Am. mainland. R: AHB 18(1958)127. Almost stemless, palm-like, source of fibres esp. ***C. palmata*** Ruíz & Pavón (C Am. to Bolivia) grown for hat manufacture (Panamá hat plant or palm, toquilla) – 6 young lvs per hat, of which 4 M exported (form. via Panamá) from Ecuador per annum); older lvs used for mats, baskets etc., also cult. orn. Other spp. for thatching & brooms

Carmenocania Wernham = *Pogonopus*

Carmichaelia R. Br. Leguminosae (III 24). (Incl. *Chordospartium, Corallospartium, Notospartium*) 23 NZ, Lord Howe Is. (1). R: NZJB 36(1998)55. Photosynthetic flat branches without green lvs. ***C. crassicaulis*** Hook. f. (*Corallospartium c.*, coral broom, NZ); ***C. stevensonii***

(Cheeseman) Heenan (*Chordospartium s.*, NZ) – not reprod. by seed in wild but seedlings look dead for 2+ yrs. Some cult. orn. esp. ***C. enysii*** Kirk & dwarfer ***C. orbiculata*** Colenso (NZ)

Carminatia Moçiño ex DC. Compositae (Eup.-Alom.). 3 SW US to El Salvador. R: PSE 110(1988)169

Carmona Cav. = *Ehretia*

Carnarvonia F. Muell. Proteaceae (V). 1 Queensland: ***C. araliifolia*** F. Muell. R: FA 16(1995)343

carnation *Dianthus caryophyllus*

carnauba wax *Copernicia prunifera*

Carnegiea Britton & Rose. Cactaceae (III 1). 1 SW US, Mex.: ***C. gigantea*** (Engelm.) Britton & Rose (saguaro, giant cactus) – largest of all cacti, reaching 4–6 m in 75–100 yrs, then branching, eventually 20 m tall, 60 cm thick, with candelabriform branching, 12 t in wt. & alleged to live for 200 yrs, not thriving in cult., mature pl. absorbing c. 750 l of water after a storm; poll. by birds & insects by day, bats by night, hollowed by woodpeckers to make cool nests, then occupied by owls etc.; Native American house rafters in Arizona, fr. to 7.5 cm diam., ed., form. v. imp. as food & drink, still coll. & used in cerem. (see DP 2(1)(1980))

Carnegieodoxa Perkins = *Hedycarya*

carnival plant *Justicia scheidweileri*

caroá fibre *Neoglaziovia variegata*

carob *Ceratonia siliqua*

Carolina allspice *Calycanthus floridus*

Carolus W.R. Anderson (~ *Mascagnia*). Malpighiaceae. 6 trop. Am. R: Novon 16(2006)186

carom seeds *Trachyspermum ammi*

Caropodium Stapf & Wettst. = *Grammosciadium*

Caropsis (Rouy & Camus) Rauschert (*Thorella* Briq.). Umbelliferae (III 8). 1 Eur.: ***C. verticillato-inundata*** (Thore) Rauschert

Caroxylon Thunb. (~ *Salsola*). Amaranthaceae (Chenop.). c. 100 SW & C As. to S Afr. ***C. nitrarium*** (Pallas) Akhani & Roalson (*S. n.*, Russia, SW & C As.) – used as boron indicator in Russia

Carpacoce Sonder. Rubiaceae (IV 13). 7 S Afr. R: Strel. 9(2000)618

Carpanthea N.E. Br. Aizoaceae (V). 1 SW Cape: ***C. pomeridiana*** (L.) N.E. Br. R: BJ 111(1990)478. Ann., cult. orn.

Carparomorchis M.A. Clem. & D.L. Jones = *Bulbophyllum*

Carpentaria Becc. Palmae (V 14i). 1 N Aus.: ***C. acuminata*** (H.A. Wendl. & Drude) Becc. R: FA 39(2011)197

Carpenteria Torrey = *Philadelpus*

Carpesium L. (~ *Inula*). Compositae (Inul.-Inul.). 25 Euras. (Eur. 2), Indomal. to Aus. Pappus 0

carpet grass *Axonopus* spp.

Carpha Banks & Sol. ex R. Br. Cyperaceae (II 7). Excl. *Asterochaete*, c. 15 S Afr. C Afr. mts, Madag., S Jap., NG, Aus. (4), Chile

Carphalea Juss. Rubiaceae (IV 1). Excl. *Paracarphalea*, 12 trop. Afr. (8), Socotra (1), Madag. (3). R: BJBB 58(1988)271. K-lobes poll.-attractants, later for wind-disp.

Carphephorus Cass. Compositae (Eup.-Liat.). Incl. *Litrisa, Trilisa* 7 SE US. R: FNA 21(2006)535

Carphochaete A. Gray. Compositae (Eup.-Ager.). 7 SW N Am. R: Phytol. 64(1987)145

Carpinaceae Vest = Betulaceae

Carpinus Tourn. ex L. Betulaceae (II, Carpinaceae). 41 N temp. (Eur. 2) to C Am. mts (Pliocene of Alabama), esp. E As. R: JAA 71(1990)51. Decid. monoec. trees – good timber for turnery, tools etc. ***C. betulus*** L. (hornbeam, Eur. to Iran) – form. much used for mill cogwheels, ox-yokes etc., still imp. as mechanism between key & hammer in a piano, butchers' chopping-blocks, skittles, pulleys, riding-boot trees, draughtsmen, chessmen, dominoes, cult. orn. (several cvs) incl. as pleached hedging; ***C. caroliniana*** Walter (American h., blue beech, water b., E N Am.) – local medic., cult. orn.; ***C. turczaninowii*** Hance var. ***coreana*** (Nakai) W. Lee (Korea) – cult. orn., creeping

Carpobrotus N.E. Br. Aizoaceae (V). Excl. *Sarcozona*, 11 W Cape to Natal (7; R: CBH 15(1993)82), Aus. (4), Calif., Chile. Sour-figs (but fr. used for jam). Cult. orn. (most in Calif. hybrids involving up to 5 spp.), some natur. esp. on cliffs in Eur. incl. ***C. edulis*** (L.) L. Bolus (pigface, Sally-my-handsome (= corruption of mesembryanthemum,

Cornwall), Cape, invasive W Eur., Medit., Aus., W US, S Atlantic Is.) – C yellow to purple, seeds disp. mammals, acidifies soils, leaf-juice for dysentery, sore throats & smeared over newborn Khoikhoi (S Afr.) & sunburn (Scilly Is.), ed. fr. ('Hottentot fig') fresh, dried or in jam, though fr. of other Cape spp., e.g. ***C. deliciosus*** (L. Bolus) L. Bolus (*C. dulcis*) ed. fresh & ***C. muirii*** (L. Bolus) L. Bolus ed. dried, tastier; ***C. glaucescens*** (Haw.) Schwantes (S Afr.) – juice rubbed on jellyfish stings & midge bites (alk. with effect like cocaine); ***C. rossii*** (Haw.) Schwantes (karkalla, S Aus.) – lvs & fr. eaten by Aborigines, now comm.

Carpodetaceae Fenzl = Rousseaceae

Carpodetus Forst. & Forst. f. Rousseaceae (Grossulariaceae s.l.). 2 NG, Solomons, NZ (1). R: SB 10(1997)861

Carpodinopsis Pichon = *Pleiocarpa*

Carpodinus R. Br. ex G. Don = *Landolphia*

Carpodiptera Griseb. (~ *Berrya*). Malvaceae (Brown.; Tiliaceae). c. 6 E Afr. & Comores (1), trop. Am.

Carpolepis (Dawson) Dawson (~ *Metrosideros*). Myrtaceae (II 7). 3 New Caledonia

Carpolobia G. Don. Polygalaceae (III). 4 trop. Afr. & Madag. R: MLW 77 – 18(1977)22. Troll's Model. Some timber & ed. fr., ***C. lutea*** G. Don. stem a chewing-stick in Nigeria

Carpolyza Salisb. = *Strumaria*

Carpotroche Endl. Achariaceae (Flacourtiaceae). 11 trop. Am. R: FN 22(1980)31. Champagnat's Model. Some parasiticidal oils used in skin disease

Carpoxylon H. Wendl. & Drude. Palmae (V 14d). 1 Vanuatu: ***C. macrospermum*** H. Wendl. & Drude – form. thought almost extinct, fr. ed., lvs for brooms etc. R: Principes 33(1989)68.

Carptotepala Mold. = *Comanthera*

Carramboa Cuatrec. See *Espeletia*

Carria V. Castro & Lacerda = *Gomesa*

Carrichtera DC. (~ *Vella*). Cruciferae (12). 1 Macaronesia to Iran, incl. Eur.: ***C. annua*** (L.) DC. – invasive in Aus. (Ward's [Mr. W. of S Aus.] weed)

Carriella Castro & Lacerda = *Gomesa*

Carrierea Franch. Salicaceae (Flacourtiaceae). 3 S & SW China, SE As. ***C. calycina*** Franch. (China) fl. after 90–100 yrs in Co. Down, Ireland

Carrissoa Bak. f. (~ *Rhynchosia*). Leguminosae (III 18). 1 Angola: ***C. angolensis*** Bak.f.

Carronia F. Muell. Menispermaceae (I). 4 NG to NSW. R: KB 30(1975)94

carrot *Daucus carota*; **Candy c.** *Athamanta cretensis*

Carruanthus (Schwantes) Schwantes. Aizoaceae (V). 2 S Cape & Karoo. R: H. E. K. Hartmann, *Ill. Handb. Succ. Pls*, Aiz. A–E (2002)101. Cult. orn.

Carruthersia Seemann. Apocynaceae (II b). 4 Philippines (2) to Solomons (1), Fiji to Tonga (2)

Carsonia Greene = *Cleome*

Carterella Terrell (~ *Hedyotis*). Rubiaceae (IV 1). 1 Baja Calif.: ***C. alexanderae*** (A. Carter) Terrell. R: Britt. 39(1987)248

Carterothamnus R. King = *Oaxacania*

Carthamus Tourn. ex L. Compositae (Card.-Cent.). Excl. *Carduncellus*, *Femeniasia*, *Phonus*, 20 Medit. to C As. R: AJBM 47(1990)28. ***C. lanatus*** L. (saffron thistle, Medit. to C As.) – cosmop. weed; ***C. tinctorius*** L. (safflower, saffron thistle, cultigen closest ally *C. palaestinus* Eig ex Rech.f., Near E) – fls form. used in dyeing food & rouge (yellow & red) in decline after use of aniline dyes, potential oilseed crop (kurdee) with polyunsaturates incl. linoleic acid for soft margarines & monounsaturate for frying – oil already used in easing Crohn's disease, in cosmetics (lip-colour for Japanese geisha) & aromatherapy, poss. effective in relief of premenstrual tension, fr. used for poultry etc., now genetically modified to produce human insulin (correct folding cf. that from fungi etc.) & carp growth hormone for aquaculture feed for farmed shrimp

Cartiera Greene = *Streptanthus*

Cartonema R. Br. Commelinaceae (I 1; Cartonemataceae). 11 trop. Aus. ext. to SW NG

Cartonemataceae Pichon = Commelinaceae

Cartrema Raf. (~ *Osmanthus*). Oleaceae (4d). 2 S N Am. R: Phytoneuron 2012–96:1

Carum L. Umbelliferae (III 8). Excl. *Lomatocarum*, *Trocdaris*, 4 Caucasus & Near East incl. ***C. carvi*** L. (caraway) – cult. for fr. used as flavouring in bread, sauerkraut, cheese & 'seedcake' & liqueur (Kümmel), since time of Pliny (carvone now synth. for flavouring)

Carvalhoa K. Schum. Apocynaceae (I d). 1 E & SE Afr.: ***C. campanulata*** K. Schum. R: AUWP 85–2(1985)47

Carvia Bremek. = *Strobilanthes*

Carya Nutt. Juglandaceae. 18 E N Am. (11) to C Am., E As. (few). R: AMBG 65(1978)1080. Heterodichogamy recorded; many nat. hybrids, even crosses between diploids & tetraploids with viable seed. Hickories – timber (wood for most drumsticks) & nuts (pecans), strips of outer bark form. used in chair-seat-making & lacrosse sticks (now usu. plastic), milky liquid from seeds extracted by Native Americans, local. medic. ***C. cordiformis*** (Wangenh.) K. Koch (bitternut, E N Am.); ***C. glabra*** (Mill.) Sweet (pig-nut, hognut, N Am.) – kernels usu. astringent; ***C. illinoinensis*** (Wangenh.) K. Koch (pecan, S US) – common dessert nut, cult. US (80% world prod., esp. Georgia), Mex., Braz., Israel, S Afr., coll. Ilinwek, first 'Eur.' cv. selected 1846, now over 500 named, esp. thin-shelled ones, used like hazelnuts or walnuts in food (18% protein), oil (70%, unsaturated) used in cosmetics etc., timber poor, shells used in plastic manuf., Texas state tree (first to be designated – 1919) poss. living to 1000 yrs; ***C. ovata*** (Mill.) K. Koch (shagbark hickory, N Am.), & ***C. laciniosa*** (Michaux f.) W. Barton (E N Am.) & ***C. tomentosa*** (Lam.) Nutt. (mockernut, E N Am.) – nuts also used

Caryocar F. Allam. Caryocaraceae. 18 trop. Am. R: FN 12(1973), OB 92(1987)179. Bat-poll. Some timbers for ship-building. Drupes of some ed., others fish-poisons (saponins, prob. also effective against termites), seeds an oil-source, many disp. agoutis. ***C. brasiliense*** Cambess. (pequí, pequiá, piquiá, piquí, Braz.) – rheas form. imp. dispersers, fr. & seeds local soap & cooking oil sources, comm. liquor from fr.; ***C. amygdaliferum*** Mutis (suarí, swarri nut) – pleasant-tasting fat; ***C. glabrum*** (Aubl.) Pers. (soapwood, NE S Am.) – inner bark used for washing; ***C. microcarpum*** Ducke (Amaz.) – lvs repellent to leaf-cutting ants, fish-poison (saponins & tannins); ***C. nuciferum*** L. (Panamá to NE S Am.) – names & uses like *C. amygdaliferum*; ***C. villosum*** (Aubl.) Pers. (NE S Am.) – names like *C. brasiliense*, also oilseed

Caryocaraceae Voigt. Magnoliidae – Malpighiales. 2/27 trop. Am. esp. Amazonia. R: FN 12(1973). Everg. trees, rarely shrubs with triterpenoid saponins. Lvs opp. (*Caryocar*) or in spirals, trifoliolate, ± dentate; stip. 0, 2 or 4, caducous. Fls in term. racemes with artic. pedicels, bisex., reg., K 5(6), imbr., basally connate in *Caryocar*, or reduced & lobed (*Anthodiscus*), C 5(6), imbr., s.t. connate basally, connate above in *Anthodiscus*, forming calyptra, A 55–750, usu. shortly basally connate into ring or 5 bundles opp. C, inner ones s.t. without anthers, anthers with longit. slits, $\underline{G}$ (4 – 20) with as many locules & styles, each carpel with 1 basal, anatr. to orthotropous, bitegmic ovule. Fr. a drupe, the stone separating into 1-seeded pyrenes or merocarps; seeds reniform with thin or 0 endosperm, embryo with large oily & proteinaceous spirally twisted hypocotyl & 2 small cotyledons; germ. hypogeal. n = 23

Genera: *Anthodiscus, Caryocar*

Form. thought close to Theaceae (Ericales)

Caryocar bat-poll. despite diff. habitats (savanna, rain forest etc.), also sphingids. Some ed. seeds (some toxic – fish-poisons)

Caryodaphnopsis Airy Shaw (~ *Persea*). Lauraceae (I). 16 trop. As. (8) & Am. (8). R: ABY 13(1991)1. ***C. theobromifolia*** (A. Gentry) van der Werff & H. Richter (*P. t.*, Ecuador) – form. *the* 'mahogany' of Ecuador, now c. 12 trees left

Caryodendron Karsten. Euphorbiaceae (Acal.-Car.). 3–4 trop. S Am. ***C. orinocense*** Karsten (inchi, tacay nut, Colombia) – seeds ed. roasted, poss. oil-crop

Caryolobis Gaertn. = *Shorea*

Caryomene Barneby & Krukoff. Menispermaceae (II). 4 trop. Am. R: MNYBG 22, 2(1971)52

Caryophyllaceae Juss. Magnoliidae – Caryophyllales. 96/2500 cosmop. (*Colobanthus* only 'dicot' to penetrate Antarctic Circle) esp. temp. & warm N hemisph. Herbs, rarely shrubs, lianes (e.g. *Schiedea*) or even small trees (*Sanctambrosia*), s.t. xeromorphic, usu. monoec., oft. with swollen nodes & s.t. with unusual sec. thickening with concentric rings of xylem & phloem; anthocyanins present, sieve-tube plastids typical of order. Lvs opp., rarely in spirals, simple, entire; stip. 0 (exc. some Paronychioideae). Fls usu. reg., hypogynous (rarely not, as in *Scleranthus*, where perigynous), in dichasial cymes or solit. K ((4)5), completely free in some Alsinoideae, C 0 (many Paronychioideae), (4)5 (small & bifid in many Alsinoideae, clawed & usu. with large blade in Caryophylloideae), A (1–4) 5–10 in 1 or 2 whorls, s.t. basally adnate to C to form short tube that may be adnate to gynophore, or inserted at edge of nectary-disk around ovary, or adnate to K, anthers with longit. slits, $\underline{G}$ (2 – 5(+)) with ± united styles, oft. surmounting gynophore, 1-loc. distally but ± partitioned proximally, placental column reaching apex or not, ovules usu. ∞, s.t. >1, bitegmic, hemitropous to campylotropous (usu.). Fr. a capsule with as many or twice as many valves or apical teeth as styles, rarely indehiscent (berry, nutlet or achene); seeds small, usu. ornamented on testa, embryo usu. peripheral, curved around copious starchy perisperm, less oft. straight, endosperm little or 0. x = 5 – 19

Trad. classification (not supported by recent molecular work) & chief genera:

I. **Paronychioideae** (stip.; C small or 0; a basal 'grade')
  1. **Polycarpeae** (lvs opp.; capsule): *Drymaria, Polycarpaea, Spergularia*
  2. **Paronychieae** (lvs opp.; nutlets; this & 3. s.t. separated as Illecebraceae): *Herniaria, Paronychia*
  3. **Corrigioleae** (lvs alt.; prob. 'sister' to rest of fam.): *Corrigiola*

I. **Alsinoideae** (stip. 0; K free)
  1. **Alsineae** (episepalous A usu. with nectariferous gland at base, styles free; capsule usu. with many seeds or nutlet): *Arenaria, Cerastium, Minuartia, Sagina, Stellaria*
  2. **Pycnophylleae** (styles at least basally connate, fr. indehiscent 1-seeded): *Pycnophyllum* (only)
  3. **Geocarpeae** (C 0, 5 vestigial episepalous 'staminodes'; capsule): *Geocarpon* (only)
  4. **Habrosieae** (C minute, styles free, G 2-ovulate; fr. indehiscent, 1-seeded): *Habrosia* (only)
  5. **Scleranthеae** (C 0, styles 2 free, G 1-ovulate, fr. indehiscent): *Scleranthus*

I. **Caryophylloideae** (stip. 0; (K))
  1. **Caryophylleae** (fls reg., styles 2(3); capsule with 4(6) teeth): *Acanthophyllum, Dianthus, Gypsophila, Saponaria*
  2. **Drypideae** (fls ± irreg., styles 3): *Drypis* (only)
  3. **Sileneae** (fls reg., styles 3 – 5, capsule with teeth = A or 2 x A): *Silene*

With Molluginaceae, differ from rest of order in anthocyanin prod. C supposed to be of staminodal orig. (Cronquist)

Many cult. orn. herbs esp. *Dianthus, Gypsophila, Saponaria, Silene*; many weeds esp. *Cerastium, Sagina, Stellaria*

Caryopteris Bunge. Labiatae (III). 7 E As. Cult. orn. esp. ***C.*** × ***clandonensis*** Rehder (*C. incana* (Houtt.) Miq. (China, Jap.) × *C. mongholica* Bunge (N China) 1930s England) – blue-flowered shrub oft. dying back in winter

Caryota L. Palmae (III 6). 13 Indomal. to trop. Aus. Only palms with bipinnate lvs, with fish-tail-like leaflets, flowering from top to bottom & dying (Holttum's Model), fr. filled with irritant crystals but disp. by animals, some cult. orn. e.g. ***C. maxima*** Blume (*C. aequatorialis*, Malay Pen.) – hapaxanthic to 35 m tall & fls with A over 100, ***C. mitis*** Lour. (Indomal.), 'fishtail palm', but prob. hybridize (taxonomy unclear). ***C. urens*** L. (kitul (kittool), toddy palm, Indomal.) – widely planted, somewhat 'weedy' in wild, a source of sago (famine-food in e.g. Myanmar), palm-sugar, toddy (7–14 l sap per day from 1 infl., up to 27 l per tree if more than 1 infl. tapped), arrack, timber & kitul (kittool) fibre (Ceylon piassava) derived from leaf-bases & used as a brush-fibre, fodder for domestic elephants, young lvs ed., mature ones exported to Eur. for cut foliage

Caryotophora Leistner = *Skiatophyllum*

casabanana *Sicana odorifera*

Casabitoa Alain = *Picramnia*

casana *Solanum cajanumense*

Casasia A. Rich. Rubiaceae (II 1). 11 C Am., WI. G 1-loc. R: JAA 68(1987)176

Cascabela Raf. = *Thevetia*

Cascadia A.M. Johnson (~ *Saxifraga*). Saxifragaceae (I 1). 1 NW Calif. to Washington: ***C. nuttallii*** (Small) A.M. Johnson – 'sister' to *Saxifragoides* (Chile!), seeds spiny

cascalote *Libidibia coriaria*

cascara (sagrada) *Frangula purshiana*

cascarilla (bark) *Croton eluteria*

Cascarilla (Endl.) Wedd. = *Ladenbergia*

Cascaronia Griseb. Leguminosae (III 11). 1 Bolivia, Argentina: ***C. astragalina*** Griseb.

Casearia Jacq. Salicaceae (Samydaceae, Flacourtiaceae). Incl. *Laetia, Samyda*, 215 trop. (94 Am. – R: FN 22(1980)225,237,280). Roux's Model. Some timbers; main branches with scale lvs, short-lived twigs with foliage lvs; pellucid glands in lvs. ***C. praecox*** Griseb. (*Gossypiospermum p.*, WI, Colombian or Maracaibo boxwood, zapatero, Venez., WI) – all trees in 1 pop. fl. at once & just for 1 day, principal boxwood largely replacing *Buxus sempervirens* in comm. for rules, veneers, carving, key-boards etc.; ***C. sylvestris*** Sw. (trop. Am.) – seeds a source of an oil like chaulmoogra oil; ***C. tomentosa*** Roxb. (*C. elliptica*, Ind.) – fish-poison, wood for carving

cashew nut *Anacardium occidentale*

Casimirella Hassler. Icacinaceae. 7 trop. Am. R: Britt. 44(1992)166. Some with tubers ed. when cooked (*Humirianthera*)

Casimiroa Llave. Rutaceae (I). c. 10 C Am. highlands to Texas. Alks. ***C. edulis*** Llave (white sapote, Mex. apple, Mex.) – variable, cult. for ed. fr., bitter-sweet flavour, used in milk shakes, icecream etc.

cassabanana *Sicana odorifera*

Cassandra D. Don = *Chamaedaphne*

cassareep see *Manihot esculenta*

cassava *Manihot esculenta*

Cassebeera Kaulf. (~ *Doryopteris*). Pteridaceae (IV). 2 trop. Am.

Casselia Nees & Mart. Verbenaceae (Cass.). 6 S Am. R: JTBS 137(2010)167

Cassia Tourn. ex L. Leguminosae (I 3). Excl. *Chamaecrista* & *Senna* (segregation supported by floral ontogeny studies), q.v., 32 trop. (*Cassia* s.s. with 3 adaxial A sigmoidally curved; Afr. 10 (R: KB 43(1988)334), Am. 12–13). Medic. & cult. orn. trees (Troll's Model), some timbers: ***C. abbreviata*** Oliv. (trop. Afr.) – antibacterial, trad. medic. for diarrhoea; ***C. brewsteri*** (F. Muell.) Benth. (NE Aus.) – pink hardwood; ***C. fistula*** L. (pudding-pipe tree, purging cassia, Ind. laburnum, golden shower, amaltas, tara, trop. As.) – widely cult., fr. to 60 cm with seeds embedded in laxative pulp (c. 1200 t p.a.), used against habitual constipation; ***C. grandis*** L.f. (horse c., trop. Am.) – cult. orn. with pink fls, medic.; ***C. javanica*** L. (Mal.) – cult. orn. with red fls, timber beautifully marked & used in house-building in Java, tanbark, fr. purgative, subsp. ***nodosa*** (Buch.-Ham.) Larsen & S. Larsen (var. *indochinensis*, *C. nodosa*, Indomal.) with similar uses; ***C.*** × ***nealiae*** Irwin & Barneby (rainbow shower, *C. fistula* × *C. javanica*) – cult. orn. raised Hawaii

cassia bark *Cinnamomum aromaticum*; **golden c.** *Chamaecrista fasciculata*; **horse c.** *Cassia grandis*; **purging c.** *C. fistula*; **ringworm c.** *Senna alata*; **tanner's c.** *S. auriculata*

Cassidispermum Hemsl. = *Burckella*

cassie *Vachellia farnesiana*

Cassine L. Celastraceae (I). 3 S Afr. R: AJB 63(1997)148. Lvs opp. See also *Elaeodendron*, *Lauridia*

Cassinia R. Br. Compositae (Gnap.-Cass.). c. 40 Aus. R: OB 104(1991)90. Poss. incl. *Calomeria*, *Haeckeria*, *Ozothamnus*. Some cult. orn., some allergenic

Cassinopsis Sonder. Icacinaceae. 6 Afr., Madag. Lemon thorn. Attims's Model

Cassiope D. Don. Ericaceae (IV). (Excl. *Harrimanella*) 18 Himal., circumboreal (Eur. 2, N Am. 3). R: Plantsman 11(1989)106. Cult. orn. ***C. tetragona*** (L.) D. Don much used as fuel by Inuit

Cassipourea Aubl. Rhizophoraceae (I). Incl. *Dactylopetalum*, 62 trop. Am., trop. & S Afr., Madag., Sri Lanka. R: KB 1925(1925)241. Roux's Model. G Alks. ***C. elliottii*** (Engl.) Alston (pillarwood, E Afr.) – comm. timber

cassumar (ginger) *Zingiber montanum*

Cassupa Bonpl. = *Isertia*

Cassytha Osbeck ex L. Lauraceae (I). 23 OW trop. esp. Aus. (19, 16 endemic; R: JABG 3(1981)187). Alks. Parasites (dodder-laurels) with habit (Roux's Model) of *Cuscuta*, scale-like green lvs soon falling. ***C. filiformis*** L. (pantrop. but poss. introd. Am.) – astringent & diuretic, source of a brown dye in E Afr., stems form. for mattresses in N Queensland, fr. ed.

Cassythaceae Bartling ex Lindl. = Lauraceae (I)

Castalis Cass. = *Dimorphotheca*

Castanea Mill. Fagaceae. 8 N temp. (Eur. 1 – ***C. sativa*** Mill. (common, sweet [form. 'sugar chestnut' as on ripening starch becomes sugar – granulated sold] or Spanish chestnut, E Medit. to N Iran, introd. to GB by Romans – meal (pollenta) among relics of R. soldiers), cult. orn. (to 2000 yrs old on Mt Etna (legally prot. since 1745); girth to 16 m in Kent, greatest of any tree in GB, but at 25°C under long days merely a bush), for nuts (starchy, oft. sold roasted as in London's winter streets, the flour still used in cooking esp. Italy, candied (marrons glacés)) & timber (but 'chestnut' of Med. buildings usu. oak), esp. coppice for fencing & gates, also form. walking-sticks ('congo sticks'); bark used in tanning). R: A. Camus, *Les Châtaigniers* (1929)11. Entomophily; others planted incl.: ***C. crenata*** Sieb. & Zucc. (Jap. chestnut, C & S Jap. – mass planting 5000 yrs ago) – timber & nuts, many French chestnut cvs are crosses with *C. sativa*; ***C. dentata*** (Marshall) Borkh. (American c., N Am.) – form. imp. timber with best nuts (also used in a 'coffee' & local medic. by Iroquois) & assoc. with Ind. village-sites (planted), now almost extinct through chestnut blight introd. from As. in late 1800s, other (As.) spp. now tried in US to replace it in commerce incl. ***C. mollissima*** Blume (Chinese chestnut) – resistant to blight (as are hybrids with *C. d.*), cult. China as long as *C. sativa* in W, catkins for lamp-wicks; ***C. pumila*** (L.) Mill. (chinquapin, C & E US) – timber for railway sleepers, seeds ed.

Castanadia R. King & H. Robinson (*Castenedia*; ~ *Eupatorium*). Compositae (Eup.-Crit.). 1 Colombia: ***C. santamartensis*** R. King & H. Robinson. R: MSBMBG 22(1987)354

Castanella Spruce ex Hook.f. = *Paullinia*

castanha de cutia *Acioa edulis*

Castanopsis (D. Don) Spach. Fagaceae. 120 trop. & warm As. (esp. Borneo). For Am. spp. see *Chrysolepis*. No clear diffs from *Castanea*; seeds roasted like chestnuts; some cult. orn. ***C. acuminatissima*** Blanco (Indomal.) – seeds ed. raw or cooked NG; ***C. argentea*** (Blume) A. DC. (Indomal.) – timber, dyebark, ed. seeds; ***C. cuspidata*** (Thunb.) Schottky (Jap., Korea) – ed. seeds, much planted in Jap. parks & gardens, lvs form. used as rice 'bowls'

Castanospermum A. Cunn. ex Mudie. Leguminosae (III 2). Incl. *Alexa*, 10 trop. Am. (R: AMBG 82(1995)551), 1 NE Aus., W New Br., Vanuatu: ***C. australe*** A. Cunn. ex Mudie (black bean, Moreton Bay or Aus. chestnut) – coastal forest & beaches, seeds ed. (black beans roasted, poisonous if raw; used in prostate cancer, screened as poss. AIDS vaccine as leucocyte movement to inflammation sites from blood vessels inhibited by castanospermine, poss. by interfering with the ability to bind with heparanase, a key part of invasion), street-tree, decorative timber, pots of seedlings sold as 'lucky beans' in Eur.

Castanospora F. Muell. Sapindaceae. 1 NE Aus.: ***C. alphandii*** (F. Muell.) F. Muell.

Castela Turpin. Simaroubaceae. Excl. *Holacantha*, 12 trop. & warm Am., Galápagos. Some med. R: TSDSNH 15,4(1968)31

Castelia Cav. = *Pitraea*

Castellanoa Traub = *Chlidanthus*

Castellanosia Cárdenas (~ *Browningia*). Cactaceae (III 1). 1 Bolivia, W Paraguay: ***C. caineana*** Cárdenas

Castellia Tineo. Gramineae (XVI 10). 1 Macaronesia, Medit. (inc. Eur.), N Afr. to Pakistan: ***C. tuberculosa*** (Moris) Bor

Castelnavia Tul. & Wedd. Podostemaceae (III). 5 Braz., 1 ext. to Bolivia. R: SB 34(2009)723

Castenedia R. King & H. Robinson = *Castanedia*

Castilla Sessé. Moraceae (III). 3 trop. Am. R: FN 7(1972)92. Form. imp. rubber sources now supplanted by *Hevea*. ***C. elastica*** Sessé (C Am., Panamá or Ulé rubber, C Am.) – Roux's Model, androdioec., thrips-poll., source of the rubber balls Columbus saw, ***C. ulei*** Warb. (uli, Amazon) etc.

Castilleja Mutis ex L.f. Orobanchaceae ('Cast.'; Scrophulariaceae s.l.). c. 190 (E N Am. 3, Euras. 5 (Eur. 3), C Am. 8, Andes 5, rest W N Am.) – Ind. paintbrush. Hemiparasitic; upper lvs brightly coloured making infl. conspicuous – some local medic., contraceptives, cult. orn. ***C. densiflora*** (Benth.) Chuang & Heckard (*Orthocarpus d.*, W N Am.) – seeds disp. with fr. of introd. host-pl., *Hypochaeris glabra* L.; ***C. integra*** A. Gray (Mex.) – iridoids sequestered by *Euphydryas anicia* butterflies

Castilloa Endl. = *Castilla*

castor oil *Ricinus communis*

Castratella Naudin. Melastomataceae. 2 NE S Am.

Castrilanthemum Vogt & Oberprieler (~ *Pyrethrum*). Compositae (Anth.-Leucanthem.). 1 SE Spain: ***C. debeauxii*** (Degen & al.) Vogt & Oberprieler. R: AJBM 54(1996)342

Castroa Guiard = *Gomesa*

Castroviejoa Galbany & al. Compositae (Gnap.). 2 Balearics

Casuarina L. Casuarinaceae. 17 SE As. to W Pac. (see also *Allocasuarina, Gymnostoma*). Timbers & fuelwoods – ironwood, she-oak, swamp oak, actinomycete r.-symbionts (*Actinobacteria*) fixing nitrogen; dioec. exc. ***C. equisetifolia*** L. (Aus. pine, beefwood, jau, whistling pine, yar, Indomal. & widely planted e.g. long-established on E Afr. coast, invasive in SE As., S Afr., Carib.) – pioneer seashore tree (Attims's Model) good for hedging incl. topiary & windbreaks, timber used for shingles, fencing etc., burns with great heat ('best firewood in the world'), tanbark (Madag.). Hybrids in Florida but not in Aus. ***C. cunninghamiana*** Miq. (river oak, E Aus.) – prot. sp., form. large pieces of bark used on Aboriginal canoes; ***C. glauca*** Sieber ex Spreng. (E Aus.) – widely cult., r.-suckering invasive Florida, Hawaii; ***C. oligodon*** L. Johnson (NG) – N-fixing, transplanted by NG highlanders, v. early silviculture

Casuarinaceae R. Br. Magnoliidae – Fagales. 4/95 Indomal., Aus. (esp.), W Pac. R: Telopea 3(1988)133. Monoec. or dioec. everg. trees & shrubs with drooping equisetoid twigs, tannins & r. oft. with nitrogen-fixing bacterial nodules. Lvs scale-like, up to 20 in sets of 4 in whorls, ± connate, forming toothed sheaths at each node; stip. 0. Fls anemophilous, P 1 or 2 in males (falling at anthesis) but 0 in females, so A 1 subtended by bract & 2 bracteoles, these males in whorled axillary spikes, G 2 subtended by an eventually woody bract & 2 bracteoles, 2-loc. but only anterior fertile, 2-branched style (winged in fr.) & 2(4) orthotropous, bitegmic ovules, these females in heads on short (usu.) lat. branches. Infr. a 'cone' of 1-seeded, winged nuts (samaras), initially encl. in accrescent woody bracteoles, which sep. at maturity appearing like a dehisced capsule; seed pend., endosperm 0, embryo (oft. more than 1) large, straight, oily. x = 8 – 14 (? orig. 9)
Genera: *Allocasuarina, Casuarina, Ceuthostoma, Gymnostoma*
Fossils from Tertiary of S Afr. & Arg. Simple fls rep. reduction rather than primitive simplicity. 'Cones' of *Casuarina equisetifolia* familiar on trop. beaches. Timbers esp. firewood

cat grass *Dactylis glomerata*; **c. thyme** *Teucrium marum*

Catabrosa P. Beauv. Gramineae (XVI 14). 3 N temp. (Eur. 1), Lesotho, Chile. R: Darw. 23(1981)181. ***C. aquatica*** (L.) P. Beauv. (N temp.) – waterhair

Catabrosella (Tzvelev) Tzvelev (~ *Colpodium*). Gramineae (XVI 14). 7 Eur. to Himal.

Catacolea Briggs & L. Johnson. Restionaceae. 1 SW W Aus.: ***C. enodis*** Briggs & L. Johnson. R: Telopea 7(1998)346

Catadysia O. Schulz = *Weberbauera*

Catalepidia P. Weston (~ *Macadamia*). Proteaceae (V 4). 1 Queensland: ***C. heyana*** (Bailey) P. Weston. R: FA 16(1995)499

Catalepis Stapf & Stent. Gramineae (XXVII 2). 1 S Afr.: ***C. gracilis*** Stapf & Stent. R: Strel. 10(2000)686

Catalonian jasmine *Jasminum grandiflorum*

Catalpa Scop. Bignoniaceae (1). 10 E As. (4), SE N Am. (2), WI (4). R: Candollea 13(1952)241. Cult. orn. Trees (Koriba's Model) & some timber. ***C. bignonioides*** Walter (Ind. bean, SE US) – catalpol (iridoid) sequestered by sphingid *Ceratomia catalpae*, cult. orn. incl. golden '**Aurea**', timber for railway sleepers etc.; ***C. speciosa*** (Warder ex Barney) Engelm. (catawba, cigar-tree, US) – similar uses, insect-damaged lvs produce more extrafl. nectar attracting insects which attack or remove eggs or larvae of the first herbivores; hybrids between As. & Am. spp. & between *C. bignonioides* & *Chilopsis linearis* in cult.

Catamixis Thomson. Compositae (Perty.). 1 NW Himal.: ***C. baccharoides*** Thomson

Catananche Vaill. ex L. Compositae (Lact.-Cat.). 6 Medit. (Eur. 2). ***C. caerulea*** L. (cupid's dart, S Eur.) – cult. orn.

Catanthera F. Muell. Melastomataceae. 17 Sumatra, Borneo (8), NG. R: Reinw. 10(1982)35. Ivy-like climbers

Catapodium Link = *Desmazeria*

cataria *Nepeta cataria*

Catasetum Rich. ex Kunth. Orchidaceae (V 12c). c. 170 trop. Am. R: FR 30(1932)257, 31(1933)99. Usu. epiphytes with 2 or 3 plicate lvs per pseudobulb & dimorphic male & female fls on sep. infls (spp. with bisex. fls = *Clowesia*); poll. male euglossine bees, pollinia violently ejected when column appendages touched. Female fls seldom produced in cult. & not always distinctive, though more produced in high light intensities. Closely related spp. with different fragrances attracting different poll. Many cult. orn.; pseudobulb mucilage used in book-binding

Catatia Humbert. Compositae (Gnap.-Gnap.). 2 Madag. R: OB 104(1991)138

catawba *Catalpa speciosa*

catberry *Ribes grossularioides*

catbriar *Smilax* spp.

catchfly *Silene* spp.

catechu *Senegalia catechu, Uncaria gambir*

Catenularia Botsch. = *Catenulina*

Catenulina Soják. Cruciferae (26). 1 C As.: ***C. hedysaroides*** (Botsch.) Soják

caterpillar plant *Spathicarpa hastifolia*

Catesbaea Gronov. ex L. Rubiaceae (III 4). 20 WI to Florida Keys. Spiny shrubs. ***C. spinosa*** L. (lily thorn, Spanish guava, Cuba) – cult., ed. fr.

catgut *Tephrosia virginiana*

Catha Forssk. ex Scop. Celastraceae (I). 1 SW Arabia to S Afr.: ***C. edulis*** (Vahl) Endl. (khat, ghat, miraa, qat, cafta, 'bubble') – cult. Ethiopia, Somalia, Yemen (in 1990 80 000 ha in

N Yemen, most imp. cash crop as in Ethiopia, but using 40% water-supply in irrigation) etc. for lvs chewed fresh by Muslims (esp. males; Somali khat-houses 'mafrish') as daily stimulant banned in much of Eur. & N Am. (mephedrone a cathinone in young lvs, effect similar to amphetamines & MDMA – addictive with suicidal after-effects), form. made into tea & still taken in coffee, fresh supplies airfreighted from E Afr. to Muslim countries. See also *Lydenburgia*

Catharanthus G. Don (~ *Vinca*). Apocynaceae (I b). 8 Madag. (7), Ind. & Sri Lanka (1). R: WAUP 96–3(1996)12. Differs from *Vinca* in 34 ways incl. sessile A without term. appendage, habit (Champagnat's Model) etc. ***C. roseus*** (L.) G. Don (Madag. periwinkle, old maid, Cayenne jasmine, pantrop. weed, orig. Madag.) – cult. orn. with at least 80 named alks notably vincristine & others which have retarding effect on progress of leukaemia discovered when tested for alleged effects in diabetes(!), also in Hodgkin's Disease, but 2 t of lvs needed for 1 g alks (= 6 wks' treatment for 1 child), see W.I. Taylor & N. Farnsworth (1975) *The C. alkaloids*

Cathariostachys S. Dransf. Gramineae (V 6). 2 Madag. R: KB 53(1998)388. Lemur food

Cathaya Chun & Kuang (~ *Tsuga*). Pinaceae. 1 W China: ***C. argyrophylla*** Chun & Kuang – embryo & pollen like *Pinus*, wood anatomy & external morphology of female cones like *Picea*, wood like *Pseudotsuga*; discovered 1938 Jinfu Shan (by 1989 3266 wild trees known in 4 provinces), fossil in Ger. & pollen from Canad. Arctic Eocene. R: NRBGE 45(1988) 385

Cathayanthe Chun. Gesneriaceae (III 2j). 1 Hainan: ***C. biflora*** Chun

Cathayeia Ohwi = *Idesia*

Cathcartia Hook.f. (~ *Meconopsis*). Papaveraceae. 4 Himal. Cult. orn.

Cathedra Miers. Olacaceae (Aptandraceae). 5 trop. S Am. R: FN 38(1984)105

Cathestecum J. Presl = *Bouteloua* (but see JWAS 27(1937)495)

Cathetostemma Blume (~ *Hoya*). Apocynaceae (V a). 1 Mal.: ***C. laurifolium*** Blume

Cathissa Salisb. = *Ornithogalum*

Cathormion (Benth.) Hassk. (~ *Albizia*). Leguminosae (II 4). 1 Ind. to Aus.: ***C. umbellatum*** (Vahl) Kosterm. – seeds sea-disp. R: FM I,11(1992)143

Catila Ravenna = *Calydorea*

cativo *Prioria copaifera*

catjang *Cajanus cajan*

catmint, catnep, catnip *Nepeta cataria* & other *N.* spp.

Catoblastus H. Wendl = *Wettinia*

Catocoryne Hook.f. Melastomataceae. 1 Peru: ***C. linnaeaoides*** Hook.f.

Catoferia (Benth.) Benth. Labiatae (VII 3d). 4 C Am. R: KB 41(1986)299. A strongly exserted

Catolesia Hind. Compositae (Eup.-Gyp.). 1 Bahia, Braz.: ***C. mentiens*** Hind. R: KB 55(2000)942

Catolobus (C. Meyer) Al-Shehbaz (~ *Arabis*). Cruciferae (15). 2 E Eur. to E As. R: Novon 15(2005)520

Catopheria Benth. = *Catoferia*

Catophractes D. Don. Bignoniaceae 1). 1 trop. & S Afr.: ***C. alexanderi*** D. Don. R: Strel. 10(2000)175

Catopsis Griseb. Bromeliaceae (2). 18 trop. Am. R: FN 14(1977)1366. Cult. orn. incl. ***C. berteroana*** (Schultes f.) Mez (S Florida to Braz.) – carnivore (absorbing organic matter through 'tank') at top of *Rhizophora* mangroves in Florida, infl. 90 cm above lvs so poll. insects clearly separated from prey (cf. *Cephalotus*)!

Catospermum Benth. = *Goodenia*

Catostemma Benth. Malvaceae (Bomb.; Bombacaceae). 11 N S Am. R: AMBG 74(1987)636

Catostigma Cook & Doyle = *Catoblastus*

cat's claw *Senegalia greggii*; **c. c. creeper** *Dolichandra unguis-cati*; **c. ear** *Hypochaeris* spp.; **c. eyes** *Dimocarpus longan* var. *malesianus*; **c. foot** *Antennaria dioica*; **c.paw** *Anigozanthos humilis*; **c. tail** *Amaranthus caudatus*, *Bulbinella* spp., *Typha* spp.; **c.t. grass** *Phleum* spp., *Trisetaria cristata*; **c.t. millet** *Setaria pumila*; **c. whiskers** *Cleome gynandra*

Cattleya Lindl. Orchidaceae (V 13b). Incl. *Sophronitis*, excl. *Guarianthe*, 114 C & S Am. R: Phytotaxa 186(2014)75. Epiphytes (mostly) with 1 – 3 thick lvs per pseudobulb & showy fls, much cult. & hybridized, the most familiar orchid of button-holes & bouquets being derived from large-flowered spp. with 1 leaf per pseudobulb. ***C. warscewiczii*** Reichb.f. (Colombia) – scented fl. to 20 cm across

Cattleyella van den Berg & M.W. Chase = × *Brassocattleya*

Cattleyopsis Lem. = *Broughtonia*

× **Cattleytonia** Moir & Sander. Orchidaceae. *Broughtonia* × *Cattleya* – 285 grexes

× **Cattlianthe** J.M.H. Shaw. Orchidaceae. *Cattleya* × *Guarianthe* – c. 2700 grexes

catuaba herbal med. based on diff. pls in diff. parts trop. Am., e.g. spp. of *Anemopaegma, Erythroxylum, Ilex, Micropholis, Secondatia, Tetragastris* (see KB 45(1990)186)

Catunaregam Wolf. Rubiaceae (II 1). 8 trop. Afr. to As. Incl. geoxylic suffrutex. ***C. spinosa*** (Thunb.) Tirv. (*Xeromphis s.*, C & S Afr., Indomal., China) – molluscicidal saponins

× **Catyclia** J.M.H. Shaw. Orchidaceae. *Cattleya* × *Encyclia*

cauassu *Calathea lutea*

Caucaea Schltr. (~ *Oncidium*). Orchidaceae (V 12h). 9 trop. S Am.

Caucaliopsis H. Wolff = *Agrocharis*

Caucalis Tourn. ex L. Umbelliferae. (III 3). 1 S Eur.: ***C. platycarpos*** L. (bur parsley) – casual in GB. R: Fl. Iraq 5,1(2013)271

Caucanthus Forssk. Malpighiaceae. 3 E & NE Afr. to Arabia

Caucasalia R. Nordenstam (~*Adenostyles*). Compositae (Sen.-Sen.). 4 Turkey, Cauc. R: PSB 206(1997)22

cauchao *Amomyrtus luma*

Caudanthera Plowes = *Caralluma*

Caulanthus S. Watson = *Streptanthus* (but see AMBG 9(1922)283)

Caularthron Raf. Orchidaceae (V 13b). 4 trop. Am. R: C.L. Withner, *The Cattleyas & their relatives* V(1998)69. Cult. orn. epiphytes incl. ***C. bilamellatum*** (Reichb.f.) R. Schultes (C Am. to Colombia) – facultative myrmecophyte with extrafl. nectar making up almost half diet of ants at some times of yr.

cauliflower *Brassica oleracea* Botrytis Group

Caulipsolon Klak = *Mesembryanthemum* (but see BJ 120 (1998)364)

Caulocarpus Bak. f. = *Tephrosia*

Caulokaempferia Larsen (*Monolophus*). Zingiberaceae (II 1). 27 Himal. to SE As. Tiny pls of wet places

Caulophyllum Michaux. Berberidaceae (II 1; Leonticaceae). 3 NE As., E N Am. (2). R: Rhodora 87(1985)463. ***C. thalictroides*** (L.) Michaux (blue cohosh, papoose r., squaw r., E N Am.) – dried rhiz. diuretic etc.; cult. orn. Ovary wall splits early to expose 2 large blue drupe-like seeds borne on sturdy funicles

Caulostramina Rollins = *Hesperidanthus* (but see CGH 204(1973)155)

Causonis Raf. (~ *Cayratia*). Vitaceae. 25 Indomal. ***C. japonica*** (Thunb.) Raf. (Indomal.) – 'fr.' = insect galls, sterile (meiotic irregularities); ***C. trifolia*** (L.) Mabb. & J. Wen (*Cayratia trifolia*, Indomal.) – r. ed. boiled N Queensland

Caustis R. Br. Cyperaceae (II 7). c. 6 Aus. R: FR 53(1944)91. Prot. spp. ***C. blakei*** Kükenthal ex S.T. Blake (NE Aus.) – dried & coloured to be sold in NSW as 'koala fern'

Cautleya (Benth.) Hook.f. Zingiberaceae (II 1). 2 Himal. R: KB 36(1982)747. Some cult. orn. like *Roscoea*

Cavacoa Léonard. Euphorbiaceae (Al.-Cross.). 3 trop. & S Afr. R: BJBB 25 (1955)320

Cavalcantia R. King & H. Robinson (~ *Eupatorium*). Compositae (Eup.-Ager.). 2 Braz. R: MSBMBG 22(1987)154

cavalo nero *Brassica oleracea* Acephala Group

Cavanillesia Ruíz & Pavón. Malvaceae (Bomb.; Bombacaceae). 4 trop. Am. R: BJBBuit. III,6(1924)214. ***C. platanifolia*** (Bonpl.) Kunth (cuipo, quipo, C Am.) – fr. mucilage improves water uptake & speeds germ., soft pith-like wood used for canoes & floating rafts of heavier timbers, washed up in Azores, poss. balsa subs.

Cavea W.W. Sm. & Small = *Saussurea*

Cavendishia Lindl. Ericaceae (VIII 5). c. 130 trop. Am. R: FN 35(1983). Hummingbird-poll.; some cult. orn.

Caxamarca Dillon & Sagást. Compositae (Sen.-Sen.). 1 N Peru: ***C. sanchezii*** Dillon & Sagást. R: Novon 9(1999)156

Cayaponia A. Silva Manso. Cucurbitaceae (XV). Incl. *Selysia*, c. 55 warm Am., 2 trop. W Afr. Some bat-poll., independently others derived in having male fls visited by bees becoming drunk on alc. nectar so do not groom off pollen before visiting female fls (cf. *Epipactis*). Some med. ***C. glandulosa*** (Poeppig & Endl.) Cogn. (Colombia) – crushed lvs & stems an insect-repellent; ***C. kathematophora*** R. Schultes (? cultigen in Colombian Amaz.) – seeds for necklaces & anklets; ***C. ophthalmica*** R. Schultes (? cultigen in NW Amaz.) – treatment of conjunctivitis

cay-cay fat *Irvingia malayana*

Cayenne incense *Protium* spp.; **c. pepper** *Capsicum annuum* Longum Group

Caylusea A. St-Hil. Resedaceae. 3 Cape Verde Is., N & E Afr. to Ind. R: MLW 67–8(1967)36

Cayratia Juss. Vitaceae. Excl. *Causonis*, c. 35 OW trop. ***C. geniculata*** (Blume) Gagnepain (W Mal.) – ropes; ***C. roxburghii*** Gagnepain (Ind.) – endangered

Ceanothus L. Rhamnaceae (inc. sed.). 55 N Am. esp. W. R: M. Rensselaer & H.E. McMinn (1942) *C.*, (subg. *Cerastes* 25) SB 40(2015)952. Calif. lilacs. Imp. constituents of chaparral, hybridizing freely in wild & cult. Alks. Decid. & everg. shrubs & small trees with nitrogen-fixing Actinomycetes (*Actinobacteria*) in r., s.t. spiny, the best blue-flowered shrubs for N Eur. gardens incl. hybrids esp. scented ***C.*** × ***delilianus*** Spach '**Gloire de Versailles**' (*C. americanus* × *C. coeruleus* Lag., Mex. & Guatemala); ***C. americanus*** L. (E N Am.) – lvs fresh or dried used as a tea (New Jersey tea) by Native Americans, also med. suggested remedy for lung-bleeding

ceara rubber *Manihot glaziovii*

Cearanthes Ravenna. Amaryllidaceae. 1 NE Braz.: ***C. fuscoviolacea*** Ravenna

Ceballosia Kunkel ex Förther = *Heliotropium*

Cebipira Juss. ex Kuntze = *Bowdichia*

Cecarria Barlow. Loranthaceae (5 2). 1 Philipp., Flores & Timor to N Queensland: ***C. obtusifolia*** (Merr.) Barlow – commemorates C.E. Carr, botanist

Cecchia Chiov. = *Oldfieldia*

Cecropia Loefl. Urticaceae (Cecropiaceae). 61 trop. Am. R: FN 94(2005)32. Dioec. (some fish-disp. in Amaz.) fast-growing pioneer trees (Corner's & Rauh's Models) with hollow septate twigs usu. inhabited by ants & with stilt-r.; lvs favoured by sloths. The ants (*Azteca* spp. prob. with carnivorous ancestry) feed on food-bodies at petiole-bases (food is glycogen!) & attack other grazers esp. leaf-cutting ants. The hollow internodes are burrowed into by pregnant female, which raises brood therein. In spp. with waxy stems (difficult for leaf-cutters to climb) there are neither food-bodies nor a thin area at top of internode for *Azteca* penetration; moreover ***C. peltata*** L. (trumpet tree) in Puerto Rico has 98% of its trees without these symbiotic traits whereas in Trinidad & trop. Am. mainland all trees have them (cf. *Musanga*). *C. peltata* – pulp & cult. orn., replacing *M. cecropioides* in SW Cameroun (? relatively pest-free), ash chewed with coca & used in gunpowder; ***C. insignis*** Liebm. (C Am.) – establishes only in gaps at least 215m2 in forest; ***C. sciadophylla*** Mart. (trop. S Am.) – oft. left when forest felled for agric. as lvs a source of alkali for coca in Amaz. where limestone rare

Cecropiaceae C. Berg = Urticaceae

cedar *Cedrus* spp.; **c. acacia** *Falcataria toona*; (**E**) **Afr.** c. *Juniperus procera*; **Alaska c.** *Xanthocyparis nootkatensis*; **Atlantic c.** *Cedrus atlantica*; **Aus. c.** *Toona ciliata*; **c. balls** *Juniperus virginiana*; **bastard c.** *Chukrasia tabularis, Guazuma ulmifolia, Soymida febrifuga*; **Bermuda c.** *J. bermudiana*; **Burma** or **Moulmein c.** *T. ciliata*; **Chilean c.** *Austrocedrus chilensis*; **Chinese c.** *Cunninghamia lanceolata*; **Clanwilliam c.** *Widdringtonia cedarbergensis*; **Cyprus c.** *Cedrus brevifolia*; **c. elm** *Ulmus crassifolia*; **incense c.** *Calocedrus decurrens*; **Jap. c.** *Cryptomeria japonica*; **c. of Lebanon** *Cedrus libani*; **M(u)lanje c.** *W. whytei*; **Port Orford** or **Oregon c.** *Chamaecyparis lawsoniana*; **red c.** *Juniperus virginiana* (US), *Toona ciliata* (Aus.); **Siberian c.** *Pinus cembra*; **southern white c.** *C. thyoides*; **Virginian pencil c.** or **red c.** *J. virginiana*; **WI c.** *Cedrela odorata*; **western red c.** *Thuja plicata*; **white c.** *Melia azedarach, Calocedrus decurrens, Chamaecyparis thyoides, Chukrasia tabularis, Thuja occidentalis*, etc.; **c. wood oil** (US) *Juniperus* spp. esp. *J. ashei, Cedrus* spp. *Cupressus* [s.l.] spp.; **yellow c.** *Rhodosphaera rhodanthema* (Aus.), *Xanthocyparis nootkatensis* (**y. Alaska c.**, N Am.)

cedrat *Citrus medica*

Cedrela P. Browne. Meliaceae (I). 18 trop. Am. R: T.D. Pennington & A.N. Muellner (2010) *A monograph of C.* OW spp. now referred to *Toona*. Good timber esp. ***C. odorata*** L. (cedro, WI cedar, natur. Medit., e.g. Stromboli, invasive trop. Afr., Galápagos but CITES-listed) used in cigar-boxes, moth-proof chests etc.; domatia, moth-poll. in Costa Rica, wood source of antimalarial gedunin

Cedrelinga Ducke. Leguminosae (II 4). 1 Braz.: ***C. cateniformis*** (Ducke) Ducke – agroforestry, timber a mahogany subs. R: MNYBG 74,1(1996)251

Cedrelopsis Baill. Rutaceae (III, Ptaeroxylaceae). 8 Madag. R: F Mad 107 bis (1991)97

cedro *Cedrela odorata*; **c. negro** *Juglans neotropica*

cedron *Simaba cedron*

Cedronella Moench. Labiatae (VII 2b). 1 Macaronesia: ***C. canariensis*** (L.) Webb & Berth. (*C. triphylla*, Canary balm) – shrubby, lvs 3-foliolate, form. used as a tea, cult. orn. R: BZ 78,2(1993)111

Cedronia Cuatrec. = *Picrolemma*

Cedrus Trew. Pinaceae. 4 (or 1 – 2) mts N Afr. to As. R: P. Maheshwari & C. Biswas (1970) *Cedrus*, RV 21(1990)111. Cedars. Everg. trees with long & short shoots of needle lvs, the short with capacity to develop into long. Infls solit. in position of short shoots; cone of closely enwrapped scales, each with 2 seeds, developing in 2–3 yrs. Some cedarwood oil (mostly *Juniperus* spp.). Cult. orn. esp. as specimen trees, when rapidly attaining an air of antiquity: ***C. atlantica*** (Endl.) Carr. (Atlantic or Atlas c., N Afr. – endangered in wild) & esp. **Glauca Group** incl. '**Glauca Pendula**' all derived from orig. tree still at Vallée aux Loups, Paris, ***C. brevifolia*** (Hook.f.) Henry (Cyprus c., Cyprus), both s.t. considered as geographical subspp. (subsp. *atlantica* (Endl.) Battand. & Trabut & subsp. *brevifolia* (Hook.f.) Meikle, usu. differing in somewhat downy shoots & smaller lvs resp.) of ***C. libani*** A. Rich. (cedar of Lebanon, As. Minor, to 3000 yrs old, timber used in Solomon's temple, Nebuchadnezzar's palace in Babylon & Cheops's ship at base of Gt Pyramid (c. 2500 BC) but only 14 remnant groves from Tripolis to Sidon left, 'tar' used like creosote & in medic. S Turkey) – all useful timber (allergenic to some) & some oils for scent etc. ***C. deodara*** (D. Don) G. Don (*C. libani* ssp. *deodara* (D. Don) Sell, deodar, Himal.) – reaching 12 m girth, imp. timber tree in Ind., nat. tree Pakistan

Ceiba Mill. Malvaceae (Bomb.; Bombacaceae). Incl. *Chorisia*, 18 trop. Am., 1 ext. to Afr. R: AJBM 60(2003)265. Massart's Model; bat-poll. & ***C. pentandra*** (L.) Gaertn. poll. by non-flying mammals in SE Peru rainforest (young fr. ed., shoots c. 25% protein dry wt. & imp. food for red spider monkeys in C Am.). Seeds in hairs from carpel walls, the kapok of commerce (105 K t p.a.) esp. from *C. pentandra* (silk-cotton tree, trop. Am. & Afr., where at 70 m the tallest angiosperm in the Continent & revered as habitat of spirits, sacred in both Afr. & Am.) – the cult. plant app. derived from a savanna form crossed with a rainforest one in Afr. (lemur-poll. in Madag.) & taken thence to Indonesia (prob. by AD 500) etc., wind-disp., kapok used in stuffing mattresses (dangerous in life-jackets as absorbs oily seawater, so many wearing them WW II drowned), in insulation etc., seeds an oil-source (soap, illuminant), wood used for matches, canoes etc.; ***C. speciosa*** (A. St. Hil.) Ravenna (~ *C. insignis, Chorisia speciosa*, paina de seda, yuchán, trop. Am.) – kapok used e.g. for arrow-proof jackets by Matico Indians, cult. orn. tree with swollen oft. spiny trunk & white to yellow or red fls before lvs; ***C. trischistandra*** (A. Gray) Bakh. (trop. Am.) – 'mushroom' smell of bat-poll. fls due to fatty acid derivatives with C^8 skeleton

Celaenodendron Standl. = *Piranhea*

celandine *Ficaria verna*; **greater c.** *Chelidonium majus*; **tree c.** *Bocconia frutescens*

Celastraceae R. Br. Magnoliidae – Celastrales. Incl. Brexiaceae, Hippocrateaceae, Lepuropetalacaeae, Parnassiaceae, Plagiopteraceae, Stackhousiaceae, excl. Goupiaceae 96/1300 trop. & temp. (fewer). R: SB 31(2006)132. Trees, shrubs & lianes, few herbs, usu. glabrous & with laticifers. Lvs in spirals or opp. (both even in a single genus, e.g. *Cassine*) or distich., rarely much reduced (*Canotia*); stip. small, rarely 0. Fls usu. small, in cymes, rarely racemes or solit. & axillary, rarely unisexual, regular, hypogynous to occ. semi-epigynous, K (2 –)5, imbr. (rarely valvate), s.t. with a basal tube, C (2 –)5, usu. imbr. or valvate, rarely 0, A on, without, or within a nectary-disk, usu. alt. with C, seldom in 2 whorls or with set of staminodes opp. K, (2)3(4) or 5, when 3 aligned with sides of ovary, filaments s.t. ± connate at base, anthers extrorse or introrse, with longit. (transverse in *Hippocratea* & *Salacia*) slits, pollen oft. in tetrads or polyads, $\underline{G}$ (2 – 5), rarely half-inf. with as many locules as G (rarely all save 1 abortive) with axile placentation & as many stigmas from 1 style, ovules (1)2 – 10, anatropous, bitegmic (exotegmic). Fr. a berry, capsule, drupe or samara; seeds with arils, wings or angular or compressed, embryo with large cotyledons, endosperm ± oily or 0. x = 8–10, 12, 14, 15, 17, 23

Classification & principal genera (Simmons; *Parnassia* & *Lepuropetalon* to be added):

I. **Celastroideae** (trees or shrubs): *Cassine, Catha, Celastrus, Crossopetalum, Euonymus, Gymnosporia, Maytenus*

II. **Stackhousioideae** (herbs with lvs. s.t. reduced to scales, Aus.): *Stackhousia*

III. **Hippocrateoideae** (Hippocrateaceae; usu. lianes): *Hippocratea*

IV. **Salacioideae** (usu. lianes): *Salacia, Tontelea*

Many mono- or oligo-specific genera. Various genera have been recog. as sep. fams at diff. times, esp. II. (= Stackhousiaceae), III. & IV. with transverse anther dehiscence (Hippocrateaceae), though *Campylostemon* (III.) & *Sarawakodendron* (I.) intermediate in many characters; *Canotia* (Canotiaceae; minute lvs). *Goupia* (ovules ∞, styles free) now considered distinct (*Bhesa* excl. to Centroplacaceae) but other fams (see above), also Parnassiaceae with small seeds & herb. habit) & Lepuropetalaceae (minute herbs form. in Saxifragaceae), now incl.

Some medic. esp. *Celastrus, Gymnosporia, Maytenus, Tripterygium,* some timbers & dyes; khat (*Catha*); cult. orn. esp. *Euonymus, Parnassia*

Celastrus L. Celastraceae (I). 31 trop. to warm temp. R: AMBG 42(1955)215. Cult. orn. lianes for pergolas etc. with fr. like *Euonymus,* some medic. ***C. angulatus*** Maxim. (NW & C China) & ***C. glaucophyllus*** Rehder & E. Wilson (W China) – insecticidal seed-oil & r.-bark; ***C. orbiculatus*** Thunb. (*C. articulatus,* oriental bitter-sweet, E As.) – invasive NZ, E US.; ***C. paniculatus*** Willd. (trop. As. to Pacific) – medic., seed-oil exported for rheumatism etc.; ***C. scandens*** L. (N Am.) – imp. local medic.

celeriac *Apium graveolens* var. *rapaceum*

Celerina Benoist. Acanthaceae (III 2c). 1 Madag.: ***C. seyrigii*** Benoist

celery *Apium graveolens* var. *dulce;* **c. cabbage** *Brassica rapa* Pekinensis Group; **Chinese c.** *Oenanthe javanica;* **c. pine, white** *Callitris glaucophylla;* **c.-top pine** *Phyllocladus aspleniifolius;* **sea c.** *A. prostratum*

Celianella Jablonsk. Phyllanthaceae (Ant.; Euphorbiaceae s.l.). 1 Venez.: ***C. montana*** Jablonsk. R: A. Radcliffe-Smith, *Gen. Euphorb.* (2001) 68

Celiantha Maguire. Gentianaceae. 3 Guayana Highland. R: MNYBG 32(1981)382

cellophane orig. proprietary name of transparent material made of regenerated cellulose; **c. plant** *Echinodorus berteroi*

Celmisia Cass. Compositae (Ast.-Hin.). 69 Aus. (10), NZ (59). R: NZJB 7(1969)400. Cult. orn., seeds not long-lived. Lvs form. used in Maori waterproof raincapes, sandals etc., also smoked like tobacco

Celome Greene = *Cleome*

Celosia L. Amaranthaceae (I 1). 65 warm Am. & Afr. Ovules 2 – ∞. ***C. argentea*** L. (trop.) – weedy polyploid complex largely octoploid (tetraploid in C Ind.), one of the 'amaranths' of the Ancients (name later transferred to *Amaranthus* spp.). Lvs used as spinach in Sri Lanka & E Afr. (mfungu); ***C. cristata*** L. (*C. argentea* var. *cristata,* cockscomb), a tetraploid cultigen of cvs app. derived from *C. argentea* with white to yellow, purple or red variously plumed (e.g. **Plumosa Group**) or fasciated (**Cristata Group**) etc. infls, much grown as pot-pls

Celsia L. = *Verbascum*

Celtica Vázquez & Barkw. (~ *Stipa*). Gramineae (X). 1 W Medit.: ***C. gigantea*** (Link) Vázquez & Barkw. (*S.g.*) – cult. orn.

Celtidaceae Endl. = Cannabaceae

Celtis Tourn. ex L. Cannabaceae (Celtidaceae; Ulmaceae II). c. 60 trop. (most), temp. (Eur. 4). Hackberries, nettle-trees, sugar- berries, oft. with ed. fr., a yellow dye from bark & timber good for charcoal. Roux's & Troll's Models. Double-forked style. ***C. australis*** L. (lote-tree, Medit.) – widely planted street-tree in Medit. for shade, timber valuable; ***C. occidentalis*** L. (beaver wood, N Am.) – wood for fences & fuel; ***C. sinensis*** Pers. (temp. E As., invasive Aus.) – insect galls ed., cult. bonsai subject; ***C. zenkeri*** Engl. (trop. Afr.) – powerful fetish tree in Ghana

celtuce *Lactuca sativa*

Cenarrhenes Labill. Proteaceae (IV). 1 Tasmania: ***C. nitida*** Labill. R: FA 16(1995)131

Cenchrus L. Gramineae (XXIV 4). Excl. *Echinaria,* incl. *Odontelytrum, Pennisetum,* c. 120 warm subcosmop. (Mal. 16 – R: Blumea 59(2014)64). R (s.s.): ISUJS 37(1963)259. Spikelet surrounded by involucre of sterile spikelets which are hardened & spiky in some spp. at maturity & act as burs in animal-disp. (bur grasses, hedgehog grass), esp. troublesome in sheep-rearing areas of Aus., N Am., ***C. biflorus*** Roxb. invasive in S Afr. Some apospory, pseudogamy. Fodders, lawn-grasses & some grains, e.g. *C. biflorus,* ***C. brownii*** Roem. & Schultes (trop. Am.); many noxious weeds. ***C. abyssinicus*** (Hackel) Morrone (*O. a.,* Yemen to S Afr.) – in running water; ***C. americanus*** (L.) Morrone (*C. spicatus, P. a., P. glaucum, P. typhoides,* pearl millet, bulrush or Afr. or Ind. or spiked m., bajri, gero) – OW cultigen domesticated c. 4500 yrs BP, food-crop also used in beer-making, game-cover plots in S Br.; ***C. caudatus*** (Schrad.) Kuntze (*P. macrourum,* S Afr.) – invasive Aus., NZ; ***C. ciliaris*** L. (*P. c.,* buffel grass, Afr. to Ind.) – drought-tolerant forage e.g. NSW, soil stabilizer; ***C. clandestinus*** (Chiov.) Morrone (*P. c.,* Kikuyu grass, E & NE Afr., invasive S Afr., trop. S Am., Galápagos, Hawaii) – apomictic tetraploid, pasture grass used for erosion control & as lawns, anthers, large, emerge overnight on long filaments making grey-white haze over sward; ***C. compressus*** (R. Br.) Morrone (*P. alopecuroides*) – cult. orn.; ***C. hohenackeri*** (Steud.) Morrone (*P. h.,* moya grass, E Afr. to Ind.) – mooted for paper-making; ***C. longisetus*** M.C. Johnst. (*P. villosum,* feathertop) & ***C. setaceus*** (Forssk.) Morrone (*P. s.,* fountain grass) – both N Afr., noxious weeds S Afr., *C. s.* also in Aus., Hawaii, W US;

C. polystachios (L.) Morrone (*P. p.*, Afr., invasive Mal., Aus.) – fodder, hay, dam-bank stabilization (Ind.); ***C. purpurea*** (Schum.) Morrone (*P. p.*, elephant or Napier grass, Afr., invasive SE US, Hawaii) – fodder & paper (short fibres good for pulp)

Cenia Comm. ex Juss. = *Cotula*

Cenocentrum Gagnepain = *Hibiscus*

Cenolophium Koch. Umbelliferae (III 8). 1 Euras.: ***C. denudatum*** (Hornem.) Tutin (Baltic parsley) – cult. orn. perenn.

Cenolophon Blume = *Alpinia*

Cenostigma Tul. Leguminosae (I 4). 14 trop. Am. R: PhytoKeys 71(2016)84

Centaurea L. Compositae (Card.-Cent.). Incl. *Cnicus, Cyanus,* excl. *Cheirolophus, Crocodylium, Plectocephalus, Rhaponticoides, Stizolophus,* c. 250 Med. (Iberia 94) & SW As. (esp. Armenia), N Euras. (few), trop. Afr., N Am. (2), Aus. (1). Herbs & some subshrubs; all fls tubular & bisex. or outer ones enlarged, ray-like & sterile. Many cult. orn. (bluebottle, knapweed; see also *Amberboa*); used in Neanderthal burial head-wreaths with *Ephedra* etc. ***C. babylonica*** (L.) L. (Turkey & Syria) – to 4 m, unbranched; ***C. benedicta*** (L.) L. (*Cnicus b.*, blessed thistle, Medit.) – form. imp. drug due to glucoside cnicin used in treatment of gout & as tonic, fr. an oil-source, with caruncle; ***C. calcitrapa*** L. (star-thistle, Medit., invasive Aus., W US) – long spiny involucral bracts, young stems ed. Egypt; ***C. cyanus*** L. (cornflower, bachelor's buttons, Euras.) – pappus rays hygroscopically moving scales allowing fr. to 'creep' over soil, ann., form. common cornfield weed with many colour-forms (blue, white, purple or pink) in cult., infusion from capitula use as eye-lotion, with alum by watercolourists; ***C. depressa*** Bieb. (SW As., natur. Medit.) – in wreath on Tutankhamun's mummiform coffin, prob. cult. Anc. Egypt; ***C. iberica*** Trev. ex Spreng. (Medit.) – spiny, the 'thistle' of Genesis; ***C. montana*** L. (C Eur.) – common perenn. cult.; ***C. nigra*** L. (hardheads, knapweed, Eur. natur. N Am.) – diuretic; ***C. pulchella*** Ledeb. (SW As.) – house-brooms; ***C. solstitialis*** L. (St. Barnaby's thistle, yellow star t., S Eur.) – subcosmop. allelopathic weed noxious in Aus., W US; ***C. stoebe*** L. (*C. maculosa*, spotted knapweed, Euras.) – invasive NW US

Centaurium Hill. Gentianaceae. (Excl. *Gyrandra, Schenkia, Zeltnera*) 20 Euras. (Eur. 14). R: T 53(2004)740. Some cult. orn. esp. ***C. scilloides*** (L.f.) Samp. (Azores, W Eur.) in rock gardens, others mostly ann. or bienn., some med.

Centaurodendron Johow (~ *Plectocephalus*). Compositae (Card.-Cent.). Incl. *Yunquea*, 2 Juan Fernandez. Pachycaul trees (Holttum's Model). ***C. dracaenoides*** Johow, ***C. palmiforme*** Skottsb. – both with fewer than 25 left (2004)

Centauropsis Bojer ex DC. Compositae (Vern.-Cent.). 8 Madag. Shrubs

Centaurothamnus Wagenitz & Dittr. Compositae (Card.-Cent.). 1 Arabia: ***C. maximus*** (Forssk.) Wagenitz & Dittr.

Centella L. Umbelliferae (Mack.). 50 warm, mostly S Afr., ***C. asiatica*** (L.) Urb. (gotu cola, pantrop. to Chile, NZ etc.) grown as cover crop, lvs ed., extracts (largely from Madag.) used in W skin ointments to promote healing & in folk med. to treat leprosy, active agents being pentacyclic triterpenoid derivatives esp. asiaticoside, though depressant of C nervous system

Centema Hook.f. Amaranthaceae (I 2). 2 trop. Afr.

Centemopsis Schinz. Amaranthaceae (I 2). 3 trop. Afr. R: KB 36(1982)681

Centipeda Lour. Compositae (Athro.). 10 Madag., Indomal., Pac. (Aus. 9), Chile. R: Muelleria 15(2001)36. Ground as snuff subs. (sneezeweed) in Aus. ***C. cunninghamii*** (DC.) A. Br. & Asch. (Aus.) – local treatment for skin & fever; ***C. minima*** (L.) A. Br. & Asch. (*C. orbicularis*, trop. As. to Aus., widely introd.) – med. in China (antibacterial triterpenes)

centipede grass *Eremochloa ophiuroides*

Centosteca Desv. = *Centotheca*

Centotheca Desv. Gramineae (XXI). 4 OW trop. Lemma bristles disp. mechanism

Centradenia G. Don. Melastomataceae. 6 Mex., C Am. R: JAA 58(1977)73. Opp. lvs of uneven sizes. Some cult. orn.

Centradeniastrum Cogn. Melastomataceae. 2 W trop. S Am. R: Biollania ed. Esp. 6(1997)153

Centranthera R. Br. Orobanchaceae (Buch.; Scrophulariaceae s.l.). 5–6 China to Aus. R: Ann. vol. ARBGC (1942)53

Centrantheropsis Bonati = *Phtheirospermum*

Centranthus DC. (~ *Valeriana*). Caprifoliaceae (Valerianaceae). 10 Medit. & Eur. (8). R: BJLS 71(1976)211. C with spurred base with nectar, the C-tube with a partition separating the style from a passage of downward-pointing hairs leading to spur only penetrable by long-tongued insects. ***C. ruber*** (L.) DC. (red valerian, Medit., Eur.) – cult. orn. natur. on walls in GB & Calif.

Centrapalus Cass. (~ *Vernonia*). Compositae (Vern.-Centrap.). 2 trop. Afr. ***C. pauciflorus*** (Willd.) H. Robinson (*C. galamensis, V. g., V. pauciflora*) – seed-oil low viscosity, rich in epoxy acid for plastics, coatings better than solvent-based paints causing smog. etc.

Centratherum Cass. Compositae (Vern.-Lychn.) 3 trop. NW, Aus., Philipp. R: Rhodora 83(1981)14

Centrilla Lindau = *Justicia*

Centrochloa Swallen = *Axonopus*

Centrogenium Schltr. = *Eltroplectris*

Centroglossa Barb. Rodr. Orchidaceae (V 12h). 5 Braz., Peru, Paraguay

Centrolepidaceae Endl. = Restionaceae

Centrolepis Labill. Restionaceae (Centrolepidaceae). 26 SE As., Mal. (mts), Aus. (20, endemic; R: JABG 15(1992)5), NZ (3)

Centrolobium Mart. ex Benth. Leguminosae (III 11). 7 trop. Am. Legume spiny to 30 cm with large wing. ***C. microchaete*** (Benth.) Lima & ***C. robustum*** (Vell.) Benth. (zebrawood) Braz. timbers (putumuju, form. sold as porcupine wood) for furniture, boat-building etc.

Centromadia Greene (~ *Hemizonia*). Compositae (Mad.-Mad.). 4 Calif., NW Mex. R: Novon 9(1999)466

Centronia D. Don (~ *Meriania*). Melastomataceae. 15 C & W trop. Am., 1 Guianas

Centropappus Hook.f. (~ *Brachyglottis*). Compositae (Sen.-Tuss.). 1 Tasmania: ***C. brunonis*** Hook.f.

Centropetalum Lindl. = *Fernandezia*

Centroplacaeae Doweld & Reveal (~ Pandaceae). Magnoliidae – Malpighiales 2/7 W Afr. Everg. trees, s.t. dioec. (*Centroplacus*). Lvs in spirals, s.t. serrate (*Centroplacus*), stipulate. Infls racemose with articulated pedicels; fls small, 5-merous. opp. K, anthers with oblique-apical slits; female fls with C s.t. 0 (*Centroplacus*), $\underline{G}$ (2, 3), ovule 1 per loc., subapical. Fr. a loculicidal capsule opening from base, K persistent; seed arillate, with small embryo & copious endosperm

Genera: *Bhesa, Centroplacus*

Centroplacus Pierre. Centroplacaceae (~ Pandaceae). 1 W Afr.: ***C. glaucinus*** Pierre

Centropodia (R. Br.) Reichb. Gramineae (XXV). 4 Afr. to C As. R: KB 37(1982)658. Kranz anatomy

Centropogon C. Presl (~ *Siphocampylus*). Campanulaceae (III 3). 212 trop. Am. R: Pflanzenr. 276b(1943)161, (1957/8)768, 276c(1968)838. Heterogeneous; fleshy fr. derived from *Siphocampylus* several times. Herbs & shrubs, usu. hummingbird-poll.; some cult. hanging baskets, some ed. fr. ***C. nigricans*** Zahlbr. (Ecuador) – sole poll. a bat (*Anoura fistulata*, first named 2005) with 85-mm tongue (150% body-length), only 1:1 bat-flower known

Centrosema (DC.) Benth. Leguminosae (III 18). 36 warm Am. Pedicel twists through 180° so that standard points downwards (cf. *Clitorea*). Some used as green manures (butterfly peas) esp. ***C. plumieri*** (Pers.) Benth. under rubber & coconuts. ***C. carajasense*** Cavalc. (Braz.) – fish-poison in Amazon

Centrosolenia Benth. (~ *Nautilocalyx*). Gesneriaceae (II 5c). 35 Guiana Shield esp. tepuis. R: SB 41(2016)93

Centrospermae Eichler = Magnoliidae – Caryophyllales

Centrostachys Wall. Amaranthaceae (I 2). 1 N Afr., Ind., Java: ***C. aquatica*** (R. Br.) Wall.

Centrostegia A. Gray ex Benth. (~ *Chorizanthe*). Polygonaceae (I 1). 1 SW N Am.: ***C. thurberi*** A. Gray ex Benth. R: Phytol. 66(1989)207

Centrostemma Decne = *Hoya*

Centrostigma Schltr. (~ *Habenaria*). Orchidaceae (IV 4d). 3 trop. Afr. R: FZ 11(1995)48

Centunculus Dill. ex L. (~ *Anagallis*). Primulaceae. Spp. incl. ***C. minimus*** L.

century plant *Agave americana*

Ceodes Forst. & Forst. f. = *Pisonia*

Cephaelis Sw. = *Sabicea*

Cephalacanthus Lindau. Acanthaceae (III 2c). 1 Peru: ***C. maculatus*** Lindau

Cephalanthera Rich. Orchidaceae (V 1). 19 N temp. (Eur. 5). Helleborines. Rostellum 0; pollen germinates *in situ*. ***C. austiniae*** (A. Gray) Reichb.f. (N Am.) & ***C. calcarata*** S.C. Chen & K.Y. Lang (Yunnan, coll. once) – mycotrophs; ***C. longifolia*** (L.) Fritsch (Medit.) – poll. solit. bees in Israel, orange papillae on lip mimicking pollen of *Cistus salviifolius*, Estonian pops with green & achlorophyllous pls (last unchanging over 14 yrs, receiving carbon from surrounding trees via Thelephoraceae fungi (mycorrhizae)); ***C. rubra*** (L.) Rich. (red helleborine, Euras.) prot. in GB, poll. solit. bees in Sweden, mimicking *Campanula persicifolia* by flowering first, fooling early hatched males

Cephalantheropsis Guillaumin (~ *Calanthe*). Orchidaceae (V 14). 4 Indomal. R: OD 62(1998)155. Turn blue when bruised

Cephalanthus L. Rubiaceae (III 6). 6 trop., N Am. R: Blumea 23(1976)179. Some rheophytes. ***C. occidentalis*** L. (button bush, SE US) – cult. orn., medic. (laxative etc.)

Cephalaralia Harms. Araliaceae. 1 E Aus.: ***C. cephalobotrys*** (F. Muell.) Harms

Cephalaria Schrad. ex Roemer & Schultes. Caprifoliaceae (Dipsacaceae). c. 80 Medit. (Eur. 14), to C As., Ethiopia, S Afr. Some cult. orn. ***C. syriaca*** (L.) Roemer & Schultes (Turkey to Iran & N Afr.) – imp. cornfield weed

Cephalipterum A. Gray (~ *Rhodanthe*). Compositae (Gnap.-Ang.). 1 W & S Aus.: ***C. drummondii*** A. Gray. R: BM n.s. 16(1999)256

Cephalobembix Rydb. = *Schkuhria*

Cephalocarpus Nees. Cyperaceae (III 1). c. 3 trop. S Am. R: MNYBG 12,3(1965)17. Habit of *Dracaena*

Cephalocereus Pfeiffer. Cactaceae (III 1). 3 Mex. Cult. orn. (? Incl. *Neobuxbaumia*) esp. ***C. senilis*** (Haw.) Pfeiff. (old man cactus, C Mex. where reaching 15 m). See also *Pilosocereus*

Cephalocleistocactus F. Ritter = *Cleistocactus*

Cephalocroton Hochst. Euphorbiaceae (Epi.-Epi.). Incl. *Cephalocrotonopsis*, c. 7 Socotra (1), trop. Afr., Madag., Sri Lanka (1). ***C. cordofanus*** Hochst. (trop. Afr.) – seeds ed., oil-source (Sudan)

Cephalocrotonopsis Pax = *Cephalocroton*

Cephalodendron Steyerm. = *Remijea*

Cephalohibiscus Ulbr. (~ *Thespesia*). Malvaceae (Malv.-Goss.). 1 Papuasia: ***C. peekelii*** Ulbr. R: NBGB 12(1935)495

Cephalomanes C. Presl. Hymenophyllaceae (Hym.). 6 trop. ***C. javanicum*** (Blume) Bosch (Mal.) – aquarium pl., local medic.

Cephalomappa Baill. Euphorbiaceae (Epi.-Ceph.). 5 Mal. R: Reinw. 11(1998)161

Cephalopappus Nees & Mart. Compositae (Mut.-Nass.). 1 NE Braz.: ***C. sonchifolius*** Nees & Mart. – herb, pappus 0

Cephalopentandra Chiov. Cucurbitaceae (XIV). 1 NE trop. Afr.: ***C. ecirrhosa*** (Cogn.) C. Jeffrey. R: U. Eggli, *Ill. Handb. Succ. Pls*, Dicots (2002)76

Cephalophilon (Meisn.) Spach = *Persicaria*

Cephalophis Vollesen. Acanthaceae (III 2c). 1 E Afr.: ***C. lukei*** Volleson. R: FTEA Acanth. (2010)676

Cephalophyllum (Haw.) N.E. Br. Aizoaceae (V). 38 S Namibia & nearby. R: H. E. K. Hartmann, *Ill. Handb. Succ. Pls*, Aiz. A–E (2002)113. Cult. orn. leaf-succs. See also *Jordaaniella*

Cephalopodum Korovin. Umbelliferae (III 8). 3 C As. R: Candollea 57(2002)268

Cephalorhizum Popov & Korovin. Plumbaginaceae (II). 2 C As.

Cephalorrhynchus Boiss. = *Cicerbita*

Cephaloschefflera (Harms) Merr. = *Schefflera*

Cephalosorus A. Gray (~ *Angianthus*). Compositae (Gnap.-Ang.). 1 W Aus.: ***C. carpesioides*** (Turcz.) Short – ann. with ectomycorrhizae. R: OB 104(1991)129

Cephalosphaera Warb. (~ *Brochoneura*). Myristicaceae. 1 trop. E Afr. mts: ***C. usambarensis*** (Warb.) Warb. (mtambara) – timber (plantations in E Afr.)

Cephalostachyum Munro (~ *Schizostachyum*). Gramineae (V 5). 12 NE Ind. to SE As. & China (6). R: KB 52(1997)700. Madag. (2) spp. to be segregated, incl. ***C. vigueri*** A. Camus (Madag.) – bamboo with cyanide eaten at levels toxic to humans by *Hapalemur aureus* (golden bamboo lemur) discovered 1987

Cephalostemon Schomb. Rapateaceae (II 1). (Incl. *Duckea*) 9 trop. S Am.

Cephalostigma A. DC. = *Wahlenbergia*

Cephalotaceae Dumort. Magnoliidae – Oxalidales. 1/1 SW Aus. Everg. carnivorous herb with short rhiz. Lvs rosetted, the inner flat, simple, the outer being ground-level pitchers at first closed by lid, its lower surface & the distal inner surface of the pitcher slippery with overlapping, downwardly-directed projections from epidermal cells; multicellular glands on pitcher surfaces, petiole & lower surface of other lvs; flask-shaped ones within the pitcher esp. in bright-coloured cushion-like projections; stip. 0. Infl. arising from centre of rosette with apical racemes of dichasia of small reg. perigynous fls, hypanthium appearing as a K-tube with K 6, valvate, as 'lobes', C 0, A 6 + 6, unequal, at summit of hypanthium above glandular setose disk, $\underline{G}$ 6 with circinnate styles, each G with 1(2) basal, erect, anatropous, bitegmic ovule. Fr. a hairy follicle, seeds with v. small straight embryo surrounded by copious fleshy endosperm; n = 10

Genus: *Cephalotus*; form. thought to be related to Crassulaceae & Saxifragaceae, though A structure diff.

Cephalotaxaceae Neger = Taxaceae

Cephalotaxus Sieb. & Zucc. ex Endl. Taxaceae (Cephalotaxaceae). 7 E Himal. to Jap. R: Phytotaxa 84(2013)3. Plum yews. Cult. orn. (Massart's Model). Lvs with 2 broad glaucous-green lines beneath; arils like *Taxus*. ***C. harringtonia*** (Forbes) K. Koch ('*C. harringtonii*'; var. *harringtonia* known only in cult.) & var. ***drupacea*** (Sieb. & Zucc.) Koidz. (China, Jap.) – most widely cult.

Cephalotomandra Karsten & Triana. Nyctaginaceae (5). 1 – 3 C Am.

Cephalotus Labill. Cephalotaceae. 1 SW Aus.: ***C. follicularis*** Labill. Discovered by Robert Brown in 1801; restricted to swampy coastal tracts between Donelly R. & Cheyne Beach E of Albany. Pitchers develop from July to January; though pls. can survive without insect food, the parallel with *Nepenthes* is remarkable even as far as to having a digestive-juice-resistant insect (gadfly, cf. mosquitoes in *N.*) & algae living in the pitchers; such pitchers arise by single-gene mutants in e.g. *Codiaeum variegatum* (L.) Blume, but these do not have digestive glands; infl. to 60 cm so poll. insects separated from prey; testa carried out of fr. by cotyledon, but hypocotyl expands into remnant fr. filling it, poss. as reserve for young seedling or protection from herbivores or dehydration

Cepobaculum M.A. Clem. & D.L. Jones = *Dendrobium*

Ceradenia L.E. Bishop. Polypodiaceae (V; Grammitidaceae). 54 trop. Am. & Afr. (few). R: AFJ 78(1988)1

Ceraia Lour. = *Dendrobium*

Ceranthera Elliott = *Dicerandra*

Ceraria Pearson & Stephens = *Portulacaria*

Cerastium L. Caryophyllaceae (II 1). Excl. *Dichodon*, c. 100 almost cosmop. (Eur. 58, N Am. 27). Many weeds (chickweed, mouse-ear c. esp. ***C. fontanum*** Baumg. subsp. ***vulgare*** (Hartman) Greuter & Burdet (common m.e.)), some cult. orn. esp. ***C. tomentosum*** L. (snow-in-summer, Sicily, Italy) – rampant rock-pl.

Cerasus Mill. = *Prunus*

Ceratandra Ecklon ex Lindl. Orchidaceae (IV 4b). 6 SW to E Cape. R: Strel. 9(2000)158. Oil-secreting callus in some (poll. oil-bees), lost in ***C. grandiflora*** Lindl. (beetle-poll.)

Ceratandropsis Rolfe = *praec.*

Ceratanthus F. Muell. ex G. Taylor = *Platostoma*

Ceratiola Michaux. Ericaceae (V 4; Empetraceae). 1 SE US: ***C. ericoides*** Michaux – dioec., cult. orn.

Ceratiosicyos Nees. Achariaceae. 1 S Afr.: ***C. laevis*** (Thunb.) Meeuse. R: Strel. 9(2000)222

Ceratobium (Lindl.) M.A. Clem. & D.L. Jones = *Dendrobium*

Ceratocapnos Durieu (~ *Corydalis*). Papaveraceae (Fumariaceae I 2). 3 W Eur. (1) & Medit. R: OB 88(1986)38. Scrambling annuals with leaf tendrils. ***C. heterocarpa*** Durieu – cleistogamous fls prod. short 1-seeded indehiscent fr., open fls in same pop. prod. long 2-seeded dehiscent fr, the lowermost seed in which is most viable

Ceratocarpus Buxb. ex L. Amaranthaceae (Axyr.; Chenopodiaceae I 3). 1 E Eur., temp. As.: ***C. arenarius*** L.

Ceratocaryum Nees. Restionaceae. 7 SW & S Cape. R: Bothalia 15(1985)479, KB 56(2001)471

Ceratocentron Senghas. Orchidaceae (V 16c). 1 Philippines: ***C. fesselii*** Senghas

Ceratocephala Moench (~ *Ranunculus*). Ranunculaceae (II 3). 3 C Eur. (2), Medit. to NW China, NZ (1), natur. N Am.

Ceratocephalus Pers. = *praec.*

Ceratochilus Blume = *Trichoglottis*

Ceratochloa DC. & P. Beauv. = *Bromus*

Ceratocnemum Cosson & Bal. Cruciferae (12). 1 Morocco: ***C. rapistroides*** Cosson & Bal.

Ceratodon Bridel. Ditrichaceae (Musci). 17 cosmop. ***C. purpureus*** (Hedw.) Brid. (cosmop.) – weedy, tolerant of bird-dung & city-smoke, dioec., females (more organic volatiles than males) attractive to springtails effecting fertilization, analogous to poll. in Magnoliidae

Ceratogyne Turcz. Compositae (Ast.-Bra.). 1 W temp. Aus.: ***C. obionoides*** Turcz.

Ceratoides Gagnebin = *Ceratocarpus*

Ceratolacis (Tul.) Wedd. Podostemaceae (III). 2 Braz. R: Novon 14(2004)108

Ceratolimon Crespo & Lledó (~ *Limoniastrum*). Plumbaginaceae (II). 4 Medit. R: BJLS 132(2000)169

Ceratolobus Blume = *Calamus* (but see KB 34(1979)1)

Ceratominthe Briq. = *Xeropoma*

Ceratonia L. Leguminosae (I 4). 2 Arabia & Somalia. Form. in I 3; some mimosoid features. Relicts of Indomal. flora (flowers in autumn): ***C. oreothauma*** Hillc., J. Lewis & Verdc. (Oman & Somalia), notable for 3-colpate pollen & ***C. siliqua*** L. (carob, locust-bean (prob. of the Bible, John the Baptist's 'husks that the swine did eat'), cultigen orig. from Arabia, introd. Iberia C12) – variable infl. type, ±bracteoles, fls. semen-scented (poll. lace-wings), variable nos of fl. parts & sex expression (1 sex usu. suppressed during development), variable pollen grains, 'inverted' fr. etc.; fr. eaten & seeds disp. by bats (*Rousettus aegyptiacus*), pods (algarroba) full of juicy pulp (prod. 315 K t. p.a.) cont. sugar & gum (tragasol, a tragacanth subs.) used as fodder (e.g. for Wellington's cavalry in Peninsular War) & alcohol source, seeds a coffee subs., form sold as 'sweets' in GB & used as weights (189–205 mg, the original carats (now 200 mg) of jewellers), also yielding a diabetic flour & suitable for baby food, timber for furniture etc. See ITCJ 1(1980)15

Ceratopetalum Sm. Cunoniaceae (III). 8 NG, E Aus. ***C. apetalum*** D. Don (coachwood, lightwood) – light timber for furniture, veneers etc., char. caramel scent; ***C. gummiferum*** Sm. (Christmas bush, NSW) – cult. orn. (K red in fr.; used instead of holly in colonial NSW), prot. sp.; ***C. succirubrum*** C. White (NE Aus, NG, New Br.) – furniture, cabinetwork

Ceratophyllaceae Gray. Magnoliidae – Ceratophyllales. 1/2 – 6 cosmop. Rootless floating submerged monoec. herbs without mycorrhiza or vessels, s.t. with r.-like branches anchoring pl., vegetatively glabrous; stems usu. branched (only 1 per node). Fern-type reserve starch grains unique in angiosperms. Lvs 1 – 4 times forked, rigid & oft. brittle in whorls of 3 – 10, with apical minute teeth & bristles but no xylem or phloem; stomata & stip. 0. Fls unisexual, solit., extra-axillary, males & females usu. on alt. nodes; bracts 6–13, linr. (form. interpreted as P), united at base, P 0, A 1 (form. groupings of 3 –)10–20(–46), in spirals on flat receptacle considered to be a fl.), not clearly differentiated into filament & anther but connective extended into 2 points, extrorse, pollen smooth, inaperturate with no or reduced exine (pollen tubes branching); $\underline{G}$ 1 with 1 orthotropous, unitegmic ovule. Fr. an achene with persistent spiny style & oft. other spines; seed with thin testa, 2 'cotyledons' (app. not homologous with those in other pls) from an annular common primordium, ± linr., radicle vestigial. n = 12

Only genus: *Ceratophyllum*

Orthotropus unitegmic ovules unique in basal angiosperms. Aquatic *Montsechia vidalii* (130–125 M-yr-old Cretaceous fossil from Spain) resembling C.

Ceratophyllum L. Ceratophyllaceae. 2 – 6 variable (up to 30 recog. by some authors), cosmop. R: KB 40(1985)243. Hornworts, so-called from old translucent horny lvs. Plant decays as it grows apically, releasing laterals; winter buds not formed, the pl. sinking in autumn, rising in spring. Water-poll., the anthers breaking off & floating up through water, pollen with same density as water. Floating veg. provides shelter for young fish but also for bilharzia-carrying snails & malarial mosquito larvae; it can rapidly choke waterways but, as in Arkansas, can be controlled using Chinese grass-carp. Both ***C. demersum*** L. (lvs 1- or 2-times forked, fr. usu. spiny) & ***C. submersum*** L. (lvs 3- or 4-times forked, fr. usu. smooth to warty) used as aquarium pls; oft. confused with *Myriophyllum* (Haloragaceae), *Najas* (Hydrocharitaceae) or *Chara* (Characeae), when sterile, but readily recog. by usu. forked lvs

Ceratophytum Pittier = *Tanaecium*

Ceratopteris Brongn. Pteridaceae (II; Parkeriaceae). 5 trop. & warm. R: Britt. 26(1974)139. Only homosporous aquatic (floating) ferns; fronds ed.: sterile ones simple to 3-pinnate, fertile ones larger & 4 – 5-pinnate; sporangia in 1 – 4 rows along veins. Some cult. aquarium pls. ***C. cornuta*** (P. Beauv.) Lepr. (OW) – cult. & ed. Liberia; ***C. pteridoides*** (Hook.) Hieron. (trop. Am.) – covered 17 000 ha in a Surinam lake in 1966; ***C. thalictroides*** (L.) Brongn. (trop. Afr. to Pac. & S Jap.) – much cult. in flooded rice-fields etc. as spring veg. (esp. Jap.)

Ceratopyxis Hook.f. Rubiaceae (III 4). 1 W Cuba: ***C. verbenacea*** (Griseb.) Hook.f.

Ceratosanthes Burm. ex Adans. Cucurbitaceae (XIII). 4 C Am. to N Arg.

Ceratosepalum Oliv. = *Triumfetta*

Ceratostema Juss. Ericaceae (VIII 5). 23 S Am. mts, mostly E Ecuador. Some ed. fr.

Ceratostigma Bunge. Plumbaginaceae (I). 8 NE trop. Afr. (1), Tibet, China, SE As. R: GH 8(1954)410. Cult. orn. shrubby, oft. dying back in winter, esp. ***C. plumbaginoides*** Bunge (*Plumbago larpentiae*, W China, glabrous) & ***C. willmottianum*** Stapf (W China & Tibet, hairy lvs)

Ceratostylis Blume. Orchidaceae (V 15). c. 140 Indomal., W Pac. (not Aus.). ***C. latifolia*** Blume (Mal. to S Jap.) – lvs ed. Jap.

Ceratotheca Endl. Pedaliaceae (3). 5 trop. & S Afr. R: MSB 25(1975)1. ***C. sesamoides*** Endl. (trop. Afr.) – cult. for seeds used like sesame; ***C. triloba*** (Bernh.) Hook.f. (S Afr.) – cult. orn., natur. N Am.

Ceratozamia Brongn. Zamiaceae (III). 21 Mex. to Belize. R: MNYBG 57(1990)201, BR 70(2004)276. ***C. longifolia*** Miq. (Mex.) – male cone heats to 11.7°C above ambient. Cult. orn. esp. ***C. mexicana*** Brongn. (Chiapas)

Cerbera L. Apocynaceae (I g). 2–6 trop. coasts Ind. & W Pac. Oceans. R: TFSS 5(2004)23. Poisonous trees (Koriba's Model) & shrubs, lvs in spirals, frs common in drift, form. used as ordeal-poisons & for suicide. ***C. floribunda*** K. Schum. (NG, NE Aus.) – blue fr. ed. cassowaries (resistant to alks); ***C. manghas*** L. (Seychelles to Pac.) – cult. orn. with fragrant white fls, wood used for vividly painted masks in S Sri Lanka (cf. *Alstonia*)

Cerberiopsis Vieill. ex Pancher & Sébert. Apocynaceae (I g). 2 New Caled. ***C. candelabra*** Vieill. ex Pancher & Sébert a tall branched tree of regular architecture (Scarrone's Model) but hapaxanthic; mass-flowering but damaged younger pls can be found in flower otherwise

Cercestis Schott. Araceae (VII 14). Incl. *Rhektophyllum*, 8 W Afr. Thermogenesis in spadix with 1 peak. ***C. afzelii*** Schott – basket-making; cult. orn. esp. ***C. mirabilis*** (N.E. Br.) Bogner with irreg. perforated lvs

Cercidiphyllaceae Engl. Magnoliidae – Saxifragales. 1/2 E As. Dioec. trees with decid., simple lvs, palmately veined & in spirals on short shoots, pinnately-veined & opp. on long; stip. small, decid. Fls wind-poll. in term. infl. on short shoots maturing before or at same time as lvs; P 0, males in short raceme, 4 lower each subtended by 4-lobed bract, inner (upper) without, heads (individual fls diff. to recog.) with A 16 – 35 with latrorse anthers & longit. slits, females in pseudanthia with (2 –)4(– 8) K-like bracts, each subtending a naked carpel with decurrent 2-ridged stigma, ovules 15 – 30, in 2 rows, anatropous, bitegmic. Fr. of sep. follicles; seeds flattened, winged, endosperm scanty, oily, embryo large, spatulate. 2n = 38

Genus: *Cercidiphyllum*

'Fls' poss. pseudanthia, condensed infls seen in allegedly allied Palaeocene fossils (*Joffrea*); widely distrib. in N hemisph. in Tertiary. Seed anatomy & pollen etc. suggested placement in monogeneric order

Cercidiphyllum Sieb. & Zucc. Cercidiphyllaceae. 2 China, Jap. R: JAA 60(1975)367. ***C. japonicum*** Sieb. & Zucc. ex J.J.Hoffm. & J.H. Schultes bis – largest decid. tree in Jap. (katsura), strong timber for house interiors, furniture, etc., cult. orn. (though oft. inferior shrubby forms)

Cercidium Tul. = *Parkinsonia*

Cercis L. Leguminosae (I 1). 6 N temp. (Eur 1) to NE Mex. (1). Orn. decid. trees with fls on branches & trunk (exc. ***C. racemosa*** Oliv., C China, with axillary racemes) before lvs expand; seeds with vestigial arils & lvs with pulvini at junction of lamina & petiole suggesting that leaf is terminal leaflet of an ancestral pinnate leaf; disjunct distr. (N Am., Medit. & E As.) of v. homogeneous group of spp. not closely allied to any other genus, cauliflory (almost exclusively trop. trait) & arils etc. suggest C. a Tertiary relic cf. *Ceratonia siliqua;* fls superficially like Papilionoideae but standard ('back') C lies *inside* wings. ***C. canadensis*** L. (redbud, SE Canada to NE Mex.) – fls used in salads & pickles; ***C. siliquastrum*** L. (Judas tree, W Med. to E Bulgaria, Lebanon & Turkey) – cult. orn., legend that Iscariot hanged himself on it prob. a confusion between *Arbor Judae* & *Arbor Judaeae*, i.e. Judaea tree, as it was commonly cult. around Jerusalem), for in Medit. story assoc. with the fig (Br. legend has elder as the gibbet, 'Judas tree' in Kent, & edible fungus found on it called Jew's ear, *Auricularia*), serial buds in leaf axils

Cercocarpus Kunth. Rosaceae (Ros.-Dryad.). 8 W & SW N Am. R: Britt. 7(1950)91. Heteroblastic; r. symbionts (*Actinobacteria*) fix nitrogen. Wood for tool-handles etc., some cult. orn. ***C. ledifolius*** Nutt. – imp. local medic.

Cerdia Moçiño & Sessé ex DC. Caryophyllaceae (I 1). 1 Mex.: ***C. virescens*** Moçiño & Sessé ex DC. R: BotJLS 152(2006)5

Cereus Mill. Cactaceae (III 3). 29 WI, E S Am. Formerly incl. many other ribbed columnar cacti esp. 'night-blooming cereus', now placed in segregate genera, e.g. *Hylocereus, Selenicereus* etc. Fls large, white; extrafl. nectaries; cult. orn., e.g. ***C. hildmannianus*** K. Schum. **'Monstrosus'** (= *C. abnormis* (Willd.) Sweet, '*C. peruvianus*'), long cult., some ed. esp. ***C. repandus*** (L.) Mill. (cadushi, WI) – despined young stems ed. Curaçao

ceriman *Monstera deliciosa*

Cerinthe Tourn. ex L. Boraginaceae (B.2.2). 7+ Eur. (5), Medit. R: T 58(2009)1315. ***C. major*** L. (honeywort, Medit.) – cult. orn.

Ceriops Arn. Rhizophoraceae (III). 2 trop. coasts of Ind. & W Pac. Oceans (fossils like them in Eocene London Clay, cf. *Nypa*), inner mangrove: ***C. decandra*** (Griff.) Theob. (Indomal.) – tannin, charcoal, wood burns when wet, so good at sea; ***C. tagal*** (Perr.) C. Robinson – Attims's Model, timber the most durable of all mangroves, bark (45% tannin) used in tanning (e.g. *Crotalaria* nets in Sri Lanka) & a constituent of soga batik-dye

Ceriosperma (O. Schulz) Greuter & Burdet = *Rorippa*

Ceriscoides (Hook.f.) Tirvengadum. Rubiaceae (II 1). 11 Indomal. R: HPB 7(2003)444. Local medic.

Cerochlamys N.E. Br. Aizoaceae (V). 4 W Cape. R: H. E. K. Hartmann, *Ill. Handb. Succ. Pls*, Aiz. A–E (2002)113. ***C. pachyphylla*** (L. Bolus) L. Bolus – cult. orn.

Cerolepis Pierre. Older name for *Camptostylus*

Ceropegia L. Apocynaceae (V b; Asclepiadaceae III 5). c. 190 Arabia (10, R: NRBGE 45(1988)287) warm Afr. incl. Canary Is., to Aus. (1). R: F. Albers & U. Meve, *Ill. Handb. Succ. Pls*, Asclep. (2002)63. Heterogeneous. Usu. succ. twiners or subshrubs, some with tubers, others leafless & *Stapelia*-like. Fls held erect, the C often swollen at base, the whole acting as a poll. trap as in *Aristolochia* (q.v.): the C-tube is lined with downward-pointing hairs, in some heat is produced (cf. *Arum*) & flies are further attracted by smell, colour & s.t. long hairs at C-lobe tips flickering in breeze; once inside, they cannot escape until the tube hairs wither, when they leave with pollinia on their proboscides. Many cult. orn.; tubers of some eaten by humans (e.g. ***C. affinis*** Vatke (trop. E Afr.) in NE & E Afr. (all parts ed.) & ***C. bulbosa*** Roxb. (E Afr. to Ind.) in Arabia & Ind.) & other animals. ***C. linearis*** E. Meyer subsp. ***woodii*** (Schltr.) H. Huber (*C. woodii*, E & S Afr.) – much cult. in hanging baskets, lvs marbled, fls blackish poll. biting midges detained within 1–2 days, easily prop. from aerial tubers

Cerosora (Bak.) Domin. Pteridaceae (III). 3 Himal., Sumatra, Borneo (& Afr. '*Pityrogramma*'). R: KB 13(1959)450

Ceroxylon Bonpl. ex DC. Palmae (IV 2). 12 Andes. R: NBGB 10(1929)841, Phytotaxa 34(2011)1. Wax palms, once called *Beethovenia*, wax on trunk form. used for candles, wax matches, gramophone records; lvs overexploited for Christian ceremonies. ***C. alpinum*** Bonpl. ex DC. (*C. andicola*, Colombia) – lower alt. than seqq., named after a farm (Los Alpes, not a mt.!), ***C. quindiuense*** (Karsten) H. Wendl. (Colombia) – at 60(–75) m the tallest palm, national tree of Colombia, 10 kg wax per tree per annum; ***C. utile*** (Karsten) H. Wendl. (Colombia & Ecuador) found at 4000 m – highest record for a palm

Ceruana Forssk. Compositae (Ast.-Gran.). 1 Egypt, trop. Afr.: ***C. pratensis*** Forssk. – used in brooms, found in Egyptian tombs

Cervantesia Ruíz & Pavón. Santalaceae (Cervantesiaceae). 4 Andes (Peru, Ecuador)

Cervantesiaceae Nickrent & Der. See Santalaceae

Cervaria Wolf = *Peucedanum*

Cervia Rodriguez ex Lag. = *Rochelia*

Cespedesia Goudot. Ochnaceae (III). 1[–6] trop. S Am.: ***C. spathulata*** (Ruíz & Pavón) Planch.

Cestrum L. Solanaceae (II 1). Excl. *Sessea*, 150+ trop. Am., (? Aus.). R: R.G. van den Berg & al., *Solanaceae V* (2001)109,153. Alks. Berries. Many cult. orn. (Roux's Model) esp. for fragrant fls, e.g. ***C. nocturnum*** L. (lady of the night, rat-ki-rani, trop. Am.), some Mex. spp. incl. ***C. roseum*** Kunth day-fls poll. hummingbirds; some med.; ***C. diurnum*** L. (WI, S Am., invasive SE US) – kills grazing horses (Florida), toxin poss. useful in human osteoporosis; ***C. laevigatum*** Schldl. (S Am., invasive S Afr.) – cannabis subs. in coastal Braz.; ***C. parqui*** L'Hérit. (S S Am.) – toxic to sheep & cattle

Ceterach Willd. = *Asplenium*

Ceterachopsis (J. Sm.) Ching = *Asplenium*

Ceuthocarpus Aiello. Rubiaceae (I 7). 1 E Cuba (serpentine): ***C. involucratus*** (Wernham) Aiello

Ceuthostoma L. Johnson (~ *Casuarina*). Casuarinaceae. 2 Palawan & Borneo to NG. R: Telopea 3(1988)133

cevadilla *Schoenocaulon officinale*

Cevallia Lag. Loasaceae (III). 1 SW N Am.: ***C. sinuata*** Lag. – connective with long process, G 1, ovule 1, pend. R: Madroño 19(1967)7

Ceylon cedar *Melia azedarach*; **C. ebony** *Diospyros ebenum*; **C. gooseberry** *Dovyalis hebecarpa*; **C. gurjun** *Dipterocarpus zeylanicus*; **C. mahogany** *M. azedarach*; **C. olive** *Elaeocarpus serratus*; **C. satinwood** *Chloroxylon swietenia*; **C. spinach** *Basella alba*

Chaboissaea Fourn. = *Muhlenbergia*

Chabrea Raf. (~ *Peucedanum*). Umbelliferae (III 10). 1 Eur.: ***C. carvifolia*** (Jacq.) Raf.

Chacaya Escal. = *Ochetophila*

Chacoa R. King & H. Robinson (~ *Eupatorium*). Compositae (Eup.-Crit.). 1 Paraguay, Arg.: ***C. pseudoprasiifolia*** (Hassler) King & H. Robinson. R: MSBMBG 22(1987)332

Chadsia Bojer. Leguminosae (III 16). 9 Madag. Fls scarlet, some cauliflorous

Chaenactis DC. Compositae (Chaen.). 27 W N Am. (17 – R: FNA 21(2006)400), Mex. R: CDH 3(1940)89. Some cult. orn. incl. ***C. glabriuscula*** DC. (n = 6, Calif.) which has given rise by aneuploidy to ***C. fremontii*** A. Gray (n = 5) & ***C. stevioides*** Hook. & Arn. (n = 5)

Chaenanthe Lindl. = *Comparettia*

Chaenomeles Lindl. Rosaceae (Ros.-Mal.). 3 E As. R: JAA 45(1964)302. CJB 68(1990)2232. 'Japonica' – spp. hybridized, these & parents but esp. ***C. speciosa*** (Sweet) Nakai (Jap. quince, China) & ***C.* × *superba*** (Frahm) Rehder (*C. speciosa* × *C. japonica* (Thunb.) Spach) grown as spring-flowering shrubs, oft. against walls, favourite bonsai subjects in Jap.; fr. made into preserves, that of *C. japonica* used to scent rooms in China

Chaenorhinum (DC.) Reichb. Plantaginaceae (Ant.; Scrophulariaceae s.l.). Incl. *Albraunia*, *Holzneria*, 21 Medit. (esp. Eur., 12), natur. temp. R: D.A. Sutton, *Rev. Antirrhineae* (1988)97; differs from *Antirrhinum* in spurred fls. Some disp. as tumbleweeds with seeds retained in capsules; some cult. orn.

Chaenostoma Benth. (~ *Sutera*). Scrophulariaceae (Man.). c. 45 trop. & S Afr. ***C. cordatum*** (Thunb.) Benth. (*S. c.*, *Bacopa* 'Snowflake', SE Cape) – cult. orn. esp. hanging baskets

Chaerophyllopsis H. Boissieu. Umbelliferae (III 8). 1 W China: ***C. huai*** H. Boissieu

Chaerophyllum Tourn. ex L. Umbelliferae (III 2). Incl. *Oreomyrrhis*, c. 65 N temp. (Eur. 12). R: EJB 58(2001)339. ***C. bulbosum*** L. (turnip-rooted chervil, Eur., natur. US) – ed. carrot-like taproot, s.t. cult.

Chaetacanthus Nees = *Dyschoriste*

Chaetachme Planch. Cannabaceae (Ulmaceae II). 1 trop. & S Afr., Madag.: ***C. aristata*** Planch. – wood s.t. used for musical instruments

Chaetacme Planch. = *Chaetachme*

Chaetadelpha A. Gray ex S. Watson. Compositae (Cich.-Cich.). 1 SW US: ***C. wheeleri*** A. Gray ex S. Watson – zig-zag twigs. R: FNA 19(2006)368

Chaetanthera Ruíz & Pavón. Compositae (Mut.-Mut.). Excl. *Oriastrum*, 30 S Peru, Chile, Andes of Arg. R: A. Davies, *C. & O.* (2010)122. Cushion-pls incl. ***C. ramosissima*** D. Don (*C. tenella*, Chile)

Chaetanthus R. Br. (~ *Leptocarpus*). Restionaceae. 3 SW Aus.

Chaetium Nees. Gramineae (XXIV 5). 3 trop. Am. R: AMBG 85(1998)416

Chaetobromus Nees. Gramineae (Danth.). 1 SW Cape to Namaqualand: ***C. involucratus*** (Schrad.) Nees (*C. dregeanus*). R: NJB 18(1998)70

Chaetocalyx DC. Leguminosae (III 13). 13 trop. Am. Twiners

Chaetocarpus Thw. Peraceae (Chaet.; Euphorbiaceae s.l.). 13 trop. Am. (8 WI), W Afr. (1), Madag., As., W Mal. (1: ***C. castanicarpus*** (Roxb.) Thwaites (trop. As., W Mal.) – good timber, young lvs ed.)

Chaetocephala Barb. Rodr. = *Myoxanthus*

Chaetochlamys Lindau = *Justicia*

Chaetolepis (DC.) Miq. Melastomataceae. 1 W Afr. (*Nerophila*), 10 trop. Am. R: JBRIT 7(2013)233. Only amphi-Atlantic M. genus; no appendages to connective

Chaetolimon (Bunge) Lincz. (~ *Acantholimon*). Plumbaginaceae (II). 3 C As.

Chaetonychia (DC.) Sweet (~ *Paronychia*). Caryophyllaceae (I 2). 1 W Medit.: ***C. cymosa*** (L.) Sweet

Chaetopappa DC. Compositae (Ast.-Chaet.). 11 (excl. *Pentachaeta*) SW N Am. R: Phytol. 64(1988)448. ***C. ericoides*** (Torrey) Nelson – local medic.

Chaetopoa C.E. Hubb. Gramineae (XXIV 1). 2 Tanzania

Chaetopogon Janchen = *Agrostis*

Chaetoptelea Liebm. = *Ulmus*

Chaetosciadium Boiss. (~*Torilis*). Umbelliferae. 1 E Medit.: ***C. trichospermum*** (L.) Boiss. R: Fl. Iraq 5,2(2013)282

Chaetoseris Shih = *Cicerbita* (but see APS 29(1991)398)
Chaetospira S.F. Blake = *Pseudelephantopus*
Chaetostachydium Airy Shaw (~ *Psychotria*). Rubiaceae (IV 7). 3 NG. Pachycaul
Chaetostichium C.E. Hubb. = *Oropetium*
Chaetostoma DC. Melastomataceae. 11 Braz. G 3-loc.
Chaetosus Benth. = *Parsonsia*
Chaetothylax Nees = *Justicia*
Chaetotropis Kunth = *Polypogon*
Chaeturus Link = *Agrostis*
Chaetymenia Hook. & Arn. Compositae (Bah.). 1 Mex.: ***C. peduncularis*** Hook. & Arn.
chaff-flower *Achyranthes* spp., *Alternanthera* spp.
chaguar fibre *Bromelia serra*
chahomilia *Salvia fruticosa*
Chailletia DC. = *Dichapetalum*
chairmaker's rush *Schoenoplectus pungens*
Chaiturus Willd. (~ *Leonurus*). Labiatae (VI). 1 W Eur. to C As.: ***C. marrubiastrum*** (L.) Reichb.
Chalarothyrsus Lindau. Acanthaceae (III 2c). 1 Mex.: ***C. amplexicaulis*** Lindau
Chalcanthus Boiss. (~ *Eutrema*). Cruciferae (28). 1 Iran mts, Afghanistan & C As.: ***C. renifolius*** (Boiss. & Hohen.) Boiss.
Chalema Dieterle = *Sicydium*
Chalepophyllum Hook.f. Rubiaceae (I 6). 5 Venez., Guyana. 2 bracteoles; K unequal
chalice vine *Solandra maxima*
Chalybea Naudin (~ *Pachyanthus*). Melastomataceae. Incl. *Huilaea*, 5 trop. Am.
Chamabainia Wight. Urticaceae (III). 1–2 Indomal., Taiwan
Chamaeacanthus Chiov. = *Campylanthus*
Chamaealoe A. Berger = *Aloe*
Chamaeangis Schltr. = *Diaphananthe*
Chamaeanthus Schltr. Orchidaceae (V 16c). 3 Thailand to Borneo
Chamaeanthus Ule = *Geogenanthus*
Chamaebatia Benth. Rosaceae (Ros.-Dryad.). 2 Calif. & Baja Calif. R: FNA 9(2014)343. Morph. similar to *seq*. Glandular, aromatic, everg. shrubs (mountain misery) forming dense undergrowth. ***C. foliolosa*** Benth. (Sierra Nevada) – cult. orn. with nitrogen-fixing nodules
Chamaebatiaria (Brewer & Watson) Maxim. Rosaceae (Ros.-Sorb.). 1 W US: ***C. millefolium*** (Torrey) Maxim. – lvs 2-pinnate (cf. *Spiraea*), cult. orn. R: FNA 9(2014)394
Chamaecereus Britton & Rose = *Echinopsis*
Chamaechaenactis Rydb. Compositae (Bah.). 1 SW US: ***C. scaposa*** (Eastw.) Rydb. R: FNA 21(2006)395
Chamaeclitandra (Stapf) Pichon. Apocynaceae (1 a). 1 trop. Afr.: ***C. henriquesiana*** (Warb.) Pichon. R: BJBB 58(1988)165
Chamaecostus C. Specht & Stevenson (~ *Costus*). Costaceae. 8 S Am. R: T: 55(2006)157
Chamaecrista Moench (~ *Cassia*). Leguminosae (I 3). c. 330 trop. to temp. Am., E As. (Afr. 36 – R: KB 43(1988)335; Am. esp. Braz. 266 – R: MNYBG 35(1982)636). Differs from *Cassia* s.s. in A all straight & *Senna* in its 2 bracteoles. Trees, shrubs & herbs; local medic., some cult. orn. incl. ***C. fasciculata*** (Michaux) Greene (*Cassia f.*, golden cassia, prairie senna, N Am.); ***C. rotundifolia*** (Pers.) Greene (S Am.) – introd. Aus. as pasture herb
Chamaecrypta Schltr. & Diels = *Diascia*
Chamaecyparis Spach. Cupressaceae. 5 E As., N Am., rather local. R: A. Farjon, *Monogr. Cupressaceae & Sciadopitys* (2005)160. Monoec.; juvenile lvs s.t. needle-like whereas all adult lvs scale-like, adpressed; seeds winged, 2–5 per scale, maturing in 1 yr cf. *Cupressus* where many per scale taking 2 yrs. Comm. timbers & cult. orn. esp. ***C. lawsoniana*** (A. Murray bis.) Parl. (Lawson's cypress, Port Orford or Oregon cedar, NW N Am.) – one of most imp. timbers of Pac. NW & v. imp. cult. orn. (Attims's Model; introd. GB 1854) with many cvs, incl. blue & golden forms, crushed foliage smells like parsley, lvs with white markings beneath; timber for house interiors incl. floors, boats, fences, sleepers, matches etc.; ***C. obtusa*** (Sieb. & Zucc.) Endl. (hinoki or Jap. cypress, Jap., Taiwan) – some allegedly 3000 yrs old in Taiwan, boles used as solid board-tables, plantation logs with swellings at intervals (tawaro-shibo) due to abnormal cambial activity prized for alcove posts in Jap. drawingrooms, bonsai subject; ***C. pisifera*** (Sieb. & Zucc.) Endl. (sawara cypress, Jap.); ***C. thyoides*** (L.) Britton, Sterns & Pogg. (S white cedar, E N Am.). See also *Xanthocyparis*

Chamaecytisus Link = *Cytisus*

Chamaedaphne Moench. Ericaceae (VIII 4). 1 N temp. (incl. Eur.): ***C. calyculata*** (L.) Moench – cult. orn., form. used as tea by Native Americans. R: FNA 9(2009)507

Chamaedorea Willd. Palmae (V 2). 77–108 trop. Am. R: D.R. Hodel (1992) *C. palms*. Dioec., unarmed, understorey palms, many tufted. ***C. cataractarum*** Mart. (C Am.) – Schoute's Model. Some ed. incl. fr., many cult. orn. with bamboo-like stems, surviving in dim light, ***C. elegans*** Mart. (Mex., Guatemala) being the most common house-palm ('*Neanthe bella*') in SE US etc. (fls smell of blood; fr. ed. by Native Americans; unopened spathes may be prep. like asparagus as may those of ***C. graminifolia*** H. Wendl. (Costa Rica), ***C. sartorii*** Liebm. (Mex. to Honduras) & ***C. tepejilote*** Liebm. (pacaya, Mex. to Colombia), a sp. cult. for this delicacy esp. in Guatemala), while trade in ***C. seifrizii*** Burret (Mex. to Honduras) worth $100 M a yr)

Chamaegastrodia Makino & F. Maek. (~ *Odontochilus*). Orchidaceae (IV 2b). 3 E As.

Chamaegeron Schrenk. Compositae (Ast.-Hom.). 4 C As.

Chamaegigas Dinter (~ *Lindernia*). Linderniaceae (Plantaginaceae (Grat.-Lind.); Scrophulariaceae s.l.). 1 Namibia: ***C. intrepidus*** Dinter – poikilohydric. R: TSP 81(1992)328

Chamaegyne Süsseng. = *Eleocharis*

Chamaeiris Medik. = *Iris*

Chamaelaucium DC. = *Chamelaucium*

Chamaele Miq. Umbelliferae (III 8). 1 Jap.: ***C. decumbens*** (Thunb.) Mak.

Chamaeleon Cass. = *Carlina*

Chamaeleorchis Senghas & Lueckel = *Oncidium*

Chamaelirium Willd. Melanthiaceae. 1 E N Am.: ***C. luteum*** (L.) A. Gray (blazing star, fairy wand, unicorn r.) – dried tubers medic., diuretic etc., cult. orn. R: FNA 26(2002)68

Chamaemeles Lindl. Rosaceae (17). 1 Madeira: ***C. coriacea*** Lindl. With *Musschia*, M's only endemic genera; G 1

Chamaemelum Mill. (~ *Anthemis*). Compositae (Anth.-Sant.). (Excl. *Cladanthus*) 2 Eur. (2), Medit., Canary Is. R: FNA 19(2006)496. ***C. nobile*** (L.) All. (chamomile, S & W Eur., Medit.) – source of Oil of Roman Chamomile, a light blue (when fresh) oil distilled from heads particularly of a double-flowered form (single or 'Ger.' c. considered inferior), used in flavouring liqueurs, as a tea, & esp. for hair shampoos (esp. blonde hair) & many other cosmetics; form. cult. as lawns with minimum maintenance & good drought tolerance (cf. grass) before mowing simplified through mechanization, still mixed with grass under heavy pressure (as at Buckingham Palace, London) esp. the sterile **'Treneague'**, the scented lvs giving off perfume when crushed, seats also planted with this, though allegedly allergenic to some; fls used medic. (oil mitigates psoriasis), incl. ointment for boils

Chamaemespilus Medik. = *Sorbus*

Chamaenerion Séguier (~ *Epilobium*). Onagraceae. 8 N temp. R: FGP 7(1972)81. ***C. angustifolium*** (L.) Holub (*E. a.*, *Chamerion a.*, rose-bay willow-herb, fireweed, wickup) – distr. greatly increased in GB (highland populations poss. indig.; lowland from Am. or Eur. in C20, poss. through increase in habitats e.g. bomb-sites), autogamy almost impossible (highly protandrous: dichogamy first described by Sprengel from this pl.), r. live for 20 yrs, lvs ed. as greens, local medic. & mosquito repellent & for smoking fish by Native Americans & used by them & in Russia for tea, pollen concentrates gold, excellent honey

Chamaepentas Bremek. Rubiaceae (IV 1). 1 trop. E Afr.: ***C. greenwayi*** Bremek.

Chamaepericlymenum Hill = *Cornus*

Chamaepus Wagenitz. Compositae (Gnap.-Gnap.). 1 Afghanistan: ***C. afghanicus*** Wagenitz. R: OB 104(1991)171

Chamaeranthemum Nees = *Chameranthemum*

Chamaeraphis R. Br. Gramineae (XXIV 4). 1 N Aus.: ***C. hordeacea*** R. Br.

Chamaerhodiola Nakai = *Rhodiola*

Chamaerhodos Bunge. Rosaceae (Ros.-Pot.). 5 C & E As., 1 W N Am.

Chamaerops L. Palmae (III 4a). 1 W Med.: ***C. humilis*** L. – only mainland Eur. palm, clump-forming in maquis or ± arborescent in light woodland, fragrance from leaf glands attractive to poll. weevils; cult. orn. (under which Goethe had inspiration) & lvs source of veg. 'horse' hair (Algerian fibre, *crin végétal*) used in upholstery; young buds ed. R: J. Dransfield & al., *Gen. Palm.* (2008)247

Chamaesaracha (A. Gray) Benth. = *Physalis* (but see Rhodora 75(1973)325)

Chamaesciadium C. Meyer (~ *Trachydium*). Umbelliferae (III 8). 1 W As.: ***C. acaule*** (M.Bieb.) Boiss.

Chamaescilla F. Muell. ex Benth. Asphodelaceae. 4 Aus. R: Nuytsia 13(2001)476
Chamaesium H. Wolff. Umbelliferae (III 8). 5 Himal., Tibet, W China
Chamaespartium Adans. = *Genista*
Chamaesphacos Schrenk ex Fischer & C. Meyer. Labiatae (VI). 1 Iran to Kazakhstan & W China, Afghanistan: ***C. ilicifolius*** Schrenk ex Fischer & C. Meyer
Chamaesyce Gray = *Euphorbia*
Chamaexeros Benth. Asparagaceae (Lomandraceae (Xanthorrhoeaceae s.l.)). 4 SW Aus. R: Nuytsia 2(1976)118
Chamaexiphion Hochst. ex Steud. = *Ficinia*
Chamarea Ecklon & Zeyher. Umbelliferae (III 8). 5 S Afr. R: EJB 48(1991)200, 261
chambala *Solanum cajanumense*
Chambeyronia Vieill. Palmae (V 14a). 2 New Caled. R: Allertonia 3(1984)330
Chamelaucium Desf. Myrtaceae (II 15). Excl. *Homoranthus*, 13 W Aus. Heath-like. 'Waxflowers'. Insect- & bird-poll. spp.; extrafl. nectaries in leaf axils (first found in M.). ***C. uncinatum*** Schauer (Geraldton wax (flower)) & others cut-fls in Eur., incl. hybrids (even with *Verticordia plumosa* (Desf.) Druce = **'Eric John'**), with ***C. megalopetalum*** F. Muell. ex Benth. = '**Mega White**' used in Sydney Olympics bouquets (2000)
Chamelophyton Garay. Orchidaceae (V 13c). 1 Venez., Guyana: ***C. kegelii*** (Reichb.f.) Garay
Chamelum Philippi = *Olsynium*
Chameranthemum Nees. Acanthaceae (III 2c). 4 trop. Am. Cult. orn. with varieg. lvs esp. ***C. gaudichaudii*** Nees (Braz.)
Chamerion (Raf.) Holub = *Chamaenerion*
chamfuta *Afzelia quanzensis*
Chamguava Landrum (~ *Psidium*). Myrtaceae (II 10). 3 C Am. R: SB 16(1991)21
Chamira Thunb. Cruciferae (inc. sed.). 1 S Afr.: ***C. circaeoides*** (L.f.) A. Zahlbr. – basal 'lvs' (cotyledons) opp. R: Strel. 10(2000)187
Chamissoa Kunth. Amaranthaceae (I 2). 2 warm Am. Aril
Chamissoniophila Brand = *Antiphytum*
chamomile (tea) *Chamaemelum nobile*; **corn c.** *Anthemis arvensis*; **dyer's** or **golden c.** *Cota tinctoria*; **German c.** *Matricaria chamomilla*; **scentless c.** *Tripleurospermum inodorum*; **stinking c.** *A. cotula*; **sweet c.** *Chamaemelum nobile*; **wild c.** *M. chamomilla*; **yellow c.** *Cota tinctoria*
Chamomilla Gray = *Matricaria*
Chamorchis Rich. (~ *Herminium*). Orchidaceae (IV 4d). 1 N Scandinavia, Alps, Carpathians: ***C. alpina*** (L.) Rich. R: Gen. Orch. 2(2001)271
champa, champak *Magnolia champaca*
Champereia Griff. Opiliaceae. 1 Indomal.: ***C. manillana*** (Blume) Merr. – root-parasite, veg., fr. ed., medic.
Championella Bremek. = *Strobilanthes*
Championia Gardner. Gesneriaceae (III 2d?). 1 Sri Lanka: ***C. reticulata*** Gardner – K(5), C(4)
chan *Shorea* spp.
chanal or **chañar** *Geoffroea decorticans*
Chandrasekharania Nair & al. Gramineae (?XVII). 1 Kerala: ***C. keralensis*** Nair & al. R: Proc. Ind. Acad. Sci. (Pl. Sci.) 91(1982)79
Changiodendron Miau = *Sabia*
Changiostyrax C.T. Chen (~ *Sinojackia*). Styracaceae. 1 SE China: ***C. dolichocarpus*** (Qi) C.T. Chen. R: Guihaia 15(1995)289
Changium H. Wolff (~ *Conopodium*). Umbelliferae (III 5). 2 Tibet, E China
Changnienia Chien. Orchidaceae (V 13e). 1 C & E China: ***C. amoena*** Chien. R: KM 10 (1993)52
Changruicaoia Z.Y. Zhu = *Heterolamium*
channel millet *Echinochloa inundata, E. turneriana*
cha-om *Senegalia pennata*
chaparral *Larrea tridentata* (but also a vegetation-type)
Chapelieria A. Rich. ex DC. (~ *Feretia*). Rubiaceae (Octotrop.). 2 Madag. Roux's Model
chaplash *Artocarpus chama*
Chapmannia Torrey & A. Gray. Leguminosae (III 11). Incl. *Arthrocarpum, Pachecoa*, 5 Horn of Afr. (4 Socotra, Somalia), 1 Florida, 1 C Am. & Venez. (? introd.). R: NJB 19(1999)598
Chapmanolirion Dinter = *Pancratium*
Chaptalia Vent. (~ *Gerbera*). Compositae (Mut.-Mut.). c. 70 warm Am. R: Darw. 6(1944) 505
Chara Vaill. ex L. Characeae. 116+ cosmop. (Aus. 17). Known since Upper Cretaceous

Characeae Agardh. Charophyta-Charales. 6/c. 320 (37 genera extinct) cosmop. Stoneworts. Freshwater (to 60 m depth) algae s.t. encrusted in lime when mature. Axis comprising giant multinucleate cells; branchlets in whorls
Genera: *Chara* (116+), *Lamprothamnium* (5+), *Lychnothamnus* (1), *Nitellopsis* (1), *Nitella* (180+), *Tolypella* (5+)
Most complex structure of any green alga. Extant part of sister group to land pls (see Appendix). Large cells used in electrical experiments etc. in pl. physiology; cytoplasmic streaming readily observed

Charadranaetes Janovec & H. Robinson. Compositae (Sen.-Sen.). 1 Costa Rica: ***C. durandii*** (Klatt) Janovec & H. Robinson. R: Novon 7(1997)165

Charadrophila Marloth. Stilbaceae (Scrophulariaceae s.l.). 1 S Afr.: ***C. capensis*** Marloth – long referred to Gesneriaceae. R: BJ 111(1989)83

Charales Lindl. Charophyta. 6 fams, 1 extant: Characeae. Sister to land pls.; oldest fossils from Late Ordovician (470 M yrs ago). See Appendix

charas *Cannabis sativa*

Chardinia Desf. Compositae (Card.-Card.). 1 W As.: ***C. orientalis*** (L.) Kuntze

chards *Cynara scolymus* (blanched summer shoots), *Tragopogon porrifolius* (young flowering shoots); **rhubarb, Swiss c.(s)** *Beta vulgaris*

Chareis auctt. = *Felicia*

Charia C. DC. = *Ekebergia*

Charianthus D. Don = *Miconia* (but see SB 30(2005)572)

Charidion Bong. = *Luxemburgia*

Charieis Cass. Older name for *Felicia*

Chariessa Miq. = *Citronella*

charlock *Sinapis arvensis*

Charpentiera Gaudich. Amaranthaceae (I 2). 6 Hawaii (5), Austral Is., Cook Is. R: Britt. 24(1972)283. Trees! Austral Is. c. 4500 km from Hawaii! Highly flammable dry timber lit & thrown over cliffs in pyrotechnical displays

Chartolepis Cass. = *Centaurea*

Chartoloma Bunge. Cruciferae (33). 1 C As.: ***C. platycarpum*** (Bunge) Bunge

Chartreuse Liqueur first made 1735, perfected to yellow form c. 1840, an infusion of over 100 herbs incl. *Angelica archangelica, Campanula alpestris, C. carnica, Gentiana bavarica, Hyssopus officinalis, Saxifraga* sp. etc. manufactured by 3 monks & aged in 100 000 litre barrels

Charybdis Speta = *Drimia*

Chasallia Comm. ex Poir. = *Chassalia*

Chascanum E. Meyer (~ *Bouchea*). Verbenaceae (Dur.). 25 Afr., Madag., Arabia to W India. R: FR 45(1938)114

Chascolytrum Desv. (~ *Briza*). Gramineae (XVI 5). Incl. *Rhombolytrum*, 28 S Am.

Chascotheca Urb. (~ *Securinega*). Phyllanthaceae (Phyll.-Wiel.; Euphorbiaceae s.l.). 1 Cuba, Hispaniola: ***C. neopeltandra*** (Griseb.) Urb.

Chasechloa A. Camus = *Echinolaena*

Chaseella Summerh. Orchidaceae (V 15). 1 C & E Afr.: ***C. pseudohydra*** Summerh. R: OM 2 (1987)164

Chaseopsis Szlach. & Sitko = *Maxillaria*

Chasmanthe N.E. Br. Iridaceae (VI 5). 3 SW Cape. R: SAJB 51(1985)253. Seeds bright orange (usu. brown in Irid.). Cult. orn. like *Gladiolus*. ***C. floribunda*** (Salisb.) N.E. Br. – invasive in Aus.

Chasmanthera Hochst. Menispermaceae (III). 2 trop. Afr. Some ed. r.

Chasmanthium Link. Gramineae (XIX). Incl. *Bromuniola*, 7 trop. Afr., E N Am., Mex. R: JAA 71(1990)176. Only temp. genus in tribe

Chasmatocallis R. Foster = *Lapeirousia*

Chasmatophyllum Dinter & Schwantes. Aizoaceae (V). 8 SW Afr., Cape. R: H. E. K. Hartmann, *Ill. Handb. Succ. Pls*, Aiz. A–E (2002)115. Cult. orn. shrubby succ.

Chasmopodium Stapf. Gramineae (XXII). 3 W trop. Afr.

Chassalia Comm. ex Poir. Rubiaceae (IV 7). c. 40 OW trop. esp. Madag. (c. 28). Chamberlain's Model

chaste tree *Vitex agnus-castus*

chats *Solanum tuberosum* (undersized tubers usu. fed to stock)

Chaubardia Reichb.f. Orchidaceae (V 12j). 3 trop. S Am., Trinidad

Chaubardiella Garay. Orchidaceae (V 12j). 8 trop. Am.
Chauliodon Summerh. Orchidaceae (V 16d). 1 trop. W Afr.: ***C. deflexicalcaratum*** (De Wild.) L. Jonss. – leafless
chaulmoogra oil *Hydnocarpus kurzii*
Chaunanthus O. Schulz (~ *Iodanthus*). Cruciferae (46). 4 Mex. R: NJB 32(2014)133
Chaunochiton Benth. Olacaceae (Aptandraceae). 3 trop. Am.
Chaunostoma J.D. Sm. = *Lepechinia* (but see BG 20(1895)9)
Chautemsia Araujo & Souza. Gesneriaceae (II 5b). 1 Braz. (Minas Gerais): ***C. calcicola*** Araujo & Souza. R: T 59(2010)207
Chavanessia A. DC. = *Urceola*
chay (root) *Hedyotis indica*
chaya *Cnidoscolus chayamansa*
Chayamaritia Middleton & Mich. Möller (~ *Henckelia*). Gesneriaceae (III 2j). 2 Laos, Thailand. R: PSE 301(2015)1961
Chaydaia Pitard = *Rhamnella*
chayote *Sicyos edulis*
chayotilla *Hanburia mexicana*
Chazaliella Petit & Verdc. = *Eumarchia*
cheat or **chess** *Bromus secalinus*
checkerberry *Gaultheria procumbens*
cheeseberry *Rubus ellipticus*
Cheesemania O. Schulz = *Pachycladon*
cheeseplant *Monstera deliciosa*
cheeses *Malva* spp. (frs resemble flat c.)
cheeseweed *Ambrosia salsola*
cheesewood *Pittosporum undulatum*
Cheilanthes Sw. Pteridaceae (IV). Excl. *Allosorus, Cheiloplecton, Gaga, Myriopteris*, 150 subcosmop. (Aus. 15 (R: Telopea 4(1991)509)), esp. Andes, Mex., S Afr. Usu. dry rocky sites – ferns with xeromorphic characters; some cult. orn. ***C. tenuifolia*** (Burm.f.) Sw. (Indomal.) – hair tonic
Cheilanthopsis Hieron. = *Woodsia*
Cheiloclinium Miers. Celastraceae (IV). 11 trop. Am.
Cheilocostus C. Specht = *Hellenia*
Cheilophyllum Pennell. Plantaginaceae (Grat.-Stem.; Scrophulariaceae s.l.). 8 WI
Cheiloplecton Fée (~ *Cheilanthes*). Pteridaceae (IV). 1 Braz.: ***C. rigidum*** (Sw.) Fée
Cheilosa Blume. Euphorbiaceae (Per.-Cheil.). 1 W Mal.: ***C. montana*** Blume – fr. ed. & fermented. R: Blumea 38(1993)161
Cheilosoria Trevis. = *Pellaea*
Cheilotheca Hook.f. Ericaceae (II 3; Monotropaceae). 2 Assam to W Mal.
Cheiradenia Lindl. Orchidaceae (V 12j). 1 NE S Am.: ***C. cuspidata*** Lindl.
Cheiranthera A. Cunn. ex Brongn. Pittosporaceae. 10 SW & SE Aus. R: AusSB 20(2007)342
Cheiranthus L. = *Erysimum*
Cheiridopsis N.E. Br. Aizoaceae (V). 29 S Afr. R: H. E. K. Hartmann, *Ill. Handb. Succ. Pls*, Aiz. A–E (2002)117. Clump-forming succ., cult. orn.
Cheirodendron Nutt. ex Seemann. Araliaceae. 5 Hawaii, 1 Marquesas (***C. bastardianum*** (Decne) Frodin)
Cheiroglossa C. Presl = *Ophioglossum*
Cheirolaena Benth. Malvaceae (Malv.-Dom.; Sterculiaceae). 1 Madag.: ***C. linearis*** Benth.
Cheirolophus Cass. (~ *Centaurea*). Compositae (Card.-Cent.). 25 SW Eur., N Afr., Canary Is.
Cheiropleuria C. Presl. Dipteridaceae (Cheiropleuriaceae). 3 SE As. & Honshu to E Mal. R: Blumea 46(2001)521
Cheiropleuriaceae Nakai = Dipteridaceae
Cheirorchis Carr = *Cordiglottis*
Cheirostemon Bonpl. = *Chiranthodendron*
Cheirostylis Blume. Orchidaceae (IV 2b). 53 OW trop. (Aus. 1). Some autogamy
Chelidonium Tourn. ex L. Papaveraceae (I). 1 temp. & subarctic Euras. (natur. E US): ***C. majus*** L. (greater celandine, swallow-wort). R: FGP 17(1982)237. Differs from *Papaver* in stalked stigmas, arillate seeds etc. Alks. In GB in last interglacial (Ipswichian) but generally considered intr. by man in this one (Mediaeval, but poss. Romans) for medic. properties of corrosive orange latex long used in eye disorders & for cancers in

Russia, classically mixed with fennel, wormwood, honey & a dash of human milk, certainly efficacious in treatment of warts, corns, suntan, freckles & other skin disorders; mutant forms with epiphyllous infls or branches, double fls & the ragged 'var. *laciniatum* (Mill.) Syme', a single-gene mutant with the effect of inhibiting intercalary growth of laminae & petals which arose at Heidelberg c. 1590, s.t. cult.

Chelonanthera Blume = *Pholidota*

Chelonanthus (Griseb.) Gilg = *Helia*

Chelone Dill. ex L. Plantaginaceae (Ant.-Chel.; Scrophulariaceae s.l.). 4 E N Am. Cult. orn. (shellflowers, turtlehead) allied to *Penstemon*, esp. ***C. glabra*** L. (balmony, E N Am.) – medic., vermifuge, iridoids (catalpol sequestered by *Euphydryas phaeton* butterflies, some Coleoptera & some Hymenoptera, which also sequester aucubin)

Chelonespermum Hemsl. = *Burckella*

Chelonistele Pfitzer. Orchidaceae (V 10b). 13 W Mal. (c. 10 restricted to Borneo). R: OM 1(1986)23

Chelonopsis Miq. Labiatae (VI). Incl. *Bostrychanthera*, 15 E As. R: NJB 26(2008)31

Chelyella Szlach. & Sitko = *Maxillaria*

Chelyocarpus Dammer. Palmae (III 2). 4 trop. S Am. R: Principes 16(1972)67, Palms 55(2011)73

Chelyorchis Dressler & N. Williams = *Rossioglossum* (but see JTBS 136(2009)181)

Chelystachya Mytnik & Szlach. = *Polystachya*

chempedak *Artocarpus integer*

chen pi *Citrus reticulata*

chenar *Platanus orientalis*

chengal *Neobalanocarpus heimii*

Chengiopanax Shang & J.Y.Huang (~ *Eleutherococcus*). Araliaceae. 3 China. R: BBR 13(1993)47

Chennapyrum Löve = *Aegilops*

Chenolea Thunb. (~ *Bassia*). Amaranthaceae (Chenopodiaceae). 2 S Afr. R: T 60(2011)71

Chenopodiaceae Vent. = Amaranthaceae

Chenopodiastrum S. Fuentes & al. Amaranthaceae (Atrip.). 6–7 N temp. R: Willd. 42(2013)14

Chenopodiopsis Hilliard. Scrophulariaceae (Manul.). 3 S Afr. R: EJB 47(1990)339

Chenopodium Tourn. ex L. Amaranthaceae (Atripl.; Chenopodiaceae I 2). Incl. *Einadia*, *Rhagodia*, excl. *Blitum*, *Chenopodiastrum*, *Lipandra*, *Oxybasis*, c. 100 temp. Small trees, shrubs but mostly weedy herbs, some grains (selection for increased fr. size, low dormancy & non-shattering infls (cf. cereals); saponins washed out before prep.), cult. orn. & medic., imp. sheep forage (*R.* spp., Aus.), etc. ***C. album*** L. (goosefoot, fat-hen, lambsquarters, N temp.) – form. a veg. (still used by Native Americans), ousted by spinach, fr. taken by poultry, grains ground into bread-flour, local medic. N Am.; ***C. berlandieri*** Moq. (C Am.) – veg. (huauzontle) Mex. City; ***C. giganteum*** D. Don (N Ind.) – ann. veg. to several m tall; ***C. macrospermum*** Hook.f. Am. (bipolar; poss. N Am. pops introd.); ***C. oahuense*** (Meyen) Aellen (Hawaiian goosefoot) – potherb; ***C. pallidicaule*** Aellen (canihua, Andes) – imp. grain; ***C. parabolicum*** (R. Br.) S. Fuentes & Borsch (*R. p.*, (Aus.)) – red, yellow or white fr. or combinations of these, foraged at random by *Zosterops* sp., principal avian consumer but seeds from diff. morphs respond differently, red germ. fastest, then yellow but white greatest response to passage through bird-gut, these diff. factors poss. explaining polymorphism; ***C. quinoa*** Willd. (quinoa, quinua, red Inca quinoa, cultigen allied to *C. berlandieri*, Andes) – similar, both high in amino-acids but ousted by cereals intr. by colonists, but now fashionable (no gluten). See also *Dysphania*

Chenorchis Z.J. Liu = *Holcoglossum*

chequers *Sorbus torminalis*

cherimoya *Annona cherimola*

Cherleria Haller ex L. (~ *Minuartia*). Caryophyllaceae. 23 Euras., W N Am.

Cherokee rose *Rosa laevigata*

cheronjee *Buchanania lanzan*

cherry *Prunus* spp., the ed. ones derived from *P. avium* & *P. cerasus*; **African c.** *Tieghemella heckelii*; **amarelle c.** *P. cerasus*; **American** or **black c.** *P. serotina*; **Barbados c.** *Malpighia emarginata*; **bird c.** *P. padus*; **Bohemian c.** *Cornus mas*; **Braz. c.** *Eugenia* spp.; **cabinet c.** *P. serotina*; **capulin c.** *P. salicifolia*; **Cayenne c.** *Eugenia uniflora*; **choke c.** *P. virginiana*; **Cornelian c.** *Cornus mas*; **Duke c.** *P.* × *gondouinii*; **finger c.** *Rhodomyrtus macrocarpa*; **flowering c.** several *P.* spp. & hybrids but esp. *P. serrulata* cvs & *P.* × *yedoensis*; **Fuji c.**

P. incisa; **Jamaican c.** *Muntingia calabura*; **c. laurel** *P. laurocerasus*; **Liberian c.** *Sacoglottis gabonensis*; **maraschino c.** *P. cerasus* 'Marasca'; **morello c.** *P. cerasus*; **Oshima c.** *P. serrulata*; **c. pie** *Heliotropium arborescens*; **Pitanga c.** *Eugenia uniflora*; **c. plum** *P. cerasifera*; **prairie c.** *P. gracilis*; **St Lucie c.** *P. mahaleb*; **sand c.** *P. pumila*; **Surinam c.** *E. uniflora*; **sweet** or **wild c.** *P. avium*; **winter c.** *Alkekengi officinarum, Solanum pseudocapsicum*; **Yoshino c.** *Prunus × yedoensis*

Chersodoma Philippi. Compositae (Sen.-Sen.). 9 Andes. R: Britt. 48(1997)591. Dioec.

chervil *Anthriscus cerifolium*; **turnip-rooted c.** *Chaerophyllum bulbosum*

Chesneya Lindl. ex Endl. Leguminosae (III 24). Incl. *Spongiocarpella*, excl. *Chesniella*, c. 24 SW & C As. R: PhytoKeys 70(2016)27

Chesniella Boriss. (~ *Chesneya*). Leguminosae (III 24). c. 10 As.

chess *Bromus secalinus*; **downy c.** *B. tectorum*

chestnut *Castanea sativa* (**common, Eur., Spanish** or **sweet c.**); **Amer. c.** *C. dentata*; **Aus. c.** *Castanospermum australe*; **Cape c.** *Calodendrum capense*; **China c.** *Sterculia monosperma*; **Chinese c.** *Castanea mollissima*; **golden c.** *Chrysolepis chrysophylla*; **horse c.** *Aesculus hippocastanum*; **Jap. c.** *Castanea crenata*; **golden c.** *Chrysolepis chrysophylla*; **Guiana c.** *Pachira aquatica*; **Moreton Bay c.** *Castanospermum australe*; **c. oak** *Quercus montana*; **Polynesian** or **Tahiti c.** *Inocarpus fagifer*; **c. vine** *Tetrastigma voinierianum*; **water c.** *Trapa natans*; **Chinese water c.** *Eleocharis dulcis*; **wild c.** (S Afr.) *Brabejum stellatifolium*

Chevaliera Gaudich. ex Beer (~ *Aechmea*). Bromeliaceae (3). 22 trop. Am. R: Phytol. 66(1989)77. ***C. magdalenae*** (André) André (*A. m.*, pita, C & S Am.) – fibre from lvs for rope, twine & thread for sewing leather

Chevalierella A. Camus. Gramineae (XIX). 1 Congo: ***C. dewildemanii*** (Vanderyst) Compère – false petioles

Chevreulia Cass. Compositae (Gnap.-Gnap.). 6 S Am., Falkland Is., Tristan da Cunha. R: BotJLS 106(1991)186, SB 36(2011)784

chewing-gum Orig. *Manilkara zapota* with sugar & flavourings, now patented subs. of resins, rubber etc.

Chewing's fescue *Festuca rubra* subsp. *commutata*

Cheyniana Rye (~ *Balaustion*). Myrtaceae (II 15). 2 SW Aus. R: Nuytsia 19(2009)141

chia seeds *Salvia* spp. esp. *S. columbariae* & *S. hispanica*

Chian turpentine *Pistacia terebinthus*

Chiangiodendron Wendt. Achariaceae (Flacourtiaceae s.l.). 1 Mex., Costa Rica: ***C. mexicanum*** Wendt – dioec. R: SB 13(1988)435. Only non-OW Pangieae

Chiapasophyllum Doweld = *Selenicereus*

Chiarinia Chiov. = *Lecaniodiscus*

Chiastophyllum (Ledeb.) A. Berger = *Umbilicus*

chibasa *Juncus* sp.

chibou *Bursera simaruba*

chica *Fridericia chica, Ramorinoa girolae*

Chichicaste Weigend (~ *Loasa*). Loascaceae (I 1). 1 Costa Rica to NW Colombia: ***C. grandis*** (Standl.) Weigend

chick pea *Cicer arietinum*

chicken claws *Salicornia europaea*

chickrassy *Chukrasia tabularis*

chickweed *Stellaria* & *Cerastium* spp.; **common c.** *S. media*; **greater c.** *S. neglecta*; **mouse-ear c.** *Cerastium* spp.; **c. wintergreen** *Trientalis europaea*

chicle *Manilkara zapota, Dyera costulata*

Chiclea Lundell = *Manilkara*

chicory *Cichorium intybus* (US = *C. endivia*)

Chidlowia Hoyle. Leguminosae (I 4). 1 trop. W Afr.: ***C. sanguinea*** Hoyle – explosively dehiscent pods

Chienia W.T. Wang = *Delphinium*

Chieniodendron Tsiang & P.T. Li = *Meiogyne*

Chigua Stevenson = *Zamia* (but see MNYBG 57(1990)169)

chihli *Brassica rapa* Pekinensis Group

Chihuahuana Urbatsch & R. Roberts (~ *Ericameria*). Compositae (Ast.-Sol.). 1 N Mex.: ***C. purpusii*** (Brandegee) Urbatsch & R. Roberts

chikanda Ed. tubers of terr. orchids esp. spp. of *Disa, Satyrium* (Zambia)

chiku *Manilkara zapota*

Chikusichloa Koidz. (~ *Leersia*). Gramineae (III 3 ii). 3 China, Jap., Ryukyu Is., Sumatra

Chile, Chilean or **Chili cedar** *Austrocedrus chilensis;* **C. crocus** *Tecophilaea cyanocrocus;* **C. jasmine** *Mandevilla laxa;* **C. needle grass** *Nassella neesiana;* **C. nut** *Gevuina avellana;* **C. pine** *Araucaria araucana;* **C. wine palm** *Jubaea chilensis*

Chileorchis Szlach. = *Chloraea*

Chileranthemum Oersted. Acanthaceae (III 2c). 2 Mex.

Chiliadenus Cass. = *Jasonia* (but see Webbia 94(1979)298)

Chilianthus Burch. (~ *Buddleja*). Scrophulariaceae (Buddlejaceae). 3 S Afr.

chilicote *Cucurbita foetidissima*

chil(l)ies *Capsicum annuum* Longum Group

Chiliocephalum Benth. (~ *Helichrysum*). Compositae (Gnap.-Gnap.). 1 Ethiopia: ***C. schimperi*** Benth. R: OB 104(1991)149

Chiliophyllum Philippi. Compositae (Ast.-Hin.). 1 C Arg.: ***C. densiflorum*** Philippi. R: SCB 92(2009)21

Chiliotrichiopsis Cabrera. Compositae (Ast.-Hin.). Excl. *Haroldia* 3 Peru, Argentina. R: SCB 92(2009)27

Chiliotrichum Cass. Compositae (Ast.-Hin.). 2 Arg., Chile. R: SCB 92(2009)37. Everg. shrubs. ***C. diffusum*** (Forst.f.) Kuntze – commonly dominant in Fuegia, cult. orn.

chilito *Mammillaria* spp.

Chillania Roiv. = *Eleocharis*

Chilocardamum O. Schulz (~ *Sisymbrium*). Cruciferae (46). 4 S Arg. R: Darw. 44(2006)343

Chilocarpus Blume. Apocynaceae (I i). 13 Indomal. R: SGP 72(2002)130

Chiloglottis R. Br. Orchidaceae (IV 3e). 23 E Aus., NZ (3). Chiloglottone sexually attractive to male thynnid wasp-poll., single scent component of ***C. trapeziformis*** Fitzg. (E Aus.) with exactly the same chem. (cf. *Ophrys*) as female sex pheromone of poll. *Neozeleboria cryptoides;* some cult. orn.

Chilopogon Schltr. = *Appendicula*

Chilopsis D. Don. Bignoniaceae (1). 1 SW N Am.: ***C. linearis*** (Cav.) Sweet – cult. orn. rheophyte, the branches used for baskets ('flowering willow'), hybridized with *Catalpa bignonioides* (see × *Chitalpa*)

Chiloschista Lindl. Orchidaceae (V 16c). c. 20 trop. As. to Fiji (not Borneo, NG). Many leafless

chilte *Cnidoscolus elasticus* & other *C.* spp.

Chimaerochloa Linder (~ *Danthonia*). Gramineae (Danth.) 1 NG: ***C. archboldii*** (Hitchc.) Pirie & Linder

Chimantaea Maguire & al. Compositae (Mut.-Wund.). 10 Venez., Guyana. R: MNYBG 9(1957)428. Trees & shrubs, some unbranched pachycauls like alpine ***C. mirabilis*** Maguire & al. (Venez.)

Chimaphila Pursh. Ericaceae (II 1; Pyrolaceae). 5 Euras. (Eur. 1), N & trop. Am. Pollen shed in monads, tetrads or polyads. Some cult. orn.; some med. esp. ***C. umbellata*** (L.) W. Barton for bladder problems

Chimarrhis Jacq. Rubiaceae (Cond.). 13 trop. Am. Petit's Model

Chimborazoa H. Beck = *Serjania*

Chimonanthus Lindl. Calycanthaceae. 6 China. R: JNTCFP 1984,2: 78. ***C. praecox*** (L.) Link (*C. fragrans*, wintersweet, (Jap.) allspice) – A movements due to differential cell growth-rates, cult. orn. (Champagnat's Model) for fragrant winter fls produced before lvs (cf. *Calycanthus*) in China long used with linen, like lavender, though pls grown from seed may take 12–14 yrs to produce fls; beetle-poll. with A 5

Chimonobambusa Makino. Gramineae (Iv). 37 Himal. (2), China (34, 31 endemic – R: FOC 22(2006)152), Jap. Bamboos, cult. orn., esp. ***C. quadrangularis*** (Franceschi) Makino (China, Jap.) with culms square in t.s. & a flowering-cycle longer than 100 yrs; ***C. tumidissinoda*** Hsueh & Yi ex Ohrnberger (*Qiongzhuea t.*, SW China) – swollen nodes make canes char. walking-sticks & umbrella-handles, shoots ed.

Chimonocalamus Hsueh & Yi (~ *Sinarundinaria*). Gramineae (IV). 11 E Himal. to Yunnan (China 9, 8 endemic – R: FOC 22(2006)103)

China or **Chinese anise** *Illicium verum;* **C. apple** *Malus prunifolia;* **C. artichoke** *Stachys affinis;* **C. aster** *Callistephus sinensis;* **C. banana** *Musa acuminata* 'Dwarf Cavendish'; **C. bellfl.** *Platycodon grandiflorus;* **C. berry** *Melia azedarach;* **C. box** *Murraya paniculata;* **C. briar** *Smilax bona-nox;* **C. cabbage** *Brassica rapa* Pekinensis Group & Chinensis Group; **C. cedar** *Cunninghamia lanceolata;* **C. chestnut** *Castanea mollissima, Sterculia monosperma;* **C. coir** *Trachycarpus fortunei;* **C. crab apple** *Malus hupehensis;* **C. date** *Diospyros kaki;* **C. date-plum** *Ziziphus jujuba;* **C. foxglove** *Rehmannia elata;* **C. grass** *Boehmeria nivea;* **C. galls** insect galls on *Rhus chinensis;* **C. gooseberry** *Actinidia deliciosa;* **C. hat plant** *Holmskioldia sanguinea;* **C. hibiscus** *Hibiscus rosa-sinensis;* **C. houses** *Collinsia bicolor;* **C. jute** *Abutilon theophrasti;* **C. kale** *Brassica oleracea* Alboglabra Group; **C. key** *Boesenbergia rotunda;* **C. lantern** *Alkekengi officinarum;* **C. melon** *Cucumis melo* Inodorus Group; **C. mustard** *B. juncea* 'Crispifolia'; **C. olive** *Canarium* spp.; **C. parsley** *Coriandrum sativum;* **C. pepper** *Zanthoxylum simulans;* **C. pink** *Dianthus chinensis;* **C. plum** *Prunus salicina;* **C. potato** *Dioscorea polystachya;* **C. raisin** *Hovenia dulcis;* **C. root** *Smilax china;* **C. sacred bamboo** *Dracaena sanderiana;* **C. spinach** *Amaranthus tricolor;* **C. sumac** *Rhus chinensis;* **C. tallow tree** *Triadica sebifera;* **C. tree** *Melia azedarach;* **C. tupelo** *Nyssa sinensis;* **C. water bamboo** *Dracaena sanderiana;* **C. white pine** *Pinus armandii;* **C. wood oil** *Aleurites moluccanus;* **C. yam** *Dioscorea polystachya;* **C. yew** *Taxus mairei*

chincherinchee *Ornithogalum thyrsoides*

Chingia Holttum = *Cyclosorus*

Chingiacanthus Hand.-Mazz. = *Isoglossa*

chingma *Abutilon theophrasti*

Chingyungia Ai = *Melampyrum*

chinquapin *Castanea* spp. esp. *C. pumila*

chioca *Ullucus tuberosus*

Chiococca P. Browne. Rubiaceae (III 4). 6 trop. Am. incl. S Florida. Shrubs & lianes. ***C. alba*** (L.) A. Hitchc. (S Florida to Paraguay) – cult. orn. s.t. used for snakebite, aphrodisiacs

Chiogenes Salisb. ex Torrey = *Gaultheria*

Chionachne R. Br. = *Polytoca*

Chionanthus L. Oleaceae (4d). (Incl. *Linociera*) c. 60 trop. (Afr. 9, Madag. 3 = *Noronhia*?) & subtrop., E As. (1), E N Am. (1). Some temp. spp. cult. orn. (fringe-flowers), esp. ***C. virginicus*** L. (old man's beard, E N Am.) – medic. bark

Chione DC. Rubiaceae (III 4). Excl. *Colleteria,* 1 trop. Am.: ***C. venosa*** (Sw.) Urb. R: SGP 73(2003)180

Chionocharis I.M. Johnston (~ *Eritrichium*). Boraginaceae (B.3.2). 1 Himal.: ***C. hookeri*** (Clarke) I.M. Johnston

Chionochloa Zotov. Gramineae (Danth.). 25 SE Aus. (2 incl. Lord Howe Is.), 23 NZ. Tussock grasses. ***C. rigida*** (Raoul) Zotov (NZ) – fire increases no. of tussocks in fl. & no. of infls per tussock

Chionodoxa Boiss. = *Scilla*

Chionogentias L.G. Adams = *Gentianella* (but see AusSB 8(1995)949)

Chionographis Maxim. Melanthiaceae. 4 S China to Jap. Allied to *Chamaelirium,* P unequal

Chionohebe B. Briggs & Ehrend. = *Veronica*

Chionolaena DC. Compositae (Gnap.-Gnap.). Incl. *Gnaphaliothamnus,* 20 Mex., S Am. R: AMBG 80(1993)405. Shrubs with revolute lvs

Chionopappus Benth. Compositae (Liab.). 1 Peru: ***C. benthamii*** S.F. Blake

Chionophila Benth. Plantaginaceae (Ant.-Chel.; Scrophulariaceae s.l.). 2 Rocky Mts. ***C. jamesii*** Benth. - cult. rock-pl.

× **Chionoscilla** J. Allen ex Nicholson = *Scilla*

Chionothrix Hook.f. Amaranthaceae (I 2). 3 Somalia

chipilin *Crotalaria longirostrata*

chiquito *Combretum butyrosum*

chir pine *Pinus roxburghii*

Chiranthodendron Sessé ex Larréat. Malvaceae (Bomb.; Sterculiaceae). 1 Mex. & Guatemala: ***C. pentadactylon*** Larréat. K(5), C 0, A basally united into curved tube but apically sep. as 5 exserted lobes, each with 2 linr., 1-celled anthers, the whole resembling a hand & thus a source of awe; poll. by perching birds & bats; fls used for eye disorders & piles; hybridized with *Fremontodendron* 'Pacific Sunset' to give × ***Chiranthomontodendron lenzii*** Dorr

chirata, chiretta *Swertia* spp.

Chirita Buch.-Ham. ex D. Don = *Henckelia;* see also *Damrongia, Liebigia, Microchirita, Primulina*

Chiritopsis W.T. Wang = *Primulina* (but see EJB 49(1992)48)

Chironia L. Gentianaceae. 30 subSaharan Afr., Madag. Some cult. orn. incl. ***C. baccifera*** L. (Christmas berry, S Afr.)

chironja citrus fr. believed to be a cross between grapefruit & an orange

Chiropetalum A. Juss. (~ *Argythamnia*). Euphorbiaceae (Acal.-Chro.-Bit.). c. 22 trop. Am.

Chirripoa Süsseng. = *Guzmania*

chiso *Perilla frutescens*

Chisocheton Blume. Meliaceae (II-Trich.). 53 Indomal. to trop. China & Vanuatu R: BBMNHB 6(1979)301. Range of form (Corner's, e.g. ***C. tomentosus*** (Roxb.) Mabb. (Malay Pen.), to Champagnat's Models) from unbranched pachycauls to weeping leptocaul shrublets & timber trees, some with myrmecophilous shoots; most spp. with leaves (pinnate) with indeterminate growth (cf. *Tachigali*), ***C. pohlianus*** Harms & ***C. tenuis*** P. Stevens (NG) with infls borne on new apical growths of lvs; 1–2 whorls of C; seeds arillate or sarcotestal. ***C. cumingianus*** (C. DC.) Harms (Indomal. to China) – fish-poison in NG, seeds source of an oil for med. & lighting in Philipp.; similar oil from ***C. macrophyllus*** King (W Mal.) & that from ***C. pentandrus*** (Blanco) Merr. used as hair-oil in Philipp.

× **Chitalpa** Elias & Wisura. Hybrids between *Catalpa* & *Chilopsis* spp., several crosses & cvs

Chitonanthera Schltr. = *Octarrhena*

Chitonochilus Schltr. = *Agrostophyllum*

chitra *Berberis aristata*

Chittagong wood *Chukrasia tabularis*

chittam *Cotinus obovatus*

chives *Allium schoenoprasum*; **Chinese** or **garlic c.** *A. tuberosum*

Chlaenaceae 'Thouars' = Sarcolaenaceae

Chlaenandra Miq. Menispermaceae (III). 1 NG: ***C. ovata*** Miq.

Chlaenosciadium Norman. Umbelliferae (I?). 1 W Aus.: ***C. gardneri*** Norman

Chlamydacanthus Lindau (*Theileamea*). Acanthaceae (III). 1 trop. E Afr., 1 Madag.

Chlamydites J.R. Drumm. = *Aster*

Chlamydoboea Stapf = *Paraboea*

Chlamydocardia Lindau. Acanthaceae (III 2c). 2 trop. W Afr.

Chlamydocarya Baill. = *Pyrenacantha*

Chlamydocola (K. Schum.) M. Bod. = *Cola*

Chlamydogramme Holttum = *Tectaria* (but see R: FM II,2(1991)37)

Chlamydojatropha Pax & K. Hoffm. Euphorbiaceae (tribe?). 1 trop. W Afr.: ***C. kamerunica*** Pax & K. Hoffm. – only ♀ fls known

Chlamydophora Ehrenb. ex Less. Compositae (Anth.-Leuc.). 1 Eur., N Afr.: ***C. tridentata*** (Del.) Less. R: BBMNHB 23(1993)142

Chlamydophytum Mildbr. Balanophoraceae. 1 trop. W Afr.: ***C. aphyllum*** Mildbr. R: BJ 106(1986)367. Found on *Tessmannia* at 3 localities 1000 km apart

Chlamydostachya Mildbr. = *Anisotes*

Chlamydostylus Bak. = *Nemastylis*

Chlidanthus Herb. Amaryllidaceae. 6 S Peru to Bolivia, NW Arg. ***C. fragrans*** Herb. (Peruvian Andes) – cult. orn.

Chloachne Stapf = *Poecilostachys*

Chloanthaceae Hutch. (Dicrastylidaceae) = Labiatae (III)

Chloanthes R. Br. Labiatae (IV 1; Chloanthaceae). 4 Aus. R: JABG 1(1977)84

Chloothamnus Buese (~ *Nastus*). Gramineae (V 6). Some SE As. '*Nastus*' spp.

Chloracantha Nesom & al. (~ *Boltonia*). Compositae (Ast.-Bolt.). 1 S N Am. & C Am.: ***C. spinosa*** (Benth.) Nesom (*Aster* s.) – rheophyte with spines. R: Phytol. 70(1991)371, 382

Chloraea Lindl. Orchidaceae (IV 2a). 52 temp. S Am. R: Orquideologia 6(1971)231. Terrestrial, some autogamous; heterogeneous

Chloranthaceae R. Br. ex Sims. Magnoliidae – Chloranthales. 4/67 trop. & warm. Aromatic trees, shrubs or herbs (even annuals) oft. with strongly swollen nodes; wood soft, vessels with scalariform end-plates (up to 200 crossbars!). Lvs opp., simple, toothed, petioles ± connate basally; stip. interpetiolar. Fls reduced, unisexual or not with 0–3 bracts, in crowded infls or (*Hedyosmum*) resembling a spike, P 0 or weakly 3-fid K, A 1–5, usu. ± connate, lat. ones with only 1/2-anthers, in *Sarcandra* A1 laminar with 2 separated pollen-sacs, anthers with longit. slits, pollen monosulcate (*Hedyosmum*) to multiaperturate, $\underline{G}$, $\overline{G}$ or half-inf. 1, ovule 1, orthotropous, bitegmic. Fr. a berry or drupe, seeds with much oily, starchy endosperm, tiny embryo with 2 cotyledons. x = 8, 14, 15

Genera: *Ascarina, Chloranthus, Hedyosmum, Sarcandra*

Widespread fossils incl. pollen (*Asteropollis*, ? = *Hedyosmum*) from early Cretaceous. Fl. constr. compared with gnetopsid *Archaeostrobilus* (Taylor & Hickey), though app. reduced, in *Hedyosmum* each stamen corresponding to 1 ebracteolate fl. (but the pigeon-holing of the infl. type into pre-conceived categories (cf. Araceae, Cercidiphyllaceae, Pandanaceae) perhaps a sterile pursuit in any case; see BJ 113[1991]339)

Some teas (*Chloranthus*) & medic. locally imp.

Chloranthus Sw. Chloranthaceae. 10 Indomal., E As. R: AJB 89(2002)940. Leeuwenberg's & Tomlinson's Models; thrips-poll. ***C. spicatus*** (Thunb.) Makino (*C. inconspicuus*, E & SE As.) – fls to flavour tea in SE As., also medic.; ***C. erectus*** (Buch.-Ham.) Verdc. (*C. officinalis*, Mal.) – lvs etc. used as tea in Java before *Camellia sinensis*, also febrifuge

Chloris Sw. Gramineae (XXIX 5). Excl. *Stapfochloa*, c. 60 trop. & warm (N Am. 11). R: BYUSBB 19, 2(1974)1. Windmill grasses. Heterogeneous? Some good pasture grasses esp. ***C. gayana*** Kunth (Rhodes grass, Afr., natur. Aus., Am.) – encouraged in Aus. citrus orchards as pollen a food for pest-predating phytoseiid mites, & ***C. truncata*** R. Br. (windmill grass, Aus.) – toxic to some stock. ***C. virgata*** Sw. (trop. Am.) – invasive in Aus., Hawaii

Chlorocalymma W. Clayton. Gramineae (XXIV 1). 1 Tanzania: ***C. cryptacanthum*** W. Clayton

Chlorocardium Rohwer & al. (~ *Ocotea*). Lauraceae (inc. sed.). 2 trop. S Am. ***C. rodiei*** (Schomb.) Rohwer & al. (*Nectandra r.*, *Ocotea r.*, *O. venenosa*, greenheart) – v. heavy timber, form. only one exploited in the forest, resistant to termites & borers so used for wharves, locks of Panamá Canal etc., but v. difficult to work, arrow poison (alks close to D-tubocurarine; see *Chondrodendron*)

Chlorocarpa Alston. Achariaceae (Flacourtiaceae s.l.). 1 Sri Lanka: ***C. pentaschista*** Alston

Chlorochorion Puff & Robbrecht (~ *Pentanisia*). Rubiaceae (III 8). 2 trop. Afr. R: BJ 110(1989)547. Name a translation of [Bernard] 'Verdcourt'

Chlorocrambe Rydb. Cruciferae (46). 1 Oregon, Idaho, Utah: ***C. hastata*** (S. Watson) Rydb. R: FNA 7(2010)685

Chlorocyathus Oliv. (~ *Raphionacme*). Apocynaceae (Peripl.). 2 trop. Afr.

Chlorogalum Kunth. Asparagaceae (Hyacinthaceae). 5 W N Am. R: Madroño 5(1940)137. Crystals like 'Agavaceae' but leaf anatomy like 'Hyacinthaceae'. Cult. orn. bulbs incl. ***C. pomeridianum*** (DC.) Kunth (N Calif.) – bulbs yield a lather usable as soap subs., outer scales v. fibrous

Chloroleucon (Benth.) Britton & Rose (~ *Albizia*). Leguminosae (II 4). 10 trop. Am. R: MNYBG 74,1(1996)136. Axillary spines (sterile peduncles)

Chloroleucum Britton & Rose = praec.

Chloroluma Baill. = *Chrysophyllum*

Chloromyrtus Pierre = *Eugenia*

Chloropatane Engl. = *Erythrococca*

Chlorophora Gaudich. = *Maclura*. See also *Milicia*

Chlorophytum Ker-Gawler. Asparagaceae (Anthericaceae). 150 OW trop. esp. Afr. (S Afr. 36, Madag. 1, Arabia 1, Socotra 1). Seeds thin, folded or flat, cf. angular small seeds of closely allied *Anthericum*. Cult. orn. esp. ***C. comosum*** (Thunb.) Jacques (trop. & S Afr.) with white fls oft. replaced by young pls. which weigh infl. axis down & take r.; form usu. grown is **'Picturatum'** (spider plant) with C yellow stripe on lvs to 30 cm, **'Variegatum'** with creamy margins, decontaminates air with formaldehyde dehydrogenase. ***C. laxum*** R. Br. (OW trop.) – tubers ed. S Arabia; ***C. suffruticosum*** Bak. (E Afr.) – stem to 50 cm; ***C. tuberosum*** (Roxb.) Bak. (OW trop.) – local (Ayurvedic) med.

Chloropyron Behr (~ *Cordylanthus*). 4 W N Am. R: SB 34(2009)188. Halophytes

Chlorosa Blume = *Cryptostylis*

Chlorospatha Engl. Araceae (VII 11). 68 trop. Am. esp. Colombia (45, 43 endemic). R: AMBG 101(2015)70

Chloroxylon DC. Rutaceae (I ?10; Flindersiaceae). 2 Madag., 1 S Ind., Sri Lanka: ***C. swietenia*** DC. ((E Ind. or Ceylon) satinwood) – timber for furniture & veneers, though alks may irritate skin, gum useful

Choananthus Rendle = *Scadoxus*

cho-cho, choco *Sicyos edulis*

chocolate *Theobroma cacao*; **c. lily** *Dichopogon* spp.; **c. mint** *Mentha* × *piperita* 'Chocolate'

Chodaphyton Minod = *Stemodia*

Chodsha-Kasiana Rauschert = *Catenulina*

Choerospondias B.L. Burtt & A.W. Hill. Anacardiaceae (II). 1 NE Ind. to N Thailand, SE China & Jap.: ***C. axillaris*** (Roxb.) B.L. Burtt & A.W. Hill – useful wood, fr. ed & used for wine, bark fibre for ropes. R: FOC 11(2008)341

choi sum *Brassica rapa* Chinensis Group

Choisya Kunth. Rutaceae (I ?5). 6 SW N Am. R: AMN 24(1940)730. ***C. ternata*** Kunth (SW Mex.) – hardy everg. shrub with scented fls, poss. self-sterile & clonal in cult. (no fr. recorded, though ***C.* × *dewitteana*** Geerinck **'Aztec Pearl'** allegedly hybrid with ***C. dumosa*** A. Gray (incl. *C. arizonica*, Arizona))

chokeberry *Photinia* spp.; **black c.** *Aronia melanocarpa*; **red c.** *A. arbutifolia*

chokecherry *Prunus virginiana*

choko *Sicyos edulis*

cholla *Cylindropuntia* spp.; **c. gum** from *C. fulgida* etc.

Chomelia Jacq. (~ *Tarenna*). Rubiaceae (III 3). Incl. *Anisomeris*, 20 trop. Am.

Chondradenia Maxim. ex F. Maek. = *Galearis*

Chondrilla Tourn. ex L. Compositae (Cich.-Cich.). 25 temp. Euras. (Eur. 5 excl. *Willemetia*). ***C. juncea*** L. (skeleton weed) – lvs arr. in plane of the meridian, wild-coll. blanched shoots used like chicory (Spain), bad weed (controlled by *Puccinia chondrollina* rust) esp. Aus., wiry stems damaging machinery

Chondrococcus Steyerm. = *Coccochondra*

Chondrodendron Ruíz & Pavón. Menispermaceae (I). 3 C & 7 trop. S Am. R: MNYBG 22,2(1971)5. Oft. large lianes. R. of ***C. tomentosum*** Ruíz & Pavón (Braz. & Peru) source of D-tubocurarine, a muscle-relaxant used in surgery not synth. artificially because now superseded in W med. by atracurium besilate, a constituent of curare arrow poisons, r. also medic. (pareira r., p. brava)

Chondropetalum Rottb. = *Elegia* (But see Bothalia 15(1985)427)

Chondropyxis D. Cooke. Compositae (Gnap.-Ang.). 1 S Aus.: ***C. halophila*** D. Cooke. R: Fl. S Aus. 3(1986)1612

Chondrorhyncha Lindl. Orchidaceae (V 12j). Excl. *Chondroscaphe*, 7 trop. Am.

Chondroscaphe (Dressler) Senghas & Gerl. Orchidaceae (V 12j). 14 trop. Am.

Chondrostylis Boerl. (~ *Agrostistachys*). Euphorbiaceae (Acal.-Agr.). 2 SE As., W Mal. R: Blumea 46(2001)89

Chondrosum Desv. = *Bouteloua*

Chonemorpha G. Don. Apocynaceae (II c). Incl. *Rhynchodia*, 10 Indomal. R: KB 1947:47, 1948:68. ***C. fragrans*** (Moon) Alston (*C. macrophylla*) – cult. orn. liane with fragrant white fls to 8 cm diam.; bark source of water-resistant fibre used for fishing-nets

Chonocentrum Pierre ex Pax & K. Hoffm. Phyllanthaceae (Euphorbiaceae s.l.). 1 Amazonia: ***C. cyathophorum*** (Muell. Arg.) Pax & K. Hoffm. R: A. Radcliffe-Smith, *Gen. Euphorb.* (2001)12

Chonopetalum Radlk. Sapindaceae. 1 trop. W Afr.: ***C. stenodictyum*** Radlk. – coll. once

Chontalesia Lundell = *Hymenandra*

Chordifex Briggs & L. Johnson. Restionaceae. 20 SW & E Aus. R: Telopea 7(1998)356

Chordospartium Cheeseman = *Carmichaelia*

Choretrum R. Br. Santalaceae (Amphorogynaceae). 6 Aus. R: Fl. NSW 2(1991)150

Choriantha Riedl = *Onosma*

Choricarpia Domin = *Backhousia*

Choriceras Baill. (~ *Longetia*). Picrodendraceae (Cal.-Diss.; Euphorbiaceae s.l.). 1–2 NE Aus, 1 ext. S NG

Chorigyne Eriksson. Cyclanthaceae. 7 Panamá & Costa Rica. R: NJB 9(1989)31

Chorilaena Endl. Rutaceae (I 3). 1 SW Aus.: ***C. quercifolia*** Endl. R: FA 26(2013)456

Choriptera Botsch. (~ *Lagenantha*). Amaranthaceae (Chenopodiaceae III 3). 3 Somalia

Chorisandrachne Airy Shaw (~ *Leptopus*). Phyllanthaceae (Wiel.-Wiel.). 1 SW Thailand: ***C. diplosperma*** Airy Shaw

Chorisepalum Gleason & Wodehouse. Gentianaceae. 5 Guayana Highland. R: MNYBG 32(1981)342

Chorisia Kunth = *Ceiba*

Chorisis DC. = *Ixeris*

Chorisiva (A. Gray) Rydb. = *Euphrosyne* (but see FNA 21(2006)31)

Chorisochora Vollesen (~*Angkalanthus*). Acanthaceae (III 2c). 4 S Afr. (1), Somalia (1), Socotra (2). R: KB 49(1994)474

Chorispora R. Br. ex DC. Cruciferae (17). 11 E Med. (1 ext to Eur., weedy elsewhere: ***C. tenella*** (Pallas) DC.), C As. Fr. indehiscent falling as 1-seeded corky units. Heterogeneous. ***C. sabulosa*** Cambess. (Himal.) – fr. ed.

Choristemon Williamson = *Leucopogon*

Choristylis Harv. = *Itea*

Choritaenia Benth. Umbelliferae (I 2c). 1 S Afr.: ***C. capensis*** (Sonder) Burtt Davy. R: Strel. 10(2000)66

Chorizandra R. Br. Cyperaceae (I 2). 6–8 Aus.(5), New Caled. (1–3)

Chorizanthe R. Br. ex Benth. (~ *Eriogonum*). Polygonaceae (Erio.-Erio.). 50 dry W Am. (41 annuals – R: Phytol. 66(1989)100). Some with ocreae

Chorizema Lab. Leguminosae (III 13). 27 SW & E Aus. (1). R: AusSB 5(1992)249. Cult. orn. esp. ***C. ilicifolium*** Lab. (*C. cordatum*, flame pea, SW Aus.) – orange-red & purplish fls

Chortolirion A. Berger = *Aloe* (but see Bothalia 25(1995)31)

Chosenia Nakai = *Salix*

Chouardia Speta = *Scilla*

Choulettia Pomel = *Plocama*

Chouxia Capuron. Sapindaceae. 6 N Madag. R: Adansonia III,21(1999)52

chow chow *Sicyos edulis*

chowlee *Vigna unguiculata*

choy sum *Brassica rapa* Chinensis Group

Chresta Vell. ex DC. (~ *Eremanthus*). Compositae (Vern.-Chr.). 11 C Braz. R: SB 10(1985)465

Christella H. Lév. = *Cyclosorus*

Christensenia Maxon. Marattiaceae (Christenseniaceae). 2 Indomal. R: AFJ 83(1993)3. Lvs palmate; veins anastomosing; synangia circular. ***C. aesculifolia*** (Blume) Maxon (Assam, S China to Solomon Is. (not NG)) – cult. orn.

Christenseniaceae Ching = Marattiaceae

Christensonella Szlach. & al. = *Maxillaria*

Christensonia Haager = *Vanda* (but see Lankesteria 2(2001)19)

Christia Moench. Leguminosae (III 19). c. 10 SE As., Indomal., Aus. Some cult. orn.

Christiana DC. Malvaceae (Brown.; Tiliaceae). (Incl. *Asterophorum, Tahitia*) 3 trop. S Am. – 1 (***C. africana*** DC.) ext to Afr., 1 Tahiti (extinct). Rauh's Model

Christianella W.R. Anderson (~ *Mascagnia*). Malpighiaceae. 5 trop. Am. R: Novon 16(2006)190

Christiopteris Copel. = *Christopteris*

Christisonia Gardner. Orobanchaceae (Orob.). 17 SW China, SE As., Indomal. R. parasitic on Acanthaceae & bamboos; infls emerge, mature & die in 2 weeks

Christmas bell *Blandfordia nobilis, B. punicea* (Aus.), *Sandersonia aurantiaca* (S Afr.); **C. berry** *Chironia baccifera* (S Afr.), *Schinus terebinthifolius;* **C. bush** *Bursaria spinosa* (S Aus.), *Ceratopetalum gummiferum* (E Aus.), *Prostanthera lasianthos* (Aus.), *Alchornea cordifolia* (W Afr.); **C. cactus** *Schlumbergera × buckleyi;* **C. rose** *Helleborus niger;* **C. tree** *Picea abies* (GB, but *Abies alba* also used on Cont.), *Nuytsia floribunda* (Aus.), *Metrosideros excelsa* (NZ); **Nordmann C. t.** *Abies nordmanniana*

Christolea Cambess. Cruciferae (26). 2 C & E As., Himal., China

Christopher, Herb *Actaea spicata*

Christopheria J.F. Sm. & J.L. Clark (~ *Episcia*). Gesneriaceae (II 5c). 1 Guianas: ***C. xantha*** (Leeuwenb.) J.F. Sm. & J.L. Clark. R: SB 38(2013)453

christophine *Sicyos edulis*

Christopteris Copel. Polypodiaceae (II). 2 Indomal. R: BJ 105(1984)1

Christ's thorn *Ziziphus spina-christi* or poss. *Paliurus spina-christi;* name also used for *Euphorbia milii, Carissa carandas* etc.

Chroesthes Benoist. Acanthaceae (III 2d). 3 S China to SE As. ***C. longifolia*** (Wight) B. Hansen (*Lepidagathis l.*, W Mal.) – common forest pl. never found in fr., lvs fall if stem shaken

Chroilema Bernh. = *Haplopappus*

Chromatotriccum M.A. Clem. & D.J. Jones = *Dendrobium*

Chromolaena DC. (~ *Eupatorium*). Compositae (Eup.-Prax.). 165 trop. & warm Am. R: MSBMBG 22(1987)383. x = 10. ***C. odorata*** (L.) R. King & H. Robinson (Siam weed, paraffin bush, triffid weed) – invasive in OW (introd. Ind. late C19 & in Afr. by 1940s, controlled in Java by dipteran galler, *Procecidochares connexa*), allelopathic effects on maize, toxic to stock, local med. (lvs anti-coagulant); ***C. squalida*** (DC.) R. King & H. Robinson – weedy trop. As.

Chromolepis Benth. Compositae (Helia.-Chr.). 1 Mex.: ***C. heterophylla*** Benth.

Chromolucuma Ducke. Sapotaceae (IV). 4 NE S Am. R: Britt. 64(2013)29

Chronanthos (DC.) K. Koch = *Cytisus*

Chroniochilus J.J. Sm. Orchidaceae (V 16c). 4 Thailand to Borneo

Chronopappus DC. Compositae (Vern.-Lych.). 1 Braz.: ***C. bifrons*** (Pers.) DC.

Chrozophora Neck. ex A. Juss. Euphorbiaceae (Acal.-Chro.-Chro.). 10 Med. (Eur. 2), trop. Afr. to Thailand. ***C. plicata*** (Vahl) Spreng. (Afr., Ind.) – fr. poss. oil-source for soap, also dye-source & purgative; ***C. tinctoria*** (L.) A. Juss. (Med.) – source of turn-sole dye (*bezetta*

rubra, tournesol) used for colouring liqueurs, wine, pastries, linen & Dutch cheeses, properties known since antiquity

Chrysactinia A. Gray. Compositae (Helia.-Pect.). 6 SW N Am. R: Phytol. M 10(1996)5. *Harnackia* & *Lescaillea* poss. referable here. Shrubs. ***C. mexicana*** A. Gray (Mex.) – medic.

Chrysactinium (Kunth) Wedd. (~ *Liabum*). Compositae (Liab.). 8 Ecuador, Peru. R: SCB 54(1983)49. Some oils

Chrysalidocarpus H. Wendl. = *Dypsis*

Chrysallidosperma H. Moore = *Syagrus*

Chrysanthellum Rich. Compositae (Cor.). 13 S N & C Am. esp. Mex., Galápagos incl. 1 pantrop. weed, ***C. americanum*** (L.) Vatke. R: Phytol. 64(1988)417

Chrysanthemoides Fabr. (~ *Osteospermum*). Compositae (Cal.). 2 E & S Afr. R: Strel. 9(2000)311. 'Drupaceous' fr. (also in *O.*) eaten & disp. by birds. ***C. monilifera*** (L.) Norl. (boneseed, bitou bush) – form. planting required in land-reclamation in Aus. (introd. 1908), now aggressive weed in Aus. (900 km of NSW coast by 2006), NZ, controlled biologically (e.g. *Tortrix* moths from S Afr.)

Chrysanthemum Tourn. ex L. (*Dendranthema*). Compositae (Anth.-Art.). 37 Eur. (2), C & E As. R: APG 29(1978)168. Garden chrysanthemums (vulgarly 'mums', 'chrysanths') derived from ***C.*** × ***morifolium*** Ramat. (*D.* × *grandiflora*) prob. a complex hybrid group (now 10 cv groups) raised in China c. C 4 from ***C. indicum*** L. (*D. indica*, E As., fls yellow, form. used as insect-fumigant in China & still for sore eyes in herbalism) & other spp., poss. orig. *C. zawadzkii*; in China first yellow, then white then purple cvs – (monograph written AD 1104; 500 cvs by 1630), introd. Eur. 1688 (first cv, '**Old Purple**') but not established until 1789 (3 cvs to Marseille, France), s.t., as in 1000-bloom cs. (up to 2300 known), trained as standard on *Artemisia annua* stock; capitula in Chinese herbal tea (ju hua) cont. active anti-HIV compounds. Korean chrysanthemums are hybrids (?backcrosses) between this & ***C. zawadzkii*** Herbich (*C. erubescens*, *D. zawadskii*, Euras.), orig. with 'single' heads with 1 row of ligulate florets. All are imp. cut-fls (cut-fl. trade in US worth $17M+, potted pl. $68M+, by 2005) & form. assoc. with nobility in China (Mah Jong tiles; the sign of the Jap. Mikado – sun symbol, royal stock descended from Sun Goddess) with 16-floreted 'double' (16 outer, 16 inner) exclusive to royal household, 14-floreted 'single' used by royal princes. ***C. yoshinaganthum*** Mak. (Jap.) – rheophyte. See also *Argyranthemum*, *Glebionis*, *Ismelia*, *Leucanthemum*, *Tanacetum*

chrysanthemum greens *Glebionis coronaria*

Chrysanthoglossom Wilcox, Bremer & Humphries. Compositae (Anth.-Leuc.). 2 N Afr. R: BBMNHB 23(1993)143

Chrysitrix L. Cyperaceae (I 2). 4 SW Cape, SW Aus.

Chrysobalanaceae R. Br. Magnoliidae – Malpighiales. 20/525 trop. esp. Am. R: FOW 9,10(2003). Trees or shrubs, s.t. geoxylic; alks & cyanogenic compounds unknown. Wood rich in silica. Twigs strongly lenticellate. Lvs in spirals, appearing 2-ranked, simple, usu. entire, with stip. Fls in term. or axillary infls, rarely solit., ± strongly zygomorphic, small, bisex. or pls polygamous, perigynous with annular nectary in hypanthium below A; K (5), imbr., C 5, imbr., rarely 0, A 2–100(–c. 300), s.t. some staminodal, filaments long, distinct, connate or connate in groups, in the more zygomorphic all on 1 side of hypanthium, anthers with longit. slits. G usu. 2 ± reduced, united by gynobasic style, oft. appearing $\underline{G}$ 1 eccentrically positioned in hypanthium, stigma simple or 3-lobed, ovules 2 per locule or single G subdivided to appear as 2 1-ovulate locules, erect, anatropous. Fr. a 1-seeded drupe; endocarp oft. hairy within, seeds with large embryo & 0 endosperm. n = 10, 11

Classification [probably to be abandoned] *& principal genera:*

1. **Chrysobalaneae** (fls small, reg.): *Licania*
2. **Parinarieae** (fls irreg., A 7–17 posterior): *Parinari*
3. **Couepieae** (fls irreg., A (15)20+ usu. in ring): *Couepia*, *Maranthes*
4. **Hirtelleae** (fls strongly irreg., A 3–75 posterior): *Atuna*, *Hirtella*, *Magnistipula*

Poss. incl. Dichapetalaceae, Euphroniaceae & Trigoniaceae, & even Balanopaceae. *Stylobasium*, once incl. here now referred to Surianaceae. Some remarkable intercontinental distr. in *Hirtella* (all Am. save 1 in each of Afr. & Madag.), *Maranthes* (all Afr. save 1 Am. closely allied to 1 As.) while *Chrysobalanus icaco* occurs in both Am. & Afr. Some oils, ed. seeds & fr. esp. *Acioa*, *Afrolicania*, *Chrysobalanus*, *Couepia*, *Licania* & *Neocarya*; timber

Chrysobalanus L. Chrysobalanaceae (1). 3 trop. Am., 1 (***C. icaco*** L., icaco, cocoplum) ext. to trop. Afr. & natur. Seychelles, Tanzania, Vietnam & Fiji. R: FOW 9(2003)4. *C. icaco* grown for ed. fr. (e.g. Cuba) usu. preserved (esp. NE S Am.: 'icacos'); seed-oil used for candles etc. in W Afr.; in Benin inhabited by ants making nests from lvs & feeding on fr. & secretions, keeping off other intruders

Chrysobraya H. Hara = *Lepidostemon*

Chrysocephalum Walp. (~ *Helichrysum*). Compositae (Gnap.-Ang.). c. 10 Aus. R: OB 104(1991)119

Chrysochamela (Fenzl) Boiss. Cruciferae (15). 3 E Med. (Eur. 1), Russia

Chrysochlamys Poeppig. Guttiferae (1.). Excl. *Tovomitopsis*, c. 55 trop. Am.

Chrysochloa Swallen. Gramineae (XXIX 5). 4 trop. Afr.

Chrysochosma (J. Sm.) Kümmerle = *Notholaena*

Chrysocoma L. Compositae (Ast.-Hom.). 20 S Afr. esp. SW Cape. R: MBSM 17(1981)259. Some cult. orn.

Chrysocoryne Endl. = *Gnephosis*

Chrysocoryne Zoellner = *Leucocoryne*

Chrysocycnis Linden & Reichb.f. = *Maxillaria* (*Mormolyca*)

Chrysodracon (Jankalski) P.L. Liu & Morden (~ *Dracaena*). Asparagaceae. 6 Hawaii. R: SB 39(2014)101

Chrysoglossum Blume. Orchidaceae (V 14). 4 trop. As. to Samoa

Chrysogonum L. Compositae (Helia.-Eng.). 1 SE US: ***C. virginianum*** L. – cult. orn., myrmechorous. R: Sida 19(2001)813

Chrysogrammitis Parris. Polypodiaceae (V; Grammitidaceae). 2 SE As. to Melanesia. R: KB 53(1998)909

Chrysolaena H. Robinson (~ *Vernonia*). Compositae (Vern.-Lep.). 9 trop. S Am.

Chrysolepis Hjelmq. (~ *Castanopsis*). Fagaceae. 2 W N Am. ***C. chrysophylla*** (Hook.) Hjelmq. (golden chestnut) – timber, ed. fr.

Chrysoma Nutt. (~ *Solidago*). Compositae (Ast.-Sol.). 1 SE US: ***C. pauciflosculosa*** (Michaux) Greene. R: SBM 20(2000)29

Chrysophae Kozo-Polj. = *Chaerophyllum*

Chrysophthalmum Schultz-Bip. (~ *Inula*). Compositae (Inul.). 3 W As. R: BJLS 137(2001)214

Chrysophyllum L. Sapotaceae (IV). Incl. *Achrouteria, Donella, Gambeya, Gambeyobotrys, Prieurella, Ragala, Zeyherella*, 43 trop. Am., c. 15 Afr., c. 10 Madag., 2–3 Indomal. to Aus. R: T.D. Pennington, Sapotaceae (1991)216. Roux's & Troll's Models; serial buds in leaf axils of some spp. subseq. give rise to fls on old wood. Many ed. fr. esp. ***C. cainito*** L. (star-apple, fr. star-shaped in t.s., prob. derived from 'C. *argenteum* Jacq.' in C Am.) & ***C. oliviforme*** L. (damson-plum, trop. Am.). Others with good timber, latex for bird-lime etc., ***C. sanguinolentum*** (Pierre) Baehni subsp. ***balata*** (Ducke) T.D. Penn. (*Ecclinusa b.*, Braz.) source of an inferior balata

Chrysopogon Trin. Gramineae (XXII). Incl. *Vetiveria*, 48 warm esp. OW (Cuba & Florida 1, Eur. 1). R (Thailand, Mal.): Austrobaileya 5(1999)506. Local medic. Aus. ***C. aciculatus*** (Retz.) Trin. (As.) – lawngrass in trop., sharp calluses can pierce animals' stomachs; ***C. fallax*** S.T. Blake (Aus.) – nut-like 'r.' dug out & eaten by wallabies; ***C. gryllus*** (L.) Trin. (French whisk, Medit.) – used for brushes; ***C. zizanioides*** (L.) Roberty (*V. z.*, vetiver (oil), khuskhus, cuscus, sevendara (grass), Ind.) – planted for erosion control in Sri Lanka tea estates, added to tinned asparagus to enhance flavour, fragrant 'r.' used in scent-making (allegedly a constituent of Chanel No. 5) & woven into mats, baskets, screens, ('sandalwood') fans (used by Moghul emperors on wetted punkas) etc. which give off scent when sprinkled with water, cont. insect-repellent zizanol & epizizanol (terpenes)

Chrysoprenanthes (Schultz-Bip.) Bramw. = *Sonchus* (but see Bot. Macar. 24(2003)182)

Chrysopsis (Nutt.) Elliott (~ *Heterotheca*). Compositae (Ast.-Chr.). 11 SE US, esp. Florida, to Mex. & Bahamas. R: FNA 20(2006)213. Cult. orn. (golden aster)

Chrysoscias E. Meyer (~ *Rhynchosia*). Leguminosae (III 18). 3–4 W Cape

Chrysosplenium Tourn. ex L. Saxifragaceae (I 3). c. 55 Eur. (5), NE As., N Am. (6), few in N Afr. & temp. S Am. R: JFSUTB 7(1957). Molecular data support split between spp. with opp. & spiral lvs. K 4, C 0. Some cult. orn. incl. ***C. oppositifolium*** L. (golden saxifrage, Eur.) – emergency foodpl.

Chrysothamnus Nutt. Compositae (Ast.-Sol.). 9 W N Am. R: FNA 20(2006)187. See also *Euthamia*

Chrysothecium (Jaub. & Spach) Hendrych (~ *Thesium*). Santalaceae. 4 Turkey, C As. R: Preslia 65(1993)319

Chrysothemis Decne. Gesneriaceae (II 5c). 9 trop. Am. R: SB 41(2016)94. Cult. orn. tuberous herbs incl. ***C. friedrichsthaliana*** (Hanst.) H. Moore (*Tussacia f.*, C Am. & Andes) –

ant-disp. (elaiosomes), fls bathed in rainwater-filled glandular K, protecting young C from insects

Chthamalia Decne = *Matelea*

Chthonocephalus Steetz. Compositae (Gnap.-Ang.). 6 temp. Aus. R: OB 104(1991)128

Chuanminshen Sheh & Shan (~ *Peucedanum*). Umbelliferae (III 10). 1 China (cult.): ***C. violaceus*** Sheh & Shan

chuchupate *Ligusticum porteri*

Chucoa Cabrera = *Onoseris* (but see T 54(2005)86 & *Paquirea*)

chufa *Cyperus esculentus*

chuglam, white *Terminalia bialata*

Chukrasia A. Juss. Meliaceae (I). 1 S China, Indomal.: ***C. tabularis*** A. Juss. – timber good (Chittagong wood, chickrassy, yinma, yonhin, Ind. redwood, bastard cedar, white c.) & source of a gum; epiphyllous fls recorded in Thailand (cf. *Chisocheton*). R: FM I,12(1995)354

chumprak *Heritiera cochinchinensis*

Chumsriella Bor = *Germainia*

Chunechites Tsiang = *Urceola*

chungah *Caralluma* spp.

Chunia H.T. Chang. Hamamelidaceae (II). 1 Hainan: ***C. bucklandioides*** H.T. Chang

Chuniophoenix Burret. Palmae (III 5). 3 S China, Hainan, Vietnam. R: Phytotaxa 218(2015)163

chupadilla *Cyrtocarpa procera*

Chuquiraga Juss. Compositae (Barn.). 22 Andes & Patagonia. R: Darw. 26(1985)219. Xeromorphic shrubs; thorns in axils

churnwood *Citronella moorei*

churras *Cannabis sativa* 'subsp. *indica*'

Chusan palm *Trachycarpus fortunei*

Chusquea Kunth. Gramineae (V 1). Incl. *Neurolepis*, *Swallenochloa*, c. 160 [+ ?50 unnamed] trop. Am., to snowline. R: SB 34(2009)679. Tree-like, shrubby & cli. bamboos with monopodial, sympodial or mixed shoots, usu. solid pith. Char. of cloud forest, oft. forming impenetrable thickets, e.g. ***C. aristata*** Munro (*N. a.*) to snowline in Andes, at 4500 m forming 'carrizal'. Cult. orn. esp. ***C. culeou*** E. Desv. (Chile) – to 6 m, hardy in GB. ***C. magnifolia*** L.G. Clark (*N. pittieri*, Venez.) – mass-flowering on 5-yr cycle

Chusua Nevski. Orchidaceae (IV 2). 8 N Ind. to China

Chydenanthus Miers. Lecythidaceae (II). 1 Andamans to NG: ***C. excelsus*** (Blume) Miers – seeds used as fish-poison. R: FM I,21(2013)92.

Chylismia (Torrey & A. Gray) Raim. (~ *Camissonia*). Onagraceae. 16 W N Am. deserts. R: UCPB 34(1962)174, CUSNH 37(1969)202

Chylismiella (Munz) W.L. Wagner & Hoch (~ *Camissonia*). Onagraceae. 1 W N Am.: ***C. pterosperma*** (S. Watson) W.L. Wagner & Hoch. R: CUSNH 37(1969)374

Chymsydia Albov (~ *Agasyllis*). Umbelliferae (III 9). 1–2 Transcaucasia

Chysis Lindl. Orchidaceae (V 13a). 10 trop. Am. Cult. orn. epiphytes

Chytranthus Hook.f. Sapindaceae. 30 trop. Afr. Corner's Model, ***C. villiger*** Radlk. (W Afr.) with basiflory. Some ed. fr.

Chytroglossa Reichb.f. Orchidaceae (V 12h). 4 Braz.

Chytroma Miers = *Lecythis* + *Eschweilera*

Chytropsia Bremek. = *Eumarchia*

Cibirhiza Bruyns. Apocynaceae (Va; Asclepiadaceae III 2). 3 Ethiopia, Oman (***C. dhofarensis*** Bruyns – local ed. esp. tubers), Tanz., Zambia. R: F. Albers & U. Meve, *Ill. Handb. Succ. Pls*, Asclep. (2002)107

Cibotarium O. Schulz = *Sphaerocardamum*

Cibotiaceae Korall (~ Dicksoniaceae). Polypodiidae – Cyatheales. 1/11 Indopacific, C Am. R: T 55(2006)712. Terrestrial with massive erect to creeping rhizomes with soft yellow hairs at apices & persistent leaf-bases. Fronds bipinnate or more dissected; veins free, simple or forked to pinnate. Sori marginal at vein ends; indusia bivalvate; spores globose tetrahedral with prominent angles & well-dev. equatorial flange (unique in C.). x = 68
Genus: *Cibotium*

Cibotium Kaulf. Cibotiaceae (Cyatheaceae s.l.). 11 trop. As. to Mal. (3), Am. (2), Hawaii (6), C Am. R: M.F. Large & J.Braggins, *Tree Ferns* (2004)63. Cult. orn. tree ferns & cut trunks used as flower-pots; stem-hairs used as styptic in As.; starch form used for laundry & food in Hawaii, where extracts used in embalming. ***C. barometz*** (L.) J. Sm. (Ind. & S China to W Mal.) – prostrate sp., the end of the trunk with bud covered in hairs passed off as the 'Vegetable Lamb of Tartary' or Scythian or Tartarian lamb in C17 (the barometz of C14 Sir John Mandeville), comm. med. Vietnam; ***C. glaucum*** (Sm.) Hook. & Arn. (Hawaii) – pith eaten like breadfruit, scales form. used & exported as 'pulu fibre' for packing & stuffing pillows etc.

Cicca L. = *Phyllanthus*

cicely, sweet *Myrrhis odorata*

Cicendia Adans. Gentianaceae. 1 Eur. & Med.; 1 Calif., W S Am.

Cicer Tourn. ex L. Leguminosae (III 26). 44 C & W As. (37), Greece (1), Canary Is. & Morocco (1), Ethiopia (1) & ***C. arietinum*** L. (chick pea, garbanzo bean, poss. derived from ***C. reticulatum*** Ladiz. (s.t. treated as subsp.), SE Turkey c. 6500 BP). R: MLW 72–10(1972). World's third pulse crop (c. 20% protein; 11M t p.a.), after beans & peas, eaten fresh or dried (a diet with cereals provides full complement of amino-acids), made into flour (Bengal gram, besan flour) for e.g. pasta (p. di ceci), coffee subs., fodder, the 'salted provender' of Isaiah & today's hummus; orig. of Lat. name Cicero. R: Blumea 52(2007)394

Cicerbita Wallroth (~ *Lactuca*). Compositae (Cich.-Cich.). Incl. *Cephalorhynchus*, *Chaetoseris*, *Mycelis* c. 38 N temp. Some cult. orn. esp. spp. of *C. s.s.* with blue fls, e.g. ***C. alpina*** (L.) Wallroth (*L. a.*, alpine sowthistle, Eur.) – prot.; ***C. muralis*** (L.) Wallroth (*L. m.*, *M. m.*, wall lettuce, temp. Euras., N Afr. (rare)) – weed

Ciceronia Urb. Compositae (Eup.-Crit.). 1 E Cuba (serpentine): ***C. chaptalioides*** Urb. R: MSBMBG 22(1987)1

Cichorium Tourn. ex L. Compositae (Cich.-Cich.). 6 Eur. (3), Med., Ethiopia. R: Gorteria suppl. 5(2000). 2 salad pls: ***C. endivia*** L. (poss. derived from ***C. pumilum*** Jacq. (*C. e.* subsp. *divaricatum*, coastal Medit.)) – long cult. (Pliny mentions it; one of the 'bitter herbs' of the Passover) for lvs, 3 cv groups: **Scarole Group** (escarole), **Fraisé Group** (fraisé), **Endivia Group** (endive, chicory in US) – usu. blanched; ***C. intybus*** L. (chicory, succory, witloof, blue sailor (Am.), Medit.) – fls open at 8 a.m., close at 4 p.m., lvs ed., usu. blanched, also used for skin complaints & locust antifeedants, 4 cv groups: **Root Chicory Group** – r. as an adulterant or subs. of coffee, **Witloof Group** (witloof; puntarelle), **Pain de Sucre Group, Radicchio Group** (radicchio) – cvs with white-veined, red-purple lvs

Ciclospermum Lag. = *Cyclospermum*

Cicuta L. Umbelliferae (III 8). 8 N temp. (Eur. 1: ***C. virosa*** L., cowbane, water hemlock). All v. toxic, poss. the most violently poisonous (when eaten) of all N temp. pls., esp. in N Am. ***C. maculata*** L. (beaver poison)

cider *Malus domestica*; **c. gum, c. tree** *Eucalyptus gunnii*

cidra *Cucurbita ficifolia*

Cienfuegosia Cav. Malvaceae (Malv.-Goss.). R: AMBG 56(1969)191. c. 26 trop. & warm Am. & Afr.

Cienkowskiella Kam = *Siphonochilus*

cigar flower *Cuphea ignea*; **c. tree** *Catalpa speciosa*

Cigarrilla Aiello. Rubiaceae (I 1/IV 24). 1 Mex.: ***C. mexicana*** (DC.) Aiello

cilantro *Coriandrum sativum*, *Eryngium foetidum*

Ciliosemina Antonelli (~ *Remijia*). Rubiaceae (I 1). 2 NW S Am. R: T 54(2005)25. Bark a source of quinine

Cimicifuga L. ex Wernisch. = *Actaea*

Ciminalis Adans. = *Gentiana*

Cinchona L. Rubiaceae (I 1). 23 C Bolivia to N Colombia & Venez., 1 ext. to Costa Rica. R: MNYBG 80(1998). Trees & shrubs. Bark (Jesuits' bark, Peru b., druggists' b.) source (300–500 t quinine from 5–10 K t bark (2003) esp. Congo) of alks esp. antimalarial quinine (suppresses glucose & prot. synth. of trophozoites in red blood cells, analgesic), but when artemisins available, no longer recommended though still not superseded by synthetics (1944; the first synth. organic dye, mauveine (1856) introd. while attempting synth. quinine) & used in tonic water (see also *Remijia*; bubbles speed entry of alcohol to blood so that gin & t. the most efficient means (short of injection) of so doing), 60% prod. used in food industry. Form. felled in forest before plantations established by colonial powers in As. esp. ***C. calisaya*** Wedd. ('*C. officinalis*', E Andes) – yellow bark, calisaya, Ledger bark, the main source of the high alk. yielding cvs of Indonesia, & ***C. pubescens*** Vahl (brown or red bark, invasive in Pacific, displacing *Miconia robinsoniana* Cogn. in Galápagos) – high in quinidine, 4–13% quinine; true ***C. officinalis*** L. (S Ecuador) not imp.

Cinchonopsis L. Andersson (~ *Cinchona*). Rubiaceae (I 1). 1 C & W Amaz.: ***C. amazonica*** (Standl.) L. Andersson. R: AMBG 82(1995)424

Cincinnobotrys Gilg. Melastomataceae. 7 trop. Afr. R: Adansonia 16(1976)355

Cineraria L. (~ *Senecio*). Compositae (Sen.-Sen.). Excl. *Bolandia*, *Oresbia*, 35 trop. & (esp.) S Afr., Madag., SW Arabia. Garden 'cinerarias' now referred to *Pericallis* × *hybrida*

Cinna L. Gramineae (XVI 15). Excl. *Limnodea*, 4 Am., temp. Euras. (Eur. 1). R: Sida 14(1991)581. A 1 or 2

Cinnabarinea F. Ritter = *Echinopsis*

Cinnadenia Kosterm. Lauraceae (inc. sed.). 2 Bhutan, Assam, Myanmar, Malay Pen.

Cinnagrostis Griseb. = *Calamagrostis*

Cinnamodendron Endl. Canellaceae. 5 trop. S Am., WI. ***C. corticosum*** Miers (WI) – bark a spice & a tonic

Cinnamomum Schaeffer. Lauraceae (I). Incl. *Phoebe* p.p., c. 250 E & SE As. (Borneo 26; R: Blumea 56(2011)241) to Aus., Fiji & Samoa, trop. Am. (48). R (p.p.): Ginkgoana 6(1986)1. Aromatic trees (Massart's Model) & shrubs (some heterodichogamous), bark yielding prod. for flavouring, scent & medic. ***C. aromaticum*** Nees (*C. cassia*, cassia bark, c. lignea, Chinese or bastard cinnamon, rou gui, Myanmar) – long cult. China (1 of 50 fundamental herbs), bark used like cinnamon (one of the oldest spices, malabathrum of the Romans – though *C. malabatrum* (Burm.f.) G. Don a diff sp.), lvs distilled for oil used as flavouring as in some gin, ingredient of 'natural viagra'; ***C. burmanni*** (Nees & T. Nees) Blume (*C. nitidum*, Indomal.) – sold as spice 'Korintji cinnamon' in US; ***C. camphora*** (L.) J. Presl (camphor (laurel), ho wood, China, Taiwan, Jap., invasive SE As., Aus., S Afr.) – timber for cabinet work & distilled to give camphor (now synth. artificially), used to keep off moths in wardrobes etc. & in liniments, aromatherapy oils etc.; ***C. culitlawan*** (L.) Kosterm. (*C. culilaban*, culilawan, China, Mal.) – bark gives spice, medic. etc.; ***C. iners*** Reinw. ex Blume (Indomal.) – bark a food-flavouring & tonic drink, in joss-sticks in Malay Pen.; ***C. porosum*** (Nees & Mart.) Kosterm. (emboya, imbuia, imbuya, Braz.) – comm. imp. hardwood; ***C. verum*** J. Presl (*C. zeylanicum*, cinnamon, Sri Lanka & SW Ind.) – cinnamon of commerce, though oft. adulterated with or replaced by *C. aromaticum* (coarser bark), cult. as coppiced low shrub & widely natur. (invasive in Seychelles), the bark sold as 'quills' (quillings when broken) & used to flavour food (esp. stewed apples & pears in GB, biscuits in Netherlands) & toothpaste, also as incense & medic. (incl. added to laudanum; high doses used in cancer treatment), in men's scent in anc. Rome

cinnamon *Cinnamomum verum*; **Chinese c.** *C. aromaticum*; **c. fern** *Osmundastrum cinnamomeum*; **Korintji c.** *C. burmanni*; **c. rose** *Rosa cinnamomea*; **white c.** *Alyxia* spp.

Cinnamosma Baill. Canellaceae. 5–6 Madag. ***C. fragrans*** Baill. (taggar) – Mangenot's Model, scented wood exported via Zanzibar to Mumbai for religious cerem.

cinquefoil *Potentilla* spp. esp. *P. reptans*

Cintia Kníze & Říha = *Weingartia*

Cionidium T. Moore = *Tectaria*

Cionomene Krukoff = *Elephantomene*

Cionosicyos Griseb. Cucurbitaceae (XV). 4–5 C Am., WI

Cionura Griseb. = *Marsdenia*

Cipadessa Blume. Meliaceae (II). 1 Indomal. (natur. Aus.): ***C. baccifera*** (Roth) Miq. R: FM I,12(1995) 57

Cipocereus F. Ritter. Cactaceae (III 3). 5 Braz. R: Bradleya 9(1991)86

Cipoia Philbrick & al. Podostemaceae (III). 2 Braz. R: SB 29(2004)113, 33(2006)825

Cipum A. Rich. = ? Sapindaceae

Cipura Aubl. Iridaceae (VII 5). c. 9 trop. Am. Lvs pleated. Some bulbs locally medic.

Cipuropsis Ule = *Vriesea*

Circaea Tourn. ex L. (~ *Fuchsia*). Onagraceae. 7 N hemisph., woods (Eur. 2). R: AMBG 69(1982)804. Poll. syrphids & small bees, the first the more imp. the shadier the habitat; hybrids common but reprod. vegetatively & resemble good spp. Fl. 2-merous with 1 whorl of A; fr. with hooked bristles; ***C. lutetiana*** L. (enchanter's nightshade, E US, Euras.) – weed

Circaeaster Maxim. Circaeasteraceae. 1 NW Himal. to NW China: ***C. agrestis*** Maxim.

Circaeasteraceae Hutch. Magnoliidae – Ranunculales. 1/1 NW Himal. to NW China. Ann. herb with rosette of simple subdecussate lvs with open dichot. venation at top of stem-like hypocotyl, stip. 0. Fls bisex., reg., hypog., minute in term. fascicles; K 2 or 3, C 0, A (1)2(3), anthers with longit. slits bisporangiate, G (1)2(3) each with 2 apical ovules, pend. from ventral margin of G, orthotropous, unitegmic, only 1 maturing. Fr. an achene, seed with no testa but copious endosperm with suberized outer layer. 2n = 30
Genus: *Circaeaster*

Kingdonia form. here has been put in a monogeneric family or (as here) in Ranunculaceae. Pollen studies (AJB 69(1982)990) show similarities with herbaceous Berberidaceae & certain R. like *Trollius* & the affinities of this app. much reduced pl. still unclear. The open dichot. venation has excited much morphological interest in the past even though such is found in petals of some *Ranunculus* spp.

Circaeocarpus C.Y. Wu = *Zippelia*

Circandra N. E. Br. (~ *Erepsia*). Aizoaceae (V). 1 SW S Afr.: ***C. serrata*** (L.) N.E. Br. – extinct? R: BBP 64(1989)473

Circassian seeds *Adenanthera pavonina*

cirio *Fouquieria columnaris*

Cirrhaea Lindl. Orchidaceae (V 12i). 7 Braz.

Cirrhopetalum Lindl. = *Bulbophyllum*

Cirsium Mill. Compositae (Card.-Card.). c. 250 N temp. (Eur. 60). Feathery pappus (cf. *Carduus*). Thistles, ***C. interpositum*** Petrak (SW China) – pachycaul to 5 m tall. Few cult. orn.; local medic. N Am.; some bad weeds, esp. ***C. arvense*** (L.) Scop. (swamp thistle, Eur., invasive N Am.) – usu. dioec., & ***C. vulgare*** (Savi) Ten. (spear or bull t., Eur. & Med., invasive Aus., S Afr., N Am.) – prob. 'true' t. of Scotland (though now applied to *Onopordum acanthium*); Native Americans eat inner stem of several spp. while young shoots of ***C. palustre*** (L.) Scop. (Eur., Medit., W As.) eaten in salads & r. of ***C. tuberosum*** (L.) All. (Eur.) form. harvested & stored as food for winter

ciruela *Spondias purpurea*

Cischweinfia Dressler & N. Williams. Orchidaceae (V 12h). 11 trop. S Am. R: Selbyana 25(2004)8. Poll. euglossine beres

Cissampelopsis (DC.) Lindl. Compositae (Sen.-Sen.). 10 trop. As. R: KB 63(2008)214

Cissampelos L. Menispermaceae (V). 20 trop. R: Phytologia 30(1975)415. Alks. ***C. pareira*** L. (false pareira r., range of genus) – medic. esp. against snakebite & used as fertility control in Urug.

Cissus L. Vitaceae. Excl. *Cyphostemma*, c. 350 trop. & warm. Heterogeneous. Usu. lianes with tendrils; fls bisex., C 4; berry usu. ined. Some cult. climbers & housepls (grape ivy), esp. ***C. antarctica*** Vent. (kangaroo vine, E Aus.), ***C. discolor*** Blume (*C. javana*, Mal.) & ***C. verticillata*** (L.) Nicolson & Jarvis (*C. sicyoides*, princess vine, trop. Am.); ***C. quadrangularis*** L. (OW trop., S Afr.) – succ. liane with 4-winged stems, oft. ± leafless; lvs of many spp. ed. & medic. e.g. ***C. gongylodes*** (Bak.) Planch. (Braz.) – cult. Kayapó (Amaz.) incl. cvs high in vitamins & nutrients, planted in 'abandoned' slash-&-burn, giving 40 yrs of ed. lvs & fr. before new clearance

Cistaceae Juss. Magnoliidae – Malvales. 9/175 temp. & warm esp. Medit. Aromatic shrubs (*Pakaraimaea* a tree) or herbs; hairs oft. clustered as to seem stellate. R.-hairs 0, at least in young pls (mycorrhiza). Lvs opp. more rarely in spirals or whorled, simple, s.t. with stip. Fls bisex., reg. (exc. K), hypog.; K 5 (2 outer oft. narrower than & s.t. adnate to inner 3) or 3, convolute, C 5 (3 in *Lechea*), convolute in opp. direction to K, seldom imbr., oft. crumpled in bud & evanescent, A (3–) ± ∞ on or just outside annular nectary-disk, centrifugal, s.t. sensitive to touch, anthers with longit. slits, G ((2)3–5(–12), 1-loc. with parietal placentation, the placentas ± deeply intruded & s.t. (as in *Cistus*) meeting to give discrete locules, style solit. to 0, stigma 1(3) s.t. lobed, ovules (1–)4–∞ on each placenta, orthotropous, rarely (as in *Fumana*) anatropous, bitegmic. Fr. a loculicidal capsule, seeds (1–)3–∞, usu. v. small, usu. with starchy endosperm & embryo usu. with flattened cotyledons & commonly curved into hook or ring, circinately coiled, or rarely straight. x = 5–11

Genera: *Atlanthemum, Cistus, Crocanthemum, Fumana, Helianthemum, Hudsonia, Lechea, Pakaraimaea, Tuberaria*

Long-believed related to Bixaceae but form. placed in Violales (= Malpighiales); v. close to Dipterocarpaceae (subfam. Pakaraimaeoideae already referred here), Sarcolaenaceae & poss. incl. those. Many cult. orn. esp. spp. of *Cistus* & *Helianthemum*; fragrant resin (*Cistus*)

Cistanche Hoffsgg. & Link. Orobanchaceae (Orob.). 10 Med. (Eur. 2), Ethiopia to W India. & NW China. Local medic., e.g. ***C. deserticola*** Ma (rou cong rong, Chinese deserts) – overharvested (70 t per annum from Mongolia alone). ***C. phelypaea*** (L.) Cout. (SW Eur. & N Afr.) – eaten as asparagus by Tuareg

Cistanthe Spach (~ *Calandrinia*). Montiaceae (Portulacaceae s.l.). Excl. *Calyptridium, Philippiamra, Spraguea*, 20 Am. Some cult. orn. incl. ***C. grandiflora*** (Lindl.) Schldl. (*Calandrinia g.*, *C. glauca*, Chile) – perenn. grown as ann. 1 m tall

Cistus Tourn. ex L. Cistaceae. Incl. *Halimium*, 29 Canary Is., Medit. (Eur. 25) E to Cauc. R: ABG 153(2006)323. Rock roses. Although self-incompatible, pollen can germ. & obstruct stigma, but *C. ladanifer* & *C. salviifolius* with sensitive A moving from G to P reducing self-poll., ***C. albidus*** L. (alkaline soils) & ***C. crispus*** L. (acid) with style elongating above A by p.m. (fls last 1 day). Almost all cult. orn. shrubs incl. ***C. ladanifer*** L. (W Med.) – early colonist after fire (flammable as 0.1–0.2% leaf fresh wt. is alpha-pinene with flashpoint at 38°C) & source of the resin ladanum coll. by dragging a kind of rake through shrubs (&, according to Pliny, goats' beards!) & used in scenting soap, deodorants, form. medic., seeds ed. & ground into flour in Morocco & Spain, & ***C. creticus*** L. (*C. incanus* subsp. *c.*, E Med.) – resin similar, prob. 'myrrh' of Genesis; resin from both obtained by boiling twigs when resin can be skimmed from water-surface; ***C. salviifolius*** L. (S Eur.) – form. lvs tea subs. in Greece & mulberry leaf subs. for silk worms in Cyprus

Cithareloma Bunge. Cruciferae (4). 2 Iran, C As.

Citharexylum B. Juss. Verbenaceae (Cith.). c. 130 trop. Am. to Arg. Dioec. trees & shrubs superficially resembling cherries; fr. a berry-like drupe separating into 2 nutlets. Timber good (fiddlewood (meaning of Lat. name), corruption of *bois-fidèle*) esp. ***C. spinosum*** L. (*C. fruticosum*, WI) – not spiny! ***C. caudatum*** L. – invasive in Hawaii

citrange, Troyer c. *Citrus* × *insitorum*

citrangequat *Citrus* × *georgiana*

Citriobatus A. Cunn. ex Loudon = *Pittosporum*

× **Citrofortunella** J. Ingram & H. Moore = *Citrus*

citron *Citrus medica*

× **Citroncirus** J. Ingram & H. Moore = *Citrus*

Citronella D. Don. Cardiopteridaceae. 21 Mal., Pac., trop. Am. R: CGH 142(1942)61. Nozeran's Model. ***C. gongonha*** (Mart.) R. Howard (Braz.) used like maté; ***C. moorei*** (Benth.) R. Howard (churnwood, Aus.) – flanged trunks

citronella oil *Cymbopogon nardus, C. winterianus*

Citropsis (Engl.) Swingle & Kellerman. Rutaceae (II 2). 8 trop. Afr. R: W.T. Swingle, *Botany of Citrus* (1943)302. Stock for citrus fr. ***C. articulata*** (Spreng.) Swingle & Kellerman (omuboro) – roots chewed for alleged 'Viagra' qualities

+ **Citroponcirus** H. Wu & al. = *Citrus*

Citrullus Schrad. Cucurbitaceae (XIV). Incl. *Acanthosicyos* p.p., 6 trop. & S Afr., prob. As. Monoec. or dioec. lianes with branched tendrils. ***C. colocynthis*** (L.) Schrad. (colocynth, bitter apple, 'vine of Sodom', 'gall', 'gourd' of Bible, cult. & natur. Medit. & Ind., wild ally = *C. mucosospermus* (Fursa) Fursa, W trop. Afr.) – green throughout Iraq summer (deep r. system), dried pulp purgative etc., cult. since time of Assyrians, rodent control since time of Columella; ***C. lanatus*** (Thunb.) Matsum. & Nakai (watermelon, Afr., natur. Am.) – poss. selected from *C. colocynthis* in early Afr. agric., refreshing red-fleshed fr. sold in Med. etc., small white-fleshed cvs used for preserves, seeds used in soups & snack-food etc. in China, excess fr. made into syrup in E Eur., ingredient of sun-lotions & other cosmetics; ***C. naudinianus*** (Sonder) Hook.f. (*A. n.*, S trop. Afr.) – imp. food & water source (Kalahari)

Citrus L. Rutaceae (II 2). Incl. *Clymenia, Eremocitrus, Feroniella, Fortunella, Microcitrus, Oxanthera, Poncirus*, c. 25 S & SE As, to N China (1) & E Aus. (6, R: Telopea 7(1998)333), New Caled. R: H.J. Webber & al. (1943–8) *The C. Industry*. 'Gold fr.' ('aurantiae'), confused with Apples of the Hesperides (= quince). Everg. (decid. in *C. trifoliata*) trees & shrubs, usu. spiny with simple lvs (trifoliolate in *C. t.*, pinnate in ***C. lucida*** (Scheff.) Mabb. (*Feroniella l.*, SE As.)) often with a joint on petiole showing derivation from trifoliolate & pinnate (as *Merrillia*) ones like those in many Rut.; 'subg. *Papeda*' with winged 'petiole', s.t. as long as 'lamina', subg. *Citrus* without. Fls usu. white, strongly fragrant, fr. a leathery-skinned berry (hesperidium), the rind insecticidal (source of bittering agents used in tonic water instead of quinine, esp. naringin half as bitter as q. but simply manipulated chemically to a dihydrochalcone 500 times as sweet as sucrose, related compounds now being dev. as comm. sweeteners in US) with (3–)5–15(–18) locules (segments) filled with inflated hair-cells (fibre used in gut disease) full of juice (diff. stereoisomers of limonene responsible for diffs between orange- & lemon juice), keeping seeds moist (& alleged to raise blood pressure, octopamine s.t. causing migraines), rich in vitamin C – 'limejuice' (actually usu. lemon, which is 3 times richer) form. used as antiscorbutic by Br. seamen (known as limers or 'limeys' in US). Acid oil droplets in juice vesicles of some make those ined. but other. spp. (R: Telopea 7(1997)167) gives most imp. fr. industry (100 M t by 2009) in warm countries esp. S US, Medit., Braz., WI, Aus. etc. based on *C.* spp. & hybrids, and in cleaning products – citrus scent deemed 'fresh'. Long cult., imp. to Romans (fr. orig. grown for scenting qualities (*C. medica*), name 'citrus' orig. applied to fragrant wood of *Tetraclinis articulata*), the orig. of some still obscure, but most in cult. are anc. apomictic diploid hybrids (polyembryony first observed by Leeuwenhoek!) & selected cvs of these, e.g. limelo (*C.* × *aurantiifolia* × *C.* × *limon*) etc., app. involving 4 wild spp. (see below); hybrids still being synth. for new fr. crops & rootstocks (grafting in *C.* the earliest documented of all); many cult. orn. (form. status symbols esp. in Holland) incl. Florentine 'bizzaria', graft hybrid of *C. medica* on *C.* × *aurantium* Sour Orange Group. ***C.* x *aurantiifolia*** (Christm.) Swingle (*C. medica* × ?*C. hystrix*, (Persian) lime) – for cooking & juice, oil from rind & from seeds used in soap etc., some cvs sweeter & much used in Ind.; ***C.* × *aurantium*** L. (*C. maxima* × '*C. reticulata*') – honey used in true Spanish nougat, many backcrosses (incl. chironja), most in following cv. groups: **Grapefruit Group** (*C. paradisi*, a backcross with female *C. maxima* made in C18 Barbados) – pop. breakfast fr. & source of juice, faddish (but app. effective because controlling sugar metabolism) slimming aid, cvs incl. pink-fleshed mutants (usu. **'Marsh Seedless'** or **'Star Ruby'**) & NZ grapefruit with winged petioles & green cotyledons, also used in shampoos & other cosmetics, **Sour Orange Group** (incl. **'Bouquet'** ('bergamot' orange in US), bigarade & bitter or Seville oranges) – upper surface of petals yield neroli oil for scent (chief ingredient of eau-de-Cologne & used in 'bisex.' scents like CK One), fr. used for marmalade, candied peel & in liqueurs (e.g. Curaçao, Cointreau), sudachi (Jap.) juice used like vinegar on mushrooms, **Sweet Orange Group** (*C. sinensis*, raised in China) – most imp. citrus crop (66% of total), pulp used fresh or for juice, rind for oil, seed-oil used in soap, extract from lvs, twigs & young fr. (petitgrain oil) used in aromatherapy, chewing sticks in Nigeria, many cvs incl. blood (red or red-streaked pulp) & navel (sec. fr. at stylar end, e.g. **'Baia'** ('Washington Navel')) oranges, famous ones incl. **'Shamouti'** ('Jaffa'), **'Valencia'** with thin rind & c. 6 seeds, ortaniques (backcrosses with '*C. reticulata*', Jamaica 1920s), **Tangelo Group** (*C.* × *tangelo*, grapefruit crossed with '*C. reticulata*') – several cvs, in GB known incorrectly as uglis, such as **'Minneola'**, **Tangor Group** (*C. nobilis*, tangor, a backcross with '*C. reticulata*') – easily removed rind; ***C. australasica*** F. Muell. (finger lime, E Aus.) – vesicles spherical, separating, so-called 'vegetable caviar' for pickles etc., parent (with ? '*C.* × *reticulata*' of 'Red Centre lime'); ***C. australis*** (Mudie) Planch. (*Microcitrus a.*, E Aus.) – fr. a passable 'lime'; ***C.* × *floridana*** (J. Ingram & H. Moore) Mabb. (× *Citrofortunella floridana*, *Citrus* × *aurantiifolia* × *C. japonica*, limequat) – ed. fr., raised Florida 1909; ***C.* × *georgiana*** Mabb. (*C.* × *insitorum* × *C. japonica*, citrangequat) – some v. hardy cvs; ***C. glauca*** (Lindl.) Burkill (*Eremocitrus g.*, desert kumquat, NE to S Aus.) – desert pl. resistant to cold, lvs isobilateral, used as rootstock, fr. ed. used in drinks, preserves & handwashes; ***C. hystrix*** DC. (*C. micrantha*, kaffir, leech or makrut lime, C Mal.) – 'lime leaves' of (esp. Thai) cuisine, effective leech repellent; ***C.* × *insitorum*** Mabb. (× *Citroncirus webberi*, (Troyer) citrange, *Citrus* × *aurantium* × *C. trifoliata*) – rootstock for other citrus fr. (resistant to tristeza virus); ***C. japonica*** Thunb. (*Fortunella j.*, *C. margarita*, *F. m.*, kumquat, S China) – fr. ed. complete with peel raw, preserved or candied; ***C.* × *junos*** Sieb. ex Tanaka ('*C. reticulata*' x *C. cavaleriei* Lév. ex Cavalerie (W China), yuzu) – imp. vinegar subs. (esp. sudachi), fashionable 'drizzle' on food,; ***C.* × *latifolia*** (Yu. Tanaka) Tanaka (*C.* × *aurantiifolia* × ?*C.* × *limon*, Tahitian lime) – usu. seedless (cf. limelo above); ***C* × *limon*** (L.) Osb. (*C medica* × *C.* × *aurantium*, lemon, prob. incl. ***C. limetta*** Risso (limetta, mosami sweetie) – weak sweet or acid flavour, dried fr. sold Ind.) – rind yields lemon oil (flavour due to citral, less than 5% by wt.), pulp gives lemon juice & citric acid used culinarily & med. (scurvy, efficient contraceptive as it kills sperm [citric acid in ejaculate], so used by prostitutes since 18C, many cvs incl. bergamot (*C.* × *bergamia*, *C. aurantium* subsp. *b.*) – rind oil used in scent, hair-oil, tanning oil & Earl Grey tea (though allegedly carcinogenic); ***C. maxima*** (Burm.) Merr. (*C. decumana*, *C. grandis*, pomelo, pummelo, shaddock (after a Capt. Chaddock, who introd. it to WI), pompelmous, pamplemousse, SE As.) – fr. to 8 kg with thick rind, source of bitter narinjin used in drinks & sweets, parent or many imp. frs; ***C. medica*** L. (citron, ? NE Ind.) – early cult. for scented fr. & spread to Medit. after Alexander but not in Bible (? seed in Cyprus 1200 BC), first discussed monoecy (Theophrastus), local medic. for skin conditions, antiseptic, major source of candied peel, (male) parent of many imp. hybrids, **'Ethrog'** & other cvs (etrog, cedrat) used in Feast of Tabernacles 136 BC onwards, **'Fingered'** (var. *sarcodactylis*, Buddha's hand) – fr. with finger-like processes resembling one affected by mites, used to scent rooms & clothes, flavour sweets & tea in China; ***C.* × *microcarpa*** Bunge (*C* × *mitis*, × *Citrofortunella m.*, × *C. microcarpa*, *Citrus japonica* × '*C. reticulata*', calamondin, calamansi) – imp. for drinks in Philippines, housepl. with fr. held all winter; ***C.*** × ***oliveri*** Mabb. (*C.* × *microcarpa* × *C. australasica*, sunrise lime) – comm. 'bush tucker' crop Aus.; ***C. reticulata*** Blanco (*C. unshiu*, mandarin, tangerine, satsuma, subtrop. China) – rather hardy (satsumas the hardiest citrus) with small fr. introd. via Kew 1805, the segments separating from one another & peeling readily, oil in shampoos & other cosmetics, TCM (chen pi), ants used to capture insect pests in China (ant nests (*Oecophylla smaragdina*) sold for purpose) since AD 304 perhaps earliest use of biol. control (imported to US 20 cent.) with bamboo pole bridges between trees, cvs incl. **'Clementine'** (clementine, poss. backcross with *C.* × *aurantium*, so name of original parent of oranges unclear as most '*C. reticulata*' prob. *C.* × *aurantium* cvs), **'Dancy'** (common red tangerine of Florida), **'Owari'** (seedless); ***C.*** × ***taitensis*** Risso (*C.* × *jambhiri*, *C. medica* × ?'*C. reticulata*', rough lemon) incl. **'Otaheite'**, a dwarf potted 'orange' esp. US Christmas flower trade; ***C. trifoliata*** L. (*Poncirus t.*, C & N China) – cult. orn. spiny, decid. trifoliolate lvs, fl. buds overwintering on leafless shoots, used as hedging & stock for other citrus, fr. fragrant but acid & with little flesh (suitable for marmalade). See also pomander

Cladanthus Cass. (~ *Chamaemelum*). Compositae (Anth.-Sant.). c. 5 Medit.

Claderia Hook.f. Orchidaceae (V 12b). 2 Thailand to NG. R: OB 72(1983)17. Strange green, prominently veined fls

Cladium P. Browne. Cyperaceae (II 7). 4 N Am. (3), 1 cosmop. exc. Am.: ***C. mariscus*** (L.) Pohl – paper-making in Danube, form. for thatching in GB ('elk sedge' of Anglo-Saxons, the s. of Sedgemoor) but now rare there. R: FR 51(1942)1,183. ***C. jamaicensis*** Crantz (SE US to W I) – dominant in Everglades

Cladocarpa (H. St. John) H. St John = *Sicyos*

Cladoceras Bremek. Rubiaceae (II 2). 1 trop. E Afr.: ***C. subcapitata*** (K. Schum. & Krause) Bremek. R: BRSBB 117(1984)247; Afr. spp. of '*Tarenna*' to be added

Cladochaeta DC. (~ *Helichrysum*). Compositae (Gnap.-Gnap.). 2 Cauc. R: OB 104(1991) 150

Cladocolea Tieghem. Loranthaceae (4 4). 25 C Am. R: JAA 56(1975)272, Novon 2(1992)351

Cladogelonium Leandri. Euphorbiaceae (Crot.-Gel.). 1 Madag.: ***C. madagascariense*** Leandri. R: A. Radcliffe-Smith, *Gen. Euphorb.* (2001)284

Cladogynos Zipp. ex Span. Euphorbiaceae (Epi.-Epi.). 1 SE As., Mal.: ***C. orientalis*** Zipp. ex Span. R: A. Radcliffe-Smith, *Gen. Euphorb.* (2001)182

Cladomyza Danser = *Dendrotrophe*

Cladopus H. Möller. Podostemaceae (III). Incl. *Torrenticola*, c. 9 E As., Mal., Queensland

Cladoraphis Franch. (~ *Eragrostis*). Gramineae (XXVII). 2 S Afr. R: Strel. 9(2000)182

Cladostachys D. Don = *Deeringia*

Cladostemon A. Braun & Vatke. Capparaceae. 1 S & SE Afr.: ***C. kirkii*** (Oliv.) Pax & Gilg. R: Strel. 10(2000)205

Cladostigma Radlk. = *Hildebrandtia*

Cladothamnus Bong. = *Elliottia*

Cladrastis Raf. Leguminosae (III 2). 1 N Am., 5 E As. (R: Rhodora 105(2003)213). R: NRBGE 9(1913)96. Yellow-wood esp. ***C. kentukea*** (Dum. Cours.) Rudd (*C. lutea*, Kentucky yellow-wood, yellow ash, SE US) – close-grained wood for gun-stocks, heartwood provides a yellow dye

Clandestinaria (DC.) Spach = *Rorippa*

Clanwilliam cedar *Widdringtonia cedarbergensis*

Claoxylon A. Juss. Euphorbiaceae (Acal.-Acal.-Claox.). Incl. *Claoxylopsis*, c. 75 Madag. (13) to Hawaii. A 10–200

Claoxylopsis Leandri = *praec.* (but see KB 43(1988)642)

Clappertonia Meissn. Malvaceae (Grew.-Apeib.; Tiliaceae). 2–3 trop. W Afr. R: FR 99(1988)267. Scarrone's Model. Fibre pls esp. ***C. ficifolia*** (Willd.) Decne. – jute-like fibre

Clappia A. Gray. Compositae (Tag.-Pect.). 1 Texas, Mex.: ***C. suaedifolia*** A. Gray. R: FNA 21(2006)251

Clara Kunth = *Herreria* (but see Bradea 9(2003)17)

claret ash *Fraxinus angustifolia* 'Raywood'

Clarisia Ruíz & Pavón. Moraceae (II). (Incl. *Sahagunia*) 3 trop. Am. R: SB 35(2010)780. Timber & ed. fr.

Clarkeasia J.R.I. Wood = *Strobilanthes* (but see EJB 51(1994)187)

Clarkella Hook.f. Rubiaceae (inc. sed.). 1 Himal., Thailand: ***C. nana*** (Edgew.) Hook.f. R: FOC 19(2011)89

Clarkia Pursh. Onagraceae. 41W N Am., S S Am. R: UCPB 20(1955)241. Cult. orn. ann. herbs esp. ***C. amoena*** (Lehm.) Nelson & Macbr. (coastal N Calif.) – the 'godetia' of gardens & ***C. unguiculata*** Lindl. (Calif.) – the 'clarkia' of gardens with cvs of many colours incl. doubles; ***C. breweri*** (A. Gray) Greene (Calif.) – only scented sp., cult. orn.

claro walnut *Juglans hindsii*

clary *Salvia sclarea*

Clastopus Bunge ex Boiss. Cruciferae (2). 2 Iran, Iraq

Clathrotropis (Benth.) Harms. Leguminosae (III 2). 7 trop. S Am. R: BZ 96(2011)1555. Timber

Clausena Burm.f. Rutaceae (II 1). 23 OW trop., S Afr. R: BMNHNAdans. 4,16(1994)107. Some ed. fr. esp. ***C. dentata*** (Willd.) M. Roemer (Ind.) – taste like blackcurrants & ***C. lansium*** (Lour.) Skeels (wampi, S China) – cult. for lime-like fr., citrus rootstock. Some local medic.: ***C. anisata*** (Willd.) Benth. (OW trop.) – mosquito repellent, molluscicide; ***C. heptaphylla*** Wight & Arn. (SE As.) – carbazole alk. with antifungal activity

Clausenellia Löve & D. Löve = *Sedum*

Clausia Trotzky (~ *Hesperis*). Cruciferae (20). 5 E Eur. (1) to C As.

Clausospicula Lazarides. Gramineae (XXII). 1 N Terr., Aus.: ***C. extensa*** Lazarides. R: Aus SB 4(1991)391. Cleistogamous

Clavija Ruíz & Pavón. Primulaceae (Theophrastaceae). 55 S Nicaragua to Braz., Hispaniola. R: OB 107(1991)1. Palm-like pachycauls (Corner's Model) oft. with cauliflory, most dioec., some gynodioec., androdioec. or polygamous

Clavinodum T.H. Wen = *Oligostachyum*

claw, devil's *Harpagophytum procumbens*; **dragon c.** *Corallorhiza odontorhiza*

Claytonia Gronov. ex L. Montiaceae (Portulacaceae s.l.). Excl. *Montia*, 27 N Am., ext. to E As., 2 natur. Eur. R: SBM 78(2006)37. Cotyledons 1 or 2; 'stemlvs' 1 pr (*Montia* several); some with ant-disp. seeds. Local food (tubers = fairy potatoes) & medic. ***C. lanceolata*** Pursh (W US) – corms ed. raw or cooked by Native Americans; ***C. perfoliata*** Donn ex Willd. (*Montia p.*, Cuban spinach, miner's lettuce, winter purslane, W N to C Am. – decaploids in Guatemala highlands, natur. GB) – infls subtended by a perfoliate disk-like organ in place of pr of lvs, ed. Some cult. orn. esp. ***C. virginica*** L. (spring beauty, E N Am.) – 2n = 12–190 with diff. nos in same pl.

Claytoniella Yurtsev = *Montia*

Cleanthe Salisb. ex Benth. = *Aristea*

clearing nut *Strychnos potatorum*

clearweed *Pilea* spp.

cleavers *Galium aparine*

Cleghornia Wight. Apocynaceae (II c). 4 Sri Lanka & SE As. to W. Mal. R: AUWP 88–6(1988)11

Cleidiocarpon Airy Shaw. Euphorbiaceae (Epi.-Epi.). 2 Myanmar, W China. Drupes. ***C. cavaleriei*** (Léveillé) Airy Shaw – promoted in China as oil-crop

Cleidion Blume. Euphorbiaceae (Acal.-Acal.-Cleid.). 25 trop. (W Afr. 1, Madag. 1, Mal. 6, New Caled. 12). Some with Corner's Model. ***C. spiciflorum*** (Burm.f.) Merr. (Indomal. to Aus.) – trad. medic.

Cleisocentron Brühl. Orchidaceae (V 16c). 6 Sikkim to Vietnam & Borneo

Cleisomeria Lindl. ex G. Don. Orchidaceae (V 16c). 2 SE As., Mal. R: OB 95(1988)131

Cleisostoma Blume (~ *Sarcochilus*). Orchidaceae (V 16c). c. 90 Nepal to New Caled. (1) & Fiji. Cult. orn. incl. ***C. striatum*** (Reichb.f.) N.E. Br. (? incl. *C. javanicum*, Assam to W Mal.)

Cleisostomopsis Seidenf. Orchidaceae (V 16c). 2 SE As.

Cleistachne Benth. (~ *Sorghum*). Gramineae (XXII 5). 1 trop. Afr., Ind.: ***C. sorghoides*** Benth. R: Strel. 10(2000)688

Cleistanthus Hook. f. ex Planch. (~ *Bridelia*). Phyllanthaceae (Brid.-Pseud.; Euphorbiaceae). c. 140 OW trop. (Afr. 23, Madag. 6). Heterogeneous. Dried fr. of ***C. collinus*** (Roxb.) Benth. (Ind., Sri Lanka) used in criminal poisonings

Cleistes Rich. Orchidaceae (II 1). Excl. *Cleistesiopsis*, c. 60 trop. Am.

Cleistesiopsis Pansarin & F. Barros (~ *Cleistes*). Orchidaceae (II 1). 2 SE N Am. R: KB 63(2008)444. ***C. divaricata*** (L.) Pansarin & F. Barros – hinged anther dispenses several (not just 1) pollen tetrad-masses to diff. insects

Cleistocactus Lem. Cactaceae (III 4). Excl. *Borzicactus*, 15 C Peru to Bolivia & N Arg., Paraguay & Urug. Hummingbird-poll. Cult. orn. esp. ***C. strausii*** (Heese) Backeb. ('*C. straussii*', S Bolivia) with 30–40 bristle-like spines to 2 cm long & 4 yellowish stouter spines per areole; ***C. sepium*** (Kunth) Roland-Gosselin (Peru) – ed. fr.

Cleistocalyx Blume = *Syzygium*

Cleistochlamys Oliv. Annonaceae (III 7). 1 trop. E. Afr.: ***C. kirkii*** (Benth.) Oliv.

Cleistochloa C.E. Hubb. Gramineae (XXIV 3). 8 C Mal. to NG & NE Aus. (dry sandstone ridges). Cleistogamy

Cleistogenes Keng (*Kengia*). Gramineae (XXIX). 14 S Eur. (2), Turkey to temp. E As. (China 10, 5 endemic – R: FOC 22(2006)460)

Cleistopetalum Okada = *Monoon* (but see APG 47(1996)4)

Cleistopholis Pierre ex Engl. Annonaceae (II). 3–4 trop. Afr. ***C. patens*** (Benth.) Engl. (otu) – Troll's Model, comm. timber, bark for rope & mats, lvs & r. medic.

Cleisostomopsis Seidenf. = *Cleisocentron*

Clelandia J. Black = *Hybanthus*

Clematepistephium Hallé (~ *Epistephium*). Orchidaceae (II 2). 1 New Caled.: ***C. smilacifolium*** (Reichb.f.) Hallé – liane to 8 m long. R: Gen. Orch. 3(2003)299

Clematicissus Planch. Vitaceae. 2 Aus. (R: Telopea 11(2006)391), ? 4 S Am.

Clematis Dill. ex L. Ranunculaceae (II 2). Incl. *Clematopsis*, *Naravelia*, c. 323 N temp. (Eur. 10), S Am., Madag. (few), Oceania & trop. Afr. mts. R: APS 43(2005)454; M. Johnson, *The genus C.* (2001); M. Toomey & E. Leeds (2001) *Illus. Encycl. C.* Lianes, shrubs or herbs; lvs usu. opp. compound or simple (in some Aus. spp. varieg. when young), K petaloid, C 0, fr. an achene usu. with long feathery style. Imp. cult. orn. (150+ spp. grown) esp. lianes, only woody Ranunculaceae of any size, some without interfascicular cambium (the sec. rays arise in the primary bundles), some with regular wood formation & some with extra thickenings in pith & cortex cells; petioles wrap round support & lignify. Several medic. (some spp. with acrid juice inflaming skin) & superstition (e.g. smoke a Blackfoot antidote to having been too close to a ghost) & many hybrids in cult., the common garden pls being derived from ***C. florida*** Thunb. (E As.) – fls on old wood in summer, ***C. patens*** Morren & Decne. (China, Jap.) – fls on old wood in spring, & esp. ***C. × jackmanii*** T. Moore (*C. lanuginosa* Lindl. (? China, not known in wild) × *C. viticella* L. (Medit. & nearby As.), arose 1858 at Jackman's nurseries at Woking, England) – fls on new wood in summer & autumn, & ***C.* 'Hendersonii'** (*C. viticella* × *C. integrifolia* L. (Euras.) crossed with *C. lanuginosa* & *C. vitalba*). ***C. afoliata*** J. Buch. (NZ) – lvs reduced to petioles & petiolules, laminae developing only in young pl. or in shade; ***C. armandii*** Franchet (SW China) – everg. cult. orn. liane, 1 clasping petiole of a pair in alt. nodes; ***C. filamentosa*** Dunn (China) – med. (kan-mu-tun) for hypertension; **C. flammula** L. (Euras.) – ed. shoots (cooked); ***C. heracleifolia*** DC. (China) – commonly cult. herbaceous sp. with blue fls, hybrid with *C. vitalba* = ***C. × jouiniana*** Schneider; ***C. ligusticifolia*** Nutt. (N Am.) – imp. local medic; ***C. montana*** Buch.-Ham. ex DC. (montana, Himal., W China) – rampant liane much planted for spring fls; ***C. rehderiana*** Craib (W China) – cult. orn. liane with yellow, cowslip-scented, nodding fls; ***C. vitalba*** L. (old man's beard, 'traveller's joy', Euras., N Afr., invasive in Aus., NZ, NW N Am.) – may attain 30 m, lengths of stem a tobacco subs., young shoots ed. Eur., diuretic

Clematoclethra (Franch.) Maxim. Actinidiaceae. 1 W & C China: ***C. scandens*** (Franch.) Maxim. R: APS 27(1989)81

Clematopsis Bojer ex Hutch. = *Clematis*

Clemensiella Schltr. = *Hoya* (but see EJB 66(2009)449)

clementine *Citrus reticulata* 'Clementine'

Cleobulia Mart. ex Benth. Leguminosae (III 18). 3–5 Mex., Braz. R: Phytol. 38(1977)51

Cleomaceae Bercht. & J. Presl. (~ Cruciferae; Capparaceae s.l.). Magnoliidae – Brassicales. 1/300 trop. & warm. Shrubs and herbs with methyl glucosinolates, s.t. armed; root-hairs 0. Lvs in spirals, often palmate; stip. 3–8-fid, usu.
caducous, s.t. 0. Infl. usu. racemose, s.t. corymbose or fls 1, s.t. monoec.; K 4, C 4, imbr. s.t. with intrastaminal nectary-discs, A (4–)6(–27); anthers linear, coiled at dehiscence, G 1(2), 2-loc. with parietal placentation & 1–18(–26+) anatr., bitergmic ovules. Fr. usu. capsulart, elongate; seeds with 0 or scant endosperm
Genus: *Cleome*, *?Dipterygium*

Cleome L. Cleomaceae. Incl. *Adinocleome*, *Arivela*, *Carsonia*, *Cleomella*, *Cleoserrata*, *Corynandra*, *Gynandropsis*, *Hemiscola*, *Isomeris*, *Mitostylis*, *Oxystylis*, *Peritoma*, *Physostemon*, *Podandrogyne*, *Polanisia*, *Tarenaya*, *Wislizenia*, c. 300 trop. & warm (Eur. 3). Most adapted to seasonal drought but ***C. chapalaensis*** Iltis (SW Mex.) grows in shallow water. Some medic. locally, others with ed. seeds, some cult. orn. ***C. anomala*** Kunth (N S Am.) – nitrogen/sulphur compounds in fl. scent attractive to poll. bats; ***C. dodecandra*** L. (*Polanisia d.*, N Am.) – young pls boiled & dried for winter use; ***C. gynandra*** L. (*G. g.*, cat whiskers,

OW trop. natur. Am.) – potherb; *C. **houtteana*** Schldl. (*C. hassleriana, C. sesquiorgyalis, T. h.*, spider flower, SE Braz. to Arg.) – cult. orn. esp. cut-fl., many cvs; *C. **monophylla*** L. (OW trop.) – lvs ed., seeds for a mustard; *C. **refracta*** (Engelm.) Mabb. (*W. r.*, jack-ass clover, SW N Am.) – polymorphic; *C. **rutidosperma*** DC. (trop. Afr.) – weedy in SE As. to trop. Aus.; *C. **viscosa*** L. (*Arivela v., Corynandra v.*, OW trop.) – seeds a cumin subs. in Sri Lanka

Cleomella DC. = *Cleome* (but see UWPSB 1(1992)29)

Cleonia L. Labiatae (VII 2b). 1 W Med. incl. Eur.: *C. **lusitanica*** (L.) L. – on gypsiferous soils

Cleophora Gaertn. = *Latania*

Cleoserrata Iltis = *Cleome* (but see Novon 17(2007)447)

Cleretum N.E. Br. Aizoaceae (V). Incl. *Dorotheanthus*, 14 S Afr. (winter rainfall). R: T 61(2012)304. Succ. ann., some cult. orn. esp. *C. **bellidiforme*** (Burm.f.) G. Rowley (*D. b., Mesembryanthemum b., M. criniflorum*, Livingston[e] [i.e. living stone] daisy) –fr. opens on wetting, fls purple to white, those cvs with yellow etc. poss. hybrids with *C. **clavatum*** (Haw.) Klak (*D. gramineus*)

Clermontia Gaudich. Campanulaceae (III 4). 22 Hawaii. R: SBM 32(1991)12. Prob. derived from *Lobelia* subg. *Tupa* of As. Woody, some epiphytic, with ed. bird-disp. fr.; latex used as bird-lime. *C. **arborescens*** (H. Mann) Hillebrand – visited by introd. white-eyes (*Zosterops japonica*, Jap.), though orig. poll.prob. extinct

Clerodendranthus Kudô = *Orthosiphon*

Clerodendrum L. Labiatae (III, Verbenaceae). Excl. *Kalaharia, Ovieda, Rotheca (Cyclonema), Volkameria*, c. 150 trop. & warm OW. Trees or shrubs (Chamberlain's etc. Models), perenn. or even ann. herbs, *C. **laciniatum*** Balf.f. (Rodrigues) – marked heterophylly; fls white, yellow, red, blue or violet; A project as a landing-stage for insects, later replaced by style rising from below as A wither & fall; K oft. brightly coloured & in most spp., e.g. *C. **trichotomum*** Thunb. (China, Jap.), accrescent (red & fleshy contrasting with blue-black fr. attractive to dispersing birds). Some medic. but name (Gk. for 'chance' & 'tree') refers to variable reports of efficacy. *C. **fistulosum*** Becc. (Borneo) has pithy stems inhabited by ants & the hollow ones of *C. **capitatum*** (Willd.) Schum. & Thonn. used as pipes & for tapping palm wine (W Afr.); those of *C. **myrmecophilum*** Ridl. (W Mal.) with hollow internodes & thin-walled layers pierced & penetrated by ants. Many cult. orn. hardy & tender esp.: *C. **bungei*** Steud. (*C. foetidum*, China) – hardy shrub; *C. **chinense*** (Osb.) Mabb. (*C. fragrans, C. philippinum*, glory bower, Honolulu or Lady Nugent's rose, China to C Mal., natur. S US & pantrop.) – extrafl. nectaries with ants building caves over them, fls fragrant offered to Buddha in Cambodia), pinkish, in cult. usu. double (wild pl. = 'var. *simplex* (Mold.) S.L. Chen'), serious weed in WI, Samoa; *C. **paniculatum*** L. (pagoda flower, SE As.) – tender with scarlet fls, not prod. fr. (aborted pollen in NG) in Mal., C Am.); *C. **speciosissimum*** Van Geert ex Drapiez (*C. fallax*, Indomal.) – cult. orn., natur. widely; *C. **thomsoniae*** Balf. (bleeding heart, bag flower, trop. W Afr.) – cult. greenhouses for white K & red C; *C. **petasites*** (Lour.) S. Moore (SE As.) – efficacious bronchial compound extracted

Clethra Gronov. ex L. Clethraceae. 30–64 trop. Am., As. to Mal., N Am. (1), Madeira (1). R: BJ 87(1967)36. Some cult. orn. esp. *C. **alnifolia*** L. (bush pepper, E N Am.) & *C. **arborea*** Aiton (folhado, lily-of-the-valley tree, Madeira) – part of original laurel forest

Clethraceae Klotzsch (~ Cyrillaceae). Magnoliidae – Ericales. 2/c. 60 Madeira (1), E As. to Mal., SE US, C Am. to Cuba. Trees & shrubs. Lvs conduplicate-subplicate, in spirals; stip. 0. Infls term., composed of racemose branches; pedicels articulated. Fls usu. reg. & bisex. K 5(6) quincuncial, free or larely united by a tube; C 5(6), imbr., free or basally connate. A 5(6) + 5(6), s.t. adnate to C, anthers ± sagittate with pores or short slits; $\underline{G}$ (3–5) with axile placentation & 1-∞ anatropus or orthotropous unitegmic ovules per locule, style entire or 3(4)-branched. Fr. indehiscent, 1–5-seeded or 3-loculicidal capsule, with persistent K. Seeds s.t. winged. n = 8

Genera: *Clethra, Purdiaea*

Clevelandia Greene = *Castilleja*

Cleyera Thunb. Pentaphylacaceae (3; Theaceae s.l.). 8 Himal. to Jap. (1), Mex. to Panamá & WI. R: JAA 18(1937)118, 22(1941)395. Some cult. orn. incl. *C. **japonica*** Thunb. (*C. ochnacea*, Himal. to Jap.) – sacred tree of Shintoism (sakaki), a tea in Himal.

Clianthus Sol. ex Lindl. Leguminosae (III 24). 1–2 NE NZ. R: NZJB 38(2000)363. *C. **puniceus*** (G. Don) Lindl. (parrot's bill), long cult. by Maoris & poss. native only in 2 areas in Urewera Nat. Park. See also *Swainsona*

Clibadium F. Allam. ex L. Compositae (Helia.-Ecl.). 29 trop. Am. R: Britt. 55(2003)253. Some, e.g. ***C. laxum*** S.F. Blake (Ecuador), with fleshy cypselas. Fish-poisons (guaco) esp. ***C. sylvestre*** (Aubl.) Baill. – extract stops the human heart reversibly

Clidemia D. Don = *Miconia*

cliff brake *Pellaea* spp.

Cliffordiochloa B. Simon = *Otachyrium*

Cliffortia L. Rosaceae (8). 132 S Afr. esp. Cape mts (124, 109 endemic), 1 ext. to Angola, Kenya. R: H. Weimarck (1934) *Monograph of the genus C.* Parallelism of leaf form with *Aspalathus*. Most spp. monoec. & dioec.

Cliftonia Banks ex Gaertn. f. Cyrillaceae. 1 SE US: ***C. monophylla*** (Lam.) Sarg. – shrub of swamps of coastal plain, fls the forage for comm. honeybees. R: CGH 186(1960)73

Climacoptera Botsch. (~ *Salsola*). Amaranthaceae (Chenopod.). 6+ SW & C As.

climbing bellflower *Gloriosa modesta;* **c. fern** *Lygodium* spp.

Clinacanthus Nees. Acanthaceae (III 2c). 3 S China to Mal. ***C. nutans*** (Burm. f.) Lindau (SE As.) – young lvs ed. Vietnam

Clinanthus Herb. (~ *Stenomesson*). Amaryllidaceae. 22 Andes. R: SB 25(2000)723. ***C. elwesii*** (Bak.) Meerow – pollen shed in tetrads

Clinelymus (Griseb.) Nevski = *Elymus*

Clinopodium Tourn. ex L. Labiatae (VII 2b). Incl. *Acinos*, ? *Bancroftia*, *Calamintha*, NW *Micromeria* & *Satureja* – R (N Am.): Britt. 18(1966)244, c. 100 subcosmop. (Eur. 11). ***C. douglasii*** Benth. (*Micromeria chamissonis*, *S. douglasii*, yerba buena, W US) – fragrant lvs a local medic. tea. Cult. orn. incl. ***C. acinos*** (L.) Kuntze (*Acinos arvensis*, basil thyme, Eur., W As.), ***C. ascendens*** (Jordan) Samp. (*C. nepeta* subsp. *a*, common c., Eur.) & ***C. menthifolium*** (Host) Stace (*Calamintha sylvatica*, wood calamint, Eur.)

Clinosperma Becc. Palmae (Ve). Incl. *Brongniartikentia*, *Lavoixia*, 4 New Caled. R: KB 63(2008)69. ***C. macrocarpum*** (H.E. Moore) Pintard & W.J. Baker (*L. c.*) – only 4 left (Mont Panié) 1980

Clinostemon Kuhlm. & Samp. (~ *Mezilaurus*). Lauraceae (II). 2 trop. Am.

Clinostigma H. Wendl. Palmae (V ?). 11 New Br., Vanuatu, Fiji, Samoa, Micronesia to Bonin Is.

Clinostigmopsis Becc. = *Clinostigma*

Clintonia Raf. Liliaceae. 4 N Am., 1 E As. Wood lilies. Unique megasporogenesis with selective elimination of genomes before double fert. so that embryo & endosperm are genetically identical diploids. Cult. orn. esp. ***C. andrewsiana*** Torrey (C & N Calif.) & ***C. borealis*** (Aiton) Raf. (E N Am.) – young lvs a potherb & salad

Cliococca Bab. = *Linum*

Clistax Mart. Acanthaceae (III 2c). 2 Braz.

Clistoyucca (Engelm.) Trel. = *Yucca*

Clitandra Benth. Apocynaceae (1 c). 1 trop. Afr.: ***C. cymulosa*** Benth. – source of inferior (? medic.) rubber, fr. ed. R: BJBB 58(1988)159

Clitandropsis S. Moore = *Melodinus*

Clitoria L. Leguminosae (III 18). Excl. *Barbieria*, 62 trop. esp. Am. (48). Fls inverted so that standard points downwards (cf. *Centrosema*) & A & pistil touch backs of visiting insects. Some trees; cult. orn. climbers esp. blue-flowered ***C. ternatea*** L. (butterfly pea, blue p. or vine, prob. native trop. Am., now pantrop.) – fls for colouring rice, also like litmus paper, local medic.

Clitoriopsis R. Wilczek. Leguminosae (III 18). 1 Congo, Sudan: ***C. mollis*** R. Wilczek

Clivia Lindl. Amaryllidaceae. 5 S Afr. R: H. Koopowitz (2002) *Cs.*: 39. Alks. Transition from bird- to butterfly-poll. Cult. orn. (kaffir lilies) esp. ***C. miniata*** (Lindl.) Bosse – stemless with fleshy r. & somewhat swollen leaf-bases like a bulb, many cvs incl. ones with coloured or varieg. lvs raised & prized in Jap. & ***C. caulescens*** R.A. Dyer – stem to 50 cm, & hybrids (also in wild) between them

cloak fern *Notholaena* spp.

clock-vine *Thunbergia grandiflora*

Cloezia Brongn. & Gris. Myrtaceae (II 5). 5 New Caled.

Cloiselia S. Moore = *Dicoma* (but see SB 31(2006)424)

Clonodia Griseb. Malpighiaceae. 2–3 trop. S Am. R: MNYBG 32(1981)203. Extrafl. nectaries

Clonostylis S. Moore = *Spathiostemon*

Closia Rémy = *Perityle*

clotbur *Xanthium spinosum*

cloudberry *Rubus chamaemorus*

clove gilliflower or **pink** *Dianthus caryophyllus, D. plumarius*

clover *Trifolium* spp.; **alsike c.** *T. hybridum*; **alyce c.** *Alysicarpus vaginalis*; **Aztec c.** *T. amabile*; **Bokhara c.** *Melilotus albus*; **Dutch c.** *T. repens*; **Hungarian c.** *T. pannonicum*; **Italian c.** *T. incarnatum*; **jack-ass c.** *Cleome refracta*; **Japanese c.** *Kummerowia striata*; **Persian c.** *T. resupinatum*; **purple** or **red c.** *T. pratense*; **running buffalo c.** *T. stoloniferum*; **strawberry c.** *T. fragiferum*; **sub, subterranean c.** *T. subterraneum*; **Swedish c.** *T. hybridum*; **white c.** *T. repens*; **Uganda c.** *T. burchellianum* subsp. *johnstonii*; **zig-zag c.** *T. medium*

cloves *Syzygium aromaticum*; **Madagascar c.** *Cryptocarya agathophylla*

Clowesia Lindl. (~ *Catasetum*). Orchidaceae (V 12c). 7 C Am. (Mex. 5). R: Selbyana 1(1975)134. Segregated from *Catasetum* by bisex. fls; poll. male euglossine bees

clubmoss *Lycopodium* spp.

clubrush *Schoenoplectus lacustris*

Clusia Plum. ex L. Guttiferae (1). Incl. *Oedematopus, Renggeria*, c. 300+ trop. & warm Am. Colour of sap useful in infrageneric classification. Mostly dioec. trees (only genus of dicot. trees with crassulacean acid metabolism) & shrubs, some apomictic; c. 85 ± epiphytic, or stranglers with anastomosing aerial r. (cf. *Ficus*), e.g. ***C. rosea*** Jacq. (balsam fig, warm & trop. Am.) – to 20 m, at anthesis staminodes deliquesce into a mass attractive to bees, sticky resin coll. for their nests, that from seeds used as bird-lime & to caulk boats, used as iron indicator in Venez. & firebreak in Sri Lanka, tough street-tree (as in Honolulu, though invasive). Some medic. (rock balsam) & resin used for incense; in Guianas roots (kufa) of several spp., esp. ***C. grandiflora*** Splitg. & ***C. palmicida*** Rich. ex Planch. & Triana, used in basket- & furniture-making; in Mex. lvs of some spp. used as playing-cards. Some cult. orn. incl. ***C. major*** (L.) Jacq. (copey, trop. Am.)

Clusiaceae Lindl. = Guttiferae

Clusiella Planch. & Triana. Calophyllaceae (Guttiferae (I1)). 7 Panamá to N S Am. R: Novon 9(1999)349. Dioec. epiphytes

cluster bean *Cyamopsis tetragonoloba*

Clutia Boerh. ex L. Peraceae (Clut.; Euphorbiaceae s.l.). 75 Afr. esp. S, Arabia (2). Some medic.

Clybatis Philippi = *Leucheria*

Clymenia Swingle = *Citrus*

Clypeola L. (~ *Alyssum*). Cruciferae (2). 9 Medit. (Eur. 2). R: KB 1935:1

Clytostoma Miers ex Bur. = *Bignonia*

Clytostomanthus Pichon = *Cydista*

Cnemidaria C. Presl = *Cyathea*

Cnemidiscus Pierre = *Glenniea*

Cneoraceae Vest = Rutaceae

Cneoridium Hook.f. Rutaceae (I 10). 1 SW N Am.: ***C. dumosum*** (Nutt.) Hook.f. (bushrue)

Cneorum L. Rutaceae (III; Cneoraceae). Incl. *Neochamaelea*, 2 W Med. inc. Eur., Canary Is. R: T 59(2010)1132. ***C. tricoccon*** L. (spurge olive, early introd. Cuba) – violent purgative, in Balearics form. disp. by *Podarcis* lizards now extinct since introd. of carnivores of which pine-martens are now new disp. agents, cult. orn.

Cnesmocarpon Adema (~ *Jagera*). Sapindaceae. 4 NG, Aus. R: Blumea 38(1993)195. Fr. hairs irritant

Cnesmone Blume (*Cnesmosa*). Euphorbiaceae (Acal.-Pluk.-Trag.). 11 Assam to W Mal. R: JAA 22(1941)427. Stinging climbers

Cnestidium Planch. Connaraceae (1V). 2 trop. Am.

Cnestis Juss. Connaraceae (IV). 12 trop. Afr., 1 Indomal. R: AUWP 89–6(1989)174. Lianes, rarely small trees, some poisonous

Cnicothamnus Griseb. Compositae (Gochn.). 2 Bolivia, Argentina. Small trees & shrubs

Cnicus Tourn. ex L. = *Centaurea*

Cnidiocarpa Pim. = *Selinum* (but see BZ 95(2010)75)

Cnidium Cusson (~ *Selinum*). Umbelliferae (III 8). c. 10 Euras.

Cnidoscolus Pohl. Euphorbiaceae (Crot.-Man.). c. 70 Am. Monoec. or rarely dioec. herbs, s.t. with stinging hairs. ***C. aconitifolius*** (Mill.) I.M. Johnston (*Jatropha a.*, C Am.) – lvs boiled & eaten by Mayans; ***C. chayamansa*** McVaugh (trop. Am.) & other spp. (chaya) – lvs eaten like spinach (rich in Vitamin C); ***C. elasticus*** Lundell (chilte, Mex.) – source of rubber, latex 44–50% rubber; ***C. stimulosus*** (Michaux) A. Gray (~ *C. urens*, bull nettle, E N Am.) – stinging hairs; ***C. urens*** (L.) Arthur (trop. Am.) – urticating hairs & sticky latex deter grazers but in Costa Rica sphingid larvae graze hairs & constrict petiole preventing latex flow incl. ***C. marcgravii*** Pohl (*Jatropha oligodon*) – ed. fr. & oil for cooking etc.

coachwhip *Fouquieria splendens*

coachwood *Ceratopetalum apetalum*

coal fossil fuel largely derived from compressed & metamorphosed plant remains of swamp-forests of the past

coat, Jacob's *Alternanthera ficoidea* 'Bettzichiana'

Coatesia F. Muell. (~ *Geijera*). Rutaceae (I 2). 1 NE Aus.: ***C. paniculata*** F. Muell. (*G. p.*, axe-breaker) – comm. timber. R: FA 26(2013)72

Coaxana J. Coulter & Rose. Umbelliferae (III 8). 2 Mex. R: CUMH 11(1975)13

cob, Kentish *Corylus maxima*

Cobaea Cav. Polemoniaceae (II). 18 trop. Am., esp. Mex. R: SBM 57(1999)28. Lianes; cult. orn. as annuals in gardens esp. ***C. scandens*** Cav. (Mex.) growing to 8 m in 1 season. Lvs pinnate term. in branched tendril tipped with hooks, which prevent the spiral movements of the tendril from dragging a branch away before it has time to clasp its support. Fls protandrous, at first greenish cream with unpleasant smell (fly-poll.), later purplish with pleasant honey smell (bee-poll.) though prob. bat-poll. in wild; after anthesis, pedicel becomes twisted; seeds winged

Cobaeaceae D. Don = Polemoniaceae

Cobana Ravenna (~ *Tigridia*). Iridaceae (VII 5). 1 Guatemala, Honduras: ***C. guatemalensis*** (Standl.) Ravenna – lvs pleated

Cobananthus Wiehler (~ *Alloplectus*). Gesneriaceae (II 5c). 1 Guatemala: ***C. calochlamys*** (J.D. Sm.) Wiehler – fls reg., dry capsule cf. fleshy fr. of *Alloplectus* & *Columnea*. R: Selbyana 2(1997)94

cobbler's pegs *Bidens pilosa*

cobnut *Corylus avellana*

coca, cocaine *Erythroxylum coca*

Coccineorchis Schltr. (~ *Stenorrhynchos*). Orchidaceae (IV 2h). 7 trop. Am.

Coccinia Wight & Arn. Cucurbitaceae (XIV). 25 trop. & S Afr., 1 ext. to Mal.: ***C. grandis*** (L.) J. Voigt (rashmato, tindora) – poss. sex chromosomes, shoots & fr. eaten (introd. Hawaii 1969, now invasive). R: PhytoKeys 54(2015)1. ***C. abyssinica*** (Lam.) Cogn. (anchote) – cult. Ethiopia for ed. tubers

Coccochondra Rauschert (*Chondrococcus*). Rubiaceae (IV 7). 4 Guayana Highland. R: PEE 144(2011)117

Coccocypselum P. Browne. Rubiaceae (IV 3). 20 trop. Am. Heterostyly

Coccoloba P. Browne. Polygonaceae (Erio.-Cocc.). 120 trop. & warm Am. (N Am. 2). Trees, shrubs & lianes. Some timber (pigeon-wood); ed. fr.; some cult. esp. ***C. uvifera*** (L.) L. (seaside grape, Jamaican kino, trop. Atlantic Am.) – pachycaul tree (Roux's Model) typical of littoral, 'fr.' (fleshy hypanthium encl. achene) used for jelly, & wine-like drink, red sap for dyeing & tanning. ***C. caracasana*** Meissn. (trop. Am.) – assoc. with *Azteca* ants in Costa Rica

Cocconerion Baill. Euphorbiaceae (Ricinoc.-Cocc.). 2 New Caled. Lvs whorled

Coccosperma Klotzsch = *Erica*

Coccothrinax Sarg. Palmae (III 2). 49 WI, esp. Cuba. R: Selbyana 12(1991)91. Limestone & serpentine. Cult. orn.; lvs used for hats & basketry

Cocculus DC. Menispermaceae (V). 8 trop. & warm excl. S Am. & Aus. R (Mal.): KB 15 (1962)479. Alks. Dioec., usu. lianes, some medic. (homoeopathic remedy for travel-sickness incl. jet-lag), some ed. fr., some cult. orn. esp. ***C. carolinus*** (L.) DC. (coral beads, SE US). ***C. hirsutus*** (L.) Theob. (Egypt, Ind.) – diuretic, laxative, famine-food

cocculus indicus *Anamirta cocculus*

coccus wood *Brya ebenus*

Cochemiea (M. Brandegee) Walton (~ *Mammillaria*). Cactaceae. 3 SW N Am.

cochineal red dye form. obtained from dried female c. insects grown on *Opuntia cochenillifera*, now prod. synthetically

Cochinochloa H.N. Nguyen & Tran. Gramineae. 1 S Vietnam: ***C. braiana*** H.N. Nguyen & Tran. R: Blumea 58(2013)31

Cochleanthes Raf. Orchidaceae (V 12j). Excl. *Warczewiczella*, 4 trop. Am. R: Lankesteriana 2(2005)96. Cult. orn. epiphytes

Cochlearia Tourn. ex L. Cruciferae (18). Excl. *Ionopsidium*, 20 N temp. (Eur. 15). Polyploidy & hybridization based on x = 6 in Eur., x = 7 in N Am. leading to spp. complexes. Two Turkish spp. accumulate nickel on serpentine. ***C. danica*** L. – spreading along motorway central reservations in GB (from seashore ballast); ***C. officinalis*** L. s.l. (scurvy grass, Eur.) – polyploid series (incl. ***C. o.*** subsp. ***anglica*** (L.) Asch. & Graebn. – octoploid, decaploid, W Eur.), rich in vitamin C, form. medic. (fashion for drinking extract in mornings in 1650s, cf. orange-juice today), salad (tarry flavour)

Cochleariella Zhang & Vogt = *Yinshania*

Cochleariopsis Löve & D. Löve = *Cochlearia*

Cochleariopsis Zhang = *Yinshania*

Cochlianthus Benth. (~ *Apios*). Leguminosae (III 18). 2 Himal.

Cochliasanthus Trew (~ *Vigna*). Leguminosae (III 18). 1 trop. Am.: ***C. caracalla*** (L.) Trew (*Phaseolus c., V. c.*, snail fl.) – cult. orn. perenn. with fragant fls & keel coiled (5 revolutions) like snail-shell, poll. strong insects like *Bombus morio* bee depressing lower left wing with legs which promotes emergence of stigma & pollen

Cochlidium Kaulf. (~ *Grammitis*). Polypodiaceae (V; Grammitidaceae). 16 trop. Am., ***C. serrulatum*** (Sw.) L.E. Bishop ext to Afr. & Ind. Ocean

Cochlidosperma (Reichb.) Reichb. = *Veronica*

Cochlioda Lindl. = *Oncidium* (but see Selbyana 22(2001)141)

Cochliostema Lem. Commelinaceae (II 1e). 2 Nicaragua to Ecuador. Epiphytic, bromeliad-like pachycauls; anthers coiled, encl. by filaments united in a stalked hood. Cult. orn.

Cochlospermaceae Planch. = Bixaceae

Cochlospermum Kunth. Bixaceae (Cochlospermaceae). 12 trop. R: BJ 101(1980)215. Mostly xeromorphic trees (Koriba's Model) or shrubs, some with tuberous stems underground, oft. flowering when leafless in dry season. ***C. gillivraei*** Benth. (N Aus.) – fr. ed., high in vitamin C; ***C. religiosum*** (L.) Alston (*C. gossypium*, silk-cotton tree, Myanmar, Ind.) – flowers as an unbranched pole with term. infl., source of an insoluble gum (karaya, kutira), a subs. for tragacanth, fr. hairs stuffed in pillows said to induce sleep, cult. orn. for yellow fls esp. nr. temples in Ind., fig-like engulfing ruins as at Angkor (Cambodia); ***C. vitifolium*** (Willd.) Spreng. (trop. Am.) – succ. stem, lignotuber

cockatoo bush *Myoporum insulare;* **c. grass** *Alloteropsis semialata*

Cockaynea Zotov = *Hystrix*

Cockerellia (R.T. Clausen & Uhl) Löve & D. Löve = *Sedum*

cockle, corn *Agrostemma githago;* **cow c.** *Vaccaria hispanica;* **white c.** *Silene latifolia*

cocklebur *Xanthium strumarium*

cockroach berry *Solanum capsicoides;* **c. plant** *Haplophyton crooksii*

cockscomb *Celosia cristata;* **c. mint** *Elsholtzia ciliata*

cocksfoot *Dactylis glomerata*

cockspur grass *Echinochloa crus-galli;* **c. thorn** *Crataegus crus-galli, Vachellia eburnea*

cocky apple *Planchonia careya*

Cocleorchis Szlach. = *Cyclopogon*

coco or **cocoyam** *Colocasia esculenta;* **c. de macaco** *Attalea* spp.; **c. de mer** *Lodoicea maldivica;* **c. grass** *Cyperus rotundus;* **c. plum** *Chrysobalanus icaco*

cocoa *Theobroma cacao*

cocobolo *Dalbergia* spp. esp. *D. retusa*

cocoña *Solanum sessiliflorum*

coconut *Cocos nucifera*

cocos palm *Syagrus romanzoffiana;* **c. wood** *Brya ebenus*

Cocos L. Palmae (V 8a). 1 poss. C Mal. (e.g. Samar, Philipp.) or Barrier Reef (still unclear; all allies trop. Am.) now widely cult. & natur. pantrop.: ***C. nucifera*** L. (coconut) – v. imp. source of saturated oil for margarine, soap, etc.; in Pac., whole cultures hinge on it – lvs for shelter, weaving, 'grass skirts' of Hawaii (introd. from Gilbert Is.) etc., timber (porcupine wood, cocowood) for building, cabinetwork & spoons etc., fr. for food & drink. Monoec. insect-poll. (wind-supplementary) with (usu.) 1–seeded water-disp. drupe (inviable after 14 days in seawater, which retards germ.) with fibrous mesocarp (husk) which yields the fibre coir used for doormats, coconut matting, cord (sennit) & rope (used by early Polynesians for flying kites), coir dust a by-prod. used for mulching & soil-less germ. medium etc.; endocarp hard with 3 pores, the seed adherent to it, endosperm hollow with spongy ed. embryo nr. base, when young with c. 500 ml of refreshing coconut milk used in cleaning prods (e.g. Lux flakes) & as a medium in plant physiology experiments; copra is the dried endosperm, oil (many cosmetics) extracted by boiling or pressure, the refuse cake being used as stockfeed (70% world trade from Philipp.); desiccated c. is sliced & dried endosperm much used in confectionery (flavour due to alpha-nonalactone); glycerine a by-prod. so c. of strategic imp. until dynamite-based explosives superseded 1945. Apical bud of geriatric trees used for tinned palm-hearts (leads to tree death) & infl. axis is tapped for toddy, when evaporated prod. jaggery (sugar), when fermented giving arrack

& fermented further vinegar. Cult. in plantations (oft. imp. pollen source for honeybees e.g. Sri Lanka) esp. nr. coasts, the polyphyletic dwarf cvs being highest yielding; orig. obscure but forms in Atlantic basically uniform (& thus prone to catastrophic disease), the Am. ones prob. introd. from Cape Verde Is. 1499 & those poss. orig. via Mozambique. Macaques (*Macaca nemestrina*) used to pick fr. (e.g. 80% of coconut gardens in Sumatra), trained as plant-collectors by E.J.H. Corner in pre-WW II Singapore. [N.B. Coconut pearls ('mestica calappa') actually fragments of mollusc shells.] R: IBM 56 (1987)118, BM n.s. 16(1999)2. See also *Butia, Lytocaryum, Syagrus*

cocowood *Cocos nucifera*

cocozelle *Cucurbita pepo*

cocus wood *Brya ebenus*

Codariocalyx Hassk. (~ *Desmodium*). Leguminosae (III 19). 2 SE As. to trop. Aus. ***C. motorius*** (Houtt.) Ohashi (*Desmodium gyrans*, semaphore or telegraph plant, trop. As.) – long term. leaflet & 2 laterals moving by jerks in warmth or response to noise (Hallé), grown as curiosity under glass, cattle-fodder

Coddia Verdc. Rubiaceae (II 1). 1 S Afr.: ***C. rudis*** (Harv.) Verdc. R: Strel. 10(2000)484

Coddingtonia S. Bowd. = ? *Psychotria*

Codia Forst. & Forst. f. Cunoniaceae (VII). Excl. *Pullea*, 12 New Caled. R: BSBF 87(1940)254

Codiaeum Rumph. ex A. Juss. Euphorbiaceae (Cod.-Cod.). 17 Mal. to Pac. ***C. variegatum*** (L.) A. Juss. (SW Pacific, ?Borneo & E Java) cult. for coloured lvs ('crotons'), long selected by people of Papuasia & W Pac. for ornament, now used for hedging also; many named cvs, some with twisted lvs, pitcher-lvs, or 2 blades separated by a length of midrib, or variously lobed, varieg. or spotted, 'vogue' foliage pls of mid–late 19 cent. in Eur.

Codiocarpus R. Howard. Stemonuraceae (Icacinaceae s.l.). 1–3 Indomal. R: Blumea 17(1969)188

codlin an early maturing, unstriped cooking apple; **c.s and cream** *Epilobium hirsutum*

Codon L. Boraginaceae (Codonaceae; Hydrophyllaceae). 2 S Afr. 10–12-merous

Codonacanthus Nees. Acanthaceae (III 2c). 2 NE Ind., S China & Jap.

Codonaceae Weigend & Hilger = Boraginaceae

Codonanthe (Mart.) Hanst. Gesneriaceae (II 5c). 20 trop. Am. R: Baileya 19(1973)5. Epiphytic subshrubs or lianes usu. assoc. with aerial ant nests, oft. with extrafl. nectaries (red spots on lvs), seeds same size as ant eggs; fls small, poll. by hummingbirds (?) attracted by coloured lvs. Some cult. orn. incl. ***C. crassifolia*** (Focke) C. Morton (C Am. to Venez. & Peru) – liane, with ants, attracted by floral & extrafl. nectar, fr. pulp & arils, place seeds in walls of their nests: plant growth-rates slower away from nests

Codonanthopsis Mansf. (~ *Codonanthe*). Gesneriaceae (II 5c). 5 Amaz. – some on ant nests

Codonechites Markgraf = *Odontadenia*

Codonoboea Ridl. (~ *Henckelia*). Gesneriaceae (III 2j). c. 120 W Mal. R: GBS 62(2011)255, EJB 70(2013)387. ***C. geitleri*** (A. Weber) C.L. Lim (*H. g.*, Pahang) – bright yellow style an anther dummy; ***C. platypus*** (Clarke) C.L. Lim (*H. p.*, W Malay Penin.) – prod. 4 lvs a yr, each lasting 22 months

Codonocarpus Cunn. ex Endl. Gyrostemonaceae. 3 Aus. At least some toxic to stock

Codonocephalum Fenzl = *Inula*

Codonochlamys Ulbr. Malvaceae. 2 Braz. ? = *Hibiscus* (*Pavonia*)

Codonopsis Wall. Campanulaceae (I 1). Incl. *Campanumoea*, *Leptocodon*, excl. *Himalacodon*, *Pankycodon*, *Pseudocodon*, 46 C & E As. (China 26) to Mal. R: D.Y. Hong, *Monogr. C.* (2015)61. Usu. with tubers (some medic. esp. ***C. pilosula*** (Franch.) Nannf. subsp. ***tangshen*** (Oliv.) Hong (dan shen dang, tang-shen, W China) – ginseng subs. (tea etc.), or ed. e.g. ***C. ussuriensis*** (Rupr. & Maxim.) Hemsl. (*C. lanceolata* (Sieb. & Zucc.) Trautv. p.p., Jap., Manchuria)); berries evolved from capsules at least twice. Many cult. orn., esp. twining spp. with prettily marked C interior

Codonorchis Lindl. Orchidaceae (IV 1). 1 S trop. & temp. S Am.: ***C. lessonii*** (Urv.) Lindl.

Codonorhiza Goldbl. & Manning (~ *Lapeirousia*). Iridaceae (III 1). 7 W Cape. R: Strelitzia 35(2015)88

Codonosiphon Schltr. = *Bulbophyllum*

Codonostigma Klotzsch = *Scyphogyne*

Codonura K. Schum. = *Baissea*

Coelachne R. Br. Gramineae (Micr.). 12 (closely allied) OW trop.

Coelachyropsis Bor (~ *Coelachyrum*). Gramineae (XXIX 5). 1 S Ind., Sri Lanka: ***C. lagopoides*** Bor

Coelachyrum Hochst. & Nees. Gramineae (XXIX 5). 4 trop. & S Afr. through Arabia to Pakistan

Coelandria Fitzg. = *Dendrobium*

Coelanthum E. Meyer ex Fenzl. Molluginaceae. 3 SW Cape to Natal. R: JSAB 24(1958)48

Coelebogyne J. Sm. = *Caelebogyne*

Coelia Lindl. Orchidaceae (V 13e). 5 C Am., WI. Cult. orn.

Coelidium J. Vogel ex Walp. = *Amphithalea*

Coeliopsis Reichb.f. Orchidaceae (V 12e). 1 C Am.: ***C. hyacinthosma*** Reichb.f. – cult. orn. R: Gen. Orch. 5(2009)40

Coelocarpum Balf.f. Verbenaceae (inc. sed.). 5 Socotra (2 endemic), Somalia, Madag.

Coelocaryon Warb. Myristicaceae. 4 trop. Afr.

Coelococcus H. Wendl. = *Metroxylon*

Coeloglossum Hartman = *Dactylorhiza*

Coelogyne Lindl. Orchidaceae (V 12). 190 Indomal., trop. China, W Pacific. R: D. Clayton (2002) *The genus C.: a synopsis.* Extrafl. nectaries; showy fls; allied to *Pleione* but epiphytic & without plicate lvs. Many cult. orn. esp. ***C. cristata*** Lindl. (Himal.) form. much grown for winter fls for bouquets, pls to 2 m diam. recorded

Coelonema Maxim. = *Draba*

Coeloneurum Radlk. Solanaceae (I). 1 Hispaniola: ***C. ferrugineum*** (Spreng.) Urb.

Coelophragmus O. Schulz = *Dryopetalon*

Coelopleurum Ledeb. = *Angelica*

Coelopyrena Valeton. Rubiaceae (IV 7). 1 E Mal.: ***C. salicifolia*** Valeton

Coelorachis Brongn. = *Rottboellia* (but see KB 24(1970)309)

Coelospermum Blume = *Caelospermum*

Coelostegia Benth. Malvaceae (Hel.; Bombacaceae). 6 W Mal. R: GBS 63(2001)126

Coelostelma Fourn. = *Matelea*

Coenadenium (Summerh.) Szlach. = *Angraecopsis*

Coenoemersa R. González & Lizb. Hern. = *Platanthera*

Coespeletia Cuatrec. See *Espeletia*

Coffea L. Rubiaceae (II 3). Incl. *Psilanthus*, c. 124 OW trop. (Madag. c. 60) to N Aus. (1: ***C. brassii*** (Leroy) A.P. Davis). R: BJLS: 152(2006)475. Some pachycaul treelets, e.g. ***C. magnistipula*** Stoffelen & Robbrecht (W Afr.) – large stip. & leaf-bases catch debris into which adventitious r. grow & ***C. macrocarpa*** A. Rich. (Mauritius) – Corner's Model. Domatia; fr. a drupe with 2 seeds, most spp. with caffeine. Cult. for seeds (coffee beans), mostly produced in trop. Am. esp. Braz., Colombia, WI etc., esp. ***C. arabica*** L. (Arabian or arabica coffee, Ethiopia, colobus-disp. allotetraploid (2n = 44), Roux's Model), the best, usu. grown at alt., flowering in response to temp. drop in Mal., timber used for furniture, ***C. canephora*** Pierre ex Fröhner (robusta or Congo c., trop. W Afr.) – diploid oft. used in instant coffee (& pills to reduce glucose uptake), some cvs rust-resistant hybrids with *C. arabica*, & ***C. liberica*** W. Bull ex Hiern (Liberian or Abeokuta c., trop. W Afr.) – bitter flavour, oft. added to robusta in blends. Stimulatory effects (up to c. 3 espressos improving dexterity by c. 10%, calming & reducing depression) due to alks incl. caffeine (1,3,7-trimethylxanthine (nat. pesticide) blocking effect of adenosine (sleep promotor), addictive & fatal in excess, 150 mg per cup (tea 80), side effects after 200 mg, excess leading to shaking, migraines etc.) & theobromine manipulating dopamine prod., i.e. activating pleasure centre in brain (as do opium & cocaine); over 700 compounds isolated but basis of flavour still unclear though alkylpyrazines, sulphur compounds & aliphatics imp.; caffeine can relieve headaches, hay fever & allegedly (with green tea) stop skin cancer, but with meals coffee can increase blood glucose & insulin, also making word-recall more difficult; 2–4 cups a day alleged to reduce gout & incidence of colon cancer (theophylline), boosting muscle-recovery in athletes & allegedly longevity at 4–5 cups a day (chlorogenic acid, an anti-oxidant, poss. reducing DNA damage), but increases oestrogen in women, while cafestol & kahweol (removed in filtered coffee) raise cholesterol levels; orig. only chewed but in C13 cleaned & roasted (Arabs) with first coffee-house in Istanbul (1554; first in Eur. 1650 ('Angel', High St., Oxford, UK) opened by a Jew from Turkey; by 1700 in London 2000, some leading to London Stock Exchange, Lloyds & newspapers like Spectator), now ann. export trade (c. 7 M t in 2010) worth $US 15 bn+ (£1.6 bn to UK alone), 600 espressos drunk per person per yr Italian coffee bars alone (= world's greatest consumption); prized 'kopi luak' (civet coffee) seeds (worth $US 750 per kg) gathered from civet faeces in As. (esp. Vietnam), 'Black Ivory coffee' being that passed through elephants (bitterness destroyed by digestion) – $50 a cup in Bangkok;

coffee powder in comm. hair-dyes, coffee-grounds mixed with slag proposed for road-foundations in Victoria. Coffee planting in Braz. led to reduction of 1.5 M km^2 Atlantic Rainforest to 4K; c. rust in Sri Lanka (1869, leading to tea-planting there and rubber-planting in Malaysia), Brazil by 1970. M.N. Clifford & K.C. Wilson (eds. 1985) *Coffee*, R. von Hünersdorf & H.G. Hasenkamp (2002, 2 vols) *Coffee: a bibliography*, G. Wrigley (1988) *Coffee*

coffee *Coffea* spp.; **Abeokuta c.** *C. liberica*; **Arabian** or **arabica c.** *C. arabica*; **Congo c.** *C. canephora*; **Kentucky c. (tree)** *Gymnocladus dioica*; **Liberian c.** *C. liberica*; **c. plum** *Flacourtia rukam*; **robusta c.** *C. canephora*; **c. weed** *Senna occidentalis*

coffin nail *Anacardium occidentale*; **c. tree** or **wood** *Juniperus recurva, Persea nanmu*

cognac See *Vitis*

Cogniauxia Baill. Cucurbitaceae (IX). 2 trop. Afr.

Cogniauxiocharis (Schltr.) Hoehne = *Pteroglossa*

cogon grass *Imperata cylindrica*

cogwood *Sarcomphalus chloroxylon, Ceanothus* spp.

Cohnia Kunth = *Cordyline*

Cohniella Pfitzer = *Trichocentrum* (but see Britt. 62(2010)155)

cohoba *Anadenanthera peregrina*

cohosh, black *Actaea racemosa, A. spicata*; **blue c.** *Caulophyllum thalictroides*

cohune nut *Attalea cohune*

coigue *Nothofagus dombeyi*

Coilocarpus F. Muell. ex Domin = *Sclerolaena*

Coilochilus Schltr. Orchidaceae (IV 3c). 1 New Caled.: ***C. neocaledonicus*** Schltr. R: Gen. Orch. 2(2001)118

Coilonox Raf. = *Ornithogalum*

Coilostigma Klotzsch = *Erica*

Coincya Porta & Rigo ex Rouy *(Hutera, Rhynchosinapis)*. Cruciferae (12). 6 Eur (6), Medit. R: BJLS 102(1990)353. ***C. wrightii*** (O. Schulz) Stace (Lundy cabbage, Lundy Is., Bristol Channel) – 1 of v. few Br. endemics, sole host for 2 spp. of beetle

Coinochlamys T. Anderson ex Benth. = *Mostuea*

Coix L. Gramineae (XXII). 4 trop. As. R: RBA 31(1951)185. ***C. aquatica*** Roxb. ('*C. gigantea*', Indomal.) – aquatic, salt-source in NG; ***C. lacryma*-jobi** L. (Job's tears, SE As.) – the sheath of the bract of the infl. a hollow pear-shaped organ cont. the 1-flowered female spikelet, the 2 males projecting from it, this 'false fr.' with thin shell ed. (adlay) esp. Ind. & Myanmar, or hard one used as beads (grey & shiny); aquatic forms grow up to 30 m long

Cojoba Britton & Rose (~ *Pithecellobium*). Leguminosae (II 4). 12 trop. Am. R: MNYBG 74,2(1997)36

coke *Erythroxylum coca*

Cola Schott & Endl. Malvaceae (Sterc.; Sterculiaceae). c. 125 trop. Afr. Corner's & Rauh's Models. ***C. lizae*** Hallé (Gabon) – obligately gorilla-disp. Cola (kola) cult. trop. esp. OW for stimulant caffeine-cont. seeds ('nuts', first recorded by Leo Africanus in 1526), chewed (prob. habit adopted from coca-chewing during slave trade) or used in cola drinks (largely supplanted by synthetics), notably ***C. acuminata*** (P. Beauv.) Schott & Endl. (Abata cola, c. 180 K t p.a.) – extract an anti-trypanosomal drug (sleeping sickness), & to a lesser extent ***C. anomala*** K. Schum. (Bamenda c.), ***C. nitida*** (Vent.) Schott & Endl. (taken to Jamaica & Braz. by 1630) – dyes, ceremonies (only by men), also treatment of erectile dysfunction, & ***C. verticillata*** (Thonn.) A. Chév. (Owé c.) – also cont. heart-stimulant kolanin. Used in 'Pemberton's French Cola Wine' (a forerunner of Coca-Cola) as caffeine source, now replaced by synthetic citrate caffeine; shape of 'classic' C.-C. bottle inspired by confusion with fr. of cacao!

Colania Gagnepain = *Aspidistra*

Colanthelia McClure & E.W. Sm. (~ *Aulonemia*). Gramineae (V 2). 7 Braz., N Arg. R: SCB 9(1973)77. Bamboos

Colax Lindl. = *Pabstia*

Colchicaceae DC. Magnoliidae – Liliales. Excl. Petermanniaceae, 15/210 Aus. & S Afr. to W Eur. & W As., N Am. R: T 56(2007)177. Geophytes without raphides but with starch-rich corms to tubers, s.t. with twining ann. stems, rarely woody; some alks. Lvs dorsiventral, ovate to linr., parallel-veined & oft. with distinct midrib, s.t. with tendrils, distich. to subopposite or verticillate. Fls reg., usu. bisex. in term. racemes, cymes, umbels, heads or solit.; P usu. 3 + 3, s.t. basally connate; A 3 + 3 dehiscing by longit. slits; $\underline{G}$ usu. 3, each loc.

with few to ∞ anatr. or campylotropous ovules on axile placentas. Capsule loculicidal or septicidal, rarely a berry (*Disporum*), with usu. globose seeds (ovoid in some *Wurmbea*), s.t. arillate; embryo linr., straight

Classification & genera:

Burchardieae (*Burchardia*); **Uvularieae** (*Disporum, Uvularia*); **Tripladenieae** (*Kuntheria, Schelhammera, Tripladenia*); **Iphigenieae** (*Camptorrhiza, Iphigenia*); **Anguillarieae** (*Baeometra, Wurmbea*); **Colchiceae** (*Colchicum, Gloriosa, Hexacyrtis, Ornithoglossum, Sandersonia*)

Iphigenia & *Wurmbea* with disjunct distribs. Many cult. orn.

Colchicum Tourn. ex L. Colchicaceae. Incl. *Androcymbium*, c. 100 Eur. (30 incl. *Bulbocodium* & *Merendera*), Medit. to Ethiopia, Somalia, C As. & N Ind., S Afr. (28). R: BJ 127(2007)167,284. Many alks. Fls crocus-like but A 6 (3 in *Crocus*), many appearing in autumn after lvs wither; P-tube long (cf. *Crocus*) with G at ground level but in spring pedicel elongates & brings capsule above ground; many cult. orn. incl. hybrids & double-flowered (tepaloid A) forms. ***C. autumnale*** L. (autumn crocus, meadow saffron, naked boys, Eur. to N Afr.) – dried corms & seeds the source of medic. colchicum (*tinctura colchici* mentioned in Assyrian medic. texts & still painkiller esp. in gout) & alk. colchicine (also in some Melanthiaceae, etc.) used in plant breeding as it causes a doubling of chromosomes by disorganizing the spindle-mechanism at mitosis when they usu. sep. into 2 daughter nuclei

Coldenia L. Boraginaceae. 1 OW trop. & warm: ***C. procumbens*** L.

Colea Bojer ex Meissn. Bignoniaceae (6). 21 Madag., Mauritius (1), Seychelles (1). R: FMad. 178(1938)32. Some pachycaul treelets with cauliflory (Corner's Model)

Coleactina Hallé = *Leptactina*

Coleanthera Stschegl. Ericaceae (VII 7; Epacridaceae). 3 W Aus.

Coleanthus Seidl. Gramineae (XVI 14). 1 temp. Euras., N Am.: ***C. subtilis*** (Tratt.) Seidl – umbel-like spikelet clusters. R: FNA 24(2007)618

Coleataenia Griseb. (*Sorengia*). Gramineae (XXIII 1). 7 trop. Am. R: JBRIT 4(2010)691

Colebrookea Sm. Labiatae (VI). 1 Pakistan to W China: ***C. oppositifolia*** Sm. – dioec., oft. gregarious, local medic., cult. orn.

Coleocarya S.T. Blake. Restionaceae. 1 NSW, Queensland: ***C. gracilis*** S.T. Blake

Coleocephalocereus Backeb. (~ *Cephalocereus*). Cactaceae (III 3). 6 E & SE Braz. R: Bradleya 6(1988)91, 7(1989)35

Coleochloa Gilly. Cyperaceae (III 2). 8 trop. & warm Afr. (esp. E), Madag. Some epiphytic. Compressed stem, distich. decid. lvs, 'ligule' of hairs & ventrally open leaf-sheaths superficially like Gramineae

Coleocoma F. Muell. Compositae (Inul.-Pluch.). 1 NW Aus.: ***C. centaurea*** F. Muell. R: FA 37(2015)419

Coleogeton (Reichb.) Les & Haynes = *Stuckenia*

Coleogyne Torrey. Rosaceae (Ros.-Kerr.). 1 SW US.: ***C. ramosissima*** Torrey (black brush). R: FNA 9(2014)390

Coleonema Bartling & H.L. Wendl. Rutaceae (I 4). 8 SW to E Cape. R: JSAB 47(1981)401. Some cult. orn. like *Diosma*, esp. ***C. pulchellum*** I. Williams ('*C. pulchrum*') natur. S Aus.

Coleophora Miers = *Daphnopsis*

Coleostachys A. Juss. Malpighiaceae. 1 N trop. S Am.: ***C. genipifolia*** A. Juss. R: Phytotaxa 277(2016)77

Coleostephus Cass. Compositae (Anth.-Leuc.). 3 W Eur. (2), N Afr. R: BBMNHB 23(1993) 143

Coleotrype C.B. Clarke. Commelinaceae (II 1d). 10 SE Afr. (4), Madag. (6)

coleslaw *Brassica oleracea*

coleus *Coleus scutellarioides*

Coleus Lour. (~ *Plectranthus*). Labiatae (VII 3e). Incl. *Capitanya, Leocus, Solenostemon*, 271 trop. & warm OW. ***C. amboinicus*** Lour. (*P. a.*, Cuban oregano, five-in-one, Spanish sage, ? E Afr., spread to As.) – flavouring, medic. & shampoo; ***C. barbatus*** (Andrews) G. Don (*P. b.*, OW trop.) – hedge-pl. in Kenya, infusion a lice-remover in Uganda, diterpene (forskolin) potential drug for hypertension, glaucoma, asthma etc.; ***C. caninus*** (Roth) Vatke (*P. c.*, scaredy cat pl., trop. OW) allegedly repellent cats & dogs; ***C. esculentus*** (N.E. Br.) G. Taylor (*P. e.*, Livingstone potato, trop. & S Afr.) – ed. tubers; ***C. rotundifolius*** (Poir.) A. Chev. & Perrot (*P. r.*, *S. r.*, Hausa potato, OW) – tubers eaten like potatoes; ***C. scutellarioides*** (L.) Benth. (*C. blumei*, *P. s.*, *S. s.*, coleus, ? NG) – cult. orn. foliage pot-pl., varieg. lvs with purple dominant over green, & both over patterned expressed only in homozygotes, some true chimaeras with green-yellow-green etc. layers, others with dark-edged yellowish lvs app. due to reversible hormonal bleaching initiated from mesophyll (another dominant char.)

colic root *Aletris farinosa*

Colicodendron Mart. (~ *Capparis*). Capparaceae. 4 S Am. R: JBRIT 2(2008)76. ***C. valerabellum*** Iltis & al. (Venez.) – pachycaul treelet

Colignonia Endl. Nyctaginaceae (3). 6 Andes of Colombia to Arg. Shrubs, herbs & lianes to 15 m. R: NJB 8(1988)231

Colima (Rav.) Aarón Rodr. & Ortiz-Catedral = *Tigridia*

Collabiopsis S.S. Ying = *Collabium*

Collabium Blume. Orchidaceae (V 14). 14 China to Mal. & Polynesia. R: OM 8(1997)148

Collaea DC. (~ *Galactia*). Leguminosae (III 18). 7 S Am.

collards *Brassica oleracea* Acephala Group

Collare-stuartense Senghas & Bockem. = *Oncidium*

colleja *Silene vulgaris*

Colleteria D.W. Taylor = *Wandersong* (but see SGP 73(2003)203)

Colletia Comm. ex Juss. Rhamnaceae (2). 5 S S Am. R: Parodiana 5(1989)279. In each axil 2 serial buds, upper develops into triangular thorn, lower into fls or a branch. Nitrogen-fixing actinobacterial nodules on r.; some cult. orn. & timber. In ***C. hystrix*** Clos (*C. armata*, S Chile & Arg. to 2000 m) hawthorn-scented fls on thorny growths, lvs evanescent; ***C. paradoxa*** (Spreng.) Escal. (*C. cruciata*, S Braz., Urug., Arg.) has flattened branches but thorny growths in juvenile & occ. adult phases; ***C. spinosissima*** J. Gmel. (Arg., Bolivia, Ecuador, Peru, Urug.) – r. saponins used as soap

Colletoecema E. Petit. Rubiaceae (IV 10). 3 trop. Afr. R: Blumea 53(2008)533. Roux's Model

Colletogyne Buchet. Araceae (VII 19). 1 Madag.: ***C. perrieri*** Buchet

Colliguaja Molina. Euphorbiaceae (Euph.-Hipp.). 3 C Chile, 1 S Braz.: ***C. brasiliensis*** Klotzsch ex Baill. – rheophyte; ***C. odorifera*** Molina (Chile) bark (colliguaji b.) – used as soap

colliguaji bark *Colliguaja odorifera*

collimamol *Luma apiculata*

Collinia (Liebm.) Oersted = *Chamaedorea*

Collinsia Nutt. Plantaginaceae (Ant.-Chel.; Scrophulariaceae). 4 N Am., esp. W US. R: SB 31(2006)404. C 2-lipped functioning like a legume fl. Cult. orn. annuals esp. ***C. bicolor*** Benth. (*C. heterophylla*, Chinese houses, Calif.)

Collinsonia L. Labiatae (VII 1). Excl. *Keiskea* 3–4 E N Am. ***C. canadensis*** L. (horse balm, stone r.) – form. medic.

Collomia Nutt. Polemoniaceae (III). 15 W N Am., Bolivia to Patagonia. R: AMN 31(1944) 217. Differs from *Gilia* in that mature capsule does not rupture K; testa of some spp. mucilaginous when wet. Some cult. orn. esp. ***C. grandiflora*** Douglas ex Lindl. (W N Am., natur. in Eur., Aus.)

Collospermum Skottsb. = *Astelia*

Colobanthera Humbert. Compositae (Ast.-Gran.). 1 Madag.: ***C. waterlotii*** Humbert

Colobanthium Reichb. = *Avellinia*

Colobanthus Bartling. Caryophyllaceae (II 1). 20 S Pac. (Aus., NZ, Kerguélen, New Amsterdam, temp. S Am. incl. Andes). C 0, A in 1 whorl. ***C. quitensis*** (Kunth) Bartling (*C. crassifolius*, trop. & S Am.) – with *Deschampsia antarctica* only angiosperms to penetrate Antarctic Circle

Colobocarpus Esser & Welzen = *Croton*

Colobogyne Gagnepain = *Acmella*

Colocasia Schott. Araceae (VII 25). 16 trop. As. Tuberous herbs with peltate lvs. Cult. ed. & orn. pls esp. ***C. esculenta*** (L.) Schott (cocoyam, taro (cognate with 'calo', 'cala' (e.g. *Caladium*) = dark or blue; *-casia*, *kachu* = tuber, though orig. *kolokasion* was *Nelumbo nucifera*), dasheen, kalo (Pac.), keladi, talas (SE As.), aivi (Ind.), imo (Jap.)) – cult. for 10 000 yrs As. (12 M t prod. p.a. by 2010; ? first in Ind., though app. native in New Caled. where cvs later established), first irrigation terraces in As. with rice perhaps orig. a weed there, known in Medit. by classical times & prob. NW soon after 1492 (now invasive in SE US), young lvs (oft. blanched) & tubers ed. (boiled), starch (grains v. small) added to improve breakdown of 'biodegradable' plastics, many selected forms & cvs (84 in Hawaii alone) incl. 'var. *antiquorum*' (eddoes) – ed. small tubers, **'Fontanesii'** (*C. violacea*) – triploid orn. with violet petioles & veins

colocynth *Citrullus colocynthis*

Cologania Kunth (~ *Amphicarpaea*). Leguminosae (III 18). 12 trop. Am. esp. Mex. R: Phytol. 73(1992)281

Cololobus H. Robinson (~ *Vernonia*). Compositae (Vern.-Vern.). 3 Braz.

colomba root See calumba r.

Colombian boxwood *Casearia praecox*; **C. mahogany** *Cariniana pyriformis*

Colombiana Osp. = *Pleurothallis*

Colombobalanus Nixon & Crepet = *Trigonobalanus*

Colona Cav. Malvaceae (Grew.-Grew.; Tiliaceae). c. 25 S China, SE As., Indomal. R: NBGB 9(1926)796. ***C. javanica*** (Blume) Burret (SE As to C Mal.) – bast for fishing-nets & rope, e.g. buffalo tethers in Java

colophony rosin, a distillate from crude oleo-resin of *Pinus* spp.

Colophospermum Kirk ex Léonard (~ *Hardwickia*). Leguminosae (I 2). 1 S trop. Afr.: ***C. mopane*** (Benth.) Léonard – dominates mopane woodlands of hot dry low-rainfall areas, wind-poll., lvs fold together in hottest time of day; resin makes it susceptible to fire; larvae of mopane worms (*Gonimbrasia* spp.) sought by Bushmen

Colorado grass *Urochloa texana*

Coloradoa Boissev. & C. Davidson = *Sclerocactus*

Colpias E. Meyer ex Benth. Scrophulariaceae (Hemim.) 1 S Afr.: ***C. mollis*** E. Meyer ex Benth. – elaiophores attractive to bees. R: Strel. 10(2000)517

Colpodium Trin. Gramineae (XVI 14). 13 Turkey & Cauc. to Nepal & E Siberia (China 5), Mts Kenya & Kilimanjaro. R: KBAS 13(1986)103. High alt. segregate of *Poa* with fewer florets etc. ***C. versicolor*** (Steven) Woronow (Cauc.) – 2n = 4

Colpogyne B.L. Burtt = *Streptocarpus*

Colpoon P. Bergius (~ *Osyris*). Santalaceae. 1 S Afr.: ***C. compressum*** P. Bergius – root-parasite

Colpothrinax Schaedtler. Palmae (III 46). 3 C Am., Cuba. R: Palms 45(2001)186. ***C. wrightii*** Schaedtler (Cuba) – trunk greatly swollen in middle at maturity

Colquhounia Wall. Labiatae (VI). c. 63 E Himal., SW China, Vietnam. Erect or twining herbs allied to *Stachys* but upper C lip entire or emarginate & shorter than lower. Some cult. orn. esp. ***C. coccinea*** Wall. – fls red, field-hedge in parts of Bhutan

coltsfoot *Tussilago farfara*

Colubrina Rich. ex Brongn. Rhamnaceae (inc. sed.). 33 trop. & warm esp. Am. R: Britt. 23(1971)2. Some cult. orn. (Roux's Model) incl. ***C. arborescens*** (Mill.) Sarg. (C Am., WI) – source of snakebark, medic. (incl. aphrodisiac) & timber & ***C. asiatica*** (L.) Brongn. (trop. As., invasive SE US) – seeds long-viable in seawater, lvs ed., fr. a fish-poison & medic., bark & r. a soap subs.; ***C. elliptica*** (Sw.) Briz. & Stern (*C. reclinata*, WI) – basis of a drink (mabee or mabi) in Puerto Rico & Haiti; ***C. glandulosa*** Perkins ('*C. rufa*', trop. Am.) – bark (saguaragy) used in treatment of fever in Braz.; ***C. oppositifolia*** Brongn. ex H. Mann (Hawaii) – hard wood used like metal by old Hawaiians

columba root See calumba r.

Columbiadoria Nesom (~ *Hesperodoria*). Compositae (Ast.-Sol.). 1 NW US: ***C. hallii*** (A. Gray) Nesom. R: Phytol. 71(1991)248

columbine *Aquilegia* spp., esp. *A. vulgaris* & hybrids

Columellia Ruíz & Pavón. Columelliaceae. 1–4 N Andes

Columelliaceae D. Don. Magnoliidae – Bruniales. Incl. Desfontainiaceae, 2/2–5 Colombia to Bolivia. Everg. trees & shrubs. Lvs opp., simple, serrate or spinulose; stip. 0. Fls in few-flowered cymes or solit., bisex., slightly irreg; K (4)5(8), valvate or weakly imbr., C ((4)5(8)), tube v. short, lobes imbr., A 2(3) or 5 (*Desfontainia*) attached to C nr. base, anthers basifixed with longit. slits, nectary-disk 0, G (2, 5) with thick style & lobed stigma & parietal placentae usu. or almost meeting to make 2-loc. ovary, ovules numerous, anatropous, unitegmic. Fr. a capsule with 4 valves or thin-walled berry (*D.*); seeds minute with small straight embryo & copious fleshy endosperm

Genera: *Columellia, Desfontainia*

Wood & pollen features etc. support incorporation of Desfontainiaceae

Columnea Plum. ex L. Gesneriaceae (II 5c). Incl. *Bucinellina, Dalbergaria, Pentadenia, Trichantha*, 270+ trop. Am. esp. C Am. Epiphytes & lianes, hummingbird-poll.; many spp. & hybrids cult. Lvs of a pr oft. unequal

Coluria R. Br. = *Geum* (but see NRBGE 15(1925)48)

Colutea Tourn. ex L. Leguminosae (III 24). 28 Med. (Eur. 3) to China, Himal., E & NE Afr., mostly in dry mts. R: MB(P) 14(1963)3, AK 12(1967)33. Cult. orn. esp. ***C. arborescens***

L. (bladder senna, Med., natur. GB) – lvs with properties like *Senna italica* & used to adulterate it; fr. inflated

Coluteocarpus Boiss. (~ *Noccaea*). Cruciferae (19). 1 SW As. mts: ***C. vesicaria*** (L.) Holmboe

Colvillea Bojer. Leguminosae (I 4). 1 Madag.: ***C. racemosa*** Bojer – A exserted ('shaving-brush syndrome'), C orange-red, much visited by parrots, timber for building & pirogue hulls, cult. orn.

Colymbada Hill = *Centaurea*

Colysis C. Presl = *Leptochilus*

colza *Brassica napus*

Comaclinium Scheidw. & Planch. (~ *Dyssodia*). Compositae (Tag.-Pect.). 1 C Am.: ***C. montanum*** (Benth.) Strother. R: Sida 11(1986)374

Comandra Nutt. Santalaceae (Comandraceae). 1 N Am. (3 subspp.), Med. inc. Eur. (1 subsp.) – ***C. umbellata*** (L.) Nutt., parasitic on r. of other pls; fr. sweet, ed. R: MTBC 22(1965)1

Comandraceae Nickrent & Der. See Santalaceae

Comanthera L.B. Sm. (~ *Syngonanthus*). Eriocaulaceae. 35 S Am. (Braz. 34). R: T 59(2010)1136, (subg. *Thysanocephalus*) SB 40(2015)141. Some comm. 'everlastings' often dyed, esp. ***C. elegans*** (Bong.) Parra & Giul. (*S. e.*, Braz.) – by 1984 40 K kg exported p.a.

Comanthosphace S. Moore. Labiatae (VI). 3–4 E As. R: Fontqueria 41(1995)3

Comarostaphylis Zucc. (~ *Arctostaphylos*). Ericaceae (III). 10 Mex. R: SB 12(1987)582

× **Comagaria** Büscher & Loos. Rosaceae. *Comarum* × *Fragaria*. × ***C. rosea*** (Mabb.) Büscher & Loos (*Potentilla r.*; *C. palustre* × *F.* × *ananassa*) esp. '**Serenata**' – cult. orn.

Comarum L. (~ *Potentilla*). Rosaceae (Ros.-Pot.). 1 N temp.: ***C. palustre*** L.

Comastoma (Wettst.) Toyok. (~ *Gentiana*). Gentianaceae. 25 boreal

Combera Sandw. Solanaceae (Benth.). 2 temp. S Am.

Comborhiza Anderb. & Bremer. Compositae (Gnap.-Rel.). 2 S Afr. R: AMBG 78(1991)1070. Geoxylic suffrutices

Combretaceae R. Br. Magnoliidae – Myrtales. 14/500 trop. & warm esp. Afr. Trees or shrubs (s.t. geoxylic), oft. scandent. Internal phloem & intraxylary phloem oft. present. Lvs in spirals, opp. or whorled, simple, entire, leaf-base oft. with 2 gland-cont. flask-shaped cavities at base; stip. minute or 0. Fls in racemes, spikes or heads, usu. small & bisex. & reg. & epigynous (half so in *Strephonema*), hypanthium oft. nectariferous within; K 4 or 5(–8) appearing as lobes of hypanthium, persistent, valvate or s.t. imbr. or v. small, C 4 or 5(–8) alt. with K, imbr. or valvate, oft. 0, A oft. twice K & bicyclic, outer s.t. reduced, 0, or rarely in prs or triplets, rarely 3-cyclic, anthers usu. versatile, with longit. slits, epigynous disk oft. present, $\overline{G}$(2–5), 1-loc. with term. style & (1)2(–20) ovules, pend., anatropous, bitegmic with zig-zag micropyle, an elaborate obturator oft. produced on funicle. Fr. 1-seeded, usu. indehiscent & water-disp. or drupaceous, generally ribbed, the ribs oft. wing-like, rarely dry & dehiscent; seeds without endosperm, embryo oily with 2 (or 3 in some *Terminalia* spp. of SE As.) folded or spirally twisted cotyledons (massive & hemispheric in *Strephonema*, united in some Afr. spp. of *Combretum*). n = 7, 11–13

Classification & principal genera:

I. **Strephonematoideae** (C present; G semi-inf., Afr.): *Strephonema* (only)

II. **Combretoideae** (G inferior) – 2 tribes:

1. **Laguncularieae** (hypanthium with 2 adnate prophylls): *Laguncularia*
2. **Combreteae** (hypanthium without adnate prophylls): *Buchenavia*, *Conocarpus*, *Terminalia* (a. Terminaliinae), *Combretum* (b. Combretinae)

Many in savannas; *Conocarpus*, *Laguncularia* & *Lumnitzera* mangroves (latter 2 closely allied). Unicellular hairs of a type only otherwise found in some Myrtaceae. Timbers in above genera & *Anogeissus*, ed. seeds & fr. for tanning (*Terminalia*), several cult. orn. esp. spp. of *Combretum*

Combretocarpus Hook.f. Anisophylleaceae. 1 wetter W Mal.: ***C. rotundatus*** (Miq.) Danser in peat-swamp-forests (gregarious in Borneo) with aerenchymatous r. floating limply in small pools, timber recommended for furniture. R: Phytotaxa 229(2015)173

Combretodendron A. Chev. = *Petersianthus*

Combretum Loefl. Combretaceae (II 2b). Incl. *Calopyxis*, *Meiostemon*, *Quisqualis*, *Thiloa*, c. 255 trop. (excl. most of Aus.). Climbers (some with eccentric xylem, alks) or erect. Lvs in spirals or opp. (usu.). ***C. fruticosum*** (Loefl.) Stuntz (trop. S Am.) – fls visited by non-flying mammals in SE Peru: 7 spp. primate by day & 3 at night; ***C. molle*** R. Br. ex G. Don (NE Afr.) – cytotoxic but eaten by healthy chimpanzees. Cult. orn. lianes esp. ***C. indicum***

(L.) DeFilipps (*Quisqualis i.*, madhumalati, Rangoon creeper, Myanmar to NG) – fragrant white fls becoming pink with age, anthelminthic (Malay name *'udani'* punned via Dutch *'hoedanig'* (what kind of?) to *Q.*) & ***C. paniculatum*** Vent. (trop. Afr.) – great length, oft. produces red fls when leafless; ***C. butyrosum*** (Bertol.f.) Tul. (trop. Afr.) – butter-like substance (chiquito) from fr.; ***C. caffrum*** (Ecklon & Zeyher) Kuntze (S Afr.) – Zulu arrow poison, with combretasins that target blood vessels in tumours, so now used in cancer treatment; ***C. imberbe*** Wawra (trop. Afr.) – leadwood, ashes for white paint (with charcoal & cattle dung) trad. for Botswana houses; others medic., dyes, poisons, scents & gums (Ind. g.) imp. locally

Comesperma Labill. (~ *Bredemeyera*). Polygalaceae (IV). 40 Aus. ***C. volubile*** Labill. (love(creeper)) – occ. cult. orn.

Cometes Burm. ex L. Caryophyllaceae (I 2). 2 NE Afr. & Ethiopia to NW Ind., deserts

comfrey *Symphytum* spp.; **blue** or **Russian c.** *S.* × *uplandicum*

comino *Aniba perutilis*

Cominsia Hemsl. = *Phrynium*

Comiphyton Floret (~ *Cassipourea*). Rhizophoraceae (I). 1 Gabon to E Zaire, not N or S of 2°: ***C. gabonense*** Floret

Commelina Plum. ex L. Commelinaceae (II 2). c. 170 trop. & warm. Infls subtended by large leafy sheathing bract; fls short-lived, the upper 3 A sterile with cross-shaped anthers with juicy lobes pierced by bees for nectar; fr. dehiscent or not. 2n = 22, 24, 28, 30, 42, 44, 56, 60, 66, 90, 120, 150. Several cult. orn. as trop. ground-cover, fls usu. blue; some with ed. tubers; some medic. esp. as poultices (mucilage), others used (crushed paste) to repair pottery (Socotra). ***C. benghalensis*** L. (OW trop.) – weedy (as in cotton) SE US; ***C. caroliniana*** Walter (*C. hasskarlii*, Ind.) – weedy pl. introd. SE US with rice seed; ***C. communis*** L. (China, Jap., natur. Eur., US) – '**Hortensis**' with large fls source of blue dye used in blue 'Awobana' paper (Jap.); ***C. erecta*** L. (trop. & warm) – andromonoec.

Commelinaceae Mirb. Magnoliidae – Commelinales. 40/650 trop. to subtemp. R: T 40(1991)19. Herbs, usu. perenn., s.t. robust, or succ., or climbers (*Aetheolirion*, *Palisota thollonii*, *Tripogandra* etc.), almost always with 3-celled glandular microhairs, oft. with mucilage cells each cont. bundle of calcium oxalate raphides (not *Cartonema*); stems swollen at nodes; vessel elements in all parts (only in r. in *Cartonema*). Lvs in spirals, simple with closed sheath with parallel-veined oft. ± succ. lamina s.t. separated from sheath by slender petiole, the blade halves rolled separately against midrib in bud, rarely plicate. Infls cymose, oft. breaking through subtending sheathing bract, rarely solit. or app. in racemes. Fls hypogynous, usu. bisex., reg. to irreg.; K 3 usu. green (petaloid & coloured in e.g. *Dichorisandra*), rarely connate basally, C 3 ephemeral usu. blue or white, s.t. clawed or basally connate to form tube, alike or 1 diff. colour &/or ± reduced, A usu. 3 + 3 but s.t. 3 staminodal or not dev., rarely 1 (e.g. *Callisia*), filaments usu. slender oft. long-hairy, anthers basifixed or versatile oft. with expanded connective, with longit. slits (rarely apical (& basal) pores), $\underline{G}$ (3), 3-loc. or apically 1-loc., or 1 or 2 locs undeveloped or 0, style term., ovules 1 – ∞ per locule on axile placentas, orthotropous to anatropous, bitegmic. Fr. usu. loculicidal capsule, seldom indehiscent or fleshy; seeds with copious mealy endosperm & compound starch grains, usu. arillate, rarely winged (*Aetheolirion*), embryo s.t. with vestigial 2nd cotyledon opp. other. x = (4–)6–16(–29).

Classification & chief genera:

I. **Cartonematoideae** (shoots glandular-pubescent (glandular microhairs lacking), raphide-canals 0 or only next to veins; fls reg., yellow)

1. **Cartonemateae** (perennials, raphide-canals 0): *Cartonema* (s.t. placed in own fam.); incl. **Triceratelleae** (annuals, raphide-canals next to veins): *Triceratella* (only)

II. **Commelinoideae** (shoots v. rarely glandular-pubescent (glandular microhairs almost always present), raphide-canals present but never nr. veins; fls var. but v. rarely both reg. & yellow. Tribes & subtribes separable on micro-characters in the main) with *Floscopa* prob. 'sister' to all

1. **Tradescantieae** (7 subtribes): a (*Palisotinae*, Afr.) – *Palisota* (only); b (*Streptoliriinae*, usu. climbers, As.) – *Spatholirion*; c (*Cyanotinae*, OW trop.) – *Cyanotis*; d (*Coleotrypinae*, OW trop.) – *Amischotolype*; e (*Dichorisandrinae*, trop. Am.) – *Dichorisandra*; f (*Thyrsantheminae*, trop. & warm esp. C Am.) – *Tinantia*; g (*Tradescantiinae*, Am.) – *Callisia*, *Gibasis*, *Tradescantia*, *Tripogandra*
2. **Commelineae**: *Aneilema*, *Commelina*, *Murdannia*, *Pollia*

Many cult. orn. esp. *Tradescantia* (incl. *Zebrina*, *Rhoeo*) but also some *Cochliostema* (bromeliad-like tank-epiphytes); some medic. & ed. (local spinach subs. in Pac.)

Commelinantia Tharp = *Tinantia*

Commelinidium Stapf = *Acroceras*

Commelinopsis Pichon = *Commelina*

Commersonia Forst. & Forst. f. Malvaceae (Bytt.-Lasio.; Sterculiaceae). Incl. *Rulingia*, excl. *Androcalva*, 25 SE As. to Aus. (14, 12 endemic), New Caled. R: AusSB 24(2011)231. Troll's Model; fibre from bark. ***C. bartramia*** (L.) Merr. (SE As. to W Pacific) – termite-proof but soft building timber (Bismarcks)

Commersorchis Thouars = *Phaius*

Commicarpus Standl. = *Boerhavia*

Commidendrum DC. Compositae (Ast.-Homoc.). 4 (1 extinct) St Helena. R: Q.C.B. Cronk, *Endemic Fl. St Helena* (2000)74. Allied to *Felicia*. Form. common on island: pachycaul trees (Leeuwenberg's Model), extant spp. scarce; ***C. burchellii*** Hemsl. extinct but ***C. rotundifolium*** (Roxb.) DC. form. reduced to 3 pls 1880s & extinct in wild 1986 but prop. by seed at Kew & 1000 trees re-established on is.; ***C. robustum*** DC. (gumwood) – form. woodland 400–600 m, building timber, national tree; ***C. spurium*** (Forst.f.) DC. with only 6 trees in wild (2000)

Commiphora Jacq. Burseraceae (IV). c. 190 warm Afr. (Somalia 52) & Madag. (44 endemics), Arabia to Sri Lanka, Mex. (2) & S Am. (1). Mosquito-poll. at night? Fleshy outgrowths of stone not seed = pseudoarils. Resin exudes, the lumps used in medic., incense etc. (oleo-resins known as bdellium); live fences (long cuttings r.); seedling r. ed., wood chewed as water source in Uganda. ***C. myrrha*** (Nees) Engl. (*C. molmol*, molmol, NE Afr. to Arabia) – gum = principal myrrh today imp. in Arab medic. (but that of the Bible *C. guidottii*), non-toxic slug-repellent; ***C. foliacea*** Sprague (Arabia, Somalia) – dye, charcoal for cleaning teeth, bark prep. in skin disease treatment; ***C. gileadensis*** (L.) C. Chr. (*C. opobalsamum*, SW Arabia, Ethiopia & Sudan) – source of balm of Gilead (Mecca myrrh), form. medic., incense, scent, allegedly spontaneously igniting with oil of jasmine in 'miracle of the Sacred Fire' forbidden by Pope in 1238; ***C. guidottii*** Chiov. ex Guid. (Somalia, Ethiopia) – scented myrrh; ***C. kataf*** (Forssk.) Engl. (N Kenya to S Arabia) – with allied spp. source of gum opopanax (300 t per annum), tick-repellent; ***C. madagascariensis*** Jacq. (Abyssinian myrrh, Tanzania) – form. males cult. Mauritius & Ind. for scent; ***C. merkeri*** Engl. (E Afr.) – oleo-resin poss. wound treatment; ***C. wightii*** (Arn.) Bhand. (*C. mukul*, Arabia to Ind. desert) – oleo-resin from trunk (guggul) imp. local medic. (arthritis, to reduce blood cholesterol, obesity & acne etc.) due to guggulsterones

Commitheca Bremek. = *Pauridiantha*

Comocladia P. Browne. Anacardiaceae (I). 16+ trop. Am.

Comolia DC. Melastomataceae. 22 trop. S Am.

Comoliopsis Wurdack. Melastomataceae. 1 Venez.: ***C. neblinae*** Wurdack. R: ABV 14,3(1984)23

Comoranthus Knobloch (~ *Schrebera*). Oleaceae (4b). 3 Madag., Comoro Is. R: FMad. 166(1952)68

Comospermum Rauschert (*Alectorurus*; ~ *Anthericum*). Asparagaceae (Anthericaceae). 1 Jap.: ***C. yedonense*** (Franch. & Sav.) Rauschert

Comparettia Poeppig & Endl. Orchidaceae (V 12h). Incl. *Diadenium*, *Scelochiloides*, *Scelochilus*, *Stigmatorthos*, c. 80 trop. Am. Cult. orn. epiphytes

compass plant *Silphium laciniatum*; *Lactuca serriola*

Comperia K. Koch = *Himantoglossum* (but see Gen. Orch. 2(2001)274)

Complaya Strother = *Sphagneticola*

Compositae Giseke (Asteraceae). Magnoliidae – Asterales. 1568/25 000 cosmop. (exc. Antarctica; second largest pl. fam.). R: V.A. Funk *et al.* (2009) *Systematics, evolution & biogeogr. of C.* Trees (to 50+ m), shrubs & herbs (esp. rhiz., many with Tomlinson's Model) & climbers, oft. storing carbohydrate as polyfructosans (esp. inulin); articulated laticifers (mostly Cich.) or ± extensive resin-duct system present; sec. thickening well-dev. even in many herbaceous spp., s.t. in unusual configurations or with medullary &/or cortical bundles. Lvs in spirals, less oft. opp., rarely whorled, simple, dissected or ± compound, oft. marcescent; stip. usu. 0. Infls of 1 – ∞ dense heads (capitula) with 1–1000+ usu. sessile fls on common receptacle, nearly always subtended by an involucre of 1 to several series of bracts (phyllaries; 0 in *Psilocarphus*), opening in racemose sequence (mixed in *Espeletia*); capitula s.t. solit. but usu. in often corymbiform cymose synflorescences, or occ. aggregated into cymose sec. heads (over 40 genera in several tribes) s.t. with a sec. involucre; receptacle flat to conical or cylindrical, s.t. with a bract (palea) subtending each fl. (esp. Helia.) or bristly (esp. Card.). Fls epigynous, bisex. or some female, sterile or functionally male (radiate heads with marginal female or sterile ray-florets & C bisex. or functionally male disk-florets, ray-florets with tubular C prolonged into ± strap-shaped ligule oft. tipped with vestigial 3 C-lobes (other 2 ± absent), disk-florets with reg. tubular (3–)5(6)-lobed or toothed C; discoid heads with only disk-florets; disciform heads with central disk-florets & marginal female florets with eligulate C, or with only the latter; ligulate heads (mostly Cich.) of bisex. florets with C of 5 lobes; in Mutis. some or all florets with 2-lipped C, outer larger but marginal florets s.t. as in Cich. & some or all of central ones like disk-florets); K forming pappus on top of ovary or 0, of (1)2–∞ scales, awns, bristles or connate to form crown, A as many as C-lobes, alt. with them & inserted in tube, usu. distinct, anthers usu. with short apical appendage & s.t. basal tails & pollen-sacs connate into tube, releasing pollen into tube through longit. slits to be pushed out by growth of style, $\overline{G}$ (2(3)), 1-loc. with term. style, bifid, branches commonly separating after passage through anther-tube; nectary commonly a thickened scale or cup on top of G (nectar usu. amino-acids & hexoses), ovule 1, basal, erect, anatropous, unitegmic. Fr. a cypsela usu. with persistent pappus of (1)2–many awns, scales or hairs, s.t. winged or beaked, rarely a drupe; embryo oily, straight with hypocotyl, endosperm 0 or v. thin peripheral. n = 2–19+ (? orig. 9)

Classification & chief genera:

I. **Barnadesioideae** (trees, shrubs usu. with axillary spines (reduced lvs) to herbs; latex ducts 0; capitula [see *Arnaldoa* for primitive features] homogamous or heterogamous, discoid, disciform or radiate; florets reg., with char. long 1-cellular hairs, deeply 5-lobed or pseudobilabiate with a ± 4-lobed limb or rarely ligulate with a deeper split between 2 lobes, red to white or purple or yellow; anthers tailed; pollen variable; style ± shortly bilobed, never pilose; sesquiterpenes 0 – 9/91 S Am. esp. Andes): *Barnadesia, Dasyphyllum*

II. **Stifftioideae** (~ III.; trees, shrubs, lianes; latex ducts 0; capitula discoid; styles glabrous; pappus with c. 100 capillary bristles in 4 or 5 series – 4 genera Caribb., S Am.): *Stifftia*

III. **Mutisioideae** (trees (some unbranched pachycauls), lianes & herbs; latex ducts 0; lvs in spirals, rarely opp.; capitula with bilabiate or ligulate fls, rarely not; disk fls with C deeply lobed; anthers usu. tailed (R: OB 109(1991)5; Funk & al.: 236) – trop. esp. S Am.):

1. **Mutisieae** (ray C usu. bilabiate, disk C bilabiate or tubular – S S Am., Afr., As.): *Chaetanthera, Chaptalia, Gerbera, Mutisia*
2. **Onoserideae** (ray C bilabiate, disk C tubular – S S Am., Andes; *Famatinanthus* (NW Arg.) now excl. as **Famatinanthoideae** sister to all Compositae save I.): *Lycoseris, Onoseris*
3. **Nassauvieae** (all C bilabiate – mostly S S Am.): *Jungia, Leucheria, Nassauvia, Perezia, Trixis*

Cult. orn. – *Gerbera, Mutisia*

IV. **Wunderlichioideae** (~ III.; trees or shrubs; latex ducts 0):

1. **Wunderlichieae** (term. leaf rosettes; C reg. – 4/36 NE S Am.): *Stenopadus*
2. **Hyalideae** (ray C bilabiate, disk C tubular – 4/6 S Am., As.): *Hyalis, Leucomeris*

V. **Gochnatioideae** (~ III.; trees, shrubs or herbs; latex ducts 0; style-arms with rounded tips, pappus bristly – 8/103 S US to Arg.; R: Funk & al.: 250): *Gochnatia, Richterago*

VI. **Hecastocleiodeae** (~ III.; shrub; latex ducts 0; lvs holly-like, capitula 1-flowered, in secondary heads subtended by spiny bract; style-arms with rounded tips, pappus bristly, 1/1 SW US mts; R: Funk & al.: 261): *Hecastocleis*

VII. **Carduoideae** (trees to usu. herbs (many biennial)); latex ducts usu. 0; capitula discoid, homogamous or with sterile outer florets; anthers acute, oft. tailed; pollen spiny not ridged, lvs in spirals:

1. **Dicomeae** (~ III.; trees, shrubs to herbs – 7 genera, trop. & S Afr. to S As.; R: Funk & al.: 269, T 62(2013)532): *Dicoma, Pleiotaxis*
2. **Oldenburgieae** (~ III.; woody; head with <1000 fls; style-branches rounded – 1/4 S Afr.; R: Funk & al.: 288): *Oldenburgia*
3. **Tarchonantheae** (~ III.; dioec. trees & shrubs; capitula discoid with <30 fls – 2/13 trop. Afr. to Arabia (1); R: Funk & al.: 280): *Brachylaena, Tarchonanthus*
4. **Cardueae** (Cynareae; herbs, s.t. hapaxanthic, shrubs or pachycaul trees; involucral bracts usu. in 5 rows & spiny – Euras. esp. Medit.); 5 subtribes (R: Funk & al.: 294): *Arctium, Carduus, Cirsium, Cousinia, Cynara, Jurinea, Onopordum, Saussurea*

(Carduinae); *Atractylis, Carlina* (Carlininae); *Carduncellus, Carthamus, Centaurea, Centaurodendron* (tree), *Serratula* (Centaureineae); *Echinops* (Echinopsidinae)

Some ed. – *Carthamus* (oil, also dye), *Cynara* (cardoon, globe artichoke); timber – *Brachylaena, Tarchonanthus*; cult. orn. – *Amberboa, Centaurea, Echinops, Onopordum, Silybum* (also med.); weeds – *Acroptilon, Carduus, Cirsium*

VIII. **Pertyoideae** (~ III.; shrubs to herbs; latex ducts 0; fls not 2-labiate, bristly pappus – 4/c. 80 SE As.; R: Funk & al.: 315): *Ainsliaea, Pertya*

IX. **Gymnarrhenoideae** (dwarf desert ann.; amphicarpic C 3- or 4-lobed, A 3 or 4 – 1/1 N. Afr., SW As.; R: Funk & al.: 329): *Gymnarrhena* (only)

X. **Cichorioideae** (Lactucoideae; capitula usu. homogamous, ligulate, bilabiate or discoid; disk-florets usu. reg., deeply 5-lobed (if outer ones diff., then bilabiate or rarely true rays), purplish, pinkish or white, less oft. yellow; anthers usu. caudate; pollen variable; style-arms usu. with single stigmatic area on inner surface; shrubs, herbs & rarely trees – cosmop.; R: Funk & al.: 339):

1. **Cichorieae** (Lactuceae; mostly herbs & some pachycaul trees, e.g. some *Sonchus*; lvs in spirals; latex ducts present; capitula ligulate; pollen usu. ridged & spiny or spiny – cosmop. esp. N hemisp.); 11 subtribes (R: Funk & al.: 380; *Prenanthes* unplaced): *Catananche, Scolymus* (Scolyminae); *Chondrilla* (Chondrillinae); *Cichorium* (Cichoriiinae); *Crepis, Taraxacum, Youngia* (Crepidinae); *Cicerbita, Lactuca* (Lactucinae); *Launaea, Sonchus* (Hyoseridinae [Sonchinae]); *Agoseris, Malacothrix, Stephanomeria* (Microseridinae); *Hieracium, Pilosella* (Hieraciinae); *Hypochaeris, Leontodon, Picris* (Hypochaeridinae); *Scorzonera, Tragopogon* (Scorzonerinae)
2. **Arctotideae** (Arctoteae; excl. 3.; herbs or shrubs; lvs in spirals; capitula radiate, rarely discoid; latex ducts 0; anthers obtuse, acute, or shortly tailed; pollen spiny – SW As., S Afr.); 2 subtribes (R: Funk & al.: 388, 395; some unassigned): *Arctotis* (Arctotidinae); *Berkheya, Gazania* (Gorteriinae; R: MBSM 3(1959) 71)
3. **Eremothamneae** (shrubs; latex ducts 0; pollen spinulose – 2/3 SW Afr.; R: Funk & al.: 415): *Eremanthus, Hoplophyllum*
4. **Liabeae** (trees to herbs; lvs opp. or whorled, capitula radiate or discoid; latex ducts usu.; anthers acute or shortly tailed; pollen spiny; style-arms elongate; trop. Am. esp. Peru & Ecuador (R: SCB 54(1983)1; Britt. 63(2011)77)): *Liabum, Munnozia, Sinclairia*
5. **Vernonieae** (trees (s.t. large or pachycaul), shrubs, lianes or herbs; lvs in spirals (or opp.) often with T-shaped or stellate hairs; latex ducts 0; capitula discoid, rarely ligulate, homogamous, anthers obtuse to acute, rarely tailed; pollen ridged &/or spiny; style-arms elongate; (50 monospecific genera!) mostly trop.; R: Funk & al.: 447); 19 subtribes: *Baccharoides* (Linziinae); *Bothriocline* (Erlangeinae); *Centratherum, Lychnophora* (Lychnophorinae; R: SB 39(2014)658); *Distephanus* (Disephaninae); *Elephantopus* (Elephantopinae); *Gymnanthemum* (Gymnantheminae); *Lepidaploa, Lessingianthus* (Lepidaploinae); *Piptocarpha* (Piptocarphinae); *Rolandra* (Rolandrinae); *Vernonanthura, Vernonia* (Vernoniinae)
6. **Platycarpheae** (~ 2.; acaulescent perennial herbs with stolons emerging from below sec. head; latex ducts 0; capitula discoid, 1–many fld – 2/3 S Afr.; R: SB 36(2011)194): *Platycarpha, Platycarphella*
7. **Moquinieae** (~ 5.; shrubs; latex ducts 0; hairs simple; homogamous or gynodioec. fls – 2/2, Braz.; R: Funk & al.: 477): *Moquinia, Pseudostifftia*

Some veg. – *Cichorium* (chicory, endive), *Lactuca* (lettuce), *Pacourina, Scorzonera, Taraxacum, Tragopogon* (salsify); timber – *Eremanthus, Gymnanthemum*; rubber – *Taraxacum*; cult. orn. – *Arctotis, Catananche, Gazania, Stokesia*; weeds – *Chondrilla, Crepis, Cyanthillium, Elephantopus, Ethulia, Hypochaeris, Lactuca, Sonchus, Taraxacum*

XI. **Corymbioideae** (~ X; perenn. herbs; latex ducts 0; lvs parallel-veined, usu. in rosettes; capitual discoid, 1-fl; involucre of 2 bracts – 1/9 Cape; R: Funk & al.: 488): *Corymbium*

XII. **Asteroideae** (latex ducts, sesquiterpenoid lactones; capitula heterogamous, radiate or disciform, less oft. discoid; disk-florets usu. with broad short lobes, usu. yellow; anthers basifixed (free in wind-poll. spp.); style-branches usu. with 2 distinct stigmatic areas; pollen spiny; cosmop.):

1. **Senecioneae** (shrubs, herbs (some succ.), trees & lianes; lvs usu. in spirals; pyrrolizidine alks; latex ducts 0; involucral bracts usu. in 1 row, oft. with outer series of reduced bracts; receptacle usu. naked; style-branches usu. truncate,

apically minutely hairy, less oft. variously appendaged – cosmop., esp. S Afr.; R: Funk & al.: 513); 4 subtribes: *Abrotanella* (Abrotanellinae); *Brachyglottis, Cremanthodium, Doronicum, Gynoxys, Ligularia, Parasenecio, Psacalium, Roldana, Sinosenecio, Tephroseris* (Tussilagininae); *Cineraria, Crassocephalum, Dendrophorbium, Emilia, Euryops, Gynura, Kleinia, Monticalia, Packera, Pentacalia, Senecio, Synotis* (Senecioninae); *Crassothonna, Othonna* (Othonninae)

Many pachycaul trees with similar candelabriform branching (Leeuwenberg's Model) in diverse genera: *Dendrosenecio* (Afr. mts), *Brachyglottis* (Aus., NZ), *Kleinia* (dry Afr.), *Pittocaulon* (C Am.), *Pladaroxylon* (St Helena), *Senecio* (Juan Fernandez), *Telanthophora* (C Am.), etc.; similar branching in Astereae (*Apodocephala, Psiadia, Vernoniopsis* in Madag.), in Cardueae (*Centaurodendron* on Juan Fernandez), in Heliantheae (*Podachaenium* in C Am.) & in Cichorieae (*Sonchus* on Juan Fernandez) but also in small shrubby spp. in these tribes

2. **Calenduleae** (small trees, shrubs & herbs; lvs in spirals or opp.; involucral bracts in 1–3 rows; receptacle naked; anthers acute, ± tailed; style-arms truncate with apical hairs; pappus 0; cypselas oft. curiously shaped – Afr., SW As., Eur.): *Calendula, Osteospermum*
3. **Gnaphalieae** (~ 6.; herbs (many rather dreary weeds), shrubs, shrublets (oft. ericoid); latex ducts 0; lvs, entire, in spirals; involucral bracts papery, s.t. showy ('everlastings') – cosmop. esp. S Afr. & Aus.; 5 subtribes (many genera unassigned) trad. recog.; R: OB 104(1991)5): *Loricaria* (Loricariinae); *Metalasia, Oedera, Stoebe* (Relhaniinae); *Anaphalis, Antennaria, Ozothamnus* (Cassiniinae); *Angianthus, Craspedia, Rhodanthe* (Angianthinae); *Filago, Gamochaeta, Gnaphalium, Helichrysum, Leontopodium, Pseudognaphalium, Xerochrysum* (Gnaphaliinae)
4. **Astereae** (shrubs, some trees (e.g. *Apodocephala, Commidendrum, Melanodendron, Vernoniopsis*), lianes & herbs; lvs usu. in spirals; latex usu. 0; involucral bracts in (2)3–5(–9) series, usu. imbricate; receptacle naked, rarely scaly; anthers obtuse, not tailed; style-arms with a shortly hairy triangular to lanceolate apical appendage – cosmop. (R: Phytol. 76(1994)201, T 61(2012)438): [subtribes in flux] *Archibaccharis, Aster, Baccharis, Brachycome, Celmisia, Diplostephium, Ericameria, Erigeron, Felicia, Grangea* (Grangeinae); *Gutierrezia, Haplopappus, Lagenophora, Machaeranthera, Microglossa, Olearia, Psiadia, Pteronia, Solidago, Symphyotrichum, Vittadinia*
5. **Anthemideae** (herbs & some shrubs (e.g. *Argyranthemum*); latex ducts 0; lvs in spirals (rarely opp.), oft. much-divided, strongly scented; involucral bracts in 1–several series, usu. with thin dry transparent tips or margins; receptacle naked or scaly; anthers obtuse to acute, not tailed; style-arms truncate, fringed with short hairs – cosmop. esp. S Afr., Medit., C As.; R: BBMNHB 23(1993)84, Funk & al.: 637); 14 subtribes (but several genera unplaced): *Achillea, Anacyclus, Matricaria* (Matricariinae); *Ajania, Artemisia, Chrysanthemum* (Artemisiinae); *Anthemis, Tanacetum, Tripleurospermum* (Anthemidinae); *Argyranthemum, Glebionis* (Glebionidinae); *Athanasia* (Athanasiinae); *Cotula, Leptinella* (Cotulinae); *Inulanthera*(?), *Ursinia* (Ursiniinae); *Leucanthemum* (Leucantheminae); *Osmitopsis* (Osmitopsidinae); *Pentzia* (Pentziinae); *Santolina* (Santolininae); *Sclerorhachis, Tanacetopsis, Trichanthemis* (Handeliinae)
6. **Inuleae** (incl. Plucheeae; perenn. herbs & shrubs (*Duhaldea* trees); latex ducts 0; lvs in spirals; involucral bracts without transparent margins; anthers with tails oft. branched; style-branches with acute hairs): Inulinae usu. with yellow radiate capitula – OW esp. Medit. to C As., E Afr.: *Anisopappus, Blumea, Buphthalmum, Inula, Pulicaria*; Plucheinae (R: T 64(2015)111) usu. with purple discoid capitula – trop. & warm incl. arid: *Epaltes, Laggera, Pluchea, Sphaeranthus*
7. **Athroismeae** (trees or shrubs; latex ducts 0; lvs in spirals; involucre cup-shaped to cylindrical, bracts without transparent margins – 5/55 Afr. Aus.; R: Funk & al.: 685): *Anisopappus, Blepharispermum*

Cult. herbs – *Tanacetum* (costmary, pyrethrum), *Artemisia* (absinthe & tarragon; also med.), *Chamaemelum* (chamomile); *Limbarda* ed.; many cult. orn. – *Achillea, Anacyclus, Anaphalis, Antennaria, Anthemis, Artemisia, Aster, Bellis, Boltonia, Brachycome, Brachyglottis, Calendula, Callistephus, Chrysanthemum, Curio* (succ.), *Doronicum, Erigeron, Felicia, Glebionis, Helichrysum, Inula, Kleinia* (succ.), *Leontopodium* (edelweiss), *Leucanthemum, Ligularia, Olearia, Osteospermum, Othonna* (succ.), *Ozothamnus, Pericallis* ('cineraria'), *Petasites, Raoulia, Santolina, Senecio, Solidago,*

Symphyotrichum, Tanacetum, Telekia, Ursinia, Xerochrysum ('helichrysum'); weeds – *Chrysanthemoides, Erigeron, Crassocephalum, Senecio* (some poisonous)

XIII. **'Heliantheae alliance'** (latex ducts 0; involucral bracts usu. in 1–3 series); anther thecae often black, without spurs or tail; usu. phytomelanin layer in cypsela

1. **Feddeeae** (liane; heads discoid with multiseriate involucre with resiniferous ducts & 9–12 white bisex. florets – 1/1 Cuba): *Feddea*
2. **Helenieae** (excl. 5., 7., 11., 12.; usu. herbs with lvs in spirals; involucral bracts in 2+ rows; radiate fls usu. yellow; anthers pale or at least not black; receptacle usu. without scales; Am. esp. SW N Am.); 4 subtribes: *Gaillardia, Helenium* (Gaillardiinae); *Hymenoxys* (Tetraneurinae)
3. **Coreopsideae** (~ 9.; shrubs & herbs; outer involucral bracts diff. in shape & colour from inner – cosmop. esp. Am.; R: Funk & al.: 720): *Bidens, Coreopsis, Cosmos, Dahlia*
4. **Neurolaeneae** (~ 9.; stems fisulose; receptacle usu. with scales; anthers blackened; cypselas with phytomelanin – trop. esp. Am.): *Neurolaena*
5. **Tageteae** (~ 2.; trees (few) to ann. herbs; receptacle with dark spots, no scales; cypselas with phytomelanin; char. smell due to monoterpenes – 32/c. 270 Am., esp. SW N Am., Aus. (1 sp.)): *Flaveria, Hydropectis* (aquatic), *Pectis, Porophyllum, Tagetes, Thymophylla*
6. **Chaenactideae** (~ 2.; herbs; lvs often lobed, in spirals; capitula discoid; receptacle naked; anthers not blackened; cypsela with phytomelanin – 3/20 SW N Am., esp. Calif.; R: Funk & al.: 696): *Chaenactis*
7. **Bahieae** (~ 2.; usu. herbs; receptacle naked; anthers not blackened; cypselas with phytomelanin; pappus usu. of basally thickened scales – trop. Afr., S. Pacif., Am. esp. SW N Am.; R: T 65(2016)1074): *Bahia, Hymenopappus*
8. **Polymnieae** (~ 9.; perenn. herbs; lvs opp.; receptacle with scales; disk fls functionally male – 1/3 E N Am.): *Polymnia*
9. **Heliantheae** (excl. 3., 4., 8., 10.); herbs, shrubs, trees & lianes; lvs strigose, usu. opp.; receptacle scaly; involucral bracts in few rows; anthers usu. blackened; cypselas with phytomelanin; 113/1461 cosmop. esp. Am.; 14 subtribes: *Acmella* (Spilanthinae); *Ambrosia, Parthenium, Xanthium* (Ambrosiinae); *Echinacea, Zinnia* (Zinniinae); *Calea, Melampodium* (Melampodiinae); *Encelia, Flourensia* (Enceliinae); *Helianthus, Pappobolus, Scalesia, Simsia, Viguiera* (Helianthinae); *Heptanthus* (Pinillosiinae); *Montanoa* (Montanoinae); *Perymenium, Wedelia* (Ecliptinae); *Rudbeckia* (Rudbeckiinae); *Silphium* (Engelmanniinae); *Verbesina* (Verbesininae)
10. **Millerieae** (~ 9.; trees to ann. herbs; lvs usu. opp.; capitula usu. radiate; receptacle with scales; anthers usu. blackened; cypselas with phytomelanin): 34/380 trop. & warm esp. Mex., N Andes); 8 subtribes recog.: *Espeletia* (Espeletiinae); *Galinsoga* (Galinsoginae); *Milleria, Smallanthus* (Milleriinae); *Tridax* (Dyscritothamninae)
11. **Madieae** (~ 2.; trees (e.g. *Dubautia*), lianes, shrubs & herbs; lvs opp. or in spirals or whorls, oft. glandular; invilucral bracts usu. subequal; receptacle scales 0 or peripheral; cypselas with phytomelanin – 24/120 mainly W N Am., esp. Calif.; R: Funk & al.: 698); 8 subtribes: *Arnica* (Arnicinae); *Dubautia, Madia* (Madiinae); *Lasthenia* (Baeriinae)
12. **Perityleae** (~ 2.; shrubs or herbs of rocky habitats; lvs usu. opp., glandular; involucral bracts subequal in 2 series; receptacle naked; cypselas with phytomelanin – 4/73 Am., esp. SW N Am. deserts; R: Funk & al.: 702); 2 subtribes: *Perityle* (Peritylinae)
13. **Eupatorieae** (shrubs & herbs incl. aquatics; hairs simple; lvs usu. opp.; pyrrolizidine alks secreted by nectaries; capitula discoid, homogamous (never yellow); anthers obtuse to acute, not tailed; pollen spiny; style-arms elongate, club-shaped, papillose – mostly Am.); 17 subtribes (R: MSBMBG 22(1987)1)): *Hofmeisteria* (Hofmeisteriinae); *Ageratina* (Oxylobinae); *Mikania* (Mikaniinae); *Trichocoronis* (Trichocoroninae); *Adenostemma* (Adenostemmatinae); *Ageratum* (Ageratinae); *Fleischmannia* (Fleischmanniinae); *Chromolaena* (Praxelinae); *Campuloclinium* (Gyptidinae); *Liatris* (Liatrinae); *Eupatorium* (much reduced but now being rebuilt; Eupatoriinae); *Symphyopappus* (Disynaphiinae); *Ayapana* (Ayapaninae); *Brickellia* (Alomiinae); *Critonia, Koanophyllum, Ophryosporus* (Critoniinae); *Bartlettina, Neomirandea* (Hebecliniinae); *Stevia* (Piqueriinae); *Trichogonia* (Trichogoniinae; R: Phytotaxa 260(2016)296)

Some ed. – *Acmella, Galinsoga, Guizotia* (oil), *Helianthus* (oil & tubers), *Madia* (oil), & *Smallanthus* (tubers); sweetener – *Stevia*; medic. tea – *Ayapana*; timber – *Montanoa*; rubber – *Parthenium*; cult. orn. – *Ageratum, Chrysogonum, Coreopsis, Cosmos, Dahlia, Echinacea, Encelia, Gaillardia, Helenium, Helianthus, Layia, Liatris, Ratibida, Rudbeckia, Sanvitalia, Tagetes, Thymophylla, Tithonia, Wedelia, Zinnia*; weeds – *Adenostemma, Ageratina, Ageratum, Bidens, Blainvillea, Chrysanthellum, Chromolaena, Gymnocoronis, Tridax, Xanthium*

Fam. most consp. in diversity in montane subtrop. & trop. areas (e.g. paramo of Andes with 101/858, 25% of total flora) with a tendency for tribal & subtribal specialization in ecology & habit, though within particular genera there may be great variation e.g. *Coreopsis* & *Erigeron* with aquatics to xerophytes; *Blepharispermum, Brachylaena, Eremanthus, Fulcaldea* & '*Vernonia*' include timber trees & *Piptocarpha* dominates the seasonally burnt landscapes of EC Braz. The capitulum in animal-poll. spp. acts in many ways like a single flower (cf. Aizoaceae), the long-tubed fls of Cardueae being visited by bees & Lepidoptera, the yellow & white fls so common in the rest of the family being attractive to flies, beetles etc. At anthesis the pollen is forced out by the stigma, the floret being functionally male, later the stigmatic arms sep. such that the floret is functionally female; oft. the stigmas then curl back to touch the pollen on the style thus effecting self-fert. A number of genera esp. *Hieracium* & *Taraxacum* have large numbers of apomictic lines ('microspecies'). Wind-poll. occurs in *Ambrosia, Artemisia* etc. It has been customary to argue that the ecological success in terms of spp. & individual nos. of the family is due to the capitulum system, the involucre acting like a K, the fr. wall like a testa, etc., but the pseudanthial head is 'duplicated in the small & unsuccessful family Calyceraceae' (Cronquist) while a similar pollen-presentation mechanism is found in *Brunonia* (Goodeniaceae s.l.; fls in cymose heads) & dev. to varying degrees in Campanulaceae, other Goodeniaceae & part of Rubiaceae. Cronquist attributes the success of the C. to the defensive combination of polyacetylenes & the bitter sesquiterpene lactones followed by the development of other chem. repellents like the alks in Senecioneae, the latex system of Cichorieae, which do not have the polyacetylene-bearing resin system of the other tribes; the bad smell of Tageteae & the char. smells of Anthemideae are such repellents: many of these compounds make the pls imp. as sources of flavourings & insecticides. Contrary to received doctrine, the fam. is primitively woody (cf. NP 73(1974)967) & prob. of S Am. (cf. Calyceraceae, though fossils from mid-Jurassic China claimed) origin followed by great radiation in Afr., the early members being pachycaul with large discoid capitula (condensed thyrses) of yellow fls & involucral bracts in several rows. Patagonian fossils 47.7 M yrs old resemble Stifftieae & poss. bird-poll.

Compared with other large fams, e.g. Gramineae, Leguminosae, the C. are of little value to humans exc. as ornamentals, the ed. ones with low levels of toxins or, as in lettuce, having had them selected out: some are insecticides & fish-poisons but many are noxious weeds, their fr. spread by wind (pappus) or animals (sticky pappus in *Adenostemma*, pappus of barbed bristles in *Bidens*, involucral bracts with hooked tips in *Arctium*, sticky bracts in *Sigesbeckia*, receptacle with hooks in *Xanthium* etc.). With increasing clearance of native veg. throughout the world, these aggressive toxic pls will inherit it – until our own species's demise

Compsoneura (A. DC.) Warb. Myristicaceae. 14 trop. Am. ***C. mexicana*** (Hemsl.) Janovec ('*C. sprucei*', C Am.) – thrips-poll. in Costa Rica

Comptonanthus R. Nordenstam = *Ifloga*

Comptonella Bak. f. Rutaceae (I 2). 8 New Caled. R: BMMHN Adans. 5(1983)394. Usu. dioec.

Comptonia L'Hérit. Myricaceae. 1 E N Am.: ***C. peregrina*** (L.) J. Coulter sole survivor of (?) 12 Eocene spp. (some Eur.). Actinobacteria r.-symbionts fixing nitrogen. Differs from *Myrica* in pinnatifid lvs & usu. monoec. Lvs a tea for E Canadian Indians; local medic., cult. orn.

Comularia Pichon = *Hunteria*

Conamomum Ridl. = *Amomum*

Conandrium (K. Schum.) Mez. Primulaceae (Myrsinaceae). 2 E Mal. R: Blumea 33(1988)109

Conandron Sieb. & Zucc. Gesneriaceae (III 2j). 1 E China, Jap.: ***C. ramondoides*** Sieb. & Zucc. – cult. orn.

Conanthera Ruíz & Pavón. Tecophilaeaceae. 5 Chile. Some ed. bulbs

Conceveiba Aubl. Euphorbiaceae (Alc.-Conc.). Excl. *Aubletiana*, 14 trop. Am. R: Boissiera 60(2005)37

Conceveibastrum (Muell. Arg.) Pax & K. Hoffm. = *praec.*

Conceveibum A. Rich. ex A. Juss. = *Aparisthmium*
concha satinwood *Zanthoxylum caribaeum*
Conchidium Griff. (~ *Eria*). Orchidaceae (V 15). 10 Indomal.
Conchocarpus Mikan (~ *Angostura*). Rutaceae (I 8). 45 trop. Am. R: KB 53(1998)266. C sympetalous, reg. to irreg.
Conchopetalum Radlk. Sapindaceae. 2 Madag.
Conchophyllum Blume = *Dischidia*
Concocidium Romowicz & Szlach. = *Gomesa* (but see PJB 51(2006)44)
Condalia Cav. Rhamnaceae (6). 18 warm Am. R: Britt. 14(1962)340. Some ed. fr., timber gives dye
Condaliopsis (Weberb.) Süsseng. = *Ziziphus*
Condaminea DC. Rubiaceae (Cond.). 3 Andes
Condea Adans. (~ *Hyptis*). Labiatae (VII 3c). 27 trop. & warm Am. R: Phytotaxa 58(2012)13
condurango *Gonolobus cundurango*
Condylago Luer. Orchidaceae (V 13). 1 Colombia: ***C. rodrigoi*** Luer. R: MSB 24(1987)21
Condylidium R. King & H. Robinson (~ *Eupatorium*). Compositae (Eup.-Ayap.). 2 C. Am., WI Am. R: MSBMBG 22(1987)206
Condylocarpon Desf. Apocynaceae (I i). 7 C Am. (1) to trop. S Am. R: AMBG 70(1983)149. Febrifuges
Condylopodium R. King & H. Robinson. Compositae (Eup.-Alom.). 4 Colombia. R: MSBMBG 22(1987)266
Condylostylis Piper (~ *Vigna*). Leguminosae (III 18). 4+ trop. Am. R: AJB 98(2011)1704
cone flower *Echinacea* spp., *Ratibida* spp., *Rudbeckia* spp.; **c.sticks** *Petrophile* spp.
conessi *Holarrhena pubescens*
Confederate daisy *Helianthus porteri*; **C. rose** *Hibiscus mutabilis*
congaberry *Carissa* spp.
Congdonia Muell. Arg. = *Declieuxia*
Congea Roxb. Labiatae (I; Symphoremataceae). c. 7 SE As., W Mal. R: GBS 21(1966)259. Climbers, cult. orn. esp. ***C. tomentosa*** Roxb. (Myanmar, Thailand) with white to lilac tomentose leaf-like bracts subtending heads of white fls
Congo bean *Cajanus cajan*; **C. copal** *Guibourtia demeusei*; **C. jute** *Hibiscus americanus*; **C. pea** *C. cajan*; **C. rubber** *Ficus lutea*; **C. stick** *Castanea sativa* coppice walking-stick; **C. wood** (Am.) *Lovoa trichilioides*
Congolanthus A. Raynal. Gentianaceae. 1 trop. Afr.: ***C. longidens*** (N.E. Br.) A. Raynal. R: Adans. 8(1968)56
congorosa *Maytenus ilicifolia*
Conicosia N.E. Br. Aizoaceae (V). Incl. *Herrea*, 2 Cape Prov. R: BJ 111(1990)482. Capsule opens hygroscopically & some seeds washed out by raindrops; it remains open when dry & many seeds shaken out & finally it breaks up into wind-disp. segments, each cont. up to 2 seeds held in pocket-like folds (Rowley). Cult. orn., ***C. pugioniformis*** (L.) N.E. Br. invasive W US
Coniferae Juss., **Coniferales, Coniferopsida** See Pinidae
Conimitella Rydb. (~ *Mitella*). Saxifragaceae (I 5). 1 W US: ***C. williamsii*** (D. Eaton) Rydb. R: FNA 9(2009)105
Coniogramme Fée. Pteridaceae (I). c. 40 OW trop. to Hawaii
Conioselinum Hoffm. Umbelliferae (III 8). 18 temp. Euras. (Eur. 1), N Am. (3). R: Willd. 33(2003)357. ***C. pacificum*** (S. Watson) J. Coulter & Rose (W N Am.) – ed. 'wild carrot'
Conium L. Umbelliferae (III 5). 5 temp. Euras. (Eur. 1), S Afr. (1). ***C. maculatum*** L. (hemlock, Euras.), biennial, only umbellifer with alks which may act as parts of coenzymes in oxid.-reduction processes, form. medic., v. poisonous (said to have killed Socrates, poss. 'gall' of Bible, rodent control since C2) due to polyacetylenes esp. coniine paralyzing resp. system, invasive Micronesia (1 pl. prod. up to 38 000 seeds)
conkerberry *Carissa* spp.
conkers *Aesculus hippocastanum*
Connaraceae R. Br. Magnoliidae – Oxalidales. 12/170 trop. esp. OW. R: AUWP 89-6(1989)131. Trees (s.t. pachycaul & unbranched), shrubs or lianes commonly with solit. crystals of calcium oxalate in parenchyma cells & v. oft. with mucilage-canals &/or tanniniferous secretory cavities; bark, fr. & seeds oft. v. poisonous though agent unknown. Lvs in spirals, pinnate to unifoliolate, pulvinate, s.t. reduced & hook-like, woody, petioles with transverse ridges; stip. 0. Fls in racemes or panicles, small, usu. bisex. (pl. rarely dioec. as in *Ellipanthus*) & usu. heterostylous (some tristyly; *Connarus* in Afr. & As. with

only medium- & long-styled fls), reg., ± hypogynous; K (4)5, s.t. basally connate, imbr. or valvate, v. oft. persistent around fr. base. C (4)5, s.t. basally connate, imbr. or rarely valvate, A 5 + 5, inner s.t. staminodes, anthers with longit. slits & 3-colporate or 3-colpate (4-colpate in *Jollydora*) pollen, disk 0 or small, usu. extrastaminal but receptacle s.t. nectariferous, G 1 (3), 5 (7 or 8), oft. 5 with 4 abortive, s.t. ± connate basally, oft. not completely closed, each with term. style & capitate stigma & 2 marginal, collateral (cf. Leguminosae) ovules, ascending, anatr. to (oft.) hemitropous, bitegmic, usu. 1 abortive. Fr. of 1–5 follicles, usu. dry-dehiscent, opening along ventral or rarely both sutures or an indehiscent nut or drupe; seeds oft. black with red/orange aril (cf. Sapindaceae) &/or sarcotestal with oily or 0 endosperm. n = 13, 14, 16

Classification & principal genera:

I. **Connareae** (trees, shrubs, lianes; G1; pollen tricolporate; OW): *Connarus, Ellipanthus*

II. **Jollydoreae** (usu. unbranched treelets; G1; pollen tetracolpate; Afr.): *Jollydora* (only)

III. **Manoteae** (lianes; G5; pollen tricolporate; seed ventrally attached; Afr.): *Manotes* (only)

IV. **Cnestideae** (shrubs & lianes; G5; pollen tricolporate; seed basally attached): *Agelaea, Cnestis, Rourea*

Seeds like Sapindaceae but free carpels unlike Sapindales

Some timbers, local med. & poisons, tannins & fibres

Connarus L. Connaraceae (I). 77 trop. R: AUWP 89–6(1989)239. Some fish-poisons, anthelminthics, useful wood

Connellia N.E. Br. Bromeliaceae (1). 5 Venez. highlands. R: AMBG 73 (1986)690

Connorochloa Barkworth & al. = *Elymus* (but see Telopea 13(2011)51)

Conobea Aubl. Plantaginaceae (Grat.-Stem.; Scrophulariaceae s.l.). 7 trop. Am.

Conocalyx Benoist. Acanthaceae (III 2c). 1 Madag.: ***C. laxus*** Benoist

Conocarpus L. Combretaceae (II 2a). 2 trop. Am. & Afr.: ***C. erectus*** L. (buttonwood, button mangrove, trop. Am., W Afr.) – mangrove, tanbark & wood for charcoal, hedge-pl. in Florida; ***C. lancifolius*** Engl. (NE Afr., S Yemen) – on sandy soils, windbreak & topiary (e.g. Kuwait)

Conocephalum Hill. Conocephalaceae (Hepaticae). 2 subcosmop. ***C. conicum*** (L.) Dumort. – dioec., highly scented, local medic., terrarium pl.

Conocephalus Blume = *Poikilospermum*

Conocliniopsis R. King & H. Robinson (~ *Eupatorium*). Compositae (Eup.-Gyp.). 1 Colombia, Venez., Braz.: ***C. prasiifolia*** (DC.) R. King & H. Robinson. R: MSBMBG 22(1987)97

Conoclinium DC. (~ *Ageratum*). Compositae (Eup.-Ager.). 4 E US, Mex. R: MSBMBG 22(1987)130

Conomitra Fenzl. Apocynaceae (V b; Asclepiadaceae III 5). 1 Sudan: ***C. linearis*** Fenzl. R: FTEA Apoc., 2(2012)212

Conomorpha A. DC. = *Cybianthus*

Conopharyngia G. Don = *Tabernaemontana*

Conopholis Wallroth. Orobanchaceae (Rhin.). 3 SE US to Panamá. R: SB 38(2013)802. ***C. americana*** (L.) Wallroth (cancer-r., squawroot, US) – med.

conophor nut *Plukenetia conophora*

Conophyllum Schwantes = *Mitrophyllum*

Conophytum N.E. Br. Aizoaceae (V). Excl. *Ophthalmophyllum*, 86 S Afr. R: H. E. K. Hartmann, *Ill. Handb. Succ. Pls*, Aiz. A–E (2002)134. Cult. orn. succ. with globose to oblong photosynthetic growths, new ones developing inside old, which dry to a thin shell prot. new in dry season

Conopodium Koch. Umbelliferae (III 8). 6 Euras., Medit. R: EJB 58(2001)339. Cotyledon 1. Tuberous r. of *C.* ***majus*** (Gouan) Loret (earth-nut, pig-nut, Eur.) ed. roasted

Conosapium Muell.-Arg. = *Excoecaria*

Conospermum Sm. Proteaceae (IV 1). c. 53 Aus. R: FA 16(1995)131. Smoke bushes; mooted for treatment of HIV (conocurovone), cult. orn.

Conostalix (Kraenzlin) Brieger = *Dendrobium*

Conostegia D. Don = *Miconia* (but see PhytoKeys 67(2016)1)

Conostephium Benth. Ericaceae (VII 7; Epacridaceae). 11 S Aus. R: Nuytsia 23(2013)315

Conostomium (Stapf) Cuf. Rubiaceae (IV 1). 4 trop. E Afr.

Conostylidaceae Takht. = Haemodoraceae

Conostylis R. Br. Haemodoraceae (II). (Excl. *Blancoa*) 45 SW Aus. R: FA 45(1987)57. Insect- & bird-poll.

Conothamnus Lindl. = *Melaleuca* (but see Muelleria 16(2002)41)

Conradina A. Gray. Labiatae (VII 2b). 6 SE US. Some cult. orn.

Conringia Heister ex Fabr. Cruciferae (20). 6 Medit., Eur. (3) to C As. R: JAA 66(1985)348. ***C. orientalis*** (L.) Dumort. (Eur., Med.) – subcosmop. weed, seeds source of cooking oil

Consolea Lem. (~ *Opuntia*). Cactaceae (II). 2–3 Florida Keys (***C. corallicola*** Small – only 20 pls left in wild, or = *C. moniliformis* (L.) Berger?), WI. R: Britt. 53(2001)105

Consolida Gray = *Delphinium*

Constancea Baldwin (~ *Eriophyllum*). Compositae (Mad.-Bae.). 1 Calif.: ***C. nevinii*** (A. Gray) Baldwin. R: FNA 21(2006)362

Constantia Barb. Rodr. Orchidaceae (V 13b). 6 Braz.

consumption weed *Baccharis halimifolia*

contrayerva *Aristolochia odoratissima*; **c. root** *Dorstenia contrajerva*

Convallaria L. Asparagaceae (Convallariaceae). 1 N temp. incl. Eur. ***C. majalis*** L. (*C. keiskei*, lily-of-the-valley, muguet; N Am. = subsp. ***majuscula*** (Greene) Gandhi & al.) – cult. orn. (incl. cvs with pink or double fls), cold spells in 2 seasons needed for emergence of radicle, then growth of epicotyl, clones to 670+ yrs old, rhiz. medic., fls used in scent & in certain snuffs, poisonous (38 cardenolide glycosides: azetidine 2-carboxylic acid, non-protein amino-acid, interfering with proline chem. in usu. being incorporated in proteins but proline t-RNA-synthetase discriminates against it), Eur. clones natur. N Am. R: FR 86(1975)543.

Convallariaceae Horan. = Asparagaceae

Convolvulaceae Juss. Magnoliidae – Solanales. 55/1850 cosmop. esp. warm. R: SB 28(2003)795. Herbaceous climbers (always twining to right), s.t. parasitic, lianes to 30 m, herbs or shrubs or rarely trees (*Humbertia*); stems oft. with unusual sec. growth & usu. with articulated non-anastomosing latex-canals or cells & internal phloem, in *Cuscuta* without chlorophyll & attached to host by haustoria, the terr. r.-system soon withering & internal phloem absent. Lvs in spirals, simple, entire to lobed, scale-like in *Cuscuta*, usu. palmately veined; stip. 0. Fls usu. lasting just 1 day, usu. in heads, dichasia or solit., oft. subtended by a pr of bracts, these s.t. enlarged & involucriform, bisex. (not in *Hildebrandtia*), usu. 5-merous ((3–)5 in *Cuscuta*, 4 in *Hildebrandtia*); K imbr., ± connate, s.t. unequal, (C) commonly funnel-shaped, hardly lobed (obliquely irreg. in *Humbertia*), oft. induplicate-valvate & oft. convolute in bud, imbr. in *Cuscuta*, A attached at tube-base with filaments oft. unequal & anthers with longit. slits, usu. annular nectary-disk around ovary-base, $\underline{G}$ (2(3–5)) with as many locules (rarely 1) & free or united styles or 1, rarely united only by common style, ovules usu. 2 per carpel (∞ in *Humbertia*), basal, erect, anatropous, unitegmic. Fr. a loculicidal, circumscissile or irreg. dehiscing capsule, less oft. baccate, drupe or nut. Seeds usu. with black shiny seed-coat; embryo straight or curved with 2 plicate, oft. bifid cotyledons or these scarcely recognizable (*Cuscuta*), endosperm with oil, protein & carbohydrate. x = 7–15+

Classification & principal genera:

1. **Aniseieae** (leaf venation pinnate; pollen prolate, style 1, stigma not globose): *Aniseia* (only)
2. **Cardiochlamyeae** (leaf venation palmate, style 1): *Cardiochlamys, Poranopsis*
3. **Convolvuleae** (leaf venation pinnate, pollen spheroidal, style 1, fr. wiith 2–4 valves): *Convolvulus, Polymeria* (only)
4. **Cresseae** (style ± bifid or 2; fr. dehiscent): *Bonamia, Evolvulus, Wilsonia*
5. **Cuscuteae** (parasites with scale lvs; Cuscutaceae – deeply embedded in fam.): *Cuscuta* (only)
6. **Dichondreae** (style ± bifid or 2; fr. a utricle; Dichondraceae): *Dichondra, Falkia, Metaporana, Nephrophyllum*
7. **Erycibeae** (leaf venation pinnate, C-lobes bifid, style 0): *Erycibe* (only)
8. **Humbertieae** (arborescent; Humbertiaceae – 'sister' to rest of fam.): *Humbertia* (only)
9. **Ipomoeeae** (incl. Argyreieae; leaf venation pinnate, spiny pollen, style 1; poss. best treated as single genus): *Ipomoea*
10. **Jacquemontieae** (leaf venation pinnate, pollen spheroidal, style 1, fr. with 8 valves): *Jacquemontia* (only)
11. **Maripeae** (style ± bifid or 2; fr. woody-baccate): *Maripa*
12. **Merremieae** (leaf venation pinnate; pollen prolate, style 1, stigma globose): *Merremia, Operculina, Xenostegia*

Many cult. orn. esp. spp. of *Ipomoea, Convolvulus & Poranopsis*, dried fr. of *Argyreia & Merremia* sold as 'wood roses'; *Hildebrandtia* has some ed. fr., the tubers of *I. batatas* are sweet potatoes, other *I.* spp. are leaf veg. & sources of hallucinogenic drugs (also *Argyreia*) & purgatives (also *Convolvulus*); some bad weeds, esp. *C.* & form. *Cuscuta*

Convolvulus Tourn. ex L. Convolvulaceae (3). Incl. *Calystegia* (solit. fls., usu. large bracts, fr. 1-loc., cf. *Convolvulus s.s.* 199 – R: PhytoKeys 51(2015)1) c. 220 cosmop. esp. temp. (Eur. 26). Pollen smooth (that of *Ipomoea* (q.v.) spiny). Bindweed; alks. Some pernicious weeds esp. ***C. arvensis*** L. (common or field bindweed, cornbind, Euras., widely natur.) – sweetly-scented fls visited by insects, introd. US 1730s as valued medic. pl., now pestilential, **'Stonestreetii'** (var. *stonestreetii*) with 5-lobed C (cf. *Rhododendron stenopetalum*) & ***C. sepium*** L. (*Calystegia s.*, temp. & warm) – purgative, rhiz. ed. boiled in China, young lvs ed. Ind. Others cult. orn. esp. ***C. tricolor*** L. (S Eur.) & ***C. gharbensis*** Battand. & Pitard (Morocco) – ann. herbs, ***C. sabatius*** Viv. (*C. mauritanicus*, Italy, N Afr.) – perenn., **'Two Moons'** with white, lilac or s.t. bicolored fls, ***C. japonicus*** Thunb. **'Flore Pleno'** (*Calystegia hederaceus*, E As.) – double-flowered form favoured in China & ***C. soldanella*** L. (*Calystegia s.*, seashores worldwide) – seeds sea-disp., scurvy-grass subs.; ***C. floridus*** L.f. (Canary Is.) – woody, r. source of an ess. oil; ***C. scammonia*** L. ((Levant) scammony, SW As.) – source of a drastic purgative; ***C. sylvaticus*** Spreng. (*Calystegia s.*, E Med.) – poss. orig. of lotus-fl. motif of anc. Greeks. See also *Cynanchum acutum*

Conyza Less. = *Erigeron*; for Afr. spp. see *Nidorella*

Conyzanthus Tamamschjan = *Symphyotrichum*

Conzattia Rose. Leguminosae (I 4). 1 Mex.: ***C. multiflora*** (Robinson) Standl. – medic. etc.

cooba *Acacia salicina*

coohoy nut *Floydia prealta*

Cook pine *Araucaria columnaris*

Cooktown ironwood *Erythrophloeum chlorostachys*; **C. orchid** *Dendrobium bigibbum*

Cooktownia D.L. Jones (~ *Habenaria*). Orchidaceae (IV 2). 1 N Queensland: ***C. robertsii*** D.L. Jones – peloric? R: Austrobaileya 5(1997)74

coolabar, coolibah *Eucalyptus microtheca*; **c. grass** *Eragrostis advena*

coolwort *Tiarella cordifolia*

Coombea P. Royen = *Medicosma*

coondi *Carapa guianensis*

Cooperia Herb. = *Zephyranthes*

Coopernookia Carolin. Goodeniaceae. 6 SW & SE Aus. R: FA 35(1992)80

copaiba balsam *Copaifera officinalis* etc.; **c. oil** *C. multijuga*

Copaifera L. Leguminosae (I 2). c. 35 trop. Am., Afr. (s.s. 4 Afr.), Mal. (1). Source of hard resins, timbers & oleo-resins (copals), used industrially, & timbers (ironwood) esp. ***C. officinalis*** (Jacq.) L. (trop. Am.) – copaiba balsam; ***C. multijuga*** Hayne (Amaz.) – oil (25 l over 6 months) from trunk can be used directly in diesel engines; ***C. salikounda*** Heckel (W Afr.) – veneer (bubinga)

copal hard resins form. much used in varnish & paint manufacture, esp. **Congo c.** dug up from ground (*Guibourtia demeusei*) & other semi-fossil c. from *Copaifera* spp. (W Afr.), *Hymenaea verrucosa* (**Zanzibar c.**, E Afr.); **Manila** or **East Ind. c.** is from *Agathis dammara*, **kauri c.** (NZ) from *A. australis*; **Madag. c.** *Hymenaea verrucosa*

copalchi *Hintonia latiflora*; **c. bark** *Croton niveus*

copalquín *Hintonia latiflora*, *Pachycormus discolor*

Copedesma Gleason = *Miconia*

Copelandiopteris Stone = *Pteris*

Copernicia Mart. ex Endl. Palmae (III 4b). 21 Cuba, Hispaniola (2) & S Am. (3) savannas. R: GH 9(1961–3)1–232. Some cult. orn. incl. ***C. prunifera*** (Mill.) H. Moore (*C. cerifera*, wax palm, carnauba w. p., NE Braz. where there are some 100 M trees) – source of carnauba wax used in shoe-polish, coating chocolates, apples, pears & citrus, in lipstick & (form.) carbon paper & candles, gramophone records etc. (some other spp. less exploited), coll. by beating off wax particles from young lvs; seeds ed.

copey *Clusia major*

Copiapoa Britton & Rose. Cactaceae (III 5). 19 Chile coastal deserts. R: G. Charles (1998) *C.*

copihue *Lapageria rosea*

copper-burr *Sclerolaena* spp.

copperleaf *Acalypha amentacea* subsp. *wilkesiana*

copperweed *Iva acerosa*

coppice poles from stumps (stools) cut back on a regular cycle of 5–15 yrs, long practised from Eur. to Himal., esp. with *Castanea sativa* & *Corylus avellana* in W Eur.; now *Tectona* *grandis* treated thus in trop. Poles used for stakes, fencing, hurdles etc. & form. in wattle-&-daub houses

copra *Cocos nucifera*

Coprosma Forst. & Forst. f. Rubiaceae (IV 13). Excl. *Nertera*, c. 100 S China to Hawaii, S. Am., Trista da Cunha. Dioec. herbs, shrubs & trees with extrafl. nectaries & domatia. R: BBPBM 132(1935)1. 15 in NZ are divaricate shrubs (51 spp. of div. shrubs in 23 families in NZ), the most twiggy called 'mickeymick' (Maori mingimingi), ***C. propinqua*** A. Cunn. effectively disp. by geckos. Some cult. orn. (v. variable), some dyes & med., Aus. spp. with ed. fr.; ***C. foetidissima*** Forst. & Forst. f. (NZ) – a stinkwood; ***C. repens*** Hook.f. (NZ) – invasive in Aus.

Coptidipteris Nakai & Momose = *Dennstaedtia*

Coptidium Nyman (~ *Ranunculus*). Ranunculaceae (II 3). 2 N N temp.

Coptis Salisb. Ranunculaceae (III 3). 15 N temp. R: JJB 24(1949)73. Alks. Cult. orn. herbs with rhiz. prod. yellow dye & some medic. (tonic, febrifuge) esp. ***C. teeta*** Wall. (Himal.) – form. traded from Myanmar, now over-exploited & only a few populations left, & ***C. trifolia*** (L.) Salisb. (*C. groenlandica*, cankerberry, veg. gold or gold thread, NE As., Alaska)

Coptocheile Hoffsgg. = ? Gesneriaceae

Coptophyllum Korth. Rubiaceae (Cinch.). Incl. *Jainia*, *Pomazota*, 16 Ind., W Mal.

Coptosapelta Korth. Rubiaceae (?I 1). 13 SE China to Mal. Some med., esp. ***C. tomentosa*** (Blume) K. Heyne (Myanmar to Borneo) – vermifuge, febrifuge etc.

Coptosperma Hook.f. (~ *Tarenna*). Rubiaceae (II 2). 19 trop. Afr. (11; R: SGP 71(2001)374), Madag. Heterogeneous

coquilla palm *Attalea funifera*

coquino squash *Cucurbita moschata*

coquito palm *Jubaea chilensis*

coracan *Eleusine coracana*

coral beads *Cocculus carolinus*; **c. bells** *Heuchera sanguinea*; **c. berry** *Symphoricarpos orbiculatus*; **c. broom** *Carmichaelia crassicaulis*; **c. bush** or **plant** *Russelia equisetiformis*, *Templetonia retusa*; **c. creeper** *Antigonon leptopus*, *Kennedia* spp.; **c. fern** *Gleichenia* spp.; **c. flower** *Erythrina* spp.; **Indian c.** *Smilax* spp.; **c. pea** *Kennedia* spp.; **roble c.** *Terminalia amazonia*; **c. root** *Corallorhiza odontorhiza*, *Cardamine bulbifera*; **c. tree** *Erythrina* spp.; **c. vine** *Antigonon leptopus*; **c. wood** *Adenanthera pavonina*

corallita *Antigonon leptopus*; **white c.** *Porana paniculata*

Corallocarpus Welw. ex Hook.f. Cucurbitaceae (XIII). 13 trop. Afr., Madag., Ind.

Corallodiscus Batalin. Gesneriaceae (III 2b). 3–5 Himal. to NW China & SE As.

Corallorhiza Rupp. ex Gagnebin. Orchidaceae (V 13e). 11 N & C Am., with 1 circumboreal. R: HPB 10(1997)8. Chlorophyll-less, rootless mycotrophs involving several genera of fungi. Some cult. orn. esp. ***C. odontorhiza*** (Willd.) Poir. (coral r., dragon claw, E N Am. to Guatemala) – fls commonly cleistogamous. ***C. trifida*** Châtel. – endomycorrhiza also ectomycorrhiza of *Pinus contorta* ('linked liaison')

Corallorrhiza Gagnebin = praec.

Corallospartium Armstr. = *Carmichaelia*

Corbassona Aubrév. = *Pycnandra*

Corbichonia Scop. Lophiocarpaceae (Corbichoniaceae; Molluginaceae). 1–3 SW Afr., trop. Afr. to As. R: T 65(2016)790

Corbichoniaceae Thulin = Lophiocarpaceae (but see T 65(2016)790)

Corchoropsis Sieb. & Zucc. Malvaceae (Dom.; Sterculiaceae). 1 E As. & Jap.: ***C. tomentosa*** (Thunb.) Makino – ann. R: APS 32(1994)252

Corchorus Tourn. ex L. Malvaceae (Grew.-Apeib.; Tiliaceae s.l.). c. 70 trop. (Aus. 18 endemic). Jute (gunny) obtained from phloem fibres extracted by retting stalks, the fibre used in sacking, twine, carpeting & paper (3 M t prod. per yr), much from Bangladesh, for sacks, hessian (oft. mixed with or used instead of hemp), etc.; young shoots ed. like spinach. Main source: ***C. capsularis*** L. (China, widely cult.) – can be grown inundated; ***C. olitorius*** L. (melokhia, molokhia, tossa jute, Jew's mallow, Ind.) – grown more in uplands, medic. infusions used by Chinese in Hawaii, C resembling lvs leading to first 'foliar theory' of the flower by Pehr Forsskål seeing it in Egypt (1762)

cord grass *Sporobolus* spp. & hybrids

Cordanthera L.O. Williams = *Telipogon*

Cordeauxia Hemsl. Leguminosae (I 4). 1 NE Afr.: ***C. edulis*** Hemsl. (ye'eb or jeheb nut, Somalia, Ethiopia) – source of purplish dye, seeds ed. (taste somewhat like *Castanea sativa*)

Cordemoya Baill. = *Hancea*

Cordia L. Boraginaceae (Cord.; Cordiaceae, Ehretiaceae). Incl. *Auxemma, Patagonula, Saccellium*, excl. *Varronia* (pollen grains porate (cf. colporate in *C.*), c. 250 trop. R: JAA suppl. 1(1991)43. Champagnat's, Fagerlind's & Prévost's Models; homostyly, distyly, dioec.; style double-forked. Hollow ant-infested domatia; trunk hissing when slashed; lvs used for sandpaper; branches used as self-tindering firesaws in Maikal Hills, Ind.; cult. fast-growing timber (Spanish elm) in trop., dye (esp. ***C. americana*** (L.) Gottschling & J.S. Mill. (*Patgonula a.*, trop. Am.)) to stain *Cedrela* wood to resemble *Swietenia*, & orn.; some ed. fr. ***C. alliodora*** (Ruíz & Pavón) Oken (cyp, cypre, Ecuador laurel, salmwood, trop. Am.) – moth-poll., assoc. with *Azteca* ants in Costa Rica, plantation tree, fr. ed. Mex.; ***C. africana*** Lam. (*C. abyssinica*, mukumari, muringa, trop. Afr.) – timber; ***C. collococca*** L. (manjack, WI) – cherry-like fr., ed. & fed to fowls; ***C. curassavica*** (Jacq.) Roemer & Schultes ('*C. cylindrostachya*', trop. Am.) – hedge-pl. in Mal. but weedy though rapidly controlled by beetle (*Schematiza cordiae*), feeding exclusively on lvs, & wasp (*Eurytoma attiva*) laying eggs on young fr.; ***C. dichotoma*** Forst. f. (Mal. to New Caled.) – local medic., esp. C; ***C. dodecandra*** DC. (ziricote, C Am.) – fine dark timber, rough lvs used like sandpaper, cult. for ed. fr.; ***C. gerascanthus*** L. (prince wood, Spanish elm, C Am., Colombia, WI) – timber; ***C. goeldiana*** Huber (freijo, Braz.) – comm. timber; ***C. myxa*** L. (Sudan teak, sebesten plum, Ind. to Aus., natur. Afr.) – planted in Med., Calif. & trop. Am. for mucilaginous fr. used med. & as bird-lime, wood for cabinetwork; ***C. sebestena*** L. (geiger tree, trop. Am.) – cult. orn. with red fls., medic. fr.; ***C. subcordata*** Lam. (kou, Indopacific strand) – buoyant (ed.) fr. sea-disp., wood for bowls in Hawaii

Cordiaceae R. Br. ex Dumort. = Boraginaceae

Cordiera A. Rich. ex DC. (~ *Alibertia*). Rubiaceae (Ic.-Cord.). 10 trop. Am.

Cordiglottis J.J. Sm. = *Thrixspermum*

Cordisepalum Verdc. Convolvulaceae (2). 2 SE As. R: Blumea 51(2006)422

Cordobia Niedenzu. Malpighiaceae. 2 S Am.

corduroy *Gossypium* spp.; *Sarcopteryx stipata*

Cordyla Lour. Leguminosae (III 1). 7 trop. Afr., Madag. (2). C 0. ***C. pinnata*** (A. Rich.) Milne-Redh. (bush mango, W Afr.) – over-exploited Senegal etc. for ed. fleshy pods, wood & medic r. (J. Florence)

Cordylanthus Nutt. ex Benth. Orobanchaceae ('Cast.'; Scrophulariaceae s.l.). Excl. *Dicranostegia*, 13 W N Am. (s.s. 13). R: SBM 10(1988). Annuals. ***C. wrightii*** A. Gray – lvs used for bleaching by Hopi Indians

Cordyline Comm. ex R.Br. Asparagaceae (Dracaenaceae). c. 20 Australasia, Pac., trop. Am. (1). Like *Dracaena* but 2 or more ovules per locule & no sec. tissue in r. Cult. orn. 'cabbage-trees' (Chamberlain's & Leeuwenberg's Models), fragments of trunk will regrow. ***C. australis*** (Forst f.) Endl. (palm lily, NZ) – form. eaten by moas, sudden decline in wild 1987 ('phytoplasma' spread by passion-vine hopper (*Scolyopoa australis*) from Aus.), lvs a source of fibre, cult. orn. esp. in tubs & at seaside, source of high fructose syrup comparable with that used in food-processing; ***C. fruticosa*** (L.) A. Chev. (*C. terminalis*, happy plant, tanget (NG), ti, trop. E As. to Polynesia) – many cvs with coloured lvs, oft. used for hedging, ritual, medic. & magic, trad. dress in NG, wrapping food for cooking, crushed for soap for clothes in Amaz., tubers baked E Oceania

Cordyloblaste Henschel ex Moritzi (~ *Symplocos*). Symplocaceae. 2 E As. R: T 57(2008)841

Cordylocarpus Desf. Cruciferae (12). 1 Algeria, Morocco: ***C. muricatus*** Desf.

Cordylogyne E. Meyer. Apocynaceae (V c; Asclepiadaceae III 1). 1 S Afr.: ***C. globosa*** E. Meyer

Cordylostigma Groeninckx & Dessein (~ *Kohautia*). Rubiaceae (IV 1). 9 warm Am. R: T 59(2010)1466

Coreanaomecon Nakai = *Chelidonium*

Corema D. Don. Ericaceae (V 4; Empetraceae). 2: ***C. album*** (L.) D. Don (Azores & Iberia), ***C. conradii*** (Torrey) Torrey (NE N Am.)

Coreocarpus Benth. Compositae (Cor.). 7 SW N Am. esp. Mex. R: SB 14(1989)448

Coreopsis L. Compositae (Cor.). 86 Am. (US & Canada 28 – R: FNA 21(2006)185). R: FMNHB 11(1936)279; CN 27(1995)1. Afr. spp. = *Bidens*. Some med. locally. Cult. orn. esp. ***C. basalis*** (A. Dietr.) S.F. Blake (*C. drummondii*, S & SE US) & 'doubles' derived from ***C. auriculata*** L. (E US), ***C. grandiflora*** Hogg ex Sweet (E N Am.), ***C. lanceolata*** L. (E N Am.

natur. W N Am., S Am., Afr., China, Aus., NZ) & ***C. tinctoria*** Nutt. (N Am.); ***C. gigantea*** (Kellogg) H.M. Hall (Calif.) – pachycaul with stems to 12.5 cm thick & ***C. tinctoria*** Nutt. (*Calliopsis t.*, E N Am.) – form. used in a hot drink, orange dye from fls

Corethamnium R. King & H. Robinson (~ *Eupatorium*). Compositae (Eup.-Crit.). 1 Colombia: ***C. chocoense*** R. King & H. Robinson. R: MSBMBG 22(1987)354

Corethrodendron Fisch. ex Basiner (~ *Hedysarum*). Leguminosae (III 25). 4 C As. to E Siberia. R: T 72(2003)573

Corethrogyne DC. (~ *Lessingia*). Compositae (Ast.-Mac.). 2 W N Am. R: SBM 20(2000)34

coriander *Coriandrum sativum*; **Bolivian c.** *Porophyllum ruderale*; **Thai c.** *Eryngium foetidum*; **Vietnamese c.** *Persicaria odorata*

Coriandropsis H. Wolff = *Coriandrum*

Coriandrum Tourn. ex L. Umbelliferae (III 4). 2 SW As. ***C. sativum*** L. (coriander, cilantro, Chinese parsley, dhani) – long cult. for fr. ('seeds'; *coriandrum* one of the earliest words recog. in the deciphering of Linear B (Crete, C17–15 BC)), preserved in Tutankhamun's tomb 1325 BC, one of the 'bitter herbs' prescribed by Jews at Feast of the Passover, form. coated with sugar & sold as coriander comfits; form. added to weak beers, now most heavily used herb (Ind. & Morocco biggest producers) – to flavour gin, confectionery, bread, curry powder & in some scents & soap

Coriaria Nissole ex L. Coriariaceae. 17 Mex. to Chile, W Med. (Eur. 1), Himal. to Jap. & NG, NZ & S Pac. R: Rhodora 74(1972)242. Wind-poll. self-incompatible (S hemisph. everg. (planar foliage resembling fern-frond) with protogynous bisex. fls on new wood; N hemisph. decid., andromonoecious or monoecious with fls on old wood); fr. poisonous, hallucinogenic. Some cult. orn. & dyes. ***C. myrtifolia*** L. (W Med.) – mostly mammal-disp., lvs & bark for tanning, crushed fr. used as fly-poison; ***C. ruscifolia*** L. subsp. ***microphylla*** (Poir.) L. Skog (shansi, Peru) – narcotic hallucinogen (flying sensation)

Coriariaceae DC. Magnoliidae – Cucurbitales. 1/17 Euras., NZ, C & S Am. Trees to subshrubs (rarely perenn. herbs), oft. with nitrogen-fixing r. nodules (*Frankia*); stems quadrangular, with corky lenticels. Branches & lvs frond-like. Lvs small, opp. or whorled, simple with palmate venation; stip. minute, caducous. Fls ± reg., in racemes, bisex. or pls polygamous; K 5, quincunc., C 5 becoming ± fleshy in fr., A 5 + 5, filaments of antepetalous A adnate to keel of C, anthers with longit. slits, $\underline{G}$ 5(10), s.t. basally united, each with long slender style & 1 pend., anatropous, bitegmic ovule. Fr. a head of nuts ± encl. in fleshy C; seed rather compressed, embryo straight, oily; endosperm scant or 0. n = 10, 15

Genus: *Coriaria*

After A structure & chromosome evidence suggested Sapindales, though seed-coat anatomy & other features suggested affinity with Ranunculaceae, with wood structure suggesting affinities with both, though C. accumulate ellagic acid unlike Ranunculales, DNA evidence gives the current placing

Coridaceae J. Agardh = Primulaceae (Myrsinaceae)

Coridothymus Reichb.f. = *Thymbra*

Coriflora W. Weber = *Clematis*

Coris Tourn. ex L. Primulaceae (Coridaceae, Myrsinaceae). 1 (variable; recog. as 2 in Eur.) Med., Somalia

Corispermum B. Juss. ex L. Amaranthaceae (Chenopodiaceae I 6). 65 N temp. (Eur. 11, natur. Am.). Bugseed

Coristospermum Bertol. = *Ligusticum*

cork *Quercus suber*; **c. elm** *Ulmus thomasii*; **c. tree** *Ochroma pyramidale*, *Phellodendron* spp.; **c. wood** *Duboisia myoporoides* (Aus.), *Entelea arborescens* (NZ), *Hakea* spp. (Aus.), *Leitneria floridana* (N Am.), *Musanga cecropioides* (W Afr.), *Myrianthus arboreus* (Congo)

corkscrew grass *Stipa* spp.; **c. hazel** *Corylus avellana* 'Contorta'; **c. rush** *Juncus effusus* 'Spiralis'

Cormonema Reisseck ex Endl. = *Colubrina*

Cormus Spach = *Sorbus*

corn grain commonly used in any particular territory, e.g. wheat in GB, maize in US; **baby c.** *Zea mays*; **c.bind** *Convolvulus arvensis*; **blue c.** *Z. mays*; **c. cockle** *Agrostemma githago*; **c. flakes, c.flour, Indian c., c. on the cob, popc., sweet c.** *Zea mays*; **c. flower** *Centaurea cyanus*; **c. marigold** *Glebionis segetum*; **c. poppy** *Papaver rhoeas*; **c.salad** *Valerianella locusta*; **c. spurrey** *Spergula arvensis*; **squirrel c.** *Dicentra canadensis*

Cornaceae Bercht. & J. Presl. Magnoliidae – Cornales. Incl. Alangiaceae, excl. Aralidiaceae & Melanophyllaceae (= Torricelliaceae), Aucubaceae (= Garryaceae), Curtisiaceae,

Davidiaceae & Mastixiaceae & Nyssaceae (= N.), Griseliniaceae, Helwingiaceae, 2/80 N temp., rare in trop. & S temp. (not S Am.). Trees, shrubs (s.t. thorny – *Alangium*) or rarely rhiz. herbs, laticifers in *A.* Lvs opp., or in spirals or distich. (*A.*), simple, usu. entire; stip. 0. Fls usu. bisex., reg., epigynous, small, in cymes or cymose heads, these s.t. with consp. whorl of large petaloid bracts; K 4(5–10) small teeth or 0, s.t. forming a tube in male fls, C 4 or 5(–10) valvate, 0 in female fls, A as many as & alt. with C, usu. attached to or around edge of epigynous disk, anthers with longit. slits, $\overline{G}$ (2–4(–9)) with as many locules or 1-loc. & pseudomonomerous, style 1, term., or styles ± distinct, 1 ovule per locule, apical, pend., anatr. & bitegmic or rarely (*A.*) unitegmic. Fr. usu. a drupe with 1–5-loc. longit. grooved endocarp with 1 seed per locule; seeds with small elongate embryo embedded in copious oily endosperm. n = 9–11

Genera: *Alangium, Cornus*

As predicted from wood anatomy, many of form. incl. fams now found to be only distantly related, though several genera now in allied Nyssaceae, incl. *Diplopanax* until lately in Araliaceae. Cult. orn., timber & ed. fr. (esp. *Cornus*)

cornel *Cornus* spp.

Cornelian cherry *Cornus mas*

Cornera Furt. = *Calamus*

Corneria Bobrov & Melikian = *Dacrydium*

cornichon *Cucumis sativus* 'Gherkins Pariser'

Cornish elm *Ulmus minor* 'Stricta'; **C. moneywort** *Sibthorpia europaea*

Cornopteris Nakai. Athyriaceae (Woodsiaceae; Dryopteridaceae II 1). 9 trop. As. R: APG 30(1979)101

Cornucopiae L. Gramineae (XVI 15). 2 E Med. (Eur. 1) to Iraq. Fr. adhere to animals & alleged to burrow into soil (cf. *Heteropogon contortus*)

Cornuella Pierre = *Chrysophyllum*

Cornukaempferia Mood & Larsen (~ *Kaempferia*). Zingiberaceae (II 1). 2 Thailand – cult. orn.

Cornulaca Del. Amaranthaceae (Chenopodiaceae III 3). 6 Egypt to C As. ***C. monacantha*** Del. imp. camel fodder

Cornus Tourn. ex L. Cornaceae. c. 60 N temp. (Eur. 4), rare S Am., Afr. (s.s. 4 Euras., Calif.; here incl. *Afrocrania, Benthamidia, Bothrocaryum, Chamaepericlymenum, Dendrobenthamia* (*Cynoxylon*), *Discocrania, Swida* (*Thelycrania*)), 15 red-fruited clearcut spp., c. 45 blue- (or white-) fruited less so. R: BR 54(1988)233; P. Cappiello & D. Shadow (2005) *Dogwoods: the genus C.* Dogwoods (from 'dogs', i.e. skewers form. made from wood of C. spp.), cornels; trees (Fagerlind's & Leeuwenberg's Models), shrubs & rhiz. herbs (*Chamaepericlymenum*) with torn lvs hanging together with pulled-out xylem-thickenings ('magnetic lvs' cf. *Maytenus, Wimmeria*), some domatia, many local medic., much cult. for bract-surrounded infls resembling simple fls, winter-bark colour etc. ***C. alba*** L. (C & E As. – cult. for red branches, esp. **'Sibirica'** ('Westonbirt'); ***C. alternifolia*** L.f. (N Am.) – pagoda form, lvs in spirals; ***C. amomum*** Mill. subsp. ***obliqua*** (Raf.) J.S. Wilson (*C. obliqua*, E N Am.) – poss. polyphenol source; ***C. canadensis*** L. (*Chamaepericlymenum c.*, bunchberry, crackerberry, E As., N Am., Greenland) – tetraploid (2n = 44), rhiz., fl. opens in less than 0.5 msecs (fastest plant-movement known), catapulting pollen into air by 'elbow springs' on filaments (not found in *Cornus* s.s.), cult. orn. with conspic. bracts; ***C. controversa*** Hemsl. – (Himal., China, Jap.) – cult. orn. pagoda-shaped; ***C. florida*** L. (E N Am.) – cult. orn., molluscicidal steroid saponins, subsp. ***urbiniana*** (Rose) Rickett (Mex.) – bracts apically adhering; ***C. hongkongensis*** Hemsl. (China) – fr. sold as snacks in Hunan; ***C. kousa*** Miq. ex Hance (E As.) – cult. orn. with consp. bracts & ed. infrs; ***C. mas*** L. (Bohemian or Cornelian cherry, C & S Eur., SW As.; '*C. femina*' of the herbals is *C. sanguinea*) – cult. orn. & for fr. (more vitamin C than oranges) used in jam or rob (syrup) & basis of alcoholic Vin de Cornouille in France, fr. favoured by pigs (Odysseus's followers changed into ps by Circe after eating them); ***C. nuttallii*** Audubon (W N Am.) – timber for tools & cabinet work, cult. orn. with large bracts; ***C. officinalis*** Siebold & Zucc. (shan zhu yu, E As.) – fr. in TCM esp. for menstrual bleeding; ***C. sanguinea*** L. (dogwood, swamp d., pegwood, Eur.) – wood for skewers, bobbins etc., poss. 'whipple-tree' of Chaucer

Cornutia Plum. ex L. Labiatae (Premn.). 12 trop. Am. R: FR 40(1936)154. ***C. pyramidata*** L. (C Am., WI) – fr. used for blue ink (red with lime) & dyeing cloth

coro *Trichocline* spp.

Corokia Cunn. Argophyllaceae (Grossulariaceae s.l.). 6 NZ, Rapa Is. Cult. everg. shrubs (Attims's Model; incl. hybrids) with stomata on inner epidermis of inferior ovary esp. **C.**

cotoneaster Raoul (NZ) – leaf colour in autumn (& size) like fr. so perhaps 'fr. flags' to attract dispersers

Corollonema Schltr. = *Oxypetalum*

Coromandel wood *Diospyros quaesita*

Coronanthera Vieill. ex C.B. Clarke. Gesneriaceae (II 4b). 10 New Caled., 1 Solomon Is. R: Selbyana 6(1983)159. Trees (to 15 m) & shrubs

Coronaria Guett. = *Silene*

Coronidium Paul G. Wilson (~ *Helichrysum, Xerochrysum*). Compositae (Gnaph.). 17 E & SE Aus. R: Nuytsia 18(2008)300. Heterogeneous

Coronilla Tourn. ex L. Leguminosae (III 22). (Excl. *Securigera* – stems angled; *C. s.s.* stems round) 9 Atlantic is., Medit., Eur. R: Willd. 19(1989)59. Cult. orn. with strongly scented fls, esp. ***C. valentina*** L. (bastard senna, Medit.), also used for erosion control. See also *Hippocrepis*

Coronopus Zinn = *Lepidium*

Corothamnus (Koch) C. Presl = *Cytisus*

Coroya Pierre = ? *Dalbergia*

Corozo Jacq. ex Giseke = *Elaeis*

Corpuscularia Schwantes (~ *Delosperma*). Aizoaceae (V). 8 SE Afr. R: H. E. K. Hartmann, *Ill. Handb. Succ. Pls*, Aiz. A–E (2002)175

Correa Andrews. Rutaceae (I 3). 11 temp. Aus. R: FA 26(2013)337. C(4). Cult. orn. (Aus. fuchsia); all spp. hybridize. Lvs of ***C. alba*** Andrews have been used as tea (Cape Barren tea)

Correllia A.M. Powell = *Perityle*

Correlliana D'Arcy = *Cybianthus*

Correorchis Szlach. = *Chloraea* (but see ASBP 77(2008)115)

Corrigiola L. Caryophyllaceae (I 3). 11 cosmop. (Eur. 1–2). R: MBMHRU 285(1968)34. ***C. litoralis*** L. (strapwort, subcosmop.) – r. used in scent & medic.

Corryocactus Britton & Rose. Cactaceae (III 1). 12 Bolivia, S Peru, N Chile. Columnar, cult. orn.

Corsia Becc. Corsiaceae. c. 25 Papuasia & Aus. (1). Heterogeneous?

Corsiaceae Becc. (~ Burmanniaceae). Magnoliidae – Liliales. 3/28 S China, Papuasia, NE Aus., Chile (*Arachnitis*). Chlorophyll-less mycotrophs with rhiz. or tubers. Lvs scale-like, in spirals or 2-ranked. Fls solit., term., irreg., bisex. or not, P petaloid, tubular basally, 3 + 3, posterior member of outer large & coloured encl. other 5 linr.-spathulate, A 6 on tube, filaments short, anthers with longit. slits, pollen monosulcate, $\overline{G}$ (3), 1-loc, with ± intruded 2-lobed placentas, style with 3 thick stigmas & ∞ tiny ovules. Fr. a capsule with 3 valves; seeds small, winged, with undifferentiated embryo & scanty endosperm. x = 9
Genera: *Arachnitis* (unisexual fls, poss. to be moved to Burmanniaceae), *Corsia* (bisex.), ?*Corsiopsis*

Corsican mint *Mentha requienii*; **C. pine** *Pinus nigra* subsp. *salzmannii*

Corsiopsis D.X. Zhang & al. ?Corsiaceae. 1 China: ***C. chinensis*** D.X. Zhang & al. R: SB 24(1999)313. Collected once

Corstorphine plane *Acer pseudoplatanus* 'Corstorphinensis'

Cortaderia Stapf. Gramineae (Danth.). Incl. *Lamprothyrsus*, excl. *Austroderia*, 20 S Am. Gynodioec. clump-forming coarse grasses, *C. selloana* is subdioec. & ***C. bifida*** Pilger apomictic. Cult. orn. (pampas grass) esp. ***C. selloana*** (Schultes & Schultes f.) Asch. & Graebner (*Gynerium argenteum*, Braz., Arg., Chile) as specimen lawn pl., also comm. for dried infls (female larger) & in S Am. for paper, noxious weed (each plume with 7 M frs) in S Eur., S Afr., SW N Am., Aus. & NZ. ***C. jubata*** (Carr.) Stapf (W trop. S Am.) – introd. S Afr. to control erosion on mine-dumps – invasive there & Calif.

Cortesia Cav. (~ *Ehretia*). Boraginaceae (Ehretiaceae). 1 Arg.: ***C. cuneifolia*** Cav.

Cortia DC. Umbelliferae (III 8). 9 C & S As.

Cortiella Norman. Umbelliferae (III 8). 3 C & S As.

Cortusa Boerh. ex L. = *Primula*

Corunastylis Fitzg. = *Genoplesium*

Coryanthes Hook. Orchidaceae (V 12i). c. 50 trop. Am. R: TSP 83(1993). Epiphytes with extrafl. nectaries; massive pendent waxy fls with part of lip forming bucket into which drops of water are secreted from knobs on column; male bees of genus *Eulaema* attracted by strong scent (app. specific to each sp. preventing hybridization) scratch on area of tissue at the base of the lip to collect the liquid scent, which intoxicates their front tarsi such that the bees lose hold & fall into the bucket; they cannot scale the sides but must leave through a tunnel in which they deposit or collect the pollinia; a bee which took 45 mins to escape could not be induced to return as the scent was not replenished until next morning, suggesting that this would prevent self-fert. Some cult. orn.

Corybas Salisb. Orchidaceae (IV 3a). c. 130 Indomal. to NZ (Aus. 20, 19 endemic), Polynesia & Jap. with ***C. macranthus*** (Hook.f.) Reichb.f. (Macquarie Is., subAntarctic) most S distrib. of any orchid. R: PM 16 (1983) (74 E of Wallace's Line), KB 41(1986)575 (27 As. & W Mal.). Predicted to be poll. by fungus-gnats; some self-poll. Some cult. orn. (helmet orchids)

Corycium Sw. Orchidaceae (IV 4b). 15 S Afr. R: Fl. Cap. 5,3(1913)281. *Pterygodium* prob. referable here. Elaiophores attractive to bees

Corydalis DC. Papaveraceae (Fumariaceae I 1). Excl. *Capnoides, Ceratocapnos, Pseudofumaria,* q.v., c. 400 N temp. (Eur. 11; Sino-Himal. c. 280), trop. Afr. mts (1). R: AHB 17(1955)115, OB 88(1986)21; M. Lidén & H. Zetterlund (1997), *C.: a gardener's guide.* Herbs with tubers, 1 cotyledon & usu. pinnate (simple in ***C. ludlowii*** Stearn, Tibet) lvs; tuber in ***C. cava*** (L.) Schweigger & Koerte (C Eur.) the main axis with depression on underside, each ann. shoot arising from axil of scale leaf; that of ***C. solida*** (L.) Clairv. (*C. bulbosa*, Euras.) is a swollen current ann. shoot. Fls transverse zygomorphic twisting through 90° to become vertical, 1 petal spurred & cont. nectar from staminal nectary; inner petals united at tip enclose stigma & anthers, which are caused to emerge by bees alighting on & pushing down inner C (cf. Leguminosae). Many alks; ***C. yanhusuo*** (Y.H. Chou & C.C. Xu) Z.Y. Su & C.Y. Wu (S China) – TCM, a primary painkiller, Parkinson's disease. Some ed. tubers in As. Many cult. orn. esp. the freq. confused *C. solida* & *C. cava* (used in comm. sedative tea mixtures) with pink to purple fls

Corylaceae Mirb. = Betulaceae

Corylopsis Sieb. & Zucc. Hamamelidaceae (I 2). 7 Bhutan to Jap. R: JAA 58(1977)382. Fls before lvs in spring, usu. in pendent spikes or racemes; cult. orn. esp. ***C. pauciflora*** Sieb. & Zucc. & ***C. spicata*** Sieb. & Zucc. (Jap.)

Corylus Tourn. ex L. Betulaceae (II; Corylaceae). 17 N temp. (Eur. 3). R: JAA 71(1990)61. Hazels [orig. *haesel* = cap; generic name from Gk for cap, i.e. 'leafy' cup around fr.] or filberts; monoec. trees & shrubs (heterodichogamy recorded) with edible seeds in nuts; domatia, local medic. (source of taxol). Anemophilous, the male fls in pendent catkins, the females in small red bud-like infls; poll. to fert. takes up to 4 months. Cult. ed. & orn. esp.: ***C. americana*** Walter (American hazel or filbert, E N Am.) – leafy involucre around nuts v. long; ***C. avellana*** L. (hazel, Eur., Cauc., Turkey) – seriously threatened in GB by intr. (Am.) grey squirrels, form. much grown for coppice (q.v.), esp. for hurdles, firewood, legume-poles, wattle-&-daub, in GB 500 000 acres in 1905 reduced to c. 94 000 in 1965, diff. cvs of value for seeds esp. 'Barcelona' nuts selected from cobnuts (though '**Kentish Cob**' ('Lambert's Filbert') much grown actually a filbert (i.e. poss. 'full beard' = long-husked nut, or after St Philibert's Day, 22 Aug., Old Style, when ripe), ***C. maxima*** Mill., SE Eur., poss. a cv. of *C. avellana*), the seeds used for oil for cooking, form. paint, soap etc., & for confectionery, esp. nut-chocolate, wood form. principal source of charcoal for gunpowder, can cause urticaria, **'Contorta'** (corkscrew hazel, orig. plant (1863, ? somatic mutant) from Frocester, England) with twisted stems; hazel imp. in Eur. mythology, St Patrick supposed to have purged Ireland of poisonous snakes by brandishing a hazel wand & since time of Pliny favoured as a water-divining rod & in rural Eng. used as subs for 'palm' on Palm Sunday; ***C. colurna*** L. (Turkish hazel, SE Eur. to N Iran) – cult. for nuts, wood form. used for spinning-wheels; ***C. × colurnoides*** C. Schneider (trazel, *C. avellana × C. colurna*) – non-suckering; ***C. heterophylla*** Fisch. ex Trautv. (E As.) & ***C. sieboldiana*** Blume (temp. As.) – seeds sold in Chinese markets

Corymbia K. Hill & L. Johnson (~ *Angophora, Eucalyptus*). Myrtaceae (II 11.). 113 Aus. (esp. N) & S NG. R: Telopea 6(1995)185. Heterogeneous? Bloodwoods & ghost gums recog. by compound infls usu. ± urceolate fr. & usu. ± tessellate bark. Timbers & cult. orn. ***C. calophylla*** (Lindl.) K. Hill & L. Johnson (*E. c.*, marri, SW Aus.) – v. large fr., ectomycorrhizae spread by red kangaroos (spores germ. after digestion); ***C. citriodora*** (Hook.) K. Hill & L. Johnson (*E. c.*, lemon-scented gum, Queensland) – source of a lemon-scented oil, structural timber, fuel crop; ***C. ficifolia*** (F. Muell.) K. Hill & L. Johnson (*E. f.*, SSW Aus) – widely cult. for showy panicles of red fls, '**Summer Red**' (*C. f. × C. ptychocarpa* (F. Muell.) K. Hill. & L. Johnson, N Aus.) – pot-pl. & cut-fl.; ***C. gummifera*** (Gaertn.) K. Hill & L. Johnson (e.g., bloodwood, E. Aus.) – sap red, one of first Aus. pls cult. GB (1771); ***C. maculata*** (Hook.) K. Hill & L. Johnson (*E. m.*, spotted gum, E Aus.) – to 68 m in NSW, grown in plantation for plywood, furniture & constr.; ***C. opaca*** (D. Carr & S. Carr.) L.

Johnson (N Aus.) – larvae from insect galls ed., kino for salves, capsules for Aborigines' decoration; ***C. papuana*** (F. Muell.) K. Hill & L. Johnson (*E. p.*, S NG) & allied spp. in Aus. = ghost gum – bark white

Corymbium Gronov. Compositae (Corymbieae). 9 SW Cape. R: MIABH 23b(1990)631. Lvs narrow with parallel veins

Corymborkis Thouars. Orchidaceae (V 3). 6 Indomal. to Pac. R: BT 71(1977)161. ***C. veratrifolia*** (Reinw.) Blume (Indomal.) – cult. as emetic, one of few pls on Christmas Is. avoided by red land crabs

Corymbostachys Lindau = *Justicia*

Corynabutilon (K. Schum.) Kearney (~ *Abutilon*). Malvaceae (Malv.-Malv.). 6 temp. Arg., Chile. R: Novon 11(2001)193

Corynaea Hook.f. Balanophoraceae. 1 trop. Am. montane forests: ***C. crassa*** Hook.f. – vestigial r. on tuber. R: FN 23(1980)41

Corynandra Schrad. ex Spreng. = *Cleome*

Corynanthe Welw. Rubiaceae (I 1). Incl. *Pausinystalia*, 6 trop. Afr. R: SB 39(2014)311. ***C. johimbe*** K. Schum. (*P. j.*) – aphrodisiac, etc. (alks), yohimbine shown to increase sexual motivation in rats & use in western 'Anabolic nutrient', alleged to increase testosterone levels

Corynanthera J. Green (~ *Micromyrtus*). Myrtaceae (II 15). 1 W Aus.: ***C. flava*** J. Green. R: Nuytsia 2(1979)368

Corynella DC. = *Poitea*

Corynemyrtus (Kiaerskov) Mattos = *Psidium*

Corynephorus P. Beauv. Gramineae (XVI 7). 5 Eur. (4), Medit. to Iran. On dunes & other sands. Hair grass; basal awns with twisted column & clavate limb & ring of hairs at junction

Corynephyllum Rose = *Sedum*

Corynocarpaceae Engl. Magnoliidae – Cucurbitales. 1/6 SW Pac. R: BJLS 95(1987)9. Everg. trees with bitter glucosides in bark & seeds. Branches often in pseudowhorls. Lvs in spirals, simple, entire, leathery, shiny; stip. intrapetiolar, decid. Fls in usu. term. umbellate thyrses, bisex. (or pls gynodioec.), reg.; K 5 quincuncial, C 5 imbr., A 5 opp. & basally adnate to C all atop a short hypanthium, alt. with 5 nectaries (staminodes) each bearing a petaloid scale (? C; ? staminode connective), anthers with longit. slits, 5 nectaries opp. A outside G, $\underline{G}$ 1 (rarely with second style) with 1 pend., anatropous, bitegmic ovule. Fr. a drupe; seeds with straight oily & starchy embryo, v. poisonous, endosperm 0. 2n = 44, 46

Genus: *Corynocarpus*

Affinities long disputed (Goldberg suggested Rosales, wood anatomy suggested alignment with Berberidaceae, while bitter principle (karakin) said to be identical with hiptagin of *Hiptage* (Malpighiaceae), pollen like *Itea* (Iteaceae), but DNA evidence gives present disposition

Corynocarpus Forst. & Forst. f. Corynocarpaceae. 6 NG, NE Aus. (2), New Caled., Vanuatu, NZ (N Is.; introd. S Is. by Maoris: ***C. laevigatus*** Forst. & Forst. f. (karaka nut, NZ etc.) – gynodioec., seeds ed. after roasting, a staple food of Maoris, cult. orn., fleshy part of fr. ed. raw, trunks used for canoes; natur. Hawaii (after seeding Kauai by air in 1929); fr. reptile-disp.)

Corynopuntia Knuth = *Grusonia*

Corynostylis Mart. = *Calyptrion*

Corynotheca F. Muell. ex Benth. Asparagaceae (Anthericaceae). 6 Aus. R: FA 45(1987)299

Corynula Hook. f. = *Leptostigma*

Corypha L. Palmae (III 7). 6 trop. As. to Aus. Hapaxanthic (Holttum's Model); palmate lvs. Cult. esp. ***C. utan*** Lam. (*C. elata*, gebang) – infl. with 3–15 M functional fls & c. 250 000 fr. (biomass some 22% of a 44 yr-old tree recorded in cult.), source of toddy, sugar, arrack, vinegar, starch, seeds ed. & used for buttons & rosaries, raffia, petiole-fibre used in hats, ropes, coarse ('kajang') mats, 'Bangkok' hats (exported from Philippines!), etc., midribs used in furniture-making in Philippines; ***C. umbraculifera*** L. (talipot, cultigen (?) poss. derived from *C. utan*) – lvs (to 8 m across, petiole to 5 m long) used for thatching, umbrellas etc. & also for sacred Buddhist books (olas) written on strips (ready-ruled!) with metal stylus, infl. (after 20–30 yrs; up to 79 yrs in Singapore Bot. Gardens) to 8 m tall (with 60 M fls), the largest of any plant, followed 8 months later by fr. & further 4 months by death

Coryphantha (Engelm.) Lem. (~ *Mammillaria*). Cactaceae (III 9). Incl. *Escobrittonia*, 48 SW N Am. Cult. orn.

Coryphomia N. Rojas = *Copernicia*

Coryphopteris Holttum = *Thelypteris*

Coryphothamnus Steyerm. Rubiaceae (IV 7). 1 SE Venez.: ***C. auyantepuiensis*** (Steyerm.) Steyerm.

Corysanthes R. Br. = *Corybas*

Corythea S. Watson = *Acalypha*

Corytholoma (Benth.) Decne = *Sinningia*

Corythophora Knuth. Lecythidaceae (I). 4 Braz.

Corytoplectus Oersted (~ *Alloplectus*). Gesneriaceae (II 5c). 11 Mex., NW S Am. R: Selbyana 29(2008)96

Coscinium Colebr. Menispermaceae (IV). 2 Indomal., SE As. ***C. fenestratum*** (Gaertn.) Colebr. (Ind., Sri Lanka) – seeds disp. bats & polecats; wood source of a turmeric-like dye, medic. (false calumba r., subs. for *Jateorhiza palmata*; tablets sold for dysentery in Vietnam; imp. in Ayurvedic med. but over-exploited), rope for elephant-logging

Cosentinia Tod. (~ *Cheilanthes*). Pteridaceae (III). 1 W Medit.: ***C. vellea*** (Aiton) Tod.

Cosmantha Y. Itô = *Echinopsis* (*Soehrensia*)

Cosmea Willd. = *Cosmos*

Cosmelia R. Br. Ericaceae (VII 3; Epacridaceae). 1 SW Aus.: ***C. rubra*** R. Br.

Cosmianthemum Bremek. Acanthaceae (III 2c). 8 W Borneo

Cosmibuena Ruíz & Pavón. Rubiaceae (I 1). 4 trop. Am. R: AMBG 79 (1992)886. *Hillia* poss. referable here. Usu. epiphytes

Cosmocalyx Standl. Rubiaceae (IV 5). 1 Mex.: ***C. spectabilis*** Standl. R: Britt. 50(1998)312

Cosmos Cav. (~ *Bidens*). Compositae (Cor.). c. 36 trop. & warm Am. esp. Mex. (***C. caudatus*** Kunth commonly natur. OW). R: BS 92(2014)381. Cult. orn. ('Mex. aster') esp. ***C. bipinnatus*** Cav. ('cosmea' of gardens, SW N Am.), natur. in As., Madag., C Am., WI, & ***C. atrosanguineus*** (Hook.) Voss (Mex.) – fls colour & smell of chocolate, extinct in wild (not seen since 1860s) in cult. only 1 sterile clone from which fert. **'Pinot Noir'** selected

Cosmostigma Wight. Apocynaceae (Va; Asclepiadaceae III 4). 3 Hainan, Indomal.

Cossinia Comm. ex Lam. (*Cossignia*, *Cossignya*). Sapindaceae (III 1). 3 Mauritius (2), New Caled. (1). R: Austrobaileya 1(1982)485

Costaceae Nakai (~ Zingiberaceae). Magnoliidae – Zingiberales. 7/110 trop. R: SB 31 (2006)89, T 55(2006)154. Perenn. non-aromatic herbs with rhiz. Stems terete, usu. unbranched (when branched, branches breaking through sheaths – *Tapeinochilos*). Lvs in spirals, sheaths closed; ligule present; pulvinus 0; lamina rolled from 1 side to other. Infl. a strobiloid spike term. (s.t. on short leafless shoots) or fls solit., axillary (*Monocostus*); fls (1 or pr) subtended by imbr. bract. Fls bisex., zygomorphic, epigynous; (K) with 2 or 3 lips, imbr., (C) 3-lobed, imbr.; A 1 oft. petaloid. Labellum petaloid, staminodal opp. A, ± 3-lobed, as long as or longer than C, base with A-base forming papillate tube. $\overline{G}$ (2)3, placentation axile; ovules ∞, anatropous. Fr. a (2) 3 – loc. capsule with persistent K, usu. loculicidal. Seeds ∞, angular or ellipsoid, with white to yellow aril; embryo straight; endosperm little, perisperm copious with starch in simple grains. n = 9, 14

Genera: *Chamaecostus, Costus, Dimerocostus, Hellenia, Monocostus, Paracostus, Tapeinochilos*

Form. subfam. Costoideae of Zingiberaceae, differing from Z. s.s. in not being aromatic etc.

Some medic., cult. orn.

Costaea A. Rich. = *Purdiaea*

Costantina Bullock = *Lygisma*

Costarica L. Gómez = *Sicyos*

Costaricia Christ = *Dennstaedtia*

Costera J.J. Sm. Ericaceae (VIII 5). 9 W Mal.

costmary *Tanacetum balsamita*

Costularia C.B. Clarke. Cyperaceae (II 7). 20 S Afr., Ind. Ocean, Mal. & New Caled. (where some woody)

Costus L. Costaceae (Zingiberaceae s.l.). Excl. *Chamaecostus, Hellenia, Paracostus*, c. 80 trop. to Aus. (c. 5). R (NW): FN 8(1972)27, 18(1977)168. Staminodal labellum large, K & P rather small; floral mechanism like *Iris*; seeds arillate. Cult. orn., some local med.; ***C. spectabilis*** (Fenzl) K. Schum. (E Afr.) – rosette herb; ***C. woodsonii*** Maas (C Am.) – presence of ants attracted by extrafl. nectaries increases seed set 3-fold

costus root *Saussurea costus*

Cota Gay (~ *Anthemis*). Compositae (Anth.-Anth.). 40 Eur., Medit. ***C. tinctoria*** (L.) Gay (*A. t.*, dyer's, golden or yellow chamomile, C & S Eur., natur. GB, N Am.) – source of yellow dye

cotignac *Mespilus germanica*

Cotinus Mill. (~ *Rhus*). Anacardiaceae (IV). 1 SE US, 1 S Eur. to China, 1 SW China. Cult. orn. esp. ***C. coggygria*** Scop. (*Rhus cotinus*, smoke tree or bush, wig tree, Hungarian, Ind., Turkish, Tyrolean or Venetian sumac, S Eur. to China) – lvs used for tanning, sterile parts of infl. elongate & become hairy, the whole infl. falling & may be blown about, wood gives a dye ('young fustic', yellow), cvs incl. commonly cult. purple-lvd **'Purpureus'**; ***C. obovatus*** Raf. (*C. americanus*, SE US) – chittamwood prod. orange dye

coto bark *Aniba coto*; **false c. b.** *Ocotea pseudocoto*

Cotoneaster Medik. Rosaceae (Ros.-Mal.). c. 80 temp. OW esp. Himal., W China (China 65, c. 20 esp. Chinese natur. Eur., 34 N Am.), or 400+ incl. many apomictic aggregates incl. tetraploids. R: CJB 68 (1990)2211; J. Fryer & B. Hylmö (2009) *Cs*. Troll's Model. ***C. cambricus*** Fryer & Hylmö ('*C. integerrimus*') – apomict once reduced to 4 pls at Llandudno, N Wales (6 in 1978, 33 in 1995), likely introd. Many cult. orn. (but smelly due to trimethylamine (cf. *Ligustrum*)) everg. & decid. with brightly coloured fr. (many retained through winter app. unattractive to birds in N temp.), esp. ***C. horizontalis*** Decne (W China) – apomictic with herring-bone sprays of foliage; most cult. spp. from China or Himal., e.g. ***C. bacillaris*** Wall. ex Lindl. (Himal.) – walking-sticks, ***C. conspicuus*** (Messel) Messel (SE Tibet) – fr. untouched by birds in wild until April, ***C. racemiflorus*** (Desf.) Schldl. (Med. to China) – apomictic, source of a sweet manna-like subs. high in dextrose used in Iran & Ind. Several invasive bird-disp. weeds e.g. ***C. divaricatus*** Rehder & E. Wilson, ***C. glaucophyllus*** Franch., ***C. pannosus*** Franch. (all temp. As.) in Aus., *C. p.* also in W US, & ***C. simonsii*** Bak. (trop. As.) in NZ

Cotopaxia Mathias & Constance. Umbelliferae (III 8). 2 Colombia, Ecuador. R: Caldasia 14(1984)21

Cottea Kunth. Gramineae (XXVII 1). 1 S US to C Mex., Ecuador to Arg.: ***C. pappophorides*** Kunth

Cottendorfia Schultes f. Bromeliaceae (1). 1 NE Braz.: ***C. florida*** Schultes f. R: FN 14(1974)212 (spp. 1–24 = *Lindmania*)

cotton *Gossypium* spp. & hybrids; **c. candy** (US) spun sugar; **devil's c.** *Abroma augustum*; **French c.** *Calotropis procera*; **c. grass** *Eriophorum* spp., *Imperata cylindrica*; **c. gum** *Nyssa aquatica*; **lavender c.** *Santolina chamaecyparissus*; **sea island c.** *Gossypium barbadense*; **c. sedge** *Eriophorum angustifolium*; **silk c.** *Ceiba pentandra*; **c. thistle** *Onopordum* spp.; **c.weed** *Achillea maritima*; **c. wood** *Populus* spp. esp. *P. deltoides*, *P. trichocarpa*

Cottonia Wight. Orchidaceae (V 16c). 1 S Ind., Sri Lanka: ***C. macrostachya*** Wight

Cotula Vaill. ex L. Compositae (Anth.-Cot.). Excl. *Leptinella*, 55 S hemisph. (esp. S Afr.) to N Afr. & Mex. some natur. N (e.g. Eur. 2 incl. ***C. coronopifolia*** L., S Afr., also invasive in Aus.). R: BBMNHB 23(1993)157. ***C. myriophylloides*** Harv. (Cape) – obligate aquatic; some cult. orn. carpeting pls

Cotylanthera Blume = *Exacum*

Cotyledon Dill. ex L. Crassulaceae (II). 11 S & E Afr. to Arabia (1). Succ. shrubs. Some cult. orn. (many pls described as C. belong in other genera like *Adromischus*, *Dudleya*, *Echeveria*, *Tylecodon* & *Umbilicus*) esp. ***C. orbiculata*** L. (Cape & SW Afr.) – fresh leaf-juice allegedly beneficial in treatment of epilepsy but it & other spp. toxic

Cotylelobium Pierre. Dipterocarpaceae (Dipt.-Dipt.). 5 Sri Lanka, W Mal.

Cotylodiscus Radlk. = *Plagioscyphus*

Cotylolabium Garay (~ *Stenorrhynchos*). Orchidaceae (1V 2h). 1 E Braz.: ***C. lutzii*** (Pabst) Garay. R: Gen. Orch. 3(2003)182

Cotylonychia Stapf = *Pentadiplandra*

couch (grass) *Elymus repens*; (Aus.) *Cynodon dactylon*; **blue c.** *Digitaria didactyla*; **onion c.** *Arrhenatherum elatius* var. *bulbosum*; **c. potato** *Homo sapiens* (sedentary form)

Couepia Aubl. Chrysobalanaceae (3). Excl. *Gaulletia*, 58 trop. Am. exc. WI. R: FOW 10(2003)5. Troll's Model. Some rheophytes; hawkmoth-poll. (cf. allied *Hirtella*) with white fls open at night & copious nectar; some distrib. agoutis. Some ed. fr. See also *Acioa*

cough root *Lomatium dissectum*

Coula Baill. Erythropalaceae (Coulaceae; Olacaceae s.l.). 1 trop. W. Afr.: ***C. edulis*** Baill. (coula) – Roux's Model, timber a comm. mahogany subs., seeds (Gaboon nuts) ed. fresh, cooked or fermented

Coulaceae Tiegh. = Erythropalaceae

Coulterella Vasey & Rose. Compositae (Tag.-Var.). 1 Baja Calif.: ***C. capitata*** Vasey & Rose. R: U. Eggli, *Ill. Handb. Succ. Pls*, Dicots (2002)20

Coulteria Kunth (~ *Caesalpinia*). Leguminosae (I 4). 7 Mex. to trop. Am. R: PhytoKeys 71(2016)51. Dioec. (?pseudocopulation). ***C. cubensis*** (Greenm.) S. Sotuyo & G. Lewis (*Caesalpinia violacea*, C Am., WI) – cult. orn.

Coulterophytum Robinson. Umbelliferae (III 9). 5 Mex.

Couma Aubl. Apocynaceae (I c). 6 NE S Am. Source of couma rubber (Braz.), that of ***C. macrocarpa*** Barb. Rodr. (sorva, leche caspi, lechi-caspi) a chewing-gum base; ***C. guianensis*** Aubl. (Rauh's Model) – ed. fr.

Coumarouna Aubl. = *Dipteryx*

counter wood *Milicia excelsa*

Couratari Aubl. Lecythidaceae (I). 20 trop. S Am. Hood (ligule – see fam.) coiled under itself, concealing nectar which can be reached only by long-tongued female euglossine bees; wind-disp. winged seeds. Barkcloth

courbaril *Hymenaea courbaril*

Courboria Brongn. = *Maerua*

courgette *Cucurbita pepo*

Couroupita Aubl. Lecythidaceae (I). 3 trop. Am. 1-sided ligule (see fam.) overlying A (with sterile pollen) & stigmas, *Xylocopa* bees entering for pollen & their backs rubbing stigmas & fertile A. ***C. guianensis*** Aubl. (cannonball tree, Guianas) – Rauh's Model, infls on branches & trunk, fleshy, fruity, followed by woody caps. to 20 cm diam. with evil-smelling pulp, timber good

Coursetia DC. Leguminosae (III 23). 38 warm Am. R: SBM 21(1988)

Coursiana Homolle = *Payera*

Courtoisia Nees = *Courtoisina*

Courtoisina Soják = *Cyperus*

cous root *Lomatium ambiguum*

couscous *Triticum turgidum* Durum Group (s.t. mixed with *Hordeum vulgare*)

cousin mahoe *Hibiscus americanus*

Cousinia Cass. (~ *Arctium*). Compositae (Card.-Card.). c. 650 E Medit. (Eur. 1) to C As. & W Himal. Some spp. moved to *Arctium*

Cousiniopsis Nevski. Compositae (Card.-Card.). 1 C As.: ***C. atractyloides*** (Winkler) Nevski

Coussapoa Aubl. Urticaceae (Cecropiaceae). 50 trop. S Am. R: FN 51(1990)16. Some stranglers (cf. *Ficus*). ***C. asperifolia*** Trécul (NE S Am.) – fr. heads secrete wax coll. (with diaspores) by mason bees & taken to their nests

Coussarea Aubl. Rubiaceae (IV 11). c. 110 trop. Am.

Coutaportla Urb. (~ *Portlandia*). Rubiaceae (inc. sed.). 2 Mex.

Coutarea Aubl. Rubiaceae (I 1). 7 Mex. to Arg. Bark locally medic. esp. for malaria, notably ***C. hexandra*** (Jacq.) K. Schum.

Couthovia A. Gray = *Neuburgia*

Coutinia Vell. Conc. = *Aspidosperma*

Coutoubea Aubl. Gentianaceae. 5 trop S Am., WI. R: MNYBG 51(1989)19. Some medic.

Coveniella Tindale. Dryopteridaceae. 1 NE Queensland: ***C. poecilophlebia*** (Hook.) Tindale. R: GBS 39(1987)169

Covillea Vail = *Larrea*

cow bane *Cicuta virosa*; **c.berry** *Vaccinium vitis-idaea*; **c. cockle** *Vaccaria hispanica*; **c. itch** *Mucuna pruriens*; **c. parsley** *Anthriscus sylvestris*; **c. parsnip** *Heracleum sphondylium*; **c. pea** *Vigna unguiculata*; **c.slip** *Primula veris*; **c.slip, Amer.** *Primula* (*Dodecatheon*) spp.; **c.slip, Cape** *Lachenalia* spp.; **c.slip, Virginian** *Mertensia virginica*; **c. tree** *Brosimum utile*; **c. wheat** *Melampyrum* spp.

Cowania D. Don ex Tilloch & Taylor = *Purshia*

Cowiea Wernham (~ *Hypobathrum*). Rubiaceae (II 5). 2 C Mal.

Coxella Cheeseman & Hemsl. = *Aciphylla*

coyoli (palm) *Acrocomia aculeata*

crab apple *Malus sylvestris*, but also applied to any natur. seedling apples as well as some cult. for fls rather than fr.; **c.eye bean** *Phaseolus vulgaris*; **c.'s eyes** *Abrus precatorius*; **c. grass** *Digitaria*, *Eleusine* & *Panicum* spp.; **c. oil and wood** *Carapa guianensis*

Crabbea Harv. Acanthaceae (III 2d). 16 trop. & S Afr. R: NJB 24(2007)502

Cracca Benth. = *Coursetia*

crackerberry *Cornus canadensis*
Cracosna Gagnepain. Gentianaceae. 3 SE As. R: Blumea 48(2003)19
Craibella R. Saunders & al. = *Pseuduvaria* (but see SB 29(2004)42)
Craibia Harms & Dunn. Leguminosae (III 16). 10 trop. Afr. Some good timber
Craibiodendron W.W. Sm. Ericaceae (VIII 2). 5 SE As. R: JAA 67(1986)441
Craigia W.W. Sm. & W.E. Evans. Malvaceae (Til.; Tiliaceae). 2 China, N Vietnam. On limestone
Crambe Tourn. ex L. Cruciferae (12). 35 Euras. (Eur. 8), Medit., Macaronesia, trop. Afr. mts. R: Pfl. 79(4, 105)(1919) 228. Cult.: ***C. cordifolia*** Steven (Cauc.) – lvs to 60 cm across, infl. to 2 m; ***C. hispanica*** L. (Med. to trop. Afr.) – comm. oilseed, subsp. ***abyssinica*** (R.E. Fr.) Prina (*C. abyssinica*, trop. Afr.) – source of industrial oil; ***C. maritima*** L. (sea kale, coasts W Eur. to SW As.) – grown for succ. blanched shoots, boiled & cooked, comm. Kent & Lincolnshire (only comm. native veg.)
Crambella Maire. Cruciferae (12). 1 Morocco: ***C. teretifolia*** (Batt.) Maire
cramp bark *Viburnum opulus*
cranberry *Vaccinium oxycoccos*, (Am.) *V. macrocarpon*, (NZ) *Ugni molinae*; **c. bean** *Phaseolus vulgaris*
crane flower *Strelitzia reginae*
cranesbill *Geranium* spp.; **bloody c.** *G. sanguineum*; **meadow c.** *G. pratense*; **shining c.** *G. lucidum*
Cranichis Sw. Orchidaceae (IV 2b). c. 50 warm Am. (Florida 1) esp. Andes
Craniolaria L. Martyniaceae. 3 S Am. ***C. annua*** L. (N S Am.) – fleshy r. form. eaten, medic.
Craniospermum Lehm. Boraginaceae (B.3.6). 1–5 temp. As.: ***C. subvillosum*** Lehm. R: BZ 85,12(2000)79
Craniotome Reichb. Labiatae (VI). 1 Himal. to Vietnam: ***C. furcata*** (Link) Kuntze
Cranocarpus Benth. Leguminosae (III 11). 3 Braz.
Crantzia Scop. = *Alloplectus*
crape fern *Todea barbara*; **c. myrtle** *Lagerstroemia indica*
Craspedia Forst. f. Compositae (Gnap.-Ang.). Excl. *Pycnorus*, 23 temp. Aus., NZ. R: OB 104(1991)111. Cult. orn. esp. ***C. uniflora*** Forst. f. (billy buttons)
Craspedolobium Harms. Leguminosae (III 16). 1 W China: ***C. schochii*** Harms. R: BZ 86,6(1998)119
Craspedophyllum (C. Presl) Copel. = *Hymenophyllum*
Craspedorhachis Benth. Gramineae (XXIX 3). 3 S trop. Afr.
Craspedosorus Ching & W.M. Chu = *Cyclosorus*
Craspedospermum Airy Shaw = *Craspidospermum*
Craspedostoma Domke = *Gnidia*
Craspidospermum Bojer ex A.DC. Apocynaceae (I e). 1 Madag.: ***C. verticillatum*** Bojer ex A. DC. – Scarrone's Model. R: WAUP 97–2(1997)11
Crassocephalum Moench (~ *Senecio*). Compositae (Sen.-Sen.). 24 warm Afr. to Yemen & Masc., ***C. crepidioides*** (Benth.) S. Moore (thickhead (Aus.)) an aggressive weed in Mal., now pantrop., local veg. (also other spp.)
Crassothonnna R. Nordenstam (~ *Othonna*). Compositae (Sen.-Sen.). 13 S Afr. R: CN 50(2012)71. Succ. terete lvs. ***C. capensis*** (L. Bailey) R. Nordenstam (*O. c.*, *'O. crassifolia'*) – cult. orn. esp. hanging baskets & (Calif.) ground-cover
Crassula Dill. ex L. Crassulaceae (III). c. 195 almost cosmop. (Eur. 4, NW 13 (R: KB 39(1984)699)), esp. trop. & S (144, R: CBH 8(1977), Cape alone with 95 (29 endemic)) Afr. Succ. herbs, shrubs & treelets, divisible into 6 sections; many cult. orn. (incl. *Rochea*) esp. ***C. arborescens*** (Mill.) Willd. (lvs obovate to orbicular) & ***C. ovata*** (Mill.) Druce S Afr. (lvs elliptic-oblanceolate) – familiar shrubby housepls (jade plant, money-plant m.- tree) rarely flowering in cult. (cf. *Epipremnum pinnatum*) s.t. with plantlets in infl.; ***C. muscosa*** L. (*C. lycopodioides*, S Afr.) – lvs arr. like clubmoss; ***C. perfoliata*** L. var. ***falcata*** (Wendl.) Toelken (Cape Prov., 'rochea') – lvs grey, sickle- shaped, 2-ranked, fls bright red. Many xeromorphic features, e.g. *C. perfoliata* with some epidermal cells swollen above rest into large bladders which meet over whole surface dead & air-filled when leaf mature, walls infiltrated with silica but ***C. aquatica*** (L.) Schönl. (N temp.) aquatic ann., while ***C. pageae*** Toelken (*Pagella archeri*, S Afr.) a liverwort-like ann. & ***C. cloisiana*** (Gay) Reiche (Arg.) also described in Podostemaceae; ***C. helmsii*** (Kirk) Cockayne (pygmyweed, Aus., NZ) – outcompeting (introd.!) *Elodea* spp. in some UK sites (introd. 1927, natur. 1956, by 1996 pestilential in 320 parks & nature reserves in England alone & thence invading the Continent, sale banned UK (2014)

Crassulaceae J. St-Hil. Magnoliidae – Saxifragales. 34/1350 almost cosmop. esp. S Afr., rare in Aus. & W Pac. R: U. Eggli, *Ill. Handb. Succ. Pls*, Crass. (2005). Succ. shrubs & herbs (s.t. ann.), s.t. treelets, with crassulacean acid metabolism & oft. with red (anthocyanin) r.-tips & crystals of calcium oxalate in parenchyma cells & unusual stem vasculature incl. cortical &/or medullary bundles. Lvs in spirals (oft. rosettes) to opp. or whorled, simple & usu. entire, oft. with hydathodes; stip. 0. Fls solit. or usu. in term. thyrses, usu. bisex., hypogynous to weakly perigynous usu. (3–)5(–32)-merous; K ± free, C free to tubular, A usu. 2 × C in 2 whorls, less oft. same as & alt. with C, rarely basally connate, anthers with longit. slits, G as many as K or P, usu. free, with nectariferous appendage nr. base, ovules (1 –)∞ per carpel, on parietal to marginal placentas, anatropous, bitegmic. Fr. usu. head of follicles, nut-like or capsule (*Diamorpha*); seeds small with usu. scant oily proteinaceous endosperm. x = 4–22+ (2n to 540+)

Classification & principal genera:

I. **Sempervivoideae** (fls (4)5(–32)-merous, Usu. N Hemisph.; generic limits unclear, with much hybridity; 5 tribes): *Orostachys* (1. Telephieae); *Umbilicus* (2. Umbiliceae); *Sempervivum* (3. Semperviveae); *Aeonium* (4. Aeonieae); *Dudleya*, *Rosularia* (5[a]. Sedeae), *Echeveria*, *Sedum* (5[b])

II. **Kalanchoideae** (usu. woody, C-tube): *Adromischus*, *Kalanchoe*, *Tylecodon*

III. **Crassuloideae** (lvs usu. decussate): *Crassula*

Oft. in arid habitats but also seen in rain forest & other moist sites (some *Crassula* spp. aquatic); closely allied to Saxifragaceae

Many cult. orn. esp. spp. of *Adromischus*, *Aeonium*, *Crassula*, *Echeveria*, *Kalanchoe* & *Villadia* under glass or as housepls in temp., *Sedum*, *Sempervivum* & *Umbilicus* outside; some local medic., veg. etc.

+ **Crataegomespilus** Simon-Louis. Graft-hybrids between spp. of *Crataegus* & *Mespilus germanica* known as Bronvaux medlars: layers of cells from 1 sp. overlying those of another but these not as sharply distinguishable as form. held

Crataegus Tourn. ex L. Rosaceae (Ros.-Mal.). Excl. *Mespilus*, c. 230 N temp. (Eur. 21; N Am. 152 – R: FNA 9(2014)491), incl. many apomictic clones (pseudogamy) & hybrids (some triploid) form. considered spp. (in N Am. diploid usu. sexual outbreeders, triploid obligate apomicts & tetraploid self-compatible facultative apomicts, sp. complexes undergoing raciation with some crossing between incipient spp. rather than wholesale interspecific hybridization). R: CJB 68(1990)2220; J.B. Phipps, *Hawthorns & medlars* (2003)62. Usu. thorny decid. shrubs, hawthorns or 'thorns', the thorns (used until C19 as fish-hooks in GB; adventitious indeterminate on trunks, determinate aphyllous on shoots (all N Am. spp.) or indeterminate), the haws being pomes; some domatia. Fr. & lvs used in herbal teas (medic. effects in hypertension, heart disease etc. largely due to flavonoids & proanthocyanidins), handcreams, fr. in Chinese sweets. In GB ***C. monogyna*** Jacq. (white thorn, open country, hedges) hybridizes (= ***C.* × *media*** Bechstein; cvs incl. red '**Paul's Scarlet**') with ***C. laevigata*** (Poir.) DC. (woods), fls of both called may (smell of latter held to be like that of Great Plague, trimethylamine also being one of first prods of rotting flesh (corpses form. kept in home for up to a week), poss. orig. in C19 of attribution of bad luck to bringing blossom indoors, though also assoc. with Virgin Mary), the first widely used for quickset or quickthorn hedging (invasive in Aus., NZ, W US), young buds ed. ('bread & cheese' in country districts), lvs a medic. tea, dried fr. added to flour in Eur., wood a subs. for box, stake considered suitable for heart of a vampire (cf. *Acer*, *Populus*), **'Biflora'** ('Praecox', Glastonbury Thorn, poss. *C.* × *media*) flowering in winter as well as spring (winter-flowering forms of *C. monogyna* known from N Afr.) – 'holy thorn' (1520), legend of orig. from staff of Joseph of Arimathea dating from 1714, sprays sent to UK monarch at Christmas; ***C. punctata*** Jacq. (NE N Am.) – nest-tree for butcher-bird hanging prey on thorns until needed. Others cult. incl. ***C. azarolus*** L. (azarole, E Med.) – ed. apple-flavoured haws, cvs to 6 cm diam. ('Medit. medlars'); ***C. crus-galli*** L. (cockspur thorn, N Am.); ***C. douglasii*** Lindl. (black haw, W N Am.) – fr. used in jellies; ***C. mexicana*** Moçiño & Sessé ex DC. ('*C. pubescens*', *C. stipulosa*, manzanilla, Mex. hawthorn, Mex.) – fr. ed. esp. for stock; ***C. pentagyna*** Waldst. & Kit. ex Willd. (Chinese haw, E Eur. to Iran) – cult. fr. tree in China; ***C. phaenopyrum*** (L.f.) Medik. (*C. cordata*, Washington thorn, NE N Am.); ***C. pinnatifida*** Bunge (As.) – grown for fr. preserves & sold like toffee-apples in Beijing

× **Crataemespilus** Camus. Sexual hybrids between *Mespilus germanica* & spp. of *Crataegus*, e.g. with *C. brachyacantha* Sarg. & Engelm. in Arkansas to give triploid × ***Crataemespilus canescens*** (J. Phipps) J. Phipps (*Crataegus* × *c.*)

Crataeva L. See *Crateva*
Crateranthus Bak. f. Lecythidaceae (V; Napoleonaeaceae). 4 trop. W Afr. R: KB 70[1–6]2015
Craterispermum Benth. Rubiaceae (III 7). 16 trop. Afr. & Madag. to Seychelles R: KB 28(1974)434. Massart's Model
Craterocapsa Hilliard & B.L. Burtt. Campanulaceae (I 2). 5 C & S Afr. R: NRBGE 32(1973)314. ***C. tarsodes*** Hilliard & B.L. Burtt – local med. for epilepsy (verbascoside poss. neurosedative)
Craterogyne Lanj. = *Dorstenia*
Craterosiphon Engl. & Gilg. Thymelaeaceae (Thym.-Daph.). 9 trop. Afr.
Craterostemma K. Schum. = *Brachystelma*
Craterostigma Hochst. Linderniaceae (Plantaginaceae (Grat.-Lind.); Scrophulariaceae s.l.). 25 OW trop. R: TSP 81(1992)85, Willd. 43(2013)221. Surviving dry season as poikilohydrics; some cult. orn. e.g. ***C. plantagineum*** Hochst. (E & S Afr. to Ind., 'blue gem') – cold treatment in Uganda. See also *Crepidorhopalon*
Crateva L. Capparaceae. 10 trop. (Madag. c. 5 endemic, Am. 4 – R: HPB 13(2008)121). R: Blumea 12(1964)186. Garlic pear. Some leaf veg. & cult. orn. esp. ***C. religiosa*** Forst.f. (barna, OW) in trop., medic. ***C. benthamii*** Eichler – obligate fish-disp. in Amaz.
Cratoxylum Blume. Hypericaceae (III; Guttiferae I). 6 Indomal. R: Blumea 15(1967)453. Distyly in ***C. formosum*** (Jack) Dyer (medic.); ***C. arborescens*** (Vahl) Blume (geronggang, geronggong) – soft wood for clogs, dayak drums, shingles
crattock *Ficus racemosa*
Cratylia Mart. ex Benth. Leguminosae (III 18). 7 S Am.
Cratystylis S. Moore. Compositae (Inul.-Pluch.). 4 arid/semi-arid W & S Aus. R: Nuytsia 14(2002)447, FA 37(2015)388. Dioec. to subdioec. ***C. conocephala*** (F. Muell.) S. Moore – imp. fodder pl.
craw-craw (plant) *Senna alata*
Crawfurdia Wall. (~ *Gentiana*). Gentianacae. 16 Himal. Twining perennials
crazyweed *Oxytropis* spp.
cream of wheat [US] *Triticum turgidum* Durum group; **c.-cups** *Platystemon californicus*
creat *Andrographis paniculata*
Creatantha Standl. = *Isertia*
creeper, canary *Tropaeolum peregrinum*; **Virginia c.** *Parthenocissus quinquefolia*
creeping Jenny *Lysimachia nummularia*
Cremanthodium Benth. (~ *Ligularia*). Compositae (Sen.-Tuss.). 70 Himal., S China. R: JLSBot48(1929)259. Cult. orn., fragrant fls in usu. nodding heads
Cremaspora Benth. Rubiaceae (Ix.-Octo.). 3–4 trop. Afr., Comoro Is. ***C. triflora*** (Thonn.) K. Schum. (*C. africana*) – frs source of blue-black body-dye
Cremastogyne (H. Winkler) Czerep. = *Betula*
Cremastopus P. Wilson = *Cyclanthera*
Cremastosciadium Rech.f. = *Eriocycla*
Cremastosperma R.E. Fries. Annonaceae (IV 2). 19 trop. S Am. R: AHB 10(1930)46
Cremastra Lindl. Orchidaceae (V 13e). 5 E As. R: SB 38(2013)67. ***C. appendiculata*** (D. Don) Mak. (*C. variabilis*, *C. wallichiana*) – r. used by Ainu for toothache; ***C. aphylla*** Yukawa (Jap.) – leafless
Cremastus Miers = *Cuspidaria*
Cremersia Feuillet & Skog. Gesneriaceae (II 5c). 1 French Guiana: ***C. platula*** Feuillet & Skog. R: Britt. 54(2003)347
Cremnophila Rose = *Sedum* (but see CSJ 50(1978)139)
Cremnophyton Brullo & Pavone = *Atriplex*. (but see Cand. 42(1987)621)
Cremnothamnus Puttock (~ *Ozothamnus*). Compositae (Gnap.). 1 NW Aus.: ***C. thomsonii*** (F. Muell.) Puttock. R: AusSB 7(1994)569
Cremocarpon Baill. (~ *Psychotria*). Rubiaceae (IV 7). 1 Comoro Is.
Cremolobus DC. Cruciferae (21). 7 Andes. R: CGH 195(1965)142. Some lianes
Cremosperma Benth. Gesneriaceae (II 3a). c. 25 N Andes. R: JWAS 25(1935)286
Cremospermopsis Skog & Kvist (~ *Cremosperma*). Gesneriaceae (III 2b). 2 Colombia. R: Novon 12(2002)264
Crenea Aubl. Lythraceae. 2 trop. S Am., Trinidad. R: Caldasia 15(1986)121. Salt water
Crenias Spreng. (~ *Podostemum*). Podostemaceae (III). 5 SE Braz. R: T 50(2001)1164
Crenidium Haegi (~ *Cyphanthera*). Solanaceae (III 1). 1 W Aus.: ***C. spinescens*** Haegi. R: FA 29(1982)34

Crenosciadium Boiss. & Heldr. (~ *Opopanax*). Umbelliferae (III 10). 1 Turkey: ***C. siifolium*** Boiss. & Heldr.

Crenuluma Plowes = *Caralluma*

Creochiton Blume. Melastomataceae. 6 C & E Mal.

creole tea *Sauvagesia erecta*

creosote bush *Larrea divaricata* subsp. *tridentata*

crepe flower *Lagerstroemia indica;* **c. jasmine** *Tabernaemontana divaricata;* **c. myrtle** *L. indica*

Crepidiastrum Nakai (~ *Ixeris*). Compositae (Lact.-Crep.). 7 E As.

Crepidifolium Sennikov. Compositae (Cic.-Cich. (Crep.)). 4 C & E As.

Crepidium Blume = *Malaxis*

Crepidomanes (C. Presl) C. Presl (~ *Trichomanes*). Hymenophyllaceae (Hym.). (Excl. *Abrodictyum, Callistopteris, Vandenboschia*) 35 trop. OW, some Am. Buds on rachis of some spp.

Crepidopteris Copel. = praec.

Crepidorhopalon E. Fischer (~ *Craterostigma, Lindernia*). Linderniaceae (Plantaginaceae (Grat.-Lind.); Scrophulariaceae s.l.). 30 trop. & S Afr., Madag. (1). R: TSP 81(1992)126, Willd. 43(2013)223. Some on copper- or cobalt-rich soils

Crepidospermum Hook.f. Burseraceae (II). 7 trop. Am. R: Britt. 39(1987)51, KB 57(2002)471

Crepinella Marchal = *Schefflera*

Crepis L. (~ *Hieracium*). Compositae (Cich.-Cich.(-Crep.)). Excl. *Askellia*, c. 200 N hemisph. (Eur. 70; N Am. 24), S Afr., S Am. R: UCPB 21(1947)199. Heterogeneous. Mostly weedy with yellow fls (hawk's beard) incl. ***C. capillaris*** (L.) Wallr. (Eur.) & ***C. vesicaria*** L. (Eur.) – extracts inhibit *Staphylococcus aureus* growth; few cult. esp. ***C. rubra*** L. (E Eur.) – pink fls

Crescentia L. Bignoniaceae (7). 6 trop. Am. R: FN 25(1980)82. ***C. cujete*** L. (calabash tree) – Champagnat's Model, cauliflorous with sulphur compounds in flower-scent attractive to bats, gourd-like berries (poss. adapted to disp. by extinct megafauna), bearing nectaries thought to be ant-attractants, the ants warding off herbivores; woody pericarp used liked gourds as bowls, scoops, maracas, etc. & form. with eye-holes used to camouflage swimming hunters who pulled down individual birds without disturbing the flock (Columbus); young fr. pickled (like walnuts), seeds ed. cooked & used to make a drink in Nicaragua ('semilla de jícaro'); ***C. alata*** Kunth (*Parmentiera a.*, C Am.) – bat-poll. (sulphides smell) though pollen & nectar stolen by social bees, horse- (& poss. extinct megafauna-) disp., timber, seeds prep. as cooling drink, pericarp as cups; ***C. amazonica*** Ducke (Amaz.) – seeds rot in fr. unless disp. by characin fish; ***C. portoricensis*** Britton (W Puerto Rico) – known from 4 adults & some juveniles

cress *Lepidium sativum* (**common** or **garden c.**, c. of '**mustard and c.**', cvs incl. **Aus.** or **golden c.**), though oft. replaced by faster-growing *Brassica napus;* **American c.** *Barbarea verna;* **Brazilian c.** *Acmella oleracea;* **brown c.** *Nasturtium × sterile;* **Greek c.** *L. sativum;* **green c.** *N. officinale;* **hoary c.** *L. draba;* **Indian c.** *Tropaeolum majus;* **land c.** *B. verna;* **meadow c.** *Cardamine pratensis;* **Pará c.** *Acmella oleracea;* **penny c.** *Thlaspi arvense;* **rock c.** *Arabis* spp.; **shepherd's c.** *Teesdalia nudicaulis;* **violet c.** *Ionopsidium acaule;* **wall c.** *Arabis* spp.; **waterc.** *N. officinale* & hybrids; **winterc.** *B. verna*

Cressa L. Convolvulaceae (4). 1–4 trop. & warm incl. Eur.: ***C. cretica*** L. – in arid saline sites, allelopathic, a tonic in Sudan. R: BotJLS 133(2000)28

crested dog's-tail *Cynosurus cristatus*

Creusa P.V. Heath = *Crassula*

Cribbia Senghas (~ *Rangaeris*). Orchidaceae (V 16d). 4 trop. Afr. R: KB 53(1997)743

Criciuma Soderstrom & Londoño = *Eremocaulon*

crila *Crinum latifolium*

crin végétal *Chamaerops humilis*

Crinipes Hochst. Gramineae (Arund.). 1 Sudan, Ethiopia, Uganda: ***C. abyssinicus*** (A. Rich.) Hochst. R: KB 12(1957)54

Crinitaria Cass. (~ *Galatella*). Compositae. Ast.-Bell. 13 Euras.

Crinitina Soják = *Galatella*

crinkle bush *Lomatia silaifolia*

Crinodendron Molina. Elaeocarpaceae. 5 temp. S Am. R: SB 16(1991)77. Bird-poll. cult. orn. esp. ***C. hookerianum*** C. Gay (Chile, 'Chile lantern-tree') – with red fls, the buds beginning to expand the season prior to anthesis

Crinonia Blume = *Pholidota*

Crinum L. Amaryllidaceae. c. 65 trop. & warm (N Am. 4), esp. subSaharan Afr. (c. 40). R: Herbertia 9(1942)63. Relationships with *Ammocharis* still unclear. Bulbous pls with lvs in spirals & consp. fls maturing together. Alks. Many cult. orn. (Cape lilies), c. 10

spp. aquatic; some locally medic. (esp. emetics). ***C. asiaticum*** L. (trop. As.) – widely cult. trop., the seed with layer of cork over endosperm aiding water-disp. (fr. floating for 1–2 weeks), lvs a rheumatism compress in Sumatra; ***C. flaccidum*** Herb. (Darling lily, Aus.) & ***C. kirkii*** Bak. (pyjama l., E Afr.) – cult. orn.; ***C. latifolium*** L. (crila, SE As.) – Vietnam trad. med. for prostate & menopause problems, mooted comm. drug; ***C.*** × ***powellii*** Bak. (*C. bulbispermum* (Burm.f.) Milne-Redh. & Schweick. (S Afr.) × *C. moorei* Hook.f. (S Afr.)) – many cvs.; ***C. pedunculatum*** R. Br. (Aus.) – used by Aborigines for jellyfish stings

Crioceras Pierre. Apocynaceae (I d). 1 Gabon to Angola: ***C. dipladeniiflorus*** (Stapf) K. Schum. – Leeuwenberg's Model. R: Fontqueria 41(1995)12

Criogenes Salisb. = *Cypripedium*

Criosanthes Raf. = *Cypripedium*

Criscia Katinas (~ *Onoseris*). Compositae (Mut.-Nass.). 1 E S Am.: ***C. stricta*** (Spreng.) Katinas

Crispiloba Steenis. Alseuosmiaceae. 1 Queensland: ***C. disperma*** (S. Moore) Steenis

Cristaria Cav. Malvaceae (Malv.-Malv.). 75 temp. S Am.

Cristatella Nutt. = *Cleome* (*Polanisia*)

Cristonia J. Ross (~ *Bossiaea*). Leguminosae (III 4). 3 Aus. R: Muelleria 28(2010)67

Critesion Raf. = *Hordeum*

Crithmum Tourn. ex L. Umbelliferae (III 8). 1 Eur., maritime: ***C. maritimum*** L. (sea samphire) – lizard-poll. in Balearics, lvs fleshy, form. much pickled (salty spicy taste)

Crithopsis Jaub. & Spach. Gramineae (XV). 1 Morocco & Crete to Afghanistan: ***C. delileana*** (Schultes) Roshev. R: NJB 13(1993)484

Critonia P. Browne (~ *Eupatorium*). Compositae (Eup.-Crit.). 43 trop. Am. R: MSBMBG 22(1987)295

Critoniadelphus R. King & H. Robinson (~ *Eupatorium*). Compositae (Eup.-Crit.). 2 C Am. R: MSBMBG 22(1987)299

Critoniella R. King & H. Robinson (~ *Eupatorium*). Compositae (Eup.-Crit.). 6 N Andes. R: MSBMBG 22(1987)341

Critoniopsis Schultz-Bip. Compositae (Vern.-Pip.). c. 45 S Am. esp. N Andes. R: Phytol. 46(1980)437

Croatella Gonç. (~ *Asterostigma*). 1 Andes (Ecuador): ***C. integrifolia*** (Madison) Gonç. R: Willd. 35(2005)323

Crobylanthe Bremek. Rubiaceae (I 9). 1 Borneo: ***C. pellacalyx*** (Ridl.) Bremek.

Crocanthemum Spach (~ *Helianthemum*). Cistaceae. 24 Am. (N Am. 21 esp. SE US, C. Mex. – R: Rhodora 67(1965)201,255; S S Am. 3). Lvs alt. (cf. *H.*). ***C. canadense*** (L.) Britton (*H. c.*, frostweed, E US) – cult. orn.

Crocidium Hook. Compositae (Sen.-Tuss.). 1 NW N Am.: ***C. multicaule*** Hook. R: FNA 20(2006)641

Crockeria Greene ex A. Gray = *Lasthenia*

Crocodeilanthe Reichb.f. & Warsc. = *Pleurothallis*

Crocodylium Hill (~ *Centaurea*). Compositae (Card.-Cent.). 3 E Medit. to Middle E

Crocopsis Pax = *Stenomesson*

Crocosmia Planch. (~ *Tritonia*). Iridaceae (VI 4). 8 trop. & S Afr., Madag. R: JSAB 50(1984) 463. Like *Tritonia* but fr. globose, not oblong, each locule with 3 or more (not 1 or 2) seeds. Dried fls in water give strong smell of saffron, ***C. aurea*** (Hook.) Planch. source of yellow dye. Cult. orn. (incl. *Curtonus*) esp. ***C.*** × ***crocosmiiflora*** (Lem.) N.E. Br. (montbretia, *C. aurea* × *C. pottsii* (Bak.) N.E. Br., raised 1880 in France)

Crocoxylon Ecklon & Zeyher = *Elaeodendron*

Crocus Tourn. ex L. Iridaceae (VI 5). 88 Medit. (Eur. 43) to W China. R: B. Mathew (1982) *The C.*, T 57(2008)487, J. Ruksans (2010) *Crocuses*. Corms with cormlets forming in axils of scales; peduncle & ovary subterr. (cf. *Colchicum*), A 3, style with 3 stigmatic branches (*C. speciosus* with multifid), P closing at night & in dull weather, visited by bees & Lepidoptera allegedly for nectar produced by ovary (some authorities suggest nectar 0 & fls deceitful); anthers dehisce outwards so that insects contact pollen while probing for nectar. Many cult. orn. esp. yellow-flowered ***C. flavus*** Weston (*C. aureus*, Balkans, long but now rarely cult., much in gardens sterile & big yellow-flowered **'Dutch Yellow'** or **'Golden Yellow'**, hybrid with ***C. angustifolius*** Weston (*C. susianus*, SW Russia) = triploid ***C.* × *luteus*** Lam. (*C.* × *stellaris* Haw., diploid) – natur. Netherlands) & purple ***C. neapolitanus*** (Ker Gawl.) Loisel. ('*C. vernus*', Dutch c., Italy); ***C. chrysanthus*** (Herb.) Herb. (E Med.) – common winter-flowering sp. with many cvs (some being hybrids with ***C. biflorus*** Mill. (S Eur., W As.)); ***C. longiflorus*** Raf. (S Italy, Malta) – stigmas used like saffron; ***C. nudiflorus*** Sm. (Pyrenees, natur. in GB on former properties of Knights of St John of Jerusalem where introd. as saffron subs., form cult. as at Saffron Walden, Essex, UK) & ***C. speciosus*** M. Bieb. (E Med. to Iran) – commonly seen autumn-flowering spp.; ***C. sativus*** L. (saffron, a sterile autotriploid cultigen prob. selected for long stigmas from *C. cartwrightianus* Herb. (Greece)) – source of saffron (Arabic za'fran, 10^6 fls to give 10 kg dried spice) form. imp. dye (alpha crocin, crocetin – a carotenoid) in anc. Greece derived from stigmas, now largely prep. in Spain (5000 ha yielding 4.7×10^9 stigmas, 47 t spice per annum, world total 300 t, 100–170 t. from Iran, best from Kashmir) for cooking (bitter taste due to picrocrocin) & colouring foods esp. Spanish rice, bouillabaisse, while saffron cakes & loaves trad. in Cornwall, v. imp. constituent of Fernet-Branca (It. digestif; prod. alleged to consume 75% world's saffron), 'karkom' (orig. of word crocus) of Song of Solomon, most prob. that featured on Minoan pottery & frescoes (1600 BC), used in c. 90 med. conditions (steroids; but 2 g harmful), app. reducing arteriosclerosis poss. explaining low levels of cardiovascular disease in Spain where consumption is high (richest known source of vitamin B_2), constituent of laudanum

crocus, autumn *Colchicum* spp. esp. *C. autumnale*; **Chilean c.** *Tecophilaea cyanocrocus*; **Dutch c.** *Crocus neapolitanus*; **saffron c.** *C. sativus*

Crocyllis E. Meyer ex Hook.f. = *Plocama*

Croftia Small = *Carlowrightia*

Crofton weed *Ageratina adenophora*

Croizatia Steyerm. Phyllanthaceae (Brid.-Sav.; Euphorbiaceae s.l.). 4 Panamá, Venez. R: SB 12(1987)1

Cromapanax Grierson. Araliceae. 1 Bhutan: ***C. lobatus*** Grierson. R: EJB 48(1991)19

Cromidon Compton. Scrophulariaceae (Man.). 12 S Afr. R: EJB 47(1990)320

Croninia J. Powell (~ *Leucopogon*). Ericaceae (VII 7; Epacridaceae). 1 W Aus.: ***C. kingiana*** (F. Muell.) J. Powell. R: Nuytsia 9(1993)125

Cronquistia R. King = *Carphochaete*

Cronquistianthus R. King & H. Robinson (~ *Eupatorium*). Compositae (Eup.-Crit.). 20 N Andes. R: Phytol. 70(1991)158

crookneck *Cucurbita pepo* etc.

Croomia Torrey. Stemonaceae (Croomiaceae). 5 E China & Jap., 1 SE US. R: JJB 87(2012)83. Vascular system of discontinuous cylinders

Croomiaceae Nakai. See Stemonaceae

Croptilon Raf. (~ *Haplopappus*). Compositae (Ast.-Chr.). 3 SE US. R: Sida 9(1981)59

crosnes *Stachys affinis*

cross of Jerusalem or **Maltese c.** *Silene chalcedonica*; **c.wort** *Cruciata laevipes*

Crossandra Salisb. Acanthaceae (III 1). 52 trop. Afr. & Madag. (25; R: KB 52(1997)383), 1 ext. to Ind., Arabia (1). Seeds flat, covered with hairs or fringed scales making seed sticky when wet. Some cult. orn. greenhouse shrubs esp. ***C. infundibuliformis*** (L.) Nees (*C. undulifolia*, kanakambaram, Afr., S Ind., Sri Lanka) – fls bright orange, allegedly aphrodisiac, worn with raw tamarind & betel by women guests in S Ind.; ***C. stenostachya*** C.B. Clarke (trop. Afr.) – pollen grains 520×19 μm, largest in terr. angiosperms

Crossandrella C.B. Clarke. Acanthaceae (III 1). 3 trop. Afr. R: PEE 143(2010)184

Crosslandia W. Fitzg. Cyperaceae (II 4). 1 N Aus.: ***C. setifolia*** W. Fitzg.

Crossoglossa Dodson = *Microstylis*

Crossoliparis Marg. = *Liparis*

Crossonephelis Baill. = *Glenniea*

Crossopetalum P. Browne. Celastraceae (I). Incl. *Myginda* c. 26 trop. Am.

Crossopteryx Fenzl. Rubiaceae (III 2). 1 trop. & S Afr.: ***C. febrifuga*** (G. Don) Benth. – timber & febrifuge. R: Strel. 10(2000)485

Crossosoma Nutt. Crossosomataceae. 3 SW N Am. R: FNA 9(2014)11

Crossosomataceae Engl. Magnoliidae – Crossosomatales. 4/8 SW N Am. R: SB 28(2003) 104. Glabrous xeromorphic shrubs or small trees with intricate often spiny branches, mostly growing on rhyolite. Lvs decid. or s.t. marcescent, simple, in spirals (opp. & distally 2-lobed in *Apacheria*); stip. minute or 0. Fls solit. (2, 3 in *Glossopetalon*), bisexual, shortly perigynous, the disk forming a nectary to which A attached or annular within A; K (3)4 or 5(6), C (3)4 or 5(6), white & imbr., A 4 to c. 50 in 3 or 4 whorls assoc. with c. 10 trunk bundles arising centrifugally or centripetally, in 2 whorls in *Apacheria* etc. but ± reduced to 1 whorl (ante-C whorl lost) in others, anthers with longit. slits, $\underline{G}$ 1–5(–9) each with (1)2–22+ ovules on marginal placenta, amphitropous or campylotropous, bitegmic. Fr. follicular; seeds arillate with thin to copious oily endosperm. n = 6

Genera: *Apacheria, Crossosoma, Glossopetalon, Velascoa*

Crossosperma Hartley (~ *Melicope*). Rutaceae (I 2). 2 New Caled.

Crossostemma Planch. ex Benth. Passifloraceae. 1 trop. Afr.: ***C. laurifolium*** Planch. ex Benth.

Crossostephium Less. = *Artemisia* (but see BBMNHB 23(1993)120)

Crossostylis Forst. & Forst. f. Rhizophoraceae (II). 10 W Pac. Attims's Model

Crossothamnus R. King & H. Robinson (~ *Eupatorium*). Compositae (Eup.-Alom.). 4 Colombia, Ecuador, Peru. R: MSBMBG 22(1987)262

Crossyne Salisb. (~ *Boophone*). Amaryllidaceae. 2 Cape. R: FR 105(1994)355

Crotalaria Dill. ex L. Leguminosae (III 8). c. 700 trop. & subtrop. (Afr. & Madag. with 511, R: R.M. Polhill (1982) *C. in Afr. & Mad.*). Alks; fr. an inflated dehiscent legume (rattlepod (Aus.)). Fodders & fibres esp. ***C. juncea*** L. (Bombay, Madras, sann or sunn hemp, tag (hemp), orig. unknown) – fibre (c. 200 K t p.a.) for cordage, canvas, fishing-nets (more durable than jute) & cigarette-papers; ***C. burkeana*** Benth. (S Afr.) – major cause of crotalism; ***C. cunninghamii*** R. Br. (Aus.) – fibres used by Aborigines to make 'sandals' for hot sand of desert W Aus.; ***C. longirostrata*** Hook. & Arn. (chipilin, chiplin, C Am.) – soporific veg. cult. SW US; ***C. micans*** Link (*C. anagyroides*, trop. Am.) – cult. OW as green manure; ***C. spectabilis*** Roth (Indomal.) – cult. for fodder in NW; ***C. usaramoensis*** Bak. f. (S Am.) – green manure & cover crop in China

croton *Codiaeum* spp. esp. *C. variegatum*; **c. oil** *Croton tiglium*

Croton L. Euphorbiaceae (Crot.). Excl. *Astraea*, incl. *Cubacroton, Moacroton*, c. 700 trop. & warm (Aus. 27 + 3 weeds – R: Austrobaileya 6(2003)354; Afr. 50; Madag. 150 endemic). R: T 60(2001)801. 31 sections. Monoec. & dioec. herbs, shrubs, some rheophytes, & trees, usu. with stellate hairs; some may cause contact dermatitis while seeds of others suggested to promote tumours. Many alks. Some timbers, teas but esp. medic. (fever bark). ***C. alamosanus*** Rose (Mex.) – dioec., wind-poll.; ***C. bonplandianus*** Baill. (Braz.) – weedy in As. (introd. Chittagong 1897–8 & all over Ind. in 90 yrs), leachate allelopathic to weedy associates; ***C. caudatus*** Geiseler (Ind.) – promising in cancer treatment; ***C. eluteria*** (L.) Wright (WI) – source of cascarilla bark used as tonic & bitters, flavouring for liqueurs & scenting tobacco (1–5% volatile oil incl. eugenol, vanillin etc.); ***C. laccifer*** L. (S & SE As.) – host-pl. for lac-prod. insects of imp. in varnish-making; ***C. malambo*** Karsten (malambo, Venez.) – bark med.; ***C. megalocarpus*** Hutch. (musine, E Afr.) – timber & shade-tree, local medic. (abdomen); ***C. niveus*** Jacq. (trop. Am.) – copalchi bark (Mex.) subs. for *C. eluteria*; ***C. scouleri*** Hook.f. (Galápagos) – capsules crushed by 2 spp. of finch & at least 1 seed always dropped, 2 other finch spp. forage but miss some, which can then germinate; ***C. setiger*** Hook. (W N Am.) – natur. Aus. (dove-weed); ***C. stellatopilosus*** H. Ohba (? = *C. longissimus* Airy Shaw, Thailand) – anti-ulcer drug (plaunotol); ***C. texensis*** (Klotzsch) Muell. Arg. (skunkweed, N Am.) – tea & insecticide; ***C. tiglium*** L. (trop. As.) – source of croton oil, one of most purgative substances known, itself not tumour-inducing but when applied with a subeffective carcinogen it is & may account for high level of oesophageal cancer in China (*C. flavens* L. may do same in WI), phorbol myristate acetate the most active skin-irritant

Crotonogyne Muell. Arg. Euphorbiaceae (Al.-Crot.). 16 W trop. Afr.

Crotonogynopsis Pax. Euphorbiaceae (Acal.). 4 trop. Afr. R: Novon 24(2015)246. Litter-box pls. Herbarium specimens smelly

Crotonopsis Michaux = *Croton*

crow('s) ash *Flindersia australis*; **c.berry** *Empetrum nigrum*; **c.foot** *Ranunculus* spp., *Dactyloctenium* spp. (N Am.); **c.f. elm** *Heritiera trifoliolata*; **c.f. grass** *Eleusine coracana* subsp. *indica*; **c. poison** *Stenanthium densum*

crowa *Ananas comosus*

Crowea Sm. Rutaceae (I 3). 3 S Aus. R: Nuytsia 1(1970)15. Waxflowers; explosive fr. dehiscence; prot. spp.

crown beard *Verbesina* spp.; **c. daisy** *Glebionis coronaria*; **c. gum** *Manilkara chicle*; **c. imperial** *Fritillaria imperialis*; **c. of thorns** that of Christ prob. *Ziziphus spina-christi*, poss. *Poterium spinosum*, but now applied to *Euphorbia milii*

Crucianella L. Rubiaceae (IV 16). c. 30 Eur. (8), Medit. to Iran & C As.

Cruciata Mill. (~ *Galium*). Rubiaceae (IV 16). 9 Eur. (5) to E Med., ***C. laevipes*** Opiz (crosswort) in GB. R: NSL 32(2000)152

Crucicaryum Brand = *Cynoglossum*

Cruciferae Juss. (Brassicaceae). Magnoliidae – Brassicales. 343/3500 cosmop. esp. temp. (particularly Med. to C As., W N Am.). R: J.G. Vaughan *et al.* (1976) *The biology & chem. of* *the C.*; PSE 259(2006)89. Herbs, rarely lianes, shrubby or small trees when oft. pachycaul, with mustard-oil glucosides (glucosinolates effective in defence against bacteria, fungi, insects & mammals (refused by horses), sinigrin (glycoside) liberating into soil antipathigenic allyl isothiocyanate, an injurious vesicant oil inimicable to vesicular-mycorrhizal activity) & oft. cyanogenic, stem oft. with unusual vasculature; mycorrhizae rare or 0. Lvs in spirals, rarely opp., simple to pinnately dissected, rarely with articulated leaflets; stip. 0. Fls in usu. bractless racemes, rarely spikes or solit., bisex. (rarely dioec. or gynodioec.: see *Erucastrum, Lepidium, Pachycladon*) & usu. regular, hypogynous, receptacle rarely prolonged into gynophore (cf. Capparaceae) but oft. with nectaries (s.t. forming ring around A or G); K 2 + 2 (v. rarely (K)), decussate, outer oft. gibbous basally, C 4 diagonal to K, imbr. or convolute, usu. with elongate claw, rarely 0 (e.g. *Pringlea*), A 2 short + 4 longer (tetradynamous), inner derived from only 2 primordia & s.t. basally connate in prs (2–4 in *Lepidium*, up to 16 in *Megacarpaea*), anthers with longit. slits, $\underline{G}$ (2), ± style, ovary with 2 locules sep. by a thin replum connecting 2 parietal placentas, each with (1–) ∞ ovules in 2 rows sep. by replum, anatr. to campylotropous, bitegmic, endotestal (cf. exotegmic fibres of Capparaceae); nectar hexose-rich. Fr. dry-dehiscent, usu. a siliqua (elongate) or silicula (short), valves falling to reveal replum; seeds with large oily embryo, folded (straight in 5 spp. of *Leavenworthia*), with the cotyledons oft. lying against radicle & 0 or little (oft. 1 layer of cells) endosperm. Germ. epigeal. n = (4) 5–12+

Classification & principal genera (R: T 61(2012)933) – 49 molecularly distinct tribes (here in alphabetical order, though many genera (e.g. *Ricotia*) unplaced):

1. **Aethionemeae** (shrubs to ann. herbs; n = 7, 8, 11, 12, 14, 16, 18, 21, 22, 24, 30!); app. 'sister' to rest of fam., 1/45 Euras.): *Aethionema*
2. **Alysseae** (stellate hairs, few-seeded fr., seeds oft. winged, x = 8; 17/262): *Alyssum, Aurinia, Hormathophylla*
3. **Alyssopsideae** (4/9 C As.): *Olimarabadopsis*
4. **Anastaticeae** (13/74): *Anastatica, Farsetia, Lobularia, Malcomia*
5. **Anchonieae** (multicellular-multiseriate glands, erect K, x = 7; 9/73 mostly Euras.; excl. 13.): *Matthiola, Parrya*
6. **Aphragmeae** (1/11): *Aphragmus*
7. **Arabideae** (branched hairs, entire or dentate lvs, x = 8; 17/488): *Arabis, Aubrieta, Draba*
8. **Asteae** (1/1): *Asta*
9. **Lunarieae** (Biscutelleae; 3/49): *Biscutella, Lunaria*
10. **Bivonaeeae** (1/1; R: T 61(2012)89): *Bivonaea*
11. **Boechereae** (x = 7, N Am., 1 sp. to E As.; R: SB 38(2013)203; 8/127 esp. N Am.): *Boechera*
12. **Brassiceae** (great fr. variation; 47[to be reduced]/227 esp. Med.; excl. 14.): *Brassica, Eruca, Moricandia, Raphanus, Sinapidendon, Sinapis*
13. **Buniadeae** (~ 5.; 1/2): *Bunias*
14. **Calepineae** (~ 12.; 3/9 Euras.): *Goldbachia*
15. **Camelineae** (usu. annuals; n = 8, s.t. red. to 5 (4); 8/32 Euras., N Am., excl. 37., 48.): *Arabidopsis, Camelina, Capsella*
16. **Cardamineae** (usu. with deeply lobed or compound lvs, x = 8, oft. moist places; 12/337): *Armoracia, Barbarea, Cardamine, Nasturtium, Rorippa*
17. **Chorisporeae** (multicellular-multiseriate glands, erect K, lomenta, x = 7; 4/56 As.; R: PSE 294(2011)65): *Chorispora, Parrya*
18. **Cochlearieae** (glabrous, entire rosette-lvs, x = 6 or 7; 2/29 N temp.): *Cochlearia, Ionopsidium*
19. **Coluteocarpeae** (Noccaeeae; glabrous perennials, smooth seeds, oft. auricled scape-lvs, x = 7; 3/127 almost all Euras., N Afr.): *Noccaea*
20. **Conringieae** (2/9 Euras.): *Conringia*
21. **Cremolobeae** (2/32 S Am.; R: T 62(2013)1228): *Menonvillea*
22. **Crucihimalayeae** (3/13 C As. ext. to N Am.): *Crucihimalaya*
23. **Descurainieae** (usu. dendritic hairs, 1–3-pinnatisect lvs, x = 7; 6/45): *Descurainia*
24. **Dontostemoneae** (2/15 C As. ext. to Eur.): *Dontostemon*
25. **Erysimeae** (1/200): *Erysimum*
26. **Euclidieae** (multicellular glands 0, erect K, x = 7): *Braya, Solms-laubachia*
27. **Eudemeae** (7/31 S Am.): *Eudema, Xerodraba*
28. **Eutremeae** (simple or 0 hairs, white fls, x = 7; 3/34 mostly As.): *Eutrema*

29. **Halimolobeae** (x = 8; 5/39 Am.): *Halimolobos, Pennellia*
30. **Heliophileae** (simple or 0 hairs, usu. appendaged C &/or A, extrafl. nectaries, n = 10; 1/90 S Afr.): *Heliophila*
31. **Hesperideae** (stalked glands, apically unicellular, x = 6–10; 2/35 Medit.): *Hesperis*
32. **Iberideae** (simple or 0 trichomes, corymbs of oft. irreg. fls; 2/30 Euras.): *Iberis, Teesdalia*
33. **Isatideae** (simple or 0 hairs, indehiscent 1- or 2-seeded fr., x = 7; 5/90 Euras.): *Isatis*
34. **Kernereae** (2/2 S Eur.): *Kernera*
35. **Lepideae** (1 ovule per loc.; seeds oft. mucilaginous, x = 8; 3/252 cosmop.): *Lepidium*
36. **Megacarpaeeae** (2/11 esp. C As.): *Megacarpaea*
37. **Microlepidieae** (~ 15.;16/59 Aus., NZ): *Pachycladon, Stenopetalum*
38. **Notothlaspideae** (1/2 NZ): *Notothlaspi*
39. **Oreophytoneae** (2/6 SW Eur., Afr.): *Murbeckiella*
40. **Physarieae** (usu. stellate hairs, pollen with 4 or more colpi (3 in rest of fam.), x = 8; 7/133 almost all US; R: SB 38(2013)178): *Physaria*
41. **Schizopetaleae** (2/16 S Am.): *Schizopetalon*
42. **Scoliaxoneae** (1/1 Mex.): *Scoliaxon*
43. **Sisymbrieae** (simple or 0 hairs, bifid stigmas, x = 7; 1/41): *Sisymbrium*
44. **Smelowskieae** (perenn. with pinnatisect lvs, several-∞-seeded fr., x = 6; 1/25 C As. to N Am.): *Smelowskia*
45. **Stevenieae** (3/11 As. ext. to Eur.): *Macropodium, Stevenia*
46. **Thelypodieae** (~ 41.; 26/244): *Mostacillastrum, Pringlea, Stanleya, Streptanthus, Thelypodium, Weberbauera*
47. **Thlaspideae** (simple or 0 hairs, entire scape-lvs, x = 7; 12/34 Eur., SW As.): *Alliaria, Thlaspi*
48. **Turritideae** (~ 15.; 1/2 Euras.): *Turritis*
49. **Yinshanieae** (1/13 China, Vietnam): *Yinshania*

Two thirds of genera with 1–4 spp., but more than half spp. in 11 genera. Although common in open dry habitats, C. include aquatic spp. of *Nasturtium* & *Subularia* & world alt. record (*Solms-laubachia*) for seed-pls. Glucosinolates & myrosinase brought together by damage releasing insect-deterrent compounds, though cabbage aphids' eating does not do this, though when attacked, they release a form of myrosinase, thus generating mustard gas. Cabbage white butterfly caterpillars tolerate glucosinolates by transforming them to non-toxic nitriles. Spirolobal cotyledons arose in 3 tribes, lianes in *Cremolobus, Heliophila* & *Lepidium*; chromosome reduction to 4 in *Physaria, Stenopetalum*

Arabidopsis thaliana (a dreary little weed) is THE botanical model organism for study of genetics & development. Many salads (cresses (*Barbarea, Lepidium, Sinapis*), rocket (*Eruca*)) & other veg. incl. *Brassica, Raphanus* & of lesser imp. *Armoracia, Crambe, Eutrema* (wasabi) & *Nasturtium* (watercress), oilseeds (*Brassica, Descurainia, Physaria*) & many cult. orn.: *Aethionema, Alyssum, Arabis, Aubrieta, Aurinia, Cochlearia, Draba, Erysimum* (wallflowers), *Heliophila, Hesperis, Hormathophylla, Iberis, Ionopsidium, Lobularia, Lunaria, Malcolmia, Matthiola* (stock), *Parrya, Schizopetalon; Isatis* gives woad & *Camelina* a fibre; weeds incl. spp. of *Arabidopsis, Capsella, Cardamine* & *Carrichtera*

crucifix orchid *Epidendrum* spp.

Crucihimalaya Al-Shehbaz & al. (~ *Arabidopsis*). Cruciferae (22). 11 C As. to Himal., Mongolia. R: Novon 9(1999)298

Cruckshanksia Hook. & Arn. Rubiaceae (IV 1). (Excl. *Oreopolus*) 7 N Chile & adjacent Arg. R: AMBG 83(1996)470

Cruddasia Prain (~ *Ophrestia*). Leguminosae (III 18). 2 trop. As.

Crudia Schreb. Leguminosae (I 2). c. 50 trop. (Afr. 10 – R: SGP 78(2008)87), esp. riverine forests, some water-disp. in Amazon (air-filled cavity between cotyledons). C ± 0. R: BJBBuit. 18(1950)407

cruel plant *Araujia sericifera*

Crumenaria Mart. Rhamnaceae (4). 4 C Am. to Argentina. R: Cand. 68(2013)268

crummock *Sium sisarum*

Crunocallis Rydb. = *Montia*

Crupina (Pers.) DC. Compositae (Card.-Cent.). 3 Med. (Eur. 2) & SW As. to China

Crusea Cham. & Schldl. Rubiaceae (IV 15). 13 SW N & C Am. R: MNYBG 22,4(1972)1

Cruzia Philippi = *Scutellaria*

cry-baby *Erythrina crista-galli*

Cryanthemum Kamelin = *Chrysanthemum*

Crybe Lindl. = *Bletia*

Cryophytum N.E. Br. = *Mesembryanthemum*

Cryosophila Blume. Palmae (III 2). 10 W Mex. to N Colombia. R: SBM 46(1996)33. Branched palms; some cult. orn. oft. under name *Acanthorrhiza* – r.-spines covering base of trunk. *C. **stauracantha*** (Heynh.) R. Evans (*C. argentea*, escoba palm, C Am.) – oft. an indicator of *Swietenia macrophylla* presence

Cryphia R. Br. = *Prostanthera*

Cryphiacanthus Nees = *Ruellia*

Crypsinopsis Pichi-Serm. = *Selliguea*

Crypsinus C. Presl = *Selliguea*

Crypsis Aiton = *Sporobolus* (but see BRCI 11D(1962)91)

Cryptadenia Meissn. = *Lachnaea*

Cryptandra Sm. Rhamnaceae (5). c. 55 temp. Aus. (esp. SW, SE)

Cryptangium Schrad. ex Nees = *Lagenocarpus*

Cryptantha Lehm. ex G. Don. Boraginaceae (B.3.8.4). Incl. *Eremocarya, Greeneocharis, Johnstonella, Oreocarya* 160 W N Am. Cult. orn.

Cryptanthemis Rupp = *Rhizanthella*

Cryptanthopsis Ule = *Orthophytum*

Cryptanthus Osb. = *Clerodendrum*

Cryptanthus Otto & A. Dietr. Bromeliaceae (3). 45 E Braz. R: FN 14(1979) 1586. Cult. orn. dwarf pls (earth-stars), esp. ***C. zonatus*** (Vis.) Beer & ***C. bivittatus*** (Hook.) Regel (only known in cult.), also hybrids incl. with spp. of *Billbergia*

Cryptarrhena R. Br. Orchidaceae (V 12j). 3 trop. Am. R: Lindleyana 8(1993)163

Crypteronia Blume. Crypteroniaceae. 4–7 SE As., Mal. Young shoots of ***C. paniculata*** Blume (Mal.) eaten with rice, timber useful

Crypteroniaceae A. DC. Magnoliidae – Myrtales. Incl. Alzateaceae, Oliniaceae, Penaeaceae, Rhynchocalycaceae, 13/45 trop. As., S Afr. & S Am. Trees and shrubs (s.t. ericoid) oft. with quadrangular twigs & s.t. accum. aluminium. Lvs opp. or whorled, simple, entire with a continuous marginal vein; stip. minute or 0. Fls in axillary racemes to thyrses, v. small, bisex. (rarely trees dioec.), reg., oft. perigynous; K (4)5(6), valvate & oft. persistent, C 0 or (4)5(6), A (4 or 8)5(6) or 10, $\underline{G}$ or $\overline{G}$ (2–4(5, 6)), 1–6-loc. with term. style, ovules 1–3 or 40–60 (*Alzatea*) per loc., anatr. to hemipterous, bitegmic, on usu. axile placentas. Fr. a loculicidal capsule or drupe (*Olinia*), the valves oft. held together apically by persistent style; seeds usu. small, flat with membranous wing & 0 endosperm. n = 11, 12, 14

Principal genera: *Alzatea, Axinandra, Crypteronia, Dactylocladus, Olinia, Penaea, Rhynchocalyx*

Alzatea, Olinia, Penaea and allies, *Rhynchocalyx* s.t. removed to own fams

Gums (*Penaea*), timber etc. (*Crypteronia*)

Cryptobasis Nevski = *Iris* (but see BZ 88[10](2003)50)

Cryptocapnos Rech.f. Papaveraceae (Fumariaceae I 2). 1 Afghanistan: ***C. chasmophytica*** Rech.f. R: OB 88(1986)91

Cryptocarpus Kunth. Nyctaginaceae (2). 1 W S Am., Galápagos: ***C. pyriformis*** Kunth

Cryptocarya R. Br. Lauraceae (I). (Incl. *Ravensara*) c. 300 trop. & warm esp. As. (Aus. 47 – R: FA 2(2007)140). Roux's Model. Alks (***C. pleurosperma*** C. White & Francis (*C. glabella*, poison walnut, NE Aus.) – sap causes severe blistering so avoided by lumberjacks). ***C. agathophyllum*** van der Werff (*Ravensara aromatica*, Madag. cloves or nutmeg, Madag.) – seeds a spice, bark used in local rum-making; ***C. massoy*** (Oken) Kosterm. (*C. aromatica*, NG) – ess. oil from aromatic bark (massoy b.); ***C. fusca*** Gillespie (Fiji) – introd. Tonga for scenting coconut oil; ***C. glaucescens*** R. Br. (brown beech, jackwood, E Aus.) – good timber; ***C. hornei*** Gillespie (Polynesia) – timber for handicrafts in Tonga; ***C. latifolia*** Sonder (S Afr.) – source of ntonga nuts used locally for their oil; ***C. moschata*** Nees & Mart. (Braz.) – fr. (Braz. nutmegs) used as spice

Cryptocentrum Benth. = *Maxillaria* (but see BJ 97(1977)562)

Cryptocereus Alexander = *Selenicereus*

Cryptochilus Wall. Orchidaceae (V 15). 5 Himal. Bird-poll.?

Cryptochloa Swallen. Gramineae (VI 3). 8 trop. Am. R: CUSNH 39(2000)53. Bamboos, some ant-disp. (elaiosomes)

Cryptocodon Fed. (~ *Asyneuma*). Campanulaceae (I 9). 1 C As.: ***C. monocephalus*** (Trautv.) Fed.

Cryptocoryne Fischer ex Wydler. Araceae (VII 9). c. 70 Indomal. Marsh & water pl., oft. stoloniferous, with spathe-length governed by water-depth (open to air apically & entered by poll. beetles detained overnight), fr. underwater, seeds with cotyledons as floats dropped after 2 mins, some, e.g. ***C. ciliata*** (Roxb.) Wydler, viviparous; many cult. orn. aquarium pls esp. ***C. spiralis*** (Retz.) Wydler (Ind. ipecacuanha, Ind.) – trad. medic. Sri Lanka, & ***C. walkeri*** Schott (Sri Lanka)

Cryptodiscus Schrenk ex Fischer & C. Meyer = *Prangos*

Cryptogramma R. Br. Pteridaceae (I). 6 alpine & boreal (incl. Eur.). ***C. crispa*** (L.) Hook. (parsley fern) – cult. orn.

Cryptogrammaceae Pichi-Serm. = Pteridaceae

Cryptogyne Hook.f. = *Sideroxylon*

Cryptolepis R. Br. Apocynaceae (III; Periplocaceae). 31 Afr. (27), As (3), Aus. (1). ***C. nigrescens*** (Afzel.) L. Joubert & Bruyns (*Parquetina gabonica*, W Afr.) – local medic., stuns fish

Cryptolluma Plowes = *Caralluma*

Cryptomeria D. Don. Taxodiaceae. 1 Jap. (introd. China): ***C. japonica*** (L.f.) D. Don (Jap. cedar) – principal conifer in Jap. forestry, much grown for timber (sugi), esp. used for boats, much cult. orn. (Rauh's Model) with many cvs incl. dwarf forms derived from witches' brooms, the type natur. in Azores, where it (unlike trees in Jap.) produces suckers; old wood buried in ground in Jap. becomes dark green (jindai-sugi). R: A. Farjon, *Monogr. Cupressaceae & Sciadopitys* (2005)118

Cryptophoranthus Barb. Rodr. = *Pleurothallis*

Cryptophragmium Nees = *Gymnostachyum*

Cryptophysa Standl. & J.F. Macbr. = *Conostegia*

Cryptopus Lindl. Orchidaceae (V 16d). 4 Madag. & Mascarenes

Cryptopylos Garay. Orchidaceae (V 16c). 1 SE As., Sumatra: ***C. clausus*** (J.J. Sm.) Garay. R: OB 95(1988)266

Cryptorhiza Urb. = *Pimenta*

Cryptosepalum Benth. Leguminosae (I 2). 11 trop. Afr. C 1, A 3. Timber, bark fibre, local medic.

Cryptospora Karelin & Kir. Cruciferae (26). 4 C As.

Cryptostegia R. Br. Apocynaceae (III; Periplocaceae). 2 Madag. R: Adansonia III,23 (2001)212. ***C. grandiflora*** R. Br. (rubber vine) – cult. orn. liane with (Madag.) rubber comparable with that of *Hevea*, widely natur. trop., toxic to stock, bad weed in Aus.

Cryptostemma R. Br. = *Arctotheca*

Cryptostephanus Welw. ex Bak. Amaryllidaceae. 3 trop. Afr. R: BM 27(2010)18. Berry

Cryptostylis R. Br. Orchidaceae (IV 3c). 23 Indomal. to W Pac. (Aus. 5) & Taiwan. Fls not resupinate. Aus. spp. with pseudocopulation (same ichneumon, *Lissopimpla excelsa*). Cult. orn. incl. ***C. subulata*** (Lab.) Reichb.f. (Aus.) – recently self-introd. NZ. ***C. erecta*** R. Br. (Aus.) – tubers form. eaten

Cryptotaenia DC. Umbelliferae (III 8). 4 N temp. & trop. Afr. mts. ***C. canadensis*** (L.) DC. (*C. japonica*, E As., N Am., natur. Austria) – cult. Jap. for salad (Jap. hornwort or parsley, mitsuba or mitzuba) & fried r.

Cryptotaeniopsis Dunn = *Pternopetalum*

Cryptothladia (Bunge) M. Cannon = *Morina*

Ctenanthe Eichler (~ *Myrosma*). Marantaceae. c. 10 Braz., Costa Rica (1). Cult. orn. like *Calathea*

Ctenardisia Ducke. Primulaceae (Myrsinaceae). (Incl. *Yunckeria*) 5 trop. Am. R: Wrightia 7(1982)42

Ctenitis (C. Chr.) C. Chr. (*Atalopteris*). Dryopteridaceae (I). c. 120 trop. & warm (not Aus.). R: Blumea 31(1985)1

Ctenium Panzer. Gramineae (XXIX 3). 20 Am., Afr. (8 – R: KB 69[9541](2014)2), Madag. Savannas. ***C. aromaticum*** (Walter) Wood (*C. gangito*, toothache grass, Virginia to Florida) – forage (r. spicy when fresh), fls only after fire

Ctenocladium Airy Shaw = *Dorstenia*

Ctenolepis Hook.f. Cucurbitaceae (XIV). Incl. *Zombitsia*, 3 trop. Afr., Madag. & Ind.

Ctenolophon Oliv. Ctenolophonaceae (Linaceae s.l.). 1 W trop. Afr., 1 Mal. (very similar). Some timber

Ctenolophonaceae Exell & Mendonça (~ Linaceae). Magnoliidae – Malpighiales. 1/2 OW trop. Buttressed trees with simple & stellately tufted hairs, oxalate crystals. Lvs opp., entire; stip. interpetiolar, caducous. Infl. thyrsoid; fls 5-merous, in bud elongate-ellipsoid.

K quincuncial imbr., rounded, accrescent in fr. C contorted. Disk extrastaminal. A 10 adnate to base of disk, of 2 lengths, anthers dorso-versatile with protruding connective. G (2) with 2 ovules per locule. Fr. a capsule, ribbed, splitting into 2 valves. Seed 1 with hairy aril, pendulous from persistent filiform columella; endosperm copious; embryo straight, cotyledons large, folded
Genus: *Ctenolophon*
Marginal stomata on disk, anthers with broad connective, both like Humiriaceae

Ctenomeria Harv. (~ *Tragia*). Euphorbiaceae (Acal.-Pluk.-Trag.). 2 S Afr.

Ctenopaepale Bremek. = *Strobilanthes*

Ctenophrynium K. Schum. = *Saranthe*

Ctenopsis De Notaris = *Festuca*

Ctenopterella Parris (~ *Ctenopteris*). Polypodiaceae (V). 12 OW trop. R: GBS 58(2007)234

Ctenopteris Blume ex Kunze = *Prosaptia*

Cuatrecasanthus H. Robinson (~ *Vernonia*). Compositae (Vern.-Pip.). 6 Ecuador, Peru. R: PhytoKeys 14(2012)24. 1-flowered capitula

Cuatrecasasiella H. Robinson (~ *Luciliopsis*). Compositae (Gnap.-Gnap.). 2 Andes. R: BJLS 106(1991)185

Cuatrecasasiodendron Standl. & Steyerm. Rubiaceae (I 5). 2 Colombia

Cuatrecasea Dugand = *Iriartella*

Cuatresia Hunz. Solanaceae (IV 5d). 11 trop. Am. R: OB 92(1987)73. Shrubs & trees

Cuba(n) bast *Hibiscus elatus*; **C. hemp** *Furcraea hexapetala*; **C. lily** *Scilla peruviana*; **C. mahogany** *Swietenia mahagoni*; **C. oregano** *Coleus amboinicus*; **C. oysterwood** *Gymnanthes lucida*; **G. spinach** *Claytonia perfoliata*

Cubacroton Alain = *Croton*

Cubanola Aiello. Rubiaceae (III 4). 2 Cuba, Hispaniola

Cubanthus (Boiss.) Millsp. = *Euphorbia*

cubebs *Piper cubeba*; **African c.** *P. clusii*; **Guinea c.** *P. guineense*

Cubelium Raf. (~ *Hybanthus*). Violaceae. 1 E N Am.: ***C. concolor*** (Spreng.) Raf.

Cubilia Blume. Sapindaceae. 1 C Mal.: ***C. cubili*** (Blanco) Adelb. (*C. blancoi*) – lvs a veg. & seeds (kubili nuts) ed., cult. Java. R: FM I,11(1994)490

Cubitanthus Barringer (~ *Anetanthus*). ? Linderniaceae. 1 Braz.: ***C. alatus*** (Cham. & Schldl.) Barringer. R: JAA 65(1984)145

cucamelon *Melothria scabra*

Cuchumatanea Seidenschnur & Beaman = *Alepidocline*

cuckold *Bidens* spp.

cuckoo flower *Cardamine pratensis*; **c. pint** *Arum maculatum*

Cucubalus Tourn. ex L. = *Silene*

cucumber *Cucumis sativus*; **apple c.** *C. sativus* 'Crystal Apple'; **bitter c.** *Momordica charantia*; **bur c.** *Sicyos angulatus*; **c. root** *Medeola virginiana*; **Lebanese c.** *C. sativus* Lebanese Group; **Persian c.** *C. s.* cvs; **squirting c.** *Ecballium elaterium*; **c. tree** *Averrhoa bilimbi*, *Magnolia acuminata*

Cucumella Chiov. = *Cucumis*

Cucumeria Luer = *Specklinia*

Cucumeropsis Naudin. Cucurbitaceae. 1 trop. W Afr. (? native trop. Am.): ***C. mannii*** Naudin (*Posadaea sphaeocarpa*) – seeds (egusi) source of oil & protein

Cucumis Tourn. ex L. Cucurbitaceae (XIV). Incl. *Cucumella* (R: Britt. 46(1994)163), *Dicoelospermum*, *Mukia*, *Myrmecosicyos* (R: KB 15(1962)357), *Oreosyce*, 52 OW trop. R: J.H. Kirkbride (1993) *Biosystematic monograph of the genus C.*; Blumea 52(2007)166 (5 cross-sterile spp. groups in 2 subg.), PNAS 107(2010)14,269. Cult. since earliest times. ***C. anguria*** L. (tindori, WI gherkin, cultigen derived from 'subsp. *longipes*' (Afr.)) – young fr. boiled or pickled; ***C. melo*** L. (melon, orig. As.) – 'subsp. *agrestis*' (wild forms with inedible fr., incl. senat seed once exported from Sudan as oilseed) & 7 major interfertile cv. groups incl. **Cantalupensis Group** (cantaloupe (true) with rough-warty but not netted skin, not in comm. prod. in Am.), **Flexuosus Group** (var. *flexuosus*, snake melon to 2m long, pickled in Middle E), **Inodorus Group** (incl. Chinese m., honeydew m. with smooth rind, casaba m. with wrinkled, flesh crisp), **Reticulatus Group** (mash or musk m., 'cantaloupe' (not true), rock m. (NSW), with ± strongly netted rind & musky orange flesh, the most imp. in comm.), & others largely grown as 'ornamental gourds' & cooking forms resembling small marrows; ***C. humifructus*** Stent (trop. & S Afr.) – fr. underground excavated & disp. by water-seeking aardvarks avoiding dangerous waterholes, also disp. by Bushmen; ***C. messorius*** (C. Jeffrey) H. Schaefer (*Myrmecosicyos m.*, Kenya) –

found around nests of harvester ants in Rift Valley; ***C. metuliferus*** E. Mey. ex Naudin (kiwano, horned or jelly melon, trop. & S Afr.) – fashionable dessert fr. grown in NZ (cf. *Actinidia deliciosa*); ***C. sativus*** L. (cucumber, Sino-Himal., v. rare in wild) – prob. cult. by 'Hoabhinian' culture 8000–3000 BP, now several cv. groups eaten raw (e.g. **'Crystal Apple'** (apple cucumber), small-fruited **Lebanese Group** (Lebanese c.) & Persian cs., & larger **Standard Group** (frame, ridge, white cs, those in N Eur. usu. parthenocarpic so poll. discouraged so as to prevent seed-formation), or pickled (**Pickling Cucumber Group** e.g. **'Gherkins Pariser'**, cornichon), most 'gherkins' being immature small-fruited cvs), also in face-cleansing cosmetics, flavour due to an aldehyde (nona-2,6-dienal) with odour threshold of 0.0001 p.p.m.!

Cucurbita Tourn. ex L. Cucurbitaceae (XV). c. 15 spp. or groups trop. & warm Am. R: EB 44(3) suppl. (1990)58. Formed part of the squash/beans/maize culture of pre-Columbian Am. Monoec.; 2 genera of solit. bees (*Peponapis* & *Xenoglossa*) derive all food from nectar & pollen of *C.* spp. Genus prob. native in S as well as C Am., *C. moschata* most like original sp. & domesticated independently in C & S Am. while the rest as follows: *C. maxima* (S Am., from *C. andreana* Naudin, Arg. & Urug.), *C. ficifolia* (C Am. S of Mex.), *C. argyrosperma* (Mex. S of M. City, perhaps derived from *C. sororia* L. Bailey, C Am.) & *C. pepo* (Mex. N of M. City, perhaps derived from var. *texana* (Scheele) Decker &/or *C. fraterna* L. Bailey perhaps domesticated independently in E US (subsp. *ovifera*) & Mex. (subsp. *pepo*)). The terms 'squash' & 'pumpkin' are used for more than 1 sp.: ***C. argyrosperma*** Hort. Huber (*C. mixta*, winter squash, pumpkin, cushaw, silverseed gourd); ***C. ficifolia*** Bouché (cidra, sidra (Spain), Malabar gourd) – made into kind of marmalade (stringy) esp. Mallorca, strands of candied fr. being 'angel's hair' or 'cabello de angel'; ***C. foetidissima*** Kunth (buffalo gourd, chilicote) – poss. disp. by extinct megafauna, drought-tolerant, seeds source of oil & protein, r. laxative; ***C. maxima*** Duchesne (*Pepo macrocarpus*, autumn & winter squash & pumpkin) – fr. globose, largest of all frs (to 2 m round & 90(– 691 in Calif. 2007) kg) to oblong, many cvs (incl. **'Turbaniformis'**, cult. orn. Turk's Cap gourd), anthelminthic in Malta, used against benign prostate hypertrophy; ***C. moschata*** Duchesne (*Pepo indicus*, pumpkin, butternut or coquina or winter squash, a(h)uyama, ayote, calabaza, zapallo) – a keeping-squash like *C. maxima*, anthelminthic (a carboxypyrrolidine called cucurbitine); ***C. pepo*** L. 'subsp. *pepo*' (veg. marrow, veg. spaghetti, summer & autumn squash & pumpkin (non-keeping), pomion) – one of oldest domesticated pls (9000 yrs BP Mex.) poss. derived from *C. melopepo* L., to 47.85 kg (UK record), seeds ('pumpkin nuts', pepitas) ed. (found in mastodon intestines) & rich in zinc & omega-3 fats (health-food spread), many cvs (8 groups incl. **'Zucchini'** (zucchini, courgette) ed. immature, **'Summer Crookneck'** – club-shaped with curved neck & 'subsp. *ovifera*' – the hard-shelled coloured orn. gourds), also vermifuge

Cucurbitaceae Juss. Magnoliidae – Cucurbitales. 97/990 trop. & warm with few temp. R: C. Jeffrey in BJLS 81(1980)233; D. Bates & al. (eds, 1990) *Biology & Utilization of the C.*: 449; T 60(2011)129. Usu. juicy climbers or trailers with hypocotylar tubers, rarely somewhat woody or arborescent (*Dendrosicyos*), usu. with coiled tendrils, 1 at each node, these rarely spines or 0 (*Ecballium*), oft. with bristly (glandular) hairs with calcified walls, freq. with bitter purgative cucurbitacins, tannins 0; stem usu. angular, vascular bundles usu. in 2 cycles & nearly always bicollateral, s.t. otherwise unusual too. Lvs in spirals, usu. palmately veined or -lobed, oft. with extrafl. nectaries, margins with hydathodes; stip. 0. Fls axillary, solit. or racemes to thyrses, unisexual (pls. monoec. or dioec.), v. rarely bisex., usu. reg. & epigynous; K (3–)5(–7), imbr. or open, C (3–)5(–10) or ((3–)5(–7)) usu. yellow or white ± valvate & oft. diff. in males & females, A essentially 5 attached to hypanthium (rarely around summit of ovary), alt. with C, but usu. reduced or displaced so that oft. app. A 3 (2 with dithecal anthers, 1 with monothecal), free or connate, anthers with longit. slits, $\overline{G}$ ((1–)3(–5)) with intruded parietal placentas, s.t. joined to form pluriloc. ovary, style solit. with 1–3(–5) usu. bilobed stigmas, or 2 or 3, each with bilobed stigma, ovules (1–)∞, anatropous, bitegmic. Fr. usu. a berry (when hard-walled termed a pepo), less oft. a capsule (s.t. fleshy & explosive), rarely samaroid or with pyrenes (*Hodgsonia*); seeds (1–)∞, large, oft. flattened & even winged, with oily straight embryo & ± 0 endosperm, cotyledons large. x = 7–14.

Classification & chief genera (R: T 60(2011)129; many monospecific; 15 tribes):

I. **Gomphogyneae** (tendrils usu. apically 2-fid, A 3, 5, caps. or berry): *Hemsleya*

II. **Triceratieae** (Fevilleeae (tendrils usu. apically 2-fid, A 1–5, pepo, samara or achene, seeds 1 or 10–15 s.t. winged): *Fevillea, Sicydium*

III. **Zanonieae** (tendrils usu. 2-fid, A 4, 5, caps., seeds few, winged): *Gerrardanthus*

IV. **Actinostemmateae** (tendrils usu. 2-fid, A 5, 6, pyxidium, seeds few): *Actinostemma* (only)
V. **Indofevilleeae** (tendrils 2-fid, A 5, fr. indehisc.), seeds many: *Indofevillea* (only)
VI. **Thladiantheae** (tendrils simple or 2-fid, A 5, fr. indehisc. fleshy R (Thladianthiinae): BZ 91(2006)767)): *Thladiantha*
VII. **Siraitieae** (tendrils 2-fid, A 5 s.t. 2 prs + 1, fr. indehisc.): *Siraitia* (only)
VIII. **Momordiceae clade** (tendrils simple or 2-fid; A 2, 3; fr. usu. spiny): *Momordica* (only)
IX. **Joliffieae** (Telfairiae; tendrils simple or 2-fid, A 3(5), fr. fleshy): *Telfairia*
X. **Bryonieae** (tendrils simple or 0, A 3, berry): *Bryonia, Ecballium*
XI. **Schizopeponeae** (tendrils 2- or 3-fid, A 3, fr. indehisc. or 3-valved): *Schizopepon*
XII. **Sicyoeae** (tendrils simple or 2–8-fid, A 2–5): *Cyclanthera, Hodgsonia, Luffa, Marah, Sicyos, Trichosanthes*
XIII. **Coniandreae** (tendrils (0) simple, 2- or 3-fid, A 2, 3, 5, fr. fleshy): *Dendrosicyos, Gurania, Kedrostis*
XIV. **Benincaseae** (tendrils (0), simple to 5-fid, A (2)3(–5), fr. usu. indehisc.): *Benincasa, Citrullus, Coccinia, Cucumis, Lagenaria, Peponium, Zehneria*
XV. **Cucurbiteae** (tendrils simple to 7-fid, A (2)3(4)): *Cayaponia, Cucurbita*

All have frost-sensitive aerial parts & perenn. spp. in temp. regions have tubers (e.g. *Bryonia*); 1 of 10 fams with elaiophores attractive to bees (*Momordica, Thladiantha*)

Presence of cucurbitacins supports affinity with Begoniaceae & Datiscaceae; seed structure differing markedly from that in Passifloraceae or Caricaceae with which they have been allied also predictive: they have no members 'intermediate' between C. & another family

Of enormous economic importance as foodpls (gourds, melons, marrows, squashes, pumpkins etc.): *Benincasa, Citrullus, Coccinia, Cucumis, Cucurbita, Lagenaria, Momordica, Sechium, Trichosanthes* etc., also loofah (*Luffa*), oilseeds & medic. (*Apodanthera, Cayaponia, Citrullus, Cucumeriopsis, Ecballium, Fevillea, Hemsleya, Solena, Telfairia, Trichosanthes*); fls of many ed. too while others provide 'ornamental gourds' in cult.

Cucurbitella Walp. Cucurbitaceae (VIII). 1 S Am.: ***C. asperata*** (Hook. & Arn.) Walp. R: AMBG 85(1998)436

cudjoe wood *Jacquinia keyensis*

Cudrania Trécul = *Maclura*

cudweed *Filago* spp., *Gamochaeta* spp., *Gnaphalium* spp.

Cuenotia Rizz. Acanthaceae. 1 NE Braz.: ***C. speciosa*** Rizz.

Cuervea Triana ex Miers (~ *Hippocratea*). Celastraceae (III). 5 trop. Am., W Afr. Roux's Model

Cufodontia Woodson = *Aspidosperma*

cuichunchulli *Pombalia parviflora*

cuipo *Cavanillesia platanifolia*

Cuitlauzina Lex. (~ *Odontoglossum*). Orchidaceae (V 12h). Incl. *Palumbina*, 7 C Am. Cult. orn.

cilantro *Eryngium foetidum*

Culcasia P. Beauv. Araceae (VII 14). 24 trop. Afr. esp. W & C. ***C. panduriformis*** Engl. & K. Krause – 'litter-box' pl.

Culcita C. Presl. Culcitaceae (Dicksoniaceae). Excl. *Calochlaena*, 2 trop. Am., Azores to Iberia (1), Mal., Aus., New Caled. R: M.F. Large & J.E. Braggins, *Tree Ferns* (2004)79. ***C. macrocarpa*** C. Presl (Atlantic Is.) – rhiz. hairs for stuffing in Madeira

Culcitaceae Pichi-Serm. Polypodiidae – Cyatheales. 1/ 2 SW Eur., to Azores, trop. Am. Terr. ferns with creeping or erect solenostelic rhiz. with articulate hairs. Petioles in t. s. with gutter-shaped vasc. bundle; frond 4–5-pinnate-pinnatifid; veins free, oft. forked. Sori term. on veins; spores tetrahedral-globose. $x = 66$

Genus: *Culcita*

Culcitium Bonpl. = *Senecio*

culilawan *Cinnamomum culitlawan*

Cullen Medik. (~ *Psoralea*). Leguminosae (III 20). 34 warm OW (Aus. 32, 24 endemic) to Med. R: AusSB 10(1997)581. ***C. corylifolium*** (L.) Medik. (*P. c.*, bawchan seed, Ind. & Sri Lanka) – principal cult. medic. pl. of S Arabia, trad. Chinese & Hindu medic., used in psoriasis, leprosy; ***C. patens*** (Lindl.) J.W. Grimes (*P. p.*, SE Aus.) – Aboriginal string source

Cullenia Wight (~ *Durio*). Malvaceae (Hel.). 3 Ind., Sri Lanka. R: BJBB 40 (1970)244. ***C. exarillata*** Robyns (SW Ghats, Ind.) – during fr. scarcity fls visited & eaten by poll. bats & other mammals ('predator- poll.')

Cullumia R. Br. Compositae (Arct.-Gort.). 15 Cape to Karoo. R: MBSM 3(1959)271

Culver's root *Veronicastrum virginicum*

Cumarinia F. Buxb. (~ *Coryphantha*). Cactaceae (III 9). 1 Mex.: ***C. odorata*** (Boedecker) F. Buxb.

cumbungi (reed) *Typha* spp.

cumin *Cuminum cyminum*; **black c.** *Nigella sativa*

Cumingia S. Vidal = *Camptostemon*

Cuminia Colla. Labiatae (VII 2b). 1 Juan Fernandez: ***C. eriantha*** (Benth.) Benth.- woody. R: KM 3(1986)154

Cuminum Tourn. ex L. Umbelliferae (III 2). 2 Medit. to Sudan & C As. R: T 65(2016)585. ***C. cyminum*** L. (cumin, Medit. – cult. since time of Minoans (C13 BC), the frs ('seeds'; 200 K t p.a.) used to flavour cheese, cakes, liqueurs & curry powder, largely replaced by caraway

Cumulopuntia F. Ritter (~ *Opuntia*). Cactaceae (II). 4 Andes

cundurango *Gonolobus cundurango*

Cuniculotinus Urbatsch & al. Compositae (Ast.-Sol.). 1 W US: ***C. gramineus*** (H.M. Hall) Urbatsch & al. R: FNA 20(2006)100

Cunila Royen ex L. (*Hedyosmos*). Labiatae (VII 2b). 15 E N Am. to Urug. R: FRB 85(1936)138. Heterogeneous. ***C. origanoides*** (L.) Britton (*C. mariana*, American dittany, E N Am.) – culinary herb, cunila oil med.

cunjevoi *Alocasia macrorrhizos*

Cunninghamia R. Br. ex Rich. Taxodiaceae. 1–2 E As. (all lineages of '*C. konishii* Hayata' derived from *C. lanceolata*; Cretaceous fossils, Tertiary of C Eur., N Am.). R: A. Farjon, *Monogr. Cupressaceae & Sciadopitys* (2005)84. Timber of ***C. lanceolata*** (Lamb.) Hook. (*C. sinensis*, Chinese cedar, China) used for house-building, boats etc. & much in reafforestation

Cunonia L. Cunoniaceae (VIII). 1 S Afr., 24 New Caled. (poss. best united with *Weinmannia*). Buds prot. by stip. ***C. capensis*** L. (S Afr.) – cult. orn., wood used for furniture etc.

Cunoniaceae R. Br. Magnoliidae – Oxalidales. Incl. Baueraceae, Davidsoniaceae & Eucryphiaceae, 27/310 mostly S hemisph. (Mal. 10/40 – R: FM 16(2002)53) esp. Aus., NG, New Caled. R: SB 17(1992)181, 26(2001)372. Trees, shrubs (s.t. unbranched) or climbers, oft. accumulating aluminium. Lvs pinnate or trifoliolate, rarely simple, opp. or whorled, margins gland-toothed; stip. oft. large & consp. (0 in *Bauera*), oft. interpetiolar & prs connate, commonly with small colleters. Fls small in heads, racemes to thyrses, rarely solit. in axils, reg., bisex. (rarely pls dioec. or polygamodioec. as in *Pancheria*), usu. hypogynous; K (3)4, 5(–10), imbr. or valvate, s.t. basally connate, C alt. with K, usu. smaller, s.t. 0 or more than K (*Bauera*), A usu. in 2 whorls, 8–10 or in 1 opp. K, rarely >20, with slender filaments longer than C & anthers s.t. versatile, with longit. slits & v. small pollen grains, shallow nectary-disk s.t. around $\underline{G}$, G (rarely half-inf.) ((1) 2(3–5)), with sep. styles, rarely carpels ± distinct, ovules (1)2 – ∞ per locule on axile to apical-axile placentas, usu. anatropous, bitegmic with zig-zag micropyle. Fr. usu. a capsule, the carpels s.t. separating & opening ventrally at least apically, rarely drupe-like, nut-like or follicular; seeds small, winged or hairy with thin testa & usu. small straight embryo embedded in copious oily (?) or starchy endosperm, rarely 0 as in *Davidsonia*. n = 12, 15, 16

Classification & principal genera:

I. **Spiranthemeae**: *Acsmithia* (only)

II. **Bauereae** (Baueraceae, Davidsoniaceae): *Bauera* (0 stip., axillary fls; has been placed in its own fam. or assigned on fr. chars. to Saxifragaceae, Grossulariaceae, etc. but pollen & embryo characters confirm its being here), *Davidsonia*

III. **Schizomerieae**: *Ceratopetalum, Schizomeria*

IV. **Eucryphieae** (Eucryphiaceae): *Eucryphia*

V. **Geissoieae**: *Geissois*

VI. **Caldcluvieae**: *Caldcluvia*

VII. **Codieae**: *Callicoma*

VIII. **Cunonieae**: *Cunonia, Pancheria, Weinmannia*

Brunelliaceae may belong here (DNA data); *Aphanopetalum* excl. to Saxifragales (as Aphanopetalaceae)

Few cult. orn. esp. *Eucryphia*; some timber (*Ceratopetalum, Platylophus, Schizomeria, Weinmannia*) & ed. fr. (*Davidsonia*)

cunure *Calliandra guildingii*

Cunuria Baill. = *Micrandra*

cup and saucer plant *Holmskioldia sanguinea*

Cupania Plum. ex L. Sapindaceae. c. 45 warm Am. Some timber (loblolly)

Cupaniopsis Radlk. Sapindaceae. 60 C Mal. to N & E Aus. (incl. ***C. anacardioides*** (A. Rich.) Radlk., tuckeroo, also NG, invasive SE US), New Caled., Fiji & Samoa. R: LBS 15 (1991). Some pachycaul treelets

Cuphea P. Browne. Lythraceae. c. 240 Am. Lvs opp. or whorled, oft. with 1 fl. at each node but origin is in axil of leaf below. Source of medium-chain triglycerides, semi-domesticated oilseed crops subs. for palm kernel & coconut oil, poss. jet-fuel; several cult. orn. incl. hybrids, esp. ***C. hyssopifolia*** Kunth (Mex., Guatemala) – edging pl. used in trop. like *Buxus*, ***C. ignea*** A. DC. (cigar flower, Mex., WI) – used in *lei* in Hawaii, & ***C. llavea*** Lex. (Mex.). ***C. spermacoce*** A. St.-Hil. (Braz.) – xylopodium in cerrado

Cupheanthus Seemann = *Syzygium*

Cuphocarpus Decne & Planch. = *Polyscias*

Cuphonotus O. Schulz. Cruciferae (37). 2 Aus. R: CGH 205(1974)154

cupid's dart *Catananche caerulea*; **c. flower** *Ipomoea quamoclit*

cuprea bark *Ciliosemina* spp.

Cupressaceae Gray. Pinidae – Pinales. Incl. Taxodiaceae excl. *Sciadopitys*, 29/150 subcosmop. R: A. Farjon, *Monogr. C. & Sciadopitys* (2005)81. Monoec. or dioec. resinous trees & shrubs. Lvs decussate or in whorls of 3 or 4, in young pl. needle-like, usu. small & scale-like in mature pl. Fls small, solit., axillary or term. on short shoots (rarely males in axillary groups); cones term., woody, leathery or berry-like, cone-scales opp. or in whorls of 3, ovules usu. several per scale, erect; pollen without air-bladders, the nucleus of the grain acting as generative nucleus, there being no male prothallial cells, the first division delayed in many genera until after poll. Seeds winged or not. Seedlings usu. with 2(– 15) cotyledons. n = 11

Genera: *Athrotaxis, Austrocedrus, Callitris, Calocedrus, Chamaeyparis, Cryptomeria, Cunninghamia, Cupressus, Diselma, Glyptostrobus, Fitzroya, Fokienia, Juniperus, Libocedrus, Metasequoia, Microbiota, Neocallitropsis, Papuacedrus, Pilgerodendron, Platycladus, Sequoia, Sequoiadendron, Taiwania, Taxodium, Tetraclinis, Thuja, Thujopsis, Widdringtonia, Xanthocyparis (Callitropsis)*. 20 are basically N hemisph., S; 16 are monospecific, most with relict distribs. Known from Jurassic (Arg.), though ancestors would be classified as Voltziaceae (cf. T 33(1984)77–79). Incl. only known polyploid conifers. Many imp. timbers, resins, flavourings & cult. orn.

× **Cupressocyparis** Dallimore. Cupressaceae. *Chamaecyparis* × *Cupressus* hybrids. See also × *Cuprocyparis*

Cupressus Tourn. ex L. Cupressaceae. Incl. *Hesperocyparis*, 14 Medit. (Eur. 2) to Middle E, Himal. to China, SW N Am. to Honduras (Am. spp. poss. to be seg. as *Hesperocyparis* etc.). Cypresses. R: A. Farjon, *Monogr. Cupressaceae & Sciadopitys* (2005)176. Monoec., cone-scales woody, seeds maturing in 2 yrs. Many cult orn. (Attims's Model), for timber (incl. shavings for hats in Jap.) & poss. some 'cedarwood oil': ***C. arizonica*** Greene (*Callitropsis a.* (Greene) D.P. Little, *H. a.*, Arizona c., SW N Am.); ***C. dupreziana*** A. Camus (Algeria, 231 pls left 2001) – diploid pollen on *C. sempervirens* recipient grows into embryo; ***C. goveniana*** Gordon (*Callitropsis g.* (Gordon) D.P. Little, *H. g.*, Calif. c., Calif.); ***C. lusitanica*** Mill. (*Callitropsis g.* (Mill.) D.P. Little, *H. l.*, Mex. c., C Am., long natur. in Eur.); ***C. macrocarpa*** Hartweg (*Callitropsis m.* (Gordon) D.P. Little, *H. m.*, 'macrocarpa', Monterey c., M. County, Calif., now restricted to 2 groves in wild) – widely planted, a parent of × *Cuprocyparis leylandii*, '**Horizontalis**' (*C. lambertiana*) common windbreak SE Aus.; ***C. sempervirens*** L. (Italian c., pencil pine (Aus.), S Eur., Libya, SW As., commem. Cyparissus, Apollo's male lover turned by A. into the tree now much planted in Christian cemeteries) – cult. erect form (wild plant with spreading branches = f. ***horizontalis*** (Mill.) Voss) familiar in Continental pictures, coppices (rare in conifers), young trees & new growth female but stressed trees more male, timber used for sarcophagi of Egyptians, Odysseus's house, statues of Gk. gods (poss. gopher wood of Bible) & infusion as foot-bath long used to combat smelly feet & still used to scent soaps, apex of cone prob. inspiration for Gaudi's architectural crosses (Barcelona etc.); ***C. tonkinensis*** Silba (Vietnam) – almost exterminated, rs used for incense; ***C. torulosa*** D. Don (Himal. c., Himal.). See also *Xanthocyparis*

× **Cuprocyparis** Farjon (~ *Callitropsis, Cupressus s.l.*). Cupressaceae. *Cupressus* × *Xanthocyparis* hybrids. R: Novon 12(2002)188. Fast-growing conifers esp. × ***C. leylandii*** (A.B. Jackson & Dallimore) Farjon (*Callitropsis* × *leylandii* (A.B. Jackson & Dallimore) D.P. Little, × *Cupressocyparis leylandii*, 'leylandii', Leyland cypress grown from cones of *Cupressus macrocarpa* coll. 1888, 5 of the 6 resultant pls. being named clones, pollen parent *Xanthocyparis nootkatensis*; cross now repeated) – sets seeds, prob. most widely planted shelter-belt tree in GB, reaching 20 m in 25 yrs & cause of neighbour disputes, incl. colour-forms esp. golden **'Castlewellan'**

cupuaçu *Theobroma grandiflorum*

Cupulanthus Hutch. = *Gastrolobium*

Cupularia Gordon & Gren. = *Dittrichia*

Cupuliferae A. Rich. = Betulaceae + Fagaceae, later restricted to F.

curagua *Ananas lucidus*

Curanga Juss. = *Picria*

curare form. arrow poisons esp. of *Strychnos toxifera* with *Chondrodendron tomentosum* (q.v.); see also *seq*.

Curarea Barneby & Krukoff. Menispermaceae (I). 4 trop. S Am. R: MNYBG 22,2(1971)7. ***C. toxicofera*** (Wedd.) Barneby & Krukoff & ***C. candicans*** (Rich.) Barneby & Krukoff sources of curare; ***C. tecunarum*** Barneby & Krukoff an oral contraceptive in Braz.

Curatella Loefl. Dilleniaceae (II). 1 trop. Am.: ***C. americana*** L. R: MBSM 9(1971)25

Curculigo Gaertn. Hypoxidaceae. c. 20 trop. esp. S. Sessile palm-like pls with plicate lvs; ovary loculi imperfect; some propagules epiphyllous; ? rat-disp. in Mal.; ***C. pilosa*** (Schum. & Thonn.) Engl. (trop. Afr.) – seeds ed. See also *Molineria*

Curcuma L. Zingiberaceae (II 1). Incl. *Hitchenia, Laosanthus, Paracautleya, Smithatris, Stahlianthus*, c. 104 trop. As. R: T 64(2015)366. Spices & starch, some used like arrowroot & in glue. ***C. amada*** Roxb. (mango ginger, Ind.) – pickles; ***C. angustifolia*** Roxb. (Ind.) – tubers yield Bombay or (East) Ind. arrowroot; ***C. aromatica*** Salisb. (Ind.) – dye; ***C. caesia*** Roxb. (Ind.) – cosmetics & medic.; ***C. longa*** L. (*C. domestica*, turmeric, triploid cultigen orig. SE As.?) – dried & ground rhiz. (1 M t p.a. by 2013) used in curry powder & orange or yellow dyes form. much used with silk & wool, incl. in carpets, anti-oxidant, disinfectant (comm. plasters in Ind.), blocks 'tumour necrosis factor' that contributes to cancers & arthritis), pre-European introd. Polynesia & powder (mena) still integral part of med. & cerem. on Rotuma (Fiji); ***C. roscoeana*** Wall. (S As.) – imp. Thai cut-fl. export; ***C. zedoaria*** (Christm.) Roscoe ('*C. zerumbet*', zedoary, NE Ind.) – condiment or tonic, used in Ind. scents; others cult. orn.

Curcumorpha A. Rao & Verma = *Boesenbergia*

Curio Heath (~ *Kleinia*). Compositae (Sen.-Sen.). 21 S Afr. Succ. ***C. articulata*** (L.f.) Heath (*K. a.* (L.f.) Haw., *Senecio a.*, candlepl.) – familiar window-sill pl. with jointed stems & lobed term. lvs; ***C. rowleyana*** (H.J. Jacobsen) Heath (*K. r.* (H.J. Jacobsen) Kunkel, *S. r.*, E Cape) – mat-forming pl. with slender stems & bead-like lvs with no apical or marginal growth, but radial expansion from an adaxial meristem (marked by a narrow 'window', which is thus not the junction of leaf margins)

Curitiba Salywon & Landrum (~ *Mosiera*). Myrtaceae. 1 S Braz., Atlantic forest: ***C. prismatica*** (D. Legrand) Salywon & Landrum. R: Britt. 69(2007)302

curlewberry *Empetrum nigrum*

currant *Vitis vinifera*; **blackc.** *Ribes nigrum*; **American blackc.** *R. americanum*; **buffalo c.** *R. aureum*; **bush c.** *Miconia* spp.; **c. bush** *Leptomeria* spp.; **flowering c.** *R. sanguineum*; **golden c., Missouri c.** *R. aureum*; **native c.** (Aus.) *Acrotriche depressa*; **redc.** *R. rubrum*; **skunk c.** *R. glandulosum*; **whitec.** *R. rubrum* cv.

Curroria Planch. ex Benth. = *Cryptolepis*

curry leaf *Bergera koenigii*

curse, Koster's *Miconia crenata*; **Paterson's c.** *Echium plantagineum*

Curtia Cham. & Schldl. Gentianaceae. 80 Guianas to Urug. R: Rodriguesia 60(2009)424

Curtisia Aiton. Curtisiaceae (Cornaceae s.l.). 1 S Afr. (Eocene of Eur.): ***C. dentata*** (Burm.f.) C.A. Sm. (*C. faginea*) – timber for spokes, furniture etc. (assagai wood). R: Bothalia 39(2009)91

Curtisiaceae Takht. (~ Cornaceae, Grubbiaceae). Magnoliidae – Cornales. 1/1 S Afr. Everg. tree. Lvs simple, serrate, opp.; hairs unicellular. Infl. a term. thyrse. Fls minute, 4-merous; K-tube turbinate; C (sub)valvate; A opp. & = K; disk epigynous, 4-angled; $\overline{G}$ (4) with 1 ovule per loc.; stigma 4-lobed. Fr. a drupe, 4-seeded, with apical K; embryo elongate in copious endosperm. n = 13

Genus: *Curtisia*
Curtonus N.E. Br. = *Crocosmia*
curua *Attalea spectabilis*
curuba *Passiflora mollissima*
Curupira G.A. Black. Olacaceae (Ximeniaceae). 1 Braz.: ***C. tefeensis*** G.A. Black – oilseed long-coll. from Amaz. forests but not scientifically described until 1948
Cuscatlania Standl. Nyctaginaceae (III). 1 C Am.: ***C. vulcanicola*** Standl.
cuscus *Chrysopogon zizanioides*
Cuscuta Tourn. ex L. Convolvulaceae (5; Cuscutaceae). c. 200 cosmop. (Eur. 17). R: MTBC 18(1932)113, SB 40(2015)275. Dodder, devil's guts, scald. Parasites with short-lived r.-systems & oft. brightly coloured chlorophyll-less thread-like twining stems (those of 1 pl. in Costa Rica totalling 500 m) linked by haustoria to host. Form. reduced yield of many crops, now largely controlled though some not restricted to 1 host sp. Local medic. esp. ***C. chinensis*** Lam. (Indomal., tu si zi; parasite of soybeans) – seeds (alks) used for acne, dandruff, TCM for kidney ailments
Cuscutaceae Dumort. = Convolvulaceae (5)
cushaw *Cucurbita argyrosperma*
cush-cush *Dioscorea trifida*
cushion bush *Leucophyta brownii*
Cusickia M.E. Jones = *Lomatium*
Cusickiella Rollins (~ *Draba*). Cruciferae (11). 2 W US. R: JJB 63(1988)65
cusparia bark *Angostura* & *Conchocarpus* spp.
Cuspidaria DC. Bignoniaceae (3). 19 trop. Am. R: AMBG 98(2014)423
Cuspidia Gaertn. Compositae (Arct.-Gort.). 1 S Afr.: ***C. cernua*** (L.f.) B.L. Burtt. R: MBSM 3(1959)315
Cussonia Thunb. Araliaceae. c. 20 trop. & S Afr. to Masc. Pachycaul trees (Leeuwenberg's Model) with soft wood, s.t. cult. ('cabbage-trees', 'umbrella-trees'), some molluscicidal saponins. ***C. arborea*** Hochst. ex A. Rich. (*C. barteri*, trop. Afr.) – hypocotyl becomes a fire-resistant lignotuber (cf. *Eucalyptus*)
custard apple *Annona reticulata, A. squamosa;* **c. powder** *Zea mays*
Cutandia Willk. Gramineae (XVI 12). 6 Medit. (Eur. 4) to Middle E. R: BJLS 76(1978)351
cutch *Senegalia catechu*
Cuthbertia Small = *Callisia*
Cutsis Burns-Balogh & al. = *Dichromanthus*
Cuttsia F. Muell. Rousseaceae (Grossulariaceae s.l.). 1 E Aus.: ***C. viburnea*** F. Muell.
Cuviera DC. Rubiaceae (III 2). Excl. *Globulostylis*, 10 trop. Afr. R: BJLS 173(2013)415. Roux's Model; some myrmecophilous with hollow swellings above nodes
Cwangayana Rauschert = *Aralia*
Cyamopsis DC. Leguminosae (III 15). 4 dry Afr. & Arabia. ***C. tetragonoloba*** (L.) Taubert (*C. psoraloides*, cluster bean, guar, guvar, gwar, cultigen poss. derived from *C. senegalensis* Guill. & Perr., W Afr.) – much cult. esp. Ind. for young pods as forage & seed-gum (500 K t p.a.), in shampoos etc.
Cyanaeorchis Barb. Rodr. Orchidaceae (V 12c). 3 Braz.
Cyanandrium Stapf = *Phyllagathis*
Cyananthus Wall. ex Benth. Campanulaceae (I 1). 18 Himal. R: APS 35(1997)397. Ovary sup.! Some cult. rock-pls
Cyanastraceae Engl. = Tecophilaeaceae
Cyanastrum Oliv. Tecophilaeaceae (Cyanastraceae). 3 trop. Afr. R: KB 53(1998)775
Cyanea Gaudich. (~ *Delissea*). Campanulaceae (III 4). 78 Hawaii. R: W.L. Wagner & al., *Manual Fl. Hawaii* 1(1990)437. Prob. derived from *Lobelia* subg. *Tupa* of As.; many palm-like pachycauls (Corner's Model), some with pinnatisect lvs; some with prickles poss. adapted to defence against flightless geese later exterminated by Polynesians. ***C. asarifolia*** St John discovered 1970, wild pop. (12) destroyed by hurricane (1992), but returned to wild from cult. pl. micropropagation
Cyanella Royen. Tecophilaeaceae. 9 S Afr. esp. SW Cape. R: Bothalia 42(2012)27. Some cult. orn. incl. ***C. hyacinthoides*** L. – ed. corms poss. comm.
Cyanicula Hopper & A.P. Brown (~ *Caladenia*). Orchidaceae (IV 3b). 10 S (esp. SW) Aus. R: AusSB 17(2004)211. Fls usu. blue
Cyanixia Goldbl. & Manning (~ *Babiana*). Iridaceae (VI 2). 1 Socotra: ***C. socotrana*** (Hook.f.) Goldbl. & Manning. R: EJB 60(2004)529

Cyanoneuron Tange (~ *Myrioneuron*). Rubiaceae (IV-Cyan.). 5 Borneo, Sulawesi. R: NJB 18(1998)147

Cyanopsis Cass. = *Volutaria*

Cyanorchis Thouars = *Phaius*

Cyanoseris (Koch) Schur = *Lactuca*

Cyanostegia Turcz. Labiatae (IV 1; Dicrastylidaceae). 5 C & W Aus. R: Brunonia 1(1978)45

Cyanothamnus Lindl. = *Boronia*

Cyanothyrsus Harms = *Daniellia*

Cyanotis D. Don. Commelinaceae (II 1c). (Incl. *Amischophacelus*) 50 OW trop. (Afr. 25). Anther dehiscence has 1 valve slipping over other, pressing pollen out through small gap. Some aquatics; some cult. orn. ***C. cristata*** (L.) D. Don (Indomal.) – juicy stems used to clean slates in S Ind.; ***C. longifolia*** Benth. (trop. Afr.) – accum. Co

Cyanthillium Blume (~ *Vernonia*). Compositae (Vern.-Erl.). 7 OW trop., ***C. cinereum*** (L.) H. Robinson now pantrop. weed

Cyanus Mill. = *Centaurea*

Cyathanthus Engl. = *Scyphosyce*

Cyathea Sm. Cyatheaceae. c. 650 (s.s., Pacific, trop. Am.) trop & warm (Afr. 14), incl. *Alsophila* (trop. to subAntarctic) & *Sphaeropteris* (Indomal., Pacific, trop. Am.) as subg. (see KB 38(1983)167), *Cnemidaria, Gymnosphaera, Hymenophyllopsis*, & *Schizocaena*. R: M.F. Large & J.E. Braggins *Tree Ferns* (2004)69,81. Tree ferns esp. consp. in montane forest where oft. aggressive early colonists of sec. growth, some epiphytic, some spiny, marcescent fronds forming skirts in some spp. poss. preventing climbers & epiphytes damaging apical buds; trunks for building, when hollow for cannon & bee-hives, starchy pith of some (as for pigs on Norfolk Is.) used as food, r.-chips exported from Mal. to Jap. & Taiwan as orchid-growing medium; all Aus. spp. now prot. Some local medic. (astragalin, a flavonoid, inhibits histamine release, cf. *Arnica, Humulus*), some cult. orn. (Corner's Model) esp. ***C. arborea*** (L.) Sm. (trop. Am.) – scales used in nest of Puerto Rico's endemic hummingbird, those of ***C. mexicana*** Schldl. & Cham. (*A. firma*, Mex.) also used in nests, ***C. brownii*** Domin (Norfolk Is.) – at 20 m +, poss. biggest tree-fern, & ***C. cooperi*** (F. Muell.) Domin (E Aus.) – invasive Usambaras (Tanz.), Hawaii; ***C. dealbata*** (Forst.) Sw. (NZ) – emblem of All Blacks rugby team, spores (ponga powder) febrifuge; ***C. macgregorii*** F. Muell. (*A. m.*) & other spp. (NG) withstand fires in otherwise treeless montane grasslands; ***C. manniana*** Hook. (*A. m., C. usambarensis*, trop. Afr.) – anthelminthic much used by Ger. troops in World War I though excess leads to blindness; ***C. medullaris*** (Forst.f.) Sw. (*S. m.*, bungy, ponga, W Pac.) – mature pls each prod. more than 15 billion spores (1.5 kg) per annum, bungy huts (W NZ) made of trunks, 'pongoware' boxes (from stem cross-sections, also of *C. dealbata*) etc. sold NZ, cult. orn.; ***C. tuyamae*** H. Ohba (Volcano Is., Jap.) – branches above ground level whereas other spp. have them underground

Cyatheaceae Kaulf. Polypodiidae – Cyatheales. 1/650 trop. & warm. Tree ferns to 24 m tall with polycyclic dictyosteles, scaly young parts (uniseriate hairs in Dicksoniaceae) & lvs to 3-pinnate & 5 m long; veins simple to forked, free, rarely anastomosing. Sori superficial variously indusiate or naked; gametophytes green, cordate; n = 69
Genus: *Cyathea*
Fossils from Jurassic or early Cretaceous

Cyathella Decne = *Cynanchum*

Cyathobasis Aellen. Amaranthaceae (Chenopodiaceae III 3). 1 Turkey: ***C. fruticulosa*** (Bunge) Aellen

Cyathocalyx Champ. ex Hook.f. & Thomson. Annonaceae (II). Excl. *Drepananthus*, 7 Indomal. R: T 59(2010)1730. Seeds tritegmic

Cyathochaeta Nees. Cyperaceae (II 7). 3 SW & E Aus. BJ 75(1952)490

Cyathocline Cass. Compositae (Ast.-Gran.). 3 trop. As. R: MBSM 15(1979)513

Cyathocoma Nees (~ *Tetraria*). Cyperaceae (II 7). 3 S Afr. R: SAJB 63(1997)167

Cyathodes Labill. (~ *Styphelia*). Ericaceae (VII 7; Epacridaceae). 3 Tasmania. R: AusSB 9(1996)494. ***C. divaricata*** Hook.f. – dioec., bird-poll.

Cyathogyne Muell.-Arg. = *Thecacoris*

Cyathomone S.F. Blake (? = *Ericentrodea*). Compositae (Cor.). 1 Ecuador: ***C. sodiroi*** (Hieron.) S.F. Blake

Cyathophylla Bocquet & Strid. Caryophyllaceae (III 1). 1 Greece: ***C. chlorifolia*** (Poir.) Bocquet & Strid

Cyathopsis Brongn. & Gris (~ *Styphelia*). Ericaceae (VII 7; Epacridaceae). 3 New Caled. R: AusSB 18(2005)451

Cyathopus Stapf. Gramineae (XVI 5). 1 Himal.: ***C. sikkimensis*** Stapf – woods. R: FOC 22(2006)363

Cyathorhachis Nees ex Steud. = *Polytoca* (but see Blumea 47(2002)569)

Cyathoselinum Benth. (~ *Seseli*). Umbelliferae (III 8). 1 SE Eur.: ***C. tomentosum*** (Vis.) Benth.

Cyathostegia (Benth.) Schery. Leguminosae (III 1). 1 Peru, Ecuador: ***C. mathewsii*** (Benth.) Schery

Cyathostelma Fourn. = *Jobinia*

Cyathostemma Griff. = *Uvaria* (but see Blumea 45(2000)377)

Cyathostemon Turcz. (~ *Astartea*). Myrtaceae. 7+ SW Aus. R: Nuytsia 24(2014)9

Cyathula Blume. Amaranthaceae (I 2). 25 trop. Some drugs in China, incl. ***C. prostrata*** (L.) Blume (anthelminthic) with some cult. orn cvs

Cybebus Garay. Orchidaceae (IV 2h). 1 Colombia: ***C. grandis*** Garay. R: Gen. Orch. 3(2003)185

Cybianthopsis (Mez) Lundell = *Cybianthus*

Cybianthus Mart. Primulaceae (Myrsinaceae). Incl. *Grammadenia* 167 trop. Am. R: ABiV 10(1980)129. Rauh's Model

Cybistax Mart. ex Meissn. (~ *Tabebuia*). Bignoniaceae (1). 1 Amaz. Braz. & Peru: ***C. antisyphilitica*** (Mart.) DC. – lvs used as blue dye for cloth. See also *Roseodendron*

Cybistetes Milne-Redh. & Schweick. = *Ammocharis*

cycad any member of Cycadidae

Cycadaceae Pers. Cycadidae – Cycadales. 1/100 E Afr. to Jap. & Aus. R: Blumea 43(1998)353. Dioec. palm-like trees to 15 m with trunk clothed in leaf-bases. Pinnae with prominent mid-vein but no laterals, vernation circinately involute. Sporophylls in spirals in definite cones (male), the females leafy, toothed to deeply lobed, with large naked seeds terminally, the male cones are dropped & a lat. meristem takes over as leader while females are lat.

Only genus: *Cycas*

Cycadales Pers. ex Bercht. & J. Presl = only order of Cycadidae, q.v.

Cycadidae Pax. (Cycadatae). 10/285. R: BR 70(2004)274. T. Walters & R. Osborne (2004) *Cycad classification*. Gymnosperms, dioec., pachycaul, slow-growing with sec. thickening, tap-rooted & usu. unbranched (many males with Chamberlain's Model); axillary buds 0. Wood centripetal, suggesting a relict condition from a protostelic ancestor. Most with a single persistent cambium but a succession of cambia give co-axial cylinders of sec. xylem & phloem in *Cycas* & some spp. of *Encephalartos* & *Macrozamia*; wood of *Zamia* & *Stangeria* comprises scalariform tracheids. Apical meristems to 3000 μm across. R. with cyanobacteria in special 'coralloid' much-branched lat. r. in at least 30 spp.; r.-hairs 0. Lvs large, pinnate. Reproductive organs in cones (exc. female *Cycas*), term. or lat.; megasporophylls with sterile tips & 8–2 orthotropous ovules, seeds large; microsporophylls scale-like or peltate with pollen-sacs on abaxial side, sperm to 300 μm long (largest sperm known) with spiral band of flagella. In some at least sex is determined by X & Y chromosomes. Cotyledons 2

An arch. group recog. from the Triassic onwards & rep. now by 2 fams: Cycadaceae (pinnae with midrib & no lat. veins, Zamiaceae (incl. Stangeriaceae; midrib & lat. veins or ∞ parallel or wavy, simple or forked veins running longitudinally)

Wind- & beetle-poll. (some thermogenesis), weevils attracted by micropylar exudates (sugars & amino-acids), cf. nectar in angiosperms. Outer layer of seed fleshy. A group in decline since the Jurassic & now with relict distr. All are cult. orn. & endangered through overcollecting; toxic due to carcinogens & neurotoxins but some yield sago (esp. *Cycas*)

Cycas L. Cycadaceae. Incl. *Epicycas* c. 100 E Afr. to (SE As. 40) Jap. & Aus. (27). R: BR 70(2004)277. Corner's Model (females); lenticels on ovules & seeds; three-layered seedcoat but *C. circinalis* & allies have additional layer of spongy tissue allowing buoyancy, those of ***C. thouarsii*** R. Br. (E Afr.) reaching Sri Lanka. Cult. orn. & some ed. ('sago-palm', trop. As., esp. ***C. rumphii*** Miq. (Mal.) & allies, '*C. circinalis*') – sago from pith, but destroy central nervous system leading to dementia, paralysis & death due to free amino-acid (methylamino-L-alanine, poss. prod. by cyanobacteria in coralloid r.), which scavenges Cu & Zn in CNS (called Guam disease because flying-foxes eaten there accum. toxin from seeds); cycasin (glycoside removed in cooking) converted to methylazomethanol, one of most toxic carcinogens known. ***C. circinalis*** L. (*C. beddomei*, SW Ind., Sri Lanka) –

lvs ed., sago for flour & bread; ***C. media*** R. Br. (NE Aus.) – seeds ed. boiled, an Aboriginal staple in Arnhemland; ***C. revoluta*** Thunb. (Ryukyu Is.) – dwarf, varieg. & cristate clones commonly cult. Jap.

Cyclacanthus S. Moore. Acanthaceae (III 2c). 2 SE As.

Cyclachaena Fresen. = *Euphrosyne*

Cycladenia Benth. Apocynaceae (II e). 1 SW US: ***C. humilis*** Benth.

Cyclamen Tourn. ex L. Primulaceae. 23 Eur., Medit. to Iran & NE Somalia (1). R: B. Mathew (2013) *Genus C.* Swollen hypocotyls ('corms') to 100 yrs old; 1 cotyledon. R. emerge from base of hypocotyl in e.g. *C. persicum*, from the top in e.g. *C. hederifolium* & all over in other spp. Reflexed C-lobes in all spp. (cf. '*Dodecatheon*'); fruiting scape of *C. persicum* extends & falls to ground, that of *C. hederifolium* & other spp. coils up spring-like before seed-release; seeds sticky, at least s.t. ant-disp. Cult. orn. esp. florist's cyclamen derived from ***C. persicum*** Mill. (SE Eur., Tunisia & Aegean (not Iran!)), orig. with scented fls (buzz-poll. by bees but largely replaced by thrips etc. & small moth (*Micropteris elegans*) reliant on it), now with double, fringed or large, even yellow (1997) fls, treated as annuals in the trade (over 200 M pls sold per yr in Eur. alone); other spp. grown outside esp. ***C. hederifolium*** Aiton (sowbread, S Eur. to SW As., natur. GB) – tuber to 20 cm diam., form. considered a cure for baldness when used as snuff

Cyclandrophora Hassk. = *Atuna*

Cyclanthaceae Poit. ex A. Rich. Magnoliidae – Pandanales. 12/225 trop. Am. R: AHB 18(1958)1. Monoec. perenn. herbs to erect shrubs or lianes (± epiphytic). Lvs in spirals or distich. with sheath, petiole & expanded blade, s.t. plicate, with parallel or parallel-pinnate venation & cross-veins, usu. bilobed or bifid, s.t. palmate or simple. Infls usu. axillary with 2–8 large decid. spathes subtending spadix. Fls v. small, weevil-poll., much reduced in structure, unisexual, males & females (± embedded in axis) in same spadix, males with P 4–24, minute, s.t. connate in 1 or 2 series or merely an entire or toothed cup to 0 when fls ± confluent, A 6 – ∞ with filaments basally connate & anthers with longit. slits, females with P 4, small, s.t. connate & as many staminodes opp. & partly adnate to them, those oft. thread-like, $\underline{G}$ (4), 1-loc. with 4 parietal or almost apical placentas & as many stigmas, rarely pseudomonomerous with 1 placenta & stigma (in *Cyclanthus* females confluent into rings round spadix, each ring with common ovular chamber & ∞ placentas & ovules), ovules ± ∞, anatropous, bitegmic. Fr. fleshy berry-like, oft. coalescent into multiple fr.; seeds up to ∞ with straight embryo & endosperm rich in oil, protein, oft. hemicellulose, rarely starch. x = 9 – 16

Classification & genera:

I. **Carludovicoidae** (fls in groups in spirals; fr. spadix not screw-like; lvs bifid, palmate or simple; poll. weevils use infls for feeding, shelter, mating & oviposition): *Asplundia, Carludovica, Dianthoveus, Dicranopygium, Evodianthus, Ludovia, Pseudoludovia, Schultesiophytum, Sphaeradenia, Stelestylis, Thoracocarpus*

II. **Cyclanthoideae** (male & female fls in sep. alt. whorls or part spirals; fr. spadix screw-like; lvs deeply bilobed): *Cyclanthus*

Insect-poll. recorded in *Asplundia, Carludovica, Cyclanthus* & *Evodianthus*; some ant-disp. fr. Cult. orn. & fibre for hats (*Carludovica*)

Cyclanthera Schrad. Cucurbitaceae (XII). Incl. *Pseudocyclanthera, Rytidostylis*, c. 40 trop. Am. Anther-locules of A united into 2 ring- shaped locules around pistil; fr. usu. explosive but fleshy, valves rolling back rapidly & expelling seeds in pulpy endocarp esp. in ***C. brachystachya*** (Ser.) Cogn. (*C. explodens*, N S Am.) – cult. orn.; ***C. pedata*** (L.) Schrad. (achocha, acoccha, caigua, korila, stuffing gourd, trop. Am.) – fr. ed. Peru & Bolivia, grown Eur.

Cyclantheropsis Harms. Cucurbitaceae (II). 3 trop. Afr., Madag.

Cyclanthus Poit. Cyclanthaceae. 1 trop. Am.: ***C. bipartitus*** Poit. – poll. by dynastine scarab beetles

Cyclea Arn. ex Wight. Menispermaceae (V). 29 China to Philipp. R: KB 14(1960)68, 34(1980)565. Alks. Some medic. locally esp. ***C. barbata*** Miers (Assam to Java) – (medic.) green jelly (tjintjau) from lvs ed. Java & ***C. peltata*** (Lam.) Hook.f. & Thomson (Mal.)

Cyclobalanopsis Oersted = *Quercus*

Cyclocarpa Afzel. ex Urb. Leguminosae (III 11). 1 OW trop.: ***C. stellaris*** Afzel. ex Urb.

Cyclocarya Iljinsk. (~ *Pterocarya*). Juglandaceae. 1 E China (form. widespread N temp.): ***C. paliurus*** (Batal.) Iljinsk. R: BM 30(2013)229

Cyclocheilaceae Marais = Orobanchaceae

Cyclocheilon Oliv. Orobanchaceae (Cyclocheilaceae). 3 NE Afr. R: KB 35(1981)806. Shrubs

Cyclocodon Griff. ex Hook.f. & Thomson (~ *Codonopsis*). Campanulaceae (I 1). 3 Indomal. R: D.Y. Hong, *Monogr. Codonopsis* (2015)210

Cyclocotyla Stapf. Apocynaceae (I c). 1 trop. W Afr.: ***C. congolensis*** Stapf. R: AUWP 85–2(1985)59

Cyclodium C. Presl. Dryopteridaceae (II). 10 trop. Am. R: AFJ 76(1986)56. 1 sp. a rheophyte

Cyclogramma Tag. = *Cyclosorus*

Cyclolepis Gillies ex D. Don. Compositae (Gochn.). 1 temp. S Am.: ***C. genistoides*** Gillies ex D. Don – char. spiny, almost leafless gynodioec. shrub of salty soils in N Patagonia

Cyclolobium Benth. Leguminosae (III 4). 1 trop. S Am.: ***C. brasiliense*** Benth. R: EJB 59(2002)248

Cycloloma Moq. Amaranthaceae (Chenopodiaceae I 4). 1 W & C N Am.: ***C. atriplicifolium*** (Spreng.) J. Coulter, seeds ed. by diff. Native American groups, natur. Eur., S Am.

Cyclonema Hochst. = *Rotheca*

Cyclopeltis J. Sm. Lomariopsidaceae (Dryopteridaceae s.l.). c. 6 Indomal., 1 trop. Am.

Cyclophyllum Hook.f. (~ *Canthium*). Rubiaceae (III 2). c. 40 Mal. to Fiji (Aus. 9; R: Austrobaileya 6(2001)42)

Cyclopia Vent. Leguminosae (III 7). 23 SW & S Cape. R: EJB 54(1997)136. Stipules 0. Several spp., orig. ***C. genistoides*** (L.) Vent., used for 'Cape' or bush ('honeybush') teas, some comm.

Cyclopogon C. Presl. Orchidaceae (IV 2h). 75 warm & trop. Am., ***C. obliquus*** (J.J. Sm.) Szlach. natur. As.

Cycloptychis E. Meyer ex Sond. = *Heliophila*

Cyclorhiza Shen & Shan (~ *Vicatia*). Umbelliferae (III 8). 2 SW China

Cyclosorus Link (*Meniscium*). Thelypteridaceae. s.l. (incl. *Chingia*, *Christella*, *Goniopteris*, *Mesophlebeion*, *Pneumatopteris*, *Pronephrium*, *Pseudocyclosorus*) c. 600 trop.; s.s. 3 trop. Some with trunks, some rheophytes. ***C. contiguus*** (Ros.) Ching (Borneo) – ed.; ***C. proliferus*** (Retz.) Tard. & C. Christ (*Ampelopteris p.*, OW) – young fronds ed., medic. incl. laxative

Cyclospermum Lag. (*Ciclospermum*). Umbelliferae (III 7). 3 C Am., WI. R: Britt. 42(1990)277. ***C. leptophyllum*** (Pers.) Britton & P. Wilson (*Apium l.*) – now pantrop. ann. weed leading to carrot-smelling milk from cows grazing it

Cyclostachya Reeder & C. Reeder = *Bouteloua*

Cyclotrichium (Boiss.) Manden. & Scheng. Labiatae (VII 2b). 6 SW As. to Iran

Cycniopsis Engl. Orobanchaceae (Buch.; Scrophulariaceae s.l.). 2 trop. Afr. R: NJB 13(1993)185, 14(1994)64

Cycnium E. Meyer ex Benth. (~ *Escobedia*). Orobanchaceae (Buch.; Scrophulariaceae s.l.). 16 warm Afr. R: DBA 32,3(1978)1

Cycnoches Lindl. Orchidaceae (V 12c). 34 trop. Am. R: Orchid J. 1(1952)178, 225,273,349,397. Epiphytes with floral mechanism & dimorphism like *Catasetum*, prod. more female infls in high light intensity, poll. male euglossine bees. Some cult. orn. (swan orchids)

Cycnogeton Endl. (~ *Triglochin*). Juncaginaceae. 8 Aus. R: O. Seberg, Div.Phylog.Monocots (2010)73. C. ***procerum*** (R. Br.) Buchenau (*T. p.*) – rhiz. ed. SE Aus. Aborigines

Cydista Miers = *Bignonia*

Cydonia Mill. (~ *Pyrus*). Rosaceae (Ros.-Mal.). Excl. *Pseudocydonia*, 1 Cauc. & Kurdistan: ***C. oblonga*** Mill. (quince, long cult., now natur. S Eur.) – differs from *Chaenomeles* in that latter has serrate lvs (not entire), decid. (not persistent) sepals & A 40–60 (not 15–25). Oft. used as r.-stock for pears; fr. (600 K t p.a. by 2011) tart (flavour due to ethyl 2-methyl-2-butanoate) & used as jelly (high pectin content) or cooked with other fr., in antiquity a love-token assoc. with Venus & prob. 'golden apples' of the Hesperides; *marmelo* in Portuguese & Spanish & thus a contender for the 'original' marmalade. R: FNA 9(2014)486

Cydoniorchis Senghas = *Bifrenaria*

Cylicodiscus Harms. Leguminosae (II 1). 1 Guineo-Congolian forests: ***C. gabunensis*** (Taubert) Harms (denya, okan, Afr. greenheart) – comm. timber

Cylicomorpha Urb. Caricaceae. 2 trop. Afr. R: V.M. Badillo (1971) *Caric.*: 34. ***C. parviflora*** Urb. (E Afr.) – hollow spiny trunk oft. tusked by elephant & with bees' nests within

Cylindrilluma Plowes = *Caralluma*

Cylindrocarpa Regel. Campanulaceae (I 9). 1 C As.: ***C. sewerzowii*** (Regel & Herder) Regel

Cylindrocline Cass. Compositae (Inul.-Pluch.). 2 Mauritius. Shrubs, small trees. ***C. lorencei*** A.J. Scott – extinct in wild

Cylindrokelupha Kosterm. = *Archidendron*

Cylindrolobovia Y Itô = *Echinopsis*

Cylindrolobus Blume = *Eria*

Cylindrophyllum Schwantes. Aizoaceae (V). 5 Cape. R: H. E. K. Hartmann, *Ill. Handb. Succ. Pls*, Aiz. A–E (2002)179. Cult. orn. with 4-ranked succ. lvs

Cylindropsis Pierre. Apocynaceae (I c). 1 trop. W Afr.: ***C. parvifolia*** Pierre. R: BJBB 63(1994)316

Cylindropuntia (Engelm.) F. Knuth (~ *Opuntia*). Cactaceae (II). 33 SW N & C Am., introd. OW. ***C. fulgida*** (Engelm.) F. Knuth (SW N Am.) – source of cholla gum; ***C. imbricata*** (Haw.) F. Knuth (SW N Am.) – invasive in S Afr., Aus.

Cylindropyrum (Jaub. & Spach) Löve = *Triticum*

Cylindrosolenium Lindau = *Stenostephanus*

Cylindrosperma Ducke = *Microplumeria*

Cymaria Benth. Labiatae (Cymar.). 2 Indomal.

Cymatocarpus O. Schulz. Cruciferae (26). 3 Transcaucasia to C As.

Cymbalaria Hill (~ *Linaria*). Plantaginaceae (Ant.-Ant.; Scrophulariaceae s.l.). 9 W Eur. (7), Medit. to Iran. R: D.A. Sutton, *Rev. Antirrhineae* (1988)158; differs from *Linaria* s.s. in axillary fls & palmately veined lvs. Some cult. orn. creepers esp. ***C. muralis*** Gaertn. f., Meyer & Scherb. (Kenilworth or Oxford ivy, ivy-leaved toadflax, mother-of-thousands, pennywort, wandering sailor, Eur. natur. GB (C17, said to have been introd. with sculpture from Italy & first known as Oxford weed) & N Am.) – fls negatively phototropic after poll., developing fr. inserted in dark crannies; peloric fls produced in tissue-culture, not recorded in wild (cf. *Linaria*); f. ***toutonii*** (A. Chev.) Cuf. (France) with deeply lobed lvs (? virus)

Cymbaria L. Orobanchaceae (Cymb.; Scrophulariaceae s.l.). 4 Eur. (1), C & E As.

Cymbidiella Rolfe. Orchidaceae (V 12b). 3 Madag.: 2 epiphytic, 1 terr. (***C. flabellata*** (Thouars) Rolfe, wasp-poll.)

Cymbidiopsis Chowdhery = *Cymbidium*

Cymbidium Sw. Orchidaceae (V 12a). c. 70 NW Ind. to Jap. & Aus. (3). R: D. du Puy & P. Cribb (2007) *The genus C.* Epiphytes (some mycotrophic) with extrafl. nectaries & large fls much cult. esp. for button-holes (many complex hybrids, some sold for up to 1 M yuan) – ***C. ensifolium*** (L.) Sw. (trop. As. to NG) is the Chien Lan or Fukien orchid, a pot-pl. in China for centuries, its fls infused as eye-treatment – also to flavour curries (Bhutan); ***C. faberi*** Rolfe (Nepal to N Vietnam & China) – fl. extract use in 'Tentatrice' scent sold in Jap. ('feminine' scent due to methyl epijasmonate); ***C. goeringii*** (Reichb.f.) Reichb.f. (*C. virescens*, Ind. to Jap.) – fls salted & in hot water used as a drink & also preserved in plum vinegar. Some local medic. esp. for dysentery & diarrhoea

Cymbiglossum Halbinger = *Rhynchostele*

Cymbispatha Pichon = *Tradescantia*

Cymbocarpa Miers. Burmanniaceae. 2 trop. S Am., WI

Cymbocarpum DC. ex C. Meyer. Umbelliferae (III 11). Incl. *Kalakia*, 3 SW As.

Cymbochasma (Endl.) Klokov & Zoz = *Cymbaria*

Cymbolaena Smoljan. = *Filago* (but see OB 104(1991)173)

Cymbonotus Cass. Compositae (Arct.-Arct.). 3 temp. Aus. R: FA 37(2015)165

Cymbopappus R. Nordenstam. Compositae (Anth.-Pen.). 3 S Afr. R: BJLS 96(1988)308

Cymbopetalum Benth. Annonaceae (III 1). 27 Mex. to trop. S Am. R: SBM 40(1993)5. Poll. scarabs (*Cyclocephala* spp.); pollen of app. apomictic (first known in fam.) ***C. brasiliense*** (Vell.) Baill. (*C. odoratissimum*, Braz.) almost 350 μm diam. Petals of ***C. penduliflorum*** (Dunal) Baill. used with vanilla by Aztecs to flavour chocolate

Cymbopogon Spreng. Gramineae (XXII 7). 59 OW trop. & warm (China 24). R: Reinwardtia 9(1977)225, (1980)390. Many yield aromatic ess. oils used in scent, 'herbal' insect-repellents, medicine & flavouring. ***C. bhutanicus*** Noltie (Bhutan) – local lemon-grass oil; ***C. citratus*** (DC.) Stapf (lemon grass, cultigen from ?As., cult. Florida) – fls rare, serai (sereh) of As. cooking, scent (citral) for soaps; ***C. flexuosus*** (Steud.) J. Watson (Malabar oil, orig. S Ind.?); ***C. jwarancusa*** (Jones) Schultes (oil grass, As.) – grown for scent & medic.; ***C. martini*** (Roxb.) J. Watson (ginger-grass, palma-rosa, rosha, rusha, Ind., cult. Mal.) – 'geranium oil', used to flavour tobacco, '**Mota**' yielding palma-rosa (65% geraniol) – inhibits yeast growth, '**Sofia**' ginger-grass oil; ***C. nardus*** (L.) Rendle (citronella, mana grass, S Ind., Sri Lanka) – used as tea, oil the source of insect-repellent; ***C. schoenanthus*** (L.) Spreng. (camel grass, N Afr. to N Ind.) – oil medic. & adulterant of otto of roses in Middle E; ***C. validus*** (Stapf) Davy (S Afr.) – imp. thatching grass; ***C. winterianus*** Jowitt ex Bor (cultigen) – trad. pesticide, cult. Ind. for cheap scent (citronella better than that of *C. nardus*)

Cymbosema Benth. (~ *Dioclea*). Leguminosae (III 18). 1 trop. Am.: ***C. roseum*** Benth. – fls a medic. tea for menstrual disorders in NW Amaz.

Cymbosetaria Schweick. = *Setaria*

Cymodocea C. Koenig. Cymodoceaceae. 7 coasts of W Afr. & Canary Is. to Med. (Eur. 1), Indopacific. R: C. den Hartog (1970) *Sea grasses of the World* (1970)161. Heterogeneous. Bell's Model

Cymodoceaceae Vines (~ Juncaginaceae). Magnoliidae – Alismatales. 5/15 trop. & warm coastal shallows esp. Aus. R: C. den Hartog (1970) *Sea grasses of the World*. Glabrous, dioec. marine rhiz. herbs without vessel elements; stem sympodial, the app. axillary infls term. Lvs in spirals, distich. or app. opp., linr., 3 – ∞-veined with open basal sheath, serrulate apex, & ligule at its junction with blade. Fls small, water-poll., solit., paired, in cymes or thyrses; P 0, anthers paired on common filament, with longit. slits & thread-like pollen grains to 1 mm long without exine, $\underline{G}$ 2, styles 2- or 3-fid, ovule 1, pend. from locule apex, orthotropous, bitegmic. Fr. an achene, endosperm 0. x = 7 (*Cymodocea*)
Genera: *Amphibolis, Cymodocea, Halodule, Syringodium, Thalassodendron*
Prob. best included in Potamogetonaceae but differ in filamentous pollen & marine habitat

Cymophora Robinson (~ *Tridax*). Compositae (Mill.-Dysc.). 4–5 Mex., Guatemala. R: Madroño 24(1977)1

Cymophyllus Mackenzie = *Carex*

Cymopterus Raf. Umbelliferae (III 9). Excl. *Vesper*, c. 27 W N Am. R: AMBG 17(1930)213. Heterogeneous. Several spp. (r., lvs etc.) ed. (N Am. Indians)

Cynanchum L. Apocynaceae (Vc; Asclepiadaceae III 1). Incl. *Glossonema, Holostemma, Glossonema, Metaplexis, Sarcostemma*, excl. *Scyphostelma* c. 250 trop. & warm (Eur. 1). Some Am. spp. perhaps to be excl. as *Funastrum*. Lianes, some v. large, some cult. orn; latex for fly-killers. ***C. acutum*** L. (Medit., S Euras.) – the 'convolvulus' (c. 1593–1080 BC) of Egyptologists, latex assoc. with motherhood & breast-feeding; ***C. rostellatum*** (Turcz.) Liede & Khanum (*M. japonica*, E As.) – stems for rope, seeds & lvs medic., floss a cotton-floss subs. (China); ***C. sarcomedium*** Meve & Liede (*S. intermedium*, Ind.) – ? a major ingredient of 'soma' of anc. Ind.; ***C. viminale*** (L.) Bassi (*S. v.*, OW trop.) – outer stem layers ed. cooked

Cynapium Nutt. ex Torrey & A. Gray (~ *Ligusticum*). Umbelliferae (III 8). 1 N Am.: ***C. apiifolium*** Nutt. ex Torrey & A. Gray

Cynara Vaill. ex L. Compositae (Card.-Card.). 8 Medit. (Eur. 7), Canary Is. R: BJLS 109(1992)87. Coarse thistle-like herbs cult. as veg.: ***C. cardunculus*** L. (*C. scolymus*, cardoon, Medit.) – lvs blanched (chards) & ed., also r., now invasive Aus., N & S Am. in pampas (where fls used to curdle milk), potential pulp crop; globe or French artichokes are cvs where young fl. buds eaten (bases of involucral bracts & receptacle, the florets discarded), infusion of lvs used in cynar liqueur

Cynarospermum Vollesen (~ *Blepharis*). Acanthaceae (III 1). 1 W Ind.: ***C. asperrimum*** (Nees) Vollesen. R: KB 54(1999)1271

Cyne Danser. Loranthaceae (3). 6 Philipp. to NG. R: Blumea 38(1993)101

Cynoctonum J. Gmelin = *Mitreola*

Cynodendron Baehni = *Chrysophyllum*

Cynodon Rich. Gramineae (XXIX 5). 10 trop. & warm (Eur. 1). R: T 19(1970)565. Hybrids with *Chloris* spp. ***C. plectostachyus*** (K. Schum.) Pilg. (E Afr.) – 1 pl. in 5 1/2 months covered c. 800 m^2 with runners to c. 16 m. Some with HCN, e.g. ***C. convergens*** F. Muell. (*Brachyachne c.*, E Aus.), so toxic to stock, others pasture & lawn-grasses (stargrasses) incl. hybrids, esp. ***C. dactylon*** (L.) Pers. (Bermuda grass, Bahama or kweek grass, dhob, dhub or doob, warm & widely natur., invasive in Aus. [where some natural hybrids with *Chloris* spp.], Hawaii) – lawngrass in Aus. ('couch') sacred to Hindus (cattlefeed), rhiz. ed. Spain, used for bonesetting in Greece since time of Dioscorides, ***C. incompletus*** Nees (S Afr.) – cvs for lawngrass in US, ***C. × magennisii*** Hurcombe (*C. dactylon × C. transvaalensis*) – fine triploid lawngrass for golf-courses in S US & Hawaii, & ***C. transvaalensis*** Davy (masindi, Uganda grass, S Afr., natur. US)

Cynoglossopsis Brand = *Cynoglossum* (but see PSE 138(1981)283)

Cynoglossum Tourn. ex L. Boraginaceae (B.3.8.1). Incl. *Paracaryum, Paracynoglossum Pardoglossum, Rindera, Solenanthus*, & poss. *Suchtelenia*, excl. *Adelinia, Andersonglossum*, c. 180 OW temp. (Eur. 11) & warm. R: Pfl 78(1921)114. Fr. of 4 nutlets covered with barbed prickles forming an animal-disp. bur. Alks. Few cult. orn. esp. ***C. amabile*** Stapf & J.R. Drumm. (Chinese forget-me-not, E As.) & ***C. officinale*** L. (hound's-tongue, Eur., W As.) – form. medic. & young lvs used as salad

Cynoglottis (Guşul.) Vural & Kit Tan (~ *Anchusa*). Boraginaceae (B.2.1.1). 2 SE Eur., SW As. R: NRBGE 41(1983)71

Cynometra L. Leguminosae (I 2). c. 85 trop. Some water-disp. in Amaz.; some timbers esp. ***C. alexanderi*** C.H. Wright (ironwood, muhimbi, C Afr.); ***C. cauliflora*** L. (cultigen orig. E Mal.) – unripe pods ed. raw, cooked or pickled (nam-nam)

Cynomoriaceae Endl. ex Lindl. Magnoliidae – Saxifragales. 1/1 Medit. to Mongolia. monoec. Chlorophyll-less r. parasite with r.-hairs. Lvs in spirals. Infl. capitate, fls minute. P (1)4 or 5(-8), s.t. basally connate; A 1 adnate to P; $\overline{G}$ with 1 pend. unitegmic ovule. Fr. an achene. n = 12
Genus: *Cynomorium*
Form. in Balanophoraceae

Cynomorium Micheli ex L. Cynomoriaceae. 1 Med. incl. Eur. to S Arabia & Mongolia (also coll. NE Somalia): ***C. coccineum*** L. – whole plant reddish brown to purplish black, parasitizing a range of salt-marsh pls; so unlike flowering pls as to be called *Fungus melitensis* (Maltese mushroom) in Middle Ages; ed. Bedouin, r. a condiment (Tuareg), valuable dysentery cure in Malta until 1860 with armed guards for pls, styptic & other local medic., main ingredient (Doctrine of Signatures?) in 'natural viagra', ed. in Iraq. R: BJ 106(1986)374

Cynorchis Thouars = *Cynorkis*

Cynophalla (DC.) J. Presl (~ *Capparis*). Capparaceae. 16 trop. Am. R: HPB 13(2008)117

Cynorhiza Ecklon & Zeyher (~ *Peucedanum*). Umbelliferae (III 11). 3 S Afr. R: T 57(2008) 358

Cynorkis Thouars. Orchidaceae (IV 2). c. 150 trop. & S Afr., Madag. (most), Masc. Some apomicts, ***C. uncata*** Kraenzlin (E Afr.) with axillary bulbils; some cult. orn.

Cynosciadium DC. Umbelliferae (III 8). 2 N Am.

Cynosurus L. Gramineae (XVI 9). 10 Eur. (4), Medit. R: NBP 1964: 23. Pasture & hay grasses (silky bent) esp. ***C. cristatus*** L. (crested dog's-tail, Eur., natur. N Am.) – fodder, form. used for weaving mats & baskets

Cynoxylon (Raf.) Small = *Cornus*

cyp or **cypre** *Cordia alliodora*

Cypella Herb. Iridaceae (VII 5). Incl. *Kelissa, Onira*, c. 30 Mex. to Arg. Corms, lvs plicate; elaiophores attractive to bees. ***C. aquatilis*** Ravenna (Braz.) – submerged; other spp. cult. orn. esp. yellow-flowered ***C. herbertii*** (Lindl.) Herb. (S Am.)

Cyperaceae Juss. Magnoliidae – Poales. 98/5300 cosmop. esp. temp. Perenn. usu. rhiz. herbs (many with Tomlinson's Model; rarely annuals), s.t. lianoid (*Scleria*), shrubby & dracaenoid or vellozioid (*Afrotrilepis, Cephalocarpus*, some *Costularia, Gahnia, Microdracoides, Trilepis*), *Websteria* a leafless aquatic, usu. with vessel elements throughout veg. parts; stems triangular (less oft. terete), usu. solid; r. with r.-hairs (exc. *Eleocharis*), mycorrhiza rare. Lvs in spirals, oft. 3-ranked with a (usu.) closed sheath & usu. long narrow blade (s.t. terete or even 0) with parallel veins; ligule at junction with sheath s.t. present. Infls term. Fls small, usu. wind-poll. (some *Cyperus, Ficinia, Rhynchospora* (*Dichromena*) etc. insect-poll.), usu. unisexual (pls monoec., occ. dioec.), sessile in axils of spirally arr. or distich. bracts (scales), forming spikes or spikelets usu. in sec. infls, rarely solit. & term., v. rarely a small bract between fl. & axis; P (1–)6(–∞) scales, bristles or 0, A (1–)3(6), anthers with longit. slits, pollen in pseudomonads (3 of the 4 nuclei formed at meiosis soon degenerating), $\underline{G}$ ((2)3(4)) with terminal style & stigma branches not always same no. as G, ovule 1, basal, anatropous, bitegmic. Fr. an achene; seed free from pericarp, with oily, ± starchy endosperm with outer proteinaceous layer. x = 5–60+
Classification & chief genera:

I. **Mapanioideae** (fls pseudanthia with A in axils of bracts surrounding female fls; app. sister to rest of fam.): 2 tribes (1. Hypolytreae – *Mapania*; 2. Chrysitricheae – *Chorizandra*)

II. **Cyperoideae** (usu. at least 1 fl. per spikelet bisex.): 7 tribes (1. Scirpeae – *Eriophorum, Scirpus*; 2. Fuireneae – *Fuirena, Schoenoplectus*; 3. Eleocharideae – *Eleocharis*; 4. Abildgaardieae – *Abildgaardia, Bulbostylis, Fimbistylis*; 5. Cypereae – *Cyperus, Isolepis*; 6. Dulichieae – *Blysmus*; 7. Schoeneae – *Cladium, Gahnia, Lepidosperma, Rhynchospora, Schoenus*)

III. **Scleroideae** (spikelets with few glumes, 1-few of which subtend a unisexual fl; heterogeneous?): 4 tribes with some genera unplaced (1. Cryptangieae – *Lagenocarpus*; 2. Trilepideae – *Microdracoides*; 3. Sclerieae – *Scleria* (only); 4. Bisboeckelereae – *Calyptrocarya*)

IV. **Caricoideae** (fls unisexual, naked, usu. in many-flowered spikes, females surrounded by a perigynium (utricle)): *Carex*

Controversy over generic limits & evolution of the infl., some contending that the bisex. fls in some genera are aggregates of unisexual. The ovules are unusual in being term. prod. of a meristem. C4 photosynthesis with multiple orig. (& some reversals to C3). C. are app. related to Gramineae, but despite their frequency, particularly in cool wet habitats, are by no means so diversified: they are scarcely palatable to animals, incl. humans & are of almost negligible economic importance (but see KB 56(2001)257) when compared with G. Some are used for thatching, basketwork etc., *Cladium*, *Cyperus* & *Lepidosperma* for paper-making, while spp. of *Actinoscirpus*, *Eleocharis* & *Cyperus* have ed. tubers; few cult. orn.

Cyperochloa Lazarides & L. Watson. Gramineae (XXI). 1 SW Aus.: ***C. hirsuta*** Lazarides & L. Watson. R: Brunonia 9(1987)215

Cyperus Micheli ex L. Cyperaceae (II 5). Incl. *Alinula*, *Ascolepis*, *Kyllinga*, *Lipocarpha*, *Mariscus*, *Pycreus*, *Remirea*, *Torulinium* etc., c. 950 cosmop. R [s.s.]: Pflr. IV,20 (1935), BSRBB 122(1989)103. Ann. or perenn. herbs with rhiz./tubers, some insect-poll., e.g. ***C. niveus*** Retz. (OW trop.) – oil-coll. honey bees in Nepal, & ***C. sphaerocephalus*** Vahl (S Afr.) both with pollenkitt; some atrocious weeds, some e.g. ***C. cristatus*** (Kunth) Mattf. & Kük. (*K. alba*, S Afr.), used as stopper for ostrich-egg water-containers, & ***C. brevifolius*** (Rottb.) Hassk. (*K. brevifolia*, trop.) local med. ***C. articulatus*** L. (adrue, warm) – roasted rhiz. medic. & air-freshener (Kenya), erosion control; ***C. alternifolius*** L. (*C. involucratus*, umbrella-plant, OW trop., widely natur.) – common housepl., local medic.; ***C. bulbosus*** Vahl (yalka, OW trop. to Aus.) – ed. tubers; ***C. canus*** J. Presl & C. Presl (tule, Mex.) – cult. Honduras to make sleeping-mats; ***C. corymbosus*** Rottb. (OW trop.) – rhiz. a strong contraceptive in Braz.; ***C. difformis*** L. (OW, natur. Am.) – trop. weed of rice e.g. in Aus. ('dirty Dora'); ***C. eragrostis*** Lam. (Am.) – imp. agric. weed, invasive in S Eur.; ***C. esculentus*** L. (yellow nutsedge, ?W As. & Afr., now subcosmop.) – weedy pest in US but var. ***sativus*** Boeck. (tiger nut, earth almond, chufa, rush or Zulu nut) – rarely flowering, tubers ed. (rich in starch, sugar & fat) roasted, one of oldest crops in Egypt, made into flour or juice served as drink (Horchata de Chufa in Spain); ***C. giganteus*** Vahl (Mex. papyrus, trop. Am.) – cult. & natur. S US; ***C. iria*** L. (warm OW, natur. US) – grasshoppers fed it have underdeveloped ovaries (presence of juvenile hormone III); ***C. laevigatus*** L. (makaloa, Euras., Afr., Am.) – form. much used in Hawaii for 'makaloa' mats for cloaks etc. esp. on Nihau with red sheaths of *Eleocharis erythropoda*; ***C. longus*** L. ((sweet) galingale, Euras, incl. GB) – mooted for 'bioenergy' prod. on poor soils, rhiz. (violet-scented) used in scent; ***C. malaccensis*** Lam. (Indomal.) – used for brushes & matting; ***C. papyrus*** L. (papyrus, C Afr. & Nile Valley (rare in upper parts), natur. E Sicily) – dominates swamps of N Uganda (clones in cult. oft. app. sterile) forming a sp.-poor trop. environment, 'sudd', poss. native in N Israel, form. leaf pith sliced into thin strips, laid side by side & another set over these at right angles then tapped so that a sheet (papyrus, same r. as 'paper') was formed – used until 8 cent. AD, examples 4000 yrs. old extant, also used for sandals, ropes & boats (Moses in the bulrushes), recently the Ra series (transAtlantic), tuberous rhiz. also ed.; ***C. prolifer*** Lam. (trop. E Afr.) – long cult. in fish-ponds in US; ***C. pulchellus*** R. Br. (trop. Aus.) – poll. flies & bees; ***C. rotundus*** L. (coco grass, nutgrass, trop., widely natur.) – 'the world's worst weed' through tubers, form. used in perfumery (Scythians embalmed corpses with it), local medic. in Ger. & As. (antifebrile cyperene (also said to inhibit prostaglandin synthesis) & cyperinol in tubers), leaf extract allelopathic; ***C. tegetiformis*** Roxb. (As.) – cult. for fibre in China for mats etc.

Cyphacanthus Leonard. Acanthaceae (III 1). 1 Colombia: ***C. atopus*** Leonard – known only from orig. specimen

Cyphanthera Miers (~ *Anthocercis*). Solanaceae (III 1). 9 S Aus. (2 spp. may be hybrids between ***C. albicans*** (A. Cunn.) Miers & *Duboisia* spp.)

Cyphia Bergius. Campanulaceae (V). 64 Afr. (esp. S), Cape Verde Is. R: Pflr. IV,276c(1968)935. Tubers ed.

Cyphiaceae A. DC. = Campanulaceae

Cyphisia Rizz. = *Justicia*

Cyphocalyx Gagnepain = *Trungboa*

Cyphocardamum Hedge. Cruciferae (35). 1 Afghanistan: ***C. aretioides*** Hedge – known only from orig. specimen

Cyphocarpa (Fenzl) Lopr. = *Kyphocarpa*

Cyphocarpus Miers. Campanulaceae (IV). 3 N Chile. R: BSAB 7(1959)250

Cyphochilus Schltr. = *Appendicula*

Cyphochlaena Hackel. Gramineae (34b). 2 Madag. R: Adansonia 5(1965)411

Cyphokentia Brongn. Palmae (V 14e). Incl. *Moratia*, 2 New Caled. R: KB 63(2008)71. ***C. cerifera*** (H. Moore) Pintaud & W. Baker (*M. c.*) – adaxial cuticle 4 times as thick as epidermal cells, seeds take 1 1/2 yrs to germinate

Cypholepis Chiov. = *Coelachyrum*

Cypholophus Wedd. Urticaceae (III). 15 C Mal. to W Pac. Heterogeneous. Some bark fibres for mats

Cypholoron Dodson & Dressler. Orchidaceae (V 12h). 2 Ecuador. R: Phytol. 24(1972)285. ***C. frigidum*** Dodson & Dressler – only 1.5 cm tall

Cyphomandra Mart. ex Sendtner = *Solanum*

Cyphomeris Standl. = *Boerhavia*

Cyphonanthus Zuloaga & Morrone (~ *Panicum*). Gramineae (XXIII 1). 1 trop. Am.: ***C. discrepans*** (Döll) Zuloaga & Morrone. R: T 56(2007)526

Cyphophoenix H. Wendl. ex Hook.f. Palmae (V 14c). 2 New Caled. (1), Loyalties (1). R: KB 63(2008)66

Cyphosperma H. Wendl. ex Hook.f. Palmae (V 14c). 4 New Caled. (1), Fiji (2). R: Allertonia 3(1984)387

Cyphostemma (Planch.) Alston (~ *Cayratia*). Vitaceae. c. 150 warm (Afr. 106). R: NS 16(1960)113. Some cult. orn. succ. esp. ***C. juttae*** (Dinter & Gilg) Descoings (elephant's foot, SW Afr.) – stem bottle-like to 1 m diam.

Cyphostigma Benth. Zingiberaceae (II 4). 1 Sri Lanka: ***C. pulchellum*** (Thw.) Benth.

Cyphostyla Gleason. = *Allomaieta*

Cyphotheca Diels. Melastomataceae. 1 Yunnan: ***C. montana*** Diels. R: NJB 10(1990)21

cypress *Cupressus, Chamaecyparis, Xanthocyparis* spp.; **African c.** *Widdringtonia* spp.; **Arizona c.** *Cupressus arizonica*; **bald c.** *Taxodium distichum*; **black c.** *Callitris endlicheri*; **Calif. c.** *Cupressus goveniana*; **Himalaya c.** *C. torulosa*; **hinoki or Jap. c.** *Chamaecyparis obtusa*; **Lawson's c.** *C. lawsoniana*; **Leyland c.** × *Cuprocyparis leylandii*; **Medit. c.** *Cupressus sempervirens*; **Mexican c.** *C. lusitanica*; **Monterey c.** *C. macrocarpa*; **Nootka c.** *Xanthocyparis nootkatensis*; **Patagonian c.** *Fitzroya cupressoides*; **c. pine** *Callitris* spp.; **c. pine, white** *C. glauca*; **Port Jackson c.** *C. rhomboidea*; **c. powder** French cosmetic with starch from *Arum maculatum*; **sawara c.** *Chamaecyparis pisifera*; **Sitka c.** *X. nootkatensis*; **southern c.** *Taxodium distichum*; **summer c.** *Bassia scoparia*; **swamp c.** *T. distichum*; (**Alaska**) **yellow c.** *X. nootkatensis*

Cypringlea Strong (~ *Scirpus*). Cyperaceae (II 2). 3 Mex. R: ABM 83(2008)15

Cyprinia Browicz = *Periploca*

Cypripediaceae Lindl. = Orchidaceae (II)

Cypripedium L. Orchidaceae (III). 51 N temp. (Eur. 3). R: P. Cribb (1997) *The genus C.* 'Lady's slipper' orchids, moccasin flowers (US), terr. with plicate lvs; contact irritant (cypripedin) from glands on shoots; labellum inflated, sac-like, column with 2 fert. anthers flanking gland-like staminode, pollen glutinous but not in pollinia; insects enter labellum but cannot return that way & are forced to pass out through opening at base, thereby brushing against stigma & then anthers. Some medic. N Am. & E As., some cult. orn., though most 'cypripediums', i.e. greenhouse epiphytes without plicate lvs, belong to *Paphiopedilum* & *Phragmipedium*. ***C. acaule*** Aiton (N Am.) – poll. queen bumble-bees; ***C. calceolus*** L. (Euras.; diff. 'races' in N Am. referred to other spp.) – grows c. 16 yrs before flowering, prot. in GB though allegedly reduced to a single plant in Yorkshire through overcollecting for herbarium & garden, now increased by ex-situ conservation & established in other sites, poll. female solit. bees attracted by compounds like bee pheromones; ***C. macranthos*** Sw. ('*C macranthus*', C Russia to NE As.) – nectarless fls (poss. mimicking *Pedicularis schistostegia* Vved.) poll. *Bombus* queen bees; ***C. reginae*** Walter (N Am.) – causes dermatitis

Cypselea Turpin = *Sesuvium*

Cypselocarpus F. Muell. Gyrostemonaceae. 1 SW Aus.: ***C. haloragoides*** (Benth.) F. Muell.

Cypselodontia DC. = *Dicoma*

Cyrilla Garden ex L. Cyrillaceae. 1 SE US to N S Am.: ***C. racemiflora*** L. (leatherwood) – tree or shrub of wet ground reprod. largely veg., with infls on last season's wood, imp. bee-tree for comm. honey. R: CGH 186(1960)76

Cyrillaceae Lindl. Magnoliidae – Ericales. Excl. Clethraceae, 2/2 N S Am. to S USA. R: CGH 186(1960)1. Glabrous to stellate-hairy trees or shrubs, s.t. accum. cobalt. Lvs in spirals, simple, entire; stip. 0. Fls in racemes or panicles, each with a bract & oft. 2 bracteoles, bisex., reg., hypogynous; K 5(-7), connate basally, imbr., persistent & oft. accrescent in fr., C as many as & alt. with K, whitish, basally connate, imbr. or convolute, A twice C or same as & alt. with C (*Cyrilla*), anthers ± versatile, with longit. slits & pollen grains in monads, nectary-disk around G-base, G ((2–)5), pluriloc. with axile placentas & ± style & 1–3 ovules per locule, each pend. from nr. tip, anatropous, unitegmic. Fr. indehiscent, drupaceous or 1-seeded 2–5-winged samara; seed oft. with testa lost & straight embryo embedded in copious oily endosperm. n = 10

Genera: *Cliftonia, Cyrilla*

Clethraceae (∞ ovules) kept sep. once more. Prob. closest to Ericaceae

Cyrillopsis Kuhlm. Ixonanthaceae. 2 NE Braz.

Cyrilwhitea Ising = *Sclerolaena*

Cyrtandra Forst.& Forst. f. Gesneriaceae (III 2j). 800+ China (1), S Jap., Nicobars to Pacific Is., esp. Borneo (150) & NG (150). Many hybrids in Hawaii. Some tree-like, ***C. dilatata*** C.B. Clarke (Borneo) only rheophytic gesneriad

Cyrtandroidea F. Br. = *Cyrtandra*

Cyrtandromoea Zoll. Plantaginaceae (?Grat.; Scrophulariaceae s.l.). 11 SE As., Mal. Iridoids present (unlike Gesneriaceae to which it has been referred)

Cyrtandropsis Lauterb. = *Cyrtandra*

Cyrtanthera Nees = *Justicia*

Cyrtanthus Aiton. Amaryllidaceae. Incl. *Anoiganthus, Vallota* 55 trop. & S Afr. R: Herbertia 1939:65. ***C. ventricosus*** (Jacq.) Willd. (S Afr.) – flowering promoted by smoke (not ethylene). Cult. orn. (fire-lilies), like *Crinum* spp., esp. ***C. elatus*** (Jacq.) Traub (*V. speciosa*, Scarborough, George or Knysna lily) & ***C. mackenii*** Hook. (ifafa lily)

Cyrtidiorchis Rauschert = *Maxillaria* (but see T 31(1982) 560)

Cyrtidium Schltr. = *Maxillaria*

Cyrtocarpa Kunth. Anacardiaceae (II). 5 SW N Am. to NE Braz. R: AMBG 78(1991)184. ***C. procera*** Kunth (chupadilla) – fr. ed., bark a soap subs.

Cyrtochiloides N. Williams & M.W. Chase (~ *Oncidium*). Orchidaceae (V 2h). 3 trop. Am. R: Lindleyana 16(2001)284

Cyrtochilum Kunth (~ *Oncidium*). Orchidaceae (V 2h). c. 130 Andes, Carib. (1). R: Lindleyana 16(2001)56

Cyrtochloa S. Dransf. (~ *Dinochloa*). Gramineae (V 4). 5 Philippines (most endemic Luzon). R: KB 53(1998)861

Cyrtococcum Stapf (~ *Panicum*). Gramineae (XXIV 2). 15 OW trop.

Cyrtocymura H. Robinson (~ *Vernonia*). Compositae (Vern.-Vern.). 6 trop. Am. esp. Braz.

Cyrtogonone Prain. Euphorbiaceae (Al.-Crot.). 1 W Afr.: ***C. argentea*** (Pax) Prain. R: A. Radcliffe-Smith, *Gen. Euphorb.* (2001)342

Cyrtogonellum Ching. Dryopteridaceae (I). 10 China. SE As.

Cyrtomidictyum Ching = *Polystichum*

Cyrtomium C. Presl = *Polystichum*

Cyrtophyllum Reinw. (~ *Fagraea*). Gentianaceae. 5 Indomal. R: GBS 64(2012)498. ***C. fragrans*** (Roxb.) DC. (*F. f.*) – cult. orn.

Cyrtopodium R. Br. Orchidaceae (V 12d). 47 trop. Am. (Florida 2). R: HPB 13(2008)191. Cult. orn. epiphytic or terr. poll. male euglossine bees. ***C. longibulbosum*** Dodson & G. Romero (Ecuador) – pseudobulbs to 3.5 m tall!

Cyrtorchis Schltr. Orchidaceae (V 16d). 18 trop. & S Afr. R: KB 14(1960)143. Hawkmoth-poll. Cult. orn. incl. ***C. crenata*** (B.S. Williams) Garay

Cyrtorhyncha Nutt. (~ *Ranunculus*). Ranunculaceae (II 3). 1 W N Am.: ***C. ranunculina*** Nutt.

Cyrtosia Blume (~ *Galeola*). Orchidaceae (II 2). 5 Indomal. to Jap. Leafless terr. orchids; fr. a fleshy berry. ***C. septentrionalis*** (Reichb.f.) Garay (*Galeola s.*, Jap.) – endozoochory

Cyrtosperma Griff. Araceae (V). 13 Indomal., Oceania (similar seeds from Eocene of Br. Columbia). R: Blumea 33(1988)428. Some rheophytes, cult. orn. & ed. tubers esp. ***C. merkusii*** (Hassk.) Schott (*C. chamissonis, C. edule*, babai, swamp taro, W Pac.) – tubers to 60 kg (10 yrs old), slow-growing but good for stagnant & brackish swamps

Cyrtostachys Blume. Palmae (V – tribe unclear). 7 Mal. esp. NG to Melanesia. R: KB 64(2009)78. Monoec. cult. orn. esp. ***C. renda*** Blume (*C. lakka*, sealing-wax palm, W Mal.) – widely planted in trop.

Cyrtostylis R. Br. (~ *Acianthus*). Orchidaceae (IV 1). 5 Aus. (5), NZ

Cyrtoxiphus Harms = *Cylicodiscus*

Cystacanthus T. Anderson = *Phlogacanthus*

Cysticapnos Mill. Papaveraceae (Fumariaceae I). 4 S Afr. R: OB 88(1986)105

Cysticorydalis Fedde ex Ikonn. = *Corydalis*

Cystoathyrium Ching (~ *Athyrium*). Cystopteridaceae. 1 China: ***C. chinense*** Ching – extinct?

Cystodiaceae Croft (~ Lindsaeaceae). 1/1 C Mal. to Solomon Is. Terr. ferns, young parts with long golden hairs. Lvs bipinnate, veins free, simple or forked. Sori submarginal, term. on veins, protected by a reflexed lobe of leaf (false indusium) & smaller, indusium
Genus: *Cystodium*

Cystodiopteris Rauschert = seq.

Cystodium J. Sm. Cystodiaceae (Lindsaeaceae s.l.). 1 Borneo to Solomon Is.: ***C. sorbifolium*** (Sm.) J. Sm. - stem prostrate. R: M.F. Large & J.E. Braggins, *Tree Ferns* (2004)280. Form. referred to Dicksoniaceae (Cyatheaceae s.l.)

Cystopteridaceae Shmakov (~ Woodsiaceae). 4/38 temp., trop. mts. R: T 61(2012)519. Lvs pinnate to 3(4)-pinnate-pinnatifid, veins free; sori abaxial, on veins, indusia 0 or attached proximally to receptacle. x = 40, 42.
Genera: *Acystopteris, Cystoathyrium, Cystopteris, Gymnocarpium*

Cystopteris Bernh. Cystopteridaceae (Woodsiaceae s.l.). 27 temp. & warm (esp. N Eur. 6; GB 3) incl. trop. alpine. R: MTBC 21, 4(1963)1. ***C. bulbifera*** (L.) Bernh. (E N Am.) – bulblets formed on fronds. Cult. orn. esp. ***C. fragilis*** (L.) Bernh. (bladder fern, N hemisph. to Chile & Kerguelen Is.)

Cystorchis Blume. Orchidaceae (IV 2b). 21 Thailand to Micronesia. Some mycotrophs incl. ***C. aphylla*** Ridl. (Mal.)

Cystostemma Fourn. = *Funastrum*

Cystostemon Balf.f. Boraginaceae (B.2.2). 15 trop. Afr. to SW Arabia. R: NRBGE 40(1982)1

Cytinaceae Link (~ Rafflesiaceae). Magnoliidae – Malvales. 2/11 Medit., S Afr., Madag., Mex. Monoec. or dioec. chlorophlyll-less parasites. Lvs in spirals. Infl. racemose, capitate or spicate. P 4–9, basally connate; A 6–10 connate, extrorse with nectariferous cavities between A; $\overline{G}$ (8–14) with intrusive parietal placentation & ∞ uni- or bitegmic ovules per carpel
Genera: *Bdallophytum, Cytinus*

Cytinus L. Cytinaceae. 8: 2 (subg. *Cytinus*) Canary Is. to Middle E (Eur. 2) on Cistaceae, monoec.; 6 (subg. *Hypolepis*) S Afr., Madag., dioec. ***C. hypocistis*** (L.) L. (Medit.) – ant-poll., ***C. visseri*** Burgoyne (S Afr.) – poll. striped fieldmouse & (carnivorous!) short-snouted elephant-shrew, attractants incl. 3-hexanone; ***C. ruber*** (Fourr.) Fritsch (Canary Is., W Medit.) – ed., local medic.

Cytisanthus Lang = *Genista*

Cytisophyllum Lang. Leguminosae (III 9). 1 W Med.: ***C. sessilifolium*** (L.) Lang

Cytisopsis Jaub. & Spach (~ *Anthyllis*). Leguminosae (III 18). 2 E Med. & N Afr.

Cytisus Tourn. ex Desf. Leguminosae (III 9). Incl. *Calicotome, Chamaecytisus c.* 60 Eur., N Afr., Canary Is., W As. R: T 55(2006)739. Brooms. Explosive fls with more than 1 chance of cross-poll.: short A deposit pollen on undersurface of insect, long ones on back but preceded by style which may contact pollen deposited by another fl., then style grows round, such that stigma occupies position just above short A & is thus ready for pollen on undersurface of a later visitor. Lvs usu. reduced, stems photosynthetic; widely planted & natur., some forced for cut-flower trade under glass. ***C. multiflorus*** (L'Hérit.) Sweet (W Med.) – invasive in Aus.; ***C. proliferus*** L.f. (*Chamaecytisus p.* subsp. *palmensis*, tagasaste, tree lucerne, Canary Is.) – fodder tree introd. Pacific, reafforesting salinized land in Aus.; ***C. scoparius*** (L.) Link (*Sarothamnus s.*, common, Eur. or Scotch broom, Eur., invasive Aus. [introd. 1800s as hop subs.], NZ, W US) – sand-binder, bee-forage, form. for fibre & dyes, prob. THE planta-genista (see *Genista*), young shoots used for dropsy etc. & now to regulate heart-rate, alk. (sparteine a diuretic used in WW II, tripling 'renal elimination' but hallucinogenic in excess); ***C. striatus*** (Hill) Rothm. (SW Eur.), widely planted & now natur. along motorways in GB. See also *Argyrocytisus, Cytisophyllum*

Cytogonidium Briggs & L. Johnson (~*Restio*). Restionaceae. 1 SW Aus.: ***C. leptocarpoides*** (Benth.) Briggs & L. Johnson. R: Telopea 7(1998)362

Cyttaranthus Léonard. Euphorbiaceae (Acal.-Agr.). 1 trop. Afr.: ***C. congolensis*** Léonard. R: BJBB 25(1955)286

Czeikia Ikonn. = *Acanthophyllum* (but see BZ 89(2004)114)

Czernaevia Turcz. = *Angelica*

D

da xao *Ziziphus jujuba*

dabai *Canarium odontophyllum*

dabéma, daboma *Piptadeniastrum africanum*

Daboecia D. Don. Ericaceae (V 5). 1 Ireland to Spain & Azores: ***D. cantabrica*** (Hudson) K. Koch (St Dabeoc's (*sic*) heath, fossils as far N as Shetland). G; bee-poll. cult. orn. everg., **'Charles Nelson'** – several C whorls, A 0, G aborted

Dacrycarpaceae A. Bobrov & Melikian = Podocarpaceae

Dacrycarpus (Endl.) Laubenf. (~ *Podocarpus*). Podocarpaceae. 9 Myanmar to NZ. R: JAA 50(1969)315. Timber esp. ***D. dacrydioides*** (Rich.) Laubenf. (*Podocarpus d.*, kahikatea, NZ) – pulp for paper, fr. eaten by Maoris; ***D. imbricatus*** (Blume) Laubenf. (SE As. to W Pac.)

Dacrydiaceae A. Bobrov & Melikian = Podocarpaceae

Dacrydium Sol. ex Forst. f. Podocarpaceae. 25 SE As. to NZ. Usu. dioec.; seeds arillate. R: Phytol.M 7(1984)25. Timber esp. ***D. cupressinum*** Sol. ex Forst. f. (rimu, red pine, NZ) – masting every 3–5 yrs, to which breeding of kakapo (*Strigops habroptilus*), flightless parrot (reduced to 125 birds), restricted as imp. in diet, ***D. elatum*** (Roxb.) Loudon (sempilor, Mal.). ***D. xanthandrum*** Pilg. (C & E Mal.) – primary forest tree with r. suckers (rare phenomenon). See also *Lagarostrobos, Lepidothamnus*

Dacryodes Vahl. Burseraceae (V). c. 35 trop. (c. 2 Am., 18 Afr., 15 As. – T: KB 63(2008)386). R: Blumea 7(1954)500. Rauh's Model. Some timbers, ed. fr. pulp & resins. ***D. belemensis*** Cuatrec. (S Am.) – fr. used in drink by Colombian Indians; ***D. edulis*** (G. Don) H.J. Lam (trop. Afr.) – oily seeds (eben, safu) ed.; ***D. excelsa*** Vahl (*D. hexandra*, WI) – source of WI elemi

Dacryotrichia Wild. Compositae (Ast.-Gran.). 1 Zambia: ***D. robinsonii*** Wild. R: GOB 1(1973)67

Dactyladenia Welw. (~ *Acioa*). Chrysobalanaceae (4). 31 trop. Afr. R: FOW 10(2003)76

Dactylaea Fedde ex H. Wolff = *Sinocarum*

Dactylaena Schrad. ex Schultes f. = *Cleome*

Dactylanthaceae Takht. = Balanophoraceae

Dactylanthus Hook.f. Balanophoraceae. 1 NZ (N Is.): ***D. taylorii*** Hook.f. – only wholly r.-parasitic fl. pl. (20 angiosperm hosts) in NZ (fossils 28 M yrs old, form. in S Is. too), inconstantly dioec. with 5 male to 1 female infls. of smelly fls with copious nectar, poll. short-tailed bats but attacked by introd. (Aus.) possums, seed fleshy (? aril). R: Englera 22(2001)1. When boiled, parasite tissue is removed to expose host r. – 'wood-rose'

Dactyliandra (Hook.f.) Hook.f. Cucurbitaceae (XIV). 1 Kenya, 1 disjunct distrib. (***D. welwitschii*** Hook.f.): Namib (SW Afr.) & Rajasthan (Ind.) deserts

Dactylicapnos Wall. (~ *Dicentra*). Papaveraceae (Fumariaceae I 1). 11+ Himal. to SE As. R: OB 88(1986)20, BM 25(2008)200. Tendrilled climbers, ***D. scandens*** (D. Don) Hutch. (*Dicentra s.*, Sino-Himal.) – to 4.5 m

Dactyliophora Tieghem. Loranthaceae (5 3). 2 Ceram, Papuasia, N Queensland. R: AusJB 22(1974)558

Dactylis Royen ex L. Gramineae (XVI 11). 1–5 Euras. ***D. glomerata*** L. (cock's-foot) – cult as meadow- & pasture-grass with seedlings sold as 'cat grass', widely natur. N Am., S Afr., invasive in Aus. & Hawaii, imp. cause of hay-fever; complex of tetraploid & diploid forms, latter incl. subsp. ***aschersoniana*** (Graebner) Thell. (*D. polygama* Horvat., C Eur.) prob. derived from subsp. ***glomerata*** by haplodiploidy

Dactylocardamum Al-Shehbaz. Cruciferae (27). 1 Peru: ***D. imbricatifolium*** Al-Shehbaz – known only from orig. specimen

Dactylocladus Oliv. Crypteroniaceae. 1 Borneo (+ ?1 NG): ***D. stenostachys*** Oliv. (jonkong) – imp. timber export of freshwater peatswamps used for boat-building & form. slats of Venetian blinds

Dactyloctenium Willd. Gramineae (XXIX). 13 warm. Crowfoot. Grains ground to paste & baked by Aus. Aborigines. ***D. aegyptium*** (L.) Willd. (Egyptian grass, Sahara & Sudan, widely natur. incl. N Am.) – usu. ann., sand-binder, fr. ed. (minor cereal), locally medic. since time of Dioscorides; ***D. australe*** Steud. (Durban grass, E & S Afr.) – perenn. sand-binder, shade-tolerant lawn-grass

Dactylopetalum Benth. = *Cassipourea*

Dactylopsis N.E. Br. = *Mesembryanthemum* (but see KB 61(2006)396)

Dactylorhiza Necker ex Nevski (*Satorkis*). Orchidaceae (IV 4d). Incl. *Coeloglossum*, c. 40 temp. Euras. (Eur. 13), Alaska, Medit., Macaronesia. Spp. hybridize & reprod. isolation not clear-cut. Some tubers eaten Iran (Persian salep). Most poll. esp. by bumble-bees by deceit but ***D. viridis*** (L.) R. Bateman & al. (*C. v.*, frog orchid, N temp.) – some nectar, poll. wasps, beetles, mosquitoes. Some, incl. hybrids, e.g. ***D. majalis*** (Reichb.) P. Hunt & Summerh. (Euras.) app. a cross between *D. incarnata* (L.) Soó *&* *D. fuchsii* (Druce) Soó – stigmatic exudate oily, with glucose & amino-acids, food for some bees; cult. orn.

Dactylorhynchus Schltr. = *Bulbophyllum*

Dactylostalix Reichb.f. Orchidaceae (V 13e). 1 Jap.: ***D. ringens*** Reichb.f.

Dactylostegium Nees = *Dicliptera*

Dactylostelma Schltr. = *Oxypetalum*

Dactylostigma D. Austin = *Hildebrandtia*

dadap *Erythrina subumbrans*

Dadjoua Parsa = ? Cruciferae

Daedalacanthus T. Anderson = *Eranthemum*

Daemia R. Br. = *Pergularia*

Daemonorops Blume = *Calamus* (but see ARBGC 12,1(1911)1)

Daenikera Hürl. & Stauffer. Santalaceae (Amphorogynaceae). 1 New Caled.: ***D. corallina*** Hürl. & Stauffer – root-parasite

daffodil *Narcissus pseudonarcissus* but also any large-flowered *N.* sp. or hybrid; **Peruvian d.** *Ismene* spp.; **Tenby d.** *N. pseudonarcissus*

dagame *Calycophyllum candidissimum*

dagga *Cannabis sativa*

dahl split pulses esp. *Cajanus cajan, Cicer arietinum*

Dahlberg daisy *Thymophylla tenuiloba*

Dahlgrenia Steyerm. = *Dictyocarpum*

Dahlgrenodendron J. Merwe & Wyk (~ *Cryptocarya*). Lauraceae (I). 1 S Afr.: ***D. natalense*** (J. Ross) J. Merve & Wyk – like *C.* but pollen unusual, endangered. R: SAJB 54(1988)80

Dahlia Cav. Compositae (Cor.). 35 mts of Mex. (esp.) to Colombia. R: Rhodora 71(1969)309, 367; BotJLS 133(2000)229. Tuberous r.; stems to 8 m, usu. unbranched, s.t. scrambling & epiphytic. Orig. imp. Aztec crop, first grown for animal fodder (tubers, inulin) & medic. (urinary problems; cf. Eur. Doctrine of Signatures!) but also orn. incl. 'double-flowered' cvs (incl. hybrids) selected, introd. to Eur. where back-crosses between diploid ***D. coccinea*** Cav. (2n = 32, Mex. – fls yellow to red) & tetraploid ***D.* × *pinnata*** Cav. (*D.* × *hortensis, D* × *rosea, D. coccinea* × *D. sorensenii* H. Hansen & Hjert. (Mex.), 2n = 64, orig. Mex. – fls purple) have given rise to garden dahlias, a mania in 1830s & 1840s; garden ds divided into 14 classes on capitulum form e.g. pompon d. with small heads of only ray-florets, 'single-flowered' with disk-florets present (e.g. **'Coltness Gem'** – much used dwarf bedder), 'cactus-flowered' with only long ray-florets. ***D. excelsa*** Benth. (incl. *D. imperialis* Roezl ex Ortgies, *D. arborea*, tree-dahlia, range of genus) – to 8 m, unbranched (suckering) exc. infl., where up to 300 capitula, lvs 2–3-pinnate, app. sterile in cult., hollow stems used as water-pipes

Dahliaphyllum Constance & Breedlove. Umbelliferae. 1 Mex.: ***D. almedae*** Constance & Breedlove – woody pachycaul to 4 m. R: ABM 26(1994)84

Dahlstedtia Malme. Leguminosae (III 16). 16 S Am. esp. Braz., Arg. R: T 61(2012)104. Smelly; bird- and Lepidoptera-poll.

dahoma *Piptadeniastrum africanum*

Dahomey rubber *Ficus lutea*

dahoon *Ilex cassine*

daikon *Raphanus sativus* 'Longipinnatus'

Daiotyla Dressler (~ *Chondrorhyncha*). Orchidaceae (V 12j). 4 C & N S Am. R: Lankesteria 5(2005)92

Dais L. (~ *Gnidia*). Thymelaeaceae (Thym.-Daph.). 2 S Afr., Madag. Bark-fibres used for thread, string etc.

Daiswa Raf. = *Paris*

daisy *Bellis perennis* (but also used of many Compositae with radiate capitula); **Barberton d.** *Gerbera jamesonii*; **burr d.** *Calotis* spp.; **d.bush** *Olearia* spp.; **Confederate d.** *Helianthus porteri*; **crown d.** *Glebionis coronaria*; **Dahlberg d.** *Thmyophylla tenuiloba*; **dog d.** *Leucanthemum vulgare*; **kingfisher d.** *Felicia bergeriana*; **Livingstone d.** *Cleretum bellidiforme*; **Michaelmas d.** *Symphyotrichum* spp., see also *Aster*; **moon d.** *L. vulgare*; **Moroccan d.** *Rhodanthemum hosmariense*; **ox-eye d.** *L. vulgare*; **paper d.** (Aus.) *Xerochrysum* spp. esp. *X. bracteatum*,

Rhodanthe manglesii; **Paris d.** *Argyranthemum frutescens*; **Shasta d.** *L.* × *superbum*; **Singapore d.** *Sphagneticola trilobata*; **Swan River d.** *Brachycome iberidifolia*; **tree d.** *Olearia* spp. (Aus.), *Montanoa* spp. (Am.); **yam d.** *Microseris* spp.

Daknopholis W. Clayton (~ *Chloris*). Gramineae (XXIX 5). 1 E Afr., Madag., Aldabra: ***D. boivinii*** (A. Camus) W. Clayton

dakua *Agathis macrophylla*

dal see **dahl**

Dalanum Dostál = *Galeopsis*

Dalbergaria Tussac = *Columnea*

Dalbergia L.f. Leguminosae (III 11). c. 250 trop. (Madag. 48, 47 endemic). Trees, shrubs & lianes. Fr. flattened, indehiscent, some in Amaz. water-disp. Many timbers (some CITES-listed) with dark colour & dense grain for furniture, musical instruments etc. (rosewood, Nicaragua wood, palisander (Braz.)). ***D. cearensis*** Ducke (kingwood, princes wood, tulipwood, Braz.) – most expensive cabinet wood in late C17; ***D. cochinchinensis*** Pierre ex Laness. (trac, SE As.); ***D. congestiflora*** Pittier (camatillo, Mex. kingwood, SW N Am.) – carving timber; ***D. cultrata*** R. Graham ex Ralph (Myanmar) – esp. for ploughs, resin red; ***D. decipularis*** Rizz. & Mattos (Braz. tulipwood, Bahia) – valuable timber ('sebastião-de-arruda') known since C19 but tree not described until 1967; ***D. foliosa*** Ralph (*D. foliacea*, SE As.) – local medic. NE Thailand; ***D. granadillo*** Pittier (granadillo, Mex.); ***D. hainanensis*** Merr. & Chun (huanghuali, Hainan) – wood for late C16/early C17 Chinese furniture, more recently combs etc.; ***D. horrida*** (Dennst.) Mabb. (Ind.) – survives in only a few sacred groves in Kerala; ***D. latifolia*** Roxb. (Ind. or Bombay blackwood or rosewood, Malabar r., black or E Ind. r., Rosetta wood, S Ind.); ***D. melanoxylon*** Guillem. & Perr. (Afr. blackwood, Mozambique ebony, trop. Afr.) – orig. 'ebony' as in anc. Egypt (furniture in Tutankhamun's tomb), even for veneers, now much used for musical instruments (e.g. most oboes & clarinets, though now often sawdust with resin & carbon fibre) & carving e.g. Makonde; ***D. nigra*** (Vell. Conc.) Benth. (Braz. rosewood, Bahia or Rio r., jacaranda, Braz.) – esp. for furniture (e.g. Denmark 1950s, by 1854 more than 5000 t p.a. imp. to UK), radio cabinets, pianos, violin pegs, sides & backs of classical guitars etc., now endangered sp. & timber trading illegal; ***D. parviflora*** Roxb. (Mal.) – liane with scented heartwood used in joss-sticks; ***D. retusa*** Hemsl. (cocobolo, C Am.) – esp. for small work like knife-handles, chess-pieces, rosaries, buttons etc.; ***D. sissoo*** Roxb. ex DC. (sheesham, sissoo, shisham, Ind.) – used for fancy trinket-boxes, fuelwood; ***D. stevensonii*** Standl. (Honduras rosewood, Belize) – used for xylophones, marimbas etc.

dalbergia, false *Pericopsis laxiflora*

Dalbergiella Bak. f. Leguminosae (III 16). 3 trop. Afr.

Dalea L. Leguminosae (III 10). c. 160 Canada to Arg., esp. Mex. & Andes, in dry & desert areas. R: MNYBG 27(1977)135. Indigo bush. Some cult. orn., local medic., stems tied together as brooms

Dalechampia Plum. ex L. Euphorbiaceae (Acal.-Pluk.-Trag.). 120 warm (c. 95 Am., 10 Afr., 10 Madag., 6 As.). R: BotJLS 105(1991)137. Mostly lianes, some with stinging glands; 9+ male & 3 female fls + 2 bracts mimic 1 fl., some with scent (trans-carvone oxide; also in *Catasetum* & other orchids) coll. male euglossines & fl. resins for bees' nests, visitors acting as poll. (buzz-poll. in Madag., where no resin-coll. bees); poll. by male euglossines evol. several times, that by female resin-collectors only once. ***D. spathulata*** (Scheidw.) Baill. (*D. roezliana*, Mex.) – cult. orn. with 2 large pink or white outer bracts prot. fls, the females in a 3-flowered cyme prot. by another bract lying below 4 more bracts with 9–14 male fls lying in front of a yellow cushion of rudimentary males s.t. secreting a resin, after flowering all that lying above the females is dropped

Dalembertia Baill. Euphorbiaceae (Hipp.). 4–5 C Am. Male fl. of A 1 encl. in K 1

Dalenia Korth. Melastomataceae. 3 Borneo. K-tube with calyptra

Dalhousiea Wall. ex Benth. Leguminosae (III 2). 1 W trop. Afr., 1–2 NE Ind., Bangladesh. R: BJB 4(1975)33

Dalibarda Kalm = *Rubus*

Dallachya F. Muell. (~ *Rhamnella*). Rhamnaceae (6). 1 NG, W Pacific: ***D. vitiensis*** (Seem.) F. Muell.

dalli *Virola surinamensis*

Dallis grass *Paspalum dilatatum*

Dallwatsonia B. Simon = *Hymenachne* (but see Austrobaileya 3(1992)678)

Dally pine *Psoralea pinnata*

Dalrymplea Roxb. ('*Dalrympelea*'; ~ *Turpinia*). Staphyleaceae. 25 Ind. to Jap. & NG

Dalzellia Wight. Podostemaceae (II). 5 trop. As. R: BMNHN 4,10(1988)71. Rootless
Dalzielia Turrill = *Marsdenia*
damanu *Calophyllum neoebudicum*
damar mata kuching *Hopea* spp.
damask or **d. violet** *Hesperis matronalis*; **d. rose** *Rosa* × *damascena*
Damasonium Mill. Alismataceae. c. 6 W & C Eur. (3 incl. ***D. alisma*** Mill., thrumwort), S temp. Aus., Calif. (1). R: EJB 58(2001)47
Damatris Cass. (~ *Haplocarpha*). Compositae (Arct.-Arct.). 1 S Afr.: ***D. pudica*** Cass.
Damburneya Raf. (~ *Nectandra*). Lauraceae (I). 20 trop. to warm Am. R: T 65(2016)988. ***D. purpurea*** (Ruíz & Pavón) Trefimov (*N. p.*, Braz.) – 99 spp. of insect-visitors
dame's rocket or **d.'s violet** *Hesperis matronalis*
damiana *Turnera diffusa*
dammar resins from trop. trees used in varnishes. See *Canarium*, *Hopea*, *Shorea*, *Vateria*, *Vatica* & *Agathis*; **white d.** *Vateria indica*
Dammara Rumph. ex Link = *Agathis*
Dammaropsis Warb. = *Ficus*
Dammera Lauterb. & K. Schum. = *Licuala*
Damnacanthus Gaertn. f. Rubiaceae (IV 10). 13 E As. Usu. heterophyllous spiny shrubs, paired thorns = lat. shoots proximally or infl. shoots. ***D. indicus*** Gaertn.f. – cult. orn. used in penjing
Damnamenia Given (~ *Pleurophyllum*). Compositae (Ast.-Hin.). 1 Auckland & Campbell Is.: ***D. vernicosa*** (Hook.f.) Given
Damnxanthodium Strother (~ *Perymenium*). Compositae (Helia.-Ecl.). 1 Mex.: ***D. calvum*** (Greenman) Strother
Dampiera R. Br. Goodeniaceae. 66 Aus., esp. SW. R: Telopea 3(1988)183. Some cult. orn.
Damrongia Kerr ex Craib (~ *Chirita*, *Ornithoboea*). Gesneriaceae (III 2i). 10 SE As. to Sumatra. R: T 65(2016)285
damson or **d. plum** *Prunus* × *domestica* 'ssp. *institia*'; **d. plum** (WI) *Chrysophyllum oliviforme*
dan shen dang *Codonopsis pilosula* subsp. *tangshen*
Danae Medik. Asparagaceae (Convallariaceae). 1 S Turkey, NW Syria, Caspian: ***D. racemosa*** (L.) Moench (Alexandrian laurel) – phylloclades, cult. orn.
Danaea Sm. Marattiaceae. 50 trop. & warm Am. R: BJLS 163(2010)375
Danaeaceae Agardh. Older name for Marattiaceae
Danais Comm. ex Vent. Rubiaceae (IV Dan.). 32 Tanzania (1), Madag. & Masc. Some dyes & fibres
dandelion *Taraxacum officinale* etc.; **Russian d.** *T. bicorne*
Dandya H. Moore. Asparagaceae (Themidaceae). 4 Mex. R: ABM 18(1992)14
danewort *Sambucus ebulus*
dang gui *Angelica sinensis*
dange *Kleinhovia hospita*
Danguya Benoist. Acanthaceae (III 2c). 1 Madag.: ***D. pulchella*** Benoist
Danguyodrypetes Leandri = *Lingelsheimia*
Danhatchia Garay & Christenson (~ *Gonatostylis*). Orchidaceae (IV 2d). 1 NSW, NZ: ***D. australis*** (Hatch) Garay & Christenson – mycotroph. parasitic on *Lycoperdon perlatum* Pers. (puffball), that assoc. with *Beilschmiedia tarairi* in NZ. R: Gen. Orch. 3(2003)82
Daniellia Bennett. Leguminosae (I 2). 9 trop. Afr., forest & savannas. Bumbo; source of copals & furniture timbers esp. ***D. ogea*** (Harms) Holl. (faro, ogea, W Afr.) & ***D. thurifera*** Bennett (hyedua, W Afr.)
Dankia Gagnepain = ? *Camellia*
danpi *Paeonia ostii*
Dansera Steenis = *Dialium*
Dansiea Byrnes. Combretaceae (II 1). 2 Queensland. R: FA 18(1990)260
danta *Nesogordonia papaverifera*
Danthonia DC. Gramineae (Danth.). Excl. *Rytidosperma* 25 Eur. (2), Medit., Am. (7 + 1 natur. E N Am.: ***D. decumbens*** (L.) DC. (*Sieglingia d.*, heath grass, Eur., Medit.) – fls oft. cleistogamous)
Danthoniastrum (Holub) Holub (~ *Metcalfia*). Gramineae (IX). 2 SE Eur., W As.
Danthonidium C. Hubb. Gramineae (Arund.). 1 Ind.: ***D. gammiei*** (Bhide) C. Hubb.
Danthoniopsis Stapf. Gramineae (XVIII). 16 Afr. & Arabia to Pakistan. Fire-free sites
Danube grass *Phragmites australis*
Danxiaorchis Zhai & al. Orchidaceae (V 13e). 1 SE China: ***D. singchiana*** Zhai & al.

dao *Dracontomelon dao*

Dapania Korth. Oxalidaceae. 1 Madag., 2 Mal. Lianes

Daphnandra Benth. Atherospermataceae (Monimiaceae I 2). 6 E Aus. R: FA 2(2007)95. Alks

Daphne L. Thymelaeaceae (Thym.-Daph.). c. 95 Euras. (Eur. 17). R: J.J. Halda (2001) *The Genus D. Wikstroemia* prob. referable here. Shrubs oft. cult.; lvs usu. in spirals (opp. in ***D. genkwa*** Sieb. & Zucc., China, where used clinically as an effective & safe abortifacient); bark fibrous & used in Himal. for rope, paper (shaped like olas & in Tibet treated with *Aconitum* spp. extract as insect-repellent) & rayon (***D. odora*** Thunb. (also cult. for local med., scent), ***D. papyracea*** Wall. ex Steud. (form. typewriter stencil paper), *D. genkwa* etc., Nepal paper from ***D. bholua*** Buch-Ham. ex D. Don); scent carnation-like in many spp. & attractive to Lepidoptera (also found in *Dianthus* (Caryophyllaceae), *Gymnadenia* (Orchidaceae), *Narcissus* (Amaryllidaceae) – all Eur., *Viburnum* (Viburnaceae) – E As., *Ribes* (Grossulariaceae) – N Am., *Petunia* (Solanaceae) – S Am.), the insects themselves prod. similar scents: imp. constituents incl. indol & menthyl anthranilate; nectar produced at base of K-tube & accessible only to Lepidoptera & long-tongued bees; glycosides such as daphnin & an acrid resin (mezerein) give pls a bitter burning taste when chewed – they are poisonous & some insect-repellent, form. medic. e.g. seed-oil of ***D. gnidium*** L. (Medit.) purgative; ***D. laureola*** L. (spurge laurel, Euras., natur. GB, invasive NW US) – everg. poll. principally by X moths, strong purgative & emetic; ***D. mezereum*** L. (mezereon, Euras, prot. in GB) – fls attractive (diterpene esters) to butterflies esp. Red Admiral & honeybees, fr. form. used as pepper subs. oft. with fatal consequences

Daphnimorpha Nakai = *Wikstroemia*

Daphniphyllaceae Muell. Arg. Magnoliidae – Saxifragales. 1/29 E As., Mal. Dioec. everg. glabrous trees or shrubs with a unique type of alk. (daphniphylline group), oft. accumulating aluminium. Lvs in spirals but s.t. so crowded at branch-tips as to appear whorled, simple, entire, pinnately-veined; stip. 0. Fls small, reg., hypogynous, in axillary racemes, each pedicel with a decid. bract; K (0)3–6, ± imbr., C 0, A 5–14 with anthers with valves; female fls s.t. with staminodes, $\underline{G}$ (2(–4)) with as many locules, styles united only basally, short, curved to circinate, ovules (1)2 per locule, apical-axile, pend., anatropous, bitegmic. Fr. a 1(2)-seeded drupe; seed with v. small straight embryo in copious oily & proteinaceous endosperm. 2n = 32

Genus: *Daphniphyllum*. Embryology confirms exclusion from Euphorbiaceae

Daphniphyllum Blume. Daphniphyllaceae. 29 China, through Indomal. (Mal. 16 – R: FM I,13(1997)145) to trop. Aus. R: Taiwania 11(1965)57, 12(1966)137. Some cult. orn. incl. ***D. macropodum*** Miq. **Humile Group** (*D. humile* Maxim., N Jap., Korea) – lvs used like tobacco by Ainu

Daphnopsis Mart. Thymelaeaceae (Thym.-Daph.). Incl. *Coleophora*, 50–65 trop. Am. to E Arg. R: AMBG 46(1959)257, Phytol. 61(1986)361. Koriba's Model. ***D. americana*** (Mill.) J.R. Johnston (*D. tinifolia*, bonace, WI) – inner bark used for cordage

Dapsilanthus Briggs & L. Johnson (~ *Leptocarpus*). Restionaceae. 4 SE As. (1), E Mal. to N Aus. (3). R: Telopea 7(1998)369, 8(1998)25

Darbya A. Gray = *Nestronia*

Darcya Hunz. = *Darcyanthus*

Darcya B.L. Turner & C. Cowan (~ *Stemodia*). Plantaginaceae (Grat.-Stem.). 3 C Am. R: Phytol. 74(1993)268

Darcyanthus Hunz. (~ *Physalis*). Solanaceae (IV). 1 Peru, Bolivia: ***D. spruceanus*** (Hunz.) Hunz. (*Darcya* s.). R: Novon 21(2011)47

dari *Sorghum bicolor*

Darling lily *Crinum flaccidum*; **D. pea** *Swainsona galegifolia*

Darlingia F. Muell. Proteaceae (V 1). 2 Queensland. R: FA 16(1995)356

Darlingtonia Torr. Sarraceniaceae. 1 N Calif., SW Oregon: ***D. californica*** Torr. – carnivorous herb differing from *Sarracenia* in having pitcher-hood with distinct flap ('fish-tail' covered in nectaries attracting insects which slide on footholdless detachable wax into pitchers), in bracts on scape & style apically 5-branched; first pr of pitchers usu. orientated N–S & taller than subsequent, all turned outwards; digestion by assoc. fauna as digestive glands 0; seeds with air-filled projections & hydrophobic surface so air- & water-disp. R: FNA 9(2009)349

Darmera Voss (*Peltiphyllum*). Saxifragaceae. 1 N Calif., SW Oregon: ***D. peltata*** (Benth.) Voss (*P. peltatum*, umbrella plant) – cult. orn. waterside pl. with long-petioled peltate lvs. R: FNA 8(2009)75

darnel *Lolium temulentum*

Darniella Maire & Weiller = *Salsola*

Darwinia Rudge. Myrtaceae (II 15). c. 45 Aus., esp. SW. Heterogeneous (some allied to *Actinodium*, others to *Homoranthus*). Many hybrids in wild; insect- & bird-poll. heath-like shrubs with oils used in perfumery; some cult. orn.

Darwiniella Braas & Lückel = *Telipogon*

Darwiniera Braas & Lückel = *Telipogon*

Darwiniothamnus Harling (~ *Erigeron*). Compositae (Ast.-Con.). 2 Galápagos

dasheen *Colocasia esculenta*

Dasiphora Raf. (~ *Potentilla*). Rosaceae (Pot.). c. 12 N temp. ***D. fruticosa*** (L.) Rydb. (*P. f.*, widdy, circumpolar, relict in GB) – commonly cult. shrub, many cvs, spontaneous 'doubles' in Ireland & W N Am.

Dasispermum Necker ex Raf. (~ *Heteroptilis*). Umbelliferae (III 8). 1 S Afr.: ***D. suffruticosum*** (Berg.) B.L. Burtt. R: NRBGE 45(1988)93

Dasistoma Raf. (~ *Seymeria*). Orobanchaceae (Gerard.; Scrophulariaceae s.l.). 1 SE US: ***D. macrophylla*** (Nutt.) Raf. R: PANSP 80(1928)427

Dasoclema James Sincl. = *Uvaria*

Dasyandantha H. Robinson. Compositae (Vern.-Pip.). 1 Venez.: ***D. cuatrecasasiana*** (Aristeg.) H. Robinson

Dasyanthina H. Robinson Compositae (Vern.-Vern.). 2 E Braz.

Dasycephala (DC.) Hook.f. = *Diodia*

Dasycondylus R. King & H. Robinson (~ *Eupatorium*). Compositae (Eup.-Gyp.). 8 Braz. R: MSMMBG 22(1987)94

Dasydesmus Craib = *Oreocharis*

Dasyglossum Königer & Schildhauer = *Cyrtochilum*

Dasygrammitis Parris (~ *Ctenopteris*). Polypodiaceae (V). 6 Sri Lanka to Polynesia. R: GBS 58(2007)238

Dasylepis Oliv. Achariaceae (Flacourtiaceae). 6 trop. Afr. R: KB 59(2004)585

Dasylirion Zucc. Asparagaceae (Convallariaceae). 17 SW N Am. R: PAPS 50(1911)431. Stemless or tree-like dioec. pachycauls (Chamberlain's Model) with linr., usu. spiny-edged lvs used for thatching & baskets, the polished leaf-bases sold as curios; sap used as a drink (sotol). Some cult. orn. esp. for 'sub-trop.' bedding; ***D. wheeleri*** S. Watson ex Rothr. – buds etc. ed., scapes for building, lvs for matting, sap drunk (sotol), poss. alcohol source

Dasymalla Endl. (~ *Pityrodia*). Labiatae (IV 1). 4 N & W Aus. R: AusSB 24(2011)5

Dasymaschalon (Hook.f. & Thomson) Dalla Torre & Harms (*Pelticalyx*, ~ *Desmos*). Annonaceae (III 7). c. 27 trop. As. Heterogeneous?

Dasynotus I.M. Johnston. Boraginaceae (B.3.8.4). 1 NW US: ***D. daubenmirei*** I.M. Johnston

Dasyochloa Willd. ex Rydb. = *Munroa* (but see Sida 17(1997)663)

Dasyphyllum Kunth. Compositae (Barn.). 40 Chile. R: RMLP n.s. 9(1959)21. Shrubs to trees, ***D. diacanthoides*** (Less.) Cabrera (Chile) to 50 m tall

Dasypoa Pilg. = *Poa*

Dasypogon R. Br. Dasypogonaceae (Xanthorrhoeaceae s.l.). 3 SW Aus. R: FA 46(1986)142. Corner's Model; nectar taken by honey-eaters

Dasypogonaceae Dumort. (~ Xanthorrhoeaceae). Magnoliidae – Arecales. Incl. Calectasiaceae, 5/23 S & SW Aus. Pachycaul trees to shrubs or rhiz. herbs. Lvs in spirals, usu. V- or U-shaped in vertical sect. Fls solit. or in globular heads on bracteate peduncles. Fls bisex. P 3 + 3; A 6 usu. attached to P; G (3), each locule with 1 ovule from axile placenta or (*Calectasia*) 1-loc. with 3 basal ovules; ovules anatropus, bitegmic. Fr. indehiscent to explosively dehiscent (*Baxteria*) capsule. Seeds subspherical with pale yellow testa; endosperm copious, starch 0

Genera: *Acanthocarpus, Baxteria, Calectasia, Dasypogon, Kingia*

Dasypyrum (Cosson & Durieu) T. Durand. Gramineae (XV). 2 Medit. incl. Eur. R: NJB 11(1991)135

Dasysphaera Volkens ex Gilg. Amaranthaceae (I 2). 4 E Afr.

Dasystachys Bak. = *Chlorophytum*

Dasystachys Oersted = *Chamaedorea*

Dasystephana Adans. = *Gentiana*

Dasytropis Urb. Acanthaceae (III 2c). 1 E Cuba (serpentine): ***D. fragilis*** Urb.

date *Phoenix dactylifera*; **Chinese d.** *Ziziphus jujuba*; **desert d.** *Balanites* spp.; **dwarf d.** *Phoenix reclinata*; **Indian d.** *Tamarindus indica*; **Trebizond d.** *Elaeagnus angustifolia*

date-plum *Diospyros* spp. esp. *D. kaki*, *D. lotus*, *D. virginiana*

Datisca L. Datiscaceae. 1 W N Am., 1 S As. Nitrogen-fixing actinobacterial r.-symbionts. ***D. cannabina*** L. (As. Minor to Ind.) – dioec., cult. orn. foliage pl., source of a yellow dye form. much used for silk; ***D. glomerata*** (Presl) Baill. (W N Am.) – androdioec.

Datiscaceae Bercht. & J. Presl. Magnoliidae – Cucurbitales. 3/4 W As. to Aus., W N Am. (v. disjunct). R: Aliso 8(1973)49. Trees or perenn. herbs. Lvs in spirals, simple (trees) to pinnate; stip. 0. Fls usu. unisexual (pls. dioec. but s.t. polygamous & androdioec. in *Datisca*), ± reg., in axillary spikes to thyrses or slender term. leafy infls; male fls with K 3–10, s.t. connate basally, C 0 or 6–8 (*Octomeles*), A same no. as & opp. K or up to 25 with short (long in *Datisca*) anthers with longit. slits, ± pistillode; female (& bisex.) fls with K 3–8 on summit of G, C 0, ± functional A around G summit, G (3–8), 1-loc. with parietal placentas & distinct styles (bifid in *Datisca*), ovules 20–80, anatropous, bitegmic. Fr. capsular, dehiscing apically between styles; seeds ∞ with v. small straight cylindrical oily embryo & v. little or 0 endosperm. n = 11 (*Datisca*), 23 + (*Tetrameles*)
Genera: *Datisca, Octomeles, Tetrameles*
Octomeles & *Tetrameles* are monospecific genera of soft-wooded trees & have been seg. as Tetramelaceae; seed structure predicted affinity with Begoniaceae

dattock *Detarium senegalense*

Datura L. Solanaceae (IV 4). Excl. *Brugmansia*, 13 S N Am. but widely natur. R: SB 38(2013)825. Annuals with erect fls (cf. *Brugmansia*) & reg. candelabriform branching (Leeuwenberg's Model). Many alks. ***D. ceratocaula*** Ortega (Mex.) – semi-aquatic, revered hallucinogen of Aztecs still in use (tropane alks as in all spp.); ***D. innoxia*** Mill. ('*D. inoxia*', S N Am.) – sacred hallucinogen in SW N Am., cult. orn. (*D. meteloides*, excl. ***D. wrightii*** Regel also cult. orn., strongly scented fls); ***D. metel*** L. (S Am.) first introd. OW now not known 'in wild'; ***D. stramonium*** L. (thorn-apple, Jimson or Jamestown weed [named after intoxicated behaviour of soldiers who ate lvs during occupation of J., Virginia in 1705], N Am. but now widely natur. (invasive S Afr., Galápagos) incl. GB) – poll. by evening-flying moths (fls open c. 6 p.m., closing within 24 hrs), seeds remain dormant for many yrs, subject of intense genetic analysis (single gene mutants incl. pink-flowered forms (*D. tatula* – purple allele dominant over white) & ones with smooth capsules (*D. inermis* – smooth recessive to armed)), contains stramonium, a drug used in treatment of asthma until 1945 & Parkinson's Disease, comprises dried lvs (alks – hyoscamine & scopolamine) gathered for 300 yrs in Eur., also halts uncontrolled growth of certain brain tumours, intoxicant & hallucinogen used by Algonquin Indians (E US) as 'wysoccan' given to boys in adolescent rites over 18–20 days after which they are deemed adults, & offered (dhatura) by Hindus to Siva

Daturicarpa Stapf = *Tabernanthe*

dau *Dipterocarpus costatus*

Daubentonia DC. = *Sesbania*

Daubentoniopsis Rydb. = *Sesbania*

Daubenya Lindl. Asparagaceae (Hyacinthaceae). (Incl. *Amphisiphon, Androsiphon*) 8 S Afr. R: Bothalia 32(2002)138. Some poll. monkey beetles

Daucosma Engelm. & A. Gray (~ *Discopleura*). Umbelliferae (III 8). 1 N Am.: ***D. laciniatum*** Engelm. & A. Gray

Daucus Tourn. ex L. Umbelliferae (III 3). Incl. *Agrocharis, Margotia, Melanoselinum, Pseudorlaya, Turgenia*, c. 44 Eur., Medit., SW & C As., trop. Afr., Aus., NZ & Am. Carrots. R: T 65(2016)585. ***D. carota*** L. complex – biennial (wild pl. toxic, white-rooted 'subsp. *carota*') cult. for food (r. ed. raw or cooked, even in c. cake (WW II & recently revived), roasted a coffee subs.) & for animals (to 4.649 kg (GB record) & 2.8 m long), the eastern races with anthocyanin in r., the W with carotene, the eastern app. first domesticated in Afghanistan, the W ones being derived from yellow-rooted eastern ones ('subsp. *sativus* (Hoffm.) Arc.'), modern stocks deriving from a few C18 Dutch cvs, some crossed with subsp. ***capillifolius*** (Gilli) Apbiza (N Afr.) to breed in resistance to carrot fly; after poll. pedicels bend inwards but when frs ripen, spread out again, the burred mericarps (diuretic used in kidney complaints & dropsy, fr. oil used in flavouring liqueurs, cosmetics etc.) adhering to animals; ***D. decipiens*** (Schrad. & J.C. Wendl.) Spalik & al. (*Melanoselinum d.*, Madeira, Azores) – hapaxanthic pachycaul to 3 m

Daumailia Arènes = *Urospermum*

Daumalia Airy Shaw = *Urospermum*

daun salam *Syzygium polyanthum*

Dauphinea Hedge. Labiatae (VIII 3e). 1 Madag.: ***D. brevilabra*** Hedge. R: NRBGE 41(1983)119

Dauresia R. Nordenstam & Pelser (~ *Senecio*). Compositae (Sen.-Sen.). 2 Namibia. R: PDE 129(2011)2

Daustinia Buril & A.R. Simões (~ *Jacquemontia*). Convolvulaceae. 1 Braz.: ***D. montana*** (Moric.) Buril & A.R. Simões. R: Phytotaxa 186(2014)255

Davallia Sm. Davalliaceae. Incl. *Araiostegia, Davallodes, Humata, Parasorus* & *Scyphularia*, c. 30(–130) W Med. (1, incl. Eur.), Himal. & N Jap. to Aus. & Tahiti, Afr. & Madag. (2). R: Blumea 39(1994)151. Epiphytes oft. cult. in hanging baskets or on 'fern-balls' (in 1900s Jap. to construct life-size horses & other animals) esp. ***D. canariensis*** (L.) Sm. (hare's-foot fern, SW Med. & Canary Is.) – late Miocene relict endemic, ***D. mariesii*** Veitch (E As.), etc.

Davalliaceae M. Schomb. Polypodiidae – Polypodiales. 1/30 trop. & warm OW. R: Blumea 39(1994)154, T 55(2006)718. Mostly epiphytes with dorsiventral scaly dictyostelic dorsiventral rhiz. & (simple to) 1–4-pinnate, cleanly abscizing lvs; veins free, forking or pinnate. Sori ± round, at ends of veins, or abaxialdorsal, usu. with pouch-shaped indusium; spores ellipsoid. Gametophytes cordate, green. x = 40
Genera: *Davallia*
Gymnogrammitis now in Polypodiaceae, *Leucostegia* in Hypodematiaceae

Davalliopsis Bosch = *Cephalomanes*

Davallodes (Copel.) Copel. = *Davallia* (but see Blumea 37(1992)176)

Daveaua Willk. ex Mariz. Compositae (Anth.-?Leuc.). 1 Portugal, Morocco: ***D. anthemoides*** Mariz. R: BBMNHB 23(1993)153

Davejonesia M.A. Clem. = *Dendrobium*

Davenportia R. Johnson (~ *Merremia*). Convolvulaceae. 1 Aus.: ***D. davenportii*** (F. Muell.) R. Johnson. R: Austrobaileya 8(2010)171

Davidia Baill. Nyssaceae (Davidiaceae). 1 SW China (fossil spp. from Palaeocene of N Am.): ***D. involucrata*** Baill. (dove-tree, ghost-tree) – cult. orn. hardy tree (Massart's Model) with scented young lvs & large flakes in bark, rare in wild (though pop. of 100 000 found 2003 in Baoxing County, Sichan), native of *Quercus-Prunus-Corylus* forests; the 'flower' comprises a condensed infl. of C-less male fls surrounding 1 female reduced to an ovary with a ring of aborted A (cf. *Euphorbia*), the whole being subtended by 2(3) large bracts to 30 cm long; the solit. fr. has 6–10 seeds & after drop, the outer layers or pericarp decay & inner ones dehisce to reveal seeds which all germinate at once, 1 eventually crowding out the others, poss. a mechanism to pay off hungry browsers for even the cotyledons have axillary buds which can develop if the apex is grazed off

Davidiaceae Li = Nyssaceae

Davidsea Soderstrom & R. Ellis (~ *Schizostachyum*). Gramineae (V 5). 1 Sri Lanka: ***D. attenuata*** (Thwaites) Soderstrom & Ellis

Davidson's plum *Davidsonia jerseyana*

Davidsonia F. Muell. Cunoniaceae (II; Davidsoniaceae). 3 NE Aus. ***D. jerseyana*** (Bailey) G. Harden & J.B. Williams ('*D. pruriens*', Davidson's plum) – fr. ed. (preserves)

Davidsoniaceae Bange = Cunoniaceae

Daviesia Sm. (~ *Pultenaea*). Leguminosae (III 13). 135 Aus., esp. SW. R: Aus SB 8(1995) 1156. Bitter peas or 'bacon-&-eggs'; yellow bee-poll. fls in some, red bird-poll. in others. ***D. arborea*** W. Hill (queenwood, NE Aus.) – tree to 14 m

Davilanthus E. Schilling & Panero (~ *Simsia*). Compositae (Helia.-Helia.) 7 Mex. R: Britt. 62(2010)317, Phytoneuron 2011–25(2011)1

Davilia Mutis = *Llagunoa*

Davilla Vand. Dilleniaceae (II). c. 25 trop. Am. R: MBSM 9(1971)75. 2 inner K accrescent in fr., forming leathery casing for it. Most primitive spp. the most restricted

dawn redwood *Metasequoia glyptostroboides*

Dawsonia R. Br. Polytrichaceae (Bryopsida). 9 E Mal. to Aus., Solomons & NZ. ***D. grandis*** Schliep. & Geheeb – rope-making in NG; ***D. superba*** Grev. var. ***superba*** (NZ) – to 65 cm tall

day lily *Hemerocallis* spp.

Dayaoshania W.T. Wang = *Oreocharis*

Dayia J.M. Porter (~ *Gilia*). Polemoniaceae (III). 2 SW N Am. R: Aliso 19(2000)71

dead finish *Acacia tetragonophylla*; **d. man's finger** *Orchis mascula*; **d.nettle** *Lamium* spp.; **red d.n.** *L. purpureum*; **white d.n.** *L. album*

deadly nightshade *Atropa belladonna*

deal [from Dutch *deel*] orig. sawn boards of imp. conifer timber; **red** or **(Baltic) yellow d.** *Pinus sylvestris*; **white d.** *Picea abies*

Deamia Britton & Rose = *Strophocactus*

death angel *Justicia pectoralis*; **d. camash** or **camus** *Toxicoscordion nuttallii*

Debesia Kuntze = *Chlorophyllum*

Debia Neupane & N. Wikstr. Rubiaceae (IV 15). 4 SE As., Mal. R: T 64(2015)314

Debregeasia Gaudich. Urticaceae (III). 4 NE Afr., trop. & warm As. R: KB 43(1988)673, 44(1989)702. Monoec. & dioec. trees & shrubs with achenes encl. in fleshy K & aggregated into spherical syncarps. Useful fibres & ed. fr. e.g. ***D. longifolia*** (Burm.f.) Wedd. ('*D. edulis*', janatsi, yanagi, Ind. to Jap. & Philipp.) – cult. orn. & ***D. saeneb*** (Forssk.) Hepper & Wood (*D. salicifolia*, *D. hypoleuca*, NE Afr. to Tibet & Bhutan)

Decabelone Decne = *Tavaresia*

Decachaena (Hook.) Lindl. = *Gaylussacia*

Decachaeta DC. (~ *Eupatorium*). Compositae (Eup.). Incl. *Erythradenia*, 8 Mex. & Guatemala. R: MSBMBG 22(1987)406

Decagonocarpus Engl. Rutaceae (I 8). 2 Amazonia

Decaisnea Hook.f. & Thomson. Lardizabalaceae. 1–2 E Himal. to C China. Pachycaul treelets, not cli. like rest of family; primitive wood with scalariform perforation plates & pitting. ***D. fargesii*** Franch. (*D. insignis* (Griff.) Hook.f. & Thomson s.l., Nepal, China) – ed. blue fr. (*D. insignis* yellow) cult. orn. R: BM n.s. 29(2012)246

Decaisnina Tieghem. Loranthaceae (3). 25 Java & Philipp. to trop. Aus., Tahiti & Marquesas. R: Blumea 38(1993)70. ***D. forsteriana*** (Schultes & Schultes f.) Barlow (NG to Tahiti, Marquesas) – exceptional in L. for distrib. to oceanic is.; ***D. sumbawensis*** (Tiegh.) Barlow – myrmecophyte (S Indonesia)

Decalepidanthus Riedl (*Pseudomertensia*). Boraginaceae (B. 3.7). 7 Himal. R: Phytotaxa 226(2015)131. ***D. sericophyllus*** Riedl – for 70 yrs known only from orig. specimen (rediscovered 1971)

Decalepis Wight & Arn. Apocynaceae (III; Periplocaceae). 5 Ind., 1 ext. to China

Decalobanthus Oostr. Convolvulaceae (12). 1 Sumatra (Bangko area): ***D. sumatranus*** Oostr.

Decamerium Nutt. = *Gaylussacia*

Decanema Decne. = *Cynanchum*

Decanemopsis Costantin & Gallaud = *Cynanchum*

Decaneuropsis H. Robinson & Skvarla (~ *Vernonia*). Compositae (Vern.). c. 12 trop. As.

Decaphalangium Melchior = *Clusia*

Decaptera Turcz. = *Menonvillea*

Decarya Choux. Didiereaceae. 1 SW Madag.: ***D. madagascariensis*** Choux – Champagnat's Model, used in charcoal prod. R: FMad. 121(1963)4

Decarydendron Danguy. Monimiaceae (V 1). 4 Madag. R: AMBG 72(1985)77, Adans. 24(2002)108. G to 1000!

Decaryella A. Camus. Gramineae (Chlorid). 1 Madag.: ***D. madagascariensis*** A. Camus

Decaryia Choux = *Decarya*

Decaryochloa A. Camus. Gramineae (V 6). 1 Madag.: ***D. diadelpha*** A. Camus. R: KB 52(1997)594

Decaschistia Wight & Arn. (~ *Hibiscus*). Malvaceae (Malv.-Hib.). 18 Ind., SE As. to Aus. Some promising fibres

Decaspermum Forst. & Forst. f. Myrtaceae (II 10). 34 Indomal. to W Pac. (Aus. 2). ***D. parviflorum*** (Lam.) A.J. Scott (Mal.) – bee-poll., cryptically dioec. with male anthesis 20 mins. before female

Decastelma Schltr. = *Metastelma*

Decastylocarpus Humbert. Compositae (Vern.-Erl.). 1 Madag.: ***D. perrieri*** Humbert

Decatoca F. Muell. (~ *Trochocarpa*). Ericaceae (VII 7; Epacridaceae). 1 NG: ***D. spenceri*** F. Muell.

Decatropis Hook.f. Rutaceae (I 5). 2 Mex. & Guatemala

Decazesia F. Muell. Compositae (Gnap.-Ang.). 1 W Aus.: ***D. hecatocephala*** F. Muell. R: OB 104(1991)127

Decazyx Pittier & S.F. Blake. Rutaceae (I 5). 1 Honduras, 1 Mex. R: BSBM 43(1982)1

Deccania Tirveng. (~ *Randia*). Rubiaceae (II 1). 1 Ind.: ***D. pubescens*** (Roth) Tirveng.

Decemium Raf. = *Hydrophyllum*

Deceptor Seidenf. Orchidaceae (V 16c). 1 Vietnam: ***D. bidoupensis*** (Tixier & Guillaumin) Seidenf. R: OB 114(1992)361

Deckenia H. Wendl. ex Seem. Palmae (V 14h). 1 Seychelles: ***D. nobilis*** H. Wendl. ex Seem. – palm cabbage prized, massive leaf-litter preventing regeneration

Declieuxia Kunth. Rubiaceae (IV 3). 27 trop. Am. savannas. R: MNYBG 28, 4(1976)1

Decodon Gmelin. Lythraceae. 1 E N Am.: ***D. verticillatus*** (L.) Elliott (fossils in Euras.) – cult. orn. aquatic with *Hippuris* habit, tristylous (monomorphic at N of range where largely veget. clones), alks

Decorsea R. Viguier. Leguminosae (III 18). 6 Afr., Madag.

Decorsella A. Chev. Violaceae (III 1a). 1 trop. W Afr.: ***D. paradoxa*** A. Chev. – seeds still developing after capsule dehiscence

Decumaria L. = *Hydrangea*

Decussocarpus Laubenf. = *Retrophyllum*

Dedeckera Rev. & J. Howell (~ *Eriogonum*). Polygonaceae (Erio.-Erio.). 1 Calif.: ***D. eurekensis*** Rev. & J. Howell – fls in summer so monopolizing poll. insects as rest of desert pls 'dormant'. R: Phytol. 66(1989)238

deer grass *Muhlenbergia rigens*

deerberry *Vaccinium stamineum*

Deeringia R. Br. Amaranthaceae (I 1). 12 OW trop. (Madag. 6). R: T 54(2005)686. ***D. amaranthoides*** (Lam.) Merr. (trop. As.) – young lvs ed., juice with vinegar & onion inhaled to clear nose

Deeringothamnus Small. Annonaceae. 2 Florida. R: Britt. 12 (1960)273. Nat. hybrids with *Asimina*

degami *Calycophyllum candidissimum*

Degeneria I. Bailey & A. C. Sm. Degeneriaceae. 2 Fiji. R: JAA 69(1988)277. ***D. vitiensis*** I. Bailey & A.C. Sm. – poll. by Coleoptera

Degeneriaceae I. Bailey & A.C. Sm. Magnoliidae – Magnoliales. 1/2 Fiji. Large glabrous trees with vessel-elements. Lvs in spirals, simple, entire; stip. 0. Fls solit. on long supra-axillary pedicels, bisex., reg., hypogynous; K 3, C 12–25 in 3–5 whorls (? or spiral), larger, A ∞ oft. 20–30 in spiral, laminar, 3-nerved, the 4 microsporangia paired & embedded in abaxial surface between veins with pollen monosulcate, staminodes s.t. present, $\underline{G}$ 1, largely open at anthesis with stigmatic surface along margins, ovules c. 20–32, laminar, in 1 row nr. margin of G, anatropous, bitegmic with consp. funicular obturator. Fr. thick ± fleshy but with hard exocarp, poss. dehiscent; seeds ∞ with orange- red sarcotesta, s.t. on dangling funicles (cf. Magnoliaceae), embryo v. small, with 3(4) cotyledons in copious oily ruminate endosperm. 2n = 24

Genus: *Degeneria*

Primitive features incl. laminar A, unsealed G & 3(4) cotyledons but G 1 & vessels more advanced than Magnoliaceae & Winteraceae resp.

Degenia Hayek. Cruciferae (2). 1 Croatia: ***D. velebitica*** (Degen) Hayek

Degranvillea Determann. Orchidaceae (IV 2h). 1 Fr. Guiana: ***D. dermatoptera*** Determann. R: Gen. Orch. 3(2003)192

Deguelia Aubl. (~ *Derris*). Leguminosae (III 16). c. 17 Amaz. to Panamá. Insecticides (rotenones)

Dehaasia Blume (~ *Alseodaphne*). Lauraceae (I). 35 SE As. to NG. R: BJ 93(1973)427. Anthers 2-loc. (cf. 4-loc. in *A.*, which is otherwise diff. to sep.).

Deherainia Decne. Primulaceae (Theophrastaceae). 2 C Am. R: NJB 9(1989)20. ***D. smaragdina*** Decne with large green fls, foetid; at anthesis A lie close to stigma but later spring away, the anthers tipped with a fibrous deposit & between lobes are crystal deposits in connective, such also found in pollen: function of movements & crystals not understood but crystal-form a useful char. in separating several allied genera

Deianira Cham. & Schldl. Gentianaceae. 7 trop. Am. R: AJBRJ 21(1977)45

Deidamia Noronha ex Thouars. Passifloraceae. 5 Madag.

Deilanthe N.E. Br. (~ *Nananthus*). Aizoaceae (V). 3 S Afr. R: H. E. K. Hartmann, *Ill. Handb. Succ. Pls*, Aiz. A–E(2002)182

Deinacanthon Mez (~ *Bromelia*). Bromeliaceae (3). 1 Paraguay, NW Arg.: ***D. urbanianum*** (Mez) Mez – textiles in Gran Chaco

Deinandra Greene (~ *Hemizonia*). Compositae (Mad.-Mad.). 21 SW N Am. R: Novon 9(1999)467

Deinanthe Maxim. = *Hydrangea* (but see Garden 121(1996)364)

Deinbollia Schum. & Thonn. Sapindaceae. 38 warm Afr. & Madag. ***D. pinnata*** (Poir.) Schum. & Thonn. (W Afr.) – ed. pulp around seeds

Deinocheilos W.T. Wang = *Oreocharis* (but see Guihaia 6(1986)1)

Deinostema Yamaz. (~ *Gratiola*). Plantaginaceae (Grat.-Grat.; Scrophulariaceae s.l.). 2 E As.

Deinostigma W.T. Wang & Z.Y. Li. Gesneriaceae (III 2j). 7 S China, Vietnam. R: GBS 68(2016)155

Deiregyne Schltr. Orchidaceae (IV 2h). 18 mts of Mex., Guatemala. R: HBML 28(1982)311

Deiregynopsis Rauschert = *Aulosepalum*

Dekinia M. Martens & Galeotti = *Agastache*

Delaetia Backeb. = *Eriosyce*

Delairea Lem. (~ *Senecio*). Compositae (Sen.-Sen.). 1 S Afr.: ***D. odorata*** Lem. (*S. mikanioides*), cli. natur. in England, serious weed in Aus., W US, Hawaii, fls scented. R: FNA 20(2006)608

Delamerea S. Moore. Compositae (Inul.-Pluch.). 1 N Kenya: ***D. procumbens*** S. Moore

Delaportea Thorel ex Gagnepain = *Vachellia*

Delarbrea Vieill. Myodocarpaceae (Araliaceae s.l.). Incl. *Pseudosciadium,* 7 Mal. (1), Queensland (1), New Caled. R: Allertonia 4,3(1986)1. Cult. orn. pachycauls

Delavaya Franch. Sapindaceae. 1 SW China: ***D. toxocarpa*** Franch. – colonist on limestone used in reafforestation

Delia Dumort. = *Spergularia*

Delilia Spreng. Compositae (Helia.-Ecl.). 2 trop. Am. (1: ***D. biflora*** (L.) Kuntze, natur. W Afr., Cape Verde Is.), Galápagos (1: ***D. repens*** (Hook.f.) Kuntze). R: PSE 194(1995)113

Delissea Gaudich. Campanulaceae (III 4). 15 Hawaii, each sp. restricted to 1 is. R: SBM 73(2005)14. Woody, allied to *Brighamia*, prob. derived from *Lobelia* subg. *Tupa* in As.

Delonix Raf. Leguminosae (I 4). Incl. *Aprevalia*, 11 E Afr. (2, 1 ext. to Arabia & Ind.), Madag. (9, endemic). R: KB 50(1995)449. White-flowered spp. poll. moths, yellow-flowered spp. & ***D. regia*** (Hook.) Raf. (flamboyant, peacock flower, flame-tree, (gul) mohur, W & N Madag., where v. rare & rediscovered there only in 1932 – widely planted street-tree (Troll's Model) with scarlet fls & lvs with up to 1000 leaflets) poll. sunbirds

Delopyrum Small = *Polygonella*

Delosperma N.E. Br. Aizoaceae (V). 142 S Afr. to Arabia. R: H. E. K. Hartmann, *Ill. Handb. Succ. Pls*, Aiz. A–E(2002)184. Succ. shrubs or herbs; some cult. orn.

Delostoma D. Don. Bignoniaceae (1). 4 Andes. R: FN 25,2(1992)30

Delostylis Raf. = *Trillium*

Delphinacanthus Benoist = *Pseudodicliptera*

Delphinium Tourn. ex L. Ranunculaceae (I 4). Incl. *Consolida* (larkspur, ann., 2 upper C united, 0 lower C), excl. *Staphisagria*, c. 350 N temp. (Eur. 25, N Am. 61 (R: Phytol. 78(1985)74); R (As.): JAA 48(1967)249, 49(1968)73, 233) to trop. Afr. mts (3–4, R: JAA 48(1967)31, 476). Hybridization freq. in disturbed sites N Am. K 5, 1 spurred, C 2 or 4 smaller, upper pr with spurs entering K spur where there is nectar; bee-poll. Many alks & v. toxic, some medic. e.g. 'bya-rkang' (Tibet) for diarrhoea; many cult. orn. (R: C. Edwards (1990) *Ds.*). herbaceous perennials with complex ancestry involving ***D. elatum*** L. (Eur.) – lower C with white or yellow hairs seen in modern hybrids, ***D. grandiflorum*** L. (As.), etc. (= ***D.* x *cultorum*** Voss), some 'double' (increased K &/or A replaced by C – 'ranunculus-type', sterile with no spur), those with red or pink fls involving red-flowered N Am. spp., e.g. **'Pink Sensation'** (1934, 4x *D. elatum* hybrids crossed with app. polyploid ***D. nudicaule*** Torrey & Gray (2x)), while irradiated ***D. cardinale*** Hook. (W N Am.) yielded forms with 'giant' fls & these were then used in hybrids, since which the Dutch have used colchicine to promote polyploid lines of *D. c.* & *D. n.* to give 4x pls to cross with *D. elatum* hybrids. ***D. nuttallianum*** Pritzel ex Walp. (W N Am.) – toxic to cattle (delphinine) but sheep immune & those used to clear pasture; ***D. semibarbatum*** Bien. ex Boiss. (*D. zalil*, zalil, Iran) – yellow fls source of a dye for silk & cotton; ***D. staphisagria*** L. (stavesacre, S Eur., SW As.) – seeds form. used to control ectoparasites, rats & ants in W Eur. since 1788

Delphyodon K. Schum. = *Parsonsia*

Delpinophytum Speg. (~ *Lepidium*). Cruciferae (7). 1 Patagonia: ***D. patagonicum*** (Speg.) Speg.

Delpya Pierre ex Radlk. = *Sisyrolepis*

Delpydora Pierre. Sapotaceae (IV). 2 trop. W Afr. R: T.D. Pennington, *S.* (1991)227. Corner's Model

Deltaria Steenis. Thymelaeaceae (Thym.-Oct.). 1 New Caled.: ***D. brachyblastophora*** Steenis

Deltocheilos W.T. Wang = *Chirita*

Demavendia Pim. Umbelliferae (III 10). 1 SW & C As.: ***D. pastinacifolia*** (Boiss. & Hausskn.) Pim.

Demidium DC. = *Gnaphalium*

Demosthenesia A.C. Sm. Ericaceae (VIII 5). 11 Andes

den shan *Salvia miltiorrhiza*

Dendranthema (DC.) Des Moul. = *Chrysanthemum*

Dendriopoterium Svent. = *Poterium*

Dendroarabis (C. Meyer) German & Al-Shehbaz. Cruciferae (7). 1 C As.: ***D. fruticulosa*** (C. Meyer) German & Al-Shehbaz. R: HPB 13(2008)290

Dendrobangia Rusby. Metteniusaceae. 2 trop. S Am. R: Candollea 62(2007)92

Dendrobates M.A.Clem. & D.L. Jones = *Dendrobium*

Dendrobenthamia Hutch. = *Cornus*

Dendrobium Sw. Orchidaceae (V 11a). Incl. *Cadetia, Diplocaulobium, Epigeneium, Flickingeria*, c. 1500 trop. & warm As. (NG c. 560) to Aus. (71) & Pac. R (sect. *Oxyglossum*): NRBGE 46(1989)161. Epiphytes with extrafl. nectaries; alks. ***D. moniliforme*** (L.) Sw. (Korea, Taiwan, Jap.) – used by early C 17 Jap. royalty to scent clothes; others locally medic. or used for basketwork & bangles etc. (stems) as in ***D. pentanema*** Schltr. (*Diplocaulobium p.*, Papuasia) – yellow fibres from stems used in otherwise black armbands in New Ireland; gum of some Aus. spp. used to fix body colour of Aborigines in N Aus. ***D. sinense*** Tang & F.T. Wang (Hainan) – poll. hornets (*Vespa bicolor*) attracted by odour like distressed bees, their prey. Many cult. orn.; over 100 in Eur. alone, incl. ***D. bigibbum*** Lindl. (Cooktown orchid, N Aus.), ***D. canaliculatum*** R. Br. (antelope orchid, N Aus.), ***D. crumenatum*** Sw. (pigeon orchid, Indomal., invasive in Carib.) – fls 9 days after cold snap in wild, used in ear disease, the shiny stems & pseudobulb surfaces used in straw-plaiting in Philippines etc., ***D. nobile*** Lindl. (Himal. to China) – many cvs, ***D. speciosum*** Sm. (rock lily, r. orchid, E Aus.) & ***D. taurinum*** Lindl. (Philipp.) – in wild always on *Pterocarpus* spp.

Dendrocacalia (Nakai) Tuyama. Compositae (Sen.-Tuss.). 1 Bonin Is.: ***D. crepidifolia*** (Nakai) Tuyama – dioec. shrub

Dendrocalamopsis Q.H. Dai & X. L. Tao = *Bambusa*

Dendrocalamus Nees. Gramineae (V 4). c. 40 Ind. & Sri Lanka to China & Philipp. R: KBAS 13(1986)54. Huge clump-forming woody bamboos (some with extrafl. necataries on auricles), ***D. giganteus*** Wall. ex Munro (giant bamboo, Myanmar, Yunnan) to 35 m the largest grass known, growing at c. 46 cm a day at best, stems used for buckets & rafts, when split for chopsticks; ***D. asper*** (Schultes f.) Heyne (SE As., Malaya) – widely cult., as are other spp., for ed. shoots, most imp. thus in Thailand (phai-tong – 60 000 ha (40 000 ha in 1995 just 1 Chinese clone), prod. 18 t. per ha); ***D. strictus*** (Roxb.) Wight & Arn. (male bamboo, Calcutta b., Ind.) – solid stems used for constr., paper pulp, charcoal etc., cult. Hawaii, S N Am., WI etc.

Dendrocereus Britton & Rose (~ *Acanthocereus*). Cactaceae (II 1). 2 Cuba, Santo Domingo

Dendrochilum Blume. Orchidaceae (V 10b). c. 275 SE As. & Mal. R: OB 130 (1997)15. Cult. orn. epiphytes

Dendrochloa C.E. Parkinson = *Schizostachyum*

Dendrocnide Miq. Urticaceae (I). 37 Indomal., Pac. R: GBS 25(1969)1. Stinging trees (Aubréville's Model; giant nettle, gympie, 'stingers', bad weeds of regrowth in N Aus.) with painful effects (pollen also irritant; *Alocasia macrorrhizos* alleged antidote), ***D. excelsa*** (Wedd.) Chew (trop. Aus.) the tallest to 40 m with 2 m buttresses (cf. *Obetia*), the sting of ***D. moroides*** (Wedd.) Chew (trop. Aus.) causing recurrent discomfort (due to a bicyclic octapeptide) for up to 6 months or longer, esp. if affected area dampened with cold water, though fr. ed. if hairs removed. Some fibres incl. ***D. sinuata*** (Blume) Chew (Himal. to Java) for ropes

Dendroconche Copel. = *Microsorum*

Dendrocoryne (Lindl.) Brieger. See *Dendrobium*

Dendrocousinsia Millsp. (~ *Sebastiania*). Euphorbiaceae (Hipp.). 6 Jamaica. R: Britt. 67(2015)93

Dendroglossa C. Presl = *Leptochilus*

Dendrokingstonia Rauschert (*Kingstonia*). Annonaceae (IV 5). 3 W Mal. R: BJLS 168(2012)78

Dendroleandria Arènes = *Helmiopsiella*

Dendrolobium (Wight & Arn.) Benth. Leguminosae (III 19). 18 trop. As., Ind. Ocean, Aus. Local medic.

Dendromecon Benth. Papaveraceae (II). 1–2 SW N Am. Everg. shrubs with arils, cult. orn.

Dendromyza Danser = *Dendrotrophe*

Dendropanax Decne & Planch. Araliaceae. c. 90 trop. Am. (50), E As., Mal. Some cult. orn. & timber esp. ***D. arboreus*** (L.) Decne & Planch. (trop. Am.)

Dendropemon (Blume) Reichb. (~ *Phthirusa*). Loranthaceae (4 4). 32 WI. R: Fontqueria 34(1992)1, SBM 92(2011)16

Dendrophorbium (Cuatrec.) C. Jeffrey. Compositae (Sen.-Sen.). c. 75 S Am. Woody

Dendrophthoe Mart. Loranthaceae (5 5). 30 OW trop. (Aus. 6). Many aggressive parasites of cult. trees e.g. ***D. falcata*** (L.f.) Ettingsh. (SE As. to Aus.) on mango & citrus, some local medic. incl. cancers; ***D. pentandra*** (L.) Miq. (Indomal.) – parasitizes *Dendrobium crumenatum* in Singapore (only orchid host known)

Dendrophthora Eichler (~ *Phoradendron*). Santalaceae (Viscaceae). 120+ trop. Am. R: Wentia 6(1961)1, BJ 122(2000)169,448

Dendrophylax Reichb.f. Orchidaceae (V 16d). Incl. *Polyradicion*, 14 Florida, WI. R: Orchids 75(2006)742. Leafless epiphytes with chlorophyllous r. Some cult. orn. incl. ***D. lindenii*** (Lindl.) Rolfe (*P. l.*, Florida, WI) – fragrant white fls

Dendroportulaca Eggli = *Deeringia*

Dendrosenecio (Hedb.) R. Nordenstam. Compositae (Sen.-Sen.). 4(–11) C & E Afr. R: KB 34(1973)53. Pachycaul 'giant' or 'tree' groundsels (Leeuwenberg's Model) in montane forest & forming groves above treeline (some to c. 250 yrs old), dwarf & creeping spp. e.g. ***D. keniensis*** (Bak.f.) Mabb. (Mt Kenya) in swamps (cf. *Espeletia, Lobelia*)

Dendroseris D. Don = *Sonchus*

Dendrosicus Raf. = *Amphitecna*

Dendrosicyos Balf.f. Cucurbitaceae (XIII). 1 S Arabia, Socotra: ***D. socotrana*** Balf.f. – tendrils 0, only arborescent (trunk to 1 m diam.) cucurbit, now threatened by goats (& in drought felled & pulped for livestock) though prot. by the ± goat-proof *Cissus subaphylla* (Balf.f.) Planch.

Dendrosida Fryxell. Malvaceae (Malv.-Malv.). 7 Mex. (4), Colombia, Venez. R: Britt. 23(1971)237

Dendrosipanea Ducke. Rubiaceae (I 6). 3 N S Am.

Dendrostellera (C. Meyer) Tieghem = *Diarthron*

Dendrostigma Gleason = *Mayna*

Dendrothrix Esser = *Senefelderopsis* (but see Novon 3(1993)245)

Dendrotrophe Miq. Santalaceae (Amphorogynaceae). Incl. *Cladomyza, Dendromyza*, 21+ Indomal., S China, SE As., trop. Aus. Some shoot dimorphism

Denea Cook = *Howea*

Denekia Thunb. (~ *Amphidoxa*). Compositae (Ast.). 1 C & S Afr.: ***D. capensis*** Thunb. R: OB 104(1991)55

Denhamia Meissner. Celastraceae (I). 15 trop. Aus. R: FA 22(1984)153

Denisonia F. Muell. = *Pityrodia*

Denisophytum R. Viguier (~ *Caesalpinia*). Leguminosae (I 4). 8 trop. Afr., Am. R: PhytoKeys 71(2016)43

Denmoza Britton & Rose. Cactaceae (III 4). 1 W & NW Argentina: ***D. rhodacanthus*** (Salm Dyck) Britton & Rose

Dennettia Bak. f. = *Uvariopsis*

Dennstaedtia Bernh. Dennstaedtiaceae. c. 60 trop. to warm temp. Some forming thickets on poor soils, ***D. glauca*** (Cav.) Looser green manure in Peru; some cult. orn.

Dennstaedtiaceae Lotsy. Polypodiidae – Polypodiales. Incl. Monachosoraceae, 10/230 cosmop. Terrestrial, usu. with creeping rhizomes oft. with polystele or solenostele, usu. with jointed hairs; scaleless lvs small to large & much-divided; petioles oft. with epipetiolar buds; veins free, forked or pinnate. Sori marginal to submarginal, indusium linr., cuplike or reflexed over sorus; spores tetrahedral & trilete or reniform & monolete. Gametophyte cordate, green. x = 26+

Genera: *Blotiella, Dennstaedtia, Histiopteris, Hypolepis, Lepolepia, Microlepia, Monachosorum, Oenotrichia, Paesia, Pteridium*

[N.B. Saccolomatoideae (dictyostele; sori term. single veins): *Saccoloma* = Saccolomataceae, Lindsaeoideae (usu. protostele; sori usu. on commissures): *Lindsaea, Odontosoria* etc. = Lindsaeaceae]

Some cult. orn. (*Blotiella, Dennstaedtia, Hypolepis, Microlepia*), some weedy (*Hypolepis, Paesia* but esp. *Pteridium* (bracken))

Denscantia Cabral & Bacig. (*Scandentia*). Rubiaceae (IV 15). 5 Braz. R: Darw. 39(2001)30, 353

Dentaria Tourn. ex L. = *Cardamine*

Dentella Forst. & Forst. f. Rubiaceae (IV 1). 1–8 Indomal. to Pacific
Dentoceras Small = *Polygonella*
denya *Cylicodiscus gabunensis*
deodar *Cedrus deodara*
Depacarpus N.E. Br. = *Meyerophytum*
Depanthus S. Moore (~ *Coronanthera*). Gesneriaceae (II 4c). 2 New Caled. Trees to 10 m
Deparia Hook. & Grev. (~ *Athyrium*). Athyriaceae (Woodsiaceae, Dryopteridaceae s.l.). c. 70 trop. OW (Afr. 1) to Hawaii, N Am. (1). R: JFSUTB III, 13(1984)375
depgul *Lancea tibetica*
Deplanchea Vieill. Bignoniaceae (1). 5 Mal., Aus., New Caled. Trees. ***D. tetraphylla*** (R. Br.) F. Muell. (NG, NE Aus.) – bird-poll.
Deppea Cham. & Schldl. Rubiaceae (IV 5). 25 C & S Mex. (most) to C Am., SE Braz. (1). R: Allertonia 4(1988)389
Deprea Raf. (~ *Physalis*). Solanaceae (IV 5d). Incl. *Larnax* 45 trop. Am. R: Kurtziana 23(1994)108. Cryptic dioecy
Derenbergia Schwantes = *Conophytum*
Derenbergiella Schwantes = *Mesembryanthemum*
Dermatobotrys Bolus. Scrophulariaceae (Teed.). 1 S Afr.: ***D. saundersii*** Bolus – epiphytic shrub on palms & *Calodendrum capense*. R: Strel. 10(2000)517
Dermatocalyx Oersted = *Schlegelia*
Dermatophyllum Scheele (*Calia*, ~ *Sophora*). Leguminosae (III 2) 6 SW N Am. R: Phytoneuron 2011–57(2011), 2012–3(2012). ***D. secundiflorum*** (Ortega) Gandhi & Reveal (*C. s.*, mescal bean) – cult. orn. everg. tree, before peyote use principal hallucinogen in SW N Am. but fr. v. toxic (cytisine) & fatal in excess
Derosiphia Raf. Older name for *Podocaelia*
derris (root) *Derris* spp.
Derris Lour. Leguminosae (III 16). Incl. *Paraderris*, excl. *Solori*, 12 OW trop. esp. SE As., 1 (***D. trifoliata*** Lour. (*D. uliginosa*)) ext. to E Afr. & W Pac. in mangrove. Lianes; sources of rotenones, efficacious as insecticides & fish-poisons, esp *D. t.*; some cult. orn. ***D. elliptica*** (Wall.) Benth. (*P. e.*, derris root, tuba-root) – chewed in suicide in NG, derris powder; ***D. montana*** Benth. (*P. m.*, *D. malaccensis*, trop. As.) – cult. for derris powder
Derwentia Raf. = *Veronica*
Desbordesia Pierre ex Tieghem = *Irvingia* (but see BJBB 65(1996)147)
Deschampsia P. Beauv. Gramineae (XVI 7). Excl. *Avenella*, *Vahlodea*, c. 30 temp. & cool, esp. N (Eur. 6, N Am. 7). Hair-grasses. Tufted growth with rough lvs esp. ***D. cespitosa*** (L.) P. Beauv. (tussock grass, N hemisph.) – coarse fodder; ***D. antarctica*** Desv. (S S Am.) – most southerly fl. pl. known (68° 21′ S, Refuge Is)
Descurainia Webb & Berth. Cruciferae (23). 35 temp. (esp. N (17) & S Am.) & cool N (Eur. 1), S Afr. R (N Am.): AMN 22(1939)481. Some Macar. spp. woody. Minor oilseeds esp. ***D. pinnata*** (Walter) Britton (N Am.); ***D. sophia*** (L.) Prantl (flixweed, Euras., natur. N Am.) – seeds used like mustard, attract & trap mosquito larvae
Desdemona S. Moore = *Basistemon*
desert date *Balanites* spp. esp. *B. aegyptiaca*; **d. kumquat** *Citrus glauca*; **d. pea (Sturt('s))** *Swainsona formosa*; **d. rose** *Adenium obesum*; **Sturt's d. r.** *Gossypium sturtianum*
Desertia Mart.-Azorín. Asparagaceae. 2 SW Afr. R: Phytotaxa 221(2015)206
Desfontainia Ruíz & Pavón. Columelliaceae (Desfontainiaceae). 1 (variable) Andes (Costa Rica to Cape Horn): ***D. spinosa*** Ruíz & Pavón – lvs holly-like but opp., fls scarlet & yellow, poll. green-backed firecrown hummingbird; source of yellow dye for cloth; tea from lvs medic. & allegedly hallucinogenic; cult. orn. R: BJ 123(2001)291
Desfontainiaceae Endl. = Columelliaceae
Desideria Pampan. = *Solms-laubachia*
Desmanthodium Benth. Compositae (Mill.-Desm.). 8 C & S (1) Am. R: Phytol. 80(1996)259
Desmanthus Willd. Leguminosae (II 1). 24 warm Am. R: SBM 38(1993). Mostly herbs. Cover crop & fodder though some woody. ***D. pernambucanus*** (L.) Thell. – pantrop. weed
Desmaria Tieghem (~ *Loranthus*). Loranthaceae (4 2). 1 S Chile: ***D. mutabilis*** (Poepp. & Endl.) B.D. Jackson. R: PSE 151(1985)121. Shoot dimorphism, decid.
Desmazeria Dumort. Gramineae (XVI 12). Incl. *Catapodium* (lemmas glabrous), 6–7 Med. (incl. Eur.) to Iran. R: BJ 94(1974)556. ***D. rigida*** (L.) Tutin (fern grass) – natur. Aus., N Am.
Desmidorchis Ehrenb. = *Caralluma* (but see Adansonia 19(1980)322)
Desmocladus Nees = *Lepidobolus*

Desmodiastrum (Prain) Pramanik & Thoth. (~ *Alysicarpus*). Leguminosae (III 19). 4 Indomal.

Desmodium Desv. Leguminosae (III 19). Excl. *Codariocalyx, Hylodesmum, Ototropis, Ougeinia*, c. 275 warm esp. E As., Braz., Mex. R (As.): Ginkgoana 1(1973)1. Legumes fall into 1-seeded units spread on animals, trousers etc. (beggarweed). Alks; some medic. locally; some fodders & green manure. ***D. gangeticum*** (L.) DC. (OW trop.) – local medic. (salpani), fibres allegedly valuable for paper-making; ***D. triflorum*** (L.) DC. (trop.) – form. used for lawns by whites in Bismarcks

Desmogymnosiphon Guinea = *Gymnosiphon*

Desmoncus Mart. Palmae (V 8b). 24 trop. Am. R: Phytotaxa 35(2011)16. Lianes superficially like *Calamus*. Some ed. fr. & cult. orn. incl. ***D. orthacanthos*** Mart. (*D. major*, picmoc, Trinidad to E S Am.) – stems used for basketry & piano-stool seats

Desmophlebiaceae Mynssen & al. (~ Athyriaceae). Polypodiidae-Polypodiales. 1/2 trop. Am. R: T 65(2016)26. Rhiz. without clathrate scales. Lvs imparipinnate with pinnae alt. at least distally, vein-tips connected by submarginal vein
Genus: *Desmophlebium*

Desmophlebium Mynssen & al. (~ *Diplazium*). Desmophlebiaceae. 2 trop. Am. R: T 65(2016)27

Desmopsis Safford. Annonaceae (IV 7). 17 Mex. to Cuba

Desmos Lour. Annonaceae (III 7). 25–30 Indomal. to W Pac. Infl. – axes act as grapples in cli.; fr. a head of berries. Some ed. & local medic., cut stems source of drinking water. ***D. chinensis*** Lour. (Assam & S China to Philipp.) – Troll's Model, grown as living fence in Java, cordage, medic., lvs used to brew alc. drink (Hainan), fls source of fragrant ess. oil

Desmoscelis Naudin. Melastomataceae. 2 trop. S Am.

Desmoschoenus Hook.f. (~ *Scirpus*). Cyperaceae (II 5). 1 NZ: ***D. spiralis*** (A. Rich.) Hook.f. R: NZJB 34(1996)131

Desmostachya (Stapf) Stapf (*Stapfiola*). Gramineae (XXIX 6). 1 N Afr. & Middle E to Ind. & SE As.: ***D. bipinnata*** (L.) Stapf – used for cheap ropes, baskets & whips used as bird-scarers (hieroglyphic sign of anc. Egyptians), cerem. armbands in Hindu funerals. R: FOC 22(2006)480

Desmostachys Miers. Icacinaceae. 7 trop. Afr. & Madag. Mangenot's Model

Desmotes Kallunki (~ *Erythrochiton*). Rutaceae (I 8). 1 Coiba Is., Panamá: ***D. incomparabilis*** (Riley) Kallunki

Desmothamnus Small = *Lyonia*

Desplatsia Bocquillon. Malvaceae (Grew.-Grew.; Tiliaceae). c. 5 trop. W Afr. to Uganda. R: NBGB 9(1926)818. Massart's Model; seed elephant-disp. (fr. 25 x 20 cm, sought also by pygmies)

Detarium Juss. Leguminosae (I 2). 3 W Afr. forests & Sudanian savanna. ***D. senegalense*** J. Gmelin (dattock, tallow-tree, W Afr.) – grey timber mahogany subs.; fr. (round indehiscent & drupe-like) ed.

determa *Sextonia rubra*

Dethawia Endl. Umbelliferae (III 8). 1 Pyrenees: ***D. splendens*** (Lapeyr.) Kerguélen (*D. tenuifolia*)

Detzneria Schltr. ex Diels = *Veronica*

Deuterocohnia Mez. Bromeliaceae (1). Incl. *Abromeitella*, 14 S Am. R: FN 14(1974)231, Bradea 6(1992)143. Terrestrial shrubs with extrafl. nectaries

Deuteromallotus Pax & K. Hoffm. = *Hancea*

Deutzia Thunb. Hydrangeaceae (II 1). c. 60 temp. As. to Philipp., Mex. Mts (c. 4). R: T.I. Zaikonnikova (1966) *Deitsii-Dekorativnye Kustarniki* (see Baileya 19 (1975)133). Seeds small, winged. Local medic. China. Many (50 in US) cult. orn. shrubs esp. hybrids between v. similar spp. & cvs

Deutzianthus Gagnepain. Euphorbiaceae (Al.-Gron.). Incl. *Loerzingia*, 2 N Vietnam, Sumatra. R: A. Radcliffe-Smith, *Gen. Euphorb.* (2001)291

Devendraea Pusalkar (~ *Lonicera*). Caprifoliaceae. 5 Himal. R: Taiwania 56(2011)212

Deverra DC. = *Pituranthos*

Devia Goldbl. & Manning. Iridaceae (VI 4). 1 W Karoo: ***D. xeromorpha*** Goldbl. & Manning. R: Strel. 10(2000)628

devil, blue *Eryngium ovinum, E. pinnatifidum*; **Mexican d.** *Ageratina adenophora*; **mountain d.** *Lambertia formosa*

Devillea Tul. & Wedd. = *Podostemum*

devil's bit *Succisa pratensis*; **d.'s club** *Oplopanax horridus*; **d.'s cotton** *Abroma augustum*; **d.'s guts** *Cuscuta* spp., *Equisetum* spp.; **d.'s paintbrush** *Pilosella aurantiaca*

Devogelia Schuit. Orchidaceae (V). 1 Moluccas: ***D. intonsa*** Schuit. R: Blumea 49(2004)362

dewberry *Rubus* spp. esp. (GB) *R. caesius*, but applied in US to trailing blackberries & *R. vitifolius*

dewdrop, golden *Duranta erecta*

Dewevrea M. Micheli. Leguminosae (III 16). 1–2 W Afr. Veg., fish-poison

Dewevrella De Wild. Apocynaceae (IIc). 1 trop. Afr.: ***D. cochliostema*** De Wild. R: AUWP 85–2(1985)67

Dewildemania O. Hoffm. Compositae (Vern.-Centr.). 7 trop. Afr.

Dewindtia De Wild. = *Cryptosepalum*

Dewintera Jaarsveld & Wyk (~ *Rogeria*). Pedaliaceae. 1 Namibia: ***D. petrophila*** (De Winter) Jaarsveld & Wyk. R: Bothalia 37(2007)198

Dewinterella D. Müll.-Doblies & U. Müll.-Doblies = *Hessea*

Deyeuxia Clarion ex P. Beauv. (~ *Calamagrostis*). Gramineae (XVI 5). Incl. *Dichelachne*, 207 temp. All polyploids. Some pasture grasses; ***D. crinita*** (L.f.) Zotov (Aus. to Pac.) – stems for paper-making. Some cult. orn. incl. ***D. × acutiflora*** (Schrad.) P. Beauv. '**Karl Foerster**' (*D. pyramidalis* (Host) Veldk. × *D. epigejos* (L.) Mabb., both Euras.) in N Am.

dhaincha *Sesbania aculeata*

dhak *Butea monosperma*

dhal see dahl

dhalebana *Geonoma baculifera*

dhani *Coriandrum sativum*

dhatura *Datura stramonium*

dhawa *Anogeissus latifolia*

Dhofaria A. Mill. Capparaceae. 1 Oman: ***D. macleishii*** A. Mill. – allied to Aus. *Apophyllum*! R: NRBGE 45(1988)55

dhop or **dhub** *Cynodon dactylon*

dhup *Canarium euphyllum*; **red d.** *Parishia insignis*

dhupa fat *Vateria indica*

Diabelia Landrein (~ *Abelia*). Caprifoliaceae. 3 China, Jap. R: Phytotaxa 3(2010)35

Diacalpe Blume = *Dryopteris*

Diacarpa Sim = *Atalaya*

Diachyrium Griseb. = *Sporobolus*

Diacidia Griseb. Malpighiaceae. 12 trop. Am. R: MNYBG 32(1981)61

Diacranthera R. King & H. Robinson. Compositae (Eup.-Gyp.). 2 Braz. R: MSBMBG 22(1987)94

Diacrodon Sprague. Rubiaceae (IV 15). 1 Braz.: ***D. compressus*** Sprague

Diadeniopsis Szlach. = *Systelloglossum* (but see PJB 51(2006)39)

Diadenium Poeppig & Endl. = *Comparettia*

Dialiopsis Radlk. = *Zanha*

Dialium L. Leguminosae (I 3). Excl. *Uittienia* 27 trop. (Am. 1) exc. Aus. C 0–2, A 2(3), fr. ± globose indehiscent 1–2-seeded; extrafl. nectaries. ***D. guineense*** Willd. (trop. W Afr.) & ***D. ovoideum*** Thw. (Sri Lanka) – velvet tamarinds, orange-red ed. fr. pulp; other spp. with ed. fr. incl. ***D. indum*** L. (Mal.) – good timber (keranji)

Dialyanthera Warb. = *Otoba*

Dialyceras Capuron. Sphaerosepalaceae. 3 NE Madag. R: Adans. 21(1999)109

Dialypetalanthaceae Rizz. & Occh. = Rubiaceae

Dialypetalanthus Kuhlm. Rubiaceae (Cond.; Dialypetalanthaceae). 1 E Braz.: ***D. fuscescens*** Kuhlm.

Dialypetalum Benth. Campanulaceae (III 1). 5 Madag. R: FMad. 186(1953)30. Pithy shrubs

Dialytheca Exell & Mendonça. Menispermaceae (III). 1 Angola: ***D. gossweileri*** Exell & Mendonça

Diamantina Novelo & al. Podostemaceae (III). 1 Braz.: ***D. lombardii*** Novelo & al. Pollen in tetrads (unique in P.). R: SB 29(2004)109

Diamena Rav. (~ *Anthericum*). Asparagaceae (Anthericaceae). 1 Peru (nr Trujillo): ***D. stenantha*** (Rav.) Rav. – extinct in wild?

Diamorpha Nutt. = *Sedum* (but poss. to be re-segregated)

Diandranthus Liou = *Miscanthus* (but see APG 54(2003)114)

Diandriella Engl. = *Homalomena*

Diandrochloa De Winter = *Eragrostis*

Diandrolyra Stapf. Gramineae (VI 3). 3 SE Braz. R: CUSNH 39(2000)55, Novon 19(2009)211. Bamboos

Diandrostachya (C. Hubb.) Jacq.-Fél. = *Loudetiopsis*

Dianella Lam. ex Juss. Asphodelaceae (Hemerocallidaceae, Phormiaceae). 20+ trop. E Afr., Indomal., Madag., W Pac. R: MBMUZ 163(1940)5. Some cult. orn. incl. epiphytes; berries blue, a dye-source in Hawaii, ***D. nigra*** Colenso the blueberry of NZ; ***D. caerulea*** Sims (NSW) – 1 of first 6 Aus. pls introd. UK (1771), fr. high in vit. C; ***D. ensifolia*** (L.) DC. (range of genus) – germ. after 3 months (embryo immature until then), in Mal. dried rhiz. chewed as vermifuge, fresh rhiz. a rat-poison, also used in scent; ***D. revoluta*** R. Br. (E Aus.) – buzz-poll.

Dianellaceae Salisb. = Asphodelaceae

Dianthoseris Schultz-Bip. ex A. Rich. = *Crepis*

Dianthoveus Hammel & Wilder. Cyclanthaceae. 1 SW Colombia to N Ecuador: ***D. cremnophilus*** Hammel & Wilder

Dianthus L. Caryophyllaceae (III 1). c. 300 Euras. (Eur. 115, diversified faster than any other pls or vertebrates with 2.2–7.6 spp. per M yrs) to Afr. mts (few), N Am. (1 native: ***D. repens*** Willd. (NE As., NW N Am.), 5 natur.). Pinks. Strongly-scented ed. fls visited by butterflies (cf. *Daphne*). Many cult. orn. esp. rock-pls & many hybrids (c. 27 000 cvs registered), some form. used in mulling ale. ***D. arboreus*** L. (SE Eur.) – shrub to 1 m, poss. the pink in the murals at Knossos; ***D. armeria*** L. (Deptford pink, Eur. to Iran) – natur. Hawaii, N & S Am.; ***D. barbatus*** L. (Sweet William [from Fr. *oillet* then Willy], Eur. mts, natur. China, Java, N & S Am.) – many cvs incl. short, narrow-lvd Sweet Johns of colonial N Am., '**Temarisou**' with leafy (bracts) shoots in place of fls bred Jap., vogue cut-fl., with *D. caryophyllus* parent of first documented (Cambridge, 1717) garden plant hybrid ('Fairchild's Mule'); ***D. caryophyllus*** L. (carnation, clove pink, ? Medit.) – source of oil for soap & scent, cult. by Moors in Valencia 1460, now comm. cut-fls (trad. in Liguria), though some carnations are hybrids (= ***D. × allwoodii*** Hort. Allwood) with *D. plumarius*, with variously coloured fls (e.g. picotee, orig. speckled, now edged with diff. colour), subject of florists' societies C17 & 18 (pinks in C18) now genetically modified 'blue' cvs (e.g. **'Moondust'**) with gene for delphinidin introd. from a pansy; ***D. chinensis*** L. (Chinese p., C & E China) – esp. **'Heddewigii'** (Jap. p.) polyploid flowering in first yr. from seed, but forms with long fimbriate C dev. by Kishu clan in Edo period Jap. (1615–1868); ***D. deltoides*** L. (maiden p., meadow p., Eur.) – natur. N Am.; ***D. gratianopolitanus*** Villars (*D. caesius*, Cheddar p., Eur. incl. GB where prot.) – poll. butterflies, diurnal hawkmoths, diurnal & nocturnal noctuids (nectar mostly sucrose but high in amino-acids); ***D. plumarius*** L. (clove p., C Eur., natur. N Am.) – orig. of most 'pinks', involved in carnation breeding

Diapensia L. Diapensiaceae. 4 Himal. & W China, circumboreal (1: ***D. lapponica*** L., arctic-alpine not discovered in Scotland before 1951, now prot.). R: T 32(1983)419

Diapensiaceae Lindl. Magnoliidae – Ericales. 5/13 arctic & N temp. to Himal. R: T 32(1983)417. Subshrubs or perenn. herbs, mycorrhizal, accum. aluminium. Lvs in spirals, stip. 0. Fls solit. or in racemes, reg., bisex., K 5 or (5), imbr., C (5), almost distinct in *Galax*, imbr. or convolute, A usu. 5 alt. with C & attached to tube with 5 staminodes (0 in *Diapensia* & *Pyxidanthera*) opp. C, all s.t. connate, this tube adnate to C-tube & falling with it in *Galax*, anthers with longit. (transverse in *Pyxidanthera*) slits, nectary a weakly dev. ring at base of ovary or 0, $\underline{G}(3)$, 3-loc., with 3-lobed stigma, ovules several – ∞, anatr. to hemitropous or campylotropous, unitegmic. Fr. a 3-valved loculicidal or denticidal capsule; seeds small with copious fleshy endosperm around ± straight embryo. x = 6

Genera: *Berneuxia, Diapensia, Galax, Pyxidanthera, Shortia*

An app. relict group (s.t. placed in own order, Diapensiales), with low chromosome numbers & a marked inability for seedling regeneration; *Diplarche* (= *Rhododendron*) removed to Ericaceae. Some cult. orn. esp. ground cover (*Galax*)

Diaperia Nutt. (~ *Filago*). Compositae (Gnap.). 3 C US, N Mex. R: FNA 19(2006)460

Diaphananthe Schltr. Orchidaceae (V 16d). Incl. *Chamaeangis*, excl. *Rhipidoglossum*, c. 33 trop. Afr. R: KB 14(1960)140

Diaphanoptera Rech.f. (~ *Acanthophyllum*). Caryophyllaceae (III 1). 6 NE Iran to Afghanistan. Heterogeneous

Diaphractanthus Humbert. Compositae (Vern.-Erl.). 1 Madag.: ***D. homolepis*** Humbert

Diarrhena P. Beauv. Gramineae (XII). 5 E As., N Am. (2, R: BTBC 118(1991)128). ? Relics of widespread Tertiary woodland

Diarthron Turcz. Thymelaeaceae (Thym.-Daph.). 19 Eur. (1) to W As. R: NRBGE 40(1982)216

Diascia Link & Otto. Scrophulariaceae (Hemim.). c. 50 S Afr. (Cape 34 – R: Strel. 9(2000)645). ***D. barberae*** Hook.f. – cult. ann. with pink fls, melittid bees (*Rediviva* spp.) with long legs orientated by translucent 'windows' at base of upper C lip (ultraviolet-absorbing cf. surroundings) harvest oils from trichomal elaiophores at apices of long spurs of fls by rubbing forelegs against trichomes & poll. (co-evolution)

Diaspananthus Miq. = *Ainsliaea*

Diaspasis R. Br. = *Scaevola*

Diastatea Scheidw. (~ *Lobelia*). Campanulaceae (III 1). 5 Mex. (most) to Peru. R: BTBC 67(1940)784

Diastella Salisb. = *Mimetes*

Diastema Benth. Gesneriaceae (III 5b). 20 trop. Am. Heterogeneous. Herbs with scaly underground rhizomes (like *Achimenes*); some cult. orn.

Diateinacanthus Lindau = *Odontonema*

Diatenopteryx Radlk. Sapindaceae (IV 1). 2 S Am.

Diblemma J. Sm. = *Microsorum*

Dibrachionostylus Bremek. Rubiaceae (IV 1). 1 trop. E Afr.: ***D. kaessneri*** (S. Moore) Bremek.

dibs *Vitis vinifera*

Dicarpellum (Loesener) A.C. Sm. (~ *Salacia*). Celastraceae (I). 4 New Caled. R: FNC 25(2004)5. Dioec.

Dicarpidium F. Muell. Malvaceae (Bytt.-Herm.; Sterculiaceae). 4 NW Aus.

Dicarpophora Speg. = *Jobinia*

Dicella Griseb. Malpighiaceae. 6 trop. S Am.

Dicellandra Hook.f. Melastomataceae. 3 trop. W Afr. R: Adansonia 14(1974)77

Dicellostyles Benth. = *Hibiscus*

Dicentra Benth. Papaveraceae (Fumariaceae I 1). Excl. *Dactylicapnos* (tendrils), *Ehrendorfia*, 10 As. & N Am. R: Britt. 13(1961)1, OB 88(1986)8. Alks. Fls pend., each outer C with large basal pouch, inner C spoon-shaped cohering at tips so as to form a hood covering A & G; bees hang on pend. fl. & probe for nectar in the pouches & in so doing push aside the hood & touch stigma on which there is usu. pollen from same fl. Cult. orn. & local medic. ***D. canadensis*** (Goldie) Walp. (squirrel-corn, E N Am.) – many small yellow tubers, fls whitish; ***D. formosa*** (Haw.) Walp. – common in gdns, natur. GB; ***D. spectabilis*** (L.) Lem. (bleeding hearts, Dutchman's breeches, Korea, China, long cult. Jap.) – most commonly cult. 'old-fashioned' (not introd. until 1846) – red fls

Dicerandra Benth. Labiatae (VII 2b). 8–9 SE US. R: SB 14(1989)197

Diceratella Boiss. Cruciferae (4). 11 NE trop. Afr., Iran

Diceratostele Summerh. Orchidaceae (V 4a). 1 trop. Afr.: ***D. gabonensis*** Summerh.

Diceratotheca J.R.I. Wood & Scotland. Acanthaceae (III 2). 1 NW Thailand: ***D. bracteolata*** J.R.I. Wood & Scotland. R: KB 67(2012)692

Dicercoclados C. Jeffrey & Y.L. Chen. Compositae (Sen.-Tuss.). 1 China: ***D. triplinervis*** C. Jeffey & Y.L. Chen. R: KB 39(1984)213

Dicerma DC. = *Aphyllodium*

Dicerocaryum Bojer. Pedaliaceae (3). 3 E & S Afr., Madag. R: MSB 25(1975)1

Dicerospermum Bakh.f. = *Poikilogyne*

Dicerostylis Blume = *Hylophila*

Dichaea Lindl. Orchidaceae (V 12j). c. 110 trop. Am. Cult. orn.

Dichaelia Harv. = *Brachystelma*

Dichaetanthera Endl. Melastomataceae. 34 trop. Afr. (7), Madag. (27)

Dichaetaria Nees ex Steud. Gramineae (Arund.). 1 Ind., Sri Lanka: ***D. wightii*** Nees ex Steud.

Dichaetophora A. Gray. Compositae (Ast.-Astr.). 1 S US, N Mex.: ***D. campestris*** A. Gray. R: Wrightia 1(1946)90.

Dichanthelium (Hitchc. & Chase) Gould (~ *Panicum*). Gramineae (XXIV). 120 Afr., Am. (N Am. 34; R: AJB 90(2003)816). Cleistogamy common

Dichanthium Willemet. Gramineae (XXII 7). 22 OW trop. (Eur. 2 natur.). R (sect. D.): BSAB 12(1968)206. Valuable pasture grasses in Aus. ***D. aristatum*** (Poir.) C. Hubb. (S As.) – lawn-grass in S US

Dichapetalaceae Baill. (~ Chrysobalanaceae). Magnoliidae – Malpighiales. 3/170 trop. (& S Afr. – 1). Trees, shrubs or lianes, usu. v. poisonous (fluoroacetic acid & pyridine alks) & with char. unicellular hairs with warty papillae. Lvs in spirals, simple, entire; stip. usu. caducous. Fls in axillary to petiolar or epiphyllous cymes, usu. bisex. & reg.

(± irreg. in *Tapura*) with articulated pedicels; K (4)5, imbr., s.t. basally connate, C (4)5, usu. 2-lobed or bifid, imbr., rarely with basal tube, A(4)5, alt. with C, rarely 3 with 2 staminodes, anthers with longit. slits, basal nectary gland opp. each C (confluent when tube present), $\underline{G}$ to $\overline{G}$ (2–4(5)), pluriloc. with term. style, lobed or rarely distinct styles, ovules 2 per loc., apical-axile, pend., anatropous, bitegmic. Fr. a flattened densely hairy drupe (rarely a capsule) with 1–3(4))-loc. stone usu. with 1 seed per loc., exocarp s.t. splitting; seed with large straight oily embryo & no endosperm, oft. with caruncle. 2n = 20, 24

Genera: *Dichapetalum, Stephanopodium, Tapura*

Some cult. orn. & sources of poisons

Dichapetalum Thouars. Dichapetalaceae. 133 trop. (Indomal.-Pacific 16, trop. Afr. & Madag. 97, Am. 20). R: AUWP 86–3(1986). Roux's Model; many spp. with epiphyllous infls. Some hyperaccumulators; many extremely poisonous & used to kill wild pigs, monkeys & rats in Afr. (fluoroacetic acid disrupts tricarboxylic acid cycle of respiration); ***D. cymosum*** (Hook.) Engl. of high veldt of S Afr. begins growth before veldt grasses & is therefore eaten by cattle leading to 'gifblaar' poisoning & death

Dichasianthus Ovcz. & Junussov (~ *Torularia*). Cruciferae (26). 1 C As.: ***D. subtilissimus*** (Popov) Ovcz. & Junussov

Dichazothece Lindau. Acanthaceae (III 2c). 1 E Braz.: ***D. cylindracea*** Lindau

Dichelachne Endl. = *Deyeuxia* (but see Blumea 22(1974)5, NZJB 20(1982)303)

Dichelostemma Kunth (~ *Brodiaea*). Asparagaceae (Themidaceae). 5 N Am. R: Four Seasons 9,1(1991)24. ***D. pulchellum*** (Salisb.) A. Heller & other spp. with ed. bulbs. Cult. orn. esp. ***D. ida-maia*** (Alph. Wood) Greene (*Brevoortia i., Brodiaea i., Rhytidea bicolor*, Calif., Oregon) – scarlet fls; ***D. volubile*** (Kellogg) A. Heller (Calif.) – fls pink, stem flexuous & twining to 1.5 m

Dicheranthus Webb. Caryophyllaceae (I 2). 1 Canary Is.: ***D. plocamoides*** Webb

Dichilanthe Thwaites. Rubiaceae (III 3). 2 Sri Lanka, Borneo

Dichiloboea Stapf = *Trisepalum*

Dichilus DC. Leguminosae (III 9). 5 S Afr. R: SAJB 54(1988)182. Alks

Dichocarpum W.T. Wang & Hsiao (~ *Isopyrum*). Ranunculaceae (III 2). 20 Himal., E As. R: APS 26(1988)249

Dichodon (Reichb.) Reichb. (~ *Cerastium*). Caryophyllaceae. 5 Arctic, C Eur., Iran

Dichoglottis Fischer & C. Meyer = *Gypsophila*

Dichondra Forst. & Forst. f. Convolvulaceae (6). 9 trop. & warm (1 natur. Eur. incl. Cornwall). R (partial): Britt. 13(1961)346. ***D. micrantha*** Urb. (prob. orig. Am., now subcosmop.) – grown as lawn-grass subs.

Dichondraceae Dumort. = Convolvulaceae

Dichondropsis Brandegee = *Dichondra*

Dichopogon Kunth = *Arthropodium*

Dichorisandra Mikan. Commelinaceae (II 1e). 25 trop. S Am., 1 ext. to C Am. & WI. Infls racemes or thyrses; seeds arillate. Some cult. orn. housepls esp. ***D. reginae*** (L. Linden & Rodigas) W. Ludw. (*Tradescantia r.*, Peru)

Dichoropetalum Fenzl (~ *Johrenia, Peucedanum*). Umbelliferae (III 10). Incl. *Holandrea, Johreniopsis*, 35 Euras. (Turkey 14). R: Willd. 37(2007)477

Dichosciadium Domin. Umbelliferae (Azor.). 1 Aus.: ***D. ranunculaceum*** (Hook.f.) Domin

Dichostemma Pierre. Euphorbiaceae (Euph.-Anth.). 2 W Afr.

Dichotomanthes Kurz. Rosaceae (16). 1 SW China: ***D. tristaniicarpa*** Kurz – 1 entirely free carpel (cf. *Heteromeles*)

Dichroa Lour. = *Hydrangea*

Dichrocephala L'Hérit. ex DC. Compositae (Ast.-Gran.). 3 OW trop. R: MBSM 15(1979)491. ***D. integrifolia*** (L.f.) Kuntze – weedy but local medic.

Dichromanthus Garay. Orchidaceae (IV 2h). 4 Texas to mts of C Am.

Dichromena Michaux = *Rhynchospora*

Dichromochlamys Dunlop. Compositae (Ast.-Pod.). 1 Aus.: ***D. dentatifolia*** (F. Muell.) Dunlop. R: JABG 2(1980)235

Dichrospermum Bremek. = *Spermacoce*

Dichrostachys (DC.) Wight & Arn. Leguminosae (II 1). 14 trop. OW (Aus. 1) esp. Madag. C 0, basal fls in spike sterile with 10 staminodes, term. ones bisex. with 10 short A; some with stip. thorns e.g. ***D. cinerea*** (L.) Wight & Arn. (Afr., Ind., nature. pantrop.) – used as a thorny hedge but now a pest in e.g. Cuba

Dichrotrichum Reinw. ex Vriese = *Agalmyla*

Dickasonia L.O. Williams. Orchidaceae (V 10b). 1 NE Ind., Myanmar: ***D. vernicosa*** L.O. Williams. R: Fl. Bhutan 3,3(2002)338

Dickinsia Franch. Umbelliferae (Azor.). 1 SW China: ***D. hydrocotyloides*** Franch.

Dicksonia L'Hérit. Dicksoniaceae (Cyatheaceae s.l.). c. 20 trop. Am., St Helena, Mal. (mts, esp. NG), Aus., New Caled., NZ. R: M.F. Large & J.E. Braggins, *Tree Ferns* (2004)282. Fossils from Jurassic; differs from *Cyathea* in indumentum having only hairs & no scales on young parts. All Aus. spp. prot.; some cult. orn. esp. ***D. antarctica*** Labill. (Aus.) to 15 m, pith a starch source for Aborigines, fronds form. sold for Christmas decoration in Aus. ***D. squarrosa*** (Forst.) Sw. (NZ) – trunks used for Maori food-storage houses & storage-pits for sweet potato

Dicksoniaceae M. Schomb. (~ Cyatheaceae). Polypodiidae – Cyatheales. 3/c. 27 E As., Aus., St Helena, trop. Am. Tree ferns wiuth polystelic rhizomes (solenostele in *Calochlaena*); stem apices & petiole bases covered in uniseriate hairs. Blades large, 2–3-pinnate; veins simple to forked, free. Sori abaxial & indusium 0 (*Lophosoria*) or marginal with indisium. x = 56, 65
Genera: *Calochlaena, Dicksonia, Lophosoria* (s.t. placed in own fam.)
Some molecular evidence suggests amalgamation with Cyatheaceae
Known from Triassic

Dicladanthera F. Muell. Acanthaceae (III 2c). 2 W Aus. R: JABG 9(1986)171

Diclidanthera Mart. Polygalaceae (II). 3–4 trop. S Am.

Diclidantheraceae J. Agardh = Polygalaceae

Diclinanona Diels. Annonaceae (III 5). 3 E Peru, W Braz. R: T 63(2014)1249

Dicliptera Juss. Acanthaceae (III 2c). Incl. *Peristrophe*, c. 150 trop. & warm (S Afr. 15 – R: KB 51(1996)12). Some local medic.; some cult. orn. incl. ***D. dodsonii*** Wassh. (W Ecuador, only 1 wild pl. known) – lianoid; ***D. javanica*** Nees (Mal.) – normal & cleistogamous fls

Diclis Benth. Scrophulariaceae (Hemim.). 9 trop. & S Afr., Madag.

Dicoelia Benth. Phyllanthaceae (Wiel.-Wiel.; Euphorbiaceae s.l.). 2 W Mal. R: Blumea 56 (2011)209

Dicoelospermum C.B. Clarke = *Cucumis*

Dicoma Cass. Compositae (Card.-Dic.). Incl. *Cloiselia, Macledium*, excl. *Dicomopsis*, c. 45 trop. & S Afr., Madag. (4), Ind. (1). Trees to herbs poss. incl. *Pasaccardoa*

Dicomopsis S. Ortíz (~ *Dicoma*). Compositae (Card.-Dic.). 1 SW trop. Afr.: ***D. welwitschii*** (O. Hoffm.) S. Ortíz. R: T 62(2013)534

Dicoria Torrey & A. Gray. Compositae (Helia.-Amb.). 4 SW N Am. Pappus 0, winged cypselas s.t. with term. tuft of hairs

Dicorynia Benth. Leguminosae (I 3). 2 trop. S Am. A 2. ***D. guianensis*** Amshoff (angelique, basralocus) – timber

Dicoryphe Thouars. Hamamelidaceae (I 1). 12 Madag., Comoro Is. Some rheophytes

dicotyledons (**dicots**). Magnoliidae p.p., angiosperms trad. (first formalized by John Ray in C17) defined as those with seedlings usu. with 2 cotyledons; lvs usu. net-veined; fls oft. 5-merous. Monocotyledons (q.v.) much less diversified. Now realized that many 'primitive' fams form. placed here (see Nymphaeaceae) are relics of the group from which monocots & dicots in the narrow sense emerged, so that it is necessary to recog. a group of 'basal' angiosperms that are neither monocots nor dicots (s.s.). See Appendix

Dicraeanthus Engl. Podostemaceae (III). 2 W Afr. ***D. africanus*** Engl. – eaten as salad

Dicraeia Thouars = *Podostemum*

Dicraeopetalum Harms. Leguminosae (III 9). 3 SE Ethiopia, S Somalia, NE Kenya, Madag. (2)

Dicranocarpus A. Gray. Compositae (Cor.). 1 SW N Am.: ***D. parviflorus*** A. Gray – some fr. without pappus. R: FNA 21(2006)219

Dicranoglossum J. Sm. = *Pleopeltis*

Dicranolepis Planch. Thymelaeaceae (Thym.-Daph.). c. 20 trop. Afr. Mangenot's Model

Dicranopteris Bernh. Gleicheniaceae (Gleich.). c. 11 trop. esp. Mal. & S temp. Forming dense tangles particularly on poor soils & after fires. ***D. pectinata*** (Willd.) Underw. (trop. Am.) – only solenostele in G. ***D. linearis*** (Burm.f.) Underw. (OW trop.) – stems used for fish-traps (last 2 yrs in saltwater), plaiting, chair-seats, hut-walls, mats (e.g. for tea-shade) etc., & split as pens for writing Arabic, veg. in Sumatra

Dicranopygium Harling. Cyclanthaceae. 50 trop. Am. R: AHB 18(1958)274

Dicranostegia (A. Gray) Pennell (~ *Cordylanthus*). Orobanchaceae (Cast.). 1 Baja Calif.: ***D. orcuttiana*** (A. Gray) Pennell. R: SB 34(2009)191

Dicranostigma Hook.f. & Thomson. Papaveraceae (I). 3–5 Himal. W China. Alks. Cult. orn. herbs

Dicranostyles Benth. Convolvulaceae (11). 15 trop. Am. R: AMBG 60(1975)385. Lianes to 30 m

Dicraspidia Standl. Muntingiaceae. 1 C Am. to Colombia: ***D. donnell-smithii*** Standl. R: T 47(1998)38

Dicrastylidaceae J.L. Drumm. ex Harv. = Labiatae

Dicrastylis J.L. Drumm. ex Harv. (*Mallophora*). Labiatae (IV 1; Chloanthaceae). 26 Aus. R: Brunonia 1(1978)437, JABG 14(1991)85

Dicraurus Hook.f. = *Iresine*

Dicrocaulon N.E. Br. Aizoaceae (V). 7 Cape. R: H. E. K. Hartmann, *Ill. Handb. Succ. Pls*, Aiz. A–E(2002)211

Dictamnus L. Rutaceae (I). 1 C & S Eur. to N China: ***D. albus*** L. (*D. fraxinella*, burning bush, dittany) – herb. perenn. with aromatic lvs & irreg. fls with 10 upward-pointing A in term. racemes, the stems covered with glands which release a volatile inflammable oil which may be lit without harming the plant; causes allergy in some

Dictyandra Welw. ex Hook.f. = *Leptactina* (but see PSE 145(1984)105)

Dictyanthus Decne = *Matelea*

Dictymia J. Sm. Polypodiaceae (II 4). 2–3 Aus. & NG to Fiji. ***D. brownii*** (Wikström) Copel. (W Pac.) – cult. orn. with crassulacean acid metabolism

Dictyocaryum H. Wendl. Palmae (V 1). 3 trop. S Am. R: FN 53(1990)53

Dictyochloa (Murb.) Camus = *Ammochloa*

Dictyocline T. Moore = *Cyclosorus*

Dictyodroma Ching = *Deparia*

Dictyolimon Rech.f. (~ *Limonium*). Plumbaginaceae (II). 4 Afghanistan to Ind.

Dictyoloma A. Juss. Rutaceae (III). 1 Peru, Braz., Bolivia: ***D. vandellianum*** A. Juss. – Leeuwenberg's Model, lvs bipinnate, ground to stun fish

Dictyoneura Blume. Sapindaceae. 2–3 Mal. R: FM I,11(1994)507

Dictyophleba Pierre. Apocynaceae (Ic). 5 trop. Afr. R: BJBB 59(1989)207. Leeuwenberg's Model

Dictyophragmus O. Schulz. Cruciferae (46). 3 Peru, Arg. R: Novon 1(1991)71

Dictyophyllaria Garay = *Vanilla* (but see Gen. Orch. 3(2003)304)

Dictyosperma H. Wendl. & Drude. Palmae (V). 1 Masc.: ***D. album*** (Bory) R. Scheffer, now rare in wild through over-exploitation of the cabbage, but cult. orn. in trop.

Dictyospermum Wight (~ *Aneilema*). Commelinaceae (II 2). 4–5 Ind. & Sri Lanka to NG

Dictyostega Miers. Burmanniaceae. 1 Mex. to Bolivia, Braz.: ***D. orobanchoides*** (Hook.) Miers

Dictyoxiphium Hook. = *Tectaria*

Dicyclophora Boiss. Umbelliferae (III 1). 1 Iran: ***D. persica*** Boiss.

Dicymanthes Danser = *Amyema*

Dicymbe Spruce ex Benth. Leguminosae (I 2). 16 trop. S Am., esp. Amazonia. ***D. altsonii*** Sandw. & ***D. corymbosa*** Spruce ex Benth. – local dominants with large coppice-shoots leading to thickets when main trunks die

Dicymbopsis Ducke = *Dicymbe*

Dicypellium Nees & Mart. Lauraceae (I). 2 E Amaz. R: BJ 110(1988)168. ***D. caryophyllaceum*** (Mart.) Nees & Mart. – bark sold in quills (*Cassia caryophyllata*) smells like cloves (oil has 95% eugenol), used as flavouring & with lvs in a stimulant tea, timber good

Dicyrta Regel = *Achimenes*

Didelotia Baill. Leguminosae (I 2). c. 11 Guineo-Congolian forests, Afr. R: Blumea 12(1964)209. Timber

Didelta L'Hérit. Compositae (Arct.). 2 SW Afr. R: MBSM 3(1959)304. ***D. spinosa*** Aiton – small tree

Didesmandra Stapf. Dilleniaceae (IV). 1 Sarawak: ***D. aspera*** Stapf

Didesmus Desv. (~ *Cakile*). Cruciferae (12). 2 E Med. (Eur. 1)

didgeridoo, didgeridu, didjeridu *Bambusa arnhemica, Pandanus spiralis*

Didiciea King & Prain = *Tipularia*

Didierea Baill. Didiereaceae. 2 S & SW Madag. R: FMad. 121(1963)29

Didiereaceae Radlk. Magnoliidae – Caryophyllales. 6/20 E & S Afr., SW Madag. (4/11 – R: FMad. 121(1963)). Decid. pachycaul xerophytes, oft. thorny, usu. dioec. (rarely gynodioec.), with betalains (not anthocyanins); wood-formation reg. Thorny short axes in axils of lvs of long axes. Phloem sieve-tube plastids of P-type with C globular protein crystalloid & subperipheral ring of proteinaceous filaments. Lvs in spirals, entire, small; stip. 0. Fls in term. ± dichasial cymes with 2 involucral bracts; K 2, petaloid (but derived from bracts?), C 2 + 2 (derived from K?), A (5-)8(10; ∞ in *Calyptrotheca*), filaments adnate to outside of annular nectary, staminodes oft. in female fls, G ((2)3(4)) with 1 style & (2)3(4)-lobed stigma & as many loc., but only 1 fert. with 1 basal, erect, campylotropous, bitegmic ovule. Fr. dry, indehiscent, usu. 3-angled, encl. in involucral bracts; seeds with small funicular aril & large curved or folded embryo with almost 0 endosperm or perisperm. 2n = c. 150, 190–200

Genera: *Alluaudia, Alluaudiopsis, Calyptrotheca, Decarya, Didierea, Portulacaria*

Serology, sieve-tube plastid features, betalains & pollen ornamentation placed D. here, confirmed by DNA & by some being successfully grafted on Cactaceae (*Pereskia*)

Didiplis Raf. (~ *Lythrum*). Lythraceae. 1 E US: ***D. diandra*** (DC.) Alph. Wood – in ponds etc.

Didissandra C.B. Clarke. Gesneriaceae (III 2h). 8 W Mal.

Didonica Luteyn & Wilbur. Ericaceae (VIII 5). 4 Panamá & Costa Rica. R: SB 16(1991)587

Didymaea Hook.f. Rubiaceae (IV 16). 5 C Am. ? Apogamous

Didymanthus Endl. Amaranthaceae (Chenopodiaceae I 5). 1 W Aus.: ***D. roei*** Endl. R: FA 4(1984)218

Didymaotus N.E. Br. Aizoaceae (V). 1 Karoo: ***D. lapidiformis*** (Marloth) N.E. Br. – cult. orn. R: H. E. K. Hartmann, *Ill. Handb. Succ. Pls*, Aiz. A–E(2002)215

Didymelaceae Leandri = Buxaceae

Didymeles Thouars. Buxaceae (Didymelaceae). 2 Madag.

Didymia Philippi = *Cyperus*

Didymiandrum Gilly. Cyperaceae (III 1). 1 trop. S Am.: ***D. stellatum*** (Boeckeler) Gilly. R: MNYBG 12(1965)51

Didymocarpus Wall. ex Buch.-Ham. Gesneriaceae (III 2j). 70+ Himal. & S China to Sumatra. See also *Henckelia, Hovanella*. Long narrow capsules split to form gutters down which raindrops carry away tiny seeds. Some cult. orn. Sect. *Codonoboea* (Ridl.) Kiew (4 Malay Pen.; R: Blumea 35(1990)71) – fls epiphyllous (cf. *Microchirita*)

Didymocheton Blume = *Dysoxylum*

Didymochlaena Desv. Hypodematiaceae (Didymochlaenaceae). 1–2 trop. ***D. truncatula*** (Sw.) J. Sm. – rhizome erect, fronds to 2 m, cult. orn.

Didymochlaenaceae Ching ex L.B. Zhang & L. Zhang = Hypodematiaceae (but see T 64(2015)34)

Didymochlamys Hook.f. Rubiaceae (inc. sed.). 2 Panamá to Venez. Epiphytes

Didymocistus Kuhlm. Phyllanthaceae (Antid.-Hym.; Euphorbiaceae s.l.). 1 trop. S Am.: ***D. chrysadenius*** Kuhlm. R: A. Radcliffe-Smith, *Gen. Euphorb.* (2001)76

Didymodoxa Wedd. (~ *Australina*). Urticaceae (V). 2 S & E trop. Afr. to N Ethiopia. R: NJB 8(1988)45

Didymoecium Bremek. = *Rennellia*

Didymoglossum Desv. (~ *Trichomanes*). Hymenophyllaceae. c. 30 trop.

Didymogonyx (L. Clark & Londoño) Tyrrell (~ *Rhipidocladum*). Gramineae (V 2). 2 Colombia, Venez.

Didymopanax Decne & Planch. = *Schefflera*

Didymophysa Boiss. Cruciferae (47). 2 Iran to C As. & Himal.

Didymoplexiella Garay (~ *Didymoplexis*). Orchidaceae (V 7). 8 SE As. to Jap. & Borneo. Mycotrophic

Didymoplexiopsis Seidenf. (~ *Didymoplexis*). Orchidaceae (V 7). 1 Vietnam, Thailand: ***D. khiriwongensis*** Seidenf. – mycotrophic

Didymoplexis Griff. Orchidaceae (V 7). 17 trop. & S Afr. (1), Madag. (1)., SE As. & Mal. Mycotrophic

Didymopogon Bremek. Rubiaceae (I 9). 1 Sumatra: ***D. sumatranum*** (Ridl.) Bremek.

Didymosalpinx Keay. Rubiaceae (?II 1). 3 trop. Afr.

Didymosperma H. Wendl. & Drude ex Hook.f. = *Arenga*

Didymostigma W.T. Wang (~ *Chirita*). Gesneriaceae (III 2j). 2 SE China

Didymotheca Hook.f. = *Gyrostemon*

Didyplosandra Wight ex Bremek. = *Strobilanthes*

Diectomis Kunth (~ *Andropogon*). Gramineae (XXII 7). 1 trop.: ***D. fastigiata*** (Sw.) P. Beauv.

Dieffenbachia Schott. Araceae (VII 3). c. 140 (only 36 described!) trop. Am. Dumb canes. Stout herbs (Chamberlain's Model), female fls with consp. staminodes; thermogenic heat with 1 peak. ***D. longispatha*** Engl. & Krause (C Am.) – poll. by 9 scarab spp. Many cult. orn. housepls (esp. C19 vogue) most cvs derived from ***D. seguine*** (Jacq.) Schott (incl. *D. maculata* with narrower lvs, oft. irreg. heavily white-spotted), the orig. 'dumb cane' used to torture slaves – chewing a portion of stem leads to speechlessness in adults & death in children or pets due to ejection of an obscure proteinaceous poison & irritant needle-like raphides (believed to cause temp. sterility & therefore use in Russian concentration camps proposed by Nazis) from damaged idioblasts, the expulsion visible in macerated tissue under 100 x microscope, '**Wilson's Delight**' – white venation due to 1 dominant gene

Diegodendraceae Capuron = Bixaceae

Diegodendron Capuron. Bixaceae (Diegodendraceae). 1 Madag.: ***D. humbertii*** Capuron

Dielitzia Short. Compositae (Gnap.-Ang.). 1 Aus.: ***D. tysonii*** Short. R: Muelleria 7(1989) 103

Diellia Brackenr. = *Asplenium*

Dielsantha F. Wimmer. Campanulaceae (III 1). 1 trop. Afr.: ***D. galeopsoides*** (Engl. & Diels) F. Wimmer

Dielsia Gilg. Restionaceae. 1 SW Aus.: ***D. stenostachya*** (W. Fitzg.) Briggs & L. Johnson. R: Telopea 8(1998)27

Dielsiocharis O. Schulz. Cruciferae (3). 2 Iran, Tajikstan. R: Novon 13(2003)171

Dielsiochloa Pilg. = *Festuca*

Dielsiodoxa Albrecht. Ericaceae. 5 SW Aus. R: Nuytsia 21(2011)110

Dielsiothamnus R.E. Fries. Annonaceae (III 7). 1 trop. E Afr.: ***D. divaricatus*** (Diels) R.E. Fries

Dielytra Cham. & Schldl. = *Dicentra*

Dienia Lindl. (~ *Malaxis*). Orchidaceae (V 11b). 5 Indopacific

Dierama K. Koch. Iridaceae (VI 5). 44 E Afr. mts to Cape. R: O.M. Hilliard & B.L. Burtt (1991) *Dierama*. Specially prot. in Zimbabwe; some cult. orn. with pendent fls (wand-flowers)

Diervilla Mill. Caprifoliaceae (Diervillaceae). 2–3 E N Am. R: GH 2(1929)46. Cult. orn. decid. shrubs

Diervillaceae Pyck = Caprifoliaceae

Dieteria Nutt. (~ *Machaeranthera*). Compositae (Ast.-Mac.). 3 W N Am. R: Sida 20(2003)1391

Dieterlea Lott. (~ *Ibervillea*). Cucurbitaceae (XIII). 3 Mex. R: Britt. 38(1986)407

Dietes Salisb. ex Klatt. Iridaceae (VII 2). 6 E & S Afr. (5). Lord Howe Is. (1! Cf. *Cunonia*). R: AMBG 68(1981)132. Some viviparous with plantlets on old infl. (Goldblatt); some cult. orn.

Dieudonnaea Cogn. = *Gurania*

Diflugossa Bremek. = *Strobilanthes*

Digastrium (Hackel) A. Camus = *Ischaemum*

Digera Forssk. Amaranthaceae (I 2). 1 OW trop.: ***D. muricata*** (L.) Mart. (*D. arvensis*) – potherb in Sudan, fodder for stock

Digitacalia Pippen. Compositae (Sen.-Tuss.). 5 Mex. R: Phytol. 69(1990)150

Digitalis Tourn. ex L. Plantaginaceae (Dig.-Dig.; Scrophulariaceae s.l.). Incl. shrubby *Isoplexis* (3 Macaronesia – R: BJ 79(1980)218) c. 22 Medit., Eur. (12) to C As. R: BJ 79(1960)222. Foxgloves (i.e. 'folks' [? = fairies'] gloves'). Hybrids in wild incl. fertile ***D. × fucata*** Ehrh. (*D. × purpurascens*, *D. lanata × D. purpurea*). Cult. orn. & medic. esp. as source of digitalis (cardiac glycosides incl. digitoxin & digoxin), from *D. purpurea* & *D. lanata* & used as cardiac stimulant since 1785 (William Withering – who was told by a countrywoman *D. purpurea* was 'good for the dropsy') but not analysed until 1933. ***D. canariensis*** L. (*I. c.*, Canary Is.) – bird-poll. shrub, cult. orn.; ***D. grandiflora*** Mill. (*D. ambigua*, Euras.) – cult. orn., yellow fls; ***D. isabelliana*** (Webb & Berth.) Lindinger (*I. i.*, Canary Is.) – pollen found on heads of insectivorous birds (no sunbirds there); ***D. lanata*** Ehrh. (Austrian d., C & SE Eur., natur. N Am.) – principal source of digitalis, most cult. Netherlands; ***D. mertonensis*** Buxton & C. Darl. (*D. grandiflora × D. purpurea*) – true-breeding tetraploid; ***D. purpurea*** L. (common f., fairy fingers, polymorphic centred on Medit., homogeneous N Eur. populations expanded from Iberia post-glacial) – source of digitalis (foxglove tea form. taken for colds & fevers, copious draughts to induce intoxication; the minute seeds coll. by children S England as part of 'War Effort' in WW II), contains loliolide, a potent ant-repellent (pls long-used as insecticidal disinfectant for walls in Forest of Dean, England), sign of insincerity in 'Language of Fls', many cvs incl. **'Monstrosa'** (? orig. Leiden

Botanic Garden before 1842) with reg. term. fl. & **'Saltwood Summer'** with sep. petals (cf. *Rhododendron stenopetalum*)

Digitaria Haller. Gramineae (XXIV 1). c. 270 trop. & warm (Eur. 3, China 22, Aus. 44, N Am. 18 + 11 introd.). R: J.T. Henrard (1950) *Monogr. D.*; Blumea 21(1973)1. Crab or finger grasses, many weedy esp. ***D. abyssinica*** (A. Rich.) Stapf (*D. scalarum*) in E Afr. & ***D. sanguinalis*** (L.) Scop. (summer grass, Euras.) in much of world. ***D. didactyla*** Willd. (blue couch, trop. Afr.) – lawn-grass; ***D. eriantha*** Steud. (subsp. ***pentzii*** (Stent) Kok; *D. decumbens, D. pentzii*) **'Pangola'** (pangola grass, S Afr.) – pasture-grass in S US setting no seed; ***D. exilis*** Stapf (W Afr., ? cultigen) – a staple crop (acha, fonio, fundi, hungry rice) tolerant of poor soils; ***D. iburua*** Stapf (black fonio, iburu, W Afr.) – eaten like millet

Digitariella De Winter = *Digitaria*

Digitariopsis C. Hubb. = *Digitaria*

Diglyphosa Blume. Orchidaceae (V 14). 3 Indomal. R: OM 8(1997)166

Dignathe Lindl. = *Cuitlanzina*

Dignathia Stapf. Gramineae (XXIX 3). 5 trop. E Afr., NW Ind.

Digomphia Benth. Bignoniaceae (1). 3 Guayana Highland & nearby

Digoniopterys Arènes. Malpighiaceae. 1 Madag.: ***D. microphylla*** Arènes

Diheteropogon (Hackel) Stapf (~ *Andropogon*). Gramineae (XXII 7). 4 trop. & S Afr. R: KB 20(1966)73. ***D. amplectens*** (Nees) W. Clayton – imp. forage trop. Afr.

Dijon mustard *Brassica juncea*

dika (fat) *Irvingia gabonensis*

Dilatris Bergius. Haemodoraceae (I). 4 SW Cape. R: Strel. 9(2000)92

Dilepis Süsseng. & Merxm. = *Flaveria*

Dilivaria Juss. = *Acanthus*

Dilkea Masters. Passifloraceae. c. 6 trop. S Am. A 6 basally connate

dill or **(E.) Indian d** *Anethum graveolens*

Dillandia Funk & H. Robinson (~ *Liabum*). Compositae (Liab.). 3 Colombia, Ecuador, Peru. R: SB 26(2001)218

Dillenia L. Dilleniaceae (IV). c. 65 Ind. Ocean, Indomal., Aus. (1). R: Blumea 7(1952)1, 9(1959)577. Most primitive spp. the most restricted. Buzz-poll. (*Xylocopa* bees) cult. orn. (Scarrone's Model) esp. ***D. indica*** L. (Ind. to C Mal., water-disp.) – fr. ed. (curry, jellies), also shampoo, & ***D. suffruticosa*** (Hook.f. & Thomson) Martelli (*Wormia subsessilis*, Mal.) – evergrowing shoots, seldom out of flower; ***D. pentagyna*** Roxb. (China, Indomal.) – decid., charcoal source; other spp. minor timbers, e.g. mudi (Solomon Is.) – ***D. crenatifolia*** Hoogl. ex Mabb. (*D. crenata*) & larger ***D. salomoniensis*** (C.T. White) Hoogl. (girth to 7 m, prob. biggest *D.* sp. & one of largest trees in Solomon Is.), ed. fr. locally medic. etc.

Dilleniaceae Salisb. Magnoliidae – Dilleniales. 11/425 trop. & warm esp. Australasia. Trees to subshrubs or even rhiz. herbs (*Acrotrema*) or lianes, with simple conduplicate lvs, usu. in spirals, rarely lobed to pinnatisect or scale-like, strong parallel sec. veins ending in teeth; stip. 0, s.t. petiole with amplexicaul wings. Fls oft. large, solit. or in racemose or thyrsoid infls, yellow or white, usu. bisex., reg., usu. without nectar; K (2)3–5(–18), spirally imbr., persistent, C (2)3–5(–7), imbr., oft. crumpled in bud, A ± ∞, but s.t. few or 1, oft. asymmetrically placed but always assoc. with 5–15 stamen-trunks, the members of each originating centrifugally or all A so, anthers with longit. slits or apical pores, $\underline{G}$ (1–10(–20), rarely in 2 whorls, ± conduplicate & s.t. not fully sealed, rarely connate forming pluriloc. ovary with distinct styles, 1–80 anatr. to amphitropous or campylotropous bitegmic ovules with zig-zag micropyle per loc. Fr. dry-dehiscent or indehiscent & then incl. in ± fleshy K (*Dillenia*); seeds endotestal with funicular aril (s.t. laciniate or even ± 0) & v. small straight embryo in copious oily proteinaceous endosperm. x = 4, 5, 8–10, 12, 13

Classification & genera:

I. **Delimioideae** (prophylls 1 or 2 per fl.; arils fimbriate): *Tetracera*

II. **Doliocarpoideae** (prophylls 0; aril margin usu. entire): *Curatella, Davilla, Doliocarpus, Neodillenia, Pinzona*

III. **Hibbertioideae** (lvs oft. ericoid, fls oft. solit., G 1–5(–10)): *Hibbertia*

IV. **Dillenioideae** (lvs with petiolar wings): *Acrotrema, Didesmandra, Dillenia, Schumacheria*

Many app. primitive features incl. ± distinct conduplicate carpels, some not completely closed, ± ∞ A & wood with scalariform vessel-elements. Some timbers, ed. fr. & cult. orn. (esp. *Dillenia*)

dillon bush *Nitraria billardierei*

Dillwynia Sm. (~ *Pultenaea*). Leguminosae (III 13). c. 38 Aus. Some cult. orn. (parrot pea)

Dilobeia Thouars. Proteaceae (IV). 2 E Madag. Leaf dichot. divided

Dilochia Lindl. Orchidaceae (V 10b). 8 Mal.

Dilochiopsis (Hook.f.) Brieger (~ *Eria*). Orchidaceae (V 15). 1 Borneo: ***D. scortechinii*** (Hook.f.) Brieger

Dilodendron Radlk. Sapindaceae. 3 trop. Am. R: AMBG 74(1987)533. ***D. costaricense*** (Radlk.) A. Gentry & Steyerm. – ed. seeds

Dilomilis Raf. Orchidaceae (V 13c). 5 WI to Braz. R: Lindleyana 12(1997)180. See also *Tomzanonia*

Dilophia Thomson. Cruciferae (26). 2 C As. to W China

Dilophotriche (C. Hubb.) Jacq.-Fél. = *Tristachya*

Dimerandra Schltr. (~ *Epidendrum*). Orchidaceae (V 13b). 8 trop. Am. R: C.L. Withner, *The Cattleyas & their relatives* V (1998)77

Dimeresia A. Gray. Compositae (Chaen.). 1 W US: ***D. howellii*** A. Gray – heads 2-flowered, each fl. with bract. R: FNA 21(2006)182

Dimeria R. Br. (~ *Ischaemum*). Gramineae (XXII 4). 59 Madag. (3), Indopacific (China 6). R: KB 7(1953)553. Spikelets 1-flowered

Dimerocostus Kuntze. Costaceae (Zingiberaceae s.l.). 2 trop. Am. R: T 55(2006)156

Dimerostemma Cass. Compositae (Helia.-Ecl.). Incl. *Angelphytum*, c. 26 trop. S Am. R: PBSW 97 (1984)618

Dimetia (Wight & Arn.) Meisn. (~ *Hedyotis*). Rubiaceae (IV 15). 7 trop. As. R: T 64(2015)315

Dimetra Kerr. Oleaceae (5). 1 NE Thailand: ***D. craibeana*** Kerr – geoxylic suffrutex form. referred to Verbenaceae, ? extinct

Dimitria Rav. = *Chilocardamum*

Dimocarpus Lour. (~ *Litchi*). Sapindaceae. 6 S & SE As. to Aus. R: Blumea 19(1971)113. Wood useful e.g. rifle-butts. ***D. longan*** Lour. (*Euphoria l.*, *Nephelium l.*, longan or longyen, long yan rou, dragon's eyes, Ind., Sri Lanka) – cult. China & Mal. (fls initiated at will through application of potassium chlorate in fertilizer – discovered from a firework spillage) for juicy fr. (1 M t by 2000), exported dried from China, var. ***malesianus*** Leenh. being mata kuching or cat's eyes; hybrids with litchi raised

Dimorphandra Schott. Leguminosae (I 4). 26 trop. Am. Intermediate between Leg. I & II. ***D. mollis*** Benth. (Braz. cerrado) – seeds rich in galactomannans; ***D. parviflora*** Spruce ex Benth. – seeds used for intoxicating snuff in Amaz.

Dimorphanthera (Drude) F. Muell. ex J.J. Sm. Ericaceae (VIII 5). 75 Mal. esp. NG (cf. *Satyria* in trop. Am.). R: FM I 6(1967)885, 9(1982)563, EJB 60(2004)280

Dimorphocalyx Thwaites. Euphorbiaceae (Cod.-Ost.). 17 Indomal. (Mal. 8 – R: Blumea 59(2015)193) to Aus. (1)

Dimorphocarpa Rollins (~ *Dithyrea*). Cruciferae (40). 4 N Am. ***D. wislizeni*** (Engelm.) Rollins – local medic. (dermatology)

Dimorphochloa S.T. Blake = *Cleistochloa*

Dimorphocoma F. Muell. & Tate. Compositae (Ast.-Pod.). 1 C Aus.: ***D. minutula*** F. Muell. & Tate

Dimorpholepis (G. Barroso) R. King & H. Robinson = *Grazielia*

Dimorphorchis Rolfe (*Lowianthus*, ~ *Arachnis*). Orchidaceae (V 16c). 5 Borneo. Cult. orn.

Dimorphosciadium Pim. (~ *Pachypleurum*). Umbelliferae (III 8). 1 C As.: ***D. gayoides*** (Regel & Schmalh.) Pim.

Dimorphostemon Kitag. = *Dontostemon*

Dimorphotheca Vaill. ex Moench. Compositae (Calend.). Incl. *Castalis*, c. 20 S & trop. Afr. R: M. Norlindh, *Stud. Calend.* 1(1943)38. Fl.-heads with dark spots attractive to poll. bee-flies (cf. *Gorteria*). Disk cypselas straight, compressed, with 2 thick wings, ray cypselas incurved, 3-angled to nearly cylindrical, usu. sharply tubercled or wrinkled, pappus 0. Cult. orn. annuals to shrubby perenn. esp. ***D. sinuata*** DC. ('*D. aurantiaca*', *D. calendulacea*, 'sun marigolds') but many pls cult. as *D.* belong to *Osteospermum*; ***D. cuneata*** (Thunb.) Less. a poss. oilseed for dry country

Dinacria Harv. ex Sonder = *Crassula*

Dinacrusa Krebs = *Althaea* + *Malva*

Dinebra Jacq. (~ *Leptochloa*). Gramineae (XXIX 5). 27 trop. & warm. ***D. scabra*** (Nees) Peterson & N. Snow (*L. s.*, trop. Am.) – bad weed of rice

Dinema Lindl. = *Encyclia*

Dinemagonum A. Juss. Malpighiaceae. 1 Chile: ***D. gayanum*** A. Juss. R: SB 14(1989)419

Dinemandra A. Juss. ex Endl. Malpighiaceae. 1 Peru, Chile: ***D. ericoides*** A. Juss. ex Endl. R: SB 14(1989)422

Dinetopsis Roberty = seq.

Dinetus Buch.-Ham. ex Sweet (~ *Porana*). Convolvulaceae (2). 8 trop. As. R: Blumea 51(2006)428

dingo fern™ *Baloskion tetraphyllum*

Dinizia Ducke. Leguminosae (II 1). 1 Braz., Guyana: ***D. excelsa*** Ducke – monospecific stands prod. v. acidic humus supporting specialized undergrowth pls, timber

Dinklageanthus Melchior ex Mildbr. = *Dinklageodoxa*

Dinklageella Mansf. Orchidaceae (V 16d). 4 trop. W Afr.

Dinklageodoxa Heine & Sandw. Bignoniaceae (1). 1 Liberia: ***D. scandens*** Heine & Sandw.

Dinocanthium Bremek. = *Pyrostria*

Dinochloa Buese. Gramineae (V 4). c. 30 Myanmar to Philipp. R: KB 36(1981)613. Scrambling bamboos with zig-zag culms spiralling around tree-trunks, fleshy fr.

Dinophora Benth. Melastomataceae. 2 trop. W Afr.

dinosaur egg = plumcot

Dinoseris Griseb. = *Hyaloseris*

Dinosperma T. Hartley. Rutaceae (I 1). 4 E Aus. R: Adansonia 19 (1997)190

Dintera Stapf. Linderniaceae (Phrymaceae – Microc.; Scrophulariaceae s.l.). 1 Namibia: ***D. pterocaulis*** Stapf. R: Strel. 10(2000)519

Dinteracanthus C.B. Clarke ex Schinz = *Ruellia*

Dinteranthus Schwantes (~ *Lithops*). Aizoaceae (V). 6 S Afr. R: H. E. K. Hartmann, *Ill. Handb. Succ. Pls*, Aiz. A–E(2002)215. Stemless mat-forming succ., cult. orn. with fls opening p.m.

Diocirea Chinnock. Scrophulariaceae (Myoporaceae). 4 Aus. R: R. Chinnock, *Eremophila* (2007)171

Dioclea Kunth. Leguminosae (III 18). c. 40 trop. Am., a few OW incl. ***D. hexandra*** (Ralph) Mabb. (*D. reflexa*, pantrop.), a drift seed from Carib. in Carolinas, successfully sea-disp. to W Afr., r. with rotenone local. medic. incl. head-lice; ***D. grandistipula*** Queiroz (Braz.) – cauliflorous

Diodeilis Raf. = *Clinopodium*

Diodella Small = *Hexasepalum*

Diodia Gronov. ex L. Rubiaceae (IV 15). c. 50 trop. & warm Am. & Afr., ***D. serrulata*** (P. Beauv.) G. Taylor (*D. maritima*) common to both. Bell's Model

Diodonopsis Pridgeon & M.W. Chase (~ *Masdevallia*). Orchidaceae (V 13c). 5 C & N S Am. R: Lindleyana 16(2001)252

Diodontium F. Muell. (~ *Glossogyne*). Compositae (Helia.-Cor.). 1 N Aus.: ***D. filifolium*** F. Muell. – coll. once. R: FA 37(2015)449

Dioecrescis Tirveng. (~ *Gardenia*). Rubiaceae (II 1). 1 Ind.: ***D. erythroclada*** (Kurz) Tirveng. R: NJB 3(1983)456

Diogenesia Sleumer (~ *Sphyrospermum*). Ericaceae (VIII 5). 13 Andes. R: NRBGE 36(1978)251

Diogoa Exell & Mendonça. Olacaceae (Strombosiaceae). 2 trop. Afr. R: SGP 77(2007)240

Dioicodendron Steyerm. Rubiaceae (Cond.). 1 NW trop. S Am.: ***D. dioicum*** (K. Schum. & K. Krause) Steyerm.

Dion Lindl. = *Dioon*

Dionaea Sol. ex Ellis. Droseraceae. 1 SE US (pine barrens): ***D. muscipula*** Ellis (Venus' flytrap, tippitiwichet [= vagina]), carnivore attracting arthropods by blue fluorescence. Lvs in basal rosette, spathulate, the petioles winged & lamina of 2 hinged lobes fringed with 14–20 teeth, interlocking when trap closed, & with 3–7(+) sensitive hairs on upper surface; the lobes fold together on stimulation of 2 of the triggers (or 1 twice) within 20–40 secs, the mechanism app. hydraulic, the loss of turgor due to an increase in cell-wall plasticity with trap cells losing in 1–3 secs 30% of their ATP used for rapid transport of protons from motor cells, action potential predominantly dependent on extracellular concentration of calcium. Once closed, it seems that the trap actually squashes the engulfed insect prey before digestive enzymes from reddish surface glands (visible to naked eye) are secreted (prey excretions & movement (struggling against triggers) promoting this), breakdown prod. stimulating trap to narrow further, mucilage from peripheral glands sealing it like a gasket; opening is by real growth & several cycles poss. Much coll. & exported ('prot.') for the carnivorous-plant trade, though readily prop. as almost every part of the plant is capable of prod. adventitious buds seen even in infl. Fls' initiation app. stimulated by trap-activity. Intolerant of thalloid liverwort competition. T: T. Bailey & S. McPherson (2013) *D.: the Venus's flytrap*

Dionaeaceae Raf. = Droseraceae

Dioncophyllaceae Airy Shaw. Magnoliidae – Caryophyllales. 3/3 trop. Afr. Lianes or shrubs with hooked or cirrhose leaf-tips & peltate scales & char. multicellular glands secreting acid insect-trapping mucilage; stems with successively produced irreg. arr. vascular bundles. Lvs in spirals, simple, abaxially circinnate, midrib prolonged & forked into hooks or tendrils; stip. 0. Fls bisex., reg., in cymes; K 5 small, s.t. basally connate, valvate or open, persistent, C 5, convolute, A 10–30, anthers with longit. slits, $\underline{G}$ (2)or(5), 1-loc. with parietal placentas, styles 2 or 5 distinct to united, stigmas capitate or feathery, ovules ∞, anatropous, bitegmic. Fr. a loculicidal capsule, opening before ripe, ovules on thickened funicles; seeds large (5–12 cm diam.), s.t. winged all round, with ± discoid embryo & copious starchy endosperm; germ. epigeal, cryptocotylar. 2n = 24, 36 (*Triphyophyllum*)
Genera: *Dioncophyllum, Habropetalum, Triphyophyllum*
Form. incl. in Flacourtiaceae but accurately assoc. with Ancistrocladaceae by Cronquist *Triphyophyllum* is certainly carnivorous

Dioncophyllum Baill. Dioncophyllaceae. 1 trop. W Afr.: ***D. thollonii*** Baill.

Dionycha Naudin. Melastomataceae. 3 Madag. ***D. bojeri*** Naudin a source of a black dye for silk

Dionychastrum A. Fern. & R. Fern. Melastomataceae. 1 Tanzania (Uluguru Mts): ***D. schliebenii*** A. Fern. & R. Fern.

Dionysia Fenzl (~ *Primula*). Primulaceae. 49 mts C As., N Iraq, Iran (esp.), Afghanistan. R: Willd. 37 (2007)37. Tufted alpine garden pls prob. referable to *Primula* though with long C-tube & fewer ovules

Dioon Lindl. Zamiaceae (II). 13 Mex. (12), Honduras (1). R: MNYBG 57(1990)203, SB 33(2008)229. ***D. edule*** Lindl. (Mex.) – average ann. growth 0.76 mm, male cone 10 °C above ambient when ripe, seeds ed. if cooked

Diora Rav. Asparagaceae (Anthericaceae). 1 Peruvian Andes: ***D. cajamarcaensis*** (Poelln.) Rav. R: OB 92(1987)189

Dioscorea Plum. ex L. Dioscoreaceae. Incl. *Borderea, Nanarepanta, Rajania* (WI, capsule with 1 wing), *Tamus* (berries), *Testudinaria*, c. 630 trop. & warm (Eur. 4, incl. 2 Pyrenean Tertiary relics form. *Borderea*). If segregate genera recog. then *D. s.s.* has to be split into c. 20 others. Yams (see D.G. Coursey (1967) *Y.*): most primitive groups (shallow rhiz. & tubers with steroids) medic., advanced groups with deep annually replaced tubers with no chem. protection – edible yams. Ann. twining stems; alks (OW where x = 10 (NW, x = 9)); extrafl. nectaries. 'Tubers' arise variously (see also below), e.g. in *D. polystachya* by lat. hypertrophy of hypocotyl, in others from hypertrophy of internodes above cotyledon, in ***D. pentaphylla*** L. etc. from internode above cotyledon as well as hypocotyl, in ***D. villosa*** L. etc. as fleshy rhizomes. Yams have been domesticated by diff. cultures independently in diff. parts of the world. Diosgenin, a precursor of progesterone, cortisone etc. for manufacture of steroidal hormones of use as oral contraceptives, is yielded in comm. quantities by a number of spp., coll. of which in Afr. is licensed: since 1943 for progesterone prod. both ***D. composita*** Hemsl. (barbasco r., Mex.) & ***D. floribunda*** M. Martens & Galeotti (Mex., produces 10% d. by wt.). Many cult. trop. (50 M t p.a. by 2010) esp. W Afr. for starch (fufu), their culture intimately related to life & cerem. of many peoples; others are coll. or cult. ***D. alata*** L. (white yam, water y., Guyana arrowroot, trop. As. (?orig. NG), invasive in trop. Afr., SE US) – most widely grown with many cvs with 3x to 8x (all ploidy levels), tubers to 50 kg & 3.5 m long; ***D. bulbifera*** L. (air potato, acom (y.), Otaheite y. or potato, aerial y., OW trop.) – axillary bulbils, some cvs ed.; ***D. × cayenensis*** Lam. (yellow y., Guinea y., *D. abyssinica* Hochst. ex Kunth (Afr.) ×?); ***D. communis*** (L.) Caddick & Wilkin (*Tamus c.*, back bryony, murraim berries, oxberry, Eur., Medit.) – tuber formed by enlargement of first 2 internodes, young shoots ed. like asparagus (C Spain), form. medic. (plaster for gout) & eaten when boiled, fr. toxic, form. rubbed on chillblains; ***D. daemona*** Roxb. (Indomal.) – tubers used to kill fish & pigs in Sumatra; ***D. elephantipes*** (L'Hérit.) Engl. (*Testudinaria e.*, elephant's-foot, Hottentot-bread, S Afr.) – tuber to 2.1 m tall & 300 kg derived from 1st internode, projecting out of soil with thick tessellated covering of cork & ann. shoots in wet season, cult. orn., & emergency food; ***D. esculenta*** (Lour.) Burkill (potato-yam, E As.) – clones with 4x, 6x, 9x, 10x; ***D. polystachya*** Turcz. (*D. batatas*, shan yao, Chinese y. or potato, temp. E As., natur. E US) – TCM; ***D. pyrenaica*** Bub. & Bordere ex Gren. (*Borderea p.*, Pyrenees) – juvenile for 10–20 yrs, living to at least 305, one of longest-living herb. pls, ants most efficacious poll.; ***D. transversa*** R. Br. (long yam, N & E Aus.) – form. imp. tuber in SE Aus.; ***D. trifida*** L.f. (cush-cush, yampee, yampi yam, trop. Am.) – each pl. produces several tubers

Dioscoreaceae R. Br. Magnoliidae – Dioscoreales. Incl. Trichopodaceae, 3/c. 630 trop. & warm with few N temp. Herbs usu. dioec. with twining, or rarely erect or spiny, shoots arising from a fleshy starchy rhizome or a tuber derived from lower internodes &/or hypocotyl; raphides usu. present, also steroidal saponins & oft. lactone alks; vessel-elements usu. present throughout, the vascular bundles in 2 dissimilar whorls or 1 whorl of alternating types & no sec. thickening exc. in tuber. Lvs in spirals, rarely opp., usu. with distinct petiole (usu. with pulvinus at each end) & lamina, entire (oft. cordate) to less oft. lobed or compound, usu. with 3–13 curved-convergent main veins & oft. with nectaries or mucilaginous pits s.t. with nitrogen-fixing bacteria. Fls in axillary spikes to panicles, rarely bisex. (*Stenomeris*), reg., epigynous; P 3 + 3, usu. basally connate with short tube, nectaries commonly present, A 3 + 3, but inner s.t. staminodal or obs., filaments s.t. shortly connate, attached to P-tube base, anthers with longit. slits, $\overline{G}$ (3), 3-loc. with axile placentation & ± distinct styles & 2 – ∞ anatropous, bitegmic ovules per loc. Fr. a denticidal capsule, oft. 3-angled or -winged, rarely a berry or indehiscent & samaroid; seeds mostly winged with small embryo, subterm. plumule & a broad lat. cotyledon s.t. with a rudimentary 2nd one, embedded in copious endosperm with oil, protein & hemicellulose. x = 9(?), 10, 12, 14 +

Genera: *Dioscorea* (95% of the spp.), *Stenomeris, Trichopus*

Stenomeris & *Trichopus* have bisex. fls; *Trichopus* has a short stem with only 1 leaf & it, like '*Tamus*' (*D.* with berries), has been seg. into sep. fam. Taccaceae s.t. incl. here; Thismiaceae poss. referable here. *Dioscorea* is an imp. carbohydrate & steroid source

Dioscoreophyllum Engl. Menispermaceae (III). 3 trop. Afr. ***D. cumminsii*** (Stapf) Diels (W Afr.) – fr. source of a sweetener, monellin (protein) 9000 times as sweet as sucrose (poss. a sugar mimic enticing disp. agents) tried for low-calorie food & drinks

Diosma L. Rutaceae (I 4). 28 SW Cape. R: JSAB 48(1982)329. Heath-like, some cult. orn.

Diosphaera Buser = *Campanula*

Diospyros L. Ebenaceae (I). c. 550 trop. (Am. c. 100; Afr. 94, Madag. c. 100, Indopacific c. 250, Aus. 15). *Euclea* prob. referable here (unless *Royena* excl.). Ebonies (see also *Dalbergia*), some with elephant-disp. fr.; Massart's & Roux's Models. Timbers (favoured furn. wood in anc. Egypt), fr. trees, fish-poisons (bark, fr. – rich in tannin when unripe, also used to tan nets) used independently in Am. & OW, local medic., etc., ***D. ebenum*** Koenig ex Retz. (Ceylon ebony, Ind., Sri Lanka) being the ebony of commerce, though other spp. also used, e.g. ***D. abyssinica*** (Hiern) F. White (trop. & SAfr.) – for tool handles & shuttles for weaving sisal, ***D. affinis*** Thw. (Ind., Sri Lanka) – wood for carving in Sri Lanka, ***D. celebica*** Bakh. (Macassar ebony, Sulawesi), ***D. haplostylis*** Boivin (Madag. e., Madag.); ***D. montana*** Roxb. (Bombay ebony, Indomal.); other timbers incl. marblewood or zebrawood with streaked or marbled figure esp. from ***D. marmorata*** R. Parker (Andamans), ***D. oocarpa*** Thw. (Sri Lanka, Andamans) & calamander or Coromandel wood (***D. quaesita*** Thw., Sri Lanka) greyish-brown with black bands much used in Sheraton furniture. Those cult. for fr. (date-plums, velvet apples, Fuji or fuyo fr.) incl. ***D. blancoi*** A. DC. (*D. discolor, D. mabolo*, mabola, butterfruit, C Mal.), ***D. kaki*** L.f. (Jap. persimmon, Chinese date, kaki, E As.) – much grown in Jap. & China (where dried & also a sugar-source, juice from green fr. used to waterproof paper), derived by polyploidy from *D. roxburghii* Carr. & cvs selected for smooth-skinned fr. & frost-hardiness, crisp-fr. cvs used for packaged 'convenience' food in Spain, ***D. lotus*** L. (As.) – fr. eaten fresh, dried or bletted, with wild form ('var. *brideliifolia*') in Philipp., ***D. nigra*** (J.F. Gmel.) Perr. (*D. digyna*, '*D. ebenaster*', black sapote, C Am., natur. As.), ***D. virginiana*** L. (persimmon, SE US) – r.-suckers, fr. made into 'bread', beer & brandy, in Civil War seeds used as buttons & coffee subs., ink & syrup prep. from fr., antiseptic & astringent, wood useful (textile shuttles, gunstocks & golf-club heads in US). ***D. andamanica*** (Kurz) Bakh. (Mal.) – facultative myrmecophyte; ***D. decandra*** Lour. & ***D. peregrina*** (Gaertn.) Guerke (SE As.) – topiary & bonsai in Buddhist temples (Thailand); ***D. major*** (Forst.f.) Bakh. (Pac.) – ed. fr. in Tonga; ***D. melanoxylon*** Roxb. (tendu, Ind., Sri Lanka) – lvs used as cigarette-papers; ***D. malabarica*** (Desr.) Kostel. (*D. embryopteris*, gaub, trop. As.) – fr. sticky & used for caulking boats; ***D. mespiliformis*** Hochst. ex A. DC. (trop. Afr.) – medic., ed. fr., timber; ***D. mollis*** Griff. (makua or ma-plua, Thailand) – fr. a source of a black dye for silk; ***D. mweroensis*** F. White (C & SE Afr.) – fish-poison & anti-bilharzia agent; ***D. natalensis*** (Harv.) Brenan (trop. & S Afr.) – some rheophytic forms; ***D. oleifera*** Cheng (China) & other spp. – source of persimmon oil used for waterproofing

Diostea Miers. Verbenaceae (Neo.). 1 Arg., Chile: ***D. juncea*** (Hook.) Miers
Diotacanthus Benth. = *Phlogacanthus*
Diothonea Lindl. = *Epidendrum* (but see Orquideología 24(2005)44)
Diotocranus Bremek. = *Mitrasacmopsis*
Dipanax Seemann = *Tetraplasandra*
Dipcadi Medik. (~ *Ornithogalum*). Asparagaceae. 20 Medit. to S Afr. & Ind. R: T 58(2009)98
Dipelta Maxim. (~ *Linnaea*). Caprifoliaceae (Linnaeaceae). 3 China. R: FOC 19(2011)646. Cult. orn. like *Weigela* but A 4 & C 2-lipped, esp. ***D. floribunda*** Maxim. (W & C China)
Dipentodon Dunn. Dipentodontaceae. 1 NE Ind., Myanmar, S China: ***Dipentodon sinicus*** Dunn
Dipentodontaceae Merr. (~ Tapisciaceae). Magnoliidae – Huerteales. 2/18 NE Ind. to Aus., C Am. Small trees with long vessel-elements with scalariform endplates. Lvs in spirals or distich., simple, toothed; stip. small, decid. Fls in globose axillary, pedunculate umbels, at first subtended by 4 or 5 involucriform bracts, or thyrses, regular, hypogynous; pedicels jointed in middle, K 5–7, connate at base, valvate, C similar & alt. with K, valvate, A as many as & alt. with C, anthers with longit. slits, nectaries (? staminodes) opp. C, $\underline{G}$ (2, 3) with simple style & term. stigma, 1-loc., but with partial partitions basally, ovules 2 basal, erect or 6 on top of free-C placenta. Fr. a tardily dehiscent capsule with 1 seed
Genera: *Dipentodon, Perrottetia*
Form. *D.* incl. in Santalales, *P.* in Celastraceae
Diphalangium Schauer = ? *Milla*
Diphasia Pierre = *Vepris*
Diphasiastrum Holub. See *Lycopodium*
Diphasiopsis Mendonça = *Vepris*
Diphasium C. Presl ex Rothm. = *Lycopodium*
Diphelypaea Nicolson = *Phelypaea*
Dipholis A. DC. = *Sideroxylon*
Diphyes Blume = *Bulbophyllum*
Diphylax Hook.f. = *Platanthera*
Diphyllarium Gagnepain. Leguminosae (III 18). 1 SE As.: ***D. mekongense*** Gagnepain
Diphylleia Michaux. Berberidaceae (II 2). 1 E N Am. (***D. cymosa*** Michaux), 1 C & N Jap. (***D. grayi*** F. Schmidt), 1 W China (***D. sinensis*** H.L. Li – rhizomes locally medic.)
Diphysa Jacq. Leguminosae (III 11). 15 trop. Am. ***D. americana*** (Mill.) M. Sousa (*D. robinioides*) – fuelwood
Dipidax Salisb. = *Onixotis*
Diplachne P. Beauv. (~ *Leptochloa*). Gramineae (XXIX 5). 3 trop. & temp. ***D. fusca*** (L.) Roem. & Schultes (*L. f., L. malabarica*, OW trop. marshes) – salt-tolerant, exuding NaCl from leaf-glands
Diplacrum R. Br. (~ *Scleria*). Cyperaceae (III 4). 6 trop.
Diplacus Nutt. (~ *Mimulus*). Phrymaceae. c. 50 N Am. R: Phytoneuron 2012–39:27, 2013–46:1,65, -66:1. ***D. aurantiacus*** (Curtis) Jepson (Oregon, Calif.) – shrub; ***D. jepsonii*** (A.L. Grant) Nesom (*M. nanus* var. *j.*, Oregon, Calif.) – herb, s.t. merely cotyledons, 2 lvs & 1 fl.
Dipladenia A. DC. = *Mandevilla*
Diplandra Hook. & Arn. = *Lopezia*
Diplandrorchis S.C. Chen = *Neottia*
Diplarche Hook.f. & Thomson = *Rhododendron*
Diplarpea Triana. Melastomataceae. 1 Colombia: ***D. paleacea*** Triana
Diplarrhena Labill. ('*Diplarrena*'). Iridaceae (VII 1). 2 SE Aus. R: FA 46(1986)26. ***D. moraea*** Labill. – cult. orn., fibres used in Aboriginal basketry
Diplasia Rich. Cyperaceae (I 1). 1 Costa Rica to W Braz.: ***D. karatifolia*** Rich.
Diplaspis Hook.f. Umbelliferae (Azor.). 3 SE Aus. (Tasmania 2). R: Aus SB 11(1998)3
Diplatia Tieghem. Loranthaceae (5 3). 3 trop. Aus.
Diplaziopsidaceae X.C. Zhang & Christenh. (Dryopteridaceae s.l.). 2–3/c. 5. R: T 61(2012)521. Terr. or saxicolous. Lvs soft, fleshy; petiole with 2 vasc. bundles, lamina 1-pinnate, subglabrous. Sori elongate usu. along 1 side of vein, vein-endings thickened and prominent adaxially
Genera: *Diplaziopsis, Homalosorus; Hemidictyum* poss. here
Diplaziopsis C. Chr. (~ *Diplazium*). Diplaziopsidaceae (Woodsiaceae/Dryopteridaceae s.l.). 2–4 E As., Polynesia

Diplazium Sw. (~ *Athyrium*). Athyriaceae (Woodsiaceae/Dryopteridaceae s.l.). Excl. *Desmophlebium*, c. 400 trop. (Afr. few, Am. c. 150) & N temp. (Eur. 2); incl. *Allantodia*. Cult. orn. incl. ***D. esculentum*** (Retz.) Sw. (trop. & E As. to Polynesia, natur. S Afr., Florida) – young shoots a veg. in Sikkim, in curry in Sri Lanka etc.
Diplazoptilon Ling = *Saussurea*
Diplectria (Blume) Reichb. = *Dissochaeta* (but see Blumea 24(1978)405)
Diploblechnum Hayata (~ *Blechnum*). Blechnaceae. 2 Taiwan to Pacific Is.
Diplobryum C. Cusset = *Hydrobryum*
Diplocarex Hayata = *Carex*
Diplocaulobium (Reichb.f.) Kraenzlin = *Dendrobium*
Diplocentrum Lindl. Orchidaceae (V 16c). 2 Ind.
Diploclisia Miers. Menispermaceae (V). 2 Indomal., China, SE As.
Diplocyatha N.E. Br. = *Orbea*
Diplocyclos (Endl.) Post & Kuntze. Cucurbitaceae (XIV). 4 trop. Afr., 1 ext. to trop. As. (***D. palmatus*** (L.) C. Jeffrey, cult. orn.)
Diplodiscus Turcz. Malvaceae (Brown.; Tiliaceae). 7 Sri Lanka (1), W Mal. R: Reinw. 5(1960)255. ***D. paniculatus*** Turcz. (Philipp.) – Baroba nut, seeds ed.
Diplodium Sw. = *Pterostylis*
Diplofatsia Nakai = *Fatsia*
Diploglottis Hook.f. Sapindaceae (IV). 11 NE Aus., 1 ext. to NG: ***D. australis*** (G. Don) Radlk. (*D. cunninghamii*, native tamarind, Aus.) – fine grey timber, ed. arils used in jam. ***D. bracteata*** Leenh. (boonjee [tamarind] NE Aus.) – comm. fr. crop
Diplokeleba N.E. Br. Sapindaceae (III 1). 2 trop. S Am.
Diploknema Pierre. Sapotaceae (II). 10 Indomal. (Yunnan 1). ***D. butyracea*** (Roxb.) H.J. Lam (Ind. to SW China) – seeds yield fat used in soap etc.; ***D. sebifera*** Pierre (Borneo) – cooling oil (tengkawang) from seeds
Diplolabellum F. Maek. = *Oreorchis*
Diplolaena R. Br. Rutaceae (I 3). 15 [many intergrading] SW Aus. R: FA 26(2013)484
Diplolegnon Rusby = *Corytoplectus*
Diplolepis R. Br. Apocynaceae (Vc; Asclepiadaceae). Incl. *Grisebachiella*, 14 Chile, Arg. R: Darwiniana 50(2012)300
Diplolophium Turcz. Umbelliferae (III 8). 5–7 trop. Afr.
Diplomeris D. Don. Orchidaceae (IV 4d). 32 Himal. R: Gen. Orch. 2(2001)286
Diplomorpha Meissn. = *Wikstroemia*
Diploon Cronq. Sapotaceae (III). 1 trop. S Am.: ***D. cuspidatum*** (Hoehne) Cronq. R: T.D. Pennington, *S.* (1991)180
Diplopanax Hand.-Mazz. Nyssaceae (Cornaceae s.l.). 2 China, Vietnam. R: Novon 12(2002)435. Congeneric with Tertiary fossils referred to *Mastixia* (cf. *Metasequoia*)
Diplopeltis Endl. Sapindaceae (III 1). 5 NW Aus.
Diplopilosa Dvořák = *Hesperis*
Diplopogon R. Br. = *Amphipogon*
Diploprora Hook.f. Orchidaceae (V 16c). 2 trop. As. R: OB 95(1988)39
Diplopterygium (Diels) Nakai. Gleicheniaceae (Gleich.). 25 trop. & warm As. to Hawaii, 1 trop. Am. (***D. bancroftii*** (Hook.) A.R. Sm.). Thicket-formers but lvs bipinnatifid, not pseudodichotomous
Diplopterys A. Juss. (~ *Banisteriopsis*). Malpighiaceae. 31 trop. Am. R: HPB 11(2006)3. Close to *Banisteriopsis*. ***D. cabrerana*** (Cuatrec.) Gates yields an hallucinogenic drug (tryptamines) s.t. mixed with *B. caapi*; ***D. pubipetala*** (A. Juss.) W.R. Anderson & C. Davis (Braz.) – leaf-glands act as EFNs for ants
Diplora Bak. = *Asplenium*
Diplorhynchus Welw. ex Ficalho & Hiern. Apocynaceae (I e). 1 trop. & S Afr.: ***D. condylocarpon*** (Muell.-Arg.) Pichon – local medic. R: MLW 80–12(1980)28
Diplosoma Schwantes. Aizoaceae (V). 2 Cape, limestone. R: H. E. K. Hartmann, *Ill. Handb. Succ. Pls*, Aiz. A–E(2002)218
Diplospora DC. (~ *Tricalysia*). Rubiaceae (II 3). 9 Indomal. R: Blumea 35(1990)297 (Aus. spp. = *Tarenna*)
Diplostephium Kunth. Compositae (Ast.-Hin.). c. 110 trop. Andes (N Chile), 1 Costa Rica. R: Lundellia 14(2011)34
Diplostigma K. Schum. = *Vincetoxicum* (but see FTEA Apoc.2(2012)492)
Diplotaenia Boiss. (~ *Peucedanum*). Umbelliferae (III 10). 4 Turkey, Iran. R: Willd. 41(2011)69. ***D. cachrydifolia*** Boiss. – a source of jatamansi (cf. *Nardostachys*)

Diplotaxis DC. (*Dyplotaxis*, ~ *Brassica*). Cruciferae (12). 25 Eur. (11), Medit. to NW Ind. R: Pfl. 70(4,105)(1919)149. ***D. erucoides*** (L.)DC. (SW Eur.) – 'white rocket'; ***D. muralis*** (L.) DC. (S & C Eur.) – 'wall rocket', largely selfed, n = 21, *D. viminea*(L.) DC. (n = 10) × ***D. tenuifolia*** (L.) DC. (Eur., n = 11, sold as 'wild rocket' in Eur., bad weed (Lincoln weed) in Aus.)

Diplotropis Benth. Leguminosae (III 9). 12 Amazonia. R: AA 15(1985)62. Heterogeneous. Comm. timbers ('sucupira')

Diplusodon Pohl. Lythraceae. c. 75 Braz., Bolivia (1). R: Bradea 5(1989)205

Diplycosia Blume = *Gaultheria*

Dipodium R. Br. Orchidaceae (V 12b). 25 Mal. to New Caled. Some cult. orn. incl. ***D. squamatum*** (Forst. f.) Sm. (*D. punctatum*, SW Pac.)

Dipogon Liebm. Leguminosae (III 18). 1 S trop. Afr., cult. elsewhere (invasive in Aus.): ***D. lignosus*** (L.) Verdc. R: Strel. 10(2000)283

Dipoma Franch. Cruciferae. 1 SW China: ***D. iberideum*** Franch.

Diposis DC. Umbelliferae (Azor.). 3 temp. S Am.

Dipsacaceae Juss. = Caprifoliaceae

Dipsacus Tourn. ex L. Caprifoliaceae (Dipsacaceae). 20 Euras. (Eur. 8), Medit., trop. Afr. & Sri Lanka (1) mts. R: BN 129(1976)383. Lvs basally connate, forming rainwater-collecting troughs, which may prevent insects cli. up to fls. Rigid bracts act as fr. catapults when animals pass by. Roots medic., esp. ***D. asper*** Wall. ex DC. (Ind.). In ***D. sativus*** (L.) Honck. (cultigen ? derived from *D. ferox* Lois., Medit.), these are the effective part of the head (teasel) form. used in raising nap on cloth (bracts downward-pointing cf. wild pls); ***D. fullonum*** L. (*D. sylvestris*, Eur., invasive in US, as is ***D. laciniatus*** L., Eur.) much used in dried fl. arrangements like other spp. cult. orn. e.g. ***D. pilosus*** L. (shepherd's rod, Euras.)

Dipteracanthus Nees = *Ruellia*

Dipteranthemum F. Muell. = *Ptilotus*

Dipteranthus Barb. Rodr. = *Zygostates*

Dipteridaceae Seward & Dale. Polypodiidae– Gleicheniales. Incl. Cheiropleuriaceae, 2/11 Indomal. to Jap. & Samoa. R: T 55(2006)710. Terr. ferns with creeping protostelic or solenostelic stems covered with bristles or hairs. Petioles with 1 vasc. bundle proximally, polystelic distally, fronds with 2 flabellate halves & highly reticulate venation. Sori without indusia, scattered over surface (*Dipteris*), or fronds dimorphic & fertile ones covered in sporangia (*Cheiropleuria*). x = 33

Genera: *Cheiropleuria, Dipteris*

Fossils from upper Triassic, many, of cosmop. distrib., by Jurassic (*Dictyophyllum, Hausmannia*)

Dipteris Reinw. Dipteridaceae. 8 As. to Polynesia. ***D. lobbiana*** (Hook.) Moore (Mal.) – facultative rheophyte

Dipterocarpaceae Blume (~ Cistaceae). Magnoliidae – Malvales. 16/535 trop. esp. Mal. R (Mal.): FM I, 9(1982)237. Trees to 60m, usu. resinous (char. branching resin canals in Dipterocarpoideae) & usu. with buttresses & ectotrophic mycorrhizae. Lvs in spirals to distich., simple, leathery, conspic. sec. veins & scalariform tertiaries; stip. s.t. prot. bud, persistent or decid. Fls usu. bisex., usu. pointed in bud, reg., nodding, scented, in usu. axillary panicles, racemes or rarely cymes; K 5, s.t. valvate in fr. but usu. imbr., oft. with basal tube, persistent (when 2–5 usu. greatly enlarging into wing-like lobes in fr.), C 5, contorted, s.t. basally connate, A 5–110 in 1–3 whorls or ± irreg., initiated centrifugally, typically served by 10 trunk-bundles, ± hypogynous (on an androgynophore in Monotoideae), filaments distinct or ± basally connate, anthers basifixed (appearing versatile in Monotoideae), with longit. slits & connective usu. consp. prolonged, $\underline{G}$, rarely semi-inf. ((2)3(–5)), pluriloc. with axile placentation & term. style & 2(–4) anatropous, bitegmic ovules per locule. Fr. dry indehiscent, 1-seeded with woody pericarp; seeds without endosperm or dormancy, cotyledons oft. folded & encl. radicle. x = 7, 11

Classification & genera:

Monotoideae (Monotaceae; wood, G & oft. lvs without resin-ducts, anthers basiversatile, K equally accrescent in fr., papery, $\underline{G}$ 3(–5)-loc., each with 2 ovules, C longer than K, trop. Afr., Madag. (1 sp.), trop. Am. (1 sp.): *Marquesia* (*Trillesanthus*), *Monotes, Pseudomonotes*

[**Pakaraimaeoideae** (as Monotoideae but P shorter than K, $\underline{G}$ (5)-loc., each with 4 ovules, Guianas): *Pakaraimaea* = Cistaceae]

Dipterocarpoideae (wood, lvs & G with resin-ducts, anthers basifixed, 2 or 3 K becoming accrescent in fr., $\underline{G}$ 2–3-loc., each with 2 ovules, trop. As., Mal.; Tertiary of Mt Elgon, E Afr.): 2 tribes – Dipterocarpeae (fr. K valvate, x = 11; *Anisoptera, Cotylelobium,*

Dipterocarpus, Stemonoporus, Upuna, Vateria. Vateriopsis, Vatica), Shoreae (fr. K valvate, x = 11; *Dryobalanops, Hopea, Neobalanocarpus, Parashorea, Shorea*)

Poss. incl. Sarcolaenaceae. The first 2 have been referred to Tiliaceae & the 1st to a sep. fam. by some authors. Seed & floral vasculature studies corr. suggested affinity with Malvales & esp. Sarcolaenaceae. In Miocene deposits in Borneo. They are richest in Mal. (10/386 with 267 spp. in Borneo alone) & char. show gregarious flowering poss. initiated by periods of high irradiation though reserves must also be built up beforehand. Fruiting assoc. with pig (*Sus barbatus*) migrations; pigs mate when trees fl. & gestation = time to fr. ripening). They are the char. trees of Indomal., being the principal emergents & the basis of the export timber-trade, *Shorea* being the most imp. genus of all in Mal. Much is used as plywood. Resins are also used in varnishes & *Shorea* frs yield a comm. ed. fat

Dipterocarpus Gaertn. f. Dipterocarpaceae (Dipt.-Dipt.). c. 70 Indomal. (Borneo 41; Tertiary fossils in Afr.). Massart's & Roux's Models; large stip. prot. buds. Light timber, which absorbs preservative, much used for sleepers & heavy constr. (keruing (Mal.) or gurjun (Ind.), apitong or 'bagac' (Philipp.) or yang (Burma)) e.g. ***D. costatus*** Gaertn. f. (dau, Andamans to Negri Sembilan), ***D. grandiflorus*** (Blanco) Blanco (bagac, Andamans to Philipp.), ***D. tuberculatus*** Roxb. (eng, SE As.), ***D. zeylanicus*** Thw. (hora, Ceylon gurjun, Sri Lanka). Oleoresins (gurjun oil or balsam) form. much exported from Ind. for slow-drying varnishes. ***D. oblongifolius*** Blume (neram, S Thailand, Malay Pen., Borneo) – char. rheophyte of fast-flowing rivers ('neram' rivers)

Dipterocome Fischer & C. Meyer. Compositae (Card.). 1 E Med. to Afghanistan: ***D. pusilla*** Fischer & C. Meyer. R: CN 45(2007)27

Dipterocypsela S.F. Blake. Compositae (Vern.-Dip.). 1 Colombia: ***D. succulenta*** S.F. Blake

Dipterodendron Radlk. = *Dilodendron*

Dipteronia Oliv. = *Acer*

Dipteropeltis Hallier f. Convolvulaceae (?6). 3 trop. W Afr. R: BJBB 64(1995)183

Dipterostele Schltr. = *Telipogon*

Dipterygium Decne. ?Cleomaceae. 1 Egypt to Pakistan: ***D. glaucum*** Decne – methylglucosinolates present unlike Cruciferae, fr. a samara

Dipteryx Schreb. Leguminosae (III 3). Excl. *Taralea*, 9 trop. Am. ***D. odorata*** (Aubl.) Forsyth f. (Tonka bean) – Attims's Model, fls every 5 yrs, fragrant seeds cured in rum used for scenting tobacco & snuff; ***D. oleifera*** Benth. (ebor or eboe) – similar seeds

Diptychandra Tul. Leguminosae (I 4). 1 Braz., Bolivia, Paraguay: ***D. aurantiaca*** Tul.

Diptychocarpus Trautv. Cruciferae (17). 1 Eur. to C As.: ***D. strictus*** (M. Bieb.) Trautv.

Dipyrena Hook. Verbenaceae (Priv.). 1 Arg.: ***D. glaberrima*** (Hook.) Hook. – pyrenes 2, 2-loc.

Dirachma Schweinf. ex Balf.f. Dirachmaceae (Geraniaceae s.l.). 1 Socotra: ***D. socotrana*** Schweinf. ex Balf.f., a tree reduced to 30 individuals by 1967; 1 C Somalia. R: BJBB 61(1991)10.

Dirachmaceae Hutch. (~ Geraniaceae). Magnoliidae – Rosales. 1/2 NE Afr. Trees **and** shrubs. Lvs in spirals, small, toothed; stip. subulate, marcescent. Fls solit., 5–8-merous; epicalyx of 4–8 lobes; C contorted with basal nectaries (s.t. on subbasal appendages), white; A basally adnate to C, anthers extrorse, dehiscing by longit. slits; $\underline{G}$(5–8), lobed with 1 bitegmic anatropous-apotropous ovule per loc., opp. K. Fr. beaked, segments opening adaxially. Seeds laterally flattened, exotestal, endosperm scant

Genus: *Dirachma*

Exotestal seed indicated exclusion from Geraniaceae; wood anatomy & poss. aril resemble Rhamnaceae

Dirca L. Thymelaeaceae (Thym.-Daph.). 4 N Am. JBRIT 3(2009)494. Leatherwood. K almost 0, C 0, A 8, G 1-loc. Flexible shoots used for baskets, bark for rope, imp. local medic., cult. orn. esp. ***D. palustris*** L. (moosewood)

Dirhamphis Krapov. Malvaceae (Malv.-Malv.). 2 W Mex. (1), Bolivia, Paraguay

Dirichletia Klotzsch = *Carphalea*

dirty Dora *Cyperus difformis*

Disa P. Bergius. Orchidaceae (IV 4c). Incl. *Forficaria, Monadenia, Schizodium*, 181 trop. & S Afr. (esp. Cape 98 (84 endemic; R: Strel. 9(2000)159), Arabia (1), Madag., Réunion. R (p.p.): CBH 9(1981)1. Terrestrial, cult. orn. incl. hybrids, oft. with conspic. fls. Wide range of poll. types (esp. Cape): butterfly fls & carpenter-bee fls both evolved twice, long-spurred long-tongued fly fls 4 times, night-scented moth fls 3 times & self-poll. 3 times. In S Afr. ***D. uniflora*** P. Bergius – fl. 10 cm across with nectar in spur, poll. large mt. pride butterfly (*Aeropetes tulbaghia*, visits only red or orange fls, also poll. ***D. ferruginea*** Sw. (Batesian mimic of *Kniphofia uvaria* etc.)), though ***D. filicornis*** (L.f.) Thunb. nectarless & poll. mason bees); ***D. chrysostachya*** Sw. & ***D. satyriopsis*** Kraenzlin (S Afr.) – pollinia disp. on feet of sunbirds; ***D. sankeyi*** Rolfe (S Afr.) – poll. spiderhunting wasps (*Hemipepsis* spp.) attracted by sweet, spicy fragrance. Tubers a food-source (chikanda) esp. ***D. walleri*** Reichb.f (trop. & S Afr.).

Disaccanthus Greene = *Streptanthus*

Disakisperma Steud. (~ *Leptochloa*). Gramineae (XXIX 5). 4 Afr., Am. R: PhytoKeys 26(2013)29

Disanthus Maxim. Hamamelidaceae (II). 1 E China, S Jap.: ***D. cercidifolius*** Maxim. – cult. orn. (autumn colour) with small paired fls, C with 2 basal nectaries & 5 small staminodes. R: Cathaya 3(1991)1

Disaster Gilli = *Commersonia*

Discalyxia Markgraf = *Alyxia*

Discaria Hook. Rhamnaceae (2). 6 S Am. (3), Aus. (2), NZ (1). R: BSAB 22(1983)301. Spiny shrubs with *Frankia* symbionts, allied to *Colletia*, those with explosive fr. expelling seeds to 2.4 m, those without have floating fr. Some cult. orn. ***D. pubescens*** (Brongn.) Druce (anchor plant, E Aus.) – thickets of spiny shoots. see also *Ochetophila*

Dischidanthus Tsiang = *Marsdenia*

Dischidia R. Br. Apocynaceae (Va; Asclepiadaceae III 4). 80 Indomal. to Aus. & W Pac. (New Caled. 1). Epiphytes with adventitious r. & with fleshy wax-covered lvs.; some local medic. ***D. major*** (Vahl) Merr. (*D. rafflesiana* Wall., Ind. to Aus.) also has pitcher-shaped lvs (stomata inside, those outside prob. closed). c. 10 cm deep, into each of which an adventitious r. grows from stem or petiole nr. it; detritus, largely coll. by ants, accumulates in pitchers, as does rainwater thus made available to r.

Dischidiopsis Schltr. = *Dischidia*

Dischisma Choisy. Scrophulariaceae (Manul.; Globulariaceae). 11 SW Cape to Namibia. R: MBSM 15(1979)94

Dischistocalyx T. Anderson ex Benth. ('*Distichocalyx*'). Acanthaceae (III 2b). 15 trop. Afr. Terrestrial, later becoming climbers & epiphytes

Disciphania Eichler. Menispermaceae (III). c. 25 trop. Am. R: MNYBG 20(1970)124

Discipiper Trel. & Stehlé = *Piper*

Discocactus Pfeiffer. Cactaceae (III 4). 10 E S Am. R: CSJ 43(1981)37 (up to 36 spp. recog. by some growers). Hawkmoth–poll. (cf. *Gymnocalycium*); fire-tolerant (cerrado)

Discocalyx (A.DC.) Mez (~ *Tapeinosperma*). Primulaceae (Myrsinaceae). 50 C Mal. (Philipp. 31) to Polynesia. ***D. dissectus*** Kaneh. & Hatusima (NG) subherbaceous with much-divided chamomile-like lvs

Discocapnos Cham. & Schldl. (~ *Fumaria*). Papaveraceae (Fumariaceae I 2). 1 SW & S Cape: ***D. mundtii*** Cham. & Schldl. R: Strel. 10(2000)308

Discocarpus Klotzsch. Phyllanthaceae (Brid.-Sav.; Euphorbiaceae s.l.). 3 NE S Am. R: AMBG 83(1996)154

Discoclaoxylon (Muell. Arg.) Pax & K. Hoffm. (~ *Claoxylon*). Euphorbiaceae (Acal.-Acal.-Claox.). 4 trop. Afr. (3 restricted to Gulf of Guinea is.)

Discocleidion (Muell. Arg.) Pax & K. Hoffm. Euphorbiaceae (Acal.-Ber.). 2 C China, 1 ext. to Ryukyus

Discocnide Chew. Urticaceae (I). 1 Mex., Guatemala: ***D. mexicanus*** (Liebm.) Chew

Discocrania (Harms) Král = *Cornus*

Discoglypremna Prain. Euphorbiaceae (Acal.-Car.). 1 trop. Afr.: ***D. caloneura*** (Pax) Prain. R: BJBB 64(1995)201

Discolobium Benth. Leguminosae (III 11). 8 S Am.

Discophora Miers. Stemonuraceae (Icacinaceae s.l.). 3 Costa Rica to Bolivia. R: CGH 142(1942)21

Discopleura DC. = *Ptilimnium*

Discopodium Hochst. Solanaceae (IV 5a). 2 trop. Afr. mts

Discospermum Dalzell (~ *Diplospora*). Rubiaceae (II 3). 7 Indomal. R: Blumea 35(1990) 300

Discovium Raf. = *Lepidium*

Discretitheca Cantino (~ *Rotheca*). Labiatae (III). 1 Nepal: ***D. nepalensis*** (Mold.) Cantino

Discyphus Schltr. Orchidaceae (IV 2g). 1 trop. Am.: ***D. scopulariae*** (Reichb.f.) Schltr. R: Gen. Orch. 3(2003)199

Diselma Hook.f. Cupressaceae. 1 Tasmania: ***D. archeri*** Hook.f. – foliage resembling *Microcachrys tetragona*. R: A. Farjon, *Monogr. Cupressaceae & Sciadopitys* (2005)460

Disepalum Hook.f. Annonaceae (III 5). Incl. *Enicosanthellum*, 9 SE As. to W Mal. R: Britt. 41(1989)356. 2-merous

Disisorhipsalis Doweld = *Pseudorhipsalis*

Diskyphogyne Szlach. & R. González = *Brachystele*

Disocactus Lindl. Cactaceae (III 2). Incl. *Aporocactus*, *Heliocereus* (× *Heliochia*, *Nopalxochia*) 11 C Am. R: Bradleya 9(1991)86. Epiphytes, cult. orn. allied to *Epiphyllum*, esp. ***D.* × *hybridus*** (Geel) Barthlott (× *Heliochia* 'Ackermannii', '*N. ackermannii*'; *D. speciosus* (Cav.) Barthlott (*Heliocereus s.*, C Am.) × *D. phyllanthoides* (Haw.) Barthlott (*N. p.*, S Mex.)) – large red to pink fls, most of the 'epiphyllums' of comm. (R: S.E. Haselton (1946) *E. handbook*) & ***D. flagelliformis*** (L.) Barthlott (*A. f.*, rat's-tail, Mex.) – crimson fls; *D. speciosus* × *Selenicereus grandiflorus* = × ***Disoselenicereus maynardii*** (Paxton) E. Meier – synth. early C19

Disparago Gaertn. Compositae (Gnap.-Rel.). 9 S Afr. R: Bothalia 23(1993)197

Disperis Sw. Orchidaceae (IV 4a). 78 trop. & S Afr., Madag. & Masc. (22; R: Adansonia 24(2002)56, 27(2005)167), Indomal. to NG. Only OW orchids with oil-secreting fls (at least 17 spp. in S Afr.) visited by poll. *Rediviva* bees (cf. *Diascia*). Some cult. orn. epiphytes & terr. spp.

Disphyma N.E. Br. Aizoaceae (V). 5 Cape (2), Aus. & NZ (3). R: R. E. K. Hartmann, *Ill. Handb. Succ. Pls*, Aiz. A–E(2002)220. Cult. orn. incl. ***D. crassifolium*** (L.) L. Bolus (pigface, SW Cape), crossed with *Glottiphyllum longum* N.E. Br. to give × ***Disphyllum* 'Sunburn'**

Disporopsis Hance. Asparagaceae (Convallariaceae). 7 trop. E As., Philipp.

Disporum Salisb. ex D. Don. Colchicaceae. Excl. *Prosartes*, 5 E As., Indomal. Form. referred to Convallariaceae (= Asparagaceae)

diss or **dis grass** *Ampelodesmos mauritanicus*

Dissanthelium Trin. = *Poa* (but see Phytol. 11(1965)361)

Dissiliaria F. Muell. ex Baill. Picrodendraceae (Cal.-Diss.; Euphorbiaceae s.l.). 6 E Queensland. R: Austrobaileya 5(1997)10

Dissocarpus F. Muell. Amaranthaceae (Chenopodiaceae I 5). 4 Aus. R: FA 4(1984)226

Dissochaeta Blume. Melastomataceae. Incl. *Deplectria*, *Macrolenes*, c. 60 Indomal. Lvs an antidote to *Antiaris toxicaria* dart-poison. ***D. muscosa*** (Blume) G. Kadereit (*M. m.*, Mal.) – liane with ed. fr. & shoots, local med.

Dissochondrus (Hillebrand) Kuntze. Gramineae (XXIV 4). 1 Hawaii: ***D. biflorus*** (Hillebr.) Kuntze

Dissomeria Hook.f. ex Benth. Salicaceae (Flacourtiaceae). 2 trop. Afr.

Dissothrix A. Gray. Compositae (Eup.-Alom.). 1 Braz.: ***D. imbricata*** (Gardner) Robinson. R: MSBMBG 22(1987)251. Coll. once

Dissotis Benth. Melastomataceae. c. 120 trop. & S Afr. Some cult. orn. shrubs & herbs

Disteganthus Lem. Bromeliaceae (3). 3 NE S Am. R: FN 14(1979)1764

Distemonanthus Benth. Leguminosae (I 3). 1 trop. W Afr.: ***D. benthamianus*** Baill. (anyaran, ayan, bongassi, bonsamdua, Nigerian satinwood) – comm. timber

Distephanus Cass. (~ *Vernonia*). Compositae (Vern.-Dis.). c. 50 SE Afr. (2), Madag., Mauritius (1)

Disterigma (Klotzsch) Niedenzu. Ericaceae (VIII 5). 35 trop. Andes. Some ed. fr. reaching local markets

Distichella Tieghem = *Dendrophthora*

Distichia Nees & Meyen (~ *Juncus*). Juncaceae. 3 Andes. R: FOW 6(2002)15

Distichirhops Haegens = seq.

Distichirrhops Haegens (~ *Aporusa*). Phyllanthaceae (Scep.; Euphorbiaceae s.l.). 3 Borneo, NG. R: Blumea suppl. 12 (2000)193

Distichlis Raf. Gramineae (XXIX 1). Incl. *Monanthochloe*, *Reederochloa*, 9 trop. Am., 1 Aus. R: RAA 22(1955)86. Alkali grass, often in saline habitats. ***D. distichophylla*** (Lab.) Fassett (Aus.) – fodder & saline indicator; ***D. eludens*** (Soderstrom & H. Decker) H. Bell & Columbus (*R. e.*, Mex.) – dioec.; ***D. spicata*** (L.) Greene (N Am.) – for binding sandy soil

Distichocalyx Benth. = *Dischistocalyx*

Distichochlamys M. Newman (~ *Scaphochlamys*). Zingiberaceae (II 1). 3 Vietnam. R: Britt. 55(2003)207

Distichorchis M.A. Clem. & D.L. Jones = *Dendobium*

Distichoselinum García Martín & Silvestre = *Thapsia* (but see Lagascalia 13(1985)232)

Distichostemon F. Muell. = *Dodonaea* (but see Austrobaileya 2(1984)57)

Distictella Kuntze (~ *Amphilophium*). Bignoniaceae. 18 trop. Am. R: AMBG 96(2009)294

Distictis Mart. = *Amphilophium*

Distoecha Philippi = *Hypochaeris*

Distomocarpus O. Schulz = *Rytidocarpus*

Distrianthes Danser. Loranthaceae. 2 NG. R: FM I,13(1997)324

Distyliopsis Endress. Hamamelidaceae (I 4). 7 Myanmar to Mal., SE As. & Taiwan

Distylium Sieb. & Zucc. Hamamelidaceae (I 4). 10 Indomal. P0; Am. spp. with K = *Molinadendron*. Domatia; some rheophytes & cult. orn. esp. ***D. racemosum*** Sieb. & Zucc. (Jap.) – fine-grained wood used for furniture & art

Distylodon Summerh. Orchidaceae (V 16d). 2 trop. Afr. R: PhytoKeys 36(2014)32

Disynaphia Hook. & Arn. ex DC. (~ *Eupatorium*). Compositae (Eup.-Dis.). 16 S Am. R: MSBMBG 22(1987)77

Disynstemon R. Viguier. Leguminosae (III 16). 1 SW Madag.: ***D. paullinioides*** (Bak.) Peltier

dita bark *Alstonia scholaris*

Ditassa R. Br. Apocynaceae (V c; Asclepiadaceae). Excl. *Minaria*, c. 100 S Am. (Braz. c. 50)

Ditaxis Vahl ex A. Juss. (~ *Argythamnia*). Euphorbiaceae (Acal.-Chro.-Dit.). c. 45 warm Am.

Ditepalanthus Fagerl. Balanophoraceae. 1 Madag.: ***D. malagasicus*** (Jum. & Perrier) Fagerl.

Dithrix (Hook.f.) Brummitt = *Habenaria*

Dithyrea Harv. Cruciferae (40). 2 SW N Am. R: FNA 7(2010)607

Dithyridanthus Garay = *Deiregyne*

Dithyrostegia A. Gray (~ *Angianthus*). Compositae (Gnap.-Ang.). 2 SW Aus. R: Muelleria 7(1989)106, OB 104(1991)114

Ditrichospermum Bremek. = *Strobilanthes*

Ditrysinia Raf. (~ *Sebastiania*). Euphorbiaceae (Hipp.). 1 E US: ***D. fruticosa*** (Bartram) Govaerts & Frodin (*D. ligustrina*). R: A. Radcliffe-Smith, *Gen. Euphorb.* (2001) 385

Ditta Griseb. Euphorbiaceae (Aden.-Aden.). 1–2 WI

dittander *Lepidium latifolium*; (in old Herbals = dittany)

dittany *Origanum dictamnus*; *Dictamnus albus*; **American d.** *Cunila origanoides*; **false d.** *Ballota acetabulosa*

Dittoceras Hook.f. = *Heterostemma*

Dittostigma Philippi = *Nicotiana*

Dittrichia Greuter (~ *Pulicaria*). Compositae (Inul.-Inul.). 5 Med. incl. Eur. introd. S Am. R: PAB 19(2000)342. ***D. viscosa*** (L.) Greuter – anti-inflammatory used (modern Crete) for bonesetting since Dioscorides, anthelminthic

Diuranthera Hemsl. = *Chlorophytum*

Diuris Sm. Orchidaceae (IV 3d). 1 Timor, c. 70 Aus. Bee-poll. (prob. most by deception), many wild hybrids; some cult. orn. (donkey orchids). ***D. maculata*** Sm. (SE Aus.) – fls mimic those of *Daviesia* spp. & *Pultenaea scabra* R. Br. (Leguminosae).

divi-divi *Libidibia coriaria*

Diyaminauclea Ridsd. Rubiaceae (I 2). 1 Sri Lanka: ***D. zeylanica*** (Hook.f.) Ridsd. R: Blumea 24(1978)345

Dizygostemon (Benth.) Wettst. Plantaginaceae (Grat.-Stem.; Scrophulariaceae s.l.). 2 Braz.

Dizygotheca N. E. Br. = *Plerandra*

Djaloniella P. Taylor. Gentianaceae. 1 trop. W Afr.: ***D. ypsilostyla*** P. Taylor. R: T 12(1963)294

djave *Baillonella toxisperma*

Djinga Cusset. Podostemaceae (III). 2 Cameroun

Dobera Juss. Salvadoraceae. 2 trop. E Afr., S Arabia to NW Ind.

Dobinea Buch.-Ham. ex D. Don. Anacardiaceae (I; Podoaceae). 2 E Himal., S China. R: FOC 11(2008)357. Female fls epiphyllous

dock *Rumex* spp.; **patience d.** or **spinach d.** *R. patientia*; **tanner's d.** *R. hymenosepalus*; **water d.** *R. hydrolapathum*; **yellow d.** *R. crispus*

docko *Artemisia vulgaris*

Dockrillia Brieger = *Dendrobium*

Docynia Decne (~*Cydonia*). Rosaceae (15). 1 E Himal. to N Thailand & S China: ***D. indica*** (Wall.) Decne – cult. orn. tree, ed. fr. s.t. used in China to speed bletting of persimmons. R: CJB 68(1990)2233

Docyniopsis (C.K. Schneider) Koidz. = *Macromeles*

Dodartia Tourn. ex L. Mazaceae (Phrymaceae-Mim.; Scrophulariaceae s.l.). 1 S Russia, W As.: ***D. orientalis*** L.

dodder *Cuscuta* spp.; **d.-laurel** *Cassytha* spp.

doddering-dillies *Briza media*

Dodecadenia Nees (~ *Litsea*). Lauraceae (II). 1 S Himal.: ***D. grandiflora*** Nees

Dodecahema Rev. & Hardham (~ *Centrostegia*). Polygonaceae (I 1). 1 Calif.: ***D. leptoceras*** (A. Gray) Rev. & Hardham. R: Phytol. 66(1989)86

Dodecastigma Ducke. Euphorbiaceae (Cod.-Ost.). 2 NE S Am.

Dodecatheon L. = *Primula* (but see FNA 9(2009)268)

dodo cloth *Tabernaemontana pachysiphon*

Dodonaea Mill. Sapindaceae (III 1). Incl. *Distichostemon* (*q.v.*), 74 trop. & warm esp. Aus. (67, 65 endemic; R: Brunonia 7(1984)1). Wind-poll. usu. viscid shrubs & trees, some cult. orn. & medic. or fodders. ***D. viscosa*** Jacq. (*D. angustifolia*, akeake (NZ), hop-bush, trop. & warm) – molluscicidal saponins, wood (good firewood) for tool handles, cult. orn. (fr. & lvs used in lei in Hawaii), winged fr. source of red dye (Hawaii) & allegedly used as hops by early settlers in Aus.

Dodonaeaceae Kunth ex Small = Sapindaceae

Dodsonia Ackerman = *Stenia*

Doellia Schultz-Bip. ex Walp. (~ *Blumea*). Compositae (Inul.-Pluch.). 2+ Arabia, Afr. R: Willd. 25(1995)21

Doellingeria Nees (~ *Aster*). Compositae (Ast.-Sym.). 3 N Am. R: Phytol. 77(1994)252. As. spp. = *Aster*

Doerpfeldia Urb. Rhamnaceae (10). 1 Cuba: ***D. cubensis*** Urb.

dog apple *Asimina reticulata*; **d.bane** *Apocynum* spp.; **d. daisy** *Leucanthemum vulgare*; **d.'s mercury** *Mercurialis perennis*; **d. nettle** *Urtica urens*; **d. plum** *Ekebergia capensis*; **d. rose** *Rosa canina*; **d. senna** *Senna italica*; **d.'s tail grass, crested** *Cynosurus cristatus*; **d.'s tooth violet** *Erythronium dens-canis*; **d. violet** *Viola canina*; **d.wood** *Cornus* spp.; **d.wood, swamp** *C. sanguinea*

dokudami *Houttuynia cordata*

Dolianthus C.H. Wright (~ *Amaracarpus*). Rubiaceae (IV 7). 13 Papua NG highlands. R: Blumea 46(2001)421

Dolichandra Cham. Bignoniaceae (3). Incl. *Macfadyena*, 8 trop. Am. R: AMBG 98(2014)428. Cult. orn. lianes esp. ***D. cynanchoides*** Cham. (S Braz., Paraguay, Urug., Arg.) – red fls; ***D. unguis-cati*** (L.) L. Lohmann (*Bignonia u.*, *Doxantha u.*, *M. u.*, cat's claw creeper, Mex. to Arg., invasive in S Afr., Aus., SE US) – 2n = 40, 80 (only polyploidy in *D.*), mass-flowering in dry but not wet regions, yellow fls, fr. to 30 cm long

Dolichandrone (Fenzl) Seem. Bignoniaceae (1). 10 E Afr. (1), Indomal. (5), trop. Aus. (3), Indopacific (1). Trees with white fls opening at night, v. fragrant, bat-poll. (cf. allied bird-poll. *Markhamia*); cult. orn. incl. ***D. spathacea*** (L.f.) K. Schum. (As. to New Caled.) – mangrove with sea-disp. seeds with corky wings

Dolichlasium Lag. (~ *Trixis*). Compositae (Mut.-Nass.). 1 Arg. Andes: ***D. lagascae*** Gillies ex D. Don

Dolichocentrum (Schltr.) Brieger = *Dendrobium*

Dolichochaete (C. Hubb.) J. Phipps = *Tristachya*

Dolichodelphys K. Schum. & K. Krause. Rubiaceae (Cond.). 1 Colombia, Ecuador, Peru: ***D. chlorocrater*** K. Schum. & K. Krause

Dolichoglottis R. Nordenstam = *Brachyglottis*

Dolichokentia Becc. = *Cyphokentia*

Dolicholobium A. Gray. Rubiaceae (Cond.). 28 Philipp. to Fiji. R: Blumea 29(1983)251. Some rheophytes incl. ***D. rheophilum*** M. Jansen (NG)

Dolicholoma D. Fang & W.T. Wang = *Petrocodon*

Dolichometra K. Schum. Rubiaceae (IV 1). 1 Tanzania (E Usambaras): ***D. leucantha*** K. Schum.

Dolichopetalum Tsiang. Apocynaceae (V a; Asclepiadaceae). 1 China: ***D. kwangsiense*** Tsiang

Dolichopsis Hassler. Leguminosae (III 18). 2 Paraguay & Arg. (chaco)

Dolichopterys Kosterm. = *Lophopterys*

Dolichorhynchus Hedge & Kit Tan = *Douepea*

Dolichorrhiza (Pojark.) Galushko (~ *Senecio*). Compositae (Sen.-Sen.). 4 Cauc., Iran

Dolichos L. Leguminosae (III 18). 60 OW trop. Usu. climbers. Some cult. orn. & fodder; others with ed. pods & seeds. ***D. kilimandscharicus*** Taub. (SC & E Afr.) – molluscicidal saponins. See also *Lablab*

Dolichostachys Benoist. Acanthaceae (III). 1 Madag.: ***D. elongata*** Benoist

Dolichostegia Schltr. = *Dischidia*

Dolichostemon Bonati. Labiatae (?). 1 SE As.: ***D. verticillatus*** Bonati

Dolichothele Britton & Rose = *Mammillaria*

Dolichothrix Hillard & B.L. Burtt. Compositae (Gnap.-Rel.). 1 Cape: ***D. ericoides*** (Lam.) Hilliard & B.L. Burtt. R: OB 104 (1991)71

Dolichoura Brade. Melastomataceae. 2 Braz. R: Britt. 599(2007)226

Dolichovigna Hayata = *Vigna*

Doliocarpus Rolander. Dilleniaceae (III). c. 45 trop. Am. R: MBSM 9(1971)29. Some lianes with potable water if cut

Dollinera Endl. = *Ototropis*

Dolomiaea DC. Compositae (Card.-Card.). Incl. *Vladimiria*, 14 Tibet, Himal.

Dolophragma Fenzl (~ *Arenaria*). Caryophyllaceae. 4–5 Himal.

domba oil *Calophyllum inophyllum*

Dombeya Cav. Malvaceae (Dom.; Sterculiaceae). Excl. *Andringitra*, c. 210 Afr. (19), Madag. (c. 175) to Mascarenes (15). Small trees & shrubs with coloured nectar, some cult. orn. incl. *D.* × ***cayeuxii*** André (*D. burgessiae* Gerrard ex Harv. (S & E Afr.) × *D. wallichii* (Lindl.) Baill. (Madag.)); other spp. locally imp. as fibre-sources, Madag. sp. for paper (2002)

Domeykoa Philippi. Umbelliferae (Azor.). 5 Peru, Chile. R: UCPB 33(1962)173

Dominella F. Wimmer = *Lysipomia*

Domingoa Schltr. Orchidaceae (V 13c). Incl. *Nageliella*, 4 Cuba, Hispaniola. R: C.L. Withner, *The Cattleyas & their relatives* IV(1996)31. Cult. orn. epiphytes esp. ***D. purpurea*** (Lindl.) van den Berg & Soto (*N. p.*)

Dominia Fedde = *Trachymene*

Domkeocarpa Markgraf = *Tabernaemontana*

Domohinea Leandri = *Tannodia*

Donatia Forst. & Forst. f. Stylidiaceae (Donatiaceae). 2 Tasmania, NZ, subAntarctic S Am. Dominate Fuegian bogs

Donatiaceae B. Chandler = Stylidiaceae

Donax Lour. Marantaceae. 1 Indomal. to Vanuatu: ***D. canniformis*** (Forst.f.) K. Schum. – split stems used for basketwork & fish-traps etc. locally, rhiz. ed., local medic. R: T 54(2005) 1083

Donella Pierre ex Baill. = *Chrysophyllum*

Donepea Airy Shaw = *Douepea*

dong (nut) *Santalum acuminatum*; **d. quai** *Angelica sinensis*

Doniophyton Wedd. (~ *Chuquiraga*). Compositae (Barn.). 2 Arg., Chile. R: PSE 206(1997)36. Annuals

donkey orchid *Diuris* spp.

Donnellsmithia J. Coulter & Rose. Umbelliferae (III 5). 15–20 Mex., C Am.

Dontostemon Andrz. ex C. Meyer. Cruciferae (24). 11 China & adj. Russia & Mongolia. R: Novon 10(2000)96,334, 17(2007)172

do(ugh)nut peach *Prunus persica* 'Saturn'

doob grass *Cynodon dactylon*

Doodia R. Br. = *Blechnum*

doon *Shorea* spp.

Doona Thwaites = *Shorea*

Dopatrium Buch.-Ham. ex Benth. Plantaginaceae (Grat.-Grat.; Scrophulariaceae). 14 trop. As. to Aus., ***D. junceum*** (Roxb.) Benth. to S Russia, a ricefield weed natur. in N Am. R: NJB 17(1997)534. Most confined to rock-pools

Dora, dirty *Cyperus diffusus*

Doratoxylon Thouars ex Hook.f. Sapindaceae (III 2). 6 Madag. to Mascarenes

Dorcoceras Bunge (~ *Boea*). Gesneriaceae (Loxoc.). 4 SE As. to Philippines. R: T 65(2016)285

Dorema D. Don = *Ferula*

Doricera Verdc. = *Ixora*

Doritis Lindl. = *Phalaenopsis*

Dorobaea Cass. (~ *Senecio*). Compositae (Sen.-Sen.). 3 Andes. R: CN 27(1995)31

Doronicum Tourn. ex L. Compositae (Sen.-Tuss.). c. 40 temp. Euras. (Eur. 12), Medit. R: AMBG 90(2003)335. Some medic. & cult. orn. (leopard's bane), the common early-flowering pls being hybrids (nat. homoploid hybrids in Italy): ***D. × excelsum*** (N.E. Br.) Stace (*D. pardalianches* L. (incl. *D. cordatum*, W Eur.) × *D. plantagineum* L. (W Eur.) × *D. columnae* Ten. ('*D. cordatum*', SE Eur.)) & ***D.*** × ***willdenowii*** (Rouy) A.W. Hill (*D.* × *excelsum* × *D. austriacum* Jacq. (Eur.)) with less cordate basal lvs

doronoki *Populus maximowiczii*

Dorothea Wernham = *Aulacocalyx*

Dorotheanthus Schwantes = *Cleretum*

Dorstenia Plum. ex L. Moraceae (IV). 105 trop. (As. 1, Afr. c. 60, Am. c. 45). R: Ilicifolia 2(1999)5. Small trees or shrubs (***D. gigas*** Schweinf. ex Balf.f. (Socotra) a pachycaul treelet to 4 m with trunk to 1m diam.) & herbs (some succ. &/or geophytic), monoec., fls on a flat or hollowed receptacle, fr. ejected 'as one might flip away a bit of soap between finger & thumb' (Willis). Medic. esp. ***D. brasiliensis*** Lam. (S Am.) – fertility control in Urug., ***D. contrajerva*** L. (trop. Am.) – Corner's Model, febrifuge in Costa Rica, rhiz. used to flavour cigarettes

Dorvalia Hoffsgg. = *Fuchsia*

Doryalis Warb. = *Dovyalis*

Doryanthaceae Dahlgren & Clifford (~ Tecophilaeaceae). Magnoliidae – Asparagales. 1/2 E Aus. Giant rosette pls with lvs in 5/13 spirals, tips eventually with dry threads, bases massive, starchy; steroidal saponins. Infl. a term. thyrse; bracts large, oft. red. Fls ± reg.; P 6, connate basally forming a nectar-cup; A 3 + 3, epipet., with elongate anthers dehiscing longit.; $\overline{G}$ 3 with simple style, septal nectaries & several – ∞ anatr. ovules in 2 rows per loc., on axile placentas. Fr. a loculicidal capsule. Seeds usu. laterally winged; testa yellow to red-brown; endosperm rich in fats; embryo straight. After flowering veg. buds in leaf-axils grow to produce a thicket (10 yrs to flower). n = 17, 18, 22, 24

Genus: *Doryanthes*

Many similarities with Asphodelaceae (Hemerocallidaceae), morphology much like Iridaceae

Doryanthes Correa. Doryanthaceae. 2 E Aus. Pachycaul pls with strap-shaped lvs to 2.5 m long (***D. palmeri*** W. Hill ex Bull), infls to 5.5 m, P to 18cm long, anthers to 3 cm long; prot., s.t. cult. esp. ***D. excelsa*** Correa (gymea lily, E NSW) – lvs form. used in Aboriginal basketry

Dorycnium Mill. = *Lotus* (but see BJ 31(1901)314)

Dorycnopsis Boiss. = *Anthyllis*

Doryopteris J. Sm. Pteridaceae (IV). Excl. *Lytoneuron*, *Ormopteris*, 7 trop. & warm esp. SE Braz. R: CGH 143(1942)1

Doryphora Endl. Atherospermataceae. 2 NE Aus. R: FA 2(2007)92. Alks. ***D. sassafras*** Endl. has fragrant wood used for furniture, insect-proof boxes etc.

Dorystaechas Boiss. & Heldr. ex Benth. = *Dorystoechas*

Dorystephania Warb. = *Sarcolobus*

Dorystoechas Boiss. & Heldr. ex Benth. Labiatae (Vii 2a). 1 SW Turkey: ***D. hastata*** Boiss. & Heldr. ex Benth. – 'sister' to Am. '*Salvia*' spp.

Doryxylon Zoll. Euphorbiaceae (Acal.-Chro.-Chro.). 1 Luzon, Lesser Sunda Is.: ***D. spinosum*** Zoll. R: Blumea 44(1999)422

Dossifluga Bremek. = *Strobilanthes*

Dossinia Morren. Orchidaceae (IV 2d). 1 Borneo: ***D. marmorata*** Morren. R: Gen. Orch. 3(2003)85

double coconut *Lodoicea maldivica*

Douepea Cambess. Cruciferae (12). 2 Saudi Arabia, Pakistan. R: Novon 11(2001)296

Dougal grass *Isolepis cernua*

Douglas fir *Pseudotsuga menziesii*

Douglasdeweya Yen & al. = *Elymus* (but see CJB 83(2005)416)

Douglasia Lindl. = *Androsace*

doum (dom) palm *Hyphaene thebaica*

Douradoa Sleumer (~ *Curupira*). Olacaceae (Ximeniaceae). 1 Braz.: ***D. consimilis*** Sleumer. R: FN 38(1984)136

dove flower or **orchid** *Peristeria elata*; **d. tree** *Davidia involucrata*; **d.-weed** *Croton setiger*

Dovea Kunth = *Elegia* (but see Bothalia 15(1985)435)

Dovyalis E. Meyer ex Arn. Salicaceae (Flacourtiaceae). 15 warm Afr., Sri Lanka (1) to NG. R: BJ 92(1972)64. Dioec. shrubs & trees (Troll's Model), some armed with axillary spines. ***D. caffra*** (Hook.f. & Harv.) Hook.f. (kei apple, S & E Afr.) – hedgepl. with ed. fr. used in jelly, marmalade etc.; ***D. hebecarpa*** (Gardner) Warb. (Ceylon gooseberry, kitembilla, Sri Lanka) – ed. fr., added to arrack with sugar to give a sherry subs.

Dowerin rose *Eucalyptus pyriformis*

down tree *Ochroma pyramidale*

Downingia Torrey. Campanulaceae (III 2). 13 W N Am., 1 ext. to Chile. R: MTBC 19,4(1941)1. Elongated pedicel-like ovary. Some cult. orn.

downy chess *Bromus tectorum*

Doxantha Miers = *Dolichandra*

Doyerea Grosourdy. Cucurbitaceae (XIII). 1 trop. Am.: ***D. emetocatharctica*** Grosourdy. R: U. Eggli, *Ill. Handb. Succ. Pls*, Dicots (2002)82

Doyleanthus Sauquet. Myristicaceae. 1 Madag.: ***D. arillata*** Sauquet. R: AJB 90(2003) 1302

Draba Dill. ex L. Cruciferae (7). Incl. *Erophila, Schivereckia* c. 380 N temp. (Eur. 44, Himal. & C As. c. 100, N Am., esp. W, 121) & boreal, Andes (c. 70). Largest genus in fam. Hybridization & apomixis in some. Shrubby spp. in Venez. Some cult. orn. rock-garden pls incl. ***D. aizoides*** L. (Alps, SW Wales, Belgium (Ardennes)). ***D. cacuminum*** Ekman – one of few Scandinavian endemics, an octoploid that has arisen at least 3 times from ***D. norvegica*** Gunnerus (hexaploid) & a diploid sp.; ***D. verna*** L. (*E. v.*, whitlow-grass, Eur., Medit.) – comprises selfing pure lines (jordanons) form. considered spp. (Jordan), alleged to cure whitlow = inflammation around the nails

Drabastrum (F. Muell.) O. Schulz. Cruciferae (37). 1 SE Aus.: ***D. alpestre*** (F. Muell.) O. Schulz. R: TPRSSA 89(1965)223

Drabella (DC.) Fourr. (~ *Draba*). Cruciferae (7). 1 Eur.: ***D. muralis*** (L.) Fourr.

Drabopsis K. Koch = *Draba*

Dracaena Vand. ex L. Asparagaceae (Dracaenaceae). Incl. *Sansevieria*, c. 120 trop. OW to Canary Is. (2), 1 C Am., 1 Cuba. Shrubs & trees (Chamberlain's & Leeuwenberg's Models, some 'litter-box' pls) with extrafascicular cambium like *Cordyline* but ovule 1 per loc. & sec. tissues in r., or perenn. herbs with short. rhiz. & thick lvs (*Sansevieria*) Extrafl. nectaries. Some used as live fences in trop.; ed. fls sold in Madag. markets; resin from stems of some spp. a source of 'dragon's blood' used to stop flow of blood in anitiquity, also used in varnishes & photo-engraving, that of ***D. cinnabari*** Balf.f. (Socotra) prob. that known to the Ancients & still used to stain pottery (& horn to resemble tortoiseshell) besides medic. & in cosmetics, also imp. browse for stock, source of leaf-fibre (incl. bowstring hemp); some medic. incl. ***D. steudneri*** Engl. (C & E Afr.). Several cult. orn. incl. ***D. draco*** (L.) L. (dragon-tree, Canary Is., Madeira, S Morocco, Cape Verde Is.) – rare in wild, source of dragon's-blood, slow-growing, branching every 10 yrs, prob. after flowering, one 20 m tall & c. 12 m girth in 1868 alleged to have been 6000 yrs old but in 1971 none more than 365 yrs. old alive, though one in Gibraltar poss. planted 1480; ***D. fragrans*** (L.) Ker-Gawler (trop. Afr.) – commonly cult., particularly varieg. cvs.; ***D. hyacinthoides*** (L.) Mabb. (*S. h., S. guineensis, S. thyrsiflora*, S Afr.) – imp. fibre; ***D. sanderiana*** Mast. (*D. braunii*, Chinese water bamboo, W Afr.) – segments with leafy shoots piled in pyramids sold as 'lucky bamboo' in Eur. (imports to UK via Rotterdam blamed for introd. As. tiger mosquitoes, hosts to dengue fever etc.) etc., 'corkscrew' forms also cult. 'orn.'; ***D. trifasciata*** (Prain) Mabb. (*S. t.*, mother-in-law's tonge, Nigeria, Congo) – cult. pot-pl. esp. '**Laurentii**' with yellow marginal stripes on lvs; ***D. zeylanica*** (L.) Mabb. (*S. z.*, Sri Lanka) – fibre

Dracaenaceae Salisb. = Asparagaceae

Dracocephalum Tourn. ex L. Labiatae (VII 2b). c. 70 Euras. (Eur. 3), N Afr. (1), N Am. (1), usu. montane. R: BZ 72,2(1987)260. Local medic., bee-pls, cult. orn. (dragonhead) allied to *Nepeta* but K 2-lipped & 15-veined, upper A longer than lower

Dracoglossum Christenh. (~ *Tectaria*). Lomariopsidaceae. 2 trop. Am. R: Thaiszia 17(2007)3

Dracomonticola Linder & Kurzweil (~ *Platanthera*). Orchidaceae (IV 4d). 1 S Afr.: ***D. virginea*** (Bolus) Linder & Kurzweil. R: Willd. 25(1995)229

Draconanthes (Luer) Luer (~ *Lepanthes*). Orchidaceae (V 13c). 2 Ecuador

Draconopteris L.B. Zhang & L. Zhang (~ *Tectaria*). Tectariaceae. 1 trop. Am.: ***D. draconoptera*** (D. Eaton) L.B. Zhang & L. Zhang. R: T 65(2016)732

Dracontioides Engl. Araceae (V). 2 E Braz. R: Aroid. 28(2005)23

Dracontium L. Araceae (V). 24 trop. Am. R: AMBG 91(2004)605. Sympodial rhizome gives 1 leaf & infl. (in ***D. gigas*** (Seem.) Engl. 3 m & 1.5 m tall resp.); leaf with 3 major lobes, the laterals developing pseudo-dichot. initially; fls bisex. with P. ***D. polyphyllum*** L. tubers ed.; some medic. esp. for snakebite cure (Amaz.) incl. ***D. spruceanum*** (Schott) G. Zhu (*D. trianae*), cult. NW Amaz.

Dracontomelon Blume. Anacardiaceae (II). 8 SE As., Indomal. to Fiji. Some timbers (NG, Pac. or Papuan walnut) esp. ***D. dao*** (Blanco) Merr. & Rolfe (*D. mangiferum*, dao, Indomal.) – 'paldao' used for veneers & also matches, fr. ed., fls used as flavouring. ***D. vitiense*** Engl. (Vanuatu to Samoa) – ed. fr.

Dracophilus (Schwantes) Dinter & Schwantes (~ *Juttadinteria*). Aizoaceae (V). 2 S Afr. R: H. E. K. Hartmann, *Ill. Handb. Succ. Pls*, Aiz. A–E(2002)225

Dracophyllum Labill. Ericaceae (VII 6; Epacridaceae). c. 48 Aus., New Caled. & esp. NZ. Trees & shrubs (Holttum's & Leeuwenberg's Models) with lvs like monocots. Some cult. orn.

Dracopsis (Cass.) Cass. = *Rudbeckia*

Dracosciadium Hilliard & B.L. Burtt. Umbelliferae (III 8). 2 Natal. R: NRBGE 43(1986)220

Dracoscirpoides Muasya (~ *Scirpus*). Cyperaceae (II 1). 3 S Afr.

Dracula Luer (~ *Masdevallia*). Orchidaceae (V 13c). 120+ trop. Am. R: MSB 46(1993)1. Cult. orn. ***D. lafleurii*** Luer & Dalström (Ecuador) – poll. mycophilous drosophilid flies (most of lifecycle in fungi) mating in fls (labellum like a mushroom cap with similar smell) & poll. (but not laying eggs there)

Dracunculus Mill. Araceae (VII 23). Excl. *Helicodiceros*, 2 Medit. incl. Eur. R: Thaiszia 4(1994)177. Cult. orn. tuberous herbs with malodorous appendage & poll. mechanism like *Arum*. ***D. vulgaris*** Schott (dragon arum, Medit.) – poisonous & avoided by grazing animals, depicted in Minoan paintings

dragon arum *Dracunculus vulgaris*, (N Am.) *Arisaema* spp.; **d. claw** *Corallorhiza odontorhiza*; **d. fruit** *Hylocereus undatus*; **d. gum** *Astragalus* spp. esp. *A. gummifera*; **d.head** *Dracocephalum* spp.; **d.-tree** *Dracaena draco*

dragon's blood Several reddish resins used in varnishes etc., orig. from *Dracaena cinnabari* &, later, *D. draco*, more recently from *Daemonorops* spp.; **d. eye, d. fruit** *Hylocereus undatus*; **d. eyes** *Dimocarpus longan*

Drakaea Lindl. Orchidaceae (IV 3e). 10 SW Aus. R: AusSB 20(2007)256. *Ophrys*-like insect mimics poll. thynnine wasps

Drake-brockmania Stapf = *Dinebra*

Drakonorchis (Hopper & A.P. Br.) D.L. Jones & M. Clements = *Caladenia*

Drambuie Liqueur made from whisky & honey derived from bees visiting *Calluna vulgaris*

Dransfieldia W. Baker & Zona. Palmae (V 14). 1 W NG: ***D. micrantha*** (Becc.) W. Baker & Zona. R: SB 31(2006)61

Draperia Torrey. Boraginaceae (Hydrophyllaceae-Rom.). 1 Calif.: ***D. systyla*** (A. Gray) Torrey

Drapetes Banks ex Lam. Thymelaeaceae (Thym.-Daph.). (Excl. *Kelleria*) 1 Fuegia & Falkland Is.: ***D. muscosa*** Lam. R: Aus SB 3(1990)634

drautabua *Acmopyle sahniana*

Dregea E. Meyer = *Marsdenia*

Dregeochloa Conert (~ *Danthonia*). Gramineae (Arund.). 2 S Afr. R: SBi 47(1966)335

Drejera Nees = *Thyrsacanthus*

Drejerella Lindau = *Justicia*

Drepananthus Maingay ex Hook.f. & Thomson (~ *Cyathocalyx*). Annonaceae (II). 26 Indopacific. R: T 59(2010)1730

Drepanocarpus G. Meyer = *Machaerium*

Drepanocaryum Pojark. Labiatae (VII 2c). 1 C As.: ***D. sewerzowii*** (Regel) Pojark. R: BZ 77,12(1992)124

Drepanostachyum Keng f. (~ *Sinarundinaria*). Gramineae (IV). c. 10 Himal. to SW China. Heterogeneous

Drepanostemma Jum. & Perrier = *Cynanchum*

Dresslerella Luer. Orchidaceae (V 13c). 13 C & NW S Am. R: MSB 26(1988)1

Dressleria Dodson. Orchidaceae (V 12c). 13 C Am. R: Phytoneuron 2012–48:2,-103:1. Lvs foetid.; fls bisex. (cf. *Catasetum*) poll. male euglossine bees

Dressleriella Brieger = *Epidendrum*

Dresslerіopsis Dwyer = *Lasianthus*

Dresslerothamnus H. Robinson. Compositae (Sen.-Sen.). 5 C Am. to Colombia. R: SB 14(1989)382. Lianes

Driessenia Korth. Melastomataceae. 14 W (11 Borneo endemics) & C Mal. R: NJB 5(1985)335

Drimia Jacq. Asparagaceae (Hyacinthaceae). Incl. *Litanthus, Rhadamanthus, Schizobasis* (compound twining infls, cf. *Bowiea*), *Tenicroa, Urginea*, c. 100 trop. & warm OW (S Afr. 50) incl. S Eur. (3). R: EJB 60(2004)554. Many fl. in late spring or summer so in winter-rainfall areas of S Afr., flowering when leafless. Many medic. & poisonous (cardiac glycosides) esp. ***D. maritima*** (L.) Stearn (*U. maritima*, squill, sea onion, Medit.) – insect- & wind-poll., diff. forms used as cardiac stimulant etc. & as rat-poison (specific to rodents, used since time of Theophrastus; other animals vomit), grown comm. in US since 1946, hung as amulet outside houses in Greece even today, bulbs resistant to maquis fires; ***D. indica*** (Roxb.) Jessop (trop. As. & Afr.) used as subs. ***D. noctiflora*** (Batt. & Trabut) Stearn (N W Afr.) – night-flowering with tepals reflexed like *Cyclamen*

Drimiopsis Lindl. & Paxton = *Ledebouria*

Drimycarpus Hook.f. Anacardiaceae (I). 3+ Indomal. $\overline{G}$. ***D. luridus*** (Hook.f.) Ding Hou (W Mal.) – v. allergenic fr.

Drimys Forst. & Forst. f. Winteraceae. Excl. *Tasmannia* (dioec., OW), 6 trop. Am. Bisex. fls). ***D. winteri*** Forst. & Forst. f. (Winter's bark, Mex. to Tierra del Fuego) – v. variable, first known as a 'medicine v. powerful against the scurvy' introd. by Capt. John Winter in 1578, also stomachic

Droceloncia Léonard. Euphorbiaceae (Acal.-Pyc.-Pyc.). 1 Madag., Comores: ***D. rigidifolia*** (Baill.) Léonard. R: A. Radcliffe-Smith, *Gen. Euphorb.* (2001)167

Droguetia Gaudich. Urticaceae (V). 7 trop. & warm. Afr. to Java. R: NJB 8(1988)36, 10(1990)431

Droogmansia De Wild. Leguminosae (III 19). 5 SC Afr. (1 variable), W Afr. (4), merging with *Tadehagi*

dropseed *Sporobolus fimbriatus*

dropwort *Filipendula vulgaris*; **water d.** *Oenanthe crocata*

Drosanthemopsis Rauschert = *Jacobsenia*

Drosanthemum Schwantes. Aizoaceae (V). 107 SW Afr., Cape & Namaqualand (most). R: H. E. K. Hartmann, *Ill. Handb. Succ. Pls*, Aiz. A–E(2002)227. Succ. shrubs; cult. orn.

Drosera L. Droseraceae. c. 100 cosmop. (Eur. 3, trop. Am. 20 – R: FN 96(2005)21) esp. S hemisph. (SW Aus. 68) in wet places. R: T 43(1994)583. Carnivorous herbs (sundews, e.g. less than 1 ha of ***D. anglica*** Hudson (N temp.) trapped c. 6 M *Pieris rapae* butterflies in one morning!) with rhiz. or s.t. scrambling (Chamberlain's Model) stems (***D. erythrorrhiza*** Lindl. (SW Aus.) to over 50 yrs old, ***D. gigantea*** Lindl. (SW Aus.) to 1 m tall, ***D. magnifica*** Rivadavia & Gonella (Braz.) to 1.5 m), & round to linr. palisade-less leaf-blades with gland-tipped red or greenish hairs capable of movement when irritated (bending to centre of leaf in 3–20 mins, the surface voltage dropping when touched & the receptor potential correlated with intensity of stimulus, above a certain threshold a series of short electrical pulses towards the base of the tentacle giving the action potential leading to the bending) & of holding (trapping mucilage c. 4% solution of acidic polysaccharide oft. suffocating insects through spiracles) & digesting insects, which in some spp. at least promotes flowering. Some S Afr. spp. poll. monkey beetles; some with cleistogamous fls, others with epiphyllous propagules. Some locally medic. (against warts (Yorkshire) but usu. resp. complaints) due to quinones (e.g. derivatives of 1,4-naphthoquinone in 'Herbae Droserae' (***D. rotundifolia*** L., Eur.) for whooping cough & asthma); cult. as curiosities (over 100 spp. grown); rootstocks ed. SW Aus. Aborigines. ***D. cistiflora*** L. (S Afr.) – fls white, yellow, pink to purple or red

Droseraceae Salisb. Magnoliidae – Caryophyllales. Excl. Drosophyllaceae, 3/c. 102 cosmop. Carnivorous herbs oft. with basal rosette of lvs & cyanogenic. Lvs in spirals to whorled, oft. circinnate (adaxially) in bud, simple, either with irritable gland-tipped hairs (*Drosera*) or an active trap (*Aldrovanda, Dionaea*); stip. oft. present. Infl. term., fls bisex., reg., hypogynous with marcescent K, C & A, solit. (*Aldrovanda*) or in circinnate thyrses; K (4)(5–8), ± basally connate, imbr., C same no., convolute, A (4)5(10–20), connate basally in *Dionaea* with v. variable pollen in tetrads, $\underline{G}$ (3(5)), 1-loc., with distinct oft. bifid styles (united in *Dionaea*) & (3–)∞ ovules (anatropous, bitegmic) on parietal placentas or a basal one. Fr. a loculicidal capsule (rarely indehiscent); seeds (3–)∞, fusiform with short straight embryo embedded in copious endosperm rich in starch, oil & protein. n = 5–24

Genera: *Aldrovanda, Dionaea, Drosera*

Slow but perceptible movements (*Drosera*) to v. rapid (*Dionaea*), in a wide range of habitats incl. wet (*Drosera*) to total submergence (*Aldrovanda*)

Drosophyllaceae Chrtek et al. (~ Droseraceae). Magnoliidae – Caryophyllales. 1/1 W Medit. Woody carnivore with taproot. Lvs linr., in spirals, abaxially (cf. Droseraceae) circinnate with stalked glands in lines forming non-moving mucilage traps (cf. Droseraceae), remarkably similar to *Byblis*; stip. 0. Fls large, reg., in term. few-flowered thyrse. K 5, imbr., basally connate; C 5 contorted ± marcescent; A 5 + 5 with pollen as monads (cf. Droseraceae); $\underline{G}$ (5), 1-loc., ovules anatropous, bitegmic. Fr. septicidal (cf. Droseraceae). n = 6

Genus: *Drosophyllum*
Many diffs from Droseraceae

Drosophyllum Link. Drosophyllaceae (Droseraceae s.l.). 1 Portugal, S Spain, Morocco: ***D. lusitanicum*** (L.) Link – calcifuge reg. selfing colonist (incl. roadsides & favoured by fire events) with marcescent lvs; water-supply from dews absorbed by cuticular pores on lvs, which are insect-trapping like *Drosera* (mucilage); used to treat conjunctivitis

Drudeophytum J. Coulter & Rose = *Tauschia*

druggists' bark *Cinchona* spp.

Drummondita Harv. (~ *Philotheca*). Rutaceae (I 3). 9 Aus. R: FA 26(2013)427

drumsticks *Isopogon* spp., *Moringa oleifera*

Drusa DC. Umbelliferae (Azor.). 1 Canary Is. & Somalia: ***D. glandulosa*** (Poir.) Engl.

Dryadanthe Endl. = *Potentilla*

Dryadella Luer (~ *Masdevallia*). Orchidaceae (V 13c). c. 50 trop. Am.

Dryadodaphne S. Moore. Atherospermataceae. 3 NG, Queensland

Dryadorchis Schltr. (~ *Thrixspermum*). Orchidaceae (V 16c). 5 NG. R: Blumea 40(1995)421. Fls last 1 day

Dryandra R. Br. = *Banksia* (but see FA 17B(1999)251)

Dryas L. Rosaceae (Ros.-Dry.). 2(–15) Arctic-alpine. R: SBT 53(1959)507. Mountain avens. Nitrogen-fixing nodules (actinobacteria); lvs char. as fossil in cool periods. Cult. orn. rock-pls esp. ***D. octopetala*** L.

Drymaria Willd. ex Schultes. Caryophyllaceae (I 1). 48 W US (9) to Patagonia (46), Galápagos (1) with ***D. cordata*** (L.) Schultes pantrop. – local medic. R: AMBG 48(1961)173. Some with elaiophores attractive to poll. male euglossine bees. ***D. arenarioides*** Willd. ex Schultes (NW Mex.) – v. toxic to stock

Drymoanthus Nicholls. Orchidaceae (V 16c). 4 NE Aus., NZ (2)

Drymocallis Fourr. ex Rydb. (~ *Potentilla*). Rosaceae (Pot.). 25–30 temp. (N Am. 15). ***D. arguta*** (Pursh) Rydb. (N Am.) – absorbs proteins from dead arthropods & other org. matter trapped in hairs (protocarnivory)

Drymoda Lindl. Orchidaceae (V 15). 3 Myanmar, Thailand, Laos, Indonesia. R: OB 89(1986)167

Drymoglossum C. Presl = *Pyrrosia*

Drymonia Mart. Gesneriaceae (II 5c). c. 140 trop. Am. Mostly lianes with salt-shaker anthers. Some cult. orn. incl. ***D. peltata*** (Oliv.) H. Moore (Costa Rica) – named from cult. pls & not seen in wild until recently, & ***D. serrulata*** (Jacq.) Mart. – decid. liane with dichogamous fls (male 1st day, female next) with more nectar than any other bee-poll. pl., *Epicharis* bees being poll. but oil sticking pollen grains together also coll. non-poll. *Trigona bees*

Drymophila R. Br. (~ *Luzuriaga*). Alstroemeriaceae (Luzuriagaceae/Liliaceae s.l.). 2 E & SE Aus. R: FA 45(1987)156

Drymophloeus Zipp. Palmae (V 14i). Excl. *Solfia*, 2 Moluccas to Samoa. R: Blumea 44(1999)1. Monoec. graceful cult. orn.

Drymotaenium Makino = *Lepisorus*

Drynaria (Bory) J. Sm. Polypodiaceae (II). Incl. *Aglaomorpha*, c. 45 OW trop. esp. China. Epiphytes. Lvs usu. dimorphic, the sterile erect, short & broad, soon becoming dry & collecting humus, the fertile deeply lobed or pinnate, petiolate; (? ant-attracting) nectaries reported

Drynariopsis (Copel.) Ching = *Aglaomorpha*

Dryoathyrium Ching = *Deparia*

Dryobalanops Gaertn. f. Dipterocarpaceae (Dipt.-Shor.). 7 Mal. R: FM I,9(1982)371. Roux's Model. Timber (Brunei teak, kapur) a pop. light brown hardwood; ***D. aromatica*** Gaertn. f. (*D. sumatrensis*) & other spp. source of camphor (baros, barus, Borneo or Sumatra c.) form. much exported from N Sumatra & Johore since C6 to Arabs (mentioned by Marco Polo), the crystals coll. from splits in the bole; fr. boiled as veg.

Dryopetalon A. Gray. Cruciferae (46). Incl. *Rollinsia*, 9 Mex., 2 ext. to SW US. R: Novon 17(2007)399. C 5–7-lobed

Dryopoa Vickery (~ *Festuca*). Gramineae (XVI 10). 1 SE Aus.: ***D. dives*** (F. Muell.) Vickery. R: FA 44a(2009)299

Dryopolystichum Copel. Dryopteridaceae (I). 1 Papuasia: ***D. phaeostigma*** (Cesati) Copel.

Dryopsis Holttum & Edwards = *Dryopteris*

Dryopteridaceae Herter. Polypodiidae – Polypodiales. Excl. Athyrioideae [= Athyriaceae, Onocleaceae, Woodsiaceae], Tectarieae s.s. [= Tectariaceae], Hypodematiaceae, 26/1875

cosmop. esp. temp. & montane. Terr. (rarely epiphytic) pls with erect, creeping or s.t. cli. stems with non-clathrate scales at apices, dictyostelic, scaly. Fronds simply pinnate (simple) to decompound; veins pinnate or forking, free to variously anastomosing. Sporangia usu. in ± orbicular sori with round to peltate or 0 indusium (acrostichoid in some groups); spores reniform, monolete. n = (40) 41

Principal genera:

I. **Dryopteridoideae** *Arachniodes, Ctenitis, Dryopteris, Polystichum*

II. **Elaphoglossoideae** *Elaphoglossum, Rumohra*

Cult. orn. esp. *Rumohra*

Dryopteris Adans. Dryopteridaceae (I). Incl. *Dryopsis* c. 250 subcosmop. (Eur. 19) but rare in lowland trop. (Aus. 1, NZ 0). R: BBMNHB 14(1986). Shield ferns, buckler f. incl. ***D. carthusiana*** (Villars) H.P. Fuchs (narrow b.f.) & ***D. dilatata*** (Hoffm.) A. Gray (*D. austriaca*, broad b.f., N temp.). Much allopolyploidy. Some cult. orn. esp. *D. dilatata* as 'Florists' fern', 'fern' used in soap, oil in spa massage, & ***D. filix-mas*** (L.) Schott (male fern, N temp.) – allopolyploid (1 genome from diploid *D. abbreviata* (Lam. & DC.) Newman), one of oldest known vermifuges which paralyzes tapeworms, which may then be removed by purgatives (but dangerous as it paralyses voluntary muscles of patients & is now replaced by quinacrine), app. used in silk reeling in anc. China

Drypetes Vahl. Putranjivaceae (Euphorbiaceae s.l.). Incl. *Sibangea*, c. 200 trop., E As., S Afr. Glucosinolates (mustard oils) – evolved independently in Brassicales. Roux's Model; some timbers. ***D. caustica*** (Cordemoy) Airy Shaw now reduced to 2 trees on Mauritius & 12 on Réunion; ***D. floribunda*** (Muell. Arg.) Hutch. (W Afr.) – ed. pulp around seeds; ***D. gossweileri*** S. Moore (W & C Afr.) – elephant-disp.; ***D. pellegrinii*** Leandri (W Afr.) – bark local medic., r. for chewing-sticks (Ghana); ***D. pendula*** Ridl. (W Mal.) – ant-pl. with hollow twigs; ***D. roxburghii*** (Wall.) Hurusawa (Ind., Myanmar) – seeds in rosaries

Drypis Micheli ex L. Caryophyllaceae (III 2). 1 S Eur. & Lebanon: ***D. spinosa*** L. – cult. orn. herb

du zhong *Eucommia ulmoides*

Duabanga Buch.-Ham. Lythraceae (Sonneratiaceae). 2–3 Indomal. R: JAA 48(1967)89. In rain forest (cf. allied *Sonneratia* in mangrove); Massart's Model. ***D. grandiflora*** (DC.) Walp. (*D. sonneratioides*) – night-flowering, timber for tea-chests, fr. ed (boiled), lvs & fr. yield a black dye, extracts induce collagen prod. so used in anti-wrinkle skin-creams

Duabangaceae Takht. = Lythraceae

Dubardella H.J. Lam = *Pyrenaria*

Dubautia Gaudich. Compositae (Mad.-Mad.). Incl. *Raillardia*, 24 Hawaii. R: Allertonia 4,1(1985)62. Trees, shrubs & lianes; lvs with parallel veins

Duboisia R. Br. Solanaceae (III 1). 3 Aus., 1 ext. to New Caled. Heterogeneous. Alks incl. atropine, a comm. source & form. used as emu poison, mood-changing masticatory (pitcheri, pituri) of C Aus. Aborigines, s.t. put behind ears like a nicotine patch, esp. ***D. hopwoodii*** (F. Muell.) F. Muell. (Aus.) & ***D. myoporoides*** R. Br. (Aus., New Caled.) – alk. comm. source (also from hybrids) for insecticide & medic. drugs (scopolamine, pre-med. relaxant before surgery, used in WW II landing-craft) & timber (corkwood) for carving

Duboscia Bocq. (~ *Desplatsia*). Malvaceae (Grew.-Grew.; Tiliaceae). 3 trop. W Afr. ***D. macrocarpa*** Bocq. – seeds imp. in gorilla diet in Gabon

Dubouzetia Pancher ex Brongn. & Gris (~ *Crinodendron*). Elaeocarpaceae. 11 Moluccas (1), NG (4), NE Aus. (2), New Caled. (6). R: KB 42(1987)796

Dubyaea DC. Compositae (Lact.-Crep.). 14 Himal., W China. R: MTBC 19,3(1940)8

Ducampopinus A. Chev. = *Pinus*

Duchesnea Sm. = *Potentilla*

duck meat *Lemna* spp.; **d. orchid** *Caleana major*; **d. plant** *Lessertia frutescens*; **d.weed** *Lemna* spp. esp. *L. minor*

Duckea Maguire = *Cephalostemon*

Duckeanthus R.E. Fries. Annonaceae (III 3). 1 trop. S Am.: ***D. grandiflorus*** R.E. Fries

Duckeella Porto & Brade. Orchidaceae (II 1). 3 trop. S Am. R: Gen. Orch. 3(2003)289

Duckeodendraceae Kuhlm. = Solanaceae

Duckeodendron Kuhlm. Solanaceae (Duckeodendraceae). 1 Amazonian Braz.: ***D. cestroides*** Kuhlm. – wood anatomy confirms fam. placement

Duckera F. Barkley = *Melanococca*

Duckesia Cuatrec. Humiriaceae. 1 Amazonian Braz.: ***D. verrucosa*** (Ducke) Cuatrec. R: CUSNH 35(1961)76

Ducrosia Boiss. Umbelliferae (III 11). 3 Egypt to NW Ind. R: NRBGE 34(1975)190

Dudleya Britton & Rose. Crassulaceae (I 5a). 47 SW N Am. (US 45 – R: FNA 9(2009)171. Glabrous succ.; some cult. orn.

duffin bean *Phaseolus lunatus*

Dufrenoya Chatin. Santalaceae. 11 SE As., W Mal.

Dugaldia Cass. = *Hymenoxys*

Dugandia Britton & Killip = *Senegalia*

Dugesia A. Gray. Compositae (Helia.-Dug.). 1 Mex.: ***D. mexicana*** (A. Gray) A. Gray. R: Britt. 26(1974)385

Duggena Vahl = *Gonzalagunia*

Duguetia A. St-Hil. Annonaceae (III 3). Incl. *Pachypodianthum* 95 trop. Am. R: FN 88(2003)69. Fr. united with fleshy receptacle to give false fr.; leaf flushes white, chlorophyllous later. Some timbers esp. ***D. quitarensis*** Benth. (Jamaica & Cuba lancewood, trop. S Am.). ***D. sessilis*** (Vell.) Maas (*D. rhizantha*, Braz.) – subterr. rhiz. with scale lvs & aerial flowering shoots

Duhaldea DC. (~ *Inula*). Compositae (Inul.-Inul.) 14 Iran to Himal. & E Afr. Trees to herbs

Duidaea S.F. Blake. Compositae (Stifft.). 4 Venez. & Guyana

Duidania Standl. Rubiaceae (I 1). 1 Venez.: ***D. montana*** Standl.

Dukea Dwyer = *Raritebe*

dukong anak *Phyllanthus urinaria*

duku *Lansium domesticum*

Dulacia Vell. (~ *Olax*). Olacaceae. 14 trop. Am. R: FN 38(1984)116

Dulichium Rich. Cyperaceae (II 6). 1 N Am. (fossils in Eur.): ***D. arundinaceum*** (L.) Britton

Dulongiaceae J. Agardh = Phyllonomaceae

Duma T. Schust. (~ *Muehlenbeckia*). Polygonaceae (II 4). 3 Aus. R: IJPS 172(2011)1061

Dumaniana Yild. & Selvi = *Pastinaca* (but see SBD 13,2(2006)5)

Dumasia DC. Leguminosae (III 18). 10 OW trop.

dumb cane *Dieffenbachia* spp. esp. *D. seguine*

Dumori butter *Tieghemella heckelii*

Dunalia Kunth (~ *Acnistus*). Solanaceae (IV 5b). 5 Andes. Some trees; some dioec. (males & females orig. described as distinct spp.)

Dunbaria Wight & Arn. Leguminosae (III 18). 20 trop. As. to Aus. R: WAUP 98–1(1998)12. Cli. Local medic.

Dungsia Chiron & Castro = *Sophronitis*

Dunkeld larch *Larix* × *marschlinsii*

Dunnia Tutcher. Rubiaceae (Dunn.). Excl. *Neohymenopogon*, 2 Ind., China. R: Blumea 24(1978)367

Dunniella Rauschert = *Pilea*

Dunstervillea Garay. Orchidaceae (V 12h). 1 Venez.: ***D. mirabilis*** Garay. R: GO 5(2009)262

Duosperma Dayton. Acanthaceae (III 2f). 26 trop. & S Afr. R: KB 61(2006)289

Duparquetia Baill. Leguminosae (I 3). 1 trop. W Afr.: ***D. orchidacea*** Baill. – fls seem to mimic those of orchids but poll. mechanism unknown

Duperrea Pierre ex Pitard. Rubiaceae (II 1). 2 Ind., China, SE As.

Duperreya Gaudich. (~ *Porana*). Convolvulaceae (6). 3 Aus. R: Austrobaileya 8(2009)48

Duplipetala Thiv (~ *Canscora*). Gentianaceae. 2 SE As. to W Mal. R: Blumea 48(2003)25

Dupontia R. Br. (~ *Colpodium*). Gramineae (XVI 15). Incl. *Arctophila*, 3 Arctic incl. Eur.

Dupontiopsis Soreng & al. = *Dupontia*

Dupuya Kirkbr. (~ *Cordyla*). Leguminosae (II 1). 2 Madag. R: Novon 15(2005)307

Durabaculum M.A.Clem. & D.L. Jones = *Dendrobium*

Durandea Planch. = *Hugonia*

Duranta L. Verbenaceae (Dur.). 17 Carib. to S Am., natur. S & E As. R: Sida 10(1984)308. Shrubs & trees, s.t. spiny. Some cult. orn. esp. ***D. erecta*** L. (*D. repens*, pigeon-berry, golden dewdrop, Florida to Braz., natur. S US)

Durban grass *Dactyloctenium australe*

Duriala (R. Anderson) Ulbr. = *Maireana*

durian *Durio zibethinus*

Durio Adans. Malvaceae (Hel.; Bombacaceae). 30 W Mal. R: Reinw. 4,3(1958)48. Bat- (vertebrate-poll. ?orig. condition), bird- & bee-poll. trees (Roux's Model) with lvs in 2 ranks; some cauliflorous. In Borneo (21) 6 hornbill-disp. seeds (fr. indehiscent), 15 disp. on ground (fr. dehiscent). ***D. zibethinus*** L. (durian) – bat-poll. (*Eonycteris* spp., so need mangrove & other pls nearby to sustain bats when d. not in fl.), each fl. produces 0.63 ml watery nectar a night, highly esteemed fr., usu. malodorous (26 volatiles, main being propanethiol & ethyl α-methylbutyrate) but with arils tasting of caramel, banana, vanilla etc. with slight onion tang, much sought by animals (fr. passes undigested through elephants losing stench & bitterness, single fr. coll. & sold for $US 400 – cf. *Coffea*), incl. humans early in morning after fall of spiny capsule up to some kg in wt., smell reduced in '**Chantaburi No 1**', '**Montong**' chief cv exported from Thailand, seeds ed. roasted, timber useful

durma mats *Phragmites australis*

durmast oak *Quercus petraea*

Duroia L.f. Rubiaceae (Ix.-Cord.). 25 trop. Am. Fagerlind's Model; myrmecophilous: in ***D. petiolaris*** Hook.f. & ***D. hirsuta*** (Poeppig & Endl.) K. Schum. (app. allelopathic growing in monospecific stands in rain forest; caustic bark used to make short-lived tattoos), stem below infl. hollow & with 2 longit. slits as ant-doors for *Myrmelachista schumanni* living in stems & killing other pls with formic acid; ***D. saccifera*** Hook.f. has ant-houses on lvs

durra *Sorghum bicolor*

Durringtonia R. Henderson & Guymer. Rubiaceae (IV 13). 1 trop. E Aus.: ***D. paludosa*** R. Henderson & Guymer – dioec. herb. pl. of swamps. R: KB 40(1985)97

durum (**wheat**) *Triticum turgidum* Durum group

Duschekia Opiz = *Alnus*

Duseniella K. Schum. (~ *Chuquiraga*). Compositae (Barn.). 1 Arg., arid Patagonia: ***D. patagonica*** (Dusén) K. Schum. – ann.

Dussia Krug & Urb. ex Taubert. Leguminosae (III 2). c. 9 trop. Am. Timber, local medic.

dusty miller *Jacobaea maritima*

Dutailliopsis T. Hartley. Rutaceae (I 2). 1 New Caled.: ***D. gordonii*** T. Hartley. R: Adansonia 19(1997)210

Dutaillyea Baill. Rutaceae (I 2). 2 New Caled. R: BMNHN 4,6(1984)29

Dutch clover *Trifolium repens*; **D. crocus** *Crocus neapolitanus*; **D. elm** *Ulmus × hollandica*; **D. iris** *Iris × hollandica*; **D. lavender** *Lavandula × intermedia* Dutch Group; **D.man's breeches** *Dicentra spectabilis*; **D. man's pipe** *Aristolochia macrophylla*; **D. rush** *Equisetum hyemale*, *Schoenoplectus lacustris*

Duthiastrum De Vos. Iridaceae (VI 5). 1 S Afr.: ***D. linifolium*** (Phill.) De Vos. R: Strel. 10(2000)629

Duthiea Hackel. Gramineae (IX). 3 Afghanistan to China. R: KB 8(1954)547

Duthiea Speta = *Drimia*

Duthiella De Vos = *Duthiastrum*

Duvalia Haw. Apocynaceae (V b; Asclepiadaceae). 18 S Afr., 5 Somalia to Arabia. R: F. Albers & U. Meve, *Ill. Handb. Succ. Pls*, Asclep. (2002)124. Cult. orn. succ. with rudimentary lvs & minute stip.

Duvaliandra M. Gilbert (~ *Caralluma*). Apocynaceae (V b; Asclepiadaceae). 1 Socotra: ***D. dioscoridis*** (Lavranos) M. Gilbert. R: Bradleya 8(1990)29

Duval-Jouvea Palla = *Cyperus*

Duvaucellia S. Bowdich = *Kohautia*

Duvernoia E. Meyer ex Nees (*Duvernoya*) = *Justicia*

Duvigneaudia Léonard = *Anomostachys*

duzhong *Eucommia ulmoides*

Dyakia Christenson. Orchidaceae (V 16c). 1 Borneo: ***D. hendersoniana*** (Reichb.f.) Christenson. R: OD 50(1986)63.

Dybowskia Stapf = *Hyparrhenia*

Dyckia Schultes f. Bromeliaceae (1). 124 trop. S Am. esp. S. R: FN 14(1974)500. Terr. & saxicolous cult. orn. with extrafl. nectaries (presence of ants maximizes fr. set in ***D. floribunda*** Griseb.); ***D. breviflora*** Bak. (Braz.) – rheophyte

Dyera Hook.f. Apocynaceae (I e). 2 W Mal. R: TFSS 5(2004)27. ***D. costulata*** (Miq.) Hook.f. (jelutong) – comm. lightweight hardwood, chicle obtained by tapping used as chewing-gum (now superseded by synthetics)

Dyerophytum Kuntze (~ *Plumbago*). Plumbaginaceae (I). 2 Socotra, 1 S W Afr., 1 Arabia to Ind.

dyer's chamomile *Cota tinctoria*; **d. greenweed** *Genista tinctoria*; **d. rocket** or **weld** *Reseda luteola*

Dymondia Compton (~ *Arctotis*). Compositae (Arct.-Arct.). 1 S Afr.: ***D. margaretae*** Compton – ? clonal as seed-set rare. R: Strel. 9(2000)318

Dypsidium Baill. = *Dypsis*

Dypsis Noronha ex Mart. Palmae (V 14f). Incl. *Antongilia, Chrysalidocarpus, Neodypsis, Neophloga, Phloga, Vonitra,* 165 Madag., Comoro Is. (1), Pemba (1). R: J. Dransfield & H. Beentje, *Palms Madag.* (1995) 123. Schoute's Model. Incl. some of the tiniest palms like ***D. hildebrandtii*** (Baill.) Becc. only 30 cm tall & ***D. tenuissima*** Beentje 15–45 cm tall, stem 2 mm diam., only cli. Madag. palm (***D. scandens*** Dransf.); ***D. crinita*** (Jum. & Perrier) Beentje & Dransf. – rheophyte, ***D. aquatilis*** Beentje – in deeper water. ***D. marojejyi*** Beentje & ***D. perrieri*** (Jum.) Beentje & Dransf. with litter-trapping crowns. Used in constr. & thatching, palm-hearts & some ed. fr.; many cult. orn. ***D. decaryi*** (Jum.) Beentje & Dransf. (*Neodypsis d.*, triangle palm, S Madag.) – lvs in 3 ranks, threatened in wild but widely cult.; ***D. fibrosa*** (C.H. Wright) Beentje & Dransf. (*Vonitra f.*, N & E Madag.) – Madag. piassava; ***D. lutescens*** (H. Wendl.) Beentje & Dransf. (*Chrysalidocarpus l., Areca madagascariensis*, bamboo or cane palm, E Madag.) – cult. orn. slender clustered stems, anthelminthic in dogs, cosmetics exported; ***D. utilis*** (Jum.) Beentje & Dransf. (*V. u.*, C & E Madag.) reduced to c. 6 trees in rain forest through over-exploitation of palm-hearts

Dysaster H. Robinson & Funk. Compositae (Ast.). 1 Peru: ***D. cajamarcensis*** H. Robinson & Funk – shrub. R: PhytoKeys 36(2014)37

Dyschoriste Nees. Acanthaceae (III 2f). c. 60 trop. & warm. R (Am.): AMBG 15(1928)9

Dyscritogyne R. King & H. Robinson = *Steviopsis*

Dyscritothamnus Robinson. Compositae (Mill.-Dysc.). 2 C Mex.

Dysodiopsis (A. Gray) Rydb. (~ *Dyssodia*). Compositae (Tag.-Pect.). 1 SC US: ***D. tagetoides*** (Torr. & A. Gray) Rydb. R: FNA 21(2006)237

Dysolobium (Benth.) Prain (~ *Vigna*). Leguminosae (III 18). 4 SE As. R: Blumea 30(1985) 363

Dysophylla Blume = *Pogostemon*

Dysopsis Baill. Euphorbiaceae (Acal.-Acal.-Dys.). 1 Andes, Juan Fernandez: ***D. glechomoides*** (A. Rich.) Muell. Arg. R: Caldasia 23(2001)420

Dysosma Woodson (~ *Podophyllum*). Berberidaceae (II 2). 7 E As.

Dysoxylum Blume. Meliaceae (II-Trich.). c. 80 Indomal. (Mal. 50 – R: FM I, 12(1995)61); Aus. (15 – R: Telopea 10(2004)725); New Caled. (9 – R: FNC 15(1988)23) to NZ (1) & Tonga. Some like ***D. sessile*** Miq. (Moluccas) unbranched pachycauls (Corner's Model); ***D. angustifolium*** King (Malay Pen.) – rheophyte, ? fish–disp., though fr. allegedly toxic to mammals. Timber: ***D. acutangulum*** Miq. (Mal.) – coffins, ***D. bijugum*** (Lab.) Seem. (sharkwood, Vanuatu, New Caled., Norfolk Is.), ***D. fraserianum*** (A. Juss.) Benth. (Aus. mahogany or rosewood, Aus.) – fragrant wood for turnery etc., ***D. gotadhora*** (Buch.-Ham.) Mabb. (SE As.) – timber for carts & furniture; ***D. loureiroi*** Pierre (SE As.) – sandalwood-like scented timber used for coffins & joss-sticks, ***D. mollissimum*** Blume (*D. forsteri, D. muelleri*, China to Aus.) – red timber for cabinet-work etc. esp. in Tonga (red bean, Aus.), form. largest of all trees in Java, ***D. pettigrewianum*** Bailey (Cairns satinwood, scrub ironbark, E Mal. to trop. Aus. & Solomon Is.), ***D. spectabile*** (G. Forst.) Hook.f. (kohekohe, NZ) – similar, winter-flowering, poll. tuis (birds), bitter lvs form. hop subs., local medic. ***D. arborescens*** (Blume) Miq. (Indmal.) – local medic. Indonesia

Dyspemptemorion Bremek. = *Justicia*

Dysphania R. Br. (~ *Chenopodium*). Amaranthaceae (Dysph.; Chenopodiaceae I 2, Dysphaniaceae). 47 cosmop. esp. warm (Aus. 17; R: Nuytsia 4(1983)180). ***D. ambrosioides*** (L.) Mosyakin & Clemants (*C. a.*, wormseed, American w., epazote, trop. Am., natur. elsewhere, esp. S Afr., 1 pl. prod. up to 40 000 fr. per yr) – cult. for medic. oil, a vermifuge, fr. also used to season rice, beans etc., fertility control in Urug.; ***D. graveolens*** (Willd.) Mosyakin & Clemants (*C. g.*, yerba del zorillo, Mex.) – medic. & condiment

Dysphaniaceae Pax = Amaranthaceae

Dyssochroma Miers = *Markea*

Dyssodia Cav. Compositae (Tag.-Pect.). Excl. *Boeberastrum, Comaclinium,* incl. *Hymenatherum,* 5 US to Guatemala. R: Sida 11(1986)374). ***D. papposa*** (Vent.) A.S. Hitchc. – ed., local medic. See also *Thymophylla*

Dystaenia Kitag. (~ *Ligusticum*). Umbelliferae (III 8). 4 Korea, Jap.

Dystovomita (Engl.) D'Arcy. Guttiferae (1.). 4 trop. Am.

dzildzilehe *Gymnopodium floribundum*

E

eagle fern *Pteridium aquilinum*; **e. wood** *Aquilaria malaccensis, Gyrinops ledermannii*

Earina Lindl. Orchidaceae (V 13f). 7 W Pac.

early Nancy *Wurmbea dioica*

earth almond *Cyperus esculentus*; **e.apple** *Smallanthus sonchifolius*; **e. chestnut** *Lathyrus tuberosus*; **e. nut** *Arachis hypogaea, Conopodium majus*; **e. stars** *Cryptanthus* spp.

East African cedar *Juniperus procera*; **E. A. sandalwood** *Osyris lanceolata*; **E. Indian arrowroot** *Curcuma angustifolia, Tacca leontopetaloides*; **E. I. dill** *Anethum graveolens*; **E. I. rosewood** *Dalbergia latifolia*; **E. I. satinwood** *Chloroxylon swietenia*; **E. London boxwood** *Buxus macowanii*

Easter cactus *Rhipsalidopsis gaertneri*; **E. lily** *Lilium longiflorum* var. *eximium*, see also *Zantedschia aethiopica*

Eastwoodia Brandegee. Compositae (Ast.-Sol.). 1 SW N Am.: ***E. elegans*** Brandegee. R: SBM 20(2000)37

Eatonella A. Gray. Compositae (Mad.-Huls.). 1 SW N Am.: ***E. nivea*** (D. Eaton) A. Gray. R: FNA 21(2006)348

eau de Cologne (orig. C18 blend) principal ingredient *Citrus* × *aurantium*, also cont. *Rosmarinus officinalis* & *C.* × *limon*; **e. de Créole** *Mammea americana*

eba *Lophira alata*

Ebandoua Pellegrin = *Jollydora*

eben *Dacryodes edulis*

Ebenaceae Guerke. Magnoliidae – Ericales. 3/575 trop. & warm, few temp. R: ANWB 103B(2001)486. Trees, shrubs or rarely geoxylic suffrutices, oft. dioec., s.t. with extrafl. nectaries, s.t. with black heartwood. Lvs usu. in spirals, distich., simple, entire, usu. with extrafl. nectaries abaxially; stip. 0. Fls small, reg., axillary, solit. or in cymose, rarely thyrsoid clusters, oft. with well-dev. staminodes or pistillodes; K (3 – 8), persistent, oft. accrescent in fr., C (3 – 8), lobes contorted, tube s.t. with 8-lobed corona (some II.), A (3-)12 – 20(– 100+), epipet. or borne on receptacle, oft. with 2 anthers per filament, usu. with longit. slits, rarely apical pores, connective oft. larger than anthers, $\underline{G}$ (2 – 5(+)), rarely $\overline{G}$ (II.), pluriloc., each loc. with 2 pend., anatropous, bitegmic ovules & ± divided by a false septum, styles ± distinct. Fr. a berry, rarely tardily dehiscent; seeds large, with thin testa, hard, s.t. ruminate endosperm with oil & hemicellulose, & straight or slightly curved embryo with flat, leafy, usu. emergent cotyledons, photosynthetic or not. n = 15

Classification & genera:

I. **Ebenoideae** (corona 0; $\underline{G}$): *Diospyros, Euclea*

II. **Lissocarpoideae** (Lissocarpaceae; C with 8-lobed corona; $\overline{G}$): *Lissocarpa*

Timbers & fr.-trees: ebonies etc., persimmons & date-plums (*Diospyros*)

Ebenopsis Britton & Rose (~ *Havardia*). Leguminosae (II 4). 3 S Texas, Mex. R: MNYBG 64,1(1996)173

Ebenus L. Leguminosae (III 25). 18 Medit. (Eur. 2) to Balochistan

Eberhardtia Lecomte. Sapotaceae (I 3). 3 S China, SE As., Sabah. R: T.D. Pennington, *S.* (1991)145. ***E. tonkinensis*** Lecomte (SE As.) – timber termite-resistant, seed-oil for cooking

Eberlanzia Schwantes (V). Aizoaceae (V). 8 SW Afr. R: H. E. K. Hartmann, *Ill. Handb. Succ. Pls*, Aiz. A–E(2002)250. See also *Ruschia*

Ebertia Speta = *Drimia*

ebony *Diospyros* spp. esp. *D. ebenum*, though orig. *Dalbergia melanoxylon*; **American** or **Jamaican e.** *Brya ebenus*; **black e.** *Euclea pseudebenus*; **Bombay e.** *Diospyros montana* etc.; **Cape e.** *Heywoodia lucens*; **German e.** 'ebonized' wood; **green e.** *Handroanthus* spp.; **e. heart** *Elaeocarpus bancroftii*; **Macassar e.** *D. celebica*; **Madagascar e.** *D. haplostylis*; **mountain e.** *Bauhinia divaricata*; **Mozambique e.** *Dalbergia melanoxylon*

ebor or **eboe** *Dipteryx oleifera*

Ebracteola Dinter & Schwantes. Aizoaceae (V). 5 SW Afr. R: H. E. K. Hartmann, *Ill. Handb. Succ. Pls*, Aiz. A–E(2002)250

Ecastaphyllum P. Browne = *Dalbergia*

Ecballium A. Rich. Cucurbitaceae (X). 1 Medit.: ***E. elaterium*** (L.) A. Rich. (squirting cucumber; subsp. ***dioicum*** (Batt.) Costich, dioec. with smaller hairier lvs, adapted to hotter summers) – monoec. ± trailing herb with fr. which falls when ripe & pericarp contracts, reducing turgidity, such that seeds in watery fluid are ejected explosively through basal hole; fr. used as purgative (elaterium) & anti-inflammatory (cucurbitacin B), allegorical in Renaissance pictures, e.g. Botticelli's *Venus & Mars* (!)

Ecbolium Kurz (~ *Justicia*). Acanthaceae (III 2c). Incl. *Populina*, 24 trop. & S Afr. Madag.

Ecclinusa Mart. Sapotaceae (IV). 11 trop. S Am. to Trinidad. R: FN(1990)622

Eccoilopus Steud. = *Spodiopogon*

Eccoptocarpha Launert. Gramineae (XXIV 5). 1 S trop. Afr.: ***E. obconiciventris*** Launert

Eccremis Willd. ex Bak. (*Excremis*). Asphodelaceae (Hemerocallidaceae). 1 Andes: ***E. coarctata*** (Ruíz & Pavón) Bak. R: T 58(2009)1128. (A) thick

Eccremocactus Britton & Rose = *Weberocactus*

Eccremocarpus Ruíz & Pavón. Bignoniaceae (4). 3 Peru & Chile. R: AMBG 84(1997)105. Climbers with tendrils & sensitive petioles; fr. a capsule with valves remaining apically united after dehiscence. ***E. scaber*** Ruíz & Pavón (Chile, Arg.) – cult. orn. with orange (cvs with yellow or red) fls

Ecdeiocolea F. Muell. Ecdeiocoleaceae (Restionaceae s.l.). 3 SW Aus. R: Telopea 13(2011)70. Single pls change from female to male in 1 season

Ecdeiocoleaceae Cutler & Airy Shaw (~ Restionaceae). Magnoliidae – Poales. 2/4 SW Aus. Monoec. everg. rush-like pls with creeping rhiz. Culms terete, striate, simple or with 1 – 3 flowering branches. Lvs rudimentary, sheaths closed (splitting), auriculate. Infl. of 1 to few spikelets; glumes usu. fertile. Fl. subtended by longer shiny blackish-brown bract. P 2 + 4, outer conduplicate, inner usu. flat. Male fls with A 4 or 6 with anthers with longit. slits, plus minute pistillode. Female fls with staminodes, $\underline{G}$ (2), each locule with 1 pend. ovule. Fr. a capsule or nut. n = c. 24, 32, 33

Genera: *Ecdeiocolea, Georgeantha*

Ecdysanthera Hook. & Arn. = *Urceola*

Echeandia Ortega. Asparagaceae (Anthericaceae). 82 SW US (3) & Mex. (69; 62 endemic) to Peru (S Am. spp. poss. generically distinct)

Echetrosis Philippi = *Parthenium*

Echeveria DC. Crassulaceae (I 5b). c. 140 warm Am. esp. Mex. R: J. Pillbeam (2008) *The genus E.*; U. Eggli, *Ill. Handb. Succ. Pls*, Crass. (2005)103. Succ. herbs & shrubs with term. leaf-rosettes & axillary (cf. *Cotyledon*) infls. Many cult. orn. spp. & hybrids, oft. used for formal bedding-displays (floral clocks etc.) when infls removed; prop. from stem or leaf-cuttings or, in some spp., fragments of infl. scape

Echidnium Schott = *Dracontium*

Echidnopsis Hook.f. Apocynaceae (V b; Asclepiadaceae). 28 trop. E Afr. to Arabia. R: F. Albers & U. Meve, *Ill Handb. Succ. Pls*, Asclep. (2002)129, PSE 265(2007)71. Some cult. orn. succ. with minute lvs on tubercles

Echinacanthus Nees. Acanthaceae (III 2). 4 Himal. & China. R: EJB 51(1994)186

Echinacea Moench. Compositae (Helia.-Zinn.). 4 E US. R: SB 27(2002)610. Cult. orn. (cone flowers) esp. ***E. pallida*** (Nutt.) Nutt. var. ***angustifolia*** (DC.) Cronq. (*E. angustifolia*) – rhiz. the most widely used Native American med. of Plains, patent med. 1870 now shown to have antiviral effects & stimulate prod. of white blood-cells so used as temp. relief from cold symptoms (though clinical trials in 2005 allege comm. preparations, worth $155 M in 2004, do not prevent or ease colds) & to aid wound-healing, & ***E. purpurea*** (L.) Moench – rhiz. local medic., both with capitula to 15 cm diam.

Echinaria Desf. Gramineae (XVI 13). 1 Medit., middle E: ***E. capitata*** (L.) Desf.

Echinaria Heister ex Fabr. = *Cenchrus*

Echinocactus Link & Otto. Cactaceae (III 9). 4 SW N Am. Ribbed, oft. large cult. orn. (most die if apex damaged because they cannot branch), oft. with long spines (to 8 cm in ***E. polycephalus*** Engelm. & Bigelow, branching sp.), incl. ***E. horizonthalonius*** Lem. – pulp eaten & ***E. platyacanthus*** Link & Otto (C Mex.) – sacred in pre-Columbian culture, now used in confectionery, medic. etc.

Echinocaulon (Meissn.) Spach = *Persicaria*

Echinocephalum Gardner (~ *Melanthera*). Compositae (Helia.). 1 Parag, Braz.: ***E. latifolium*** Gardner. R: Nuytsia 23(2013)381

Echinocereus Engelm. Cactaceae (III 1). 67 SW N Am. R: KMMS 1(1985). All in cult. (hedgehog cacti). ***E. enneacanthus*** Engelm. & ***E. pectinatus*** (Scheidw.) Engelm. etc. with ed. (strawberry-flavoured) fr., the spine-clusters easily removed

Echinochloa P. Beauv. Gramineae (XXIV). c. 45 warm (Eur. 1). R: T. Yabuno & H. Yamaguchi (2001) *hie to iu shokubutsu* [*E.*]. Polyploid series (x = 9). Grains ground to paste & baked by Aus. Aborigines. ***E. colona*** (L.) Link (jungle rice, Shama millet, trop. & warm) & ***E. crus-galli*** (L.) P. Beauv. (barnyard grass, cockspur, Euras.) – hexaploid weeds, esp. in rice; bad weed in subtrop. & warm; ***E. esculenta*** (A. Br.) H. Scholz (*E. utilis*, billion dollar grass, Jap. [barnyard] millet, Sanwa m.) – hexaploid derived from *E. crus-galli*, cereal in China &

Jap., cropping 6 weeks after sowing (imp. E Eur.), natur. US, where grown for fodder, as in Aus., birdseed, grown for game-cover GB & US; ***E. frumentacea*** Link ([Ind.] barnyard millet, b. grass, jungle rice, Ind.) – hexaploid cereal (Ind.) derived from *E. colona*, birdseed, fodder crop NE Aus.; ***E. glabrescens*** Munro ex Hook.f. (As.) – widespread hexaploid weed of rice; ***E. obtusiflora*** Steud. (W Afr.) – diploid weed in *Oryza glaberrima*; ***E. oryzicola*** (Vasinger) Vasinger (tetraploid, E As.) & ***E. oryzoides*** (Ard.) Fritsch (hexaploid, W As. or Ind.?) – widespread paddy-rice mimic weeds; ***E. pyramidalis*** (Lam.) A. Hitchc. & Chase (antelope grass, trop. & S Afr., Madag.) – fodder, locally used flour; ***E. turneriana*** (Domin) J. Black (12-ploid) & ***E. inundata*** P. W. Michael & Vick. (8-ploid) (channel millets, Aus.) – nat. forage, once promoted as grain crops

Echinocitrus Tanaka = *Triphasia*

Echinocodon Hong. Campanulaceae (I 1). 1 China: ***E. draco*** (Pampan.) Hong (*E. lobophyllus*). R: D.Y. Hong, *Monogr. Codonopsis* (2015)223

Echinocodon Kolak. = *Campanula*

Echinocodonia Kolak. = *Campanula*

Echinocoryne H. Robinson (~ *Vernonia*). Compositae (Vern.-Lep.). 6 Braz.

Echinocystis Torrey & A. Gray. Cucurbitaceae (XII). 1 N Am.: ***E. lobata*** (Michaux) Torrey & A. Gray – cli. herb with tubers, s.t. used medic. & seeds used as beads by Native Americans, cult. orn.

Echinodorus Rich. Alismataceae. 28 trop. Am. (N Am. 4). R: KB 63(1975)529. Cult. orn. aquarium pls (burhead, sword plant) esp. ***E. berteroi*** (Spreng.) Fassett (cellophane pl., N Am. to Patagonia) – submerged lvs membranous, ribbon-like

Echinofossulocactus Lawrence = *Echinocactus*

Echinoglochin (A. Gray) Brand = *Plagiobothrys*

Echinolaena Desv. Gramineae (XXIII 3). 9 trop. Am., Madag. ***E. polystachya*** (Kunth) Hitchc. (German grass, trop. Am., natur. OW, Hawaii) – fodder

Echinomastus Britton & Rose = *Sclerocactus*

Echinopaepale Bremek. = *Strobilanthes*

Echinopanax Decne & Planch. ex Harms = *Oplopanax*

Echinopepon Naud. (~ *Echinocystis*). Cucurbitaceae (XII). Incl. *Apatzingania*, c. 20 Am. esp. Mex. R: AMBG 85(1998)262. ***E. arachnoidea*** (Dieterle) A.K. Monro & Stafford (*A. a.*) – ann., fr. ripens in ground (cf. *Arachis*)

Echinophora Tourn. ex L. Umbelliferae (III 1). 9 Medit. (Eur. 2) to Iran, 1 carpel aborts; umbel with 1 bisex. fl. surrounded by males, the spiny pedicels to the latter encl. fr.

Echinopogon P. Beauv. Gramineae (XVI 5). 7 Aus., NZ, NG. R: HIP 33(1935)t.3261. If grazed young ***E. caespitosus*** C. Hubb. & ***E. ovatus*** (Forst.f.) P. Beauv. (Aus.) cause staggers

Echinops L. Compositae (Card.-Ech.). c. 120 Eur. (12), Medit. to C As. & trop. Afr. mts. Alks. Spherical heads comprising ∞ 1-flowered capitula, each with an involucre, bee-poll. Cult. orn. robust herb. pls (globe thistles), esp. ***E. bannaticus*** Rochel ex Schrad. ('*E. ritro*', SE Eur.) – blue fls

Echinopsis Zucc. Cactaceae (III 4). Incl. *Lobivia* & *Trichocereus*, 101 S Am. Heterogeneous. Cult. orn. ribbed cacti with diurnal or nocturnal fls incl. ***E. chiloensis*** (Colla) Friedrich & Rowley (*E. chilensis*, *T. chiloensis*, Chile) – long spines (poss. prot. against mistletoe (*Tristerix aphyllus*) infection as deterring seed-disp. birds), fr. ed., made into drinks & ***E. pachanoi*** (Britton & Rose) Friedrich & Rowley (San Pedro cactus, Ecuador, Peru) – to 10 m tall, hallucinogenic (mescaline); ***E. pasacana*** (Ruempler) Friedrich & Rowley (*T. pasacana*, W Arg., S Bolivia) – vasc. system (closed) used for boxes, church beams, doors & fences

Echinopterys A. Juss. Malpighiaceae. 3 Mex. Mericarps spiny

Echinorhyncha Dressler (~ *Chondrorhyncha*). Orchidaceae (V 12j). 5 N S Am. R: Lankesteria 5(2005)94

Echinosepala Pridgeon & M.W. Chase (~ *Pleurothallis*). Orchidaceae (V 13c). 11 R: Lindleyana 17(2002)100

Echinosophora Nakai = *Sophora*

Echinospartum (Spach) Fourr. (~ *Genista*). Leguminosae (III 9). 5 SW Eur. R: Fl. Iberica 7,1(1999)119

Echinostephia (Diels) Domin. Menispermaceae (V). 1 SE Queensland: ***E. aculeata*** (Bailey) Domin. R: FA 2(2007)379

Echiochilon Desf. Boraginaceae (B.1). Incl. *Leurocline, Sericostoma, Tetraedrocarpus*, 15 arid NW Afr. to Arabia, E Afr. & NW Ind. R: BJLS 130(1999)196. Some firewood sources

Echiochilopsis Caball. = *Echiochilon*

Echioides Ortega = *Arnebia*

Echiostachys Levyns (~ *Lobostemon*). Boraginaceae (B.2.2). 3 W Cape. R: Strel. 9(2000)375

Echitella Pichon = *Mascarenhasia*

Echites P. Browne. Apocynaceae (II e). 9 Carib. R: Britt. 49(1997)329, 54(2003)310. Foliage of ***E. rubrovenosa*** Linden (Braz.) app. 'mimetic' of that of *Gymnostachyum verschaffeltii* Lem.

Echium Tourn. ex L. Boraginaceae (B.2.2). 60 Macaronesia (28, 27 endemic), Eur. (18), W As., N & S Afr. R: JAA suppl. 1(1991)137. Alks; some form. medic. incl. ***E. vulgare*** L. (viper's bugloss, Euras., pernicious weed in N Am., NZ where ***E. candicans*** L.f. (Madeira) – shrub to 2.5 m also aggressive colonist). Cult. orn., esp. shrubby & pachycaul spp. of Macaronesia, some unbranched (Holttum's Model; others Scarrone's Model) & with infls to 4 m tall, also ***E. plantagineum*** L. (*E. lycopsis*, bugloss, Euras.) – biennial, pernicious weed (Paterson's curse) but emergency fodder (Salvation Jane), though toxic for horses, & imp. honeybee-pl. (for 'Aus. bluebell honey') in E Aus. (introd. 1843, weedy by c. 1890, now prod. seed-rain up to 30 000 seeds per m^2), seed-oil rich in 'good' omega 3 fatty acids like fish oil; ***E. nervosum*** Dryander (Madeira) – visited by lizards, most other spp. bee-poll.; ***E. wildpretii*** Pearson ex Hook.f. (Canary Is.) – red fls & copious dilute nectar attractive to birds though colour in ultaviolet & shape attractive to bees

Eclecticus O'Byrne. Orchidaceae (V 16c). 1 Thailand: ***E. chungii*** O'Byrne

Eclipta L. Compositae (Helia.-Ecl.). 6 Aus., S Am., ***E. prostrata*** (L.) L. (*E. alba*) introd. OW – blackish dye for hair, tattooing etc. in Ind., source of thiophene derivatives active against nematodes. R: Nuytsia 23(2013)47. Pappus 0

Ecliptostelma Brandegee = *Marsdenia*

Ecpoma K. Schum. = *Sabicea*

Ectadiopsis Benth. = *Cryptolepis*

Ectadium E. Meyer. Apocynaceae (III; Periplocaceae). 3 S Afr. R: SAJB 56(1990)113

Ectinocladus Benth. = *Alafia*

Ectopopterys W.R. Anderson. Malpighiaceae. 1 Colombia, Peru: ***E. soejartoi*** W.R. Anderson. R: CUMH 14(1980)11

Ectotropis N.E. Br. (~ *Delosperma*). Aizoaceae (V). 1 EC S Afr.: ***E. alpina*** N.E. Br. R: H. E. K. Hartmann, *Ill Handb. Succ. Pls*, Aiz. A–E(2002)257

Ectozoma Miers = *Juanulloa*

Ectrosia R. Br. = *Eragrostis* (but see R: FA 44B(2005)426)

Ectrosiopsis (Ohwi) Jansen = *Eragrostis*

Ecua Middleton (~ *Parsonsia*). Apocynaceae (IIc). 1 Moluccas: ***E. moluccensis*** Middleton. R: Blumea 41(1996)33

Ecuadendron D. Neill. Leguminosae (I 2). 1 W Ecuador: ***E. acosta-solisianum*** D. Neill. R: Novon 8(1998)45

Ecuador laurel *Cordia alliodora*

Ecuadorella Dodson & G. Romero = *Otoglossum*

Ecuadoria Dodson & Dressler = *Microthelys*

edamame *Glycine max*

Edanyoa Copel. = *Bolbitis*

Edbakeria R. Viguier = *Pearsonia*

eddoes *Colocasia esculenta* cvs

edelweiss *Leontopodium nivale* subsp. *alpinum*; **NZ e.** *Leucogenes* spp.

Edgaria C.B. Clarke = *Herpetospermum* (but see FOC 19(2011)34)

Edgeworthia Meissn. Thymelaeaceae (Thym.-Daph.). 3 China, Jap. & (? introd.) SE US (Georgia). Shrubs with bark used for high-class paper (for currency etc.) esp. ***E. papyrifera*** Sieb. & Zucc. (*E. chrysantha*, paperbush, mitsumata, China, long cult. Jap.) – living stems tied in knots, ess. oil from fls used in folk eye med. (China) & ***E. gardneri*** (Wall.) Meissn. (Nepal, Sikkim); cult. orn

edinam *Entandrophragma angolense*

Edisonia Small = *Matelea*

Edithcolea N.E. Br. Apocynaceae (Vb; Asclepiadaceae). 1 E & NE Afr. to Socotra & Arabia: ***E. grandis*** N.E. Br. – cult. orn. R: FTEA Apoc.2(2012)323

Edithea Standl. = *Omiltemia*

Edmondia Cass. (~ *Helichrysum*). Compositae (Gnap.-Gnap.). 3 Cape. R: OB 104(1991)153

Edmundoa Leme (~ *Canistrum*). Bromeliaceae (3). 3 E Braz.

Edraianthus (A. DC.) A. DC. (~ *Campanula*). Campanulaceae (I 3). Excl. *Muehlbergella*, 13 Balkans, ***E. graminifolius*** (L.) A. DC. ext. to Italy SE Eur. (9) to Cauc. R: GBIUS 26(1973)1.

Superficially like *Wahlenbergia* but capsule with irreg. dehiscence & lvs linr., elongate. Cult. orn. rock-pls esp. *E. g.*

Edrastima Raf. (~ *Hedyotis*). Rubiaceae (IV 15). 5 trop. OW. R: T 64(2015)315

Eduardoregelia Popov = *Tulipa*

Edwardsia Salisb. = *Sophora*

eelgrass *Zostera marina*

Eenia Hiern & S. Moore = *Anisopappus*

Efulensia C.H. Wright (~ *Deidamia*). Passifloraceae. 2 trop. Afr.

efwatakala grass *Melinis minutiflora*

Eganthus Tieghem = *Minquartia*

Egenolfia Schott = *Bolbitis*

Egeria Planch. (~ *Elodea*). Hydrocharitaceae. 3 subtrop. S Am. R: Darwiniana 12(1961)293, AB 19(1984)74. ***E. densa*** Planch. – invasive in Sumatra, Aus., NZ, W US

egg fruit or **e. plant** (incl. **Thai e. p.**) *Solanum melongena;* **dinosaur e.** = plumcot; **pea e. plant** *Solanum torvum*

Eggelingia Summerh. Orchidaceae (V 16d). 3 trop. Afr.

eggs and bacon *Lotus corniculatus*

eglantine *Rosa rubiginosa*

Egleria Eiten = *Eleocharis* (but see Phytol. 9(1964)481)

Eglerodendron Aubrév. & Pellegrin = *Pouteria*

Egletes Cass. Compositae (Ast.-Gran.). 6 trop. Am. R: Lloydia 12(1949)239, 248

egusi *Melothria mannii*

Egyptian bean *Nelumbo nucifera;* **E. grass** *Dactyloctenium aegyptium;* **E. lotus** *Nymphaea lotus;* **E. onion** *Allium* × *proliferum*

Ehrendorferia Fukuhara & Lidén (~ *Dicentra*). Papaveraceae. 2 Calif.

Ehretia P. Browne. Boraginaceae (Ehr.; Ehretiaceae). Incl. *Carmona, Rotula,* c. 40 trop. & warm (Mal. 12, Am. 3 – R: AMBG 76(1989)1059), fossils from Eur. Some timbers incl. ***E. acuminata*** R. Br. (koda wood, S China to Aus.); some medic. incl. ***E. philippinensis*** A. DC. (Philipp.) – effective in treatment of diarrhoea. ***E. aquatica*** (Lour.) Gottschling & Hilger (*R. a.*, OW trop., E Braz.) – rheophyte with tough branches, psychoactive (smoked with tobacco or mixed with coconut milk in Kerala), med. for urinary problems; ***E. microphylla*** Lam. (*C. retusa, E. buxifolia, E. monopyrena,* Fukien tea, Indomal. to Aus.) – cult. orn. esp for hedging in Mal. & bonsai, natur. (bird-disp.) in Hawaii

Ehretiaceae Mart. = Boraginaceae

Ehrharta Thunb. Gramineae (IV 2). Incl. *Microlaena, Tetrarrhena, Zotovia,* 37 S Afr. (25, 1 ext. to Ethiopia), Masc., Indonesia to NZ. Some pasture grasses; some invasives (6 spp. in Aus.) like ***E. calycina*** Sm. (S Afr.) in Aus., W US & ***E. erecta*** Lam. (S Afr.) in Aus., NZ, W US. ***E. juncea*** (R. Br.) Spreng. (*T. j.*, SE Aus.) – eaten by wombats, in absence of which forms 8-m tall tangles inimical to lyrebirds; ***E. stipoides*** Labill. (*M. s.*, Aus., NZ) – basal cleistogamous florets like *Amphicarpum amphicarpon,* turf & soil-binder; ***E. villosa*** Schultes f. (S Afr.) – sand-binder in NSW

Eichhornia Kunth. Pontederiaceae. 7 trop. Am. Rhizomatous aquatics with floating or submerged lvs. ***E. azurea*** (Sw.) Kunth (Braz.) – fls dimorphic; ***E. paniculata*** (Spreng.) Solms-Laub. (trop. Am.) – tristylous though monomorphic in Jamaica (morphs self-compatible?). ***E. crassipes*** (Mart.) Solms-Laub. (water-hyacinth, trop. Am.) – tristylous (only in NE Braz.), infls elongating overnight & usu. all fls opening at once, next morning bending over so that fr. develops in water; a free-floater with swollen petioles cont. aerenchyma, the laminas raised above water-level & acting as sails; 2 parents can give 30 offspring by veg. budding in 23 days & 1200 in 4 months, a yield of 470 t/ha.; orig. introd. as cult. orn. (Louisiana Exposition 1884) now most aggressive trop. aquatic weed, choking many trop. waterways & natur. in Portugal; when grown on sewage & other 'grey' water (tolerant of 100 p.p.m phenol) with bacteria & fungi assoc. with r., can remove toxins (e.g. arsenic from drinking water) & yield 800 kg dry matter/ha/day, which in Indonesia is harvested for pigs & has been suggested as a methane source, fibre now used in basketry & furniture

Eichlerago Carrick = *Prostanthera*

Eichleria Progel = *Rourea*

Eichlerodendron Briq. = *Xylosma*

Eidothea A. Douglas & Hyland. Proteaceae (IV). 2 NE Aus. R: Telopea 9(2002)821

Eigia Soják (*Stigmatella*). Cruciferae (4). 1 Israel, NW Saudi Arabia: ***E. longistyla*** (Eig) Soják

Einadia Raf. = *Chenopodium* (but see Nuytsia 4(1983)199)

einkorn diploid *Triticum* spp. esp. *T. monococcum*
Einomeia Raf. = *Aristolochia*
Eionitis Bremek. = *Oldenlandia*
Eirmocephala H. Robinson. Compositae (Vern.-Vern.). 3 C Am., Andes to Bolivia
Eisocreochiton Quis. & Merr. = *Creochiton*
Eitenia R. King & H. Robinson (~ *Eupatorium*). Compositae (Eup.-Prax.). 2 Braz. R: MSBMBG 22(1987)395
Eizia Standl. Rubiaceae (I 5). 1 Mex.: ***E. mexicana*** Standl.
ejow *Arenga pinnata*
Ekebergia Sparrman. Meliaceae (II-Trich.). 4 trop. & S Afr. Some timbers esp. ***E. capensis*** Sparrman (Cape ash, dog plum, trop. & S Afr.)
ekhimi *Piptadeniastrum africanum*
Ekimia Duman & M. Watson (~ *Prangos*). Umbelliferae (III 2). 3 SW As. R: T 65(2016)585
ekki *Lophira* spp.
Ekmania Gleason. Compositae (Vern.-Pip.). 1 Cuba: ***E. lepidota*** (Griseb.) Gleason
Ekmanianthe Urb. (1). Bignoniaceae. 2 Cuba, Hispaniola
Ekmaniocharis Urb. = *Mecranium*
Ekmaniopappus Borhidi (~ *Herodotia*). Compositae (Sen.-Tuss.) 2 Hispaniola
Ekmanochloa A. Hitchc. Gramineae (VI 3). 2 E Cuba (serpentine). R: AMBG 80(1993)857. Caespitose; blades of ***E. subaphylla*** A. Hitchc. ± suppressed, the culms photosynthetic
ekpogoi *Berlinia* spp.
ekra, ekar *Saccharum bengalense*
Elachanthemum Ling & Y.R. Ling (~ *Artemisia*). Compositae (Anth.-Art.). 1 China: ***E. intricatum*** (Franch.) Ling & Y.R. Ling. R: BBR 23(2003)147
Elachanthera F. Muell. = *Luzuriaga*
Elachanthus F. Muell. Compositae (Ast.-Pod.). 3 arid Aus. R: Phytoneuron 2012–39:22
Elacholoma F. Muell. & Tate (~*Mimulus*). Phrymaceae (Mim.; Scrophulariaceae s.l.). 1 C Aus.: ***E. hornii*** F. Muell. & Tate
Elachyptera A.C. Sm. (~ *Hippocratea*). Celastraceae (III). 7 trop. Afr., Madag., Am.
Elaeagia Wedd. Rubiaceae (Cond.). c. 15 trop. Am. ***E. utilis*** Wedd. (Colombia) – pasto lacquer
Elaeagnaceae Juss. Magnoliidae – Rosales. 3/45 temp. & warm N hemisph. to trop. As. & Aus. Shrubs to small trees, rarely scandent, oft. shoots reduced to spines, usu. decid., with nitrogen-fixing r. nodules (actinobacteria); indumentum of scales or stellate hairs, oft. giving grey or silvery appearance. Lvs simple, entire, in spirals (opp. in *Shepherdia*); stip. 0. Fls bisex. (s.t. pls dioec. or polygamo-dioec.), reg., perigynous with hypanthium usu. constricted above G (cupulate to ± flat in male fls), solit. or in small umbels; K (2)4(6) valvate lobes on hypanthium, oft. petaloid, C 0, A in throat alt. with & same no. as K (*Elaeagnus*) or 2 × K alt. & opp. (*Hippophae* & *Shepherdia*), anthers with longit. slits, nectary-disk (lobed) also in throat, $\overline{G}$ 1 with 1 style & 1 basal, anatropous, bitegmic ovule with funicular obturator. Fr. drupe- or berry-like, the achene enveloped by, but free from, persistent hypanthium which becomes fleshy oft. with a bony inner layer; seed 1 with straight embryo & 2 fleshy oily & proteinaceous cotyledons & ±0 endosperm. n = 6, 10, 11, 13, 14

Genera: *Elaeagnus, Hippophae, Shepherdia*

Seed-structure like Rhamnaceae

Pls mostly of steppe & coasts; some cult. orn. & ed. fr.

Elaeagnus Tourn. ex L. Elaeagnaceae. c. 40 Eur. (1 natur.), As. to N Queensland, N Am. (1). Alks. Some cult. orn. shrubs (oleaster) esp. ***E. angustifolia*** L. (Russian olive, Trebizond date, SE Eur., C As. to China, disjunct, invasive US), ***E. multiflora*** Thunb. (Jap., China) – ed fr. (goumi) with several cvs, ***E. pungens*** Thunb. (China & Jap.) – common varieg. forms, ***E. × reflexa*** Morren & Decne (*E. pungens* × *E. glabra* Thunb. (China, Jap.)) – scrambling hedge-pl. widely used in NZ, now escaped & ***E. × submacrophylla*** Serv. (*E.* × *ebbingei*; *E. macrophylla* Thunb. (Korea, Jap.) × *E. pungens*), ***E. umbellata*** Thunb. (autumn olive, As.) – invasive Hawaii, N Am., fr. ed. raw or preserved; many other spp. locally imp. as fr., e.g. ***E. latifolia*** L. (*E. conferta*, S & E As.) in Sikkim, swamping creeper in forest canopy of Krakatoa
Elaeis Jacq. Palmae (V 8c). 2 trop. Am. (1) & Afr. (1). Monoec. palms which hybridize in cult. ***E. guineensis*** Jacq. (oil palm, W Afr., now widely planted in trop., firstly from 4 (exceptionally fine) pls in Bogor (last 1 fell 1993), Java 1848 (via Mauritius) the first estates (Sumatra) 1911, natur. Malay Pen. where orn. 1870, first estates 1917), now (esp. 'tenera'

cvs) have replaced much Mal. lowland rain forest – poll. weevils (now introd. to As.) suppl. by wind, most imp. source (50 M t p. a. by 2011) of (saturated) oil for margarine, orig. (W Afr.) medic.; oil from pericarp (palm-oil) used for soap, candles etc., oil from rest is palm kernel oil used in many cosmetics etc.; palm-oil 'diesel' taxis in some trop. cities, thickening agent in napalm manufacture; Malaysia with (2003) 51% maket-share (harvested by E Indonesians), oil yield c. 3475 kg/ha/yr.; trunk can be tapped for toddy; fr. peduncle fibre used for 'ecomat' mulches. ***E. oleifera*** (Kunth) Cortés (Am. oil palm, trop. Am.) – lesser imp., stem rots at base but new r. form above, so pl. potentially immortal

Elaeocarpaceae Juss. Magnoliidae – Oxalidales. Incl. Tremandraceae, 12/550 trop. & warm exc. Cont. Afr. Trees or shrubs (s.t. ericoid), oft. with alks; hairs simple. Lvs usu. in spirals, simple (oft. serrate), withering red to orange or yellow; stip. persistent or not. Fls pendent, usu. bisex., reg., without epicalyx, in racemose or cymose infls; K (3)4, 5(– 11), usu. valvate, s.t. basally connate, C (3)4 or 5(6), usu. free, oft. apically fringed, usu. valvate, rarely 0, A (1) 2 × K to 300, originating centrifugally & oft. ± in 5 antesepalous groups on a ± definite disk or enlarged receptacle forming an androgynophore, anthers with apical slits or pores or short longit. lat. slits, connective oft. ± conspic. prolonged, $\underline{G}$ ((1)2 – 8(9)) with as many locs & 1 style, each loc. with 1 – 30 pend., anatropous, bitegmic ovules with zig-zag micropyle. Fr. a capsule (oft. armed) or a drupe; seeds 1-∞ with much oily, proteinaceous endosperm, oft. arillate or hairy, & with straight or J – U-shaped embryo. n = 12, 14, 15, 21, c. 90

Genera: *Aceratium, Aristotelia, Crinodendron, Dubouzetia, Elaeocarpus, Peripentadenia, Platytheca, Sericolea, Sloanea, Tetratheca, Tremandra, Vallea*

Form. considered allied to Tiliaceae (= Malvaceae) but seed-structure militated against that; Tremandraceae form. allied with Polygalaceae here through DNA analysis. Some cult. orn. (*Crinodendron, Tetratheca* etc.) & some locally imp. timbers, dyes & fr.

Elaeocarpus Burm. ex L. Elaeocarpaceae. c. 290 trop. & warm OW exc. Afr. (Madag. 8, As. c. 25, New Caled. 29, NZ 2, Hawaii 1). Aubréville's Model; domatia; some rheophytes; heteroblasty in NZ. Some timbers esp. for flooring e.g. in NE Aus. ***E. obovatus*** G. Don (blueberry ash) & ***E. kirktonii*** F. Muell. ex F.M. Bail. (white beech, whitewood), cult. orn. & ed. fr., some local medic. ***E. bancroftii*** F. Muell. (ebony heart, Aus.) – Karanda nuts (Queensland); ***E. bifidus*** Hook. & Arn. (Hawaii) – form. bark for cordage, wood for frames of grass huts; ***E. dentatus*** (Forst. & Forst.f.) Vahl (NZ) – bark source of a blue-black dye; ***E. hookerianus*** Raoul (pokaka, NZ) – juvenile divaricate; ***E. reticulatus*** Sm. (E Aus.) – lignotubers; ***E. serratus*** L. (Indomal.) – fr. (Ceylon olive) used in curry; ***E. sphaericus*** (Gaertn.) K. Schum. (*E. ganitrus*, olive nut, Ind. to W Mal.) – seeds used as beads, Ayurvedic medic. (CNS depressant); ***E. tectorius*** (Lour.) Poir. (*E. robustus*, Indomal.) – village fr. tree

Elaeodendron Jacq. (~ *Cassine*). Celastraceae (I). c. 40 Macaronesia, OW trop. (Afr. 8; R: SAJB 64(1998)93), WI. Seeds exarillate (cf. *Euonymus*)

Elaeoluma Baill. Sapotaceae (IV). 4 trop. Am. R: FN 52(1990)240

Elaeophora Ducke = *Plukenetia*

Elaeophorbia Stapf = *Euphorbia*

Elaeopleurum Korovin = *Seseli*

Elaeoselinum Koch ex DC. (~ *Thapsia*) Umbelliferae (III 12). (Excl. *Distichoselinum, Margotia*) 4 [or 1] Medit. (Eur. 3). R: Lagascalia 13(1985)213

Elaeosticta Fenzl (~ *Scaligeria*). Umbelliferae (III 8). Incl. *Muretia*, 26 Eur. (2) to C As. R: NSL 20(1983)140

Elaphandra Strother. Compositae (Helia.-Ecl.) 13 trop. Am. R: SBM (1991)17

Elaphanthera Hallé (~ *Exocarpos*). Santalaceae. 1 New Caled.: ***E. baumannii*** (Stauffer) Hallé

Elaphoglossum Schott ex J. Sm. Dryopteridaceae (II; Lomariopsidaceae). c. 600 trop. & warm (Eur. 1: ***E. semicylindricum*** (Bowdich) Benl (Azores, Madeira), Aus. 1, NZ 0) to Jap. esp. Am. (e.g. Venez. 98). R: AFJ 70(1980)47. App. migrated from Am. to Afr. Epiphytes with simple, tongue-like fronds without indusia, some with epiphyllous bud propagules. Some cult. orn.

Elasis D. Hunt. Commelinaceae (II 1f). 1 Ecuador: ***E. hirsuta*** (Kunth) D. Hunt. R: KB 33(1978)332

Elateriopsis Ernst = *Hanburia*

Elateriospermum Blume. Euphorbiaceae (Crot.-Elat.). 1 S Thailand, W Mal.: ***E. tapos*** Blume – wood for general use, seeds ed. when prussic acid removed by boiling, also used in a game like conkers. R: Blumea 49(2004)428

elaterium *Ecballium elaterium*

Elatinaceae Dumort. Magnoliidae – Malpighiales. 2/34 temp. & (esp.) trop. Herbs to (rarely) suffrutices of wet places, oft. creeping & rooting, resinous. Lvs opp. or whorled, simple, entire to toothed; stip. minute. Fls small, reg., bisex., axillary, solit. or in cymes; K 2 – 5(6), s.t. connate, C 2 – 5(6), imbr., persistent, A same as or (2)3–6(–10) in 1 or 2 whorls, anthers with longit. slits, G ((2)3–5) with distinct styles & as many loc. but partitions not reaching apex in some *Bergia* spp. Ovules ∞, on axile placentas, anatropous, bitegmic, s.t. with zig-zag micropyle. Fr. a septicidal capsule; seeds with straight or curved embryo & 0 endosperm. x = 6, 9
Genera: *Bergia, Elatine*
Prob. allied to Guttiferae but exceptional in alliance in being largely aquatic while embryology recalls Myrtales

Elatine L. Elatinaceae. 10 trop. & temp. (Eur. 8). Waterwort. Some grown in aquaria, incl. ***E. alsinastrum*** L. (Medit., Euras.) with unbranched stems & whorls (unique in fam.) of lvs ('*Hippuris* syndrome')

Elatinoides (Chav.) Wettst. = *Kickxia*

Elatostema Forst. & Forst. f. Urticaceae (II). Incl. *Pellionia, Procris*, c. 300 trop. OW to NZ (1). R: FRBeih. 83, 1 – 2(1935–6). Prob. heterogeneous. Fr. ejected by staminodes; some apogamous. ***E. elegans*** Winkler (NG) – rheophyte. Some local medic., some cult. orn. for coloured lvs & stems esp. ***E. repens*** (Lour.) Hall.f. (Myanmar to W Mal.) – pot-pl.

Elatostematoides Robinson = *Elatostema*

Elattospermum Soler. = *Breonia*

Elattostachys (Blume) Radlk. Sapindaceae. c. 20 Mal. to Aus. (4) & W Pac.

Elburzia Hedge. Cruciferae (47). 1 NW Iran: ***E. fenestrata*** (Boiss. & Hohen.) Hedge. R: NRBGE 29(1969) 181

Elcomarhiza Barb. Rodr. = *Marsdenia*

elder *Sambucus nigra*; **American e.** *S. nigra* var. *canadensis*; **e.berry** *S.* spp.; **blue e.** *S. caerulea*; **box e.** *Acer negundo*; **dwarf e.** *S. ebulus*; **marsh e.** *Euphrosyne xanthiifolia*; **poison e.** *Rhus vernix*; **stinking e.** *S. pubens*

elecampane *Inula helenium*

Electranthera Mesfin & al. (~ *Coreopsis*). Compositae (Coreops.). 3 Mex. R: Phytoneuron 2015–68:4

Elegia L. Restionaceae. Incl. *Chondropetalum, Dovea*, 51 S Afr. R: Bothalia 15(1985)418,427,435, 40(2010)11. ***E. capensis*** (Burm.f.) Schelpe – orig. described as a horsetail; ***E. tectorum*** (L.f.) Moline & Linder – form. imp. thatch in Cape

Eleiodoxa (Becc.) Burret. Palmae (I 3b). 1 W Mal.: ***E. conferta*** (Griff.) Burret – sarcotesta a tamarind subs. (with raw prawns = 'umei' (Sarawak)). R: J. Dransfield & al., *Gen. Palm.* (2008)170

Eleiosina Raf. Older name for *Sibiraea*

Eleiotis DC. Leguminosae (III 19). 2 Ind. to Myanmar, Sri Lanka

Elekmania R. Nordenstam (~ *Senecio*). Compositae (Senec.). c. 11 Hispaniola. R: CN 44(2006) 66

elemi oleo-resins from var. pls form. much used in varnishes, printing inks & ointments. **African e.** *Canarium schweinfurthii*; **American e.** *Bursera simaruba*; **Brazilian e.** *Protium heptaphyllum*; **Carana e.** *P. carana*; **E Afr. e.** *Boswellia frereana*; **Manila e.** *C. luzonicum*; **WI e.** *Dacryodes excelsa*; **Yucatan e.** *Amyris plumieri*

Eleocharis R. Br. Cyperaceae (II 3). Incl. *Websteria*, c. 200 cosmop. (Eur. 15, N Am. 67). Spike rushes; some with bulbils. R: T 46(1997)437. ***E. caespitosissima*** Bak. (Madag.) – amphicarpic with aerial & basal (seeds asexual?) fr. buried by geotropic culms; ***E. confervoides*** (Poir.) Steud. (*W. c.*, trop.) – leafless; ***E. dulcis*** (Burm.f.) Henschel **'Tuberosa'** (*E. tuberosa*, (Chinese) water chestnut, OW trop.) – tubers or corms gathered from ground or the crops of captured magpie geese by Arnhem Land (N Aus.) Aborigines, form. imp. E Aus., cult. China etc. in flooded fields drained before harvest, the principal white crunchy veg. in Chinese food like chop suey; ***E. erythropoda*** Steud. ('*E. calva*', kohekohe, N Am., Hawaii) – red basal sheath used in Niihau mats (see *Cyperus laevigatus*); ***E. sphacelata*** R.Br. (E Aus.) – corms eaten by Sydney Aboriginal people pre-conquest; lvs of other spp. used for matting & for women's skirts in NG

Eleogiton Link = *Isolepis*

Eleorchis F. Maek. Orchidaceae (V 10a). 1 Jap.: ***E. japonica*** (A. Gray) F. Maek.

elephant apple *Limonia acidissima*; **e. climber** *Argyreia nervosa*; **e. ear** *Begonia* spp.; **e. ('s) foot** *Cyphostemma juttae, Dioscorea elephantipes*; **e. f. tree** *Beaucarnea recurvata*; **e. garlic** *Allium* *ampeloprasum* Ampeloprasum Group; **e. grass** *Pennisetum purpureum*; **e. tree** *Pachycormus discolor*

Elephantomene Barneby & Krukoff. Menispermaceae (II). 1 NE S Am.: ***E. eburnea*** Barneby & Krukoff – disjunct distrib., Fr. Guiana & Peru/Braz. border

Elephantopus Vaill. ex L. Compositae (Vern.-Ele.). c. 28 trop. & warm. ***E. scaber*** L. a bad weed in warm regions

Elephantorrhiza Benth. Leguminosae (II 1). 9 trop. & S Afr. R: Bothalia 11(1974)247. r. for tanning

Elettaria Maton. Zingiberaceae (II 4). 7 Ind. to W Mal. Infls on prostrate shoots with scale lvs arising from rhizomes. ***E. cardamomum*** (L.) Maton (cardamom, Ind.) – widely cult. in As. (& since 1920 Guatemala) for spicy seeds (36 K t p.a. by 2006) used in medic., as a masticatory & in cooking, imported to Eur. since Roman period: still used in meat dishes such as hamburgers, pizza, sausages & Swedish meatballs, also in Danish pastries, Arab coffee, ice-cream & liqueurs

Elettariopsis Bak. Zingiberaceae (II 4). c. 10 SE As., W Mal. (5; R: NRBGE 40(1982)140). Poss. pesticides

Eleusine Gaertn. Gramineae (XXIX 5). 4 ann., 5 perenn. (6 diploid, 3 polyploid) Afr. (8; R: KB 27(1972)251.), S Am. (1: ***E. tristachya*** (Lam.) Lam.). R: AMBG 84(1972)841. Many toxic (cyanogenic). ***E. coracana*** (L.) Gaertn. (finger millet, coracan, kurakkan, ragi, allotetraploid cultigen prob. orig. E Afr. from *E. c.* subsp. *africana* (Kenn.-O'Byrne) Hilu & De Wet (*E. indica* subsp. *a.*, tetraploid (2n = 36, *E. indica* (2n = 18) × *E. floccifolia* (Forssk.) Spreng.) race of pantrop. weed toxic to stock (crowsfoot grass)) – imp. grain crop in Afr. & Ind., cult. in Ethiopia (earliest Afr. agriculture) 3rd Millenium BC, fermented for alcoholic drinks, seeds viable for more than 10 yrs without weevil damage, straw used in brick-making in Ind.; others weeds (crabgrass) incl. ***E. indica*** (L.) Gaertn. (yardgrass, trop. OW, now pantrop.)

Eleutharrhena Forman. Menispermaceae (I). 1 China, Assam: ***E. macrocarpa*** (Diels) Forman. R: KB 30(1975)98

Eleutherandra Slooten. Achariaceae (Flacourtiaceae). 1 Mal.: ***E. pes-curvi*** Slooten

Eleutheranthera Poit. Compositae (Helia.-Ecl.). 2 trop. Am., natur. OW. R: FA 37(2015)503

Eleutheranthus K. Schum. = *Opercularia*

Eleutherine Herb. Iridaceae (VII 5). 2 trop. Am. R: AMBG 78(1991)945. Lvs plicate. ***E. bulbosa*** (Mill.) Urb. (*E. americana*, S Am., natur. S Afr.) – fls open 5–7 p.m., cult. for local medic., tubers sold in Borneo markets for food-flavouring

Eleutherococcus Maxim. (*Acanthopanax*). Araliaceae. Incl. *Kalopanax*, excl. *Chiengopanax*, c. 40 E As., Himal., Mal. Some cult. orn., usu. prickly trees & shrubs, few timbers. ***E. senticosus*** (Rupr. & Maxim.) Maxim. (Siberian ginseng, NE As.) – source of a tonic (used by Russian athletes in Moscow Olympics, 1984, though app. no evidence for efficacy); ***E. septemlobus*** (Thunb.) Mabb. (*A. ricinifolius*, *K. s.*, *'K. pictus'*, temp. E As.) – cult. orn. decid. tree, timber (sen); ***E. trifoliatus*** (L.) S.Y. Hu (E As.) – ginseng subs.

Eleutheroglossum (Schltr.) M.A. Clem. & D.L. Jones. See *Dendrobium*

Eleutheropetalum (H. Wendl.) H. Wendl. ex Oersted = *Chamaedorea*

Eleutherospermum K. Koch. Umbelliferae (III 5). Incl. *Tamamschjania*, 2 Cauc., S W As.

Eleutherostigma Pax & K. Hoffm. = *Plukenetia*

Eleutherostylis Burret. Malvaceae (Grew.-Grew.; Tiliaceae). 1 NG, Moluccas: ***E. renistipulata*** Burret

Eleuthranthes F. Muell. = *Opercularia*

Elgon olive *Olea welwitschii*

Elide Medik. = *Asparagus*

Eliea Cambess. Hypericaceae (III). 1 Madag.: ***E. articulata*** Cambess.

Eligmocarpus Capuron. Leguminosae (I 3). 1 SE Madag. (v. local): ***E. cynometroides*** Capuron. R: Adansonia 8(1968)205

Elingamita Baylis. Primulaceae (Myrsinaceae). 1 NZ (Three Kings Group): ***E. johnsonii*** Baylis – c. 12 left in 1970, dioec., extrafl. nectaries (only NZ 'dicot' thus)

Eliokarmos Raf. = *Ornithogalum*

Elionurus Humb. & Bonpl. ex Willd. (*Elyonurus*). Gramineae (XXII). 15 trop. Afr. to NW Ind., Am., Aus. (1). R: KB 32(1978)665. Some ess. oils (unexploited)

Elisena Herb. = *Ismene*

Elissarrhena Miers = *Anomospermum*

Elizabetha Schomb. ex Benth. Leguminosae (I 2). 11 trop. S Am. R: PKNAW C79(1976)323. ***E. princeps*** Schomb. ex Benth. – bark ash used in hallucinogenic snuffs

Elizaldia Willk. = *Nonea*
elk nut *Pyrularia pubera*; **e. sedge** *Cladium mariscus*
elkshorn fern *Platycerium* spp.
Ellangowan poison bush *Eremophila deserti*
Elleanthus C. Presl. Orchidaceae (V 2). Excl. *Epilyna*, c. 100 trop. Am. Bird-poll. Some cult. orn.
Elleimataenia Kozo-Polj. = *Osmorhiza*
Ellenbergia Cuatrec. Compositae (Eup.-Ager.). 1 Peru: ***E. glandulata*** Cuatrec. R: MSBMBG 22(1987)159
Ellertonia Wight = *Kamettia*
Elliottia Muhlenb. ex Elliott. Ericaceae (V 3). (Incl. *Cladothamnus*, *Tripetaleia*) 4 Jap., Alaska, NW & SE N Am. R: JJB 63(1988)163. Some cult. orn. shrubs
Ellipanthus Hook.f. Connaraceae (I). 6 E Afr. coast & Madag. (2, R: AUWP 89 – 6(1989)268) to Mal. ***E. tomentosus*** Kurz (Ind. to C Mal.) – appetizer, medic. (flatulence), hard wood for house-posts, bridges etc.
Ellipeia Hook.f. & Thomson = *Uvaria*
Ellipeiopsis R.E. Fries = *Uvaria*
Ellisia L. Boraginaceae (Hydrophyllaceae). 1 N Am.: ***E. nyctelea*** (L.) L. ('Aunt Lucy'). R: Rhodora 42(1946)35
Ellisiophyllaceae Honda = Plantaginaceae
Ellisiophyllum Maxim. Plantaginaceae (Dig.- Sib.; Scrophulariaceae s.l). 1 Ind. to Jap., Taiwan to E NG: ***E. pinnatum*** (Benth.) Makino
Ellisochloa Peterson & N. Barker (~ *Merxmuellera*). Gramineae (V-Centrop.). 2 S Afr. R: T 60(2011)1118
elm *Ulmus* spp.; **American e.** *U. americana*; **Camperdown e.** *U. glabra* 'Camperdown'; **cedar e.** *U. crassifolia*; **cork e.** *U. thomasii*; **Cornish e.** *U. minor* 'Stricta'; **crowsfoot e.** *Heritiera trifoliolata*; **Dutch e.** *U. × hollandica*; **English e.** *U. minor* 'Atinia'; **hickory e.** *U. thomasii*; **Huntingdon e.** *U. × vegeta*; **Japanese e.** *U. davidiana*, *Zelkova serrata*; **rock e.** *U. thomasii*, *Milicia excelsa*; **Scotch e.** *U. glabra*; **slippery e.** *U. rubra*; **Spanish e.** *Cordia* spp.; **water e.** *Planera aquatica*; **white e.** *U. americana*; **WI e.** *Guazuma ulmifolia*; **wych e.** *U. glabra*
Elmera Rydb. (~ *Heuchera*). Saxifragaceae (I 5). 1 mts of Washington: ***E. racemosa*** (S. Watson) Rydb. – cult. orn. like *Heuchera* but petals cleft & scapes leafy. R: FNA 9(2009)105
Elmerrillia Dandy = *Magnolia*
Elodea Michaux. Hydrocharitaceae. 5 temp. Am., natur. OW. R: AB 21(1985)112. Submerged aquatics. Grown in ponds etc. but oft. becoming weedy, e.g. ***E. canadensis*** Michaux (Canadian pondweed or waterweed, American w., N Am., invasive Eur., Aus. (only male), NZ) – at depths to 14 m in Canada, dioec., introd. Ireland 1836, Br. 1842 & rapidly spread by veg. reprod. (only females) until 1880s when in decline & now scarce, introd. NZ 1870s & still a pest with stems to 6 m long & dry matter per sq.m greater than any other aquatic macrophyte, male fls (P 6, A 9) break off as buds & float to surface where they open, females (P 6, 3 staminodes, G (3)) with ovaries ext. to surface where poll.; ***E. callitrichoides*** (Rich.) Caspary (*E. ernstiae*, S Urug., NE Arg.) – natur. Eur. incl. GB (1948); ***E. nuttallii*** (Planch.) St John (N Am.) – introd. Belgium 1939, Br. Is. 1966 & still spreading & displacing *E. canadensis* as stronger-growing, though not pestilential. See also *Crassula* & *Lagarosiphon*
Eloyella Ortíz (~ *Phymatidium*). Orchidaceae (V 12h). 10 trop. Am.
Elsholtzia Willd. Labiatae (VII 1). c. 40 temp. OW incl. trop. mts. R: BBMNHB 10(1982)69. ***E. ciliata*** (Thunb.) Hylander (*E. cristata*, cockscomb mint, kinh gioi, Vietnamese balm, C & E As., natur. Eur., N Am.) – potherb sold in Eur., cult. orn., used like lemon balm & for hangovers in Jap.; ***E. haichowensis*** Y.Z. Sun (China) – copper indicator
Elsiea F.M. Leighton = *Ornithogalum*
Eltroplectris Raf. (~ *Stenorrhynchos*). Orchidaceae (IV 2h). 13 trop. Am., 1 ext. to Florida
Elvasia DC. Ochnaceae (II). c. 15 trop. S Am. R: BJ 113(1991)171. Some hydrochory
Elvira Cass. = *Delilia*
Elymandra Stapf. Gramineae (XXII 7). 6 trop. Afr., 1 also in Braz. R: KB 20(1966)287
Elymus L. (~ *Triticum*). Gramineae (XV). Incl. *Festucopsis*, *Hystrix*, *Kengyilia*, *Pascopyrum*, *Sitanion*, c. 150 N temp. (s. s., China 88, 62 endemic – R: FOC 22(2006)400; N Am. 32 – R: FNA 24(2007)288) esp. As. Mostly polyploids involving *Pseudoroegneria* spp; some with genomes from *Agropyron*, *Australopyrum*, *Hordeum* (e.g. N Am. tetraploids) etc. Good forage, some cult. orn. N Am. for bouquets (bottle-brush grass). ***E. repens*** (L.) Gould (*Agropyron r.*, *Elytrigia r.*, couch, twitch or witch grass, Euras., invasive Masc.,

Aus., Canada) – usu. hexaploid (involving *H.*, *P.* & other genes), bad weed of cult. with tough mildly diuretic rhiz. (coll. WW II & component of modern 'fasting' tea); ***E. smithii*** (Rydb.) Gould (*Pascopyrum s.*, N Am.) – octoploid with *Leymus* genes, imp. forage. See also *Leymus*

Elyonurus Willd. = *Elionurus*

Elythranthera (Endl.) A.S. George (~ *Caladenia*). Orchidaceae (IV 3b). 2 W Aus.

Elytranthe (Blume) Blume. Loranthaceae (3). 10 E Ind. to Vietnam & W Mal. R: FM I,13(1997)326

Elytraria Michaux. Acanthaceae (I). 21 trop. & warm. R: Aliso 32(2014)21

Elytrigia Desv. = *Elymus*; see also *Thinopyrum*

Elytropappus Cass. Compositae (Gnap.-Rel.). 8 S Afr. R: JSAB 1(1935)89, Strel. 9(2000)319. Poss. heterogeneous. ***E. rhinocerotis*** (L.f.) Less. (rhinoceros bush) – char. pl. of Karoo, ousting more desirable fodder pls

Elytrophorus P. Beauv. Gramineae (Arund.). 2 OW trop. (Aus. 1), wet places

Elytropus Muell. Arg. Apocynaceae (II c). 1 Chile: ***E. chilensis*** Muell. Arg.

Elytrostachys McClure. Gramineae (V 2). 2 C & N S Am. R: SCB 9(1973)79. Bamboos

Emarhendia Kiew & al. Gesneriaceae (III 2i). 1 Pahang (limestone): ***E. bettiana*** (M.R. Hend.) Kiew & al. R: BBP 70(1998)398

Embadium J. Black = *Hackelia* (but see TRSAS 89(1965)285)

Embelia Burm.f. Primulaceae (Myrsinaceae). Incl. *Grenacheria*, c. 150 trop. & warm OW (Mal. c. 60). Roux's Model. Some ed. & medic. esp. ***E. ribes*** Burm.f. (Indomal.) – liane with fr. used to adulterate pepper, as tapeworm treatment etc., fed to rats induce infertility; ***E. micrantha*** A. DC. (Mauritius) – favoured kidney-stone remedy & now v. rare; ***E. philippensis*** A. DC. (Philipp.) – ed. fr. & lvs, cordage

Embergeria Boulos = *Sonchus*

embero *Lovoa trichilioides*

Emblemantha Stone. Primulaceae (Myrsinaceae). 1 Sumatra: ***E. urnulata*** Stone. R: PANSP 140(1988)275

emblic *Phyllanthus emblica*

Emblica Gaertn. = *Phyllanthus*

Emblingia F. Muell. Emblingiaceae (Polygalaceae s.l.). 1 W Aus.: ***E. calceoliflora*** F. Muell. – autogamous

Emblingiaceae Airy Shaw (~ Polygalaceae). Magnoliidae – Brassicales. 1/1 W Aus. Prostrate hispid suffrutex. Lvs simple, ± opp.; stip. minute. Fls bisex., resupinate, solit., axillary. (K) lobed, deeply divided adaxially; C 2, slipper-shaped; nectary abaxial; androgynophore curved abaxially; A 8 or 9, 4 abaxial fertile, 4 or 5 adaxial staminodes; $\underline{G}$ (2,3) with axile placentation & 1 basal ovule per carpel. Fr. 1-seeded, indehiscent. Seeds reniform, arillate; endosperm scant

Genus: *Emblingia*

Form. in Polygalaceae

Embolanthera Merr. Hamamelidaceae (I 1). 2 SE As., Philippines

Embothrium Forst. & Forst. f. Proteaceae (V 3). 8 C & S Andes. ***E. coccineum*** Forst. & Forst. f. (Chilean firebush, Arg. & Chile) – variable shrub of open places from coast to treeline, some cvs hardy & cult. esp. **'Norquinco Valley'** (Arg. Andes)

emboya *Cinnamonum porosum* but applied to other timbers in error

Embreea Dodson (~ *Stanhopea*). Orchidaceae (V 12i). 2 Colombia, Ecuador. R: R. Jenny, *Stanhopea book* (2010)74

Embryopsida proposed name for land-plants

embul *Musa × paradisiaca*

Emelianthe Danser. Loranthaceae (5 6). 1 E & NE Afr.: ***E. panganensis*** (Engl.) Danser. R: R. Polhill & D. Wiens, *Mistletoes of Afr.* (1998)96

emeri *Terminalia ivorensis*

Emerus Mill. = *Hippocrepis*

Emex Necker ex Campderá = *Rumex*

Emicocarpus K. Schum. & Schltr. Apocynaceae (V c; Asclepiadaceae). 1 SE Afr.: ***E. fissifolius*** K. Schum. & Schltr.

Emilia (Cass.) Cass. (~ *Senecio*). Compositae (Sen.-Sen.). 90 OW trop. (3 incl. ***E. sonchifolia*** (L.) DC. – local medic. & veg. now pantrop. weed). Some cult. orn. ('Flora's paintbrush'). ***E. bittersdorffii*** Beentje (SW Tanz.) with 1(2) fleshy lvs adpressed to ground, growing with *Eriospermum adpressifolium* O. Weber with almost identical lvs

Emiliella S. Moore (~ *Emilia*). Compositae (Sen.-Sen.). 5 Angola, Zambia. R: GOB 2(1975)85

Eminia Taubert. Leguminosae (III 18). 4 trop. Afr. R: BJBB 53(1983)153. R. used in brewing (amylase)

Eminium Schott. Araceae (VII 23). 9 E Med. to C As. R: Willdenowia 20(1991)49. Cult. orn. with infrs oft. developing below ground & growing to surface before seed-disp.

Emmenanthe Benth. Boraginaceae (Hydrophyllaceae-Rom.). 1–2 SW N Am.: ***E. penduliflora*** Benth. – seeds germ. only in presence of charred wood, cult. orn. ann. with pale yellow or pink fls

Emmenopterys Oliv. Rubiaceae (Cond.). 1 S & W China (Eocene of Germany, W N Am.): ***E. henryi*** Oliv. (China) – to 1000 yrs old, cult. orn. introd. 1907 but no fls until 1971 (Italy). R: Plantsman 5 (2006)30

Emmenosperma F. Muell. Rhamnaceae (inc. sed.). 3 Aus. (2), New Caled. ***E. alphitonioides*** F. Muell. (Aus.) – 'yellow ash'

Emmeorhiza Pohl ex Endl. Rubiaceae (IV 15). 1 trop. S Am.: ***E. pohliana*** C. Presl

emmer *Triticum turgidum* Dicoccon Group

Emmotaceae Tiegh. = Metteniusaceae

Emmotum Desv. ex Ham. Metteniusaceae (Emmotaceae/Icacinaceae). 13 trop. S Am. R: AMBG 98(2011)8

Emodiopteris Ching & S.K. Wu = *Dennstaedtia*

Emorya Torrey. Scrophulariaceae (Budd.; Buddlejaceae). 2 Texas & N Mex. R: Sida 18(1999)694

Emorycactus Doweld = *Echinocactus*

Empedoclesia Sleumer = *Orthaea*

Empetraceae Hook. & Lindl. = Ericaceae

Empetrum Tourn. ex L. Ericaceae (V 4; Empetraceae). 3–18 N temp. (Eur. 1) & Arctic, S Andes, Falkland Is., Tristan da Cunha. Diploids usu. dioec.; monoec. & bisex. pls app. derived. ***E. nigrum*** L. (crowberry, curlew berry, N temp.) – allelopathic, ed. black fr.

emping *Gnetum gnemon*

Emplectanthus N.E. Br. (~ *Riocreuxia*). Apocynaceae (Vb; Asclepiadaceae). 2 S Afr R: BJ 120(1998)124.

Emplectocladus Torrey = *Prunus*

Empleuridium Sond. & Harv. Celastraceae (I). 1 S Afr.: ***E. juniperinum*** Sond. & Harv. – form. referred to Rutaceae. R: AMBG 72(1985)182

Empleurum Sol. ex Aiton. Rutaceae (I 4). 2 S Afr. R: JSAB 50(1984)427. C0, wind-poll. Lvs used like buchu (*Agathosma* spp.)

Empodisma L. Johnson & Cutler. Restionaceae. 3 Aus., NZ. R: PhytoKeys 13(2012)59

Empodium Salisb. Hypoxidaceae. 9 S Afr.

Empogona Hook.f. (~ *Tricalysia*). Rubiaceae (II 3). 29 trop. Afr. R: AMBG 96(2009)206

Empusella (Luer) Luer = *Specklinia*

emu apple *Owenia acidula*; **e. bush** *Eremophila* spp.; **e. grass** *Podocarpus drouynianus*

en choy *Ipomoea aquatica*

Enallagma (Miers) Baill. = *Amphitecna*

Enantia Oliv. = *Annickia*

Enantiophylla J. Coulter & Rose. Umbelliferae (III 9). 1 C Am.: ***E. heydeana*** J. Coulter & Rose

Enarganthe N.E. Br. (~ *Ruschia*). Aizoaceae (V). 1 W S Afr.: ***E. octonaria*** (L. Bolus) N.E. Br. R: H. E. K. Hartmann, *Ill. Handb. Succ. Pls*, Aiz. A–E(2002)258

Enarthrocarpus Labill. Cruciferae (12). 5 E Med. (Eur. 2), N Afr. R: Pfr. IV,105,1(1919)210

Enaulophyton Steenis. Melastomataceae. 2 Borneo incl. ***E. lanceolatum*** Steenis, a rheophyte

Encelia Adans. Compositae (Helia.-Enc.). 15 SW N Am., Chile, Peru, Galápagos. R: PAAAS 49(1913)358. Shrubs or herbs, some cult. orn. incl. ***E. farinosa*** A. Gray ex Torrey (brittlebush, SW N Am.) – stem resin used as 'chewing gum', local medic., form. used as incense by Spanish

Enceliopsis (A. Gray) Nelson. Compositae (Helia.-Enc.). 3 W N Am. R: FNA 21(2006) 112

Encephalartos Lehm. Zamiaceae (I). 65 trop. & S Afr. R: MNYBG 57(1990)203, BR 70(2004)285. Corner's Model. Poll. curculionids. Stems a source of sago (cf. *Cycas*); cult. orn. incl. ***E. altensteinii*** Lehm. (S Afr.) – at 300 yrs in Amsterdam Botanic Garden contender for oldest pot-pl., ***E. woodii*** Sander (Natal) – extinct in wild, prop. suckers from single male known at Kirstenbosch, S Afr.

Encephalosphaera Lindau. Acanthaceae (III 1). 2 trop. S Am.

enchanter's nightshade Orig. applied to *Mandragora* spp. but transferred to *Circaea* spp.

Encheiridion Summerh. = *Microcoelia*

Encholirium Mart. ex Schultes & Schultes f. Bromeliaceae (1). 23 Braz. (rock outcrops). R: BolB 23(2005)10

Enchosanthera King & Stapf ex Guillaumin = *Creochiton*

Enchylaena R. Br. = *Maireana* (but see FA 4(1984)213)

Encopella Pennell. Linderniaceae (Phrymaceae – Microc.; Scrophulariaceae s.l.). 1 Cuba: ***E. tenuifolia*** (Griseb.) Pennell

Encyclia Hook. (~ *Epidendrum*). Orchidaceae (V 13b). Excl. *Prosthechea*, c. 160 trop. Am. R: C.L. Withner, *The Cattleyas & their relatives* V(1998), VI (2000). Some cult. orn. epiphytes, usu. with showy fls & pseudobulbs (cf. *Epidendrum*), some fragrant incl. ***E. citrina*** (Llave & Lex.) Dressler (Mex.), the 'intoxicating' ***E. fragrans*** (Sw.) Lemée (C & N S Am.) & less pleasant ***E. incumbens*** (Lindl.) Mabb. (*Epidendrum aromaticum*, C Am.)

Endadenium Leach = *Euphorbia*

Endertia Steenis & De Wit. Leguminosae (I 2). 1 Borneo: ***E. spectabilis*** Steenis & De Wit – timber

Endiandra R. Br. Lauraceae (I). (? 1 Assam,) c. 100 Mal. (Borneo 8 – 9) to Aus. (38, endemic), New Caled. (6) & Fiji. Some timbers esp. ***E. palmerstonii*** (Bailey) C. White & Francis (Queensland or Aus. walnut, walnut bean, NE Aus.) – wood varieg. black, green, brown, pink etc., valuable & form. much used for panelling, seeds viable for 1.5 yrs

endive *Cichorium endivia* Endivia Group

Endlicheria Nees. Lauraceae (I). c. 60 trop. Am. R: RTBN 34(1937)500. Some fish-disp. in Amazon

Endocaulos C. Cusset. Podostemaceae (III). 1 Madag.: ***E. mangorense*** (Perrier) Cusset

Endocellion Turcz. ex Herder (~ *Petasites*). Compositae (Sen.-Tuss.). 2 Siberia. R: FGP 7(1972)393

Endocomia Wilde (~ *Horsfieldia*). Myristicaceae. 4 Indomal. R: Blumea 30(1984)179

endod *Phytolacca dodecandra*

Endodesmia Benth. Calophyllaceae (Guttiferae I 2). 1 trop. W Afr.: ***E. calophylloides*** Benth.

Endolepis Torrey = *Stutzia*

Endomallus Gagnepain = *Cajanus*

Endonema A. Juss. Crypteroniaceae (Penaeaceae). 2 W Cape. R: Strel. 9(2000)560

Endopappus Schultz-Bip. (~ *Chrysanthemum*). Compositae (Anth.). 1 N Afr.: ***E. macrocarpus*** Schultz-Bip. R: BBMNHB 23(1993)154

Endopleura Cuatrec. Humiriaceae. 1 Amazonian Braz.: ***E. uchi*** (Huber) Cuatrec. R: CUSNH 35(1961)80

Endosamara Geesink (~ *Millettia*). Leguminosae (III 16). 1 – 2 Indomal. Lianes

Endosiphon T. Anderson ex Benth. = *Ruellia*

Endospermum Benth. Euphorbiaceae (Aden.-End.). 8 SE As. to Fiji. R: EJB 68(2011)460. Aubréville's Model. Myrmecophily with stem-domatia, ants keeping off other pls, esp. ***E. moluccanum*** (Teijsm. & Binn.) Becc. (Moluccas) therefore called 'Arbor Regis' by Rumphius. Some timbers esp. ***E. peltatum*** Merr. (SE As. to C Mal.) – hardwood plantations in Philippines

Endosteira Turcz. = *Cassipourea*

Endostemon N.E. Br. Labiatae (VII 3d). 18 trop. & S Afr. to Ind. (1). R: KB 49(1994)689

Endotheca Raf. = *Aristolochia*

Endotropis Raf. (*Ventia*; ~ *Rhamnus*). Rhamnaceae. 6 N Am. R: T 65(2016)926. Unisexual fls cf. *Frangula*

Endresiella Schltr. = *Trevoria*

Endressia Gay. Umbelliferae (III 8). 2 Pyrenees, N Spain. (*Arpitium* Necker ex Sweet app. corr. name)

Endressia Whiffin (~ *Mollinedia*). Monimiaceae. 1 NE Queensland: ***E. wardellii*** (F. Muell.) Whiffin. R: FA 2(2007)87

Endusa Miers ex Benth. = *Minquartia*

Endymion Dumort. = *Hyacinthoides*

Enekbatus Trudgen & Rye (~ *Baeckea*). Myrtaceae (Lept.). 10 SW Aus. R: Nuytsia 20(2010)241

Enemion Raf. (~ *Isopyrum*). Ranunculaceae (III 1). 6 NE As., N Am. R: NRBGE 28(1968)270. C 0 – replaced by A (homoeotic mutant)

eng *Dipterocarpus tuberculatus*

Engelhardia Leschen. ex Blume. Juglandaceae. 9 Himal. to Mal. R: AMBG 65(1978)1076. Heterogeneous. Some timber & tanbarks. ***E. roxburghiana*** Wall. (*E. chrysolepis*, Indomal.) – bark (koboku) dihydroflavonoid sweeteners suggested in tumour treatment

Engelhardtia Leschen. ex Blume = praec.

Engelmannia A. Gray ex Nutt. Compositae (Helia.-Eng.). 1 N Am.: ***E. peristenia*** (Raf.) Goodm. & C. Lawson (*E. pinnatifida*) – cult. orn. R: FNA 21(2006)87

Englerastrum Briq. = *Plectranthus*

Englerella Pierre = *Pouteria*

Engleria O. Hoffm. Compositae (Ast.-Hom.). 2 Angola, Namibia

Englerina Tieghem (~ *Tapinanthus*). Loranthaceae (5 7). 25 trop. Afr. R: R. Polhill & D. Wiens, *Mistletoes of Afr.* (1998)119

Englerocharis Muschler. Cruciferae (4). 4 Andes. R: KB 67(2012)251

Englerodaphne Gilg = *Gnidia*

Englerodendron Harms. Leguminosae (I 2). 4 trop. Afr. R: Adans. 29(2007)64

Englerophytum K. Krause (*Bequaertiodendron*). Sapotaceae (IV). 5 – 10 trop. Afr. R: T.D. Pennington, S. (1991)47

Engomegoma Breteler. Olacaceae (Strombosiaceae). 1 Gabon: ***E. gordonii*** Breteler. R: BJ 118(1996)113

Engysiphon G. Lewis = *Geissorhiza*

Enhalus Rich. Hydrocharitaceae. 1 Indomal. to W Pac.: ***E. acoroides*** (L.f.) Royle. R: FA 39(2011)32. Marine; chief food of dugong. Female fls float horiz. at low tide (assoc. with spring tides) & catch males which break off & float (cf. *Elodea*); as tide rises, fls stand vertically & pollen sinks (heavier than water) on to stigmas; testa bursts when seed ripens & frees embryo. Black fibre-strands from leaf margins used for fishing nets in E Mal., W Pacific, lasting for generations; seeds ed.

Enhydra DC. = *Enydra*

Enhydrias Ridl. = *Blyxa*

Enicosanthellum Tien Ban = *Disepalum*

Enicosanthum Becc. = *Monoon*

Enicostema Blume. Gentianaceae. 3 C Am. & WI (1), Afr. to Lesser Sunda Is. (1), Madag. (1). R: Adansonia 9(1969)57. Alks. Medic. locally

Enkianthus Lour. Ericaceae (I). 17 Himal. to Jap. R: NJB 14(1994)389. Cult. orn. decid. shrubs with pagoda form (Fagerlind's Model). ***E. quinqueflorus*** Lour. (China) – form. used in New Year celebrations in S China

Enkleia Griff. Thymelaeaceae (Thym.-Daph.). 4 Andamans, SE As., Mal. R: JAA 42(1961)384

Ennealophus N.E. Br. (~ *Trimezia*). Iridaceae (VII 5). 6 S Am. Lvs plicate; elaiophores attractive to bees

Enneapogon Desv. ex P. Beauv. Gramineae (XXVII 1). 24 warm (Eur. 1, Aus. 16, 15 endemic). R (p.p.): KB 22(1968)393. Many plumose awned lemma-lobes. Many spp. with disarticulating culms

Enneastemon Exell = *Monanthotaxis*

Enneatypus Herzog = *Ruprechtia*

Enochoria Bak. f. = *Schefflera* + *Meryta*

Enriquebeltrania Rzed. (*Beltrania*). Euphorbiaceae (Acal.). 2 Mex. R: SB 31(2006)538

enset *Ensete ventricosum*

Ensete Horan. Musaceae. Incl. *Musella*, 7 OW trop. (Eocene of NW US). R: KB 1947(1947)97. Hapaxanthic banana-like herbs (Holttum's Model). ***E. lasiocarpa*** Franch. (*M. l.*, Vietnam, Yunnan) – c. 60 cm tall, insect-poll., fr. rather dry, stem ed., fermented to wine, fibre for chairs & ropes, medic.; ***E. ventricosum*** (Welw.) Cheesman (enset, NE Afr.) – starchy food from pseudostems, fl. heads & seeds eaten cooked, seeds used for beads in N Uganda, cult. orn.

Enslenia Nutt. = *Cynanchum*

Entada Adans. Leguminosae (II 1). 28 trop. Am & Afr. to Aus. Trees & lianes, some sea-disp., seeds germ. in N Is., NZ after 21 months in sea, carved as snuff-boxes & form. for boxes for wax matches. ***E. africana*** Guill. & Perr. (W Afr.) – fls to 2 cm but fr. to 80+ cm; ***E. gigas*** (L.) Fawcett & Rendle (*E. scandens*, sea bean, nicker or Mackay b., trop Am. & Afr.) – liane with legumes to 1.8m long, seeds flat & brown to 5cm across (common drift seeds ('lucky beans') in W Br.), lvs a veg. & seeds ed. roasted, fibre for nets, sails etc., saponin source; ***E. rheedei*** Spreng. ('*E. phaseoloides*', '*E. pursaetha*', Queensland bean, trop. Afr. to Aus.) – similar liane to 75m long but legumes (to 200 × 13 cm) not spirally twisted, seeds used for hair-washing in As.

Entandrophragma C. DC. Meliaceae (I). 11 trop. Afr. Rauh's Model. Timber (sapele, subs. for true Am. mahoganies incl. ***E. angolense*** (Welw.) C. DC. (edinam, gedu nohor), ***E. candollei*** Harms (omu), ***E. cylindricum*** (Sprague) Sprague (aboudikro), ***E. utile*** (Dawe & Sprague) Sprague (utile, assié)). Seed-oil with rare isomer of oleic acid can be used in nylon manufacture

Entelea R. Br. Malvaceae (Grew.-Apeib.; Tiliaceae). 1 NZ.: ***E. arborescens*** R. Br. – Leeuwenberg's Model, timber half the wt. of cork (corkwood) used for fishing-net floats etc. by Maoris

Enterolobium Mart. Leguminosae (II 4). 11 trop. Am. R: MNYBG 74,1(1996)245. Cult. orn., timber & legumes for livestock (though some in Braz. cause lesions (photosensitive)) esp. ***E. cyclocarpum*** (Jacq.) Griseb. (guanacaste), disp. by modern (introd.) horses suggesting extinct horse-like herbivores were orig. disp. agents, but also by peccaries which eat seeds, crushing some & spitting out rest there being a variation in seed wt. (the heavier the harder) correlated with crushability which ensues that peccaries both disperse seeds yet eat some as a 'reward', bark & fr. used as soap subs., seeds also part of Aztec chocolate 'brew'

Enteropogon Nees (~ *Chloris*). Gramineae (XXIX 5). 8 trop. savanna (Aus. 6)

Enterosora Bak. (~ *Grammitis*). Polypodiaceae (V; Grammitidaceae). 8 trop. Am., 2 Afr. R: SB 17(1992)345

Enterospermum Hiern = *Tarenna*

Entolasia Stapf. Gramineae (XXIV 2). 6 trop. Afr. to E Aus. & New Caled. (1)

Entomophobia de Vogel (~ *Pholidota*). Orchidaceae (V 10b). 1 Borneo: ***E. kinabaluensis*** (Ames) de Vogel. R: OM 1(1986)41. Fls almost completely closed

Entoplocamia Stapf. Gramineae (XXVII 3). 1 Namibia: ***E. aristulata*** (Hack. & Rendle) Stapf. R: Strel. 10(2000)695

Enydra Lour. Compositae (Helia.-Mel.). 10 warm. Wet places. Bell's Model

Eokochia Freitag & G. Kadereit (~ *Kochia*). Amaranthaceae. 1 C Medit.: ***E. saxicola*** (Guss.) Freitag & G. Kadereit – endangered

Eomatucana F. Ritter = *Matucana*

Eomecon Hance. Papaveraceae (I). 1 E China: ***E. chionantha*** Hance – arils, cult. orn.

Eosanthe Urb. Rubiaceae (inc. sed.). 1 E Cuba (serpentine): ***E. cubensis*** Urb. R: Britt. 51(1999)229

Epacridaceae R. Br. = Ericaceae

Epacris Cav. Ericaceae (VII 4) Epacridaceae. 40 SE Aus. (38), NZ (2), New Caled. (1, introd.?). Aus. heath. Sweet (dianthus)-scented

Epallage DC. = *Anisopappus*

Epaltes Cass. Compositae (Inul.-Pluch.). Excl. *Ethuliopsis*, *Sphaeromorphaea*, 10 trop. Lvs usu. decurrent; pappus 0

Eparmatostigma Garay = *Vanda* (but see HBML 23(1972)178)

epazote *Dysphania ambrosioides*

Eperua Aubl. Leguminosae (I 2). 14 NE S Am. R: SCB 28(1975). Resins; timber (wallaba) ± resistant to decay in water, that of ***E. falcata*** Aubl. (flagelliflory, bat-poll.) much used for roof shingles. ***E. purpurea*** Benth. – ashes added to clay for pots; this & other spp. held to encourage growth of thick hair

Ephedra Tourn. ex L. Ephedraceae. c. 40 Medit. (Eur. 4) to China (fossils in NE 126 – 145 M yrs), 14 W US & Mex., 13 Andes. Mostly shrubby switch-pls, 1 ± arborescent, some lianoid, many with rhiz. & underground buds; minute calcium oxalate crystals in intercellular spaces of wood, as in *Welwitschia*. Am. lineage derived from OW; fleshy bracts primitive, dry ones evolved several times. Wind-poll. though some visited by insects seeking nectar on outside of 'P'; double fert. (without triploid endosperm) in ***E. nevadensis*** S. Watson (W N Am.). Alks (6) incl. ephedrine (v. close to 'crystal meth'; not all spp.) medic. in China (esp. ***E. sinica*** Stapf, ma huang; fleshy bracts also in Mongolian food) for 5000 yrs but stimulant qualities not understood in W until 1924 & still used in treatment of asthma (bronchodilator), sinusitis, weight loss by suppressing appetite, etc. & stimulant, mimicking effect of adrenalin ('herbal Ecstasy'; abused by athletes), banned in 2004 but now synth.; poss. contender for 'soma' or 'haoma' of the ancients being made effective by eating *Peganum harmala* first, found in Neanderthal burial head-wreaths. Some cult. orn. e.g. ***E. gerardiana*** Wall. (Himal.) – fleshy berry-like 'P' ed. but acid; ***E. trifurca*** Torr. ex S. Watson (Mormon tea, SW N Am.) – form. imp. tea

Ephedraceae Dumort. Gnetidae. 1/c. 40 N hemisph. & W S Am. Xeromorphic equisetoid shrubs, climbers or small trees, usu. dioec.; sec. xylem ring porous. Lvs opp. or in whorls of 3 (4), scale-like, evanescent (stems photosynthetic). Female infl. a short shoot with 2 – 8 prs (or whorls) of bracts, the lowermost sterile, s.t. becoming swollen & juicy, & 1 – 3 fls, each a nucellus encl. by 2 layers (integuments & bract envelope); male fls subtended by a bract, 'P' 2-lipped, microsporangiophore ('stamen') 1 C, forked or even 3. Female gametophyte develops from lowermost of a linr. tetrad of spores produced by megaspore mother-cell & is cellular prothallus with (1)2 or 3 archegonia with necks to 40 cells long; male gametophyte a prothallus with a tube nucleus, sperm cell, 1 sterile cell & 2 prothallial cells; pollen tube releases 2 nuclei, 1 uniting with egg nucleus, the 2nd with another but no embryo or nutritive tissue (cf. Magnoliidae) results from latter. Zygote divides 3 times & all (but usu. only 3 – 5) can become embryos, each with 2 cotyledons. Seeds solit. or paired forming syncarp with 2 prs of bracts, membranous & winged or fleshy & coloured. 2n = 14, 28

Genus: *Ephedra*

Superficially v. diff. from Gnetaceae with which it is generally assoc. Pollen from early Cretaceous of Portugal

Ephedranthus S. Moore. Annonaceae (IV 2). 4 trop. S Am. R: BMPEG 15(1999)136

Ephemerantha P. Hunt & Summerh. = *Dendrobium*

Ephippiandra Decne. Monimiaceae (V 1). 7 Madag. R: AMBG 72(1985)81. Mangenot's Model

Ephippianthus Reichb.f. Orchidaceae (V 13e). 2 Korea, Jap., Sakhalin

Ephippiocarpa Markgraf = *Callichilia*

Epiblastus Schltr. Orchidaceae (V 15). c. 20 E Mal., Polynesia

Epiblema R. Br. Orchidaceae (IV 3i). 1 SW Aus.: ***E. grandiflorum*** R. Br. – poss. food mimic of papilionoid legumes. R: Gen. Orch. 2(2001)204

Epicampes J. Presl = *Muhlenbergia*

× **Epicattleya** Rolfe. Orchidaceae. *Cattleya* × *Epidendrum* – 183 grexes

Epicharis Blume = *Dysoxylum*

Epicion Small = *Cynanchum*

Epiclastipelma auctt. = seq.

Epiclastopelma Lindau = *Mimulopsis*

Epicranthes Blume = *Bulbophyllum*

Epicycas Laubenf. = *Cycas*

Epidanthus L.O. Williams = *Epidendrum*

Epidendropsis Garay & Dunsterv. = *Epidendrum*

Epidendrum L. Orchidaceae (V 13b). c. 1400 trop. Am. (Florida 7). Epiphytes rarely with pseudobulbs, much cult. (crucifix orchids; see also *Encyclia*, *Barkeria* s.t. incl. here). Extrafl. nectaries. ***E. ibaguense*** Kunth (~ *E. secundum* Jacq.) fls all year round, nectar 0 & app. mimicking *Lantana camara* L. & *Asclepias curassavica* L. in Panamá, attracting monarch butterflies, though the widely cult. pl. is actually ***E. radicans*** Pavón ex Lindl. (Mex.); ***E. × obrienianum*** Rolfe (1888; *E. jamiesonis* Reichb.f. (Ecuador) × *E. radicans*) – first hort. hybrid to fl. (1888), invasive (plantlets on stems) in Florida, Hawaii

Epidryos Maguire. Rapateaceae (I 1). 2 Colombia, & Panamá, 1 Guianas

Epifagus Nutt. Orobanchaceae (Orob.). 1 N Am.: ***E. virginiana*** (L.) Barton (beech drops) on *Fagus grandifolia*

Epigaea L. Ericaceae (V 3). 3: ***E. asiatica*** Maxim. (Jap.), ***E. gaultherioides*** (Boiss.) Takht. (Cauc. & E As. Minor), ***E. repens*** L. (E US) – cryptic dioecy (app. gynodioec.), myrmechory, all cult. orn. creeping everg. shrubs

Epigeneium Gagnepain = *Dendrobium*

Epigynum Wight. Apocynaceae (II c). 4 Himal. to Borneo. R: HPB 10(2005)68

Epilasia (Bunge) Benth. Compositae (Lact.-Scor.). 3 W & C As. to China

Epilobium L. Onagraceae. 165 temp. (Eur. 24 native & natur.) esp. W N Am. (excl. *Chamaenerion*, incl. *Boisduvalia*, *Zauschneria*), arctic & trop. mts. R(China): SBM 34(1992)1. Willow-herbs. Herbs or subshrubs, some weedy & hybridizing e.g. ***E. parviflorum*** Schreb. (Euras., Med.) with A 4 shorter than style (pollen for cross-poll.), 4 longer, which curl back giving autogamy; seeds with chalazal tuft of hairs making seeds buoyant in air; some bird-poll. (*Zauschneria* – Calif. fuchsia). Cult. orn. incl. ***E. brunnescens*** (Cockayne) Raven & Engelhorn ('*E. nerteroides*', NZ) – natur. NW England, ***E. canum*** (Greene) Raven (*E. californicum Z. c.*, Calif.) – orange-red-flowered, bird-poll., ***E. hirsutum*** L. (codlins & cream, Euras., N Afr.) – natur. N Am.

Epilyna Schltr. (~ *Elleanthus*). Orchidaceae (V 2). 2 Ecuador

Epimedium Tourn. ex L. Berberidaceae (II 2). 54 N Afr., N Italy to Caspian (2), W Himal., NE As., Jap. R: W.T. Stearn, *The genus E.* (2002)41. K 4 (unequal & in 2 prs) + 4 petaloid, C 4 flat or extended into pouches, A 4 (but see fam. for var. interpretations of structure). Fls pend., protogynous, anthers later bending up over stigma & dehiscing, followed by style elongation carrying stigma among anthers & poss. of self-poll.; seeds with membranous arils, ant-disp. In modern Chinese herbalism rhiz. used for impotence (horny goat weed, esp. ***E. brevicornu*** Maxim. (China), ***E. grandiflorum*** Morren (E As.)), arthritis, paralysis of legs & high blood pressure in elderly women, ***E. sagittatum*** (Sieb. & Zucc.) Maxim. (yin yang huo, C China, natur. Jap.) for hypertension. Many cult. orn. spp. & hybrids much confused in cult. incl. ***E. alpinum*** L. (barrenwort, S Eur., natur. in N) & ***E. × youngianum*** Fischer & C. Meyer (*E. diphyllum* Lodd. (S Jap.) × *E. grandiflorum* Morren (E As.)) orig. Jap., other spontaneous hybrids occurring in gdns. ***E. elatum*** Morren & Decne (W Himal.) – effective mosquito-repellent

Epimeredi Adans. = *Anisomeles*

Epinetrum Hiern = *Albertisia*

Epipactis Zinn. Orchidaceae (V 1). c. 50 Eur. (9), temp. As., Himal., SE As., trop. Afr. Helleborines. ***E. consimilis*** D. Don (As.) – in Israel poll. hoverflies, the warts on lip mimicking aphids on which they usu. lay eggs; ***E. helleborine*** (L.) Crantz (Euras., introd. US 1875 & now in 28 US states & 6 Canadian provinces) & ***E. purpurata*** Sm. (Eur.) – nectar 'toxic' (ethanol), infected with bacteria & fungi transferred from fr. by poll. wasps which become slow & sluggish after drinking it & do not groom off pollinia to effect poll.; ***E. thunbergii*** A. Gray (Jap.) – poll. *Camponotus japonicus* ants as well as hoverflies

Epiphyllanthus A. Berger = *Schlumbergera*

Epiphyllum Haw. Cactaceae (III 2). 18 trop. Am. Epiphytes with large fragrant fls & extrafl. nectaries; cult. orn. incl. ***E. crenatum*** (Lindl.) G. Don (Mex. to Honduras) – diurnal fls, parent of many hybrids incl. intergeneric ones, & ***E. oxypetalum*** (DC.) Haw. (queen of the night) – headily scented night fls, though 'epiphyllums' of hort. usu. *Disocactus × hybridus*

Epipogium S. Gmelin ex Borkh. Orchidaceae (V 8b). 3 temp. Euras. (***E. aphyllum*** Sw., ghost orchid found UK 1854, thought 'extinct' there 1986, but rediscovered 2009), OW trop.to Aus. Leafless mycotrophs with branched rhiz. & 0 r.; endotrophic mycorrhiza.; usu. self-poll. ***E. roseum*** (D. Don) Lindl. (OW trop. to Aus.) – seeds disp. a few days after fls emerge

Epipremnopsis Engl. = *Amydrium*

Epipremnum Schott. Araceae (IV 4). c. 20 SE As. to W Pac. (alleged fossils in Oligocene of N Egypt). Lianes, some medic. & cult. orn. esp. ***E. pinnatum*** (L.) Engl. (*Monstera dilacerata*, Indomal. to W Pac.) – like *M. deliciosa* with perforated lvs, many cvs esp. **'Aureum'** (poss. wild sp. – *E. aureum* (André) Bunting from Society Is., 'money plant' – rarely flowering so owners of fl. pls considered 'in the money'), irreg. varieg., widely planted in trop.

Epiprinus Griff. Euphorbiaceae (Epi.-Epi.). 4 Assam to W Mal.

Epirixanthes Blume (~ *Salomonia*). Polygalaceae (IV). 5 Indomal. Chlorophyllous parasites. R: FM I,10(1988)488. ***E. elongata*** Blume (Ind. to C Mal.) – contraception & abortifacient (W Mal.) app. psychological

Epischoenus C.B. Clarke. Cyperaceae (II 7). 8 S Afr. R: SAJB 61(1995)147

Episcia Mart. Gesneriaceae (II 5c). 9 trop. Am. R: Selbyana 5(1978)25, 6(1982)183. Stolons (unique in fam.). Cult. orn. spp. & hybrids esp. from ***E. cupreata*** (Hook.) Hanst. (*E. splendens*)

Episcothamnus H. Robinson = *Lychnophoriopsis*

Epistemma D.V Field & J. Hall. Apocynaceae (III; Periplocaceae). 4 W & C Afr.

Epistephium Kunth. Orchidaceae (II 2). 21 N S Am. Some to 180 cm tall; lvs net-veined like *Smilax*; toothed cup-shaped organ at summit of G, persistent in fr.

Epitaberna K. Schum. = *Heinsia*

Epithelantha A. Weber ex Britton & Rose. Cactaceae (III 9). 1[–6] SW N Am.: ***E. micromeris*** (Engelm.) Britton & Rose – hallucinogenic 'false peyote'

Epithema Blume. Gesneriaceae (III 1d). 20 OW trop. (Afr. 1: ***E. tenue*** C.B. Clarke, W & C Afr.) – epiphyllous fls on leaf 3 (cf. *Streptocarpus*), hapaxanthic. R: GBS 67(2015) 165

Epitriche Turcz. (~ *Angianthus*). Compositae (Gnap.-Ang.). 1 SW Aus.: ***E. demissus*** (A. Gray) Short. R: Muelleria 5(1983)181

Epixiphium (A. Gray) Munz (~ *Maurandya*). Plantaginaceae (Ant.; Scrophulariaceae s.l.). 1 SW N Am.: ***E. wislizeni*** (A. Gray) Munz

Eplingia L.O. Williams = *Trichostema*

Eplingiella Harley & Pastore (~ *Hyptis*). Labiatae (VII 3c). 3 S Am. R: KB 69[9357](2014)2

epos *Perideridia* spp.

Equisetaceae Michaux ex DC. Equisetidae (Equisetales). 1/15 almost cosmop. Rhiz. herbs with aerial hollow, jointed stems impregnated with silica, s.t. unbranched. Lvs a series of teeth united by a sheath, usu. without chlorophyll (stems photosynthetic). Sporangia in term. cones, s.t. on special unbranched stems, each with whorls of sporangiophores without bracts, each peltate & bearing 5 – 10 sporangia which originate on outer surface but during ontogeny are carried underneath peltate head. Spores with 4 spathulate bands which are hygroscopic ('haptera') coiling & uncoiling when humidity changes. Prothalli male or herm. (when female first), so that selfing is poss., though proportions of male to herm. can be manipulated by differing light intensities or crowding; antherozoids spirally coiled & multiflagellate. n = 108

Only genus: *Equisetum*

Fossils from Permian/Triassic. Poss. related to Calamitaceae but ancestors ('*Equisetites*') were herbaceous as well as arborescent, the 2 growing together in the Carboniferous, but the reduced lvs app. derived from larger dichotomizing structures (Sporne). Usu. in marshes, wet woods, on riverbanks etc.

Equisetidae Warm. (Sphenopsida). (Extant) 1/15 amost cosmop. Vasc. spore-pls with r., stems & whorled lvs; some with sec. thickening. Sporangia thick-walled, homosporous (or heterosporous), usu. borne in a reflexed position on sporangiophores in whorls; antherozoids multiflagellate (Sporne). Three extinct orders incl. Calamitales, which were dominant group of swamp-forests of Carboniferous, & extant Equisetales, comprising single fam. Equisetaceae

Equisetum Tourn. ex L. Equisetaceae. Incl. *Hippochaete*, 15 almost cosmop. (exc. Australasia; Eur. 10). R: BNH 8(1963)42, NH 30(1978)413. Horsetails; fossils 150 MYA. Bell's Model; alks incl. nicotine but toxicity due to thiaminase breaking down the vitamin thiamine (like silica poss. antiherbivores (? dinosaurs)); local medic., long used to staunch blood flow & poss. treatment of Alzheimer's Disease; extracts in cosmetics as silica used in collagen synthesis; affinity for gold in solution & concentrate it more than any other pl. but indicators because only 0.25 mg of gold per kg of stems or rhizomes. Up to 13 m long (***E. giganteum*** L., but stems only 2 cm thick & no sec. thickening, supported by turgor pressure so a sprawler). Several bad weeds (devil guts) & stems s.t. used for scouring (s. rush) & polishing (form. for pewter, & lime wood-carvings by Grinling Gibbons), others medic. (form. used to remove white spots in fingernails). ***E. arvense*** L. (temp., natur. Aus.) in Jap.) – dies down in winter, cones ed. after boiling, shoots used to fatten geese in N Am., eaten like asparagus by Romans who also dried them as a tea & a thickening powder; ***E. hyemale*** L. (Dutch rush, N temp.) – used for sanding in cabinet-making; ***E. sylvaticum*** L. (N temp.) – a yellow dye extracted from dried stems in Norway; ***E. telmateia*** Ehrh. (N temp.) – cones juicy & sweet, stems yield a black fibre for Vancouver Indians' baskets

Eragrostiella Bor (~ *Tripogon*). Gramineae (XXIX 6). 6 E Afr., Sri Lanka to N Aus. R: CHA 22(1976)1. Aus. spp. 'resurrection' pls

Eragrostis Wolf. Gramineae (XXVII 2). Incl. *Ectrosia*, *Pogonarthria*, *Psammagrostis*, excl. *Thellungia*, c. 435 temp. & trop. (Eur. 9, China 32, Mal. 24 – R: Blumea 47(2002)160, Aus. 73, 63 endemic + 15 introd. – R: FA44B(2005)346, Am. 120 [N Am. 25]). R: Preslia 24(1952)281, T 43(1994)392. The Afr. spp. ***E. hispida*** K. Schum., ***E. nindensis*** Ficalho & Hiern & ***E. paradoxa*** Launert have lvs which revive after desiccation to 6 – 12% water-content; some dioec. incl. ***E. reptans*** (Michaux) Nees (Braz.); pericarp of many spp. peeled away readily when moist. ***E. australasica*** (Steud.) C. Hubb. (Aus.) – canes for fibre & fencing; ***E. nigra*** Nees ex Steud. (As.) – stalks for brooms (Bhutan); ***E. squarrosa*** (Steud.) Fourn. (*P. s.*, C Afr.) – minor broom grass. Some cult. orn. & fodder incl. ***E. curvula*** (Schrad.) Nees (Afr. or weeping lovegrass, S Afr., invasive Aus., US) – apomictic, planted to stabilize soil in tea estates, ***E. lehmanniana*** Nees (trop. & S Afr.) – invasive W US, & ***E. tef*** (Zuccagni) Trotter (*E. abyssinica*, t'ef or teff, NE Afr., natur. Aus., poss. derived from *E. pilosa* (L.) P. Beauv. [S Eur., OW trop., natur. US]) – also cult. for gluten-free ed. seeds (0.9 m t p.a. by 2000s, imp. in Ethiopia, where a staple fermented & cooked like crepes (injira), but also now e.g. Netherlands), straw used in brick manufacture (this or other *E.* spp. in Egyptian pyramid bricks [3359 BC])

Eranthemum L. Acanthaceae (III 2a). 30 trop. As. Cult. orn. foliage pls closely allied to *Pseuderanthemum*. ***E. viscidum*** Blume (W Mal.) – used for eye infections (Java)

Eranthis Salisb. Ranunculaceae (I 2). 8 Euras. R: APG 38(1987)96. Tuberous toxic herbs much cult. for early spring fls (winter aconites) esp. ***E. hyemalis*** (L.) Salisb. (S Eur., natur. N Eur. & Am.); ***E. cilicica*** Schott & Kotschy (E Medit.) oft. considered conspecific but alleged hybrid, ***E.*** × ***tubergenii*** Bowles is sterile with aborted anthers & has more carpels than either. Consp. P app. K, C being nectaries, though organs intermediate between these latter & A found in *E.* × *tubergenii* (cf. *Nymphaea*)

Erasanthe Cribb & al. (~ *Aeranthes*). Orchidaceae (IV 16d). 1 Madag.: ***E. henrici*** (Schltr) Cribb & al. R: Adans. 29(2007)28

Erato DC. (~ *Liabum*). Compositae (Liab.). 5 trop. Am. R: SB 31(2006)597. Some woody

Erblichia Seem. (~ *Piriqueta*). Passifloraceae (Turneraceae). Excl. *Arboa*, 1 C Am.: ***E. odorata*** Seem. R: Adansonia 19(1979)459

Ercilla A. Juss. Phytolaccaceae (I). 1 – 2 Chile. Temp. rain forest; adventitious clinging r. like *Hedera*; coloured P prob. K. Cult. orn. esp. ***E. volubilis*** A. Juss.

Erdisia Britton & Rose = *Corryocactus*

Erechtites Raf. (~ *Senecio*). Compositae (Sen.-Sen.). 5 Am. introd. OW (Eur. 1). R: AMBG 43(1956)1. Alks

Eremaea Lindl. = *Melaleuca* (but see Nuytsia 9(1993)137)

Eremaeopsis Kuntze = *Eremaea*

Eremalche Greene (~ *Malvastrum*). Malvaceae (Malv.-Malv.). 3 Calif., N Baja Calif.

Eremanthus Less. Compositae (Vern.-Lych.). 27 Braz., esp. arid cerrado. R: AMBG 74(1987)265. Trees & shrubs; some timber (*Vanillosmopsis*)

Eremia D. Don = *Erica*

Eremiastrum A. Gray = *Monoptilon*

Eremiella Compton = *Erica*

Eremiolirion Manning & Forest. Tecophilaeaceae. 1 C & NW Namibia: ***E. amboensis*** (Schinz) Manning & Mannheimer. R: Bothalia 42(2012)25

Eremiopsis N.E. Br. = *Erica*

Eremiris (Spach) Rodionenko = *Iris* (but see BZ 91(2006)1707)

Eremitella Yatsk. & Contreras. Orobanchaceae. 1 Mex.: ***E. mexicana*** Yatsk. & Contreras. R: Novon 19(2009)267

Eremitis Doell. Gramineae (VI 2). 3 E Braz. 3 types of infl., 1 burying itself like a peanut

Eremium Seberg & Linde-Laursen = *Leymus*

Eremobium Boiss. Cruciferae (4). 1 N Afr. & Middle E: ***E. aegyptiacum*** (Spreng.) Hochr.

Eremoblastus Botsch. Cruciferae (5). 1 W Kazakhstan to C As.: ***E. caspicus*** Botsch.

Eremocarpus Benth. = *Croton*

Eremocarya Greene = *Cryptantha* (but see SB 37(2012)75)1

Eremocaulon Soderstrom & Londoño (~ *Bambusa*). Gramineae (V 3). Incl. *Criciuma*, 4 Bahia, Braz. R: SB 27(2002)704. Some cli.

Eremocharis Philippi. Umbelliferae (Azor.). 9 Chile, Peru. R: UCPB 33(1962)153

Eremochion Gilli = *Horaninovia*

Eremochloa Buese. Gramineae (XXII 3). 12 Indomal. to Aus. R: Blumea 46(2001)399. ***E. ophiuroides*** (Munro) Hackel (centipede grass, SE As.) – used for erosion control & (SE US) as lawn-grass

Eremocitrus Swingle = *Citrus*

Eremocrinum M.E. Jones. Asparagaceae (Anthericaceae). 1 Utah & N Arizona: ***E. albo-marginatum*** (M.E. Jones) M.E. Jones. R: FNA 26(2002)216

Eremodaucus Bunge. Umbelliferae (III 5). 1 Cauc. to C As. & Afghanistan: ***E. lehmannii*** Bunge

Eremodraba O. Schulz = *Neuontobotrys* (but see AMBG 77(1990)602)

Eremogeton Standl. & L.O. Williams. Scrophulariaceae (Leuc.). 1 Mex., Guatemala: ***E. grandiflorus*** (A. Gray) Standl. & L.O. Williams. R: Sida 11(1985)167

Eremogone Fenzl (~ *Arenaria*). Caryophyllaceae (II 1). c. 90 N temp. (14 Am.)

Eremohylema Nelson = *Pluchea*

Eremolaena Baill. Sarcolaenaceae. 3 E Madag. R: Candollea 69(2014)187

Eremolepidaceae Tieghem ex Nakai. See Santalaceae

Eremolepis Griseb. = *Antidaphne*

Eremolimon Lincz. = *Limonium*

Eremomastax Lindau. Acanthaceae (III 2e). 1 trop. Afr., Madag.: ***E. speciosa*** (Hochst.) Cuf. – variable, seeds oft. with toothed scales, spreading when wetted

Eremonanus I.M. Johnson = *Eriophyllum*

Eremopanax Baill. = *Polyscias*

Eremopappus Takht. = *Centaurea*

Eremophea Paul G. Wilson. Amaranthaceae (Chenopodiaceae I 5). 2 Aus. R: FA 4(1984)224

Eremophila R. Br. Scrophulariaceae (Myoporaceae). Excl. *Calamophoreus, Diocirea, Glycocystis*, 215 Aus. esp. W, 1 ext. to NZ. R: R.J. Chinnock, *E. & allied genera* (2007)180. Some timber & tanbark (e.g. ***E. oppositifolia*** R. Br., emu bush, weeooka), trad. medic. & cerem. for Aboriginal people (esp. ***E. alternifolia*** (S. Moore) Ostenf. & ***E. longifolia*** (R. Br.) F. Muell.) but many toxic to stock, e.g. ***E. deserti*** (Benth.) Chinnock (Ellangowan poison bush), some weedy ('poverty bush' in W Aus because common on degraded or overgrazed land). ***E. cuneifolia*** Kraenzl. (W Aus.) – trad. med. for coughs; ***E. duttonii*** F. Muell. (C Aus.) – antiseptic (kangaroos rolling in it, poss. thereby removing ticks)

Eremophyton Bég. Cruciferae (12). 1 Morocco, Algeria, Libya: ***E. chevalieri*** (Baratte) Bég.

Eremopoa Rosch. = *Poa*

Eremopogon Stapf (~ *Dichanthium*). Gramineae (XXII 7). 4 OW trop.

Eremopyrum (Ledeb.) Jaub. & Spach (~ *Triticum*). Gramineae (XV). 4 C As., N Am. R: NJB 11(1991)271

Eremosemium Greene = *Grayia*

Eremosis (DC.) Gleason (~ *Critoniopsis*). Compositae (Vern.). 27 Mex., C Am. R: Phytoneuron 2016–50:10. Trees to 15 m+

Eremosparton Fischer & C. Meyer. Leguminosae (III 24). 3 SE Russia (1) to C As., sandy deserts

Eremospatha G. Mann & H. Wendl. ex Schaedtler. Palmae (I 2a). 11 trop. Afr. R: Phytotaxa 51(2012)11. Herm. rattans

Eremostachys Bunge = *Phlomoides* (but see NSL 27(1990)135)

Eremosynaceae Dandy = Escalloniaceae

Eremosyne Endl. Escalloniaceae (Eremosynaceae). 1 SW Aus.: ***E. pectinata*** Endl. – ann.

Eremothamnus O. Hoffm. Compositae (Cich.-Erem.). 1 coastal S Namibia: ***E. marlothianus*** O. Hoffm. R: T 43(1994)36

Eremothera (Raven) W.L. Wagner & Hoch (~ *Camissonia*). Onagraceae. 7 W N Am. deserts. R: CUSNH 37(1969)350

Eremotropa Andres = *Monotropastrum*

Eremurus M. Bieb. Asphodelaceae. c. 45 Eur. (2), alpine W & C As. R: MAISSP 8,23(1909)1. Foxtail lilies, giant asphodels. Typical of dry open, almost barren habitats of W As. to subalpine grasslands of Afghanistan & W China. Columnar infls to 3.5 m, of white, pink or yellow fls with P withering before A & G mature, followed by capsules of wind-disp. winged seeds. Lvs of ***E. spectabilis*** M. Bieb. (Crimea to Pakistan) ed., transplanted from wild to Turkish graves. Several other cult. orn. esp. hybrids & forms of ***E. stenophyllus*** (Boiss. & Buhse) Bak. (SW As.) with yellow fls & ***E. olgae*** Regel (Turkestan) with white (the hybrid between them = ***E.*** × ***isabellinus*** P.L. Vilm. (Shelford Hybrids))

Erepsia N.E. Br. Aizoaceae (V). 29 SW Cape. R: H. E. K. Hartmann, *Ill. Handb. Succ. Pls*, Aiz. A–E(2002)258. Shrubs, some cult. orn. (incl. *Kensitia, Semnanthe*), fls always open

Ergocarpon C. Towns. Umbelliferae (III 1). 1 Iraq/Iran border: ***E. cryptanthum*** (Rech.f.) C. Towns. R: KB 17(1964)437

Eria Lindl. Orchidaceae (V 15). Excl. *Callostylis, Campanulorchis, Conchidium, Dilochiopsis, Mycaranthes, Pinalia*, c. 230 Indopacific. Epiphytes, some cult. orn.

Eriachaenium Schultz-Bip. Compositae (Mut.-Adenoc.). 1 S S Am.: ***E. magellanicum*** Schultz-Bip.

Eriachne R. Br. Gramineae (Micr.). 48 Aus. with 6 ext. to S Mal., 2 to China & Sri Lanka. R: Aus SB 8(1995)359, FA 44B(2005)132. Prob. related to *Micraira* but not morphologically obviously so. ***E. mucronata*** R. Br. (Aus.) – used as lead indicator in Queensland

Eriadenia Miers = *Mandevilla*

Eriandra P. Royen & Steenis. Polygalaceae (II). 1 Papuasia: ***E. fragrans*** P. Royen & Steenis – tree to 32 m. R: FM I,10(1988)492

Eriandrostachys Baill. = *Macphersonia*

Erianthecium L. Parodi = *Chascolytrum*

Erianthemum Tieghem. Loranthaceae (5 6). 16 E & S Afr. R: R. Polhill & D. Wiens, *Mistletoes of Afr.* (1998)234. ***E. dregei*** (Ecklon & Zeyher) Tieghem (S Afr.) – 'wood rose' outgrowths produced at junction with host sold as curios

Erianthus Michaux = *Saccharum*

Eriastrum Wooton & Standl. Polemoniaceae (III). 14 SW US. R: Madroño 8(1945)65. Some cult. orn. like *Gilia* but K-lobes unequal

Eriaxis Reichb. f. Orchidaceae (II 2). 1 New Caled.: ***E. rigida*** Reichb.f. R: Gen. Orch. 3(2003)309

Eribroma Pierre = *Sterculia*

Erica Tourn. ex L. Ericaceae (V 5). Incl. *Acrostemon, Anomalanthus, Arachnocalyx, Blaeria, Bruckenthalia* (pollen in monads), *Coccosperma, Coilostigma, Eremia, Eremiella, Ericinella, Grisebachia, Nagellocarpus, Philippia, Platycalyx, Salaxis, Scyphogyne, Simocheilus, Stokoeanthus, Sympieza, Syndesmanthus, Thamnus, Thoracosperma*, c. 820 S Afr. (770 – D. Schumann et al. (1993) *Ericas of S Afr.*; most, esp. SW Cape 658 – 635 endemic; R: Strel. 9(2000)423), trop. Afr. mts, Madag., Medit., Macaronesia, Eur. (16). R (partial): CBH 19(2000)1. Heaths; orig. Eur., S Afr. spp. in radiation from ancestor of *E. arborea*. Shrubs & small trees (Leeuwenberg's & Rauh's Models) with endotrophic mycorrhiza; some (Eur., S Afr.) with elaiosomes, some cyanogenic. Many fynbos spp. germ. enhanced by smoke. Fls bell-shaped, pend. & visited esp. by bees searching for nectar secreted by disk, bird-poll. spp. in Cape with significantly thicker stems; ***E. cedromantana*** E. Oliv. (W Cape) – ? poll., disp. ants; ***E. hanekonii*** E. Oliv. (Cape) – poll. rodents (*Acomys subspinosus*). ***E. cinerea*** L. & *E. tetralix* (crossed-leaved h.) cover great areas of drier & wetter moorland resp. in Eur. Many cult. orn. esp. 19 cent. S Afr. spp. in greenhouses (***E. gracilis*** Wendl. (Cape) still much prop. for winter decoration), incl. ***E. arborea*** L. (tree h., Medit. to trop. Afr. mts) – distillate used for rheumatism in Spain, where wood used for musical instruments, spoons, clogs, nodules at ground level trad. used to make briar ('bruyère') pipes, ***E. australis*** L. (S Eur.) – favoured for charcoal in Spain, ***E. carnea*** L. (*E. herbacea*, Eur.) – lime-tolerant winter-flowering, ***E. ciliaris*** L. (Dorset h., W Eur.), ***E. erigena*** R. Ross (Irish h., W Eur.) – lime-tolerant, ***E. x hyemalis*** W. Watson (unknown parentage) – cult. cut-fl. Hawaii etc., ***E. verticillata*** P. Bergius (Cape) – form. extinct in wild, re-introd. from 10 pls now in cult. around world; ***E. lusitanica*** Rudolphi (SW Eur., natur. SW England) – major weed in many parts of NZ, ***E. tetralix*** L. (Eur.) – source of yellow dye, ***E. vagans*** L. (Cornish h., W Eur.) – lime-tolerant

Ericaceae Juss. Magnoliidae – Ericales. Incl. Empetraceae, Epacridaceae, Monotropaceae, Pyrolaceae, 113/4250 cosmop. exc. deserts; usu. montane in trop. R: BR 68(2002)335. Shrubs & trees (s.t. pachycaul), more rarely lianes or (sub)herbaceous, s.t. epiphytic, mycorrhizal & oft. on acid soils or even achlorophyllous (Monotropoideae). Lvs simple, oft. ericoid, in spirals, opp. or whorled; stip. 0. Fls usu. bisex., ± reg., usu. in usu. term. bracteate racemes (s.t. solit.) with 2 bracteoles; K (2 –)4, 5(– 7), valvate or imbr., persistent, C ((3)4,5(– 7)) usu. a tube with convolute or imbr. lobes, s.t. free, A in 2 whorls, usu. 2 × C, less oft. up to 20 or (some of) whorl opp. C absent, s.t. attached at tube-base & rarely united, anthers becoming inverted during ontogeny with term. pores, less oft. slits, appendages oft. present, pollen grains in tetrads (monads in e.g. *Enkianthus*), nectary-disk intrastaminal (rarely 0), surrounding & oft. attached to G, $\underline{G}$ to $\overline{G}$ ((2 –)4,5 (– 12)), pluriloc. (1-loc. in *Scyphogyne* etc. with 1 pend. ovule), style hollow, placentation axile basally, parietal apically or all axile, ovules (1 –)∞ on each placenta, anatr. to ± campylotropous, unitegmic. Fr. a capsule (loculicidal or septicidal), berry or drupe (esp. Vaccinioideae), rarely a nut; seeds ± ∞, usu. small, s.t. winged, with straight or short cylindric or spathulate embryo embedded in copious oily & proteinaceous endosperm. x = (8 –)12 or 13(– 23)

Classification & chief genera:

I. **Enkianthoideae** (Vacc.-Enkiantheae; lvs pseudoverticillate; n = 11): *Enkianthus* (only)

II. **Monotropoideae** (herbs; n = 8, 13, 19, 23 etc.; mostly N temp.): 1. **Pyroleae** (Pyroloideae; G imperf. 5-loc.; s.t. treated as sep. fam., Pyrolaceae): *Pyrola;* 2. **Monotropeae** (chlorophyll-less; s.t. treated as sep. fam., Monotropaceae; R: WJB 33(1975)1): *Cheilotheca, Monotropa*; 3. **Pterosporeae**: *Pterospora*

III. **Arbutoideae** (Vacc.-Arbuteae; C urceolate, fr. berry or drupe, n = 13): *Arbutus, Arctostaphylos*

IV. **Cassipoideae** (infls axillary): *Cassiope* (only)

V. **Ericoideae** (incl. Rhododendroideae – lvs revolute, infls usu. term., C usu. other than urceolate, decid., anther & filament appendages 0, usu. viscin threads in pollen tetrads, $\underline{G}$, fr. usu. septifragal or septicidal capsule): Lvs v. revolute with abaxial channel, infls variable in position, C gamopetalous, persistent, appendages if present flattened spurs, viscin threads absent, $\underline{G}$, fr. usu. septicidal capsule) – 1. **Ericeae** (incl. Calluneae, Daboecieae, excl. **Bryantheae** [R: Britt. 64(2012)75] – *Bryanthus, Ledothamnus*): *Erica, Calluna, Daboecia*; 2. **Empetreae** (Empetraceae):

Empetrum; 3. **Bejarieae**: *Bejaria*; 4. **Phyllodoceae** (incl. Epigaeae): *Epigaea, Kalmia, Phyllodoce*; 5. **Rhodoreae**: *Rhododendron*;

VI. **Harrimanelloideae**: *Harrimanella* (only)

VII. **Epacridoideae** (Styphelioideae; Epacridaceae) – 1. **Prionoteae**: *Prionotes*; 2. **Archerieae**: *Archeria* (only); 3. **Oligarrheneae** (incl. Needhamielleae): *Oligarrhena*; 4. **Richeeae** (Epac.- Richeoideae): *Dracophyllum, Richea*; 5. **Epacrideae**: *Epacris*; 6. **Cosmelieae**: *Andersonia*; 7. **Styphelieae**: *Acrotriche, Leptecophylla, Leucopogon, Styphelia*

VIII. **Vaccinioideae** (infl.usu. axillary; n = 12, a v. variable group without viscin threads & with sup. to inf. ovaries & fr. a berry, drupe or loculicidal capsule) – 1. **Oxydendreae**: *Oxydendrum* (only); 2. **Lyonieae**: *Agarista, Lyonia, Pieris*; 3. **Andromedeae**: *Andromeda, Zenobia* (only); 4. **Gaultherieae**: *Gaultheria, Leucothoe*; 5. **Vaccinieae** (many epiphytes, generic limits shaky): *Agapetes, Cavendishia, Dimorphanthera, Disterigma, Gaylussacia, Macleania, Psammisia, Thibaudia, Vaccinium*)

Wittsteinioideae referred to Alseuosmiaceae. Although char. of moorlands of N temp. & Aus., E. are most diversified in Cape, the Andes & (esp. *Rhododendron*) Sino-Himal. & NG, though poss. orig. Euras.

Imp. cult. orn. esp. spp. of *Rhododendron* (incl. azaleas) & *Erica*, also *Arbutus, Arctostaphylos, Cassiope, Cavendishia, Elliottia, Enkianthus, Gaultheria, Kalmia, Leucothoe, Pieris* etc. Almost all require acidic conditions (all Br. spp. exc. *Arbutus unedo* for example); some fr. trees or bushes esp. *Vaccinium*, but also *Acrotriche, Disterigma, Empetrum, Gaylussacia, Leucopogon, Macleania*, though others e.g. *Kalmia* & others in its subfam. poisonous. Some medic. (*Andromeda, Chimaphila, Gaultheria, Moneses, Pyrola* etc.), honey (*Calluna*)

Ericameria Nutt. (~ *Haplopappus*). Compositae (Ast.-Hin.). 31 SW & W N Am. R: Phytol. 68(1990)144. Local medic., poss. rubber sources

Ericentrodea S.F. Blake & Sherff. Compositae (Cor.). 6 Andes. R: Novon 3(1993)77

Erichsenia Hemsl. Leguminosae (III 13). 1 W Aus.: ***E. uncinata*** Hemsl.

Ericinella Klotzsch = *Erica*

Ericksonella Hopper & A.P. Br. (~ *Caladenia*). Orchidaceae (IV 3b). 1 SW Aus.: ***E. saccharata*** (Reichb.f.) Hopper & A.P. Br. R: AusSB 17(2004)208

Ericomyrtus Turcz. (~ *Baeckea*). Myrtaceae. 4 SW Aus. R: Nuytsia 25(2015)133

Erigenia Nutt. Umbelliferae (III 5). 1 E N Am.: ***E. bulbosa*** (Michaux) Nutt. – tuberous, one of the first spring-flowers

Erigeron L. Compositae (Ast.-Con.). Incl. *Aphanostephus, Apopyros, Conyza s.s., Darwiniothamnus, Hysterionica, Trimorpha*, excl. *Pappochroma*, c. 400 cosmop. (not Aus., though introd.) esp. N Am. (US & Canada 173 – R: FNA 20(2006)256) & C Am. (234), with OW 100 (Eur. 15 incl. introd. ***E. canadensis*** L. (*C. c.*, horseweed, N Am.) – also weedy Jap.); ***E. albidus*** (Spreng.) A. Gray (? *E. sumatrensis*), & ***E. bonariensis*** L. of S. Am. – weedy subcosmop.; ***E. karvinskianus*** DC. (*E. mucronatus*, C Am.), apomict natur. on walls trop. & temp., invasive in Azores, Aus.)). Fleabanes. Polyploidy common, agamospermy evolved at least 3 times, autogamy twice. Shrubby spp. (2, *D.*) on Galápagos. ***E. heteromorphus*** Robinson (Mex.) – aquatic. Many cult. orn. incl. ***E. annuus*** (L.) Pers. (daisy f., E N Am., natur. Eur.) & a no. of complex hybrids used as herb. perennials, many derived from ***E. speciosus*** (Lindl.) DC. (W N Am.)

erima (NG) *Octomeles sumatrana*

erimado *Ricinodendron heudelotii*

Erinacea Adans. Leguminosae (III 9). 1 W Medit. (incl. Eur.): ***E. anthyllis*** Link (*E. pungens*, hedgehog broom) – cult. orn. shrublet with branch-thorns

Erinna Philippi (~ *Leucocoryne*). Amaryllidaceae (Alliaceae Gill.). 1 Chile: ***E. gillesioides*** Philippi – A 3 with 3 staminodes

Erinocarpus Nimmo ex J. Graham. Malvaceae (Grew.-Apeib.; Tiliaceae). 1 SW Ind.: ***E. nimmonii*** J. Graham – androphore, fr. spiny, bark-fibre used for rope

Erinus L. Plantaginaceae (Dig.-Ver.; Scrophulariaceae s.l.). 1 – 2 Morocco, Pyrenees & Alps (***E. alpinus*** L.) – cult. orn. rock-pls

Erioblastus Nakai ex Honda = *Deschampsia*

Eriobotrya Lindl. Rosaceae (Ros.-Mal.). 15 – 20 Himal. to E As. & W Mal. R: CJB 68(1990)2233. Cult. orn. everg. trees & shrubs esp. ***E. japonica*** (Thunb.) Lindl. (loquat, Jap. medlar, nispero, China & Jap.) – widely cult. (Fagerlind's Model) in subtrop. for yellow acid fr. eaten raw or in jams etc., bird-disp. invasive S Braz., crossed with *Rhaphiolepis* spp., cvs grafted on *Cydonia oblonga*

Eriocarpum Nutt. (~ *Haplopappus*). Compositae (Ast.-Sol.) 10 N Am.

Eriocaulaceae Martinov. Magnoliidae – Poales. 7/1265 trop. & warm esp. S Am., few temp. (N 1/2). R: Rodriguesia 63(2012)13. Perenn. (usu.; s.t. hapaxanthic), usu. monoec. small herbs (to suffrutices with trunks to 80 cm) with oxalate crystals, growing in wet places; vessel-elements in all veg. organs; sec. thickening rare (some *Paepalanthus*, *Syngonanthus*). Lvs in a dense basal spiral, parallel-veined, grass-like but without a well differentiated basal sheath. Infl. a dense centripetally flowering, usu. grey or white, head on a scape, the receptacle naked, hairy or with chaffy bracts subtending fls. Fls small wind- or insect-poll., without nectaries exc. for glands in petal-tips of *Eriocaulon*, (2- or) 3-merous, oft. reg., usu. unisexual, the sexes mixed or females marginal, rarely pls dioec.; K 2 or 3, s.t. with basal tube or forming spathe, C (0) 2 or 3, s.t. with a tube, males with filaments adnate to tube or oft. with stipe- like androphore at top of which C & filaments diverge, A 2 or 4 (2-merous fls), 3 or 6((or 1)3 + -merous fls) opp. C when 2 or 3, anthers with longit. slits, $\underline{G}$ (2 or 3), oft. stipitate, with as many loc. & term. style with as many branches, each loc. with 1 ventral – apical, pend., orthotropous, bitegmic ovule. Fr. a loculicidal capsule; seeds with small lenticular embryo forming a cap over the copious starchy endosperm at micropylar end; cotyledon not photosynthetic. n = 9, 15, 20, 25

Classification & genera: **Eriocauloideae** (usu. aquatic: *Eriocaulon*); **Paepalanthoideae** (usu. terr.: *Comanthera, Leiothrix, Mesanthemum, Paepalanthus, Rondonanthus, Syngonanthus*)

App. closely allied to Xyridaceae

Infls of *Syngonanthus* spp. used as 'everlastings'

Eriocaulon Gronov. ex L. Eriocaulaceae. c. 400 trop. & warm incl. c. 30 in Jap. & 11 in N Am. incl. ***E. aquaticum*** (Hill) Druce (*E. septangulare*, pipewort) also in Ireland & Hebrides. Some spp. with trunks to 80 cm. ***E. australe*** R. Br. (E Aus., NG) – salt-source in NG

Eriocephalus Dill. ex L. Compositae (Anth.-Anth.). 32 S Afr. R: FSA 33,4(2001)1. Some cult. orn. & fodder, form used as diuretic, but now imp. ess. oil for scent

Eriocereus (A. Berger) Riccob. = *Harrisia*

Eriochilus R. Br. Orchidaceae (IV 3b). 9 Aus.

Eriochiton (R. Anderson) A.J. Scott = *Maireana*

Eriochlamys Sonder & F. Muell. Compositae (Gnap.-Ang.). 4 Aus. R: OB 104(1991)129

Eriochloa Kunth. Gramineae (XXIV 5). 1 S Am.: ***E. distachya*** Kunth. See also *Urocholoa*

Eriochrysis P. Beauv. Gramineae (XXII 5). 12 trop. Afr. & Am., Ind. (1)

Eriocnema Naud. Melastomataceae. 1 Braz.: ***E. acaulis*** Triana

Eriocoelum Hook.f. Sapindaceae. 10 trop. Afr.

Eriocoma Kunth (~ *Achnatherum*). Gramineae (X). 29 Am.

Eriocycla Lindl. = *Seseli*

Eriodendron DC. = *Ceiba*

Eriodes Rolfe (*Tainiopsis*). Orchidaceae (V 14). 1 NE Ind. to Vietnam: ***E. barbata*** (Lindl.) Rolfe. R: OB 89(1986)65

Eriodictyon Benth. Boraginaceae (Hydrophyllaceae-Nam.). 8 SW N Am. R: BG 60(1915)122. Some cult. orn. shrubs incl. ***E. californicum*** (Hook. & Arn.) Torrey (Calif., Oregon) – lvs used medic. (yerba santa) & as a tea. ***E. trichocalyx*** A.A. Heller (SW N Am.) – erosion control & fire-break in Calif.

Erioglossum Blume = *Lepisanthes*

Eriogonella Goodman = *Chorizanthe*

Eriogonum Michaux. Polygonaceae (Erio.-Erio.). Excl. *Nemacaulis, Johanneshowellia* c. 250 W N Am. (US 224 – R: FNA 5(2005)221) to Mex. R: Phytol. 66(1989)267. With *Penstemon* biggest N Am. angiosperm genus. Shrubs, herbs & cushion-pls from seashore to highest US mts, differing from most P. in having no ocreae but cymose umbels or heads of fls (oft. compound heads). Several cult. orn., local medic. & ed. lvs & r.; ***E. leptophyllum*** (Torrey) Wooton & Standl. – analagesic, snakebite remedy, etc.; ***E. ovalifolium*** Nutt. – silver indicator in Montana

Eriogynia Hook. = *Luetkea*

Eriolaena DC. Malvaceae (Bomb.; Sterculiaceae). 8 Indomal., China, SE As.

Eriolarynx (Hunz.) Hunz. (~ *Vassobia*). Solanaceae (IV 5b). 3 Arg., Bolivia

Eriolobus (DC.) M. Roemer (~ *Malus*). Rosaceae (16). 1 E Med.: ***E. trilobatus*** (Poir.) M. Roemer

Eriolopha Ridl. = *Alpinia*

Erioneuron Nash (~ *Tridens*). Gramineae (XXIX 1). Excl. *Dasyochloa* 3 S US, Mex., Peru, Bolivia, Arg. R: Sida 17(1997)656

Eriope Bonpl. ex Benth. Labiatae (VII 3c). c. 40 trop. & warm S Am. esp. C & E Braz. R: HIP 28,3(1978)1. ***E. crassipes*** Benth. (Braz. savannas) – explosive poll. mechanism (like *Ulex*) assoc. with bee-visits

Eriopexis (Schltr) Brieger. See *Dendrobium*

Eriophorella Holub = *Trichophorum*

Eriophoropsis Palla = seq.

Eriophorum L. (~ *Scirpus*). Cyperaceae (II 1). 18 N temp. (Eur. 7, N Am. 11) & Arctic, S Afr. (1). R: BZeit. 54(1896)156. Typically in wet moorland. Female fls each with P of bristles which grow into long hairs after fert. & act to disperse fr., s.t. used to stuff pillows (cotton grass), form. used in GB: those of ***E. angustifolium*** Honck. also in paper & candlewicks, ***E. vaginatum*** L. in yarn & fabric

Eriophyllum Lag. Compositae (Mad.-Baer.). 13 W N Am. R: FNA 21(2006)353. Some cult. orn.

Eriophyton Benth. Labiatae (VI). Incl. *Alajja*, 5–6 C As. to Himal. R: T 60(2011)482. ***E. wallichii*** Benth. (Himal.) – shaggy with waterproof hairs

Eriopidion Harley = *Eriope*

Eriopsis Lindl. Orchidaceae (V 10?). 3 trop. Am. Epiphytes, some cult. orn.

Erioscirpus Palla = *Eriophorum* (but poss. distinct)

Eriosema (DC.) Desv. Leguminosae (III 18). 150 trop. & warm. Some ed. & med.

Eriosemopsis F. Robyns. Rubiaceae (III 2). 1 S Afr.: ***E. subanisophylla*** F. Robyns. R: Strel. 10(2000)485

Eriosolena Blume = *Daphne*

Eriosorus Fée = *Jamesonia* (but see CGH 200(1970)54); see also *Tryonia*

Eriospermaceae Lem. = Asparagaceae

Eriospermum Jacq. ex Willd. Asparagaceae (Eriospermaceae). 102 subSaharan Afr. Some with epiphyllous lvs

Eriosphaera Less. = *Galeomma*

Eriostemon Sm. Rutaceae (I 3). 2 E Aus. R: Nuytsia 12(1998)242. Cult. orn. (waxflowers, wax pls). See also *Philotheca*

Eriostrobilus Bremek. = *Strobilanthes*

Eriostylos C. Towns. Amaranthaceae (I 2). 1 Somalia: ***E. stefaninii*** (Chiov.) C. Towns. R: KB 34(1979)237

Eriosyce Philippi. Cactaceae (III 5). 32 Chile, Arg. R: F. Kattermann, E. (1994). Cult. orn. allied to *Echinocactus* but with axillary spines at top of fr.

Eriosynaphe DC. Umbelliferae (III 10). 1 SE Russia to C As.: ***E. longifolia*** (Fischer) DC.

Eriotheca Schott & Endl. (~ *Pachira*). Malvaceae (Bomb.; Bombacaceae). 20 trop. Am. R: BJBB 33(1963)124. Rauh's Model. ***E. pubescens*** (Mart. & Zucc.) Schott & Endl. – polyploid, seeds polyembryonous

Eriothrix Cass. = *Eriotrix*

Eriothymus (Benth.) Reichb. (~ *Hedeoma*). Labiatae (VII 2b). 1 Braz.: ***E. rubiaceus*** (Benth.) J.A. Schmidt – ? extinct, known only from 1 specimen

Eriotrix Cass. Compositae (Sen.-Sen.). 2 Réunion. Shrubs

Erioxylum Rose & Standl. = *Gossypium*

Erisma Rudge. Vochysiaceae (II). 16 trop. S Am. R: ABN 3(1954)462, MNYBG 81(1998)21. Seeds a source of tallow for candles, soap etc., those of ***E. japura*** Spruce ex Warm. coll. & stored for lean times in Amaz.

Erismadelphus Mildbr. Vochysiaceae (II). 2 trop. W Afr. R: ABN 1(1953)594

Erismanthus Wall. ex Muell. Arg. Euphorbiaceae (Acal.-Eris.). 2 SE As. to Hainan & W Mal. R: Blumea 41(1993)379

Erithalis P. Browne. Rubiaceae (III 4). 8 Florida, WI. R: Sida 21(2005)1579

Eritrichium Schrad. ex Gaudin. Boraginaceae (3.5.3). c. 50 N temp. (Eur. 4). R: BZ 86,6(2003)79. Some cult. orn. (American forget-me-nots) esp. ***E. nanum*** (L.) Gaudin (Alps)

Erlangea Schultz-Bip. Compositae (Vern.-Erl.). c. 10 trop. Afr.

Ermania Cham. ex Botsch. = *Smelowskia*

Ermaniopsis Hara = *Solms-laubachia*

Ernestia DC. Melastomataceae. 16 trop. S Am.

Ernestimeyera Kuntze = *Alberta*

Ernodea Sw. Rubiaceae (IV 15). 9 SE US, WI

Erocallis Rydb. = *Lewisia*

Erodiophyllum F. Muell. Compositae (Ast.-Gran.). 2 W & S Aus.

Erodium L'Hérit. Geraniaceae. c. 80 Eur. (34), Medit. to C As., temp. Aus. (3) & S trop. S Am. Awn twists into corkscrew & is hygroscopic; mericarp with sharp point with backward-pointing hairs. Fr. falls & in damp awn untwists, lengthens & drives fr. into soil (cf. *Stipa*). Cult. orn. (storksbills) with cincinnal umbel (cf. *Geranium*) incl. ***E. cicutarium*** (L.) L'Hérit. (Eur., Medit., natur. N & S Am. where used as forage) – form. r. a source of a dye in Hebrides. Others with ed. r. or lvs, some weedy (filaree in Calif.), esp. ***E. botrys*** (Cav.) Bertol. (Medit.) – invasive, like *E. cicutarium*, Aus., W US

Erophaca Boiss.(~ *Astragalus*). Leguminosae (III 24). 1 Medit.: ***E. baetica*** (L.) Boiss. (*A. lusitanica*) – seeds a coffee subs.

Erophila DC. = *Draba*

Erpetion Sweet = *Viola*

Errazurizia Philippi. Leguminosae (III 10). 1 coastal Chile, 3 SW US deserts. R: MNYBG 27(1977)13

Ertela Adans. (*Monniera*). Rutaceae (I 8). 2 trop. S. Am. R: Candollea 45(1990)369. Leeuwenberg's Model

eru *Gnetum* spp.

Eruca Mill. (~ *Brassica*). Cruciferae (12). 1 Medit. incl. Eur., NE Afr. R: JAA 66(1985)324. ***E. vesicaria*** (L.) Cav. 'subsp. *sativa* (Mill.) Thell.' (*B. eruca* L., garden or salad rocket, arugala, jamba, Medit.) – salad greens, oilseed subs. for rape in Ind. (poss. mustard-seed of Bible), biodiesel potential (tamarira)

Erucaria Gaertn. (~ *Cakile*). Cruciferae (10). 6 E Med. (Eur. 1), Arabia, Iran. Spirolobal cotyledons

Erucastrum (DC.) C. Presl (~ *Brassica*). Cruciferae (12). Excl. *Hirschfeldia*, 25 Macaronesia, Medit., C & S Eur. (4). R: JAA 66(1985)305. ***E. incanum*** (L.) Koch (*H. i.*, Medit.) – widespread weed S Br., N Am.

Ervatamia (DC.) Stapf = *Tabernaemontana*

Erxlebenia Opiz = *Pyrola*

Erycibe Roxb. Convolvulaceae (7). c. 75 Indomal. to trop. Aus. & Jap. C-lobes bifid & sessile, conical or subglob. 5 – 10-rayed stigma unique in fam. ***E. ramiflora*** Hallier f. (Sumatra) cauliflorous; ***E. stenophylla*** Hoogl. (Borneo) – rheophyte

Erycina Lindl. Orchidaceae (V 12h). Incl. *Psygmorchis*, 7 trop. Am. R: Schlechteriana 2(1991)115. Some short-lived epiphytes on coffee & guava & in sec. forest; some with yellow fls poss. mimicking Malpighiaceae with oils coll. *Centris* bees, some autogamous. Some cult. orn.

Erymophyllum Paul G. Wilson (~ *Rhodanthe*). Compositae (Gnap.-Ang.). 5 W Aus. R: OB 104(1991)105

Eryngiophyllum Greenman = *Chrysanthellum*

Eryngium Tourn. ex L. Umbelliferae (II 1). c. 230 trop. & temp. (exc. trop. & S Afr.; Euras. & N Afr. 61 – R: BB 159(2011)76). R: SBNat 596(1999)2. Lvs spiny-toothed, simple & monocot-like or lobed; fls in bracteate heads with K oft. longer than C, bee-poll.; many ed. or cult. orn. esp. cvs of ***E. planum*** L. (temp. OW), ***E × zabelii*** Hort. ex Christ (*E. alpinum* L. (SE Eur.) × *E. bourgatii* Gouan (SW Eur.)), some weedy. ***E. agavifolium*** Griseb. ('*E. bromeliifolium*', Arg.) & ***E. pandanifolium*** Cham. & Schldl. (*E. lassauxii*, S Am.) – cult. orn. with pandan-like lvs from Pampas, lvs source of caraguata fibre; ***E. aquaticum*** L. (rattlesnakemaster, N Am.) & ***E. yuccifolium*** Michaux (N Am. – used against snakebite etc.; ***E. campestre*** L. (eryngo (r.), Eur., (?) natur. sandy fields in GB) – candied r. (esp. from Colchester, UK) form. used as 'kissing comfits' (considered aphrodisiac), also med. (coughs & colds); ***E. foetidum*** L. (cilantro, fitweed, ngo gai, Thai coriander, trop. Am.) – lvs ed raw or steamed Mal., pickled in Sikkim, culinary herb in WI, used like coriander (retains flavour when dried) in Vietnam, local medic.; ***E. giganteum*** M. Bieb. (Cauc.) – cult. orn. ('Miss Willmott's ghost', as Ellen W. surreptitiously scattered seeds when visiting gardens) visited by wasps; ***E. maritimum*** L. (sea holly, Eur., natur. Aus., N Am.) – young shoots form. ed., r. candied (? eringoes of Falstaff), form. valued as tonic; ***E. ovinum*** A. Cunn. ('*E. rostratum*', SE Aus.) – 'blue devil' (also *E. pinnatifidum* Bunge, W Aus.), form. conspic. roadside pl.

Erysimum Tourn. ex L. Cruciferae (25). Incl. *Cheiranthus, Syrenia*, c. 200 N hemisph. (Eur. 58, esp. E; SW As. 96). R: ANMWB 111(2009)181, 112(2011)369, 113(2012)147, 115(2013)57,75, 116(2014)87. Some weeds, e.g. ***E. cheiranthoides*** L. (treacle mustard, Eur., natur. N Am.), & cult. orn. esp. ***E. cheiri*** (L.) Crantz (*Cheiranthus c.*, wallflower, prob. hybrid orig. of 2 or more Aegean spp., cultigen natur. GB) – spring-bedding, scented perenn. (Scarrone's Model) with yellow, brown, red or pink fls form. medic. & used in scent, sign of romantic love, crossed with *E. bicolor* DC. (Macaronesia) to give ['*E.* × *kewense* Hort.'] **'Harpur Crewe'** – strong scent, double fls (A0, G0); also alpine or fairy wallflowers esp. 'Siberian wallflower', ***E.* × *marshallii*** (T. Moore) Bois ('*E.* × *allionii*', prob. *E. humile* Pers. (*E. decumbens*, SW Alps) × *E. perofskianum* Fischer & C. Meyer (Cauc. to Afghanistan)) differing in its bilobed stigma; ***E. asperum*** (Nutt.) DC. (western wallflower, W & C N Am.)

Erythea S. Watson = *Brahea*

Erythradenia (Robinson) R. King & H. Robinson = *Decachaeta*

Erythraea Borkh. = *Centaurium*

Erythranthe Spach (~ *Mimulus*). Phrymaceae. 116 As., Am. R: Phytoneuron 2012–39:33, 2013–43:1, 2014–31:1. Monkey-fl., m. musk. Stigma sensitive to contact such that it closes up after being touched by visiting insect. Bee- & bird-poll. e.g. sister spp. ***E. cardinalis*** (Benth.) Spach (*M. c.*, W N Am.) – red fls, poll. hummingbirds, & ***E. lewisii*** (Pursh) Nesom & Fraga (*M. l.*, W N Am.) – pink fls, poll. bumblebees, the colour diff. (& therefore poll. system) due to single mutation; ***E. guttata*** (DC.) Nesom (*M. g.*, Alaska to Mex.) – densely glandular pubescent lvs to 15 cm, fls to 6 cm, ***E. lutea*** (L.) Nesom (*M. l.* incl. *M. smithii*, yellow monkey-fl., Chile, Arg.) – usu. glabrous, lvs to 2.5 cm, fls to 2.5 cm & their sterile triploid hybrid, ***E.* x *robertsii*** (Silverside) Nesom – common natur. upland pl. in Brit. Is., fertile (? tetraploid) pls being called ***E. peregrina*** (Vall.-Marín) Nesom (*M. p.*); hybrid between *E. l.* & ***E. cuprea*** (Veitch) Nesom (*M. c.*, Chile) = ***E.* × *maculosa*** (T. Moore) Mabb. (*M.* × *m.*); ***E. moschata*** (Lindl.) Nesom (*M. m.*, musk pl., N Am.) – much cult. cottage pl. in C19 for fragrant lvs but since c. 1914 surviving cult. clones scentless as orig. veg. clone crossed repeatedly with scentless later introd. ones; ***E. pardalis*** (Pennell) Nesom (*M. cupriphilus*, Calif.) close to *E. guttata*, restricted to 2 coppermines

Erythranthera Zotov = *Rytidosperma*

Erythrina L. Leguminosae (III 18). c. 120 warm (As.- Aus. 12, Afr. & Madag. 38, Am. c. 70). R: Lloydia 37(1974)332. Extrafl. nectaries attractive to ants, which act as guards (some with ant-infested twigs); many alks (some hypnotic incl. those of *E. americana*, *E. berteroana* & *E. variegata*). Prob. all bird-poll. red or orange fls (coloured nectar), some Mal. spp. with long peduncles acting as perches for non-hovering sunbirds, many Am. spp. without & with smaller tubular C-tubes sticking out assoc. with long-billed hummingbird poll. (short-billed ones merely thieves) & with high sucrose & low amino-acid content in nectar, whereas spp. visited by passerines have fls twisted back towards peduncle & low sucrose & high amino-acid contents (the evolutionary transition having occurred at least 4 times); insufficient calorific value in nectar in fls open on any tree in 1 day to sustain a single hummingbird's energy requirements so outcrossing inevitable; nectar of Mal. spp. 'sweet, if somewhat bitter, watery' (Corner) & as fls open in dry season in monsoon climates an imp. sugar & water-source for birds (& squirrels), e.g. ***E. velutina*** Willd. (NE S Am.) in Fernando de Noronha Arch. (NE Braz.) poll. by 2 perching bird spp., also doves & lizards. Fls cooked & eaten; seeds used as beads. Cult. as shade trees & orn. (coral trees) incl. many hybrids (the first ***E.* x *bidwillii*** Herb. (*E. crista-galli* × *E. herbacea* L. (SE N Am.) in Aus. (long held to be only known hybrid legume), ***E.* × *malottiorum*** Mabb. & Lorence (*E. c.-g.* × *E. fusca* Lour. (riparian & estuarine S Am.) – raised Hawaii (33 hybrids grown there), closely resembling *E. dominguezii* Hassler of geographical range intermediate between the hybrid's parents but an upland dry forest tree, & ***E.* × *sykesii*** Barneby & Krukoff – cult. orn. sterile (? orig. Aus.) with nectar sugar suggesting hybrid between OW & NW bird-poll. spp. (? *E. coralloides* A. DC. (S US, Mex.) × *E. lysistemon* Hutch. (S Afr.)). ***E. abyssinica*** Lam. (E Afr.) – bark. medic. (incl. gonorrhoea); ***E. americana*** Mill. (C Mex.) – fls ed.; ***E. berteroana*** Urb. (C Am., introd. WI) – fls & young shoots a sleep-inducing veg. (pito) exported to US; ***E. caffra*** Thunb. (kaffir boom, SE Afr.) – red seeds with black hilum used as beads, wood v. light; ***E. crista-galli*** L. (warm S Am.) – fls inverted with v. small wings, keel forming nectar sac at base, bird-poll. oft. so nectariferous as to be called 'cry-baby' in Louisiana, grown under glass in N Eur., effective painkiller (up to 9 days) because protein binds to nerve-cells transmitting pain; ***E. herbacea*** L. (Cherokee bean, S N Am.) – seeds red with black line from hilum used as beads; ***E. mildbraedii*** Harms (trop. Afr.) – thorns used for making 'rubber stamps'; ***E. mitis*** Jacq. (*E. umbrosa*, mortel, Venez.) – shade tree for woody crops; ***E. subumbrans*** (Hassk.) Merr. (*E. lithosperma*, dadap, Indomal.) – coffee-shade; ***E. sandwicensis*** Degener (~ *E. tahitensis* Nad., *E. monosperma*, Hawaii) – fl. colour polymorphic in populations (orange, yellow, white or pale green), toxic to 8 introd. spp. of bruchid beetles, timber (wiliwili) for outriggers & form. surfboards & fish-floats; ***E. variegata*** L. (*E. indica*, E Afr. to Pac.) – poplar-like (esp. fastigiate '**Tropical Coral**') coffee-shade, wind-break & crop-support, face-powder prep. from wood in Thailand

Erythrocephalum Benth. Compositae (Card.-Dic.). c. 14 trop. E Afr. Herbs with scaly pappus

Erythrochiton Nees & Mart. Rutaceae (I 8). 7 trop. Am. R: Britt. 44(1992)123. K coloured, (C). ***E. hypophyllanthus*** Planch. & Linden (Colombia & Venez.) – known only from one coll., unbranched treelet with fls borne on abaxial surface of lvs

Erythrochlamys Guerke = *Ocimum*

Erythrococca Benth. Euphorbiaceae (Acal.-Acal.-Claox.). 50 trop. & S Afr., S Arabia. R: AnnB 25(1911)606

Erythrocoma Greene = *Geum*

Erythrodes Blume. Orchidaceae (IV 2d). 26 trop. As., Mal., New Caled. (1) to Samoa & Tonga. Some cult. orn. terr.

Erythronium L. (~ *Tulipa*). Liliaceae. 29 temp. N Am. (23; most W N Am.), Euras. (2). R C. Clennett (2014) *The genus E.* P reflexed. Many cult. orn. allied to tulips, with membranous-coated corms, some hybrids, incl.: ***E. americanum*** Ker-Gawler (amberbell, trout lily, E N Am.) – medic., fls yellow; ***E. californicum*** Purdy (fawn lily, N Calif.) – fls whitish; ***E. dens-canis*** L. (dog's-tooth violet, Euras.) – corms source of a starch used for vermicelli & cakes in Jap., also eaten with reindeer or cow milk in Mongolia & Siberia, lvs ed. boiled, cult. orn. with pink fls; ***E. grandiflorum*** Pursh (W US) – corms dug up & eaten by bears in Canada, staple food for Native Americans

Erythropalaceae Planch. ex Miq. (Strombosiaceae; Olacaceae s.l.). Magnoliidae – Santalales. 7/68 trop. Trees, shrubs & lianes. Lvs in spirals or distich., conduplicate. Fls 3–6-merous. K small, oft. cupular. C basally connate. A 2 or 3 × C (some staminodes). $\underline{G}$ ((2)3(-5)). Endosperm with oil or starch; embryo minute with 1 or 2 cotyledons. n = 16, 19, 20

Genera: *Coula, Erythropalum, Heisteria, Octoknema, Scorodocarpus, Strombosia, Tetrastylidium* (& prob. others from Olacaceae)

Some timbers (*Coula, Heisteria, Scorodocarpus*), ed. fr. (*Coula*) & lvs (*Strombosia*)

Erythropalum Blume. Erythropalaceae (Olacaceae s.l). 1 Indomal.: ***E. scandens*** Blume – potherb. R: FPM 3(2012)304

Erythrophleum Afzel. ex R. Br. Leguminosae (I 1). 9 OW trop. (Afr. 4, Madag. 1, E As. & Mal. 3, Aus. 1: ***E. chlorostachys*** (F. Muell.) Baill. (Cooktown ironwood) – timber). Alks. ***E. suaveolens*** (Guill. & Perrottet) Brenan (*E. guineense*, sasswood, trop. Afr.) – 3–5 axillary buds, timber (missanda), bark (sassy, s. bark) source of poison for fishes & arrows & form. much used ordeal-poison (causes heart to stop) for 'trying' criminals; other spp. used similarly

Erythrophysa E. Meyer ex Arn. Sapindaceae. 9 Ethiopia (1), NW Cape (1), Transvaal (1), W Madag. (6). R: MMNHP 19(1969)17. Petiole oft. winged

Erythrophysopsis Verdc. = praec.

Erythropsis Lindl. ex Schott & Endl. = *Firmiana*

Erythrorchis Blume (~ *Galeola*). Orchidaceae (II 2). 5 Indomal. to Aus. (1). Mycotrophs

Erythrorhipsalis A. Berger = *Rhipsalis*

Erythroselinum Chiov. = *Lefebvrea*

Erythroseris Kilian & Gemeinh. (~ *Cichorium*). Compositae (Cich.-Cich.). 2 Socotra, Somalia. R: Willd. 37(2007)292

Erythrospermum Lam. Achariaceae (Flacourtiaceae). 4 Mauritius to Fiji

Erythrostemon Klotzsch (~ *Caesalpinia*). Leguminosae (I 4). 31 trop & warm Am. R: PhytoKeys 71(2016)115. ***E. gilliesii*** (Hook.) Klotzsch (Arg, Urug.) – cult. orn. with antitumour activity in seeds

Erythroxylaceae Kunth. Magnoliidae – Malpighiales. 4/240 trop. esp. Am. Glabrous trees & shrubs oft. with alks incl. cocaine. Lvs in spirals (opp. in *Aneulophus*), simple (oft. with longit. markings), entire; stip. intrapetiolar. Fls small, reg., usu. bisex., (4)5-merous, oft. heterostylous, solit. or in axillary fascicles; K a tube with imbr. or valvate lobes, C imbr. & usu. with adaxial ± basal appendages, disk 0, A 10(–12) usu. forming a tube, anthers with longit. slits, $\underline{G}$ (2 or 3 (4)) with as many locs & styles (± connate), ovule 1(2) in 1 fert. loc., axile, pend., anatr. to hemitropous, bitegmic. Fr. a 1-seeded drupe; seed with straight embryo in copious (rarely 0) starchy endosperm. n = 12

Genera: *Aneulophus, Erythroxylum, Nectaropetalum, Pinacopodium*

Erythroxylum a source of narcotics etc.

Erythroxylum P. Browne. Erythroxylaceae. 230 trop. (Mal. 6; R: Sandakania 7(1996)67; Aus. 3) esp. Am. (187; R: Britt. 56(2004)2) & Madag. Rauh's & Troll's Models. Branches oft. covered with distich. scales (vestigial lvs); lvs oft. with longit. folds & pale broad band along centre; alks with medic. properties, cocaine (absent in OW spp.; sequestered by lymanitrid moths) form. prescribed for hay-fever & sinus problems. ***E. coca*** Lam. (coca, E Andes) – cv. ('var. *ipadu*') perhaps brought to Amaz. from Andean highlands prop. vegetatively by men (cassava by women) as hedge-pl. & for lvs, dried & powdered & mixed with lime the daily masticatory (poss. narcotic-chewing habit orig. from As. pre-Columbus) of W S Am.; maintains blood glucose levels despite poor diets (also high in calcium & vitamins so valued in areas with no dairy prods); source of comm. cocaine ('coke'; 920 t by 2010), now also grown in e.g. trop. As. (lvs 1.4% cocaine), the oldest anaesthetic (like coffee & opium manipulating dopamine prod., i.e. activating pleasure-centre of brain), a debilitating addictive narcotic causing euphoria, indifference to pain & tiredness, increased alertness & enhancing sexual desire ('girl'), sniffed in 'lines' from flat surface through straw, also dissolved & injected, effect lasting c. 15 – 40 mins; crack = freebase cocaine in raisin-sized bits ('rocks', freebasing = cocaine hydrochloride dissolved in water & heated with another agent to 'free' cocaine) with instant euphoria wearing off after c. 15 mins, investigated by Freud (1884) incidentally discovering anaesthetic action, form. with cola in non-alcoholic drinks (1885–1903; orig. Coca Cola formula an effective spermicidal douche, today's still with non-narcotic leaf extract) & 'Coca des Incas' tonic 'wine' & 'Vin Mariani' sanctioned by the Pope; comm. tea effective against alt. sickness in Andes; susceptible to *Fusarium* wilts so host-specific ones poss. mycoherbicides to curb illicit prod. ***E. novogranatense*** (Morris) Hieron. (Amaz.) – also cult. for alks, form. hedge-pl. in Malaysia (now banned); other spp. with useful timber incl. ***E. affine*** A. St.-Hil. (Honduras redwood, S Am.) & ***E. ellipticum*** R. Br. ex Benth. (trop. Aus.) – firewood burning when 'green'

Escallonia Mutis ex L.f. Escalloniaceae (Grossulariaceae s.l.). Excl. *Forgesia*, *Valdivia*, 37 S Am. esp. Andes. R: KNAWC 58,2(1968). Everg. shrubs & trees widely cult. & oft. hybridized. ***E. rubra*** (Ruíz & Pavón) Pers. (incl. *E. macrantha*, Chile) – variable sp. used as hedge-pl. in SW England & Ireland

Escalloniaceae R. Br. ex Dumort. (~ Grossulariaceae). Magnoliidae – Escalloniales. Incl. Polyosmaceae, Tribelaceae, 7/120 Réunion, SE As. to SW Aus. & New Caled., S Am. Trees & shrubs, some pachycaul, rarely annuals, oft. accum. aluminium. Lvs simple, us u. in spirals (opp. in *Polyosma*), margins usu. with broad glandular teeth; stip. 0. Fls usu. bisex., reg., (4)5(–9)-merous, usu. in racemes or panicles; K united, persistent; C free usu. imbr., A opp. K; G (2 (–5)), with parietal placentation & 1-∞ ovules per carpel. Fr. a septicidal capsule, rarely indehisc. (drupe in *Polyosma*). n = 12

Genera: *Anopterus, Eremosyne, Escallonia, Forgesia, Polyosma, Tribeles, Valdivia*

(*Eremosyne, Polyosma* & *Tribeles* s.t. referred to separate monogeneric fams)

Escallonia cult. orn.

escarole *Cichorium endivia* Scarole Group

Eschscholzia Cham. Papaveraceae (II). c. 10 W N Am. R: Pittonia 5(1905)205. Concave receptacle; (K) falling as a cap; in dull weather each petal rolls longit. encl. some A; valves of fr. curl spirally & fr. explodes; alks. Cult. as orn. ann. esp. ***E. californica*** Cham. (Calif. poppy, Calif. (state flower)) – chlorophyllous seed-coat with stomata, cvs with white or pink fls as well as yellow wild form. colourless latex mildly narcotic & used by Native Americans against toothache, forming ± pure stands in NZ; subsp. ***mexicana*** (Greene) C. Clark (*E. mexicana*) – copper indicator in Arizona

Eschweilera Mart. ex DC. Lecythidaceae (I). 85 trop. Am. incl. Mex. (1). FN 21,2(1990)158. Ligule (see fam.) pressed down on A & bearing staminodes & nectar available only to big bees which can force way in, i.e. *Xylocopa* & female euglossines; seed sessile. Timber (manbarklak) more resistant than greenheart to marine borers. ***E. itayensis*** Knuth – bark ash 'salt' used to coat pellets of hallucinogenic *Virola* paste in NW Amaz.; ***E. odorata*** Miers – seeds coll., ed.; ***E. ovalifolia*** (DC.) Niedenzu – obligate fish-disp. in Amaz.

Esclerona Raf. = *Xylia*

escoba *Cryosophila stauracantha*

Escobaria Britton & Rose = *Coryphantha*

Escobariopsis Doweld = *Mammillaria*

Escobedia Ruíz & Pavón. Orobanchaceae (Esc.; Scrophulariaceae s.l.). c. 15 trop. Am. R: Ceiba 8(1961)93. R. of esp. ***E. scabrifolia*** Ruíz & Pavón (palillo) used for dyeing

escobilla *Malpighia emarginata*

Escobrittonia Doweld = *Coryphantha*

Escocoryphantha Doweld = *Escobaria*

Escontria Rose. Cactaceae (III 1). 1 S Mex.: ***E. chiotilla*** (K. Schum.) Rose – fr. ed., sold in markets

Esenbeckia Kunth. Rutaceae (I 6). 30 trop. Am. R: FN 33(1982)24. Bark of some spp. used like angostura, others medic. (gasparillo, WI)

Esfandiaria Charif & Aellen = *Anabasis*

Eskemukerjea Malick & Sengupta (~ *Fagopyrum*). Polygonaceae (inc. sed.). 2 Nepal

Esmeralda Reichb.f. = *Arachnis*

Espadaea A. Rich. Solanaceae (I). 1 Cuba: ***E. amoena*** A. Rich.

esparto grass *Macrochloa tenacissima, Lygeum* spp.

espavé *Anacardium excelsum*

Espejoa DC. (~ *Jaumea*). Compositae (Bah.). 1 Mex., C Am.: ***E. mexicana*** DC.

Espeletia Mutis ex Bonpl. Compositae (Mill.-Esp.). 88 (s.t. split into 7 genera largely on habit etc.: *Carramboa, Coespeletia, Espeletia* s.s., *Espeletiopsis, Libanothamnus, Ruilopezia, Tamania*). Char. pachycaul pls of paramo (cf. *Dendrosenecio*) with dense rosettes of lvs in spirals (cf. tribe!), oft. densely pubescent; some wind-poll., ***E. schultzii*** Wedd. (Venez.) visited by hummingbirds

Espeletiopsis Cuatrec. See *Espeletia*

espercet *Sulla coronaria*

Espostoa Britton & Rose. Cactaceae (III 4). 9 Peru, 1 ext. to Bolivia. Cult. orn. like *Cephalocereus* but fls scaly & hairy, ***E. lanata*** (Kunth) Britton & Rose to 10 m tall

Espostoopsis Buxb. Cactaceae (III 4). 1 N Bahia: ***E. dybowskii*** (Rol.-Goss.)Buxb.

essia *Petersianthus macrocarpus*

Esterhazya Mikan. Orobanchaceae (Ger.; Scrophulariaceae s.l.). 5 Bolivia, Braz. R: Britt. 37(1985)195

Esterhuysenia L. Bolus (~ *Lampranthus*). Aizoaceae (V). 5 W S Afr. R: H. E. K. Hartmann, *Ill. Handb. Succ. Pls*, Aiz. A–E(2002)267

estragon *Artemisia dracunculus*

Estrevesia P. Braun = *Harrisia*

Esula (Pers.) Haw. = *Euphorbia*

Etaballia Benth. Leguminosae (III 11). 1 Amaz.: ***E. dubia*** (Kunth) Rudd. R: BMPEG 13(1997)90

eteng *Pycnanthus angolensis*

Etericius Desv. Rubiaceae. 1 Guyana = ?

Ethulia L.f. Compositae (Vern.-Erl.). 19 trop. Afr. (15 endemics), Indomal. R: KB 43(1988)165. ***E. conyzoides*** L.f., medic. in Afr., a ricefield weed

Ethuliopsis F. Muell. (~ *Epaltes*). Compositae (Pluch.). 1 Aus.: ***E. cunninghamii*** (Hook.) F. Muell. R: Austrobaileya 9(2013)50

Etlingera Giseke (~ *Amomum*). Zingiberaceae (II 4). c. 100 Indomal. (Borneo 42 – R: A.D. Poulsen, *E. of Borneo* (2006)26). Poll. spider-hunters; fr. dug up by pigs. Some local ed. & medic. esp. ***E. elatior*** (Jack) R.M. Sm. (torch ginger, Mal.) – fls used in curries, cult. orn.

etrog *Citrus medica* 'Ethrog' & other cvs

Euadenia Oliv. Capparaceae. 3 trop. Afr. Chamberlain's Model

Euanthe Schltr. = *Vanda*

Euaraliopsis Hutch. ex Y.R. Ling = *Trevesia*

Eubotryoides (Nakai) H. Hara (~ *Leucothoe*). Ericaceae. 1 Jap.: ***E. grayana*** (Maxim.) Hara

Eubotrys Nutt. (~ *Leucothoe*). Ericaceae (VIII 4). 2 SE US. R: Britt. 64(2012)167

Eubrachion Hook.f. Santalaceae (Eremolepidaceae). 2 WI, S Am. R: SBM 18(1988)43

eucalypt *Eucalyptus* spp.; also *Angophora* & *Corymbia* spp.

Eucalyptopsis C. White. Myrtaceae (II 11). 2 Buru, NG

Eucalyptus L'Hérit. Myrtaceae (II 11). 800+ (some intergrading) Mindanao (Mal. 16) to Aus. (R: FA 19 (1988); NSW 240) with 'eucalypt' fossils NZ. See also *Corymbia* (bloodwoods). Everg. (***E. populnea*** F. Muell. facultatively decid. in N Queensland) trees (Attims's Model) usu. with fire-resistant epicormic shoot-systems; most with distinct juvenile phase with diffs in shape, position & colour of lvs, waxy surface of juvenile lvs preventing adult tortoise beetles getting foothold & grazing; shallow-angled lvs largely in spp. of moderate to high rainfall & mild to warm climates while more vertical lvs with reduced heat-loads & increased water efficiency in drier sites but also wet montane forest (Tasmania) where they can intercept light at low angles in high latitudes; fls (some beetle-poll.) with an operculum thought to be derived from K & C through intercalary growth, leaving vestigial lobes around rim, dropped at anthesis, A ∞, oft. brightly coloured; seeds usu. ∞, small. Most char. genus (gums, marlock) of Aus. landscape (koalas alleged to eat only c. 19 spp.), varying from dwarf shrubs with lignotubers to 250 yrs old & coppice shoots (mallees) to some of the tallest trees known (***E. regnans*** F. Muell. 100m+ in Styx Valley, Tasm. (world's tallest angiosperm), & 1 felled Gippsland, Victoria (c. 1872) held to be 132.5 m tall though even 152.4 m claimed, cf. *Pseudotsuga*). Over 200 spp. introd. elsewhere & dominate the scenery of parts of Calif. (first introd. 1849), E Afr., Sri Lanka, Portugal, Israel, etc., Addis Ababa (= 'new flower') first permanent capital of Ethiopia because gums gave a permanent supply of firewood: fast-growing (***E. gunnii*** Hook.f. is the fastest tree hardy in GB (while ***E. nitens*** (Deane & Maiden) Maiden (SE Aus.) reaches 20 m in 6 yrs) & other spp. in plantation in Kenya more productive than even estuarine grasses) & most imp. dicot. plantation trees worldwide, though aggressive & invasive in e.g. Cape Province, S Afr. so now only to be grown in designated areas there. Besides timber (e.g. sleepers of Trans-Siberian Railway, early C17 [!] Engl. spice-cabinets), cardboard & paper (incl. high-quality coated p. e.g. tissue p.), sources of oils (medicinal, though 90% of that in Aus. imported from Swaziland, eucalyptol being an expectorant) & tannins & grown as orn. & for cut foliage. Five main groups recognizable by bark: gums (smooth & decid.), boxes (rough but fibrous), peppermints (finely fibrous), stringybarks (long fibrous) & ironbarks (hard, rough-fissured & dark, used e.g. as outer layer, over Douglas fir, on *St Roch*, the first boat to circumnavigate N Am.); scribbly gums with black channels ('scribbles') caused by *Ogmograptis scribula* moth larvae. L.D. Pryor (1976) *Biology of eucalypts*; S. Kelly (1983) *Eucalypts*, ed. 2. ***E. acmenoides*** Schauer (white mahogany, yellow stringybark, E Aus.) – timber; ***E. amygdalina*** Labill. (black peppermint, Tasmania) – a source of Aus. kino; ***E. astringens*** (Maiden) Maiden (W Aus.) – bark (mallet b., 40 – 50% tannin) form. much used for tanning shoe-leather; ***E. botryoides*** Sm. (bangalay, bastard mahogany, swamp m., SE Aus.) – timber imp. in ship-building; ***E. brassiana*** S.T. Blake (Cape York red gum, NE Aus., S NG) – fuel crop; ***E. bridgesiana*** R. Baker (apple box, E Aus.); ***E. camaldulensis*** Dehnh. (*E. rostrata*, (river or Murray) red gum, Aus., invasive S Afr.) – most widespread sp., now more planted than *E. globulus* elsewhere, prob. from Eur. introd. from SE Aus. (1803), to at least 950 yrs old, local medic., principal source of Aus. kino & form. much used for road woodblocks as in Melbourne & Port Adelaide-Adelaide road (9 mi) the busiest road in Aus. early C20; ***E. capitellata*** Sm. (brown stringybark, NSW) – general constr.; ***E. cinerea*** F. Muell. ex Benth. (Argyle apple, SE Aus.); ***E. cneorifolia*** A. Cunn. ex DC. (S Aus.) – comm. source of e. oil on Kangaroo Is.; ***E. copulans*** L. Johnson & K. Hill (Blue Mts, NSW) – known from only 2 wild pls (?hybrid); ***E. cornuta*** Labill. (yate, SW Aus.) – tough timber; ***E. crebra*** F. Muell. (red ironbark, NE Aus.); ***E. deglupta*** Blume (kamarere, C & E Mal.) – to 60 m in Mindanao, general constr., much planted in Mal., fuel crop; ***E. delegatensis*** F. Muell. ex R. Baker (alpine ash, Tasmanian oak, SE Aus.) – imp. timber-tree; ***E. diversicolor*** F. Muell. (karri, W Aus., invasive S Afr.) – to almost 90 m, v. hard timber ± resistant to termites & teredo; ***E. dives*** Schauer (SE Aus.) – imp. source of e. oil for e.g. soap; ***E. dumosa*** A. Cunn. ex Oxley (S & SE Aus.) – source of e. oil, also a manna called lerp or larap; ***E. fastigata*** Deane & Maiden (brown barrel, SE Aus.) – timber; ***E. forrestiana*** Diels (W Aus.) – 'fuchsia mallee'; ***E. fraxinoides*** Deane & Maiden ((Aus.) white ash, SE Aus.); ***E. globoidea*** Blakely (white stringybark, NSW, Victoria, invasive SE Eur., S Afr., W US, Hawaii) – timber for heavy constr.; ***E. globulus*** Labill. (blue gum, fever tree [because form. planted to drain malarial swamps in E Medit.], Victoria & Tasmania) – form. most widely cult. sp. throughout world (through efforts of F. Muell., e.g. Medit., Ind., S Afr., Calif.), natur. in Calif. etc., major source of e. oil in Spain, imp. firewood crop, good bee-forage, pulp etc. cut juvenile twigs in floristry, buds & fr. solit. (rare in *E.*); ***E. gomphocephala*** A. Cunn. ex DC. (tuart, SW Aus.) – one of Aus. heaviest & strongest timbers form. used for railway sleepers etc.; ***E. grandis*** W. Hill (flooded gum, rose g., NE Aus., invasive S Afr.) – constr. timber; ***E. gunnii*** Hook.f. (cider gum, c. tree, Tasmania) – potable sweet sap fermented to make alcoholic drink, hardiest in GB, growing up to 1.5 m per annum; ***E. leucoxylon*** F. Muell. (SE Aus.) – railway sleepers, poles etc.; ***E. longifolia*** Link (woollybutt, NSW) – durable timber for railway sleepers; ***E. luehmanniana*** F. Muell. (yellow ash, NSW) – wet mallee; ***E. macrorhyncha*** F. Muell. ex Benth. (red stringybark, S & E Aus.) – lvs a source of rutin; ***E. marginata*** Donn ex Sm. (jarrah, W Aus.) – principal timber of WAus. much exported; ***E. megacornuta*** C. Gardner (W Aus.) – spectacular cut-fl.; ***E. melliodora*** A. Cunn. ex Schauer (yellow box, E Aus.); ***E. microcorys*** F. Muell. (tallow wood, E Aus.) – valuable timber for telegraph poles etc.; ***E. microtheca*** F. Muell. (coolibah (as in 'Waltzing Matilda'), coolibar, desert box, N Aus.) – one of strongest & hardest of all timbers, good fuel; ***E. moluccana*** Wall. ex Roxb. (*E. hemiphloia*, grey box, E Aus.) – timber for railway sleepers etc.; ***E. muelleriana*** Howitt (yellow stringybark, SE Aus.) – timber for general outdoor use; ***E. niphophila*** Maiden & Blakely (*E. pauciflora* subsp. *n.*, snow gum, SE Aus.); ***E. nobilis*** L. Johnson & K. Hill (white gum, NSW) – at 79.22 m tallest tree in NSW (1997); ***E. obliqua*** L'Hérit. (messmate, E Aus.) – one of most imp. Aus. hardwoods for constr., pulp & furniture, a single sheet of bark tied at both ends making Aboriginal canoes; ***E. occidentalis*** Endl. (SW W Aus.) – much planted in Israel, Hawaii; ***E. oreades*** F. Muell. ex R. Baker (Blue Mt. ash, E Aus.) – timber for barrels, cabinet work etc.; ***E. paniculata*** Sm. (grey ironbark, E Aus., invasive S Afr.) – timber v. hard used for sleepers on Trans-Siberian Railway, in China, Ind. & Afr.; ***E. pauciflora*** Sieber ex Spreng. (white Sally, E & SE Aus.); ***E. pilularis*** Sm. (blackbutt, E Aus.) – timber for poles, ship-decking etc.; ***E. piperita*** Sm. (Sydney peppermint, NSW) – oil used for colds etc. by Surgeon White of First Fleet (1788); ***E. populnea*** F. Muell. subsp. ***bimbil*** L. Johnson & K. Hill (bimble box, NE Aus.); ***E. pyriformis*** Turcz. (Dowerin rose, W Aus.) – cult. Aus., Calif. for spectacular big fls; ***E. recurva*** Crisp (NSW) – ± entirely veg. reprod. (clones 13 000 yrs old?) like *E. imlayensis* Crisp & Brooker (NSW); ***E. redunca*** Schauer (wandoo, SW Aus.) – tough timber, tanbark; ***E. regnans*** F. Muell. (Aus. mt. ash or oak, SE Aus.) – major v. tall timber-tree in even-aged stands, growing up to 3 m a year, when fruiting prod. c. 12×10^5 seeds/ha annually, seeds germ. after fire when predators (esp. ants) killed & mother trees incinerated giving even-aged stand; ***E. resinifera*** Sm. (red mahogany, E Aus.) – timber & kino source; ***E. robusta*** Sm. (swamp or white mahogany, E Aus.) – timber for shingles, ship-building etc.; ***E. saligna*** Sm. (Sydney blue gum, saligna g., NSW, Queensland) – general constr.; ***E. salmonophloia*** F. Muell. (salmon gum, SW Aus.) – durable timber, comm. foliage 'cut-fl.' NE Aus.; ***E. sideroxylon*** Cunn. ex Woolls (mugga, E Aus., invasive S Afr.) – form. exploited for tannin, non-floating wood imp. as firewood (ironbark) NSW; ***E. sieberi*** L. Johnson (silvertop ash, NSW to NE Tasm.) – timber for flooring, handles, wood-chips; ***E. staigeriana*** F. Muell. ex Bailey (N Queensland) – trialled for citronella-like oils prod.; ***E. tereticornis*** Sm. (forest red gum, f. river g., E Aus., NG) – fuel crop, general consruction, imp. pl. for honey (toffee-flavoured); ***E. viminalis*** Labill. (white gum, S & SE Aus.) – timber for shingles etc., manna, widely cult.

Eucarpha (R. Br.) Spach (~ *Knightia*). Proteaceae (V 1). 2 New Caled.

Eucarya T. Mitchell ex Sprague & Summerh. = *Santalum*

Eucephalus Nutt. (~ *Aster*). Compositae (Ast.-Ast.). 10 N Am. R: FNA 20(2006)39

Euceraea Mart. Salicaceae (Samydaceae, Flacourtiaceae). 2 trop. S Am. R: FN 22(1980) 252

Euchaetis Bartling & Wendl. Rutaceae (I 4). 23 S & SW Cape. R: JSAB 47(1981)157

Eucharis Planch. & Linden (~ *Urceolina*). Amaryllidaceae. (Incl. *Caliphruria*) 17 with 2 nat. hybrids trop. Am. R: AMBG 76(1989)170. A on margin of corona; alks. Cult. orn. esp. ***E. × grandiflora*** Planch. & Linden (oft. sold as *E. amazonica*, sterile hybrid, poss. involving *E. sanderi* Bak. (W Colombia) & *E. moorei* (Bak.) Meerow (Andes of Ecuador & Peru), Amazon or star lily, Andes of Colombia & Peru) – large scented white fls; ***E. amazonica*** Hogg (NE Peru) – emetic tea prep. from pl. incl. bulbs

Euchile (Dressler & G. Pollard) Withner = *Prosthechea* (but see C.L. Withner, *The Cattleyas & their relatives* V(1998)137)

Euchilopsis F. Muell. (~ *Pultenaea*). Leguminosae (III 13). 1 SW Aus.: ***E. linearis*** (Benth.) F. Muell.

Euchiton Cass. (~ *Gnaphalium*). Compositae (Gnap.-Gnap.). 21 Aus., NZ, NG, E As. R: OB 104(1991)166. ***E. sphaericus*** (Willd.) Holub (SE Aus.) – perenn. pasture weed

Euchlaena Schrad. = *Zea*

Euchlora Ecklon & Zeyher (~ *Lotononis*). Leguminosae (III 8). 2 Cape. R: T 60(2011)170

Euchorium Ekman & Radlk. Sapindaceae (III 2). 1 W Cuba: ***E. cubense*** Ekman & Radlk.

Euchresta Bennett. Leguminosae (III 5). 4 E As., Java (1). R: APS 30(1992)43. Seeds toxic but local medic.

Euclasta Franch. Gramineae (XXII). 2 Afr., Ind., trop. Am.

Euclea Burm. ex L. (~ *Diospyros*). Ebenaceae (I). 12 – 20 trop. Afr. to Arabia & Comoro Is. Allied to *D.* sect. *Royena*. Lvs in spirals, opp. or whorled; fr. ed. ***E. divinorum*** Hiern (E & S Afr.) – bark a potential comm. tannin-souce; ***E. pseudebenus*** E. Meyer (black or Cape ebony, S Afr.) – timber for furniture etc.; ***E. racemosa*** Murray subsp. ***schimperi*** (A. DC.) F. White (*E. latidens*, *E. schimperi*, trop. Afr.) – timber, ed. fr., insecticidal smoke & dye from bark, seeds for beads in Karamoja (Uganda)

Euclidium R. Br. Cruciferae (26). 1 E Eur to Middle E, natur. Aus., N Am.: ***E. syriacum*** (L.) R. Br. R: FNA 7(2010)552

Euclinia Salisb. (~ *Randia*). Rubiaceae (II 1). 2 trop. Afr. (Madag. sp. = *Melanoxerus*) Fagerlind's Model

Eucnide Zucc. Loasaceae (II). 13 SW N Am., esp. Mex. R: JAA 48(1967)56. Like *Mentzelia* but C with basal tube & G with 4 or 5 placentas. ***E. bartonioides*** Zucc. (W Texas to Mex.) – cult. orn. biennial with yellow fls

Eucodonia Hanst. (~ *Achimenes*). Gesneriaceae (II 5b). 2 C & S Mex. R: Selbyana 1(1976)389

Eucomis L'Hérit. Asparagaceae (Hyacinthaceae). 12 trop. (1) & S Afr. R: BM n.s. 28(2011)184. Fls in term. bracteate raceme topped with cluster of sterile leafy bracts – 'pineapple lilies' incl. some cult. orn. esp. ***E. autumnalis*** (Mill.) Chitt. (*E. undulata*) – lvs undulate & ***E. comosa*** (Houtt.) Wehrh. (*E. punctata*) – lvs with purple spots (both S Afr.)

Eucommia Oliv. Eucommiaceae. 1 (?) China (Tertiary Eur.): ***E. ulmoides*** Oliv. – poss. no longer known in wild state, much cult. in China (cf. *Ginkgo*); timber for furniture & fuel; aucubin-cont. bark (du zhong, duzhong) a Chinese tonic & for arthritis, solid latex yields a gutta percha which can be seen as strands if leaf broken across, used for lining oil pipelines & insulating electric cables, also tooth-fillings. R: IDSY 2008:16

Eucommiaceae Engl. Magnoliidae – Garryales. 1/1 China (fossils from Cenozoic of Mex. & N hemisph.). Dioec. wind-poll. tree with articulated laticifers in phloem & cortex & scattered latex-cells elsewhere. Lvs in spirals, simple, toothed, decid.; stip. 0. Fls solit. & shortly pedicellate in axils of bracts racemosely arr. on proximal sect. of a distally leafy shoot, reg.; P 0, males with A 5 – 10 with short filaments & apically prolonged connective, anthers with longit. slits, females with G (2), 1-loc., flattened with short style & 2 unequal stigmas & 2 collateral, pend. anatr. unitegmic ovules, 1 aborting. Fr. a samara; seed with large embryo embedded in copious endosperm. 2n = 34

Genus: *Eucommia*

Form. assoc.with many diff. groups but aucubin in bark indicative of true affinity

Eucorymbia Stapf. Apocynaceae (II c). 1 W Mal.: ***E. alba*** Stapf

Eucosia Blume = *Goodyera*

Eucrosia Ker-Gawler. Amaryllidaceae. 7 Andes. R: SB 12(1987)460. Lvs stalked; ? butterfly-poll.; some cult. orn.

Eucryphia Cav. Cunoniaceae (IV; Eucryphiaceae). 7 SE Aus. (5, endemic), Chile. R: Plantsman 5(1983)169. Everg. trees or shrubs cult. for white fls esp. (allegedly lime-tolerant) ***E. × nymansensis*** Bausch (***E. cordifolia*** Cav. (ulmo, Chile: comm. timber & tannin source, allegedly lime-tolerant) × ***E. glutinosa*** (Poeppig & Endl.) Baill. (Chile)); ***E. lucida*** (Labill.) Baill. (leatherwood, Tasmania) – good timber, cult. orn., honey sold UK; ***E. moorei*** F. Muell. (plumwood, SE Aus.) – useful timber

Eucryphiaceae Gay = Cunoniaceae

Eucrypta Nutt. Boraginaceae (Hydrophyllaceae-Hydro). 2 SW N Am. R: Lloydia 1(1935)144

Eudema Humb. & Bonpl. Cruciferae (27). 4 Andes. R: JAA 71(1990)100

Eudianthe (Reichb.) Reichb. = *Silene*

Eugeissona Griff. Palmae (I 1). 6 Thailand, Malay Pen. (2), Borneo (4). Branching due to unequal development of prod. of dichotomizing apex; dioec.; C woody; sago, pollen ed. ***E. minor*** Becc. (Borneo) – stilt-r. (form. favoured for umbrella-handles & walking-sticks ('rajah canes')), sequentially prod. allowing pl. to 'walk' through forest; ***E. tristis*** Griff. (bertam, Thailand, Malay Pen.) – lvs used for thatch etc., the plant forming dense groves app. antagonistic to regeneration of dipterocarps; ***E. utilis*** Becc. (Borneo) – seeds rhino-disp., starch from trunk the basic sago source for Penans of Sarawak, pollen & 'cabbage' (apical bud) also ed.

Eugenia Micheli ex L. Myrtaceae (II 10). Incl. *Calycorectes*, excl. *Hexachlamys*, c. 750 trop. esp. Am. (C Am. 140; Afr. & is. 120, Mal. 60, Aus. 1, E NG 1 with woody caps., New Caled. 60). Differs from superficially similar but not closely related *Syzygium* (to which many spp. form. here now referred) in several morph. characters, *E.* usu. having cotyledons united, seed-coat smooth & free of pericarp & infls being racemes of pedicellate fls. Cryptic dioecy in Afr. spp.; some fish-disp. in Amazon. Many with ed. fr., some cult. ***E. brasiliensis*** Lam. (*E. dombeyi*, grumichama, S Braz.) – fr. eaten fresh, in pies, candied etc.; ***E. luschnathiana*** (O. Berg) B.D. Jackson (pitomba, Braz.) – aromatic fr. for jelly; ***E. pyriformis*** Cambess. (*E. uvalha*, uvalha, Braz.) – aromatic fr. ed. used in drinks; ***E. stipitata*** McVaugh (arazá, Amaz. Per) – promising fr. if processed (acidic); ***E. uniflora*** L. (Cayenne or Surinam cherry, pitanga (name also applied to other spp.), trop. Am., invasive S Afr.,

SE US, Carib.) – widely cult. trop. as hedge-pl., for ed. fr. used in jellies etc., crushed lvs insect-repellent

Euglypha Chodat & Hassler = *Aristolochia*

Euhesperida Brullo & Furnari = *Satureja*

Euklisia (Torr. & A. Gray) Rydb. = *Streptanthus*

eulalia *Miscanthus sinensis*

Eulalia Kunth. Gramineae (XXII 5). Incl. *Pseudopogonatherum*, 37 OW trop. (China 14, Aus. 4). ***E. aurea*** (Bory) Kunth (*E. fulva*, sugar grass, SE As. to Aus.) – palatable to stock

Eulaliopsis Honda. Gramineae (XXII). 2 Afghanistan, Ind., China, Taiwan, Philipp. R: JETB 25(2001)374. ***E. binata*** (Retz.) C. Hubb. (baib, Bharbur or sawai grass, Ind.) – imp. in paper-making & also local cordage

Euleria Urb. = *Picrasma*

Eulobus Nutt. ex Torr. & Gray (~ *Camissonia*). Onagraceae. 1 Calif., Ariz. (***E. californicus*** Nutt. ex Torr. & A. Gray), 3 Baja Calif. R: CUSNH 37(1969)189

Eulophia R. Br. Orchidaceae (V 12h). Excl. *Orthochilus*, c. 200 trop. (Am. 2, ***E. alata*** (L.) Fawc. & Rendle also in Afr.). R: Lindleyana 13(1998). Usu. terr., some mycotrophic; poll. carpenter bees. Tubers of some spp. provide Ind. salep. ***E. horsfallii*** (Batem.) Summerh. (trop. Afr.) – lvs to 2.5 m x 15 cm, infl stalk to 3m

Eulophidium Pfitzer = *Oeceoclades*

Eulophiella Rolfe. Orchidaceae (V 12b). 5 Madag. R: OD36(1972)120. Cult. orn.

Eulychnia Philippi. Cactaceae (III 1). 4 Chile, Peru. Cult. orn. ***E. acida*** Philippi (Chile) – long spines may prevent infection from bird-disp. mistletoe *Tristerix aphyllus*

Eumachia DC. (*Margaritopsis*). Rubiaceae (IV 7). Incl. *Chazaliella, Chytropsia, Readea*, c. 70 trop. (W Pacific 12, Afr. c. 25, Am. 27). ***E. sp.*** (*Chazaliella letouzeyi*, Cameroon & Gabon) – usu. unbranched pachycaul

Eumorphia DC. Compositae (Anth.-Phym.). 6 S Afr. R: NJB 5(1986)538

Eunomia DC. = *Noccaea*

Euodia Forst. & Forst. f. (*Evodia*). Rutaceae (I 2). 7 NG & NE Aus. to Pac. R: Allertonia 8(2000)53. ***E. hortensis*** Forst. & Forst. f. (W Pac.) – planted for lvs to scent coconut oil, local medic. incl. abortifacient, introd. to Niue by Polynesians, many selected cult. orn. leaf-forms (cf. *Melicope denhamii*). See also *Melicope, Picrella* & *Tetradium*

Euonymopsis Perrier (*Evonymopsis*). Celastraceae (I). 8 Madag. R: NS Paris 10(1942)202

Euonymus Tourn. ex L. (*Evonymus*). Celastraceae (I). Excl. '*Astrocassine*' (Madag.), *Wilczeckra* (trop. Afr.), c. 120 N temp. (Eur. 4) esp. As. (China 90, 50 endemic), Aus. R: Thaiszia 11(2001)1. ***E. globularis*** Ding Hou (Aus.) links with *Brassiantha* & *Hedraianthera*. Decid. or everg. trees & shrubs, rarely creeping & rooting, fr. a 3 – 5–valved brightly coloured capsule dehiscing to expose bird-disp. scarlet to orange arillate seeds; seeds of many spp. with cardiotoxic glycosides. Many cult. orn. esp. ones with corky outgrowths on stems & bright autumn foliage or everg. lvs. ***E. alatus*** (Thunb.) Sieb. (E As.) – branches with corky wings; ***E. atropurpureus*** Jacq. (Ind. arrow wood, C & E N Am.); ***E. benguetensis*** Merr. (Sumatra, Luzon) – barkcloth, incl. shrouds; ***E. europaeus*** L. (spindle-tree, louseberry, Eur., W As.) – indicator of hedges 100+ yrs old, host for aphid *Myzus persicae*, charcoal form. used in gunpowder, oil extracted from seeds, yellow dye from seeds form. used to colour butter, timber form. used for skewers, spindles, toothpicks & viol bows, poss. use as insecticide as form. used against head-lice; ***E. fortunei*** (Turcz.) Hand.-Mazz. (*E. hederaceus*, Ind. to C China, Jap., SE As. & Mal.) – everg. with many cvs incl. varieg.; ***E. japonicus*** Thunb. (Jap.) – everg. widely planted, many cvs incl. varieg. offered as houseplss; ***E. maackii*** Rupr. (*E. hamiltonianus* subsp. *m.*, N China to Korea) – cult. orn. used as box subs. ('paich-ha')

Eupatoriadelphus R. King & H. Robinson = *Eutrochium*

Eupatoriastrum Greenman. Compositae (Eup.-Crit.). Incl. *Matudina* 5 Mex., C Am. R: PSE 190(1995)117

Eupatorina R. King & H. Robinson (~ *Eupatorium*). Compositae (Eup.-Crit.). 1 Hispaniola: ***E. sophiifolia*** (L.) R. King & H. Robinson. R: MSBMBG 22(1988)308

Eupatoriopsis Hieron. Compositae (Eup.-Prax.). 1 Braz.: ***E. hoffmanniana*** Hieron. R: MSBMBG 22(1987)388

Eupatorium Tourn. ex L. Compositae (Eup.). 41 Euras. (Eur. 1), E N Am. (20). R: MSBMBG 22(1987)64. Form. incl. some 1200 spp. also from trop., now seg. into many genera e.g. *Ayapana*. Some cult. orn. incl. ***E. cannabinum*** L. (hemp agrimony, Eur. & Med. to C As.) – visited by Lepidoptera, form. medic. ('ague weed') as was ***E. perfoliatum*** L. (boneset, E N Am.) – 'gravel r.' for urinary problems. See also *Eutrochium*

Euphlebium (Kraenzlin) Brieger = *Dendrobium*

Euphorbia L. Euphorbiaceae (Euph.-Euph.). Incl. *Cubanthus, Elaeophorbia* (trees with drupes), *Monadenium* (cult. orn. monoec. succ. – R: P.R.O. Bally (1961) *The genus M.*), *Pedilanthus* (monoec. succ. shrubs, usu. bird-poll. – R: CGH 182(1957)1), *Poinsettia, Synadenium* (cyathium glands atop a tube) c. 1900 cosmop. (Eur. 105; Turkey 91, S Afr. 185 – R: Bothalia 42(2012)217), esp. warm. Euphorbia J. Four subgg.: ***Athymalus*** (148 OW esp. Afr. (S Afr. 80) in 7 sections – R: T. 62(2013)1188); ***Chamaesyce*** (600 – R: T 61(2012)776) – C4 pls; ***Esula*** (457 – R: T 62(2013)330); ***Euphorbia*** (661 – R: T 62(2013)303). Monoec. or dioec. herbs (c. 1300), some geophytic, succ. (esp. Afr., Madag.), shrubs & trees (Corner's, Tomlinson's, Prévost's, Leeuwenberg's, Koriba's, Champagnat's, Aubréville's, Rauh's, Attims's & Troll's Models), ***E. abyssinica*** J. Gmelin (*E. obovalifolia, E. amplophylla*, E Afr.) – tree to 30m, prob. tallest) with corrosive milky latex; stems oft. spiny, the pls s.t. cactoid (but latex & paired spines distinguish them at once), lvs in spirals, opp. or whorled; fls in cyathia (developmentally interm. between fl. & infl.) without P, the males reduced s.t. to A 1 (jointed, rep. pedicel, below), surrounding 1 female (s.t. absent) comprising G (3) mature before males, the whole surrounded by a 'whorl' of green P-like bracts & between them 4 horn-shaped glands prob. rep. stip. of bracts (in allied *Anthostema* both male & females have P); explosive fr. (occ. drupes), seeds disp. ants, wind, birds etc. Succ. spp. occupy same position in Afr. as Cactaceae in Am. & a series can be drawn up showing increasing xeromorphy assoc. with aridity: lvs normal & photosynthesizing; shoot not water-storing stems which do not photosynthesize (e.g. ***E. bupleurifolia*** Jacq., S Afr.) but lvs falling in dry season or stem fleshy & green, the lvs present only in wet season (e.g. ***E. neriifolia*** L., trop. & warm As.); where the lvs are abortive & soon fall, the principal photosynthesis being due to the stems, these may be thin & cylindrical but branched (*E. tirucalli*) or flattened (***E. xylophylloides*** Brongn. ex Lem., Madag., cf. *Schlumbergera*) or a stout C stem with a number of thinner apical branches (e.g. ***E. caput-medusae*** L., S Afr.), the branches covered with cushion-like papillae; in other spp. these scarcely sep. in development giving a ridged stem resembling the Cereeae of Cactaceae (e.g. ***E. polygona*** Haw., S Afr.) & the stem may be almost spherical like *Echinocactus* (e.g. ***E. globosa*** (Haw.) Sims, S Afr.). Infls in some leafy spp. surrounded by brightly coloured bracts, app. attractive to poll. birds as in *E. pulcherrima* ('poinsettia'). Latex of all spp. toxic, s.t. carcinogenic, & can bring out skin allergies but used in wart treatment; it has been used to stun fish. Many cult. as succ. curiosities or as pop. hardy border pls etc., while ***E. tirucalli*** L. (trop. & S Afr.) promising source of hydrocarbons for fuel (in latex, almeidina, which may be tapped) & for charcoal (incl. in fireworks in Ind.) & much planted around burial-grounds (tirucalli), also to re-vegetate asbestos tailings in Transvaal. ***E. albomarginata*** Torrey & A. Gray (N Am.) – local medic.; ***E. amygdaloides*** L. (wood spurge, Eur., SW As.) – seeds can wait more than 125 yrs for forest clearing before germ. in England, subsp. ***robbiae*** (Turrill) Stace (Mrs Robb's bonnet) – cult. ground-cover introd. (1896) from Istanbul by a Mrs Robb in her hatbox; ***E. antisyphilitica*** Zucc. (candelilla, SW N Am.) – waxy exudate (overcollected so as to be unsustainable) refined for use in polishes & creams for leather, furniture, babies, coating sweets such as Smarties, etc. & in lipstick & lip balm (form. for gramophone records), mixed with rubber for electric insulation materials, also in waterproofing fabrics, paper etc., when mixed with paraffin used as candles; ***E. arbuscula*** Balf.f. (Socotra) – local medic., insecticides, adhesive (e.g. for hand-written Qur'an); ***E. balsamifera*** Aiton (NW Afr., Canary Is. & Arabia (diff. subsp.)) – young shoots ed. boiled, latex a bird-lime, that of subsp. ***adenensis*** (Deflers) Govaerts a valuable glue in Arabia; ***E. characias*** L. (S Eur.) – commonly cult. hardy shrub, 'subsp. ***veneta*** (Willd.) Litard' (*E. wulfenii*) larger eastern subsp. poss. not distinct; ***E. cornastra*** (Dressler) R.-Sm. (*Poinsettia c.*, Mex.) – like poinsettia but bracts white, long-day pl. poss. comm.; ***E. cotinifolia*** L. (trop. Am.) – toxic tripterpenes form. used as fish poisons; ***E. cyparissias*** L. (cypress spurge, Eur., natur. N Am.) – ground-cover becoming weedy, form. used as cosmetic in Ukraine; ***E. drupifera*** Thonn. (*Elaeophorbia d.*, W Afr.) – ordeal-poison in Ivory Coast, rubbed into eyes (danger to cornea); ***E. esula*** L. (Eur., invasive N Am.) – 'hay' can yield 4 times more energy per annum than wheat straw; ***E. fischeriana*** Steud. (*E. pallasii*, E As.) – antitumour drug in Chinese med. for over 2000 yrs; ***E. fulgens*** Karw. ex Klotzsch (Mex.) – widely grown for cut-fl. trade; ***E. graminea*** Jacq. (Caribb.) – '**Inneuphe**' [Diamond Frost] vogue cult. for small white bracts; ***E. helioscopia*** L. (Euras.) – wart treatment, form. (Saturday night's pepper – Wiltshire, UK) used (until 20 cent. Isle of Man fishermen) to swell penis (same effect achieved by rhinoceros straddling other *E.* spp. – Walker); ***E. hirta*** L. (C Am., now pantrop. weed (asthma plant, a.

weed – Aus.)) – latex effective bee & scorpion sting treatment (Uganda); ***E. intisy*** Drake (Madag.) – form. imp. source of (intisy) rubber now almost exterminated; ***E. ipecacuanhae*** L. (ipecacuanha spurge, Carolina i., E US) – form. used as emetic; ***E. lancifolia*** Schldl. ('ixbut', C Am.) – natural increaser of milk-flow in women; ***E. lathyrus*** L. (caper spurge, mole plant, Eur., natur. N Am.) – fls visited by wasps in UK, caruncluate seeds toxic but fr. used as caper subs. (poor & poss. dangerous), 8% dry wt. terpenes can be 'cracked' to give fuel oil, plant incorrectly believed to be mole-repellent, latex causing blisters & allegedly used by Mediaeval beggars to elicit sympathy; ***E. marginata*** Pursh (snow-on-the-mt., N Am.) – glands with white petal-like appendages, garden-pl. but latex corrosive, seed-oil (also that of ***E. heterophylla*** L. (warm Am.)) app. sup. to linseed; ***E. milii*** Des Moul. (*E. bojeri*, Madag.) – spiny shrub with bright red or yellow bracts oft. grown as housepl. esp. var. ***splendens*** (Bojer) Ursch & Leandri (*E. splendens*, crown of thorns) with brilliant red bracts, crossed with ***E. lophogona*** Lam. (Madag.) to give stouter ***E. × lomi*** Rauh increasingly seen in florists' shops; ***E. obesa*** Hook.f. (Cape) – dioec. subspherical succ., rare in wild; ***E. paralias*** L. (Eur., Medit.) – invasive Aus.; ***E. peplus*** L. (petty spurge, radium weed, Eur.) – bad weed of cult., poss. leukaemia & non-melanoma skin cancer treatment (ingenol mebutate); ***E. pereskiifolia*** Houllet ex Baill. (*Synadenium p.*, trop. Afr.) – planted in sacred places & on graves in E Afr., persisting as forest develops on site; ***E. pilulifera*** L. (trop.) – in 19 cent. used for a tea depressing respiration & relaxing bronchioles (Aus.); ***E. pulcherrima*** Willd. ex Klotzsch (poinsettia, Mex.) – basis of multi-M pound Christmas industry (US trade in potted pls worth $241 M by 2005), rooted cuttings being flowered under short days & growth retardants, e.g. B9, to produce bright red bracts on small pl. for housepl. sale, some cvs with white bracts, used for hedging & veg. in trop., latex used as depilatory in Mex., red dye from bracts, extrafl. nectaries; ***E. tithymaloides*** L. (*Pedilanthus t.*, bird cactus, Florida to S Am.) – used as hedge-pl. in cemeteries; ***E. tortilis*** Rottler ex Ainslie (~ *E. antiquorum*, S Ind.) – hedge-pl.; ***E. trigona*** Haw. (? S Ind., ? Angola, ? hybrid) – pop. succ. pot-pl. in hotels etc., unknown in fl.

Euphorbiaceae Juss. Magnoliidae – Malpighiales. Excl. Peraceae, Phyllanthaceae, Picrodendraceae, Putranjivaceae 214/5600 cosmop. (few in Amazon basin) exc. Arctic. R: A. Radcliffe-Smith (2001) *Genera Euphorbiacearum*; R. Govaerts & al. (2000) *World Checklist . . . E.*, 1–4. [S.l.] Trees, shrubs, lianes or herbs, s.t. succ., monoec. or dioec. with stems & lvs oft. with specialized cells or tubes of milky or coloured latex; internal phloem s.t. present; many accum. aluminium. Lvs usu. simple & in spirals, s.t. opp. or whorled, s.t. compound, with pinnate or palmate venation, s.t. v. reduced; stip. large & prot. term. but s.t. reduced to merely glands or 0. Fls usu. reg. in basically cymose infls, s.t. v. reduced & in bisex. pseudanthia; P usu. reg. inconspicuous, s.t. basally connate or 0, A (1–)3–50(–400), s.t. basally connate, anthers with longit. slits or rarely apical pores, nectary-disk of discrete or united segments s.t. present without or within A, $\underline{G}$ ((1)2–5(–20)), pluriloc. with distinct styles or style with bifid or more-branched branches, rarely G pseudomonomerous, each loc. with 1 or 2 pend., apical-axile, anatr. or hemitropous bitegmic ovules, the nucellus oft. protruding through micropyle & contact with obturator (placental) roofing the micropyle & forming a passageway for pollen-tubes. Fr. oft. a capsular schizocarp, the mericarps separating from persistent columella & opening adaxially to release seeds, or a drupe or berry; seeds oft. with a caruncle around micropyle & with straight or curved embryo embedded in copious oily, rarely 0, endosperm, oft. with poisonous proteins. x = 6 – 14+

Classification & principal genera (Corner's 1976 work on seeds drew attention to the heterogeneity of Euphorbiaceae s.l., now confirmed by DNA studies):

[Phyllanthoideae – Phyllantheae, Bischofieae (*Bischofia*) = **Phyllanthaceae**; Drypeteae = **Putranjivaceae**]

[Oldfieldioideae = **Picrodendraceae**]

I. **Cheilosoideae** (Acal. – Cheil.; latex 0, exine echinate, sarcotesta; Indomal.): *Cheilosa & Neoscortechinia (only)*

II. **Acalyphoideae** (excl. Clutieae, Pereae = **Peraceae**; 1 ovule per loc., latex 0; lvs oft. with petiolar or laminar glands, seeds ± caruncle, endosperm usu. copious, x usu. = 8–11): 14 tribes, incl. Chrozophoreae (*Caperonia, Chrozophora*), Alchorneeae (*Alchornea*), Acalypheae (*Acalypha, Claoxylon, Cleidion, Macaranga, Mallotus, Mercurialis*), Ricineae (*Ricinus*), Plukenetiae (Dalechampieae; *Dalechampia, Tragia*).

III. **Crotonoideae** (as above but latex reddish or yellowish to milky, rarely 0, innocuous, laticifers articulate or not, lvs oft. palmately veined, lobed or compound, indumentum simple or oft. stellate, usu. C present): 12 tribes incl. Adenoclineae

(*Endospermum*), Manihoteae (*Manihot, Cnidoscolus*), Jatropheae (Joannesieae; *Jatropha*), Ricinodendreae (*Ricinodendron*), Codiaeae (*Codiaeum*), Aleuritideae (*Aleurites, Neoboutonia, Vernicia*), Crotoneae (*Croton*), Heveeae (*Hevea*)

V. **Euphorbioideae** (as above but latex whitish, oft. caustic or poisonous, laticifers inarticulate (rarely 0), lvs simple (rarely lobed), pinnately veined or 3-veined, indumentum of simple hairs or 0, C 0): 3 tribes incl. Hippomaneae (*Excoecaria, Hippomane, Sapium, Triadica*), Hureae (R: Britt. 65(2013)311; *Hura*), Euphorbieae (*Anthostema, Euphorbia*)

The different basic chromosome nos & seed-structure of the 1st 2 subfams long suggested exclusion. *Euphorbia* (q.v.) represents the most florally advanced group in the fam.

Imp. crops incl. rubber (*Hevea, Manihot, Micrandra, Sapium*), cassava (*Manihot*) & other veg. (*Cnidoscolus*), oils both industrial & medic. (*Aleurites, Baliospermum, Caryodendron, Cephalocroton, Cleidiocarpon, Croton, Garcia, Gymnanthes, Jatropha, Joannesia, Mallotus, Plukenetia, Ricinus, Sapium, Schinziophyton, Tetracarpidium, Triadica, Vernicia*), timber (*Endospermum, Excoecaria, Givotia, Hevea, Ricinodendron*), dye-pls. (*Chrozophora, Homalanthus, Mallotus, Mercurialis*), hydrocarbon sources (*Euphorbia, Jatropha*) & many cult. orn. esp. for foliage (*Acalypha, Breynia, Codiaeum, Euphorbia* (incl. *Monadenium, Pedilanthus*, poinsettia), *Ricinus*)

Euphoria Comm. ex Juss. = *Litchi*

Euphorianthus Radlk. (~ *Diploglottis*). Sapindaceae (III 1). 1 Indomal.: ***E. euneurus*** (Miq.) Leenh. R: FM I,11(1994)536

Euphrasia Tourn. ex L. Orobanchaceae (Rhin.; Scrophulariaceae s.l.). c. 350 incl. many microspp. N temp. (Eur. 46), NG, Aus., NZ. Hemiparasites, some reg. autogamous. ***E. officinalis*** L. (s.s. = *E. rostkoviana* Hayne, eyebright, Eur.) – form. used in eye disease

Euphronia Mart. & Zucc. Euphroniaceae (Vochysiaceae s.l.). 3 Guyana Shield, NE S Am. R: AMBG 74(1987)89

Euphroniaceae Marc.-Berti. (~ Chrysobalanaceae, Vochysiaceae). Magnoliidae – Malpighiales. 1/3. N trop. S Am. Trees. Lvs in spirals, white-tomentose adaxially, stipulate. Fls herm. in term. racemes to thyrses. K 5 unequal; C 3 contorted in bud, A = K (-7), adnate to C, basally connate; staminode 1 long + 4 or 5 short. G 3, semi-inf., each loc. with 2 apotropous, bitegmic ovules. Fr. a 3-valved septicidal capsule; 1 seed per loc., slightly winged, with thin endosperm

Genus: *Euphronia*

Euphrosyne DC. Compositae (Helia.-Amb.). Incl. *Chorisiva, Cyclachaena, Leuciva, Oxytenia*, 5 N Am. Some aquatic. ***E. xanthiifolia*** (Nutt.) A. Gray (*C. x., Iva x.*, marsh elder) – introd. Eur. where imp. hay-fever pl., noxious weed

Euplassa Salisb. Proteaceae (V 4). 20 trop. Am. R: KB 59(2004)27. Some timbers incl. veneers esp. ***E. meridionalis*** Salisb. (*Adenostephanus guyanensis*, Braz. lacewood)

Euploca Nutt. (~ *Heliotropium*). Boraginaceae (Heliotropiaceae). c. 100 cosmop., esp. Afr., Aus. trop. Am. R (as *H.* sect. *Orthostachys*): Sendtnera 5(1998)97

Eupodium J. Sm. (~ *Marattia*). Marattiaceae. 3 trop. Am. R: KB 65(2010)116. Stalked synangia

Eupomatia R. Br. Eupomatiaceae. 3 E NG, E Aus. R: Austrobaileya 6(2002)333. Heterodichogamy recorded. Staminodes with food-bodies, musky fragrance & sticky exudate attractive to poll. weevils (*Elleschodes*), overarching & prot. them from predators. ***E. laurina*** R. Br. (Aus.) – fls last less than 1 day, poll. *Elleschodes hamiltonii* T. Blackburn only, fr. ed.

Eupomatiaceae Orb. Magnoliidae – Magnoliales. 1/3 NG, E Aus. Trees & rhiz. wiry shrubs; sieve-tube plastids of P type. Lvs distich. (spirals), simple, entire, aromatic; stip. 0. Fls 1 – 3 in axils or term. on longer shoots, reg., bisex., receptacle urceolate; P 0, calyptra (bract) attached to rim of receptacle falling to expose A 20 – 100 nr. rim of receptacle, outer ones with short, broad laminar base & well-defined anther & prolonged thickened connective, staminodes c. 40 – 80, fleshy & ± petaloid, all forming continous spiral & basal united to form synandrium falling from fl. together ('littering the ground . . . with what look rather like sea anemones' (Stevens)), $\underline{G}$ 13 – 70 in spirals on receptacle, ± connate at margins, but unclosed, style 0, stigma feathery, each carpel with 2 – 11 anatropous bitegmic ovules. Fr. subglobose berry-like aggregate with carpels laterally coalescent so as to appear sunk in fleshy receptacle; seeds 1 or 2 per carpel, with v. small embryo embedded in oily ruminate endosperm. 2n = 20

Genus: *Eupomatia*

Euptelea Sieb. & Zucc. Eupteleaceae. 2 Assam to SW & C China (1), Jap. (1). R: JAA 27(1946)175. Fert. 2 months post poll. Cult. orn.

Eupteleaceae Willhelm. Magnoliidae – Ranunculales. 1/2 E As. (Upper Eocene of Kazakhstan; ? fossils in Oregon). Small subglabrous, decid. trees. Lvs in spirals, simple, toothed; stip. 0. Fls small, wind- to insect-poll., long-pedicellate, solit. in axils of 6 – 12 closely crowded bracts of veg. shoots, bisex. or s.t. some male; P 0, A 6 – 20 in 1 whorl with short filaments & long red anthers & prolonged connectives, $\underline{G}$ 6 – 31 incompletely closed with decurrent stigma, style 0, each carpel with 1 – 3(4) ± marginal, anatropous, bitegmic ovules. Fr. a head of small samaras with papery pericarp; seeds with tiny embryo & poorly differentiated cotyledons embedded in oily & proteinaceous endosperm. 2n = 28 Genus: *Euptelea*

Eureiandra Hook.f. Cucurbitaceae (XIII). 8 trop. Afr. to Socotra

Euroschinus Hook.f. Anacardiaceae (I). 9 Mal. (1), Aus. (1). New Caled. (7; R: Bot. Helv. 104(1994)124). ***E. falcatus*** Hook.f. (Aus.) – second-rate timber for cheap furniture etc.

Eurotia Adans. = *Axyris*

Eurya Thunb. Pentaphylacaceae (3; Ternstoemeriaceae; Theaceae s.l.). c. 70 trop. & warm As. & W Pac. R: BSBF 42(1895)151. Dioec. trees & shrubs (Mangenot's Model), some cult. orn. & locally used timber. ***E. acuminata*** DC. (Indomal.) – aluminium-accumulator

Euryalaceae J. Agardh = Nymphaeaceae

Euryale Salisb. Nymphaeaceae (Euryalaceae). 1 China to N Ind. & Jap. (fossils known from Eur.): ***E. ferox*** Salisb. – like *Victoria* but lvs with flat margins, smaller fls & all A fertile; cleistogamous fls over longer period & setting more seeds than chasmogenous ones; cult. for 3000 yrs by Chinese for ed. rhizomes & seeds (fox nut), now 'puffed' by roasting & sold in markets in Ind.

Eurybia (Cass.) Cass. (~ *Aster*). Compositae (Ast.-?Mac.). 28 N Am., N Euras. (1 – ***E. sibirica*** (L.) Nesom). R: FNA 20(2006)365

Eurybiopsis DC. = *Minuria*

Euryblema Dressler (~ *Chondrorhyncha*). Orchidaceae (V 12j). 2 trop. Am. R: Lankesteriana 5(2005)94

Eurycarpus Botsch. = *Solms-laubachia* (but see Novon 10(2000)346)

Eurycaulis M.A. Clem. & D.L. Jones. See *Dendrobium*

Eurycentrum Schltr. Orchidaceae (IV 2d). 7 Papuasia, Vanuatu

Eurychone Schltr. Orchidaceae (V 16d). 2 trop. Afr. R: BM n.s. 18(2001)155. Cult. orn.

Eurychorda Briggs & L. Johnson (~ *Restio*). Restionaceae. 1 E Aus.: ***E. complanata*** (R. Br.) Briggs & L. Johnson. R: Telopea 7(1998)357

Eurycles Drapiez = *Proiphys*

Eurycoma Jack. Simaroubaceae. 3 Indomal. Corner's Model; A5, staminodes 5. ***E. longifolia*** Jack (tongkat ali, Indomal.) – locally medic. (hypertension, diabetes, stomach ache) – bark with eurycomalactone poss. antimalarial, aphrodisiac (testosterone booster, r. tissue comm. for treatment of impotence)

Eurycorymbus Hand.-Mazz. Sapindaceae (III 1). 1 S China, Taiwan: ***E. cavaleriei*** (H. Lév.) Rehder & Hand.-Mazz. (*E. austrosinensis*)

Eurydochus Maguire & Wurd. (~ *Gongylolepis*). Compositae (Stifft.).1 S Am.: ***E. bracteatus*** Maguire & Wurd.

Eurylobium Hochst. = *Stilbe*

Eurynotia R. Foster = *Ennealophus*

Euryodendron H.T. Chang. Pentaphylacaceae (3; Ternstroemeriaceae). 1 S China: ***E. excelsum*** H.T. Chang

Euryomyrtus Schauer (~ *Baeckea*). Myrtaceae (II 15). 6 SW & SE Aus. R: Nuytsia 13(2001)546

Euryops (Cass.) Cass. (~ *Othonna*). Compositae (Sen.-Oth.). c. 100 S Afr. to Arabia & Socotra. R: OB 20(1968)1. Resin form. used for gum; some cult. orn. esp. ***E. acraeus*** M.D. Henderson (S Afr., oft. confused with *E. evansii* Schltr. not in cult.), also hybrids in trop.

Eurypetalum Harms. Leguminosae (I 2). 2 Cameroun to Gabon. R: Adans. 29(2007)68. Some cabinet timbers

Eurysolen Prain. Labiatae (VI). 1 NE Ind. to Yunnan, Thailand, W Mal.: ***E. gracilis*** Prain

Eurystemon Alexander (~ *Heteranthera*). Pontederiaceae. 1 N C Am.: ***E. mexicanum*** (S. Watson) Alexander

Eurystyles Wawra. Orchidaceae (IV 2h). 20 trop. Am. R: Britt. 37(1985)160. Obligate epiphytes

Eurytaenia Torrey & A. Gray. Umbelliferae (III 8). 2 N Am. R: Phytoneuron 2012–67:1

Euscaphis Sieb. & Zucc. = *Staphylea* (but see FOC 11(2008)498)

Eusideroxylon Teijsm. & Binnend. Lauraceae (I). 1 Sumatra to Borneo: ***E. zwageri*** Teijsm. & Binnend. (belian, billian, Borneo ironwood) – ± pure stands (Roux's Model) in Sumatra, seeds elephant- or rhino-disp., heavy constr. timber sinking in water, much exported. See also *Potoxylon*

Eusiphon Benoist = *Ruellia*

Eustachys Desv. (~ *Chloris*). Gramineae (XXIX 5). 16 trop. Am. (N Am. 4; S Am. 9 – R: Cand. 51(1996)225) & Afr., S Afr., savannas

Eustegia R. Br. (Apocynaceae Vc; Asclepiadaceae). 1 Namaqualand & W Cape: ***E. minuta*** (L.f.) Schultes. R: BJ 121(1999)36

Eustephia Cav. Amaryllidaceae. 4 S Peru to Bolivia. Alks; cult. orn.

Eusteralis Raf. = *Pogostemon*

Eustigma Gardner & Champ. Hamamelidaceae (I 3). 4 S China, SE As. R: PhytoKeys 65(2016)47. C minute, fleshy with basal nectaries abaxially, A subsessile, G (2) with large stigmas

Eustoma Salisb. Gentianaceae. 3 S N Am. to N S Am. R: SW Nat. 2(1957)38. ***E. grandiflorum*** (Raf.) Shinn. (N Am). – showy ann. or biennial with purple fls widely cult. pot-pl. & cut-fl. ('prairie gentian', 'lisianthus') since early 1980s, many cvs incl. 'doubles'

Eustrephus R. Br. ? Philesiaceae (Smilacaceae s.l.). 1 polymorphic S & E NG, E Aus., New Caled.: ***E. latifolius*** R. Br. (wombat berry) – cult. orn. R: Austrobaileya 4(1995)393

Eustylis Engelm. & A. Gray = *Alophia*

Eutaxia R. Br. (~ *Pultenaea*). Leguminosae (III 13). 23 Aus. (SW 22). R: Nuytsia 20(2010)111. Cult. orn. esp. ***E. microphylla*** (R. Br.) C.H. Wright & Dewar (S & E Aus.)

Euterpe Mart. Palmae (V 10). 7 trop. Am. R: FN 72(1996)22. Monoec. palms, a principal source of palm-hearts esp. ***E. edulis*** Mart. (assai, açai, Braz. to Arg.) – chocolate- & cherry-flavoured fr. eaten fresh or juiced, high in antioxidants, used in health & energy drinks (dried powder a 'superfood' in Aus.), & ***E. oleracea*** Mart. (manicole, Trinidad to S trop. Am.); some fish-disp. Amazon, cult. orn.

Eutetras A. Gray. Compositae (Per.-Per.). 2 C Mex. R: SW Nat 11(1966)118

Euthamia (Nutt.) Cass. Compositae (Ast.-Sol.). 5 N Am. R: FNA 20(2006)97. ***E. graminifolia*** (L.) Nutt. (*Solidago g.*) – poss. rubber source as is ***E. nauseosa*** (Pursh) Nesom (*Chrysothamnus n.*, W N Am.) – gum form. chewed by Native Americans

Euthemis Jack. Ochnaceae (III). 2 SE As. to Borneo. R: BJ 113(1991)180

Eutheta Standl. = *Melasma*

Euthryptochloa T. Cope = *Phaenosperma*

Euthystachys A. DC. Stilbaceae. 1 S Afr.: ***E. abbreviata*** (E. Meyer) A. DC. R: Strel. 10(2000)542

Eutrema R. Br. Cruciferae (28). Incl. *Neomartinella* (3 China – R: Novon 10(2000)337), *Platycraspedum* (2 China, E Tibet – R: Novon 10(2002)2), *Taphrospermum* (7 Russia to Himal. – R: HPB 5(2000)101), *Thellungiella*, 26 Arctic (Eur. 1), C & E As., N Am. (2 incl. ***E. edwardsii*** R. Br. – also Arctic & Altai). R: HPB 10(2005)129. ***E. japonica*** (Miq.) Koidz. (*E. wasabi, Wasabia w.*, Jap., Sakhalin) – cult. condiment (wasabi) for eating with raw fish (sashimi), used like horseradish esp. in Jap., comm. prod. NZ, worth $160 per kg by 2014, evanescent flavour, so much in Eur., N Am. is ersatz (*Armoracia rusticana* & mustard)

Eutrochium Raf. (~ *Eupatorium*). Compositae (Eup.). 5 N Am. R: FNA 21(2006)393. ***E. purpureum*** (L.) E. Lamont (*Eupatorium p.*, Joe-pye weed, E N Am.) – 'gravel r.' for urinary problems

Euxylophora Huber. Rutaceae (I 8). 1 Amaz. Peru & Braz.: ***E. paraensis*** Huber (Braz. satinwood, pau amarello) – clear yellow hard timber for flooring, furniture etc.

Euzomodendron Cosson = *Vella*

Evacidium Pomel = *Filago* (but see OB 104(1991)173)

Evandra R. Br. Cyperaceae (II 7). 2 SW Aus.

Evax Gaertn. = *Filago*

Evea Aubl. = *Psychotria*

Everardia Ridl. ex Thurn. Cyperaceae (III 1). 12 Venez., Guyana. R: MNYBG 12,3(1965) 20

everlastings usu. coloured heads of *Xerochrysum bracteatum* but many other materials also used esp. spp. of *Alkekengi, Limonium, Rhodanthe, Schoenia, Syngonanthus* etc.; **mountain e.** *Antennaria dioica;* **pearly e.** *Anaphalis margaritacea*

Everistia S. Reynolds & R. Henderson (~ *Canthium*). Rubiaceae (III 2). 1 E Aus.: ***E. vacciniifolia*** (F. Muell.) S. Reynolds & R. Henderson. R: Austrobaileya 5(1999)354

Eversmannia Bunge (~ *Onobrychis*). Leguminosae (III 25). 4 SE Russia, N Iran & C As.

Evodia Lam. = *Euodia*
Evodianthus Oersted. Cyclanthaceae. 1 trop. Am.: ***E. funifer*** (Poit.) Lindm. – Chamberlain's Model, pend. aerial r. used for basket-weaving in Ecuador
Evodiella van der Linden = *Melicope*
Evodiopanax (Harms) Nakai = *Gamblea*
Evolvulus L. Convolvulaceae (4). 98 warm & trop. Am., 2 ext. to OW incl. ***E. alsinoides*** (L.) L. – weedy, only pl. on some Sri Lanka serpentines, allegedly snail-poll., local medic. incl. smoked fror bronchial complaints. Some cult. orn.
Evonymopsis Perrier = *Euonymopsis*
Evonymus L. = *Euonymus*
Evota (Lindl.) Rolfe = *Ceratandra*
Evotella Kurzweil & Linder. Orchidaceae (IV 4b). 1 SW & S Cape: ***E. rubiginosa*** (Bolus) Kurzweil & Linder. R: PSE 175(1991)215
Evrardia Gagnepain = *Odontochilus*
Evrardiana Aver. = *Odontochilus*
Evrardianthe Rauschert = *Odontochilus*
Evrardiella Gagnepain = *Aspidistra*
Ewartia Beauverd. Compositae (Gnap.-Cass.). 4 SE Aus. R: OB 104(1991)95. Dioec.
Ewartiothamnus Anderb. (~ *Ewartia*). Compositae (Gnap.-Gnap.). 1 NZ: ***E. sinclairii*** (Hook.f.) Anderb. R: OB 104 (1991)94
Exaculum Caruel (~ *Cicendia*). Gentianaceae. 1 Eur.: ***E. pusillum*** (DC.) Caruel (*C. p.*)
Exacum L. Gentianaceae (incl. *Cotylanthera* – mycophytes; R: JJB 50(1975)321). 69 OW trop. (As. 25, Socotra & Oman 4, Afr. 2, Madag. 38). R: OB 84(1985). Style turned to 1 side: diff. fls with it turned both ways found on same plant (enantiostyly). Some cult. orn. esp. ***E. affine*** Balf. f. ex Regel (Socotra, Dhofar) – oft. seen as flowering pot-pl. ('Arabian' or 'Persian violets'), s.t. with 'double' fls
Exalaria Garay & G. Romero (~ *Cranichis*). Orchidaceae (Orch.-Cran.). 1 N Andes: ***E. parviflora*** (Presl) Garay & G. Romero. R: HPB 4(1999)479
Exallage Bremek. (~ *Oldenlandia*). Rubiaceae (IV 15). 15 Indopacific, ***E. auricularia*** (L.) Bremek. natur. Afr.
Exandra Standl. = *Simira*
Exarata A. Gentry. Schlegeliaceae (Bignoniaceae s.l.). 1 Ecuador & Colombia: ***E. chocoensis*** A. Gentry. R: SB 17(1992)503
Exarrhena R. Br. = *Myosotis*
Exbucklandia R.W. Br. Hamamelidaceae (II). 2 E Himal. & S China, Malay Pen. & Sumatra. Large stip. folded against one another & prot. young axillary bud or infl.; fls in heads in groups of 4 sunk in axis. ***E. populnea*** (R. Br. ex Griff.) R.W. Br. (E Himal. to Malay Hills) – multiple branches at nodes arising from branching of trace to primary axillary bud; timber useful, lvs quiver like aspen
Excavatia Markgraf = *Ochrosia*
excelsior short thin curled wood shavings, esp. from *Populus* spp., used for stuffing mattresses etc.
Excentradenia W.R. Anderson (~ *Hiraea*). Malpighiaceae. 4 N S Am. R: CUMH 21(1997)29
Excentrodendron H.T. Chang & Miau = *Burretiodendron*
Excoecaria L. Euphorbiaceae (Hipp.). Incl. *Conosapium*, *Spirostachys*, c. 40 OW trop. to Pac. Prévost's Model. ***E. agallocha*** L. (blind-your-eye(s), Indomal.) – latex blinding, lvs turn bright red before falling; ***E. cochinchinensis*** Lour. (SE As.) – cult. orn., medic. See also *Spirostachys*
Excremis Willd. ex Schultes f. = *Eccremis*
Exechostylus K. Schum. = *Pavetta*
Exellia Boutique (~ *Monanthotaxis*). Annonaceae (III 7). 1 trop. Afr.: ***E. scammopetala*** (Exell) Boutique. R: BJBB 21(1951)117
Exellodendron Prance. Chrysobalanaceae (2). 5 trop. S Am. R: FOW 9(2003)182
Exhalimolobos Al-Shehbaz & C. Bailey (~*Halimolobos*). Cruciferae (29). 9 Mex. to S Am. R: SB 32(2007)146
Exoacantha Labill. Umbelliferae (III 3). 1 SW As.: ***E. heterophylla*** Labilll.
Exocarpos Labill. Santalaceae. 26 SE As., Mal. to Hawaii (Aus. 10, 9 endemic; New Caled. 5). R: MBMUZ 213(1959)117. Root-parasites. Some Aus. timbers (ballart) for cabinet-making. ***E. cupressiformis*** Labill. (Aus.) – ed. fr.; ***E. phyllanthoides*** Endl. (Norfolk Is.) – tanbark
Exocarya Benth. Cyperaceae (I 1). 1 E NG, NE Aus.: ***E. sclerioides*** (F. Muell.) Benth.

Exochaenium Griseb. (~ *Sebaea*). Gentianaceae. 22 trop. Afr. R: SB 37(2012)240
Exochanthus M.A. Clem. & D.L. Jones. See *Dendrobium*
Exochogyne C.B. Clarke. Cyperaceae (III). 1 N S Am. & SE Braz.: ***E. amazonica*** C.B. Clarke. R: MNYBG 12,3(1965)52
Exochorda Lindl. Rosaceae (Ros.-Exoc.). 1 C As., China: ***E. racemosa*** (Lindl.) Rehder – cult. orn. *Spiraea*-like shrub allied to *Prunus*, fr. of 5 carpels flattened & arr. starwise. R: F.Y. Gao (1998) *E.*
Exodeconus Raf. (~ *Physalis*). Solanaceae (IV). 6 S Am., Galápagos. R: PSE 193(1994)156
Exogonium Choisy = *Ipomoea*
Exohebea R. Foster = *Tritoniopsis*
Exolobus Fourn. = *Gonolobus*
Exomiocarpon Lawalrée. Compositae (Helia.-Ecl.). 1 Madag.: ***E. madagascariense*** (Humbert) Lawalrée. R: BJBB 17(1943)62
Exomis Fenzl ex Moq. Amaranthaceae (Chenopodiaceae I 3). Excl. *Manochlamys*, 1 S Afr.: ***E. microphylla*** (Thunb.) Aellen. R: Strel. 10(2000)223
Exorhopala Steenis = *Helosis*
Exorrhiza Becc. = *Clinostigma*
Exospermum Tieghem = *Zygogynum*
Exostema (Pers.) Bonpl. Rubiaceae (IIII 4). 25 trop. Am. esp. WI. R: PSE 212(1999)216. Some rheophytes; febrifugal alks in bark (Jamaica b.). ***E. caribaeum*** (Jacq.) Roemer & Schultes (princewood, C Am., WI) – wood for turnery etc.
Exostigma Sancho (~ *Podocoma*). Compositae (Ast.). 2 S Am. R: SB 37(2012)516
Exostyles Schott. Leguminosae (III 1). 4 SE Braz. R: KB 59(2004)523. Timbers
Exothea Macfad. Sapindaceae (III 2). 4 Florida, WI, C Am., Surinam. R: KB 59(2004) 523
Exotheca Andersson. Gramineae (XXII 7). 1 trop. Afr. & Vietnam: ***E. abyssinica*** (A. Rich.) Andersson – odd distrib. poss. reflexion of coastal trading going back to antiquity
Extriplex E. Zacharias (~ *Atriplex*). Amaranthaceae (Atriplic.). 2 SW N Am. R: SB 35(2010)850
eyan *Lovoa trichilioides*
eye, dragon's *Hylocereus undatus*
eyebright *Euphrasia* spp. esp. *E. officinalis*
Eylesia S. Moore = *Buchnera*
eyong *Sterculia oblonga*
Eysenhardtia Kunth. Leguminosae (III 10). 12 – 15 C Am. R: ISJR 56(1982)395. ***E. polystachya*** (Ortega) Sarg. (*E. amorphoides*, kidneywood, lignum-nephriticum, SW N Am.) – wood-chips placed in water against a black background produce peacock blue phosphorescence, form. imported for medic.
Ezoloba B. Wyk & Boatwr. (~ *Lotononis*). Leguminosae (III 8). 1 W Cape: ***E. macrocarpa*** (Ecklon & Zeyh.) Wyk & Boatwr.
Ezosciadium B.L.Burtt (*Trachysciadium* auctt.). Umbelliferae (III 8). 1 S Afr.: ***E. capense*** (Ecklon & Zeyher) B.L. Burtt. R: PSE 276(2008)172

F

Faba Mill. = *Vicia*
Fabaceae Lindl. = Leguminosae
Faberia Hemsl. (~ *Prenanthes*). Compositae (Cich.-Cich.-Cich.). 8 C & SW China. R: T 62(2013) 1235. Poss. allopolyploid origin
Faberiopsis Shih & Y.L. Chen = *praec.* (but see APS 34(1996)438)
Fabiana Ruíz & Pavón. Solanaceae (Petun.). 15 warm temp. S Am. R: Kurtziana 22(1993)112. Moth-poll. ***F. imbricata*** Ruíz & Pavón (pichi, Chile, Arg.) – cult. orn. shrub, medic.
Fabrisinapis C. Towns. = *Hemicrambe*
Facchinia Reichb. (~ *Minuartia*). Caryophyllaceae. 7 Eur. mts. R: EJB 72(2015)355
Facelis Cass. Compositae (Gnap.-Gnap.). 4 S Am., ***F. retusa*** (Lam.) Schultz-Bip. introd. OW, N Am. R: BJLS 106(1991)191
Facheiroa Britton & Rose (~ *Espostoa*). Cactaceae (III 4). 3 NE Braz.
Factorovskya Eig = *Medicago*
Fadenia Aellen & C. Towns. = *Salsola* (but see NJB 11(1991)315)

Fadogia Schweinf. Rubiaceae (III 2). c. 45 trop. Afr. Heterogeneous. Some with large red, ? bird-poll. fls; some ed. fr.

Fadogiella F. Robyns. Rubiaceae (III 2). 3 (1 known only from (destroyed!) type specimen) trop. Afr.

Fadyenia Hook. = *Tectaria*

Fagaceae Dumort. Magnoliidae – Fagales. Excl. Nothofagaceae, 7/970 cosmop. exc. trop. & S Afr. R: R. Govaerts & D. G. Frodin, *World Checklist... Fagales* (1998)107. Strongly tanniferous monoec. (rarely dioec.) trees & shrubs, some creeping; r. oft. with ectotrophic mycorrhizae. Lvs simple to deeply lobed, in spirals, rarely opp. or whorled; stip. decid. Infls pend., rarely erect or variously reduced, even to solit fls. Fls usu. wind-poll. (insects in *Castanea*), inconspicuous, males in ± reduced dichasia in catkins or heads, P (4 –)6(7 – 9), small, scale-like, s.t. basally connate or almost 0, A (4 –) 6 – 12(– 90), anthers with longit. slits; females 1 – 7(– 15) at base of male infls or in diff. axils, individually or collectively involucrate, ± 6 – 12 staminodes, $\overline{G}$ ((2)3) or (6 (7 – 15)) with styles & locules as many as carpels but septa not reaching to apex, each loc. with 2 axile, anatropous, bitegmic ovules. Fr. usu. a nut with stony or leathery pericarp, subtended by accrescent involucre (cupule); seed 1 with large straight starchy or oily embryo & 0 endosperm. n = 12 (13, 21)

Classification & genera:

Quercoideae (infl. spike or catkin): *Castanea, Castanopsis, Chrysolepis, Lithocarpus, Quercus, Trigonobalanus*

Fagoideae (infl. capitate): *Fagus* (*Nothofagus* with unitegmic ovules & diff. infls excl.)

Prominent & oft. dominant in angiosperm forests of N hemisph. (oaks, beeches) & montane forests of Mal. Delayed fert. The char. cupule has been thought derived from the coalescence of 3-lobed extensions of the pedicel beneath each fl., though it has been more convincingly argued that it is derived from P; within the fam., increased enclosure of fr. by cupule has led to decrease in ecol. imp. of former; cupule reduction elsewhere has increased its imp. (BR 59(1993)81).

Source of some of the most imp. timbers – oak, beech, chestnut, also cork (*Quercus suber*), seeds (up to 46% oil, by wt.) for human (*Castanea*) & stock (*Fagus, Quercus*) & wild pig (*Castanopsis, Lithocarpus* – leading to spectacular pig migrations in 'masting' yrs) consumption; many cult. orn.

Fagara P. Browne ex L. = *Zanthoxylum*

Fagaropsis Mildbr. ex Siebenl. Rutaceae (I 2). 4 trop. & NE Afr. (2), Madag. (2). R: SBU 30,1(1992)67. ***F. angolensis*** (Engl.) H. Gardner (mafu) – furniture timber, veneers etc.

Fagerlindia Tirv. = *Benkara*

Fagonia Tourn. ex L. Zygophyllaceae (I). 34 Med. (Eur. 1), SW As. & NE trop. Afr. to NW Ind., SW Afr., SW N Am. (7), Chile (1). R: SandB 3(2005)226

Fagopyrum Mill. (~ *Polygonum*). Polygonaceae (II 2). 16 As. & E Afr. Fls like *Polygonum* but heterostylous. Grown for fr. ('seed'), the seeds with floury endosperm (buckwheat) esp. ***F. esculentum*** Moench (orig. Yunnan?) – flour much used in USA for pancakes, for noodles (soba) in Jap., for porridge (grechka) in Russia, imp. bee-pl. for comm. honey, pls used in b. pillows sold in USA & form. roof-insulation (e.g. Netherlands), source of rutin, seeds of other pls inhibited from germ. for period after cropping b.; ***F. tataricum*** (L.) Gaertn. (Siberian, Tartary or Kangra b., orig. Sichuan?) – flour used locally

Fagraea Thunb. Gentianaceae. Excl. *Cyrtophyllum, Limahlania, Picrophloeus, Utania*, 5 Indomal. Cult. orn. with scented fls, some medic. & locally used timber. ***F. auriculata*** Jack (Myanmar to Philipp.) – tree to epiphyte, fls to 30 cm across; ***F. berteroana*** A. Gray ex Benth. – fls in leis in Hawaii, to scent coconut oil in Fiji

Faguetia Marchand. Anacardiaceae (I). 1 Madag.: ***F. falcata*** Marchand

Fagus Tourn. ex L. Fagaceae. 10 N temp. (Eur. 1). Beeches (beech from *boc*, poss. cognate with 'book' & origin of Buckinghamshire, England, where common). Decid. monoec. trees (Troll's Model) with imp. timber, nuts & cult. orn. e.g. ***F. crenata*** Blume (Jap. b., buna, Jap.) – imp. forest tree, etc. but esp. ***F. grandifolia*** Ehrh. (American b., Nova Scotia to Mex.) – differing from Eur. b. in longer, coarsely serrate lvs with more veins, uses similar, distilled wood for creosote & ***F. sylvatica*** L. ((Eur.) b., Eur. to Cauc. incl. subsp. ***orientalis*** (Lipsky) Greuter & Burdet (oriental or Turkish b., C Greece to SE Russia)) – to 300 yrs old in Carpathians, in pure stands in mts of Eur., also (artificially) in areas where oak removed & b. encouraged for timber for chairs etc. (e.g. 'Windsor' chairs in Chilterns, England), casting dense shade with little undergrowth beyond mycotrophs like *Lathraea* spp. etc., masts at irreg. intervals with males in drooping heads & fr. of 1 or 2 brown 3-angled 'nuts' in a prickly involucre, toxic saponins, straight-grained timber used for woodworking tools like planes, kitchen utensils, turnery etc. as well as furniture, seeds produce an oil/butter (e.g. WW II Ger.) the residue a flour subs. form. in Fr. & Silesia, timber form. source of creosote, lvs used with gin & brandy in alc. drinks, cult. orn. esp. as hedges (to 30 m, world's tallest, along A93 N of Perth, Scotland) when lvs retained (marcescent) as in young trees, many cvs incl. **Atropunicea Group** (f. *purpurea*, purple b.) now replacing earlier-cult. paler **Cuprea Group** (copper b.) – first coll. in Ger. forests, **'Dawyck'** ('Fastigiata', Dawyck b.) – fastigiate, **'Aspleniifolia'**/**'Laciniata'** – deeply cut lvs etc., **'Pendula'** (weeping b.), **'Quercoides'** (f. *quercoides*) – oak-like form & bark, a specimen of the wild type grown at 3000 m in Java for 60 yrs was a densely branched everg. shrub with ann. rings but no fls, while the allied *Castanea sativa* produced fls regularly

faham tea *Jumellea fragrans*

Fahrenheitia Reichb.f. & Zoll. ex Muell. Arg. = *Paracroton*

Faidherbia A. Chev. (~ *Acacia*). Leguminosae (II 4). 1 warm Afr. to E Med.: ***F. albida*** (Del.) A. Chev. – drops lvs in wet season & gets new ones in dry, in agroforestry improving maize yield 280%, valuable fodder tree, source of gum, tannin etc., candidate for wood of Ark of the Covenant

Faika Philipson. Monimiaceae (V 2). 1 W NG: ***F. villosa*** (Kaneh. & Hatusima) Philipson. R: FM I,10(1986)284

fair maids of France (or **Kent**) *Ranunculus aconitifolius*

fairy bells *Prosartes* spp.; **f. fingers** *Digitalis purpurea*; **f. flax** *Linum catharticum*; **f. floss** spun sugar; **f. potato** *Claytonia* spp.; **f. wand** *Chamaelirium luteum*

Falcaria Fabr. Umbelliferae (III 8). 3–4 C Eur., Medit., W & C As. R: Pfr IV,228, Heft 90(1927)129

Falcata J. Gmelin = *Amphicarpaea*

Falcataria (I. Nielsen) Barneby & Grimes. Leguminosae (II 4). 3 E Mal. to Queensland. Timbers esp. ***F. moluccana*** (Miq.) Barneby & Grimes (*Paraserianthes f.*, *Albizia falcataria*, *A. moluccana*, batai wood, moluccana, sau, E Mal.) – fast-growing (to 5 m height increase (record = 10.74 m in 13 months in Sabah (1975), the world's fastest-growing tree) & 15 cm diam. increase per annum) for tea-chests, pulp, matches, chopsticks & fuel, s.t. invasive; ***F. toona*** (Bailey) I. Nielsen (*P. toona*, red siris, cedar acacia, Aus.) – resistant to termites, used for carving in Sri Lanka

Falcatifoliaceae A. Bobrov & Melikian = Podocarpaceae

Falcatifolium Laubenf. (~ *Dacrydium*). Podocarpaceae. 5 Mal. to New Caled. R: Phytol.M 7(1984)29. Dioec.

Falckia Thunb. = *Falkia*

Falconeria Hook.f. = *Kashmiria*

Falconeria Royle (~ *Sapium*). Euphorbiaceae (Euph.-Hipp.). 1 Himal. to W Mal.: ***F. insignis*** Royle. R: A. Radcliffe-Smith, *Gen. Euphorb.* (2001)372

Falkia Thunb. (*Falckia*). Convolvulaceae (6). 3 Afr.

Fallopia Adans. (~ *Polygonum*). Polygonaceae (II 4). Excl. *Reynoutria*, c. 6 N temp. ***F. baldschuanica*** (Regel) Holub (*F. aubertii*, *R. b.*, Russian vine, W China & Tibet) – rampant liane planted as a screen; ***F. convolvulus*** (L.) A. Löve (black bindweed, Eur.) – weedy, fr. ed. in Prehistory (found in stomach of Tollund Man)

Fallugia Endl. (~ *Geum*). Rosaceae (Colur.). 1 SW N Am.: ***F. paradoxa*** (Tilloch & Taylor) Torr. – cult. orn. shrub, poss. oilseed. R: Aliso 20(2001)7

Falona Adans. = *Cynosurus*

false acacia *Robinia pseudoacacia*; **f. oat** *Arrhenatherum elatius*; **f. sarsaparilla** *Hardenbergia violacea*; **f. spikenard** *Maianthemum racemosum*; **f. walnut** *Neoguillauminia cleopatra*

Famatina Rabean = *Phycella*

Famatinanthus Ariza & Freire (~ *Aphyllocladus*). Compositae (Famat.). 1 Arg. Andes: ***F. decussatus*** (Hieron.) Ariza & Freire. R: SB 39(2014)353

Fanninia Harv. Apocynaceae (Vc; Asclepiadaceae III 1). 1 S Afr.: ***F. caloglossa*** Harv. R: F. Albers & U. Meve, *Ill. Handb. Succ. Pls*, Asclep. (2002)137

fanwort *Cabomba* spp.

Faradaya F. Muell. = *Oxera* (but see Blumea 44(1999)321)

Faramea Aubl. Rubiaceae (IV 11). 130 trop. Am. Pollen dimorphic

Farfugium Lindl. Compositae (Sen.-Tuss.). c. 3 E As. R: APG 8(1939)77. Cult. orn. esp. ***F. japonicum*** (L.) Kitam. (*F. tussilagineum*, *Ligularia t.*) **'Aureomaculatum'** (leopard plant)

Fargesia Franch. (~ *Thamnocalamus*). Gramineae (IV). Incl. *Sinarundinaria*, c. 90 E Himal. to China (78, 77 endemic). Heterogeneous. Cult. orn. esp. ***F. murielae*** (Gamble) T.P. Yi

(*Arundinaria m.*, *A. sparsiflora* Rendle, *T. m.*) – hapaxanthic clone in Eur. flowering 86 yrs after introd. (1993 – 8), though low light intensity promotes reversion to veg. growth, ***F. nitida*** (Mitford) T.P. Yi (*A. n.*, *S. n.*, C China)) – culms to 6m flowering after 100 yrs in cult. & ***F. spathacea*** Franch. (*A.s.*, *T. s.*, umbrella bamboo, C China)

farkleberry *Vaccinium arboreum*

Farmeria Willis ex Hook.f. Podostemaceae (III). Incl. *Maferria*, 2 SW Ind., Sri Lanka

faro *Daniellia ogea*

Faroa Welw. Gentianaceae. 17 trop. Afr. R: GOB 1(1973)69

Farquharia Stapf. Apocynaceae (IIb). 1 S Nigeria: ***F. elliptica*** Stapf

Farrago W. Clayton. Gramineae (XXIX 3). 1 Tanzania: ***F. racemosa*** W. Clayton

Farringtonia Gleason = *Siphanthera*

farro *Triticum turgidum* Dicoccon Group

Farsetia Turra. Cruciferae (4). 27 Morocco to NW Ind., Tanz. mts. R: SBU 25,3(1986)

Fascicularia Mez. Bromeliaceae (3). 1 Chile: ***F. bicolor*** (Ruíz & Pavón) Mez ('*F. pitcairniifolia*') – usu. terr., cult. orn. natur. Eur. is. R: BJLS 129(1999)322

Fasciculochoa B. Simon & Weiller = *Otachyrium*

fat hen *Chenopodium album*

Fatoua Gaudich. Moraceae (I). 2 Madag. to E As., N Aus. & New Caled

× **Fatshedera** Guillaumin. Araliaceae. *Fatsia* × *Hedera*. **× *F. lizei*** (C.-Cochet) Guillaumin (female *Fatsia japonica* 'Moseri' × male *Hedera hibernica* (though 2n = 44 – 49 & that of *F. japonica* 12 or 24), but *Orobanche hederae* exclusively parasitic on *Hedera hibernica* (W Eur. & Medit.) in wild, attacks *Fatsia japonica* in cult.)), an hybrid raised in Lizé Frères nursery, Nantes, 1910, but not repeated, with floral structure intermediate between parents & sterile but with large stip. outgrowths found in neither though typical of other genera in A., cult. orn.

Fatsia Decne & Planch. Araliaceae (1). Incl. *Boninofatsia*, 3 E As. ***F. japonica*** (Thunb.) Decne & Planch. – cult. orn. everg. for bold effect (name misrendering of Jap. yatsude as '*fatsi*')

Faucaria Schwantes. Aizoaceae (V). 6 E Cape & Karoo. R: H. E. K. Hartmann, *Ill. Handb. Succ. Pls*, Aiz. F–Z(2002)23. Cult. orn. succ. with 4-ranked lvs, usu. long-toothed esp. ***F. tigrina*** (Haw.) Schwantes (tiger's-jaws, Cape)

Faucherea Lecomte (~ *Manilkara*). Sapotaceae (I 2). 11 Madag.

Faujasia Cass. Compositae (Sen.-Sen.). 4 Réunion. R: KB 47(1992)78

Faujasiopsis C. Jeffrey. Compositae (Sen.-Sen.). 3 Mascarenes

Faurea Harv. Proteaceae (IV 3). 18 trop. & S Afr., Madag. Some timber esp. ***F. saligna*** Harv. (Afr. beech, trop. & S Afr.), also yields tannin & medic. locally

Fauria Franch. = *Nephrophyllidium*

Favargera Löve & D. Löve = *Gentiana*

Favratia Feer (~ *Campanula*). Campanulaceae (I 6). 1 SE Alps: ***F. zoysii*** (Jacq.) Feer (*C. z.*) – cult. orn. rockpl. R: BJ 12(1890)610

Fawcettia F. Muell. = *Tinospora*

Faxonanthus Greenman = *Leucophyllum*

Faxonia Brandegee. Compositae (Mill.-Gal.) 1 Baja Calif.: ***F. pusilla*** Brandegee

faya *Morella faya*

feaberry *Ribes uva-crispa*

feather flower *Verticordia* spp.; **f. grass** *Stipa* spp., *Leptochloa* spp.; **f. hyacinth** *Muscari comosum* 'Plumosum'; **parrot's f.** *Myriophyllum aquaticum*; **prince [of Wales]'s f.** *Amaranthus hybridus*; **f.top** *Cenchrus longisetus*

Feddea Urb. Compositae (Hel.-Fed.). 1 E Cuba (serpentine): ***F. cubensis*** Urb.

Fedia Gaertn. = *Valerianella* (but see NM 54(1990)3, AJBM 48(1991)157)

Fedorovia Kolak. = *Theodorovia*

Fedorovia Yakovlev = *Ormosia*

Fedtschenkiella Kudr. = *Dracocephalum*

Feea Bory = *Trichomanes*

Feeria Buser (~ *Trachelium*). Campanulaceae (I 3). 1 Morocco: ***F. angustifolia*** (Schousb.) Buser. R: BHBoiss. 2(1894)517

Fegimanra Pierre. Anacardiaceae (I). 3 trop. Afr.

Feijoa O. Berg = *Acca*

Feldstonia Short. Compositae (Gnap.-Ang.). 1 W Aus.: ***F. nitens*** Short. R: OB 104(1991)132

Felicia Cass. Compositae (Ast.-Hom.). 85 S Afr. with few ext. to trop. Afr. & Arabia. R: MBSM 9(1973)195. Some cult. orn. esp. ***F. amelloides*** (L.) Voss & ***F. bergeriana*** (Spreng.) O. Hoffm. (kingfisher daisy) – blue fls, S Afr.

Feliciadamia Bullock (~ *Miconia*). Melastomataceae. 1 W Afr.: *F.* ***stenocarpa*** (Jacq.-Fél.) Bullock. R: KB 15(1962)393

Felipponia Hicken = *Mangonia*

Felipponiella Hicken = *Mangonia*

Femeniasia Susanna (~ *Carthamus*). Compositae (Card.-Cent.). 1 Minorca: *F.* ***balearica*** (Rodríguez Femenías) Susanna. R: Coll.B 17(1988)83

Fendlera Engelm. & A. Gray. Hydrangeaceae (I; Philadelphaceae). 5 SW N Am. R: Lundellia 4(2001)2

Fendlerella A.A. Heller. Hydrangeaceae (II 1; Hydrangeaceae). 4 SW US

Fenerivia Diels (~ *Polyalthia*). Annonaceae (IV 4). 10 Madag. R: T 60(2011)1412. C = homoeotic outer A, K a vestigial flange

Fenestraria N.E. Br. Aizoaceae (V). 1 SW Afr.: *F.* ***rhopalophylla*** (Schltr & Diels) N.E. Br. – stemless clump-forming succ. with erect club-shaped lvs, the truncate tips with transparent windows, which, in the wild, are all of the pl. above ground, light being focused by them to internal photosynthetic tissues; cult. orn. R: H. E. K. Hartmann, *Ill. Handb. Succ. Pls*, Aiz. F–Z(2002)26

Fenixia Merr. Compositae (Helia.-Ecl.). 1 Philippines: *F.* ***pauciflora*** Merr.

fennel *Foeniculum vulgare*; **Florence f.** *F. vulgare* var. *azoricum*; **f. flower** *Nigella sativa*; **giant f.** *Ferula communis*; **hog's f.** *Peucedanum officinale*; **water f.** *Oenanthe aquatica*

fenni *Anacardium occidentale*

fenugreek *Trigonella foenum-graecum*

Fenzlia Endl. = *Lithomyrtus*

Ferdinandea Pohl = *Ferdinandusa*

Ferdinandusa Pohl. Rubiaceae (Cond.). 25 trop. Am. Some local febrifuges

Feretia Del. Rubiaceae (II 5). 2 trop. Afr. R: KB 34(1979)368. Seeds a coffee subs.

Fergania Pim. (~ *Ferula*). Umbelliferae (III 10). 1 C As.: *F.* ***polyantha*** (Korovin) Pim. R: NSL 19(1982)119

Fergusonia Hook.f. Rubiaceae (?IV 7). 1 S Ind., Sri Lanka: *F.* ***tetracocca*** (Thw.) Baill.

fern (cut-fl. industry) *Asparagus setaceus*, being replaced by *Rumohra adiantiformis*; **air f.** (US) *Sertularia argentea* L., a zoophyte dyed green & resembling plastic *Ceratophyllum*; **beech f.** *Phegopteris connectilis*; **bird's nest f.** *Asplenium nidus*; **bladder f.** *Cystopteris fragilis*; **bristle f.** *Trichomanes* spp.; **cinnamon f.** *Osmunda cinnamomea*; **dingo f.**™ *Baloskion tetraphyllum*; **eagle f.** *Pteridium aquilinum*; **elk's horn f.** *Platycerium* spp.; **filmy f.** Hymenophyllaceae; **f. grass** *Desmazeria* spp.; **frankincense f.** *Anemia caffrorum*; **hart's tongue f.** *Asplenium scolopendrium*; **holly f.** *Polystichum* spp.; **kangaroo f.** *Microsorum pustulatum*; **koala f.** *Caustis blakei*; **lady f.** *Athyrium* spp.; **maidenhair f.** *Adiantum* spp.; **male f.** *Dryopteris filix-mas*; **marsh f.** *Acrostichum* spp.; **ostrich f.** *Onoclea struthiopteris*; **parsley f.** *Cryptogramma crispa*; **potato f.** *Microgramma bifrons*; **royal f.** *Osmunda* spp.; **shield f.** *Dryopteris* spp.; **stag's horn f.** *Platycerium* spp.; **f. tree** *Filicium decipiens*, *Jacaranda* spp.; **walking f.** *Asplenium rhizophyllum*; **whisk f.** *Psilotum* spp.

Fernaldia Woodson. Apocynaceae (I e). 3 Mex., C Am. R: Rhodora 104(2002)188. *F.* ***pandurata*** (A. DC.) Woodson (C Am.) – ed. fls exported to US

Fernandezia Ruíz & Pavón. Orchidaceae (V 12h). Incl. *Pachyphyllum*, c. 50 trop. Am.

Fernandoa Welw. ex Seem. Bignoniaceae (1). 15 OW trop. (Afr. 5, Madag. 3, S China & SE As. to Sumatra 7)

Fernelia Comm. ex Lam. Rubiaceae (II 5). 4 Masc. R: KB 37(1983)551

Fernseea Bak. Bromeliaceae (3). 2 Braz. R: Bradea 3(1983)343

Ferocactus Britton & Rose. Cactaceae (III 9). Incl. *Glandulicactus*, *Leuchtenbergia* (oldest name), *Parrycactus*, *Stenocactus*, *Thelocactus*, c. 44 SW N Am. R: Bradleya 2(1984)19. Stems candied; cult. orn. large ovoid ribbed cacti esp. *F.* ***wislizeni*** (Engelm.) Britton & Rose with largest spines to 10cm, hooked; *F.* ***cylindraceus*** (Engelm.) Orcutt – water source for Native Americans; *F.* ***histrix*** (DC.) G. Lindsay (N C Mex.) – fls (s.t. pickled) & fr. ed., sacred in pre-Columbian times, local medic.

Feronia Corr. Serr. = *Limonia*

Feroniella Swingle = *Citrus* (but see W. Swingle, *Bot. Citrus* (1943)468)

Ferraria Burm. ex Mill. Iridaceae (IV 3). 18 trop. (4) & S Afr. (15). R: Bothalia 41(2011)295. Poll. Coleoptera, Diptera, bees (e.g. *F.* ***ferrariola*** (Jacq.) Willd. (S Afr.) – spicy scent) & wasps. Cult. orn. though fls of some 'malodorous & fugacious', *F.* ***crispa*** Burm. (S Afr.) – now noxious weed Aus.

Ferreirea Allemão = *Sweetia*

Ferreyanthus H. Robinson & Brettell = *Ferreyranthus*

Ferreyranthus H. Robinson & Brettell. Compositae (Liab.). 8 Peru & S Ecuador (1). R: Arnaldoa 2,2(1994)10. Trees & shrubs form. common on slopes along roads

Ferreyrella S.F. Blake. Compositae (Eup.-Ager.). 2 Peru. R: MSBMBG 22(1987)163

Ferrocalamus Hsueh & Keng f. (~ *Indocalamus*). Gramineae (IV). 2 China. R: FOC 22(2006)135

Ferruminaria Garay & al. = *Bulbophyllum*

Ferula Tourn. ex L. Umbelliferae (III 10). Incl. *Dorema*, 170 Medit. (Eur. 8) to C As., trop. Afr. mts. R: E.P. Korovin (1947) *Gen. F. monogr. illus.* Drugs & gums & some statuesque garden pls with large lvs. ***F. ammoniacum*** (D. Don) Spalik & al. (*D. a.*, Iran to Ind.) – principal source of gum ammoniacum used in scent, incense & medic., obtained from insect-punctures etc. Asafoetida (hing in Ind.), gum-resin s.t. used in veterinary med. & (with tamarind & anchovies) allegedly Worcestershire sauce, derived from ***F. assafoetida*** L. (W Iran), ***F. foetida*** (Bunge) Regel (E Iran etc.) & ***F. narthex*** Boiss. (Afghanistan), form. used as condiment ('food of the gods') in Iran & more widely as medic.; gum galbanum also medic. derived from ***F. gummosa*** Boiss. (*F. galbaniflua*, Iran) & ***F. rubricaulis*** Boiss.; sagapenum another resin medic. from diff. spp., also 'silphion [silphium]' prob. now extinct but form. of great imp. in Cyrenaica as a source of an aromatic gum, & sumbul from ***F. sumbul*** Hook.f. (C As.) – scent & incense; ***F. hermonis*** Boiss. (zallouh, E Medit.) roots for comm. tea ('herbal [or Lebanese] viagra' as ferulic acid & feruloside dilate blood vessels). Cult. orn. incl. ***F. communis*** L. (giant fennel, S Eur. to Syria) – scapes form. used for furniture (Cyprus), tipped with a pine cone = thyrse of 'bacchantes', pith burns slowly leaving vasc. tissue, Prometheus bringing fire to Earth in it

Ferulago Koch. Umbelliferae (III 10). 43 Medit. (Eur. 9) to C As. R: FR 100(1989)119

Ferulopsis Kitagawa. Umbelliferae (III 10). 2 N & C As.

fescue *Festuca* spp.; **bearded f.** *Vulpia ambigua*; **Chewing's f.** *F. rubra* subsp. *commutata*; **f. foot** *Schedonorus arundinaceus*; **hair f.** *F. filiformis*; **hard f.** *F. brevipila*; **meadow f.** *Lolium pratense*; **red f.** *F. rubra*; **sheep's f.** *F. ovina*

Fessia Speta = *Scilla*

Festuca L. Gramineae (XVI 10). Incl. *Vulpia*, excl. *Leucopoa* (dioec.), *Pseudobromus* & *Schedonorus*, c. 430 temp. (s.s., China 55, 25 endemic – R: FOC 22(2006)225, Am. 209 [N 37 native – R: FNA 24(2007)389]), trop. mts. R: KBAS 13(1986)94. Fescues, bluegrass. Lvs roll inwards when dry. Pasture- & lawn-grasses, alpine spp. oft. viviparous. ***F. brevipila*** R. Tracy (*F. trachyphylla*, '*F. longifolia*', ~ *F. ovina*, hard f., W Eur.), ***F. filiformis*** Pourret (*F. tenuifolia*, hair f., Eur.) & ***F. rubra*** L. (red. f., N temp.) esp. subsp. ***commutata*** Gaudin (*F. nigrescens*, Chewing's f.) – imp. lawn-grasses. ***F. bromoides*** L. (*V. b.*, Eur.) – fodder-pl. in Aus. but awns harmful to sheep & reduce fleece value, invasive Aus., NZ, W US; ***F. costata*** Nees (S Afr.) – used in making indoor brooms; ***F. glauca*** Vill. (Euras.) – cult. orn. grey foliage (many other spp. sold as *F. g.* in N Am.); ***F. myuros*** L. (Eur.) – invasive Aus.; ***F. ovina*** L. (sheep's f., temp. N OW) – pasture-grass

Festucella Alexeev = *Hookerochloa*

Festucopsis (C. Hubb.) Meld. (~ *Elymus*). Gramineae (XV). 1 Albania: ***F. serpentini*** (C.Hubb.) Meld.

× **Festulolium** Asch. & Graebner. Gramineae. *Festuca* × *Lolium*

feterita *Sorghum bicolor* Caudatum Group

fetterbush *Pieris floribunda*

fever bark *Alstonia constricta*, *Croton* spp., *Ilex verticillata*; **f. bush** *Garrya elliptica*, *Lindera* spp.; **f.few** *Tanacetum parthenium*; **f. tree** *Vachellia xanthophloea*, *Eucalyptus globulus*, *Pinckneya bracteata*, *Zanthoxylum capense*; **f.wort** *Triosteum perfoliatum*

Fevillea L. Cucurbitaceae (II). Excl. *Anisosperma*, 8 trop. Am. R: Sida 21(2005)1975. A 5 all alike. Water-disp. purgative & emetic seeds with oil content greater than any other dicot (55%), those of ***F. cordifolia*** L. widely cult. for oil (seeds washed up in Cornwall 2000), ***F. pedatifolia*** (Cogn.) Jeffrey used as candles in Peru

Fezia Pitard ex Batt. Cruciferae (12). 1 Morocco: ***F. pterocarpa*** Pitard ex Batt.

Fibigia Medik. Cruciferae (2). 13 E Medit. (Eur. 3) to Afghanistan. Some cult. orn. herbs esp. ***F. lunarioides*** (Willd.) Sweet (Aegean)

Fibraurea Lour. Menispermaceae (IV). 2 Assam. SE As., Philipp., Borneo, Sulawesi. R: KB 40(1985)546. Alks. ***F. recisa*** Pierre yields an antibacterial used in Chinese medicine ('huangteng'), also a yellow dye

Ficalhoa Hiern. Sladeniaceae (Theaceae s.l.). 1 E Afr. mts: ***F. laurifolia*** Hiern. R: JB 36(1898)329

Ficaria Schaeff. (~ *Ranunculus*). Ranunculaceae. 5 Eur. to C As. Cotyledon 1, pronounced tubers, K3 (cf. *Ranunculus s.s.*). ***F. verna*** Hudson (*R. f.*; celandine, pilewort, Eur., W As., natur. N Am.) – fls with glistening P due to epidermal cells full of pigment & cells beneath packed with white starch grains, tubers in haemorrhoid treatment (resemblance post-hoc Doctrine of Signatures?) ed. if roasted, blanched lvs s.t. eaten, P & lvs used for cleaning teeth in England, cult. orn. with several cvs incl. 'doubles', tetraploid (*R. f.* subsp. *bulbilifer*) with consp. bulbils

Ficinia Schrad. Cyperaceae (II 5). 60 trop. & S Afr. (50 restricted to Cape). ***F. radiata*** Kunth (Cape) – insect-poll.

Ficoidaceae Juss. = Aizoaceae

Ficus Tourn. ex L. Moraceae (V). c. 850 trop. & warm esp. Indomal. to Aus. (c. 550, R: GBS 21(1965)1 – Borneo 150, New Guinea 150; Afr. 105 (R: Kirkia 13(1990)253), Am. c. 190). R: T 64(2015)589. Figs. (Gyno)dioec. or monoec. trees, shrubs & r.-clinging lianes, epiphytes & stranglers (with coalescing r., some 'individuals' (poss. also somatic mutants) comprising more than one genome!) incl. myrmecophyte ***F. obscura*** Blume var. ***borneensis*** (Miq.) Corner (Borneo), occ. podagric semi-succ. (e.g. ***F. palmeri*** S. Watson, Baja Calif.), unbranched pachycaul treelets (Corner's Model, e.g. ***F. theophrastoides*** Seem., Solomon Is.) to tall trees (Rauh's Model) with slender twigs & small lvs, or rheophytes, with milky latex; r. to 120 m deep recorded, clogging drains & engulfing buildings; lvs simple (***F. otophora*** Corner & Guillaumin (New Caled.) with sep. pinnae, ***F. hirta*** Vahl palmate), in spirals to distich. or opp. with stip. enveloping bud but soon lost; glands ('nectaries') exuding wax (Weiblen). Fls borne in a globose, oblong or pyriform receptacle (fig, syconium, 'fr.' to 15 cm diam. in ***F. punctata*** Thunb., Mal.) with a small apical opening (ostiole) closed by overlapping bracts, 2 or 3 fls to several thousand in largest figs, male, female & gall fls (short-styled sterile females): all 3 in monoec. figs or males & galls in gall pl., the females in seed pl. in gynodioec. figs (the males usu. around ostiole); males with P (0)1 – 7, A 1 – 7; infls protogynous with poll.-attractive app. sp.-specific volatiles to usu. sp.-specific gall-wasps (Hymenoptera – Agaonidae, a symbiosis poss. 65 M yrs old, fossil wasps (England) 34 M yrs old), though ***F. ottoniifolia*** (Miq.) Miq. (S trop. Afr.) occupied by diff. spp. in forest & open country & ***F. sur*** Forssk. (Arabia) with 2 wasp spp. found in a single syconium, ***F. septica*** Burm.f. (Indomal.) with 3 poll. *Ceratosolen* spp. in Taiwan alone (while ***F. lutea*** (Afr.) & ***F. microcarpa*** (As.) cult. elsewhere poll. by local wasps), the pregnant females (1 to several per syconium) forcing a passage through the ostiole, oft. losing wings & antennae (& poss. any attached fungal spores too) in the process, removing pollen from 'pollen pockets' & poll., laying eggs in short-styled fl. in which the ovules are stimulated to produce galls in which the grubs feed, then dying; CO_2 content is higher within the fig than without in some spp. & is believed in some spp. to control the hatching of adults, it in turn being controlled by the activity of bacteria, in turn controlled by populations of nematodes e.g. in *F. religiosa* early male phase has 10% CO_2, 10% O_2 & some ethylene so females inactive; as seeds mature, the emerging male wasps (wingless with reduced legs & eyes) fertilize females still in galls, then cut holes in galls & fig-wall aiding their exit, then die (in *F. religiosa* the change in atmosphere rousing the females); females pass the male fls which dust them with pollen actively or passively taken to other receptive figs. There is also a number of parasitic Hymenoptera which do not poll., ovipositing through syconium wall, some also taking pollen; 30 – 80% ovules are galled. Classification into 6 subg. (R: Blumea 48(2003)168) largely matches poll. & molecular taxonomy: subgg. *Pharmacosycea* (c. 80 poll. *Dolichoris*, trop.) & *Sphaerosuke* (*Urostigma*; c. 280 poll. diff. genera of wasps, matching sections or subsections, trop.) – monoec.; subgg. *Ficus* (c. 60 poll. *Blastophaga* spp., OW), *Terega* (*Sycidium*; c. 110 poll. *Kradibia* & *Liporrhopalum* spp., OW) & *Synoecia* (c. 75 r.-climbers poll. *Wiebesia* spp., Indomal. to Jap.) – (gyno-)dioec., subg. *Sycomorus* (c. 140 poll. *Ceratosolen* spp., OW) – (gyno-)dioec. & some monoec. Wasp spp. visit more than 1 *F.* sp. in monoec. figs, v. rarely in dioec., though hybrids between dioec. spp. found on Krakatao. Syconia (aborted if not poll.) borne in axils, on branches or trunks even at ground level in diff. spp., while others are geocarpic, the figs borne on underground stolons to 10 m long & buried up to 10 cm deep. The true frs are tiny drupes borne in the fleshy receptacle, which may be brightly coloured & is the principal food of many birds & mammals esp. bats, pigs, deer & primates (incl. human) which are the disp. agents (poss. aseasonal availability & high calcium levels in syconia contributing to fs being 'keystone' resources in rain-forests, some 1200 vertebrate spp. reliant), though some fish-disp. in Amaz., prot. by latex with ficin (a protease) until ripe. Latex of some spp. is used as birdlime & vermifuge, meat-tenderizer & chill-proofing agent in beer, while others provide bark cloth (e.g. ***F. thonningii*** Blume (trop. Afr., where oft. planted as shade), ***F. lutea*** Vahl (*F. nekbudu*, obada, trop. & S Afr. to Seychelles)), though in Mex. (since at least AD 79) now largely replaced by *Trema micrantha*. Many others cult. orn. (some artificial hybrids), sources of rubber, fibres, paper, timber, medic. & are revered in religion (considered haunts of spirits etc. in diff. parts of world) & figuring consp. in mythology (e.g. Homer). ***F. benghalensis*** L. (*F. indica*, banyan (from 'banians' the traders seen resting below such trees), Ind., Pakistan, seeding in Florida) – 2 sorts of gall fls in July as opposed to Nov. flowering, small fls with *Blastophaga* wasps, large with non-poll. gallers, initially epiphytic, crown spreading by aerial r. (r.-hairs 0, cortex & pericycle thick, periderm with chloroplasts & lenticels present) dropping down to become accessory trunks, such that a plant may occupy some hectares e.g. 220-yr-old (2009) over 1.6 ha at Calcutta, 450 m circumference with 100 subsid. trunks & 2900 proproots (Alexander 's army said to have sheltered beneath one; at least their views of 'r.' & 'stems' were shaken), timber & fibre, a sacred tree of the Hindus, **'Krishnae'** with cup-shaped lvs (10% of seedlings said to come true); ***F. benjamina*** L. (*F. nitida*, Benjamin tree, weeping or Java fig, Indomal.) – oft. initially epiphytic & s.t. a strangler, dropping r. all round 'host' tree & shading out its crown to smother it completely & become a living shell around the dead tree, cult. orn. with drooping branches, clipped 'square' as street-tree shade in Cairo; ***F. carica*** L. (common fig) – decid. tree (allied to ***F. palmata*** Forssk. of NE Afr. to Ind. but with diff. wasp poll.) prob. native in SW As. but early spread to Medit. where Egyptians were cultivating it 4000 BC, poss. present in pre-pottery A Neolithic level of Jericho, evidence now suggesting cult. there 11 300 yrs ago, i.e. pre-dating domestication of cereals & legumes: some 600 cvs (over 1 M t p.a., 26% in Turkey), much grown are Adriatic fs (esp. **'Brown Turkey'** ('Brown Naples')) with no male fls & parthenocarpic while Smyrna fls require poll. achieved by suspending infls of 'wild' caprifigs from which the poll., *Blastophaga psenes* L., emerge, now, with almonds, olives & carobs typical of the anthropogenic landscape of the W Medit., where many grown for eating fresh or drying, extract a laxative ('syrup of figs'), latex used against warts & to curdle instantly warm milk (Turkey), natur. N Eur. as along river in Sheffield, UK, witness to higher water temps from water form. used as coolant in (now redundant) steel factories; ***F. coronata*** Spin (NSW) – lvs for sandpaper, 'fr.' ed.; ***F. elastica*** Roxb. ex Hornem. (rubber plant, Ind. rubber tree, Indomal., unknown in wild) – large tree with buttresses usu. grown as juvenile in houses, form. imp. source of rubber (Ind., Ind. or Assam r.); ***F. exasperata*** Vahl (trop. Afr., Yemen to Ind.) – cytotoxic but eaten by healthy chimpanzees; ***F. globosa*** Blume (SE As., W Mal.) – strangler with coalescing r., first through union of epidermal hairs, then compression of adjacent cortex, at periphery rays prod. parenchyma eventually uniting & some becoming cambial cells linking existing cambia so that continuous ring of vasc. cambium reorganized & giving rise to more sec. tissues; ***F. granatum*** Forst. f. (Vanuatu) – imp. local 'fr.'; ***F. hirta*** Vahl (NE Ind. to W Mal. & S China) – r. used in soup in China; ***F. laevis*** Blume (Mal.) – r.-cli. without dimorphic lvs; ***F. lutea*** (*F. vogelii*, trop. & S Afr. to Seychelles) – source of Abbo, Congo or Dahomey rubber; ***F. lyrata*** Warb. ex De Wild. & Durand (fiddle-leaf f., trop. Afr.) – juveniles cult. as house-pls; ***F. macrocarpa*** Blume (Mal.) – topiary subject in Malaysia, incl. yellow-leaved form; ***F. macrophylla*** Desf. ex Pers. (Moreton Bay f., E Aus.) – to 60 m, but usu. smaller in cult. where used as street & shade tree (as at Sydney, also NZ which wasp has since reached); ***F. microcarpa*** L.f. (Indomal.) – assoc. with ghosts so not felled, introd. C Am. & now potential pest as orig. wasps poll., & invasive Hawaii where orig. wasp & alien disp. frugivorous birds introd., **'Green Island'** rarely flowering dwarf pl. used for hedges & bonsai, var. ***hillii*** (Bailey) Corner (Hill's fig, NE Aus.) – avenue tree in NSW; ***F. pseudopalma*** Blanco (Philipp.) – unbranched pachycaul cult. orn.; ***F. pumila*** L. ('*F. repens*', cli. f., Vietnam to Jap.) – liane with aerial r. secreting a gummy exudate & absorbing fluid constituent to leave r. cemented to support, at top of which it becomes more tree-like, var. ***awkeotsang*** (Mak.) Corner 'fr.' used for jelly sweetmeat in China (okgue), though some cvs e.g. **'Minima'** are juvenile forms much cult. in conservatories etc.; ***F. racemosa*** L. (crattock, Indomal.) – timber, shade, ed. 'fr.'; ***F. religiosa*** L. (peepul, pipal, pipul, bo(dhi) tree, Ind. to SE As.) – decid. fast-growing tree, usu. an epiphyte when young & splits host (not a strangler) with lvs with v. long drip-tips cont. vestigial veins rep. an ancestral large lamina here unexpanded, sacred to Hindus & Buddhists (Buddha had the true insight beneath one; the Anuradhapura, Sri Lanka tree grown from a sprig of that tree brought from Ind. 288 BC, complete with gall-wasps), wood for rosary beads in Ind., lvs used for miniature paintings, fibre form. used for paper in Myanmar, host for lac insects & some silkworms, cult. orn. incl. NZ, since reached by wasp; ***F. retusa*** L. (Indomal.) – topiary subject with blunt lvs in Mal.; ***F. rubiginosa*** Desf. ex Vent. (*F. novae-walliae*,

Port Jackson fig, NSW) – cult. street-tree, natur. NZ as wasp has now been blown from Aus. as has that of *F. macrophylla* (1993); ***F. semicordata*** Buch.-Ham. ex Sm. (Nepal to SE As.) – imp. fodder crop; ***F. septica*** Burm.f. (W Pacific) – local medic. NG, bactericidal; ***F. sycomorus*** L. (sycamore (of the Bible), mulberry fig, trop. & S Afr.), brought to Medit. & cult. for ed. figs (parthenocarpic clones, maturing when syconia slashed) & timber used for Pharaohs' sarcophagi (now extinct in wild in Egypt where first domesticated); ***F. tinctoria*** Forst. f. (Indomal. to Pac.) – cult. orn. with coalescing aerial r.; ***F. virens*** Aiton (Indomal.) – cult. for shade, young shoots pickled in Sikkim

fiddle dock *Rumex pulcher*; **f. greens** or **f. heads** *Osmundastrum cinnamomeum*; **f.-leaf** *Ficus lyrata*; **f.wood** *Citharexylum spp., Petitia domingensis, Vitex* spp.

Fiebrigia Fritsch = *Gloxinia* (*Seemannia*)

Fiebrigiella Harms. Leguminosae (III 11). 1 Bolivia: ***F. gracilis*** Harms

Fiedleria Reichb. = *Petrorhagia*

Fieldia Cunn. Gesneriaceae (II 4b). Excl. *Lenbrassia*, 1 SE Aus.: ***F. australis*** Cunn. – anisophyllous cli. usu. on tree-fern trunks

fig *Ficus* spp. esp. (ed.) *F. carica*; **climbing f.** *F. pumila*; **fiddle-leaf f.** *F. lyrata*; **Hill's f.** *F. microcarpa* var. *hillii*; **Hottentot f.** *Carpobrotus edulis*; **Indian f.** *Opuntia* spp.; **Java f.** *F. benjamina*; **mulberry** or **sycamore f.** *F. sycomorus*; **Port Jackson f.** *F. rubiginosa*; **weeping f.** *F. benjamina*

figwort *Scrophularia* spp.; **Cape f.** *Phygelius capensis*; **water f.** *S. auriculata*

Fijian kauri *Agathis macrophylla*

Filaginella Opiz = *Gnaphalium*

Filago Loefl. Compositae (Gnap.-Gnap.). Incl. *Cymbolaena, Evacidium, Evax*, excl. *Logfia*) 38 OW, some weedy Am. R: OB 104(1991)170, 171

filaree *Erodium* spp.

Filarum Nicolson. Araceae (VII 10). 1 Peru: ***F. manserichense*** Nicolson. R: Britt. 18(1967)348

filbert *Corylus* spp.

filé *Sassafras albidum*

Filetia Miq. Acanthaceae (III 2c). 8 Sumatra, Malay Pen.

Filgueirasia Guala. Gramineae (V 2). 2 Braz.

Filicium Thw. ex Hook.f. Sapindaceae (III 2). 3–4 OW trop. ***F. decipiens*** (Wight & Arn.) Thw. (fern tree (E Afr.), Ind.) – fern-like lvs, wood for furniture

Filicopsida = Polypodiidae

Filifolium Kitam. = *Artemisia*

Filipedium Raiz. & Jain = *Capillipedium*

Filipendula Mill. Rosaceae (Ros.-Ulm.). 15 Euras. (Eur. 2), NE N Am. (4). R: JJB 69(1994)290. ***F. ulmaria*** (L.) Maxim. (meadowsweet, Euras., natur. N Am.) – form. medic. (salicylic acid compounds (like willows) basis of efficacy in arthritis treatment etc.), acetyl-salicic acid first isolated 1835 leading to aspirin (A[cetyl]SPIR[*aea*, to which genus *F.* once referred]IN, Bayer in 1899) & synth. ('Aspro', Melbourne, Aus. in 1914), fragrant oil, cult. orn.; ***F. vulgaris*** Moench (dropwort, Euras.) – form. medic. esp. renal diseases, cult. orn.

Fillaeopsis Harms. Leguminosae (II 1). 1 W & SW Afr., forest margins: ***F. discophora*** Harms – easily polished timber used for wood-block floors etc.

filmy ferns Hymenophyllaceae

Fimbribambusa Widjaja (~ *Bambusa*). Gramineae (V 4). 2 C & E Mal. R: Reinw. 11(1997)80

Fimbriella Farw. ex Butzin = *Platanthera*

Fimbristemma Turcz. = *Matelea*

Fimbristylis Vahl. Cyperaceae (II 4). Excl. *Abildgaardia, Bulbostylis*, 200+ warm (Eur. 3; China 53; Aus. c. 70). Some local medic., copper indicators (Queensland) & fibres esp. ***F. umbellaris*** (Lam.) Vahl (*F. globulosa*, trop. As.) – cult. for weaving, esp. sleeping-mats

Findlaya Hook.f. = *Orthaea*

Finetia Gagnepain (~ *Anogeissus*). Combretaceae (II 2a). 1 SE As.: ***F. rivularis*** Gagnepain

Fingardia Szlach. = *Malaxis*

finger cherry *Rhodomyrtus macrocarpa*; **dead man's f.** *Orchis mascula*; **fairy fs** *Digitalis purpurea*; **f. grasses** *Digitaria* spp.; **f. lime** *Citrus australasica*

Fingerhuthia Nees. Gramineae (XXVII 3). 2 trop. & S Afr. R: Strel. 9(2000)186

finish, dead *Acacia tetragonophylla*

Finlaysonia Wall. Apocynaceae (III; Periplocaceae). c. 3 Indomal. to Aus. ***F. khasiana*** (Kurz) Venter (*Hanghomia marseillei*, SE As.) – liane with r. of cerem. significance, burnt as incense etc. in Laos

finocchio *Foeniculum vulgare* var. *azoricum*

Finschia Warb. (~ *Grevillea*). Proteaceae (V 3). 3 Papuasia, 1 ext. W Pac. *F. **chloroxantha*** Diels (Papuasia) – ed. seeds, planted Vanuatu

Fintelmannia Kunth = *Trilepis*

Fioria Mattei = *Hibiscus*

fiorin *Agrostis stolonifera*

fique *Furcraea* spp.

fir *Abies* spp., but form. used for many other coniferous woods; **alpine f.** *A. lasiocarpa*; **balsam f.** *A. balsamea*; **bristlecone f.** *A. bracteata*; **Caucasian f.** *A. nordmanniana*; **Douglas f.** *Pseudotsuga menziesii*; **Fraser f.** *A. fraseri*; **giant f.** *A. grandis*; **Greek f.** *A. cephalonica*; **Himalayan f.** *A. pindrow*; **Japanese f.** *A. firma*; **Memel f.** *Pinus sylvestris*; **noble f.** *A. procera*; **Norway f.** *P. sylvestris*; **red f.** *A. magnifica*; **St Lucia f.** *A. bracteata*; **silver f.** *A. alba*; **umbrella f.** *Sciadopitys verticillata*; **white f.** *A. grandis*

fire bush *Pyracantha coccinea*, *Hamelia patens*; **Chilean f. bush** *Embothrium coccineum*; **f. cracker plant** *Russelia equisetiformis*; **f. pink** *Silene virginica*; **f. thorn** *Pyracantha coccinea*; **f. tree** *Nuytsia floribunda*; **f.weed** *Chamaenerion angustifolium*, *Senecio madagascariensis* (Aus.); **f.wheel tree** *Stenocarpus sinuatus*

Firmiana Marsili. Malvaceae (Sterc.; Sterculiaceae). 12 OW trop., E Afr. eastwards. R: Reinwardtia 4(1957)281, 5(1960)384, Blumea 34(1989)117. Ovary wings in wind-disp. Some light timbers, seeds used for caffeine-cont. tea, & cult. orn. esp. *F. **simplex*** (L.) W. Wight (wutong, E As.) – foliage pl. with lvs to 30 cm across, v. common street-tree in Jap., planted in China for wood used in wind instruments, Phoenix supposed to alight on it, fr. ed.

Fischeria DC. Apocynaceae (V c; Asclepiadaceae). 16 trop. Am. R: SB 11(1986)229

fish berries *Anamirta cocculus*; **f. grass** *Cabomba* spp.; **f. plant** *Houttuynia cordata*

Fissendocarpa (Haines) Bennet = *Ludwigia*

Fissenia Endl. = *Kissenia*

Fissicalyx Benth. Leguminosae (III 11). 1 Venez., Guyana: *F. **fendleri*** Benth.

Fissistigma Griff. Annonaceae (III 7). 60 OW trop. (China 23). Local medic. China for arthritis & sciatica

fitches *Nigella sativa* seeds

Fitchia Hook.f. Compositae (Cor.). 6 Polynesia. R: UCPB 29, 1(1957). 'Sister' to *Oparanthus* & allied to Carib. genera

fitsroot *Astragalus glycyphyllos*

Fittingia Mez. Primulaceae (Myrsinaceae). 9 NG. R: Blumea 33(1988)94. Pachycaul treelets

Fittonia Coem. Acanthaceae (III 2c). 2 Peru. Leaf infusions used for toothache in NW Amaz.; cult. orn. house-pls with white or coloured leaf-veins, esp. *F. **albivenis*** (Veitch) Brummitt (*F. verschaffeltii*) cvs: **Argyroneura Group** (lvs with white veins) & **Verschaffeltii Group** (lvs with red or pink veins)

fitweed *Eryngium foetidum*

Fitzalania F. Muell. = *Meiogyne* (but see FA 2(2007)45)

Fitzroya Hook.f. ex Lindl. Cupressaceae. 1 S Chile & S Arg.: *F. **cupressoides*** (Molina) I.M. Johnston (*F. patagonica*, alerce, Patagonian cypress) – Attims's Model, to 3622 yrs old in Chile, CITES-listed, timber esp. for shingles. R: BM n.s. 16(1999)229

Fitzwillia Short. Compositae (Gnap.-Ang.). 1 W Aus.: *F. **axilliflora*** (Ewart & J.W. White) Short. R: OB 104(1991)132

five-in-one *Coleus amboinicus*

Flabellaria Cav. Malpighiaceae. 1 trop. Afr.: *F. **paniculata*** Cav.

Flabellariopsis R. Wilczek. Malpighiaceae. 1 trop. Afr.: *F. **acuminata*** (Engl.) R. Wilczek. R: BJBB 25(1955)304

Flacourtia Comm. ex L'Hérit. Salicaceae (2; Flacourtiaceae). c. 15 trop. & S Afr. to Fiji. Troll's Model. Fr. ± ed., some medic. esp. *F. **indica*** (Burm. f.) Merr. (*F. ramontchi*, Madag. or Governor's plum, ramontchi, OW trop.) – fr. for jelly, medic.; *F. **inermis*** Roxb. (batako plum, lovi-lovi, orig. uncertain) – fr. variable, for jelly; *F. **jangomas*** (Lour.) Räuschel (another cultigen); *F. **rukam*** Zoll. & Moritzi (coffee plum, rukam, Mal.)

Flacourtiaceae Rich. ex DC., s.s. = Salicaceae. See also Achariaceae, Aphloiaceae, Berberidopsidaceae, Gerrardinaceae & Lacistemataceae (other genera once referred to this 'dustbin' fam. now in Asteropeiaceae, Celastraceae, Crypteroniaceae, Dioncophyllaceae, Malvaceae, Muntingiaceae, Myrtaceae, Passifloraceae, Putranjivaceae, Theaceae, Thymelaeaceae etc. etc.!)

flag *Iris* spp.; **blue** or **poison f.** *I. versicolor*; **f. root** or **sweet f.** *Acorus calamus*; **water** or **yellow f.** *I. pseudacorus*

Flagellaria L. Flagellariaceae. 4 OW trop. Stems used for basketry, fish-traps etc. *F. **indica*** L. (As.) – young lvs used as shampoo, stripped stems (to 20 m long) for dilly bags (N Queensland)

Flagellariaceae Dumort. Magnoliidae – Poales. Excl. Hanguanaceae, 1/4 OW trop. Glabrous lianes with sym. rhiz. (accum. sucrose & not starch, without sec. thickening or axillary buds but poss. with apically-dichotomizing meristems (Schoute's Model); silica only in vasc. bundles; vessel-elements throughout veg. body. Lvs in spirals with closed sheath, short petiole & 0 ligule, lamina parallel-veined, circinate in bud, apically cirrhose at maturity acting as a sensitive tendril. Fls small, bisex., or pls dioec., reg., 3-merous, (?) wind-poll., in term. bracteate panicles; P 3 + 3, white or greenish, basally connate A 3 + 3 with longit. slits, $\underline{G}$ (3), 3-loc. with 3 styles, s.t. weakly connate basally & stigmatic ± throughout length, 1 orthotropous (? or anatropous) ovule per loc. on axile placenta. Fr. a drupe with 1 pyrene & usu. 1 or 2 seeds with copious endosperm with starch in simple grains; embryo small, undifferentiated, capping endosperm. 2n = 38

Genus: *Flagellaria*

Hanguana now referred to Commelinales

Flagenium Baill. (~ *Fernelia*). Rubiaceae (II 5). 6 Madag. R: BJLS 155(2007)559

flamboyant *Delonix regia*; **yellow f.** *Peltophorum pterocarpum*

flame flower *Ixora coccinea*; **f. of the forest** (Afr.) *Mussaenda erythrophylla*, (Ind.) *Butea monosperma*; **Nandi f.** *Spathodea campanulata*; **f. pea** *Chorizema ilicifolium*; **f. tree** (Aus.) *Brachychiton acerifolius*, *Nuytsia floribunda*, (Afr.) *S. campanulata*, (WI) *Delonix regia*; **f. vine** *Pyrostegia venusta*

flamingo flower *Anthurium* ×*ferrierense* & *A. scherzerianum*

Flanagania Schltr. = *Cynanchum*

flannel flower *Actinotus helianthi*; **f. plant** *Verbascum thapsus*

flatweed *Hypochaeris radicata*

flatwoods plum *Prunus umbellata*

Flaveria Juss. Compositae (Tag.-Flav.). 22 Am. (esp. SW N Am.), with 2 pantrop. weeds incl. *F. **trinervia*** (Spreng) C. Mohr (*F. australasica*, orig. S Mex.). R: AMBG 65(1978)590. Genus with both Kranz & non K. leaf-anatomy syndromes

flax *Linum usitatissimum*; **blue f.** *L. narbonense*; **bush f.** *Astelia nervosa*; **fairy f.** *L. catharticum*; **false f.** *Camelina sativa*; **flowering f.** *L. grandiflorum*; **NZ f.** *Phormium tenax*; **prairie f.** *L. lewisii*

fleabane *Pulicaria* spp., *Erigeron* spp., *Pluchea* spp.; **common f.** *Pulicaria dysenterica*; **daisy f.** *E. annuus*

fleawort *Tephroseris integrifolia*

Fleischmannia Schultz-Bip. (~ *Eupatorium*). Compositae (Eup.-Flei.). 95 N Am., W S Am. R: MSBMBG 22(1987)285

Fleischmanniopsis R. King & H. Robinson. Compositae (Eup.-Crit.). 5 Mex., C Am. R: MSBMBG 22(1987)310

Flemingia Roxb. ex Aiton f. Leguminosae (III 18). 30 OW trop. Some local medic., cover-crops, lac insect hosts & dyes esp *F. **grahamiana*** Wight & Arn. (Ind.) & *F. **macrophylla*** (Willd.) Merr. (SE As., hedge-pl. in trop.) – sources of waras or warrus = dried resinous legume hairs used for dyeing silk; *F. **vestita*** Benth. ex Bak. a r.-crop (souphlong) in N Ind.

Fleroya Y.F. Deng = *Mitragyna* (but see T 56(2007)247)

fleur-de-lis or fleur-de-lys *Iris* × *germanica* 'Florentina' or *I. pseudacorus* (cf. ASNB 13,10(1989)1)

Fleurya Gaudich. = *Laportea*

Fleurydora A. Chev. Ochnaceae (III). 1 Guinea: *F. **felicis*** A. Chev. R: BJ 113(1991)172. Only capsular O. in Afr., allied to S Am. pls (cf. *Pitcairnia*)

Flexanthera Rusby. Rubiaceae (I 7). 2 Colombia, Bolivia

Flickingeria A. Hawkes = *Dendrobium*

Flinders grass *Iseilema* spp.

Flindersia R. Br. Rutaceae (I; Flindersiaceae). 17 E Mal., E Aus. (15), New Caled. R: JAA 50(1969)481, 56(1975)243. Scarrone's Model; trees with tubercled capsules of winged seeds; alks. Timbers (yellow-wood, red beech etc.): *F. **australis*** R. Br. (crow('s) ash, Aus. teak, E Aus.); *F. **brayleyana*** F. Muell. (Queensland or Aus. maple, NE Queensland) – beetle-poll. in N Queensland, high-quality timber; *F. **collina*** Bailey & *F. **maculosa*** (Lindl.) Benth. (leopard tree, l.-wood, E Aus.); *F. **pimenteliana*** F. Muell. (silkwood, NG, NE Queensland); *F. **schottiana*** F. Muell. (*F. pubescens*, silver or bumpy ash, NE &

E Aus.); *F. **xanthoxyla*** (Hook.) Domin (*F. oxleyana*, Long Jack, E Aus.) – timber & yellow dye

Flindersiaceae C. White ex Airy Shaw = Rutaceae

flintwood *Scolopia braunii*

flixweed *Descurainia sophia*

floating pennywort *Hydrocotyle ranunculoides*

Floerkea Willd. Limnanthaceae. 1 N Am.: *F. **proserpinacoides*** Willd. – spicy stem for salads. R: FNA 7(2010)173

flooded gum *Eucalyptus grandis*

flopper *Kalanchoe daigremontiana*

floradora *Marsdenia floribunda*

Flora's paintbrush *Emilia* spp.

Florence whisk *Sorghum bicolor* Technicum Group

flores de palo Loranthaceae esp. *Phoradendron* spp.

Florestina Cass. Compositae (Bah.). 6 S US, Mex., Guatemala. R: Phytotaxa 268(2016)91. *F. **tripteris*** DC. (S N Am.) – toxic (cyanogenic), serious weed Aus. since 1989

Florida arrowroot *Zamia pumila*; **F. boxwood** *Schaefferia frutescens*; **F. moss** *Tillandsia usneoides*; **F. trema** *Trema micrantha*; **F. velvet bean** *Mucuna pruriens* var. *utilis*

florist's fern *Dryopteris dilatata*

Floscaldasia Cuatrec. Compositae (Ast.-Hin.). 2 Colombia, Ecuador (Andes). R: Novon 10(2000)145

Floscopa Lour. Commelinaceae (II). 20 trop. & warm. Local medic.

Flosmutisia Cuatrec. Compositae (Ast.-Hin.). 1 Colombia: *F. **paramicola*** Cuatrec.

floss, candy (UK), **fairy f.** (Aus.) spun sugar

Flourensia DC. Compositae (Helia.-Enc.). 13 S N Am., 18 S Am. (disjunct). R: Fieldiana Bot. n.s. 16(1984)

flowering ash *Fraxinus ornus*; **f. cherry** *Prunus serrulata* & hybrids; **f. currant** *Ribes sanguineum*; **f. maple** *Callianthe* hybrids; **f. rush** *Butomus umbellatus*

Floydia L. Johnson & B. Briggs. Proteaceae (V 1). 1 NE Aus.: *F. **prealta*** (F. Muell.) L. Johnson & B. Briggs – nut (coohoy n.) ed. roasted; timber. R: FA 16(1995)417

Flueckigera Kuntze = *Ledenbergia*

Flueggea Willd. (~ *Securinega*). Phyllanthaceae (Phyll.-Fluegg.; Euphorbiaceae s.l.). Excl. *Richeriella*, c. 15 trop. to Iberia (1: *F. **tinctoria*** (L.) G. Webster) & Turkey. R: Allertonia 3(1984)259. *F. **neowawraea*** W. Hayden (Hawaii) – reduced to 41 trees by 1982; *F. **virosa*** (Willd.) Royle (*S. v.*, trop. OW) – timber for houseposts in Solomon Is., fr. eaten on Mafia Is. (E Afr.), seeds germ. best after passage through baboon (? & human) gut

fly honeysuckle *Lonicera xylosteum*; **f. orchid** *Ophrys insectifera*; **shoo f.** *Nicandra physalodes*; **Venus's f. trap** *Dionaea muscipula*

Flyriella R. King & H. Robinson. Compositae (Eup.-Alom.). 4 SW N Am. R: Phytol. M 11(1997)143

foam, meadow *Limnanthes* spp.; **f.bark** *Jagera pseudorhus*; **f.flower** *Tiarella cordifolia*

Fockea Endl. Apocynaceae (V a; Asclepiadaceae). 6 trop. & S Afr. R: AMBG 93(2006)341. Water-storing tubers with aerial twining stems; cult. orn. esp. *F. **edulis*** (Thunb.) K. Schum. (*F. crispa*, Karoo) – one of 'oldest living potpls' (1989, potted Vienna c. 1801), but see *Encephalartos*

Foeniculum Mill. Umbelliferae (III 8). 1 As., natur. widely: *F. **vulgare*** Mill. (*F. officinale*, fennel, invasive Aus., W US), rep. by 2 vars in cult.: var. ***azoricum*** (Mill.) Thell. (Florence f., finocchio, anise (USA)) – swollen leaf-bases ed. when blanched, & var. ***dulce*** (Mill.) Fiori & Paol. – fr. larger, cult. for ess. oils (esp. *trans*-anethole) for flavouring (e.g. anisette, ouzo, pastis raki) & medic. since Mycenaean times (C13 BC)

Foetidia Comm. ex Lam. Lecythidaceae (III; Foetidiaceae). 1 Tanz., 14 Madag., 2 Masc. R: Britt. 60(2008)336

Foetidiaceae Airy Shaw = Lecythidaceae

fog, Yorkshire *Holcus lanatus*

Fokienia A. Henry & H. Thomas. Cupressaceae. 1 SE China, Laos, Vietnam: *F. **hodginsii*** (Dunn) A. Henry & H. Thomas – wood almost worked out so stumps now revisited for wood-carving & boat-building, pe-mou oil used in scent-making extracted by distillation of r. R: A. Farjon, *Monogr. Cupressaceae & Sciadopitys* (2005)156

Foleyola Maire. Cruciferae (12). 1 W Sahara: *F. **billotii*** Maire

folhado *Clethra arborea*

Folotsia Costantin & Bois = *Cynanchum*

fonio *Digitaria exilis*; **f., black** *D. iburua*
Fonkia Philippi = *Gratiola*
Fontainea Heckel. Euphorbiaceae (Cod.-Bal.). c. 9 NG (2), NE Aus. (6), New Caled. & Vanuatu (1). R: Austrobaileya 2(1985)112
Fontanesia Labill. Oleaceae (3).1 Euras.: ***F. phillyraeoides*** Labill. (subsp. ***p.*** -Sicily, Turkey & Syria; subsp. ***fortunei*** (Carr.) Hegi – EC China) – cult. orn. shrub
Fontellaea Morillo = *Philibertia*
Fontinalis Hedw. Fontinalaceae (Musci). c. 45 subcosmop. Aquatic mosses, some v. long. ***F. antipyretica*** Hedw. (N temp.) – dioec., named because used around chimneys to prevent houses burning down
Fontquera Maire = *Perralderia*
fool's parsley *Aethusa cynapium*
Foonchewia R.J. Wang (~ *Dunnia*). Rubiaceae (Foonch.). 1 E Guangdong: ***F. guangdongensis*** R.J. Wang & H.Z. When. R: JSE 50(2012)468
Forbesina Ridl. = *Eria*
Forchhammeria Liebm. Resedaceae. 10 Calif. to C Am. & WI. Form. in Euphorbiaceae, then Capparaceae (but G 2-loc. with axile placentation). ***F. pallida*** Liebm. (Mex.) – dioec., wind-poll.
Forcipella Baill. Acanthaceae (III 2c). 5 Madag.
Fordia Hemsl. Leguminosae (III 16). 18 SE As., W Mal. R: Blumea 36(1991)191. Heterogeneous? Ramiflorous & cauliflorous trees to 30 m. ***F. rheophytica*** (Buijsen) Dasuki & Schot ('*F. angustifoliola*') a rheophyte of rapids in Sarawak
Fordiophyton Stapf. Melastomataceae. Incl. *Stapfiophyton*, 16 S China, SE As.
Forestiera Poir. Oleaceae (4d). c. 16 S US, C Am., WI. Some cult. orn. incl. ***F. acuminata*** (Michaux) Poir. (N Am.) – timber good for turnery; ***F. pubescens*** Nutt. (stretchberry, S US) – large pls good indicators of water, fr. ed
Forficaria Lindl. = *Disa*
Forgesia Comm. ex Juss. (~ *Escallonia*). Escalloniaceae. 1 Réunion: ***F. racemosa*** J.F. Gmel.
forget-me-not *Myosotis* spp.; **American f.** *Eritrichium* spp.; **Cape f.** *Anchusa capensis*; **Chatham Is. f.** *Myosotidium hortensia*; **Chinese f.** *Cynoglossum amabile*; **garden f.** *Myosotis sylvatica*; **water f.** *M. scorpioides*
Formania W. Sm. & Small. Compositae (Anth.-Art.). 1 SW China: ***F. mekongensis*** W. Sm. & Small
Formanodendron Nixon & Crepet = *Trigonobalanus*
Formosa lily *Lilium formosanum*
Formosia Pichon = *Anodendron*
Forrestia A. Rich. = *Amischotolype*
Forsellesia Greene = *Glossopetalon*
Forsskaolea L. Urticaceae (V). 6 Canary Is., SE Spain (1), Afr., Arabia, Ind. R: NJB 8(1988)34. Bark of ***F. tenacissima*** L. (SE Spain to Sahara & Ind.) used for rope
Forstera L.f. Stylidiaceae. (Incl. *Phyllachne*) 9 Tasmania, NZ (5 – R: NZJB 47(2009)299), temp. S Am.
Forsteronia G. Meyer. Apocynaceae (II c). c. 40 trop. Am. Rubber of little value
Forsythia Vahl. Oleaceae (2). 9 SE Eur. (1), E As. (8). Decid. shrubs with winged seeds (& heterostylous fls before lvs emerge), much cult. esp. ***F. intermedia*** Zabel (long considered ***F. suspensa*** (Thunb.) Vahl (China, long cult. Jap.; pith solid at nodes; fr. wall medic. in China) × ***F. viridissima*** Lindl. (China, pith hollow at nodes) but app. true wild Chinese sp.) esp. **'Spectabilis'**, one of most widely planted of all shrubs, from which colchicine-induced tetraploids & back-crossed triploids have been raised, also hybrids esp. **'Arnold Dwarf'** (*F. i.* × *F. japonica* Mak. (Jap.)) & 'Farrand' hybrids ('Arnold Dwarf' × *F. ovata* Nakai (Korea))
Forsythiopsis Bak. = *Oplonia*
Fortunatia J.F. Macbr. = *Oziroe*
Fortunearia Rehder & E. Wilson. Hamamelidaceae (I 3). 1 C & E China: ***F. sinensis*** Rehder & E. Wilson – cult. orn. decid. shrub with male fls & bisex. fls in sep. racemes
Fortunella Swingle = *Citrus*
Fortuynia Shuttlew. ex Boiss. Cruciferae (12). 2 Iran, Afghanistan, Balochistan
Fosbergia Tirveng. & Sastre (~ *Randia*). Rubiaceae (II 1). 5 Myanmar, Thailand, China & Vietnam
Fosterella L. B. Sm. Bromeliaceae (1). 31 C Am. & W S Am. R: J. Peters (2009) *Rev. genus F.*
Fosteria Molseed = *Tigridia*

Fothergilla L. Hamamelidaceae (I 4). 2 E N Am. R: Arnoldia 31(1971)89. Cult. orn. (usu. *F.* × ***intermedia*** Ranney & Fantz (*F. gardenii* L. × *F. major* Lodd.)), decid. shrubs with consp. white filaments, C 0

fo-ti *Reynoutria multiflora*

foulia bean *Vicia faba*

fountain grass *Cenchrus setaceus;* **f. plant** *Russelia equisetiformis;* **f. tree** *Spathodea campanulata*

Fouquieria Kunth. Fouquieriaceae. 11 arid SW N Am. R: Aliso 7(1972)439. ***F. columnaris*** (Kellogg) Curran (*Idria c.*, boojum tree, cirio, Baja Calif.) – succ. columnar trunk to 20 m & 360 yrs old (R.R. Humphrey (1974) *The boojum & its home*); ***F. splendens*** Engelm. (coachwhip, ocotillo) – planted as spiny hedge, latex poss. rubber source

Fouquieriaceae DC. (~ Polemoniaceae). Magnoliidae – Ericales. 1/11 SW N Am. R: Aliso 7(1972)439. Woody or succ. xeromorphic spiny little-branched shrubs & small trees. Lvs in spirals, small, simple, those of long shoots each arising from a decurrent ridge, basal part of petiole marcescent as a spine, in axils of which are short shoots with clustered spineless lvs; stip. 0. Fls hypogynous, ± reg. in spikes to panicles; P in spiral: K 5 imbr., persistent, 2 outer oft. larger, C (5) forming a tube or salveriform C with imbr. lobes; A 10(– 23), in single whorl on receptacle, exserted but those opp. K s.t. larger, anthers with longit. slits; $\underline{G}$ (3) with 1 3-branched style, placentation basally axile but otherwise parietal with deeply intruded placentas, which meet as fr. matures forming a central column & may even appear axile or free-central, ovules (6)14 – 18(– 20), anatropous, bitegmic. Fr. a loculicidal capsule; seeds with membranous margins of 1-cellular trichomes & straight embryo in thin (or 0) oily proteinaceous endosperm. n = 12

Genus: *Fouquieria* (incl. *Idria*)

Fourniera Bommer = *Cyathea*

four-o'clock plant *Mirabilis jalapa*

Fourraea Greuter & Burdet (~ *Arabis*). Cruciferae. 1: ***F. alpina*** (L.) Greuter & Burdet. R: Novon 15(2005)521

Foveolaria Ruíz & Pavón = *Styrax*

Foveolina Källersjö (~ *Pentzia*). Compositae (Anth.-Pen.). 5 S Afr. R: BJLS 96(1988)316. Annuals

fox and cubs *Pilosella aurantiaca;* **f.bane** *Aconitum vulparia;* **f. berry** *Vaccinium vitis-idaea;* **f.glove** *Digitalis* spp. esp. *D. purpurea;* **f.g., Chinese** *Rehmannia elata;* **f.g., Mexican** *Tetranema roseum;* **f. grape** *Vitis labrusca;* **f. nut** *Euryale ferox;* **f.tail grass** *Alopecurus pratensis;* **f.t.g., giant** *Setaria magna;* **f.t.g., green** *S. viridis;* **f.t.g., yellow** *S. pumila;* **f.t. lily** *Eremurus* spp.; **f.t., marsh** *A. geniculatus;* **f.t. millet** *S. italica;* **f.t. palm** *Wodyetia bifurcata*

Fragaria Tourn. ex L. (~ *Potentilla*). R: CJB 40(1962)869; J.F. Hancock (1999) *Strawberries*; strawberries i.e. *streabariye* (AD 995) = straying (runners) with achenes (drupelets in mutants) over surface of swollen receptacle ('fr.'). Comm. strawberries (UK record 231 g receptacle – active flavour principle unknown; synthetics = ethyl 1-methyl 2-phenylglycate; receptacle & lvs causing urticaria in some, & sensitivity to strawberries a not infrequent 'allergy'; lvs rep. in peers' coronets – UK) some 57 K T a year, 98% N hemisphere (US 28%), usu. ***F. ananassa*** (Weston) Rozier (*P. a.*, *F. chiloensis* var. *a.*, octoploid (2n = 56), ***F. chiloensis*** (L.) Mill. (*P. c.*, coastal Alaska to Calif., S Am., cult. since pre-Columbian times but spread through S Am. by Spanish, introd. (male) Eur. 1716) × ***F. virginiana*** Mill. (E N Am., Native Americans form. made a bread (wattahimneah) from it & maize, introd. Eur. 1629) – both allopolyploids with *F. vesca* as one parent) – hybrids in Brittany & diff. bot. gardens by 1750s, early cvs incl. '**Downton**' (1817 England, later '**Keens' Seedling**' there) or earlier in France (though natural hybrids now known from NW Am.), esp. form. '**Royal Sovereign**' (UK, 1892) until mid-C20, but until recently mostly relatively tasteless '**Camarosa**' (Calif.) but 65% now '**Elsanta**' (NL), high ellagic acid (binds with DNA protecting it from carcinogens), some cvs like '**Little Scarlet**' much used for jam), in C19 picked early a.m. Kent & turned into jam for mid-day sales in London, 'pineberry' with pineapple-flavoured white 'fr.' with red 'pips'; ***F. moschata*** Weston (*P. m.*, '*F. muricata*', hautbois, Eur.) – 'fr.' small & largely replaced by *F. ananassa;* ***F. vesca*** L. (*P. v.*, wild s., N temp.) – small 'fr.' used in fancy pastries etc., esp. in France & C Eur, app. incl. continuously flowering 'alpine' s. (*F. semperflorens*), '**Monophylla**' (*F. v.* var. *m.*) with simple lvs & transitions to 3-foliolate – first spontaneous pl. mutation observed (1763), & '**Muricata**' (Plymouth s.) – small bristle-clad 'fr.'.

Fragariopsis A. St-Hil. = *Plukenetia*

Fragosa Ruíz & Pavón = *Azorella*

Frailea Britton & Rose (~ *Parodia*). Cactaceae (III 5). 9 S Am. Cult. orn.

Franchetella Pierre = *Pouteria*

Franciscodendron Hyland & Steenis (~ *Hildegardia*). Malvaceae (Sterc.; Sterculiaceae). 1 Queensland: ***F. laurifolium*** (F. Muell.) Hyland & Steenis. R: Brunonia 10(1987)211

Francoa Cav. Francoaceae (Saxifragaceae s.l.). 1 Chile (polymorphic): ***F. sonchifolia*** Cav. ('*F. ramosa*', bridal wreath, wedding fl.) – cult. orn. herb

Francoaceae A. Juss. (~ Saxifragaceae). Magnoliidae – Geraniales. Incl. Bersamaceae, Greyiaceae, Ledocarpaceae, Melianthaceae, Rhynchothecaceae, Vivianiaceae, 9/38 sub-Saharan Afr., S Am. Trees, shrubs s.t. subpachycaul or rhiz. to ann. herbs. Lvs pinnate to simple, in spirals or opp.; s.t. stip. Fls s.t. resupinate, in scapose infl. or thyrses or solit., K 4 or 5 induplicate-valvate, C(0, 3)4 or 5, free, s.t. clawed, A = & opp. C (*Rhynchocalyx*) or K (*Tetilla*), 8 (*Francoa*), or 5 + 5 with latrorse or longit. (*Rhynchocalyx*) dehiscence, nectary lobed, $\underline{G}$ (2–5) semi-inf. with axile to basal placentation with 1–many anatr. bitegmic ovules. Fr. a loculicidal or septicidal capsule; seeds large with yellow aril or small, exarillate. n = 26

Genera: *Balbisia, Bersama, Francoa, Greyia, Melianthus, Rhynchocalyx, Rhynchotheca, Tetilla, Viviania*

Francoeuria Cass. = *Pulicaria*

frangipani *Plumeria rubra;* **native f.** *Hymenosporum flavum*

Frangula Mill. (~ *Rhamnus*). Rhamnaceae. c. 50 N temp. Fls 5-merous, bisex. (cf. *R.*). Massart's Model. ***F. alnus*** Mill. (*R. frangula*, alder buckthorn, Euras., Medit., invasive N Am.) – excellent charcoal form. used for small-arms gunpowder (esp. slow fuses in WW II) & cult. for this, medic. as in bowel regulators; ***F. purshiana*** (DC.) Cooper (*R. p.*, cascara sagrada, W N Am.) – bark the source of comm. purgatives (anthraquinones), with *F. alnus* used since 1870 but not analyzed until 1975

Frankenia L. Frankeniaceae. c. 70 temp. (Eur. 6, Aus. 50, Am. 14 – R: SBM 17(1987)1) & subtrop. salty habitats. Herbs or subshrubs with wiry branches & usu. inrolled hairy lvs. Some medic., poisons or ashes a salt-source; some cult. rock-pls (sea heath esp. ***F. laevis*** L., Eur.). ***F. salina*** (Molina) I.M. Johnston (*F. grandifolia*, yerba reuma, SW N Am.) – shrub used in treatment of rheumatism

Frankeniaceae Desv. Magnoliidae – Caryophyllales. 1/70 mostly temp. & subtrop. saline habitats. Herbs or shrubs. Lvs opp., the prs united by a common sheath, simple & oft. ericoid with inrolled margins & oft. with salt-excreting glands; stip. 0. Fls axillary, solit. or cymose, bibracteolate, usu. bisex., reg.; K (4 – 7), basally connate with short induplicate-valvate lobes, C 4 – 7, imbr. with long claws, A 4 – 7(– 24 where inner whorl staminodes) usu. 3 + 3, ± basally connate, anthers with longit. slits, $\underline{G}$ ((1-)3(4)), 1-loc. with as many parietal placentas & 1 style with distinct stigmas, each placenta with (1)2 – 6(– ∞) anatropous, bitegmic ovules. Fr. a loculicidal capsule, encl. in persistent K; seeds with C straight embryo in abundant starchy endosperm. n = 10, 15

Genus: *Frankenia*

Sister to Tamaricaceae (also halophytic) poss. to come here but lvs opp.

frankincense *Boswellia* spp. esp. *B. sacra;* **f. fern** *Anemia caffrorum;* **f. pine** *Pinus taeda*

Franklandia R. Br. Proteaceae (IV). 2 SW Aus. R: FA 16(1995)316. ***F. fucifolia*** R. Br. – moth-poll.

Franklinia Bartram ex Marshall (~ *Gordonia*). Theaceae. 1 SE Georgia, USA (last seen in wild in 1803): ***F. alatamaha*** Marshall – cult. orn. decid. tree, can hybridize with *Gordonia lasianthus* (= × ***Gordlinia grandiflora*** Ranney & Fantz), *Schima* spp.

Franseria Cav. = *Ambrosia*

Frantzia Pittier (~ *Sechium*). Cucurbitaceae (XII). 5 C Am. ***F. tacaco*** (Pittier) Wunderlin – cult. veg.

Fraser fir *Abies fraseri*

Frasera Walter = *Swertia*

Fraunhofera Mart. (~ *Maytenus*). Celastraceae (I). 1 Braz.: ***F. multiflora*** Mart.

Fraxinus Tourn. ex L. Oleaceae (4a). 43 N temp. (Eur. 5) a few ext. to trop. R: Pfl. 4,243(1920)9, PSE 273(2008)33. Ash. Timber trees usu. with pinnate lvs (some domatia; ***F. excelsior*** f. ***diversifolia*** (Aiton) Lingels. with simple lvs) & 1-seeded winged samaras esp. ***F. excelsior*** L. (common or Eur. a., Eur., SW As.) – serial axillary buds, fls in short racemes before lvs with P 0, A 2, G 2, anemophilous, with male, female & bisex. fls in diff. combinations on diff. trees, elastic timber (French, Polish or Slavonian a.) form. for wheels & still for shooting-brake timbers (e.g. Morris Minor, also carbody frames of

Morgan cars), tool-handles & sports goods e.g. tennis racquets, polo mallets, billiard cues, hockey-sticks (handle = cane), cricket stumps & croquet-mallet handles, the toughest of Br. timbers (stools 1000 yrs old still sound, but threatened by ash dieback (ascomycete fungal) disease, *Hymenoscyphus fraxineus* (*Chalara f.*), ? orig. E As., in Poland 1992, UK 2012), ash-stick used to avert snakebite (Pliny, Norse sagas onwards), bark form. used for fevers, buds used as a slimming remedy in Glos., UK, lvs used for gout & to adulterate tea (C18 England) & as cattle-fodder (Scandinavia), sap form. used for earache, warts etc. (see yggadrasil), some cult. cvs incl grotesque 'weeping' '**Pendula**' grafted on wild form. Other spp. with similar uses & cult. orn. incl. ***F. americana*** L. (American, Canadian or white a., N Am., in US threatened by emerald ash borer, *Agrilus planipennis*) incl. '**Ascidia**' with lvs pitcher-shaped at base; ***F. angustifolia*** Vahl (*F. rotundifolia*, Medit.) – invasive Aus., cult. orn. esp. '**Raywood**' (claret a.); ***F. caroliniana*** Mill. (water a., E N Am.); ***F. chinensis*** Roxb. (China, Korea, Jap.) – source of 'Chinese insect white wax' extruded by insects feeding on pollarded trees (insects form. coll. from *Ligustrum lucidum* & transported by humans in 25 kg loads for over 300 km to trees, where each kg of insects gave up to 5 kg of wax in 6 mm layers on lvs after 100 days), used in Chinese candles, for coating pills & high-quality paper & for polishing jade, soapstone etc.; ***F. latifolia*** Benth. (Oregon a., W N Am.); ***F. nigra*** Marshall (Amer., black, brown, Can., alpine or white a., E N Am.), subsp. ***mandshurica*** (Rupr.) Sun (*F. mandshurica*, Jap. a., NE As.); ***F. ornus*** L. (flowering or manna a., S Eur., W As.) – white C in fragrant insect-poll. fls produced at same time as lvs, cult. in Sicily & Calabria for manna sugar or syrup exuding from branches when damaged by insects & is a mild laxative, mite-galls with organs intermediate between stems & lvs; ***F. pennsylvanica*** Marshall (*F. pubescens*, Amer., Can., green, red or white a., E N Am., natur. Buenos Aires) – imp. urban forestry in Can. prairies esp. to replace elms; ***F. quadrangulata*** Michaux (*F. tetragona*, blue a. [because of a blue dye from inner bark], N Am.); ***F. uhdei*** (Wenz.) Lingelsh. (trop. Am.) – weedy Hawaii

Freatulina Chřtek & Slavíková = *Drosera*

Fredolia (Bunge) Ulbr. = *Anabasis*

Freesia Ecklon ex Klatt. Iridaceae (VI 4). Incl. *Anomatheca*, 16 trop. & S Afr. esp. winter rainfall SW Cape. R: Strelitzia 27(2010)26. Tunicate corms & 2-ranked lvs usu. with fragrant (ionones) fls, the basis of imp. cut-flower industry (by 2001 215 ha greenhouses in Netherlands, yielding 500 M stems a year for Dutch market alone; 60 ha in Japan, 10 ha UK, US), of hybrids now with many named cvs incl. double ones (grown at high temps many revert to fertile 'singles') & esp. tetraploids e.g. '**Buttercup**' (1911); the ability to smell fl. varies & some find it difficult to detect any scent at all. Hybridization first 1878 nr Firenze between ***F. leichtlinii*** Klatt subsp. ***alba*** (G.L. Meyer) Manning & Goldbl. (*F. alba*, S Cape coast, fls white, v. strong scent) & subsp. ***leichtlinii*** (? *F. leichtliniana* Klatt. ex Rob., S Cape, fls greenish-yellow, v. fragrant, invasive Aus.), but in earnest since 1898 between *F. l.* & S Afr. ***F. corymbosa*** (Burm.f.) N.E. Br. (esp. forms with small pink fls, '*F. armstrongii*', usu. scentless), to give florists' freesias, ***F. × kewensis*** J. Wright bis (*F. × hybrida*, *F. × tubergenii*). ***F. laxa*** (Thunb.) Goldbl. & Manning (*A. cruenta*, *Lapeirousia l.*, trop. & S Afr.) – cult. orn., red fls

Fregea Reichb.f. = *Sobralia*

freijo *Cordia goeldiana*

Fremontia Torrey = *Fremontodendron*

Fremontodendron Cov. Malvaceae (Bomb.). 3 SW N Am. R: SB 16(1991)3. Cult. orn. ± everg. shrubs & trees with irritant brown hairs & petaloid K, C 0 & A united basally in a tube, esp. '**California Beauty**' (1953) but now largely '**California Glory**'; *F. californicum* (Torrey) Cov. (1 sold in 1850s for £37. 80!; source of cordage) × *F. mexicanum* Davidson, & allegedly hardier than both)

French bark *Pinus pinaster*; **F. bean** *Phaseolus vulgaris*; **F. berries** *Rhamnus saxatilis*; **F. cotton** *Calotropis procera*; **F. honeysuckle** *Sulla coronaria*; **F. jujube** *Ziziphus jujuba*; **F. lavender** *Lavandula stoechas*; **F. marigold** *Tagetes erecta*; **F. millet** *Panicum miliaceum*; **F. mulberry** *Callicarpa americana*; **F. rhubarb** *Angelica archangelica*; **F. ryegrass** *Arrhenatherum elatius*; **F. turpentine** *Pinus pinaster*; **F. whisk** *Chrysopogon gryllus*

Frerea Dalzell = *Caralluma* (but see R: Bradleya 8(1990)9)

Fresenia DC. = *Felicia*

Freya Badillo = *Sabazia* (but see Ernstia 33(1985)9)

Freycinetia Gaudich. Pandanaceae. c. 250 Sri Lanka to NZ (1) & Polynesia. R: Blumea 16(1968)361. Usu. lianes (Petit's Model), dioec. (some monoec. in Aus.) with brightly coloured fleshy bracts but fr. a syncarp of berries (drupes in *Pandanus*), some poll. bats attracted by fleshy bracts, e.g. ***F. baueriana*** Endl. (Norfolk Is., NZ) – bats extinct or rare in NZ & prob. poll. & disp. by introd. Aus. possum, ***F. arborea*** Gaudich. (Hawaii) by *Zosterops japonica* (white-eye, introd. 1929), the orig. bird-poll. (drepanids) now extinct; ***F. reineckei*** Warb. (Samoa) produces occ. bisex. spikes & fls. Prop-r. used for high-quality ropes; leaf-fibre for weaving, skirts etc.

Freyera Reichb. = *Geocaryum*

Freylinia Colla. Scrophulariaceae (Teed.). 8 trop. (1) & S Afr. Cult. orn. fragrant shrubs

Freyliniopsis Engl. = *Manuleopsis*

Freziera Willd. Pentaphylacaceae (3; Theaceae s.l.). Incl. *Patascoya*, 57 trop. Am. esp. mts. Dioec. with more females than males (unlike most trop. trees). ***F. forerorum*** A. Gentry (E Panamá) with v. asymmetric lvs

friar's balsam tincture largely derived from *Styrax* spp. (& Cape aloes); **f.'s cowl** *Arisarum vulgare*

Fridericia Mart. (~ *Arrabidaea*). Bignoniaceae (3). 67 trop. Am. R. AMBG 98(2014)431. ***F. chica*** (Bonpl.) L. Lohmannn (*A. c.*, S Am.) – cultigen, form. body cosmetic (red) from lvs, dye-pl. (Peru) for (red) barkcloth & to blacken teeth, local med.; ***F. dichotoma*** (Jacq.) L. Lohmann (*A. rotundata*, Braz.) – cult. orn.

Friedrichkarlmeyeria T. Ali & Thines (~ *Thlaspi*). Cruciferae (Colut.). 1 S Caspian: ***F. umbellata*** (DC.) T. Ali & Thines. R: T 65(2016)93

Friesodielsia Steenis. Annonaceae (III 7). 55 trop. W Afr. (7), Indomal. ? Heterogeneous. Troll's Model; some lianes

frijoles *Phaseolus vulgaris*

fringe flower or **tree** *Chionanthus* spp. esp. *C. virginicus*; **f.-myrtle** *Calytrix* spp. esp. *C. tetragona*; **f.d lily** *Thysanotus* spp.

frisé *Cichorium endivia* Frisé Group

Frithia N.E. Br. Aizoaceae (V). 2 Pretoria area, S Afr. R: Both. 30(2000)1. Stemless succ. herbs with club-shaped lvs with apical windows through which light passes to photosynthetic tissues within, cult. orn.

Fritillaria Tourn. ex L. Liliaceae. c. 100 W Eur. & Medit. (Eur. 24) to E As. (Jap. 8, 7 endemic, + 1 natur. – R: BM 22(2005)192), N Am. (20). R: HIP 39(1980)265. Bulbous with many alks & 1 or more fleshy scales & s.t. rice-grain-like bulblets; seeds many, flat. Many cult. orn. & few ed. etc. ***F. assyriaca*** Bak. (W As.) – highest known amount of chromatin per nucleus of any cell; ***F. camschatcensis*** (L.) Ker ('black sarana', E As., coastal NW Am.) – fls purple-black, bulbs ed. dried & cooked; ***F. cirrhosa*** D. Don (Himal.) – lvs with tendril-like tips (cf. *Gloriosa*), bulb pounded with orange juice & sugar for chest ailments (pei mou); ***F. imperialis*** L. (crown imperial, Iran to N Ind.) – fls in whorl below term. tuft of bracts, whole pl. with fox or skunk smell, a favourite of the Dutch Masters, once medic. & starch source, ed. cooked (imperialine a heart poison); ***F. lanceolata*** Pursh (N Am.) – imp. Native Amer. food-pl.; ***F. meleagris*** L. (snake's head, guinea fl., Eur., W As., natur. GB where prot. though not recorded before 1732) – P chequered shades of red & purple (Lat.: *fritillus* = dicebox, ds form. chequered thus), form. called leper-lily as fl. resembles bells lepers were obliged to carry, cult. orn. since Tudor times, seeds water-disp.; ***F. persica*** L. (Cyprus to Iran) – bulb with single massive scale invested with remnants of older ones; ***F. roylei*** Hook. (Himal.) – bulb medic. in China (pei-mu) used like *F. cirrhosa*; ***F. thunbergii*** Miq. (? W China) – long grown for med. China, Jap., where natur.

fritillary *Fritillaria* spp.

Fritzschia Cham. Melastomataceae. 3 Braz.

Froelichia Moench. Amaranthaceae (II 2). 16 warm Am., Galápagos (weedy in Jap., Aus., W Afr., N Am.). Lat. fls sterile & developing wings serving to aid disp.

Froelichiella R.E. Fries. Amaranthaceae (II 2). 1 Braz.: ***F. grisea*** R.E. Fries

Froesia Pires. Ochnaceae (Quiinaceae). 6 trop. S Am. R: Novon 4(1994)246

Froesiochloa G.A. Black. Gramineae (VI 3). 1 trop. S Am.: ***F. boutelouoides*** G.A. Black

Froesiodendron R.E. Fries (~ *Cardiopetalum*). Annonaceae (III 1). 3 Amaz. R: Britt. 47(1995)267

frog-bit *Hydrocharis morsus-ranae*; **American f.-b.** *Limnobium spongia*; **f. orchid** *Dactylorhiza viridis*

Frolovia (DC.) Lipsch. = *Dolomiaea* (but see Willd. 33(2003)391)

Frommia H. Wolff. Umbelliferae (III 8). 1 S C Afr.: ***F. ceratophylloides*** H. Wolff

Frondaria Luer (~ *Pleurothallis*). Orchidaceae (V 13c). 1 C Colombia to C Bolivia: ***F. caulescens*** (Lindl.) Luer

Froriepia K. Koch. Umbelliferae (III 8). 2 Turkey to Iran

frost flowers *Aster* spp.; **f. grape** *Vitis vulpina;* **f.weed** *Crocanthemum canadense*

Fructicicola (Schltr.) M.A. Clem. & D.L. Jones = *Bulbophyllum*

frumenty *Triticum aestivum*

Fryxellia D. Bates (~ *Anoda*). Malvaceae (Malv.-Malv.). 1 SW N Am.: ***F. pygmaea*** (Correll) D. Bates. R: Britt. 26(1974)95

Fuchsia Plum. ex L. Onagraceae. 106 C & S Am., Tahiti (1), NZ (3; AMBG 82(1985)501). R: PCAS IV, 25(1943)1, UCPB 53(1969)38, AMBG 69(1982)1, 209, 72(1984)222, 76(1989)546, 82(1995)501. Shrubs (some with tubers) & trees with opp. or whorled lvs & usu. pend. bird-poll. fleshy fls & ed. berries; closest ally, *Circaea*. A few locally medic. but many cult. orn. esp. hybrids grown in hanging baskets & as standards e.g. ***F. × hybrida*** Hort. ex Sieb. & Voss (prob. involving *F. magellanica* & *F. fulgens* DC. (Mex.) but *F. coccinea* Dryander (S Braz.), *F. microphylla* Kunth & *F. thymifolia* Kunth (both Mex.) prob. in ancestry of modern hybrids). ***F. excorticata*** (Forst. & Forst. f.) L.f. (NZ) – decid. gynodioec. tree a target for introd. possums; ***F. magellanica*** Lam. (S Chile, Arg.) – forms (e.g. **'Riccartonii'**) grown for hedging in Azores, Isle of Man & Ireland, where ± natur., invasive in Aus., Masc., Hawaii; ***F. procumbens*** R. Cunn. ex A. Cunn. (NZ) – prostrate, fls erect, trioecious

fuchsia, Californian *Epilobium* spp., *Ribes speciosum*

Fuernrohria K. Koch. Umbelliferae (III 4). 1 Cauc., Armenia: ***F. setifolia*** K. Koch

Fuerstia T.C.E. Fries. Labiatae (VII 3d). 8 trop. Afr.

Fuertesia Urb. Loasaceae (III). 1 Hispaniola: ***F. dominguensis*** Urb.

Fuertesiella Schltr. Orchidaceae (IV 2b). 1 mts of E Cuba & Dominican Republic: ***F. pterichoides*** Schltr. R: Gen. Orch. 3(2003)36

Fuertesimalva Fryxell. Malvaceae (Malv.-Malv.) 14 Mex. (2), Andes. R: Sida 17(1996)70

fufu (W Afr.) mashed or pounded starchy food, orig. *Dioscorea* spp.

Fuirena Rottb. Cyperaceae (II 2). 30 warm (Eur. 1, N Am. 7)

Fuji cherry *Prunus incisa;* **F. fruit** *Diospyros* spp.

fuki *Petasites japonicus*

Fukien tea *Ehretia microphylla*

Fulcaldea Poir. Compositae (Barn.). 1 NE Braz., 1 4000 km away: ***F. laurifolia*** (Bonpl.) Poir. (N Peru, Ecuador) – wood for constr. R: T 60(2011)1099

fuller's herb *Saponaria officinalis;* **f. teasel** *Dipsacus sativus*

Fumana (Dunal) Spach. Cistaceae. c. 9 Eur., N Afr. Cult. orn. like *Helianthemum* but outer A sterile

Fumaria Tourn. ex L. Papaveraceae (Fumariaceae I 2). 50 Eur. (39), Medit. to C As. & Himal. (1), trop. E Afr. highlands (1). R: OB 88(1986)42. Fumitory. Usu. autogamous annuals, mostly polyploid, with fls like *Corydalis;* many scramble with sensitive petiolules; caruncles; alks. ***F. officinalis*** L. (Eur., Medit. to Iran) – form. yellow dye source

Fumariaceae Marquis = Papaveraceae

Fumariola Korsh. Papaveraceae (Fumariaceae I 2). 1 C As.: ***F. turkestanica*** Korsh. R: OB 88(1986)42

fumitory *Fumaria* spp.

Funastrum Fourn. (~ *Cynanchum*). Apocynaceae (Vc). Am. spp. of '*Sarcostemma*' (true *S.* = *C.*)

fundi *Digitaria exilis*

Funifera Andrews ex C. Meyer. Thymelaeaceae (Thym.-Daph.). 3–4 Braz.

Funkia Spreng. = *Hosta*

Funkiaceae Horan. = Asparagaceae

Funkiella Schltr. Orchidaceae (IV 2h). c. 25 Texas, Mex., Guatemala, Costa Rica mts. R: FFG 35(1991)19, suppl. 2(1993)230

Funtumia Stapf. Apocynaceae (IIb). 2 trop. Afr. R: MLW 81 – 16(1981)10. Prévost's Model. Alks. Rubber sources esp. ***F. elastica*** (Preuss) Stapf (bastard wild, Lagos, Iré or silk r.), comm. valuable tree in regrowth of forest after felling timber trees

Furarium Rizz. = *Passovia*

Furcaria (DC.) Kostel. = *Hibiscus*

Furcraea Vent. Asparagaceae (Agavaceae). 21 trop. Am. R: U. Eggli, *Ill. Handb. Succ. Pls, Monocots* (2001)78. Pachycaul succ. (Holttum's Model) ± stem with large lvs in rosettes & term. infls even larger than in *Agave;* usu. hapaxanthic with bulblets in infls; some imp. fibres (fique) & cult. orn. esp.: ***F. foetida*** (L.) Haw. (*F. gigantea*, Mauritius hemp, N S Am.) – comm. cult. in Mauritius, St Helena etc. (firebreak in Sri Lanka) & ***F. hexapetala*** (Jacq.) Urb. (*F. cubensis*, ? poss. not distinct from *F. foetida*, Cuba hemp, Cuba, Haiti) – both for twine, cordage, sacking etc. Others less used incl. ***F. cabuya*** Trel. (cabuya, C Am.) – ropes, hammocks etc.

Furtadoa Hotta (~ *Homalomena*). Araceae (VII 6). 2 Sumatra, Malay Pen. ***F. sumatrensis*** Hotta – rheophyte

furze *Ulex* spp.

Fusaea (Baill.) Saff. Annonaceae (III 3). 2 trop. S Am. R: Britt. 51(1999)195. Petit's Model

Fusifolium Raf. = *Drimia*

Fusispermum Cuatrec. Violaceae (I). 3 Panamá, Colombia, Peru. R: PKNAW C87(1984)121

fustian (from Fostat, Cairo) orig. cotton-flax coarse fabric, later a twilled cotton; **gene** [i.e Genoa] **f.** – origin of word 'jeans'

fustic *Maclura tinctoria;* **young f.** *Cotinus coggygria*

futi, futui *Jacaranda copaia*

fuyo fruit *Diospyros* spp.

G

Gabo(o)n chocolate *Irvingia gabonensis;* **G. mahogany** *Aucoumea klaineana*

Gabonius Mackinder & Wieringa (~ *Hymenostegia*). Leguminosae (I 2). 1 Gabon: ***G. ngouniensis*** (Pellegrin) Mackinder & Wieringa. R: Phytotaxa 142(2013)15

Gabunia K. Schum. ex Stapf = *Tabernaemontana*

Gadellia Shulkina = *Campanula*

Gaertnera Lam. Rubiaceae (IV 7). 85 OW trop. R: AMBG 96(2009)592, 99(2014)691

Gaertnera Schreb. = *Hiptage*

Gaga Pryer & al. (~ *Cheilanthes*). Pteridaceae. 19 SW US to Bolivia. R: SB 37(2012)855

gage See *Prunus*

Gagea Salisb. Liliaceae. Incl. *Lloydia*, c. 100 [75–275] temp. Euras. (Eur. 24), W N Am. R: Phytotaxa 15(2011)51. Axillary bulbils develop if poll. fails. 3 in Br. Is. (***G. lutea*** (L.) Ker-Gawler (yellow star-of-Bethlehem, Eur. to Himal., where major food of Himal. snow-cock), ***G. serotina*** (L.) Ker-Gawl. (*L. s.*, Snowdon lily, range of genus, prot. GB, cult. orn.) & ***G. bohemica*** (Zauschner) Schultes & Schultes f. (C & S Eur., SW As., NW Afr.), a relict pop. discovered in Welsh mts)

Gagnebina Necker ex DC. Leguminosae (II 1). 8 Madag., 2 ext. to W Ind. Ocean. R: KB 41(1986)463

Gagnepainia K. Schum. Zingiberaceae (II 2). 3 SE As.

Gagria Král = *Pachyphragma*

gaharu (wood) *Aquilaria* spp., *Gyrinops ledermannii*

Gahnia Forst. & Forst. f. Cyperaceae (II 7). c. 30 E As., Mal., Aus. (22, 20 endemic), Pac. R: Bot. Arch. 40(1940)151. Some woody, ***G. sieberiana*** Kunth (NE Aus.) pachycaul; fr. exposed by attachment to marcescent A bases thrust outwards, their tips being retained by surrounding bracts, some nutlets red, bird-disp.

gai laan *Brassica oleracea* Alboglabra Group

Gaiadendron G. Don. Loranthaceae (2). 2 trop. Am mts. ***G. punctatum*** (Ruíz & Pavón) G. Don – terr., tree-like in S Am., epiphytic in C Am., parasitic on r. (cf. *Nuytsia*)

Gaillardia Foug. Compositae (Hele.-Gai.). 20 N & temp. S (2) Am. R: Res. Stud. Coll. Wash. 12(1944)195. Some local medic. & cult. orn. (blanket flowers, Ind. b.) esp. perenn. ***G. × grandiflora*** Van Houtte (*G. aristata* Pursh (N Am.) × *G. pulchella* Foug. (N Am.), natur. W N Am.) & ann. ***G. pulchella***

Gaillonia A. Rich. ex DC. = *Plocama* (but see NJB 18(1998)32)

Gaimardia Gaud. Restionaceae (Centrolepidaceae). 4 NG, Tasmania, NZ, Antarctic S Am. (1)

Galactodenia Sundue & Labiak (~ *Terpsichore*). Polypodiaceae (V). 5 trop. Am. R: SB 37(2012)340

Galactia P. Br. Leguminosae (III 18). c. 55 warm esp. Am. (Aus. 3). Latex (rare in L.)

Galactites Moench. Compositae (Card.-Card.). 2 Canary Is. & Medit. (Eur. 2)

Galactophora Woodson. Apocynaceae (II d). 6 trop. S Am. R: Sida 21(2005)2058

Galagania Lipsky = *Muretia*

galanga *Kaempferia* spp., *Alpinia* spp.

galangal *Alpinia galanga, A. officinarum*

Galantharum Boyce & S.Y. Wong. Araceae. 1 Kalimantan: ***G. kishii*** Boyce & S.Y. Wong. R: Aroideana 38E,2(2015)25

Galanthus L. Amaryllidaceae. 20 Eur. (4) to Iran. R: A.P. Davis (1999) *The genus G.*, M. Bishop & al. (2006) *Snowdrops*. Snowdrops (an allusion to mediaeval earring fashions, not weather). Alks; lectins effective insecticides. P 3 + 3 shorter (cf. *Acis* & *Leucojum* where all same length), inner with nectaries, visited by bees which touch stigma when grasping pend. fl. & are showered with pollen when probing for nectar. App. orig. Cauc. twice disp. to Medit. & Middle E, thrice to Balkans. Many cult. orn. (30 M exported annually from Turkey, ***G. elwesii*** Hook. f. (Balkans, S Ukraine, Turkey) much depleted along Med. coast, yet can be readily prop. by 'chipping' from bulb fragments), & esp. ***G. nivalis*** L. (Eur., natur. GB), **'Scharlokii'** with 2 fls per peduncle, others with 'double' (P 26[-32], G 0) fls, poss. orig. 'moly' increasing acetylcholine levels, known in 'wild' in GB since 1731, some commanding high prices (cf. *Tulipa*), 1 bulb of '**Flocon de Neige**' £265 in 2008; ***G. reginae-olgae*** Orph. (*G. corcyrensis*, *G. nivalis* subsp. *r.*, SE Eur.) – flowering autumn, lvs revolute; ***G. woronowii*** Losink. (Caucasus) – orig. source of galanthamine (acetylcholine esterase inhibitor), postponing Alzheimer's disease

Galatella Cass. (~ *Aster*). Compositae (Ast.-Bell.). Incl. *Linosyris*, excl. *Crintiaria*, c. 30 Euras.

Galax Sims. Diapensiaceae. 1 SE US: ***G. urceolata*** (Poir.) Brummitt ('*G. aphylla*', beetle-weed) – cult. ground-cover. R: T 21(1972)309

Galaxia Thunb. = *Moraea*

galba *Calophyllum brasiliense*

galbanum *Ferula gummosa, F. rubricaulis*

Galbulimima Bailey. Himantandraceae. 2 E Mal., NE Aus. Alks; fr. taken by fr.-pigeons. ***G. belgraveana*** (F. Muell.) Sprague (NG) when taken with lvs of *Homalomena* spp. leads to violent intoxication & deep sleep with hallucinations

Gale Duhamel = *Myrica*

gale, sweet *Myrica gale*

Galeana Llave. Compositae (Per.-Gal.). 1 Mex., C Am.: ***G. pratensis*** (Kunth) Rydb.

Galeandra Lindl. Orchidaceae (V 12c). c. 18 C & S Am. R: GC III, 12(1892)430. Cult. orn. terr. & epiphytic incl. ***G. devoniana*** Lindl. – flood-tolerant epiphyte in igapó of Amaz.

Galearia Zoll. & Moritzi. Pandaceae (Euphorbiaceae s.l.). 5 SE As. to Solomon Is. Local medic. R: FM I,20(2011)18

Galearis Raf. Orchidaceae (IV 4d). Incl. *Amerorchis*, c. 10 Himal., E As. (China 8); 1 temp. N Am., Greenland

Galeatella (F. Wimmer) Degener & I. Degener = *Lobelia*

Galega Tourn. ex L. Leguminosae (III 24). 6 Euras. (Eur. 1), E Afr. mts. Alks. ***G. officinalis*** L. (goat's rue, Eur. to Iran) – fodder (also ***G. orientalis*** Lam., Cauc.) & cult. orn., form. medic., stem-fibre used for paper

Galenia L. Aizoaceae (I). 29 S & S trop. Afr. R: H. E. K. Hartmann, *Ill. Handb. Succ. Pls*, Aiz. F–Z(2002)28. Heterogeneous?

Galeobdolon Adans. = *Lamium*

Galeoglossum A. Rich. & Galeotti (~ *Cranichis*). Orchidaceae (IV 2b). 3 Mex. R: SB 36(2011)266

Galeola Lour. Orchidaceae (II 2). 6 Indomal. Lianoid mycotrophs to several m. ***G. cathcartii*** Hook.f. (NE Ind. – Thailand) – to 9m, seeds winged

Galeomma Rauschert (*Eriosphaera*). Compositae (Gnap.-Gnap.). 2 Cape. R: OB 104(1991)135

Galeopsis L. Labiatae (VI). 11 temp. Euras. (Eur. 9). Hemp nettles. ***G. tetrahit*** L. (2n = 32, Eur.) = *G. speciosa* Mill. (2n = 16, Eur.) × *G. pubescens* Besser (2n = 16, Eur.), first allotetraploid to be resynthesized

Galeottia A. Rich. (*Mendoncella*). Orchidaceae (V 12j). 12 trop. Am. R: Lindleyana 3(1988)221. Cult. orn.

Galeottiella Schltr. (~ *Brachystele*). Orchidaceae (IV 2c). 6 trop. Am. R: Gen. Orch. 3(2003)59

Galianthe Griseb. = *Spermacoce*

galingale *Cyperus longus*

Galiniera Del. Rubiaceae (II 5). 2 trop. Afr., Madag.

Galinsoga Ruíz & Pavón. Compositae (Mill.-Gal.). 15 temp. & subtrop. C & S Am. R: Rhodora 79(1977)319. ***G. parviflora*** Cav. & ***G. quadriradiata*** Ruíz & Pavón (*G. ciliata*), diploid & polyploid resp., weeds now almost cosmop., cooked as veg. in SE As.

galip nut *Canarium indicum*

Galipea Aubl. Rutaceae (I 8). 15 trop. Am. Alks., some medic.; see also *Angostura*

Galitzkya Botsch. (~ *Alyssum*). Cruciferae (2). 3 C As to W China & Mongolia

Galium L. Rubiaceae (IV 16). c. 600 cosmop. (Eur. 145, Turkey 101, China 63). Incl. *Relbunium* (Endl.) Hook.f. (R: BJ 76(1955)516) with involucrate fls. (r. sources of dyes in Peru). Heterogeneous? Slender herbs with usu. square stems & whorled lvs & leaflike 'stip.' (bedstraws), some with coumarin when dried (allegedly not in living pls) & used to keep with linen or as mattress-stuffing, esp. ***G. odoratum*** (L.) Scop. (*Asperula o.*, woodruff, Euras., Medit.) – flavouring, a green syrup, for drinks esp. fr.-cup & beer (Ger.), incl. 'witte' beer (Netherlands, Oregon), snuff etc., locally medic., cult. orn.; ***G. album*** Mill. ('*G. mollugo*', hedge b., Eur., natur. N Am.) – occ. cult.; ***G. aparine*** L. (goosegrass, kisses, cleavers, sticky bobs, sticky willy, stickyweed (Aus., where invasive), sweethearts, Euras., S S Am.) –scrambler with reflexed hooks on stems & schizocarps which are animal-disp. & form. fed to poultry & used as coffee subs. (Ireland), lvs used in treatment of tonsilitis, adenoids & lymph system problems; ***G. circaezans*** Michaux (N Am.) – myrmechory; ***G. tricornutum*** Dandy (Eur.) – fr. form. used to cover pinheads so as to prot. seamstresses' fingers; ***G. verum*** L. (lady's b., Eur. to Iran, natur. N Am.) – yellow fls smelling of urine, attractive to flies etc., form. used as styptic & to curdle milk in cheese-making in Eng., medic., r. yield a red dye, dried pl. placed in drawers to deter moths

gall *Quercus pubescens*, (Bible) poss. *Conium maculatum* but see *Citrullus*; **takout, tamarix** or **teggaout g.** *Tamarix* spp.

Gallardoa Hicken. Malpighiaceae. 1 Argentina: ***G. fischeri*** Hicken

gallberry *Ilex glabra*

Gallesia Casar. (~ *Seguieria*). Petiveriaceae (Phytolaccaceae II). 1 Peru, Braz.: ***G. integrifolia*** (Spreng.) Harms – smells of garlic, fr. a samara; locally medic. for worms etc.

Gallienia Dubard & Dop (~ *Fernelia*). Rubiaceae (II 5). 1 Madag.: ***G. sclerophylla*** Dubard & Dop

Galopina Thunb. Rubiaceae (IV 13). 4 SW Cape to S Malawi

Galphimia Cav. Malpighiaceae. c. 20 Texas to Arg. K without oil glands but large A suggesting pollen reward for pollinators. Cult. orn. shrubs esp. ***G. gracilis*** Bartling (Mex.)

Galpinia N.E. Br. Lythraceae. 1 S Afr.: ***G. transvaalica*** N.E. Br. R: Strelitzia 10(2000)345

Galtonia Decne = *Ornithogalum*

Galvezia Dombey ex Juss. Plantaginaceae (Ant.-Ant; Scrophulariaceae s.l.). 4 coastal Peru, Ecuador, Galápagos. R: D.A. Sutton, *Rev. Antirr.* (1988)514

gama grass *Tripsacum dactyloides*

gamalu *Pterocarpus marsupium*

Gamanthera van der Werff. Lauraceae (I). 1 Costa Rica: ***G. herrerae*** van der Werff

Gamanthus Bunge = *Halimocnemis*

gamar *Gmelina arborea*

gamba grass *Andropogon gayanus*

Gambelia Nutt. (~*Antirrhinum*). Plantaginaceae (Ant.; Scrophulariaceae s.l.). 4 SW N Am. R: D.A. Sutton, *Rev. Antirr.* (1988)510. Shrubs, some juncoid; ***G. speciosa*** Nutt. – poll. hummingbirds

Gambeya Pierre = *Chrysophyllum*

Gambeyobotrys Aubrév. = *Chrysophyllum*

Gambia pods *Vachellia nilotica*

gambier or **gambir** *Uncaria gambier*

Gamblea C.B. Clarke. Araliaceae. Incl. *Evodiopanax*, 4 E Himal. to Jap. & W Mal. R: Adans. III, 22(2000)47

gamboge *Garcinia* spp. esp. *G. xanthochymus*

Gamocarpha DC. Calyceraceae. 6 temp. S Am.

Gamochaeta Wedd. (~ *Gnaphalium*). Compositae (Gnap.-Gnap.). c. 50 trop. & warm Am. Many weedy OW (cudweeds cf. *Gnaphalium*), e.g. ***G. americana*** (Mill.) Wedd., ***G. calviceps*** (Fern.) Cabrera in E Aus., NZ, S Afr., but esp. ***G. pensylvanica*** (Willd.) Cabrera (? SE N Am.) in garden crops S & SE As. to E Aus.

Gamochaetopsis Anderb. & Freire (~ *Lucilia*). Compositae (Gnap.-Gnap.). 1 subAntarctic S Andes: ***G. alpina*** (Poepp. & Endl.) Anderb. & Freire. R: BJLS 106(1992)186

Gamolepis Less. = *Steirodiscus*

Gamopoda Bak. = *Rhaptonema*

Gamosepalum Hausskn. = *Alyssum*

Gamosepalum Schltr. = *Aulosepalum*

Gamotopea Bremek. Rubiaceae (IV 7). 5 trop. S Am.

gampi *Wikstroemia sikokiana*

Gamwellia Bak. f. = *Gleditsia*

gandaria *Bouea macrophylla*
gan cao *Glycyrrhiza uralensis*
gandhini *Trachyspermum roxburghianum*
gang-flower *Polygala vulgaris*
Ganguelia Robbrecht (~ *Oxyanthus*). Rubiaceae (II 1). 1 Angola: ***G. gossweileri*** (S. Moore) Robbrecht – pyrophytic geofrutex. R: SAJB 62(1996)21
ganja *Cannabis sativa* 'subsp. *indica*'
Ganophyllum Blume. Sapindaceae (III 2). 1 trop. W Afr. (***G. giganteum*** (A. Chev.) Hauman – timber & ed. fr.), 1 Andamans, Vietnam, Malay Pen., Sumatra, Java, Philipp., NG (***G. falcatum*** Blume – timber, seeds yield an oil)
Gantelbua Bremek. = *Hemigraphis*
Ganua Pierre ex Dubard = *Madhuca*
Gaoligongshania D.Z. Li & al. (~ *Indocalamus*). Gramineae (IV). 1 Yunnan: ***G. megalothyrsa*** (Hand.-Mazz.) D.Z. Li & al. R: APS 33(1995)598
Garaventia Looser = *Tristagma*
Garaya Szlach. = *Mesadenella*
Garayanthus Szlach. = *Cleisostoma*
Garayella Brieger = *Chamelophyton*
garbanzo beans *Cicer arietinum*
Garberia A. Gray. Compositae (Eup.-Liat.). 1 SE US: ***G. heterophylla*** (Bartram) Merr. & F. Harper – shrub. R: FNA 21(2006)538
Garcia Vahl ex Rohr. Euphorbiaceae (Al.-Gar.). 2 Mex. (cult. trop. Am.). Yield an oil like tung oil; fls consp. (K (2 or 3), C 8–12 pinkish, A 60+), cult. orn.
Garciadelia Jestrow & Jiménez (~ *Adelia*). Euphorbiaceae (Acal.-Adel.). 4 Hispaniola. R: T 59(2010)1809
Garcibarrigoa Cuatrec. (~ *Pseudogynoxys*). Compositae (Sen.-Sen.). 2 Ecuador, Colombia
Garcilassa Poeppig. Compositae (Helia.-Helia.). 1 Costa Rica to Peru & Bolivia: ***G. rivularis*** Poepp. & Endl.
Garcinia L. Guttiferae (2.). Incl. *Rheedia*, *Tripetalum*, c. 260 trop. esp. As. (Am. c. 10), S Afr. Polygamous usu. slow-growing trees & shrubs (Attims's Model; some rheophytic) with usu. nocturnal highly scented fls, & berries with (oft. ed.) fleshy endocarp around seeds, some parthenocarpic; some wild spp. & also *G. mangostana* in cult. apomictic (adventive embryony); resins give pigments incl. gamboge, some waxes & timbers, some bark & seeds local medic. (W Afr.). ***G. atroviridis*** Griff. ex T. Anderson (NE Ind.) – water extracts cause wt. loss in rats due to hydroxycitric acid, an inhibitor of fatty acid & cholesterol synthesis; ***G. brasiliensis*** Mart. (*Rheedia b.*, bakupari, Braz.) – fr. sold in markets; ***G. cowa*** Roxb. ex DC. (Indomal.) – endocarp & pericarp good flavour; ***G. dulcis*** (Roxb.) Kurz (mundu, Mal.) – fr. tree; ***G. gummi-gutta*** (L.) N. Robson (*G. cambogia*, *G. gutta*, brindleberry, Ind.) – pink deeply lobed endocarp, pericarp (goraka) dried for fish curry, with guaraná used as appetite suppressant; ***G. humilis*** (Vahl) C. Adams (achacha, Bolivia) – comm. fr. (largest plantation in Queensland); ***G. indica*** (Thouars) Choisy (Ind.) – ed. fat (kokam, kokum or Goa butter) from seeds, pericarp used in flavouring curries; ***G. kola*** Heckel (W & C Afr.) – chewing-sticks; ***G. mangostana*** L. (allopolyploid female (2n = 88–90) = *G. hombroniana* Pierre (2n = 48) × *G. malaccensis* Hook.f. (2n = ?42), mangosteen, Mal.) – delicious endocarp, one of the best trop. frs only productive there, now promoted for medic. as high in anti-oxidants & xanthones, latex used in pimple cream; ***G. mannii*** Oliv. (W Afr.) – antibacterial chewing-stick in Cameroun; ***G. pedunculata*** Roxb. ex Buch.-Ham. (Bengal) – large ed. fr.; ***G. sopsopia*** (Buch.-Ham.) Mabb. (*G. paniculata*, E Himal.) – fr. cherry-sized, mangosteen-flavoured; ***G. xanthochymus*** Hook.f. ('*G. pictoria*', [anc.] tamal, N Ind.) & other spp. when tapped yield gamboge used in watercolours & dyeing, e.g. Buddhist priests' robes
gardener's garters *Phalaris arundinacea* var. *picta*
Gardenia Ellis. Rubiaceae (II 1). c. 140 trop. & warm OW. Shrubs & trees (Corner's & Scarrone's Models) with opp. or whorled lvs (3s, app. 2 prs with 1 reduced to a scale) & usu. large scented (like tuberoses) white or yellow fls. Pollen deposited on sides of stigma before anthesis, stigmas spreading on second or third day. Form. much cult. under glass for button-holes, esp. ***G. augusta*** (L.) Merr. (*G. florida*, *G. jasminoides*, Cape jasmine, China) esp. double-flowered forms, used to scent tea etc., fr. source of a yellow dye (for daikon colouring & yellowing fr. juice), used in treatment of influenza & colds in modern Chinese herbalism; ***G. tannaensis*** Guillaumin (Vanuatu) – much cult. Samoa for fls;

G. volkensii K. Schum. (E & S Afr.) – long-lived large grey fr. Other spp. medic. & app. insecticidal, minor sources of timbers & dyes

Gardeniopsis Miq. Rubiaceae (?II 4). 1 W Mal.: ***G. longifolia*** Miq.

Gardneria Wall. Loganiaceae (Strychnaceae). 5 Ind. & C Jap. to Java. R: BJBB 32(1962)431. Alks

Gardnerina R. King & H. Robinson. Compositae (Eup.-Ager.). 1 Braz.: ***G. angustata*** (Gardner) R. King & H. Robinson – known only from 1 specimen). R: MSBMBG 22(1987)157

Gardnerodoxa Sandw. = *Neojobertia*

Gardoquia Ruíz & Pavón = *Micromeria*

Garhadiolus Jaub. & Spach (~ *Rhagadiolus*). Compositae (Cich.-Cich.-Crep.). 2 S W As. to China

gari *Manihot esculenta*

Garidella Tourn. ex L. (~ *Nigella*). Ranunculaceae (I 3). 2 S Eur. (2) to C As. C longer than K unlike in *N.*

garlic *Allium sativum*; **black g.** *A. sativum*; **Canadian g.** *A. canadense*; **g. chives** *A. tuberosum*; **crow** or **false g.** *A. vineale*; **elephant g.** *A. ampeloprasum* Ampeloprasum Group; **field g.** *A. oleraceum*; **hedge g.** *Alliaria petiolata*; **g. pear** *Crateva* spp.; **society g.** *Tulbaghia* spp.; **g. weed** *Petiveria alliacea*; **wild** or **wood g.** *A. ursinum*

garnet berry *Ribes rubrum*

Garnieria Brongn. & Gris (~ *Persoonia*). Proteaceae (II 2). 1 New Caled.: ***G. spathulifolia*** (Brongn. & Gris) Brongn. & Gris

Garnotia Brongn. Gramineae (XXII 1). 29 S & E As. to Pac. R: KB 27(1972)515

Garnotiella Stapf = *Asthenochloa*

Garrettia Fletcher. Labiatae (Peronemat.). 1 Yunnan, Thailand, Java: ***G. siamensis*** Fletcher. R: KB 1937: 71

Garrya Douglas ex Lindl. Garryaceae. 15 Washington State to Panamá, WI (1). R: CGH 209(1978). Male fls with vestigial nectary-disk; females with P0 or 2(4) appendages. Cult. orn. shrubs esp. ***G. elliptica*** Douglas ex Lindl. (Calif. to Oregon), form. medic. (feverbush), bark with at least 5 alks incl. delphinine otherwise known only from *Aconitum* & *Delphinium*; ***G. wrightii*** Torr. (SW N Am.) – rubber source

Garryaceae Lindl. Magnoliidae – Garryales. Incl. Aucubaceae 2/23 Himal. to Jap., W N & C Am. Dioec. everg. trees & shrubs usu. with highly toxic alks. Lvs decussate though traces arising at diff. levels, simple. ± connate basally; stip. 0. Fls small, wind-poll., 1–3 in axils of decussate bracts in usu. term. infls; males with P 4, apically connate, bract-like, A 4 alt. with P, anthers with longit. slits, females with P 0 or 2 appendages nr. styles, $\overline{G}$ (2(3)), 1-loc. with distinct styles & 1 pend., anatropous, unitegmic ovule per carpel. Fr. a 2-seeded berry dry & tardily dehiscent at maturity; seed 1(2) with small linr. embryo in copious oily endosperm with reserves of hemicellulose. n = 8 (*Aucuba*), 11 (*Garrya*)
Genera: *Aucuba*, *Garrya*
Aucuba form. in Cornaceae s.l. but pollen diff. & petroselinic acid as major fatty acid in seeds otherwise unknown in Cornales suggested placement confirmed by DNA
Cult. orn. & medic. locally

Garuga Roxb. Burseraceae (III). 4 Himal., Indomal. to W Pac. R: Blumea 7(1953)459, 498. Some locally used timber

Garuleum Cass. Compositae (Cal.). 8 S Afr. Pappus 0

Gasoul Adans. = *Mesembryanthemum*

gasparillo *Esenbeckia* spp.

Gasteranthus Benth. (~ *Besleria*). Gesneriaceae (II 3a). 38 trop. Am. (7 extinct, 10 at risk). R: SBM 59(2000)29. One group poll. by hummingbirds, others by bees

× **Gasteraloe** Guillaumin. Asphodelaceae. *Aloe* × *Gasteria* hybrids. Comm. potpls incl. × ***G. beguinii*** (Radl) Guillaumin (*A. aristata* Haw. (S Afr.) × *G. carinata* (Mill.) Duval)

Gasteria Duval. Asphodelaceae (Aloaceae). 23 S Afr. R: U. Eggli, *Ill. Handb. Succ. Pls*, Monocots (2001)192; Aloe 44(2007)84. Mostly stemless succ. with rosettes of lvs, much cult. (incl. hybrids with *Aloe*, *Gonialoe*) & many variants form. regarded as spp.

Gastonia Comm. ex Lam. = *Polyscias*

× **Gastonialoe** J.M.H. Shaw. Asphodelaceae. *Gasteria* × *Gonialoe* hybrids. Comm. potpls incl. × ***G.* 'Goliath'** (*Gasteria brachyphylla* (Salm-Dyck) Jaarsveld × *Gonialoe variegata*)

Gastranthus Moritz ex Benth. = *Stenostephanus*

Gastridium P. Beauv. Gramineae (XVI 5). 2 Canaries, W Eur. (1), Medit. to Iran. Glumes persistent on axis

Gastrochilus D. Don (~ *Saccolabium*). Orchidaceae (V 16c). c. 55 Himal. & E As. to Jap. R: AOSB 54 (1985)1111, Guihaia 16(1996)129. Cult. orn.

Gastrococos Morales = *Acrocomia* (but see Principes 11(1967)114)

Gastrocotyle Bunge. Boraginaceae (B.2.1.1). 1–2 E Med. to C As. & NW Ind.: ***G. hispida*** (Forssk.) Bunge

Gastrodia R. Br. Orchidaceae (V 7). c. 60 Afr. (3 – R: KB 65(2010)316), E As., Indomal. to NZ (Aus. 7, 6 endemic). Some mycotrophs, with tubers roasted & eaten by Aborigines, e.g. ***G. sesamoides*** R. Br. (potato orchid, S & E Aus., NZ, natur. SW Cape) – poll. xylocopid bees gathering pseudopollen from lip, tubers eaten by Aborigines in Tasmania; ***G. elata*** Blume (SE As.) – dried & powdered a common cure for headaches in China

Gastrolepis Tieghem. Stemonuraceae (Icacinaceae s.l.). 1 New Caled.: ***G. austrocaledonica*** (Baill.) R. Howard. R: JAA 21(1940)481

Gastrolobium R.Br. Leguminosae (III 13). Incl. *Brachysema*, *Jansonia*, *Nemcia*, *Oxylobium* p.p. 109 W, esp. SW, Aus. R: AusSB 15(2002)623. Bee- (yellow fls) & few bird-poll. (red); produce fluoroacetate (in *G. bilobum* up to 2.65 g per kg) highly toxic to stock ('heart-leaf') as blocks Krebs Cycle at citrate stage, so form. an eradication programme in W Aus., though emus & some marsupials ± tolerant. ***G. bilobum*** R. Br. (SW Aus.) – ectomycorrhizae spread by red kangaroo, spores germ. after digestion; ***G. calycinum*** Benth. (York Road poison, SW Aus.) & ***G. spinosum*** Benth. (prickly poison, SW Aus.) – v. toxic to stock

Gastrolychnis (Fenzl) Reichb. = *Silene*

Gastronychia Small = *Paronychia*

Gastropyrum (Jaub. & Spach) Löve = *Triticum*

Gastrorchis Schltr. = *Phaius*

Gastrosiphon (Schltr.) M.A. Clem. & D.L. Jones = *Corybas*

gatal-gatal *Schima wallichii*

gau choi *Allium tuberosum*

gaub tree *Diospyros malabarica*

Gaudichaudia Kunth. Malpighiaceae. c. 10 Mex. to Bolivia

Gaudinia P. Beauv. = *Trisetaria*

Gaudiniopsis (Boiss.) Eig = *Ventenata*

Gaulettia Sothers & Prance (~ *Couepia*). Chrysobalanaceae. 9 S Am. R: Phytoneuron 172(2014)181

× **Gaulnettya** Marchant = *Gaultheria*

Gaultheria Kalm ex L. Ericaceae (VIII 4). Incl. *Diplycosia*, *Pernettya*, 134 Mal. (24), E As. (c. 33), Aus. & NZ (14), N Am. (6), C & S Am. (c. 58). R: BJLS 106(1991)229. Many with methyl salicylate, the basis of medic. wintergreen. Some cult. orn. shrubs incl. ***G. mucronata*** (L.f.) Hook. & Arn. (*P. mucronata*, S S Am.) – grown for game cover, potentially wind-poll., some forms bisex., most wild ones functionally but not structurally dioec.); ***G. procumbens*** L. (checkerberry, partridge berry, N Am.) – medic. & refreshing tea for E Canad. Indians, source of orig. wintergreen (now extracted from *Betula lenta*), & ***G. shallon*** Pursh (salal, shallon, N Am.) – fr. ed. esp. cooked, natur. in GB; ***G. × wisleyensis*** Marchant ex Middleton ('× *Gaulnettya wisleyensis*', the distinction from *Pernettya* (fr. of *G.* a capsule, that of *P.* a berry) not tenable) – hybrids between *G. mucronata* & *G. shallon*, spontaneous in S GB where parents natur., incl. **'Wisley Pearl'**

× **Gaulthettya** Camp = *Gaultheria*

Gaura L. = *Oenothera*

Gaurella Small = *Oenothera*

Gauropsis C. Presl = *Clarkia*

Gaussenia A. Bobrov & Melikian = *Dacrydium*

Gaussia H. Wendl. Palmae (V 2). 5 C Am., WI, on limestone. R: SB 11(1986)145. Stems with swollen 'bellies'

gauze tree *Lagetta lagetto*

Gavarretia Baill. (~ *Conceveiba*). Euphorbiaceae (Acal.-Alc.). 1 Amaz. Braz.: ***G. terminalis*** Baill. R: A.R. Sm., *Gen. Euph.* (2001)197. G 2-loc.

Gavilea Poeppig. Orchidaceae (IV 2a). 17 temp. S Am. R: Darw. 50(2012)36

gay feather *Liatris* spp.

Gaya Kunth. Malvaceae (Malv.-Malv.). 33 trop. Am. R: Bonplandia 9(1996)57. Epicalyx 0

Gayella Pierre = *Pouteria*

Gaylussacia Kunth. Ericaceae (VIII 5). 49 N & S (most) Am. R: BJ 86(1967)309. Huckleberries distinguished from *Vaccinium* in ovary with 10 divisions because of outgrowths in the 5 carpels. Fr. ed. esp. in pies, notably ***G. baccata*** (Wangenh.) K. Koch (E N Am.). Cult.

orn. incl. ***G. brachycera*** (Michaux) Torrey & Gray (E US) – 13 000-yr-old clone of 40 ha in Pennsylvania (biggest pl. clone known)

Gayophytum A. Juss. Onagraceae. 9 temp. W N Am. & W S Am. R: Britt. 16(1964)343

gaz of Khunsar see *Astragalus brachycalyx*

Gazachloa Phipps = *Danthoniopsis*

Gazania Gaertn. Compositae (Arct.-Gort.). 16 trop. (1) & S Afr. R: MSBM 3(1959)364. Milky latex; dark spots on ligules attractive to poll. bee-flies (cf. *Gorteria*). Cult. orn. esp. ***G. rigens*** (L.) Gaertn. & ***G. pectinata*** (Thunb.) Hartweg (*G. pinnata*) – both S Afr., & many hybrids incl. **Sunbathers Series** with capitula failing to close in cloudy weather. ***G. linearis*** (Thunb.) Druce (S Afr.) – tomentum twisted, soaked in fat & made into a fringe for loincloths in Zulu culture

gean *Prunus avium*

Geanthus Reinw. = *Etlingera*

gear *Papaver somniferum*

Gearum N.E. Br. Araceae (VII 4). 1 C Braz.: ***G. brasiliense*** N.E. Br. – coll. 1828, rediscovered 1978, beetle-poll. R: KB 49(1994)786

gebang (palm) *Corypha utan*

gedu nohor *Entandrophragma angolense*

geebung *Persoonia* spp.

Geesinkorchis de Vogel. Orchidaceae (V 10b). 4 W Mal. R: Blumea 50(2005)505

geiger tree *Cordia sebestena*

Geigeria Greiss. Compositae (Inul.-Pluch.). 28 trop. & S Afr. R: NJB 23(2005)318

Geijera Schott. Rutaceae (I 2). Excl. *Coatesia*, c. 10 N NG, E Aus. (3, 2 endemic), New Caled. Alks. ***G. parviflora*** Lindl. (E Aus.) – fodder-tree (sheep bush, wilga)

Geissanthus Hook.f. Primulaceae (Myrsinaceae). c. 55 W trop. S Am.

Geissaspis Wight & Arn. Leguminosae (III 11). 2 warm As.

Geissois Labill. Cunoniaceae. (V). Excl. *Karrabina*, *Lamanonia*, 19 New Caled., Vanuatu, Fiji

Geissolepis Robinson. Compositae (Ast.-?). 1 Mex.: ***G. suaedifolia*** Robinson. R: SBM 20(2000)47

Geissoloma Lindl. ex Kunth. Geissolomataceae. 1 Lageberg Mts, Cape, S Afr.: ***G. marginatum*** (L.) A. Juss. R: Strel. 19(2000)309

Geissolomataceae A. DC. Magnoliidae – Crossosomatales. 1/1 S Afr. Xeromorphic everg. shrub accumulating aluminium. Lvs opp., simple; stip. petiolar, minute. Fls solit., term. with 3 prs of bracts (the more distal the more petaloid, s.t. with vestigial fls in axils), bisex., reg., 4-merous; P 4 petaloid, basally connate, A 4 + 4 adnate to P-base, anthers with longit. slits, nectary with 4 nectar recesses opp. P, $\underline{G}$ (4), 4-loc. each with slender style, the stigmas distally connivent, & 2 pend., anatropous, bitegmic ovules. Fr. a loculicidal capsule enveloped by persistent P; seeds 1 per loc. with straight embryo & scant endosperm

Genus: *Geissoloma*

Form. assoc. with Myrtales but absence of internal phloem etc. made this unlikely; wood studies supported alliance with Grubbiaceae (Cornales) & Bruniaceae (Bruniales)

Geissomeria Lindl. Acanthaceae (III 1). 15 Mex. to trop. S Am.

Geissopappus Benth. = *Calea*

Geissorhiza Ker-Gawler. Iridaceae (VI 5). 103 S Afr. esp. Cape (winter rainfall). R: AMBG 72(1985)277, Novon 5(1995)156, Bothalia 39(2009)127. Some cult. orn. like *Ixia* but related to *Hesperantha*

Geissospermum Allemão. Apocynaceae (1a). 5 Braz. R: AMBG 71(1984)1077. Nozeran's Model. Some medic. bark used as febrifuge: alks

Geitonoplesium Cunn. ex R. Br. Asphodelaceae (Hemerocallidaceae). 1 C & E Mal. to Fiji (polymorphic): ***G. cymosum*** (R. Br.) R. Br. R: Austrobaileya 4(1987)396

Gelasia Cass. = *Scorzonera*

Gelasine Herb. Iridaceae (VII 5). c. 6 subtrop. S Am. R: NJB 4(1984)347. Lvs plicate; elaiophores attractive to bees

Geleznowia Turcz. (~ *Philotheca*). Rutaceae (I 3). 1 SW Aus.: ***G. verrucosa*** Turcz. (*G. calycina*) – cutfl. gathered in wild, so plantations proposed. R: FA 26(2013)415

Gelibia Hutch. = *Polyscias*

Gelidocalamus Wen (~ *Indocalamus*). Gramineae (IV). 9 China. R: FOC 22(2006)132. Heterogeneous

Gelrebia Gagnon & G.P. Lewis (~ *Caesalpinia*). Leguminosae (I 4). 8 Afr. R: PhytoKeys 71(2016)54

Gelsemiaceae Struwe & Albert (~ Loganiaceae). Magnoliidae – Gentianales. Incl. Pteleocarpaceae, 3/16 N to trop. Am., Afr., E As. R: BJLS 175(2014)495. Shrubs & lianes with heterostylous fls split from Loganiaceae because of imbr. C, twice-dichot. divided stigmas, latrorse anthers & flattened seeds. Fr. a capsule; endosperm horny, starchy. n = 8, 10

Genera: *Gelsemium, Mostuea, Pteleocarpa* (lvs in spirals)

Heterostyly & complex indole alks like Rubiaceae. Local stimulants etc.

gelseminum, gelsemium *Gelsemium sempervirens*

Gelsemium Juss. Gelsemiaceae (Loganiaceae s.l.). 2 S US to Guatemala, 1 SE As. to W Mal. Allspice jasmine. Double-forked style; alks, highly toxic. ***G. elegans*** (Gardner & Champ.) Benth. (SE As. to Mal.) – alk. used in murder & suicide; ***G. sempervirens*** (L.) J. St-Hil. (Am.) – cult. orn. liane, form. medic. (neuralgia, migraine etc.) as gelseminum, gelsemium

Gemmabryum J. Spence & H. Ramsay (~ *Bryum*). Bryaceae (Bryidae). c. 100 cosmop. Asexual reprod. structures incl. bulbils, rhizoidal tubers, stem tubers & filiform rhizoidal gemmae

Gemmaria Salisb. = *Strumaria*

Gemmingia Heist. ex Fabr. ?Older name for *Aristea*

Gendarussa Nees = *Justicia*

genépi absinthe-type liqueur flavoured with *Artemisia glacialis* & *A. umbelliformis*

geneva Original gin (Hollands) distilled from rye or barley

Genianthus Hook.f. = *Secamone*

Geniosporum Wall. ex Benth. = *Platostoma*

Geniostemon Engelm. & A. Gray. Gentianaceae. 4 Mex. R: Phytol. 76(1994)8

Geniostoma Forst. & Forst. f. Loganiaceae (Geniostomaceae). 24 Mal. to NZ, Tahiti & Jap. R: Blumea 26(1980)245. Attims's Model; herm. but self-incompatible

Geniostomaceae Struwe & Albert = Loganiaceae

genip, genipa, genipapo *Genipa americana*

Genipa Tourn. ex L. Rubiaceae (II 1). 3 trop. Am. Fagerlind's Model; some fish-disp. in Amaz. ***G. americana*** L. (genip, genipa, Mex. to Paraguay) – tree used as live-fences, timber useful, fr. rather unpalatable but source of a barkcloth dye & body dye (iridoid, genipin, turning black with protein, e.g. skin) for Am. Indians & used in drinks (genipapo)

Genista L. Leguminosae (III 9). 87 Eur. (58), Canary Is., Medit. to W As. R: NRBGE 27(1966)11 & BSB 2,45(1972)269 (*Teline*) – brooms, 'planta-genista' origin of 'Plantagenets' name prob. *Cytisus scoparius*; alks; explosive fls showering poll. insects with pollen. Shrubs, s.t. spiny, differing from *Cytisus* etc. in style curved only nr. apex. Cult. orn. & dyepls. ***G. aetnensis*** (Biv.) DC. (*Cytisanthus a.*, Mt Etna broom, Sicily, Sardinia) – to 9 m, fls scented; ***G. anglica*** L. (petty whin, needle w. or furze, W Eur.) – spiny cult. orn.; ***G. linifolia*** L., ***G. monspessulana*** (L.) L. Johnson (Medit.) both invasive in Aus., *G. m.* also in NZ, though v. similar ***G. ferox*** (Poir.) Dum. Cours. (N Afr.) not invasive; ***G. tinctoria*** L. (dyer's greenweed, Eur., W As., natur. N Am.) – fls yield a yellow dye &, when mixed with woad, Kendal green; ***G. tridentata*** L. (*Chamaespartium* t., W Iberia) – medic. tea in S Portugal

Genistidium I.M. Johnst. Leguminosae (III 23). 1 Texas, Mex.: ***G. dumosum*** I.M. Johnst. R: JAA 22(1941)113

Genlisea A. St-Hil. Lentibulariaceae. 29 trop. Am. (11), trop. & S. Afr. & Madag. R: A. Fleischmann (2013) *Monogr. genus G.* Rosette-pls with 2 kinds of leaf, thread-like forked & chlorophyll-less ones = traps (soil particles within suggest prey are sucked in); r. 0; fr. like a globe dehiscing along equator & to certain extent trop.; DNA levels as low as 63 M base-prs (*Arabidopsis* 155 M)

Gennaria Parl. Orchidaceae (IV 4d). 1 Macaronesia, W Med. incl. Eur.: ***G. diphylla*** (Link) Parl. R: Gen. Orch. 2(2001)292

Genoplesium R. Br. (~ *Prasophyllum*). Orchidaceae (IV 3g). c. 45 Aus., 2 ext. to NZ, New Caled. Midge orchids. ***G. baueri*** R. Br. (E Aus.) – leafless, holoparasitic, reduced to 210 pls but prob. overlooked (Weston)

gentian *Gentiana* spp.; **field g.** *Gentianella campestris*; **marsh g.** *Gentiana pneumonanthe*; **prairie g.** *Eustoma grandiflorum*; **yellow g.** *G. lutea*

Gentiana Tourn. ex L. Gentianaceae. Excl. *Comastoma, Crawfurdia*, 361 (c. 15 sects) temp. (Eur. 27, As. 312 – China 247 (SW mts 190), Aus. 1) & Arctic, usu. montane elsewhere but absent from Afr. exc. Morocco. R: D. Wilkie (1936) *Gentians*; BBMNHB 20(1990)362; T.N. Ho & S.W. Liu (2001) *Worldwide Monograph of G.*:116. Usu. perenn. herbs, oft. tufted, with blue to purple or red, yellow or white fls; r. with bitter glucosides yield medic.; cult. orn. (incl. hybrids like ***G.*** × ***macaulayi*** Chitt. (*G. sino-ornata* Balf.f. × *G. farreri* Balf.f., both China) esp. ***G. acaulis*** L. (*G. kochiana*, Alps & Pyrenees)) – fls dark blue to 5 cm tall, poll. by bees, other spp. by butterflies, while yellow-flowered ***G. lutea*** L. (yellow gentian, Eur., W As.) by short-tongued insects (nectar accessible), lvs resembling *Veratrum* spp., comm. source of g. r. used as tonic incl, appetite-stimulant in anorexia, used by anc. Egyptians, & in flavouring liqueurs & Suze; ***G. makinoi*** Kuzn. (Jap.) – cut-fl. esp. '**Marsha**' in Holland; ***G. pneumonanthe*** L. (marsh g., Eur., N As.) – fls a source of a blue dye; ***G. prostrata*** Haenke (Alps, N Am., Andes, S S Am.) – ann., bipolar distrib.; ***G. verna*** L. (Ireland, GB (Teesdale), C Eur. to As. mts) – poss. survived the glaciation in GB (pollen evidence)

Gentianaceae Juss. Magnoliidae – Gentianales. Incl. Saccifoliaceae & Loganiaceae p.p., 95/1850 cosmop. but esp. temp. & subtrop. & in trop. mts. R: L. Struwe & V.A. Albert (2002) *G.: systematics & natural history*. Small trees, some pachycaul, shrubs & (usu.) herbs, oft. with mycorrhiza (s.t. chlorophyll-less & mycotrophic), usu. accum. bitter iridoid substances & oft. with internal phloem. Lvs opp., seldom whorled or even in spirals (some *Swertia* spp.), simple (scale-like in mycotrophic *Voyria*, *Voyriella* etc.); stip. 0. Fls solit. or in cymose (rarely racemose) infls, usu. bisex. & reg.; K (4 or 5(–12)) with lobes imbr. (s.t. valvate or open), rarely reduced or 0 when tube 2-cleft, rarely K free, C (4 or 5(–12)) with ± elongate tube & usu. convolute lobes (with long thread-like projection in *Urogentias*) & oft. with scales or nectary-pits within, A as many as & alt. with lobes, attached to tube, rarely some staminodal or 0, anthers with longit. slits (rarely term. pores as in *Exacum*), nectary-disk or glands usu. around $\underline{G}$ (2), 1-loc. with parietal placentas s.t. ± deeply intruded & bifid, seldom 2-loc. with axile placentas or 1-loc. with free-C placenta, style term. with entire or 2-lobed stigma (0 in *Lomatogonium*, where stigmas decurrent along ovary-sides), ovules ± ∞, anatropous, unitegmic. Fr. a septicidal capsule, rarely a berry; seeds usu. with small, straight embryo in copious oily endosperm but in chlorophyll-less genera seeds tiny with undifferentiated embryo & scanty endosperm. x = 5–13+

Classification & chief genera (some generic limits controversial; *Voyria* unplaced): **Saccifolieae** (Saccifoliaceae; fls usu. 5-merous, placentation parietal; trop. Am.; 'sister' to rest of fam.): *Saccifolium*, *Voyriella*; **Exaceae** (8/180; R: SB 37(2012)238): *Exacum*, *Sebaea*; **Chironieae**: *Centaurium*, *Zeltnera*; **Helieae**: *Irlbachia*, *Macrocarpaea*; **Potalieae**: *Anthocleista*, *Fagraea*, *Lisianthus*; **Gentianeae**: *Gentiana*, *Gentianella*, *Halenia*, *Swertia*

Cult. orn. esp. *Centaurium*, *Eustoma* ('lisianthus'), *Exacum*, *Gentiana*, *Sabatia*, & medic. esp. *Canscora*, *Gentiana*, *Potalia*. Some timber (*Fagraea*)

Gentianella Moench (~ *Gentiana*). Gentianaceae. c. 260 N temp (Eur. 22; N Am. 30), Aus (12), NZ 30 – R: NZJB 42(2004)449, S Am. differing from *Gentiana* in lacking small lobes or pleats between C-lobes, in sessile G & rounded wingless seeds; 1 nectary per C lobe (others form. here cf. *Swertia* etc.). Tendency to rapid isolation of ecological & seasonal variants. Many short-lived, some ann. incl. ***G. lilliputiana*** (C.J. Webb) Glenny (NZ) – to 20 mm with 1 4-merous fl.; cult. orn. & some med. ***G. anglica*** (Pugsley) E.F. Warburg (Br. 'endemic') prob. merely an early-flowering morph of ***G. amarella*** (L.) Boerner (N temp.) selected & maintained by former grassland management practices; ***G. campestris*** (L.) Boerner (felwort, field gentian, Eur.); ***G. cosmantha*** (Griseb.) Pringle & ***G. splendens*** (Gilg) Fabris (S Am.) – bird-poll.; ***G. florida*** (Griseb.) Holub (S Am.) – dioec.; ***G. quinquefolia*** (L.) Small (ague weed, E N Am.)

Gentianodes Löve & D. Löve = *Gentiana*

Gentianopsis Ma (? =*Pterygocalyx*). Gentianaceae. 24 N temp. Cult. orn.

Gentianothamnus Humbert. Gentianaceae. 1 Madag.: ***G. madagascariensis*** Humbert. R: BSBF 84(1937)388

Gentingia Johansson & Wong (~ *Prismatomeris*). Rubiaceae (Prismat.). 1 NW Malay Pen.: ***G. subsessilis*** (King & Gamble) Johansson & Wong. R: Blumea 33(1988)351

Gentlea Lundell = *Ardisia*

Gentrya Breedlove & Heckard = *Castilleja*

Genyorchis Schltr. Orchidaceae (V 15). 6 trop. Afr.

Geoblasta Barb. Rodr. = *Bipinnula* (but see Gen. Orch. 3(2003)11)

Geocalpa Brieger = *Pleurothallis*

Geocarpon Mackenzie = *Mononeuria* (but see FNA 5(2005)148)

Geocaryum Cosson (incl. *Huertia*). Umbelliferae (III 8). 13–15 E Medit. R: L. Engstrand (1977) *Biosystematics & taxonomy of G.* 2 Gk. spp. already extinct

Geocaulon Fern. Santalaceae (Comandraceae). 3 Alaska, Canada, NE US: ***G. lividum*** (Richardson) Fern.

Geocharis (K. Schum.) Ridl. Zingiberaceae (II 4). 6 W Mal.

Geochloa Linder & N. Barker (~ *Danthonia*). Gramineae (Danth.) 3 S Cape. Fls after fires

Geochorda Cham. & Schldl. = *Bacopa*

Geococcus J.L. Drumm. ex Harv. Cruciferae (37). 1 semi-arid Aus.: ***G. pusillus*** J.L. Drumm. ex Harv. – geocarpic. R: TRSSA 89(1965)231

Geodorum Andrews (~ *Eulophia*). Orchidaceae (V 12b). 12 Indomal., W Pac. (Aus. 1). ***G. nutans*** (C. Presl) Ames (Taiwan to Philipp.) – source of a strong adhesive used in musical instruments; ***G. densiflorum*** (Lam.) Schltr. (*G. pictum*, Ind. to Pacific) – eaten (yeenga) by Aus. Aborigines

Geoffraya Bonati = *Vandellia*

Geoffroea Jacq. Leguminosae (III 11). 2 S Am. R: EJB 56(1999)330. ***G. decorticans*** (Hook. & Arn.) Burkart (*Gourliea d.*, chanal, chañar, Arg., Bolivia, Chile, Peru, Urug., Paraguay) – imp. fodder, pods ed., medic., dyepl., timber

Geogenanthus Ule. Commelinaceae (II 1e). 5 trop. S Am. Cult. orn. esp. ***G. poeppigii*** (Miq.) Faden (*G. undatus*, seersucker pl., Braz., Peru) – lvs with undulating surface

Geohintonia Glass & FitzMaurice. Cactaceae (III 9). 1 NE Mex.: ***G. mexicana*** Glass & FitzMaurice – on gypsum outcrops

Geomitra Becc. = *Thismia*

Geonoma Willd. Palmae (V 11). 68 trop. Am. R: Phytotaxa 17(2011)1. Monoec., unarmed forest palms to 3150 m alt. (***G. weberbaueri*** Dammer ex Burret, Peru) incl. some rheophytes incl. ***G. cuneata*** H. Wendl. ex Spruce subsp. ***linearis*** (Burret) A. Henderson (Colombia, Ecuador). ***G. baculifera*** (Poit.) Kunth (dhalebana, NE S Am.) – thatch (lasts 6–12 yrs)

Geopanax Hemsl. = *Schefflera*

Geophila D. Don. Rubiaceae (IV 7). c. 30 trop. Bell's Model. ***G. repens*** (L.) I.M. Johnst. – local medic.

Georgia bark tree *Pinckneya bracteata*

Georgeantha Briggs & L. Johnson. Ecdeiocoleaceae. 1 N of Perth, W Aus.: ***G. hexandra*** Briggs & L. Johnson. R: Telopea 7(1998)307

Geosiridaceae Jonker = Iridaceae

Geosiris Baill. Iridaceae (III; Geosiridaceae). 2 (?+1) Madag. Dust-like seeds. ***G. aphylla*** Baill. – fls open in morning

Geostachys (Bak.) Ridl. Zingiberaceae (II 4). 15 SE As., W Mal.

Geraea Torrey & A. Gray. Compositae (Helia.-Enc.). 2 SW N Am. R: FNA 21(2006)122. Cult. orn.

Geraldton wax (flower) *Chamelaucium uncinatum*

Geraniaceae Juss. Magnoliidae – Geraniales. Excl. Biebersteiniaceae, Dirachmaceae, & Francoaceae, 5/800 temp., few trop. Herbs or shrubs (s.t. pachycaul &/or geophytic), usu. with aromatic oils in multicellular capitate-glandular hairs. Lvs in spirals, less oft. opp., usu. lobed, compound or dissected pinnately or palmately; stip. usu. present (*Hypseocharis* 0). Fls in cymes, usu. bisex., reg. (irreg. in *Pelargonium*); K (4)5, imbr. or less oft. valvate, s.t. basally connate or forming a lobed tube (in *Pelargonium* adaxial sepal with spurred nectary, C (0, 2, 4)5, imbr., nectary-glands alt. with P, around A (0 in *Pelargonium*), A 5 + 5 but some or all of outer staminodal, rarely (*Monsonia*) 5 + 5 + 5, filaments ± basally connate, anthers with longit. slits, $\underline{G}$ ((4)5, pluriloc. with axile placentas & 1 style with distinct stigmas & elongating persistent column (beak), ovules anatr. to campylotropous, bitegmic, (1)2(–12) per loc., superposed & at least upper pend. Fr. usu. 5 1-seeded mericarps separating acropetally from beak & oft. opening to release seed; seeds (exotegmic like Guttiferae) with straight or usu. curved green embryo & 0 or scant endosperm. n = 4, 7–11, 14, 15

Orig. S Afr.? Geranieae (septicidal capsule: *Erodium, Geranium, Monsonia* & *Pelargonium*) all with some spp. with spiny petioles 'sister' to *Hypseocharis* (Hypseocharitaceae form. in Oxalidaceae; loculicidal capsule); other genera form. here (re-)seg. in discrete fams – the allied Ledocarpaceae & Vivianiaceae (= Francoaceae), but also Biebersteiniaceae (Sapindales) & Dirachmaceae (Rosales). Cult. orn. (*Geranium, Pelargonium, Monsonia* (incl. *Sarcocaulon*) etc.), forage (*Erodium*), ess. oils (*Pelargonium*) & locally used med., dyes etc.

Geraniopsis Chrtek = *Geranium*

geranium (of house-pls) *Pelargonium* spp.; **g. oil** ('East Ind.' or 'Turkish') *Cymbopogon martini, Geranium* or *Pelargonium* spp.

Geranium Tourn. ex L. Geraniaceae. c. 430 temp. (Eur. 38, Am. 137 (15 introd.) – R: SBM 95(2012)), montane trop. (Mal. 15, 13 endemic; Hawaii 7 endemic shrubby with parallel-veined lvs, allied to Am. spp., incl. ***G. arboreum*** A. Gray to 3950 m – red-flowered, bird-poll. (unique in fam.)). Ann. or perenn. rabbit-proof herbs & s.t. (e.g. Macaronesia with ***G. maderense*** Yeo (Madeira) to 1.5 m in fl.) pachycaul shrublets (Holttum's, Corner's & Tomlinson's Models); infls of 1 fl. or a pr as units in dichasial cymes or cincinni; fr. explosive, oft. all mericarps opening at once & ejecting seeds. Many cult. orn. (cranesbills) esp. for ground-cover; R: P. Yeo (1985) *Hardy Gs.* ***G. lucidum*** L. (shining c., Eur. & Medit. to Himal.) – weed; ***G. macrorrhizum*** L. (C S & SE Eur.) – Zdravetz oil from hairs; ***G. maculatum*** L. (N Am.) – local medic.; ***G. nepalense*** Sweet (As. mts) – dye a subs. for *Rubia cordifolia* from r. also used in tanning; ***G. phaeum*** L. (black or mourning widow, W & C Eur.) – blackish fls; ***G. pratense*** L. (meadow c., Euras.); ***G. robertianum*** L. (Herb Robert, Eur. to NW Afr. & SW China, (?natur.) E N Am.) – form. medic. (skin & kidneys) & used against red-water fever in livestock (Ireland); ***G. sanguineum*** L. (bloody c., Eur., E Medit.) – cult. orn. incl. dwarf var. ***striatum*** Weston (NW Eng.); ***G. viscosissimum*** Fischer & C. Meyer (NW Am.) – 'protocarnivorous' in that hairs can absorb proteins from arthropods etc. trapped by sticky hairs; ***G. wallichianum*** D. Don (NE Afghanistan to Kashmir) – r. with 25–32% tannin used in tanning & dyeing

Gerardia Benth. = *Agalinis*

Gerardia Plum. ex L. = *Stenandrium*

Gerardiina Engl. Orobanchaceae ('Micr.').1 trop. & S Afr.: ***G. angolensis*** Engl. R: Strel. 10(2000)520

Gerardiopsis Engl. = *Anticharis*

Gerardoa Luer = *Specklinia*

Gerascanthus P. Br. = *Cordia*

Gerbera L. Compositae (Mut.-Mut.). 29 NE to S Afr. with ***G. piloselloides*** (L.) Cass. W Afr. to China. R: OB 78(1985). Herbs, heterogeneous with some allied to *Chaptalia*. Cult. orn. greenhouse rosette-pls orig. ***G. jamesonii*** Bolus ex Adlam (Barberton or Transvaal daisy, Transvaal) but modern cutfls usu. hybrids (*'G. hybrida'*, first made England c. 1890; trade in US worth $32 M+ by 2005) with ***G. viridifolia*** (DC.) Schultz-Bip. (Ethiopia to S Afr.)

Gereaua Buerki & Callam. (~ *Haplocoelum*). Sapindaceae. 1 Madag.: ***G. perrieri*** (Capuron) Buerki & Callam. R: SB 35(2010)179

Gerlachia Szlach. = *Stanhopea*

Germainia Bal. & Poitr. Gramineae (XXII 6). 10 Assam & SE As. to NE Aus. R: TFB 6(1972)29

German chamomile *Matricaria chamomilla*; **G. grass** *Echinolaena polystachya*

germander *Teucrium* spp. esp. *T. chamaedrys*; **g. speedwell** *Veronica chamaedrys*; **wood g.** *T. scorodonia*

gero *Cenchrus spicatus*

geronggang or **geronggong** *Cratoxylum arborescens*

Geropogon L. Compositae (Lact.-Scor.). 1 Eur., Medit., Turkey: ***G. hybridus*** (L.) Schultz-Bip. R: Lazaroa 9(1986)31

Gerrardanthus Harv. ex Hook.f. Cucurbitaceae (III). 3–5 trop. & S Afr. ***G. macrorhiza*** Harv. ex Hook.f. (Natal) – spherical succ. stem to 50 cm diam.

Gerrardina Oliv. Gerrardinaceae (Flacourtiaceae s.l.). 2 E trop. & S Afr.

Gerrardinaceae Alford (~ Flacourtiaceae). Magnoliidae-Huerteales. 1/2 trop. E & S Afr. R: T 55(2006)962. Small trees & shrubs, s.t. scandent. Lvs in spirals, simple, stipulate, margins serrate. Infls axillary cymes. Fls bisex., reg., articulated, perigynous. K 5 persistent, lobes imbr.; C5 imbr., A 5 opp. C, anthers with longit. slits; disk cupular $\underline{G}$ (2) with 1 style & 1 loc. with 4 pend. anatropus ovules borne on 2 placentae. Fr. a berry with 1–4 seeds; aril 0; embryo minute

Genus: *Gerrardina*

Gerritea Zuloaga & al. Gramineae (XXIII 3). 1 Bolivia: ***G. pseudopetiolata*** Zuloaga & al. R: Novon 3(1993)213

Geschollia Speta = *Drimia*

Gesneria L. Gesneriaceae (II 5a). Incl. *Pheidonocarpa, Rhytidophyllum*, 70 trop. Am. R: SCB 29(1976)43, Selbyana 6(1982)200. Suffrutices, shrubs & trees to 10 m. Some with pseudostipules; mucilage from lvs scented & (?) attractive to bat poll. Some cult. orn.

Gesneriaceae Rich. & Juss. Magnoliidae – Lamiales. 151/3525 trop. (esp. OW – over 60% genera & spp., China (R: EJB 49(1992)5) with 56 genera (28 endemic); R(NW): Selbyana 6(1983)1), few Tertiary relics in temp. Euras. R: http:/persoon.si.edu/Gesneriaceae/ Checklist; Selbyana 31(2013)83. Usu. herbs, s.t. epiphytic, or shrubs, s.t. lianes (e.g. *Asteranthera*), rarely trees (e.g. *Lenbrassia, Sanango, Shuaria, Solenophora*). Lvs opp., rarely whorled or in spirals, s.t. members of a pr unequal, usu. simple (rarely pinnatifid); stip. 0. Fls solit., axillary or in usu. bracteose thyrses to solit., s.t. epiphyllous, bisex.; K 5 or usu. (5), lobes usu. valvate, C (5), irreg., & usu. 2-lipped & oft. spurred, or almost reg., lobes imbr., adaxial ones usu. encl., A 4 (posterior 1 absent) alt. with C & attached to tube, anthers all connivent or in prs, less oft. A 5 (*Ramonda*, some *Sinningia* spp.) or 2, usu. staminodes 1–3 in place of missing A, anthers with longit. slits, nectary-disk oft. at base of $\underline{G}$ (2) or $\overline{G}$ (2) with term. style, 1-loc. & with 2 parietal placentas ± intruded & bifurcate, s.t. meeting (e.g. *Monophyllaea*) so G 2-loc., ovules ∞, anatropous, unitegmic. Fr. a usu. loculicidal (or septicidal) capsule, less oft. a berry; seeds ∞, small with straight embryo in oily endosperm (Gesnerioideae; 0 in Didymocarpoideae) & 1 (esp. C.) or 2 cotyledons. x= 4–17+

Classification & principal genera:

I. **Sanangoideae** (small tree; term. bracteose thyrse): *Sanango* (only)

II. **Gesnerioideae** (incl. Coronantheroideae; small trees to lianes or herbs with isocotylous seedlings) – 5 tribes: 1. Titanotricheae (*Titanotrichum* only); 2. Napeantheae (*Napeanthus* only); 3. Beslerieae, a. Beslerinae (*Besleria, Gasteranthus*), b. Anetanthinae (*Anetanthus*); 4. Coronanthereae, a. Coronatherineae (*Coronanthera*), b. Mitrariinae (*Fieldia, Mitraria*), c. Negriinae (*Lenbrassia, Negria*); 5. Gesnerieae (incl. Episcieae, Gloxinieae, Sinningieae), a. Gesneriinae (*Gesneria, Rhytidophyllum*), b. Gloxiniinae (*Achimenes, Gloxinia, Kohleria, Smithiantha*), c. Columneinae (*Alloplectus, Codonanthe, Columnea, Drymonia* (lianes), *Nautilocalyx, Nematanthus*), d. Sphaerorrhizinae (*Sphaerorrhiza* only), e. Ligeriinae (*Sinningia*)

III. **Didymocarpoideae** (Cyrtandroideae; anisocotylous seedlings, & 1 (further from light) s.t. becoming the only photosynthetic organ, forming a lamina by intercalary growth, placenta lamelliform in t.s.; OW) – 2 tribes: 1. Epithemateae (anisocotylous seedlings, placenta triangular in t.s., OW exc. for 1 *Rhynchoglossum* sp.), a. Loxotidinae (*Rhynchoglossum*), b. Monophyllaeinae (*Monophyllaea*), c. Loxoniinae (*Stauranthera*), d. Epithematinae (*Epithema*); 2. Trichosporeae, a. Jerdoniinae (*Jerdonia*), b. Carallodiscinae (*Corallodiscus*), c. Tetraphyllinae (*Tetraphyllum*), d. Leptoboeinae (*Boeica*), e. Ramondinae (*Haberlea, Jancaea, Ramonda*), f. Litostigminae (*Litostigma*), g. Streptocarpinae (*Streptocarpus*), h. Didissandrinae (*Didissandra*), i. Loxocarpinae (*Boea*), j. Didymocarpinae (*Aeschynanthus, Agalmyla, Codonoboea, Cyrtandra, Didymocarpus, Henckelia, Oreocharis, Paraboea, Primulina*)

Prob. incl. Calceolariaceae (as **Calceolarioideae**), *Peltanthera* (**Peltantheroideae**)

Reg. fls arisen independently several times. Many genera esp. in NW show transitions from insect- to bird-poll. & *Columnea* (NW) is paralleled by *Aeschynanthus* in being bird-poll. epiphytes; some have extrafl. nectaries incl. *Codonanthe* epiphytic on ants' nests; 1 of the 10 fams with elaiophores attractive to bees (*Drymonia*)

Some locally imp. med. but internationally imp. as cult. orn. esp. *Sinningia* (gloxinias), *Streptocarpus* (incl. *Saintpaulia*), but also *Achimenes, Columnea, Haberlea, Nematanthus, Ramonda* etc. (H.E. Moore (1957) *African violets, gloxinias & their relatives*)

Gesnouinia Gaud. Urticaceae (IV). 2 Canary Is. Monoec. shrubs & trees. ***G. arborea*** (L.f.) Gaud. – cult. orn. tree to 7 m

Gethyllis L. Amaryllidaceae. 32 S Afr. R: Willdenowia 15(1986). Fragrance form. used to flavour brandy & to scent linen-cupboards; some ed. fr.

Gethyum Philippi = *Solaria*

Getonia Roxb. (*Calycopteris*). Combretaceae. 1 Indomal.: ***G. floribunda*** Roxb. – cli.

Geum L. Rosaceae (Ros.-Colur.). Incl. *Acomastylis, Coluria, Oncostylus, Orthurus, Parageum, Waldsteinia, Novosieversia*, excl. *Sieversia*, c. 45 temp. (Eur. 14) incl. NZ & cold. R: BJ 125(2004)414. Avens. Cult. orn. esp. forms of ***G. chiloense*** Balbis ex Ser. (Chile) – fls red, e.g. **'Mrs J. Bradshaw'** (apomictic 'double' breeding true) & ***G. ternatum*** (Stephan) Smedmark (*W. t.*, Carpathians & 5000 km away in Siberia, Jap.); ***G. aleppicum*** Jacq. (E As., N Am.) & other N Am. spp. locally medic.; ***G. japonicum*** Thunb. (E As.) – diuretic in trad. Chinese medic., anti-HIV triterpene acids in extract; ***G. triflorum*** Pursh (N Am.) – seeds crushed for scent; ***G. rivale*** L. (water a., N temp.) – glycosides medic., nodding bee-poll. fls found in less disturbed habitats than ***G. urbanum*** L. (wood a., Herb Bennet (*herba* *benedicta*), Eur., W As., Medit.) with erect yellow fls, r. clove-scented used for flavouring beer etc., locally medic., but where they meet hybrid swarms (***G. × intermedium*** Ehrh.; in N Am., natur. *G. u.* hybridizes with ***G. canadense*** Jacq.) form; both have hooked achenes promoting animal-disp., the hook an accrescent Z-shaped style, of which the upper piece lost after fert.; this fr. type has evolved twice from ancestors with plumose styles

Geunsia Blume = *Callicarpa*

Gevuina Molina. Proteaceae (V 4). 1 Chile, Arg. (OW spp. = *Bleasdalea*): ***G. avellana*** Molina (Chile nut) – cult. orn., useful timber, seeds ed., flavour reminiscent of hazel-nuts, poss. subs. for macadamia in cool climates

ghat *Catha edulis*

Ghaznianthus Lincz. (~ *Acantholimon*). Plumbaginaceae (II). 1 Afghanistan: ***G. rechingeri*** (Freitag) Lincz.

gheombi *Sindoropsis letestui*

gherkin *Cucumis sativus*; **WI g.** *C. anguria*

Ghikaea Volkens & Schweinf. Orobanchaceae (Buch.; Scrophulariaceae s.l.). 1 NE Afr.: ***G. speciosa*** (Rendle) Diels. R: FTEA Scroph.(2008)183

Ghinia Schreb. = *Tamonea*

ghost gum *Corymbia* spp.; **Miss Willmott's g.** *Eryngium giganteum*; **g. orchid** *Epipogium aphyllum*; **g. tree** *Davidia involucrata*

Giadotrum Pichon = *Cleghornia*

giam *Hopea* spp.

giant bamboo *Dendrocalamus giganteus*; **g. cactus** *Carnegiea gigantea*; **g. fennel** *Ferula communis*; **g. hogweed** *Heracleum mantegazzianum*; **g. taro** *Alocasia macrorrhizos*

Gibasis Raf. Commelinaceae (II 1g). 11 trop. Am. esp. Mex. R: KB 41(1986)107. Cult. orn. esp. ***G. pellucida*** (Martens & Galeotti) D. Hunt (Tahitian bridal veil, Mex.)

Gibasoides D. Hunt. Commelinaceae (II 1f). 1 Mex.: ***G. laxiflora*** (C.B. Clarke) D. Hunt. R: KB 33(1978)331

Gibbaeum Haw. ex N.E. Br. Aizoaceae (V). Incl. *Muiria*, 28 Cape. R: H. E. K. Hartmann, *Ill. Handb. Succ. Pls*, Aiz. F–Z(2002)35, 179. Cult. orn. succ. with shoots like *Conophytum*, incl. ***G. hortenseae*** (N.E. Br.) Thiede & Klak (*M. hortenseae*, Little Karoo)

Gibbaria Cass. (~ *Osteospermum*). Compositae (Cal.). Excl. *Nephrotheca*, 2 S Afr. R: Strel. 9(2000)328

Gibbesia Small = *Paronychia*

Gibbsia Rendle = *Leucosyke*

Gibraltar mint *Mentha pulegium*

Gibsoniothamnus L. O. Williams. Schlegeliaceae (Scrophulariaceae s.l.). 11 Mex. to C Am. R: Britt. 56(2004)215. Epiphytic, rarely terr., shrubs

gidgee, gidya *Acacia* spp. esp. *A. cambagei*

Gifola Cass. = *Filago*

Gigantochloa Kurz ex Munro. Gramineae (V 4). c. 30 Indomal. R: Reinwardtia 10(1987)291, 11(1997)83. Giant bamboos. Some with extrafl. nectaries; fls of ***G. albociliata*** (Munro) Kurz (SE As.) fragrant & visited by meliponid bees. ***G. apus*** (Schultes f.) Munro (Indomal.) & other spp. with ed. young shoots, strips of older ones of *G. apus* used to cut umbilical cords in Flores; ***G. scortechinii*** Gamble (Mal.) – used in particle-board (esp. in Java, where long cult.)

Gigasiphon Drake (~ *Bauhinia*). Leguminosae (I 1). 4–5 trop. Afr.

Gigliolia Becc. = *Areca*

Gilberta Turcz. (~ *Myriocephalus*). Compositae (Gnap.-Ang.). 1 Aus.: ***G. tenuifolia*** Turcz. R: Nuytsia 8(1992)419

Gilbertiella Boutique = *Monanthotaxis*

Gilbertiodendron Léonard. Leguminosae (I 2). Incl. *Pellegriniodendron* c. 30 trop. W Afr. Troll's Model. Some good timber. ***G. dewevrei*** (De Wild.) Léonard can form single-dominant forests in C Afr., mass-flowering with over 11 M fls & 10 000 seeds per ha

Gilead, balm of *Abies balsamea*, *Commiphora gileadensis*, *Liquidambar orientalis*, *Populus × jackii* 'Gileadensis', *P. nigra*, (orig. ?) *Pistacia terebinthus*

Gilesia F. Muell. = *Hermannia*

Gilgiochloa Pilg. Gramineae (XVIII). 1 trop. Afr.: ***G. indurata*** Pilg.

Gilia Ruíz & Pavón. Polemoniaceae (III). Excl. *Aliciella*, *Giliastrum*, 39 NW esp. W N Am. R: Aliso 19(2000)65. Some cult. orn. esp. annuals but most now removed to other genera e.g. *Ipomopsis*, *Linanthus*

Giliastrum (Brand) Rydb. (~ *Gilia*). Polemoniaceae (III). 9 W N & C Am. R: Aliso 17(1998)83

Gilibertia Ruíz & Pavón = *Dendropanax*

Gilipus Raf. = ? *Myrica/Morella*

Gillbeea F. Muell. Cunoniaceae (IV). 3 Queensland, NG. R: NJB 20(2000)437

Gillenia Moench (*Porteranthus*). Rosaceae (Ros.-Gill.). 2 E N Am.: ***G. stipulata*** (Willd.) Nutt. (American ipecacuanha, SE US) – glycosides in r. & bark form. medic.; ***G. trifoliata*** (L.) Moench (Bowman's r., E N Am.). R: FNA 9(2014)425

Gillespiea A.C. Sm. Rubiaceae (IV 7). 1 Fiji: ***G. speciosa*** A.C. Sm.

Gilletiella De Wild. & T. Durand = *Anomacanthus*

Gilletiodendron Vermoesen. Leguminosae (I 2). 5 trop. Afr. Timber, resins, ed. seeds

Gilliesia Lindl. Amaryllidaceae (Alliaceae-Gill.). c. 6 Chile. ***G. graminea*** Lindl. – insect mimicry with osmophores but no nectar, poll. by deceit (like some orchids)

gilliflower (gyllofer, gilofre, corruptions of caryophyllum, i.e. clove) *Dianthus* spp.; also *Erysimum* & *Matthiola* spp.

Gilmania Cov. (*Phyllogonum*) Polygonaceae (Erio.-Erio.). 1 Death Valley area, Calif.: ***G. luteola*** (Cov.) Cov. – in a 'good' yr 10s M pls giving edge of V. a golden ring. R: Phytol. 66(1989)243

Gilruthia Ewart. Compositae (Gnap.-Ang.). 1 W Aus.: ***G. osbornii*** Ewart & J.M. White. R: OB 104(1991)125

gin spirit usu. distilled from rye or barley, flavour largely due to *Juniperus communis*

Ginalloa Korth. Santalaceae (Viscaceae). 5 Indomal. Hyperparasites on Loranthaceae

gingelly *Sesamum indicum*

ginger *Zingiber officinale*; **g.bread palm** *Hyphaene thebaica*; **g.b. plum** *Neocarya macrophylla*; **cassumar g.** *Z. montanum*; **g. grass** *Cymbopogon martini*; **g. g. oil** *C. m.* 'Sofia'; **Japanese g.** *Z. mioga*; **kahili g.** *Hedychium gardnerianum*; **g. lily** *H.* spp.; **mango g.** *Curcuma amada*; **g. mint** *Mentha × gracilis*; **Mioga g.** *Z. mioga*; **shell g.** *Alpinia zerumbet*; **Siamese g.** *A. galanga*; **torch g.** *Etlingera elatior*; **wild g.** *Asarum canadense*

gingham *Gossypium* spp.

Gingidia Dawson (*Gingidium = Anisotome, Aciphylla* spp.?). Umbelliferae (III 8). Incl. *Lignocarpa, Scandia* (older names), 13 Aus. (3), NZ (10). R: NZJB 4(1966)84. Gynodioec.; some woody. Hybrids, also with *Aciphylla* spp.

Gingidium Forster & Forster f. = *Gingidia*

Ginkgo Kaempf. ex L. Ginkgoaceae. 1 China (SE Eur 5 M yrs ago): ***G. biloba*** L. (maidenhair tree) – Massart's Model, dioec. (male trees [occ. prod. female cones] with earlier leaf-fall & usu. less spreading form, prod. approx. 1 trillion pollen grains a year), flowering when 20 yrs old, mycorrhizal, a relic in Dalou Mts, SW China; app. identical fossils 200 M yrs old, ± continuous with fossil *G. adiantoides* (Unger) Heer, though 6 spp. in Cretaceous, so treated as living pteridosperm by some theorists. Much cult., to 1000 yrs old (840-yr-old tree in Seoul venerated; prob. introd. Eur. & N Am. via Korea) & bole to 9 m circumference in temples, few pests, tolerating pollution & salt; wild hill pops poss. to 3000 yrs old, with basal (following damage) & aerial (in old age) outgrowths (chi-chi) used for bonsai; timber & street-tree though females are objectionable as fallen seeds (poss. once disp. by carrion-feeding dinosaurs) stink of rancid butter or vomit (butyric acid) & make a mess, these (ginkgo nuts, when nauseous layer removed leaving ed. female gametophyte) roasted or consumed with bird's-nest soup & in C18 Jap. with sake), though excess is toxic through affecting Vitamin B6 activity, also an oil-source (incl. in detergents) though causing dermatitis in sensitive people; some insecticidal compounds, high in anti-oxidants, medic. in relief of tinnitus, congestion, glaucoma & headaches, allegedly retarding memory loss (though 2007 trials do not support this), v. imp. in preventing blood-flow, while ginkgoilides used in Parkinson's Disease. Falling lvs in autumn bright yellow due to 6-hydroxykynurenic acid capturing IV radiation & fluorescing it at 'yellow' wavelengths. Many cvs incl. dwarf, fastigiate, weeping & varieg. forms; **'Ohazuki'** ('Epiphylla') with peduncles adnate to petioles. R: P. Crane (2013) *G.*

Ginkgoaceae Engl. Ginkgoidae. 1/1 extant, China (at least 6 genera extinct). Decid. dioec. trees with resinous, slightly mucilaginous wood, strap-shaped (extinct) or fan-shaped lvs all with open dichot. venation on long & short shoots. Ovules on peduncles in axils of lvs or scale-lvs of short shoots, 2(–4+) per peduncle with fleshy collars around bases; meiosis leads to linr. tetrad of spores, lowermost giving gametophyte with chlorophyll & 2 or 3 archegonia. Microsporangiophores borne on catkin-like axis, the gametophytes with 2 prothallial cells, a tube nucleus, a sterile cell & a sperm cell, released as pollen which has

2 large motile sperms with spiral bands of flagella. Fert. may be up to 4 months after poll. & after the ovule has fallen; embryo with 2 cotyledons. n = 12 (incl. XY sex system, though XX allegedly being female, XY male 'unlikely determinants' (Crane))
Extant genus: *Ginkgo*

Ginkgoidae Engl. 1/1 extant (order Ginkgoales). Gymnosperms known from Triassic onwards, trees (poss. also shrubs in Mesozoic) with lvs with dichot. venation. Ovules 2–10, term. on axillary branching axes. Seed large with outer fleshy & inner stony layers. Microsporangiophores in catkins with 2–12 pend. microsporangia; sperms with spiral bands of flagella. Families: Ginkgoaceae, Trichopityaceae (extinct, ? lvs circular in cross-sect.) poss. distantly allied though *Yimaia* spp. from China more clearly related

Ginoria Jacq. Lythraceae. Incl. *Haitia*, 13 Mex. & WI. R: AMBG 97(2010)55

ginseng *Panax pseudoginseng*; **American g.** *P. quinquefolius*; **Siberian g.** *Eleutherococcus senticosus*

Giraldiella Dammer = *Gagea*

Girardinia Gaud. Urticaceae (I). 2 warm & trop. OW. R: KB 36(1981)143. 1 E & NE Afr.: ***G. bullosa*** (Steud.) Wedd. – bark-fibre for sewing. 1 warm & trop. OW: ***G. diversifolia*** (Link) I. Friis (*G. palmata*, Nilgiri nettle) – bark-fibre made into cloth, ropes, bowstrings, in dry season in Java deer eat the stinging lvs just before fighting for mates

Girgensohnia Bunge ex Fenzl. Amaranthaceae (Chenopodiaceae III 3). 5 Eur. (1) to C As. Saline semi-deserts

girigiri *Sphenostylis stenocarpa*

girl *Erythroxylum coca*

Gironniera Gaud. Cannabaceae (Ulmaceae s.l.). 6 Indomal. to Pac. Timber used locally for tea-chests & matches

Gisekia L. Gisekiaceae (Sarcobataceae; Phytolaccaceae s.l.). 1 (variable) OW trop.: ***G. pharnacioides*** L. – ed. R: KB 48(1993)345

Gisekiaceae Nakai (~ Sarcobataceae). Magnoliidae – Caryophyllales. 1/1 OW trop. Prostrate herbs. Lvs opp. Fls. 5-merous in dichasia; P quincuncial; A 5 or 10–15, alt. with P; G (3–)5(–15), 1 ovule per carpel. Infr. heterocarpic mericarps with K ± accrescent. n = 9
Genus: *Gisekia*

Gitara Pax & K. Hoffm. = *Acidoton*

Githopsis Nutt. Campanulaceae (I 8). 4 W N Am. R: SB 8(1983)436

Giulianettia Rolfe = *Glomera*

Givotia Griff. Euphorbiaceae (Ricinod.). 4 E Afr. (1), Madag. (2), Ind. & Sri Lanka (1: ***G. moluccana*** (L.) Sreem. (*G. rottleriformis*) – timber soft & light for figures & toys etc., cult. orn.). Dioec.

Gjellerupia Lauterb. Opiliaceae. 1 NG: ***G. papuana*** Lauterb. R: Willd. 9(1979)37

Glabraria L. = *Brownlowia*

Glabrella M. Möller & W.H. Chen (~ *Briggsia*). Gesneriaceae. 2 China. R: GBS 66(2014)198

gladdon *Iris foetidissima*

Gladiolimon Mobayen. Plumbaginaceae (II). 1 Afghanistan: ***G. speciosissimum*** (Aitch. & Hemsl.) Mobayen

Gladiolus Tourn. ex L. Iridaceae (VI 3). (Incl. *Acidanthera* & sunbird-poll. spp. form. called *Anomalesia*, *Petamenes* (*Homoglossum*) & *Oenostachys*) c. 270 Eur. (6), Medit., trop. Afr. mts (84 – R: P. Goldblatt (1996) *G. in trop. Afr.*) but esp. S Afr. (168 – R: P. Goldblatt & J. Manning (1998) *G. in S Afr.*); Cape 105, 86 endemic – R: Strel. 9(2000)125, Madag. (8; R: BMNHN 4,11(1990)235); (afrikanders, bluebell (S Afr.)), gladiolus being applied to *Acorus* or *Iris pseudacorus* in mediaeval GB. Seeds almost always winged; poll. bees, beetles, flies, butterflies, hawkmoths (white fls) or sunbirds, reflected in fl. structure; app. 32 shifts of poll. vector in 165 spp. in S Afr. Florists' gladiolus (***G. × hortulanus*** L. Bailey) with complex ancestry involving *G. cardinalis* Curtis, *G. carneus* Delaroche, *G. dalenii* Geel (*G. natalensis*, Afr., Madag.) – corms ed. after tannins leached out, medic., app. cult. W Afr. (Goldblatt), *G. oppositiflorus* Herb., *G. papilio* Hook.f. (*G. purpureoauratus*), *G. saundersii* Hook.f. & *G. tristis* L. (all S Afr.), early hybrids known as ***G. × colvillei*** Sweet (1823, *G. cardinalis* × *G. tristis*, leading to 'Nanus Hybrids', ***G. × insignis*** Paxton, 1834) & ***G. × gandavensis*** Van Houtte (1837, *G. dalenii* × *G. oppositiflorus*), over 30 000 cvs derived from 8 spp., with a vogue approaching tulipomania excesses by 1912, record pl. 2.55 m tall; cut-fl. trade in US worth $24 M+ by 2005; other spp. cult. incl. ***G. caryophyllaceus*** (Burm.f.) Poir. (S Afr.) – common weed in SW W Aus.; ***G. communis*** L. (2n = 60, 90, 120; incl. ***G. illyricus*** Koch (Eur. to Turkey, N Afr., widely natur. incl. GB (where prot.) e.g. Scilly

Is. – 'whistling jacks'; 2n = 60, 90), ***G. murielae*** Kelway (*Acidanthera bicolor, G. callianthus, 'G. murieliae'*, 'peacock orchid', sword lily, Ethiopia to E Afr.) – ? triploid, frangipani-scented (to most) fls, ***G. undulatus*** L. (Cape) – invasive in Aus. ***G. candidus*** (Rendle) Goldbl. (*G. ukambanensis*, Tanzania to Arabia) – corm v. imp. wild food in Dhofar, sought also by red-legged partridges; ***G. italicus*** Mill. (S Eur. to C As., Macaronesia & Afghanistan) – perhaps the 'hyacinth' of antiquity, or rose of Sharon of the Bible; ***G. watsonioides*** Bak. (trop. Afr. mts) – up to snow-line

Gladiopappus Humbert. Compositae (Card.-Dicom.) 1 Madag.: ***G. vernonioides*** Humbert – subshrub

gladwin, gladwyn (stinking) *Iris foetidissima*

Glandonia Griseb. Malpighiaceae. 3 trop. S Am. R: MNYBG 32(1981)135

Glandora D.C. Thomas & al. (~ *Lithodora*). Boraginaceae (B.2.2). 8 Medit. T 57(2008)92. Cult. orn. shrublets esp. ***G. prostrata*** (Loisel.) D.C. Thomas (*'L. diffusa', L. prostrata, Lithospermum p.*, W Medit.)

Glandularia J. Gmelin (~ *Verbena*). Verbenaceae (Verb.). c. 100 temp. & warm Am. S Am. Cult. orn. herbs esp. ***G. x hybrida*** (Groenl. & Ruempler) Nesom & Pruski (*V. × hybrida, V. × hortensis*, complex hybrid group involving S Am. spp. incl. *G. incisa* (Hook.) Tronc., *G. peruviana* (L.) Sm., *G. platensis* (Spreng.) Schnack & Covas & *G. canadensis* (L.) Nutt. from N Am.) – garden verbena. ***G. aristigera*** (S. Moore) Tronc. (*V. tenuisecta*, temp. & warm S Am.) – invasive E Aus., S Afr.

Glandulicactus Backeb. = *Ferocactus*

Glandulicereus Guiggi = *Stenocereus*

Glaphyropteridopsis Ching = *Cyclosorus*

Glaphyropteris (Fée) Fée = *Cyclosorus*

Glaribraya H. Hara = *Eutrema*

glasswort *Salicornia* spp., *Salsola* spp.

Glastaria Boiss. Cruciferae (33). 1 SW Turkey, Syria, Iraq: ***G. glastifolia*** (DC.) Kuntze

Glastonbury thorn *Crataegus monogyna* 'Biflora'

Glaucidiaceae Tamura = Ranunculaceae

Glaucidium Sieb. & Zucc. Ranunculaceae (Glauc.; Glaucidiaceae, Hydrastidaceae, Paeoniaceae s.l.). 1 Jap.: ***G. palmatum*** Sieb. & Zucc. – cult. orn. with embryology & cytology like *Paeonia* though immunologically allied with R.

Glaucium Mill. Papaveraceae (I). c. 23 Eur. (2), SW & C As. R: FR 89(1979)499. Many alks. ***G. flavum*** Crantz (sea poppy, horned p., Eur., Medit., natur. N Am.) – seed-oil for illumination & soap, form. medic., cult. orn.

Glaucocarpum Rollins = *Hesperidanthus* (but see Madroño 4(1938)233)

Glaucocochlearia (O. Schulz) Pobed. = *Cochlearia*

Glaucosciadium B.L. Burtt & P. Davis. Umbelliferae (III 10). 1 S Turkey, Cyprus: ***G. cordifolium*** (Boiss.) B.L. Burtt & P. Davis. R: KB 4(1949)229

Glaux L. = *Lysimachia*

Glaziocharis Taubert ex Warm. = *Thismia*

Glaziophyton Franch. Gramineae (V 2). 1 E Braz. mt. tops: ***G. mirabile*** Franch. – reed-like sprouting typical bamboo twigs after burning

Glaziostelma Fourn. = *Tassadia*

Glaziova Bureau = *Amphilophium*

Glazioviantnus G. Barroso = *Chresta*

Gleadovia Gamble & Prain. Orobanchaceae (Orob.). 2 W Himal., W China

Gleasonia Standl. (~ *Henriquesia*). Rubiaceae (I 4). 5 trop. S Am.

Glebionis Cass. (*Xanthophthalmum, ~ Chrysanthemum*). Compositae (Anth.-Gleb.). 2 Eur., N Afr. R: FNA 19(2006)554. Annuals, cult.: ***G. coronaria*** (L.) Spach (*C. c.*, crown daisy, Medit.) – lvs & fls ed. in China & Jap. (chrysanthemum greens, shungiku, tong ho); ***G. segetum*** (L.) Fourr. (*C. s.*, corn marigold, Eur., W As., natur. N Am.) – lvs ed. China

Glechoma L. Labiatae (VII 2c). c. 6 temp. Euras. (Eur. 2).R: BZ 77,12(1992)124. ***G. hederacea*** L. (ground ivy, alehoof, Eur., natur. N Am., where introd. by Eur. settlers for medic. tea) – form. medic. tea for colds & as tonic (acts on mucous membranes) & added to ale to 'clear the brain' & to beer on long voyages

Glechon Spreng. Labiatae (VII 2b). 6–7 Braz., Paraguay. R: FRB 115(1939)18

Gleditsia Clayton. Leguminosae (I 4). 14 (2–3 E N Am., 1 S Am., 1 Caspian, rest China & Jap. to NG). Dioec. trees (honey-locusts), 1 a shrub; spp. from NE Arg. & China v. similar though separated for 60 M yrs. Tepals scarcely different from one another, in spirals, floral parts inconstant in number; lvs pinnate or bipinnate oft. on same tree (cf. constancy of *Astragalus*), usu. with stout branched thorns in axils, arising from uppermost of serial buds; some used as hedges, others timber, shade & cult. orn., fr. & seeds medic. (diuretic etc.) esp. ***G. sinensis*** Lam. (S China, Vietnam) its fr. also used as soap in China. Cult. orn. esp. ***G. triacanthos*** L. (OW temp.) – large fr. & thorns to c. 5m suggesting form. disp. by megafauna, wood for railway sleepers etc. incl. pins to hold insulators on telegraph wires, thorns used to pin together tattered Confederate uniforms in US Civil War, local medic. N Am., many cvs, some thornless trees become thorny after 40–55 yrs, fodder-tree (esp. fr.) in Aus., though invasive there, S Afr., Chile & Arg.

Glehnia Schmidt ex Miq. (~ *Angelica*). Umbelliferae (III 9). 2 NE As., W N Am. ***G. littoralis*** Schmidt ex Miq. – cough treatment in modern Chinese herbalism

Gleichenella Ching (~ *Dicranopteris*). Gleicheniaceae. 1 trop. S Am.: ***G. pectinata*** (Willd.) Ching

Gleichenia Sm. Gleicheniaceae (Gleich.). 10 S Afr., Masc., Mal. to NZ. Leaf-segments v. small. See also *Dicranopteris*

Gleicheniaceae C. Presl. Polypodiidae – Gleicheniales. Incl. Stromatopteridaceae, 6/125 trop. & warm, S temp. R: T 55(2006)710. Terr. oft. thicket-forming ferns of open ground; rhizomes usu. protostelic, creeping, dichotomizing. Rachis (pinnate to) bipinnate or pseudodichotomously branching to 7 m, indeterminate. Veins free. Sori round, of 2–15 pear-shaped sporangia borne abaxially without indusia. Gametophytes massive, slow-growing, with endotrophic mycorrhiza when old, & v. large antheridia (to 100 μm diam.) with several hundred antherozoids. Gametophytes green. n = 22, 34, 39, 43, 56

Classification & genera:

Gleichenioideae (frond at least 1-forked): *Dicranopteris, Diplopterygium, Gleichenella, Gleichenia, Sticherus*

Stromatopteridoideae (frond 1-pinnate): *Stromatopteris*

Fossils back to Mesozoic & poss. Carboniferous. Some used for weaving

Glekia Hilliard (~ *Phyllopodium*). Scrophulariaceae (Man.). 1 Lesotho, E Cape: ***G. krebsiana*** (Benth.) Hilliard. R: NRBGE 45(1988)482. Commem. G.L.E. Krebs

Glenniea Hook.f. Sapindaceae. 8 trop. Afr., Madag., Indomal. R: Blumea 22(1977)411

Glia Sonder (~ *Annesorhiza*). Umbelliferae (III 8). 1 SW Cape: ***G. prolifera*** (Burm.f.) B.L. Burtt (*G. capensis*). R: NRBGE 45(1988)198

Glinus Loefl. ex L. Molluginaceae. c. 10 warm (Eur. 1, Aus. 3) & trop. Some potherbs

Glionnetia Tirv. (~ *Randia*). Rubiaceae ('III 2', Vang.). 1 Seychelles: ***G. sericea*** (Bak.) Tirv. R: BMNHN 4,6(1984)197

Gliopsis Rauschert = *Rutheopsis*

Gliricidia Kunth. Leguminosae (III 23). 5 trop. Am. ***G. sepium*** (Jacq.) Walp. – explosive fr. dehiscence with seeds ejected to 40 m; grown widely as cocoa shade incl. in OW, also as living fence, termite-proof (? coumarin) building timber in C Am., firewood crop (10–15 t/ha/yr dry matter in W Afr.), green manure in Sri Lanka; seeds or powdered bark a poison for rats & mice; fls fried & eaten in C Am.

Glischrocaryon Endl. (~ *Loudonia*). Haloragaceae. 5 S & SW Aus. Looks like *Haloragis* but pollen like *Gunnera*. R: BAIM 10(1975)150

Glischrocolla (Endl.) A. DC. Crypteroniaceae (Penaeaceae). 1 S Afr.: ***G. formosa*** (Thunb.) Dahlgren. R: BN 120(1967)59

Glischrothamnus Pilg. = *Mollugo*

Globba L. Zingiberaceae (II 2). c. 100 E As., Indomal. R: AJB 91(2004)107. Lower cymes usu. replaced by bulbils. ***G. marantina*** L. – cult. spice, bulbils ed. as appetizer; other spp. local medic.

globe amaranth *Gomphrena globosa*; **g. artichoke** *Cynara scolymus*; **g.flower** *Trollius* spp.; **g. mallow, scarlet** *Sphaeralcea coccinea, S. augustifolia* subsp. *cuspidata*; **g. thistle** *Echinops* spp.

Globimetula Tieghem. Loranthaceae (5 6). 13 trop. Afr. R: R. Polhill & D. Wiens, *Mistletoes of Afr.* (1998)209. Sunbirds probe edges of specialized C segments causing reflex, further probes splitting C tube & A coil inwards explosively

Globularia Tourn. ex L. Plantaginaceae (Globulariaceae). 23 Cape Verde Is., Canary Is., Eur. (15), As. Minor, NE Afr. R: BJ 69(1938)318, BAGS 35(1967)305. Herbs & subshrubs, cult. orn. esp. rock garden pls

Globulariaceae DC. = Plantaginaceae

Globulariopsis Compton. Scrophulariaceae (Man.). 7 W & SW Cape. R: O.M. Hilliard, *Tribe Selagineae* (1989)16

Globulostylis Wernham (~ *Cuviera*). Rubiaceae (III 2). 8 W trop. Afr. R: BJLS 173(2013)425. Roux's Model; no ants cf. *C.*

Glochidion Forst. & Forst. f. (~ *Phyllanthus*). Phyllanthaceae (Phyll.-Phyll.). c. 320 As. to Polynesia. Some local timbers & medic. (alks). R: PNAS 100(2003)5264

Glochidocaryum W.T. Wang = *Microula*

Glochidotheca Fenzl (*Turgeniopsis*). Umbelliferae (III 3). 1 Eur., W As.: ***G. foeniculacea*** Fenzl

Glockeria Nees = *Stenostephanus*

Gloeocarpus Radlk. Sapindaceae. 1 Philipp.: ***G. patentivalvis*** (Radlk.) Radlk. R: Blumea 35(1991)389

Gloeospermum Triana & Planch. Violaceae (III 1a). 12 trop. Am. Endosperm 0

Glomera Blume. Orchidaceae (V 10b). Incl. *Glossorhyncha*, c. 130 Mal. to W Pac. R: FGP 9(1974)82, 10(1974)81. Poll. birds (*Glomera s.s.*) & nocturnal Lepidoptera ('*Glossorhyncha*'). Some cult. orn.

Glomeropitcairnia Mez. Bromeliaceae (2). 2 Carib. R: FN 14(1977)1388. Up to 20 litres water recorded in one 'tank'

Gloneria André = *Psychotria*

Gloriosa L. Colchicaceae (Liliaceae s.l.). Incl. *Littonia*, c. 10 OW trop. & S Afr. Alks. ***G. modesta*** (Hook.) Manning & Vinnersten (*L. m.*, cli. bellflower, S Afr.) – cult. orn. with orange fls; ***G. superba*** L. (v. variable esp. Somalia, s.t. subdivided) – flame lily, scrambler (Leeuwenberg's Model) of forest edge & scrub esp. in abandoned cult. with tendrils at leaf-tips (cf. '*Littonia*') & pedicels adnate for 2 nodes above the point of emergence of their vasc. bundles from the stem, nodding fls with reflexed P, spreading A & style projecting from fl. horiz. yet close to one another so that seed-set by selfing poss. in absence of poll., alks incl. colchicine, brittle underground tubers poisonous (used for suicide in Ind.), used for gout since C1, nat. fl. Zimbabwe, adopted as 'nat.' fl. by Sri Lankan Tamils, several selected forms with specific names (& diff. ploidies) in cult., the best poss. '*G. rothschildiana*' (octoploid)

glory bower *Clerodendrum chinense*; **g. bush** *Tibouchina urvilleana*; **g. of the snow** *Scilla forbesii*; **g. of the sun** *Leucocoryne ixioides*; **g. pea** *Clianthus* & *Swainsona* spp.

Glossanthis Polj. = *Trichanthemis*

Glossarion Maguire & Wurd. Compositae (Stifft.). 2 Venez., Guyana. R: Britt. 41(1989)39. Shrubs

Glossocalyx Benth. Siparunaceae (Monimiaceae s.l.). 4 trop. W Afr. Cook's & Roux's Models; in some 1 leaf of each pr rep. by midrib only

Glossocardia Cass. Compositae (Cor.). 12 SE As. to Pac., introd. Afr. R: Blumea 35(1991)466

Glossocarya Wall. ex Griff. (~ *Rotheca*). Labiatae (III; Verbenaceae). 9 Indomal., Aus. (R: JABG 13(1990)17)

Glossochilopsis Szlach. = *Malaxis*

Glossochilus Nees. Acanthaceae (III 2c). 2 S Afr.

Glossodia R. Br. (~ *Caladenia*). Orchidaceae (IV 3b). 2 SE Aus. Wax-lip (orchids)

Glossogyne Cass. = *Glossocardia*

Glossolepis Gilg = *Chytranthus*

Glossoloma Hanst. (~ *Alloplectus*). Gesneriaceae (II 5c). c. 30 trop. S Am. R: SBM 88(2009)

Glossonema Decne = *Cynanchum* (but see KB 37(1982)344)

Glossopappus Kunze. Compositae (Anth.-Leuc.). 1 SW Eur., N Afr.: ***G. macrotus*** (Durieu) Briq. R: BBMNHB 33(1993)143

Glossopetalon A. Gray (*Forsellesia*). Crossosomataceae. 4 W US. R: FNA 9(2014)12

Glossopholis Pierre = *Tiliacora*

Glossorhyncha Ridl. = *Glomera*

Glossostelma Schltr. (~ *Asclepias*). Apocynaceae (Ve; Asclepiadaceae). 12 trop. & S Afr. R: KB 50(1995) 531

Glossostemon Desf. Malvaceae (Bytt.-Theob; Sterculiaceae). 1 Iran, Iraq, Arabia: ***G. bruguieri*** Desf. – moghat r., sold in bazaars, medic.

Glossostigma Wight & Arn. (~ *Mimulus*). Phrymaceae (Scrophulariaceae s.l.). 8 Aus. & NZ, 1 ext. to Afr. & Ind. R: Phytoneuron 2012–39: 23. Minute short-lived herbs ('mud-mats')

Glossostipula Lorence (~ *Randia*). Rubiaceae (Ix.-Cord.). 2 Mex., Guatemala. R: Candollea 41(1986)453

Glottidium Desv. = *Sesbania*

Glottiphyllum Haw. ex N.E. Br. Aizoaceae (V). 16 Karoo. R: Bradleya 11(1993)38. Cult. orn. dwarf succ. with tongue-shaped or ± cylindrical succ. lvs in 2 or 4 ranks

Gloveria Jordaan = *Gymnosporia* (but see SAJB 64(1998)299)

gloxinia *Sinningia speciosa*

Gloxinella (H.E.Moore) Roalson & Boggan (~ *Kohleria*). Gesneriaceae (II 5b). 1 trop. Am.: ***G. lindeniana*** (Regel) Roalson & Boggan

Gloxinia L'Hérit. Gesneriaceae (II 3). 15 trop. Am. R: Selbyana 1(1976)385. Heterogeneous. Bird- & bee-poll. groups of spp. (cf. *Gasteranthus* & *Sinningia*), male euglossines coll. scents

Gloxiniopsis Roalson & Boggan (~ *Gloxinia*). Gesneriaceae (II 5b). 1 trop. Am.: ***G. racemosa*** (Benth.) Roalson & Boggan

Gluema Aubrév. & Pellegrin. Sapotaceae (I 3). 2 trop. W Afr. R: KB 61(2006)179. To 40 m tall

Glumicalyx Hiern. Scrophulariaceae (Manul.). 6 S & SE Afr. R: NRBGE 35(1977)155

Gluta L. Anacardiaceae (I). Incl. *Melanorrhoea*, 30 Madag. (1), Indomal. ***G. renghas*** L. (rengas, Mal.) – timber useful though can cause dermatitis, seeds ed. roasted; ***G. laccifera*** (Pierre) Ding Hou (*M. l.*, SE As.) – lacquer & ***G. usitata*** (Wall.) Ding Hou (*M. u.*, theetsee, thitsi, trop. As.) – source of Burmese lacquer

Glutago Comm. ex Raf. = *Oryctanthus*

Glyceria R. Br. Gramineae (XI). 48 temp. (Eur. 10) esp. N Am. R: BZ 91(2006)258. Luscious pasture-grasses for cows; waterfowl favour fr. form. used by Native Americans; those of ***G. fluitans*** (L.) R. Br. (sweet or manna (because fls shower down when shaken) grass, Euras., natur. N Am., N temp., invasive NZ) the basis of 'manna croup'; ***G. maxima*** (Hartm.) Holmb. (Euras., natur. N Am.) cyanogenic, invasive in Aus. & NZ where causing stock deaths

Glycine Willd. Leguminosae (III 18). Excl. *Neonotonia*, 19 As. to Aus. (16). R: EB 35(1981)275. Cleistogamous infls in axils of lower lvs of most spp., some on rhiz. ***G. max*** (L.) Merr. (soy or soya bean, black b. (in Taiwan), palaeopolyploid prob. selected in NE China c. C11 BC (known to Europeans since C16) from '*G. soja* Sieb. & Zucc.' (C & E As., Taiwan), its weedy form being '*G. gracilis* Skvortzov') – short-day pl. with seeds with 35–40% animal-like protein, richest of all pl. foods, first cult. in US 1924 (first brought in ballast, 1804), now imp. oilseed (poly-unsaturated omega-3 oil; 35% of US oil & fats, 50% of world's oilseed prod. (the 'veg. oil' of much processed food) – 100 M ha (esp. Am.), 81% GM, 260 M t per annum by 2011), animal feed, cosmetics, candles; S Am. forests being cleared to plant glyphosate-resistant (genetically modified) cvs; flour used in shortening & confectionery, ice cream, salad oil, mayonnaise, etc. ('lecithin' emulsifier as in chocolate) & to bulk out e.g. sausages; meal given form & texture = T[exturized]S[oy]P[rotein] eaten by vegetarians as meat subs. (no cholesterol); soy milk made from seeds cooked, mashed & strained much promoted, esp. for the lactose-intolerant, with gypsum precipitated to bean curd (tofu, tou-fou, tau-foo) much used in China & SE As.; seeds cooked with roasted wheat & fermented by *Aspergillus oryzae* give miso paste (first recorded in China 1134 before which it was merely used as a green manure) & soy(a) sauce (now replacing monosodium glutamate in restaurants) used in Worcestershire sauce; seeds fermented by a form of *Bacillus subtilis* (*B. natto*) = natto (nattokinase dissolves blood clots); alleged to reduce incidence of breast cancer (low rates in As.), though excess (phyto-oestrogens – anti-oxidant isoflavones slow sperm mobility & disturb menstrual cycle) linked to that & ovarian cancer; drying oil used in paints, as plasticizer & stabilizer for vinyl plastics, rubber extender & in soaps, detergents, biodiesel (more useable energy & less greenhouse gas than ethanol from sugar-cane), (smudgeable) newspaper ink, paint etc.; forms grown as pulses = edamame; fibre with bamboo for 'wool' cardigans. ***G. wightii*** (Arn.) Verdc. – at germ. releasing canavanine inhibiting growth of e.g. lettuce

Glycocystis Chinnock (~ *Eremophila*). Scrophulariaceae (Myoporaceae). 1 SW Aus.: ***G. beckeri*** (F. Muell.) Chinnock. R: R.J. Chinnock, *Eremophila* (2007)159

Glycorchis D.L. Jones & M.A. Clem. = *Ericksonella*

Glycosmis Corr. Serr. Rutaceae (II 1). c. 50 Indomal. (1 natur. Afr. & Am.). R: PANSP 137(1985)1, GBS 46(1994)113. Alks. ***G. parviflora*** (Sims) Little (*G. citrifolia*, SE As.) introd. 1788 via England to Jamaica & now spread throughout S Am.; ***G. pentaphylla*** (Retz.) DC. (Indomal.) – ed. fr., twigs used as toothpicks in Ind., local medic. (also other spp.); ***G. perakensis*** Naray. (Malay Pen.) – rheophyte

Glycoxylon Ducke = *Pradosia*

Glycydendron Ducke. Euphorbiaceae (Crot.). 1–2 NE S Am.

Glycyrrhiza Tourn. ex L. Leguminosae (III 24). 36 Euras. (Eur. 5) with few in Aus., N Am. & temp. S Am. R: NSL 40(2008)95. Saponins; oligoglycosides incl. glycyrrhizin 50 times sweeter than sugar (but liquorice aftertaste), slows tooth decay. Many spp. local sources of liquorice incl. ***G. glabra*** L. (Medit. to C As.) – used by Roman soldiers to combat thirst (steroid causing water retention) cult. (esp. Russia, Spain, Middle E) for rhizomes, a source of liquorice used in confectionery (incl. Pontefract cakes or pomfrets, orig. medic., later confectionery, made at P., Yorkshire, UK from c. 1660 to 1960), cough mixtures, lozenges & other medic. esp. for sore throats & mouth ulcers (carbenoxalone from r.), form. for indigestion, but for food poisoning in modern Chinese herbalism, used in plug tobacco, brewing stout, r. beer, shoe-polish, soap, fire-extinguishers, fibre for plastics & fibreboard (US); ***G. lepidota*** Pursh (NW N Am.) – local medic.; ***G. uralensis*** Fisch. ex DC. (gan cao, E As.) – TCM

Glycyrrhizopsis Boiss. & Bal. = *Glycyrrhiza*

Glyphaea Hook.f. Malvaceae (Grew.-Apeib.; Tiliaceae). 3 trop. Afr. ***G. brevis*** (Spreng.) Monachino – chewing-stick in Nigeria

Glyphochloa W. Clayton. Gramineae (XXII 3). 9 C & S Ind. R: KB 35(1981)814

Glyphosperma S. Watson = *Asphodelus*

Glyphostylus Gagnepain = *Excoecaria*

Glyphotaenium (J. Sm.) J. Sm. = ? *Enterosora*

Glyptocarpa Hu = *Pyrenaria*

Glyptocaryopsis Brand = *Plagiobothrys*

Glyptopetalum Thw. (~ *Euonymus*). Celastraceae (I). 35 Indomal. R: Reinwardtia 14(2014)184

Glyptopleura Eaton. Compositae (Cich.-Cich.-Micr.). 2 W US. R: FNA 19(2006)361

Glyptostrobus Endl. Cupressaceae (Taxodiaceae). 1 C Vietnam & E Laos (almost extinct), SE China (Eur. Tertiary): ***G. pensilis*** (D. Don) K. Koch (*G. lineatus*) – like *Taxodium* but lvs trimorphic, cult. orn., fewer than 250 mature trees left by 2012, buoyant wood used in floats. R: A. Farjon, *Monogr. Cupressaceae & Sciadopitys* (2005)133

Gmelina L. Labiatae (Premn.). 31 E As., Indomal., Aus., New Caled. (5 – R: FNC 25(2004)22). R: KB 67(2012)295. Some light & medium timbers (grey teak) esp. ***G. arborea*** Roxb. ex Sm. (gamar, yemani, Ind.) – utility timber & firewood crop (up to 30 m^3/ha/annum) & ***G. leichardtii*** (F. Muell.) Benth. (white beech, NE Aus.)

Gnaphaliothamnus Kirpiczn. = *Chionolaena*

Gnaphalium Vaill. ex L. Compositae (Gnap.-Gnap.). Incl. *Omalotheca*, c. 80 cosmop. (Eur. 1+ ***G. undulatum*** L. (S Afr.) natur.; 19 Afr. & Madag.). R: OB 104(1991)155,167, Phytol. 68(1990)241. Cudweeds. ***G. obtusifolium*** L. (N Am.) – local medic.

Gnaphalodes A. Gray = *Actinobole*

Gnephosis Cass. Compositae (Gnap.-Ang.). 16 temp. Aus. R: OB 104(1991)130

Gnetaceae Blume. Gnetidae (Gnetales). 1/30 trop. Usu. lianes, less oft. trees or shrubs with sec. thickening (single cambium in trees, successive ones in lianes, like angiosperms), vessels in xylem, & sec. phloem with companion cells as well as sieve-tubes (unlike most gymnosperms, but derived from diff. initials unlike angiosperms & wood anatomy generally more like conifers). Lvs decussate, everg., with broad lamina & reticulate venation. Fls in whorls in spike-like infls, the whorls subtended by fleshy collars (pl. dioec. to bisex. s.t. with males & females in same infl.); females with nucellus with 3 coats (? 3 integuments or 2 + P), the innermost ext. beyond others as micropylar tube & at maturity middle one stony, outer one fleshy; males with 1 microsporangiophore with 1 or 2 sporangia surrounded by a tubular P, the pollen grains with 3 nuclei (? tube nucleus, sterile cell & sperm cell). Pollen drawn down micropyle by droplet mechanism; no archegonia are formed & poss. any prothallial nucleus can act as egg. Several pollen-tubes may penetrate prothallus giving many zygotes & as there are many prothalli per seed & suspensors may branch giving multiple embryos, polyembryony is of a high order but only 1 usu. reaches maturity in seed. n = 11

Genus: *Gnetum*

Gnetidae Pax (Gnetatae). 3/71 in 3 fams: Ephedraceae, Gnetaceae, Welwitschiaceae. Each of the fams of G. has attained features somewhat characteristic of Magnoliidae, e.g. vessels, angiosperm-like lvs in *Gnetum* (beyond *Drewria* (Cretaceous) & some pollen records, there is no fossil evidence of G., suggesting that other fossils may have been pigeon-holed as 'angiosperms')

Gnetum L. Gnetaceae. 30 Indomal. (21), Amazonia (7), trop. W Afr. (2). R: P. Maheshwari & R. Vasil (1961) *G.* Angiosperm-like (molecular evidence confirms convergence) lvs, vessel elements, double fertilization; some insect-poll., some water-disp., fish-disp. etc. Sources of fibre & ed. seeds esp. ***G. gnemon*** L. (melindjo, Indomal. to N Aus.) – tree (Roux's Model) cult. for ed. young lvs & seeds ('Fructus Beretinus' (Beretina, Philippines) brought to Eur. from Drake's voyage, 1580) cooked & roasted, excellent crackers (emping) made from flour, fibre from bark (inner bark from several spp. NG ('tulip') for bilum bags); ***G. africanum*** Welw. (& ***G. buchholzianum*** Engl.; eru, W. & C Afr.) – liane with locally imp. ed. lvs & tuber; ***G. costatum*** K. Schum. (Papuasia) – seeds, lvs & fls for potherb

Gnidia L. (*Lasiosiphon*). Thymelaeaceae (Daph.). 160 trop. & S Afr. to Arabia (1), Madag. to W Ind. & Sri Lanka. Heterogeneous. Leeuwenberg's Model. ***G. glauca*** Steud. (E Afr.) – source of high-quality paper

Gnomophalium Greuter (~ *Gnaphalium*). Compositae (Inul.). 1 Medit.: ***G. pulvinatum*** (Del.) Greuter. R: Willd. 33(2003)242

Goa bean *Psophocarpus tetragonolobus*; **G. butter** *Garcinia indica*; **G. ipecacuanha** *Naregamia alata*; **G. powder** *Andira araroba*

Goadbyella R. Rogers = *Microtis*

goat's beard *Tragopogon pratensis*, *Aruncus dioicus*; **g. nut** *Simmondsia chinensis*; **g. rue** *Galega officinalis*, *Tephrosia virginiana*; **g. thorn** *Astragalus* spp.; **g. weed** *Ageratum conyzoides*; **g. w., horny** *Epimedium grandiflorum*; **g. willow** *Salix caprea*

gobbo *Hibiscus esculentus*

gobō *Arctium lappa*

Gochnatia Kunth. Compositae (Gochn.). Excl. *Anastraphia*, c. 40 trop. & warm Am., SE As. mts (2). R: RMP 12(1971)1

Godetia Spach = *Clarkia*

Godmania Hemsl. Bignoniaceae (1). 2 trop. Am. Trees

Godoya Ruíz & Pavón. Ochnaceae (III). 2 W S Am. R: BJ 113(1991)173. Lvs simple

Godwinia Seemann = *Dracontium*

Goebelia Bunge ex Boiss. = *Sophora*

Goeppertia Nees (~ *Calathea*). Marantaceae. 248 trop. Am. R: SB 37(2012)626. Many cult. orn. foliage pls esp. ***G. luciani*** (Linden) Borchs. & S. Suárez (*C. l.*) – lvs varieg., ***G. splendida*** (Lemaire) Borchs. & S. Suárez (*C. s.*, Braz.) – lvs spotted yellow above, purple beneath, etc. Some, e.g. ***G. gymnocarpon*** (H. Kenn) Borchs. & S. Suárez (*C. g.*, Braz.) with mass-flowerings; others e.g. ***G. ovandensis*** (Matuda) Borchs. & S. Suárez (*C. o.*) with extrafl. nectaries & with seeds disp. by predatory *carnivorous* ants, which strip off arils (aril-less seeds grow best), but in absence of ants *Eurybia* sp. (Lepidopteran ant-attended herbivore) reduces seed prod. by 66% as opposed to 33%. ***G. allouia*** (Aubl.) Borchs. & S. Suárez (*C. a.*) – medic., baby-clothing, tubers (topee-tampo, topi-tamboo, topinambo) eaten like potatoes in WI; ***G. cyclophora*** (Bak.) Borchs. & S. Suárez (*C. c.*) – medic., wrapping-paper; ***G. elliptica*** (Roscoe) Borchs. & S. Suárez (*C. e.*) – wrapping material; ***G. standleyi*** (J.F. Macbr.) Borchs. & S. Suárez (*C. s.*, N S Am.) – blue dye for bark cloth; ***G. violacea*** (Roscoe) Borchs. & S. Suárez (*C. v.*) – young shoots & fl. buds ed.

Goerkemia Yild. = *Isatis*

Goeldinia Huber = *Allantoma*

Goerziella Urb. = *Amaranthus*

Goethalsia Pittier. Malvaceae (Grew.-Grew.; Tiliaceae). 1 C Am. to Colombia: ***G. meiantha*** (Donn. Sm.) Burret. R: AMBG 51(1964)10

Goethartia Herzog = *Pouzolzia*

Goethea Nees = *Hibiscus*

Goetzea Wydler. Solanaceae (I; Goetzeaceae). 2 Puerto Rico, Hispaniola. ***G. elegans*** Wydler reduced to c. 100 pls (Puerto Rico)

Goetzeaceae Miers = Solanaceae

goflo (Canary Is.) *Pteridium aquilinum*

goji berry *Lycium chinense*

Golaea Chiov. = *Crabbea*

gold fruit *Citrus* spp.; **g. of pleasure** *Camelina sativa*; **g. orach(e)** *Atriplex hortensis*; **g. thread** or **vegetable g.** *Coptis trifolia*

Goldbachia DC. Cruciferae (14). 7 Eur. (1) to temp. As.

gold-dust *Aurinia saxatilis*

golden apple *Spondias cytherea*, [**of the Hesperides**] *Cydonia oblonga*; **g. aster** *Chrysopsis* spp.; **g. cassia** *Chamaecrista fasciculata*; **g. chain** *Laburnum anagyroides*; **g. chestnut** *Chrysolepis chrysophylla*; **g. club** *Orontium aquaticum*; **g. currant** *Ribes aureum*; **g. dewdrop** *Duranta erecta*; **g. drop** *Onosma frutescens*; **g. feather** *Tanacetum parthenium* 'Aureum'; **g. gram** *Vigna radiata*; **g. larch** *Pseudolarix amabilis*; **g. rain** *Laburnum anagyroides*, *Koelreuteria paniculata*; **g. rod** *Solidago* spp.; **g. samphire** *Limbarda crithmoides*; **g. seal** *Hydrastis canadensis*; **g. shower** *Cassia fistula*, *Pyrostegia venusta*; **g. thread** *Coptis trifolia*; **g. walnut, Nigerian** *Lovoa trichilioides*; **g. willow** *Salix alba* 'Vitellina'; **g. yew** *Taxus baccata* 'Aurea'

goldenaster *Ionactis* spp.

goldfussia *Strobilanthes persicifolia*

Goldfussia Nees = *Strobilanthes*

goldilocks *Ranunculus auricomus*

Goldmanella Greenman. Compositae (Cor.). 1 C Am.: ***G. sarmentosa*** (Greenman) Greenman

Goldmania M. Rose ex M. Micheli = *Microlobius*

Golionema S. Watson = *Olivaea*

Gomara Ruíz & Pavón = *Sanango*

Gomaranthus Rauschert = *Sanango*

gomas(h)io *Sesamum indicum*

gombo *Hibiscus esculentus*

Gomesa R. Br. Orchidaceae (V 112h). c. 120 S Am. esp. E Braz. Cult. orn. epiphytes

Gomidesia O. Berg = *Myrcia*

Gomortega Ruíz & Pavón. Gomortegaceae. 1 S Chile: ***G. keule*** (Molina) I.M. Johnston – intoxicating fr., form. imp. narcotic in Chile

Gomortegaceae Reiche. Magnoliidae – Laurales. 1/1 S Chile. Tree, everg., aromatic with ethereal oil-cells in lvs & young stems. Lvs opp., simple, entire; stip. 0. Fls in 'racemes' (but term. fl.!), bisex., P (5–)7(–10), in spirals to 3-merous whorls, inner ones smaller; A 7–13 in spirals, outer 1–3(4) tepaloid with imperf. anthers, others with filaments & anthers dehiscing from base upwards by 2 valves, (1–)3(4) staminodes between A & $\overline{G}$ ((2)3(–5)), 2- or 3-loc., style with 2(3) branches, each loc. with 1 pend., anatr. ovule. Fr. a drupe, yellow, ed., usu. 1-loc. & 1-seeded; seed with large embryo in oily endosperm. 2n = 42
Genus: *Gomortega*
Allied to Monimiaceae

Gomphandra Wall. ex Lindl. Stemonuraceae (Icacinaceae s.l.). c. 60 SE As. to Solomon Is. ***G. quadrifida*** (Blume) Sleumer (Myanmar to W Mal.) – local medic., wood-tar form. to blacken teeth

Gomphia Schreb. = *Ouratea*

Gomphichis Lindl. Orchidaceae (IV 2b). 24 Andes, Braz., Costa Rica

Gomphocalyx Bak. Rubiaceae (IV 15). 1 Madag.: ***G. hernarioides*** Bak.

Gomphocarpus R. Br. (~ *Asclepias*). Apocynaceae (Vc). 20 Afr., Arabia. R: KB 56(2001)779. ***G. fruticosus*** (L.) R. Br. (*A.f.*, Med., Afr., Arabia, natur. N Am.) – latex used as hair-remover from hides in Yemen; ***G. physocarpus*** E. Meyer (*A. p.*, S Afr.) – introd. Hawaii as fibre crop, now natur.

Gomphogyne Griff. Cucurbitaceae (I). 6 E Himal. to C China & Papuas. R: TFB 35(2007)50

Gompholobium Sm. Leguminosae (III 13). Incl. *Burtonia*, 44 Aus. (esp. SW). R: AusSB 21(2008)68. Wedge peas

Gomphostemma Wall. ex Benth. Labiatae (VI). c. 30 SE As., Mal. R: ARBGC 3(1891)227. Rain forest. Some medic. esp. ***G. javanicum*** (Blume) Benth. (SE As., W & C Mal.) said to have antitumour activity

Gomphostigma Turcz. Scrophulariaceae (Budd.). 2 S Afr. R: MLW 77–8(1977)15. 1 a rheophyte, other in Karoo

Gomphotis Raf. = *Thryptomene*

Gomphrena L. Amaranthaceae (II 2). Incl. *Philoxerus*, c. 125 trop. & warm Am., Aus. (33 + 2 introd. – R: AusSB 11(1998)78), natur. OW trop. Heterogeneous. Cult. orn. for bedding & everlastings esp. ***G. globosa*** L. (globe amaranth, OW trop.) – fl. heads subtended by 2 or 3 purple, orange, rose, white or varieg. leafy bracts. ***G. claussenii*** Moq. (Braz.) – efficient at removing zinc & cadmium from contaminated soils

gomuti palm *Arenga pinnata*

gonagra *Rumex hymenosepalus*

Gonatanthus Klotzsch = *Remusatia*

Gonatogyne Klotzsch ex Muell. Arg. (~ *Savia*). Phyllanthaceae (Brid.-Sav.). 1 S Braz.: ***G. brasiliensis*** (Baill.) Muell. Arg. R: A.R. Sm., *Gen. Euph.* (2001)9

Gonatopus Hook.f. ex Engl. Araceae (VII 1). 5 trop. & S Afr. ***G. boivinii*** (Decne) Engl. – tubers & fr. toxic to dogs, man & birds

Gonatostylis Schltr. Orchidaceae (IV 2d). 2 New Caled. R: Gen. Orch. 3(2003)90

gonçalo-alves *Astronium fraxinifolium*

Gongora Ruíz & Pavón. Orchidaceae (V 10). c. 70 trop. Am. R: R. Jenny (1993) *Monograph of the genus G. R & P.* Cult. orn. epiphytes with extrafl. nectaries incl. ***G. batemanni*** (Lindl.) Mabb. & Jenny (*G. cassidea*, C Am.)

Gongrodiscus Radlk. Sapindaceae. 3 New Caled.

Gongronema (Endl.) Decne (~ *Marsdenia*). Apocynaceae (Va; Asclepiadaceae). 15 OW trop.

Gongrospermum Radlk. Sapindaceae. 1 Philippines: ***G. philippinense*** Radlk. R: FM I,11 (1994)548

Gongrostylus R. King & H. Robinson (~ *Eupatorium*). Compositae (Eup.-Ayap.). 1 C Am.: ***G. costaricensis*** (Kuntze) R. King & H. Robinson – epiphytic liane. R: MSBMBG 22(1987)201

Gongrothamnus Steetz = *Distephanus*

Gongylocarpus Schldl. & Cham. Onagraceae. 2 Mex., C Am. G eventually sunk in receptacle

Gongylolepis Schomb. Compositae (Stifft.). 15 trop. S Am. (14 in Guayana Highland, 11 endemics of Venez. G.). R: AMBG 76(1989)997. Trees & shrubs

Gongylosciadium Rech.f. (~ *Pimpinella*). Umbelliferae (III 8). 1 Turkey, Cauc., Iran: ***G. falcarioides*** (Bornm. & Wolff) Rech.f.

Gongylosperma King & Gamble. Apocynaceae (III; Periplocaceae). 2 Malay Pen.

Gongylotaxis Pim. & Kljuykov (~ *Scaligeria*). Umbelliferae (III 5). 1 Afghanistan: ***G. rechingeri*** Pim. & Kljuykov. R: EJB 53(1996)188

Gonialoe (Bak.) Boatwr. & Manning (~ *Aloe*). Asphodelaceae. 3 arid S Afr. R: SB 39(2014)69. ***G. variegata*** (L.) Boatwr. & Manning (*A. v.*, partridge-breasted aloe) – fav. house-pl. with varieg. lvs, incl. comm. hybrids with *Gasteria* spp. = × *Gastonialoe*

Gonioanthela Malme = *Peplonia*

Goniocaulon Cass. Compositae (Card.-Cent.). 1 trop. Afr., Ind.: ***G. glabrum*** Cass.

Goniochilus M. Chase = *Leochilus* (but see CUMH 16(1987)124)

Goniocladus Burret = *Physokentia*

Goniodiscus Kuhlm. Celastraceae (I). 1 Braz.: ***G. elaeospermus*** Kuhlm. – oil extracted from seeds

Goniolimon Boiss. Plumbaginaceae (II). 20 Russia (Eur. 11) to Mongolia, NW Afr. Like *Limonium* but styles hairy & stigmas capitate; some cult. orn.

Gonioma E. Meyer. Apocynaceae (Ie). 1 Cape, 1 SW Madag. R: WAUP 97–2(1997)16. Alks. ***G. kamassi*** E. Meyer (kamassi, Cape) – dense wood (Cape boxwood) used for engraving etc., exported as 'boxwood'

Goniophlebium C. Presl (~ *Polypodium*). Polypodiaceae (IV). c. 23 trop. As. to Fiji (? & Am.). R: Blumea 34(1990)371. Some cult. orn.

Goniopteris C. Presl = *Cyclosorus*

Goniorrhachis Taub. Leguminosae (I 2). 1 SE Braz.: ***G. marginata*** Taub. – timber

Gonioscypha Bak. = *Rohdea*

Goniostemma Wight = *Secamone*

Goniothalamus (Blume) Hook.f. & Thomson. Annonaceae (III 5). Incl. *Richella*, c. 130 Indomal. to New Caled. Some ed. fr. (grape-flavoured); medic., used in child-birth. ***G. tapis*** Miq. (W Mal.) – fragrant fls sold in markets; ***G. wightii*** Hook.f. & Thomson – bark yields a strong fibre

Gonocalyx Planch. & Linden ex Lindl. Ericaceae (VIII 5). 9+ C Am. R: SB 15(1990) 747

Gonocarpus Thunb. (~ *Haloragis*). Haloragaceae. Excl. *Trihaloragis*, 40 Aus. (35) & NZ ext. to SE As. & Jap. R: BAIM 10(1975)164. ***G. micranthus*** Thunb. (*Haloragis m.*, SE As. to Aus.) – aquatic natur. W Galway (Ireland)

Gonocaryum Miq. Cardiopteridaceae (Icacinaceae s.l.). 10 Indomal. to Taiwan. Nozeran's Model; many aluminium hyperaccumulators. Local medic.

Gonocormus Bosch = *Crepidomanes*

Gonocrypta Baill. = *Pentopetia*

Gonocytisus Spach. Leguminosae (III 9). 3 E Med. (Eur. 1). Like *Spartium* but lvs 3-foliolate

Gonolobus Michaux (~ *Marsdenia*). Apocynaceae (Vc; Asclepiadaceae III 3). 100–150 warm Am. (1 introd. W Afr.). ***G. cundurango*** Triana (condurango, cundurango, trop. Am.) – bitters, form. medic.; ***G. edulis*** Hemsl. (guayato, Costa Rica) – fr. ed.

Gonopyrum Fischer & C. Meyer = *Polygonella*

Gonospermum Less. = *Tanacetum* (but see BBMNHB 23(1993)107)

Gonostegia Turcz. = *Pouzolzia*

Gontscharovia Boriss. (~ *Satureja*). Labiatae (VII 2b). 1 C As.: ***G. popovii*** (Fedsch. & Gontsch.) Boriss.

Gonyanera Korth. = *Acranthera*

Gonypetalum Ule = *Tapura*

Gonystylaceae Tiegh. = Thymelaeaceae

Gonystylus Teijsm. & Binnend. Thymelaeaceae (Gon.). 25 Indomal., Pac. Seeds large, arillate, ***G. bancanus*** (Miq.) Kurz (ramin, W Mal.) – catfish-disp. (giving strange flavour to fish flesh) peat swamp-forest tree with knee-r. & inner bark with irritant fibres, lightweight comm. timber used for planking, furniture, mouldings, dowelling, walking canes, clothes-hangers, garden-tool handles, parquet, plywood, etc., much exported, source of agarwood (see *Aquilaria*)

Gonzalagunia Ruíz & Pavón. Rubiaceae (I ?5). c. 35 trop. Am. Roux's Model

Good King Henry *Blitum bonus-henricus*

Goodallia Benth. Thymelaeaceae (Thym.-Daph.). 1 Guyana: ***G. guianensis*** Benth.

Goodenia Sm. Goodeniaceae. Incl. *Calogyne, Catospermum, Neogoodenia, Symphyobasis*, c. 180 Australasia (Aus. 178 mainly endemic, 3 ext. to SE As. with 1 endemic Java). R: FA 35(1992)47, Nuytsia 13(2000)376, (2001)529. G 1-loc. apically, ± 2-loc. basally. ***G. ovata*** Sm. (SE Aus.) – infusion used to quieten children

Goodeniaceae R. Br. Magnoliidae – Asterales. 10/410 trop. & warm, mostly Aus. Sappy shrubs, herbs or even trees with simple to stellate hairs (multicellular with term. cell papillate in *Brunonia*), oft. storing inulin & poisonous; laticifers 0. Lvs simple, in spirals (rarely opp. or whorled), oft. with axillary tuft of hairs; stip. 0. Fls bisex., solit. (rare) or in heads, racemes or cymes, thyrses, etc.; K ((3–)5), lobed, s.t. v. reduced, C (5), 2-labiate or 1-labiate (i.e. adaxial bifid to base), s.t. fan-like, lobes valvate, A 5 alt. with & attached to tube of C or free, anthers with longit. slits, free, connivent or connate, style growing up through & presenting pollen to insects, intrastaminal nectaries s.t. present, $\overline{G}$ (2), s.t. half-inf. or (*Velleia*) $\underline{G}$, (1)2-loc. (4-loc. in *Scaevola porocarya* F. Muell., 2 locs & ovules reduced almost to common condition in other *S.* spp.), ovules 1–∞ per loc., mostly erect or ascending on axile placentas, anatr. & unitegmic. Fr. usu. a capsule dehisc. through 2 valves, s.t. a drupe or nut; seeds usu. flat, s.t. winged, with straight embryo embedded in usu. copious oily endosperm. n = 7–9

Principal genera: *Dampiera, Goodenia, Leschenaultia, Scaevola, Velleia*

Brunonia allied to *Leschenaultia-Anthotium-Dampiera* group. Fam. allied to Calyceraceae & Compositae

Some cult. orn.

Goodia Salisb. Leguminosae (III 14). 6 SW & SE Aus. R: Muelleria 29(2011)142

Goodmania Rev. & Ertter (*Gymnogonum*). Polygonaceae (Erio.-Erio.). 1 Calif. & Nevada: ***G. luteola*** (C. Parry) Rev. & Ertter. R: FNA 5(2005)433

Goodyera R. Br. Orchidaceae (IV 2d). c. 90 N temp. (Eur. 1: ***G. repens*** (L.) R. Br.), Indomal., Mozambique (1), Madag. Local medic. N Am.

Gooringia F. Williams = *Odontostemma*

goose plum, wild *Prunus hortulana, P. munsoniana, P. rivularis*

gooseberry *Ribes uva-crispa*; **American g.** *R. cynosbati*; **Barbados g.** *Pereskia aculeata*; **Cape g.** *Physalis peruviana*, **Ceylon g.** *Dovyalis hebecarpa*; **Chinese g.** *Actinidia deliciosa*; **Indian g.** *Phyllanthus emblica*; **Otaheite g.** *Phyllanthus acidus*; **prickly g.** *Ribes cynosbati*; **star g.** *Breynia androgyna*

goosefoot *Chenopodium album*

goosegog *Ribes uva-crispa*

goosegrass *Galium aparine*

gooseneck loosestrife *Lysimachia clethroides*

gooya *Owenia acidula*

gopher apple *Licania michauxii*; **g. wood** see *Pinus*, though poss. *Cupressus sempervirens*

goraka *Garcinia gummi-gutta*

Gorceixia Bak. Compositae (Vern.-Lychn.). 1 SE Braz.: ***G. decurrens*** Bak.

Gordonia Ellis. Theaceae. Excl. *Polyspora*, c. 25 warm N. Am. Cult. orn. everg. trees & shrubs esp. ***G. lasianthus*** (L.) Ellis (loblolly bay, black laurel, SE US) – cabinet wood, hybridized with *Franklinia* (q.v)

Gorgonidium Schott. Araceae (VII 4). 4 Peru, Bolivia, N Arg. R: BJ 109(1988)529

gorli oil *Caloncoba echinata*

Gormania Britton = *Sedum*

Gorodkovia Botsch. & Karav. = *Smelowskia*

gorse *Ulex europaeus;* **dwarf g.** *U. minor, U. gallii*

Gorteria L. Compositae (Arct.-Gort.). 8 SW Cape to Namibia. R: Willd. 44(2014)105. ***G. diffusa*** Thunb. – poll. bee-flies attracted by dark marks on ray-florets, mimicking resting flies

Gosela Choisy. Scrophulariaceae (Man.; Globulariaceae). 1 S Afr.: ***G. eckloniana*** Choisy. R: SAJB 56(1990)477

Gossampinus Schott & Endl. = *Ceiba*

Gossia N. Snow & Guymer (~ *Austromyrtus*). Myrtaceae (II 10). c. 30 Queensland, NG, New Caled., Fiji

Gossweilera S. Moore = *Omphalopappus*

Gossweilerochloa Renvoize = *Tridens*

Gossweilerodendron Harms (~ *Prioria*). Leguminosae (I 2). 2 Gulf of Guinea forests. ***G. balsamiferum*** (Vermoesen) Harms (agba, tola wood) – timber for furniture etc., copals

Gossypianthus Hook. (~ *Guilleminea*). Amaranthaceae (II 2). 2 N & C Am.

Gossypioides Skovsted ex J.B. Hutch. Malvaceae (Malv.-Goss.). 2 trop. Afr., Madag.

Gossypiospermum (Griseb.) Urb. = *Casearia*

Gossypium L. Malvaceae (Malv.-Goss.). 49 warm temp. to trop. (Aus. 17). R: P.A. Fryxell (n.d.) *Nat. Hist. Cotton tribe*: 37, Rheedea 2(1992)109. Petit's Model; fls visited by bees & (Am.) hummingbirds; extrafl. nectaries. Cotton (26 M t per annum (81% GM), 9.1 M ha in Ind. (main grower); most imp. cash-crop in China, genetically modified forms resistant to herbicides & bollworm caterpillars grown in US, Aus. etc.): seeds covered with long hairs (single-celled, 3000 times longer than wide), which when dried are flat & can be spun (200 M per kg) – unbleached is calico [from Calicut, Ind.], plain weave = gingham, 'Chambray' etc., also twill (with diagonal ridges in fabric as in denim etc.), satin, corduroy, muslin, terrycloth, & short hairs (fuzz) used for felt, paper, twine, viscose (for rayon), sausage-skins & cellophane & in plastics & photographic film; oil (poly-unsaturated) extracted & (C Am.) flour (high protein), incaparina; seed-cake valuable cattle-feed. Cotton used for cotton wool, thread, carpets etc., oil in cooking & soap powders; poss. male contraceptive (China) as gossypol (protective terpenoid in glands, toxic to humans (heart & liver), in small quantities inhibits sperm formation, glandless seeds now genetically engineered, allowing seeds to be used as protein sources); petals source of yellow dyes (Ind.); over 1000 diff. uses claimed for waste in former USSR, incl. growth promoters, concrete plasticizers, food preservatives, etc. Complex & controversial history; diploid ***G. arboreum*** L. (tree cotton cult. Pakistan 1800 BC & taken to Middle E in 1st millenium BC) more variable than diploid ***G. herbaceum*** L. (domesticated E Afr.) in Afr., independently selected from diff. ancestors, though their genome claimed as also appearing 1–2 mya (transoceanic disp.) in diff. NW cottons (diverging 5–10 mya) which (comm. ones) are tetraploid esp. ***G. barbadense*** L. (*G. peruvianum*, Sea Island c., Peruvian c.), the one widely planted in S US (seeds known from deposits over 5000 yrs old in coastal Ecuador), Egyptian cotton derived from a single plant in a Cairo garden (1820). Naturally coloured cvs selected in anc. times; some cult. orn. incl. ***G. sturtianum*** J.H. Willis (Sturt('s) desert rose, Aus.) – emblem of N Terr., Aus.

gotu cola *Centella asiatica*

gou qi zi *Lycium chinense*

Gouania Jacq. Rhamnaceae (4). c. 50 trop. & warm (Am. 15 – R: AMBG 99(2014)501, Afr. 2, Madag. & Ind. Ocean 5, Aus. 2, New Caled. 1). Some with watch-spring tendrils; saponins lead to use as shampoo etc. – ***G. lupuloides*** (L.) Urb. (trop. Am.) used as chewing-stick gives soapy mouthwash when chewed, considered aphrodisiac in Carib.; ***G. polygama*** (Jacq.) Urb. (Carib.) – with *Smilax domingensis* Willd., *Pimenta dioica* & sugar, basis of pru, a Cuban drink

Gouinia Fourn. ex Benth. Gramineae (XXIX 3). Excl. *Schenckochloa*, 14 trop. Am. R: ABM 23(1993)13

Gouldia A. Gray = *Kadua*

Gouldochloa Valdés R. & al. = *Chasmanthium*

goumi *Elaeagnus multiflora*

Goupia Aubl. Goupiaceae. 4–5 Guyana, N Braz. ***G. glabra*** Aubl. (Guyana) – tall buttressed tree (Roux's Model), fine but smelly timber for outdoor use incl. railway sleepers

Goupiaceae Miers (~ Celastraceae). Magnoliidae–Malpighiales. 1/4–5 NE S Am. Everg. trees accum. aluminium. Lvs in spirals, becoming distich.; stip. narrow, caducous. Fls bisex., reg., (4)5-merous; K imbr., basally connate; C long, indupl.-valv.; nectary-disk large, cupular; A5 inserted on inner edge of disk, anthers with longit. slits; G (5), each loc. with several basal, axile ovules per loc. Fr. a drupe, 1–3-loc. Seeds with straight embryo in copious endosperm
Genus: *Goupia*
Part of dismembered heterogeneous Celastraceae (q.v.)

gourd See Cucurbitaceae, *Benincasa, Cucumis, Cucurbita, Lagenaria, Trichosanthes,* also *Crescentia,* (of Bible) *Citrullus colocynthis;* **bottle g.** *L. siceraria;* **silverseed g.** *Cucurbita argyrosperma;* **stuffing g.** *Cyclanthera pedata*

gourd-tree *Adansonia gregorii*

Gourliea Gillies ex Hook. & Arn. = *Geoffroea*

goutweed *Aegopodium podagraria*

Govenia Lindl. Orchidaceae (V 13e). 24 trop. Am. (Florida 1). Some cult. orn. terr.

governor's plum *Flacourtia indica*

gowan Scottish word for var. pls with yellow fls, e.g. *Ranunculus* & *Taraxacum* spp.

Goyazia Taubert. Gesneriaceae (II 5b). 2 Braz. R: Selbyana 1(1976)392

Goyazianthus R. King & H. Robinson. Compositae (Eup.-Alom.). 1 Braz.: ***G. tetrastichus*** (Robinson) R. King & H. Robinson. R: MSBMBG 22(1987)258

Goydera Liede = *Vincetoxicum* (but see Novon 3(1993)265)

Grabowskia Schldl. = *Lycium*

gra-chai *Bosenbergia rotunda*

Graciela Rzed. = *Strotheria*

Gracelianthus R. Gonzalez & Szlach. = *Aulosepalum*

Graderia Benth. Orobanchaceae (Buch.; Scrophulariaceae). 5 Afr. & Socotra

Gradyana Athie-Souza & al. Euphorbiaceae (Hippom.). 1 NE Braz.: ***G. franciscana*** Athie-Souza & al. R: SB 40(2015)527

Graellsia Boiss. Cruciferae (47). 7 Morocco (1), Turkey to Pakistan. R: BJLS 102(1990) 17

Graffenrieda DC. Melastomataceae. 45 trop. Am.

Grafia A. Hawkes = *Phalaenopsis*

Grafia Reichb. Umbelliferae (III 5). 1 Eur.: ***G. golaka*** (Hacq.) Reichb.

Grahamia Gillies (~ *Anacampseros*). Anacampserotaceae. 1 Chile: ***G. bracteata*** Gillies – subshrub

Grajalesia Miranda. Nyctaginaceae (VI). 1 Mex.: ***G. fasciculata*** (Standl.) Miranda

gram whole pulses; **black g.** *Vigna mungo;* **golden** or **green g.** *V. radiata;* **horse g.** *Macrotyloma uniflorum;* **red g.** *Cajanus cajan*

grama or **gramma** *Bouteloua* spp.; **blue g.** *B. gracilis;* **side-oats g.** *B. curtipendula*

Gramineae Juss. (Poaceae). Magnoliidae – Poales. 752/11 300 cosmop. (Aus. 230/1300+; N Am. 236/1373) but esp. trop. & N temp. sub-arid. R: A. Arber (1934) *The Gramineae;* E.A. Kellogg in Kubitzki (1990-) **13** (2015). Usu. perenn. & oft. rhiz. herbs (many with Tomlinson's Model), annuals or (bamboos; F.A. McClure (1966) *The bs.*) ± woody & tree-like but without sec. thickening, many with McClure's or Bell's Models; cell-walls, esp. epidermis ± strongly silicified (silica bodies = phytoliths, often taxonomically and forensically useful), vessel elements usu. in all veg. organs; stems usu. terete & usu. with hollow internodes & prominent nodes; r. oft. with r.-hairs but oft. with endomycorrhizae also. Lvs distich. (spirals in *Micraira*), never 3-ranked, with usu. open sheath & usu. elongate lamina usu. with basal meristem & pr of basal auricles (narrowed to a petiolar base (pseudopetiole, an ancestral feature) above sheath as in many bamboos); ligule usu. adaxial at junction of lamina & sheath, rarely 0. Fls usu. wind-poll., usu. bisex., in 1–8-flowered spikelets in spike-like to panicle-like sec. infls; spikelets usu. with pr of subopp. bracts (glumes) & 1– several distich. florets oft. on zig-zag rhachilla, the florets usu. comprising a pr of subopp. subtending scale-like bracts (lemma & palea s.t. considered outer P), 2 or 3 small lodicules (inner P as concluded by Robert Brown (early C19) because they express genes usu. active in petals elsewhere), the upper bract (palea) s.t. interpreted as derived from P, up to 6 or more lodicules in Bambusoideae), A (1–)3 or 6 (esp. Bambusoideae, where up to >100 in *Ochlandra*), anthers elongate, basifixed but deeply sagittate so as to appear versatile, with longit. slits & nearly smooth pollen grains, $\underline{G}$ (2 (3 in

Bambusoideae)), 1-loc. with 2(3) stigmas, oft. large & feathery, ovule 1, orthotropous to almost anatr. etc., (1)2-tegmic. Fr. (caryopsis) usu. encl. in persistent lemma & palea, usu. dry-indehiscent, integuments adnate to pericarp (though free in c. 11% of genera), the seed rarely falling free of these accessory structures such as when pericarp becomes mucilaginous when wet & expelling the seed on drying out, fleshy in some bamboos (e.g. *Alvimia* & *Olmeca* (NW), *Dinochloa, Melocalamus, Melocanna, Ochlandra* (OW)); embryo straight with well-dev. plumule covered by a closed cylindrical coleoptile, radicle with a similar coleorhiza, & enlarged lat. cotyledon (scutellum), all peripheral to copious starchy endosperm usu. with proteinaceous tissue & s.t. also oily, rarely (*Melocanna*) 0. x = 2–23+

Classification & principal genera (Kellogg – N.B. bold roman numbers = her tribes):

I. **Anomochlooideae** (herbaceous forest pls resembling Marantaceae, but with bambusoid leaf anatomy, pseudopetioles, & ligular hairs (or 0); R: T 45(1996)642): trop. Am. – *Anomochloa, Streptochaeta* (only)

II. **Pharoideae** (herbaceous perenn.; blades resupinate with pseudopetioles) – trop.: *Leptaspis, Pharus, Scrotochloa* (only)

III. **Puelioideae** (herbaceous perenn. with hollow culms & broad blades with pseudopetioles & fringed ligules) – W trop. Afr.: *Guaduella, Puelia* (only)

IV. **Ehrhartoideae** (herbaceous perenn. s.t. with woody caudex & fringe-like ligules; spikelets 1-flowered with glumes scarcely dev. or 2; 3 tribes with *Suddia* unplaced):
1. **Phyllorachideae** (monoec. perenn. with hard culms; blades cordate to sagittate with pseudopetioles) – trop. Afr., Madag.: *Humbertochloa, Phyllorachis* (only)
2. **Ehrharteae** (rhiz. herbs with membranous or fringed ligule; spikelets with 1 fert. fl. + 2 empty lemmas + 2 glumes, A 2–4, 6) – OW warm: *Ehrharta*
3. **Oryzeae** (perenn. to ann. herbs ± pseudopetioles; spikelets with 1–3 fls, 1 fert. +0–2 just empty lemmas, glumes small or 0; 11 genera, damp areas, some aquatics): *Leersia, Oryza* (rice) (1. Oryzinae), *Zizania* (2. Zizaniinae)

V. **Bambusoideae** (Biotropica 11(1979)161; woody or herbaceous, blades with pseudopetioles; ligule membranous; some with 3 lodicules, A 6 & 3 stigmas; 3 tribes with similar leaf anatomy; trop. esp. forest & aquatic habitats, some with periodic flowering as rarely as every 120 yrs; mostly tetraploids, *Dendrocalamus* hexaploids):
1. **Arundinarieae** (**IV**.; culms woody, usu. hollow; with basipetal branch development; lvs with outer ligule; spikelets laterally compressed; c. 26 genera, many hybridizing) – temp. woody bamboos to trop. mts: *Arundinaria, Chimonobambusa, Fargesia, Oldeania, Phyllostachys, Sasa, Yushania*
2. **Bambuseae** (**V**; culms woody, to 40 m; with acropetral or bidirectional development; lvs with outer ligule; spikelets laterally compressed; c. 63 genera) – *Chusquea* (1. Chusqueinae), *Arthrostylidium, Aulonemia* (2. Arthrostylidiinae), *Guadua* (3. Guaduinae), *Bambusa, Dendrocalamus, Gigantochloa* (4. Bambusinae), *Melocanna, Schizostachyum* (5. Melocanninae), *Nastus* (6. Hickeliinae), *Racemobambos* (7. Racemobambusinae)
3. **Olyreae** (**VI**; incl. Parianeae; monoec. creeping herbs or 'rambling canes' with broad laminas; outer ligule 0; spikelets 1-flowered; c. 21 genera, largely S Am.) – *Buergersiochloa* (1. Buergersiochloinae); *Pariana* (2. Parianinae), *Olyra* (3. Olyrinae)

VI. **Pooideae** (incl. Stipoideae; T 29(1980)664; perenn. to ann. herbs differing from other subfams in stigmas borne on style-branches arising separately from G apex; hybridization & polyploidy common; c. 177 [too many recog.] genera, largely N temp. but also on trop. mts; 10 tribes):
1. **Brachyelytreae** (**VII**; 1-flowered spikelets) – *Brachyelytrum* (only)
2. **Nardeae** (**VIII**; incl. Lygeae; microhairs, lodicules 0, stigma 1) – *Lygeum, Nardus* (only)
3. **Phaenospermateae** (**IX** incl. Duthieinae; ligule membranous; lodicules 0, 2 or 3; 8 genera) – *Stephanachne*
4. **Stipeae** (**X**; R: T 61(2012)18; tussocky or reed-like, spikelets 1-flowered, lemmas usu. with term. hygroscopic awn; c. 28 genera): *Austrostipa, Nassella, Piptatherum, Piptochaetium, Stipa*
5. **Meliceae** (**XI**; incl. Brylkinieae; sheath margins connate; lodicules thick, usu. united for most of length) – *Glyceria, Melica*
6. **Diarrheneae** (**XII**; blades glossy, midvein usu. excentric; 2–7-flowered spikelets & ellipsoid knobbed fr. with 2 term. stigmas) – *Diarrhena* (only)

7. **Brachypodieae** (**XIII**; infl. unbranched, spikelets with (3)5–24 fls, distal ones reduced) – *Brachypodium* (only)
8. **Bromeae** (**XIV**; ~ Triticeae; auricles usu. 0; sheath margins connate; infl. a panicle, 2–30 florets per spikelet, lemmas herbaceous, awns 1 or more per lemma, dorsal, lodicules glabrous): *Bromus, Littledalea* (only)
9. **Triticeae** (**XV**; auricles usu. present; sheath margins usu. free; infl. usu. a spike, 1-several (variable) fls per spikelet, lemmas hard or herbaceous, awns term. (or 0), lodicules usu. hairy; poss. best treated as 1 genus): *Elymus, Hordeum* (barley), *Leymus, Secale* (rye), *Triticum* (wheat)
10. **Poeae** (**XVI**; incl. Hainardieae, Seslerieae; infls spikes, racemes or non-capitate panicles, spikelets laterally compressed, usu. disarticulating above glumes; endosperm with lipids; c. 106 genera (some unplaced) in 15 subtribes) – *Amphibromus* (1. Torreyochloineae), *Arrhenatherum, Avena* (oats), *Helictotrichon, Lagurus, Trisetum* (2. Aveninae), *Phalaris* (3. Phalaridinae), *Anthoxanthum* (4. Anthoxanthinae), *Agrostis, Ammophila, Briza, Calamagrostis, Deyeuxia* (5. Agrostidinae), *Scolochloa* (6. Scolochloinae), *Avenula, Deschampsia, Holcus* (7. Airinae), *Ammochloa* (8. Ammochloinae), *Cynosurus* (9. Cynosurineae), *Festuca, Lolium* (10. Loliinae), *Dactylis* (11. Dactylidinae), *Parapholis* (12. Parapholiinae), *Sesleria* (13. Sesleriinae), *Puccinellia* (14. Coleanthinae), *Alopecurus, Phleum, Poa* (15. Poinae incl. Alopecuridinae)

VII. **Aristidoideae** (panicles of 1-flowered needle-like spikelets; lemma with 3-branched awn; trop.): *Aristida, Stipagrostis*

VIII. **Panicoideae** (incl. Centothecoideae; leaf anatomy usu. with Kranz syndrome; spikelets oft. 2-flowered, the lower male or sterile; lodicules 2, A 3, stigmas 2; mainly trop., c. 212 genera in 8 tribes):

1. **Steyermarkochloeae** (**XVII**; monoec., veg. culms with 1 leaf, flowering ones with more, aerenchymatous) – *Steyermarkochloa* (only)
2. **Tristachyideae** (**XVIII** ~ Arundinelleae; spikelet cluster falling as a unit; lemma oft. tufted hairy; 8 genera) – *Loudetia, Tristachya*
3. **Chasmanthieae** (**XIX** ~ Centotheceae; blades broad, s.t. with pseudopetioles; A 1–3; 6 genera) – *Chasmanthium, Zeugites*
4. **Gynerieae** (**XX**; dioec. blades disarticulating from sheaths; lemma plumose, A 2) – *Gynerium* (only)
5. **Centotheceae** (**XXI** incl. Thysanolaeneae; superfl. similar to some Bambusoideae; 5 genera) – *Centotheca, Thysanolaena*
6. **Andropogoneae** (**XXII** incl. Arundinelleae, excl. Paspaleae; infls of fragile racemes (s.t. paniculate) with usu. paired spikelets the disp. unit; trop. savannas, c. 90 genera (many, e.g. *Chrysopogon, Coix* (Coicinae), *Polytoca* (Chionachininae R: Blumea 47(2002)553), unplaced) in 7 subtribes) – *Arundinella, Garnotia,* (1. Arundinellinae), *Tripsacum, Zea* (maize) (2. Tripsacinae), *Hemarthria, Rottboellia* (3. Rottboelliinae), *Dimeria, Ischaemum* (4. Ischaeminae incl. Dimeriinae), *Eulalia, Imperata, Miscanthus, Saccharum* (sugar-cane), *Sorghastrum, Sorghum* (5. Saccharinae incl. Sorghinae), *Apocopsis* (6. Germainiinae), *Andropogon, Bothriochloa, Cymbopogon, Heteropogon, Hyparrhenia, Schizachyrium, Themeda* (7. Andropogoninae incl. Anthistiriinae)
7. **Paspaleae** (**XXIII**; not readily characterizeable, 31 genera in 3 subtribes) – *Mesosetum* (1. Arthropogoninae), *Hymenachne, Otachyrium* (2. Otachyriinae), *Axonopus, Paspalum* (3. Paspalinae)
8. **Paniceae** (**XXIV**; spikelets falling entire, lemma & palea of upper floret encl. fr.; pantrop., c. 72 genera (some incl. *Dichanthelium, Echinochloa, Sacciolepis,* unplaced) in 6 subtribes) – *Digitaria* (1. Anthephorinae incl. Digitariinae), *Acroceras, Oplismenus* (2. Boivinellinae), *Cleistochloa, Neurachne* (3. Neurachininae (Aus., W Pac.)), *Cenchrus, Pseudoraphis, Setaria, Spinifex* (4. Cenchrinae incl. Setariinae, Spinificinae), *Eriochloa, Melinis, Urochloa* (5. Melinidinae; R: BB 138(1988)), *Panicum* (6. Paniceae)

IX. **Danthonioideae** (~ Arundinoideae; R: AMBG 97(2010)306; auricles 0, spikelets with 1–many fls, the distal oft. reduced; style-branches 2, widely separated; 17 genera) – *Cortaderia, Danthonia, Pentameris, Rytidosperma, Arundo, Phragmites*

X. **Chloridoideae** (leaf anatomy with 'Kranz syndrome' assoc. with C_4 photosynthesis app. adapted to high light intensities; spikelets shattering at maturity; mainly trop., c. 132 genera (some unplaced) in 5 tribes):

1. **Centropodieae** (**XXV** ~ Arundinoideae; lemma with 9 veins) – *Centropodia*
2. **Triraphideae** (**XXVI**; ligule a fringe of hairs; lemmas keeled) – *Triraphis*
3. **Eragrostideae** (**XXVII**; panicles or racemes with 1–several-flowered spikelets & 3(–13)-nerved lemmas; trop. with many colonist spp.; 3 subtribes) – *Enneapogon* (1. Cotteinae), *Eragrostis* (2. Eragrostidinae), *Uniola* (3. Uniolinae)
4. **Zoysieae** (**XXVIII** ~ Cynodonteae, incl. Sporoboleae; R: KB 28(1973)37; ligule usu. a fringe of hairs, spikelets with 1 fl., pericarp free from seedcoat) – *Sporobolus, Zoysia*
5. **Cynodonteae** (**XXIX**, incl. Chlorideae, Orcuttieae, Pappophoreae, excl. Zoysieae; not readily characterized; 82 genera (some, e.g. *Dactyloctenium*, unplaced) in 7 subtribes) – *Bouteloua, Distichlis, Muhlenbergia* (1. Boutelouinae incl. Monanthochloinae (halophytes)), *Orcuttia* (2. Orcuttiinae), *Ctenium, Gouinia* (3. Gouiniinae), *Triodia* (4. Triodiinae, Aus.), *Astrebla, Chloris, Cynodon, Eleusine, Leptochloa, Lepturus* (5. Eleusininae incl. Chloridinae, LepTureae), *Tripogon* (6. Tripogoninae), *Tridens* (7. Pappophorinae)

XI. **Micrairoideae** (~ Arundinoideae, incl. Eriachneae, Hubbardieae, Isachinae; R [s.s.]: SB 32(2007)77; herbs, ligule a fringe of hairs; spikelets usu. with 2 fls, styles 2; 9 genera) – *Eriachne, Isachne, Micraira*

XII. **Arundinoideae** (herbs, ligule membranous; style-branches 2; c. 19 genera) – *Arundo, Hakonechloa, Molinia, Phragmites*

On all continents & to high alts, in tropical forests to deserts, characterizing the steppes of C As., grasslands of Africa, prairies of N Am., pampas & bamboo forests of S Am., many of these maintained symbiotically with grazing herbivores; anthropogenic pastureland and therefore much of Euras. Afr., Aus. & Am. landscape similarly maintained, incl. biodiverse habitats such as chalk grassland in UK. Accum. of silica (up to 7% dry wt) and its recycling through herbivory or decay makes it available to diatoms, the app. increase in diatomite in Miocene assoc. with early expansion of grasslands

A v. natural family, with possible fossils from Palaeocene/Eocene boundary (Tennessee), though phytoliths typical of grasses claimed in Late Cretaceous dinosaur coprolites (Ind.): it is argued that grasses arose in shady moist environments invading open habitats several times independently, the living representatives of the earliest lineages being perenn. broad-leaved pls. of trop. forests. At least 80% have a polyploid ancestry & this has been analyzed in e.g. wheat, *Sporobolus* (*Spartina*) *anglicus* etc., though reduction in chromosome number via 'polyhaploidy' recorded. Vegetative spread of clones estimated to be 1000 yrs old known in *Festuca.* The combination of basal shoots (tillers) & intercalary meristems allows grasses to tolerate burning & grazing which eliminate their competitors & hold up successional sequences to forest; in turn, saliva from herbivorous mammals appears to cont. growth-factors stimulatory to the grass. Silica poss. grazing-tolerant adaptation (cf. *Equisetum*) but, after fires in mills etc., lumps of 'glass' formed form. thought to be 'thunderbolts' & the cause rather than result of fire! Few (esp. Aus.) with toxins (HCN, oxalates etc.). Accum. of glycine betaines in some spp. may be linked to tolerance of saline habitats; uniquely grasses acquire iron through chelation of ferric ions

Though no parasites, succ. or epiphytes, considerable variation from the lamina-less sheaths of *Spartochloa* to laminas 5 m long in *Chusquea* (*Neurolepis*), from the solit. 1-flowered spikelet of *Aciachne* to the 2 m plumes of *Gynerium*, from the familiar dry 'seeds' to the fleshy berries several cm long of *Melocanna*, from annuals a few mm tall to 40 m bamboos; the stereotyped pooids typical of the N temp. are misleadingly uniform. Most are wind-poll., the pollen viable for less than a day though of immense effect on hay-fever sufferers (allergenic proteins, some assoc. with antimicrobial activity). Some are apomictic or cleistogamous (e.g. wheat; cleistogamy more common than in any other fam. (5% of spp., in 19% of genera, some prod. inside sheaths of lower nodes as in *Amphicarpum, Microlaena*)), while others have bulbils in infls (esp. Arctic spp.); vivipary reported in *Melocanna*. Many are wind-disp. (ballistic disp. in *Raddia* & *Sucrea*) or transported ectozoically (awns, hooks etc.), the awns s.t. hygroscopic & screwing the fr. into the ground, some fleshy & endozoic; some whole infls acting as 'tumbleweeds'

Oft. confused with sedges & rushes by laypersons, G. differ from Juncaceae (P 6, fr. a 1–3-loc. capsule with 3 to ∞ seeds) & Cyperaceae (1 scale beneath each floret, stems usu. 3-angled & solid) in a number of clear-cut characters

Although not food of early hominids or modern related animals, most major civilizations are based on the triploid endosperm of G. (AMBG 68(1981)87, cereals occupying

7 M km^2: wheat, barley, oats, rye etc. in Euras., millets & tef in parts of Afr., rice in E As., maize in C Am. (wheat, rice & maize on 5.5 M km^2 provide more than 50% of calories consumed by humans); only the Maori have become (form.) international power without, being based on sweet potato), & on animals raised on G. as forage, as well as bamboos as building materials in many trop. societies (& now temp. flooring (laminate 'plyboo'), furniture, acupuncture needles, bicycle-frames (UK since 1890s), fuel brickettes, charcoal for calligraphy, clothing fibre, carpets, etc., by 2009 worth 70.9 billion yuan a yr in China alone). Other imp. prod. include sugar-cane & aromatic oils (esp. *Cymbopogon* spp.) for soap, cooking etc., thatching (e.g. 60 K thatched houses in UK) & weaving materials, sand-binders (*Ammophila* etc.) & toxic metal-tolerant colonist grasses (*Agrostis* spp. etc.) for reclamation of derelict land; others are sources of paper (*Arundo, Eulaliopsis, Leymus, Stipa* etc. spp.), ed. bamboo shoots, beads (*Coix, Polytoca*), reeds for wind instruments (*Arundo donax* & other spp.), fishing-rods, lawn-grasses (esp. *Festuca* spp. in temp., *Cynodon dactylon* etc. in trop.) & turf for playing-fields, parks & golf-courses, & cult. orn. e.g. *Cenchrus, Cortaderia, Hakonechloa, Melinis, Miscanthus, Phalaris, Zea* & many bamboos. Many are bad weeds (*Avena fatua, Elymus repens, Imperata cylindrica, Nassella trichotoma, Poa annua* etc.; in NZ more spp. natur. than native), & others rep. fire-climax veg. of a noxious type e.g. *Heteropogon contortus, I. cylindrica*

gramma See grama

Grammadenia Benth. = *Cybianthus*

Grammangis Reichb.f. Orchidaceae (V 12b). 2 Madag. Cult. orn. epiphytes

Grammatophyllum Blume. Orchidaceae (V 12a). 12 Myanmar to Polynesia. Some locally medic. & supposed aphrodisiac. ***G. speciosum*** Blume (tiger orchid) – spectacular cult. orn. epiphyte with infls to 3 m of up to 100 fls c. 15 cm diam., those at base of infl always abnormal (K2, C2, no lip & non-functional column)

Grammatopteridium Alderw. = *Selliguea*

Grammatotheca C. Presl (~ *Lobelia*). Campanulaceae (III 1). 1 S Afr. (natur. Aus.): ***G. bergiana*** (Cham.) C. Presl

Grammitidaceae Newman = Polypodiaceae (V)

Grammitis Sw. Polypodiaceae (V; Grammitidaceae). Incl. *Adenophorus, Ctenopteris, Prosaptia, Xiphopteris*, c. 400 trop. (NG 64, R: Blumea 29(1983)13) esp. As. & Am., warm & S temp. (s.s. 150 trop. & Aus. (NZ 9, R: NZJB 14(1976)85); s. strictissimo (excl. *A., C., P., Cochlidium*, etc. Azores 1, Afr. 7, Pacific 4, trop. Am. 14) – blackish sclerotic lamina margins. ***G. succinea*** L.D. Gomez (Oligocene amber of Dominican Republic) – only grammitid fossil known

Grammosciadium DC. Umbelliferae (III 2). 7–8 E Med.

Grammosolen Haegi. Solanaceae (III 1). 2 arid S Aus. R: Telopea 2(1981)178

Grammosperma O. Schulz = *Sarcodraba*

granadilla *Passiflora quadrangularis*; **purple g.** *P. edulis*; **sweet g.** *P. ligularis*;. **yellow g.** *P. laurifolia*

granadillo *Dalbergia granadillo, Platymiscium* spp.

Grandidiera Jaub. Achariaceae (Flacourtiaceae). 1 trop. E Afr.: ***G. boivinii*** Jaub.

Grandiphyllum Docha Neto (~ *Oncidium*). Orchidaceae (V 12h). 7 S Am.

Grangea Adans. Compositae (Ast.-Gran.). 10 trop. & warm Afr., Madag., 1 (***G. maderaspatana*** (L.) Poir. – local medic.) warm As. R: MBSM 15(1979)450

Grangeopsis Humbert. Compositae (Ast.-Gran.). 1 Madag.: ***G. perrieri*** Humbert. R: MBSM 15(1979)524

Grangeria Comm. ex Juss. Chrysobalanaceae (I). 2 Madag. (1), Réunion & Mauritius (1). R: FOW 9(2003)9

granite gooseberry *Ribes curvatum*

Granitites Rye (~ *Alphitonia*). Rhamnaceae (inc. sed.). 1 W Aus. (granite outcrops): ***G. intangendus*** (F. Muell.) Rye. R: Nuytsia 10(1996)451

granny-bonnets *Aquilegia* spp.

Grantia Boiss. = *Iphiona*

grape *Vitis vinifera*; **bush g.** *V. acerifolia*; **bullace g.** *V. rotundifolia*; **canyon g.** *V. arizonica*; **cat g.** *V. palmata*; **chicken g.** *V. vulpina*; **fox g.** *V. labrusca, V. rotundifolia*; **frost g.** *V. vulpina*; **holly g.** *Berberis repens*; **g. hyacinth** *Muscari* spp.; **g. ivy** *Cissus* spp.; **mountain g.** *V. monticola*; **Oregon** or **Rocky Mt. g.** *B. aquifolium*; **sand g.** *V. rupestris*; **seaside g.** *Coccoloba uvifera*; **skunk g.** *V. labrusca*; **tree g.** *C. uvifera*; **g.-vine** (largely US) *Vitis vinifera* (N.B. 'on the g.-v.' = 'on the g.-v. [cf. bush] telegraph' (US Civil War))

grapefruit *Citrus* × *aurantium* Grapefruit Group
Graphandra Imlay. Acanthaceae (III 2b). 1 Thailand: ***G. procumbens*** Imlay. R: KB 1939:126
Graphardisia (Mez) Lundell = *Ardisia*
Graphephorum Desv. Gramineae (XVI 2). 2 N & C Am.
Graphistemma (Benth.) Benth. = *Cynanchum*
Graphistylis R. Nordenstam (~ *Senecio*). Compositae (Sen.-Sen.). 8 S Braz.
Graphorkis Thouars. Orchidaceae (V 12b). 4 trop. Afr., Madag. & Masc. Cult. orn. incl. ***G. lurida*** (Sw.) Kuntze (Afr.) – flowering throughout yr., oft. epiphytic on palms esp. *Hyphaene* spp.
grappa *Vitis vinifera*
grapple plant *Harpagophytum procumbens*
Graptopetalum Rose. Crassulaceae (I 5b). 18 SW N Am. R: U. Eggli, *Ill. Handb. Succ. Pls*, Crass. (2005)128. Cult. orn. (incl. hybrids with *Echeveria* spp.) succ. with rosettes of succ. lvs esp. ***G. paraguayense*** (N.E. Br.) Walther (W Mex.!)
Graptophyllum Nees. Acanthaceae (III 2c). 10 Aus., SW Pac. Shrubs oft. with coloured or spotted lvs; ***G. pictum*** (L.) Griff. (? NG) – much cult. in trop. as foliage pl. with purplish or green lvs marked with yellow, local medic., used in Thai weddings as variegation suggests silver & gold (wealth)
grass See Gramineae; (slang) *Cannabis sativa*; (skirts) *Cocos nucifera*; **emu g.** *Podocarpus drouynianus*; **g. of Parnassus** *Parnassia palustris*; **g., steel** *Xanthorrhoea johnsonii*; **g. tree** *X.* spp.; **g. widow** *Olsynium douglasii*; **g. wrack** *Zostera marina*
Grastidium Blume = *Dendrobium*
Gratiola L. Plantaginaceae (Grat.-Grat.; Scrophulariaceae s.l.). Incl. *Amphianthus*, 26 Euras. (4), Australasia (3–5), N Am. (17), S Am. mts (2–3). Dried pl. of ***G. officinalis*** L. (Eur. to W As.) form. medic. ('hedge hyssop')
Gratwickia F. Muell. Compositae (Gnap.-Ang.). 1 Aus.: ***G. monochaeta*** F. Muell. R: OB 104(1991)117
Grauanthus Fayed. Compositae (Ast.-Gran.). 2 trop. Afr.
gravel root *Eupatorium purpureum*
Gravesia Naudin. Melastomataceae. 110 Afr. (5), Madag.
Gravesiella Fernandes & R. Fernandes = *Cincinnobotrys*
Gravisia Mez = *Aechmea*
Grayia Hook. & Arn. (*Eremosemium*). Amaranthaceae (Atrip.; Chenopodiaceae I 3). Incl. *Zuckia*, 4 W US. R: SB 35(2010)853
Grazelianthus Peixoto & Per. Moura. Monimiaceae. 1 E Braz.: ***G. arkeocarpus*** Peixoto & Per. Moura. R: KB 63(2008)138
Grazielia R. King & H. Robinson (~ *Eupatorium*). Compositae (Eup.-Dis.). 11 S Am. R: MSBMBG 22(1987)79
Grazielodendron Lima (~ *Pterocarpus*). Leguminosae (III 11). 1 Braz.: ***G. riodocensis*** Lima. R: Bradea 3(1983)401
greasewood *Adenostoma fasciculatum*
Great Northern bean *Phaseolus vulgaris*
grechka *Fagopyrum esculentum*
Greek basil *Ocimum basilicum* 'Minimum'; **G. cress** *Lepidium sativum*
green alder *Alnus alnobetula* subsp. *crispa*; **g. briar** *Smilax* spp.; **g. cress** *Nasturtium officinale*; **g. ebony** *Handroanthus* spp.; **g. gram** *Vigna radiata*; **Kendal g., dyer's g. weed** *Genista tinctoria*; **g. rose** *Rosa chinensis* 'Viridiflora'; **g. soju** vodka-like spirit (Korea) from *Ipomoea batatas*, *Oryza sativa* etc.; **g. sprangletop** *Disakisperma dubium*
Greenea Wight & Arn. Rubiaceae (I 5). 9 SE As. to W Mal. R: TFB 41(2013)65. ***G. corymbosa*** (Jack) Voigt – leaf infusion against fevers
Greenella A. Gray = *Gutierrezia*
greengage *Prunus* × *domestica* 'subsp. *italica*'
greenheart *Chlorocardium rodiei*; **Afr. g.** *Cylicodiscus gabunensis*
greenhood *Pterostylis* spp.
Greeneocharis Guerke & Harms = *Cryptantha* (but see SB 37(2012)751)
Greeniopsis Merr. Rubiaceae (I 5). 7 Philippines. R: T 59(2010)1556
Greenmania Hieron. = *Unxia*
Greenmaniella W. Sharp. Compositae (Helia.-Mel.). 1 Mex.: ***G. resinosa*** (W. Watson) W. Sharp
Greenovia Webb & Berth. = *Aeonium*

greens (or g.tuff, GB) *Brassica* spp. – lvs used as veg.; **Cape g.** *Berzelia* spp.; **chrysanthemum g.** *Glebionis coronaria*; **Holland g.** *B. rapa* 'Tyfon'; **Japanese g.** *B. juncea* 'Crispifolia'; **sour g.** *Rumex venosus*; **Texsel g.** *B. carinata*

Greenwayodendron Verdc. (~ *Polyalthia*). Annonaceae (IV 1). 2 trop. Afr. Petit's Model

Greenwoodia Burns-Balogh = *Kionophyton*

Greigia Regel. Bromeliaceae (3). 28 C & NW S Am. R: FN 14(1979)1629. ± caulescent. ***G. sphacelata*** (Ruíz & Pavón) Regel (Chile) – fr. ed.

Grenacheria Mez = *Embelia*

grenadine *Punica granatum*

Greslania Balansa. Gramineae (V 4). 4 New Caled.

Greuteria Amirahm. & Kaz. Osaloo (~ *Hedysarum*). Leguminosae (III 25). 2 N Afr. R: T 64(2015)57

Grevea Baill. Montiniaceae (Grossulariaceae). 3 E Afr., Madag.

Grevillea R. Br. ex J. Knight. Proteaceae (V 3). 357 Sulawesi (1, endemic), NG (3, 1 endemic), New Caled. (3, endemic), Vanuatu, Aus. (357, 350 endemic). R: D.J. McGillivray (1993) *G.*, Nuytsia 9(1993)237, FA 17A(2000)21. Prob. incl. *Finschia*, *Hakea*. Fls protandrous, stigma carrying mature pollen in position to be coll. by poll. (s.t. birds; ***G. leucopteris*** Meissn. (SW W Aus.) nocturnal scarabs; ***G. myosodes*** McGillivray (NW Aus.) poss. mammal-poll., smells mousy, also effectively poll. by humans sucking nectar!) before stigma ripe, nectar coloured; ant-disp. in ***G. pteridifolia*** J. Knight (N Aus.) & some other spp., seed-wing high in lipid & protein, rest of seed with cyanide. Some timbers & cult. orn. incl. hybrids, esp. ***G. buxifolia*** (Sm.) R. Br. (grey spider fl., E NSW); ***G. hilliana*** F. Muell. (NE Aus.) – timber; ***G. robusta*** Cunn. ex R. Br. (silky oak, E Aus.) – timber for cabinet-work etc., cult. as coffee- & tea-shade, a tree to 40 m when a seedling much used as a house-pl., street-tree as at Kathmandu; ***G. striata*** R. Br. (beefwood, Aus.) – cabinet-work (Aus.)

Grewia L. Malvaceae (Grew.-Grew.; Tiliaceae). c. 280 warm OW (Madag. 82 incl. 5 lianes). Trees, shrubs & climbers (Roux's & Troll's Models); some Afr. spp. with seeds disp. endozoically via elephants (apricot-flavoured fr. of ***G. burttii*** Exell (C Afr.) & ***G. pachycalyx*** K. Schum. (C & S Afr.) – also ed. humans); some domatia. (Aus.) twigs with chewed ends used as paint-brushes; others yield locally ed. fr. (c. 50% diet of some 'hunter-gatherers' in C Tanzania at some times of yr.) incl. for trad. wine S Afr.), wood (e.g. spears), fibres (cordage) & medic. e.g. ***G. asiatica*** L. (phalsa, Himal.) – fr. ed., juice for drinks in Ind., Philippines; ***G. carpinifolia*** A. Juss. (W Afr.) – lvs & s.t. fls used for a mucilaginous sauce; ***G. crenata*** (Forst. f.) Schinz & Guillaumin (Tonga) – fibre form. for textiles & cordage, fr. ed., stems for friction-lighting fires, spears etc.; ***G. paniculata*** Roxb. ex DC. (Indomal.) – local medic., wood distillate high in acetone; ***G. tenax*** (Forssk.) Fiori (*G. populifolia*, trop. Afr.) – fr. in Tutankhamun's tomb. Some cult. orn. greenhouse shrubs e.g. ***G. ferruginea*** Hochst. ex A. Rich. (*G. seringeana*, NE Afr.)

grey box *Eucalyptus moluccana*; **g. mangrove** *Avicennia marina*; **g. spider fl.** *Grevillea buxifolia*

Greyia Hook. & Harv. Francoaceae (Melianthaceae, Greyiaceae). 3 S Afr. Some cult. orn.

Greyiaceae Hutch. = Francoaceae

greywood, silver *Terminalia bialata*

Grias L. Lecythidaceae (I). 6 Panamá to Peru. R: FN 21(1979)197. Locally medic. (NW Amaz.); ed. fr. high in vitamins esp. ***G. cauliflora*** L. (anchovy pear, C Am., Jamaica) – lvs to 110 × 28 cm

Griegia auct. = *Greigia*

Grielaceae Martinov. See Neuradaceae

Grielum Burm.f. ex L. Neuradaceae. 5 S Afr.

Griffinia Ker-Gawler. Amaryllidaceae. 15–20 Braz. Incl. *Hyline* (fragrant nocturnal fls cf. blue & violet in *G.* s.s. – R: Herbertia 54(1999)54). Cult. orn. esp. ***G. liboniana*** Morren (C Braz.)

Griffithella (Tul.) Warm. Podostemaceae (III). 1 Ind. (W Ghats): ***G. hookeriana*** (Tul.) Warm.

Griffithia Wight & Arn. = *Benkara*

Griffithsochloa Pierce = *Bouteloua*

Griffonia Baill. Leguminosae (I 1). 4 trop. W Afr. Local medic., cordage, dyes. ***G. simplicifolia*** (DC.) Baill. – comm. source of 5-hydroxytryptophan dietary supplement

Grimaldia Schrank = *Chamaecrista*

Grimmeodendron Urb. Euphorbiaceae (Euph.-Hipp.). 2 Greater Antilles

Grimmia Hedw. Grimmiaceae (Musci). c. 200 cosmop. ***G. pulvinata*** (Hedw.) Sm. (subcosmop.) – common ± hemispherical hoary-grey cushions on walls & roof-slates in N Eur.

Grindelia Willd. Compositae (Ast.-Mac.). c. 70 WN (18 – R: FNA 20(2006)424) & S Am. (25 – Kurtziana 27(1999)329). Some cult. orn. coarse herbs & everg. shrubs e.g. ***G. glutinosa*** (Cav.) Mart. (S Am.) & local medic. esp. ***G. hirsutula*** Hook. & Arn. (*G. camporum*, Calif.) – grown for resin prod. like rosin & ***G. squarrosa*** (Pursh) Dunal (NW Am., natur. GB & Aus.) – used for burns, poison ivy dermatitis etc.

Grisebachia Klotzsch = *Erica*

Grisebachianthus R. King & H. Robinson (~ *Eupatorium*). Compositae (Eup.-Crit.). 7 E Cuba. R: MSBMBG 22(1987)328

Grisebachiella Lorentz = *Diplolepis*

Griselinia Forst. f. Griseliniaceae (Cornaceae s.l.). 7 NZ & Chile. R: Britt. 45(1993)261. Some cult. orn. & timber for boats, railway sleepers (esp. ***G. littoralis*** (Raoul) Raoul, NZ).

Griseliniaceae Forst. & Forst. f. ex A. Cunn. (~ Cornaceae). Magnoliidae – Apiales. 1/7 NZ, Chile. Dioec. shrubs & trees, oft. littoral. Lvs distich., conduplicate. K minute. Fr. a berry. n = 18

Genus: *Griselinia*. A fragment of the shattered heterogeneous Cornaceae

Griseocactus Guiggi = *Stenocereus*

Griseocereus (P.V. Heath) P.V. Heath = *Stenocereus*

Grisollea Baill. Stemonuraceae (Icacinaceae s.l.). 3 Madag., Comores, Seychelles. R: SB 38(2013)499

Grisseea Bakh.f. = *Parsonsia*

grits *Zea mays*

groats usu. oats, wheat or buckwheat

Grobya Lindl. Orchidaceae (V 12c). 5 Braz. R: BJLS 145(2004)119. Cult. orn. epiphytes poll. anthophorid bees

Groenlandia Gay (~ *Potamogeton*). Potamogetonaceae. 1 W Eur. & N Afr. to SW As.: ***G. densa*** (L.) Fourr. R: Fl.Ib. 17(2010)85

gromwell *Lithospermum officinale*, *Mertensia maritima*

Gronophyllum R. Scheffer = *Hydriastele*

Gronovia Martyn ex L. Loasaceae (III). 2 trop. Am. A 5, staminodes 0, G 1

Grosourdya Reichb.f. Orchidaceae (V 16c). 11 SE As., W Mal.

Grossera Pax. Euphorbiaceae (Al.-Gross.). 8 trop. Afr. (7), Madag. (1). R: SB 39(2014)494

Grossheimia Sosn. & Takht. = *Centaurea*

Grossularia Mill. = *Ribes*

Grossulariaceae DC. Magnoliidae – Saxifragales. 1/160 N temp. to Andes. Shrubs, s.t. armed with prickles, s.t. dioec. Lvs in spirals, conduplicate-plicate, palmately-veined, margins lobed & v. oft. with hydathodes; stip. 0 or rarely consp. & basally adnate to petiole. Fls ± reg., usu. bisex. (or pl. dioec.), with a well-dev. hypanthium, in term. or axillary racemes of (1–)5–50 fls, usu. on short shoots; K ((3–)5(–9)) persistent, imbr. or valvate as lobes on hypanthium or with tube, usu. petaloid, C as many & alt. with K but smaller, nectary-disk oft. 5-lobed, $\overline{G}$ (2), with parietal placentas & ∞ anatr. bitegmic ovules. Fr. a soft berry with term. marcescent P; seeds (3–)10–60 with outer mucilaginous layer (? aril) & oily endosperm with hemicellulose. n = 8

Genus: *Ribes*

Fr. (gooseberries, currants) & cult. orn.

Grosvenoria R. King & H. Robinson (~ *Eupatorium*). Compositae (Eup.-Crit.). 4 Andes. R: MSBMBG 22(1987)352

ground elder *Aegopodium podagraria*; **g. ivy** *Glechoma hederacea*; **g. pine** *Ajuga chamaepitys*

groundnut *Arachis hypogaea*, (Am.) *Apios americana*; **Bambara g.** *Vigna subterranea*

groundsel *Senecio vulgaris*; **g. bush** *Baccharis halimifolia*; **tree** or **giant g.** *Dendrosenecio* spp., *B. halimifolia*

Grubbia P. Bergius. Grubbiaceae. 3 SW & S Cape. R: JSAB 43(1977)115

Grubbiaceae Endl. ex Meissn. Magnoliidae– Cornales. 1/3 Cape. Ericoid shrubs. Lvs opp., simple, revolute; stip. 0 but leaf-bases joined by transverse ridge. Fls bisex., reg., small in compact axillary cymes; K 4, small, valvate, C 0, A 8, 4 weakly adnate to K & longer than rest, anthers inverted & adnate to filament & with longit. slits, pollen in monads, $\overline{G}$ (2) with apical disk & style with simple or bilobed stigma, at first ± 2-loc., later 1-loc. with C placenta with 2 apical, pend., unitegmic, anatr. ovules. Fr. a 1-seeded drupe, those of an infl. forming a compact cluster looking like a cypress cone; seed with thin testa & long straight embryo in oily proteinaceous endosperm

Genus: *Grubbia*

Grubovia Freitag & G. Kadereit (~ *Kochia*). Amaranthaceae. 3 C As. R: T 60(2011)72

gru-gru *Acrocomia aculeata*
gruie *Owenia acidula*
grumichama *Eugenia brasiliensis*
Grumilea Gaertn. = *Psychotria*
Grushvitzkya Skvortsova & Aver. = *Trevesia*
Grusonia Reichb.f. ex Britton & Rose (~ *Opuntia*). Cactaceae (II). 18 SW N Am.
Grypocarpha Greenman = *Philactis*
guaba *Inga vera*
guabiroba *Campomanesia guaviroba*
Guacamaya Maguire. Rapateaceae (I 2). 1 Colombia, Venez.: ***G. superba*** Maguire. R: MNYBG 10,1(1958)35
guacamole *Persea americana*
guachilote *Parmentiera aculeata*
guacimilla *Trema micrantha*
guaco *Clibadium* spp.
Guadua Kunth (~ *Bambusa*). Gramineae (V 3). 27 trop. Am. R: CUSNH 39(2000)58. Stems with armed stolons poss. deterrent to (extinct) megafauna. ***G. angustifolia*** Kunth (guadua) – all-purpose bamboo for construction in Colombia; ***G. sarcocarpa*** Londoño & Peterson (Peru) – fleshy fr.
Guaduella Franch. Gramineae (III). 6 trop. Afr. R: KB 16(1962)247, 37(1983)660. Resemble *Aframomum* & easily mistaken
guaiac *Bulnesia sarmientoi*
Guaiacum Plum. ex L. Zygophyllaceae (II). 6 warm Am. Everg. trees & shrubs of dry areas, sources of lignum-vitae much used in seawater etc. (hardest of comm. timbers, S.G. 1.333, with fibre-layers diagonally opposed & lots of resin so self-lubricating & therefore good for pestles etc.), from which medic. resin guaiacum is obtained by heating; timber used for pulleys, bowls for bowling, etc. esp. from ***G. officinale*** L. & ***G. sanctum*** L., the former almost exterminated (CITES-listed) for medic. use in venereal disease by natives from 1450, introd. Spain 1501 & in Eur. used with mercury in treatment of syphilis; g. lozenges ('Plummer's pills') for sore throats
Guaicaia Maguire = *Glossarion*
Guajacum L. = *Guaiacum*
guaje *Leucaena esculenta*
Guamatela J.D. Sm. Guamatelaceae. 1 C Am.: ***G. tuerckheimii*** J.D. Sm.
Guamatelaceae Oh & D. Potter (~ Rosaceae). Magnoliidae – Crossosomatales. 1/1 C Am. R: SB 31(2006)736. Shrub with villous shoots. Lvs opp., simple, ovate, margins serrulate; stip. setaceous. Fls in term. racemes, bisex., reg., with filiform bracts & bracteoles. K5 imbr., persistent; C5 lanceolate, persistent, usu. shorter than K; A 10 free on shallow hypanthium, anthers introrse, longit. dehiscent; $\underline{G}3$ with connate styles separating after anthesis, each loc. with ∞ locules. Fr. usu. 3 follicles, dehiscing ventrally; seeds ∞ with membranous aril; endosperm scant
Genus: *Guamatela*
Lvs opp. cf. Rosaceae
Guamia Merr. = *Meiogyne*
guanacaste *Enterolobium cyclocarpum*
Guanchezia G. Romero & Carnevali (~ *Bifrenaria*). Orchidaceae (V 12g). 1 Venez.: ***G. maguirei*** (C. Schweinf.) G. Romero & Carnevali. R: HPB 7(2002)79
guango *Samanea saman*
Guapeba B.A. Gomes = *Pouteria*
Guapira Aubl. (~ *Neea*). Nyctaginaceae (5). c. 70 trop. Am.
guapiruvu *Schizolobium parahyba*
guar *Cyamopsis tetragonoloba*
guaraná *Paullinia cupana*
Guardiola Cerv. ex Bonpl. Compositae (Mill.-Guar.). 10 SW N Am.
Guarea F. Allam. Meliaceae (I 6). Excl. *Leplaea*, 69 trop. Am. (R: EJB 70(2013)190). Champagnat's Model; most spp. with lvs with indefinite growth (cf. *Chisocheton*, *Tachigali*). Some moth-poll. incl. ***G. rhopalocarpa*** Radlk. (C Am.) – 5 flowering periods (4–20 wks) per yr.; ***G. macrophylla*** Vahl (Braz.) – extrafl. nectaries on fr. as well as petioles, buds etc. Some timbers esp. ***G. guidonia*** (L.) Sleumer (WI redwood, alligator w., 'acajou', trop. Am.), also medic. locally, & ***G. kunthiana*** A. Juss. (Am.) – fine timber

Guarianthe Dressler & Higgins (~ *Cattleya*). Orchidaceae (V 13b). 4 trop. Am. R: Lankesteria 7(2003)37. Cult. orn. incl. ***G. bowringiana*** (Veitch) Dressler & Higgins (*C. b.*, C Am.) – overcollected in wild but saved from extinction by successful cult., up to 47 fls per spike (over 600 fls recorded in a season from 1 pl.), its floriferousness leading to its being used in over 100 registered hybrids involving most spp. of *C.*, *Epidendrum*, *Laelia* & *Sophronitis*

× **Guaricattonia** J.M.H. Shaw. Orchidaceae. *Broughtonia* × *Cattleya* × *Guarianthe* – 255 grexes

guatambu moroti *Balfourodendron riedelianum*

Guatemala grass *Tripsacum fasciculatum*

Guatteria Ruíz & Pavón. Annonaceae (III 4). Incl. *Guatteriella*, *Guatteriopsis*, *Heteropetalum* (R: AHB 10(1930)73), 300+ trop. Am. (Peru 54). R: AHB 12(1939)108, 291, PSE 148(1985)20. With *Inga*, *Ocotea*, largest trop. Am. tree genera. Mangenot's Model; alks. Some local timbers & fibres; ***G. modesta*** Diels – oral contraceptive in Peru; ***G. scandens*** Ducke one of few cli. A. in Am.

Guatteriella R.E. Fries = *Guatteria*

Guatteriopsis R.E. Fries = *Guatteria*

guava *Psidium* spp. esp. *P. guajava*; **g.berry** *Myrciaria floribunda*; **black g.** *Guettarda argentea*; **Chilean g.** *Ugni molinae*; **Costa Rican g.** *P. friedrichsthalianum*; **Pará g.** *Campomanesia* spp.; **pineapple g.** *Acca sellowiana*; **purple g.** *P. cattleyanum*; **Spanish g.** *Catesbaea spinosa*; **strawberry g.** *P. cattleyanum*

guaxima *Hibiscus americanus*

guayacán *Libidibia paraguariensis*

Guayania R. King & H. Robinson (~ *Eupatorium*). Compositae (Eup.-Heb.). 5 Guayana Highland. R: MSBMBG 22(1987)410

Guaymasia Britton & Rose = *Coulteria*

guayato *Gonolobus edulis*

guayule *Parthenium argentatum*

guayusa *Ilex guayusa*

Guazuma Mill. Malvaceae (Bytt.-Theob.; Sterculiaceae). 3 trop. Am. Roux's Model. ***G. ulmifolia*** Lam. (bastard cedar, WI elm) – natur. OW trop, fr. passes through cattle which act as disp. agents suggesting that prob. orig. disp. by extinct megafauna of Pleistocene, light timber for boats, barrels, fuelwood etc., local medic. (form. herbal tea for losing wt.)

gubgub *Vigna unguiculata*

Gudrunia Braem = *Tolumnia*

Gueldenstaedtia Fischer. Leguminosae (III 24). Excl. *Tibetia*, 4 Sino-Himal. to Siberia. R: AnnBF 41(2004)284

guelder rose *Viburnum opulus*

Guenthera Andréansky = *Brassica*

guere palm *Astrocaryum* spp.

Guerkea K. Schum. = *Baissea*

Guernsey lily *Nerine sarniensis*

Guerreroia Merr. = *Glossocardia*

Guettarda L. Rubiaceae (III 3). c. 70 Vanuatu, New Caled., (esp.) trop. Am. with 1, ***G. speciosa*** L. (Aubréville's Model), common on trop. coasts, fls nocturnal (lasting 1 night), fr. water-disp. Some ed. fr. esp. ***G. argentea*** Lam. (black guava, Am.) & locally used timbers. See also *Tinadendron*

Guettardella Champ. ex Benth. = *Antirhea*

Guevaria R. King & H. Robinson (~ *Eupatorium*). Compositae (Eup.-Ager.). 5 Ecuador & Peru. R: MSBMBG 22(1987)161

guggul *Commiphora wightii*

Guiana chestnut *Pachira aquatica*

Guianodendron Schütz Rodrigues & Azevedo (~ *Acosmium*). Leguminosae (III 2). 1 NE S Am.: ***G. praeclarum*** (Sandw.) Schütz Rodrigues & Azevedo. R: Novon 16(2006)129

Guibourtia Bennett. Leguminosae (I 2). c. 14 trop. Afr. (14), Am. (1). Timbers & copal esp. from ***G. arnoldiana*** (De Wild. & T. Durand) Léonard (benge, C Afr.), ***G. copallifera*** Bennett (Sierra Leone copal, trop. Afr.), ***G. demeusei*** (Harms) Léonard (Afr. rosewood, bubinga, Congo c., trop. Afr.) & ***G. ehie*** (A. Chev.) Léonard (amazakoué, anokye, shedua, trop. Afr.)

Guichenotia Gay. Malvaceae (Bytt.- Lasio.; Sterculiaceae). 17 SW Aus. R: AusSB 16(2003) 330

Guiera Adans. ex Juss. Combretaceae (II 2). 1 N trop. Afr.: ***G. senegalensis*** J.F. Gmel.

Guihaia Dransf., S.K. Lee & F.N. Wei. Palmae (III 4a). 2 S China & N Vietnam (karst limestone). R: Principes 29(1985)9

Guihaiothamnus H.S. Lo. Rubiaceae. 1 SE China: ***G. acaulis*** H.S. Lo. R: FOC 19(2011)145

Guilandina L. (~ *Caesalpinia*). Leguminosae (I 4). c. 7 trop. & warm. ***G. bonduc*** L. (*C. b.*, *C. bonducella*, bonduc, nicker bean, trop.) – common drift seed on SW Br. coast etc. (seeds viable for at least 2½ yrs. in seawater), used as beads

Guildford grass *Romulea rosea*

Guilfoylia F. Muell. (~ *Cadellia*). Surianaceae. 1 NE Aus.: ***G. monostylis*** (Benth.) F. Muell. Stipules

Guilielma Mart. = *Bactris*

Guillainia Vieill. = *Alpinia*

Guillauminia Bertrand = *Aloe*

Guilleminea Kunth. Amaranthaceae (II 2). 5 C Am. (***G. densa*** (Schultes) Moq. weedy in OW). R: Sida 12(1974)325

Guillenia Greene = *Streptanthus*

Guillonea Cosson = *Thapsia*

Guindilia Gillies ex Hook. & Arn. (*Valenzuelia*). Sapindaceae (IV 1). 3 Chile, Arg.

Guinea corn *Sorghum* spp.; **g.(-)flower** *Fritillaria meleagris*, (Aus.) *Hibbertia* spp.; **G. grains** *Aframomum melegueta*; **G. grass** *Urochloa maxima*; **G. henweed** *Petiveria alliacea*; **G. peach** *Nauclea latifolia*; **G. pepper** *A. m.*, *Xylopia aethiopica*; **G. yam** *Dioscorea* × *cayenensis*

Guinetia Rico & Sousa = *Calliandra* (but see KB 54(2000)977)

Guioa Cav. Sapindaceae. 64 Thailand to Samoa. R: P.L. van Welzen (1990) *Guioa* Cav. *(S.)*. Some ed. arils

Guiraoa Cosson. Cruciferae (12). 1 Spain: ***G. arvensis*** Cosson

guisaro *Psidium guineense*

Guizotia Cass. Compositae (Mill.-Mill.). 6 warm Afr. R: BT 69(1974)1. ***G. abyssinica*** (L.f.) Cass. (ramtil, Niger seed, Ethiopia) – cult. (esp. Ind.) for oilseed fed to cagebirds & oil used for cooking & paint

gul mohur (tree) *Delonix regia*

Gularia Garay = *Schiedeella*

Gulubia Becc. = *Hydriastele*

Gulubiopsis Becc. = *Hydriastele*

gum *Eucalyptus* spp., also *Angophora* & *Corymbia* spp.; **almond g.** *Prunus dulcis*; **American red g.** *Liquidambar styraciflua*; **Amrad g.** *Vachellia nilotica*; **Amritsar g.** *Senegalia modesta*; **angico g.** *Parapiptadenia rigida*; **animé g.** *Hymenaea courbaril*; **g. arabic** *S. senegal* etc.; **Ashanti g.** *Terminalia* spp.; **black g.** *Nyssa sylvatica*; **blue g.** *Eucalyptus globulus*; **Cape g.** *V. karroo*; **Cape York red g.** *E. brassiana*; **carob seed g.** *Ceratonia siliqua*; **cashew g.** *Anacardium occidentale*; **cherry g.** *Prunus cerasus*; **cider g.** *E. gunnii*; **g. dragon** *Astragalus* spp.; **E Afr. g.** *Vachellia drepanolobium*; **flooded g.** *E. grandis*; **forest red/river g.** *E. tereticornis*; **gatty g.** *Anogeissus latifolia*; **ghost g.** *Corymbia* spp.; **hog g.** *Symphonia globulifera*; **karaya** or **kutira g.** *Cochlospermum religiosum*, *Sterculia urens*; **khayer g.** *Senegalia catechu*; **kolhol g.** *S. senegal*; **lemon-scented g.** *Corymbia citriodora*; **locust g.** *Ceratonia siliqua*; **mesquite g.** *Prosopis glandulosa*; **Murray red or river red g.** *E. camaldulensis*; **rose g.** *E. grandis*; **saligna g.** *E. saligna*; **salmon g.** *E. salmonophloia*; **snow g.** *E. niphophila*; **spotted g.** *Corymbia maculata*; **spruce g.** *Picea mariana*, *P. rubens*; **sweet g.** *Liquidambar styraciflua*; **Sydney blue g.** *E. saligna*; **S. red g.** *Angophora costata*; **talh g., g. talha** *Vachellia seyal*; **Tartar g.** *Sterculia cinerea*; **g. tragacanth** *Astragalus* spp.; **g. tree** *Eucalyptus* spp.; **water g.** *Tristania neriifolia*; **white g.** *E. nobilis*; **yellow g.** *Xanthorrhoea* spp. esp. *X. resinosa*

gumbo *Hibiscus esculentus*; **g. limbo** *Bursera simaruba*

Gumillea Ruíz & Pavón = ? *Picramnia*

gumwood *Commidendrum robustum*

gundabluey *Acacia victoriae*

Gundelia Tourn. ex L. Compositae (Cich.). 1 (–4) Cyprus & Turkey to Afghanistan: ***G. tournefortii*** L. – steppe-pl. tumbleweed (? of the Bible; pollen found in shroud of Turin) with ed. fr. & young infls (tertiary heads of 1-fld primary capitula, outer 4 or 6 with functionally male fls), akoub, molluscicidal saponins. R: T 43(1994)37

Gundlachia A. Gray. Compositae (Ast.-Sol.). 6 Texas to Venez. (arid habitats). R: Britt. 48(1997)533. Shrubs

gungo pea *Cajanus cajan*

Gunillaea Thulin. Campanulaceae (I 2). 2 trop. Afr. & Madag.

Gunnarella Senghas. Orchidaceae (V 16c). 90 New Caled.

Gunnaria S.C. Chen ex Z.J. Liu & L.J. Chen = *Vanda* (but see JSE 47(2009)602)

Gunnarorchis Brieger = *Thelasis*

Gunnera L. Gunneraceae. c. 60 trop. & S Afr., Madag., Mal., Tasmania, NZ, Antarctic is., Hawaii, S Am. Pachycaul ± stemless herbs with some of the largest lvs known to creeping herbs, e.g. tiny ann. ***G. herteri*** Osten (Urug., Braz.); ***G. petaloidea*** Gaud. (Hawaii) – mycorrhizal as well as bacterial symbionts. ***G. macrophylla*** Blume (Mal. Mts) – intercropped with brassicas, improving yield (*Nostoc* symbionts) by 50%, infr. a tonic; ***G. perpensa*** L. (S Afr.) – medic. Cult. orn. esp. form. ***G. manicata*** Linden ex Delchevalerie (*G. brasiliensis*, S Braz.) & ***G. tinctoria*** (Molina) Mirbel (*G. chilensis*, Chile, invasive GB, Ireland, Azores, NZ) – lvs to 1.5 m across, peeled young petioles ed. Chile, but most so-named in modern cult. actually hybrids with *G. m.* with lvs to 3 m across by waterside (Shaw), & other spp. creeping & grown as rock-pls

Gunneraceae Meissn. Magnoliidae – Gunnerales. 1/c. 60 S trop. & S hemisph. Terr., oft. pachycaul (rarely ann.) herbs with stems & adventitious r. harbouring symbiotic *Nostoc punctiforme* (intracellular infection starting in 2 glands below cotyledons in seedlings) or *Chlorococcus* colonies in warts entered through hydathodes or mucilage glands; stem polystelic. Lvs 7 mm to 3 m diam, in spirals, orbicular or ovate (– peltate), with palmate venation v. prominent adaxially & large median axillary scale (? stip.). Infl. axill. or pseudoterm. panicle of small epigynous fls, the basal ones oft. female, upper ones male & middle ones bisex. or pl. dioec.; K 2, P 2or almost 0, small, valvate, A (1–)2 with short filaments & anthers with longit. slits, $\overline{G}$ (2), 1-loc. with 2 term. styles & 1 (2) pend., anatropous, bitegmic ovule. Fr. a drupe; seed with v. small embryo embedded in copious oily endosperm. n = 17

Genus: *Gunnera*

Form. incl. in Haloragaceae (Saxifragales) but differing in habit, A, G, embryology & chem. etc. (Cronquist). Pollen known from Early Cretaceous onwards in S hemisph., N Am. & poss. As.

Gunnessia P. Forster. Apocynaceae (Va; Asclepiadaceae III 4). 1 Queensland: ***G. pepo*** P. Forster. R: Austrobaileya 3(1990)282

Gunniopsis Pax (~ *Aizoon*). Aizoaceae (I). 14 W, S & C Aus. R: JABG 6(1983)133

gunny *Corchorus* spp., as in **g. bag** = hessian; see also *Musa*

gunpowder plant *Pilea microphylla*

Gurania (Schldl.) Cogn. (~ *Psiguria*). Cucurbitaceae (XIII). c. 75 trop. Am. Monoec. but changing sex so as to appear dioec.; some with tough bird-poll. fls like *P.*

Guraniopsis Cogn. = *Apodanthera*

Guringalia Briggs & L. Johnson = *Chordifex*

gurjun or **g. oil** *Dipterocarpus* spp.

Gustavia L. Lecythidaceae (I). 41 trop. Am. R: FN 21(1979)128. Corner's & Leeuwenberg's Models; reg. fls (some self-sterile) poll. trigonid bees. ***G. superba*** O. Berg (W trop. S Am.) – pls regenerate from fragments of cotyledons. Some with ed. fr. sold in Colombian markets; some timbers esp. ***G. augusta*** L. (stinkwood), some medic.; bark ash 'salt' used to coat pellets of hallucinogenic (*Virola*) paste in NW Amaz.

Gutenbergia Schultz-Bip. Compositae (Vern.-Erl.). c. 20 trop. Afr.

Guthriea Bolus. Achariaceae. 1 S Afr.: ***G. capensis*** Bolus. R: Strelitzia 10(2000)46

Gutierrezia Lag. Compositae (Ast.-Sol.). (Excl. *Xanthocephalum*) 28 W N (16, R: SB 10(1985)7) & S Am

gutta percha *Madhuca pallida*, *Palaquium* spp. esp. *P. gutta*, *Eucommia ulmoides*; **g. sundek** *Payena leerii*

Guttiferae Juss. (Clusiaceae). Magnoliidae – Malpighiales. Excl. Bonnetiaceae, Calophyllaceae, Hypericaceae, 14/750 trop. Everg. trees & shrubs with yellow or otherwise brightly coloured resinous juice in schizogenous secretory canals & cavities, usu. glabrous. Lvs usu. opp., simple & usu. entire & oft. with many slender lat. veins & resin-cavities s.t. appearing as pellucid dots; stip. 0. Fls bisex. or not (dioecy), 4- or 5-merous in term. cymes, rarely solit., reg., hypogynous, oft. with bracteoles (s.t. passing into K), K 2(3)4 or 5(–20), imbr., C (0,3)4, 5(–8) free, imbr. or convolute, s.t. basally connate, A (4–)∞ centrifugally dev. from few trunk-bundles, or in 2–5 centrifugal bundles opp. & oft. adnate to C when many staminodal, or fewer, anthers with longit. slits, $\underline{G}$ ((1–)5(–20)), with as many locules or 1-loc. (intruded placentas not reaching centre) & as many styles, ± basally connate or 1 style with lobed or peltate stigma, placentation axile (rarely parietal on intruded placentas) with (1)2–∞ anatr. to hemitropous bitegmic ovules per carpel.

Fr. a drupe or berry or septicidal capsule; seeds exotegmic, oft. arillate, with straight or curved embryo & 0 endosperm. x = 7–10

Classification & genera:

[Kielmeyeroideae = Calophyllaceae; Clusioideae:]

1. **Clusieae** (inc. Tovomiteae; fr. septifragal, seeds arillate): *Chrysochlamys, Clusia, Dystovomita, Tovomita, Tovomitopsis*
2. **Garcinieae** (incl. Allanblackieae; usu. dioec., fr. baccate): *Allanblackia, Garcinia*
3. **Symphonieae** (incl. Moronobeeae; fls bisex., fr. baccate): *Lorostemon, Montrouziera, Moronobea, Pentadesma, Platonia, Symphonia, Thysanostemon*

If Hypericaceae & Calophyllaceae incl. then Podostemaceae also falls here. Timbers (*Montrouziera, Platonia, Symphonia*), gums, pigments & resins from stems (esp. *Garcinia*, gamboge), ed. fr. (*Garcinia* incl. mangosteen, *Platonia*), oilseeds (*Allanblackia, Garcinia, Pentadesma*) & cult. orn.

Gutzlaffia Hance = *Strobilanthes*

guvar *Cyamopsis tetragonoloba*

Guya Frapp. ex Cordem. = *Drypetes*

Guyania Airy Shaw = *Guayania*

Guynesomia Bonifacino & Sancho. Compositae (Ast.-Hin.). 1 C Chile: ***G. scoparia*** (Phil.) Bonifacino & Sancho. R: T 53(2004)675

Guyonia Naudin. Melastomataceae. 2 trop. W Afr.

Guzmania Ruíz & Pavón. Bromeliaceae (2). 167 trop. Am. R: FN 14(1977)1275, 1401. Terr. or epiphytic; some cult. orn. incl. hybrids with *Vriesea* spp.

gwar *Cyamopsis tetragonoloba*

Gyalanthos Szlach. & Marg. = *Pleurothallis*

Gymapsis Bremek. = *Strobilanthes*

gymea lily *Doryanthes excelsa*

Gyminda Sarg. Celastraceae (I). 4 SN & C Am., WI

Gymnacanthus Nees = *Ruellia*

Gymnachne L. Parodi = *Chascolytrum*

Gymnaconitum (Stapf) Wei Wang & Z.D. Chen (~ *Aconitum*). Ranunculaceae (I 4). 1 China: ***G. gymnandrum*** (Maxim.) Wei Wang & Z.D. Chen – ann. R: T 62(2013)719

Gymnacranthera Warb. Myristicaceae. 7 Indomal. R: Blumea 31 (1986)451. Some oilseeds used in candle-manufacture

Gymnadenia R. Br. (~ *Habenaria*). Orchidaceae (IV 4d). Incl. *Nigritella*, c. 20 NE Am. & temp. Euras. (Eur. 2, incl. ***G. conopsea*** (L.) R. Br. (scented orchid, Euras. to Jap.)). Poll. butterflies & hawkmoths

Gymnadeniopsis Rydb. = *Platanthera*

Gymnanthemum Cass. (~ *Vernonia*). Compositae (Vern.-Gym.), c. 24 OW trop. ***G. amygdalinum*** (Del.) Walp. (*V. a.*, bitterleaf, trop. Afr.) – stem pith taken by sick chimpanzees (vernoniosides effective against drug-resistant malarial parasites rich there) which avoid app. toxic lvs, though these used as veg. & in soup by humans (W Afr.), & insecticidal; ***G. arboreum*** (Buch.-Ham.) H. Robinson (*V. a.*, Indomal.) – timber tree

Gymnanthera R. Br. Apocynaceae (III; Periplocaceae). 2 Mal. to Aus. (2, 1 endemic). Alks

Gymnanthes Sw. (*Ateramnus* auctt.). Euphorbiaceae (Hipp.). Incl. *Neoshirakia, Shirakiopsis*, c. 45 trop. esp. Am. ***G.* sp.** (*S. indica*, Indomal.) – timber, drying oil (poss. crop in wet habitats), seeds ed. (pericarp toxic – latex), medic., fish-poison, dyes; ***G. lucida*** Sw. (Carib.) – fine wood (Cuban oysterwood) for cabinet-making

Gymnarrhena Desf. Compositae (Gymnarrhenoideae). 1 N Afr. to Middle E: ***G. micrantha*** Desf.

Gymnartocarpus Boerl. = *Parartocarpus*

Gymnaster Kitam. = *Miyamayomena*

Gymneia (Benth.) Harley & Pastore (~ *Hyptis*). Labiatae (VII 3c). 7 Braz. R: Phytotaxa 58(2012)53

Gymnema R. Br. = *Marsdenia*

Gymnemopsis Costantin = *Marsdenia*

Gymnocactus Backeb. = *Turbinicarpus*

Gymnocalycium Pfeiffer ex Mittler. Cactaceae (III 4). 56 S Am. R: G. Charles (2009) *G. in habitat & culture*. Alks incl. mescaline. Many cult. orn. globose cacti

Gymnocarpium Newman. Cystopteridaceae (Woodsiaceae, Dryopteridaceae s.l.). 7 N temp. (Eur. 3), Indomal. (1), Taiwan (1). R: ABF 15(1978)101. Some cult. orn. incl. ***G. dryopteris*** (L.) Newman (oak fern, N temp.)

Gymnocarpos Forssk. (~ *Paronychia*). Caryophyllaceae (I 2). 10 Canary Is. to E As. esp. Horn of Afr. T: EJB 53(1996)9, 59(2002)231. ***G. decandrus*** Forssk. (*P. d.*, Macaronesia to Pakistan) – much used as camel fodder

Gymnochilus Blume = *Cheirostylis*

Gymnocladus Lam. Leguminosae (I 4). 6 E N Am. (1), E & SE As. (5, 2 v. rarely coll.). R: JAA 57(1976)91. Dioec. or polygamous pachycaul trees with bipinnate lvs & fr. opening along parietal suture like a follicle (primitive). ***G. chinensis*** Baill. (China) – fr. pulp saponins used to wash clothes, also inhibits HIV replication; ***G. dioica*** (L.) Koch (*G. canadensis*, chicot, Kentucky coffee tree, E US) – poss. orig. disp. by now extinct megafauna, cult. orn. with durable timber & with seeds roasted (non-protein amino-acids thereby destroyed) as coffee subs., form. planted by Native Americans in New York State, local medic. esp. laxative

Gymnocondylus R. King & H. Robinson (~ *Eupatorium*). Compositae (Eup.-Ayap.). 1 Braz.: ***G. galeopsifolius*** (Gardner) R. King & H. Robinson. R: MSBMBG 22(1987)206

Gymnocoronis DC. Compositae (Eup.-Aden.). 5 C & S Am. R: MSBMBG 22(1987)62. All aquatic, ***G. spilanthoides*** (Hook. & Arn.) DC. (S Am.) – noxious weed in NSW ('Senegal tea') & NZ, cult. aquaria Eur. where now natur. (N Italy)

Gymnodiscus Less. Compositae (Sen.-Oth.). 2 SW Cape to Namaqualand

Gymnogonum Parry = *Goodmania*

Gymnogrammitidaceae Ching = Polypodiaceae (II)

Gymnogrammitis Griff. (~ *Davallia*). Polypodiaceae (II). 1 E Himal. & S China to SE As.: ***G. dareiformis*** (Hook.) Ching – cult. orn. epiphyte. R: Blumea 37 (1992)186

Gymnolaena (DC.) Rydb. Compositae (Hele.-Pect.). 3 Mex. R: Sida 3(1967)110

Gymnolomia Kunth = *Eleutheranthera*

Gymnoluma Baill. = *Elaeoluma*

Gymnomyosotis (A.DC.) Nikif. = *Myosotis*

Gymnopentzia Benth. Compositae (Anth.-Phym.). 1 S Afr.: ***G. bifurcata*** Benth. R: NJB 5(1986)538

Gymnopetalum Arn. = *Trichosanthes*

Gymnophragma Lindau. Acanthaceae. 1 NE NG: ***G. simplex*** Lindau

Gymnophyton Clos. Umbelliferae (Azor.). 6 Andes of Chile & Arg. R: UCPB 33(1962)137

Gymnopodium Rolfe. Polygonaceae (Erio.-Gym.). 1 C Am.: ***G. floribundum*** Rolfe (dzildzilche) – used in Mex. honey exported to GB, shrubs yielding good charcoal

Gymnopogon P. Beauv. Gramineae (XXIX 3). 14 warm Am., Ind. to Thailand (1). R: AMBG 98(2010)302

Gymnopoma N.E. Br. = *Skiatophytum*

Gymnopteris Bernh. = *Hemionotis*

Gymnoschoenus Nees. Cyperaceae (II 7). 2 Aus. ***G. sphaerocephalus*** (R. Br.) Hook.f. (SE Aus.) – distinctive ('button grass') plains esp. W Tasmania. R: FR 48(1940)58

Gymnosciadium Hochst. = *Pimpinella*

Gymnosiphon Blume. Burmanniaceae. Incl. *Ptychomeria*, 27 trop. rain forests (As. 7, Afr. 4 – R: KB 65(2010)83), Madag. 2, Am. 14)

Gymnosperma Less. Compositae (Ast.-Sol.). 1 S US to C Am.: ***G. glutinosum*** (Spreng.) Less. R: SBM 20(2000)51

Gymnospermae Lindl. (**gymnosperms).** 79/985 in 12 fams. A grouping used to designate those seed pls not considered to be angiosperms (Magnoliidae), differing from those in not having the seeds encl. in carpels & not having a double fert. char. of a. Fertilization oft. 4–6 + months after poll. Four extant groups: Cycadidae (cycads), Ginkgoidae, Pinidae (conifers), Gnetidae, each of 1 order) – argued by some as poss. not closely interrelated, the extinct seed-ferns (pteridosperms) having given rise to most of these & to the angiosperms independently

Gymnospermium Spach (~ *Leontice*). Berberidaceae (II 1; Leonticaceae). 6–8 Eur. (1) to E As. R: BZ 55(1970)191

Gymnosphaera Blume = *Cyathea*

Gymnospora (Chodat) Pastore (~ *Polygala*). 2 SE Braz. R: Novon 22(2013)305

Gymnosporia (Wight & Arn.) Hook.f. (~ *Maytenus*). Celastraceae (I). Incl. *Gloveria*, *Putterlickia*, c. 100 OW trop. N to Spain. R: T 55(2006)518. Spiny trees form in *M.* incl. ***G. senegalensis*** (Lam.) Loes. (*M. s.*, Spain to Bangladesh) – local medic. in Afr., extracts with cytotoxic effect on some cancers

Gymnostachys R. Br. Araceae (I). 1 E Aus.: ***G. anceps*** R. Br. – lvs form. for cordage. R: FA 39(2011)239

Gymnostachyum Nees. Acanthaceae (III 2b). 30 Ind. to C Mal.

Gymnostemon Aubrév. & Pellegrin. Simaroubaceae. 1 trop. W Afr.: ***G. zaizou*** Aubrév. & Pellegrin

Gymnostephium Less. Compositae (Ast.-Hom.). 8 S Afr. R: Phytol. 76(1994)92

Gymnosteris Greene. Polemoniaceae (III). 2 W US. R: AMN 31(1944)230

Gymnostoma L. Johnson (~ *Casuarina*). Casuarinaceae. 18 Mal. to W Pac. Rauh's Model

Gymnostyles Juss. = *Soliva*

Gymnotheca Decne. Saururaceae. 2 SW China. R: FOW 11(2005)5

gympie *Dendrocnide* spp.

Gynandriris Parl. = *Moraea*

Gynandropsis DC. = *Cleome*

Gynatrix Alef. (~ *Plagianthus*). Malvaceae (Malv.-Malv.). 2 SE Aus.R: Fl. Victoria 3(1996) 351. Not dioec. Local fibre pls

Gynerium Willd. ex P. Beauv. Gramineae (XX). 1 trop. Am.: ***G. sagittatum*** (Aubl.) P. Beauv. (uva grass, white roseau) – giant reed to 10 m, stems used for arrow-shafts (so form. a strategic resource), laths etc., lvs for weaving hats, mats etc., young shoots used for shampoo. R: KB 49(1994)313

Gynizodon Raf. (*Anneliesia*) = *Miltonia*

Gynocardia Roxb. Achariaceae (Flacourtiaceae). 1 Assam & Myanmar: ***G. odorata*** Roxb. – seeds yield an oil similar to, but less efficacious than, chaulmoogra oil (cf. *Hydnocarpus*), fr. used as fish-poison

Gynochthodes Blume. Rubiaceae (IV 10). Incl. *Tetralopha*, 93 Madag., SE As. to Pac. R: Adans. 33(2011)286

Gynocraterium Bremek. = *Staurogyne* (but see KB 1939:557)

Gynoglottis J.J. Sm. Orchidaceae (V 10b). 1 Mal.: ***G. cymbidioides*** (Reichb.f.) J.J. Sm.

Gynopachis Blume = *Aidia*

Gynophorea Gilli = *Erysimum*

Gynophyge Gilli = *Daucus*

Gynostemma Blume. Cucurbitaceae (I). c. 17 Indomal., E As. (China 14). R: APS 33(1995)405, 34(1995)207. ***G. pentaphyllum*** (Thunb.) Makino (jiaogulan) – medic. incl. for altitude sickness

Gynotroches Blume. Rhizophoraceae (II). 1 Myanmar to Mal., Caroline & Solomon Is.: ***G. axillaris*** Blume – Massart's Model

Gynoxys Cass. Compositae (Sen.-Tuss.). c. 120 C Am. to Peru. Shrubs & trees

Gynura Cass. (~ *Kleinia*). Compositae (Sen.-Sen.). 44 OW trop. R: JSE 49(2011)286. Cult. orn., medic. & some ed. ***G. aurantiaca*** (Blume) DC. (Java, Sulawesi, widely natur.) – foliage pl. with velvety-purple hairy lvs (common hanging basket pl. '**Purple Passion**' poss. hybrid with *G. procumbens* (Lour.) Merr.); ***G. bicolor*** (Willd.) DC. (Himal.) – cult. orn. 'Okinawa spinach'; ***G. japonica*** (Thunb.) Juel (*G. pinnatifida*, E As.) – cult. for foliage & tubers, comm. medic. in modern Chinese herbalism; ***G. pseudochina*** (L.) DC. (OW trop.) – source of medic. 'China r.'

Gypothamnium Philippi. Compositae (Mut.-Onis.). 1 N Chile: ***G. pinifolium*** Philippi – shrub

Gypsacanthus Lott, Jaramillo & Rzed. Acanthaceae (III 2c). 1 Mex.: ***G. nelsonii*** Lott, Jaramillo & Rzed. R: BSBM 46(1984)47

Gypsophila L. Caryophyllaceae (III 1). Excl. *Psammophiliella*, 140 temp. Euras., Egypt, Aus. & NZ (1, ? introd.). R: Wentia 9(1962)4. Heterogeneous?; allied to *Saponaria* but with many small fls with short tubes. Some medic. & cult. orn. esp. ***G. elegans*** M. Bieb. (Ukraine – Iran) & ***G. paniculata*** L. (baby's breath, C Eur. to C As.) – much planted, the latter esp. for wedding bouquets etc. with cvs incl. 'doubles' esp. '**Bristol Fairy**'; ***G. patrinii*** Ser. (E Russia to temp. As.) – used as copper indicator in Russia; ***G. rokejeka*** Del. (Egypt, E Med.) used in halva with sesame seeds & honey; ***G. struthium*** Loefl. (Spain) – form. used as soap in Medit.

gypsywort *Lycopus europaeus*

Gyptidium R. King & H. Robinson (~ *Eupatorium*). Compositae (Eup.-Gyp.). 2 Braz., Arg. R: MSBMBG 22(1987)89

Gyptis (Cass.) Cass. (~ *Eupatorium*). Compositae (Eup.-Gyp.). 7 trop. S Am. R: MSBMBG 22(1987)87

Gyrandra Griseb. (~ *Centaurium*). Gentianaceae. 3 Mex, C Am. R: T 53(2004)721

Gyranthera Pittier. Malvaceae (Bomb.; Bombacaceae). 2 Panamá, Venez. ***G. caribensis*** Pittier – extreme anisocotyly

Gyrinops Gaertn. (~ *Aquilaria*). Thymelaeaceae (Thym.-Aq.). 9 Sri Lanka (1), Laos (1), E Mal. ***G. ledermannii*** Domke & ***G. versteegii*** (Gilg) Domke (NG) – agarwood (eaglewood, gaharu), the resins (produced in response to fungal invasion) used as insecticides; ***G. walla*** Gaertn. (wallapatta, Sri Lanka) – light timber & bark-fibre used for ropes

Gyrinopsis Decne = *Aquilaria*

Gyrocarpaceae Dumort. = Hernandiaceae

Gyrocarpus Jacq. Hernandiaceae (Gyrocarpaceae). 4 trop. & warm. R: BJ 89(1969)181. Alks. ***G. americanus*** Jacq. (trop.) – light timber for toys, rafts etc.

Gyrocaryum Valdés (~ *Omphalodes*). Boraginaceae (B.3.4). 1 Spain: ***G. oppositifolium*** Valdés

Gyrocheilos W.T. Wang. Gesneriaceae (III 2j). 5 S China. R: EJB 72(2015)235

Gyrodoma Wild. Compositae (Ast.-Gran.). 1 Mozambique: ***G. hispida*** (Vatke) Wild. R: Kirkia 9(1974)294

Gyrogyne W.T. Wang (~ *Stauranthera*). Gesneriaceae (II1 1c?). 1 S China: ***G. subaequifolia*** W.T. Wang – coll. once, extinct?

Gyroptera Botsch. = *Choriptera*

Gyrostelma Fourn. = *Matelea*

Gyrostemon Desf. Gyrostemonaceae. 12 Aus. (11 W Aus.). R: BJ 106(1985)108. Wind-poll.

Gyrostemonaceae A. Juss. (~ Resedaceae). Magnoliidae – Brassicales. 5/18 Aus. R: BJ 106(1985)112. Trees or shrubs (rarely annuals) with mustard-oil glucosides but without myrosin-cells, yellow when dried. Lvs in spirals, simple, entire, oft. ± succ.; stip. minute or 0. Fls in spikes, racemes or solit. axillary, small, unisexual (pls usu. dioec.) with enlarged receptacle; P (?K) usu. discoid or cupular, ± lobed, persistent in fr., A7–∞, in 1 or more whorls developing centripetally, filaments v. short or 0, anthers with longit. slits, $\underline{G}$ (1–)±∞ adnate to C column, forming compound ovary with as many locs as carpels (1 in *Cypselocarpus*), column oft. apically expanded with short distinct styles, placentation axile with 1 campylotropous ovule per loc. Fr. dry or succ. schizocarp, each carpel opening dorsally &/or ventrally & separating from column, or indehiscent (*Cypselocarpus, Tersonia*); seeds basally arillate with peripheral embryo curved around copious oily endosperm. n = 14, 15

Genera: *Codonocarpus, Cypselocarpus, Gyrostemon, Tersonia, Walteranthus*

All genera with indehiscent fr. restricted to temp. W Aus. Embryology suggested alliance with Resedaceae

Gyrostipula J. Leroy (~ *Breonia*). Rubiaceae (I 2). 2 Madag. & Comoro Is.

Gyrotaenia Griseb. Urticaceae (II). 6 WI

H

H *Papaver somniferum* (heroin)

Haageocactus Backeb. = *seq.*

Haageocereus Backeb. Cactaceae (III 4). Excl. *Weberbauerocereus*, 19 deserts of Peru & N Chile. Nat. hybrids with *Espostoa* spp. Cult. orn. ribbed cylindrical cacti

Haarera Hutch. & E. A. Bruce = *Erlangea*

Haastia Hook.f. = *Brachyglottis*

Habenaria Willd. Orchidaceae (IV 4d). c. 800 pantrop. & subtrop. R: Gen. Orch. 2(2001)300. Heterogeneous. Terr., some moth-poll. Some cult. orn., some ed tubers, some in Ayurvedic medic. in Ind. See also *Gymnadenia, Platanthera*

Habenella Small = *Habenaria*

Haberlea Friv. Gesneriaceae (III 2e). 1 Balkans: ***H. rhodopensis*** Friv. (incl. *H. ferdinandi-coburgii*, a fine cult. form known from 1 locality in Bulgaria), a Tertiary relic like *Jancaea* & *Ramonda*, dehydration-tolerant through 'vitrification' of sucrose & raffinose

Hablitzia M. Bieb. Amaranthaceae (Chenopodiaceae I 1). 1 Cauc.: ***H. tamnoides*** M. Bieb. – perenn. subterr. stem with ann. cli. shoots with sensitive petioles. R: FT 11(2000)57

Habracanthus Nees = *Stenostephanus*

Habranthus Herb. Amaryllidaceae. c. 35 temp. S Am. Like *Zephyranthes* but P zygomorphic. Some cult. orn. esp. ***H. robustus*** Herb. (Arg., S Braz.) & ***H. tubispathus*** (L'Hérit.) Traub (Barbados snowdrop, warm S Am.)

Habrochloa C. Hubb. Gramineae (XXVI). 1 C Afr.: ***H. bullockii*** C. Hubb. R: HIP 37(1967)t. 3645

Habroneuron Standl. Rubiaceae (? I 5). 1 Mex.: ***H. radicans*** (Wernham) S. Darwin

Habropetalum Airy Shaw. Dioncophyllaceae. 1 Sierra Leone: ***H. dawei*** (Hutch. & Dalz.) Airy Shaw. R: KB 1951:334

Habrosia Fenzl. Caryophyllaceae (II 4). 1 W As.: ***H. spinulifera*** (Ser.) Fenzl

Habzelia A.DC. = *Xylopia*

Hachettea Baill. Balanophoraceae. 1 New Caled.: ***H. austrocaledonica*** Baill. – dioec., on Cunoniaceae R: APG 33(1982)95

hackberry *Celtis* spp.

Hackelia Opiz (~ *Lappula*). Boraginaceae (B.3.5.1). 45 N temp., Aus, C & S Am. R: MNYBG 26, 1(1976)121

Hackelochloa Kuntze. Gramineae (XXII 3). 2 trop. OW, introd. Am. Fr. disp. as in *Rottboellia*

hackmatack *Larix laricina, Populus balsamifera*

Hacquetia Necker ex DC. (~ *Sanicula*). Umbelliferae (II 1). 1 C Eur.: ***H. epipactis*** (Scop.) DC. – cult. orn. rock-pl. with short-stalked umbels surrounded by petal-like green bracts

Hadrodemas H. Moore = *Callisia*

Hadrolaelia (Schltr) Chiron & V.P. Castro = *Cattleya*

Haeckeria F. Muell. (~ *Calomeria*). Compositae (Gnap.-Cass.). 2 SE Aus. R: AusSB 17 (2004)460

Haegiela Short & Paul G. Wilson (~ *Epaltes*). Compositae (Gnap-Ang.). 1 Aus.: ***H. tatei*** (F. Muell.) Short & Paul G. Wilson. R: Muelleria 7(1980)259

Haemacanthus S. Moore = *Satanocrater*

Haemanthus Tourn. ex L. Amaryllidaceae. 25 Cape. R: JSAB supp. 12(1984). Cult. orn. (blood lilies (incl. ***H. albiflos*** Jacq.!); trop. spp. referred to *Scadoxus*). Alks; 2n = 16 (reduced from *Scadoxus* condition where 18)

Haematocarpus Miers (*Baterium*). Menispermaceae (I). 2 E Himal. to Sulawesi. R: KB 26(1972)419, 30(1975)81

Haematodendron Capuron. Myristicaceae. 1 Madag.: ***H. glabrum*** Capuron. R: Adansonia 12(1972)375

Haematostaphis Hook.f. Anacardiaceae (II). 1 trop. W Afr.: ***H. barteri*** Hook.f. (blood plum) – fr. ed.

Haematostemon (Muell. Arg.) Pax & K. Hoffm. Euphorbiaceae (Acal.-Pluk.-Pluk.). 2 trop. S Am. R: MNYBG 17(1967)143

Haematoxylum Gronov. Leguminosae (I 4). 5 trop. Am., Namibia (1). Timber (Nicaragua wood) & dyes esp. ***H. brasiletto*** Karsten (peachwood (modern 'brazilwood' for N Am.), Nicaragua wood, Am.) & ***H. campechianum*** L. (logwood, campeachy wood or campeche, Mex. & WI) – dark heartwood source of haematoxylin used in microscopical preparations & as dye ('logwood chips', rasped by prisoners in C17 Holland), largely replaced by *Caesalpinia echinata*, & in ink, timber for furniture, cult. orn. & used as spiny hedge, fls good bee-forage

Haemodoraceae R. Br. Magnoliidae – Commelinales. Excl. Lanariaceae, 14/100 trop. & N Am., S Afr., Aus. (7/86, most in SW), NG. R: AMBG 77(1990)722. Perenn. herbs with tubers or short rhiz. & usu. raphides & typical red pigment (phenalenones) in r. & rhiz.; vessels usu. restricted to r., s.t. also stem, rarely 0; vasc. bundles in whorls in *Lophiola* (2 in stem, 1 in rhiz.) or scattered. Lvs all basal, distich., with sheathing base & linr. parallel-veined lamina (tubular in *Tribonanthes*, plicate in *Barberetta* & *Wachendorfia*). Fls bisex., reg. to irreg., 3-merous, hypogynous to epigynous, in racemes, panicles, cymes or cymose umbels, which are usu. long-hairy; persistent P 6 to (6) in 1 (Conostyloideae) or 2 whorls, tube straight or curved, A 3 or 6, free or adnate to tube, anthers basifixed or versatile, with longit. slits, $\underline{G}$ to $\overline{G}$ (3), 3-loc. with axile placentation & septal nectaries, 1 – ∞ anatr. to orthotropous, bitegmic ovules per loc. Fr. a loculicidal capsule with (1–)several – ∞ seeds or nut-like (*Phlebocarya*); embryo small with term. cotyledon & lat. plumule embedded in copious starchy endosperm with oil, protein & hemicellulose. x = 4–8, 15+

Classification & principal genera:

I. **Haemodoroideae** (P in 2 whorls, tube ± 0, A 3 or 6; n = 12, 15, 19–21): *Haemodorum, Wachendorfia*

II. **Conostyloideae** (P in 1 whorl, tube oft. long & curved, A 6, fls always long-hairy; n = 4–8, 11; SW Aus.): *Anigozanthos, Conostylis*

Form. placed nr. Iridaceae or Liliaceae. Natural hybrids freq. The red alpha-phenylphenalenones are not found in any other organisms

Some cult. orn. incl. cut-fls (*Anigozanthos* (kangaroo-paws)), dyes (*Lachnanthes*), local medic. & ed. (Aus.)

Haemodorum Sm. Haemodoraceae (I). 20 Aus., 1 ext. to NG. R: FA 45(1987)136. ***H. corymbosum*** Vahl (E Aus.) & other spp. – roasted rhiz. ed. (Aborigines)

Haenianthus Griseb. Oleaceae (4d). 2 WI. R: CJB 69(1991)489

hagberry or **hegberry** *Prunus padus, P. avium*

Hagenbachia Nees & Mart. Asparagaceae (Anthericaceae). 6 C & S Am. R: NJB 7(1987)255. Form. referred to Haemodoraceae

Hagenia J. Gmelin. Rosaceae (8). 1 mts C Afr., Sudan, Ethiopia to Zimbabwe: ***H. abyssinica*** (Bruce) J. Gmelin – Rauh's Model, to 35m, female fls (koso or kousso) used as taenicide

hag-taper (i.e. hedge-taper, high-taper) *Verbascum thapsus*

Hagsatera González (~ *Epidendrum*). Orchidaceae (V 13b). 2 Mex. R: C.L. Withner, *The Cattleyas & their relatives* V(1998)141

haiari *Lonchocarpus* spp.

Hainania Merr. = *Pityranthe*

Hainardia Greuter (~ *Monerma*). Gramineae (XVI 12). 1 Medit. incl. Eur.: ***H. cylindrica*** (Willd.) Greuter – weedy in S Afr. R: FNA 24(2007)689

hair fescue *Festuca filiformis;* **h. grass** spp. of *Aira, Corynephorus, Deschampsia, Koeleria, Muhlenbergia* & *Trisetum* etc.

Haitia Urb. = *Ginoria*

Haitiella L. Bailey = *Coccothrinax*

Hakea Schrad. (~ *Grevillea*). Proteaceae (V 3). 149 Aus. esp. SW. R: FA 17B(1999)31. Everg. xerophytes (needlebush, needle-wood) with divided lvs only when young in most xeromorphic spp. (cf. *Acacia*), others resembling plastic foliage; like *Grevillea* but seeds with long term. wings. Insect-poll. derived from bird-poll., ***H. dactyloides*** (Gaertn.) Cav. (SE Aus.) fl. biennially in Blue Mts NSW; ***H. eyreana*** (S. Moore) McGillivray (E & S Aus.) – bark used on burns. Some timbers (corkwood) & cult. orn., others used in reclamation of arid areas, e.g. ***H. salicifolia*** (Vent.) B.L. Burtt in Spain, though invasive (Aus., S Afr.) as are ***H. decurrens*** R. Br. ('*H. sericea*') in Iberia, ***H. drupacea*** (Gaertn.) Roem. & Schultes (*H. suaveolens*, W Aus.), ***H. gibbosa*** (Sm.) Cav. (NSW) & ***H. sericea*** Schrad. & Wendl. (*H. acicularis*, SE Aus.) – major weeds in Cape fynbos, etc.

Hakoneaste F. Maek. = *Ephippianthus*

Hakonechloa Makino ex Honda (~ *Phragmites*). Gramineae (Arund.). 1 Jap.: ***H. macra*** (Munro) Honda – varieg. cvs cult. as pot-pls in Jap.

Halacsya Doerfler. Boraginaceae (B.2.2). 1 W Balkans (serpentine): ***H. sendtneri*** (Boiss.) Doerfler

Halacsyella Janchen = *Edraianthus*

Halanthium K. Koch = *Halimocnemis*

Halarchon Bunge (~ *Halimocnemis*). Amaranthaceae (Chenopodiaceae III 3). 1 Afghanistan: ***H. vesiculosus*** (Moq.) Bunge

Haldina Ridsd. = *Adina* (but see R: FOC 19(2011)146)

haldu *Adina cordifolia*

Halenbergia Dinter = *Mesembryanthemum*

Halenia Borkh. Gentianaceae. c. 70 Euras. mts (3; Eur. 1), Am. esp. Andes (C Am. 6, R: BTBC 111(1984)366). Many with cleistogamous fls, few cult. orn.

Halerpestes Greene (~ *Ranunculus*). Ranunculaceae (II 3). c. 10 As., Am.

Halesia Ellis ex L. Styracaceae. 3–5 E China & E N Am. G inf.; fr. with 2 or 4 longit. wings. As. spp. app. close to *Rehderodendron*. Cult. orn. decid. shrubs (bell trees) esp. ***H. carolina*** L. (*H. tetraptera*, E US) – snowdrop tree, silver bell, opossumwood, with white drooping fls

Halesiaceae D. Don = Styracaceae

halfa *Macrochloa tenacissima*

Halfordia F. Muell. Rutaceae (I 2). 1 NG, E Aus., Vanuatu, New Caled.: ***H. kendack*** (Montr.) Guillaumin (*H. scleroxyla*) – timber (jitta, kerosene wood) for furniture, burls green. R: FM 26(2013)83

Halgania Gaud. Boraginaceae (Ehr.; Ehretiaceae). c. 20 Aus.

× **Halimiocistus** Janchen = *Cistus*

Halimione Aellen (~ *Atriplex*). Amaranthaceae (Chenopdiaceae I 3). 3 Euras., Medit.

Halimiphyllum (Engl.) Boriss. Zygophyllaceae. 5 C As.

Halimium (Dunal) Spach = *Cistus*

Halimocnemis C. Meyer. Amaranthaceae (Chenopodiaceae III 3). 19 Eur. (1) to C As.

Halimodendron Fischer ex DC. = *Caragana*

Halimolobos Tausch. Cruciferae (29). Excl. *Exhalimolobos*, 6 SW US to C Am. R: SB 32(2007)144

Hallea J. Leroy = *Mitragyna*

hallelujah *Oxalis acetosella*

Halleorchis Szlach. & Olsz. Orchidaceae (IV 2d.). 1 Cameroun & Gabon: ***H. aspidogynoides*** Szlach. & Olsz. R: Gen. Orch. 3(2003)98

Halleria L. Stilbaceae (Scrophulariaceae.s.l.). 5 trop. & S Afr. (3), Madag. (2). Roux's Model. ***H. lucida*** L. (Afr. honeysuckle, Afr.) – cult. orn. shrub with scarlet drooping fls & allegedly ed. fr.

Hallia Thunb. = *Psoralea*

Hallieracantha Stapf = *Ptyssiglottis*

Hallianthus H. Hartman (~ *Leipoldtia*). Aizoaceae (V). 1 SW Cape: ***H. planus*** (L. Bolus) H. Hartman. R: BJ 104(1983)143

Halmoorea Dransf. & Uhl = *Orania*

Halocarpaceae A. Bobrov & Melikian = Podocarpaceae

Halocarpus Quinn (~ *Dacrydium*). Podocarpaceae. 3 NZ. R: Phytol.M 7(1984)31. ***H. biformis*** (Hook.) Quinn (manoao) – timber used locally for building, railway-sleepers etc.

Halocharis Moq. Amaranthaceae (Chenopodiaceae III 3). 7 SW & C As.

Halocnemum M. Bieb. Amaranthaceae (Chenopodiaceae II 2). 3 C Med. incl. Eur. to C As. ***H. strobilaceum*** (Pallas) M. Bieb. – alks

Halodule Endl. Cymodoceaceae. 8 shallow trop. seas. R: VKNAW II, 59(1970)1. Favoured food for dugongs and turtles. Pollen in rafts reaching stigmas (cf. *Halophila, Lepilaena, Ruppia*) but many spp. unknown in sexual state

Halogeton C. Meyer ex Ledeb. Amaranthaceae (Chenopodiaceae III 3). 5 Med. (Eur. 1) to C As. Toxic weeds in W US (sodium oxalate). ***H. glomeratus*** (M. Bieb.) Ledeb. (SE Russia) – ant-poll.; ***H. sativus*** (L.) Moq. (W Med.) – form. cult. & burnt for base-rich ash (barilla)

Halopegia K. Schum. Marantaceae. 2 Congo (1), SE As. to Java (1)

Halopeplis Bunge ex Ung.-Sternb. Amaranthaceae (Chenopodiaceae II 1). 3 S W Eur. (1), E Medit., Cauc.

Halophila Thouars. Hydrocharitaceae. 19 trop. coasts WI, Ind. & Pac. Oceans. R: C. den Hartog (1970) *Sea grasses of the world*: 238. Bell's Model; pollen in rafts reaching stigmas (cf. *Halodule, Lepilaena, Ruppia*). ***H. stipulacea*** (Forssk.) Asch. spread from Ind. Ocean & Red Sea through Suez Canal (opened 1869) to Malta etc. & still spreading

Halophytaceae Soriano (~ Basellaceae). Magnoliidae – Caryophyllales. 1/1 temp. Arg. Succ. monoec. herb. Wood without rays. Lvs in spirals; stip. 0. Fls in racemes; bracteoles 2; P 4 membranous, 0 in females; A 4; $\underline{G}$ (3) 1-loc., 1-ovulate; style with 3 stigmas. Fr. a thin-walled nutlet encl. in axial tissue which hardens to form a syncarp encl. a few nutlets; seed with annular embryo

Genus: *Halophytum*

Excl. from Amaranthaceae (Chenopodiaceae): sieve-tube plastids diff. while cuboid pollen resembles that of some *Basella* spp.

Halophytum Speg. Halophytaceae. 1 S Arg.: ***H. ameghinoi*** (Speg.) Speg.

Halopyrum Stapf. Gramineae (Chlorid.). 1 Ind. Ocean coasts: ***H. mucronatum*** (L.) Stapf

Haloragaceae R. Br. Magnoliidae – Saxifragales. Incl. Tetracarpaeaceae, 9/150 cosmop. but esp. Aus. R: BAIM 10(1975). Shrubs or small trees (*Haloragodendron*) but usu. aquatic or amphibious herbs; cortex commonly with air-cavities; in aquatics, vasc. system reduced (s.t. to single C fibro-vascular strand). Lvs in spirals, opp. or whorled, v. varied in form, serrate or deeply lobed; stip. 0. Fls solit. & axillary or in term. spikes to thyrses, usu. small & unisexual, epigynous, reg., (2–)4-merous, wind-poll., 2-bracteolate; K valvate, s.t. 0, persistent in fr., C oft. larger, imbr. or 0, A 4 + 4 or 4(3) with usu. short filaments & rather large anthers with longit. slits, $\overline{G}$ ((2–)4) with as many locules (s.t. partitions ± absent) & distinct feathery styles, 1(2) anatropous, bitegmic pend. ovule per loc. Fr. a nut or drupe (a schizocarp with (2–)4 mericarps in *Myriophyllum*, follicles in *Tetracarpaea*); seeds with straight cylindrical embryo in ± copious oily endosperm. n = (6) 7(8)

Principal genera: *Gonocarpus, Haloragis, Myriophyllum*

Pelargonidin in lvs (like Saxifragaceae). Some cult. aquatics in aquaria esp. *Myriophyllum* spp.

Haloragidaceae auctt. = praec.

Haloragis Forst. & Forst. f. Haloragaceae. Excl. *Meionectes*, 26 Aus. (23), New Caled (1), NZ (1), Rapa (1), Juan Fernandez (2). R: BAIM 10(1975)64. Desert ephemerals to obligate aquatics

Haloragodendron Orch. Haloragaceae. 6 S Aus. (1 almost extinct – ***H. lucasii*** (Maiden & Betche) Orchard (male-sterile clones) rediscovered Sydney 1980s). R: BAIM 10(1975)140. Trees & shrubs to 3 m

Halosarcia Paul G. Wilson = *Tecticornia* (but see Nuytsia 3(1980)25)

Halosciastrum Koidz. (~ *Cymopterus*). Umbelliferae (III 8). 1 E As.: ***H. melanotilingia*** (Boissieu) Pim. & Tichomirov

Halosicyos Mart. Crov. Cucurbitaceae (XIII). 1 C Arg.: ***H. ragonesei*** Mart. Crov. R: U. Eggli, *Ill. Handb. Succ. Pls*, Dicots (2002)84

Halostachys C. Meyer ex Schrenk. Amaranthaceae (Chenopodiaceae I). 1 SE Russia to C As.: ***H. caspica*** (Bieb.) Schrenk (*H. belangeriana*)

Halothamnus Jaub. & Spach (~ *Salsola*). Amaranthaceae (Chenopodiaceae III 3). 21 Middle E to Afghanistan & Somalia. R: OB 143(1993)

Halotis Bunge = *Halimocnemis*

Haloxanthium Ulbr. = *Atriplex*

Haloxylon Bunge. Amaranthaceae (Chenopodiaceae III 3). Incl. *Hammada*, c. 25 W Med. (Eur. 1) to Iran, Mongolia, Myanmar & SW China. Steppe-pls with alks & jointed twigs (saxaul/saxoul) in deserts E of Caspian & Aral Seas to China (covering 1 M sq. km in C as deserts alone); imp. fuel & for stock-pens, charcoal & fodder for camels, sheep etc. (100 km of Gt Wall of China in Mongolia made of earth + saxaul), dune-stabilizers esp. ***H. persicum*** Bunge ex Boiss. & Buhse (C As.) to 6 m tall & bole to 20 cm diam. with timber used for general carpentry & source of charcoal, introd. Mongolia for dune-stabilizing; ***H. aphyllum*** (Minkw.) Iljin (widespread) – similar & a promising firewood crop; ***H. salicornicum*** Bunge ex Boiss. (*Hammada s.*, Iran, Afghanistan) – source of soap & poss. Biblical manna

Halphophyllum Mansf. = *Gasteranthus*

halvah *Sesamum indicum*

Hamadryas Comm. ex Juss. Ranunculaceae (II 3). 6 Antarctic S Am. Dioec.

Hamamelidaceae R. Br. Magnoliidae – Saxifragales. 28/100 widespread but chiefly subtrop. esp. E As. Shrubs or trees, oft. with stellate hairs. Lvs usu. in spirals (distich.), simple, oft. palmately-lobed; stip. usu. present. Fls bisex. or unisexual, usu. reg., ± epigynous, s.t. crowded (& considered by some to be s.t. pseudanthial) but usu. in spikes or heads, wind- or insect-poll.; K (0)4 or 5(–7), small, free to connate, C usu. (0)4 or 5 small & narrow, A (1–)4 or 5(–24), alt. with C developing centripetally (*Matudaea*) or centrifugally (*Fothergilla*), anthers with valves or longit. slits, connective usu. extended, G (2 (or 3)) united at least basally, placentation usu. axile, styles distinct, each loc. with 1(2 – ∞) ± pend., anatr. (orthotropous in Altingioideae), bitegmic ovules, oft. all but 1 aborting. Fert. oft. delayed until long after poll. (cf. Pinidae). Fr. a woody capsule, septicidal (& loculicidal); seeds with thick hard testa & large straight embryo in oily & proteinaceous endosperm. x = 8, 12, 15, 16

Classification & principal genera:

I. **Hamamelidoideae** (male fls in male infls s.t. not discrete, others clearly sep.; locules with 1 or 2 ovules; 4 tribes):
   1. **Hamamelideae**: *Dicoryphe, Hamamelis, Trichocladus*
   2. **Corylopsideae**: *Corylopsis* (only)
   3. **Eustigmateae**: *Eustigma*
   4. **Fothergilleae** (incl. Disanthoideae ($\underline{G}$, n = 8) prob. best recog. as distinct): *Distylium, Sycopsis*

II. **Exbucklandioideae** (pls polygamo-monoec. with fls in capitula; A 10–14; stip. broad, encl. young shoot): *Exbucklandia*

III. **Rhodoleioideae** (fls bisex. in 5–10-flowered capitulum subtended by bracts so as to resemble single fl.; prob. best in II.): *Rhodoleia*

IV. **Altingioideae** (Liquidambaroideae; Altingiaceae; pls dioec., male infls with P 0, females globose heads with P of ∞ scales): *Liquidambar* (only)

Altingioideae oft. recog. as discrete fam., though clearly allied. App. relics of an anc. group, *Disanthus* having most arch. features

Timbers (*Distylium, Exbucklandia, Liquidambar*), extracts medic. & used in scent (*Hamamelis, Liquidambar*), cult. orn. (*Corylopsis, Disanthus, Fothergilla, Hamamelis, Liquidambar, Parrotia, Sycopsis* etc.)

Hamamelis L. Hamamelidaceae (I 1). 4 E N Am., E As. Witch hazels. R: C. Lane (2005) *W. h.* Cult. orn. shrubs (As. spp. fl. in winter, Am. spp. in autumn when leafy) esp. ***H.* × *intermedia*** Rehder (***H. japonica*** Sieb. & Zucc. (Jap.) × ***H. mollis*** Oliv. (China)) & ***H. virginiana*** L. (E N Am.) – bark & lvs source of medic. witch hazel for bruises, haemorrhoids, varicose veins etc. & grown comm. in England for eye-lotions, form. used as divining-rod (witch/wych an Old Engl. term for pliant branches), used as grafting stock for other spp. though seeds take up to 2 yrs. to germinate; all spp. with strongly-scented fls

Hamatocactus Britton & Rose = *Thelocactus*

Hamelia Jacq. Rubiaceae (IV 5). 16 trop. Am. R: MNYBG 26,4(1976)90. Interpetiolar stip. small, decid.; extrafl. nectaries continue to secrete for up to 21 days after fall of C. Cult. orn. shrubs with red or yellow fls, esp. hummingbird-poll. ***H. patens*** Jacq. (fire bush) oft. confused with *Ixora* spp. – ed. fr. used in a fermented drink, tanbark

hami melon *Cucumis melo* Inodora Group

Hamilcoa Prain. Euphorbiaceae (Euph.-Stom.). 1 Cameroun: ***H. zenkeri*** (Pax) Prain. R: A.R.-Sm., *Gen. Euph.* (2001)351

hamilla *Berrya cordifolia*

Hamiltonia Roxb. = *Spermadictyon*

Hammada Iljin = *Haloxylon* (but possibly distinct)

Hammarbya Kuntze (~ *Malaxis*). Orchidaceae (V 11b). 1 N temp.: ***H. paludosa*** (L.) Kuntze (*M. p.*, bog orchid) – poll. male fungus-gnats in Minnesota, foliar embryos at leaf-tips effect veg. reproduction

Hammatolobium Fenzl (~ *Tripodion*). Leguminosae (III 22). 2 Medit. (Eur. 1)

Hammeria Burgoyne (~ *Lampranthus*). Aizoaceae. 3 Cape

Hampea Schldl. Malvaceae (Malv.-Goss.). 21 Mex. to Colombia. R: P. Fryxell (n.d.) *Nat. Hist. Cotton tribe*: 72

Hanabusaya Nakai (~ *Adenophora*). Campanulaceae (I 6). 1 Korea: ***H. asiatica*** (Nakai) Nakai

Hanburia Seemann. Cucurbitaceae (XII). Incl. *Elateriopsis*, 7 trop. Am. Fr. explosive as in ***H. mexicana*** Seemann (chayotilla)

Hancea Seem. (*Cordemoya*, ~ *Mallotus*). Euphorbiaceae (Acal.-Acal.-Rottl.). 17 Indomal. (13), Madag. & Masc. (4). R: Blumea 52(2007)362

Hanceola Kudô. Labiatae (VII 3b). 8 China

Hancockia Rolfe. Orchidaceae (V 14). 1 E & SE As.: ***H. uniflora*** Rolfe. R: OM 6(1992)63

Hancornia B.A. Gomes. Apocynaceae (I c). 4 Braz. ***H. speciosa*** B.A. Gomes – source of mangabeira (or Pernambuco) rubber, fr. used in marmalade & to flavour ice cream, sherbet etc.

hand, Buddha's *Citrus medica* 'Fingered'

Handelia Heimerl. Compositae (Anth.-Han.). 1 Iran, C As. to China: ***H. trichophylla*** (Schrenk) Heimerl. R: BBMNHB 23(1993)110

Handeliodendron Rehder. Sapindaceae (II 2). 1 China: ***H. bodinieri*** (H. Lév.) Rehder. R: JAA 16(1935)65

Handroanthus Mattos (~ *Tabebuia*). Bignoniaceae (1). 30 trop. Am. R: SB 32(2007)663. Some excellent timbers (green ebony, ipê), poss. most durable Am. wood (dead specimens of ***H. guayacan*** (Seem.) S. Grose (*T. g.*, Mex. to Colombia) still standing in the Panamá Canal), 400 yr-old beams in Panamá still in excellent condition, poss. replacement for *Paubrasilia echinata* in bow-making; some spectacular cult. orn. & some barks medic. (allegedly cures for cancers, now patented) incl. malaria. ***H. billbergii*** (Bur. & K. Schum.) S. Grose (*T. b.*, Ecuador) – most sought after timber for carving; ***H. chrysanthus*** (Jacq.) S. Grose (*T. c.*, Mex. to Venez.) – most imp. tree of coastal Ecuador; ***H. heptaphyllus*** (Vell.) Mattos (*T. h.*, S Am.) – nat. fl. Paraguay; ***H. impetiginosus*** (DC.) Mattos (*T. i.*, inc. *T. avellanedae*, pau d'arco, N Mex. to Arg.) – cabinet timber (lapacho) form. used as ball bearings, trad. medic. tea (taheebo) now comm.; ***H. serratifolius*** (Vahl) S. Grose (*T. s.*, WI to Bolivia) – valuable timber (washiba), cult. for consp. yellow fls (nat. fl. Braz.)

Hanghomia Gagnepain & Thénint = *Finlaysonia*

Hanguana Blume. Hanguanaceae (Flagellariaceae s.l.). 10 Sri Lanka, SE As. & Mal. to N Aus. R: Willd. 40(2010)205. Some floating (aerenchyma). Some spp. have identical male pls, but females have diff. sized fr.

Hanguanaceae Airy Shaw. (~ Flagellariaceae). Magnoliidae – Commelinales. 1/10 Indomal. Perenn. dioec. terr. herbs. Lvs with petiole & midrib, in spirals. Infl. a term. panicle of small 3-merous fls. P 3 (smaller) + 3 yellow or green, marcescent, males with A filaments broad & basally connate, 6-lobed disk around pistillode; females with 6 staminodes, G(3) with axile placentation. Stigmas 3; 1 basal ovule per carpel. Fr. a 1(-3)-seeded berry; seed bowl-shaped, endosperm starchy, copious, embryo small. n = c. 24, 36, 45

Genus: *Hanguana*

Remarkable convergence with Flagellariaceae (Poales) of which form. considered part

Haniffia Holttum. Zingiberaceae (II 1). 2 Peninsular Thailand & Malaya. R: NJB 30 (2000)287

Hannafordia F. Muell. Malvaceae (Bytt.-Lasio.; Sterculiaceae). 4 C Aus.

Hannoa Planch. (~ *Quassia*). Simaroubaceae. 5–7 trop. Afr.

Hannonia Braun-Blanquet & Maire. Amaryllidaceae. 1 Morocco.: ***H. hesperidium*** Braun-Blanquet & Maire

Hansenia Turcz. (~ *Ligusticum*). Umbelliferae (III 8). Incl. *Notopterygium*, 6 E As. R: Willd. 38(2008)162

Hanseniella Cusset (~ *Hydrobryum*). Podostemaceae (III). 2 N Thailand

Hanslia Schindler (~ *Desmodium*). Leguminosae (III 19). 2 Mal. to trop. Aus.

Hansteinia Oersted = *Stenostephanus*

Hapaline Schott. Araceae (VII 11). 7 Myanmar to Borneo. R: KB 51(1996)67. Seasonally dormant geophytes

Hapalochilus (Schltr.) Senghas = *Bulbophyllum*

Hapalorchis Schltr. Orchidaceae (IV 2h). 10 warm Am. R: HBML 29(1982)326

Haplanthodes Kuntze (~ *Andrographis*). Acanthaceae (III 2b). 4 Ind. R: BBSI 23(1983) 198

Haplanthoides Li = *Andrographis*

Haplanthus Nees = *Andrographis*

Haplocalymma S.F. Blake = *Viguiera*

Haplocarpha Less. Compositae (Arct.-Arct.). Excl. *Damatris*, *Landtia*, 6 E Afr. to Ethiopia

Haplochilus Endl. = *Zeuxine*

Haplochorema K. Schum. (~ *Boesenbergia*). Zingiberaceae (II 1). 1 Sumatra, 3 Borneo. R: NRBGE 44(1987)211

Haploclathra Benth. Calophyllaceae (Guttiferae (I1). 4 Amazonia. R: MNYBG 22,4(1972) 129. Some locally used timbers (red)

Haplocoelopsis Davies. Sapindaceae (I). 1 E & SE Afr.: ***H. africana*** Davies. R: KB 52(1997) 230

Haplocoelum Radlk. Sapindaceae. Excl. *Gereaua*, 4–5 C & E Afr.

Haplodictyum C. Presl = *Cyclosorus*

Haplodypsis Baill. = *Dypsis*

Haploesthes A. Gray. Compositae (Tag.-Flav.). 3 SW N Am. R: Wrightia 5(1975)108

Haplolobus H.J. Lam. Burseraceae (V). c. 16 Mal. & Pac.

Haplolophium Cham. = *Amphilophium*

Haplopappus Cass. Compositae (Ast.-Mac.). (Excl. N Am spp. = *Ericameria*, *Tonestus*) c. 70 S Am. esp. Chile. R: CIWP 389(1928)1. Some cult. orn. ***H. tenuisectus*** (Greene) S.F. Blake (burroweed, Arizona) – see also *Isocoma*

Haplopetalon A. Gray = *Crossostylis*

Haplophandra Pichon = *Odontadenia*

Haplophloga Baill. = *Dypsis*

Haplophragma Dop = *Fernandoa*

Haplophyllophorus (Brenan) Fernandes & R. Fernandes = *Cincinnobotrys*

Haplophyllum A. Juss. (~ *Ruta*). Rutaceae (I 10). 66 Medit. (Eur. 8) to NW Afr., Arabia & E Siberia. R: HIP 40(1–3)(1986). Alks. ***H. tuberculatum*** (Forssk.) A. Juss. (N Afr. & Middle E) – imp. local medic.

Haplophyton A. DC. Apocynaceae (1a). 1 SW N Am.: ***H. cimicidum*** A. DC. (*H. crooksii*, cockroach plant) – alks (insecticidal)

Haplopteris C. Presl = *Monogramma*

Haplorhus Engl. Anacardiaceae (I). 1 high Andes of Peru, N Chile: ***H. peruviana*** Engl. – v. rare, lvs willowy

Haplormosia Harms. Leguminosae (III 2). 1 W Afr., swamp forest: ***H. monophylla*** (Harms) Harms – timber (haplormosia, idewa)

Haplosciadium Hochst. Umbelliferae (III 8). 1 E & NE Afr.: ***H. abyssinicum*** Hochst. – geocarpic rosette-pl.

Haploseseli H. Wolff & Hand.-Mazz. = *Physospermopsis*

Haplosphaera Hand.-Mazz. Umbelliferae (III 8). 2 S & E As.

Haplospondias Kosterm. Anacardiaceae (II). 1 Myanmar, S China: ***H. brandisiana*** (Kurz) Kosterm. – coll. once

Haplostachys (A. Gray) Hillebrand. Labiatae (VI). 5 (4 extinct) Hawaii. App. derived from N Am. *Stachys* spp. ***H. haplostachya*** (A. Gray) St John reduced to 1 pop.

Haplostephium Mart. ex DC. = *Lychnophora*

Haplostichanthus F. Muell. = *Polyalthia* (but see Blumea 39(1994)215)

Haplothismia Airy Shaw. Thismiaceae (Burmanniaceae s.l.). 1 S Ind. (Western Ghats): ***H. exannulata*** Airy Shaw – coll. rarely. R: KB 7(1952)277

happy plant *Cordyline fruticosa*; **h. tree** *Camptotheca acuminata*

Haptanthaceae C. Nelson = Buxaceae (but see SB 30(2005)773,779)

Haptanthus Goldberg & C. Nelson. Buxaceae (Haptanthaceae). 1 Honduras: ***H. hazlettii*** Goldberg & C. Nelson – first coll. 1980, refound 2010. R: SB 14(1989)16

Haptocarpum Ule. = ? *Cleome* (but see NP ed. 2, 17b(1936)220)

Haptotrichion Paul G. Wilson (~ *Waitzia*). Compositae (Gnap.-Ang.). 2 Carnarvon Dist., W Aus. R: Nuytsia 8(1992)422

Haradjania Rech.f. = *Myopordon*

Haraella Kudô = *Gastrochilus*

Harbouria J. Coulter & Rose. Umbelliferae (III 8). 1 SW US: ***H. trachypleura*** (A. Gray) J. Coulter & Rose

hard fern *Blechnum spicant*; **h. fescue** *Festuca brevipila*; **h. heads** *Centaurea nigra*

Hardenbergia Benth. Leguminosae (III 18). 3 Aus. Allied to *Kennedia*. Cult. orn. esp. ***H. violacea*** (Schneev.) Stearn (*H. monophylla*, false sarsaparilla, E Aus.) with 1-foliolate lvs

hardhack *Spiraea tomentosa*

Harding grass *Phalaris aquatica*

Hardingia Docha Neto & Baptista = *Gomesa*

Hardwickia Roxb. Leguminosae (I 2). 1 semi-arid Ind.: ***H. binata*** Roxb. – heaviest of Ind. timbers used for constr. & orn. work, bark used for tanning & as fibre & for sails & paper, resin a wood preservative, lvs used as fodder

harebell *Campanula rotundifolia*, (Aus. & NZ) *Wahlenbergia* spp.

hare's ear *Bupleurum* spp.

hare's-foot fern *Davallia canariensis*

harewood *Acer pseudoplatanus*

Harfordia Greene & C. Parry. Polygonaceae (Erio.-Erio.). 1 Baja Calif., Mex.: ***H. macroptera*** (Benth.) Greene & C. Parry

haricot bean *Phaseolus vulgaris*

Harlanlewisia Epling = *Scutellaria*

harlequin flower *Sparaxis* spp.

Harleya S.F. Blake. Compositae (Vern.-Pep.). 1 Mex., C Am.: ***H. oxylepis*** (Benth.) S.F. Blake

Harleyodendron R. Cowan. Leguminosae (III 1). 1 NE Braz.: ***H. unifoliolatum*** R. Cowan. R: Britt. 31(1979)72

harmal *Peganum harmala*

Harmandia Pierre ex Baill. Olacaceae (Aptandraceae). 1 SE As.: ***H. mekongensis*** Pierre ex Baill. R: FPM 3(2012)307

Harmandiella Costantin = *Marsdenia*

Harmogia Schauer (~ *Baeckea*). Myrtacee (II 15). 1 E Aus.: ***H. densifolia*** (Sm.) Schauer. R: SB 20(2007)316

Harmonia Baldwin. Compositae (Mad.-Mad.). 5 Calif. R: FNA 21(2006)297

Harmsia K. Schum. Malvaceae (Dom.; Sterculiaceae). 2 NE Afr. R: T 48(1999)5

Harmsiodoxa O. Schulz. Cruciferae (37). 3 Aus. R: TRSSA 89(1965)204

Harmsiopanax Warb. Araliaceae. 3 Mal. R: FM I, 9(1979)9. Some hapaxanthic (Holttum's Model) unique in fam.). Fr. with 2 mericarps like Umbelliferae but leaf-base, habit & petal-shape typical of Araliaceae. ***H. ingens*** Philipson (NG) – to 18 m, spiny, hapaxanthic

Harnackia Urb. (~ *Lascaillea*). Compositae (Hele-Pect.). 1 E Cuba: ***H. bisecta*** Urb. – liane on serpentine. R: Madroño 24(1977)137

Haroldia Bonif. (~ *Chiliotrichiopsis*). Compositae (Ast.-Hin.). 1 Arg.: ***H. mendoncina*** (Cabrera) Bonif. R: SCB 92(2009)50

Haroldiella Florence. Urticaceae. 2 Fr. Polynesia. R: J. Florence, *Fl. Polyn. Fr.* 1(1997) 218

Haronga Thouars = *Harungana*

Harpachne Hochst. ex A. Rich. = *Eragrostis*

Harpagocarpus Hutch. & Dandy = *Fagopyrum* (but see KB 1926:364)

Harpagonella A. Gray (~ *Pectocarya*). Boraginaceae (B.3.8.4). 2 SW N Am.

Harpagophytum DC. ex Meissn. Pedaliaceae (2). 2 S Afr. R: MSABH 13(1970)15. ***H. procumbens*** DC. ex Meissn. (devil's claw, grapple plant) has fr. with large woody grapples c. 2.5 cm long, pointed & barbed, so that it is animal-disp. (&, like *Xanthium*, a nuisance to shepherds) but it can cause grazing animals to become lame or even starve as their jaws may become locked by them; used as mouse-traps in Madag., also comm. medic., for pain relief in W Eur.

Harpalyce Sessé & Moçiño ex DC. Leguminosae (III 4). 24 Mex., Cuba & Braz.

Harpanema Decne. = *Camptocarpus*

Harpephyllum Bernh. Anacardiaceae (II). 1 S Afr.: ***H. caffrum*** Bernh. (kaffir plum) – dioec.(?) cult. orn. flowering when unbranched, fr. used in jelly & local medic. R: Strel. 10(2000) 57

Harperella Rose (~ *Ptilimnium*). Umbelliferae (II 8). 1 SE US: ***H. nodosa*** (Rose) Rose. R: T 61(2012)413

Harperia W. Fitzg. = *Desmocladus*

Harperocallis McDaniel ('*Isidrogalvia*'). Tofieldiaceae. 11 Am. esp. Andes. R: PhytoKeys 21(2013)39. ***H. flava*** McDaniel (Florida) – genetically v. uniform

Harpochilus Nees (~ *Justicia*). Acanthaceae (III 2c). 3 Braz. ***H. neesianus*** Mart. (N.E. Braz.) – bat-poll. fls smell of sour cabbage

Harpochloa Kunth (~ *Microchloa*). Gramineae (XXIX 5). 2 C & S Afr.

Harpullia Roxb. Sapindaceae (II 1). 26 Indomal., trop. Aus. (8) & New Caled. (1), Tonga. R: Blumea 28(1982)1. Usu. pinnate lvs, ***H. mabberleyana*** W. Takeuchi (NG) – treelet with simple lvs. Some cult. orn. trees incl. ***H. pendula*** Planch. ex F. Muell. (Aus. tulipwood, NE Aus.) – street-tree in Lima, Peru, timber beautifully marked black to yellow used in cabinet-making

Harrimanella Cov. (~ *Cassiope*). Ericaceae (VI). 2 Arctic & subarctic. R: FNA 9(2009)494. Cult. orn. shrublets differing from *C.* in spiral lvs & term. infls

Harrisella Fawcett & Rendle (~ *Dendrophylax*). Orchidaceae (V 16). 1 trop. Am. to Florida: ***H. porrecta*** (Rchb.f.) Fawcett & Rendle. R: FNA 26(2002)621

Harrisia Britton. Cactaceae (III 1). 20 Florida, WI (Carib. 11), S Am. R: SB 30(2013)218. Cult. orn. slender ribbed cacti with caffeine, some ed. fr. ***H. martinii*** (Labouret) Britton (Arg., Parag.) declared a noxious weed in C Afr., serious in Qld

Harrisonia R. Br. ex A. Juss. Rutaceae (III 4). 3–4 OW trop. ***H. abyssinica*** Oliv. (trop. Afr.) – local treatment of diarrhoea, r. used for swollen testicles on Mafia Is.

Harrysmithia H. Wolff. Umbelliferae (III 8). 2 China

harsinger (tree) *Nyctanthes arbor-tristis*

Harthamnus H. Robinson = *Plazia*

Hartia Dunn = *Stewartia*

Hartleya Sleumer (~ *Gastrolepis*). Stemonuraceae (Icacinaceae s.l.). 1 NG: ***H. inopinata*** Sleumer. R: Blumea 17(1967)218

Hartliella E. Fischer (~ *Lindernia*). Linderniaceae (Plantaginaceae (Grat.-Lind.); Scrophulariaceae s.l.). 4 Congo (Katanga). R: TSP 81(1992)204. On soils rich in heavy metals

Hartmannia Spach = *Oenothera*

Hartmanthus Hammer (~ *Delosperma*). Aizoaceae. 2 S Afr. R: H. E. K. Hartmann, *Ill. Handb. Succ. Pls*, Aiz. F–Z(2002)55

Hartogia Thunb. ex L.f. = *Cassine*

Hartogiella Codd = *Cassine*

Hartogiopsis H. Perrier. Celastraceae (I). 1 Madag.: ***H. trilobocarpa*** (Bak.) H. Perrier. R: NS 10(1942)194

hart's tongue (fern) *Asplenium scolopendrium*

Hartwegiella O. Schulz = *Mancoa*

Hartwrightia A. Gray ex S. Watson. Compositae (Eup.-Gyp.). 1 Georgia, Florida: ***H. floridana*** A. Gray ex S. Watson. R: MSBMBG 22(1987)282

Harungana Lam. Hypericaceae (I; Guttiferae I). Incl. *Psorospermum*, 50 trop. Afr. to Mauritius: ***H. madagascariensis*** Lam. ex Poir. (invasive in Aus.) – resin yellow turning red on exposure, wood easily worked, lvs medic.

Harveya Hook. Orobanchaceae (Orob.; Scrophulariaceae s.l.). 13 trop. & S Afr., Masc. R: SBM 80(2006)15. Holo- to hemi-parasites, drying black, form. ink-source

Haselhoffia Lindau = *Physacanthus*

hashish *Cannabis sativa* 'subsp. *indica*'

Hasseanthus Rose = *Dudleya*

Hasselquistia L. = *Tordylium*

Hasseltia Kunth. Salicaceae (Flacourtiaceae). Excl. *Neosprucea*, 3 trop. Am. R: FN 22(1980)73

Hasseltiopsis Sleumer. Salicaceae (Flacourtiaceae). 1 C Am.: ***H. dioica*** (Benth.) Sleumer

Hasskarlia Baill. = *Tetrorchidium*

Hasslerella Chodat = *Polypremnum*

Hasteola Raf. = *Senecio* (But see SB 19(1994)211)

Hastingsia (Durand) S. Watson (~ *Schoenolirion*). Asparagaceae (Hyacinthaceae). 4 W N Am. R: Madroño 38(1991)135

hasu *Nelumbo nucifera*

Hatiora Britton & Rose. Cactaceae (III 6). Excl. *Rhipsalidopsis*, 3 E & SE Braz.

Hatschbachiella R. King & H. Robinson (~ *Eupatorium*). Compositae (Eup.-Eup.). 2 S Braz. R: MSBMBG 22(1987)71

hau *Hibiscus tiliaceus*

haulm aerial stems of potatoes, beans, peas etc.

Haumania Léonard. Marantaceae. 2 trop. Afr.

Haumaniastrum Duvign. & Plancke (~ *Platostoma*). Labiatae (VII 3d). 35 trop. Afr. R: KB 52(1997)298. ***H. robertii*** (Robyns) Duvign. & Plancke (*Acrocephalus r.*, Congo) – used as copper indicator in Katanga

Hausa potato *Coleus rotundifolius*

Haussknechtia Boiss. Umbelliferae (III 10). 1 SW Iran: ***H. elymaitica*** Boiss. – umbels globose, not seen since 1860s

hautbois or **hautboy** *Fragaria muricata*

Hauya Moçiño & Sessé ex DC. Onagraceae. 2 C Am. Woody

Havardia Small. Leguminosae (II 4). 5 Texas to C Am. R: MNYBG 74,1(1996)165

Havetia Kunth = *Clusia*

Havetiopsis Planch. & Triana = *Clusia*

haw fr. of *Crataegus* spp. esp. *C. monogyna*; **black h.** *C. douglasii*, *Viburnum prunifolium*

Hawaiian goosefoot *Chenopodium oahuense*

hawkbit *Leontodon* spp.

Hawkesiophyton Hunz. = *Markea*

hawk's beard *Crepis* spp.

hawkweed *Hieracium* spp.

Haworthia Duval (~ *Aloe*). Asphodelaceae (Aloaceae). Excl. *Astroloba*, *Haworthiopsis*, *Tulista*, 42 dry SW Afr. R: U. Eggli, *Ill. Handb. Succ. Pls*, Monocots (2001)199. Stemless or short-stemmed rosette-pls with succ. warty lvs, much cult. as orn.

Haworthiopsis Rowley (~ *Haworthia*). Asphodelaceae. 18 S Afr. R: SB 39(2014)70, Phytotaxa 265,1(2016)1

hawthorn *Crataegus* spp. esp. *C. monogyna*; **Chinese h.** *C. pentagyna*; **Indian h.** *Rhaphiolepis indica*; **Mexican h.** *C. pubescens*; **water h.** *Aponogeton distachyos*

Haya Balf.f. (~ *Polycarpaea*). Caryophyllaceae (I 1). 1 Socotra: ***H. obovata*** Balf.f.

Hayata Averyanov = *Cheirostylis* (but see Taiwania 54(2009)311)

Hayataella Masam. = *Ophiorrhiza*

Haydenia M. Simmons = *seq.* (but see SB 36(2011)929)

Haydenoxylon M. Simmons (~ *Gymnosporia*). Celastraceae. 3 trop. Am. R: Novon 23(2014)224

Haydonia R. Wilczek = *Vigna*

Haylockia Herb. = *Zephyranthes*

Haymondia Egan & B. Pan (~ *Pueraria*). Leguminosae (III 18). 1 Himal.: ***H. wallichii*** (DC.) Egan & B. Pan. R: Phytotaxa 218(2015)212

Haynaldia Schur = *Dasypyrum*

hayrattle *Rhinanthus minor*

Hazardia Greene (~ *Haplopappus*). Compositae (Ast.-Mac.). 13 SW N Am. R: Madroño 26(1979)105

haze, blue *Pseudoselago spuria*

hazel *Corylus* spp. esp. *C. avellana*; **h. alder** *Alnus serrulata*; **American h.** *C. americana*; **corkscrew h.** *C. avellana* 'Contorta'

Hazomalania Capuron = *Hernandia*

Hazunta Pichon = *Tabernaemontana*

head, Medusa *Taeniatherum caput-medusae*

headache tree *Premna serratifolia*

heart, bleeding *Homalanthus populifolius*; **h-leaf** *Gastrolobium* spp.; **h.pea** or **seed** *Cardiospermum grandiflorum*
heart's ease *Viola tricolor*
heath *Erica* spp.; **Australian h.** *Epacris* spp.; **Cornish h.** *Erica vagans*; **cross-leaved h.** *E. tetralix*; **Dorset h.** *E. ciliaris*; **h. grass** *Danthonia decumbens*; **Irish h.** *E. erigena*; **St Dabeoc's h.** *Daboecia cantabrica*; **sea h.** *Frankenia* spp.; **tree h.** *E. arborea*
heather (**common** or **Scottish** or **white**) *Calluna vulgaris*
heaven, tree of *Ailanthus altissima*
Hebanthe Mart. (~ *Pfaffia*). Amaranthaceae (II 2). 7 trop. & S Am. R: Sendtnera 4(1997)18. Lianes
Hebanthodes Pedersen. Amaranthaceae. 1 Peru: ***H. peruviana*** Pedersen – coll. once
Hebe Comm. ex Juss. = *Veronica*
Hebea L. Bolus = *Tritoniopsis*
Hebecarpa (Chodat) J. Abbott (~ *Polygala*). Polygalaceae. 19+ warm Am. esp. Mex. R: KB 70[39](2015)2
Hebecladus Miers = *Jaltomata*
Hebeclinium DC. (~ *Eupatorium*). Compositae (Eup.-Heb.). 20 trop. Am. R: MSBMBG 22(1987)399
Hebecoccus Radlk. = *Lepisanthes*
Hebejeebie Heads = *Veronica*
Hebenstretia L. Scrophulariaceae (Manul.). c. 40 trop. & S Afr., N to Eritrea (1). R: MBSM 15(1979)1, 18(1982)183. Some cult. orn.
Hebepetalum Benth. (~ *Roucheria*). Linaceae (I). 3 N S Am.
Heberdenia Banks ex A. DC. Primulaceae (Myrsinaceae). 1 Macaronesia: ***H. bahamensis*** (Gaertn.) Sprague (!, *H. excelsa*) – part of laurel forest of Madeira
Hebestigma Urb. Leguminosae (III 23). 1 Cuba: ***H. cubense*** (Kunth) Urb.
Hecastocleis A. Gray. Compositae (Hecast.). 1 SW US: ***H. shockleyi*** A. Gray – shrub. R: FNA 19(2006)71
Hecatactis F. Muell. ex Mattf. = *Lagenophora*
Hecatostemon S.F. Blake. Salicaceae (Flacourtiaceae). 1 trop. S Am.: ***H. completus*** (Jacq.) Sleumer
Hechtia Klotzsch. Bromeliaceae (1). c. 55 C Am. R: FN 14(1974)577. Dioec., habit of *Agave* or *Yucca*; some cult. orn. ***H. montana*** Brandegee (Mex.) – roasted 'hearts' ed.
Hecistopteris J. Sm. Pteridaceae (V; Vittariaceae). 3 trop. Am.
Heckeldora Pierre. Meliaceae (II-Trich.). 7 trop. W Afr. R: Blumea 52(2007)180
Hectorella Hook.f. (~ *Lyallia*). Montiaceae (Hectorellaceae). 1 NZ: ***H. caespitosa*** Hook.f.
Hectorellaceae Philipson & Skip. = Montiaceae
Hecubaea DC. = *Helenium*
Hedbergia Molau (~ *Bartsia*). Orobanchaceae (Rhin.; Scrophulariaceae s.l.). 3 trop. Afr mts. R: NJB 8(1988)193
Hedeoma Pers. Labiatae (VII 2b). 42 SW N Am., S Am. R: Sida 8(1980)218. Some cult. incl. ***H. pulegioides*** (L.) Pers. (American pennyroyal, Am.) – dried lvs home medic.
Hedera Tourn. ex L. Araliaceae (1). 12 Eur. (4), Medit. to E As. Ivies. R: P.Q. Rose (1980) *Ivies*; Adans. III, 24(2002)210. Woody lianes with distinct juvenile & mature stages, discussed by Theophrastus, the first (diploid) with usu. lobed lvs & rooting stems, the mature (tetraploid) with rootless flowering shoots with elliptic lvs, this 'phase-change' assoc. with increase in nuclear size & DNA content. 2n = 24–192 (mostly 24, 48). Some sp. boundaries not clear-cut; many cvs selected for ground-cover, climbers & pot-pls (s.t. grafted on × *Fatshedera* to give 'standards'), incl. many varieg. forms. ***H. helix*** L. (Eur., Medit., W As., invasive in Aus., NZ, W US) – fls v. late, poll. by wasps & moths etc., v. nutritious fr. ripens over winter & imp. food-source for nestling birds in spring (trop. 'behaviour'), molluscicidal saponins (toxic); wood used as boxwood subs., young twigs form. source of dyes & infusion form. used like dry-cleaning fluids & to remove 'shine' from serge clothes, form. used in treatment of corns, verrucas & warts, local medic. & considered to counteract effects of alcohol so figuring in Bacchus's chaplet & as a sign for a tavern; signifying 'fidelity' in 'Language of Fls', orn. cvs incl. '**Conglomerata**' – adult form prop. vegetatively, '**Congesta**' – stems erect with 2-ranked juvenile lvs, & many varieg. ***H. algeriensis*** Hibb. ('*H. canariensis*', Canary ivy, NW Afr.) esp. '**Gloire de Marengo**' ('Variegata') – housepl., ***H. colchica*** (K. Koch) K. Koch (Persian i., Cauc. to N Iran) – lvs v. large; ***H. hibernica*** (Kirchner) Bean (*H. helix* subsp. *h.*, *H. h.* 'Hibernica', Irish or Atlantic i., Iberia, W France, W Br., Ireland) – tetraploid, poss. allotetraploid

involving *H. helix* & another sp. from Morocco); ***H. iberica*** (McAllister) Ackerfield & J. Wen (S Iberia) – hexaploid (*H. canariensis* Willd. × *H. hibernica*); ***H. rhombea*** (Miq.) Bean (Jap. i., temp. E As.) – esp. '**Variegata**'

Hederopsis C.B. Clarke = *Macropanax*

Hederorkis Thouars (~ *Bulbophyllum*). Orchidaceae (V 16a). 2 Masc. R: Adansonia 16(1976)225

hedge bedstraw *Galium album*; **h. mustard** *Sisymbrium officinale*; **h. parsley** *Torilis japonica*; **h. woundwort** *Stachys sylvatica*

hedgehog broom *Erinacea anthyllis*; **h. cactus** *Echinocereus* spp.; **h. grass** *Cenchrus* spp.; **h. holly** *Ilex aquifolium* 'Ferox'

Hedinia Ostenf. = *Smelowskia*

Hediniopsis Botsch. & Petrovsky = *Smelowskia*

Hedosyne (A. Gray) Strother (~ *Iva*). Compositae (Helia.-Amb.). 1 SW US: ***H. ambrosiifolia*** (A. Gray) Strother. R: FNA 21(2006)90

Hedraianthera F. Muell. Celastraceae (I). 1 E Aus.: ***H. porphyropetala*** F. Muell. R: FA 22(1984)167

Hedranthera (Stapf) Pichon = *Callichilia*

Hedstromia A.C. Sm. Rubiaceae (IV 7). 1 Fiji: ***H. latifolia*** A.C. Sm.

Hedyachras Radlk. = *Glenniea*

Hedycarya Forst. & Forst. f. Monimiaceae (V 1). 11 W Pac. (New Caled. 9), Aus., NZ. R: Adansonia 18(1978)25, BMNHN 4,5(1983)247. Dioec. trees & shrubs (Rauh's & Attims's Models) incl. rheophytes (e.g. ***H. rivularis*** Guillaumin, New Caled.); ***H. angustifolia*** Cunn. (E Aus.) – timber for cabinetwork

Hedycaryopsis Danguy = *Ephippiandra*

Hedychium J. Koenig. Zingiberaceae (II 1). 50 Madag. (?), Indomal., Himal. Robust herbs with rhiz. & showy fls with long tubes & narrow free P-lobes, larger staminodes (lip 2-lobed) & with stigmas projecting just beyond anther. Local medic. & many cult. orn. (ginger lilies). ***H. coronarium*** J. Koenig (butterfly lily, trop. As., widely natur. trop. Am., invasive S Afr., Hawaii) – fls v. fragrant (gardenia-jasmine-like in summer, dull in winter), stems poss. paper source; ***H. elatum*** R. Br. (Nepal) & ***H. greenii*** W.W. Sm. (W Bhutan) – bulbils in infl.; ***H. flavescens*** Carey ex Roscoe (Sri Lanka) – invasive NZ, S Afr., Masc.; ***H. gardnerianum*** Ker (kahili ginger, Himal.) – invasive in Macar., Aus., NZ, S Afr., Masc., Hawaii etc.; ***H. spicatum*** Buch.-Ham. ex Sm. (Ind.) – rhiz. much used in perfumery e.g. abir (q.v.)

Hedyosmos Mitch. = *Cunila*

Hedyosmum Sw. Chloranthaceae. 45 trop. Am. (R: FN 40(1988)1; WI 5, R: JAA 69(1988)51), SE As. (1). Infl. a spike of ebracteolate fls rep. by A1 (s.t. interpreted as a solit. strobiloid fl. with A several × 100 on spiral axis (P = bract as in *Chloranthus*); pollen monosulcate. G. Some medic. esp. ***H. translucidum*** Cuatrec. (Colombia) – sudorific tea used by S Am. shamans

Hedyotis L. Rubiaceae (IV 15). Excl. *Debia, Dimetia, Exallage, Leptopetalum, Mexotis,* incl. *Metabolos, Pleiocraterium,* c. 180 Indomal. R: T 62(2013)369. See also *Houstonia, Kadua, Oldenlandia, Stenaria*

Hedypnois Mill. Compositae (Lact.-Hyp.). 3 Macaronesia to Iran. (Eur. 2), ***H. cretica*** (L.) Dum.-Cours. natur. S US, ***H. rhagadioloides*** (L.) F.W. Schmidt widely introd.

Hedysarum Tourn. ex L. Leguminosae (III 25). c. 160 N temp. (Eur. 18, N Am. 4), Medit. R: AHP 19(1902)183. Some cult. orn. incl. ***H. alpinum*** L. (circumboreal) – r. eaten raw, boiled or roasted by Inuit & Native Americans. See also *Corethrodendron, Sulla*

Hedyscepe H. Wendl. & Drude. Palmae (V 14j). 1 Lord Howe Is: ***H. canterburyana*** (C. Moore & F. Muell.) H. Wendl. & Drude (*Kentia c.*, umbrella palm) – cult. orn. R: FA 49(1994)411

Hedythyrsus Bremek. Rubiaceae (IV 1). 3 trop. Afr. R: SB 36(2011)1031

Heeria Meissn. Anacardiaceae (I). Excl. *Ozoroa*, 1 SW Cape: ***H. argentea*** (Thunb.) Meissn. R: Strel. 9(2000)270

Hegemone Bunge ex Ledeb. = *Trollius*

Hegnera Schindler (~ *Desmodium*). Leguminosae (III 9). 1 Indomal.: ***H. obcordata*** (Miq.) Schindler

Heimerliodendron Skottsb. = *Pisonia*

Heimia Link. Lythraceae. 3 (closely allied) S US to Arg. Alks. Cult. orn. shrubs used in fertility control & crushed lvs, fermented & drunk, as mildly intoxicating hallucinogen (everything seems to be yellow) esp. ***H. salicifolia*** (Kunth) Link (sinicuichi)

Heinsenia K. Schum. (~ *Aulacocalyx*). Rubiaceae (II 1). 1 trop. Afr.: ***H. diervilleoides*** K. Schum.

Heinsia DC. Rubiaceae (II 6). 4–5 trop. Afr. Some myrmecophily; some ed. fr., timber

Heisteria Jacq. Erythropalaceae (Olacaceae s.l.). c. 33 trop. Am., Afr. (3). R: FN 38(1984)42. Massart's, Roux's & Troll's Models. Some timber. ***H. latifolia*** Standl. (*H. olivae*, trop. Am.) – psychoactive (scopolamine); ***H. parvifolia*** Sm. (trop. Afr.) – local oilseed; ***H. spruceana*** Engl. (trop. Am.) – leaf infusions used to relieve swollen limbs

Hekistocarpa Hook.f. Rubiaceae (II 7). 1 Cameroun, Nigeria: ***H. minutiflora*** Hook.f. R: PSE 229(2001)65

Hekkingia H. Ballard & Munzinger. Violaceae (III 1d). 1 NE S Am.: ***H. bordenavei*** H. Ballard & Munzinger

Heladena A. Juss. Malpighiaceae. 6 trop. & warm Am.

Helanthium (Hook.f.) J.G. Sm. = *Echinodorus* (but perhaps distinct)

Helcia Lindl. = *Trichopila*

Heldreichia Boiss. Cruciferae. 1 Turkey to Afghanistan: ***H. bupleurifolia*** Boiss. R: T 59(2010)197

Heleastrum DC. = *Eurybia*

Helenium L. Compositae (Hele.-Gai.). 30 (18 N) Am. Local medic.; cult. orn. esp. cvs of ***H. autumnale*** L. (sneezeweed, N Am.) – powdered dried lvs used to induce sneezing. ***H. amarum*** (Raf.) Rock (bitterweed, N Am.) – sesquiterpene lactones toxic to herbivores & used as antifeedant against Colorado beetle. See also *Hymenoxys*

Heleochloa Host ex Roemer = *Crypsis*

Helia Mart. Older name for *Irlbachia*

Heliabravoa Backeb. = *Polaskia*

Heliamphora Benth. Sarraceniaceae. 23 (closely allied) Guayana Highland. R: Preslia 64(1992)219. Subshrubby to herbaceous pitcher pls like *Sarracenia* but G 3-loc., C 0. Ants & other arthropods main food, attracted by sarracenin secreted by special glands, though spiders s.t. take them before they fall in; in low light, only weakly pouched laminas produced

Heliamphoraceae Chrtek et al. = Sarraceniaceae

Helianthella Torrey & A. Gray. Compositae (Helia.-Enc.). 9 W N Am. R: AMN 48(1952)8. Extrafl. nectaries. Some cult. orn.

Helianthemum Mill. Cistaceae. c. 80 Eur. (30) to Sahara, NE Afr. to C As. For Am. spp. see *Crocanthemum* (lvs alt.). Rock-roses. Cult. orn. shrublets esp. forms & hybrids of ***H. nummularium*** (L.) Mill. (*H. chamaecistus*, Eur.)

Helianthopsis H. Robinson = *Pappobolus*

Helianthostylis Baill. Moraceae (IV). 2 Amazonia. R: FN 83(2001)243

Helianthus L. Compositae (Helia.-Helia). 51 N Am. R: MTBC 22, 3(1969)1, FNA 21(2006)141. Sunflowers. ***H. anomalus*** S.F. Blake, ***H. deserticola*** Heiser & ***H. paradoxus*** Heiser (all SW US; 3 of only 8 confirmed homoploid spp.) all derived from *H. annuus* × *H. petiolaris* Nutt. (W N Am.); ***H. annuus*** L. (common s., US) – only N Am. pl. to become a major economic crop, introd. Eur. 1510, second only to soyabean as oil crop by 1970s with 2/3 prod. in former USSR, now fourth, oil (6 M t by 2010; seeds 25–32% oil: high in polyunsaturates esp. linoleic acid) used in cooking oil, margarine (peanut butter subs), paints, biodiesel, bioplastics, etc., local medic., cult. orn. ann. to 3 m (UK record = 7.17 m; 'bonsai' specimens flowering at 5.6 cm!) with capitulum to 30 cm diam. or more (ray-florets sterile; signifying 'haughtiness' because of size in 'Language of Fls'), seeds ed. usu. salted, oil-cake used as fodder, pericarp made into fuel-logs in Canada; allelopathic (cypsela-walls) so that dead patches found on lawns where 'seeds' used in bird-feeders; sterile hybrids made with *Tithonia rotundifolia* (Mill.) S.F. Blake (Carib.). Other (rabbit-proof) cult. orn. incl. ***H. microcephalus*** Torrey & A. Gray (? = ***H. parviflorus*** Hornem., E US), *H.* × ***multiflorus*** L. (2n = 51, allotriploid between *H. annuus* (2n = 34) & *H. decapetalus* L. (2n = 68, C & SE US) arising in Eur. cult.) & ***H. porteri*** (A. Gray) Pruski (*Viguiera p.*, Confederate daisy, SE US); ***H. tuberosus*** L. (Jerusalem (poss. corruption of girasole, i.e. sunflower in Italian, though introd. UK from Terneuzen (?corrupted to Jerusalem), SW Netherlands) artichoke, N Am., invasive in C & E Eur.) – ed. tubers with sweet taste due partly to fructose from inulin (also cause of flatulence), good carbohydrate food for diabetics (sweeter on molar basis than glucose from starch), eaten by Native Americans before Eur. contact, introd. OW before 1613, boiled, also livestock feed, bred with *H. annuus* increasing latter's disease resistance

Helicanthes Danser. Loranthaceae (5 3). 1 Ind.: ***H. elastica*** (Desr.) Danser

Helichrysopsis Kirpiczn. (~ *Gnaphalium*). Compositae (Gnap.-Gnap.). 1 trop. E Afr.: ***H. septentrionale*** (Vatke) Hilliard. R: OB 104(1991)153

Helichrysum Mill. Compositae (Gnap.-Gnap.). Incl. *Laphangium*, c. 600 warm OW (Eur. 14) esp. S Afr. (Cape 81, 35 endemic – R: Strel. 9(2000)329), not Aus. (= *Coronidium*; for 'everlastings' see *Xerochrysum*). Polyphyletic & prob. incl. *Anaphalis* & Afr. spp. of '*Achyrocline*'? Many cult. orn. herbs & shrubs incl. ***H. forsskaolii*** (J. Gmelin) Hilliard & B.L. Burtt (*H. fruticosum*, E & S Afr.), ***H. petiolare*** Hilliard & B.L. Burtt ('*Gnaphalium lanatum*', liquorice (licorice) pl., S Afr., natur. in hedges in Portugal) – orn. fol. esp. in hanging baskets. ***H. italicum*** (Roth) G. Don (Medit.) – oil said to have antiviral activity; ***H. luteoalbum*** (L.) Reichb. (*G. l., L. l.*, temp. OW) – subcosmop. weed; ***H. serpyllifolium*** (Berg.) Pers. (S Afr.) – lvs used as tea (Hottentot t.)

Helicia Lour. Proteaceae (V 1). c. 100 Indomal., Pac. ***H. diversifolia*** C. White (Queensland) – source of helicia nuts

Helicilla Moq. = *Suaeda*

Heliciopsis Sleumer. Proteaceae (IV 4). 14 Myanmar to C Mal. Timber with oaklike figure used in furniture etc.

Helicodiceros Schott (~ *Dracunculus*). Araceae (VII 23). 1 W Med. is.: ***H. muscivorus*** (L.f.) Engl. – cult. orn. with stinking (due to dimethyl disulphide & trisulphide) hairy infl. likened to dead animal's anus (Bown), app. attractive to poll., though incoming insects s.t. eaten by lizards basking in thermogenetic infl. R: Thaiszia 4(1994) 179

Heliconia L. Heliconiaceae (Musaceae s.l.). c. 100–200 trop. Am. (Colombia c. 180; R (subg. *Stenochlamys* – 42): OB 82(1985)) & Moluccas to Fiji & Samoa (6, R: Allertonia 6,1(1990)). R: F. Berry & W.J. Kress (1991) *H.: an identification guide*. Tomlinson's Model; hummingbird-poll. in Am., some poll. bats in OW. Cult. orn. banana-like pls (lobster-claw). ***H. bihai*** (L.) L. (balisier, trop. Am.) – wild pl. prob. not cult., though many cvs of ***H. indica*** Lam. (OW) referred to as it, lvs used for wrapping food, poss. paper source; ***H. chartacea*** Barreiros (Braz.) – self-splitting lvs

Heliconiaceae Vines (Musaceae subfam. Heliconioideae). Magnoliidae – Zingiberales. 1/100–200 W Pacific & trop. Am. Stems with distinct internodes or reduced so that leaf-sheaths form pseudostem, hapaxanthic, arising from symp. rhiz. Lvs distich. with sheath, usu. petiole & lamina with midrib & fine parallel laterals looped at margin. Infl. term. thyrse, usu. on leafy shoots but s.t. on specialized ones (cf. Zingiberaceae); fls in fascicles in axils of coloured spathaceous bracts, each subtended by bracteoles, bisex., 3-merous. P petaloid, zygomorphic, basally forming a tube, laterals united with median K free; A 2 (+ 1 staminode) + 3, filaments adnate to P tube; anthers basifixed. $\overline{G}$ (3) with 1 filiform hollow style & axile placentation; 1 anatropus ovule per locule. Fr. a drupe on elongated pedicel, with 3 pyrenes; seeds with thin perisperm & copious endosperm rich in starch & oil. n = (11) 12

Genus: *Heliconia*

Staminode in same position as that in Marantaceae & Zingiberaceae

Helicostylis Trécul. Moraceae (III). 7 trop. Am. R: FN 7(1972)75, 83(2001)266. Hallucinogenic bark used in cerem. witchcraft in Guianas

Helicotropis Delgado (~ *Vigna*). Leguminosae (III 18). 4+ trop. Am. R: AJB 98(2011)1709

Helicteres L. Malvaceae (Hel.; Sterculiaceae). 60 trop. As. & Am. (38 – R: Bonpl. 11(2001)24). Shrubs & trees (some bat-poll.) with useful bark fibre e.g. kaivum (***H. isora*** L., As. – fr. medic.)

Helicteropsis Hochr. = *Hibiscus* (but see FMad 129(1955)126)

Helictochloa Romero-Zarco = ? *Helictotrichon* (but see Cand. 66(2011)96)

Helictonema Pierre (~ *Hippocratea*). Celastraceae (III). 1 W Afr.: ***H. velutinum*** (Afzel.) Hallé. R: BMNHN 4,5(1983)20

Helictotrichon Besser. Gramineae (XVI 2). Excl. *Amphibromus* (glabrous G), *Avenula*, *Trisetopsis*, c. 100 N temp. esp. Euras. (Am. 1). Oatgrass

Helietta Tul. Rutaceae (I 7). 8 trop. Am. to Calif. (N Am. 3 – R: Britt. 36(1984)455). R: Britt. 50(1998)360. ***H. parvifolia*** A. Gray (Texas, Mex.) – effective insecticide against Mex. flies

Helinus E. Meyer ex Endl. Rhamnaceae (4). 5 trop. & S Afr. (3), Madag. (1), NW Ind. (1)

Heliocarpus L. Malvaceae (Grew.-Apeib.; Tiliaceae). 1 trop. Am.: ***H. americanus*** L. (*H. donnell-smithii*) – variable, source of fibre for hammocks etc., bark beaten to give paper, soft wood used for floats, bottle-stoppers etc. R: AMBG 36(1949)511

Heliocarya Bunge = *Caccinia*

Heliocauta Humphries. Compositae (Anth.-Mat.). 1 Morocco: ***H. atlanticum*** (Litard & Maire) Humphries. R: BBMNHB 23(1993)106

Heliocereus (A. Berger) Britton & Rose = *Disocactus*

× **Heliochia** G. Rowley = *Disocactus*

Heliohebe Garnock-Jones = *Veronica*

Heliomeris Nutt. Compositae (Helia.-Helia). 5 W N Am. R: PIAS 88(1979)366

Heliophila Burm. f. ex L. Cruciferae (30). Incl. *Brachycarpaea, Cycloptychis*, etc., c. 90 S Afr. R: Fl. S Afr. 13(1970)17, Novon 15(2005)387. Spirolobal cotyledons. Some cult. orn. herbs & shrubs esp. ***H. longifolia*** DC. with blue fls reminiscent of flax. ***H. glauca*** Burch. ex DC. is a shrub to 2 m, ***H. scandens*** Harv. a woody cli. to 3 m

Heliopsis Pers. Compositae (Helia.-Zinn.). c. 15 upland trop. Am., esp. Mex. R: OJS 57(1957)171. ***H. helianthoides*** (L.) Sweet (N Am.) cult. orn., like ***H. longipes*** (A. Gray) S.F. Blake source of a promising insecticide (scabrin)

Heliosperma (Reichb.) Reichb. = *Silene*

Heliostemma Woodson = *Matelea*

heliotrope *Heliotropium* spp. esp. *H. arborescens*; **winter h.** *Petasites fragrans*

Heliotropium Tourn. ex L. Boraginaceae (Heliotropiaceae). Incl. *Argusia, Ceballosia, Nogalia, Tournefortia*, excl. *Euploca, Ixorhea, Myriopus*, c. 300 trop. & temp. R: JAA suppl. 1(1991)74. Small trees (*Argusia*), lianes (*Tournefortia*) to herbs; alks; some not eaten by locusts in E Afr. even though common in swarming places of young stages. Cult. orn. & locally imp. medic., incl. purgatives; some ed. fr. (*Tournefortia*). ***H. amplexicaule*** Vahl (S Am., natur. OW trop.) – cult. orn. & fert. control; ***H. arborescens*** L. (*H. peruvianum*, heliotrope, cherry pie, Peru) – cult. orn., in S Eur. used in scent; ***H. arboreum*** (Blanco) Mabb. (*H. foertherianum, T. argentea*, IndoPacif. coasts) – local medic., useful wood; ***H. balfourii*** Guerke (Socotra) – most widely used tanning pl. on S.

Helipterum DC. = *Helichrysum*; for Aus. spp. see *Rhodanthe, Syncarpha*

Helixanthera Lour. Loranthaceae (5 5). c. 45 trop. Afr. (12 – R: R. Polhill & D. Wiens, *Mistletoes of Africa* (1998)17) to Sulawesi. ***H. parasitica*** Lour. (Indomal.) – fr. ed. Philippines

Helixyra Salisb. ex N.E. Br. = *Moraea*

Hellalia Král = *Sedum*

hellebore *Helleborus* spp.; **black h.** *H. niger*; **false h.** *Veratrum* spp.; **green h.** *H. viridis*; **stinking h.** *H. foetidum*; **white h.** *V. album*

helleborine *Epipactis* spp.; *Cephalanthera* spp.

Helleborus Tourn. ex L. Ranunculaceae (I 1). 21 Eur. (11), Medit., As., limestones. R: B. Mathew (1989) *Hellebores*. Cardiac glycosides v. poisonous with burning taste. Rhiz. with aerial shoots taking several yrs to flower; carpels slightly connate basally, C prob. rep. by nectaries, K coloured; seeds with elaiosome along raphe, attractive to ants (? & snails) which act as disp. agents. Many cult. orn. for winter & spring fls esp.: ***H. foetidus*** L. (setterwort, stinkwort, stinking h., bear's-foot, W & S Eur.) – green fls seen as yellow by bees, alleged to find smelly fls (cf. *Crocus & Cytisus*) attractive, form. used as dangerous cathartic & veterinary medic.; ***H. lividus*** Aiton subsp. ***corsicus*** (Briq.) Fourn. (*H. argutifolius*, Corsica, Sardinia); ***H. niger*** L. (Christmas rose, black (referring to cut surface of rhiz.) h., Alps & Apennines) – fls white to pink becoming green after fert., rodent & bird pest control since 1 cent. in W Eur.; ***H. orientalis*** Lam. (Lenten rose, Greece & Turkey) – fls cream, some hybrids with *H. niger*, some ('***H. × hybridus*** Hort. ex Vilm.') combining several spp., some 'doubles' incl. petaloid nectaries ('anemone-centred') & petaloid A, e.g. '**Hidcote Double**' (examples of both types found in wild); ***H. thibetanus*** Franch. (Tibet) – germ. hypogeal (rest of genus epigeal); ***H. vesicarius*** Aucher (S Turkey) – mature follicle 3-winged inflated to 8 cm diam., ? wind-disp.; ***H. viridis*** L. (green h., W & C Eur., natur. N Am.)

Hellenia Retz. (*Cheilocostus*). Zingiberaceae. 6 Indomal. ***H. speciosa*** (Koenig) Govaerts (*C. s., Costus s.*) – diosgenin in rhiz. (comm.)

Hellenocarum H. Wolff (~ *Carum*). Umbelliferae (III 8). 3 Eur., SW As.

Helleria Fourn. = *Festuca*

Helleriella A. Hawkes (~ *Platyglottis*). Orchidaceae (V 13d). 2 C Am.

Hellerochloa Rauschert = *Festuca*

Hellmuthia Steud. (~ *Scirpus*). Cyperaceae (II 5). 1 Cape: ***H. membranacea*** (Thunb.) Haines & Lye. R: BN 129(1976)61

helmet flower *Scutellaria* spp.; **h. orchid** *Corybas* spp.; **policeman's h.** *Impatiens glandulifera*

Helmholtzia F. Muell. Philydraceae. Excl. *Orthophylax*, 2 Moluccas, NG, NE Aus.

Helminthocarpon A. Rich. = *Dorycnopsis*

Helminthostachys Kaulf. Ophioglossaceae. 1 Sri Lanka, Himal. to Queensland & New Caled.: ***H. zeylanica*** (L.) Hook. – dorsiventral rhiz. & 2-ranked lvs on upper surface, fert.

spikes with lat. sporangiophores of globose sporangia, ed. in salads (Philipp.), alleged antimalarial

Helminthotheca Zinn (~ *Picris*). Compositae (Lac.-Hyp.). 4 SE Eur. to Iran, ***H. echioides*** (L.) Holub (*P. echioides*, oxtongue, Medit. widely introd.) – wide range of insects trapped by grapple-hook hairs. R: T 24(1975)111

Helmiopsiella Arènes. Malvaceae (Dom.; Sterculiaceae). 4 Madag. R: BMNHN 4,10(1988) 69

Helmiopsis Perrier. Malvaceae (Dom.; Sterculiaceae). 8 Madag.

Helmontia Cogn. Cucurbitaceae (XIII). 2–4 Braz., Guyana, Venez.

Helogyne Nutt. Compositae (Eup.-Alom.). 8 Andes. R: MSBMBG 22(1987)264

Helonema Süsseng. = *Eleocharis*

Helonias L. Melanthiaceae. Excl. *Heloniopsis, Ypsilandra*, 1 E US: ***H. bullata*** L. – cult. orn. with fragrant fls, blue anthers. R: FNA 26(2002)70

Heloniopsis A. Gray (~ *Helonias*). Melanthiaceae. 9 Taiwan, Korea, Jap. R: AMBG 96 (2009)525. Cult. orn.

Helonoma Garay (~ *Beloglottis*). Orchidaceae (IV 2h). 4 Guayana Highlands, Peruvian Andes

Helosaceae Bromhead = Balanophoraceae

Helosciadium Koch (~ *Apium*). Umbelliferae (III 8). 5 Eur. R: PSE 287(2010)11

Helosis Rich. Balanophoraceae (Helosaceae). Incl. *Exorhopala*, 1 Malay Penin., 2 trop. Am. R: SB 40(2015)603. ***H. cayennensis*** (Sw.) Spreng. (*H. guyanensis*) – used NW Amaz. as styptic (blood-red! Cf. 'Doctrine of Signatures')

Helwingia Willd. Helwingiaceae (Cornaceae s.l.). 4 Himal., Thailand to Jap. ***H. japonica*** (Thunb.) Dietr. (China & Jap.) – cult. orn., lvs ed.

Helwingiaceae Decne. (~ Aquifoliaceae). Magnoliidae – Aquifoliales. 1/4 E As. Dioec. glabrous shrubs & trees. Lvs simple, in spirals; stip. 2 ± fimbriate, decid. Infls cymose-fasciculate, epiphyllous, initiated adjacent to leaf-axil on base of leaf primordium. P (prob. C, though others favour K) 3–5, valvate. A = & alt. P; $\underline{G}$ (2–4) alt. with P, ovule 1 per loc. Fr. a drupe with 3–5-lobed endocarp; seeds with straight embryo in weakly ruminate endosperm. n = 19

Genus: *Helwingia*

Despite earlier dispositions (e.g. Cornaceae), close to Phyllonomaceae (also epiphyllous) on DNA evidence. Epiphylly due largely to intercalary growth.

Helxine Req. = *Soleirolia*

Hemandradenia Stapf. Connaraceae (I). 2 W & C Afr. R: AUWP 89–6(1989)275

Hemarthria R. Br. Gramineae (XXII 3). 14 warm OW (Eur. 1, China 6, Aus. 1). R: Blumea 45(2000)450. Some fodders

Hemerocallidaceae R. Br. = Asphodelaceae

Hemerocallis L. Asphodelaceae (Hemerocallidaceae). c. 15 C Eur. (1) to China & (esp.) Jap. R: A.B. Stout (1986, ed. 3) *Daylilies*, R.W. Munson (1989) *H. The daylily*. Cult. orn. (daylilies, spiderlilies) in China for millennia, now over 38 000 cvs incl. 13 000 named clones of ***H. fulva*** (L.) L. (As., early natur. in Eur. (introd. by C1 A.D.), later US) – fls orange, scentless, dried used as food-flavouring in China & Jap., form. medic. (hemerocallin antidote to arsenic poisoning), commonly a self-sterile triploid ('**Europa**'), & many hybrids esp. derived from & ***H. lilioasphodelus*** L. (*H. flava*, E Siberia to Jap.) – fls yellow, fragrant. Used as bio-indicators (airborne fluoride pollution) in Braz.; fls eaten in salads. *The Daylily J.*

Hemiadelphis Nees = *Hygrophila*

Hemiandra R. Br. Labiatae (IV 2). 14 SW Aus.

Hemiangium A.C. Sm. = *Semialarium*

Hemianthus Nutt. (~ *Micranthemum*). Linderniaceae (Phrymaceae – Microc.; Scrophulariaceae s.l.). 3–4 C Am.

Hemiarrhena Benth. (~ *Lindernia*). Linderniaceae (Plantaginaceae (Grat.-Lind.); Scrophulariaceae s.l.). 1 NW Aus.: ***H. plantaginea*** (Benth.) F. Muell. R: Willd. 43(2013)224

Hemiarthron (Eichler) Tieghem = *Psittacanthus*

Hemibaccharis S.F. Blake = *Archibaccharis*

Hemiboea C.B. Clarke. Gesneriaceae (III 2j). Incl. *Metabriggsia*, 25 S China & S Jap. to SE As. R: APS 25(1987)81, 220

Hemiboeopsis W.T. Wang = *Henckelia* (but see ABY 6(1984)397)

Hemicarpha Nees = *Lipocarpha*

Hemichaena Benth. Phrymaceae (Leuc.; Scrophulariaceae s.l.). 5 C Am. R: Phytoneuron 2012–39:24

Hemichlaena Schrad. = *Ficinia*

Hemichroa R. Br. Amaranthaceae (Chenopodiaceae IV 1). Excl. *Surreya*, 1 SW & S Aus.: ***H. pentandra*** R. Br. R: T 62(2013)109

Hemicicca Baill. = *Phyllanthus*

Hemicrambe Webb (~ *Brassica*). Cruciferae (12). Incl. *Nesocrambe* 3 Morocco, Socotra. R: Novon 14(2004)156. ***H. fruticosa*** (C. Towns.) Gómez-Campo (*Fabrisinapis f.*, Socotra) – known only from 2 pls

Hemicrepidospermum Swart = *Crepidospermum*

Hemicyatheon (Domin) Copel. = *Hymenophyllum*

Hemidesmus R. Br. Apocynaceae (III; Asclepiadaceae). 1 S Ind., SE As., Mal.: ***H. indicus*** (L.) Sm.– local medic.

Hemidictyaceae Christenh. & H. Schneid. (~ Diplaziopsidaceae, Aspleniaceae, Woodsiaceae). 1/1 trop. Am. R: T 61(2012)522. Terr. Lvs 1-pinnate with thickened prominent vein-endings like D., app. closest to Aspleniaceae
Genus: *Hemidictyum*

Hemidictyum C. Presl. Hemidictyaceae (Woodsiaceae; Dryopteridaceae s.l.). 1 trop. Am.: ***H. marginatum*** (L.) C. Presl (*Diplazium limbatum*)

Hemidiodia K. Schum. = *Diodia*

Hemieva Raf. = *Suksdorfia*

Hemifuchsia Herrera = *Fuchsia*

Hemigenia R. Br. Labiatae (IV 2). c. 50 Aus. Heterogeneous. Lvs decussate or in whorls of 3 or 4

Hemigramma Christ = *Tectaria*

Hemigraphis Nees = *Strobilanthes*

Hemilophia Franch. Cruciferae. 5 SW China. R: Adans. III,21(1999)240, EJB 59(2002)443

Hemimeris L.f. Scrophulariaceae (Hemim.). 6 N & W Cape. R: Strel. 9(2000)651. Elaiophores attractive to bees

Hemimunroa L. Parodi = *Munroa*

Hemionitidaceae Pichi-Serm. = Pteridaceae

Hemionitis L. Pteridaceae (IV). (As. sp. excl.) 6 trop. Am. Lvs dimorphic. Cult. orn.

Hemiorchis Kurz. Zingiberaceae (II 2). 3 C Himal. to Myanmar. Orchid-like

Hemipappus K. Koch = *Tanacetum*

Hemiphora (F. Muell.) F. Muell. Labiatae (IV 1; Dicrastylidaceae). 5 W Aus. R: AusSB 24(2011)6

Hemiphragma Wall. Plantaginaceae (Dig.-Hem.; Scrophulariaceae s.l.). 1 W Himal. to Assam: ***H. heterophyllum*** Wall. – cult. orn. tufted herb with berries

Hemiphylacus S. Watson. Asparagaceae. 5 Mex.

Hemipilia Lindl. Orchidaceae (IV 4d). Incl. *Amitostigma*, *Ponerorchis*, c. 60 Himal., E As., Thailand. ***H. flabellata*** Bureau & Franch. (China) – poll. bees attracted by nectar source in fls. of associated *Ajuga forrestii* Diels

Hemipiliopsis Luo & S.C. Chen = *Hemipilia*

Hemipogon Decne. Apocynaceae (Vc; Asclepiadaceae III 1). 10 S Am.

Hemiptelea Planch. (~ *Zelkova*). Ulmaceae. 1 N China & Korea: ***H. davidii*** (Hance) Planch. – cult. orn. spiny decid. tree or shrub with fr. only half-encircled by wing

Hemiscola Raf. = *Cleome*

Hemiscleria Lindl. = *Epidendrum*

Hemiscolopia Slooten. Salicaceae (Flacourtiaceae). 1 Indomal.: ***H. trimera*** Slooten

Hemisiphonia Urb. = *Micranthemum*

Hemisorghum C. Hubb. ex Bor = *Sorghum*

Hemisphaera Kolak. = *Campanula*

Hemisphaerocarya Brand = *Cryptantha*

Hemisteptia Fisch. & C. Meyer (~ *Saussurea*). Compositae (Card.-Card.). 1 S, E & SE As., E Aus.: ***H. lyrata*** (Bunge) Fisch. & C. Meyer. R: FA 37(2015)50

Hemistylus Benth. Urticaceae (IV). 4 trop. S Am.

Hemitelia R. Br. = *Cyathea*

Hemithrinax Hook.f. (~ *Thrinax*). Palmae (III 2). 3 Cuba

Hemitomes A. Gray. Ericaceae (II 3; Monotropaceae). 1 W US: ***H. congestum*** A. Gray. R: FNA 9(2009)395

Hemitria = *Phthirusa*

Hemizonella (A. Gray) A. Gray (~ *Madia*). Compositae (Mad.-Mad.). 1 SW N Am.: ***H. minima*** (A. Gray) A. Gray. R: FNA 21(2006)296

Hemizonia DC. Compositae (Mad.-Mad.). Excl. *Centromadia, Deinandra, Holozonia*, 1 W N Am.: ***H. congesta*** DC. R: FNA 21(2006)291

Hemizygia (Benth.) Briq. = *Syncolostemon* (but see Bothalia 12(1976)2)

hemlock *Conium maculatum*; **water h.** *Cicuta virosa*

hemlock (spruce) *Tsuga* spp.; **Canada h.** *T. canadensis*; **Carolina h.** *T. caroliniana*; **eastern h.** *T. canadensis*; **Japanese h.** *T. sieboldii*; **western h.** *T. heterophylla*; **white h.** *T. canadensis*

Hemmantia Whiffin. Monimiaceae. 1 NE Queensland: ***H. webbii*** Whiffin. R: FA 2(2007)72

hemp *Cannabis sativa*; **African h.** *Sparrmannia africana*; **Ambari h.** *Hibiscus cannabinus*; **Bahama h.** *Agave sisalana*; **Bombay h.** *Crotalaria juncea*; **bowstring h.** *Dracaena* spp.; **Chinese h.** *Abutilon theophrasti*; **Cuba h.** *Furcraea hexapetala*; **Deccan h.** *H. cannabinus*; **Indian h.** *Cannabis sativa*; **Madras h.** *Crotalaria juncea*; **Manila h.** *Musa textilis*; **Mauritius h.** *Furcraea foetida*; **h. nettle** *Galeopsis* spp.; **NZ h.** *Phormium tenax*; **Queensland h.** *Sida rhombifolia*; **Russian h.** *Cannabis sativa*; **sisal h.** *Agave sisalana*; **sunn, sann** or **tag h.** *Crotalaria juncea*

Hemsleya Cogn. ex F.B. Forbes & Hemsl. Cucurbitaceae (I). 27 E As. (China 25, 21 endemic). Allied to *Gomphogyne*. Cucurbitacins poss. effective in tumour control

hen-and-chickens *Bellis perennis* 'Prolifera', proliferating captula forms of *Calendula officinalis*

henbane *Hyoscyamus niger*

henbit *Lamium amplexicaule*

Henckelia Spreng. (~ *Didymocarpus*). Gesneriaceae (III 2j). Incl. *Chirita* s. s., excl. *Codonoboea*, 56 Ind. to Thailand (not on limestone). R: T 60(2011)773

henequén *Agave angustifolia*; **Salvador h.** *A. vivipara* var. *letonae*

Henleophytum Karsten. Malpighiaceae. 1 Cuba: ***H. echinatum*** (Griseb.) Small

henna *Lawsonia inermis*; **black h.** *Indigofera tinctoria*; **neutral h.** *Ziziphus jujuba*

Hennecartia Poisson. Monimiaceae (V 3). 3 S Braz., Paraguay, NE Arg. Pollen received on a 'hyperstigma'

Henonia Moq. Amaranthaceae (I 1). 1 Madag.: ***H. scoparia*** Moq.

Henoonia Griseb. (*Bissea*). Solanaceae (I; Goetzeaceae). 1 Cuba: ***H. myrtifolia*** Griseb.

Henophyton Cosson & Durieu (~ *Oudneya*). Cruciferae (12). 1 Morocco to Libya: ***H. deserti*** (Cosson & Durieu) Cosson & Durieu

Henrardia C. Hubb. Gramineae (XV). 2 Turkey & Iran to C As. R: NJB 13(1993)488

Henricia Cass. = *Psiadia*

Henricksonia B. Turner. Compositae (Cor.). 1 Mex.: ***H. mexicana*** B. Turner. R: AJB 64 (1977)78

Henriettea DC. Melastomataceae. Incl. *Henriettella, Llewelynia*, 67 trop. S Am.

Henriettella Naud. = praec.

Henriquezia Spruce ex Benth. Rubiaceae (I 4; Henriqueziaceae). 7 Amaz. Braz.

Henriqueziaceae Bremek. = Rubiaceae

Henrya Nees ex Benth. (~ *Tetramerium*). Acanthaceae (III 2c). 2 C Am. R: CUMH 17(1990) 99

Henryettana Brand = *Antiotrema*

Hensmania W. Fitzg. Asphodelaceae (Hemerocallidaceae). 3 SW Aus. R: FA 45(1987)249

henweed, Guinea *Petiveria alliacea*

Hepatica Mill. = *Anemone*

Heppiella Regel. Gesneriaceae (II 5b). 4 Andes. R: SB 15(1990)720

Heptacodium Rehder. Caprifoliaceae. 1 C & E China: ***H. miconioides*** Rehder – capitula of 2 rows of 3 fr. around a central bud capable of prod. later sets of fls, cult. orn. R: IDSY 2012(2013)30

Heptanthus Griseb. Compositae (Helia.-Pin.). 7 Cuba

Heptaptera Margot & Reuter. Umbelliferae (III 5). 8 E Medit. (Eur. 4), SW As. R: NRBGE 31(1971)91

Heracleum L. Umbelliferae (III 11). 65 N temp. (Eur. 8), trop. mts. Coarse biennials & perennials. ***H. mantegazzianum*** Sommier & Levier (?*H. pubescens* (Hoffm.) Bieb., or *H. speciosum* Weinmann, giant hogweed, Cauc., natur. Eur. (GB 1893), US) – to 3 m with umbels to 1 m across, causing photodermatitis, sensitizing skin to ultraviolet radiation, in bright sunlight (same but lesser effect in carrots & parsnips etc.), cult. orn., invasive in N & E Eur. (also tall invaders are ***H. persicum*** Desf. ex Fischer & ***H. sosnowskyi*** Manden. –

introd. Russia 1947 as fodder-crop), hybridizing in UK & Ger. with ***H. sphondylium*** L. (hogweed, N temp.) – form. pigfood, medic. & used in liqueurs & basis of alcoholic drink in E Eur., subsp. ***montanum*** (Gaudin) Briq. (*H. lanatum*, cow parsnip) – r., fls & stems ed. N Am., fr. used as spice in Sikkim; ***H. maximum*** Bartram (*H. lanatum*, N Am.) – imp. local medic. esp. for rheumatism, shoots ed.

Herb Bennet *Geum urbanum*; **H. Christopher** *Actaea spicata*; **H.-of-Grace** *Ruta graveolens*; **H. Paris** *Paris quadrifolia*; **H. Patience** *Rumex patientia*; **H. Robert** *Geranium robertianum*; **willowh.** *Epilobium* spp.

Herbertia Sweet. Iridaceae (VII 5). 5 temp. S Am., 1 (***H. lahue*** (Molina) Goldbl.) ext. to S US. R: Britt. 56(2004)363. Lvs plicate; cult. orn.

Herbstia Sohmer. Amaranthaceae (I 2). 1 Braz.: ***H. brasiliana*** (Moq.) Sohmer. R: Britt. 28(1976)448

Hercules' club *Aralia spinosa*, *Zanthoxylum clava-herculis*

Herderia Cass. Compositae (Vern.-Erl.). 1 trop. W Afr.: ***H. truncata*** Cass.

Hereroa (Schwantes) Dinter & Schwantes. Aizoaceae (V). 28 S Afr. R: H. E. K. Hartmann, *Ill. Handb. Succ. Pls*, Aiz. F–Z(2002)56. Cult. orn. succ. hummock-formers or shrubs

Hererolandia Gagnon & G.P. Lewis (~ *Caesalpinia*) Leguminosae (I 4). 1 Namibia: ***H. pearsonii*** (L. Bolus) Gagnon & G.P. Lewis. R: PhytoKeys 71(2016)29

Hericinia Fourr. = *Ranunculus*

Herissantia Medik. (~ *Abutilon*). Malvaceae (Malv.-Malv.). 6 trop. Am., ***H. crispa*** (L.) Brizicky ± pantrop.

Heritiera Dryand. Malvaceae (Sterc.; Sterculiaceae). Excl. *Argyrodendron*, incl. *Tarrietia* 30 trop. Afr., Indomal. to Aus. & New Caled. (1). R: Reinw. 4(1959)465. Monoec. trees with large buttresses. Some comm. timbers incl. mengkulang (Mal.); ***H. cochinchinensis*** (Pierre) Kosterm. (chumprak, SE As.), ***H. littoralis*** Dryand. (OW) – timber for shipbuilding e.g. masts for dhows in E Afr., ***H. trifoliolata*** (F. Muell.) Kosterm. (crowsfoot elm, NE Aus.) – interior panelling in old railway carriages, ship cabins etc. & ***H. utilis*** (Sprague) Sprague (niangon, W Afr.). ***H. macrophylla*** (Wall.) Kurz (E Ind., Myanmar) – changes lvs every 2 yrs 8 months

Hermannia Tourn. ex L. Malvaceae (Bytt.-Herm.; Sterculiaceae). 250 trop. & warm esp. S Afr. Some cult. orn. with honey-scented fls, form. medic. Cape (syphilis)

Hermanschwartzia Plowes = *Stapeliopsis*

Hermas L. Umbelliferae (Azor.). 9 Cape. R: SB 40(2015)353. ***H. villosa*** (L.) Thunb. – pachycaul treelet

Hermbstaedtia Reichb. Amaranthaceae (I 1). 5 trop. & S Afr. R: KB 37(1982)83

Hermidium S. Watson = *Mirabilis*

Herminiera Guillemin & Perrottet = *Aeschynomene*

Herminium L. Orchidaceae (IV 4d). 19 temp. Euras. (Eur. 1: ***H. monorchis*** (L.) R. Br., musk orchid – general tonic in China), Thailand, C Mal.

Hermodactylus Mill. = *Iris*

Hernandia Plum. ex L. Hernandiaceae. 23 trop. esp. Indopacific. R: BJ 89(1969)122. Many alks. Primitive spp. the most restricted in distr. Monoec. trees (Attims's & Rauh's Models); some timbers for canoes etc. esp. ***H. nymphaeifolia*** (C. Presl) Kubitzki (*H. peltata*, *H. ovigera* auctt., jack-in-the-box, OW trop. coasts) – some trees with male fls opening a.m., females p.m., others with reverse, leaf-extract a painless depilatory, street-tree

Hernandiaceae Blume. Magnoliidae – Laurales. 4/60 trop. R: BJ 89(1969)78. Trees, shrubs or lianes with alks & scattered spherical ethereal oil-cells that may unite as mucilage cavities. Lvs in spirals, simple (s.t. 3-lobed) or palmate; stip. 0. Fls small, reg., epigynous, bisex. or not (pls then polygamous, monoec., rarely dioec.) in cymes; P 4–8 or 3 or 4(–6) + 3 or 4(–6), usu. imbr., A 3–5(–7), filaments oft. with nectary appendages (cf. Lauraceae), anthers with longit. valves, $\overline{G}$ 1 with elongate style & 1 pend., anatropous, bitegmic ovule. Fr. dry-indehiscent, oft. winged or in accrescent involucre derived from 2 or 3 connate bracteoles; seeds without endosperm, embryo with large, folded, wrinkled or lobed oily cotyledons

Classification & genera:

I. **Hernandioideae** (infl. thyrsoid, n = 18, 20): *Hernandia*, *Illigera*

II. **Gyrocarpoideae** (Gyrocarpaceae; infl. dichasial, n = 15): *Gyrocarpus*, *Sparattanthelium*.

Some timbers

Herniaria Tourn. ex L. Caryophyllaceae (I 2). 48 Eur. (17) & Afr. to Ind., 1 N Arg. & Bolivia. C 0 or rudimentary. ***H. glabra*** L. (rupturewort, herniary, Eur. to C As.) – ant-poll., medic. (diuretic etc.)

herniary *Herniaria glabra*

Herodotia Urb. & E. Ekman. Compositae (Sen.-Tuss.). Excl. *Ekmaniopappus*, 1 Hispaniola: ***H. haitensis*** Urb. & E. Ekman

Herpestis Gaertn. f. = *Bacopa*

Herpetacanthus Nees. Acanthaceae (II 2c). c. 25 Panamá to Braz.

Herpetophytum (Schltr.) Brieger = *Dendrobium*

Herpetospermum Wall. ex Hook.f. Cucurbitaceae (XI). 1(–3) Himal., China: ***H. pedunculosum*** (Ser.) C.B. Clarke. R: Blumea 59(2014)2

Herpolirion Hook.f. Asparagaceae (Anthericaceae). 1 SE Aus., NZ: ***H. novae-zelandiae*** Hook.f. R: FA 45(1987)242

Herpysma Lindl. Orchidaceae (IV 2d). 1 S & SE As. to Sumatra: ***H. longicaulis*** Lindl. R: Fl. Bhutan 3,3(2002)95

Herpyza C. Wright. Leguminosae (III 18). 1 W Cuba: ***H. grandiflora*** (Griseb.) C. Wright. R: AJBM 61(2004)64

Herrania Goudot. Malvaceae (Bytt.-Theob.; Sterculiaceae). 17 trop. S Am. R: JAA 39(1958) 227. Monkey-disp. in Amaz. ***H. albiflora*** Goudot (N S Am.) – Corner's Model; ***H. camargoana*** R. Schultes (Amazonia) – ground seeds used as a condiment on meat

Herrea Schwantes = *Conicosia*

Herreanthus Schwantes = *Conophytum*

Herreranthus R. Nordenstam (~ *Senecio*). Compositae (Sen.-Sen.). 1 Cuba: ***H. rivalis*** (Greenman) R. Nordenstam. R: CN 44(2006)62

Herreria Ruíz & Pavón. Asparagaceae (Herreriaceae). 27 S Am. ***H. montevidensis*** Klotzsch ex Griseb. – medic. but imp. in basket-weaving (Braz.)

Herreriaceae Kunth = Asparagaceae

Herreriopsis Perrier. Asparagaceae (Herreriaceae). 1 Madag.: ***H. elegans*** Perrier

Herrickia Wooton & Standl. (~ *Eurybia*). Compositae (Ast.-Mach.). 4 W US. R: FNA 20(2006)361

Herschelia Lindl. = *Disa*

Herschelianthe Rauschert = *Disa*

Hertia Less. (~ *Othonna*). Compositae (Sen.-Oth.). 10 SW As., N & S Afr.

Herya Cordemoy = *Pleurostylia*

Hesiodia Moench = *Sideritis*

Hesperalbizia Barneby & Grimes (~ *Albizia*). Leguminosae (II 4). 1 SW Mex.: ***H. occidentalis*** (Brandegee) Barneby & Grimes. R: MNYBG 74(1996)112

Hesperaloe Engelm. Asparagaceae (Agavaceae). 5 SW N Am. R: CSM 23(1978)56. Cult. orn. stemless herbs forming grassy clumps

Hesperantha Ker-Gawler. Iridaceae (VI 5). Incl. *Schizostylis*, 82 subSaharan Afr. R: AMBG 90(2003)396, Both. 37(2007)177. Some cult. orn. esp. ***H. coccinea*** (Backh. & Harv.) Goldbl. & Manning (*S. c.*, kaffir lily, S Afr.) – red fls, some cvs with washed-out pink

Hesperelaea A. Gray. Oleaceae (4d). 1 NW Mex.: ***H. palmeri*** A. Gray – exterminated by goats (3 trees last seen 1875)

Hesperethusa M. Roemer = *Naringi*

Hesperevax (A. Gray) A. Gray (~ *Filago*). Compositae (Gnap.-Gnap.). 3 W US. R: SB 17(1992)293

Hesperhodos Cockerell = *Rosa*

Hesperidanthus (Robinson) Rydb. (~*Thelypodium*). Cruciferae (46). 5 W N Am. R: HPB 10(2005)49

Hesperis Tourn. ex L. Cruciferae (31). 34 Eur. (13), Medit. to Iran, C As., W China. R: FR 84(1973)259. Biennial or short-lived perenn. herbs; ***H. matronalis*** L. (dame's violet or d.'s rocket, (sweet) r., damask (v.), C & S Eur., natur. N Eur. & N Am. – invasive) – old-fashioned garden pl. with fragrant fls, seeds crushed to give an oil

Hesperocallidaceae Traub = Asparagaceae

Hesperocallis A. Gray. Asparagaceae (Agavaceae). 1 SW US deserts: ***H. undulata*** A. Gray – cult. orn. with fragrant fls, garlicky ed. bulbs. R: FNA 26(2002)221

Hesperochiron S. Watson. Boraginaceae (Hydrophyllaceae-Rom.). 2 SW N Am. Cult. orn. stemless herbs

Hesperochloa (Piper) Rydb. = *Festuca*

Hesperocnide Torrey = *Urtica*

Hesperodoria Greene = *Chrysothamnus*

Hesperocyparis Bartel & Price = *Cupressus*

Hesperogreigia Skottsb. = *Greigia*

Hesperolaburnum Maire. Leguminosae (III 9). 1 Morocco: ***H. platycarpum*** (Maire) Maire
Hesperolinon (A. Gray) Small = *Linum*
Hesperomannia A. Gray. Compositae (Vern.-Hesp.). 3 Hawaii. Trees
Hesperomecon Greene = *Platystigma* (but see UKSB 47(1987)25)
Hesperomeles Lindl. Rosaceae (17). 11 C Am. to Bolivia (almost all Andean; Peru 7). R: CJB 68(1990)2230
Hesperonia Standl. = *Mirabilis*
Hesperopeuce (Engelm.) Lemmon (~ *Tsuga*). Pinaceae. 1 W N Am.: ***H. mertensiana*** (Bong.) Rydb. – to 1238 yrs old, 75m tall with bole to 2.75 m diam. R: NRBGE 45(1988)387
Hesperoscordum Lindl. = *Muilla*
Hesperoseris Skottsb. = *Dendroseris*
Hesperostipa (Elias) Barkworth (~ *Stipa*). Gramineae (X). 5 N Am. R: Phytol. 74(1993)15. ***H. comata*** (Trin. & Rupr.) Barkw. (*S. c.*, N Am.) – cult. for forage
Hesperothamnus Brandegee (~ *Millettia*). Leguminosae (III 16). 5 Mex.
Hesperoxiphion Bak. Iridaceae (VII 5). 4 Andes, Colombia. R: BN 132(1979)466. Lvs plicate
Hesperoyucca (Engelm.) Bak. (~ *Yucca*). Asparagaceae (Agavaceae). 3 SW N Am. R: Sida 19(2001)842.Capsule loculicidal, pollen in glutinous masses. ***H. whipplei*** (Torrey) Trel. (*Y. w.*) – s.t. hapaxanthic, infl. to 3.65 m in 14 days (world record pl. growth), strong fibre, fls ed.
Hesperozygis Epling. Labiatae (VII 2b). 8 Mex. (1), S Braz. R: FR 115(1939)12
Hessea Herb. Amaryllidaceae. 13 S Afr. R: CBH 16(1994)44
hessian (US = burlap), fabric orig. made from hemp, later from jute or a mix of both
Hestia S.Y. Wong & Boyce (~ *Schismatoglottis*). Araceae (VII 8). 1 W Mal.: ***H. longifolia*** (Ridl.) S.Y. Wong & Boyce
Hetaeria Blume. Orchidaceae (IV 2d). c. 25 Sri Lanka & SE As. to Tahiti
Heterachne Benth. Gramineae (XXVII 2). 3 N Aus. R: HIP (1935) t.3283
Heteracia Fischer & C. Meyer. Compositae (Lact.-Crep.). 2 SW As. to China
Heteradelphia Lindau. Acanthaceae (III 2e). 2 trop. W Afr.
Heteranthelium Hochst. ex Jaub. & Spach. Gramineae (XV). 1 Turkey to Pakistan: ***H. piliferum*** Hochst. ex Jaub. & Spach. R: NJB 13(1993)486
Heteranthemis Schott (~ *Chrysanthemum*). Compositae (Anth.-Gleb.). 1 SW Eur., N Afr.: ***H. viscidehirta*** Schott – cult. orn. ann. R: FNA 19(2006)551
Heteranthera Ruíz & Pavón. Pontederiaceae. Excl. *Eurystemon*, *Zosterella*, 11 trop. & warm Afr. (1), Am. ext. to N Am. (7). Some with merely submerged linr. lvs, others with orbicular floating ones, some with both. Cult. orn. in aquaria
Heteranthia Nees & Mart. Solanaceae (II). 1 Braz.: ***H. decipiens*** Nees & Mart.
Heteranthocidium Szlach. & al. = *Oncidium* (but see PJB 51(2006)54)
Heteranthoecia Stapf. Gramineae (Micr.). 1 trop. Afr.: ***H. guineensis*** (Franch.) Robyns
Heteraspidia Rizz. = *Justicia*
Heteroaridarum Hotta = *Aridarum*
Heteroarisaema Nakai = *Arisaema*
Heterocalycium Rauschert = *Cuspidaria*
Heterocarpha Stapf & C. Hubb. = *Dinebra*
Heterocaryum A. DC. (~ *Lappula*). Boraginaceae (B.3.5.2). 6 SW to C As.
Heterocentron Hook. & Arn. Melastomataceae. 28 Mex. & C Am. A dimorphic, some merely attractants for insects. Some cult. orn. esp. ***H. elegans*** (Schltdl.) Kuntze (Spanish shawl, C Am.) – ground-cover, but incl. some aggressive spp. e.g. ***H. subtriplinervium*** (Link & Otto) A. Braun (*H. macrostachyum*, Mex.), widely natur. in trop.
Heterochaenia A. DC. Campanulaceae (I 2). 4 Masc. R: ABG 155(2008)247. Pachycaul treelets
Heterochiton Graebner & Mattf. = *Herniaria*
Heterocodon Nutt. Campanulaceae (I 8). Excl. *Homocodon*, 1 W N Am.: ***H. rariflorus*** Nutt.
Heterocoma DC. Compositae (Vern.-Lychn.). Incl. *Bishopalea*, *Sipolisia*, *Xerxes*, 6 Braz.
Heterocondylus R. King & H. Robinson (~ *Eupatorium*). Compositae (Eup.-Ayap.). 13 C & S Am., esp. Braz. R: MSBMBG 22(1987)204
Heterocypsela H. Robinson. Compositae (Vern.-Dip.). 1 E Braz.: ***H. andersonii*** H. Robinson. R: Phytol. 44(1979)442
Heterodendrum Desf. = *Alectryon*
Heteroderis (Bunge) Boiss. Compositae (Lact.-Crep.). 1 SW & C As. to Pakistan: ***H. pusilla*** (Boiss.) Boiss.
Heterodraba Greene = *Athysanus*

Heteroflorum M. Sousa (~ *Peltophorum*). Leguminosae (I 4). 1 Mex.: ***H. sclerocarpum*** M. Sousa. R: Novon 15(2005)213

Heterogaura Rothr. = *Clarkia*

Heterogonium C. Presl (~ *Tectaria*). Tectariaceae (Dryopteridaceae s.l.). 22 Mauritius, SE As., Mal. R: Kalikasan 4(1975)205. Form hybrids with *T.* spp.

Heterolamium C.Y. Wu (~ *Meehania*). Labiatae (VII 2c). 1 China: ***H. debile*** (Hemsl.) C.Y. Wu

Heterolepis Cass. Compositae (Cich.). 3 Cape to Karoo. ***H. aliena*** (L.f.) Druce – cult. orn. shrub

Heterolobium Peter = *Gonatopus*

Heteromeles M. Roemer (~ *Photinia*). Rosaceae (Ros.-Mal.). 1 Calif., NW Mex.: ***H. arbutifolia*** (Lindl.) M. Roem. (*H. salicifolia*, toyon, tollon) – char. of chaparral, used like holly in decorations & orig. of name Hollywood, Calif., hybrids with *Photinia* sp. known. R: FNA 9(2014)447,488

Heteromera Pomel (~ *Chrysanthemum*). Compositae (Anth.-?Leuc.). 2 N Afr. R: BBMNHB 23(1993)153

Heteromma Benth. Compositae (Ast.-Gran.). 3 S Afr. mts

Heteromorpha Cham. & Schldl. Umbelliferae (III 8). 7 trop. & S Afr. to Yemen. R: KB 51(1996)234. Usu. trees. ***H. arborescens*** (Spreng.) Cham. & Schldl. (*H. trifoliata*) – tree, shrubby & subherbaceous forms, locally medic.

Heteropanax Seemann. Araliaceae. 8 Ind., S China

Heteropappus Less. = *Aster*

Heteropetalum Benth. = *Guatteria*

Heteropholis C. Hubb. (~ *Hackelochloa*). Gramineae (XXII 3). 6 C Afr. to Aus. R: GBS 36(1983)137

Heterophragma DC. Bignoniaceae (I). 2 Ind., SE As.

Heterophyllaea Hook.f. Rubiaceae (I V 1). Incl. *Teinosolen*, 8 Bolivia, Arg.

Heteroplexis C.C. Chang. Compositae (Ast.-Bac.). 3 China. R: Guihaia 5(1985)337

Heteropogon Pers. Gramineae (XXII 7). 6 trop. & warm Afr., S Eur. (1) etc. ***H. contortus*** (L.) Roemer & Schultes (tanglehead, trop. & warm) – good grazing when young but awns with hygroscopic action painful to stock & man, fire-climax veg. of much of Madag., thatch (pili) in Hawaii, strips of split stems used in basketry & armlets etc. by Aus. Aborigines

Heteropolygonatum Tamura & Ogiso. Asparagaceae (Convallariaceae). 10 China, Vietnam. R: Plantsman n.s. 9(2010)175, Phytotaxa 188(2014)218. Epiphytes

Heteropsis Kunth. Araceae (IV 3). 20 trop. S Am. R: SB 38(2013)939. ***H. flexuosa*** (Kunth) Bunting (nibi, N S Am.) & ***H. spruceana*** Schott (N S Am.) – aerial r. for lashing poles in house constr., furniture, basketry etc.

Heteropteris Fée = *Neurodium*

Heteropterys Kunth (*Heteropteris* Kunth). Malpighiaceae. 130 Mex. to Arg., W Afr. (1). Fr. a samara. Some locally used fibres

Heteroptilis E. Meyer ex Meissn. = *Dasispermum*

Heteropyxidaceae Engl. & Gilg = Myrtaceae (I 2)

Heteropyxis Harv. Myrtaceae (I 2; Heteropyxidaceae). 3 C & S Afr. R: MBSM 10(1971)222

Heterorhachis Schultz-Bip. ex Walp. Compositae (Arct.-Gort.). 1 Cape: ***H. aculeata*** (Burm.f.) Roessler – cult. orn. shrub. R: Strel. 9(2000)336

Heterosamara Kuntze (~ *Polygala*). Polygalaceae. 18 OW trop. R: Fontqueria 50(1998)123

Heterosavia (Urb.) Petra Hoffm. (~ *Gonatogyne*). Phyllanthaceae (Phyll.-Flueg.). 4 Carib. R: Britt. 60(2008)152

Heterosciadium Lange ex Willk. = *Daucus*

Heterosmilax Kunth = *Smilax* (but see Britt. 36(1984)184)

Heterospathe R. Scheffer. Palmae (V 14). 39 C & E Mal., Papuasia (NG 16) to Fiji. Monoec., unarmed feather-palms. Some cult. orn. incl. ***H. elata*** R. Scheffer (C Mal. to NW Pac.) – fr. chewed like betel, petioles & lvs used for basketry & hat-making etc.

Heterosperma Cav. Compositae (Cor.). 5+ SW US to S Am.

Heterostachys Ung.-Sternb. Amaranthaceae (Chenopodiaceae II 2). 2 C & S Am.

Heterostemma Wight & Arn. Apocynaceae (Vb; Asclepiadaceae III 5). Incl. *Dittoceras*, 15 Indomal., W Pac.

Heterostemon Desf. Leguminosae (I 2). 7 trop. Am. esp. Upper Amazon. R: PKNAW, C 79(1976)42. Extrafl. nectaries

Heterotaxis Lindl. = *Maxillaria*

Heterothamulopsis Deble & al. Compositae (Ast.-Bac.). 1 S Braz.: ***H. wagenitzii*** (Hellwig) Deble & al. – dioec. shrub. R: Balduinia 1(2005)4

Heterothalamus Less. Compositae (Ast.-Bac.). 3 C. Arg., Urug., S Braz. Shrubs. R: Balduinia 1(2005)6

Heterotheca Cass. Compositae (Ast.-Chr.). 28 S N Am. R (sect. *Phyllotheca*): Univ. Waterloo Biol. Ser. 37(1996)1. Cult. orn. but some referred to *Chrysopsis* & *Pityopsis*. ***H. inuloides*** Cass. often sold in US as *Arnica montana*

Heterotis Benth. = *Dissotis*

Heterotoma Zucc. (~ *Lobelia*). Campanulaceae (III 1). 1 Mex. & C Am.: ***H. lobelioides*** Zucc. – cult. orn. R: SB 15(1990)296

Heterotrichum DC. Melastomataceae. 10 trop. Am. Cult. orn. shrubs, some with ed. fr.

Heterotristicha Tobler = *Tristicha*

Heterotropa Morren & Decne = *Asarum*

Heterozeuxine Hashimoto = *Zeuxine*

Heterozostera (Setch.) Hartog (~ *Zostera*). Zosteraceae. 3 Aus. 1 Chile. R: AB 81(2005)101

Heuchera L. Saxifragaceae (I 5). 35 S Mex. to Arctic esp. W US (E, 7; N Am. 32 – R: FNA 9(2009)85). R: MSPS 2(1936)1; D. Heims & G. Ware (2005) *Hs & heucherellas*: 72. Poss. hybridization leading to multiple origin of autotetraploid lineages from diploid ancestors with bigger fls. Cult. orn. tufted pls (alum- r.) for ground-cover, esp. ***H. sanguinea*** Engelm. (coralbells, SW N Am.) & its hybrids. Some local medic.

× **Heucherella** Wehrh. Saxifragaceae. *Heuchera* × *Tiarella*. R: D. Heims & G. Ware (2005) *Heucheras & hs:* 165. Sterile hybrids known only in cult., the first × ***H. tiarelloides*** (Lemoine) Wehrh. ex Stearn (*Heuchera* × *brizoides* Hort. ex Lemoine (*H. sanguinea* Engelm. × *H. americana* L.) × *Tiarella cordifolia* L.)

Hevea Aubl. Euphorbiaceae (Crot.-Hev.). 9 Amazon basin. Rauh's Model. Fr. explodes disp. seeds to 20+ m, seeds water-disp. but mainly destroyed by fish, e.g. piranhas at floodtime. ***H. brasiliensis*** (A. Juss.) Muell. Arg. (natural or Pará rubber) – domatia; seeds (oilseeds poss. source of biodiesel) stay afloat up to 2 months; trunk the source of best natural rubber (so-named because it would rub out pencil-marks, first rubbers for sale 1770 at 3 shillings) & most of that planted in OW esp. form. Mal. but now Thailand biggest producer (95% of global supply [8.4 M t by 2005] in SE As.). Latex tapped by making sloping incisions in bark & exudate coll. in suspended cups, coagulated with acid & pressed into sheets etc. with literally thousands of uses (good motor tyres with c. 40 % natural rubber, & still best for medic. gloves); the last 'tappings' or waste called almeidina in Angola; old logs used for chipboard, plywood & furniture (now planted for this in Malaysia) in Thailand & Sri Lanka, now exported to Eur. & US as beds, chopping-boards etc. The OW provenances selected from Amazonia some of the best known but reduced to a small number of successful trees when intr. via Kew & Sri Lanka to Mal., where Tamils imported from Ind. as tappers

Hewardia J. Sm. = *Adiantum*

Hewittia Wight & Arn. Convolvulaceae (12). 1 OW trop.: ***H. malabarica*** (L.) Suresh (*H. sublobata*) – natur. Carib. R: FPM 5(2015)136

Hexachlamys O. Berg (~ *Eugenia*). Myrtaceae (II 10). 15 S Am.

Hexacyrtis Dinter. Colchicaceae (Liliaceae s.l.). 1 Namibia: ***H. dickiana*** Dinter. R: Strel. 10(2000)589

Hexadesmia Brongn. = *Scaphyglottis*

Hexaglottis Vent. = *Moraea*

Hexalectris Raf. Orchidaceae (V 13a). 10 S US, Mex. Mycotrophs

Hexalobus A. DC. Annonaceae (III 6). 5 trop. & S Afr. R: SB 36(2011)38

Hexaneurocarpon Dop = *Fernandoa*

Hexapora Hook.f. = *Micropora*

Hexaptera Hook. = *Menonvillea*

Hexapterella Urb. Burmanniaceae. 2 trop. S Am. R: AMBG 76(1989)956

Hexasepalum Bartl. ex DC. (*Diodella*). Rubiaceae (IV 15). 1 Afr., 4 trop. Am. R: SB 41(2016)409

Hexaspermum Domin = *Phyllanthus*

Hexaspora C. White. Celastraceae (I). 1 N Queensland: ***H. pubescens*** C. White. R: FA 22(1984)168

Hexastemon Klotzsch = *Erica*

Hexastylis Raf. = *Asarum*

Hexatheca C.B. Clarke. Gesneriaceae (III 2j). 4 Borneo. R: NRBGE 46(1989)54

Hexinia Yang = *Launaea*
Hexisea Lindl. = *Scaphyglottis*
Hexopetion Burret = *Astrocaryum*
Hexuris Miers = *Peltophyllum*
Heynea Roxb. (~ *Trichilia*). Meliaceae (II). 2 Ind. & S China to W Mal. R: FM 1,12(1995)41, FOC 11(2008)120. ***H. trijuga*** Roxb. – local medic. bark
Heynella Backer. Apocynaceae (Va; Asclepiadaceae III 4). 1 Java: ***H. lactea*** Backer – rare. R: Blumea 6(1950)381
Heywoodia Sim. Phyllanthaceae (Wiel.-Astroc.; Euphorbiaceae s.l.). 1 E Afr., S Afr. (disjunct): ***H. lucens*** Sim (Cape ebony). R: A.R.-Sm., *Gen. Euph.* (2001)6
Heywoodiella Svent. & Bramw. = *Hypochaeris*
hiba *Thujopsis dolabrata*
Hibbertia J. Kenn. ex Andrews. Dilleniaceae (III). Incl. *Adrastaea, Pachynema*, c. 225 Madag. (1), Mal. (2), Aus. (c. 200), New Caled., Fiji (1). Mostly ericoid or cli. shrubs, some with phylloclades. A varied, from 200 not in obvious groups to ∞ in 15 bundles to A 1. Some cult. orn. with yellow fls (guinea-fl.). ***H. conspicua*** (Harv.) Gilg (W Aus.) – buzz-poll. by bees; ***H. scandens*** (Willd.) Gilg (snake cli., Aus.) – cult.
Hibiscadelphus Rock (~ *Hibiscus*). Malvaceae (Malv.-Hib.). 8 Hawaii, extinct (4) or endangered. R: Novon 5(1995)183. Trees & shrubs. ***H. wilderianus*** Rock – extinct & only 1 tree of it ever known; ***H. giffardianus*** Rock – reduced to 1 tree by 1930 & known now only in cult.
hibiscus, Chinese *Hibiscus rosa-sinensis*
Hibiscus L. Malvaceae (Malv.-Hib.). c. 700 warm temp. (Eur. 2) to trop. (Incl. *Abelmoschus, Anotea, Cenocentrum, Decaschistia, Dicellostyles, Goethea, Helicteropsis, Julostylis, Jumellianthus, Kosteletzkya, Kydia, Malachra, Malvaviscus, Nayariophyton, Pavonia, Peltaea, Phragmocarpidium, Rojasimalva, Senra, Talipariti* (22 trop. to Jap. – R: CUMH 23(2001)231; Scarrone's Model) etc.), *Urena*. Trees to herbs, fls white to red, yellow or even bluish, usu. with basal maroon spots on petals & extrafl. nectaries (even in orig. ant-less Hawaii), fr. capsules to berries (*Malvaviscus*; R: AMBG 80(1993)441); alks. Fibres, medic. & many cult. orn., ((rose) mallows, some as *Decaschistia, Malvaviscus, Pavonia*). ***H. abelmoschus*** L. (*A. moschatus*, trop. OW) – grown for musky seeds (ambrette); ***H. americanus*** (L.f.) Mabb. (*Urena lobata*, Congo jute, cousin mahoe, guaxima, aramina, trop.) – jute subs., local medic.; ***H. borneensis*** Airy Shaw (Borneo) – tree to 30 m with trunk to 80 cm diam.; ***H. cannabinus*** L. (kenaf, Ambari or Deccan hemp, Bimlipatum jute, trop. Afr., long cult. Ind. & SE Eur.) – fibre like jute used for paper (S US), door-panel insulation in Toyota Prius cars, seed-oil for illumination in Afr.; ***H. capitatus*** (L.) Mabb. (*Malachra c.*, trop. Am.) – source of excellent jute-like fibre; ***H. elatus*** Sw. (*Talipariti e.*, blue mahoe, Cuban bast, Jamaica & Cuba) – bark fibre for ropes & hat-making, timber for gunstocks, cabinet-making etc.; ***H. esculentus*** L. (*A. e.*, bandakai, bindi, gobbo, gombo, gumbo, okra, lady's fingers) – ed. young fr. (slimy), allopolyploid (2n = 130), one genome poss. from diploid *H.* sp. (*A. tuberculatus* Pal & Singh, 2n = 58); ***H. fryxellii*** Mabb. (NW Aus.) – prickly; ***H. hamabo*** Sieb. & Zucc. (Jap., Korea) – fibre source; ***H. heterophyllus*** Vent. (rosella, E Aus.) – the 'soft' wood used with a harder one to make fire (Aborigines), rosella jam made from inflated red K; ***H. macrophyllus*** Roxb. ex Hornem. (Ind. to Java) – fibre-pl. in SE As., timber for building; ***H. malvaviscus*** L. (*Malvaviscus arboreus*, trop. Am., widely natur.) – hummingbird-poll. red fls, berries; ***H. manihot*** L. (*A.m.*, aibika, ibika, tree spinach, Mal., domesticated ? NG) – cordage like jute, lvs ed., 'var. *caillei* A. Chev.' (*A. caillei*) a cultigen (the okra of W Afr.) prob. derived from it; ***H. mutabilis*** L. (Confederate rose, China) – fls open white & fade purple-pink in 1 day, bark for textiles & paper, imp. in Chinese art; ***H. rosa-sinensis*** L. (Chinese hibiscus, China rose, shoe-flower, unknown in wild & poss. anc. hybrid involving several spp., but widely cult. trop. & warm & under glass in temp.) – many cvs incl. hybrids with Hawaiian endemic spp., fls used for shining shoes in Ind. & colouring eyebrows black; ***H. roxburghianus*** (Wight) Mabb. (*Kydia calycina*, Ind.) – monoec., bark used for coarse ropes; ***H. rudis*** (Benth.) Mabb. (*Malachra r.*, trop. Am.) – root cooked with sugar for treatment of bloody diarrhoea; ***H. sabdariffa*** L. (roselle, Jamaica or red sorrel, ? trop. Afr., but now natur. pantrop.) – fibre used for rope, fleshy red K used in drinks, jellies, lvs used like spinach; ***H. schizopetalus*** (Boulger) Hook.f. (trop. E Afr.?) – petals deeply laciniate, poss. an old cv. or parent of *H. rosa-sinensis*, though usu. sterile; ***H. syriacus*** L. (rose of Sharon (Am.), E As.) – many cvs, the common hibiscus of gardens (temp.); ***H. tiliaceus*** L. (*Talipariti t.*, hau (Hawaii), trop.) – lvs get redder with time (malvidin prod.), leaf starch like sago, imp. fibre for cordage, mats, sails, nets,

exported to Eur., wood for bows & canoe outriggers, firewood & erosion control, local medic., lvs to wrap food, several useful cvs selected Polynesia; ***H. waimeae*** A. Heller (Hawaii) – scented fls

Hickelia A. Camus. Gramineae (V 6). Incl. *Pseudocoix*, 4 Tanzania (1), Madag. (3). R: KB 49(1994)438. Scrambling bamboos

Hickenia Lillo = *Araujia*

hickory *Carya* spp.; **Australian h.** *Acacia bakeri*, *A. implexa*; **h. elm** *Ulmus thomasii*

Hicksbeachia F. Muell. Proteaceae (V 4). 2 NE Aus. R: Telopea 3(1988)231. ***H. pinnatifolia*** F. Muell. (red boppel nut, rose nut) – Corner's Model, seed ed.

Hicoria Raf. = *Carya*

Hicriopteris C. Presl = *Dicranopteris*

Hidalgoa Llave. Compositae (Cor.). 6 Mex., C Am. Climbing (petioles). ***H. ternata*** Llave (*H. wercklei*) – cli. 'dahlia' form. pop. cult. early C20

Hieracium Tourn. ex L. Compositae (Lact.-Hier.). Excl. *Pilosella*, c. 770 (diploid sexually reprod. spp.; c. 5200 triploid or tetraploid apomictic microspp.), temp. (excl. Aus. exc. as weeds), trop. mts. Hawkweeds. Allied to *Crepis* & linked to it via S Am. spp.

Hieris Steenis. Bignoniaceae (2). 1 Penang (Malay Pen.): ***H. curtisii*** (Ridl.) Steenis. R: FM I,8(1977)127

Hiernia S. Moore. Orobanchaceae (Buch.; Scrophulariaceae s.l.). 1 Angola, Namibia: ***H. angolensis*** S. Moore. R: Strel. 19(2000)523

Hierobotana Briq. Verbenaceae (Verb.). 1 Colombia, Ecuador, Peru: ***H. inflata*** (Kunth) Briq.

Hierochloe R. Br. = *Anthoxanthum*

Hieronyma Allemão (*Hieronima*, *Hyeronyma*). Phyllanthaceae (Ant.-Hier.; Euphorbiaceae s.l.). 21 trop. Am. (S Am. 10; R: BJ 111(1990)297). Some timbers for cabinetwork

Hieronymiella Pax. Amaryllidaceae. 4 Arg., Bolivia (1)

Hieronymusia Engl. (~ *Suksdorfia*). Saxifragaceae (I 6). 1 Arg., Bolivia: ***H. alchemilloides*** (Griseb.) Engl.

Hijmania Vianna (*Maria*). Moraceae. 4 trop. Afr. R: Phytotaxa 247(2016)97

Hilaria Kunth. Gramineae (XXIX 1). 10 S US to Guatemala. R: JWAS 46(1956)311. Some cult. incl. ***H. mutica*** (Buckley) Benth. (tobosa grass, SW US)

hildaberry *Rubus* 'Hildaberry'

Hildaea C. Silva & R.P. Oliveira (~ *Panicum*). Gramineae. 5 trop. Am.

Hildebrandtia Vatke ex A. Braun. Convolvulaceae (4). 13 Arabia, Afr. (R: KB 51(1996)526), Madag. K accrescent in fr. Some ed. fr. (*Cladostigma*)

Hildegardia Schott & Endl. Malvaceae (Sterc.; Sterculiaceae). 11 Cuba (1), Afr. (3), Madag. (3), Indomal. (3), Aus. (1). Rauh's Model

Hildewintera Ritter ex Rowley = *Cleistocactus*

Hilgeria Foerther = *Heliotropium*

Hillebrandia Oliv. Begoniaceae. 1 Hawaii: ***H. sandwicensis*** Oliv. C 5, poss. homoeotic mutants of A, as seen in allied As. *Begonia* spp. (cf. 'double' fls in cult. *B.*)

Hilleria Vell. Conc. Petiveriaceae (Phytolaccaceae II). 3 S Am., 1 ext. to Afr., Madag. & Masc. (***H. latifolia*** (Lam.) H. Walter) – potherb in W Afr.

Hillia Jacq. (~ *Cosmibuena*). Rubiaceae (I 3). 24 trop. Am. R: AMBG 81(1994)582. Some epiphytic shrubs, solit. term. fls, hair-tuft at end of seeds (cf. Apocynaceae) unique in R., alks; some cult. orn (*Ravnia*) looking like gesneriads

Hilliardia R. Nordenstam (~ *Matricaria*). Compositae (Anth.-Cot.). 1 Natal: ***H. zuurbergensis*** (Oliv.) R. Nordenstam. R: OB 93(1987)147

Hilliardiella H. Robinson (~ *Cyanthillium*). Compositae (Vern.-Centrop.). 8 E & S Afr.

Hilliella (O. Schulz) Zhang & Li = *Yinshania*

Hill's fig *Ficus microcarpa* var. *hillii*

Hilsenbergia Tausch ex Meissn. = *Bourreria* (but see Adansonia III,25(2003)157)

Himalacodon Hong & Q. Wang (~ *Codonopsis*). Campanulaceae (I). 1 E Himal.: ***H. dicentrifolius*** (Clarke) Hong & Q. Wang. R: JSE 52(2014)548

Himalaiella Raab-Straube = *Jurinea* (but see Willd. 33(2003)390)

Himalayacalamus Keng f. (~ *Drepanostachyum*). Gramineae (IV). 8 Himal. to SW China

Himalayan blackberry *Rubus armeniacus*; **H. honeysuckle** *Leycesteria formosa*

Himalayopteris Shao & Lu (~ *Selliguea*). Polypodiaceae. 1 Himal.: ***H. erythrocarpa*** (Kuhn) Shao & Lu. R: Novon 21(2011)91

Himalrandia Yamaz. Rubiaceae (II 4). 3 Himal.

Himantandraceae Diels. Magnoliidae – Magnoliales. 1/2 E Mal. to N Aus. Large aromatic trees with alks, young parts densely covered with fimbriate peltate scales. Lvs in spirals (distich.), simple, entire, gland-dotted; stip. 0. Fls large, bisex., solit. (2 or 3) in axils; K (2), calyptrate, 1 encl. other (prob. bracts), C 3–23 (prob. staminodes) in spirals, linr., A 13–130, in spirals, not differentiated into filament & anther, 1 pr of sporangia on each side, each with single longit. slit, c. 13–22 staminodes in spirals between A & G, pollen-grains monosulcate, G (6)7–10(–30) in spirals, closed, each with ovary & style, weakly connate (more fully in fr.), with 1(2) pend., anatr. ovules. Fr. gall-like, a syncarp (?drupe) of coalesced carpels; seeds small with copious, oily non-ruminate endosperm. 2n = 24 Genus: *Galbulimima*

Himantochilus Anderson ex Benth. = *Anisotes*

Himantoglossum Koch. Orchidaceae (IV 4d). Incl. *Comperia*, 11 Eur. N Afr., E Med. R: BotJLS 142(2003)17. Many threatened because of salep used in icecream. ***H. hircinum*** (L.) Spreng. (lizard orchid, Eur.) – visited by solit. bees but otherwise unattractive to insects, distr. slowly spreading northwards in GB (early evidence of global warming)

Himantostemma A. Gray = *Matelea*

Himatanthus Willd. ex Schultes (~ *Plumeria*). Apocynaceae (I g). 9 S Am. R: T 62(2013)1305. Koriba's Model

Hindsia Benth. Rubiaceae (IV 3). 11 Braz. R: AJBRJ 34(1996)56. ***H. violacea*** Benth. – cult. orn., coll. once & extinct in wild

hing *Ferula* spp.

hinoki *Chamaecyparis obtusa*

Hinterhubera Schultz-Bip. ex Wedd. Compositae (Ast.-Hin.). 8 Andes

Hintonella Ames. Orchidaceae (V 12h). 1 Mex.: ***H. mexicana*** Ames. R: Gen. Orch. 5(2009)278

Hintonia Bullock. Rubiaceae (inc. sed.). 4 C Am. ***H. latiflora*** (DC.) Bullock (copalchi, copalquín) – used for malaria & other fevers (quinine)

Hionanthera Fernandes & A. Diniz = *Ammannia*

Hippeastrum Herb. Amaryllidaceae. Excl. *Rhodophiala*, c. 55 Mex. & WI to Arg., Bolivia, W Afr. (1, natur.?). Cult. orn. bulbous pls ('amaryllis') esp. hybrids of complicated parentage involving Am. spp. esp. ***H. aulicum*** (Ker) Herb., ***H. elegans*** (Spreng.) H. Moore, ***H. puniceum*** (Lam.) Kuntze (*H. equestre*, Barbados or fire lily, perhaps 'the true' *Amaryllis belladonna*), ***H. reginae*** (L.) Herb. (also in W Afr., natur.?), ***H. reticulatum*** (L'Hérit.) Herb., ***H. striatum*** (Lam.) H. Moore (*H. rutilum*), much cult. for selling at Christmas as dormant bulbs to grow. Early hybrids incl. ***H. × johnsonii***(Bury) Herb. (*A. × j.*, *H. reginae × H. vittatum* (L'Hérit.) Herb.), 1799) & '**Reginae**' strains (*H. striatum × H. vittatum* with *H. psittacinum* (Ker) Herb. etc., Eur., 1870s), then these crossed with *H. leopoldii* Bak. & *H. pardinum* Veitch ex T. Moore (Peru). ***H. calyptratum*** (Ker Gawl.) Herb. – fls smell of burning rubber!

Hippeophyllum Schltr. Orchidaceae (V 11b). 10 Mal.

Hippia L. Compositae (Anth.-Cot.). 8 SW & S Cape. R: Strel. 9(2000)336

Hippobroma G. Don (~ *Lobelia*). Campanulaceae (III 2). 1 Jamaica, natur. pantrop.: ***H. longiflora*** (L.) G. Don (*L. l.* L.) – toxic latex. R: BTBC 67(1940)782

Hippobromus Ecklon & Zeyher. Sapindaceae. 1 S Afr.: ***H. pauciflorus*** (L.f.) Radlk. R: Strel. 9(2000)643

Hippocastanaceae A. Rich. = Sapindaceae

Hippochaete Milde = *Equisetum*

Hippocratea L. Celastraceae (III). 3 trop. Am., Afr. (s.l. incl. *Elachyptera*, *Prionostemma*, *Pristimera*, *Reissantia* etc. 120 trop.). Twining shrubs s.t. used as 'rope' bridges in Afr., medic. & insecticidal in Mex. ***H. comosa*** Sw. (WI) – seeds ed.

Hippocrateaceae Juss. = Celastraceae

Hippocrepis L. Leguminosae (III 22). 34 Eur. (10), W As., Medit. R: Willdenowia 19(1989)59. Some fodders incl. ***H. comosa*** L. (horseshoe vetch, Eur.) – cult. orn. with scented fls esp. ***H. emerus*** (L.) Lassen (*Emerus major*, scorpion senna [jointed legumes suggesting scorpion tail & therefore sting cure – 'Doctrine of Signatures'], Eur., Turkey)

Hippodamia Decne = *Solenophora*

Hippolytia Polj. (~ *Tanacetum*). Compositae (Anth.-Ant.). 19 C As. to N China. R: APS 17, 4(1978)70, BBMNHB 23(1993)105

Hippomane L. Euphorbiaceae (Hipp.). 3 Florida & WI to Venez, Galapagos (1). G 6–9-loc.; fr. a drupe; alks. ***H. mancinella*** L. (manchineel) – notoriously (much cited in sensational literature) poisonous coastal tree cult. as windbreak, latex with phorbol on skin gives severe dermatitis, in eyes can blind (form. thought of like upas – death to sleepers below), form. arrow-poison, useful timber

Hippomarathrum Link = *Cachrys*

Hippophae L. Elaeagnaceae. 5 temp. Euras. (Eur. 1). R: NJB 22(2002)371. Alks. ***H. rhamnoides*** L. (sea buckthorn, sallow thorn, Eur. to N China) – small dioec. tree or shrub of sandy coasts & shingle-banks etc. in mts, some rheophytic forms; bracteoles form hood over A in wet weather & sep. on drying out so that pollen may be blown away; domesticated in China & Eur. (plantation crop on Baltic coast & around Berlin), fr. ed., rich in vitamins A, C (7 times as much as lemons), & E – juice, preserves, sauce for meat or fish in Eur., with milk or cheese in C As.; wood suitable for turning; yellow dye; oil in cosmetics. Some spp. cult. as medic. pls Nepal

Hippotis Ruíz & Pavón. Rubiaceae (Cond.). 11 trop. Am.

Hippuridaceae Vest = Plantaginaceae

Hippuris L. Plantaginaceae (Hipp.; Hippuridaceae). 1 cosmop., with ecological races in Arctic & Baltic: ***H. vulgaris*** L. (mare's tail) – young lvs eaten by Inuit. The char. habit (*Hippuris* syndrome) seen in unrelated genera: *Decodon* & *Rotala* (Lythraceae), *Elatine* (Elatinaceae), *Pogogyne* (Labiatae) etc. etc.

Hiptage Gaertn. Malpighiaceae. 20–30 trop. As. to Fiji. Fr. a 3-winged samara. Some cult. orn. incl. ***H. benghalensis*** (L.) Kurz (Indomal.) – cult. in trop. for fragrant fls, bonsai in Thailand, locally medic. (insecticidal), invasive in Masc.

Hiraea Jacq. Malpighiaceae. 40 trop. Am. ***H. schultesii*** Cuatrec. (Colombia) – leaf infusion for conjunctivitis

Hirania Thulin. Sapindaceae (?). 1 SC Somalia: ***H. rosea*** Thulin. R: NJB 24(2007)510

Hirpicium Cass. Compositae (Arct.-Gort.). 11 trop. & S Afr. R: MBSM 3(1959)333

Hirschfeldia Moench (~ *Erucastrum*). Cruciferae (12). 1 Medit.: ***H. incana*** (L.) Lagr.-Foss. (*E. i.*) – gynodioec., widespread weed in N Am.

Hirschia Bak. = *Iphiona*

Hirtella L. Chrysobalanaceae (4). 108 trop. Am., E Afr. (1) & Madag. (1). R: FOW 10(2003)95. Mangenot's & Troll's Models; pink or purple butterfly-poll. dayfls (cf. *Couepia*). Bark of ***H. americana*** L. (C & N S Am.) baked with clay (cf. *Licania*) gives heat-resistance to pottery cooking-pots; ***H. carbonaria*** Little (Ecuador, Colombia) – charcoal; ***H. pendula*** Lam. – fresh fr. used as ear pendants on St Lucia

Hirtellina Cass. = *Staehelina* (but see Boissiera 51(1996)75)

Hirtzia Dodson = *Pterostemma*

Hispaniella Braem = *Oncidium*

Hispaniolanthus Cornejo & Iltis (~ *Capparis*). Capparaceae. 1 W Haiti: ***H. dolichopodus*** (Helwig) Cornejo & Iltis. R: HPB 14(2009)9

Hispidella Barnadez ex Lam. Compositae (Lact.-Hier.). 1 Iberian Pen.: ***H. hispanica*** Barnadez ex Lam.

Histiopteris (J. Agardh) J. Sm. Dennstaedtiaceae. 8 trop. esp. Mal. Terr. ***H. incisa*** (Thunb.) J. Sm. (pantrop.) – cult. orn. in wild, lvs scramble or climb, the lamina to 3 m long on petiole to 2 m

Hitchcockella A. Camus. Gramineae (V 6). 1 Madag.: ***H. baronii*** A. Camus

Hitchenia Wall. = *Curcuma*

Hitcheniopsis (Bak.) Ridl. = *Scaphochlamys*

Hitoa Nad. = *Ixora*

Hladnikia Reichb. (~ *Grafia*). Umbelliferae (III 5). 1 Slovenia: ***H. pastinacifolia*** Reichb.

ho wood *Cinnamomum camphora*

hoary alison *Berteroa incana*; **h. cress** *Erucastrum incanum*

hobblebush *Viburnum lantanoides*

Hochreutinera Krapov. Malvaceae (Malv.-Malv.). 2 Mex., temp. S Am.

Hochstetteria DC. = *Dicoma*

Hockinia Gardner. Gentianaceae. 1 E Braz.: ***H. montana*** Gardner

Hodgkinsonia F. Muell. = *Margaritopsis*

Hodgsonia Hook.f. & Thomson. Cucurbitaceae (XII). 2 Indomal. R: Blumea 46(2001) 169. 'Seeds' = pyrenes (unique in fam.). ***H. heteroclita*** (Roxb.) Hook.f. & Thomson ('*H. macrocarpa*', Sikkim to Thailand) – seeds eaten by Sikkimese, opium roasted in seed-oil

Hodgsoniola F. Muell. Asphodelaceae (Hemerocallidaceae; Johnsoniaceae). 1 SW Aus.: ***H. junciformis*** (F. Muell.) F. Muell. R: FA 45(1987)306

Hoehnea Epling. Labiatae (VII 2b). 4 S Braz., Paraguay. R: FRB 85(1936)130, 85, 115(1939)8

Hoehneella Ruschi. Orchidaceae (V 12j). 1–2 SE Braz.

Hoehnelia Schweinf. = *Ethulia*

Hoehnella Szlach. & Sitko = *Maxillaria*

Hoehnephytum Cabera. Compositae (Sen.-Sen.). 3 Braz.

Hoffmannanthus H. Robinson & al. (~ *Vernonia*). Compositae (Vern.-Erl.). 1 trop. E Afr.: ***H. abbotianus*** (O. Hoffm.) H. Robinson & al. R: PhytoKeys 39(2014)58

Hoffmannia Sw. Rubiaceae (IV 5). c. 100 Mex. to Arg. Some cult. orn. foliage pls under glass

Hoffmanniella Schltr. ex Lawalrée. Compositae (Helia.-Ecl.). 1 C Afr., Cameroun: ***H. silvatica*** Schltr. ex Lawalrée

Hoffmannseggella H. Jones = *Laelia*

Hoffmannseggia Cav. Leguminosae (I 4). 23 SW N Am. (10) to Chile (S Am. 11). R: Lundellia 9(2006)8; Afr. spp. = *Pomaria*. Tubers of some SW US spp. (esp. ***H. glauca*** (Ortega) Eifert – also S Am., s.t. invasive) ed. roasted

Hofmeisterella Reichb.f. Orchidaceae (V 12h). 2 Colombia, Ecuador. R: Gen. Orch. 5(2009)278

Hofmeisteria Walp. Compositae (Eup.-Hofm.). Excl. *Oaxacania*, 10 Mex. R: MSBMBG 22(1987)453, Phytol. M 11(1997)145

hog brake *Pteridium aquilinum, Ambrosia artemisiifolia;* **h. fennel** *Peucedanum officinale;* **h. gum** *Metopium toxiferum, Moronobea* spp., *Symphonia globulifera;* **h. nut** *Carya glabra;* **h. plum** *Spondias* spp. esp. *S. mombin, Symphonia globulifera, Ximenia americana;* **h.weed** *Heracleum sphondylium;* **h.w., giant** *H. mantegazzianum*

Hohenackeria Fischer & C. Meyer. Umbelliferae (III 6). 2 E Eur. (2), Cauc., N Afr.

Hohenbergia Schultes & Schultes f. Bromeliaceae (3). 47 trop. Am. R: FN 14(1979)1731. Some cult. orn. terr. or epiphytic

Hohenbergiopsis L.B. Sm. & Read. Bromeliaceae (3). 1 Guatemala, S Mex.: ***H. guatemalensis*** (L.B. Sm.) L.B. Sm. & Read. R: Phytol. 33(1976)440

Hoheria Cunn. Malvaceae (Malv.-Malv.). 6 NZ. R: Plantsman 5(1983)178, NZJB 38(2000)375. Cult. orn. small trees with heteroblasty & white fls esp. ***H. lyallii*** Hook.f. & ***H. populnea*** Cunn. (bark for cordage, wood for cabinet-making, locally medic.)

Hoita Rydb. (~ *Orbexilum*). Leguminosae (III 20). 3 SW N Am. R: MNYBG 61(1990)49

hoja santa (leaves) *Piper auritum*

Holacantha A. Gray (~ *Castela*). Simaroubaceae. 2 SW N Am. R: Britt. 5(1944)137

Holalafia Stapf = *Alafia*

Holandrea Reduron & al. = *Dichoropetalum* (but see PSE 243(2004)205)

Holarrhena R. Br. Apocynaceae IIb). 4 trop. Afr., Indomal. R: MLW 81–2(1981). Many alks. Decid. trees & shrubs, imp. locally medic., cult. orn. esp. ***H. pubescens*** (Buch.-Ham.) G. Don (*H. antidysenterica*, conessi, kurchi, Indomal.) – bark (Tellichery b.) effective against dysentery, (*H. febrifuga*, kumbanzo, trop. Afr.) – febrifuge, timber for spoons etc.

Holboellia Wall. = *Stauntonia* (but see Cathaya 8–9(1997)86)

Holcoglossum Schltr. Orchidaceae (V 16c). Incl. *Neofinetia*, 14 SE As. to Taiwan, S Jap. R: NRBGE 44(1987)251. ***H. amesianum*** (Reichb.f.) Christenson – anther grows through 360° against gravity, inserting pollen into stigma cavity; ***H. falcatum*** (Thunb.) Garay & H. Sweet (*N. f.*, E As.) – scented fls (like lily-of-the-valley), cult. orn., orig. by Samurai & nobility

Holcolemma Stapf & C. Hubb. (~ *Setaria*). Gramineae (XXIV). 4 E Afr. & Ind., Aus. R: KB 32(1978)773

Holcosorus T. Moore = *Selliguea*

Holcus L. Gramineae (21b). 9 Eur. (XVI 7), Medit. to Middle E., S Afr. (1). Intergrading with *Deschampsia*. ***H. lanatus*** L. (Yorkshire fog, creeping soft grass, Euras., invasive Aus., Hawaii, Masc., N Am.) – pasture-grass, 2n = 14, ancestor of ***H. mollis*** L., a polyploid complex, backcrossing in Eur. to give triploid sterile veget. clones

hold-me-tight *Achyranthes indica*

Holigarna Buch.-Ham. ex Roxb. Anacardiaceae (I). 7+ Indomal. G inf. Some timber

Holland greens *Brassica rapa* 'Tyfon'

Hollandaea F. Muell. Proteaceae (V 1). 4 NE Aus. R: Austrobaileya 8(2012)673

Hollermayera O. Schulz. Cruciferae (46). 1 Chile: ***H. valdiviana*** (Phil.) Ravenna. R: NJB 1(1981)142

Hollisteria S. Watson. Polygonaceae (Erio.-Erio.). 1 C Calif.: ***H. lanata*** S. Watson. R: Phytol. 66(1989)210

Hollrungia K. Schum. = *Passiflora*

holly *Ilex* spp. esp. *I. aquifolium* & *I.* × *altaclerensis*; **American h.** *I. opaca*; **h. fern** *Polystichum* spp.; **h. grape** *Berberis repens*; **Japanese h.** *I. crenata*; **h. oak** *Quercus ilex*; **sea h.** *Eryngium maritimum*; **Singapore h.** *Malpighia coccigera*. See also *Heteromeles*

hollyhock *Alcea rosea*

holm (e) = holly

Holmbergia Hicken. Amaranthaceae (Atrip.; Chenopodiaceae I 2). 1 Urug., Paraguay, Arg.: ***H. tweedii*** (Moq.) Speg. – berryoid. R: SB 35(2010)852

Holmesia Cribb = *Angraecopsis*

Holmgrenanthe Elisens. Plantaginaceae (Ant.-Ant.; Scrophulariaceae s.l.). 1 SW N Am.: ***H. petrophila*** (Cov. & C. Morton) Elisens. R: SBM 5(1985)54

Holmgrenia W.L. Wagner & Hoch = *Neoholmgrenia*

Holmskioldia Retz. Labiatae (V; Verbenaceae). 1 Himal.: ***H. sanguinea*** Retz. (Chinese hat plant, cup-&-saucer plant) – orn. with red fls. now pantrop. cult. See also *Karomia*

Holocalyx M. Micheli. Leguminosae (III 1). 1 trop. S Am.: ***H. balansae*** M. Micheli – timber. R: Britt. 62(2010)111

Holocarpa Bak. = *Pentanisia*

Holocarpha (DC.) Greene. Compositae (Mad.-Mad.). 4 Calif. R: FNA 21(2006)287

Holocheila (Kudô) S. Chow (~ *Pogostemon*). Labiatae (VI). 1 SW China: ***H. longipedunculata*** S. Chow

Holocheilus Cass. Compositae (Mut.-Nass.). 7 S Braz., Paraguay, Urug., N & C Arg. R: RMLP n.s. 11(1968)1. Perenn. herbs

Holochlamys Engl. Araceae (IV 1). 2 NG, New Br.

Holodictyum Maxon = *Asplenium*

Holodiscus (K. Koch) Maxim. Rosaceae (Ros.-Spir.). 1(-5) W N Am. to Colombia: ***H. discolor*** (Pursh) Maxim. – local medic. N Am., wood for knitting-needles, roasting tongs etc., cult. orn. shrub. R: BTBC 70(1943)275.

Holographis Nees. Acanthaceae (III 1). 10 Mex., arid & semi-a. R: JAA 64(1983)129

Hologyne Pfitzer = *Coelogyne*

Hololachna Ehrenb. (~ *Reaumuria*). Tamaricaceae. 1 C As.: ***H. songarica*** (Pallas) Ehrenb.

Hololeion Kitam. (~ *Hieracium*). Compositae (Cich.-Cich.-Crep.). 3 E As.

Hololepis DC. (~ *Vernonia*). Compositae (Vern.-Lychn.). 2 SE Braz.

Holopogon Komarov & Nevski = *Neottia*

Holoptelea Planch. Ulmaceae. 2 trop. Afr. (1), Ind. (1). Timber

Holopyxidium Ducke = *Lecythis*

Holoregmia Nees (~ *Craniolaria*). Martyniaceae. 1 NE Braz.: ***H. viscida*** Nees. R: KB 58(2003)207

Holoschkuhria H. Robinson. Compositae (Bah.). 1 N Peru: ***H. tetramera*** H. Robinson. R: CN 38(2002)47

Holoschoenus Link = *Scirpoides*

Holostachyum (Copel.) Ching = *Aglaomorpha*

Holostemma R. Br. = *Cynanchum*

Holosteum L. Caryophyllaceae (II 1). 3–4 temp. Euras. (Eur. 1). Mouse-ear

Holostyla DC. = *Caelospermum*

Holostylis Duchartre = *Aristolochia*

Holostylon F. Robyns & Lebrun = *Plectranthus*

Holothrix Rich. ex Lindl. Orchidaceae (IV 4d). c. 45 trop. & S Afr., Arabia

Holozonia Greene (~ *Lagophylla*). Compositae (Mad.-Mad.). 1 W US: ***H. filipes*** (Hook. & Arn.) Greene. R: FNA 21(2006)294

Holstia Pax = *Tannodia*

Holstianthus Steyerm. Rubiaceae (I 5). 1 Guayana Highland: ***H. barbigularis*** Steyerm. R: AMBG 73(1986)495. 11 other R. genera restricted to G. H.

Holtonia Standl. (~*Elaeagia*). Rubiaceae (Cond.). 1 Colombia: ***H. myriantha*** (Standl.) Standl.

Holttumiella Copel. = *Taenitis*

Holttumochloa Wong. Gramineae (V 4). 3 Malay Penin. R: KB 48(1993)518

Holtzea Schindler = *Desmodium*

Holubia Löve & D. Löve = *Gentiana*

Holubia Oliv. Pedaliaceae (2). 1 S Afr.: ***H. saccata*** Oliv. R: HIP t. 1475 (1884)

Holubiella Koda = *Botrychium*

Holubogentia Löve & D. Löve = *Gentiana*

holy basil *Ocimum tenuiflorum;* **h. clover** *Onobrychis viciifolia;* **h. flax** *Santolina* spp.; **h. grass** *Anthoxanthum nitens;* **h. rose** *Rosa × richardii;* **h. thistle** *Silybum marianum;* **h. thorn** *Crataegus monogyna* 'Biflora'

Holzneria Speta = *Chaenorhinum* (but see BJ 103(1982)16)

Homalachne Kuntze = *Holcus*

Homalanthus A. Juss. (*Omalanthus*). Euphorbiaceae (Hipp.). 23 trop. As. (Mal. 13 – R: Blumea 42(1997)429) to Aus. ***H. nutans*** (Forst. f.) Guillemin (mamala, Polynesia) – used for hepatitis in Samoa, bark source of prostratin for HIV/AIDS treatment; ***H. populifolius*** Graham (bleeding heart, mamala, NG, trop. Aus.) – Koriba's Model, widely cult. & natur., source of black dye for staining rattan goods etc., bark used in synthesis of prostratin in AIDS & cancer research

Homalium Jacq. Salicaceae (Flacourtiaceae). c. 180 trop. & warm (Afr. (R: BJBB 43(1973)239) & Madag. 59, Mal. 23, Am. 3). K & /or C accrescent & forming wings in fr.; ***H. grandiflorum*** Benth. (W Mal.) – fls once every 10–15 yrs; ***H. guillainii*** (Vieill.) Briq. (New Caled.) – accumulates nickel to 14% leaf ash dry wt. & carries a moss which accumulates chromium to 5000 μg/g (20 times that in itself). Some timbers for building, boats etc., ***H. tomentosum*** (Vent.) Benth. (SE As., Mal.) – source of Moulmein lancewood

Homalocalyx F. Muell. Myrtaceae (II 15). Incl. *Wehlia*, 11 Aus. R: Brunonia 10(1987)139

Homalocarpus Hook. & Arn. (~ *Bowlesia*). Umbelliferae (Azor.). 6 Chile. R: UCPB 38(1965)58

Homalocheilos J.K. Morton = *Isodon*

Homalocladium (F. Muell.) L. Bailey = *Muehlenbeckia*

Homalodiscus Bunge ex Boiss. = *Ochradenus*

Homalomena Schott. Araceae (VII 6). Excl. *Adelonema*, c. 500 trop. As. Some rheophytes, most with char. scented tissues (cf. *Pelargonium*) esp. terpenoids, e.g. ***H. davidiana*** A. Hay (NG) – liquorice, ***H. lauterbachii*** Engl. (NG) – anise, ***H. melanesica*** A. Hay (Bismarck & Solomon Is.) – lemon balm, ***H. soniae*** A. Hay (NG) – parsley. Thermogenic heat with 2 peaks. Some cult. orn. (incl. varieg. cvs, e.g. *H. davidiana*) & locally used poisons & medic. etc. ***H. peekelii*** Engl. (Papuasia) – scented pl. worn around neck in New Ireland

Homalopetalum Rolfe. Orchidaceae (V 13b). 8 trop. Am. R: C.L. Withner, *The Cattleyas & their relatives* V(1998)151

Homalosciadium Domin. Umbelliferae (I ?). 1 SW Aus.: ***H. homalocarpum*** (F. Muell.) H. Eichler

Homalosorus Small ex Pichi-Serm. (~ *Diplaziopsis*). Diplaziopsidaceae. 1 E N Am.: ***H. pycnocarpos*** (Spreng.) Pichi-Sermolli

Homalospermum Schauer (~ *Leptospermum*). Myrtaceae (II 14). Myrtaceae. 1 SW Aus. coastal swamplands: ***H. firmum*** Schauer. R: Telopea 2(1983)381

Homeria Vent. = *Moraea*

hominy *Zea mays*

Homochroma DC. = *Zyrphelis*

Homocodon Hong (~ *Heterocodon*). Campanulaceae (I 8). 2 Bhutan, China. R: FOC 19(2011)551

Homocollecticon (Summerh.) Szlach. & Olsw. = *Cyrtorchis*

Homoglossum Salisb. = *Gladiolus*

Homognaphalium Kirpiczn. = *Gnaphalium* (but see OB 104(1991)149)

Homogyne Cass. (~ *Petasites*). Compositae (Sen.-Tuss.). 3 Eur. mts. ***H. alpina*** (L.) Cass. (C Eur.) – cult. orn., natur. in GB, lactones insect-antifeedant

Homolepis Chase. Gramineae (XXIII 1). 5 trop. Am.

Homollea Arènes (~ *Tarenna*). Rubiaceae (? II 2). 3 Madag.

Homolliella Arènes = *Paracephaelis*

Homonoia Lour. Euphorbiaceae (Acal.-Acal.-Lasio.). 2 SE As., Mal. R: Blumea 43(1998)137. Rheophytes. ***H. riparia*** Lour. – planted as erosion retardant in Sumatra, rope made from bark in S China, skin-medic. in Mal.

Homopholis C. Hubb. Gramineae (XXIV). Incl. *Walwhalleya*, 4 Aus. ***H. proluta*** (F. Muell.) R. Webster (*W. p., Panicum p.*) – drought-resistant forage, seeds ground to paste & baked by Aborigines

Homopogon Stapf = *Trachypogon*

Homoranthus Cunn. ex Schauer. Myrtaceae (II 15). Incl. *Rylstonea*, 31 E & S Aus., some v. localized. R: AusSB 24(2001)355. Ess. oil suggested for scent-making; some cult. orn. shrubs

Homozeugos Stapf. Gramineae (XXII 7). 6 trop. Afr. R: GOB 1(1973)11

Honckenya Ehrh. (*Honkenya*). Caryophyllaceae (II 1). 1 N temp. incl. Eur., circumpolar, S Patagonia: ***H. peploides*** (L.) Ehrh. (sea purslane) – sandy coasts, Bell's Model, eaten pickled by Alaskan Inuit, whose diet otherwise almost exclusively animal. R: FNA 5(2005)137

Honduras mahogany *Swietenia macrophylla*; **H. redwood** *Erythroxylum affine*

Hondurodendron Ulloa & al. Olacaceae (Aptandraceae). 1 NW Honduras: ***H. urceolatum*** Ulloa & al. – dioec. R: AMBG 97(2010)459

honesty *Lunaria annua*

honewort *Trinia glauca*

honey larval food prep. by bees from floral nectar etc. (see hydromel); **h.berry** *Melicoccus bijugatus*; **h. flower** *Melianthus major*, *Lambertia formosa*; **h. locust** *Gleditsia* spp.; **h. palm** *Jubaea chilensis*; **h.suckle** *Lonicera* spp.; **African h.s.** *Halleria lucida*; **Cape h.s.** *Tecoma capensis*; **coral h.s.** *L. sempervirens*; **fly h.s.** *L. xylosteum*; **French h.s.** *Sulla coronaria*; **Himalayan h.s.** *Leycesteria formosa*; **Jamaica h.s.** *Passiflora laurifolia*; **panagol h.** *Pogostemon benghalensis*; **swamp h.s.** *Rhododendron viscosum*; **trumpet h.s.** *L. sempervirens*; **h.wort** *Cerinthe major*

Honorius Gray = *Ornithogalum*

Honolulu rose *Clerodendrum chinense*

Hoodia Sweet ex Decne. Apocynaceae (Vb; Asclepiadaceae). Incl. *Trichocaulon*, 14 SW trop. & S Afr. R: F. Albers & U. Meve, *Ill. Handb. Succ. Pls*, Asclep. (2002)142. San trad. appetite suppressant (xhoba) dev. esp. from ***H. gordonii*** (Masson) Decne (S Afr.) as wt.-loss aid not comm. as now considered not efficacious; cult. orn. succ. with flat or cup-shaped, almost lobeless C

× **Hoodiopsis** Lückh. Apocynaceae (Vb). Natural hybrid between *Hoodia* & *Stapelia* spp.

Hooglandia McPherson & Lowry. Cunoniaceae (II). 1 New Caled.: ***H. ignambiensis*** McPherson & Lowry. R: AMBG 91(2004)261

hook grass *Carex* (*Uncinia*) spp. (NZ)

Hookerochloa Alexeev (~ *Poa*). Gramineae (XVI 15). 2 Aus., NZ

hoop pine *Araucaria cunninghamii*

hop *Humulus lupulus*; **h.-bush** *Dodonaea* spp.; **h. clover** *Medicago lupulina*; **h. hornbeam** *Ostrya* spp.; **h. tree** *Ptelea trifoliata*; **h. trefoil** *Trifolium campestre*

Hopea Roxb. (~ *Shorea)*. Dipterocarpaceae (Dipt.-Shor.). 104 Indomal. Roux's Model. Poll. by thrips; some spp. triploid & apomictic (emergent & understorey spp.). ***H. centipeda*** Ashton (Borneo) – rheophyte; ***H. ponga*** (Dennst.) Mabb. (*H. wightiana*, ilapongu, Ind.) – shoots oft. with echinate galls initiated by *Mangalorea hopeae* Takagi so that larvae are prot. by a spiny 'fr.' reminiscent of those in other fams (no spines in Dipterocarpaceae otherwise!), similar forming in other insect-attacked genera in this & allied fams. Resin of ***H. papuana*** Diels (NG) coll. megachilid bees & inhibits growth of pollen-assoc. fungi. Some form. imp. as sources of resin (damar mata kuching) used in linoleum & paints; some timbers (giam) esp. ***H. mengarawan*** Miq. (merawan, W Mal.) & ***H. odorata*** Roxb. (thingan, SE As.) etc., this sp. also a street-tree (e.g. Kuala Lumpur, Singapore, Vietnam)

Hopi tea *Thelesperma megapoticum*

Hopia Zuloaga & Morrone (~ *Panicum*). Gramineae (XXIII 3). 1 SW N Am.: ***H. obtusa*** (Kunth) Zuloaga & Morrone – forage. R: T 56(2007)150

Hopkinsia W. Fitzg. Restionaceae (Anarthriaceae, Hopkinsiaceae). 2 SW Aus.

Hopkinsiaceae Briggs & L. Johnson = Restionaceae

Hoplestigma Pierre. Boraginaceae (Hoplestigmataceae). 2 W C Afr.

Hoplestigmataceae Gilg = Boraginaceae

Hoplophyllum DC. Compositae (Cich.-Eremo.). 2 NW Cape, S Afr. R: Taxon 43(1994)36

Hoplophytum Beer = *Aechmea*

Hoppea Willd. Gentianaceae. 2 Myanmar, Ind., Sri Lanka, ***H. dichotoma*** Willd. introd. W Afr., Ethiopia, Philippines, Aus. R: Blumea 48(2003)29

hora *Dipterocarpus zeylanicus*

Horaninovia Fischer & C. Meyer. Amaranthaceae (Chenopodiaceae III 3). 6 SW & C As.

horchata (de chufa) *Cyperus esculentus*

Hordelymus (Jessen) Harz = *Leymus*

Hordeum Tourn. ex L. (~ *Triticum*). Gramineae (XV). 43 N temp. (20 diploid, rest polyploid; Eur. 8 + cultigens, China 10 – R: FOC 22(2006)306), S Afr. (*H. capense*). Barley (app. source of one genome in N Am. *Elymus*). Spikelets in 3s on axis forming dense spike, fls of C or lat. spikelets oft. aborted. ***H. vulgare*** L. 'subsp. *spontaneum* (K. Koch) Koern.' (*H. spontaneum*, W As.) with brittle rachis & husked grains first harvested c. 9000 BC, domesticated Fertile Crescent & E of Zagros Mts, giving rise to 2-rowed 'subsp. *distichon* (L.) Koern.' (*H. distichon*) with non-brittle rachis, which mutated to 'subsp. *vulgare*' (6-rowed or '4-rowed' b. with C spikelet sterile – 'bere' (= bear), 'big' (Scotland)) all spikelets of each triad fertile, a single recessive gene giving 6-rowed ears from 2-rowed; predominantly selfed (2n = 14), hardiest forms cult. to 70° N (Norway), somewhat tolerant of saline soils; cereal grown by anc. Egyptians (& allegedly recorded on tablet 3800 yrs old in Iraq) & Sumerians (by 1750 BC 40% of yield used in ale with a gallon being a dignitary's daily ration, poss. used as currency but later replaced by date wine as soil became unsuitable for b.; porridge-like beer a candidate for biblical 'manna') & found in Swiss lake-dwellings, now (123 M t by 2010) mostly used in malting when germinated in water, kiln-dried & used as substrate for yeasts in beer – (first successfully bottled c. 1736, £8347M spent on it (cf. £4051M on bread) in GB by 1985, 29M pints per day drunk in GB by 1988, i.e. 108 l per head per annum (10 times wine consumption), 132 l in Ger. 1997), known to increase 'good' cholesterol, boost immune defences & prevent blood-clotting, high in anti-oxidants, 'bitter' beer (end C19) being served after only a few days' cellarage, 'mild' with lower hop content & sweeter, 'porter' a hoppy ale made with roasted barley & a predecessor of 'stout' (London c. 1730), though oft. 'extended' with maize or rice in US, diff. flavours added esp. Belgium (e.g. kriek with sour cherries) & (malt) whisky-making (invented C6 by Irish monks with distilleries flourishing by C13; blended now mixed with other grain w.), Irish (*uisce beatha* = whisky) now triple-distilled, Scotch usu. twice (now 4M bottles a day made in Scotland (its biggest export, worth £4.2 bn by 2011, world's biggest producer, followed by USA and then Jap.)) with diff. cvs (form. '**Maris Otter**' in old Brit. ales now replaced by bigger-yielding but lesser-tasting ones; '**Golden Promise**' in whisky), malt-extract used in proprietary spreads (e.g. Marmite, Vegemite), malt vinegar, med. etc.; pearl b. used in soups & stews is grain worn to ± spherical shape; b. water a watery solution used as soft drink & medic., b. sugar made from that since C17; toasted b. flour (tsampa) with salted buttery tea an imp. dish in Tibet; alk. (gramine) led to synthesis of lignocaine, univ. used anaesthetic. Some cult. orn. – ***H. jubatum*** L. (N temp.), some weeds esp. ***H. marinum*** Huds. (Euras.) e.g. in Aus., ***H. murinum*** L. (Euras.) in W US (subsp. ***leporinum*** (Link) Arc. (*H. leporinum*, Medit.) – seeds 200 yrs old from adobe in SW N Am. found to be viable), ***H. secalinum*** Schreb. (meadow b., Eur., N Afr.) – tetraploid (poss. *H. marinum* × *H. brevisubulatum* (Trin.) Link (W As.)), as is ***H. capense*** Thunb. (S Afr.)

horehound *Marrubium vulgare;* **black h.** *Ballota nigra;* **water h.** *Lycopus europaeus*

Horichia Jenny. Orchidaceae (V 12i). 1 Panamá: ***H. dressleri*** Jenny. R: Gen. Orch. 5(2009)416

horizontal scrub *Anodopetalum biglandulosum*

Horkelia Cham. & Schldl. = *Potentilla* (but see FNA 9(2014)246)

Horkeliella (Rydb.) Rydb. = *Potentilla* (but see FNA 9(2014)270)

Hormathophylla Cullen & T. Dudley (~ *Alyssum*). Cruciferae (2). 10 W Medit. ***H. spinosa*** (L.) P. Küpfer (*A. s., Ptilotrichum s.*, S France, E & S Spain) – spiny shrublet, cult. orn.

Hormidium (Lindl.) Heynh. = *Prosthechea*

Horminum L. Labiatae (VII 2b). 1 S Eur. mts. ***H. pyrenaicum*** L. – cult. orn. with showy purplish-blue fls. R: Fl. Iber. 12(2010)453

Hormocalyx Gleason = *Miconia*

Hormuzakia Guasul. (~ *Anchusa*). Boraginaceae (B.2.1.1). 2 SE Med.

hornbeam *Carpinus betulus;* **American h.** *C. caroliniana;* **hop h.** *Ostrya* spp.

horn-nut *Trapa natans;* **h.-of-plenty** *Valerianella cornucopiae;* **h.wort** *Ceratophyllum* spp.; **Jap. h.w.** *Cryptotaenia canadensis*

Hornea Bak. Sapindaceae. 1 Mauritius: ***H. mauritiana*** Bak.

horned melon *Cucumis metalliferus;* **h. poppy** *Glaucium flavum;* **h. pondweed** *Zannichellia palustris;* **h. rampion** *Phyteuma* spp.

Hornschuchia Nees. Annonaceae (III 1). 10 E Braz. R: Britt. 47(1995)296

Hornstedtia Retz. Zingiberaceae (II 4). c. 50 Indomal. (Borneo 8) to Aus. ***H. rumphii*** Valeton (Mal.) – fr. ed.; ***H. scyphifera*** (König) Steud. (Mal.) – runners 'walk' on 1 m high adventitious r. in Malay Pen.

Hornungia Reichb. Cruciferae (23). 3 W Eur. to Med., ***H. procumbens*** (L.) Hayek cosmop. weed. R: Novon 7(1997)339

horny goat weed *Epimedium* spp., esp. *E. grandiflorum*

Horovitzia Badillo (~ *Carica*). Caricaceae. 1 Oaxaca, Mex.: ***H. cnidoscoloides*** (Lorence & Torres) Badillo – covered in stinging hairs

Horridocactus Backeb. = *Eriosyce*

horse brush *Tetradymia* spp.; **h. chestnut** *Aesculus* spp. esp. *A. hippocastanum;* **h. gram** *Macrotyloma uniflorum;* **h. hair, vegetable** *Chamaerops humilis;* **h. mint** *Mentha longifolia;*

h. radish *Armoracia rusticana;* **h.r. tree** *Moringa oleifera;* **h. shoe vetch** *Hippocrepis comosa;* **h. tail** *Equisetum* spp.; **h. weed** *Erigeron canadensis*

Horsfieldia Willd. Myristicaceae. 103 Indomal. to Aus. R: GBS 37(1985)115, 38(1985)55,185, 39(1986)1. ***H. iryaghedhi*** (Gaertn.) Warb. (Sri Lanka to Mal.) – fine timber, fls v. fragrant & suggested for scent-making, oilseed used for candle-making

Horsfordia A. Gray. Malvaceae (Malv.-Malv.). 4 SW N Am.

Horstrissea Greuter & al. Umbelliferae (III 8). 1 Crete: ***H. dolinicola*** Greuter & al. R: Willdenowia 19(1990)389

Horta Vell. Conc. = *Clavija*

hortensia *Hydrangea macrophylla*

Hortia Vand. Rutaceae (I 80. 10 trop. S Am. esp. Braz. R: SB 37(2012)199). Alks; local medic. ***H. excelsa*** Ducke (Braz.) – timber (pau amarelo); ***H. regia*** Sandw. (Guyana) – ed. fr.

Hortonia Wight Monimiaceae (I). 3 Sri Lanka

Horvatia Garay. Orchidaceae (V 12g). 1 Peru: ***H. andicola*** Garay. R: Gen. Orch. 5(2009)152

Horwoodia Turrill. Cruciferae. 1 Arabia to Iraq: ***H. dicksoniae*** Turrill. R: JB 77(1939)117

Hosackia Benth. ex Lindl. (~ *Lotus*). Leguminosae (III 22).11 SW Canada (N Am. 9) to Guatemala

Hosea Ridl. Labiatae (III). 1 Sarawak, Brunei: ***H. lobbii*** (C.B. Clarke) Ridl.

Hoseanthus Merr. = *Hosea*

Hoshiapuria Hajra, Daniel & Philcox = *Rotala*

Hosiea Hemsl. & E. Wilson. Icacinaceae. 2 W & C China, Jap.

Hoslundia Vahl. Labiatae (VII 3d). 1 trop. Afr.: ***H. opposita*** Vahl (trop. Afr.) – source of vanilla-scented oil, fr. ed. R: FTEA Lam. (2009)191

Hosta Tratt. Asparagaceae (Agavaceae, Hostaceae). c. 40 China, Korea & Jap. (15). R: Plantsman 3(1981)20; W.G. Schmid (1991) *The Genus H; Hosta J.* Lvs of some cooked & ed. Korea, Jap. Cult. orn. perenn. herbs (1000+ cvs) esp. ***H. plantaginea*** (Lam.) Asch. (China) – fragrant white fls, incl. hybrids & varieg. pls, with short rhizomes (plantain lilies), long cult. Jap. & many cvs not assignable to wild spp., many not prod. viable seeds, ***H. ventricosa*** Stearn (China) – natural tetraploid with pseudogamous apomixis

Hostaceae Mathew = Asparagaceae

hot water plant *Achimenes grandiflora*

Hottarum Bogner & Nicolson (~ *Piptospatha*). Araceae (VII 8). 5 W Mal.

Hottea Urb. Myrtaceae (Myrt.). 5 Hispaniola

Hottentot bread *Dioscorea elephantipes;* **H. fig** *Carpobrotus edulis;* **H. tea** *Helichrysum serpyllifolium*

Hottonia Boerh. ex L. Primulaceae. 2 W N Am. (1), Eur. & W As. (1). R: AnnB 25(1911)253. Floating aquatics with finely dissected submerged lvs; heterostylous aerial fls. ***H. palustris*** L. (water violet, Euras.) – cult. in aquaria

Houlletia Brongn. Orchidaceae (V 12i). 7 trop. Am. Cult. orn. epiphytes

Houmiri, Houmiriaceae See *Humiria*, Humiriaceae

houmus *Cicer arietinum*

hound's-tongue *Cynoglossum officinale*

house leek *Sempervivum* spp. esp. *S. tectorum;* **h. lime** *Sparrmannia africana*

houses, Chinese *Collinsia bicolor*

Houssayanthus Hunz. Sapindaceae (IV 1). 5 Mex. (2), Venez. (1), trop. S Am. (2). R: Candollea 42(1987)805

Houstonia Gronov. (~ *Hedyotis*). Rubiaceae (IV 15). 20 N Am. R: SBM 48(1996)19. ***H. caerulea*** L. (*Hedyotis c.*, bluets, innocence, C & E N Am.) – cult. orn.

Houttuynia Thunb. Saururaceae. 1 Jap. S to mts of Nepal & Java: ***H. cordata*** Thunb. (dokudami, fish plant) – parthenogenetic & almost completely male-sterile as microspores degenerate, cult. orn., shoots (crushed lvs with fishy smell) eaten as veg. with fish in China & Vietnam. R: FOW 11(2005)7

Hovanella A. Weber & B.L. Burtt = *Streptocarpus* (but see BBP 70(1998)333)

Hovea R. Br. Leguminosae (III 4). 38 Aus. (all States). R: Aus SB 14(2001)34. Alks. Some cult. orn. (purple pea)

Hovenia Thunb. Rhamnaceae (1). 7 E As. Decid. trees incl. ***H. dulcis*** Thunb. (Chinese or Jap. raisin (tree), China, Korea, Jap.) – pedicels fleshy & sweet, turning red after frost, bird-disp., invasive S Braz.: coll. from wild & also used in med. esp. for hangovers, timber valuable

Hoverdenia Nees. Acanthaceae (III 2c). 1 Mex.: ***H. speciosa*** Nees

Howardia Klotzsch = *Aristolochia*

Howea Becc. (*Howeia*). Palmae (V 14g). 2 Lord Howe Is. (Aus.). Cult. orn. monoec. (?) wind-poll. palms esp. ***H. forsteriana*** (F. Muell.) Becc. (kentia) – to 18 m but commonly grown as a sessile housepl., app. derived from ***H. belmoreana*** (C. Moore & F. Muell.) Becc. but now isolated in flowering 7 weeks earlier & occupying sedimentary outcrops (an example of allegedly 'sympatric' speciation)

Howeia Becc. = *praec.*

Howellanthus (Constance) Walden & R. Patt. Boraginaceae (Hydro.-Rom.). 1 N Calif.: ***H. dalesianus*** (J. Howell) Walden & R. Patt. – on ultramafics. R: Madroño 57(2010)270

Howellia A. Gray. Campanulaceae (III 2). 1 W N Am.: ***H. aquatilis*** A. Gray – aquatic s.t. with *Hippuris* 'syndrome'. R: Cons. Biol. 2(1988)275

Howelliella Rothm. Plantaginaceae (Ant.-Ant.; Scrophulariaceae s.l.). 1 E Calif.: ***H. ovata*** (Eastw.) Rothm.

Howethoa Rauschert = *Lepisanthes*

Howittia F. Muell. Malvaceae (Malv.-?Malv.). 1 SE Aus.: ***H. trilocularis*** F. Muell. 2 collateral ovules per locule

Hoya R. Br. Apocynaceae (Va; Asclepiadaceae III 4). c. 250 Indomal. to Pac. Many assoc. with ants, some with specialized lvs housing them, some growing on carton, some rooting in ant-inhabited cavities in trees, etc.; some with accrescent infl. axes; nectar coloured. Some local medic.; cult. orn. (as house-pls removing airborne pollutants) r.-climbers, twiners or sprawling shrubs (wax flowers, w. plant) esp. ***H. carnosa*** (L.f.) R. Br. (S China to Aus.) – used in Hawaiian leis & ***H. lanceolata*** Wall. ex D. Don subsp. ***bella*** (Hook.) Kent (*H. bella*, Ind. to Myanmar) & ***H. parasitica*** (Hornem.) Wight (Ind., SE As.)

hoya santa See hoja santa

Hoyella Ridl. = *Dischidia*

Hsenhsua X.H. Jin & al. (~ *Habenaria*). Orchidaceae (IV 4d). 1 Tibet: ***H. chrysea*** (W.W. Sm.) X.H. Jin & al.

Hua Pierre ex De Wild. Huaceae. 1 trop. Afr.: ***H. gabonii*** Pierre ex De Wild.

Hua Wang *Paeonia × suffruticosa*

Huaceae A. Chev. Magnoliidae – Oxalidales. 2/3 trop. Afr. Everg. shrubs & trees with garlic smell & usu. stellate or peltate hairs. Lvs in spirals (distich.), simple, entire, with basal glands on margin or abaxially; stip. present, caducous in *Afrostyrax*. Fls solit. or in cymose clusters, axillary, small, bisex., reg.; K 5, valvate (*Hua*), or (3–5) with irreg. lobes (*Afrostyrax*), C (4)5, induplicate-valvate, A (8)10, $\underline{G}$ (5), 1-loc. with term. style & 1 (*Hua*) or (4–)6 (*Afrostyrax*) basal, erect, anatropous, bitegmic ovules. Fr. indehiscent (*Afrostyrax*) or dehiscent, 5-valved (*Hua*); seed 1(2) with straight embryo & garlic-scented copious endosperm
Genera: *Afrostyrax*, *Hua*

Hualania Philippi (~ *Bredemeyera*). Polygalaceae. 1 C Arg.: ***H. colletioides*** Philippi

huanghuahaosu *Artemisia annua*

huanghuali *Dalbergia hainanensis*

Huanaca Cav. Umbelliferae (Azor.). 4 S Arg. & Chile. R: Kurtziana 6(1971)7. Poss. incl. *Stilbocarpa*

huang qi *Astragalus membranaceus*

huangteng *Fibraurea recisa*

huanita *Bourreria huanita*

Huarpea Cabrera. Compositae (Barn.). 1 S Andes (Arg.): *H. andina* Cabrera – perenn. herb

huasango *Loxopterygium huasango*

huauzontle *Chenopodium berlandieri*

Hubbardia Bor. Gramineae (Micr.). 2 W Ind. R: KB 67(2012)533. ***H. heptaneuron*** Bor of waterfalls – once thought extinct because of damming rivers, refound 2010

Hubbardochloa Auq. (~ *Muehlenbergia*). Gramineae (XXIX). 1 C Afr. mts: ***H. gracilis*** Auq.

Hubera Chaowasku = *Huberantha*

Huberantha Chaowasku (~ *Polyalthia*). Annonaceae. 27 trop. OW. R: KB 70,23(2015)1

Huberia DC. Melastomataceae. 16 Andes of Ecuador & Peru, SE Braz. R: RBB 27(2004)546

Huberodendron Ducke. Malvaceae (Bom.; Bombacaceae). 4 trop. Am.

Huberopappus Pruski. Compositae (Vern.-Pip.). 1 Venez.: ***H. maigualidae*** Pruski. R: Novon 2(1992)19

Hubertia Bory (~ *Senecio*). Compositae (Sen.-Sen.). 25 Madag., Masc.

huckleberry *Gaylussacia* spp., *Vaccinium* spp., *Solanum scabrum*

Hudsonia L. Cistaceae. 1 N Am., esp. dunes: ***H. ericoides*** L. (beach heath) – cult. orn.

Hueblia Speta = *Chaenorhinum*

Huernia R. Br. Apocynaceae (Vb; Asclepiadaceae III 5). c. 50 trop. & S Afr., S Arabia. R: F. Albers & U. Meve, *Ill. Handb. Succ. Pls*, Asclep. (2002)159. Cult. orn. dwarf succ.
Huerniopsis N.E. Br. = *Piaranthus*
Huertea Ruíz & Pavón. Tapisciaceae (Staphyleaceae s.l.). 4 WI, Andes
Huetia Boiss. = *Geocaryum*
Hughesia R. King & H. Robinson (~ *Eupatorium*). Compositae (Eup.-Crit.). 1 Peru: ***H. reginae*** R. King & H. Robinson. R: MSBMBG 22(1987)361
Hugonia L. Linaceae (I). 40 OW trop. (sect. *Durandea* with mericarps poss. to be excl.). Lower branches of infl. rep. by conspic. cli.-hooks
Hugoniaceae Arn. = Linaceae
Hugueninia Reichb. = *Descurainia*
Huidobria Gay (~ *Loasa*). Loasaceae (I1). 2 N Chile. R: Sendtnera 4(1997)82. Heterogeneous
Huilaea Wurd. = *Chalybea*
huixincao *Rhodobryum giganteum*
Hulemacanthus S. Moore. Acanthaceae (III 2d). 1–2 NG. R: Blumea 57(2013)215
Hullettia King ex Hook.f. Moraceae (II). 2 S Myanmar & Thailand, Malay Pen., Sumatra. R: JAA 41(1960)334
Hullsia Short (~ *Brachycome*). Compositae (Ast.-Bra.). 1 N Aus.: ***H. argillicola*** Short. R: JABG 28(2014)168
Hulsea Torrey & A. Gray. Compositae (Mad.-Hul.). 7 W N Am. R: Britt. 27(1975)228. Cult. orn. aromatic lvs
Hulteniella Tzvelev (~ *Arctanthemum*). Compositae (Anth.-Art.). 1 Arctic: ***H. integrifolia*** (Richardson) Tzvelev. R: FNA 19(2006)534
Hulthemia Dumort. = *Rosa*
Hultholia Gagnon & G.P. Lewis (~ *Caesalpinia*). Leguminosae (I 4). 1 SE As.: ***H. mimosoides*** (Lam.) Gagnon & G.P. Lewis. R: PhytoKeys 71(2016)58
hulver (arch.) *Ilex aquifolium*
Humata Cav. = *Davallia*
Humbertacalia C. Jeffrey. Compositae (Sen.-Sen.). 8 Madag., Masc. R: KB 47(1992)82
Humbertia Comm. ex Lam. Convolvulaceae (8; Humbertiaceae). 1 Madag.: ***H. madagascariensis*** Lam. – large tree, hard sandalwood-scented timber, secretory cells only in fl., ovules ∞; 'sister' to rest of C.
Humbertiaceae Pichon = Convolvulaceae
Humbertianthus Hochr. Malvaceae (Malv.-Hib.). 1 Madag.: ***H. cardiostegius*** Hochr. R: FMad. 129(1955)121
Humbertiella Hochr. Malvaceae (Malv.-Hib.). 6 SW Madag. R: BMNHN 4,12(1990)7
Humbertina Buchet = *Arophyton*
Humbertiodendron Leandri. Trigoniaceae. 1 E Madag.: ***H. saboureaui*** Leandri
Humbertioturraea J. Leroy. Meliaceae (II-Trich.). 5 Madag. R: KB 44(1989)369
Humbertochloa A. Camus & Stapf. Gramineae (IV 1). 2 trop. E Afr., Madag.
Humblotiella Tard. = *Lindsaea*
Humblotiodendron Engl. = *Vepris*
Humboldt willow *Salix humboldtiana*
Humboldtia Vahl. Leguminosae (I 2). 6 S Ind., Sri Lanka (1). Flowering shoots with extrafl. nectaries & hollow obconical internodes, in the top of each of which, opp. leaf, is a slit to an ant-inhabited cavity
Humboldtiella Harms = *Coursetia*
Humea Sm. = *Calomeria*
Humeocline A. Anderb. (~ *Calomeria*). Compositae (Gnap.-Gnap.). 1 Madag.: ***H. madagascariensis*** (Humbert) A. Anderb. R: OB 104(1991)139
Humiria Aubl. Humiriaceae. 4 trop. S Am. R: CUSNH 35(1961)87. Bat-disp. Some locally used med. & timbers, some of which (cf. *Aquilaria*) beautifully scented & used as incense once attacked by fungi; bark a source of 'umiry-balsam'; exocarp of some spp., e.g. ***H. balsamifera*** Aubl., ed.
Humiriaceae A. Juss. (Houmiriaceae). Magnoliidae – Malpighiales. 8/50 trop. S Am. to Costa Rica & W Afr. (1). R: CUSNH 35(1961)25. Trees & shrubs. Lvs everg., in spirals (s.t. distich.), simple, ± entire, drying black; stip. caducous, tiny or 0. Fls bisex., ± reg., in cymes, oft. corymbiform; K (5), lobes imbr., 2 outer oft. smaller than others, rarely 0, C 5, thick, convolute or imbr., usu. 3–5- veined, whitish (rarely red), A 10–30 (s.t. 5 groups of 3 opp. K & 5 opp. C; ∞ in bundles in *Vantanea*) with filaments forming a tube, s.t. some without anthers (staminodes), anthers with expanded prolonged connective, nectary-

disk intrastaminal, free or adnate to G, usu. cupulate to tubular, with lobes, or s.t. of 10–20 distinct scales, G ((4)5(–7)), pluriloc. with axile placentas & 1 style, s.t. 1-loc. apically (partitions not reaching summit), 1 or 2 (superposed) pend., anatropous, bitegmic ovules per loc. Fr. a drupe with usu. pluriloc. stone, s.t. with resinous secretory cavities & adapted to water-disp., the stone with as many valves as G, 1 or more being pushed off at germ.; seeds 1 or 2 with slightly curved embryo & copious oily endosperm. n = 12 Genera: *Duckesia, Endopleura, Humiria, Humiriastrum, Hylocarpa, Sacoglottis, Schistostemon, Vantanea*

Fr. disp. by water (empty cavities giving buoyancy) or bats. Some timber etc. (*Humiria*)

Humirianthera Huber = *Casimirella*

Humiriastrum (Urb.) Cuatrec. Humiriaceae. 16 C Am. to SE Braz. R: CUSNH 35(1961)122. Local medic.

Humularia Duvign. Leguminosae (III 11). c. 35 trop. Afr. Some r., smoked like tobacco, stimulatory

Humulus L. Cannabaceae (Celtidaceae). 3 N temp. Twining dioec. climbers, male fls in loose axillary panicles, females in short bracteate spikes, cone-like at maturity, each bract with 2 fls.; astragalin (flavonoid) inhibits histamine release (cf. *Arnica*). ***H. japonicus*** Sieb. & Zucc. (? ***H. lupulus*** (Lour.) Merr., Jap. hop, temp. E As.) – cult. orn.; ***H. lupulus*** L. (hop [from hoppan, AS to climb], N temp.) – fibre for coarse cloth in Sweden (with jute = hessian) & paper, Romans ate shoots like asparagus (France, Italy, Spain, Turkey today), hop–pillows sleep-inducing (CNS-depressant, cured insomnia of George III of England), phyto-oestrogens causing breast enhancement, but now (first in C9 Eur., England 1524) most used in brewing beer (in US 75% in Washington State alone, usu. seedless; in UK e.g. '**Fuggle**' usu. with seeds), due to (bacteriostatic, effective against caries so poss. use in mouthwashes) resinous & bitter substances (incl. alks like codeine & morphine) in female infls (10 000–15 000 lupulin glands per infl.), when grown trained up wires as in Kent, where the stems (bines) are cut down when the hops coll., prop. by cuttings of rhiz., introd. from Flanders early C15 as beer keeps better than ale (though strongly opposed at first), lager (orig. Bavaria) brewed at cool temps (lager yeasts active then)

Hunaniopanax C.J. Qi & T.R. Cao = *Aralia*

hun(g)choy *Ipomoea aquatica*

Hunga Pancher ex Prance. Chrysobalanceae (2). 11 NG (3), New Caled. & Loyalty Is. (8, 3 on serpentine). R: FOW 9(2003)186

hungry rice *Digitaria exilis*

Hunnemannia Sweet. Papaveraceae (II). 2 E Mex. R: Phytol. 73(1992)330. ***H. fumariifolia*** Sweet – alks, cult. orn. like *Eschscholzia* but sepals sep.

Hunteria Roxb. Apocynaceae (If). 12 trop. Afr., ***H. zeylanica*** (Retz.) Thw. (local medic.) ext. to W Mal. R: WAUP 96–1(1996)88. Alks

Huntleya Bateman ex Lindl. Orchidaceae (V 12j). 14 trop. Am. Cult. orn. epiphytes

huntsman's cup or **horn** *Sarracenia purpurea*

Hunzikeria D'Arcy. Solanaceae (Petun.). 3 SW US to Venez. R: AMBG 65(1978)705

Huodendron Rehder. Styracaceae. 4 S China, SE As.

Huon pine *Lagarostrobos franklinii*

Huperzia Bernh. (~ *Lycopodium*). Lycopodiaceae. Incl. *Phylloglossum*, c. 200 subcosmop. (Eur. 1: ***H. selago*** (L.) Schrank & Mart. (*L. selago*) – powerful (dangerous) emetic). R: OB 92(1987)163. Like *Lycopodium* but sporophylls not apical; many hybrids in N Am., reprod. by gemmae (? wind-disp.). ***H. drummondii*** (Kunze) Christenh. & H. Schneider (*P. d.*, Aus., NZ) – dies back to a tuber in summer; ***H. serrata*** (Thunb.) Trevis. (Indomal., C Am.) – Chinese herbal medic., huperzine A now in clinical trials for Alzheimer's Disease (traded in US as 'Cerebra') as it blocks an enzyme in brain

Hura L. Euphorbiaceae (Euph.-Hur.). 2trop. Am. R: Britt. 65(2013)322. Monoec. bat-poll. trees (Koriba's Model) with irritant latex. ***H. crepitans*** L. (huru) – widely cult. trop., fr. 5–20-loc. with explosive dehiscence, expelling seeds to 14 m, form. used wired together as sand-boxes before blotting-paper used, latex a fish-poison alleged to kill even anacondas, timber (assacu)

huru *Hura crepitans*

Husnotia Fourn. = *Ditassa*

Hutchinsia R. Br. = *Hornungia*

Hutchinsiella O. Schulz = *Hornungia*

Hutchinsonia Robyns (~ *Rytigynia*). Rubiaceae (III 2). 2 trop. W Afr.

Hutera Porta & Rigo = *Coincya*

Huthamnus Tsiang = *Jasminanthes*
Huthia Brand = *Cantua*
Huttonaea Harv. Orchidaceae (IV 4c). 5 S Afr. Elaiophores attractive to bees
Huttonella Kirk = *Carmichaelia*
Huxleya Ewart = *Clerodendrum*
Huynhia Greuter (~ *Arnebia*). Boraginaceae (B.2.2). 2 W As. R: Willd. 11(1981)37
hu-zhang *Reynoutria japonica*
hyacinth *Hyacinthus orientalis*; see also *Gladiolus italicus*; **h. bean** *Lablab purpureus*; **feather h.** *Muscari comosum* 'Plumosum'; **grape h.** *M.* spp.; **Roman h.** *Bellevalia romana*; **summer h.** *Ornithogalum candicans*; **tassel h.** *M. comosum*; **water h.** *Eichhornia crassipes*
Hyacinthaceae Batsch ex Borckh. = Asparagaceae
Hyacinthella Schur (~ *Hyacinthus*). Asparagaceae (Hyacinthaceae). 18 SE Eur. (3), SW As. R: Candollea 36(1981)513, 37(1982)157. Some cult. orn.
Hyacinthoides Heister ex Fabr. (*Endymion*). Asparagaceae (Hyacinthaceae). 11 W Eur., N Afr. R: T 59(2010)78. Bluebells. Splash-cup disp. mechanism. Cult. orn. ('Constancy' in 'Language of Fls') esp. ***H. hispanica*** (Mill.) Rothm. (Spanish bluebell, SW Eur.) & ***H. non-scripta*** (L.) Rothm. (*Scilla nutans*, (English) bluebell, W Eur.; '**Bracteata**' with long leafy bracts) – bulbs form. source of glue for book-binding & fixing arrow-flights (since Stone Age) & of starch for linen etc.; these form hybrid swarms, ***H. × massartianac*** Geerinck (*H. × variabilis*) in cult., spreading in UK
Hyacinthus Tourn. ex L. Asparagaceae (Hyacinthaceae). 3 W & C As. R: BN 127(1974)297. Hyacinth (cf. *Gladiolus italicus*) v. anc. name, pre-Gk. (cf. *Crocus*) & non-Indoeuropean, orig. a name of a god assoc. with spring. ***H. orientalis*** L. (hyacinth, NE Medit. limestone, natur. Eur.) – introd. to W Eur. via Turkey & Venice to Padua flowering 1562, by 1734–9 a hyacinthomania (cf. *Tulipa*), 351 cvs offered by one nurseryman in 1753, 2000 known in 19 cent. but by 1996 only 220 (60 in comm.); cvs of diff. colours & fasciated rachis, diploids 1560–1850, triploids 1700–1900, higher ploidies 1900–1950, 'double' fls selected end 17 cent. Haarlem with some cvs realizing £200 a bulb (cf. *Tulipa*); subject of 18 cent. Florists' societies ('sorrow' in 'Language of Fls'); may be prop. by gouging out base of (ed.) bulbs which stimulates growth of new bulblets therein; fls used in scent-making in S France (competing with synthetic phenyl acetaldehyde)
Hyaenanche Lamb. Picrodendraceae (Picr.-Hyaen.). 1 S Afr.: ***H. globosa*** (Gaertn.) Lamb. – lvs verticillate. R: A.R.-Sm., *Gen. Euphorb.* (2001)89
Hyalea Jaub. & Spach = *Centaurea*
Hyalis D. Don ex Hook. & Arn. (~ *Plazia*). Compositae (Mut.-Hyal.). 2 S Bolivia, Paraguay & Arg. Shrubs
Hyalisma Champ. = *Sciaphila*
Hyalocalyx Rolfe. Passifloraceae (Turneraceae). 1 trop. E Afr., Madag.: ***H. setifer*** Rolfe
Hyalochaete Dittrich & Rech. f. = *Jurinea*
Hyalochlamys A. Gray (~ *Angianthus*). Compositae (Gnap.-Ang.). 1 SW Aus.: ***H. globifera*** A. Gray. R: OB 104(1991)129
Hyalocystis Hallier f. Convolvulaceae (12). 2 trop. Afr.
Hyalolaena Bunge. Umbelliferae (III 8). 10 SW & C As.
Hyalopoa (Tzvelev) Tzvelev (~ *Colpodium*). Gramineae (XVI 14). 4 C As.
Hyalosepalum Troupin = *Tinospora*
Hyalosema (Schltr.) Rolfe = *Bulbophyllum*
Hyaloseris Griseb. Compositae (Stifft.). Excl. *Dinoseris*, 7 Bolivia, Arg. R: Kurtziana 7(1973)195. Shrubs
Hyalosperma Steetz. Compositae (Gnap.-Ang.). 9 Aus. R: OB 104(1991)124
Hyalotricha Copel. = *Campyloneurum*
Hyalotrichopteris Wagner = *Campyloneurum*
Hybanthopsis Paula-Souza (~*Hybanthus*). Violaceae (III 2a). 1 E Braz.: ***H. bahiensis*** Paula-Souza. R: Britt. 55(2003)210
Hybanthus Jacq. Violaceae (III 2a). Excl. *Afrohybanthus*, *Pombalia*, c. 41 trop., SW N Am. (3), E N Am. (1); see also *Cubelium*, *Pigea*. R: BJ 67(1936)437. Alks.
Hybochilus Schltr. = *Leochilus* (but see CUMH 16(1987)120)
Hybosema Harms = *Gliricidia*
Hybosperma Urb. = *Colubrina*
Hybridella Cass. Compositae (Helia.-Zal.). 1 Mex.: ***H. globosa*** Cass. R: Madroño 24(1977)30
Hydatella Diels = *Trithuria*

Hydatellaceae Hamann. Magnoliidae – Nymphaeales. 1/10 Ind., Aus., NZ. R: T 57(2008)179. Small tufted ann. (rarely perenn.) aquatics, glabrous save minute multicellular hairs in leaf axils, with vessels only in r. which are unbranched. Lvs in spirals, slender, subterete to flattened but centric internally. Infl. a term. head with 2–∞ bracts each with 1 or several fls in axis; fls water-poll., naked, unisexual of A 1 or pseudomonomerous G with 2 or 3 or 5–10 filamentous structures (? styles) & 1 pend., anatropous, bitegmic ovule. Fr. small dry, indehiscent or opening by 3 valves; seed 1 with starchy perisperm but almost 0 endosperm & minute scarcely differentiated embryo
Genus: *Trithuria*
Form. placed in Centrolepidaceae or allied to Commelinales. *Archaeofructus* (early Cretaceous of China) has some similarities

Hydnocarpus Gaertn. Achariaceae (Flacourtiaceae). 40 Indomal. (Borneo 13). R: BJ 69(1936)1. Roux's Model; main branches with only scale lvs; short-lived twigs with foliage lvs. Several spp. sources of oil (chaulmoogra o.) form. much used in treatment of eczema, leprosy & other skin conditions esp. ***H. castanea*** Hook.f. & Thomson (kalaw, Myanmar), ***H. kurzii*** (King) Warb. (SE As.), ***H. pentandra*** (Buch.-Ham.) Oken (*H. laurifolia, H. wightiana*, Ind.), ***H. venenata*** Gaertn (lucraban, Ind., Sri Lanka)

Hydnophytum Jack. Rubiaceae (IV 7). 52 Indomal. (esp. NG). Epiphytes, some with ant-inhabited 'tubers', like *Myrmecodia*, but c. 40% not inhabited by ants though other animals bring in nutrients (Jebb), havens for small frogs; fr. spread by birds on to bark; local medic.

Hydnora Thunb. Aristolochiaceae (Hydnoraceae). c. 7–8 Arabia, trop. & S Afr., Madag., Masc., arid & semi-a. Poll. dung beetles. ***H. abyssinica*** A. Br. ex Schweinf. (*'H. africana', H. johannis*, Arabia, Afr.) – infls break through even asphalt to flower, imp. fuel in Sudan, medic. & ***H. triceps*** Drège ex Meyer (S Afr.) – entirely subterr., restricted to *Euphorbia* spp., others on legume spp. Ed. fr. esp. ***H. abyssinica*** – subterr., sought by foxes, rhinoceros, warthog, monkeys & humans, ***H. esculenta*** Jum. & Perrier (Madag.)

Hydnoraceae C. Agardh = Aristolochiaceae

Hydrangea Gronov. Hydrangeaceae (II 2). Incl. *Broussaisia* (fr. berry-like), *Cardiandra* (rhiz. herbs, lvs in spirals, A ∞), *Decumaria, Deinanthe* (coarse herbs), *Dichroa, Pileostegia* (evergreen liane), *Platycrater, Schizophragma* (decid. lianes with clinging rootlets), c. 50 Himal. to Jap. & Philipp., Am. R: T 64(2015)749. Erect or cli., everg. or decid. shrubs with small fertile bisex. fls., oft. with showy sterile fls with C-like K surrounding them (in cult. forms, these s.t. comprise all the infl.). Dried r. source of hydrangin, a diaphoretic & diuretic alk. – some spp. medic. & steamed lvs of ***H. macrophylla*** (Thunb.) Ser. subsp. ***serrata*** (Thunb.) Makino (Jap., Korea) used in a drink (amacha) in Jap., though *H.* spp. can poison humans & stock. Many cult. orn. forms esp. derived from *H. macrophylla* subsp. ***macrophylla*** (hortensia) – typical pl. with all sterile fls but wild 'var. *normalis* E. Wilson' (Jap.) with both fertile & sterile, colour of fls dependent on capacity to absorb Al^{3+} which cause pigments to go from red to blue (this is made difficult on limy soils where 'blue' forms go pink though the addition of iron reverses this; during development, the stomata of K disintegrate as lobes green & lose char. colours). Others cult. incl. ***H. anomala*** D. Don, a cli. (aerial rootlets) sp. of which subsp. ***petiolaris*** (Sieb. & Zucc.) McClint. (Jap., Taiwan) is oft. cult. as wall pl., sterile fls inverting as inner fls mature into fr.; ***H. febrifuga*** (Lour.) Y. DeSmet & Samaia (*Dichroa f.*, SE As.) – trad. fever med., quinine subs. in WWII (tolyl derivative of beta-dichroine hypnotic used in methaqualone sleeping-tablets (Mandrax) used as recreational drug in 1960s), r. poss. source of anti-inflammatories; ***H. hydrangeoides*** (Sieb. & Zucc.) B. Schulz (*S. h.*, Korea, Jap.) – decid. root-cli; ***H. paniculata*** Sieb. (E As.) – wood for umbrella-handles; ***H. viburnoides*** (Hook.f. & Thomson) Y. DeSmet & Granados (*Pileostegia v.*, E As.) – everg. liane

Hydrangeaceae Dumort. Magnoliidae – Cornales. 9/185 N temp. to Mal. Trees, shrubs, lianes or rhiz. herbs (some *Hydrangea, Kirengeshoma*), oft. accum. aluminium. Lvs opp. (rarely whorled or in spirals), simple, joined by a line across stem; stip. 0. Fls usu. bisex., reg. or the marginal ones sterile, irreg. with enlarged K (v. rarely polygamo-dioec. in '*Broussaisia*'), in cymes (oft. corymbiform to paniculiform); K (4 or 5(–12)), lobes valvate or imbr., C 4 or 5(–12), valvate, imbr. to convolute, A 1–several × C (to 200 in *Carpenteria*), when ∞ arising centripetally, filaments s.t. basally connate, G ((2)3–5(–12)), half to fully inf. (sup. in *Carpenteria, Fendlera, Fendlerella, Jamesia, Whipplea*), pluriloc. or 1-loc. with intruded parietal placentas & distinct (s.t. basally connate) styles (rarely 1) & intrastaminal nectary-disk at apex of G, (1–)∞ anatropous, unitegmic ovules on each placenta. Fr. a capsule or berry; seeds with straight embryo in fleshy endosperm. n = 13–18+

Classification & principal genera: I. **Jamesioideae** – *Jamesia* & *Fendlera* only (W N Am.); II. **Hydrangeoideae** – *Deutzia, Philadelphus* (1. Philadelpheae, s.t. seg. as distinct fam.), *Hydrangea* (2. Hydrangeae)

Many cult. orn. shrubs, *Hydrangea* medic.

Hydranthelium Kunth = *Bacopa*

Hydrastidaceae Augier ex Martinov = Ranunculaceae

Hydrastis Ellis. Ranunculaceae (Hydr.; Hydrastidaceae). 1 C & E N Am.: ***H. canadensis*** L. (golden seal, yellow or turmeric r.) – rhiz. (berberine hydrochloride) used in prep. of a tonic & form. a yellow dye, the sp. much reduced through over-exploitation. R: Aliso 13(1993)551

Hydriastele H. Wendl. & Drude. Palmae (V 14). Incl. *Gronophyllum, Gulubia, Siphokentia,* 48 E Mal. to NE Aus. & Fiji. R: KB 59(2004)62. Slender, monoec., weevil-poll. palms incl. rheophytes

Hydrilla Rich. Hydrocharitaceae. 1 OW, introd. to SE US & C Am.: ***H. verticillata*** (L.f.) Royle – allied to *Elodea*. R: FA 39(2011)21. Submerged aquatic, male fls floating to surface as buds, opening & floating (or through air) to females (opening at surface after hypanthium lengthening); s.t. cult. in aquaria (invasive NZ & US, introd. Miami, US 1959 as 'star-vine' & 'oxygen plant' – tilapia & 2 snail spp. introd. to control it but ate other pls, & now tilapia a pest itself), pops in US monoec. & female

Hydrobryopsis Engl. = *Zeylanidium*

Hydrobryum Endl. Podostemaceae (III). Excl. *Zeylanidium*, 17 S Ind., E Nepal, Assam, China & S Jap., Thailand. R: APG 55(2004)133

Hydrocera Blume ex Wight & Arn. (*Tytonia*). Balsaminaceae. 1 Indomal.: ***H. triflora*** (L.) Wight & Arn. – semi-aquatic, poss. with water-disp. seeds; K 5, C 5, fr. a 5-seeded berry-like drupe (cf. *Impatiens*), split by swelling mucilage into 5 valves; fls used to dye fingernails in Ind.

Hydrocharis L. Hydrocharitaceae. 3 OW (Eur. 1). R: AB 14(1982)177. Floating aquatics oft. (erroneously) reported as rootless. ***H. morsus-ranae*** L. (frog-bit, Eur., E As., natur. N Am.) – cult. orn. dioec. pl. with horiz. runners which form new pl. at tips, overwintering as buds from stolons

Hydrocharitaceae Juss. Magnoliidae – Alismatales. Incl. Najadaceae, 17/150 cosmop. Aquatics (marine or freshwater, usu. perenn.), submerged or partly emergent, s.t. free-floating; vessels only in r. Lvs in spirals, opp. or whorled, oft. ± sheathing, subtending axillary scales & s.t. with distinct lamina, the margins oft. with thick-walled prickle-hairs (pl. usu. otherwise glabrous). Fls reg. (slightly irreg. in *Vallisneria*), usu. unisexual (monoec., dioec. or trioec.; bisex. fls oft. cleistogamous), solit. or in few-flowered cymes, infls subtended by (1)2 distinct ± connate bracts forming a spathe; K 3 (rarely 0), C 3 (rarely 0) attached to G or hypanthium, or C (P) 0, A 1- 3 to ∞ in 1 – ∞ 3- merous whorls, s.t. paired opp. K, when ∞ developing centripetally, some s.t. nectariferous staminodes, anthers with longit. slits, pollen grains globose, united in thread-like chains in marine genera (*Halophila, Thalassia*), male fls oft. released from submerged infls & floating on surface (*Elodea, Enhalus, Hydrilla, Lagarosiphon, Vallisneria*), though pollen alone does this in some *Elodea* spp., $\underline{G}$ 1 or $\overline{G}$ ((2)3–6(–20)), 1-loc. oft. with ± deeply intruded partial partitions, styles shortly basally connate, ovule(s) anatr. (orthotropous), bitegmic. Fr. a septicidal capsule, usu. opening irreg., submerged, or indehiscent with thin pericarp (*Najas*); seeds (1) several – ∞, endosperm 0 (scanty in *Ottelia*), embryo straight. n = 6–12

Classification & principal genera: **Hydrocharitoideae** (Lvs aerial, margins entire: *Hydrocharis, Limnobium* only); **Stratioideae** (lvs aerial, distichous, spiny: *Stratiotes* only); **Anacharidioideae** (lvs submerged, freshwater: *Blyxa, Elodea, Ottelia*); **Hydrilloideae** (Najadaceae, Halophiloideae, Hydrilloideae, Thalassioideae, Vallisnerioideae; lvs submerged, freshwater & marine, water-poll. common: *Enhalus & Halophila* (marine), *Najas, Thalassia* (marine))

Many monospecific genera. Of freshwater genera, *Najas* (cosmop.), *Ottelia* & *Vallisneria* pantrop., most of others OW. Female fls pushed to surface by hypanthium in *Elodea, Hydrilla* & *Lagarosiphon*, by peduncle in *Enhalus* & *Vallisneria*; male fls have just K in *Elodea* & *Vallisneria*, K + C in others. All germ. underwater

Najas & Ottelia ed.; green manures (*Najas, Stratiotes, Thlassia*); some cult. orn. esp. *Vallisneria*; some introd. invasive (see *Egeria, Elodea, Hydrilla, Lagarosiphon, Najas*)

Hydrochloa P. Beauv. = *Luziola*

Hydrochorea Barneby & Grimes (~ *Pithecellobium*). Leguminosae (II 4). 3 S Am. R: MNYBG 74(1996)23

Hydrocleys Rich. Alismataceae (Limnocharitaceae). 5 S Am. R: Phytol. 57(1985)421. Aquatics with milky latex, ***H. nymphoides*** (Willd.) Buchenau (*Limnocharis humboldtii*, water poppy) – cult. orn. with large yellow fls

Hydrocotylaceae Bercht. & J. Presl = Araliaceae

Hydrocotyle Tourn. ex L. Araliaceae (Hydr.; Umbelliferae s.l.). c. 200 cosmop. (Eur. 1, Aus. 55 mostly endemic). Creeping perenn. herbs oft. with peltate lvs; some cult. orn. groundcover (navelwort, pennywort) esp. ***H. sibthorpioides*** Lam. (As., but widely natur.), ***H. bonariensis*** Lam. (S Am.) – lvs to 120 mm across, aggressive lawn-weed in Aus., ***H. ranunculoides*** L.f. (floating pennywort, N Am.) – pest in UK since 1990, shoots growing up to 20 cm a day!, sale banned UK in 2014, invasive W Aus.; ***H. javanica*** Thunb. (Mal.) – lvs a fish-poison in Indonesia

Hydrodea N.E. Br. = *Mesembryanthemum*

Hydrodiscus Koi & Kato (~ *Diplobryum*). Podstemaceae. 1 N Laos: ***H. koyamae*** (Kato & Fukuoka) Koi & Kato R: AJB 97(2010)387

Hydrodyssodia B. Turner = *Hydropectis*

Hydrogaster Kuhlm. Malvaceae (Grew.-Grew.; Tiliaceae). 1 E Braz.: ***H. trinervis*** Kuhlm. – dioec., trunk stores water (released if injured)

Hydroidea Karis (~ *Atrichanthum*). Compositae (Gnap.). 1 SW Cape: ***H. elsiae*** (Hilliard) Karis. R: OB 104(1991)74

Hydrolea L. Hydroleaceae (Hydrophyllaceae s.l.). 11 trop. R: Rhodora 90(1988)169. Semi-aquatics; lvs in spirals, some with axillary thorns; autogamous. ***H. zeylanica*** (L.) Vahl (OW) – local medic., leafy shoots potherb in Indonesia

Hydroleaceae R. Br. (~ Hydrophyllaceae). Magnoliopsida – Solanales. 1/11 trop. Shrubs & herbs, oft. helophytes with aerenchyma. Lvs simple, in spirals; stip. 0. Fls (4)5-merous in cymes or racemes, rarely solit.; K basally connate, persistent, C imbr., basally connate, A adnate to C, usu. basally swollen, $\underline{G}$ (2(4)), diagonal with axile placentation & many ovules. Fr. a septicidal (or loculicidal) capsule; seeds longitudinally ridged & ruminate. n = (9)10(12)
Genus: *Hydrolea*

Hydrolythrum Hook.f. = *Rotala*

hydromel honey & water, when fermented = mead

Hydromystria G. Meyer = *Limnobium*

Hydropectis Rydb. Compositae (Tag.-Pect.). Incl. *Hydrodyssodia*, 3 Mex. R: Phytol. 78(1995)213. Aquatic (±), annuals

Hydrophilus Linder (~ *Leptocarpus*). Restionaceae. 1 SW Cape: ***H. rattrayi*** (Pillans) Linder. R: Bothalia 15(1985)484

Hydrophylax L.f. Rubiaceae (IV 15). 1–3 Afr., Madag., Ind., Thailand. ***H. maritima*** L.f. (As.) – dune-colonist, Bell's Model, fr. corky, indehiscent

Hydrophyllaceae R. Br. = Boraginaceae

Hydrophyllum Tourn. ex L. Boraginaceae (Hydrophyllaceae-Hydro.). 11 N Am. (W & E spp. separated). R: AMN 27(1942)710. Fls protandrous with the A scales united to C-tube forming nectar-tubes which bees probe. Some form. eaten by Native Americans as greens, esp. ***H. virginianum*** L. (E N Am.) – cult. orn.

Hydrorchis M.A. Clem. & D.L. Jones = *Microtis*

Hydrostachyaceae Engl. (~ Hydrangeaceae). Magnoliidae – Cornales. 1/22 S trop. & S Afr., (esp.) Madag. Submerged (ann. to) perenn. dioec. (monoec.) aquatics with tubers & no vessels. Lvs basal, entire to 3-pinnatifid, ligulate basally & oft. covered with small scaly or fringed appendages, a small membranous intrapetiolar stip. present; stomata 0. Fls sessile in bract axils of dense spike term. a C scape, small, unisexual; P 0, males with A 1 with short filament & extrorse anther, females with $\underline{G}$ (2), 1-loc. with 2 parietal placentas & 2 elongate persistent styles, s.t. basally connate, with ∞ anatropous, unitegmic ovules. Fr. a fissuricidal capsule with ∞ tiny seeds with 0 endosperm. n = 10–12
Genus: *Hydrostachys*

Hydrostachys Thouars. Hydrostachyaceae. 22 S trop. & S Afr., (esp.) Madag. R: Adansonia 13(1973)76. Flower when water level drops; bases attached to rocks; leaf morphology v. variable within spp.

Hydrostemma Wall. = *Barclaya*

Hydrothauma C. Hubb. Gramineae (Panicoid.). 1 Zambia: ***H. manicatum*** C. Hubb. – aquatic ann. with air-canals

Hydrothrix Hook.f. Pontederiaceae. 1 NE Braz.: ***H. gardneri*** Hook.f. – submerged ann. with cleistogamous fls (outer stamen fertile, inner 1 or 2 staminodal)

Hydrotriche Zucc. Plantaginaceae (Grat.-Grat.; Scrophulariaceae). 4 Madag. R: Adansonia 19(1979)145. ***H. hottoniiflora*** Zucc. – 'Hippuris' syndrome

hyedua *Daniellia thurifera*

Hyeronima Allemão = *Hieronyma*

Hygea Hanst. = ? Scrophulariaceae (s.l.)

Hygrochilus Pfitzer = *Phalaenopsis* (but see Phytotaxa 159,4(2014)256)

Hygrochloa Lazarides. Gramineae (XXIV 4). 2 N Aus. Aquatics

Hygrophila R. Br. Acanthaceae (III 2d). c. 25 trop., wet places. Some with sticky seeds disp. on birds' feet; some cult. orn. as submerged aquatics in aquaria, esp. ***H. difformis*** (L.f.) Blume (*Synnema triflorum*, water wisteria, Ind. to Thailand) – oft. a weed of ricefields & ***H. polysperma*** (Roxb.) T. Anders. (trop. As.) – now invasive SE US; ***H. costata*** Nees & T. Nees (yerba de hicotea, trop. S Am.) – invasive NE Aus.; ***H. obovata*** (Hornem.) Buch.-Ham. (*H. schulli*, Ind., Sri Lanka) – used like spinach, local medic.

Hygroryza Nees. Gramineae (III 3ii). 1 S & SE As. to China: ***H. aristata*** (Retz.) Nees – floating. R: FOC 22(2006)180

Hylaea J. Morales (~ *Prestonia*). Apocynaceae (II e). 2 NE S Am. R: Novon 9(1999)83

Hylaeanthe A. Jonker & Jonker. Marantaceae. 5–6 trop. S Am.

Hylaeorchis Carnevali & Romero (~ *Bifrenaria*). Orchidaceae (V 10). 1 S Venez. & Amaz. Braz.: ***H. petiolaris*** (Schltr.) Carnevali & Romero. R: HPB 6(2001)503

Hylandia Airy Shaw. Euphorbiaceae (Cod.-Bal.). 1 Queensland: ***H. dockrillii*** Airy Shaw (blushwood). R: KB 29(1974)329

Hylandra Löve = *Arabidopsis*

Hylebates Chippindale. Gramineae (XXIV). 2 trop. Afr.

Hylenaea Miers. Celastraceae (III). 3 trop. Am. ***H. comosa*** (Sw.) Miers (WI) – seeds ed.

Hyline Herb. = *Griffinia*

Hylocarpa Cuatrec. Humiriaceae. 1 Amazonian Braz.: ***H. heterocarpa*** (Ducke) Cuatrec. R: CUSNH 35(1961)84

Hylocereus (A. Berger) Britton & Rose (~ *Selenicereus*). Cactaceae (III 2). 14 trop. Am. Usu. cli. with aerial r., stems usu. 3-angled, fls nocturnal & white, rarely red. Fr. of some ed. (pitaya); some widely cult. esp. ***H. undatus*** (Haw.) Britton & Rose (dragon('s) fr., orig. unknown) – one of the 'night-blooming cereus' with fragrant white fls & ed. fr. (sold in markets) much cult. Vietnam & Malaysia, used on Chinese altars (red & long-lasting)

Hylocharis Miq. = *Oxyspora*

Hylodendron Taub. Leguminosae (I 2). 1 Gulf of Guinea: ***H. gabunense*** Taub. – trunk thorny

Hylodesmum Ohashi & R. Mill (*Papilionopsis*, *Podocarpium*, ~ *Desmodium*). Leguminosae (III 19). 14 Indomal. R: EJB 57(2000)173. Rim aril 0, cf. *D.*

Hylomecon Maxim. (~ *Chelidonium*). Papaveraceae (I). 1 temp. E As.: ***H. japonica*** (Thunb.) Prantl & Kündig – arils, alks, cult. orn. incl. var. ***dissecta*** (Franch. & Sav.) Fedde (cf. *Chelidonium*)

Hylomyza Danser = *Dufrenoya*

Hylophila Lindl. Orchidaceae (IV 2d). Incl. *Dicerostylis*, 7 Thailand to Solomons

Hylotelephium H. Ohba. See *Sedum*

Hymenachne P. Beauv. (~ *Panicum*). Gramineae (XXIII 2). 12 trop., swamps. R: AJB 90(2003)817. Some fodders. ***H. amplexicaulis*** (Rudge) Nees (Mex. to Arg.) – forms part of floating islands in Amazon, now weed of nat. significance in Aus. & invasive SE US

Hymenaea L. Leguminosae (I 2). Incl. *Trachylobium*, 16 trop. Am. & E Afr. (1). Troll's Model. Fr. ed. & sold in markets; copals form. much used for varnish etc. up to 3 kg, the best quality subfossil, esp. from ***H. courbaril*** L. (WI locust, Braz. cherry or copal, anamé gum, trop. Am.) – large fr. suggesting lost megafaunal disp. agents, heavy timber (courbaril, jatoba, locust wood) like mahogany; ***H. oblongifolia*** Huber var. ***palustris*** (Ducke) Lee & Lang (jutaí, Amaz.) – used in pottery-making (cf. *Licania octandra*); ***H. stigonocarpa*** Mart. ex Hayne (Braz.) = seeds 40% xyloglucan used to strengthen paper; ***H. verrucosa*** Gaertn. (*T. v.*, anime, E Afr. to Seychelles) – source of Zanzibar & Madag. copal

Hymenandra (A. DC.) Spach. Primulaceae (Myrsinaceae). 8 Indomal. (R: GBS 43(1991)1), 9 trop. Am. (R: Sida 18(1999)713)

Hymenanthera R. Br. = *Melicytus*

Hymenasplenium Hayata = *Asplenium*

Hymenatherum Cass. = *Thymophylla*

Hymeneria (Lindl.) M. Clements & D.L. Jones = *Pinalia*

Hymenidium Lindl. (~ *Pleurospermum*). Umbelliferae (III 5). 40 Pakistan to China. R: FR 111(2000)539, BZ 89(2004)1652

Hymenocallis Salisb. Amaryllidaceae. Excl. *Ismene*, 50 SE US (N Am. 15, esp. Florida) to NE S Am. R: KB 1954: 201, PL 18(1962)57. Filaments basally united by a consp. tube (corona) larger than P. Some with chlorophyllous seed integuments; alks – some used locally medic., pancratistatin an antineoplastic & antiviral isocarbostyril

Hymenocardia Wall. ex Lindl. Phyllanthaceae (Antid.-Hym.; Hymenocardiaceae). 6 trop. & S Afr., 1 SE As. to Sumatra. Some dyes

Hymenocardiaceae Airy Shaw = Phyllanthaceae

Hymenocarpos Savi (~ *Anthyllis*). Leguminosae (III 22). 4 Medit. (incl. Eur.), W As. incl. ***H. circinnatus*** (L.) Savi

Hymenocephalus Jaub. & Spach = *Psephellus*

Hymenochlaena Bremek. = *Strobilanthes*

Hymenoclea Torrey & A. Gray = *Ambrosia*

Hymenocnemis Hook.f. = *Gaertnera*

Hymenocoleus Robbrecht. Rubiaceae (IV 7). 12 trop. Afr. R: BJBB 45(1975)273, 47(1977)8

Hymenocrater Fischer & C. Meyer. Labiatae (VII 2c). c. 10 Iran, Afghanistan. R: BZ 77,12(1992)126. Some medic., cult. orn.

Hymenodictyon Wall. Rubiaceae (I 1). 24 OW (esp. Madag. – 13) E to Sulawesi. R: BJLS 152(2006)337. Alks; some epiphytes, fls +/– open simultaneously on any 1 pl. ***H. orixense*** (Roxb.) Mabb. (*H. excelsum*, trop. As.) – soft timber for tea-boxes, school slates etc., bark a febrifuge

Hymenoglossum C. Presl = *Hymenophyllum*

Hymenogyne Haw. Aizoaceae (V). 2 Cape. R: T 64(2015)520

Hymenolaena DC. (~ *Pleurospermum*). Umbelliferae (III 5). 3 C As. to Himal. R: FR 111(2000)532

Hymenolepis Cass. (~ *Athanasia*). Compositae (Anth.-Ath.). 7 S Afr. R: NJB 5(1986)517

Hymenolobium Benth. Leguminosae (III 11). c. 17 trop. S Am. V. tall trees; some timbers

Hymenolobus Nutt. = *Hornungia*

Hymenolophus Boerl. = *Urceola*

Hymenolyma Korovin = *Hyalolaena*

Hymenonema Cass. Compositae (Lact.-Cat.). 2 Greece

Hymenopappus L'Hérit. Compositae (Bah.). 11 S N Am. R: Rhodora 58(1956)163. R. of some spp. form. chewed as medic. by Native Americans

Hymenophyllaceae Mart. Polypodiidae – Hymenophyllales. 9/1000 trop. & temp. Filmy ferns. Usu. epiphytic with scale-less usu. slender creeping rhiz., vasc. tissue usu. in protostele, only 1 tracheid (in sect.) or 0 in some spp., some rootless. Lvs 5 mm – 60 cm long but lamina only 1 cell thick (exc. '*Cardiomanes*' = *Hymenophyllum* p.p.) with circinnate vernation & open venation & 0 stomata. Sporangia on receptacles continuous with vein-tips; indusia tubular or 2-lobed; spores green, globose, trilete. Gametophytes filamentous or ribbon-like, ofen reprod. by gemmae or mere fragmentation. n = 11, 12, 18, 28, 32–34, 36

Genera: *Abrodictyum, Callistopteris, Cephalomanes, Crepidomanes, Didymoglossum, Hymenophyllum* (incl. *Cardiomanes*), *Polyphlebium, Trichomanes, Vandenboschia*

Pls of ever-humid habitats esp. cloud forests of trop. mts, streamsides etc.

Hymenophyllopsidaceae Pichi-Serm. = Cyatheaceae

Hymenophyllopsis Goebel = *Cyathea*

Hymenophyllum Sm. Hymenophyllaceae (Hym.). Incl. *Amphipterum, Cardiomanes, Craspedophyllum, Hemicyatheon, Mecodium, Meringium, Myriodon*, c. 275 trop. & S temp. ('basal' group), N temp. only in Eur. (2) & Jap. Differs from *Trichomanes* (s.l.) in thread-like (not 2–4 mm across), glabrous (not hairy) rhizome & 2-lipped (not tubular) indusium. ***H. reniforme*** Hook. (*Cardiomanes reniforme, Mecodium r.*, NZ) – lamina some cells thick

Hymenophysa C. Meyer = *Lepidium*

Hymenopogon Wall. = *Dunnia*

Hymenopyramis Wall. ex Griff. Labiatae (Peronemat.). 6 Ind. to China & SE As.

Hymenorchis Schltr. Orchidaceae (V 16c). 12 Philipp. to New Caled. (1) esp. NG

Hymenosporum R. Br. ex F. Muell. Pittosporaceae. 1 E Aus., NG: ***H. flavum*** (Hook.) F. Muell. – cult. orn. tree (Fagerlind's Model) with fragrant fls ('native honeysuckle') & winged seeds, wood (wollum wollum) for small cabinetwork

Hymenostachys Bory = *Trichomanes*

Hymenostegia (Benth.) Harms. Leguminosae (I 2). Excl. *Annea, Gabonia*, 4 Gulf of Guinea

Hymenostemma (Kunze) Willk. Compositae (Anth.-Leucanthemops.). 1 Spain, Morocco: ***H. pseudanthemis*** (Kunze) Willk. R: BBNHB 23(1993)140

Hymenostephium Benth. (~ *Viguiera*). Compositae (Helia.-Helia.) c. 26 Mex. to Arg.

Hymenothrix A. Gray (~ *Bahia*). Compositae (Bah.). Incl. *Amauriopsis*, c. 10 SW N Am. R: Britt. 14(1962)101. ***H. dissecta*** (A. Gray) Baldwin (*B. d.*) – local medic.

Hymenoxys Cass. Compositae (Hele.-Tet.). Excl. *Tetraneuris*, incl. *Dugaldia*, *Plummera*, 25 W N Am. to Arg. Some cult. orn. but others toxic weeds of rangeland (sesquiterpene lactones killing many millions of sheep & goats per annum) esp. ***H. hoopesii*** (A. Gray) Bierner (*Helenium h.*, sneezeweed, W US) – causes 'spewing sickness' in stock, ***H. odorata*** DC. & ***H. richardsonii*** (Hook.) Cockerell (pingue) in Texas; *H. r.* & other spp. form. used as chewing-gum by Native Americans

Hyobanche L. Orobanchaceae (Buch.; Scrophulariaceae s.l.). 8 S Afr.

Hyophorbe Gaertn. Palmae (V 2). 5 Masc. Unarmed, monoec. R: GH 11(1978)212. Some cult. orn. incl. ***H. lagenicaulis*** (L. Bailey) H. Moore, reduced (1996) to 8 adult trees in wild (Round Is., Mauritius) but now common street-tree in e.g. Hainan; ***H. amaricaulis*** Mart. reduced (1988) to 1 tree in a botanic garden

Hyoscyamus Tourn. ex L. Solanaceae (IV 2). 17 W Eur. (5) & N Afr. to Somalia, SW & C As. to China. R: BBMNHB 27(1997)26. Highly toxic alks. ***H. muticus*** L. (E Medit.) – imp. in witches' brews; ***H. niger*** L. (henbane, temp. Euras., natur. N Am.) – seed dormancy up to at least 100 yrs, form. cult. (in Neolithic as source of recreational alkaloidal drug; used by Assyrians for toothache, for which lvs smoked in rural England until C20; assoc. with gods Jupiter & Thor) as hypnotic & hallucinatory narcotic from dried lvs, used to control pl.-pests in Eur. since C2 & to stun fowls to facilitate stealing ('henbane'), one of poisons administered by Dr Crippen (killed wife 1910), scopolamine now used for motion sickness; other spp. used locally

Hyoseris L. Compositae (Cich.-Cich.Hyos.). 2 Medit. incl. Eur.

Hyospathe Mart. Palmae (V 10). 6 trop. Am. R: NJB 9(1989)189, AJB 91(2004)953

Hypacanthium Juz. = *Arctium*

Hypagophytum A. Berger. Crassulaceae (III). 1 Ethiopia: ***H. abyssinicum*** (A. Rich.) A. Berger. R: OB 121(1993)47

Hyparrhenia Andersson ex Fourn. Gramineae (XXII 7). 58 Afr., Madag., (with few in) trop. Am. & As., 1 ext. to Medit. R: KB Add. ser. 2(1969)1. ***H. filipendula*** (Hochst.) Stapf (Indo-mal.) – source of paper pulp of moderate quality, imp. thatching grass C Afr.; ***H. hirta*** (L.) Stapf (OW) – grown in SW N Am. for erosion control; ***H. rufa*** (Nees) Stapf (trop. Afr.) – good fodder (Jaragua grass), invasive trop. S Am.

Hypecoaceae Willk. & Lange = Papaveraceae

Hypecoum Tourn. ex L. Papaveraceae (Fumariaceae II; Hypecoaceae). 20 Med. (Eur. 6) to C As. & N China. R: NJB 10(1990)129. Alks. Fls 2-merous; inner C 3-sect, middle lobe encl. A. In ***H. procumbens*** L. (Medit.) pollen shed into these lobes which close up before stigma is receptive; when pressed by insects the lobes open & dust the visitors with pollen

Hypelate P. Browne. Sapindaceae (III 2). 1 Florida, WI: ***H. trifoliata*** Sw. – white timber

Hypelichrysum Kirpiczn. = *Pseudognaphalium*

Hypenanthe (Blume) Blume = *Medinilla*

Hypenia (Benth.) Harley (~ *Hyptis*). Labiatae (VII 3c). 23 trop. & warm Am. R: BJLS 98(1988)91. Internodes with waxy scales so ants cannot climb to rob nectar ('greasy pole syndrome')

Hyperacanthus E. Meyer ex Bridson (~ *Gardenia*). Rubiaceae (II 1). 4 SE Afr., 1 Madag. R: KB 40(1985)275, 57(2002)959

Hyperaspis Briq. = *Ocimum*

Hyperbaena Miers ex Benth. Menispermaceae (I). 19 warm Am. R: Britt. 33(1981)81

Hypericaceae Juss. (~ Guttiferae). Magnoliidae – Malpighiales. 8/560 cosmop. Trees (e.g. trop. Afr. *Hypericum*), (usu.) shrubs or herbs. Lvs opp., margins entire with gland lines or dots; stip. 0. Fls 4- or 5-merous. C oft. contorted; A (5-) ∞, centrifugal, oft. in bundles; anthers with small glands; $\underline{G}$ (3–5) with 1 style or several basally united. Fr. a capsule, rarely a berry or drupe. Seeds (5-) ∞, oft. winged

Classification & genera:

I. **Vismieae** (stellate hairs, term. bud scales 0, berry): *Harungana*, *Vismia*

II. **Hypericeae** (glabrous or s.t. simple hairs, term. bud scales 0, septicidal caps.): *Hypericum*, *Lioanthus*, *Triadenum*, *Thornea*

III. **Cratoxyleae** (glabrous, term. bud scales, loculicidal caps.): *Cratoxylum*, *Eliea*

Medic. & cult. orn. esp. *Hypericum*; *Cratoxylum*, *Harungana* with useful wood

Hypericophyllum Steetz. Compositae (Bah.). 10 trop. Afr. R: KB 69(2014)9500:2

Hypericopsis Boiss. = *Frankenia*

Hypericum Tourn. ex L. Hypericaceae (II; Guttiferae I). Incl. *Santomasia*, 488 temp., trop. mts (Eur. 56 with 5 from N Am. natur.); subg. *H.* mainly OW, subg. *Brathys* mainly NW. R: BBMNHB 5(1977)293, 8(1981)55, 12(1985)163, 16(1987)1, 20(1990)1, 26(1996)75, 31(2001)37, 32(2002)61, Syst. Div. 4(2006)28, Phytotaxa 3(2010)5, 72(2012)5. Trees, shrubs or herbs with opp. gland-dotted lvs; some apomictic (evolved 3 times; apospory & pseudogamy); extrafl. nectaries; fr. usu. a capsule but berries in 4 distinct parts of genus. Some locally medic. (diuretic etc.) in OW (comm. as 'Hypercal Ointment' (with *Calendula officinalis*) & some cosmetics) & Am., temp. & trop., but toxic (quinones) esp. to sheep; many cult. orn. (St John's wort) esp. ***H. androsaemum*** L. (tutsan, Eur., W As., N Afr.) – fr. a berry, fire- & drought-tolerant diuretic shrub invasive in Aus. & NZ, ***H. calycinum*** L. (rose of Sharon, SE Bulgaria, N Turkey) – cult. for ground-cover in shade but invasive in Aus., ***H.* × *hidcoteense*** Hilling ex Geer. **'Hidcote'** (prob. *H. calycinum* × *H.* × *cyathiflorum* N. Robson 'Gold Cup' (*H. addingtonii* N. Robson (SW China) × *H. hookerianum* Wight & Arn. (Ind. to Nepal & Thailand)) prob. arising Hidcote, Glos., UK) – familiar sterile garden shrub hypericum now virus-infected, ***H.* × *inodorum*** Mill. (*H. elatum*, *H. hircinum* L. (Medit. to SW Saudi Arabia) × *H. androsaemum* but without goat odour of *H. hircinum*); ***H. perforatum*** L. (Eur. & Medit. to C China, natur. W N Am. (Klamath weed), where controlled biologically) – facultative apomict, poisonous to stock through photo-sensitization esp. in sunny countries e.g. Aus., SW US, Iraq, first noted to affect white but not black sheep by Cirillo (1787), form. medic. to staunch bleeding, in fractures etc. & used since Classical times as a 'feel good' herb, now cult. Aus. etc. for extracts (St John's wort; hypericin, an anthroquinone alleged to boost serotonin levels) used in mild depression in Eur. & US, incl. as a 'preloader' to reduce after-effects of Ecstasy tablets

Hypertelis E. Meyer ex Fenzl (~ *Mollugo*). Molluginaceae. Excl. *Kewa*, 5 S Afr., ext. to Eur., Aus., Am. R: T 65(2016)786

Hyperthelia W. Clayton. Gramineae (XXII 7). 6 trop. Afr., ***H. dissoluta*** (Steud.) W. Clayton (thatch in S Afr.) introd. S Am. R: KB 20(1967)438. Grains of ***H. edulis*** (C. Hubb.) W. Clayton coll. for food

Hyphaene Gaertn. Palmae (III 8a). 8 Afr., Madag., Arabia, Ind., Sri Lanka. Dioec., oft. forked by true dichotomy (Schoute's Model); bee-poll.; seeds disp. by elephants & baboons. All used by humans incl. for toddy, ***H. petersiana*** Klotzsch ex Mart. (S trop. Afr.) basis of imp. basket-weaving industry in Botswana; some cult. orn. incl. ***H. thebaica*** (L.) Mart. (doum or dom palm, gingerbread palm, Nile region) – fibrous part of fr. considered to taste like gingerbread, endocarp used for buttons (veg. ivory subs.), lvs used for mats, ropes, paper & fuel

Hypnum Hedw. Musci (Hypnaceae). 50–200 cosmop. Mosses with antibacterial, antifungal properties; used in sleep-inducing (hence generic name) pillows in Fiji. ***H. cupressiforme*** Hedw. (cosmop.) – salt-tolerant

Hypobathrum Blume. Rubiaceae (II 5). c. 35 Indomal. (Borneo 24). Some rheophytes. Local medic., sour fr. used in chutneys

Hypocalymma (Endl.) Endl. Myrtaceae (II 17). c. 15 SW Aus. R: NJB 22(2002)536. Some cult. orn. shrubs

Hypocalyptus Thunb. Leguminosae (III 12). 3 S & SW Cape. R: Strel. 9(2000)485

Hypochaeris Vaill. ex L. Compositae (Lact.-Hyp.). c. 60 Eur. (9), As., N Afr., (esp.) S Am. ***H. glabra*** L. (Eur., Medit.) – C florets have wind-disp. cypselas, marginal ones beakless cypselas disp. by animals; ***H. oligocephala*** (Svent. & Bramwell) Lack (*Heywoodiella o.*, NW Tenerife) – suffrutex. Some weeds (cat's-ear), like ***H. radicata*** L. (Eur., N Afr.) invasive in Aus. (flatweed), & cult. orn.

Hypochoeris L. = praec.

Hypocylix Woloszczak = *Salsola*

Hypocyrta Mart. = *Nematanthus*

Hypodaphnis Stapf. Lauraceae (I). 1 trop. W Afr.: ***H. zenkeri*** (Engl.) Stapf

Hypodematiaceae Ching (~ Dryopteridaceae). Polypodiidae. 3/6. R: T 55: 705
Genera: *Didymochlaena, Hypodematium, Leucostegia*

Hypodematium Kunze. Hypodematiaceae (Dryopteridaceae s.l.). 3 OW warm (Afr. 1) esp. limestone

Hypoderris R. Br. Tectariaceae (Dryopteridaceae s.l.). 3 trop. Am. R: SB 39(2014)389. ***H. brownii*** J. Sm. ex Hook. – cult. orn., s.t. with epiphyllous plantlets

Hypodiscus Nees. Restionaceae. 15 S & SW Cape, Namaqualand. R: Bothalia 15(1985)488

Hypoestes Sol. ex R. Br. Acanthaceae (III 2c). c. 40 OW trop. Some cult. orn. foliage pot-pls esp. ***H. phyllostachya*** Bak. ('*H. sanguinolenta*', polka-dot plant, Madag.). ***H. aristata*** (Vahl) Roemer & Schultes (*H. verticillatus*, trop. & warm Afr., Arabia) – source of 2 anti-neoplastic alks (hypoestestafins)

Hypogomphia Bunge. Labiatae (VI). 1–3 Iran to Kazakhstan

Hypogon Raf. = *Collinsonia*

Hypogynium Nees (~ *Andropogon*). Gramineae (XXII). 2 trop. Afr. & Am.

Hypolaena R. Br. Restionaceae. 8 SW Aus.

Hypolepis Bernh. Dennstaedtiaceae. c. 55 trop. & warm. Sori prot. by small reflexed margins of frond (no indusium). Some cult. orn., ***H. sparsiora*** (Schrad.) Kuhn (Afr., Madag.) – weedy

Hypolobus Fourn. Apocynaceae (Vc; Asclepiadaceae III 3). 1 Braz.: ***H. infractus*** Fourn.

Hypolytrum Rich. Cyperaceae (I 1). 60 trop. & warm (Am. 29). Corner's Model

Hypophyllanthus Regel = *Helicteres*

Hypopitys Crantz (~ *Monotropa*). Ericaceae (II 3). 1 N temp., C Am.: ***H. monotropa*** Crantz (*M. h.*)

Hypoxidaceae R. Br. Magnoliidae – Asparagales. 7/160 S hemisph. to trop. Afr. mts & N Am. Herbs (some pachycaul) with corms or rhiz. Lvs usu. tristichous, linr. to lanceolate, oft. plicate, hairy, bases marcescent, in rosettes; r. with vessels. Infls spikes to umbel-like clusters or solit. fl. Fls reg.; P 3 + 3 s.t. with short tube, usu. yellow or white (red in *Rhodohypoxis*); nectaries 0; A 3 + 3 (A 3 in *Pauridia*), anthers with longit. dehiscence; $\overline{G}$ (3), 3-loc. (1-loc. in *Empodium*), each loc. with several anatr. to hemianatr. ovules. Fr. a capsule with remains of P or fleshy; seeds small, globose, testa usu. black; embryo small; endosperm copious. n = 6–9, 11

Genera: *Curculigo, Empodium, Hypoxidia, Hypoxis, Molineria, Pauridia, Rhodohypoxis*

Some medic. (*Hypoxis*), cult. orn. esp. *Molineria, Rhodohypoxis*; some fibres (*Curculigo*)

Hypoxidia Friedmann (~ *Hypoxis*). Hypoxidaceae. 2 Seychelles. R: BMNHN 4,6(1984)453. Fls to 12 cm diam.

Hypoxis L. Hypoxidaceae. c. 90 warm (N Am. 7) & trop. (Afr. 69 – R: Bothalia 36(2006)14), esp. S hemisph., incl. many polyploid apomicts. Some S Afr. spp. used in prostate, lung & other cancers (hypoxiside, a diglucoside); some cult. orn. with white or yellow fls. ***H. aurea*** Lour. (Indomal.) – tonic & aphrodisiac in Chinese trad. medicine; ***H. marginata*** R. Br. (N Aus.) – tuber ed. N Queensland

Hypsela C. Presl = *Lobelia* (Aus. spp. = *Isotoma*)

Hypselandra Pax & K. Hoffm. = *Boscia*

Hypselodelphys (K. Schum.) Milne-Redh. (= *Trachyphrynium*). Marantaceae. 5 trop. Afr.

Hypseloderma Radlk. = *Camptolepis*

Hypseocharis Remy. Geraniaceae (Hypseocharitaceae). 6 Andes. Loculicidal capsule

Hypseocharitaceae Wedd. = Geraniaceae

Hypseochloa C. Hubb. Gramineae (XVI 5). 2 Mt Cameroun & Tanzania

Hypserpa Miers. Menispermaceae (V). c. 7 Indomal. to Polynesia. R: KB 52(1997)981. ***H. nitida*** Miers (Indomal.) – local medic., mountaineers' ropes in SE As.

Hypsophila F. Muell. Celastraceae (2). 2 NE Aus. R: FA 22(1984)168

Hyptianthera Wight & Arn. Rubiaceae (II 5). 2 N Ind., Thailand

Hyptidendron Harley (~ *Hyptis*). Labiatae (VII 3c). c. 19 NE Braz. to Bolivia. R: BJLS 98(1988)93. ***H. arboreum*** (Benth.) Harley (Guayana Higland, Colombia/Peru) – to 28 m tall

Hyptis Jacq. Labiatae (VII 3c). Excl. *Hypenia* & *Hyptodendron*, c. 280 warm & trop. Am. with few weedy spp. OW. R: RMLP 7(1949)153. Explosive poll. mech. with middle of C lobes holding A & style under tension until set off by insects (leading to cross-poll.) or wind (to self-poll.). Some locally used medic. (e.g. ***H. crenata*** Pohl ex Benth. (Bolivia, Braz.) – effective painkiller) & seeds for food (e.g. ***H. emoryi*** Torrey, SW N Am.) or oil (e.g. ***H. spicigera*** Lam. (black sesame or beni seed, Am. & Afr.), stems also provide a fibre); ***H. suaveolens*** (L.) Poit. (S Am.) – locally medic. & poss. anticancer agent, dye-source in Ind., black frogspawn-like jelly a sweetmeat prep. from seeds in Taiwan

Hyrtanandra Miq. = *Pouzolzia*

Hyssaria Kolak. = *Campanula*

hyssop *Hyssopus* spp. esp. *H. officinalis*, (of the Bible) *Origanum syriacum, Agastache* spp.; **anise h.** *A. foeniculum*; **hedge h.** *Gratiola officinalis;* **water h.** *Bacopa monnieri*

Hyssopus Tourn. ex L. Labiatae (VII 2b). 2 S Eur. (1), Medit. to C As. ***H. officinalis*** L. (hyssop, S Eur.) – medic., oil for liqueurs (incl. Chartreuse), antibacterial

Hysterionica Willd. = *Erigeron* (but see Phytol. 76(1994)173)

Hysterobaeckea (Niedenzu) Rye (~ *Baeckea*). 3 S Aus. R: Nuytsia 25(2015)213

Hystrichophora Mattf. Compositae (Vern.-Erl.). 1 Tanzania: ***H. macrophylla*** Mattf., known only from 1 specimen. R: KB 43(1988)249

Hystrix Moench = *Elymus* (but see NJB 17(1997)454)

I

i *Inocarpus fagifer*

Ianhedgea Al-Shehbaz & O'Kane = *Descurainia* (but see EJB 56(1999)323)

Ianthe Salisb. = *Pauridia*

Ianthopappus Roque & Hind. Compositae (Wund.-Hyal./Mut.-Hyal.). 1 S Am.: ***I. corymbosus*** (Less.) Roque & Hind. R: Novon 11(2001)97

Ibarraea Lundell = *Ardisia*

Ibatia Decne. = *Matelea*

Iberidella Boiss. = *Noccaea*

Iberis Dill. ex L. Cruciferae (32). c. 27 Eur. (19), Medit. Candytuft. Fls in umbelliform clusters which elongate with age, outer 2 C longer than inner. Usu. calcicoles. Some locally medic. but imp. cult. orn. esp. ***I. amara*** L. (W Eur.) – best scent, & ***I. umbellata*** L. (florist's c., S Eur.) – annuals & ***I. sempervirens*** L. (Medit.) – everg. rock-pl. & ***I violacea*** R. Br. (*I. pruitii*, Medit.) – perenn.

Ibervillea Greene. Cucurbitaceae (XIII). Incl. *Tumamoca*, c. 8 SW N Am. R: Madroño 41(1994)15. ***I. sonorae*** (S. Watson) Greene – swollen caudex to several kg

Ibetralia Bremek. = *Alibertia*

Ibicella (Stapf) Eselt. Martyniaceae (Pedaliaceae s.l.). 3 warm & trop. S Am. ***I. lutea*** (Lindl.) Eselt. (natur. Calif., Aus., S Afr.) – carnivorous with viscid lvs entangling gnats & flies

ibika *Hibiscus manihot*

iboga *Tabernanthe iboga*

Iboza N.E. Br. = *Tetradenia*

iburu *Digitaria iburua*

Icacina A. Juss. Icacinaceae. 6 trop. Afr. Mangenot's Model. ***I. oliviformis*** (Poir.) Raynal (*I. senegalensis*, W Afr.) – pyrophyte, tubers (to 50 kg) & seeds can provide a flour

Icacinaceae Miers. Magnoliidae – Asteridae. Excl. Metteniusaceae, Pennantiaceae, 33/200 trop., few temp. Trees & shrubs, s.t. scrambling, or lianes, s.t. accum. aluminium & prod. alks; stems oft. with unusual anatomy, s.t. with interxylary phloem. Lvs in spirals (opp. in *Iodes*), simple, entire to toothed, weakly revolute; stip. 0. Fls usu. reg., bisex. (rarely pl. polygamous to dioec.), (3–)5(–7)-merous in usu. axillary infls; pedicel articulated with fl., K a tube with imbr. (rarely valvate) lobes, C usu. valvate, free or ± connate, rarely 0, A same no. as C or K with filaments free or on C-tube alt. with lobes, anthers with longit. slits, disk s.t. present, $\underline{G}$ ((1–)3) with term. style, pseudomonomerous in *Phytocrene*, usu. only 1 loc. fertile with (1)2 anatropous, unitegmic ovules pend., back-to-back, from top of G, with funicular thickening nr. micropyle. Fr. a 1-seeded drupe; seed with usu. minute embryo & well-dev. oily endosperm or 0. x = 10, 11

Principal genera: *Alsodeiopsis, Iodes, Phytocrene, Pyrenacantha*. Many unispecific. *Citronella, Gonocaryum, Leptaulus, Pseudobotrys* referred to Cardiopteridaceae, *Gomphandra* etc. to Stemonuraceae (both Aquifoliales)

Some local foods (*Icacina; Miquelia* & *Phytocrene* lianes with potable water) & medic.

Icacinopsis Roberty = *Dichapetalum*

icaco, icacos *Chrysobalanus icaco*

Icacorea Aubl. = *Ardisia*

ice-plant *Mesembryanthemum crystallinum, Tetragonia tetragonoides*

icecream bean, i. plant *Inga edulis*

Iceland poppy *Papaver nudicaule*

Ichnanthus P. Beauv. Gramineae (XXIII 3). 36 trop. Am. with 1 pantrop. R: SB 7(1982)85

Ichnocarpus R. Br. Apocynaceae (II c). Excl. *Micrechites*, 3 Indomal. to trop. Aus. R: Blumea 39(1994)73. Some rubber, fibres & local medic.

Ichthyostoma Hedrén & Vollesen. Acanthaceae (III 2c). 1 SE Ethiopia & Somalia: ***I. thulinii*** Hedrén & Vollesen. R: NJB 16(1996)441

Ichthyothere Mart. Compositae (Mill.-Mill.). 18 trop. S Am. Some fish-poisons (active principle a polyacetylene – ichthyothereol)

Ichtyoselmis Lidén & Fukuhara (~ *Dicentra*). Papaveraceae. 1 China, Myanmar: ***I. macrantha*** (Oliv.) Lidén & Fukuhara
ichu *Jarava ichu*
Icianthus Greene = *Streptanthus*
Icma Philippi = ? (Compositae)
Icomum Hua = *Aeollanthus*
Icuria Wieringa (~ *Bikinia*). Leguminosae (I 4). 1 C Mozambique: ***I. dunensis*** Wieringa – ± monospecific dune forest. R: WAUP 00–3(1999)241
Ida A. Ryan & Oakeley = *Sudamerylycaste*
Idahoa Nelson & Macbr. Cruciferae. 1 W US: ***I. scapigera*** (Hook.) Nelson & Macbr. R: FNA 7(2010)566
Idanthisa Raf. = *Anisacanthus*
Idenburgia Gibbs = *Sphenostemon*
Idertia Farron (~ *Ouratea*). Ochnaceae. 1 trop. Afr.: ***I. axillaris*** (Oliv.) Farron. R: PEE 146(2013)354
Idesia Maxim. (*Cathayeia*). Salicaceae (Flacourtiaceae). 1 China, Jap.: ***I. polycarpa*** Maxim. – cult. orn. dioec. or polygamous tree
idewa *Haplormosia monophylla*
idigbo *Terminalia ivorensis*
Idiopappus H. Robinson & Panero (~ *Verbesina*). Compositae (Helia-Ecl.). 1 Ecuador: ***I. saloyensis*** (Domke) H. Robinson & Panero
Idiopteris T. Walker = *Pteris*
Idiospermaceae S.T. Blake = Calycanthaceae
Idiospermum S.T. Blake. Calycanthaceae (Idiospermaceae). 1 Queensland: ***I. australiense*** (Diels) S.T. Blake – Roux's Model; mature twigs with vessel elements with simple perforation plates, those of young twigs with scalariform; S populations have trees with up to G 5; seedlings arising from separated cotyledons grow better than those from intact seeds; embryos to 6.5 cm across & 100 g (largest known?) found in stomachs of poisoned cattle in Queensland & thus rediscovered (1971) but still a threatened rain-forest tree; pollen reward & fls brood-sites for beetles, thrips etc. R: FA 2(2007)105
Idiothamnus R. King & H. Robinson (~ *Eupatorium*). Compositae (Eup.-Crit.). 4 S Am. R: MSBMBG 22(1987)335
Idria Kellogg = *Fouquieria*
Ifafa lily *Cyrtanthus mackenii*
Ifloga Cass. Compositae (Gnap.-Gnap.). Incl. *Trichogyne*, 16 Medit. (Eur. 1), Macaronesia, S Afr. R: T 60 (2011)1073
Ighermia Wiklund (~ *Asteriscus*). Compositae (Inul.-Inul.). 1 Morocco: ***I. pinifolia*** (Maire & Wilczek) Wiklund
Igidia Speta. Asparagaceae (Hyacinthaceae). 1 Madag.: ***I. volubilis*** (Perrier) Speta – twining infl. to 2m (cf. *Bowiea*). R: EJB 60(2004)553
Ignatius bean *Strychnos ignatii*
Ignurbia R. Nordenstam (~ *Senecio*). Compositae (Sen.-Sen.). 2 Hispaniola. R: Willd. 36(2006)464
Iguanura Blume. Palmae (V 14). c. 32 W Mal. R: GBS 28(1976)191, 48(1998)1, KB 34(1979)143
igusa *Juncus effusus*
Ihlenfeldtia Hartmann (~ *Cheiridopsis*). Aizoaceae (V). 2 Cape. R: BJ 114(1992)29
Ihsanalshehbazia T. Ali & Thines (~ *Thlaspi*). Cruciferae (Colut.). 1 W Medit.: ***I. granatensis*** (Boiss. & Reut.) T. Ali & Thines. R: T 65(2016)93
Ikonnikovia Lincz. Plumbaginaceae (II). 1 C As., NW China: ***I. kaufmanniana*** (Regel) Lincz.
ilama *Annona diversifolia*
ilang-ilang = ylang-ylang
ilapongu *Hopea ponga*
ilb *Ziziphus spina-christi*
Ildefonsia Gardner = *Bacopa*
Ileostylus Tieghem (~ *Loranthus*). Loranthaceae. 1 NZ, Norfolk Is.: ***I. micranthus*** (Hook.f.) Tieghem
ilex *Quercus ilex*
Ilex L. Aquifoliaceae. Incl. *Nemopanthus*, 600+ cosmop. (China 204 (170 endemic); Eur. 3) esp. trop. & temp. As. & Am. (fossil pollen 80 M yrs old from SE Aus.). Hollies. R: F.C. Galle (1997) *Hollies, the genus I.*; H. Soc. Am., *The Holly handbook*. Dioec., oft. everg. trees

(Massart's & Rauh's Models) & shrubs (some epiphytes, e.g. ***I. baasiana*** Stone & Kiew (Borneo) – tubers) with spirals of (opp. in some Bornean spp.) lvs & drupes; some domatia, alks incl. caffeine & theobromine; embryo oft. scarcely dev. at disp., slowly maturing later. ***I. canariensis*** Poir. & ***I. perado*** Aiton (Macaronesia) – part of the laurel forest of Madeira. Timber, stimulants & cult. orn. (many hybrids & other cvs), most commonly cult. *I. aquifolium* & *I. × altaclerensis*, which differs in spines pointing towards leaf-apex. ***I. × altaclerensis*** (Loudon) Dallimore (*I. × altaclarensis*, hybrids ('altacs') between *I. aquifolium* & *I. perado*) incl. varieg. forms like '**Golden King**' (female) & '**Camelliifolia**' with usu. entire spineless lvs; ***I. anomala*** Hook. & Arn. (Hawaii, Tahiti) – wood form. for saddle-trees, canoe decoration & anvils for kapa beating; ***I. aquifolium*** L. ((Eur.) holly, Eur. & Medit., invasive Aus, W N Am.) – vern. name (arch. *hulver, holm(e)* from Anglo-Saxon *holegn* (i.e. holy tree when tradition of decorating houses with it in winter, prob. assoc. with Romans' Saturnalia, was taken over by Christians for Christmas) – thorns & red drupes added to attraction in Christian eyes, much cult., with many cvs incl. '**Ferox**' (hedgehog h., male) with rows of spines on adaxial surfaces of lvs & varieg. incl. **'Golden Queen'** & **'Silver Queen'** (both male), & form. the most used hedgepl. for formal gardening since Tudor times, some cvs with entire thornless lvs like adult forms of the type – app. usu. only manifest in maturity (cf. *Hedera* & poss. indicative of protection from megaherbivores (cf. *Gleditsia*)), timber white (colour preserved if timber cut in winter & dried before hot weather) used for veneers, inlay (often stained red, green, or blue), musical instruments incl. bagpipes, etc. (150 000 trees felled 1802 in Needlewood Forest, Staffs, England for bobbins for Lancashire cotton mills) or stained black as ebony subs. as in handles of silver teapots, burns when felled (little water), foliage also flammable, form. pollarded for winter fodder (v. high calorific value browse) in GB, sprigs used to bleed chillblains (also treated with lard & powdered holly fr.), bark also basis of a birdlime, fr. essence used in liqueurs (Ger.); ***I. cassine*** L. (dahoon, SE US, Mex., Cuba) – lvs used as a tea, cult. orn.; ***I. crenata*** Thunb. (Jap. h., Himal. to Jap., Philippines) – much cult. in Jap. incl. as bonsai; ***I. glabra*** (L.) A. Gray (gallberry, inkberry, C & E N Am.) – imp. bee-plant; ***I. guayusa*** Loes. (guayusa, E Colombia, Ecuador, W Peru) – lvs a stimulant tea (caffeine; highest concentration in any pl.); ***I. kaushue*** S.Y. Hu (*I. kudingcha*, kudingcha, China) & other spp. – tea subs., medic.; ***I. mucronata*** (L.) M. Powell & al. (*N. m.*, E N Am.) – cult. for orn. fr.; ***I. opaca*** Aiton (American h., C & E US) – uses similar to *I. aquifolium*, many cvs; ***I. paraguariensis*** A. St-Hil. ((yerba) mate, Braz. or Paraguay tea, Paraguay & adjacent parts of Argentina & Braz., Urug.) – locally imp. tea (900 K t per annum by 2002) made from lvs of cult. & wild pls; ***I. purpurea*** Hassk. (China, Jap.) – much planted China, fr. branches sold at Chinese New Year; ***I. verticillata*** (L.) A. Gray (feverbark, N Am.) – lvs a tea subs., cult. orn. with many cvs; ***I. vomitoria*** Aiton (yaupon, SE US., Mex., natur. Bermuda) – topiary, dried lvs a tea for Native Americans

Iliamna Greene (~ *Sphaeralcea*). Malvaceae (Malv.-Malv.). 7 W N Am. R: CDH 1(1936)217

iliau *Wilkesia gymnoxiphium*

ilima *Sida fallax*

Iljinia Korovin ex Komarov (~ *Haloxylon*). Amaranthaceae (Chenopodiaceae III 3). 1 C As.: ***I. regelii*** (Bunge) Komarov

Illawara palm *Archontophoenix cunninghamiana*; **I. pine** *Callitris rhomboidea*; **I. plum** *Podocarpus elatus*

Illecebraceae R. Br. = Caryophyllaceae

Illecebrum Ruppius ex L. Caryophyllaceae (I 2). 1 Canary Is., W Eur., Medit.: ***I. verticillatum*** L.

Illiciaceae Bercht. & J. Presl. = Schisandraceae

Illicium L. Schisandraceae (Illiciaceae). 34 Ind. to Korea & W Mal., SE N Am. to Hispaniola. R: BBR 21(2001)164, 322. Thermogenesis, poll. gall-midges (brood-sites), heating after fert. so benefiting larval dev. Fr. with explosive dehiscence. First insecticidal fumigants known (China, C2 BC); some cult. orn. (beetle-poll. – some thermogenesis) & comm. oils etc. ***I. anisatum*** L. (Jap. star anise, Korea & Jap.) – toxic seeds used to kill fish, locally medic., branches used to decorate Buddhist graves; ***I. floridanum*** Ellis (purple or tree a., SE US) – cult. orn.; ***I. verum*** Hook.f. (star a., Chinese a., SE China, NE Vietnam) – unripe fr. a culinary spice (main sweet constituent = *trans*-anethol) esp. in liqueurs when distilled, also medic. (oseltamvir, tamiflu) against avian influenza, though poss. promoting suicide in Jap. teenagers, also stopping blood supply to tumours & coronary artery growing into stents

Illigera Blume. Hernandiaceae. 18 OW trop. R: BJ 89(1969)157. Usu. lianes with palmately lobed lvs & sensitive petioles

illipe *Shorea macrophylla* & other spp.; *Madhuca* spp.

ilomba *Pycnanthus angolensis*

Iltisia S.F. Blake = *Microspermum*

Ilysanthes Raf. = *Lindernia*

Imantina Hook.f. = *Gynochthodes*

Imbralyx Geesink = *Fordia*

Imbricaria Comm. ex Juss. = *Mimusops*

imbu *Spondias* spp.

imbuia, imbuya *Cinnamomum porosum*

Imeria R. King & H. Robinson (~ *Eupatorium*). Compositae (Eup.-Crit.). 1 Venez., Braz.: ***I. memorabilis*** (Maguire & Wurd.) R. King & H. Robinson. R: MSBMBG 22(1987)357

Imerinaea Schltr. Orchidaceae (V 12b). 1 Madag.: ***I. madagascarica*** Schltr.

Imitaria N.E. Br. = *Gibbaeum*

immortelles = everlastings

imo *Colocasia esculenta*

Impatiens Riv. ex L. Balsaminaceae. c. 1000 trop. & N temp. esp. Ind. (Madag. 270, N & C Am. 6, Eur. 1). R (Afr. 109: C. Grey-Wilson (1980) *I. of Afr.*). Prob. orig. S China. Balsams, jewel-weed, touch-me-not (allusion to explosive 5-valved capsule with fleshy pericarp, outer layers of which are v. turgid &, when ripe, a touch can set off the valves, rolling up inwards, starting at base, & scattering seeds) – tender succ. herbs much grown as bedding (trade worh $100 M+ per annum in US by 2005) & housepls with resupinate fls, insect or bird-poll., & extrafl. nectaries, some Ind. spp. with 3-sporangiate anthers; chromosomes can be counted in pollen-grains on herbarium sheets app. at a resting stage at first mitotic interphase until released from pl.; some epiphytes. ***I. balsamina*** L. (garden balsam, C & S Ind., – cult., fls used to dye fingernails; ***I. capensis*** Meerb. (orange b., N Am.(!)) – local medic., cult., natur. Eur.; ***I. glandulifera*** Royle (policeman's helmet, Himal.) – fr. ed., natur. in N Am. & Eur. incl. GB where v. aggressive streamside herb (germ. early so swamping competitors, spreading 35 km a yr.) growing to 3m in 1 season & ousting native veg., rust-control now being used; ***I. hawkeri*** W. Bull (Papuasia) – orig. cult. NG for varieg. lvs (cf. *Codiaeum*, *Polyscias*), v. variable & now fashionable as pot-pl. (**New Guinea Group** ('NG Hybrids', F1 hybrids sterile) = *I. h.* × *I. linearifolia* Warb. (? Sulawesi & Java), poss. with other spp.), salt-source in NG; ***I. mirabilis*** Hook. f. (Thailand, Langkawi) – treelet with swollen storage-trunk; ***I. niamniamensis*** Gilg (W & C Afr.) – cult. orn., eaten like épinards in Congo; ***I. pachycaulon*** M. Newman (Laos) – pachycaul to 1.3 m; ***I. parviflora*** DC. (Siberia) – scruffy ann. first natur. 1830s Eur. (Geneva), ± natur. GB (spreading 24 km a yr.) where form. a favourite subject for pl. physiologists; ***I. tinctoria*** A. Rich. (*I. elegantissima*, NE Afr.) – cult. orn., fragrant white fls, tubers source of cosmetic red dye; ***I. walleriana*** Hook.f. (busy lizzie, sultan's fl., Tanzania to Mozambique) – ubiquitous pot-pl. but not natur. in GB as no seeds set, cf. bee-poll. *I. glandulifera*, many cvs, some hybrids with *I. auricoma* Baill. (Comoro Is.) giving yellow fls; ***I. winkleri*** Hook.f. (Borneo, esp. limestone) – succ. to 2 m, bole to 6 cm diam.

Impatientella Perrier = *Impatiens*

Imperata Cirillo. Gramineae (XXII 5). 13 trop. & warm. ***I. cylindrica*** (L.) P. Beauv. (lalang, alang-alang, blady, cogon or cotton grass, kunai (NG), OW trop. to Medit.) – fr. ed. emus & kangaroos, v. bad weed (invasive trop. As., Hawaii, SE US) of burnt-over forest, abandoned pasture etc., suggested for reclamation & for thatching, hat-making etc. (ropes in anc. Egypt), palatable to few animals incl. water buffalo

Imperatoria L. = *Peucedanum*

imphee *Sorghum bicolor*

Inca peanut *Plukenetia volubilis*; **I. wheat** *Amaranthus caudatus*

Incaberry *Physalis peruviana*

Incaea Luer = *Dryadella*

incaparina *Gossypium* spp.

Incarum Gonç. (~ *Asterostigma*). Araceae (VII 4). 1 Andes: ***I. pavonii*** (Schott) Gonç. R: Willd. 35(2005)319

Incarvillea Juss. Bignoniaceae (1). 16 Himal., C & E As. (Eur. in Oligocene). R: NRBGE 23(1961)303. Perenn. to ann. herbs with tuberous or woody r. & alks. Cult. orn. hardy pls esp. ***I. delavayi*** Bureau & Franch. (purple fls) & ***I. mairei*** (A. Léveillé) Grierson (*I. grandiflora*, crimson fls) from China

incense *Boswellia* spp., *Commiphora* spp., *Aquilaria* spp.; **Cayenne i.** *Protium heptaphyllum;* **i. cedar** *Calocedrus decurrens;* **i. plant** *Calomeria amaranthoides;* **i. tree** *P. heptaphyllum*

inchi *Caryodendron orinocense*

Indagator Halford (~ *Helmiopsis*). Malvaceae (Malv.-Brownl.). 1 N Queensland: ***I. fordii*** Halford. R: Austrobaileya 6(2002)337

India Nageswara Rao = *Robiquetia* (but see JETB 22(1998)701)

India rubber tree *Ficus elastica*

Indian almond *Terminalia catappa;* **I. arrow wood** *Euonymus atropurpureus;* **I. barnyard millet** *Echinochloa frumentacea;* **I. bean-tree** *Catalpa bignonioides;* **I. beech** *Millettia pinnata;* **I. berry** *Anamirta cocculus;* **I. blanket** *Gaillardia* spp.; **I. coral** *Smilax* spp.; **I. corn** *Zea mays;* **I. cress** *Tropaeolum majus;* **I. dill** *Anethum graveolens;* **I. fig** *Opuntia ficus-indica;* **I. gooseberry** *Phyllanthus emblica;* **I. gum** spp. of *Anogeissus, Combretum, Limonia;* **I. hawthorn** *Rhaphiolepis indica;* **I. hemp** *Apocynum cannabinum;* **I. horse-chestnut** *Aesculus indica;* **I. kale** *Caladium lindenii;* **I. laburnum** *Cassia fistula;* **I. laurel** *Terminalia elliptica;* **I. lavender** *Bursera penicillata;* **I. lettuce** *Lactuca indica;* **I. liquorice** *Abrus precatorius;* **I. madder** *Oldenlandia umbellata, Rubia cordifolia;* **I. mallow** *Abutilon theophrasti;* **I. millet** *Cenchrus spicatus;* **I. mulberry** *Morinda citrifolia;* **I. nard** *Nardostachys jatamansi;* **I. olive** *Olea europaea* subsp. *cuspidata;* **I. paint-brush** *Castilleja* spp.; **I. patchouli** *Pogostemon indicus;* **I. pipe** *Monotropa uniflora;* **I. plum** *Oemleria cerasiformis;* **I. potato** *Ipomoea pandurata;* **I. redwood** *Polyalthia cerasoides;* **I. root** *Asclepias curassavica;* **I. shot** *Canna indica;* **I. silk cotton tree** *Bombax ceiba;* **I. spinach** *Basella alba;* **I. strawberry** *Potentilla indica;* **I. tobacco** *Lobelia inflata;* **I. willow** *Monoon longifolium* (fastigiate form)

Indianthus Suksathan & Borchs. (~ *Schumannianthus*). Marantaceae. 1 S Ind., Sri Lanka: ***I. virgatus*** (Roxb.) Suksathan & Borchs. R: BJLS 159(2009)393

Indigastrum Jaub. & Spach (~ *Indigofera*). Leguminosae (III 15). 9 S trop. Afr. (8), 1 pantrop. R: Bothalia 22(1992)167

indigo *Indigofera* spp., (China, Jap.) *Persicaria tinctoria;* **Assam i.** *Strobilanthes cusia;* **bastard i.** *Amorpha fruticosa;* **i. bush** *Dalea* spp.; **Chinese green i.** *Rhamnus* spp.; **false i.** *Amorpha* & *Baptisia* spp.; **Java i.** *I. arrecta, I. tinctoria;* **Natal i.** *I. arrecta*

Indigofera L. Leguminosae (III 15). Incl. *Vaughania*, c. 730 trop. & warm (Afr.-Madag. c. 490, As.-Pacific 115). Dendroid hairs. Indigo (from Gk. for Indian). Local medic.; form. imp. dye-pls (bleu de Nimes orig. of 'denim') esp. in WI (C18) & India (C19) cult. in Ind. & Sumatra esp. Java i. (***I. arrecta*** Hochst. ex A. Rich. (trop. Afr.) & ***I. tinctoria*** L. (SE As.) – cultigen, natur. pantrop.), the natural prod. (still in hair dyes – black henna, Hmong batik, food-colour in e.g. M&M's) almost replaced by aniline dyes (Perkin attempted synthesis of quinine from orig. aniline, but prod. instead synthetic 'mauveine' dye patented 1856), though some spp. still grown as green manure & fodders, esp. ***I. spicata*** Forssk. (trop. OW) but dogmeat made from horses fed on it hepatotoxic (indospicine); ***I. pentaphylla*** L. (Ind.) – lvs sold in market as sour component for curry; some cult. orn. incl. ***I. heterantha*** Wall. ex Brandis (*I. gerardiana*, NW Himal.) – decid. shrub, fls boiled in milk as a tonic in Kashmir

Indobanalia A.N. Henry & Roy. Amaranthaceae (I 2). 1 SW Ind.: ***I. thyrsiflora*** (Moq.) A.N. Henry & Roy

Indocalamus Nakai. Gramineae (IV). Excl. *Ferrocalamus, Gelidocalamus*, 23 China (22, endemic – R: FOC 22(2006)135), Japan (1) – heterogeneous, s.s. 1 China (***I. sinicus*** (Hance) Nakai). Culms used for chopsticks, lvs for weaving. ***I. tessellatus*** (Munro) Keng f. – orig. described from packing around tea sent to UK from China, flowering cycle of at least 115 yrs

Indochloa Bor = *Euclasta*

Indocourtoisia Bennet & Raiz. = *Courtoisina*

Indocypraea Orchard (~ *Wollastonia*). Compositae (Helia.). 1 Indomal.: ***I. montana*** (Blume) Orchard. R: Nuytsia 23(2013)436

Indofevillea Chatterjee. Cucurbitaceae (V). 2 Himal. R: BZ 91(2006)768

Indokingia Hemsl. = *Polyscias*

Indomelothria W. de Wilde & Duyfjes (~ *Zehneria*). Cucurbitaceae (XIV). 2 SE As., W Mal. R: Blumea 51(2006)6

Indoneesiella Sreemad. = *Andrographis*

Indopiptadenia Brenan. Leguminosae (II 1). 1 Ind., Nepal: ***I. oudhensis*** (Brandis) Brenan – endangered due to timber over-exploitation. R: Phytotaxa 164(2014)61

Indopoa Bor. Gramineae (Chlorid.). 1 Ind.: ***I. paupercula*** (Stapf) Bor – caryopsis needle-like in narrow pocket along keel of lemma. R: KB 13(1958)225

Indopolysolenia Bennet = *Leptomischus*
Indorouchera Hallier f. (~ *Hugonia*). Linaceae (I). 2 Nicobar & Andaman Is., SE As. to W Mal. R: FM I,10(1988)615. Lianes
Indosasa McClure (~ *Sinobambusa*). Gramineae (IV). 15 S China (15, 13 endemic – R: FOC 22(2006)143) & Vietnam
Indoschulzia Pimenov & Kljuykov = *Kedarnatha* (but see KB 50(1995)639)
Indosinia J.E. Vidal. Ochnaceae. 1 S Vietnam: ***I. involucrata*** (Gagnep.) J. Vidal. R: BJ 113(1991)183
Indotristicha P. Royen (~ *Dalzellia*). Podostemaceae (II). 2 Ind. R: BMNHN 4,10(1988)173
Indovethia Boerl. = *Sauvagesia*
indungulu *Siphonochilus aethiopicus*
Indurgia Speta = *Drimia*
Inezia E. Phillips. Compositae (Anth.-Cot.). 2 S Afr. R: BBMNHB 23(1993)146
Inga Mill. Leguminosae (II 4). c. 300 trop. & warm Am. esp. Andean foothills. T.D. Pennington (1997), *The genus I. Botany*. With *Guatteria* & *Ocotea*, largest neotrop. genera; only pinnate-leaved II. Some fish-, some monkey-disp. Amazon; some with foliar nectaries visited by ants which remove other insects. Some planted as crop-shade or orn., growing well on impoverished soils, others with ed. sweet pulp (sarcotesta) around seeds, esp. ***I. edulis*** Mart. (icecream bean, i. tree, trop. Am.) – cult. for fr. (to 2m long in Amaz. Peru), those of ***I. feuillei*** DC. (pacay) offered by Atahualpa (last Inca Emperor) to Pizarro, who killed him anyway. Some timbers, fuelwood etc. e.g. ***I. laurina*** (Sw.) Willd. (Spanish oak, WI), ***I. vera*** Willd. (guaba) – fls open afternoon & poll. lepidoptera & birds (sucrose-rich nectar), at night nectar hydrolyzed (? micro-organisms) to hexoses & with sour smell attractive to bats; ***I. acrocephala*** Steud. (Amaz.) – red latex used to fix dyes for tourist souvenirs
Ingonia Pierre ex M. Bod. = *Cola*
Inhambanella (Engl.) Dubard. Sapotaceae (I 3). 1 W & 1 SE trop. Afr. R: T.D. Pennington, *S.* (1991)140
injira *Eragrostis tef*
inkberry *Phytolacca americana*, *Ilex glabra*
inkweed *Phytolacca octandra*
innocence *Houstonia caerulea*
Inobulbon (Schltr.) Schltr. & Kränzlin = *Dendrobium*
Inocarpus Forst. & Forst. f. Leguminosae (III 11). 3 Mal., Pac. Leaf, fl. & fr. v. 'unlegeminous' in aspect; sap red. R: Blumea 52(2007)401. ***I. fagifer*** (Parkinson) Fosb. (*I. edulis*, i, Polynesian, Tahiti or O'taheite chestnut, Pac.) – seeds ed. raw or roasted
Inodes O.F. Cook = *Sabal*
inoi, inoy *Poga oleosa*
inside-out-flower *Vancouveria hexandra*
Inti M. Blanco = *Maxillaria* (but see Lankesteria 7(2007)524)
Intsia Thouars. Leguminosae (I 2). 3 trop. As., coasts of Ind. & Pac. Oceans. Timber esp. ***I. bijuga*** (Colebr.) Kuntze (Borneo teak, kwila, esp. abundant on Kabara so centre of canoe-building in Fiji & much of Tonga) & ***I. palembanica*** Miq. (*Afzelia p.*, merbau, Borneo or Malacca teak, Mal.)
Intybellia Cass. = *Crepis*
Inula L. Compositae (Inul.-Inul.). c. 100 temp. & warm OW (Eur. 18 (excl. *Limbarda*)). Herbs (oft. statuesque) with alks. Some medic., cult. orn., dyes etc.: ***I. britannica*** L. (Euras.) – cult. orn. with lactone antifeedant effective on flour-beetles; ***I. conyzae*** (Griess.) Meikle (*I. conyza*, ploughman's spikenard, Eur., Med.); ***I. helenium*** L. (elecampane (enula campana), C As., natur. Eur., N Am., Jap. etc.) – r. (*Radix helenii*, Enulae) trad. medic. in skin & chest disease now accepted as useful in asthma treatment, also candied & used to flavour absinthe, cult. as r. veg. at time of Pliny; ***I. magnifica*** Lipsky (E Cauc.) & ***I. racemosa*** Hook.f. (W Himal., cult. as medic. pl. in Ind.; a lactone (isoalantodiene) potent plant-growth regulator) – robust herb. perennials to 2–3 m
Inulanthera Källersjö (~ *Athanasia*). Compositae (Anth.-Urs.?). 10 S Afr. R: NJB 5(1986)539
Inuloides R. Nordenstam (~ *Osteospermum*). Compositae (Cal.). 1 SW Cape: ***I. tomentosa*** (L.f.) R. Nordenstam. R: CN 44(2006)44
Inulopsis (DC.) O. Hoffm. (~ *Podocoma*). Compositae (Ast.-Pod.). 4 S Am. R: Phytol. 76(1994)115
Inversodicraea Engl. ex R. Fries = *Ledermanniella*
Involucrella (Hook.f.) Neupane & N. Wikstr. Rubiaceae (IV 15). 2 SE As. R: T 64(2015)316

Io Nordenstam (~ *Senecio*). Compositae (Sen-Sen.) 1 Madag.: ***I. ambondrombeensis*** (Humbert) Nordenstam. – lvs opp. R: CN 40(2003)47

Iocenes R. Nordenstam = *Senecio*

Iochroma Benth. Solanaceae (IV 5b). Incl. *Acnistus*, c. 26 trop. S Am. R: New Plantsman 5(1998)154. Shrubs & trees, some local medic. & hallucinogenic in Amaz. & cult. orn. esp. ***I. cyaneum*** (Lindl.) Green (*I. tubulosum*), NW S Am.) with deep blue tubular fls (other spp. with yellow, scarlet or white fls), change from blue to red due to deletion of gene coding for enzyme in anthocyanin pathway

Iodanthus (Torrey & A. Gray) Steud. (*Oclorosis*). Cruciferae (16). 1 C & E US: ***I. pinnatifidus*** (Michaux) Steud. R: JAA 69(1988)116

Iodes Blume (*Ioedes*). Icacinaceae. Incl. *Polyporandra*, c. 20 OW trop. ***I. scandens*** (Becc.) Utteridge & Byng (*P. s.*, E Mal. to Polynesia) – liane with lvs cooked with taro

Iodina Miers = *Jodina*

iodine bush *Mallotonia gnaphalodes*

Iodocephalus Thorel ex Gagnepain. Compositae (Vern.-Centrop.). 1 SE As.: ***I. gracilis*** Thorel ex Gagnepain

Ioedes Blume = *Iodes*

Iogeton Strother. Compositae (Helia.-Ecl.). 1 Panamá: ***I. nowickeanus*** (D'Arcy) Strother

Ionacanthus Benoist = *Mellera*

Ionactis Greene (~ *Aster*). Compositae (Ast.). 5 N Am. R: Britt. 44(1992)247, Phytol. 77(1994)262. Goldenasters

Ione Lindl. = *Sunipia*

Ionidium Vent. = *Hybanthus*

Ionopsidium Reichb. (~ *Cochlearia*). Cruciferae (18). 9 W Medit. ***I. acaule*** (Desf.) Reichb. (violet cress, Portugal) – minute cult. ann. for paving

Ionopsis Kunth. Orchidaceae (V 12h). 6 trop. & warm Am. (Florida 1). Cult. orn. epiphytes

Iostephane Benth. Compositae (Helia.-Helia.). 4 Mex. R: Madroño 30(1983)34. ***I. madrensis*** (S. Watson) Strother (cachana) – r. medic.

Iotasperma Nesom (~ *Erigeron*). Compositae (Ast.-Pod.). 2 Aus. R: Phytol. 76(1994)144,274

ipê *Handroanthus* spp.

ipecacuanha *Carapichea ipecacuanha*; **American i.** *Gillenia stipulata*; **bastard i.** *Asclepias curassavica*; **Carolina i.** *Euphorbia ipecacuanhae*; **false i.** *Richardia scabra*; **Goa i.** *Naregamia alata*; **Indian i.** *Cryptocoryne spiralis*; **i. spurge** *E. ipecacuanhae*; **white i.** *Pombalia calceolaria*

Ipheion Raf. (~ *Tristagma*). Amaryllidaceae (Alliaceae-Gill.). 3 S Am. ***I. uniflorum*** (Lindl.) Raf. (*T. u.*, *Triteleia u.*, Arg., Urug.) – cult. orn. spring bulb

Iphigenia Kunth. Colchicaceae (Liliaceae s.l.). 9 OW trop. (Socotra 1), 1 ext. to Aus. Alks

Iphigeniopsis F. Buxb. = *Camptorrhiza*

Iphiona Cass. Compositae (Inul.-Inul.). 12 Middle E & NE Afr. to C As. R: NJB 5(1985)169

Iphionopsis A. Anderb. (~ *Iphiona*). Compositae (Inul.-Pluch.). 2 NE & E Afr., 1 Madag. R: NJB 5(1985)51

ipoh *Antiaris toxicaria*

Ipomoea L. Convolvulaceae (9). Incl. *Mina*, *Pharbitis*, c. 600 trop. & warm temp. (Am. 327 (R: T 45(1996)3); Eur. 2 incl. ***I. sagittata*** Poir., trop. Am. & Eur.). Paraphyletic unless *Argyreia*, *Astripomoea*, *Lepistemon*, *Stictocardia*, *Turbina* incl.). Cli. herbs or shrubs (even trees incl. some pioneers in Mex. arid areas), some with tubers, others caudiciform succ. (e.g. ***I. bolusiana*** Schinz, S & trop. Afr., Madag.), distinguishable from *Convolvulus* in stigma not linearly divided & spiny pollen; extrafl. nectaries. Alks. Some ed. tubers, fls, shoots etc., drugs & cult. orn. (morning glory). ***I. alba*** L. (*I. bona-nox*, moonflower, trop. Am., widely natur.) – cult. for large white scented night fls, calyces a curry veg. in Sri Lanka; ***I. aquatica*** Forssk. (hung choy, en c., kangkong, ong tsoi, water spinach, OW, invasive SE US) – cult. for ed. shoots esp. white-stemmed form in e.g. Hong Kong; ***I. arborescens*** (Willd.) G. Don (Mex.) – Champagnat's Model; ***I. batatas*** (L.) Lam. (sweet potato, 'yam' (US), Braz. arrowroot, kau kau (NG), cultigen spread from C Am. ('kumar' in Peru) to Polynesia ('kumara') in pre-Columbian times & thence to NZ by Maoris (staple crop, no grains), now widely cult. in trop. & warm (world prod. 100 M t per annum by 2009) incl. Jap. where v. imp. – hexaploid (closest to *I. tabascana* McDonald & Austin (tetraploid) & *I. trifida* (Kunth) G. Don (diploid, with feral tetraploid forms known in wild, poss. derived from *I. leucantha* Jacq. (diploid, ? *I. trichocarpa* (Kunth) G. Don × *I. lacunosa* L., trop. Am.)), swollen tubers ed. starch (as for dang myun noodles in Korea, also with natural fibres used in biodegradable plastic for Toyota cars – polylactic acid bioplastic) & alcohol source, fifth world r.-crop (Henry VIII considered it aphrodisiac –

Doctrine of Signatures?; 'Let the sky rain potatoes' – Falstaff), many cvs, early Hawaiians having 230 but all exc. 24 now lost, genes from '*I. trifida*' being intr. to stocks to improve nematode resistance, some cult. orn cvs for hanging-baskets, e.g. '**Blackie**' & bilious pale ones; ***I. cairica*** (L.) Sweet (railway creeper, trop. OW) – natur. Am., invasive Aus.; ***I. coccinea*** L. (*Quamoclit c.*, US) – fls scarlet; ***I. costata*** F. Muell. ex Benth. (yala, N & C Aus.) – ed. tuber coll. Aborigines; ***I. imperati*** (Vahl) Griseb. (*I. stolonifera*, trop. & warm) – sand-binder; ***I. indica*** (Burm.) Merr. (trop. S Am.) – cult. pantrop.; ***I. lobata*** (Cerv.) Thell. (trop. Am.) – lvs palmately lobed, cult. orn. with red bird-poll. fls fading to yellow; ***I. leptophylla*** Torrey (W N Am.) – ants attracted to extrafl. nectaries on lvs & K permit seed prod. increase × 100; ***I. nil*** (L.) Roth (Jap. morning glory, trop. Am., introd. As. by Portuguese early 1500s) – much used in plant physiology (*Pharbitis n.*), cult. orn., esp. fashionable in Jap. Edo period being much depicted on screens & many cvs incl. those with highly divided C, seeds allegedly purgative; ***I. obscura*** (L). Ker (OW trop.) – potherb in Sri Lanka; ***I. orizabensis*** (J. Pellett.) Steud. (Mex.) – scammony r., the source of a resin, ipomoea, a drastic purgative; ***I. pandurata*** (L.) G. Meyer (Ind. potato, p. vine, N Am.) – large tuberous r., pedicel nectaries attractive to *Crematogaster* & other ants so fr. survival increased 3–4 times, cult. orn. with white fls; ***I. pes-caprae*** (L.) R. Br. (trop. beaches) – sand-binder; ***I. purga*** (Wender.) Hayne (jalap, Mex.) – dried tubers with high levels of purging resin, ipomoea; ***I. purpurea*** (L.) Roth (morning glory, trop. Am. (? orig. Mex.), natur. N Am. etc.) – cult. orn., ***I. quamoclit*** L. (cupid flower, trop. Am.) – fls scarlet, lvs deeply dissected pinnately, cult. orn., weed of Aus. sugar-cane; ***I. sloteri*** (House) Oostr. (cardinal cli.) – tetraploid derived from diploid hybrid *I. coccinea* × *I. quamoclit* (*I.* × *multifida* (Raf.) Shinners), cult. orn.; ***I. spathulata*** Hall. f. (E Afr.) – r. infusion used against eye disease; ***I. tricolor*** Cav. (*I. rubrocaerulea*, Mex. & C Am.) – ergoline alks the effective hallucinogenic agent in seeds, cult. orn. esp. '**Heavenly Blue**' (reddish buds but increasing pH turns C blue); ***I. tuboides*** Degener & Oostr. (Hawaii) – extrafl. nectaries but no native ants in Hawaii (phylogenetic inertia?); ***I. violacea*** L. (*I. macrantha*, *I. tuba*, coastal & pantrop.) – floating seeds with air cavity, cult. orn.

Ipomopsis Michaux. Polemoniaceae (III). c. 30 W N Am. & Florida, S Am. (1: Arg. & Chile). R: Aliso 3(1956)351. Like *Gilia* but stems with well-dev. lvs. Some cult. orn. herbs, ***I. aggregata*** (Pursh) V. Grant (W N Am.) – foetid, & other spp. local medic.

Ipsea Lindl. (~ *Pachystoma*). Orchidaceae (V 14). 3 Ind., Sri Lanka, Thailand. R: KB 42(1987)937

Iranecio R. Nordenstam (~ *Senecio*). Compositae (Sen.-Sen.). 16 SE Eur. (1) to Iran. R: Fl. Iranica 164(1989)53

Irania Hadač & Chrtek = *Fibigia*

Iré rubber *Funtumia elastica*

Irenea Szlach. & al. = *Cyrtochilum*

Irenella Süsseng. (~ *Iresine*). Amaranthaceae (II 2). 1 Ecuador: ***I. chrysotricha*** Süsseng.

Irenepharsus Hewson. Cruciferae (37). 3 SE Aus. R: JABG 6(1982)1

Irenodendron Alford & Dement (~ *Laetia*). Salicaceae. 3 S Am. R: JBRIT 9(2015)332

Iresine P. Browne. Amaranthaceae (II 2). c. 45 trop. & warm Am. (Mex. 29). Trees, shrubs, lianes & herbs, usu. dioec. Some locally medic., others cult. for orn. foliage (blood-leaf) esp. ***I. herbstii*** Hook. (S Am.) – lvs purplish-red or green, with yellow veins. ***I. vermicularis*** (L.) Moq. with *Sesuvium portulacastrum* pollution-tolerant, replacing mangroves nr Rio de Janeiro

Iriartea Ruíz & Pavón. Palmae (V 1). 1 trop. Am: ***I. deltoidea*** Ruíz & Pavón (*I. ventricosa*, paxiuba) – stem (used for dugouts) with thickened 'belly' halfway up trunk withers basally to be supported by aerial r., allowing tree to move away from obstacles (cf. *Socratea*), some r. spiny, bee-poll. R: FN 53(1990)61

Iriartella H. Wendl. Palmae (V 1). 2 trop. S Am. R: FN 53(1990)69

Iridaceae Juss. Magnoliidae – Asparagales. 72/2200 cosmop. but esp. S Afr., E Medit., C & S Am. R: P. Goldblatt & J.C. Manning (2008) *The Iris family*: 14, 89. Usu. geophytic herbs (symp. rhiz., corms or bulbs), less oft. (S Afr.) everg. (*Aristea*, *Bobartia*, *Dietes*, *Dierama*, *Pillansia*) or even shrubby (*Klattia*, *Nivenia*, *Witsenia* – R: P. Golblatt (1993) *The Woody I.*), rarely ann. (3 spp. of *Sisyrinchium* – Cronquist) or achlorophyllous mycotroph (*Geosiris*); usu. with crystals of calcium oxalate in some cells, alks 0 but s.t. pl. poisonous, storing starch or fructosans; vessels in r. (in aerial organs also, in *Sisyrinchium*), commonly mycorrhizal & without r.-hairs. Lvs usu. distich., parallel-veined, with sheathing base & narrow blade (rarely with petiole & expanded blade). Fls bisex. (heterostyly [v. rare in monocots] in some *Nivenia*), reg. or irreg. in term. infls or solit., subtended by a bract, the infl. by 1 or 2 expanded bladeless sheaths forming a spathe; P 3 + 3 alike or not, oft. united by basal tube, nectaries septal or at base of P or on inner P (e.g. *Tigridia*) or base of A (*Iris*) or 0 (*Sisyrinchium*), A 3 (2 in *Diplarrhena*) opp. outer P, filaments oft. united by basal tube, anthers extrorse with longit. slits (apical pores in *Cobana*, some *Aristea*), $\overline{G}$ (3), 3-loc. with axile placentation ($\underline{G}$ in *Isophysis*, 1-loc. with parietal placentation in some *Iris* (*Hermodactylus*)) & term. 3-lobed style, the branches s.t. subdivided or oft. expanded & petaloid with stigma on outer side of branch, (1–) ±∞, campylotropus (IV) or anatropous, bitegmic ovules per loc. Fr. a loculicidal capsule; seeds s.t. arillate or sarcotestal with rather small linr. embryo & fleshy endosperm with reserves of hemicellulose, protein & oil (usu. 0 starch). n = 3–19+

Classification & principal genera:

I. **Isophysidoideae** ($\underline{G}$, Tasmania): *Isophysis* (only)

II. **Geosiridaceae** (incl. Nivenioideae; some woody or mycotrophic, fugacious blue fls, Afr., Madag., NG, Aus.): *Aristea, Nivenia, Patersonia*

III. **Iridoideae** (fugacious fls, perigonal nectaries, style long-branched below A level; 4 tribes):

1. **Sisyrinchieae** (style-branches ext. between rather than opp. A; esp. S hemisph.): *Bobartia, Sisyrinchium*
2. **Irideae** (P divided into limb & claw, flattened style-branches & crests tepaloid; esp. OW): *Galaxia, Iris, Moraea*
3. **Mariceae** (like 2. but style-branches merely thickened; C & S Am.): *Neomarica, Trimezia*
4. **Tigridieae** (like 3. but lvs plicate; C & S Am.): *Cypella, Tigridia*

IV. **Crocoideae** (Ixioideae; corms of (1–)3(more) internodes with rs from lower half, cf. *Ferraria* & *Moraea* (III.) with new rs from base of apical shoot, infls of sessile fls in spikes or panicles, P-tubes, OW esp. S Afr.; 3 tribes):

1. **Pillansieae** (lvs with pseudo-midrib, panicle, Cape): *Pillansia* (only)
2. **Watsonieae** (spike, deeply-divided style-branches, trop. & S Afr.): *Lapeirousia, Watsonia*
3. **Ixieae** (style diff., OW esp. S Afr.): *Babiana, Crocus, Dierama, Freesia, Geissorhiza, Gladiolus, Hesperantha, Ixia, Romulea, Tritonia, Tritoniopsis*

Many v. imp. garden pls esp. in commerce as cut-fls etc. (C. Innes (1985) *The World of I.*): *Crocosmia* (montbretia), *Crocus, Freesia, Gladiolus, Iris, Ixia, Sparaxis, Tigridia* & limited uses as spices (*Crocus*), dyes (*Crocosmia*) & scents (*Iris*); some S Afr. spp. (esp. *Romulea, Sparaxis, Watsonia*) now serious weeds in Aus.

Iridodictyum Rodionenko = *Iris*

Iridosma Aubrév. & Pellegrin. Simaroubaceae. 1 W trop. Afr.: ***I. letestui*** (Pellegrin) Aubrév. & Pellegrin

iris, Algerian *Iris unguicularis*; **black i.** *Iris nigricans*; **Cape i.** *Moraea* spp.; **Dutch i.** *I.* × *hollandica*; **English i.** *I. latifolia*: **Japanese i.** *I. ensata*; **snake's-head i.** *I. tuberosa*; **Spanish i.** *I. xiphium*; **stinking i.** *I. foetidissima*

Iris Tourn. ex L. Iridaceae. (VII 2). Incl. *Belamcanda, Chamaeiris, Hermodactylus, Juno, Limniris, Pardanthopsis, Xiphion*, c. 280 Euras. (Eur. 31), N Afr., N Am. Outer P 3 the falls narrowed basally, s.t. bearded, inner 3 the standards usu. erect, A 3 free at base of falls, style-branches 3, petal-like covering A; just above the anther on the outer side of the style is a little flap whose upper surface is the stigma. Bees, attracted by the flag-like falls & s.t. scent, enter the fl. for nectar when pollen from other fls is brushed against the stigma; when they retreat, carrying newly deposited pollen, they close the flap, thereby separating the anther from the stigma & preventing self-poll. (similar is found in *Viola* but here the fl. comprises 3 functional units). Seeds oft. arillate or flattened (wind-disp.). Fls of many colours & much cult. (flags; cut-fl. trade alone in US worth $120 M per annum by 2005), some v. anc. hybrids; 9–10 subgg. recog. (s.t. even as distinct genera): *Iris* (I.) with rhiz. & distinctly bearded falls (incl. bearded (pogon i., many cvs), oncocyclus (poll. by night-sheltering *Eucara* bees), & regelia i. & their hybrids (Regeliocyclus i.), *Limniris* (Li.) with rhiz. & beardless falls (incl. 'Pac. Coast Irises' (PCIs) or Calif. i., many hybrids), heterogeneous, & *Hexagonae* (Louisiana i.), waterside spp. & hybrids of S US for warm countries), *Nepalensis* (e.g. ***I. decora*** Wall. (*I. nepalensis*, Himal.) with dahlia-like tuberous r., *Xiphium* (X., incl. *Hermodactylus*) with bulbs & fibrous r. (incl. xiphion (English, Dutch & Spanish i.) usu. on alk. soils & *Hermodactyloides* (R: Davis & Hedge Festschrift (1989)83) with bulbs with fibrous netted tunics (reticulata i.)), *Scorpiris* with bulbs & usu. thick fleshy r. when dormant (comprises juno i.), B. Mathew (1981) *The Iris*; F. Köhlein

(1987) *The I.*, ed. 2. Bulbs derived from rhiz. several times. The common flag iris (I.) of gardens of complex history involving Eur. spp., *I. pumila* L., *I. lutescens* Lam. (S Eur.), *I. variegata* L. (C & SE Eur.) etc. & referred to as '*I. germanica*'. (orig. ***I.* × *germanica*** L. = *I. pallida* Lam. (Eur.) × *I. variegata*), of which group '**Florentina**' (*I. florentina*) poss. used to decorate Sphinx & known to Thutmose III (1501–1447 BC), rhiz. ground & used as scent of violets, to scent royal bedclothes (1480s), now in e.g. Chanel No 19 (1971) & to flavour certain gins, or medic. (orris, i.e. iris, r.), as fixative in potpourri & for powdering wigs & hair in C18, the fl. poss. the orig. fleur-de-lis (i.e. -Louis), from 1889 other spp. from E. Medit. bred in (tetraploids = ***I.* × *conglomerata*** Henderson); the whole group has 2n = 24–c. 64 with increasing pl. size. Many other hybrids, some between distantly related spp. Other cult. spp. incl.: ***I. albicans*** Lange (I., Yemen) – cult. in Muslim graveyards, emblem of Upper Egypt, fls white, bred into *I.* × *conglomerata*; ***I. atrofusca*** Bak. (I., Israel) – ordinary sword-lvs on deep soils but in heavily grazed areas linr. lvs like the toxic *Asphodelus aestivus* in the area (? mimicry); ***I. clarkei*** Hook.f. (Li., Himal.) – dried lvs fodder for yaks & horses; ***I. confusa*** Sealy (Li., W China) & ***I. wattii*** Bak. ex Hook.f. (Li., Assam & China) – stems elongate, carrying up lvs before flowering: ***I. danfordiae*** (Bak.) Boiss. (X., Turkey) – yellow-flowered reticulata of early spring; ***I. domestica*** (L.) Goldbl. & Mabb. (subg. *Pardanthopsis*; *Belamcanda chinensis*, leopard lily, E Russia, N Ind., China, Jap., natur. N Am.) – spotted orange fls, seeds black (Irid. usu. brown)., medic., cult. orn. incl. hybrids with other spp.; ***I. ensata*** Thunb. (Li., *I. kaempferi*, Jap. i., E As.) – red-purple fls, long cult. Jap., many cvs incl. tetraploids & hybrids with *I. pseudacorus*; ***I. foetidissima*** L. (subg. *Xyridion*, gladdon, (stinking) gladwin, gladwyn or iris, S & W Eur. to N Afr.) – bruised lvs smell of meat ('roast beef plant'), rhiz. steeped in ale form. an efficient purge; ***I. graminea*** L. (subg., *Xyridion*, NE Spain to Cauc.) – fls smell of ripe plums; ***I. latifolia*** (Mill.) Voss (X., *I. xiphioides*, English i., Spain & Pyrenees) – falls deep blue-purple; ***I. missouriensis*** Nutt. (S US.) – medic., seeds form. used as arrowhead poison; ***I. nelsonii*** Randolph (Louisiana) – genomes of 3 spp., 1 of only 8 confirmed homoploid hybrid spp.; ***I. nigricans*** Dinsm. (black iris, Middle E) – floral emblem of Jordan; ***I. paradoxa*** Steven – transplanted to Turkish graves; ***I. pseudacorus*** L. (Li., yellow flag, water f. (orig. 'gladiolus' in Mediaeval GB; prob. orig. fleur-de-lis), W Eur., Medit. to Iran, invasive Aus., NZ, NW US) – rhiz. & seeds form. medic. (strong purgative used in c. 1700 Skye for enemas, colds, toothache etc.), rhiz. form. a source of black dye & ink, seeds a coffee subs.; ***I. pumila*** L. (I; E Eur. to Urals) – natural amphidiploid (*I. pseudopumila* Boiss. & Heldr. × *I. attica* Tineo), cult. orn., now crossed with others incl. *I.* × *germanica*; ***I. reticulata*** M. Bieb. (X., Cauc.) – violet-scented fls in early spring, many hybrids with *I. histrioides* (G. Wilson) S. Arn. (C N Turkey); ***I. sibirica*** L. (Li., C Eur. to Siberia) – cult. ('Siberian') hybrids with *I. sanguinea* Hornem. ex Donn (C As. to Jap.); ***I. spuria*** L. (subg. *Xyridion*, Eur., Medit.) – v. variable in wild, over 70 cvs (many hybrids); ***I. tenax*** Dougl. (Li, W N Am.) – fine cord from lvs used for fishing-nets & snares strong enough to capture bears & elk; ***I. tuberosa*** L. (X., *Hermodactylus t.*, snake's head iris, S France to Middle E) – usu. dingy cult. orn.; ***I. unguicularis*** Poir. (Li. (subg. *Siphonostylis*), *I. stylosa*, Algerian i., Algeria to E Medit.) – winter-flowering, seeds covered with glistening sessile glands attractive to disp. agents (ants); ***I. versicolor*** L. (Li. blue f., E US) – amphidiploid, ***I. virginica*** L. (E US) × *I. setosa* Pall. (NE N Am.), dried rhizomes medic.; ***I. xiphium*** L. (X., Spanish i., W Medit.) – a parent of the bulbous Dutch i. (***I.* × *hollandica*** Wehr.) grown as cut-fls, derived also from *I. filifolia* Boiss. & *I. tingitana* Boiss. & Reuter (both W Medit.) etc. (poss. *I. latifolia*)

Irish ivy *Hedera hibernica*; **I. yew** *Taxus baccata* 'Fastigiata'

Irlbachia Mart. (*Helia*). Gentianaceae. 17 trop. S Am. R: PKNAW C88(1985)406. ***I. alata*** (Aubl.) Maas (Amaz.) – bat-poll *herb*, pops fl. all yr. round; ***I. purpurascens*** (Aubl.) Maas (NE S Am.) – medic.

Irmischia Schldl. = *Metastelma*

iroko *Milicia excelsa*

iron grass *Lomandra* spp.; **i. oak** *Quercus stellata*

ironbark *Eucalyptus* spp. esp. *E. sideroxylon* (firewood); **grey i.** *E. paniculata*; **red i.** *E. crebra*; **scrub i.** *Dysoxylum pettigrewianum*

ironweed (US) *Vernonia* spp.

ironwood *Backhousia subargentea, Cynometra alexanderi; Eusideroxylon zwageri, Mesua ferrea, Nestegis apetala, Ostrya virginiana, Parrotia persica, Xylia xylocarpa* & also spp. of *Casuarina, Copaifera, Olea*; **desert i.** *Olneya tesota*

Irvingbaileya R. Howard. Stemonuraceae (Icacinaceae s.l.). 1 Queensland: ***I. australis*** (C. White) R. Howard. R: Britt. 5(1943)50

Irvingia Hook.f. Irvingiaceae (Ixonanthaceae s.l.; Simaroubaceae s.l). Incl. *Desbordesia*, 7 trop. Afr. (Guineo-Congolian forests), 1 Mal. R: FOW 1(1999)8. Troll's Model. Fr. usu. ed. (a samara in ***I. glaucescens*** Engl. (*D. g.*, trop. W Afr.)), seeds used for soap & as wax source, those of ***I. gabonensis*** (O'Rorke) Baill. (dika nuts, W trop. Afr.) giving d. butter, the paste from the mashed kernels being d. bread or Gaboon chocolate, fr. smelly & disp. elephants; ***I. malayana*** Oliv. ex A. Benn. (SE As., W Mal.) – hard-wooded relic in cleared forest, seeds yielding cay-cay fat

Irvingiaceae Exell & Mendonça (Ixonanthaceae s.l.; Simaroubaceae s.l.). Magnoliidae – Malpighiales. 3/12 trop. OW. R: FOW 1(1999). Trees. Lvs distich., revolute, margins entire; stip. large, encircling term. bud. Infls term. or axillary panicles. Fls 5-merous; K imbricate, free; C imbricate; A (9)10; intrastaminal nectary-disk consp.; $\underline{G}$ (2,5) with 1 pend. ovule per carpel. Fr. a drupe or samara

Genera: *Allantospermum, Desbordesia, Irvingia, Klainedoxa*

Excl. from Simaroubaceae (q.v.) because of mucilage cavities in stem, stip., leaf & fr. anatomy, phytochemistry

Irwinia G. Barroso. Compositae (Vern.-Pip.). 1 NE Braz.: ***I. coronata*** G. Barroso

Iryanthera Warb. Myristicaceae. c. 23 trop. S Am. to Panamá. Bark of some sources of hallucinogenic pastes & medic.

Isabelia Barb. Rodr. Orchidaceae (V 13b). 3 Braz.

Isachne R. Br. Gramineae (Micr.). c. 100 trop. & warm (China 18) esp. trop. As. ***I. globosa*** (Thunb.) Kuntze (*I. australis*, As. to Aus.) – fodder

Isaloa Humbert = *Barleria*

Isalus J. Phipps = *Tristachya*

Isandra F. Muell. = *Symonanthus*

Isandraea Rauschert = *Symonanthus*

isano oil *Ongokea gore*

Isanthus Michaux = *Trichostema*

isañu *Tropaeolum tuberosum*

Isatis Tourn. ex L. Cruciferae (30). Incl. *Boreava, Pachypterygium, Sameraria, Tauschera*, c. 85 Eur. (5–9), Medit. to Afghanistan. ***I. tinctoria*** L. (woad, SW As., ? SE Eur.) – form. cult. for lvs to be dried, powdered & fermented (smelly) to give v. fast blue dye (indigotin) from lvs favoured as Med. taste for red declined (thereafter Virgin Mary always figured wearing blue), a speciality of Toulouse (France) in 14 cent., form. used for sailors' clothes, policemen's uniforms & students' gowns at Christ's Hospital (a London school, hence 'Blue-coat Boys'), displaced by indigo (1631, though revived in France when Am. indigo supplies cut off in Napoleonic Wars) & synthetics (1890) but factory in England until 1930s

Ischaemum L. Gramineae (XXII 4). 87 warm & trop. esp. As. ***I. rugosum*** Salisb. (trop., orig. Indopacific) – bad weed of ricefields, natur. S As., noxious weed in USA

Ischnea F. Muell. Compositae (Sen.-Tuss.). 4 NG. R: PSE 191(1994)254. Ray florets tubeless, disk florets male, pappus 0

Ischnocarpus O. Schulz = *Pachycladon*

Ischnocentrum Schltr. = *Glomera*

Ischnochloa Hook.f. = *Microstegium*

Ischnogyne Schltr. Orchidaceae (V 10b). 1 China: ***I. mandarinorum*** (Kraenzl.) Schltr.

Ischnolepis Jum. & Perrier. Apocynaceae (III; Asclepiadaceae I). 1 Madag.: ***I. graminifolia*** (Costantin & Gallaud) Klack (*I. tuberosa*). R: Candollea 54(1999)332

Ischnosiphon Koern. Marantaceae. 35 trop. Am. R: OB 43(1977). Stalks used as pipes for blowing snuff up noses in Amaz. ***I. arouma*** (Aubl.) Koern. (tirite) – lvs used for basketry, stems split for sifters, fans & matapi (a tube used for processing cassava)

Ischnostemma King & Gamble = *Pentatropis*

Ischnurus Balf.f. = *Lepturus*

Ischyrolepis Steud. = *Restio* (but see Bothalia 15(1985)397)

Iseia O'Don. = *Aniseia*

Iseilema Andersson. Gramineae (XXII 7). 24 Indomal. to Aus. R: HIP 33(1935)t.3286. Barcoo or Flinders grass

Isertia Schreb. Rubiaceae (I 8). 14 trop. Am. Some cult. orn. shrubs; soap. ***I. laevis*** (Triana) B. Boom – moth poll. syndrome but also visited by hummingbirds during day

Isidodendron Fernàndez-Alonso & al. Trigoniaceae. 1 Colombia: ***I. tripterocarpum*** Fernàndez-Alonso & al.

Isidorea A. Rich. ex DC. Rubiaceae (I 7). 20 WI

Isidrogalvia Ruíz & Pavón = *Tofieldia* but see *Harperocallis*
Isinia Rech.f. = *Lavandula*
Iskandera N. Busch. Cruciferae (5). 2 C As.
Islaya Backeb. = *Eriosyce*
Ismelia Cass. (~ *Glebionis*). Compositae (Anth.-Gleb.). 1 Morocco: ***I. carinata*** (Schousb.) Sch. Bip. (*I. versicolor*, *Chrysanthemum c.*) – stems simple or forked; cult. orn. R: FNA 19(2006) 552
Ismene Salisb. ex Herb. (~ *Hymenocallis*). Amaryllidaceae. 10–15 Andes of S Ecuador to Bolivia. Cult. orn. incl. hybrids, esp. ***I. narcissiflora*** (Jacq.) M. Roemer (*H. n.*, basket fl., Andes of Peru & Bolivia) & ***I. × festalis*** Worsley (spider lily, *H. × f.*, *I. longipetala* (Lindl.) Meerow (Peru) × *I. narcissiflora*)
Isoberlinia Craib & Stapf ex Holland. Leguminosae (I 2). 5 trop. Afr. Imp. components of miombo woodlands in S trop. Afr. Timber, local medic. & ed. (lvs)
Isocarpha R. Br. Compositae (Eup.-Ayap.). 5 trop. & warm Am. R: MSBMBG 22(1987)215. ***I. oppositifolia*** (L.) Cass. used as a tonic in Cuba
Isochilostachya Mytnik & Szlach. = *Polystachya* (but see ASBP 80(2011)80)
Isochilus R. Br. Orchidaceae (V 13d). 10 trop. Am. Cult. orn. epiphytes
Isochoriste Miq. = *Asystasia*
Isocoma Nutt. (~ *Haplopappus*). Compositae (Ast.-Mac.). 16 SW N Am. R: Phytol. 70(1991)70. Some cult. orn. ***I. tenuisecta*** Greene (*H. t.*, burroweed, Arizona)
Isodendrion A. Gray. Violaceae (III 1c). 4 Hawaii (1 extinct: ***I. pyrifolium*** A. Gray (form. on all is.), 3 rare). Shrubs
Isodesmia Gardner = *Chaetocalyx*
Isodictyophorus Briq. = *Plectranthus*
Isodon (Benth.) Spach (*Rabdosia*, ~ *Plectranthus*). Labiatae (VII 3b). 96 trop. & warm As., Afr. (2). R: JAA 69(1988)289.
Isoetaceae Reichb. Lycopsida-Isoetales. 1/192 cosmop. exc. Pac. is. Terr. or aquatic pls with lvs ± linr., flat to terete, arising from corm-like rhiz., apically tapered, basally spathulate, in bases of outermost of which few large megaspores with the many microspores at bases of inner ones
Genus: *Isoetes*
Isoetes L. Isoetaceae. Incl. *Stylites*, 192 cosmop. (Eur. 11) exc. Pac. is. Quillworts. R: NHBeih. 3(1962)3, Phytotaxa 277(2016)101. Rosette-pls (some decid. terr., most aquatic) with r. dichot. branched or tussock-forming with elongate dichot. branching stems & unbranched r. ('*Stylites*'). Spp. difficult to identify, usu. requiring microscopic examination of spores. Many allopolyploid hybrids incl. ***I. lacustris*** L. (decaploid, Eur.). Spores said to be disp. in excreta of earthworms. Crassulacean acid metabolism like xerophytes as daytime carbon limitation in oligotrophic aquatic habitats, but terr. temp. spp. without, even when submerged, though trop. alpine ones e.g. ***I. andicola*** (Amstutz) Gomez (*S. a.*, Andean Peru to 4750 m, discovered 1940) have variable levels of CAM & are unique in terr. pls as stomata 0 & overall structure comparable with Cretaceous *Nathorstia* Richter & ultimately Lepidodendrales (Sporne). Some cult. orn.
Isoetopsis Turcz. Compositae (Ast.). 1 temp. Aus.: ***I. graminifolia*** Turcz. – mycorrhizal ann., female florets 3-lobed with zygomorphic throat. R: OB 104(1991)123
Isoglossa Oersted. Acanthaceae (III 2c). 50 OW trop. to Arabia (As. 8, R: NJB 5(1985)1)). Some W Afr. spp. on 9-yr flowering cycles, ***I. woodii*** C.B. Clarke (SE Afr.) on 4–7-yr
Isolepis R. Br. (~ *Scirpus*). Cyperaceae (II 5). 69 temp. esp Aus. (29, 15 endemic) & Afr., trop. (esp. mts). R: KB 57(2002)273. Heterogeneous? ***I. cernua*** (Vahl) Roemer & Schultes (*Scirpus c.*, Euras., Medit., Aus., NZ, W N Am.) – grown in tube & sold as 'optical fibre pl.' or 'Dougal grass'
Isoleucas Schwartz. Labiatae (VI). 2 SE Yemen, Somalia. R: SGP 77(2007)234
Isoloma J. Sm. = *Lindsaea*
Isoloma (Benth) Decne. = *Kohleria*
Isolona Engl. Annonaceae (III 6). 19 trop. Afr., Madag. R: SBM 87(2009)30.Roux's Model. ***I. cauliflora*** Verdc. (Tanz.) – infls in leaf-litter
Isomacrolobium Aubrév. & Pellegrin (~ *Anthonotha*). Leguminosae (I 2). 12 W & C Afr. R: PEE 144(2011)65
Isomeris Nutt. ex Torrey & A. Gray = *Cleome*
Isometrum Craib = *Oreocharis* (but see ABY 8(1986)36)
Isonandra Wight (~ *Palaquium*). Sapotaceae (II). 10 S Ind., Sri Lanka
Isonema R. Br. Apocynaceae (IIa). 3 W Afr.

Isopappus Torrey & A. Gray = *Croptilion*

Isophysis T. Moore. Iridaceae (I). 1 Tasmania.: ***I. tasmanica*** (Hook.f.) T. Moore – G! R: FA 46(1986)4

Isoplexis (Lindl.) Loudon = *Digitalis*

Isopogon R. Br. ex J. Knight. Proteaceae (IV 4). c. 35 Aus. esp. SW (25). R: FA 16(1995)194. Like *Petrophile* but bracts fall before fr. Some cult. orn. (drumsticks)

Isopyrum L. Ranunculaceae (III 1). 4 Euras. (Eur. 1). R: NRBGE 28(1968)272. Alks. Some cult. orn. delicate herbs

Isostigma Less. Compositae (Cor.). 13 subtrop. S Am., on campos. R: BG 81(1926)241, Willd. 34(2004)535

Isotheca Turrill. Acanthaceae (2c). 1 Trinidad: ***I. alba*** Turrill

Isotoma (R.Br.) Lindl. (~ *Solenopsis*). Campanulaceae (III 1). 14 Aus. R: Pfl. IV,276b(1953)398

Isotrema Raf. = *Aristolochia*

Isotria Raf. Orchidaceae (II 1). 2 E US. R: FNA 26(2002)511

Isotropis Benth. Leguminosae (III 13). 14 Aus.

isphaghul seeds *Plantago ovata*

istle = ixtle

ita palm *Mauritia flexuosa*

Itaculumnia Hoehne = *Habenaria*

Italian alder *Alnus cordata*; **I. basil** *Ocimum basilicum* 'Fino Verde Compatto'; **I. cypress** *Cupressus sempervirens*; **I. millet** *Setaria italica*; **I. poplar** *Populus nigra*; **I. rye** *Lolium multiflorum*; **I. senna** *Senna italica*; **I. whisk** *Sorghum bicolor* Technicum Group

Itaobimia Rizz. = *Riedeliella*

Itasina Raf. (*Thunbergiella*). Umbelliferae (III 8). 1 S Afr.: ***I. filifolia*** (Thunb.) Raf. R: NRBGE 45(1988)93

Itatiaia Ule = *Tibouchina*

Itaya H. Moore. Palmae (III 2). 1 Amaz. Colombia, Ecuador, Peru, Braz.: ***I. amicorum*** H.E. Moore. R: Principes 16(1972)85

ité *Mauritia flexuosa*

Itea L. Iteaceae. Incl. *Choristylis*, 27 Himal. to Jap. & W Mal., E & S Afr., E N Am. (1). Cult. orn. everg. trees & shrubs esp. ***I. ilicifolia*** Oliv. (China) & ***I. virginica*** L. (E US); ***I. riparia*** Collett & Hemsl. (Myanmar) – rheophyte

Iteaceae J. Agardh (~ Grossulariaceae). Magnoliidae – Saxifragales. Incl. Pterostemonaceae, 2/29 E & SE As., N Am. Trees & shrubs, s.t. scandent. Lvs in spirals, simple with spiny or glandular teeth; stip. small or 0. Infl. racemes to panicles or corymbs, axillary (or term.). Fls 5-merous, K basally connate, lobes valvate, C valvate, A opp. K (s.t. with staminodes opp. C), G (2–5) to subinferior, each loc. with 4-many bitegmic ovules. Fr. a capsule with persistent P; seeds with large curved or elongate embryo in fleshy endosperm. n = 11

Genera: *Itea, Pterostemon*

One of the fragments of the shattered heterogeneous Grossulariaceae s.l.

Iteadaphne Blume (~ *Litsea*). Lauraceae (II). 1 Sumatra, Java: ***I. caudata*** (Nees) Li. R: ABY 7(1985)132

Iti Garnock-Jones & P. Johnson = *Cardamine*

Itoa Hemsl. Salicaceae (Flacourtiaceae). 2 S China, trop. As.

Ituridendron De Willd. = *Omphalocarpum*

Itysa Ravenna = *Calydorea*

Itzaea Standl. & Steyerm. Convolvulaceae (?4). 1 C Am.: ***I. sericea*** (Standl.) Standl. & Steyerm.

Iva L. Compositae (Helia.-Amb.). 9 N Am. (US & Canada 7 – R: FNA 21(2006)25) to WI. R: CN 42(2005)35. ***I. annua*** L. (sumpweed) – cypsela kernels eaten by Native Americans since prehistoric times, poss. oil source. See also *Cyclachaena, Euphrosyne, Hedosyne, Oxytenia*

Iva wine or **liqueur** flavoured with *Achillea erba-rotta* subsp. *moschata*

Ivania O. Schulz. Cruciferae (46). 2 Chile. R: HPB 15(2010)343

Ivanjohnstonia Kazmi = *Cynoglossum*

Ivesia Torrey & A. Gray = *Potentilla* (but see FNA 9(2014)219)

Ivodea Capuron. Rutaceae (I 2). 30 Madag. (28, 27 endemic), Comores. R: Adans. 37(2015)63

ivory nut *Phytelephas macrocarpa*; **vegetable i.** *Ammandra decasperma, P. aequatorialis* (best quality) etc.; **i.wood** *Siphonodon australis*

ivy *Hedera* spp. esp. *H. helix*; **Atlantic i.** *H. hibernica*; **Boston i.** *Parthenocissus tricuspidata*; **Canary i.** *H. algeriensis*; **grape i.** *Cissus* spp.; **Irish i.** *H. hibernica*; **Japanese i.** *H. rhombea*; **Kenilworth or Oxford i.** *Cymbalaria muralis*; **Persian i.** *H. colchica*; **poison i.** *Rhus radicans*; **Swedish i.** *Plectranthus verticillatus*

Ixanthus Griseb. Gentianaceae (~ *Blackstonia*). 1 Canary Is.: ***I. viscosus*** (Aiton) Griseb.

ixbut *Euphorbia lancifolia*

Ixchelia H. Ballard & Wahlert (~ *Hybanthus*). Violaceae. 2 C Am. R: Britt. 67(2015)275

Ixerba Cunn. Strasburgeriaceae (Ixerbaceae). 1 N NZ: ***I. brexioides*** Cunn.

Ixerbaceae Griseb. ex Doweld & Reveal = Strasburgeriaceae

Ixeridium (A. Gray) Tzvelev (~ *Ixeris*). Compositae (Lact.-Crep.). c. 13 E & SE As. to NG

Ixeris (Cass.) Cass. Compositae (Lact.-Crep.). 10 E & SE As.

Ixia L. Iridaceae (VI 5). 78 S Afr. (winter rainfall zone), natur. W Aus. R: JSAB 28(1962)45, SAJB51(1985)66, Bothalia 41(2011)83. Some poll. monkey beetles. Cult. orn. corms ((Afr.) corn lilies) esp. hybrids involving ***I. maculata*** L. (Cape). ***I. paniculata*** Delaroche (W S Afr.) – natur. subcosmop.

Ixianthes Benth. Stilbaceae (Scrophulariaceae s.l.). 1 S Afr.: ***I. retzioides*** Benth. – rheophyte, lipid-secreting trichomes suggesting poll. by large oil-collecting bees but only poss. sp. thought outside pl. range, poss. explaining low fr.- & seed-set, but in one pop. bee *Rediviva gigas* with 7-fold increase in seed-set. R: Strel. 10(2000)523

Ixiochlamys F. Muell. & Sonder (~ *Podocoma*). Compositae (Ast.-Pod.). 4 Aus. R: JABG 2(1980)241

Ixiolaena Benth. Compositae (Gnap.). 1 Aus.: ***I. viscosa*** Benth. See also *Leiocarpa*

Ixioliriaceae Nakai (~ Tecophilaeaceae). Magnoliidae – Asparagales. 1/3 E Med. to C As. Perennials with tunicated corms; alks 0. Lvs linr., flat. Infl. subumbellate, term.; fls bisex., reg., 3-merous. P 3 + 3, blue; A 3 + 3 inserted at base of P, anthers basifixed, longit.-dehiscent; $\overline{G}$ (3) with ∞ anatr. ovules in axile placentation, style 3-branched. Capsule loculicidal, apically dehiscent; seeds ∞, ovate to pyriform, reticulate, black, embryo straight in starch-free endosperm

Genus: *Ixiolirion*

Ixiolirion Fischer ex Herb. Ixioliriaceae (Tecophilaeaceae s.l.). 3 Egypt to C As. Cult. orn. bulbs esp. ***I. tataricum*** (Pallas) Herb. (*I. ledebourii*, *I. montanum*, SW & C As., Kashmir) – fls blue

Ixoca Raf. = *Silene*

Ixocactus Rizz. = *Phthirusa* (but see SB 16(1991))

Ixodia R. Br. Compositae (Gnap.-Cass.). 2 Aus. R: OB 104(1991)87

Ixodonerium Pitard. Apocynaceae (IIc). 1 SE As.: ***I. annamense*** Pitard

Ixonanthaceae Planch. ex Miq. Magnoliidae – Malpighiales. 3/21 trop. Trees & shrubs with simple entire to toothed lvs in spirals; stip. small or 0. Fls bisex., ± reg., ± hypogynous, in racemose to cymose axillary infls, oft. 5-merous; K imbr., s.t. with basal union, marcescent, C imbr. or convolute, marcescent, A 5–20 with widened filaments, free or adnate to annular to cupular intrastaminal nectary-disk, anthers with longit. slits, $\underline{G}$ 5), pluriloc. with axile-apical placentas & term. style, s.t. divided into locelli (cf. Linaceae) & apically 1-loc.; style & filaments folded in bud; 2 pend., anatr. bitegmic ovules per loc. Fr. a septicidal (s.t. also loculicidal) capsule with or without persistent C column; seeds arillate or basally winged, endosperm little or 0

Genera: *Cyrillopsis*, *Ixonanthes*, *Ochthocosmus*

Some genera now excl. as Irvingiaceae (via Simaroubaceae)

Ixonanthes Jack. Ixonanthaceae. 3 SE As., Mal. R: Blumea 26(1980)191. Bark of ***I. icosandra*** Jack used for tanning fishing-nets in Malay Pen.

Ixophorus Schldl. Gramineae (XXIV 4). 1 Mex.: ***I. unisetus*** (J. Presl) Schldl. R: IJPS 165(2004)1102

Ixora L. Rubiaceae (II 2). Incl. *Captaincookia*, *Doricera*, *Myonima*, *Versteegia*, c. 500 trop. (As. 200+, Afr. 37 – R: OBB 9(1998)67, Am. 60). Fls long-tubular, white, yellow, orange, pink or red, poll. by (?) Lepidoptera; some cauliflory. Shrubs or trees, some rheophytic, some cult. orn. greenhouse shrubs (incl. many hybrids) esp. ***I. coccinea*** L. (flame flower, S Ind.) – fls red. ***I. klanderiana*** F. Muell. (Queensland) – ed. fr.; ***I. margaretae*** (Hallé) Mouly & B. Bremer (*C. m.*, New Caled.) – remarkable pachycaul (Corner's Model) of v. restricted distr.; ***I. trilocularis*** (Balf.f.) Mouly & B. Bremer (*D. t.*, Rodrigues) – consp. foliar dimorphism

Ixorhea Fenzl (~ *Heliotropium*). Boraginaceae (Heliotropiaceae). 1 NW Arg.: ***I. tschudiana*** Fenzl

Ixtlania M.E. Jones = *Justicia*

ixtle or **ixtli fibre** *Agave* spp.; **Jaumave i.** *A. funkiana*; **tula i.** *A. lechuguilla*. See also *Yucca*

Ixyophora Dressler (~ *Chondrorhyncha*). Orchidaceae (V 12j). 5 E Andes slopes. R: Lankesteriana 5(2005)95

Izabalaea Lundell = *Agonandra*

Izozogia Navarro. Zygophyllaceae. 1 Bolivia: ***I. nellii*** Navarro. R: Novon 7(1997)1

J

jaba *Raphanus sativus*

Jablonskia Webster (~ *Securinega*). Phyllanthaceae (Jablonsk.); Euphorbiaceae s.l.). 1 N trop. S Am.: ***J. congesta*** (Muell. Arg.) Webster. R: SB 9(1984)232

jaborandi Name for several pungent aromatic S Am. pls which promote saliva when chewed, esp. Piperaceae & Rutaceae (notably *Pilocarpus* spp. which provide the drug jaborandi (dried lvs))

Jaborosa Juss. Solanaceae (IV). 22 temp. S Am. (Chile 21, 11 endemics). R: Kurtziana 19(1987)77

jaboticaba *Myrciaria* spp. esp. *M. cauliflora*

jaca = jak

Jacaima Rendle = *Matelea*

jacaranda *Dalbergia nigra*

Jacaranda Juss. Bignoniaceae (1). c. 50 trop. Am. R: FN 25(1992)51. Shrubs & trees (Scarrone's Model) usu. with bipinnate lvs (fern-trees); 2 or 3 buds per axil. Some timber (palisander), e.g. ***J. copaia*** (Aubl.) D. Don (futi, futui, NE S Am.) & pulp; some medic. esp. bark; cult. orn. esp. as street-trees, esp. ***J. mimosifolia*** D. Don (*J. ovalifolia*, NW Arg.) – invasive in S Afr. so new plantings banned

Jacaratia A. DC. Caricaceae. 7 trop. Am. R: V.M. Badillo (1971) *Caric.*: 38. Locally eaten fr. ***J. dolichaula*** (Donn.Sm.) Woodson (S Mex. to Panamá) – loses lvs in lower part of crown during flowering, sphingid-poll., dioec. with females resembling male infl. as petals (5) resemble male fl. buds & lobed stigma like a female fl., but no nectar reward!

jacareuba *Calophyllum brasiliense*

Jacea Mill. = *Centaurea*

jack = jak; **j.ass clover** *Cleome refracta*; **j.ass, laughing** *Arethusa bulbosa*; **j.** (i.e. jakes = latrine)**-by-the-hedge** *Alliaria petiolata*; **j.-in-the-box** *Hernandia nymphaeifolia*; **j.-in-the-pulpit** *Arum maculatum, Arisaema* spp.; **j. fruit** = jak; **long j.** *Triplaris* spp.; **j. pine** *Pinus banksiana*; **j.wood** *Cryptocarya glaucescens*

Jackia Wall. = *Jackiopsis*

Jackiopsis Ridsd. Rubiaceae (Vang.). 1 W Mal.: ***J. ornata*** (Wall.) Ridsd. R: Blumea 25(1979)295

jacks, whistling *Gladiolus communis*

Jacksonia R. Br. ex Sm. (~ *Pultenaea*). Leguminosae (III 13). 74 Aus. R: AusSB 20(2007)476. Cult. orn. incl. ***J. scoparia*** R. Br. (stinkwood, E Aus.) – burning wood foetid

Jacmaia R. Nordenstam (~ *Senecio*). Compositae (Sen.-Sen.). 1 Jamaica: ***J. incana*** (Sw.) R. Nordenstam

jacobaea *Jacobaea elegans*

Jacobaea Mill. (~ *Senecio*). Compositae (Sen.-Sen.). c. 35 OW. Many alks toxic to stock; weeds & some cult. orn. ***J. aquatica*** (Hill) Gaertn. f. & al. (Eur.) – forms hybrid swarms (***J. × ostenfeldii*** (Druce) B. Bock) with *J. vulgaris* in GB; ***J. elegans*** (L.) Moench (*S. e.*, jacobaea, S Afr., natur. Calif.) – cult. orn. with purple (or white) ray-florets; ***J. maritima*** (L.) Pelzer & Meijden (*S. bicolor, S. cineraria*, dusty miller, W & C Medit., natur. GB) – cult. orn., alleged efficacious in eye complaints, forms hybrid swarms (***J. × albescens*** (Burb. & Colgan) Verloove & Lambinon) with *J. vulgaris* in UK; ***J. vulgaris*** Gaertn.f. (*S. j.*, ragwort, stinking Willie [i.e. Duke of Cumberland] (Scotland), Eur., Medit., natur. widely) – when attacked by cinnabar moth prod. 'regrowth' shoots with small fr. giving less competitive seedlings, used against sore throats & rheumatism

Jacobean lily *Sprekelia formosissima*

Jacobinia Nees ex Moricand = *Justicia*

Jacob's coat *Alternanthera ficoidea* 'Bettzichiana', *Acalypha amentacea* subsp. *wilkesiana*; **J.'s ladder** *Polemonium* spp. esp. *P. caeruleum*

Jacobsenia L. Bolus & Schwantes (~ *Drosanthemum*). Aizoaceae (V). Incl. *Drosanthemopsis* 3 W S Afr. R: MIABH 27(1997)121. Heterogeneous? Cult. orn. shrublets

Jacquemontia Choisy. Convolvulaceae (10). c. 120 trop. & warm, esp. Am. (Mal. 1). Some cult. orn. lianes. ***J. ovalifolia*** (Choisy) H. Hallier (trop. Afr., Am., Hawaii) – form. medic. in Hawaii

Jacquesfelixia J. Phipps = *Danthoniopsis*

Jacqueshuberia Ducke. Leguminosae (I 4). 7 Braz., Colombia. Fls (red, purple or yellow) expose partly joined filaments & pollen sticky with viscin threads at night – bat-poll.

Jacquinia Dill. ex L. Primulaceae (Theophrastaceae). Excl. *Bonellia* c. 10 Carib., ***J. armillaris*** Jacq. ext. to N S Am. Some fr. used to poison fish; some cult. orn. shrubs incl. ***J. armillaris*** (*J. arborea, J. barbasco*) – fish-poison, yellow & brown seeds used as beads; ***J. keyensis*** Mez (Florida, WI) – source of cudjoe wood; ***J. pungens*** A. Gray (Mex.) – leafing in dry season, lvs spine-tipped

Jacquiniella Schltr. Orchidaceae (V 13b). 12 trop. Am. Cult. orn. epiphytes

jade flower *Viburnum macrocephalum* f. *keteleeri*; **j. plant** *Crassula arborescens, C. ovata*; **j. vine** *Strongylodon macrobotrys*

Jadunia Lindau (~ *Strobilanthes*). Acanthaceae (III 2c). 2 NG

Jaegeria Kunth. Compositae (Mill.-Jaeg.). 9 Mex. to Urug., Galápagos. R: Phytol. 55(1984)243

Jaeschkea Kurz. Gentianaceae. 3 Himal.

Jagera Blume. Sapindaceae. 2 E Mal. to NE Aus. R: FM I,11(1994)614. Some unbranched pachycauls (Corner's Model); fr. hairs irritant. Some fish-poisons, e.g. ***J. pseudorhus*** (A. Rich.) Radlk. (foambark, NE Aus., NG) – saponins (also foaming agents)

jaggery crude sugar from palms or sugar-cane

Jahnia Pittier & S.F. Blake = *Turpinia*

Jailoloa Heatubun & W. Baker (~ *Ptychosperma*). Palmae (V 14i). 1 E Halmahera: ***J. halmaherensis*** (Heatubun) Heatubun & W. Baker. R: KB 69[9525](2014)5

jak or **jack (fruit)** *Artocarpus heterophyllus*

jalap *Ipomoea purga*; **Brazilian j.** *Merremia tuberosa*; **false j.** *Mirabilis jalapa*; **Indian j.** *Operculina turpethum*

Jaimehintonia B. Turner. Amaryllidaceae (Alliaceae-Gill.). 1 Mex.: ***J. gypsophila*** B. Turner. R: Novon 3(1993)86

Jainia Balakr. = *Coptophyllum*

Jalcophila Dillon & Sagást. Compositae (Gnap.-Gnap.). 3 Andes of Bolivia, Peru, Ecuador. R: BotJLS 106(1991)185

Jaliscoa S. Watson. Compositae (Eup.-Oxy.). 3 Mex. R: MSBMBG 22(1987)440

Jaltomata Schldl. (~ *Saracha*). Solanaceae (IV 7). Incl. *Hebecladus*, c. 60 trop. & warm Am. Coloured nectar. Fr. sold in Mex., lvs used like spinach

jam tree *Muntingia calabura*

Jamaica bark *Exostema* spp.; **J. cherry** *Muntingia calabura*; **J. dogwood** *Piscidia piscipula*; **J. ebony** *Brya ebenus*; **J. honeysuckle** *Passiflora laurina*; **J. horse bean** *Canavalia ensiformis*; **J. pepper** *Pimenta dioica*; **J. plum** *Spondias mombin*; **J. sarsaparilla** *Smilax regelii*; **J. sorrel** *Hibiscus sabdariffa*

Jamaicella Braem = *Tolumnia*

jamba *Eruca vesicaria*

jamberry *Physalis philadelphica*

jambolan *Syzygium cumini*

Jambosa Adans. = *Syzygium*

jambu *Syzygium* spp. esp. *S. jambos*; **wax j.** *S. samarangense*

jambú *Acmella oleracea*

Jamesbrittenia Kuntze (~ *Sutera*). Scrophulariaceae (Manuleeae). 84 trop. & S Afr., 1 N Afr. to Ind. R: O. Hilliard (1994) *The M. a tribe of S.* ***J. atropurpurea*** (Benth.) Hilliard (*S. a.*, S Afr.) – fls a saffron subs.; ***J. fodina*** (Wild) Hilliard (*S. f.*, Zimbabwe) – ash has 15.3% nickel & 4.8% chromium by wt.

Jamesia Torrey & A. Gray. Hydrangeaceae (I). 2 W N Am. R: Britt. 41(1989)335. ***J. americana*** Torrey & A. Gray – cult. orn. shrub with pink fls

Jamesianthus S.F. Blake & Sherff. Compositae (Tag.-Pect.). 1 Alabama: ***J. alabamensis*** S.F. Blake & Sherff. R: FNA 21(2006)377

Jamesonia Hook. & Grev. Pteridaceae (III). Incl. *Eriosorus*, c. 55 C Am., WI, N Andes, SE Braz., Tristan da Cunha. R: CGH 191(1962)109. Xerophytes with fronds of indefinite apical growth

Jamestown weed *Datura stramonium*

jamun *Syzygium cumini*

Janakia Joseph & Chandra = *Decalepis*

janatsi *Debregeasia longifolia*

Jancaea Boiss. (*Jankaea*; ~ *Ramonda*). Gesneriaceae (III 2e). 1 Mt Olympus, Greece: ***J. heldreichii*** (Boiss.) Boiss. – Tertiary relic (cf. *Haberlea*, *Ramonda*), cult. orn.; hybrids synth. with *R. myconi* (Pyrenees)

Jankaea Boiss. = praec.

Janotia J. Leroy (~ *Breonia*). Rubiaceae (I 2). 1 Madag.: ***J. macrostipula*** (Capuron) J. Leroy. R: Adans. 14(1975)682

Jansenella Bor. Gramineae (XVIII). 1 Ind. & Sri Lanka: ***J. griffithiana*** (C. Muell.) Bor. R: KB 10(1955)96

Jansonia Kippist = *Gastrolobium*

Janusia A. Juss. ex Endl. Malpighiaceae. (Incl. *Schwannia*) 18 Calif. to Arg.

Japan or **Japanese alder** *Alnus japonica*; **J. anemone** *Anemone* × *hybrida*; **J. apricot** *Prunus mume*; **J. arrowroot** *Pueraria montana* var. *thomsonii*; **J. artichoke** *Stachys affinis*; **J. ash** *Fraxinus nigra* subsp. *mandshurica*; **J. barnyard millet** *Echinochloa esculenta*; **J. birch** *Betula maximowicziana*; **J. cedar** *Cryptomeria japonica*; **J. chestnut** *Castanea crenata*; **J. clover** *Kummerowia striata*; **J. cypress** *Chamaecyparis obtusa*; **J. galls** *Rhus chinensis*; **J. ginger** *Zingiber mioga*; **J. greens** *Brassica juncea* 'Crispifolia'; **J. honeysuckle** *Lonicera japonica*; **J. hop** *Humulus japonicus*; **J. hornwort** *Cryptotaenia canadensis*; **J. hyacinth** *Ophiopogon* spp.; **J. iris** *Iris ensata*; **J. knotweed** *Reynoutria japonica*; **J. lacquer** *Rhus verniciflua*; **J. larch** *Larix kaempferi*; **J. laurel** *Aucuba japonica*; **J. lilac** *Syringa reticulata*; **J. lime** *Tilia japonica*; **J. maple** *Acer palmatum*; **J. medlar** *Eriobotrya japonica*; **J. millet** *Echinochloa esculenta*; **J. peppermint** *Mentha canadensis*; **J. morning glory** *Ipomoea nil*; **J. parsley** *C. canadensis*; **J. pear** *Pyrus pyrifolia*; **J. pepper** *Zanthoxylum piperitum*; **J. persimmon** *Diospyros kaki*; **J. privet** *Ligustrum japonicum*; **J. raisin-tree** *Hovenia dulcis*; **j. red pine** *Pinus densiflora*; **J. toad-lily** *Tricyrtis hirta*; **J. walnut** *Juglans ailanthifolia*; **J. wax tree** *Rhus succedanea*; **J. white pine** *P. parviflora* var. *pentaphylla*; **J. yew** *Taxus cuspidata*

Japonasarum Nakai = *Asarum*

Japonoliriaceae Takht. = Petrosaviaceae

Japonolirion Nakai (~ *Tofieldia*). Petrosaviaceae (Japonoliriaceae). 1 Jap.: ***J. osense*** Nakai – vulnerable relic on serpentine

japonica *Chaenomeles speciosa*

jaragua grass *Hyparrhenia rufa*

Jaramilloa R. King & H. Robinson (~ *Eupatorium*). Compositae (Eup.-Oxy.). 2 N Colombia. R: MSBMBG 22(1987)450

Jarandersonia Kosterm. Malvaceae (Brown.; Tiliaceae). 6 Borneo. R: SB 37(2012)217

Jarava Ruíz & Pavón (~ *Stipa*). Gramineae (X). Excl. *Pappostipa*, 30 S Am. R: Gayana (Bot.) 59(2002)29. ***J. ichu*** Ruíz & Pavón (*S. i.*, ichu, Mex. to Arg. highlands) – fodder; ***J. plumosa*** (Spreng.) S. Jacobs & Everett (Arg., Chile, Urug.) – invasive S Aus.

Jardinea Steud. (~ *Phacelurus*). Gramineae (XXII 3). 3 trop. Afr.

Jarilla Rusby. Caricaceae. 3 Mex., Guatemala. R: ABM 20(1992)78. Tuberous herbs of montane decid. forest

jarrah *Eucalyptus marginata*

Jasarum Bunting. Araceae (VII 11). 1 Guyana, Venez.: ***J. steyermarkii*** Bunting – only submerged aroid in S Am.

Jasione L. Campanulaceae (I 4). 15 Eur. (9), Medit., SW As. R: BBC 48,2(1931)8. Some cult. orn. incl. ***J. montana*** L. (sheep's-bit, policeman's buttons, Eur., N Afr.) – fls in dense heads, blue

Jasionella Stoy. & Stefanoff = *Jasione*

jasmin, jasmine *Jasminum* spp.; **allspice j.** *Gelsemium* spp.; **Arabian j.** *Jasminum sambac*; **Cape j.** *Gardenia augusta*; **Catalonian j.** *J. grandiflorum*; **Cayenne j.** *Catharanthus roseus*; **common j.** *J. officinale*; **crepe j.** *Tabernaemontana divaricata*; **Italian j.** *J. humile* 'Revolutum'; **Madag. j.** *Marsdenia floribunda*; **rock j.** *Androsace* spp.; **Spanish j.** *J. grandiflorum*; **star j.** *Trachelospermum jasminoides*; **white j.** *J. officinale*; **winter j.** *J. nudiflorum*; **yellow j.** *J. humile*, *J. mesnyi*

Jasminocereus Britton & Rose. Cactaceae (III 1). 1 Galápagos: ***J. thouarsii*** (A. Weber) Backeb. – cult. orn. tree to 8 m with bole to 30 cm diam.

Jasminochyla (Stapf) Pichon = *Landolphia*

Jasminum Tourn. ex L. Oleaceae (1). Incl. *Menodora*, c. 200+ trop. (Mal. 52 – R: Sandakania 5(1994)3) to (few) temp. OW (China 46 – R: BBR 4,1(1984)88). P. Green & D. Miller (2009) *The genus J. in cultivation*. Jasmin(e), jessamine (arch.). Decid. & everg. shrubs & lianes with berries (dry in *J. mesnyi* & *J. nudiflorum*) & scented white, yellow or pink fls (many heterostylous) used in scent-making & perfuming tea, the oil (c. £5000 per kg in 1993) called malatti (Ind.); many cult. orn. esp. ***J. auriculatum*** Vahl (Himal. to Sri Lanka) – cult. Ind. for scent & assoc. (yuthika) with Krishna; ***J. beesianum*** Forrest & Diels (W China) – fls pale red; ***J. fluminense*** Vell. (trop. Afr., Arabia, Seychelles) – natur. S Am., invasive SE US, Hawaii). ***J. grandiflorum*** L. (*J. officinale* f. *grandiflorum*, Catalonian or Spanish j., S Arabia, NE Afr., Himal.) – more robust but less hardy than *J. o.*, taken by Moors to Spain, white fls (C lobes 15 – 20 mm) used in scent-making (8000–10000 fls per kilo, 8 kg giving 1 g (25 drops) of oil) bulk of comm. j. oil, 300 compounds incl. benzylacetate (readily artif. synth.) & indole (!), for scent, e.g. Chanel N° 5 (1921, Grasse, Fr.) & methyl jasminate (now synth.) first in Eau Sauvage (Dior, 1966), esp. Egypt, & in tea; ***J. humile*** L. (yellow j., Himal., invasive NZ) esp. '**Revolutum**' (Italian j.) – fragrant yellow fls; ***J. laurifolium*** Roxb. ex Hornem. (*J. nitidum*, Ind.) – fls white, fragrant; ***J. mesnyi*** Hance (*J. primulinum*, yellow j., W China) – fls yellow; ***J. nudiflorum*** Lindl. (winter j., W China) – yellow fls in winter; ***J. odoratissimum*** L. (Macaronesia) – yellow fls used in scent-making; ***J. officinale*** L. (common j., Caucasus, Himal. & SW China, long cult. China, natur. SE Eur., SW As.) – white fls (C lobes 0–11 mm.) used in scent-making, much cult. orn. esp. f. ***affine*** (Lindl.) Rehder ('*J. grandiflorum*'); ***J. parkeri*** Dunn (NW Ind.) – dwarf shrub endangered in wild; ***J. polyanthum*** Franch. (Myanmar, China, invasive NZ) – much cult. pot-pl. with v. fragrant fls pinkish without; ***J. sambac*** (L.) Aiton (Arabian j., bela, beli, zambac, ? Ind.) – nat. fl. Philippines (1930s), Indonesia (1990), fls (mohle flowers, moli) used to scent tea (1 M kg fls per annum), incl. double-flowered '**Grand Duke of Tuscany**'

Jasonia (Cass.) Cass. (~ *Pulicaria*). Compositae (Inul.-Inul.). Incl. *Chiliadenus* 9 Medit. R: ABotMalac. 29(2004)222, 31(2005)82. ***J. glutinosa*** (L.) DC. (SW Eur., NW Afr.) – imp. local medic.; ***J. tuberosa*** (L.) DC. – herbal tea

jatamansi *Nardostachys jatamansi, Diplotaenia cachrydifolia*

Jateorhiza Miers. Menispermaceae (III). 2 trop. Afr. ***J. palmata*** (Lam.) Miers (*J. columba*, calumba, colomba, columba) – liane the source of *Radix Columbae* (c. root), a tonic

jatoba *Hymenaea courbaril*

Jatropha L. Euphorbiaceae (Crot.-Jat.). 180+ trop. & warm (Afr. 70), N Am. R: UCPB 74(1979). Monoec. or dioec. trees (Chamberlain's & Leeuwenberg's Models), shrubs & herbs incl. annuals; alks. ***J. cuneata*** Wiggins & Rollins (SW N Am.) – stems for basket-making; ***J. curcas*** L. (physic nut, purging nut, pulza, trop. Am.) – cult. as boundary hedge, (foaming) chewing-stick in Nigeria, seed-oil for candle- & soap-making, mooted drought-tolerant, high-octane biofuel, purgative, seed provides short-lived taper when lodged in cleft stick, shown to have antitumour activity; ***J. dioica*** Sessé ex Cerv. (*J. spathulata*, SW N Am.) – bark for tanning & dyes; other spp. cult. orn. esp. ***J. podagrica*** Hook. (C Am.) – stem grotesquely swollen, fls red, or locally medic. (e.g. ***J. unicostata*** Balf.f. (Socotra)). See also *Cnidoscolus*

jau *Casuarina equisetifolia*

Jaubertia Guillemin = *Plocama*

jaumave fibre *Agave funkiana*

Jaumea Pers. Compositae (Tag.-Jaum.). 2 W N Am., S S Am.

Jaundea Gilg = *Rourea*

Java almond *Canarium luzonicum*; **J. cardamom** *Amomum maximum*; **J. fig** *Ficus benjamina*; **J. grass** *Polytrias amaura*; **J. indigo** *Marsdenia tinctoria*; **J. moss** *Taxiphyllum barbieri*; **J. plum** *Syzygium cumini*

Javieria Archila & al. = *Brassavola*

Javorkaea Borh. & J.-Komlódi (~ *Rondeletia*). Rubiaceae (I 5). 1 Honduras: ***J. hondurensis*** (Donn. Sm.) Borh. & J.-Komlódi. R: ABH 29(1983)16

Jedda Clarkson. Thymelaeaceae (Thym.-Daph.). 1 Cape York, NE Aus.: ***J. multicaulis*** Clarkson – cryptogeal germ. (only T. & only Aus. dicot thus), resistant to fire & grazing. R: Austrobaileya 2(1986)203

Jefea Strother. Compositae (Helia.-Ecl.). 5 S US to Guatemala. R: SBM 33(1991)22

Jeffersonia Barton. Berberidaceae (II 2; Podophyllaceae). 1 NE As., 1 E N Am. R: KB 1920:242. Cult. orn. herbs incl. ***J. diphylla*** (L.) Pers. (E N Am.) – local med.

Jeffreya Wild. Compositae (Ast.-Hom.). 2 trop. Afr. R: FTEA Comp.,2(2002)476

Jeffreyeia H. Robinson & al. (~ *Vernonia*). Compositae (Vern.-Erl.). 5 E Afr. R: PhytoKeys 39(2014)54
jeheb nut *Cordeauxia edulis*
Jehlia Rose = *Lopezia*
Jejewoodia Szlach. (~ *Ceratochilus*). Orchidaceae (V 16c). 6 Borneo
Jejosephia A.N. Rao & Mani = *Trias* (but see JETB 7(1985)216)
jellico *Berula bracteata*
jelutong *Dyera costulata*
jengkol *Archidendron jiringa*
Jenkinsia Griff. = *Miquelia*
Jenmaniella Engl. Podostemaceae (III). 7 NE S Am. R: Med.Bot.Mus.Utrecht 107(1951)119,137
Jennyella Lueckel & Fessel = *Houlletia*
Jensenobotrya Herre. Aizoaceae (V). 1 SW Afr.: ***J. lossowiana*** Herre – cult. orn. dwarf shrub. R: H. E. K. Hartmann, *Ill. Handb. Succ. Pls*, Aiz. F–Z(2002)67
Jensia Baldwin (~ *Madia*). Compositae (Mad.-Mad.). 2 Calif. R: FNA 21(2006)301
Jepsonia Small. Saxifragaceae (I 6). 3 S Calif. R: Britt. 21(1969)286. ***J. parryi*** (Torrey) Small – cult. orn. herb with lvs in spring, withering before fls appear in autumn; corm-like rhiz.
jequerity seeds *Abrus precatorius*
jequitiba *Cariniana* spp.
Jerdonia Wight. Gesneriaceae (III 2a). 1 S Ind.: ***J. indica*** Wight. Form. in Scrophulariaceae s.l.
jereton *Schefflera morototoni*
Jericho, rose of *Anastatica hierochuntica*, *Selaginella lepidophylla*
jerry-jerry *Ammannia baccifera*
Jersey long jacks *Brassica oleracea* Acephala Group
Jerusalem artichoke *Helianthus tuberosus*; **J. cherry** *Solanum pseudocapsicum*; **J. cross** *Silene chalcedonica*; **J. rye** *Triticum turgidum* Polonicum Group; **J. sage** *Phlomis fruticosa*; **J. thorn** *Parkinsonia aculeata*
jessamine *Jasminum* spp.
Jessea H. Robinson & Cuatrec. Compositae (Sen.-Sen.). 4 Costa Rica, Panamá. R: Novon 4 (1994)49, BJ 118(1996)147
Jessenia Karsten = *Oenocarpus*
Jesuits' bark *Cinchona* spp.; **J. nut** *Trapa natans*; **J. tea** *Otholobium glandulosum*
jewel orchid *Anoectochilus* spp., *Ludisia discolor*; **j.weed** *Impatiens* spp.
Jew's apple *Solanum melongena*; **J. mallow** *Corchorus olitorius*; **J. plum** *Spondias cytherea*
jiaogulan *Gynostemma pentaphyllum*
jiaomu *Zanthoxylum* spp. esp. *Z. bungeanum*
jicama *Pachyrhizus erosus*
jiggle-joggles *Briza media*
Jimson weed *Datura stramonium*
jiquitiba *Cariniana* spp.
jitta *Halfordia kendack*
Jirawongsea Picheans. = *Boesenbergia* (but see Fol. Malays. 9(2008)2)
Joannesia Vell. Euphorbiaceae (Crot.-Jat.). 1 Venez., 1 Amaz. Braz., 1 coastal Braz. Monoec. trees. ***J. princeps*** Vell. (araranut tree, S Braz. coast) – source of anda-assy oil, purgative & for skin disease, timber good
Jobinia Fourn. Apocynaceae (Vc; Asclepiadaceae III 4). 25 trop. S Am. R: AMBG 99(2013)57
jobo *Spondias mombin*
Job's tears *Coix lacryma-jobi*
jocote *Spondias purpurea*
Jodina Hook. & Arn. ex Meissn. (*Iodina*). Santalaceae (Cervantesiaceae). 1 S Braz., Urug., Arg.: ***J. rhombifolia*** (Hook. & Arn.) Reissek
Jodrellia Baijnath = *Bulbine* (but see FTEA Asphod.(2002)13)
Joe-pye weed *Eutrochium purpureum*
Johanneshowellia Reveal (~ *Eriogonum*). Polygonaceae (I 1). 2 W US. R: FNA 5(2005) 443
Johannesteijsmannia H. Moore. Palmae (III 4b). 4 W Mal. R: GBS 26(1972)63. Some cult. orn. (esp. ***J. altifrons*** (Reichb.f. & Zoll.) H. Moore (vogue pot-pl. in hotels etc.) with large scarcely-divided lvs used as thatch
Johnny-go-to-bed [at noon] *Tragopogon pratensis*; **J.-jump-up** *Viola tricolor*
Johnson grass *Sorghum halepense*

Johnsonia R. Br. Asphodelaceae (Hemerocallidaceae). 5 SW Aus. R: FA 45(1987)242

Johnsoniaceae Lotsy = Asphodelaceae

Johnstonalia Tortosa (*Johnstonia*). Rhamnaceae (4). 1 Peru: ***J. axilliflora*** (M. Johnston) Tortosa. R: Novon 16(2006)433

Johnstone River hardwood *Backhousia bancroftii*

Johnstonella Brand = *Cryptantha* (but see SB 37(2012)754)

Johnstonia Tortosa = *Johnstonalia*

Johrenia DC. Umbelliferae (III 10). 5 E Medit. (Eur. 1). R: Willd. 37(2007)491

Johreniopsis Pim. = *Dichoropetalum*

joint *Cannabis sativa* 'subsp. *indica*'

Joinvillea Gaudich. ex Brongn. & Gris. Joinvilleaceae. 2 W Mal., Solomon & Caroline Is. to Hawaii. Shoots to 5 m long

Joinvilleaceae A.C. Sm. & Toml. (~ Flagellariaceae). Magnoliidae – Poales. 1/2 W Mal. & Pac. is. R: T 19(1970)887. Coarse erect herbs, with unbranched hollow (exc. nodes) stems & sympodial rhiz. without sec. growth; cell-walls ± strongly silicified like Gramineae. Lvs in spirals, with open sheath, grassy lamina, ligule & pr of auricles at base of lamina, which is plicate in bud & parallel-veined. Fls bisex., 3-merous, reg., in term. much-branched bracteate panicles with caducous bracteoles; P 3 + 3, imbr., marcescent, s.t. basally connate, A 6 with basifixed anthers with longit. slits, $\underline{G}$ (3), 3-loc. with term. styles s.t. basally connate & 1 orthotropous pend. ovule per loc. on axile placenta. Fr. ± 3-quetrous drupe with 1 endocarp & 1–3 seeds with copious starchy endosperm capped by small undifferentiated embryo
Genus: *Joinvillea*, form. in Flagellariaceae (q.v.)

jo-jo weed *Soliva sessilis*

jojoba *Simmondsia chinensis*

Jollydora Pierre ex Gilg. Connaraceae (II). 4 Liberia, E Nigeria to Angola. R: AUWP 89–6(1989)284. Usu. unbranched treelets (Corner's Model)

Jonesiopsis Szlach. = *Caladenia*

Jonesyella Szlach. = *Caladenia*

jonkong *Dactylocladus stenostachys*

Jonopsidium Reichb. = *Ionopsidium*

jonquil *Narcissus jonquilla*, (NSW) *N. papyraceus*

Joosia Karsten. Rubiaceae (I 1). 11 W S Am. R: Britt. 49(1997)30

Jordaaniella H. Hartman (~ *Cephalophyllum*). Aizoaceae (V). 4 SW Cape. R: H. E. K. Hartmann, *Ill. Handb. Succ. Pls*, Aiz. F–Z(2002)67. Cult. orn. mat-formers

Jorena Adans. = ?

Joseanthus H. Robinson (~ *Vernonia*). Compositae (Vern.-Pip.). 5 Ecuador, Colombia

Josephinia Vent. Pedaliaceae (3). 3–4 Kenya & Somalia (1), trop. & arid Aus. (2–3), 1 ext. to Mal.

Joseph's coat *Amaranthus tricolor*

Joshua tree *Yucca* spp. esp. *Y. brevifolia*

Jossinia Comm. ex DC. = *Eugenia*

jostaberry *Ribes* × *nidigrolaria*

Jostua Luer = *Masdevallia*

Jouvea Fourn. Gramineae (XXIX 1). 2 Baja Calif. to Panamá. R: BTBC 66(1939)315

Jouyella Szlach. = *Bipinnula*

Jovellana Ruíz & Pavón. Calceolariaceae (Scrophulariaceae s.l.). 4 NZ, Chile. Some cult. orn. like *Calceolaria*

Jovetia Guédès. Rubiaceae (II 5). 2 Madag.

Jovibarba (DC.) Opiz = *Sempervivum*

jowar *Sorghum bicolor*

Joycea Linder = *Rytidosperma* (but see Telopea 6(1996)611)

joyweed *Alternanthera* spp.

ju hua *Chrysanthemum morifolium*

Juania Drude. Palmae (IV 2). 1 Juan Fernandez: ***J. australis*** (Mart.) Hook.f. – cult. orn. dioec. palm, reduced to c. 1000 in wild (2004). R: GH 10(1969)386

Juanulloa Ruíz & Pavón. Solanaceae (IV 3). c. 11 trop. Am. R: Kurtziana 21(1991)209. Lianes, shrubs, s.t. epiphytes, s.t. myrmecophilous, ? bird-poll.

Jubaea Kunth. Palmae (V 8a). 1 coastal C Chile (excl. *Paschalococos* q.v.): ***J. chilensis*** (Molina) Baill. (*J. spectabilis*, Chilean wine-palm, coquito, honey-palm) – massive bole (widest primary growth – 1 m – known) tapped or felled for sap (up to 300 litres per tree) for treacle (palm honey) so that tree now rare in wild, but widely cult.; seeds ed., sold in UK supermarkets. R: IBM 56(1987)120

Jubaeopsis Becc. Palmae (V 8a). 1 Transkei: ***J. caffra*** Becc. – dichot. stem, 2n = 16–200. R: IBM 56(1987)122

Jubelina A. Juss. (~ *Diplopterys*). Malpighiaceae. 5 trop. Am. R: CUMH 17(1990)21

Judas tree *Cercis siliquastrum*; **J.'s bag** *Adansonia digitata*

Juelia Aspl. = *Ombrophytum*

jug orchid *Acanthephippium* spp.

Jugastrum Miers = *Eschweilera*

Juglandaceae DC. ex Perleb. Magnoliidae – Fagales. Incl. Rhoipteleaceae, 9/62 temp. & warm N hemisph. to Arg. & Mal. R: AMBG 65(1978)1058, 88(2001)231. Monoec. or dioec. wind-poll. trees & shrubs, aromatic (glands). Lvs in spirals (opp. (spirals in seedlings) in *Alfaroa* & *Oreomunnea*), (trifoliolate or) pinnate with aromatic peltate glands, leaflets subopp., oft. serrate; stip. usu. 0 (present in *Rhoiptelea*). Fls small, unisexual, in axils of bracts of catkins, s.t. in term. panicles, these unisexual or not; C 0, males with K (1–)4(5) ± adnate to bractlets, or K (& s.t. bractlets) 0, A (2–)5–40(–100+) in 1 or 2 or more whorls, with short filaments & anthers with longit. slits, females with bract & bracteoles oft. united & forming an involucre becoming a husk in fr., K 4 teeth or 0 (e.g. *Carya*), $\underline{G}$ (*Rhoiptelea*) or $\overline{G}$ (2(3)) with free styles s.t. basally united, rarely stigmas sessile, 2(3)-loc. below (4–8 with extra partitions in some), 1-loc. apically, with 1 orthotropous (morphologically truly apical) unitegmic (anatr. bitegmic in *Rhoiptelea*) ovule. Fr. a nut or samara, or drupe-like ('tryma'), the soft husk (involucre) splitting to release bony pericarp; seed 1, pachychalazal, ± 0 endosperm & oily embryo with usu. 4-lobed massive cotyledons. x = 16

Genera: *Alfaroa, Carya, Cyclocarya, Engelhardia, Juglans, Oreomunnea, Platycarya, Pterocarya, Rhoiptelea*

Samaroid fr. with epigeal germ. seeds, drupe-like with hypogeal. Fossils of several genera (some extinct, form. N Am.) in Eur. etc. Timber: *Carya* (hickories), *Juglans* (walnut); ed. nuts: *Carya* (pecan) & *Juglans*; cult. orn.

Juglans L. Juglandaceae. 20 Medit. (Eur. 1) to E As., N Am. to Andes. R: AMBG 65(1978)1071. Walnuts, differing from *Carya* in chambered pith in twigs. Heterodichogamy recorded. Fr. drupe-like with hard pericarp splitting into 2 'boats' along midribs of carpels. Produce juglone, an allelopathic red crystalline quinone active against tobacco mosaic virus, in lvs, bark & fr.; some domatia. Ed. seeds, timber & cult. orn. ***J. ailanthifolia*** Carr. (*J. sieboldiana*, Jap. w., Jap.) – seeds ed.; ***J. cinerea*** L. (butternut, NE US) – bark medic., ed. seeds, timber for furniture; ***J. hindsii*** (Jepson) R. E. Sm. (claro w., N & C Calif.) – cabinet timber; ***J. mandshurica*** Maxim. (Manchurian w., NE China, Korea) – timber; ***J. neotropica*** Diels (*J. honorei*, cedro negro, S Am.) – ed. seeds, cabinet timber form. imp. Colombia; ***J. nigra*** L. (black or Am. w., E US, natur. C Eur.) – seeds ed. (used in confectionery in US), timber for furniture, gunstocks, WW I aeroplane propellors etc.; ***J. regia*** L. ((English, Black Sea, European or Persian) w., SE Eur., W–C As. & W China (? orig., cult. under date-palms in pre-Biblical Palestine), natur. US) – walnuts of commerce (form. a cordial prep., Queen Anne decanters engraved 'Noyer', as was until end 18 cent. a milk from blanched pulverized seeds soaked in vats), now allegedly improving cardiovascular health (oil rich in alpha-lineoleic acid & vitamin E poss lowering blood pressure), timber 'the foremost cabinet wood of N Am.' much used in 18 cent. for furniture (ousted by mahogany less vulnerable to woodworm), gunstocks & as veneer, seed-oil (poly-unsaturated omega-3 oil) used in salad oil (usu. from France), paint (esp. for white & bright 'oils' instead of linseed that yellows with time) & soap, dye from husk used to stain floors & dye wool & hair red-yellow (as by Anc. Gk. women), fertility symbol in Pliny's time, plant-pest control since time of Evelyn (1664), some hardier forms known as Carpathian w., hybrid (***J. x intermedia*** Jacques) with *J. nigra* breeds true

jujube *Ziziphus jujuba, Z. mauritiana* (**Indian j.**)

Julbernardia Pellegrin. Leguminosae (I 2). c. 11 trop. Afr. (c. 5 Guineo-Congolian forest, c. 5 major constituents of Sudano-Zambesian woodlands – miombo). timber, dyes etc. ***J. globiflora*** (Benth.) Troupin (*Berlinia g.*) – infusion used in ordeals in C Afr.

julep, mint see *Zea mays*

Julianiaceae Hemsl. = Anacardiaceae (I)

Julocroton Mart. = *Croton*

Julostylis Thw. = *Hibiscus* (but see BBAS 34(1993)280)

jumbie beans Various leguminous seeds esp. *Abrus precatorius, Adenanthera pavonina, Erythrina* spp., *Leucaena* spp., *Ormosia* spp.

Jumellea Schltr. (~ *Angraecum*). Orchidaceae (V 16d). 57 Afr. (2), Madag. (39), Masc. ***J. fragrans*** (Thouars) Schltr. (*A. f.*, Mauritius, Réunion) – flavouring (faham) for a fine tea (thé de Bourbon), icecream, rum etc.

Jumelleanthus Hochr. = *Hibiscus*

jumping beans See *Sebastiania*

Juncaceae Juss. Magnoliidae – Poales. Excl. Prioniaceae, 7/430 temp. & cold + trop. mts. R: FOW 6–8 (2002). Herbs glabrous (exc. *Luzula*), oft. with starchy symp. rhiz. (rarely ann.) rarely with mycorrhiza, silica bodies 0; vessels in all veg. organs. Stems oft. not ext. above ground exc. in fl., photosynthetic. Lvs in spirals, usu. basal, simple, parallel-veined with sheath & usu. flat to channelled terete, folded, or centric lamina (rarely 0), sheath oft. with apical auricles oft. ± confluent to form adaxial ligule. Fls usu. bisex. (rarely pl. dioec.) & wind-poll. (no nectaries when insect-poll.), small, (solit.) in heads or cymes; P 3 + 3 (or P 3) usu. greenish to blackish (rarely white or yellow), A 3 + 3 (or 3 + 0) with basifixed anthers with longit. slits & pollen in tetrads, $\underline{G}$ (3), 1- or 3-loc. (parietal or axile placentas (3 basal ovules in *Luzula*)), ovules 3–∞, anatropous, bitegmic, morphologically term. Fr. a loculicidal capsule (rarely indehisc.); embryo small, straight, embedded in starchy endopserm. n = 3–36

Genera: *Distichia, Juncus, Luzula, Marsippospermum, Oxychloe, Patosia, Rostkovia*, with greatest generic diversity in S hemisph. esp. Andes

App. closer to Gramineae & Cyperaceae than form. held, confirming laypersons' impressions! Some matting & chair-seats (*Juncus*); cult. orn.

Juncaginaceae Rich. Magnoliidae – Alismatales. Excl. Maundiaceae, 4/43 temp. & cold. Perenn. (usu.), oft. rhiz. herbs of wet places with bifurcating axes & elongate secretory canals, oft. cyanogenic; vessels confined to r. & rhiz. Lvs in spirals, mostly basal with ligule at junction of open sheath & slender lamina, s.t. not distinct. Fls small, wind-poll., bisex. or not, in term. bractless spikes or racemes (in *Lilaea*, each leaf-axil with 1 or 2 female fls & spike with male & bisex. fls); P 3 + 3 (2 + 2; 3; *Lilaea* with P 1 in males & bisex., 0 in females), A (3, 4)6(8; 1 in *Lilaea*), anthers with longit. slits, $\underline{G}$ 6 (alt. ones sterile or 0; 4 in *Tetroncium*), usu. adnate to C axis but distinct at maturity (free in *Cycnogeton*, 1-loc. (but prob. $\underline{G}$ 3 in *Lilaea* where style filiform to 30 cm), 1 erect, bitegmic, anatr. ovule per loc. Fr. a follicle (dry-indehiscent in *Lilaea*) or schizocarp with 1-seeded mericarps (*Triglochin*); seed 1 without endosperm, embryo straight. n = 6, 8, 9, 15

Genera: *Cycnogeton, Lilaea* (oft. placed in its own fam.), *Tetroncium, Triglochin*

Poss. incl. other allied aquatic fams (see p. 1000). Marshy habitats, largely coastal. Some *Triglochin* spp. ed., some poisonous (hydrogen cyanide)

Juncellus (Griseb.) C.B. Clarke = *Cyperus*

Juncus Tourn. ex L. Juncaceae. Excl. *Oreojuncus*, c. 310 cosmop. (Eur. 53 (*J. planifolius* R. Br. (S Pac. disjunct) natur. Ireland, Oregon & Hawaii), Aus. 47 (31 endemic) with 21 natur., N Am. 95) but rare in trop. R: FOW 7, 8 (2002). Rushes (*Distichia, Luzula, Oxychloe* prob. referable here). Usu. symp. rhiz. prod. 1 leafy shoot a yr.; lvs grassy, needle-like or centric, infls oft. appearing lat. on a leaf-like cylindrical stem, but actually bract takes over term. position. Some medic. e.g. chibasa (Colombia) – r. used in skin infections; pith of several spp. used as rush-lights before candles common. Some small e.g. ***J. bufonius*** L. (toad rush, cosmop.), others robust & used for matting (in Egypt since Neolithic) & chair-seats esp. ***J. effusus*** L. (Euras., N Am., S Am. mts, invasive in Aus. & NZ) in Anc. Rome for basket-weaving & mod. Jap. where cult. for weaving into standard floor-matting of Jap. house (igusa, tatami), some variants cult. orn. incl. '**Spiralis**' with spiral stems found in Ireland, W Scotland & Surrey. ***J. acutus*** L. (Medit., S Afr., Atlantic Is., Am.) – used for bone-setting in Crete since time of Dioscorides, invasive in Aus.; ***J. allioides*** Franch. (SW China) – large white P, insect-poll.; ***J. articulatus*** L. (N temp.) – invasive in Aus., NZ; ***J. balticus*** Willd. (*J. arcticus* var. *balticus*, N temp.) – plaited chaplet of 'Christ's' crown of thorns preserved in Notre Dame, Paris; ***J. b.*** subsp. ***andicola*** (Hook.) Snogerup (*J. arcticus* var. *andicola*, totorilla, trop. Am mts) – cult. Ecuador for handicrafts, also cattle fodder; ***J. microcephalus*** Kunth (trop. Am. mts) – ed. r. tubers; ***J. rigidus*** Desf. (*J. arabicus*, Medit., W & C As., Afr.) – used for making writing instruments in Egypt to 300 BC

June grass *Koeleria pyramidata*

Juneberry *Amelanchier* spp.

Junellia Moldenke (*Thryothamnus*). Verbenaceae (Verb.). Incl. *Urbania*, excl. *Mulguraea*, c. 40 S Am. R: Darw. 49(2011)48

Jungia L.f. (*Trinacte*). Compositae (Mut.-Nass.). 28 Mex. to Andes. R: ARSSL 4(1995)27

jungle rice *Echinochloa colona, E. frumentacea*

juniper *Juniperus* spp. esp. *J. communis*, (in Bible) *Retama raetam*; **African j.** *J. procera*; **plum j.** *J. drupacea*; **Rocky Mt. j.** *J. scopulorum*; **Sierra j.** *J. occidentalis*

Juniperus Tourn. ex L. Cupressaceae. 67 N hemisph. (Eur. 9) to trop. Afr. mts & WI. Junipers. R: R.P. Adams (2011) *Junipers of the world*, ed. 3. Monoec. or dioec. trees & shrubs, some with needle-like lvs, some with scale-like ones at maturity; females cones with 3–8 fleshy coalescing scales becoming berry-like & bird-disp., the seeds hard & unwinged. Timbers, cult. orn. & flavourings, some medic., source (esp. J. ashei Buchholz, SE US) of cedarwood oil for soap, disinfectant, insecticide (incl. termites) etc., ***J. bermudiana*** L. (Bermuda or Barbados cedar, Bermuda) – critically endangered, timber form. for pencils; ***J. californica*** Carr. (Calif., Baja C.) – 'fr.' ed.; ***J. chinensis*** L. (temp. E As.) – v. many orn. cvs; ***J. communis*** L. ((common) juniper, N temp.) – shrub with sweet aromatic fr. (200 t coll. wild pls in C Eur. imported to GB annually), medic. Ireland (diuretic) & N Am., 'fr.' distillate form. considered abortifacient (GB), used to flavour gin (= genever, Dutch for juniper), liqueurs & eaten with meat, many orn. cvs incl. creeping forms & slow-growing ones esp. **'Compressa'**; ***J. drupacea*** Labill. (plum j., E Med.) – cone ed., extract a kind of pekmez in S Turkey; ***J. excelsa*** M. Bieb. (E Medit. to C As.) – fr. used medic.; ***J. horizontalis*** Moench (N Am.) – many creeping cvs; ***J. monosperma*** (Engelm.) Sarg. (SW N Am.) – medic., ed. 'fr.', wood for Navajo dance-sticks; ***J. occidentalis*** Hook. (Sierra j., W US) – wood for fencing, ed. 'fr.'; ***J. oxycedrus*** L. (Medit.) – heartwood distilled to give parasiticidal oil of Cade; ***J. phoenicea*** L. (Medit., Canary Is.) – said to have been used to build Mogador in Morocco, oil from cones used in anc. Egypt; ***J. procera*** Hochst. ex Endl. (~ *J. excelsa*, E Afr. cedar, E Afr. mts, SE Arabia) – timber for pencils etc.; ***J. recurva*** Buch.-Ham. ex D. Don (coffin tree, c. wood, Himal.) – cult. orn.; ***J. sabina*** L. (savin, C & S Eur. to W As.) – young twigs medic. & insecticidal; ***J. scopulorum*** Sarg. (Rocky Mt j. or cedar, W & C N Am.) – imp. local medic., many orn cvs; ***J. squamata*** Buch.-Ham. ex D. Don (Ind. to Taiwan) – lvs & twigs burnt in Sikkim temples; ***J. virginiana*** L. (red cedar, (Virginian) pencil c., NE N Am.) – imp. local medic., wood for insect-proof chests & balls placed among clothes, pencils, oil for scenting soap, woodchips used as insecticidal mulch, cult. orn. (many cvs) but alt. host for apple rust

Juno Tratt. = *Iris*

Junopsis W. Schulze = *Iris* (subg. *Nepalenses*)

Juno's tears *Verbena officinalis*

junsai *Brasenia schreberi*

Jura turpentine *Picea abies*

Jurinea Cass. Compositae (Card.-Card.). c. 200 C & S Eur. (17), NW Afr.(1), SW & esp. C As. Some cult. orn. incl. ***J. mollis*** (L.) Reichb. (Medit.) – when lepidopterans attack rosette, multiple rosettes & therefore more fls & seeds produced but mammalian attack leads to only lat. heads which are less successful at prod. seeds

Jurinella Jaub. & Spach = *Jurinea*

Juruasia Lindau. Acanthaceae (III 2c). 2 Braz.

Jussiaea L. = *Ludwigia*

Justago Kuntze = *Cleome*

Justenia Hiern = *Bertiera*

Justicia Houston ex L. Acanthaceae (III 2c). Incl. *Adhatoda, Beloperone, Jacobinia, Monechma, Rungia*, c. 700 trop. (S Am. 38 – R: AMBG 89(2002)234) & warm (China 43), temp. N Am. R: KB 43(1988)551, 592. ?Heterogeneous. Herbs & shrubs, some small trees; ***J. heterocarpa*** T. Anders. (trop. Afr.) – heterocarpy, otherwise unknown in fam.; ***J. klossii*** (S. Moore) Mabb. (*R. klossii*, mushroom plant, NG) – cult. potherb. Cult. orn. esp. ***J. adhatoda*** L. (*A. vasica*, adhatoda, NW Himal. to Bhutan) – lvs medic., wood used for beads, good fuel; ***J. brandegeeana*** Wassh. & L.B. Sm. (*B. guttata, Drejerella g.*, shrimp-plant, lobster-plant, Mex., natur. Florida) – form. v. pop. housepl.; ***J. gendarussa*** Burm.f. (Ind.) – local medic., hedge-pl. veg. prop. in Indomal., prob. introd. Afr. by Arab traders; ***J. ladanoides*** Lam. (*J. schimperi*, trop. Afr.) – local veg.; ***J. pectoralis*** Jacq. (death angel, trop. S Am.) – hallucinogenic snuff; ***J. scheidweileri*** V. Graham (*Porphyrocoma pohliana*, carnival pl., Braz.)

jutaí *Hymenaea oblongifolia* var. *palustris*

jute *Corchorus* spp.; **American, Chinese** or **Manchurian j.** *Abutilon theophrasti*; **Bimlipatum j.** *Hibiscus cannabinus*; **Congo j.** *Hibiscus americanus*

Juttadinteria Schwantes. Aizoaceae (V). 5 S Afr. R: H. E. K. Hartmann, *Ill. Handb. Succ. Pls*, Aiz. F–Z(2002)70. Cult. orn. succ.

Juzepczukia Chrshan. = *Rosa*

K

Kabulia Bor & C. Fischer = *Polygonum*

Kabulianthe (Rech.f.) Ikonn. (~ *Gypsophila*). Caryophyllaceae. 1 Afghanistan: ***K. honigbergeri*** (Fenzl) Ikonn. R: BZ 89(2004)114

Kabuyea Brummitt (~ *Cyanastrum*). Tecophilaeaceae. 1 Kenya, Tanzania: ***K. hostifolia*** (Engl.) Brummitt. R: KB 53(1998)771

kachnar *Bauhinia* spp.

kacip fatimah *Labisia pumila*

kadam(b), kadamba *Neolamarckia cadamba*

Kadenia Lavrova & Tikh. = *Selinum*

Kadenicarpus Doweld = *Turbinicarpus*

Kadsura Kaempf. ex Juss. Schisandraceae. 16 Ind. to Jap., W Mal., Moluccas. R: FOW 4(2003)31. Twining shrubs with unisexual fls, parts in spirals. ***K. japonica*** (L.) Dunal (E As.) – cult. dioec. with red fr., trad. medic. ('Fructus Kadsurae'); ***K. longipedunculata*** Finet & Gagnepain (China) – monoec., prod. methyl butyrate, poll. female pollen-eating dipterans (*Megommata* sp.) deceived as female fls no reward; ***K. scandens*** (Blume) Blume (Mal.) – fr. ed. (sour)

Kadua Cham. & Schldl. (~ *Hedyotis*). Rubiaceae (IV 15). 28 Pacific (Hawaii 21). R: SB 30(2005)831. ***K. affinis*** DC. (*H. terminalis*, Hawaii) – prob. most polymorphic pl. sp. in H.

Kaempferia L. Zingiberaceae (II 1). c. 40 Ind. to S China & Mal. R: NRBGE 38(1980)1. Cult. orn. pot-pls incl. ***K. galanga*** L. (SE As) – rhiz. a spice & scent (galanga, name referring to other pls too). See also *Boesenbergia*

Kafirnigania Kamelin & Kinzik. Umbelliferae (III 10). 1 C As.: ***K. hissarica*** (Korovin) Kamelin & Kinzik.

kaffir boom *Erythrina caffra*; **k. lily** *Hesperantha coccinea*; **k. lime** *Citrus hystrix*; **k. plum** *Harpephyllum caffrum* (N.B. all these names are offensive to many people and their use is to be discouraged)

Kageneckia Ruíz & Pavón. Rosaceae (1). 2–3 Peru, Chile. Dioec. trees. ***K. lanceolata*** Ruíz & Pavón (Peru) – black dye from lvs

kagné butter *Allanblackia oleifera*

kahikatea *Dacrycarpus dacrydiodes*

kahili ginger *Hedychium gardnerianum*

kahua bark *Terminalia arjuna*

kai choy *Brassica juncea*; **k. lan** *Brassica oleracea* Alboglabra Group

Kaieteuria Dwyer = *Ouratea*

Kailarsenia Tirvengadum (~ *Randia*). Rubiaceae (II 1). Incl. *Larsenaikia* 6–7 trop. As., 3 Aus. R: NJB 14(1993)525

Kailashia Pimenov & Kljuykov (~ *Pachypleurum*). Umbelliferae (III 8). 2 Himal. R: FR 116(2005)82

Kairoa Philipson. Monimiaceae (V 2). 3 NG. R: HPB 14(2009)72

Kairothamnus Airy Shaw. Picrodendraceae (Cal.-Pseud.; Euphorbiaceae s.l.). 1 NG: ***K. phyllanthoides*** (Airy Shaw) Airy Shaw. R: FM 20 (2011)57

Kaisupeea (~ *Boea*). Gesneriaceae (III 2i). 3 Myanmar, S Laos, Thailand. R: NJB 21(2001)116

kaivum fibre *Helicteres isora*

kajang (mats) *Corypha utan*

Kajewskia Guillaumin = *Veitchia*

Kajewskiella Merr. & Perry (~ *Xanthophytum*). Rubiaceae (? I 7). 2 Solomon Is. R: Blumea 25(1979)283

Kakadu plum *Terminalia ferdinandiana*

kaki *Diospyros kaki*

kaku oil *Lophira* spp.

Kalaharia Baill. (~ *Clerodendrum*). Labiatae (II; Verbenaceae). 2 trop. & S Afr. R: PEE 146(2013)136

Kalakia Alava = *Cymbocarpum*

kalamansi *Citrus × microcarpa*

kalamet *Mansonia gagei*

Kalanchoe Adans. Crassulaceae (II). Incl. *Bryophyllum*, 144 S & E Afr. & Madag. (60 – R: P. Boiteau & C. Allorge-Boiteau (1995) *K. de Madag.*) ? to As. R: U. Eggli, *Ill. Handb. Succ. Pls*, Crass. (2005)143. Succ. shrubs (Leeuwenberg's & Rauh's Models), scramblers & herbs s.t. hapaxanthic; 3 sections incl. sect. *Bryophyllum* with plantlets borne in leaf-margins & infls; some invasive, toxic to stock (cardiac glycosides). Favoured for pl. physiology experiments; ***K. petitiana*** A. Rich. (Ethiopia) – facultative crassulacean acid metabolism, shifting from C_3 as leaves age or there is water stress. Cult. orn. ***K. blossfeldiana*** Poelln. (Madag.) occ., but commonly grown pot-pls for red fls in winter are hybrids (poss. with ***K. glaucescens*** Britten (*K. flammea*, E & NE Afr.), ***K. pumila*** Bak. (Madag.) etc.) e.g. '**Dorothy**' (pinnately lobed lvs, big fls); ***K. beharensis*** Drake (Madag.) – tree-like to 6 m; ***K. daigremontiana*** Hamet & Perrier (flopper, Madag.) – commonly cult. sp. with plantlets at margin of blotched lanceolate lvs, hybrids with next = ***K. × houghtonii*** D.B. Ward; ***K. delagoensis*** Ecklon & Zeyher (*K. tubiflora*, mother of millions, Madag., S Afr.) – lvs linr. with apical plantlets, widely cult. & weedy, invasive Aus.; ***K. pinnata*** (Lam.) Pers. (? orig. Madag., but widely natur. in trop., invasive Aus., Galapagos) – commonly cult. sp. with pinnate lvs & plantlets ('air plant'); ***K. synsepala*** Bak. (Madag.) – long stolons, cult. orn.

Kalappia Kosterm. Leguminosae (I 3). 1 Sulawesi: ***K. celebica*** Kosterm. – locally valuable timber now rare through over-exploitation. R: FM I,12(1996)625

kalaw *Hydnocarpus castanea*

Kalbfussia Schultz-Bip. = *Leontodon*

Kalbreyera Burret = *Geonoma*

Kalbreyeracanthus Wassh. = *Stenostephanus*

Kalbreyeriella Lindau. Acanthaceae (III 2c). 3 Panamá, Colombia

kale *Brassica oleracea* Acephala Group; **Chinese k.** *B. o.* Alboglabra Group; **Indian k.** *Caladium lindenii*; **sea k.** *Crambe maritima*; **Siberian k.** *B. napus* Pabularia Group; **Tuscan k.** *B. o.* Acephala Group (cavalo nero)

Kali Mill. (?= *Salsola*). Amaranthaceae (Chenopodiaceae). c. 20 temp. OW, ? N Am. ***K. komarovii*** (Iljin) Akhani & Roalson (*S. k.*, land seaweed, E As.) – local veg. (okahijiki, C Jap.), in wet conditions long-winged 'frs' with green seeds & short-winged with yellow, but only latter in dry (or when abscisic acid applied), yellow seeds more cold-tolerant & with longer dormancy than green

Kalidiopsis Aellen = *Kalidium*

Kalidium Moq. Amaranthaceae (Chenopodiaceae II 1). 5 Medit. (Eur. 2) to C As.

Kalimantanorchis Tsukaya & al. (~ *Tropidia*). Orchidaceae. 1 W Borneo: ***K. nagamasui*** Tsukaya & al. R: SB 36(2011)52

Kalimeris (Cass.) Cass. = *Aster*

Kalimpongia Pradhan = *Dickasonia*

Kalinia H. Bell & Columbus (~ *Eragrostis*). Gramineae. 1 SW N Am.: ***K. obtusiflora*** (Fourn.) H. Bell & Columbus. R: Aliso 30(2012)91

Kaliphora Hook.f. Montiniaceae (Cornaceae s.l., Melanophyllaceae, Torricelliaceae). 1 Madag.: ***K. madagascariensis*** Hook.f. – Roux's Model

Kallstroemia Scop. Zygophyllaceae (IV). 17 trop. & warm Am. R: CGH 198(1969)41. ***K. grandiflora*** Torrey & A. Gray (Arizona poppy, SW N Am.) – conspic. desert ann.

Kalmia L. Ericaceae (V 3). Incl. *Leiophyllum*, *Loiseleuria*, 10 N Am. (8, ***K. procumbens*** (L.) Galasso & al. (*L. p.*, alpine azalea) circumboreal, cult. orn. shrublet – R: FNA 9(2009)480), 2 natur. Eur.), Cuba. R: Rhodora 76(1974)315; R.A. Jaynes (1997) *K.*: 37. C usu. with 10 pouches in which anthers held under tension like bows until insect probing for nectar releases them & is showered with pollen. American laurel; lvs poisonous. Cult. orn. usu. everg. shrubs esp. ***K. angustifolia*** L. (sheep laurel, pig l., lambkill), ***K. buxifolia*** (Berg.) Gift & Kron (*Leiophyllum b.*, sand myrtle), ***K. latifolia*** L. (mt. l., calico bush) & ***K. polifolia*** Wangenh. (bog l.) – all E N Am.

Kalmiella Small = *Kalmia*

Kalmiopsis Rehder (~ *Phyllodoce*). Ericaceae (V 3). 2 Oregon. R: FNA 9(2009)478. Only Oregon endemic genus; heterostyly (? unique in fam.). ***K. leachiana*** (L. Henderson) Rehder – cult. orn. rock-pl., prot. in wild, hybridizes with *P.* spp. R: JAA 13(1932)31

kalo *Colocasia esculenta*

kalonji *Nigella sativa*

Kalopanax Miq. = *Eleutherococcus*

Kalopternix Garay & Dunsterv. = *Epidendrum*
kamal *Nelumbo nucifera*
kamala *Mallotus philippensis*
kamarere *Eucalyptus deglupta*
kamassi *Gonioma kamassi*
Kamelinia Khassanov & Malzev. Umbelliferae (III 5). 2 C As.
Kamettia Kostel. (*Ellertonia*). Apocynaceae (Ic). 1 S Ind., 1 Thailand. R: TFB 33(2005)76
Kamiesbergia Snijman = *Hessea*
kamini *Murraya paniculata*
Kampmannia Steud. = *Cortaderia*
Kampochloa W. Clayton. Gramineae (Chlorid.). 1 S trop. Afr.: ***K. brachyphylla*** W. Clayton
kampyo *Lagenaria siceraria*
Kanahia R. Br. Apocynaceae (Vc; Asclepiadaceae III 1). 2 trop. E Afr., Arabia. R: NJB 6(1986)787
kanakambaram *Crossandra infundibuliformis*
Kanakomyrtus N. Snow. Myrtaceae (II 10). 6 New Caledonia
Kanaloa Lorence & K. Wood. Leguminosae (II 1). 1 Hawaii: ***K. kahoolawensis*** Lorence & K. Wood. R: Novon 4(1994)137
kanda *Prunus africana*
Kandaharia Alava. Umbelliferae (III 11). 1 Afghanistan: ***K. rechingerorum*** Alava. R: Candollea 31(1976)92
Kandelia (DC.) Wight & Arn. Rhizophoraceae (III). 2 Ind. to S Jap., W Mal. Mangoves. ***K. candel*** (L.) Druce – tanbark
kanga butter *Pentadesma butyracea*
kangaroo apple *Solanum aviculare, S. laciniatum*; **k. fern** *Microsorum pustulatum*; **k. grass** *Themeda* spp. esp. *T. australis*; **k. paw** *Anigozanthos* spp.; **k. thorn** *Acacia paradoxa*; **k. vine** *Cissus antarctica*
kangkong *Ipomoea aquatica*
Kania Schltr. (~ *Metrosideros*). Myrtaceae (II 5). 6 Philippines (2), NG (4). R: Blumea 28(1982)177, KB 45(1990)205
Kanimia Gardner = *Mikania*
Kanjarum Ramam. = *Strobilanthes*
kanluang *Nauclea orientalis*
kan-mu-tun *Clematis filamentosa*
kan-non-chiku *Rhapis excelsa*
kanna *Mesembryanthemum tortuosum*
Kansas gay-feather *Liatris spicata*; **K. thistle** *Solanum rostratum*
Kantou Aubrév. & Pellegrin = *Inhambanella*
Kaokochloa De Winter. Gramineae (XXVII 1). 1 SW Afr.: ***K. nigrirostris*** De Winter. R: Strel. 10(2000)700
kaolang *Sorghum* spp. (China)
Kaoue Pellegrin = *Stachyothyrsus*
kapa *Broussonetia papyrifera*
kaphal bokra, k. pothu *Morella esculenta*
kapok *Ceiba pentandra*; see also *Bombax* spp.
kapong *Tetrameles nudiflora*
Kappia Venter = *Chlorocyathus*
kapur *Dryobalanops* spp. esp. *D. aromatica*
karaka *Corynocarpus laevigata*; **k. berry** *Rubus* 'Karaka Black'
karamanni wax *Moronobea* spp.
Karamyschewia Fischer & C. Meyer = *Oldenlandia*
karanda *Carissa carandas*; **k. nuts** *Elaeocarpus bancroftii*
karanja *Millettia pinnata*
Karatas Mill. = *Bromelia*
Karatavia Pim. & Lavrova. Umbelliferae (III 8). 1 C As.: ***K. kultiassovii*** (Korovin) Pim. & Lavrova
karaya gum *Cochlospermum religiosum*
Kardanoglyphos Schldl. = *Rorippa*
Kardomia Peter G. Wilson (~ *Babingtonia*). Myrtaceae (II 15). 6 E Aus. R: AusSB 20(2007)316
karela, karella *Momordica charantia*
Karelian birch wood *Betula pendula, B. pubescens*

Karelinia Less. (~ *Pluchea*). Compositae (Pluch.). 1 Russia & NE Iran to Mongolia: ***K. caspia*** Less.

Kariba weed *Salvinia molesta*

Karimbolea Descoings = *Cynanchum*

Karina Boutique. Gentianaceae. 1 Congo: ***K. tayloriana*** Boutique. R: BJBB 41(1971)262

Karinia Rezn. & McVaugh = ?*Scirpoides*

karité *Vitellaria paradoxa*

kariyat *Andrographis paniculata*

karkalla *Carpobrotus rossii*

Karnataka P.K. Mukherjee & Constance. Umbelliferae (III 8). 1 S Ind.: ***K. benthamii*** (C.B. Clarke) P.K. Mukherjee & Constance. R: Britt. 38(1986)145

karnikar *Pterospermum acerifolium*

karo *Pittosporum crassifolium*

Karomia Dop. Labiatae (III; Verbenaceae). 1 Vietnam, 8 E & S Afr. & Madag. R: GOB 7(1985)36

Karorchis M.A. Clem. & D.L. Jones = *Bulbophyllum*

Karrabina Rozefelds & H. Hopkins (~ *Geissois*). Cunoniaceae (V). 2 E Aus. R: AusSB 26(2013)180

karri *Eucalyptus diversicolor*

Karroochloa Conert & Türpe = *Tribolium* (but see R: SBi 50(1969)290)

karuka nut *Pandanus* sp. (?*P. bowersiae*)

Karvandarina Rech.f. Compositae (Card.-Cent.). 1 Iran, Pakistan: ***K. aphylla*** Rech.f. & al.

Karwinskia Zucc. Rhamnaceae (6). 16 SW US to Bolivia & WI. ***K. calderonii*** Standl. (Salvador) – fine timber; ***K. humboldtiana*** (Roemer & Schultes) Zucc. (SW N Am.) – fr. ed. but seeds toxic, paralyzing motor nerves, used in treatment of tetanus in Mex.

Kaschgaria Polj. Compositae (Anth.-Art.). 2 C As. to W China. R: BBMNHB 23(1993)117

Kashmiria Hong. Plantaginaceae (Dig.-Ver.; Scrophulariaceae s.l.). 1 Himal.: ***K. himalaica*** (Hook.f.) Hong. R: BN 133(1980)565

kataka *Strychnos potatorum*

kath *Uncaria gambir*

Katherinea A.D. Hawkes = *Epigeneium*

Katinasia Bonif. (~ *Nardophyllum*). Compositae (Ast.-Hin.). 1 W Arg.: ***K. cabrerae*** (Bonif.) Bonif. R: SCB 92(2009)53

katon *Sandoricum koetjape*

katsura *Cercidiphyllum japonicum*

katuk *Breynia androgyna*

kau kau *Ipomoea batatas*

Kaufmannia Regel (~ *Primula*). Primulaceae. 1–2 C As.

kauila *Alphitonia ponderosa*

Kaulfussia Nees = *Felicia*

kaunghmu *Anisoptera scaphula*

Kaunia R. King & H. Robinson (~ *Eupatorium*). Compositae (Eup.-Oxy.). 14 S Am. R: MSBMBG 22(1987)448

kauri *Agathis* spp.; **Fijian k.** *A. macrophylla*

kava *Piper methysticum*, (NZ) *P. excelsum*

Kaviria Akhani & Roalson (~ *Salsola*). Amaranthaceae (Chenopodiaceae). c. 10 trop. & warm OW

kawaka *Libocedrus plumosa*

kawa-kawa *Piper excelsum*

kawal *Senna obtusifolia*

Kayea Wall. (~ *Mesua*). Calophyllaceae (Guttiferae I1). 75 Indomal.

keaki *Zelkova serrata*

Kearnemalvastrum D. Bates. Malvaceae (Malv.-Malv.). 2 Mex. to Colombia. R: SBM 25(1988)240

Keayodendron Leandri. Phyllanthaceae (Phyll.-Brid.-Keay.); Euphorbiaceae s.l.). 1 trop. Afr.: ***K. bridelioides*** (Hutch. & Dalziel) Leandri. R: A.R.-Sm., Gen. Euph. (2001)29

Kebirita Kramina & Sokoloff (~ *Lotus*). Leguminosae (III 22). 1 NW Afr.: ***K. roudairei*** (Bonnet) Kramina & Sokoloff (*L. r.*). R: BMSN 106(2001)61

kechapi *Sandoricum koetjape*

Keckiella Straw (~ *Penstemon*). Plantaginaceae (Ant.-Chel.; Scrophulariaceae s.l.). 7 W N Am. R: Britt. 19(1967)203. Some cult. orn.

kedam *Neolamarckia cadamba*

Kedarnatha P.K. Mukherjee & Constance. Umbelliferae (III 8). 5 Himal., Myanmar. R: FR 115(2004)233

Kedhalia C.K. Lim (~ *Haniffia*). Zingiberaceae (II 1). 1 Kedah: ***K. flaviflora*** C.K. Lim. R: Folia Mal. 10(2009)2

kedondong *Canarium* spp.

Kedrostis Medik. Cucurbitaceae (XIII). c. 20 OW trop. Oft. with swollen caudiciform bases

Keenania Hook.f. Rubiaceae (I 8). 5 Assam, SE As.

Keerlia DC. = *Aphanostephus*

Keetia E. Phillips (~ *Canthium*). Rubiaceae (III 2). 40 trop. & S Afr. R: KB 41(1986)965

Kefersteinia Reichb.f. Orchidaceae (V 12j). c. 70 trop. Am.

Kegeliella Mansf. Orchidaceae (V 12i). 4 trop. Am.

kei apple *Dovyalis caffra*

Keiria Bowdich = ? *Cissus*

Keiskea Miq. (~ *Collinsonia*). Labiatae (VII 1). 5 E As. R: BBMNHB 10(1982)70

Keithia Benth. = *Rhabdocaulon*

Keithia Spreng. = ? (not Capparaceae)

keladi *Colocasia esculenta*

Kelissa Ravenna = *Cypella*

Kelita A. Bean (~ *Ptilotus*). Amaranthaceae. 1 Queensland: ***K. uncinella*** A. Bean. R: Muelleria 28(2010)105

Kelleria Endl. (~ *Drapetes*). Thymelaeaceae (Thym.-Daph.). 11 Mt Kinabalu (1), NG, Aus. (2), NZ. R: AusSB 3(1990)609. ? Beetle-poll.

Kelleronia Schinz (~ *Tribulus*). Zygophyllaceae (IV). 3 NE Afr., S Arabia

Kellochloa Lizarazu & al. (~ *Panicum*). Gramineae. 2 N Am. R: PSE 301(2015)2256

Kelloggia Torrey ex Benth. Rubiaceae (IV ? 16). 2 China (1), SW US (1)

Kelly grass *Rottboellia cochinchinensis*

Kelseya Rydb. Rosaceae (2). 1 W US: ***K. uniflora*** (S. Watson) Rydb. – cult. rock-pl. R: FNA 9(2014)414

Kelussia Mozaff. (~ *Opopanax*). Umbelliferae (III 10). 1 Iran: ***K. odoratissima*** Mozaff. R: BZ 88,2(2003)88

kemanji *Ocimum* × *africanum*

kemiri (nut) *Aleurites moluccanus*

kempas *Koompassia malaccensis*

Kemulariella Tamaschjan (~ *Aster*). Compositae (Ast.-Ast.). 6 Cauc.

kenaf *Hibiscus cannabinus*

kenda *Macaranga peltata*

Kendal green dye made from fls of *Genista tinctoria* with woad

Kendrickia Hook.f. Melastomataceae. 1 S Ind., Sri Lanka: ***K. walkeri*** (Gardner) Triana – r.-cli.

kendyr fibre *Apocynum venetum*

Kengia Packer = *Cleistogenes*

kenguel seed *Silybum marianum*

Kengyilia C. Yen & J.-L. Yang = *Elymus* (but see FOC 22(2006)431)

Kennedia Vent. Leguminosae (III 18). 15 Aus., NG. Showy lianes with red to almost black fls, some cult. orn. (coral peas) incl. ***K. nigricans*** Lindl. (black bean, W Aus.) – fls almost black

Kenopleurum Candargy = *Thapsia*

Kensitia Fedde = *Erepsia*

kentia *Howea forsteriana*

Kentiopsis Brongn. Palmae (V 14a). Incl. *Mackeea*, 4 New Caled. R: KB 63(2008)642. ***K. magnifica*** (H. Moore) Pintaud & Hodel (*M. m.*) – to 25m, tallest palm in N.C.

Kentish cob *Corylus maxima*

Kentranthus Raf. = *Centranthus*

Kentrochrosia Schumann & Lauterb. = *Kopsia*

Kentrosiphon N.E. Br. = *Gladiolus*

Kentrothamnus Süsseng. & Overk. Rhamnaceae (2). 1 Bolivia, Arg. *Frankia* symbionts

Kentucky blue grass *Poa pratensis*; **K. coffee tree** *Gymnocladus dioica*; **K. yellowwood** *Cladrastis kentukea*

kepayang *Pangium edule*

keppel *Stelechocarpus burahol*

Keracia (Cosson) Calest. = *Hohenackeria*

keranji *Dialium* spp. esp. *D. indum*

Keratochlaena Morrone & Zuloaga (*Sclerochlamys*). Gramineae (XXIII 1). 1 N Braz.: ***K. rigidifolia*** (Filg. & al.) Morrone & Zuloaga. R: Darw. 47(2009)231

keratto *Agave* spp.

Keraudrenia Gay. Malvaceae (Bytt.-Lasio.; Sterculiaceae). 13 Aus.

Keraunea Cheek & Sim-Bianch. (~ *Neuropeltis*). Convolvulaceae. 2 E Braz. R: NJB 31(2013)454

Keraymonia Farille. Umbelliferae (III 5). 4 Himal. R: FR 111(2000)525

Kerbera Fourn. = *Melinia*

Kerguélen cabbage *Pringlea antiscorbutica*

Kerianthera Kirkbride. Rubiaceae (I 8). 2 Braz. R: Britt. 37(1985)109

Kerigomnia P. Royen = *Octarrhena*

Kermadecia Brongn. & Gris. Proteaceae (V 4). 4 New Caled. R: AJB 62(1975)135. ***K. rotundifolia*** Brongn. & Gris – timber for general use

kermek *Limonium gmelinii* & *L. latifolium*

Kernera Medik. (~ *Cochlearia*). Cruciferae (34). 1 C & S Eur. mts. ***K. saxatilis*** (L.) Sweet – cult. orn. rock-pl.

kerong *Millettia pinnata*

kerosene weed *Ozothamnus ledifolius*; **k. wood** *Halfordia kendack*

Kerria DC. Rosaceae (Ros.-Kerr.). 1 temp. E As: ***K. japonica*** (L.) DC. – cult. orn. suckering shrub, usu. double-flowered form (reverts to simple form if overwintered in warm greenhouse). R: FNA 9(2014)389

Kerriochloa C. Hubb. Gramineae (XXII). 1 SE As.: ***K. siamensis*** C. Hubb.

Kerriodoxa Dransf. Palmae (III 5). 1 peninsular Thailand: ***K. elegans*** Dransf. R: Principes 27(1983)3. One of 12 most endangered pl. spp. on Phuket but spreading in cult.

Kerriothyrsus C. Hansen = *Phyllagathis* (but see Willdenowia 17(1988)153)

Kerry lily *Simethis mattiazzii*

Kerstingiella Harms = *Macrotyloma*

keruing *Dipterocarpus* spp.

ketaki *Pandanus tectorius*

Keteleeria Carr. Pinaceae. 3 S China, Taiwan, SE As. (Eur. in Tertiary). R: RV 121(1990)129. Miocene fossils from E As., W N Am. Some timber & cult. orn.

Keumkangsania Kim = *Hanabusaya*

Kewa Christenh. (~ *Hypertelis*). Kewaceae. 8 S Afr., Madag., St Helena. R: Phytotaxa 181(2014)240

Kewaceae Christenh. (~ Molluginaceae). Magnoliidae – Caryophyllales. 1/8 S Afr., Madag., St Helena. R: Phytotaxa 181(2014)240. Herbs or subshrubs + anthocyanins (betalains. 0). Lvs fleshy, linear, ± fasciculate; stip. sheathing. Infl. umbel-like, pedunculate. P 5, 3(4) becoming petaloid; A (3–)5–15(–20), anthers (& stigmas) brightly coloured; $\underline{G(3–5)}$, style 0, placentation axile, with many ovules per loc. Fr. a membranous loculicidal capsule or schizocarp. n = 8.

Genus: *Kewa*

A fragment of shattered Molluginaceae. Presence of anthocyanins considered a reversal

kewda *Pandanus fascicularis*

key, Chinese *Boesenbergia rotunda*; **K. lime** *Citrus × aurantiifolia*

keyaki *Zelkova serrata*

Keyserlingia Bunge ex Boiss. = *Sophora*

Keysseria Lauterb. (~ *Lagenophora*). Compositae (Ast.-Lag.). 9 C & E Mal., 3 Hawaii

Khadia N.E. Br. Aizoaceae (V). 6 Transvaal. R: H. E. K. Hartmann, *Ill. Handb. Succ. Pls*, Aiz. F–Z(2002)72. Drink made from r. (khadi)

khai-nam *Wolffia arrhiza*

khaki weed *Alternanthera pungens*

Khaosokia D. Simpson & al. Cyperaceae. 1 S Thailand (limestone cliffs): ***K. caricoides*** D. Simpson & al. R: BJLS 149(2005)358

Khasiaclunea Ridsd. (~ *Adina*). Rubiaceae (I 2). 1 NE Ind., Bhutan: ***K. oligocephala*** (Havil.) Ridsd. R: Blumea 24(1978)347

khat *Catha edulis*

Khaya A. Juss. Meliaceae (I). c. 5 trop. Afr. & Madag. Rauh's model. Bark locally medic. Wood used as subs. for *Swietenia*, esp. imp. in comm. ***K. anthotheca*** (Welw.) C. DC. (Afr. mahogany, trop. Afr.), also ***K. grandifoliola*** C. DC. (Benin mahogany, B.wood, W & C

Afr.), ***K. ivorensis*** A. Chev. & ***K. senegalensis*** (Desr.) A. Juss. (bisselon, W Afr.) & ***K. madagascariensis*** Jum. & Perrier (Madag. m., E Madag.)

khayer *Senegalia catechu*

khella *Visnaga daucoides*

khesara *Lathyrus sativus*

Khmeriosicyos W. J. de Wilde & Duyfjes. Cucurbitaceae (XIV). 1 Cambodia: ***K. harmandii*** W.J. de Wilde & Duyfjes – coll. once. R: Blumea 49(2004)440

khus-khus *Chrysopogon zizanioides*

kiaat *Pterocarpus angolensis*

Kibara Endl. Monimiaceae (V 2). Excl. *Wilkiea*, 39 Nicobar Is., Mal. R: Blumea 30(1985)389. Some ed. fr., lvs used for flavouring. ***K. moluccana*** Perkins (E Mal.) – pericarp rubbed in hair to colour it black

Kibaropsis Vieill. ex Jérémie (~ *Hedycarya*). Monimiaceae (V 1). 1 New Caled.: ***K. caledonica*** (Guillaumin) Jérémie – cotyledons 4. R: Adans. 17(1977)80

Kibatalia G. Don. Apocynaceae (IIb). 15 SE As. to Philipp. R: AUWP 86–5(1986)36. Leeuwenberg's Model; steroidal alks. Wood for shoes etc. ***K. arborea*** (Blume) G. Don (W & C Mal.) – local medic.

Kibera Adans. = *Erucastrum*

Kibessia DC. = *Pternandra*

Kickxia Dumort. (~ *Linaria*). Plantaginaceae (Ant.-Ant.; Scrophulariaceae s.l.). Excl. *Nanorrhinum* 9 temp. OW. R: NJB 20(1999)664. Fluellen. Oft. weedy: ***K. elatine*** (L.) Dumort. (Eur., Medit.) caused abandonment of c. 1500 acres barley SW of Dayton, Calif.

kidney bean *Phaseolus vulgaris*; **k. fern** *Trichomanes* spp.; **k. vetch** *Anthyllis vulneraria*; **k. wood** *Eysenhardtia polystachya*

Kielmeyera Mart. Calophyllaceae (Guttiferae II). 47 S Braz., Peru. Char. of campos. ***K. coriacea*** Mart. (pau santo) – ground bark a subs. for ground cork

kif *Cannabis sativa* 'subsp. *indica*'

Kigelia DC. Bignoniaceae (6). 1 (variable; forest pls poss. distinct sp.) trop. Afr. (fossil wood in Libya): ***K. africana*** (Lam.) Benth. (*K. pinnata*, sausage-tree) – infls of claret, bat-poll. fls. hanging on long peduncles, fr. gourd-like, locally medic. (purgative) & comm. skin-creams (antibacterial iridoids & naphthoquinones poss. cancer cytotoxins)

Kigelianthe Baill. = *Fernandoa*

Kiggelaria L. Achariaceae (Kiggelariaceae, Flacourtiaceae s.l.). 1 trop. & S Afr.: ***K. africana*** L. (*K. dregeana*, wild peach) – good timber (Natal mahogany). R: Strel. 9(2000)530

Kiggelariaceae Link = Achariaceae

Kihansia Cheek. Triuridaceae. 2 SE Cameroun, Tanzania. R: KB 70,7(2015)1

Kiharopyrum Löve = *Aegilops*

Kikuyu grass *Cenchrus clandestinus*

Kikuyuochloa H. Scholz = *Cenchrus*

Killickia Bräuchler & al. (~ *Micromeria*). Labiatae. 4 E S Afr. R: BJLS 157(2008)576

Killipia Gleason = *Miconia*

Killipiella A.C. Sm. = *Disterigma*

Killipiodendron Kobuski = *Freziera*

Kilmarnock willow *Salix caprea* 'Pendula'

kimchi Korea's national dish of fermented vegetables (orig. *Brassica rapa* Pekinensis Group), cf. sauerkraut

Kinabaluchloa Wong. Gramineae (V 4). 2 W Mal. R: KB 48(1993)523. Bamboos

Kinepetalum Schltr. = *Tenaris*

Kinetochilus (Schltr.) Brieger = *Dendrobium*

king cup *Caltha palustris*; **k.fisher daisy** *Felicia bergeriana*; **K. William (Billy) pine** *Athrotaxis selaginoides*; **k. wood** *Astronium fraxinifolium*, *Dalbergia cearensis*

Kingdonia Balf.f. & W. Sm. Ranunculaceae (II 2; Circaeasteraceae; Kingdoniaceae). 1 W & N China: ***K. uniflora*** Balf.f. & W. Sm.

Kingdoniaceae A.S. Foster ex Airy Shaw. See Ranunculaceae (II)

Kingella Tieghem = *Trithecanthera*

Kinghamia C. Jeffrey (~ *Gutenbergia*). Compositae (Vern.-Erl.). 5 trop. W Afr. R: KB 43(1980)274

Kingia R. Br. Dasypogonaceae (Xanthorrhoeaceae s.l.). 1 SW Aus.: ***K. australis*** R. Br. (black gin) – pachycaul tree (Corner's Model) with mantle of concealed aerial r. under marcescent leaf-bases; as the stem dies from the base after 300+yrs, r. keep contact with ground (cf. *Iriartea*); oldest specimen known c. 650 yrs old; flowering stimulated by fire; nectar taken by honey-eaters. R: FA 46(1986)146

Kingianthus H. Robinson. Compositae (Helia.-Ecl.). 2 Ecuador. R: Phytol. 58(1978)415

Kingidium P. Hunt = *Phalaenopsis*

Kingiodendron Harms (~ *Prioria*). Leguminosae (I 2). 6 Indomal. R: Blumea 18(1970)46, KB 32(1977)244

Kingstonia Hook.f. & Thomson = *Dendrokingstonia*

kingwood *Dalbergia cearensis*; **Mex. k.** *D. congestiflora*

kinh gioi *Elsholtzia ciliata*

Kinia Raf. = ? *Ottelia, Blyxa*

kinnikinni(c)k *Arctostaphylos uva-ursi*

kino astringent resin-like substance from tapped trees, used medic. & for tanning locally: **African k.** *Pterocarpus erinaceus*; **Australian k.** *Corymbia* & *Eucalyptus* spp. esp. *E. camaldulensis* & *E. resinifera*; **Bengal k.** *Butea monosperma*; **E. Indian** or **Malabar k.** *P. marsupium*; **Jamaican k.** *Coccoloba uvifera*

Kinostemon Kudô = *Teucrium*

Kinugasa Tatew. & Sûto = *Paris*

Kionophyton Garay (~ *Stenorrhynchos*). Orchidaceae (IV 2h). 2 Mex., Guatemala. R: Gen. Orch. 3(2003)215

Kippistia F. Muell. (~ *Minuria*). Compositae (Ast.-Pod.). 1 S Aus.: ***K. suaedifolia*** F. Muell. R: Nuytsia 3(1980)215

Kirengeshoma Yatabe. Hydrangeaceae (II 1). 2 E China, Korea & Jap. ***K. palmata*** Yatabe – cult. orn. herb

kiri wood *Paulownia tomentosa*

Kirilowia Bunge = *Bassia*

Kirkbridea Wurd. Melastomataceae. 2 Colombia

Kirkia Oliv. Kirkiaceae. (Simaroubaceae s.l.). Incl. *Pleiokirkia*, 6 trop. & S Afr. R: KB 62(2007)152. Some cult. orn. incl. ***K. acuminata*** Oliv. (trop. Afr.) – timber & ***K. wilmsii*** Engl. (pepper-tree, S Afr.)

Kirkiaceae Takht. (Simaroubaceae s.l.). Magnoliidae – Sapindales. 1/6 trop. & S Afr., Madag. Monoec. trees & shrubs. Lvs pinnate, in spirals to ± opp.; leaflets serrate; stip. 0. Fls usu. 4-merous, in dichasia. A inserted outside & beneath disk; $\underline{G}$ (4(8)) with 1 (2) pend. ovules per loc., styles free, distally connate. Fr. a schizocarp with 1-seeded mericarps; seeds with curved embryo & 0 endosperm

Genus: *Kirkia*

Form. subfam. of Simaroubaceae excl. on absence of quassinoids & by DNA work

Kirkianella Allan = *Sonchus*

kisidwe *Allanblackia floribunda*

Kissenia R. Br. ex Endl. Loasaceae (I 1). 2: 1 S Arabia, Somalia, Ethiopia, 1 SW Afr.

kisses *Galium aparine*

Kissodendron Seemann = *Polyscias*

Kita A. Chev. = *Hygrophila*

Kitagawia Pim. (~ *Peucedanum*). Umbelliferae (III 10). 5 C As.

Kitaibela Willd. Malvaceae (Malv.-Malv.). 1 Lower Danube: ***K. vitifolia*** Willd. – cult. orn.; 1 E Med. R: BJ 109(1987)59

Kitamuraea Rauschert = *Aster*

Kitchingia Bak. = *Kalanchoe*

kite tree *Nuxia floribunda*

kitembilla *Dovyalis hebecarpa*

Kitigorchis Maek. (~ *Oreorchis*). Orchidaceae (V 8). 1 Jap.: ***K. itoana*** Maek. – almost extinct. R: EJB 54(1997)290

kittool or **kitul** *Caryota urens*

kiwano *Cucumis metuliferus*

kiwi (fruit) *Actinidia deliciosa*; **k. berry** *A. arguta*; **golden k.** *A. chinensis*

Kjellbergia Bremek. = *Strobilanthes*

Kjellbergiodendron Burret. Myrtaceae (Lept.). 1+ Sulawesi, NG: ***K. celebicum*** (Koord.) Merr. R: JAA 33(1952)162

Klackenbergia Kissling (~ *Sebaea*). Gentianaceae. 2 Madag. R: T 58(2009)910

Klaineanthus Pierre ex Prain. Euphorbiaceae (Crot.-Aden.). 1 trop. W Afr.: ***K. gaboniae*** Pierre ex Prain. R: A. R.-Sm., *Gen. Euph.* (2001)277

Klaineastrum Pierre ex A. Chev. = *Memecylon*

Klainedoxa Pierre ex Engl. Irvingiaceae (Simaroubaceae s.l.). 2 trop. Afr. R: FOW 1(1999)5. ***K. gabonensis*** Pierre ex Engl. (trop. W Afr.) – fr. eaten by elephants (seeds pass through & germ. in dung), seeds ed., timber valuable

Klamath weed *Hypericum perforatum*

Klaprothia Kunth. Loasaceae (I 2). 2 trop. S Am. R: SB 15(1990)671

Klarobelia Chatrou (~ *Malmea*). Annonaceae (IV 2). 12 Amaz. (Peru 6)

Klasea Cass. (~ *Serratula*). Compositae (Card.-Cent.). 46 Euras. esp. C As. R: BJLS 152(2006)452

Klaseopsis L. Martins = *Rhaponticum* (but see T 55(2006)974)

Klattia Bak. Iridaceae (V). 3 S & SW Cape. R: P. Goldblatt, *Woody I.* (1993)99. Shrubby, poll. sunbirds

Kleberiella V. Castro & Catharino = *Gomesa*

Kleinhovia L. Malvaceae (Bytt.-Bytt.; Sterculiaceae). 1 trop. As. to Aus.: ***K. hospita*** L. (dange) – commonly cult. (in trop.) orn. tree with red fls, stick carried as talisman across crocodile-infested streams in E Indonesia. R: TFSS 7(2011)356

Kleinia Mill. (~ *Senecio*). Compositae (Sen.-Sen.). c. 50 Canary Is., trop. & S Afr., Madag., Arabia (incl. *Notonia*), S Ind., Sri Lanka. R (partial): HIP 39,4(1988)1. *Gynura* & *Solanecio* poss. referable here. Cult. orn. succ. with red, yellow or white fls esp. ***K. neriifolia*** Haw. (*S. kleinia*, Canary Is.) – pachycaul treelet with pseudodichotomous branching

Kleinodendron L.B. Sm. & Downs = *Savia*

Klemachloa R. Parker = *Dendrocalamus*

Klingia Schönl. = *Gethyllis*

klinki pine *Araucaria hunsteinii*

Klossia Ridl. Rubiaceae (? IV 2). 1 Malay Pen.: ***K. montana*** Ridl.

Klotzschia Cham. Umbelliferae (?Azor.). 3 Braz.

Klugia Schldl. = *Rhynchoglossum*

Klugiodendron Britton & Killip = *Abarema*

Kmeria (Pierre) Dandy = *Magnolia*

knapweed *Centaurea nigra*; **Russian k.** *Rhaponticum repens*

Knautia L. Caprifoliaceae (Dipsacaceae). 60 Eur. (48), Medit. R: BJ 36(1905)435. Some cult. orn. like *Scabiosa*, incl. ***K. arvensis*** (L.) Coulter (field scabious, Eur. to N Afr.) with heads of fls, the more towards the periphery the more the C is extended outwards; stigmas of a head ripen together; fatty outgrowth (elaiosome) at fr.-base attractive to dispersing ants, seed germ. quickest in those where they have removed it

knawel *Scleranthus annuus*

Knema Lour. Myristicaceae. 93 Indomal. R: Blumea 25(1979)321. 27(1981)223, 32(1987)115. Some medic. seed-oils (some e.g. ***K. glauca*** (Blume) Petermann (W Mal.) form. for lighting) & locally used timbers

Knersia H. Hartmann & Liede (~ *Drosanthemum*). Aizoaceae. 1 Cape: ***K. diversifolia*** (L. Bolus) H. Hartmann & Liede

Knightia Sol. ex R. Br. Proteaceae (V 1). Excl. *Eucarpha*, 1 NZ: ***K. excelsa*** Sol. ex R. Br. (rewarewa) – beautifully figured timber for cabinet-making

Kniphofia Moench. Asphodelaceae (Aloaceae). 70 Yemen (1), trop. (22) & S (48, esp. Drakensberg) Afr. mts, Madag. (2). Red-hot pokers, torch lilies. R: (S Afr.) Bothalia 9(1968)363, (trop. Afr.) KB 28(1973)465; C. Whitehouse (2016) *K.: the complete guide*. Cult. orn. spp. & hybrids (fls in winter in NZ) with heads of pendent red or yellow fls visited by sunbirds; bees entering some fls in Brit. gardens may be trapped & unable to leave cylindrical C. ***K. caulescens*** Bak. ex Hook.f. (E S Afr.) – rosettes of lvs term. stout stems to 30 cm; ***K. uvaria*** (L.) Oken (S Afr.) – 1 of first Cape pls introd. Eur. though the commonly cult. pls today are complex hybrids & cvs (***K. × praecox*** Bak. etc.), natur. Victoria

knobcone pine *Pinus attenuata*

knobthorn or **knobwood** *Zanthoxylum* spp. esp. *Z. capense*

knol-kohl = kohlrabi

Knorringia (Czuk.) Tsvelev (~ *Persicaria*). Polygonaceae (II 5). 2 C As.

knotweed or **knotgrass** *Reynoutria*, *Persicaria* & *Polygonum* spp. esp. *P. aviculare*; **Japanese k.** *R. japonica*

Knowltonia Salisb. = *Anemone*

Knoxia L. Rubiaceae (III 8). 7 Indomal., 2 Afr.

koala fern *Caustis blakei*

Koanophyllon Arruda (~ *Eupatorium*). Compositae (Eup.-Crit.). c. 120 trop. Am. R: MSBMBG 22(1987)314

koboku *Engelhardia roxburghiana*

Kobresia Willd. = *Carex* (but see Pfl. IV,20(1909)33)

Kochia Roth = *Bassia*

Kochummenia Wong (~ *Randia*). Rubiaceae (II 1). 2 Malay Pen. R: MNJ 38(1984)31

koda wood *Ehretia acuminata*

Kodalyodendron Borh. & Acuña = *Amyris*

kodo millet *Paspalum scrobiculatum*

Koeberlinia Zucc. Koeberliniaceae (Capparaceae s.l.). 1 SW N Am. (***K. spinosa*** Zucc. (allthorn) – leafless xerophyte of dry areas), 1 Bolivia. R: Britt. 60(2008)172

Koeberliniaceae Engl. (~ Capparaceae). Magnoliidae – Brassicales. R: FNA 7(2010)184. 1/2 SW N Am., Bolivia. Thorny shrub or small tree with sparse 1-cellular hairs; glucosinolates 0. Lvs minute, in spirals, ephemeral on thorn-tipped green twigs; stip. 0. Fls bisex. K 4(5), imbr.; C 4(5), imbr.; A 8(10) with flattened filaments & basal nectaries; disk 0; G (2(3)), 2-loc. with axile placentation & ∞ anatropous bitegmic ovules per loc. Fr. a (1)2–4-seeded berry with persistent term. style. Endosperm 0 or scant, s.t. persistent perisperm. x = 11.
Genus: *Koeberlinia*

Koechlea Endl. = *Ptilostemon*

Koehneola Urb. Compositae (Helia.-Pin.). 1 E Cuba (serpentine): ***K. repens*** Urb.

Koehneria S.A. Graham, Tobe & Baas (~ *Pemphis*). Lythraceae. 1 S Madag.: ***K. madagascariensis*** (Bak.) S.A. Graham, Tobe & Baas. R: AMBG 73(1986)805. Pollen diff. from *Pemphis*

Koeiea Rech.f. = *Rhammatophyllum*

Koeleria Pers. = *Trisetaria*

Koellensteinia Reichb.f. Orchidaceae (V 12j). 18 trop. Am.

Koellikeria Regel. Gesneriaceae (II 3). 1 trop. Am.: ***K. erinoides*** (DC.) Mansf. R: Selbyana 6(1982)174

Koelpinia Pallas. Compositae (Lact.-Scor.). 5 N Afr. to E As. (Eur. 1)

Koelreuteria Laxm. Sapindaceae. 4 China (2), Taiwan (1), Fiji (1!). R: JAA 57(1976)137. Capsule large & bladdery & may be blown as a seed-disp. mechanism. ***K. elegans*** (Seem.) A.C. Sm. (Fiji) – allegedly native (but cf. *Broussonetia*); ***K. henryi*** Dümmer (*K. e.* subsp. *formosana*, Taiwan) – autumn-fls cf. seq.; ***K. paniculata*** Laxm. (golden rain, pride of Ind. & China, natur. Jap., Korea & US – introd. Thomas Jefferson 1809), cult. orn. tree esp. as street-tree in C Eur., etc., fls medic. & source of yellow dye in China, seeds used as beads, poss. oil source

Koelzella Hiroe = *Prangos*

Koenigia L. (~ *Persicaria*). Polygonaceae (II 1). Incl. *Aconogon*, 60 Arctic & N Eur. mts, temp. E As., Himalayas, ***K. islandica*** L. circumpolar to Tierra del Fuego, 'bipolar' like *Anemone multifida*, *Carex macloviana*, *Osmorhiza berteroi*, 0.5 – 20 cm tall, 1 of few annuals in arctic-alpine flora. R: BotJLS 124(1997)305

Koernickanthe Andersson. Marantaceae. 1 trop. Am.: ***K. orbiculata*** (Koern.) Andersson. R: NJB 1(1981)240

Kogelbergia Rourke (~ *Stilbe*). Stilbaceae. 2 W Cape. R: Both. 30(2000)12

Kohautia Cham. & Schldl. (~ *Oldenlandia*). Rubiaceae (IV 15). Excl. *Cordylostigma*, 27 OW trop. Primitively moth-poll., then bee- & butterfly-poll. Some weedy like ***K. tenuis*** (S. Bowdich) Mabb. (*K. senegalensis*, W Afr.)

kohekohe *Dysoxylum spectabile*, (Hawaii) *Eleocharis erythropoda*

Kohleria Regel. Gesneriaceae (II 5b). Incl. *Moussonia* (R: Selbyana 1(1975)2); & *Capanea* (some bat-poll. fls 5 cm diam., green with purple spots), c. 35 trop. Am. (Colombia 14, 9 endemic). R: SCB 79(1992). Shrubs & herbs, some with scaly runners, or aerial rhiz. in place of infls. Many cult. orn. incl. ∞ hybrids

Kohlerianthus Fritsch = *Columnea*

kohlrabi *Brassica oleracea* Gongylodes Group

Kohlrauschia Kunth (~ *Petrorhagia*). Caryophyllaceae (III 1). 5 Medit. & W As.

koie-yan *Oxera splendida*

Koilodepas Hassk. Euphorbiaceae (Epi.-Epi.). 9 S Ind. to Hainan & NG. R: AMBG 97(2010)222

kokam or **kokum butter** *Garcinia indica*

kokerite *Attalea maripa*

Kokia Lewton. Malvaceae (Malv.-Goss.). 4 Hawaii. R: P.A. Fryxell, *Nat. Hist. Cotton Tribe* (n.d.) 79. Some extinct (1) or only surviving in botanic gardens, e.g. ***K. cookei*** Degener reduced to 1 pl. in Eur., now being prop. by leaf-fragments & callus. ***K. drynarioides*** (Seemann) Lewton with extrafl. nectaries, though no native ants in Hawaii (? 'phylogenetic inertia')

kok(k)o *Albizia lebbeck*

kokoon *Kokoona zeylanica*

Kokoona Thwaites. Celastraceae (I). 10 Indomal. (Mal. 6 – R: FM I,6(1962)258). ***K. zeylanica*** Thw. (kokoon, Sri Lanka) – seed-oil effective as leech-deterrent

kokrodua *Pericopsis alata*

koksaghyz *Taraxacum bicorne*

kokum *Garcinia indica*

kola *Cola* spp.

kolhol gum *Vachellia nilotica*

Kolkwitzia Graebner (~ *Linnaea*). Caprifoliaceae (Linnaeaceae). 1 C China: ***K. amabilis*** Graebner (beautybush) – floriferous shrub like *Abelia*, rare in wild, cult. esp. in US

Kolobochilus Lindau = *Razisea*

Kolobopetalum Engl. Menispermaceae (III). 4 trop. Afr.

Kolowratia C. Presl = *Alpinia*

Kolpakowskia Regel = *Ixiolirion*

Komaroffia Kuntze (~ *Nigella*). Ranunculaceae (I 3). 2 Iran & C As. Differs from *N.* in entire or palmate lvs

Komarovia Korovin. Umbelliferae (? tribe). 1 C As.: ***K. anisosperma*** Korovin

komatsuma, komatsuna *Brassica rapa* Perviridis Group

kombo *Pycnanthus angolensis*

Kompitsia Costantin & Gallaud = *Pentopetia*

Konantzia Dodson & N. Williams = *Ionopsis*

konjaku *Amorphophallus konjac*

Kontumia S.K. Wu & Phan. Polypodiaceae (IV). 1 Vietnam: ***K. heterophylla*** S.K. Wu & Phan. R: Novon 15(2005)245

Koompassia Maingay. Leguminosae (I 3). 3 Mal. R: FM I,12(1996)631. ***K. excelsa*** (Becc.) Taubert (tualang, W Mal.) – at 88 m a tree* in Sarawak the tallest trop. angiosperm – timber splits too easily to be valuable (phloem inclusions), though buttresses used for dining-tables; ***K. malaccensis*** Maingay (kempas, W Mal.) – timber good

Koordersiochloa Merr. = *Streblochaete* (but see Reinw. 13(2012)300)

Koordersiodendron Engl. Anacardiaceae (II). 1 Philipp. to NG: ***K. pinnatum*** (Blanco) Merr. – timber good

Kopsia Blume. Apocynaceae (I b). 24 SE As. to W Mal., Carolines. R: HPB 9(2004)92. Alks. Some cult. orn. trees & shrubs

Kopsiopsis (G. Beck) G. Beck = *Boschniakia*

korakaha *Memecylon umbellatum*

korarima *Aframomum corrorima*

Korean grass *Zoysia tenuifolia*

korila *Cyclanthera pedata*

Korintji cinnamon *Cinnamomum burmanni*

Kornasia Szlach. = *Malaxis*

Korolkowia Regel = *Fritillaria*

Korovinia Nevski & Vved. = *Galagania*

Korshinskia Lipsky. Umbelliferae (III 5). 5 SW & C As. R: EJB 52(1995)342

Korthalsella Tieghem. Santalaceae (Viscaceae). c. 30 NE & E Afr. (1), Madag., Masc., Himal. to Jap., NZ & Hawaii (exceptional in being on oceanic is.). R: Novon 7(1997)269. Leaf-sheath ocrea swollen & ant-inhabited in 10 of 27 spp. examined; explosive fr. (not bird-disp.)

Korthalsia Blume. Palmae (I 3a). 26 Indomal. (25 Mal.). R: KB 36(1981)163. Rattans. Herm.; hapaxanthic shoots branching (Schoute's Model) high in canopy; suckers

Korupodendron Litt & Cheek. Vochysiaceae (II). 1 W C Afr.: ***K. songweanum*** Litt & Cheek. R: Britt. 54(2002)14

Kosmosiphon Lindau. Acanthaceae (III 2a). 1 trop. W Afr.: ***K. azureus*** Lindau

koso or **kousso** *Hagenia abyssinica*

Kosopoljanskia Korovin = *Schrenkia*

Kosteletzkya C. Presl = *Hibiscus*

* Very recently overtaken by *Shorea faguetiana* Heim (Dipterocarpaceae) at 94.1 m in Borneo

Kostermansia Soeg.-Reks. Malvaceae (Hel.; Bombacaceae). 1 Malay Pen.: ***K. malayana*** Soeg.-Reks. – fls every 8–9 yrs. R: Reinw. 5(1959)1

Kostermanthus Prance. Chrysobalanaceae (4). 3 W Mal. R: FOW 10(2003)172

Koster's curse *Miconia crenata*

kosumba *Schleichera oleosa*

Kotchubaea Regel ex Hook.f. = *Kutchubaea*

kotibé *Nesogordonia papaverifera*

koto *Pterygota* spp.

kotofihy *Prunus africana*

Kotschya Endl. Leguminosae (III 11). 31 trop. Afr.

kou *Cordia subcordata*

Kovalevskiella Kamelin = *Notoseris*

kowhai *Sophora tetraptera*

Koyamacalia H. Robinson & Brettell = *Parasenecio*

Koyamaea W. Thomas & Davidse. Cyperaceae (III inc. sed.). 1 Venez. & Braz.: ***K. neblinensis*** W. Thomas & Davidse. R: SB 14(1989)189

Koyamasia H. Robinson (~ *Vernonia*). Compositae (Vern.-Centr.). 1 Thailand: ***K. calcarea*** (Kitamura) H. Robinson

Kozlovia Lipsky. Umbelliferae (III 2). 4 C As. to Ind. R: EJB 58(2001)340

kozo *Broussonetia kazinoki*

krabak *Anisoptera* spp. esp. *A. curtisii*

Kraenzlinella Kuntze (~ *Pleurothallis*). Orchidaceae (V 13c). 9 trop. Am.

Kraenzlinorchis Szlach. = *Habenaria*

Krameria Loefl. Krameriaceae. 18 SW US to Arg. & Chile, arid & semi-arid. Local medic. (rhatany r.) & dyes. ***K. lanceolata*** Torrey (SW US) – hemiparasite with haustoria

Krameriaceae Dumort. (~ Zygophyllaceae). Magnoliidae – Zygophyllales. 1/18 warm Am. Hemiparasitic shrubs or trees or rhiz. herbs. Lvs in spirals, simple (–trifoliolate), entire; vessels & stip. 0. Fls bisex., pea-like, solit. & axillary or in term. panicles; K (4)5, imbr., showy, the 3 outer oft. larger than 2 inner & ± encl. flower, C (4)5, the 3 adaxial long-clawed, 2 lower smaller, broad, thick, sessile oft. lipid-secreting glands (nectary-disk 0), A (3)4 alt. with upper C (5th rarely below & sterile) with thick filaments s.t. basally connate or adnate to C & anthers with 1 or 2 term. pores or short slits, $\underline{G}$ pseudomonomerous (2), 1 carpel reduced & empty, the other with 2 collateral pend. anatr. bitegmic ovules. Fr. dry-dehiscent, 1-seeded, usu. armed with barbed bristles or spines; seeds with straight embryo & 0 endosperm. n = 6

Genus: *Krameria*

Form. thought closest to Polygalaceae; female *Centris* bees coll. saturated fatty-acids from the lipid-glands for larvae (they also visit Malpighiaceae for same)

Krapfia DC. (~ *Ranunculus*). Ranunculaceae (II 3). 8 Andes

Krapovickasia Fryx. Malvaceae (Malv.-Malv.). 4 C & S Am. R: Britt. 30(1978)454

Krascheninnikovia Gueldenst. ('*Eurotia*'). Amaranthaceae (Axyr.; Chenopodiaceae I 3). 1 Med., temp. As., W N Am.: ***K. ceratoides*** (L.) Gueldenst. – boron indicator in Russia; subsp. ***lanata*** (Pursh) Haklau (winterfat, W N Am.) – imp. food for stock esp. sheep, lice-killing shampoo. R: T 57(2008)572

Krasnovia Popov = *Kozlovia*

Krassera Schwartz = *Anerincleistus*

Krausella H.J. Lam = *Planchonella*

Krauseola Pax & K. Hoffm. Caryophyllaceae (I 1). 1 N Kenya & S Ethiopia, 1 Mozambique

Kraussia Harv. Rubiaceae (II 5). 4 trop. & S Afr. R: KB 50(1995)775

Krebsia Harv. = *Stenostelma*

Kreczetoviczia Tsvelev = ? *Trichophorum*

Kremeria Durieu = *Coleostephus*

Kremeriella Maire. Cruciferae (12). 1 NW Afr.: ***K. cordylocarpus*** (Coss. & Durieu) Maire

Kreodanthus Garay. Orchidaceae (IV 2d). 14 trop. Am.

kretek clove-flavoured cigarette

Kreysigia Reichb. = *Schelhammera*

kriek beer flavoured with sour cherries

Krigia Schreb. Compositae (Lact.-Mic.). 7 N Am. R: FNA 19(2006)362. Some cult. orn. with yellow or orange fls

krim-saghyz *Taraxacum megalorhizon*

krobonko *Telfairia occidentalis*
Kroenleinia Lodé = *Echinocactus*
Krokia Urb. = *Pimenta*
Krubera Hoffm. (~ *Capnophyllum*). Umbelliferae (III 10). 1 Medit. incl. Eur.: ***K. peregrina*** (L.) Hoffm.
Krugia Urb. = *Marlierea*
Krugiodendron Urb. Rhamnaceae (6). 2 Mex., Costa Rica, WI (1: ***K. ferreum*** (Vahl) Urb. – bark & r. medic.). V. hard wood
Krukoviella A.C. Sm. Ochnaceae (III). 1 Peru, Braz.: ***K. disticha*** (Tieghem) Dwyer (*K. scandens*). R: BJ 113(1991)174
Krylovia Schischkin = *Aster* (*Rhinactinidia*)
Kryptostoma (Summerh.) Geer. = *Habenaria*
Kubitzkia van der Werff. Lauraceae (I). 1 NE S Am.: ***K. macrantha*** (Kosterm.) van der Werff ('*Systemonodaphne geminiflora*'). R: BJ 110(1988)161
kudingcha *Ilex* spp. esp. *I. kaushue*
Kudoacanthus Hosok. (~ *Leptostachya*). Acanthaceae (III). 1 Taiwan: ***K. albonervosus*** Hosok.
Kudrjaschevia Pojark. = *Nepeta*
kudzu vine *Pueraria montana* var. *thomsonii*
kümmel *Carum carvi*
Kuepferia Favre (~ *Gentiana*). Gentianaceae. 14 Himal. to Tibet. R: T 63(2014)349
kufa *Clusia* spp.
Kuhitangia Ovcz. = *Acanthophyllum*
Kuhlhasseltia J.J. Sm. Orchidaceae (IV 2d). 9 Korea & Jap. to NG
Kuhlmannia J. Gómes = *Pleonotoma*
Kuhlmanniella Barroso = *Dicranostyles*
Kuhlmanniodendron Finschi & Groppo (~ *Carpotroche*). Achariaceae (Lind.). 2 E Braz. R: SB 38(2013)169
Kuhnia L. = *Brickellia*
kukui nut *Aleurites moluccanus*
Kulinia Briggs & L. Johnson = *Desmocladus* (but see Telopea 7(1998)349)
kullam nut *Balanites rotundifolia*
kumara *Ipomoea batatas*
Kumara Medik. (~ *Aloe*). Asphodelaceae (Aloeaceae). 2 SW S Afr. R: SB 39(2014)67
kumato trade-name for *Solanum lycopersicum* 'Olmeca'
kumbanzo *Holarrhena pubescens*
kumbuk *Terminalia arjuna*
Kumlienia Greene (~ *Ranunculus*). Ranunculaceae (II 3). 1 SW N Am.: ***K. hystricula*** (A. Gray) Greene
Kummerowia Schindler. Leguminosae (III 19). 2 As., N Am. ***K. striata*** (Thunb.) Schindler (*Lespedeza s.*, Jap. clover, E As., widely natur. SE US) – fodder
kumquat *Citrus japonica*; **desert k.** *C. glauca*
kunai *Imperata* spp.
Kundmannia Scop. Umbelliferae (III 8). 2 S Eur., Medit.
Kungia Fu (~ *Sinocrassula*). Crassulaceae (I 1). 2 SW China
Kunhardtia Maguire. Rapateaceae. (I 2). 2 Venez.
Kuniwatsukia Pichi-Serm. = *Anisocarpium*
Kunkeliella Stearn = *Thesium*
Kunstleria Prain. Leguminosae (III 16). 8 Kerala (1), W & C Mal. (7). R: Blumea 38(1994)465. Lianes, rarely coll.
Kuntheria Conran & Clifford (~ *Schelhammera*). Colchicaceae. 1 N Queensland: ***K. pedunculata*** (F. Muell.) Conran & Clifford. R: FA 45(1987)417
Kunzea Reichb. Myrtaceae (II 14). 38 Aus. (38, 37 endemic), NZ (1: ***K. ericoides*** (A. Rich.) J. Thompson (burgan) – poll. chafer beetles, some Aus. forms with lignotubers; timber & medic.). Some cult. orn. heath-like shrubs like *Leptospermum* but A longer than C. ***K. ambigua*** (Sm.) Druce (tick bush, E Aus.) – allegedly found only in tick-infested areas nr. Sydney; ***K. baxteri*** (Klotzsch) Schauer (SW Aus.) – poll. honey possum (*Tarsipes rostratus*); ***K. pomifera*** F. Muell. (monterry, muntries, SE Aus.) – fr. ed.
Kupea Cheek & S. Williams. Triuridaceae. 1 W Cameroun: ***K. martinetugei*** Cheek & S. Williams, 1 Tanzania: ***K. jonii*** Cheek. R: KB 58(2003)225, 940
kurakkan *Eleusine coracana*

Kuramosciadium Pimenov & al. Umbelliferae (I 8). 1 Uzbekistan: ***K. corydalifolium*** Pimenov & al. R: SB 36(2011)492

kurara, kuraru *Andira inermis*

kurchi bark *Holarrhena pubescens*

kurdee *Carthamus tinctorius*

kurrajong *Brachychiton populneus*

Kurramiana Omer & Qaiser (~ *Gentiana*). Gentianaceae. 1 Pakistan: ***K. micrantha*** (Aitch. & Hemsl.) Omer & Qaiser. R: PakJB 24(1992)101

kurrat *Allium ampeloprasum*

Kurrimia Wall. ex Thw. = *Bhesa*

Kuruna Attigala & al. (~ *Arundinaria*). Gramineae (IV). 7 S Ind., Sri Lanka. R: SB 41(2016)178

Kurzamra Kuntze. Labiatae (VII 2b). 1 temp. S Am.: ***K. pulchella*** (Clos) Kuntze

Kurziodendron Balakr. = *Trigonostemon*

kusamaki *Podocarpus macrophyllus*

kussum oil *Schleichera oleosa*

Kutchubaea Fischer ex DC. (*Kotchubaea*). Rubiaceae (Ix.-Cord.). 11 trop. S Am. R: MNYBG 10,1(1963)212

kuth *Saussurea costus*

kutira gum *Cochlospermum religiosum*

kutjera *Solanum centrale*

kutki *Picrorhiza kurrooa*

kuwini *Mangifera × odorata*

kweek grass *Cynodon dactylon*

kwei *Osmanthus fragrans*

kweme nut *Telfairia pedata*

kwila *Intsia bijuga*

Kydia Roxb. = *Hibiscus*

kyetpaung *Urceola esculenta*

Kyhosia Baldwin. Compositae (Mad.-Mad.). 1 Calif., Oregon: ***K. bolanderi*** (A. Gray) Baldwin. R: FNA 21(2006)295

Kyllinga Rottb. = *Cyperus*

Kyllingiella R. Haines & Lye = *Cyperus*

kyor *Sagittaria sagittifolia*

Kyphocarpa (Fenzl) Lopr. (*Cyphocarpa*). Amaranthaceae (I 2). 3–4 S trop. & S Afr.

Kyrsteniopsis R. King & H. Robinson (~ *Eupatorium*). Compositae (Eup.-Alom.). 5 Mex., Guatemala. R: MSBMBG 22(1987)243

L

Labatia Sw. = *Pouteria*

labdanum = ladanum

Labiatae Juss. (Lamiaceae). Magnoliidae – Lamiales. S.l. incl. Chloanthaceae, Verbenaceae p.p. 240/6700 cosmop. exc. high alts & lats, esp. Medit., SW As., China, Aus. & S Am. Trees, shrubs or herbs, rarely lianes. Young stems oft. 4-angled. Lvs opp. (s.t. whorled or even in spirals as in *Icomum*), simple (rarely pinnate or digitate); stip. 0. Fls bisex. (pl. s.t. gynodioec., rarely dioec.), usu. bracteolate & in cymes (oft. in a terminal thyrse or verticillasters (compact axillary cymes)) or single fls in axils; K usu. persistent & with 4 or 5 (-9) teeth or lobes or bilabiate, fleshy & berry-like in fr. in *Hoslundia*, C ((4)5(-16)), (1) 2-labiate or less oft. ± reg., A 4 or 2 (+ 2 staminodes) or 5–8(-16), attached to tube, anthers usu. with longit. slits, connective s.t. such that pollen-sacs are sep. (e.g. *Salvia*), usu. nectary-disk (nectar sucrose-rich) annular to anterior at base of $\underline{G}$ (2) s.t. on gynophore, oft. each carpel longit. divided in 2, style oft. bifid apically (both pointed cf. Verbenaceae), 1 erect, anatr. to hemitropous (rarely orthotropous) unitegmic ovule with funicular obturator in each of 4 lobes. Fr. of(1–)4 1(2)-seeded mericarps (nutlets) with tough pericarp, dry-indehiscent or a drupe with pyrenes; seed with straight embryo & little or 0 oily endosperm. n = 10–240 (polyploidy common)

Classification & chief genera [see also Nature Sci. Reps 34343(2016)9]:

I. **Symphorematoideae** (non-aromatic lianes with simple or stellate hairs, A 4–18, style term., ovules orthotropus, pend.; Indomal.): *Congea*, *Sphenodesme*

II. **Viticoideae** (Verbenaceae p.p.; non-aromatic trees, shrubs (rarely geoxylic), lianes, lvs s.t. compound, A (2)4(5), style term., drupes; trop. esp. As., few temp. (*Vitex*)): *Vitex*; *Gmelina, Premna* & *Cornutia* to be excl. as **Premnoideae**

III. **Ajugoideae** (incl. Teucrioideae, Verbenaceae p.p.; woody or herbaceous, s.t. aromatic, hairs usu. simple, lvs usu. simple, A (2)4(5), style term., drupes to schizocarps or dry-indehiscent): *Aegiphila, Ajuga, Caryopteris, Clerodendrum, Oxera, Rotheca, Teucrium*; *Acrymia, Cymaria* excl. as **Cymarioideae**

IV. **Prostantheroideae** (Chloanthoideae; usu. non-aromatic small trees to shrubs, lvs simple, A 4–8, style term., fr. dry, seeds with endosperm; Aus.):
   1. **Chloantheae** (Chloanthaceae, Dicrastylidaceae) – *Dicrastylis, Pityrodia*. R: AusSB 24(2011)1
   2. **Westringieae** – *Hemigenia, Prostanthera* (aromatic)

V. **Scutellarioideae** (usu. non-aromatic shrubs & herbs, lvs simple, A4, style term., schizocarps): *Holmskioldia, Scutellaria, Tinnea*

VI. **Lamioideae** (Stachyoideae incl. Pogostemonoideae; shrubs or herbs, rarely trees, mostly non-aromatic, lvs simple, pollen usu. tricolpate, style gynobasic, schizocarps; R: T 60(2011)471): *Ballota, Lamium, Leonotis, Leucas, Melittis, Moluccella, Phlomis, Physostegia, Pogostemon, Prasium, Sideritis, Stachys* (over 30% monospecific)

VII. **Nepetoideae** (usu. aromatic, pollen usu. hexacolpate, style gynobasic):
   1. **Elsholtzieae** (K 10-nerved): *Elsholtzia, Perilla*
   2. **Mentheae:** (K usu. 11-nerved; several generic limits hazy): *Perovskia, Rosmarinus, Salvia* (a. Salviinae); *Clinopodium, Hyssopus, Lycopus, Monarda, Origanum, Mentha, Micromeria, Prunella, Satureja, Thymus* (b. Menthinae); *Agastache, Dracocephalum, Nepeta* (c. Nepetinae); *Melissa* (inc. sedis)
   3. **Ocimeae** (A declinate): *Lavandula* (only) (a. Lavandulinae; Lavanduleae); *Isodon* (b. Hanceolinae); *Eriope, Hypenia, Hyptis* (c. Hyptidinae); *Basilicum, Hoslundia, Ocimum, Orthosiphon, Platostoma* (d, Ociminae); *Aeollanthus, Coleus, Plectranthus* (e. Plectranthinae)

Trad. distinction from Verbenaceae s.l. (usu. without ess. oils & 4-lobed G) untenable; some genera, e.g *Callicarpa, Tectona* (teak), are not readily placed here, though *Garrettia, Hymenopyramis, Peronema* & *Petraeovitex* comprise **Peronematoideae**, & some form. here now in own fam., e.g. *Tetrachondra*.

Bilat. symmetrical K evolved many times, C actinomorphy at least 9 times, A 2 derived in diff. ways, lobed G many times, gynobasic style twice, fleshy fr. at least 8 times. Labiatae s.s. usu. insect-poll. pls of open country (*Gomphostemma* in rainforest), lipped fls acting as landing-stages for poll.: in most temp. spp. lower lip (C 3) acting as such, the upper usu. hooded & prot. A, while in trop. spp. upper lip of C 4 with A lying along lower (C 1) lip or ascendant from it; in *Teucrium* lower lip is 5-lobed & upper 0, A oft. completely exposed. Pollinators incl. birds as well as insects (incl. Lepidoptera) – see *Salvia* for most intricate mech.; explosive systems in *Aeollanthus* & *Hyptis*, where A held under tension by enfolding lobes of C such that arrival of insect on lower lip releases A & poll. is dusted with grains. Some disp. by modifications of K – persistent & bladdery or hooks formed by teeth (wind- & animal-disp. respectively). Volatile oils incl. terpenes which are allelopathic in some chaparral spp. but also the basis of many of their uses to humans

Fam. long-recog. because of its medic. (esp. coughs, colds) & culinary value, many still of great importance as flavourings & scents (e.g. CK's 'Escape for Men', but refused by cattle): *Lavandula* (lavender), *Mentha* (mint), *Ocimum* (basil), *Origanum* (oregano, marjoram), *Rosmarinus* (rosemary), *Salvia* (sage), *Satureja* (savory), *Thymus* (thyme), also *Pogostemon* (patchouli) & *Perilla* (used in printing inks & paints) etc.; tubers of *Stachys* (Chinese artichoke) & *Coleus* (Hausa potato) ed.; many cult. orn. in above genera & *Ajuga, Callicarpa, Caryopteris, Clerodendrum, Clinopodium, Congea, Holmskioldia, Horminum, Lamium, Moluccella, Monarda, Nepeta, Perovskia, Physostegia, Plectranthus, Prostanthera, Rotheca, Scutellaria, Thymbra, Vitex, Westringia* etc. etc. Genera form. in Verbenaceae incl. imp. timbers: *Gmelina, Premna, Tectona* (teak), *Vitex*

Labichea Gaudich. ex DC. Leguminosae (I 3). 15 Aus. R: Muelleria 6(1985)22

Labidostelma Schltr. = *Polystemma*

Labisia Lindl. Primulaceae (Myrsinaceae). 7 SE As., Mal. R: Blumea 50(2005)580. Subherbaceous (Rauh's Model) with creeping stems; blue fr. poss. attractive to pheasants. Some medic. locally (childbirth, gonorrhoea) esp. ***L. pumila*** (Blume) Fern.-Vill. (kacip fatimah) capsules (prep. from r.) sold in Malaysia

lablab *Lablab purpureus*; also used for 'greens' eaten with rice in trop. As.

Lablab Adans. (~ *Dolichos*). Leguminosae (III 18). 1 (prob.) trop. Afr., widely cult.: ***L. purpureus*** (L.) Sweet (*L. niger, Dolichos l.*, lablab, hyacinth bean, bonavist, prob. derived from wild 'subsp. *uncinatus* Verdc.') – pods & seeds ed., green manure & fodder

Labordia Gaudich. = *Geniostoma*

Labourdonnaisia Bojer. Sapotaceae (I 2). 3 Madag. Timber hard

Labrador tea *Rhododendron tomentosum*

Labramia A. DC. (~ *Manilkara*). Sapotaceae (I 2). 8 Madag.

+**Laburnocytisus** Trel. Leguminosae. Graft hybrids between *Laburnum* & *Cytisus*. +***L.* 'Adamii'** derived (once, 1826) accidentally from graft on *Laburnum anagyroides* of *C. purpureus* Scop. (Austria), the chimaera (epidermis *C.*, rest *L.*) oft. breaking down so that there are branches of 'pure' *L.* or (rarely) *C.*, cult. curiosity

Laburnum Fabr. Leguminosae (III 9). 2 SC & SE Eur. Cult. orn. trees & shrubs esp. ***L. anagyroides*** Medik. (golden chain or rain, C & S Eur.) – long-lived (tree planted at Leiden in 1601, 17 m tall in 1937, suckers still alive 2015), all parts poisonous (cytisine (alk.), seeds can be fatal to children), timber hard, used as ebony subs. esp. in inlays, musical instruments etc., also ***L. alpinum*** (Mill.) J. Presl (Scotch l., S Eur.) & their hybrid ***L. × watereri*** (Wettstein) Dippel ('*L. × vossii*') with longer racemes

laburnum, Indian *Cassia fistula*; **Scotch l.** *Laburnum alpinum*

lac insect resin, secreted by *Laccifer lacca* esp. (in Ind.) on *Butea monosperma, Cajanus cajan, Ficus religiosa, Schleichera oleosa & Ziziphus mauritiana*, the original shellac

Lacaena Lindl. Orchidaceae (V 12i). 2 C Am. R: Die Orchidee 30(1979)55. Cult. orn.

Lacaitaea Brand = *Trichodesma*

Lacandonia E. Martínez & C.H. Ramos = *Triuris*

Lacandoniaceae E. Martínez & C.H. Ramos = Triuridaceae

Laccodiscus Radlk. Sapindaceae. 4 W Afr.

Laccopetalum Ulbr. (~ *Ranunculus*). Ranunculaceae (II 3). 1 Peruvian Andes (4000–4800 m): ***L. giganteum*** (Wedd.) Ulbr. – coarse herb with pachycaul stems & large yellowish green fls with (?) 10 000 carpels, local medic., critically endangered

Laccospadix H. Wendl. & Drude (~ *Calyptrocalyx*). Palmae (V 14g). 1 NE Queensland: ***L. australasicus*** H. Wendl. & Drude. R: FA 39(2011)197

Laccosperma (G. Mann & H. Wendl.) Drude (*Ancistrophyllum*). Palmae (I 2a). 6 trop. Afr. R: Phytotaxa 51(2012)43. Rattans – hapaxanthic, herm.

lace, Queen Anne's *Anthriscus sylvestris*

lace-bark *Lagetta lagetto*; **l.-b. pine** *Pinus bungeana*

lace-leaf *Aponogeton madagascariensis*

lacewood *Platanus × hispanica* veneers; **Brazilian l.** *Euplassa meridionalis*

Lachanodes DC. (~ *Senecio*). Compositae (Sen.-Sen.). 1 St Helena: ***L. arborea*** (Roxb.) R. Nordenstam (*S. redivivus*, she cabbage) – tree (Attims's Model), form. building timber, now reduced to c. 50 pls in wild but also in cult., allies in S Am. & Australasia. R: Q.C.B. Cronk, *Endemic Flora S. H.* (2000)83

Lachemilla (Focke) Rydb. = *Alchemilla*

Lachenalia Jacq.f. ex Murray. Asparagaceae (Hyacinthaceae). Incl. *Polyxena*, 134 Namibia to E Cape (mostly winter rainfall region). R: Plantsman 8(1986)129; G. Duncan (2012) *The genus L.* Cape cowslip esp. ***L. aloides*** (L.f.) Engl. – cult. orn., many cvs, & hybrids with other spp. – pendent orange or yellow fls bird.-poll. while ***L. unifolia*** Jacq. has pink or purplish fls visited by bees

Lachnaea Royen ex L. Thymelaeaceae (Thym.-Daph.). Incl. *Cryptadenia* 40 Cape. R: Strel. 9(2000)680

Lachnagrostis Trin. (~ *Agrostis)*. Gramineae (XVI 5). c. 30 S temp. esp. Australasia. R: Telopea 9(2001)445. Heterogeneous

Lachnanthes Elliott. Haemodoraceae (I). 1 Massachusetts to Florida, Cuba: ***L. caroliniana*** (Lam.) Dandy (*L. tinctoria*) – weedy in comm. cranberry bogs, r. form. medic. (chest disease) & source of red dye. R: FNA 26(2002)48

Lachnocapsa Balf.f. Cruciferae (4). 1 Socotra: ***L. spathulata*** Balf.f.

Lachnocaulon Kunth = *Paepalanthus*

Lachnoloma Bunge. Cruciferae (26). 1 Iran to China: ***L. lehmannii*** Bunge

Lachnophyllum Bunge. Compositae (Ast.-Hom.). 2 W to C As.

Lachnopylis Hochst. = *Nuxia*

Lachnorhiza A. Rich. (~ *Vernonia*). Compositae (Vern.-Linz.). 1 W Cuba: ***L. piloselloides*** A. Rich. R: ABH 37(1992)89

Lachnosiphonium Hochst. = *Catunaregam*

Lachnospermum Willd. Compositae (Gnap.-Rel.). 3 SW Cape to Namaqualand. R: OB 104(1991)74, Strel. 9(2000)337

Lachnostachys Hook. Labiatae (IV 1; Dicrastylidaceae). 6 S Aus. R: Nuytsia 11(1996) 81

Lachnostoma Kunth = *Matelea*

Lachnostylis Turcz. Phyllanthaceae (Brid.-Sec.; Euphorbiaceae s.l.). 3 S Cape. R: Strel. 9(2000)458

Lacistema Sw. Lacistemataceae (Flacourtiaceae s.l.). 1–2 trop. Am. R: FN 22(1980)183. Species delimitation difficult, fls v. small

Lacistemataceae Mart. (~ Flacourtiaceae). Magnoliidae – Malpighiales. 2/4 trop. Am. Everg. trees accum. aluminium. Lvs usu. serrate, distchous, stipulate. Infl. a spike or dense raceme. P inconspicuous; A 1 with well-separated & even stipitate thecae, $\underline{G}$ (2 or 3) with 1 or 2 apical ovules per carpel. Fr. a 1(-3)-seeded capsule. endosperm copious
Genera: *Lacistema, Lozania*
A fragment of the shattered heterogeneous Flacourtiaceae

Lackeya Fortunato & al. Leguminosae (III 18). 1 SE US: ***L. multiflora*** (Torrey & A. Gray) Fortunato & al. R: KB 51(1996)365

Lacmellea Karsten. Apocynaceae (I c). c. 20 trop. S Am. Massart's Model. Some rubber & ed. fr. ***L. lactescens*** (Kuhlm.) Markgraf – powdered lvs chewed as coca subs. in NW Amaz.

Lacomucinaea Nickrent & M. García (~ *Thesium*). Santalaceae (Thesiaceae). 1 S Afr.: ***L. lineata*** (L.f.) Nickrent & M. García – succ. lvs. R: Phytotaxon 224(2015)179

lacoocha *Artocarpus lacucha*

Lacostea Bosch = *Trichomanes*

Lacosteopsis (Prantl) Nakaike = *Crepidomanes*

lacquer (SE As.) *Gluta laccifera;* **Burmese l.** *G. usitata;* **Chinese** or **Japanese l.** *Rhus verniciflua;* **pasto l.** *Elaeagia utilis*

Lacroixia Szlach. = *Dinklagiella* (But see ABF 40(2003)69)

Lactoridaceae Engl. = Aristolochiaceae

Lactoris Philippi. Aristolochiaceae (Lactoridaceae). 1 Masatierra, Juan Fernandez: ***L. fernandeziana*** Philippi – Troll's Model, wind-poll., gynomonoec. shrub with ethereal oil-cells in parenchyma & pendent green fls, only c. 1000 individuals left in montane cloud forest (1999, though only 12 in 1962) due to depredations by goats & more recently rampant *Rubus* spp.; wood anatomy corr. predicted present disposition. Fossils in Patagonia; Cretaceous pollen from S Afr. referred here

Lactuca Tourn. ex L. Compositae (Lact.-Lact.). Incl. *Lactucella, Lagedium, Mulgedium, Pterocypsela, Scariola* (s.s. with 4+ ribs on cypsela c. 75; Eur. 17), *Steptoramphus*, excl. *Cephalorrhynchus, Cicerbita* (incl. *Mycelis*), c. 60 cosmop. esp. N temp. (Eur. c. 30). Bitter sesquiterpene lactones lactucin & lactupicrin with sedative properties, ***L. virosa*** L. (opium lettuce, C & S Eur.) form. cult. for this (lactucarium), used as home remedy for coughs etc. Some cult. veg. & weeds. ***L. indica*** L. (*P. i.*, Ind. lettuce, E & SE As.) – cult. for ed. lvs; ***L. sativa*** L. (lettuce, cultigen ? orig. E Medit. from *L. serriola*) – bitter principles bred out, aphrodisiac to anc. Egyptians but opp. for Anc. Greeks, cos (upright heads of lvs with broad midribs) & cabbage (globose heads of lvs, e.g. ubiquitous tasteless iceberg l.)) types known to anc. Persians, Pliny mentioning 11 distinct sorts, now *the* salad green of temp. regions with many cvs incl. forms with thick main stem ed. (asparagus lettuce, celtuce (stem l.)), '**Grand Rapids**' a cv in which the red/far-red reactions of phytochrome orig. investigated; ***L. serriola*** L. (*L. scariola*, prickly l., Eur. now subcosmop. weed) – lvs tending to be held edgewise upwards (compass plant), ed. locally, seeds source of culinary Egyptian lettuce-seed-oil

lactucarium *Lactuca virosa*

Lactucella Naz. = *Lactuca*

Lactucosonchus (Schultz-Bip.) Svent. = *Sonchus*

Lacunaria Ducke. Ochnaceae (Quiinaceae). 7 trop. Am. R: SB 37(2012)167. ***L. jenmanii*** (Oliv.) Ducke (NE S Am.) – hairy seeds used as beads

Ladakiella German & Al-Shehbaz (~ *Alyssum*). Cruciferae (22). 1 Himal.: ***L. klimesii*** (Al-Shehbaz) German & Al-Shehbaz. R: NJB 28(2010)647

ladanum *Cistus ladanifer* & *C. creticus*

Ladeania A. Egan & Reveal (~ *Psoralidium*). Leguminosae (III 20). 2 W N Am. R: Novon 19(2009)311

Ladenbergia Klotzsch. Rubiaceae (I 1). (Incl. *Cascarilla*) 34 trop. Am. R: NJB 17(1997)269. Alks in bark used like quinine but less efficacious

Ladino clover *Trifolium repens* f. *lodigense*

lad's love *Artemisia abrotanum*

lady apple *Syzygium suborbiculare*; **l. fern** *Athyrium filix-femina*; **L. Nugent's rose** *Clerodendrum chinense*; **l. of the night** *Brunfelsia americana, Cestrum nocturnum*; **l. palm** *Rhapis excelsa*

Ladyginia Lipsky. Umbelliferae (III 10). 3 SW & C As. R: EJB 49(1992)215

lady's bedstraw *Galium verum*; **l. comb** *Scandix pecten-veneris*; **l. fingers** *Anthyllis vulneraria, Hibiscus esculentus*; **l. mantle** *Alchemilla vulgaris*; **l. slipper** *Cypripedium* spp.; **l. smock** *Cardamine pratensis*; **l. tresses** *Spiranthes spiralis*

Laelia Lindl. (~ *Cattleya*). Orchidaceae (V 13b). Incl. *Schomburgkia*, 33 trop. Am. R: C.L. Withner (1990) *The Cattleyas & their relatives* II, (1993) III: 169. Braz. spp. = *Cattleya*. Cult. orn. epiphytes with extrafl. nectaries, esp. ***L. anceps*** Lindl. (Mex., Honduras) – many cvs & hybrids (incl. some with *C.* spp. etc.) & ***L. purpurata*** Lindl. (Braz.) – fls to 20 cm across; ***L. speciosa*** (Kunth) Schltr. (Mex.) – mucilaginous paste from pseudobulbs used to make images of animals, skulls etc. on All Saints Day & 'Day of the Dead' (Uphof); ***L. thomsoniana*** (Reichb.f.) L.O. Williams (*S. t.*, WI) – pseudobulbs carved into pipe-bowls locally; ***L. tibicinis*** (Lindl.) L.O. Williams (*S. t.*, Mex. to Panamá) – hollow ant-filled pseudobulbs to 55 cm tall absorbing nutrients brought in by ants

× **Laeliocatanthe** J.M.H. Shaw. Orchidaceae. *Cattleya* × *Guarianthe* × *Laelia* – 155 grexes

× **Laeliocattleya** Rolfe. Orchidaceae. *Cattleya* × *Laelia* – 565 grexes

Laeliopsis Lindl. & Paxton = *Broughtonia*

Laennecia Cass. (~ *Conyza*). Compositae (Ast.-Pod.). 18 SW N Am. to Bolivia. R: Phytol. 68(1990)205

Laestadia Kunth ex Less. Compositae (Ast.-Hin.). 6 trop. Andes, WI. Pappus 0

Laetia Loefl. ex L. Older name for *Casearia*

Lafoensia Vand. Lythraceae. 5–6 S Mex. to S Braz. R: Mem.Soc.Ci.Nat.La salle 45(1986)115. Trees & shrubs with bat-poll. 8–16-merous fls. Some construction timber esp. ***L. acuminata*** DC. (Ecuador); ***L. punicifolia*** Bert. ex DC. (Mex. to N S Am.) – tree with yellow fls turning to red, cult. orn., source of yellow dye

Lafuentea Lag. Plantaginaceae (Dig.-Ver.; Scrophulariaceae s.l.). 2 S Spain (***L. rotundifolia*** Lag. – shrubby), Morocco

Lagarinthus E. Meyer = *Schizoglossum*

Lagarosiphon Harv. Hydrocharitaceae. c. 16 trop. Afr. & Madag. R: BJBB 53(1983)441. Male fls released; A 3, staminodes 2 or 3. ***L. major*** (Ridl.) Moss (S Afr.) – grown in aquaria ('*Elodea crispa*'), pest in Masc., NSW, NZ

Lagarosolen W.T. Wang = *Petrocodon* (but see ABY 6(1984)11)

Lagarostrobos Quinn (~ *Dacrydium*). Podocarpaceae. Incl. *Manoao*, 1 NZ: ***L. colensoi*** (Hook.) Quinn (*M. c.*, silver pine), 1 Tasmania: ***L. franklinii*** (Hook.f.) Quinn (*D. franklinii*, Huon pine, Tasmania) – lax seed-cones (cf. *Dacrydium*), trees to 1500 yrs old so poss. use in dendrochronology, timber (used for endgrain blocks in printmaking by Margaret Preston), H.p. oil in medic. soaps, cult. orn. R: AusJB 30(1982)316, NZJB 33(1995)196

Lagascea Cav. Compositae (Helia.-Helia.). 9 Mex., C Am., R: Fieldiana 38(1978)75. Synflorescence of 1(or 2[8])-flowered capitula, unique in Heliantheae

Lagedium Soják = *Lactuca*

Lagenandra Dalz. Araceae (VII 9). 14 S Ind., Sri Lanka. R: MLW 78–13(1978). Some aquarium pls (allied to *Cryptocoryne*)

Lagenantha Chiov. Amaranthaceae (Chenopodiaceae III 3). 2 NE Afr.

Lagenanthus Gilg = *Lehmanniella*

Lagenaria Ser. Cucurbitaceae (XIV). 6 trop. Afr., Madag. (1) with 1 ext. to rest of trop. Night-flowering. ***L. siceraria*** (Molina) Standl. (calabash or bottle gourd) – 1 subsp. in Afr. & Am., 1 in As. & Pac. perhaps domesticated independently though that in E Polynesia poss. disp. from Am., that in OW known to Pliny & prob. cult. 'Hoabhinian' culture 8000–3000 yrs ago; young fr. ed. (dried & peeled = kampyo used in sushi, 'New Guinea beans'), locally medic. (lauki; purgative), mature fr. with tough pericarp used for flasks, cups, dippers, penis-sheaths (NG), helmets for Polynesian rowers, etc., many diff. shaped forms cult.

Lagenia E. Fourn. (~ *Araujia*). Apocynaceae (V c). 2 S Am.

Lagenias E. Meyer (~ *Sebaea*). Gentianaceae. 1 W Cape: ***L. pusillus*** (Cham.) E. Meyer. R: SB 37(2012)252

Lagenifera Cass. = *Lagenophora*

Lagenithrix Nesom = *Pappochroma* (*Erigeron* s.l.)

Lagenocarpus Nees. Cyperaceae (III 1). 30 trop. Am.

Lagenocypsela Svenson & Bremer. Compositae (Ast.-Lag.). 2 NG. R: Aus. SB 7 (1994)265

Lagenopappus Nesom = *Pappochroma* (*Erigeron* s.l.)

Lagenophora Cass. (*Lagenifera*). Compositae (Ast.-Lag.). Excl. *Keysseria*, 14 Indomal., Pac., trop. Am. (6). R: Blumea 14(1966)285. Heterogeneous? ***L. mikadoi*** (Koidz.) H. Koyama (Ryukyus) – rheophyte

Lagerstroemia L. Lythraceae. 55 trop. As. (Mal. 14) to Aus. R: GBS 24(1969)185. Trees (Champagnat's Model) with A 15–200. Cult. orn. (some hybrids) incl topiary & some good timber; alks. ***L. calyculata*** Kurz (Myanmar to Vietnam) – street-tree as in Phnom Penh; ***L. hypoleuca*** Kurz (Andaman pyinma, Andamans) – timber; ***L. indica*** L. (crape or crepe myrtle, crepe flower, China) – widely cult. trop. & warm, many cvs with white, pink or purple fls with A 36–42 & potent psychoactive properties; ***L. microcarpa*** Wight (*L. lanceolata*, Ind.) – reddish timber for many uses incl. bridges, coffee-boxes, furniture, tanbark good; ***L. speciosa*** (L.) Pers. (*L. flos-reginae*, pride-of-India, queen-flower, pyinma, Ind. & China to Aus.) – cult. orn. with purple or white fls with A 130–200, medic., timber (banglang) for railway sleepers

Lagetta Juss. Thymelaeaceae (Thym.-Daph.). 3 WI. ***L. lagetto*** (Sw.) Nash (*L. lintearia*, lace-bark, lagetto, gauze tree) – stretched inner bark a reticulated lace-like material used ornamentally & as textile in Jamaica

lagetto *Lagetta lagetto*

Laggera Schultz-Bip. ex Benth. (~ *Blumea*). Compositae (Inul.-Pluch.) 10 OW trop. Heterogeneous. Stems winged (decurrent leaf-bases)

Lagoa T. Durand. Apocynaceae (V c; Asclepiadaceae III 1). 1 Braz.: ***L. calcarata*** (Decne) Baill.

Lagochilopsis Knorr. = *seq*.

Lagochilus Bunge ex Benth. Labiatae (VI). c. 40 C As. to Iran & Afghanistan to NW China. Alks; some local medic. (e.g. ***L. inebrians*** Bunge (C As.) – dried lvs a stimulant tea), cult. orn.

Lagoecia L. Umbelliferae (III 8). 1 Medit. (incl. Eur.): ***L. cuminoides*** L. – 1 of the 2 loc. aborts, fr. a cumin subs. R: Fl. Iraq 5,2(2013)119

Lagonychium M. Bieb. = *Prosopis*

Lagophylla Nutt. Compositae (Mad.-Mad.). 4 W N Am.

Lagopsis (Benth.) Bunge (~ *Marrubium*). Labiatae (VI). 4 C As. to Jap.

Lagos rubber *Funtumia elastica*

Lagoseriopsis Kirpiczn. (~ *Launaea*). Compositae (Cich.-Cich (Crep.)). 1 C As.: ***L. popovii*** (Krash.) Kirpiczn.

Lagoseris M. Bieb. = *Crepis* (but prob. distinct)

Lagotis Gaertn. Plantaginaceae (Dig.-Ver.; Scrophulariaceae s.l.). 20 E Eur (2), N & C As. to Cauc., Himal. & W China

Lagrezia Moq. (~ *Celosia*). Amaranthaceae (I 2). 12 Madag., Ind. Ocean

Laguna Cav. = *Hibiscus*

Lagunaria (DC.) Reichb. Malvaceae (Malv.). 1 Queensland, 1 Norfolk Is., Lord Howe Is.: ***L. patersonia*** (Andrews) G. Don – cult. orn. everg. tree, seeds amongst irritant bristles inside fr. R: Blumea 51(2006)349. Extrafl. nectaries

Laguncularia Gaertn. f. Combretaceae (II 1). 1 trop. Am. & W Afr. mangroves: ***L. racemosa*** (L.) Gaertn. f. (button wood, trop. Am.) – tanbark, timber for external use. R: AMBG 45(1958)162

lagundi *Vitex negundo*

Lagurus L. Gramineae (XVI 2). 1 Medit. (incl. Eur.): ***L. ovatus*** L. (rabbit's foot; natur. Channel Is.) – cult. orn. for dried displays, invasive Aus., S Afr. R: FNA 24(2007)670

Lagynias E. Meyer ex F. Robyns. Rubiaceae (III 2). 5 trop. E & S Afr. R: BJBB 11(1928)312

Lahia Hassk. = *Durio*

lahpet *Camellia sinensis*

Lakshmia Veldk. (~ *Andropogon*). Gramineae (XXII). 1 S Ind., Sri Lanka: ***L. venusta*** (Thwaites) Veldk. R: Rheedea 18(2008)81

lalang *Imperata cylindrica*

Lalldhwojia Farille. Umbelliferae (III 10). 4 Himal.

Lallemantia Fischer & C. Meyer. Labiatae (VII 2b). 5 Turkey to C As. & Himal. Some oils, potherbs & local medic.

Lamanonia Vell. Conc. (~ *Geissois*). Cunoniaceae (V). 5 S Braz., Arg., Paraguay. R: RBB 16(1993)75

Lamarchea Gaud. = *Melaleuca*

Lamarckia Moench. Gramineae (XVI 11). 1 Medit. incl. Eur.: ***L. aurea*** (L.) Moench – cult. orn., weedy in Aus., S Afr. R: FNA 24(2007)484

lamb, Scythian, vegetable l. of Tartary *Cibotium barometz;* **l.'s succory** *Arnoseris minima*

Lambertia Sm. Proteaceae (V 1). 9 SW & 1 E Aus. R: FA 16(1995)425. Some cult. orn. incl. ***L. formosa*** Sm. (honey flower, mt. devil, E Aus.)

lamb-kill *Kalmia angustifolia*

lamb's ears *Stachys byzantina;* **l. lettuce** *Valerianella locusta;* **l.quarters** *Chenopodium album;* **l.-tail** *Umbilicus oppositifolius*

lambswool (i.e. lamasool) Ale with roasted apples, sugar & spice drunk on 31 October (also in wassail in e.g. Sussex)

Lamechites Markgraf = *Ichnocarpus*

Lamellisepalum Engl. = *Sageretia*

Lamiacanthus Kuntze = *Strobilanthes*

Lamiaceae Martinov. See Labiatae

Lamiastrum Heister ex Fabr. = *Lamium*

Lamiodendron Steenis (~ *Fernandoa*). Bignoniaceae (1). 1 NG: ***L. magnificum*** Steenis. R: FM I,8(1977)161

Lamiophlomis Kudô = *Phlomoides*

Lamiostachys Krestovsk. = *Stachys* (but see BZ 91(2006)1258)

Lamium Tourn. ex L. Labiatae (VI). Incl. *Wiedemannia* (*Lamiastrum* with spiny bracts & K-lobes poss. distinct) 20 N Afr., Euras. (Eur. 12). R: J. Mennema (1989) *Taxonomic revision of L.* Heterogeneous? Dead nettles, non-flowering stems superficially resembling *Urtica* spp. & referred to as 'urtica' in med. ***L. album*** L. (white dead nettle, Euras. (?anc. introd. GB), natur. E N Am.) – fls large, white, bee-poll., lvs cooked like spinach, local medic.; ***L. amplexicaule*** L. (henbit, Eur., Med. to Iran, natur. N Am.) – prod. cleistogamous fls in spring & autumn, can cause 'staggers' in stock; ***L. galeobdolon*** (L.) Crantz (*Lamiastrum g.*, *G. luteum*, yellow archangel, W Eur. to Iran) – cult. orn. incl. varieg. cvs, invasive NW US, & ***L. maculatum*** (L.) L. (Eur., Medit.) – over 20 cvs esp. varieg. for ground-cover; ***L. purpureum*** L. (red. d. n., Eur., Med.) – locally medic.

Lamottea Pomel = *Carthamus*

Lamourouxia Kunth. Orobanchaceae (Ger.; Scrophulariaceae s.l.). 28 Mex. to Peru. R: SCB 6(1971)

Lampadaria Feuillet & Skog. Gesneriaceae (II 5c). 1 French Guiana: ***L. rupestris*** Feuillet & Skog. R: Britt. 54(2003)344

Lampas Danser. Loranthaceae (5). 1 N Borneo: ***L. elmeri*** Danser. R: Blumea 38(1993)108

Lampaya Philippi = *Lampayo*

Lampayo Philippi ex Murillo. Verbenaceae (Neo.). 2 Bolivia, Chile, Argentina deserts

Lamprachaenium Benth. = *Phyllocephalum*

Lampranthus N.E. Br. Aizoaceae. Excl. *Oscularia*, 194 Namibia to E Cape (Cape 124, 118 endemic – R: Strel. 9(2000)246), Aus. (1). R: H. E. K. Hartmann, *Ill. Handb. Succ. Pls*, Aiz. F–Z(2002)75. Cult. orn. subshrubs, esp. hybrids (many complex)

Lamprocaulos Masters = *Elegia*

Lamprocephalus R. Nordenstam (~ *Senecio*). Compositae (Sen.-Sen.). 1 SW Cape: ***L. montanus*** R. Nordenstam. R: Strel. 9(2000)338

Lamprococcus Beer = *Aechmea*

Lamproconus Lem. = *Pitcairnia*

Lamprolobium Benth. Leguminosae (III 4). 2 NE Aus. R: Austrobaileya 3(1991)394. Alks

Lampropappus (O. Hoffm.) H. Robinson (~ *Vernonia*). Compositae (Vern.-Gym.). 3 trop. Afr.

Lamprophragma O. Schulz = *Pennellia*

Lamprothamnus Hiern. Rubiaceae (II 5). 1 trop. E Afr.: ***L. zanguebaricus*** Hiern

Lamprothyrsus Pilg. = *Cortaderia*

Lamy butter *Pentadesma butyracea*

Lamyra (Cass.) Cass. = *Ptilostemon*

Lamyropappus Knorr. & Tamamschjan. Compositae (Card.-Card.). 1 C As.: ***L. shacaptaricus*** (B. Fedtsch.) Knorr. & Tamamschjan. R: BZ 38(1954)909

Lamyropsis (Charadze) Dittr. Compositae (Card.-Card.). 6 Medit. (Eur. 2), SW As.

Lanaria Aiton. Lanariaceae (Haemodoraceae s.l.). 1 S Afr.: ***L. lanata*** (L.) Durand & Schinz. R: Strel. 10(2000)640

Lanariaceae H. Huber ex Dahlgren (~ Haemodoraceae). Magnoliidae – Asparagales. 1/1 S Afr. Pl. with vertical rhiz. Lvs linr., basal, sheathing, in spirals (appearing distich.). Infls shortly branched, corymbose panicle clothed in dendritic hairs. Fls bisex. P 6 connate, tube half length P; A 3 + 3 adnate in mouth of P, anthers basifixed, introrse; G 3, ± inf. with 2 anatr. ovules per loc. Fr. a 1-seeded capsule; seed shiny black with phytolmelan in coat. n = 18
Genus: *Lanaria*
Phenalenone typical of H. absent. Poss. to be incl. in Hypoxidaceae

Lancea Hook.f. & Thomson. Mazaceae (Phrymaceae; Scrophulariaceae s.l.). 2 Tibet, China. ***L. tibetica*** Hook.f. & Thomson (depgul, Tibet) – r. roasted & smoked with tobacco as narcotic in Ladakh

lancewood Elastic woods used for fishing-rods, bows etc. e.g. *Duguetia quitarensis*, *Oxandra lanceolata*; **degame l.** *Calycophyllum candidissimum*; **Moulmein l.** *Homalium tomentosum*; **NZ l.** *Pseudopanax crassifolius*

Lancisia Fabr. = *Cotula*

land cress *Barbarea verna*; **l. seaweed** *Kali komarovii*

Landiopsis Capuron ex Bosser. Rubiaceae (?). 1 Madag.: ***L. capuronii*** Bosser. R: Adans. 20(1998)132

Landolphia P. Beauv. Apocynaceae (I c). 56 trop. Am., Afr., Madag. (6 – R: SGP 69(1999)93). Many lianes (Leeuwenberg's Model) with hook-tendrils like *Strychnos* spp.; fr. (some ed.) a berry of acid pulp (seed-hairs), many yield a rubber (Madag. or E Afr. r.) e.g. ***L. gummifera*** (Lam.) K. Schum. (vahy, Madag.)

Landoltia Les & Crawford = *Spirodela* (but see Novon 9(1999)532)

Landtia Less. (~ *Haplocarpha*). Compositae. 3 E Afr. to Ethiopia. ***L. rueppellii*** (Sch. Bip.) Vatke (*H. r.*) – in alpine frost-heaved soil, recurved peduncle facilitating germ.

Lanessania Baill. = *Trymatococcus*

Langebergia A. Anderb. (~ *Petalacte*). Compositae (Gnap.-Cass.). 1 S Afr.: ***L. canescens*** (DC.) A. Anderb. – shrub. R: OB 104(1991)93.

Langlassea H. Wolff = *Prionosciadium*

Langloisia Greene (~ *Gilia*). Polemoniaceae (III). Incl. *Loeseliastrum*, 3 SW N Am. deserts. R: Madroño 33(1986)167, 170

langsat *Lansium domesticum*

Langsdorffia Mart. Balanophoraceae. 4 Madag. (1), NG (1), S Am. (2, incl. ***L. hypogaea*** Mart. – wax made into candles). R: SB 36(2011)427

Lanium (Lindl.) Benth. = *Epidendrum*

Lankesterella Ames (~ *Stenorrhynchos*). Orchidaceae (IV 2h). 11 trop. Am.

Lankesteria Lindl. Acanthaceae (III). 7 trop. Afr., Madag.

Lankesteriana Karremans (~ *Anathallis*). Orchidaceae (V 13c). 19 trop. Am. R: Lankesteriana 13(2014)321

Lannea A. Rich. Anacardiaceae (II). 40 trop. Afr., Indomal. (1: ***L. coromandelica*** (Houtt.) Merr. (*L. grandis*, *Odina wodier*) – timber for external use, furniture etc., gum used for sizing paper & cloth, confectionery, tanbark also powdered as tooth-powder, roadside tree oft. pollarded), r. roasted or boiled (v. starchy) in Uganda; ***L. schweinfurthii*** (Engl.) Engl. (E & S Afr.) – ed. fr.; ***L. stuhlmannii*** (Engl.) Engl. (trop. Afr.) – r. a 'wool' used for life-belts etc. ('flotite' in NZ); ***L. welwitschii*** (Hiern) Engl. (trop. Afr.) – elephants strip off bark in pieces to 10 m long (Ghana)

Lanonia A.J. Hend. & Bacon (~ *Licuala*). Palmae (III 4b). 8 Vietnam (6), Hainan (1), Java (1). R: SB 36 (2011)887

Lansium Corr. Serr. Meliaceae (II-Trich.). 3 Mal. R: Blumea 31(1985)140. ***L. domesticum*** Corr. Serr. (duku, langsat, long kong [Thailand]) – ed. arils, a market fr. with a no. of distinctive cvs, some, at least, apomictic; pericarp used for incense in Java

Lantana L. Verbenaceae (Lant.). c. 100 trop. Am., trop. & S Afr. (few). R (sect. L.): JBRIT 6(2012)405. Shrubs & herbs, oft. armed, some with ed. fr.; spp. with 2 pyrenes per fr. prob. = *Lippia*. Alks. Some hedgepls, others aggressive weeds esp. ***L. camara*** L. (trop. Am.) – diploid to hexaploid obligate outbreeder, fls open yellow when visited by thrips (in Ind.), changing to orange once poll., serious weed (toxic triterpenoids (lantadenes), verbascoside, an inhibitor of protein kinase C), poss. a hybrid cultigen (***L. strigocamara*** R. Sanders) in S Eur., Azores, trop. As., Aus., NZ, New Caled., trop. & S Afr., Masc.,

Seychelles, SE US, Galápagos, Hawaii (where biologically controlled); ***L. montevidensis*** (Spreng.) Briq. (S Am.) – cult. orn. with yellow-centred fls

Lantanopsis C. Wright ex Griseb. Compositae (Helia.-Ecl.). 3 Cuba, Hispaniola

lantern(s), Chinese *Alkekengi officinarum*; **l. tree, Chile** *Crinodendron hookerianum*

Lanugia N.E. Br. = *Mascarenhasia*

Laosanthus Larsen & Jenjittikul = *Curcuma* (but see NJB 21(2001)135)

lapacho(l) *Handroanthus* spp. esp. *H. impetiginosus*

Lapageria Ruíz & Pavón. Philesiaceae (Smilacaceae s.l.). 1 S Chile & nearby Arg.: ***L. rosea*** Ruíz & Pavón (Chilean bellflower) – national fl. of Chile ('copihue'), cult. orn. liane with pend. red (or white) fls, berry allegedly ed., hybrids formed with *Philesia magellanica*

Lapanthus Louzada & Versieux (~ *Orthophytum*). Bromeliaceae (3). 2 SE Braz. R: SB 35(2010)497

Lapeirousia Pourret. Iridaceae (III 1). Excl. *Afrosolen*, *Codonorhiza*, *Schizorhiza*, 27 subSaharan Afr. esp. SW Coast. R: Strelitzia 35(2015)24. Some (starchy corms) locally ed. (Goldblatt), some cult. orn. See also *Freesia*

Laphamia A. Gray = *Perityle*

Laphangium (Hilliard & B.L. Burtt) Tzvelev = *Helichrysum*

Lapidaria (Dinter & Schwantes) N.E. Br. (~ *Lithops*). Aizoaceae (V). 1 SW Afr.: ***L. margaretae*** (Schwantes) N.E. Br. – cult. orn. succ. herb. R: H. E. K. Hartmann, *Ill. Handb. Succ. Pls*, Aiz. F–Z(2002)75

Lapiedra Lag. Amaryllidaceae. 1 W Med. incl. Eur.: ***L. martinezii*** Lag. R: Lagascalia 8(1978)13

Lapithea Griseb. = *Sabatia*

Laplacea Kunth = *Gordonia*

Laportea Gaudich. (*Fleurya*). Urticaceae (I). 22 trop. (3 pantrop.) & warm, temp. E As., E N Am. R: GBS 25(1969)111. Heterogeneous; see also *Dendrocnide*. Some with stings e.g. ***L. mooreana*** (Hiern) Chew (trop. Afr.), though those of ***L. alatipes*** Hook.f. (trop. & S Afr.) not a deterrent to gorillas which eat shoots. Some lvs used like spinach, some local medic.; some with strong fibres esp. ***L. canadensis*** (L.) Wedd. (*Urtica whitlowii*, wood nettle, E N Am.) – fibre comparable with ramie, v. strong

Lappula Gilib. Boraginaceae (B.3.5.1). c. 55 temp. Euras. (Eur. 6 incl. ***L. squarrosa*** (Retz.) Dumort. (*L. myosotis*) – fls change from white to red & blue), 5 N Am. R: JAA suppl. 1(1991)108. Fr. hooked

Lapsana L. (~ *Crepis*). Compositae (Lact.-Crep.). 1 Eur. & SW As.: ***L. communis*** L. (nipplewort) – used to treat nipple soreness of nursing mothers in Scottish Highlands – cf. 'Doctrine of Signatures', shape of infl. buds. R: T 44(1995)5

Lapsanastrum J.-H. Pak & Bremer (~ *Lapsana*). Compositae (Lact.-Crep.). 4 E As., N Am. R: T 44(1995)19

laran *Neolamarckia cadamba*

larap *Eucalyptus dumosa*

larch *Larix* spp., see also *Abies procera*; **common** or **European l.** *L. decidua*; **Dunkeld l.** *L. × marschlinsii*; **golden l.** *Pseudolarix amabilis*; **hybrid l.** *L. × marschlinsii*; **Japanese l.** *L. kaempferi*; **tamarack** or **American l.** *L. laricina*

Lardizabala Ruíz & Pavón. Lardizabalaceae. 1 Arg., Chile: ***L. funaria*** (Molina) Looser (*L. biternata*) – monoec. everg. liane, cult. orn., fr. (aquiboquil, aquibuquil) ed., good bark fibre. R: CBM n.s. 29(2012)253

Lardizabalaceae R. Br. Magnoliidae – Ranunculales. (Incl. Sargentodoxaceae) 7/c. 40 Himal. to SE As. & Taiwan, C Chile. R: CBM 29(2012)242. Pachycaul shrubs (*Decaisnea*) & lianes (monoec. or dioec. [S Am.]; *D.* polygamous). Lvs in spirals, palmate (pinnate in *Decaisnea*), rarely simple (*Sargentodoxa*); stip. usu. 0. Fls small, 3-merous, reg., in usu. drooping racemes from scaly axillary buds; K 3 + 3(–8) (3 in *Akebia*), usu. petaloid & imbr. or outer valvate, C 3 + 3, smaller, nectariferous or 0, A (3)6(–8) opp. C, filaments ± basally connate, anthers with longit. slits, the pollen-sacs ± embedded in thickened connective with short term. appendage, $\underline{G}$ 3 (or 6–12) in 1(–5) whorls of 3 ± 6 staminodes (pistillodes s.t. in male fls), carpels with term. ± sessile stigma, s.t. (at least *Akebia*) conduplicate & unclosed, with (few–)∞ anatr. to campylotropous or orthotropous bitegmic ovules. Fr. a head of berries or fleshy follicles, oft. ed., the pericarp (at least in *Decaisnea*) with latex-system; seeds with short straight embryo in copious oily endosperm s.t. also with carbohydrate. 2n = 28, 30, 32

Genera: *Akebia*, *Boquila*, *Decaisnea*, *Lardizabala*, *Sargentodoxa*, *Sinofranchetia*, *Stauntonia*

Major disjunction in distr. Most arch. genus the pachycaul *Decaisnea* (Cronquist), the most advanced the dioec. Chilean genera. Cult. orn. & some locally ed. fr.

Larentia Klatt (~ *Alophia*). Iridaceae (VII 5). 3 C & N S am. R: 20(2010)412. Lvs plicate

Laretia Gillies & Hook. (~ *Azorella*). Umbelliferae (Azor.). 2 Chilean Andes

Laricorchis Szlach. = *Maxillaria*

Larix Mill. Pinaceae. 10 cool N hemisph. (Eur. 2). R: NSPV 9(1972)4; RV 121(1990) 193. Larches. Much like *Cedrus* but lvs decid. & cones mature in 1 yr, protogynous. Timber & turpentine. ***L. decidua*** Mill. (common or European l., Alps & Carpathians) – timber for telegraph poles, pit-props, boat-building, shingles, fencing & gates, also nurse for broadleaf trees, distilled resin the source of Venice turpentine, sugar-like manna form. medic. in bronchial conditions, bark with 8–9% tannin & used for tanning leather esp. in former USSR, cut branches used as 'palm' in N Eur.; ***L. kaempferi*** (Lamb.) Carr. (*L. leptolepis*, Jap. l., Jap.); ***L. laricina*** (Du Roi) K. Koch (tamarack (l.), hackmatack, Am. l., N Am.) – uses similar to those of *L. decidua*; ***L. × marschlinsii*** Coaz (*L. × eurolepis*, *L. decidua × L. kaempferi*, Dunkeld or hybrid l.) – much planted larch resistant to l. canker, seed from seed orchards of parental spp. best coll. from *L. decidua* as *L. kaempferi* fls first & with more pollen so has higher precentage of pure *L. kaempferi* offspring; ***L. occidentalis*** Nutt. (W N Am.) – source of Venice turpentine

larkspur *Delphinium* spp.

Larnax Miers = *Deprea* (but see Phytol. 78(1995)354)

Larrea Cav. (*Covillea*). Zygophyllaceae (II). 6 SW N Am., S Am., desert disjuncts esp. ***L. divaricata*** Cav. (S Am.) & *L.* ***tridentata*** (DC.) Cov. (*L. d.* subsp. *t.*, chapparal, creosote bush) in SW N Am., where suckering clones to 7.8 m radius estimated at 11 700 yrs old (Mojave desert); twigs steeped in boiling water yield antiseptic lotion (no support for trad. medic. use; causes liver disease), fl. buds pickled & eaten like capers

Larryleachia Plowes (~ *Lavrania*). Apocynaceae (Vb; Asclep. III 5). 5 S Afr. R: F. Albers & U. Meve, *Ill. Handb. Succ. Pls*, Asclep. (2002)176

Larsenaikia Tirv. = *Kailarsenia*

Larsenia Bremek. = *Strobilanthes*

Larsenianthus Kress & Mood (~ *Hedychium*). Zingiberaceae (II 1). 4 Bangladesh, N. Ind., Myanmar. R: Phytokeys 1(2010)21

Lasallea Greene = *Symphyotrichum*

Laseguea A. DC = *Mandevilla*

Laser Borkh. ex P. Gaertner & al. Umbelliferae (III 2). 7 C & S Eur., W As. R: T 65(2016)585

Laserpitium Tourn. ex L. Umbelliferae (III 2). 6 Euras. R: T 65(2016)585. Fr. & r. locally medic., liqueur flavourings etc.

Lasersisia Liben = *Synsepalum*

Lasia Lour. Araceae (V). 2 Indomal. R: Blumea 33(1988)459. ***Lasia spinosa*** (L.) Thw. (*L. aculeata*) – potherb like spinach, rhizome cooked, curried in Sri Lanka

Lasiacis (Griseb.) A. Hitchc. Gramineae (XXIV 2). 16 Madag., warm Am. (N Am. 2). R: AMBG 65(1978)1133. At maturity glumes & lower lemma turn black & epidermis fills with oil globules attractive to frugivorous birds, the tough upper floret not being digested by disp. agents

Lasiadenia Benth. Thymelaeaceae (Thym.-Daph.). 2 trop. S Am. R: Britt. 38(1986)114. ***L. rupestris*** Benth. used to remove warts in Amaz.

Lasiagrostis Link = *Stipa*

Lasianthaea DC. (~ *Zexmenia*). Compositae (Helia.-Ecl.). 12 SW US to (esp.) Mex. R: MNYBG 31, 2(1979)1, Phytol. 65(1988)359, 66(1989)496. Cult. orn.

Lasianthera P. Beauv. Stemonuraceae (Icacinaceae s.l.). 1 trop. W Afr.: ***L. africana*** P. Beauv. – Prévost's Model

Lasianthus Jack. Rubiaceae (IV 10). c. 180 trop. Afr. (c. 15, many restricted to Uluguru Mts, Tanzania [R: NJB 23(2006)656]), Indomal. (China 30, Mal. 131 [R: Blumea 57(2012)5]) to Aus., WI (1)

Lasiarrhenum I.M. Johnston = *Lithospermum* (but see Phytol. 77(1994)38)

Lasimorpha Schott. Araceae (V). 1 W & C trop. Afr.: ***L. senegalensis*** Schott. R: Blumea 33(1988)465

Lasiobema (Korth.) Miq. = *Phanera*

Lasiocarpus Liebm. Malpighiaceae. 4 Mex.

Lasiocaryum I.M. Johnston. Boraginaceae (B.3.2). 3 C As., Himal.

Lasiocephalus Willd. ex Schldl. = *Senecio*

Lasiocereus Britton & Rose. Cactaceae. 2 Peru

Lasiochlamys Pax & K. Hoffm. Salicaceae (Flacourtiaceae). 13 New Caled. R: Blumea 22(1974)124

Lasiochloa Kunth = *Tribolium*

Lasiocladus Bojer ex Nees. Acanthaceae (III). 5 Madag.

Lasiococca Hook.f. Euphorbiaceae (Acal.-Acal.-Lasioc.). 3 E Himal., Vietnam, Hainan, Malay Pen.

Lasiocoma Bolus = *Euryops*

Lasiocorys Benth. = *Leucas*

Lasiocroton Griseb. Euphorbiaceae Acal.-Adel.). 6 WI

Lasiodiscus Hook.f. (~ *Colubrina*). Rhamnaceae (inc. sed.). 7 trop. Afr., 1 Madag. R: KB 50(1995)495. Roux's Model

Lasiolaena R. King & H. Robinson (~ *Eupatorium*). Compositae (Eup.-Gyp.). 6 E Braz. R: KB 54(2000)916

Lasiopetalum Sm. Malvaceae (Bytt.-Lasio.; Sterculiaceae). 30–35 Aus.

Lasiopogon Cass. Compositae (Gnap.-Gnap.). 8 S Afr., 1 ext. to N Afr. & Middle E. R: OB 104(1991)155

Lasiorrhachis Stapf (~ *Saccharum*). Gramineae (XXII 5). 3 Madag.

Lasiosiphon Fres. = *Gnidia* (but perhaps distinct)

Lasiospermum Lag. Compositae (Anth.-Anth.). 4 S Afr., Egypt. R: FSA 33,4(2000)64. Some toxic to stock

Lasiospora Cass. Compositae (Cich.-Cich.(Scorz.)). 2 Medit.

Lasiostelma Benth. = *Brachystelma*

Lasiurus Boiss. Gramineae (XXII). 1 Mali to NE Ind.: ***L. scindicus*** Henrard – valuable fodder in subdesert

Lastarriaea Rémy (~ *Chorizanthe*). Polygonaceae (Erio.-Erio.). 3 Calif., N & C Chile. R: Phytol. 66(1989)213

Lasthenia Cass. Compositae (Mad.-Baer.). 18 W N Am. (17 – R: FNA 21(2006)336), Chile (1). R: UCPB 40(1966)1. Some cult. orn. (*Baeria*), some seeds eaten as porridge

Lastrea Bory = *Thelypteris*

Lastreopsis Ching. Dryopteridaceae (II). Excl. *Parapolystichum*, c. 28 trop. & S temp. (Am. 6) esp. Aus.

Latace Philippi = *Nothoscordum*

Latania Comm. ex Juss. Palmae (III 8b). 3 Masc. R: Fl. Masc. 189(1984)6. Dioec. fan-palms. Cult. orn. in trop. esp. ***L. lontaroides*** (Gaertn.) H. Moore (*L. borbonica*, Réunion)

Lateristachys Holub = *Lycopodiella*

Lateropora A.C. Sm. = *Symphysia*

Lathraea L. Orobanchaceae (Rhin.; Scrophulariaceae s.l.). 7 temp. Euras. (Eur. 3). R: E. Heinricher (1931) *Monogr. der Gattung L.* ***L. squamaria*** L. (toothwort, Eur. to Himal.) parasitic on r. of *Fagus*, *Corylus* etc. with thick rhiz. bearing 4 rows of tooth-like scale lvs, each hollow with glands in side-chambers in which dead insects s.t. found, in S France app. s.t. mycotrophic; ***L. clandestina*** L. (S & W Eur.) – cult. parasite with brilliant fls growing on *Populus*, *Salix* etc. & natur. in GB (esp. in C Cambridge), NZ, nectar pH 11.5 due to dissolved ammonia tolerated by poll. bees but not by nectar-robbing ants

Lathraeocarpa Bremek. Rubiaceae (IV 1). 2 S Madag. R: T 58(2009)209

Lathriogyna Ecklon & Zeyher = *Amphithalea*

Lathrocasis L.A. Johnson (~ *Allophyllum*). Polemoniaceae (III). 1 W N Am.: ***L. tenerrima*** (A. Gray) L.A. Johnson. R: Aliso 19(2000)67

Lathrophytum Eichler. Balanophoraceae. 1 Braz. (nr. Rio): ***L. peckoltii*** Eichler. R: FN 23(1980)54

Lathyrus Tourn. ex L. Leguminosae (III 28). 160 N temp. (most; Eur. 54), trop. E Afr. mts, temp. S Am. R: NRBGE 41(1983)209. Vetchlings; fls yellow, blue, red etc. Close to *Vicia* but most distinguishable by their winged stems & parallel veins in lvs but one sect. only separable on characters of style-pubescence: a tuft of these hairs brushes the pollen out of apex of keel where the anthers shed it. Cli. spp. usu. with branched tendrils; in ***L. aphaca*** L. (yellow v., Eur. & Med. to Afghanistan) leaf rep. by tendril, photosynthesis carried out by v. large stip.; in ***L. nissolia*** L. (Eur. & Med., natur. N Am.) there are phyllodes with parallel veins (cf. *Acacia*). Many cult. orn. (incl. *Orobus*) & some fodders etc. ***L. cicera*** L. (Medit.) & ***L. clymenum*** L. (Medit.) – fodders, but eating seeds can lead to lathyrism (paralysis of legs); ***L. grandiflorus*** Sm. (S Eur.) – perenn. 'everlasting' pea with 2 or 3 rose-purple fls per peduncle; ***L. japonicus*** Willd. (*L. maritimus*, beach pea, sea p., circumpolar) – seeds viable for 4–5 yrs in seawater, a coffee subs. for Inuit; ***L. latifolius*** L.

(*L. sylvestris* subsp. *l.*, C & S Eur., natur. N Am.) – perenn. 'everlasting' pea with several to many white to purple fls per peduncle; ***L. linifolius*** (Reichard) Baessler (*L. macrorrhizus*, *L. montanus*, W & C Eur.) – tubers ed. like potatoes (dried used to allay travellers' hunger), local medic., form. used to flavour some whiskies; ***L. odoratus*** L. (sweet pea, Crete, Sicily & S Italy) – v. many cvs (autogamous), cult. since anc. times for large scented fls, large-flowered '**Cupani**' (1980s) app. similar to that grown in a Palermo monastery garden 1695, 'self' colours by mid-19 cent., modern cvs with frilled or waved standard but weaker scent (most 'scent' for soap etc. now synthetic) leading to 'Spencer' sweet peas since 1900, *the* cut flower of the Edwardians, dwarf cvs incl. tendril-less '**Supersnoop Group**'; ***L. pratensis*** L. (common or meadow v., Eur. & Medit. to Afghanistan); ***L. sativus*** L. (Ind., Riga or dogtooth pea, khesari) – cultigen, poss. orig. a weed of lentils, cult. early Neolithic & perhaps first crop domesticated in Eur., fodder & used like *Cajanus cajan* but beta-N-oxalyl-amino-alanine causing motorneurone disease in C Ind.; ***L. tingitanus*** L. (Tangier pea, W Medit.) – green manure; ***L. tuberosus*** L. (earth chestnut, Euras., Medit.) – tubers ed., fls form. distilled for scent

Latipes Kunth = *Leptothrium*

Latouchea Franch. Gentianaceae. 1 E China: ***L. fokienensis*** Franch.

Latourorchis Brieger = *Dendrobium*

Latrobea Meissn. (~ *Pultenaea*). Leguminosae (III 13). 9 SW Aus.

Latua Philippi. Solanaceae. (IV). 1 S Chile: ***L. pubiflora*** (Griseb.) Baill. – heteroblastic, spiny, woody hummingbird-poll. pl. to 10 m, with alks incl. hyoscyamine & scopolamine, fish-poison, form. used as malevolent hallucinogen, leading to permanent madness

lauan lightweight timbers of *Parashorea* & *Shorea* spp.; **red l.** *Shorea negrosensis* etc.; **white l.** *S. contorta*, *Parashorea malaanonan*

Laubenfelsia A. Bobrov & Melikian = *Dacrycarpus*

Laubertia A. DC. Apocynaceae (II e). 3 trop. Am. R: Rhodora 104(2002)172

laudanum alcohol tincture of opium

laughing jackass *Arethusa bulbosa*

lauki *Lagenaria siceraria*

Laumoniera Nooteboom = *Brucea*

Launaea Cass. Compositae (Cich.-Cich.(Hyos.)). 54 Canaries to S Afr. & E As. (Eur. 6), WI. R: Englera 17(1997)1. ***L. sarmentosa*** (Willd.) Kuntze a pl. char. of trop. sandy beaches

Lauraceae Juss. Magnoliidae – Laurales. 56/2700 trop. & warm esp. SE As. & Braz. Aromatic usu. everg. trees & shrubs (*Cassytha* a parasitic twiner), usu. with spherical ethereal oil-cells in parenchymatous tissues, many accum. aluminium. Lvs in spirals (rarely whorled or opp.), simple (lobed in e.g. *Sassafras*, scale-like in *Cassytha*), oft. coriaceous, oft. glaucous adaxially, venation usu. pinnate; stip. 0. Infls thyrsoid to umbellate. Fls bisex. (or pl. polygamous to dioec.) with well-dev. hypanthium like a K-tube (epigyny in *Hypodaphnis*), reg., usu. 3-merous & small, in axillary usu. thyrsoid infls (rarely solit.); pedicel oft. swollen; P (?derived from A) usu. 3 + 3, sepaloid, A in 4 whorls of 3, the innermost oft. sterile or 0 & s.t. 1 or 2 of the other 3 also, whorl 3 usu. with pr of glands at base, anthers usu. opening by (1)2 or 4 valves from base upwards (small pores in *Micropora*), $\underline{G}$ ($\overline{G}$ in *Hypodaphnis*) 1 (?(3)), 1-loc. with 1 large ± apical, anatropous, bitegmic ovule. Fr. a 1-seeded berry (rarely dry-indehiscent), oft. encl. in persistent accrescent fleshy to woody hypanthium; seeds endotestal with large straight oily embryo (s.t. also starchy), endosperm 0. n = (11) 12 (15)

Classification & chief genera (after Rohwer):

I. **Perseeae** (infl. thyrsoid, umbels without involucres): *Alseodaphne, Aniba, Beilschmiedia, Cassytha, Cinnamomum, Cryptocarya, Dehaasia, Endiandra, Endlicheria, Eusideroxylon, Licaria, Ocotea, Persea, Phoebe*

II. **Laureae** (umbels surrounded by involucre of bracts): *Actinodaphne, Laurus, Lindera, Litsea, Neolitsea.*

Uncertain position: *Cinnadenia, Chlorocardium*

Generic limits debatable. The dodder-like *Cassytha*, despite its macro-morphological features was corr. placed in L. by Robert Brown in 1810; unlike other parasitic groups, it has a normal well-dev. embryo. Several genera in Macaronesia represent relict laurel forest trees: *Laurus, Ocotea, Persea.* Fam. with imp. relationships with birds (food/disp.) Many spices & flavourings (*Aniba, Cinnamomum* (camphor, cinnamon), *Cryptocarya, Laurus, Licaria, Lindera, Litsea, Sassafras*), oilseeds (*Cryptocarya, Licaria, Lindera, Nectandra, Neolitsea*; fr. of *Persea americana* the avocado (pear)), some medic. (esp. *Sassafras*); timbers

(*Aniba, Beilschmiedia, Caryodaphnopsis, Chlorocardium, Dicypellium, Endiandra, Eusideroxylon, Lindera, Nectandra, Ocotea, Persea, Sextonia, Umbellularia*)

Lauradia Vand. = *Sauvagesia*

laurel orig. *Laurus nobilis* but the l. crown of Caesar *Ruscus hypoglossum* & the l. of shrubberies usu. *Prunus laurocerasus;* **Alexandrian l.** *Calophyllum inophyllum, Danae racemosa;* **bay l.** *L. nobilis;* **bog l.** *Kalmia polifolia;* **Californian l.** *Umbellularia californica;* **camphor l.** *Cinnamomum camphora;* **cherry l.** *P. laurocerasus;* **Chilean l.** *Laurelia sempervirens;* **Chinese l.** *Antidesma bunius;* **Ecuador l.** *Cordia alliodora;* **Indian l.** *Terminalia elliptica;* **Japanese l.** *Aucuba japonica;* **mountain l.** *K. latifolia;* **pig** or **sheep l.** *K. angustifolia;* **Portugal l.** *P. lusitanica;* **spurge l.** *Daphne laureola;* **l. wood** *Albizia lebbeck*

Laurelia Juss. Atherospermataceae (Monimiaceae II 2;). 1 NZ, 1 Chile & Peru. Polygamous to dioec. trees with alks. ***L. novae-zelandiae*** Cunn. (pukatea, NZ) – timber & medic.; ***L. sempervirens*** (Ruíz & Pavón) Tul. (*L. aromatica, L. serrata*, Chilean laurel, Chile & Peru) – fr. a spice (Peruvian nutmegs)

Laureliopsis Schodde. Atherospermataceae (Monimiaceae II 1). 1 S Arg., S Chile: ***L. philippiana*** (Looser) Schodde – fr. a spice (tepa). R: Parodiana 2(1983)298

Laurembergia P. Bergius. Haloragaceae. 4 trop. & warm

Laurentia Adans. = *Lobelia;* see also *Hippobroma, Isotoma, Palmerella, Porterella, Wimmerella*

Laurida Ecklon & Zeyher (~ *Cassine*). Celastraceae (I). 2 E S Afr. R: SAJB 63(1997)227. ***L. tetragona*** (L.f.) R. Archer – fr. ed.

Laurocerasus Duhamel = *Prunus*

Laurophyllus Thunb. Anacardiaceae (I). 1 W Cape: ***L. capensis*** Thunb. R: Strel. 9(2000)271

Laurus Tourn. ex L. Lauraceae (II). 1–2 Medit. (1), Macaronesia (1: ***L. azorica*** (Seub.) Franco (*L. canariensis*) – part of Madeira laurel forest). R: BotJLS 68(1974)51. Alks. ***L. nobilis*** L. ((true or bay) laurel, sweet bay or b. tree) – domatia, lvs used to flavour food, fr. in veterinary medic., aromatherapy, lvs the original crown of l. (resting on your ls.; baccalaureat (Bachelor of Arts etc.), poet laureate) of Greece, superseded in UK wreaths by *Prunus laurocerasus*

laurustinus *Viburnum tinus*

Lautembergia Baill. = *Orfilea*

Lauterbachia Perkins. Monimiaceae (V ?2). 1 NG: ***L. novoguineensis*** Perkins – known only from one gathering, now destroyed. R: FM I,10(1986)326

lavandin *Lavandula × intermedia*

Lavandula Tourn. ex L. Labiatae (VII 3a). 39 Atlantic Is., Medit. (Eur. 7) to Somalia & Ind. Lavenders. R: T. Upson & S. Andrews (2004) *The genus L.*: 107. Protandrous bee-poll. fls sources of nectar for good honey & used in sachets (l. bags – scent from K) to scent linen (also insecticidal), though lavender = 'distrust' in 'Language of Fls'. Aromatic oil (1350 t per annum (2002) in comm. from France alone) for scent in cosmetics & in porcelain-painting, to flavour food, & drinks, lavender water & medic. (antiseptic so used to control food spoliation; for headaches & depression (generalized anxiety disorder), allegedly increasing blood-flow to penis by 40%, though may cause breast growth in boys, oil-burners soporific) used empirically since anc. times; extracted today in GB & S France largely from ***L. × intermedia*** Emeric ex Lois. (lavandin, esp. '**Dutch Group**' ('English' or Dutch lavender), sterile spontaneous tetraploid hybrid (SE France) = ***L. angustifolia*** Mill. (*L. officinalis, L. vera*, (orig.) English l., finest oil, many cvs incl dwarfs e.g. '**Munstead**') × less hardy ***L. latifolia*** Medik. (both Medit.), much cult., oil of lesser quality (Bulgaria, France), many orn. cvs), also ***L. stoechas*** L. (French l., Spanish l., Medit., Portugal) – invasive in Aus., cult. orn. hybrids

Lavatera Tourn. ex L. = *Malva*

Lavauxia Spach = *Oenothera*

lavender *Lavandula* spp. esp. *L. angustifolia, L. × intermedia* (**Dutch l., 'English l.'**); **French** or **Spanish l.** *L. stoechas;* **Indian l.** *Bursera penicillata;* **sea l.** *Limonium* spp. esp. *L. vulgare*

laver (bread) *Porphyra* spp. esp. *P. umbilicalis*

Lavigeria Pierre. Icacinaceae. 1 trop. W Afr.: ***L. macrocarpa*** Pierre

Lavoisiera DC. Melastomataceae. 46 Braz.

Lavoixia H. Moore = *Clinostigma* (but see Allertonia 3(1984)334)

Lavradia Roemer = *Sauvagesia*

Lavrania Plowes. Apocynaceae (V b; Asclepiadaceae III 5). 1 Namibia: ***L. haagnerae*** Plowes. R: F. Albers & U. Meve, *Ill. Handb. Succ. Pls*, Asclep. (2002)178

Lawia Griff. ex Tul. = *Dalzellia*

Lawrencella Lindl. (~ *Helichrysum*). Compositae (Gnap.-Ang.). 2 Aus. R: Nuytsia 8(1992)369

Lawrencia Hook. (~ *Plagianthus*). Malvaceae (Malv.-Malv.). 12 Aus. (W Aus. 11). R: Nuytsia 5(1984)201. Some dioecy. ***L. squamata*** Nees ex Miq. – thorny

Lawsonia L. Lythraceae. 1 N Afr., trop. OW: ***L. inermis*** L. – pulverized lvs (henna) a green powder staining fingernails (hands & feet of brides in Sumatra – lasts up to 6 months) etc. red, with indigo hair glossy blue-black, much used in E, also nails & hair of Egyptian mummies swathed in cloth dyed with it; fls used to scent oil etc. & distilled for similar; bark medic.; poss. source of synth. insecticides, cult. orn. R: FM I,22(2016)48

Lawson's cypress *Chamaecyparis lawsoniana*

Laxmannia R. Br. Asparagaceae (Anthericaceae). 13 Aus. R: FA 45(1987)254

Laxmanniaceae Bubani = Asparagaceae

Laxoplumeria Markgraf. Apocynaceae (1a). 3 E Peru, Braz.

Layia Hook. & Arn. ex DC. Compositae (Mad.-Mad.). 14 W N Am., NW Mex. R: FNA 21(2006)262. ***L. platyglossa*** (Fischer & C. Meyer) A. Gray (*L. elegans*, tidy-tips) – consp. early summer field flower with yellow ray-florets tipped with white, much-cult. ann.

Lazarum A. Hay (~ *Typhonium*). Araceae. 36 Aus. (14), trop. Am. (22)

Leachia Plowes = *Larryleachia*

Leachiella Plowes = *Lavrania*

lead tree *Leucaena leucocephala*; **l.wood** *Combretum imberbe*; **l.wort** *Plumbago* spp.

Leandra Raddi = *Miconia* (but see Phytotaxa 262(2016)1)

Leandriella Benoist. Acanthaceae (III). 2 Madag.

Leaoa Schltr. & Porto = *Scaphyglottis*

leatherwood *Cyrilla racemiflora*, *Dirca* spp., *Eucryphia lucida*

Leavenworthia Torrey. Cruciferae (16). 8 SE US. R: JAA 69(1988)118. Limestone glades. 5 spp. with straight embryos

Lebanese cucumber *Cucumis sativus* Lebanese Group

Lebeckia Thunb. Leguminosae (III 8). c. 35 S Afr.

Lebetanthus Endl. (~ *Prionotes*). Ericaceae (VII 1; Epacridaceae). 1 S S Am.: ***L. myrsinites*** (Lam.) Dusén (*L. americanus*)

Lebetina Cass. = *Dyssodia*

Lebronnecia Fosb. Malvaceae (Malv.-Goss.). 1 Marquesas: ***L. kokioides*** Fosb. – known from only 1 tree & seedlings when described. R: Adans. 6(1966)509

Lebrunia Staner. Calophyllaceae. 1 trop. Afr.: ***L. bushaie*** Staner. R: BJBB 13(1934)105

Lebruniodendron Léonard. Leguminosae (I 2). 1 Gulf of Guinea: ***L. leptanthum*** (Harms) Léonard. R: BJBB 21(1951)421

Lecananthus Jack. Rubiaceae (IV 6). 3 W Mal. R: Blumea 43(1998)337

Lecaniodiscus Planch. ex Benth. Sapindaceae. 2 trop. Afr. ***L. cupanioides*** Planch. ex Benth. – chewing-stick in Nigeria, aril ed.

Lecanium C. Presl = *Trichomanes*

Lecanolepis Pichi-Serm. = *Trichomanes*

Lecanophora Speg. (~ *Cristaria*). Malvaceae (Malv.-Malv.). 5 temp. Arg. R: Darw. 9(1950)254

Lecanopteris Reinw. Polypodiaceae (IV). 13 Mal. R: GBS 45(1994)293. Epiphytes with fleshy rhiz. inhabited by *Crematogaster* & *Iridomyrex* ants bringing in nutrients to be absorbed, but can survive without. ***L. spinosa*** Jermy & Walker (Sulawesi) – spines used by ants in runways

Lecanorchis Blume. Orchidaceae (II 2). c. 20 Indomal. & (esp.) Jap.

Lecanosperma Rusby = *Heterophyllaea*

Lecanthus Wedd. Urticaceae (II). 3 OW trop., esp. China

Lecardia Poisson ex Guillaumin = *Salaciopsis*

Lecariocalyx Bremek. Rubiaceae (IV 7). 1 Borneo: ***L. borneensis*** Bremek.

leche caspi *Couma macrocarpa*

Lechea Kalm ex L. Cistaceae. 17 Am. R: CGH 121(1938)

Lechenaultia R. Br. (*Leschenaultia*). Goodeniaceae. 30 Aus. (most SW endemics), 1 ext. to NG. R: FA 35(1992)17, Nuytsia 16(2006)164. Some cult. orn.

lechi-caspi *Couma macrocarpa*

Lecocarpus Decne. Compositae (Mill.-Mel.). 3 Galápagos. R: Madroño 20(1969)255

lecheguilla *Agave lechuguilla*

Lecointea Ducke. Leguminosae (III 1). 8 trop. Am. R: BZ 91(2006)1077. Timber

Lecokia DC. Umbelliferae (III 5). 1 Crete to Iran: ***L. cretica*** DC.

Lecomtea Koidz. = *Cladopus*

Lecomtedoxa (Engl.) Dubard. Sapotaceae (I 3). 6 Gabon. R: T.D. Pennington, *S.* (1991)143. Seeds a source of cooking oil

Lecomtella A. Camus. Gramineae (Panicoid.). 1 Madag.: ***L. madagascariensis*** A. Camus

Lecontea A. Rich. ex DC. = *Paederia*

Lecosia Pedersen. Amaranthaceae. 2 SE Braz.

Lecythidaceae A. Rich. Magnoliidae – Ericales. Incl. Scytopetalaceae, 24/360 trop. esp. S Am. rain forests. R: T 47(1998)817. Trees or shrubs, oft. pachycaul. Lvs in spirals oft. at branch-tips, simple, entire or toothed; stip. small & caducous or 0. Fls bisex., reg. or not, epigynous (or half so) with hypanthium s.t. extended beyond G, oft. large, showy & ephemeral, insect- or bat-poll. with much nectar, axillary, solit. or in term. or axillary panicles, fascicles or racemes s.t. from old wood; K (2–)4–6(–12), imbr. (valvate in some OW genera, rarely connate & forming calyptra), C 3–6 (8, 12, 18), imbr. or 0 (*Foetidia*), A (10–)∞(- c. 1210) symmetrical in several centrifugal series basally connate on a basal ring, or this ring asymmetrical & extended on 1 side into a flat ligule s.t. curved over G as a hood (the A thereon staminodal), in fls with C 0 outer A sterile & connate forming a corona, intrastaminal disk scarcely dev. in NW genera, enlarged & ± covering G in OW, anthers opening by longit. slits (or apical pores), $\overline{G}$ (2–6) with a term. style & as many loc. as G with axile placentation (basal in *Eschweilera*) & 2–115 anatropous, bitegmic ovules per loc. Fr. a capsule with distal operculum, oft. v. large (monkey-pots) or a drupe or berry; seeds oft. nut-like, winged or oft. with funicular aril, endosperm usu. 0 (well dev. in *Asteranthos*), embryo large oily & proteinaceous, the cotyledons s.t. rudimentary & hypocotyl much-thickened. n = 11, 18, 21 (Scytopetaloideae), 13 (Planchonioideae), 16 (Napoleonaeoideae), 17 (Lecythidoideae)

Classification & chief genera:

Four subfams s.t. treated as sep. but closely related fams:

I. **Lecythidoideae** (C present, A in several whorls, ± basally united, fr. a berry or capsule, trop. Am.): 11 genera incl. *Bertholletia, Cariniana, Couroupita, Eschweilera, Gustavia, Lecythis*

II. **Barringtonioideae** (Planchonioideae, Barringtoniaceae; C present, fr. a usu. 1-seeded berry or dry 4-winged indehiscent capsule, OW): 5 genera incl. *Barringtonia, Petersianthus, Planchonia*

III. **Asteranthoideae** (Foetidioideae, Foetidiaceae; C 0, staminal disk large, fr. a drupe, OW): *Foetidia*

IV. **Scytopetaloideae** (Scytopetalaceae, Asteranthaceae; stip. minute, C s.t. 0, staminodes connate, corolla-like, $\underline{G}$, fr. indehiscent): 6 genera incl. *Asteranthos, Scytopetalum*

V. **Napoleonaeoideae** (Napoleonaeaceae; C 0, outer A sterile forming corona or disk, fr. indehiscent; W Afr.): *Crateranthus, Napoleonaea* (only)

Trad. placed in Myrtales, L. differ in embryology, bitegmic ovules, lvs in spirals etc.; ovule characters suggested present disposition (Cronquist) where Scytopetalaceae (= IV) were already assigned. The pachycaul genus *Gustavia* is considered the most arch. (Cronquist). The range of fl. & fr. form assoc. with diff. poll. & disp. mechanisms, the last including water, air & a range of animals. In Amaz., poll. from open-flowered *Gustavia* (small & middle-sized bees) to those with increasing ligule (*Couroupita* (q.v.) etc.) to *Couratari* with hidden nectar available only to female euglossines. Ed. fr. or seeds (*Bertholletia* – Braz. nuts, *Grias, Gustavia, Lecythis*), timber (*Bertholletia, Careya, Cariniana, Couroupita, Eschweilera, Gustavia, Lecythis, Petersianthus*), cult. orn.

Lecythis Loefl. Lecythidaceae (I). 26 trop. Am. R: FN 21(1979)197. Like *Couroupita* but A of ligule (see fam.) sterile; fr. a 'monkey-pot' form. used to trap monkeys, which grab sugar inside cannon-ball-like caps. (secured) & cannot withdraw extended fist. Some timbers & ed. oily seeds (paradise nuts) esp. ***L. zabucajo*** Aubl. (Braz., Guianas) – sapucaia nuts, with delicate flavour suitable for chocolates, & other spp.; ***L. minor*** Jacq. & ***L. ollaria*** L. have toxic seeds which if eaten lead to temporary loss of hair & nails (& nausea) caused by the selenium analogue of the amino-acid cystathionine

Leda C.B. Clarke = *Isoglossa*

Ledebouria Roth (~ *Scilla*). Asparagaceae (Hyacinthaceae). c. 45 trop. & S Afr., with c. 2 in both Madag. & Ind. R: JSAB 36(1970)233. Cult. orn., esp. ***L. socialis*** (Bak.) Jessop (*S. violacea*, S Afr.) with blotched or striped lvs, some with toxic bulbs

Ledebouriella H. Wolff. Umbelliferae (III 8). 2 C As. R: NS 24(1987)150

Ledenbergia Klotzsch ex Moq. (*Flueckigera*). Petiveriaceae (Phytolaccaceae III). 3 Mex. to Venez. Trees & shrubs

Ledermanniella Engl. Podostemaceae (III). Excl. *Inversodicraea*, c. 44 trop. Afr. R: BMNHN 4,5(1983)361, 6(1984)249. Some restricted to single waterfalls or rapids

Ledger bark *Cinchona calisaya*

Ledocarpaceae Meyen = Francoaceae

Ledothamnus Meissn. Ericaceae (Er.-Bry.). 7 Guayana Highland. R: MNYBG 29(1978)141, ABV 14,1(1983)170

Ledum L. = *Rhododendron*

Ledurgia Speta = *Drimia*

Leea Royen ex L. Vitaceae (Leeaceae). 34 OW trop. (Afr. & Madag. 2, Indomal. 32). Chamberlain's Model. Some locally medic. & cult. orn. foliage pls

Leeaceae Dumort. = Vitaceae

leek *Allium ampeloprasum* Porrum Group, (Bible) *A. a.* Kurrat Group; **lady's l.** *A. cernuum*; **l. orchid** *Prasophyllum* spp.; **round-headed l.** *A. sphaerocephalon*; **sand l.** *A. scorodoprasum*

Leersia Sw. Gramineae (III 3 i). 18 trop. & warm temp. (Eur. 1). R: SBi 46(1965)129, ISJS 44(1969)215. Marsh grasses allied to *Oryza* but sterile lemmas 0, used as fodder in As. esp. ***L. hexandra*** Sw. (bareet grass, trop.). ***L. oryzoides*** (L.) Sw. (N temp., invasive in Aus.) – cleistogamous fls

Leeuwenbergia Letouzey & Hallé. Euphorbiaceae (Ricinod.). 2 trop. Afr. R: Adans. 14(1974)380

Lefebvrea A. Rich. Umbelliferae (III 11). 10 trop. & warm Afr. R: T 57(2008)361. Hapaxanthic

Legazpia Blanco (~ *Torenia*). Linderniaceae (Plantaginaceae (Grat.-Lind.); Scrophulariaceae s.l.). 1 E As. to W Pacific: ***L. polygonoides*** (Benth.)Yamaz. (*T. p.*) – local medic., esp. skin disease. R: JJB 30(1955)359

Legenere McVaugh. Campanulaceae (III 2). 1 Calif., S Chile: ***L. valdiviana*** (Philippi) F. Wimmer. R: BSAB 17(1976)176. Name an anagram of E.L. Greene

Legnephora Miers. Menispermaceae (V). 5 NG, NE Aus. R: KB 27(1972)275. Alks

Legocia Livera = *Christisonia*

Legousia Durande (*Specularia*). Campanulaceae (I 8). 7 Eur. (5), Medit. to C As. Some cult. orn. annuals incl. ***L. hybrida*** (L.) Delarbre (Venus's looking-glass)

Legrandia Kausel. Myrtaceae (II 10). 1 Chile: ***L. concinna*** (Philippi) Kausel. R: FN 45 (1986)131

Leguminosae Juss. (Fabaceae). Magnoliidae – Fabales. 741/20 200 cosmop. (I & II mostly trop.). Trees (esp. Troll's Model), shrubs incl. lianes, & herbs, s.t. thorny, oft. with r. nodules cont. nitrogen-fixing bacteria (25–30% of I, 60–70% of II, c. 95% of III; cf. *Parasponia*, with leghaemoglobin regulating oxygen environment around *Rhizobium* in nodules) & freq. with non-protein amino-acids in seeds &/or veg. parts as well as alks. Lvs pinnate (rarely palmate), bipinnate, unifoliolate, trifoliolate or simple, s.t. phyllodic or reduced to a tendril (some *Lathyrus*), usu. in spirals, the petiole & leaflets with basal pulvini oft. controlling orientation & 'sleep movements'; stip. present, s.t. large or rep. by thorns (s.t. ant-inhabited in *Vachellia*) or prickles, leaflets s.t. with stipellules. Fls usu. bisex., reg. or not, hypogynous to perigynous, usu. in racemes, spikes or heads, K 5 or ((3–)5(6)), when a tube with valvate (rarely imbr.) lobes (II), or oft. ± bilabiate (III), C (0–)5 irreg. (adaxial small & lying within its laterals in I or large (standard) & outside them (resupinate occ.) in III) or (3–)5(6) s.t. basally connate, reg. & usu. valvate (II), A usu. twice P (to ∞ in I & II), distinct or ± connate, alike or not, s.t. some staminodal, forming a sheath around G in III, oft. coloured & long-exserted in II, anthers usu. with longit. slits (pores in some I), nectary oft. a ring on receptacle around G, $\underline{G}$ 1 (2–16 in some II), each with (1)2–∞ anatr. to campylotropous, bitegmic ovules oft. with zig-zag micropyles. Fr. usu. dry & dehiscent down both sutures (a legume), occ. breaking up into 1-seeded sections (lomentum) or indehiscent & samaroid or a drupe; seeds hard, exotestal, s.t. with an elongate funicle & an aril (oft. merely a vestige), rarely sarcotestal or winged, embryo straight to curved with 0 or little endosperm (much in *Prosopis*). n = 5–14. R.M. Polhill & P.H. Raven (eds, 1981) *Advances in legume systematics* 1; M. Crisp & J. Doyle (eds, 1995) *idem* 7; G. Lewis & al. (eds, 2005) *Legumes of the World*

Classication & chief genera:

Form. s.t. treated as 3 fams, the legumes now usu. arr. as below but I is paraphyletic & basal to II & III (for a classification with monophyletic groups so as to retain II & III,

II would include some subgroupings of I while the rest of I would prob. be rep. as 3 subfams, of which that being sister to rest of fam. would be I 1; see T 62(2013)217):

I. **Caesalpinioideae** (L. Watson & M.J. Dalwitz (1983) *The genera of L. C.*; fls usu. irreg., C imbr. in bud, free or some united; adaxial C overlapped by laterals, K free (exc. in 3), radicle usu. straight; lvs bipinnate or pinnate, rarely unifoliolate or simple, usu. trees (s.t. forming pure stands in Afr. forests) & shrubs of trop.): 4 tribes (diff. to recog. & with both II & III nested within):
   1. **Cercideae** (lvs with palmate venation & 2 halves with independent nyctitropic movement, whether free or not, on a single pulvinus unique in fam.; trees, shrubs or lianes; R: Phytoneuron 2010–48(2010)2): *Bauhinia, Cercis*
   2. **Detarieae** (incl. Amherstieae; stip. intrapetiolar, K ± 0 in *Aphanocalyx*, C ± 0 in *Brachystegia, Crudia*; ectotrophic mycorrhiza typical but rare elsewhere in L.; trees (Troll's Model) to suffrutices esp. consp. in Afr. woodlands (miombo) e.g. *Brachystegia, Isoberlinia, Julbernardia*): *Afzelia* (timber), *Amherstia* (cult. orn.), *Baikiaea* (timber), *Brownea, Copaifera* (copals), *Cynometra* (ed.), *Daniellia* (copals), *Detarium, Guibourtia* (copals), *Intsia* & *Peltogyne* (timber), *Saraca, Schotia* & *Tamarindus* (ed. fr.). *Macrolobium* + 22 genera incl. *Brachystegia* & allies of trop. Afr. & Am. form. excl. as Macrolobieae (bracteoles prot. fl. buds, K well-dev.)
   3. **Cassieae** (stip. lat. or 0, anthers with slits or pores): trop. & warm, form. divided into 5 subtribes, but heterogenous in that Cassiinae (Cassieae s.s.) app. referable to 4., in which Sclerolabieae & *Ceratonia* now placed – *Cassia, Chamaecrista, Dialium, Koompassia* (timber; tallest dicot. in trop.), *Senna* (medic.)
   4. **Caesalpinieae** (stip. interpetiolar or 0, rarely herbs; paraphyletic as II is nested within it, the tribe defined by negative characters & the base group for the whole family, comprising relics & complexes undergoing rapid speciation): widely distrib. esp. trop., incl. *Acrocarpus* (plantation), *Biancaea* (timber, cult. orn.), *Caesalpinia* (cult. orn.), *Ceratonia* (carobs), *Colvillea* & *Coulteria* (cult. orn.), *Cordeauxia* (ed. seeds), *Delonix* (cult. orn.), *Gleditsia* (timber), *Gymnocladus* (timber, cult. orn.), *Haematoxylum* (dyes), *Melanoxylon, Paubrasilia, Pterogyne* & *Recordoxylon* (timber), *Tachigali* (some hapaxanthic, some with 'evergrowing' lvs)

II. **Mimosoideae** (fls reg., C valvate in bud (excl. *Dinizia*), oft. with basal tube; lvs bipinnate or, less oft., pinnate or phyllodic; trees, shrubs & some herbs incl. aquatics (*Neptunia*) esp. trop. & warm): 4 tribes oft. with uncommon amino-acids in seeds (but not canavanine); Journal: Bull. Int. Group Study M.:
   1. **Mimoseae** (incl. Parkieae – trees, K-lobes with valvate aestivation: *Parkia* (ed.) & *Pentaclethra* (only), both with disjunct distribs.; paraphyletic as 3. & 4. nested in it; more than half spp. in *Mimosa*, some 15 genera uni- or bi-specific; trop. & warm esp. S Am. & trop. Afr.): *Adenanthera* (beads), *Anadenanthera* (hallucinogens), *Cylicodiscus* (timber), *Desmanthus, Dinizia* (v. tall), *Entada* (v. long), *Leucaena* (ed. fr., pulp), *Mimosa, Neptunia, Piptadenia, Prosopis* (fodder etc.), *Xylia* (timber)
   2. **Mimozygantheae** – 1 sp. (of *Mimozyganthus*) prob. referable to 1.
   3. **Acacieae** (gum exudations common, filaments usu. free (cf. 4.); trees or shrubs incl. climbers, rarely herbs; paraphyletic): *Acacia* etc. referable to 4. for which oldest name thus Acacieae), *Acaciella, Senegalia, Vachellia* (timber, gums, tanbarks, fodder)
   4. **Ingeae** (like 3. (to be moved there) but filaments united; trop. esp. Am. & As. to Aus.): *Albizia* (timber), *Archidendron, Calliandra, Enterolobium, Faidherbia, Inga, Paraserianthes* (timber), *Pithecellobium*

III. **Papilionoideae** (Faboideae; fls like I but adaxial petal (standard) outside laterals (wings; exc. Swartzieae & Sophoreae p.p.), K basally united; radicle usu. curved; no bipinnate lvs or compound pollen-grains as found in other subfams; the most familiar legumes with the standardized papilionoid flower assoc. with insect- (& bird-) poll., able to synthesize quinolizidine alks & isoflavones as well as non-protein amino-acids incl. canavanine, found from rain forests to edges of deserts with greatest diversity of form in Braz. planalto, Mex., E Afr., Madag. & Sino-Himal.; spectacular radiations of a few basic stocks in Mediterranean, Cape & Aus.); 28 tribes:
   1. **Swartzieae** (polyphyletic; usu. trees; adaxial C not outside wings, A usu. free; fr. var., usu. trop.): *Cordyla, Swartzia*
   2. **Sophoreae** (paraphyletic?: trees, shrubs or rarely herbs; fls reg. to papilionoid, A free or not; fr. var.; canavanine 0; mostly trop., ext. to higher latitudes in *Sophora*; many 'non-papilionoid-flowered' genera form. here now moved,

radial symmetry evolving in parallel): *Camoensia* (largest L. fl.), *Castanospermum* (medic., ed. seeds), *Cladrastis*, *Myroxylon* (balsam), *Ormosia*, *Pericopsis* (afrormosia), *Sophora*

3. **Dipterygeae** (trees; lvs paripinnate; K with 2 enlarged upper lobes; balsams & gums, trop. S Am.): *Dipteryx*
4. **Brongniartieae** (trees, shrubs or subshrubs; Am.): *Brongniartia*, *Harpalyce*, *Hovea*, *Poecilanthe*
5. **Euchresteae** (shrubs with quinolizidine alks; Indomal.): *Euchresta* (only)
6. **Thermopsideae** (shrubs or perenn. herbs usu. with 3-foliolate lvs & oft. with lupine alks; N temp.): *Baptisia*, *Thermopsis*
7. **Podalyrieae** (incl. Liparieae – s.t. with canavanine: *Amphithalea*, *Liparia*; shrubs or small trees; oft. with quinolizidine alks; Cape, Afromontane to Ind.; R: Crisp & Doyle, *op. cit.*: 304): *Cadia*, *Cyclopia* (teas), *Podalyria*
8. **Crotalarieae** (shrubs or herbs s.t. with quinolizidine & pyrrolizidine alks; mostly Afr.; R: Crisp & Doyle, *op. cit.*: 305; CBH 13(1991)265): 11 genera incl. *Aspalathus* (teas), *Crotalaria*, *Lebeckia*, *Lotononis*, *Rafnia*
9. **Cytiseae** (Genisteae; shrubs, herbs or small trees, some armed or switch pls; mostly Eur. & Afr. + *Lupinus*): *Cytisus*, *Genista*, *Laburnum*, *Lupinus*, *Retama*, *Ulex*
10. **Amorpheae** (small trees or shrubs or herbs; canavanine 0, NW): *Amorpha*, *Dalea*
11. **Dalbergieae** (trees, lianes & shrubs with hard timber, rarely suffrutices; gum & phenolics abundant; trop. (all rep. Am. exc. *Dalbergiella* & *Inocarpus* incl. Adesmieae (herbs or shrubs of montane & temp. S Am.: *Adesmia* (only) & Aeschynomeneae (mostly shrubs & herbs; warm: *Aeschynomene*, *Arachis* (ground nut), *Kotschya*, *Ormocarpum*, *Smithia*, *Stylosanthes* (fodder), *Zornia*): *Andira*, *Brya* (timber), *Dalbergia* (rosewood), *Machaerium* & *Pterocarpus* (all timber), *Inocarpus* (ed. seeds)
12. **Hypocalypteae** (form. in Liparieae = 7.): *Hypocalyptus* (only)
13. **Mirbelieae** (usu. shrubs s.t. with canavanine; Aus. (esp. SW) with *Gompholobium* ext. to NG; 19 genera poss. united as *Pultenaea*): *Daviesia*, *Dillwynia*, *Gastrolobium*, *Jacksonia*, *Pultenaea*, *Oxylobium*
14. **Bossiaeeae** (shrubs; quinolizidine alks or canavanine; Aus.): *Bossiaea*
15. **Indigofereae** (herbs, less oft. shrubs or small trees, s.t. with flattened or even leafless branches; mainly trop. Afr. but *Indigofera* widespread warm; R: Crisp & Doyle, *op. cit.*: 217): 7 genera incl. *Cyamopsis* (ed.), *Indigofera*, *Phylloxylon* (timber)
16. **Millettieae** (Tephrosieae; R. Geesink LBS 8(1984); paraphyletic & overlapping in some characters with other tribes esp. Galegeae where gradual change in vasculature assoc. with herbaceous habit therein; trees, lianes or shrubs, mainly trop.): *Derris*, *Lonchocarpus* (fish-poisons & insecticides – rotenones), *Millettia*, *Tephrosia*, *Wisteria* (cult. orn.; shows similarities (loss of a chloroplast-DNA inverted repeat) with some temp. tribes)
17. **Abreae** (woody lianes or shrubs, lvs paripinnate, A 9, pantrop.): *Abrus* (only)
18. **Phaseoleae** (polyphyletic; dextrarotatory twining, prostrate or erect herbs, occ. shrublets, rarely trees; lvs usu. 3-foliolate; cosmop. esp. warm, the most imp. economically; R[Phaseolinae – 27/340+]: AJB 98(2011)1712): *Cajanus* (pigeon pea), *Canavalia* (sword bean), *Dioclea*, *Erythrina*, *Flemingia* (r.-crop), *Glycine* (soya), *Lablab*, *Mucuna*, *Phaseolus* (beans), *Rhynchosia*, *Vigna*
19. **Desmodieae** (shrubs or oft. herbs, s.t. trees; canavanine freq.; mainly trop. esp. Ind. to China): *Alysicarpus*, *Codariocalyx*, *Desmodium*, *Uraria*
20. **Psoraleeae** (small trees or shrubs, rarely herbs; rarely trop.): *Cullen*, *Otholobium*, *Pediomelum*, *Psoralea*
21. **Sesbanieae** (form. in 23.): *Sesbania* (only)
22. **Loteae** (herbs or small shrubs with canavanine; mainly N temp. incl. Coronilleae (lomenta, orig. 5 times; OW esp. Medit. & Afr.: *Coronilla* & *Hippocrepis*): *Anthyllis*, *Lotus*
23. **Robinieae** (excl. Sesbanieae; trees, shrubs & herbs with canavanine in seeds, centred in trop. Am. but ext. to rest of Am.; R: Crisp & Doyle, *op. cit.*: 158): *Gliricidia*, *Robinia* (cult. orn.)
24. **Galegeae** (Astragaleae; polyphyletic; herbs or shrubs usu. with canavanine, principally N temp. but incl. Carmichaelieae (small trees, shrubs or subshrubs mostly with ribbed or flattened branches; NZ region: *Carmichaelia*): *Astragalus*, *Clianthus*, *Colutea*, *Glycyrrhiza* (liquorice), *Swainsona*

25. **Hedysareae** (herbs or subshrubs with 1-seeded fr. or lomenta, s.t. with canavanine; N temp. to Horn of Afr.; R: T 64(2015)49): *Caragana* (moved to **Caraganeae**), *Ebenus*, *Hedysarum* (fodder), *Onobrychis* (sainfoin)
26. **Cicereae** (perenn. & ann. herbs, oft. spiny; E Medit. to W As.): *Cicer* (chick pea) only
27. **Trifolieae** (ann. or perenn. herbs usu. with canavanine; largely Euras.): *Medicago*, *Ononis*, *Trifolium* (clover)
28. **Fabeae** (Vicieae; ann. or perenn. herbs with canavanine, largely temp., imp. economically): *Lathyrus* (sweet pea), *Lens* (lentil), *Pisum* (pea), *Vicia* (vetch)

Over a quarter of all spp. in 5 genera: *Acacia*, *Astragalus*, *Crotalaria*, *Indigofera* & *Mimosa*, all char. of open or disturbed veg. Form. thought closely allied to Sapindales, some genera, esp. *Ceratonia*, *Dialium* & *Gymnocladus*, with features reminiscent of S., but now known to be close to Polygalaceae – hardly surprising to laypeople when fls examined

Some aquatics (*Neptunia*, *Aeschynomene*) with aerenchyma useful in making floats etc., some xerophytes & many climbers: leaf-tendrils in *Lathyrus*, *Vicia* etc., stem-tendrils in *Bauhinia*, others with twining stems (*Phaseolus*); lianes oft. with unusual anatomy. Thorns branches as in *Gleditsia* or stip. as in *Vachellia*, where s.t. ant-inhabited. Lvs v. small in e.g. *Ulex*, *Stauracanthus*, though seedlings with typical 3- foliolate or 3-lobed lvs, s.t. phyllodic (*Acacia*, *Lathyrus*, *Phylloxylon*), when small the stems photosynthetic (*Cytisus*, *Carmichaelia* – where flattened). Reg. lvs usu. with 'sleep' movements, the leaflets folding together upwards, downwards etc.; in *Mimosa*, they are sensitive to touch & assume sleep position rapidly, while in *Codariocalyx* the lat. leaflets move continuously (& noisily?) at high temperatures. Poll. in III principally by insects, the adaxial standard being the 'flag', the lat. wings encl. the other 2 petals forming a keel in which A & G lie. Nectar accumulates around base of G & on either side of the base of the free 10th stamen is an opening leading to it: moderately long-tongued insects like bees can reach it. Bees land on the wings depressing them with their wt. while probing for nectar under the standard: the keel is depressed & the stigmas emerge, collecting any pollen already on the insect, rapidly followed by the A. The stigmas & A may return to the keel & the whole process can be repeated as in *Trifolium*; or, they are under tension such that a visit leads to explosive poll. which cannot be repeated as in *Medicago* & *Ulex*; or there is a piston mechanism squeezing pollen in small quantities out of the keel-tip, requiring several insect visits as in *Lotus*, *Lupinus*, *Ononis*; or this latter may be achieved by a brush of hairs as in *Lathyrus* & *Vicia*. Cleistogamy quite freq. esp. in cult. crops (cf. *Triticum*). Some fls also brought to ground level or below where fr. ripens (geocarpy) as in *Arachis*, *Macrotyloma*, *Medicago*, *Tephrosia*, *Vigna*. Legumes oft. opening explosively, s.t. inflated as in *Colutea*; seed-coats usu. hard, making good beads & weights (*Abrus*, *Adenanthera*, *Afzelia*, *Delonix*, *Erythrina*, *Guilandina*)

Defence mechanisms of L. varied & complex, showing an increasing sophistication in the more advanced tribes: I & II with ants, tannins & terpenoids while III with many chemicals incl. diverse alks, non-protein amino-acids & isoflavonoids, the most advanced tribes with phytoalexins rather than stored compounds. Some of these compounds imp. in tanning, insecticides etc. (*Acacia*, *Derris* etc.). Presence (derived several times?) of *Rhizobium* bacterial nodules makes many imp. in land improvement & as green manures (*Leucaena*, *Medicago*, *Onobrychis* etc.) & valuable fodder (*Trifolium*), though stachyose & verbascose not digested exc. by intestinal bacteria leading to globally warming methane expulsion. V. imp. fam. for food-pls esp. pulses (beans, peas, gram) & oil (soya, groundnut) but also tanbarks, v. many imp. trop. timbers, copals, gums, insecticides & cult. orn. (see above), form. dyes & kino & still medic. (*Cassia*, *Senna*), rivalling Gramineae in world economy, grain & forage legumes grown over 12–15% of world's arable lands & providing 33% of human protein nitrogen needs

Lehmanniella Gilg. Gentianaceae. 4 Panamá, Colombia, Peru. R: PKNAW, C 88(1985)411

Leibergia J. Coulter & Rose = *Lomatium*

Leibnitzia Cass. Compositae (Mut.-Mut.). 4 S & E As., 2 N Am. R: NJB 8(1988)67. Herbs. ***L. anandria*** (L.) Turcz. (E As.) – cleistogamous in autumn

Leiboldia Schldl ex Gleason (~ *Vernonia*). Compositae (Vern.-Liab.). 2 S Mex R: BJ 108(1987)225

Leichardtia R. Br. = *Marsdenia*

Leichhardt (pine) *Nauclea orientalis*

Leichhardtia F. Muell. = *Phyllanthus*

Leidesia Muell. Arg. (~ *Seidelia*). Euphorbiaceae (Acal.-Acal.-Merc.). 1 S Afr.: ***L. procumbens*** (L.) Prain. R: A.R.-Sm., *Gen. Euph.* (2001)206
Leioanthum M.A. Clem. & D.L. Jones = *Dendrobium*
Leiocarpa Paul G. Wilson. Compositae (Gnap.). 10 Madag. (?), Aus. R: Nuytsia 13(2001)597
Leiocarpaea (C. Meyer) German & Al-Shehbaz (~ *Bunias*). Cruciferae (14). 1 Siberia: ***L. cochlearioides*** (Murray) German & Al-Shehbaz. R: NJB 28(2010)648
Leionema (F. Muell.) Paul G. Wilson (~ *Eriostemon*). Rutaceae (I 3). 24 E Aus. (R: FA 26(2013)431), 1 NZ. R: Nuytsia 12(1998)270
Leiophaca Lindau = *Whitfieldia*
Leiophyllum (Pers.) Hedwig f. = *Kalmia*
Leiopoa Ohwi = *Festuca*
Leiospora (C. Meyer) F. Dvorák. Cruciferae (26). 6 Russia to E As.
Leiostemon Raf. Older name for *Pennellianthus*
Leiothamnus Griseb. = *Symbolanthus*
Leiothrix Ruhl. Eriocaulaceae. 37 S Am. (Braz. 35 endemic)
Leiothylax Warm. Podostemaceae (III). 3 trop. Afr. R: Adansonia 20(1980)202
Leiotulus Ehrenb. = *Malabaila* (but see FR 105(1994)152)
Leiphaimos Schldl. & Cham. = *Voyria*
Leipoldtia L. Bolus. Aizoaceae (V). 8 SW Cape, Karoo to Namibia. R: H. E. K. Hartmann, *Ill. Handb. Succ. Pls*, Aiz. F–Z(2002)109. Cult. orn. succ. shrublets
Leitgebia Eichler = *Sauvagesia*
Leitneria Chapman. Simaroubaceae (Leitneriaceae). 2 SE US (also Eocene of London). R: Castanea 76(2011)332. ***L. floridana*** Chapman (corkwood) – light timber (lighter than cork) used for floats for fishing-nets
Leitneriaceae Benth. = Simaroubaceae
lelayang *Parishia* spp.
Leleba Rumph. ex Nakai = *Bambusa*
Lellingeria A.R. Sm. & R.C. Moran. Polypodiaceae (V; Grammitidaceae). Excl. *Leucotrichum*, *Stenogrammitis*, 34 trop. R: AFJ 81(1991)72
Lelya Bremek. Rubiaceae (IV 1). 1 trop. Afr.: ***L. prostrata*** (R. Good) W. Lewis. R: AMBG 52(1965)189
Lemaireocereus Britton & Rose (~ *Pachycereus*). Cactaceae. 1 Mex.: ***L. hollianus*** (J. Coulter) Britton & Rose. See also *Stenocereus*
lemba *Molineria latifolia*
Lembertia Greene = *Monolopia*
Lembocarpus Leeuwenb. Gesneriaceae (II 5c). 1 Guiana: ***L. amoenus*** Leeuwenb. – tuber prod. 1 leaf & 1 infl. a season. R: ABN 7(1958)318
Lemboglossum Halb. = *Rhynchostele*
Lembotropis Griseb. (~ *Cytisus*). Leguminosae (III 9). 2 C Eur. to C Russia
Lemeea P.V. Heath = *Aloe*
Lemmaphyllum C. Presl (~ *Lepisorus*). Polypodiaceae (IV). 6 Mal. Cult. orn. epiphytes
Lemmonia A. Gray = *Nama*
Lemna L. Araceae (Lemnaceae). 13 cosmop. esp. Arctic regions. R: IBM 34(1965)16. Duckweed or duckmeat esp. ***L. minor*** L., eaten by wildfowl & tried as fodder for farm animals, floating (***L. trisulca*** L. (temp.) at depths to 14 m in Canada) plant-body, photosynthetic with single r. &. conspic. rootcap. ***L. minuta*** Kunth (Am.) – natur. France 1965, UK 1977, v. invasive everg. in canals etc.
Lemnaceae Gray = Araceae
lemon *Citrus* × *limon*; **l. aspen** *Acronychia acidula*; **l. balm** *Melissa officinalis*; **l. basil** *Ocimum* × *africanum*; **l. grass** *Cymbopogon citratus*; **l. mint** *Mentha* × *piperita* 'Citrata'; **l. myrtle** *Backhousia citriodora*; **rough l.** *Citrus* × *taitensis*; **l.-scented gum** *Corymbia citriodora*; **l. thorn** *Cassinopsis* spp.; **l. thyme** *Thymus* × *carolipaui* 'Culinary Lemon', also forms of *T. pulgeioides*; **l. verbena** *Aloysia citrodora*; **water l.** *Passiflora laurifolia*; **l. wood** *Aspidosperma tomentosum*, *Calycophyllum candidissimum*, *Pittosporum eugenioides*, *Xymalos monospora*
Lemoorea Short. Compositae (Gnap.-Ang.). 1 Aus.: ***L. burkittii*** (Benth.) Short. R: Muelleria 7(1989)112
Lemphoria O. Schulz = *Andrzeiowskia*
Lemurella Schltr. Orchidaceae (V 16d). 4 Madag.
Lemurodendron Villiers & Guinet. Leguminosae (II 1). 1 Madag.: ***L. capuronii*** Villiers & Guinet. R: BMNHN 4,11(1989)3

Lemurophoenix Dransf. Palmae (V 14f). 1 NE Madag.: ***L. halleuxii*** Dransf. R: KB 46(1991)61

Lemuropisum Perrier (~ *Caesalpinia*). Leguminosae (I 4). 1 SW Madag.: ***L. edule*** Perrier – seeds ed. when immature, toxic later

Lemurorchis Kraenzlin. Orchidaceae (V 16d). 1 Madag.: ***L. madagascariensis*** Kraenzlin

Lemurosicyos Keraudren. Cucurbitaceae. 1 Madag.: ***L. variegata*** (Cogn.) Keraudren. R: BSBF 110(1964)405

Lemyrea (A. Chev.) A. Chev. & Beille (~ *Fernelia*). Rubiaceae (II 5). 4 Madag. R: Novon 14(2004)122

Lenbrassia G. Gillett (~ *Fieldia*). Gesneriaceae (I 4c). 1 Queensland: ***L. australiana*** (C. White) G. Gillett – tree to 13 m. R: JAA 55(1974)431

Lencymmoea C. Presl. Myrtaceae (?). 1 Myanmar: ***L. salicifolia*** C. Presl

Lennea Klotzsch. Leguminosae (III 23). 5 C Am.

Lennoa Lex. Boraginaceae (Lennoaceae). 1 C Mex., Colombia & Venez.: ***L. madreporoides*** Lex. R: SB 11(1986)539. A of 2 lengths

Lennoaceae Solms-Laub. = Boraginaceae

Lenophyllum Rose. Crassulaceae (I 5b). 7 SW N Am. R: U. Eggli, *Ill. Handb. Succ. Pls*, Crass. (2005)181. Cult. orn. succ.

Lens Mill. Leguminosae (III 28). 4 Medit., W As., Afr. (1: ***L. ervoides*** (Brign.) Grande) incl. 2 cultigens: ***L. nigricans*** (M. Bieb.) Godron & ***L. culinaris*** Medik. (*L. esculenta, Ervum l.*, lentil, masur, cultigen with cleistogamous fls & readily germ. seeds app. derived from '*L. orientalis* (Boiss.) Schmalh.' (*L. culinaris* subsp. *o.*, E Medit. to Iraq) with long-dormant seeds – one of earliest crops cult. (form. with oats & barley, threshed together & separated by throwing – cereals go further) for seeds, split into cotyledons for sale (4.5 Mt by 2011), & used in soups (the 'mess of pottage' for which Esau traded his birthright), incl. fast-boiling cvs, or as flour, foliage a valuable forage for animals. R: BotJLS 133(2000) 55

Lent lily *Narcissus pseudonarcissus*

Lenten rose *Helleborus orientalis*

Lentibulariaceae Rich. (~ Scrophulariaceae). Magnoliidae – Lamiales. 3/300 cosmop. Carnivorous pls of wet places (some epiphytes), s.t. rootless & free-floating, with stalked &/or sessile glands. Lvs simple, in spirals, in rosettes in *Pinguicula* & *Genlisea*, which has tubular trap-lvs also; stip. 0; stem in *Utricularia* oft. with spirals or whorls of simple or dissected photosynthetic appendages bearing bladders with a trap-mechanism capturing small animals (see *Utricularia*). Fls bisex., in a bracteate raceme (rarely 1-flowered) or solit. & term. without bracts (*Pinguicula*); K (2–4 or 5), lobed or ± 2-cleft, C (5), 2-labiate & ± 5-lobed, lobes imbr., the lower lip basally spurred, A 2 (anterior) borne on C-tube, nectary-disk 0, $\underline{G}$ (2), 1-loc. with free-C placentation & ± sessile unequally 2-lobed stigma & 2–∞ anatropous, unitegmic ovules. Fr. usu. a capsule opening by 2–4 valves or irreg., or circumscissile (indehiscent & 1-seeded in some *Utricularia* (*Biovularia*) spp.); embryo scarcely differentiated, endosperm 0. x = 7 . . . 32

Genera: *Genlisea, Pinguicula, Utricularia*

Carnivory achieved variously: *Pinguicula* with glandular hairs & inrolling lvs, *Genlisea* with bottle-like pitchers with bands of hairs & digestive glands, *Utricularia* with bladder-traps. Some *Utricularia* spp. weedy in ricefields; some cult. orn.

lentil *Lens culinaris*

lentisco, lentisk *Pistacia lentiscus*

Lenwebbia N. Snow & Guymer (~ *Austromyrtus*). Myrtaceae (II 10). 2 Aus. R: SBM 65(2003)25

Lenzia Philippi. Montiaceae (Portulacaceae s.l.). 1 Chile: ***L. chamaepitys*** Philippi

Leocordia Delile (~ *Lotononis*). Leguminosae (III 8). 51 trop & S Afr., Madag.

Leocereus Britton & Rose. Cactaceae (III 4). 1 E Braz.: ***L. bahiensis*** Britton & Rose. R: Bradleya 8(1990)107. Caffeine synth., cult. orn.

Leochilus Knowles & Westc. Orchidaceae (V 12h). Incl. *Goniochilus, Hybochilus, Papperitzia*, 12 trop. Am. R: SBM 14(1986). Cult. orn. epiphytes

Leocus A. Chev. = *Coleus*

Leonardendron Aubrév. = *Anthonotha*

Leonardoxa Aubrév. Leguminosae (I 2). 1 Guineo-Congolian forests: ***L. africana*** (Baill.) Aubrév. – foliar nectaries, swollen petioles excavated & diff pops app. independently occupied by diff. genera of ants (*Aphomomyrmex* & *Petalomyrmex* spp.) which are app. restricted to tree & patrol young lvs & prot. from other herbivores necessary for survival of shoot to maturity, but *Catalaucus mckeyi* ants exlude them & do not patrol so parasitic on system. R: Adansonia III,22(2000)95

Leonia Ruíz & Pavón. Violaceae (II). 5–6 trop. S Am. ***L. glycycarpa*** Ruíz & Pavón – used for bird-lime in Braz.

Leonis R. Nordenstam (~ *Senecio*). Compositae (Sen.-Sen.). 1 Cuba, Hispaniola: ***L. trineura*** (Griseb.) R. Nordenstam (*S. t.*). R: CN 44(2006)55

Leonohebe Heads = *Veronica*

Leonotis (Pers.) R. Br. (~ *Leucas*). Labiatae (VI). 9 trop. Afr., 1 ext. to Am. & As. (***L. nepetifolia*** (L.) R. Br., natur. in US – pantrop. weed, young infls cooked Uganda, local medic. in As., Afr. & Am., infusion a lice-remover), cult. orn. R: KB 58(2003)600. Heterogeneous. ***L. ocymifolia*** (Burm.f.) Iwarsson (trop. & S Afr.) – locally medic. as are other spp. incl. ***L. leonurus*** (L.) R. Br. (S Afr.) also cult. orn. hedge-pl.

Leonticaceae Airy Shaw = Berberidaceae

Leontice L. Berberidaceae (II 1; Leonticaceae). 3 SE Eur. (1) to N Afr. & C As. Alks. Fr. a papery wind-disp. bladder. Locally medic. & countering opium, ***L. leontopetalum*** L. an agric. weed

Leontochir Philippi = *Bomarea*

Leontodon L. (~ *Picris*). Compositae (Cich.-Hypo.). Excl. *Scorzoneroides*, c. 40 temp. Euras. (Eur. 27) to Medit. & Iran. Subg. *Oporinia* prob. generically distinct. Oft., esp. ***L. saxatilis*** Lam. (*L. taraxacoides*), weedy (hawkbit) in lawns like *Taraxacum* spp.

Leontonyx Cass. = *Helichrysum*

Leontopodium R. Br. ex Cass. Compositae (Gnap.-Gnap.). Excl. *Sinoleontopodium* 58 Eur. (1) to Myanmar & China, esp. mts. R: OB 104(1991)134. Some dioec. ***L. nivale*** (Ten.) Hand.-Mazz. subsp. ***alpinum*** (Cass.) Greuter (*L. alpinum*, edelweiss, Eur. mts) – the familiar tufted woolly herb with small fl. heads (insect-poll., esp. flies) crowded into dense cymes surrounded by consp. bract-like lvs, beloved of alpinists, used in 'whitening' cosmetics, source of 'anti-free radicals' compounds, fibres in hairs same size as UV wavelength so absorbing it & reproduced as glass optical fibres for poss. use in solar panels, satellites etc., cult. orn. as are some other spp.

Leonuroides Rauschert = *Panzerina*

Leonurus L. Labiatae (VI). Excl. *Chaiturus*, c. 25 Euras. (Eur. 1). Heterogeneous. Alks. ***L. cardiaca*** L. (motherwort, Eur., Medit.) – form. locally imp. medic. & source of green dye, cult. orn; ***L. japonicus*** Houtt. ('*L. sibiricus*', As., now pantrop. weed) – local medic.

leopard bane *Doronicum pardalianches*; **l. flower, l. lily** *Iris domestica*; **l. plant** *Farfugium japonicum* 'Aureomaculatum'; **l. tree** *Libidibia ferrea*, *Flindersia collina*; **l. wood** *Brosimum guianense*, *F. collina*, *F. maculosa*

Leopoldia Parl. = *Muscari*

Leopoldinia Mart. Palmae (V 12). 3 Colombia, Venez., Braz. R: Phytotaxa 32(2011)1. Some fish-disp. in Amaz. ***L. piassaba*** Wallace ex Archer (piassaba palm) – source of Pará piassava, a valuable fibre coll. in the wild, used like *Musa textilis* in ropes, also for brushes & brooms

Lepanthes Sw. Orchidaceae (V 13c). c. 1000 trop. Am. Some pseudocopulation (fungus-gnats). Few cult. orn. usu. dwarf epiphytes

Lepanthopsis (Cogn.) Ames. Orchidaceae (V 13c). c. 40 trop. Am. R: MSB 39(1991)1

Lepargochloa Launert = *Loxodera*

Lepechinella Airy Shaw = *Lepechiniella*

Lepechinia Willd. Labiatae (VII 2a). Incl. *Sphacele*, c. 40 warm Am., Hawaii (1) Some dioec.; some medic., cult. orn. ***L. chamaedryoides*** (Balbis) Epling (Chile) – natur. Réunion

Lepechiniella Popov. Boraginaceae (B.3.5.1). c. 6 C As.

Lepeostegeres Blume. Loranthaceae (3). 9 W Mal. to NG. R: Blumea 38 (1993)175, FM I,13(1997)340

Lepianthes Raf. = *Piper*. See also *Pothomorphe*

Lepidacanthus C. Presl = *Aphelandra*

Lepidagathis Willd. Acanthaceae (III 2d). Incl. *Lophostachys*, c. 110 trop. & warm. Local medic. See also *Chroesthes*

Lepidaploa (Cass.) Cass. (~ *Vernonia*). Compositae (Vern.-Lep.). c. 140 trop. Am.

Lepidaria Tieghem. Loranthaceae (3). 12 S Thailand, N & W Mal. R: FM I,13(1997)348

Lepiderema Radlk. Sapindaceae. 8 N Aus. (6, R: Austrobaileya 1(1982)488), NG (2, R: Blumea 36(1991)235). Some caulifl. pachycauls (Corner's Model)

Lepidesmia Klatt (~ *Ayapana*). Compositae (Eup.-Ayap.). 1 Cuba, Colombia, Venez. coastal: ***L. squarrosa*** Klatt. R: MSBMBG 22(1987)213

Lepidium Tourn. ex L. Cruciferae (35). Incl. *Cardaria, Coronopus, Stroganowia, Stubendorffia, Winklera*, c. 250 cosmop. esp. temp. (Eur. 20, Aus. 35 endemic + 8 natur., N Am. 42, S Am. 50 + 12 natur. – R: Darw. 48(2010)142). R: NDSNG 41(1906)1. Pepperwort, peppercress, swine cress. A 2; seeds slimy; alks. Many weedy herbs like ant-poll. ***L. coronopus*** (L.) Al.-Shehbaz (*Coronopus squamatus*, wart cress, ? orig. S Am.) but ***L. arbuscula*** Hillebr. & ***L. serra*** H. Mann (Hawaii) shrubby, some Aus. spp. lianes; ***L. sisymbrioides*** Hook.f. (S NZ) & ***L. solandri*** Kirk (NZ) – dioec. Some cult. for salads etc. ***L. draba*** L. (*Cardaria d.*, hoary cress, Medit., Euras.) – pungent seeds form. a pepper subs., invasive GB (introd. 1809 in straw bedding from Netherlands) & US, spreading by r. suckers (cf. *Armoracia*); ***L. latifolium*** L. (dittander, Eur., Medit., N Afr., invasive W US) – cult. salad pl. of anc. Greeks, form. medic. (leprous sores) & veterinary (camels); ***L. meyenii*** Walp. (maca, Andes) – form. cult. salad veg., allegedly aphrodisiac, r. eaten as porridge & fermented to alcoholic drinks etc., dried now a 'superfood'; ***L. sativum*** L. (Egypt & W As., cress, Gk. c.) – usu. eaten at cotyledon (char. 3-lobed) or seedling stage with mustard sown 4 days later

Lepidobolus Nees. Restionaceae. Incl. *Desmocladus*, c. 27 S Aus.

Lepidobotryaceae Léonard (Oxalidaceae s.l.). Magnoliidae – Celastrales. 2/3 trop. Afr. & Am. R: Novon 3(1993)408. Dioec. trees with unifoliolate, articulate 2-ranked lvs with stip. & stipels, lately removed from Oxalidaceae because of bitter (not sour or acidic) substances in lvs & bark, unclawed petals, disk, $\underline{G}$ 2 or 3 (rather than 5) septicidal or irreg. (not loculicidal) dehiscent capsules, collateral rather than superposed ovules, obturators & 0 endosperm

Genera: *Lepidobotrys, Ruptiliocarpon*

'Similarities' to Sapindaceae & Euphorbiaceae superficial

Lepidobotrys Engl. Lepidobotryaceae (Oxalidaceae s.l.). 1 trop. Afr.: ***L. staudtii*** Engl.

Lepidocaryum Mart. Palmae (I 2c). 1 W Amazonia: ***L. tenue*** Mart. (*L. tessmannii*) – dioec., lvs for thatch. R: R. Henderson, *Palms Amaz.* (1995)77

Lepidoceras Hook.f. Santalaceae (Eremolepidaceae). 2 C Peru (2 colls), S Chile. R: SBM 18(1988)48

Lepidocordia Ducke. Boraginaceae (Ehretiaceae). 2 N trop. S Am. R: AJB 77(1990)548

Lepidogrammitis Ching = *Lemmaphyllum*

Lepidogyne Blume. Orchidaceae (IV 2d). 1 Mal.: ***L. longifolia*** (Blume) Blume

Lepidolopha Winkler. Compositae (Anth.-?Han.). 9 C As. R: BBMNHB 23(1993)105

Lepidolopsis Polj. Compositae (Anth.-Han.). 1 C As., Iran, Afghanistan: ***L. turkestanica*** (Regel & Schmalh.) Polj. R: BBMNHB 23(1993)109

Lepidomicrosorium Ching & Shing = *Microsorum*

Lepidonia S.F. Blake (~ *Vernonia*). Compositae (Vern.-Leib.). 9 S Mex. R: BJ 108(1987)225

Lepidopetalum Blume. Sapindaceae. 6 Andamans & Nicobars, Sumatra, Philipp. to Bismarck Arch. R: Blumea 36(1992)439

Lepidopharynx Rusby = *Hippeastrum*

Lepidophorum Necker ex Cass. Compositae (Anth.). 1 Portugal, Spain: ***L. repandum*** (L.) DC. R: BBMNHB 23(1993)139

Lepidophyllum Cass. Compositae (Ast.-Hin.). 1 Patagonia: ***L. cupressiforme*** (Lam.) Cass. R: SCB 92(2009)56

Lepidorrhachis (H. Wendl. & Drude) Cook. Palmae (V 14). 1 Lord Howe Is. (2 mts above 750 m): ***L. mooreana*** (F. Muell.) Cook. R: FA 49(1994)408

Lepidospartum (A. Gray) A. Gray. Compositae (Sen.-Tuss.). 3 SW US. R: FNA 20(2006) 632

Lepidosperma Labill. Cyperaceae (II 7). 81 Mal., Aus., New Caled., NZ. R: AusSB 25(2012)231. ***L. gladiatum*** Labill. (sword sedge, Aus.) – weakly ant-disp., sand-binder & material for paper-making; ***L. squamatum*** Labill. (Aus.) – fibre for table-mats

Lepidostemon Hook.f. & Thomson. Cruciferae (26). 6 Himal. R: Novon 10(2000)329, EJB 59(2002)444

Lepidostephium Oliv. (~ *Athrixia*). Compositae (Gnap.). 2 S Afr. R: OB 104(1991)49

Lepidostoma Bremek. Rubiaceae (I 9). 1 Sumatra: ***L. polythyrsum*** Bremek.

Lepidothamnaceae A. Bobrov & Melikian = Podocarpaceae

Lepidothamnus Philippi (~ *Dacrydium*). Podocarpaceae. 2 NZ, 1 S Chile. R: AusJB 30(1982)316. ***L. laxifolius*** (Hook.f.) Quinn (mt. rimu, NZ mts) – planted to check erosion

Lepidotis P. Beauv. = *Lycopodiella*

Lepidotrichilia (Harms) T.D. Penn. Meliaceae (II). 4 Madag. (3), E Afr. (1)

Lepidotrichum Velen. & Bornm. (~ *Aurinia*). Cruciferae (2). 1 SW Black Sea: ***L. uechtritzianum*** (Bornm.) Velen. & Bornm.

Lepidozamia Regel. Zamiaceae (I). 2 NE Aus. R: FA 48(1998)638. ***L. hopei*** (W. Hill) Regel (NE Queensland) – at 20 m the tallest cycad; ***L. peroffskyana*** Regel – poll. host-specific *Tranes* weevils

Lepilaena J. L. Drumm. ex Harv. (~ *Althenia*). Potamogetonaceae (Zannichelliaceae). 6 Aus., NZ. Pollen in rafts reaching stigma, cf. *Halodule, Halophila, Ruppia*

Lepinia Decne. Apocynaceae (I i). 4 Caroline (1) & Solomon Is. & NG (1), Marquesas (1), Tahiti & Moorea (1). R: Allertonia 7(1997)256. Postgenital apical union of 3–5 mericarps to give basket-like fr.

Lepiniopsis Valeton. Apocynaceae (I i). 2 C & E Mal. (1), Micronesia (1)

Lepionurus Blume. Opiliaceae. 1 Indomal.: ***L. sylvestris*** Blume – r.-parasite, local medic.

Lepironia Rich. Cyperaceae (I 2). 1 Madag. to Polynesia: ***L. articulata*** (Retz.) Domin (*L. mucronata*) – cult. esp. China for fibre used in junk sails & mats for packing tobacco, rubber, kapok etc.

Lepisanthes Blume. Sapindaceae. 24 OW trop. (Aus. 1). R: Blumea 17(1969)33. Some with Chamberlain's Model. Some cult. orn. trees, some with ed. fr. grown as *Aphania* or *Erioglossum* spp. e.g. ***L. fruticosa*** (Roxb.) Leenh. (luna nut, Mal.)

Lepismium Pfeiffer (~ *Rhipsalis*). Cactaceae (III 6). 15 E Bolivia & Arg., a few ext. to Braz. R: Bradleya 5(1987)99, F. Supplie (2007) *R. & L.* Cult. orn. like *Rhipsalis* but without char. branching pattern & usu. spiny. See also *Pfeiffera*

Lepisorus (J. Sm.) Ching (~ *Pleopeltis*). Polypodiaceae (IV). Incl. *Belvisia, Paragramma*, c. 45 trop. & warm OW (Mal. 4 – R: Blumea 43(1998)109). Some cult. orn.

Lepistemon Blume (~ *Ipomoea*). Convolvulaceae (9). 6 OW trop. (Afr. 1)

Lepistemonopsis Dammer (~ *Ipomoea*). Convolvulaceae (9). 1 trop. E Afr.: ***L. volkensii*** Dammer

Leplaea Vermoesen (~ *Guarea*). Meliaceae (II-Trich.). 7 trop. Afr. R: PEE 145(2012)212. Some timbers esp. ***L. cedrata*** (A. Chev.) Koenen & J. de Wilde (*G. c.*) & ***L. thompsonii*** (Sprague & Hutch.) Koenen & J. de Wilde (*G. t.*) – pale mahogany subs.

Leporella A.S. George. Orchidaceae (IV 3f). 1 Aus.: ***L. fimbriata*** (Lindl.) A.S. George – pheromones attract male flying ants (*Myrmecia urens*) for pseudocopulation poll.

leprosy gourd *Momordica charantia*

Leptacanthus Nees = *Strobilanthes*

Leptactina Hook.f. Rubiaceae (II 2). 19 trop. & S Afr. R: Adans. 36(2014)122. Petit's Model. ***L. rheophytica*** Sonké & Neuba (Equatorial Guinea) – rheophyte

Leptadenia R. Br. Apocynaceae (V b; Asclepiadaceae III 5). 4 OW trop. ***L. pyrotechnica*** (Forssk.) Decne (Afr. to Ind.) – fr. & twigs eaten by Bedouin

Leptagrostis C. Hubb. Gramineae (Arund.). 1 Ethiopia: ***L. schimperiana*** (Hochst.) C. Hubb. R: KB 1937:63

Leptaleum DC. Cruciferae (26). 1 E Med. incl. Eur. to C As.: ***L. filifolium*** (Willd.) DC.

Leptaloe Stapf = *Aloe*

Leptarrhena R. Br. Saxifragaceae. 1 W N Am.: ***L. pyrolifolia*** (D. Don) Ser. (*L. amplexifolia*) – cult. orn. rhiz. herb, local medic. R: FNA 9(2009)131

Leptaspis R. Br. Gramineae (II). Excl. *Scrotochloa*, 3 OW trop. Panicle tough or shed intact

Leptatherum Nees (~ *Microstegium*). Gramineae (XXII 5). 3 trop. & warm OW. R: Blumea 54(2009)179

Leptaulus Benth. Cardiopteridaceae (Icacinaceae s.l.). 6 trop. Afr., Madag. Petit's Model

Leptecophylla C. Weiller (~ *Cyathodes*). Ericaceae (VII 7). 13 NG, Aus., NZ to Hawaii. R: Muelleria 12(1999)196. ***L. tameiameiae*** (Cham. & Schldl) Weiller (Marquesas, Hawaii) – used in leis

Lepterica N.E. Br. = *Scyphogyne*

Leptinella Cass. (~ *Cotula*). Compositae (Anth.-Cot.). 34 NG, Aus., NZ, subAntarctic is., S Am. R: NZJB 25(1987)99. Cult. orn. carpeting pls esp. ***L. squalida*** Hook.f. (*C. squalida*, NZ)

Leptoboea Benth. Gesneriaceae (III d). 3 S & SE As.

Leptocallisia (Benth.) Pichon = *Callisia*

Leptocanna Chia & Fung = *Schizostachyum*

Leptocarpha DC. Compositae (Helia.-Ecl.). 1 Chile: ***L. rivularis*** DC.

Leptocarpus R. Br. Restionaceae. Incl. *Meeboldina, Stenotalis*, excl. *Calopsis*, c. 15 SW Aus., ***L. tenax*** (Labill.) R. Br. ext. to SE. See also *Apodasmia*

Leptocarydion Hochst. ex Stapf. Gramineae (XXIX 3). 1 E & S Afr.: ***L. vulpiastrum*** (De Not.) Stapf. R: Strel. 10(2000)701

Leptoceras Fitzg. = *Leporella*

Leptoceras (R. Br.) Lindl. (~ *Caladenia*). Orchidaceae (IV 3b). 1 S Aus.: ***L. menziesii*** (R. Br.) Lindl. R: Gen. Orch. 2(2001)110

Leptocereus (A. Berger) Britton & Rose. Cactaceae (III 1). 11 Cuba, Hispaniola, Puerto Rico

Leptochilus Kaulf. Polypodiaceae (IV). 9 Jap. & China to Queensland & Solomons. R: Blumea 42(1997)274

Leptochiton Sealy. Amaryllidaceae. 2–3 Andes

Leptochloa P. Beauv. Gramineae (XXIX 5). Excl. *Diplachne, Disakisperma, Trigonochloa,* 5 Aus., S US to S Am.

Leptochloopsis Yates = *Uniola*

Leptocionium C. Presl = *Hymenophyllum*

Leptoclinium Gardner ex Benth. Compositae (Eup.-Alom.). 1 Braz.: ***L. trichotomum*** (Gardner) Bak. R: MSBMBG 22(1987)258

Leptocodon (Hook.f.) Lem. = *Codonopsis*

Leptocoryphium Nees = *Anthenantia*

Leptodactylon Hook. & Arn. = *Linanthus*

Leptodermis Wall. Rubiaceae (IV 12). 40 Himal. to Jap. (China 34, 30 endemic). Some cult. orn.

Leptoderris Dunn. Leguminosae (III 6). 20 trop. Afr.

Leptodesmia (Benth.) Benth. Leguminosae (III 19). 3 Madag., 1 ext. to Ind.

Leptofeddea Diels = *Leptoglossis*

Leptoglossis Benth. Solanaceae (Petun.). 7 Peru, Arg. R: HBML 27(1980)12. Xerophytes

Leptoglottis DC. = *Schrankia*

Leptogonum Benth. Polygonaceae (Erio.-Lept.). 1 Hispaniola: ***L. domingense*** Benth. R: NJB 10(1990)487. Small tree

Leptogramma J. Sm. = *Stegnogramma*

Leptolaena Thouars. Sarcolaenaceae. Excl. *Xerochlamys,* 10 Madag. R (subg. *L.*): Adans. III,23(2001)173

Leptolepia Mett. ex Diels (~ *Microlepis*). Dennstaedtiaceae. 1 NG, Aus., NZ: ***L. novaezelandiae*** (Colenso) Kuhn

Leptolepidium Shing & S.K. Wu = *Cheilanthes*

Leptolobium Vogel (~ *Acosmium*). Leguminosae (III 9). 12 trop. Am. R: T. 57(2008)981. ***L. panamense*** (Benth.) Sch. Rodr. & Azevedo (*A. p.*) – introd. C Afr. as firebreak

Leptoloma Chase = *Digitaria*

Leptomeria R. Br. Santalaceae (Amphorogynaceae). 17 Aus. esp. SW. R: AusSB 12(1999)62. ***L. acida*** R. Br. (26 mg vitamin C per 100 g fr.) & other spp. (currant bushes) with fr. used in preserves; ***L. drupacea*** (Labill.) Druce – homoeotic mutants with embryosac-like bodies in place of microspores (unique in angiosperms)

Leptomischus Drake. Rubiaceae (IV 1). 7 SE As. R: APS 31(1993)273

Leptonema A. Juss. Phyllanthaceae (Ant.-Lept.; Euphorbiaceae s.l.). 2 Madag.

Leptonychia Turcz. Malvaceae (Bytt.-Bytt.; Sterculiaceae). c. 30 OW trop. (SE As. 3, R: Blumea 32(1987)443). Troll's Model. Local medic.

Leptonychiopsis Ridl. = *Leptonychia*

Leptopetalum Hook. & Arn. (~ *Oldenlandia*). Rubiaceae (IV 15). 8 Indopacific

Leptopharyngia (Stapf) Boit. = *Tabernaemontana*

Leptopharynx Rydb. = *Perityle*

Leptophoenix Becc. = *Hydriastele*

Leptophyllochloa Caldéron = *Trisetaria*

Leptoplax O. Schulz (~ *Peltaria*). Cruciferae (2). 1 Greece: ***L. emarginata*** (Boiss.) O. Schulz

Leptopteris C. Presl (~ *Osmunda*). Osmundaceae. 6 NG, Aus. (1), New Caled., Polynesia, NZ. R: M.F. Large & J.E. Braggins, *Tree Ferns* (2004)295. ***L. wilkesiana*** (Brack.) Christ (W Pac.) – tree-fern to 133 yrs old in Fiji. Cult. orn. esp. ***L. superba*** (Colenso) C. Presl (NZ) with plume-like fronds to 1.2 m

Leptopus Decne (~ *Andrachne*). Phyllanthaceae (Phyll.-Por.; Euphorbiaceae s.l.). 9 Cauc. to Mal. R: KB 63(2008)46, 64(2010)627

Leptopyrum Reichb. Ranunculaceae (III 1). 1 W Siberia to E As.: ***L. fumarioides*** (L.) Reichb. – cult. orn. ann., natur. Eur.

Leptorhabdos Schrenk. Orobanchaceae (Orob.-'Micr.'; Scrophulariaceae s.l.). 1 Cauc. & Iran to C As. & Himal.: ***L. parviflora*** (Benth.) Benth.

Leptorhynchos Less. Compositae (Gnap.-Ang.). 8 temp. Aus. R: OB 104(1991)120, Nuytsia 13(2001)608

Leptorumohra (H. Itô) H. Itô = ? *Dryopteris*

Leptosaccharum (Hackel) A. Camus = *Eriochrysis*

Leptoscela Hook.f. Rubiaceae (IV 1). 1 E Braz.: ***L. ruellioides*** Hook.f. R: HIP 12(1872) 44

Leptosema Benth. (~ *Brachysema*). Leguminosae (III 13). 13 Aus. R: Aus SB 12(1999)9. Bird-poll.; ***L. chambersii*** F. Muell. (trop. Aus.) – much nectar so fls sucked by Aborigines

Leptosiphon Benth. = *Linanthus*

Leptosiphonium F. Muell. (~ *Ruellia*). Acanthaceae (III 2a). 10 Papuasia

Leptosolena C. Presl (~ *Alpinia*). Zingiberaceae (II 4). 1 Philippines: ***L. haenkei*** C. Presl. R: APG 56(2005)45

Leptospermopsis S. Moore = *Leptospermum*

Leptospermum Forst. & Forst. f. Myrtaceae (II 14). 79 SE As. (Mal. 3) to NZ (1, Aus. 77), R: Telopea 3(1989)301. Heterogeneous. Everg. shrubs (some rheophytes) & trees, some beetle-poll., flowering & fruiting coincides with emergence of possum young fom pouch; cult. orn. (incl. hybrids) & lvs used for tea (as by Cook's crew; tea trees, Aus.). ***L. laevigatum*** (Gaertn.) F. Muell. (SE Aus.) – invasive in Aus., Hawaii & S Afr. (fynbos, even replacing introd. *Acacia saligna*); ***L. liversidgei*** R. Bak. & H.G. Sm. ('mozzie blocker', E Aus.). releases citronella into air, deterring mosquitoes; ***L. petersonii*** Bailey (*L. citratum*, E Aus.) – source of lemon-scented oil & comm. grown in Kenya & Guatemala, aldehydes incl. citral used in soft drinks, scents & synth. of vitamin A, though oil may cause breast growth in boys; ***L. scoparium*** Forst. & Forst. f. (manuka, E Aus., NZ) – most commonly cult. sp. (natur. in Scilly Is.), timber & tea, honey medic. (incl. skin problems, stops growth of antibiotic-resistant *Staphylococcus aureus*)

Leptospron (Benth.) Delgado (~ *Vigna*). Leguminosae (III 18). 2+ trop. Am. R: AJB 98(2011)1709

Leptostachya Nees (~ *Justicia*). Acanthaceae (III 2c). 1 trop. As.: ***L. wallichii*** Nees

Leptostelma D. Don (~ *Erigeron*). Compositae (Ast.-Con.). 6 S Am. R: CN 46(2008)2

Leptostigma Arn. (~ *Nertera*). Rubiaceae (IV 13). (Incl. *Corynula*) 6: SE As. 1, NZ 1, W S Am. 4. R: APG 33(1982)73

Leptostylis Benth. = *Pycnandra* (but see FNC 1(1967)20)

Leptotaenia Nutt. ex Torrey & A. Gray = *Lomatium*

Leptoterantha Louis ex Troupin. Menispermaceae (III). 1 trop. Afr.: ***L. mayumbense*** (Exell) Troupin. R: BJBB 19(1949)426

Leptotes Lindl. Orchidaceae (V 13b). 9 trop. S Am. Cult. orn. like *Cattleya* with lvs terete or subterete; fr. of ***L. bicolor*** Lindl. (Braz.) locally used to flavour icecream

Leptothrium Kunth. Gramineae (XXIX 3). 3 Carib. & S Am., Senegal to Pakistan

Leptothyrsa Hook.f. Rutaceae (I 8). 1 Amazon: ***L. sprucei*** Hook.f. R: Candollea 45(1990)376

Leptotriche Turcz. = *Myriocephalus* (but see OB 104(1991)128)

Leptunis Steven = *Asperula*

Lepturella Stapf = *Oropetium*

Lepturidium A. Hitchc. & Ekman (~ *Brachyachne*). Gramineae (XXIX). 1 Cuba: ***L. insulare*** A. Hitchc. & Ekman – salt flats

Lepturopetium Morat. Gramineae (Chlorid.). 2 New Caled., Marshall Is., Cocos Is. R: Adans. 20(1981)377

Lepturus R. Br. Gramineae (XXIX 5). 16 coasts of E Afr. to Polynesia & Aus.

Lepuropetalaceae Nakai = Celastraceae

Lepuropetalon Elliott. Celastraceae (Lepuropetalaceae). 1 SE US, Mex., C Chile, Urug.: ***L. spathulatum*** Elliott – winter ann. c. 2 cm tall, one of the smallest terr. herbs, growing on powerlines & in cemeteries etc.

Lepyrodia R. Br. Restionaceae. c. 30 SW & SE Aus. For NZ sp. see *Sporadanthus*

Lepyrodiclis Fenzl. Caryophyllaceae (II 1). 3 W As. R: ANMW 61(1937)74. ***L. holosteoides*** (C. Meyer) Fischer & C. Meyer – introd. Ger., serious weed of wheat & peas NW US

Lerchea L. Rubiaceae (IV 2). c. 10 SE As. (China 2), Mal. R: Blumea 32(1987)91

Lereschia Boiss. (~ *Cryptotaenia*). Umbelliferae (III 8). 1 S Eur.: ***L. thomasii*** (Tern.) Boiss.

Leretia Vell. Conc. (~ *Mappia*). Icacinaceae. 2 trop. S Am. R: JAA 23(1942)58

Leroya Cavaco = *Pyrostria*

Leroyia Cavaco = *Pyrostria*

lerp manna from *Eucalyptus dumosa*

Lescaillea Griseb. (~ *Chrysactinia*). Compositae (Hele.-Pect.). 1 W Cuba (serpentine): ***L. equisetiformis*** Griseb. – liane. R: Madroño 24(1977)138

Lesia J.L. Clark & J.F. Sm. (~ *Nematanthus*). Gesneriaceae (II 5c). 1 NE S Am.: ***L. savannarum*** (C. Morton) J.L. Clark & J.F. Sm. R: SB 38(2013)456

Lesliea Seidenf. = *Phalaenopsis* (but see OB 95(1988)190)

Lespedeza Michaux. Leguminosae (III 19). 44 temp. N & S Am., trop. & E As., Aus. R: JJB 89(2014)4. C s.t. 0 & fls cleistogamous; herbs & shrublets with v. small 1-seeded indehiscent legumes cult. for forage & as green manures; alks. ***L. bicolor*** Turcz. (E As.) & ***L. thunbergii*** (DC.) Nakai (? hybrid cultigen, E As.) – cover crops & fuel

Lesquerella S. Watson = *Physaria*

Lesquereuxia Boiss. & Reuter (~ *Siphonostegia*). Orobanchaceae (Cymb.; Scrophulariaceae s.l.). 1 E Med.: ***L. syriaca*** Boiss. & Reuter – r.- parasite on *Hedera helix, Castanea sativa* etc.

Lessertia DC. Leguminosae (III 24). Incl. *Sutherlandia*, 55 trop. E (few) & S Afr. ***L. frutescens*** (L.) Goldbl. & Manning (*S. f.*, balloon pea, duck pl., S Afr.) – form. considered cancer cure, cult. orn. shrub with red fls & large bladder-like frs, when floated in water reminiscent of ducks

Lessingia Cham. Compositae (Ast.-Mac.). (Excl. *Benitoa, Corethrogyne*) 12 SW N Am. R: FNA 20(2006)452. Some cult. orn.

Lessingianthus H. Robinson (~ *Vernonia*). Compositae (Vern.-Lep.). c. 102 S Am. esp. Braz.

Letestua Lecomte. Sapotaceae (I 2). 1 W trop. Afr.: ***L. durissima*** (A. Chev.) Lecomte

Letestudoxa Pellegrin. Annonaceae (III 3). 2 W trop. Afr. Lianes

Letestuella G. Taylor. Podostemaceae (III). 1 SW Afr.: ***L. tisserantii*** G. Taylor

Lethedon Spreng. Thymelaeaceae (Gon.). 15 New Caled., Vanuatu (1), NE Aus. (1)

Lethia Ravenna = ? *Calydorea*

letterwood *Brosimum guianense*

Lettowia H. Robinson & Skvarla (~ *Vernonia*). Compositae (Vern.-Erl.). 1 E Afr.: ***L. nyassae*** (Oliv.) H. Robinson. R: PhytoKeys 25(2013)48

Lettowianthus Diels. Annonaceae (II). 1 trop. E Afr.: ***L. stellatus*** Diels

lettuce, asparagus l. *Lactuca sativa*; **Ind. l.** *L. indica*; **lamb's l.** *Valerianella locusta*; **miner's l.** *Claytonia perfoliata*; **opium l.** *L. virosa*; **prickly l.** *L. serriola*; **stem lettuce** *L. sativa*; **l. tree** *Pisonia grandis*; **wall l.** *Cicerbita muralis*; **water l.** *Pistia stratiotes*

Leucactinia Rydb. (~ *Urbinella*). Compositae (Hele.-Pect.). 1 Mex.: ***L. bracteata*** (S. Watson) Rydb. R: Phytol. M 10(1996)21

Leucadendron R. Br. Proteaceae (IV 4). 83 S Afr. (Cape 82, 79 endemic – R: Strel. 9(2000)576). R: CBH 3(1972). Dioec.; when fr. matures, P splits into 4 parts united around stigma & acts as wing in disp. ***L. argenteum*** (L.) R. Br. – lvs covered with soft silky hairs much exploited for bookmarks, mats, s.t. painted with landscapes. Cult. orn. incl. ***L.* 'Safari Sunset'** (*L. salignum* P.J. Berg. × *L. laureola* (Lam.) Fourc. ['*L. laureolum*']) – frost-hardy (raised NZ)

Leucaena Benth. Leguminosae (II 1). 22 Texas to Peru. R: SBM 55 (1998)79. Some polyploids; spp. grown beyond natural ranges & hybridizing with indigenous ones. Livestock fodder, green manure, firewood crops, soil conservation; at least 13 spp. & hybrids with ed. fr. (unripe), seeds (also fl. buds & galls, esp. Mex.) in local markets. ***L. esculenta*** (Moçiño & Sessé) Benth. (guaje, Mex.) – seeds eaten with salt but in others toxic mimosine leads to loss of hair, infertility etc.; ***L. leucocephala*** (Lam.) De Wit (*L. glauca, L. latisiliqua*, lead tree, allopolyploid (female parent ***L. pulverulenta*** (Schldl) Benth., Mex.) arising in Mex., invasive in S Eur., As., Aus., Hawaii, trop. & S Afr., Masc., Seychelles, SE US, trop. Am., Galápagos) – firewood crop, green manure, pulp & forage (toxic to non-ruminants causing hair loss, e.g. horses' manes & tails) in trop., seeds (flat, brown) used as beads (Jumbie beans), now mooted as charcoal source & for energy (1 M barrels of oil per annum from 12 000 ha), anthelminthic in Sumatra

Leucampyx A. Gray = *Hymenopappus*

Leucanthemella Tzvelev (~ *Tanacetum*). Compositae (Anth.-Art.). 2 SE Eur. (1: ***L. serotina*** (L.) Tzvelev – cult. orn. with white or red ray-florets), E As. R: BBMNHB 23(1993)139

Leucanthemopsis (Giroux) Heyw. (~ *Tanacetum*). Compositae (Anth.-Leucanthemops.). 9 S Eur., Morocco (1). R: AIBC 32,2(1975)175

Leucanthemum Mill. (~ *Chrysanthemum*). Compositae (Anth.-Leuc.). Excl. *Mauranthemum* (R: T 44 (1995)377) 43 Eur. esp. mts, N As. R: BBMNHB 23(1993)141. Cult. orn. esp. ***L.* × *superbum*** (J. Ingram) Kent (Shasta daisy, poss. *L. maximum* (Ramond) DC. (*L. vulgare* subsp. *m.*, Pyrenees) × *L. lacustre* (Brot.) Samp. (SW Eur.) though poss. involving *Nipponanthemum nipponicum* (Maxim.) Kitam. or merely a fine cv of *L. lacustre*, first noticed natur. at Mt Shasta, Washington, USA), esp. 'double-flowered' '**Esther Read**' (poss. with 'double' *L. vulgare* in ancestry) & ***L. vulgare*** Lam. (*C. l.*, moon, dog or ox-eye daisy, marguerite, Euras., natur. N Am.) – home remedy for catarrh, ed. (just about)

Leucas Burm. ex R. Br. Labiatae (VI). c. 100 Afr. & Arabia to Indomal. [s.s. As. 38, Afr. spp. = ? *Leonotis*], some weedy elsewhere esp. ***L. martinicensis*** (Jacq.) R. Br. (Afr. Arabia) – in China burnt to repel mosquitoes, but also ***L. biflora*** (Vahl) Sm. (Indomal.), ***L. glabrata***

(Vahl) Sm. (Arabia) & ***L. urticifolia*** (Retz.) Sm. (NE Afr.). R: SBNat 341(1980)1. Some locally medic.

Leucaster Choisy. Nyctaginaceae (I). 1 SE Braz.: ***L. caniflorus*** (Mart.) Choisy

Leucelene Greene = *Chaetopappa*

Leucheria Lag. Compositae (Mut.-Nass.). 47 Andes, Chile, Patagonia, Falkland Is. R: Darwiniana 20(1976)9. Herbs

Leuchtenbergia Hook. Oldest name for *Ferocactus*

Leuciva Rydb. = *Euphrosyne* (but see FNA 21(2006)29)

Leucobarleria Lindau = *Neuracanthus*

Leucoblepharis Arn. (~ *Blepharispermum*). Compositae (Athro.). 1 Ind.: ***L. subsessilis*** (DC.) Arn. R: BJ 112(1990)183

Leucobryum Hampe. Leucobryaceae (Musci). c. 122 subcosmop. (N Am. 2) esp. trop. Some free-floating in periodically inundated areas esp. ***L. glaucum*** (Hedw.) Ångstr. (N temp., WI, Andes, Hawaii) – dioec. Subs. for sphagnum moss & for stuffing cushions in trop. As.

Leucocalantha Barb. Rodr. = *Pachyptera*

Leucocarpus D. Don. Phrymaceae (Leuc.; Scrophulariaceae s.l.). 1 trop. Am.: ***L. perfoliatus*** (Kunth) Benth. – cult. orn. like *Mimulus*. R: Phytoneuron 2012–39: 33

Leucochloron Barneby & Grimes (~ *Pithecellobium*). Leguminosae (II 4). 4 Braz. R: MNYBG 74,1(1996)130

Leucochrysum (DC.) Paul G. Wilson. Compositae (Gnap.-Ang.). 5 S Aus. R: Nuytsia 8(1992)439

Leucocodon Gardner. Rubiaceae (IV 6). 1 Sri Lanka: ***L. reticulatum*** Gardner. One of few Sri Lankan endemic genera

Leucocorema Ridl. = *Trichadenia*

Leucocoryne Lindl. Amaryllidaceae (Alliaceae). 15–20 Chile. R: AMHNV 5(1972)9. Bulbous pls cult. orn. like *Ixia* esp. ***L. ixioides*** (Hook.) Lindl. (glory of the sun)

Leucocrinum Nutt. ex A. Gray. Asparagaceae (Anthericaceae). 1 SW US: ***L. montanum*** Nutt. ex A. Gray – cult. orn. with fragrant white fls. R: FNA 26(2002)217

Leucocroton Griseb. Euphorbiaceae (Acal.-Adel). Excl. *Garciadelia*, 26 Cuba (serpentine)

Leucocyclus Boiss. = *Achillea* (but see BBNHB 23(1993)130)

Leucogenes Beauverd (~ *Leontopodium*). Compositae (Gnap.-Gnap.). 4 NZ. R: OB 104(1991)134. Hybrids with *Raoulia* spp. in wild. Cult. orn. silvery tomentose herbs (NZ edelweiss) resembling *Leonotopodium*

Leucoglossum Wilcox & al. = *Leucanthemum*

Leucohyle Klotzsch = *Trichopilia*

Leucojum L. Amaryllidaceae. Excl. *Acis*, 2 Eur. to Iran. Snowflakes. R: F.C. Stern (1956) *Snowdrops & snowflakes*, Plantsman 14(1992)72. Alks. App. derived from west of a complex forced south in Pleistocene, the eastern part becoming *Galanthus*, differing in having inner P segments shorter than outer. Cult. orn.: ***L. aestivum*** L. (summer s., Eur. to Iran, natur. in valleys of Shannon & Thames) – harvested in Balkans for galanthamine used in Alzheimer's disease treatment, seeds with outer layer (? aril) of air-filled tissue allowing water-disp.; ***L. vernum*** L. (spring s., C Eur., natur. GB) – seeds with caruncle (? aril) attractive to ants, allegedly disp. agents

Leucolophus Bremek. = *Urophyllum*

Leucomeris D. Don (~ *Gochnatia*). Compositae (Wund./Mut.-Wund.). 2 SE As. Small trees

Leucomphalos Benth. ex Planch. Leguminosae (III 2). Incl. *Baphiastrum*, *Bowringia* 6 E As. (1), Afr. & Madag. (5). R: WAUP 94–4(1994)10

Leuconotis Jack. Apocynaceae (I c). 4 Mal. R: FM I,18(2007)264. Trees, poss. rubber sources

Leucopholis Gardner = *Chionolaena*

Leucophrys Rendle (~ *Urochloa*). Gramineae (XXIV 5). 1 S Afr.: ***L. mesocoma*** (Nees) Rendle. R: Strel. 10(2000)702

Leucophyllum Bonpl. Scrophulariaceae (Leuc.). 12 SW N Am. R: Sida 11(1985)107. ***L. frutescens*** (Berland.) I.M. Johnson (ashplant, Texas & Mex.) – grown as low hedge-pl. or as a lawn locally

Leucophysalis Rydb. (~ *Physalis*). Solanaceae (IV 5c). 2 N Am. R: Rhodora 111(2009)210

Leucophyta R. Br. (~ *Calocephalus*). Compositae (Gnap.-Ang.). 1 Aus.: ***L. brownii*** Cass. (*C. b.*, cushion bush) – cult. orn. R: OB 104(1991)125

Leucopoa Griseb. (~ *Festuca*). Gramineae (XVI 10). 37 temp. Dioec. (cf. *F.*). Heterogeneous

Leucopogon R. Br. (~ *Styphelia*). Ericaceae (VII 7; Epacridaceae). c. 23 Mal., Aus. esp. SW, New Caled. Heterogeneous? Fls term. or in upper axils; other (c. 70) spp. referred to diff.

genera. Some with ed. fr. ('Aus. currant') disp. by birds, ***L. malayanus*** Jack (Mal.) – inner bark waterproofs canoes, lvs & r. medic.

Leucopsis (DC.) Bak. = *Noticastrum*

Leucoptera R. Nordenstam. Compositae (Anth.-Ath.). 3 SW Cape to Namaqualand. R: BN 129(1976)140

Leucopterum Small = *Rhynchosia*

Leucorchis E. Meyer = *Pseudorchis*

Leucosalpa Scott-Elliott. Orobanchaceae ('Butt.'; Scrophulariaceae s.l.). 4 Madag. R: Boissiera 7(1943)287

Leucosceptrum Sm. Labiatae (VI). 1 Himal. to Vietnam: ***L. canum*** Sm.

Leucosidea Ecklon & Zeyher. Rosaceae (8). 1 S Afr.: ***L. sericea*** Ecklon & Zeyher. R: Strel. 10(2000)473

Leucospermum R. Br. Proteaceae (IV 4). Excl. *Vexatorella* = *Paranomus*, 48 SW Cape to Zimbabwe. R: JSAB supp. 8(1972). Seeds over 200 yrs old germ. Some cult. orn. shrubs incl. hybrids, esp. ***L. cordifolium*** (J. Knight) Fourc. (Cape) for cut-fls. ***L. conocarpodendron*** (L.) Buek (Cape) – ant-disp. (elaiosomes)

Leucosphaera Gilg. Amaranthaceae (I 2). 1 S trop. Afr.: ***L. bainesii*** (Hook.f.) Gilg

Leucospora Nutt. (~ *Stemodia*). Plantaginaceae (Grat.-Stem.; Scrophulariaceae s.l.). 1 E N Am.: ***L. multifida*** (Michaux) Nutt. (*S.m.*) – cult. orn. ann.

Leucostegane Prain. Leguminosae (I 2). 2 Malay Pen., Borneo. R: FM I,12(1996)635

Leucostegia C. Presl. Hypodematiaceae (Dryopteridaceae s.l.). 2 C Himal. & S China, S Ind. to E Polynesia. R: Blumea 37(1992)184. Terr. & epiphytic cult. orn.

Leucosyke Zoll. & Moritzi. Urticaceae (III). (Incl. *Gibbsia*, *Maoutia*) c. 50 Mal. to Polynesia. R: BJ 73(1943)191. ***L. capitellata*** (Poir.) Wedd. (Taiwan to Java & NG) – fibre for ropes, medic. locally; ***L. puya*** (Hook.) den Bakker & Mabb. (*M. puya*, Himal. to Myanmar) – fibre for cloth incl. sails

Leucosyris Greene (~ *Aster*). Compositae (Ast.-Sol.). Incl. *Arida*, 9 W N Am. R: Phytoneuron 2012–98:2

Leucothoe D. Don. Ericaceae (VIII 4). Excl. *Agarista*, *Eubotryoides*, *Eubotrys*, 5 Himal., Jap., N Am. R: Britt. 65(2013)422. Cult. orn. shrubs esp. ***L. fontanesiana*** (Steud.) Sleumer ('*L. catesbaei*', SE US) – locally med.

Leucothrinax C. Lewis & Zona (~ *Thrinax*). Palmae (III 2). 1 N Carib.: ***L. morrisii*** (H. Wendl.) C. Lewis & Zona – much cult. R: Palms 52(2008)87

Leucotrichum Labiak (~ *Lellingeria*). Polypodiaceae (V). 5 trop. Am. R: T 59(2010)915

Leuenbergeria Lodé (~ *Pereskia*). Cactaceae. 8 C Am, to N S Am.

Leunisia Philippi. Compositae (Mut.-Nass.). 1 C Chile: ***L. laeta*** Philippi – shrub

Leurocline S. Moore = *Echiochilon*

Leutea Pim. (~ *Ferula*). Umbelliferae (III 10). 8–10 SW & C As.

Leuzea DC. = *Rhaponticum*

Levant galls *Quercus pubescens*; **L. garlic** *Allium ampeloprasum*; **L. madder** *Rubia peregrina*; **L. scammony** *Convolvulus scammonia*; **L. storax** *Liquidambar orientalis*

Levenhookia R. Br. Stylidiaceae. 10 S Aus. Shoe-shaped labellum embracing column springs downwards if touched

lever wood *Ostrya virginiana*

Levieria Becc. Monimiaceae (V 1). 7 E Mal., N Aus. R: Blumea 26(1980)373

Levisticum Hill. Umbelliferae (III 8). 1 E Medit.: ***L. officinale*** Koch (lovage, bladderseed, natur. Eur. & N Am.) – cult. since time of Pliny, lvs may be blanched like celery, fr. & r. used as flavouring in liqueurs etc.

Levya Bureau ex Baill. = *Bignonia*

Lewisia Pursh. Montiaceae (Portulacaceae s.l.). Excl. *Lewisiopsis*, 16 W N Am. R: B. Mathew (1989) *The genus L.*; FNA 4(2003)476. Thick starchy r. of some spp. ed. esp. ***L. rediviva*** Pursh (bitter r.) – form. staple & imp. article of trade, imp. local medic., cult. orn. rock-pl. surviving 2 yrs drought or even boiling

Lewisiopsis Govaerts = *Cistanthe*

Lexarzanthe Diego & Calderón = *Romanschulzia* (but see ABM 68(2004)74)

Leycephyllum Piper = *Rhynchosia*

Leycesteria Wall. Caprifoliaceae. 5 W Himal. to SW China. R: KB 1932:161. ***L. formosa*** Wall. (Himal. honeysuckle, W China, Himal. & E Tibet) – cult. orn. thicket-forming shrub with verticillate simple lvs & consp. bracts, fr. attractive to gamebirds, natur. Eur. incl. GB, invasive in Azores & upland NSW

leylandii × *Cuprocyparis leylandii*

Leymus Hochst. (~ *Triticum*). Gramineae (XV). Incl. *Hordelymus*, 54 N temp. (Eur. 7; China 24, 11 endemic – R: FOC 22(2006)387; N Am. 11 native – R: AJB 71(1984)609, FNA 24(2007)353), 1 Arg. Steppes, oft. saline, alkaline or dunes. Incl. polylpoids with genome from *Psathyrostachys*. Fibres, grains & local medic. N Am.; some useful sand-binders esp. ***L. arenarius*** (L.) Hochst. (*T. a.*, lyme grass, Euras.) – cult. in Jap. for ropes, mats & paper

Leysera L. (~ *Asteropterus*). Compositae (Gnap.-Rel.). 3 S Afr. R: BN 131(1978)369

Lhotskya Schauer = *Calytrix*

Lhotzkyella Rauschert = *Matelea*

li tree *Spondias dulcis*

Liabellum Rydb. (~ *Sinclairea*). Compositae (Liab.). 5 Mex.

Liabum Adans. Compositae (Liab.). Excl. *Sampera*, 22 C Am., WI, Andes (esp. Peru). R: SBM 97(2015)29

Lianthus N. Robson. Hypericaceae (II). 1 Yunnan: ***L. ellipticifolius*** (H.L. Li) N. Robson. R: BNHMB 31(2001)38

Liatris Gaertn. ex Schreb. Compositae (Eup.-Lint.). 41 N Am., Mex., Bahamas. R: FNA 21(2006)512. Spiciform cymes unique in fam. Local medic.; cult. orn. perenn. herbs (gayfeather) esp. ***L. spicata*** (L.) Willd. (Kansas g., E N Am.), locally medic.

Libanothamnus Ernst. See *Espeletia*

Libanotis Haller ex Zinn = *Seseli*

Liberatia Rizz. = *Lophostachys*

Liberbaileya Furt. = *Maxburretia*

Libertia Spreng. Iridaceae (VII 3). 16 NG, Aus., NZ, Andes. R: Plantsman 14(2015)104. 2n = 14–228. Cult. orn. herbs with fibrous r. & white or blue fls

Libidibia (DC.) Schldl. (~ *Caesalpinia*). Leguminosae (I 4). Incl. *Stahlia*, 7 trop. Am. R: PhytoKeys 71(2016)94. Tanbarks, wood for guitars etc. ***L. coriaria*** (Jacq.) Schldl. (*C. c.*, divi-divi, cascalote, trop. Am.) – pods with 40–45% tannin, giving light-coloured leather; ***L. ferrea*** (Tul.) Queiroz (*C. f.*, leopard tree, E. Braz.) – street-tree with spotted bark; ***L. monosperma*** (Tul.) Gagnon & G.P. Lewis (*S. m.*, Hispaniola, Puerto Rico) – timber; ***L. paraguariensis*** (Parodi) G. P. Lewis (*C. p.*, *C. melanocarpa*, S Am.) – lvs used for tanning (guayacán) – 21% tannin, forage with fr. available all yr. round

Libocedrus Endl. Cupressaceae. Excl. *Austrocedrus*, *Papuacedrus*, 5 New Caled. (3), NZ (2). R: A. Farjon, *Monogr. Cupressaceae & Sciadopitys* (2005)438. Mangenot's Model. Timber esp. ***L. bidwillii*** Hook.f. (pahautea, NZ) & ***L. plumosa*** (D. Don) Sarg. (*L. doniana*, kawaka, NZ)

Librevillea Hoyle. Leguminosae (I 2). 1 Gabon: ***L. klainei*** (Harms) Hoyle

Libyella Pampan. = *Poa*

Licania Aubl. Chrysobalanaceae (1). Excl. *Afrolicania*, *Angelesia*, c. 220 trop. Am. (Peru 43). R: FOW 9(2003)218. Troll's Model.; ***L. michauxii*** Prance (gopher apple, SE US) – geoxylic suffrutex with spreading underground 'trunk' to 30 m & aboveground shoots to 30 cm, imp. Seminole drug; some fish-disp. Amaz. Timbers (strong & resistant to borers but diff. to work due to silica), oil (oiticica o.), esp. for illumination, & ed. fr., used for pots (silica grains). ***L. arborea*** Seemann (Am.) – oil; ***L. cecidophora*** Prance (NC Peru) – leaf galls used in necklaces & capes (the only insect galls used for ornament!); ***L. octandra*** (Roemer & Schultes) Kuntze (*L. utilis*, caripé, pottery tree of Pará, NE S Am.) – powdered silica-rich bark mixed with clay makes heat-resistant pots; ***L. tomentosa*** (Benth.) Fritsch (Am.) – cult. for fr.

Licaria Aubl. Lauraceae (I). 50 trop. Am. R: MIABH 28–9(2000)118. Heterodichogamy recorded. Some oilseeds & local medic. esp. ***L. pucheri*** (Ruíz & Pavón) Kosterm. (puchurin nut, Peru)

Lichtensteinia Cham. & Schldl. Umbelliferae (II?). 7 S Afr., 1 St Helena. R: EJB 48(1991)221

licorice See liquorice

Licuala Wurmb. Palmae (III 4b). Excl. *Lanonia*, c. 130 SE As. (Malay Penin. 41 – R: Sandakania 19(1997)1) to Vanuatu & Aus. Palms with bisex. fls & palmate lvs much grown in trop. as cult. orn.; lvs used to improve opium burning; stems for walking-sticks ('Penang lawyers', i.e. loyak or loyar (Malay names for the palms))

Lidbeckia Berg. Compositae (Anth.-Cot.). 3 S Afr. R: Strel. 9(2000)339

Liebigia Endl. (~ *Chirita*). Gesneriaceae (III 2j). 12 Sumatra, Java & Bali. R: T 60(2011)779

Lietzia Regel = *Sinningia*

Lifago Schweinf. & Muschler. Compositae (Inul.-Inul.). 1 Morocco, Algeria: ***L. dielsii*** Schweinf. & Muschler

Ligaria Tieghem (~ *Loranthus*). Loranthaceae (4 3). 2 C Braz. R: Britt. 42(1990)66. ***L. cuneifolia*** (Ruíz & Pavón) Tiegh. (Peru) – infests *Corryocactus brevistylus* (K. Schum.) Britton & Rose

Ligeophila Garay = *Aspidogyne*

Lightfootia L'Hérit. = *Wahlenbergia*

Lightia Schomb. = *Euphronia*

Lightiodendron Rauschert = *Euphronia*

lightwood *Acacia implexa, Ceratopetalum apetalum*

Ligia Fasano = *Thymelaea*

lign-aloes *Aquilaria malaccensis*

Lignariella Baehni = *Aphragmus* (but see HPB 4(2000)115)

Lignocarpa Dawson. Older name for *Gingidia*

lignum *Muehlenbeckia* spp. esp. *M. florulenta;* **l. rhodium** *Amyris balsamifera*

lignum-nephriticum *Eysenhardtia polystachya*

lignum-vitae *Guaiacum* spp.; **Maracaibo l.** *Bulnesia arborea;* **Paraguay l.** *B. sarmientoi;* **Queensland l.** *Vitex lignum-vitae*

Ligularia Cass. Compositae (Sen.-Tuss.). (Excl. *Cremanthodium*) 125 temp. Euras. (China most). Herb. perenn. cult. orn. with revolute lvs, esp. ***L. dentata*** (A. Gray) H. Hara (*L. clivorum*, China & Jap.), its cvs & hybrids. See also *Farfugium* (lvs involute)

Ligulariopsis Y.L. Chen = *Ligularia* (but see APS 34(1996)631)

Ligusticella J. Coulter & Rose. = *Podistera*

Ligusticopsis Leute. Umbelliferae (III 8). 14 temp. E As.

Ligusticum Tourn. ex L. Umbelliferae (III 8). Incl. *Arafoe, Coristospermum, Mutellina, Pachypleurum, Tamanschjanella*, 30–40 circumboreal. R: ANMW 74(1970)457. ? Heterogeneous. Some ed. & medic. (US = osha). ***L. canbyi*** J. Coulter & Rose (N Am.) – still used by Flathead Indians for colds & sore throats; ***L. porteri*** J. Coulter & Rose (chuchupate, Mex.) – r. tea medic.; ***L. scoticum*** L. (E N Am., Eur.) – potherb like celery; ***L. sinense*** Oliv. (China) – trad. drug pl.; ***L. officinale*** (Makino) Kitag. (*Cnidium o.*) – imp. drug pl. Jap., poss. hybrid

Ligustrina Rupr. = *Syringa*

Ligustrum Tourn. ex L. (~ *Syringa*). Oleaceae (4c). c. 40 Eur. (1) & N Afr. (1), E & SE As. to Aus. (1). Privets. R: BJ Beib. 132 (1924)19. Decid. & everg. shrubs & small trees, oft. invasive, e.g. ***L. robustum*** (Roxb.) Blume (Indomal.) in SE US & Masc.; fleshy bird-disp. fr. cf. *Syringa* but prob. to be included in *S.*, ***L. sempervirens*** (Franch.) Lingels. (SW China) showing a reversal in its dehiscent 'berry'; heavy scent, offensive to many due to ammonia undertone due to trimethylamine (esp. after rain as in *Cotoneaster* & *Sorbus*, Rosaceae) giving overall fishy smell not found in *S.* or *Jasminum* spp. but tainting honey of bees feeding on *L.* nectar; extrafl. nectaries. Commonly planted as hedges, form. ***L. vulgare*** L. (common or European p., Eur., Medit., invasive Aus., SE US) since at least 1548 (fr. toxic (glycoside), though form. used as home remedy for mumps in Wilts, UK – some child fatalities recorded, form. source of dyes), but chiefly now ***L. ovalifolium*** Hassk. (Jap.), the common 'golden' p. is **'Aureum'**, all tolerant of city pollution; others incl. ***L. japonicum*** Thunb. (Jap. p., Korea, Jap.) – everg. & ***L. lucidum*** Aiton f. (glossy p., China, Korea, invasive Aus., NZ) – common street-tree in S Eur., comm. insect wax coll. in China; ***L. sinense*** Lour. (*L. indicum, L. wallichii*, Chinese p., Himal. to China & SE As., invasive in Aus., NZ, SE US) – with honey, extract a pop. cold cure in modern Chinese herbalism

Lijndenia Zoll. & Moritzi (~ *Memecylon*). Melastomataceae. 17 OW trop. (Madag. 6, Indomal. 3, Afr. 3). R: NJB 2(1982)121

lilac *Syringa* spp. esp. *S. vulgaris;* **Calif. l.** *Ceanothus* spp.; **Cape l.** (Aus.) *Melia azedarach;* **Himalayan l.** *S. emodi;* **Indian** or **Persian l.** *M. azedarach;* **Rouen l.** *S.* × *persica*

Lilaea Bonpl. (~ *Triglochin*). Juncaginaceae (Lilaeaceae). 1 Rocky Mts, Mex., Andes: ***L. scilloides*** (Poir.) Hauman (*L. subulata*) – lvs for thatch & brooms in S Am. (natur. Portugal, Spain & Aus.). R: Fl. Iberica 17(2010)52

Lilaeaceae Dumort. = Juncaginaceae

Lilaeopsis Greene. Umbelliferae (III 8). 15 N & S Am. (2 also in NZ), Aus. (2), Mauritius (1), 1 natur. Portugal. R: SBM 6(1985)

lili (flowers) *Soymida febrifuga*

Liliaceae Juss. Magnoliidae – Liliales. Incl. Calochortaceae, 15/550 N hemisph. Perennials with bulbs or rhiz.; aerial stem unbranched. Lvs oval to filiform, usu. parallel-veined. Infl. usu. raceme, s.t. umbel, thyrse or 1-flowered. fls herm., usu. reg. P 3 + 3 (0, *Scoliopus*), oft. with tube, oft. nectariferous. A 3 + 3, free, extrorse. Style 3-lobed. $\underline{G}$ (1-)3 with (2-)several-∞ anatropus bitegmic ovules. Fr. a capsule or berry. Seeds oft. flattened &

then oft. winged, or ellipsoidal/angular with aril (elaiosome) or globose; endosperm without starch

Classification & genera: Molecular work has stabilized the limits of L., with most genera form. here assigned to unrelated but morphologically superficially similar Asparagaceae & Asphodelaceae (Asparagales) besides allied fams in Liliales (Melanthiaceae, Colchicaceae etc.). Three subfams remain:

1. **Lilioideae** (capsule loculicidal; seeds oft. flattened): *Amana, Cardiocrinum, Clintonia, Erythronium, Fritillaria, Gagea, Lilium, Medeola, Notholirion, Tulipa*
2. **Calochortoideae** (capsule septicidal): *Calochortus, Tricyrtis*
3. **Streptopoideae** (seeds striate): *Prosartes, Scoliopus, Streptopus*

All cult. orn., *Tulipa* (& *Lilium*) being of great comm. importance as cut-fls etc.

Liliopsida = Magnoliidae (monocotyledons)

Lilium Tourn. ex L. Liliaceae. Incl. *Nomocharis* 122 N temp. (Eur. 10, N Am. 20) to Philippines. Lilies. R: H.B.D. Woodcock & W.T. Stearn (1950) *Lilies of the world*. Most self-incompatible, poll. insects & (N Am.) hummingbirds; bulbs scaly, s.t. stoloniferous or rhiz., many ed.; aerial stems unbranched, s.t. with r. above the bulbs ('stem-rooting l.'). Allied to *Fritillaria* & linked by *Notholirion*, much cult. with many hybrids (cut-fl. trade in US worth $77M+ by 2005), *L. candidum* in GB in C15 but only sp. known to Shakespeare, some ed. as accompaniment to meat in S China. ***L. auratum*** Lindl. (golden-rayed l. of Jap., Honshu) – strongly scented fls to 30 cm diam., bulbs eaten in Jap.; ***L. brownii*** Miellez. (Myanmar, China) – ed. crop, cough treatment in modern Chinese herbalism; ***L. bulbiferum*** L. (orange l., C & E Eur.) – only Eur. l. with upward-pointing fls, long cult., an emblem of the Orangemen in N Ireland, parent (with *L. maculatum* Thunb., Jap.) of ***L × hollandicum*** Bergmans – imp. cut-fl.; ***L. canadense*** L. (Canada l., E N Am.); ***L. candidum*** L. (Madonna l., Bourbon l., Balkans (? natur.), ? Crete, Israel & Lebanon) – overwintering basal lvs prod. autumn, no fr. in cult. though found in Gk. pops, grown since at least 1500 BC in part for white fls used in scent-making (500 kg yield 300 g pure essence, a vase of scent in Tutankhamun's 1327 BC tomb), long cult. for treatment of corns & bunions, a phallic symbol (bulb & straight scape) of fert., figured in Cretan frescos 5000 yrs old, poss. the Rose of Sharon of the Bible & more recently in mediaeval paintings Christ figure crucified on it, also oft. assoc. with the Virgin Mary Annunciation since 1311; ***L. catesbaei*** Walter (leopard l., pine l., SE US); ***L. columbianum*** Hort. ex Leichtlin (W N Am.) – bulbs imp. food for Native Americans; ***L. formosanum*** A. Wallace (Formosa lily, Taiwan) – flowering within a few months from seed, weedy Aus., NZ, E & S Afr.; ***L. lancifolium*** Thunb. (*L. tigrinum*, tiger lily, E China, Korea, Jap., natur. N Am.) – bulbs eaten in Jap.; ***L. leichtlinii*** Hook.f. var. ***maximowiczii*** (Regel) Bak. (E As.) – comm. ed. bulbs; ***L. longiflorum*** Thunb. var. ***eximium*** (Courtois) Bak. (Easter l., Bermuda l., November l. (Aus.), Jap.) – fragrant white fls; ***L. martagon*** L. (Turk's-cap l., Eur. to Mongolia, natur. GB) – most widespread sp., bulbs ed.; ***L. michauxii*** Poir. (Carolina l., SE US); ***L. philippinense*** Bak. (Philippines, natur. Florida) – tepals to 25 cm long; ***L. regale*** E. Wilson (Christmas l. (Aus.), royal l., W China); ***L. superbum*** L. (Turk's-cap l., E US)

lilli-pilli, lilly(-)pilly *Syzygium smithii*

lily *Lilium* spp. but **l. of the field** was *Anemone coronaria* or poss. *Narcissus tazetta* or *Sternbergia lutea*; **African l.** *Agapanthus* spp.; **arum l.** *Zantedeschia aethiopica*; **atamasco l.** *Zephyranthes atamasca*; **Amazon l.** *Eucharis × grandiflora*; **beetle l.** *Baeometra uniflora*; **belladonna l.** *Amaryllis belladonna*; **Berg l.** *Ornithogalum candicans*; **Bermuda l.** *Lilium longiflorum* var. *eximium*; **blood l.** *Haemanthus* spp.; **boat l.** *Tradescantia spathacea*; **Bourbon l.** *L. candidum*; **Brisbane l.** *Proiphys cunninghamii*; **Canadian l.** *L. canadense*; **Carolina l.** *L. michauxii*; **Chinese sacred l.** *N. tazetta*; **chocolate l.** *Dichopogon* spp.; **Christmas l.** (Aus.) *L. regale*; **climbing l.** *Gloriosa superba*; **cobra l.** *Arisaema* spp.; **corn l.** *Ixia* spp., *Clintonia borealis*; **Darling l.** *Crinum flaccidum*; **day l.** *Hemerocallis* spp.; **Easter l.** *L. longiflorum* var. *eximium*, see also *Zantedeschia aethiopica*; **Eucharist l.** *Eucharis × grandiflora*; **fawn l.** *Erythronium californicum*; **fire l.** *Cyrtanthus* spp.; *Hippeastrum puniceum*; **flame l.** *G. superba*; **Formosa L.** *L. formosanum*; **fringed l.** *Thysanotus* spp.; **George l.** *C. elatus*; **glory l.** *G. superba*; **golden(-rayed) l. of Japan** *L. auratum*; **Guernsey l.** *Nerine sarniensis*; **gymea l.** *Doryanthes excelsa*; **ifafa l.** *C. mackenii*; **impala l.** *Adenium multiflorum*; **Jacobean l.** *Sprekelia formosissima*; **Kaffir l.** *Hesperantha coccinea*; **Kerry l.** *Simethis mattiazzii*; **Knysna l.** *C. elatus*; **Lent l.** *Narcissus pseudonarcissus*; **leopard l.** *L. catesbaei*, *Iris domestica*; **Madonna l.** *L. candidum*; **Malta l.** *S. formosissima*; **mariposa l.** *Calochortus* spp.; **May l.** *Maianthemum bifolium*; **November l.** *L. longiflorum*; **orange l.** *L. bulbiferum*; **orchid l.** *Barclaya longifolia*; **palm l.** *Cordyline australis*; **peace l.** *Spathiphyllum* spp.; **Peruvian l.** *Alstroemeria aurea*; **pine l.** *L. catesbaei*; **plantain l.**

Hosta spp.; **Poor Knight's l.** *Xeronema callistemon*; **pyjama l.** *Crinum kirkii*; **queen l.** *Phaedranassa* spp.; **rain l.** *Zephyranthes minima*; **rock l.** *Arthropodium cirrhatum, Dendrobium speciosum*; **royal l.** *L. regale*; **spider l.** *Hymenocallis* spp.; **St Bernard's l.** *Anthericum liliago*; **St Bruno's l.** *Paradisea liliastrum*; **St John's l.** *Clivia miniata*; **Scarborough l.** *Cyrtanthus elatus*; **sego l.** *Calochortus* spp.; **Snowdon l.** *Gagea serotina*; **spider l.** *Ismene × festalis*; **star l.** *Eucharis × grandiflora*; **swamp l.** *Zephyranthes* spp.; **sword l.** *Gladiolus murielae*; **l. thorn** *Catesbaea spinosa*; **tiger l.** *L. lancifolium*; **torch l.** *Kniphofia* spp.; **trout l.** *Erythronium americanum*; **Turk's-cap l.** *L. martagon, L. superbum*; **vanilla l.** *Arthropodium* spp.; **veld l.** *Crinum* spp.; **voodoo l.** *Sauromatum venosum*; **weevil l.** *Molineria capitulata*; **wood l.** *Clintonia* spp., *Trillium grandiflorum*; **zephyr l.** *Zephyranthes atamasca*

lily of the valley *Convallaria majalis*; **l. o. t. v. tree** *Clethra arborea*

Lima bean *Phaseolus lunatus*

Limacia Lour. Menispermaceae (V). 3 Myanmar to W Mal. R: KB 1957:447. Local medic. (alks) esp. ***L. scandens*** Lour., trad. med. in Malaysia for depression; ***L. oblonga*** Hook.f. & Thomson (Thailand to Borneo) – sweet ed. fr.

Limaciopsis Engl. Menispermaceae (V). 1 trop. Afr.: ***L. loangensis*** Engl.

Limahlania Wong & Sugumaran (~ *Fagraea*). Gentianaceae. 1 SE As. to Borneo: ***L. crenulata*** (Clarke) Wong & Sugumaran – cult. orn., Fagerlind's Model. R: GBS 64(2012)491

Limatodis Blume = *Calanthe*

limay Several dipterocarp woods (trade name in Philippines)

limba *Terminalia superba*

Limbarda Adans. (~ *Inula*). Compositae (Inul.-Inul.). 1 Eur. & Medit.: ***L. crithmoides*** (L.) Dumort. (*Inula c.*, golden samphire) – lvs s.t. used as greens

lime (fr.) *Citrus × aurantiifolia*; **l. basil** *Ocimum americanum*; **l. berry** *Triphasia trifolia*; **blood l.** *C.* cv.; **Chinese l.** *T. trifolia*; **finger l.** *C. australasica*; **kaffir l.** *C. hystrix*; **Key l.** *C. × aurantiifolia*; **makrut l.** *C. hystrix*; **myrtle l.** *T. trifolia*; **Persian l.** *C. × aurantiifolia*; **sunrise l.** *C. × oliveri*; **Tahitian l.** *C. × latifolia*

lime (tree) *Tilia* spp.; **American l.** or basswood *T. americana*; **European** or **common l.** *T. × europaea*; **house l.** *Sparrmannia africana*; **Japanese l.** *T. japonica*; **silver** or **weeping l.** *T. tomentosa* 'Petiolaris'; **small-leaved l.** *T. cordata*

Limeaceae Shipunov. Magnoliidae – Caryophyllales. Excl. Macarthuriaceae, 1/20 S As., Aus., Afr. Subshrubs & herbs. Lvs in spirals; stip. 0. A connate basally. G (2–7) opp. K with axile placentation & 1–3 ovules per carpel. Fr. a schizocarp or membranous capsule; seeds s.t. arillate. n = 9

Genus: *Limeum*

Anthocynanins – app. secondarily regained

limelo *Citrus × aurantiifolia × C. × limon*

limequat *Citrus × floridana*

limetta *Citrus limetta* (? *C. × limon*)

Limeum Burm. ex L. Limeaceae (Molluginaceae). c. 20 S Afr. (most) to trop. Afr., Arabia & Pakistan. R: MBSM 2(1956)133

Limnalsine Rydb. = *Montia*

Limnanthaceae R. Br. Magnoliidae – Brassicales. 2/8 temp. N Am. R: FNA 7(2010)172. Small delicate sub-succ. annuals of moist places, usu. with myrosin cells & mustard oils. Lvs in spirals, pinnatisect to pinnate; stip. 0. Fls reg., bisex., solit. or in racemes, on long axillary pedicels, (4)5-merous (*Limnanthes*) or 3-merous (*Floerkea*); K ± distinct, valvate, C distinct, convolute in bud, A twice C in 2 whorls or as many as & alt. with C (*Floerkea* s.t.), filaments opp. K with basal nectary gland, G 2 or 3 (*Floerkea*), (4)5 (*Limnanthes*), united by gynobasic style, deeply lobed into globular segments, style cleft or with lobed stigma, each locule with 1 basal, anatropous, unitegmic ovule. Fr. separating into indehiscent 1-seeded mericarps; embryo straight with unusual fats, endosperm 0. n = 5

Genera: *Floerkea, Limnanthes*

Form. allied with Geraniaceae but many morph. differences. Embryology like Boraginaceae but chem. places them here

Limnanthes cult. orn. & poss. oilseed

Limnanthemum S. Gmelin = *Nymphoides*

Limnanthes R. Br. Limnanthaceae. 7 W N Am. esp. Calif. R: FNA 7(2010)173. Potential oilseeds with long-chain fatty acids esp. ***L. alba*** Hartweg ex Benth. with wax like jojoba, but also for cosmetics from ***L. douglasii*** R. Br. – conspic. spring fl. in the wild, cult. orn. (meadow foam, poached egg fl.) with white-tipped yellow C

Limnas Trin. (~ *Alopecurus*). Gramineae (XVI 15). 3 C As. to NE Siberia

Limniboza R. Fries = *Platostoma*

Limniris (Tauch) Reichb. = *Iris*

Limnobium Rich. Hydrocharitaceae. Incl. *Hydromystria*, 1 SE US: ***L. spongia*** (Bosc) Steud. (*H. laevigatum*, American frogbit) – floating herb with spongy layer in abaxial leaf-surface & stolons, cult. orn. (only female pl. in Eur.), r.-hairs used to demonstrate protoplasmic streaming. R: Rhodora 94(1992)124

Limnocharis Bonpl. Alismataceae (Limnocharitaceae). 2 trop. Am. ***L. flava*** (L.) Buchenau grown & natur. in SE As. & Mal., pest in Aus. – eaten like spinach, also pig-fodder

Limnocharitaceae Takht. ex Cronq. = Alismataceae

Limnocitrus Swingle (~ *Pleiospermium*). Rutaceae (II 2). 1 SE As. (tidal swamps): ***L. littoralis*** (Miq.) Swingle

Limnodea L. Dewey. Gramineae (XVI 15). 1 S US: ***L. arkansana*** (Nutt.) L. Dewey – prairie. R: FNA 24(2007)776

Limnophila R. Br. Plantaginaceae (Grat.-Grat.; Scrophulariaceae s.l.). 37 OW trop. (Aus. 5 – R: AusJB 33(1985)367). R: KB 24(1970)101. Aromatic marsh herbs or aquatics s.t. grown in aquaria e.g. ***L. aromatica*** (Lam.) Merr. (rice paddy herb, Indomal.), some locally medic. & for flavouring food. ***L. chinensis*** (Osb.) Merr. (trop. & warm As.) – lvs eaten fresh or cooked

Limnophyton Miq. Alismataceae. 3 trop. Afr., ***L. obtusifolium*** (L.) Miq., ext. to Madag., Mal.

Limnopoa C. Hubb. Gramineae (Micr.). 1 trop. As.: ***L. meeboldii*** (C. Fischer) C. Hubb. – forming mats on water surface. R: HIP 35(1943)t.3432

Limnosciadium Mathias & Constance. Umbelliferae (III 8). 2 S C US. R: AJB 28(1941)162

Limnosipanea Hook.f. Rubiaceae (I 6). 7 Panamá to trop. S Am. ***L. spruceana*** Hook.f. (Amazon) only truly aquatic Rubiacea, with habit of *Hippuris*

Limodorum Boehmer. Orchidaceae (V 1). 3 Medit. incl. Eur. to Iran. ***L. abortivum*** (L.) Sw. – leafless mycotroph, 4 lat. A s.t. fertile

Limonia L. (*Feronia*). Rutaceae (II 2). 1 Ind. to Java: ***L. acidissima*** L. (*F. elephantum*, elephant apple, wood a.) – elephants eat lvs, useful timber, gum (Ind. g.) used in glue, watercolours etc., fr. ed. esp. when strained to remove seeds (also in jam) & used as soap subs. in Jap., *Citrus* may be grafted on to it

Limoniaceae Ser. = Plumbaginaceae

Limoniastrum Fabr. (~ *Limonium*). Plumbaginaceae. (II, Limoniaceae). Excl. *Ceratolimon*, *Saharanthus*, 2 Medit. (Eur. 1)

Limoniopsis Lincz. Plumbaginaceae (II). 2 Turkey to Cauc.

Limonium Mill. (*Statice*). Plumbaginaceae (II, Limoniaceae). c. 350 cosmop. esp. maritime & arid N hemisph. (excl. *Myriolepis*, Eur. 85). Herbs or subshrubs with persistent K around capsule; pollen finely or coarsely reticulate, stigmas with rounded or prominent papillae (those with coarse p. & rounded s. or with fine p. & prominent s. self-incompatible); monomorphic (coarse p. & prominent s.) self-compatible or apomicts e.g. ***L. binervosum*** (G.E. Sm.) Salmon agg. (Atlantic Eur.); hybridization & 'microspp.' ***L. carolinianum*** (Walter) Britton (E US) – 9–44% dry wt. soluble phenolics deterrent to grazing Canada geese. Cult. orn. esp. for cut-fls & everlastings (statice, sea lavender, marsh rosemary; see also *Psylliostachys*), some esp. ***L. gmelinii*** (Willd.) Kuntze (E Eur. to Siberia) & ***L. latifolium*** (Sm.) Kuntze (E Eur.; stellate indumentum) with r. (kermek) used for tanning & others, in Urug., used for fertility control. ***L. arborescens*** (Brouss.) Kuntze (Canary Is.) – shrubby, rare in wild (where to 8 m tall) but commonly cult.; ***L. lobatum*** (L.f.) Chaz. (*L. thouinii*, Medit.) – cult. orn.; ***L. sinuatum*** (L.) Mill. (Medit.) – K blue-violet, C white & ***L. vulgare*** Mill. (W Eur., N Afr.) – K pale purple, C blue-purple – commonly seen 'everlastings'; ***L. suffruticosum*** (L.) Kuntze – used as a boron indicator in former USSR

Limosella L. Scrophulariaceae (Man.; Plantaginaceae Grat.). c. 15 cosmop. (Eur. 3, Afr. 10). R: BJ 66(1934)488. ***L. aquatica*** L. (mudwort, N temp.) – multiplies by runners

Linaceae DC. ex Perleb. Magnoliidae – Malpighiales. Excl. Ctenolophonaceae, 10/250 cosmop. Trees, lianes (s.t. with climbing hooks), shrubs & herbs, s.t. cyanogenic. Lvs in spirals (to opp. or whorled), simple, entire, usu. involute; stip. decid., oft. small, s.t. glands or 0. Fls bisex., usu. reg. & 5(4)-merous in thyrsoid infls, rarely solit.; K quincunc. imbr., distinct to basally connate, s.t. ± dissimilar, C distinct, convolute, oft. clawed, 2–5 extrastaminal nectary glands, A alt. with C or opp. (*Anisadenia*) or 10 (15), when s.t. unequal, or alt. with staminodes when 5, filament bases connate into a tube, anthers with longit. slits, $\underline{G}$ ((2)3–5(–8)), pluriloc. with axile or apical-axile placentas & s.t.

1-loc. apically (partitions not reaching apex), styles distinct or 1 deeply cleft, each locule with 2 pend., anatropous, bitegmic ovules, s.t. separated by incomplete septa from ovary wall & oft. with placental obturators. Fr. a septicidal capsule, drupe or pr of 1-seeded mericarps (*Anisadenia*); seeds with straight oily embryo & little or 0 endosperm (thick in *Indorouchera*). n = 6–11+

Classification & principal genera:

I. **Hugonioideae** (woody, usu. lianes with grapnels (modified shoots); lvs in spirals or 2-ranked; K oft. unequal; A 10 – of 2 lengths; fr. a drupe or with mericarps; OW): *Hugonia, Roucheria*

II. **Linoideae** (usu. herbs; lvs opp. or in spirals; K ± equal; A 5 + 5 staminodes; fr. a septicidal capsule): *Anisadenia, Linum*

Some timbers & locally eaten fr., cult. orn. esp. *Linum*, which provides linseed oil & flax

Linanthastrum Ewan = *Linanthus*

Linanthus Benth. Polemoniaceae (III). Incl. *Leptodactylon, Leptosiphon*, 55 W N Am. R: Aliso 19(2000)81. Shrubs to herbs incl. annuals, some cult. orn., esp. ***L. grandiflorus*** (Benth.) Greene (*Leptosiphon g.*, SW US)

linaloe or **linaloa oil** *Bursera* spp.

Linaria Mill. Plantaginaceae (Ant.-Ant.; Scrophulariaceae s.l.). 150 Euras. (Eur. 77 – Iberia 54), Medit., widely natur. temp. R: D.A. Sutton, *Rev. Antirrh.* (1988)260. Toadflax; some locally medic. esp. in treatment of haemorrhoids; cult. orn. (fls blue, purple or yellow) incl. ***L. vulgaris*** Mill. (common t., butter-&-eggs, Euras., natur. N Am.) – fls yellow & orange, '**Peloria**' being a form with reg. fls & 5 spurs due to cytosine methylation (reversible) of a single gene (cf. *Cymbalaria*).

Linariantha B.L. Burtt & R.M. Sm. Acanthaceae (III 2c). 1 Borneo: ***L. bicolor*** B.L. Burtt & R.M. Sm. R: NRBGE 26(1965)328

Linariopsis Welw. Pedaliaceae (3). 2–3 W & SW trop. Afr.

Lincoln weed *Diplotaxis tenuifolia*

Linconia L. Bruniaceae (1). 3 W Cape. R: T 60(2011)1144

Lindackeria C. Presl. Achariaceae (Flacourtiaceae s.l.). 13 trop. Afr. (R: BJ 94(1974)311), trop. Am. (6, R: FN 22(1980)11)

Lindauea Rendle = *Lepidagathis*

Lindbergella Bor = *Poa*

Lindelofia Lehm. = *Cynoglossum*

linden *Tilia* spp.

Lindenbergia Lehm. Orobanchaceae (Scrophulariaceae s.l.). 12 OW trop. R: BotJLS 110(1995)285. Some cult. orn.

Lindenbergiaceae Doweld = Orobanchaceae

Lindenia Benth. Rubiaceae (I 5). 3 C Am. (1), Fiji (1), New Caled. (1). R: JAA 57(1976)426. Rheophytes in middle of shallow watercourses; transpacific distrib.! Cult. orn. with long white fls

Lindeniopiper Trel. = *Piper*

Lindera Thunb. (~ *Litsea*). Lauraceae (II). 100 trop. & temp. As., Aus. (1), E N Am. (3, fever bush, wild allspice). Aromatic dioec. trees; 2-loc. anthers (*Litsea* 4-loc.). ***L. benzoin*** (L.) Blume (spice bush, E N Am.) – bush with lvs form. used as a tea & fr. as an allspice subs., bark a febrifuge, poss. oil source; ***L. praecox*** (Sieb. & Zucc.) Blume (*Parabenzoin p.*, Jap.) – seed-oil used as illuminant

Lindernia All. Linderniaceae (Plantaginaceae (Grat.-Lind.), Scrophulariaceae s.l.). Incl. *Bryodes, Ilysanthes, Psammetes*, excl. *Vandellia*, 30 trop. & warm. R: Willd. 43(2013)225

Linderniaceae Borsch & al. (~ Byblidaceae; Plantaginaceae (Grat.-Lind.) + Phrymaceae (Micro.); Scrophulariaceae s.l.). Magnoliidae – Lamiales. 25/260 trop. to warm. esp. Am. R: Willd. 43(2013)219. Subshrubs to annuals, oft. with square stems, some poikilohydric. Lvs opp. Infl. a raceme or axillary racemes. C 2–lipped; A 4 (5) or 2 (+ 2 staminodes); G (2), stigma sensitive. Capsule septicidal or septifragal. Seeds with ruminate endosperm

Prinicipal genera: *Craterostigma, Crepidorhopalon, Lindernia, Micranthemum, Stemodiopsis* (poss to be excl.), *Torenia*

Many monospecific genera; many (semi-)aquatics cf. Phrymaceae. Cult. orn. (esp. *Torenia*) incl. aquarium pls

Lindemiella Eb. Fisch. & al. (~ *Lindernia*). Linderniaceae. 16 trop. Afr., Madag. (usu. rocky outcrops). R: Willd. 43(2013)227

Lindheimera A. Gray & Engelm. Compositae (Helia.-Eng.). 1 SW N Am.: ***L. texana*** A. Gray & Engelm. (Texas star) – cult. orn. ann. R: Sida 15(1993)533

Lindleya Kunth. Rosaceae (1). 1 Mex.: ***L. mespiloides*** Kunth – capsule. R: JBRIT 6(2012)349

Lindmania Mez (~ *Cottendorfia*). Bromeliaceae (1). 38 Guayana Highland. R: AMBG 73(1986)690

Lindneria T. Durand & Lubbers = *Ornithogalum*

Lindsaea Dryander ex Sm. Lindsaeaceae (Dennstaedtiaceae s.l.). Incl. *Ormoloma*, c. 180 trop. to Aus. (15), New Caled. (10) & Jap. ***L. integra*** Holttum (N Mal.) – rheophyte

Lindsaeaceae C. Presl ex M. Schomb. (~ Dennstaedtiaceae). Polypodiidae – Polypodiales. Excl. Cystodiaceae, Lonchitidaceae, 7/220 trop. R: T 55(2006)714. Usu. terr. with protostelic (rarely solenostelic) rhiz. Blades 1–3-pinnate or more divided, usu. glabrous; veins usu. free. Sori (sub)marginal, opening towards margin. Spores tetrahedral, trilete, rarely bilateral, monolete. Gametophyte cordate

Genera: *Lindsaea, Neolindsaea, Odontosoria, Osmolindsaea, Sphenomeris, Tapeinidium, Xyropteris*

Lindsayella Ames & C. Schweinf. = *Sobralia*

Lindsayomyrtus B. Hyland & Steenis. Myrtaceae (II 13). 1 E Mal. & Queensland: ***L. racemoides*** (Greves) Craven. R: AusSB 3(1990)731

linen cloth made from *Linum usitatissimum*; **China** or **Canton l.** that from *Boehmeria nivea*

ling *Calluna vulgaris*; **l.berry** *Vaccinium vitis-idaea*; **l. nut** *Trapa natans*

Lingelsheimia Pax. Phyllanthaceae (Phyll.-Flueg.). 7 Afr., Madag.

Lingnania McClure = *Bambusa*

lingonberry *Vaccinium vitis-idaea*

lingue *Persea lingue*

Linguella D.L. Jones & M.A. Clem. = *Pterostylis*

Linkagrostis Romero García, Blanca & Morales Torres = *Agrostis*

Linnaea Gronov. Caprifoliaceae (Linnaeaceae). 1 circumpolar (incl. Eur.): ***L. borealis*** L. (twinflower) – trailing everg. cult. orn. (Bell's Model). Prob. incl. *Abelia, Dipelta, Kolkwitzia* (R: Phytotaxa 125(2013)25)

Linnaeaceae Backlund = Caprifoliaceae

Linnaeobreynia Hutch. = *Capparis*

Linnaeopsis Engl. = *Streptocarpus*

Linnaeosicyos Schaef. & Kocyan (~ *Trichosanthes*). Cucurbitaceae (XII). 1 Hispaniola: ***L. amara*** (L.) Schaef. & Kocyan. R: SB 33(2008)350

Linocalix Lindau = *Justicia*

Linociera Sw. ex Schreb. = *Chionanthus*

Linodendron Griseb. Thymelaeaceae (Thym.-Daph.). 3 Cuba

linoleum Floor-covering (invented 1860, Scotland) composed of wood flour or cork dust with linseed oil on a jute backing

Linospadix H. Wendl. Palmae (V 14g). 7 NE & E Aus. (5 – R: Principes 41(1997)196,211), NG (2 – R: Blumea 46(2001)243). Small cult. orn. palms with pinnate or bifid lvs incl. ***L. monostachyos*** (Mart.) H. Wendl. (NE Aus.) – stems used for walking-sticks

Linostoma Wall. ex Endl. Thymelaeaceae (Thym.-Daph.). 3 Indomal. to trop. Aus. R: JAA 42(1961)306

Linosyris Cass. = *Crinitaria*

linseed oil *Linum usitatissimum*

Lintonia Stapf = *Chloris*

Linum Tourn. ex L. Linaceae (II). c. 190 temp. & subtrop. esp. Medit. (Eur. 36). Flax. Fls red, yellow, blue ('sister' to yellow & '*Cliococca*', '*Hesperolinon*', *Radiola*, '*Sclerolinon*') or white, some heterostyled; seeds with mucilaginous testa, swelling on wetting. ***L. pratense*** (Norton) Small (S US) selfed – after C falls, K moves to force A on to stigma. Comm. flax derived from retting stems of ***L. usitatissimum*** L. (cultigen app. domesticated just once, with seeds coll. 8th millenium BC, domesticated 7th millennium in Syria & in NW Eur by 5th, Ind. by 3rd) with non-dehiscent capsules & big seeds derived from 'subsp. *bienne* (Mill.) Thell.' (*L. angustifolium, L. bienne*), W Eur., Medit.) & fibre used for textiles (as to wrap mummies, but declined due to cotton imports (700 K t per annum), biggest producer Russia), thread, carpets, twine, canvas, paper etc., the fibres with great tensile strength, that left after removal = tow form. used for ropes & sacks; seeds of other cvs yield linseed oil (known in archaeology from 8000 BC, incl embalming mummies, 3 M t per annum with main producers now Canada – 500 000 t. in 2006, then China, of 'linola oil'), a drying oil (polyunsaturated omega-3-oils, rich in alpha linolenic acid) used in food-processing (GM cvs with omega-3 long-chain polyunsaturated fatty acids – fish oils), paints (but yellowing with time, cf. *Juglans*), varnish, linoleum (q.v.), printing inks, water-proofing, soap etc., also medic. (some cvs high in seed mucilage used for digestive probs), seed-cake good stock feed; increasing fertilizer can alter permanently amounts of DNA in cells (Lamarckian evolution!). Other locally imp. fibres & medic. incl. ***L. catharticum*** L. (purging or fairy f., Eur. & Medit., natur. subcosmop.) – form. purgative & diuretic, emetic & for menstrual irregularities; ***L. lewisii*** Pursh (prairie f., W N Am.) – medic., fibre, also ed. seeds like ***L. marginale*** A. Cunn. (E Aus.) – ed. Aborigines, though suspected of cyanide poisoning of stock. Many cult. orn. esp. ***L. flavum*** L. (C & S Eur. to Russia) – fls yellow, ***L. grandiflorum*** Desf. (flowering f., N Afr.) – fls red, ***L. perenne*** L. (Euras. & N Am.) & allied ***L. narbonense*** L. (blue f., Medit.) – fls blue

Linzia Sch.-Bip. ex Walp. (~ *Vernonia*). Compositae (Vern.-Lin.) c. 7 Afr.

Lipandra Moq. (~ *Chenopodium*). Amaranthaceae (Atrip.; Chenopodiaceae). 1 Euras. (+ widely natur.): ***L. polysperma*** (L.) S. Fuentes & al. R: Willd. 42(2012)14

Liparia L. Leguminosae (III 7). Incl. *Priestleya*, 20 SW Cape fynbos. R: NJB 17(1997)16. Sunbird-, bee- & rodent-poll. (***L. parva*** Vogel ex Walp.). L. ***villosa*** L. – seeds germ. after 200 yrs

Liparis Rich. Orchidaceae (V 11b). c. 400 cosmop. (not NZ; Eur. 1: ***L. loeselii*** (L.) Rich. (fen orchid, N temp.) – regularly autogamous, the success of seed-set increased 4 times by rain through pressure on anther cap pushing pollinia on to stigma or by cohesion of water droplet). R: JLSB 22(1886)244. Terr. orchids; some local medic. esp. Chinese trad., (dingy) cult. orn., ***L. prazeri*** King & Pantl. (? *L. diphyllos* J. Graham, Ind.) poss. comm.

Liparophyllum Hook.f. (~ *Villarsia*). Menyanthaceae. 8 Tasmania & NZ. R: Novon 19(2009)407. Some with ant-disp. seeds (caruncles); ***L. gunnii*** Hook.f. – homostylous

Lipoblepharis Orchard (~ *Wollastonia*). Compositae (Helia.). 5 Indopacific. R: Nuytsia 23(2013)438. ***L. urticifolia*** (Blume) Orchard – cult. orn.

Lipocarpha R. Br. = *Cyperus* (but see WAUP 89–1(1989)6)

Lipochaeta DC. (~ *Wollastonia*). Compositae (Helia.-Ecl.). 6 Hawaii. R: Britt. 53(2001)544

Lipostoma D. Don = *Coccocypselum*

Lipotriche R. Br. (~ *Melanthera*). Compositae (Helia.). 12 Afr. R: KB 69[9528](2014)2

lippia *Phyla nodiflora*

Lippia Houst. ex L. (~ *Lantana*). Verbenaceae (Lant.). Excl. *Burroughsia*, c. 120 trop. Afr. & Am., widely natur. Prob. incl. '*Lantana*' spp. with 2 pyrenes per fr. ***L. carviodora*** Meikle (NE Afr.) – dioec. (rare in fam.). ***L. plicata*** Bak. (E Afr.) – insecticidal, eaten by healthy chimps. Some with lvs used as tea, others as 'oregano' in C & S Am. (esp. ***L. graveolens*** Kunth (S N to C Am.) & ***L. micromera*** Schauer (WI to N S Am.)). Cult. orn. incl. ***L. dulcis*** Trev. (C Am.) – hernandulcin 800 times as sweet as sucrose, known to Aztecs

Lipschitziella Kamelin = *Jurinea* (but see Willdenowia 33(2003)391)

Lipskya (Koso-Polj.) Nevski. Umbelliferae (III 4). 1 C As.: ***L. insignis*** (Lipsky) Nevski

Lipskyella Juz. = *Cousinia*

Liquidambar L. Hamamelidaceae (IV; Altingiaceae). Incl. *Altingia*, *Semiliquidambar*, 15 E Med. (1), E As. (5), Indomal. (8), SE N & C Am (1). R: Phytokeys 31(2013)30. Monoec. trees; C 0. Valuable timber & aromatic balsam (storax) used medic. & in scent; cult. orn. decid. spp. with spectacular autumn colours. ***L. excelsa*** (Noronha) Oken (*A. e.*, rasamala, SE As. to Mal.) – to 60 m, heavy wood for beams, yellow scented resin used in perfume, lvs ed. Java; ***L. formosana*** Hance (S China, Taiwan) – silkworms fed on it yield 'Marvello hair'; ***L. orientalis*** Mill. (As. Minor) – source of Levant storax, the balm (of Gilead) of the Bible; ***L. styraciflua*** L. (sweet gum, American red g., Connecticut to C Am.) – street-tree as in NSW & US, timber (satin walnut, bilsted) for cabinet-work & veneers etc., (Am.) storax medic., antiseptic, tapped for disinfectant sap, used in skin disease (coll. from cloud forests by Aztecs)

liquorice (licorice) *Glycyrrhiza glabra*; **l. plant** *Helichrysum petiolare*

Liriodendron L. Magnoliaceae (II). 2 E N Am. (1), China (1). R: D. Hunt (ed.) *Magnolias & their allies* (1998)135. Scarrone's Model; alks; lactone antifeedant affects gypsy moth. ***L. tulipifera*** L. (tulip tree or t. poplar, E N Am.) – food-bodies at bases of some tepals, cult. orn. (a tree at Wheathampstead Herts, GB planted 1581 still alive 1987), timber (yellow poplar, (American or canary) whitewood) used for cabinet-work, shingles, clapboards, riding-boot lasts etc. form. for Ind. canoes (18 m one made from single trunk for Daniel Boone) & for cigar- & bible-boxes for white settlers, root-bark used as a 'vermouth' in colonial USA; ***L. chinense*** (Hemsl.) Sarg. (Chinese t. t., C China, N Indochina) – cult. orn. differing in smaller yellow-green (rather than orangeish) fls, separated for 10–16M yrs but still interfertile (hybrids (1970) = ***L.*** × ***sinoamericanum*** Yieh ex Shang & Zhang R. Wang)

Liriope Lour. Asparagaceae (Convallariaceae). 8 Jap., China & Vietnam. Cult. orn. ground-cover esp. ***L. muscari*** (Decne) L. Bailey (Jap., China) incl. fasciated forms; some locally medic., ***L. graminifolia*** (L.) Bak. for diabetes, the candied tubers tonic & supposed aphrodisiac

Liriosma Poeppig = *Dulacia*

Liriothamnus Schltr. = *Trachyandra*

Lisaea Boiss. (~ *Turgenia*). Umbelliferae (III 3). 3 E Medit., SW As.

Lisianthius P. Browne = seq.

Lisianthus P. Browne. Gentianaceae. 30 trop. Am. R: JAA 53(1972)76.Woody & semi-woody, some cult. orn. ***L. nigrescens*** Cham. & Schldl. (C Am.) – fls dark blue-black. See also *Eustoma*

Lisowskia Szlach. = *Malaxis*

Lissanthe R. Br. (~ *Styphelia*). Ericaceae (VII 7; Epacridaceae). 9 Aus. R: AusSB 18(2005)555

Lissocarpa Benth. Ebenaceae (II; Lissocarpaceae). 8 N & W S Am. R: ANMW 105B(2004)523

Lissocarpaceae Gilg = Ebenaceae

Lissochilus R. Br. = *Eulophia*

Lissospermum Bremek. = *Strobilanthes*

Listera R. Br. = *Neottia*

Listia E. Meyer (~ *Lotononis*). Leguminosae (III 8). 7 trop. & S Afr. R: T 60(2011)171

Listrobanthes Bremek. = *Strobilanthes*

Listrostachys Reichb. f. Orchidaceae (V 16d). 1 trop. Afr.: ***L. pertusa*** (Lindl.) Reichb.f.

Litanthus Harv. = *Drimia*

Litchi Sonn. Sapindaceae. 1 trop. China to W Mal.: ***L. chinensis*** Sonn. with (R: Blumea 24(1978)398, FM I,11(1994)653) 3 recog. subspp.: ***philippinensis*** (Radlk.) Leenh. (Philippines, wild) poss. ancestor of cult. ***javensis*** Leenh. (Java) & ***chinensis*** (litchi, lychee, *Nephelium litchi*) of which 2 forms grown – 'mt.' used as a stock or grown in mts (small prickly fr. more like wild pl.) & 'water' grown best in monsoon climates (incl. Madag.), the litchi of commerce (a Chinese monograph of C11; 2 M t per annum by 2000) eaten fresh, tinned or dried ('l. nuts', fragments used in tea), the edible flesh being an aril, on Ming porcelain = wish for a son, trees living to 1000 yrs, the wood useful for constr. incl. saltwater piles & keels, hybrids with longan raised

Lithachne P. Beauv. (~ *Olyra*). Gramineae (VI 3). 4 trop. Am. R: CUSNH 39(2000)68. ***L. humilis*** Soderstrom (Honduras) – 'sleep' movements with lvs hanging vertically & not held upwards as in most bamboos

Lithobium Bong. Melastomataceae. 1 Braz.: ***L. cordatum*** Bong.

Lithocarpus Blume. Fagaceae. c. 300 Indomal. (***L. corneus*** (Lour.) Rehder (S China) – sub-pachycaul treelet with huge fr.), 1 SW US (***L. densiflorus*** (Hook. & Arn.) Rehder (*Quercus d.*) – prob. excl. as *Q.*, acorns for food incl. bread, local medic.), Miocene cupule fossils in W N Am. R: A. Camus (1952–4) *Les chênes* 3: 511. Paper pulp, some ed. fr.

Lithocaulon Bally = *Pseudolithos*

Lithococca Small ex Rydb. = *Heliotropium*

Lithodora Griseb. Boraginaceae (B.2.2). Excl. *Glandora*, 3–5 Medit. R: T 57(2008)92

Lithodraba Boelcke (~ *Xerodraba*). Cruciferae (35). 1 Arg.: ***L. mendocinensis*** (Haum.) Boelcke – dense cushion-pl. R: Darw. 9(1951)349

Lithomyrtus F. Muell. (~ *Myrtella*). Myrtaceae (II 10). 11 Aus. esp. N Terr. R: Austrobaileya 5(1999)182

Lithophila Sw. Amaranthaceae (II 2). 1 WI, 2 Galápagos

Lithophragma (Nutt.) Torrey & A. Gray. Saxifragaceae. 10 W N Am. R: FNA 8(2009)77. Some with subsuperior G; some cult. orn. like *Tellima*

Lithophytum Brandegee = *Plocosperma*

Lithops N.E. Br. Aizoaceae (V). 37 S Afr. R: Excelsa 3(1973)41.; D.T. & N. Cole (2005) *L. Living stones*. Glabrous succ. with ann. growths of 1 or more obconical bodies composed of a pr of fleshy lvs on top of a dual column; leaf apices flattened & oft. with transparent 'windows' where aqueous tissue of the body of the leaf meets the surface; the solit. fl. produced between the lvs terminates season's growth after which a new shoot develops in the axil of 1 or both lvs; new shoots withdraw moisture from old lvs which remain as withered sheaths. Pls well camouflaged in their stony surroundings, the colour variations being due to orange-red chromoplasts in the epidermal & subepidermal cells as well as betalain pigments in the latter; in those with 'windows', light penetrates to chloroplasts inside lvs; in areas of v. high light intensity, pl. has app. protective layer of calcium oxalate. Many cult. orn. incl. hybrids

Lithospermum Tourn. ex L. Boraginaceae (B.2.2). Incl. *Lasiarrhenum, Macromeria, Nomosa, Onosmodium, Psilolaemus*, excl. *Aegonychon, Buglossoides, Glandora*, c. 80 temp. exc. Aus. R: JAA suppl. 1(1991)119. Heterostyly evolved several times. Some dyes & cult. orn. incl. ***L. canescens*** (Michaux) Lehm. (yellow puccoon, N Am.) – red dye used as face-paint by Native Americans; ***L. erythrorhizon*** Siebold & Zucc. (E As.) – purple dye esp. for lip-stick; ***L. incisum*** Lehm. (N Am.) – blue dye, medic. & oral contraceptive; ***L. macromeria*** J. Cohen (*M. viridiflora*, SW N Am.) – mixed with tobacco by Hopi in rain-making ceremonies; ***L. officinale*** L. (gromwell, Euras., natur. N Am.) – lvs used as a tea; ***L. ruderale*** Douglas ex Lehm. (W N Am.) – medic. & contraceptive (inspiration for perfecting oral contraceptives). See also *Lithodora*

Lithostegia Ching. Dryopteridaceae (I). 1 E Himal. to SW China: ***L. foeniculacea*** (Hook.) Ching

Lithraea Miers ex Hook. & Arn. (~ *Rhus*). Anacardiaceae (I). 3 S Am. R: Phytol. 8(1962)329. ***L. caustica*** (Molina) Hook. & Arn. (Chile) – severe dermatitis with painful swellings lasting many days

Litogyne Harv. = *Epaltes* (but see Strelitzia 10(2000)148)

Litosanthes Blume = *Lasianthus* (but see Candollea 44(1989)209)

Litostigma Y.G. Wei & al. (~ *Petrocosmea*). Gesneriaceae (III 2f). 2 S China. R: EJB 67(2010)178

Litothamnus R. King & H. Robinson (~ *Eupatorium*). Compositae (Eup.-Gyp.). 2 Braz. R: Phytol. 81(1996)385

Litrisa Small = *Carpephorus*

Litsea Lam. Lauraceae (II). 300+ warm & trop. As. & Aus., Am. Dioec. trees & shrubs (Massart's Model) with alks; ***L. rheophytica*** Kosterm. (Sarawak) – rheophyte. Some timbers. ***L. cubeba*** (Lour.) Pers. (SE As. to Philipp.) – fr. ed. & med., silkworms reared on lvs, effective insecticide in seed-oil, used in aromatherapy; ***L. glutinosa*** (Lour.) Robinson (Himal., SE As. to Aus., invasive S Afr., Masc.) – bark & fr. medic.; ***L. monopetala*** (Roxb.) Pers. (Himal.) – imp. fodder tree; ***L. verticillata*** Hance (Vietnam, Cambodia, China) – 21 compounds active against HIV

Littaea Tagl. = *Agave*

Littledalea Hemsl. Gramineae (XIV). 4 C As. to W China (4, 3 endemic – R: FOC 22(2006)370. Close to *Bromus* but leaf-sheaths with free margins

Littonia Hook. = *Gloriosa*

Littorella P. Bergius = *Plantago*

Litwinowia Woronow (~ *Euclidium*). Cruciferae (17). 1 SW As. to China: ***L. tenuissima*** (Pall.) Pavlov

live forever *Sedum telephium*; **l. long** *S.* spp.

living stones *Lithops* spp.

Livingston(e) daisy *Cleretrum bellidiforme*; **L. potato** *Coleus esculentus*

Livistona R. Br. Palmae (III 4b). Excl. *Saribus*, c. 30 NE Afr., Arabia, Ryukyus, Indomal. to Aus. (15, 13 endemic). R: GBS 60(2009)185. Fan-palms with bisex. fls & v. varied ecol., fibrous 'bandages' around leaf-bases a natural fabric suggested as inspiration for weaving in China & Ind. (Corner). Cult. orn. & locally eaten bud (cabbage palm) esp. ***L. australis*** (R. Br.) Mart. (cabbage tree, E Aus.) – trunks used to build huts for first Eur. settlers (Sydney, 1788), cabbage eaten, now prot. in wild; ***L. mariae*** F. Muell. (*L. rigida*, N Aus.) – a pop. taken by Aborigines to outback C Aus., not remants of a Gondwana forest

lizard orchid *Himantoglossum hircinum*, (Aus.) *Burnettia cuneata*; **l. plant** *Kalanchoe delagoensis*

Llagunoa Ruíz & Pavón. Sapindaceae (III 1). 3–42 W trop. S Am.

llareta *Azorella compacta*

Llavea Lag. Pteridaceae (I). 1 Mex., Guatemala: ***L. cordifolia*** Lag. – cult. orn.

Llerasia Triana. Compositae (Ast.-Hin.). 14 Colombia, Bolivia. R: Biotropica 2(1970)39. Trees, lianes, shrubs

Llewelynia Pittier = *Henriettea*

Lloydia Salisb. ex Reichb. = *Gagea*

Loasa Adans. Loasaceae (I 1). c. 36 Mex. to S Am. esp. mts. R (Arg.): BJ 76(1955)423; Sendtnera 3(1996)219. Heterogeneous? Herbs & subshrubs usu. with stinging hairs & yellow (usu.), white or red fls facing downwards; petals boat-shaped, prot. A, nectaries large & conspic. Some cult. orn.

Loasaceae Juss. Magnoliidae – Cornales. 18/300 Am. & (*Kissenia*) Afr., Arabia, (*Plakothira*) Marquesas (!). Herbs, s.t. cli., shrubs or even small trees with coarse silicified (& oft. calcified) hairs, s.t. stinging, oft. gland-tipped. Lvs simple & oft. lobed, in spirals or opp.; stip. 0. Fls bisex., reg., solit. or usu. in thyrses; K (4)5(–8), convolute or imbr., persistent & oft. accrescent, C (4)5(–8) or 10 (when incl. 5 petaloid staminodes), distinct or lobes of a tube, induplicate-valvate in bud, A (10–)∞ centripetal (II.), centrifugal (I.) or 5 (III., IV.), distinct or with basal tube or in antepetalous bundles, s.t. the anthers almost sessile on C-tube, some oft. petaloid or nectariferous staminodes, anthers with longit. slits, $\underline{G}$ to $\overline{G}$(3–5(–7)), 1-loc. with parietal placentas, oft. ± deeply intruded (rarely pluriloc. with axile placentas or (III., IV.) pseudomonomerous with 1 pend., apical ovule), each placenta with 1–∞ anatr. to hemitropous, unitegmic ovules. Fr. a capsule, rarely dry-indehiscent (cypsela), seeds, s.t. winged, with straight or curved embryo in copious oily endosperm or endosperm ± 0. n = 7–15 +

Classification & principal genera:

I. **Loasoideae** (s.t. stinging hairs; infls thyrsoid; fls pendent or erect; A centrifugal): 1. Loaseae (fls 5(-8)-merous) – *Blumenbachia, Cajophora, Loasa, Nasa*; 2. Klaprothieae (fls 4-merous) – *Klaprothia, Plakothira*

II. **Mentzeliodeae** (s.t. stinging hairs; infls thyrsoid; fls erect; A centripetal) – *Eucnide, Mentzelia*

III. **Gronovioideae** (s.t. stinging hairs; infls thyrsoid; fls erect, 5-merous; $\overline{G}$ with 1 pendent ovule) – *Gronovia*

IV. **Petalonychoideae** (stinging hairs 0; infls term. racemes) – *Petalonyx* (only)

Unitegmic seeds suggested this affinity. Some cult. orn. herbs esp. *Cajophora, Mentzelia*

Lobanilia R.-Sm. (~ *Claoxylon*). Euphorbiaceae (Acal.-Acal.-Claox.). 8 Madag. R: KB 44(1989)334

Lobelia Plum. ex L. Campanulaceae (III 1). Incl. *Hypsela* c. 415 trop. & warm esp. Am., few temp. (Eur. 2, ***L. anceps*** L.f. circumAustral); *s. s.* c. 20 E N Am. R: Pfl. IV 267b(1943)408, (1953)775, AMBG 98(2011)42 (18 sects.; heterogeneous so to be split or all subfam. incl. here). Incl. *Pratia* (fleshy berry) as there are intermediates e.g. ***L. angulata*** Forst. f. (Am., As.). in Sulawesi showing transition to typical capsule, natur. Scotland. Herbs, shrubs & pachycaul trees (Giant L.), some aquatics (***L. dortmanna*** L., water l., W Eur., N N Am. – r. take up free CO_2 (cf. *Isoetes*)) & succ., some dioec. (e.g. ***L. irrigua*** R. Br. – islands nr. Tasmania, almost extinct but saved by cult.). Fls twisted through 180°; style pushes through the anther-tube forcing out the pollen, after which stigmas sep., exposing receptive surface free of 'home' pollen (cf. Compositae). Many alks. The pachycaul spp. (Holttum's & Tomlinson's Models) of the Afr. mts (closest allies incl. ***L. organensis*** Gardner (Braz.), like *Dendrosenecio* spp. there, adapted to diurnal fluctuation in temp. above tree-line: their massiveness acts as thermal buffer & hollow infl. axes have chemicals causing ice crystals to form at highish temperatures thus prot. rest of pl.; lvs close over bud at night & provide a haven for many invertebrates; pith eaten by gorillas which make nests in the pl.; fls bird-poll. Many cult. orn. incl. ***L. bridgesii*** Hook. & Arn. (*L. blanda*, Chile); ***L. cardinalis*** L. (cardinal fl., N Am. to N S Am.) – fls scarlet, more alks than *L. inflata*; ***L. erinus*** L. (S Afr. to Somalia) – the common bedding sp. with many cvs esp. '**Pendula**' used in hanging baskets, medic. snuff in S Afr.; ***L. giberroa*** Hemsl. (E Afr. mts) – pachycaul, gorilla food, trad. medic. (angina, poisoning), latex for nettle-stings; ***L. inflata*** L. (Ind. tobacco, pokeweed, E N Am.) – cult. for lvs used medic. esp. in chest conditions; ***L. oligophylla*** (Wedd.) Lammers (*Hypsela reniformis*, S Am.) – creeping cult. pl.; ***L. siphilitica*** L. (blue cardinal fl., C & E N Am.), hybrids with *L. cardinalis* – ***L. × speciosa*** Sweet, several cvs; ***L. tupa*** L. (Chile) – psychoactive & poss. hallucinogenic, certainly narcotic & medic. esp. in treatment of toothache but overpowering smell can cause sickness

Lobeliaceae Juss. = Campanulaceae (V)

Lobivia Britton & Rose = *Echinopsis*

loblolly bay *Gordonia lasianthus*; **l. magnolia** *Magnolia grandiflora*; **l. pine** *Pinus taeda*; **l. tree** *Cupania* spp.

Lobostemon Lehm. Boraginaceae (B.2.2). Excl. *Echiostachys*, 28 S Afr. esp. W Cape. R: Phytotaxa 37(2011). Cult. orn. incl. ***L. fruticosus*** (L.) Buek (Cape) – dressings for wounds & sores

Lobostephanus N.E. Br. = *Emicocarpus*

lobster claw *Heliconia* spp.; **l. plant** *Justicia brandegeeana*

Lobularia Desv. Cruciferae (4). 4 Cape Verde & Canary Is., Medit. (Eur. 2) to Arabia. R: OB 91(1987)5. ***L. maritima*** (L.) Desv. ((sea or sweet) alyssum, S Eur.) – commonly grown white-flowered edging pl., ants effective pollinators

Loch Ness monster See *Pinus sylvestris*

Lochia Balf.f. = *Gymnocarpos*

Lockhartia Hook. Orchidaceae (V 12h). c. 30 trop. Am. Cult. orn. (R: AOSB 43(1974)399) epiphytes with small fls

Lockhartiopsis Archila = *Lockhartia*

Lockia Averyanov = *Luisia* × *Vanda*

locoweed *Astragalus* spp. (also *Oxytropis* spp.)

locust (bean) Several legumes etc. with large mealy pods esp. spp. of *Astronium, Byrsonima, Ceratonia, Gleditsia, Hymenaea, Parkia, Robinia* etc. (l. of Bible = *Ceratonia siliqua*); **African l.** *Parkia filicoidea* etc.; **black l.** *R. pseudoacacia*; **l. bean gum** *C. siliqua*; **black l.** *Robinia pseudoacacia*; **honey l.** *Gleditsia* spp.; **WI l.** *Hymenaea courbaril*; **l. wood** *Astronium fraxinifolium, Hymenaea courbaril, Robinia pseudoacacia*

lodgepole pine *Pinus contorta* var. *latifolia*

lodh bark *Symplocos racemosa*

Lodia Mosco & Zanov. = *Rapicactus*

Lodoicea Comm. ex DC. Palmae (III 8b). 1 Seychelles: ***L. maldivica*** (J. Gmelin) Pers. (coco-de-mer, double coconut) – dioec. palm to 30m (male) & 200–350 yrs old, with v. large lvs (petiole to 8m long, lamina c. 4 × 2 m) & male infl. to 2 m long, fls to 10 cm diam., slow-growing & now restricted to reserves, its conservation first mooted 1864; fr. takes 6 yrs to develop & contains 1(–3) 2-lobed seeds in a bony endocarp, the largest seeds (50 cm long, to 18 kg+) of any pl., killed by seawater & first known washed up on other islands & believed to be from the sea (Seychelles then unknown), their suggestive shape (like female hindquarters) indicating the devil's work or, at least, aphrodisiacal qualities; an evolutionary dead-end on a small granite fragment of Gondwanaland, rarely cult. but endocarp used as bowls etc. & form. extravagant silver-mounted centrepieces etc. R: J. Dransfield & al., *Gen. Palm.* (2008)323

Loefgrenianthus Hoehne. Orchidaceae (V 13b). 1 Braz.: ***L. blanche-amesii*** (Loefgren) Hoehne

Loeflingia L. Caryophyllaceae (I 1). 7 N Am. (1), Medit. (Eur. 3)

Loerzingia Airy Shaw = *Deutzianthus* (but see KB 16(1963)365)

Loeselia L. Polemoniaceae (III). 14 Calif., ***L. glandulosa*** (Cav.) G. Don ext. to Venez. R: Phytol. 77(1994)319. Fl. ± irreg.

Loeseliastrum (Brand) Timbrook = *Langloisia*

Loesenera Harms. Leguminosae (I 2). 4 trop. W Afr.

Loeseneriella A.C. Sm. (~ *Hippocratea*). Celastraceae (III). 16 OW trop. ***L. apocynoides*** (Oliv.) J. Raynal (trop. Afr.) – for tea-baskets in Uganda

Loewia Urb. Passifloraceae (Turneraceae). 3 E & NE trop. Afr.

loganberry *Rubus loganobaccus*; **S African l.** = 'Youngberry'

Logania R. Br. Loganiaceae. Excl. *Orianthera*, c. 20 Aus. R: FA 28(1986)29

Loganiaceae R. Br. ex Mart. s.s. (see below). Magnoliidae – Gentianales. Excl. Desfontainiaceae, Gelsemiaceae, Plocospermataceae, etc. 16/400 trop. to temp. (few). R: Pfl. ed. 2, 28b1(1980, incl. Buddlejaceae [= Scrophulariaceae], Retziaceae = Stilbaceae, etc.). [S.l.:] Trees, shrubs, lianes & herbs, s.t. accum. aluminium & oft. with alks & internal phloem. Lvs opp. (rarely in spirals or ternate etc.), simple, entire to lobed; stip. present, or lines linking petiole-bases or 0, petioles s.t. with an ocrea. Fls usu. bisex., s.t. heterostylous (V), reg. or not, solit. or usu. in cymes; K 4 or 5 free or not, valvate or imbr., C (4 or 5(–16 (*Anthocleista, Potalia*)), variously shaped, the lobes valvate, imbr. or contorted in bud, A inserted on C-tube & usu. alt. with lobes, s.t. fewer (*Usteria*), anthers basifixed, versatile with longit. slits, $\underline{G}$ ((1)2(–4)) s.t. half-inf. with as many loc., the partitions s.t. incomplete apically, with term. style s.t. apically lobed & seldom branched & 2–∞ anatr. to hemitropous or amphitropous unitegmic ovules on axile placentas. Fr. a capsule, berry or drupe (*Neuburgia*); seeds 1–∞ s.t. winged, with fleshy, starchy or horny endosperm surrounding straight (slightly curved in *Gardneria*) embryo. n = 6–12, 19

Classification & genera:

I. **Spigelieae** (herbaceous, connate leaf-sheaths & stip., usu. half-inf. ovary): *Mitrasacme, Mitreola, Phyllangium, Schizacme, Spigelia*

II. **Loganieae** (Geniostomaceae; ochreae freq., fls s.t. unisexual, C imbr. or contorted): *Geniostoma, Logania* (wood anatomy like Oleaceae)

III. **Strychneae** (C fleshy, valvate): *Gardneria, Neuburgia, Spigelia, Strychnos* – some similarities with Apocynaceae & Rubiaceae

[IV. **Plocospermeae** (capsule with few large narrow hair-tufted seeds): *Plocospermum* = Plocospermataceae]

[V. **Gelsemieae** (heterostyly): *Gelsemium, Mostuea* = Gelsemiaceae]

VI. **Antonieae** (C valvate, seeds winged): *Antonia, Bonyunia, Norrisia, Usteria*

[VII. **Buddlejeae** (indumentum s.t. stellate, lvs s.t. not opp., n = 7, 19): *Androya, Buddleja, Emorya, Gomphostigma* = Scrophulariaceae; *Nuxia* = Stilbaceae]

[VIII. **Retzieae** (lvs acicular, usu. opp.): *Retzia* = Stilbaceae]

[IX. **Potalieae** (glabrous, fleshy oft. large fls, berries): *Anthocleista, Fagraea, Potalia* = Gentianaceae]

[X. **Desfontainieae** (lvs holly-like, berry): *Desfontainia* = Columelliaceae]

Although fam. 'may represent a relict from common ancestors' (Leeuwenberg *et al.*), it is now shattered as above with yet other genera in e.g. Gesneriaceae

Imp. poisons – *Strychnos* (also poss 'apple' in Garden of Eden)

Logfia Cass. (~ *Filago*). Compositae (Gnap.- Gnap.). 4 Eur., Medit. to Afghanistan, N Am. R: T 60(2011)575

logwood *Haematoxylum campechianum*

Loheria Merr. Primulaceae (Myrsinaceae). 6 Philippines (4), NG (2). R: Micronesica 24(1991)65. Pachycaul treelets

Loiseleuria Desv. = *Kalmia*

Lojaconoa Bobrov = *Trifolium*

lolagbola *Oxystigma oxyphyllum*

Loliolum Krecz. & Bobrov (~ *Vulpia*). Gramineae (XVI 10). 1 E Medit. to C As.: ***L. subulatum*** (Sol.) Eig

loliondo *Olea welwitschii*

Lolium L. Gramineae (XVI 10). Incl. *Schedonorus*, 26 temp. Euras. Infl. s.t. unbranched (ryegrass) with spikelets in 2-ranked heads. Valuable fodder & lawn-grasses; spp. interfertile & intergrade. ***L. arundinaceum*** (Schreb.) Darbysh. (*F. a., S. a.*, Eur., Medit., invasive Aus., NZ, US) – ergovaline, vasorestricting alk. (fescue foot, NZ) from infecting fungus (*Epichloe* [*Neotyphodium*] *coenophiala*) of hybrid origin; ***L. multiflorum*** Lam. (*L. italicum, L. perenne* subsp. *multiflorum*, Italian ryegrass, introd. to GB c. 1830 from C & S Eur.) – fodder; ***L. perenne*** L. (perenn. r., Eur., natur. N Am., invasive Aus., NZ) – contains loliolide a potent ant-repellent, much used for playing-fields & lawns, artificial autotetraploids more competitive than diploids, prob. the first grass deliberately sown for pastures but stock can suffer 'ryegrass staggers' from fungal infection (*Epichloe festucae* var. *lolii* [*Neotyphodium l.*]); their hybrid = ***L. × hybridum*** Hausskn. – fodder grass; ***L. pratense*** (Huds.) Darbysh. (*F. p., S. p.*, meadow fescue, Euras.) – pasture grass; ***L. temulentum*** L. (darnel, the 'tares' of the Bible, Euras.) – fr. infected with *Epichloe* fungal mycelia form. mixed with barley to make an intoxicating beer (loline alks)

lolot *Piper lolot*

Lomagramma J. Sm. Dryopteridaceae (II; Lomariopsidaceae s.l.). c. 20 Indomal. to Pac. (not Aus.); ***L. guianensis*** (Aubl.) Ching perhaps separable

Lomandra Labill. Asparagaceae (Lomandraceae). c. 50 Aus., 2 ext. to NG & 1 of them to New Caled. R: FA 46(1986)100. Dioec. 'iron grasses'; P sepaloid or inner petaloid. ***L. banksii*** (R. Br.) Lauterb. (NE Aus., NG, New Caled.) – shrubby or treelet to 3 m; ***L. longifolia*** Labill. (E Aus.) – flour from fr., leaf-bases etc.; ***L. multiflora*** (R. Br.) Britten (trop. Aus.) – yellow dye from rhiz.

Lomandraceae Lotsy = Asparagaceae

Lomanodia Raf. = *Astronidium*

Lomanthus R. Nordenstam (~ *Senecio*). Compositae (Senec.). 17 S Am. R: CN 47(2009)34

Lomaphlebia J. Sm. (~ *Grammitis*). Polypodiaceae (Grammitidaceae). 2 trop. Am.

Lomariopsidaceae Alston (~ Dryopteridaceae). Polypodiidae – Polypodiales. Incl. Nephrolepidaceae, 4/66 trop. & warm. R: T 55(2006)718. Terr. with creeping to cli. rhiz. Fronds 1-pinnate with pinnae entire to crenate; veins free. Sori discrete, round, with round to reniform indusium or 0 or sporangia acrostichoid (fronds dimorphic). n = 41 (fewer in some *Lomariopsis* spp.)

Genera: *Cyclopeltis, Lomariopsis, Nephrolepis* (poss. distinct fam.; cult. orn., fibre & veg.), *Thysanoria*

Lomariopsis Fée. Lomariopsidaceae (Dryopteridaceae s.l.). 40 trop. (Aus. 1). Thick rhizomes cli. trees to15 m; some with gametophytes liverwort-like, aquatic

Lomatia R. Br. Proteaceae (V 3). 12 Aus. (9), S Am. Like *Grevillea* but fr. with several seeds. ***L. myricoides*** (Gaertn. f.) Domin (SE Aus.) – rheophyte. Some cult. orn. incl. ***L. hirsuta*** (Lam.) Diels (Peru, Chile, Ecuador, Arg.) – timber for furniture; ***L. silaifolia*** (Sm.) R. Br. (crinkle bush, E Aus.) – prot. in wild; ***L. tasmanica*** W. Curtis (Tasmania) – seed unknown, triploid clones to 43 000 yrs old!

Lomatium Raf. Umbelliferae (III 10). 74 W N Am. R: AMBG 25(1938)225. Heterogeneous. Many with ed. r. esp. ***L. ambiguum*** (Nutt.) J. Coulter & Rose (cous r.) – eaten raw or pounded & made into cakes or biscuits; ***L. dissectum*** (Nutt.) Mathias & Constance (cough r.) – boiled r. medic.

Lomatocarpa Pim. Umbelliferae (III 8). 4 SW & C As., Afghanistan

Lomatocarum Fisch. & C. Meyer (~ *Carum*). Umbelliferae (III 8). 1 C. As.: ***L. alpinum*** (M. Bieb.) Fisch. & C. Meyer

Lomatogoniopsis T.N. Ho & S.W. Liu (~ *Lomatogonium*). Gentianaceae. 3 China. R: T.N. Ho & S.W. Liu, *Worldwide Monog. Swertia* (2015)309

Lomatogonium A. Braun. Gentianaceae. 24 N Euras. R: T.N. Ho & S.W. Liu, *Worldwide Monog. Swertia* (2015)319

Lomatophyllum Willd. (~ *Aloe*). Asphodelaceae (Aloaceae). 14 Madag., Mascarenes

Lomatozona Bak. Compositae (Eup.-Prax.). 4 Braz. R: MSBMBG 22(1987)390. Lvs opp., pappus basally connate

Lombardochloa Roseng. & Arrill. = *Briza*

Lombardy poplar *Populus nigra* 'Italica'

Lomelosia Raf. (*Pycnocomon*; ~ *Scabiosa*). Caprifoliaceae (Dipsacaceae). Incl. *Scabiosiopsis*, *Tremastelma*, 50+ Medit. to Cauc. R: Willdenowia 15(1985)72. Cult. orn. incl. ***L. caucasica*** (M. Bieb.) Greuter & Burdet (*Scabiosa c.*, Cauc.) – cult. orn. herbaceous perenn. scabious with pale blue fls; ***L. argentea*** (L.) Greuter & Burdet (*S. a.*, E Med.) – yellow dye (Turkey); ***L. cretica*** (L.) Greuter & Burdet (*S. c.*, W Medit.) – shrubby

Lonas Adans. Compositae (Anth.). 1 SW Medit. incl. Eur.: ***L. annua*** (L.) Vines & Druce – cult. orn. ann. with yellow fls

Lonchitidaceae C. Presl ex M. Schomb. (~ Lindsaeaceae). 1/2 trop. Like Lindsaeaceae but sori covered by reflexed segment-margin
Genus: *Lonchitis*

Lonchitis L. Lonchitidaceae (Lindsaeaceae; Dennstaedtiaceae s.l.). 1 trop. Am., 1 Afr. & Madag.

Lonchocarpus Kunth. Leguminosae (III 16). Incl. *Willardia*, 120 trop. Am., ***L. sericeus*** (Poir.) Kunth reaching W Afr. Many used as fish-poisons (barbasco, haiari) in S Am. (3 p.p.m. eliminates piranha but not other spp.) & sources of rotenone (2–4% by wt.) for biodegradable insecticides esp. ***L. nicou*** (Aubl.) DC. (timbo, Guyana); some folk medic. exported to Eur.; some timbers esp. *L. sericeus* (savonette); some cult. orn. small trees with lilac fls ('*Willardia*')

Lonchophora Durieu = *Matthiola*

Lonchostephus Tul. = *Mourera*

Lonchostoma Wikström = *Brunia* (but see Strel. 9(2000)385)

Loncomelos Raf. = *Ornithogalum*

Londesia Fischer & C. Meyer = *Bassia*

London plane *Platanus* × *hispanica*; **L. pride** *Saxifraga* × *urbium*

long bean *Vigna unguiculata*; **l. jack** *Flindersia xanthostyla*, *Triplaris* spp.; **l.jacks, Jersey** *Brassica oleracea* Acephala Group; **l. kong** *Lansium domesticum*; **l. purples** *Orchis mascula* (poss. *Arum maculatum*); **l. yam** *Dioscorea transversa*; **l. yan rou** *Dimocarpus longan*

longan or **longyen** *Dimocarpus longan*

Longetia Baill. (~ *Austrobuxus*). Picrodendraceae (Cal.; Euphorbiaceae s.l.). 1 New Caled.: ***L. buxoides*** Baill. R: A.R.-Sm., *Gen. Euph.* (2001)94

longoza *Aframomum angustifolium*

Lonicera L. Caprifoliaceae. 180 N hemisph. (Eur. 16) to Mex. & Philipp. (China 57). R: RMBG 14(1903)27. Honeysuckle (motif in C18 ceilings etc.). Usu. decid. shrubs & lianes, opp. lvs s.t. connate. Paired fls s.t. discrete with sep. fr., s.t. at summit of a common ovary or even more united. Spp. with short-tubed fls bee-poll., long-tubed (e.g. ***L. hildebrandiana*** Collett & Hemsl. (N Myanmar & SW China) with fls to 15(–18) cm long) v. fragrant poll. by hawkmoths: fls open in evening just after anthers have dehisced & moved into horiz. position, acting as landing-stage for moths with the style projecting beyond it;

later A wither & droop, style taking up horiz. position when fl. has turned from white to yellow. Some medic. N Am., esp. ***L. involucrata*** Banks ex Spreng. – fr. ed. bears; many cult. orn. spp. & hybrids incl. ***L. caprifolium*** L. (Euras.) – fls white or purplish, fragrant, liane locally medic. (infusion drunk for asthma), though increasingly replaced in cult. by hybrid with ***L. etrusca*** Santi (Medit.) = ***L. × italica*** Schmidt ex Tausch ('*L.* × *americana*') – decid., true ***L × americana*** (Mill.) K. Koch (*L. etrusca* × *L. implexa* Sol.) being everg.; ***L. ciliosa*** (Pursh) Poir. (N Am.) – plant boiled to make shampoo (Flathead Indians); ***L. caerulea*** L. (incl. var. *edulis*, E As.) – fr. like blueberries in look & taste, some forms (**'Blue Belle'** & **'Cinderella'** in N Am.) cult.; ***L. fragrantissima*** Lindl. & Paxton (China, not found wild) – ± decid. shrub with white fragrant fls in winter; ***L. japonica*** Thunb. (Jap. h., E As., invasive in C Eur., Aus., NZ, US [like several other spp.] incl. Hawaii) – ± everg. liane, trad. Vietnamese medic. (diuretic), many cvs incl. varieg.; ***L. ligustrina*** Wall. var. ***yunnanensis*** Franch. (*L. nitida*, China) – everg. shrub with small lvs, much used as hedging (usu. **'Ernest Wilson'**) & game-cover, rarely fertile exc. **'Fertilis'** etc.; ***L. maackii*** (Rupr.) Maxim. (E As.) – allelopathic invasive Upper Midwest US; ***L. nigra*** L. (C & S Eur. mts) – shrub with molluscicidal saponins; ***L. periclymenum*** L. (woodbine, Eur. & Medit.) – form. used to treat thrush; ***L. xylosteum*** L. (fly h., Euras., locally natur. in GB)

Loniceroides Bullock = *Marsdenia*

lontar *Borassus flabellifer*

loofah *Luffa aegyptiaca*

loosestrife, gooseneck *Lysimachia clethroides*; **purple** *Lythrum salicaria*; **yellow l.** *Lysimachia* spp. esp. *L. vulgaris*

Lopez root *Toddalia asiatica*

Lopezia Cav. Onagraceae. 22 Mex. & C Am. R: AMBG 60(1973)478. C 4, upper 2 ascending with 1 or 2 nectar-drop-like tubercles at base, A 2 with the upper fertile & enfolded by expanded petaloid one; at first insects alight on A, which later grow up out of the way leaving style in their position; in e.g. ***L. racemosa*** Cav. (*L. coronata*, cult. orn.) explosive pollen-shedding in that the fertile stamen is under tension & the arrival of insect releases it. Interstaminal nectaries in bird-poll. spp.; chem. stimulants & the pseudonectaries above in fly-poll. Ovary grows back into pedicel, poss. protection against salt-spray

Lophacme Stapf. Gramineae (XXIX 3). 2 S C Afr.

Lophanthera A. Juss. Malpighiaceae. 4 Braz. R: MNYBG 32(1981)36. ***L. lactescens*** Ducke – cult. orn. tough street-tree with panicles of 300–500 fls

Lophanthus Adans. (~ *Nepeta*). Labiatae (VII 2c). c. 20 C As. mts to China. R: BZ 77,9(1992)71

Lophatherum Brongn. Gramineae (XIX). 2 E As., Indomal. to trop. Aus. R: FOC 22(2006)445. R. tubers; false petioles; awns of sterile lemmas involved in disp. ***L. gracile*** Brongn. (E & SE As. to Polynesia) – diuretic & febrifuge (Vietnam), forage

Lophiarella Szlach. & al. = *Trichocentrum* (but see SB 38(2013)53)

Lophiocarpaceae Doweld & Reveal (Phytolaccaceae III). Incl. Corbichoniaceae, 2/5–7 OW trop. to S Afr. Subshrubs to herbs. Lvs simple. Infls spikes or cymes; P(K?) 5, A 4 (*Lophiocarpus*) or several often petaloid (*Corbichonia*); G(2), 1-loc. with 1 (*L.*) to several basal ovule per loc. Fr. a capsule (*C.*) or achenes (*L.*); seeds arillate (*C.*), n = 9

Genera: *Corbichonia*, *Lophiocarpus* (s.t. treated as separate monogeneric familes but still kept next to one another)

Lophiocarpus Turcz. Lophiocarpaceae (Phytolaccaceae III). 4 S Afr. Sieve-tube plastid type confirms moving from Amaranthaceae (Chenopodiaceae)

Lophiola Ker-Gawler. Nartheciaceae (Lophiolaceae). 2 E N Am. R: FNA 26(2002)48. Form. placed in Haemodoraceae, Melanthiaceae

Lophiolaceae Nakai = Nartheciaceae

Lophira Banks ex Gaertn. f. Ochnaceae (II). 2 trop. W Afr. R: BJ 113(1991)170. Koriba's Model. Antibacterial flavonoids (lophirones) & inhibitors of tumour promotors. Seeds yield an oil (meni oil, kaku, zawa) used for cooking, on hair & as soap, chewing-sticks; timber (ekki, ironwood, Afr. oak) esp. (rain forest, high frequency indicating abandoned cultivation) ***L. alata*** Banks ex Gaertn. f. (azobé, bongossi, eba, crown to 30 m diam.) sup. to reinforced concrete for wharves, dams & locks & used (untreated) as railway sleepers (***L. lanceolata*** Tieghem ex Keay in savanna)

Lophocarpinia Burkart. Leguminosae (I 4). 1 Paraguay, Arg.: ***L. aculeatifolia*** (Burkart) Burkart. R: PhytoKeys 71(2016)32

Lophocereus (A. Berger) Britton & Rose (~ *Pachycereus*). Cactaceae. 2 SW N Am. L: ***L. schottii*** (Engelm.) Britton & Rose (*P. s.*) – toxic to most *Drosophila* spp. living in rotten trunks but not *D. pachea*, resistant to alks & reliant on a steroid precursor of moulting hormone

Lophochlaena Nees = *Pleuropogon*

Lophochloa Reichb. = *Rostraria*

Lophocolea (Dumort.) Dumort. Lophocoleaceae (Musci-Jungermanniales). c. 80 cosmop. Leafy liverworts. ***L. bidentata*** (L.) Dumort. – common in dank town-lawns; ***L. heterophylla*** (Schrad.) Dumort. – flour beetle affected by lactone antifeedant

Lophogyne Tul. Podostemaceae (III). 1 E C Braz.: ***L. lacunosa*** (Gardner) Bove & Philbrick. R: Britt. 63(2011)157

Lopholaena DC. Compositae (Sen.-Oth.). 18 trop. & S Afr. R: TRSSA 21(1934)221, NJB 11(1991)79

Lopholepis Decne = *Perotis*

Lophomyrtus Burret. Myrtaceae (II 10). 2 NZ. Cult. orn. esp. ***L. x ralphii*** (Hook.f.) Burret (*L. bullata* Burret × *L. obcordata* (Raoul) Burret)

Lophopappus Rusby. Compositae (Mut.-Nass.). 5 Andes. R: Phytotaxa 103(2013)25. Shrubs

Lophopetalaceae auctt. = Celastraceae

Lophopetalum Wight ex Arn. Celastraceae (I; Lophopetalaceae). 18 Indomal. to trop. Aus. ***L. multinervium*** Ridl. (Mal.) – tree with consp. kneeroots in swamp forests; ***L. toxicum*** Loher (Philippines) – arrow poisons (digitaloid glycosides) from bark; other spp. with bark (oily) making good firelighters

Lophophora J. Coulter (~ *Echinocactus*). Cactaceae (III 9). 23 S Texas, N & E Mex. R: Britt. 21 (1969)299. Spineless cacti with turnip-like r. & alks, the dried flat crowns mescal buttons cont. the hallucinogenic peyote. ***L. williamsii*** (Salm-Dyck) J. Coulter – used for over 7000 yrs, 30 alks, mescaline the active one, leading to brilliantly coloured hallucinations & a feeling of weightlessness, also used for treatment of alcohol addiction. See also *Epithelantha*

Lophophytaceae Bromhead = Balanophoraceae

Lophophytum Schott & Endl. Balanophoraceae. 3 warm S Am. R: FN 23(1980)45. ***L. mirabile*** Schott & Endl. – pollen tricolporate & hexacolporate (tetrahedral)

Lophopogon Hackel. Gramineae (XXII 6). 2 Ind. R: HIP 37(1967)t.3648. Glumes 3-toothed

Lophopterys A. Juss. Malpighiaceae. 7 trop. S Am. R: CUMH 23(2001)

Lophopyxidaceae H. Pfeiffer (~ Celastraceae). Magnoliidae – Malpighiales. 1/1 Mal., W Pacific. Monoec. straggly liane with woody tendrils (axillary branchlets). Lvs small, serrate, in spirals; stips. small, knob-like. Fls 5-merous, small, sessile, in axillary, paniculate infls with basal bracts often tendrils. K valvate, basally connate; C v. small; disk yellowish, A = & opp. K, alt. with cordate glands; $\underline{G}$ ((4)5) with 2 apical anatropus, epitropous, bitegmic ovules per loc. Fr. a 5-winged samara; seed 1, with endosperm & erect embryo
Genus: *Lophopyxis*
Sister to Putranjivaceae

Lophopyxis Hook.f. Lophopyxidaceae (Celastraceae). 1 Mal., W Pac.: ***L. maingayi*** Hook. f. – applied to stings from poison fish, split stems used for tying thatch

Lophoschoenus Stapf = *Costularia*

Lophosciadium DC. = *Ferulago*

Lophosoria C. Presl. Dicksoniaceae (Lophosoriaceae). 1–3 trop. Am. R: M.F. Large & J.E. Braggins, *Tree Ferns* (2004)297. ***L. quadripinnata*** (J. Gmelin) C. Chr. – small tree-fern, the stipe-base hairs & rhiz. hairs found as linings in bird nests made of *Cyathea* in Mex.

Lophosoriaceae Pichi-Serm. = Dicksoniaceae

Lophospatha Burret = *Salacca*

Lophospermum D. Don ex R.Taylor (~ *Asarina*). Plantaginaceae (Ant.-Ant.; Scrophulariaceae s.l.). 6 Mex. to Guatemala. R: D.A. Sutton, *Rev. Antirr.* (1988)497. Some cult. orn. scramblers

Lophostachys Pohl = *Lepidostachys*

Lophostemon Schott (~ *Tristania*). Myrtaceae (II 2). 4 N & E Aus., S NG. Cult. orn. street-trees in trop. esp. ***L. confertus*** (R. Br.) Peter G. Wilson & Waterhouse (red, brush or Brisbane box, E Aus.) – to 1500 yrs old, timber, varieg. form in cult.

Lophostigma Radlk. Sapindaceae (IV 1). 2 Bolivia, Peru, Ecuador. R: SB 18(1993)379

Lophostoma (Meissn.) Meissn. Thymelaeaceae (Thym.-Daph.). 4 N trop. S Am. R: JAA 44(1963)155

Lophothecium Rizz. = *Justicia*

Lophotocarpus T. Durand = *Sagittaria*

Lopimia Mart. = *Hibiscus* (*Pavonia*)

Lopriorea Schinz. Amaranthaceae (I 2). 1 E Afr.: ***L. ruspolii*** (Lopr.) Schinz

loquat *Eriobotrya japonica*; **wild l.** (trop. Afr.) *Uapaca kirkiana*

Lorandersonia Urbatsch & al. (~ *Chrysothamnus*). Compositae (Ast.-Sol.). 7 N Am. R: FNA 20(2006)177

Loranthaceae Juss. Magnoliidae – Santalales. 77 (1 extinct)/1000 trop. & temp. (esp. S incl. Aus. but not Tasmania). R: T 53(2010)545); (Afr.): R. Polhill & D. Wiens (1998) *Mistletoes of Africa*: 77. Photosynthetic hemiparasites (& hyperparasites), typically brittle shrublets on tree-branches, less oft. terr. shrubs (*Atkinsonia*, *Gaiadendron*, *Macrosolen*), lianes or even (*Nuytsia*) trees on host r. Haustorium 1, or several at ends of epicortical r., plant rarely *Cuscuta*-like, the haustoria usu. promoting gall-like host growth. Stem oft. dichasial but nodal constrictions 0. Lvs usu. perenn. & opp. or ternate, simple, entire, rarely scale-like; stip. 0. Fls usu. bisex., usu. reg., consp. & freq. red or yellow, insect- or bird-poll., in dichasia s.t. resembling heads, racemes, umbels etc., rarely solit.; K a toothed or lobed rim or cup (calyculus) at summit of G, C (3–)5, 6(–9) oft. with a basal tube equally or unequally cleft, valvate & s.t. nectariferous basally (to 25 cm long in *Aetanthus*), A as many as & opp. & adnate to C, anthers with longit. slits, disk s.t. present, $\overline{G}$ (0 or) (3 or 4), usu. 1-loc. with c. 4–12 'ovules' comprising 8-nucleate egg-sacs (to 48 mm long in *Moquiniella* – longest known in Magn.) without obvious nucellus or integument. Fr. usu. a berry or drupe with latex (cf. Viscaceae) & 1(–3) seeds, rarely dry-indehiscent (*Nuytsia*); seed without testa but ± covered with viscous material & oft. with more than 1 embryo, at least s.t. without obvious radicle but with 2 (9–12 in some *Psittacanthus*) cotyledons eventually becoming united, endosperm copious, ?starchy (derived from primary endosperm nuclei). n = 8–12 (reducing series from 12, polyploidy v. rare)

Principal genera (poss. best arranged in 5 tribes: Nuytsieae (*Nuytsia* only), Gaiadendreae (*Atkinsonia*, *Gaiadendron* only), Elytrantheae (14), Psittacantheae (c. 17 Am, 1 NZ – *Tupeia*), Lorantheae (OW c. 40)): *Agelanthus*, *Amyema*, *Cladocolea*, *Decaisnina*, *Dendrophthoe*, *Helixanthera*, *Macrosolen*, *Phthirusa*, *Psittacanthus*, *Scurrula*, *Struthanthus*, *Tapinanthus*, *Taxillus*, *Tripodanthus* (many form. referred to *Loranthus*). App. closest to Olacaceae, but Viscaceae (= Santalaceae) form. incl.; an old S fam. (like Proteaceae & Restionaceae) with *Nuytsia* 'sister' to rest. Hyperparasitism rare (cf. Viscaceae). Some with explosive poll. Some are troublesome in plantation crops in trop., esp. spp. of *Dendrophthoe* & *Phthirusa*; some used as bird-lime, some dried for orn.

Loranthus Jacq. Loranthaceae (5 2). 10 Euras., Sumatra & Sulawesi. ***L. europaeus*** Jacq. (Eur.) – parasitic on Fagaceae (established naturally England [Kew]), decid., trioec., fls small

Lordhowea R. Nordenstam (~ *Phaneroglossa*). Compositae (Sen.-Sen.). 1 Lord Howe Is.: ***L. insularis*** (Benth.) R. Nordenstam – treelet allied to S Afr. taxa (cf. *Dietes*). R: OB 44(1978)38

lords and ladies *Arum maculatum*

Lorentzia Griseb. = *Pascalia*

Lorentzianthus R. King & H. Robinson (~ *Eupatorium*). Compositae (Eup.-Crit.). 1 Bolivia, Arg.: ***L. viscidus*** (Hook. & Arn.) R. King & H. Robinson. R: MSBMBG 22(1987)330

Lorenzia E.G. Gonç. Araceae (VII 4). 1 N Braz.: ***L. umbrosa*** E.G. Gonç. R: SB 37(2012)48

Lorenzochloa Reeder & C. Reeder (~ *Aciachne*). Gramineae (X). 9 Am.

Loretoa Standl. = *Capirona*

Loreya DC. = *Bellucia*

Loricalepis Brade. Melastomataceae. 1 Braz.: ***L. duckei*** Brade

Loricaria Wedd. Compositae (Gnap.-Lor.). 19 Andes. R: OB 104(1991)62

Lorinseria C. Presl = *Woodwardia*

Loropetalum R. Br. ex Reichb. Hamamelidaceae (I 1). 3 Himal., China. R: T 48(1999)697. ***L. chinense*** (R. Br.) Oliv. – cult. orn. (esp. purple-lvd cvs) everg. (Troll's Model), used in penjing

Lorostelma Fourn. = *Tassadia*

Lorostemon Ducke. Guttiferae (3.). 5 Braz.

lote-tree *Celtis australis*

Lotononis (DC.) Ecklon & Zeyher. Leguminosae (III 8). Incl. *Buchenroedera*, excl. *Euchlora*, *Ezoloba*, *Leobordea*, *Listia*, 91 S Afr. (R: CBH 14(1991) – most), ext. to Medit. (Eur. 2) & Ind.

lotus *Ziziphus lotus*; (Anc. Egypt) *Nymphaea lotus*, *N. nouchali* var. *caerulea*; **night l.** *N. pubescens*; **sacred l.** *Nelumbo nucifera*

Lotus Tourn. ex L. Leguminosae (III 22). Incl. *Dorycnium*, *Tetragonolobus*, excl. *Acmispon*, *Barnebyella*, *Hosackia*, *Ottleya*, *Syrmatium*, c. 125 N OW temp. (Eur. 30, Macaronesia 27). R:

BJ 25(1898)166, 31(1901)314. Piston poll. mechanism (see fam.) with style receptive only after abrasion, promoting cross-poll. Some cult. orn. & for forage. ***L. berthelotii*** Lowe ex Masf. (Canary Is.) – fls scarlet; ***L. corniculatus*** L. (bird's foot trefoil, eggs & bacon, Euras., Medit.) – fodder but S populations oft. highly cyanogenic (pack-animals of the Sudan Campaign (1896) died from eating it), invasive in Aus., US; ***L. jolyi*** Batt. (Sahara) – imp. toxic pl.; ***L. tetragonolobus*** L. (*T. purpureus*, asparagus pea, winged p., Medit.) – ed. pods, seeds a coffee subs.; ***L. uliginosus*** (L.) Schkuhr (Eur., Medit.,) – invasive in Aus.

Loudetia Hochst. ex Steud. Gramineae (XVIII). 25 trop. & S Afr., Madag., S. Am. (1). ***L. esculenta*** C. Hubb. (Sudan) – grain ed.; ***L. simplex*** (Nees) C. Hubb. (S Afr.) – minor grass for brooms

Loudetiopsis Conert = *Tristachya*

Loudonia Lindl. = *Glischrocaryon*

Louiseania Carrière = *Prunus*

Louisiella C. Hubb. & Léonard. Gramineae (XXIV 6). 1 Sudan, Congo: ***L. fluitans*** C. Hubb. & Léonard, 1 Am.: ***L. elephantipes*** (Trin.) Zuloaga. R: SB 39(2014)1113. Aquatics

louro *Ocotea* spp.; **l. inamui** *O. cymbarum*; **l. preto** *O.* spp.; **red l.** *Sextonia rubra*

Lourteigia R. King & H. Robinson (~ *Eupatorium*). Compositae (Eup.-Gyp.). 11 Colombia & Venez. R: MSBMBG 22(1987)132

Lourtella S. Graham & al. Lythraceae. 1 Peru, S Bolivia: ***L. resinosa*** S. Graham & al. R: SB 12(1987)519

louseberry *Euonymus europaeus*

lousewort *Pedicularis* spp. esp. *P. sylvatica*

Louteridium S. Watson. Acanthaceae (III 2c). 6 Mex. to C Am.

Louvelia Jum. & Perrier = *Ravenea*

lovage *Levisticum officinale*; **Scotch l.** *Ligusticum scoticum*

Lovanafia M. Pelt. = *Dicraeopetalum*

love apple *Solanum lycopersicum*; **l. (creeper)** *Comesperma volubile*; **African** or **weeping l.grass** *Eragrostis curvula*; **l.-in-idleness** *Viola tricolor*; **l.-in-a-mist** *Nigella damascena*; **l.-lies-bleeding** *Amaranthus caudatus*

lovi-lovi *Flacourtia inermis*

Lovoa Harms. Meliaceae (I). 2 trop. Afr. ***L. trichilioides*** Harms (Afr. or Nigerian golden walnut, Benin w., apopo, bibolo, embero, eyan, tigerwood, Congo wood) – much used for furniture as subs. for mahogany

Lowiaceae Ridl. Magnoliidae – Zingiberales. 1/16 Indomal. Glabrous perenn. herbs with sympodial rhizomes & vessels only in r. Lvs distich., with sheathing base, petiole & expanded simple lamina, rolled in bud, with parallel-pinnate venation though some veins not reaching tip. Fls bisex., orchid-like, bracteate, in axillary cymes, foul-smelling. lasting 1 day; P 6 petaloid, the lat. 2 inner small, anterior 1 an elliptic or spatulate labellum, the 3 outer narrow & free, A 5 (0 opp. labellum), anthers with longit. slits, $\overline{G}$ (3), ext. into slender hypanthium-like neck, 3-loc. with term. style & 3 stigmas & ∞ anatr. (?) bitegmic ovules on axile placentas. Fr. a loculicidal capsule with ∞ seeds with 3-lobed arils & starchy reserves. n = 9

Genus: *Orchidantha* (cult. orn.)

Lowianthus Becc. Older name for *Dimorphorchis*

Lowryanthus Pruski. Compositae (Athro.). 1 Madag.: ***L. rubens*** Pruski. R: Phytoneuron 2014–51: 1

Loxanthera (Blume) Blume. Loranthaceae (3). 1 W Mal.: ***L. speciosa*** Blume. R: Blumea 38(1993)114

Loxanthocereus Backeb. = *Haageocereus*

Loxocalyx Hemsl. Labiatae (VI). 3 China, Jap.

Loxocarpus R. Br. (~ *Henckelia*). Gesneriaceae (III 2i). 23 Indomal.

Loxocarya R. Br. Restionaceae. 5 SW Aus.

Loxococcus H. Wendl. & Drude. Palmae (V 14). 1 Sri Lanka: ***L. rupicola*** (Thw.) H. Wendl. & Drude

Loxodera Launert. Gramineae (XXII 3). 5 trop. Afr.

Loxodiscus Hook.f. Sapindaceae (III 1). 1 New Caled.: ***L. coriaceus*** Hook.f.

Loxogrammaceae Ching ex Pichi-Serm. = Polypodiaceae (I)

Loxogramme (Blume) C. Presl. Polypodiaceae (I; Loxogrammaceae). Incl. *Anarthropteris* 34 Indomal., Pac. (1), Afr. (4), C Am. (1). ***L. scolopendrium*** (Bory) C. Presl (Mal.) – lvs for cigarette papers (local)

Loxoma Garay = *Smithsonia*

Loxoma auctt. See *Loxsoma*

Loxomorchis Rauschert (*Loxoma*) = *Smithsonia*

Loxonia Jack. Gesneriaceae (III 1c). 3 W Mal. R: PSE 127(1977)201

Loxoptera O. Schulz = *Cremolobus*

Loxopterygium Hook.f. Anacardiaceae (I). 3 trop. S Am. (disjunct). ***L. huasango*** Spruce ex Engl. (huasango, SW Ecuador to Peru) – timber, esp. for parquet; ***L. sagotii*** Hook.f. (Venez., Guianas) – general purpose timber

Loxoscaphe T. Moore = *Asplenium*

Loxostemon Hook.f. & Thomson = *Cardamine*

Loxostigma C.B. Clarke. Gesneriaceae (III 2j). 7+ E Himal. to Vietnam. R: NRBGE 34(1975)103

Loxostylis A. Spreng. ex Reichb. Anacardiaceae (I). 1 S Afr.: ***L. alata*** A. Spreng. ex Reichb. – cult. orn. R: Strel. 10(2000)58

Loxothysanus Robinson. Compositae (Bahieae). 2 E Mex. R: Wrightia 5(1974)45

Loxsoma R. Br. ex A. Cunn. Loxsomataceae. 1 N NZ: ***L. cunninghamii*** R. Br. ex A. Cunn.

Loxsomataceae C. Presl. Polypodiidae – Cyatheales. 2/2 NZ, trop. Am. R: T 55(2006)712. Ferns with creeping hairy rhizomes with solenosteles & hairs with circular multicellular base; fronds bi- or more-pinnate, veins free, forked. Sori marginal, terminating veins, each with urceolate indusium; sporangia on thick short stalks. x = 46
Genera: *Loxsoma, Loxsomopsis*
Jurassic fossils (*Stachypteris*) referred here

Loxsomopsis Christ. Loxsomataceae. 1 trop. Am.: ***L. costaricensis*** Christ – isolated populations in Costa Rica & Colombia to Bolivia

loyak, loyar *Licuala* spp.

lozane *Tephrosia macropoda*

Lozanella Greenman. Cannabaceae (Ulmaceae II). 2 trop. Am.

Lozania S. Mutis. Lacistemataceae (Flacourtiaceae s.l.). 3 trop. Am. R: FN 22(1980)201

luan See lauan

Lubaria Pittier. Rutaceae (I 8). 1 Venez., Costa Rica: ***L. aroensis*** Pittier

lucerne *Medicago sativa*; **Paddy's l.** *Sida rhombifolia*; **tree l.** *Cytisus proliferus*

Lucilia Cass. Compositae (Gnap.-Gnap.). Excl. *Belloa*, 12 S Am. R: OB 104(1991)158

Luciliocline A. Anderb. & Freire (~ *Lucilia*). Compositae (Gnap.-Gnap.). 13–16 Andes of Peru, Bolivia, Arg. R: BJLS 106(1991)187

Luciliopsis Wedd. = *Chaetanthera*. See also *Cuatrecasasiella*

Lucinaea DC. = *Schradera*

Luckhoffia A. White & B. Sloane = *Hoodia* × *Stapelia* natural hybrid

lucky bamboo *Dracaena sanderiana*; **l. bean(s)** *Abrus precatorius, Afzelia quanzensis, Castanospermum australe, Entada gigas, Erythrina* spp.; **l. nut** *Thevetia peruviana*

Lucombe oak *Quercus* × *crenata*

lucraban *Hydnocarpus venenata*

Luculia Sweet. Rubiaceae (Cinch.). 5 Himal. & Yunnan. Cult. orn. everg. shrubs

lucuma *Pouteria lucuma*

Lucuma Molina = *Pouteria*

Lucya DC. Rubiaceae (IV 1). 1 WI: ***L. tetrandra*** (L.) K. Schum.

Ludekia Ridsd. (~ *Neonauclea*). Rubiaceae (I 2). 2 Borneo, Philippines. R: Blumea 24(1978)335

Ludia Comm. ex Juss. Salicaceae (Flacourtiaceae). 23 E Afr., Madag., Masc. R: Adansonia 12(1972)79

Ludisia A. Rich. Orchidaceae (IV 2d). 1 SE As., W Mal.: ***L. discolor*** (Ker) Lindl. (jewel orchid) – cult. orn. R: Gen. Orch. 3(2003)116

Ludovia Brongn. Cyclanthaceae. 3 trop. Am. Climbers with female fls (P 0) sunken

Ludwigia L. (*Ludvigia*). Onagraceae. 82 cosmop. (Eur. 1) esp. Am. R: Reinwardtia 6(1963) 327; AMBG 76(1989)221 & 92(2005)307; SB 17(1992)481. Herbaceous & floating to woody (Attims's Model) & erect water pls (***L. anastomosans*** (DC.) Hara (Braz.) – tree) of wet places, many prod. aerenchyma facultatively, ***L. adscendens*** (L.) Hara ('*L. repens*', trop. As.) in water prod. erect spongy r. reaching to water surface as well as ordinary r. Some cult. orn. esp. in aquaria e.g. ***L. sedoides*** (Humb. & Bonpl.) Hara; some invasive incl. ***L. grandiflora*** (Michx.) Greuter & Burdet (*L. hexapetala*, water primrose, trop. Am.) – invasive N temp., sale banned in UK since 2014; ***L. hyssopifolia*** (G. Don) Exell (OW trop. weed) – seeds buoyant but also adhere to birds etc., those nr. base of capsule shed with corky endosperm but those nr. apex free, pl. stocked by Chinese herbalists, source of black dye

Lueckelia Jenny (~ *Polycycnis*). Orchidaceae (V 12i). 1 trop. S Am.: ***L. breviloba*** (Cooper) Jenny. R: Gen. Orch. 5(2009)427

Lueddemannia Reichb.f. Orchidaceae (V 12i). 3 W trop. S Am.

Luehea Willd. Malvaceae (Grew.-Grew.; Tiliaceae). 25 trop. Am. A few cult. orn. ***L. divaricata*** Mart. (Braz.) – timber for general constr.; ***L. seemannii*** Planch. & Triana (C & N S Am.) – moth-poll.

Lueheopsis Burret. Malvaceae (Grew.-Grew.; Tiliaceae). 6–7 trop. S Am.

Lueranthos Szlach. & Marg. = *Pleurothallis*

Luerella Braas = *Masdevallia*

Luerssenia Kuhn ex Luerssen = *Tectaria*

Luerssenidendron Domin = *Acradenia*

Luetkea Bong. (*Eriogynia*). Rosaceae (Ros.-Spir.). 1 NW N Am.: ***L. pectinata*** (Pursh) Kuntze (*E. p.*) – high mt. cult. orn. rock-pl., local medic. esp. menstrual pains. R: FNA 9(2014)421

Luetzelburgia Harms. Leguminosae (III 11). 13 Braz. R: BJLS 174(2014)341. Timber etc.

Luffa Mill. Cucurbitaceae (XII). 8 trop. (5 OW, 3 Am.). R: SB 39(2014)209. Fr. & seeds medic. Am. ***L. acutangula*** (L.) Roxb. (As.) – young fr. (sing-kwa, sinqua (melon)) ed.; ***L. aegyptiaca*** Mill. ('*L. cylindrica*', loofah, veg. sponge, OW) – bleached vasc. system of mature fr. the loofah of bathrooms, also used in engine filters, linings for steel helmets, insulation and bath mats, besides as a carrier for immobilization of microbial cells in bioreactors

Lugoa DC. = *Tanacetum*

Lugonia Wedd. = *Philibertia*

Luina Benth. (~ *Tetradymia*). Compositae (Sen.-Tuss.). 2 NW N Am. R: FNA 20(2006)627

Luisia Gaudich. Orchidaceae (V 16c). c. 35 trop. As. to Jap. & Polynesia (Mal. 9, R: Reinw. 10(1988)383)

Luisma M. Murillo & A.R. Sm. Polypodiaceae (V; Grammitidaceae). 1 Colombia: ***L. bivascularis*** M. Murillo & A.R. Sm. R: Novon 13(2003)313

Lulia Zardini. Compositae (Mut.-Mut.). 1 Braz.: ***L. nervosa*** (Less.) Zardini. R: BSAB 19(1982)254

lulo *Solanum quitoense*

Luma A. Gray. Myrtaceae (II 10). 2 Peru, Chile, Arg. Cult. orn. shrubs incl. ***L. apiculata*** (DC.) Burret (arrayán, collimamol, palo colorado)

lumbang oil *Aleurites moluccana*

Lumnitzera Willd. Combretaceae (II 1). 2 E Afr. to Pac. R: JEconTaxBot 21(1997)325. Mangroves with sea-disp. fr. Timber. ***L. racemosa*** Willd. with white fls borne throughout tree (Attims's Model) – butterfly-poll.; ***L. littorea*** (Jack) J. Voigt with red somewhat zygomorphic fls borne on outside of canopy (Scarrone's Model), bird-poll.

Lunania Hook. Salicaceae (Samydaceae; Flacourtiaceae). 14 trop. Am. R: FN 22(1980)207

Lunaria Tourn. ex L. Cruciferae (9). 3 C & SE Eur. R: JAA 68(1987)190. Silicula with satiny paper-white septum. Cult. orn. esp. ***L. annua*** L. (honesty, penny flower, S Eur., natur. N Eur., N Am.) – dried infrs for winter decoration, young r. allegedly ed.

Lunasia Blanco. Rutaceae (I 1). 1 C Mal. to N Aus.: ***L. amara*** Blanco – many alks, local medic. R: JAA 48(1967)460

Lunathyrium Koidz. = *Deparia*

Lundellia Leonard = *Holographis*

Lundellianthus H. Robinson (~ *Lasianthaea*). Compositae (Helia.-Ecl.). 8 C Am. R: SB 14(1989)544

Lundia DC. Bignoniaceae (3). 13 trop. Am. R: AMBG 98(2014)447. Some with calyptrate K

Lundinia R. Nordenstam (~ *Senecio*). Compositae (Sen.-Sen.). 1 Cuba, Hispaniola: ***L. plumbea*** (Griseb.) R. Nordenstam. R: CN 44(2006)66

lungwort *Pulmonaria officinalis*; **sea l.** *Mertensia maritima*

lunumidella *Melia azedarach*

lupin(e) *Lupinus* spp.; **Egyptian l.** *L. albus*; **pearl l.** *L. mutabilis*; **Russell l.** *L.* × *regalis*; **tree l.** *L. arboreus*; **white l.** *L. albus*; **yellow l.** *L. luteus*

Lupinophyllum Hutch. = *Tephrosia*

Lupinus Tourn. ex L. Leguminosae (III 9). 220 Andes, Rocky Mts (Calif. 82), Medit. (Eur. 6), trop. Afr. highlands, E S Am. Lupins. Poll. mechanism with piston action (see fam.); many alks. Cult. orn., green manure & fodder (domesticated independently in OW (e.g. ***L. albus*** L. (S Eur.)) & Am. (***L. mutabilis***)) with palmate lvs, most commonly seen 'Russell lupins' (***L.* × *regalis*** Bergmans, introd. 1937) prob. hybrids between *L. arboreus* &

L. polyphyllus (? & other spp.), with 1- or 2-coloured fls. ***L. albus*** L. (*L. termis*, Egyptian or white l., derived from subsp. ***graecus*** (Boiss. & Spruner) Franco & Sylva, Aegean) – cattle food, form. used as flour by Egyptians, Greeks & Romans, seeds ed. boiled to remove alks or roasted & s.t. ground as coffee subs.; ***L. angustifolius*** L. (Medit.) – some strains with alks to 2.5% dry wt., some with 0 (single gene diff.), distinguishable by sheep; ***L. arboreus*** Sims (tree l., Calif., invasive in Aus., NZ, W US) – shrubby pl. used in land restoration, 180 kg N /ha/yr fixed on China clay tips in Cornwall; ***L. arcticus*** S. Watson (N Am.) – seeds from 8000 (? 13 000) BC germ. 1966); ***L. benthamii*** A.A. Heller (Calif.) – nectarless mimic of *Delphinium parryi* A. Gray; ***L. latifolius*** Lindl. ex J. Agardh (W US) – carpeted devastated forest-floors after Mt St Helens (Washington, USA) eruption (1980); ***L. luteus*** L. (yellow l., Medit.) – green manure, seeds a coffee subs.; ***L. mutabilis*** Sweet (tarwi, tarhui, pearl lupin, cultigen Andes) – high alt. pulse crop (cvs without toxic quinolizidine alks); ***L. perennis*** L. (E N Am.) – fodder pl.; ***L. polyphyllus*** Lindl. (W N Am., invasive NZ) – form. cult. orn. cvs

Luronium Raf. Alismataceae. 1 Eur.: ***L. natans*** (L.) Raf. – cult. orn. floating herb, some stages rootless. R: Fl. Iberica 17(2010)22

Luteidiscus H. St. John = *Tetramolopium*

lutqua *Baccaurea ramiflora*

Lutzia Gand. = *Alyssoides*

Luvunga Buch.-Ham. ex Wight & Arn. Rutaceae (II 2). c. 10 Indomal., 1 ext to Aus. (***L. monophylla*** (DC.) Mabb.). R: PANSP 137(1985)221. Local med.

Luxemburgia A. St-Hil. Ochnaceae (I). 18 Braz. R: BJ 113(1991)176

Luziola Juss. Gramineae (III 3ii). Incl. *Hydrochloa*, 11 Am. R: AMBG 52(1965)472. Wet places; A 6–12!

Luzonia Elmer = *Dioclea*

Luzula DC. (~ *Juncus*). Juncaceae. 108 cosmop. (Aus. 15, 12 endemic; N Am. 23) esp. temp. Euras. (Eur. 31). R: FOW 6(2002)18. Usu. hairy with flat lvs (wood rushes). Some incl. ***L. campestris*** (L.) DC. (Euras., natur. N Am.) with ant-disp. seeds with juicy outgrowths

Luzuriaga Ruíz & Pavón. Alstroemeriaceae (Luzuriagaceae, Philesiaceae s.l.). 3 Peru to Tierra del Fuego, Falklands, NZ. R: Willdenowia 17(1988)170. Some cult. orn.

Luzuriagaceae Lotsy = Alstroemeriaceae

Lyallia Hook.f. Montiaceae (Portulacaceae s.l.). Excl. *Hectorella*, 1 Kerguélen Is.: ***L. kerguelensis*** Hook.f. – Tertiary relict

Lyauteya Maire = *Cytisopsis*

Lycapsus Philippi. Compositae (Per.-Lyc.). 1 Desventuradas Is. (Chile): ***L. tenuifolius*** Philippi

× **Lycamerlycaste** J.M.H. Shaw. Orchidaceae. *Lycaste* × *Sudamerlycaste* – 121 grexes

Lycaste Lindl. Orchidaceae (V 12g). Excl. *Ida*, c. 32 trop. (esp. C) Am. R: H.F. Oakeley, *L., Ida & Anguloa* (2008)22. Cult. orn. epiphytes esp. ***L. virginalis*** (Scheidw) Linden (*L. skinneri*, Mex. to Belize) – many cvs

lychee *Litchi sinensis*

Lychniothyrsus Lindau = *Ruellia*

Lychnis Tourn. ex L. = *Silene*

Lychnodiscus Radlk. Sapindaceae. 7 trop. Afr.

Lychnophora Mart. Compositae (Vern.-Lych.). 30 Braz. Some coniferoid xerophytes

Lychnophoriopsis Schultz-Bip. (~ *Lychnophora*). Compositae (Vern.-Lych.). 4 SE Braz.

Lycianthes (Dunal) Hassler (~ *Capsicum*). Solanaceae (IV 6). 150 trop. Am., E As. (c. 20). Buzz-poll. ***L. moziniana*** (Poir.) Bitter (Mex. mts) – weedy (?) cultigen of Aztecs ('tlanochtle') with fr. rich in vitamin C; ***L. rantonnetii*** (Lesc.) Bitter (S Am.) – cult. orn. shrub

Lycium L. Solanaceae (IV 1). Incl. *Grabowskia, Phrodus*, c. 92 warm temp. (Eur. 3, Aus. 1, S Afr. 17, Am. (R: AMBG 19(1932)179). Some cryptic dioec. Shrubs, oft. thorny (boxthorn), some with ed. fr., s.t. used as hedging, some becoming noxious weeds e.g. ***L. ferocissimum*** Miers (S Afr.) in Aus., NZ, SE US; ***L. barbarum*** L. (Duke of Argyll's tea-tree, SE Eur. to China) – medic. in modern Chinese med., natur. GB esp. on walls; ***L. chinense*** Mill. (goji berry, gou qi zi, wolfberry, China) – fr. eaten with meat, promoted as healthy food in Taiwan; ***L. nodosum*** Miers (S Am.) – ash a salt-source in Paraguayan chaco; ***L. pallidum*** Miers (W N Am.) – fr. a delicacy in SW US

Lycocarpus O. Schulz = *Sisymbrium*

Lycochloa Samuelsson. Gramineae (XI). 1 Syria: ***L. avenacea*** Samuelsson – ? derived from *Schizachne*

Lycomormium Reichb.f. Orchidaceae (V 12e). 5 trop. S Am. R: Orquideologia 9(1974)186. Poll. male euglossine bees

Lycopersicon Mill. = *Solanum*

Lycopodiaceae P. Beauv. ex Mirb. Lycopodiidae – Lycopodiales. 4/300 cosmop. R: OB 92(1987)153. Everg. herbs, s.t. epiphytic. Stems dichot. branched, protostelic. Lvs scale- or needle-like with 1 vasc. strand. Sporangia in term. strobili or in axils of lvs; spores all alike. Prothalli monoec. with mycorrhiza

Genera: *Huperzia* (incl. *Phylloglossum*), *Lycopodiella*, *Lycopodium*, *Pseudolycopodiella* (clubmosses)

Lycopodiastrum Holub = *Lycopodium*

Lycopodiella Holub (~ *Lycopodium*). Lycopodiaceae. 40 moist temp. (Eur. 2) & trop. R: OB 92(1987)174. Fertile interspecific hybrids in wild. ***L. cernua*** (L.) Pichi-Serm. (trop.) – local medic., kapok subs., cult. orn.

Lycopodiidae Bek. (Lycopodiatae). Lycophytes (clubmosses & allies). 6/1200 in 3 fams (usu. put in own Orders). Sporophytes with r., microphylls usu. in spirals. Sporangia thick-walled, homo- or heterosporous, borne on a sporophyll or assoc. with one; antherozoids bi- or multiflagellate. Prob. split from rest of vascular pls at least 350 M yrs ago, maybe in the Devonian. Five Orders (2 extinct – Protolepidodendrales appearing in Devonian & Lepidodendrales typically arborescent pls. of the coal-forests of the Upper Carboniferous) with the extant Lycopodiales (Lycopodiaceae), Isoetales (Isoetaceae) & Selaginellales (Selaginellaceae)

Lycopodium Dill. ex L. Lycopodiaceae. Incl. *Diphasiastrum*, 40 trop. & temp. (Eur. 5; see fam. for segregate genera). R: OB 92(1987)170. Clubmosses. Some with alks (only pteridophytes thus). Some cult. orn., others used for stuffing upholstery, basket-, bag- & fishing-net-making; spores (l. powder) form. used in sound experiments in physics, those of ***L.*(*D.*) *alpinum*** L. (N temp. & arctic) dye wool yellow & those of ***L. clavatum*** L. (N temp.) v. flammable & form. used in fireworks & stage-lighting, also medic. (incl. jet-lag remedies), hair powder & dusting agent on some condoms (veg. sulphur) though some humans are allergic & develop granulosis resembling syphilis! ***L. (D.) complanatum*** L. (temp., trop mts) – medic., spores for rubber gloves, in Viking remains (York, England), introd. because an aluminium-accumulator (to 1% dry wt) so used as a mordant instead of alum. ***L. magellanicum*** (P. Beauv.) Sw. (S Am.) – spores on S. Georgia 2005 with 3 times conc. of paracoumaric acids shielding DNA from UV radiation compared with 1965 (ozone 'hole')

Lycopsis L. (~ *Anchusa*). Boraginaceae (B.2.1.1). 2 Eur., W As.

Lycopus Tourn. ex L. Labiatae (VII 2b). c. 14 wet, esp. N temp. (Eur. 2), Aus. (1). R: AMN 68(1962)95. ***L. europaeus*** L. (gipsywort, so named as r. source of a fast black dye gypsies believed to stain faces brown)

Lycoris Herb. Amaryllidaceae. c. 30 (some sterile triploids like ***L. flavescens*** M. Kim & S. Lee (S Korea, 2n = 19, *L. chinensis* Traub (2n = 16) × *L. sanguinea* Maxim. (2n = 22) arising several times) China & Jap. to Myanmar. R: Sida 16(1994)305. Cult. orn. bulbous herbs

Lycoseris Cass. Compositae (Mut.-Onis.). 11 trop. Am. R: NJB 11(1991)555. Dioec. shrubs

Lycurus Kunth = *Muhlenbergia*

Lydenburgia N. Robson (~ *Catha*). Celastraceae (I). 2 S Afr.

Lygeum Loefl. ex L. Gramineae (VIII). 1 Medit. incl. Eur. ***L. spartum*** L. (albardine) – fibre used for mats, wine-bottle-covers in Italy, paper (esp. N Afr.), sails & ropes etc., one of the 'esparto' grasses (cf. *Ampelodesmos*, *Stipa*). R: BM n.s. 19(2002)38

Lyginia R. Br. Restionaceae (Anarthriaceae, Lyginiaceae). 3 SW Aus. Filaments ± connate

Lyginiaceae Briggs & L. Johnson = Restionaceae

Lygisma Hook.f. Apocynaceae (Va; Asclepiadaceae III 4). 3 SE As.

Lygodesmia D. Don. Compositae (Cich.-Cich.-Micr.). 5 N Am. esp. W. R: FNA 19(2006)369. ***L. juncea*** (Pursh) Hook. (W N Am.) – imp. local medic.

Lygodiaceae M. Roem. (~ Schizaeaceae). Polypodiidae – Schizaeales. 1/25 trop. & warm. Cli. ferns with protostelic rhiz. & fronds of determinate growth in young stages & homologous ones with dichotomizing apices with indeterminate, giving rise to twining axes; veins free or anastomosing. Sori on lobes of ultimate segments; sporangia abaxial, 1 per sorus, each covered with indusium-like flange; spores 128–256 per sporangium, tetrahedral & trilete. Gametophytes cordate. n = 29, 30

Genus: *Lygodium*

Lygodisodea Ruíz & Pavón = *Paederia*

Lygodium Sw. Lygodiaceae (Schizaeaceae s.l.). 25 trop. & warm (E US 1). Twining axes used for basketry, fish-traps, mats & yarn; local medic. ***L. circinnatum*** (Burm.f.) Sw. ((paku) ata, Mal.) – exceptionally high nickel hyperaccumulator, petioles for mat-making etc. (Bali); ***L. japonicum*** (Thunb.) Sw. & ***L. microphyllum*** (Cav.) R. Br. (As. to Aus.) – invasive SE US

Lygos Adans. = *Retama*

Lymanbensonia Kimnach (~ *Lepismium*). Cactaceae (III 6). 3 S Ecuador to S Peru, Bolivia

Lymania Read (~ *Araeococcus*). Bromeliaceae (3). 9 trop. Am. R: BJLS 157(2008)48

lyme grass *Leymus arenarius*

Lyonia Nutt. Ericaceae (VIII 2). 36 E & SE As., E N Am. (5), Mex., WI. R: JAA 62(1981)129,315. Some cult. orn. shrubs

Lyonothamnus A. Gray. Rosaceae (Amygd.-Lyon.). 1 is. off S Calif.: ***L. floribundus*** A. Gray – lvs simple to divided, opp. R: FNA 9(2014)346

Lyonsia R. Br. = *Parsonsia*

Lyperanthus R. Br. Orchidaceae (IV 3f). 2 Aus., NZ. R: Gen. Orch. 2(2001)162. New Caled. spp. = *Megastylis*

Lyperia Benth. (~ *Sutera*). Scrophulariaceae (Man.). 6 S Afr. R: O.M. Hilliard (1994) *The Manuleeae*: 212

Lyrocarpa Hook. & Harv. Cruciferae (40). 3 Calif., Mex. R: CDH 3(1941)169

Lyrochilus Szlach. = *Pteroglossa*

Lyroglossa Schltr. (~ *Stenorrhynchos*). Orchidaceae (IV 2h). 2 trop. Am.

Lyrolepis Rech.f. = *Carlina*

Lysiana Tiegh. Loranthaceae (3). 6 Aus. G 4-loc. with axile placentas. ***L. casuarinae*** (Miq.) Tiegh. – fr. ed. Aborigines

Lysicarpus F. Muell. Myrtaceae (II 5). 1 Queensland: ***L. angustifolius*** (Hook.) Druce

Lysichiton Schott. Araceae. (II). 1 Kamchatka, Sakhalin & Kuriles: ***L. camtschatcensis*** (L.) Schott – disp. water & bears; 1 W N Am.: ***L. americanus*** Hultén & H. St John (skunk cabbage, invasive Scotland) – poll. *Pelecomalius testaceum* (staphylinid beetle) seeking food & mating site, locally medic. & eaten; both (and their hybrid ***L. × hortensis*** J.D. Arm. & B. Phillips) cult. orn. waterside pls, the first with white, the second yellow spathes but poss. only subspp.

Lysichlamys Compton = *Euryops*

Lysiclesia A.C. Sm. = *Orthaea*

Lysidice Hance. Leguminosae (I 2). 2 S China, Vietnam. ***L. rhodostegia*** Hance – cult. orn. shrub

Lysiloma Benth. Leguminosae (II 4). 8 Florida, Mex., WI. R: MNYBG 74,1(1996)257. Some mahogany-like timbers esp. sabicú (***L. sabicu*** Benth., WI) used in boat-building

Lysimachia Tourn. ex L. Primulaceae (Myrsinaceae). Incl. *Glaux*, c. 150 temp. (esp. Himal.; China 135, Eur. 14, excl. *Asterolinon* (= *Anagallis*), poss. incl. *Anagallis*, *Trientalis*) & warm. R: NRBGE 16(1928)51, Willd. 39(2009)50. Herbs, rarely shrubs, elaiophores attractive to bees, some locally medic., some cult. orn. ***L. clethroides*** Duby (gooseneck loosestrife, NE to SE As.) – lvs a local condiment, cult. orn. pendent infls; ***L. minorcensis*** Rodr. (Minorca) – extinct in wild; ***L. maritima*** (L.) Manns & Anderb. (*Glaux m.*, saltwort, sea milkwort, N temp. coasts) – hibernating shoot with r. prod. in axil of seedling, which dies, gives rise vegetatively to new pl. for some yrs before flowering, shoots fleshy, C0, K petaloid, pink; ***L. nummularia*** L. (creeping Jenny, moneywort, Eur., natur. E N Am.) – form. medic. & tea; ***L. vulgaris*** L. (yellow loosestrife, Euras.) – garden pl., invasive NW US

Lysinema R. Br. Ericaceae (VII 5; Epacridaceae). Excl. *Woollsia*, 5 SW Aus.

Lysionotus D. Don. Gesneriaceae (III 2j). 30 Himal. to Jap. (China 28). Epiphytic shrubs, some cult. orn.

Lysiosepalum F. Muell. Malvaceae (Bytt.-Lasio.; Sterculiaceae). 5 SW Aus. R: Nuytsia 13(2001)579

Lysiostyles Benth. Convolvulaceae (11). 1 trop. S Am.: ***L. scandens*** Benth.

Lysiphyllum (Benth.) de Wit (~ *Bauhinia*). Leguminosae (I 1). 8 Mal. to trop. Aus. (4)

Lysipomia Kunth. Campanulaceae (III 3). 30 high Andes. R: Britt. 8(1955)71. ***L. mitsyae*** Sylvester & Quandt poss. smallest eudicot (less than 5.5 mm tall) with persistent cotyledons

Lythraceae J. St Hil. Magnoliidae – Myrtales. Incl. Punicaceae, Sonneratiaceae, Trapaceae, 28/500 trop. with few temp. T 47(1998)436. Herbs (incl. ann. aquatics – *Trapa*), less oft. shrubs or trees (bark flaky, some mangroves), oft. with alks & usu. with 4-angled young stems & internal phloem. Lvs opp. (rarely whorled or in spirals), simple; stip. vestigial or 0. Fls bisex., oft. heterostylous, solit. (rare), fascicled in axils or term. racemes to thyrses, reg. or not, with consp. hypanthium s.t. spurred or with epicalyx, 4–6 (8–16)-merous; K valvate lobes of hypanthium, C free attached at summit of or within hypanthium, crumpled in bud (or 0), oft. clawed, A usu. twice K or C & in 2 whorls inserted in hypanthium (rarely fewer (1 in *Rotala*) or ∞ & centrifugal as in *Lagerstroemia*), anthers with longit. slits, $\underline{G}$ (2–4(–6)), pluriloc., but partitions s.t. not reaching apex, rarely pseudomonomerous with 1 ovule, usu. surrounded by annular nectary-disk & term. by style, 2–4(many)-loc. with axile placentation & (2–)±8 anatr. bitegmic ovules with zig-zag micropyles per loc. Fr. usu. a capsule, dehiscing variously; seeds (1–)±8 s.t. winged, ± 0 endosperm, embryo straight, oily. n = 5–11–24

Principal genera: *Ammannia, Cuphea, Diplusodon, Lagerstroemia, Lythrum, Punica, Rotala, Sonneratia, Trapa*

Distyly (e.g. *Lythrum, Pemphis*) app. derived from tristylous condition; some aquatics (e.g. *Decodon, Rotala*) with '*Hippuris* syndrome'

Punica is pomegranate, *Trapa* water chestnut. Many dye-pls esp. henna (*Lawsonia*) & *Lafoensia*, timber from *Duabanga, Lagerstroemia* & *Physocalymma* spp.; many cult. orn. esp. *Cuphea, Lagerstroemia* & *Lythrum*; some hallucinogens (*Heimia, Lagerstroemia*)

Lythrum L. Lythraceae. 36 cosmop. (Eur. 13, incl. *Peplis*). Herbs with C 6, A 12, in tristylous fls – styles of 3 lengths on diff. pls promoting outcrossing, Darwin having demonstrated that fewest viable seeds formed by selfing; the longest A have largest pollen-grains which are deposited by insects on longest styles with largest stigmatic papillae. This is condition in ***L. salicaria*** L. (purple loosestrife, OW (diploid Aus., tetraploid Eur.), invasive N Am.) – cult. orn. In some, tristyly has broken down, distyly believed to have evolved polyphyletically for some have populations with fls equivalent to the long- & short-styled forms, others to the long- & mid-styled. ***L. portula*** (L.) D. Webb (*Peplis p.*, Eur. to Cauc.) – water purslane

Lytocaryum Tol. = *Syagrus* (but see Palms 54(2010)8)

Lytoneuron (Klotzsch) Yesilyurt (~ *Doryopteris*). Pteridaceae (IV). 12 trop. & warm Am. esp. Braz. R: Phytotaxa 22(2015)116

M

ma huang, ma-huang *Ephedra* spp. esp. *E. sinica, E. vulgaris*

Maackia Rupr. Leguminosae (III 2). 8 E As. R: NRBGE 8(1913)99. Some cult. orn. decid. trees

Maasia Mols & al. (~ *Polyalthia*). Annonaceae (IV 3). 6 trop. As. R: SB 33(2008)493

Maba Forst. & Forst. f. = *Diospyros*

Mabea Aubl. Euphorbiaceae (Euph.-Hipp.). c. 40 trop. Am. Nozeran's Model. Water-disp. seeds nearly destroyed by fish. ***M. occidentalis*** Benth. app. poll. by red woolly opossum & bats (only E. thus)

mabee or **mabi bark** *Colubrina elliptica*

mabola *Diospyros blancoi*

Mabrya Elisens. Plantaginaceae (Ant.-Ant.; Scrophulariaceae s.l.). 5 SW N Am. R: D.A. Sutton, *Rev. Antirrhineae*(1988)490

Maburea Maas. Olacaceae. 1 Guyana: ***M. trinervis*** Maas. R: BJ 114(1992)276

maca *Lepidium meyenii*

macachi *Arjona tuberosa*

Macadamia F. Muell. (~ *Brabejum*). Proteaceae (V 4). 9 E Mal., Aus. (most; New Caled. spp. = *Virotia*). Seeds ed. (macadamia nuts, Queensland nuts, bopplenuts (Aus.), c. 100 K t per annum 'in shell') esp. ***M. integrifolia*** Maiden & Betche (Queensland) – principal sp. grown, & ***M. tetraphylla*** L. Johnson (Queensland, NSW) – fr. dehiscing on tree, both confused with ***M. ternifolia*** F. Muell. (rarely cult.) & grown with one another, also their hybrid, much cult. Calif., Hawaii (4th most imp. agric. export) & Malawi. Sold either in endocarp & then cracked like almonds, or shelled, roasted & salted, poss. the most delicious of all nuts (20 cals each) but expensive; seeds rich in vitamin B, 9% protein, 10% carbohydrate, 70–80% oil (mono-unsaturated; highest in any comm. food oils) used in cosmetics & poss. lowering blood pressure. ***M. hildebrandtii*** Steenis (Sulawesi) – planted in Sumatra as fire-lane tree in *Pinus* plantations

Macairea DC. Melastomataceae. 22 trop. S Am. R: MNYBG 50(1989)54

Macaranga Thouars (~ *Mallotus*). Euphorbiaceae (Acal.-Acal.-Rottl.). c. 260 OW trop. (Afr. 26, Madag. 10). R: T.C. Whitmore (2008) *The genus M.* Dioec. (e.g. thrips-poll. ***M. hullettii*** King ex Hook.f. (Malay Peninsula)) or rarely monoec. trees (Aubréville's etc. Models), climbers & shrubs, many typical of sec. forest & some with extrafl. nectaries & ant-inhabited twigs (ants eaten by squirrels) – myrmecophytism with mult. origins, long-spiny in W Afr.: seedlings of ***M. triloba*** (Thunb.) Muell. Arg. (W Mal.) 7–10 cm tall already with hollows in stipules & stems; some with lvs to 60 cm diam. e.g. ***M. mappa*** (L.) Muell. Arg. (*M. grandifolia*, C Mal., cult. orn.); some locally medic. ***M. peltata*** (Roxb.) Muell. Arg. (kenda, S India, Sri Lanka to Thailand) – lvs used for steaming jaggery etc.

Macarenia P. Royen. Podostemaceae (III). 1 Colombia: ***M. clavigera*** P. Royen. R: MedBotUtrecht 107(1951)137

Macarisia Thouars. Rhizophoraceae (I). 2–7 Madagascar

macaroni *Triticum turgidum* Durum Group

Macarthuria Huegel ex Endl. Macarthuriaceae (~ Limeaceae, Molluginaceae). 10 SW (esp.) & SE Aus. Rush-like, shrubby; seeds ant-disp.

Macarthuriaceae Christenh. 1/10 S Aus. R: Phytotaxa 181(2014)238. Rush-like shrubs. Lvs in spirals, some reduced to scales; stip. 0. P (?K) 5, s.t. + (?)C attached to base of A; A 8 basally connate; G (3(–7)), 1 (3)-loc., each loc. with 1 ovule. Fr. a loculicidal caps.; seeds arillate

Genus: *Macarthuria*

Macartney rose *Rosa bracteata*

macary bitter *Picramnia antidesma*

macassà *Aeollanthus suaveolens*

Macassar ebony *Diospyros celebica*; **M. oil** *Schleichera oleosa*

macaw bush *Solanum mammosum*; **m. palm** *Aiphanes erosa*

Macbridea Elliott. Labiatae (VI). 2 SE US. Antedated by *M.* Raf.

Macbrideina Standl. Rubiaceae (Cond.). 1 Peru, Ecuador: ***M. peruviana*** Standl.

Maccraithea M.A. Clem. & D.L. Jones = *Dendrobium*

Macdougalia Heller = *Hymenoxys*

mace *Myristica fragrans*

Macfadyena DC. = *Dolichandra*

Macgregoria F. Muell. Celastraceae (II; Stackhousiaceae). 1 E Aus.: ***M. racemigera*** F. Muell.

Machaeranthera Nees (~ *Aster*). Compositae (Ast.-Mac.). Excl. *Psilactis*, *Xylorhiza*, 30 W N Am. R: SB 28(2003)180 (s.s. 2 – R: Sida 20(2003)1390). Some cult. orn. herbs incl. ***M. gymnocephala*** (DC.) W. Thompson

Machaerina Vahl. Cyperaceae (II 7). c. 50 trop. & warm (not Afr.) esp. Aus. Some fibres, esp. ***M. gunnii*** (Hook.f.) Kern (New Guinea, Aus.) – stems for women's skirts (New Guinea)

Machaerium Pers. Leguminosae (III 11). 130 trop. Am. (Braz. c. 80), 1 (***M. lunatum*** (L.f.) Ducke) extending to W Afr. coast. Many lianes with recurved stipular thorns; some timbers (palisander) & locally medic. ***M. isadelphum*** (E. Meyer) Standl. (*M. angustifolium*, Brazil) – ash a salt-source

Machaerocarpus Small = *Damasonium*

Machaerocereus Britton & Rose = *Stenocereus*

Machairophyllum Schwantes. Aizoaceae (V). 4 Cape. R: Bothalia 33(2003)33. Cult. orn. cushion-formers to 120 cm diam.

Machaonia Bonpl. Rubiaceae (III 3). 30 trop. Am.

mache *Valerianella locusta*

Machilus Rumph. (~ *Persea*). Lauraceae (I). 90 As.

Mackay bean *Entada gigas*

Mackaya Harv. Acanthaceae (III 2c). 2 As., 1 S Afr.: ***M. bella*** Harv. – cult. orn. shrub

Mackeea H. Moore = *Kentiopsis*

Mackenziea Nees = *Strobilanthes*

Mackinlaya F. Muell. Umbelliferae (Mack.; Araliaceae). 5 E Mal., W Pacific. R: BBMNHB 1(1951)3. 3-pinnate lvs, 2-celled ovary; alks

Mackinlayaceae Doweld = Umbelliferae-Mack.

Maclaudia Venter & Verhoeven. Apocynaceae (III). 1 Guinea: ***M. felixii*** Venter & Verhoeven. R: BotJLS 115(1994)58

Macleania Hook. Ericaceae (VIII 5). 37 trop. Am. R: Biollania esp. ed. 6(1997)455. Some cult. orn. & ed. fr. sold in local markets

Macleaya R. Br. (~ *Bocconia*). Papaveraceae (I). 2 temp. E As. Alks. Cult. orn. herbs esp. ***M. cordata*** (Willd.) R. Br. (plume poppy, China & Jap.) to 2.5 m with small fls (C 0) & ***M. × kewensis*** Turrill, its hybrid with ***M. microcarpa*** (Maxim.) Fedde (C China)

Macledium Cass. = *Dicoma* (but see T 50(2001)740)

Maclura Nutt. Moraceae (I). Incl. *Cudrania, Plecospermum*, excl. *Chlorophora* p.p., 11 Indomal. to Aus., Afr. Am. R: PKNAWC 89(1986)243. ***M. pomifera*** (Raf.) C. Schneider (Osage orange, bow wood, Arkansas, Oklahoma & Texas, natur. NW, C & E US) – apomictic dioec. spiny tree with male fls in racemes & females in heads, fr. a yellow syncarp of achenes enclosed in fleshy P on a fleshy receptacle (poss. adapted to extinct megafauna), timber used for gym-equipment, bows, war-clubs, railway sleepers, street-surfaces, fence-posts, widely planted in U.S. as living fence in C19 with wood extract used in pesticide synthesis & dye (e.g. khaki in WW I, leather), roots source of yellow dye, lvs used to feed silkworms, fr. insect-repellent; ***M. tinctoria*** (L.) Steud. (*Chlorophora t.*, fustic, trop. Am.) – dioec., wind-poll., heartwood yields yellow, brown or green dye (fustic), bark medic.; ***M. tricuspidata*** Carr. (*Cudrania t.*, China, Korea) – fr. ed., cult. as hedgepl. in U.S.

Maclurochloa Wong. Gramineae (V 4). 1 Malay Penin.: ***M. montana*** (Ridl.) Wong. R: KB 48(1993)528

Maclurodendron Hartley. Rutaceae (I 2). 6 SE As. to Mal. R: GBS 35(1982)1

Maclurolyra C. Calderón & Soderstrom (~ *Olyra*). Gramineae (VI 3). 1 Panamá: ***M. tecta*** C. Calderón & Soderstrom. R: SCB 11(1973)6

Macnabia Benth. ex Endl. = *Erica*

Macodes (Blume) Lindl. Orchidaceae (IV 2d). 10 Vietnam, Jap. & Mal. to Vanuatu

Macoubea Aubl. Apocynaceae (I d). 3 trop. Am. R: MMNHN 30(1985)170, Novon 9(1999)88. Scarrone's Model. Some ed. fr.

Macowania Oliv. Compositae (Gnap.). 12 S Afr. (10), Ethiopia & Yemen (2). R: NRBGE 34(1975)260

Macphersonia Blume. Sapindaceae. 8 trop. E Afr., Madag. R: MMNHNP 19(1969)117

macqui *Aristotelia chilensis*

Macrachaenium Hook.f. Compositae (Mut.-Nass.). 1 S Am. Andes: ***M. gracile*** Hook.f. – herb in *Nothofagus* forest

Macradenia R. Br. Orchidaceae (V 12h). 11 Florida & C Am. to trop. S Am. Cult. orn. epiphytes

Macraea Hook.f. = *Trigonopterum*

Macranthera Nutt. ex. Benth. Orobanchaceae (Gerard.; Scrophulariaceae s.l.). 1 SE US: ***M. flammea*** (Bartram) Pennell. R: PANSP 80(1928)436

Macranthisiphon Bureau ex K. Schum. = *Bignonia*

Macroberlinia (Harms) Hauman = *Berlinia*

Macrobia (Webb & Berth.) Kunkel = *Aichryson*

Macrobriza (Tzvelev) Tzvelev = *Briza*

macrocarpa *Cupressus macrocarpa*

Macrocarpaea (Griseb.) Gilg. Gentianaceae. 106 trop. Am. Shrubs & trees, s.t. pachycaul, to 5m, same fls (Grant) day- (hummingbird-, bee- & butterfly-poll.) & night-flowering (moth- & bat-poll.)

Macrocaulon N.E. Br. = *Carpanthea*

Macrocentrum Hook.f. Melastomataceae. 15 N trop. S Am.

Macrochaetium Steud. = *Cyathocoma*

Macrochlaena Hand-Mazz. = *Nothosmyrnium*

Macrochloa Kunth (~ *Stipa*). Gramineae (X). 1–2 Medit. Midveins with multiple vasc. bundles (unique in 16.). ***M. tenacissima*** (L.) Kunth (*S.t.*, esparto, Algerian grass, alfa, halfa, W Medit.) – stems used in paper-making, ropes, sails, mats etc.

Macrochordion de Vriese = *Aechmea*

Macroclinidium Maxim. (~ *Pertya*). Compositae (Pert.). 3 Jap. R: SB 37(2012)560. Herbs; hybrids, incl with *P.* spp.

Macroclinium Barb. Rodr. (~ *Notylia*). Orchidaceae (V 12h). c. 40 trop. Am.

Macrocnemum P. Browne. Rubiaceae (Cond.). c. 6 C Am. to Colombia. Aubréville's Model

Macrococculus Becc. Menispermaceae (I). 1 New Guinea: ***M. pomiferus*** Becc. R: KB 26(1972)418

Macrodiervilla Nakai = *Weigela*

Macroditassa Malme = *Peplonia*

Macroglena (C. Presl) Copel. = *Abrodictyum*

Macroglossum Copel. = *Angiopteris*

Macrohasseltia L.O. Williams. Salicaceae (Flacourtiaceae). 1 C Am.: ***M. macroterantha*** (Standl. & L.O. Williams) L.O. Williams

Macrolenes Naud. ex Miq. = *Dissochaeta*

Macrolepis A. Rich. = *Bulbophyllum*

Macrolobium Schreb. Leguminosae (I 2). c. 70 trop. Am. Extrafl. nectaries. Timber, fish-poisons

Macromeles Koidz. (~ *Malus*). Rosaceae (16). 3–4 E & SE As.

Macromeria D. Don = *Lithospermum* (but see Phytol. 77(1994)394)

Macropanax Miq. Araliaceae. 17 Himal. & China to W Mal. R: JNTCFP 1985, 1:29

Macropelma K. Schum. = *Sacleuxia*

Macropeplus Perkins. Monimiaceae (V 2). 4 E Brazil. R: Rodr. 81(2001)85

Macropetalum Burchell ex Decne. = *Brachystelma*

Macropharynx Rusby. Apocynaceae (II e). 5 trop. S Am. R: Rhodora 99(1997)253

Macropidia J.L. Drumm. ex Harv. (~ *Anigozanthos*). Haemodoraceae (II). 1 N Aus.: ***M. fuliginosa*** (Hook.) Druce

Macropiper Miq. = *Piper*

Macroplectrum Pfitzer = *Angraecum*

Macropodanthus L.O. Williams. Orchidaceae (V 16c). 8 Mal.

Macropodiella Engl. Podostemaceae (III). 6 W trop. Afr. R: Adansonia 17(1977)293

Macropodina R. King & H. Robinson (~ *Eupatorium*). Compositae (Eup.-Gyp.). 3 S Brazil, Arg. R: MSBMBG 22(1987)28

Macropodium R. Br. Cruciferae (45). 2 C As. & Sakhalin, Jap.

Macropsychanthus Harms ex K. Schum. & Lauterb. = *Dioclea* (but see Blumea 43(1998) 235)

Macropteranthes F. Muell. Combretaceae (II 1). 5 N Aus. R: Austrobaileya 4(1994)152. ***M. kekwickii*** F. Muell. (bulwaddy) forms dense thickets in N Aus.

Macroptilium (Benth.) Urb. Leguminosae (III 18). c. 20 trop. Am. R: Boissiera 28(1978)151. ***M. lathyroides*** (L.) Urb. (phasey bean, WI) – trop. cult. pioneer legume esp. '**Murray**'

Macrorhamnus Baill. = *Colubrina*

Macrorungia C.B. Clarke = *Anisotes*

Macrosamanea Britton & Rose (~ *Albizia*). Leguminosae (II 4). 11 trop. S Am. R: MNYBG 74,1(1996)182

Macroscepis Kunth. Apocynaceae (Vc; Asclepiadaceae III 3). 7 trop. Am.

Macrosciadium Tikh. & Lavrova = *Ligusticum*

Macroselinum Schur (~ *Peucedanum*). Umbelliferae (III 10). 1 S Eur., Cauc.: ***M. latifolium*** (DC.) Schur

Macrosiphonia Muell. Arg. = *Mandevilla*

Macrosolen (Blume) Reichb. (~ *Elytranthe*). Loranthaceae (3). 25 S & SE As. to New Guinea esp. Borneo. local medic. ***M. parasiticus*** (L.) Danser (S India, Sri Lanka) – bushes to several m tall, trunks to 15 cm diam.

Macrosphyra Hook.f. Rubiaceae (II 1). 3 trop. Afr. R: BJBB 28(1958)27

Macrostegia Nees = *Vitex*

Macrostelia Hochr. = *Hibiscus*

Macrostigmatella Rauschert = *Eigia*

Macrostylis Bartling & Wendl. (~ *Euchaetis*). Rutaceae (I 4). 10 S & SW Cape. R: JSAB 47(1981)373

Macrosyringion Rothm. (~ *Odontites*). Orobanchaceae (Rhin.; Scrophulariaceae s.l.). 2 Medit. R: Willd. 26(1996)64

Macrothelypteris (H. Itô) Ching. Thelypteridaceae. 10 Natal & Masc. to Hawaii. R: Blumea 17(1969)25. Mostly with 3-pinnatifid fronds & small indusia. Cult. orn. esp. ***M. torresiana*** (Gaudich.) Ching (Masc. to Jap. & Polynesia) – natur. Am.

Macrothumia Alford (~ *Neosprucea*). Salicaceae. 1 Braz.: ***M. kuhlmannii*** (Sleumer) Alford. R: Novon 16(2006)294

Macrotomia DC. ex Meissn. = *Arnebia*

Macrotorus Perkins. Monimiaceae (V 2?). 1 SE Brazil: ***M. utriculatus*** (Mart.) Perkins. R: Phytotaxa 234(2015)201

Macrotyloma (Wight & Arn.) Verdc. Leguminosae (III 18). 24 OW trop. R: HIP 38,4(1982). Pulses esp. ***M. uniflorum*** (Lam.) Verdc. (*Dolichos pubescens*, horse gram, OW trop.) – fodder etc. ***M. geocarpum*** (Harms) Maréchal & Baudet (*Kerstingiella g.*, ground bean, W Afr.) – ed. seeds, fr. buried like *Arachis*

Macroule Pierce. See *Ormosia*

Macrozamia Miq. Zamiaceae (I). 40 Aus. esp. E). R: FA 48(1998)639, BR 70(2004)289. Dioec. cult. orn. with alks; linalool dominating cone-volatiles of weevil-poll. spp., beta-myrcene that of thrips-poll. spp. e.g. ***M. macdonnellii*** (Miq.) A. DC. (C Aus.). Seeds ed. locally (Queensland nut) if soaked & pounded or baked. ***M. lucida*** L. Johnson (NE Aus.) – male cones attractive to thrips but thermogenesis increases so driving them off to settle on females; ***M. riedlei*** (Gaudich.) Gardner (SW Aus.) – possums effect disp., seed form. trad. food 'by-yu'; ***M. spiralis*** (Salisb.) Miq. (NSW) – source of good quality arrowroot & ed. seeds (burrawang)

Maculigilia V. Grant = *Linanthus*

Macvaughiella R. King & H. Robinson (~ *Eupatorium*). Compositae (Eup.-Ager.). 2 C Am. R: SB 16(1991)639, Phytol. M 11(1997)152

Madagascar bean *Phaseolus lunatus*; **M. cardamom** *Aframomum angustifolium*; **M. clove** *Cryptocarya* sp. (*Ravensara aromatica*); **M. copal** *Hymenaea verrucosa*; **M. ebony** *Diospyros haplostylis*; **M. jasmine** *Marsdenia floribunda*; **M. mahogany** *Khaya madagascariensis*; **M. nutmeg** *C. agathophyllum*; **M. palm** *Pachypodium lamerei*; **M. periwinkle** *Catharanthus roseus*; **M. plum** *Flacourtia indica*; **M. rubber** *Cryptostegia, Landolphia, Marsdenia* & *Mascarenhasia* spp.

Madagasikaria C. Davis. Malpighiaceae. 1 S Madag.: ***M. andersonii*** C. Davis. R: AJB 89(2002)702

Madagaster Nesom (~ *Aster*). Compositae (Ast.-Hin.). 5 Madag. R: Phytol. 75(1993)94

madake bamboo *Phyllostachys reticulata*

Madangia P.I. Forst. & al. = *Hoya* (but see Austrobaileya 5(1997)53)

madar *Calotropis* spp.

Madaractis DC. (~ *Senecio*). Compositae (Senec.). c. 5? C As.

Madarosperma Benth. = *Tassadia*

Maddenia Hook.f. & Thomson = *Prunus*

madder *Rubia tinctorum*; **field m.** *Sherardia arvensis*; **Ind. m.** *Hedyotis indica, R. cordifolia*; **Levant** or **wild m.** *R. peregrina*

Madeira marrow *Sicyos edulis*; **M. vine** *Anredera cordifolia*

Madhuca Buch.-Ham. ex J. Gmelin. Sapotaceae (II). Incl. *Ganua*, 100 Indomal., esp. W Mal., to Aus. R: Blumea 7(1953)364, 10(1966)1. Timbers (nyatoh) & oilseeds (the original illipe nuts used for margarine etc.) esp. ***M. longifolia*** (L.) Macbr. (mahwa, mahua, mowa, moa, mi, India) – fls also ed. (rich in nectar), seed-cake ('mahwa meal') used as worm-killer in lawns; ***M. motleyana*** (de Vriese) Macbr. (Thailand to Borneo) – seed-oil for cooking & lighting; ***M. pallida*** (Burck) Baehni (*Ganua p.*, Mal.) – yields a gutta-percha; ***M. utilis*** (Ridl.) H.J. Lam (W Mal.) – timber

madhumalati *Combretum indicum*

Madia Molina. Compositae (Mad.-Mad.). 10 W N Am. (R: FNA 21(2006)303), Hawaii, Chile. Tarweeds. Oilseeds esp. ***M. elegans*** D. Don ex Lindl. (W N Am.) – seeds a staple for some Native Americans; ***M. sativa*** Molina (W N Am., Chile) – used as olive oil subs., cult. in Eur.

Madlabium Hedge. Labiatae (VII 3e). 1 N Madag.: ***M. magenteum*** Hedge. R: FMad 175(1998)260

Madras hemp *Crotalaria juncea*; **M. thorn** *Pithecellobium dulce*

madroña, madrone, madroño (laurel) *Arbutus menziesii*

madwort *Asperugo procumbens*

Maerua Forssk. Capparaceae. 60 OW trop. esp. Afr. Some geoxylic suffrutices (*Courbonia*), cyanogenic; fr. a lomentum-like berry, that of ***M. crassifolia*** Forssk. ed. Sahara. Some locally used timbers (incl. toothpicks) & ed. lvs.

Maesa Forssk. Primulaceae (Maesaceae, Myrsinaceae). c. 150 OW trop. to Jap. & Aus. (Mal. 78). Attims's & Champagnat's Models. Some locally medic.; some cult. orn. e.g. ***M. indica*** (Roxb.) Sweet (India) – lvs used in curry. ***M. lanceolata*** Forssk. (trop. & S Afr., Madag.) – toxic, the fr. an effective bactericide

Maesaceae Anderb. & al. = Primulaceae

Maesobotrya Benth. Phyllanthaceae (Scep.; Euphorbiaceae s.l.). 20 trop. Afr.

Maesopsis Engl. Rhamnaceae (7). 1 trop. Afr.: ***M. eminii*** Engl. (musizi) – comm. timber (Roux's Model)

Maeviella Rossow (~ *Bacopa*). Plantaginaceae (Grat.-Grat.; Scrophulariaceae s.l.). 1 NE Brazil: ***M. cochlearia*** (Huber) Rossow. R: AJBRJ 27(1985)172

Maferria Cusset = *Farmeria* (but see Adans. 14(1992)34)

mafoureira or **mafurra** *Trichilia emetica*
mafu *Fagaropsis angolensis*
Maga Urb. = *Montezuma*
Magadania Pim. & Lavrova. Umbelliferae (III 8). 2 NE As. R: BZ 70(1985)530
Magallana Cav. = *Tropaeolum*
Magdalenaea Brade. Orobanchaceae (Esc.; Scrophulariaceae s.l.). 1 SE Brazil: ***M. limae*** Brade
Magnistipula Engl. Chrysobalanaceae (4). 13 trop. Afr. (11) & Madagascar (2). R: FOW 10(2003)178. Some with ant-infested stem domatia. ***M. butayei*** de Wild. (trop. Afr.) – imp. in rain-making ceremonies
Magnolia Plum. ex L. Magnoliaceae (I). Incl. *Elmerrillia, Kmeria, Manglietia, Michelia, Pachylarnax, Talauma,* 219 Himal. to Jap. & W Mal., E N Am. to trop. Am. (Colombia c. 20, more than China; subg. *M.* (*Kmeria, Manglietia, Talauma*) c. 132 As., Am. – branches usu. sylleptic, lvs conduplicate, A usu. caducous; subg. *Yulania* (Spach) Reichb. c. 79 E As., N Am. – branches proleptic, lvs conduplicate; subg. *Gynopodium* (Dandy) Figlar & Nooteb. 8 SE As., Mal. – lvs not conduplicate). R: Blumea 49(2004)88; N. Treseder (1978) *Magnolias* (temp. spp.); D. Hunt (ed. 1998) *Ms & their allies*. Trees & shrubs with alks much cult. (incl. hybrids) for spectacular white, pink, purple or yellow fruity-scented, beetle-poll. fls (term. in *M. s.s.*, lateral with gynophore between A & G in '*Michelia*') with congested fr. & often sarcotestal seeds (some '*Talauma*' with ± indehiscent fr.); some timbers (canary whitewood; e.g. for trad. Jap. shoes) & medic. barks. ***M. acuminata*** (L.) L. (cucumber tree, E N Am.) – fr. purplish red to 10 cm long, timber for flooring etc., local medic.; ***M. champaca*** (L.) Pierre (champa, champak, sapu, Himal. to W Mal.) – cult. around Hindu & Jain temples, ess. oil from fls used in scent-making, timber for tea-boxes, furniture, carving & fuel, bark a febrifuge, lvs used to feed silkworms; ***M. denudata*** Desr. (*M. conspicua, M. heptapeta,* yulan, E & S China) – cult. orn. decid. tree, fls ed.; ***M. doltsopa*** (DC.) Figlar (*Michelia d.*, E Himal. to W China) – timber good; ***M. grandiflora*** L. (bull bay, loblolly m., SE U.S.) – evergreen (Fagerlind's Model) usu. grown against a wall in GB, flowering when 20 yrs old from seed (faster from cuttings), several cvs incl. '**Goliath**' with fls to 30 cm diam. & hybrids with *M. virginiana* ('Freeman hybrids' incl. '**Exmouth**'); ***M. kobus*** DC. (*M. praecocissima*, Jap.) – cult. orn. decid. tree, timber used for engraving, matches etc. in Jap.; ***M. macrophylla*** Michaux (N Am.) – biggest native U.S. tree leaf (to 75 cm long, 25 cm wide) & fls (to 30 cm across); ***M. megaphylla*** (Hu & W.C. Cheng) Kumar (SW China) – construction timber, critically endangered; ***M. officinalis*** Rehder & E. Wilson (cultigen, ? orig. C China) – bark used as tonic in Chinese medic., lowers cortisol (stress hormone) due to magnolol & honokiol anti-oxidants; ***M. pleiocarpa*** (Dandy) Figlar & Nooteb. (*Pachylarnax p.*, Assam) – timber for cabinet-making; ***M × soulangeana*** Soul.-Bod. (*M. denudata × M. liliiflora* Desr. (*M. quinquepeta*, pollen parent, temp. China), a deliberate cross c. 1820) – most commonly planted m. with many cvs; ***M. stellata*** (Sieb. & Zucc.) Maxim. ('*M. tomentosa*', Jap.) – free-flowering small sp., poss. old cv of *M. kobus*; ***M. virginiana*** L. (*M. glauca*, E N Am.) – decid. or evergreen cult. orn., timber used for broom-handles etc.
Magnoliaceae Juss. Magnoliidae – Magnoliales. 2/221 trop. to warm temp. esp. N. R: D.G. Frodin & R. Govaerts (1996) *World checklist & bibliography of M.* Trees & shrubs usu. with alks & always with oil-cells in parenchyma. Lvs in spirals (s.t. distichous on same shoots), simple, entire (lobed in *Liriodendron*), often glaucous adaxially; stipules large, enclosing term. bud, decid. Fls large, usu. term. & solit., bisexual (rarely unisexual as in *Magnolia* (*Kmeria*)), reg., usu. with long receptacle; P spiral or in 3 or more whorls, 6–18, often ± all petaloid, A ∞ in spirals, originating centripetally, often ± strap-shaped with 4 paired microsporangia embedded in surface (adaxial; abaxial in *Liriodendron*) & ± elongate connective, pollen monosulcate, $\underline{G}$ (1–)∞, conduplicate & s.t. not completely closed but with ± distinct style & term. stigma & (1) 2(–±∞) anatropous, bitegmic ovules on marginal placenta. Fr. a follicle or indehiscent & berry-like or samaroid (*Liriodendron*), with carpels growing together s.t. forming fleshy syncarp; seeds usu. large, endotestal with sarcotesta, suspended by fibrils from vasc. bundle of raphe in spp. with dehiscent fr., embryo v. small with suspensor in copious oily, proteinaceous endosperm. n = 19

Genera: *Liriodendron, Magnolia*

Both genera show E As./E N Am. (trop. Am.) distribution. *Archaeanthus* of C Kansas Late Cretaceous with bilobed lvs & fls. 10 cm diam. poss. here. Modern pls with many primitive features but no staminodes intermediate between P & A; sarcotestal seed achieves same ecological effect as a primitive arillate seed dangling from the funicle rather than the vasc. bundle of the raphe. Thermogenesis detected in some fls. Many magnificent cult. orn. & some timbers

Magnoliidae Novák ex Takht. (Angiospermae, Anthophyta). 13 405/285 500 in 413 fams in 64 orders. See D.E. Soltis & al. (2005) *Phylogeny & evolution of angiosperms*. Seed pls usu. with closed carpels (unclosed in certain basal angiosperms, Resedaceae, Malvaceae etc.; though truly 'angio-ovuly' as pointed out by Robert Brown in 1825!) & app. always with 'double fert.' where cells other than the egg unite during fertilization to give (s.t. short-lived) triploid endosperm, though diploid in at least some Onagraceae

Prob. diverged from seed-ferns some 130 M yrs ago. The dominant group of pls on which civilization relies, trad. split into dicotyledons and monocotyledons, though d. (s.s.) and m. nested in a more encompassing M. (see Appendix), & comprising the bulk of Spermatophyta (q.v.), themselves the bulk of vasc. pls (13 300/282 500). The origin of M. has vexed many but attempts to project the modern circumscription of the group into the past so as to judge fossils as 'true' angiosperms or not are muddleheaded: clearly the 'first' angiosperm would have been seen merely as an interesting 'gymnosperm' at the time (see T 33(1984)77, 43(2004)3, EJB 51(1994)117). Similarly the ancestor of the group of plants to dominate the future world is presently seen as an 'angiosperm'. Pls resembling modern angiosperms at least in major part are known from the Triassic of Texas (*Sanmiguelia*), semi-aquatic pachycauls with plicate lvs & vessel-less xylem, app. monocot-like infls of fls with monosulcate pollen & closed carpels with app. bitegmic ovules, though aquatic P-less *Archaefructus* sp. (125 M yrs old) said to be 'sister' to living M. (Science 296(2002)899), & *S.* thought 'enigmatic' (see also Ceratophyllaceae)

Magnoliophyta Cronq. & al. See Magnoliidae

Magnoliopsida Brongn. See Magnoliidae

Magodendron Vink. Sapotaceae (IV). 2 E New Guinea: R: Blumea 40(1995)91

Magonia A. St-Hil. Sapindaceae (III 1). 1 Paraguay, Brazil: ***M. pubescens*** A. St-Hil.

Magoniella A. Sanchez (~ *Ruprechtia*). Polygonaceae. 2 trop. S Am. R: SB 36(2011)708. Lianes

maguey *Agave cantala*

Maguireanthus Wurd. Melastomataceae. 1 Guyana: ***M. ayangannae*** Wurd. R: MNYBG 10, 5(1964)155

Maguireocharis Steyerm. Rubiaceae (I 6). 1 Guayana Highland: ***M. neblinae*** Steyerm. R: MNYBG 23(1972)230

Maguireothamnus Steyerm. Rubiaceae (I 5). 2 Venezuela. R: MNYBG 10, 5(1964)220

Magydaris Koch ex DC. Umbelliferae (III 5). 2 Medit. (inc. Eur.)

Mahafalia Jum. & Perrier = *Cynanchum*

mahaleb (**cherry**) *Prunus mahaleb*

Maharanga A. DC. (~ *Onosma*). Boraginaceae (B.2.2). 9 E Himal. to SW China, Thailand (1). R: JAA 35(1954)78

Mahawoa Schltr. Apocynaceae (V c; Asclepiadaceae III 1). 1 Sulawesi: ***M. montana*** Schltr – known only from type specimen now destroyed

× **Mahoberberis** C. Schneider = *Berberis*

mahoe (WI) *Thespesia populnea*, (NZ) *Melicytus ramiflorus*

mahogany *Swietenia* spp. esp. *S. mahagoni* (**Cuban m.**), (Afr.) *Khaya* spp. Many medium coloured medium-weight timbers have been called m. (see KB 1936:193) incl.: **Australian m.** *Dysoxylum fraserianum*; **bastard m.** *Eucalyptus botryoides*; **Bataan** or **Philippine m.** *Shorea polysperma*; **Benin m.** *Khaya* spp.; **Borneo m.** *Calophyllum inophyllum*; **Burma m.** *Pentace burmanica*; **Cape m.** *Trichilia emetica*; **Ceylon m.** *Melia azedarach*; **cherry m.** *Tieghemella heckelii*; **Colombian m.** *Cariniana pyriformis*; **East India m.** *Pterocarpus dalbergioides*; **Gaboon m.** *Aucoumea klaineana*; **Honduras m.** *Swietenia macrophylla*; **Indian white m.** *Canarium euphyllum*; **Natal m.** *Kiggelaria africana*, *Trichilia emetica*; **pod** or **red m.** *Afzelia quanzensis*, *Eucalyptus resinifera*; **white m.** *E. acmenoides*, *E. robusta*

maholtine *Wissadula amplissima*

Mahonia Nutt. = *Berberis*

ma-huang See ma huang

Mahurea Aubl. Calophyllaceae (Guttiferae (I1). 2 trop. S Am. R: MNYBG 29(1978)134

mahwa or **mahua** *Madhuca longifolia*

Mahya Cordemoy = *Lepechinia*

Maianthemum G. Weber ex Wigg. Asparagaceae (Convallariaceae). (Incl. *Smilacina*) 28 N temp. (N Am. 5), Himal., C Am. R (Am.): JAA 67(1986)371. Fls 2- or 3-merous; ***M. paludicola*** LaFrankie (Costa Rica) – rhiz. upright. ***M. dilatatum*** (A. Wood) A. Nelson & Macbr.

(N Am.) – imp. weed in cranberry swamps in US; cult. orn. esp. ***M. bifolium*** (L.) F.W. Schmidt (May lily, Euras.) & ***M. racemosum*** (L.) Link (*Smilacina r.*, false, American or wild spikenard, N Am.) – fr. ed. & ***M. stellatum*** (L.) Link (N Am.) – both imp. local medic.

maiden pink *Dianthus deltoides*; **m's blush** *Sloanea australis*

maidenhair fern *Adiantum* spp. esp. *A. capillus-veneris*, see also *Asplenium trichomanes*; **m. tree** *Ginkgo biloba*

Maidenia Rendle = *Vallisneria* (but see Cowie & al. (2000)196)

maidu *Pterocarpus macrocarpus*

Maieta Aubl. Older name for *Miconia*

maigyee *Strobilanthes cusia*

Maihuenia (F. Weber) K. Schum. Cactaceae (I). 2 S Chile, S Arg. R: BJ 119(1997)58. Cult. orn. highly mucilaginous, frost-hardy

Maihueniopsis Speg. (~ *Opuntia*). Cactaceae. 7 S Am.

maile *Alyxia oliviformis*

Maillardia Frappier & Duchartre = *Trophis*

Maillea Parl. = *Phleum*

Maingaya Oliv. Hamamelidaceae (I 1). 1 Penang, Perak (Malay Pen.): ***M. malayana*** Oliv. R: KM 10(1993)83

maire *Nestegis* sp.

Maireana Moq. Amaranthaceae (Chenopodiaceae I 5). 57 Aus. R: FA 4(1984)179. Char. of dry plains of C & S Aus. esp. ***M. pyramidata*** (Benth.) Paul G. Wilson (blackbush, S Aus.) & ***M. sedifolia*** (F. Muell.) Paul G. Wilson (bluebush) – common on Nullabor

Mairetis I.M. Johnston. Boraginaceae (B.2.2). 1 Canary Is., Morocco: ***M. microsperma*** (Boiss.) I.M. Johnston. R: JAA 34(1953)4

Mairia Nees. Compositae (Ast.-Hin.). 3 S & SW Cape. R: Phytol. 76(1994)90

maize *Zea mays*

majestic palm *Ravenea rivularis*

Majidea J. Kirk ex Oliv. Sapindaceae (III 1). 4–5 E Afr., Madag.

Majorana Mill. = *Origanum*

makalao *Cyperus laevigatus*

makoré *Tieghemella heckelii*

makrut lime *Citrus hystrix*

makua *Diospyros mollis*

Malabaila Hoffm. (~ *Pastinaca*). Umbelliferae. Incl. *Leiotulus*, 10 Euras. R: FR 105(2008)141

Malabar gourd *Cucurbita ficifolia*; **M. kino** *Pterocarpus marsupium*; **M. nightshade** *Basella alba*; **M. oil** *Cymbopogon flexuosus*; **M. rosewood** *Dalbergia latifolia*; **M. spinach** *B. alba*; **M. tallow** *Vateria indica*

malabathrum *Cinnamomum aromaticum*

Malacantha Pierre = *Pouteria*

Malacca cane *Calamus* spp. esp. *C. scipionum*; **M. teak** *Afzelia rhomboidea*, *Intsia palembanica*

Malaccotristicha Cusset & G. Cusset = *Terniopsis*

Malaceae Small = Rosaceae

Malachra L. = *Hibiscus* (but see R: BJ 16(1892)345)

Malacocarpus Fischer & C. Meyer (~ *Peganum*). Nitrariaceae (Zygophyllaceae s.l.). 1 C As.: ***M. crithmifolius*** (Retz.) C. Meyer

Malacocera R. Anderson. Chenopodiaceae (I 5). 4 Aus. R: JABG 2(1980)139

Malacomeles (Decne) Engl. = *Amelanchier*

Malacothamnus Greene. Malvaceae (Malv.-Malv.). 11 Calif., N Baja Calif. R: LWB 6(1951)113. Cult. orn. tree-like shrubs

Malacothrix DC. Compositae (Cich.-Cich.(-Micr.)). 21 W N Am. R: AMN 58(1957)494, Madroño 16(1962)258

Malacurus Nevski = *Elymus*

Malagasia L. Johnson & B. Briggs. Proteaceae (V 4). 1 Madag.: ***M. alticola*** (Capuron) L. Johnson & B. Briggs. R: BotJLS 70(1975)175

malagueta pepper *Aframomum melegueta*

Malaifilix L.B. Zhang & Schuettp. (~ *Tectaria*). Tectariaceae. 1 W Mal.: ***M. grandidentata*** (Cesati) L.B. Zhang & Schuettp. R: T 65(2016)733

Malaisia Blanco = *Trophis*

malambo *Croton malambo*

Malanea Aubl. Rubiaceae (III 3). 35 trop. Am. Lianes

malanga *Xanthosoma* spp.

Malania Chun & S.K. Lee. Olacaceae (Ximeniaceae). 1 S China: ***M. oleifera*** Chun & S.K. Lee – oil-pl.

malanye *Strobilanthes cusia*

malatti *Jasminum* spp.

Malaxis Sol. ex Sw. Orchidaceae (V 11b). Excl. *Dienia, Hammarbya*, c. 180 subcosmop. (Eur. 1 (*Microstylis*), Afr. 12, N Am. 10), not NZ. R: JLSB 24(1888)308. Terr. herbs (adder's mouth, US); fls. not resupinate

Malay apple *Syzygium malaccense*

Malcolmia R. Br. Cruciferae (4). Excl. *Marcus-kochia, Maresia, Strigosella, Zuvanda*, 6 Euras. ***M. maritima*** (L.) R. Br. (Virginia stock, Albania & Greece) – cult. orn. annual with reddish to white fls

male bamboo *Bambusa bambos, Dendrocalamus strictus*; **m. fern** *Dryopteris filix-mas*

Malea Lundell = *Vaccinium*

Malephora N.E. Br. Aizoaceae (V). 16 S Afr. R: H. E. K. Hartmann, *Ill. Handb. Succ. Pls*, Aiz. F–Z(2002)138. Cult. orn. succ. shrubs used for erosion control in Calif.

Malesherbia Ruíz & Pavón. Passifloraceae (Malesherbiaceae). 24 S Peru, N Chile, W Arg. R: Gayana 16(1967)1

Malesherbiaceae D. Don = Passifloraceae

Malinvaudia Fourn. = *Matelea*

Malleastrum (Baill.) J. Leroy. Meliaceae (II-Trich.). 23 Madag., Comoro Is., Aldabra

mallee *Eucalyptus* spp. with pl. forming a thicket of stems; **fuchsia m.** *E. forrestiana*

Malleola J.J. Sm. & Schltr. = *Robiquetia*

Malleostemon J. Green. Myrtaceae (II 15). 7 SW Aus. R: Nuytsia 4(1983)295

mallet bark *Eucalyptus astringens*

Mallinoa J. Coulter = *Ageratina*

Mallophora Endl. = *Dicrastylis*

Mallophyton Wurd. Melastomataceae. 1 Venezuela: ***M. chimantense*** Wurd. R: MNYBG 105(1964)145

Mallotonia (Griseb.) Britton (~ *Heliotropium*). Boraginaceae. 1 Florida, Mex., WI: ***M. gnaphalodes*** (L.) Britton (iodine bush). R: AMBG 2(1915)47

Mallotopus Franch. & Savat. = *Arnica*

Mallotus Lour. (*Trevia*). Euphorbiaceae (Acal.-Acal.-Rottl.). Incl. *Neotrewia, Octospermum* & *Trevia*, excl. *Hancea*, c. 110 OW trop. Herb. specimens to 100 yrs old smell of fenugreek. Some light timbers, medic. & dyes of local imp.: ***M. discolor*** F. Muell. ex Benth. (E Aus.) – yellow dye; ***M. nudiflorus*** (L.) Kuljo & Welzen (*T. n.*, Ind. to C Mal.) – disp. rhinoceros in Nepal (cattle elsewhere), soft timber for tea-chests, packing-cases etc., seed-oil like tung; ***M. paniculatus*** (Lam.) Muell. Arg. (*M. cochinchinensis*, SE As., Mal.) – wood for matches, packing-cases; ***M. philippensis*** (Lam.) Muell. Arg. (kamala, NW Himal. to Aus.) – red dye from fr., seed-oil a tung subs.

mallow spp. of *Abutilon, Hibiscus, Lavatera, Malva* & other Malvaceae; **common m.** *M. sylvestris*; **curled m.** *M. verticillata* 'Crispa'; **Egyptian m.** *M. parviflora*; **Indian m.** *A. theophrasti*; **Jew's m.** *Corchorus olitorius*; **marsh m.** *Althaea officinalis*; **musk m.** *M. moschata*; **rose m.** *Hibiscus* spp.; **scarlet globe m.** *Sphaeralcea coccinea, S. angustifolia* subsp. *cuspidata*; **tree m.** *Malva arborea*

Malmea R. Fries. Annonaceae (IV 2). Excl. *Klarobelia, Mosannona, Pseudomalmea*, 6 trop. Am. R: AHB 10(1930)37, (1931)318

Malmeanthus R. King & H. Robinson (~ *Eupatorium*). Compositae (Eup.-Crit.). 3 Brazil, Uruguay, Parag., Arg. R: MSBMBG 22(1987)359

maloga bean *Vigna lanceolata*

Malope L. Malvaceae (Malv.-Malv.). 3 Medit. (Eur. 2). Annuals with 3 leafy involucral bracts & many 1-seeded mericarps app. in superposed whorls. ***M. trifida*** Cav. (Spain, N Afr.) – cult. orn. with white to purple fls

Malortiea H. Wendl. = *Reinhardtia*

Malosma (Nutt.) Abrams. (~ *Rhus*). Anacardiaceae (I). 1 SW N Am.: ***M. laurina*** (Nutt.) Abrams

Malouetia A. DC. Apocynaceae (II b). c. 20 trop. Am. & Afr. (4, R: AUWP 85–2(1985)70). Alks. Some fish-poisons & allegedly hallucinogenic narcotics sometimes fatal. ***M. tamaquarina*** A. DC. common in Amaz., toxic but eaten by the pajuíl bird (*Nothocrax urumutum* (Spix)), which is often domesticated & flesh eaten all year, though bones in fruiting season (Mar–June) poison dogs

Malouetiella Pichon = *Malouetia*

Malperia S. Watson. Compositae (Eup.-Alom.). 1 Calif. & Baja Calif.: ***M. tenuis*** S. Watson. R: MSBMBG 22(1987)235

Malpighia Plum. ex L. Malpighiaceae. c. 130 trop. Am. esp. Carib. R: F.M.Meyer (2000) *Revision der Gattung M.* (PM 23). Trees & shrubs, s.t. with stinging hairs; some with cleistogamous fls. Some cult. orn. incl. ***M. coccigera*** L. (Singapore holly, WI), & for ed. fr. esp. ***M. emarginata*** Moçiño & Sessé ex DC. ('*M. glabra*', acerola, Barbados or WI cherry, Mayan cultigen, first modern plantations in Puerto Rico) & smaller-fruited ***M. glabra*** L. (*M. punicifolia*, escobilla, trop. Am. to Texas) – both with fr. high in vitamin C (to 4%), used in breakfast-juice, syrups, jam etc.

Malpighiaceae Juss. Magnoliidae – Malpighiales. 74/1300 trop. & warm esp. S Am. Small trees, shrubs or lianes often with unusual sec. growth & s.t. with alks. Lvs usu. opp., simple & entire, very often with 2 large fleshy glands on petiole or abaxial surface; stipules s.t. large & united. Fls usu. bisexual & bilaterally symmetrical to ± reg., 5-merous, borne on jointed 2-bracteolate pedicels in racemes or cymes; receptacle ± convex, K imbricate s.t. basally connate, very often with pair of conspic. abaxial glands at base, C imbricate, crumpled in bud, often clawed, with ciliate to fringed margins, A (1)2(3) whorls (some often sterile), filaments ± basally connate, anthers with longit. slits (rarely term. pores), disk 0, $\underline{G}$ ((2)3(–5)), pluriloc. with axile placentas & ± distinct styles & 1 anatropous to hemitropous bitegmic ovule per loc. Fr. often a schizocarp of winged to nut-like mericarps, s.t. a nut or drupe; seeds with large straight to circinate oily embryo with ± 0 endosperm. x = 6, 9–12+

Principal genera: *Acridocarpus, Banisteriopsis, Bunchosia, Byrsonima, Heteropterys, Hiptage, Hiraea, Malpighia, Mascagnia, Stigmaphyllon, Tetrapterys*

Latex & laticifers in *Galphimia* & *Verrucularia* suggested affinity with Euphorbiaceae. App. with most archaic genera in Guayana Highland, the NW genera visited by oil-gathering bees such as *Centris* spp. absent from OW, where char. oil-glands of K are vestigial or 0 (Anderson in Cronquist) but in 41 genera (only 10 fams with them) & known from mid-Eocene fossils. Bird-disp. seems to have arisen independently in unrelated genera (*Bunchosia, Byrsonima, Malpighia*)

Ed. fr. (*Bunchosia, Malpighia*), hallucinogens (*Banisteriopsis, Diplopterys, Tetrapterys*), timber (*Byrsonima*) & some cult. orn.

Malpighiantha Rojas. Malpighiaceae. 2 Argentina (= ?)

Malpighiodes Niedenzu (~ *Mascagnia*). Malpighiaceae. 4 N S Am. R: Novon 16(2006)191

malt (extract) *Hordeum vulgare*

Maltebrunia Kunth. Gramineae (III 3i). 4 Gabon, Tanz., Madag. R: HIP 36(1962)t.3595

Maltese cross *Silene chalcedonica*

malukang fat *Polygala butyracea*

Malus Mill. Rosaceae (Ros.-Mal.). c. 40 N temp. (Eur. 4, N Am. 9). R: CJB 68(1990)2234; B.E. Juniper & D.J. Mabberley (2006) *The story of the apple.* Apples – fr. (pome) an enlarged receptacle, containing inf. ovary (core) with seeds (pips); flavour due to ethyl 2-methyl butyrate (synthetic 'apple' fragrance = methyl acetate). Sometimes incl. in *Pyrus* but differing in ± pubescent lvs, entire or weakly lobed as in ***M. iosensis*** (Wood) Britton (C N Am.) – cult. orn., ***M. toringo*** (Siebold) de Vriese (E As.; for distinctly 3-lobed *M. trilobata* see *Eriolobus t.*), acute leaf-margin teeth, yellow pollen (purple in *Pyrus*), gritty sclereid tissue absent from fr., etc. Fr. & orn. trees; millions of 'wild' apples known from Neolithic & Bronze Age Switzerland, the Lake-dwellers storing them dried; ***M. domestica*** (Suckow) Borkh. (*M. communis, M. pumila, 'M. sylvestris* subsp. *mitis'*, orchard apple, derived from extremely variable wild populations form. called *M. sieversii* (Lindl.) M. Roemer, C As.) – fls protogynous, visited by bees, extrafl. nectaries, self-incompatible, the most imp. temp. fr. tree with large bear- (and later horse-)disp. pomes ('megafauna' syndrome, cf. primitive bird-disp. of many other *M.* spp.) to 1.357 kg – UK record (Eur. consumption (1993) c. 20 kg per person per annum now falling, world prod. (2011) c. 75 M t) with c. 2 500 cvs UK (those from pips = 'pippins', orig. small green-fruited cvs = 'codlins', a pop. C13 (& long after) apple being '**Costard**', hence term 'costermonger') & poss. 6000 in former USSR (perhaps c. 20 000 overall), used as dessert ('**Red Delicious**', orig. 'Hawkeye' (Iowa c. 1880), prob. still most widely grown), dipped in molten toffee (toffee apples), cooked, dried or for juice, s.t. fermented to cider (hard c. in US), some 70% in GB, or distilled to apple brandy (calvados, (US) applejack) esp. in Normandy whence cider-making introd. GB c. 1630 ('**Redstreak**' in Herefordshire, which county was producing 1.5 M gallons per annum in 1890s), apple wine (Germany), apple vinegar culinary & a cure-all, flavouring incl. bubble-pipes in Middle E., wood chips used for smoking salmon; good eating cvs (some triploids) incl. '**Bramley's Seedling**' ('cooker', chance seedling c. 1810, Notts, England, orig. tree still alive 2003), '**Cox's Orange Pippin**' (c. 1825 chance seedling from 'Ribston Pippin', form. UK favourite overtaken 2009 by '**Gala**' ['Golden Delicious' × 'Kidd's Orange Red' ('Cox's Orange Pippin' × 'Delicious')] – 22 K t), '**Golden Delicious**' (orig. 'Mullins' Yellow Seedling', c. 1890, W. Virginia, USA) & '**Granny Smith**' (orig. NSW c. 1830) – green when ripe; '**Flower of Kent**' allegedly that observed by Newton triggering his gravitational theory; modern breeding introd. fire-blight resistance gene from gut of silk-moth; orn. cvs incl. '**Niedwetzkyana**' (C As. pre-1900) – lvs purple-red, fls. red; escaped forms often known as crab apples (& confused with spiny ***M. sylvestris*** (L.) Mill. (Eur.)), a common name also used for cult. orn. spp. & hybrids. ***M. baccata*** (L.) Borkh. (NE As. to N China) – fr. ed. fresh, dried or preserved, used for breeding in cold resistance (alleged hybrids with *M. domestica* = ***M. × adstringens*** Zabel; ***M. × floribunda*** Siebold ex van Houtte (*M. sieboldii* (Regel) Rehder (Jap.) × *M. baccata*) – pop. early-flowering tree raised Jap. before 1862; ***M. fusca*** (Raf.) Schneider (W N Am.) – local medic., wood for tool-handles; ***M. hupehensis*** (Pampan.) Rehder (*M. theifera*, Chinese crab apple, China, Assam) – cult. orn. with early fls & small orn. fr., lvs used as tea; ***M. prunifolia*** (Willd.) Borkh. (Chinese a., NE As.) – alleged hybrids with *M. domestica* = ***M. × astracanica*** Dum.-Cours.

Malva Tourn. ex L. Malvaceae (Malv.-Malv.). Incl. cosmop. weeds etc. form. in *Lavatera*, c. 20 Eur. (12), Medit., temp. As., Aus., trop. Afr. mts, Am. Mallows, the fr. known as cheeses (shape reminiscent of some types of c.). Nectar secreted in receptacle covered with hairs preventing rain-wash & access by short-tongued insects, the fl. v. protandrous with A in centre at first, the styles later lengthening to occupy this position as A curve back & down; in small-flowered spp. the styles later curve down & twist among A effecting self-poll. Cult. orn. & some leaf vegetables, mucilage (sometimes mixed with goose-grease) used in poultices for wounds & sores. ***M. arborea*** (L.) Webb & Berth. (*L. arborea, M. dendromorpha, M. eriocalyx*, Eur., Medit.) – woody, lvs form used as lavatory paper on Jersey; ***M. × clementii*** (Cheek) Stace (*L. × c., M. thuringiaca* (L.) Vis. (*L. t.*, C & S Eur.) × *M. olbia* (L.) Alef. (*L. o.*, W Medit.)) & ***M. moschata*** L. (musk mallow, Eur., N Afr., natur. N Am.) – cult. orn.; ***M. nicaeensis*** All. (Euras., Medit., natur. N Am.) & ***M. parviflora*** L. (Egyptian m., Medit. to Afghanistan) – imp. ed. mallows used since 6000 BC, viable seeds of latter recovered from 200-yr.-old adobe in Calif. & Mex.; ***M. pusilla*** Sm. (*M. rotundifolia*, p.p., Euras.) – locally medic., fibre-crop for small farms on Amaz. varzéa, seeds germ. after sealing in cannister for 120 yrs (Dr Beal's experiment); ***M. sylvestris*** L. (common m., Eur., Medit., natur. N Am.) – fr. ed.; ***M. trimestris*** (L.) Salisb. (*L. t.*, (Medit.)) – ann., cult. orn.; ***M. verticillata*** L. '**Crispa**' (curled m., Euras.) – cult. salad pl.

Malvaceae Juss. Magnoliidae – Malvales. Incl. Bombacaceae, Sterculiaceae, Tiliaceae, 229/5100 cosmop. esp. trop. Trees (often with fibrous bark), shrubs & herbs, rarely scandent, usu. with tufted or stellate hairs & parenchyma typically with scattered mucilage-cells & -cavities or -canals. Lvs in spirals, simple to ± dissected & usu. palmately-veined; petiole usu. pulvinate at both ends; stipules usu. present. Fls usu. bisexual & reg., often with epicalyx of 3(or more) sterile bracts, solit. & axillary or in cymes, s.t. cauliflorous; K usu. 5 valvate, s.t. basally united, C 5 often adnate to base of A-tube, convolute (or imbricate), A (5–)∞ initiated centrifugally & assoc. with limited no. of trunk-bundles, connate in a tube for most of length, anthers with longit. slits & usu. spinulose pollen, $\underline{G}$ (1–)5(–∞) with as many (twice in Ureneae = Hibisceae p.p.) styles as carpels, ± basally united, & as many locules with axile placentation & 1–∞ anatropous to campylotropous bitegmic ovules with zig-zag micropyles. Fr. a loculicidal capsule, schizocarp, rarely berry or samara; seeds exotegmic, sometimes winged or arillate; cotyledons folded, endosperm oily & proteinaceous, copious to 0. x = 6–17, 20+

Classification & principal genera:

I. **Byttnerioideae** (indumentum stellate; epicalyx usu. 0; androgynophore usu. 0; A usu. basally united, pollen usu. not spiny; usu. with staminodes); 4 tribes: a. Theobromateae – *Guazuma, Theobroma*; Byttnerieae – *Byttneria, Leptonychia*; Lasiopetaleae – *Commersonia, Lasiopetalum*; Hermannieae – *Hermannia, Melochia, Waltheria*

II. **Grewioideae** (indumentum stellate; epicalyx usu. 0; K almost free; androgynophore often present, A usu. free; (G); R: SB 37(2012)708): *Colona, Grewia, Luehea, Microcos* (Grewieae); *Corchorus, Sparrmannia, Triumfetta* (Apeibeae)

III. **Tilioideae** (trees; indumentum stellate; epicalyx usu. 0; K free, with nectaries on abaxial surface, C present; androgynophore 0; A more than 5, free; (G); aril 0): *Craigia, Mortoniodendron, Tilia* (only)

IV. **Brownlowioideae** (trees, rarely shrubs, some dioec.; indumentum stellate to lepidote; epicalyx 0; K campanulate; A free; G s.t. free): *Berrya, Brownlowia, Pentace*

V. **Helicteroideae** (trees, rarely shrubs; indumentum stellate; K a tube; androgynophore; A 10–30 forming a tube): *Helicteres, Triplochiton*, (Durioneae – trees (Roux's Model), indumentum lepidote, seeds arillate – *Durio*; prob. distinct)

VI. **Sterculioideae** (trees, often tall; indumentum stellate; epicalyx 0, C 0, staminodes 0; fls. usu. unisexual): *Brachychiton, Cola, Firmiana, Heritiera, Scaphium, Sterculia*

VII. **Dombeyoideae** (indumentum stellate or lepidote; epicalyx; P often marcescent; androgynophore; A usu. forming a tube, pollen usu. spiny; (G)): *Dombeya, Melhania*

VIII. **Bombacoideae** (trees, s.t. spiny; indumentum stellate; K usu. with basal tube over a third length; androgynophore 0; A usu. forming tube, pollen usu. not spiny; G (2–5); fr. usu capsule with hairy endocarp): *Adansonia, Bombax, Ceiba, Fremontodendron, Matisia, Ochroma, Pachira*

IX. **Malvoideae** (Herbs, shrubs or trees; stipules usu.; indumentum usu. stellate; epicalyx usu.; K usu. 5 with tube more than third length; C5, convolute. adnate to A-tube; androgynophore 0; A forming a tube, pollen usu. spiny, staminodes 0; G((3)5-∞); fr. usu. schizocarps; R: AusSB 18(2005)57); 3 tribes: [a. Kydieae = b.]; b. Hibisceae (incl. Decaschistieae, Kydieae, Ureneae) – *Hibiscus, Humbertiella, Megistostegium, Perrierophytum* (only); c. Gossypieae (R: BG 129(1968)303) – *Cienfuegosia, Gossypium, Thespesia*; d. Malveae: *Abutilon, Alcea, Althaea, Callianthe, Iliamna, Malva, Sida, Sphaeralcea, Wissadula*

Form. several genera assigned to family with difficulty (see ed. 2: 93, 435, 684), though the narrowerminded would have the subfams above as fams

A v. imp. fam. economically with tree crops incl. cocoa (*Theobroma*), cola nuts (*Cola*), but also fr. trees (*Durio*), & timbers (*Apeiba, Argyrodendron, Berrya, Guazuma, Heritiera, Sterculia, Tilia, Triplochiton* (obeche), *Hoheria, Mansonia, Nesogordonia, Ochroma* (balsa), *Pachira, Pterospermum* & *Thespesia* etc.). Many cult. orn. & fibre-pls esp. cotton (*Gossypium*) & jute (*Corchorus*), but also *Abroma, Abutilon, Bombax, Brachychiton, Ceiba, Dombeya, Firmiana, Fremontodendron, Grewia, Helicteres, Hibiscus, Kleinhovia, Lagunaria, Malachra, Plagianthus, Sida, Sterculia, Thespesia, Waltheria* & *Wissadula*, the most familiar cult. orn. being mallows (*Malva*), hollyhocks (*Alcea*), *Abutilon, Callianthe, Hibiscus, Malope, Sidalcea* etc. Some food-pls typically v. mucilaginous like okra (*Hibiscus*) & *Malva*, some medic. (*Adansonia* (baobab), *Scaphium, Sterculia*) & *Sphaeralcea* used in hair conditioners

Spp. of *Hibiscadelphus, Kokia* & *Lebronnecia*, woody island pls of Pacific & *Trochetiopsis* (St Helena), extinct or almost so

Malvastrum A. Gray. Malvaceae (Malv.-Malv.). 15 trop. & warm. R: Rhodora 84(1982)1. ***M. coromandelica*** (L.) Garcke (trop. Am., now pantrop.) – fibre for brooms etc.

Malvaviscus Fabr. = *Hibiscus*

Malvella Jaub. & Spach (~ *Sida*). Malvaceae (Malv.-Malv.). 4 Am., Medit. (Eur. 1). R: SW Nat. 19(1974)97

mamala *Homalanthus nutans*

mamey *Pouteria campechiana*

Mamillopsis C.J. Morren ex Britton & Rose = *Mammillaria*

Mammea L. Calophyllaceae (Guttiferae (I1). c. 75 trop. (C Am. 2, Afr. 3, Madag. (most), Indomal. to W Pacific). Trees; cryptic dioec. as app. herm. fls have (sterile) inaperturate pollen; some monkey-disp. in Amaz. ***M. americana*** L. (mammee apple, San Domingo apricot, WI) – cult. for ed. fr. & fls used in liqueur-making (eau de Créole), seeds toxic to fish, chicks & some insects (insecticidal substituted coumarins, also in ***M. africana*** Sabine, trop. Afr.); ***M. suriga*** (Roxb.) Kosterm. (India) – cult. orn., timber, fls to dye silk, fr. ed.

mammee (apple) *Mammea americana*; **m. zapote** *Pouteria sapota*

Mammillaria Haw. Cactaceae (III 9). 135 SW US to Colombia & Venezuela esp. Mex. R: CSJ 43(1981)41 (incl. *Dolicothele*); J. M. Soc. Low, often tuft-forming cacti with ed. fr. (chilitos), v. commonly cult. orn., some with hooked spines (e.g. ***M. bocasana*** Poselger, C Mex.) or sometimes unisexual fls (e.g. ***M. dioica*** M. Brandegee (occ. bisexual ones), SW Calif. & Baja Calif.) or with long soft hair-like spines (e.g. ***M. hahniana*** Werderm., C Mex.); some with apices with true dichotomies (Schoute's Model) e.g. ***M. parkinsonii*** Ehrenb. & ***M. perbella*** Hildm. ex K. Schum. (C Mex.)

Mammilloydia F. Buxb. = *Mammillaria*
mammoth tree *Sequoiadendron giganteum*
mamoncillo *Melicoccus bijugatus*
Mamorea Sota = *Thismia*
man orchid *Orchis anthropophora*
mana grass *Cymbopogon nardus*
manacá *Brunfelsia uniflora*
Mananthes Bremek. = *Justicia*
Manaosella J. Gómes. Bignoniaceae (3). 1 Brazil, Venez., Bolivia: ***M. cordifolia*** (A. DC.) A. Gentry. R: AMBG 98(2014)450
manbarklak *Eschweilera* spp.
manchineel *Hippomane mancinella*
Manchurian jute *Abutilon theophrasti*; **M. oak** *Quercus mongolica*; **M. walnut** *Juglans mandshurica*; **M. water rice** *Zizania palustris*
Mancoa Wedd. Cruciferae (29). 8 C & S Am. R: SB 22(2007)149
mandarin (orange) *Citrus reticulata*
Mandenovia Alava = *Heracleum* (but see NRBGE 32(1973)191)
Mandevilla Lindl. Apocynaceae (II d). Incl. evening-flowering *Macrosiphonia*, c. 130 trop. Am. R: AMBG 20(1933)645. Usu. lianes, many form. cult. orn. greenhouse pls (dipladenia) with some hybrids (esp. ***M. × amabilis*** (Backhouse) Dress = *M. splendens* (Hook.f.) Woodson × ?, c. 1868, UK – many cvs seen in trop.), incl. Chilean jasmine (***M. laxa*** (Ruíz & Pavón) Woodson, Bolivia & N Arg., natur. NSW). Local medic., latex against warts in NW Amazonia, that of ***M. vanheurckii*** (Muell. Arg.) Markgraf antifungal (Colombian Amaz.)
mandioca *Manihot esculenta*
Mandirola Decne. Gesneriaceae (II 5b). 3 trop. Am.
Mandragora Tourn. ex L. Solanaceae (IV). 3 Medit. (Eur. 1) C As., Himal. R: BBMNHN 28(1998)29. Mandrake. Alks; thick tuberous roots with fancied resemblance to human form, esp. ***M. officinarum*** L. (the mandrake of western literature assoc. with many myths (see ASNB 13,11(1991)49), allegedly screaming when pulled from ground & deafening human gatherer so that dogs were supposed to extract them, S Eur.) – mind-changing since antiquity (hallucinations between consciousness & sleep), containing hyoscyamine, grown anc. Egyptian gardens, form. medic., e.g. as anaesthetic up to 1846 (introd. of ether) & scopolamine used in 'Kwells' travel sickness pills, in Bible imp. because smell assoc. with fertility, fr. = golden apples of Aphrodite; ***M. turcomanica*** Mizg. – now restricted to SW Turkmenistan ('white mandragora') mistakenly thought to be the 'soma' (khaoma) of ancient India & Iran (see *Cynanchum*)
mandrake *Mandragora officinarum*
Manekia Trel. Piperaceae (3; Peperomiaceae). Incl. *Sarcorhachis*, 3 trop. Am. R: SB 37(2012)593. Lianes (most with juvenile monopodial veg. state, then adventitious rs) climbing to canopy with pendent sympodial infls
Manettia Mutis. Rubiaceae (IV 15). c. 80 trop. Am. R: JB 57 suppl. (1918)1. Lianes & twining herbs, some cult. orn. esp. ***M. cordifolia*** Mart. (Bolivia to Arg. & Peru) – ipecacuanha adulterant, fls bright red, & ***M. luteorubra*** (Vell.) Benth. (*M. inflata*, Paraguay & Uruguay) – fls yellow-tipped. In Peru lvs of some spp. chewed, blackening (? & preserving) teeth, also febrifugal
Manfreda Salisb. = *Agave* (but see CSM 30(1985)56)
mangel-wurzel *Beta vulgaris*
Mangenotia Pichon = *Cryptolepis*
Mangenotiella Schmid. Primulaceae. 1 New Caled.: ***M. stellata*** Schmid. R: Adans. 34(2012)338
mange-tout *Pisum sativum* var. *macrocarpon*
Mangifera L. Anacardiaceae (I). 69 Indomal. Mango. R: Lloydia 12(1949)73; A.J.G.H. Kostermans & J.-M. Bompard (1993) *The mangoes*. Scarrone's Model. Seeds (diff. spp.) disp. elephants, rhino, bats, hornbills. ***M. indica*** L. (mango, cultigen (? wild Indomal., introd. Afr. by AD 1000, Somalia 1331, Philippines early 1400s) allied to *M. sylvatica* Roxb., with v. many cvs (e.g. '**Kensington Pride**' ('Bowen', Aus. – 90% of Aus ms), some polyploid – L.B. Singh (1960) *The Mango*) – cross-poll. (between diff. cvs) required even in parthenocarpic cvs (pseudogamy), the drupe, poss. disp. by extinct megafauna (cf. *Mallotus*), an excellent fr. (flavour due to car-3-ene & dimethylstyrene), also used in chutney, pickles & squashes, immature windfalls sliced or powdered (amchur) for curry, the

ground seed a source of a flour, urine of cows fed exclusively lvs form. source of 'India yellow' dye, the timber for floor-boards, tea-chests etc. (35 M t per annum by 2009, most in India), irritant to some (mangiferine, a ketone); 25 other spp. with ed. fr., e.g. ***M. caesia*** Jack (bauno, binjai, W Mal.), ***M. pajang*** Kosterm. (bambangan, Borneo) with skin peeled like a banana, & some diploid hybrids like ***M.* × *odorata*** Griff. (apple m., kuwini, *M. foetida* Lour. (Mal.) × *M. indica*) – grows better than *M. indica* in everwet sites, flour used in Javanese delicacies (do dol); some grown as shade-trees etc.

mangium *Acacia mangium*

Manglietia Blume = *Magnolia*

Manglietiastrum Law = *Magnolia*

mango *Mangifera* spp. esp. *M. indica*; **apple m.** *M.* × *odorata*; **bush m.** *Cordyla pinnata*; **plum m.** *Bouea macrophylla*

mangold *Beta vulgaris*

Mangonia Schott. Araceae (VII 4). 2 Brazil, Uruguay

mangosteen *Garcinia mangostana*

mangrove Woody pls growing in muddy swamps inundated by tides, e.g. spp. of *Aegiceras, Avicennia, Bruguiera, Ceriops, Conocarpus, Kandelia, Laguncularia, Lumnitzera, Pelliciera, Rhizophora, Sonneratia, Xylocarpus*; **grey m.** *Avicennia marina*; **m. bark** bark of these pls used in comm. leather-tanning, though leather turned intense red by it

Manicaria Gaertn. Palmae (V 9). 2 trop. Am. R: Palms 54(2010)124. ***M. saccifera*** Gaertn. (bussu palm) – Schoute's Model, source of sago in Venez., lvs used as sails & thatch (troolie in Guyana), the spathe used as a hat, the temiche cap of the Lower Amazon, fr. used as spinning-tops, seeds an oil source

manicole *Euterpe oleracea*

Manihot Mill. Euphorbiaceae (Crot.-Man.). c. 100 trop. & warm Am. R: FN 13(1973)19. Monoec. trees, shrubs & herbs; Leeuwenberg's Model. ***M. esculenta*** Crantz (*M. utilissima*, cassava, manioc, mandioca, tapioca, gari (W Afr.), yuca, cultigen domesticated from 'subsp. *flabellifolia*' S Amaz., cult. 2000 yrs) – shrubby tree with large tuberous roots a major trop. staple (228 M t 2010) rather immune to insect attack because of high levels of cyanide (eating unprepared cassava leads to cyanide-assoc. disease, konzo), a reliable crop (cult. by women in Amaz., cf. coca) on somewhat impoverished soils with over 100 cvs with differing amounts of cyanide ('sweet' ones with glycosides only in bark of tubers), removed by squeezing the ground tuber in water & by evaporation during drying; cassava meal (Brazilian, Pará or Rio arrowroot) & tapioca (sediment from boiled extract) used in soups, puddings etc., a glue form. used on postage stamps, sugar, alcoholic drinks (comm. beer in Afr.; tucupi in Braz.) & acetone all derived from it, the toxic juice evaporated being cassareep used for preserving meat & in certain table sauces, wood suitable for chip- & particle-board, extrafl. nectaries attract red ants which prevent pl. being climbed by beans which only then climb tougher maize in the maize – beans – cassava system of the Kayapó people of S Am.; ***M. glaziovii*** Muell. Arg. (Brazil) – source of Ceara or Manicoba rubber & oilseeds

Manihotoides D. Rogers & Appan = *Manihot*

Manila copal *Agathis dammara*; **M. elemi** *Canarium luzonicum*; **M. hemp** *Musa textilis*; **M. maguey** *Agave cantala*

Manila grass *Zoysia matrella*

Manilkara Adans. Sapotaceae (I 2). 65 trop. (As. to Pacific 15, Afr. & Madag. 20, Am. 30). Aubréville's Model. Milky latex & fr. of commercial imp. esp. ***M. bidentata*** (A. DC.) A. Chev. (balata, trop. Am.) – in some Tobago forests almost half of all large trees, source of non-elastic rubber used in machine-belts, boot-soles, largely from wild trees; ***M. chicle*** (Pittier) Gilly (crown gum, trop. Am.) – subs. for *M. zapota*; ***M. hexandra*** (Roxb.) Dubard (palu, India) – hard timber; ***M. obovata*** (Sabine & G. Don) J.H. Hemsl. (African pear, W Afr.); ***M. zapota*** (L.) P. Royen (sapodilla (plum), chiku, chicle, naseberry, beef apple, Mex. & C Am.) – tapped like rubber (6 K t per annum by 1930) but only once every 2–3 yrs, used in cable construction, golfballs, dental surgery, the chewing-gum of the Aztecs, the tree being encouraged by the Mayas such that its presence in many sites a reflection of the practices of that lost culture, fr. ed. (? disp. by extinct megafauna) also used in juice, syrup, wine & vinegar, insect-resistant timber for Mayan beams, musical instruments, bark medic. (dysentery, diarrhoea)

Manilkariopsis (Gilly) Lundell = *Manilkara*

Maniltoa R. Scheffer. Leguminosae (I 2). 20–25 Indomal. to Aus. esp. New Guinea. ***M. lenticellata*** C. White (New Guinea) – new lvs pale & limp (cf. *Amherstia, Brownea, Saraca*)

manio *Podocarpus nubigenus, P. salignus*
manioc *Manihot esculenta*
Manisuris L. (~ *Rottboellia*). Gramineae (XXII 3). 1 Ind.: ***M. myuros*** L.
manjack *Cordia collococca*
Manjekia W. Baker & Heatubun (~ *Adonidia*). Palmae (V 14i). 1 Biak: ***M. maturbongsii*** (W. Baker & Heatubon) W. Baker & Heatubon. R: KB 69[9525](2014)9
manketti (nut oil) *Ricinodendron rautanenii*
Mankyua B.Y. Sun & al. Ophioglossaceae. 1 Jeju Is., Korea: ***M. chejuense*** B.Y. Sun & al.
manna Various ed. materials, some of pl. origin & usu. sweet, often exudations following insect attack, like honeydew but set hard; that of Bible could have been lichen, but see also *Hordeum*; modern comm. sources incl. *Fraxinus ornus* (**m. ash**) & it is coll. from *Tamarix* spp., *Haloxylon salicornicum* (poss. Biblical m.), *Larix decidua, Olea europaea*, etc.; **m. croup** *Glyceria fluitans*
Mannagettaea H. Sm. Orobanchaceae (Orob.). 3 E Siberia to W China
Manniella Reichb.f. Orchidaceae (IV 2e). 2 trop. W Afr.
Manniophyton Muell. Arg. Euphorbiaceae (Al.-Crot.). 1 C trop. Afr.: ***M. africanum*** Muell. Arg. (gasso nut) – liane with fibres used for ropes & nets. R: A. R-Sm., *Gen. Euph.* (2001)343
manoao *Halocarpus biformis*
Manoao Molloy = *Lagarostrobos* (but see NZJB 33(1995)196)
Manochlamys Aellen (~ *Exomis*). Amaranthaceae (Atrip.; Chenopodiaceae I 3). 1 S Afr.: ***M. albicans*** (Aiton) Aellen. R: Strel. 10(2000)224
Manoelia Bowdich = *Withania*
Manongarivea Choux = *Lepisanthes*
Manostachya Bremek. Rubiaceae (IV 15). 3 trop. Afr.
Manotes Sol. ex Planch. Connaraceae (III). 4–5 humid trop. Afr. R: AUWP 89–6(1989)294
Manothrix Miers = *Mesechites*
Mansoa DC. Bignoniaceae (3). 11 trop. S Am. R: AMBG 98(2014)450. ***M. alliacea*** (Lam.) A. Gentry etc. – widely used indig. medic., cult. orn.; ***M. standleyi*** (Steyerm.) A. Gentry (trop. Am.) – garlicky leaf decoction used as febrifuge
Mansonia J.R. Drumm. Malvaceae (Hel.; Sterculiaceae). 5 W, C, E Afr., Assam, Myanmar – v. disjunct. ***M. altissima*** (A. Chev.) A. Chev. (W Afr.) – timber (aprono, bété) used for furniture; ***M. gagei*** J.R. Drumm. (kalamet, Burma) – fragrant wood used as cosmetic
Mantalania Capuron ex J. Leroy. Rubiaceae (?II 1). 2–3 Madag. Rauh's Model
Mantisalca Cass. (~ *Centaurea*). Compositae (Card.-Cent.). 1(–5) Medit. (Eur. 1: ***M. salamantica*** (L.) Briq. & Cavill.)
Mantisia Sims = *Globba*
manuka *Leptospermum scoparium*
Manulea L. Scrophulariaceae (Manul.). 73 S Afr. R: O.M. Hilliard (1994) *The Manuleeae*: 291. Some cult. orn. annuals & shrublets
Manuleopsis Thell. Scrophulariaceae (Manul.). 1 SW Afr.: ***M. dinteri*** Thell. R: O. Hilliard (1994) *The Manuleeae*: 78
Manyonia H. Robinson. Compositae (Vern.-Dip.). 1 Tanz.: ***M. peculiaris*** (Verdc.) H. Robinson
manzanilla *Crataegus mexicana*
manzanita *Arctostaphylos pungens*
manzanote *Olmediella betschleriana*
Maoutia Wedd. = *Leucosyke*
Mapania Aubl. Cyperaceae (I 1). 85 trop. (exc. Madag.). D.A. Simpson (1992) *Revision of the genus M.* Some resemble pandans (***M. baldwinii*** Nelmes (Ivory Coast) Corner's Model), 'litter-box pls'; some in Malay Pen. allegedly rat-disp.; some local medic. ***M. palustris*** (Steud.) Fernandez-Villar (Indomal.) – used for weaving mats & baskets
Mapaniopsis C.B. Clarke. Cyperaceae (I 1). 2 SE Venez. to N Brazil
Mapinguari Carnevali & R. Singer = *Maxillaria* (but see Lankesteriana 7(2007)525)
maple *Acer* spp.; **Australian m.** *Flindersia brayleyana*; **field** or **hedge m.** *A. campestre*; **flowering m.** *Callianthe* hybrids; **great** or **Scottish m.** *A. pseudoplatanus*; **Japanese m.** *A. palmatum*; **Norway m.** *A. platanoides*; **Pacific m.** *A. macrophyllum*; **m. peas** *Pisum sativum*; **Queensland m.** *F. brayleyana*; **red m.** *A. rubrum*; **rock, sugar** or **striped m.** *A. saccharum* (**fiddle-back** or **bird's-eye m.** from its burrs); **silver m.** *A. saccharinum*; **tiger m.** *A.* spp. (US) esp. *A. macrophyllum* (W), *A. rubrum* (E)
Mapouria Aubl. = *Psychotria*
Mappia Jacq. Icacinaceae. Excl. *Leretia* 4 trop. Am. R: Phytotaxa 116,1(2013)1

Mappianthus Hand.-Mazz. Icacinaceae. 2 S China, Borneo

Maprounea Aubl. Euphorbiaceae (Hipp.). 5 trop. Am. (3), W Afr. (2)

mapuche *Araucaria araucana*

maqui *Aristotelia chilensis*

Maquira Aubl. Moraceae (III). 4 trop. Am. R: FN 7(1972)64, 83(2001)263. Arrow poisons (cardenolides). ***M. sclerophylla*** (Ducke) C. Berg (Amaz. Brazil) – hallucinogenic snuff from fr.

Maracanthus Kuijt (~ *Oryctina*). Loranthaceae. 2 N. Venez., Costa Rica. R: Britt. 28(1976)231

maracuja, maracuyá *Passiflora quadrangularis*

Marah Kellogg. Cucurbitaceae (XII). 7 W N Am. R: Madroño 13(1955)113. Tendrilled monoec. climbers with large tubers. Some local skin medic., cult. orn. incl. ***M. macrocarpus*** (Greene) Greene (S Calif. & Baja Calif.) – seeds source of a red dye

Marahuacaea Maguire (~ *Amphiphyllum*). Rapateaceae (I 1). 2 Venez. R: ABV 14,3(1984)16

marama bean *Tylosema esculentum*

maramara *Schefflera morototoni*

marang *Artocarpus odoratissimus*

Maraniona C. Hughes & al. Leguminosae (III 11). 1 N Peru: ***M. lavinii*** C. Hughes & al. R: SB 29(2004)371

Maranta Plum. ex L. Marantaceae. 32 trop. Am. R (subg. M.: 16): NJB 6(1986)729).? Heterogeneous. ***M. arundinacea*** L. (arrowroot, Bermuda or St Vincent a., C Am.) – rhizome the source of arrowroot (so-called because used as a poultice to remove poison from arrow wounds), a readily digestible (small-grained) starch used for infants & invalids esp. in treatment of diarrhoea, also used as face-powder & form. coating on carbon-less paper used for computer printout, poss. fuel-alcohol source & fibre for tear-resistant paper bags, not cult. pre-Columbian Am., mostly grown at St Vincent, WI; other spp. cult. orn. housepls esp. ***M. leuconeura*** C. J. Morren (prayer plant, Brazil) notably var. ***kerchoveana*** C. J. Morren with light green lvs marked with a row of dark brown blotches on either side of midrib

Marantaceae R. Br. Magnoliidae – Zingiberales. 27/525 trop. esp. Am. (not Aus.). Rhizomatous perennial herbs, some with sublianoid stems, rhizomes sympodial usu. starchy, vessels usu. only in roots. Lvs ± distichous with open sheath, distinct & sometimes winged petiole & simple lamina rolled from 1 side in bud & often patterned, with distinct pulvinus (allowing laminar movement vis-à-vis sun) at lamina base, venation pinnate-parallel with prominent midrib. Fls bisexual, basically 3-merous in thyrses comprising (1)2-flowered cymules, those of a pair mirror-image asymmetric), s.t. on a separate shoot from rhizome; K 3, not petaloid, C 3 with a basal tube, usu. white, with 1 petal often hood-like & larger, 1 functional stamen (posterior member of inner whorl), petaloid bearing 1 pollen-sac on 1 edge, staminodes (2)3 or 4, petaloid but small (2 from inner A whorl, other (1)2 the laterals of outer), 1 of the inner forming a labellum over pistil before anthesis, other often a landing-stage for insects, $\overline{G}$ (3), 3-loc. (2 often empty or obsolete) with term. style & septal nectaries at ovary summit, each loc. with 1 almost basal, anatropous to campylotropous bitegmic ovule. Fr. a capsule, caryopsis-like or berry; seeds with micropylar operculum & basal aril (in capsules), embryo linear & usu. curved to plicate in starchy perisperm. n = 4–14+

Principal genera: *Calathea, Ctenanthe, Goeppertia, Ischnosiphon, Maranta, Monotagma, Phrynium*

Pollen released from the anther is deposited in a subapical cavity on the outside of the style; when an insect forces its head between the 2 inner staminodes searching for nectar, it pushes the labellum & releases the style until then under tension, receiving the pollen on its back (cf. Leguminosae); on euglossine bees deposited where they cannot groom & remove it. Some cleistogamy but also pseudocleistogamy (some *Calathea* fls forcibly opened by pollinator)

Arrowroot (*Maranta, Myrosma*), wax (*Calathea*), basket-weaving (*Donax, Ischnosiphon, Phrynium*), v. sweet polypeptides (*Thaumatococcus*) & some cult. orn. esp. *Calathea* (also ed. tubers & fls), *Ctenanthe, Goeppertia* & *Maranta*

Maranthes Blume. Chrysobalanaceae (3). 12 trop. (Afr. 10, trop. As. & Pacific 1, Panamá 1 (closely allied to As. sp.)). R: FOW 10(2003)56. ***M. aubrevillei*** (Pellegrin) Prance (W Afr.) – toothed lvs (v. rare in C.), bubbling & hissing slash; ***M. corymbosa*** Blume (As., Pacific) – imp. timber in Solomon Is. & Philippines (parquet, plywood); ***M. polyandra*** (Benth.) Prance (W Afr.) – bat-poll., form. favoured for charcoal by Nigerian blacksmiths

Marantochloa Brongn. ex Gris. Marantaceae. c. 15 trop. Afr. to Comoros, Réunion. R: BJBB 65(1996)369. ***M. cordifolia*** (K. Schum.) Koechl. – pith & young lvs ed. gorillas

marara *Pseudoweinmannia lachnocarpa*

maraschino cherry or **marasco** *Prunus cerasus* 'Marasca'

Marasmodes DC. Compositae (Anth.-Pen.). 4 SW Cape. R: BJLS 96(1988)306

Marathrum Bonpl. Podostemaceae (III). Incl. *Vanroyenella*, 10 trop. Am. R: MedBotUtrecht 107(1951)70,131

Marattia Sw. Marattiaceae. Excl. *Eupodium, Ptisana*, 7 trop. Am., Hawaii

Marattiaceae Kaulf. Marattiidae – Marattiales. Incl. Angiopteridaceae, Christenseniaceae, Danaeaceae, 6/c. 80 trop. & warm. R: T 55(2006)709, 57(2008)740. Large scaly terr. ferns with erect but short stout stems (with polycyclic dictyostele), s.t. markedly dorsiventral. Roots, stems & lvs with mucilage canals. Fronds often v. large, simple to 1 or more pinnate (xylem polycyclic) with pneumathodes (lenticels), large fleshy starchy stipules at base & swollen pulvini along petioles & rachises. Sori intramarginal on lower surface with free sporangia or these in synangia encl. 1000–7000 spores; indusium 0. Spores all of 1 type giving rise to monoec. prothalli, potentially rather long-lived resembling anthoceroid liverwort. n = 40

Genera: *Angiopteris* (Angiopteridaceae – fronds pinnate, veins free, sporangia separate), *Marattia* (*Eupodium, Ptisana*; Marattiaceae s.s. – fronds pinnate, veins free, synangia paired, stem erect, radially symmetrical), *Danaea* (Danaeaceae – as M. s.s. but synangia single, stem dorsiventral), *Christensenia* (Christenseniaceae (Kaulfussiaceae) – fronds palmate, veins reticulate), the fam. sometimes treated as an order comprising the above segregate fams

Similar fossils from Upper Carboniferous

Some cult. orn. & ed. stems etc. (*Angiopteris, Marattia*)

Marattiidae Klinge (~ Polypodiidae). 6/c. 80 trop. & warm.

Only fam.: Marattiaceae (q.v.)

Marattiopsida Doweld. See Marattiidae

marble wood *Diospyros* spp. esp. *D. marmorata* & *D. oocarpa*

marc *Vitis vinifera*

Marcania Imlay. Acanthaceae (III 2c). 1 Thailand: ***M. grandiflora*** Imlay

Marcelliopsis Schinz. Amaranthaceae (I 2). 3 S Afr.

Marcetella Svent. = *Poterium*

Marcetia DC. Melastomataceae. 31 trop. S Am., esp. Bahia (25)

Marcgravia Plum. ex L. Marcgraviaceae. c. 60 trop. Am. Climbing epiphytes, the climbing shoots with small 2-ranked, round lvs adpressed to surface climbed & covering adventitious roots, the pl. later producing (cf. *Hedera*) pendulous shoots with leathery lanceolate lvs & the capacity to revert to the climbing condition; the pendulous shoots tipped with dense racemes of green fls with stalked pitcher-like nectaries (bracts); poll. hummingbirds, bats (e.g. ***M. myriostigma*** Triana & Planch. (Brazil)) & non-flying mammals ((e.g. ***M. nepenthoides*** Seem. (C Am.) – opossums), others allegedly poll. by bees or lizards; infls of ***M. rectiflora*** Triana & Planch. (WI, cult. orn. liane to 10 m) erect with pendulous nectaries, those in other spp. pendulous; some fls cleistogamous

Marcgraviaceae Bercht. & J. Presl. Magnoliidae – Ericales. 8/130 trop. Am. R: Caldasia 29(2007)215. Glabrous lianes or epiphytes with clinging roots, rarely erect shrubs or trees. Lvs simple, entire, often dimorphic, those on juvenile rooting shoots sessile & oriented distichously, those on flowering ones petiolate & in spirals; stipules 0. Fls bisexual, reg. in term., often pendulous, racemes, spikes or umbels, often bird-poll. though some cleistogamous, bracts pitcher-like, saccate or spurred nectaries; pedicels usu. with 2 K-like prophylls; K (4) 5, s.t. basally connate, C 3–5, imbricate, ± basally connate & s.t. (*Marcgravia*) distally so, forming a calyptra (decid.), A 3–40, the filaments ± basally connate & sometimes adnate to C, anthers with longit. slits, $\underline{G}$ (2–20) with simple or lobed stigma, 1-loc. becoming pluriloc. by intrusion of placental partitions with numerous anatropous, bitegmic ovules on resultant axile placentas. Fr. tardily dehiscent capsule (s.t. berry-like) with few-∞ small seeds with straight or weakly curved embryo in little or 0 endosperm. Starch app. 0. n = 18

Genera: *Marcgravia* (Marcgravioideae); *Marcgraviastrum, Norantea, Ruyschia, Sarcopora, Schwartzia, Souroubea* (Noranteioideae)

Marcgraviastrum (Szyszyl.) de Roon & S. Dressler (~ *Norantea*). Marcgraviaceae. 15 trop. Am. (E Brazil 2)

Marcus-kochia Al-Shehbaz (~ *Malcolmia*). Cruciferae (Anast.). 4 Medit., SW As. R: HPB 19(2014)56

mare's tail *Hippuris vulgaris*

Maresia Pomel. Cruciferae (4). 5 Medit. (Eur. 1) to Caspian & S Iran. R: HPB 19(2014)53

Mareya Baill. Euphorbiaceae (Acal.-Acal.-Claox.). 3 trop. Afr. (Guineo-Congolian). R: BJBB 65(1996)4, 66(1997)138

Mareyopsis Pax & K. Hoffm. Euphorbiaceae (Acal.-Acal.-Mar.). 2 W trop. Afr. R: BJBB 65(1996)15, 66(1997)134. Dioec. (cf. *Mareya*)

Margaranthus Schldl. = *Physalis*

Margaretta Oliv. (~ *Asclepias*). Apocynaceae (V c; Asclepiadaceae III 1). 1 trop. Afr.: ***M. rosea*** Oliv. R: KB 51(1996)724

Margaritaria L.f. Phyllanthaceae (Phyll.-Fluegg.; Euphorbiaceae). 14 trop. R: JAA 60(1979)407. ***M. discoidea*** (Baill.) Webster (trop. Afr.) – fr. ed. Mafia Is.

Margaritolobium Harms = *Muellera*

Margaritopsis C. Wright = *Eumachia*

Margbensonia A. Bobrov & Melikian = *Podocarpus*

Margelliantha Cribb. Orchidaceae (V 16d). 6 E Afr. R: KB 34(1979)329

Marginaria Bory = *Pleopeltis*

Marginariopsis C. Chr. = *Pleopeltis*

margosa *Azadirachta indica*

Margotia Boiss. = *Thapsia* (but see Lagascalia 13(1985)228)

marguerite *Leucanthemum* spp. esp. *L. vulgare*, *Argyranthemum* spp.

Margyricarpus Ruíz & Pavón. Rosaceae (8). Incl. *Tetraglochin*, c. 8 Andes, S Brazil, Uruguay. Hybrids with *Acaena* spp. recorded. ***M. pinnatus*** (Lam.) Kuntze (*M. setosus*, pearl fruit, Andes) – cult. orn. rock-pl. with white berries, used in fertility control in Uruguay

Marianthus Huegel ex Endl. (~ *Billardiera*). Pittosporaceae. 13 SW Aus., 1 S Aus. to Vic. R: AusSB 17(2004)128

marigold, common or **pot** *Calendula officinalis*; **African m.** *Tagetes erecta*; **bur m.** *Bidens tripartita*; **corn m.** *Glebionis segetum*; **French m.** *T. erecta*; **marsh m.** *Caltha palustris*; **Scotch m.** *Calendula officinalis*

marihuana, marijuana *Cannabis sativa* 'subsp. *indica*'

Marila Sw. Calophyllaceae (Guttiferae (I 1). c. 40 trop. Am.

Marina Liebm. Leguminosae (III 10). 38 Mex., ext. to SW US & C Am., Venez. R: MNYBG 27(1977)55. ***M. scopa*** Barneby (C Am. to Venez.) – stems tied together for brooms

marionberry *Rubus* 'Marion'

Mariosousa Seigler & Ebinger (~ *Acacia*). Leguminosae. 13 SW US to C Am. R: Novon 16(2006)415. Prickles 0

Maripa Aubl. Convolvulaceae (11). 19 warm & trop. Am. esp. N S Am. Lianes to 30 m. R: AMBG 60(1973)357. Some ed. fr.

Mariposa (Alph. Wood) Hoover = *Calochortus*

mariposa lily *Calochortus* spp.

Mariscopsis Chermezon = *Queenslandiella*

Marisculus Goetgh. = *Alinula*

Mariscus Vahl = *Cyperus*

marita *Pandanus conoideus*

marjoram, sweet m. *Origanum majorana*; **pot m.** *O. onites*; **'Spanish' m.** see *Urtica*; **winter m.** *O. dictamus* subsp. *viride*

Markea Rich. Solanaceae (IV 3). 9 trop. Am. Epiphytic or scandent, some myrmecophilous. R: Kurtziana 25(1997)71. See also *Merinthopodium*

Markhamia Seem. ex Baill. Bignoniaceae. 6 trop. Afr. (4) & As. (2). Some cult. orn. & timber

marking nut *Semecarpus anacardium*

Markleya Bondar = *Attalea*

marlberry *Ardisia* spp.

Marlierea Cambess. = *Myrcia*

Marlieriopsis Kiaerskov = *Blepharocalyx*

marlock *Eucalyptus* spp.

Marlothia Engl. = *Helinus*

Marlothiella H. Wolff. Umbelliferae (III 8). 1 S Afr.: ***M. gummifera*** H. Wolff. R: EJB 48(1991)225

Marlothistella Schwantes (~ *Ruschia*). Aizoaceae (V). 2 W & E Cape. R: H. E. K. Hartmann, *Ill. Handb. Succ. Pls*, Aiz. F–Z(2002)142

marmalade plum *Pouteria sapota*

Marmaroxylon Killip (~ *Zygia*). Leguminosae (II 4). 9 Am. R: KB 46(1991)515. Timbers

marmelo *Bunchosia* spp.

Marmoritis Benth. (*Phyllophyton, Pseudolophanthus;* ~ *Nepeta*). Labiatae (VII 2c). 4–5 Pakistan to China. R: BZ 77,12(1992)125

Marojejya Humbert. Palmae (V 14f). 2 NE Madag. Litter-trapping crowns

maroola plum *Sclerocarya birrea* subsp. *caffra*

Marquesia Gilg (*Trillesanthus*). Dipterocarpaceae (Monot.). 4 trop. Afr.

marram grass *Ammophila arenaria*

marri *Corymbia calophylla*

marrow *Cucurbita pepo;* **Madeira m.** *Sicyos edulis*

Marrubium Tourn. ex L. Labiatae (VI). c. 40 Eur. (12). Medit., As. Some cult. orn. incl. ***M. vulgare*** L. ((white) horehound, Euras., Medit., Macaronesia) – form. much-used medic. herb (sore throats, tonic) as tea, in sweets & in liqueurs, invasive in Aus. where controlled with horehound plume moth (*Wheeleria spilodactylus*)

Marsdenia R. Br. Apocynaceae (V a; Asclepiadaceae III 4). Incl. *Dregea, Gymnema, Leichardtia, Loniceroides, Stephanotis, Thozetia, Wattakaka,* c. 200 trop. & warm. (Aus. 33, Afr. 15). Usu. lianes. Some medic. (incl. emetics), poisonous; some Madag. spp. a source of Madagascar rubber; some cult. orn. ***M. abyssinica*** (Hochst.) Schltr (*D. a.*, trop. Afr.) – lvs cooked Uganda; ***M. australis*** (R. Br.) Druce (*L. a.*, alunqua, bush banana, Aus.) – fr. ed. Aborigines; ***M. castillonii*** Lillo ex Meyer (Arg., Paraguay) – roots ed.; ***M. erecta*** R. Br. (SE Eur. to Iran) – latex blisters skin; ***M. flavescens*** A. Cunn. ex Hook. & ***M. viridiflora*** R. Br. (Aus.) – tubers ed. Aborigines; ***M. floribunda*** (Brongn.) Schltr. (*S. f.*, stephanotis, Madagascar jasmine, floradora, Madag.) – commonly cult. liane grown for large white waxy fls used in wedding corsages etc., woody fr. matures over 2 yrs; ***M. hamiltonii*** Wight (India) – ed. fr.; ***M. sylvestris*** (Retz.) P. Forst. (*G. s.*, OW) – used (Ayurvedic medic.) to control hyperglycaemia, acting on tastebuds (gymnemic acid occupying glucose receptors blocking sweet-taste sensation, even overcoming miraculin (*Synsepalum* spp.)) & in intestine (preventing glucose absorption); ***M. tenacissima*** (Roxb.) Moon (Himal., NE India) – source of fibre for rope etc.; ***M. tinctoria*** R. Br. (Java indigo, Mal.) – stimulates appetite, form. cult. for indigo-like dye; ***M. volubilis*** (L.f.) Cooke (*D. v., W. v.*, Ind. to W Mal.) – eaten with curry, also febrifuge & emetic (!)

marsh betony *Stachys palustris;* **m. elder** *Euphrosyne xanthiifolia;* **m. fern** *Acrostichum* spp.; **m. grass** *Sporobolus* spp.; **m. mallow** *Althaea officinalis;* **m. marigold** *Caltha palustris;* **m. rose** *Mimetes (Orothamnus) zeyheri;* **m. rosemary** *Limonium* spp., *Rhododendron* spp.

Marshallfieldia J.F. Macbr. = *Adelobotrys*

Marshallia Schreb. Compositae (Hele.-Mar.). 7 US. R: CGH 181(1957)41. Cult. orn. (Barbara's buttons) esp. ***M. grandiflora*** Beadle & Boynton (E N Am.)

Marshalljohnstonia Henrickson. Compositae (Lact.). 1 Mexico: ***M. gypsophila*** Henrickson. R: SB 1(1977)169

Marsilea L. Marsileaceae. c. 70 trop. (esp. Afr.) & temp. (Eur. 4). R: SBi 49(1968)273, (Am.) SBM 11 (1986)1. Lvs petiolate with 4 clover-like lobes floating in deep water, held above water in shallow. Some Aus. spp. with v. drought-resistant sporocarps living up to 100 yrs; on water uptake, gelatinous interior swells, splitting sporocarp & worm-like mass exudes carrying sori with it leading to germ. of spores & fertilization. Sporocarps (nardoo) of Aus. spp. esp. ***M. drummondii*** A. Braun ground & eaten by Aborigines & early white settlers (Burke & Wills died from eating it – thiaminase, preventing absorption of thiamine); ***M. crenata*** C. Presl (SE As. to Aus.) – young fronds a veg., local medic.

Marsileaceae Mirb. Polypodiidae – Salviniales. Incl. Pilulariaceae, 3/c. 75 trop. & temp. R: T 55(2006)711. Small amphibious or aquatic heterosporous pls with creeping rhizomes with solenosteles. Fronds circinate, borne in 2 dorsal rows; veins dichotomously branched. Micro- & mega-sporangia together in hard sporocarps (not produced when pls submerged) attached to frond-base or in axil, the sporocarps perhaps tightly folded pinnae with a number of elongate sori, each covered with a membranous indusium. Spores can be dormant for up to 100 yrs. Male gametophyte of only 9 cells incl. 1 prothallial cell. $n = 10, 20$

Genera: *Marsilea* (frond 4-lobed), *Pilularia* (frond without lobes), *Regnellidium* (frond with 2 lobes). *Pilularia* form. placed in own fam. but young *Marsilea* pls have unlobed fronds like *Pilularia*

Marsippospermum Desv. Juncaceae. 4 temp. S Am., Falkland Is., NZ & is. (1). R: FOW 6(2002)5

Marssonia Karsten = *Napeanthus*
Marsypianthes Mart. ex Benth. Labiatae (VII 3c). 5–6 trop. Am. R: FRB 85(1936)184
Marsypopetalum R. Scheffer. Annonaceae (IV 7). 6 SE As. to C Mal. R: SAB 9(2011)24
Martellidendron (Pichi-Serm.) Callm. & Chassot (~ *Pandanus*). Pandanaceae. 6 Madag., Seychelles. R: T 52(2003)755
Marthella Urb. Burmanniaceae. 1 Trinidad: ***M. trinitalis*** (Johow) Urb.
Marticorenia Crisci. Compositae (Mut.-Nass.). 1 Chile: ***M. foliosa*** (Philippi) Crisci – shrub. R: JAA 55(1974)38
Martinella Baill. Bignoniaceae (3). 2 trop. Am. R: AMBG 98(2014)453. ***M. obovata*** (Kunth) Bureau & K. Schum. used & cult. by Am. Indians for eye troubles
Martinezia Ruíz & Pavón = *Prestoea* + *Aiphanes*
Martiodendron Gleason. Leguminosae (I 3). 4 trop. S Am.
Martretia Beille. Phyllanthaceae (Antid.-Mart.; Euphorbiaceae s.l.). 1 W & C Afr.: ***M. quadricornis*** Beille. R: BJBB 59(1989)319
Martynia Martyn ex L. Martyniaceae (Pedaliaceae s.l.). 1 Mex.: ***M. annua*** L. – fr. s.t. pickled; sticky lvs used to remove lice from fowls
Martyniaceae Horan. (~ Pedaliaceae). Magnoliidae – Lamiales. 5/15 trop. & warm Am. R: T 18(1969)527. Sticky-hairy herbs (rarely shrubs, *Holoregmia*), usu. annual; roots often tuberous. Lvs simple, opp., in spirals, with toothed margins. Fls bisexual, in term. racemes. K5, s.t. connate, C(5); A didynamous, 4 with 1 staminode, or 2 with 3, inserted in tube; $\underline{G}(2)$, placentation parietal, 2-∞ ovules per carpel. Fr. an elongated incompletely loculicidal capsule with apical spurs or hooks; endocarp woody; seeds 4-∞ with ± gelatinous testa, embryo straight in thin endosperm. n = 15,16
Genera: *Craniolaria, Holoregmia, Ibicella, Martynia, Proboscidea*
App. not close to Pedaliaceae, from which differing in pollen & placentation
Insects get trapped in sticky indumentum (cf. Lentibulariaceae)
Veg. & poss. comm. oilseeds (*Proboscidea*)
marula *Sclerocarya birrea* subsp. *caffra*
marupa *Quassia amara*
Maruta (Cass.) Gray = *Anthemis*
marvel of Peru *Mirabilis jalapa*
Marvello hair Gut from Chinese silkworms fed on *Liquidambar formosana*
Mary Jane *Cannabis sativa*
Maryland pinkroot *Spigelia marilandica*
Mary's bean *Merremia discoidesperma*
marzipan confection of almond meal with sugar or honey
Mascagnia (DC.) Bertero. Malpighiaceae. Incl. *Triopterys*, excl. *Adelphia, Aenigmatanthera, Alicia, Amorimia, Carolus, Christianella, Malpighiodes, Niedenzuella*, c. 45 Mex. to Argentina
Mascarena L. Bailey = *Hyophorbe*
Mascarenhasia A. DC. Apocynaceae (II b). 8 Madag. with 1 ext. to E & S Afr. R: WAUP 97–2(1997)21. Prévost's Model. Rubber sources (Madagascar r.)
Maschalocephalus Gilg & K. Schum. Rapateaceae (II 2). 1 Sierra Leone, Liberia: ***M. dinklagei*** Gilg & K. Schum. Only Afr. R.
Maschalocorymbus Bremek. = *Urophyllum*
Maschalodesme K. Schum. & Lauterb. Rubiaceae (II 5). 2 New Guinea
Masdevallia Ruíz & Pavón. Orchidaceae (V 13c). c. 580 trop. Am. highlands. R: MSB 16(1986)1, M.E. Gerritsen & R. Parsons (2005), *Ms*. Cult. orn. epiphytes with K with elongated tails. Some attract poll. flesh-flies through colour & smell of bad meat but there is no 'reward'. Sect. Chimaeroideae predicted to be poll. by fungus-gnats
Masdevalliantha (Luer) Szlach. & Marg. = *Andinia*
mash melon *Cucumis melo* Reticulatus Group
masindi *Cynodon transvaalensis*
maslin mixed grain esp. rye with wheat; bread of mixed grains
Masoala Jum. Palmae (V 14f). 2 NE Madag. Litter-trapping crowns. ***M. madagascariensis*** Jum. – now reduced to 3 trees in wild
Massangea C.J. Morren = *Guzmania*
Massartina Maire = *Elizaldia*
Massia Bal. = *Eriachne*
Massonia Thunb. ex Houtt. Asparagaceae (Hyacinthaceae). Incl. *Whiteheadia*, c. 10 dry S Afr. R: JSAB 42(1976)406, FR 108(1997)66. ***M. depressa*** Thunb. ex Houtt. – fls poll. small rodents incl. 2 gerbil spp., jelly-like nectar poss. discouraging robbers

massoy bark *Cryptocarya massoy*

Massularia (K. Schum.) Hoyle. Rubiaceae (II 1). 2 trop. Afr. R: Phytotaxa 203(2015)263. ***M. acuminata*** (G. Don) Hoyle – Petit's Model, pollen in pollinia; chewing-stick in Nigeria

mast fallen fr. of beech, oak etc. form. much used as food for pigs etc. Right to be permitted to allow animals into woodland to eat it known as pannage; some woodlands (e.g. in Domesday Book) measured by amount of mast rather than area

Mastersia Benth. Leguminosae (III 18). 2 Assam (1), C Mal. (1). R: Blumea 30(1984)77. ***M. bakeri*** (Koord.) Backer (Mal.) – a cover-pl. in plantations

Mastersiella C. Benedict. Restionaceae. 3 SW & S Cape. R: Bothalia 15(1985)481

masterwort *Astrantia major* & *A. maxima*

mastic *Pistacia lentiscus*; **American m.** *Schinus molle*; **Barbados m.** *Sideroxylon foetidissimum*; **Bombay m.** *P. atlantica*

Mastichodendron (Engl.) H.J. Lam = *Sideroxylon*

Mastigion Garay & al. = *Bulbophyllum*

Mastigosciadium Rech.f. & Kuber. Umbelliferae (III 5). 1 Afghanistan: ***M. hysteranthum*** Rech.f. & Kuber

Mastigostyla I.M. Johnston. Iridaceae (VII 5). Incl. *Cardenanthus*, c. 20 Peru, Arg., Bolivia. R: Rhodora 64(1962)291. Lvs plicate

Mastixia Blume. Nyssaceae (Cornaceae s.l.; Mastixiaceae). 19 Indomal. R: Blumea 23(1976)51. Some used as plywood in India

Mastixiaceae Calestani = Nyssaceae

Mastixiodendron Melchior. Rubiaceae (Cond.). 7 E Mal. to Fiji. R: JAA 58(1977)349. Polypetalous corollas; some spp. with semi-sup. ovaries

masur *Lens culinaris*; **m. wood** *Betula pendula* & *B. pubescens*

mat grass *Nardus stricta*

mata kucing, m. kuching *Dimocarpus longan* var. *malesianus*

matac *Asclepias curassavica*

matai *Prumnopitys taxifolia*

Matalbatzia Archila = *Oncidium*

Matara tea *Senna auriculata*

matarique *Psacalium decompositum*

Matayba Aubl. (~ *Cupania*). Sapindaceae. c. 50 trop. Am. Some timbers & medic. oilseeds

mate *Ilex paraguariensis*

Matelea Aubl. Apocynaceae (Vc; Asclepiadaceae). 180 trop. S Am. Incl. many form. segregate small genera

mathai, mathe (seeds) *Trigonella foenum-graecum*

mathers *Vaccinium ovalifolium*

Mathewsia Hook. & Arn. Cruciferae (41). 6 S Peru, N Chile. R: ABN 15(1966)105

Mathiasella Constance & C. Hitchc. Umbelliferae (III 10). 1 Mex.: ***M. bupleuroides*** Constance & C. Hitchc. R: AJB 41(1954)56

Mathieua Klotzsch = *Eucharis*

Mathurina Balf.f. Passifloraceae (Turneraceae). 1 Rodrigues: ***M. penduliflora*** Balf.f. – heterophylly like many Rodrigues endemics, cult. orn. (Estonia)

Matisia Bonpl. Malvaceae (Bomb.; Bombacaceae). c. 30 trop. S Am. ***M. cordata*** Bonpl. (*Quararibea c.*, sapote, trop. S Am.) – poll. by non-flying mammals in SE Peru rainforest, fr. ed., pulp rich in vitamins A & C

Matonia R. Br. Matoniaceae. 2 Mal., exposed ridges & high mts. R: Blumea 38(1993)167. Frond with initial dichotomy, followed by each half undergoing reg. series of unequal dichotomies such that growing point curved back parallel to petiole; the pinnae forming a fan-like frond on surface of rachis away from petiole; pinnae pinnatifid. Fossil *Matonidium* with similar char. frond v. widespread in Triassic

Matoniaceae C. Presl. Polypodiidae – Gleicheniales. 2/4 Mal. R: T 55(2006)711. Terr. ferns with creeping rhizome with 2 co-axial cylinders of vasc. tissue surrounding central solid stele. Fronds 3-chotomous-pedate or pseudodichotomous; veins free or ± anastomosing. Sori superficial, consisting of a small number of sporangia in a ring around a receptacle which is continuous with the stalk of an umbrella-shaped indusium. n = 25, 26

Genera: *Matonia, Phanerosorus*. Relict group, fossils back to Lower Mesozoic referred here

Matricaria Tourn. ex L. (*Chamomilla*). Compositae (Anth.-Mat.). Excl. *Tripleurospermum*, 6 Euras. (Eur. 3 incl. ***M. chamomilla*** L. (*M. recutita*, wild chamomile, German c.) – widely used supplement but little clinical evidence of efficacy, & ***M. discoidea*** DC. ('*M. matricarioides*', pineapple weed, NW N Am., whence introd. E N Am., Euras. (GB by 1869),

N Afr. – resistant to some herbicides, fr. spread on vehicle tyres, in Am. medic. tea & insecticide, in GB poultice for boils). R: BBMNHB 23(1993)153

Matsumurella Mak.(~ *Lamium*). Labiatae (VI). 5 E As.

Matteuccia Tod. (~ *Onoclea*). Onocleaceae (Dryopteridaceae s.l.). 1 circumboreal: ***M. struthiopteris*** (L.) Tod. (ostrich fern) – fronds dimorphic, sterile ones forming leafy cup around deeply bipinnatifid fertile ones with pinnule edges contracted around sori, cult. orn., the young fronds ed. if steamed or tinned. '***M. intermedia*** C. Chr.' = *M. struthiopteris* × *Onoclea orientalis* (Hook.) Hook.

Mattfeldanthus H. Robinson & R. King. Compositae (Vern.-Lep.). 2 Brazil. R: Bonpl. 14(2005)75

Mattfeldia Urb. Compositae (Sen.-Sen.). 1 Hispaniola: ***M. triplinervis*** Urb.

Matthaea Blume. Monimiaceae (V 2). 6 Mal. R: Blumea 28(1972)77. Mangenot's Model. Stems & lvs used for headaches

Matthiola R. Br. Cruciferae (5). 48 Macaronesia, W Eur., Medit. Stock. R: MHB 1,18(1900)9. Cult. orn. herbs (sometimes subshrubby), esp. ***M. incana*** (L.) R. Br. (Brompton s., S Eur., natur. Calif.) – highly-scented fls grown for cutting, 'doubles' with no. of cells in floral apex increased & all organs after K develop as C, known since 1538, one strain 'eversporting' prod. c. 50% 'doubles' (seed viability longer) every generation (seedlings with pale lvs 'doubles', darker = 'singles'), '**Annua**' (ten week stock), a fast-maturing form used for bedding; ***M. longipetala*** (Vent.) DC. (S Eur.) esp. subsp. ***bicornis*** (Sm.) P. Ball (*M. bicornis*, night-scented s., E Medit., natur. SW N Am.) – heavy scent in evening

Mattiastrum (Boiss.) Brand = *Cynoglossum*

mattipaul *Ailanthus malabarica*

Matucana Britton & Rose (~ *Oreocereus*). Cactaceae (III 4). 12 Peru. R: R. Bregman, *The genus M.* (1996): 37.

Matudacalamus Maekawa = *Aulonemia*

Matudaea Lundell. Hamamelidaceae (I 4). 2 C Mex.to Honduras

Matudanthus D. Hunt. Commelinaceae (II 1f). 1 Mex.: ***M. nanus*** (Martens & Gal.) D. Hunt. R: KB 33(1978)333

Matudina R. King & H. Robinson = *Eupatoriastrum*

Maughaniella L. Bolus = *Diplosoma*

mauka *Mirabilis expansa*

Mauloutchia (Baill.) Warb. (~ *Brochoneura*). Myristicaceae. 10 E Madag. R: Adans. 13(1973)209, BJLS 143(2004)359. Aril rudimentary

Maundia F. Muell. Maundiaceae (Juncaginaceae s.l.). 1 E Aus.: ***M. triglochinoides*** F. Muell. R: FA 39(2011)56

Maundiaceae Nakai (~ Juncaginaceae). Magnoliidae – Alismatales. 1/1 E Aus. Herb. Lvs distichous with closed sheath; ligule 0. Fls sessile in spike on scape; bracts 0. P 2(–4 in term. fls), clawed; A (4–)6, anthers with sep. thecae; G ((3)4) with orthotropous, apical, pend. bitegmic ovules. Fr. a schizocarp
Genus: *Maundia*

Maurandella (A. Gray) Rothm. (~ *Maurandya*). Plantaginaceae (Ant.-Ant.; Scrophulariaceae s.l.). 1 SW N Am: ***M. antirrhiniflora*** (Willd.) Rothm. R: D.A. Sutton, *Rev. Antirrhineae* (1988)487

Maurandya Ortega (~ *Asarina*). Plantaginaceae (Ant.-Ant.; Scrophulariaceae s.l.). 2 W & S Am. R: D.A. Sutton, *Rev. Antirrhineae* (1988)484. ***M. scandens*** (Cav.) Pers. – cult. orn.

Mauranthemum Vogt & Oberprieler (~ *Leucanthemum*). Compositae (Anth.-Leuc.). 4 W Medit. ***M. paludosum*** (Poir.) Vogt & Oberprieler (*L. p.*) – cult. orn.

Mauria Kunth. Anacardiaceae (I). c. 12 Andes

Mauritia L.f. Palmae (I 2c). 2 trop. S Am. Dioec.; often in vast stands. ***M. flexuosa*** L.f. (ita palm, ité, N S Am., Trinidad) – water-disp. in Amaz., sources of food (sago, oil, wine), fibre from young lvs (= tibirisi used for hammocks & mats), thatch, etc., found in archaeological sites c. 9000 yrs old

Mauritiella Burret (~ *Mauritia*). Palmae (I 2c). 3 N S Am. R: Palms 54(2010)121

Mauritius hemp *Furcraea foetida*

Maurocenia Mill. (~ *Cassine*). Celastraceae (I). 1 S Afr.: ***M. frangula*** Mill. (*M. capensis*, *M. frangularia*) – fr. ed. R: Bothalia 26(1998)7

Mausolea Polj. = *Artemisia*

maw seed *Papaver somniferum*

mawah *Pelargonium* spp. & hybrids

Maxburretia Furt. Palmae (III 4a). 3 Thailand & Malay Pen., limestone hills. R: GH 11(1978)187

Maxia O. Nilsson = *Montia*

Maxillaria Ruíz & Pavón. Orchidaceae (V 12g). Incl. *Chrysocycnis, Cryptocentrum, Cyrtidiorchis, Maxillariella, Mormolyca, Trigonidium*, c. 650 trop. & warm Am. (Florida 2). R: Phytoneuron 225(1)(2015)7. Some cleistogamy; some pseudocopulation. Some with 0 pseudobulbs e.g. ***M. valenzuelana*** (A. Rich.) Nash – iris-like lvs in fans; several spp. with elaiophores attractive to bees. Cult. orn. epiphytes (incl. *Ornithidium*) with false pollen on lip (protein &/or starch) in ***M. rufescens*** Lindl. & other spp.; ***M. tenuifolia*** Lindl. – coconut-scented fls

Maxillariella M. Blanco & Carnevali = *Maxillaria* (but see Lankesteriana 7(2007)527)

Maximiliana Mart. = *Attalea*

Maximowiczia Khokhr. = *Scirpus*

Maximowicziella Khokhr. = *Scirpus*

Maxonia C. Chr. Dryopteridaceae (II). 1 WI, S Am.: ***M. apiifolia*** (Sw.) C. Chr.

Maxwellia Baill. Malvaceae (Bytt.-Lasio.; Sterculiaceae). 1 New Caledonia: ***M. lepidota*** Baill.

may *Crataegus* spp., (Aus.) *Spiraea cantoniensis*; **m. apple** *Podophyllum peltatum*; **m. blob** *Caltha palustris*; **m. weed** *Anthemis cotula*

Mayaca Aubl. Mayacaceae. 3 trop. Am., 1 S Congo, Angola, Zambia. R: NS Paris 14(1952)234

Mayacaceae Kunth (~ Xyridaceae). Magnoliidae – Poales. 1/4 trop. Am., Afr. *Lycopodium*-like freshwater aquatic herbs with vessels in roots (adventitious) & stems; longit. air-channels throughout vegetative body, hairy only in leaf-axils. Lvs sessile, without sheaths, in spirals, narrow & often apically bifid. Fls bisexual, reg., term. but app. axillary due to sympodial stem-growth, aerial but without nectaries; K 3 green, valvate, C 3 whitish, imbricate, A 3 alt. with C, anthers with apical pores or short slits & monosulcate pollen, $\overline{G}$ (3), 1-loc. with term. style & 3 parietal placentas with several–∞ bitegmic, orthotropous ovules. Fr. a loculicidal capsule (often immersed on recurved pedicel at maturity. Seeds globose to ovoid); embryo small, forming a cap on starchy proteinaceous endosperm beneath an operculum at micropylar end. n = 8

Genus: *Mayaca*

Mayanaea Lundell (~ *Orthion*). Violaceae (II 2a). 1 Guatemala: ***M. caudata*** (Lundell) Lundell. R: Wrightia 5(1974)58

Mayariochloa Salariato & al. (~ *Scutachne*). Gramineae (XXIV 2). 1 Cuba: ***M. amphistemon*** (C. Wright) Salariato & al. R: SB 37(2012)110

Mayna Aubl. Achariaceae (Flacourtiaceae). 6 trop. Am. R: FN 22(1980)22. Local med.

Mayodendron Kurz = *Radermachera*

mayten *Maytenus boaria*

Maytenus Molina. Celastraceae (I). Excl. *Denhamia, Gymnosporia, Haydenia*, incl. *Moya, Tricerma*, c. 200 trop. to warm (Madeira 1: ***M. umbellata*** (R. Br.) Mabb. (*M. dryandri*), exc. Pacific). Alks incl. caffeine. ***M. acuminata*** (L.f.) Loes. (trop. & S Afr.) – fine silky threads from broken leaf edges due to polyisoprenes from laticifer-like cells (not xylem thickenings as in *Cornus*, cf. *Wimmeria*). ***M. boaria*** Molina (mayten, Chile) – lvs a febrifuge; ***M. ilicifolia*** Mart. ex Reissek (*M. officinalis*, congorosa, S Am.) – medic. (? antitumoural pristimerin) & fert. control; ***M. vitis-idaea*** Griseb. (Peru etc.) – ash a salt-source in Paraguayan chaco

Mazaceae Reveal (~ Phrymaceae, Orobanchaceae). Magnoliidae – Lamiales 3/33 C As., China, SW Pacif. Herbs. lvs opp. (or in spirals), margins dentate. Fls with conspic. lower lip; staminodes 0. n = 19

Genera: *Dodartia, Lancea, Mazus*

Part of the questionable accelerating disintegation of Scrophulariaceae s.l.

Mazaea Krug & Urb. (*Neomazaea*). Rubiaceae (I 5). 2 Cuba (serpentine). R: Brittonia 51(1999)221

Mazus Lour. Mazaceae (Phrymaceae (Mim.), Scrophulariaceae s.l.). c. 20 E & SE As. to Aus., NZ. R: Brittonia 8(1954)29. Mat-forming ground-cover pls esp. ***M. pumilus*** (Burm.f.) Steenis (*M. reptans*, '*M. japonicus*', '*M. rugosus*', E & SE As) – natur. in lawns in Scotland, U.S., Jamaica, used for snakebite in Java

mazzard *Prunus avium*

Mazzettia Iljin = *Vladimiria*

mbocaya *Acrocomia aculeata*

mboga *Sesuvium portulacastrum*
mbura *Parinari curatellifolia*
Mcneillia Dillenb. & Kadereit (~ *Arenaria*). Caryophyllaceae. 5 Medit. R: T 63(2014)83
Mcvaughia W.R. Anderson. Malpighiaceae. 2 NE Braz. R: SB 40(2015)534
mead fermented honey & water (hydromel)
meadow crane's-bill *Geranium pratense*; **m. fescue** *Lolium pratense*; **m. foam** *Limnanthes* spp.; **m. foxtail** *Alopecurus pratensis*; **m.grass** *Poa* spp.; **annual m.g.** *P. annua*; **common m.g.** *P. pratensis*; **rough m.g.** *P. trivialis*; **wood. m.g.** *P. nemoralis*; **m. pink** *Dianthus deltoides*; **m. rue** *Thalictrum* spp.; **m. saffron** *Colchicum autumnale*; **m. saxifrage** *Saxifraga granulata*; **m.sweet** *Filipendula ulmaria*
mealberry *Arctostaphylos uva-ursi*
mealies *Zea mays*
Mearnsia Merr. (~ *Metrosideros*). Myrtaceae (II 7). c. 18 Philipp. to New Caled. & NZ
Meborea Aubl. = *Phyllanthus*
Mecardonia Ruíz & Pavón (~ *Bacopa*). Plantaginaceae (Grat.-Grat.; Scrophulariaceae s.l.). 10 warm Am. R: Candollea 42(1987)431. ***M. tenella*** (Cham. & Schldl) Pennell (S Braz.) – attracts oil-, scent- & pollen-gathering bees. Some cult. orn. esp. '**Early Yellow**' perenn. grown as ann.
Mecca balsam *Commiphora gileadensis*; **M. galls** *Quercus pubescens*; **M. myrrh** *C. gileadensis*
Mechowia Schinz. Amaranthaceae (I 2). 2 S trop. Afr.
Mecodium (Copel.) Copel. = *Hymenophyllum*
Mecomischus Cosson ex Benth. Compositae (Anth.-Sant.). 2 Morocco, Algeria. R: BBNHB 23(1993)130
Meconella Nutt. Papaveraceae (III). 3 W N Am.
Meconopsis Viguier. Papaveraceae (IV). Excl. *Cathcartia*, c. 60 Himal. to W China. R: G. Taylor (1934) *An account of the genus M.*, J.L.S. Cobb (1989) *M.* Alks. Cult. orn. with blue or reddish fls, esp. 'blue poppies' incl. ***M. betonicifolia*** Franch. (*M. baileyi*, N Myanmar to SW China) & ***M. × sheldonii*** G. Taylor (*M. betonicifolia* × *M. grandis* Prain (Himal.)), some hapaxanthic incl. ***M. horridula*** Hook.f. & Thomson (to 6000 m in Himal. & China). See also *Papaver*
Mecopus Bennett. Leguminosae (III 19). 1 Indomal.: ***M. nidulans*** Bennett
Mecranium Hook.f. = *Miconia* (but see SBM 39(1993))
mecrusse *Androstachys johnsonii*
Medemia Wuerttemb. ex H. Wendl. Palmae (III 8a). 1 N trop. Afr.: ***M. argun*** (Mart.) H. Wendl. (Egypt, Sudan) – first known from fr. (cf. *Lodoicea*), endangered through overuse as matting (refound Sudan 1995 after not seen living since 1964); fr. used as offerings in Ancient Egyptian tombs, ed., esp. toothsome after burying. R: J. Dransfield & al., *Gen. Palm.* (2008)317
Medeola L. Liliaceae. 1 E N Am.: ***M. virginiana*** L. (cucumber root) – rhizomes crisp & ed. like cucumber. R: FNA 26(2002)150
Medeolaceae Takht. = Liliaceae
Mediasia Pim. (~ *Seseli*). Umbelliferae (III 8). 1 SW & C As.: ***M. macrophylla*** (Regel & Schmalh.) Pim.
Medicago Tourn. ex L. Leguminosae (III 27). (Excl. *Trigonella*) 83 Eur. (43), Medit., Ethiopia, S Afr. R: CJB 67(1989)3260. Medick, bur clover, burweed. Fodder & green manures, some cult. orn. ***M. arborea*** L. (tree m., moon trefoil, S Eur.) – shrubby to 3.5 m, cult. orn.; ***M. hypogaea*** E. Small (*Factorovskya aschersoniana*, SE Med. to Iraq) – geocarpic annual, the 1 mm gynophore extending to 10 cm underground; ***M. intertexta*** (L.) Mill. (*M. echinus*, calvary clover, Medit.) – fr. twisted & spiny like crown of thorns; ***M. lupulina*** L. (black medick, hop clover, nonsuch, yellow trefoil, Euras., natur. E Afr., N Am.) – fodder pl., the (?) original shamrock, hairy forms more resistant to whitefly oviposition; ***M. polymorpha*** L. (*M. nigra*, OW) – viable seeds 200 yrs old from adobe in Calif. & Mex.; ***M. sativa*** L. (lucerne, alfalfa, SW As., natur. Eur., N Am.) – successfully domesticated autotetraploid (2n = 4x = 32, few diploids known) selected by horse-raising culture ('alfalfa' Arabic from old Irani 'an aspo-asti' (= horsefodder) introd. Eur. 2400 BC (Persian-Greek Wars)), long-cult. for fodder & silage, seeds grown like beansprouts, poss. oilseed, lvs a comm. source of chlorophyll, blood tonic, temp. relief from arthritis
medick *Medicago* spp.; **black m.** *M. lupulina*
Medicosma Hook.f. Rutaceae (I 2). 25 New Guinea (1), E Aus. (6, endemic), New Caled. 15. R: AusJB 33(1985)27. Ant-disp. through persistent placental endocarp

Medinilla Gaudich. ex DC. Melastomataceae. c. 375 trop. Afr. (3–4), (esp.) Madag. (c. 70 & New Guinea, to Pacific (Borneo 48, R: Blumea 35(1990)5; Philippines 80, R: Blumea 40(1995)113; Aus. 1, New Caledonia 0). Some cult. orn. pot-pls with large lvs esp. ***M. magnifica*** Lindl. (Philippines) – 4-winged stems; ***M. radicans*** (Blume) Blume (C Mal.) – lvs used for dysentery

Mediocalcar J.J. Sm. Orchidaceae (V 15). 17 Sulawesi to Samoa, esp. New Guinea. R: OM 8(1997)21

Mediusella (Cavaco) Dorr (~ *Leptolaena*). Sarcolaenaceae. 2 Madag. R: Adans. 31(2009)315

medlar *Mespilus germanica;* **Bronvaux m.** *Crataegus.* hybrids with *M. germanica;* **Japanese m.** *Eriobotrya japonica;* **Medit. m.** *C. azarolus*

Medranoa Urbatsch & R. Roberts. Compositae (Ast.). 5 SW N Am.

Medusa head *Taeniatherum caput-medusae*

Medusagynaceae Engl. & Gilg = Ochnaceae

Medusagyne Bak. Ochnaceae (Medusagynaceae). 1 Mahé (Seychelles): ***M. oppositifolia*** Bak. – G (17–25) with as many locs & styles forming a 'crown', each loc. with 1 ascending & 1 descending bitegmic anatropous ovule, fr. a capsule opening septicidally from base with carpels diverging from central column-like spokes of an opening umbrella, thought extinct (not seen between 1908 & 1970) but c. 12 pls (2 populations) found at c. 250 m, the sp. now spread in cult. by seed. R: KM 6(1989)166

Medusandra Brenan. Peridiscaceae (Medusandraceae). 2 trop. W Afr. Fr. relished by baboons & parrots

Medusandraceae Brenan = Peridiscaceae

Medusanthera Seemann. Icacinaceae. 8 Mal. to W Pacific. Roux's Model. R: KB 66(2011)53

Medusorchis Szlach. = *Habenaria*

Meeboldia H. Wolff. Umbelliferae (inc. sed.). Incl. *Sinodielsia* (R: FR 97(1986)753), 5 Himalaya

Meeboldina Süsseng. = *Leptocarpus* (but see APS 43(2005)552)

Meehania Britton. Labiatae (VII 2c). 7 E As., E U.S. (1: ***M. cordata*** (Nutt.) Britton. – cult. ground-cover). R: BZ 77,12(1993)123

Meehaniopsis Kudô = *Glechoma*

Megacarpaea DC. Cruciferae (36). 9 Eur. (1) to C As., Himal. & China. ***M. megalocarpa*** Schischkin & B. Fedtsch. – monoec. Some statuesque hapaxanthic pachycauls with A 18–24 in ***M. polyandra*** Benth. (Himal.) – young lvs ed.

Megacaryon Boiss. = *Echium*

Megacodon (Hemsl.) H. Sm. Gentianaceae. 2 Himal. ***M. stylophorus*** (C.B. Clarke) H. Sm. – cult. orn. perennial to 2 m

Megacorax S. González & W.L. Wagner (~ *Lopezia*). Onagraceae. 1 Mex. (1 mt. in Durango): ***M. gracielanus*** S. González & W.L. Wagner. R: Novon 12(2002)360

Megadenia Maxim. Cruciferae (9). 1 Siberia, China (disjunct): ***M. pygmaea*** Maxim.

Megahertzia A.S. George & Hyland. Proteaceae (V 1). 1 Queensland: ***M. amplexicaulis*** A.S. George & Hyland. R: FA 16(1995)497

Megalachne Steud. Gramineae (XVI). 2 Juan Fernandez

Megalastrum Holttum (~ *Ctenitis*). Dryopteridaceae (II). 91 trop. Am., Afr. to Madag.(3). R: AFJ 104(2014)112, 181

Megaleranthis Ohwi = *Trollius*

Megalochlamys Lindau. Acanthaceae (III 2c). 10 E & S Afr., 2 ext. to Arabia

Megalodonta Greene = *Bidens*

Megalonium (A. Berger) Kunkel = *Aeonium*

Megalopanax Ekman = *Aralia*

Megaloprotachne C.E. Hubb. Gramineae (XXIV 1). 1 S & S trop. Afr.: ***M. albescens*** C.E. Hubb. R: Strel. 10(2000)703

Megalorchis Perrier. Orchidaceae (IV 4d). 1 E Madag. mts: ***M. regalis*** (Schltr.) Perrier. R: Gen. Orch. 2(2000)315

Megalostoma Leonard. Acanthaceae (III 2c). 1 C Am.: ***M. viridescens*** Leonard

Megalostylis S. Moore = *Dalechampia*

Megalotheca F. Muell. = *Loxocarya*

Megalotus Garay = *Robiquetia* (but see HBML 23(1972)184)

Megaphrynium Milne-Redh. (~ *Sarcophrynium*). Marantaceae. 4 trop. Afr. ***M. macrostachyum*** (Benth.) Milne-Redh. – used for roofing

Megaphyllaea Hemsl. = *Chisocheton*

Megasea Haw. = *Bergenia*

Megaskepasma Lindau. Acanthaceae (III 2c). 1 Venez.: ***M. erythrochlamys*** Lindau – cult. for brilliantly coloured bracts

Megastachya P. Beauv. (~ *Eragrostis*). Gramineae (XXI). 1 Afr. & Madag.: ***M. mucronata*** (Poir.) P. Beauv. R: Strel. 11(2000)704

Megastigma Hook.f. Rutaceae (I ?5). 2 Mexico, C Am.

Megastoma Cosson & Durieu = *Ogastemma*

Megastylis (Schltr.) Schltr. Orchidaceae (IV 2f). 7 New Caled., 1 ext. to Vanuatu

Megathyrsus (Pilg.) B. Symon & S. Jacobs = *Urochloa*

Megatritheca Cristóbal (~ *Byttneria*). Malvaceae (Bytt.-Bytt.; Sterculiaceae). 2 Gabon, Congo. R: Adans. 5(1965)370

Megistostegium Hochr. Malvaceae (Malv.-Hib.). 3 Madag. R: FMad 129(1955)83. Bird-poll. shrubs

Megistostigma Hook.f. (~ *Sphaerostylis*). Euphorbiaceae (Acal.-Pluk.-Trag.). 5 SE As. to W Mal.

Meiandra Markgraf = *Alloneuron*

Meineckia Baill. Phyllanthaceae (Phyll.-Par.; Euphorbiaceae s.l.). Incl. *Zimmermannia*, *Zimmermanniopsis*, 30 SW & NE trop. Afr., Socotra, S Arabia, Madag. (12), S India & Sri Lanka, Assam, trop. Am. R: KB 63(2008)53. ***M. paxii*** Radcl.-Sm. (*Zimmermannia capillipes*, Tanzania) – roots anthelminthic

Meiocarpidium Engl. & Diels. Annonaceae (II). 1 W trop. Afr.: ***M. lepidotum*** Engl. & Diels

Meiogyne Miq. Annonaceae (IV 7). Incl. *Fitzalania*, 24 Indomal. to Pacific. R: SB 39(2014)400

Meiomeria Standl. = *Dysphania*

Meionandra Gauba = *Valantia*

Meionectes R. Br. (~ *Haloragis*). Haloragaceae. 2 Aus.

Meiostemon Exell & Stace = *Combretum* (but see BSB II,40(1964)18)

Meiracyllium Reichb.f. Orchidaceae (V 13b). 2 C Am. R: C.L. Withner, *The Cattleyas & their relatives* V(1998)165. Cult. orn. epiphytes

Meizotropis J. Voigt (~ *Butea*). Leguminosae (III 18). 2 Himal. R: BBSI 29('1987')216

mel grass *Ammophila arenaria*

Melachone Gilli = *Dolianthus*

Meladenia Turcz. = *Cullen*

Meladerma Kerr. Apocynaceae (III; Periplocaceae). 3 Thailand

Melaleuca L. Myrtaceae (II 4). Incl. *Beaufortia*, *Callistemon* (A free, some with small collar), *Calothamnus* (infls 1-sided), *Conothamnus*, *Eremaea*, *Lamarchea*, *Petraeomyrtus*, *Phymatocarpus*, *Regelia*, c. 300 Indomal. (few), Aus. (c. 280), Pacific. Fls in heads or spikes, the infl. axis often extending as a leafy shoot, the fls with consp. long stamens in 5 bundles opp. petals; fr. often persistent & enlarging over many yrs. Cult. orn. (bottlebrushes (incl. hybrids), paperbarks – Aus.), timber & medic. oil. ***M. alternifolia*** (Maiden & Betche) Cheel (tea tree, E Aus.) – oil trad. Aboriginal antiseptic, now for acne, athlete's foot, arthritis; ***M. cajuputi*** Maton & Sm. ex Powell (Myanmar to Aus.) – distilled for cajuput oil used like eucalyptus oil in cough sweets etc., timber good; ***M. citrina*** (Curtis) Dum-Cours. (*Callistemon c.*, SE Aus.) – cult. orn. (several cvs), allelopathic (leptospermone), a chem. analogue (mesotrione) comm. herbicide; ***M. cruenta*** Craven & R.T. Edwards (*Calothamnus sanguinea*, blood flower., SW Aus.) – cult. orn.; ***M. lanceolata*** Otto (moonah, E Aus.); ***M. quadrifida*** (R. Br.) Craven & R.T. Edwards (*C. q.*, SW Aus.) – honeyeaters imp. poll.; ***M. quinquenervia*** (Cav.) S.T. Blake ('*M. leucadendra*', E Aus., SE New Guinea, New Caled., invasive SE U.S. 50 acres a day in Everglades (1 acre being cleared) – orig. sown from aeroplanes to dry Everglades out & now controlled biologically, Carib., in Braz. by introduced myrtle rust (*Uredo rangelii*, S Am.) but that now affecting *Eucalyptus* there) – source of niaouli oil like cajuput, the trees forming almost pure stands after the native vegetation of New Caled. is destroyed by fire; ***M. uncinata*** R. Br. agg. (broombush, S Aus.) – brush fencing in SE Aus., lobster-pots etc., used for lowering water-tables, ess. oil, biomass, some morphs assoc. with *Rhizanthella gardneri*; ***M. viminalis*** (Gaertn.) Byrnes (*Callistemon v.*, NE Aus.) – rheophyte, oil anthelminthic *in vitro*, cult. orn.

Melampodium L. Compositae (Mill.-Mel.). 39 trop. & warm Am. esp. Mex. R: Rhodora 74(1972)1. ***M. americanum*** L. (Mex. to Guatemala) – insecticidal lactone effective against army worm

Melampyrum Tourn. ex L. Orobanchaceae (Rhin.; Scrophulariaceae s.l.). 35 N temp. (Eur. 24). R: FR 23(1926)159, 385, 24(1927)127. Cow-wheats. Hemiparasites. Seeds of ***M. arvense*** L. (Euras.) taint wheat so form. pls & even seeds were removed from fields in S England

Melananthus Walp. Solanaceae (Schw.). 5 trop. Am. G with 1 ovule

Melancium Naudin = *Melothria*

Melandrium Roehl. = *Silene*

Melanocenchris Nees. Gramineae (XXIX 6). 3 Chad & NE trop. Afr. to India & Sri Lanka. R: BBSI 16(1974)141

Melanochyla Hook.f. Anacardiaceae (I). 30 Mal. R: FM I, 8(1978)490

Melanococca Blume (~ *Rhus*). Anacardiaceae (I). 1 C Mal to Tahiti: ***M. tomentosa*** Blume

Melanocommia Ridl. = *Semecarpus*

Melanodendron DC. Compositae (Ast.). 1 St Helena: ***M. integrifolium*** DC. (black cabbage) – allied to *Felicia*, a tree (Leeuwenberg's Model), form. building timber. R: Q.C.B. Cronk (2000) *Endemic Flora St Helena*: 80

Melanodiscus Radlk. = *Glenniea*

Melanolepis Reichb. ex Zoll. Euphorbiaceae (Acal.-Chro.-Chro.).2 Taiwan & SE As. to Pacific. R: Blumea 44(1999)437. ***M. multiglandulosa*** (Blume) Reichb. & Zoll. – imp. local medic., cult. Taiwan

Melanoloma Cass. = *Centaurea*

Melanophylla Bak. Torricelliaceae (Melanophyllaceae). 7 Madag.

Melanophyllaceae Takht. ex Airy Shaw = Torricelliaceae

Melanopsidium Cels ex Colla (~ *Billotia*). Rubiaceae (II Cond.). 1 Brazil: ***M. nigrum*** Colla. R: Brittonia 52(2000)331

Melanorrhoea Wall. = *Gluta*

Melanortocarya Selvi & al. (~ *Nonea*). Boraginaceae (B.2.1.1). 1 SE Medit.: ***M. obtusifolia*** (Willd.) Selvi & al. R: T 55(2006)915

Melanosciadium Boissieu. Umbelliferae (III 8). 1 W China: ***M. pimpinelloides*** Boissieu

Melanoselinum Hoffm. = *Daucus*

Melanospermum Hilliard (~ *Phyllopodium*). Scrophulariaceae (Manul.). 6 C & S Afr. R: NRBGE 45(1988)482

Melanostachya Briggs & L. Johnson (~ *Restio*). Restionaceae. 1 SW Aus.: ***M. ustulata*** (Ewart & Sharman) Briggs & L. Johnson. R: Telopea 7(1998)361

Melanoxerus Kainul. & B. Bremer (~ *Euclinia*). Rubiaceae (II 1). 1 Madag.: ***M. suavissimus*** (Cavaco) Kanul. & B. Bremer. R: T 63(2014)828

Melanoxylon Schott = *Melanoxylum*

Melanoxylum Schott. Leguminosae (I 4). 1 Amazonia: ***M. brauna*** Schott – timber (brauna) used for bridges etc.

Melanthera J.P. Rohr. Compositae (Helia.-Verb.). Excl. *Lipotriche*, *Wollastonia*, 3–5 Carib. & N Andes. R: Brittonia 53(2001)550

Melanthiaceae Batsch. (~ Liliaceae). Magnoliidae – Liliales. Incl. Trilliaceae, excl. Campynemataceae, Nartheciaceae, Petrosaviaceae, Tofieldiaceae, 17/190 N hemisph. to SE As. & S Am. Perennial herbs, s.t. pachycaul (*Veratrum*), with rhizomes & spirals (pseudoverticillate in *Trillium* etc.) of or distichous (often evergreen) lvs. Infls spikes or racemes (to panicles but solit. in *Trillium* etc.) of usu. bisexual ± hypogynous (cf. C.) fls. P 3 + 3, s.t. basally connate, s.t. marcescent; A 3 + 3; $\underline{G}$ (3(–10)), 3-loc., s.t. apically free, each loc. with ∞ ovules. Fr. loculicidal or septicidal capsule or berry; seeds usu. rounded & winged or with term. appendages. n = 5, 8, 15, 17

Principal genera (many E As. – E N Am. affinities): *Helonias, Paris, Schoenocaulon, Trillium, Veratrum*

Some cult. orn. & local medic. (many with v. toxic alks): *Anticlea, Chamaelirion, Helonias, Heloniopsis, Melanthium, Schoenocaulon, Trillium, Veratrum, Xerophyllum*

Melanthium Clayton ex L. (~ *Zigadenus, Veratrum*). Melanthiaceae. 5 N Am. Some cult. orn. rhizomatous herbs incl. ***M. virginicum*** L. (bunchflower, E U.S.)

Melasma P. Bergius. Orobanchaceae (Esc.; Scrophulariaceae s.l.). 5 S Afr. (3), Am. (2). R: NBGB 15(1940)121

Melasphaerula Ker-Gawler. Iridaceae (VI 3). 1 S Afr.: ***M. ramosa*** (L.) N.E. Br. (*M. graminea*) – cult. orn. like *Sparaxis* but G acutely 3-angled. R: Strel. 9(2000)136

Melastoma Burm. ex L. Melastomataceae. Incl. *Otanthera* 22 Indomal. to Pacific (Aus. 1). R: Blumea 46(2001)351. Dendroid hairs; dry capsules to fleshy indehiscent fr. Some cult. orn. esp. ***M. malabathricum*** L. (Indomal.) – locally medic., fr. sweet, stains mouth black like *Vaccinium*, a common plant of regrowth (eventually overcoming even *Imperata cylindrica*) app. everflowering, one form rheophytic

Melastomastrum Naud. Melastomataceae. 6 trop. Afr. R: BMNHN 3°, 270 Bot. 14(1974) 49

Melastomataceae Juss. Magnoliidae – Myrtales. 163/5600 trop. & warm esp. S Am. Shrubs & herbs, less often trees or lianes, often with 4-angled stems, s.t. myrmecophilous, often accumulating aluminium & with internal phloem as well as cortical and/or pith vascular bundles. Lvs opp. (s.t. anisophyllous or 1 suppressed), rarely whorled, simple, often with 3–9 prominent sub-parallel veins; stipules rare (e.g. *Astronidium*). Fls usu. bisexual & without nectar (present in *Blakea*), insect-poll., in cymes, usu. ± perigynous, reg. exc. A, (3)4 or 5(–10)-merous; K lobes valvate or a rim on hypanthium, s.t. a calyptra, C usu. free, convolute in bud, A usu. 2 whorls, often dimorphic, filaments often twisted at anthesis bringing anthers to 1 side of fl., these with single term. pore or less often 2 pores or longit. slits, connective often with appendages, G ((2)3–5(–15)) with as many loc. (s.t. 1- loc. through partitions not developing), with term. style & (1–)∞ anatropous (or campylotropous) bitegmic ovules with zig-zag micropyle on usu. axile placentas per loc. Fr. a loculicidal or septifragal capsule or berry (s.t. both in 1 genus e.g. *Aciotis*); seed usu. small, endosperm 0, the cotyledons often unequal. x = 7–18+

Classification & principal genera:

App. *Pternandra* sister to rest of fam., which can be arranged in

Melastomatoideae (inc. Astronioideae in part; fr. many-seeded, embryo small): *Astronia, Astronidium, Blakea, Brachyotum, Dissotis, Gravesia, Medinilla, Melastoma, Miconia, Microlicia, Monochaetum, Osbeckia, Phyllagathis, Sonerila, Tibouchina, Topobea*

Memecyloideae (inc. Astronioideae in part, s.t. treated as sep. fam.; some with buzz-poll.; nodes swollen; fr. a 1–5-seeded berry, embryo large): *Memecylon, Mouriri*

Memecyloideae often considered to approach Myrtaceae; wood anatomy (fibres with bordered (not simple) pits; included phloem etc.) like Myrt., also Lythraceae & Onagraceae. Details of A app. associated with poll. mechanisms imp. in generic & specific distinction. 15 genera with elaiophores attractive to bees (only 10 fams thus). Fleshy fr. distrib. mainly by birds but also marsupials, monkeys, bats & other mammals, turtles & other reptiles, & fish

Despite large size, a fam. of little econ. interest, though some timber (*Astronia, Miconia*), ed. fr. (*Bellucia, Heterotrichum, Miconia* (cult.)), dyes (*Dionycha, Memecylon* etc.) & cult. orn. esp. *Bertolonia, Dissotis, Medinilla, Melastoma, Memecylon, Miconia, Monolena, Rhexia, Sonerila, Tibouchina, Triolena*, some becoming weedy, e.g. *Heterocentron, Melastoma, Memecylon, Miconia*

Melchiora Kobuski = *Balthasaria*

Melhania Forssk. Malvaceae (Dom.; Sterculiaceae). c. 50 trop. OW. Epicalyx with 3 large segments

Melia L. Meliaceae. (II-Mel.). 3 OW trop. R: GBS 37(1984)49. ***M. azedarach*** L. (As. to Aus.) – extrafloral nectaries on petioles of young pls, timber tree (*M. dubia, M. composita*, white cedar, bakain, lunumidella, Ceylon cedar or mahogany) used for construction, bark & lvs medic. & insecticidal (antifeedant triterpenoids), fr. used as beads, early-flowering forms domesticated in China & India (Persian or Indian or Cape (Aus.) lilac, China tree, chinaberry, 'syringa', Pride of India) known to Arabs by 1080, & widely natur. in Am., Afr., SE U.S., Hawaii & Medit. where cult. orn. for lilac-scented fls., '**Floribunda**' a precocious cv (breeding true) used in 'subtrop.' bedding

Meliaceae Juss. Magnoliidae – Sapindales. 51/700 trop. & subtrop. (few). R: Blumea 22(1975)419. Trees, often pachycaul (Corner's, Champagnat's, Aubréville's (*Vavaea*) & Rauh's (*Xylocarpus*) models), rarely shrubs or suckering shrublets, dioec., polygamous, monoec. or with only bisexual fls; bark bitter & astringent. Lvs pinnate to bipinnate (*Melia*), unifoliolate or simple, in spirals (rarely decussate) with usu. entire leaflets & basally swollen petiole, s.t. (*Chisocheton, Guarea*), with indefinite growth from apical 'pseudogemmula' (cf. *Tachigali*), s.t. spiny; stipules 0. Fls if unisexual often with rudiments of opp. sex, in spikes to thyrses, axillary to supra-axillary, cauliflorous or even epiphyllous (*Chisocheton*), reg.; K (2)3–5(–7), s.t. transitional to bracteoles, usu. atop a tube, imbricate or s.t. almost closed when basally circumscissile, C 3–7(–14) in 1(–2, spiral in some *Chisocheton*) whorls, s.t. basally connate, imbricate or convolute or valvate, A usu. atop a tube with 3–19(–30) anthers in 1(2) whorls, nectary usu. a disk usu. around ovary, $\underline{G}$ ((1)2–6(–20)) with as many locules & usu. axile placentation (1–loc. with intruded parietal placentas in *Heckeldora* etc.) with 1–∞ bitegmic anatropous, campylotropous or orthotropous ovules per loc. Fr. a capsule, berry or drupe; seeds winged & then attached to woody columella, or with corky outer layers, or with fleshy sarcotesta or aril or a combination of both, or none of these, usu. without endosperm. n = 8- c. 180 (x = 6, 7)

Classification [likely to be superseded] & principal genera:

I. **Cedreloideae** (Swietenioideae; buds usu. with scale lvs, fr. a woody capsule with central columella & winged seeds or columella rudimentary & seeds with woody or corky sarcotesta): *Capuronianthus, Cedrela, Entandrophragma, Khaya, Lovoa, Swietenia, Toona,* [**Xylocarpeae** (A 8–10 atop a tube, seeds unwinged with corky or woody outer layer):] *Carapa, Xylocarpus*

II. **Melioideae** (incl. Quivisianthoideae; buds usu. naked; fr. a capsule, berry or drupe with usu. unwinged seeds): [**Melieae** (lvs pinnate or bipinnate, fr. a drupe), only:] *Azadirachta & Melia, Turraea,* [**Trichilieae** (lvs 1-foliolate to pinnate, fr. a caps., berry or drupe)] *Aglaia, Chisocheton, Dysoxylum, Guarea, Lansium, Owenia, Sandoricum, Trichilia*

Char. limonoids allied to those in Rutaceae. Poss. orig. West Gondwana. *Neomangenotia* form. assigned to own subfam. referable to *Commiphora*

Imp. high-quality timbers esp. *Cedrela, Dysoxylum, Entandrophragma, Khaya, Lovoa, Melia, Swietenia* (mahogany), *Toona*, often over-exploited; fr. trees (*Lansium, Sandoricum*), insecticides (*Azadirachta, Melia*), tanbark (*Xylocarpus*), medic. (*Munronia*), oilseeds etc.

Melianthaceae Horan. = Francoaceae

Melianthus Tourn. ex L. Francoaceae (Melianthaceae). 8 S Afr. Shrubs with rather pachycaul stems, foetid when bruised, posterior sepal saccate or spurred. ***M. comosus*** Vahl – invasive Aus.; ***M. major*** L. (honeyflower, invasive NZ) – cult. orn. (Tomlinson's Model) for 'subtrop.' bedding, fls rich in nectar (black, taken by sunbirds), locally medic.

Melica L. Gramineae (XI). 90 temp. (Eur. 10; China 23 – R: FOC 22(2006)216; N Am. 19 – FNA 24(2007)88) exc. Aus. R: KBAS 13(1986)113. Melick. some cult. orn.; some forest grasses incl. ***M. uniflora*** Retz. (wood melick, Eur., Medit.)

melick *Melica* spp.; **wood m.** *M. uniflora*

Melichrus R. Br. (~ *Styphelia*). Ericaceae (VII 7; Epacridaceae). 4 Aus.

Melicocca L. = *Melicoccus*

Melicoccus P. Browne. Sapindaceae (IV 2). 10 trop. Am. R: FN 87(2003)28. ***M. bijugatus*** Jacq. (honeyberry, mamoncillo, quenette) – fr. tree, seeds ed. roasted, bark decoction used against dysentery

Melicope Forst. & Forst. f. Rutaceae (I 2). Incl. *Platydesma*, 234 Madag. to NZ, Hawaii (*Pelea*, 47; Mal. to Aus. 128). R: Allertonia 8(2000)65. Mostly dioec. Alks; some antibacterial activity (local medic.). ***M. anisata*** (H. Mann) Hartley & Stone (Hawaii) – for scenting clothes (anethole), fr. used in lei; ***M. denhamii*** (Seem.) Hartley (C Mal. to Tonga) – selected leaf-form cvs (cf. *Euodia hortensis*)

Melicytus Forst. & Forst. f. Violaceae (III 16). Incl. *Hymenanthera*, 8–9 Solomon Is., E Aus., NZ, Norfolk Is., Fiji. Dioec. trees & shrubs; fls. almost reg.; ovules ∞ per loc.; berry. ***M. alpinus*** (Kirk) Garnock-Jones (*H. a.*, NZ) – white fr. on underside of prostrate stems (? reptile-disp.); ***M. dentatus*** (DC.) Molloy & Mabb. (*H. d.*, SE Aus.) – spiny cult. orn. shrub; ***M. ramiflorus*** Forst. & Forst. f. (mahoe, NZ etc.) – cult. orn., timber form. used for charcoal in gunpowder

Melientha Pierre. Opiliaceae. 1 Yunnan & SE As. to Philipp.: ***M. suavis*** Pierre – r.-parasite, caulifl., wood for charcoal, fr. ed., young shoots a veg. in Thailand

melilot *Melilotus* spp.; **white m.** *M. albus*; **yellow m.** *M. officinalis*

Melilotus Mill. (~ *Trigonella*). Leguminosae (III 27). 20 temp. & subtrop. Euras. (Eur. 16), N Afr., Ethiopia. R: BJ 29(1901)660, CJPS 49(1969)1. Melilot. Fragrant herbs grown as green manure, forage crops (though a glucoside of o-hydroxycinnamic acid gets converted to dicoumarol, an anticlotting factor if it reaches blood of sheep or cattle) or bee-pls esp. ***M. albus*** Medik. (white m., Bokhara clover, Euras., natur. N Am.,? white-flowered form of *M. officinalis*) & ***M. officinalis*** (L.) Pallas (yellow m., Euras., natur. N Am.) – anticoagulant warfarin developed from it; ***M. messanensis*** (L.) All. (*M. siculus*, messina, Medit.) – salt-tolerant, introd. Aus. to improve salted land

melindjo *Gnetum gnemon*

Melinia Decne = *Philibertia* (but see Darw. 30(1990)279)

Melinis P. Beauv. Gramineae (XXIV 5). Incl. *Rhynchelytrum*, 22 trop. & S Afr. R: BB 138(1988)50. ***M. minutiflora*** P. Beauv. (molasses grass, efwatakala g., trop. Afr.) – widely introd. for forage (invasive in U.S., trop. Am., Galapagos, Hawaii), whole pl. sweet-smelling, said to be insect-repellent; ***M. repens*** (Willd.) Zizka (*R. r., Tricholaena rosea*, Natal grass, trop. & S Afr.) – cult. pasture grass natur. U.S., cult. orn., basket- & hat-weaving S Afr.

Meliosma Blume. Sabiaceae (Meliosmaceae). c. 25 trop. As. (15) & Am. (10; Cretaceous fossils in Euras. & N Am.). Explosively dehiscent anthers held under tension by staminodes. Some timbers & cult. orn. with fragrant fls

Meliosmaceae Meissn. See Sabiaceae

Melissa Tourn. ex L. Labiatae (VII 2). 4 Eur. (1), Macaronesia, N Afr. to Iran & C As. ***M. officinalis*** L. (balm, lemon or sweet or tea b., S Eur., natur. N Am. etc.) – cult. for lemon-scented lvs used as seasoning & medic. (mild sedative & antidepressant (alleged to reduce anxiety, also bloating on long flights), terpenoids acting on involuntary nervous system & relieving spasm in smooth muscle), in scents & liqueurs, lvs in empty skeps attract bee-swarms as same terpenoids as in Nasonov glands of honeybees

Melissitus Medik. = *Trigonella*

Melittacanthus S. Moore. Acanthaceae (III 2c). 1 Madagascar: ***M. divaricatus*** S. Moore

Melittis L. Labiatae (VI). 1 Euras.: ***M. melissophyllum*** L. (bastard balm). R: Fl. Iber. 12(2010)178

Mellera S. Moore. Acanthaceae (III 2a). 4–5 warm Afr.

Mellichampia A. Gray ex S. Watson = *Cynanchum*

Melliniella Harms (~ *Alysicarpus*). Leguminosae (III 19). 1 trop. W Afr.: ***M. micrantha*** Harms

Melliodendron Hand.-Mazz. Styracaceae. 1 S China: ***M. xylocarpum*** Hand.-Mazz. R: BM 30(2013)205

Mellissia Hook.f. = *Withania*

Melloa Bureau = *Dolichandra*

Melocactus (L.) Link & Otto. Cactaceae (III 3). 33 W Mex. to S Peru (E Brazil 14, 11 endemic) with 2 nat. hybrids. R: Bradleya 9(1991)1. Term. fl.-bearing structure permanently distinct from rest; effectively hapaxanthic. Cult. orn. (melon cactus) esp. Braz. spp. & ***M. intortus*** (Mill.) Urb. (*M. communis*, Turk's-cap c., WI) – to 1 m tall, seeds removed from fr. & disp. by fireants

Melocalamus Benth. (~ *Dinochloa*). Gramineae (V 4). 11 India & S China to SE As. Climbers with fleshy fr.

Melocanna Trin. (~ *Schizostachyum*). Gramineae (V 5). 2 Indomal. Some vivipary. ***M. baccifera*** (Roxb.) Kurz (*M. bambusoides*, Terai bamboo, NE India, Myanmar) – stems used for allegedly white-ant resistant building material, fr. a berry size of an avocado, ed. baked; flowering on a 7–51 yr cycle

Melocarpum (Engl.) Beier & Thulin (~ *Zygophyllum*). Zygophyllaceae (I). 2 Horn of Afr. R: PSE 240(2003)37

Melochia Dill. ex L. Malvaceae (Bytt.-Herm.; Sterculiaceae). c. 60 trop. esp. Am. R: CUSNH 34(1967)191. Some weedy, some locally medic. & veg.; ***M. corchorifolia*** L. (trop.) – in N Aus. fibre for string esp. fishing-nets; ***M. pyramidata*** L. (trop. & warm Am.) – locally used fibres

Melodinus Forst. & Forst. f. Apocynaceae (I e). c. 25 Indomal. & Pacific. R: SGP 73(2003)4. Alks. ***M. monogynus*** Roxb. ex Lindl. (India) – local medic., cult. orn.

Melodorum Lour. Annonaceae (III 7). Excl. *Sphaerocoryne*, c. 5 Indomal. (Aus. spp. = *Uvaria*). Troll's Model. Some ed. fr.

melokhia *Corchorus olitorius*

Melolobium Ecklon & Zeyher. Leguminosae (III 9). 15 S Afr.

Melomphis Raf. = *Ornithogalum*

melon *Cucumis melo*; **m. cactus** *Melocactus* spp.; **Chinese m.** *C. melo* Inodorus Group; **horned** or **jelly m.** *C. metuliferus*; **m.-loco** *Apodanthera undulata*; **mash** or **musk m.** *C. melo* Reticulatus Group; **m. pear** *Solanum muricatum*; **rock m.** *C. melo* Reticulatus Group; **snake m.** *C. melo* Flexuosus Group; **m. tree** *Carica papaya*; **tree m.** *S. muricatum*; **water m.** *Citrullus lanatus*

Melosperma Benth. Plantaginaceae (Grat.-Melo.; Scrophulariaceae s.l.). 1 Chile, Arg.: ***M. andicola*** Benth. – seeds large. R: Parodiana 3(1985)373

Melothria L. Cucurbitaceae (XIV). Incl. *Cucumeriopsis*, c. 12 NW (***M. mannii*** (Naudin) (*C. m.*) – seeds (egusi) source of oil & protein, ext. to W Afr., ***M. pendula*** L. weedy As.). Some ed. fr. & local medic., ***M. scabra*** Naudin (cucamelon, C Am.) – raw or pickled in Aus., UK

Melothrianthus Mart. Crov. Cucurbitaceae (XIII). 1 Brazil: ***M. smilacifolius*** (Cogn.) Mart. Crov. R: NS 15(1954)58

Melpomene A.R. Sm. & R.C. Moran (~ *Grammitis*). Polypodiaceae (V; Grammitidaceae). 29 trop. Afr. (few) & Am. R: Novon 2(1992)426

Memecylaceae DC. = Melastomataceae
Memecylanthus Gilg & Schltr. = *Periomphale*
Memecylon L. Melastomataceae (Memecylaceae). c. 320 OW trop. (As. 153 [Malay Penin. 32, Borneo 27 – R: OB 69(1983)5], Afr. 57 – R: T 63 (2014)551). R: DC. Mon. Phan. 7(1891)1130. Afr. spp. plagiotropic, As. with erect branches; some rheophytes; elaiophores attractive to bees. Some spp. of sect. *Afzeliana* (W Afr.) with 'star' fls (C much reduced; A connectives enlarged, blue). Some timbers (v. dense, used for anchors), & cult. orn., some ed. fr. ***M. caeruleum*** Jack (*M. floribundum*, SE As., Mal.) – invasive Seychelles; ***M. umbellatum*** Burm.f. (korakaha, Indomal.) – fr. ed., lvs yield yellow dye
Memel fir *Pinus sylvestris*
Memora Miers = *Adenocalymma*
Memoremea Otero & al. (~ *Omphalodes*). Boraginaceae (B.3.3). 1: E Eur.: ***M. scorpioides*** (Haenke) Otero & al. R: Phytotaxa 173 (2014)266
mena *Curcuma longa*
Menabea Baill. = *Pervillea*
Mendoncella A. Hawkes = *Galeottia*
Mendoncia Vell. ex Vand. Acanthaceae (II; Mendonciaceae). c. 65 trop. Am., 4 Afr., 3 Madag.
Mendonciaceae Bremek. = Acanthaceae (II)
Mendoravia Capuron. Leguminosae (I 3). 1 SE Madag., local: ***M. dumaziana*** Capuron – wood for construction, carpentry. R: Adansonia 8(1968)208
menduro *Balanites maughamii*
Menendezia Britton = *Tetrazygia*
Menepetalum Loes. Celastraceae (I). 4 New Caledonia
Menezesiella Chiron & V. Castro = *Gomesa*
mengkulang *Heritiera* spp.
meni oil *Lophira* spp.
Meniscium Schreb. Older name for *Cyclosorus*
Meniscogyne Gagnepain (~ *Elatostema*). Urticaceae (II). 2 SE As.
Menisorus Alston = *Cyclosorus*
Menispermaceae Juss. Magnoliidae – Ranunculales. 72/450 trop. & warm, few temp. Usu. dioec. lianes & scandent shrubs, rarely trees or herbs, usu. with bitter sesquiterpenoids (poisonous) & alks, the stems often with unusual sec. thickening & becoming flattened. Lvs simple (3-foliolate in *Syntriandrum* etc.), in spirals, often peltate, rarely lobed but with palmate venation; stipules usu. 0. Infls axillary racemes, panicles or cymose heads, fls rarely solit., rarely on old wood. Fls ± reg., usu. dull, 3-merous often with 2 whorls of K, C & A; K usu. free, imbricate or valvate (1–)6(–12+), C often 6 but s.t. more or fewer or even 0, A (2–3, 6(–40) often opp. C, filaments s.t. ± connate, anthers with longit. (rarely transverse) slits & usu. introrse, females with $\underline{G}$ (1)3(6–30) in 1 or more whorls, often held on a gynophore, with postgenital fusion, each with 2 (1 soon aborting) submarginal, pendulous, hemitropous to amphitropous bi- or unitegmic ovules. Fr. a head of drupelets or nutlets, usu. ± curved, the term. style often appearing ± basal; seed often horseshoe-shaped with straight to curved embryo & oily proteinaceous endosperm, s.t. ruminate, or scanty or 0. n = (9-) 11–13, 19, 25

Classification & principal genera:

I. **Pachygoneae** (Tiliacoreae, Triclisieae incl. Hyperbaeneae & Peninantheae); endosperm 0): *Chondrodendron, Hyperbaena, Tiliacora, Triclisia*
II. **Anomospermeae** (endosperm strongly ruminate; cotyledons not foliaceous, appressed): *Abuta*
III. **Tinosporeae** (endosperm weakly ruminate; cotyledons foliaceous, divaricate): *Disciphania, Tinospora*
IV. **Fibraureae** (incl. Coscinieae; endosperm not ruminate; cotyledons thin, foliaceous, divaricate): *Coscinium, Fibraurea*
V. **Menispermeae** (endosperm not ruminate; cotyledons subcarnose, appressed): *Cissampelos, Cocculus, Stephania*

App. close to Lardizabalaceae but L. have compound lvs & copious endosperm

Many medic., esp. in preparation of curare or fish-poisons, sweeteners or contraceptives: *Abuta, Anamirta, Chondrodendron, Cissampelos, Curarea, Dioscoreophyllum, Fibraurea, Jateorhiza, Pachygone, Pycnarrhena, Stephania, Tinospora*; dyes: *Coscinium*; few cult. orn. lianes: *Cocculus, Menispermum*, while some rainforest spp. gigantic

Menispermum Tourn. ex L. Menispermaceae (V). 2–4 E N Am. & E As. ***M. canadense*** L. (moonseed, yellow parilla, E N Am.) – cult. orn. liane with medic. rhizome

Menitskia (Krestovsk.) Krestovsk. = *Stachys* (but see BZ 91(2006)1893)

Menkea Lehm. Cruciferae (37). 6 Aus. R: CGH 200(1970)175

Menodora Bonpl. = *Jasminum* (but see AMBG 19(1932)87)

Menonvillea DC. Cruciferae (21). Excl. *Aimara*, 24 Chile, W Arg. R: Phytotaxa 162,5(2014)241. Shrublets to annuals; gynophore present

Menstruocalamus Yi = *Chimonobambusa*

Mentha Tourn. ex L. Labiatae (VII 2b). 18–19 temp. (Eur. 10 with 4 common hybrids, Aus. 6, NZ 1, Am. 1). Mints. C subreg., 4-merous. V. aromatic herbs, many local medic. (e.g. esp. ***M. canadensis*** L. (relict amphidiploid, ***M. arvensis*** L. (field mint, now Euras., 2n = 72) × *M. longifolia* (2n = 24)), N Am.) & long-cult. as flavourings (oil c. 30 K t by 2007), the Romans using ***M. aquatica*** L. (waterm., Euras., Medit., 2n = 96), most common now being ***M. spicata*** L. (spearm., *M. longifolia* × *M. suaveolens*, 2n = 48) – folk medic. heart disease (Crete), anti-oxidants, incl. '**Alopecuroides**' (backcross with *M. s.*, *M.* × *villosa* nm. *alopecuroides*, Bowles's m.) – rumoured to be the best, & ***M. suaveolens*** Ehrh. (apple or woolly m., S & W Eur., incl. '**Variegata**', pineapple m.), a hairier pl., used in mint sauce & jelly, ***M.* × *piperita*** L. (*M. spicata* × *M. aquatica*, C17 England, peppermint), sterile (irradiated cvs used in U.S.), the flavouring for chocolate, crème-de-menthe, tea, icecream, toothpaste etc. (menthol binds to protein-receptor in mammals sensitive to cold), incl. '**Chocolate**' (chocolate m.), '**Citrata**' (nm. *citrata*, eau-de-cologne m., bergamot, lemon or orange m.), '**Officinalis**' (nm. *officinalis*, white p.) & '**Piperita**' (nm. *piperita*, black p. used in tea), while an imp. source of menthol for cigarettes etc. is *M. canadensis* (*M. arvensis* var. *piperascens*, Japanese m.). Others cult. incl. ***M.* × *gracilis*** Sole ('*M.* × *gentilis*', *M. spicata* × *M. arvensis*, ginger or Scotch (spear)m.) – cult. for spearmint oil; ***M. longifolia*** (L.) Hudson (horse m., Eur., Medit. to Himal. & S Afr.) – variable, often confused with *M. spicata*, but all hairs simple; ***M. pulegium*** L. (pennyroyal, Eur., W As., invasive Aus., W U.S.) – used in soap etc., form. to sweeten sailors' drinking water & medic. (pulegone form. abortifacient in GB), a form being Gibraltar m.; ***M. requienii*** Benth. (Corsican m., W Medit.) – cult. orn. mat-former, natur. Br. Is.

Mentocalyx N.E. Br. = *Gibbaeum*

Mentodendron Lundell = *Pimenta*

Mentzelia Tourn. ex L. Loasaceae (II). 80+ warm Am. (S Am. 8 – R: AMBG 94(2007)658). R: AMBG 21(1934)103. Small trees, shrubs & herbs, stinging hairs 0; K + C shed as a unit; outer A s.t. staminodes. ***M. albicaulis*** (Hook.) Torrey & A. Gray (W N Am.) – ed. seeds, local medic. & tobacco subs. Some cult. orn. esp. ***M. laevicaulis*** (Douglas) Torrey (blazing star, W N Am.) – local medic., & ***M. lindleyi*** Torrey & A. Gray (bartonia, C Calif.) – scented yellow fls opening in evening

Menyanthaceae Dumort. (~ Gentianaceae). Magnoliidae – Asterales. 6/58 cosmop. R: Novon 19(2009)410. Glabrous aquatics or helophytes with intercellular canals & spaces in stems, where vasc. bundles often scattered. Lvs simple (3-foliolate in *Menyanthes*), cordate or reniform to linear, in spirals; petiole sheathing at base, stipules 0 or wing-margins of petiole. Fls bisexual, reg., usu. 5-merous & heterostylous, solit. or in various infl. types, K basally connate, C with tube & valvate to imbricate lobes (margins fringed or crested), filaments attached to tube alt. with lobes, anthers sagittate with longit. slits, s.t. scales (? staminodes) alt. with filaments, nectary-disk often around G (2), superior to half-inf., 1-loc. with term. style & ∞ anatropous, unitegmic ovules on the 2 parietal placentas. Fr. a capsule, variously dehiscent or a berry; seeds, s.t. winged, with linear embryo in copious firm oily endosperm. n = 9, 17

Genera: *Liparophyllum, Menyanthes, Nephrophyllidium, Nymphoides, Ornduffia, Villarsia*

Form. assoc. with Gentianaceae but G. have opp. lvs, internal phloem etc. & Cronquist places M. in Solanales but cladistic work (& inulin storage) places M. in an enlarged Asterales

Some locally medic. or ed., some cult. orn. & bad weeds

Menyanthes Tourn. ex L. Menyanthaceae. 1 circumboreal: ***M. trifoliata*** L. (bogbean, buckbean) – self-incompatible heterostylous fls, rhizome medic. (trad. cure for arthritis in Germany, for weight-loss in N Am., rheumatism, asthma) & for treatment of bovine TB etc., lvs used as hop subs. in beer, rhizome powdered for a bread by Inuit & in N Euras., mitsugashiwalactone attractive to cats

Menziesia Sm. = *Rhododendron*

Meopsis (Calest.) Kozo-Polj. = *Daucus*

mephedrine *Catha edulis*

meranti *Shorea* spp.

merawan *Hopea mengerawan*

merbau *Intsia bijuga, I. palembanica*

Merciera A. DC. Campanulaceae (I 2). 6 SW Cape. R: Bothalia 36(2006)1. G 1-loc. with basal ovule

Mercurialis Tourn. ex L. Euphorbiaceae (Acal.-Acal.-Merc.). 8 Medit. & temp. Euras. (Eur. 7) to N Thailand. Mercury ('mercurial' as females with paired cocci resemble testes, the *mercurialis mascula* of Pliny therefore being female!). Dioec. (*M. annua* diploid, polyploid monoec./andromonoec.) with extrafl. nectaries; lvs opp.; G (2). 'Roots' impart purple tinge to water; some dye-sources incl. ***M. perennis*** L. (dog's m., Euras., Medit.) – carpeting rhiz. herb (highly toxic), often char. of disturbed woodland; ***M. annua*** L. (annual m., Eur., Medit.) –weed of gardens, seeds with caruncles

mercury *Mercurialis* spp.; **annual m.** *M. annua*; **dog's m.** *M. perennis*

Merendera Ramond = *Colchicum*

Meresaldia Bullock = *Tassadia*

Meriandra Benth. Labiatae (VII 2a). 2 Ethiopia, Arabia, Himalaya. R: NRBGE 31(1971)55

Meriania Sw. Melastomataceae. c. 100 trop. Am. Connective spurred

Merianthera Kuhlm. Melastomataceae. 7 E Braz. (rocky outcrops). R: T 61(2012)1046

Mericarpaea Boiss. Rubiaceae (IV 16). 1 W As.: ***M. vaillantioides*** Boiss.

Mericocalyx Bamps = *Otiophora*

Meringium C. Presl = *Hymenophyllum*

Meringogyne H. Wolff = *Angoseseli*

Meringurus Murb. = *Gaudinia*

Merinthopodium J.D. Sm. (~ *Markea*). Solanaceae (IV 3). 3 trop. Am. ***M. neurantha*** (Hemsl.) J.D. Sm. (*Markea n.*) – bat-poll. in Costa Rica

Merinthosorus Copel. = *Aglaomorpha*

Merismostigma S. Moore = *Caelospermum*

Meristotropis Fischer & C. Meyer = *Glycyrrhiza*

Merope M. Roemer (~ *Atalantia*). Rutaceae (II 2). 1 trop. As.: ***M. angulata*** (Willd.) Swingle – tide-disp. fr.

Merostachys Spreng. Gramineae (V 2). 48 trop. Am. R: CUSNH 39(2000)71. Bamboos flowering in 11–34 yr cycles; some climbers

Merremia Dennst. ex Endl. Convolvulaceae (12). c. 100 trop. Heterogeneous. Some serious weeds in plantations, some cult. orn. & locally medic. & ed. ***M. discoidesperma*** (J.D. Sm.) O' Don. (Mary's bean, C Am.) – most widespread of all drift-seeds, local medic. C Am., amulet in Outer Hebrides; ***M. gemella*** (Burm.f.) Hall.f. (SE As. to trop. Aus.) – stems for string (Aus.), esp. for tieing up kangaroos for roasting; ***M. peltata*** (L.) Merr. (E Afr. is. to Pacific) – lvs to 40 by 30 cm; ***M. tuberosa*** (L.) Rendle (Brazilian jalap, C Am. but now natur. pantrop.) – form. grown for medic. (resins laxative), cult. orn. with yellow fls, dried fr. with accrescent K = 'wood rose' of dried floral decorations

Merrillanthus Chun & Tsiang. Apocynaceae (V c; Asclepiadaceae III 1). 1 Hainan: ***M. hainanensis*** Chun & Tsiang

Merrillia Swingle (~ *Murraya*). Rutaceae (II 2). 1 Myanmar to W Mal.: ***M. caloxylon*** (Ridl.) Swingle – pinnate lvs with rachis joints like petiole of *Citrus* spp., fr. with large amounts of eupatorin (also in *Eupatorium* spp.) with high antitumour activity

Merrilliodendron H.L. Kaneh. Icacinaceae. 1 Philippines & W Pacific: ***M. megacarpum*** (Hemsl.) Sleumer (*M. rotense*) – cauliflory

Merrilliopanax H.L. Li. Araliaceae. 3 E Himal. to SW China. R: BMNHN 4,5(1983)289

Merrittia Merr. = *Blumea*

mersawa *Anisoptera* spp. esp. *A. laevis*

Mertensia Roth. Boraginaceae (B.3.3). 40 N temp. (Eur. 1, N Am. 24 – R: JAA suppl. 1(1991)88) to Mex. & Afghanistan. R: AMBG 24(1937)17. Bluebells (N Am.). Cult. orn. rock-pls incl. ***M. maritima*** (L.) Gray (gromwell, sea lungwort, oysterleaf, N temp. coasts) – rhizome eaten by Inuit, & ***M. virginica*** (L.) Link (Virginian b. or cowslip, E N Am.)

Merumea Steyerm. Rubiaceae (I 1/IV 24). 2 Guayana Highland. R: MNYBG 23(1972)232

Merwia B. Fedtsch. = *Ferula*

Merwilla Speta. Asparagaceae (Hyacinthaceae). 3 Zimbabwe to S Afr. R: Phyton 38(1998)107

Merwiopsis Safina = *Pilopleura*

Merxmuellera Conert (~ *Rytidosperma*). Gramineae (Danth.). Excl. *Ellisochloa*, 9 trop. Afr., Madag.

Meryta Forst. & Forst. f. Araliaceae. 30 NZ, Aus., New Caled. (11, endemic), Pacific is. Dioec. pachycauls (Chamberlain's Model), lvs simple (comb. unique in A.). Some cult. orn.

Mesadenella Pabst & Garay (~ *Stenorrhynchos*). Orchidaceae (IV 2h). 7 trop. Am.

Mesadenus Schltr. (~ *Brachystele*). Orchidaceae (IV 2h). 7 Florida (1), C. Am., Carib., SE Brazil

Mesanthemum Koern. Eriocaulaceae. 18 trop. & warm Afr., Madagascar (2)

Mesanthophora H. Robinson. Compositae (Vern.-Mes.). 2 C Paraguay, Bolivia. R: Novon 2(1992)169

mescal *Lophophora williamsii*; see also *Agave vivipara*; **m. bean** *Dermatophyllum secundiflorum*

mesclun salad leaves, e.g. mix of lettuce, rocket, spinach etc.

Mesechites Muell. Arg. Apocynaceae (II d). 8 trop. Am. R: Candollea 61(2007)220

Mesembryanthemaceae Philib. = Aizoaceae

Mesembryanthemum Tourn. ex L. Aizoaceae (IV). Incl. *Aptenia, Aspazoma, Brownanthus, Caulipsolon, Cryophytum, Dactylopsis, Phyllobolus, Psilocaulon, Sceletium, Synaptophyllum*, 103 Medit. (Eur. 2) to Iran, Arabia, drier S Afr., S Aus., Calif., W S Am., Atlantic Is. R: Bothalia 43(2013)201. Many succ. & cult. orn. form. referred here now in *Carpobrotus, Cleretum, Conophytum* etc. Dried fermented lvs of some (*Phyllobolus*) spp. chewed, leading to intoxication due to mesembrine, like cocaine; local soporific, used for alcohol- and drug-dependence by San (S Afr.), now tablets prescribed by doctors. ***M. crystallinum*** L. (*Cryophytum c.*, ice-pl., Cape, natur. in Aus., Calif., Medit.) – lvs covered with glistening papillae, cult. bedding pl. with lvs eaten like spinach, like ***M. nodiflorum*** L. (Medit., Iran) accum. & releasing salt in soil so dominating disturbed habitats; ***M. tortuosum*** L. (*Sceletium t.*, kanna, S Afr.) – alks incl. mesembrine, appetite-suppressant, chewed to reduce stress, now comm., poss. recreational drug

Mesocapparis (Eichler) Cornejo & Iltis (~ *Capparis*). Capparaceae. 1 Braz.: ***M. lineata*** (Pers.) Cornejo & UIltis. R: HPB 13(2008)113

Mesoglossum Halb. = *Rhynchostele*

Mesogramma DC. (~ *Senecio*). Compositae (Senec.-Senec.). 1 S Afr.: ***M. apiifolium*** DC. – annual. R: CN 42(2005)78

Mesogyne Engl. Moraceae (III). 1 trop. Afr.: ***M. insignis*** Engl.

Mesomelaena Nees (~ *Gymnoschoenus*). Cyperaceae (II 7). 5 SW Aus. R: Telopea 2(1981) 181

Mesona Blume = *Platostoma*

Mesophlebion Holttum = *Cyclosorus*

Mesoptera Hook.f. = *Psydrax*

Mesopteris Ching (~ *Cyclosorus*). Thelypteridaceae. 1 W China, Vietnam: ***M. tonkinensis*** (C. Chr.) Ching. R: AFJ 105(2015)18

Mesosetum Steud. Gramineae (XXIII 1). 27 trop. Am. (Braz. 20). R: AA 19(1989)51

Mesosphaerum P. Browne (~ *Hyptis*). Labiatae (VII 3c). 25 trop. Am. mts. R: Phytotaxa 58(2012)29

Mesospinidium Reichb.f. = *Brassia* (but see Orquideologia 8(1973)165)

Mesostemma Vved. = *Stellaria*

Mespilodaphne Nees = *Ocotea*

mespilus, snowy *Amelanchier ovalis*

Mespilus Tourn. ex L. (~ *Crataegus*). Rosaceae (Ros.-Mal.). 1 SE Eur. to C As.: ***M. germanica*** L. (*C. g.*, ? orig. Transcaucasia) – poss. cult. Assyrians & Babylonians, poss. Roman introd. to GB (seed found at Silchester), fr. ed. (carnivore frugivory) esp. after frost & slightly rotten (bletted) when malic acid reduced & sugar increased, 'cotignac' a Med. preserve made at Orléans, France, & offered to sovereigns, Joan of Arc etc. R: J.B. Phipps (2003) *Hawthorns & Medlars*: 104

mesquite *Prosopis glandulosa*

Messerschmidia Hebenstr. = *Heliotropium*

messina *Melilotus messanensis*

messmate *Eucalyptus* spp. with stringy bark, esp. *E. obliqua*

Mestoklema N.E. Br. ex Glen. Aizoaceae (V). 6 Cape. R: Bothalia 13(1981)454

Mesua L. Calophyllaceae (Guttiferae (I1)). Excl. *Kayea*, 5 Sri Lanka & W Ghats (4, endemic) to Sumatra. ***M. ferrea*** L. (ironwood, na) – A <1000, the tree sacred in India & planted around temples in Vietnam, timber v. hard used for railway sleepers in S India, form. for lances, fls used in medic. & cosmetics & to scent the stuffing of pillows, seeds with 79% oil principal oil source for Dai people (trop. China)

Mesyniopsis W. Weber = *Linum*

Mesynium Raf. = *Linum*

Metabolos Blume = *Hedyotis* (but see BJBB 63(1994)256)

Metabriggsia W.T. Wang = *Hemiboea* (but see EJB 49(1992)30)

Metadacrydium Baum.-Bodenh. ex A. Bobrov & Melikian = *Dacrydium*

Metadina Bakh.f. = *Adina* (but see FOC 19(2011)215)

Metaeritrichium W.T. Wang = *Microula*

Metagentiana T.N. Ho & S.W. Liu (~ *Gentiana*). Gentianaceae. Excl. *Sinogentiana*, 14 Myanmar, Thailand. China. R: BBAS 43(2002)87

Metalasia R. Br. Compositae (Gnap.-Rel.). 57 S Afr. esp. SW Cape. R: OB 99(1989)1. Lvs s.t. twisted spirally. Long-lasting fl. stems often dyed brightly for graveside tributes

Metalepis Griseb. = *Cynanchum*

Metanarthecium Maxim. (~*Aletris*). Nartheciaceae. 1 Japan: ***M. luteoviride*** Maxim.

Metanemone W.T. Wang. Ranunculaceae (II 2). 1 Yunnan: ***M. ranunculoides*** W.T. Wang

Metapanax Frodin (~ *Macropanax*). Araliaceae. 2–4 China, N Vietnam. R: Brittonia 53(2001)117. Cult. orn.

Metapetrocosmea W.T. Wang (~ *Petrocosmea*). Gesneriaceae (III 2j). 1 Hainan: ***M. peltata*** (Merr. & Chun) W.T. Wang

Metaplexis R. Br. = *Cynanchum*

Metapolypodium Ching = *Goniophlebium*

Metaporana N.E. Br. (~ *Bonamia*). Convolvulaceae (IV). 6 Socotra (1), E Afr. (1), Madag. (4). R: JAA 71(1990)252

Metarungia Baden = *Anisotes* (but see NJB 1(1981)143, KB 34(1980)638)

Metasasa W.T. Lin = *Acidosasa*

Metasequoia Miki ex Hu & W.C. Cheng. Cupressaceae (Taxodiaceae). 1 C China (fossil Eur., W N Am.): ***M. glyptostroboides*** Hu & W.C. Cheng (dawn redwood) – cult. orn. (60 km avenue in China!) decid. (branchlets) tree to 45m or more, if topped branching out into shade-tree, first coll. 1943 & seed coll. 1947 introd. to Western gardens in 1948; c. 6000 left in wild. R: BR 42(1976)215. Genus first described from Cretaceous & Tertiary fossils (diff. spp.) as *Taxites* Brongn. (1828) & as *Metasequoia* (1941), from modern pls in 1948

Metasocratea Dugand = *Socratea*

Metastachydium Airy Shaw ex C.Y. Wu & Li. Labiatae (VI). 1 C As. to W China: ***M. sagittatum*** (Regel) C.Y. Wu & Li. R: APS 13(1975)73

Metastelma R. Br. = *Cynanchum* (but possibly distinct)

Metastevia Grashoff = *Stevia*

Metathelypteris (H. Itô) Ching = *Thelypteris*

Metatrophis F. Br. Moraceae. 1 Rapa: ***M. margaretae*** F. Br. – not seen since 1934

Metaxya C. Presl. Metaxyaceae. 2 trop. S Am.

Metaxyaceae Pichi-Serm. Polypodiidae – Cyatheales. 1/2 trop. S Am. R: T 55(2006). Terr. ferns. Rhizome creeping, solenostelic, apices hairy. Fronds 1-pinnate; petiole with omega-shaped vasc. bundle. Sori round, abaxial on the veins (sometimes 2 or 3 on same vein – unique in modern ferns); indusium 0. n = 95, 96
Genus: *Metaxya*

Metcalfia Conert. Gramineae (IX). Excl. *Danthoniastrum*, 1 Mex.: ***M. mexicana*** (Scribner) Conert. R: Willd. 2(1960)417

Meteoromyrtus Gamble = *Eugenia*

Meterostachys Nakai (~ *Orostachys*). Crassulaceae (I). 1 S Jap. & S Korea: ***M. sikokiana*** (Mak.) Nakai. R: U. Eggli, *Ill. Handb. Succ. Pls*, Crass. (2005)182

Metharme Philippi ex Engl. Zygophyllaceae (II). 1 N Chile: ***M. lanata*** Philippi ex Engl.

methi *Trigonella foenum-graecum*

Methysticodendron R. Schultes = *Brugmansia*

Metopium P. Browne. Anacardiaceae (I). 3 Florida, Mex., WI. Purging resins esp. hog gum from ***M. toxiferum*** (L.) Krug & Urb. (Florida)

Metoxypetalum Morillo = *Oxypetalum* (but see Ernstia 3(1994)145)

Metrodorea A. St-Hil. Rutaceae (I 6). 5 NE S Am. to Peru & Bolivia. R: FN 33(1982)116

Metrosideros Banks ex Gaertn. Myrtaceae (Lept.). Incl. *Carpolepis*, *Tepualia*, excl. *Mearnsia*, 60+ E Mal., Pacific esp. NZ, S Afr. (1: ***M. angustifolia*** (L.) Sm., only capsular Myrtacea in Afr.), Chile. Trees, shrubs (incl. rheophytes) & woody climbers (some stranglers) app.

merging with *Callistemon* in New Caled. Some cult. orn. & timbers esp.: ***M. excelsa*** Sol. ex Gaertn. (Christmas tree (NZ), pohutukawa, NZ) – timber form. for ship-building, locally medic., floral emblem of La Coruña, NW Spain (from old specimen growing there), invasive Cape fynbos; ***M. polymorpha*** Gaudich. (*M. collina*, Hawaii) – wood for construction & form. for idols, (exc. for young lvs) resistant to volcanic fumes (SO_2 to 100 p.p.m.) as stomata close; ***M. robusta*** Cunn. (rata, NZ) – sometimes epiphytic, timber for telegraph-poles etc.

Metroxylon Rottb. Palmae (I 3c). c. 7 E Mal. Sago palms. R: Principes 30(1986)170. Trunks hapaxanthic (Holttum's Model) but, in most, suckers develop from base; fr. takes 3 yrs to ripen. Trunks felled when infl. appears & sago removed from stem by crushing & washing (felled trunks for rearing highly esteemed sago grubs, *Rhynchophorus ferrugineus*), esp. ***M. sagu*** Rottb. (natur. W Pacific to SE As.) spineless or ('*M. rumphii* (Willd.) Mart.') prickly – petioles used for walls in Indonesia. Some 'seeds' used as buttons

Mettenia Griseb. = *Chaetocarpus*

Metteniusa Karsten. Metteniusaceae. 7 NW trop. S Am. R: JAA 21(1940)485. ***M. edulis*** Karsten – cooked seeds ed.

Metteniusaceae H. Karst. ex Schnizl. (~ Icacinaceae). Magnoliidae – Metteniusales. Incl. Oncothecaceae (R: BMNHN Adansonia 3(1981)305), 7/42 trop. Am. Trees & shrubs. Lvs simple, in spirals; stip. 0. Fls reg., shortly pedicellate in axillary racemes to thyrses, pedicels with 1 bract & 2 sepaloid bracteoles; K5 ± quincuncial before anthesis, persistent in fr., C (5) basally adnate, lobes quincuncial in bud, A5, attached to C, alt. with lobes, anthers basifixed, extrorse with prolonged connective & longit. slits, $\underline{G}$ (5), 5-loc. & -grooved with 5 short free styles & apical-axile placentation or 1-loc. with parietal, (1)2 collateral, pendulous, anatropous, unitegmic ovules per loc. Fr. a drupe. with 1- or 5-loc. stone, each loc. with 1 or 0 seed; seed with curved embryo in copious endosperm. n = 25 Genera: *Apodytes* (timber), *Dendrobangia*, *Emmotum*, *Metteniusa*, *Oncotheca* (to Icacinaceae?), *Poraqueiba* (ed.), *Rhaphiostylis*. Resolution of I., M., Cardiopteridaceae & Stemonuraceae continuing

Metternichia Mikan. Solanaceae (I; Goetzeaceae). 1 E Brazil: ***M. principis*** Mikan – seeds winged (unique in S.)

meu *Meum athamanticum*

Meum Mill. Umbelliferae (III 8). 3 N Afr. & Eur. incl. ***M. athamanticum*** Jacq. (baldmoney, meu, spignel) – cult. orn., form. cult. for ed. roots, medic. (flatulence) & scent for snuff

Mexacanthus T. Daniel. Acanthaceae (III 2c). 1 W Mexico: ***M. macvaughii*** T. Daniel. – poll. hummingbirds, carpenter bees (pollen on wings). R: SB 6(1981)288

Mexerion Nesom (~ *Gnaphalium*). Compositae (Gnap.-Gnap.). 2 Mex. R: Phytol. 68(1990)247

Mexianthus Robinson. Compositae (Eup.-Crit.). 1 Mex.: ***M. mexicanus*** Robinson. R: Sida 13(1989)340

Mexican apple *Casimiroa edulis*; **M. devil** *Ageratina adenophora*; **M. foxglove** *Tetranema roseum*; **M. hat** *Rudbeckia* spp.; **M. kingwood** *Dalbergia congestiflora*; **M. mahogany** *Swietenia humilis*; **M. poppy** *Argemone mexicana*; **M. tarragon** *Tagetes lucida*

Mexicoa Garay = *Oncidium*

Meximalva Fryx. (~ *Sida*). Malvaceae (Malv.-Malv.). 2 Mexico

Mexipedium V. Albert & M. Chase (~ *Phragmipedium*). Orchidaceae (III). 1 Mex.: ***M. xerophyticum*** (Soto & al.) V. Albert & M. Chase. R: Lindleyana 7(1992)172

Mexotis Terrill & H. Robinson (~ *Hedyotis*). Rubiaceae (IV 1). 5 Mex., Guatemala. R: JBRIT 3(2009)60

Meyenia Nees (~ *Thunbergia*). Acanthaceae (II). 1 India, Sri Lanka: ***M. hawtayneana*** (Wall.) Nees

Meyerocactus Doweld = *Echinocactus*

Meyerophytum Schwantes. Aizoaceae (V). 2 Cape. R: H. E. K. Hartmann, *Ill. Handb. Succ. Pls*, Aiz. F–Z(2002)170

Meyna Roxb. ex Link = *Canthium*

mezcal = mescal

mezereon *Daphne mezereum*

Mezia Schwacke ex Niedenzu. Malpighiaceae. 1 trop. Am.: ***M. includens*** (Benth.) Cuatrec. – imp. local medic. vine

Meziella Schindler = *Myriophyllum* (form. thought to be in Haloragaceae)

Mezilaurus Kuntze ex Taubert. Lauraceae (I). Excl. *Clinostemon*, 18 trop. S Am. to Costa Rica (1). R: AMBG 74(1987)158

Mezleria C. Presl = *Monopsis*; see also *Lobelia*

Mezobromelia L.B. Sm. Bromeliaceae (2). 9 Colombia, Ecuador. R: FN 14(1977)1364

Mezochloa Butzin = *Alloteropsis*

Mezoneuron Desf. (*Mezonevron*, ~ *Caesalpinia*). Leguminosae (I 4). 24 As to Hawaii, Afr. (few). R: Phytotaxa 274 (2016). ***M. enneaphyllum*** (Roxb.) Voigt (Indomal.) – cult. orn. tree

Mezzettia Becc. Annonaceae (II). 4 W Mal. to Moluccas. R: Blumea 35(1990)217. ***M. macrocarpa*** Heyden & Kessler (N Borneo) – obligate elephant-disp. seeds

Mezzettiopsis Ridl. = *Orophea*

mfukufuku *Brexia madagascariensis*

mfungu *Celosia argentea*

mi *Madhuca longifolia*

Mibora Adans. Gramineae (XVI 13). 2 W Eur., NW Afr. ***M. minima*** (L.) Desv. – cult. orn. annual, sometimes flowering when less than 1 cm tall

mibuna *Brassica rapa* Japonica Group

Michaelmas daisy *Symphyotrichum* spp. & hybrids

Michauxia L'Hérit. Campanulaceae (I 3). 7 SW As. Fls 6–10-merous. ***M. campanuloides*** L'Hérit. (E Medit.) – cult. orn. biennial with white fls

Michelia L. = *Magnolia*

Micheliella Briq. = *Collinsonia*

Michelsonia Hauman. Leguminosae (I 2). 1 E Congo: ***M. microphylla*** (Troupin) Hauman – forming almost pure stands, timber (over-exploited mid C 20). R: WAUP 99–3(1999) 249

Micholitzia N.E. Br. = *Hoya*

Mickelia R.C. Moran & al. Dryopteridaceae (II). 10 trop. Am. R: Britt. 62(2010)338

mickymick (Maori *mingimingi*) *Coprosma* spp.

Micklethwaitia G. Lewis & Schrire (~ *Cynometra*). Leguminosae (I 2). 1 Mozambique: ***M. carvalhoi*** (Harms) G. Lewis & Schrire. R: KB 59(2004)166

Miconia Ruíz & Pavón. Melastomataceae. Incl. *Anaectocalyx, Calycogonium, Charianthus, Clidemia, Conostegia, Hormocalyx, Killipia, Leandra, Maieta, Mecranium, Necramium, Ossaea, Pachyanthus, Pleiochiton, Sagraea, Tetrazygia, Tococa*, excl. *Rupestrea*, c. 1900 trop. Am. (largest genus of woody pls there; Colombia 338). R: Phytotaxa 106(2013)1; (sect. *Chaenopleura*) SBM 81 (2007). Many related spp. in same habitat, 20 in Trinidad with complementary fruiting times throughout yr & same dispersing birds. Leeuwenberg's & Mangenot's Models; some with hollow ant-infested stems or with bladdery outgrowths of lvs inhabited by ants (in ***M. mayeta*** (D. Don) Michelang. (*Maieta guianensis*) 80% of host N intake from waste deposited by *Pheidole minutula* ants); some poll. hummingbirds. Some timber (rodwood), ed. fr. (bush currants) & local medic.; some yellow dyes & some cult. orn. Oft. in sec. veg. & some weedy in Pac. (esp. ***M. crenata*** (Vahl) Michelang. (*Clidemia hirta*) – 'Koster's [correctly Köster's] curse' (introd. from Braz., but by a W.F. Parr!) in Fiji, also invasive in trop. As., Seychelles, Madag. (introd. 1948), Hawaii). ***M. albicans*** (Sw.) Steud. (trop. Am.) – disp. (Braz. savanna) by vole-like rat, *Necromys* (*Bolomys*) *lasiurus*, only small rodent-disp. known; ***M. argentea*** (Sw.) DC. (trop. Am.) – pioneer establishing in gaps greater than 102 m^2 & becoming canopy emergent; ***M. bicolor*** (Mill.) Triana (*Tetrazygia b.*, Carib.) – cult. orn. with small fls but conspic. yellow A; ***M. calvescens*** DC. (*M. magnifica*, trop. Am., invasive Tahiti (introd. cult. orn. 1937, penetrating native forest, by 1996 covering over two-thirds of is.), Hawaii & Sri Lanka) – lvs to 70 cm long; ***M. fadyenii*** (Hook.) Judd & Skean (*T. f.*, Jamaica) – hummingbird-poll. treelet, cult. orn.; ***M. hammelii*** (Almeda) Almeda (*Clidemia h.*, C Am.) – *Ololaelaps* mites in domatia; ***M. physophora*** (Vahl) DC. (*Tococa guianensis, T. occidentalis*, trop. Am.) – ants clearing (competing) veg. around it; ***M. rubra*** (Aubl.) Mabb. (*C. r.*, Amaz.) – cult. orn. shrub; ***M. salicina*** (DC.) Mabb. (*Leandra s.*, Braz.) – rheophyte; ***M. subciliata*** Benth. (Amaz.) – fr. fish-disp.; ***M. xalapensis*** (D. Don) M. Gómez (*Conostegia x.*, C Am.) – delicious flavour like blueberry, cult.

Micrachne Peterson (~ *Cynodon*). Gramineae (XXIX 5). 5 C & E Afr. T: T 64(2015)459. Incl. most '*Brachyachne*' spp. ***M. patentiflora*** (Stent & Rattray) Peterson – lvs revive after reduction of water content to 5%

Micractis DC. (~ *Sigesbeckia*). Compositae (Mill.-Mill.). 4 trop. Afr., Madag. R: Gleditschia 18(1990)211

Micraeschynanthus Ridl. = *Aeschynanthus*

Micraira F. Muell. Gramineae (Micr.). 15 trop. Aus. R: FA 44b(2005)120. Moss-like pls with spiral phyllotaxis, 'resurrection' pls with lvs reviving after desiccation

Micrandra Benth. Euphorbiaceae (Crot.-Mic.). Incl. *Cunuria* 12 trop. S Am. Some rubber ('M.' r.), local medic.

Micrandropsis Rodr. Euphorbiaceae (Crot.-Mic.). 1 Amazon: ***M. scleroxylon*** (Rodr.) Rodr. R: AA 3,2(1973)5

Micrantha Dvořák (~ *Hesperis*). Cruciferae (5). 1 Iran: ***M. multicaulis*** (Boiss.) Dvořák

Micranthemum Michaux. Linderniaceae (Phrymaceae (Microc.); Scrophulariaceae s.l.). Incl. *Amphiolanthus, Hemisiphonia,* 14 Am., WI. Aquatics s.t. cult. in aquaria

Micranthes Haw. (~ *Saxifraga*). Saxifragaceae (I 2). c. 85 N temp. N Am. 45 – R: FNA 8(2009)49, Arctic. R: BJLS 178(2015)63. Seeds usu. unitegmic (*S.* usu. bitegmic)

Micrantheum Desf. Picrodendraceae (Cal.-Pseud.; Euphorbiaceae s.l.). 3 Aus.

Micranthocereus Backeb. Cactaceae (III 3). 9 C & E Brazil

Micranthus (Pers.) Ecklon. Iridaceae (VI 2). 7 SW Cape. R: Bothalia 43(2013)130

Micrargeria Benth. Orobanchaceae ('Micr.'; Scrophulariaceae s.l.). 4–5 trop. Afr., India

Micrargeriella R. Fries. Orobanchaceae ('Micr.'; Scrophulariaceae s.l.). 1 Congo, Zambia: ***M. aphylla*** R. Fries

Micrasepalum Urb. Rubiaceae (IV 15). 2 Cuba, Hispaniola

Micrechites Miq. (~ *Ichnocarpus*). Apocynaceae. 10 Himal. to NG (Mal 8 – R: FM I,18(2007)281)

Microberlinia A. Chev. Leguminosae (I 2). 2 Gulf of Guinea. Timber (zebrawood, zebrano, zingana) esp. ***M. brazzavillensis*** A. Chev.; ***M. bisulcata*** A. Chev. – crown 75 m diam. (? biggest known), fr. expelled 72 m from canopy

Microbiota Komarov (~ *Thuja*). Cupressaceae. 1 E Siberia: ***M. decussata*** Komarov – dwarf cult. orn. R: A. Farjon, *Monogr. Cupressaceae & Sciadopitys* (2005)420

Microbriza Parodi ex Nicora & Rugolo = *Chascolytrum*

Microcachrydaceae Doweld & Reveal = Podocarpaceae

Microcachrys Hook.f. Podocarpaceae. 1 Tasmania (fossils from Cape Tertiary): ***M. tetragona*** (Hook.) Hook.f. – seeds encl. in scarlet aril

Microcala Hoffsgg. & Link = *Cicendia*

Microcalamus Franch. Gramineae (XXIV 2). 1 trop. W Afr.: ***M. barbinodis*** Franch. – superficially like a bamboo

Microcalia A. Rich. = *Lagenophora*

Microcardamum O. Schulz = *Hornungia*

Microcarpaea R. Br. Phrymaceae (Microc.; Scrophulariaceae s.l.). 2 trop. As. R: Phytoneuron 2012–39: 21

Microcaryum I.M. Johnston. Boraginaceae (B.3.2). 1 E Himal.: ***M. pygmaeum*** (Clarke) I.M. Johnston

Microcasia Becc. = *Bucephalandra*

Microcephala Pobed. Compositae (Anth.-Han.). 5 Iran, C As. R: MBSM 12(1976)655

Microchaete Benth. = *Monticalia*

Microcharis Benth. (~ *Indigofera*). Leguminosae (III 15). 36 Arabia, Afr., Madag. R: Bothalia 22(1992)165

Microchilus Presl (~ *Erythrodes*). Orchidaceae (IV 2d). c. 140 trop. Am.

Microchirita (Clarke) Y.Z. Wang (~ *Chirita*). Gesneriaceae (III 2j). 18 India to Java (limestone cf. *Henckelia*). R: T 60(2011)778. Fls often borne on petiole. ***M. elphinstonia*** (Craib) A. Weber & Middleton (*C. e.*, Thailand) – ann. with epiphyllous infls, cult. orn.

Microchlaena Ching = *Anisocarpium*

Microchloa R. Br. Gramineae (XXIX 5). Incl. *Rendlia,* 6 Afr., with 1 pantrop. R: SBi 47(1966)291. Lvs of ***M. caffra*** Nees (Afr.) & ***M. kunthii*** Desv. (Afr., natur. Arizona to Mex.) revive after desiccation to 3–9% of water content

Microcitrus Swingle = *Citrus*

Microcnemum Ung.-Sternb. Amaranthaceae (Chenopodiaceae II 2). 1 Medit. (incl. Eur.) to Cauc.: ***M. coralloides*** (Loscos & Pardo) Buen

Micrococca Benth. (~ *Claoxylon*). Euphorbiaceae (Acal.-Acal.-Claox.). 12 OW trop. (Mal. 3 – R: Blumea 47(2002)150)

Microcodon A. DC. Campanulaceae (I 2). 3 W Cape. R: Strelitzia 9(2000)392

Microcoelia Lindl. Orchidaceae (V 16d). c. 30 trop. & S Afr., Madag. R: SBU 23,4(1981)1. Leafless epiphytes with photosynthetic roots

Microcoelum Burret & Potztal = *Lytocaryum*

Microconomorpha (Mez) Lundell = *Cybianthus*

Microcorys R. Br. Labiatae (IV 2). c. 20 SW Aus. Heterogeneous

Microcos Burm. ex L. (~ *Grewia*). Malvaceae (Grew.-Grew.; Tiliaceae). 53 Indomal., Fiji. Forest pls, cf. most *G.* spp.in open habitats. ***M. nervosa*** (Lour.) S.Y. Hu (S China) – imp. local medic., exported to U.S.

Microculcas Peter = *Gonatopus*

Microcybe Turcz. (~ *Phebalium*). Rutaceae (I 3). 4 Aus. R: FA 26(2013)480

Microcycas A.DC. Zamiaceae (III). 1 W Cuba: ***M. calocoma*** (Miq.) A. DC. – almost extinct, several spermatozoids per male gametophyte & poss. hundreds of ovules on each female

Microdactylon Brandegee = *Matelea*

Microderis DC. = *Leontodon*

Microdesmis Hook.f. Pandaceae. 10 trop. Afr. (8), SE As. & W Mal. (2). Roux's Model. ***M. puberula*** Hook.f. (trop. Afr.) – flexible hard timber for tool-handles, combs etc.

Microdon Choisy. Scrophulariaceae (Man.; Globulariaceae). Incl. *Agathelpis*, 7 W & SW Cape. R: O.M. Hilliard, *The tribe Selagineae* (1999)11

Microdracoides Hua. Cyperaceae (III 2). 1 trop. W Afr.: ***M. squamosus*** Hua – pachycaul with woody trunk & term. panicle of fls.

Microepidendrum Brieger ex W. Higgins (~ *Epidendrum*). Orchidaceae (V 13b). 1 Mex.: ***M. subulatifolium*** (A. Rich. & Galeotti) W. Higgins

Microgilia J.M. Porter & L.A. Johnson (~ *Gilia*). Polemoniaceae (III). 1: ***M. minutiflora*** (Benth.) J.M. Porter & L. A. Johnson. R: Aliso 19(2000)79

Microglossa DC. Compositae (Ast.). 19 Afr., Madag., Masc., trop. As. Usu. scandent or twining shrubs. ***M. afzelii*** O. Hoffm. (trop. Afr.) – local medic. esp. for chests; ***M. pyrifolia*** (Lam.) Kuntze (OW trop.) – imp. local medic.

Microgonium C. Presl = *Didymoglossum*

Microgramma C. Presl (~ *Polypodium*). Polypodiaceae (V). Incl. *Solanopteris*, c. 18 trop. Am., 1 Afr., Madag. & Masc. Cult. orn. epiphytes. ***M. bifrons*** (Hook.) Lellinger (*S. b.*, potato fern, trop. Am.) & other spp. with complex lateral rhizomatous sacs infested with *Azteca* & other ants which bring in debris later exploited by roots entering sacs

Microgyne Less. Compositae (Ast.-Pod.). 1 Arg., Brazil, Uruguay: ***M. trifurcata*** Less. – local medic. R: SB 33(2006)852

Microgynella Grau = *Microgyne*

Microgynoecium Hook.f. Amaranthaceae (Atrip.; Chenopodiaceae I 3). 1 Tibet: ***M. tibeticum*** Hook.f.

Microholmesia Cribb = *Angraecopsis*

Microlaelia (Schltr.) Chiron & Castro = *Sophronitis*

Microlaena R. Br. = *Ehrharta*

Microlagenaria (C. Jeffrey) A. M. Lu & J. Q. Li = *Siraitia* (but see R: APS 31(1993)53)

Microlecane Schultz-Bip. ex Benth. = *Bidens*

Microlepia C. Presl (~ *Dennstaedtia*). Dennstaedtiaceae. c. 70 trop. esp. OW (Afr. 2) to Jap. & NZ. Some with fronds several m long, supported by other pls; some rheophytes. Some cult. orn. incl. ***M. speluncae*** (L.) Moore (OW trop., natur. Am.)

Microlepidium F. Muell. (~ *Capsella*). Cruciferae (37). 2 Aus. R: CGH 205(1974)158

Microlepis (DC.) Miq. Melastomataceae. 4 S Brazil

Microliabum Cabrera. Compositae (Liab.). 5 C Bolivia to Arg. R: SB 15(1990)738

Microlicia D. Don. Melastomataceae. c. 130 trop. S Am.

Microlobius C. Presl (*Goldmania*; ~ *Stryphnodendron*). Leguminosae (II 1). 1 Mex. to Honduras, S Brazil, Paraguay & Arg. (disjunct): ***M. foetidus*** (Jacq.) M. Sousa & G. Andrade – garlic-scented, trad. medic. R: AIBU Mex, Bot 63(1992)104

Microloma R. Br. Apocynaceae (V c; Asclepiadaceae III 1). 10 S Afr. R: BJ 112(1991)453

Microlonchoides Candargy = *Jurinea*

Microlonchus Cass. = *Mantisalca*

Microlophopsis Czerep. = ? *Serratula*

Micromeles Decne = *Sorbus* (*Aria*)

Micromelum Blume. Rutaceae (II 1). 10 Indomal., Pacific. R: W.T. Swingle, *Bot. Citrus* (1943)139. ***M. minutum*** (Forst.f.) Wight & Arn. (Mal.) – local medic.

Micromeria Benth. (~ *Satureja*). Labiatae (VII 2b). 54 Eur. (21), Medit., Macar. (Canary Is. 15 endemic incl. ***M. teneriffae*** (Poir.) G. Don), As., S Afr.; Am. spp. [sect. *Pseudomelissa*] = *Clinopodium*). R: Willd. 38(2008)365

Micromonolepis Ulbr. (~ *Chenopodium*). 1 W N Am.: ***M. pusilla*** (S. Watson) Ulbr.

Micromyrtus Benth. Myrtaceae (II 15). 43 Aus.

Micromystria O. Schulz = *Andrzeiowskia*

Micronoma H. Wendl. Palmae. 1 E Peru = ?

Micronychia Oliv. (~ *Protorhus*). Anacardiaceae (I). 10 Madag. R: Adans. III,22(2000)5, 31(2009)31

Micropapyrus Süsseng. = *Rhynchospora*

Microparacaryum (Riedl) Hilger & Podlech (~ *Lepechiniella*). Boraginaceae (B.3.8.1). 3 SW As.

Micropeplis Bunge = *Halogeton*

Micropera Lindl. (*Camarotis*). Orchidaceae (V 16c). c. 20 Indomal. to Aus. (New Caled. 1)

Microphlebodium Gómez = *Pleopeltis*

Micropholis (Griseb.) Pierre. Sapotaceae (IV). 38 trop. Am. R: FN 52(1990)172. Some timbers

Microphyes Philippi. Caryophyllaceae (I 1). 3 Chile

Microphysa Schrenk (~ *Galium*). Rubiaceae (IV 16). 1 C As.: ***M. elongata*** (Schrenk) Pobed. R: FOC 19(2011)216

Microphysca Naudin = *Tococa*

Microphytanthe (Schltr.) Brieger = *Dendrobium*

Micropleura Lag. Umbelliferae (Mack.). 2 Colombia, Chile

Microplumeria Baill. Apocynaceae (I a). 1 Amazonian Brazil: ***M. anomala*** (Muell. Arg.) Markgr.

Micropolypodium Hayata (~ *Xiphopteris*). Polypodiaceae (V; Grammitidaceae). Excl. *Moranopteris*, 24 trop. Am. & Mal. R: Novon 2(1992)419

Micropora Hook.f. (*Hexapora*). Lauraceae (I). 1 Malay Pen.: ***M. curtisii*** (Hook.f.) Hook.f. – coll. once on Penang Hill

Micropsis DC. Compositae (Gnap.-Gnap.). 5 Brazil, Arg., Uruguay. R: OB 104(1991)158

Micropterum Schwantes = *Cleretum*

Micropus L. Compositae (Gnap.-Gnap.). Excl. *Bombycilaena*, 1 Medit.: ***M. supinus*** L. R: OB 104(1991)173,174

Micropyropsis Romero-Zarco & Cabezudo = *Lolium* (but see Lagascalia 11(1983)95)

Micropyrum (Gaudin) Link = *Festuca*

Microrphium C.B. Clarke. Gentianaceae. 1 W Mal. to Philippines: ***M. pubescens*** C.B. Clarke. R: Blumea 48(2003)30

Microsaccus Blume. Orchidaceae (V 16c). 12 Indomal.

Microschoenus C.B. Clarke = *Juncus*

Microsciadium Boiss. Umbelliferae (III 8). 1 As. Minor: ***M. minutum*** (Urv.) Briq.

Microsechium Naud. = *Sicyos*

Microsemia Greene = *Streptanthus*

Microseris D. Don. Compositae (Cich.-Cich. (Micro.)). Incl. *Stebbinsoseris*, excl. *Nothocalais*, *Uropappus*, 14 W N Am., Chile & Peru (1), Aus. (1) & NZ (1). Australasian spp. app. derived from W N Am. R (partial): CDH 4(1955)207. ***M. laciniata*** (Hook.) Schultz-Bip. (W N Am.) – latex a chewing-gum; ***M. lanceolata*** (Walp.) Schultz-Bip. ('*M. scapigera*', murnong, myniong, yam daisy, Aus.) – tuber ed. roasted

Microsisymbrium O. Schulz = *Streptanthus*

Microsorum Link (*Microsorium*, ~ *Polypodium*). Polypodiaceae (IV). Incl. *Neocheiropteris*, *Phymatosorus* (R: Webbia 28(1978)457). c. 70 OW trop. to NZ. R: Blumea 42(1997)294. Some local ed. & medic.; cult. orn. epiphytes esp. ***M. pustulatum*** (Forst.f.) Copel. (*P. p.*, *M. diversifolium*, *Phymatodes d.*, kangaroo fern, E Aus., NZ); ***M. scandens*** (Forst.f.) Tindale (E Aus., NZ) – climbs ivy-like; ***M. scolopendria*** (Burm.f.) Copel. (*P.s.*, OW) – used to scent clothes & flavour coconut-oil

Microspermum Lag. Compositae (Eup.-Ager.). Incl. *Iltisia*, 9–10 C Am. R: MSBMBG 22(1987)184, Phytol M 11(1997)155

Microstachys A. Juss. (~ *Sebastiania*). Euphorbiaceae (Euph.-Hipp.). 15 trop. (As. 1, Aus. 1, Afr. 3–4). ***M. chamaelea*** (L.) Hook. f. (As.) – local medic.

Microstaphyla C. Presl = *Elaphoglossum*

Microstegium Nees. Gramineae (XXII). Excl. *Leptatherum*, 16 OW trop. & warm esp. China. R: Blumea 57(2012)161. ***M. vimineum*** (Trin.) A. Camus (As.) – invasive USA etc.

Microsteira Bak. Malpighiaceae. 25 Madag. Polygamo-dioec.

Microstelma Baill. = *Gonolobus*

Microstemma R. Br. = *Brachystelma*

Microstephanus N.E. Br. = *Pleurostelma*

Microsteris Greene = *Phlox*

Microstigma Trautv. (~ *Matthiola*). Cruciferae (5). 3 Siberia, Mongolia, China

Microstrobilus Bremek. = *Strobilanthes*

Microstrobos J. Garden & L. Johnson = *Pherosphaera*

Microstylis (Nutt.) Eaton = *Malaxis*

Microtatorchis Schltr. = *Taeniophyllum*

Microtea Sw. Microteaceae (Phytolaccaceae III). 12 trop. Am.

Microteaceae Schäferhoff & Borsch (~ Phytolaccaceae). Magnoliidae – Caryophyllales. 1/12 trop. Am. R: Willd. 39(2009)223. Ann. herbs. Lvs in spirals. Infl. of 1–3- flowered units. P (4); A (2–)5–9 with spherical anthers; G (2–4), 1-loc. Fr. a spiny or muricate achene
Genus: *Microtea*
Form. often placed in Amaranthaceae (Chenopodiaceae)

Microterangis (Schltr.) Senghas = *Aerangis*

Microthelys Garay = *Funkiella*

Microthlaspi F.K. Meyer (~ *Thlaspi*). Cruciferae (Colut.). 5 Euras., N Afr. R: T 65(2016)94

Microtidium M.A. Clem. & D.L. Jones = *Microtis*

Microtis R. Br. Orchidaceae (IV 3g). 19 China & Jap. to Aus. (10 – all in W Aus., R: JABG 13(1990)49, 8 endemic) & NZ. Onion orchids (single onion-like leaf); fls. green, some fragrant, stems & tubers emitting strong scent too; if not visited then self-poll., many apomicts. ***M. parviflora*** R. Br. (Aus., NZ) & other spp. ant-poll.; ***M. unifolia*** (Forst.f.) Reichb.f. (E Aus.) – tubers form. ed. Aborigines in Sydney area

Microtoena Prain. Labiatae (VI). 24 Himal. to W China, natur. in Mal. ***M. patchouli*** (C.B. Clarke) C.Y. Wu & Hsuan (*M. cymosa*, Chinese patchouli, trop. As.) – oil used like patchouli

Microtrichia DC. = *Grangea*

Microtrichomanes (Prantl) Copel. = *Hymenophyllum*

Microtropis Wall. ex Meissn. Celastraceae (I). 66 trop. As. (Am. spp. = *Quetzalia*)

Microula Benth. Boraginaceae (B.3.8.3). Incl. *Actinocarya*, 30 Himal. to W China (most)

Mida A. Cunn. ex Endl. Santalaceae (Nanodeaceae). 1 NZ: ***M. salicifolia*** A. Cunn.; 1 Juan Fernandez: ***M. fernandeziana*** (Philippi) Sprague & Summerh. (*Santalum f.*) – extinct

Middletonia Puglisi (~ *Paraboea*). Gesneriaceae (Loxoc.). 4 Indomal. R: T 65(2016)286

midge orchid *Genoplesium* spp.

midyim *Austromyrtus dulcis*

Miersia Lindl. Amaryllidaceae (Alliaceae). 3 Chile

Miersiella Urb. Burmanniaceae. 1 trop. S Am.: ***M. umbellata*** (Miers) Urb.

Miersiophyton Engl. = *Rhigiocarya*

Migandra Cook = *Chamaedorea*

mignonette *Reseda odorata*; **m. vine** *Anredera cordifolia*; **wild m.** *R. lutea*

Mikania Willd. Compositae (Eup.-Mik.). c. 400 trop. (OW 9, R: BJ 103(1982)211). R: MSBMBG 22(1987)419. Some dioec. (Greater Antilles). Lianes (Stone's Model), ***M. dentata*** Spreng. (*M. ternata*, S Brazil) grown as hanging-basket pl.; ***M. cordata*** (Burm.f.) Robinson – used as antimalarial in Afr., headache cure in Sumatra; ***M. micrantha*** Kunth (trop. Am.) – invasive trop. As. & Aus., Hawaii, Masc.

Mikaniopsis Milne-Redh. (~ *Cissampelopsis*). Compositae (Sen.-Sen.). c. 15 trop. Afr.

Mila Britton & Rose (~ *Echinopsis*). Cactaceae (III 4). 1 C Peru: ***M. caespitosa*** Britton & Rose – cult. orn. variable dwarf sp.

Mildbraedia Pax = *Plesiatropha*

Mildbraediochloa Butzin = *Melinis*

Mildbraediodendron Harms. Leguminosae (III 1). 1 WC Afr.: ***M. excelsum*** Harms – timber

Mildella Trevis. (~ *Pellaea*). Pteridaceae (IV). 3 trop.

mile-a-minute *Persicaria perfoliata*

milfoil *Achillea millefolium*; **water m.** *Myriophyllum* spp.

Milicia Sim (~ *Chlorophora*). Moraceae (I). 2 trop. Afr. Troll's Model. ***M. excelsa*** (Welw.) C. Berg (*Chlorophora e.*, iroko, odum, mvule, Afr. teak, rock elm, counter wood, yellow-wood) – timber good for furniture etc. (but allergenic to some), termite-proof, used as oak & teak subs., now endangered through logging, plantations with bananas in Uganda, best not grown in pure stands (galls), bast a 'fallback' food for gorillas, local medic., incl. eczema

Milium Tourn. ex L. Gramineae (XVI 15). 5 N temp. OW (Eur. 2), E N Am. Spikelets 1-flowered; x = 4, 5, 7, 9. ***M. effusum*** L. (millet grass) – cult. orn. perenn.

Miliusa Leschen. ex A. DC. Annonaceae (IV 7). c. 40 Mal. to Aus. R: Blumea 48(2003)432

milk thistle *Silybum marianum, Sonchus* spp.; **m. tree** *Brosimum* & *Manilkara* spp., *Sapium glandulosum* etc.; **m. vetch** *Astragalus* spp.; **m.weed** *Asclepias* spp.; **m. wood** *Alstonia* spp.; **m. wort** *Polygala* spp.; **sea m. wort** *Lysimachia maritima*

Milla Cav. Asparagaceae (Alliaceae s.l.). 10 SW N Am. to Guatemala. R: GH 8(1953)278, Herbertia 54(1999)232. Some cult. orn.

Milleria Martyn ex L. Compositae (Mill.-Mill.). 2 Mex. to C Am. & Peru. R: Phytol. 81(1996)350

millet, African *Cenchrus spicatus*; **Am. m.** *C. americanus*; **[Ind.] barnyard m.** *Echinochloa frumentacea*; **bulrush m.** *C. spicatus*; **cat-tail m.** *Setaria italica*; **channel m.** *E. inundata, E. turneriana*; **common or French m.** *Panicum miliaceum*; **foxtail** or **German m.** *S. italica*; **m. grass** *Milium effusum*; **great m.** *Sorghum bicolor*; **Hungarian m.** *Setaria italica*; **Ind. m.** *C. spicatus*; **Italian m.** *S. italica*; **Jap. [barnyard] m.** *E. esculenta*; **Kodo m.** *Paspalum scrobiculatum*; **little m.** *Panicum sumatrense*; **milo m.** *Sorghum bicolor*; **pearl m.** *C. spicatus*; **proso** or **Russian m.** *P. miliaceum*; **Sanwa m.** *E. esculenta*; **Shama m.** *E. colona*; **spiked m.** *C. spicatus*; **m. spray** (birdseed) *Setaria italica*; **Texas m.** *Urochloa texana*

Millettia Wight & Arn. Leguminosae (III 16). Incl. *Pongamia* c. 150 OW trop. (Afr. – Madag. c. 100). Heterogeneous. Some fish & bilharzia-snail poisons, cult. orn. & comm. timbers, esp. ***M. laurentii*** De Wild. (wenge, trop. Afr.) & ***M. stuhlmannii*** Taubert (panga-panga, SE trop. Afr.). ***M. grandis*** (E. Meyer) Skeels (*M. caffra*, S Afr.) – wood for walking-sticks (umzimbeet); ***M. pinnata*** (L.) Panigr. (*P. p.*, Indian beech, karanja, kerong, saw, thinwin, Indomal.) – poon(ga) oil in skin treatment etc., lamps, poss. biofuel, fuelwood; ***M. thonningii*** (Schum. & Thonn.) Bak. (W Afr.) – chewing-stick in Nigeria. See also *Callerya*

Milligania Hook.f. Asteliaceae. 5 Tasmania. R: FA 45(1987)169

Millingtonia L.f. Bignoniaceae (2). 1 SE As. & Mal.: ***M. hortensis*** L.f. – cult. orn. with fragrant fls & fr. to 30 cm long, lvs a poor subs. for opium in cigarettes, timber suggested for tea-chests. R: FM I,8(1977)133

Millotia Cass. Compositae (Gnap.). Excl. *Scyphocoronis, Toxanthes*, 11 temp. Aus. R: AusJB 8 (1995)1

milo *Sorghum bicolor* Subglabrescens Group, *Thespesia populnea*

Miltianthus Bunge = *Zygophyllum*

Miltitzia A. DC. = *Phacelia*

Miltonia Lindl. Orchidaceae (V 12h). 12 trop. Am. esp. Braz. Cult. orn. epiphytes, fls favourite late C19 button-holes

× **Miltonidium** G. Wilson. Orchidaceae. *Miltonia* × *Oncidium* – 238 grexes

Miltoniodes Brieger & Lückel = *Oncidium*

Miltoniopsis God.-Leb. (~ *Miltonia*). Orchidaceae (V 12h). 5 trop. Am. Cult. orn. (pansy orchids), incl. intergeneric hybrids

Milula Prain = *Allium*

Mimetanthe Greene (~ *Mimulus*). Phrymaceae (Mim.; Scophulariaceae s.l.). 1 N Am.: ***M. pilosa*** (Benth.) Greene. R: Phytoneuron 2012–39:25

Mimetes Salisb. Proteaceae (IV 4). Incl. *Diastella, Orothamnus* 21 SW & S Cape. R: JSAB 42(1976)185, 50(1984)171. Complex compound infls of bird-poll. fls. ***M. zeyheri*** Meissn. (*O. z.*, marsh rose)

Mimetophytum L. Bolus = *Mitrophyllum*

Mimophytum Greenman = *Omphalodes* (but see Phytoneuron 2013–64:9)

mimosa *Acacia dealbata*. Florist's mimosa & 'mimosa bark' of tanners derived from several *Acacia* spp.; **water m.** *Neptunia prostrata*

Mimosa Tourn. ex L. Leguminosae (II 1). c. 530 trop. & warm esp. Am. (461). R: MNYBG 65(1991). Herbs, shrubs, lianes & trees often with stipular thorns. Some bad weeds of sugar-cane etc. esp. ***M. diplotricha*** C. Wright ('*M. invisa*', trop. Am.), ***M. pigra*** Humb. & Bonpl. ex Willd. (trop., invasive in trop. As. & Aus. (introd. Darwin Bot. Garden late C19, by 1999 800 sq. km of N Terr. flood-plain infested; forming monospecific stands), trop. & S Afr., SE U.S.) & ***M. pudica*** L. (sensitive plant, trop. Am. but widely natur.) – lvs bipinnate with 4 sec. petioles & marked 'sleep movements' at night where leaflets fold together & petioles droop, a condition reached v. rapidly if pl. touched or shaken, subjected to sudden temp. change, high hydrostatic pressure, chemical agents etc., poss. a mechanism reducing transpiration or deterring grazers etc. the stimulus app. passed through phloem of pulvinus, which has living wood-fibres, & implemented by contractile ('muscle') proteins (form. considered due to migration of ions – action potential poss. a chloride spike through membranes leading to drop in turgor of pulvinus cells), cult. orn. curiosity &

sand-binder; ***M. lewisii*** Barneby (N.E. Brazil) – bat-poll. (unique in genus); ***M. scabrella*** Benth. (*M. bracaatinga*, bracaatinga, Brazil) – coffee-shade, paper & plywood, locally imp. fuelwood crop; ***M. splendida*** Barneby (Braz.) – pachycaul treelet in cerrado; ***M. tenuiflora*** (Willd.) Poir. (*M. hostilis*, trop. & warm Am.) – potent drink rich in tryptamines form. prepared from roots & taken before battles, source of tepescohuite (Mex.) used in treatment of burns

Mimosaceae R. Br. = Leguminosae (II)

Mimosopsis Britton & Rose = *Mimosa*

Mimozyganthus Burkhart. Leguminosae (II 2). 1 SE Bolivia, SW Paraguay to Arg.: ***M. carinatus*** (Griseb.) Burkart. R: Darw. 3(1939)448

Mimulicalyx Tsoong. Phrymaceae (Mim.; Scrophulariaceae s.l.). 2 China

Mimulopsis Schweinf. Acanthaceae (III 2e). c. 15 trop. Afr., Madag. Some with gregarious flowering, ***M. solmsii*** Schweinf. (trop. Afr.) – hapaxanthic after 9 yrs

Mimulus L. Phrymaceae (Mim.; Scrophulariaceae s.l.). Excl. *Diplacus*, *Erythranthe*, *Thyridia*, 7 warm. R: Phytoneuron 2012–39:18

Mimusops L. Sapotaceae (I 1). 41 trop. Afr. (20), Madag. (15), Masc. (4), Seychelles (1), Indomal. (1: ***M. elengi*** L. ((anc.) vakul) – cult. for fragrant fls). ***M. laurifolia*** (Forssk.) Friis (persea, Ethiopia) – tallest tree in Arabia but endangered through overgrazing and wood-cutting, fr. ed. & cult. Egypt (now extinct there), lvs with olive lvs tied to reeds for Tutankhamun's tomb. See also *Baillonella*, *Manilkara*, *Tieghemella*

Mina Cerv. = *Ipomoea*

Minaria Konno & Rapini (~ *Ditassa*). Apocynaceae (Vc; Asclepiadaceae). 21 trop. S Am., esp. E Braz. R: T 55(2006)424

Minasia H. Robinson. Compositae (Vern.-Lych.). 5 E Braz. Some pachycaul rosette-trees

mincemeat orig. meat cut up small, later anything thus, esp. mixture of suet, apples, almonds, candied peel, currants, raisins etc. with sugar used in mince-pies at Christmas (GB)

mind-your-own-business *Soleirolia soleirolii*

Mindium Adans. = *Michauxia*

miner's lettuce *Claytonia perfoliata*

Ming aralia *Polyscias fruticosa*

Minicolumna Brieger = *Epidendrum*

Minjaevia Tzvelev = *Silene* (*Atocion*)

minjiri *Senna siamea*

Minkelersia M. Martens & Galeotti = *Phaseolus*

minneola *Citrus* × *aurantium* Tangelo Group

Minquartia Aubl. Olacaceae (Coulaceae; ?Erythropalaceae). 1 N trop. S Am.: ***M. guianensis*** Aubl. – Nozeran's Model; good wood, black dye. R: FN 38(1984)38

mint *Mentha* spp. & hybrids esp. *M. spicata*; **apple m.** *M. suaveolens*; **bergamot m.** *M.* × *piperita* 'Citrata'; **Bowles's m.** *M. spicata* 'Alopecuroides'; **cat m.** *Nepeta* spp.; **chocolate m.** *M.* × *piperita* 'Chocolate'; **cockscomb m.** *Elsholtzia ciliata*; **Corsican m.** *M. requienii*; **eau-de-cologne m.** *M.* × *piperita* 'Citrata'; **field m.** *M. arvensis*; **Gibraltar m.** *M. pulegium*; **ginger m.** *M.* × *gracilis*; **horse m.** *M. longifolia*; **Japanese m.** *M. canadensis*; **m. julep** see *Zea mays*; **lemon** or **orange m.** *M.* × *piperita* 'Citrata'; **pepperm.** *M.* × *piperita*; **black p.m.** *M.* × *p.* 'Piperita'; **white p.m.** *M.* × *p.* 'Officinalis'; **pineapple m.** *M. suaveolens* 'Variegata'; **Scotch m.** *M.* × *gracilis*; **spearm.** *M. spicata*; **water m.** *M. aquatica*; **woolly m.** *M. suaveolens*

mintbush *Prostanthera* spp.

Minthostachys (Benth.) Spach. Labiatae (VII 2b). 12 Andes. Poss. comm. mint flavourings esp. oil from ***M. mollis*** (Kunth) Griseb (tipo, N Andes) – local condiment & medic. tea

Minuartia Loefl. ex L. Caryophyllaceae (II 1). Excl. *Cherleria*, *Eremegone*, *Facchinia*, *Mcneillia*, *Minuartiella*, *Mononeuria*, *Pseudocherleria*, *Rhodalsine*, *Sabulina*, *Triplateia*, c. 54 Euras. Some cult. orn.

Minuartiella Dillenb. & Kadereit (~ *Minuartia*). Caryophyllaceae (II 1). 4 Turkey, Iran (mts). R: T 63(2014)84

Minuopsis W. Weber = *Minuartia*

Minuria DC. Compositae (Ast.-Pod.). 10 C & SE Aus. R: Nuytsia 3(1980)221, 6(1987) 63

Minuriella Tate = *Minuria*

Minurothamnus DC. = *Heterolepis*

Mionandra Griseb. Malpighiaceae. 1 Bolivia, Paraguay, Arg.: ***M. camareoides*** Griseb.

Miquelia Meissn. Icacinaceae. 8 Indomal. ***M. caudata*** King (Malay Pen., Borneo) – climber with potable water in stem

Miqueliopuntia Frič ex F. Ritter (~ *Opuntia*). Cactaceae (II). 1 Chile: ***M. miquelii*** (Monville) F. Ritter

miraa *Catha edulis*

mirabelle *Prunus cerasifera*, also *P.* × *domestica* 'subsp. *institia*'

Mirabilis L. Nyctaginaceae (6). c. 60 warm Am. (esp. SW N), Himal. (1!). R: Sida 60(1995)614. Fls 1–several in K- like involucre of 5 bracts, K C-like, persistent in fr., some spp. (*Oxybaphus*) with enlarged papery involucre in fr. Some medic. N Am.; ***M. expansa*** (Ruíz & Pavón) Standl. (mauka, Andes) – anc. crop with ed roots & stems, now endangered; some orn. cult. esp. night-flowering ***M. jalapa*** L. (four-o'clock, marvel of Peru, false jalap, Mex., widely natur.) – fls open 16:00–18:00 hrs, visible fluorescence emitted by a yellow betaxanthin absorbed by a violet betacyanin giving a contrasting fluorescent pattern attracting poll. insects, cult. as annual but tuberous root to c. 20 kg containing purgative trigenollin when perennial, fls in water give a crimson dye used for tinting seaweed cakes & jellies in China, cosmetic powder from ground seeds used in Jap.; ***M. multiflora*** (Torrey) A. Gray (S U.S.) – root coll. for food since Prehistoric times, antitumour activity in extract

miraculous berry *Synsepalum dulcificum*; **m. fruit** *Thaumatococcus daniellii*

Miraglossum Kupicha (~ *Schizoglossum*). Apocynaceae (V c; Asclepiadaceae III 1). 7 S Afr. R: KB 38(1984)625

Mirandaceltis Sharp = *Aphananthe*

Mirandea Rzed. Acanthaceae (III 2c). 6 Mexico. R: ABM 62(2003)12

Mirandopsis Szlach. & Marg. = *Pleurothallis*

Mirbelia Sm. (~ *Pultenaea*). Leguminosae (III 13). 32 Aus.

Miricacalia Kitam. (~ *Ligularia*). Compositae (Sen.-Tuss.). 1 Jap.: ***M. makinoana*** (Yatabe) Kitam.

mirin *Oryza sativa*

miro *Prumnopitys ferruginea*

misai kuching *Orthosiphon aristatus*

Misbrookea Funk (~ *Werneria*). Compositae (Sen.-Sen.). 1 Peru & Bolivia, mts: ***M. strigosissima*** (A. Gray) Funk. R: Brittonia 49(1997)111

Miscanthidium Stapf = *Miscanthus*

Miscanthus Andersson (~ *Sorghum*). Gramineae (XXII 5). c. 30 OW trop., S Afr., E As. (China 7). ? Heterogeneous; hybridize with *Saccharum* spp. Some local uses for thatch or brooms esp. ***M. ecklonii*** (Nees) Mabb. (*Miscanthidium capense*, S Afr.) & ***M. junceus*** (Stapf) Pilg. (C Afr.); mooted as biofuel esp. ***M. × giganteus*** Greef & Deuter ex Hodkinson & Renvoize (*M. sinensis* × *M. sacchariflorus* (Maxim.) Hack. (E As.)) – sterile triploid to 3.5 m, straw for 'biopots', poss. fibre-board; some cult. orn. esp. ***M. sinensis*** Andersson (eulalia, E As.) – commonly grown varieg. cvs, also yellow dye (Jap.)

Mischarytera (Radlk.) H. Turner (~ *Arytera*). Sapindaceae. 3 New Guinea, N Aus. R: Blumea suppl. 9(1995)

Mischobulbum Schltr. = *Tainia* (but see OM 6(1992)64)

Mischocarpus Blume. Sapindaceae. 15 SE As. to Aus. (9, 7 endemic). R: Blumea 23(1977)251. Some ant-infested domatia

Mischocodon Radlk. = *Mischocarpus*

Mischodon Thw. Picrodendraceae (Pic.-Misch.; Euphorbiaceae s.l.). 1 S India, Sri Lanka: ***M. zeylanicus*** Thw. – lvs verticillate. R: A. R.-Sm., *Gen. Euphorb.* (2001)108

Mischogyne Exell. Annonaceae (III 6). 1 trop. Afr.

Mischophloeus R. Scheffer = *Areca*

Mischopleura Wernham ex Ridl. = *Sericolea*

misery, mountain *Chamaebatia* spp.

Misodendraceae J. Agardh (Myzodendraceae). Magnoliidae – Santalales. 1/8 temp. S Am. Dioec. or monoec. hemiparasitic shrublets almost exclusively on *Nothofagus* spp., with thickened haustorial region & stout twigs, the apex aborting annually & laterals growing out to give pseudo-2- or 3-chotomous shoots. Lvs small, s.t. scale-like, in spirals; stipules 0. Fls small in racemes or spikes, males with P 0, A 2 or 3 around small nectary-disk, s.t. reduced to sessile anther, anthers with term. slits, females with P 3, basally connate & adnate, as are alt. accrescent staminodes, to $\underline{G}$ (3), 1-loc. with 3 stigmas & 3 pendulous ovules, not differentiated into nucellus & integuments, hanging from free-central

placenta. Fr. an achene or nut with accrescent feathery staminodes (wind-disp.); seed without testa, embryo straight in oily green endosperm. n = 8
Genus: *Misodendrum*

Misodendron auctt. = *Misodendrum*

Misodendrum Banks ex DC. Misodendraceae. 8 S Am. S of 33°, mainly on Andean *Nothofagus* spp. R: Parodiana 1(1982)245, SB 32(2007)560. 'Woodroses' sold as curios

Misopates Raf. (~ *Antirrhinum*). Plantaginaceae (Ant.-Ant.; Scrophulariaceae s.l.). 7 Medit. (Eur. 3) to Cape Verde Is., Ethiopia & NW India. R: D.A. Sutton, *Rev. Antirrh.* (1988)145. ***M. orontium*** (L.) Raf. (weasel's snout, Euras.) – natur. N Am., GB

Miss Wilmott's ghost *Eryngium giganteum*

missanda *Erythrophleum suaveolens*

Missouri currant *Ribes aureum*

mistletoe *Viscum album*, name applied to any Loranthaceae or tree-living Santalaceae; **American m.** *Phoradendron* spp. esp. *P. leucarpon*; **m. cactus** *Rhipsalis* spp. esp. *R. baccifera*; **dwarf m.** *Arceuthobium* spp.; **WI m.** *Phthirusa caribaea*

mistol *Ziziphus mistol*

Mitchell grass *Astrebla* spp. esp. *A. pectinata*

Mitchella L. Rubiaceae (IV 10). 1 N Am., 1 S Korea & Jap. Heterostyly. ***M. repens*** L. (partridge berry, N Am.) – lvs medic., esp. genito-urinary complaints

Mitella Tourn. ex L. Saxifragaceae. Excl. *Bensoniella, Conimitella*, 9 N Am., E As. R: FNA 9(2009)108. Mitrewort, bishop's cap. Some cult. orn. herbs

mitnan *Thymelaea hirsuta*

Mitolepis Balf.f. = *Cryptolepis*

Mitophyllum Greene = *Streptanthus*

Mitostemma Masters. Passifloraceae. 3 trop. S Am.

Mitostigma Decne = *Philibertia*

Mitostylis Raf. = *Cleome* (but see Novon 23(2014)53)

Mitracarpum auctt. = *Mitracarpus*

Mitracarpus Zucc. Rubiaceae (IV 15). 30 trop. Am., 1 introd. OW: ***M. hirtus*** (L.) DC.

Mitragyna Korth. Rubiaceae (I 1). Incl. *Hallea*, 7 OW trop. (Afr. 1, As. 6). R: Blumea 24(1978)56. Afr. spp. with knee-like pneumatophores; many alks. Some timbers (nazingu, Uganda) esp. ***M. parvifolia*** (Roxb.) Korth. (S As.) – subs. for *Chloroxylon swietenia*, & ***M. stipulosa*** (DC.) Kuntze (abura, W Afr.) – furniture, barrels etc., lvs medic.; some, e.g. ***M. speciosa*** (Korth.) Haviland (Mal.), with lvs smoked like, & alleged to be more dangerous than, opium

Mitranthes O. Berg. Myrtaceae (II 10). 1 Cuba, 6 Jamaica. All rare

Mitrantia Peter G. Wilson & Hyland. Myrtaceae (II 5). 1 N Queensland: ***M. bilocularis*** Peter G. Wilson & Hyland. R: Telopea 3(1988)264

Mitraria Cav. Gesneriaceae (II 4b). 1 Chile, Arg.: ***M. coccinea*** Cav. – cult. orn. liane allied to *Fieldia* (Aus.)

Mitrasacme Labill. Loganiaceae. 54 Indomal., E As., Aus. (48, 43 endemic), New Caled., NZ. Reminiscent of *Hedyotis* (Rubiaceae)

Mitrasacmopsis Jovet. Rubiaceae (IV 1). 1 E Afr. & Madagascar: ***M. quadrivalvis*** Jovet

Mitrastemma Makino = seq.

Mitrastemon Makino. Mitrastemonaceae (Rafflesiaceae s.l.). 1 SE As. to Mal. & Jap.; 1 Mex. to NW S Am. R: Blumea 38 (1993)221

Mitrastemonaceae Makino (~ Rafflesiaceae). Magnoliidae – Ericales. 1/2 SE As. to Mal., Mex. to NW S Am. Root-parasites on Fagaceae. Lvs scale-like, opp. Fls solit., term. P 4 connate; A extrorse, connate & surrounding G exc. apical pore; G with 8–20 parietal placentas with ∞ unitegmic ovules (integument 2 cells thick). Fr. berry-like circumscissile capsule with ∞ minute seeds. Seed with undifferentiated embryo. n = 20
Genus: *Mitrastemon*

Mitrastylus Alm & T.C.E. Fries = *Erica*

Mitratheca K. Schum. = *Oldenlandia*

Mitrella Miq. (~ *Fissistigma*). Annonaceae. 6–7 Mal. (Aus. 1)

Mitreola L. Loganiaceae. 6 trop. R: MLW 74–23(1974)

Mitrephora (Blume) Hook.f. & Thomson. Annonaceae (IV 7). 47 SE As. to Aus. R: SBM 90(2010)33

mitrewort *Mitella* spp.; **false m.** *Tiarella* spp.

Mitriostigma Hochst. Rubiaceae (II Sherb.). 5 trop. & S Afr. R: NJB 27(2009)305

Mitrocereus (Backeb.) Backeb. = *Pachycereus*

Mitrophyllum Schwantes. Aizoaceae (V). 6 Cape. R: BJ 97(1976)339. Cult. orn. succ. pls
Mitropsidium Burret = *Psidium*
mitsuba or **mitzuba** *Cryptotaenia canadensis*
mitsumata *Edgeworthia papyrifera*
mituan *Thymelaea hirsuta*
Mitwabachloa Phipps = *Zonotriche*
Miyakea Miyabe & Tatew. = *Anemone* (*Pulsatilla*)
Miyamayomena Kitam. = *Aster*
Mizonia A. Chev. = *Pancratium*
mizuna *Brassica rapa* Japonica Group
mkani fat *Allanblackia stuhlmannii*
Mkilua Verdc. Annonaceae (III 1). 1 trop. E Afr.: ***M. fragrans*** Verdc. – allied to *Toussaintia* (pollen) or *Cymbopetalum*, fls used in making scent for Swahili women & Arabs. R: KB 24(1970) 449
mkweo *Beilschmiedia kweo*
Mlanje cedar *Widdringtonia whytei*
Mnesithea Kunth. Gramineae (XXII 3). 3 Ind., SE As.
Mniochloa Chase. Gramineae (VI 3). 1 Cuba: ***M. pulchella*** (Griseb.) Chase. R: AMBG 80(1993)854
Mniodes (A. Gray) Benth. Compositae (Gnap.-Lor.). 4 Peru. R: Fieldiana Bot. n.s. 26(1991)52. Dioec. cushion-formers
Mniopsis Mart. = *Podostemum*
Mniothamnea (Oliv.) Niedenzu = *Brunia*
moa *Madhuca longifolia*
moabi *Baillonella toxisperma*
Moacroton Croizat = *Croton*
Mobilabium Rupp. Orchidaceae (V 16c). 1 Queensland: ***M. hamatum*** Rupp
mobola (plum) *Parinari curatellifolia*
moccasin flower *Cypripedium* spp.
Mocinnodaphne Lorea-Hernández = ? Lauraceae (Rogers), but see ABM 32(1995)26
mock orange *Philadelphus* spp. esp. *P. coronarius*
mocker nut *Carya tomentosa*
Mocquerysia Hua. Salicaceae (Flacourtiaceae). 2 trop. Afr. Epiphyllous infls
Modesciadium P. Vargas & Jim. Mejías (~ *Stoibrax*). Umbelliferae. 1 Morocco: ***M. involucratum*** (Maire) P. Vargas & Jim. Mejías. R: Phytotaxa 212(2015)75
Modestia Charadze & Tamamschjan = *Jurinea*
Modiola Moench. Malvaceae (Malv.-Malv.). 1 Am. (widely natur.): ***M. caroliniana*** (L.) G. Don – gargle
Modiolastrum K. Schum. Malvaceae (Malv.-Malv.). 7 S Am. R: Brittonia 32(1980)484
Moehringia L. (~ *Arenaria*). Caryophyllaceae (II 1). c. 25 N temp. (Eur. 25, N Am. 3). Elaiosomes
Moenchia Ehrh. Caryophyllaceae (II 1). 3 W & C Eur. (3), Medit.
Moerenhoutia Blume = *Platylepis*
Moghania J. St-Hil. = *Flemingia*
moghat root *Glossostemon bruguieri*
Mogoltavia Korovin. Umbelliferae (III 8). 2 C As.
Mohavea A. Gray. Plantaginaceae (Ant.; Scrophulariaceae s.l.). 2 SW U.S. R: D.A. Sutton, *Rev. Antirrh.* (1988)482. Fls (little nectar) mimic *Mentzelia* spp. (copious), deceiving pollinators. Cult. orn. annuals
mohikana *Melicope anisata*
mohle, mohli (flowers) *Jasminum sambac*
Mohria Sw. = *Anemia*
mohur *Delonix regia*
mokru *Ischnosiphon arouma*
molasses uncrystallizable sugars (syrup) from raw sugar; **m. grass** *Melinis minutiflora*
molvae (Philippines) *Vitex* spp.
Moldenhawera Schrad. Leguminosae (I 4). 10 E Brazil, Venezuela. R: KB 54(2000)830
Moldenkea Traub = *Hippeastrum*
Moldenkeanthus Morat = *Paepalanthus*
mole plant *Euphorbia lathyrus*
moli *Jasminum sambac*

Molinadendron Endress (~ *Distylium*). Hamamelidaceae (I 4). 3 Mex., C Am. K present (cf. *D.*)

Molinaea Comm. ex Juss. Sapindaceae. 8 Madag. (5), Masc.

Molineria Colla (~ *Curculigo*). Hypoxidaceae. 7 Indomal. Cult. orn. esp. ***M. capitulata*** (Lour.) Herb. (*C. c.*, weevil lily, Indomal. to Aus.) – housepl. in Aus., fr. ed.

Molineriella Rouy (~ *Periballia*). Gramineae (XVI 7). 1 Medit.: ***M. minuta*** (L.) Rouy

Molinia Schrank. Gramineae (Arund.). 3 temp. Euras. (Eur. 1: ***M. caerulea*** (L.) Moench (purple moor grass)) – cult. orn. esp. '**Variegata**', catches insects between paleae in a trap like *Dionaea* (lodicules shrink slowly but if punctured by insect snap shut, though app. not absorbing nutrients). Oft. diff. to distinguish from allopatric *Eragrostis*

Moliniopsis Hayata = *Molinia*

Mollera O. Hoffm. = *Calostephane*

Mollia Mart. Malvaceae (Grew.-Grew.; Tiliaceae). c. 15 trop. S Am. R: MNYBG 29(1978)68

Mollinedia Ruíz & Pavón. Monimiaceae (V 2). c. 20 trop. Am. (Peru 15). G initially unsealed

Molloybas M.A. Clem. & D.L. Jones = *Corybas*

Molluginaceae Bartl. (~ Caryophyllaceae). Magnoliidae – Caryophyllales. Excl. Kewaceae, Limeaceae, 12/120 trop. & warm esp. S Afr. R: T 65(2016)783. Herbs (often weedy; rarely shrubs), usu. glabrous, app. always with anthocyanins & not betalains (? reversal). Lvs simple, scarcely succ., opp., in spirals or whorled; stipules small & decid. or 0 (larger in *Pharnaceum*). Fls usu. bisexual, reg., often small, solit. or in cymes; K (4 – *Polpoda*) 5, rarely basally connate (*Coelanthum*), C (? staminodes) 5 small (*Glinus*) or 0, A 3–5(15 or more in *Glinus*), usu. basally connate, anthers with longit. slits, s.t. nectary a ring around, or on lower part of, G̲ (2–5) or 1 (*Adenogramma*), usu. with distinct styles & at least basally pluriloc., placentation usu. axile with (1–)∞ campylotropous (almost anatropous) bitegmic ovules. Fr. usu. loculicidal capsule, often with persistent K (nutlet in *Adenogramma*); seeds with embryo curved around perisperm (endosperm 0). $x = 9$

Genera: *Adenogramma, Coelanthum, Glinus, Hypertelis, Mollugo, Paramollugo, Pharnaceum, Polpoda, Psammotricha, Suessenguthiella, Trigastrotheca*

Form. incl. in Aizoaceae but scarcely succ.; G̲, & free sepals & anthocyanins present – a classic example of chemotaxonomic studies leading to recognition of family status

Mollugo L. Molluginaceae. Excl. *Paramollugo, Trigastrotheca*, c. 15 trop. & warm (Eur. 2), esp. Am. Some potherbs esp. ***M. pentaphylla*** L. (As.) – medic.

molokhia *Corchorus olitorius*

Molongum Pichon. Apocynaceae (I d). 3 trop. S Am. R: AUWP 87–1(1987)

Molopanthera Turcz. Rubiaceae (I 1). 1 E Brazil: ***M. paniculata*** Turcz.

Molopospermum Koch. Umbelliferae (III 5). 1 Pyrenees, Cevennes, Alps: ***M. peloponnesiacum*** (L.) Koch. R: Bauhinia 10(1992)75

Moltkia Lehm. Boraginaceae (B.2.2). 6 N Italy to N Greece, 3 SW As. Like *Lithospermum* but C-throat not crested & nutlets bent. Cult. orn. rock-pls esp. ***M. × intermedia*** (Froebel) J. Ingram (*M. petraea* (Tratt.) Griseb. (C Greece to Dalmatia) × *M. suffruticosa* (L.) Brand (N Italy))

Moltkiopsis I.M. Johnston. Boraginaceae (B.2.2). 1 N Afr. to Iran: ***M. ciliata*** (Forssk.) I.M. Johnston. R: JAA 34(1953)2

moluccana *Falcataria moluccana*

Moluccella L. Labiatae (VI). Incl. *Sulaimania*, 8 Medit. to NW India. R: T 60(2011)483. ***M. laevis*** L. (bells of Ireland, shell flower, E Medit.) – cult. orn. with large green bell-like K, grown as an everlasting

moly *Allium moly*, though in original sense poss. *Galanthus nivalis* or merely mythological

mombin, red *Spondias purpurea*; **yellow m.** *S. mombin*

Mommsenia Urb. & Ekman. Melastomataceae. 1 Hispaniola: ***M. apleura*** Urb. & Ekman. R: ArkB 20A, 5(1926)31

Momordica Tourn. ex L. Cucurbitaceae (VIII). 45 OW trop. (As. 10 – R: BZ 87,3(2002)133). R: BZ 91(2006)772. Usu. monoec. (7 independent shifts from dioec.) scramblers with elaiophores attractive to bees & bitter fr., ed. when cooked esp. ***M. balsamina*** L. (balsam apple, Afr. to W As., Aus., natur. Am.) – fr. foul-smelling but 'red jelly' & seeds ed. (cult. Mal.), lvs & stems camel fodder; ***M. charantia*** L. (balsam pear, bitter cucumber, leprosy gourd, OW trop., natur. SE U.S.) – immature fr. (karel(l)a) a favourite of Pakistanis in UK, extract has effect like insulin; ***M. cochinchinensis*** (Lour.) Spreng. (NE India & S China to

N Aus.) – widely cult. trop., roots produce a lather for clothes-washing; ***M. rostrata*** A. Zimm. (Kenya) – caudiciform sp. cult. orn.

Mona O. Nilsson = *Montia*

Monachather Steud. (~ *Danthonia*). Gramineae (Arund.). 1 Aus.: ***M. paradoxa*** Steud. – if dominant, indicator of good range condition. R: FA 44B(2005)21

Monachosoraceae Ching = Dennstaedtiaceae

Monachosorum Kunze. Dennstaedtiaceae (Monachosoraceae). c. 3 Himal. to S China & Mal.

Monachyron Parl. = *Melinis*

Monactis Kunth. Compositae (Helia.-Ecl.). c. 9 Ecuador, Peru. R: Phytologia 34(1976)34

Monadenia Lindl. = *Disa*

Monadenium Pax = *Euphorbia*

Monandriella Engl. = *Ledermanniella*

Monanthes Haw. Crassulaceae (I 4). 9 Canary & Salvage Is., (?) Morocco. R: Bradleya 10(1992)59. Cult. orn. dwarf succ.

Monanthochloe Engelm. = *Distichlis* (but see Kurtziana 5(1969)369)

Monanthochilus (Schltr.) Rice = *Sarcochilus*

Monanthocitrus Tanaka. Rutaceae (II 2). 4 Borneo, New Guinea. R: PANSP 140(1988)272

Monanthos (Schltr.) Brieger = *Dendrobium*

Monanthotaxis Baill. Annonaceae (III 7). 56 trop. Afr. & Madag. Some lianes

monarch of the East *Sauromatum venosum*

Monarda L. Labiatae (VII 2b). 16 N Am. R: UCPB 20(1942)147. A 2. Cult. orn., teas & herbs etc. esp. ***M. didyma*** L. (Oswego tea, bee balm, bergamot, E N Am.) – eaten with meat, oil used as pomade (***M. citriodora*** Cerv. ex Lag. (C U.S. to N Mex.) with greatest oil yield by wt.), ***M. fistulosa*** L. (N Am.) – analgesic, diuretic etc., culinary herb, scent for horses & men, insecticide, & their hybrids; ***M. menthifolia*** Graham (*M. fistulosa* var. *m.*, N Am.) – poss. comm. source of geraniol; others medic.

Monardella Benth. Labiatae (VII 2b). c. 30 W N Am. R: AMBG 12(1925)1. Some cult. orn. & medic. teas esp. *M. odoratissima* Benth.

Monarrhenus Cass. (~ *Pluchea*). Compositae (Inul.-Pluch.). 2 Mauritius. Shrubs

Monarthrocarpus Merr. (~ *Desmodium*). Leguminosae (III 19). 2 Ind., E Mal

Mondia Skeels. Apocynaceae (III Periplocaceae). 2 trop. Afr. R: FTEA Apoc. 2(2012)143

mondo grass *Ophiopogon planiscapus*; **black m. g.** *O. p.* 'Nigrescens'

Monechma Hochst. = *Justicia*

Monelytrum Hackel. Gramineae (XXIX 1). 1 SW Afr.: ***M. luederitzianum*** Hackel. R: Strelitzia 10(2000)705

Monenteles Labill. = *Pterocaulon*

Monerma P. Beauv. = *Lepturus*; see also *Hainardia*

Moneses Salisb. ex Gray. Ericaceae (II 1). 1 cool N temp.: ***M. uniflora*** (L.) A. Gray – medic. N Am., esp. skin problems. R: FNA 9(2009)384

money plant *Crassula ovata, Epipremnum pinnatum* 'Aureum'; **m. tree** *C. ovata*; **m.wort** *Lysimachia nummularia*; **Cornish m.w.** *Sibthorpia europaea*

mongongo *Ricinodendron rautanenii*

Moniera Loefl. = *Monniera*

Monilaria (Schwantes) Schwantes. Aizoaceae (V). 5 Cape. R: MSABH 14(1973)49. Some cult. orn. dwarf succ.

Monilicarpa Cornejo & Iltis (~ *Capparis*). Capparaceae. 2 S Am. R: JBRIT 2(2008)67

Monimia Thouars. Monimiaceae (VI 2). 3 Mauritius & Réunion. R: AMBG 72(1985)68

Monimiaceae Juss. Magnoliidae – Laurales. Excl. Atherospermataceae, Siparunaceae 28/210 trop. & warm, esp. S. R: NJB 7(1987)25, 8(1988)25. Evergreen trees, shrubs or lianes often accumulating aluminium & with alks.; twigs often flattened at nodes. Lvs usu. opp., simple, often with gland-dots; stipules 0. Fls small, usu. unisexual, reg. to oblique, usu. perigynous with concave hypanthium, (solit. or) in cymose infls., s.t. cauliflorous; K 2 + 2, fleshy, decussate, C 7–20 (–more) or P not diff. into K & C, reduced or 0, A 8 in 1 or 2 series with short filaments & basal (? staminodes) nectaries, anthers with longit. slits or valves, staminodes around G (1–)few–∞ (*Tambourissa* to 2000!) with short styles & term. stigmas, carpels s.t. sunk in receptacle, with 1 anatropous bitegmic or (Siparunoideae) unitegmic apical & pendulous (Hortonioideae, Monimioideae) or basal & erect (Atherospermatoideae, Siparunoideae) ovule. Fr. a head of drupes or nuts, often enclosed in hypanthium; seeds endotestal with copious oily endosperm, cotyledons 2(4). x = 18–22, 39, 43

Classication & principal genera:

I. **Hortonioideae** (fls bisexual, inner P petaloid; fr. a drupe): *Hortonia* (only)

II. **Atherospermatoideae** (fls bisexual or unisexual, P sepaloid or petaloid; fr. a plumose nutlet; 2 tribes = Atherospermataceae):

1. **Atherospermateae** (leaf-hairs centrifixed, inner staminodes elongating in fr.): *Atherosperma, Laureliopsis* (only)
2. **Laurelieae** (leaf-hairs basifixed, inner staminodes not enlarging): *Daphnandra, Laurelia*]

III. **Siparunoideae** (fls unisexual, P sepaloid; anthers opening by valves; = Siparunaceae): *Siparuna* (only)

IV. **Glossocalycoideae** (fls unisexual, P obscure, fr. a drupe; = Siparunaceae): *Glossocalyx* (only)

V. **Mollinedioideae** (fls unisexual, P sepaloid, filaments with appendages, anthers opening by slits, fr. a drupe; 3 tribes):

1. **Hedycaryeae** (anthers with longit. slits; receptacle splitting): *Hedycarya, Tambourissa*
2. **Mollinedieae** (anthers with longit. slits; receptacle opening by abscission of upper part): *Kibara, Mollinedia, Steganthera*
3. **Hennecartieae** (anthers with equatorial slit): *Hennecartia* (only)

VI. **Monimioideae** (pls dioec.; P sepaloid (inner petaloid in *Peumus*); filaments without appendages (except *Palmeria*), anthers opening by slits, fr. a drupe; 3 tribes):

1. **Palmerieae** (lianes): *Palmeria* (only)
2. **Monimieae** (trees & shrubs, P sepaloid): *Monimia* (only)
3. **Peumeae** (tree, P petaloid): *Peumus* (only)

12 genera monospecific. Some with almost closed hypanthium (cf. *Ficus*), pollen germinating on a mucilaginous plug closing the opening

Some timbers (*Hedycarya, Peumus, Xymalos*), scents & medic.

Monimiastrum Guého & A.J. Scott = *Eugenia* (but see KB 34(1980)483)

Monimopetalum Rehder (~ *Euonymus*). Celastraceae (I). 1 China: ***M. chinense*** Rehder – endangered. R: JAA 7(1926)233

Monium Stapf = *Anadelphia*

Monizia Lowe = *Daucus* (but see Webbia 69(2014)19)

monkey apple *Annona* spp., *Anisophyllea laurina, Strychnos* spp.; **m. bread** *Adansonia digitata*; **m. cocoa** *Theobroma angustifolium*; **m. comb** *Pithecoctenium* spp.; **m. flower, m. musk** *Erythranthe* spp.; **m. nut** *Arachis hypogaea*; **m. plant** *Ruellia makoyana*; **m. pot** *Lecythis* spp., *Cariniana* spp.; **m. puzzle (tree)** *Araucaria araucana*

monk's rhubarb *Rumex pseudoalpinus*

monkshood *Aconitum* spp.

Monniera Loefl. = *Ertela*

Monnina Ruíz & Pavón. Polygalaceae (IV). Incl. *Pteromonnina* (poss. discrete), c. 150 New Mex. to Chile, esp. Andes. One of G 2 usu. rudimentary; fr. a drupe, rarely samaroid. ***M. salicifolia*** Ruíz & Pavón (Peru) – fr. source of a blue dye

Monocardia Pennell = *Bacopa*

Monocarpia Miq. Annonaceae (IV 6). 4 SE As., W Mal. (Borneo 3 – R: BJ 122(2000)234). Petit's Model

Monocelastrus Wang & Tang = *Celastrus*

Monochaetum (DC.) Naudin. Melastomataceae. 45 upland trop. Am. A dimorphic

Monochasma Maxim. ex Franch. & Savat. Orobanchaceae (Cymb.; Scrophulariaceae s.l.). 4 E As.

Monochilus Fischer & C. Meyer (~ *Clerodendrum*). Labiatae (III). 2 Brazil. R: Novon 9(1999)324

Monochoria C. Presl. Pontederiaceae. 4–8 NE Afr. to Jap. & Aus. (4). Annuals. ***M. hastata*** (L.) Solms-Laub. (Indomal.) & ***M. vaginalis*** (Burm.f.) C. Presl (Indomal. to Jap.) – all save roots relished as veg., locally medic., weed of rice fields, e.g. Calif.; ***M. korsakowii*** Regel & Maack (C & E As.) – natur. N Black Sea by 1932, N Italian rice fields by 1985

Monocladus Chia, Fung & Y.L. Yang = *Bonia*

Monococcus F. Muell. Petiveriaceae (Phytolaccaceae II). 1 E Aus., New Caled., Vanuatu: ***M. echinophorus*** F. Muell.

Monocosmia Fenzl = *Calandrinia*

Monocostus K. Schum. Costaceae (Zingiberaceae). 1 E Peru: ***M. uniflorus*** (Petersen) Maas. R: FN 8(1972)17, T 55(2006)156

monocotyledons (**monocots;** Liliopsida). Magnoliidae. 2723/70 300 here arr. in 79 fams. Angiosperms usu. without reg. sec. thickening, with 1 'cotyledon' (better a prophyll as no trace of 'second' one), parallel-veined lvs & 3-merous (exc. *Pentastemona*) fls. No parasites. Range of seed-size from dust-like in orchids to the biggest seeds known, in *Lodoicea* (Palmae). See Appendix & R. Dahlgren *et al.* (1985) *The families of the Monocotyledons*

Monoculus R. Nordenstam (~ *Osteospermum*). Compositae (Cal.). 2 S Afr. R: CN 44(2006)39

Monocyclanthus Keay. Annonaceae (III 6). 1 trop. W Afr.: ***M. vignei*** Keay. R: KB 8(1953)69

Monocymbium Stapf. Gramineae (XXII 7). 3 trop. & S Afr. R: RBA 30(1950)175

Monodia S. Jacobs. Gramineae (XXIX 4). 1 W Aus.: ***M. stipoides*** S. Jacobs. R: KB 40(1985)659

Monodiella Maire = *Centaurium*

Monodora Dunal. Annonaceae (III 6). c. 14 trop. Afr. R: SBM 87(2009)88. Troll's Model; fleshy 'syncarps', though in ***M. crispata*** Engl. & Diels, at least, a single term. carpel & not the syncarp it resembles. ***M. myristica*** (Gaertn.) Dunal (calabash or West African nutmeg) – seeds used like nutmeg, as rosary beads & locally medic.

Monogereion G. Barroso & R. King. Compositae (Eup.-Ayap.). 1 NE Brazil: ***M. carajensis*** G. Barroso & R. King. R: MSBMBG 22(1987)210

Monogramma Comm. ex Schkuhr. Pteridaceae (V; Vittariaceae). Incl. *Haplopteris*, c. 40 Madag. to Mal. Epiphytes, incl. minutest of all ferns, mature fronds 3 mm–2 cm long

Monogramme *auctt.* = praec.

Monolena Triana. Melastomataceae. c. 15 trop. S Am. Forest-floor herbs, seeds distrib. heavy rain. Cult. orn. esp. ***M. primuliflora*** Hook.f. (Costa Rica to Ecuador & Braz.)

Monolepis Schrad. = *Blitum*

Monolluma Plowes (~ *Edithcolea*). Apocynaceae. 5 Afr., Arabia

Monolophus Wall. ex Delafosse & al. Older name for *Caulokaempferia*

Monolopia DC. Compositae (Mad.-Bae.). Incl. *Lembertia*, 5 Calif. R: FNA 21(2006)349. ***M. major*** DC. – cult. orn. annual

Monomeria Lindl. (~ *Bulbophyllum*). Orchidaceae (V 15). Excl. *Acrochaene*, 2 SE As.

Mononeuria Reichb. (~ *Minuartia*). Caryophyllaceae. 9 E N Am., Greenland. R: T 63(2014)84

Monoon Miq. (~ *Polyalthia*). Annonaceae. Incl. *Cleistopetalum*, *Enicosanthum*, *Woodelliantha*, c. 60 Indomal. to Aus. R: T 61(2012)1029. ***M. hypogaeum*** (King) B. Xue & R. Sanders (*P. h.*, Malay Pen.) – fls on runners at ground-level; ***M. longifolium*** (Sonn.) B. Xue & R. Saunders (*P. l.*, S Ind., Sri Lanka) – riparian, fastigiate form **'Pendula'** ('Temple Pillar', Ind. willow) – fastigiate, common garden & roadside tree in Ind. & Sri Lanka

Monopera Barringer (~ *Angelonia*). Plantaginaceae (Grat.-Ang.; Scrophulariaceae s.l.). 2 S Am. R: Brittonia 35(1983)111

Monopetalanthus Harms = *Aphanocalyx*

Monopholis S.F. Blake = *Monactis*

Monophrynium K. Schum. = *Phrynium*

Monophyllaea R. Br. Gesneriaceae (III 1b). 30+ S Thailand & Mal. R: NRBGE 37(1978)1. Pls resembling single lvs rooted at foot of petioles & bearing infls at junction of 'lamina' & petiole, the 'lamina' to 60 cm long; poss. derived (cf. *Whytockia*) from a system with caulescent growth & anisophylly as in some Acanthaceae; dendroid hairs

Monophyllanthe K. Schum. (~ *Maranta*). Marantaceae. 1–2 N S Am.

Monophyllorchis Schltr. Orchidaceae (V 4b). 1 Colombia & Ecuador: ***M. microstyloides*** (Reichb.f.) Garay

Monoplegma Piper = *Oxyrhynchus*

Monoporus A. DC. Primulaceae (Myrsinaceae). 9 Madagascar

Monopsis Salisb. Campanulaceae (III 1). 15 trop. & S Afr. R: Pfl. IV,276b(1953)698,783, 276c(1968)889. Some cult. orn., like *Lobelia* but stigma-lobes filiform & recurved, esp. ***M. debilis*** (L.f.) C. Presl ('*M. simplex*', S Afr.) natur. W Aus.

Monopteryx Spruce ex Benth. Leguminosae (III 2). 3–4 N S Am. R: Brittonia 36(1984)48. Large trees with big buttresses of mythological imp. in NW Amaz.; timbers. ***M. angustifolia*** Spruce ex Benth. (Amazonia) – seeds coll. & stored for lean times

Monoptilon Torrey & A. Gray. Compositae (Ast.-Chaet.). 2 SW N Am. deserts. R: FNA 20(2006)210

Monopyle Moritz ex Benth. Gesneriaceae (II 5b). 27 Guatemala to Bolivia

Monopyrena Speg. = *Verbena*

Monorchis Ség. Older name for *Herminium*

monos plum *Pseudanamomis umbellulifera*

Monosalpinx Hallé. Rubiaceae (?II 1). 1 trop. W Afr.: ***M. guillaumetii*** Hallé – Roux's Model. R: Adansonia 8(1968)369
Monoschisma Brenan = *Pseudopiptadenia*
Monosepalum Schltr. = *Bulbophyllum*
Monosis DC. (~ *Vernonia*). Compositae (Vern.). 7 As.
Monospatha W.T. Lin = *Yushania*
Monostachya Merr. = *Rytidosperma*
Monostylis Tul. = *Apinagia*
Monotaceae Kosterm. = Dipterocarpaceae
Monotagma K. Schum. Marantaceae. 37 trop. S Am. R: Phytotaxa 20(2011)1. ***M. spicatum*** (Aubl.) Macbr. – wrapping material
Monotaxis Brongn. Euphorbiaceae (Acal.- Amp.). 8 Aus. R: Austrobaileya 6(2002)274
Monotes A. DC. Dipterocarpaceae (Monot.). c. 30 trop. Afr., Madag. (1) & Tertiary Eur. ***M. elegans*** Gilg (C S Afr.) – timber
Monotheca A. DC. = *Sideroxylon*
Monothecium Hochst. Acanthaceae (III 2c). 4 trop. Afr. to S India
Monothrix Torrey = *Perityle*
Monotoca R. Br. Ericaceae (VII 7; Epacridaceae). 18 Aus.
Monotrema Koern. Rapateaceae (II 2). 5 Colombia, Venez., Brazil
Monotropa L. Ericaceae (II 3). Excl. *Hypopitys*, 1 N temp.: ***M. uniflora*** L. (Indian pipe, Himal., Jap., N & C Am.) – locally medic.
Monotropaceae Nutt. See Ericaceae
Monotropanthum Andres = seq.
Monotropastrum Andres (~ *Cheilotheca*). Ericaceae (II 3). 2 E As., Sumatra. R: WJB33(1975)1
Monotropsis Schwein. Ericaceae (II 3). 2 SE U.S. R: SB 39(2014)586
Monroa orig. spelling for *Munroa*
Monrosia Grondona = *Polygala*
Monsanima Liede & Meve (~ *Cynanchum*). Apocynaceae (Ascl.). 2 E Braz. R: AMBG 99(2013)66
Monsonia L. Geraniaceae. Incl. *Sarcocaulon* (R: Bothalia 12(1979)581; fleshy shrublets with spines representing toughened petioles & with lvs in their axils; A 15) 40 SW As., Afr., Madag. R: MLW 79–9(1979)
monster, Loch Ness See *Pinus sylvestris*
Monstera Adans. Araceae (V 4). c. 60 trop. Am. R: CGH 207(1977)3. Large epiphytic lianes with entire, perforated or pinnatifid lvs & often long corky aerial roots; holes in lvs caused by local slowing of growth of lamina, drying & splitting as rest expands; some with dopamine; some cult. as housepls when young, esp. ***M. deliciosa*** Liebm. (ceriman, (Swiss) cheeseplant, Mex. to Panamá) – introd. 1840 but rare in wild, with ed. spadix (to 15 °C above ambient when receptive) with banana & pineapple flavour though eaten unripe disagreeable (calcium oxalate crystals); ***M. maderaverde*** Grayum & Karney (Honduras) – pendent rs to 10 m used in hat-making etc.
Montagueia Bak. f. = *Polyscias*
Montamans Dwyer = *Notopleura*
montana *Clematis montana*
Montanoa Cerv. Compositae (Helia.-Mont.). 25 trop. Am. R: MNYBG 36(1982). Tree daisy. Pachycaul treelets & trees; some cult. orn., some invasive S Afr. ***M. quadrangularis*** Schultz-Bip. (*M. moritziana*, arboloco, C Am.) – timber used locally (telegraph-poles, rafters), lvs to 90 cm; others medic. esp. ***M. tomentosa*** Cerv. (Mex.) for female ailments, also ed. Aztecs
montbretia *Crocosmia* × *crocosmiiflora*
Montbretiopsis L. Bolus = *Tritonia*
Monteiroa Krapov. Malvaceae (Malv.-Malv.). 10 mts SE & S Brazil, ***M. glomerata*** (Hook. & Arn.) Krapov. ext. to Uruguay & Arg. R: Bonplandia 12(2003)50
monterry *Kunzea pomifera*
Monterey pine *Pinus radiata*
Montezuma Moçiño & Sessé ex DC. = *Thespesia*
Montia Micheli ex L. Montiaceae (Portulacaceae s.l.). 12 temp. (Eur. 1, Aus. 2), wet places. ***M. fontana*** L. ((water) blinks, N temp., trop. mts, SE Aus.) – annual of wet places, fls cleistogamous in dull weather or when inundated, lvs eaten as salad esp. in France

Montiaceae Raf. (~ Portulacaceae). Magnoliidae – Caryophyllales. Incl. Hectorellaceae, 15/235 temp. esp. Am. R: T 59(2010)235. Glabrous herbs oft. with swollen roots & lvs in basal rosettes. Lvs simple. Fls solit., axillary, or in cymes, s.t. up to 9 K-like 'bracteoles'. P 4, 5(–19), s.t. basally united; A equal & opp. P or 1 less, alt. with P (*Hectorella, Lyallia*), or -100, s.t. basally united; G (2–8), with 1 ovule per loc. Fr. circumscissile, or 1-seeded indehisc.; seeds oft. minutely papillate, oft. with strophiole or elaisome (aril). n = 5–13 etc.
Principal genera: *Cistanthe, Claytonia, Lewisia, Montia, Montiopsis*

Montiastrum (A. Gray) Rydb. = *Montia*

Monticalia C. Jeffrey (~ *Pentacalia*). Compositae (Sen.-Sen.). c. 70 C & S Am. (Andes)

Montigena Heenan = *Swainsona* (but see NZJB 36(1998)42)

Montinia Thunb. Montiniaceae (Grossulariaceae s.l.). 1 S Afr.: ***M. caryophyllacea*** Thunb. R: Strelitzia 10(2000)414

Montiniaceae Nakai (~ Grossulariaceae). Magnoliidae – Solanales. 3/5 S Afr., Madag. Trees & shrubs with a peppery scent. Lvs simple, bracteoles 0. Fls small, unisexual (males 3- or 4-merous, females 4-). C free, s.t. valvate. A extrorse. $\overline{G}$ with 1–12 ovules per carpel & persistent short style. Fr. a capsule & seeds winged (*Montinia*), indehiscent or a drupe. n = 16, 34
Genera: *Grevea, Kaliphora, Montinia*

Montiopsis Kuntze (~ *Calandrinia*). Montiaceae (Portulacaceae s.l.). Incl. *Calandriniopsis*, 40 W S Am. esp. Chile. R: Phytol. 74(1993)273

Montrichardia Crüger. Araceae (VII 15). 2 Mex., trop. S Am. R: Aroideana 28(2005)86. ***M. arborescens*** (L.) Schott – pachycaul to 7 m, v. fragrant beetle-poll. infls with large starchless pollen (in water v. thick intine swells & splits exine explosively releasing an intine- (rather than pollen-) tube cementing pollen to hairless pollinators, water-disp. in Amaz., locally medic., spadix ed.

Montrouziera Pancher ex Planch. & Triana. Guttiferae (3.). 5 New Caled. Leeuwenberg's & Rauh's Models. Principal exploited timber of New Caled.

Monttea C. Gay. Plantaginaceae (Grat.- Melo.; Scrophulariaceae s.l.). 3 Chile, Arg. R: Parodiana 3(1985)380. Inside anterior lip oils provide poll. reward for centridine bees (cf. oil-collecting bees in *Diascia*)

Monvillea Britton & Rose = *Acanthocereus*; see also *Cereus*

mooley plum *Owenia acidula*

mooli *Raphanus sativus* 'Longipinnatus'

moon beam *Tabernaemontana divaricata*; **m. daisy** *Leucanthemum vulgare*; **m. flower** *Ipomoea alba*; **m. seed** *Menispermum canadense*; **m. trefoil** *Medicago arborea*; **m.wort** *Botrychium lunaria*

moonah *Melaleuca lanceolata*

Moonia Arn. Compositae (Cor.). 1 Sri Lanka, S India: ***M. heterophylla*** Arn. R: Brittonia 27(1975)97. Allied to *Dahlia*

moorgrass, blue *Sesleria caerulea*; **purple m.** *Molinia caerulea*

Mooria Montr. = *Cloezia*

Moorochloa Veldk. (*Brachiaria*). Gramineae (XXIV 5). 3 warm. R: Reinw. 12(2004)138. See also *Urochloa*

mooseberry *Viburnum lantanoides*; **moosewood** *Acer pensylvanicum, Dirca palustris, V. lantanoides*

mopane *Colophospermum mopane*

Mopania Lundell = *Manilkara*

Moquinia DC. Compositae (Cich.-Moq.). 1 E Brazil: ***M. racemosa*** (Spreng.) DC. R: T 43(1994)41. Dioec. shrub or tree

Moquiniastrum (Cabrero) G. Sancho (~ *Gochnatia*). Compositae (Gochn.; Mut.-Mut.). 21 trop. & warm Am. esp. Braz. Usu. gynodioec. R: Phytotaxa 147(2013)29

Moquiniella Balle. Loranthaceae (5 6). 1 S Afr.: ***M. rubra*** (A. Spreng.) Balle – embryo-sac longest in angiosperms (48 mm). R: Strelitzia 10(2000)342

Mora Schomb. ex Benth. Leguminosae (I 4). 6 trop. Am. V. tall emergent trees (to 50 m) with large seeds, 1 per fr., those of ***M. oleifera*** (Hemsl.) Ducke (to 1 kg, cf. *Lodoicea*) with the largest embryo of any pl., drifting in currents & ***M. megistosperma*** (Pittier) Britton & Rose (18 × 12 cm) the largest dicot. seeds; ***M. paraensis*** (Ducke) Ducke (Brazil) – air-filled cavity between cotyledons promoting water-disp. in Amaz. Some timbers for shipbuilding etc., some red dyes

Moraceae Link. Magnoliidae – Rosales. 40/1200 trop. & warm, few temp. Monoec. or dioec. trees, shrubs, lianes (incl. stranglers) & herbs, usu. with laticifers with milky latex

(0 in *Fatoua*) & s.t. alks. Lvs in spirals or opp., usu. simple, often with cystoliths & cell-walls with silica or calcium carbonate; stipules present, s.t. minute. Fls small, wind-poll. (insect-poll. in *Ficus* etc.), in axillary infls with the axis often thickening to form a head or invaginated receptacle (almost closed in syconia of *Ficus*); P (0–4) or 5(–10) ± basally connate, s.t. in 2 whorls, A (1–4(–6) as many as & opp. P, $\underline{G}$ or $\overline{G}$ (2(3)) with 1 often rudimentary so 1(2)-loc. with usu. 2 style(arm)s & 1 ± apical, anatropous to hemitropous or campylotropous bitegmic ovule. Fr. a drupe (s.t. with dehiscent exocarp), P &/or receptacle often becoming fleshy; seeds with straight or curved embryo with often unequal cotyledons in fleshy & oily or 0 endosperm. n = 12+. R: GBS 19(1962)187

Classification & principal genera:

I. **Moreae** (infl. unisexual or, at least, not discoid, female 1-flowered or racemose, male fls with P 3–5, A 3–5 inflexed in bud; ? heterogeneous): *Broussonetia, Fatoua, Maclura, Streblus, Trophis*

II. **Artocarpeae** (like I., female infls with several to many fls, A straight in bud; prob. divisible into 3 tribes); *Artocarpus, Sorocea*

III. **Castilleae** (like I., but female infl. a stout spike or head, male similar with involucre, male fls usu. without pistillodes): *Antiaris, Castilla, Perebea, Naucleopsis, Pseudolmedia*

IV. **Dorstenieae** (incl. Brosimeae; infls mostly bisexual, not enclosed): *Brosimum, Dorstenia*

V. **Ficeae** (fls in invaginated receptacles, sterile female fls with gall-wasps; app. sister to thrips-poll. III.): *Ficus*

Cannabaceae & Urticaceae kept distinct

Imp. fr. trees (*Artocarpus, Brosimum, Ficus, Morus, Treculia*), rubber (*Castilla, Ficus*), timber (*Artocarpus, Brosimum, Ficus, Maclura, Milicia, Streblus*), dyes (*Maclura*), fibre incl. paper (*Broussonetia, Ficus, Poulsenia*), arrow poisons – cardenolides (*Antiaris, Maquira, Naucleopsis*); cult. orn. (*Broussonetia, Ficus, Maclura*)

Moraea Mill. Iridaceae (VII 2). Incl. *Galaxia, Gynandriris, Helixyra, Hexaglottis, Homeria, Rheome, Roggeveldia, Sessilistigma*, 226 Medit. (Eur. 2), trop. (25) & S Afr. (Cape 115, 79 endemic; R: Strelitzia 9(2000)137). R: Bothalia 43(2013)36,153. Some with bulbils in axils of lower lvs, some (S Afr.) poll. monkey beetles; seeds dry (some thin & flat wind-disp.) – Goldblatt; many toxic to stock (cardiac glycosides similar to those in *Drimia*), some noxious weeds (esp. Aus., e.g. ***M. flaccida*** (Sweet) Steud. (*H. f.*, Cape; homeridin with digitalis-like effects on heart); some ed. (esp. form. ***M. fugax*** (Delaroche) Jacq. (S Afr.)) & leaf-fibre for cordage (Goldblatt), some cult. orn. (peacock lilies) with fugacious fls (first Cape 'bulbs' cult. successfully UK), incl. ***M. ramossisima*** (L.f.) Druce (*M. ramosa*, S & W Cape) – roots metamorphosed into thorns; ***M. sisyrinchium*** (L.) Ker (*Gynandriris s.*, Barbary nut, Medit. to Pakistan, poss. spread by agric.) – fls open p.m.

morama bean *Tylosema esculentum*

Morangaya G. Rowley (~ *Echinocereus*). Cactaceae. 1 Mex.: ***M. pensilis*** (M. Brandegee) G. Rowley

Moranopteris Hirai & Prado (~ *Micropolypodium*). Polypodiaceae (V). 29 trop. Am. R: T 60(2011)1127

Moratia H. Moore = *Cyphokentia*

Morelia A. Rich. ex DC. Rubiaceae (II 1). 1 trop. Afr.: ***M. senegalensis*** A. Rich. ex DC.

Morella Lour. (~ *Myrica*). Myricaceae. c. 46 cosmop. Shrubs usu. with nitrogen-fixing actinomycetes (diff. *Frankia* spp. from those on *Myrica*) in roots; some wax sources, ed. fr. & cult. orn. incl. ***M. californica*** (Cham. & Schldl.) Wilbur (*Myrica c.*, bay berry, coastal W U.S.) & ***M. cerifera*** (L.) Small (*Myrica c.*, wax myrtle, candleberry, tallow shrub, E U.S.) – medic. ('Thompsonian powder') esp. analgesic, wax on fr. surface rich in palmitic acid removed in boiling water made into candles by early settlers & used in scented bayberry soap; ***M. esculenta*** (D. Don) I. Turner (*Myrica e.*, Indomal.) – fr. ed., bark source of drug kaphal bokra & k. pothu for diarrhoea, asthma etc., also fish-poison & yellow dye; ***M. faya*** (Aiton) Wilbur (*Myrica f.*, Macaronesia, natur. S Portugal) – part of Madeiran laurel forest, introd. Hawaii for afforestation or for wine-making by Portuguese labourers in 1880s, an invasive pest since 1944 though fixing 10–20 kg nitrogen/ha/yr (no N-fixing native pls in Hawaii); ***M. javanica*** (Blume) I. Turner (*Myrica j.*, Java) – fr. ed., wood excellent as fuel & charcoal, planted for land reclamation; ***M. rubra*** Lour. (*Myrica r.*, yang-mei, yumberry, E As.) – cult. China for ed. fr. fresh, candied or tinned, comm. anti-oxidant juice in US

Morelotia Gaudich. Cyperaceae (II 7). 2 Hawaii, NZ

Morenia Ruíz & Pavón = *Chamaedorea*

Moreton Bay chestnut *Castanospermum australe;* **M. B. fig** *Ficus macrophylla;* **M. B. pine** *Araucaria cunninghamii*

Morettia DC. Cruciferae (4). 4 N Afr. to Arabia. R: BSB II,53(1980)253

Morgania R. Br. (~ *Stemodia*). Plantaginaceae (Grat.-Stem.; Scrophulariaceae s.l.). 4 Aus. R: JABG 13(1990)90

Moricandia DC. Cruciferae (12). 7 Medit. (Eur. 3) to Pakistan

Moriera Boiss. = *Aethionema*

Morierina Vieill. Rubiaceae (I 7). 2 New Caledonia

Morina Tourn. ex L. Caprifoliaceae (Morinaceae). Incl. *Cryptothladia,* 10 Balkans to Sino-Himal. R: BBMNHB 12(1984)12, 15. ***M. persica*** L. (Balkans) – seeds eaten like rice

Morinaceae Raf. = Caprifoliaceae

Morinda Vaill. ex L. Rubiaceae (IV 10). Excl. *Rennellia,* 39 trop. (New Caledonia 14 – R: OB 122(1993)20), Am. (26). R: Adans. 33(2011)297. Petit's Model; fr. a head of ± coherent drupes in succ. enlarged calyces; some medic. & dye-sources esp. ***M. citrifolia*** L. (al, bai-yor, Indian mulberry, orig. ? Micronesia) – fls & fr. borne at alt. nodes so always outside canopy, domatia present, bark yields a permanent red dye (Turkey red; *Symplocos* sp. used as mordant), roots a yellow one, for batik & ikat, source of canarywood, lvs (sold Aus.) used in N Thai curry, fr. allegedly toxic, source of foetid insecticidal hair-oil, medic. (xeronine) Hawaii, anti-oxidant used in West to support immune response & aid digestion; ***M. lucida*** Benth. (trop. Afr.) – chewing-stick in Nigeria, diabetes treatment in Zaire; ***M. officinalis*** How (ba ji tan, S China) – TCM incl. sexual problems, anti-oxidant, poss. anti-cancer, comm. medic. (high blood pressure) Vietnam; ***M. titanophylla*** Petit (Zaire, Uganda) – pachycaul

Morindopsis Hook.f. Rubiaceae (II 5). 2 SE As.

Moringa Rheede ex Adans. Moringaceae. 13 semi-arid Afr. to As. R: KB 40(1985)1. Trunks often swollen. Seeds imp. water purifiers. ***M. hildebrandtii*** Engl. (SW Madag., extinct in wild) – long-cult. W coast around graves etc. (cf. *Ginkgo biloba*); ***M. oleifera*** Lam. (horse-radish tree, sajna, NW India) – alks, fls. bird-poll., spinning seeds with 3 equidistant wings, cult. 3000 yrs, oilseed (Ben nut, oil of Ben, 40% seed weight, used in artist's paints, salad oil, in delicate machinery, soap & comm. skin prods, form. to embalm mummies) now 6–7 M t. per annum with some cvs fruiting in 6–8 months, lvs a veg. (7 times more vitamin C than citrus, 4 times more calcium & twice the protein of milk, 3 times more potassium than banana), pods ed. (drumsticks), roots (& bark medic.) smell of horse-radish, timber suited for pulp for cellophane & rayon, fuelwood crop, seed-cake used to purify water (removing 90.0–99.99% bacteria); ***M. ovalifolia*** Dinter & A. Berger (S Afr.) – trunk water-storing; ***M. peregrina*** (Forssk.) Fiori (Middle East) – oil used like *M. oleifera* in anc. Egypt esp. for scent

Moringaceae Martinov. Magnoliidae – Brassicales. 1/13 semi-arid Afr. to As. Decid. trees to subshrubs with tuberous rootstocks, with mustard-oil glucosides & 1-cellular hairs. Lvs in spirals, 1–3-imparipinnate with opp. entire leaflets; stipules 0, conspic. glands at base of petioles & pinnae. Fls bisexual, ± irreg., in axillary panicles or thyrses, with cupular hypanthium lined with a nectary-disk apically free; K 5, spreading or reflexed, unequal, imbricate, C 5 imbricate with the outermost usu. largest & inner 2 smallest, A 5 opp. C & alt. with staminodes, filaments around disk margin, anthers with longit, slits, $\underline{G}$ ((2)3(4)), 1-loc. on short gynophore with term. hollow style, parietal placentas, each with 2 rows of pendulous, anatropous, bitegmic ovules with zig-zag micropyles. Fr. woody loculicidal capsule, elongate pod without replum, explosively dehiscent; seeds 3-winged (less often wingless) with straight oily embryo ± without endosperm & 2(3) cotyledons. n = 11, 14

Genus: *Moringa*

Morisia Gay. Cruciferae (12). 1 Corsica & Sardinia: ***M. monanthos*** (Viv.) Asch. (*M. hypogaea*) – cult. orn. sessile herb which buries fr.

Morisonia Plum. ex L. Capparaceae. 5 Carib. & trop. S Am.

Morithamnus R. King, H. Robinson & G. Barroso (~ *Eupatorium*). Compositae (Eup.-Gyp.). 2 E Brazil. R: MSBMBG 22(1987)115

Moritizia DC. ex Meissn. Boraginaceae (B.2.1.2). 3 NW & SW S Am. R: Iheringia 17(1973)31

Morkillia Rose & Painter. Zygophyllaceae (V). 2 Mex.

Mormodes Lindl. Orchidaceae (V 12c). c. 80 trop. Am. R: Selbyana 2(1978)149. Column twisted to 1 side of claw of lip; pollinia violently ejected if insect (male euglossine bee) touches joint of anther to column. Cult. orn. epiphytes

Mormolyca Fenzl = *Maxillaria*

Mormon tea *Ephedra trifurca*
morning glory *Ipomoea purpurea*
Moroccan daisy *Rhodanthemum hosmariense*
Morolobium Kosterm. = *Archidendron*
Morongia Britton = *Mimosa*
Moronobea Aubl. Guttiferae (3). 7 trop. S Am. Latex a source of adhesive & flammable resins (hog gum), karamanni wax to affix arrow-heads
Morrenia Lindl. = *Araujia* (but see KB 58(2003)714)
Morrisiella Aellen = *Atriplex*
morrison *Verticordia* spp.
Morronea Zuloaga & Scataglini (~ *Panicum*). Gramineae (Panic.). 6 Mex. to Arg. R: SB 38(2013)1081
Morsacanthus Rizz. Acanthaceae (III). 1 Brazil: ***M. nemoralis*** Rizz.
mortel *Erythrina mitis*
mortiña *Vaccinium floribundum*
Mortonia A. Gray. Celastraceae (I). 5 S N Am.
Mortoniella Woodson. Apocynaceae (I g). 1 C Am.: ***M. pittieri*** Woodson. R: AMBG 26(1939)257
Mortoniodendron Standl. & Steyerm. Malvaceae (Til.). c. 18 Mex. to N Colombia. Arils
Mortoniopteris Pichi-Serm. = *Crepidomanes*
Morus Tourn. ex L. Moraceae (I). 13 temp. & warm, trop. Afr. Decid. monoec. to dioec. trees – mulberries. Male fls in catkins, females in pseudo-spikes, wind-poll.; fr. a bird-disp. juicy syncarp of fleshy P with the same *ecological* effect as *Rubus* (q.v.). Form. used for barkcloth in Mex. (superseded by *Trema micrantha*); cult. orn. & fr. trees esp. ***M. alba*** L. (*M. australis*, white m., C & E China, natur. Eur. & N Am., invasive S Afr.) – used for coughs, colds & sore eyes in modern Chinese herbalism, the food-pl. (as are other As. spp. incl. *M. indica* L., *M. laevigata* Wall. ex Brandis & *M. serrata* Roxb.) of silkworm *Bombyx mori* in China (1 coccoon yields 600+ m of thread; James I's 1610 plan for silk sufficiency in GB mistaken in that *M. nigra* planted (incl. 4 acres at what is now Buckingham Palace, London) instead of *M. alba*!), coppiced to produce readily harvestable lvs, bark used for paper money in Ming Period & allegedly by Mongols, '**Macrophylla**' (*M. bombycis*, *M. kagayamae*) – street tree in Medit.; ***M. celtidifolia*** Kunth (trop. Am.) – trad. Mex. paper, folded without breaking; ***M. nigra*** L. (black or common m.,? derived from *M. alba*) – Camerarius in 1694 noted seedless fr. set when male trees far from females, cult. for millennia (juice used for red & purple dyes by anc. Egyptians), introd. GB C16 for juicy fr. (causing illness if eaten unripe, also source of rouge), C12 wine & timber used for furniture, snuff-boxes, inlay, hockey-stick heads, etc., can be propagated from cuttings 2.5 m long; ***M. rubra*** L. (red m., E U.S.) – cult. orn., timber, medic. (urinary problems)
mosambi *Citrus limetta*
Mosannona Chatrou (~ *Malmea*). Annonaceae (IV 2). 14 trop. Am.
Moscharia Ruíz & Pavón. Compositae (Mut.-Nass.). 2 Chile. R: CGH 205(1974)163. Annuals, cult. orn.
moschatel *Adoxa moschatellina*
Moschopsis Philippi = *Boopis*
Mosdenia Stent. Gramineae (XXIX 3). 1 Transvaal: ***M. leptostachys*** (Ficalho & Hiern) W. Clayton. R: Strelitzia 10(2000)766
Mosheovia Eig = *Scrophularia*
Mosiera Small (~ *Myrtus*). Myrtaceae (II 10). 24 C Am. R: RJBN 6,3('1985')5. 4-merous fls.
Mosla (Benth.) Buch.-Ham. ex Maxim. Labiatae (VII 1). 10 E As., Mal. ***M. chinensis*** Maxim. (China) – drug pl.
mosquito bush *Ocimum* spp.; **m. wood** *Mosquitoxylum jamaicense*
Mosquitoxylum Krug & Urb. (~ *Rhus*). Anacardiaceae (I). 1 Jamaica: ***M. jamaicense*** Krug & Urb. (mosquito wood) – timber for building
moss any sp. of Musci; **m. campion** *Silene acaulis*; **clubm.** Lycopodiaceae; **Java m.** *Taxiphyllum barbieri*; **m. pink** *Polemonium* spp.
Mossia N.E. Br. Aizoaceae (V). 1 Transvaal, Lesotho: ***M. intervallaris*** (L. Bolus) N.E. Br. R: H. E. K. Hartmann, *Ill. Handb. Succ. Pls*, Aiz. F–Z(2002)178
Mostacillastrum O. Schulz (~ *Sisymbrium*). Cruciferae (46). 30 Peru, Bolivia, Arg.R: HPB 17(2012)4

Mostuea Didr. Gelsemiaceae (Loganiaceae s.l.). c. 12 trop. Afr. (7), S Am. (1). R: MLW 61–4(1961). ***M. batesii*** Bak. (*M. stimulans*, W Afr.) – stimulant keeping revellers awake through all-night dances

Motandra A. DC. Apocynaceae (II c). 3 trop. W Afr. R: MLW 83–7(1983)3

moth mullein *Verbascum blattaria*; **m. orchid** *Phalaenopsis* spp. & hybrids; **m. vine** *Araujia sericifera*

mother-in-law's tongue *Dracaena trifasciata*

mother-of-millions *Kalanchoe delagoensis*

mother-of-thousands *Saxifraga stolonifera*, *Cymbalaria muralis*, *Soleirolia soleirolii*

Motherwellia F. Muell. Araliaceae. 1 NE Aus.: ***M. haplosciadea*** F. Muell.

motherwort *Leonurus cardiaca*

Motleyia J.T. Johansson. Rubiaceae (Prismat.). 1 NW Borneo: ***M. borneensis*** J.T. Johansson. R: Blumea 32(1987)149

motusay *Philodendron radiatum*

Moullava Adans. Leguminosae (I 4). 4 S & SE As., Afr. (1). R: PhytoKeys 71(2016)65. ***M. spicata*** (Dalzell) Nicolson (W Ind.) – climbs with recurved prickles on stem & leaf-rhachis, seed-oil form. used in lamps

Moultonia Balf.f. & W.W. Sm. = *Monophyllaea*

Moultonianthus Merr. Euphorbiaceae (Acal.-Eris.). 1 Sumatra, Borneo: ***M. leembruggianus*** (Boerl. & Koord.) Steenis. R: Blumea 40(1995)384

Mount Etna broom *Genista aetnensis*

mountain ash *Sorbus aucuparia*; **Australian m. a.** *Eucalyptus regnans*; **m. avens** *Dryas octopetala*; **m. devil** *Lambertia formosa*; **m. ebony** *Bauhinia divaricata*; **m. misery** *Chamaebatia* spp.; **m. pawpaw** *Vasconcellea pubescens*; **m. plantain** *Musa troglodytarum*; **m. pride** *Spathelia sorbifolia*; **m. rice** *Piptatherum* spp.; **m. rimu** *Lepidothamnus laxifolius*

Mourera Aubl. Podostemaceae (III). 8 N S Am.

Mouretia Pitard. Rubiaceae (IV 4). 5 SE As. R: NJB 17(1997)123

Mouriri Aubl. Melastomataceae. (Memecylaceae). 81 trop. Am. R: FN 15(1976)33. Mangenot's Model; elaiophores attractive to *Centris* bees with 'buzz'-poll.; some fish-disp. in Amaz. Some ed. fr. sold in markets

mourning widow *Geranium phaeum*

mouse ear *Myosotis* spp. (Am.), *Pilosella officinarum*, (**chickweed**) *Cerastium* spp., *Holosteum* spp.; **common m. e.** *C. fontanum* subsp. *vulgare*; **m. plant** *Arisarum proboscideum*; **m. tail** *Myosurus minimus*

Moussonia Regel (~ *Kohleria*). Gesneriaceae (II 5b). 12 C Am. R: Selbyana 1(1975)22

Moutabea Aubl. Polygalaceae (II). c. 11 trop. Am.

moutai *Oryza sativa*

moutan *Paeonia* × *suffruticosa*

moxa *Artemisia chinensis*

moya grass *Cenchrus hohenackeri*

Moya Griseb. = *Maytenus*

Mozaffariania Pimenov & Maassoumi. Umbelliferae (inc. sed.). 1 Iran: ***M. insignis*** Pimenov & Maassoumi. R: BZ 87,11(2002)96

Mozartia Urb. = *Myrcia*

mpesi *Trema orientalis*

Mrs Robb's bonnet *Euphorbia amygdaloides* subsp. *robbiae*

Msuata O. Hoffm. Compositae (Vern.-Centrop.). 1 trop. Afr.: ***M. buettneri*** O. Hoffm. R: FFG 37(1992)292

mtambara *Cephalosphaera usambarensis*

Mtonia Beentje. Compositae (Ast.-Gran.). 1 Tanzania: ***M. glandulifera*** Beentje. R: KB 54(1999)97

mu xiang *Saussurea costus*

Muantijamvella J. Phipps = *Tristachya*

Muantum Pichon = *Beaumontia*

mube *Stauntonia hexaphylla*

Mucinaea Pinter & al. = *Drimia*

Mucizonia (DC.) A. Berger = *Sedum*

Mucoa Zarucchi (~ *Ambelania*). Apocynaceae (I d). 2 trop. S Am. R: AUWP 87–1(1987)40

Mucronea Benth. (~ *Chorizanthe*). Polygonaceae (Erio.-Erio.). 2 Calif. R: Phytol. 66(1989)202

Mucuna Adans. Leguminosae (III 18). 105 trop. (Pacific is. 11, R: KB 45(1990)1). Lianes often with legumes covered in irritant hairs; bat-poll. spp. in both O & NW. Alks; L-dopa in seeds used in treatment of Parkinson's disease; some fodders (buffalo bean) & cult. orn. ***M. bennettii*** F. Muell. (New Guinea) – fls scarlet, pergola pl.; ***M. pruriens*** (L.) DC. (cow itch, trop. As., widely natur.) – irritant hairs form. used as vermifuge, var. ***utilis*** (Wight) Burck (*M. deeringiana*, Bengal bean, (Florida) velvet b.) with glabrous pods & used as fodder & cover-crop

mudar fibre *Calotropis* spp.

Mudgee wattle *Acacia spectabilis*

mudi *Dillenia crenatifolia, D. salomoniensis*

mud-mat *Glossostigma* spp.

muduga oil *Butea monosperma*

mudwort *Limosella aquatica*

Muehlbergella Feer (~ *Edraianthus*). Campanulaceae (I 7). 1 Caucasus: ***M. oweriniana*** (Rupr.) Feer

Muehlenbeckia Meissn. Polygonaceae (II 5). Incl. *Homalocladium*, excl. *Duma*, 20 New Guinea, Aus. (14), NZ, C & S Am. (9 – R: BJ 114(1992)375). Dioec. climbers or creeping pls with wiry stems, ***M. florulenta*** Meissn. (*M. cunninghamii*, lignum) forming dense thickets in Aus. Some cult. orn. esp. ***M. axillaris*** (Hook.f.) Endl. (Aus., NZ) – creeping shrublet; ***M. adpressa*** (Labill.) Meissn. (Aus.) – climber with ed. 'berries' (fleshy calyx) used in pies etc.; ***M. australis*** (Forst. f.) Meissn. & ***M. complexa*** (A. Cunn.) Meissn. (both NZ) – form. eaten by moas; ***M. platyclada*** (F. Muell.) Meissn. (*Homalocladium p.*,? Solomon Is.) – myrmecophilous, cult. orn. shrub (Troll's Model) with jointed stems usu. leafless in fl.

Muellera L.f. (~ *Lonchocarpus*). Leguminosae (III 16). Incl. *Bergeronia, Margaritolobium*, 26 trop. Am. R: T 61(2012)102. Fish-poisons

Muelleranthus Hutch. Leguminosae (III 14). Excl. *Paragoodia*, 4 Aus. R: Muelleria 29(2011)117

Muellerargia Cogn. Cucurbitaceae (XIV). 2 Madag. (1), E Mal. to Torres Strait (1)

Muellerina Tieghem (~ *Phrygillanthus*). Loranthaceae (5 1). 4 E Aus.

Muellerolimon Lincz. (~ *Goniolimon*). Plumbaginaceae (II). 1 W Aus.: ***M. salicorniaceum*** (F. Muell.) Lincz. R: BZ 67(1982)676

mugga *Eucalyptus sideroxylon*

mugongo *Ricinodendron rautanenii*

muguet *Convallaria majalis*

mugwort *Artemisia* spp.

muhimbi *Cynometra alexanderi*

Muhlenbergia Schreb. Gramineae (XXIX 1). Incl. *Epicampes*, 175 trop. & warm Am. (esp S N; N Am. 69 + 1 natur.), S As. (8). Some fodders ('hair grass'); some with roots used for exported brooms, esp. ***M. macroura*** (Kunth) A. Hitchc. (*E. m.*, zakaton, Mex.); ***M. paniculata*** (Nutt.) Columbus (*Schedonnardus p.*, tumble grass, W & C N Am.) – a tumbleweed; ***M. sericea*** (Michaux) Peterson (*M. filipes*, SE U.S.) – on sand-dunes, cult. orn., used for sweetgrass baskets for tourists; ***M. rigens*** (Benth.) A. Hichc. (deer grass, W U.S.) – grains ed., fibre for basket-making, hats etc., formerly Native Americans using fire to manage its growth

muhugwe or **muhuhu** *Brachylaena huillensis*

Muilla S. Watson ex Benth. Asparagaceae (Alliaceae s.l.). 3 SW N Am. R: Aliso 10(1984)623. Cult. orn. with fibre-covered corms

muira puama *Ptychopetalum olacoides*

Muiria N.E. Br. = *Gibbaeum*

Muiriantha C. Gardner. Rutaceae (I 3). 1 SW Aus.: ***M. hassellii*** (F. Muell.) C. Gardner. R: FA 26(2013)447

Mukdenia Koidz. (~ *Bergenia*). Saxifragaceae (I 4). 1 N China, Manchuria, Korea: ***M. rossii*** (Oliv.) Koidz. – cult. orn., crossed with *B.* spp. = × **Mukgenia** Gress. R: Plantsman n.s. 10(2011)244

Mukia Arn. = *Cucumis*

mukumari *Cordia africana*

mukunawanna *Alternanthera sessilis*

mukwa *Pterocarpus angolensis*

mula or **muli** *Raphanus sativus* 'Longipinnatus'

mulberry *Morus* spp. esp. *M. nigra* (**common** or **black m.**); **m. bush** (of nursery rhymes) *Rubus* spp.; **m. fig** *Ficus sycomorus*; **Indian m.** *Morinda citrifolia*; **paper m.** *Broussonetia papyrifera*; **red m.** *Morus rubra*; **white m.** *M. alba*

mulga *Acacia aneura*; **m. wire** = bush telegraph, 'grapevine'

Mulgedium Cass. = *Lactuca*

Mulguraea N. O'Leary & Peralta (~ *Junellia*). Verbenaceae. 11 arid S Am. R: SB 34(2009)782

Mulinum Pers. (~ *Azorella*). Umbelliferae (Azor.). 20 S Andes

mulla mulla *Ptilotus* spp.

mullein *Verbascum* spp.; **common m.** *V. thapsus*; **dark** or **black m.** *V. nigrum*; **moth m.** *V. blattaria*; **m. pink** *Silene coronaria*; **white m.** *V. lychnitis*

Mullerochloa Wong (~ *Bambusa*). Gramineae (V 4). 1 N Queensland: ***M. moreheadiana*** (Bailey) Wong – climbing bamboo to 30m. R: Blumea 50(2005)434

Multidentia Gilli. Rubiaceae (III 2). 10 trop. (E 8) Afr. R: KB 42(1987)645

Muluorchis J.J. Wood = *Tropidia*

mume *Prunus mume*

mums *Chrysanthemum* × *morifolium*

Munbya Pomel = *Psoralea*

mundu *Garcinia dulcis*

Mundulea (DC.) Benth. Leguminosae (III 16). 12 Madag. (1 ext. to OW trop.: ***M. sericea*** (Willd.) A. Chev. – fish-poison (rotenone))

mung (beans) *Vigna radiata*

muninga *Pterocarpus angolensis*

Muniria Streiber & Conn (~ *Pityrodia*). Labiatae (IV 1). 4 N Aus. R: AusSB 24(2011)7

munjeet *Rubia cordifolia*

Munnozia Ruíz & Pavón. Compositae (Liab.). 46 Andes. R: SCB 54(1983)54

Munroa Torrey (*Monroa*). Gramineae (XXIX 1). 5 Am. Disjunct – (1 W U.S., 4 SW S Am.). R: BANCC 52(1978)229

Munrochloa M. Kumar & Remesh (~ *Oxytenanthera*). Gramineae (2c). 1 S Ind.: ***M. ritchiei*** (Munro) M. Kumar & Remesh. R: JBRIT 2(2008)374

Munroidendron Sherff. Araliaceae. 1 Hawaii: ***M. racemosum*** (C. Forbes) Sherff

Munronia Wight. Meliaceae (II). 5 Indomal. R: Blumea 53(2008)614. Shrublets, some medic. esp. ***M. pinnata*** (Wall.) Theob. ***M. breviflora*** (Ridl.) Mabb. & Muellner (Malay Pen.) – fr. never found

Muntafara Pichon = *Tabernaemontana*

Muntingia Plum. ex L. Muntingiaceae (Tiliaceae s.l.). 1 trop. Am.: ***M. calabura*** L. (calabura, jam tree (Sri Lanka), Jamaican cherry, Panamá berry (Hawaii)) – Troll's Model, flowers continuously with each fl. lasting 1 day (C falling p.m.), ovules do not develop until poll., pedicels elongating to present fls for bee-poll., later lowered below lvs, the ed. fr. disp. by birds & bats, so natur. in trop. As.; A 10–100, the pistil reduced at high nos & possibility of fr. formation proportionately reduced; firewood crop

Muntingiaceae C. Bayer & al. (~ Tiliaceae). Magnoliidae – 3/3 trop. Am. R: T 47(1998)38. Trees or shrubs with stellate & glandular hairs. Lvs with asymmetric bases, in spirals, distichous on plagiotropic branches; stipule-like prophylls heteromorphic. Fls in supra-axillary fascicles or solit. K (4)5(-7), valvate; C (4)5(-7) shortly clawed & crumpled in bud, imbricate; A ∞, anthers with longit. slits; $\underline{G}$ to $\overline{G}$ (5(-multi))-loc. with ∞ anatropous ovules. Fr. baccate with ∞ seeds

Genera: *Dicraspidia*, *Muntingia*, *Neotessmannia*

muntry *Kunzea pomifera*

Munzothamnus Raven (~ *Stephanomeria*). Compositae (Cich.-Cich.(-Micr.)). 1 San Clemente (Channel Is. off Calif.): ***M. blairii*** (Munz & I.M. Johnst.) Raven. R: FNA 19(2006)349

Muraltia DC. Polygalaceae (IV). 119 S Afr. (106 Cape, 100 endemic), 1 (***M. flanaganii*** Bolus) ext. to trop. Afr. R: JSAB supp. 2(1954), Strel. 9(2000)563

Murbeckiella Rothm. Cruciferae (39). 5 SW Eur. (4) & Algeria, 1 Caucasus

Murchisonia Brittan = *Thysanotus* (but see FA 45(1987)340)

Murdannia Royle. Commelinaceae (II 2). c. 50 trop. & warm. ***M. edulis*** (Stokes) Faden (Indomal.) – roots used in trad. Indian med.

Muretia Boiss. = *Elaeosticta*

Murianthe (Baill.) Aubrév. = *Manilkara*

Muricaria Desv. Cruciferae (12). 1 Morocco, Libya: ***M. prostrata*** (Desf.) Desv.

Muricococcum Chun & How = *Cephalomappa*
Muriea Hartog = *Manilkara*
muringa *Cordia africana*
murnong *Microseris lanceolata*
murraim berries *Dioscorea communis*
Murray red gum *Eucalyptus camaldulensis*

Murraya Koenig. Rutaceae (II 2). Excl. *Bergera*, c. 5 Indomal. to Pacific. R: W.T. Swingle, *Bot. Citrus* (1943)192. Unarmed trees & shrubs. Alks. ***M. paniculata*** (L.) Jack (*M. exotica*, Chinese box, kamini, ? S China) – fls. several times a year, leaf coumarin inhibits rat thyroid activity, trad. Chinese med. anti-implantation agent (yuehchukene) an indole derivative, wood for cutlery handles & walking-sticks, wood & roots powdered to give sweet-scented thanaka powder used by Thai & Burmese women, yellow dye from shoots, cult. orn. (incl. bonsai) pantrop. & warm, asymptomatic host for citrus greening (huanglongbing, while allied ***M. lucida*** (G. Forst.) Mabb. (Mal. to W Pacif.) app. not)

Murtonia Craib = *Desmodium*
murumuru *Astrocaryum murumuru*

Musa L. Musaceae. c. 74 trop. As. Bananas and plantains. R: N.W. Simmonds (1966) *The bananas*, ed. 2, GBS 55(2003)107. Tomlinson's Model; like *Ensete* but with distinct petiole between blade & sheath & not hapaxanthic, while A usu. 5 (not 6 exc. ***M. nanensis*** Swangpol & Traiperm, Thailand). Spp. with pendent infls bat-poll., the fls. functional for only 1 night, those with erect infls self-poll. or poll. sunbirds; wild forms with arillate seeds surrounded by simple trichomes in mucilage, ed. bananas with degenerate ovules in a yellow mass derived from pericarp; some with seeds germ. only after 6 months (cf. *Ravenala*), unusual in rainforest pls. Perhaps first grown in China as fibre-pls for cloth (still in S China for gunny sacks); seen by Alexander the Great (327 BC), cult. Medit. c. AD 650, Kenya by 1300, W Afr. C15 (before Europeans; Canary Is. 1402), Carib. 1516 (introd. by Portuguese), widely exported since use of refrigerated ships (1901). Prod. 1 leaf c. every 10 days for 7–9 months, then flowers. Fr. a berry, flavour due to mixture of 2 aliphatic esters (amyl acetate, amyl propionate) with eugenol (an aromatic phenol), the immature fr. c. 7% tannin, though, like cheese, can promote migraines in some (tyramine affecting blood vessels to brain) but reducing blood pressure & poss. stroke, cont. compounds converted into serotonin so reducing depression; pulp used in shampoos (synth. amyl acetate also used as banana fragrance) etc. in Eur., fermented to 'beer', vinegar & distilled (e.g. waragai, Uganda). Clonal parthenocarpic forms ed. & a major carbohydrate source (145.4 M t by 2011; 30% int. trade prod. in Ecuador, biggest exporter by 2012, worth $1.9 billion), ed. fresh or cooked, or as crisps, dessert bananas derived from triploid forms (occ. viable gametes so hybrids with diploid or tetraploid bs poss.) of ***M. acuminata*** Colla (Indomal.) – deliberately planted NG 6950–6440 yrs ago, now esp. '**Dwarf Cavendish**' (Cavendish or Chinese b.) – widely cult. by 2011 half world prod., app. resistant to Panamá disease [*Fusarium oxysporum*, a wilt orig. SE As.] in rs, so replacing susceptible '**Gros Michel**' in 1950s, but new wilt strains making it prone, the bat-poll. (0.63 ml watery nectar per fl. per night) wild pl. (seeding form known also from Pemba, E Afr.) from wetter areas than ***M. balbisiana*** Colla (Indomal.), a more disease-resistant sp. (useful fibre) with which hybrids have been formed – ***M. × paradisiaca*** L. (plantain) – recog. as hybrid by Linnaeus, the major cooking bananas used in trop. (c. 4500 yrs ago Ind.) incl. sour embul (apple banana) exported by air from Sri Lanka to UK. Other spp. locally eaten, their lvs also wax & fibre-sources, many cult. orn. ***M. basjoo*** Sieb. & Zucc. ex Iinuma (orig. China) – cult. orn. (Japanese favourite); ***M. ingens*** Simmonds (New Guinea) to 15 m tall with 'trunk' to 89 cm diam.; ***M. textilis*** Née (abacá, Manila hemp, ? Borneo, long-cult. Philipp. & now S Am.) – fibre (75 K t by 2010) for clothing & footwear in pre-Spanish Philipp., cables incl. marine cordage, twine, paper money (80% cotton), sausage-skins, tea-bags, 'paper walls' of Japanese houses, on silk & cotton in ikat, 'Canton fibre' less good & allegedly from hybrid with *M. balbisiana*; ***M. troglodytarum*** L. (mountain plantain, ? orig. New Caledonia) – major food-pl. Tahiti, figuring prominently in Gauguin's paintings

Musaceae Juss. Magnoliidae – Zingiberales. Excl. Heliconiaceae, Strelitziaceae, 2/81 trop. Glabrous large pachycaul herbs with massive subterranean corm & hapaxanthic shoots without sec. growth; roots with scattered vessel-elements & phloem strands. Aerial shoots (pseudostems) largely rolled sheathing leaf-bases. Lvs simple & in spirals with long petiole & expanded lamina rolled from 1 side to other in bud, & with prominent midrib & pinnate-parallel venation, lateral veins incurved only at margin & between which tears may form under windy conditions. Fls unisex., irreg. with much nectar from septal nectaries (visited by birds, bats & insects) in bracteate, term. or lateral infl. arising from corm and growing up through pseudostem; leathery keeled bracts in spirals, each subtending few-flowered cymes (apical males). P 6, petaloid in 2 distinct whorls but outer & 2 inner merely teeth or lobes of a P-tube split along 1 side, the 3rd (adaxial) inner free, A 5 or 6, the 6th opp. free P but oft. a staminode or 0, anthers with longit. slits, $\underline{G}$ (3), 3-loc., with term. style & ∞ anatr., bitegmic ovules per loc. Fr. fleshy berry with few–∞ seeds with rudimentary arils & straight or (*Ensete*) curved embryo in copious starchy & mealy endosperm & perisperm. x = 7–11

Genera: *Ensete, Musa* (bananas) – imp. fr. & fibre-pls, cult. orn.

Musanga R. Br. Urticaceae (Cecropiaceae). 2 trop. Afr. R: BJBB 46(1976)496. Rauh's model; v. like *Cecropia* but without ants & poss. derived thus. ***M. cecropioides*** R. Br. (umbrella tree) – cult. orn. fast-growing (stilt-roots allow germ. on temporary humps or tree-trunks rich in nutrients in clearings, disappearing to leave bole on stilts, reaching 24 m in 15–20 yrs & soon dying) but being ousted by introd. allied *C. peltata* (? fewer pests), with light timber (corkwood) used for floats & rafts, isothermic ceilings, musical instruments, baskets etc., long thin fibres suitable for paper & twine, charcoal for floor-polish, imp. medic. (alleged to improve milk-flow in women; contains oestrogen)

muscadine *Vitis rotundifolia*

Muscadinia (Planch.) Small = *Vitis*

Muscari Mill. Asparagaceae (Hyacinthaceae). 50 Eur. (13), Medit., W As. R: Lily Yearbook 29(1965)126. Grape hyacinths. Alks incl. colchicine. In Neanderthal burial head-wreaths with *Ephedra* etc. Cult. orn. & fls used in scent-making incl. ***M. armenaicum*** Leichtlin ex Bak. (SE Eur. to Cauc.) – lvs developing in autumn, but very commonly ***M. neglectum*** Guss. ex Ten. ('*M. racemosum*', Eur., Medit., natur. Aus., N Am.) & ***M. botryoides*** (L.) Mill. (C & SE Eur.); others incl. ***M. comosum*** (L.) Mill. (tassel hyacinth, Medit., natur. Aus., N Am.) – bulbs ed., '**Plumosum**' ('Monstrosum', feather h.) – branched infls of sterile fls, P thread-like, ***M. racemosum*** Mill. (*M. muscarimi, M. moschatum*, SW Turkey) – long-cult. for scent-making

Muscarimia Kostel. ex Losinsk. = *Muscari*

muscatel *Vitis vinifera*

Muschleria S. Moore. Compositae (Vern.-Erl.). 1 Angola: ***M. angolensis*** S. Moore

Musci. Mosses. Bryophytes (landplants with dominant haploid generation) with multicellular rhizoids. Few parasites; strongly defended with antibiotics. For useful & other notable spp. see *Bryum, Climacium, Dawsonia, Fontinalis, Hypnum, Leucobryum, Polytrichum, Pseudoscleropodium, Rhodobryum, Rhytidiadelphus, Sphagnum, Thamnobryum*

muscovado raw or unrefined cane-sugar

Musella (Franch.) Li = *Ensete* (but see APS 16, 3(1978)56)

Museniopsis (A. Gray) J. Coulter & Rose = *Tauschia*

Musgravea F. Muell. Proteaceae (V 2). 2 Queensland. R: FA 17B(1999)170

mushroom plant *Justicia klossii*

mushyberry *Acronychia wilcoxiana*

musine *Croton megalocarpus*

Musineon Raf. Umbelliferae (III 5). 3 W N Am. Roots ed.

musizi *Maesopsis eminii*

musk (plant) *Mimulus moschatus*; **m. basil** *Basilicum polystachyum*; **m. mallow** *Malva moschata*; **m.melon** *Cucumis melo* Reticulatus Group; **monkey m.** *Erythanthe* spp.; **m. pl.** *E. moschata*; **m. rose** *Rosa* **x** *moschata*; **m. weed** *Myagrum perfoliatum*; **m.wood** *Alangium villosum*

muskit = mesquite

muslin *Gossypium* spp.

Mussaenda Burm. ex L. Rubiaceae (II 6). c. 130 OW trop. (As. c. 97; not Aus., Madag. – see *Bremeria*). Erect or scrambling shrubs, some locally medic. & several spectacular cult. orn. esp. ***M. erythrophylla*** Schum. & Thonn. (Ashanti blood, flame of the forest, trop. W Afr.) – the enlarged sepal bright red, such sepals making the infls of many *M.* spp. conspicuous (cf. *Euphorbia*); ***M. kingdon-wardii*** Jayaw. (N Myanmar) – rheophyte

Mussaendopsis Baill. Rubiaceae (Cond.). 2 W Mal. R: Flora 189(1994)161. Durable timber

Mussatia Bureau ex Baill. = *Bignonia*

Musschia Dumort. (~ *Campanula*). Campanulaceae (I 5). 3 Madeira (cf. *Chamaemeles*). R: AJBM 64(2007)143. Pachycaul treelets with dull fls, allied to *Campanula lactiflora* M. Bieb.

(Caucasus); fr. a capsule with transverse slits between vasc. ribs. ***M. aurea*** (L.f.) Dumort. – visited by nectar-seeking lizards

must(um) unfermented (or part-fermented) grape-juice

mustard *Brassica nigra* (**m. seed** of Bible, though poss. *Eruca vesicaria* or *Salvadora persica*; **black m.**), *Sinapis alba, B. napus* (m. of 'm. & cress'); **ball m.** *Neslia paniculata*; **brown m.** *B. nigra*; **buckler m.** *Biscutella laevigata*; **Chinese m.** *Brassica juncea* 'Crispifolia', *B. oleracea* Chinensis Group; **Dijon m.** *B. juncea*; **garlic m.** *Alliaria petiolata*; **hedge m.** *Sisymbrium officinale*; **hoary m.** *Erucastrum incanum*; **Indian m.** *B. juncea*; **Mithridate m.** *Thlaspi arvense*; **m. spinach** *B. campestris*; **tower m.** *Turritis glabra*; **treacle m.** *Erysimum cheiranthoides*; **tumble m.** *S. altissimum*; **white m.** *B. hirta*

Mutellina Wolf = *Ligusticum*

Mutisia L.f. Compositae (Mut.-Mut.). 63 Andes from Colombia to S Arg. & Chile, SE Brazil, Paraguay, Uruguay & NE Arg. R: OL 13(1965)1. Shrubs & lianes, lvs often term. by a tendril. Some cult. orn. ***M. acuminata*** Ruíz & Pavón (Peru to Bolivia) – hummingbird-poll. in altiplano of Peru though nectar hexose-rich (typical of Compositae)

Mutisiopersea Kosterm. = *Persea*

mutton wood *Myrsine variabilis*

mvule *Milicia excelsa*

Mwasumbia Couvreur & D.M. Johnson. Annonaceae (IV 1). 1 Tanz.: ***M. alba*** Couvreur & D.M. Johnson – allied to *Sirdavidia* (W Afr.). R: SB 34(2009)270

Myagrum Tourn. ex L. Cruciferae (33). 1 Medit. & C Eur. to India: ***M. perfoliatum*** L. – introd. Aus., where troublesome (musk weed), N Am. R: FNA 7(2010)568

myall *Acacia pendula*

Myanmaria H. Robinson (~ *Vernonia*). Compositae (Vern.-Gym.). 1 Myanmar: ***M. calycina*** (DC.) H. Robinson

Mycaranthes Blume (~ *Eria*). Orchidaceae (V 15). 36 Mal.

Mycelis Cass. = *Cicerbita*

Mycerinus A.C. Sm. Ericaceae VIII 5). 5 Guayana Highland. R: MNYBG 29(1978)175

Mycetia Reinw. Rubiaceae (IV 4). Incl. *Myrioneuron*, c. 54 Indomal.

Mycopteris Sundue (~ *Terpsichore*). Polypodiaceae (V). 17 trop. Am. R: Britt. 66(2014)175. Always assoc. with *Acrospermum* spp. (Ascomycota)

Myginda Jacq. = *Crossopetalum*

Myladenia Airy Shaw = *Xylosma*

Myllanthus R. Cowan = *Raputia*

myniong *Microseris lanceolata*

Myodocarpaceae Doweld (~ Araliaceae). Magnoliidae – Apiales. 2/19 E Mal. to NE Aus., New Caledonia. Woody pls. Lvs simple to pinnate. Infls of racemes or panicles of usu. umbels. K valvate; C imbricate; A inflexed in bud. Fr. fleshy or (*Myodocarpus*) dry, winged with flat seeds); endocarp with large oil-ducts
Genera: *Delarbrea, Myodocarpus*

Myodocarpus Brongn. & Gris. Myodocarpaceae (Araliaceae s.l.). 12 New Caledonia

Myonima Comm. ex Juss. = *Ixora* (but see KB 37(1983)555)

Myoporaceae R. Br. = Scrophulariaceae

Myopordon Boiss. Compositae (Card.-Card.). 5 SW & C As. R: BDBG 71(1958)271

Myoporum Sol. ex Forst. f. Scrophulariaceae (Myoporaceae). 30 Aus. (most), Mauritius, E As., E Mal., NZ, Hawaii. R: R.J. Chinnock, *Eremophila & allied genera* (2007)95. Some toxic to stock, some locally coll. ed. fr. & manna; some drought-tolerant pls (boobialla, boobyalla) cult. orn. e.g. ***M. insulare*** R. Br. (cockatoo bush, Aus.) & used as shelter-belts in S Eur. e.g. ***M. laetum*** Forst. f. (*'M. acuminatum', 'M. tenuifolium'*, ngaio, NZ); ***M. sandwicense*** A. Gray (Hawaii) – timber

Myoschilos Ruíz & Pavón. Santalaceae. 1 Chile, Arg.: ***M. oblongus*** Ruíz & Pavón

Myosotidium Hook. (~ *Omphalodes*). Boraginaceae (B.3.4). 1 Chatham Is. (NZ): ***M. hortensia*** (Decne) Baill. (*M. nobile*, Chatham Is. forget-me-not) – cult. orn., roots form. eaten by Moriori people who prob. smoked lvs

Myosotis Dill. ex L. Boraginaceae (B.3.7). c. 90 temp. (Eur. 41) esp. Medit. & NZ, incl. trop mts. R: JAA suppl. 1(1991)159. Forget-me-nots (name popularized by Coleridge; scorpion grass). C with scales almost closing throat, where s.t. a coloured ring nectar-guide; fl. often changing from pink to blue at anthesis; pollen grains c. 5 x 2.4 μm (poss. smallest in angiosperms). Cult. orn. small herbs esp. ***M. scorpioides*** L. (water f., Euras., natur. N Am.) – K without hooked hairs, s.t. (K with hooked hairs) ***M. arvenis*** (L.) Hill (Euras.), &

less greyish ***M. sylvatica*** Ehrenb. ex Hoffm. (garden f., Euras., invasive Aus.) – that usu. seen, many cvs incl. pink ones

Myosoton Moench. Caryophyllaceae (II 1). 1 temp. Euras.: ***M. aquaticum*** (L.) Moench – natur. N Am.

Myosurus L. (~ *Ranunculus*). Ranunculaceae (II 3). 6 temp. esp. Am. (Eur. 2 incl. ***M. minimus*** L. (mouse tail, Euras., Afr., Aus., N Am.) – receptacle elongate)

Myoxanthus Poepp. & Endl. (~ *Pleurothallis*). Orchidaceae (V 13c). 48 trop. Am. R: MSB 44(1992)1

myrabolans = myrobalans

Myracrodruon Allem. (~ *Astronium*). Anacardiaceae (I). 2 S Am. R: RBB 14(1991)133

Myrceugenella Kausel = *Luma*

Myrceugenia O. Berg. Myrtaceae (II 10). 44 Juan Fernandez, Chile & Arg. & (1000 km away) SE Brazil. R: FN 29(1981), Brittonia 36(1984)161

Myrcia DC. ex Guillemin. Myrtaceae (II 10). Incl. *Calyptranthes, Gomidesia, Marlierea*, c. 780 trop. Am., esp. SE Brazil. Some fish-disp. in Amaz. ***M. luquillensis*** (Alain) E. Lucas & A.R. Lourenço (*C. l.*, Luquillo Mts, Puerto Rico) – 5 trees left; ***M. splendens*** (Sw.) DC. (Amaz.) – bark extract used to paint insides of gourds black; ***M. tomentosa*** (Aubl.) DC. – ed. fr. (cabelluda). See also *Plinia*

Myrcialeucus N. Rojas = *Eugenia*

Myrcianthes O. Berg. Myrtaceae. (II 10). Excl. *Pseudanamomis*, 35 trop. Am. (Peru 14). ***M. fragrans*** (Sw.) McVaugh (twinberry, trop. Am.) – blowpipe darts

Myrciaria O. Berg. Myrtaceae (II 10). Incl. *Paramyrciaria*, 22 trop. Am. R: Napaea 9(1993)14. Like *Eugenia* but K-tube extended above ovary. Many ed. fr. esp. ***M. cauliflora*** (Mart.) O. Berg (jaboticaba, S Brazil, where much cult.), ***M. dubia*** (Kunth) McVaugh (camu, camu-camu) – high vitamin C (dried powder a 'superfood'), also fermented drinks, ***M. floribunda*** (Willd.) O. Berg (guava berry, rumberry, trop. Am.). See also *Plinia*

Myrciariopsis Kausel = *Myrciaria*

myrhh See myrrh

Myriactis Less. Compositae (Ast.-Lag.). 22 Cauc. to Jap. & New Guinea (OW 16), C Am. (6). Heterogeneous. Pappus 0

Myrialepis Becc. Palmae (I 3e). 1 Sumatra, Malay Pen.: ***M. paradoxa*** (Kurz) Dransf. – hapaxanthic rattan. R: KB 37(1982)242

Myrianthemum Gilg = *Medinilla*

Myrianthus P. Beauv. Urticaceae (Cecropiaceae). 7 trop. Afr. R: BJBB 46(1976)472. ***M. arboreus*** P. Beauv. with ed. fr. & timber (corkwood) lighter than cork; ***M. holstii*** Engl. (trop. & S Afr.) – fr. ed., esp chimpanzees

Myriaspora DC. = *Bellucia*

Myrica L. Myricaceae. Excl. *Morella*, 2 N temp. (Eur. 1:. ***M. gale*** L. (*Gale belgica*, bog myrtle, (sweet) gale, N Am., NW Eur., NE Siberia) – morphologically term. ovules, common pl. of wet heaths, able to change sex from yr to yr, insecticidal (increases no. of oil-glands when attacked; comm. midge-repellent on Skye), lvs used to flavour & improve foaming of beer, 'gale beer' a trad. drink, form. for fragrant candles, medic. & source of yellow dye; Calif. 1: ***M. hartwegii*** S. Watson (*G. h.*)). R: T 48(1999)367

Myricaceae Rich. ex Kunth. Magnoliidae – Fagales. 4/50 subcosmop. Aromatic trees & shrubs with indumentum of long colourless unicellular hairs & peltate usu. yellow multicellular glands; roots usu. with nitrogen-fixing bacteria (? not *Canacomyrica*). Lvs simple, in spirals; stipules in *Comptonia*. Fls small, wind-poll., usu. unisexual in spikes (simple or compound); P 0 (present in *Canacomyrica*), males often with 2 bracteoles as well as bract, A (2–)4(5, 6 in *Canacomyrica* where at summit of G in bisexual fls, –20), fewest in most acropetal fls, anthers with longit. slits, females with 2 bracteoles (sometimes K-like) & G (2), 1-loc., with ± distinct styles & 1 basal erect orthotropous, unitegmic ovule. Fr. a drupe or nutlet, sometimes with accrescent bracteoles; seed with straight embryo & ± 0 endosperm. n = 8, 12

Genera: *Canacomyrica, Comptonia, Morella, Myrica*

Some waxes, timber & ed. fr. (*Morella*)

Myricanthe Airy Shaw. Euphorbiaceae (Ricinoc.-Cocc.). 1 NW New Caledonia: ***M. discolor*** Airy Shaw. R: KB 35(1980)390

Myricaria Desv. Tamaricaceae. Incl. *Tamaricaria*, 13 temp. Eur. (1) to C As.

Myriocarpa Benth. Urticaceae (?II). 18 trop. Am.

Myriocephalus Benth. Compositae (Gnap.-Ang.). 14 temp. Aus. R: Nuytsia 14(2002)438. Compound heads of 1–9-flowered capitula

Myriocladus Swallen. Gramineae (V 2). 12 sandstone tablelands of Venez. R: Brittonia 50(1998)432. Bamboos resembling bromeliads on poles

Myriodon (Copel.) Copel. = *Hymenophyllum*

Myriolepis (Boiss.) Lledó & al. = *Myriolimon*

Myriolimon Lledó & al. (~ *Limonium*). Plumbaginaceae. 2 C & W Medit. R (*Myriolepis*): T 52(2003)67

Myrioneuron R. Br. ex Hook.f. = *Mycetia*

Myriophyllum Vaill. ex L. Haloragaceae. Incl. *Meziella, Vinkia*, c. 60 cosmop. (Eur. 3) esp. Aus. (36, 31 endemic). R: Brunonia 8(1985)173. Water milfoil. Mostly submerged aquatics like *Ceratophyllum* but lvs pinnate & not dichot.; aerial wind-poll. fls & overwintering buds. Some medic. N Am.; some cult. in aquaria esp. ***M. aquaticum*** (Vell. Conc.) Verdc. (*M. brasiliense*, parrot's feather, trop. Am., invasive Aus., NZ, S Afr., U.S., GB (1960; sale banned from 2014) – latter 3 only veget.). ***M. spicatum*** L. (N temp., Eur. pops introd. U.S. v. aggressive) – form. imp. emergency food for Native Amer.; ***M. trifida*** (Nees) Moody & Les (*Meziella t.*, SW Aus.) – known only from 1 specimen until 1990 when pop. discovered

Myriopteris (~ *Cheilanthes*). Pteridaceae. 47 Am. (Mex. 34), S Afr. (1). R: PhytoKeys 32(2013)54

Myriopteron Griff. Apocynaceae (III; Periplocaceae). 1 NE India to W Mal.: ***M. extensum*** (Wight) K. Schum. (*M. paniculatum*)

Myriopus Small (~ *Heliotropium*). Boraginaceae (Heliotropiaceae). 25 trop. Am. Usu. lianes

Myriostachya (Benth.) Hook.f. Gramineae (Chlorid.). 1 S India, Sri Lanka to SE As.: ***M. wightiana*** (Steud.) Hook.f.

Myripnois Bunge. Older name for *Pertya*

Myristica Gronov. Myristicaceae. 175 trop. As. (Mal. 152 – R: FM I,14(2000)359) to Aus. (3, R: Blumea 36(1991)183). Dioec.; some domatia. ***M. argentea*** Warb. (W New Guinea) & other E Mal. spp. – minor sources of mace; ***M. fragrans*** Houtt. (nutmeg, Moluccas, where monopoly sought by Dutch – some 14 000 people massacred or enslaved in 1621, while R(h)un is. first English colony, to be exchanged for Manhattan by Dutch 1674) – comm. nutmeg (c. 10–12 K t per annum) cult. in Mal. & Grenada (WI), seedlings sexed by colour reaction with ammonium molybdate, when ground the seed used as flavouring for milk puddings, biscuits etc. though excess is toxic (myristicin), hallucinatory & addictive with reputation as aphrodisiac (in GB carried in pocket as heath talisman), fimbriate aril is mace used in flavouring for fish, doughnuts etc., oil medic. also used to scent toothpaste, cigarettes, candles, soap, etc.; ***M. inspida*** R. Br. (E Mal., trop. Aus.) – beetle-poll., useful timber

Myristicaceae R. Br. Magnoliidae – Magnoliales. 20/500 trop. – usu. lowland rainforest (Mal. 335 – R: FM I,14(2000)). Dioec. or monoec. evergreen trees (Massart's Model unique to fam.) usu. with aromatic tissues containing spherical ethereal oil-cells often with the phenolic, myristicin; bark typically with red sap when slashed. Shoots with long term. buds. Lvs simple, entire, in spirals, & often distichous, often gland-dotted & glaucous abaxially; stipules 0. Fls in usu. axillary cymes or racemes; P ((2)3(–5)) with valvate lobes, A 2–∞ with ± united filaments & often laterally connate anthers with longit. slits & monosulcate or inaperturate pollen grains, $\underline{G}$ 1 with unclosed carpel, stigma rarely with style, & 1 nearly basal, ± anatropous, bitegmic ovule. Fr. fleshy to leathery, usu. dehiscent along 2 sutures to reveal pendulous endotestal seed with conspicuous aril, copious oily & usu. ruminate endosperm & small embryo, the cotyledons sometimes basally connate. n = 19, 21+

Principal genera: *Gymnacranthera, Horsfieldia, Iryanthera, Knema, Myristica, Virola*; a very uniform archaic group poss. orig. Afr./S Am.

Male fls have no pistil, females no staminodes. According to Corner the seeds are the most primitive surviving; they are very large & the attractive arils are taken by birds, the toxic seed dropped. Some oils & spices esp. nutmeg & mace (*Myristica*), timber (esp. *Horsfieldia, Virola*) traded as penarhan, & hallucinogens (esp. *Virola*)

Myrmechis (Lindl.) Blume. Orchidaceae (IV 2d). Incl. *Tubilabium*, c. 15 Jap., Indomal.

Myrmecodia Jack. Rubiaceae (IV 7). 26 Mal. to Fiji. R: Blumea 37(1993)271. Epiphytes with swollen stems ('tubers') penetrated by numerous interconnecting galleries & chambers inhabited by ants, the system forming independent of them in the hypocotyl & deriving from the activity of phellogens producing cork layers which split. Ants, which patrol the stems collecting nectar, bring nutrient-rich material (decay facilitated by fungi) into tuber enhancing mineral regime for plant in a nutrient-poor habitat but deter pollinators (so pls largely selfed) & disp. agents. Seeds spread on to bark by birds, also removed by ants & planted in their runways. ***M. beccarii*** Hook.f. (NE Queensland) – poll. Apollo Jewel butterfly (*Hypochrysops apollo*), laying eggs on pl., emergent larvae carried in by ants, exploiting honeydew as well as newly-made cavities

Myrmeconauclea Merr. (~ *Neonauclea*). Rubiaceae (I 2). 4 Mal. R: Blumea 24(1978)342. Some rheophytes incl. ***M. strigosa*** (Korth.) Merr., unique in being myrmecophilous too

Myrmecophila (Christ) Nakai = *Lecanopteris*

Myrmecophila Rolfe = *Laelia*

Myrmecopteris Pichi-Serm. = *Lecanopteris*

Myrmecosicyos C. Jeffrey = *Cucumis*

Myrmedoma Becc. = seq.

Myrmephytum Becc. Rubiaceae (IV 7). 8 Philippines, Sulawesi, W New Guinea. R: Blumea 36(1991)43. Epiphytic ant-pls

Myrmidone Mart. = *Tococa*

myrobalans *Terminalia* spp. esp. *T. chebula*, *Prunus cerasifera*

Myrocarpus Allemão. Leguminosae (III 2). 5 trop. S Am. R: Phytologia 23(1972)401. Some timber & oil used in scent-making, esp. ***M. frondosus*** Allemão (Brazil) – source of cabreuva oil

Myrosma L.f. Marantaceae. 1 trop. Am.: ***M. cannifolia*** L.f. (marble arrowroot). R: Blumea 57(2012)126

Myrosmodes Reichb.f. (~ *Aa*). Orchidaceae (IV 2b). 12 Andes, esp. among *Azorella* tussocks

Myrospermum Jacq. Leguminosae (III 2). 3 trop. Am. Lvs dotted; fr. winged, indehiscent

Myrothamnaceae Niedenzu. Magnoliidae – Gunnerales. 1/2 S Afr. & Madag. Small dioec. xeromorphic aromatic glabrous shrubs. Lvs opp., flabellate, plicate, venation not reticulate; resin- or oil-ducts & stipules 2. Fls reg., wind-poll. in erect bracteates spikes; P 0–4 esp. in term. fls, A 3–8 alt. with P when present, anthers with prolonged connective & longit. slit & pollen shed in tetrads, $\underline{G}$ (3 or 4), 3- or 4-loc. with distinct recurved styles & axile placentation & rather numerous anatropous bitegmic ovules per loc. Fr. a capsule, the carpels separating apically & opening ventrally; seeds ∞ with thin testa & copious oily endosperm. n = 10
Genus: *Myrothamnus*

Myrothamnus Welw. Myrothamnaceae. 2 Kenya, S trop. Afr., Madag. ***M. flabellifolia*** Welw. (S Afr.) – resurrection pl. rapidly reviving after rain when the folded blackish fragile lvs green up, for the mitochondria (? & plastids) separated from rest of cell contents during desiccation (up to 2 yrs) by barriers perforated on rehydration, oil antibacterial, antifungal

Myroxylon L.f. Leguminosae (III 2). 2 trop. Am., natur. OW. ***M. balsamum*** (L.) Harms (Venez. to Peru) – good timber for furniture, source of balsam of Tolu used in ointments & for flavouring cough syrups etc., var. ***pereirae*** (Royle) Harms – source of balsam of Peru, form. much valued medic. now with similar uses

myrrh *Commiphora* spp. esp. *C. myrrha*, (Genesis) *Cistus creticus*; **Abyssinian m.** *Commiphora madagascariensis*; **garden m.** *Myrrhis odorata*; **Mecca m.** *C. gileadensis*; **scented m.** *C. guidottii*

Myrrhidendron J. Coulter & Rose. Umbelliferae (III 10). 5 C Am. to Colombia. Some woody

Myrrhinium Schott. Myrtaceae (II 10). 1 trop. S Am.: ***M. atropurpureum*** Schott – A 4–8, sweet juicy C attractive to frugivorous passerines, fr. ed. R: FN 45(1986)142

Myrrhis Mill. Umbelliferae (III 2). 1 Eur.: ***M. odorata*** (L.) Scop. (sweet cicely, garden myrrh)– only male fls. at end of flowering so that last (protandrous) bisexual fls pollinated, form. potherb, prob. widely introd. for strewing on church floors in mediaeval GB, home remedy & flavouring for brandy (*trans*-anethole the main sweet constituent), lvs form. used to polish oak panelling

Myrrhoides Heister ex Fabr. (~ *Chaerophyllum*). Umbelliferae (III 2). 1 Eur., Medit. to C As.: ***M. nodosa*** (L.) Cannon. R: Fl. Iraq 5,2(2013)176

Myrsinaceae. R. Br. = Primulaceae

Myrsine L. Primulaceae (Myrsinaceae). (s.s. 5 Azores, Afr. & As.) incl. *Rapanea*, *Suttonia* c. 200 trop. & warm. Rauh's & Attims's Models. Some timbers incl. ***M. melanophloeos*** (L.) Sweet (*R. m.*, Cape beech, trop. & S Afr.) & ***M. variabilis*** R. Br. (*R. v.*, mutton wood, E Aus.); ***M. africana*** L. (Azores to China) – cult. orn.; ***M. lessertiana*** A. DC. (Hawaii) –

wood surface used for beating kapa dyed with wood's red sap or black dye from its charcoal

Myrsiphyllum Willd. = *Asparagus*

Myrtaceae Juss. Magnoliidae – Myrtales. 134/5500 trop. & warm + temp. Aus. R: R. Govaerts & al. (2008) *World Checklist of M.* Trees & shrubs with abundant, scattered secretory cavities &, characteristically, internal phloem in pith; usu. with ectotrophic mycorrhizae. Lvs usu. opp., simple & often leathery; stipules rudimentary or 0. Fls usu. bisexual, reg. with hypanthium extended beyond ovary, rarely perigynous, with conspic. bracts, K, C or A attractive to animals esp. birds, in complex infls (solit. in *Myrtus communis*); K (3)4 or 5(6), often imbricate & free or not, s.t. v. reduced or splitting at anthesis or forming calyptra, C similar (a calyptra in *Eucalyptus*), A usu. ∞ developing centripetally on rim of hypanthium, free or basally united into 4 or 5 groups each supplied by 1 vasc. trunk-bundle, less often 2 × K or C in 2 whorls or single whorl opp. K (*Heteropyxis*), anthers with longit. slits or term. pores, nectary-disk lining prolonged hypanthium or at apex of ± inf. G (2–5(–16)) with as many loc. or rarely pseudomonomerous & term. style (G, stigma sessile in *Psiloxylon*), placentation axile, each loc. with 2–∞ anatropous to campylotropous, usu. bitegmic ovules per loc. though embryo often apomictic (developing from nucellus after degeneration of zygote). Fr. a (1–)few(–∞)- seeded berry, loculicidal capsule, or drupe-like or nut; seeds mesotestal, often polyembryonous initially, with usu. ± 0 endosperm, the embryos various with cotyledons s.t. curved or spiralled. x = (6–9–)11(12)

Classification & chief genera (see PSE 251(2005)13):

I. **Psiloxyloideae** (Heteropyxidoideae; lvs in spirals, dioec., x = 12; E Afr., Masc.): 1. Psiloxyleae (berry; *Psiloxylon* only); 2. Heteropyxideae (capsule, winged seeds; *Heteropyxis* only)

II. **Myrtoideae** (lvs opp. or in spirals, fls bisex., x = 11) – 14 tribes: 1. Xanthostemoneae (*Xanthostemon*); 2. Lophostemoneae (*Lophostemon*); 3. Osbornieae (*Osbornia* only); 4. Melaleuceae (*Melaleuca*); 5. Kanieae (*Tristaniopsis*); 6. Backhousieae (*Backhousia*); 7. Metrosidereae (*Mearnsia, Metrosideros, Tepualia* (Am.)); 8. Tristanieae (*Xanthomyrtus*); 9. Syzygieae (*Syzygium*); 10. Myrteae (*Acca, Campomanesia, Decaspermum, Eugenia, Myrceugenia, Myrcia, Myrcianthes, Myrtus, Pimenta, Plinia, Psidium, Rhodamnia*); 11. Eucalypteae (*Angophora, Corymbia, Eucalyptus*); 12. Syncarpieae (*Syncarpia* only); 13. Lindsayomyrteae (*Lindsayomyrtus* only); 14. Leptospermeae (*Agonis, Leptospermum*); 15. Chamelaucieae (*Babingtonia, Baeckea, Calytrix, Chamelaucium, Darwinia, Homaranthus, Kunzea, Thryptomene, Verticordia*)

Fleshy fr. in *Syzygium* group, derived in parallel in *Eugenia* group (often confused & form. even considered congeneric!)

Timber incl. paper pulp (*Corymbia, Eucalyptus,* which characterize much of Aus. veg., *Lophostemon, Syncarpia*), ed. fr. (*Acca, Eugenia, Psidium* (guava), *Myrciaria, Myrteola, Syzygium, Ugni* etc.), spices & medic. or scent oils (*Darwinia, Eucalyptus, Melaleuca* (tea tree), *Pimenta* (allspice, bay rum), *Leptospermum, Melaleuca, Myrtus, Syzygium*(cloves)) & cult. orn. esp. *Agonis, Chamelaucium* (cut-fl.), *Corymbia, Eucalyptus, Kunzea, Leptospermum, Melaleuca* (*Callistemon*), *Myrtus, Verticordia*

Myrtama Ovcz. & Kinzik. (~ *Myricaria*). Tamaricaceae. 1 Pakistan, Kashmir, Tibet, to 6500m: ***M. elegans*** (Royle) Ovcz. & Kinzik. (*Myricaria e., Tamaricaria e.,* C & S As.) – rheophyte

Myrtastrum Burret. Myrtaceae. (II 10). 1 New Caledonia: ***M. rufopunctatum*** (Brong. & Gris) Burret. R: NBGB 15(1941)494

Myrtekmania Urb. = *Pimenta*

Myrtella F. Muell. Myrtaceae (II 10). 2 New Guinea, 1 ext. to Guam. R: Austrobaileya 5(1999)178. Aus. spp. = *Lithomyrtus*

Myrteola O. Berg. Myrtaceae (II 10). 3 S S Am. & trop. Am. mts. R: SB 13(1988)120. Ed. fr.

Myrtillocactus Console. Cactaceae (III 1). 4 Mex., Guatemala. Tree-like or shrubby. Fr. ed. like *Vaccinium myrtillus*

myrtle *Myrtus communis*; **aniseed m.** *Anetholea anisata*; **m. beech** *Nothofagus cunninghamii*; **bog m.** *Myrica gale*; **crape m.** *Lagerstroemia indica*; **lemon m.** *Backhousia citriodora*; **m. lime** *Triphasia trifolia*; **sand m.** *Kalmia buxifolia*; **strawberry m.** *Ugni molinae*; **Tasmanian m.** *Nothofagus cunninghamii*; **wax m.** *Morella cerifera*; **willow m.** *Agonis* spp.; **m. wood** *Umbellularia californica*

Myrtopsis Engl. Rutaceae (I 2). 9 New Caledonia

Myrtus Tourn. ex L. Myrtaceae (II 10). (S.s.) 2 Medit. & N Afr. ***M. communis*** L. (myrtle, Medit., origin unclear) – lvs, fragrant fls & fr. rich in oil, the 'Eau d'Ange' of scent-making, also medic., wood for walking-sticks, furniture etc., bark & roots yield tannin used on finest leathers of Russia & Turkey, cult. orn. with several cvs, long assoc. with ritual & ceremony & garlands in Classical Rome (app. dwarf **'Tarentina'** worn by judges & victors at orig. Olympic Games)

Mysanthus G. Lewis & Delgado. Leguminosae (III 18). 1 Brazil: ***M. uleanus*** (Harms) G. Lewis & Delgado. R: KB 49(1994)343

Mysore thorn *Biancaea decapetala*

Mystacidium Lindl. Orchidaceae (V 16d). 10 E & S Afr. Cult. orn.

Mystacorchis Szlach. & Marg. = *Pleurothallis*

Mystropetalon Harv. Balanophoraceae. 1 SW Cape: ***M. thomii*** Harv. – on Proteaceae, monoec., pollen unique in being triangular, square or pentagonal end-on, but square from the side, elaiosomes around fr. attractive to ants which disperse seeds. R: BJ 106(1986)370

Mystroxylon Ecklon & Zeyher (~ *Cassine*). Celastraceae (I). 1 E & S Afr., Madag., Masc.: ***M. aethiopicum*** (Thunb.) Loes. – drupes ed. R: Strelitzia 10(2000)218

Mytilaria Lecomte. Hamamelidaceae (II). 1 Kwangsi & Laos: ***M. laosensis*** Lecomte. G (2) semi-inf., partly sunk in fleshy axis

Myuropteris C. Chr. = *Leptochilus*

Myxochlamys Takano & Nagam. (~ *Scaphochlamys*). Zingiberaceae. 2 Borneo. R: APG 58(2007)21

Myxopappus Källersjö (~ *Pentzia*). Compositae (Anth.-Pen.). 2 S Afr. R: BotJLS 96(1988)314. Annuals

Myxopyrum Blume. Oleaceae (5). 4 Indomal. R: Blumea 29(1984)499. ***M. nervosum*** Blume (Mal.) – local medic.

Myzodendraceae J. Agardh = Misodendraceae

Myzodendron Banks & Sol. ex R. Br. = *Misodendrum*

Myzorrhiza Philippi = *Orobanche*

mzimbeet *Androstachys johnsonii*

Mzymtella Kolak. = *Campanula*

N

na *Mesua ferrea*

Nabaluia Ames. Orchidaceae (V 10b). 3 N Borneo. R: OM 1(1986)47

Nabalus Cass. (~ *Prenanthes*). Compositae (Cich.-Cich.(Crep.)). 18 N Am., E As. Some cult. orn.

Nabea Lehm. ex Klotzsch (*Macnabia*) = *Erica*

Nablonium Cass. = *Ammobium*

Nacrea Nelson = *Anaphalis*

Nageia Gaertn. (*Decussocarpus*, ~ *Podocarpus*). Podocarpaceae. 5 Indomal. R: NRBGE 45(1988)381. Dioec. Some cult. orn. incl. ***N. wallichiana*** (C. Presl) Kuntze, only conifer native in S Ind.

Nageiaceae Fu = Podocarpaceae

Nageliella L.O. Williams = *Domingoa* (but see C.L. Withner, *The Cattleyas & their relatives* V(1998)169)

Nagelocarpus Bullock = *Erica*

nagoon berry *Rubus arcticus*

Naias Adans. = *Najas*

Naiocrene (Torrey & A. Gray) Rydb. = *Montia*

Najadaceae Juss. = Hydrocharitaceae

Najas L. Hydrocharitaceae (Najadaceae). c. 40 cosmop. (Eur. 4 + 2 natur., Aus. 8). Acarpellate (naked seed inside fr. wall); seeds needed to identify spp. Some bad weeds in rice-fields but good green fertilizer & valuable fish-food (e.g. in water 5 m deep for tilapia, to 14 m in Canada), also as packing material. ***N. tenuifolia*** R. Br. (Aus.) – found in mudpools at 60°C in Java

naked boys or **n. ladies** *Colchicum autumnale*

nal *Phragmites australis*

nali nut *Canarium harveyi*

Naletonia Bremek. = *Psychotria*

Nama L. Boraginaceae (Nam.; Hydrophyllaceae-Nam.). 56 SW US & trop. Am., Hawaii (1). ***N. hispida*** A. Gray (SW N Am.) – lotion for spider-bites

Namacodon Thulin. Campanulaceae (I 2). 1 SW Afr.: ***N. schinzianum*** (Markgr.) Thulin – capsule with septicidal dehiscence (unique in C.). R: Strelitzia 10(2000)201

Namaquanthus L. Bolus. Aizoaceae (V). 1 S Afr.: ***N. vanheerdei*** L. Bolus. R: H. E. K. Hartmann, *Ill. Handb. Succ. Pls*, Aiz. F–Z(2002)180

Namaquanula D. Mueller-Doblies & U. Mueller-Doblies = *Hessea*

Namataea D.W. Thomas & D.J. Harris. Sapindaceae. 1 Cameroun: ***N. simplicifolia*** D.W. Thomas & D.J. Harris. R: KB 54(2000)951

Namation Brand = *Petunia*

Namibia (Schwantes) Dinter & Schwantes (~ *Juttadinteria*). Aizoaceae (V). 2 SW Namibia. R: H. E. K. Hartmann, *Ill. Handb. Succ. Pls*, Aiz. F–Z(2002)180

Namibithamnus H. Robinson & al. (~ *Vernonia*). Compositae (Vern.). 2 S Afr. R: PhytoKeys 60(2016)93

nam-nam *Cynometra cauliflora*

Namophila U. Mueller-Doblies & D. Mueller-Doblies. Asparagaceae (Hyacinthaceae). 1 S Namibia: ***N. urotepala*** U. Mueller-Doblies & D. Mueller-Doblies. R: FR 108(1997)77

Nananthea DC. Compositae (Anth.-Anth.). 1 Corsica, Sardinia: ***N. perpusilla*** (Lois.) DC. R: BBMNHB 23(1993)134

Nananthus N.E. Br. Aizoaceae (V). 7 S Afr. R: H. E. K. Hartmann, *Ill. Handb. Succ. Pls*, Aiz. F–Z(2002)181. Cult. orn. dwarf succ.

Nanarepenta Matuda = *Dioscorea*

Nancy, early *Wurmbea dioica*

Nandi flame *Spathodea campanulata*

Nandina Thunb. Berberidaceae (I; Nandinaceae). 1 Ind. to Jap. (orig. ? C China): ***N. domestica*** Thunb. (heavenly bamboo) – evergreen shrub (Chamberlain's Model), lvs 2- or 3-pinnate, much cult. orn. esp. Jap. (60 named cvs), invasive in SE US

Nandinaceae Horan. = Berberidaceae

nangai **(nut)** *Canarium* spp.

Nani Adans. = *Xanthostemon*

Nanking cherry *Prunus tomentosa*

nanmu wood *Persea nanmu*

Nannoglottis Maxim. Compositae (Ast.). 9 S to C China

Nannorrhops H. Wendl. Palmae (III 5). 1 Arabia to Pakistan: ***N. ritchieana*** (Griff.) Aitch. – bushy hapaxanthic (Schoute's Model), unarmed, ed. fr. & bud, cult. orn. with fibre used for rope, lvs for basketry etc.

Nannoseris Hedb. = *Crepis*

Nannothelypteris Holttum = *Cyclosorus*

nannyberry *Viburnum lentago*

Nanobubon A. Magee (~ *Peucedanum*). Umbelliferae (III 11). 2 W Cape. R: T 57(2008)356

Nanochilus K. Schum. (~ *Hedychium*). Zingiberaceae (II 4). 2 Sumatra, Moluccas

Nanocnide Blume. Urticaceae (I). 2 E As.

Nanodea Banks ex Gaertn. f. Santalaceae (Nanodeaceae). 1 temp. S Am.: ***N. muscosa*** Banks ex Gaertn.f. – poll. small flies

Nanodeaceae Nickrent & Der. See Santalaceae

Nanodes Lindl. = *Epidendrum*

Nanolirion Benth. = *Caesia*

Nanooravia Kiran Raj & Sivad. = *Dimeria* (but see NJB 31(2013)162)

Nanophyton Less. Amaranthaceae (Chenopodiaceae III 3). c. 10 Eur. to SW & C As. ***N. erinaceum*** (Pallas) Bunge – piperidine derivatives used in treatment of hypertension

Nanorrhinum Betsche (*Pogonorrhinum*, ~ *Kickxia*). Plantaginaceae (Ant.-Ant.). 16 trop. & subtrop. OW. R: NJB 20(1999)668

Nanostelma Baill. = *Vincetoxicum*

Nanothamnus Thomson = *Blumea*

Nanozostera Toml. & Posluszny = *Zostera* (but see T 50(2001)432)

Nanuza L.B. Sm. & Ayensu = *Vellozia*

Napaea Clayton ex L. Malvaceae (Malv.-Malv.). 1 E to C US: ***N. dioica*** L. – cult. orn. with useful fibre from bark

Napeanthus Gardner. Gesneriaceae (II 2). 19 trop. Am. R: ABN 7(1958)340

Napier grass *Cenchrus purpureus*

Napoleonaea P. Beauv. Lecythidaceae (V; Napoleonaeaceae). c. 16 trop. W Afr. R: KB 70,6(2015)13. Massart's Model. Ed. fr., some prob. spread by humans; some locally medic. e.g. ***N. beninensis*** Jongkind (*N. vogelii s.l.*) – prob. thrips-poll., chewing-stick in Nigeria, ed. pulp around seeds, cult. orn. ('*N. imperialis*'); ***N. gossweileri*** Bak.f. – geoxylic suffrutex

Napoleonaeaceae A. Rich. = Lecythidaceae

Napoleon's willow *Salix babylonica*

nara or **narras** *Citrullus horridus*

naranjilla *Solanum quitoense*

Naravelia Adans. = *Clematis* (but see APG 37(1986)106)

Narcissus Tourn. ex L. Amaryllidaceae. 27 Eur. (26), Medit. R: J.W. Blanchard (1990) *N.: a guide to wild daffodils*. Daffodils; corona well developed & free of A, mono-, di- or trimorphic. Some bee-poll. e.g. *N. pseudonarcissus*, some butterfly-poll. e.g. *N. jonquilla*, some both e.g. *N. triandrus*. Cult. orn. spp. & hybrids with many phenanthridine alks (incl. lycorine & narcissine), increased galanthamine prod. selected to reduce pressure on e.g. *Leucojum vernum* (higher levels in stressed pls. so grown on 'marginal' land); imp. as cut-fls (5 300 ha in UK, more than anywhere else) & in scent-making; many garden pls (10 000 cvs) of complex orig. & frequently ± natur., there being a botanical classification for wild forms & a horticultural one (13 Divisions) based on fl. proportions, e.g. Trumpet narcissi (Division 1, corona as long as or longer than P etc.), Large- (Division 2, corona 1/3 to as long as P) e.g. '**Carlton**' the most commonly grown cv. (registered 1927 & now some 9450 million bulbs (c. 350 000 t) & perhaps the biggest genetic 'individual' in the world, but cf. *Saccharum*) & '**King Alfred**' (orig. 'Rembrandt'), & Small-cupped (Division 3, less than 1/3 as long) narcissi derived from '*N.* × *incomparabilis*', etc. Many 'double' forms due to additional P, petaloid A, or split or duplicated corona. Cult. spp. & hybrids (signifying egotism in 'Language of Fls') incl. ***N. bulbocodium*** L. (W Medit.) – v. variable even within wild populations, some v. tall swamp pops in S Portugal (= *N. quintanilhae* (A. Fern.) Fern. Casas), P v. reduced, corona the consp. part of fl., nat. hybrids with *N. cyclamineus* in Portugal = ***N.* × *caramulensis*** Ribeiro & al.; ***N.* × *compressus*** Haw. (*N. jonquilla* × *N. tazetta*); ***N. cyclamineus*** DC. (NW Spain & Portugal) – P strongly reflexed; ***N.* × *cyclazetta*** Chater & Stace (*N. cyclamineus* × *N. tazetta*) esp. '**Tête à Tête**'; ***N. jonquilla*** L. (jonquil, SW Eur. natur. elsewhere) – lvs rush-like, fls v. fragrant used in scent-making; ***N.* × *medioluteus*** Mill. (*N.* × *biflorus*, *N. poeticus* × *N. tazetta*) – natural hybrid (S France) usu. with 2 fls per infl., P white; ***N. minor*** L. subsp. ***asturiensis*** (Jord.) Barra & López (*N. asturiensis*, *N. cuneiflorus*, '*N. minimus*', C & N Spain, N Portugal) – only 6–12 cm tall; ***N.* × *monochromus*** Sell (*N. cyclamineus* × *N. pseudonarcissus*) – commonly cult. early-flowering esp. '**February Gold**'; ***N.* × *odorus*** L. (*N. jonquilla* × *N. pseudonarcissus*); ***N. papyraceus*** Ker-Gawler (paperwhite, 'jonquil' (NSW), W Medit.) – grown forced for winter fls, poss. a form of *N. tazetta*; ***N. poeticus*** L. (pheasant's-eye, France to Greece) – fragrant white fl. with v. short red-tipped corona, used in scent-making; ***N. pseudonarcissus*** L. (*N. incomparabilis*, *N. obvallaris*, wild daffodil, Lent lily, Tenby d., SW Eur.) – 'wild' in parts of GB, parent of many hybrids & commonly natur.; ***N. serotinus*** Loefl. ex L. (Medit.) – fls white & ***N. viridiflorus*** Schousboe (SW Spain, Morocco) – fls green, night-flowering, moth-poll., both autumn-flowering without lvs, the scapes photosynthetic & doubling in length after flowering; ***N. tazetta*** L. (Medit.) – fls 4–8, white with pale yellow corona, long cult. (poss. biblical 'lily of the field'), e.g. '**Grand Soleil d'Or**' late yellow, scented cut-fl. in UK & a sterile form early taken to China (? c. AD 700 via Iran), the 'Chinese sacred lily' (poss. back-cross with '*N.* × *incomparabilis*') v. popular at New Year ('singles' – close to '**Grand Monarque**', & doubles); ***N. triandrus*** L. (Angel's [after a tired pl. collector, A. Gancedo] tears, SW Eur.) – fls nodding, white or corona yellowish, some pops tristylous

nard *Nardostachys jatamansi*; **n. grass** *Nardus stricta*

nardoo *Marsilea* spp. (Aus.) esp. *M. drummondii*

Nardophyllum (Hook. & Arn.) Hook. & Arn. Compositae (Ast.-Hin.). 5 S Andes. R: SCB 92(2009)62

Nardosmia Cass. = *Petasites*

Nardostachys DC. (~ *Patrinia*). Caprifoliaceae (Valerianaceae). 1 Himal.: ***N. jatamansi*** (D. Don) DC. (*N. grandiflora*, jatamansi, Ind. nard, spikenard) – rhiz. with oil form. prized in salves in Roman society, medic. & in scent-making etc. (cf. *Diplotaenia*, *Valeriana*), over-exploited

Narduretia Villar = *Vulpia*

Narduroides Rouy = *Festuca*

Nardurus (Bluff, Nees & Schauer) Reichb. = *Festuca*

Nardus L. Gramineae (VIII). 1 Eur. & W As.: ***N. stricta*** L. (mat grass, nard g.) on drier moors, poor grazing. R: FNA 34(2007)62

Naregamia Wight & Arn. (~ *Turraea*). Meliaceae (II-Trich.). 1 Angola, 1 W Ind.: ***N. alata*** Wight & Arn. (Goa ipecacuanha) – local medic. (alks)

Narenga Bor = *Miscanthus*

Nargedia Beddome. Rubiaceae (II 5). 1 Sri Lanka: ***N. macrocarpa*** (Thw.) Bedd.

nargusta *Terminalia amazonia*

Naringi Adans. Rutaceae (II 2). 2 S & SE As.

narinjin *Citrus maxima*

narra *Pterocarpus indicus*

Narrawa burr *Solanum cinereum*

Nartheciaceae Fries ex Bjurzon (~ Melanthiaceae). Magnoliidae – Dioscoreales. Excl. Petrosaviaceae, Tofieldiaceae, 5/30 N temp. esp. E As. & E N Am. Perenn. herbs, usu. rhiz. (rarely corm) with fibrous r. Lvs basal, usu. distich. Infl. usu. a raceme; peduncle erect usu. with bracts. Fls 3-merous, reg., bisexual, bracteate & usu. bracteolate. P 3 + 3 petaloid, basally connate, A 3 + 3, $\underline{G}$ 3 to half epigynous with ∞ ovules. Fr. a capsule
Genera: *Aletris, Lophiola, Metanarthecium, Narthecium, Nietneria*
In their (dreary) simplicity superficially similar to unrelated genera now referred to Alismatales (Tofieldiaceae), Commelinales (Haemodoraceae), Liliales (Melanthiaceae) & Petrosaviales (Petrosaviaceae). Many E As. – E N Am. disjunctions. Some cult. orn., local medic.

Narthecium Hudson. Nartheciaceae (Melanthiaceae s.l.). 6 N temp. (Eur. 3, E As. 1, N Am. 1), disjunct. ***N. ossifragum*** (L.) Hudson (bog asphodel, Eur.) – iris-like pl. with yellow nectarless fls, filament hairs trapping pollen & raindrops on them allowing pollen to float to stigmas, toxic (steroidal saponins) to sheep (liver disease) & cattle, subs. for saffron in Shetland & C17 hair-dye in Lancashire

Narvalina Cass. Compositae (Cor.). 2 Hispaniola. R: Phytoneuron 2015–31:4. Shrubs

Nasa Weigend (~ *Loasa*). Loasaceae I (1). c. 100 trop. Am.

naseberry *Manilkara zapota*

nashi (pear) *Pyrus pyrifolia*

Nashia Millsp. (~ *Lantana*). Verbenaceae (Lant.). 7 Cuba, Bahamas. Lvs used as a tea

Nassauvia Comm. ex Juss. Compositae (Mut.-Nass.). Excl. *Triptilion*, 38 S Andes to Patagonia. R: Darw. 24(1982)283. Shrubs & herbs

Nassella (Trin.) E. Desv. (~ *Jarava*). Gramineae (X). 117 warm & trop. Am. (Arg. 72) esp. Andes. R: T39(1990)597. ***N. neesiana*** (Trin. & Rupr.) Barkw. (Chilean needle grass, S Am.) – noxious weed in SE Aus.; ***N. tenuissima*** (Trin.) Barkw. (SW US, NW Mex.) – invasive S Afr.; ***N. trichotoma*** (Nees) Arechav. (S Am.) – natur. S Afr., trop. As. to Aus. & NZ, noxious weed in NSW ('serrated tussock'), US (where all pops eliminated)

Nastanthus Miers (~ *Boopis*). Calyceraceae. 6 SW & S S Am. R: Gayana 67(2010)159

Nasturtiicarpa Gilli = *Calymmatium*

Nasturtiopsis Boiss. Cruciferae (12). 2 N Afr. to Israel

nasturtium *Tropaeolum majus*

Nasturtium R. Br. (~ *Rorippa*). Cruciferae (16). 5 N temp. R: Novon 8(1998)125. Watercress. ***N. microphyllum*** Boenn. ex Reichb. (2n = 64 [octoploid?], *R. m.*, *N. officinale* × ?, W Eur.) – not cult., & smaller-fld ***N. officinale*** R. Br. (2n = 32, *R. nasturtium-aquaticum*, Eur.) – salad pl. grown in inundated beds, local folk medic. Ireland; both natur. cosmop., their hybrid (i.e. back-cross, '*N.* × *sterile* (Airy Shaw) Oefelein', '*R.* × *s.*', brown cress, 2n = 48) usu. in anthropogenic habitats, some seed set

Nastus Juss. Gramineae (V 6). Excl. *Chloothamnus, Oreiostachys*, 20 Madag., Réunion

nasubi *Solanum melongena*

Natal grass *Melinis repens*; **N. indigo** *Indigofera arrecta*; **N. mahogany** *Kiggelaria africana, Trichilia emetica*; **N. plum** *Carissa macrocarpa*

Nathaliella B. Fedtsch. Scrophulariaceae (Scroph.). 1 C As.: ***N. alaica*** B. Fedtsch. R: BZ 17(1932)327

Natsiatopsis Kurz. Icacinaceae. 1 Myanmar, Yunnan: ***N. thunbergiifolia*** Kurz

Natsiatum Buch.-Ham. ex Arn. Icacinaceae. 1 E Himal. to SE As.: ***N. herpeticum*** Buch.-Ham. ex Arn. R: FOC 11(2008)514

natto *Glycine max*

Nauclea L. Rubiaceae (I 2). Incl. *Burttdavya, Sarcocephalus*, 12 OW trop. R: Blumea 24(1978)325. ***N. diderrichii*** (De Wild. & T. Durand) Merr. (badi, bilinga, opepe, W Afr.) –

timber resistant to borers & used for harbour work & yam mortars, medic. but the alk. a cumulative heart-poison; ***N. latifolia*** Sm. (*S. l.*, *S. esculentus*, Afr. or Guinea peach, trop. Afr.) – fr. ed. with apple taste, seeds germ. best when passed through baboons, used in diabetes treatment in Congo & as chewing-stick in Nigeria; ***N. orientalis*** (L.) L. (*N. cordata*, kanluang, Leichhardt (pine), Indomal.), etc. – termite-resistant trade timber, fr. ed.

Naucleaceae Wernham = Rubiaceae

Naucleopsis Miq. Moraceae (III). 22 trop. Am. R: FN 7(1972)104, 83(2001)269. Arrow-poisons (cardenolides)

Naudinia Planch. & Linden. Rutaceae (I 8). 1 Colombia: ***N. amabilis*** Planch. & Linden – (C)

Naufraga Constance & Cannon. Umbelliferae (I?). 1 Majorca: ***N. balearica*** Constance & Cannon – closest relations in NZ & Chile

Nauplius (Cass.) Cass. = *Asteriscus*

Nautilocalyx Linden. Gesneriaceae (II 5c). Excl. *Centrosolenia*, 35 trop. Am. (not SE Brazil). R: Selbyana 5(1978)29. Cult. orn.

Nautochilus Bremek. = *Ocimum*

Nautonia Decne. Apocynaceae (Va; Asclepiadaceae III 1). 1 S Braz.: ***N. nummularia*** Decne

Nautophylla Guillaumin = *Logania*

naval stores all prods derived from resins of *Pinus* spp.: soap, paint, varnish, shoe-polish etc.; orig. those for ships being rosin, turpentine, pitch & tar

Navarretia Ruíz & Pavón. Polemoniaceae (III). 31 W N Am., Chile & Arg. (1). Annuals with var. capsule dehiscence, some irreg., ***N. filicaulis*** (A. Gray) Greene (W N Am.) with both loculicidal & septicidal dehiscence. Cult. orn. incl. ***N. squarrosa*** (Eschsch.) Hook. & Arn. (skunkweed, W N Am., natur. GB) – foetid

navelwort *Umbilicus rupestris; Hydrocotyle* spp.

Navia Mart. ex Schultes & Schultes f. Bromeliaceae (1). 98 N S Am. R: FN 14(1974)451, AMBG 73(1986)703. Wind-poll.

Nayariophyton Paul = *Hibiscus* (but see BJ 110(1988)43)

nazingu *Mitragyna* spp.

Nealchornea Huber. Euphorbiaceae (Euph.-Stom.). 2 Upper Amazon. ***N. yapurensis*** Huber – fish-poison

Neamyza Tieghem = *Peraxilla*

Neanotis W. Lewis. Rubiaceae (IV 15). 31 trop. As. to Aus. R: AMBG 53(1966)32. Some smelling of faeces but lvs used as vegetable

Neanthe P. Browne = ? Leguminosae

Neanthe Cook = *Chamaedorea*

Neatostema I.M. Johnston. Boraginaceae (B.2.2). 1 Macaronesia to Medit. (inc. Eur.) & Iraq: ***N. apulum*** (L.) I.M. Johnston. R: JAA 34(1953)5

Nebelia Necker ex Sweet = *Brunia* (but see Strelitzia 9(2000)385)

Neblinaea Maguire & Wurd. Compositae (Stifft.). 1 Venez., Guyana: ***N. promontorium*** Maguire & Wurd. – shrub. R: MNYBG 9(1957)391

Neblinantha Maguire. Gentianaceae. 2 Guayana Highland. R: MNYBG 51(1989)29

Neblinanthera Wurd. Melastomataceae. 1 Venez.: ***N. cumbrensis*** Wurd. R: MNYBG 10, 5(1964)153

Neblinaria Maguire = *Bonnetia*

Neblinathamnus Steyerm. Rubiaceae (I 6). 3 Venez.

Necepsia Prain. Euphorbiaceae (Acal.-Pyc.-Necep.). 3 trop. Afr., Madag. R: BJBB 56(1986)179

Nechamandra Planch. (~ *Lagarosiphon*). Hydrocharitaceae. 1 Ind., SE As., natur. Sudan: ***N. alternifolia*** (Wight) Thw.

Neckera Hedw. Musci (Neckeraceae). 50 cosmop. Moss used to caulk Bronze & Iron Age boats

Neckia Korth. = *Sauvagesia*

Necramium Britton = *Miconia*

Necranthus Gilli = *Orobanche*

Nectandra Rolander ex Rottb. Lauraceae (I). Excl. *Damburneya*, c. 95 trop. Am. R: FN 60(1993). Some hard, heavy timbers (silverballi in Guyana) like *Ocotea*; ***N. elaiophora*** Barb.-Rodr. (Amaz.) – volatile oil a kerosene subs., also used in skin disease; ***N. pichurim*** (Kunth) Mez (pichurim or purchury bean, Braz.) – seeds medic. See also *Aniba*

nectarberry See *Rubus*

nectarine *Prunus persica* var. *nucipersica*

Nectaropetalum Engl. Erythroxylaceae. 8 trop. & S Afr., Madag. (1)

Nectaroscilla Parl. (~ *Scilla*). Asparagaceae (Hyacinthaceae). 1–2 E Medit.

Nectaroscordum Lindl. = *Allium*

Nectouxia Kunth (~ *Salpichroa*). Solanaceae. (IV). 1 Mex.: ***N. formosa*** Kunth. R: A.T. Hunziker, *Gen. Solan.* (2001)350

nedun *Pericopsis mooniana*

Neea Ruíz & Pavón. Nyctaginaceae (5). c. 80 S Florida to Bolivia. Alks incl. caffeine: ***N. parviflora*** Poeppig & Endl. (Peru) – dental preservative (blackening teeth), ***N. theifera*** Oersted (Braz.) – lvs used as tea (caparrosa), source of comm. black dye

Needhamiella L. Watson. Ericaceae (VII 5; Epacridaceae). 1 SW Aus.: ***N. pumilio*** (R. Br.) L. Watson. R: KB 18(1965)272

needle, Adam's *Yucca* spp. esp. *Y. filamentosa;* **n.bush** or **n. wood** *Hakea* spp.; **n. furze** or **whin** *Genista anglica;* **n. grass, Chilean** *Nassella neesiana;* **n.wood** *Schima wallichii*

neelakurinji *Strobilanthes kunthiana*

neem *Azadirachta indica*

Neeopsis Lundell. Nyctaginaceae (5). 1 Guatemala: ***N. flavifolia*** (Lundell) Lundell. R: Wrightia 5(1976)241

neep(s) *Brassica napus* Napobrassica Group, s.t. applied to *B. rapa* Rapifera Group

Neeragrostis Bush = *Eragrostis* (but see SCB 87(1997)35)

Neesenbeckia Levyns. Cyperaceae (II 7). 1 S Afr.: ***N. punctoria*** (Vahl) Levyns. R: JSAB 13(1947)74

Neesia Blume. Malvaceae (Hel.-Dur.; Bombacaceae). 8 W Mal. R: Reinw. 5(1961)483. Fr. with irritant hairs lost only when the skin is sloughed off. Light timbers. ***N. altissima*** (Blume) Blume – dried fr. hung above doors in Sumatra to ward off spirits

Neesiochloa Pilg. Gramineae (XXIX 7). 1 NE Braz.: ***N. barbata*** (Nees) Pilg. R: SCB 87(1997)35

Negria F. Muell. Gesneriaceae (II 4c). 1 Lord Howe Is.: ***N. rhabdothamnoides*** F. Muell. – tree to 9 m. R: FA 49(1994)341

Negripteridaceae Pichi-Serm. = Pteridaceae

Negripteris Pichi-Serm. = *Cheilanthes*

negrita *Sphaeralcea* spp.

negro pepper *Xylopia aethiopica*

Negundo Boehmer ex Ludwig = *Acer*

Neillia D. Don. Rosaceae (3). Incl. *Stephanandra,* c. 15 E Himal. to Jap. & W Mal. R: JAA 52(1971)142, Novon 16(2006)92. Some cult. orn. shrubs like *Spiraea,* esp. ***N. incisa*** (Thunb.) Oh (*S. i.*, E As.) – weeping branches of deeply incised lvs

Neisosperma Raf. = *Ochrosia*

Neja D. Don = *Erigeron*

Nekemias Raf. (~ *Ampelopsis*). Vitaceae. 9 E & SE As., E N Am. R: PhytoKeys 42(2014)12

Nelia Schwantes. Aizoaceae (V). 2 Namaqualand. R: H. E. K. Hartmann, *Ill. Handb. Succ. Pls,* Aiz. F–Z(2002)184. Fls permanently open

Nelmesia Veken. Cyperaceae (II 4). 1 N Congo: ***N. melanostachya*** Veken. R: BJBB 25(1955)143

Nelsia Schinz. Amaranthaceae (I 2). 2 S trop. & S Afr.

Nelsonia R. Br. Acanthaceae (I). 2 OW trop. (***N. canescens*** (Lam.) Spreng. weedy NW, salt subs.). R: Aliso 32(2014)25

Nelsonianthus H. Robinson & Brettell (~ *Senecio*). Compositae (Sen.-Tuss.). 1 Mex., 1 Guatemala – epiphytic shrubs

Nelumbium Juss. = *Nelumbo*

Nelumbo Adans. Nelumbonaceae. 1 S & E N Am., WI to Colombia (*N. lutea*), 1 Lower Volga, S & SE As. to trop. Aus. (*N. nucifera*), more widespread in Cretaceous; sometimes treated as subspp. R: BBP 68(1994)421, M. Griffiths (2009) *The lotus quest.* ***N. lutea*** (Willd.) Pers. (*N. pentapetala,* water chinquapin, American lotus) – fls yellow, seeds & rhiz. ed.; ***N. nucifera*** Gaertn. (*N. speciosa,* (anc.) kamal, sacred lotus, Egyptian bean) – fls red, sacred in Ind., Tibet & China, being the 'padma' from which lotus motif of As. derived, in Hindu religion considered to have sprung from navel of the god Vishnu & to have given birth to Brahma (sacred colour red) creator of the world, introd. to Egypt c. 500 BC but no longer in Nile, receptacle (oft. seen in Chinese food) heating (to 20° above ambient) & volatilizing odour attractive to pollinators held overnight (cf. Araceae), when dried a source (in China) of antihaemorrhagic quercetin & used in birth-control (suppresses quercetin), 'seeds' (fr.), used in 'moon cakes', viable for up to 1300 years in river-muds, much grown in E for rhiz. (source of Chinese arrowroot, hasu) which can grow to 20 m in 1 season, lvs (water absorbed & gas-exchange through omphalos) used to wrap food for steaming, fibre from petioles for textiles, also cult. orn. (c. 600 cvs in China alone, incl. hybrids with *N. lutea*) with red- or white-flowered (incl. 'doubles' much used in temples) scented cvs. Water-resistant leaf-surface (due to bumps spiked with tiny hairs such that water droplets buoyed up by airpockets below) imitated ('biomimicry') in graffiti-resistant, self-cleaning paint (Lotusan) & monsoon mackintoshes

Nelumbonaceae A. Rich. Magnoliidae – Proteales. 1/2 E As. & E N Am. Aquatic rhizomatous herbs with aporphin alks incl. quercetin (sesquiterpene alks in Nymphaeaceae) alks; articulated laticifers present, vascular bundles 'scattered', vessels only in r. Phyllotaxy unique, leaf-primordia in 3s with 1 scale-leaf on underside of rhiz. 1 on top next to 1 foliage-leaf, the lower orig. wrapped around term. bud but split on growth, the upper scale-leaf wrapped around base of petiole & a basal stip.; branches arise from axils of foliage lvs. Lvs peltate, concave, usu. held above water on long petioles, venation dichotomous. Fl. solit., ebracteolate, from axil of upper scale-leaf, held above water, beetle-poll., bisexual; P c. 22–30, in spirals, outer 2 K-like, rest in ± 2 series, the outer less consp. than inner, A c. 200–300, in spirals with slender filaments, 4 introrse-latrorse pollen-sacs on narrow laminar connective & term. thermogenic appendage, staminodes 0, pollen grains 3-colpate, $\underline{G}$ (2-)10–30 in 2–4 ± distinct whorls, individually sunken in large spongy receptacle, each with 1(2) ventral-apical, anatropous, bitegmic ovule(s). Fr. comprising separate hard-walled nuts loose in accrescent receptacle; seed 1 without perisperm & ± 0 endosperm, cotyledons 2 (arising as separate lobes of annular primordium) basally connate, forming a sheath around green plumule; radicle non-functional (r. adventitious). n = 8

Genus: *Nelumbo*

Form. incl. in Nymphaeales but electron-microscopic work suggested no affinity with N. or Magnoliales, but aporphin alks in cuticular waxes suggested alliance with Ranunculales, though DNA (v. surprisingly) places fam. in Proteales. Seeds ed.; epicuticular wax crystalloids allow water droplets to remove dust etc, now mimicked in surfaces for shop windows, cars etc.

Nemacaulis Nutt. (~ *Eriogonum*). Polygonaceae (Erio.-Erio.). 1 SW N Am.: ***N. denudata*** Nutt. R: Madroño 27(1980)101

Nemacianthus M.A. Clem. & D.L. Jones = *Acianthus*

Nemacladus Nutt. Campanulaceae (II). Excl. *Parishella*, 13 SW N Am. R: AMN 22(1939)522

Nemaconia Knowles & Westc. (~ *Ponera*). Orchidaceae (V 13d). 6 trop. Am.

Nemaluma Baill. = *Pouteria*

Nemastylis Nutt. Iridaceae (VII 5). 4 S US to C Am. Lvs plicate. Some cult. orn.

Nematanthus Schrad. Gesneriaceae (II 5c). 31 S & SE Braz. R: Candollea 39(1984)297, Selbyana 25(2005)222. Epiphytic shrubs, cult. orn. housepls (*Hypocyrta* spp.) esp. ***N. wettsteinii*** (Fritsch) H. Moore (SE Braz.) with many cvs

Nematolepis Turcz. Rutaceae (I 3). 7 Aus. R: Nuytsia 12(1998)277

Nematopoa C. Hubb. Gramineae (Arund.). 1 Zimbabwe: ***N. longipes*** (Stapf & C. Hubb.) C. Hubb.

Nematopteris Alderw. = *Scleroglossum*

Nematosciadium H. Wolff = *Arracacia*

Nematostemma Choux = *Cynanchum*

Nematostylis Hook.f. (~ *Alberta*). Rubiaceae (III 5). 1 Madag.: ***N. anthophylla*** (A. Rich.) Baill. R: BJBB 54(1984)348

Nematuris Turcz. = *Ampelamus*

Nemcia Domin = *Gastrolobium*

Nemesia Vent. Scrophulariaceae (Hemim.). 60 trop. (few) & S Afr. (Cape 27 – R: Strelitzia 9(2000)656). Cult. orn. annuals derived from ***N. strumosa*** Benth. (SW Cape) & ***N. versicolor*** E. Meyer ex Benth. (? = *N. affinis* Benth., S Afr.), ***N. caerulea*** Hiern introducing scent, many cvs

Nemopanthus Raf. = *Ilex*

Nemophila Nutt. Boraginaceae (Hydrophyllaceae-Hydro.). 19 W & SE (2) N Am. R: UCPB 19(1941)341. Cult. orn. annuals esp. ***N. menziesii*** Hook. & Arn. (baby-blue eyes, Calif., Oregon) – fls blue with white centres

Nemosenecio (Kitam.) R. Nordenstam (~ *Senecio*). Compositae (Sen.- Tuss.). 6 China, Jap. R: KB 39(1984)262

Nemuaron Baill. Atherospermataceae. 1 New Caled.: ***N. vieillardii*** (Baill.) Baill.
Nemum Desv. (~ *Scirpus*). Cyperaceae (II 4). 8 trop. Afr. R: BSRBB 141(2008)175
Nenax Gaertn. Rubiaceae (IV 13). 11 S Afr. esp. SW Cape
Nenga H. Wendl. & Drude. Palmae (V 14b). 5 SE As. to Mal. R: Principes 27(1983)55
Nengella Becc. = *Hydriastele*
Neoabbottia Britton & Rose = *Leptocereus*
Neoacanthophora Bennet = *Aralia*
Neoachmandra W. de Wilde & Duyfjes = *Zehneria* (but see Blumea 51(2006)12)
Neoalsomitra Hutch. Cucurbitaceae (I). 12 Indomal. to Aus. & W Pac. R: Blumea 48(2003)100
Neoancistrophyllum Rauschert = *Laccosperma*
Neoapaloxylon Rauschert (*Apaloxylon*). Leguminosae (I 2). 3 Madag. Bark fibre for rope, r. ed.
Neoastelia J. Williams (~ *Astelia*). Asteliaceae (Liliaceae s.l.). 1 NE NSW: ***N. spectabilis*** J. Williams. R: FA 45(1987)173
Neoathyrium Ching & Z.R. Wang = *Cornopteris*
Neoaulacolepis Rauschert = *Aniselytron*
Neobaclea Hochr. Malvaceae (Malv.-Malv.). 1 temp. Arg.: ***N. spirostegia*** Hochr.
Neobakeria Schltr. = *Daubenya*
Neobalanocarpus Ashton (? *Hopea* × *Shorea*),. Dipterocarpaceae (Dipt.-Shor.). 1 Peninsular Thailand, Malay Pen.: ***N. heimii*** (King) Ashton (chengal) – durable timber for boats & houses, hardest & finest in Malaysia. R: FM I,9(1982)388
Neobartlettia R. King & H. Robinson = *Bartlettina*
Neobartlettia Schltr. = *Palmorchis*
Neobassia A.J. Scott = *Eremophora* + *Sclerolaena* spp. (but see FA 4(1984)221)
Neobathiea Schltr. Orchidaceae (V 16d). 5 Madag. Cult. orn.
Neobaumannia Hutch. & Dalz. = *Knoxia*
Neobeguea J. Leroy. Meliaceae (I). 3 Madag. R: Adansonia 16(1976)170
Neobennettia Senghas = *Lockhartia*
Neobenthamia Rolfe. Orchidaceae (V 13). 1 Uluguru & Nguru Mts, Tanzania: ***N. gracilis*** Rolfe – cult. orn. straggly
Neobertiera Wernham. Rubiaceae (I 6). 4 NE S Am. R: Phytotaxa 206(2015)118
Neobesseya Britton & Rose = *Escobaria*
Neoblakea Standl. Rubiaceae (?III 3). 1 Venez.: ***N. venezuelensis*** Standl.
Neobolusia Schltr. (~ *Brachycorythis*). Orchidaceae (IV 4d). 3 trop. E & S Afr.
Neobouteloua Gould (~ *Chondrosum*). Gramineae (XXIX). 2 Argentina & Chile
Neoboutonia Muell.-Arg. Euphorbiaceae (Al.-Neo.). 3 trop. Afr. Gap colonists
Neobracea Britton. Apocynaceae (II e). 4 Cuba, Bahamas
Neobreonia Ridsd. = *Breonia*
Neobrittonia Hochr. Malvaceae (Malv.-Malv.). 1 Mex. to Panamá: ***N. acerifolia*** (G. Don) Hochr.
Neobuchia Urb. Malvaceae (Bomb.; Bombacaceae). 1 Haiti: ***N. paulinae*** Urb.
Neoburttia Mytnik & al. = *Polystachya* (but see PBJ 56(2011)46)
Neobuxbaumia Backeb. (~ *Cephalocereus*). Cactaceae (III 1). 8 Mex. Some bat-poll. incl. ***N. mezcalaensis*** (Bravo) Backeb. – androdioec. (rare in fam.)
Neobyrnesia J.A. Armstr. Rutaceae (I 3). 1 N Aus.: ***N. suberosa*** J.A. Armstr. R: FA 26(2013)336
Neocabreria R. King & H. Robinson (~ *Eupatorium*). Compositae (Eup.-Crit.). 5 trop. S Am. R: MSMMBG 22(1987)371
Neocaldasia Cuatrec. = ? (Compositae)
Neocallitropsis Florin. Cupressaceae. 1 E New Caled.: ***N. pancheri*** (Carrière) Laubenf. – oil for scenting soaps etc. marketed as 'oil of araucaria'. R: A. Farjon, *Monogr. Cupressaceae & Sciadopitys* (2005)483
Neocalyptrocalyx Hutch. (~ *Capparis*). Capparaceae. 7 S Am. R: HPB 13(2008)109
Neocarya (DC.) Prance (~ *Parinari*). Chrysobalanaceae (2). 1 W Afr.: ***N. macrophylla*** (Sabine) Prance (gingerbread plum) – fr. ed. (potential fr. tree in N Aus.), timber good. R: FOW 9(2003)195
Neocaspia Tzvelev = *Salsola*
Neocentema Schinz. Amaranthaceae (I 2). 1 Somalia, 1 Tanzania
Neochamaelea (Engl.) Erdtman = *Cneorum*
Neocheiropteris Christ = ? *Lepisorus*

Neochevalierodendron Léonard. Leguminosae (I 2). 1 Gabon: ***N. stephanii*** (A. Chév.) Léonard
Neocinnamomum H. Liu (~ *Cinnamomum*). Lauraceae (I). 6 S China to SE As.
Neoclemensia Carr = *Gastrodia*
Neocodon Kolak. & Serdyuk. = *Campanula*
Neocogniauxia Schltr. Orchidaceae (V 13c). 2 WI. R: C.L. Withner, *The Cattleyas & their relatives* IV(1996)85
Neocollettia Hemsl. Leguminosae (III 18). 1 Myanmar, C Java: ***N. wallichii*** (Kurz) Schindler (*N. gracilis*) – geocarpic with strong r. on gynophore poss. pulling fr. down into earth
Neoconopodium (Kozo-Polj.) Pim. & Kljuykov = *Kozlovia*
Neocouma Pierre. Apocynaceae (I d). 2 N S Am. R: AUWP 87–1(1987)
Neocracca Kuntze = *Coursetia*
Neocribbia Szlach. = *Solenangis*
Neocryptodiscus Hedge & Lamond = *Prangos*
Neocuatrecasia R. King & H. Robinson (~ *Eupatorium*). Compositae (Eup.-Gyp.). 12 Peru, Bolivia. R: Novon 12(2002)388
Neocupressus Laubenf. = *Cupressus*
× **Neocupropsis** Laubenf. = × *Cuprocyparis*
Neocussonia Hutch. = *Schefflera*
Neodeutzia Small = *Deutzia*
Neodielsia Harms = *Astragalus*
Neodillenia Aymard. Dilleniaceae (II). 3 Amazonia. R: HPB 10(1997)121
Neodissochaeta Bakh.f. = *Dissochaeta*
Neodistemon Babu & A.N. Henry. Urticaceae (III). 1 Indomal.: ***N. indicum*** (Wedd.) Babu & A.N. Henry. R: T 19(1970)651
Neodonnellia Rose = *Tripogandra*
Neodregea C. Wright = *Wurmbea* (but see Strel. 10(2000)589)
Neodriessenia Nayar. Melastomataceae. 6 Borneo. R: BJ 106(1985)1
Neodryas Reichb.f. = *Cyrtochilum*
Neodunnia R. Viguier = *Millettia*
Neodypsis Baill. = *Dypsis*
Neoeplingia Ramam. & al. = *Lepechinia* (but see BSBM 43(1982)61)
Neoescobaria Garay = *Trichopilia*
Neofabricia J. Thompson (~ *Leptospermum*). Myrtaceae (II 14). 3 N Queensland. R: Telopea 3(1989)291
Neofinetia Hu = *Vanda*
Neofranciella Guillaumin = *Atractocarpus*
Neogaerrhinum Rothm. (~ *Antirrhinum*). Plantaginaceae (Ant.-Ant.; Scrophulariaceae s.l.). 2 SW N Am. R: D.A. Sutton, *Rev. Antirrh.* (1988)479
Neogaillonia Lincz. = *Plocama*
Neogardneria Schltr. ex Garay. Orchidaceae (V 12j). 1 NE trop. Am.: ***N. murrayana*** (Hook.) Garay. R: Gen. Orch. 5(2009)513
Neoglaziovia Mez. Bromeliaceae (3). 3 E Braz. R: FN 14(1979)2036. ***N. variegata*** (Arruda) Mez – source of caroa fibre used for nets & suggested for paper & artificial silk
Neogleasonia Maguire = *Bonnetia*
Neogoezia Hemsl. Umbelliferae (III 5). 5 Mex. R: OB 92(1987)59
Neogontscharovia Lincz. (~ *Acantholimon*). Plumbaginaceae (II). 3 C As., Afghanistan
Neogoodenia C. Gardner & A.S. George = *Goodenia*
Neogriseocereus Guiggi = *Stenocereus*
Neoguillauminia Croizat (~ *Euphorbia*). Euphorbiaceae (Euph.-Neo.). 1 New Caled.: ***N. cleopatra*** (Baill.) Croizat – P-like structures develop late in male fls
Neoguarea (Harms) Koenen & J. de Wilde (~ *Guarea*). Meliaceae (II-Trich.). 1 C Afr.: ***N. glomerulata*** (Harms) Koenen & J. de Wilde. R: PEE 145(2012)233
Neogunnia Pax & K. Hoffm. = *Gunniopsis*
Neogyna Reichb.f. = *Coelogyne*
Neohallia Hemsl. = *Justicia*
Neoharmsia R. Viguier. Leguminosae (III 2). 2 W & NW Madag.
Neohemsleya Penn. Sapotaceae (III). 1 Tanz.: ***N. usambarensis*** Penn. R: T.D. Pennington, *S.* (1991) 175
Neohenricia L. Bolus. Aizoaceae (V). 2 S Afr. R: H. E. K. Hartmann, *Ill. Handb. Succ. Pls, Aiz. F–Z*(2002)186

Neohenrya Hemsl. = *Vincetoxicum*
Neohintonia R. King & H. Robinson = *Koanophyllum*
Neoholmgenia W.L. Wagner & Hoch (*Holmgrenia*). Onagraceae. 2 W N Am. R: Novon 19(2009)131
Neoholstia Rauschert = *Tannodia*
Neohouzeaua A. Camus (~ *Schizostachyum*). Gramineae (V 5). 7 trop. As. ***N. dullooa*** (Gamble) A. Camus (*S. d.*, SE As.) – fls on 14–17-yr cycle
Neohuberia Ledoux = *Eschweilera*
Neohumbertiella Hochr. = *Humbertiella*
Neohusnotia A. Camus = *Acroceras*
Neohymenopogon Bennet (~ *Dunnia*). Rubiaceae (IV 4). 3 SE As.
Neohyptis J.K. Morton = *Plectranthus*
Neojatropha Pax = *Mildbraedia*
Neojeffreya Cabrera = *Pterocaulon* (but see Hickenia 1(1978)160)
Neojobertia Baill. Bignoniaceae 3). 2 NE Braz. R: AMBG 99(2014)454
Neokochia (Ulbr.) G.L. Chu & S. Sand. (~ *Kochia*). Amaranthaceae (Chenopodiaceae). 1–2 N Am.
Neokoehleria Schltr. = *Comparettia*
Neolabatia Aubrév. = *Pouteria*
Neolamarckia Bosser ('*Anthocephalus*'). Rubiaceae (I 2). 2 Indomal. to trop. Aus. R: BMNHN 4,6(1984)247. Fast-growing colonist trees (Roux's Model). ***N. cadamba*** (Roxb.) Bosser ('*Anthocephalus chinensis*', cadamba, kadam(b)(a), kedam, laran) – first planted Indonesia 1933 but growth sensitive to soil moisture, timber used for disposable chopsticks in Malaysia, matchboxes, tea-chests & considered for pulp
Neolauchea Kraenzlin = *Isabelia*
Neolaugeria Nicolson. Rubiaceae (III 3). 35 WI. R: Brittonia 31(1979)119
Neolehmannia Kraenzlin = *Epidendrum*
Neolemonniera Heine. Sapotaceae (I 3). 5 W trop. Afr.
Neolepia W. Weber = *Lepidium*
Neolepisorus Ching = *Microsorum*
Neoleptopyrum Hutch. = *Leptopyrum*
Neoleroya Cavaco = *Pyrostria*
Neolindenia Baill. = *Louteridium*
Neolitsea Merr. (~ *Litsea*). Lauraceae (II). c. 80 Indomal. to E As. & Aus. (3). Dioec.; alks. Some medic. & oil-sources esp. ***N. sericea*** (Blume) Koidz. (*L. glauca*, China & Jap.) – oil for burning & soap; ***N. zeylanica*** (Nees) Merr. (Indomal.) – fr. used as 'peas' in peashooters by Malay boys
Neolloydia Britton & Rose. Cactaceae (III 9). 1 Texas & Mex.: ***N. conoidea*** (DC.) Britton & Rose – cult. orn. R: Bradleya 4(1986)1.
Neololeba Widjaja (~ *Bambusa*). Gramineae (V 4). 5 C Mal. to Aus. R: Reinw. 11(1997)112
Neolophocarpus Camus = *Schoenus*
Neolourya L. Rodrigues = *Peliosanthes*
Neoluederitzia Schinz. Zygophyllaceae (IV). 1 S Namibia: ***N. sericeocarpa*** Schinz. R: Strelitzia 10(2000)564
Neoluffa Chakrav. = *Siraitia*
Neomacfadya Baill. = *Fridericia*
Neomandonia Hutch. = *Tradescantia*
Neomangenotia J. Leroy = *Commiphora*
Neomarica Sprague (~*Trimezia*). Iridaceae (VII 4). c. 30 trop. Am. R: Novon 11(2001)377, HPB 14(2009)97. Cult. orn. herbs with fugacious fls & s.t. with scarlet seeds (usu. brown as in I. generally – Goldblatt), sometimes viviparous with plantlets in old infls
Neomartinella Pilg. = *Eutrema* (but see Novon 10(2000)337)
Neomazaea Krug & Urb. = *Mazaea*
Neomezia Votsch (~ *Deherainia*). Primulaceae (Theophrastaceae). 1 Cuba: ***N. cubensis*** (Radlk.) Votsch
Neomicrocalamus Keng f. (~ *Racemobambos*) Gramineae (V 4). 5 E As.
Neomillspaughia S.F. Blake (~ *Podopterus*). Polygonaceae (Erio.-Cocc.). 2 C Am.
Neomirandea R. King & H. Robinson (~ *Eupatorium*). Compositae (Eup.-Heb.). 27 Mex. to Ecuador. R: MSBMBG 22(1987)415. Heterogeneous. Small trees, shrubs & perenn. herbs, oft. epiphytes, ***N. araliifolia*** (Less.) R. King & H. Robinson habit of strangling fig
Neomitranthes Legrand (~ *Calyptrogenia*). Myrtaceae (II 10). 17 E Braz.

Neomolina Hellwig = *Baccharis*

Neomolinia Honda = *Diarrhena*

Neomoorea Rolfe. Orchidaceae (V 12g). 1 Panamá to Ecuador: ***N. wallisii*** (Reichb.f.) Schltr. – cult. orn. epiphyte. R: Gen. Orch. 5(2009)180

Neomortonia Wiehler. Gesneriaceae (II 5c). 3 C Am. R: Selbyana 6(1982)194. ?Heterogeneous

Neomuellera Briq. = *Plectranthus*

Neomussaenda Tange (~ *Greenea*). Rubiaceae (II 6). 2 Borneo. R: NJB 14(1994)495

Neomyrtus Burret. Myrtaceae (Myrt.). 1 NZ: ***N. pedunculata*** (Hook.f.) Allan

Neonauclea Merr. Rubiaceae (I 2). c. 65 Indomal. to China (1: ***N. tsaiana*** S.Q. Zow – timber). R: Blumea 34(1989)177. Prob. incl. *Ladekia, Myrmeconauclea, Ochreinauclea*. Some rheophytes, some myrmecophytes (W of New Guinea)

Neonavajoa Doweld = *Pediocactus*

Neonelsonia J. Coulter & Rose. Umbelliferae (III 5). 2 Mex. to N S Am.

Neonesomia Urbatsch & R. Roberts. Compositae (Ast.). 2 Texas, NW Mex. R: Sida 21(2004)252

Neonicholsonia Dammer (~ *Prestoea*). Palmae (V 10). 1 C Am.: ***N. watsonii*** Dammer. R: FN 72(1996)79

Neonotonia Lackey (~ *Glycine*). Leguminosae (III 18). 2 OW trop.

Neooreophilus Archila = *Lepanthes*

Neopalissya Pax = *Necepsia*

Neopallasia Polj. = *Artemisia* (but see APS 18(1980)86)

Neopanax Allan = *Pseudopanax*

Neoparrya Mathias. Umbelliferae (III 8). 2 SW US

Neopatersonia Schönl. = *Ornithogalum*

Neopaulia Pim. & Kljuykov = *Paulita*

Neopaxia O. Nilsson = *Montia*

Neopectinaria Plowes = *Stapeliopsis*

Neopentanisia Verdc. Rubiaceae (III 8). 2 S trop. Afr. R: BJ 110(1989)548

Neopetalonema Brenan = *Gravesia*

Neophloga Baill. = *Dypsis*

Neopicrorhiza Hong = *Picrorhiza*

Neopilea Leandri = *Pilea*

Neoplatytaenia Geld. = *Semenovia*

Neopometia Aubrév. = *Pradosia*

Neoporteria Britton & Rose = *Eriosyce*

Neopreissia Ulbr. = *Atriplex*

Neopringlea S. Watson. Salicaceae (Flacourtiaceae). 3 Mex., Guatemala. R: SB 8(1983)430

Neoptychocarpus Buchheim. Salicaceae (Flacourtiaceae). 2 trop. S Am. R: FN 22(1980)254

Neoraimondia Britton & Rose. Cactaceae (III 1). 2 Peru, N Chile, Bolivia. Cult. orn. columnar cacti

Neorapinia Mold. = *Vitex*

Neoraputia Emmerich ex Kallunki (~ *Raputia*). Rutaceae (I 8). 6 Peru (1), Braz. R: Britt. 61 (2009)28. Not cauliflorous like *R.*

Neorautanenia Schinz. Leguminosae (III 18). 3 S trop. Afr. ***N. brachypus*** (Harms) C.A. Sm. (*Dolichos seineri*) – giant swollen caudex to 150 kg; ***N. mitis*** (A. Rich.) Verdc. – alleged to kill bilharzia-carrying snails

Neoregelia L.B. Sm. Bromeliaceae (3). 97 trop. & warm S Am. R: FN 14(1979)1533. Subg. *Hylaeaicum* prob. distinct genus. Infls sunk in centre of rosette with brightly coloured inner lvs. Many cult. orn. spp. & hybrids esp. ***N. farinosa*** (Ule) L.B. Sm. (Braz.) with crimson inner lvs

Neoregnellia Urb. (~ *Helicteres*). Malvaceae (Hel.; Sterculiaceae). 1 Cuba, Hispaniola: ***N. cubensis*** Urb.

Neorhine Schwantes = *Rhinephyllum*

Neorites L.S. Sm. Proteaceae (V 1). 1 NE Aus.: ***N. kevedianus*** L.S. Sm. R: FA 16(1995)352

Neoroepera Muell. Arg. & F. Muell. Picrodendraceae (Cal.-Pseud.; Euphorbiaceae). 2 NE Aus. R: Austrobaileya 3(1992)618

Neorosea Hallé = *Tricalysia*

Neorudolphia Britton (~ *Rhodopis*). Leguminosae (III 18). 1 Puerto Rico: ***N. volubilis*** (Willd.) Britton

Neoruschia Catharino & V. Castro = *Gomesa*

Neosabicea Wernham = *Manettia*
Neoschimpera Hemsl. = *Amaracarpus*
Neoschischkinia Tzvelev = *Agrostis*
Neoschmidia T. Hartley (~ *Eriostemon*). Rutaceae (I 2). 2 New Caled. R: Adans. III,25(2003)7
Neoschroetera Briq. = *Larrea*
Neoschumannia Schltr. Apocynaceae (Vb; Asclepiadaceae III 5). 3 (disjunct) trop. Afr. R: Phytotaxa 77(2013)19
Neosciadium Domin. Araliaceae (Hydroc.; Umbelliferae s.l.). 1 SW Aus.: ***N. glochidiatum*** (Benth.) Domin. R: OBZ 115(1968)28
Neoscortechinia Pax. Euphorbiaceae (Cheil.). 6 Myanmar, Nicobar Is., Mal., Solomon Is. R: Blumea 39(1994)301
Neosepicaea Diels. Bignoniaceae (1). 4 Moluccas, New Guinea, Queensland
Neoshirakia Esser = *Gymnanthes* (but see Blumea 43(1998)129)
Neosinacalamus Keng f. = *Bambusa*
Neosloetiopsis Engl. = *Streblus*
Neosparton Griseb. Verbenaceae (Neo.). 3 temp. S Am. R: Darw. 11(1957)167
Neosprucea Sleumer (~ *Hasseltia*). Salicaceae (Flacourtiaceae). 4 trop. Am. R: SBM 85(2008)19
Neostapfia Davy (~ *Anthocloa*). Gramineae (XXIX 2). 1 Calif.: ***N. colusana*** (Davy) Davy – covered in brown viscid glands, endangered pl. of spring pools
Neostapfiella A. Camus. Gramineae (Chlorid.). 3 Madag. R: NS 11(1943)189
Neostenanthera Exell. Annonaceae (III 5). 5 trop. Afr. R: SB 39(2014)20
Neostrearia L.S. Sm. Hamamelidaceae (I 1). 1 NE Aus.: ***N. fleckeri*** L.S. Sm.
Neostricklandia Rauschert = *Phaedranassa*
Neotainiopsis Bennet & Raiz. = *Eriodes*
Neotatea Maguire (~ *Bonnetia*). Calophyllaceae (Bonnetiaceae; Guttiferae I1.). 4 NE S Am.
Neotchihatchewia Rauschert = *Tchihatchewia*
Neotessmannia Burret. Muntingiaceae (Tiliaceae s.l.). 1 E Peru: ***N. uniflora*** Burret – coll. once. R: T 47(1998)39
Neothorelia Gagnepain. Resedaceae. 1 SE As.: ***N. laotica*** Gagnepain – excl. from Capparaceae (K6, C6, G 3-loc. etc.). R: TFB 33(2005)229
Neothymopsis Britton & Millsp. = *Thymopsis*
Neotina Capuron = *Tina*
Neotinea Reichb.f. Orchidaceae ((IV 4d). 4 Macaronesia to W Eur. & Medit. incl. GB (2). R: H. Kretzschmar & al. *The orchid.....N.* ed. 2(2007)191. Usu. on alk. soils. ***N. maculata*** (Desf.) Stearn (*N. intacta*) – prob. autogamous
Neotorularia Hedge & Léonard (*Torularia*, ~ *Braya*). Cruciferae (26). 6 Medit. (Eur. 2) to C As. & Afghanistan
Neotreleasea Rose = *Tradescantia*
Neotrewia Pax & K. Hoffm. = *Mallotus*
Neottia Guett. Orchidaceae (V 1). Incl. *Listera*, c. 60 N temp. (Eur. 3) & boreal. Twayblades (*L.*), some mycotropic throughout life (*N. s.s.*), incl. ***N. nidus-avis*** (L.) Rich., bird's-nest orchid (Eur.); ***N. listerodes*** Lindl. (China) – ant-poll.; ***N. ovata*** (L.) Bluff & Fingerh. (*L. o.*, Eur. to Siberia) – higher percentage of B chromosomes at higher latitudes & in wetter habitats, grows 13–15 yrs before flowering, rostellum when touched splits open explosively, ejecting viscous fluid that glues pollinia to visiting insect
Neottianthe (Reichb.) Schltr. = *Hemipilia*
Neotuerckheimia J.D. Sm. = *Amphitecna*
Neoturczaninowia Kozo-Polj. Umbelliferae. Spp.?
Neotysonia Dalla Torre & Harms. Compositae (Gnap.-Ang.). 1 SW Aus. (Mt Narryer): ***N. phyllostegia*** (F. Muell.) Paul G. Wilson. R: OB 104(1991)114
Neourbania Fawcett & Rendle = *Maxillaria*
Neo-uvaria Airy Shaw. Annonaceae (IV 7). c. 6 SE As. to C Mal.
Neoveitchia Becc. Palmae (V 14d). 1 Vanuatu, 1 Fiji (***N. storckii*** (H.A. Wendl.) Becc. – almost extinct)
Neowawraea Rock = *Flueggea*
Neowerdermannia Frič (~ *Gymnocalycium*). Cactaceae (III 5). 2 S Am. Cult. orn.
Neowilliamsia Garay = *Epidendrum*
Neowimmeria Degener & I. Degener = *Lobelia*
Neowollastonia Wernham = *Melodinus*

Neowormia Hutch. & Summerh. = *Dillenia*
Neoxythece Aubrév. & Pellegrin = *Pouteria*
Neozenkerina Mildbr. = *Staurogyne*
Nepa Webb = *Stauracanthus*
Nepal paper *Daphne bholua*
Nepenthaceae Dumort. Magnoliidae – Caryophyllales. 1/c. 140 Madag. & Seychelles to Aus. & New Caled. Shrubby dioec. carnivorous pls, oft. climbing &/or epiphytic; cortical &/or medullary vasc. bundles oft. present. Lvs in spirals, abaxially circinate, comprising ± distinct winged petiole, strap-shaped lamina & term. tendril by which pl. climbs, its apex usu. in form of a pitcher with a lid projecting over mouth; multicellular nectar-glands & peltate hydathodes on stem & lvs with digestive glands in pitchers partly filled with digestive fluid; stip. 0. Bracts & bracteoles 0. Fls small, reg., in racemes or thyrses; K (3)4, imbricate & usu. free, with nectar-glands within, C 0, A (4–)8–25, the filaments united into central column, pollen grains in tetrads, $\underline{G}$ ((3)4) with as many loc. & ± sessile stigma & ∞ anatropous, bitegmic ovules in many rows on axile placentas. Fr. a loculicidal capsule; seeds ∞ filiform with straight embryo in starchy as well as oily & proteinaceous endosperm. n = 40
Genus: *Nepenthes*

Nepenthes L. Nepenthaceae. c. 140 Madag. (2), Seychelles (1), Sri Lanka (1), Assam (1), SE As. (8), to N Queensland (3), Solomon Is. (1) & New Caled. (1; Mal. – R: FM 15(2000), Sumatra 36+, Borneo 31+; pollen from Eur. Tertiary). R: Blumea 42(1997)1. Pitcher plants; sea level (some ± halophytic), to 3500 m (Mt Kinabalu, Borneo), 1 Mal. sp. myrmecophytic. Hybrids frequent in wild & cult., some involving up to 6 spp. Pitchers develop as intercalary invaginations below tendril tips, lips growing out below them, basal pitchers capturing creeping arthropods (mainly ants, though ***N. albomarginata*** Lobb ex Lindl. (W Mal.) up to 22 termites a minute), the later-formed upper ones (s.t. diff.-shaped) flying insects. Arthropods attracted by nectar, amines & bright pitcher colour (fluorescence), either fall from the lip (peristome) which has anisotropic cuticular cells and hygroscopic 'nectar' that cause tarsi to aquaplane, or down inner walls which have slippery cuticles (several thousand wax scales per cell like tiles snapping off stalks when touched) & slide into base of fluid-filled pitcher (fluid slightly salty taste) where they decompose & their nutrients absorbed, 10 pitchers found to cont. remains of 1994 arthropods (150 spp.). Some pitchers to 35 cm long & 18 cm across (***N. rajah*** Hook.f., Mt Kinabalu ultramafics, even cont. rats) large enough to hold 2 litres of fluid (potable & ? medic. before pitcher opens), as well as fungi, slime-moulds, protozoans, desmids, diatoms, rotifers, oligochaetes, crustaceans, larvae of mosquitoes & flies, even tadpoles (*Microhyla borneensis* [*M. nepenthicola*] in ***N. ampullaria*** Jack (W Mal.); ***N. bicalcarata*** Hook.f. (Borneo) with 33 spp. of metazoans, preyed on by a red carpenter ant (*Camponotus schmitzi*) restricted to inhabiting tendril, diving into fluid, herding & slaughtering mosquito larvae); above fluid-level spiders catch falling prey; ***N. hemsleyana*** Macfarl. (Borneo) – Hardwicke's woolly bats sleep & defaecate in pitchers; ***N. lowii*** Hook.f. (Borneo) – tree-shrews take nectar from pitcher lid & defaecate within. Some stems esp. ***N. ampullaria*** locally used for twine etc., some medic.; many cult. orn. (incl. many complex hybrids raised in C19, first being ***N. × dominyana*** Veitch ex T. Moore & Mast., 1858), trade worth $US 15 M a year by 2000

Nepeta L. Labiatae (VII 2c). 200+ temp. Euras. (Eur. 24), N Afr., trop. Afr. mts. R: BZ 78,1(1993)95. Usu. dry habitats; gynodioecy (sect. *Oxynepeta* dioec.). Catmint, some spp. irresistible to cats; rabbit-proof cult. orn. esp. ***N. cataria*** L. (catnip, catnep, Eur., SW & C As.) – lvs medic. (imp. N Am.) & used as a tea, also psychedelic effects like cannabis ('cataria'), nepetalactone (an iridoid sequestered by *Romalea guttata* grasshoppers) sensitizing genetically inherited response in Felidae so that cats behave as with cannabis (cf. *Actinidia, Boschniakia, Menyanthes, Teucrium, Valeriana*) but similar activity found in urine of tomcats so plant prod. poss. mimicking a pheromone assoc. with courtship behaviour, oil also attracting male aphids; ***N. × faassenii*** Bergmans ex Stearn (*N. racemosa* Lam. (*N. mussinii*, Caucasus, Iran) × *N. nepetella* L. (W Med.)) – common catmint of gardens, sterile pl. oft. confused with *N. racemosa*

Nephelaphyllum Blume. Orchidaceae (V 14). 11 Indomal.
Nephelea Tryon = *Cyathea*
Nephelium L. Sapindaceae. 16 Indomal. Some fr. trees esp. ***N. lappaceum*** L. (rambutan, Mal.) – apomictic, wild seeds only germ. after passage through monkey, pericarp form. source of black dye, seeds allegedly affecting glucose absorption so poss. use in diabetes

control, & ***N. ramboutan-ake*** (Labill.) Leenh. (*N. mutabile*, pulasan, W Mal.) – many cvs, pulp (sarcotesta) ed. (2.3 M t per annum). See also *Dimocarpus*, *Litchi*

Nephelochloa Boiss. (~ *Eremopoa*). Gramineae (XVI 15). 1 Turkey: ***N. orientalis*** Boiss.

Nephopteris Lellinger. Pteridaceae (III; Adiantaceae s.l.). 1 C Colombia: ***N. maxonii*** Lellinger – at 3000 m. R: AFJ 56(1966)180

Nephradenia Decne. Apocynaceae (V c; Asclepiadaceae III 4). 10 trop. Am.

Nephrangis (Schltr.) Summerh. Orchidaceae (V 16d). 2 trop. Afr. R: KB 3(1948)301

Nephrocarpus Dammer = *Basselinia*

Nephrocarya Candargy = *Nonea*

Nephrodesmus Schindler (~ *Arthroclianthus*). Leguminosae (III 19). 6 New Caled.

Nephrodium Michaux = *Dryopteris*

Nephrolepidaceae Pichi-Serm. = Lomariopsidaceae

Nephrolepis Schott. Lomariopsidaceae (Nephrolepidaceae). 19 trop. & warm. R: Blumea 50(2005)286. Young fronds a veg., local medic., fibre for hats, mats & basketry; cult. orn. (sword ferns) esp. ***N. cordifolia*** (L.) C. Presl (trop., widely natur., invasive in Aus., SE US), ***N. exaltata*** (L.) Schott (trop. Am.) – many cvs poss. incl. '**Bostoniensis**' (Boston fern, poss. = ***N. x hippocrepicis*** Miyam., *N. c.* × *N. biserrata* (Sw.) Schott (trop.)) – sterile, from which 50+ cvs derived, ***N. falcata*** (Cav.) C. Chr. (Philippines) esp. '**Furcans**' – pinnae bifurcate & sometimes even subdivided, cult. Pac. for leis as in Hawaii. ***N. brownii*** (Desv.) Hovenkamp & Miyam. (*N. multiflora*, OW trop.) – invasive in Masc., SE US; ***N. hirsutula*** (Forst.f.) C. Presl (Indomal., Pac.) – dominant in sec. grasslands in e.g. Niue

Nephromeria (Benth.) Schindler = *Desmodium*

Nephropetalum Robinson & Greenman = *Ayenia*

Nephrophyllidium Gilg (*Fauria*). Menyanthaceae. 1 N Jap., NW N Am.: ***N. crista-galli*** (Hook.) Gilg – cult. orn. bog pl.

Nephrophyllum A. Rich. Convolvulaceae (6). 1 Ethiopia: ***N. abyssinicum*** A. Rich.

Nephrosperma Balf.f. Palmae (V 14k). 1 Seychelles: ***N. vanhoutteanum*** (Van Houtte) Balf.f. – cult. orn.

Nephrotheca R. Nordenstam & Källersjö. Compositae (Cal.). 1 Cape: ***N. ilicifolia*** (L.) R. Nordenstam & Källersjö. R: CN 44(2006)33

Nephthytis Schott. Araceae (VII 12). 1 Borneo, 5 trop. Afr.(!)

Nepsera Naud. Melastomataceae. 1 trop. Am.: ***N. aquatica*** Naud.

Neptunia Lour. Leguminosae (II 1). 12 trop. & warm esp. Aus. & Am. R: AusJB 14(1966)379. Sensitive lvs like *Mimosa*, those of ***N. plena*** (L.) Benth. (trop. & warm Am., weedy in pineapple plantations) with living wood-fibres in pulvinus; ***N. prostrata*** (Lam.) Baill. (*N. aquatica*, *N. oleracea*, water mimosa, trop.) – floating stems with aerenchyma, locally eaten as potherb

neram *Dipterocarpus oblongifolius*

Neraudia Gaudich. Urticaceae (III). 5 Hawaii

Neriacanthus Benth. Acanthaceae (III 1). 4 trop. Am.

Nerine Herb. Amaryllidaceae. 23 S Afr. R: Plant Life 23(1967)suppl. Cult. orn. bulbs with many alks, one group flowering while lvs photosynthesizing, other when they have withered. ***N. bowdenii*** Will. Watson – fls after lvs die, hardiest sp.; ***N. sarniensis*** (L.) Herb. (Guernsey lily) – described from Guernsey & alleged to have arrived in ballast of ship from Jap. but prob. presented to islanders by shipwrecked sailors from Cape in 1650s or introd. from cult. in Eng. 1660s; many hybrids & a N. Society

Nerisyrenia Greene. Cruciferae (40). 8 S N Am. R: Rhodora 80(1978)159, Phytol. 75(1993)231

Nerium Tourn. ex L. Apocynaceae (II a). 1 Medit. (incl. Eur.) to Cape Verde Is. & Jap.: ***N. oleander*** L. (oleander, rose-bay) – rheophyte (Leeuwenberg's Model) with bright but nectarless fls poll. by deceit, seeds water-disp., bone setting in Crete (since Dioscorides), cult. orn. (c. 400 cvs, some scented), v. toxic due to cardiac glycosides incl. oleandrin (rat poison but gazelles & hyrax immune), in humans 1 leaf a potentially lethal dose, death within 24 hrs. R: AUWP 87–2(1987)7

Nernstia Urb. Rubiaceae (I 7). 1 Mex.: ***N. mexicana*** (DC.) Urb.

neroli *Citrus* × *aurantium*

Nerophila Naud. = *Chaetolepis*

Nertera Banks ex Gaertn. (~ *Coprosma*). Rubiaceae (IV 3). 6 Indomal. to NZ, Am., subAntartic is. ***N. granadensis*** (L.f.) Druce (*C. g.*, *N. depressa*, bead pl., Aus., NZ, S Am.) – cult. orn. ground-cover or pot-pl. with orange beadlike fr.

Nervilia Comm. ex Gaudich. Orchidaceae (V 8a). c. 60 trop. & warm OW (Arabia, Afr., Madag. 16 (R: OM 5, 1991)). Terr., infl. withers before lvs appear. Some cult. orn., local medic. esp. ***N. fordii*** (Hance) Schltr. (China) now overexploited

Nesaea Comm. ex Kunth = *Ammannia* (but see Bothalia 21(1991)35)

Nesampelos R. Nordenstam (~ *Senecio*). Compositae (Sen.-Sen.). 3 Hispaniola. R: CN 44(2006)58, 45(2007)37

Nesiota Hook.f. Rhamnaceae (3). 1 St Helena: ***N. elliptica*** (Roxb.) Hook.f. – 99% self-incompatible, extinct in wild (1994) & last of 3 pls raised from last tree found alive (1978; 12 in 1875) dead by 2004, allies in Afr., Masc.

Neslia Desv. (~ *Camelina*). Cruciferae (15). 2 SE Eur., Medit., SW As. R: FR 64(1961)11. ***N. paniculata*** (L.) Desv. (ball mustard)

Nesocaryum I.M. Johnston (~ *Cryptantha*). Boraginaceae (B.3.8.4). 1 Desventuradas Is. (Chile): ***N. stylosum*** I.M. Johnston

Nesocodon Thulin (~ *Heterochaenia*). Campanulaceae (I 2). 1 Mauritius: ***N. mauritianus*** (I. Richardson) Thulin – infl. term., nectar coloured. R: KB 34(1980)813

Nesocrambe A. Mill. = *Hemicrambe*

Nesogenaceae Marais = Orobanchaceae

Nesogenes A. DC. Orobanchaceae (Nesogenaceae). 8 Tanz., Madag., Ind. & Pac. Oceans. R: KB 35(1981)799, 38(1983)37

Nesogordonia Baill. Malvaceae (Dom.; Sterculiaceae). c. 20 trop. Afr. (3), Madag. ***N. papaverifera*** (A. Chév.) Capuron (*Cistanthera p.*, danta, kotibé, W Afr.) – hard reddish timber used for veneers, plywood & flooring

Nesohedyotis (Hook.f.) Bremek. Rubiaceae (IV 1). 1 St Helena: ***N. arborea*** (Roxb.) Bremek. – allies in Indomal. R: Q.C.B. Cronk, *Endemic Fl. St Helena* (2000)69

Nesolindsaea Lehtonen & Christenh. (~ *Lindsaea*). Lindsaeaceae. 2 Seychelles, Sri Lanka. R: BJLS 163(2010)336

Nesoluma Baill. Sapotaceae (III). 3 Polynesia

Nesomia B. Turner (~ *Ferreyrella*). Compositae (Eup.-Ayap.). 1 Mex.: ***N. chiapensis*** B. Turner. R: Phytol. 7(1991)208

Nesopteris Copel. = *Cephalomanes*

Nesothamnus Rydb. = *Perityle*

Nesphostylis Verdc. Leguminosae (III 18). 2 trop. Afr., 2 trop. As.

Nestegis Raf. Oleaceae (4d). 5 Norfolk Is. & NZ (1), NZ (3), Hawaii (1). R (NZ): JAA 44(1963)378. Some timber (maire); ***N. apetala*** (Vahl) L. Johnson (ironwood, Norfolk Is.) – valuable timber; ***N. sandwicensis*** (A. Gray) Degener, I. Degener & L. Johnson (Hawaii) – form. imp. for tool-handles & fuel

Nestlera Spreng. (~ *Relhania*). Compositae (Gnap.-Rel.). 1 S Afr.: ***N. biennis*** (Jacq.) Spreng. R: AMBG 78(1991)1069

Nestoria Urb. = *Pleonotoma*

Nestotus R. Roberts & al. (~ *Haplopappus*). Compositae (Ast.). 2 NW N Am. R: FNA 20(2006)169

Nestronia Raf. (~ *Buckleya*). Santalaceae. 1 E US: ***N. umbellula*** Raf. – r.-parasite

nettle *Urtica* spp. & other pls with stinging hairs; **bull n.** *Cnidoscolus stimulosus*; (**common**) **stinging n.** *U. dioica*; **deadn.** *Lamium* spp.; **dog n.** *U. urens*; **false n.** *Boehmeria* spp.; **giant n.** *Dendrocnide* spp.; **hedge n.** *Stachys* spp.; **hemp n.** *Galeopsis* spp.; **Nilgiri n.** *Girardinia diversifolia*; **Roman n.** *U. pilulifera*; **small n.** *U. urens*; **n. tree** *Celtis* spp. esp. *C. australis*; **wood n.** *Laportea canadensis*

Nettoa Baill. = *Corchorus*

Neuburgia Blume. Loganiaceae (Strychnaceae). 10–12 E Mal. to Pac. Fr. a drupe

Neumanniaceae 'Tieghem' = Aphloiaceae

Neuontobotrys O. Schulz. Cruciferae (46). Incl. *Eremodraba*, 14 Chile, Argentina. R: Darw. 44(2006)349

Neuracanthus Nees. Acanthaceae (III). 30 trop. Afr. & Arabia to Ind. R: KB 53(1998)10

Neurachne R. Br. Gramineae (XXIV 3). Incl. *Paraneurachne*, 8 Aus. R: CQH 13(1972)

Neuractis Cass. = *Glossocardia*

Neurada B. Juss. Neuradaceae. 1 E Medit. to Ind. desert: ***N. procumbens*** L.

Neuradaceae Kostel. (Grielaceae). Magnoliidae – Malvales. 3/9 Afr. deserts & Medit. to Ind. desert. Annual to subshrubby prostrate tomentose herbs with mucilage-ducts in pith. Lvs toothed to pinnatifid, in spirals, stip. 0. Fls bisexual, reg., solit. on axillary pedicels, ± epigynous; K 5 valvate, accrescent, C 5 imbr. or convolute, A 10 on hypanthium with unique bipolar pollen with 3(4) pores at each end, G (10, ? derived from 5),

± inferior, pluriloc. but 2–4 ± reduced or ovules not maturing, with distinct styles & 1 (?2) apical-axile, pendulous, anatropous, bitegmic ovule. Fr. dehiscing ventrally (tardily), dry; seeds without endosperm, germ. in fr. n = 6
Genera: *Grielum, Neurada, Neuradopsis*

Neuradopsis Bremek. & Oberm. Neuradaceae. 3 SW Afr.

Neurocallis auctt. = *Pteris*

Neurocalyx Hook. Rubiaceae (IV 2). 5 Sri Lanka, 1 ext. to S Ind.

Neurodium Fée = *Pleopeltis*

Neurolaena R. Br. Compositae (Neur.). 11 trop. Am. R: PSE 140(1982)119, Phytologia 58(1985)497

Neurolakis Mattf. Compositae (Vern.-Linz.). 1 Cameroun, Chad: ***N. modesta*** Mattf.

Neurolepis Meissn. = *Chusquea* (but see CUSNH 39(2000)79)

Neurolobium Baill. = *Diplorhynchus*

Neuroloma Andrz. ex DC. = *Parrya*

Neuromanes Trevis. = *Trichomanes*

Neuropeltis Wall. Convolvulaceae (?4). 4 trop. As., 10 trop. Afr. R: PEE 143(2010)176. Epiphyllous infls

Neuropeltopsis Oostr. Convolvulaceae (?4). 1 Borneo: ***N. alba*** Oostr. R: Blumea 12(1964)365. Epiphyllous infls

Neurophyllodes (A. Gray) Degener = *Geranium*

Neuropoa Clayton = *Poa* (but see KB 40(1985)728)

Neurosoria Mett. = *Cheilanthes*

Neurotheca Salisb. ex Benth. Gentianaceae. 3 trop. Afr., 1 ext. to S Am. (***N. loeselioides*** (Progel) Baill.). R: Adansonia 8(1968)57

Neustanthus Benth. (~ *Pueraria*). Leguminosae (III 18). 6 trop. As. R: Phytotaxa 218(2015)203

Neustruevia Juzep. = *Pseudomarrubium*

neutral henna *Ziziphus jujuba*

Neuwiedia Blume. Orchidaceae (I; Apostasiaceae). 8 Mal.

Nevada Holmgren (~ *Boechera*). Cruciferae (11). 1 Nevada: ***N. holmgrenii*** (Rollins) Holmgren. R: FNA 7(2010)414

Neves-armondia K. Schum. = *Pithecoctenium*

Nevillea Esterh. & Linder (~ *Restio*). Restionaceae. 2 SW Cape. R: Bothalia 15(1985)482

Neviusia A. Gray. Rosaceae (Ros.-Kerr.). 1 Alabama, 1 N Calif. (fossils in Canadian Eocene). R: Novon 2(1992)285. ***N. alabamensis*** A. Gray (cliff-faces above Black Warrior River, Alabama) – cult. orn. decid. shrub

Nevrocallis Fée = *Pteris*

Nevrodium Fée = *Neurodium*

Nevskiella Krecz. & Vved. = *Bromus*

New Guinea beans *Lagenaria siceraria*; **N. G. rosewood** *Pterocarpus indicus*; **N. G. teak** *Vitex cofassus*

New Jersey tea *Ceanothus americanus*

New Zealand cranberry *Ugni molinae*; **N. Z. flax** *Phormium tenax*; **N. Z. pygmyweed** *Crassula helmsii*; **N. Z. spinach** *Tetragonia tetragonoides*

Newbouldia Seemann ex Bureau. Bignoniaceae (1). 1 trop. W Afr.: ***N. laevis*** (P. Beauv.) Bureau – cult. orn. used as living fence, medic.

Newcastelia F. Muell. Labiatae (IV 1; Chloanthaceae). 9 trop. Aus. R: Nuytsia 11(1996)90

Newmania Lý & Škorničk. (~ *Haniffia*). Zingiberaceae (Zing.-Zing.). 3 Vietnam. R: T 60(2011)1390, GBS 67(2015)354

Newtonia Baill. Leguminosae (II 1). 15 trop. Afr. (Am. spp. referable to new genus). R: BJBB 60(1990)119. Timber

Neyraudia Hook.f. Gramineae (XXVI). 5 OW trop. (China 4). R: JAA 71(1990)164. ***N. arundinacea*** (L.) Henr. (*N. reynaudiana*, As.) – invasive SE US; ***N. curvipes*** Ohwi (Bhutan, Mt Kinabalu) – disjunct distrib.

Nezahualcoyotlia R. González (~ *Cranichis*). Orchidaceae (Orch.-Cran.). 1 Mex.: ***N. gracilis*** (L.O. Williams) R. González

ngai (camphor) *Blumea balsamifera*

ngaio *Myoporum laetum*

ngali nut *Canarium indicum*

ngapi nut *Archidendron jiringa*

ngo gai *Eryngium foetidum*

Nialel Adans. = *Aglaia*

niangon *Heritiera utilis*

Nianhochloa H.N. Nguyen & Tran (~ *Kinabaluchloa*). Gramineae. 1 S Vietnam: ***N. bidoupensis*** H.N. Nguyen & Tran. R: Adansonia 34(2012)258

niaouli oil *Melaleuca quinquenervia*

nibi *Heteropsis flexuosa, Thoracocarpus bissectus*

Nicandra Adans. Solanaceae (IV). 1 Peru: ***N. physalodes*** (L.) Gaertn. – cult. orn. annual (shoo-fly) alleged to keep off flies (steroid lactones insect antifeedants), with alks, antimicrobial, weedy & natur. in Euras., trop. Am. & US (some seeds lack 1 chromosome & these can remain dormant for 30 yrs)

Nicaragua wood *Dalbergia* & *Haematoxylum* spp. esp. *H. brasiletto*

Nichallea Bridson. Rubiaceae (II 2). 1 trop. Afr.: ***N. soyauxii*** (Hiern) Bridson. R: KB 33(1978)288

Nicipe Raf. = *Ornithogalum*

nicker bean *Guilandina bonduc, Entada gigas*

Nicobariodendron Vasudeva & Chakrab. Celastraceae. 1 Nicobar Is.: ***N. sleumeri*** Vasudeva & Chakrab. – infls racemes, A2. R: JETB 7(1985)513.

Nicodemia Ten. (~ *Buddleja*). Scrophulariaceae (Buddlejaceae). 8 S Afr., Madag. Some cult. orn. esp. winter flowering ***N. madagascariensis*** (Lam.) R. Parker (*B. m.*, Madag.)

Nicolaia Horan. = *Etlingera*

Nicolasia S. Moore. Compositae (Inul.-Pluch.). 7 SW trop. Afr. Heterogeneous. R: MBSM 2 (1954)1

Nicolletia A. Gray. Compositae (Tag.-Pect.). 3 S N Am. R: Sida 7(1978)369

Nicolsonia DC. = *Desmodium*

Nicoraella Torres = ? older name for *Anatherostipa*

Nicoraepoa Soreng & L. Gillespie (~ *Poa*). Gramineae (XVI 15). 6 Arg., Chile, Falklands. R: AMBG 94(2007)842

Nicotiana Tourn. ex L. Solanaceae (III 2). 75 Am., S Pac., Aus. (17, R: JABG 3(1981)1), SW Afr. (***N. africana*** Merxm.), incl. homoploid hybrids. R: T.H. Goodspeed (1954) *The genus N.* (CB 16), T 53 (2004)77. Tobacco (7 M t by 2010); herbs & a few shrubs (40% of spp. allotetraploid); alks esp. nicotine (10% of carbon metabolism directed to their prod.), sequestered by tobacco hornworm (*Manduca sexta*) as its defence; *N. africana* & *N. langsdorffii* Schrank (Braz.)) bird-poll., *N. otophora* Griseb. (Arg.) bat-poll., *N. alata* & *N. sylvestris* poll. hawkmoths, *N. forgetiana* Hort. Sander ex W. Watson (Braz.) poll. butterflies. Tobacco-chewing by Aus. Aborigines prob. antedating that in Am. Smoking tobacco mostly from ***N. tabacum*** L. (cultigen, prob. natural allotetraploid, *N. sylvestris* Speg. (Arg.) × *N. tomentosiformis* Goodspeed (Bolivia)) – in Amaz. form. only medic. & rarely smoked though 'cigars' there now 90 cm long, grown as annual esp. in US, trop. Am., Medit., S trop. Afr. & China (largest producer & consumer), lvs coll. when yellowing, dried & fermented ('cured') in diff. ways to give tobacco for cigars (4.9 billion smoked in US in 2004), cigarettes (c.-machine US 1880; 6.25 trillion smoked in 2012 (will kill 1 billion people this century), c. trade worth £9916 M in GB alone in mid-1980s; 320 M smokers in China, on average 1800 cs a yr) etc. & variously flavoured; rolled lvs for outer surface of cigars & bidi (beedi) cigarettes; ground & flavoured for snuff; mixed with molasses etc. for chewing tobacco (Cavendish plug or cake tobacco); different cvs for diff. purposes e.g. Virginia t. for cigarettes, Havana t. – large-lved form for cigars; also source of comm. nicotine, the stimulatory alk. increasing efficiency & reducing stress, used for insecticides etc. since 1664; a principal source of taxation since time of James I, Braz. crop used to pay for W Afr. slaves in C18, & cause of lung cancer (5.4 M people die each year, 60 M in developed countries 1950–2000, now 10% of all deaths annually) as nicotine converted to nitrosamines causing methylation of DNA leading to tumours, also increasing hydrochloric acid leading to peptic ulcers, carbon monoxide leading to carboxyhaemoglobin (thereby reducing oxygen) & heart attack, tar constricting blood-vessels, so annual net economic toll (2000s) prob. some $200 billion; GM for antibiotic prod. (tooth-decay, hepatitis B) & forms turning red on high nitrate contaminated soils, so land-mine indicator; glands oft. trapping insects. ***N. alata*** Link & Otto (temp. S Am.) but usu. ***N. × sanderae*** Hort. Sander ex W. Watson (*N. alata* × *N. forgetiana*) – cult. orn. flowering t. or t. plant, usu. with fragrant fls in evening; ***N. attenuata*** Torr. ex S. Watson (W N Am.) – smoked, medic. (skin diseases); ***N. glauca*** Graham (S Bolivia to N Arg., natur. US) – cult. orn. shrubby tree, invasive S Eur., Cape Verde Is., S Afr. & Aus. where hybridizing with native spp.; ***N. quadrivalvis*** Pursh (W US) – known only from orig.

specimen but (1804) smoked by Native Americans; ***N. rustica*** L. (wild or Aztec t., *N. paniculata* L. (S Am.) × *N. undulata* Ruíz & Pavón (Peru)) – pl. smaller & hardier than *N. t.*, cult. in pre-Columbian times in Mex. & E N Am., orig. tobacco introd. to Eur. & that smoked by Raleigh etc., first Eur. t. cult. Virginia 1612 (John Rolfe), now grown for insecticides in Euras., source of citric acid, etc.

nicuri *Syagrus coronata*

Nidema Britton & Millsp. (~ *Epidendrum*). Orchidaceae (V 13b). 2 trop. Am. Cult. orn.

Nidorella Cass. (~ *Conyza*). Compositae (Ast.-Gra.). 15 trop. & S Afr. R: BSB 43(1969)209. Some small trees incl. ***N. vernonioides*** Sch. Bip. ex A. Rich. (*C. v.*, E Afr.)

Nidularium Lem. Bromeliaceae (3). Excl. *Canistropsis*, 50 E S Am. R: FN 14(1979)1604. Cult. orn. epiphytes with prickly lvs & sessile infls

Niebuhria DC. = *Maerua*

Niedenzuella W.R. Anderson (~ *Mascagnia*). Malpighiaceae. 16 trop. Am. R: Novon 16(2006)194

Niederleinia Hieron. = *Frankenia*

Niedzwedzkia B. Fedtsch. = *Incarvillea*

Niemeyera F. Muell. (~ *Pouteria*). Sapotaceae (IV). Excl. *Sebertia*, 4 NE Aus.

Nierembergia Ruíz & Pavón. Solanaceae (Petun.). 21 Mex. (1, also in Chile) to Chile, esp. Arg. R: Darw. 5(1941)494. Some with elaiophores attractive to bees. Cult. orn. esp. ***N. linariifolia*** Graham (*N. hippomanica* var. *violacea*, '*N. caerulea*', S Am.) – C violet-blue; ***N. repens*** Ruíz & Pavón (*N. rivularis*, Andes & warm temp. S Am.) – fls white

Nietneria Klotzsch ex Benth. Nartheciaceae (Melanthiaceae s.l.). 1 Venez., Guyana: ***N. corymbosa*** Klotzsch & Schomb. ex B.D. Jackson

Nigella Tourn. ex L. Ranunculaceae (I 3). Excl. *Garidella*, 20 Euras. (Eur. 14), Medit. R: PSE 142(1983)71. Annuals with usu. finely pinnate lvs & fls with coloured K & 2-lipped petals with hollow nectariferous claws; fr. of ± united follicles making a capsule. Cult. orn. esp. ***N. damascena*** L. (love-in-a-mist, Medit.) – some 'double' cvs; ***N. sativa*** L. (fennel flower, Medit.) – cult. since Assyrian times for seeds (fitches, kalanji, black cumin), preserved in Tutankhamun's tomb, used to flavour bread (e.g. 'nan' bread, N Ind.) & cakes, medic. (oil antibacterial)

Nigellicereus (P. Heath) P. Heath = *Stenocereus*

Niger seed *Guizotia abyssinica*

Nigerian golden walnut *Lovoa trichilioides*

night-flowering cactus *Selenicereus grandiflorus*

night-scented stock *Matthiola longipetala* subsp. *bicornis*

nightshade, black *Solanum nigrum*; **deadly n.** *Atropa belladonna*; **enchanter's n.** *Circaea lutetiana*; **Malabar n.** *Basella alba*; **woody n.** *S. dulcamara*

Nigritella Rich. = *Gymnadenia*

Nigromnia Carolin = *Scaevola*

Nihon Otero & al. (~ *Omphalodes*). Boraginaceae (B.3.8.2). 5 Japan. R: Phytotaxa 173(2014)266

nikau palm *Rhopalostylis sapida*

Nikitinia Iljin = *Klasea*

Nile grass *Acroceras macrum*

Nilgirianthus Bremek. = *Strobilanthes*

nim *Azadirachta indica*

Nimiria Prain ex Craib = *Vachellia*

ninde *Aeollanthus myrianthus*

ninebark *Physocarpus opulifolius*

niopo snuff *Anadenanthera peregrina*

niové *Staudtia kamarunensis*

Nipa Thunb. = *Nypa*

Niphaea Lindl. Gesneriaceae (II 5b). 3–5 S Mex., Guatemala. R: Selbyana 29(2008)172

Niphantha Luer = *Stelis*

Niphidium J. Sm. Polypodiaceae (V). 10 trop. Am. R: AFJ 62(1972)101

Niphogeton Schldl. Umbelliferae (III 8). 18 N Andes

nipple fruit *Solanum mammosum*; **n. wort** *Lapsana communis*

Nipponanthemum Kitam. Compositae (Anth.-Art.). 1 Jap.: ***N. nipponicum*** (Maxim.) Kitam. – cult. orn. R: FNA 19(2006)555

Nipponobambusa Muroi = *Sasa*

Nipponocalamus Nakai = *Pleioblastus*

Nirarathamnos Balf.f. Umbelliferae (III 8). 1 Socotra: ***N. asarifolius*** Balf.f.

nirre *Nothofagus antarctica*

nispero *Eriobotrya japonica*

Nispero Aubrév. = *Manilkara*

Nissolia Jacq. Leguminosae (III 11). 14 trop. Am. esp. Mex. R: CUSNH 32(1956)173. Fish poisons

Nistarika Nayar = *Leptochilus*

Nitella Agardh. Characeae. c. 180 cosmop. (Aus. 24). Known since Jurassic with v. similar *Palaeonitella* Kidston & Lang from Lower Devonian

Nitellopsis Hy. Characeae. 1 Euras.: ***N. obtusa*** (Desv.) J. Groves (starry stonewort) – introd. US 1978, now aggressive

Nitidobulbon Ojeda & al. = *Maxillaria* (but see Novon 19(2009)98)

Nitraria L. Nitrariaceae (Zygophyllaceae s.l.). 5–8(12) salt deserts of Sahara & S Russia (Eur. 1) to Afghanistan & E Siberia, SE Aus. (1: ***N. billardierei*** DC. (dillon bush) – salt-rich pericarp eaten by emus, germ. improved (mammals less effective)). R: ABY 21(1999)288. Some locally eaten fr. & soda sources

Nitrariaceae Bercht. & Presl (Zygophyllaceae s.l.). Magnoliidae – Sapindales. Incl. Tetradiclidaceae (Peganaceae), 4/c. 15 deserts of Euras., SE Aus., Sahara, S N Am. Shrubs, oft. pubescent & spiny, to herbs. Lvs simple or multifid, semi-succ., in spirals or fascicles (opp. in *Tetradiclis*); stip. v. small or leafy. Fls reg., app. bisexual but oft. functionally unisexual, in dichasial cymes, 3- or 4-merous (*Tetradiclis*) or K ((4)5), persistent, C 4, 5(6), s.t. hooded, induplicate-valvate, A 4, (10–)15 without appendages, inserted at margin of disk, anthers ± basifixed; $\underline{G}$ (2–4(–6)), with 1 or 6-many pendulous, apotropous, bitegmic ovules per loc. Fr. a capsule, berry or drupe with fleshy exocarp & bony endocarp, 1-seeded; endosperm usu. 0. n = 24, 48, 60

Genera: *Malacocarpus, Nitraria, Peganum, Tetradiclis*

Dyes, local medic. incl. hallucinogens (*Peganum*)

Nitrophila S. Watson. Amaranthaceae (Polycn.; Chenopodiaceae IV 1). 4(–8) SW N & temp. S Am. R: T 62(2013)109

Nitrosalsola Tzvelev = *Caroxylon*

Nivellea Wilcox & al. Compositae (Anth.-Leuc.). 1 Morocco: ***N. nivellei*** (Br.-Blanq. & Maire) Wilcox & al. R: BotJLS 122(1996)125

Nivenia Vent. Iridaceae (V). 11 SW Cape esp. mts. R: P. Goldblatt, *Woody I.* (1993)51. Shrubby; 7 with heterostyly (? some androdioecy). Some cult. orn.

Niveophyllum Matuda = *Hechtia*

Noaea Moq. Amaranthaceae (Chenopodiaceae III 3). 3 SW & C As., N Afr. R: SBD 9,2(2002)3

Noahdendron Endress & al. Hamamelidaceae (I 1). 1 N Queensland: ***N. nicholasii*** Endress & al. R: BJ 107(1985)369

noble fir *Abies procera*

Nocca Cav. = *Lagascea*

Noccaea Moench (~ *Thlaspi*). Cruciferae (19). Incl. *Eunomia*, 128 Euras., N Afr., N Am. (3), Mex. (1), Patagonia (1). R: HPB 19(2014)25

Noccidium F.K. Meyer. Cruciferae (15). 2 SW As.

Nodocarpaea A. Gray. Rubiaceae (IV 15). 1 Cuba: ***N. radicans*** (Griseb.) A. Gray

Nodonema B.L. Burtt = *Streptocarpus*

Nogalia Verdc. = *Heliotropium*

Nogo Baehni = *Lecomtedoxa*

Nogra Merr. Leguminosae (III 18). 3 As. (prob. heterogeneous)

Nohawilliamsia M. Chase & Whitten. Orchidaceae (V 10). 1 N S Am.: ***N. pirarensis*** (Reichb.f.) M. Chase & Whitten

Noisettia Kunth. Violaceae (III 2b). 1 Peru, Braz., Guyana: ***N. orchidiflora*** (Rudge) Ging.

Nolana L. ex L.f. Solanaceae (IV; Nolanaceae). 18+ W S Am. from N Chile to S Peru, Galápagos (1). R: FN 26(1981). Oft. fleshy-leaved shore-pls, some cult. orn. grown as annuals

Nolanaceae Bercht. & J. Presl = Solanaceae

Nolina Michaux. Asparagaceae (Dracaenaceae). Excl. *Beaucarnea*, 21 SW N & C Am. R: PAPS 50(1911)412. Dioec. pachycaul xerophytes (Corner's & Leeuwenberg's models) with linear lvs but no swollen trunk-base (cf. *B.*). Some cult. orn. incl. ***N. longifolia*** (Schultes & Schultes f.) Hemsl. (zacate, Mex.) – lvs used for brooms, basketry & thatching, ***N. microcarpa*** S. Watson (SW N Am.) – used similarly

Nolinaceae Nakai = Asparagaceae

Nolletia Cass. Compositae (Ast.-Hom.). 14 S & N Afr. to Eur. (1). R: Phytotaxa 112(2013)1

Noltea Reichb. Rhamnaceae (3). 1 S Afr:. ***N. africana*** (L.) Endl. – cult. orn. R: Strelitzia 10(2000)467

Nomaphila Blume = *Hygrophila*

Nomismia Wight & Arn. = *Rhynchosia*

Nomocharis Franch. = *Lilium* (but see BotJLS 87(1983)285)

Nomopyle Roalson & Boggan (~ *Kohleria*). Gesneriaceae (II 5b). 2 Ecuador, Peru

Nomosa I.M. Johnson = *Lithospermum* (but see JAA 35(1954)24)

Nonatelia Aubl. = *Palicourea*

nonda *Parinari nonda*

Nonea Medik. Boraginaceae (B.2.1.1). Incl. *Elizaldia*, excl. *Melanortocarya*, 35 Medit. (Eur. 8 (1 extinct), dry places)

none-so-pretty *Saxifraga* × *urbium*

nongo *Albizia grandibracteata*

noodles *Ipomoea batatas, Oryza sativa, Triticum turgidum* (but millet in China 4000 yrs ago); **soba n.** *Fagopyrum esculentum*; **udon n.** *T. turgidum*

Noogoora burr *Xanthium occidentale* (= *X. strumarium* s.l.)

Nootka cypress *Xanthocyparis nootkatensis*

Nopalea Salm-Dyck = *Opuntia*

Nopalxochia Britton & Rose = *Disocactus*

Norantea Aubl. Marcgraviaceae. 2 trop. Am. See also *Marcgraviastrum, Sarcopera, Schwartzia*. *Philodendron*-like pls with fls all fertile. ***N. guianensis*** Aubl. – red nectar-spurs looking like fr., visited by birds which poll., the rest of the fl. dull so potential beak damage reduced

Nordenstamia Lundin (~ *Aequatorium*). Compositae (Sen.-Tuss.). c. 20 W Peru & Ecuador to N Arg., esp. cloud forest. R: CN 44(2006)15,20. Trees to 18 m, shrubs

Nordmann (Christmas tree) *Abies nordmanniana*

Norfolk Island pine *Araucaria heterophylla*

nori *Pyropia tenera, P. yezoensis*

Norlindhia R. Nordenstam (~ *Osteospermum*). Compositae (Cal.). 3 S Afr. R: CN 44(2006)41

Normanbokea Klad. & F. Buxb. = *Turbinicarpus*

Normanboria Butzin = *Acrachne*

Normanbya F. Muell. ex Becc. (~ *Ptychosperma*). Palmae (V 14i). 1 Queensland: ***N. normanbyi*** (W. Hill) L. Bailey (black palm). R: FA 39(2011)198

Normandia Hook.f. = *Coprosma*

Normandiodendron J. Léonard. Leguminosae (I 2). 2 W C Afr.

Normandy cress *Barbarea verna*

Normania Lowe = *Solanum*

Noronhia Stadman ex Thouars. Oleaceae (4d). 66 E & S Afr. (12), Madag. (46 R: FMad 166(1953)15), Masc. (8)

Norrisia Gardner. Loganiaceae (Strychnaceae). 2 W Mal.

Northea Hook.f. = seq.

Northia Hook.f. Sapotaceae (I 2). 1 Seychelles: ***N. hornei*** (Hartog) Pierre (capucin) – endosperm 0

Northiopsis Kaneh. = *Manilkara*

Norway maple *Acer platanoides*; **N. spruce** *Picea abies*

Nosema Prain = *Platostoma*

Nostolachma T. Durand. Rubiaceae (II 1). 10 Indomal.

Notanthera (DC.) G. Don (~ *Loranthus*). Loranthaceae (4 2). 1 temp. S Am.: ***N. heterophylla*** (Ruíz & Pavón) G. Don

Notaphoebe Griseb. = *Alseodaphne*

Notechidnopsis Lavranos & Bleck (~ *Echidnopsis*). Apocynaceae (V b; Asclepiadaceae III 5). 2 S Afr. R: KB 54(1999)341

Notelaea Vent. Oleaceae (4d). 11 E Aus. R: JAA 49(1968)333. Hard timbers. ***N. longifolia*** Vent. – lignotubers

Nothaphoebe Blume = *Alseodaphne*

Nothapodytes Blume. Icacinaceae. 11 E As., Indomal. R: BMNHN 4,7(1985)82. ***N. nimmoniana*** (J. Grah.) Mabb. (*N. foetida*, Indomal.) – alks, some significantly cytotoxic

Notheria O'Byrne & J.J. Verm. (~ *Eria*). Orchidaceae (V 15). 15 Mal. R: GBS 52(2000)285

Nothoalsomitra Telford (~ *Alsomitra*). Cucurbitaceae (XII). 1 SE Queensland: ***N. suberosa*** (Bailey) Telford. R: FA 8(1982)388

Nothobaccaurea Haegens (~ *Baccaurea*). Phyllanthaceae (Scep.). 2 W Pacific. R: Blumea suppl. 12(2000)198

Nothobaccharis R. King. & H. Robinson. Compositae (Eup.-Crit.). 1 Peru: ***N. candolleana*** (Steud.) R. King & H. Robinson. R: MSBMBG 22(1987)324

Nothobartsia Bolliger & Molau. Orobanchaceae (Rhin.). 2 W Medit. R: PSE 179(1992)60

Nothocalais (A. Gray) Greene (~ *Microseris*). Compositae (Lact.-Mic.). 4 C & W N Am. R: FNA 19(2006)335

Nothocestrum A. Gray. Solanaceae (IV 5a). 5–6 Hawaii, all endangered spp. R: Man. Fl. Pl. Hawaii 2(1990)1262. Trunks to 15 cm diam.

Nothochelone (A. Gray) Straw. Plantaginaceae (Ant.-Chel.; Scrophulariaceae s.l.). 1 NW N Am.: ***N. nemorosa*** (Lindl.) Straw – cult. orn. R: Brittonia 18(1966)85

Nothochilus Radlk. Orobanchaceae (Esc.; Scrophulariaceae s.l.). 1 Braz.: ***N. coccineus*** Radlk.

Nothocissus (Miq.) Latiff (~ *Ampelocissus*). Vitaceae. 5 Mal.

Nothocnide Blume ex Chew (*Pseudopipturus*). Urticaceae. 5 Mal. to Solomon Is. R: GBS 24(1969)361. ***N. repanda*** (Blume) Chew (Mal.) – stems sold in Indonesia & pounded for shampoo

Nothodoritis Tsi = *Phalaenopsis* (but see APS 27(1989)58)

Nothofagaceae Kuprian. (~ Fagaceae). Magnoliidae – Fagales. 1/34 W Pacific, temp. S Am. Monoec. trees, usu. everg., with peltate glandular hairs. Lvs distichous; tip. usu. peltate, encl. colleters. P uniseriate (connate in males); A 10–15(–many), connective usu. extended; G (2,3) with unitegmic ovules. Cupule. Fr. 2(3),4-valved. n = 13

Genus: *Nothofagus*

Trees excl. from F. because of unitegmic ovules, diff. pollen exine & infl. diffs. Antarctic fossils first recorded on Seymour Is. 1901–4 and within the Antarctic Circle (85° S) at 1.5 km alt. on Scott's ill-fated Terra Nova expedition (1910–12)

Nothofagus Blume. Nothofagaceae (Fagaceae s.l.). 34 New Guinea, New Caled., temp. Aus., NZ, temp. S Am. (Pliocene Antarctica, poss. like *Salix arctica* Pall. woodland today). R: BotJLS 105(1991)68. Southern beeches. At anthesis ovules merely bulges in ovary & do not develop until fert. 9–10 weeks later. Imp. timber trees, in S hemisph. second only to eucalypts; some fast-growing spp. promoted in Eur. for timber or to 'replace elm in the landscape', after the depredations of Dutch Elm disease, or as cult. orn. incl. ***N. alpina*** (Poepp. & Endl.) Oersted (*N. nervosa*, *N. procera*, raulí, Chile & Arg.), ***N. antarctica*** (Forst. f.) Oersted (nirre, Antarctic beech, Chile, Arg.) – decid.; ***N. cunninghamii*** (Hook.) Oersted (myrtle beech, Tasmanian b., T. myrtle, SE Aus.) – evergreen, timber for furniture, flooring etc.; ***N. dombeyi*** (Mirb.) Oersted (coigue, Arg., Chile) – decid., timber imp.; ***N. fusca*** (Hook.f.) Oersted (red b., NZ) – evergreen, timber for railway sleepers; ***N. menziesii*** (Hook.f.) Oersted (silver b., Southland b., NZ) – evergreen; ***N. obliqua*** (Mirb.) Oersted (roble b., Chile & Arg.) – decid., the hybrid with *N. alpina*, ***N. × dodecaphleps*** M. Grant & E. Clement, the most promising in GB; ***N. solandri*** (Hook.f.) Oersted (black b., NZ) – evergreen, timber for general construction

Notholaena R. Br. (~ *Cheilanthes*). Pteridaceae (IV). c. 50 warm (esp. SW US to Mex.) to trop. Am. Cloak ferns. Heterogeneous. Usu. xerophytes. Some cult. orn.

Notholirion Wall. ex Boiss. Liliaceae. 5 Afghanistan to W China. Cult. orn. bulbous pls

Nothomyrcia Kausel = *Myrceugenia*

Nothopanax Miq. = *Polyscias*

Nothopegia Blume. Anacardiaceae (I). 7 Sri Lanka, Ind.

Nothopegiopsis Lauterb. = *Semecarpus*

Nothoperanema (Tag.) Ching = *Dryopteris*

Nothophlebia Standl. = *Pentagonia*

Nothorhipsalis Doweld = *Lepismium*

Nothoruellia Bremek. & Nannenga-Bremek. = *Ruellia*

Nothosaerva Wight. Amaranthaceae (I 2). 1 trop. Afr. & Mauritius to trop. As.: ***N. brachiata*** (L.) Wight

Nothoschkuhria Baldwin (~ *Schkuhria*). Compositae (Bah.). 1 Bolivia, N Arg.: ***N. degenerica*** (Kuntze) Baldwin. R: Phytoneuron 2015–56:1

Nothoscordum Kunth. Amaryllidaceae (Alliaceae-All.). 19 Am. *Allium*-like but odourless & style term.; adventitious embryony from nucellar tissue; few cult. orn. ***N. borbonicum*** Kunth ('*N. inodorum*', poss. *N. gracile* (Aiton) Stearn × *N. entrerianum* Ravenna arising nr Buenos Aires) – almost cosmop. weed ('onion weed')

Nothosmyrnium Miq. Umbelliferae (III 8). 2 E As.

Nothospondias Engl. Simaroubaceae. 1 W trop. Afr.: ***N. staudtii*** Engl.

Nothostele Garay. Orchidaceae (IV 2h). 2 Braz. R: Gen. Orch. 3(2003)40

Nothotalisia W. Thomas. Picramniaceae. 3 C & NW S Am. R: Britt. 63(2011)50

Nothotsuga Hu ex Page (~ *Tsuga, Keteleeria*). Pinaceae. 1 China: ***N. longibracteata*** (W.C. Cheng) Page – endangered. R: NRBGE 45(1988)390, RV 121(1990)141

Nothovernonia H. Robinson & Funk (~ *Vernonia*). 2 trop. Afr. R: Phytokeys 3(2011)26

Noticastrum DC. Compositae (Ast.-Chr.). 19 trop. S Am. R: RMP 13(1985)313

Notiosciadium Speg. Umbelliferae (I). 1 Argentina: ***N. pampicola*** Speg.

Notobasis Cass. (~ *Cirsium*). Compositae (Card.-Card.). 1 Medit. (incl. Eur.) to C As.: ***N. syriaca*** (L.) Cass. R: FA 37(2015)60

Notobulbon B. Wyk (~ *Peucedanum*). Umbelliferae (III 11). 12 Cape, 1 ext. to E Afr. R: SB 34(2009)228. Shrubby, ***N. galbanum*** (L.) A. Magee (*Bubon g.*) – pachycaul treelet to 3 m

Notobuxus Oliv. (~ *Buxus*). Buxaceae. 5 trop. & S Afr.

Notocactus (K. Schum.) Frič = *Parodia*

Notoceras R. Br. Cruciferae (4). 1 Canary Is., Medit., incl. Eur., to NW Ind.: ***N. bicorne*** (Aiton) Amo

Notochaete Benth. = *Phlomoides*

Notochloe Domin. Gramineae (Danth.). 1 Blue Mts, NSW: ***N. microdon*** (Benth.) Domin. R: FA 44B(2005)34

Notodanthonia Zotov = *Rytidosperma*

Notodon Urb. = *Poitea*

Notodontia Pierre ex Pitard = *Spiradiclis*

Notoleptopus Voronts. & Petra Hoffm. (~ *Andrachne*). Phyllanthaceae (Par.). 1 Mal. to Aus.: ***N. decaisnei*** (Benth.) Voronts. & Petra Hoffm. R: KB 63(2008)50

Notonerium Benth. = *Heliotropium*

Notonia DC. = *Kleinia*

Notoniopsis R. Nordenstam = *Kleinia*

Notopleura (Hook.f.) Bremek. (~ *Psychotria*). Rubiaceae IV 7). 100 trop. Am. R: AMBG 88(2001)481, Novon 13(2003)228,260. Succ. herbs & subshrubs, some epiphytic

Notopora Hook.f. Ericaceae (VIII 5). 5 E Venez. R: MNYBG 29(1978)158

Notoptera Urb. = *Otopappus*

Notopterygium Boissieu = *Hansenia* (but see APS 38(2000)430)

Notoseris Shih (~ *Prenanthes*). Compositae (Lact.-Lact.). Incl. *Paraprenanthes, Stenoseris*, 33 E As. (R (China): APS 25(1987)196) to Mal.

Notospartium Hook.f. = *Carmichaelia*

Notothixos Oliv. Santalaceae (Viscaceae). 8 Sri Lanka to SE Aus. R: Brunonia 6(1983)25

Notothlaspi Hook.f. Cruciferae (38). 2 NZ. ***N. rosulatum*** Hook.f. (penwiper pl.) – lvs at first covered with hairs, the flannel-textured rosette of them supposedly much like Victorian penwiper

Nototriche Turcz. Malvaceae (Malv.-Malv.). 100 S Am. R: TLS II,7(1909)214. Usu. cushion pls; infls epiphyllous

Nototrichium (A. Gray) Hillebrand (~ *Achyranthes*). Amaranthaceae (I 2). 2 Hawaii. R: Man. Fl. Pl. Hawaii 1(1990)193. Small trees & shrubs

Notoxylinon Lewton = *Gossypium*

Notylia Lindl. Orchidaceae (V 12h). Excl. *Macroclinium*, c. 60 trop. Am. Small. cult. orn. epiphytes with extrafl. nectaries & fragrant fls

Notyliopsis P. Ortiz (~ *Notylia*). Orchidaceae (V 12h). 1 Colombia: ***N. beatricis*** P. Ortiz. R: Orquideologia 20(1996)187

Nouelia Franch. Compositae (Wund./Mut.-Wund.). 1 SW China: ***N. insignis*** – shrub

Nouettea Pierre = *Epigynum*

Novaguinea Hind. Compositae (Ast.-Lag.). 1 W NG: ***N. rudalliae*** Hind. R: KB 59(2004) 177

Novelloa Philbrick (~ *Oserya*). Podostemaceae. 2 C Am. R: SB 36(2011)114

November lily *Lilium longiflorum*

Novenia Freire (~ *Oritrophium*). Compositae (Ast.-Hin.). 1 Andes: ***N. acaulis*** (Benth.) Freire & Hellwig (*N. tunariense*). R: BSAB 24(1986)295

Novopokrovskia Tzvelev = *Erigeron*

Novosieversia F. Bolle = *Geum*

Nowickea Martínez & McDonald = *Phytolacca* (but see R: Brittonia 41(1989)399)

Noyera Trécul = *Perebea*

ntonga nut *Cryptocarya latifolia*

Nucularia Battand. Amaranthaceae (Chenopodiaceae III 3). 1 Algeria, Sahara: ***N. perrinii*** Battand.

Nudilus Raf. = *Forestiera*

Nuihonia Dop = *Craibiodendron*

Nujiangia X.H. Lin & D.Z. Li = *Habenaria* (but see JSE 50(2012)68)

Numaeacampa Gagnepain = ? Solanaceae

Nuñas bean *Phaseolus vulgaris*

Nuphar Sm. Nymphaeaceae. 9 N temp. & cold (Eur. 2). R: Rhodora 109(2007)22. Waterlilies with alks, yellow or purplish hypogynous fls usu. held above water & large berries maturing above water, breaking off & splitting into carpels, the seeds without arils (cf. *Nymphaea*) but slimy pericarp with air-bubbles aiding dispersal to 80m an hr in *N. lutea*. Some cult. orn. esp. ***N. lutea*** (L.) Sm. (brandy-bottle, yellow waterlily, W Eur. – C As.) – herbivory by waterlily beetle (eats only floating lvs) promotes submerged leaf growth, fls with alcoholic smell attractive flies (Eur.), though largely bee-poll. in S Norway, carved C13 roof-bosses in Westminster Abbey, London; ***N. advena*** (Aiton) Aiton f. (*N. l.* subsp. *a.* (E N Am.)) – smell attractive to beetles, medic. & rhiz. ed.; ***N. polysepala*** Engelm (*N. l.* subsp. *p.*, N N Am.) – seeds used for flour by Klamath

Nurmonia Harms = *Turraea*

nut technically a hard 1-seeded brittle fr., in common parlance any edible 'kernel' or even a pod with edible seeds (e.g. monkey n.) – E.A. Menninger (1977) *Edible nuts of the world*; **arara n.** *Joannesia princeps*; **Areca n.** *Areca catechu*; **Australian chestn.** *Castanospermum australe*; **awusa n.** *Plukenetia conophora*; **Barbary n.** *Moraea sisyrinchium*; **Barcelona n.** *Corylus avellana*; **Baroba n.** *Diplodiscus paniculatus*; **Ben n.** *Moringa oleifera*; **betel n.** *Areca catechu*; **bladder n.** *Staphylea* spp.; **bread n.** *Brosimum alicastrum*; **Brazil n.** *Bertholletia excelsa*; **butter n.** *Juglans cinerea*; **candle n.** *Aleurites moluccanus*; **cashew n.** *Anacardium occidentale*; **Chile n.** *Araucaria araucana, Gevuina avellana*; **chufa n.** *Cyperus esculentus*; **clearing n.** *Strychnos potatorum*; **cocon.** *Cocos nucifera*; **cohune n.** *Attalea cohune*; **coohoy n.** *Floydia prealta*; **coquilla n.** *Attalea funifera*; **dika n.** *Irvingia gabonensis*; **double cocon.** *Lodoicea maldivica*; **earthn.** *Conopodium majus*; **fox n.** *Euryale ferox*; **Gaboon n.** *Coula edulis*; **galip n.** *Canarium indicum*; **gasso n.** *Manniophyton africanum*; **ginger n.** *Zingiber officinale*(-flavoured biscuit); **n. grass** *Cyperus rotundus*; **ground n.** *Arachis hypogaea*; **grugru n.** *Acrocomia totai*; **helicia n.** *Helicia diversifolia*; **illipe n.** *Shorea, Madhuca* spp.; **inoi n.** *Poga oleosa*; **ivory n.** *Phytelephas macrocarpa*; **jack n.** *Artocarpus heterophyllus*; **jojoba n.** *Simmondsia chinensis*; **Jamaica cobn.** *Omphalea triandra*; **Jesuit's n.** *Trapa natans*; **karaka n.** *Corynocarpus laevigata*; **karanda n.** *Elaeocarpus bancroftii*; **karuka n.** *Pandanus bowersiae*; **kaya n.** *Torreya nucifera*; **kemiri n.** *Aleurites moluccanus*; **kola n.** *Cola* spp.; **kubili n.** *Cubilia cubili*; **kukui n.** *A. m.*; **ling n.** *Trapa natans*; **luna n.** *Lepisanthes fruticosa*; **macadamia n.** *Macadamia* spp.; **manketti n.** *Ricinodendron rautanenii*; **marking n.** *Semecarpus anacardium*; **monkey n.** *Arachis hypogaea*; **nali n.** *Canarium harveyi*; **nangai n.** *C.* spp.; **ngali n.** *C. indicum*; **ngapi n.** *Pithecellobium lobatum*; **nicuri palm n.** *Syagrus coronata*; **olive n.** *Elaeocarpus sphaericus*; **owusa n.** *Plukenetia conophora*; **oyster n.** *Telfairia pedata*; **paradise n.** *Lecythis* spp.; **pean.** *Arachis hypogaea*; **physic n.** *Jatropha curcas*; **pign.** *Conopodium majus*; **pili n.** *Canarium* spp.; **pistachio n.** *Pistacia vera*; **quandong n.** *Santalum acuminatum*; **Queensland n.** *Macadamia* spp., *Macrozamia* spp.; **red boppel n.** or **rose n.** *Hicksbeachia pinnatifolia*; **rush n.** *Cyperus esculentus*; **sapucaia n.** *Lecythis* spp.; **n.sedge, yellow** *C. esculentus*; **shea n.** *Vitellaria paradoxa*; **singhara n.** *Trapa natans*; **snake n.** *Ophiocaryon paradoxum*; **soap n.** *Sapindus* spp.; **Solomon n. oil** *Canarium indicum*; **Spanish n.** *Moraea sisyrinchium*; **swarri n.** *Caryocar nuciferum*; **tacay n.** *Caryodendron orinocense*; **tagua n.** *Phytelephas macrocarpa*; **tallow n.**; *Ximenia americana*; **tiger n.** *Cyperus esculentus*; **union n.** *Bouchardatia neurococca*; **n.wood** *Terminalia arostrata*; **yeheb n.** *Cordeauxia edulis*; **Zulu n.** *Cyperus esculentus*

nutmeg *Myristica fragrans*; **Brazilian n.** *Cryptocarya moschata*; **calabash n.** *Monodora myristica*; **false n.** *Pycnanthus angolensis*; **Madagascar n.** *Cryptocarya agathophylla*); **Peruvian n.** *Laurelia sempervirens*; **W African n.** *Monodora myristica*

Nuttallanthus D. Sutton = *Linaria* (but see D.A. Sutton, *Rev. Antirrh.* (1988)455)

nux vomica *Strychnos nux-vomica*

Nuxia Comm. ex Lam. Stilbaceae (Loganiaceae s.l.). 15 S Arabia to trop. Afr., Masc. & S Afr. R: MLW 75–8(1975). ***N. floribunda*** Benth. (kite tree, trop. & S Afr.) – cult. orn. tree; ***N. verticillata*** Comm. ex Lam. (Madag.) – used as flavouring for betsa-betsa

Nuytsia R. Br. ex G. Don. Loranthaceae (I). 1 SW Aus.: ***N. floribunda*** (Labill.) G. Don (fire-tree, flame-tree, Christmas tree) – largely clonal tree (underground stems to 110 m) to 12 m, parasitic on r. of grass, 'garotting' r. (& underground cables), with brilliant yellow-orange bird- & insect-poll. (beetles & wasps prob. most efficient) fls, dry wind-disp. fr. with broad wings, cotyledons 2–4 unequal

nyala tree *Xanthocercis zambesiaca*

nyatoh *Madhuca* spp., *Palaquium* spp.

Nyctaginaceae Juss. Magnoliidae – Caryophyllales. 27/390 trop. & warm esp. Am., few temp. R: T 59(2010)908. Trees, shrubs & herbs (with swollen nodes) producing betalains but not anthocyanins, oft. with unusual sec. growth with concentric rings of vasc. bundles or alt. rings of xylem & phloem. Lvs usu. opp., simple, oft. unequal; stip. 0. Fls usu. bisexual, oft. in cymes, s.t. subtended by large & s.t. coloured involucre & when reduced giving a 1-fld pseudanthium with K-like involucre & C-like K; P tubular with (3–)5(–8) valvate to plicate lobes, oft. C-like, A (1–) as many as K (–40) with filaments of unequal length, s.t. basally connate, anthers with longit. slits, annular nectary-disk oft. around $\underline{G}$ 1 with long slender style & 1 basal campylotropous (-hemitropous) (uni- or) bitegmic ovule. Fr. an achene or nut, oft. encl. in persistent base of P-tube (anthocarp, oft. glandular & animal-disp.); seed with large usu. curved embryo with ± copious starchy perisperm, endosperm 0 or a cap over radicle. n = (8-)11(-13+)

Classification & principal genera:

1. **Leucastereae** (lvs in spirals; involucre 0, ±bracteoles; style thick or 0): *Reichenbachia*
2. **Boldoeae** (lvs in spirals; involucre 0, bracteoles 0, style filiform): *Boldoa*
3. **Colignonieae** (~ 6.; lvs opp. or whorled, unarmed): *Colignonia* (only)
4. **Bougainvilleeae** (lvs in spirals; involucre of coloured bracts): *Bougainvillea*
5. **Pisonieae** (lvs usu. opp.; P inconspic.): *Guapira, Neea, Pisonia*
6. **Nyctagineae** (incl. Abronieae; lvs opp. to whorled; cotyledons 1 or 2): *Abronia, Acleisanthes, Boerhavia, Mirabilis*

Some with pollen c. 200 μm long, poss. largest in Magnoliidae (Stevens)

Some ed. & medic. (*Mirabilis, Neea, Pisonia*), dyes (*Neea*) but imp. cult. orn. esp. *Bougainvillea* & *Mirabilis*

Nyctaginia Choisy. Nyctaginaceae (6). 1 SW N Am.: ***N. capitata*** Choisy – cult. orn. herb with deep red fls

Nyctanthaceae J. Agardh = Oleaceae

Nyctanthes L. Oleaceae (5). 2 Ind. to Java. R: NSL 41(2009)203. ***N. arbor-tristis*** L. (harsinger (tree), (anc.) parijat) – holy tree of Ind. where the fallen corollas are swept up each morning, fragrant fls opening at night source of scent & a fugitive orange dye (cheap saffron subs.)

Nycticalanthus Ducke. Rutaceae (I 8). 1 Amazonia: ***N. speciosus*** Ducke

Nyctocalos Teijsm. & Binnend. Bignoniaceae (2). 3 Indomal.

Nyctocereus (A. Berger) Britton & Rose (~ *Peniocereus*). Cactaceae (III 1). 1 C Am.: ***N. serpentinus*** (Lag. & Rodr.) Britton & Rose

Nylandtia Dumort. = *Muraltia*

Nymania Lindb. (*Aitonia*; ~ *Turraea*). Meliaceae (II-Trich.). 1 S Afr.: ***N. capensis*** (Thunb.) Lindb. – shrub with inflated pink to purple capsules, which are blown about (cf. *Alkekengi*), break up & scatter seeds

Nymphaea L. Nymphaeaceae. c. 60 cosmop. (Eur. 4; Aus. 17 + 4 natur. – R: Telopea 13(2011)240). R: PCIW 4(1905), (subg. *Hydrocallis* (G united) 14) SMB 16(1987)), P.D. Slocum *Waterlilies & lotuses* (2005) 79, IJPS 168(2007)655. Waterlilies. Herbs with rhiz. or tubers, the fls solit. on pedicels with phyllotactic spirals independent of those of lvs, which usu. float on water-surface & have stomata, cuticle & palisade on adaxial surface. Fls oft. floating, some night-flowering (subg. *H.* (Am.), subg. *Lotos* (OW)), poll. by beetles, Hymenoptera & syrphid flies; K 4, C ∞ – outer 4 alt. with K & next 4 with those, the 8 being basal members of app. spirals (whorls) of C eventually passing into A ∞ continuing spirals, C & A inserted on 10–20-loc. G (to 47) sunk in receptacle with radiating apical stigmas; fr. a large berry with ∞ seeds with spongy aril entrapping air-bubbles, the fr. maturing under water, dehiscing to release seeds which float until aril decays, when they sink & germinate (cf. *Nuphar*); *N. caerulea* & other spp. have coleoptiles & mesocotyls. Rhiz. & seeds of some spp. ed., many cult. orn. & (many of those) hybrids of obscure parentage. ***N. alba*** L. (Euras., Medit.) – common white-flowered (open most of day), in shallow water in Ukraine with smaller lvs & fls ('*N. minoriflora*', app. environmentally induced), black dye from rhiz.; ***N. ampla*** (Salisb.) DC. (warm & trop. Am.) – narcotic of Mayans; ***N. micrantha*** Guillemin & Perrottet (W Afr.) – plantlets form in umbo of decaying lvs, prob. a parent of ***N. × daubenyana*** Baxter ex Daubeny, which arose in Oxford Botanic Garden, the other prob. ***N. nouchali*** Burm.f. (*N. stellata*, OW trop.) – rhiz. ed. & medic., seeds in rosaries in Ind., nat. fl. Bangladesh; ***N. maculata*** Schum. & Thonn. (*N. caerulea*, *N. nouchali* var. *caerulea*, N & C Afr.), a narcotic in Ancient Egypt, & ***N. lotus*** L. (Egypt, trop. & S Afr., Madag., hotsprings of NW Romania) – found as wreaths on mummies of Rameses II etc. back to 2000 BC, the original Egyptian lotus, *Nelumbo nucifera* not introd. before 500 BC; ***N. odorata*** Aiton (N Am., invasive NW US) – rhiz. medic.; ***N. prolifera*** Wiersama (Arg.) – usu. inside fls are tubers giving rise to other tuberiferous fls; ***N. pubescens*** Willd. (Indomal. to trop. Aus.) – night lotus with white fls, devoted to Ciwa (sacred colour white, cf. *Nelumbo*); ***N. tetragona*** Georgi (*N. pygmaea*, Finland & N Russia to Jap., N Am.) – smallest sp. in cult., much used in hybridization; ***N. thermarum*** Eb. Fischer (Rwanda, hot springs) – smallest of all waterlilies, extinct in wild, saved by succ. germ. at Kew

Nymphaeaceae Salisb. Magnoliidae – Nymphaeales. Incl. Barclayaceae, 5/75 cosmop. JPR 119(2006)561. Aquatic herbs with rhiz. or tubers, oft. with alks; vessels 0; root-hairs from specialized cells, like many monocots, & scattered vascular bundles like them or a single ring. Lvs in spirals on rhiz., usu. rounded cordate & floating; stip. median-axillary or 0. Fls solit. in axils or extra-axillary, bisexual, reg., aerial, insect-poll., hypo- to epigynous; K 4–6 (–12), s.t. petaloid (*Nuphar*), C (0–)8–∞ inserted around top of ovary, usu. free, oft. passing into A 14–∞ (700), in spirals & usu. free & laminar (3-veined) with elongate microsporangia or transitional to forms with differentiated filaments & anthers with longit. slits, pollen grains usu. monosulcate, G (3–) 5–35, ± united into compound sup. to inferior ovary with ∞ scattered anatr. or rarely orthotropous bitegmic ovules. Fr. berry-like ± (irreg.) dehiscent through swelling of mucilage within; seeds arillate, with hooked hairs or nude with 2 cotyledons, wholly distinct or arising as lobes from common primordium (that appearing to be 1 is a true leaf); endosperm scanty, perisperm copious. n = 10–29+

Genera: *Nuphar* (= **Nupharoideae**) – r. with 10–18 xylem poles & large pith, K 5–14 cf. **Nymphaeoideae** (rest) – 5–9 poles, pith 0 or small, K 4 or 5), *Barclaya*, *Euryale*, *Nymphaea*, *Victoria*

N. demonstrate a range of interesting disp. features incl. air-filled arils (*Nymphaea*), pericarp with air-bubbles (*Nuphar*), minute hooked hairs (*Barclaya*) – associated with water or pig-dispersal, etc. Fam. specialized ecologically & has penetrated the temp. zones through a 'loop-hole' in that the rhiz. of perennial spp. persist in the buffered habitat of pond- & river-bottoms; they have irregular morphology, not 'conforming' to morphological 'rules'. Criteria used to keep other homogeneous fams separate are inappropriate in that they would shatter this group, much in the same way that the characters which separate fams in modern classifications prob. represent differences at the specific or even population level in the ancestral groups of angiosperms (Taxon 33(1984)77): inferior or sup. ovaries, C present or 0, laminar or 'orthodox' A, orthotropous or anatr. ovules, 'scattered' (unlike that in modern monocotyledons) or ringed vasc. bundles & varied chemical features. Moreover, spirals of floral parts typical of early 'angiosperms' (how an early angiosperm is recognizable from its 'gynmosperm' ancestor is a moot point) fossils (Nature 319(1986)723) & long assumed to represent a primitive condition, as well as arillate seeds, also held to be primitive (Corner), are here. Despite their advanced aquatic habit, N. have retained beetle-(some thermogenesis) & other insect-poll., another allegedly primitive feature, while their pollen is also of a primitive type. Overall, there is compelling evidence that the fam. is a successful, specialized relic of the stock which existed before the monocotyledons were recognizably distinct from other angiosperms (early Cretaceous fossils referred to Nymphaeales incl. *Pluricarpellatia* Mohr & al. of Braz.) – it is tempting, then, to imagine that that stock, presumably a terr. one, had large insect-poll. fls with. parts in spirals & monosulcate pollen, animal-disp. arillate seeds & 'irregular' morphology & anatomy

Euryale, *Nuphar*, *Nymphaea* & *Victoria* yield ed. rhiz. or seeds & all are cult. orn.

Nymphoides Séguier (*Limnanthemum*). Menyanthaceae. 38 cosmop. (Eur. 1: ***N. peltata*** Kuntze (water-fringe, Euras.), Aus. c. 19). Floating lvs are part of infl. axis (cf. *Nymphaea*); some with air-filled seed-hairs for disp., others (Aus.) disp. by ants. Some ed. tubers & medic. seeds; cult. orn.

Nypa Steck. Palmae (II). 1 Ind. to Ryukyu & Solomon Is. (natur. [weedy] W Afr., Panamá [water-disp. from Afr.?]): ***N. fruticans*** Wurmb (nipa palm) – char. pl. of mangrove swamps with creeping rhiz. with truly dichotomizing apices (Schoute's Model) & fronds to 10 m tall, leaf-bases persisting for 4 yrs & providing aeration through aerenchyma (no pneumatophores), infls with 'irreg.' positions, infr. heads breaking up into char. obovoid angular fibrous sea-disp. fr., the seeds app. germinating before release, the plumule assisting in detachment; fossils known from Cretaceous & Eocene of much of OW incl. England & poss. showing the distrib. of pl. around Tethys Sea, though poss. a drift-seed from further S; lvs used for thatch, cigarette papers, basketry etc., the drosophilid-poll. (also bees, beetles) infls (heating to 10°C above ambient) tapped for sugar (gula malacca, an ingredient of 'three palms pudding' with coconut milk & sago) & toddy, immature seeds ed. R: KSVH IV,10(1964)5

Nypaceae Brongn. ex Le Maout & Decne = Palmae

Nyssa Gronov. ex L. Nyssaceae (Cornaceae). 7 S N Am. (3, R: Sida 15(1992)323), Costa Rica (1), China (3). Some timbers & cult. orn. decid. trees (tupelo) with bright autumn colour, esp. ***N. aquatica*** L. (cotton gum, water t., SE US) – swamp tree with loop-like pneumatophores, standing floods of 2 m or more, water-disp., good bee-pl., wood for broom-handles, clogs & rootwood for net-floats; ***N. ogeche*** Bartram ex Marshall (ogeechee lime, SE US) – fr. ed. when preserved; ***N. sinensis*** Oliv. (Chinese tupelo, C China); ***N. sylvatica*** Marshall (black gum, cotton g., pepperidge, SE N Am.) – cobalt indicator, the principal tupelo timber used for wharves, veneers, pulp etc., anthelminthic

Nyssaceae Juss. ex Dumort. (Cornaceae s.l.). Magnoliidae – Cornales. Incl. Davidiaceae, Mastixiaceae, 5/30 Indomal. to China, S N Am. to C Am. Trees & shrubs. Lvs usu. in spirals. Fls 4- or 5-merous. A 4–26; G (5–10). Fr. 1–5-seeded. n = 11, 13, 21, 22
Genera: *Camptotheca, Davidia, Diplopanax, Mastixia, Nyssa*
Diplopanax form. in Araliaceae. Fossils widespread N hemisphere
Timber (*Mastixia, Nyssa*), medic. (*Camptotheca*), cult. orn. (*Davidia, Nyssa*)

Nyssanthes R. Br. Amaranthaceae (I 2). 4 E Aus. R: Austrobaileya 8(2011)268. P 4, 2 inner smaller

O

oak *Quercus* spp., in GB usu. *Q. robur* (**English, French, Polish** or **Slavonian o.**) or *Q. petraea*; **African o.** *Oldfieldia africana, Lophira* spp.; **Algerian o.** *Q. canariensis*; **(American) red o.** *Q. rubra, Q. falcata* etc.; **(A.) white o.** *Q. alba, Q. montana*; **o. apple** gall on *Q.* spp.; **Australian o.** *Eucalyptus regnans*; **bear o.** *Q. ilicifolia*; **black o.** *Q. emoryi*; **Botany Bay o.** *Allocasuarina* spp.; **Brazilian o.** *Posoqueria latifolia*; **bull o.** *Allocasuarina luehmannii*; **bur(r) o.** *Q. macrocarpa*; **Californian o.** *Q. lobata*; **Ceylon o.** *Careya arborea, Schleichera oleosa*; **chestnut o.** *Q. montana*; **cork o.** *Q. suber*; **durmast** or **sessile o.** *Q. petraea*; **evergreen** or **holly o.** *Q. ilex*; **o. fern** *Gymnocarpium dryopteris*; **holm o.** *Q. ilex*; **iron o.** *Q. stellata*; **jack o.** *Q. marilandica*; **Japanese o.** *Q. mongolica* etc.; **Kermes o.** *Q. coccifera*; **live o.** *Q. virginiana*; **Lucombe o.** *Q.* × *crenata*; **Manchurian o.** *Q. mongolica*; **manna o.** *Q. cerris*; **Oregon white o.** *Q. garryana*; **patana o.** *Careya arborea*; **pin o.** *Q. palustris*; **poison o.** *Rhus radicans*; **possum o.** *Q. nigra*; **post o.** *Q. stellata*; **Quebec o.** *Q. alba*; **river o.** *Casuarina cunninghamiana*; **satin o.** *Alloxylon* spp.; **scarlet o.** *Q. coccinea*; **she-o.** *Casuarina* spp.; **shin o.** *Q. gambelii*; **shingle o.** *Q. imbricaria*; **silky o.** *Grevillea robusta, Cardwellia sublimis*; **Spanish o.** *Inga laurina*; **swamp o.** *Casuarina* spp.; **Tasmanian o.** *Eucalyptus delegatensis*; **tulip o.** *Argyrodendon* spp.; **Turkey o.** *Q. cerris*

Oakes-amesia C. Schweinf. & P. Allen = *Ornithocephalus*

Oakesiella Small = *Uvularia*

oakum untwisted rope used to caulk boats or as surgical dressings, today from linen or hemp

oat *Avena sativa*; **black o. grass** *Stipa avenacea*; **false o.** *Arrhenatherum elatius*; **o. grass** *Avena, Arrhenatherum* & *Helictotrichon* spp.; **swamp o.** *Sphenopholis pensylvanica*; **water o.** *Zizania* spp.; **wild o.** *Avena sativa*; **yellow o.** *T. flavescens*

Oaxacania Robinson & Greenman (~ *Hofmeisteria*). Compositae (Eupt.-Oaxac.). 2 Baja Calif.

obada *Ficus lutea*

obeche *Triplochiton scleroxylon*

obedient plant *Physostegia virginiana*

Oberna Adans. = *Silene*

Oberonia Lindl. Orchidaceae (V 11b). c. 300 OW trop. (Afr. to Masc. 1; Ind. 41, R: OM 4(1990)). Some locally medic., few cult. orn.

Oberonioides Szlach. = *Malaxis*

Obetia Gaudich. Urticaceae (I). 8 trop. & S Afr., Madag., Masc. R: KB 38(1983)221. Trees with stinging hairs

Obione Gaertn. = *Atriplex*

oblionker *Aesculus hippocastanum*

Oblivia Strother (~ *Otopappus*). Compositae (Helia.-Ecl.). 3 trop. Am. R: SB 14(1989)541

Obolaria L. Gentianaceae. 1 E N Am.: ***O. virginica*** L. – mycotroph (cf. *Voyria*), purplish green with scale lvs

Obolinga Barneby = *Zygia*

Obregonia Frič. Cactaceae (III 9). 1 NE Mex.: ***O. denegrii*** Frič – cult. orn. like *Leuchtenbergia* but tubercles short

Obtegomaria Doroszenko & Cantino (~ *Satureja*). Labiatae (VII 2b). 1 NE Colombia: ***O. caerulescens*** (Benth.) Doroszenko & Cantino. R: Novon 8(1998)2

oca *Oxalis tuberosa;* **o. quina** *Ullucus tuberosus*

Oceanopapaver Guillaumin = *Corchorus*

Ocellochloa Zuloaga & Morrone (~ *Panicum*). Gramineae (XXIII 3). 12 trop. Am. R: SB 34(2009)688

Ochagavia Philippi. Bromeliaceae (3). 4 Chile incl. Juan Fernandez. R: Willd. 32(2002)336. ***O. carnea*** (Beer) L.B. Sm. & Looser (*O. lindleyana*, coastal C Chile) – cult. orn., natur. Scilly Is.

Ochanostachys Masters. Olacaceae (Coulaceae). 1 W Mal.: ***O. amentacea*** Masters – fr. ed., timber for furniture etc. R: FPM 2(2012)309

Ochetophila Poepp. ex Endl. (~ *Discaria*). Rhamnaceae (2). 2 S Am. R: NZJB 43(2005)866

Ochlandra Thw. Gramineae (V 5). 9 Sri Lanka (1), S Ind. R: Rheedea 5(1995)64, 9(1999)31. Bamboos; lodicules 1–15 & to 1.5 cm long, A 15–120, fr. fleshy; some used for pulp in Ind. ***O. scriptoria*** (Dennst.) C. Fischer (flowering annually) & ***O. travancorica*** (Bedd.) Gamble take over from cardamom under trees, preventing forest regeneration leading to 'brakes' in S Ind., *O. t.* favoured by elephants

Ochna L. Ochnaceae (II). 86 OW trop. Roux's Model; K coloured, persistent; G 3–12 with common style, which falls after fert., when receptacle becomes fleshy & each carpel becomes an oil-rich drupelet attractive to birds. Cult. orn. trees & shrubs, some with ed. fr. ***O. pulchra*** Hook. (S Afr.) – oil for candles, soap; ***O. subcordata*** (Stapf) Engl. (Afr.) – flagelliflory

Ochnaceae DC. Magnoliidae – Malpighiales. Excl. *Diegodendron & Strasburgeria*, incl. Medusagynaceae & Quiinaceae (R: AMBG 89(2002)65), 33/585 trop. esp. Braz. R: BJ 113(1991)105. Everg. trees & shrubs, few lianes or herbs, usu. glabrous, young stems usu. with cortical & s.t. medullary vasc. bundles. Lvs usu. simple, in spirals, & with v. many parallel lat. veins; petioles with siphonosteles; stip. present (0 in *Medusagyne*). Fls bisex., ± reg., in panicles etc., rarely solit. fls in axils; K (3–)5(–12), imbr., oft. persistent, some s.t. larger than rest, C (4)5(–10) convolute (rarely imbr.), A (1, 5,10–)∞ (s.t. in 3–5 whorls), assoc. with 5 trunk-bundles with members arising centripetally, oft. eccentrically positioned & s.t. on long androgynophore, anthers usu. with term. pores, staminodes s.t. internal to A or a tube or lobed disk around $\underline{G}$ (1)2–15(-25, *Medusagyne*), united at least by a common style, the ovary ± partitioned with ± distinctly axile placentation or so deeply lobed that carpels appear distinct, when receptacle oft. enlarging in fr., each loc. with 1–∞ anatropous to almost campylotropous bitegmic (unitegmic in *Lophira*) ovules. Fr. var., oft. of distinct 1-seeded drupelets, s.t. a capsule, nut or drupe; seeds oft. winged, the endosperm oily & proteinaceous or 0. x = 12, 14 (?7)

Classification & principal genera:

I. **Luxemburgioideae** (fls obliquely zygom., A filaments united) – *Luxemburgia*

II. **Ochnoideae** (fls reg., filaments free) – *Brackenridgea, Lophira, Ochna, Ouratea*

III. **Sauvagesioideae** (fls zygom., filaments free) – *Sauvagesia*

Also from Medusagynaceae (fls reg., filaments free, G 16–25), *Medusagyne* & Quiinaceae (fr. berryoid, ribbed), *Quiina*

Buzz-poll. Some timber & oil (*Lophira, Ouratea*), dyes (*Brackenridgea*), ed. fr. (*Ochna*), & cult. orn. (*Ochna, Ouratea*)

Ochoterenaea F. Barkley. Anacardiaceae (I). 1 Colombia: ***O. colombiana*** F. Barkley. R: BTBC 69(1942)442

Ochotonophila Gilli = *Acanthophyllum* (but see R: FR 59(1956)169)

Ochradenus Del. Resedaceae. 6 Middle E to Socotra, Libya, Pakistan. R: NRBGE 41(1984)491

Ochreata (Lojac.) Bobrov = *Trifolium*

Ochreinauclea Ridsd. & Bakh.f. = *Neonauclea* (but see Blumea 24(1978)331)

Ochrocarpos Noronha ex Thouars = *Garcinia*

Ochrocephala Dittr. (~ *Rhaponticum*). Compositae (Card.-Cent.). 1 N trop. Afr. savanna: ***O. imatongensis*** (Philipson) Dittr. – capitulum v. large. R: BJ 103(1983)476

Ochroma Sw. Malvaceae (Bomb.; Bombacaceae). 1 trop. Am.: ***O. pyramidale*** (Lam.) Urb. (*O. lagopus*, balsa, down tree, cork t.) – fast-growing tree (Koriba's Model, cf. *Ceiba*) of clearings, seeds requiring v. high temps for germ., fls to 12 × 8 cm, bat-poll. ('mushroom' smell due to fatty acid derivatives with C_8 skeleton; also effectively coati-poll.), seed embedded in hairs (cf. *Ceiba*), world's lightest comm. timber used for insulation, model aeroplanes, architect's models, form. for life-belts etc. (Kon-Tiki voyage rafts)

Ochropteris J. Sm. Pteridaceae (VI). 2 Madag., Masc.

Ochrosia Juss. Apocynaceae (I b). (Incl. *Neisosperma*) c. 40 Masc. & Seychelles to Aus. (Mal. 13 – R: Blumea 49(2004)104) & W Pac. R: AUWP 87–5(1987)47. Koriba's Model. Some timber (yellow wood), medic. & dyes. ***O. elliptica*** Labill. (pokosola, trop. Aus. to New Caled.) – ed. seeds

Ochrosperma Trudgen (~ *Baeckea*). Myrtaceae (II 15). 6 E Aus. R: Nuytsia 6(1987)9.

Ochrothallus Pierre ex Baill. = *Pycnandra*

Ochthephilus Wurd. Melastomataceae. 1 Guyana: ***O. repentinus*** Wurd. R: MNYBG 23(1972)197

Ochthocharis Blume. Melastomataceae. 2 trop. Afr., 5 trop. As. R: KB 36(1981)13

Ochthochloa Edgew. = *Chloris* (but see KB 36(1981)560)

Ochthocosmus Benth. Ixonanthaceae. 16 trop. S Am. (7, esp. Guayana Highland, R: Britt. 32(1980)128), trop. Afr. (8)

Ochthodium DC. Cruciferae. 1 E Medit.: ***O. aegyptiacum*** (L.) DC.

Ochyrella Szlach. & R. González = *Eltroplectris*

Ocimum Tourn. ex L. Labiatae (VII 3d). 65 trop. (& subtrop.) esp. Afr. (incl. *Becium*). Some indicators of heavy metals, also ***O. obovatum*** E. Meyer ex Benth. (*B. o.*, C Afr.) for copper (also cult. orn.). Many aromatic herbs & shrubs with thymol-containing ess. oils used in flavouring etc. (basil), esp. ***O. × africanum*** Lour. (*O. × citriodorum*, *O. americanum × O. basilicum*, sweet or lemon b., kemanji) – cult. for oil; ***O americanum*** L. (*O. canum*, hoary b., lime b., partminger, trop. OW!) – lime flavour, good with seafood, potherb in Ind., medic. in Sudan; ***O. basilicum*** L. (trop. As.) – known to Pharaohs, used as fly-repellent, snuff, scent, c. 100 t oil now prod. a year, chopped lvs used in casseroles, sauces, pizza & liqueurs esp. in Medit., local tonic (flavour linalool, main sweet constituent estragol, a phenylpropanoid), juvenile hormone analogue in lvs, seeds used in frogspawn-like puddings, '**Minimum**' (*O. minimum*, bush or Greek b.); ***O. campechianum*** Mill. (Peruvian b., trop. & warm Am.); ***O. gratissimum*** L. (*O. suave*, *O. urticifolium*, S As., trop. Afr.) – oil contains eugenol, trad. medic. in Afr., infusion a louse-remover in N Uganda; ***O. tenuiflorum*** L. (*O. sanctum*, holy or Thai b., tulsi, Indomal.) – mosquito-repellent herb sacred to Hindus & found nr every Hindu house in Ind., stem-sections used in rosaries there

Oclemena Greene (~ *Aster*). Compositae (Ast.) 3 N Am. R: FNA 20(2006)78

Oclorosis Raf. Older name for *Iodanthus*

Ocotea Aubl. Lauraceae (I). c. 200 trop. & warm Am. (Peru 60) with few Macaronesia (1), trop. & S Afr. (7) to Madag. (30) & Masc. R: MIABH 20(1986)87. With *Guatteria*, *Inga*, largest trop. Am. tree genus. Massart's Model.; alks. ***O. insularis*** (Meissn.) Mez (*O. pedalifolia*, trop. Am.) – hollow ant-infested stems. Timbers (louro in S Am.) esp. ***O. bullata*** (Burchell) Baill. ((black) stinkwood, S Afr.), ***O. cymbarum*** Kunth (*O. barcellensis*, l. inamui, Braz.) – fragrant bark (Orinoco sassafras) form. comm. & ***O. usambarensis*** Engl. (E Afr. camphorwood, trop. E Afr.) – endangered as overexploited. ***O. foetens*** (Aiton) Baill. – part of Madeiran laurel forest; ***O. quixos*** (Lam.) Schmidt (Ecuadorian Andes) – cinnamaldehyde presence inspired members of 1514 voyage of Orellana so that they thought they saw female warriors whom they called Amazonians, hence the name of the river. See also *Chlorocardium*, *Sextonia*

ocotillo *Fouquieria splendens*

Octamyrtus Diels. Myrtaceae (II 10). 6 Moluccas, NG. R: GBS 56(2004)148

Octarrhena Thw. Orchidaceae (V 15). Incl. *Chitonanthera*, c. 50 Sri Lanka, Mal., New Caled. (2), Polynesia

Octerium Salisb. = *Deidamia*
Octoceras Bunge. Cruciferae (26). 1 C As. to Iran & Afghanistan: ***O. lehmannianum*** Bunge
Octoclinis F. Muell. = *Callitris*
Octoknema Pierre. Erythropalaceae (Olacaceae s.l.). 14 trop. Afr. R: KB 66(2011)368. Roux's Model
Octoknemaceae Tiegh. = Erythropalaceae
Octolepis Oliv. Thymelaeaceae (Oct.). 6 trop. Afr. (1 monoec.), Madag. (5 dioec.). R: Adans. 27(2005)91
Octolobus Benth. & Hook. f. Malvaceae (Sterc.; Sterculiaceae). 3 trop. Afr. R&: KB 1937: 394
Octomeles Miq. Datiscaceae (Tetramelaceae). 1 Mal. (not Malay Pen., Java): ***O. sumatrana*** Miq. (binuang, erima (NG)) – Massart's Model, soft pale timber good for canoes, weatherboarding, plywood, veneers etc. but large crown diff. silviculturally though in plantation in NG
Octomeria R. Br. (~ *Eria*). Orchidaceae (V 13c). Excl. *Pleurothallopsis*, c. 100 trop. Am. esp. Braz. Some cult. orn. epiphytes
Octomeron F. Robyns = *Platostoma*
Octopoma N.E. Br. (~ *Ruschia*). Aizoaceae (V). 3 Little Karoo, S Afr. R: T 65(2016)259
Octospermum Airy Shaw = *Mallotus*
Octotheca R. Viguier = *Schefflera*
Octotropis Beddome. Rubiaceae (? III 5). 2 S Ind., Myanmar
Ocyroe Philippi (~ *Nardophyllum*). Compositae. 1 SW N Am.: ***O. armata*** (Wedd.) Bonif. R: Britt. 60(2008)207
Oddoniodendron De Wild. Leguminosae (I 2). 6 Gulf of Guinea. R: Adans. 26(2004)242
Odicardis Raf. = *Veronica*
Odina Roxb. = *Lannea*
Odixia Orch. (~ *Cassinia*). Compositae (Gnap-Cass.). 2 Tasmania. R: OB 104(1991)87
odiyal flour *Borassus flabellifer*
odoko *Scottellia coriacea*
Odonellia K. Robertson (~ *Jacquemontia*). Convolvulaceae (1). 2 trop. Am. R: Britt. 34(1982)417
Odontadenia Benth. Apocynaceae (II c). 20 trop. Am. R: BJBB 67(1999)389. Some cult. lianes; flea-, louse- & mosquito-repellents in Colombia
Odontanthera Wight = *Cynanchum*
Odontelytrum Hackel = *Cenchrus* (but see Strelitzia 10(2000)706)
Odontitella Rothm. (~ *Odontites*). Orobanchaceae (Rhin.). 1 Iberia: ***O. virgata*** (Link) Rothm.
Odontites Ludwig. Orobanchaceae (Rhin.; Scrophulariaceae s.l.). 34 W & S Eur. (14), Medit. to Himal. R: Willd. 26(1996)79. Hemiparasites esp. ***O. vernus*** (Bellardi) Dumort. (red bartsia, Euras.)
Odontocarya Miers. Menispermaceae (III). 30 trop. Am. R: MNYBG 202(1970)82. Some locally med. NW Amazonia, ***O. asarifolia*** Barneby – ed. in Chaco
Odontochilus Blume (~ *Anoectochilus*). Orchidaceae (IV 2d). Excl. *Chamaegastrodia* c. 25 Himal., Indomal. to Hawaii
Odontocline R. Nordenstam (~ *Senecio*). Compositae (Sen.-Sen.). 6 Jamaica. Trees & climbers
Odontoglossum Kunth = *Oncidium* (but see L. Bockemühl (1989) *Odontoglossum*)
Odontonema Nees (~ *Justicia*). Acanthaceae (III 2c). 26 trop. Am. Some cult. orn. (*Thyrsacanthus*)
Odontonemella Lindau = *Mackaya*
Odontophorus N.E. Br. Aizoaceae (V). 4 Namaqualand. R: H. E. K. Hartmann, *Ill. Handb. Succ. Pls*, Aiz. F–Z(2002)191
Odontophyllum Sreemad. = *Aphelandra*
Odontorrhynchus Correa. Orchidaceae (IV 2h). 6 Andes. R: HBML 28(1982)340
Odontosoria Fée. Lindsaeaceae (Dennstaedtiaceae s.l.). Excl. *Sphenomeris*, 15 trop. extending to Korea & Florida. Heterogeneous? Lvs of indefinite growth forming spiny thickets (cf. *Dicranopteris*). Cult. orn. incl. ***O. chinensis*** (L.) J. Sm. (*S. chinensis*, E As. to Polynesia) – can be woody, local medic., red dye obtained from fronds
Odontospermum Necker ex Schultz-Bip. = *Asteriscus*
Odontostelma Rendle. Apocynaceae (V c; Asclepiadaceae III 1). 1 S trop. Afr.: ***O. welwitschii*** Rendle. R: F. Albers & U. Meve, *Ill. Handb. Succ. Pls*, Asclep. (2002)186

Odontostemma Benth. ex G. Don (~ *Arenaria*). Caryophyllaceae. c. 65 Himal., W China (59, 57 endemic). R: PhytoKeys 63(2016)77

Odontostomum Torrey. Tecophilaeaceae. 1 California: ***O. hartwegii*** Torrey. R: FNA 26(2002)204

Odontotrichum Zucc. = *Psacalium*

Odosicyos Keraudren = *Tricyclandra*

odum *Milicia excelsa*

Odyendea Pierre ex Engl. (~ *Quassia*). Simaroubaceae. 1 Gabon-Cameroun: ***O. gabonensis*** (Pierre) Engl.

Odyssea Stapf. Gramineae (XXIX). 2 trop. & SW Afr. & Red Sea coasts. Xerophytes

Oeceoclades Lindl. (~ *Eulophia*). Orchidaceae (V 12b). c. 40 trop. Am. (1), Afr. & W Ind. Ocean. R: HBML 24(1976)249. Some cult. orn. incl. ***O. maculata*** (Lindl.) Lindl. (trop. Afr.) – spreading weed (introd. Florida (?naturally) 1974, now widespread, poss. on slaving-ships before that), autogamous with movement of pollinia to stigma effected by rain

Oecopetalum Greenman & C. Thompson. Icacinaceae. 2 Mex. & C Am.

Oedematopus Planch. & Triana = *Clusia*

Oedera L. Compositae (Gnap.-Rel.). 18 S Afr. R: AMBG 78(1991)1071. Shrubs

Oedibasis Kozo-Polj. Umbelliferae (III 8). 4 C As.

Oedina Tieghem (~ *Dendrophthoe*). Loranthaceae. 4 Tanzania, N Malawi. R: R. Polhill & D. Wiens, *Mistletoes of Afr.* (1998)225

Oedochloa C. Silva & R.P. Oliveira. Gramineae. 9 trop. S Am. R: MPE 93(2015)230

Oemleria Reichb. (*Osmaronia*). Rosaceae (Ros.-Exoch.). 1 W N Am.: ***O. cerasiformis*** (Hook. & Arn.) Landon (Ind. plum, oso-berry) – like *Prunus* but G 5 free, dioec., with fragrant fls & ed. fr., cult. orn. early-flowering shrub. R: FNA 9(2014)385

Oenanthe L. Umbelliferae (III 8). 40 N temp. (Eur. 13), Indomal. to Aus. & trop. Afr. mts. Toxic polyacetylene hydrocarbons; some with tubers e.g. ***O. pteridifolia*** Lowe (Madeira). Water dropwort esp. ***O. crocata*** L. (water hemlock, Eur.) – poisonous (most toxic UK pl.) & used to stupify fish & ***O. aquatica*** (L.) Poir. (water fennel) – poisonous. ***O. javanica*** (Blume) DC. (Indomal.) – imp. vegetable in Taiwan & Mal., eaten with rice. Cult. orn. incl. ***O. pimpinelloides*** L. (Eur. to SW As.) – natur., forming almost pure stands, in NZ

Oenocarpus Mart. Palmae (V 10). (Incl. *Jessenia*) 9 trop. S Am. ***O. bataua*** Mart. (*J. bataua, J. polycarpa*, pataua, seje) – human disp. from NW Amaz., form. (archaeological sites 9000 ys old) imp. oil like olive o., 18–24% of pericarp (only 1% of kernel) but 40% more protein than soya, spines used as darts in blowpipes; ***O. distichus*** Mart. (bacaba palm) – fr. an oil source

Oenosciadium Pomel = *Oenanthe*

Oenostachys Bullock = *Gladiolus*

Oenothera Tourn. ex L. Onagraceae. Incl. *Calylophus* (R: AMBG 64(1977)67), *Gaura* (R: MTBC 23(1972)1), *Stenosiphon*, 145 Am. (esp. temp.), many natur. elsewhere (e.g. GB where hybridizing). R: SBM 24(1988)1, 50(1997)1, 83(2007)138. Evening primroses; many sphingid-poll. Endosperm diploid; chromosomes of some species in rings, with the rings passed through to next generation without any recombination, combinations of rings giving new genotypes sometimes with startling morphological features: an ignorance of the mechanism led de Vries to argue that these were mutations (he also had polyploids as well) & that mutations of great magnitude were the stuff of evolution, where Darwin had argued for the accumulation of small ones. Some medic. & ed. seeds (also roots boiled or as hors d'oeuvres etc.), seeds a source of gamma-linolenic acid imp. in production of fatty acids & prostaglandins & used (evening primrose oil; £36M- worth sold in 1993) in treatment of premenstrual tension, eczema etc. though efficacy questioned as placebos found to be as good (more gamma-linolenic acid in borage anyway), so some grown as crops incl. ***O. glazioviana*** Micheli (*O. erythrosepala*, orig. Eur.). Cult. orn. with day fls e.g. ***O. lindheimeri*** (Engelm. & A. Gray) W.L. Wagner & Hoch (*G. l.*, SW N Am.) – cut-fl., natur. Aus., or opening & scented in evening, poll. moths, e.g. ***O. macrocarpa*** Pursh (*O. missouriensis*, W N Am.), ***O. salicifolia*** Desf. ex Lehm. (Mex.), ***O. speciosa*** Nutt. (N Am.) – esp. pink cvs. ***O. biennis*** L. (N Am.) – seeds germinated after 80 but not 90 years of Dr Beal's experiment set up in 1879, roots & lvs ed., sup. to borage (low in saturated fats)

Oenotheraceae C.C. Robin = Onagraceae

Oenotheridium Reiche = *Clarkia*

Oenotrichia Copel. (~ *Microlepia*). Dennstaedtiaceae. 1 NG, Aus., New Caled.: ***O. maxima*** (E. Fourn.) Copel.

Oeonia Lindl. Orchidaceae (V 16d). 5 Madag., 1 ext. to Masc. R: BMNHN 4,11(1989)157

Oeoniella Schlechter. Orchidaceae (V 16). 5 Mascarenes

Oerstedella Reichb.f. (~ *Epidendrum*). Orchidaceae (V 13). 32 trop. Am. Cult. orn.

Oerstedianthus Lundell = *Ardisia*

Oerstedina Wiehler. Gesneriaceae (II 5c). 3 C Am.

Oestlundia W. Higgins (~ *Encyclia*). Orchidaceae (V 13b). 4 trop. Am. R: Selbyana 22(2001)1

Oestlundorchis Szlach. = *Deiregyne*

Ofaiston Raf. Amaranthaceae (Chenopodiaceae III 3). 1 S Russia to C As.: ***O. monandrum*** (Pallas) Moq.

ofram *Terminalia superba*

Oftia Adans. Scrophulariaceae (Teed.). 3 S Afr. R: BN 124(1971)451; see *Ranopisoa* for Madag. sp.

Ogastemma Brummitt (*Megastoma*). Boraginaceae (I). 1 Canary Is. & N Afr.: ***O. pusillum*** (Bonnet & Barratte) Brummitt – red dye used as cosmetic in N Afr. R: BotJLS 130(1999)251

Ogcodeia Bur. = *Naucleopsis*

ogea *Daniellia ogea*

ogeechee lime *Nyssa ogeche*

Oglifa (Cass.) Cass. = *Logfia*

Ohbaea Byalt & Sokolova (*Balfouria* (H. Ohba) H. Ohba) = *Sedum* (but see KB 54(1999)476)

'ohelo berry' *Vaccinium reticulatum*

ohia *Syzygium malaccense*

Ohwia Ohashi (~ *Desmodium*). Leguminosae (III 19). 2 Indomal., E As.

Oianthus Benth. = *Heterostemma*

oil hydrocarbon deposits almost entirely derived from plant & bacteria remains decaying at bottom of lakes, seas etc.; **o. grass** *Cymbopogon* & *Chrysopogon* spp., esp. *Cymbopogon jwarancusa*; **o. palm** *Elaeis guineensis*

oioi *Apodasmia similis*

Oiospermum Less. Compositae (Vern.-Cent.). 1 NE Braz.: ***O. involucratum*** (Spreng.) Less.

Oistanthera Markgraf = *Tabernaemontana*

Oistonema Schltr. = *Dischidia*

oiticica *Licania* spp.

okahijiki *Kali komarovii*

okan *Cylicodiscus gabunensis*

okari *Terminalia kaernbachii*

Okenia Schldl. & Cham. Nyctaginaceae (6). 1–2 Florida to Mex. & Nicaragua

ok-gue *Ficus pumila*

Okinawa spinach *Gynura bicolor*

Okoubaka Pellegrin & Normand. Santalaceae (Cervantesiaceae). 1–2 trop. Afr. Trees (Mangenot's Model) to 40 m, parasitizing all around; seeds to 43 g (biggest of any parasite)

okoumé *Aucoumea klaineana*

okra *Hibiscus esculentus*

okwen *Brachystegia* spp.

Olacaceae R. Br. Incl. Aptandraceae, Ximeniaceae, excl. Erythropalaceae, Schoepfiaceae; ~ Santalaceae). Magnoliidae – Santalales. 23/160 trop., S Afr. Usu. everg. root-parasites, trees, lianes & shrubs. Lvs simple, entire, alt.; stip. 0. Fls small, bisex. (or pl. dioec.), reg., hypogynous (epi- or perigynous), in axillary panicles, racemes or heads; K ± cupular & ± 3–6-toothed, oft. accrescent in fr., C 3–6, usu. valvate & s.t. basally united, disk annular or glands alt. with C outside A or around G, A same no. & opp. C or 2–5 times as many in 1 whorl with some staminodes, filaments sometimes basally united or adnate to C, $\underline{G}$ ((2)3(–5)) with term. style & usu. basally 2–5-loc. & apically 1-loc., each loc. with 1 pendulous, usu. anatropous, bitegmic or unitegmic ovule from top of free-central (or axile) placenta. Fr. a usu. 1-seeded drupe or nut oft. encl. by accrescent K; seeds with small embryo near tip of copious oily (& s.t. starchy) endosperm, thin testa, cotyledons 2–6. x = 19, 20

Principal genera: *Anacolosa, Dulacia, Olax* (11 genera monospecific); this is heterogeneous & has been split into 8 fams but as yet there is no general agreement on a stable classification (BotJLS 181(2016)7)

Some timbers & ed. fr. (*Ochanostachys, Ximenia*), ed. lvs (*Olax*), oils (*Aptandra, Curupira, Ongokea, Ximenia*), local spices & medic.

Olax L. Olacaceae. c. 50 OW trop., S Afr. Prob. includes *Dulacia*. Troll's Model, r.-parasites. Some lvs & fr. smell of garlic (cf. *Scorodocarpus*) – local condiments, lvs of ***O. zeylanica*** L. (Sri Lanka) ed. ***O. nana*** Wall. ex Benth. (W Himal.) – ann. shoots from woody stock; ***O. psittacorum*** Vahl – haustoria on a range of host trees; ***O. subscorpioidea*** Oliv. (W Afr.) – chewing-stick in Nigeria

old maid *Catharanthus roseus*; **o. man** *Artemisia abrotanum*; **o. m.'s beard** *Clematis* spp. esp. *C. vitalba, Chionanthus virginicus, Tillandsia usneoides*; **o. m. cactus** *Cephalocereus senilis*; **o. m. saltbush** *Atriplex nummularia*

Oldeania Stapleton (~ *Yushania*). Gramineae (IV). 1 trop. Afr. mts (?+ Madag. spp.): ***O. alpina*** (K. Schum.) Stapleton (*Arundinaria a., Y. a.*) – 15–20-yr life-cycle (Rwanda), forming altitudinal belts (e.g. 650 km^2 on Aberdares, Kenya) with pls to 19.5 m high. R: Phytokeys 25(2013)100

Oldenburgia Less. Compositae (Card.-Old.). 4 Cape. R: SAJB 53(1987)493. Subshrubs (2), cushion-pls (2)

Oldenlandia Plum. ex L. (~ *Hedyotis*). Rubiaceae (IV 15). Excl. *Exallage, Involucrella, Leptopetalum, Scleromitrion*, c. 250 trop. esp. Afr. Shrubs & herbs. Dyes esp. from ***O. umbellata*** L. (*H. puberula* (G. Don) Arn., chay, Ind. madder, Indomal.) – roots boiled yield yellow colour turning red in alkali & used for cloth esp. turbans

Oldenlandiopsis Terrell & W. Lewis (~ *Oldenlandia*). Rubiaceae (IV 1). 1 Carib.: ***O. callitrichoides*** (Griseb.) Terrell & W. Lewis. R: Britt. 42(1990)185

Oldfeltia R. Nordenstam & Lundin (~ *Senecio*). Compositae (Senec.-Paiv.). 1 Cuba: ***O. polyphlebia*** (Griseb.) R. Nordenstam & Lundin – tree. R: CN 38(2002)66

Oldfieldia Benth. & Hook. f. Picrodendraceae (Pic.; Euphorbiaceae s.l.). 4 trop. Afr. R: BJBB 26(1956)338. Lvs opp. or in spirals; seeds reminiscent of Meliaceae. Timber from ***O. africana*** Benth. & Hook. f. (Afr. oak or teak)

Olea Tourn. ex L. Oleaceae (4d). 33 OW trop. & warm temp. R: KB 57(2002)91. Olives. Everg. trees & shrubs with extrafl. nectaries. ***O. europaea*** L. (olive, complex from Macaronesia (subsp. ***cerasiformis*** Kunkel & Sunding, Madeira; subsp. ***guanchica*** Vargas & al., Canary Is.) to W Ind., Himal. & S Afr.) – subsp. ***europaea*** (incl. var. *sylvestris*, oleaster, spontaneous from seeds of cvs in Med.) long-lived (to c. 2000 yrs; those up to 1000 yrs now stolen in Spain for landscape designs in Eur.) cultigen of Medit. with large drupe with oily mesocarp, poss. derived from subsp. ***cuspidata*** (G. Don) Ciferri (subsp. *africana, O. cuspidata*, Ind. o., Arabia, As., invasive in Aus., Hawaii) with small drupes with thin mesocarp, certainly in cult. N of Dead Sea since 3700–3600 BC for mono-unsaturated oil & fresh or preserved drupes, the fr. used in cooking or preserved in brine (now 1.5 M t a yr) & eaten as appetizer sometimes stuffed with red pepper etc., black or green, many cvs (large or small etc.), expressed oil (1 M t oil per annum) used in cooking, as salad oil (best = virgin oil) & for preserving tinned sardines, lubrication, lighting & soap, a green form in Greece being made of pyrene oil from the 'stones [endocarp]', also medic. (anti-inflammatory polyphenols highest in fresh oil; bowel regulation & poss. insulin subs., a spoonful before an evening's drinking helping to prevent drunkenness; oleuropein (bitter glycoside usu. removed in processing) & squalene stop cholesterol oxidizing & becoming toxic), also medic. 'olive leaf extract' (for cold prevention), twigs used as toothbrushes to promote healthy gums in Saudi Arabia, source of manna in Dhofar, foliage an anc. sign of good will – 'olive branch' (a dove brought Noah one), crown of o. lvs hung on door when boy born in anc. Rome, crown of olives presented to Olympic victors & poets, though dried lvs now an oregano adulterant (to 30%); all (Italy 2013, Fr. 2015) now threatened by Olive Quick Decline Syndrome due to *Xylella fastidiosa* bacteria (orig. Am.). Good timber esp. *O. europaea* subsp. *cuspidata*, ***O. capensis*** L. (*O. laurifolia*, S Afr.) & ***O. welwitschii*** (Knobl.) Gilg & Schellenb. (*O. c.* subsp. *welwitschii*, Elgon olive, loliondo, S & E trop. Afr.) & other spp. (ironwood) – heaviest wood known with specific gravity 1.49

Oleaceae Hoffsgg. & Link. Magnoliidae – Lamiales. 25/565 subcosmop. esp. As. Trees & shrubs, sometimes lianoid, usu. with peltate secretory hairs. Nodes somewhat swollen (but no line as in Gentianales or many other opp.-leaved Lamiales). Lvs opp. (spirals in some *Jasminum*), simple, pinnate, 1- or 3-foliolate; stip. 0. Fls reg., usu. bisex., in basically cymose infls or solit.; K 4(–15)-lobed, valvate (0 in *Fraxinus*), C (4(–12, *Jasminum*)), lobes imbr., valvate or convolute, s.t. ± distinct or 0 (*Fraxinus*), A 2(4) attached to C-tube, anthers with longit. slits, s.t. a disk around $\underline{G}$ (2), 2-loc. with term. style, each loc. with

(1)2(–4, ∞) anatropous or amphitropous unitegmic ovules in axile placentas. Fr. a capsule, berry, drupe or samara; seed with straight embryo in oily or 0 endosperm. n = 10, 11, 13–4, 23–4

Classification & chief genera:

1. **Jasmineae** (K 5–15-lobed; fr. a capsule or berry): *Jasminum* (only)
2. **Forsythieae** (K 4-lobed; fr. a tough capsule or samara): *Abeliophyllum*, *Forsythia* (only)
3. **Fontanesieae** (K 4-lobed; C-tube deeply 4-lobed; fr. a samara): *Fontanesia* (only)
4. **Oleeae** (K & C 4-lobed or 0; fr. a drupe or berry or 2-loc. capsule) – 4 subtribes: a. Fraxininae (fr. a samara) – *Fraxinus* (only), b. Schreberinae (fr. a woody capsule) – *Schrebera*, c. Ligustrinae (fr. a drupe to capsule) – *Ligustrum*, *Syringa*, d. Oleinae (fr. a drupe) – *Chionanthus*, *Noronhia*, *Olea*, *Osmanthus*, *Phillyrea*
5. **Myxopyreae** (fr. berry, capsule or schizocarp): *Dimetra*, *Myxopyrum*, *Nyctanthes* (only) Trad. assoc. with Gentianales, but A 2 & absence of internal phloem anomalous there, so Cronquist placed fam. near here

Many timbers (*Olea*, *Fraxinus*, *Nestegis*, *Notelaea*), olives (*Olea*), scents (*Jasminum*, *Nyctanthes*, *Osmanthus*, *Syringa*) & many cult. orn. shrubs & hedging pls (*Abeliophyllum*, *Chionanthus*, *Fontanesia*, *Forestiera*, *Forsythia*, *Fraxinus*, *Jasminum*, *Ligustrum* (privet), *Nyctanthes*, *Osmanthus*, *Phillyrea*, *Syringa* (lilac)): several hardy genera show disjunct As.–Eur. distributions – *Forsythia*, *Ligustrum*, *Osmanthus*, *Syringa* (KB 26(1972)487)

oleander *Nerium oleander*; **yellow o.** *Thevetia peruviana*

Oleandra Cav. Oleandraceae. c. 20 trop. (As. 9 – R: PhytoKeys 11(2012)1). Rauh's Model; fronds simple with sori near midrib. Some with erect stem-branches, forming thickets (almost unique in ferns)

Oleandraceae Ching ex Pichi-Serm. (~ Davalliaceae). Polypodiidae – Polypodiales. 1/20 trop. R: T 55(2006)718. Terr. to epiphytic, sometimes scandent, ferns. Rhiz. dictyostelic, with peltate scales. Frond simple, abscising cleanly. Sori usu. term. on veins & round with reniform to suborbicular indusium. n = 41

Genus: *Oleandra* (*Arthropteris* & *Psammiosorus* moved to Tectariaceae)

Oleandropsis Copel. = *Selliguea*

Olearia Moench. Compositae (Ast.-Hin.). c. 180 Aus. (130), NG (25) & NZ. Tree daisy, daisy bush. Heterogeneous. Trees, shrubs (incl. *Pachystegia* – R: NZJB 25 (1987)144) & herbs. A few timber trees & many cult. orn. (R: JRHS 90(1965)207,245) incl. ***O. albida*** Hook.f. (NZ) – 2n = 400 +!, ***O. avicenniifolia*** (Raoul) Hook.f. (akeake, S Is., NZ) & ***O. insignis*** Hook.f. (*P. i.*, NZ)

oleaster *Elaeagnus* spp., *Olea europaea*

Oleicarpon Airy Shaw = *Dipteryx*

Oleiocarpon Dwyer = *Dipteryx*

Olfersia Raddi (~ *Polybotrya*). Dryopteridaceae (II). 2 trop. Am. R: AFJ 76(1987) 161. ***O. cervina*** (L.) Kunze – cult. orn. terr.

Olgaea Iljin (~ *Alfredia*). Compositae (Card.-Card). 16 C As. to N China

olibanum *Boswellia* spp.

Oligactis (Kunth) Cass. (~ *Liabum*). Compositae (Liab.). 7 trop. Am. R: SCB 54(1983)37

Oligandra Less. = *Lucilia*

Oliganthemum F. Muell. = *Allopterigeron*

Oliganthes Cass. Compositae (Vern.-Centap.). 10 Madag. R: FMad 189(1960)182. Treelets & shrubs

Oligarrhena R. Br. = *Monotoca*

Oligobotrya Bak. = *Maianthemum*

Oligocarpus Less. (~ *Osteospermum*). Compositae (Cal.). 1 S Afr., 1 St Helena

Oligoceras Gagnepain. Euphorbiaceae (Al.-Gross.). 1 Vietnam: ***O. eberhardtii*** Gagnepain – fr. ed.

Oligochaeta (DC.) K. Koch (~ *Myopordon*). Compositae (Card.-Cent.). 4 SW & C As., Ind. R: VGIETHZ 37(1962)315

Oligochaetochilus Szlach. = *Pterostylis*

Oligocladus Chodat & Wilczek. Umbelliferae (II). 2 Arg.

Oligocodon Keay. Rubiaceae (II 1). 1 trop. W Afr.: ***O. cunliffeae*** (Wernham) Keay. R: BJBB 28(1958)36

Oligolobos Gagnepain = *Ottelia*

Oligomeris Cambess. Resedaceae. 1 (***O. linifolia*** (Vahl) Macbr.) Canary Is., N Afr., W Ind. & SW N Am., 2 SW Afr. R: MLW 67–8(1967)69

Oligoneuron Small = *Solidago*

Oligophyton Linder. Orchidaceae (IV 4d). 1 Zimbabwe: ***O. drummondii*** Linder & Williamson. R: KB 41(1986)314

Oligospermum Hong = *Veronica*

Oligostachyum Wang & Ye (~ *Arundinaria*). Gramineae (IV). 15 China. R: FOC 22(2006)127. Heterogeneous

Oligothrix DC. Compositae (Sen.-?Oth.). 1 Cedarberg Mts, S Afr.: ***O. gracilis*** DC. – v. rare. R: Strelitzia 9(2000)345

Olimarabidopsis Al-Shehbaz & al. (~ *Arabidopsis*). Cruciferae (31). 3 E Medit. to W China. R: Novon 9(1999)302

Olinia Thunb. Crypteroniaceae (Oliniaceae). 10 E & S Afr., St Helena. R: KB 68(2013)433

Oliniaceae Arn. ex Sond. = Crypteroniaceae

Olivaea Schultz-Bip. ex Benth. (~ *Grindelia*). Compositae (Ast.-Mac.). 2 Mex. R: Britt. 15(1963)86. Aquatic annuals

olive *Olea europaea;* **Afr. o.** *O. e.* subsp. *cuspidata;* **Autumn o.** *Elaeagnus umbellata;* **Californian o.** *Umbellularia californica;* **Ceylon o.** *Elaeocarpus serratus;* **Chinese o.** *Canarium* spp.; **Elgon o.** *O. welwitschii;* **Indian o.** *O. europaea* subsp. *cuspidata;* **Java o.** *Sterculia foetida;* **Russian o.** *Elaeagnus* spp.; **spurge o.** *Cneorum tricoccum*

Oliveranthus Rose = *Echeveria*

Oliverella Tieghem (~ *Tapinanthus*). Loranthaceae (5 6). 3 E & SC Afr. R: R. Polhill & D.L. Wiens, *Mistletoes of Afr.* (1998)114

Oliveria Vent. Umbelliferae (III 8). 1 Syria to Iran: ***O. decumbens*** Vent. R: Fl. Iraq 5,2(2003)156

Oliveriana Reichb.f. Orchidaceae (V 12h). 6 Colombia

olivillo *Aextoxicon punctatum*

Olmeca Soderstrom. Gramineae (V 3). 5 Mex., Honduras. Bamboos with fleshy or dry fr. R: T 60(2011)93

Olmedia Ruíz & Pavón = *Trophis*

Olmediella Baill. Salicaceae (Flacourtiaceae). 1 C Am.: ***O. betschleriana*** (Goeppert) Loes. (manzanote) – grown as a park tree in C Am. & in Eur. botanic gardens, (?) extinct in wild

Olmedioperebea Ducke = *Maquira*

Olmediophaena Karsten = *Maquira*

Olneya A. Gray (~ *Coursetia*). Leguminosae (III 23). 1 SW N Am.: ***O. tesota*** A. Gray (desert ironwood, tesota) – old dried wood (v. heavy) used for carving, seeds form. eaten by Native Americans

olona *Touchardia latifolia*

Oloptum Röser & Hamasha (~ *Piptatherum*). Gramineae (X). 2 cosmop. R: PSE 298(2012)365

Olsynium Raf. (~ *Sisyrinchium*). Iridaceae (VII 3). c. 12 W N Am. (1), temp. S Am. R: SB 15(1990)507. Nectaries (cf. *S.*). Some cult. orn. incl. ***O. douglasii*** (A. Dietr.) Bicknell (grass widow, W N Am.)

olulu *Brighamia insignis*

Olymposciadium H. Wolff = *Aegokeras*

Olyra L. Gramineae (VI 3). 24 trop. Am. R: SCB 69(1989)2. Prob. incl. *Lithachne* etc. Forest grasses, some fls visited by insects. ***O. latifolia*** L. a weed spread to Afr. & Madag.

Omalanthus A. Juss. = *Homalanthus*

Omalocarpus Choux = *Deinbollia*

Omalotes DC. = *Tanacetum*

Omalotheca Cass. = *Gnaphalium*

Omania S. Moore = *Lindenbergia*

Ombrocharis Hand.-Mazz. Labiatae (VII 1). 1 China: ***O. dulcis*** Hand.-Mazz.

Ombrophytum Poeppig ex Endl. Balanophoraceae. 4 warm S Am., Galápagos. R: FN 23(1980)55. Largely subterr. & ? apomictic

Omegandra Leach & C. Townsend. Amaranthaceae. 1 N Aus.: ***O. kanisii*** Leach & C. Townsend. R: KB 48(1993)787

Omiltemia Standl. Rubiaceae (IV 5). 4 Mex. R: SB 9(1984)410

omixochitl *Agave polianthes*

Omoea Blume. Orchidaceae (V 16c). 2 Java, Philippines

omoto *Rohdea japonica*

Omphacomeria (Endl.) A. DC. Santalaceae. 1 SE Aus.: ***O. acerba*** (R. Br.) A. DC.

Omphalea L. Euphorbiaceae (Crot.-Aden.-Aden.). 17 trop. (Afr. 1, Aus. 3 – 2 endemic). ***O. queenslandiae*** Bailey (Queensland) – fr. ed. musky rat-kangaroos; ***O. triandra*** L. (Jamaican cobnut, trop. Am.) & other spp. with seeds ed. after cooking

Omphalocarpum P. Beauv. Sapotaceae (V). 6 W & C Afr. R: T.D. Pennington, *S.* (1991)260. Local oilseeds. ***O. mortehanii*** De Wild. (Congo) – elephant-disp.; ***O. procerum*** P. Beauv. (trop. Afr.) – trunciflory

Omphalodes Mill. Boraginaceae (B.3.4). Excl. *Memoremea*, *Nihon*, 23 temp. Euras. (Eur. 8), N Am. incl. Mex. (8, R: Phyoneuron 2013:9). Cult. orn. herbs esp. ***O. verna*** Moench (blue-eyed Mary, SE Alps to Roumania, widely escaped Eur.)

Omphalogonus Baill. = *Cryptolepis*

Omphalogramma (Franch.) Franch. Primulaceae. 15 Himal., W China. R: NRBGE 20(1949)125. Cult. orn. like *Primula* but fls bractless & seeds winged

Omphalolappula Brand = *Lappula*

Omphalopappus O. Hoffm. Compositae (Vern.-Erl.). 1 Angola: ***O. newtonii*** O. Hoffm.

Omphalophthalma Karsten = *Matelea*

Omphalopus Naud. Melastomataceae. 1 Sumatra, Java, NG

Omphalotrix Maxim. Orobanchaceae (Rhin.; Scrophulariaceae s.l.). 1 NE As.: ***O. longipes*** Maxim.

Omphalotrigonotis W.T. Wang (~ *Omphalodes*). Boraginaceae (B.3.7?). 1–2 China

omu *Entandrophragma candollei*

omuboro *Citropsis articulata*

Ona Ravenna = *Olsynium*

Onagraceae Juss. Magnoliidae – Myrtales. 21/650 cosmop. esp. temp & warm Am. R: SBM 83(2007)1. Herbs & shrubs, rarely trees to 30 m oft. with epidermal oil-cells, usu. with internal phloem. Lvs whorled, opp. or in spirals, simple, entire to pinnatifid; stip. s.t. present. Fls usu. bisex. & reg. & oft. 4-merous, solit. or in spikes to panicles, usu. with long hypanthium nectariferous within (bird or insect (rarely beetle)-poll. or reg. selfed (apomixis 0)); K oft. valvate lobes on hypanthium, C usu. same as K, valvate, imbr. or convolute & oft. clawed (rarely 0), A within hypanthium or on disk, oft. in 2 whorls, s.t. reduced to A 2, anthers with longit. slits, pollen in monads or tetrads with viscin threads in groups, $\overline{G}$ a compound ovary with as many locs as K or partitions imperfect so placentation axile or parietal, each loc. with (1–)several–∞ anatropous, bitegmic ovules. Fr. a loculicidal capsule, berry or nut; seeds usu. ∞ with straight oily embryo & endosperm 0 (diploid initially). n = (5-) 7 (–18, orig. ?8)

Principal genera: *Camissoniopsis, Chylismia, Clarkia, Epilobium, Fuchsia, Lopezia, Ludwigia* (= **Ludwigioideae** (Jussiaeaoideae) – fls 4- or 5-merous, hypanthium 0; cf. rest (**Onagroideae** – fls (2- or) 4-merous, hypanthium long or occ. 0), *Oenothera*

Fuchsia with woody habit, fleshy fr. & unspecialized placentation form. held to most resemble ancestral O.

Orig. S Am. Many cult. orn. esp. *Clarkia*, *Epilobium*, *Fuchsia*, *Ludwigia*, *Oenothera* (incl. *Gaura*), though species of *E.*, *O.*, *Chamaenerion* & *Circaea* can be weedy

Oncaglossum Sutorý (~ *Cynoglossum*). Boraginaceae (B.3.8.4). 1 C Mex.: ***O. pringlei*** (Greenm.) Sutorý. R: Novon 20(2010)464

Oncella Tiegh. Loranthaceae (5 7). 4 trop. E Afr. R: R. Polhill & D. Wiens, *Mistletoes of Afr.* (1998)230

× **Oncidiopsis** J.M.H. Shaw. Orchidaceae. *Miltoniopsis* × *Oncidium* – 722 grexes

Oncidium Sw. Orchidaceae (V 12h). Incl. *Cochlioda*, *Odontoglossum*, *Sigmatostalix*, *Solenidiopsis*, *Symphyglossum*, excl. *Cyrtochilum*, *Trichocentrum*, c. 300 trop. Am. (***O. ensatum*** Sw. ext. to Florida) to temp. S Am. R: Bradleya 1(1974)398. Epiphytes with extrafl. nectaries, some bird-poll., shiny surface of others mimicking oil-fls of e.g. Malpighiaceae (deceit). Cult. orn. epiphytes (1150 guineas paid for 1 form of ***O. alexandrae*** (Bateman) M. Chase & N. Williams (*Odontoglossum crispum*, Colombia) in 1906), now 1000 named cvs, esp. *O.* grex **Gower Ramsay** ('Goldiana' × 'Guinea Gold') as cul-tfl.; some parents of intergeneric hybrids with spp. of *Brassia*, *Macradenia* etc.

Oncinema Arn. Apocynaceae (Vc; Asclepiadaceae). 1 Cape: ***O. lineare*** (L.f.) Bullock (*O. roxburghii*). R: Strelitzia 9(2000)286

Oncinocalyx F. Muell. = *Teucrium* (but see JABG 14(1991)77)

Oncinotis Benth. Apocynaceae (II c). 7 Afr. (6), Madag. (1). R: AUWP 85–2(1985)5

Oncoba Forssk. Salicaceae (Flacourtiaceae). 4 trop. Afr. Some cult. orn. incl. ***O. spinosa*** Forssk. (trop. Afr.) – ed. pulp, fr. a rattle in Tanz.

Oncocalamus (G. Mann & H. Wendl.) H. Wendl. Palmae (I 2a). 4 trop. Afr. R: Phytotaxa 51(2012)62. Hapaxanthic rattans

Oncocalyx Tieghem (~ *Loranthus*). Loranthaceae (5 7). 13 E & S Afr., Arabia. R: R. Polhill & D. Wiens, *Mistletoes of Africa* (1998)101. Birds probing for nectar split C-tube releasing A, projecting pollen on to bird's beak (cf. *Globimetula*)

Oncocarpus A. Gray = *Semecarpus*

Oncodostigma Diels = *Meiogyne*

Oncophyllum D.L. Jones & M.A. Clem. = *Goodyera*

Oncorachis Morrone & Zuloaga (~ *Panicum*). Gramineae (XXIII 1). 2 C & E Braz. R: T 58 (2009)372

Oncosiphon Källersjö (~ *Pentzia*). Compositae (Anth.-Pen.). 8 S Afr. esp. Atlantic coast. R: BotJLS 96(1988)310. Annuals. Some cult. orn., ***O. suffruticosum*** (L.) Källersjö invasive S Aus.

Oncosperma Blume. Palmae (V 14h). 5 Sri Lanka to Moluccas. V. spiny monoec. palms with true dichotomous branching, s.t. cult. orn. incl. ***O. tigillarium*** (Jack) Ridl. (*O. filamentosum*, SE As., Mal.) – young lvs a veg., timber for flooring, underwater piles etc.

× **Oncostele** J.M.H. Shaw. Orchidaceae. *Oncidium* × *Rhynchostele* – 560 grexes

Oncostema Raf. = *Scilla*

Oncostemma K. Schum. = *Vincetoxicum*

Oncostemum A. Juss. (~ *Badula*). Primulaceae (Myrsinaceae). 90 Madag., Mascarenes. Some Madag with Corner's Model, several with detritus-collecting crowns

Oncostylus (Schldl.) F. Bolle = *Geum* (but see FRB 72(1933)27)

Oncotheca Baill. Metteniusaceae (Oncothecaceae). 2 New Caled.

Oncothecaceae Kobuski ex Airy Shaw = Metteniusaceae

Ondetia Benth. Compositae (Inul.-Pluch.). 1 Namibia: ***O. linearis*** Benth. R: Strelitzia 10(2000)152

Ondinea Hartog = *Nymphaea*

ong tsoi *Ipomoea aquatica*

Ongokea Pierre. Olacaceae (Aptandraceae). 1 W trop. Afr.: ***O. gore*** (Hua) Pierre (*O. klaineana*) – seeds yield a drying oil (isano o.)

onion *Allium cepa*; **o. couch** *Arrhenatherum elatius* var. *bulbosum*; **Egyptian o.** *A.* × *proliferum*; **o. grass** *Romulea* spp.; **Japanese bunching o.** *A. fistulosum*; **multiplier** or **potato o.** *A. cepa* Aggregatum Group; **o. orchid** *Microtis* spp.; **sea o.** *Drimia maritima*; **tree o.** *A.* × *proliferum*; **o. weed** (Aus.) *Asphodelus fistulosus*, *Nothoscordum borbonicum*, *Romulea* spp.; **Welsh o.** *Allium fistulosum*

Onira Ravenna = *Cypella* (but see NJB 3(1983)204)

Onixotis Raf. = *Wurmbea* (but see Strelitzia 9(2000)76)

Onobrychis Mill. Leguminosae (III 25). 130 Euras. (Eur. 23), Ethiopia. R: PFSUM 56 (1925) 1, 57 (1926)1. ***O. viciifolia*** Scop. (sainfoin, holy clover, As., natur. Eur.) – cult. for fodder, good bee-pl.

Onoclea L. Onocleaceae (Dryopteridaceae s.l.). Excl. *Matteuccia*, 1 N temp.: ***O. sensibilis*** L. (*O. orientalis*) – cult. orn., hybridizes with *M. struthiopteris*

Onocleaceae Pichi-Serm. (~ Blechnaceae). Polypodiidae – Polypodiales. 4/5 N temp. to Mex. R: T 55(2006)716, 61(2012)528. Terr. with creeping branched to ascending, unbranched (s.t. stoloniferous) rhiz.; scales not clathrate. Fronds dimorphic; petiole with 2 vasc. bundles distally joining in gutter-shape; blades pinnatifid or pinnate-pinnatifid, veins free or anastomosing. Sori encl. by reflexed lamina margin & also membranous evanescent true indusium; spores reniform. n = 37, 39, 40

Genera: *Matteuccia*, *Onoclea*, *Onocleopsis*, *Pentarhizidium*

Cult. orn., *Matteuccia* ed.

Onocleopsis Ballard (~ *Onoclea*). Onocleaceae (Dryopteridaceae s.l.). 1 Mex., C Am.: ***O. hintonii*** Ballard. R: AFJ 35(1945)1

Onohualcoa Lundell = *Mansoa*

Ononis L. Leguminosae (III 27). 75 Eur. (49), Medit., Canary Is., Ethiopia & Iran. Restharrow. R: BBC 49, 2(1932)381. Shrubs & herbs, oft. with thorny lat. branches; fls with piston mechanism at first, later like *Trifolium* (see fam.). Some cult. orn. incl. ***O. spinosa*** L. (*O. campestris*, Eur.) – locally medic.

Onopordum Vaill. ex L. Compositae (Card.-Card.). c. 60 Eur. (13). Medit. & W As. Coarse prickly bienn. (to triennial) herbs with spiny decurrent lvs & usu. spiny involucral bracts. Some cult. orn. incl. ***O. acanthium*** L. (cotton thistle, the modern-day 'Scotch t.', Eur. (prob.

introd. Br. Is.) to C As.) & ***O. nervosum*** Boiss. ('*O. arabicum*', Spain, Portugal) to 3 m tall, fls of the first form. used to adulterate saffron

Onoseris Willd. Compositae (Mut.-Onos.). 31 Mex. to Andes. R: JAA 25(1944)349. Herbs & shrubs

Onosma L. Boraginaceae (B.2.2). Excl. *Podonosma*, 150 Medit. (Eur. 33, Turkey 88) to Himal. & China. Some dyes esp. red dye used like alkanet from ***O. echioides*** L. (Medit. to W Himal.), some cult. orn. incl. ***O. frutescens*** Lam. (golden drop, E Medit.) etc.

Onosmodium Michaux = *Lithospermum* (but see Phytol. 78(1995)40)

Onuris Philippi. Cruciferae (27). 6 Chile, Patagonia. R: Parodiana 3(1984)53

Onus Gilli = *Mellera*

Onychium Kaulf. Pteridaceae (III). c. 10 NE Afr. & Iran to NG (esp. China). Some cult. orn. incl. ***O. siliculosum*** (Desv.) C. Chr. (Indomal.) – juice from crushed lvs alleged to prevent baldness

Onychopetalum R. Fries. Annonaceae. 3 Braz., Peru

Onychosepalum Steud. = *Desmocladus*

Oocephala (S.B. Jones) H. Robinson (~ *Vernonia*). Compositae (Vern.-Erl.). 2 trop. Afr.

Oocephalus (Benth.) Harley & Pastore (~ *Hyptis*). Labiatae (VII 3c). 18 trop. S Am. R: Phytotaxa 58(2012)33

Ooia S.Y. Wong & Boyce (~ *Hottarum*). Araceae. 3 Borneo. R: Webbia 68(2013)89

Oonopsis (Nutt.) Greene (~ *Haplopappus*). Compositae (Ast.-Mac.). 4 C US. R: FNA 20(2006)410

Oophytum N.E. Br. (~ *Conophytum*). Aizoaceae (V). 2 W S Afr. R: H. E. K. Hartmann, *Ill. Handb. Succ. Pls*, Aiz. F–Z(2002)192

Oosterdyckia Boehmer = *Cunonia*

Oparanthus Sherff. Compositae (Cor.). 6 Rapa, Marquesas. R: Allertonia 7(1997)281. Monoec. trees allied to *Fitchia* & Carib. genera

opepe *Nauclea diderrichii*

Opercularia Gaertn. Rubiaceae (IV 13). 18 Aus. (11 in W Aus.)

Operculicarya Perrier (~ *Lannea*). Anacardiaceae (II). 5 Madag., Aldabra. R: Adans. III,28(2006)359. Some used as bonsai

Operculina Silva Manso. Convolvulaceae (12). 15 trop. ***O. turpethum*** (L.) Silva Manso (*Ipomoea t.*, Ind. jalap, turpeth root, turpethum, E Afr. to Pac.) – roots drastic purgative

Ophellantha Standl. = *Acidocroton*

Ophidion Luer. Orchidaceae (V 13). 4 C Am. to Andes. R: Selbyana 7(1982)79

Ophiobotrys Gilg (~ *Osmelia*). Salicaceae (Samydaceae, Flacourtiaceae). 1 W trop. Afr.: ***O. zenkeri*** Gilg

Ophiocarpus (Bunge) Ikonn. = *Astragalus*

Ophiocaryon Endl. (~ *Meliosma*). Sabiaceae (Meliosmaceae). 7 trop. S Am. incl. ***O. paradoxum*** Schomb. ex Hook. (snakenut) – imported to Eur. as curiosity, embryo coiled & visible like snake

Ophiocephalus Wiggins = *Castilleja*

Ophiochloa Filgueiras & al. = *Axonopus* (but see Novon 3(1993)360)

Ophiocolea Perrier. Bignoniaceae (6). 9 Madag., Comoro Is. R: FMad 118(1938)25

Ophioderma (Blume) Endl. = *Ophioglossum*

Ophioglossaceae Martinov. Ophioglossidae – Ophioglossales. 4/55 cosmop. (esp. temp. & boreal). R: T 55(2006)709. Small homosporous terr. herbs, some trop. spp. epiphytic. Rhiz. & petiole fleshy; roots mycorrhizal, without root-hairs. Lvs solit. or few, without circinate vernation, comprising succ. sterile oft. entire blade & a long-stalked spike or panicle with sorus-less sporangia sunk in it. Prothalli subterr., colourless, mycotrophic. Polyploidy to v. high levels (x = 45(46–130 [*Mankuya*]))

Genera: *Botrychium* (but see GBS 40(1987)1 for splitting), *Helminthostachys*, *Mankyua*, *Ophioglossum*

Form. thought to be allied to the progymnosperm line (periderm, circular bordered-pitted tracheids & non-circinnate vernation etc. suggested alliance with cycads). Some locally eaten & medic.

Ophioglossella Schuit. & Ormerod (~ *Grosourdya*). Orchidaceae (V 16c). 1 NG: ***O. chrysostoma*** Schuit. & Ormerod. R: KB 53(1998)742

Ophioglossidae Klinge (incl. Psilotidae). 6/73 cosmop. Two orders, Ophioglossales + Psilotales, each of of 1 fam. – Ophioglossaceae, Psilotaceae, q.v.

Ophioglossum Tourn. ex L. Ophioglossaceae. Incl. *Cheiroglossa*, 25–30 subcosmop. (Eur. 3). R: MTBC 19, 2(1938)111. Adder's-tongue, snake-tongue. Root-buds; fertile blade usu. a spike with 2 rows of sporangia. 2n = 140–1440 (96-ploid: ***O. reticulatum*** L. (trop.) – highest chromosome no. known). Some ed. & locally medic. incl. ***O. petiolatum*** Hook. (Pacific) – ground to medic. powder Taiwan; ***O. vulgatum*** L. (N temp.) – mid 19 cent. English Home Counties use in snake-bite potion ('Adder's-spear ointment'; ?Doctrine of Signatures); few cult. orn. long-lived & slow-growing incl. ***O. palmatum*** L. (*C. p.*, SE As., Madag., Réunion, trop. & warm Am.) – bizarre epiphyte with coarse palmatifid frond bearing pendent spikes of sporangia

Ophiomeris Miers = *Thismia*

Ophionella Bruyns (~ *Pectinaria*). Apocynaceae (Vb). 2 E Cape. R: BotJLS 131(1999)393

Ophiopogon Ker-Gawler. Asparagaceae (Convallariaceae). 54 Indomal. to Himal. & Jap. Jap. hyacinth. Cult. orn. everg. turf-forming pls esp. ***O. japonicus*** (L.f.) Ker-Gawler (Jap., Korea) – tuberous roots ed., fls white to lilac not prod. in trop. lowlands where oft. grown, fr. blue, pea-sized, & ***O. planiscapus*** Nakai (mondo grass, Jap.) esp. '**Nigrescens**' (black m. g.) – overplanted

Ophiorhipsalis Doweld = *Lepismium*

Ophiorrhiza L. Rubiaceae (IV 2). c. 250 Indomal. (China 70, Ind. 47 – R: BBSI 39(1997)20). Fr. acts as splash-cup for seed-disp. Medic. esp. ***O. mungos*** L. (Indomal.); ***O. tomentosa*** Jack (Mal.) – s.t. viviparous

Ophiorrhiziphyllon Kurz = *Staurogyne*

Ophiuros Gaertn. f. Gramineae (XXII 3). 4 NE trop. Afr. to S China & Aus.

Ophrestia H.M. Forbes. Leguminosae (III 18). c. 16 trop. OW

Ophryococcus Oersted = *Hoffmannia*

Ophryosporus Meyen. Compositae (Eup.-Crit.). 37 S Am. R: MSBMBG 22(1987)363

Ophrypetalum Diels. Annonaceae (III 6). 1 trop. E Afr.: ***O. odoratum*** Diels

Ophrys L. Orchidaceae (IV 4b). 10 (251 recog. by some workers) Eur. (10), W As., N Afr. R: E. Nelson (1962) *Gestaltwandel . . . Monographie . . . Gattung O.* Terrestrial herbs (usu. on alkaline soils) with swollen tubers replaced annually. Lip oft. resembling large insect entering fl. Insects attracted by cyclic sesquiterpene alcohols & hydrocarbons (aliphatic acetates esp. octyl a. in ***O. speculum*** Link (*O. vernixia*, Medit.) – on lip, attractive to scoliid wasps, prob. chem. mimetic of female bee sex hormones in ***O. lutea*** Cav. (Medit.) at least) are males, which emerge before females & attempt to copulate with the fl., leaving sperm & taking pollinia for a second fl. (cf. *Caladenia, Chiloglottis, Cryptochilus, Leporella*), diff. spp. with diff. relative proportions of attractants (cf. *Chiloglottis*). Some tubers form. used as salep in Turkish delight; some cult. orn. & named (e.g. spider orchids) after fancied resemblance to insects or spiders, not their pollinators: ***O. apifera*** Hudson (bee orchid, W & C Eur.) – poll. in S by *Eucera* & *Tetralonia* bees, self-poll. in N, usu. hapaxanthic after 5–8 yrs veg. growth, accounting for great fluctuations in ann. nos. of flowering pls, mutant form with deformed lip known as wasp o.; ***O. insectifera*** L. (fly o., Eur.) – in UK largely poll. by male digger wasps; ***O. sphegodes*** Mill. (Eur.) – 24 active compounds (14 common in pl. cuticles found in same mix in bee female pheromones; cf. novel compounds in *O. speculum* like female pheromones) in odour 'bouquets' of lip with variation between active esters & aldehydes in diff. fls, each 'remembered' by poll. solit. bee (male *Andrena nigroaenea*), so not revisited

Ophthalmoblapton Allemão. Euphorbiaceae (Euph.-Hur.). 4 E Braz. R: Britt. 65(2013) 324

Ophthalmophyllum Dinter & Schwantes = *Conophytum*

Opilia Roxb. Opiliaceae. 3 OW trop. R: Willdenowia 12(1982)161. R.-parasites

Opiliaceae Valeton (~ Olacaceae). Magnoliidae – Santalales. 10/32 trop. (Am. only *Agonandra*). Usu. everg. root-parasitic trees & shrubs, sometimes lianoid. Lvs simple, alternate, with cystoliths; stip. 0. Fls small, usu. bisex. (dioec. in *Agonandra* & *Gjellerupia*), in axillary or cauliflorous spikes to panicles or umbels; K small, cupular ± 4 or 5 small lobes or teeth (not accrescent), C (3)4 or 5, sometimes basally connate, valvate (s.t. 0 in female fls), A opp. C, s.t. on C or C-tube, anthers with longit. slits, disk of free or ± connate nectaries around $\underline{G}$ (2–5) with simple or 0 style, sunk in disk, 1-loc. with 1 pendulous (erect in *Agonandra*), anatropous, unitegmic ovule (integument s.t. not recognizable). Fr. a drupe; seeds with small embryo with (2)3(4) cotyledons in copious oily starchy endosperm. n = 10

Genera: *Agonandra* (Am., basal ovule, dioec.), *Cansjera, Champereia, Gjellerupia, Lepionurus, Melientha, Opilia, Pentarhopalopilia, Rhopalopilia, Urobotrya*

Anthobolus (Santalaceae) may belong here. Some oilseeds (*Agonandra*), ed. fr. (*Champereia, Melientha*) & local medic.

Opisthiolepis L.S. Sm. Proteaceae (V 3). 1 Queensland: ***O. heterophylla*** L.S. Sm. R: FA 16(1995)373

Opisthocentra Hook.f. Melastomataceae. 1 N Braz.: ***O. clidemioides*** Hook.f.

Opisthopappus Shih (~ *Chrysanthemum*). Compositae (Anth.-Art.). 2 NE C China. R: APS 17,3(1979)110

Opithandra B.L. Burtt = *Oreocharis*

opium *Papaver somniferum*

Opizia J. Presl = *Bouteloua*

Oplismenopsis L. Parodi. Gramineae (XXIII 1). 1 Uruguay, Arg., Braz.: ***O. najada*** (Hackel & Arech.) L. Parodi – floating

Oplismenus P. Beauv. Gramineae (XXIV 2). 11 trop. & warm. R: PM 13(1981). Forest shade, some with sticky awns for disp. ***O. hirtellus*** (L.) P. Beauv. (trop. Am.) – fr. in comm. bird food in US, '**Variegatus**' ('*Panicum variegatum*') – cult. greenhouse hanging-basket pl.

Oplonia Raf. Acanthaceae (III 2c). 19 trop. Am. (Peru 1, WI 12) & Madag. (5), presumed extinct in Afr. R: BBMNHB 4(1971)259, ABASH 23(1977)303. Heterostyly

Oplopanax (Torrey & A. Gray) Miq. Araliaceae. 3 NW N Am., Jap. Prickly decid. treelets, cult. orn. esp. ***O. horridus*** (Sm.) Miq. ('devil's club', N Am.) – v. imp. medic. pl. for Native Americans (analgesic etc.)

Opocunonia Schltr. (~ *Caldcluvia*). Cunoniaceae (VI). 1 NG, New Britain: ***O. nymanii*** (K. Schum.) Schltr. – variable. R: FM I,16(2002)107

Opoideia Lindl. (~ *Peucedanum*). Umbelliferae (inc. sed.). 1 Iran: ***O. galbanifera*** Lindl.

opopanax *Opopanax chironium, Commiphora* spp., esp. *C. kataf, Vachellia farnesiana*

Opopanax Koch. Umbelliferae (III 10). 3 Balkans (Eur. 2) to Iran. ***O. chironium*** Koch (Medit.) – source of gum opopanax used in scent-making & form. medic.

Opophytum N.E. Br. = *Mesembryanthemum*

opossum wood *Halesia carolina*

Opsiandra Cook = *Gaussia*

Opsicarpium Mozaff. Umbelliferae (III 10). 1 Iran: ***O. insignis*** Mozaff. R: BZ 88,2(2003)89

optical fibre plant *Isolepis cernua*

Opuntia Mill. Cactaceae (II). Incl. *Nopalea*, excl. *Austrocylindropuntia, Brasiliopuntia, Consolea, Cylindropuntia, Corynopuntia, Cumulopuntia, Tephrocactus*, c. 75 Massachusetts & Br. Columbia to Galápagos & Straits of Magellan (range of fam. in Am.). Flattened joints, with usu. early decid. lvs & many minute irritant glochids oft. with larger spines; alks incl. mescaline. Some poll. hummingbirds, winter-flowering assoc. with birds' migrations. Some ed. fr. (prickly pears, 'tunas') & fls, others for living fences, spineless forms for forage or ed. ('nopalitos') esp. in bad times; some introd. to OW & natur., esp. ***O. aurantiaca*** Lindl. (tiger pear, poss. hybrid as no seeds) to S Afr. & E Aus. (& S Am.), now controlled by a cochineal insect in Aus.; ***O. cochenillifera*** (L.) Mill. (*N. c.*, cochineal cactus or pl., cultigen orig. Mex.) – long cult. trop. Am. (but rarely fls in cult.), cochineal (dried bodies of mealy bugs feeding on pl.) form. coll. for preparation of carmine (scarlet dye now synth. artificially); ***O. ficus-indica*** (L.) Mill. (Ind. or Barbary fig, cultigen orig. Mex.) – widely cult. (incl. some parthenocarpic clones) for fr. exported to Eur. etc., extract alleged to alleviate hang-overs, flavouring for tea (Pickwick t.) in US., natur. Medit., S Afr., Aus.; ***O. stricta*** (Haw.) Haw. (incl. widespread spiny form (*O. dillenii*), Mex. to S Am.) – natur. Medit., temp. As., S Afr., Madag., Aus. (c. 25 M ha in E Aus. by 1925, advancing 100 ha per hr, occupying range-land to detriment of grazing, now controlled by larvae of moths introd. from orig. habitats – early success of 'biological control'). Many cult. orn. incl. ***O. humifusa*** (Raf.) Raf. (C & E US) – hardy England, natur. Switzerland, Aus.; ***O. fragilis*** (Nutt.) Haw. (brittle cactus, SW US to 56°N) – rarely fls, segments detached & spread by animals & (?) water; ***O. microdasys*** (Lehm.) Pfeiffer (C & N Mex.) – commonly cult. sp. with many closely set easily detached irritant yellow glochids

orache (**gold**) *Atriplex hortensis*

orange *Citrus × aurantium*; **bergamot o.** *C. × limon*; **blood o.** *C. × aurantium* Sweet Orange group; **mandarin o.** *C. reticulata*; **mock o.** *Philadelphus* spp.; **navel o.** *C. × aurantium* Sweet Orange group; **Osage o.** *Maclura pomifera*; **Quito o.** *Solanum quitoense*; **satsuma o.** *C. reticulata*; **Seville** or **sour o.** *C. × aurantium* Sour Orange Group; **sweet o.** *C. × aurantium* Sweet Orange group

Orania Zipp. Palmae (V 2). Incl. *Halmoorea*, 28 Madag. (3), S Thailand to (esp.) NG, Aus. R: KB 67(2012)134. Poisonous (rodenticide in NG). ***O. disticha*** Burret (NG) – lvs distich.

Oraniopsis (Becc.) Dransf. & al. Palmae (IV 2). 1 N Queensland: ***O. appendiculata*** (Bailey) Dransf. & al. R: Principes 29(1985)57

Orbea Haw. (~ *Stapelia*). Apocynaceae (V b; Asclepiadaceae III 5). Incl. *Angolluma, Orbeanthus, Orbeopsis, Pachycymbium* & *Stapeliopsis*, 56 SW Arabia, trop. & S Afr. R: SBM 62(2002)28. Cult. orn. succ. with stinking mottled fls attractive to flies & succ. fanciers esp. ***O. variegata*** (L.) Haw. (*Stapelia v.*, Cape, natur. S Aus.) – 1 of first Cape pls known in Eur. (drawing sent back in 1624)

Orbeanthus Leach = *Orbea*

Orbeopsis Leach = *Orbea*

Orbexilum Raf. Leguminosae (III 20). Excl. *Pediomelum, Psoralidium*, 11 US., Mex. R: Lundellia 11(2008)1

Orbignya Mart. ex Endl. = *Attalea*

Orbivestus H. Robinson (~ *Vernonia*). Compositae (Vern.-Erl.). c. 4 trop. & S Afr. R: T 65(2016)287

Orchadocarpa Ridl. (~ *Loxocarpus*). Gesneriaceae (III 2i). 1 Malay Peninsula: ***O. lilacina*** Ridl.

orchid, bee *Ophrys apifera*; **bird's-nest o.** *Neottia nidus-avis*; **bog o.** *Hammarbya paludosa*; **butterfly o.** *Platanthera* spp.; **donkey o.** *Diurus* spp.; **duck o.** *Caleana major*; **early purple o.** *Orchis mascula*; **fen o.** *Liparis loeselii*; **fly o.** *Ophrys insectifera*; **frog o.** *Dactylorhiza viridis*; **green-winged o.** *Anacamptis morio*; **Jersey o.** *Orchis laxiflora*; **jug o.** *Acanthephippium* spp.; **lady's slipper o.** *Cypripedium* spp. esp. *C. calceolus*; **l.'s tresses o.** *Spiranthes* spp.; **leek o.** *Prasophyllum* spp.; **o. lily** *Barclaya longifolia*; **lizard o.** *Himantoglossum hircinum*, (Aus.) *Burnettia cuneata*; **man o.** *Orchis anthropophora*; **midge o.** *Genoplesium* spp.; **military o.** *O. militaris*; **monkey o.** *O. simia*; **moth o.** *Phalaenopsis* spp. & hybrids; **musk o.** *Herminium monorchis*; **onion o.** *Microtis* spp.; **pansy o.** *Miltoniopsis* spp.; **peacock o.** *Gladiolus murielae*; **poor man's o.** *Schizanthus* spp.; **potato o.** *Gastrodium sesamoides*; **pyramid o.** *Anacamptis pyramidalis*; **rock o.** *Dendrobium speciosum*; **soldier o.** *O. militaris*; **spider o.** *Ophrys* spp.; **swan o.** *Cycnoches* spp.; **sweet-scented o.** *Gymnadenia conopsea*; **o. tree** *Amherstia nobilis*, *Bauhinia* spp.; **twayblade o.** *Neottia* spp.; **wax-lip o.** *Glossodia* spp.

Orchidaceae Juss. Magnoliidae – Asparagales. Incl. Apostasiaceae & Cypripediaceae, 762/26 000 cosmop. (e.g. Mal. 4500 – biggest fam. there (& in world, larger than Compositae), NG alone 133/2300). R: A.M. Pridgeon et al. (eds, 1999–2014) *Genera Orchidacearum*. Perenn. mycotrophic epiphytic (the great majority) or terr. herbs, rarely lianes (e.g. *Clematepistephium, Vanilla*) or annuals (see *Zeuxine*), s.t. without chlorophyll, v. rarely (Aus., *Rhizanthella*) completely subterr. or rheophytic (e.g. *Appendicula*), always with raphides in some cells & oft. with mucilage-cells & alks, frequently with crassulacean acid metabolism & generally with roots (always adventitious, oft. aerial, s.t. photosynthetic) with multi-layered velamen, in terr. spp. oft. swollen into tubers, or stems forming corms or rhiz.; stems of epiphytic spp. oft. thickened to form a pseudobulb. Lvs usu. entire & glabrous, plicate to convolute, in spirals, distich., rarely opp. or whorled, s.t. scale-like, oft. ± fleshy & basally sheathing. Fls usu. bisex., 3-merous, epigynous, irreg., in racemes or panicles to solit., usu. resupinate; P usu. petaloid though outer 3 s.t. greenish, the median app. adaxial (truly abaxial) s.t. diff. from others or 2 or 3 basally connate, inner 3 with app. abaxial (truly adaxial) one (exc. Apostasioideae) usu. larger & diff. colour from laterals forming labellum (lip), the laterals oft. like outer P, nectaries var. (s.t. hollow spur from base of labellum, or a cup on or embedded in G, or extrafl. etc.), A 1(–3) all truly abaxial opp. labellum, when 1 united with style forming gynostemium (column) & truly median stamen of outer whorl, other 2 laterals of inner s.t. staminodal with vasc. strands of adaxial A in gynostemium, anthers with longit. slits, pollen grains solit. in Apostasioideae & Cypripedioideae, in tetrads & pollinia in rest with 1–8 pollinia per pollen-sac, each oft. with slender tip or caudicle, $\overline{G}(3)$, 1-loc. (3-in Apostasioideae) with marginal placentas & usu. gynostemium subtended by enlarged stigma-lobe (rostellum) to which caudicles oft. attached & from which a sticky viscidium is removed when pollinia taken by pollinators, ovules anatropous, (uni-) bitegmic, minute, ∞, development triggered by poll. & fertilization oft. delayed (up to 6 months). Fr. usu. a capsule (rarely baccate) with 3(6) longit. slits but apically & basally closed (fissuricidal); seeds (s.t. winged) minute & ∞ (to several million) with minute undeveloped embryo, endosperm formation arrested at 2–4(–16)-nucleate stage, only testa usu. persisting. Seeds usu. germinating only in presence of appropriate fungus, when forming a protocorm with basal rhizoids, no radicle & usu. no cotyledon, the protocorm eventually giving rise to apical lvs. n = 6–29+

Classification & chief genera (BotJLS 177(2015)169):

I. **Apostasioideae** (lvs in spirals, plicate; fls weakly irreg., s.t. resupinate (*Neuwiedia*), A 2 or 3, pollen in monads & pollinia 0, some selfing, n = 24; Indomal.; R: Gen. Orch. 1(1999)94): *Apostasia* & *Neuwiedia* (only)

II. **Vanilloideae** (mycotrophic &/or lianoid, n = 9–18; sister to all orchids save I., so that reduction to A1 occurred at least twice)
   1. **Pogonieae** (R: KB 63(2008)446): *Cleistes*
   2. **Vanilleae:** *Epistephium, Galeola, Vanilla*

III. **Cypripedioideae** (usu. terr.; lvs in spirals or distich., s.t. plicate; fls resupinate, labellum slipper-shaped, A 2 (inner whorl) present with median outer A a staminode, true pollinia rare, n = 9; N temp. to trop. exc. Afr.; R: Gen. Orch. 1(1999)105): *Cypripedium, Mexipedium, Paphiopedilum, Phragmipedium, Selenipedium* (only)

IV. **Orchidoideae** (incl. Spiranthoideae; usu. survive adverse periods as dormant tuberoids, lvs soft herbaceous, pseudocopulation frequent, (?) orig. S but now the common O. of N; R: Gen. Orch. 2(2000)6) – 4 tribes):
   1. **Codonorchideae:** *Codonorchis* (only)
   2. **Cranichideae** (7 subtribes): *Chloraea* (a. Chloraeinae); *Cranichis, Ponthieva* (b. Cranichidinae); *Galeottiella* (c. Galeottiellinae); *Anoectochilus, Aspidogyne, Cheirostylis, Erythrodes, Goodyera, Zeuxine* (d. Goodyerinae); *Manniella* (Manniellinae); *Pterostylis* (f. Pterostylidineae); *Discyphus* (g. Discyphinae); *Cyclopogon, Pelexia, Sarcoglottis, Spiranthes* (h. Spiranthinae)
   3. **Diurideae** (trop. As. to (esp. Aus.), S Am.; 8 subtribes): *Corybas* (a. Acianthinae); *Caladenia* (b. Caladeniinae; R: AusSB 17(2004)177); *Cryptostylis* (c. Cryptostylidinae); *Diuris* (d. Diuridinae); *Chiloglottis, Paracaleana* (e. Drakaeinae); *Megastylis* (f. Megastylinae); *Genoplesium, Prasophyllum* (g. Prasophyllinae); *Rhizanthella* (h. Rhizanthellinae); *Thelymitra* (i. Thelymitrinae)
   4. **Orchideae** (Afr., N hemisph.; 4 subtribes): *Disperis* (a. Brownleeinae); *Pterygodium* (b. Coryciinae); *Disa* (c. Disinae); *Anacamptis, Cynorkis, Dactylorhiza, Habenaria, Holothrix, Ophrys, Orchis, Peristylus, Platanthera, Satyrium* (d. Orchidinae)

V. **Epidendroideae** (usu. epiphytes with pseudobulbs, oft. with distich. fleshy lvs & lat. infls but exceptions to all these; R: Gen. Orch. 4–6(2005–2014) – all but *Devogelia* arr. in 16 tribes):
   1. **Neottieae** (incl. Palmorchideae; usu. terr. without pseudobulbs): *Cephalanthera, Epipactis, Neottia*
   2. **Sobralieae:** *Elleanthus, Sobralia*
   3. **Tropidieae:** *Tropidia*
   4. **Triphoreae** (incl. Diceratosteleae; terr. or mycotrophic; 2 subtribes): *Diceratostele* (a. Diceratostelinae); *Triphora* (b. Triphorinae)
   5. **Xerorchideae:** *Xerorchis* (only)
   6. **Wullschlaegelieae:** *Wullschlaegelia* (only)
   7. **Gastrodieae** (leafless mycotrophs): *Didymoplexis, Gastrodia*
   8. **Nervilieae** (terr. with globose corm; 2 subtribes): *Nervilia* (only; *Nerviliinae*); *Epipogium* (mycotroph; 2. Epipogiinae)
   9. **Thaieae:** *Thaia* (only)
   10. **Arethuseae** (incl. Coelogyneae; 2 subtribes): *Anthogonium, Arundina* (a. Arethusinae); *Bletilla, Coelogyne, Dendrochilum, Glomera, Pholidota, Pleione* (b. Coelogyninae)
   11. **Malaxideae** (incl. Dendrobieae; 2 subtribes): *Bulbophyllum, Dendrobium* (a. Dendrobiinae); *Liparis, Malaxis, Oberonia* (b. Malaxidinae)
   12. **Cymbidieae** (incl. Maxillarieae; many cult. orn. incl. intergeneric hybrids; 10 subtribes): *Cymbidium, Grammatophyllum* (a. Cymbidiinae); *Eulophia, Oeceoclades* (b. Eulophiinae); *Catasetum, Mormodes* (c. Catasetinae); *Cyrtopodium* (only; d. Cyrtopodiinae); *Peristeria* (e. Coeliopsidinae); *Eriopsis* (only; f. Eriopsidinae); *Lycaste, Maxillaria, Sudamerlycaste* (g. Maxillariinae); *Brassia, Cyrtochilum, Fernandezia, Miltonia, Notylia, Oncidium, Ornithocephalus, Rhynchostele, Rodriguezia, Telipogon, Tolumnia* (h. Oncidiinae); *Coryanthes, Gongora, Stanhopea* (i. Stanhopeinae); *Batemannia, Dichaea, Kefersteinia, Zygopetalum* (j. Zygopetalinae)
   13. **Epidendreae** (incl. Calypsoeae; 6 subtribes): *Bletia* (a. Bletiinae); *Cattleya, Encyclia, Epidendrum, Laelia, Prosthechea, Scaphyglottis* (b. Laeliinae); *Brachionidium, Dracula, Dryadella, Lepanthes, Lepanthopsis, Masdevallia, Octomeria,*

Platystele, Pleurothallis, Restrepia, Specklinia, Stelis, Trichosalpinx (c. Pleurothallidinae); *Isochilus* (d. Ponerinae); *Corallorrhiza, Govenia* (e. Calypsoinae); *Agrostophyllum* (f. Agrostophyllinae – elaters)

14. **Collabieae:** *Calanthe, Phaius, Plocoglottis, Spathoglottis*
15. **Podochileae:** *Appendicula, Ceratostylis, Eria, Octarrhena, Phreatia, Podochilus*
16. **Vandeae** (4 subtribes): *Bromheadia* (a. Adrorhizinae); *Polystachya* (b. Polystachyinae); *Aerides, Cleisostoma, Gastrochilus, Luisia, Phalaenopsis, Taeniophyllum, Thrixspermum, Trichoglottis, Vanda* (c. Aeridinae); *Aerangis, Angraecum, Jumellea, Tridactyle* (d. Angraecinae)

Distinct from all other monocots in intricate poll. biology, reduction of adaxial A & presence of ∞ endosperm-less seeds & gynostemium (cf. Apocynaceae-Asclepiadoideae & Stylidiaceae) but otherwise not an advanced group & prob. early derived from trop. terr. Asparagalean type with sympodial rhiz., unbranched stem, plicate lvs, A 6, G(3) (fl. prototype as deduced by Robert Brown early 19 cent.) & capsules. Alleged fossil pollen on bee from Upper Cretaceous

Found in driest deserts to highest mts, from few mm to several m tall, some chlorophyll-less mycotrophs (the habit prob. evolved some 20 times!) but no aquatics save rheophytes, nor halophytes (but see *Bletia, Brassavola*), poll. (night-flowering in 1 *Bulbophyllum*) by bees, wasps, hornets (see *Dendrobium*), flies (see *Acianthus*), mosquitoes (see *Platanthera*), ants (e.g. *Microtis*), crickets (see *Angraecum*), beetles, birds (e.g. *Cryptochilus?, Sacoila, Stenorrhynchos*), bats & poss. frogs, which transfer pollinia, oft. with quick-setting 'glue' (? polyisoprene) or explosive systems projecting them up to 60 cm from fl.; the pollinators oft. attracted by scent (from elaiophores as found in only 9 other fams) & males even carrying out pseudocopulation (*Ophrys* (q.v.), *Caladenia, Chiloglottis, Cryptostylis, Drakaea, Lepanthes, Leporella, Trichoceros, Trigonidium*), though it is unlikely that any sp. is poll. by just 1 animal sp., while the fam. is notorious for occurrence of natural hybrids & their synthesis in horticulture (up to 20 spp. in 5 genera combined in 1 pl.), comm. hybrids entering trade at rate of c. 150 a month (Hunt) so some 110 000 noted by 2000. Sometimes insects imprisoned (see *Coryanthes, Porroglossum*) & s.t. intoxicated, but pollen never offered (a third of orchids offer no reward at all, merely deception), though pollen-like 'pseudopollen' (some *Polystachya* spp.) or nutritive oils offered, though some spp. mimic other fls & there is no 'reward' – see e.g. *Cephalanthera, Epiblema, Epidendrum, Orchis, Traunsteinera,* while others attract egg-laying Diptera by imitating carrion (e.g. *Bulbophyllum* spp.), fungal fr. bodies (*Corybas*) or even aphids; some insects coll. pheromones, hold territory or roost in fls & incidentally poll. them. Some autogamy (*Neotinea, Oeceoclades, Stigmatodactylus, Thelymitra, Zeuxine* – weedy). Fr. usu. dry (indehiscent with seeds escaping as it rots in *Galeola* etc.), with hygroscopic hairs oft. between seeds, which are expelled by irreg. movements of hairs when wetted. The small seeds have allowed O. to compete with spore-pls (esp. ferns) in the rain forest canopy. Mycorrhizas (non-specific) incl. strains of *Armillaria* & *Rhizoctonia* pathogenic in other pls, suggesting they were originally thus in orchids too (Smith & Douglas)

Despite the huge size of fam., it is of almost no significance to the Common Man in most of the world, though millions of pounds are expended in cultivating the more showy tropical spp., hybridizing them & selling them as cut-fls (by 2005 US potted pl. trade alone worth $139M), while they have an aura of the exotic [see *Oncidium*, but 'They lack all grace, they have no vitality, rarely have they even fragrance. There is something altogether sinister, a cold & ruthless hate about them' – Frank Kingdon Ward, 1924] & even temp. spp. are sought out & dug up, such that many are endangered (all epiphytic & lithophytic spp. in Aus. now protected): fortunately many are now propagated by microtechniques from meristems. Hybrids first made by Jap. in 18 cent., in the W by Veitch Nursery (Exeter, UK) in 1850s, first to germinate being a *Cattleya*, first to flower being a *Calanthe*; first germ. on mycorrhiza-free medium 1922. Most oft. seen in florists' are spp. & hybrids of *Arachnis, Cattleya, Cymbidium, Guarianthe, Oncidium, Paphiopedilum, Papilionanthe, Phalaenopsis* & *Vanda*. Salep (10 – 20 M tubers per annum in Turkey alone) used in Turkish delight, icecream etc. gathered from *Dactylorhiza, Eulophia, Ophrys* & *Orchis* spp. while tubers (chikanda) of esp. spp. of *Disa* & *Satyrium* pounded & cooked with groundnuts in C Afr. (cf. *Gastrodia* in Aus.), few local medic., flavourings from *Jumellea, Leptotes* & esp. comm. *Vanilla*, while *Dendrobium* spp. used in Aus. Aborigines' body-paint & *Geodorum* spp. provide a strong gum for musical instruments

Orchidantha N.E. Br. Lowiaceae. 16 S China to W Mal. (Borneo 7). R: GBS 25(1970)239. ***O. fimbriata*** Holttum (W Mal.) – fls smell of coconut-oil & bugs, lvs used to wrap food & for back & chest pains; ***O. inouei*** Nagamasu & Saki (Borneo) – poll. dung-beetles

Orchipedum Breda. Orchidaceae (IV 2b). 3 W Mal.

Orchis Tourn. ex L. Orchidaceae (IV 4b). Incl. *Aceras*, 21 N temp. to SW China & Ind. R: H. Kretzschmar & al. *The orchid....O.* ed.2(2007)225. See also *Anacamptis*, *Neotinea*. Terr. orchids sometimes cult., dried starchy tubers form. wild-coll. ones used medic. & culinarily (bassorin, salep – easily digested) as in hot drinks, Turkish delight & T. icecreams, their shape reminiscent of testicles (Gk. *orchis*) so assoc. with potency, ***O. mascula*** (L.) L. (early purple orchid, palma-Christi (Med. England), dead man's finger, Eur.) prob. being Shakespeare's 'long purples', though prudishly replaced by *Lythrum salicaria* in Millais's picture of Ophelia (1851). Many Br. spp. rare & endangered, ***O. militaris*** L. (military or soldier orchid, Eur., Medit.) – orchinol (a phytoalexin) in tubers & ***O. simia*** L. (monkey o., Eur. Medit.) being protected spp.; other Br. spp. incl. ***O. anthropophora*** (L.) All. (*Aceras a.*, man orchid (Eur., N Afr.) & ***O. laxiflora*** Lam. (Jersey o., Channel Is. & Belgium to Medit.). ***O. caspia*** Trautv. (E Medit.) – nectarless sp. attracting insects by deceit in mimicking nectariferous fls of *Asphodelus* & *Bellevalia* spp. etc. in same habitat in Israel

Orcuttia Vasey. Gramineae (XXIX 2). 5 Calif., Mex. R: BTBC 68(1941)149. Dormant in drought yrs but flowering in spring pools when flooding adequate

Oreacanthus Benth. = *Brachystephanus* (but see KB 37(1982)467)

Oreanthes Benth. Ericaceae (VIII 5). 7 Andes of Ecuador

Orectanthe Maguire. Xyridaceae (Abolbodaceae). 2 Venez. R: AMBG 79(1992)879. Pollen to 185 μm in diam., poss. largest in angiosperms

Oregandra Standl. = *Chione*

oregano *Origanum* spp. esp. *O. vulgare*, (Am.) *Lippia* spp. esp. *L. graveolens*; **Cuban o.** *Coleus amboinicus*

Oregon (pine) *Pseudotsuga menziesii*

oregonia *Buxus sempervirens* 'Variegata'

Oreiostachys Gamble (~ *Nastus*). Gramineae (V 6). 4 Indomal.

Oreithales Schldl. = *Anemone*

Oreobambos K. Schum. Gramineae (V 4). 1 trop. E Afr.: ***O. buchwaldii*** K. Schum. – bamboo

Oreoblastus Susl. = *Solms-laubachia*

Oreobliton Durieu. Amaranthaceae (Chenopodiaceae I 1). 1 Algeria, Tunisia: ***O. thesioides*** Durieu

Oreobolopsis Koyama & Guag. Cyperaceae (II 1). 3 Calif. (1), Bolivia, Ecuador, to 4000 m. R: Novon 12(2002)342

Oreobolus R. Br. Cyperaceae (II 7). 14 Pac. wet alpine & subantarctic (Aus. 5, 4 endemic). R: BotJLS 96(1988)119

Oreocalamus Keng = *Chimonobambusa*

Oreocallis R. Br. Proteaceae (V 3). 1 Peru, Ecuador: ***O. grandiflora*** (Lam.) R. Br. (OW spp. = *Alloxylon*)

Oreocarya Greene = *Cryptantha*

Oreocereus (A. Berger) Riccob. (~ *Borzicactus*). Cactaceae (III 4). 6 W S Am. Cult. orn. esp. ***O. celsianus*** (Salm Dyck) Riccob. (*B. celsianus*, NW Arg., Bolivia) – largely poll. *Patagona gigas* (giant hummingbird) & ***O. trollii*** (Kupper) Backeb. (*B. trollii*, S Bolivia, N Arg.) with long matted hairs

Oreochaenactis Coville (~ *Chaenactis*). Compositae (Hele.-Cha.). 1 Calif.: ***O. thysanocarpha*** (A. Gray) Coville. R: FNA 21(2006)414

Oreocharis Benth. Gesneriaceae (V 4). Incl. *Ancylostemon*, *Bournea*, *Briggsia*, *Isometrum*, *Opithandra*, *Paraisometrum*, 80+ E Himal. & China to SE As. R: APS 25(1987)264. Some cult. orn. esp. ***O. muscicola*** (Diels) M. Möller & A. Weber (*Briggsia m.*, Bhutan to E Tibet, Yunnan). ***O. mileensis*** (W.T. Wang) M. Möller & A. Weber (*P. m.*, Yunnan) – coll. 2006 after 100 yrs

Oreochloa Link. Gramineae (XVI 13). 4 S Eur. R: OBC 3(1946)239

Oreochorte Kozo-Polj. = *Anthriscus*

Oreochrysum Rydb. (~ *Solidago*). Compositae (Ast.-Sol.). 1 WN Am.: ***O. parryi*** (A. Gray) Rydb. R: Phytol. 75(1993)334

Oreocnide Miq. Urticaceae (III). Incl. *Villebrunea*, 15 China & Jap., Indomal. Shrubs, s.t. scandent, & trees. ***O. integrifolia*** (Gaud.) Miq. (trop. As.) – useful fibre; ***O. rubescens*** (Blume) Miq. (Indomal.) – bark yields red dye for basketry

Oreocome Edgew. (~ *Selinum*). Umbelliferae (III 8). 6 Pakistan to W China. R: Willdenowia 31(2001)115

Oreocomopsis Pim. & Kljuykov (~ *Oreocome*). Umbelliferae (III 8). 3 Himal., S China

Oreodendron C. White = *Phaleria*

Oreogrammitis Copel. (~ *Grammitis*). Polypodiaceae (V). 10 trop. As. to Polynesia. R: GBS 58(2007)252

Oreograstis K. Schum. = *Carpha*

Oreoherzogia W. Vent = *Atadinus*

Oreojuncus Záveská Drábková & Kirschner (~ *Juncus*). Juncaceae. 2 N temp., Arctic. R: Preslia 85(2013)498

Oreoleysera Bremer (~ *Leysera*). Compositae (Gnap.-Rel.). 1 SW Cape: ***O. montana*** (Bolus) Bremer. R: OB 194(1991)67

Oreoloma Botsch. = *Sterigmostemum* (but see BZ 65(1980)425)

Oreomitra Diels = *Pseuduvaria*

Oreomunnea Oersted. Juglandaceae. 2 C Am. R: AMBG 59(1972)298

Oreomyrrhis Endl. = *Chaerophyllum* (but see UCPB 27(1955)347)

Oreonana Jepson. Umbelliferae (III 5). 3 California mts. R: Madroño 26(1979)133

Oreonesion A. Raynal. Gentianaceae. 1 trop. W Afr.: ***O. testui*** A. Raynal. R: Adansonia 5(1965)271

Oreopanax Decne. & Planch. Araliaceae. c. 150 trop. Am. Some cult. orn. hairy trees & shrubs esp. ***O. capitatus*** (Jacq.) Decne. & Planch. grown as juvenile epiphyte

Oreophilus Higgins & Archila = *Lepanthes* (but see Selbyana 29(2008)202)

Oreophysa (Boiss.) Bornm. Leguminosae (III 24). 1 Iran mts: ***O. microphylla*** (Jaub. & Spach) Browicz (*O. triphylla*). R: KB 16(1963)495

Oreophyton O. Schulz. Cruciferae (39). 1 E & NE Afr. mts: ***O. falcatum*** (A. Rich.) O. Schulz

Oreopoa H. Scholz & Parolly = *Poa* (but see Willd. 34(2004)146)

Oreopolus Schldl. (~ *Cruckshanksia*). Rubiaceae (IV 1). 1 Andes: ***O. glacialis*** (Poepp.) Ricardi. R: AMBG 83(1996)468

Oreoporanthera Hutch. = *Poranthera*

Oreopteris Holub = *Thelypteris*

Oreorchis Lindl. Orchidaceae (V 13e). Incl. *Diplolabellum* (mycotroph), 16 W Himal. to Taiwan & Jap. R: EJB 54(1997)292. ***O. patens*** (Lindl.) Lindl. – poll. Diptera

Oreosalsola Akhani (~ *Salsola*). Amaranthaceae. 9 SW As. R: Phytotaxa 249(2016)162

Oreoschimperella Rauschert (*Schimperella*). Umbelliferae (III 8). 3 Yemen, Ethiopia, Kenya mts

Oreosedum Grulich = *Sedum*

Oreoselinum Mill. = *Peucedanum*

Oreosolen Hook.f. Scrophulariaceae (Scroph.). 1 Himalaya: ***O. wattii*** Hook.f. – used in Tibetan medic. (spermicidal saponins)

Oreosparte Schltr. (~ *Hoya*). Apocynaceae (V a; Asclepiadaceae III 4). 1 Sulawesi: ***O. celebica*** Schltr. R: Webbia 68(2013)92

Oreosphacus Leyb. = *Satureja*

Oreostemma Greene (~ *Aster*). Compositae (Ast.-?Mac.). 3 W US. R: FNA 20(2006)359

Oreostylidium S. Berggren = *Stylidium*

Oreosyce Hook.f. = *Cucumis*

Oreothyrsus Lindau = *Ptyssiglottis*

Oreoxis Raf. Umbelliferae (III 8). 4 W N Am. ***O. alpina*** (A. Gray) J. Coulter & Rose – cult. orn. alpine

Oresbia Cron & R. Nordenstam (~ *Cineraria*). Compositae (Sen.-Sen.) 1 W Cape: ***O. heterocarpa*** Cron & R. Nordenstam. R: Novon 16(2006)216

Oresitrophe Bunge. Saxifragaceae (I 4). 1 NE China: ***O. rupifraga*** Bunge

Orestias Ridl. Orchidaceae (V 11b). 4 trop. Afr. R: ANMW 107B(2005)210

Orfilea Baill. (*Lautembergia*). Euphorbiaceae (Alch.-Alch.). 2–3 Madag., 1 Mauritius

orgeat *Prunus dulcis* (orig. *Hordeum vulgare*)

orham wood *Ulmus americana*

Orianthera C. Foster & Conn (~ *Logania*). Loganiaceae. 13 Aus. R: Telopea 16(2014)152

Orias Dode = *Lagerstroemia*

Oriastrum Poepp. & Endl. (~ *Chaetanthera*). Compositae (Mut.-Mut.). 18 Andes. R: A. Davies, *C. & O.* (2010)223

Oricia Pierre = *Vepris*

Oriciopsis Engl. = *Vepris*

oriental beech *Fagus sylvatica* subsp. *orientalis*; **o. bittersweet** *Celastrus orbiculatus*; **o. spruce** *Picea orientalis*

Origanum Tourn. ex L. Labiatae (VII 2b). 38 Euras. (Eur. 13), Medit. R: LBS 4(1980)1. Dwarf shrubs or herbs with ess. oils, some cult. as potherbs esp. ***O. dictamnus*** L. (dittany, Greece & Crete) – cult. orn., still med. in Crete (since Hippocrates, for dyspepsia, effective against bacteria), flavour in vermouth, & subsp. ***viride*** (Boiss.) Hayek (*O. heracleoticum*, winter marjoram, Medit.); ***O. majorana*** L. (*Majorana hortensis*, (sweet) marjoram, Medit., Turkey but now widely natur.) – cult. herb for flavouring (methyl chavicol) meat & sausages, ess. oil used in tinned meat, form. medic.; ***O. onites*** L. (pot m., SE Eur., E Medit.) – inferior; ***O. syriacum*** L. (za'atar, Near E) – hyssop of the Bible (e.g. at Crucifixion); ***O. vulgare*** L. (oregano, Eur. to C As., natur. E US) – dried lvs much used in pizza etc., source of red dye, hybrids with *O. majorana* (***O.*** × ***majoricum*** Cambess.) sold as oregano in Aus.; a Turkish sp. used against rheumatism as containing corvacrone, a prostaglandin synthetase inhibitor & thus a painkiller

Orinoco sassafras *Ocotea cymbarum*

Orinus A. Hitchc. Gramineae (XXIX). 5 Himal. to W China (4 – R: FOC 22(2006)464). Desert dunes at high alts

Orites R. Br. Proteaceae (V 1). 7 temp. E Aus., 1 S Am. Some timbers & cult. orn. e.g. ***O. excelsa*** R. Br. (prickly ash, E Aus.)

Oritrephes Ridl. = *Anerincleistus*

Oritrophium (Kunth) Cuatrec. Compositae (Ast.-Hin.). 21 Andes, Mex. (2). R: Biollania esp. ed. 6(1997)287, ABV 31(2008)84

Orixa Thunb. Rutaceae (I 2). 1 China, Korea, Jap.: ***O. japonica*** Thunb. – cult. orn. dioec. shrub with alks & scented lvs used for hedging in Jap.

Orlaya Hoffm. Umbelliferae (III 3). 5 SE Eur. (3) to C As. ***O. grandiflora*** (L.) Hoffm. (Medit.) – vogue garden-pl.

Orleanesia Barb. Rodr. Orchidaceae (V 13c). 9 trop. S Am.

Ormenis (Cass.) Cass. = *Cladanthus*

Ormerodia Szlach. = *Cleisostoma* (but see ABF 40(20003)68)

Ormocarpopsis R. Viguier. Leguminosae (III 11). Incl. *Peltiera*, 7 Madag. R: Adans. 35(2013)66

Ormocarpum P. Beauv. Leguminosae (III 11). Excl. *Zygocarpum*, 18–20 trop. & warm OW. ***O. cochinchinense*** (Lour.) Merr. (*O. orientale*, SE As., Mal.) – 'greens' in NG

Ormoloma Maxon = *Lindsaea*

Ormopteris J. Sm. (~ *Pellaea*). Pteridaceae. 6 trop. S Am. R: Phytotaxa 221(2015)118

Ormopterum Schischkin. Umbelliferae (III 8). 2 C As., Pakistan

Ormosciadium Boiss. Umbelliferae (III 11). 1 E As. Minor: ***O. aucheri*** Boiss. R: Fl. Iraq 5,2(2013)252

Ormosia Jackson. Leguminosae (III 2). c. 130 C & E S Am. (c. 80), E As. (China 35 – R: APS 22(1984)117) to NE Aus. Five segregates based on diff. dispersal mechanisms: *Fedorovia*, *Macroule*, *Placolobium*, *Ruddia*, *Trichocyamos*. Alks; seeds orn. & used as beads, some resembling *Abrus precatorius*, & others cult. orn. or useful timber esp. ***O. krugii*** Urb. (WI) & ***O. monosperma*** (Sw.) Urb. (WI to NE Venez.). ***O. nobilis*** Tul. (Amaz.) – fine roots grow under damaged bark & coalesce, repairing it

Ormosiopsis Ducke = *Ormosia*

Ormosolenia Tausch (~ *Peucedanum*). Umbelliferae (III 10). 1 E Medit.: ***O. alpina*** (Schultes) Pim. R: EJB 49(1992)214

Ornduffia Tippery & Les (~ *Villarsia*). 7 Aus. R: Novon 19(2009)409

Ornichia Klack. (~ *Chironia*). Gentianaceae. 3 Madag. R: BMNHN 4,8 (1986) 195

Ornithidium Salisb. ex R. Br. = *Maxillaria*

Ornithoboea Parish ex C.B. Clarke. Gesneriaceae (III 2i). Excl. *Damrongia*, 16 SE As. to Mal. R: GBS 66 (2014)78. Some with fls like (?mimicking) orchid fls

Ornithocarpa Rose. Cruciferae (16). 2 Mex.

Ornithocephalus Hook. Orchidaceae (V 12h). c. 50 trop. Am. Some cult. orn.

Ornithochilus (Lindl.) Benth. = *Phalaenopsis* (but see OB 95(1988)42)

Ornithogalum Tourn. ex L. Asparagaceae (Hyacinthaceae). Incl. *Galtonia*, excl. *Albuca*, *Dipcadi*, *Pseudogaltonia*, c. 160 Eur., Medit. & Near E. Afr. esp. W Cape, Madag. (1). R: T 58(2009)100. Many generic splits form. recognized based on poll. syndrome characters; some poll. monkey-beetles, some moths etc. Alks incl. colchicine, cardiac glycosides, cardiotoxic cardenolides. Cult. orn. (some dreary) bulbous pls, some toxic, others ed., incl. ***O. angustifolium*** Bor ('*O. umbellatum*', star of Bethlehem, GB etc. with 2n = 18, 27,

36, the true ***O. umbellatum*** L. (*O. divergens*, lowland Eur., Middle E, invasive Aus., US) with 2n = 36, 45, 54) – bulb s.t. eaten, though 2 digitalis-like glycosides concentrated in bulbs (& fls) poisonous to stock & humans; ***O. candicans*** (Bak.) Manning & Goldl. (*Galtonia c.*, summer hyacinth, Berg lily, Drakensberg, S Afr.) – cult. orn.; ***O. pyrenaicum*** L. (Bath asparagus, Medit., natur. N Eur.) – young infls eaten like asparagus; ***O. thyrsoides*** Jacq. (chincherinchee – name said to sound like the scapes being rubbed against one another, S Afr.) – populations of diploids, triploids, hexaploids & aneuploids, winter cut-fl. exported to Eur. (in C19 by sea, in bud), toxic (even seeds) to stock, noxious weed in S Aus.

Ornithoglossum Salisb. Colchicaceae (Liliaceae s.l.). 9 trop. & S Afr. R: OB 64(1982) – 'bird's- tongue', with alks toxic to stock

Ornithophora Barb. Rodr. = *Gomesa*

Ornithopus L. Leguminosae (III 22). 5 temp. S Am. (1), Atlantic is., Eur. (4), Medit. to W As. Bird's foot (esp. ***O. perpusillus*** L., Eur.); ***O. sativus*** Brot. (serradella, Medit.) – good fodder

Ornithostaphylos Small (~ *Arctostaphylos*). Ericaceae (III). 1 S & Baja California: ***O. oppositifolia*** (C. Parry) Small – threatened by US/Mex. border-fence construction. R: FNA 9(2009)403

Orobanchaceae Vent. (Scrophulariaceae s.l.). Magnoliidae – Lamiales. Incl. Cyclocheilaceae, Lindenbergiaceae, Nesogenaceae, Rehmanniaceae, 99/2275 N hemisph. to trop. Autotrophic herbs to facultative & obligate herbaceous root-parasites without chlorophyll (when stems usu. fleshy, the radicle becoming a haustorium penetrating host-root), rarely shrubby (*Radamaea*, former Cyclocheilaceae, Nesogenaceae) or lianoid; many turning black on drying. Lvs in spirals to opp., oft. toothed to deeply lobed (in holoparasites scaly with oft. disorganized stomata); stip. 0. Fls bisex., solit. as in *Phelipaea* or in term. bracteate racemes or spikes; K ((0, 1–)4 or 5), segments open or valvate, C (5), ± 2-labiate, oft. curved, lobes imbr. (adaxial ones internal), A (2)4 with adaxial 1 staminodal or 0, attached to C & alt. with lobes, anthers with longit. slits, $\underline{G}$ (2(3)) with thin style & 2–4-lobed stigma, 1-loc. with (2)4(6) intruded parietal placentas or axile placentation, with 1-∞ anatropous unitegmic ovules. Fr. usu. a loculicidal capsule (rarely a drupe etc.) each of the 2(3) valves typically with 2 placentas; seeds minute (c. 1 M/g), with undifferentiated embryo in oily or 0 endosperm. n = 7+

Classification & principal genera: **Gerardieae** (*Agalinis*); **Orobancheae** (*Aeginetia, Harveya, Orobanche*); **Escobedieae** (*Alectra*); 'Micrargerieae' (*Micrargeria*); **Buchnereae** (*Buchnera, Sopubia, Striga*); 'Xylocalyceae' (*Xylocalyx* – only); **Cymbarieae** (*Monochasma*); 'Castillejeae (R[Castillejinae]: 34(2009)184)' (*Castilleja*); **Rhinantheae** (*Bartsia, Euphrasia, Lathraea, Melampyrum, Odontites, Pedicularis, Rhinanthus*)

Incl. hemiparasitic 'Scrophulariaceae' (= Plantaginaceae) form. segregated merely by tradition, *Lathraea* being placed in one or other; some genera e.g. *Harveya* have range from hemi- to holo-parasites, showing that trend happened in parallel many times

Parasitism rarely restricted to 1 host sp.; seeds wind-disp. Mazaceae, Paulowniaceae, Phrymaceae poss. here

Some locally eaten or pests (*Aeginetia, Orobanche, Striga*); cult. orn. (*Agalinis, Lindenbergia*), dyes (*Alectra, Orthocarpus, Striga*), local medic.

Orobanche Tourn. ex L. Orobanchaceae (Orob.). Excl. *Phelipanthe*, 150 temp. (Eur. 45) & warm esp. N. Broomrapes. Some restricted to particular fams as hosts, ***O. crenata*** Forssk. (Arabia) – major pest of legumes, ***O. cumana*** Wallr. (*O. cernua* subsp. *c.*, Russia) of sunflowers, rarely to 1 sp. as in ***O. hederae*** Vaucher ex Duby on *Hedera hibernica* (W Eur. & Medit., but also on *Fatsia japonica* in cult.), known as broomrapes because form. thought to be outgrowths of brooms etc., 'rapum' being a knob or tuber. Seeds germ. giving spiral filaments carrying testa apically, 'searching' habitat in a spiral movement for hosts; pl. oft. ann. with tuber developing as swelling behind tip lying in contact with host-root, the tuber becoming nodulated & covered with papillae, one of which penetrates host as far as xylem; thereafter host & parasite tissue difficult to distinguish but parasite certainly has own vessel-elements; from near junction, infl. arises as bud & pushes above ground (even through tarmac). Some infls eaten by Amerindians (cf. *Pholisma*), some local medic.

Orobus Tourn. ex L. = *Lathyrus*

Orochaenactis Cov. Compositae (Cha.). 1 California: ***O. thysanocarpha*** Cov. R: FNA 21(2006)914

Orogenia S. Watson. Umbelliferae (III 5). 2 W N Am.

Orontium L. Araceae (II). 1 E N Am.: ***O. aquaticum*** L. (golden club) – (?) bee-poll. cult. orn. aquatic, starch in rhiz. & seeds ed. once boiled

Oropetium Trin. Gramineae (XXIX 6). 6 OW trop. KB 30(1975)467. ***O. capense*** Stapf (Afr.) lvs revive after reduction to 8% water content

Orophea Blume. Annonaceae (IV 7). 37 Ind. & S China to Moluccas (subg. *Sphaerocarpon* 22 S Ind. to W Mal. – R: Blumea 46(2001)147). R: Blumea 33(1988)1. ***O. katschallica*** Kurz (Andamans) – sap used to repel *Apis dorsata* bees during honey-collecting

Orophochilus Lindau. Acanthaceae (III 1). 1 Peru: ***O. stipulaceus*** Lindau

Orostachys Fischer (~ *Sedum*). Crassulaceae (I 1). 11 Eur. (2) & temp. As. R: U. Eggli, *Ill. Handb. Succ. Pls*, Crass. (2005)186. Heterogeneous; prob. some *Sedum* (*Hylotelephium* etc.) spp. referable here. Some cult. orn. with hapaxanthic leaf-rosettes oft. with offsets withering to compact winter-buds of callose lvs which grow basally in spring to give callus-tipped lvs (s.t. spiny)

Orothamnus Pappe ex Hook. = *Mimetes*

Oroxylum Vent. Bignoniaceae (2). 1 Sri Lanka to Sulawesi, Timor: ***O. indicum*** (L.) Kurz – sparsely branched pachycaul tree (Chamberlain's Model) of clearings & roadsides, with 2–3-pinnate lvs, stinking bat-poll. fls with copious nectar opening at night & fr. to 1 m long; young lvs cooked as vegetable, bitter bark medic., dye used in rattan basketry in Sarawak

Oroya Britton & Rose (~ *Oreocerus*). Cactaceae (III 4). 2 Peru. R: Ashingtonia 1(1975)136

Orphanidesia Boiss. & Bal. = *Epigaea*

Orphanodendron Barneby & Grimes. Leguminosae (I 4). 2 NW Colombia. R: Britt. 42(1990)249

Orphium E. Meyer. Gentianaceae. 1 S Afr.: ***O. frutescens*** (L.) E. Meyer – cult. orn. shrub with pink buzz-poll. fls. R: Strelitzia 10(2000)311

orpine *Sedum telephium*

Orrhopygium Löve = *Aegilops*

orris root *Iris* × *germanica* 'Florentina'

Ortachne Nees ex Steud. Gramineae (X). 2 Costa Rica to Peru, Patagonia. Dwarf grasses of montane woodland glades. R: KBAS 13(1986)86

ortanique *Citrus* × *aurantium*

Ortegia Loefl. ex L. Caryophyllaceae (I 1). 1 Spain, Portugal, Italy: ***O. hispanica*** L.

Ortegocactus Alexander (~*Mammillaria*). Cactaceae (III 9). 1 SE Mex.: ***O. macdougallii*** Alexander

Ortgiesia Regel = *Aechmea*

Orthaea Klotzsch. Ericaceae (VIII 5). 35 trop. Am. R: NJB 7(1987)31

Orthandra Burret = *Mortoniodendron*

Orthantha (Benth.) Wettst. = *Odontites*

Orthanthera Wight. Apocynaceae (V b; Asclepiadaceae III 5). 4 Ind. & Nepal (1), S trop. Afr. (3)

Orthechites Urb. = *Secondatia*

Orthilia Raf. Ericaceae (II 1; Pyrolaceae). 1 circumboreal: ***O. secunda*** (L.) House. R: FNA 9(2009)388

Orthion Standl. & Steyerm. Violaceae (III 2a). Excl. *Mayanaea*, 6 C Am.

Orthiopteris Copel. = *Saccoloma* (but see PhytoKeys 53(2015)44)

Orthocarpus Nutt. Orobanchaceae (Cast.; Scrophulariaceae s.l.). 9 W Am. R: SB 17(1992)560. ***O. luteus*** Nutt. – lvs yield a red dye, whole pl. a yellow one

Orthoceras R. Br. Orchidaceae (IV 3d). 2 E Aus., New Caled., NZ. Autogamous

Orthochilus Hochst. ex A. Rich. (*Pteroglossaspis*; ~ *Eulophia*). Orchidaceae (V 9). 33 trop. Afr., Madag., Am. R: T 63(2014)17

Orthoclada P. Beauv. Gramineae (XIX). 1 trop. Am., 1 SE trop. Afr. Lvs with 'false petioles'

Orthodon Benth. = *Mosla*

Orthogoneuron Gilg. Melastomataceae. 1 trop. Afr.: ***O. dasyanthum*** Gilg

Orthogynium Baill. Menispermaceae (tribe?). 1 Madag. (?): ***O. gomphloides*** Baill. – female unknown

Orthomene Barneby & Krukoff. Menispermaceae (II). 4 trop. Am. R: MNYBG 22,2(1971)79

Orthopappus Gleason (~ *Elephantopus*). Compositae (Vern.-Ele.). 1 trop. Am.: ***O. angustifolius*** (Sw.) Gleason

Orthophytum Beer. Bromeliaceae (3). 24 E Braz. R: Plantsman 13(1991)180. Some cult. orn.

Orthopichonia H. Huber. Apocynaceae (Ic). 6 trop. W Afr. R: AUWP 89–4(1989)29

Orthopogon R. Br. = *Oplismenus*

Orthopterum L. Bolus (~ *Faucaria*). Aizoaceae (V). 2 E Cape. R: H. E. K. Hartmann, *Ill. Handb. Succ. Pls*, Aiz. F–Z(2002)200. Unlike *F.* fl. repeatedly opening & closing

Orthopterygium Hemsl. Anacardiaceae (I; Julianiaceae). 1 Peru: ***O. huaucui*** Hemsl. – winged pedicel aids disp.

Orthoraphium Nees (~ *Stipa*). Gramineae (X). 1 Himal.: ***O. roylei*** Nees. R: FOC 22(2006)211

Orthosia Decne. Apocynaceae (V c; Asclepiadaceae III 1). 31 trop. Am. R: AMBG 99(2013)66

Orthosiphon Benth. Labiatae (VII 3d). 40 OW trop., Colombia (1!). Champagnat's Model. ***O. aristatus*** (Blume) Miq. (misaim kuching, Mal.) – cult. diuretic with high levels of potassium salts & a glycoside

Orthosphenia Standl. Celastraceae (I). 1 Mex.: ***O. mexicana*** Standl. R: Ciencia 16(1956) 141

Orthotactus Nees = *Justicia*

Orthotheca Pichon = *Xylophragma*

Orthothylax (Hook.f.) Skottsb. (~ *Helmholtzia*). Philydraceae. 1 E Aus. montane rain forest: ***O. glaberrimus*** (Hook.f.) Skottsb. R: KB 1934: 97

Orthrosanthus Sweet. Iridaceae (VII 3). 9 SW Aus. (4 – R: FA 46(1986)10), S & C Am. (R: AMBG 74(1987)578)

Orthurus Juz. = *Geum*

Ortizacalia Pruski. Compositae (Sen.-Sen.). 1 Costa Rica: ***O. austin-smithii*** (Standl.) Pruski – liane. R: Phytoneuron 2012–50:1

Orumbella J. Coulter & Rose = *Podistera*

Orvala L. = *Lamium*

Orychophragmus Bunge. Cruciferae (12). 7 E As. R: Novon 10(2000)349. ***O. violaceus*** (L.) O. Schulz – favourite pl. in Chinese gdns

Oryctanthus (Griseb.) Eichler (*Glutago*). Loranthaceae (4 4). Excl. *Oryctina*, 13 trop. Am. R: BJ 95(1976)478, 114(1992)182

Oryctes S. Watson (~ *Physalis*). Solanaceae (IV 5c). 1 SW US: ***O. nevadensis*** S. Watson. R: A.T. Hunziker, *Genera Solanacearum* (2001)227

Oryctina Tieghem (~ *Cladocolea*). Loranthaceae (4 4). 6 S Am. R: Novon 10(2000)396

Orygia Forssk. = *Corbichonia*

Oryxis Delgado & G. Lewis. Leguminosae (III 18). 1 Braz.: ***O. monticola*** (Benth.) Delgado & G. Lewis. R: KB 52(1997)221

Oryza Tourn. ex L. Gramineae (III 3i). 20 trop. (Mal. 7, R: Blumea 32(1987)169). R: SCB 91(2001)11. Rice, esp. ***O. sativa*** L. (genome sequenced 2005), 'japonica' prob. first derived from ***O. rufipogon*** Griff. (of which *'O. nivara'* annual self-fert., photoperiod-insensitive), As.; noxious weed in N Aus., US ricefields, & arising as a selected weed (c. 5000 BC Lower Yangtze in flooded *Colocasia* fields, with non-shattering infl. due to single amino-acid subst., asparagine for lysine, leading to failure to mature abscision layers); 'indica' being crosses between 'japonica' & local wild races in S & SE As., so prob. several centres of its domestication; now principal carbohydrate source of As. (biggest producer China, biggest exporter Ind. (2013), biggest importer (14%) Indonesia), unhusked grain known as paddy; up to 50 K landraces, many cvs grouped as lowland (grown under inundation; 200–400 kg per ha per annum) & upland (dry; 100 kg per ha per annum) r., or by grain size (Basmati & Patna = long-grained, rose = medium, pearl = short), some glutinous & used for puddings, others not so & poss. the world's most imp. food-pl. (by 2013 worldwide 745 M t a yr), eaten with e.g. curries, & in Eur. as paella & risotto, but oft. without the husks when much protein & vitamin removed (that with = brown r.; black r. is due to colour leaching from unmilled husks in cooking), & oft. 'polished', genetically modified (gene from *Narcissus*, later mouse) 'golden rice' with β carotene, a vitamin A precursor, in endosperm until lately too controversial to be comm., though now modified to prod. human breast-milk proteins to be used in treatment of diarrhoea etc.; husks & polishings (shude(s)) an imp. cattle food; flour used in cooking (incl. vermicelli noodles & edible rice paper) & breakfast foods etc. or as comm. starch & face-powder for Geisha; fermented to give beer (& used as 'extender' of barley in e.g. Budweiser beer), sweet white wine (mirin, also vinegar) & eventually sake (moutai in China); oil (rice bran oil cont. 'oryzanol' with much-promoted med. benefits) from endosperm comm. cooking oil; husks used as pot-plant growing-medium; straw imp. fodder, packing & building & used for 'sea-grass matting' (Aus.); local medic. (iron deficiency, diarrhoea); fls used in toothpaste in China. Native Aus. spp. ground to paste & baked by Aborigines; ***O. glaberrima*** Steud. ('red rice', W Afr.) – locally cult. (in Niger Delta for 3500 yrs), weedy (app. dervied from

'*O. barthii* A. Chev.', coll. not cult. for 3500 yrs) in rice but reduced by early flooding after seeding

Oryzetes Salisb. = *Hygrophila*

Oryzidium C. Hubb. & Schweick. Gramineae (Panicoid.). 1 Zambia to SW Afr.: ***O. barnardii*** C. Hubb. & Schweick. – floating. R: Strelitzia 10(2002)708

Oryzopsis Michaux (~ *Piptatherum*). Gramineae (X). 1. N Am.: ***O asperifolia*** Michaux

orzo rice-shaped pasta (*Triticum turgidum*)

Osa Aiello (~ *Hintonia*). Rubiaceae (I 7). 1 Costa Rica: ***O. pulchra*** (D.R. Simpson) Aiello

Osage orange *Maclura pomifera*

Osbeckia L. Melastomataceae. 50 OW trop. (As. 31, R: Ginkgoana 4(1977)1). Local medic.

Osbertia Greene (~ *Haplopappus*). Compositae (Ast.-Chr.). 3 WN & C Am. R: Phytol. 71(1991)132

Osbornia F. Muell. Myrtaceae (II 3). 1 C Mal. to NE Aus.: ***O. octodonta*** F. Muell. in mangrove, fr. leathery, indehiscent

Oschatzia Walp. (~ *Azorella*). Umbelliferae (Azor.). 2 SE Aus.

Oscularia Schwantes (~ *Lampranthus*). Aizoaceae (V). 23 N & W Cape. R: H. E. K. Hartmann, *Ill. Handb. Succ. Pls*, Aiz. F–Z(2002)195

Oserya Tul. & Wedd. Podostemaceae (III). Excl. *Noveloa*, 5 trop. Am. R: ABN 3(1954)216

osha *Ligusticum* spp. (N Am.)

Oshima cherry *Prunus serrulata*

osier *Salix* spp.

Osmadenia Nutt. (~ *Hemizonia*). Compositae (Mad.-Mad.). 1 Calif., NW Mex.: ***O. tenella*** Nutt. R: FNA 21(2006)269

Osmanthus Lour. Oleaceae (4d). c. 30 As. (SW 1, W Mal. 1) esp. China (Am. spp. = *Cartrema*). R: NRBGE 22(1958)439; Q. Xiang & Y. Lin (2008) *An. illus. monogr. sweet O.* Cult. orn. everg. shrubs & trees with extrafl. nectaries, esp. ***O. × burkwoodii*** (Hort. Burkw. & Skipw.) P. Green (× *Osmarea burkwoodii*, *Osmanthus decorus* (Boiss. & Bal.). Kasapl. (*Phillyrea d.*, Caucasus, Lazistan) × *O. delavayi* Franch. (W China) – oft. considered an 'intergeneric hybrid' (faulty taxonomy!)); ***O. fragrans*** Lour. (kwei, prob. E As., but long cult. in As.) – 166+ cvs, male infls used to flavour tea (& soups in Shanghai restaurants), sprinkle on food & in confectionery; ***O. heterophyllus*** (G. Don) P. Green (Taiwan, Jap.) – cult. orn. with holly-like foliage. See also *Nestegis*

× **Osmarea** C. Curtis. See *Osmanthus*

Osmaronia Greene = *Oemleria*

Osmelia Thw. Salicaceae (Samydaceae; Flacourtiaceae). 3 Mal., 1 Sri Lanka

Osmiopsis R. King & H. Robinson (~ *Eupatorium*). Compositae (Eup.-Prax.). 1 Haiti: ***O. plumieri*** (Urb. & Ekman) R. King & H. Robinson. R: MSBMBG 22(1987)397

Osmites L. = *Relhania*

Osmitopsis Cass. Compositae (Anth.-Osmit.). 9 SW Cape. R: BN 125(1972)9, 129(1976)21. local medic.

Osmoglossum (Schltr.) Schltr. = *Cuitlauzina*

Osmolindsaea (Kramer) Lehtonen & Christenh. (~ *Lindsaea*). Lindsaeaceae. 7 Madag., Indomal. to Jap. R: SB 38(2013)888

Osmorhiza Raf. Umbelliferae (III 2). 10 Am., E As. R: AMBG 71(1984)1128, 89(2002)414. ***O. berteroi*** DC. (*O. chilensis*, Am., bipolar) – cf. *Anemone multifida*, *Carex macloviana*, *Koenigia islandica*. Some ed. roots & medic., esp. ***O. occidentalis*** (Torrey & A. Gray) Torrey (N Am.) – antiseptic, analgesic, febrifuge etc.; some cult. orn. incl. ***O. longistylis*** (Torrey) DC. (W N Am.) – main sweet constituent *trans*-anethol

Osmoxylon Miq. Araliaceae. 50 C & E Mal. (R: Blumea 23(1976)99), Taiwan & W Pac. ***O. borneense*** Seemann (Borneo) – rheophyte; ***O. lineare*** (Merr.) Philipson (Philipp.) – cult. orn. (e.g. Singapore streets) with 'false' & real frs

Osmunda Tourn. ex L. Osmundaceae. Excl. *Osmundastrum*, 5–6 temp. (Eur. 1), trop. E As. Poss. incl. *Leptopteris*, *Todea*. Fibre used for orchid-growing, hairs around young fronds used with wool in Jap. to make a textile for raincoats. Cult. orn. (royal ferns) & locally eaten, esp. ***O. regalis*** L. (N temp. to Afr. & S Am.) with stock like small tree-fern, used in brewing Celtic heather ale (spores contain a thiaminase destroying vitamin B, & therefore yeast so stopping fermentation, a prehistoric 'Camden tablet'!)

Osmundaceae Martinov. Polypodiidae – Osmundales. 4/14 trop. & temp. R: T 55(2006) 710. Stems erect & enveloped in persistent frond-bases & coarse roots; vasc. system of separate xylem strands. Fronds 1–2-pinnate or -pinnatifid with stip. at petiole-bases. Sporangia not arr. in sori, on underside of fronds or on specialized fronds or

pinnae (*Osmunda*) s.t. forming a kind of panicle at frond-tip; sporangia short-stalked with poorly-developed lat. annulus & many spores. n = 22
Genera: *Leptopteris, Osmunda, Osmundastrum, Todea*
Fossils back to Permian referred here. Cult. orn.

Osmundastrum C. Presl (~ *Osmunda*). Osmundaceae. 1 ***O. cinnamomea*** (L.) C. Presl (cinnamon fern, fiddleheads, Am. & E As.) – 'living fossil' as v. like *Osmunda claytoniites* C. Phipps & al. (Antarctic Triassic), local medic. N Am.

oso-berry *Oemleria cerasiformis*

Ossaea DC. = *Miconia*

Ossiculum Cribb & van der Laan. Orchidaceae (V 16d). 1 Cameroun: ***O. aurantiacum*** Cribb & van der Laan. R: KB 41(1986)823

Ostenia Buchenau = *Hydrocleys*

Osteocarpum F. Muell. = *Sclerolaena* (but see FA 4(1984)231)

Osteomeles Lindl. Rosaceae (17). 1 (variable) China to Hawaii: ***O. anthyllidifolia*** Lindl. – cult. orn. shrub with pinnate lvs & pyrenes. R: CJB 68(1990)2230

Osteophloeum Warb. Myristicaceae. 1 Amazonia: ***O. platyspermum*** (A. DC.) Warb. – bark a source of hallucinogenic paste, also medic. R: NJB 20(2001)445

Osteospermum L. Compositae (Cal.). Excl. *Inuloides, Monoculus, Norlindhia, Tripteris*, c. 45 S Afr. to Arabia (St Helena sp. = *Oligocarpus*). Heterogeneous; poss. incl. *Chrysanthemoides* etc. Shrubs & herbs; some cult. orn. as '*Dimorphotheca*', esp. S Afr. ***O. ecklonis*** (DC.) Norlindh & ***O. jucundum*** (E. Phillips) Norlindh ('*O. barberae*') & many hybrids, some with spoon-shaped ray-florets

Ostericum Hoffm. = *Angelica*

Ostlundia W. Higgins = *Encyclia*

Ostodes Blume. Euphorbiaceae (Cod.-Ost.). 1 E Himal. to Borneo: ***O. paniculata*** Blume – lvs used for wrapping, latex for gum in Sikkim

Ostrearia Baill. Hamamelidaceae (I 1). 1 Queensland: ***O. australiana*** Baill. R: FA 3(1989)1

ostrich fern *Matteuccia struthiopteris*

Ostrowskia Regel. Campanulaceae (I 1). 1 Turkestan: ***O. magnifica*** Regel – thicket-forming to 2.5m with tubers & whorled lvs, C to 15 cm diam., difficult cult. orn.

Ostrya Scop. Betulaceae (II; Carpinaceae). 9 N temp. to C Am. R: JAA 71(1990)57. Hop-hornbeam (nutlets encl. in bladder-like green involucres). Timbers esp. ***O. carpinifolia*** Scop. (S Eur., As. Minor) – irritant hairs on fr. scales, & ***O. virginiana*** (Mill.) K. Koch (ironwood, leverwood, E N Am.) – local medic.

Ostryocarpus Hook.f. Leguminosae (III 16). Excl. *Aganope* 1–2 trop. OW. Fish nets & poisons

Ostryoderris Dunn = *Aganope*

Ostryopsis Decne. Betulaceae (II; Carpinaceae). 2 E Mongolia & SW China. Shrubs; A 4–6

Osvaldoa J. Grande (~ *Panicum*). Gramineae (XXIII 3). 1 trop. Am.: ***O. valida*** (Mez) J. Grande. R: T 63(2014)270. Allied to *Paspalum*

Oswego tea *Monarda didyma*

Osyridicarpos A. DC. Santalaceae (Thesiaceae). 1(–6) trop. & S Afr.: ***O. schimperianus*** (A. Rich.) A. DC. R: FTEA Sant. (2005)21

Osyris L. Santalaceae. Excl. *Colpoon*, 2 Medit. to trop. Afr. & China. Dioec. hemiparasites; some tanbark. ***O. alba*** L. (Medit.) – dioec., female fls with antherodes deceiving pollen-seeking pollinators as minority of fls male, used in bone-setting in Crete (since time of Dioscorides); ***O. lanceolata*** Hochst. & Steud. (*O. tenuifolia*, E Afr. sandalwood, Medit. to trop. OW) – tanbark, timber like true sandalwood, oil used in scent-making

Otacanthus Lindl. Plantaginaceae (Grat.-Stem.; Scrophulariaceae s.l.). 7 Braz. R: Britt. 53(2001)143. ***O. caeruleus*** Lindl. – aromatic cult. orn. (resembling Acanthaceae), K foliose, natur. Madag., Masc.

Otachyrium Nees. Gramineae (XXIII 2). Incl. *Steinchisma*, 16 Aus., SE US to S Am. R: SCB 57(1984)1

Otaheite apple *Syzygium malaccense, Spondias* spp.; **O. chestnut** *Inocarpus fagifer*; **O. gooseberry** *Phyllanthus acidus*; **O. myrtle** *Securinega durissima*; **O. potato** *Dioscorea bulbifera*; **O. walnut** *Aleurites moluccanus*

Otanthera Blume = *Melastoma*

Otanthus Hoffsgg. & Link = *Achillea*

Otatea (McClure & E.W. Sm.) Calderón & Soderstrom. Gramineae (V 3). 8 Mex. to N Colombia. R: SB 36(2011)318

Oteiza Llave (~ *Calea*). Compositae (Mill.-Gal.). 3 Mex., Guatemala

Othake Raf. = *Palafoxia*
Otherodendron Makino = *Microtropis*
Othocallis Salisb. = *Scilla*
Otholobium Stirton (~ *Psoralea*). Leguminosae (III 20). 61 E & S Afr. (53), S Am. (8 – R: MNYBG 61(1990)16). ***O. glandulosum*** (L.) Grimes (*P. g.*, Jesuits' tea, Chile) – local medic.
Othonna L. Compositae (Sen.-Sen.). Excl. *Crassothonna, Hertia*, c. 120 trop. & S Afr., Aus. Xeromorphic shrubs & herbs with tubers; ray-florets fertile, disk-florets sterile
Othonnopsis Jaub. & Spach = *Hertia*
Otilix Raf. = *Lycianthes*
Otion Crisp & P. Weston (~ *Pultenaea*). Leguminosae (III 13). 8 N, C & SW Aus.
Otiophora Zucc. Rubiaceae (IV 1). 20 trop. Afr. & Madag. R: GOB 1(1973)25
Otoba (A. DC.) Karsten. Myristicaceae. c. 7 trop. Am. Source of otoba fat used in soap-making
Otocalyx Brandegee = *Rondeletia*
Otocarpus Durieu. Cruciferae (12). 1 Algeria: ***O. virgatus*** Durieu
Otocephalus Chiov. = *Calanda*
Otochilus Lindl. Orchidaceae (V 10b). 5 E Himal. to SE As. R: BT 71(1976)8. Pseudobulbs growing on top of one another, forming chains
Otoglossum (Schltr.) Garay & Dunsterv. (~ *Oncidium*). Orchidaceae (V 12h). 13 N S Am.
Otomeria Benth. Rubiaceae (IV 1). 8 trop. Afr., Madag.
Otonephelium Radlk. Sapindaceae. 1 Ind.: ***O. stipulaceum*** (Bedd.) Radlk.
Otopappus Benth. Compositae (Helia.-Ecl.). 14 Mex. & C Am., WI. R: SB 8(1983)185
Otophora Blume = *Lepisanthes*
Otoptera DC. (~ *Vigna*). Leguminosae (III 18). 2 Afr.
Otospermum Willk. Compositae (Anth.-?Leuc.). 1 SW Eur., NW Afr.: ***O. glabrum*** (Lag.) Willk. R: BBMNHB 23(1993)153
Otostegia Benth. Labiatae (VI). Excl. *Rydingia*, 11 Egypt, Arabia to W Afr. R: SGP 77(2007)233
Otostylis Schltr. Orchidaceae (V 12j). 4 trop. S Am., Trinidad. R: Sida 21(2004)842
Ototropis Nees (~ *Desmodium*). Leguminosae (III 19). 13 China, Indomal. R: JJB 87(2012)110
Ottelia Pers. Hydrocharitaceae. c. 25 trop. & warm, esp. OW (NW 1). R: AB 18(1984)263. Submerged aquatics incl. ***O. alismoides*** (L.) Pers. (China & Jap. to Aus. & NE Afr.) – cult. orn., green veg. in Mal., natur. Italy
Ottleya Sokoloff = *Acmispon* (but see FR 110(1999)93)
otto of roses See *Rosa*
Ottoa Kunth. Umbelliferae (III 5). 1 Mex.: ***O. oenanthoides*** Kunth
Ottochloa Dandy. Gramineae (XXIV 2). 3 China (1), Indomal. to N Aus. R: Blumea 4(1941)530
Ottonia Spreng. = *Piper*
Ottoschmidtia Urb. Rubiaceae (III 3). 2 Cuba, Hispaniola
Ottoschulzia Urb. Icacinaceae. c. 3 WI, Guatemala
Ottosonderia L. Bolus (~ *Ruschia*). Aizoaceae (V). 1 W S Afr.: ***O. monticola*** (Sond.) L. Bolus. R: H. E. K. Hartmann, *Ill. Handb. Succ. Pls*, Aiz. F–Z(2002)200
otu *Cleistopholis patens*
Oubanguia Baill. Lecythidaceae (IV; Scytopetalaceae). 3 trop. W Afr. R: KB 70,6(2015)41
oud *Aquilaria* spp.
Oudneya R. Br. = *Moricandia*
Ougeinia Benth. (~ *Desmodium*). Leguminosae (III 19). 1 Ind., W Nepal: ***O. oojeinensis*** (Roxb.) Hochr. (*D. o.*) – light timber
Ouratea Aubl. Ochnaceae (II). Incl. *Gomphia*, excl. *Campylospermum, Idertia, Rhabdophyllum*. c. 150 trop. Am. Roux's Model. Some timbers, oils & local med. esp. ***O. margaretae*** Sastre (Braz.) – abortifacient, ***O. parviflora*** (A. St-Hil.) Engl. (batiputa, Braz.) – seed-oil medic.
ouricuri (palm) *Syagrus coronata*
Ourisia Comm. ex Juss. Plantaginaceae (Dig.; Scrophulariaceae s.l.). 28 Andes (15), NZ (12), Tasmania (1). R: SBM 77(2006)44. Some natural hybrids (NZ). Some cult. orn. rock-pls
Ourisianthus Bonati = *Artanema*
Outreya Jaub. & Spach = *Jurinea*
ouzo Flavour due to some or all of *Foeniculum vulgare, Pimpinella anisum, Pistacia lentiscus*
Ovidia Meissn. Thymelaeaceae (Thym.-Daph.). 2 temp. S Am. Gynodioec. R: Darw. 13(1964)77. Cult. orn. shrubs like *Daphne*

Ovieda L. (~ *Clerodendrum*). Labiatae (III). 8 Cuba, Hispaniola. L. R: Willd. 46(2016) 262

ovo *Spondias purpurea*

owala *Pentaclethra macrophylla*

Owenia F. Muell. Meliaceae (II-Trich.). 5 E Aus. Disp. emus. ***O. acidula*** F. Muell. (gooya, gruie, emu apple, mooley or sour plum) – fr. with refreshing pulp; ***O. cepiodora*** F. Muell. – timber tree reduced to 31 trees by 1984

owusa nut *Plukenetia conophora*

Oxalidaceae R. Br. Magnoliidae – Oxalidales. Excl. Lepidobotryaceae, 5/565 trop. to temp. (few). Small trees, shrubs & esp. herbs with tubers or bulbs, usu. accum. oxalates, s.t. succ., rarely lianoid. Lvs in spirals, pinnate, palmate but oft. 3-foliolate, rarely 1-foliolate or phyllodic, oft. with pulvinate leaflets folded together at night ('sleep movements'); stip. usu. 0. Fls bisex., reg., 5-merous, usu. with tristyly & s.t. cleistogamous when C 0, (solit. or) in axillary cymes on peduncles; K 5, quincuncial (3 outer oft. larger), all persistent in fr., C 5 s.t. basally weakly connate, convolute (rarely imbr.) & oft. clawed, A 10 in ± 2 whorls with outer usu. with shorter filaments, all of which basally connate & s.t. 5 without anthers, outer oft. with basal nectaries, anthers with longit. slits, $\underline{G}$ ((3–)5) with axile placentas & discrete styles & capitate stigmas, each loc. with (1) 2(–several) ± pendulous, anatropous or hemitropous, bitegmic ovules. Fr. a ± ribbed or angled loculicidal capsule or berry (*Averrhoa, Sarcotheca*); seeds oft. with mucilaginous testa involved in expulsion from capsule, embryo straight, embedded in usu. copious oily endosperm. n = (5–)7(–12)

Genera: *Averrhoa, Biophytum, Dapania, Oxalis, Sarcotheca*

Hypseocharis now in Geraniaceae; the woody *Averrhoa* & *Sarcotheca* with the lianoid *Dapania* form. made another fam., reflecting an obsession going back to the anc. Greeks of the essential nature of the diffs between woody & herbaceous pls: this has done much to retard the progress of botany & was the crux of Hutchinson's system

Some fr. trees (*Averrhoa*), ed. tubers (*Oxalis*), cult. orn. & weeds (*Oxalis, Biophytum*)

Oxalis L. Oxalidaceae. c. 500 cosmop. esp. S Am. & Cape (Cape shamrock (s.t. distrib. as true s. in US; S Afr. c. 200 – R: JSAB suppl. 1(1944), Cape 118 (94 endemic) – R: Strelitzia 9(2000)552)). 4 subgenera: *Thamnoxys* (lvs pinnate, 71 trop. Am. – R: Bradea 7(1994)5), *Oxalis* (lvs digitate), *Trifidus* (lvs 3-fid, style bifid, 2 S Am. – R: Bradea 7(2000)594), *Monoxalis* (lvs simple, style simple, 2 Andes – R: Bradea 7(2000)203). Herbs, subshrubs (***O. gigantea*** Barnéoud (Chile) to 3m), or lianes (e.g. ***O. scandens*** Kunth, S Am.)), s.t. succ. or even aquatic (e.g. ***O. natans*** L.f. (S Afr.) – endangered sp.). Some with distyly derived from tristyly; nectar of some spp. with pH 1.6 (most usu. pH 5.6–5.9); some fly-poll., e.g. ***O. acetosella*** L. (wood sorrel, Euras.) – form. used instead of *Rumex* spp. as 'sorrel', also against lice & ticks on sheep; many with tubers making them weedy pests esp. ***O. pes-caprae*** L. (*O. cernua*, Bermuda buttercup, soursob, S Afr.) – tristylous with diploids & tetraploids in S Afr., short-styled pentaploids common in Medit. (introd. Sicily 1796, Iran by 1928, Pakistan 1940s) & Aus. but rare in S Afr., bulbs ± ed. (distrib. by polecats etc.) though toxic to stock in Aus. & NZ (continued grazing by sheep leading to kidney damage), widely natur. in trop. & warm (incl. W US) with spontaneous double fls incl. Eur., where only 2 native spp. but 10 S Afr. & S Am. spp. widely natur. Some locally medic., tubers or lvs ed.; some cult. orn. but rapidly spreading vegetatively or by seed (arils turn inside-out expelling seeds from capsule explosively). ***O. corniculata*** L. (orig. unknown, ? E As.) – cosmop. weed with yellow fls, oft. seen on greenhouse floors, lvs ed. Ind. as scurvy cure; ***O. debilis*** Kunth ('*O. corymbosa*', S Am.) – sparsely hairy lvs with translucent dots a tamarind subs. & ***O. latifolia*** Kunth (Mex. to Peru, widely natur. (esp. Aus.)) – hairless lvs without dots, both cosmop. weeds with pinkish fls; ***O. magellanica*** Forst. f. (NG, Aus., NZ, Chile, Arg.) – cult. orn. (s.t. pest), fls white; ***O. noctiflora*** R. Macfarl. & al. (S Afr.) – (?)moth-poll. white fls open at night; ***O. purpurea*** L. (S Afr.) – invasive Aus.; ***O. tetraphylla*** Cav. (*O. deppei*, Mex.) – cult. orn. with 4-foliolate lvs & red fls, tubers ed. (for which form. cult. in Eur.); ***O. tuberosa*** Molina (*O. crenata*, oca, Andes) – octoploid (app. derived from *O. picchensis* Knuth (S Peru) & a tuberous sp. from Bolivia), long cult. as root veg. in Peru with tubers white, yellow or red (the latter 2 not flowering)

Oxandra A. Rich. Annonaceae. 30 trop. Am. R: AHB 10(1931)153. ***O. lanceolata*** (Sw.) Baill. (WI) – timber (asta, lancewood)

Oxanthera Montr. = *Citrus*

oxberry *Dioscorea communis*

Oxera Labill. Labiatae (III). Incl. *Faradaya*, 24 N Borneo, NG, Aus., New Caled., Vanuatu, Polynesia. R: KB 44(1999)265, 321. Lianes (Leeuwenberg's Model) & pachycaul treelets (Corner's Model). ***O. splendida*** (F. Muell.) Gâteblé & Barrabé (*F. s.*, N Borneo to N Aus.) – fish-poison in NE Queensland (koie-yan)

ox-eye *Buphthalmum* spp.; **o. daisy** *Leucanthemum vulgare*

Oxford & Cambridge bush *Rotheca myricoides* 'Ugandensis'; **O. ivy** or **weed** *Cymbalaria muralis*; **O. ragwort** *Senecio squalidus*

oxlip *Primula elatior*

oxtongue *Hemlinthotheca echioides*, *Picris* spp.

Oxyanthera Brongn. = *Thelasis*

Oxyanthus DC. Rubiaceae (Ix.-Sherb.). 40 Afr. Some timber

Oxybaphus L'Hérit. ex Willd. = *Mirabilis*

Oxybasis Kar. & Kir. (~ *Chenopodium*). Amaranthaceae (Atrip.). 5 temp. R: Willd. 42(2012) 15

Oxycarpha S.F. Blake. Compositae (Helia.-Spil.). 1 Venez.: ***O. suaedifolia*** S.F. Blake. R: CGH 53(1918)52

Oxycaryum Nees = *Cyperus* (but see BN 124(1971)281)

Oxyceros Lour. (~ *Randia*). Rubiaceae (II 1). 12 Indomal. R: Reinw. 12(2008)292

Oxychlamys Schltr. = *Aeschynanthus*

Oxychloe Philippi (~ *Juncus*). Juncaceae. Excl. *Patosia*, 5 C & S Andes. R: FOW 6(2002)9. Cushion-forming pls to several m. diam.

Oxychloris Lazarides (~ *Chloris*). Gramineae (XXIX 5). 1 Aus.: ***O. scariosa*** (F. Muell.) Lazarides – dry savanna, adventive in Switzerland. R: Nuytsia 5(1984)283

Oxycoccus Hill = *Vaccinium*

Oxydendrum DC. Ericaceae (VIII 1). 1 E US: ***O. arboreum*** (L.) DC. (sourwood, tree sorrel) – decid. tree to 25 m with acid-tasting lvs alleged to slake thirst. R: Castanea 61(1996)131

oxygen plant *Hydrilla verticillata*

Oxyglossellum M.A. Clem. & D.L. Jones = *Dendrobium*

Oxygonum Burchell ex Campderá. Polygonaceae (II Oxyg.). c. 30 trop. & S Afr., Madag. (1). R: KB 1(1957)145

Oxygraphis Bunge. Ranunculaceae (II 3). 4 temp. As.

Oxygyne Schltr. Thismiaceae (Burmanniaceae s.l.). Excl. *Saionia*, 1 W trop. Afr: ***O. triandra*** Schltr

Oxylaena Benth. ex Anderb. = *Gibbaria*

Oxylobium Andrews (~ *Pultenaea*). Leguminosae (III 13). (Excl. *Podolobium*) 15 Aus. esp. SW. Shaggy peas. See also *Gompholobium*

Oxylobus (DC.) A. Gray. Compositae (Eup.-Oxy.). 5 Mex. to Venez. R: MSBMBG 22(1987)436

Oxymyrrhine Schauer (~ *Baeckea*). Myrtaceae (Lept.). 4 SW Aus. R: Nuytsia 19(2009)155

Oxyosmyles Speg. = *Ixorhea*

Oxypappus Benth. Compositae (Tag.- Pect.). 1 Mex.: ***O. scaber*** Benth.

Oxypetalum R. Br. Apocynaceae (Vc; Asclepiadaceae III 1). Excl. *Rojasia*, *Tweedia* c. 120 trop. & S Am.

Oxyphyllum Philippi. Compositae (Mut.-Nass.). 1 N Chile: ***O. ulicinum*** Philippi – shrub

Oxypolis Raf. Umbelliferae (III 10). 4 N Am. R: T 61(2012)414

Oxyrhachis Pilg. Gramineae (XXII 2). 1 trop. E Afr., Madag.: ***O. gracillima*** (Bak.) C.E. Hubb. – upland bogs. R: Strelitzia 10(2000)708

Oxyrhynchus Brandegee. Leguminosae (III 18). 1 E Mal., 3 C Am. R: Britt. 62(2010)241

Oxyria Hill. Polygonaceae (II 3). 1(-4) Arctic to mts of Euras. & California: ***O. digyna*** (L.) Hill – like *Rumex* but K 4, cult. orn. rock-pl. with ed. lvs

Oxysepala Wight = *Bulbophyllum*

Oxyspora DC. Melastomataceae. 24 Indomal. to S China. R: GBS 35(1982)216. Some cult. orn.

Oxystelma R. Br. (~ *Cynanchum*). Apocynaceae (V c; Asclepiadaceae). 2 trop. OW. ***O. esculentum*** (L.f.) Sm. (*Sarcostemma secamone*, OW) – local medic., fr. eaten in times of scarcity

Oxystigma Harms (~ *Prioria*). Leguminosae (I 2). 5 trop. Afr. ***O. msoo*** Harms – giant legume; ***O. oxyphyllum*** (Harms) Léonard (*Pterygopodium o.*, lolagbola, tchitola, W Afr.) – comm. timber

Oxystophyllum Blume = *Dendrobium*

Oxystylis Torrey & Frémont = *Cleome* (but see FNA 7(2010)215)

Oxytenanthera Munro. Gramineae (V 4). 1 trop. Afr.: ***O. abyssinica*** (A. Rich.) Munro – bamboo with 7–21-yr flowering cycle, culms used in boat-making & winnower frames, tapped for wine in Tanzania. R: Strelitzia 10(2000)708

Oxytenia Nutt. = *Euphrosyne* (but see FNA 21(2006)29)

Oxytheca Nutt. Polygonaceae (I 1). Excl. *Acanthoscyphus, Sidotheca*, 3 W N Am. (3), Chile & Arg. Andes (subspp. diff.). R: FNA 5(2005)434

Oxytropis DC. Leguminosae (III 24). 350 N temp. (Eur. 24) esp. C As. R: MAISP VII, 22–1(1874)1, (N Am.) PCAS IV, 27 (1952)177. Crazyweed or locoweed (cf. *Astragalus*), harmful to stock; some cult. orn. ***O. deflexa*** (Pallas) DC. (N temp.) – pops (Finmark, Norway (500 pls), As. (5000 km away) & N Am.) separated by glaciations

Oyedaea DC. Compositae (Helia.-Ecl.). 18 trop. Am. R: CUSNH 20(1924)412

Oyster Bay pine *Callitris rhomboidea;* **o. leaf** *Mertensia maritima;* **mock o.** *Tragopogon porrifolius;* **o. nut** *Telfairia pedata;* **o. plant** *Tragopogon porrifolius*, (Aus.) *Acanthus spinosus;* **Spanish o. p.** *Scolymus hispanicus;* **vegetable o.** *T. porrifolius;* **o. wood, Cuban** *Gymnanthes lucida*

Oziroe Raf. (*Fortunatia;* ~ *Camassia*). Asparagaceae (Hyacinthaceae). 5 S Am. R: Darw. 40(2002)63

Ozoroa Del. (~ *Heeria*). Anacardiaceae (I). 40 trop. Afr. R: GOB 14(1966)19. Some locally ed. fr. & medic.

Ozothamnus R. Br. (~ *Cassinia*). Compositae (Gnap.- Cass.). 50 Aus., New Caled. (***O. pinifolius*** (Forst.f.) DC.), NZ. R: AusJB 6(1958)229, OB 104(1991)87. Many hybrids with poss. congeneric *Cassinia* spp. ***O. ferrugineus*** (Labill.) Sweet (SE Aus.) – tree to 5 m. Cult. orn. everg. shrubs & perenn. herbs e.g. ***O. ledifolius*** (DC.) Hook.f. (kerosene weed, Tasmania) – scented exudate highly flammable & ***O. rosmarinifolius*** (Labill.) Sweet (SE Aus.)

P

paan *Areca catechu*

Pabellonia Quezada & Martic. = *Leucocoryne*

Pabstia Garay (*Colax*). Orchidaceae (V 12j). 5–6 SE Braz. Cult. orn.

Pabstiella Brieger & Senghas (~ *Pleurothallis*). Orchidaceae (V 13c). c. 30 trop. Am.

pacay *Inga feuillei*

pacaya *Chamaedorea* spp. esp. *C. tepojilote*

Pachecoa Standl. & Steyerm. = *Chapmannia*

Pachira Aubl. Malvaceae (Bomb.; Bombacaceae). Incl. *Bombacopsis, Rhodognaphalon*, c. 50 trop. Am. & Afr. (6). Aubréville's Model. Some light timbers & cult. orn. incl. ***P. aquatica*** Aubl. (Guiana chestnut, saba nut, trop. Am. estuaries) – seeds ed., sold as pot-pl. (oft. with stems plaited together) as feng shui pl., Hawaiian money tree, etc. in Eur.

Pachites Lindl. Orchidaceae (IV 4c). 2 SW Cape. R: Strelitzia 9(2000)170

Pachyacris Schltr. ex Bullock = *Xysmalobium*

Pachyanthus A. Rich. = *Miconia* (but see Britt. 64(2012)185)

Pachycarpus E. Meyer (~ *Asclepias*). Apocynaceae (V c; Asclepiadaceae III 1). 37 trop. (15 – R: KB 53(1998)338) & S Afr. ***P. asperifolius*** Meissn. (S Afr.) – self-incompat., poll. spider-hunting wasps, sec. compounds in nectar deterring other potential pollinators

Pachycaulos J.L. Clark & J.F. Sm. (~ *Hypocyrta*). Gesneriaceae (II 5c). 1 Mex. to N Peru: ***P. nummularia*** (Hanst.) J.L. Clark & J.F. Sm. R: SB 38(2013)458

Pachycentria Blume (~ *Medinilla*). Melastomataceae. 8 Myanmar, Mal. R: Blumea 45(2000)341. ***P. constricta*** (Blume) Blume (W Mal.) & ***P. glauca*** Triana (Borneo) – ants nest in hollow root swelling; seeds carried into nests by ants to grow in 'ant gardens'

Pachycereus (A. Berger) Britton & Rose (~ *Carnegiea*). Cactaceae (III 1). Excl. *Lemaireocereus, Lophocereus*, 5 SW US & Mex. R: SB 34(2009)69. Cult. orn. tree-like cacti incl. ***P. pecten-aboriginum*** (S. Watson) Britton & Rose, ***P. pringlei*** (S. Watson) Britton & Rose – male, female & hermaphrodite when near roosts of poll. nectar-feeding bats (no males when roosts more than 50 km away), fr. ed., seeds ground into flour

Pachycladon Hook.f. Cruciferae (37). Incl. *Cheesemania*, 11 NZ (S Alps). R: NZJB 40(2002)557. ***P. wallii*** (Carse) Heenan & A. D.Mitchell (*C. w.*) – gynodioec.

Pachycormus Cov. Anacardiaceae (I). 1 Baja Calif.: ***P. discolor*** (Benth.) Standl. (elephant tree, copalquin) – stem swollen, thin brown flaking bark etc. looking like Burseraceae. R: CJSAm 63(1991)35

Pachycornia Hook.f. = *Tecticornia* (but see FA 4(1984)308)
Pachyctenium Maire & Pampan. = *Daucus*
Pachycymbium Leach = *Orbea*
Pachydesmia Gleason = *Miconia*
Pachyelasma Harms. Leguminosae (I 4). 1 W Afr. rainforest: *P.* ***tessmannii*** (Harms) Harms – pods abortifacient, fish-poison
Pachygenium (Schltr.) Szlach. & al. = *Pelexia*
Pachygone Miers. Menispermaceae (1). 10 China & Indomal. to Pac. Some used to stupefy fish & against vermin in Aus.
Pachylaena D. Don ex Hook. & Arn. Compositae (Mut.-Mut.). 1 Andes of Chile & Arg.: *P.* ***atriplicifolia*** D. Don ex Hook. & Arn. – herb. R: BJLS 157(2008)375
Pachylarnax Dandy = *Magnolia*
Pachylecythis Ledoux = *Lecythis*
Pachyloma DC. Melastomataceae. 6 Braz. Appendages behind connective
Pachymitus O. Schulz. Cruciferae (37). 1 Aus.: *P.* ***cardaminoides*** (F. Muell.) O. Schulz. R: TRSSA 89(1965)226
Pachynema R. Br. ex DC. = *Hibbertia* (but see Aus SB 5(1992)477)
Pachyneurum Bunge. Cruciferae (7). 1 C As.: *P.* ***grandiflorum*** (C. Meyer) Bunge
Pachypharynx Aellen = *Atriplex* (galled)
Pachyphragma (DC.) Reichb. (~ *Thlaspi*). Cruciferae (17). 1 Caucasus: *P.* ***macrophyllum*** (Hoffm.) N. Busch
Pachyphyllum Kunth = *Fernandezia* (but see R: JBRIT 2(2008)285)
Pachyphytum Link, Klotzsch & Otto (~ *Echeveria*). Crassulaceae (I 5b). 15 EC Mex. Cult. orn. succ. like *Echeveria* (intergeneric hybrids = × *Pachyveria*) but each petal with 2 scale-like appendages within
Pachyplectron Schltr. Orchidaceae (IV 2d). 3 New Caled.
Pachypleuria (C. Presl) C. Presl = *Davallia*
Pachypleurum Ledeb. = *Ligusticum*
Pachypodanthium Engl. & Diels = *Duguetia*
Pachypodium Lindl. Apocynaceae (II b). 25 Madag. (20), S & SW Afr. R: S.H.J.V. Rapanarivo & al. (1999) *P. (Apocynaceae)*. Thorny pachycaul succ. treelets of dry forest to mound-forming spp. of mts (Leeuwenberg's Model); lvs in spirals; fls red, yellow or white. Many cult. orn. esp. *P.* ***lamerei*** Drake (Madag. palm, Madag.), *P.* ***succulentum*** (L.f.) Sweet (S Afr.) etc. *P.* ***lealii*** Welw. (*P. giganteum*, S Afr.) – arrow-poison (glucoside with action like digitalin); *P.* ***rutenbergianum*** Vatke (Madag.) – to 12m, bark mixed with *Raphia* leaf-pinnae to give a textile
Pachyptera DC. ex Meissn. (~ *Mansoa*). Bignoniaceae (Bign.). 4 trop. Am. R: AMBG 99(2014)455
Pachypterygium Bunge = *Isatis*
Pachyrhachis A. Rich. = *Bulbophyllum*
Pachyrhizus Rich. ex DC. Leguminosae (III 18). 5 NW trop. R: NJB 8(1988)167. Cult. for ed. tubers & starch esp. *P.* ***erosus*** (L.) Urb. (jicama, yam bean, C Am.) cult. in pre-Columbian times for tubers eaten cooked or raw (as in Mal. where eaten with fermented prawn sauce), seeds toxic
Pachyrhynchus DC. = *Lucilia*
Pachysandra Michaux. Buxaceae. 3 E As. (2), E US (1). R: BBAS 33(1992)201. Alks. Monoec. (male fls with nectar; females 0) shrubby herbs cult. as ground-cover esp. *P.* ***terminalis*** Sieb. & Zucc. (Jap.) in shade
Pachystachys Nees. Acanthaceae (III 2c). 18 trop. Am. R: NJB 34(2016)522. Cult. orn. greenhouse shrubs esp. *P.* ***lutea*** Nees (Peru) with consp. yellow bracts. See also *Mirandea*
Pachystegia Cheeseman = *Olearia*
Pachystela Pierre ex Radlk. = *Synsepalum*
Pachystele Schltr. = *Scaphyglottis*
Pachystelis Rauschert = *Scaphyglottis*
Pachystelma Brandegee = *Matelea*
Pachystigma Hochst. Rubiaceae (III 2). 10 warm Afr.
Pachystoma Blume. Orchidaceae (V 14). 3 China & Indomal. to W Pac. R: OB 89(1986)
Pachystrobilus Bremek. = *Strobilanthes*
Pachystroma Muell. Arg. Euphorbiaceae (Hipp.). 1 S Braz., Bolivia: *P.* ***longifolium*** (Nees) I.M. Johnston. R: A. R.-Sm., *Gen. Euph.* (2001)393

Pachystylidium Pax & K. Hoffm. (~ *Cnesmone*). Euphorbiaceae (Acal.-Pluk.-Trag.). 1 Ind. to C Mal.: ***P. hirsutum*** (Blume) Pax & K. Hoffm. R: A.R.-Sm., *Gen. Euph.* (2001)260

Pachystylus K. Schum. Rubiaceae (II 2). 2 NG

Pachythamnus (R. King & H. Robinson) R. King & H. Robinson = *Ageratina*

Pachytrophe Bureau = *Streblus*

× Pachyveria Haage & Schmidt. Crassulaceae. Hybrids between spp. of *Echeveria* & *Pachyphytum*. Cult. orn. succ.

Pacific maple *Acer macrophyllum*

Pacifigeron Nesom. Compositae (Ast.-Hin.). 1 Rapa: ***P. rapensis*** (F. Br.) Nesom – pseudodichotomously branched shrub. R: Phytol. 76(1994)160

Packera Löve & D. Löve (~ *Senecio*). Compositae (Sen.- Sen.). c. 75 N Am. (54 – R: FNA 20(2006)570), Siberia. Some cult. orn. esp. alpines

paco-paco *Wissadula spicata*

Pacouria Aubl. (~ *Landolphia*). Apocynaceae (Ic). 2 trop. Am.

Pacourina Aubl. Compositae (Vern.-Pac.). 1 trop. Am.: ***P. edulis*** Aubl. – aquatic with sessile heads & ed. lvs

padauk, padouk *Pterocarpus* spp.; **Andaman p.** *P. dalbergioides*; **Burma p.** *P. macrocarpus*; **W African p.** *P. soyauxii*

Padbruggea Miq. = *Callerya*

paddlewood *Aspidosperma excelsum*

paddy *Oryza sativa*; **P.'s lucerne** *Sida rhombifolia*

padma *Nelumbo nucifera*

padri *Stereospermum colais*

Padus Mill. = *Prunus*

Paederia L. Rubiaceae (IV 12). 30 trop. R: OBB 3(1991)1. Lianes with faecal smell (sulphur group released on damage) but lvs of ***P. foetida*** L. (*P. scandens*, Jap. & SE As. to W Mal., natur. elsewhere, invasive SE US, Hawaii) used as a vegetable, local medic.

Paederota L. = *Veronica*

Paederotella (Wulff) Kem.-Nat. = *Veronica*

Paedicalyx Pierre ex Pitard = *Xanthophytum*

Paenula Orchard. Compositae (Gnap.). 1 C NSW: ***P. storyi*** Orchard. R: Telopea 11(2005)5

Paeonia Tourn. ex L. Paeoniaceae. 32 temp. Euras. (Eur. 13), W N Am. (2 incl. ***P. brownii*** Dougl. ex Hook.). Peonies. R: D-Y. Hong, *Peonies of the world* 1(2010)49. Perenn. herbs with rhiz. & thickened tuberous roots (sects *Paeonia* (Spain to Jap.) & *Onaepia* (2 W N Am.)) or subpachycaul shrubs (sect. *Moutan*, tree ps, W China & Tibet) mainly in calcareous woodland. Some medic. esp. ***P. ostii*** T. Hong & J.X. Zhang (China) – root-bark (danpi); many cult. orn. (signifying shame & bashfulness in 'Language of Fls'), herb. forms largely derived from ***P. lactiflora*** Pallas (Tibet to China & Siberia) though some cvs involving other spp., cult. since 900 BC, tree ps from ***P. × suffruticosa*** Andrews (moutan, several wild spp. in ancestry) – nat. fl. China (1994), in cult. since C7 becoming a 'rage' a cent. later when grafted on to herbaceous stocks, its red roots much prized but flowering pl. discarded once petals fell (Hua Wang, the King of Flowers), fermented petals used in wine-making, medic. esp. modern Chinese herbalism (period pains, high blood pressure etc.), but many other spp. now involved (publ. of American P. Society). ***P. emodi*** Wall. ex Royle (Himal.) – 1 of only 8 confirmed homoploid hybrid spp.; ***P. delavayi*** Franch. (*P. lutea*, W China) & esp. ***P. ludlowii*** (Stern & Taylor) D.Y. Hong (*P. lutea* 'Ludlowii', Tibet) – common yellow-fld tree p.; ***P. mascula*** (L.) Mill. (S & E Eur. to As. Minor) – natur. Steep Holm in Bristol Channel since at least 1803, capsules with red (inviable, 'lure') & black (viable) seeds; ***P. officinalis*** L. (Eur.) – spicy seeds allegedly eaten in Lent in Med. England

Paeoniaceae Raf. Magnoliidae – Saxifragales. 1/32 temp. Euras., W N Am. Usu. glabrous herbs & subpachycaul decid. shrubs with vessels with scalariform to simple endplates (cf. Ranunculaceae) & calcium oxalate crystals. Lvs in spirals & binately lobed or dissected; stip. 0. Fls solit., term., large, at least s.t. beetle-poll., bisex., ± reg. with ± concave receptacle & usu. continuous phyllotactic spiral from lvs & bracts to K, C, A-trunks & G; K (3-)5(–7), leathery, persistent, C 5–10(–13) with 3 or more vasc. strands like K, A ∞ with dichotomizing vascular strands derived from 5 basal trunks, members of each group usu. maturing centrifugally, anthers with longit. slits, $\underline{G}$ oft. surrounded by nectary-disk, (2)3–8(–15) with expanded ± subsessile stigma & several–∞ marginal, anatropous, bitegmic ovules with massive outer integument. Fr. a head of follicles; seeds black-purple to black, mesotestal, large with funicular aril (with copious oily endosperm). x = 5
Genus: *Paeonia*

Position of fam. form. controversial (Rosidae suggested), some characters (centrifugal A, pollengrain sculpturing, arils etc.) like Dilleniaceae but other features like Ranunculaceae, DNA fixing present position. Though *Glaucidium* (K C-like, C 0, seeds compressed & winged) has embryological & cytological features like P., s.t. treated as a fam. of its own, it is now referred (back) to Ranunculaceae. Cult. orn.

paeony *Paeonia* spp.

Paepalanthus Mart. Eriocaulaceae. Incl. *Actinocephalus, Blastocaulon, Lachnocaulon, Tonina*, c. 460 trop. Am., esp. S Am., Afr., Madag. Some to 180 cm in Amaz. savannas

Paesia J. St-Hil. Dennstaedtiaceae. 12 Mal. to Polynesia, NZ & trop. Am. Like *Hypolepis* but with true indusium. ***P. scaberula*** (A. Rich.) Kuhn (NZ) – troublesome weed in disturbed habitats (cf. *Pteridium*)

Pagaea Griseb. = *Irlbachia*

Pagamea Aubl. Rubiaceae (IV 7). 24 trop. S Am. R: MNYBG 12,3(1965)270, Britt. 4 1(1989)129. Local medic. esp. ***P. coriacea*** Spruce ex Benth. – bark used to restore mobility resulting from attacks of unknown cause

Pagameopsis (Standl.) Steyerm. Rubiaceae (inc. sed.). 2 Venez. R: AMBG 74(1987)106

Pagella Schönl. = *Crassula*

Pagesia Raf. = *Mecardonia*

Pagetia F. Muell. = *Bosistoa*

Pagiantha Markgraf = *Tabernaemontana*

pagoda flower *Clerodendrum paniculatum*; **p. tree** *Alstonia* spp., *Plumeria rubra, Styphnolobium japonicum, Terminalia* spp.

Pagothyra (Leeuwenb.) J.F. Sm. & J.L. Clark (~ *Episcia*). Gesneriaceae (II 5c). 1 NE S Am.: ***P. maculata*** (Hook.f.) J.F. Sm. & J.L. Clark. R: SB 38(2013)461

pahautea *Libocedrus bidwillii*

paich-ha *Euonymus maackii*

paigle *Primula veris, Ranunculus* spp.

paina de seda *Ceiba speciosa*

paintbrush, devil's *Pilosella aurantiaca*; **Flora's p.** *Emilia* spp.; **Indian p.** *Castilleja* spp.

Painteria Britton & Rose (~ *Havardia*). Leguminosae (II 4). 3 trop. Am. esp. Mex. R: MNYBG 74,1(1996)178

Paivaea O. Berg = *Campomanesia*

Pajanelia DC. Bignoniaceae (1). 1 Indomal.: ***P. longifolia*** (Willd.) K. Schum. – good timber

Pakaraimaea Maguire & Ashton. Cistaceae (Dipterocarpaceae-Pak.). 1 Guayana Highland: ***P. dipterocarpacea*** Maguire & Ashton. R: T 26(1977)354

pak-choi *Brassica rapa* Chinensis Group

pakwan *Breynia androgyna*

Paladelpha Pichon = *Alstonia*

Palaeocyanus Dostál = *Cheirolophus*

Palafoxia Lag. Compositae (Bah.). 12 S US, Mex. R: Rhodora 78(1976)567. Some cult. orn. ann. herbs

palamut *Quercus ithaburensis* subsp. *macrolepis* & other spp.

Palandra Cook = *Phytelephas*

Palaquium Blanco. Sapotaceae (II). 110 Taiwan & Indomal. to Samoa. Timber (nyatoh) & gutta-percha obtained by ringing or felling trees – rubbery substance which softens on heating, form. used in dentistry, C19 golf-balls (& first Atlantic cables), ***P. gutta*** (Hook.) Baill. usu. used, but other spp. exploited (see also *Payena*) but now ± superseded by synthetics. Some oilseeds esp. ***P. hexandrum*** (Griff.) Baill. (W Mal.) – imp. veg fat in Sumatra

palas *Butea monosperma*

Palaua Cav. Malvaceae (Malv.-Malv.). 15 Andes. R: BJ 42(1908)104

paldao *Dracontomelon dao*

Paleaepappus Cabrera = *Nardophyllum*

Palenia Philippi = *Heterothalamus*

Paleocyanus Dostál = *Cheirolophus*

Paleodicraeia C. Cusset. Podostemaceae (III). 1 Madag.: ***P. imbricata*** (Tul.) C. Cusset. R: Adans. 12(1973)563

Palhinhaea Franco & Vasc. = *Lycopodiella*

Paliavana Vell. ex Vand. (~ *Sinningia*). Gesneriaceae (II 5e). 8 Braz.

Palicourea Aubl. Rubiaceae (IV 7). c. 250 trop. Am. Hummingbird-poll. Local medic. esp. emetics in Amaz.; some cult. orn.

palillo *Escobedia scabrifolia*

Palimbia Besser ex DC. Umbelliferae (III 10). 3 S & E Russia to C As.

palisade grass *Panicum brizanthum*

palisander (? orig. pau santo), timbers of Braz. spp. of *Dalbergia, Jacaranda* & *Machaerium*

Palisota Reichb. Commelinaceae (II 1a). 18 trop. Afr. Stout subpachycaul herbs (***P. thollonii*** Hua a liane to 15 m, biggest in C.) with brightly coloured berries; some cult. orn. esp. ***P. barteri*** Hook. (W & C Afr.) – fls open 4–6 a.m. in wild & ***P. hirsuta*** (Thunb.) K. Schum. (W & C Afr.) – andromonoec.; ***P. flagelliflora*** Faden (Cameroun) – infls on flagelliform rooting shoots at ground-level

Paliurus Mill. Rhamnaceae (1). 5 S Eur. (1) to Jap. (Tertiary N Am.). R: BJ 116(1994)340. Roux's Model. ***P. spina-christi*** Mill. (Christ's thorn, S Eur. to N China) – tree to 7 m with stip. thorns, 1 straight, 1 curved (other spp. with 2 straight) & fr. with horizontal wing, hedge-pl.

Pallasia Klotzsch = *Wittmackanthus*

Pallenis (Cass.) Cass. (~ *Asteriscus*). Compositae (Inul.-Inul.). 3 Medit.

palm Any sp. of Palmae or similar-looking pls, but in rural dists of GB = *Salix* spp. (N England also *Buxus, Larix* & *Taxus* spp.; N Am. *Tsuga* spp.); **American oil p.** *Elaeis oleifera;* **Alexandra p.** *Archontophoenix alexandrae;* **bacaba p.** *Oenocarpus distichus;* **bamboo p.** *Dypsis lutescens;* **bangalow p.** *A. cunninghamiana;* **betel p.** *Areca catechu;* **black p.** *Normanbya normanbyi;* **b. roseau p.** *Bactris major;* **bottle p.** *Beaucarnea recurvata;* **Buddha p.** *Saribus rotundifolius;* **p. cabbage = p. hearts**; **cabbage p.** *Livistona australis; Roystonea oleracea;* **c. palmetto palm** *Sabal palmetto;* **cane p.** *Dypsis lutescens;* **carnauba wax p.** *Copernicia prunifera;* **coco-de-mer p.** *Lodoicea maldivica;* **coconut p.** *Cocos nucifera;* **cohune p.** *Orbignya cohune;* **coquito p.** *Jubaea chilensis;* **corozo p.** *Elaeis oleifera;* **date p.** *Phoenix dactylifera;* **double coconut p.** *Lodoicea maldivica;* **doum p.** *Hyphaene thebaica;* **ejow p.** *Arenga pinnata;* **fishtail p.** *Caryota mitis;* **foxtail p.** *Wodyetia bifurcata;* **gingerbread p.** *H. thebaica;* **gomuti p.** *A. pinnata;* **gru-gru p.** *Acrocomia aculeata;* **p. hearts –** buds of spp. of *Acrocomia, Bactris, Euterpe, Roystonea* & *Sabal* spp. etc. but also geriatric *Cocos nucifera;* **honey p.** *Jubaea chilensis;* **ivory p.** *Phytelephas macrocarpa;* **Japanese peace p.** *Rhapis excelsa;* **kentia p.** *Howea forsteriana;* **kitul (kittool) p.** *Caryota urens;* **lady p.** *Rhapis excelsa;* **macaw p.** *Aiphanes minima;* **Madagascar p.** *Pachypodium lamerei;* **majestic p.** *Ravenea rivularis;* **nikau p.** *Rhopalostylis sapida;* **nipa p.** *Nypa fruticans;* **oil p.** *Elaeis guineensis;* **palmetto p.** *Sabal* spp.; **palmyra p.** *Borassus flabellifer;* **Panama-hat p.** *Carludovica palmata;* **Parag. p.** *Acrocomia aculeata;* **paxiuba p.** *Socratea exorrhiza;* **peach p.** *Bactris gasipaes;* **piassaba p.** *Attalea funifera, Leopoldinia piassaba;* **ponytail p.** *Beaucarnea recurvata;* **raffia p.** *Raphia farinifera;* **rattan p.** *Calamus* spp. etc.; **royal p.** *Roystonea regia;* **sago p.** *Metroxylon sagu;* **saw palmetto p.** *Serenoa repens;* **sea p.** *Postelsia palmiformis;* **sealing-wax p.** *Cyrtostachys renda;* **sugar p.** *Arenga pinnata;* **tagua p.** *Phytelephas macrocarpa;* **talipot p.** *Corypha umbraculifera;* **thatch p.** *Thrinax parviflora;* **toddy p.** *Caryota urens;* **traveller's p.** *Ravenala madagascariensis;* **triangle p.** *Dypsis decaryi;* **vegetable ivory p.** *Phytelephas macrocarpa;* **wax p., (Colombian)** *Ceroxylon* spp., **(carnauba)** *Copernicia prunifera;* **windmill p.** *Trachycarpus fortunei;* **wine p.** *Jubaea chilensis*

palma-Christi *Ricinus communis* (Medit.: *Orchis* spp. esp. *O. mascula*); **palma fibre** *Samuela carnerosana;* **p. rosa** *Cymbopogon martini* 'Mota'

Palmae Juss. (Arecaceae). Magnoliidae – Arecales. 184/2550 trop. (NG 32/273; Afr. few) & warm. R: J. Dransfield & al. (2008) *Genera palmarum.* Everg. trees, unbranched (or dichotomously branched), usu. erect, s.t. armed, s. t. miniature &/or suckering or slender & lianoid (rattans) to 150 m with clear internodes; sec. thickening diffuse without new vasc. tissue (no cambium), vasc. bundles closed, numerous, usu. with silicified fibres; raphides, polyphenols, s.t. alks present; roots (s.t. rep. by spines) with mycorrhizae & 0 root-hairs; vessels throughout. Lvs in spirals (rarely distich. or tristichous), usu. in term. rosettes & oft. v. large, with basal sheath, tubular but oft. splitting at maturity, petiole, lamina pinnate (feather palms) or palmate (fan p.), less oft. entire (s.t. bifid) or 2-pinnate (*Caryota*), simple initially & oft. splitting during development into V-shaped (induplicate) or ∧-shaped (reduplicate) leaflets from plicate condition like Cyclanthaceae; ligule a prolongation of sheath, s.t. surrounding stem, or 0. Infls usu. axillary (simple to) paniculate (to 6 orders of branching), the branchlets oft. thick & spadix-like, the peduncle with prophyll & 1–several spathes. Fls usu. small & ± sessile, usu. insect-poll. & unisexual (monoec. & dioec. common) or bisex., 3-merous, ± reg.; P oft. 3 + 3, leathery or fleshy, green to yellow, red or white, the outer oft. smaller usu. imbr., inner usu. valvate in males, imbr. in females, rarely 2 + 2 or in spirals (10) or 0, A (3)6–950+, usu. 3 + 3, filaments s.t. connate &/or adnate to P, anthers latrorse with longit. slits & oft. monosulcate pollen grains; staminodes free, forming a cup or 0, (G(1–)3(–10), pluriloc., or pseudomonomerous with only 1 fertile loc., s.t. with septal nectaries, stigmas sessile or atop free or ± united styles, each loc. with 1 anatropous to hemitropous, campylotropous or orthotropous, bitegmic ovule. Fr. usu. a fleshy or fibrous drupe, rarely ± dehiscent (*Astrocaryum, Socratea* spp.); seed 1(–10), endosperm usu. v. oily, with protein & hemicellulose, s.t. ruminate. n = 13–18 (polyploidy in 1 *Areca* sp., also *Voaniola* (2n = 596+))

Classification & chief genera: 5 subfams:

I. **Calamoideae** (Lepidocaryoideae; erect or scandent, s.t. suckering; lvs reduplicate, (entire-) pinnate or (rarely) palmate; fls singly or in prs, pls with bisex. fls, polygamous, monoec. or dioec., A 6(–70), G (3) covered with reflexed imbr. scales; fr. usu. with 1 sarcotestal seed; mostly As., 3 tribes):
   1. **Eugeissoneae** (lvs pinnate, OW) *Eugeissona* (only)
   2. **Lepidocaryeae** (3 subtribes): *Eremospatha, Laccosperma, Oncocalamus* (a, Ancistrophyllinae incl. Oncocalaminae); *Raphia* (only; b, Raphiinae); *Lepidocaryum, Mauritia* (c, Mauritiinae)
   3. **Calameae** (lvs pinnate; OW, 6 subtribes): *Korthalsia* (a, Korthalsiinae), *Salacca* (b, Salaccinae), *Metroxylon* (only; c, Metroxylinae); *Pigafetta* (only; d, Pigafettinae); *Plectocomia* (e, Plectocomiinae); *Calamus, Daemonorops* (f, Calaminae)

II. **Nypoideae** (dichot. branched creeping stem in mangrove; lvs pinnate with reduplicate leaflets; infl. with term. head of female fls & lat. branches of males, A 3 with filaments united in a column, G 3(4), 1 or more becoming water-disp. fr. with fibrous mesocarp, As.): *Nypa fruticans* (only) – s.t. referred to separate fam.

III. **Coryphoideae** (lvs usu. induplicately palmate, less oft. pinnate); fls pedicellate to sessile, borne singly or in clusters, or in triads of 2 male & 1 female (6.), A 6–24, oft. apocarpous; some in dry country or temp., 8 tribes):
   1. **Sabaleae**: *Sabal* (only)
   2. **Cryosophileae** (Thrinacinae): *Coccothrinax, Thrinax*
   3. **Phoeniceae** (lvs pinnate): *Phoenix* (only)
   4. **Trachycarpeae** (Livistoneae; 2 subtribes + unplaced): *Chamaerops, Rhapis, Trachycarpus* (a. Rhapidinae); *Licuala, Livistona* (b, Livistoninae); *Copernicia, Pritchardia, Serenoa, Washingtonia* (unplaced)
   5. **Chuniophoeniceae**: *Nannorrhops*
   6. **Caryoteae** (lvs bipinnate or pinnate, leaflets praemorse, OW): *Arenga, Caryota*
   7. **Corypheae** (lvs palmate, bisex. or fls not dimorphic): *Corypha* (only)
   8. **Borasseae** (lvs palmate; dioec.; 2 subtribes): *Bismarckia, Hyphaene* (a, Hyphaeninae); *Borassus, Lodoicea* (b, Lataniinae)

IV. **Ceroxyloideae** (lvs pinnate, reduplicate, infls with several peduncular bracts, fls solit., G 3-ovulate; 3 tribes):
   1. **Cyclospatheae** (proximal fls bisex.): *Pseudophoenix* (only)
   2. **Ceroxyleae** (dioec., fls stalked): *Ceroxylon, Ravenea*
   3. **Phytelepheae** (Phytelephoideae; dioec. with reduplicately paripinnate lvs with persistent petioles, male infls usu. dense spikes, females usu. globose heads, A ∞, female fls with P 7 + 7 & ∞ staminodes in spirals, G (5–10); Am.; R: OB 105(1991)5): *Phytelephas* (vegetable ivory)

V. **Arecoideae** (monoec. or dioec., usu. with reduplicately paripinnate lvs, fls in triads of central female flanked by males; 14 tribes + some unplaced):
   1. **Iriarteeae** (infls with more than 2 peduncular bracts, fls not sunken in pits): *Iriartea, Socratea, Wettinia*
   2. **Chamaedoreeae** (Hyophorbeae; monoec. or dioec., fls sessile): *Chamaedorea, Hyophorbe*
   3. **Podococceae** (as 1. but fls sunk in pits): *Podococcus* (only)
   4. **Oranieae** (Oraniinae): *Orania* (only)
   5. **Sclerospermeae** (Sclerospermatinae): *Sclerosperma* (only)
   6. **Roystoneae** (Roystoneinae): *Roystonea* (only)
   7. **Reinhardtieae** (Malortieinae): *Reinhardtia* (only)
   8. **Cocoseae** (G 3-loc.; 3 subtribes): *Attalea, Beccariophoenix, Butia, Cocos, Jubaea, Lytocaryum, Syagrus* (a, Attaleinae incl. Beccariophoenicinae, Butiinae); *Acrocomia, Aiphanes, Astrocaryum, Bactris, Desmoncus* (b, Bactridinae); *Elaeis* (c, Elaeidinae)
   9. **Manicarieae** (Manicariinae): *Manicaria* (only)
   10. **Euterpeae** (Euterpeinae): *Euterpe, Oenocarpus, Prestoea*

11. **Geonomateae** (like 4. but fls sunk in pits): *Geonoma*
12. **Leopoldinieae** (Leopoldiniinae): *Leopoldinia* (only)
13. **Pelagodoxeae**: *Sommieria*
14. **Areceae** (usu. 1 peduncular bract, G usu. pseudomonomerous; the bulk of palms arr. in 9 subtribes, but many form. recog. = tribes above): *Archontophoenix* (SW Pacific esp. New Caled., a, Archontophoenicinae); *Areca, Nenga, Pinanga* (only; b, Arecinae); *Burretiokentia, Cyphophoenix, Cyphosperma, Physokentia* (SW Pacific esp. New Caled., c, Basseliniinae); *Carpoxylon, Neoveitchia* (both almost extinct, d, Carpoxylinae); *Brongniartikentia, Clinosperma* (New Caled., e, Clinospermatiinae); *Dypsis* (Madag., f, Dypsidinae); *Calyptrocalyx, Howea* (g, Linospadicinae); *Oncosperma* (h, Oncospermatinae); *Ptychosperma, Veitchia* (i, Ptychospermatinae); *Rhopalostylis* (j, Rhopalostylidinae); *Phoenicophorium* (Seychelles, k, Verschaffeltiinae); *Cyrtostachys, Heterospathe, Hydriastele, Iguanura* (unplaced)

Largest seeds (*Lodoicea*), leaves (*Raphia*), infls (*Corypha*), lianes (*Calamus*) & widest primary growth (*Jubaea*) known. Pls oft. char. of habitats being typical understorey pls of rainforest in As. (at G. Matang, Sarawak 100 spp., poss. most diverse place in world) & Am. (few in Afr. but remarkable generic diversity in New Caled., Seychelles & Madag.) & in canopy as lianes (rattans) but also high in Andes (*Ceroxylon*), in dry country & OW mangrove (*Nypa*), to 44° N (*Chamaerops*). Most sessile until apical meristem reaches max. size & trunk develops (In V.1. elongation before max. size reached so prop-roots stabilize apically widening structure): in all (cf. woody pls in general, vasc. bundles can remain active for hundreds of years. Poll. insects (some thermogenesis – 20/46, assoc. with beetle-poll., esp. weevils). The apical bud is usu. well-protected from predators by leaf-bases, spines or poss. toxic tissues: this is the palm-heart of commerce – its removal leads to death of the trunk

App. an arch. group with the Afr. & Am. genera retaining the most primitive characters (Whitmore). Palms are intimately assoc. with human societies in trop. & warm countries esp. *Cocos* in Pac., *Phoenix* in N Afr., *Metroxylon, Arenga, Borassus* & *Caryota* in As. etc. A no. of spp. so long cult. as to be unknown wild: *Cocos nucifera* (coconut), *Corypha umbraculifera* (talipot), *Phoenix dactylifera* (date). Many imp. international commodities esp. oils (*Acrocomia, Astrocaryum Attalea, Bactris, Syagrus* but esp. *Cocos* & *Elaeis*), sago (*Metroxylon*), fibres (*Astrocaryum, Attalea, Caryota, Chamaerops, Cocos, Leopoldinia, Raphia, Sabal, Serenoa, Trachycarpus*), waxes (*Ceroxylon, Copernicia*), fr. (*Cocos, Phoenix, Salacca*) & palm-hearts (e.g. *Euterpe*), drugs (*Areca, Serenoa*), rattan furniture (esp. *Calamus*), vegetable ivory (*Phytelephas*), some timber & many cult. orn. (easily transplanted as readily regenerate rs) street-trees & house-pls (*Rhapis*-growing rivalling tulipomania). Many (e.g. *Arenga, Borassus, Caryota, Corypha, Nypa*) tapped for sap evaporated to sugar or fermented to toddy & distilled (arrack). E.J.H. Corner (1966) *Natural history of Palms*; Principes (later Palms, journal)

Palmerella A. Gray (~ *Solenopsis*). Campanulaceae (III 2). 1 W N Am.: ***P. debilis*** A. Gray. R: Novon 16(2006)72

Palmeria F. Muell. Monimiaceae (VI 1). 15 NG (12), Aus. (3, end.). R: Blumea 2 8(1982)85

Palmervandenbroekia Gibbs = *Polyscias*

palmetto *Sabal palmetto*; **saw p.** *Serenoa repens*

palmiet *Prionium serratum*

palmilla *Yucca schidigera*

palmiste *Roystonea oleracea*

palmito palm-heart

Palmolmedia Ducke = *Naucleopsis*

Palmorchis Barb.-Rodr. Orchidaceae (V 1). c. 20 trop. Am.

palmyra *Borassus flabellifer*

palo amarillo *Calycophyllum multiflorum*; **p. blanco** *Arctostaphylos* spp.; **p. colorado** *Luma apiculata*; **p. madroño** *Amomyrtus luma*; **p. santo** *Bulnesia sarmientoi*; **p. verde** *Parkinsonia* spp.

palosapis *Anisoptera thurifera* & other spp.

Paloue Aubl. Leguminosae (I 2). Incl. *Paloveopsis*, 5 trop. Am.

Paloveopsis Cowan = *Paloue* (but see AA 19(1989)150)

palsywort *Primula veris*

palta *Persea americana*

Paltonium C. Presl = *Pleopeltis*

palu *Manilkara hexandra*

Paludorchis P. Delforge = *Anacamptis*

Palumbina Reichb.f. = *Cuitlauzina*

Pamburus Swingle (~ *Atalantia*). Rutaceae (II 2). 1 S Ind., Sri Lanka: ***P. missionis*** (Wight) Swingle

pameroon bark *Trichilia moschata*

Pamianthe Stapf. Amaryllidaceae. 2 N Andes. R: Britt. 36(1984)22. Epiphytes; bulb with long false neck. ***P. peruviana*** Stapf (Peru) – cult. orn. epiphyte with scented fls

pampas grass *Cortaderia selloana*

Pamphalea DC. = *Panphalea*

Pamphilia Mart. ex A. DC. = *Styrax*

pamplemousse *Citrus maxima*

Pamplethantha Bremek. = *Pauridiantha*

Panama berry *Muntingia calabura*; **bois de P.** *Quillaja saponaria, Saponaria officinalis*; **P-hat palm, plant** *Carludovica palmata*; **P. passionfruit** *Passiflora* hybrids

Panamanthus Kuijt (~ *Struthanthus*). Loranthaceae (4 4). 1 Panamá: ***P. panamensis*** (Rizz.) Kuijt – allied to *Gaiadendron*. R: AMBG 78(1991)172

Panax L. Araliaceae (2). 6 N Am., E As. Glabrous herbs with rhiz. & thick roots used medic. e.g. colds (ginseng; active principles glycosides) in E & thought to affect membrane transport of steroids, esp. ***P. pseudoginseng*** Wall. (sanchi, E As., extinct in wild), used in cancer treatment in China, but also ***P. ginseng*** C. Meyer (ren shen, sang, Korea, E As.) now v. rare in wild, roots fetching up to $US 10 000, ingredient of 'natural Viagra', anti-oxidant & immune system stimulant, & ***P. quinquefolius*** L. (American g., E N Am.) – imp. local medic.

Pancheria Brongn. & Gris. Cunoniaceae (VIII). 26 New Caled. Accumulate nickel; some with Corner's Model

Panchezia Montr. = *Pancheria*

Pancicia Vis. (~ *Pimpinella*). Umbelliferae (III 8). 1 SE Eur.: ***P. serbica*** Vis.

Pancovia Willd. Sapindaceae. 12 trop. Afr. Some timber & ed. fr.

Pancratium Dill. ex L. Amaryllidaceae. c. 10 Canary Is., W Afr., Medit. (Eur. 2), Namibia. ***P. maritimum*** L. (sea daffodil, Medit.) – figured in Minoan bronze (1560 BC) & poss. 'Rose of Sharon', with other spp. (e.g. ***P. longiflorum*** Roxb. ex Ker (Ind.)) cult. orn. like *Hymenocallis* but seeds angular (not globose or oblong)

panda bamboo, baby *Pogonatherum paniceum*

Panda Pierre. Pandaceae (Euphorbiaceae s.l.). 1 trop. W Afr.: ***P. oleosa*** Pierre – Cook's Model, orthotropous ovules, alks, large pyrene distrib. only by elephants, seed-oil used locally for cooking. R: A. R.-Sm., *Gen. Euph.* (2001)128

Pandaca Noronha ex Thouars = *Tabernaemontana*

Pandacastrum Pichon = *Tabernaemontana*

Pandaceae Engl. & Gilg (~ Euphorbiaceae). Magnoliidae – Malpighiales. 3/16 OW trop. Dioec. evergreen trees & shrubs. Lvs simple, in spirals on orthotropic axes, distich. on plagiotropic axes of determinate growth, oft. without axillary buds though such a bud between branch & axis; stip. small. Fls small, reg. usu 5-merous, in axillary clusters (*Microdesmis*) or term. (*Galearia*) or cauliflorous (*Panda, Galearia*) thyrses, or solit.; K free or connate, C valvate or imbr., A 5, 5 + 5, 10 or 15, anthers with longit. slits, disk 0, $\underline{G}$ (2–5) with 1 anatropous (orthotropous in *Panda*) bitegmic ovule per loc. Fr. a drupe with bony endocarp with as many loc. & seeds as G; seeds without caruncles, oily endosperm copious. n = 15

Genera: *Galearia, Microdesmis, Panda*

Stony fr. not found in Euphorbiaceae, though 'throw-away' branches found in Phyllanthaceae. Close to Irvingiaceae. Oilseeds (*Panda*), timber, local medic.

pandan wangi *Pandanus amaryllifolius*

Pandanaceae R. Br. Magnoliidae – Pandanales. 5/575 OW trop. to NZ. Dioec. (bisex. infls or fls in *Freycinetia*) usu. pachycaul trees & shrubs or (*Freycinetia*) lianes with clasping aerial roots, s.t. epiphytic; prop-roots (if aborted early = spines or protuberances) usu. at base of stem; primary thickening, oft. with compound vasc. bundles; vessels throughout vegetative body; branching sympodial, apical meristems forming infls (s.t. infls only on lat. branches). Lvs simple, glabrous, in 3 (4 in *Sararanga*) ranks appearing as spirals ('screw-pines') because of spiral growth of stem, bases sheathing, blades usu. elongate (to 5 m), usu. xeromorphic with parallel veins & marginal spines (also on adaxial midrib). Fls v. small, ∞, in usu. term. panicles (*Sararanga*) or in 1–several spadices subtended by coloured spathes, in racemes, homologies obscure – pedicellate & bracteate with irreg. 3–4-lobed cup (?P), A ∞ with fleshy filaments, G 10–80, 1-seeded with sessile stigmas, in forked double row, in *Sararanga*; in others P 0 & distinction between A & G of individual fls obscured in development from app. reg. initials (in *F.*, A grouped around vestigial pistillode, G (1–12), 1-loc. with several ovules & oft. with basal staminodes; in others A arr. in phalanges, oft. with branched filaments & elongated connectives, G 2–30 in phalanges, oft. incompletely closed, each with 1(– few) ovules); ovules anatropous, bitegmic. Fr. – berries (*Freycinetia, Sararanga*) or drupes, in heads, drupes with 12–80 pyrenes in *Sararanga*, monocarpellary with 1 seed or 'polydrupes' (connate carpels of phalanges) with united or separate endocarps; seeds small with copious oily (starchy in *Freycinetia*) endosperm, strophiole (from raphe) s.t. present. n = 25, 28, 30 (some aneuploidy)

Classification & genera (Candollea 67(2012)326):

1. **Pandanoideae** (arborescent, G 1-ovulate): *Benstonea, Pandanus, Martellidendron, Sararanga*
2. **Freycinetioideae** (lianoid, G multi-ovulate): *Freycinetia*

Trad. assoc. with Palmae but strikingly diff.; fam. in all continents (exc. Aus.) in Upper Cretaceous (pollen though poss. N Am. referable to Araceae), now OW only

Some ed. fr., scent-making, flavouring, thatch & cult. orn. (*Pandanus*)

pandani, pandanni, pandanny *Richea pandanifolia*

Pandanus Parkinson. Pandanaceae. Excl. *Benstonea, Martellidendron*, c. 450 OW trop. (Mal. c. 340, Madag. c. 85, Afr. only 26). R: BJ 94(1974)466. Screw-pines (all Aus. spp. protected). Usu. erect with stilt-roots, some maritime, others rheophytes, in mountain forests or epiphytes; Corner's (e.g. ***P. princeps*** Stone (Madag.)), Stone's, Leeuwenberg's & Scarrone's Models, in subg. *Vinsonia* sect. *Acanthostyla* (13 spp. Madag. swamps) monopodial trunk to 15 m with huge term. rosette of lvs & lat. horizontal forked branches with infls & smaller lvs, a growth-form unique in angiosperms, the apical lvs of saplings to 10 m long & 36 cm wide; wind- or insect-poll. fr. fibrous, disp. by sea, freshwater, turtles, fish, birds & bats, cooked & eaten. Figured in C5 Buddhist paintings (Madhya Pradesh, Ind.). Lvs used for basketry & hat-making, mats (Tonga), thatch, umbrellas (Solomon Is.) etc., fibres from stilt-roots used for chair-seats, cordage etc.; some flavourings for food (e.g. *Nelumbo* 'moon cakes') esp. ***P. amaryllifolius*** Roxb. ex Lindl. (pandan wangi) – usu. sterile (? orig. Moluccas), scented lvs used to flavour rice, jellies etc. in Mal.; ***P. bowersiae*** St John (NG) – karuka nut; ***P. conoideus*** Lam. (marita, Mal., cultigen) – no viable seeds (veg. prop.), red sauce from fr. eaten with sweet potato, taro etc. (NG); ***P. julianettii*** Martelli (marita, NG) – cult. form of '*P. brosimos* Merr. & Perry', imp. local fr. tree, lvs for thatch; ***P. kaida*** Kurz (? cultigen, SE As.) – planted on paddy-field 'bunds' in Sri Lanka, pollen insect-repellent; ***P. leram*** Jones (Nicobar breadfruit, E Ind. Ocean) – fr. ed.; ***P. simplex*** Merr. (Philippines) – lvs for mats, baskets, suitcases, exported to Eur. & US; ***P. spiralis*** R. Br. (trop. Aus.) – one of most used spp. in trad. Aboriginal society (seeds & base of young lvs ed., roots a dye-source, stems used for carrying fire, didgeridoos, lvs for fibre & trad. medic.); ***P. tectorius*** Parkinson (*P. fascicularis, P. odoriferus, P. odoratissimus* (poss. distinct), ketaki, OW maritime) – males form 30 branches & fl. annually, females c. 16 & biennially (Fiji), on Pacific atolls disp. by landcrabs eating fr. & discarding seed, oft. cult. for ed. fr., male spadices distilled for kewda (perfume) esp. in Orissa (30 M infls annually) to flavour food, tobacco, soap, hair-oil etc., lvs for sails, matting & basketry, as living hedge & cult. orn. (incl. **'Veitchii'** (*P. veitchii*, Polynesia) – most commonly cult. as house-pl., lvs varieg., fls & fr. unknown), a staple on some Pac. atolls, where black dye prepared from roots; ***P. utilis*** Bory (Madag., cultigen) – lvs much used for mats etc.

Panderia Fischer & C. Meyer = *Bassia*

Pandiaka (Moq.) Hook.f. Amaranthaceae (I 2). 12 trop. & S Afr. R: KB 34(1980)425

Pandorea (Endl.) Spach. Bignoniaceae (1). 6 E Mal., Aus., New Caled. Evergreen lianes, cult. orn. esp. ***P. jasminoides*** (Lindl.) K. Schum. (bower plant, Aus.) & ***P. pandorana*** (Andrews) Steenis (wonga-wonga vine, Mal. & W Pac.) – wiry branches straightened over fire by C Aus. Aborigines for spear-shafts

Paneroa E. Schilling (~ *Ageratum*). Compositae (Eup.). 1 Mex.: ***P. stachyofolia*** (Robinson) E. Schilling. R: Novon 18(2008)520

panga-panga *Millettia stuhlmannii*

Pangium Reinw. Achariaceae (Flacourtiaceae). 1 Mal.: ***P. edule*** Reinw. (kepayang, seeds washed up on shores of Netherlands (poss. aided by humans!)) – Aubréville's Model, fr. to 30cm, elephant-disp. seeds (buah keluak) with cyanogenic glycoside (gynocardine) removed by boiling ed. (selected cvs low in glycoside), a fermented sauce eaten with taro in NG, oil for cooking, illumination & soap, bony seed-coats used for rattles, door-'curtains' etc., bark for string bags ('bilum') in NG & an antimalarial paste in Sumatra

pangola grass *Digitaria eriantha* 'Pangola'

panic grass *Panicum* spp.

Panicum L. Gramineae (XXIV 6). Incl. *Yakirra*, excl. *Apochloa, Cyphonanthus, Dichanthelium, Kellochloa, Morronea, Ocellochloa, Oncorachis, Renvoizea, Sorengia, Stephostachys, Urochloa*, c. 100 trop. to warm temp. (China 21 – R: FOC 22(2006)504). R: JAA suppl. 1(1991)224. Panic or crab grass & millets. Some apospory, pseudogamy. Some fodders, grains, weeds & cult. orn. esp. OW. ***P. brizanthum*** Hochst. ex A. Rich. (*Brachiaria b.*, palisade grass, Afr.) – fodder; ***P. coloratum*** L. (buffalo grass, Afr.) – pasture; ***P. gilvum*** Launert (Afr.) – weed in NSW suspected of photosensitizing sheep; ***P. hemitomon*** Schultes (pifine grass, N Am.) – prairie fodder; ***P. luzonense*** J. & C. Presl (Philippines) – ann. weed of ricefields, fr. with irritant prickly decid. hairs; ***P. miliaceum*** L. (proso millet, common, French or Russian m., cultigen (?) domesticated C As. 10 K yrs ago) – flour, alcoholic drinks, pig-, poultry- & caged-bird food, game-cover in GB, cult. E & C Eur. 5th millenium BC, with *Setaria italica* in perhaps the earliest known noodles (4000 yrs old, China); ***P. molle*** Sw. (water grass, trop. Am.) – bad weed; ***P. sonorum*** L. (sauwi, Afr.) – domesticated; ***P. sumatrense*** Roth ex Roemer & Schultes (little millet, sama, Mal.) – minor grain imp. in S Ind., 'subsp. *psilopodium* (Trin.) de Wet' (*P. p.*) alleged progenitor; ***P. virgatum*** L. (switch grass, N Am.) – touted as biofuel source, orn. cvs

panirband *Withania coagulans*

Panisea (Lindl.) Lindl. Orchidaceae (V 10b). 11 NE Ind. to SE As. R: NJB 7(1987)511

Pankycodon Hong & X.T. Ma (~ *Codonopsis*). Campanulaceae (I). 1 Sinohimal.: ***P. purpureus*** (Wall.) Hong & X.T. Ma. R: JSE 52(2014)549

Panopsis Salisb. (~ *Brabejum*). Proteaceae (V 4). 25 trop. Am. Some timbers

Panphalea Lag. Compositae (Mut.-Nass.). 9 subtrop. & temp. S Am. R: Notas Mus. Eva Perón Bot. 16(1953)225. Herbs; pappus 0

pansy *Viola* spp. esp. *V.* × *wittrockiana*; **field p.** *V. arvensis*; **p. orchid** *Miltoniopsis* spp.

Pantacantha Speg. Solanaceae (Benth.). 1 Patagonia: ***P. ameghinoi*** Speg.

Pantadenia Gagnepain. Euphorbiaceae (Cod.-Ost.). Incl. *Parapantadenia*, 1 Madag, 1 Vietnam

Pantathera Philippi = *Megalachne*

Pantlingia Prain = *Stigmatodactylus*

Panulia (Baill.) Kozo-Polj. = *Apium*

Panurea Spruce ex Benth. Leguminosae (III 2). 2 Colombia, N Braz.

Panzeria Moench = *Panzerina*

Panzerina Soják (*Leonuroides*). Labiatae (VI). 6 W Siberia to Mongolia. R: Taxon 31(1982)559

Panzhuyuia Z.Y. Zhu = *Alocasia*

Paolia Chiov. = *Coffea*

papain *Carica papaya*

papapsco *Acer* spp.

Papaver Tourn. ex L. Papaveraceae (IV). Incl. *Stylomecon*, 80 Eur. (27), As., Cape Verde Is. (1), S Afr. (1: ***P. aculeatum*** Thunb., allied to E Med. spp., introd. Aus.), W N Am. Poppies; emblem of goddess Demeter. V. many alks. Fls in bud nodding through asymmetric growth of pedicel, later rectified at anthesis; fr. a capsule roofed by persistent stigmas, seeds protected thus & shaken out through pores beneath them. Many cult. orn. & for alks esp. ***P. somniferum*** L. (opium p. = 'subsp. *somniferum*' (2n = 22) cultigen (with poss. 1000 cvs) poss. derived from 'subsp. *setigerum* (DC.) Arc.' (*P. setigerum*, 2n = 44, SW As., but these two effectively 'triploid' & hexaploid of pl. similar to *P. glaucum* Boiss. & Hausskn. (2n = 14) & *P. gracile* Boiss. (2n = 28)), E Med.) – 'subsp. *somniferum*' (C 20–80 mm long!) early cult. W Med. (capsules/seeds in Switzerland 5500–8000 BCE) but now esp. Iran to China, most imp. drug pl. unmatched by synthetics, opium (7 K t in 2011, only 20% of 1906 total!) being dried latex obtained from immature capsules by lancing them & containing c. 25 diff. alks esp. morphine (9–17%, available in pure form by 1806) – powerful analgesic & potentially addictive narcotic (like *Coca* & *Coffea* manipulating dopamine prod., suppressing emotional activity & making pain, hunger, discomfort, fear & anxiety tolerable; blood vessel dilation giving pleasant warm feeling), thebaine (basis of oxycodon, a painkiller used in cancer treatment), codeine, narceine, narcotine & papaverine (alleged to promote prolonged penile erections (when injected!)), source of heroin ('gear', 'smack', 'scag', 'H'; 'chasing the Dragon' = inhaling smoke from heated powder through tube), form. used in many patent medicines (laudanum = extract in alcohol with saffron & cinnamon introd. 1527), incl C19 cigarettes used to treat asthma (!), gums & syrups, etc., heroin trade (472 t by 2011) worth $US 30 billion (93% ('illegal') from Afghanistan (half GDP by 2010 – $2 billion), used as local currency with opium 'futures' traded), 'legal' (now esp. France & Tasmania) with cvs high in thebaine for US market, though GM yeast producing the alks; cult. orn. incl. teratologous '**Hen and Chickens**' with multiple capsules sold as 'cut-fl.'; seeds (maw s.) almost without opium, used on bread etc. in baking (high incidence of glaucoma in p. seed eaters) & as birdseed; 'subsp. *hortense* (Hussenot) Syme' – source of poppy oil used in artists' paints, woodwork & salad oil, soap etc. ***P. alpinum*** L. *s.l.* (alpine poppy, Eur. mts) – cult. orn.; ***P. bracteatum*** Lindl. (W As.) – poss. comm. source of codeine (prepared from thebaine in this sp. allied to *P. orientale* which has no morphine); ***P. cambricum*** L. (*Meconopsis c.*, Welsh p., W Eur.) – cult. orn.; ***P. nudicaule*** L. (Iceland p., Arctic & N temp. As.) – some opium, many cvs with diff. fl. colours; ***P. orientale*** L. (SW As. & hybrids with *P. bracteatum* = ***P. setiferum*** Goldbl. (*P. pseudo-orientale*), hexaploid) – the unruly coarse perenn. 'oriental' ps of herbaceous borders; ***P. radicatum*** Rottb. (N Eur., W As.) – with *Salix arctica* the most northerly plant (83°N); ***P. rhoeas*** L. (red, common, corn, Flanders or field p., Euras., N Am.) – orig. E Med. (C used in funeral garlands in anc. Egypt) & poss. selected by anthropogenic habitats up to NW Eur., prob. 'flowers of the field' of Isaiah, the p. of Flanders in World War I & Poppy Day in GB, petals a source of red dye used in some wines & medicines, also local medic. (earache, toothache, neuralgia – though smelling fls widely believed to *cause* headaches), selected cult. orn. cvs being 'Shirley ps' (Rev. W. Wilks C19)

Papaveraceae Juss. Magnoliidae – Ranunculales. Incl. Fumariaceae, Pteridophyllaceae 43/750 mostly N temp., also Aus., trop. Afr. mts, S Afr., S Am. etc. C. Grey-Wilson (1993) *Poppies*. Herbs, scramblers (Fumarioideae), subshrubs or pachycaul treelets with many isoquinoline alks in articulated laticifers (or latex-cells); vasc. bundles in 1, 2 or more rings. Lvs usu. in spirals, (entire–) lobed or dissected; stip. 0. Fls oft. large, reg. to irreg., bisex., hypogynous (perigynous in *Eschscholzia*), solit. (less oft. in cymes etc.); K 2(3), sometimes basally connate, rather asymmetrical, oft. caducous, C usu. 2 × K in 2 whorls (0 in *Macleaya*), oft. 2 + 2 when 1 or both s.t. with basal spur or pouch – Fumarioideae), 3 + 3(–16), imbr. & crumpled in bud, A (4–6 in *Meconella*, Fumarioideae, *Hypecoum*) ± ∞, developing centripetally in multiples of K, nectaries 0, $\underline{G}$ (2+), 1-loc. ± style, the stigmas oft. connate to form ± discoid roof but separate atop ± ∞ carpels in 1 whorl in *Platystemon*, ovules (1–)∞, anatropous to amphitropous or ± campylotropous, bitegmic, oft. on parietal placentas s.t. meeting to form pluriloc. ovary with axile placentation. Fr. a capsule with longit. valves or pores (nut-like in some Fumarioideae, follicle with lomentum-like units in *Platystemon*); seeds s.t. arillate, with elongate embryo in copious oily endosperm. Seedlings s.t. with 1 cotyledon. n = (5)6 or 7(8–11, 19)

Classification & chief genera:

Pteridophylloideae (Pteridophyllaceae; lvs fern-like, A4): *Pteridophyllum*

[I. Chelidonioideae (= **IVb.** – Chelidonieae; hairs multicellular; pollen grains tricolpate or polyporate; G (+), seeds usu. arillate): *Bocconia, Glaucium*]

[II. Eschscholzioideae (= **IVc.** – Eschscholzieae; hairs unicellular; pollen grains polycolpate; G 2): *Eschscholzia*]

[III. Platystemonoideae (= **IVd.** – Platystemoneae; hairs multicelluar-multiseriate; pollen grains tricolpate; G many): *Meconella*]

IV. **Papaveroideae** (a, Papavereae); hairs multicelluar-multiseriate; pollen grains usu. tricolpate; G (many): *Argemone, Meconopsis, Papaver*

Fumariaceae here as subfam. **Fumarioideae** (R: OB 88(1986)5) – latex 0, K bract-like, not encl. bud): Fumarieae (Fumariaceae I – fls irreg., A 6): *Corydalis, Dicentra, Fumaria, Rupicapnos*) & Hypecoeae (Fumariaceae II – fls ± reg., A 4): *Hypecoum* (only)

Most of above & *Chelidonium, Eomecon, Macleaya, Pseudofumaria, Roemeria, Romneya & Sanguinaria* cult. orn. & oft. medic. esp. *Papaver somniferum* (opium) but also *Eschscholzia*; some oilseeds

papaw *Carica papaya, Asimina triloba*; **mountain p.** *Vasconcellea pubescens*

papaya *Carica papaya*; **strawberry p.** *C. papaya* cv.

paper birch *Betula papyrifera*; **p. bush** *Edgeworthia papyrifera*; **p. daisy** *Xerochrysum* spp. esp. *X. bracteatum, Rhodanthe manglesii*; **p. mulberry** *Broussonetia papyrifera*

paperbark *Melaleuca* spp., *Streblus asper*

Paphia Seemann (~ *Agapetes*). Ericaceae (VIII 5). 20 E NG (16) to Queensland, New Caled. & Fiji. R: EJB 60(2004)271

Paphinia Lindl. Orchidaceae (V 12i). 16 trop. Am. R: Die Orchidee 29(1978)207. Cult. orn. epiphytes

Paphiopedilum Pfitzer. Orchidaceae (III). c. 85 trop. As. to Papuasia. R: P. Cribb, *The genus P.*, ed. 2 (1998)48. Epiphytes or terr. herbs without pseudobulbs; some poll. syrphids laying eggs on staminode, in e.g. ***P. rothschildianum*** (Reichb.f.) Stein (N Borneo) mimicking aphid colony on which they usu. lay eggs, in ***P. villosum*** (Lindl.) Stein (Thailand) attracted by winy smell & glittering staminode releasing 'honeydew' but with a slippery wart in middle, so insect slips into pouch & escapes up tunnel past stigma, effecting poll., before pressing against anther (pollen viability 8 weeks, flowers lasting 2–3 months). Cult. orn. (almost 25K grexes) incl. many hybrids (esp. imp. **Winston Churchill** grex) oft. grown as 'cypripedium' but lvs folded & not plicate as in bee-poll. *C.*: ***P. lawrenceanum*** (Reichb.f.) Pfitzer (Borneo) – speckled foliage; ***P.*** × ***maudiae*** (Rolfe) McQuade (*P. callosum* (Reichb.f.) Stein (SE As.) × *P. lawrenceanum*) – one of most commonly cult. of all orchids; ***P. sanderianum*** (Reichb.f.) Stein (Sarawak) – lat. P to 90 cm+ long

Papilionaceae Giseke = Leguminosae (III)

× **Papilionanda** R. Schultes & Pease. Orchidaceae. *Papilionanthe* × *Vanda* – 871 grexes

Papilionanthe Schltr. (~ *Vanda*). Orchidaceae (V 16c). 11 Himal. to Mal. Cult. orn. scramblers esp. **Miss Joaquim** grex (*P. hookeriana* (Reichb.f.) Schltr. (SE As. to W Mal.) × *P. teres* (Roxb.) Schltr. (Himal. to SE As.) – 1890) – nat. fl. Singapore (only country with a hybrid emblem)

Papilionopsis Steenis = *Hylodesmum* – trick-pl. of *H.* infl. inserted in a Burmanniacea (cf. *Actinotinus*, *Stalagmitis*)

Papillilabium Dockr. Orchidaceae (V 16c). 1 E Aus.: ***P. beckleri*** (Benth.) Dockrill

Papistylus Kellerman & al. Rhamnaceae. 2 SW Aus. R: Nuytsia 16(2007)306

papoose root *Caulophyllum thalictroides*

Pappagrostis Rosch. = *Stephanachne*

Pappea Ecklon & Zeyher. Sapindaceae. 1 Dhofar, trop. E to S Afr.: ***P. capensis*** Ecklon & Zeyher (wild plum) – monoec., ed. fr., seed-oil & timber. R: Strelitzia 10(2000)505

Papperitzia Reichb.f. = *Leochilus*

Pappobolus S.F. Blake Compositae (Helia.-Helia.). Incl. *Helianthopsis*, 38 Colombia, Ecuador, Peru, Bolivia. R: SBM 36(1992)

Pappochroma Raf. (~ *Erigeron*). Compositae (Ast.-Lag.). 9 Aus. R: Phytol. 85(1998)277

Pappophorum Schreb. Gramineae (XXIX 7). 9 S US to Arg. R: SB 14(1989)356

Pappostipa (Speg.) Romaschenko & al. (~ *Jarava*). Gramineae (X). 31 trop. & warm Am. R: JBRIT 2(2008)181

Pappothrix (A. Gray) Rydb. = *Perityle*

paprika *Capsicum annuum* Longum Group

Papuacalia Veldk. Compositae (Sen.-Tuss.). 14 NG mts. R: Blumea 36(1991)168. Trees & shrubs

Papuacedrus Li (~ *Libocedrus*). Cupressaceae. 1 Moluccas & NG: ***P. papuana*** (F. Muell.) Li – timber for housing. R: BM 12(1995)66

Papuaea Schltr. (~ *Macodes*). Orchidaceae (IV 2d). 1 NG: ***P. reticulata*** Schltr. R: Gen. Orch. 3(2003)134

Papualthia Diels = *Polyalthia*

Papuanthes Danser. Loranthaceae (5 3). 1 NG: ***P. albertisii*** (Tieghem) Danser. R: FM I,13(1997)379

Papuapteris C. Chr. = *Polystichum*

Papuasicyos Duyfjes. Cucurbitaceae (XIV). Incl. *Urceodiscus*, 8 NG R: Blumea 48 (2003)124, 51(2006)38. Double-forked style

Papuastelma Bullock = *Marsdenia*

Papuechites Markgraf. Apocynaceae (II c). 1 Moluccas & NG: ***P. aambe*** (Warb.) Markgraf – latex used on sores. R: Blumea 40(1995)446

Papulipetalum (Schltr.) M.A. Clem. & D.L. Jones = *Bulbophyllum*

Papuodendron C. White = *Hibiscus*

Papuzilla Ridl. = *Lepidium*

papyrus *Cyperus papyrus*; **Mexican p.** *C. giganteus*

Paquirea Panero & Freire (~ *Chucoa*). Compositae (Mut.-Onos.). 1 Peruv. Andes: ***P. lanceolata*** (Beltrán & Ferreyra) Panero & Freire – coll. once. R: Phytoneuron 2013–11:2

Pará cress *Acmella oleracea*; **P. (or para) grass** *Panicum muticum*; **P. nut** *Bertholletia excelsa*; **P. rubber** *Hevea brasiliensis*

Parabaena Miers. Menispermaceae (III). 6 Indomal. R: KB 39(1984)103

Parabambusa Widjaya. Gramineae (V 4). Incl. *Pinga*, 2 NG. R: Reinw. 11(1997)121, 123

Parabarium Pierre ex C. Spire = *Urceola*

Parabeaumontia Pichon = *Vallaris*

Parabenzoin Nakai = *Lindera*

Paraberlinia Pellegrin = *Julbernardia*

Parabignonia Bur. ex K. Schum. = *Dolichandra*

Paraboea (C.B. Clarke) Ridl. Gesneriaceae (III 2i). Incl. *Phylloboea, Trisepalum*, 130 SE As. to W Mal., esp. limestone. R: EJB 65(2008)171, T 65(2016)287. Calcium salts accumulate in lvs leading to necrotic blisters

Parabouchetia Baill. = *Heliotropium*

Paracaleana Blaxell (~ *Caleana*). Orchidaceae (IV 3e). 13 S (esp. SW) Aus. R: AusSB 19(2006)214. Poll. thynnid wasps (sawflies in C.) by sexual deception (cf. *Ophrys*)

Paracalia Cuatrec. Compositae (Sen.-Tuss.). 2 Peru, Bolivia. R: Britt. 12(1960)183. Scandent shrubs

Paracalyx Ali. Leguminosae (III 18). 6 OW trop. R: USK 5(1968)93

Paracarpaea (K. Schum.) Pichon = *Cuspidaria*

Paracarphalea Razafim. & al. (~ *Carphalea*). Rubiaceae (Knox.). 3 Madag. R: Phytotaxa 263(2016)107

Paracaryopsis (H. Riedl) R. Mill = *Adelocaryum* (but see EJB 48(1991)56)

Paracaryum (A. DC.) Boiss. = *Cynoglossum* (but see PSE 148(1985)296)

Paracautleya R.M. Sm. = *Curcuma*

Paracephaelis Baill. (~ *Tarenna*). Rubiaceae (II 2). Incl. *Homolliella*, 7 E Afr., Madag.

Paraceterach (F. Muell.) Copel. (~ *Gymnopteris*). Pteridaceae (IV). Excl. *Paragymnopteris*, c. 2 warm OW. R: AFJ 76(1987)186

Parachampionella Bremek. = *Strobilanthes*

Parachimarrhis Ducke. Rubiaceae (Cond.). 1 Amazonia: ***P. breviloba*** Ducke

Parachionolaena Dillon & Sagást. = *Chionolaena*

Paracladopus Kato (~ *Cladopus*). Podostemaceae. 2 Thailand, Laos. R: T 57(2008)202

Paracoffea (Miq.) J. Leroy = *Psilanthus*

Paracolpodium (Tzvelev) Tzvelev (~ *Colpodium*). Gramineae (XVI 14). 4 Caucasus to Himal.

Paracorynanthe Capuron. Rubiaceae (I 1). 2 Madag. R: BJLS 152(2006)377

Paracostus C. Specht (~ *Costus*). Costaceae. 2 W Afr., Mal. R: T 55(2006)162

Paracroton Miq. (*Fahrenheitia*). Euphorbiaceae (Al.-Par.). 4 S Indomal.

Paracryphia Bak. f. Paracryphiaceae. 1 New Caled.: ***P. alticola*** (Schltr.) Steenis. R: BJBBuit. 18(1950)459

Paracryphiaceae Airy Shaw. Magnoliidae – Paracryphiales. Incl. Quintiniaceae, Sphenostemonaceae, 3/35 C Mal. to trop. Aus. & New Caled. Trees to 40 m or shrubs, some lianoid, with unicellular hairs; vessel-elements with up to 200 cross-bars in perforation plates. Lvs simple, usu. toothed, in spirals to subverticillate; stip. minute or 0. Fls bisex. or male, ∞ in compound spikes; P 4 decussate resembling larger outer bract encl. inner 3 bracteoles, caducous (*Paracryphia*), or K 4 or 5 & C 4 or 5 (*Quintinia*) or C 0 (*Sphenostemon*). A 4, 8(-12), filaments in male fls somewhat flattened, anthers basifixed with longit. slits, $\underline{G}$ (2, 8–15), s.t. inf. (*Quintinia*), with as many distinct stigmas & 1 (2) or 4 or many anatropous, unitegmic ovules in single row on axile placenta of each loc. Fr. a capsule (*Paracryphia* with carpels separating from a central column exc. at its apex & opening ventrally, cf. *Medusagyne*) or berry (*Sphenostemon*); seeds small, usu. winged, the straight embryo in copious endosperm

Genera: *Paracryphia, Quintinia, Sphenostemon*

Q. form. in Grossulariaceae, *S.* in Aquifoliaceae. S-type plastids like those in Theaceae & Actinidiaceae

Paractaenum P. Beauv. (*Parectenium*). Gramineae (XXIV 4). 2 Aus.

Paracynoglossum Popov = *Cynoglossum*

Paradavallodes Ching = *Davallia*

Paraderris (Miq.) Geesink = *Derris*

Paradina Pierre ex Pitard = *Mitragyna*

Paradisanthus Reichb.f. Orchidaceae (V 12j). 4 trop. S Am.

paradise flower *Caesalpinia pulcherrima*; **grains of p.** *Aframomum melegueta*; **p. nut** *Lecythis* spp.; **p. tree** *Quassia glauca*

Paradisea Mazzucc. Asparagaceae (Anthericaceae). 2 S Eur. Cult. orn. esp. ***P. liliastrum*** (L.) Bertol. (St Bruno's lily, mts of S Eur.) though often in fact ***P. lusitanica*** (Cout.) Samp.
Paradolichandra Hassler = *Dolichandra*
Paradombeya Stapf. Malvaceae (Dom.; Bombacaceae). 3 Myanmar, SW China
Paradrymonia Hanst. Gesneriaceae (II 5c). 10 trop. Am. (not SE Braz.). R: SB 41(2016)95
Paradrypetes Kuhlm. Rhizophoraceae. 2 W Amaz. & E Braz. R: SB 17(1992)74. Echinate pollen & unisexual apetalous fls! Form. in Euphorbiaceae
Paraeremostachys Adylov & al. = *Phlomoides*
Parafaujasia C. Jeffrey. Compositae (Sen.-Sen). 2 Mascarenes
Parafestuca Alexeev = *Trisetaria*
paraffin bush *Chromolaena odorata*
Paragelonium Leandri = *Aristogeitonia*
Paragenipa Baill. (~*Fernelia*). Rubiaceae (II 5). 1 Madag.: ***P. lancifolia*** (Bak.) Tirv. & Robbrecht. R: NJB 5(1985)458
Parageum Nakai & H. Hara = *Geum*
Paraglycine F.J. Herm. = *Ophrestia*
Paragoldfussia Bremek. = *Strobilanthes*
Paragonia Bur. = *Tanaecium* (but see AMBG 85(1998)465)
Paragonis J.R. Wheeler & N. Marchant. Myrtaceae (II 14). 1 SW Aus.: ***P. grandiflora*** (Benth.) J.R. Wheeler & N. Marchant. R: Nuytsia 16(2007)430
Paragoodia I. Thompson (~ *Muelleranthus*). Leguminosae (III 14). 1 W Aus.: ***P. crenulata*** (A. Lee) I. Thompson. R: Muelleria 29(2011)174
Paragophyton K. Schum. = *Spermacoce*
Paragramma (Blume) T. Moore = *Lepisorus*
Paragrewia Gagnepain ex R. Rao = *Leptonychia*
Paraguay tea *Ilex paraguariensis*; **P. palm** *Acrocomia aculeata*
paraguayo *Prunus persica*
Paragulubia Burret = *Hydriastele*
Paragymnopteris Shing (~ *Paraceterach*). Pteridaceae (IV). 5 warm OW
Paragynoxys (Cuatrec.) Cuatrec. Compositae (Sen.-Tuss.). 12 NW trop. S Am. R: Britt. 55(2003)157. Pachycaul trees to 15 m tall
Parahancornia Ducke. Apocynaceae (I c). 6 trop. S Am. R: Novon 1(1991)42
Paraharveya E. Fischer & Siedentop. Orobanchaceae (Orob.). 1 C & E Afr.: ***P. alba*** (Hepper) E. Fischer & Siedentop
Parahebe W. Oliver = *Veronica*
Parahemionitis Panigr. (~ *Hemionitis*). Pteridaceae (IV). 1 Ind. to C Mal.: ***P. arifolia*** (Burm.f.) Panigr. R: AFJ 83(1993)90
Paraholcoglossum Z.J. Liu & al. = *Holcoglossum*
Parahyparrhenia A. Camus. Gramineae (XXII 7). 6 trop. W Afr., Ind., Thailand. R: KB 20(1967)434
Paraia Rohwer, Richter & van der Werff. Lauraceae (I). 1 Amaz.: ***P. bracteata*** Rohwer, Richter & van der Werff. R: AMBG 78(1991)392
Paraisometrum W.T. Wang = *Oreocharis* (but see Novon 7(1997)431)
Paraixeris Nakai = *Crepidiastrum*
Parajaeschkea Burkill = *Gentianella*
Parajubaea Burret. Palmae (V 8a). 3 Ecuador, Bolivia & Colombia, high alts (some 10 months dry). R: Britt. 42(1990)92. Unarmed monoec. Some fibre for ropes, seeds ed., cult. orn.
Parajusticia Benoist = *Justicia*
Parakaempferia A. Rao & Verma. Zingiberaceae (II 1). 1 Assam: ***P. synantha*** A. Rao & Verma – coll. once. R: BBSI 11(1971)206
parakeelya *Parakeelya* spp.
Parakeelya Hershk. (~ *Calandrinia*). Montiaceae. 40 Aus. R: Phytol. 84(1998)101
Parakibara Philipson. Monimiaceae (V 2). 1 Moluccas: ***P. clavigera*** Philipson. R: FM I,10(1986)286
Paraknoxia Bremek. Rubiaceae (III 8). 1 trop. Afr.: ***P. parviflora*** (Stapf & Verdc.) Bremek. R: BJ 110(1989)550
Parakohleria Wiehler = *Pearcea*
Paralabatia Pierre = *Pouteria*
Paralagarosolen Y.G. Wei = *Petrocodon* (but see APS 42(2004)528)

Paralamium Dunn. Labiatae (VI). 1 Yunnan, Vietnam: ***P. griffithii*** (Hook.f) Suddee & A. Paton (*P. gracile*). R: KB 59(2004)316

Paralasianthus H. Zhu (~ *Lasianthus*). Rubiaceae (IV 10). 5 SE As. R: Phytotaxa 202(2015)274

Paralbizzia Kosterm. = *Archidendron*

Paralepistemon Lejoly & Lisowski (~ *Ipomoea*). Convolvulaceae (9). 1 S trop. Afr.: ***P. shirensis*** (Oliv.) Lejoly & Lisowski. R: BTBC 118(1991)267

Paraleptochilus Copel. = *Leptochilus*

Paraligusticum Tichom. (~ *Ligusticum*). Umbelliferae (III 8). 1 Altai Mts: ***P. discolor*** (Ledeb.) Tichom.

Paralinospadix Burret = *Calyptrocalyx*

Paralophia Cribb & Hermans (~ *Eulophia*). Orchidaceae (V 12b). 2 Madag. R: CBM 22(2005)48

Paralstonia Baill. = *Alyxia*

Paralychnophora MacLeish (*Sphaerophora*; (~ *Eremanthus*). Compositae (Vern.- Lych.). 6 Braz. R: Britt. 64(2012)294. Pachycaul treelets

Paralyxia Baill. = *Aspidosperma*

Paramachaerium Ducke. Leguminosae (III 11). 5 trop. Am. R: Britt. 33(1981)435

Paramacrolobium Léonard. Leguminosae (I 2). 1 trop. Afr.: ***P. coeruleum*** (Taub.) Léonard – timber incl. railway sleepers. R: BJBB 24(1954)348

Paramammea Leroy = *Mammea*

Paramansoa Baill. = *Fridericia*

Paramapania Uittien (~ *Mapania*). Cyperaceae (I 1). 7 Mal., ***P. radians*** (C.B. Clarke) Uittien ext. to W Pacific

Parameconopsis Grey-Wilson = *Papaver*

Paramelhania Arènes. Malvaceae (Dom.; Sterculiaceae). 1 SE Madag.: ***P. decaryana*** Arènes

Parameria Benth. Apocynaceae (II c). 3 Ind. to China to W Mal. R: FM I,18(2007)304. Some medic., fibres & rubber

Parameriopsis Pichon = praec.

Paramichelia Hu = *Magnolia*

Paramicropholis Aubrév. & Pellegrin = *Micropholis*

Paramicrorhynchus Kirpiczn. = *Launaea*

Paramiflos Cuatrec. = *Espeletia*

Paramignya Wight (~ *Luvunga*). Rutaceae (II 2). 12 Indomal. R: Swingle, *Bot. Citrus* (1943)253. ***P. monophylla*** Wight (Ind.) – citrus stock; ***P. scandens*** (Griff.) Craib (Assam to W Mal.) – local medic. & tonic, improving ability to withstand low temps

Paramitranthes Burret = *Siphoneugena*

Paramollugo Thulin (~ *Mollugo*). Molluginaceae. 5 trop. R: T 65(2016)784

Paramoltkia Greuter (~ *Moltkia*). Boraginaceae (B.2.2). 1 SW Balkans: ***P. doerfleri*** (Wettst.) Greuter & Burdet. R: Willd. 11(1981)38

Paramomum Tong = *Amomum*

Paramongaia Velarde. Amaryllidaceae. 1 Peru: ***P. weberbaueri*** Velarde – cult. orn. like *Pamianthe* but without false neck on bulb, fls yellow. R: BM n.s.14(1997)142

Paramyrciaria Kausel = *Myrciaria* (but see Cand. 46(1991)512)

Paramyristica de Wilde = *Myristica*

Paraná pine *Araucaria angustifolia*

Paranecepsia R.-Sm. Euphorbiaceae (Acal.-Pyc.-Necep.). 1 SE Afr.: ***P. alchorneifolia*** R.-Sm. R: KB 30(1976)684

Paranephelium Miq. Sapindaceae. 4 SE As., W Mal. R: Blumea 29(1984)425. ***P. macrophyllum*** King (SE As. to Malay Pen.) – form. cult. for seed-oil used in lamps & skin problems

Paranephelius Poeppig. Compositae (Liab.). 7 Peru, Bolivia, Arg. R: SCB 54(1983)45

Paraneurachne S.T. Blake = *Neurachne* (but see CQH 13(1972)21)

Paranneslea Gagnepain = *Anneslea*

Paranomus Salisb. Proteaceae (IV 4). (Incl. *Vexatorella*) 23 S Afr. R: CBH 2(1970), SAJB 50(1984)373

Parantennaria Beauverd. Compositae (Gnap.-Cass.). 1 E Aus. mts: ***P. uniceps*** (F. Muell.) Beauverd. R: OB 104(1991)95

Parapachygone Forman (~ *Pachygone*). Menispermaceae. 1 Queensland: ***P. longifolia*** (Bailey) Forman. R: FA 2(2007)375

Parapantadenia Capuron = *Pantadenia*

Parapentapanax Hutch. = *Aralia*

Parapentas Bremek. Rubiaceae (IV 1). 3–4 trop. Afr., (?) 1 Madag.

Paraphalaenopsis A. Hawkes (~ *Phalaenopsis*). Orchidaceae (V 16c). 4 W Borneo. R: H.R. Sweet, *Genus Phalaenopsis* (1986)118. Cult. orn.

Paraphlomis Prain. Labiatae (VI). c. 20 E As., Mal.

Parapholis C. Hubb. (~ *Pholiurus*). Gramineae (XVI 12). 6 Eur. to S As. R: BN 115(1962)1. ***P. incurva*** (L.) C. Hubb. (Eur., Medit.) – salt-tolerant fodder & ***P. strigosa*** (Dumort.) C. Hubb. (Medit.) – both invasive in Aus.

Paraphyadanthe Mildbr. = *Caloncoba*

Parapiptadenia Brenan. Leguminosae (II 1). 3 trop. S Am. Comm. timbers. ***P. rigida*** (Benth.) Brenan source of angico gum used like g. arabic

Parapiqueria R. King & H. Robinson (~ *Eupatorium*). Compositae (Eup.- Ayap.). 1 Braz.: ***P. cavalcantei*** R. King & H. Robinson. R: MSBMBG 22(1987)210

Parapolydora H. Robinson. Compositae (Vern.-Erl.). 2 trop. & S Afr.

Parapolystichum (Keyserl.) Ching (~ *Lastreopsis*). Dryopteridaceae. 27 trop. to NZ. R: Britt. 67(2015)81

Parapodium E. Meyer. Apocynaceae (Vc; Asclepiadaceae III 1). 3 S Afr.

Paraprenanthes Chang ex Shih = *Notoseris* (but see APS 26(1988)418)

Paraprotium Cuatrec. = *Protium*

Parapteroceras Averyanov = *Tuberolabium*

Parapteropyrum A.J. Li = *Fagopyrum* (but see APS 19(1981)330)

Parapyrenaria H.T. Chang = *Pyrenaria*

Paraquilegia J.R. Drumm. & Hutch. Ranunculaceae (III 1). 5 W Iran to Himal. & W China. Like *Isopyrum* but follicles several, *Semiaquilegia* but staminodes 0. ***P. anemonoides*** (Willd.) Ulbr. (*P. grandiflora*) – cult. orn.

Pararchidendron I. Nielsen (~ *Pithecellobium*). Leguminosae (II 4). 1 Java & Aus.: ***P. pruinosum*** (Benth.) I. Nielsen – street-tree. R: BMNHN 4, 6(1984)379

Parardisia Nayar & Giri = *Ardisia*

Pararistolochia Hutch. & Dalziel (~ *Aristolochia*). Aristolochiaceae. 31 trop. OW. R: APS 30(1992)510

Parartocarpus Baill. (~ *Artocarpus*). Moraceae (II). 2 Mal. R: FM I, 17(2006)128. Some ed. seeds (those of ***P. venenosus*** (Zoll. & Mor.) Becc. ed. after soaking in seawater for some days)

Pararuellia Bremek. Acanthaceae (III 2a). 10 SE As. to Mal. ***P. napifera*** (Zoll.) Bremek. (C Mal.) – treatment of kidney stones

Parasamanea Kosterm. = *Albizia*

Parasarcochilus Dockr. = *Pteroceras*

Parasassafras Long (~ *Litsea*). Lauraceae (II). 2 Himal. to W China & Myanmar

Parascheelea Dugand = *Attalea*

Parascopolia Baill. = *Lycianthes*

Paraselinum H. Wolff. Umbelliferae (III 8). 1 Peru: ***P. weberbaueri*** H. Wolff. R: BTBC 84(1957)197

Paraselliguea Hovenkamp. Polypodiaceae (II). 1 Borneo: ***P. leucophora*** (Bak.) Hovenkamp. R: Blumea 42(1997)485

Parasenecio W.W. Sm. & Small. Compositae (Sen.-Tuss.). 60+ Russia, E As. ('*Cacalia*' of As.), Alaska. R (as *Koyamacalia*): Phytologia 27(1973)270

Paraserianthes I. Nielsen (~ *Albizia*). Leguminosae (II 4). Excl. *Falcataria*, 1 W Mal.: ***P. lophantha*** (Willd.) I. Nielsen – invasive Sicily, Aus., S Afr.

Parashorea Kurz. Dipterocarpaceae (Dipt.-Shor.). 14 S China, SE As., Mal. Timbers ((white) lauan (luan) or seraya) esp. ***P. malaanonan*** (Blanco) Merr. (bagtikan, Borneo, Philippines) – major veneer; ***P. stellata*** Kurz (thingadu, Myanmar) – imp. timber

Parasicyos Dieterle = *Sicyos*

Parasilaus Leute. Umbelliferae (III 5). 2 SW & C As.

Parasitaxaceae A. Bobrov & Melikian = Podocarpaceae

Parasitaxus Laubenf. (~ *Podocarpus*). Podocarpaceae (Parasitaxaceae). 1 New Caled.: ***P. usta*** (Vieill.) Laubenf. – parasitic (root-graft, no haustoria but vasc. bundles penetrating to host cambium) on *Falcatifolium taxoides* (Brongn. & Gris) Laubenf., only parasitic gymnosperm known. R: FNC 4(1972)44

Paraskevia W. Sauer & G. Sauer = *Pulmonaria* (but see Phyton 20(1980)285)

Parasopubia H.-P. Hofm. & Eb. Fischer (~ *Sopubia*). Orobanchaceae (Buch.). 2 SE As. R: BJ 125(2004)357

Parasorus Alderw. = *Davallia*

Parasponia Miq. = *Trema*
Parastemon A. DC. Chrysobalanaceae (1). 3 Nicobar Is. to NG. R: FOW 9(2003)177. ***P. urophyllus*** (A. DC.) A. DC. – hard timber for construction, boats etc.
Parastrephia Nutt. Compositae (Ast.-Hin.). 3 C Andes. R: Phytol. 75(1993)347
Parastriga Mildbr. = *Harveya* (but see JAA 11(1930)52)
Parastrobilanthes Bremek. = *Strobilanthes*
Parastyrax W.W. Sm. Styracaceae. 4 China, Myanmar
Parasympagis Bremek. = *Strobilanthes*
Parasyncalathium J.W. Zhang & al. (~ *Syncalathium*). Compositae (Cich.-Cich.). 1 Himal.: ***P. souliei*** (Franch.) J.W. Zhang & al. R: T 60(2011)1680
Parasyringa W.W. Sm. = *Ligustrum*
Paratecoma Kuhlm. Bignoniaceae (1). 1 coastal Braz.: ***P. peroba*** (Rec.) Kuhlm. (peroba) – imp. building & furniture timber, now threatened with extinction
Paratephrosia Domin = *Tephrosia*
Parathelypteris (H. Itô) Ching = *Thelypteris*
Paratheria Griseb. Gramineae (XXIV 4). 2 W Afr., Madag., trop. Am.
Parathesis (A. DC.) Hook.f. Primulaceae (Myrsinaceae). 95 trop. Am. R: CTRFBS 5 (1966), Phytol. 55(1984)237, Sida 20(2003)914. Scarrone's Model
Parathyrium Holttum = *Deparia*
Paratriaina Bremek. (~ *Triainolepis*). Rubiaceae (IV 8). 1 Madag.: ***P. xerophila*** Bremek.
Paratrophis Blume = *Streblus*
Paravallaris Pierre = *Kibatalia*
× Paravanda A.D. Hawkes. Orchidaceae. *Paraphalaenopsis* × *Vanda* – 147 grexes
Paravitex Fletcher = *Vitex* (but see KB 1937:74)
Pardanthopsis (Hance) Lenz = *Iris*
Pardoglossum Barbier & Mathez = *Cynoglossum* (but see Candollea 2 8(1973)281)
Parduyna Salisb. = *Schelhammera*
Parectenium auctt. = *Paractaenum*
pareira root or **p. brava** *Chondrodendron tomentosum*; **false p. r.** *Cissampelos pareira*; **white p. r.** *Buta rufescens*; **yellow p. r.** *Aristolochia glaucescens*
Parenterolobium Kosterm. = *Albizia*
Parentucellia Viv. = *Bellardia*
Parepigynum Tsiang & P.T. Li. Apocynaceae (II c). 1 Yunnan: ***P. funingense*** Tsiang & P.T. Li. R: APS 11(1973)395
Parhabenaria Gagnepain = *Pecteilis*
Pariana Aubl. Gramineae (VI 2). 35 Amazon, Costa Rica, Trinidad. R: CUSNH 39(2000)93. Insect-poll. Lvs of ***P. lunata*** Nees used to pack gold & platinum dust
Parianella Hollowell & al. (~ *Pariana*). Gramineae (Bamb.). 2 Braz. R: Phytotaxa 7(2013)30
Parietaria Tourn. ex L. Urticaceae (IV). c. 10 (diff. to identify) subcosmop. (Eur. 6). Fls bisex. in axillary cymes though first fl. female & last are males, the bisexuals v. protogynous with style emerging from bud, A developing explosively later by which time style dropped so fls appear male; achenes of some spp. attractive to ants. ***P. judaica*** L. (*P. diffusa*, asthma weed, pellitory of the wall, W & S Eur.) – wall weed in GB, serious in Sydney, Aus., used to clean glass in Med. Palestine; ***P. officinalis*** L. (C & S Eur.) – form. used as laxative & for urinary problems
parijat *Nyctanthes arbor-tristis*
parilla, yellow *Menispermum canadense*
Parinari Aubl. (*Parinarium*). Chrysobalanaceae (2). 39 trop. (Am. 18, Afr. 6, As. & Pac. 15). R: FOW 9(2003)198. See also *Neocarya*. Frs disp. by bats, elephants (e.g. ***P. excelsa*** Sabine, trop. Afr.), baboons, pigeons, rheas, emus, agoutis & fish. Some timber & ed. fr. ***P. capensis*** Harv. (sand apple, trop. & S Afr.) – geoxylic suffrutex in poor grasslands of C Afr.; ***P. curatellifolia*** Planch. ex Benth. (trop. Afr.) – ed. fr. (mbura, mobola) & timber for poles etc. (Livingstone buried beneath one); ***P. elmeri*** Merr. (Mal.) – supports for Iban longhouses (N Borneo); ***P. nonda*** F. Muell. ex Benth. (nonda, N Aus.) – fr. ed.
Paripon Voigt = ? *Bactris*
Paris L. Melanthiaceae (Trilliaceae). Incl. *Daiswa* (R: Britt. 35(1983)255), *Kinugasa*, *Trillidium*, c. 20 OW temp. (Eur. 1: ***P. quadrifolia*** L. (herb Paris), form. medic. with whorl of 4 (or more) net-veined lvs on aerial shoots produced irreg. (not annually) from monopodial rhiz. & 4-merous fls attractive to pollinating carrion-flies, though A grow on to stigmas if crossing fails, fr. a poisonous berry with seeds without sarcotesta). R: Plantsman 9(1987)81, 10(1988)167. ***P. japonica*** (Franch. & Sav.) Franch. (*K. j.*, *T. j.*, Jap.) – largest

eukaryote genome, DNA in 1 cell 100 m long (in *Homo sapiens* 2 m). Some cult. orn. esp. ***P. polyphylla*** Sm. (*D. p.*, Himal.) – largest pl. chromosomes known

Paris daisy *Argyranthemum frutescens*

Parishella A. Gray (~ *Nemacladus*). Campanulaceae (II). 1 S Calif.: ***P. californica*** A. Gray. R: Pfl. IV,276c(1968)923

Parishia Hook.f. Anacardiaceae (I). 7 Indomal. timber (lelayang) for plywood, light construction, etc. ***P. insignis*** Hook.f. (dhup) – trade timber in Andamans

Pariti Adans. = *Thespesia*. See also *Hibiscus*

Parkeriaceae Hook. Older name for Pteridaceae

Parkia R. Br. Leguminosae (II 1). 34 trop. (Indopacific 12 (R: KB 49(1994)192), Afr. 3, Madag. 1, Am. 18). Fls in dense heads lasting 1 night; entomophilous spp. (Am.) with all fls fertile (no nectar), others with term. nectariferous ones or with fertile ones separated from basal staminodal ones by zone of nectariferous; heads pendulous or erect, staminodal fls sometimes forming a fringe (? landing-stage for bats, OW spp. of which landing head upwards, Am. ones head downwards); red fl. heads seen as black against sky in most spp., yellow ones as white against foliage (bats colour-blind); seeds disp. by mammals & birds; mealy pods ed. (locusts) esp. ***P. biglobosa*** (Jacq.) G. Don (Afr., introd. WI) & ***P. filicoidea*** Welw. ex Oliv. (African locust, Afr.), seeds of first boiled & fermented in W Afr., tannin from pods used on floor-surfaces; ***P. speciosa*** Hassk. (petai, W Mal.) – imp. 'jungle fr.' in Mal., ed. pods imparting garlic scent to eaters due to cyclic polysulphides

Parkinsonia Plum. ex L. Leguminosae (I 4). (Incl. *Cercidium*) 11–12 drier Am., S Afr. (1), NE Afr. (3). Consp. along water-courses in desert SW US (palo verde). Some ed. seeds & fr. ***P. aculeata*** L. (Jerusalem thorn, trop. & warm Am., serious invasive in trop. Aus. controlled by beetle (*Penthobruchus germaini*) restricted to eating its seeds) – cult. orn. spiny tree, phloem fibre suitable for mixed paper pulp; ***P. microphylla*** Torrey (SW N Am.) – some seeds viable after 250 yrs as herbarium material!

Parlatoria Boiss. Cruciferae (47). 3 SW As.

Parma violets *Viola alba* cv.

Parmentiera DC. Bignoniaceae (7). 9 C Am. to NW Colombia. R: FN 25(1980)97. See also *Crescentia*. Cult. orn. trees incl. ***P. cereifera*** Seemann (candle-tree, Panamá) – fodder tree with fr. to 1.3 m long, borne on trunk; ***P. aculeata*** (Kunth) Seemann (*P. edulis*, aguachilote, C Am.) – fr. ed. (poor)

Parnassia Tourn. ex L. Celastraceae (Parnassiaceae). c. 70 N temp. (Eur. 1: ***P. palustris*** L. (grass of Parnassus, N temp.); China 61) & Arctic S to Ind., Sumatra, Morocco, Mex. Some cult. orn. herbs with solit. term. white fls, nectarless staminodes forming glistening knobs (false nectaries) attractive to poll. insects

Parnassiaceae Martinov = Celastraceae

Parochetus Buch.-Ham. ex D. Don. Leguminosae (III 27). 1 mts of trop. Afr. & As. to Java: ***P. communis*** Buch.-Ham. ex D. Don – some cleistogamous fls, cult. orn. rock-pl. & in hanging baskets

Parodia Speg. Cactaceae (III 5). Excl. *Blossfeldia*, *Frailea*, c. 50 S Am. Cult. orn. small globose cacti

Parodianthus Tronc. Verbenaceae (Cass.). 2 Arg.

Parodiochloa C. Hubb. = *Poa*

Parodiodendron Hunz. Picrodendraceae (Pic.-Pic.-Pic.; Euphorbiaceae s.l.). 1 Arg.: ***P. marginivillosum*** (Speg.) Hunz. R: Kurtziana 5(1969)331

Parodiodoxa O. Schulz. Cruciferae. 1 N Arg. mts: ***P. chionophila*** (Speg.) O. Schulz

Parodiolyra Soderstrom & Zuloaga (~ *Olyra*). Gramineae (VI 3). 5 trop. Am. R: SCB 69(1989)64, Novon 9(1999)590

Parodiophyllochloa Zuloaga & Morrone (~ *Panicum*). Gramineae (XXIV 2) 6 trop. Am. R: SB 33(2008)69

Parolinia Webb. Cruciferae (4). 5 Canary Is. R: BN 123(1970)395

Paronychia Mill. Caryophyllaceae (I 2). Excl. *Chaetonychia*, *Gymnocarpos*, 110 cosmop. esp. Medit., Turkey, SE US (N Am. 26) & Peru to Bolivia (not native Aus.). R: MBMHRH 285(1968)64. Tufted herbs with small axillary fls concealed by stip., some ant-poll.; some cult. orn. rock-pls incl. ***P. argentea*** Lam. (Medit.) – form. medic.

Paropsia Noronha ex Thouars. Passifloraceae. 11 trop. Afr. & Madag., 1 E Mal. R: BJBB 40(1970)50

Paropsiopsis Engl. (~ *Paropsia*). Passifloraceae. 2 trop. W Afr. R: EJB 66(2009)32

Paropyrum Ulbr. = *Isopyrum*

Parosela Cav. = *Dalea*

Paroxygraphis W.W. Sm. Ranunculaceae (II 3). 1 E Himalaya: ***P. sikkimensis*** W.W. Sm. – dioec.

Parquetina Baill. = *Cryptolepis*

Parramatta grass *Sporobolus indicus*

Parrotia C. Meyer. Hamamelidaceae (I 4). Incl. *Shaniodendron*, 1 E China, 1 SW Caspian: ***P. persica*** (DC.) C. Meyer (ironwood) – C 0, A 5–7, fls bisex.; cult. orn. with flaking bark & bright autumn colour, hybrid with *Sycopsis sinensis* formed in Switzerland c. 1950. R: IDSY 2007:6

Parrotiopsis (Niedenzu) C. Schneider. Hamamelidaceae (I 4). 1 Himalaya: ***P. jacquemontiana*** (Decne.) Rehder – cult. orn. like *Parrotia* but A c. 15–24 & fl. heads surrounded by large white bracts, alleged to prevent regeneration of coniferous forest, twigs used in basket-making

parrot's bill *Clianthus speciosus*; **p.'s feather** *Myriophyllum aquaticum*

Parrya R. Br. Cruciferae (43). 42 N temp. R: KB 68(2013)458. Cult. orn. rock-pls esp. ***P. nudicaulis*** (L.) Regel with scented violet fls & ed. roots

Parrycactus Doweld = *Ferocactus*

Parryella Torrey & A. Gray. Leguminosae (III 10). 1 Mex.: ***P. filifolia*** Torrey & A. Gray – basketry, seeds toothache relief

Parryodes Jafri (~ *Arabis*). Cruciferae (7). 1 Bhutan, W Tibet: ***P. axilliflora*** Jafri

Parryopsis Botsch. = *Solms-laubachia*

Parsana Parsa & Maleki = *Laportea*

parsley *Petroselinum crispum*; **Baltic p.** *Cenolophium denudatum*; **bur p.** *Caucalis platycarpos*; **Chinese p.** *Coriandrum sativum*; **cow p.** *Anthriscus sylvestris*; **p. fern** *Cryptogramma crispa*; **fool's p.** *Aethusa cynapium*; **Hamburg p.** *P. crispum* 'Tuberosum'; **hedge p.** *Torilis japonica*; **Italian p.** *P. crispum* var. *neapolitanum*; **p. piert** *Alchemilla arvensis*; **sea p.** *Apium prostratum*; **turnip-rooted p.** *P. crispum* 'Tuberosum'; **water p.** *Oenanthe sarmentosa*

parsnip *Pastinaca sativa*; **cow p.** *Heracleum sphondylium*; **Peruvian p.** *Arracacia xanthorhiza*; **water p.** *Sium* spp.

Parsonsia R. Br. Apocynaceae (II e). Incl. *Lyonsia*, 82 E As., Indomal. to W Pac. (Mal. 27 – R: Blumea 42(1997)193; Aus. 35, 33 endemic). Some locally medic. incl. ***P. alboflavescens*** (Dennst.) Mabb. (Indomal.); ***P. edulis*** (G. Benn.) Guillaumin (*P. esculenta*, New Caled.) – fr. ed.

Parthenice A. Gray. Compositae (Helia.-Amb.). 1 SW N Am.: ***P. mollis*** A. Gray. R: FNA 21(2006)23

Parthenium L. Compositae (Helia.-Amb.). 16 N Am., WI. R: CGH 172(1950)1. Cypselas released with an involucral bract & 2 adjacent disk florets (sterile). ***P. argentatum*** A. Gray (guayule, Texas & N Mex.) – 20% by weight rubber, cult. emergency rubber pl. as in World War II in US; ***P. fruticosum*** Less. (S Am.) – lactones inhibiting growth of insect larvae; ***P. hysterophorus*** L. (N Am., serious invasive Aus., Ethiopia) – pollen inhibits fr. set in other spp. through allelopathy on stigmas, serious dermatitis through sesquiterpene lactone (parthenin) can be fatal

Parthenocissus Planch. Vitaceae. 15 temp. As. (1 to W Ghats, Ind.), N Am. (3, 1 ext. to Carib.). Decid. lianes with tendrils oft. tipped with adhesive disks, much cult. ('Virginia creeper', 'ampelopsis') esp. ***P. quinquefolia*** (L.) Planch. (true V. c., NE US to Mex.) – leaflets 5, bark medic., phloem eaten, boiled stalk water used as cooking syrup by E Canadian Indians, less oft. seen in Eur. than ***P. tricuspidata*** (Sieb. & Zucc.) Planch. (Boston ivy, C China to Jap.) – lvs simple to 3-lobed, many cvs

partminger *Ocimum americanum*

partridge berry *Gaultheria procumbens*; **p.-breasted aloe** *Gonialoe variegata*; **p. wood** *Andira inermis*

Parvatia Decne. = *Stauntonia* (but see Cathaya 8–9(1997)77)

Parvisedum R.T. Clausen = *Sedella*

Parvotrisetum Chrtek = *Trisetaria*

Pasaccardoa Kuntze (~ *Dicoma*). Compositae (Card.-Dicom.). 3 trop. & S Afr. Annuals

Pascalia Ortega (~ *Wedelia*). Compositae (Helia.-Verb.). 2 S Am. R: SBM 33(1991)40

pascalia weed *Wedelia glauca*

Paschalococos Dransf. (~ *Jubaea*). Palmae (V 5b). 1 Easter Is.: ***P. disperta*** Dransf. – extinct (poss. due to introd. rodents) when described from subfossil endocarps, loss poss. linked to decline of the *moai* culture as prob. a major food-source, & trunks used for rollers for the famous monuments there

Pascopyrum A. Löve = *Elymus* (but see FNA 24(2007)349)

Pasithea D. Don. Asphodelaceae (Hemerocallidaceae). 1 Chile: ***P. caerulea*** (Ruíz & Pavón) D. Don. R: T 58(2009)1129

Paspalidium Stapf = *Setaria* (but see JAA suppl. 1(1991)305)

Paspalum L. Gramineae (XXIII 3). Incl. *Reimarochloa, Thrasya*, c. 370 trop. & warm esp. Am. (N Am. 24 + 19 introd.). R: KBAS 13(1986)287, JAA suppl. 1(1991)277. Char. of pampas, campos etc., ***P. pyramidale*** Nees growing to 15 m in Amazonia. Apospory, pseudogamy. Many imp. fodders (some invasive) & some grains: ***P. conjugatum*** Bergius (trop. Am.) – invasive Aus. (in lawns & sugar-cane), Hawaii; ***P. dilatatum*** Poir. (Dallis grass, S Am., natur. US & S Afr., invasive S Eur., Aus. (fungus causing ergot), Hawaii) – fodder; ***P. distichum*** L. (*P. vaginatum*, warm coasts, invasive S Eur., Aus.) – stabilizing saltmarsh; ***P. notatum*** Fluegge (Bahia grass, trop. Am., introd. US) – fodder, some forms as lawn grass, erosion control in Afr.; ***P. scrobiculatum*** L. (Kodo millet, rice grass, Ind.) – livestock feed

pasque-flower *Anemone pulsatilla*

Passacardoa Wild = *Pasaccardoa*

Passaea Adans. = *Ononis*

Passerina L. Thymelaeaceae (Thym.-Daph). 20 C to S Afr. (Cape 15). R: Bothalia 23(2003) 61. C 0, G 1-loc. Cult. orn. heath-like shrubs

Passiflora L. Passifloraceae. c. 500 trop. & warm Am., 24 Indomal. & Pac. (Afr. 0). R: FMBot. 19(1938)1; T. Ulmer & J.M. MacDougal (2004) *P.*: 27; sect. *Dyosmia* (21) – CBM n.s. 30(2013)324. Passionflowers: lianes (***P. arborea*** Spreng. (Colombia) shrub or tree to 15 m) with axillary tendrils & entire or lobed lvs (sometimes central lobe not developing so lvs crescentic or bilobed), usu. extrafl. nectaries on petioles, stipules or tendrils, mimicking insect eggs; alks (fertility control in Uruguay). Co-evolved (in S Am.) with heliconiid butterflies, species of which have 'partitioned' host resources allowing coexistence of several spp. per plant; some bat-poll. e.g. ***P. ovalis*** Vell. (*Tetrastylis o.*, S Braz.) & ***P. mucronata*** Lam. (E Braz.). Vernacular name bestowed by Catholic missionaries in S Am. as 'Calvary Lesson' – 3 styles = nails of the Crucifixion, A 5 = 5 wounds, corona = crown of thorns, K 5 + C 5 = apostles (less Peter & Judas), lobed lvs & tendrils = hands & scourges of Christ's persecutors. Many cult. orn. incl. hybrids (***P.*** × ***violacea*** Lois. (*P. caerulea* × *P. racemosa* Brot. (SE Braz.) – first (Fulham, London, 1819)) & some imp. trop. fr. (passion fruit, incl. hybrids e.g. comm. ed. 'Panama passion fruit') – ed. part being arillate pulp used in drinks & ices; susceptible to nematodes. ***P. adenopoda*** DC. (trop. Am.) – lvs with sharp recurved trichomes puncturing cuticle of heliconiid caterpillars & killing them through haemorrhage; ***P.*** × ***belotii*** Pépin (*P.* × *alatocaerulea, P. alata* Curtis (E Braz., NE Peru) × *P. caerulea*) – fls used in scent-making; ***P. caerulea*** L. (blue p.f., Braz. to Arg.) – ± hardy in GB; ***P. coccinea*** Aubl. (trop. S Am.) – used as contraceptive; ***P. edulis*** Sims (purple granadilla, Braz. to Arg., invasive S Afr., Hawaii, Galápagos) – cult. for fr. in Aus., Mex., Hawaii (f. ***edulis*** with fls open a.m., fr. purple, & f. ***flavicarpa*** Degener with fls open p.m., fr. yellow, self-incompatible (cf. f. *e.*), all stock app. derived from 1 fr. in Covent Garden Market, London (1912 – × ?), US for juice etc.; ***P. foetida*** L. (running pop, trop. Am., widely natur. OW trop.) – ed. fr., but lvs (cyanogenic) refused even by starving horses; ***P. incarnata*** L. (apricot vine, E N Am.) – ed. fr. long cult. by Native Americans; ***P. laurifolia*** L. (water lemon, yellow granadilla, Jamaica honeysuckle, trop. Am.) – widely cult. for ed. fr.; ***P. ligularis*** Juss. (sweet granadilla, trop. Am.) – fr. sup. to that of *P. laurifolia*; ***P. maliformis*** L. (curuba, sweet calabash, trop. Am.) – cult. for grape-flavoured juice; ***P. mixta*** L.f. (trop. S Am.) – invasive NZ; ***P. mollissima*** (Kunth) L. Bailey (banana passion fruit, trop. Am.) – fr. ed.; ***P. quadrangularis*** L. (granadilla, maracuja, maracuyá, ? orig. NW S Am.) – large yellow fr. to 30 cm, eaten as veg. when immature, flavour (barbadine) for yoghurt etc.; ***P. tarminiana*** Coppens & V. Barney (banana poka, NW S Am.) – serious weed in Hawaii (almost 4000 ha, controlled by ascomycete fungus, *Septoria passiflorae*), also invasive Aus., NZ

Passifloraceae Juss. ex Roussel. Magnoliidae – Malpighiales. Incl. Malesherbiaceae, Turneraceae, 29/920 trop. & warm temp. esp. Am. Lianes with axillary tendrils (? derived from infls), (*Paropsia* & allies) shrubs & trees, or. s.t. herbaceous, oft. with unusual secondary growth & alks; s.t. smelly when bruised. Lvs in spirals, entire or lobed palmately (compound in *Deidamia*), usu. with nectaries on petiole; stip. usu. small decid., or 0; axillary bud oft. aborting or growing into infl. or tendril, vegetative branches forming from accessory bud. Fls reg., usu. bisex. (solit. or) in cymes, s.t. heterostylous, with flat to tubular hypanthium (fls hypogynous in *Paropsia* etc.), oft. with elongate androgynophore; K (3–) 5(–8) sometimes basally connate, imbr. (rarely valvate), persistent, C as many as &

alt. with K (or 0), imbr., oft. with corona of 1 or more rows of filaments or scales around A (4)5(–∞), usu. alt. with C, free or on gynophore (connate in tube around G in *Androsiphonia*), anthers with longit. slits, oft. (staminodal) disk around G, G (2)3(–6), 1-loc. with parietal placentas but styles usu. only basally connate (1 in *Barteria* & *Crossostemma*), ± ∞ anatropous (or orthotropous) bitegmic ovules usu. with long funicles. Fr. a berry or capsule; seeds oft. with fleshy apical aril (0 in *Malesherbia*) & bony testa, straight embryo embedded in copious oily endosperm. n = 5, 6, 9–11, 13
Principal genera: *Adenia, Basananthe, Passiflora, Piriqueta, Turnera*
Malesherbiaceae & Turneraceae long-known close allies
Some fr. (*Passiflora*) & cult. orn.

passionflower *Passiflora* spp.; **blue p.** *P. caerulea*

passionfruit *Passiflora* spp. esp. *P. edulis*; **banana p.** *P. mollissima*

Passovia Karsten (~ *Phthirusa*). Loranthaceae. 21 trop. Am. R: PDE 129(2011)179. ***P. pyrifolia*** (Kunth) Tiegh. (*Phthirusa p.*, S Am.) – acarpellate

passum raisin wine (*Vitis vinifera*) first made in Carthage

pasta *Triticum turgidum*; **p. di ceci** *Cicer arietinum*

pastel *Isatis tinctoria*

Pastinaca Tourn. ex L. Umbelliferae (III 11). Excl. *Malabaila*, 14 temp. Euras. (Eur. 4). Alks; furanocoumarin toxic to insects. ***P. sativa*** L. (parsnip, invasive N Am.) – bienn. with ed. taproot (UK record: 3.62 m long) with best flavour when frosted (starch converted to sugars – ? 'antifreeze'), source of vitamins C & E, allegedly useful in arthritis treatment

Pastinacopsis Golosk. Umbelliferae (III 11). 1 C As.: ***P. glacialis*** Golosk.

pastis *Pimpinella anisum* (also *Foeniculum vulgare*)

pasto lacquer *Elaeagia utilis*

Pastorea Tod. ex Bertol. = *Cochlearia*

Patagonula L. = *Cordia*

patana oak *Careya arborea*

Patascoya Urb. = *Freziera*

pataua *Oenocarpus bataua*

patchouli or **patchouly** *Pogostemon cablin*; **Chinese p.** *Microtoena patchouli*; **Indian p.** *P. indicus*

Patellaria J. Williams & Ford-Lloyd = *Beta*

Patellifolia A.J. Scott & al. (~ *Hablitzia*). Amaranthaceae (Chenopodiaceae I). 1 Macaronesia to Medit. & NE Afr.: ***P. procumbens*** (C. Sm.) A.J. Scott & al.

Patellocalamus W.T. Lin = *Ampelocalamus*

Patersonia R. Br. Iridaceae (II). 24 Sumatra, Borneo & NG (6 – R: AMBG 98(2012)517), Aus. (17 endemic), New Caled. (1). R: BJBB 44(1974)41. Secondary thickening in rhiz. (cf. woody I. subfam. Nivenioideae). Some cult. orn.

Paterson's curse *Echium plantagineum*

patience *Rumex patientia*

Patima Aubl. (~ *Sabicea*). Rubiaceae (I 8). 1 E Guyana Shield: ***P. guianensis*** Aubl. – myrmecophilous. R: AMBG 92(2005)109

Patinoa Cuatrec. Malvaceae (Bomb.; Bombacaceae). 4 trop. S Am. Fish-poison

Patis Ohwi (~ *Piptatherum*). Gramineae (X). 3 E As., E N Am. R: T 60(2011)1713

Patosia Buchenau (~ *Oxychloe*). Juncaceae. 1 Bolivian, Chilean & Arg. Andes: ***P. clandestina*** (Philippi) Buchenau – cushions to several m. diam. R: FOW 6(2002)13

Patrinia Juss. Caprifoliaceae (Valerianaceae). c. 25 Eur. (1) to Himal. & E As. Some cult. orn. rock-pls esp. ***P. triloba*** Miq. (*P. palmata*, E As.)

Pattalias S. Watson = *Seutera*

pattern wood *Alstonia congensis*

pau amarel(l)o *Euxylophora paraensis, Hortia* spp. esp. *H. excelsa*; **p. d'arco** *Handroanthus impetiginosus*; **p. marfim** *Balfourodendron riedelianum*; **p. santo** *Kielmeyera coriacea*

Paua Caball. = *Andryala*

Paubrasilia Gagnon & al. (~ *Caesalpinia*). Leguminosae (I 4). 1 E Braz.: ***P. echinata*** (Lam.) Gagnon & al. (Bahia wood, Braz. redwood, Braz. wood, peach wood, Pernambuco wood, trop. Am.) – CITES-listed, heartwood (orig. confused with *Biancaea sappan*) used for violin bows (since mid C18, wood with lower vessel freq. best) etc., also red dye-source ('pau brasil' [*brasa* = glowing coals or corruption of local name for *B. sappan*], the orig. of 'Braz.' because of amounts of dye exported thence to Portugal, replacing 'Santa Cruz', the terr.'s first European name). R: PhytoKeys 71(2016)36

Pauia Deb & Dutta = *Atropa*

Pauldopia Steenis. Bignoniaceae (1). 1 NE Ind. to SE As.: ***P. ghorta*** (D. Don) Steenis. R: ABN 18(1969)427

Paulia Korovin = *Paulita*

Paulita Soják. Umbelliferae (III 8). 3 C As.

Paullinia L. Sapindaceae (IV 1). c. 220 trop. Am. (Braz. 99), *P. pinnata* also in Afr. R: Pfl. IV 165,1(1931)219. Lianes with watch-spring tendrils & alks incl. caffeine, theophylline & theobromine; some fish-disp. in Amazon. ***P cupana*** Kunth (guaraná, S Am.) – seed ground & dried as sticks ('g. bread', currency in Amaz., allaying hunger, building stamina), basis of comm. fizzy soft drink (4 K t a yr by 2015; c. 5 M bottles a day in Braz. by 1980, 'Tai' in US), also dried as a tea (4.3% caffeine, i.e 3–5 times that in coffee) &, with cassava, principal alcohol of Mato Grosso, locally medic., tonic (allegedly aphrodisiac) imported to UK & used in chewing gum there, with *Garcinia gummi-gutta* a comm. appetite suppressant; ***P. pinnata*** L. (Am. & Afr.) – arrow & fish-poison, stems for cordage & roots for chewing-sticks (Nigeria); ***P. yoco*** R. Schultes & Killip (S Am. esp. Colombia) – caffeine-high drink (yoco)

Paulownia Sieb. & Zucc. Paulowniaceae (Scrophulariaceae s.l.). 7–10 E As. R: QJTM 12(1959)1. Some plantation trees for pulp; cult. orn. esp. ***P. tomentosa*** (Thunb.) Steud. (*P. imperialis*, China) – fls strongly scented to some but faint to others (cf. *Freesia*), cabinet wood much used in Jap. (kiri) for musical instruments, long cult. orn. China, form. coppiced Eur. for 'tropical bedding'; ***P. × taiwaniana*** T. Hu & H.T. Chang (*P. kawakamii* Itô (S China, Taiwan) × *P. fortunei* (Seem.) Hemsl. (China, Laos, Vietnam)) – fast-growing timber tree in S Am.

Paulowniaceae Nakai (~ Scrophulariaceae, Orobanchaceae). Magnoliidae – Lamiales. 4/25 E As. to Indomal. Decid. trees (*Wightia* epiphytic) to lianes. Lvs entire, opp. Infls term. thyrse. K densely brown-hairy; C 2-lipped with elongated inflated tube; A4; $\underline{G}(2)$. Fr. a capsule with winged seeds; endosperm 0? n = 19, 20

Genera: *Brandisia, Paulownia, Shiuyinghua, Wightia*

V. similar in aspect to *Catalpa* (Bignoniaceae) but ovary & seed anatomy diff.

Paulseniella Briq. = *Elsholtzia*

Pauridia Harv. Hypoxidaceae. Incl. *Saniella, Spiloxene* (both form. Amaryllidaceae), 30 S Afr. esp. SW Cape. R: Phytotaxa 182(2014)1. Some poll. monkey beetles

Pauridiantha Hook.f. Rubiaceae (I 9). Incl. *Pamplethantha, Poecilocalyx, Stelechantha*, c. 50 trop. Afr. R: OBB 15(2008)S9. Cook's & Roux's Models

Paurolepis S. Moore (~ *Gutenbergia*). Compositae (Vern.-Erl.). 3 Afr.

Pausandra Radlk. Euphorbiaceae (Cod.-Ost.). 6 trop. Am.

Pausinystalia Pierre = *Corynanthe* (but see BotJLS 120(1996)307)

Pavetta L. Rubiaceae (II 2). c. 360 OW trop. R: FR 37(1934)1. Lvs with bacterial nodules prob. secreting growth-substances (& not nitrogen-fixing); some 'litter-bin' spp. W Afr. Some cult. orn. shrubs & small trees. ***P. indica*** L. (Indomal.) – local medic.

Pavieasia Pierre. Sapindaceae. 2 SE As., China. R: APS 17(1979)34

Pavonia Cav. = *Hibiscus*

pawpaw = papaw; **mountain p.** *Vasconcellea pubescens*

Paxia Gilg = *Rourea*

Paxistima Raf. Celastraceae (I). 2 N Am. R: Sida 14(1990)231. Cult. orn. everg. ground-cover shrubs. ***P. myrsinites*** (Pursh) Raf. – medic., fr. ed.

paxiuba palm *Iriartea deltoidea*

Paxiuscula Herter = *Argythamnia*

Payena A. DC. Sapotaceae (II). 19 W Mal. (Borneo 12), Mindanao. ***P. leerii*** (Teijsm. & Binnend.) Kurz (W Mal.) – source of a gutta-percha (g. sundek)

Payera Baill. Rubiaceae (IV- Dan.). Incl. *Coursiana*, 10 Madag. R: Adans. 15(1993)68

Paypayrola Aubl. Violaceae (III 1d). c. 8 trop. S Am. Trees with A-tube

Paysonia O'Kane & Al-Shehbaz (~ *Physaria*). Cruciferae (40). 8 SE US. R: Novon 12(2002)380

pea *Pisum sativum*; **asparagus p.** *Lotus tetragonolobus*; **balloon p.** *Lessertia frutescens*; **p.-bean** *Phaseolus vulgaris*; **bitter p.** *Daviesia* spp.; **black-eyed p.** *Vigna unguiculata*; **blue p.** *Hovea* spp., *Clitoria ternatea*, *Psoralea pinnata*; **bush p.** *Pultenaea* spp.; **butterfly p.** *C. ternatea*; **chick p.** *Cicer arietinum*; **Congo p.** *Cajanus cajan*; **coral p.** *Abrus precatorius*, *Kennedia* spp.; **cow p.** *Vigna unguiculata*; **Darling p.** *Swainsona galegifolia*; **(Sturt's) desert p.** *S. formosa*; **dogtooth p.** *Lathyrus sativus*; **dun p.** *Pisium sativum* 'var. *arvense*'; **p. eggplant** *Solanum torvum*; **everlasting p.** *L. latifolius*, *L. grandiflorus*; **field p.** *P. sativum* 'var. *arvense*'; **flame p.** *Chorizema ilicifolium*; **glory p.** *Clianthus puniceus*, *Swainsona formosa*; **grass p.** *L. sativus*;

grey **p.** *P. sativum* 'var. *arvense*'; **gungo p.** *Cajanus cajan;* **Indian p.** *L. sativus;* **maple, mutter** or **partridge p.** *P. sativum* 'var. *arvense*'; **marrowfat p.** *P. sativum* 'var. *medullare*'; **parrot p.** *Dillwynia* spp.; **pigeon p.** *C. cajan;* **purple p.** *Hovea* spp.; **Riga p.** *L. sativus;* **rosary p.** *Abrus precatorius;* **sea p.** *L. japonicus;* **shaggy p.** *Oxylobium* spp.; **snow p.** = mange-tout; **Spanish p.** *Cicer arietinum;* **sugar p.** *Pisum sativum* 'var. *macrocarpon*'; **sweet p.** *L. odoratus;* **s. p. sprouts** *P. sativum* + *Triticum aestivum;* **Tangier p.** *L. tingitanus;* **winged p.** *Lotus tetragonolobus*

peace-lily *Spathiphyllum* spp.

peach *Prunus persica;* **African or Guinea p.** *Nauclea latifolia;* **do[ugh]nut** or **Saturn p.** *P. persica* cv.; **p. palm** or **nut** *Bactris gasipaes;* **p. tomato** *Solanum topiro;* **wild p.** *Kiggelaria africana;* **p. wood** *Haematoxylum brasiletto, Paubrasilia echinata*

peacharine *Prunus persica* cv.

peacock flower *Caesalpinia pulcherrima, Delonix regia, Moraea* spp.; **p. orchid** *Gladiolus murielae;* **p. tiger flower** *Tigridia pavonia*

peanut *Arachis hypogaea;* **hog p.** *Amphicarpaea bracteata;* **Inca p.** *Plukenetia volubilis*

pear *Pyrus* spp. esp. *P. communis;* **African p.** *Manilkara obovata;* **avocado, alligator** or **aguacate p.** *Persea americana;* **balsam p.** *Momordica charantia;* **Bollwiller p.** × *Sorbopyrus auricularis;* **Bradford p.** *Pyrus calleryana* 'Bradford'; **Chinese p.** *P. pyrifolia;* **corella p.** *P. communis;* **garlic p.** *Crateva* spp.; **Japanese** or **nashi p.** *P. pyrifolia;* **melon p.** *Solanum muricatum;* **prickly p.** *Opuntia* spp.; **thorn p.** *Scolopia ecklonii;* **tiger p.** *O. aurantiaca;* **vegetable p.** *Sicyos edulis;* **white p.** *Apodytes dimidiata;* **Whitty p.** *Sorbus domestica;* **woody p.** *Xylomelum* spp.

Pearcea Regel. Gesneriaceae (II 5b). Incl. *Parakohleria*, 17 N Colombia to NW Bolivia. R: SCB 84(1996)17. Fleshy capsules dehiscing to expose sticky seed-mass

pearl barley *Hordeum vulgare;* **p. fruit** *Margyricarpus pinnatus;* **p. lupin** *Lupinus mutabilis;* **p. millet** *Cenchrus spicatus;* **p.wort** *Sagina* spp. esp. *S. procumbens*

Pearsonia Dümmer. Leguminosae (III 8). 13 trop. & S Afr. (12), Madag. (1) R: KB 29(1974)383. ***P. metallifera*** Wild – accumulates up to 15.3% nickel & 4.8% chromium (in ash)

peat partially decayed plant-matter, accum. in wetlands & used as fuel & plant-growing medium

pecan (nut) *Carya illinoinensis*

Pechuel-loeschea O. Hoffm. (~ *Pluchea*). Compositae (Inul.-Pluch.). 1 SW Afr. to Zimbabwe: ***P. leubnitziae*** (Kuntze) O. Hoffm. R: Strelitzia 10(2000)153

Peckoltia Fourn. = *Matelea*

Pecluma Price. Polypodiaceae (V). 29 warm Am. R: AFJ 73(1983)109

Pecteilis Raf. (~ *Habenaria*). Orchidaceae (IV 2d). 8 trop. & E As. Cult. orn. terr.

Pectinaria Haw. Apocynaceae (V h; Asclepiadaceae III 5). 3 S Afr. Cult. orn. dwarf succ. herbs. R: CSJ 4 3(1981)62

Pectis L. Compositae (Tag.-Pect.). c. 85 warm & trop. Am. (Mex. 43), Galápagos. ***P. papposa*** Harv. & A. Gray (SW US, N Mex.) – fls used to flavour meat etc.

Pectocarya DC. ex Meissn. Boraginaceae (B.3.8.4). 15 W Am.

Pedaliaceae R. Br. Magnoliidae – Lamiales. Excl. Martyniaceae, 10/34 trop. & warm OW esp. coasts & arid, esp. Afr. (As. to Aus. only *Josephinia*). Decid. trees shrubs & herbs with indumentum of short-stalked hairs with head of 4 or more mucilage-filled cells & usu. storing stachyose rather than starch. Lvs usu. opp. (upper s.t. in spirals), simple, s.t. rather succ.; stip. 0. Fls bisex., solit. or in dichasia with 1 or 2 'extrafloral nectaries' (abortive fls) in axils of bracts at pedicel base; K 5 small, usu. forming lobed tube, C (5), irreg. s.t. with basal spur, the limb oft. oblique or ± 2-labiate, A 4, oft. with 1 staminode, inserted near C base, anthers with longit. slits, nectary-disk oft. around $\underline{G}$ (2) with term. style, 2-loc. with axile placentas, oft. subdivided by partitions ± reaching centre or 8 1-ovulate locelli (*Josephinia*), with 1–∞ anatropous, unitegmic ovules. Fr. a loculicidal capsule or drupe or nut, oft. armed with horns, hooks or prickles, or winged; seeds with straight embryo in 0 or little oily endosperm. n = 8, 13

Classification & chief genera (Martynioideae long considered distinct on pollen evidence = Martyniaceae; Trapelloideae = Plantaginaceae):

1. **Sesamothamneae** (distinct long & short shoots, pollen in tetrads, fr. with 0 processes): *Sesamothamnus* (only)
2. **Pedalieae** (fls s.t. solit., anther thecae divergent, fr. usu. with spines or wings): *Harpagophytum, Pterodiscus, Rogeria, Uncarina*
3. **Sesameae** (fls solit., anther thecae not divergent; fr. usu. with spines, horns etc.): *Sesamum*

Oilseeds (*Ceratotheca*, but esp. *Sesamum*, sesame); local medic., while animal-disp. armed fr. of *Harpagophytum* & *Uncaria* used as mouse-traps; some cult. orn.

Pedaliodiscus Ihlenf. (~ *Pedalium*). Pedaliaceae (2). 1 Kenya, Tanzania: ***P. macrocarpus*** Ihlenf. R: BDBG 81(1968)147

Pedalium Royen ex L. Pedaliaceae (2). 1 OW trop. (prob. orig. NE Afr.): ***P. murex*** L. – indicator of saline soil, lvs a veg., seeds locally medic. R: FMad. 179(1971)32

Peddiea Harv. Thymelaeaceae (Thym.-Daph.). c. 10 trop. (Tanz. 6, in E arc mts) & S Afr., Madag. (1)

Pedersenia Holub (*Trommsdorffia*, ~ *Iresine*). Amaranthaceae. 11 trop. Am. R: Preslia 70(1998)181. Lianes

Pedicellarum Hotta (~ *Pothos*). Araceae (III 1). 1 Borneo: ***P. paiei*** Hotta – coll. only twice, fls pedicellate. R: Telopea 9(2001)554

Pedicularis Tourn. ex L. Orobanchaceae (Rhin.; Scrophulariaceae s.l.). c. 750 N hemisph. (esp. mts of C & E As. (Bhutan 76, China allegedly 350); Eur. 54), 1 in Andes. Louseworts esp. ***P. sylvatica*** L. (Eur.) – presumed to become lice on sheep contacting them. Hemiparasites with alks. ***P. canadensis*** L. (N Am.) – shoots ed., medic. & animal delouser; ***P. lanata*** Cham. & Schldl. (N Am.) – roots ed.; ***P. palustris*** L. (N temp.) – glycoside toxic to insects, poultice in C18 Hebrides (Scotland) cured humans crippled by 'reddish worms' under skin of knees & ankles

Pedilanthus Necker ex Poit. = *Euphorbia*

Pedilochilus Schltr. Orchidaceae (V 15). 15 Papuasia

Pedilonum (Blume) Blume = *Dendrobium*

Pedinogyne Brand = *Trigonotis*

Pedinopetalum Urb. & H. Wolff. Umbelliferae (III 8). 1 Hispaniola: ***P. domingense*** Urb. & H. Wolff

Pediocactus Britton & Rose. Cactaceae (III 9). 6 W US. R: FNA 4(2003)211. Cult. orn. small cacti

Pediomelum Rydb. (~ *Orbexilum*). Leguminosae (III 20). 21 N Am. R: MNYBG 61(1990)56. Medic. incl. ***P. esculentum*** (Pursh) Rydb. (breadnut, breadroot) – root imp. local food

Pedistylis Wiens (~ *Emelianthe*). Loranthaceae (5 7). 1 S Afr.: ***P. galpinii*** (Sprague) Wiens – 'wood rose' outgrowths prod. at junction with host, sold as curios. R: R. Polhill & D. Wiens, *Mistletoes of Africa* (1998)98

Peekelia Harms = *Cajanus*

Peekeliopanax Harms = *Gastonia*

peel, candied *Citrus medica* & other spp.

peelu extract *Salvadora persica*

peepul *Ficus religiosa*

Peersia L. Bolus (~ *Rhinephyllum*). Aizoaceae (V). 3 S Afr. R: H. E. K. Hartmann, *Ill. Handb. Succ. Pls*, Aiz. F–Z(2002)201

Pegaeophyton Hayek & Hand.-Mazz. Cruciferae (28). 7 C As. & Himal. to W China. R: EJB 57(2000)158

Peganaceae Tieghem ex Takht. = Nitrariaceae

Peganum L. Nitrariaceae (Peganaceae; Zygophyllaceae s.l.). 5–6 Medit. (Eur. 1) to Mongolia, S N Am. Alks. ***P. harmala*** L. (harmal, Medit. to As., noxious weed S Aus.) – alks incl. harmine (used as 'truth drug' by Nazis; also found in *Banisteriopsis*) & harmalol so avoided by all stock save camels, locally medic. esp. in eye disease, rheumatism, Parkinson's disease etc., burnt as intoxicant in C As. since (?) 5th millenium BC, seeds hallucinogenic (images reflected in local art-styles (?) incl. Persian carpets & concept of 'flying carpets' cf. 'flying broomsticks' in Eur. perhaps due to henbane) & sexually stimulatory (carbolines), source of an oil & of a dye (Turkey red) used for dyeing carpets & the hats known as tarbooshes

Pegia Colebr. Anacardiaceae (II). 2 E As., Mal.

Pegolettia Cass. Compositae (Inul.-Pluch.). 9 Afr., Arabia & Middle E with ***P. senegalensis*** Cass. to Ind. (? Java). R: Cladistics 2(1986)158

pegwood *Cornus sanguinea*

Pehria Sprague. Lythraceae. 1 Colombia, Venez.: ***P. compacta*** (Rusby) Sprague

pei-mu *Fritillaria cirrhosa*, *F. roylei*

Peixotoa A. Juss. Malpighiaceae. 29 Braz. R: CUMH 15(1982)1. Some apomixis

pejibaye, pejivalle *Bactris gasipaes*

Pelagatia O. Schulz = *Weberbauera*

Pelagodendron Seemann (~ *Aidia*). Rubiaceae (II 2). 3 Pacific

Pelagodoxa Becc. Palmae (V 13). 1 Melanesia (anthropogenic habitats), Marquesas: ***P. henryana*** Becc. – large bifid lvs, now v. rare. R: Principes 15(1971)45

Pelargonium L'Hérit. Geraniaceae. 280 trop. (few) & S (most) Afr. (Cape 148, 79 endemic – R: Strelitzia 9(2000)517) with 2 in E Medit. to Iraq (allied to S Afr. spp.), & 1 in each of S Arabia (Socotra), St Helena, Tristan da Cunha & S Ind., Aus. 6 (5 endemic), NZ 1 (***P. inodorum*** Willd. also Aus.). R: D. Clifford (1970) *Ps*; Phytotaxa 159(2014)31; J.J. van der Walt & al. (1977–88) *Ps of southern Africa* 1–3; (sect. *P.*) Bothalia 15(1985)345. 'Geraniums' of greenhouses but differing from *Geranium* in irreg. fls & 5–7 of A 10 fertile, rest just filaments; shrubs & herbs (some ann. e.g. ***P. apetalum*** P. Taylor, C Afr.), s.t. with tuberous roots, or succ. when oft. with stipular spines & reduced lvs; dark spots on C attractive to poll. bee-flies. V. imp. house-pls & bedding-pls (fav. Victorian ones '**General Tom Thumb**', '**Old Frogmore Scarlet**' & *P. peltatum* (from 1844); US prod. worth $7M+ by 2005) esp. now *P.* × *hortorum*, *P.* × *domesticum*, *P. peltatum* & spp. with scented lvs due to essential oils thought to be deterrent to animal-grazers, though some used as basis for scent & soaps (geranium oil, see Arnoldia 34(1974)104 for spp. & scents), the oils reminiscent of lemon, peppermint, pennyroyal, nutmeg, strawberry, mint, camphor, apple, ginger, rue etc., those of ***P.* × *asperum*** Ehrb. ex Willd. (*'P. graveolens'*, rose-oil geranium, *P. g.* L'Hérit. × *P. radens* H. Moore (*P. radula*), both S Afr.), ***P. odoratissimum*** (L.) L'Hérit. & ***P. quercifolium*** (L.) L'Hérit. (all S Afr.) being coll. comm. (mawah oil) in France, Algeria & Réunion (where '**Rosé**' (*P. capitatum* (L.) L'Hérit. (S Afr.) × *P. radens*) much grown). ***P. crispum*** (Bergius) L'Hérit. (SW Cape) – lemon-scented; ***P.* × *domesticum*** L. Bailey (regal p. or geranium, group of cvs derived from *P. cucullatum* (L.) L'Herit. (*P. angulosum*), *'P. grandiflorum'* (S Afr. shrubs) & other spp.) crossed with *P. crispum* = 'angels'; ***P. endlicherianum*** Fenzl (Turkey) – hardy in GB; ***P.* × *hortorum*** L. Bailey (zonal p. or g., complex hybrid cultigen involving S Afr. *P. inquinans* (L.) L'Hérit. & *P. zonale* (L.) L'Hérit., their hybrid, *P.* × *hybridum* (L.) L'Hérit. (poss. correct name for *P.* × *hortorum*) back-crossed with parents) – familiar 'geranium' with many ('zonal') cvs incl. varieg. & even haploid ('**Kleine Liebling**' with n = 9, roots diploid) ones, much used for bedding e.g. scarlet-fld '**Paul Crampel**' (raised by Lemoine); ***P. monoliforme*** E. Meyer ex Harv. (S Afr.) – succ. still sprouting 7 months after being pressed; ***P. peltatum*** (L.) L'Hérit. (ivy-leaved p. or g., S Afr.) – familiar hanging basket geranium, modern cvs hybrids involving *P. zonale* or *P.* × *hortorum*); ***P. pinnatum*** (L.) L'Hérit. (S Afr.) – taproot-like tuber to 25 cm, lvs pinnate; ***P. triste*** (L.) Hérit. (Cape) – fleshy r., first sp. cult. Eur. (early C17), 101 more cult. in Eur. by 1789

Pelatantheria Ridl. (~ *Cleisostoma*). Orchidaceae (V 16c). 8 Ind. to Taiwan & W Mal. R: OB 95(1988)115

Pelea A. Gray = *Melicope*

Pelecostemon Leonard. Acanthaceae (III 2c). 1 Colombia: ***P. trianae*** Leonard. R: CUSNH 31(1958)648

Pelecyphora C. Ehrenb. Cactaceae (III 9). 2 NE Mex. R: EGF 3(1989)279

Pelexia Poit. ex Lindl. Orchidaceae (IV 2h). Incl. *Pachygenium*, excl. *Veyretia*, c. 75 warm & trop. Am. R: HBML 28(1982)342

Peliosanthes Andrews. Asparagaceae (Convallariaceae). 3 Indomal. ***P. teta*** Andrews – blue fr. attractive to (?)pheasant dispersers

Peliostomum E. Meyer ex Benth. Scrophulariaceae (Apt.). 7 trop. & S Afr.

Pellacalyx Korth. Rhizophoraceae (II). 7–8 Indomal.

Pellaea Link. Pteridaceae (IV). c. 50 trop. & warm (Eur. 1) to Canada, esp. SW N Am. R (sect. *P.*): AMBG 44(1957)125. Heterogeneous. Ferns of rocks (cliff brake, US) – local medic. (blood problems), cult. orn.

Pellegrinia Sleumer. Ericaceae (VIII 5). 5 Andes

Pellegriniodendron Léonard = *Gilbertiodendron* (but see BJBB 25(1955)203)

Pelletiera A. St-Hil. = *Anagallis*

Pellia Raddi. Pelliaceae (Hepaticae). 5 N temp. Thalloid liverworts. ***P. endiviifolia*** (Dicks.) Dumort. calcareous vicariant of ***P. epiphylla*** (L.) Corda on acidic substrates

Pelliceria Planch. & Triana = *Pelliciera*

Pelliciera Planch. & Triana ex Benth. Tetrameristaceae (Pellicieraceae). 1 Costa Rica to Colombia (Pac. shores) & Carib. coast (rare), form. circum-Carib. mangrove: ***P. rhizophorae*** Planch. & Triana – contraction of range poss. due to salinity changes (intolerant of more than 3.7%). R: JAA 32(1951)260

Pellicieraceae Beauvis. ex Bullock = Tetrameristaceae

Pellionia Gaudich. = *Procris*

pellitory *Anacyclus pyrethrum*; **p. of the wall** *Parietaria judaica*

Pelma Finet = *Bulbophyllum*

Peloria L. = *Linaria*

Pelozia Rose = *Lopezia*

Peltaea (C. Presl) Standl. = *Hibiscus* (but see Kurtziana 2(1965)161)

Peltandra Raf. Araceae (VII 20). 2 N Am. Allied to *Typhonodorum*. R: FNA 22(2000)135. ***P. virginica*** (L.) Schott – starchy rhiz. form. imp. Native Amer. food (roasted)

Peltanthera Benth. (? Gesneriaceae). 1 Peru: ***P. floribunda*** Benth. – form. in Loganiaceae, Scrophulariaceae

Peltapteris Link = *Elaphoglossum*

Peltaria Jacq. Cruciferae (47). 3 E Medit. (Eur. 2) to Iran & C As. Some cult. orn. ***P. emarginata*** (Boiss.) Hausskn. (Greece, serpentines) has nickel up to 1% dry weight of lvs

Peltariopsis (Boiss.) N. Busch. Cruciferae (47). 2 Caucasus, N Iran. R: NRBGE 36(1978)32

Peltastes Woodson. Apocynaceae (II e). 10 trop. Am. R: Candollea 60(2005)293

Pelticalyx Griff. Older name for *Dasymaschalon*

Peltiera Labat & Dupuy = *Ormocarpopsis*

Peltiphyllum (Engl.) Engl. = *Darmera*

Peltoboykinia (Engl.) H. Hara (~ *Boykinia*). Saxifragaceae (I 3). 1 Jap.: ***P. tellimoides*** (Maxim.) H. Hara – cult. orn. coarse herb with peltate lvs. R: BotJLS 90(1985)31

Peltobractea Rusby = *Peltaea*

Peltocalathos Tamura. Ranunculaceae (II 3). 1 S Afr.: ***P. baurii*** (MacOwan) Tamura. R: Bothalia 43(2013)180

Peltodon Pohl = *Hyptis* (but see FRB 85(1936)195)

Peltogyne Vogel. Leguminosae (I 2). 23 trop. Am. esp. Amazonia. R: AA 6(1) suppl. (1976)1. Some water-disp. fr. in Amaz. Timber (amaranth(e)) for panelling going black on contact with water (purple heart, purplewood) esp. ***P. paniculata*** Benth. ***P. pubescens*** Benth. (*P. paniculata* subsp. *pubescens*) – crown to 30 m+ diam.

Peltophoropsis Chiov. = *Parkinsonia*

Peltophorum (Vogel) Benth. Leguminosae (I 4). Excl. *Heteroflorum* 5 trop. Scarrone's Model. Timber (brasiletto) & shade trees; ***P. dubium*** Vog. (Braz.) – wood, medic., seeds 43.8% protein, so poss. crop pl. Some cult. orn. esp. ***P. pterocarpum*** (DC.) K. Heyne (*P. ferrugineum*, yellow flamboyant, Indomal. seashores) – bark medic. & a constituent of soga, a yellow-brown dye used in batik

Peltophyllum Gardner (*Hexuris, Soredium*; ~ *Triuris*). Triuridaceae. 2 trop. S Am. So-named because orig. specimen had leaf from a Menispermacea!

Peltopus (Schltr.) Szlach. & Marg. = *Bulbophyllum*

Peltostigma Walp. Rutaceae (I 5). 2 C Am., WI

Pelucha S. Watson. Compositae (Hele.-Psat.). 1 is. Sea of Cortes, Mex.: ***P. trifida*** S. Watson

peluskins *Pisum sativum* 'var. *arvense*'

Pembertonia P. Short (~ *Brachycome*). Compositae (Ast.-Bra.). 1 W Aus.: ***P. latisquamea*** (F. Muell.) P. Short. R: JABG 28(2014)169

pe-mou oil *Fokienia hodginsii*

Pemphis Forst. & Forst. f. Lythraceae. 1 E Afr. & trop. As. to Ryukyus & Marshall Is.: ***P. acidula*** Forst. & Forst. f. – on coasts esp. eroding beaches, Attims's Model, allied to *Punica*, distylous having lost 'mid-type', wood for knife handles (Cocos Is.). R: FM I,22(2016)50. See also *Koehneria*

Penaea L. Crypteroniaceae (Penaeaceae). 4 SW Cape. R: OB 29(1971), SAJB 55(1989)410. Endosperm tetraploid. Sources of gums tasting like liquorice, locally medic.

Penaeaceae Sweet ex Guillemin = Crypteroniaceae

Penang lawyer *Licuala* spp.

penarhan Myristicaceae spp.

pencil cedar *Juniperus virginiana*; **p. pine** (Aus.) *Polyscias murrayi, Cupressus sempervirens*

penda *Xanthostemon oppositifolius*

Pendulorchis Z.J. Liu & al. = *Holcoglossum*

Penelopeia Urb. Cucurbitaceae (XV). 2 Hispaniola. R: FR 17(1921)8

Penianthus Miers. Menispermaceae (I). 4 W & C Afr. R: BJBB 53(1983)41. Corner's Model

Peniocereus (A. Berger) Britton & Rose. Cactaceae (III 1). Excl. *Nyctocereus*, 19 SW N Am. to C Am. Cult. orn. day- or night-flowering cacti (aroma of moth-poll. ***P. greggii*** (Engelm.) Britton & Rose (queen of the night, Mex.) detectable by humans at 30 m – tubers to 20 kg, fr. ed.

Peniophyllum Pennell = *Oenothera*

Penkimia Phukan & Odyuo = *Holcoglossum*

Pennantia Forst. & Forst. f. Pennantiaceae (Icacinaceae s.l.). 2 NE Aus., Norfolk Is., NZ. R: JRSNZ 32(2002)669. ***P. corymbosa*** Forst. & Forst. f. (NZ) – hard timber used for tool-handles, cabinet-work & form. (Maoris) for fire-making by friction; ***P. endlicheri*** Reissek (incl. *P. baylisiana* (Norfolk Is., Three Kings Is. – NZ) – rarely prod. pollen, effectively only female

Pennantiaceae J. Agardh (~ Icacinaceae). Magnoliidae – Apiales. 1/3 NE Aus, Norfolk Is., NZ. Trees & shrubs. Fls unisexual. K minute, free; C connate, valvate; disk 0; G (3). Fr. a drupe. n = 25
Genus: *Pennantia*

Pennellia Nieuw. Cruciferae (29). 8 SW US, Mex. (esp.), Guatemala, Bolivia, N Arg. R: SB 22(2007)151

Pennellianthus Crosswh. (*Leiostemon*; ~ *Penstemon*). Plantaginaceae (Ant.- Chel.). 1 N Jap. & Kamchatka: ***P. frutescens*** (Lamb.) Crossw. (*L. f.*). R: AMN 83(1970)362

Pennilabium J.J. Sm. Orchidaceae (V 16c). 15 Thailand to Philippines

Pennisetum Pers. = *Cenchrus* (but see JAA suppl. 1(1991)205)

Pennsylvanian Dutch thyme *Thymus pulegioides*

penny cress *Thlaspi arvense*; **p. flower** *Lunaria annua*; **p. pies** *Umbilicus rupestris*

pennyroyal *Mentha pulegium*; **American p.** *Hedeoma pulegioides*

pennywort *Hydrocotyle* spp., *Cymbalaria muralis*, *Sibthorpia europaea*, *Umbilicus rupestris*; **floating p.** *H. ranunculoides*

Penstemon Schmidel. Plantaginaceae (Ant.-Chel.; Scrophulariaceae s.l.). 250 N Am. (esp. W). R: CBLUP 3(1908)98; R. Nold (1999) *Ps.* Biggest N Am. endemic genus (cf. *Eriogonum*). Shrubs & perenn. herbs with posterior staminode bent down to lower side of C, naked or bearded. Sucrose-rich or -dominant nectars in primarily hummingbird-poll. spp., hexose-rich or -dominant ones in insect-poll. spp. even in closely related spp. pairs. Local medic. esp. ***P. fruticosus*** (Pursh) Greene (NW Am.); many cult. orn. (beard tongue) esp. in US with over 130 named cvs, some treated as annuals incl. ***P. barbatus*** (Cav.) Roth (SW N Am.) – fls pink, catalpol (an iridoid) sequestered by *Neoterpes graefiaria* moths, ***P. spectabilis*** Thurber ex A. Gray (SW N Am.) – fls blue & purple, poll. pseudomasarid wasps & ***P. virgatus*** A. Gray (SW US) – fls violet, catalpol sequestered by *Poladryas arachne* butterflies & *Meris alticola* moths. See also *Keckiella*, *Pennellianthus*

Pentabothra Hook.f. = *Vincetoxicum*

Pentabrachion Muell.-Arg. (~ *Microdesmis*). Phyllanthaceae (Brid.-Pseud.). 1 Cameroun, Gabon: ***P. reticulatum*** Muell.-Arg. R: A. R.-Sm., *Gen. Euph.* (2001)15

Pentacalia Cass. (~ *Senecio*). Compositae (Sen.-Sen.). c. 200 C & S Am. R: Phytol. 40(1978)37, 49(1981)241. Lianes & epiphytes

Pentacarpaea Hiern = *Pentanisia*

Pentace Hassk. Malvaceae (Brown.; Tiliaceae). 25 Myanmar to W Mal. R: PLPH 87(1964). Some timbers esp. ***P. burmanica*** Kurz (thitka, Burmese mahogany) – light timber

Pentaceras Hook.f. Rutaceae (I 2). 1 E Aus.: ***P. australe*** (F. Muell.) Benth. – some alks with anti-tumour activity. R: FA 26(2013)81

Pentachaeta Nutt. (~ *Chaetopappa*). Compositae (Ast.-Pent.). 6 SW N Am. R: UCPB 65(1973)32. ***P. aurea*** Nutt. – pollen used as cosmetic

Pentachlaena Perrier. Sarcolaenaceae. 3 Madag. R: Adansonia III,22(2000)16

Pentachondra R. Br. Ericaceae (VII 7; Epacridaceae). 4 S Aus., 1 ext. to NZ (***P. pumila*** (Forst. & Forst. f.) R. Br. – cult. orn. dwarf shrub)

Pentaclethra Benth. Leguminosae (II 1). 3 trop. Am. (1) & Afr. (2). Troll's Model. Useful timber & locally medic., oilseeds: ***P. macroloba*** (Willd.) Kuntze (*P. filamentosa*, pracaxi fat, S Am.) – explosive fr. ejecting toxic (alks & free amino-acids), water-disp. seeds to 10 m, wood rotting after only 10–20 yrs, & v. similar ***P. macrophylla*** Benth. (owala oil, Afr.) – multipurpose tree for agroforestry

Pentacme A. DC. = *Shorea*

Pentacoelium Sieb. & Zucc. (~ *Myoporum*). Scrophulariaceae (Myoporaceae). 1 NE As.: ***P. bontioides*** Sieb. & Zucc. R: FOC 19(2011)492

Pentacrostigma K. Afzel. = *Ipomoea*

Pentactina Nakai (~ *Spiraea*). Rosaceae (Ros.-Spir.). 1 ***P. rupicola*** Nakai (Korea) – allied to *Petrophytum*

Pentacyphus Schltr. Apocynaceae (V c; Asclepiadaceae III 1). Incl. *Tetraphysa* 5 N S Am.

Pentadenia (Planch.) Hanst. = *Columnea*

Pentadesma Sabine. Guttiferae (3.). 5 trop. Afr. R: BJBB 41(1971)430. ***P. butyracea*** Sabine (tallow tree, W Afr.) – Rauh's Model, seeds source of Sierra Leone, Kanga or lamy butter, used for cooking, soap, margarine & candles

Pentadiplandra Baill. Pentadiplandraceae (Capparaceae s.l.). 1 Nigeria to Angola: ***P. brazzeana*** Baill. – ed. fr., with thermostable (cf. thaumatein) brazzein 20 000 times sweeter than sucrose, fish-poison, aphrodisiac

Pentadiplandraceae Hutch. & Dalz. (~ Capparaceae). Magnoliidae – Brassicales. 1/1 W Afr. Polygamous shrubby glabrous cli. Lvs simple, entire, in spirals; stip. minute. Infls short axillary subcorymbose racemes. Fls bracteate with long pedicels, reg. K 5, valvate; C 5; androgynophore short, A 9–13, $\underline{G}$ (3–5) opp. K with axile placentation, each loc. with c. 10 ovules in 2 ranks. Fr. a berry with 1 seed per loc. Seeds reniform, pubescent; embryo strongly curved, endosperm 0

Genus: *Pentadiplandra*

Pentadynamis R. Br. = *Crotalaria*

Pentaglottis Tausch. Boraginaceae (B.2.1.1). 1 SW Eur.: ***P. sempervirens*** (L.) L. Bailey (alkanet) – cult. orn. herb with blue fls

Pentagonanthus Bullock = *Raphionacme*

Pentagonia Benth. Rubiaceae (Cond.). 34 trop. Am. Lvs v. large, oft. bright red; local medic. ***P. gigantifolia*** Ducke (Amaz. Peru) – unbranched pachycaul (Corner's Model)

Pentagramma Yatsk. & al. (~ *Pityrogramma*). Pteridaceae (IV). 6 N Am. R: SB 40(2015)639

Pentalepis F. Muell. (~ *Chrysogonum*). Compositae (Helia.- Ecl.). 6 N Aus. R: FA 37(2015)504

Pentalinon Voigt (*Urechites*). Apocynaceae (II e). 2 Florida & C Am. to WI. R: JAA 70(1989)383

Pentaloncha Hook.f. Rubiaceae (Urophyll.). 2 trop. W Afr.

Pentameris P. Beauv. Gramineae (Danth.). Incl. *Pentaschistis*, *Prionanthium*, 84 Afr., Madag.

Pentamerista Maguire. Tetrameristaceae. 1 N S Am.: ***P. neotropica*** Maguire. R: MNYBG 23(1972)187

Pentanema Cass. (~ *Inula*). Compositae (Inul.-Inul.). 20 Turkey & C As. to Sri Lanka, natur. elsewhere

Pentanisia Harv. Rubiaceae (III 8). 15 trop. Afr. ***P. ouranogyne*** S. Moore (NE & E Afr.) – arrow-poison in N Uganda

Pentanopsis Rendle. Rubiaceae (IV 15). 2 NE trop. Afr. R: PSE 247(2004)237

Pentanura Blume (~ *Phyllanthera*). Apocynaceae (III; Periplocaceae). 2 Myanmar, Sumatra

Pentapanax Seemann = *Aralia*

Pentapeltis (Endl.) Bunge = *Xanthosia*

Pentapera Klotzsch = *Erica*

Pentapetes L. Malvaceae (Dom.; Sterculiaceae). 1 Indomal.: ***P. phoenicea*** L.

Pentaphalangium Warb. = *Garcinia*

Pentaphorus D. Don (~ *Gochnatia*). Compositae (Gochn.). 5 trop. Am.

Pentaphragma Wall. ex G. Don. Pentaphragmataceae. 30 Mal. ***P. horsfieldii*** (Miq.) Airy Shaw – produces 2 lvs a year, each lasting 33 months; lvs used like spinach

Pentaphragmataceae J. Agardh. Magnoliidae – Asterales. 1/30 Mal. Perenn. ± succ. herbs oft. with branched hairs; alks & latex-system 0. Lvs simple, in spirals, usu. serrate, asymmetric at base; stip. 0. Fls usu. bisex., in dense sympodial ± axillary helicoid cymes with usu. conspic. bracts; K 5 unequal, imbr., persistent, C ((4)5) or rarely (4)5, valvate, usu. fleshy, A alt. with C attached to tube or when C free to top of G, anthers extrorse with longit. slits, $\overline{G}$ (2 or 3), 2- or 3-loc. separated from hypanthium by nectariferous pits, style without collecting-hairs, axile placentas with ∞ anatropous, unitegmic ovules. Fr. a berry with apical persistent P; seeds minute with copious starchy endosperm. n = 54–56

Genus: *Pentaphragma*

Differing from Campanulaceae in absence of pollen-presenting mechanism & poss. diff. pollen-types

Pentaphylacaceae Engl. (~ Theaceae). Magnoliidae – Ericales. Incl. Ternstroemiaceae, 12/340 trop. & warm. Small trees, s.t. accum. aluminium. Lvs simple, serrate, entire, in spirals becoming distich.; stip. 0. Fls bisex., small, 5-merous with 2 persistent bracteoles adpressed to K in racemes oft. with sterile leafy tips; K free, imbr., persistent, C free, imbr., A alt. with C, filaments expanded esp. at middle, anthers with term. pores with valves, ($\underline{G}$), 5-loc. with stout persistent style & 2 pendulous, anatropous to ± campylotropous bitegmic ovules on axile placentas. Fr. usu. a berry or ± dehiscent woody capsule with persistent central axis (like Theaceae); seeds ± winged with horseshoe-shaped embryo in scanty endosperm

Classification & genera:

1. **Pentaphylaceae**: *Pentaphylax*
2. **Ternstroemieae**: *Ternstroemia*, Anneslea
3. **Frezierieae**: *Adinandra, Archboldiodendron, Balthasaria, Cleyera, Eurya, Euryodendron, Freziera, Symplocarpon, Visnea*

Sladeniaceae poss. here. Some late Cretaceous fossils referred here. Timber & cult. orn. (*Anneslea, Cleyera, Eurya*)

Pentaphylax Gardner & Champ. Pentaphylacaceae (1). 1 S China to Sumatra: ***P. euryoides*** Gardner & Champ. R: FM I,5(1955)121

Pentaplaris L.O. Williams & Standl. Malvaceae (Bomb./Malv.; Tiliaceae). 3 Costa Rica, Ecuador, Bolivia, Peru. R: Britt. 51(1999)134

Pentapleura Hand-Mazz. Labiatae (VII 2b). 1 Kurdistan: ***P. subulifera*** Hand.-Mazz.

Pentapogon R. Br. Gramineae (XVI 5). 1 SE Aus.: ***P. quadrifidus*** (Labill.) Baill. R: FA 44A(2009)233

Pentaptilon E. Pritzel. Goodeniaceae. 1 SW Aus.: ***P. careyi*** (F. Muell.) E. Pritzel. R: FA 35(1992)297

Pentarhaphia Lindl. = *Gesneria*

Pentarhizidium Hayata = *Matteuccia*

Pentarhopalopilia (Engl.) Hiepko (~ *Rhopalopilia*). Opiliaceae. 2 W & S C Afr. & E Afr. coast, 2 Madag. R: BJ 108(1987)280

Pentarrhaphis Kunth = *Bouteloua*

Pentarrhinum E. Meyer = *Cynanchum* (but see KB 47(1992)480)

Pentas Benth. Rubiaceae (IV 1). 34 Afr. to Arabia & Madag. R: BJBB 23(1953)237. Shrubs & herbs, some cult. orn. esp. ***P. lanceolata*** (Forssk.) Deflers (E Afr. to S Arabia)

Pentasacme Wall. ex Wight & Arn. Apocynaceae (V b; Asclepiadaceae III 5). 4 Indomal. R: Blumea 36(1991)109. Rheophytes. Lvs smoked like marijuana in Mal.

Pentaschistis (Nees) Spach = *Pentameris* (but see CBH 12(1990)1)

Pentascyphus Radlk. Sapindaceae. 1 Guyana: ***P. thyrsiflorus*** Radlk.

Pentaspadon Hook.f. Anacardiaceae (I?). 6 SE As. to Papuasia. Fagerlind's Model. Utility timbers. ***P. motleyi*** Hook.f. – stem a source of oil used in treating skin-disease, esp. ringworm

Pentaspatella Gleason = *Sauvagesia*

Pentastelma Tsiang & P.T. Li. Apocynaceae (V c; Asclepiadaceae III 1). 1 Hainan: ***P. auritum*** Tsiang & P.T. Li. R: FOC 16(1995)253

Pentastemona Steenis. Stemonaceae (Pentastemonaceae). 2 Sumatra. R: Blumea 36(1991) 245. Only 5-mery in monocots. ***P. sumatrana*** Steenis – vivipary (bulbils) in apical part of infl.

Pentastemonaceae Duyfjes = Stemonaceae

Pentastemonodiscus Rech.f. Caryophyllaceae (II 5). 1 Afghanistan: ***P. monochlamydeus*** Rech.f.

Pentasticha Turcz. = *Fuirena*

Pentatherum Náb. = *Agrostis*

Pentathymelaea Lecomte = *Wikstroemia*

Pentatrichia Klatt. Compositae (Gnap.). 4 S Afr. R: OB 104(1991)43

Pentatropis R. Br. ex Wight & Arn. Apocynaceae (V c; Asclepiadaceae III 1). 3 Afr., Madag.

Penthoraceae Rydb. ex Britton (~ Saxifragaceae). Magnoliidae-Saxifragales. 1/2 E As., E N Am. R: FNA 8(2009)230. Perenn. herbs with rhiz. Lvs simple, in spirals; stip. 0. Infls scorpioid or corymbiform cymes. Fls reg., small, bisex.; K 5(–8), unequal, connate at base, reflexed in fr.; C 1–8 on hypanthium rim, or 0; A 10(–16) in 2 whorls on edge of hypanthium, anthers with longit. slits; G (4)5(–8), partly inf. at anthesis but sup. at maturity, each carpel with style & 1 marginal, pendulous placenta with 30–100 anatr., bitegmic ovules distally. In fr. carpels with circumscissile dehiscence distal to syncarpous base; seeds ellipsoid to obovoid with large straight embryo & scant endosperm

Genus: *Penthorum*

Sister to Haloragaceae

Penthorum Gronov. ex L. Penthoraceae (Saxifragaceae s.l.). 1 E As., 1 E N Am. (***P. sedoides*** L. – medic.)

Pentisia (Lindl.) Szlach. = *Cyanicula*

Pentodon Hochst. Rubiaceae (IV). 2 warm & trop. Am., Afr., Arabia, Seychelles

Pentopetia Decne. Apocynaceae (III; Periplocaceae). Incl. *Gonocrypta*, 21 Madag. R: Candollea 54(1999)276. Some rubber sources

Pentopetiopsis Costantin & Gallaud = praec.

Pentossaea W. Judd = *Ossaea*

Pentstemon auctt. = *Penstemon*

Pentstemonacanthus Nees. Acanthaceae. 1 Braz.: ***P. modestus*** Nees

Pentzia Thunb. Compositae (Anth.-Pen.). 23 Morocco & Algeria (2), S Afr. R: BBMNB 23(1993)151. ***P. incana*** (Thunb.) Kuntze (*P. virgata*, (African) sheep bush, S Afr.) – shipped to Adelaide, Aus. by Kew Gardens 1864 as sheep fodder, also to improve mutton flavour. See also *Oncosiphon*

penwiper plant *Notothlaspi rosulatum*

peony *Paeonia* spp.; **tree p.** *P.* sect. *Moutan* esp. *P. suffruticosa*

Peperomia Ruíz & Pavón. Piperaceae (2; Peperomiaceae). c. 1600 trop. & warm esp. Am. (As. c. 100, Aus. c. 20, Afr. c. 20, Madag. c. 40) in 14 subgg. R: T 64 (2015)427. Mostly small ± succ. herbs (Tomlinson's, Leeuwenberg's, Petit's & Roux's Models), many epiphytic with water-storing tissue in upper leaf-epidermis (facultative crassulacean acid metabolism dependent on developmental & environmental conditions); lvs (peltate evolved several times) in spirals to decussate or verticillate; endosperm octoploid. Many cult. orn. (100+ spp. in cult.) house-pls tolerant of little light (rock balsam), esp. ***P. argyreia*** (Miq.) C.J. Morren (trop. Am.) – lvs silver-striped, ***P. caperata*** Yuncker (Braz.) – lvs strongly plicate-bullate (several cvs) & ***P. velutina*** Linden (Ecuador) – lvs silver & red. ***P. berteroana*** Miq. subsp. ***berteroana*** on Juan Fernandez, subsp. ***tristanensis*** (Christoph.) Valdebenito on Tristan da Cunha 5000km away; ***P. blanda*** (Jacq.) Kunth (*P. leptostachya*, trop.) – locally medic., cult. orn.; ***P. foliiflora*** Ruíz & Pavón & ***P. haenkeana*** Opiz (trop. Am.) – epiphyllous infls; ***P. galioides*** Kunth (Andes) – extracts active against Chagas disease; ***P. pellucida*** (L.) Kunth (trop., orig. Am.) – salad veg. Philippines, oil antagonistic to growth of *Helminthosporium oryzae* (brown spot) in rice, bacteria (e.g. *B. subtilis, E. coli*); ***P. peltigera*** C. DC. (Ecuador) – medic. (heart disease)

Peperomiaceae A.C. Sm. = Piperaceae

Pepinia Brongn. ex André = *Pitcairnia*

pepino *Solanum muricatum*

pepita *Cucurbita pepo*

Peplidium Del. Phrymaceae (Mim.; Scrophulariaceae s.l.). 14 Aus., 1 ext. to OW trop. R: Phytoneuron 2012–39:22

Peplis L. = *Lythrum*

Peplonia Decne. Apocynaceae (V c; Asclepiadaceae III 1). Incl. *Macroditassa*, 8 Braz. R: KB 59(2004)533

Peponidium Baill. ex Arènes = *Pyrostria*

Peponium Engl. Cucurbitaceae (XIV). 20 Afr., Madag.

Peponopsis Naud. Cucurbitaceae (XV). 1 Mex.: ***P. adhaerans*** Naud. – v. rare

pepper *Piper nigrum*; **African p.** *Xylopia aethiopica*; **Ashanti** or **Benin p.** *P. guineense*; **bell p.** *Capsicum annuum* Grossum Group; **p.berry** *Drimys* spp.; **betel p.** *P. betle*; **bird p.** *C. a.* var. *glabriusculum*; **black p.** *P. nigrum*; **bush p.** *Clethra alnifolia*; **cayenne p.** *Capsicum annuum* Longum Group; **cherry p.** *C. a.* Cerasiforme Group; **chilli p.** *C. a.* Longum Group; **Chinese p.** *Zanthoxylum simulans*; **cone p.** *C. a.* Conoides Group; **p.corns, pink** or **red** *Schinus terebinthifolius*; **Ethiopian p.** *X. aethiopica*; **green p.** *C. a.* Grossum Group; **Guinea p.** *Aframomum melegueta, X. aethiopica*; **Indian long p.** *P. longum*; **Jamaican p.** *Pimenta dioica*; **Japanese p.** *Z. piperitum*; **Java p.** *Piper cubeba*; **Kawa p.** *P. methysticum*; **long p.** *P. longum*; **Madagascar p.** *P. nigrum*; **malagueta** or **melegueta p.** *A. melegueta*; **negro p.** *X. aethiopica*; **red** or **sweet p.** *C. a.* Grossum Group; **Saturday-night p.** *Euphorbia helioscopia*; **Sichuan p.** *Z. simulans*; **p. tree** *S. molle, Kirkia wilmsii, P. excelsum*; **wall p.** *Sedum acre*; **water p.** *Persicaria hydropiper*; **W African black p.** *Piper clusii*; **white p.** *P. nigrum*

peppercress *Lepidium* spp.

pepperidge *Nyssa sylvatica*

peppermint *Mentha* × *piperita*; name used for some *Eucalyptus* spp. e.g. *E. amygdalina* (**black p.**), *E. piperita* (**Sydney p.**); **Japanese p.** *M. canadensis*

pepperwood *Umbellularia californica*

pepperwort *Lepidium* spp.

pequí, pequiá *Caryocar brasiliense, C. villosum*

Pera Mutis. Peraceae (Per.; Euphorbiaceae (Acal.-Per.)). 30–35 trop. Am.

Peracarpa Hook.f. & Thomson. Campanulaceae (I 8). 1 Indomal.: ***P. carnosa*** (Wall.) Hook.f. & Thomson. R: BBAS 38(1997)53

Peraceae Klotzsch (~ Euphorbiaceae s.l.). 5/140 trop. Usu. dioec. trees & shrubs. Lvs in spirals, vernation involute, margins entire; stips usu. small or 0; P (0 in *Pera*, which has coloured bracts) of K 4, 5 (–8 in males) & C 5 persistent; A 5–15 usu. basally connate, stigma bilobed (trumpet-shaped – *Pera*); G (2)3(4)-loc., each loc. with 1 anatr. bitegmic ovule. Fr. septa membranous, dehisc. septicidally & s.t. loculicidally; seeds black, carunculate (arillate), very shiny with usu. copious endosperm. n = 18
Genera: *Chaetocarpus, Clutia, Pera, Pogonophora, Trigonopleura*

Perakanthus F. Robyns. Rubiaceae (III 2). 1 Malay Peninsula: ***P. velutinus*** (Ridl.) Ridl. R: Sandakania 16(2005)42

Perama Aubl. Rubiaceae (inc. sed.). 9 trop. Am. R: Britt. 29(1977)191

Peranema D. Don = *Dryopteris*

Peranemataceae Ching = Dryopteridaceae

Peraphyllum Nutt. = *Amelanchier*

Peratanthe Urb. = *Nertera*

Peraxilla Tieghem. Loranthaceae (3). 2 NZ. R: AusJB 14(1966)429. Usu. on *Nothofagus* spp.; explosively bird-poll. fls

Perdicium L. Compositae (Mut.-Mut.). 2 SW Cape. R: NJB 5(1986)543. Herbs

Perebea Aubl. Moraceae (III). 9 trop. Am. R: FN 7(1972)60, 83(2001)253. Roux's Model; copious brilliant yellow latex

Peregrina W.R. Anderson (~ *Janusia*). Malpighiaceae. 1 Braz. & Paraguay: ***P. linearifolia*** (A. St.-Hil.) W.R. Anderson – geoxylic suffrutex? R: SB 10(1985)303

Pereilema J. Presl = *Muhlenbergia*

pereira, yellow *Aristolochia glaucescens*

Perella (Tieghem) Tieghem = *Peraxilla*

Perenideboles Goyena. Acanthaceae (III). 1 Nicaragua: ***P. ciliatum*** Goyena

Pereskia Mill. Cactaceae (I). 17 trop. Am. (disjunct: Carib. & SE trop. S Am.). R: MNYBG 41(1986)1. Heterogeneous? Leafy trees, shrubs or lianes (Leeuwenberg's Model) oft. used as stocks for cacti-grafting. Some cult. orn. esp. ***P. aculeata*** Mill. (Barbados gooseberry, trop. Am., invasive S Afr., NE Aus.) – liane to 10 m with recurved spines, cult. for ed. fr., lvs ed. as are those of ***P. grandifolia*** Haw. ('*P. bleo*', Braz.)

Pereskiopsis Britton & Rose. Cactaceae (II). 6 Mex. & C Am. Allied to *Opuntia* but habit & lvs like *Pereskia*

Perezia Lag. Compositae (Mut.-Nass.). 32 Andes from Colombia to S Patagonia, S Braz., Paraguay, Uruguay & NE Arg. R: CGH 199(1970)1. Herbs. N Am. spp. referred to *Acourtia*

Pereziopsis J. Coulter = *Onoseris*

Pergularia L. Apocynaceae (V c; Asclepiadaceae III 1). 2 Afr. to Ind.: ***P. daemia*** (Forssk.) Chiov. (trop. & S Afr. to Arabia, Ind., Sri Lanka) & ***P. tomentosa*** L. (N & trop. Afr. to Ind.) – latex smeared on hides which are buried so removing hair in Arabia, locally medic. R: KB 61(2006)246

Periandra Mart. ex Benth. Leguminosae (III 18). 6 Braz. & Hispaniola. R: RBB 22(1999)343. Some with roots used as liquorice subs.

Perianthomega Bur. ex Baill. Bignoniaceae (1). 1 SE Braz., NE Paraguay, SE Bolivia: ***P. vellozoi*** Bur. – petiole of compound leaf twining. R: AMBG 99(2014)456

Periarrabidaea Samp. = *Tanaecium*

Periballia Trin. (~ *Deschampsia*). Gramineae (XVI 7). Excl. *Molineriella*, 2 Medit.

Periboea Kunth = *Lachenalia*

Pericalia Cass. = *Roldana*

Pericallis D. Don (~ *Cineraria*). Compositae (Sen.-Sen.). 15 Macaronesia. R: OB 44(1978)15. Primitively woody. Florist's cinerarias (***P.*** × ***hybrida*** (Regel) R. Nordenstam) derived in GB from *P. cruenta* (L'Hérit.) Bolle, *P. lanata* (L'Hérit.) R. Nordenstam (*S. heritieri*) & poss. other spp., now naturalizing in Calif., many cvs

Pericalymma (Endl.) Endl. (~ *Leptospermum*). Myrtaceae (II 4). 4 SW Aus. R: Nuytsia 13(1999)12

Pericalypta Benoist. Acanthaceae (III). 1 Madag.: ***P. biflora*** Benoist. R: BSBF 109(1962)131

Pericampylus Miers. Menispermaceae (V). 2–3 Indomal. ***P. glaucus*** (Lam.) Merr. – basketry, eye medic.

Perichasma Miers = *Stephania*

Perichlaena Baill. (~ *Fernandoa*). Bignoniaceae (1). 1 Madag.: ***P. richardi*** Baill. R: FMad 178(1938)16

Pericome A. Gray. Compositae (Per.-Per.). 2 SW N Am. R: SW Nat 18(1973)335. ***P. caudata*** A. Gray – medic. incl. hair-loss prevention

Pericopsis Thw. Leguminosae (III 2). 3 trop. Afr. (*Afrormosia*), 1 Sri Lanka to Micronesia. R: BJBB 32(1962)213. Valuable timbers esp. ***P. elata*** (Harms) van Meeuwen (afrormosia, asamela, kokrodua, CITES-listed) esp. for parquet, & ***P. laxiflora*** (Bak.) van Meeuwen (false dalbergia) in W Afr.; ***P. mooniana*** (Thw.) Thw. (nedun, As.) – form. imp. cabinet-wood

Perictenia Miers = *Odontadenia*

Perideridia Reichb. Umbelliferae (III 8). 13 N Am. R: UCPB 55(1969)1. Herbs of wet places with ed. tubers (epos) esp. ***P. gairdneri*** (Hook. & Arn.) Mathias (yampah) – flour form. much eaten by Native Americans & early settlers

Peridictyon Seberg & al. (~ *Festucopsis*). Gramineae (XV). 1 Balkans: ***P. sanctum*** (Janka) Seberg, Frederiksen & Baden. R: Willd. 21(1991)96

Peridiscaceae Kuhlm. (~ Bixaceae). Magnoliidae – Saxifragales. Incl. Medusandraceae, 4/13 trop. Afr., Am. R: T 56(2007)65. Trees accum. aluminium. Lvs simple, large, in spirals becoming distich., venation palmate; stip. intrapetiolar, decid. Fls bisex., small, reg., with large persistent bracteoles, in axillary fascicles or groups of racemes; K 4–7 imbr., C 0 or 5, A 5–∞ on or around outside of large fleshy lobed annular to cupular disk with ± connate filaments & small anthers with longit. slits, $\underline{G}$ (3 or 4(5)), 1-loc. sunk in disk (*Peridiscus*) or not, with distinct styles & 6–8 pendulous ovules. Fr. a 1-seeded drupe or capsule; small embryo lying alongside copious horny endosperm
Genera: *Medusandra, Peridiscus, Soyauxia, Whittonia*

Peridiscus Benth. Peridiscaceae. 1 Amazonian Braz., Venez.: ***P. lucidus*** Benth.

Periestes Baill. = *Hypoestes*

Periglossum Decne. = *Cordylogyne*

Perilepta Bremek. = *Strobilanthes*

Perilimnastes Ridl. = *Anerincleistus*

Perilla Ard. ex L. Labiatae (VII 1). 1–6 Ind. to Jap. ***P. frutescens*** (L.) Britton (Himal. to E As, natur. Ukraine) – much cult. in E As. & SE Eur. for orn. (incl. garnish in Jap. – chiso, shiso, though wild forms high in lung-toxic ketones) & oil (yegoma) like linseed o. used to waterproof paper, as in umbrellas, in paints & printing inks etc., medic. in modern Chinese herbalism

Perillula Maxim. Labiatae (VII 1). 1 Jap.: ***P. reptans*** Maxim.

Periomphale Baill. (~ *Wittsteinia*). Alseuosmiaceae. 1 New Caled.: ***P. balansae*** Baill. FNC 20(1996)101

Peripentadenia L.S. Sm. Elaeocarpaceae. 2 N Aus. R: KB 42(1987)809

Peripeplus Pierre. Rubiaceae (IV 7). 1 W trop. Afr.: ***P. bracteosus*** (Hiern) Petit. R: BBJB 34(1964)23

Periphanes Salisb. = *Hessea*

Peripleura (Burbidge) Nesom = *Vittadenia* (but see Phytol. 76(1994)131)

Periploca Tourn. ex L. Apocynaceae (III; Periplocaceae). Excl. *Parquetina*, 12 Medit. (Eur. 2), E As., trop. Afr. R: SAJB 63(1997)123. Some locally medic. incl. ***P. aphylla*** Decne. (Egypt & Iran to Ind.) – fls ed., cordage, fodder & firewood; ***P. graeca*** L. (silk vine, SE Eur., W As.) – cult. orn. decid. twiner to 10 m; ***P. visciformis*** (Vatke) K. Schum. (NE Afr., Arabia) – fodder, fibre & dye-pl.

Periplocaceae Schltr. = Apocynaceae (III)

Periptera DC. Malvaceae (Malv.-Malv.). 5 W Mex., Guatemala (once-collected). R: BSBM 33(1974)39

Peripterygia (Baill.) Loes. Celastraceae (I). 1 New Caled.: ***P. marginata*** (Baill.) Loes.

Peripterygium Hassk. = *Cardiopteris*

Perispermum Degener = *Bonamia*

Perissocarpa Steyerm. & Maguire. Ochnaceae (II). 3 N Braz., Venez. R: ANMW 100B(1998)683

Perissocoeleum Mathias & Constance. Umbelliferae (III 10). 4 Colombia

Perissolobus N.E. Br. = *Machairophyllum*

Peristeranthus Hunt. Orchidaceae (V 16c). 1 NE Aus.: ***P. hillii*** (F. Muell.) Hunt

Peristeria Hook. Orchidaceae (V 12e). 13 trop. Am. Poll. male euglossine bees. Cult. orn. epiphytes esp. ***P. elata*** Hook. (Costa Rica to Venez.) with white waxy fls, national fl. of Panamá

Peristethium Tiegh. (~ *Struthanthus*). Loranthaceae. 15 C & NSW S Am. R: AMBG 98(2012)544

Peristrophe Nees = *Dicliptera*

Peristylus Blume (~ *Habenaria*). Orchidaceae (IV 4d). c. 100 China to Pac.

Peritassa Miers. Celastraceae (IV). 14 trop. S Am., Tobago. R: Brittonia 3(1940)502

Peritoma DC. = *Cleome* (but see FNA 7(2010)205)

Perittostema I.M. Johnston = *Lithospermum*

Perityle Benth. Compositae (Perit.-Perit.). 67 SW N Am. (esp. SW US (35 – R: FNA 21(2006)317)) with 1 in Peru & Chile. R: MNYBG 21(1970)1

periwinkle, greater *Vinca major*; **lesser p.** *V. minor*; **Madagascar p.** *Catharanthus roseus*

Pernambuco wood *Paubrasilia echinata*

Pernettya Gaudich. = *Gaultheria*

Pernettyopsis King & Gamble = *Diplycosia*

peroba *Paratecoma peroba*; **p. rosa** *Aspidosperma* spp.

Peronema Jack. Labiatae (Peronem.). 1 S Myanmar to Sumatra & Borneo: ***P. canescens*** Jack (sungkai) – timber for house-building

Peronocactus Doweld = *Parodia*

Perotis Aiton. Gramineae (XXIX 3). Incl. *Lopholepis*, 15 OW trop. (China 3). Glumes awned

Perotriche Cass. = *Stoebe*

Perovskia Karelin (~ *Salvia*). Labiatae (VII 2a). 7 NE Iran to NW Ind., Tibet. Cult. orn. shrubby herbs esp. ***P. atriplicifolia*** Benth. (Afghanistan, Pakistan) & '**Blue Spire**' (*P. a.* cv. or *P. a.* × *P. abrotanoides* Karelin (Afghanistan, W Himal.), '*P.* Hybrida')

Perplexia Iljin = *Jurinea*

Perralderia Cosson. Compositae (Inul.-Inul.). 3 NW Afr. R: BotJLS 102(1990)166

Perralderiopsis Rauschert = *Iphiona*

Perriera Courchet. Simaroubaceae. 2 Madag.

Perrieranthus Hochr. = *Perrierophytum*

Perrierastrum Guillaumin = *Plectranthus*

Perrierbambus A. Camus. Gramineae (V 6). 2 Madag.

Perrieriella Schltr. (~ *Oeonia*). Orchidaceae (V 16). 1 Madag.: ***P. madagascariensis*** Schltr.

Perrierodendron Cavaco. Sarcolaenaceae. 5 Madag. R: Adansonia III,22(2000)19

Perrierophytum Hochr. Malvaceae (Malv.-Hib.). 9 Madag. R: FMad 129(1955)87

Perrierosedum (A. Berger) H. Ohba (~ *Sedum*). Crassulaceae (inc. sed.). 1 Madag.: ***P. madagascariense*** (Perrier) H. Ohba. R: U. Eggli, *Ill. Handb. Succ. Pls*, Crass. (2005)196

Perrottetia Kunth. Dipentodonaceae (Celastraceae s.l.). 17 China, Mal., Aus. & C Am. Anomalous in C. in exotegmic fibres in seeds (cf. Passifloraceae (Turneraceae))

Perryodendron T. Hartley (~ *Melicope*). Rutaceae (I 2). 1 NG: ***P. parviflorum*** (C. White) T. Hartley. R: Adans. 19(1997)198

persea *Mimusops laurifolia*

Persea Mill. Lauraceae (I). Incl. *Apollonias*, excl. *Machilus*, 80–90 trop. Am. to subtrop. N hemisphere. R (NW): MNYBG 14(1966)1). Poss. incl. *Phoebe*. Rauh's Model; alks; some heterodichogamy. Cult. orn., some ed. fr. & timber. ***P. americana*** Mill. (avocado (pear), aguacate, alligator pear, palta,? C Am.) – large fr. adapted to disp. by ? megafauna (extinct 10 K yrs), cult. since 8000 BCE for highly nutritious fr., many cvs (Guatemalan, Mexican & W Ind.; 'cocktail as' = fr. of unfertilized fls), 25% mono-unsaturated oil (some comm. aromatherapy oils) good for cooking as smoke point 250°C (olive oil 208°C)) & vitamins A, B & E & potassium, considered the most delicious of all salad veg., also eaten as dessert fr., much cult. S Afr., Israel, Calif. (75% there = '**Fuerte**', a cross between Guatemalan & Mexican cvs); ***P. barbujana*** (Cav.) Mabb. & Nieto Fel. (*A. b.*, *P. canariensis*, Macaronesia) – part of orig. Madeiran laurel forest, domatia ant-occupied; ***P. borbonia*** (L.) Spreng. (SE US swamps) – timber, imp. Seminole drug pl.; ***P. indica*** (L.) Spreng. (Macaronesia) – component of Madeiran laurel forest, cult. orn. S US; ***P. lingue*** (Bertero) Nees (lingue, Chile) – tanbark & comm. timber; ***P. nanmu*** Oliv. (*Machilus n.*, *Phoebe n.*, coffin wood, nanmu, China) – timber esp. for coffins & funerary art

Persian berries *Rhamnus saxatilis*; **P. clover** *Trifolium resupinatum*; **P. cucumber** *Cucumis sativus*; **P. insect powder** *Tanacetum coccineum*; **P. lilac** *Melia azedarach*; **P. lime** *Citrus* × *aurantiifolia*; **P. rose** *Rosa* × *damascena*; **P. shield** *Strobilanthes auriculata* var. *dyeriana*; **P. violet** *Exacum affine*

Persicaria (L.) Mill. (~ *Polygonum*). Polygonaceae (II 5). Excl. *Aconogonon*, *Bistorta*, *Rubrivena*, c. 100 subcosmop. Some cult. orn. but some cause dermatitis in stock in Aus. Usu. herbs, s.t. aquatic (***P. amphibia*** (L.) Delarb. – yellow dye (Shetlands) & ***P. hydropiper*** (L.) Delarb. (arsesmart, smartweed, tade (Java), water pepper)). ***P. japonica*** (Meissn.) Gross (Jap.) – heterostyly; ***P. maculosa*** Gray (*P. mitis* Delarb., *Polygonum persicaria*, redleg, redshank, N temp.) – allotetraploid (*P. lapathifolia* (L.) Delarb. × ?), bad agric. weed; ***P. mollis*** (D. Don) Gross (Himal.) – young shoots a veg. in Sikkim; ***P. odorata*** (Lour.) Soják (*Polygonum o.*, laksa, rau ram, Asian or Vietnamese mint, V. coriander, SE As.) – potherb, other spp. also ed. & medic.; ***P. perfoliata*** (L.) Gross (mile-a-minute, Indomal.) – prickly, invasive weed in E N Am.; ***P. tinctoria*** (Aiton) Spach (Vietnam & China to Jap.) – source of an indigo dye; ***P. vacciniifolia*** (Wall.) Ronse Decraene (Himal.) – cult. orn. ground-cover

persimmon (Am.) *Diospyros virginiana*; **Japanese p.** *D. kaki*

Persoonia Sm. Proteaceae (II 2). c. 100 Aus. R: FA 16(1995)50. Some cult. orn. shrubs & trees (geebung, snodgollion, snot-goblin, snottygobble). ***P. falcata*** R. Br. (trop. Aus.) & other spp. – ed. fr.; ***P. longifolia*** R. Br. (Barkerbush, SW Aus.) – comm. cut foliage exported to US. See also *Acidonia, Garnieria, Toronia*

Pertusadina Ridsd. = *Adina* (but see Blumea 24(1978)353)

Pertya Schultz-Bip. Compositae (Perty.). Incl. *Myripnois*, 23 Afghanistan to Jap. Shrubs & herbs

Peru or **Peruvian balsam** *Myroxylon balsamum* var. *pereirae*; **P. bark** *Cinchona* spp.; **P. basil** *Ocimum campechianum*; **P. cotton** *Gossypium barbadense*; **P. daffodil** *Ismene* spp.; **marvel of P.** *Mirabilis jalapa*; **P. nutmeg** *Laurelia sempervirens*

Perulifera A. Camus = *Pseudechinolaena*

Peruviopuntia Guiggi = *Austrocylindropuntia*

Pervillea Decne. (~ *Toxocarpus*). Apocynaceae (II). 4 Madag. R: NJB 16(1996)172

Perymeniopsis H. Robinson (~ *Perymenium*). Compositae (Helia.-Ecl.). 1 Mex.: ***P. ovalifolia*** (A. Gray) H. Robinson

Perymenium Schrad. Compositae (Helia.-Ecl.). 43 Mex. to Peruvian Andes. R: (Mex. & C Am., 33): Allertonia 1,4(1978)1

Pescatoria Reichb.f. Orchidaceae (V 12j). Incl. *Bollea*, 23 trop. Am. R: OD 32(1968)86. Cult. orn. epiphytes without pseudobulbs; lvs plicate

Peschiera A. DC. = *Tabernaemontana*

Pessopteris Underw. & Maxon = *Niphidium*

Petagnaea Caruel (*Petagnia*). Umbelliferae (III). 1 Sicily: ***P. gussonii*** (Spreng.) Rauschert

Petagnia Guss. = *Petagnaea*

Petagomoa Bremek. = *Psychotria*

petai *Parkia speciosa*

Petalacte D. Don (*Billya*). Compositae (Gnap.-Cass.). Excl. *Anderbergia* 1 Cape: ***P. coronata*** (L.) D. Don. R: Strelitzia 9(2000)351

Petaladenium Ducke. Leguminosae (III 2). 1 Rio Negro (Braz.): ***P. urceoliferum*** Ducke

Petalidium Nees. Acanthaceae (III 2f). 35 trop. & S Afr., W Himal. & W Ind. (1)

Petalocentrum Schltr. = *Oncidium*

Petalochilus R. Rogers = *Caladenia*

Petalodiscus (Baill.) Pax = *Wielandia*

Petalolophus K. Schum. = *Pseuduvaria*

Petalonyx A. Gray. Loasaceae (IV). 5 SW N Am. R: Madroño 19(1967)11. Hairs barbed

Petalostelma Fourn. (~ *Cynanchum*). Apocynaceae (V c). 7 S Am.

Petalostemon Michaux = *Dalea*

Petalostigma F. Muell. Picrodendraceae (Cal.-Petal.; Euphorbiaceae). 5 Aus., NG. R: KB 31(1976)366. Dioec. ***P. pubescens*** Domin (quinine bush, q. tree, Queensland) – disp. by emus, voided & endocarp explodes in sun, the seeds taken by ants attracted by elaiosomes

Petalostylis R. Br. Leguminosae (I 3). 2 arid Aus. R: Muelleria 6(1986)212. Alks

Petamenes Salisb. ex J.W. Loudon = *Gladiolus*

Petasites Mill. Compositae (Sen.-Tuss.). 19 N temp. (Eur. 9; N Am. 1: ***P. frigidus*** (L.) Fries medic., ed lvs). Butterbur. R: FGP 7(1972)382. Usu. ± dioec. stoloniferous perenn. herbs with large lvs. Some cult. orn. esp. ***P. fragrans*** (Villars) C. Presl (? correctly *P. pyrenaicus* (Loefl.) G. López, winter heliotrope, C Medit., natur. GB from Italy 1803) – lvs last until second season, vanilla-scented fls in spring, only male clones known; ***P. hybridus*** (L.) Gaertn. & al. (Euras.) – females (heads of 150 female fls with 1–3 males which produce nectar) not uncommon in N & N Midlands of GB, rare elsewhere in GB as preferentially planted for heads with more fls for bees & larger lvs to pack butter (elsewhere in Eur. male exists alone only where sp. introduced) used since Middle Ages as anticonvulsive, the active principle (petasin, a terpene) 14 times as effective as papaverine, discovered in 1950s, also febrifuge, petasol butenoate poss. hay-fever relief, etc.; ***P. japonicus***

(Sieb. & Zucc.) Maxim. (China, Korea, Jap., females rarely natur. GB) – petioles to 2 m tall with laminas to 1.5 m across on Sakhalin Is. (subsp. ***giganteus*** Kitam.) but clones in cult. smaller, local medic., petioles used like rhubarb or as vegetable (fuki) by Jap. & fl.-buds used as condiment etc.

Petastoma Miers = *Fridericia*

Petchia Liv. Apocynaceae (I b). Incl. *Cabucala*, 8 Cameroun (1), Madag. (6, 1 ext. to Comoro Is.), Sri Lanka (1). R: WAUP 97–2(1997)53. ***P. erythrocarpa*** (Vatke) Leeuwenb. (Madag.) – local medic.

× **Petchoa** Boker & J.M.H. Shaw. See *Petunia*

Petelotiella Gagnepain. Urticaceae (II). 1 SE As.: ***P. tonkinensis*** Gagnepain – coll. only once

Petenaea Lundell. Petenaeaceae (Tiliaceae s.l.). 1 S Mex., Honduras: ***P. cordata*** Lundell. R: Wrightia 3(1962)22

Petenaeaceae Christenh. & al. Magnoliidae-Huerteales. 1/1 Mex., Guatemala, Honduras. R: BotJLS 164(2010)23. Trees or shrubs. Lvs in spirals, palmately veined; stip. minute, caducous. Fls in axillary, cymose, long-pedunculate infls. K valvate, densely covered by pink, beaded hairs, bases with 2 or 3 obovoid subsessile glands; C 0; disc annular, glandular; A 8–12, anthers with apical pore-like slit; $\underline{G}$, stigma discoid. Fr. a berry, 4-or 5-lobed with persistent style; seeds numerous, oblong-pyramidal or irregular

Genus: *Petenaea*

Peteniodendron Lundell = *Pouteria*

Peteravenia R. King & H. Robinson (~ *Eupatorium*). Compositae (Eup.-Crit.). 5 Mex., C Am. R: Phytol. 21(1971)394, MSBMBG 22(1987)339

Peteria A. Gray. Leguminosae (III 23). 4 SW N Am. R: Rhodora 58(1956)348. ***P. glandulosa*** (S. Watson) Rydb. – ed. tubers

Petermannia F. Muell. Petermanniaceae (Colchicaceae s.l.). 1 NSW, Queensland: ***P. cirrosa*** F. Muell. – cli. to 6 m by leaf-opposed tendrillar sterile infls, rare. R: FA 46(1986)182

Petermanniaceae Hutch. (~ Colchicaceae, Philesiaceae). Magnoliidae – Liliales. 1/1 E Aus. Perenn. rhiz. prickly woody cli. with fibrous roots. Lvs petiolate, entire with reticulate venation, in spirals. Infl. a term. cyme partly forming tendril & overtopped by sympodial growth; bracts 0. Fls bisex., reg., small. P 3 + 3, 1-veined with basal nectaries; A 3 + 3, filaments free, anthers extrorse, dehiscing longitudinally; $\overline{G}$ (3), 1-loc. with 3 parietal placentas, each with ∞ anatropous ovules in 2 rows, stigma 3-lobed. Fr. a berry. Seeds angular, ∞; endosperm copious; embryo linear. n = 5

Genus: *Petermannia*

Peterodendron Sleumer (~ *Poggea*). Achariaceae (Flacourtiaceae). 1 trop. E Afr.: ***P. ovatum*** (Sleumer) Sleumer

Petersianthus Merr. (*Combretodendron*). Lecythidaceae (II). 1 trop. W Afr., 1 Philippines. R: KB 70,6(2015)9. Koriba's Model. ***P. macrocarpus*** (P. Beauv.) Liben (essia, W Afr.) – imp. comm. timber

petha *Benincasa hispida*

Petilium Ludw. = *Fritillaria*

Petimenginia auctt. = *Petitmenginia*

Petiniotia Léonard = *Sterigmostemum* (but see EJB 62(2006)116)

petitgrain oil *Citrus × aurantium*

Petitia Jacq. = *Vitex* (but see FR 42(1937)230)

Petitiocodon Robbrecht (~ *Didymosalpinx*). Rubiaceae (II 5). 1 Nigeria & Cameroun: ***P. parviflorus*** (Keay) Robbrecht. R: BJBB 58(1988)116

Petitmenginia Bonati. Orobanchaceae (Buch.; Scrophulariaceae s.l.). 2 S China, SE As. R: BJ 125(2004)349

Petiveria Plum. ex L. Petiveriaceae (Phytolaccaceae II). 1 trop. & warm Am.: ***P. alliacea*** L. (garlic weed, Guinea henweed) – spiny achenes, locally medic. (incl. analgesic mouthwash for toothache, herbal bath against witchcraft), crushed lvs onion-scented, taints milk & meat of stock, dried lvs basis of capsules (anamú) promoted to replace coca (Colombia), fr. insecticidal (astilbin, a flavonoid) toxic to whitefly & put among clothes against moths, also antifungal coumarin. R: JAA 66(1985)30

Petiveriaceae C. Agardh (~ Phytolaccaceae). Magnoliidae – Caryophyllales. Incl. Rivinaceae, 9/20 trop. & warm Am. Trees to lianes & herbs. Lvs simple s.t. smelling of garlic. Fls in racemes or spikes; P 4 (5 – *Seguieria*); A alt. with P, in centrifugal whorls; nectary 0; G 1 with 1 bitegmic ovule, style ± 0, across. Fr. with accrescent P as wings or a drupe. n = 18, 54

Principal genera: *Ledenbergia*, *Petiveria*, *Rivina*

Petkovia Stefanoff = *Campanula*

Petopentia Bullock (~ *Tacazzea*). Apocynaceae (III; Periplocaceae). 1 Natal & Transkei: ***P. natalensis*** (Schltr.) Bullock. R: SAJB 56(1990)393

Petradoria Greene (~ *Chrysothamnus*). Compositae (Ast.-Sol.). 1 SW US: ***P. pumila*** (Nutt.) Greene. R: SBM 20(2000)74

Petraeomyrtus Craven = *Melaleuca* (but see AusSB 12(1999)678)

Petraeovitex Oliv. Labiatae (Peronem.). 8 wetter Mal. (Borneo 5), NZ, Pac. R: GBS 21(1965)215. Leeuwenberg's Model. ***P. multiflora*** (Sm.) Merr. (Mal.) – liane to 30 m with infls to 70 cm

Petrea Houst. ex L. Verbenaceae (Petr.). 11 trop. Am. R: AMBG 81(1994)624. Cult. orn. decid. trees, shrubs & lianes esp. ***P. volubilis*** L. – liane to 10 m with pale lilac to purple fls

Petrina J. Phipps = *Danthoniopsis*

Petrobium R. Br. Compositae (Helia.-Cor.) 1 St Helena: ***P. arboreum*** (Forst. & Forst. f.) Spreng. – pachycaul tree, dioec. or (?) gynodioec., allies in Polynesia. R: Q.C.B. Cronk, *Endemic Flora St. Helena* (2000)81

Petrocallis R. Br. Cruciferae. 1 mts S Eur.: ***P. pyrenaica*** (L.) R. Br. – cult. orn. rock-pl. See also *Elburzia*

Petrocodon Hance. Gesneriaceae (III 2j). Incl. *Lagarosolen*, 23 C & S China

Petrocoma Rupr. = *Silene*

Petrocoptis A. Braun ex Endl. = *Silene*

Petrocosmea Oliv. Gesneriaceae (III 2j). 27 S China (24), NE Ind., Myanmar, Thailand, Vietnam, mts. Some cult. orn. R: ABY 7(1985)49

Petroedmondia Tamamsch. Umbelliferae (III 5). 1 SW As.: ***P. syriaca*** (Boiss.) Tamamsch. R: Fl. Iraq 5,2(2013)139

Petrogenia I.M. Johnston = *Bonamia*

petroleum crude oil used for gas(oline; petrol in UK) & fuel-oil

Petrollinia Chiov. = *Inula*

Petromarula Vent. ex Hedwig f. Campanulaceae (I 9). 1 Crete: ***P. pinnata*** (L.) A. DC.

Petronymphe H. Moore. Asparagaceae (Anthericaceae). 1 Mex.: ***P. decora*** H. Moore – cult. orn., variously referred to Alliaceae, Amaryllidaceae & Liliaceae. R: GH 8(1951)258

Petrophila R. Br. = seq.

Petrophile R. Br. ex Knight (~ *Isopogon*). Proteaceae (IV 2). c. 53 S Aus. (SW c. 47). R: FA 16(1995)149. Some cult. orn. shrubs (cone-sticks)

Petrophyton Rydb. = seq.

Petrophytum (Nutt.) Rydb. Rosaceae (Ros.-Spir.). 3 W N Am. R: FNA 9(2014)411. Cult. orn. rock-pls

Petroravenia Al-Shehbaz. Cruciferae (46). 3 Arg., Chile. R: Novon 4(1994)191

Petrorhagia (Ser.) Link (~ *Velezia*). Caryophyllaceae (III 1). 33 Canary Is. & Medit. (Eur. 18) to Kashmir (introd. S Afr., Aus., Hawaii, N Am. (4)). R: BBMNHB 3(1964)119. Some cult. orn., ***P. prolifera*** (L.) P. Ball & Heywood (Eur.) invasive in Aus.

Petrosavia Becc. Petrosaviaceae (Melanthiaceae s.l.). 2 W Mal. to Jap. R: Britt. 55(2003)223

Petrosaviaceae Hutch. (~ Nartheciaceae). Magnoliidae – Petrosaviales. Incl. Japonoliriaceae, 2/3 W Mal. to Jap. R: Britt. 55(2003)223. Rhiz. glabrous herbs. Lvs in spirals. P 3 + 3 (larger), each with 1 trace; A inserted at base of P or free; G sup. to semi-inf., partially united with 4-∞ ovules per carpel. n = 12, 13, 30 (x = 15)
Genera: *Japonolirion, Petrosavia*

Petrosedum Grulich (~ *Sedum*). Crassulaceae (I 3). 7 Eur., Medit.

Petroselinum Hill. Umbelliferae (III 8). 2 Eur. (2), Medit. Parsley. ***P. crispum*** (Mill.) Fuss (garden p., Eur., W As.) – infls & green stems used for garlands in anc. Rome, one of richest sources of vitamin C, cult. for lvs used as garnish, salads & in meat-dishes, sausages etc. (flavour due to terpineol; dried fr. medic.), usu. cult. a form with curled lvs compared with wild one, '**Tuberosum**' ('Atika', Hamburg or turnip-rooted p.) – ed. root, lvs hardier than type, var. ***neapolitanum*** Danert (Italian p.) – lvs not curled

Petrosimonia Bunge. Amaranthaceae (Chenopodiaceae III 3). 12 SE Eur. (7) to C As.

Petrusia Baill. = *Tetraena*

pe-tsai *Brassica rapa* Chinensis Group

Petteria C. Presl. Leguminosae (III 9). 1 Balkans: ***P. ramentacea*** (Sieber) C. Presl – seeds toxic

pettigree or **pettigrue** *Ruscus aculeatus*

petty spurge *Euphorbia peplus*; **p. whin** *Genista anglica*

Petunga DC. = *Hypobathrum*

Petunia Juss. Solanaceae (Petun.). Incl. *Calibrachoa* (27 S Am. – R: T 61(2012)126), c. 40 trop. (esp. Braz.) & warm S Am. R: KSVH 46,5(1911)1. Cult. orn. bedding-pls (trade in US worth $94M a yr by 2005) esp. many (over 200 incl. striped, picotee & double) cvs of ***P.* × *atkinsiana*** (Sweet) Baxter (*P.* × *hybrida*, cultigen app. derived from *P. axillaris* (Lam.) Britton & al. (white fls) × *P. integrifolia* (Hook.) Schinz & Thell. ('*P. violacea*'), scented cvs poss. due to influence of *P. a.*); at least 1 sp. in Ecuador source of hallucinogen inducing sense of flying or levitation. Some small-flowered cult. pls referred to *C.*, their hybrids with *P. s. s.* then × *Petchoa*

Peucedanum Tourn. ex L. (~ *Angelica*). Umbelliferae (III 10). Incl. *Cervaria, Imperatoria, Macroselinum, Oreoselinum, Pteroselinum, Taeniopetalum, Thyselium, Tommasinia*, excl. *Afroligusticum, Afrosciadium, Lefebvrea* etc., 60 Euras. Many diff. spp. used medic., in Eur. ***P. officinale*** L. (hog fennel) used veterinarily; ***P. palustre*** (L.) Moench (Eur. to C As.) – food of swallowtail butterfly

Peucephyllum A. Gray. Compositae (Bah.). 1 SW US, N Mex.: ***P. schottii*** A. Gray. R: FNA 21(2006)378

Peumus Molina. Monimiaceae (VI 3). 1 Chile: ***P. boldus*** Molina (boldo) – Champagnat's Model, wood hard, bark a dye-source, fr. ed., lvs used as a digestive tea after meals like coffee (alks incl. boldine)

Pevalekia Trinajstić = *Acuston*

peyote or **peyotl** *Lophophora williamsii*

Peyritschia Fourn. = *Trisetaria*

Peyrousea DC. = *Schistostephium*

Pezisicarpus Vernet = *Urceola*

Pfaffia Mart. Amaranthaceae (II 2). Excl. *Hebanthe* 26 trop. S Am. Mostly shrubby some with basis for patented anti-tumour compounds. ***P. iresinoides*** Spreng. (Braz.) – ecdysteroids

Pfeiffera Salm-Dyck (~ *Lepismium*). Cactaceae (III 1). 6 Ecuador, Bolivia, Peru, Arg. R: Willd. 40(2010)164

Pfitzeria Senghas = *Comparettia*

Pfosseria Speta = *Scilla*

Phacelia Juss. Boraginaceae (Hydrophyllaceae-Rom.). c. 200 W N Am. (most), E US, S Am. Bee-poll. with anthers turning inside-out at maturity. Many cult. orn. esp. in US. ***P. tanacetifolia*** Benth. (W N Am.) – cult. as bee-fodder in Eur. & by organic gardeners to attract hoverflies for aphid control, green manure (Jersey) planted after potatoes to deter eelworm

Phacellanthus Sieb. & Zucc. Orobanchaceae (Orob.). 1 E As.: ***P. tubiflorus*** Sieb. & Zucc. R: FOC 18(1998)30

Phacellaria Benth. Santalaceae (Amphorogynaceae). 4 SE As. Hyperparasites on Loranthaceae, Viscaceae

Phacellothrix F. Muell. Compositae (Ast.-Ast.). 1 E Mal. & E trop. Aus.: ***P. cladochaeta*** (F. Muell.) F. Muell.

Phacelophrynium K. Schum. = *Phrynium* (but see GBS 13(1951)294)

Phacelurus Griseb. Gramineae (XXII 3). 7 warm OW (Eur. 1, China 3). R: KB 33(1978) 175

Phacocapnos Bernh. = *Cysticapnos*

Phaeanthus Hook.f. & Thomson. Annonaceae (IV 7). 8 Indomal. R: Blumea 45(2000)205. Alks. Local medic.

Phaedranassa Herb. Amaryllidaceae. 11 Andes. R: PL 25(1969)55. Cult. orn. (queen lilies)

Phaedranthus Miers = *Amphilophium*

Phaenanthoecium C. Hubb. Gramineae (Arund.). 1 NE trop. Afr. mts: ***P. koestlinii*** (A. Rich.) C. Hubb.

Phaenocoma D. Don. Compositae (Gnap.-Rel.). 1 SW Cape: ***P. prolifera*** (L.) D. Don – cult. orn. R: OB 104(1991)70, CBM 13(1996)56

Phaenohoffmannia Kuntze = *Pearsonia*

Phaenosperma Munro ex Benth. Gramineae (IX). 1 Assam to Jap.: ***P. globosa*** Munro ex Benth. R: FOC 22(2006)188

Phaeocephalus S. Moore = *Hymenolepis*

Phaeoneuron Gilg = *Ochthocharis*

Phaeonychium O. Schulz = *Solms-laubachia* (but see NJB 20(2000)158)

Phaeopappus (DC.) Boiss. (*Tomanthea*) = *Centaurea*

Phaeoptilum Radlk. Nyctaginaceae (4). 1 SW Afr.: ***P. spinosum*** Radlk. – *Lycium* habit. R: Strelitzia 10(2000)425

Phaeosphaerion Hassk. = *Commelina*

Phaeostemma Fourn. = *Matelea*

Phaeostigma Muldashev (~ *Ajania*). Compositae (Anth.-Art.). 3 China. R: BBMNHB 23(1993)115

Phagnalon Cass. Compositae (Gnap.). 43 Canary Is. & Medit. (Eur. 6) to Arabia & Middle E. R: OB 104(1991)53

Phainantha Gleason. Melastomataceae. 4 trop. S Am.

Phaiophleps Raf. = *Olsynium*

phai-tong *Dendrocalamus asper*

Phaius Lour. (~ *Calanthe*). Orchidaceae (V 14). c. 40 Indomal. to S China, trop. Afr. & Aus. (3), New Caled. (4). Turn blue if bruised. Cult. orn. terr. orchids esp. ***P. tankervilleae*** (Sowerby) Blume (Himal. to Aus., invasive Carib. incl. Florida, Hawaii) with infl. of brown & yellow fls to 1 m

Phalacrachena Iljin = *Plectocephalus*

Phalacraea DC. (~ *Piqueria*). Compositae (Eup.-Ager.). 4 Colombia, Ecuador & Peru. R: MSBMBG 22(1987)147

Phalacrocarpum (DC.) Willk. (~ *Chrysanthemum*). Compositae (Anth.- Leucanthemops.). 2 Spain & Portugal. R: BBMNHB 23(1993)140

Phalacroloma Cass. = *Erigeron*

Phalacroseris A. Gray. Compositae (Cich.-Cich.(-Cich.)). 1 California: ***P. bolanderi*** A. Gray. R: FNA 19(2006)374

Phalaenopsis Blume. Orchidaceae (V 16c). Incl. *Doritis, Hygrochilus, Kingidium, Ornithochilus*, 67 Himal. to Aus. R: E.A. Christenson (2001) *P. A monograph*. Alks. Cult. orn. (most imp. economic orchid 'crop') epiphytes or chasmophytes (moth orchids) incl. hybrids with fls lasting up to 4 months with sec. infl. arising from cut-off peduncle, esp. overused white cvs (incl. naturally occurring tetraploids & synth. triploids) mostly derived from ***P. amabilis*** (L.) Blume (Mal. to Aus.) – poll. carpenter bees, young lvs ed. Java. ***P. taenialis*** (Lindl.) Christenson & U.C. Pradhan (*Doritis t., K. t.*, NW Himal. to Thailand) & others with few, decid. or 0 lvs

Phalaris L. Gramineae (XVI 3). 21 Eur., Medit., N As., N (5 native) & S Am. Alks. R: Webbia 49(1995)266. Fodders etc., *P. aquatica* & *P. arundinacea* with tryptamine-based alks gramine & hordemine, gramine at 0.01% a feeding stimulant but at higher conc. repellent: ***P. aquatica*** L. (*P. tuberosa*, Toowoomba canary grass, Medit., invasive Aus., W US) – fodder esp. in Aus. but alks s.t. poisoning stock ('Phalaris Staggers') there, incl. 'var. *stenoptera*' (Harding grass, origin obscure) – forage; ***P. arundinacea*** L. (reed canarygrass, N temp., invasive in Aus., N Am. [where a mix of native & Eur. strains, cvs etc.]) – valuable hay & fodder when young, erosion control, '**Picta**' (var. *picta*, ribbon grass) – varieg. cult. orn.; ***P. canariensis*** L. (canary grass, W Medit., widely natur.) – canary seed of commerce; ***P. minor*** Retz. (Medit., S Afr.) – invasive Aus., SE US

Phaleria Jack. Thymelaeaceae (Thym.-Daph.). c. 30 Indomal., W Pac. Some cauliflory. ***P. capitata*** Jack (Indomal.) – fibre, ed. fr. (cotyledons toxic); ***P. macrocarpa*** Boerl. (NG) – medic., bark for string bags

Phalocallis Herb. = *Cypella*

phalsa *Grewia asiatica*

Phanera Lour. (~ *Bauhinia*). Leguminosae (I 1). Incl. *Lasiobema*, excl. *Schaella*, c. 90 As. to Aus. Some local medic., cult. orn.

Phanerodiscus Cavaco. Olacaceae (Aptandraceae). 3 Madag. R: Adansonia III,25(2003)121

Phaneroglossa R. Nordenstam (~ *Senecio*). Compositae (Sen.-Sen.). 1 SW Cape mts: ***P. bolusii*** (Oliv.) R. Nordenstam, allied to *Lordhowea* (cf. *Dietes*). R: Strelitzia 9(2000)352

Phanerogonocarpus Cavaco = *Tambourissa*

Phanerophlebia C. Presl = *Polystichum*. But see AMBG 83(1996)184

Phanerophlebiopsis Ching = *Arachniodes*

Phanerosorus Copel. Matoniaceae. 1 Sarawak, 1 is. S of Moluccas & off W NG, on limestone. R: FM II,3(1998)292. Fronds pendent, lat. branches oft. tipped with dormant apex

Phanerostylis (A. Gray) R. King & H. Robinson = *Brickellia*

Phania DC. Compositae (Eup.-Ager.). 5 Cuba. R: MSBMBG 22(1987)145

Phanopyrum (Raf.) Nash (~ *Panicum*). Gramineae (XXIII 1). 1 SE US: ***P. gymnocarpon*** (Elliott) Nash. R: AJB 90(2003)796

Pharbitis Choisy = *Ipomoea*

Pharnaceum L. Molluginaceae. 25 S Afr. R: JSAB 24(1959)18

Pharus P. Browne. Gramineae (II). 7 warm Am. R: KBAS 13(1986)67

Phaseolus Tourn. ex L. Leguminosae (III 18). 70+ trop. & warm Am. (R (N & C Am.): SBM 23(2002)9). Beans (see also *Vigna* – stip. with basal appendages, etc.). Many cult. (5 domesticated N Am.) for ed. seeds &/or pods esp.: ***P. acutifolius*** A. Gray (S N Am.) esp. 'var. *latifolius* G. Freeman' (tepary bean) – drought-resistant; ***P. coccineus*** L. (scarlet runner b., C Am.) – runner beans much cult. GB; ***P. dumosus*** Macfad. (C Am. WI) – promising but perenn. in Andes (*P. flavescens*); ***P. lunatus*** L. (incl. *P. limensis*, butter or Lima or Burma or duffin or Madag. or Rangoon b., Andes) – imp. ed. beans; ***P. vulgaris*** L. (kidney or French or flageolet or dwarf or string or snap or wax or haricot b., frijoles, pea-b., trop. Am., 'wild' forms being '*P. aborigineus* Burkart', Andes) – predominantly self-poll., one of most imp. legumes (20 M t a yr by 2011) giving ed. young pods (e.g. cannellini beans) as well as varied range of dried & tinned seeds (e.g. borlotti, crabeye, cranberry, Great Western, pinto, red, rosecoco, turtle & white beans, Nuñas beans being cooked in hot oil like popcorn) allegedly promoting hypoglycaemia in humans, with tomato sauce the ubiquitous baked beans (navy beans – mainly US navy) of commerce (first tinned 1911 when a luxury but now 1M tins consumed a day (5 kg per person per annum) in UK) allegedly effective in lowering cholesterol levels, & forage, resistant to cowpea weevils & adzuki bean weevils as 1% tissue by weight is alpha-amylase inhibitor inhibiting insect-gut digestion leading to insect starvation (gene introduced thence to peas via *Agrobacterium* genetic engineering) & also used in human weight-loss programmes

phasey bean *Macroptilium lathyroides*

Phaulopsis Willd. Acanthaceae (III 2f). 22 trop. Afr., ***P. talbotii*** S. Moore ext. (? introd.) to NW S Am. & WI, Ind. (1: ***P. dorsiflora*** (Retz.) Santapau (octoploid sister of tetraploid *P. talbotii* – both weedy)). R: SBU 31,2(1996)78. Some potherbs, local medic.

Phaulothamnus A. Gray. Achatocarpaceae. 1 Calif. & Texas to N Mex.: ***P. spinescens*** A. Gray. R: FNA 4(2003)13

pheasant's eye *Adonis annua*; **p. e. narcissus** *Narcissus poeticus*; **p. tail's grass** *Anemanthele lessoniana*

Phebalium Vent. Rutaceae (I 3). 28 Aus. R: FA 26(2013)458. Heterogeneous? Alks

Phedimus Raf. (~ *Sedum*). Crassulaceae (I 2). 18 E Eur., As.

Phegopteris (C. Presl) Fée. Thelypteridaceae. 4 N temp., Mal. R: Blumea 17(1969)9. ***P. connectilis*** (Michaux) Watt (*Thelypteris p.*, beech fern, N temp.) – cult. orn.

Pheidochloa S.T. Blake. Gramineae (Micr.). 1 NG, 1 trop. Aus. R: Blumea 19(1971)61

Pheidonocarpa Skog (~ *Gesneria*). Gesneriaceae (II 5a). 1 Cuba, Jamaica: ***P. corymbosa*** (Sw.) Skog. R: SCB 29(1976)40

Pheladenia D.L. Jones & M.A. Clem. (~ *Caladenia*). Orchidaceae (IV 3b). 1 SW & SE Aus.: ***P. deformis*** (R. Br.) D.L. Jones & M.A. Clem. R: AusSB 17(2004)234

Phelipaea Desf. = *Phelypaea*

Phelipanche Pomel (~ *Orobanche*). Orobanchaceae. Spp.? ***P. arenaria*** (Borkh.) Pomel (*O. a.*, *O. ionantha*, Euras.) – 1 seed weighs 0.0001 mg. Noxious paras. weeds esp. ***P. ramosa*** (L.) Pomel (Euras.) on hemp, potato, tobacco & tomato

Phellinaceae Takht. (~ Aquifoliaceae). Magnoliidae – Asterales. 1/11 New Caled. Dioec. small, everg. trees with alks. Lvs in pseudowhorls at branch-tips, glabrous. Fls 4- or 5(6)-merous in axillary racemes to panicles; K basally connate; C valvate, free; A alt. with C; $\underline{G}$ (2–5) with 1 apical hemitropous to campylotropous ovule per carpel. Fr. a drupe with 1–5 pyrenes. Seeds with copious endosperm. n = 17

Genus: *Phelline*

Phelline Labill. Phellinaceae. 11 New Caled.

Phellocalyx Bridson. Rubiaceae (II 1). 1 Malawi, Mozambique & Tanz.: ***P. vollesenii*** Bridson. R: KB 35(1980)316

Phellodendron Rupr. Rutaceae (I 2). 2 E As. R: EJB 63(2006)137. Alks. Dioec. decid. trees – some cult. orn. (cork trees), locally medic. & useful timber. ***P. amurense*** Rupr. (E As.) – invasive New York City

Phellolophium Bak. Umbelliferae (III 8). 1 Madag.: ***P. madagascariense*** Bak.

Phellopterus Benth. = *Glehnia*

Phelpsiella Maguire. Rapateaceae (I 1). 1 Venez.: ***P. ptericaulis*** Maguire. R: MNYBG 10(1958)29

Phelypaea Tourn. ex L. (*Diphelypaea*, *Phelipaea*). Orobanchaceae (Orob.). 3 Medit. (Eur. 2). R: CBM 26(2010)383. Parasitic on *Centaurea* spp. Some emergency foods

Phemeranthus Raf. (~ *Talinum*). Montiaceae (Portulacaceae s.l.). 25–30 Am.

Phenakospermum Endl. Strelitziaceae. 1 E trop. S Am.: ***P. guyanense*** (Rich.) Miq. – Tomlinson's Model with term. infl., bat-poll., arils bright red & hairy

Phenax Wedd. Urticaceae (III). 12 trop. Am., some natur. trop. As.

phenomenal berry see *Rubus*

Pherolobus N.E. Br. = *Cleretum*

Pherosphaera Archer (*Microstrobos*). Podocarpaceae. 1 NSW, 1 Tasmania. R: T 53(2004)537. ***P. fitzgeraldii*** (F. Muell.) Hook.f. (*M. f.*, Blue Mts, NSW) – dioec. (app. monoec. in cult.). ? water-poll., only 755 pls (2014; protected) in 7 waterfalls left, but cult.

Pherosphaeraceae Nakai = Podocarpaceae

Pherotrichis Decne. Apocynaceae (V c; Asclepiadaceae III 3). 2 Mex.

Phialacanthus Benth. Acanthaceae (III 2c). 5 Himal. to Malay Pen.

Phialanthus Griseb. Rubiaceae (III 4). 18 WI

Phialiphora Groeninckx. Rubiaceae (IV 15). 2 NW Madag. R: T 59(2010)1821

Phialodiscus Radlk. = *Blighia*

Phiambolea Klak (~ *Amphibolia*). Aizoaceae. 11 Cape. R: Bradleya 21(2003)112

Phidiasia Urb. = *Odontonema*

Philacra Dwyer. Ochnaceae (I). 4 Venez. & N Braz. R: BJ 113(1991)177

Philactis Schrad. Compositae (Helia.-Zinn.). 1(–4) C Am. R: Britt. 21(1969)322

Philadelphaceae Martinov = Hydrangeaceae

Philadelphus L. Hydrangeaceae (II 1; Philadelphaceae). Incl. *Carpenteria*, c. 60 N temp. (Eur. 1), orig. Am. R: JAA 35(1954)275, 36(1955)52, 325, 37(1956)15. Mock orange; shrubs with cucumber-scented lvs & strongly protogynous white, usu. strongly scented, fls, regrettably known as syringa (*Syringa* = lilac) from (Fr.) seringat; water scented with the fls independently used by anc. Parthians (OW) & in Mex. (***P. mexicanus*** Schldl.) in pre-Columbian times. ***P. lewisii*** Pursh (W N Am.) – medic., stems for basketry, arrows & combs; many cult. orn. incl. hybrids, esp. ***P. coronarius*** L. (Eur., SW As.) – poss. oil-source, v. fragrant, many cvs & ***P. microphyllus*** A. Gray (SE US to C Mex.) & their hybrid ***P.* × *lemoinei*** Lemoine as well as its hybrid with *P. coulteri* S. Watson (Mex.) = ***P.* × *purpureomaculatus*** Lemoine with many cvs such as '**Sybille**'; ***P.* × *nivalis*** Jacques (*P. coronarius* × ? *P. pubescens* Lois. (SE US)) – tree-like in UK; ***P.* × *virginalis*** Rehder (prob. a *P. p.* hybrid) incl. 'double-flowered' cvs like **'Minnesota Snowflake'**

Philbornea H. Hallier (~ *Hugonia*). Linaceae (I). 1 Sumatra, Philippines, Borneo: ***P. magnifolia*** (Stapf) H. Hallier. R: FM I,10(1988)614

Philcoxia P. Taylor & Souza. Plantaginaceae (Grat.-Grat.; Scrophulariaceae s.l.). 3 Braz. R: KB 55(2000)159. Somewhat resembling Lentibulariaceae (but placentation diff.) in reduced root-system, glandular-pubescent lvs with circinnate vernation & A2

Philenoptera Hochst. ex A. Rich. (~ *Lonchocarpus*). Leguminosae (III 16). 12 trop. Afr., Madag. R: KB 55(2000)82. some fish-poisons. ***P. nelsii*** (Schinz) Schrire (trop. Afr.) – wood form. used for wagon-wheels etc.

Philesia Comm. ex Juss. Philesiaceae (Smilacaceae s.l.). 1 S Chile: ***P. magellanica*** J. Gmelin – everg. shrub with petioled 1-veined revolute lvs & red fls, cult. orn.

Philesiaceae Dumort. (excl. Behniaceae (= Asparagaceae), Petermanniaceae & Luzuriagaceae; ~ Smilacaceae). Magnoliidae – Liliales. 2/2 S Am. Shrubs & lianes with fibrous roots. Lvs in spirals, entire with parallel veins & net-like transverse venation. Infls of 1–3 bisex., reg., term. or axillary, pendulous fls. P 3 + 3, the outer s.t. sepaloid (*Philesia*), with basal nectaries; A 3 + 3, s.t. basally united; G (3), 1-loc., ovules anatropous in 2 rows. Fr. a red berry with few to ∞ globose seeds with oily endosperm. n = 15, 19

Genera: *Lapageria, Philesia*

Eustrephus poss. here, *Geitonoplesium* now in Asphodelaceae

The shifting sands of placements of these & allied genera suggest a broadened Liliaceae (= Liliales) may (once more) be preferable. Cult. orn.

Philgamia Baill. Malpighiaceae. 4 Madag.

Philibertia Kunth (~ *Cynanchum*). Apocynaceae. Incl. *Amblystigma*, 43 S Am. R: KB 59(2004)425

Philippia Klotzsch = *Erica*

Philippiamra Kuntze (~ *Cistanthe*; ? *Diazia*). Montiaceae. 8 S Am.

Philippiella Speg. Caryophyllaceae (I 2). 1 Patagonia: ***P. patagonica*** Speg. R: MBMHRU 285(1968)398

Philippinaea Schltr. & Ames = *Orchipedum*

Philippine violet *Barleria cristata*

Phillyrea Tourn. ex L. Oleaceae (4d). 2 Madeira, Medit. (Eur. 2) to N Iran. Some locally used timber; some cult. orn. everg. trees & shrubs esp. ***P. latifolia*** L. (S Eur., SW As.) – oft. mistaken for everg. oak

Philodendron Schott. Araceae (VII 5). c. 700 trop. Am. R: KB 45(1990)37; SBM 47(1996)1. Chamberlain's Model. Usu. epiphytic lianes with entire to deeply lobed lvs, some scarab-poll. with resin gluing pollen to beetles; thermogenic heat with 2 peaks. ***P. acutatum*** Schott (trop. Am.), ***P. fragrantissimum*** (Hook.) G. Don (trop. Am.) & ***P. solimoesense*** A.C. Sm. (trop. Am.) – bisex. fls (homoeotic replacement by sterile A of carpels in same whorl – unique in angiosperms. Some locally ed. fr. & medic. (berberine-type alks reputedly bactericidal & fungicidal); many cult. orn. incl. hybrids but many v. poisonous to children & pets (fermented lvs of ***P. craspedodromum*** R.E. Schultes used as fish-poison in NW Amazon), the most familiar being ***P. bipinnatifidum*** Schott ex Endl. (SE Braz. to Arg., Paraguay, Bolivia) – stems to 10 cm diam., spadix to 17 °C above ambient (lipid metabolism) & ***P. hederaceum*** (Jacq.) Schott (*P. scandens*, trop. Am.) – poss. most common house-pl. in world; ***P. radiatum*** Schott (motusay, Mex.) – imp. for 'wickerwork' so that pops in decline; ***P. saxicola*** K. Krause (Braz.) – aerial rs to 1.5 m, cortex form. used to store & carry diamonds

Philodice Mart. = *Syngonanthus*

Philoglossa DC. Compositae (Liab.). 5 S Am. R: Phytologia 26(1973)381, SCB 54(1983)59

Philogyne Salisb. ex Haw. = *Narcissus*

Philonotion Schott (~ *Schismatoglottis*). Araceae (VII 8). 3 NE S Am. R: T 59(2010)121

Philotheca Rudge. Rutaceae (I 3). Excl. *Eriostemon*, 53 Aus. R: FA 26(2013)366. Prob. incl. *Geleznowia*. Waxflowers

Philoxerus R. Br. = *Gomphrena*

Philydraceae Link. Magnoliidae – Commelinales. 4/5 Aus. to SE As. & S Jap. Perenn. herbs with rhizome, tuber or thickened stem-base; vessels (with scalariform endplates) only in roots, no sec. growth. Lvs parallel-veined, basal ones distich., rest in spirals, usu. with basal sheath. Fls bisex., zygomorphic, in axils of bracts in simple or branched spike, open 1 day, P outer 2 (lat.), small, other (upper) united with inner upper 2 to give upper lip with 3 prominent veins (& sometimes teeth), lower inner also forming a lip, A 1 s.t. with filament united with inner P, anther introrse to extrorse with longit. slits, pollen s.t. in tetrads, $\underline{G}$ (3), 3-loc. with axile placentas or 1-loc. with intruded parietal placentas & term. style, ovules ∞ anatropous, bitegmic. Fr. usu. a loculicidal capsule; seeds usu. ± ∞, embryo with term. cotyledon & lat. plumule embedded in starchy, oily, proteinaceous endosperm. n = 8, 16, 17

Genera: *Helmholtzia, Orthothylax, Philydrella, Philydrum*

Philydrella Caruel. Philydraceae. 1 SW Aus.: ***P. pygmaea*** (R. Br.) Caruel. R: FA 45(1987)42

Philydrum Banks ex Gaertn. Philydraceae. 1 E & SE As., Mal to Aus.: ***P. lanuginosum*** Banks ex Gaertn. R: FA 45(1987)41

Philyra Klotzsch. Euphorbiaceae (Acal.-Chro.-Dit.). 1 S Braz., Paraguay: ***P. brasiliensis*** Klotzsch. R: A. R.-Sm., *Gen. Euph.* (2001)141

Philyrophyllum O. Hoffm. Compositae (Athro.). 2 S Afr. R: Bothalia 33(2003)118. Shrubs

Phinaea Benth. Gesneriaceae (II 2b). 3 trop. Am. R: Selbyana 29(2008)170. G semi-inf.; hairy herbs with scaly rhiz. ***P. multiflora*** C. Morton (Mex.) – cult. orn. with white fls. See also *Amalophyllon*

Phippsia (Trin.) R. Br. Gramineae (XVI 14). 3 circumpolar (Eur. 2), ***P. algida*** (Sol.) R. Br. also in Rockies & Andes. R: PSE 220(2000)241. Hybrids with *Puccinellia* spp.

Phitopis Hook.f. = *Schizocalyx*

Phitosia Kamari & Greuter (~ *Crepis*). Compositae (Cich.-Cich.(Chron.)). 1 Greece: ***P. crocifolia*** (Boiss. & Heldr.) Kamari & Greuter

Phlebiophragmus O. Schulz = *Mostacillastrum*

Phlebocarya R. Br. Haemodoraceae (II). 3 SW Aus. G 1-loc. in fr. R: FA 45(1987)128

Phlebodium (R. Br.) J. Sm. = *Polypodium*

Phlebochilus (Benth.) Szlach. = *Caladenia*

Phlebolobium O. Schulz. Cruciferae (46). 1 Falkland Is.: ***P. maclovianum*** (d'Urv.) O. Schulz. R: NBGB 11(1932)641

Phlebophyllum Nees = *Strobilanthes*

Phlebotaenia Griseb. = *Polygala*

Phlegmarius (~ *Huperzia*). Lycopodiaceae. ? 300 + trop.

Phlegmatospermum O. Schulz. Cruciferae (37). 4 Aus. R: CGH 205(1974)150

Phleum L. Gramineae (XVI 15). 16 N temp. (Eur. 11), temp. S Am. R: KBAS 13(1986)142. Cat-tail grass. Cult. forage grasses esp. ***P. pratense*** L. (timothy, Euras., invasive Aus., W US) – grazing & hay, many cvs, first widely grown in US (introd. *Timothy* Hansen (1720) & later reintrod. UK), common cause of hay-fever

Phloeophila Hoehne & Schltr. (~ *Pleurothallis*). Orchidaceae (V 13c). 11 trop. Am. R: Lindleyana 16(2001)253

Phloga Noronha ex Hook.f. = *Dypsis*

Phlogacanthus Nees. Acanthaceae (III 2b). 15 Indomal. A 2. Some cult. greenhouse shrubs esp. ***P. thyrsiformis*** (Hardw.) Mabb. (*P. thyrsiflorus*, N Ind.) – C orange

Phlogella Baill. = *Dypsis*

Phlojodicarpus Turcz. ex Ledeb. Umbelliferae (III 8). 4 Siberia

Phlomidoschema (Benth.) Vved. (~ *Stachys*). Labiatae (VI). 1 Iran to W Himal.: ***P. parviflora*** (Benth.) Vved.

Phlomis Tourn. ex L. Labiatae (VI). Excl. *Phlomoides*, c. 65 Medit. (Eur. 12) to C As. & China, usu. dry stony habitats. R: J. M.Taylor (1998) *P*. Some cult. orn. coarse herbs esp. ***P. fruticosa*** L. (Jerusalem sage, Medit. W to Sardinia)

Phlomoides Moench (~ *Phlomis*). Labiatae (VI). Incl. *Eremostachys*, c. 160 C Eur. to Russian Far E. R: T 61(2012)175. ***P. rotata*** (Hook.f.) Mathiesen (*Phlomis r.*, Himal.) – hapaxanthic pachycaul

phlox, water *Wurmbea* (*Onixotis*) spp.

Phlox L. Polemoniaceae (III). 69 N Am., NE As. (1). R: MAM 3(1955). J.H. Locklear (2011) *P*. Ann. or perenn. herbs or shrubs with A of unequal length or unequally inserted in C-tube. Some local medic.; many handsome cult. orn. (signifying 'agreement' in 'Language of Fls') esp. ***P. drummondii*** Hook. (E Texas) – ann. with many cvs of diff. fl. colour, ***P. paniculata*** L. (*P. decussata*, E N Am.) – common perenn. garden phlox with scented fls (also hybrids with *P. divaricata* L. (*P. canadensis*, E US) = ***P. × arendsii*** H.B. May), ***P. subulata*** L. (NE US) – mat-forming rock-pl.

Phlyctidocarpa Cannon & W.L. Theob. Umbelliferae (III 8). 1 SW Afr.: ***P. flava*** Cannon & W.L. Theob. R: Strelitzia 10(2000)69

Phoebanthus S.F. Blake. Compositae (Helia.-Helia.). 2 SE US. R: FNA 21(2006)113

Phoebe Nees (~ *Persea*). Lauraceae (I). 100 Indomal. Alks. Am. spp. referred to *Cinnamomum*

Phoenicanthus Alston. Annonaceae (IV 7). 2 Sri Lanka. R: Rev. Handbk Fl. Ceylon 5(1985)23

Phoenicaulis Nutt. Cruciferae (11). 1 W N Am.: ***P. cheiranthoides*** Nutt. – cult. orn. rock-pl. R: FNA 7(2010)415

Phoenicophorium H. Wendl. Palmae (V 14k). 1 Seychelles: ***P. borsigianum*** (K. Koch) Stuntz

Phoenicoseris (Skottsb.) Skottsb. = *Dendroseris*

Phoenicosperma Miq. = *Sloanea*

Phoenix Kaempf. ex L. Palmae (III 3). 14 trop. & warm Afr. & As., 1 Eur.: ***P. theophrasti*** Greuter (~*P. dactylifera*) – Crete (& SW Turkey), figured in Minoan frescoes, endangered sp. allied to ***P. dactylifera*** L. (date palm, cultigen, though poss. wild in Pakistan & NW Ind. or NE Sahara & Arabia where trees with small ined. fr., form spontaneous hybrids with *P. reclinata* & *P. sylvestris*). R: KB 53(1998)527. Dioec. (males heterozygous, females homozygous recessive) trees (only palms with induplicate pinnate lvs, lower leaflets spines), imp. sources of fr., sugar, lvs for raincoats (Philippines) & cult. orn. (many hybridize in cult.), most imp. *P. dactylifera* cult. since 4000 BC, ? insect- or wind-poll. (artificial poll. practised since 2300 BC as sex differentiation understood in anc. Babylon & Assyria), some 90 M trees worldwide, each producing up to 700 kg fr. annually (60–70% sugar), by 2009 7.4 M t a yr (20% in Egypt), staple of nomadic peoples in Arabia & N Afr. (soft drupes eaten locally, semi-dry exported, e.g. Deglet Noor, dry used locally & stored), comm. crop in Iraq (form. top producer & exporter), N Afr. & Calif., dried or preserved with sugar & used as dessert fr., in puddings etc. (so sweet that bacterial delay inhibited, so date cakes can last several decades), endocarps form. used for charcoal & (with crocodile dung etc.) for asthma treatment, lvs for matting, thatch etc., in Israel old leaf-bases harbour an epiphytic angiosperm flora of 32 spp., ants long used to control pests in Arabia (noted by Forssk. (1755) as first 'biological control' in Eur. literature), seeds (germ. after 2000 yrs Israel (Masada fortress), i.e. older than *Nelumbo nucifera*) & pollen contain oestrone. ***P. acaulis*** Roxb. (Himal. to S China) – fr. chewed like betel in Sikkim; ***P. canariensis*** Wildpret (Canary Is.) – widely cult. in S Eur. towns, Aus. etc., the trunk broader than in date palm; ***P. paludosa*** Roxb. (Assam to Malay Pen.) – mangrove; ***P. reclinata*** Jacq. (dwarf date palm, trop. Afr.) – large hedge-pl. because of vigorous suckering habit, overexploited in NE Kenya thus endangering Tana River crested mangabey which is heavily reliant on it; ***P. roebelenii*** O'Brien (Laos, Vietnam, China) – elegant pot-pl.; ***P. sylvestris*** (L.) Roxb. (Pakistan to Bangladesh) – imp. source of palm sugar & toddy

Pholidia R. Br. = *Eremophila*

Pholidocarpus Blume. Palmae (III 4b). 6 Mal. R: ARBGC 13(1931). ***P. macrocarpus*** Becc. to 40 m tall in Johore, corky fr. prob. elephant-disp. as is oblig. in ***P. majadum*** Becc.

Pholidostachys H. Wendl. ex Hook.f. (~ *Calyptrogyne*). Palmae (V 11). 7 trop. Am. R: Phytotaxa 43(2012)9. ***P. synanthera*** (Mart.) H.E. Moore – thatch lasts 10–12 yrs

Pholidota Lindl. ex Hook. (~ *Coelogyne*). Orchidaceae (V 10b). c. 30 Indomal.

Pholisma Nutt. ex Hook. Boraginaceae (Lennoaceae). Incl. *Ammobroma*, 3 SW N Am. R: SB 11(1986)543. Perennials with 4–10-merous fls, in sand on roots esp. of *Croton* or *Hymenoclea* spp. ***P. sonorae*** (A. Gray) Yatskievych (SW US) – succ. underground stems used as food by Arizona Indians

Pholistoma Lilja (~ *Nemophila*). Boraginaceae (Hydrophyllaceae-Hydro.). 3 SW N Am. R: BTBC 66(1939)344. ***P. auritum*** (Lindl.) Lilja (Calif., Nevada) – cult. orn. ann. with lavender to blue or violet fls

Pholiurus Trin. Gramineae (XVI 15). 1 SE Eur. & C As. (saline soils): ***P. pannonicus*** (Host) Trin.

Phonus Hill (~ *Carthamus*). Compositae (Card.-Cent.). 2 W Medit. R: AJBM 47(1990)26

Phoradendron Nutt. Santalaceae (Viscaceae). c. 240 Am. esp. trop. R: SBM 66(2003)36. Alks. Dried growths of *P.* spp. (& other Loranthaceae) sold as 'flores de palo' (wooden fls) in C Am. ***P. californicum*** Nutt. (Calif.) – fr. ed.; ***P. leucarpum*** (Raf.) Reveal & M. Johnston (*P. flavescens*, *P. serotinum*, N Am.) – the principal mistletoe of commerce in US, toxic amines, local medic. (also other spp.).

Phoringopsis M.A. Clem. & D.L. Jones = *Arthrochilus*

Phormiaceae J. Agardh = Asphodelaceae

Phormium Forst. & Forst. f. Asphodelaceae (Hemerocallidaceae; Phormiaceae). 2 NZ, Norfolk Is. ***P. tenax*** Forst. & Forst. f. (NZ, Norfolk Is., natur. Spain, pestilential weed St Helena) – poll. honey-eaters (& poss geckos), lvs source of NZ flax or hemp (bush flax), long used by Maori for textiles, cordage & nets (many cvs), source of antimicrobial & blood-clotting enzymes for injuries (& 'bullet-proof' vests), fermented & sweetened with nectar, cult. orn. incl. cvs with red lvs; ***P. cookianum*** Le Jolis (*P. colensoi*, NZ) – similiar cult. orn. but smaller, also many varieg. cvs being hybrids with *P. t.*

Phornothamnus Bak. = *Gravesia*

Photinia Lindl. (~ *Aronia*). Rosaceae (Ros.-Mal.). Incl. *Stranvaesia*, c. 40 Himal. to Thailand & Jap., C Am. (!). Cult. orn. shrubs & trees to 15 m, most planted being ***P. × fraseri*** Dress (***P. glabra*** (Thunb.) Franch. & Savat. (Jap. – imp. hedge-pl.) × *P. serratifolia* (Desf.) Kalkman (E As.)) esp. **'Red Robin'**, & ***P.* 'Redstart'** (*P. × fraseri* × *P. davidiana* (Decne) Cardot (*S. d.*, Vietnam, China))

Photinopteris J. Sm. = *Aglaomorpha*

Phragmanthera Tieghem (~ *Tapinanthus*). Loranthaceae (5 6). 34 trop. Afr., Arabia. R: R. Polhill & D. Wiens, *Mistletoes of Afr.* (1998)246. Some pests of plantation crops, e.g. ***P. capitata*** (Spreng.) Balle (W Afr.) – ant-nests in stems

Phragmipedium Rolfe. Orchidaceae (III). c. 25 trop. Am. R: C. Cash, *Slipper orchids* (1991)137. Cult. orn. terr. orchids incl. ***P. caudatum*** (Lindl.) Rolfe with P lobes to 60 cm long!

Phragmites Adans. Gramineae (25). 4 or 5 cosmop. (Eur. 1). R: JAA 71(1990)156. ***P. australis*** (Cav.) Steud. (*P. communis*, reed, nal (Ind.), Danube grass, cosmop., though E N Am. genotypes largely replaced by aggressive Europeans, also invasive Aus., NZ) – forms floating fens as at mouth of Danube, where harvested, the pulp used for paper, cellophane, cardboard & synthetic textiles, fibre-board, fuel, alcohol, insulation & fertilizer, its growth some of the most rapid of any pl., grains consumed by Native Americans, rhiz. by Aus. Aborigines, young shoots by Jap., lvs &/or culms used for mats (durma m.) in Ind., fish-traps & basket-chairs in Afr. & thatch (now from Lake Baikal) in Netherlands, culms used like quill pens by anc. Egyptians

Phragmocarpidium Krapov. = *Hibiscus* (but see Bonpl. 3(1969)14)

Phragmorchis L.O. Williams. Orchidaceae (V 16). 1 Philippines: ***P. teretifolia*** L.O. Williams

Phragmotheca Cuatrec. Malvaceae (Bomb.; Bombacaceae). 11 Panamá to Peru. R: Britt. 43(1991)73, Caldasia 18(1996)258. Some ed. fr.

Phravenia Al-Shehbaz & Warwick. Cruciferae (46). 1 SW N Am.: ***P. viereckii*** (O. Schulz) Al-Shehbaz & Warwick. R: T 60(2011)1161

Phreatia Lindl. Orchidaceae (V 15). c. 200 NE Ind. to Taiwan & Aus. (esp. NG)

Phrissocarpus Miers = *Tabernaemontana*

Phrodus Miers = *Lycium* (but see Kurtziana 19(1987)69)

Phryganocydia Mart. ex Bur. = *Bignonia*

Phrygilanthus Eichler = *Notanthera*

Phryma L. (~ *Mimulus*). Phrymaceae (Mim.; Scrophulariaceae s.l.). 1 Ind. to Jap., E N Am.: ***P. leptostachya*** L. R: Phytoneuron 2012–39:24

Phrymaceae Schauer (~ Orobanchaceae). Magnoliidae – Lamiales. Excl. Mazaceae, Microcarpaeae, 13/210 trop. & warm. Usu. herbs, with simple hairs. Stems apically 4-angled. Lvs simple, toothed, opp. infl. a raceme. Fls bisex. K(5); C 2-lipped, anterior 3-lobed; A2–4 inserted in corolla, anterior pair longer; (other s.t. staminodes or 0); G(2), 1-loc. with 1 sub-basal hemitropous ovule, stigma-lobes 2 unequal. Fr. a capsule, achene or berry; endosperm scanty

Principal genera: **Leucocarpeae** (cymes): *Hemichaena, Leucocarpus* (only); **Mimuleae** (racemes): *Diplacus, Erythranthe, Glossostigma, Mimulus, Peplidium, Phryma*

Some genera placed here (Microcarpaeae) now in Linderniaceae (form. Scrophulariaceae s.l. via Plantaginaceae-Grateoleae); other genera now Mazaceae. Cult. orn. esp. *Diplacus, Erythranthe, Mimulus*

Phryna (Boiss.) Pax & K. Hoffm. = *Phrynella*

Phryne Bubani = *Sisymbrium*

Phrynella Pax & K. Hoffm. Caryophyllaceae (III 1). 1 Turkey: ***P. ortegioides*** (Fischer & C. Meyer) Pax & K. Hoffm.

Phrynium Willd. Marantaceae. Incl. *Cominsia, Monophrynium, Phacelophrynium*, c. 40 Indomal. to Pacific. Local basketry. ***P. giganteum*** Scheffer (*C. g.*) – lvs used like plates for taro & sweet potato in Polynesia; ***P. maximum*** Blume (*Phacelophrynium m.*, W Mal.) – grown in Borneo for lvs to wrap food. See also *Ataenidia*

Phtheirospermum Bunge ex Fischer & C. Meyer. Orobanchaceae (Rhin.; Scrophulariaceae). 4 E As.

Phthirusa Mart. Loranthaceae (4 4). Incl. *Ixocactus*, excl. *Dendropemon, Passovia*, 7 Mex. (2), N Andes (3), E Braz. (2). R: PDE 129(2011)159. ***P. caribaea*** (Krug & Urb.) Britton & P. Wilson (WI mistletoe, Carib.) – pest on woody crops

Phuodendron (Graebner) Dalla Torre & Harms = *Valeriana*

Phuopsis (Griseb.) Hook.f. Rubiaceae (IV 16). 1 Caucasus, E Turkey, NW Iran: ***P. stylosa*** (Trin.) B.D. Jackson (*Crucianella s.*) – cult. orn. ann. R: FOC 19(2011)291

Phuphanochloa Sungkaew & Teerawat. = *Bambusa* (but see KB 63(2008)669)

Phycella Lindl. (~ *Hippeastrum*). Amaryllidaceae. Incl. *Famatina*, c. 6 Chile, Arg., Venez.

Phygelius E. Meyer ex Benth. Scrophulariaceae (Frey.). 2 S Afr. R: Plantsman 9(1988)233. Cult. orn. subshrubs esp. ***P. capensis*** E. Meyer ex Benth. (Cape figwort) – pendent scarlet fls & ***P. × rectus*** Coombes (*P. capensis × P. aequalis* Harv. ex Hiern (fls pink)) – many cvs

Phyla Lour. (~ *Lantana*). Verbenaceae (Lant.). 5 trop. & warm. R: AMBG 98(2012)578. Bell's Model. Some cult. orn. ground-cover esp. ***P. nodiflora*** (L.) Greene (lippia) – bee-fodder, local medic., now trop. 7 temp., v. invasive Aus.

Phylacium Benn. Leguminosae (III 18). 2 SE As. to NE Aus. R: Blumea 24(1978)485

Phylactis Schrad = *Philactis*

Phylica L. Rhamnaceae (3). 188 S Afr. (133 (126 endemic) in Cape – R: Strelitzia 9(2000)13,599), Madag. & Tristan da Cunha (incl. Gough Is. where ant-disp. ***P. arborea*** Thouars not regenerating as ants gone & introd. mice eat seeds). R: JSAB 8(1942)3. Mostly heath-like shrubs (Leeuwenberg's Model) with revolute lvs

Phyllacantha Hook.f. = *Phyllacanthus*

Phyllacanthus Hook.f. Rubiaceae (inc. sed.). 1 W Cuba: ***P. grisebachianus*** Hook.f. – prob. extinct. R: HIP 11(1871)77

Phyllachne Forst. & Forst. f. = *Forstera*

Phyllactis Pers. = *Valeriana*

Phyllagathis Blume. Melastomataceae. Incl. *Brittenia, Cyanandrium, Enaulophyton, Tylanthera*, 58 S China to W Mal. R (p.p.): Blumea 47(2002)463. Apical part of 'leaf' of ***P. scortechinii*** King & other spp. (Malay Pen.) falls, the basal part bears infls & buds. Some local medic. ***P. steenisii*** Cellin. (*E. lanceolatum*, Borneo) – rheophyte; ***P. subacaulis*** (Cogn.) Cellin. (*B. a.*, Sarawak) – acaulescent herb

Phyllangium Dunlop (~ *Mitrasacme*). Loganiaceae. 5 temp. Aus. R: FA 28(1996)59

Phyllanoa Croizat. Violaceae. 1 Colombia: ***P. colombiana*** Croizat – coll. once, long referred to Euphorbiaceae. R: A. R.-Sm., *Gen. Euph.* (2001)68

Phyllanthaceae Martinov (~ Euphorbiaceae). Magnoliidae – Malpighiales. 57/2000 trop. & warm (N to Iberia – *Flueggea*). R: Mol. Phyl. Evol. 36(2005)112. Trees to herbs accum. aluminium, usu. monoec.; milky latex 0. Lvs in spirals, oft. becoming distich., stipulate. Fls in usu. axillary thyrsoid infls, s.t. reduced to glomerules, or solit.; K (2–)4–6(–12), usu. imbr., oft. basally connate, C (0, 3–)5(–9), disk usu. present, A (2)3–8(–60) oft. ± connate, extrorse, $\underline{G}$ ((1)2–5(–20)), each loc. with 2 anatr. or hemitr. ovules (cf. E.). Fr. usu. an explosively dehiscent septicidal capsule or schizocarp, rarely a berry or drupe. Seeds usu. without caruncles & usu. copious endosperm. n = (8,10, 11)12 or 13

Classification & principal genera:

Euphorbiaceae – Phyllanthoideae (exc. Drypeteae = Putranjivaceae) long considered distinct in terms of seed structure; a number of tribes in E. – Phyll. form. recog., prob. best arr. as 2 subfams

I. **Phyllanthoideae** (infls usu. fasciculate): *Bridelia, Phyllanthus, Sauropus*

II. **Antidesmatoideae** (usu. tanniferous): *Antidesma, Aporosa, Baccaurea, Bischofia, Cleistanthus, Uapaca*

Picrodendraceae poss. to be incl.

Fruit trees (*Antidesma, Baccaurea, Phyllanthus*), timber (*Antidesma, Aporosa, Bischofia, Bridelia, Heywoodia, Hieronyma, Richeria, Uapaca*), local. medic. (*Phyllanthus*) & cult. orn. (*Breynia*)

Phyllanthera Blume = *Cryptolepis*

Phyllanthodendron Hemsl. = *Phyllanthus* (but poss. distinct)

Phyllanthopsis (Scheele) Voronts. & Petra Hoffm. (~ *Phyllanthus*). Phyllanthaceae. 2 Mex. R: KB 63(2008)47

Phyllanthus L. Phyllanthaceae (Phyll.-Phyll.; Euphorbiaceae s.l.). 800+ trop. & warm (Afr. c. 150, Madag. c. 70, Am. c. 200). R (WI): JAA 37–39(1956–8) var. pp. To be split up excl. subg. *Phyllanthodendron* etc., or incl. *Glochidion*. Monoec. or dioec. trees, shrubs & herbs (Roux's & Troll's Models) oft. with flattened leaf-like organs bearing fls on margins or branches with lvs arr. so that branches appear like pinnate lvs; alks; some nickel accumulators (New Caled.; ***P. balgooyi*** Petra Hoffm. & A. Baker (Sabah, Palawan) to 1.6% dry weight) & rheophytes. Some ed. fr., cult. orn. etc. (e.g. for fish-poisons & insecticides in Amazon). ***P. acidus*** (L.) Skeels (*P. distichus*, Otaheite gooseberry, orig. S Am.?, natur. widely) – trimonoecy, fr. used in pickles & preserves; ***P. amarus*** Schum. & Thonn. (? orig. trop. Am., now pantrop. weed) – extracts reduce or eliminate detectable hepatitis B virus surface antigen in humans; ***P. buxifolius*** (Blume) Muell. Arg. (orig.?) – trop. cult. orn.; ***P. emblica*** L. (emblic, ambal, amioki, Ind. gooseberry, trop. As.) – animal-disp. the drying endocarp splitting explosively to release seeds, ed. fr. (acid at first then sweet, saliva breaking down glycoside but rich in vitamin C & minerals), medic., tanbark, fuelwood & dyes, hair-oil (amla); ***P. fluitans*** Benth. ex Muell. Arg. (S Am.) – floating aquatic like *Salvinia*, unique in fam.; ***P. mirabilis*** Muell. Arg. (NW Thailand) – subsucc. bottle-shaped trunk, on limestone; ***P. muellerianus*** (Kuntze) Exell (trop. Afr.) – chewing-stick in W Afr.; ***P. myrtifolius*** (Wight) Muell. Arg. (Sri Lanka) – cult. trop. As. for hedging, edging; ***P. niruri*** L. (trop. Am.) & other spp. (esp. ***P. urinaria*** L. dukung anak, trop. As.) locally medic. (malaria in Sabah) in As.; ***P. nummulariifolius*** Poir. (trop. & S Afr. to Seychelles) – stems for basket-making in Uganda; ***P. piscatorum*** Kunth (NE S Am.) – fish-poison in Colombia; ***P. tenellus*** Roxb. (Madag.?) – now cosmop. weed in trop. & subtrop. gardens

Phyllapophysis Mansf. = *Catanthera*

Phyllarthron DC. Bignoniaceae (6). 15 Madag., Comoro Is. R: FMad. 178(1938)68. Leeuwenberg's Model; lvs resembling linear series of articulated segments, app. rachis & petiole wings of ancestral compound leaf (cf. *Citrus*)

Phyllis L. Rubiaceae (IV 13). 2 Macaronesia

Phyllitis Hill = *Asplenium*

Phyllitopsis Reichst. = *Asplenium*

Phyllobaea Benth. = seq.

Phylloboea Benth. = *Paraboea*

Phyllobolus N.E. Br. = *Mesembryanthemum* (but see BJ 119(1997)148)

Phyllobotryon Muell. Arg. Achariaceae (Flacourtiaceae). Incl. *Phylloclinium*, 5 trop. Afr. R: BJBB 51(1981)426. Corner's Model (some 'litterbox' pls); infls epiphyllous

Phyllobotryum Muell. Arg. = praec.

Phyllocactus Link = *Epiphyllum*

Phyllocara Guşul. (~ *Anchusa*). Boraginaceae (B.2.1.). 1 Turkey: ***P. aucheri*** (DC.) Guşul.

Phyllocarpus Riedel ex Tul. = *Barnebydendron*

Phyllocephalum Blume. Compositae (Vern. – Centrop.). c. 9 Indomal. R: Rhodora 83(1981)9

Phyllocharis Diels = *Ruthiella*

Phyllocladaceae Bessey = Podocarpaceae

Phyllocladus Rich. ex Mirb. Podocarpaceae (Phyllocladaceae). 4 Borneo & N Luzon to NG, Tasmania, NZ. Fossils in S S Am., W Antarctica, NZ, S Aus.; migration to Mal. prob. recent (Keng). R: JAA 59(1978)249. Celery pines. Rauh's Model; sylleptic growth giving photosynthetic units with 3 orders of branching; lvs needle-like in seedlings but ephemeral & non-photosynthetic in adults; poll. drop receives pollen directly & its presence promotes metabolic re-absorption of fluid. Cult. orn. v. slow-growing evergs. & timber; bark of NZ spp. esp. ***P. trichomanoides*** D. Don (tanekaha) form. used for red dye by Maoris, timber valuable. ***P. aspleniifolius*** (Lab.) Hook.f. (celery-top pine, Tasmania) – imp. timber, trees up to 780 yrs old so poss. use in dendrochronology

Phylloclinium Baill. = *Phyllobotryon*

Phyllocomos Masters = *Anthochortus*

Phyllocosmus Klotzsch = *Ochthocosmus*

Phyllocrater Wernham (~ *Oldenlandia*). Rubiaceae (IV 1). 1 Borneo: ***P. gibbsiae*** Wernham

Phylloctenium Baill. Bignoniaceae (6). 3 Madag. R: FMad 178(1938)65

Phyllocyclus Kurz (~ *Canscora*). Gentianaceae. 5 Myanmar to S China. R: Blumea 48(2003)33. Bracts perfoliate

Phyllodes Lour. Older name for *Stachyphrynium*

Phyllodium Desv. (~ *Desmodium*). Leguminosae (III 19). 8 S As., N Aus. Local medic.

Phyllodoce Salisb. Ericaceae (V 3). 8 circumpolar, N temp. (Eur. 1, N Am. 5). R: Plantsman 10(1988)88. Some cult. orn. everg. heathers esp. ***P. caerulea*** (L.) Bab. (circumpolar, N temp.) – protected sp. in GB, rest in mts W N Am. W to Jap. Hybrids with spp. of *Kalmiopsis* & *Rhodothamnus*

Phyllogeiton (Weberb.) Herzog = *Berchemia*

Phylloglossum Kunze = *Huperzia* (but see OB 92(1987)170)

Phyllogonum Cov. Older name for *Gilmania*

Phyllolepidum Trinajstić (~ *Alyssum*). Cruciferae (2). 2 Eur. R: PB 145(2011)827

Phyllomelia Griseb. (~ *Mazaea*). Rubiaceae (I 5). 1 W Cuba (serpentine): ***P. coronata*** Griseb. R: Britt. 51(1999)227

Phyllonoma Willd. ex Schultes. Phyllonomaceae (Grossulariaceae s.l.). 4 Mex. to NW Bolivia. Some Al hyperaccumulators. R: Britt. 29(1977)69. Infls epiphyllous, those of ***P. ruscifolia*** Willd. ex Schultes (*P. integerrima*), at least, initiated on leaf primordia & not 'adnate' etc.; ***P. laticuspis*** (Turcz.) Engl. – imp. local medic. Mex., reputedly remedy for smallpox

Phyllonomaceae Small (~ Aquifoliaceae). Magnoliidae – Aquifoliales. 1/4 trop. Am. Glabrous (exc. stip.) trees & shrubs, s.t. epiphytic, accum. aluminium. Lvs simple, in spirals; stip. fimbriate. Infls monochasial cymes of small fls on adaxial surface of lvs, bisex. K (4) 5 free or weakly connate at base, C (3–)5, valvate, A free alt. with C, marcescent, anthers with lat. slits, disk, $\overline{G}$ (2(3)), 1-loc. with parietal placentation & ∞ ? unitegmic ovules, style usu. bifid. Fr. a 3–10-seeded berry; seeds with v. small straight embryo & fleshy endosperm

Genus: *Phyllonoma*

Form. part of extremely heterogeneous Grossulariaceae

Phyllopentas (Verd.) Kårehed & B. Bremer (~ *Pentas*). Rubiaceae c. 12 trop. Afr. R: T 56(2007)1076, SB 36(2011)1025

Phyllophyton Kudô = *Marmoritis*

Phyllopodium Benth. (~ *Polycarena*). Scrophulariaceae (Man.). 26 S Afr. esp. W & SW Cape. R: O. Hilliard, *Manuleeae* (1994)429

Phyllorachis Trimen. Gramineae (IV 1). 1 S trop. Afr.: ***P. sagittata*** Trimen

Phylloscirpus C.B. Clarke. Cyperaceae (II 1). 3 Andes

Phyllosma H. Bolus ex Schltr. Rutaceae (I 4). 2 SW Cape mts. R: Strelitzia 9(2000)635, JSAB 47(1981)755

Phyllospadix Hook. Zosteraceae. 5 Jap., W N Am. R: C. den Hartog (1970) *Seagrasses of the world*, 5. Some locally eaten roots, also lvs when herring eggs on them

Phyllostachys Sieb. & Zucc. Gramineae (IV). 51+ Himal. to Jap. esp. China (51, 49 endemic – R: FOC 22(2006)163). Heterogeneous. Bell's Model. Largest & most frequently cult. hardy bamboos forming thickets by rhizomatous spread, some used for paper, clothing fibre (T-shirts, socks, nappies, towels, knitting yarn etc.), timber, fishing-rods & ed. shoots; stripped stems of ***P. nigra*** (Lindl.) Munro (black bamboo, China, long cult. Jap.) & other spp. the w(h)angee canes used as walking-sticks, umbrella-handles, musical instruments, furniture (purplish-black culms) etc. ***P. aurea*** Carr. ex M. Riv. & Riv. (fish-pole bamboo, SE China, natur. Jap.) – stems for fishing-rods, plant-stakes etc., young shoots ed.; ***P. dulcis*** McClure (C China) – best-flavoured ed. b. of C China; ***P. edulis*** (Carr.) Houz. (*P. pubescens*, China introd. Jap., now 50% of Chinese bamboo forests) – major source of ed. bamboo shoots esp. those tinned for export from China & Jap.; ***P. nigra*** (Lindl.) Munro (China) – common cult. orn.; ***P. reticulata*** (Rupr.) K. Koch (*P. bambusoides*, madake b., China, long cult. Jap.) – most imp. timber b. in E, in plantation pulp prod. twice that of pine, Jap. flutes (shakuhachi) cult. orn. to 25 m tall (fl. cycle 120 yrs)

Phyllostegia Benth. Labiatae (VI). 1 Tonga, 1Tahiti, 32 Hawaii. R: Novon 9(1999)265. App. derived from N Am. *Stachys* spp.

Phyllostelidium Beauverd = *Baccharis*

Phyllostemonodaphne Kosterm. Lauraceae (I). 1 SE Braz.: ***P. geminiflora*** (Mez) Kosterm. R: BJ 110(1988)167

Phyllostylon Capanema ex Benth. Ulmaceae. 2 trop. Am. R: Sida 15(1992)263. Fine-grained yellow wood oft. ebonized esp. ***P. brasiliensis*** Capanema ex Benth. (baitoa, San Domingo boxwood)

Phyllota (DC.) Benth. (~ *Pultenaea*). Leguminosae (III 13). 11 SW & E Aus. R: PLSNSW II,9(1966)341

× **Phyllothamnus** Schneider. Ericaceae (V 3). Hybrids between *Phyllodoce* & *Rhodothamnus* spp.

Phyllotrichum Thorel ex Lecomte. Sapindaceae. 1 SE As.: ***P. mekongense*** Lecomte

Phylloxylon Baill. Leguminosae (III 15). 7 Madag. R: KB 50(1995)479. Most (critically) endangered. Hard timber

Phylogyne Salisb. ex Haw. = *Narcissus*

Phylohydrax Puff (~ *Hydrophylax*). Rubiaceae (IV 15). 2 E Afr., Madag. R: PSE 154(1986)343. Dune pioneers

Phymaspermum Less. Compositae (Anth.-Phym.). 17 S Afr. R: SB 41(1986)435

Phymatarum Hotta (~ *Schismatoglottis*). Araceae (VII 8). 1 Borneo: ***P. borneense*** Hotta – rheophyte. R: Telopea 9(2000)198

Phymatidium Lindl. Orchidaceae (V 12h). 10 Braz. R: KB 62(2007)530. ***P. tillandsiodes*** Barb. Rodr. – bromeliad-like, growing with bromeliads

Phymatocarpus F. Muell. = *Melaleuca* (but see Muelleria 12(1999)133)

Phymatochilum Christenson = *Miltonia*

Phymatopteris Pic. Serm. (~ *Phymatopsis*). Polypodiaceae (II). Spp.?

Phymatosorus Pichi-Serm. = *Microsorum*

Phymosia Desv. Malvaceae (Malv.- Malv.). 8 C Am. R: Madroño 21(1971)153. Some cult. orn. trees & shrubs

Phyodina Raf. = *Callisia*

Physacanthus Benth. Acanthaceae (III 2a). 5 trop. Afr.

Physaliastrum Makino = *Withania*

Physalidium Fenzl = *Graellsia*

Physalis L. Solanaceae (IV 5c). Excl. *Alkekengi*, c. 90 warm Am. esp. Mex. Cult. orn. & some ed. fr. R (partial): Rhodora 69(1967)82, 203, 319. Alks; some local medic. ***P. peruviana*** L. (Cape gooseberry, Incaberry, pichuberry, trop. S Am., invasive Aus.) – fr. small, ed., rich in vitamin C; ***P. philadelphica*** Lam. (*P. ixocarpa*, tomatillo, jamberry, Mex., natur. E N Am.) – fr. yellow to purple, ed.; ***P. pruinosa*** L. (strawberry tomato, E N Am.) – common cult. 'husk tomato', fr. yellow, ed.; ***P. viscosa*** L. (Am.) – natur. Aus., noxious weed in Victoria

Physandra Botsch. = ?*Halimocnemis*

Physaria (Torr. & A. Gray) A. Gray. Cruciferae (40). Incl. *Lesquerella*, 106 NE Russia, W N Am., S S Am. R: Rhodora 41(1939)392; R.C. Rollins & E.A. Shaw (1973) *The genus L. in N. America*. 11 with chromosomes reduced to 2n = 4. Local medic.; some cult. orn. rock-pls (bladder pod), ***P. fendleri*** (A. Gray) O'Kane & Al-Shehbaz (*L. f.*, SW N Am.) – cult. instead of castor oil for hydroxy fatty acids

Physena Noronha ex Thouars. Physenaceae (Capparaceae s.l.). 2 Madag.

Physenaceae Takht. (~ Capparaceae). Magnoliidae – Caryophyllales. 1/2 Madag. Dioec. trees or shrubs. Lvs entire, glabrous, distich.; stip. 0. Fls in axillary racemes, reg,

hypogynous. K 5–9, imbr.; C 0; A (8-)10–14(-25) with long anthers opening by slits; G(2) with 2 sub-basal campylotropous bitegmic ovules per loc. Fr. inflated; seed 1, endosperm 0
Genus: *Physena*

Physetobasis Hassk. = *Holarrhena*

Physeterostemon Goldenberg. Melastomataceae. 4 Bahia, Braz. R: T 55(2006)966

physic nut *Jatropha curcas*

Physinga Lindl. = *Epidendrum*

Physocalymma Pohl. Lythraceae. 1 trop. S Am.: ***P. scaberrimum*** Pohl – oil used in scent-making

Physocalyx Pohl. Orobanchaceae (Esc.; Scrophulariaceae s.l.). 2 Braz.

Physocardamum Hedge. Cruciferae (2). 1 E Turkey: ***P. davisii*** Hedge. R: NRBGE 28(1968)293

Physocarpus (Camb.) Raf. Rosaceae (Ros.-Neill.). 8–10 N Am., NE As. Cult. orn. shrubs with inflated follicular fr. esp. ***P. opulifolius*** (L.) Maxim. (ninebark, C & E N Am.) – bark scales off in narrow strips, local medic.

Physocaulis (DC.) Tausch = *Myrrhoides*

Physoceras Schltr. Orchidaceae (IV 4d). 12 Madag.

Physochlaina G. Don. Solanaceae (IV 2). 8 Caucasus to Ind. & China. R: BBMNHB 27(1997)27. Alks. Some cult. orn. perenn. herbs

Physodium C. Presl = *Melochia*

Physogyne Garay = *Schiedeella*

Physokentia Becc. Palmae (V 14c). 7 New Britain, Solomon Is., Vanuatu, Fiji. R: Principes 13(1969)120

Physoleucas (Benth.) Jaub. & Spach = *Leucas*

Physominthe Harley & Pastore (~ *Hyptis*). labiatae (VII 3c). 2 Braz. R: KB 70,1(2015)1

Physoplexis (Endl.) Schur (~ *Phyteuma*). Campanulaceae (I 9). 1 Alps: ***P. comosa*** (L.) Schur (*Phyteuma c.*) – cult. orn. rock-pl.

Physopsis Turcz. Labiatae (IV 1). 5 W Aus. R: Nuytsia 11(1996)101

Physoptychis Boiss. Cruciferae (2). 2 E Turkey, NW Iran. R: Phytotaxa 258(2016)75

Physopyrum Popov. = *Persicaria*

Physorhynchus Hook. Cruciferae (12). 2 S Iran to NW Ind.

Physosiphon Lindl. = *Pleurothallis*

Physospermopsis H. Wolff. Umbelliferae (III 5). 15 C to E As. R: FR 111(2000)535

Physospermum Cusson ex Juss. Umbelliferae (III 5). 2 temp. Euras. (Eur. 2). Bladderseed

Physostegia Benth. Labiatae (VI). 12 N Am. R: CGH 211(1982)1. ? Amphidiploid origin (n = 19). Cult. orn. herb. pls esp. ***P. virginiana*** (L.) Benth. (obedient plant) – if fls moved sideways, they stay put – catalepsy, the fls not springing back because of rigidity of bracts & friction between trichomes of bract & those of calyx & pedicel

Physostelma Wight = *Hoya*

Physostemon Mart. = *Cleome*

Physostigma Balf. Leguminosae (III 18). 4 trop. Afr. Alks. ***P. venenosum*** Balf. (Calabar or ordeal bean, trop. W Afr., introd. Am. & As.) – ordeal poison (physostigmine) an alk. used in ophthalmic med. (1934 first – but no longer – used for myasthenia gravis paralysis), inhibiting acetylcholine esterase interfering with nerve impulse transmission, still for squint & glaucoma

Physothallis Garay = *Pleurothallis*

Physotrichia Hiern. Umbelliferae (III 8). 10 trop. Afr.

Phytelephas Ruíz & Pavón. Palmae (IV 3). (Incl. *Palandra*) 6 trop. Am. R: OB 105(1991)48. Dioec. unarmed short-stemmed palms with A 900+ & aggregated fr., the 'cellulose' (2 mannans) endosperm v. hard (vegetable ivory) used for billiard-balls, buttons, chessmen, dice, etc., esp. that of ***P. macrocarpa*** Ruíz & Pavón (ivory nuts, tagua), though the best quality that of ***P. aequatorialis*** Spruce (*Palandra a.*) – male infls ed. humans & cattle, fr. ed.; ***P. seemannii*** Cook – stem rots at base & new roots arise above, so pls potentially immortal, crown traps litter (adventitious roots between lvs absorbing nutrients) but also colonized & then 'strangled' by *Cecropia obtusifolia*

Phyteuma L. Campanulaceae (I 9). 22 Eur. (esp. Alps), 1 ext. to N. Am. R: R. Schulz (1904) *Monogr. P.* Horned rampion. Fls in heads, tips of C forming tube with anthers inside, the style pushing pollen out where insects take it; later style emerges & stigmas open while C separate & recurve (cf. Compositae). Some cult. orn. rock- & border-pls incl. ***P. orbiculare*** L. (Eur.) – root s.t. ed. in salad etc. (rampion)

Phytocrene Wall. Icacinaceae. 11 SE As., Mal. Lianes with large vessel-elements – sources of potable water when cut

Phytolacca Tourn. ex L. Phytolaccaceae (I). 25 trop. & warm. R: AMBG 55(1968)303. Trees, shrubs & herbs (Leeuwenberg's & Koriba's Models) with alks, some poisonous, some used as veg., cult. orn. etc., medic. (antiviral proteins used against leukaemia): ***P. acinosa*** Roxb. (incl. *P. esculenta*, E As., natur. trop. As.) – young lvs & shoots a veg.; ***P. americana*** L. (*P. decandra*, pokeweed, pigeonberry, inkberry, N Am.) – cult. orn., dye from berries used to colour ink, wine, sweets, potherb ('poke salad'; boiled water to be discarded – toxic), kills snails which carry bilharzia, medic. (antirheumatic); ***P. dioica*** L. (trop. S Am.) – dioec. everg. tree with thick trunk & wide-spreading surface roots, cambia successive formed in layers external to earlier ones, planted in S Eur. etc. as shade-tree (bella umbra); ***P. dodecandra*** L'Hérit. (endod, trop. & S Afr., Madag.) – v. poisonous, molluscicidal saponins (triterpenes) used in field trials; '***P. electrica***' Lévy (Nicaragua) – alleged to have given electric shock to its collector (1876); ***P. octandra*** L. (inkweed, NZ) – introd. Aus.

Phytolaccaceae R. Br. Magnoliidae – Caryophyllales. Excl. Lophiocarpaceae, Microteaceae, Petiveriaceae, Sarcobataceae, 4/30 trop. & warm esp. Am. R: Pfl. 4(1960)83. Trees, shrubs, lianes or usu. herbs, oft. glabrous & somewhat succ., producing betalains & not anthocyanins; sec. growth with concentric rings of vasc. bundles; sieve-tubes with P-type plastids incl. central globular protein crystalloid & a sub-peripheral ring of proteinaceous filaments. Lvs simple, in spirals; stip. minute or 0. Fls usu. reg., usu. bisex., small in spikes to panicles, rarely in cymes or solit.; P 4 or 5(–10), s.t. basally connate, A (2–)4(–∞), when ∞ developing centrifugally, anthers with longit. slits, $\underline{G}$ (1)2–17, ± connate but with distinct styles & as many loc. as G, each loc. with 1 usu. basal, campylotropous bitegmic ovule. Fr. a drupe or nut, carpels oft. separating; seeds with embryo curved around ± copious hard or starchy perisperm. n = 9 (true endosperm 0) & sometimes an aril (*Barbeuia*)

Classification & chief genera:

I. **Phytolaccoideae** (P 5, G 3–17): *Ercilla, Phytolacca*

[II. **Rivinoideae** (P 4(5 in *Seguieria*)), G 1: *Hilleria, Petiveria, Rivina*; = Petiveriaceae]

[III. **Microteoideae** (P usu. 5, G (2)): *Microtea*; = Lophocarpaceae, Microteaceae]

IV. **Agdestidioideae** (Agdestidaceae; P 4, G (3)4): *Agdestis* (only) – poss. referable to Sarcobataceae

[V. **Barbeuioideae** (P 5, G 2, fr. a capsule): *Barbeuia* (only) = Barbeuiaceae]

Some cult. orn. (*Agdestis, Ercilla, Phytolacca*), locally medic., potherbs etc. (esp. *Phytolacca*)

Phytomorula Kofoid = *Acacia*. Described as an alga, *P. regularis* Kofoid actually *A.* pollen grains!

Piaggiaea Chiov. = *Wrightia*

pianograss *Themeda arguens*

Piaranthus R. Br. Apocynaceae (V b; Asclepiadaceae III 5). 8 Namibia, S Afr. Cult. orn. succ. R: SB 24(1999)394

piassava or **piassaba** Fibre from var. palms used as bristles of scrubbing-brushes, brooms, mechanical road-sweepers (bass broom fibre) etc. **African p.** *Raphia hookeri, R. palma-pinus*; **Bahia p.** *Attalea funifera*; **Ceylon p.** *Caryota urens*; **Madag. p.** *Dypsis fibrosa*; **Pará p.** *Leopoldinia piassaba*

Picardaea Urb. Rubiaceae (Cond.). 2 Cuba, Hispaniola

piccabeen *Archontophoenix cunninghamiana*

Picconia A. DC. Oleaceae (4d). 2 Macaronesia. ***P. excelsa*** (Aiton) DC. – part of orig. laurel forest of Madeira

Picea A. Dietr. Pinaceae. 34 cool N hemisph. (Eur. 2). R: RV 121(1990)221. Spruce (from *Pruce*, i.e. Prussia). Monoec. evergreens differing from *Abies* in persistent raised leaf-bases & pendulous woody female cones with persistent scales; cones mature in 1 yr. Imp. softwoods (e.g. outer case frame in Steinway pianos) used in chipboard, hardboard, as pulp (cotton subs., Cellucotton developed USA WW I for wound-dressings, later tampons & tissues (1925)), cellophane & rayon; local medic. (esp. N Am.) for resp. problems; resin chewed as confection. ***P. abies*** (L.) Karsten (*P. excelsa*, Norway spruce, N & C Eur., native in GB in last interglacial, now reached 46.7 m in surrey) – common Christmas tree in GB (introd. thus by Queen Charlotte), pulp & timber ('white deal', Baltic whitewood) form. much used for telegraph poles & (because of resonance) violins & cellos, bark used in tanning (Germany), source of Jura turpentine & Burgundy pitch used in wound-dressings, young shoots & lvs basis of spruce beer, many cult. orn. cvs incl.

dwarfs esp. '**Clanbrasiliana**' c. 2 m tall; *P.* ***breweriana*** S. Watson (S Oregon & N Calif.) – striking drooping branches; *P.* ***glauca*** (Moench) Voss (*P. alba*, white s., N N Am.) – timber & pulp, imp. local medic., fine roots used for sewing bark canoes, lvs burnt as insecticide etc.; *P.* ***jezoensis*** (Sieb. & Zucc.) Carr. (NE As. to Jap.)– needles mixed with those of *Abies sachalinensis* to yield Jap. pine needle oil; *P.* ***mariana*** (Mill.) Britton, Sterns & Pogg. (bog or (Canadian) black s., N N Am.) – timber & pulp, local medic., resin used as chewing gum (spruce g.), boiled branches giving s. beer, trunk & twigs accum. gold & platinum, the bark arsenic; *P.* ***orientalis*** (L.) Petermann (oriental s., Caucasus, NE Turkey) – many cvs incl. dwarfs; *P.* ***rubens*** Sarg. (red s., N N Am.) – timber for pulp & planks (form. ships), major source of spruce gum; *P.* ***sitchensis*** (Bong.) Carr. (Sitka s., W N Am.) – to 1350 yrs old, 95 m tall with bole to 5.25 m diam., imp. local medic., esp. resp. problems & disinfectant, introd. GB 1834, much planted in Scottish highlands, timber (form. used in aeroplane wings & fuselage) & pulp, celebrated mutant gold-foliaged '**Bentham's Sunlight**' ('Aurea') on Queen Charlotte Is. felled 1997 as an environmental protest

pichi *Fabiana imbricata*

Pichinia S.Y. Wong & Boyce. Araceae (VII 8). 1 Sarawak: *P.* ***disticha*** S.Y. Wong & Boyce. R: GBS 61(2010)544

Pichisermollia H. Monteiro = *Areca*

Pichleria Stapf & Wettst. = *Zosima*

Pichonia Pierre (~ *Sersalisia*). Sapotaceae (IV). 12 E NG, Solomon Is., New Caled. (7; R: AusSB 25(2012)34). R: T.D. Pennington, *Sapotaceae* (1991)228

pichuberry *Physalis peruviana*

pichurim *Nectandra pichurim*

pick-a-back plant *Tolmiea menziesii*

pickerel weed *Pontederia cordata*, (E Anglia) *Potamogeton natans*

Pickeringia Nutt. ex Torrey & A. Gray. Leguminosae (III 6). 1 SW N Am.: *P.* ***montana*** Nutt. ex Torrey & A. Gray – xerophytic shrub of chapparal spreading by underground stems & rapidly re-establishing after fire

picmoc *Desmoncus orthacanthus*

Picnomon Adans. (~ *Cirsium*). Compositae (Card.-Card.). 1 Medit. (incl. Eur.), SW As.: *P.* ***acarna*** (L.) Cass. – invasive S Aus. R: FA 37(2015)57

picotee *Dianthus caryophyllus*

Picradeniopsis Rydb. ex Britton (~ *Bahia*). Compositae (Hele.-Cha.). 8 N Am. R: T 65(2016)1074

Picraena Lindl. = *Picrasma*

Picralima Pierre. Apocynaceae (I f). 1 W trop. Afr.: *P.* ***nitida*** (Stapf) T. & H. Durand (*P. klaineana*) – bark a febrifuge, seeds used as quinine subs. (alks), imp. for malaria in Cameroun. R: WAUP 96–1(1996)128

Picramnia Sw. Picramniaceae (Simaroubaceae s.l.). 45 trop. Am. *P.* ***antidesma*** Sw. (macary bitter, WI); others locally medic., dyes (e.g. *P.* ***sellowii*** Planch. (Colombia to Paraguay) for bark-cloth) etc.

Picramniaceae Fernando & Quinn (~ Simaroubaceae). Magnoliopsida – Picramniales. 3/54 trop. Am. R: T Britt. 63(2011)54. Dioec. usu. everg. trees & shrubs with bitter bark (unique sugar-linked anthracenone derivatives). Lvs pinnate, in spirals; stip. 0. Infls usu. long axillary (term. or cauliflorous) racemes or thyrses. Fls (3–)5(6)-merous; K usu. connate basally, lobes imbr. or valvate; C s.t. 0 in males, small & imbr. in females; A alt. with K; $\underline{G}$ (1–3(4)), each with 2 or 3 apical ovules, styles short. Fr. a berry or samaroid capsule; seed 1 planoconvex to narrowly ellipsoid with 0 endosperm

Genera: *Alvaradoa* (fr. a samaroid capsule), *Nothotalisia*, *Picramnia* (fr. a berry) – s.t. placed in separate subfams

Picrasma Blume. Simaroubaceae. 7 trop. Am. (Cuba 3), Indomal.(2). R: Britt. 5(1944)139. Stip. (rare in fam.), alks. *P.* ***excelsa*** (Sw.) Planch. (*Aeschrion e.*, WI) – source of quassia chips, form. used in bitters & as insecticide

Picrella Baill. (*Zieridium*, ~ *Euodia*). Rutaceae (I 2). 3 New Caledonia. R: Adansonia III, 25(2003)252

Picria Lour. (*Curanga*). Linderniaceae (Plantaginaceae (Grat.-Lind.); Scrophulariaceae s.l.). 1 Indomal. to China: *P.* ***felterrae*** Lour. – local medic. R: FOC 18(1998)29

Picridium Desf. = *Reichardia*

Picris L. Compositae (Lact.-Hypo.). c. 50 Eur. (12), Medit. (most), As., N Aus., Afr. mts. Ox-tongue (esp. *P.* ***hieracioides*** L. (Euras., widely natur.)). *P.* ***radicata*** (Forssk.) Less. (N Afr., Arabia) – used in funerary garlands in anc. Egypt

Picrodendraceae Small (~ Phyllanthaceae; Euphorbiaceae – Oldfieldioideae). Magnoliidae – Malpighiales. 25/105 trop. (to SW Aus.), esp. SW Pacific. Monoec. or dioec. trees & shrubs without milky latex. Lvs simple to palmate, usu. in spirals, s.t. verticillate, s.t. stipulate. K 3(-13) in males, (3)4–8)(-13) in females, C 0, A 2-∞ oft. extrorse, G (2)3(-5) with 2 ovules per loc. Fr. capsules like E.; seeds oft. with caruncles, endosperm usu. copious. n = 12

Principal genera: *Austrobuxus, Petalostigma, Pseudanthus, Stachystemon* (10 monospecific). Some timber (*Androstachys, Oldfieldia*)

Poss. to be combined with Phyllanthaceae

Picrodendron Griseb. Picrodendraceae (Pic.-Pic.-Pic.; Euphorbiaceae s.l.). 1 WI: ***P. baccatum*** (L.) Krug & Urb. R: JAA 65(1984)105

Picrolemma Hook.f. Simaroubaceae. 2 E Peru, Amazonian Braz. Some with hollow ant-infested stem domatia; some fish-poisons

Picrophloeus Blume (~ *Fagraea*). Gentianaceae. 4 Mal. R: GBS 64(2012)511

Picrorhiza Royle ex Benth. Plantaginaceae (Dig.-Ver.; Scrophulariaceae s.l.). Incl. *Neopicrorhiza* 4 Himal. R: OB 75(1984)56. ***P. kurrooa*** Royle ex Benth. (*P. scrophulariiflora*, kutki) – locally medic. esp. febrifuge

Picrosia D. Don. Compositae (Lact.-Mic.). 2 warm S Am.

Picrothamnus Nutt. = *Artemisia* (but see FNA 19(2006)498)

picrotoxin *Anamirta cocculus*

Pictetia DC. Leguminosae (III 11). 8 WI. R: SBM 56(1999)36. Weedy colonists

Pieris D. Don. Ericaceae (VIII 2). 7 E As., E N Am. R: JAA 63(1982)103. Cult. orn. everg. shrubs & trees incl. ***P. floribunda*** (Pursh) Benth. (fetterbush, SE US) – Leeuwenberg's Model, & ***P. japonica*** (Thunb.) G. Don (E As.) – over 40 cvs

Pierranthus Bonati. Linderniaceae (Plantaginaceae (Grat.-Lind.); Scrophulariaceae s.l.). 1 SE As.: ***P. capitatus*** (Bonati) Bonati. R: Willd. 43(2013)229

Pierrebraunia Esteves = *Arrojadoa*

Pierreodendron Engl. (~ *Quassia*). Simaroubaceae. 2 trop. Afr.

Pierrina Engl. Lecythidaceae (?IV; Scytopetalaceae). 12 W trop. Afr. R: KB 70(2015)6:53

piert, parsley *Alchemilla arvensis*

Pietrosia Nyár. = *Andryala*

pifine grass *Panicum hemitomon*

pig balsam *Tetragastris balsamifera*; **p. laurel** *Kalmia angustifolia*; **p. nut** *Conopodium majus, Carya glabra, Simmondsia chinensis*; **p.weed** *Amaranthus albus, A. retroflexus* etc.

Pigafetta (Blume) Becc. Palmae (I 3d). 2 E Mal. R: Principes 42(1998)39. ***P. filaris*** (Giseke) Becc. – fast-growing colonist to 50 m with small seeds

Pigea Ging. (~ *Hybanthus*). Violaceae. Spp.? ***P. floribunda*** Lindl. (*H. f.* Aus.) – nickel hyperaccumulator

pigeon bean *Vicia faba* var. *equina*; **p. berry** *Phytolacca americana, Duranta erecta*; **p. grass** *Setaria pumila* & other spp.; **p. orchid** *Dendrobium crumenatum*; **p. pea** *Cajanus cajan*; **p. wood** *Coccoloba* spp.

pigface *Carpobrotus edulis, Disphyma* spp. esp. *D. crassifolium*

piggyback plant *Tolmiea menziesii*

pignoli, pignon *Pinus pinea*

Pilbara Lander. Compositae (Ast.) 1 NW Aus.: ***P. trudgenii*** Lander. R: Nuytsia 23(2013)118

Pilea Lindl. Urticaceae (II). Incl. *Sarcopilea*, c. 650 trop. & warm excl. Australasia. Monoec. & dioec. herbs (Troll's Model) without stinging hairs; some greenhouse pls (clearweed) esp.: ***P. cadierei*** Gagnepain & Guillaumin (aluminium plant, Vietnam) – lvs with silvery stripes, ***P. involucrata*** (Sims) C.H. Wright & Dewar (trop. Am.) – many cvs & ***P. microphylla*** (L.) Liebm. (*P. muscosa*, artillery pl., gunpowder pl., trop. Am., widely natur. Afr. & As. & Balkans) – anthers eject pollen explosively, trad. med. S Am. (e.g. eyewash Venez.), mordant in New Britain; ***P. fairchildiana*** Jestrow & Jiménez (*S. domingensis*, Hispaniola) – *Aeonium*-like pachycaul; ***P. melastomoides*** (Poir.) Wedd. (Indomal.) – veg. in Java; ***P. peperomioides*** Diels (Yunnan, rare & endangered in wild) – elongated stems with succ. lvs like *Peperomia*, widely cult. pot-pl. introd. to W via Norway

Pileanthus Labill. Myrtaceae (II 15). 6 SW Aus. R: Nuytsia 15(2002)38

Pileostegia Hook.f. & Thomson = *Hydrangea* (but see BJLS 165(2011)368)

Pileus Ramírez = *Jacaratia*

pilewort *Ficaria verna*

Pilgerina Z. Rogers & al. (~ *Staufferia*). Santalaceae (Cervantesiacae). 1 Madag.: ***P. madagascariensis*** Z. Rogers & al. R: AMBG 95(2008)398

Pilgerochloa Eig = *Ventenata*

Pilgerodendron Florin (~ *Libocedrus*). Cupressaceae. 1 S S Am. N to 40° S: ***P. uviferum*** (D. Don) Florin – Attims's Model, timber useful. R: A. Farjon, *Monogr. Cupressaceae & Sciadopitys* (2005)452

pili *Heteropogon contortus*; **p. nut** *Canarium luzonicum, C. ovatum*

Pilidiostigma Burret. Myrtaceae (II 10). 5 NE Aus. R: SB 29(2004)394

Piliocalyx Brongn. & Gris (~ *Syzygium*). Myrtaceae. 8 New Caled.

Piliostigma Hochst. (~ *Bauhinia*). Leguminosae (I 1). 5 trop. (Afr. 2). Local medic.

Pillansia L. Bolus. Iridaceae (VI 2). 1 SW Cape: ***P. templemannii*** (Bak.) L. Bolus – up to 35 corm 'skeletons' beneath current corm. R: Strelitzia 10(2000)633

pillarwood *Cassipourea elliottii*

pillwort *Pilularia globulifera* & other spp.

Piloblephis Raf. (~ *Satureja*). Labiatae (VII 2b). 1 SE US: ***P. rigida*** (Benth.) Raf. – medic. Seminole

Pilocarpus Vahl. Rutaceae (I). 16 trop. Am. R: FN 33(1982)132. Alks. Source of medicinal jaborandi, active alk. being pilocarpine used for glaucoma (Braz. plantations of ***P. microphyllus*** Stapf ex Wardleworth (Surinam, Braz.))

Pilocereus Lem. = *Cephalocereus*

Pilocopiapoa F. Ritter = *Copiapoa*

Pilocosta Almeda & Whiffin. Melastomataceae. 5 trop. Am. R: Novon 3(1993)311

Pilogyne Schrad. = *Zehneria*

Pilophyllum Schltr. (~ *Chrysoglossum*). Orchidaceae (V 14). 1 Mal.: ***P. villosum*** (Blume) Schltr. R: OM 8(1997)170

Pilopleura Schischkin. Umbelliferae (III 8). 2 C As.

Pilorea Raf. = *Edraianthus*

Pilosella Hill (~ *Hieracium*). Compositae (Cich.-Hier.). c. 110 (sexually reprod., c. 700 apomictic 'microspp.') Euras., N Afr. Sexual spp. (cf. *Hieracium*) or partially apomictic, most hybridizing freely. Some cult. orn. but invasive (stolons) esp. ***P. aurantiaca*** (L.) F.W. Schultz & Schultz-Bip. (devil's paintbrush, fox & cubs, Eur., natur. N Am.) – florets orange-red. ***P. officinarum*** F.W. Schultz & Schult.-Bip. (mouse ear, Eur., Medit., invasive NZ) – lvs give a medic. tea (coughs incl. whooping c.)

Piloselloides (Less.) C. Jeffrey ex Cufod. = *Gerbera*

Pilosocereus Byles & G. Rowley (~ *Cephalocereus*). Cactaceae (III 3). 35–65 S US (1), Mex. to trop. S Am. esp. Braz. Cult. orn. shrubs & trees to 10 m. ***P. lanuginosus*** (L.) Byles & G. Rowley (*P. tweedyanus*, N S Am.) – sulphur compounds in scent (bat-poll.)

Pilosperma Planch. & Triana = *Clusia*

Pilostemon Iljin = *Jurinea*

Pilostigma Tieghem = *Amyema*

Pilostyles Guillemin. Apodanthaceae (Rafflesiaceae s.l.). 1 Iran, 3 SW Aus., 1 trop. Afr., c. 4 S Calif. to trop. Am. R: PhytoKeys 36(2014)50. The pls parasitize all 3 subfams of Leguminosae & are internal save fls; ***P. haussknechtii*** Boiss. (SW As.) on *Astragalus* produces paired fls at base of host lvs

Pilothecium (Kiaerskov) Kausel = *Myrtus*

Pilularia Vaill. ex L. Marsileaceae. 3–6 N temp. (Eur. 2), Aus., NZ, Ethiopian mts, W S Am. Pillwort esp. ***P. globulifera*** L. (Eur.) – grows at lake-edges, lvs filiform, entire, with pea-shaped 4-chambered sporocarp, each chamber with a sorus bearing both macro- & microsporangia, rhizome & roots with 'vessels'. ***P. minuta*** Dur. ex A. Br. (SW Eur.) – one of smallest of all ferns

Pimelea Banks ex Gaertn. Thymelaeaceae (Thym.-Daph.). (Incl. *Thecanthes*) c. 110 NZ (17) & Aus. (90 endemic; W Aus. 45, R: Nuytsia 6(1988)129) with few ext. to Philippines & Lord Howe Is. Fls in term. heads oft. surrounded by involucre of leafy & oft. coloured bracts (***P. physodes*** Hook. (W Aus.) = Qualup bell). Some cult. orn. everg. shrubs (riceflowers); some with bark used as sources of twine by Aborigines & early settlers; some toxic to stock in Aus.

Pimelodendron Hassk. Euphorbiaceae (Euph.-Stom.). 6–8 Mal. to Aus. R: Blumea 49(2004)412. ***P. amboinicum*** Hassk. (Sulawesi to Solomon Is.) – ed. seeds, bark a purgative, milky latex used as varnish

Pimenta Lindl. Myrtaceae (II 10). 15 trop. Am. R: FN 45(1986)78. Aromatic trees. ***P. dioica*** (L.) Merr. (*P. officinalis*, allspice, pimento, Jamaica pepper, C Am., WI) – cryptically dioec., spice reminiscent of cinnamon, nutmeg & cloves etc. derived from unripe fr. used as flavouring (as in Benedictine & Chartreuse) & medic. etc. (c. 5 K t a yr); ***P. racemosa***

(Mill.) J. Moore (*P. acris*, bay rum tree, trop. Am.) – introd. Pac., oil form. distilled with rum, used in scent- & soap-making (flavour due to eugenol)

Pimentelia Wedd. Rubiaceae (I 1). 1 Peru: ***P. glomerata*** Wedd. R: AMBG 82(1995)423

pimento *Pimenta dioica*

Pimia Seemann = *Commersonia*

pimpernel (**common** or **scarlet**) *Anagallis arvensis*; **blue p.** *A. monelli*; **bog p.** *A. tenella*; **yellow p.** *A. nemorum*

Pimpinella L. Umbelliferae (III 8). c. 150 Euras. (Eur. 16), Afr., W N Am. (1), S Am. Some spices & cult. orn. esp. ***P. anisum*** L. (anise, Greece to Egypt) – cult. since 2000 BC, food- & drink-flavouring subs. for *Artemisia absinthium* (absinthe, anis, anisette, arak, ouzo, pastis, raki), distilled oil medic., familiar as aniseed balls; ***P. major*** (L.) Hudson & ***P. saxifraga*** L. – burnet saxifrage, Eur.

pin cushion *Brunonia australis*; **p. oak** *Quercus palustris*

Pinacantha Gilli. Umbelliferae (inc. sed.). 1 Afghanistan: ***P. porandica*** Gilli. R: FR 61(1959)207

Pinaceae Spreng. ex Rudolphi. Pinidae. 12/225 N temp. S to C Am., WI, Sumatra & Java. R: RV 121(1990). Usu. everg., resinous monoec. trees (few prostrate shrubs) with ectomycorrhizae & usu. opp. or whorled branches & scaly, oft. resinous (resin ducts in wood & phloem), term. buds. Lvs linear, in spirals. Male cones (poss. transformed compound strobili) small, herbaceous, with microsporophylls in spirals, each with 2 pollen-sacs abaxially; female cones usu. woody, with scales in spirals, each usu. with 2 ovules adaxially & subtended by a ± united bract; seeds usu. 2 per scale & usu. winged. Embryo with (2-)4–11(-20) cotyledons. Germination usu. epigeal (hypogeal in *Keteleeria*). n = 12 (13, 22 – *Pseudolarix*)

Genera: *Abies, Cathaya, Cedrus, Hesperopeuce, Keteleeria, Larix, Nothotsuga, Picea, Pinus, Pseudolarix, Pseudotsuga, Tsuga*

Fossils from upper Jurassic but as considered 'sister' to all other extant conifers. Most imp. sources of timber & pulp, turpentine, resins, cult. orn., ed. seeds etc.

Pinacopodium Exell & Mendonça. Erythroxylaceae. 2 trop. Afr.

Pinalia Lindl. (~ *Eria*). Orchidaceae (V 15). c. 100 Indomal.

pinang *Areca catechu*

Pinanga Blume. Palmae (V 14b). 130 Indomal., SE As. Usu. undergrowth palms with up to 9 spp. growing sympatrically on limestone in Borneo; beetle-poll. (pollen v. variable) though in ***P. cleistantha*** Dransf. (Malay Pen.) & ***P. simplicifrons*** (Miq.) Becc. (Sumatra) infls enclosed in prophyll so no obvious pollinator access. Some rheophytes in Borneo; some masticatories like betel, a few cult. orn. small pls. ***P. manii*** Becc. (Andamans, Nicobars) – local palm-hearts

Pinaropappus Less. Compositae (Cich.-Cich. (Micr.)). 8 S N & C Am. R: CUMH 9(1972)371

Pinarophyllon Brandegee. Rubiaceae (IV 5). 2 C Am.

pinaster *Pinus pinaster*

Pinckneya Rich. Rubiaceae (Cond.). 1 SE US: ***P. bracteata*** (Bartram) Raf. (*P. pubens*, fever tree, Georgia bark tree) – 1 or more K lobes rose-coloured 'flags' to 7 cm long, bitter bark medic., used against malaria in Am. Civil War (cinchonin). R: PSE 201(1996)251

Pinda P.J. Mukherjee & Constance (~ *Heracleum*). Umbelliferae (III 11). 1 W Ghats, S Ind.: ***P. concanensis*** (Dalzell) P.K. Mukherjee & Constance. R: KB 41(1986)224

pine *Pinus* spp.; **Aleppo p.** *P. halepensis*; **American pitch p.** *P. palustris*; **Athel p.** *Tamarix aphylla*; **Australian p.** *Casuarina equisetifolia*; **Austrian p.** *P. nigra* subsp. *nigra*; **Bhutan** or **blue p.** *P. wallichiana*; **Brazilian p.** *Araucaria angustifolia*; **bristle-cone p.** *P. longaeva*; **bunya-bunya p.** *A. bidwillii*; **Canadian red p.** *P. resinosa*; **Caribbean pitch p.** *P. caribaea*; **celery p.** *Phyllocladus* spp.; **celery-top p.** *P. aspleniifolius*; **Chile p.** *A. araucana*; **chir p.** *Pinus roxburghii*; **Cook p.** *A. columnaris*; **Corsican p.** *P. nigra* subsp. *salzmannii*; **cypress p.** *Callitris* spp.; **Dade County p.** *P. elliotii* var. *densa*; **Dally p.** *Psoralea pinnata*; **digger p.** *Pinus sabiniana*; **golden p.** *Pseudolarix amabilis*; **hoop p.** *A. cunninghamii*; **Huon p.** *Lagarostrobos franklinii*; **Illawara p.** *C. rhomboidea*; **jack p.** *Pinus banksiana*; **Japanese red p.** *P. densiflora*; **J. white p.** *P. parviflora* var. *pentaphylla*; **King Billy** or **William p.** *Athrotaxis selaginoides*; **klinki p.** *Araucaria hunsteinii*; **knobcone p.** *P. attenuata*; **Leichhardt p.** *Nauclea orientalis*; **p. lily** *Lilium catesbaei*; **loblolly p.** *P. taeda*; **lodge-pole p.** *P. contorta* subsp. *latifolia*; **longleaf p.** *P. palustris*; **maritime p.** *P. pinaster*; **Monterey p.** *P. radiata*; **Norfolk Is. p.** *A. heterophylla*; **Oregon p.** *Pseudotsuga menziesii*; **Oyster Bay p.** *C. rhomboidea*; **Paraná p.** *A. angustifolia*; **parasol p.** *Sciadopitys verticillata*; **pencil p.** (Aus.) *Cupressus sempervirens, Polyscias murrayi*; **pitch p.** *Pinus palustris*; **plum p.** *Podocarpus elatus*; **ponderosa p.** *Pinus ponderosa*;

Port Jackson p. *Callitris rhomboidea*; **radiata p.** *P. radiata*; **red. Am. p.** *P. resinosa*; **Scots p.** *P. sylvestris*; **screwp.** *Pandanus* spp.; **silver p.** *Lagarostrobos colensoi*; **slash p.** *Pinus elliotii*; **southern p.** *P. echinata, P. elliotii, P. palustris, P. taeda*; **stone p.** *P. pinea*; **sugar p.** *P. lambertiana*; **Swiss p.** *P. cembra* (see also *Abies alba*); **umbrella p.** *P. pinea*; **western white p.** *P. monticola*; **Weymouth p.** *P. strobus*; **white (Am.) p.** *P. strobus, Podocarpus* spp.; **w. cypress p.** *C. glauca*; **Wollemi p.** *Wollemia nobilis*; **yellow p.** *Pinus echinata*; **y. Am. p.** *P. strobus*

pineapple *Ananas comosus*; **p. guava** *Acca sellowiana*; **p. lily** *Eucomis* spp.; **p.(-scented) sage** *Salvia elegans*; **p. weed** *Matricaria matricarioides*

pineberry *Fragaria×ananassa*

Pineda Ruíz & Pavón. Salicaceae (Flacourtiaceae). 2 Andes of Bolivia, Ecuador & Peru, above tree-line. ***P. incana*** Ruíz & Pavón – lvs a source of black dye. R: KB 61(2006)207

Pinelia Lindl. = *Homalopetalum*

Pinelianthe Rauschert = *Homalopetalum*

Pinellia Ten. Araceae (VII 24). 9 China, Jap. R: Willd. 37(2007)504. TCM (ban xia) esp. ***P. ternata*** (Thunb.) Makino. Some cult. orn. incl. ***P. tripartita*** (Blume) Schott (Hong Kong, S Jap.) – poss. self-poll.

piney varnish *Vateria indica*

Pinga Widjaja = *Parabambusa* (but see Reinw. 11(1997)123)

Pingraea Cass. = *Baccharis*

pingue *Hymenoxys richardsonii*

Pinguicula Tourn. ex L. Lentibulariaceae. c. 55 Am., Medit. & few circumboreal (Eur. 13). R: BB 127/8(1966)1. Small carnivorous herbs with lvs in rosettes & usu. viscid above (secretory glands), trapping small insects as in ***P. vulgaris*** L. (butterwort – lvs used to coagulate milk & vegetable rennet (tämjölk in Scandinavia), due to bacteria (also found in snail slime & *Drosera* mucilage) assoc. with mucilage rather than digestive enzymes of pl.), bog violet, N N Am., Euras. exc. NE) where leaf-margins curl over insects & sessile glands secrete enzymes which digest prey after mucilage glands collapse & epidermal cells below them lose turgor bringing insects in touch with sessile glands

pinguin fibre *Bromelia pinguin*

Pinillosia Ossa. Compositae (Helia.-Pin.). 1 Cuba, Hispaniola: ***P. berteroi*** (Spreng.) Urb.

pink *Dianthus* spp. esp. *D. plumarius* (app. from *pinkster*, i.e. Whitsun (Dutch), now used for the colour); **Cheddar p.** *D. gratianopolitanus*; **Chinese p.** *D. chinensis*; **clove p.** *D. caryophyllus*; **Deptford p.** *D. armeria*; **p.-eye** *Tetratheca* spp.; **Japanese p.** *D. chinensis* 'Heddewigii'; **maiden p.** *D. deltoides*; **mullein p.** *Silene coronaria*; **rose p.** *Sabatia* spp.; **sea p.** *Armeria maritima*

pinnay oil *Calophyllum inophyllum*

Pinochia M. Endress & B. Hansen (~ *Forsteronia*). Apocynaceae (II c). 4 C Am. R: EJB 64(2007)270

Pinophyta. See Pinidae

Pinidae Cronq. & al. (Coniferae, Pinatae, Pinophyta). Gymnospermae. 65/627 in 6 fams (1 order), trop. to cold. Dioec. or monoec. branching woody pls, oft. with long & short shoots; sec. wood usu. of tracheids, occ. vessels. Oft. with resin-canals in lvs, cortex & s.t. wood. Lvs in spirals or opp. (whorls in *Sciadopitys*), usu. needle-like or scale-like. Female cones with bract-scales united with ovuliferous scales each bearing usu. several to 2 ovules; males usu. with many scale-like microsporophylls with 2–∞ pollen-sacs. Embryo with 2–∞ cotyledons

Fams: Araucariaceae, Cupressaceae,, Pinaceae, Podocarpaceae, Sciadopityaceae, Taxaceae

Imp. timber trees & cult. orn. incl. 'dwarf conifers' (mutants or diseased clones)

Pinosia Urb. = *Drymaria*

pinto beans *Phaseolus vulgaris*

Pintoa C. Gay. Zygophyllaceae (II). 1 Chile: ***P. chilensis*** C. Gay

Pinus Tourn. ex L. Pinaceae. c. 110 N temp. (Eur. 11) & S to C Am. (47 – R: FN 75 (1997)62), Sumatra & Java. Pines. R: N.T. Mirov (1967) *The genus P.*, A. Farjon, *World Checklist … Conifers* (2001)168, T 54(2005)37. Everg. monoec. trees (Rauh's Model) with scale lvs soon lost & long needle lvs in clusters of (1)2–5, each year's growth rep. by unbranched long shoot term. in a series of buds, 1 carrying on the axial growth, the rest forming a whorl of branches in the next season. Male cones with many fertile scales, each with 2 pollen-sacs abaxially, each grain with 2 air-bladders; female cones take 2–3 yrs to mature following arrival of pollen grain which is brought into contact with nucellus by drying up of mucilage & then does not develop complete pollen tubes for another yr.

Pines among the most imp. timbers in temp. (incl. merchant ships of classical Medit. & perhaps the 'gopher wood' of the Ark ('kapher' = resinous wood)) & trop. regions – in the latter, *P. patula* & *P. caribaea* the most planted of all trees; cult. for timber, pulp (eco-friendly nappies, baby-wipes) & resinous products (incl. constituents of comm. cough mixtures in Eur.) like resin, turpentine (solvent being terpenes from distilled resins; fuel for first [1946] Honda motorcycles; 225 K t p.a. by 1994) & rosin (used on bow-strings of stringed instruments; together = naval stores, when crude = colophony, used in cigarette flavourings & as soldering flux but allergenic fumes cause asthma in some 20% of operatives) esp. from *P. palustris* & *P. pinaster*, rosin being used to size paper etc. & in ballet-dancing shoes; wood tar & pitch obtained by destructive distillation, esp. from *P. sylvestris* ('Stockholm tar'); some with ed. seeds (pignons) & evolved with bird-dispersers, or manna & many cult. orn. Although helpful in identification, the no. of needles in a cluster is not a good phylogenetic marker & the genus is divided into 2 groups marked by no. of vasc. bundles in needles, viz. sect. ***Strobus***, soft pines (timber soft, resin little) having 1, & sect. ***Pinus***, hard pines (relatively hard wood & frequently much resin) having 2. ***P. albicaulis*** Engelm. (N Am.) – seeds ed. Native Americans; ***P. armandii*** Franch. (Chinese white p., temp. C & W China) – cheap furniture, pinyon seeds in China, though in some people makes food taste bitter (pine nut syndrome, pine mouth); ***P. attenuata*** Lemmon (knobcone p., NW US) – fire opens cones left on tree for 50 yrs + & s.t. engulfed by bark first; ***P. banksiana*** Lamb. (*P. divaricata*, jack p., E N Am., invasive NZ) – local medic., pulp; ***P. bungeana*** Zucc. ex Endl. (lace-bark p., C & N China) – cult. orn. with flaky bark; ***P. canariensis*** Sweet ex Spreng. (Canary p., Canary Is.) – app. closely allied to *P. roxburghii* 8000 km away in Himal.; ***P. caribaea*** Morelet (Caribbean or Bahamas pitch p., C Am. & WI) – hard timber much planted in lowland trop., high incidence of pls ± unbranched ('foxtails') in plantation esp. in aseasonal climates; ***P. cembra*** L. (Siberian cedar, Swiss p., Euras.) – lvs for turpentine, timber for building, seeds ed. (mast-fruiting over 1000s of km^2; ***P. cembroides*** Zucc. (SW N Am.) – seeds locally imp. as food; ***P. contorta*** Douglas ex Loudon subsp. ***latifolia*** (Engelm.) Critchf. (lodge-pole p., W N Am., invasive Aus., NZ) – medic. esp. TB, coughs etc., resin chewed like gum, timber for tepee poles, imp. plantation tree introd. GB 1853/4; ***P. densata*** Masters (China, *P. tabuliformis* Carr. × *P. yunnanensis* Franch.) – one of only 8 confirmed homoploid hybrid pl. spp.; ***P. densiflora*** Sieb. & Zucc. (Jap. red p., Korea, Jap.) – typical Jap. sp. with many cvs; ***P. echinata*** Mill. (yellow or southern p., SE US) – hard wood, used to make canoes to 12 m+ long; ***P. edulis*** Engelm. (SW US) – distrib. piñon jays, prod. irregularly abundant nutritious food in which seasons bird testes grow & remain big throughout season so breeding in late winter & late summer, pitch for chewing gum, when heated used to remove facial hair (Apache), medic. (disinfectant), black & red dyes, waterproofing baskets etc., seeds traded; ***P. elliottii*** Engelm. (slash or southern p., SE US, invasive Aus., S Afr.), var. ***densa*** Little & Dorman – fire-resistant juvenile phase, v. hard & termite-proof timber; ***P. halepensis*** Mill. (Aleppo p., Medit., invasive Aus., S Afr.) – flavours retsina, turpentine; ***P. kesiya*** Royle ex Gordon (*P. khasya*, mts of SE As.) – turpentine; ***P. koraiensis*** Sieb. & Zucc. (NE As. to Jap.) – to 350 yrs old, pine nuts in China; ***P. lambertiana*** Douglas ex Taylor & Phillips (sugar p., W N Am.) – cones to 50 cm (longest in genus), timber, damaged heartwood yields sweet laxative material, seeds ed.; ***P. longaeva*** D. Bailey (Calif., Nevada & Utah) – anc. 'bristle-cone pines' confused with shorter-lived ***P. aristata*** Engelm. (Rocky Mts) to 2000 yrs old, *P. longaeva* to 5063 yrs old known (oldest living individual; one 4900 yrs old felled in 1964) & by extrapolation on dead specimens (4000 yrs to decay), material known to be 8200 yrs old now available for correcting carbon-dating; ***P. monophylla*** Torrey & Frémont (*P. cembroides* var. *m.*, SW US) – 1-needled, imp. local medic., seeds ed.; ***P. monticola*** Douglas ex D. Don (W white p., NW Am.) – valuable timber; ***P. mugo*** Turra (C Eur., Balkans) – thin (cf. other alpine *P.* spp.) wax layer with chromophores absorbing u-v. light (i.e. u-v. filter), source of pumilio pine oil used in inhalants, soaps etc.; ***P. nigra*** J.F. Arnold (black p., Eur. & SW As.), subsp. ***nigra*** (Austrian p., C & S Eur.) & esp. subsp. ***laricio*** Maire (var. *corsicana* (Loudon) Hyl. (var. *maritima*), Corsican p., W Medit.) – widely planted, invasive Aus., NZ; ***P. palustris*** Mill. (longleaf or (American) pitch or southern p., SE US) – remains in dwarf 'grass stage' until exposed to heat of a forest fire, source of timber, pulp, & turpentine, imp. in preparation of dressings in medic., varnishes, printing inks, soap, polish, sealing-wax, fireworks, plastics etc. etc.; ***P. parviflora*** Sieb. & Zucc. (Jap. white p., S & C Jap.) – typical sp. in Jap. landscape, timber for building, shingles, sculptures, var. ***pentaphylla*** (Mayr) A. Henry (*P. pentaphylla*, Jap. white p., Jap.) – bonsai (300-yr-old sold 2012 for 100M yen ($US 1.3M); ***P. patula*** Schiede ex Schldl. & Cham.

(Mex. to Nicaragua) – widely planted in trop., invasive Afr., Madag., Hawaii; ***P. pinaster*** Aiton (maritime p., pinaster, Medit.) – much grown in Les Landes, imp. source of turpentine ('French t.'), extracts taken 'to strengthen immune system' & alleged to improve memory, bark ('Fr. b.') used in orchid potting-compost, natur. in S GB, invasive Aus., NZ, S Afr., S S Am., Hawaii; ***P. pinea*** L. (stone or umbrella p., N Medit., invasive S Afr.) – forms with thin testa ('var. *fragilis* Duhamel') selected & cult. for ed. seeds (pine nuts, pignons, pignoli) much used in salads, pesto & baklava, found in Roman middens in GB where it cannot be grown, sugar-coated like almonds at Easter in Portugal, cones used as altar fuel by R.; ***P. ponderosa*** Douglas ex Lawson (ponderosa p., W & S N Am.) – valuable timber incl. dugout canoes, imp. local medic., inner bark & seeds ed., blue dye; ***P. pungens*** Lamb. (E N Am.) – cones retained 15–20 yrs; ***P. radiata*** D. Don (Monterey or radiata p., SW N Am., invasive Aus. (spread by cockatoos), NZ, S Afr.) – almost extinguished by disease in Calif. (endangered) but forming biggest cult. forests in world (NZ) & planted in Medit. climates for timber, 1% of all seedlings with truly dichotomizing apices; ***P. resinosa*** Aiton ((Am. or Canadian) red p., NE N Am.) – timber & tanbark; ***P. rigida*** Mill. (NE N Am.) – epicormics; ***P. roxburghii*** Sarg. (*P. longifolia*, chir, Himalaya) – turpentine, charcoal for Chinese fireworks; ***P. sabiniana*** Douglas ex D. Don (digger p., California) – seeds ed.; ***P. strobus*** L. ((Am.) white or Weymouth p., E N Am.) – imp. local medic. esp. for colds, male catkins cooked, imp. timber tree (the most imp. in NE N Am.), form. for ship-masts supplied from New England, so lack of supply in American War of Independence a factor in conflict as GB ship-masts weaker; ***P. sylvestris*** L. (Scots pine, S. or Norway fir, Euras.) – relict short-leaved trees in old Caledonian forests of Scotland (subsp. ***scotica*** (P. Schott) E. Warb. (var. *s.*)), trees of lowlands being long-leaved introductions from Eur., escaped from plantations, a tree producing 1000 M pollen grains a yr, one of hardest timbers in pines ((Archangel or Baltic – v. straight trunk, 'var. *rigensis*') yellow or red deal, Archangel or B. redwood, Memel fir) used for sleepers, telegraph poles, windmill sail-stocks & other outside work besides joists & pitprops, most common carcase for veneered furniture 1670–1800 UK, for hardboard, pulp, cellophane & plastics, the pine tar antiseptic, the oil from lvs used in men's colognes etc. (pollen contains traces of human sex hormones!), form. planted in England as landmarker, submerged decaying logs in deep Scottish lochs filling with gas & exploding as surfacing poss. some 'Loch Ness Monster' sightings; ***P. taeda*** L. (loblolly p., frankincense or southern p., SE US) – timber & pulp; ***P. teocote*** Schiede ex Schldl. & Cham. (Aztec p., Mex.) – turpentine, resin & ointment (Aztecs) etc.; ***P. torreyana*** C. Parry ex Carr. (S Calif.) – seeds ed.; ***P. wallichiana*** A.B. Jackson (? = *P. dicksoniana* Forbes, Bhutan or blue p., Himal.) – 5-leaved

Pinzona Mart. & Zucc. (~ *Curatella*). Dilleniaceae (II). 1 trop. Am.: ***P. coriacea*** Mart. & Zucc. – scandent (cf. *C.*). R: MSBMBG 85(2001)806

Piofontia Cuatrec. = *Diplostephium*

Pionocarpus S.F. Blake = *Iostephane*

Piora J. Koster. Compositae (Ast.-Lag.). 1 NG: ***P. ericoides*** J. Koster

pipal *Ficus religiosa*

pipe, Dutchman's *Aristolochia macrophylla*; **p. tree** *Syringa vulgaris*; **p.wort** *Eriocaulon aquaticum*

Piper L. Piperaceae (2). Incl. *Macropiper*, *Trianaeopiper* (reduced axillary branches resembling axillary infls), c. 2000 trop. (Philippines 20 – R: Blumea 51(2006)569), Am. c. 500, Peru c. 240). R (NW): BBMNHB 19(1989)117, 20(1990)193. Dioec. or monoec. (NW hermaphr.) shrubs incl. rheophytes, lianes & small trees (Attims's, Bell's, Mangenot's, Petit's & Troll's Models) oft. with swollen nodes; odour pungent; fr. a 1-seeded drupe with thin mesocarp. Fls of some Costa Rican spp. visited by insects; others there produce food bodies in ant-hollowed petioles – ants keep off other spp. of liane & bring nutrients absorbed by pl., in ***P. cenocladium*** C. DC. the food bodies produced in response to ant presence (unique) but larvae of *Phyllobaenus* beetles eat ants & can occupy stems when food bodies still produced; many bat-disp. NW. Extracts shown to have antifertility & insecticidal effects. Cult. orn. & many spices esp. ***P. nigrum*** L. (pepper, Madag. p., S Ind., Sri Lanka) – more consumed worldwide than any other spice (c. 290 K t a yr, 17% world market) though having hypoglycaemic effect in humans (piperine an alk. (readily convertible to MDMA) but pungency from piperidine, chavicine & piperittine (amides) confined to & with limited distrib. in genus), ground unripe fr., 'peppercorns', giving black pepper (immature = green), with pericarp removed giving white, a liane (monoec. or dioec., hermaphr. forms selected) much cult. Mal. up poles or wires, mentioned by Theophrastus (C4 BC), peppercorns in nostrils of mummy of Rameses II (1324 BC), & much of former wealth of Venice & Genoa due to its trade. Other imp. spp. at least locally incl. ***P. aduncum*** L. (S Am.) – fr. ed. in Puerto Rico, introd. C19 to W Mal. where a consp. roadside weedy shrub, also invasive As., Melanesia, Polynesia; ***P. amalago*** L. (C Am., WI) – used to flavour Aztec chocolate; ***P. angustifolium*** Lam. (matico, trop. S Am.) – treatment of skin complaints in Ecuador; ***P. auritum*** Kunth (hoja santa, C Am.) – lvs to 30 cm in which fish/meat cooked (tamales) served with green sauce (mole verde); ***P. betle*** L. (betel p., cultigen, orig. ? Indomal.) – dioec., widely cult. (since at least 'Hoabinhian' culture 8000–3000 BP) for fresh lvs made up into a packet or plug containing betel nut & slaked lime as masticatory, roots etc. locally medic. (antiseptic essential oils); ***P. clusii*** C. DC. (W Afr. black p., Afr. cubebs) – used like ***P. cubeba*** L.f. (cubebs, Java p., Indonesia) – cult. in E & WI for dried unripe fr. used medic. & to flavour cigarettes & gin; ***P. guineense*** Thonn. & Schum. (Guinea c., Ashanti or Benin p., trop. Afr.) – used similarly; ***P. lolot*** C. DC. (lolot, SE As.) – flavour for satay etc.; ***P. longum*** L. ((Ind.) long p., trop. E Himal.) – seasoning & medic., seeds fed to rats induce infertility; ***P. methysticum*** Forst.f. (kava, kawa-kawa, kawa p., pepper-tree, yang(g)ona) ?cultigen derived from '*P. wichmannii* C. DC.', NG to Vanuatu) – dioec. shrub, cult. orn. to 6 m, with roots a source of narcotic sedative drink (effect due to a lactone binding to receptors assoc. with euphoria & wellbeing) cult. esp. Fiji & W Pac. used for 3000 yrs, form. prepared by children & women chewing them & spitting out pulp to be diluted & fermented, locally imp. medic. & supposed aphrodisiac (juvenile hormone analogue in rs) & now comm. valium subs. used for stress esp. on long flights; ***P. sarmentosum*** Roxb. (Indomal.) – lvs used in Thai cuisine; ***P. schultesii*** Yuncker (Colombia) – eases tubercular coughs; ***P. umbellatum*** L. (trop. Am., natur. OW) – local medic., potherb

Piperaceae Giseke. Magnoliidae – Piperales. 5/3600 trop. Shrubs, lianes or epiphytes, herbs or small trees, usu. with swollen nodes; aromatic, usu. with spherical ethereal oil cells, oft. with alks; vasc. bundles usu. scattered like those of monocots but with intrafascicular cambium, the outermost oft. becoming continuous by cambial growth. Lvs simple, usu. in spirals, petiolate. Fls v. small, unisexual or not, in axils of small peltate bracts on dense fleshy spikes; P 0, A 1–10, oft. 2 or 3 + 3, filaments oft. free, the anthers with longit. slits (1 in *Peperomia*, 2 in *Piper*), $\underline{G}$ 1 (*Peperomia*), (3) or (4) (*Piper*), 1-loc. with 1 orthotropous, ± basal, erect bitegmic (unitegmic in *Peperomia*) ovule. Fr. a berry or drupe; seed 1, endotegmic, with scanty endosperm & copious starchy perisperm, the embryo minute & hardly differentiated when seed ripe. 2n = 8–c. 128 (x = ?11 (*Peperomia*) & ?12 (*Piper*))

Classification & genera (T 57(2008)586):

1. **Verhuelliodeae:** *Verhuellia*
2. **Piperoideae:** *Peperomia, Piper*
3. **Zippelioideae:** *Manekia, Zippelia*

Spices & stimulants (*Piper*) incl. some jaborandi, house-pls (*Peperomia*)

Piperanthera C. DC. = *Peperomia*

Piperia Rydb. = *Platanthera*

Pippenalia McVaugh. Compositae (Sen.-Tuss.). 1 Mex.: ***P. delphiniifolia*** (Rydb.) McVaugh. R: CUMH 9(1972)470

Piptadenia Benth. Leguminosae (II 1). Excl. *Pityrocarpa*, c. 24 trop. Am. Some timber etc., see also *Anadenanthera, Parapiptadenia* & *Piptadeniastrum*

Piptadeniastrum Brenan. Leguminosae (II 1). 1 C Afr. forests: ***P. africanum*** (Hook.f.) Brenan (agboin, dabéma, daboma, dahoma, ekhimi) – Troll's Model, decid., trade timber used for planking. R: KB 10(1955)179

Piptadeniopsis Burkart. Leguminosae (II 1). 1 Paraguay: ***P. lomentifera*** Burkart. R: Darw. 6(1944)478

Piptanthus Sweet. Leguminosae (III 6). 2 Himal. R: Britt. 32(1980)281. Alks. ***P. nepalensis*** (Hook.) Sweet (*P. laburnifolius*) – cult. orn. sappy shrub with yellow fls

Piptatheropsis Romasch. & al. (~ *Piptatherum*). Gramineae (X). 5 N Am. R: T 60(2011) 1712

Piptatherum P. Beauv. (~ *Oryzopsis*). Gramineae (16). Excl. *Patis, Piptatheropsis*, 22 Euras. Some grains for Native Americans (mountain rice). [To be excl.:] ***P. miliaceum*** (L.) Coss. (*O. m.*, Medit.) – germ. much improved after heating seed to 90°C (fire-adapted), used to stabilize mine-dumps (Aus.)

Piptocalyx Oliv. ex Benth. = *Trimenia*

Piptocarpha R. Br. Compositae (Vern.-Pip.). 43 trop. Am.

Piptochaetium J. Presl (~ *Stipa*). Gramineae (X). 35 US to Arg. steppes. R: AMBG 89(2002)317

Piptocoma Cass. Compositae (Vern.-Pip.). 18 trop. Am. R: Rhodora 83(1981)77, 398 (*Pollalesta*). Shrubs & trees to 30m

Piptolepis Schultz-Bip. Compositae (Vern.-Lych.). c. 6 SE Braz.

Piptophyllum C. Hubb. Gramineae (Arund.). 1 Angola: ***P. welwitschii*** (Rendle) C. Hubb. R: KB 12(1957)53

Piptoptera Bunge (~ *Halimocnemis*). Amaranthaceae (Chenopodiaceae III 3). 1 C As.: ***P. turkestanica*** Bunge

Piptospatha N.E. Br. (~ *Schismatoglottis*). Araceae (VII 8). Excl. *Hottarum*, 15 W Mal. R: Telopea 9(2000)201. Rheophytes

Piptostachya (C. Hubb.) J. Phipps = *Zonotriche*

Piptostigma Oliv. Annonaceae (IV 1). 14 trop. Afr.

Piptothrix A. Gray = *Ageratina*

Pipturus Wedd. Urticaceae (III). 30 Masc. to Polynesia. Some bark used as fibre, string or bark-cloth in Pac. ***P. asper*** Wedd. (*P. arborescens*, Mal.) – poultices for boils (Philippines)

pipul *Ficus religiosa*

Piqueria Cav. Compositae (Eup.-Piq.). 5–6 trop. R: Phytol. 11(1997)165. ***P. trinervia*** Cav. (tabardillo, Mex., C Am., Haiti) – pioneering allelopathic (monoterpenoids) weed, cult. orn. for winter fls ('stevia')

Piqueriella R. King & H. Robinson (~ *Eupatorium*). Compositae (Eup.-Ager.). 1 Braz.: ***P. brasiliensis*** R. King & H. Robinson. R: MSBMBG 22(1987)166

Piqueriopsis R. King = *Microspermum*

piquí, piquiá *Caryocar brasiliense*, *C. villosum*

Piranhea Baill. Picrodendraceae (Pic.-Pic.-Pic.; Euphorbiaceae s.l.). Incl. *Celaenodrendron* 4 trop. Am. R: KB 51(1996)546. ***P. mexicana*** (Standl.) R.-Sm. (Mex.) – dioec., wind-poll.

Piratinera Aubl. = *Brosimum*

Piresia Swallen (*Reitzia*). Gramineae (VI 3). 5 Venez. & Trinidad to Braz. R: Britt. 34(1982)203, SB 37(2012)134. Fern-like; infls borne in leaf-litter layer

Piresiella Judz. & al. (~ *Piresia*). Gramineae (VI 3). 1 W Cuba: ***P. strephioides*** (Griseb.) Judz. & al. R: AMBG 80(1993)856

Piresodendron Aubrév. ex Le Thomas & Leroy = *Pouteria*

Piriadacus Pichon = *Fridericia*

Pirinia Král. Caryophyllaceae (I 1). 1 SW Bulgaria: ***P. koenigii*** Král. R: Preslia 56(1984)162

Piriqueta Aubl. Passifloraceae (Turneraceae). Excl. *Afroqueta*, 44 trop. Am. Some with epiphyllous fls

Pironneava Gaudich. ex Regel = *Hohenbergia*

Piscaria Piper = *Croton*

Piscidia L. Leguminosae (III 16). 7 C Am. to WI & Florida. ***P. piscipula*** (L.) Sarg. (*P. erythrina*, dogwood, Mex., Florida, C Am., WI) – hard timber for boat-building etc., charcoal, roots yield a fish-poison & analgesic

Pisonia Plum. ex L. Nyctaginaceae (5). 40 trop. & warm esp. Am. Trees & shrubs (Leeuwenberg's Model) with uni- or bi-sexual fls. Glandular anthocarps ensnare or disable birds which may die as a result. Local medic. ***P. aculeata*** L. (trop. Am.) – dioec. cli.; ***P. grandis*** R. Br. (E Afr. coast, Madag.–Polynesia, esp. atolls & islets with guano) figures in wayang (shadow-puppet) shows in Java (where v. rare) & form. coll. only for coronation of Sultan of Solo, some cvs cult. ed. lvs (lettuce tree) incl. some albino ones – Moluccan cabbage, '**Alba**' (poss. cv of *P. umbellifera* (Forst. & Forst.f.) Seemann (W Pacific))

Pisoniella (Heimerl) Standl. Nyctaginaceae (5). 1 warm Am.: ***P. arborescens*** (Lag. & Rodrigues) Standl.

pistachio *Pistacia vera*; **p. galls** *P. terebinthus* etc.

Pistacia L. Anacardiaceae (I). 12 Medit. (Eur. 3), As. & Mal., S US & C Am. R: PJBJ 5(1952)187. Resins (early type of turpentine), oils, blue dye from lvs (Turkey) & ed. seeds: ***P. atlantica*** Desf. (*P. mutica*, Macaronesia to Pakistan) – tan-galls, gum (Bombay mastic) used in varnishes, ripe fr. coll. Near E in Neolithic & Bronze Age & still in some markets; ***P. chinensis*** Bunge (Afghanistan to China & Philippines) – young shoots a Chinese veg.; ***P. lentiscus*** L. (lentisco, lentisk, mastic, Medit.) – Sardinian warbler only found near fruiting shrubs of this sp., resin used for chewing since time of Theophrastus, also used for varnishes esp. for oil-pictures, in quelling halitosis & as filler for caries, ingredient in ouzo, 160–170 t per annum from male pls on Chios; ***P. terebinthus*** L. (terebinth [cognate with turpentine], Medit.) – prod. parthenocarpic fr. poss. a 'diversion' for seed-predatory chalcidoid wasps (Spain), source of tan-galls & form. (Chian) turpentine, resin, poss orig. 'balm of Gilead'; ***P. vera*** L. (pistachio, Iran to C As., much cult. in Med. & US) – domesticated C As., ed. seeds high in vitamins A & E (also poss. lowering cholesterol), used (550 K t a yr by 2009) as dessert, peanut-butter subs., in ice cream & confectionery

Pistaciaceae Martinov. See Anacardiaceae (I)

Pistaciovitex Kuntze = *Vitex*

Pistia L. Araceae (VII 26). 1 trop. (? orig. Lake Victoria): ***P. stratiotes*** L. (water lettuce) – free-floating stoloniferous rosette-pl. with feathery roots & lvs with 'sleep-movements' (moving together at night) & covered with short depressed hairs giving a water-repellent surface; larva of a mosquito sp. lives anchored to root & obtains air by piercing roots with tail; known since time of Pliny (AD 77) & oft. a serious pest in reservoirs etc. (invasive in trop. As., trop. & S Afr., SE US, Carib., surviving in thermal springs in Slovenia), the fragmenting colonies soon occupying open water, used locally for pig-fodder, against warts in Amazonia, etc.

Pistorinia DC. (~ *Sedum*). Crassulaceae (I 5a). 3 Iberia, NW Afr. R: U. Eggli. *Ill. Handb. Succ. Pls*, Crass. (2005)203. Sympetaly evolved independently in *Cotyledon* etc.

Pisum Tourn. ex L. Leguminosae (III 28). Excl. *Vavilovia*, 2 Medit. (Eur. 1), W As. Makash (1983) *The pea*. ***P. sativum*** L. (garden pea, selected from 'steppe' type ('*P. humile*'), Medit. to Iran, poss. also through hybridization with *P. elatius* Bieb. (E Medit.)) – self-poll. in bud, cult. since c. 7000 BC (as long as wheat & barley, 'var. *pumilio* Meikle' one of extant crops of Neolithic agriculture), seeds (17 M t a yr by 2011) used as veg., fresh, tinned, frozen or dried or as flour; 'var. *arvense* (L.) Poir.' (field, dun, grey, maple, mutter or partridge p., peluskins) – seeds usu. for animal feed, pls as fodder & green manure; 'var. *macrocarpon* Ser.' (sugar p., mange-tout, snow p.) – pods without fibrous inner layer, the whole unripe fr. ed., shoots ed. salads (pea shoots; with wheat = 'sweet pea shoots', tau miu), 'var. *medullare*' (marrowfat p.) – allowed to dry before harvest, used in 'mushy ps' & wasabi ps. Peas used by Mendel to demonstrate the laws of inheritance (though seed colour dominance was first shown by T.A. Knight in 1787), his 'wrinkled' peas being due to a lack of one form of 'starch-branching' enzyme & his dwarf pls due to failure of 3β hydroxylase (last enzyme in gibberellic acid synth.)

pita fibre *Chevaliera magdalenae*

pitanga *Eugenia uniflora*

Pitardia Battand. ex Pitard = *Nepeta*

Pitavia Molina. Rutaceae (I). 1 Chile: ***P. punctata*** Molina

Pitaviaster T. Hartley (~ *Euodia*). Rutaceae (I 2). 1 Aus.: ***P. haplophyllus*** (F. Muell.) T. Hartley. R: FA 26(2013)121

pitaya ed. cactus fr. esp. of spp. of *Hylocereus*, *Stenocereus* & *Opuntia*

Pitcairnia L'Hérit. Bromeliaceae (1). (Incl. *Pepinia* – R: SB 13(1988)297) 285 trop. Am., W Afr. (1: ***P. feliciana*** A. Chev., only non-Am. bromeliad, saxicolous). R: FN 14(1974)244, 605, AMBG 73(1986)700. Mostly terr., some epiphytic or saxicolous. Some cult. orn. usu. with spiny lvs

pitch, Burgundy *Picea abies*; **p. pine** *Pinus palustris* & other spp.

pitcher plants See *Cephalotus*, *Darlingtonia*, *Heliamphora*, *Nepenthes*, *Sarracenia*

pitcheri *Duboisia* spp.

pitchfork *Bidens* spp.

pith plant *Aeschynomene* spp.

Pithecellobium Mart. Leguminosae (II 4). 18 trop. & warm Am. R: MNYBG 74,2(1997)2. ***P. dulce*** (Roxb.) Benth. (Madras thorn, Manila tamarind, C Am.) – introd. As. as shade-tree & thorny hedge, fr. ed. (pulp made into a 'lemonade'), seed-oil for soap etc., firewood crop, tanbark; ***P. unguis-cati*** (L.) Benth. (black Jessie, S Florida to N S Am.) – spiny shrub or tree with ed. fr. See also *Archidendron*, *Zygia*

Pithecoctenium Mart. ex Meissn. = *Amphilophium*

Pithecoseris Mart. ex DC. Compositae (Vern. – Chr.). 1 N Braz.: ***P. pacurinoides*** Mart. ex DC. – *Echinops*-like

Pithocarpa Lindl. Compositae (Gnap.-Ang.). 4 SW W Aus. R: Nuytsia 23(2013)106

pito *Erythrina berteroana*

pitomba *Eugenia luschnathiana*

pitpit *Saccharum edule* (lowland NG), *Setaria palmifolia* (upland)

Pitraea Turcz. (~ *Priva*). Verbenaceae (Priv.). 1 temp. S Am.: ***P. cuneato-ovata*** (Cav.) Caro – sometimes weedy, tubers (unique in V.). R: Kurtziana 1(1961)273

Pittierothamnus Steyerm. = *Amphidasya*

Pittocaulon H. Robinson & Brettell (~ *Senecio*). Compositae (Sen. -Tuss.). 6 Mex. & C Am. R: Phytol. 26(1973)451. Trees & shrubs incl. ***P. praecox*** (Cav.) H. Robinson & Brettell (*S. p.*, Mex.) – Leeuwenberg's Model, water-storing pith & cortex, cult. orn.

Pittoniotis Griseb. (~ *Antirhea*). Rubiaceae (III 3). 2 trop. Am.

Pittosporaceae R. Br. Magnoliidae – Apiales. 9/210 trop. & warm OW, esp. Aus. (7 endemic genera, another ext. to Mal., 1 OW trop. – *Pittosporum*). R: Nuytsia 5(1984)405. Trees, shrubs, lianes, s.t. spiny; schizogenous secretory canals present. Lvs simple, subentire, leathery, in spirals; stip. 0. Fls usu. bisex. & reg., solit. or in corymbs or thyrses, each with 2 bracteoles; K 5 sometimes basally connate, imbr., decid., C 5 usu. basally connate forming ± distinct tube with imbr. 3–5-veined lobes, A 5 alt. with C, s.t. weakly connate basally, anthers with longit. slits or term. pores, $\underline{G}$ (2(3–5)), usu. 1-loc. with simple style & several to usu. ± ∞ anatropous to ± campylotropous, unitegmic ovules on each usu. parietal placenta. Fr. a loculicidal capsule or berry; seeds oft. in viscid pulp (aril; wing in *Hymenosporum*) & with 2–5 cotyledons at base of copious oily proteinaceous endosperm. n = 12

Genera: *Auranticarpa, Bentleya, Billardiera, Bursaria, Cheiranthera, Hymenosporum, Pittosporum*

Form. in Rosales, misleading floral features obscuring true affinities

Cult. orn. esp. *Billardiera, Bursaria, Hymenosporum, Pittosporum*

Pittosporopsis Craib. Icacinaceae. 1 SE As.: ***P. kerrii*** Craib. R: FOC 11(2008)508

Pittosporum Banks ex Gaertn. Pittosporaceae. c. 140 trop. & S Afr. (Afr. & Arabia 5, R: KB 42(1987)319) to NZ (R (S & E As.): JAA 32(1951)263, 303; (Aus. (20) & NZ): AMBG 43(1956)87, AusSB 13(2000)845) & Pac. (R: Allertonia 1(1977)73). Everg. trees (Leeuwenberg's Model) & shrubs (oft. with sickly-sweet fls) much cult. in trop. & Medit. climates, esp. as street-trees; some local timbers etc. ***P. coriaceum*** Dryander (Macaronesia) – part of Madeiran laurel forest; ***P. crassifolium*** Banks & Sol. ex Cunn. (karo, NZ natur. Scilly Is.), ***P. eugenioides*** Cunn. (lemonwood, tarata, NZ); ***P. napaulense*** (DC.) Rehder & Wilson (*P. verticillatum* Wall., Ind.) – local med.; ***P. pullifolium*** Burkill (Papuasia) – seeds ed.; ***P. resiniferum*** Hemsl. (Borneo & Philippines) – seed-oil (c. 30% terpenes) for illumination, fuel-oil & medic.; ***P. stenopetalum*** Bak. (*P. verticillatum* Bojer, Madag.) – evil-smelling sap said to ward off spirits; ***P. tenuifolium*** Gaertn. (NZ) – cult. orn. with many cvs, foliage pl. for fl. arranging; ***P. tobira*** (Thunb.) Aiton f. (tobira, China, Jap.) – cult. esp. near sea; ***P. undulatum*** Vent. (E Aus.) – oil extracted from fragrant fls, wood used in golf-clubs (cheesewood), aggressive weed-tree in disturbed forests of Jamaica (Healey), Azores, Aus., NZ, S Afr., Mex., Hawaii

Pituranthos Viv. (*Deverra*). Umbelliferae (III 8). 10 OW desert & semi-desert. ***P. triradiatus*** (Hochst.) Aschers. & Schweinf. (*D. t.*, Arabia) – lvs with 0.6–1.7% dry wt. furanocoumarins leading to photosensitization in grazers

pituri *Duboisia* spp.

Pitygentias Gilg = *Gentianella*

Pityopsis Nutt. (~ *Heterotheca*). Compositae (Ast.-Chr.). 7 E US to C Am. R: Univ. Waterloo Biol. Ser. 29(1985)1. Some cult. orn. rock-pls

Pityopus Small. Ericaceae (II 3; Monotropaceae). 1 W US: ***P. californica*** (Eastw.) H. Copel. R: FNA 9(2009)294

Pityphyllum Schltr. = *Maxillaria* (but see Lankesteriana 7(2007)533)

Pityranthe Thw. (~ *Diplodiscus*). Malvaceae (Brown.). 2 Sri Lanka, Taiwan

Pityrocarpa (Benth.) Britton & Rose (~ *Piptadenia*). Leguminosae. 3 trop. Am. R: SB 32(2007)573

Pityrodia R. Br. Labiatae (IV 1, Dicrastylidaceae). Excl. *Dasymalla, Muniria, Quoya*, c. 25 Aus. R: JABG 2(1979)1

Pityrogramma Link. Pteridaceae (III). 11 trop. Am. (Andes 10), SW & W US (1). R: CGH 189(1962)52. Excl. Afr. & Madag. spp. = *Cerosora*. Terr. ferns with white or yellow waxy powder on abaxial surface of fronds, some cult. orn. esp. ***P. calomelanos*** (L.) Link (trop. Am., natur. pantrop.) – first colonizer of erupted volcanoes in Mex., spore-prints used as face-paint in NG, local medic. incl. malaria, & ***P. triangularis*** (Kaulf.) Maxon (W N Am.) – black petioles used in Native Americans' basketware

Placea Miers. Amaryllidaceae. 6 Chile

Placocarpa Hook.f. Rubiaceae (III 4). 1 Mex.: ***P. mexicana*** Hook.f.

Placodiscus Radlk. Sapindaceae. 10 trop. Afr. (esp. W). Corner's Model

Placolobium Miq. See *Ormosia*

Placopoda Balf.f. Rubiaceae (IV 1). 1 Socotra: ***P. virgata*** Balf.f.

Placospermum C. White & Francis. Proteaceae (II 1). 1 Queensland: ***P. coriaceum*** C. White & Francis. R: FA 16(1995)47

Pladaroxylon (Endl.) Hook.f. (~ *Senecio*). Compositae (Sen.-Sen.). 1 St Helena: ***P. leucadendron*** (Forst.f.) Hook.f. (he cabbage) – pachycaul tree (Leeuwenberg's Model), allies in S Am. & Australasia, form. building timber, now c. 50 left

Plaesianthera (C.B. Clarke) Liv. = *Hygrophila*

Plagiantha Renvoize = *Otachrium* (but see KB 37(1982)323)

Plagianthus Forst. & Forst. f. Malvaceae (Malv.-Malv.). 2–3 NZ. R: KB 20(1967)512. Cult. orn. trees & shrubs incl. ***P. regius*** (Poit.) Hochr. (*P. betulinus*)– bark a fibre-source & raffia subs.

Plagiobasis Schrenk. Compositae (Card.-Cent.). 1 C As.: ***P. centaureoides*** Schrenk

Plagiobothrys Fischer & C. Meyer. Boraginaceae (B.3.8.4). c. 70 W Am. (Calif. 39, 18 endemic), E As. (1), Aus. (5). R: JAA suppl. 1(1991)95

Plagiocarpus Benth. Leguminosae (III 4). 7 N Aus. R: Muelleria 28(2010)41

Plagioceltis Mildbr. ex Baehni = *Celtis*

Plagiocheilus Arn. ex DC. Compositae (Ast.-Gran.). 7 S Am. R: Phytol. 26(1973)159

Plagiochloa Adamson & Sprague = *Tribolium*

Plagiocladus J. Brunel (~ *Phyllanthus*). Phyllanthaceae (Phyll.-Fluegg.). 1 W trop. Afr.: ***P. diandrus*** (Pax) J. Brunel

Plagiogyria (Kunze) Mett. Plagiogyriaceae. 12 trop. & subtemp. (Am. 2, not Aus.). R: Blumea 43(1998)414. Forest ferns on mountain ridges

Plagiogyriaceae Bower. Polypodiidae – Cyatheales. 1/12 E As., Am. R: T 55(2006)712. Terr. ferns with usu. erect dictyostelic rhiz. without indumentum (v. rare in ferns). Lvs pinnate, when young bathed in mucilage through which pneumatothodes protrude; veins simple or forked. Sporangia on enlarged distal part of veins of v. narrow pinnae (indusium 0), so fertile fronds appearing acrostichoid, annulus complete, oblique. n = ?66
Genus: *Plagiogyria*

Plagiolirion Bak. (~ *Urceolina*). Amaryllidaceae. 1 Cauca Valley, Colombia: ***P. horsmannii*** Bak.

Plagiolophus Greenman. Compositae (Helia.-Ecl.). 1 Yucatán (Mex.): ***P. millspaughii*** Greenman

Plagiopetalum Rehder = *Anerinocleistus*

Plagiopteraceae Airy Shaw = Celastraceae

Plagiopteron Griff. Celastraceae (III; Plagiopteraceae). 1 China, S Myanmar & Thailand: ***P. suaveolens*** Griff. – long referred to Flacourtiaceae. R: TFB 32(2004)124

Plagiorhegma Maxim. = *Jeffersonia*

Plagioscyphus Radlk. Sapindaceae. 10 Madag.

Plagiosetum Benth. = *Paractaenum*

Plagiosiphon Harms. Leguminosae (I 2). 5 Guineo-Congolian rainforest

Plagiospermum Oliv. = *Prinsepia*

Plagiostachys Ridl. Zingiberaceae (II 4). c. 25 W Mal. (esp. Borneo) to S China

Plagiostyles Pierre. Euphorbiaceae (Stom.). 1 S Nigeria to C Afr.: ***P. africana*** (Muell. Arg.) Prain. R: A. R.-Sm., *Gen. Euph.* (2001)348

Plagiotheca Chiov. = *Isoglossa*

Plagius L'Hérit. ex DC. (~ *Chrysanthemum*). Compositae (Anth.- Leuc.). 3 Balearics (***P. flosculosus*** (L.) Alavi & Heyw. – cult. orn.), Tunisia, Morocco. R: Willd. 36(2006)51

Plakothira Florence (~ *Klaprothia* [anagram]). Loasaceae (I 2). 3 Marquesas. Shrubs & herbs. R: Allertonia 7(1997)238

Planaltoa Taubert. Compositae (Eup.-Alom.). 2 Braz. R: MSBMBG 22(1987)260

Planchonella Pierre (~ *Pouteria*). Sapotaceae (IV). 110 E Mal. to W Pac. (10; NG 5, Aus. 13, New Caled. 32). R: T 56(2007)337. Some gymnomonoecy (unique in S.). ***P. australis*** (R. Br.) Pierre (*Pouteria a.*, black or brush apple, trop. Aus.) – timber, esp. for carving, fr. for jam

Planchonia Blume. Lecythidaceae (II; Barringtoniaceae). 9 Andamans to N Aus. (2; Mal. 7 – R: FM I,21(2013)97). ***P. careya*** (F. Muell.) Knuth (trop. Aus., NG, cocky apple) – ed., local medic. (wound-healing antibacterials); ***P. timorensis*** Blume (Lesser Sunda Is.) – beaten bark in river/sea used to stupefy 'lobsters' (Sumba); ***P. valida*** (Blume) Blume (W Mal.) – lvs turn red before falling

plane *Platanus* spp., (of Isaiah) *Viburnum tinus*, (arch., Scotland) *Acer pseudoplatanus*; **American p.** *P. occidentalis*; **common** or **London p.** *P.* × *hispanica* 'Acerifolia'; **Corstorphine p.** *A. p.* (golden form); **oriental p.** *P. orientalis*

Planea Karis (~ *Metalasia*). Compositae (Gnap.-Rel.). 1 Cape: ***P. schlechteri*** (L. Bolus) Karis. R: OB 104(1991)73

planer tree *Planera aquatica*

Planera J. Gmelin. Ulmaceae. 1 SE US: ***P. aquatica*** J. Gmelin (water elm, planer tree). R: FNA 3(1997)376

Planichloa Simon = *Ectrosia*

Planocarpa Weiller (~ *Cyathodes*). Ericaceae (VII 7). 3 Tasmania. R: AusSB 9(1996)510

Planodes Greene (~ *Sibara*). Cruciferae (16). 2 S US & N Mex. R: FNA 7(2010)492

Planotia Munro = *Neurolepis*

Plantaginaceae Juss. (Antirrhinaceae, Veronicaceae). Magnoliidae – Lamiales. Incl. Callitrichaceae, Globulariaceae p.p, Hippuridaceae, Scrophulariaceae p.p., Trapellaceae (form. in Pedaliaceae), but excl. Linderniaceae, 97/1850 cosmop. Herbs or less oft. shrubs, s.t. pachycaul, s.t. aquatics, s.t. with alks &/or medullary vasc. bundles. Lvs usu. simple, in spirals to opp. or verticillate (*Hippuris*), s.t. venation parallel; hairs in leaf axils, stip. 0. Fls oft. 2-lipped but s.t. small & reg., bisex. (some *Plantago* monoec. (also *Callitriche*) or gynomonoec.), usu. in pedunculate bracteate heads, thyrses, racemes or spikes or solit. ± bracteoles & s.t. wind-poll.; K ((3) 4(5)), C ((3) 4(5)), s.t. 2-lipped, both imbr., A (1 or 2, 3) 4, 5(-8) alt. with C & attached to it, anthers versatile, introrse with longit. or transverse slits, $\underline{G}$ to $\overline{G}$ (*Hippuris*, *Trapella*) (1, 2) usu. with capitate or 2-lobed stigma, 2-loc. with 1–40 anatropous to hemitropous unitegmic ovules. Fr. a capsule (oft. septicidal), a berry (*Hemiphragma*), with 4 drupelet-like mericarps (*Callitriche*) or achene in persistent K; seeds (1-) ∞, oft. flattened & s.t. winged, with usu. straight embryo in translucent endosperm. n = 4–12+

Principal genera (arr. in several tribes incl. ***Antirrhineae*** (R: D.A. Sutton (1988) *Rev. A.*) & ***Veroniceae*** (R: T 53(2004)434) but some genera (e.g. Cheloneae) here likely to be excl.; position of *Cyrtandromoea*, [*Lindenbergia*, *Rehmannia* & *Triaenophora* = Orobanchaceae] particularly unclear): *Angelonia*, *Antirrhinum*, *Aragoa*, *Bacopa*, *Callitriche*, *Digitalis*, *Globularia*, *Gratiola*, *Limnophila*, *Penstemon*, *Plantago*, *Stemodia*, *Veronica*

P. s.l. incl. v. reduced forms (*Callitriche*, *Hippuris*, *Plantago*, *Trapella*) assoc. with aquatic habit & wind-poll. etc. (cf. Salicaceae for similar trends)

Medic. (*Adenosma*, *Digitalis*, *Gratiola*, *Plantago*, *Veronicastrum*); cult. orn. esp. *Antirrhinum*, *Chelone*, *Collinsia*, *Digitalis*, *Erinus*, *Globularia*, *Linaria*, *Ourisia*, *Penstemon*, *Russelia*, *Veronica*) & weeds (*Dopatrium*, *Kickxia*, *Plantago*, *Scoparia*)

Plantago Tourn. ex L. Plantaginaceae (Plant.). Incl. *Bougueria*, *Littorella*, c. 200 cosmop. (Eur. 36, Aus. 24 native). Plantains, ribwort. R: Pfl. IV, 269(1937)39, 440. Herbs & few shrubs e.g. ***P. sempervirens*** Crantz (*P. cynops*, C & S Eur.) & pachycaul shrublets in e.g. Hawaii (***P. robusta*** Roxb. (St Helena) to 1 m tall) but many weeds esp. ***P. lanceolata*** L. (ribwort p., Euras., natur. temp., invasive Aus.) – ? orig. a pl. of sea-cliffs, iridoids sequestered by some butterflies, stops bleeding in mins (many local medic. uses incl diarrhoea & piles), ***P. major*** L. (common p. Euras., widely natur.) – ? orig. a marsh pl., fr. & lvs used for influenza & colds in modern Chinese herbalism, '**Rosularis**' (rose plantain) a cult. orn. teratologous mutant with head of lvs instead of fls, & ***P. media*** L. (hoary p., Euras.) – ± insect-poll., as well as herbs of saltmarsh (***P. maritima*** L. – sea p., N temp., self-incompatible) & sea-cliffs (***P. coronopus*** L. – buck's-horn p., Euras. – form. cult. as salad veg. in Eur., invasive Aus.); alks. Seeds of some spp. mucilaginous when wetted – efficient laxatives (8 K t a yr) esp. ***P. afra*** L. ('*P. psyllium*', psyllium, Medit. to Ind.) & ***P. indica*** L. (*P. arenaria*, *P. psyllium*, Medit. to C As.) – lowers cholesterol, added to cereals & pasta; ***P. ovata*** Forssk. (isphagul, Medit. to Ind.) – used in treatment of dysentery, bowel regulation etc. (isphagul seeds); ***P. uniflora*** L. (*L. u.*, *L. lacustris*, shoreweed, Eur.) – land form with rosettes of ± flattened lvs with protogyny, 2 sessile fls flanking 1 male (wind-poll.), submerged form with larger centric lvs (oft. confused with *Isoetes*) & runners (fls 0), taking up carbon through lvs & roots (cf. *I.*)

plantain *Plantago* spp., later applied to *Platanus orientalis* & then to *Musa* × *paradisiaca* etc.; **buck's-horn p.** *P. coronopus*; **common p.** *P. major*; **hoary p.** *P. media*; **p. lily** *Hosta* spp.; **mountain p.** *M. troglodytarum*; **ribwort p.** *P. lanceolata*; **rose p.** *P. major* 'Rosularis'; **sea p.** *P. maritima*; **water p.** *Alisma plantago-aquatica*

Plarodrigoa Looser = *Cristaria*

Platanaceae Lestib. Magnoliopsida – Proteales. 1/7 N hemisph. Decid., monoec. trees with branched hairs & bark consp. scaling irreg. Lvs simple, palmately-lobed & -veined (toothed & pinnately veined in 1 sp.) in spirals becoming distich., cuticle morph. like Proteaceae; stip. usu. conspic. & united around twig; petiole base encircling axillary bud.

Fls small, reg. in dense unisexual wind-poll. heads in pendulous strings; K 3 or 4(–7) without vasc. tissue, s.t. basally united, C 3 or 4(–7) minute, alt. with K in males, 0 in females, A as many as & opp. K, anthers ± sessile with longit. slits & connective with prolonged apical appendage (3 or 4 staminodes oft. in females), G (3–)5–8(9) in 2 or 3 whorls, distally unclosed (s.t. vestigial G in males) with 1(2) pendulous, orthotropous or slightly hemitropous, bitegmic ovule. Frs hairy (accrescent P) achenes or nutlets in globose heads; seeds wind-disp., with slender straight embryo in scanty oily, proteinaceous endosperm. n = 16–21

Genus: *Platanus* (planes) – timber

Heads app. derived from main branches of a panicle; fossils with more prominent C in (?insect-poll.) 5-merous fls known. DNA work revealed hitherto unsuspected affinity

Platanocephalus Crantz = *Nauclea*

Platanthera Rich. (~ *Habenaria*). Orchidaceae (IV 4d). Incl. *Piperia* c. 135 N hemisph. (Eur. 5, E As. 50, N Am. 32) to Mal mts. Butterfly orchids. Many poll. Lepidoptera, primitively night-flying moths. Local medic. (rheumatism etc., N Am.). ***P. chlorantha*** Cust. ex Reichb. (*P. montana*, Eur.) – green ointment form. used in Dorset, England, for ulcers; ***P. obtusata*** (Pursh) Lindl. (N temp.) – poll. by mosquitoes in Alaska

Platanus Tourn. ex L. Platanaceae. 7 N hemisph. (planes, plantains (arch.), sycamore (N Am. 5 – R: Lundellia 6(2003)116)): 1 SE As. – ***P. kerrii*** Gagnepain, lvs unlobed, venation pinnate (also seen in first lvs of spring shoots in other spp.), female fls in heads of 10–12 (primitive wood features); 1 SE Eur. to N Iran – ***P. orientalis*** L. (oriental p., chen(n)ar, 'var. *cretica*' = everg. mutants (Crete), '**Mirkovec**' a dwarf); 1 NE N Am. – ***P. occidentalis*** L. (aliso, Am. p., buttonwood, sycamore); SW N Am. R: AJB 60(1973)678. Some timbers for furniture & pulp but esp. cult. orn. esp. ***P.*** × ***hispanica*** Mill. ex Münchh. '**Acerifolia**' (*P.* × *acerifolia*, *P. hybrida*, common or London p., allegedly an hybrid of *P. orientalis* & *P. occidentalis* raised at Oxford Botanic Garden from seed received from Montpellier, as re-synth. hybrids are fertile, but poss. only a cv. of *P. orientalis* from Turkey) – much planted in temp. towns & cities (most common tree in City of Westminster & many London squares e.g. Berkeley Square (1789), Cavendish Square (1810)) as withstands pollution, though hairs from lvs & fr. can cause bronchial problems, veneer sold as 'lacewood', many in US are backcrosses with *P. occidentalis*

Platea Blume. Icacinaceae. 5 Mal.

Plateilema (A. Gray) Cockerell. Compositae (Hele.-Gai.). 1 Texas, Mex.: ***P. palmeri*** (A. Gray) Cockerell. R: FNA 21(2006)444

Plathymenia Benth. Leguminosae (II 1). 1 Braz., Arg.: ***P. reticulata*** Benth. (vinhatico) – yellowish timber prized for cabinet-work & parquet flooring. R: EJB 60(2003)112

Platonia Mart. Guttiferae (III). 1 Guyana, Braz.: ***P. insignis*** Mart. (*P. esculenta*, bacury, bakury) – Massart's Model, parrot-poll.?, large yellow fr. ed., seeds source of oil for candles & soap, timber good

Platostoma P. Beauv. Labiatae (VIII 4b). Incl. *Acrocephalus, Ceratanthus, Geniosporum, Mesona*, 45 OW trop. R: KB 52(1997)272. ***P. palustre*** (Blume) A. Paton (*P. chinense, M. p., M. procumbens* China, Mal.) – cooling drink & flavouring, frogspawn-like black jelly from lvs used as sweetmeat

Platyadenia B.L. Burtt = *Henckelia*

Platyaechmea (Bak.) L.B. Sm. & Kress = *Aechmea*

Platycalyx N.E. Br. = *Erica*

Platycapnos (DC.) Bernh. Papaveraceae (Fumariaceae I 2). 3 Macaronesia, W Medit. esp. Spain (Eur. 3). R: OB 88(1986)39

Platycarpha Less. Compositae (Plat.). Excl. *Platycarphella*, 1 S Afr.: ***P. glomerata*** (Thunb.) less. R: SB 36(2011)196

Platycarphella Funk & H. Robinson (~ *Platycarpha*). Compositae (Plat.). 2 S Afr. R: SB 36(2011)200

Platycarpum Bonpl. Rubiaceae (I 4; Henriqueziaceae). 12 N trop. S Am.

Platycarya Sieb. & Zucc. Juglandaceae. 3 C & E China, Korea, Jap., Vietnam. R: AMBG 65(1978)1069. ***P. strobilacea*** Sieb. & Zucc. (C & S China) – cone-like infrs of winged nutlets in axils of stiff lanceolate bracts, the wing allegedly derived from combination of 1 bracteole + lateral P lobe; cult. orn., bark & fr. sources of black dye used in E for textiles, nets, etc.; ***P. simplicifolia*** G.-R. Long (China) – lvs simple!

Platycaulos Linder. Restionaceae. 12 S & SW Cape. R: Bothalia 15(1985)435, 40(2010)6

Platycelyphium Harms. Leguminosae (III 2). 1 drier E & NE Afr.: ***P. voense*** (Engl.) Wild

Platycentrum Naud. = *Leandra*

Platycerium Desv. Polypodiaceae (III). 18 trop. esp. OW (Afr. & Madag. 6, As. 9, Andes of Peru & Bolivia 1). Elk's horn or stag's horn fern. Ferns of rocks, or epiphytes with short rhizome & dimorphic fronds: sterile ± entire, shield-like, marcescent fronds clasping support & in some spp. e.g. ***P. grande*** (Fée) Kunze (Mal. to Aus.) & ***P. superbum*** De Joncheere & Hennipman (E Aus.) accumulating humus within where roots ramify, while in ***P. bifurcatum*** (Cav.) C. Chr. (SE As. to Aus. & Pac.), commonly cult. sp., no such accumulation, the only humus being supplied by decaying host bark & old sterile fronds; fertile fronds pendulous & much branched, covered with stellate hairs & bearing sporangia in clusters but not sori abaxially. Root-buds from which most spp. propagated (not from spores). All Aus. spp. protected; some cult. orn. house-pls incl. ***P. elephantotis*** Schweinf. (*P. angolense*, trop. Afr.) with hygroscopic movements of marcescent lvs protecting growing-points during drought

Platychaete Boiss. = *Pulicaria*

Platychorda Briggs & L. Johnson (~ *Restio*). Restionaceae. 2 SW W Aus. R: Telopea 9(2001)254

Platycladus Spach (~ *Thuja*). Cupressaceae. 1 NE Iran, China & Korea: ***P. orientalis*** (L.) Franco (*Biota o.*, *T. o.*, Chinese arbor-vitae, thuja) – distinct from *T.* in female cone-scales fleshy when young & in wingless seeds, medic. (exported from As.), cult. orn. (allegedly to 2000+ yrs old) esp. around Chinese temples & on Emperors' graves, with many cvs incl. dwarf ones. R: BM n.s. 16(1999)185

Platycodon A. DC. Campanulaceae (I 1). 1 NE As.: ***P. grandiflorus*** (Jacq.) A. DC. (balloon flower, Chinese bellflower) – like *Campanula* but with apically dehiscent 5-valved capsule, roots (toraji) ed. Korea, used for influenza & colds in modern Chinese herbalism, dried in medic. herb mixture for flavouring liqueurs (Jap.), many cvs (incl. 'doubles'), esp. dwarf '**Apoyama**' (Jap.). R: FOC 19(2011)528

Platycoryne Reichb.f. (~ *Habenaria*). Orchidaceae (IV 4d). c. 20 trop. Afr., Madag. R: KB 13(1958)58

Platycraspedum O. Schulz = *Eutrema* (but see Novon 10(2000)2)

Platycrater Sieb. & Zucc. = *Hydrangea*

Platycyamus Benth. Leguminosae (III 16). 2 Braz. & Peru

Platydesma H. Mann = *Melicope*

Platyglottis L.O. Williams = *Scaphyglottis*

Platygyna Merc. (~ *Tragia*). Euphorbiaceae (Acal.-Pluk.-Trag.). 7 Cuba. R: AMH 64(1972)89

Platygyria Ching & S.K. Wu = *Lepisorus*

Platykeleba N.E. Br. = *Cynanchum*

Platylepis A. Rich. Orchidaceae (IV 2d). 17 trop. Afr. to Seychelles, Moluccas, Tahiti (disjunct)

Platylobium Sm. (~ *Bossiaea*). Leguminosae (III 14). 9 SE Aus. R: Muelleria 29(2011)155

Platylophus D. Don. Cunoniaceae (III). 1 Cape: ***P. trifoliatus*** (Thunb.) D. Don – white timber for furniture, picture-framing etc., honey

Platymiscium Vogel. Leguminosae (III 11). 19 trop. Am. Tall trees, some with ant-infested stem domatia, CITES-listed. Timber (granadillo) for kitchen-knife handles, brush-backs, turnery etc. esp. ***P. pinnatum*** (Jacq.) Dugand (roble, Colombia)

Platymitra Boerl. Annonaceae (IV 7). 2 W Mal. R: Blumea 33(1988)471

Platyopuntia (Eng.) Kreuz. = *Opuntia*

Platypholis Maxim. Orobanchaceae (Orob.). 1 Bonin Is. (Jap.): ***P. boninsimae*** Maxim.

Platypodanthera R. King & H. Robinson (~ *Eupatorium*). Compositae (Eup.-Tri.). 1 E Braz.: ***P. melissifolia*** (DC.) R. King & H. Robinson. R: MSBMBG 22(1987)106

Platypodium Vogel. Leguminosae (III 11). 1–2 trop. Am. Timber

Platypterocarpus Dunkley & Brenan. Celastraceae (I). 1 trop. E Afr. (W Usambaras), ? extinct: ***P. tanganyikensis*** Dunkley & Brenan – winged dehiscent fr. (only Afr. C. thus)

Platyrhiza Barb. Rodr. Orchidaceae (V 12h). 1 Braz.: ***P. quadricolor*** Barb. Rodr.

Platyrhodon Hurst = *Rosa*

Platyruscus Khokr. & Tikhom. = *Ruscus*

Platysace Bunge. Umbelliferae (Azor.?). 25 Aus.

Platyschkuhria (A. Gray) Rydb. Compositae (Bah.). 1 SW US: ***P. integrifolia*** (A. Gray) Rydb. R: Britt. 23(1971)269

Platysepalum Welw. ex Bak. Leguminosae (III 16). 7–8 trop. Afr.

Platyspermation Guillaumin. Alseuosmiaceae. 1 New Caled.: ***P. crassifolium*** Guillaumin – $\overline{G}$ or semi-inferior, form. referred to Rutaceae or Escalloniaceae

Platystele Schltr. Orchidaceae (V 13c). c. 100 trop. Am. R: MSB 38(1990)1, 39(1991)147, 44(1992)112. Fls less than 1 mm diam.; ***P. jungermannioides*** (Schltr.) Garay (C Am.) – fls when less than 1 cm tall; ***P. enervis*** Luer (Ecuador) – even smaller pl. Some cult. orn.

Platystemma Wall. Gesneriaceae (III 2d). 1 Himal.: ***P. violoides*** Wall. R: FOC 18(1998)284

Platystemon Benth. Papaveraceae (III). 1 W N Am.: ***P. californicus*** Benth. (cream-cups) – cult. orn. with mainly opp. lvs & alks. R: UKSB 47(1967)25

Platystigma Benth. (*Hesperomecon*, ~ *Meconella*). Papaveraceae (III). 1 Calif.: ***P. lineare*** Benth. T: UKSB 47(1967)62

Platytaenia Kuhn = *Taenitis*

Platytaenia Nevski & Vved. = *Semenovia*

Platytheca Steetz. Elaeocarpaceae (Tremandraceae). 3 SW Aus.

Platythelys Garay = *Aspidogyne* (but see Bradea 2(1977)196)

Platythyra N.E. Br. = *Aptenia*

Platytinospora (Engl.) Diels. Menispermaceae (III). 1 W trop. Afr.: ***P. buchholzii*** (Engl.) Diels

Platyzoma R. Br. = *Pteris*

Platyzomataceae Nakai = Pteridaceae

Plazia Ruíz & Pavón. Compositae (Mut.-Onis.). Incl. *Harthamus*, 4 S Andes, Arg. Shrubs. R: PhytoKeys 34(2014)2

Plecosorus Fée = *Polystichum*

Plecospermum Trécul = *Maclura*

Plecostachys Hillard & B.L. Burtt (~ *Helichrysum*). Compositae (Gnap.-Gnap.). 2 SW Cape to Natal. R: OB 104(1991)140

Plectaneia Thouars. Apocynaceae (I i). 3 Madag. R: WAUP 97–2(1997)81. Some rubber

Plectis Cook = *Euterpe*

Plectocephalus D. Don (~ *Centaurea*). Compositae (Card.-Cent.). 7 Russia & Ukraine (2), Ethiopia (1), N Am. (2), S Am. (2)

Plectocomia Mart. ex Blume. Palmae (I 3e). 16 trop. As. to W Mal. R: Kalikasan 10(1981)1. Hapaxanthic rattans (Holttum's Model) of little economic imp.

Plectocomiopsis Becc. Palmae (I 3e). 5 S Myanmar & Thailand, W Mal. R: KB 37(1982)244. Hapaxanthic dioec. rattans

Plectorrhiza Dockr. Orchidaceae (V 16c). 2 E Aus., 1 Lord Howe Is.

Plectrachne Henrard = *Triodia*

Plectranthus L'Hérit. Labiatae (VII 3e). Excl. *Coleus* (incl. *Capitanya, Leocus, Solenostemon*), 65 trop. & warm OW esp. S Afr., Madag. Herbs & shrubs, s.t. succ.; K in fr. of some acts as disp. aid. ***P. caninus*** Roth (scaredy cat pl., trop. OW) – allegedly repellent to cats & dogs; ***P. oertendahlii*** T.C.E. Fries (Natal) & ***P. verticillatus*** (L.f.) Druce (Swedish ivy, S Afr.) – cult. orn. in hanging baskets

Plectrelminthus Raf. Orchidaceae (V 16d). 1 trop. W Afr.: ***P. caudatus*** (Lindl.) Summerh. – cult. orn.

Plectritis DC. = *Valeriana* (but see CDH 5(1959)119)

Plectrocarpa Gillies ex Hook. & Arn. Zygophyllaceae (II). 3 temp. S Am.

Plectroniella F. Robyns. Rubiaceae (III 2). 2 trop. Afr.

Plectrophora H. Focke. Orchidaceae (V 12h). 10 NE S Am., Trinidad

Pleea Michaux. Tofieldiaceae (Nartheciaceae s.l.; Melanthiaceae s.l.). 1 SE US: ***P. tenuifolia*** Michaux. R: FNA 26(2002)60

Plegmatolemma Bremek. = *Justicia*

Pleiacanthus (Nutt.) Rydb. (~ *Lygodesmia*). Compositae (Cich.-Cich.(-Micr.)). 1 W US: ***P. spinosus*** (Nutt.) Rydb. R: FNA 19(2006)361

Pleiadelphia Stapf = *Elymandra*

Pleioblastus Nakai (~ *Arundinaria*). Gramineae (IV). 40 Vietnam to Jap. (China 17 – R: FOC 22(2006)121). Heterogeneous. Culms used for furniture, umbrella-handles; some with ed. shoots; some cult. orn.

Pleiocardia Greene = *Streptanthus*

Pleiocarpa Benth. Apocynaceae (I f). 5 trop. Afr. R: WAUP 96–1(1996)134. Alks

Pleiocarpidia K. Schum. (~ *Urophyllum*). Rubiaceae (I 9). 27 W Mal. R: RTBN 37(1940)198

Pleioceras Baill. Apocynaceae (II a). 5 trop. Afr. R: MLW 83–7(1983)26. Aestivation to left!

Pleiochiton Naud. ex A. Gray = *Miconia* (but see Britt. 65(2013)22)

Pleiococca F. Muell. = *Acronychia*

Pleiocoryne Rauschert (*Polycoryne*). Rubiaceae (II 1). 1 trop. W Afr.: ***P. fernandensis*** (Hiern) Rauschert

Pleiocraterium Bremek. = *Hedyotis*

Pleiogynium Engl. Anacardiaceae (II). 2 C Mal. to Pac. ***P. timoriense*** (DC.) Leenh. (*P. cerasiferum*, *Owenia c.*, Burdekin plum, C Mal. to Pac.) – imp. timber in Tonga where seeds strung into skirt-like garments, cult. orn. street-tree in Afr. etc., fr. ed. cassowaries & bowerbirds (Aus.) & used in jam & jellies

Pleiokirkia Capuron = *Kirkia* (but see Adansonia 11(1961)88)

Pleioluma (Baill.) Baehni (~ *Pouteria*). Sapotaceae (IV). 30+ Mal. to W Pacif.

Pleiomeris A. DC. Primulaceae (Myrsinaceae). 1 Macaronesia: ***P. canariensis*** (Willd.) A. DC.

Pleione D. Don. Orchidaceae (V 10b). 21 Nepal to Taiwan & Thailand. R: P. Cribb & I. Butterfield (1999) *The genus P., ed. 2.* Terr. orchids with corm-like pseudobulbs, prob. poll. bumblebees; cult. orn. almost hardy incl. hybrids (some recorded in wild) esp. ***P. bulbocodioides*** (Franch.) Rolfe (China) & ***P. hookeriana*** (Lindl.) Rollison (Himal.)

Pleioneura (C. Hubb.) J. Phipps = *Danthoniopsis*

Pleioneura Rech.f. Caryophyllaceae (III 1). 1 C As. to W Himal.: ***P. griffithiana*** (Boiss.) Rech.f.

Pleiosepalum Hand.-Mazz. = *Aruncus*

Pleiospermium (Engl.) Swingle (II 2). Rutaceae. 5 S Ind. & Sri Lanka (1), Sumatra (1), Java (1), Borneo (2). R: W.T. Swingle, *Bot. Citrus* (1943)274

Pleiospilos N.E. Br. Aizoaceae (V). 4 Cape. R: BJ 106(1986)472. Cult. orn. stemless succ. resembling stones & with fls oft. coconut-scented

Pleiostachya K. Schum. (~ *Ischnosiphon*). Marantaceae. 2 C Am., Ecuador. R: NJB 1(1981)238

Pleiostachyopiper Trel. = *Piper*

Pleiostemon Sonder = *Flueggea*

Pleiotaenia J. Coulter & Rose = *Polytaenia*

Pleiotaxis Steetz. Compositae (Card.-Dicom.). 26 trop. Afr. R: KB 21(1967)180. Shrubs & herbs

Pleisorbus L.H. Zhou & C.Y. Wu = *Sorbus*

Plenasium C. Presl = *Osmunda*

Plenckia Reisseck (~ *Maytenus*). Celastraceae (I). 4 S Am.

Pleocarphus D. Don (~ *Jungia*). Compositae (Mut.-Nass.). 1 N Chile: ***P. revolutus*** D. Don – shrub

Pleocaulus Bremek. = *Strobilanthes*

Pleocnemia C. Presl. Dryopteridaceae. 19 Myanmar & Mal. to Fiji. R: KB 29(1974)341. Local veg. & medic.

Pleodendron Tieghem. Canellaceae. 2 Costa Rica, Hispaniola, Puerto Rico

Pleogyne Miers. Menispermaceae (I). 1 trop. E Aus.: ***P. australis*** Benth. – alks. R: FA 2(2007)368

Pleomele Salisb. = *Dracaena*

Pleonotoma Miers. Bignoniaceae (3). 17 trop. Am. R: AMBG 99(2014)457

Pleopeltis Humb. & Bonpl. ex Willd. (~ *Polypodium*). Polypodiaceae (V). c. 90 trop. R: BS 92(2014)45. Fronds oft. with peltate scales

Pleradenophora Esser (~ *Sebastiania*). Euphorbiaceae (Hipp.). 5 C Am. R: Phytotaxa 81(2013)34

Plerandra A. Gray (~ *Schefflera*). Araliaceae. 49 Melanesia. R: Britt. 65(2013)44. ***P. elegantissima*** (Masters) Lowry & al. (*S. e.*, New Caled.) – pot. pl., juvenile forms with deeply lobed lvs, adult ones more entire

Pleroma D. Don (~ *Tibouchina*). Melastomataceae. c. 175 S Am. esp. Braz. (all but 5)

Plesiatropha Pierre (*Mildbraedia*). Euphorbiaceae (Crot.-Crot). 2 trop. Afr. R: Adans. 27(2005)329

Plesioneuron (Holttum) Holttum = *Cyclosorus*

Plesmonium Schott = *Amorphophallus*

Plethadenia Urb. Rutaceae (I 5). 2 Cuba, Hispaniola. R: Willd. 30(2000)116

Plethiandra Hook.f. (~ *Medinilla*). Melastomataceae. 8 W Mal. (Borneo 7). R: EJB 62(2006)131. A 16–40 (8–122 in *M.*)

Plettkea Mattf. (~ *Pycnophyllopsis*). Caryophyllaceae (II 1). 4 Andes (Peru at 4000–5000 m!)

Pleurandropsis Baill. = *Asterolasia*

Pleuranthemum (Pichon) Pichon = *Hunteria*

Pleuranthodendron L.O. Williams. Salicaceae (Flacourtiaceae). 1 trop. Am.: ***P. lindenii*** (Turcz.) Sleumer. R: FN 22(1980)117

Pleuranthodes Weberb. Rhamnaceae (4). 2 Hawaii

Pleuranthodium (K. Schum.) R.M. Sm. (~ *Alpinia*). Zingiberaceae (II 4). 23 NG, trop. Aus. R: EJB 48(1991)63

Pleuraphis Torrey = *Hilaria*

Pleuricospora A. Gray. Ericaceae (II 3; Monotropaceae). 1 W N Am.: ***P. fimbriolata*** A. Gray – C 0. R: FNA 9(2009)396

Pleurisanthes Baill. Icacinaceae. c. 6 trop. S Am. R: CGH 142(1942)41

pleurisy-root *Asclepias tuberosa*

Pleuroblepharis Baill. = *Crossandra*

Pleurobotryum Barb. Rodr. = *Pleurothallis*

Pleurocalyptus Brongn. & Gris. Myrtaceae (II 1). 2 New Caled.

Pleurocarpaea Benth. Compositae (Vern.-Linz.). 3 trop. Aus. R: Nuytsia 23(2013)110

Pleurocitrus Tanaka = *Citrus*

Pleurocoffea Baill. = *Coffea*

Pleurocoronis R. King & H. Robinson (~ *Eupatorium*). Compositae (Eup.-Alom.). 3 SW N Am. R: MSBMBG 22(1987)237

Pleuroderris Maxon = *Tectaria*

Pleurogyna Eschsch. ex Cham. & Schldl. = *Lomatogonium*

Pleurogynella Ikonn. = *Swertia*

Pleuromanes (C. Presl) C. Presl = *Hymenophyllum*

Pleuropappus F. Muell. (~ *Angianthus*). Compositae (Gnap.-Ang.). 1 S Aus.: ***P. phyllocalymmeus*** F. Muell. R: OB 104(1991)130

Pleuropetalum Hook.f. Amaranthaceae (I 1). 3 trop. Am., Galápagos. Shrubs & treelets

Pleurophora D. Don. Lythraceae. 7 S Am.

Pleurophragma Rydb. = *Thelypodium*

Pleurophyllum Hook.f. Compositae (Ast.-Hin.). 3, islands S of NZ. Megaherbs forming *P.* meadows, ***P. speciosum*** Hook.f. (Campbell Is.) with rosettes to 1.2 m diam. ***P. hookeri*** J. Buch. (Macquarie Is.) – vertical contractile stem keeping lvs at ground-level by reduction of cell vol. & increase of wall-thickness in pith cells, leading to wrinkling of stem outer surface (unique?)

Pleuropogon R. Br. (~ *Glyceria*). Gramineae (XI). 5 W US with 1 circumpolar. R: AJB 28(1941)358

Pleuropterantha Franch. Amaranthaceae (I 2). 3 NE trop. Afr.

Pleuropteropyrum Gross = *Koenigia*

Pleurosoriopsis Fomin. Polypodiaceae (V). 1 China, Jap.: ***P. makinoi*** (Maxim.) Fomin

Pleurosorus Fée = *Asplenium*

Pleurospa Raf. = *Montrichardia*

Pleurospermopsis Norman = *Hymenidium*

Pleurospermum Hoffm. Umbelliferae (III 5). Excl. *Notopterygium*, *Pterocyclus*, 2 temp. Euras. R: FR 111(2000)517

Pleurostachys Brongn. Cyperaceae (II 7). c. 32 S Am. R: BJ 75(1952)456

Pleurostelma Baill. = *Vincetoxicum*

Pleurostima Raf. (~ *Barbacenia*). Velloziaceae. 25 trop. S Am. R: RBB 3(1980)37, B de B 8(1980)65

Pleurostylia Wight & Arn. Celastraceae (III). 3–4 trop. & S Afr., S As., N Aus., New Caled. Timber good for joists

Pleurothallis R. Br. Orchidaceae (V 13c). Excl. *Acianthera*, *Echinosepala*, *Kraenzlinella*, *Pabstiella*, *Phloeophila*, *Specklinia*, *Zootrophion*, c. 550 trop. Am. (Florida 1). R: MSB 20(1986)1. Some cult. orn. epiphytes. ***P. marthae*** Luer & Escobar (Colombia) – poll. nocturnal fungus-gnats

Pleurothallopsis Porto & Brade (~ *Octomeria*). Orchidaceae (V 13c). 18 trop. Am. R: Lindleyana 16(2001)255

Pleurothyrium Nees (~ *Ocotea*). Lauraceae (I). 46 trop. Am. R: AMBG 80(1993)52. Some with ant-infested stem domatia

Plexipus Raf. = *Chascanum*

Plicosepalus Tieghem. Loranthaceae (5 7). Incl. *Tapinostemma*, 12 Middle E, E & S Afr., arid, mostly on Burseraceae & Leguminosae. R: R. Polhill & D. Wiens, *Mistletoes of Africa* (1998)87

Plinia Plum. ex L. Myrtaceae (II 10). c. 40 trop. Am. ***P. edulis*** (Vell.) Sobral (*Marlierea e.*, cambuca, Braz.) – fr. ed., flyingsaucer-shaped. See also *Myrciaria*

Plinthanthesis Steud. (*Blakeochloa*, ~ *Danthonia*). Gramineae (Danth.). 3 SE Aus. R: T 30(1978)478

Plinthus Fenzl. Aizoaceae (I). 6 S Afr. R: H. E. K. Hartmann, *Ill. Handb. Succ. Pls*, Aiz. F–Z(2002)221

Plocama Aiton. Rubiaceae (Putorieae). Incl. *Aitchisonia, Crocyllis, Gaillonia, Jaubertia, Pseudogaillonia, Pterogaillonia, Putoria*, 34 S Eur. (1), SW As., ***P. pendula*** Aiton (Canary ls.) – subdioec., saurochory. R: T 56(2007)322, 516. Foetid when crushed

Plocaniophyllon Brandegee (~ *Deppea*). Rubiaceae (IV 5). 1 Mex., Guatemala: ***P. flavum*** Brandegee

Plocoglottis Blume. Orchidaceae (V 14). c. 40 Mal. to Pac. Terr. Lip connected to lower part of column by 2 elastic flanges. On fl. opening lip is forced down by lat. K stretching elastic 'bands'; light touch of lip by insect releases trigger & lip springs up to column

Plocosperma Benth. Plocospermataceae (Loganiaceae s.l.). 1 Mex., Guatemala, Costa Rica: ***P. buxifolium*** Benth. R: ABN 16(1967)57

Plocospermataceae Hutch. (~ Loganiaceae). Magnoliidae – Lamiales. 1/1 C Am. (Cryptically) dioec. shrubby tree. Lvs small., (sub)opp., articulated near petiole base; stip. 0. Infls axillary of 1–7 flowered congested racemes or dichasia; bracteoles 0. Fls 5(or 6)-merous, reg. campanulate-rotate; (C) imbr., A introrse or extrorse, versatile, longit. dehiscent, inserted on C-tube alt. C; nectary in females; G(1), stipitate, placentation parietal with 2 ovules per carpel, style twice-forked. Fr. a capsule with 1–4 narrow hair-tufted seeds
Genus: *Plocosperma*

Ploiarium Korth. Bonnetiaceae (Guttiferae II). 2–3 SE As. to NG. R: JAA 31(1950)201. Some timber, cult. orn.

ploughman's spikenard *Inula conyzae*

Plowmania Hunz. & Subils (~ *Brunfelsia*). Solanaceae (Petun.). 1 Mex., Guatemala: ***P. nyctaginoides*** (Standl.) Hunz. & Subils. R: Kurtziana 18(1986)127

Plowmanianthus Faden & C. Hardy. Commelinaceae (II 1e). 5 trop. Am. R: SB 29(2004)316

Pluchea Cass. Compositae (Inul.-Pluch.). c. 50 warm (OW 29 – R: Englera 23(2001)). Heterogeneous. Fleabane. Herbs & shrubs, some weedy, e.g. ***P. camphorata*** (L.) DC. (camphor weed, N Am.); ***P. carolinensis*** (Jacq.) G. Don (trop. Am.) – comm. medic. in Caribb. for sore throats; ***P. indica*** (L.) Less. (Indomal.) – lvs ed., locally medic., hedge-pl.; ***P. sericea*** (Nutt.) Cov. – local medic. (diarrhoea etc.), thatch

Plukenetia Plum. ex L. Euphorbiaceae (Acal.-Pluk.-Pluk.). Incl. *Tetracarpidium*, 17 trop. (Am. 12 – R: SB 18(1993)575; Afr. & Madag. 4; As. 1). Spp. (sacha inchi), esp. ***P. volubilis*** L. (Inca peanut, Braz.) oil-rich in omega-3 fatty acids, a 'superfood'. ***P. conophora*** Muell. Arg. (*T. c.*, conophor, awusa (owusa) nut, trop. Afr.) – liane cult. for oilseeds used in cooking etc.

plum *Prunus × domestica* (see also *Vitis vinifera* for **p. duff, p. pudding**); **Alleghany p.** *P. umbellata*; **apricot p.** *P. simonii*; **bakato p.** *Flacourtia inermis*; **beach p.** *P. maritima*; **bill[y]goat p.** *Terminalia ferdinandiana*; **p.blossom** *P. mume*; **Bokhara p.** *P. bokhariensis*; **Burdekin p.** *Pleiogynium timoriense*; **Canada p.** *Prunus nigra*; **chickasaw p.** *P. angustifolia*; **coco p.** *Chrysobalanus icaco*; **coffee p.** *F. rukam*; **date p.** *Diospyros kaki*; **Davidson's p.** *Davidsonia jerseyana*; **flatwoods p.** *P. umbellata*; **gingerbread p.** *Neocarya macrophylla*; **governor p.** *F. indica*; **hog p.** *Spondias* spp.; **Illawara p.** *Podocarpus elatus*; **Indian p.** *Oemleria cerasiformis*; **Jamaica p.** *S. mombin*; **Japanese p.** *Prunus salicina*; **Java p.** *Syzygium cumini*; **Kakadu p.** *T. ferdinandiana*; **kaffir p.** *Harpephyllum caffrum*; **Madagascar p.** *F. indica*; **p. mango** *Bouea macrophylla*; **marmalade p.** *Pouteria sapota*; **maroola p.** *Sclerocarya birrea* subsp. *caffra*; **mobola p.** *Parinari curatellifolia*; **monkey p.** *Ximenia caffra*; **monos p.** *Pseudanamomis umbellulifera*; **mooley p.** *Owenia acidula*; **Natal p.** *Carissa macrocarpa*; **Oklahoma p.** *Prunus gracilis*; **p. pine** *Podocarpus elatus*; **Russian p.** *Prunus × rossica*; **sapodilla p.** *Manilkara zapota*; **p. sauce** *P. mume*; **Sebesten p.** *Cordia myxa*; **sour p.** *Owenia acidula*; **Spanish p.** *Spondias purpurea*; **spiny p.** *X. americana*; **sugar p.** *P. × domestica* 'Prune d'Agen'; **wild p.** *Pappea capensis*; **w. goose p.** *Prunus hortulana, P. rivularis*; **p. wood** *Santalum lanceolatum*; **p. yew** *Cephalotaxus* spp.

Plumatichilos Szlach. = *Pterostylis*

Plumbagella Spach (~ *Plumbago*). Plumbaginaceae (I). 1 C As.: ***P. micrantha*** (Ledeb.) Spach – cult. orn. ann. R: FOC 15(1996)191

Plumbaginaceae Juss. Magnoliidae – Caryophyllales. 30/710 cosmop. esp. maritime. Shrubs, lianes or, usu., (perenn.) herbs; stems oft. with cortical &/or medullary bundles or alt. rings of concentric xylem & phloem; anthocyanins not betalains present. Lvs simple, entire to lobed, in spirals; stip. usu. 0; foliage usu. with scattered chalk-glands exuding water & calcium salts. Fls reg., bisex., 5-merous, oft. with heterostyly & in panicles or cymose heads (Staticoideae) or racemes (Limonioideae); K (5) forming 5- or 10-ribbed tube, oft. petaloid & membranous, C usu. (5), oft. persistent, lobes convolute, A 5 opp. C, adherent to C in *Armeria*, anthers with longit. slits, pollen oft. dimorphic in Armerioideae, G (5), 1-loc. with distinct styles (*Armeria*) or 1 apically lobed, & 1 basal, anatropous, bitegmic ovule (no synergids in *Plumbago* & 3 other genera) on slender funicle. Fr. (usu.) an achene, s.t. a circumscissile capsule or one with apical valves, encl. in persistent K; seed with straight embryo in copious or 0 starchy (4n or 5n!) endosperm. n = usu. 6–9

Classification & chief genera:

I. **Plumbaginoideae** (infl. a spike, raceme or head): *Ceratostigma, Plumbago*

II. **Limonioideae** (Staticoideae, Armerioideae; thyrses with circinnate partial infls): *Acantholimon, Armeria, Limonium* (from which many small genera have lately been split)

Conspic. saltmarsh or maritime genera (some with betaines, quaternary ammonium compounds involved in salt secretion) incl. *Limonium & Armeria*; *Aegialitis* grows in mangrove. Heteromorphic incompatibility systems broken down in some monomorphic *Limonium* spp.; some apomicts. Cult. orn. incl. spp. of *Acantholimon, Armeria, Ceratostigma, Limonium* ('statice' – 'everlasting'), *Plumbago*

Plumbago Tourn. ex L. Plumbaginaceae (I). c. 25 trop. & warm (Eur. 1). Leadwort (***P. europaea*** L. (W Medit. to C As.) form. used to treat eye disease, a side-effect being the skin going lead-coloured). Endosperm tetraploid. Some locally medic., e.g. rheumatism in Mal., several cult. orn. esp. ***P. auriculata*** Lam. (*P. capensis*, S Afr.) – everg. greenhouse scandent shrub with blue fls; ***P. indica*** L. (*P. rosea*, SE As.) – fruit unknown, pl. always in man-made habitats (? mutant of an As. sp.)

plumcot plum × apricot cross

Plumeria Tourn. ex L. Apocynaceae (Ig). Excl. *Himatanthus*, 8 trop. Am. R: AMBG 25(1938)189; L. Ross & al. (2008) *Frangipani*. Pachycaul trees & shrubs (Leeuwenberg's Model), lvs in spirals, propagated by cuttings when the large apex & lvs typical of young seedlings are regained – in the mature tree, these get smaller with each branching of the axis (apoxogenesis); fls fragrant. Cult. orn. (often sterile) esp. ***P. rubra*** L. (frangipani, pagoda tree, temple t. or fl., Mex. to Panamá) – mass-flowering & poll. by hawkmoths fooled by strong scent & colour (no nectar!), Aztec nobles carried it in streets & those visiting such had to carry it, fls offered in Buddhist temples, bark purgative, & ***P. obtusa*** L. (WI) – differing in lvs with rounded tips etc., always planted in Muslim cemeteries in Indonesia (national fl.).

Plummera A. Gray = *Hymenoxys*

Plumosipappus Czerep. = *Centaurea*

plumwood *Eucryphia moorei*

pluot = plumcot

Plutarchia A.C. Sm. Ericaceae (VIII 5). 11 Ecuador & Colombia

Plutonopuntia P. Heath = *Opuntia*

Plymouth strawberry *Fragaria vesca* 'Muricata'

Pneumatopteris Nakai = *Cyclosorus*

Pneumonanthe Gled. = *Gentiana*

Poa L. Gramineae (XVI 15). Incl. *Austrofestuca, Dissanthelium, Neuropoa, Parodiochloa, Tovarochloa*, excl. *Nicoraepoa*, 500+ temp. & cold (Eur. 45 + 4 stabilized hybrids, China 81, 14 endemic – R: FOC 22(2006)257, Mal. 34 (endemic), Aus. 46 (40 endemic, 6 introd.) – R: FA 44A(2009)301), NZ 34, Am. 175 [N 70, 61 native – R: FNA 24(2007)486]), trop. mts. Perenn. (c. 15 ann.), s.t. dioec., some apomixis (apospory, pseudogamy), e.g. ***P. alpina*** L.– 2n = 22–44, etc., & ***P. arctica*** R. Br. – 2n = 36–106 (both circumboreal); v. uniform usu. distinguishable from *Festuca* in having keeled lemma & round hilum. Meadow grass, (US) blue g. – imp. forage grasses in pasture & some lawn gs. ***P. annua*** L. (ann. m. g., winter g. (Aus.), Eur.) – weed poss. a tetraploid hybrid between *P. infirma* Kunth (Eur., Medit.) & *P. supina* Schrad. (Euras.) or either & another sp., now cosmop. cool climates, seeds sense broad temp. ranges (i.e. open ground) & germinate; ***P. arachnifera*** Torrey (Texas b.g., W US) – cult. winter fodder & lawns; ***P. compressa*** L. (Canada b.g., ?orig. Eur.) – fodder in N Am., soil-stabilizer; ***'P.' labradorica*** Steud. (× *Dupoa l.*; *Dupontia fisheri* R. Br. (female) × *P. eminens* J. Presl sterile, veg. reprod. in coastal marshes of recently deglaciated N Quebec & Labrador; ***P. nemoralis*** L. (wood m.g., Euras., Medit., introd. N Am.) – pasture & hay; ***P. pratensis*** L. (common m.g., Euras., Medit., invasive Aus. & N

Am. (Kentucky blue grass) late C17 or early C18) – lawns & pasture; *P. trivialis* L. (rough m.g., Euras., Medit., introd. N Am.) – meadows & pastures

Poaceae Barnh. = Gramineae

poached egg flower *Limnanthes douglasii*

Poacynum Baill. = *Apocynum*

Poaephyllum Ridl. Orchidaceae (V 15). 6 SE As., Mal.

Poagrostis Stapf = *Pentameris*

Poarium Desv. = *Stemodia*

Pobeguinea (Stapf) Jacq.-Fél. = *Anadelphia*

Pochota Goyena (~ *Pachira*). Malvaceae (Bomb.). 1 trop. Am.: ***P. fendleri*** (Seem.) Alverson & Duarte. R: Novon 24(2015)117

Poculodiscus Danguy & Choux = *Plagioscyphus*

pod mahogany *Afzelia quanzensis*

Podachaenium Benth. Compositae (Helia.-Zinn.). 2 Mex., Costa Rica. R: SB 7(1982)481. Trees & shrubs. ***P. eminens*** (Lag.) Schultz-Bip. – cult. orn. with candelabriform branching

Podadenia Thw. (~ *Ptychopyxis*). Euphorbiaceae (Acal.-Acal.-Blum.). 1 Sri Lanka: ***P. thwaitesii*** (Baill.) Muell. Arg. R: A. R.-Sm., *Gen. Euphorb.* (2001)170

Podaechmea (Mez) L.B. Sm. & Kress = *Aechmea*

Podagrostis (Griseb.) Scribner & Merr. (~ *Agrostis*). Gramineae (XVI 5). 6–7 Am. (N 2)

Podalyria Willd. Leguminosae (III 7). 17 Cape, 1 ext. to Natal. R: SB 36(2011)639. Alks; some cult. orn. shrubs

Podandra Baill. = *Philibertia* (but see AJBM 52(1994)37)

Podandriella Szlach. & Olsz. = *Habenaria*

Podandrogyne Ducke = *Cleome*

Podangis Schltr. Orchidaceae (V 16d). 1 trop. Afr.: ***P. dactyloceras*** (Reichb.f.) Schltr – cult. orn.

Podanthus Lag. Compositae. (Helia.-Ecl.). 2 Chile, Arg.

Podistera S. Watson. Umbelliferae (III 8). 4 W N Am.

Podlechiella Maassoumi & Kaz. Osaloo (~ *Astragalus*). Leguminosae (III 24). 1 Cape Verde Is.: ***P. vogelii*** (Webb) Maassoumi & Kaz. Osaloo (*A. v.*)

podo timber of E Afr. spp. of Podocarpaceae

Podoaceae Baill. ex Franch. See Anacardiaceae (I)

Podocaelia (Benth.) Fernandes & R. Fernandes (= *Derosiphia*). Melastomataceae. 1 trop. W Afr.: ***P. tubulosa*** (Sm.) Fernandes & R. Fernandes

Podocalyx Klotzsch. Picrodendraceae (Pod.; Euphorbiaceae s.l.). 1 Amazonia: ***P. loranthoides*** Klotzsch. R: A. R.-Sm., *Gen. Euphorb.* (2001)84

Podocarpaceae Endl. Pinidae. Incl. Phyllocladaceae, 17/172 S hemisph. to Jap. & C Am., trop. Afr. mts. R: Phytol.M 7(1984)4; AusJB 30(1982)319. Everg., usu. dioec. resinous trees (root-parasite in *Parasitaxus*), shoots proleptic. Lvs in spirals (decussate in *Microcachrys*), linear to scale- like, in *Phyllocladus* subtending complex branch-systems poss. similar to those of earliest gymnosperms, oft. lobed or dentate ('phylloclades') & bearing scales on edges. Male cones catkin-like with many bracts, each bearing 2 pollen-sacs, pollen grains with (0)2(3) air-bladders; female cones with 1–many bracts, each or only 1 with 1 ovule, & s.t. a peduncle (receptacle) which may become dry or fleshy; mature cones drupe-like, rarely cone-like or with usu. 1 protruding seed s.t. seated on an aril-like fleshy growth attractive to bird-dispersers (0 in *Pherosphaera*); cotyledons 2. n = 9–19

Genera: *Acmopyle, Afrocarpus, Dacrycarpus, Dacrydium, Falcatifolium, Halocarpus, Lagarostrobos, Lepidothamnus, Microcachrys, Nageia, Parasitaxus, Pherosphaera, Phyllocladus, Podocarpus, Prumnopitys, Retrophyllum, Saxegothaea*

Fossils from Triassic. Pollen drop receives pollen indirectly & retraction of drop due to evaporation (exc. *Phyllocladus*)

Cult. orn. & imp. timbers *(Dacrycarpus, Dacrydium, Halocarpus, Phyllocladus, Podocarpus, Prumnopitys)*

Podocarpium (Benth.) Y.C. Yang & S.H. Huang (non Unger) = *Hylodesmum*

Podocarpus L' Hérit. ex Pers. Podocarpaceae. 94 S temp. through trop. highlands to WI & Jap. R: Blumea 30(1985)251, EJB 72(2015) var. pp. Rauh's Model. Lvs usu. narrow & flat; female cones with only 2–4 scales, of which only 1 or 2 bearing ovules; seeds drupe-like on a fleshy red or purple 'receptacle'; some S Am. spp. with ant-infested stem domatia. Cult. orn. & many imp. timbers (white pine, podo, (African) yellow wood). ***P. drouynianus*** F. Muell. (SW Aus.) – slow-growing shrub, cult. for orn. foliage (emu grass); ***P. elatus*** R. Br. ex Endl. (Illawara plum, plum pine, NG, E Aus.) – ed seeds;

P. macrophyllus (Thunb.) Sweet (kusamaki, S China, Jap.) – old trees in Chinese temples, some forms much used for hedges in Jap. or topiary; ***P. nivalis*** Hook. (NZ) – hardy in GB as rock-pl.; ***P. nubigenus*** Lindl. (S Chile, SW Arg.) & ***P. salignus*** D. Don (S Chile) – manio; ***P. totara*** G. Benn. ex D. Don (totara, NZ) – war canoes (hulls) & general construction; ***P. urbanii*** Pilger (yacca, Jamaica). See also *Afrocarpus, Dacrycarpus, Nageia, Prumnopitys, Retrophyllum*

Podochilopsis Guillaumin = *Adenoncos*

Podochilus Blume. Orchidaceae (V 15). c. 60 Indomal., SW China, W Pac.

Podochrosia Baill. = *Rauvolfia*

Podococcus G. Mann & H. Wendl. Palmae (V 3). 2 trop. Afr. R: KB 63(2008)252

Podocoma Cass. Compositae (Ast.-Pod.). 5 Braz., Arg. R: BJLS 163(2010)495

Podocytisus Boiss. & Heldr. Leguminosae (III 9). 1 Balkans, Turkey: ***P. caramanicus*** Boiss. & Heldr. – cult. orn., unarmed

Podogynium Taubert = *Zenkerella*

Podolasia N.E. Br. Araceae (V 3). 1 W Mal.: ***P. stipitata*** N.E. Br. R: Blumea 33(1988)463

Podolepis Labill. Compositae (Gnap.-Ang.). 18 Aus. R: Muelleria 33(2015)22. Cult. orn. 'everlastings'

Podolobium R. Br. (~ *Pultenaea*). Leguminosae (III 13). 6 SW Aus. R: M.D. Crisp & J.J. Doyle, Adv. Legume Syst. 7(1995)279

Podolotus Royle (~ *Astragalus*). Leguminosae (III 22). 1 Oman to NW Ind.: ***P. hosackioides*** Benth.

Podonephelium Baill. Sapindaceae. 9 New Caled. R: SB 38(2013)1107

Podonosma Boiss. (~ *Onosma*). Boraginaceae (B.2.2). 3 E Medit. R: JAA 35(1954)70

Podopetalum F. Muell. = *Ormosia*

Podophania Baill. = *Hofmeisteria*

Podophorus Philippi. Gramineae (XVI). 1 Juan Fernandez: ***P. bromoides*** Philippi –. extinct. R: CUSNH 48(2003)581

Podophyllaceae DC. = Berberidaceae

Podophyllum L. Berberidaceae (II 2; Podophyllaceae). 14 E N Am. (1) & Himal. to E As. R: W.T. Stearn, *The genus Epimedium* (2002)258. Rhiz. long-known as medic. (e.g. podophyllum resin in proprietary treatment of verrucas), now used in production of etopside used in lung & other cancers; flesh of 'berry' mainly derived from placenta: ***P. hexandrum*** Royle (*P. emodi*, Himal.) & ***P. peltatum*** L. (may apple, American mandrake, E N Am.) – major disp. in Delaware forests is eastern box turtle, ripe fr. ed. (toxic unripe; 'slightly acid, mawkish, eaten by pigs & boys' – Asa Gray), long used medic. by Native Americans for warts & now against testicular cancer

Podopterus Bonpl. Polygonaceae (Erio.-Cocc.). 3 Mex. & Guatemala

Podorungia Baill. Acanthaceae (III 2c). 5 Madag. R: FMad. 182(1967)126

Podosorus Holttum = *Microsorum*

Podosperma Labill. = *Podotheca*

Podospermum DC. (~ *Scorzonera*). Compositae (Cich.-Cich.(Scorz.)). 19 Sw to C As.

Podostelma K. Schum. = *Pleurostelma*

Podostemaceae Rich. ex Kunth. Magnoliidae – Malpighiales. Incl. Tristichaceae, 50/280 trop. (esp. Am. & As.), few temp. Moss-like herbs of stony rivers, usu. submerged & oft. ann., producing aerial fls & fr. at low water, thallus oft. lichenoid, with usu. 0 vessels or even 0 xylem; sieve-plates recorded only in *Tristicha* & *Marathrum*; primary root 0 (plumule & radicle recorded in *Tristicha*, hypocotyl in *Hydrobryum*); lvs (when discernible) entire to dissected with 0 axillary buds, the thallus held by root-like haptera attached to cyanobacterial 'biofilm' on substrate. Fls bisex., reg. or not, small, solit. or in (oft. spiciform) cymes, wind- or insect-poll. (exposed in dry seasons), or cleistogamous (under water), subtended or encl. by 2 bracteoles (spathe-like & encl. up to 20 fls in Podostemoideae); P 2 or 3(–5) ± connate (? K, Tristichoideae) or 2–∞ free or a small annular scale or 0 (Podostemoideae), A – rarely 1, in up to several whorls with filaments usu. basally connate, pollen grains in diads or monads, $\underline{G}$ ((1)2(3)) with as many loc. & ± basally connate styles & (2–)±∞ anatropous, bitegmic ovules on thickened axile placentas. Fr. a septicidal capsule with usu. ∞ v. small seeds, oft. with mucilaginous testa, straight embryo & 0 endosperm (nor double fert.). n = 10

Classification & principal genera (many small – 17 monospecific): **Tristichoideae** (G(3) – *Dalzellia*); **Weddellinoideae** (G(2), capsule not ribbed – *Weddellina* (only)); **Podostemoideae** (G(2), capsule ribbed – *Apinagia, Hydrobryum, Ledermanniella, Marathrum, Podostemum, Rhyncholacis*)

Some developmental processes in the thallus lead to photosynthetic organs which are leaf-stem intermediates ('fuzzy morphology'). Form. thought nearest to Crassulaceae (cf. aquatic tendency in *Crassula*) but raised to Podostemopsida by some, P. have close relationship (surprisingly) with Guttiferae as shown by xanthones & DNA. Some locally eaten as salad

Podostemopsida Cusset & G. Cusset. See Podostemaceae

Podostemum Michaux. Podostemaceae (III). Incl. *Crenias*, *Devillea*, 11 trop. to N Am. (1). R: SBM 70(2004)30. ***P. ceratophyllum*** Michaux (E N Am.) – functional pl. from meristem at base of hypocotyl reaching flowering size in 2 months, some cleistogamous populations; poss. indicator of clean streams in US

Podostigma Elliott = *Asclepias*

Podotheca Cass. Compositae (Gnap.). 6 temp. Aus. R: Muelleria 7(1989)39

Podranea Sprague. Bignoniaceae (1). 1 trop. & S Afr. ***P. ricasoliana*** (Tanf.) Sprague – cult. orn. liane like *Pandorea*. R: FN 25,2(1992)111

Poecilandra Tul. Ochnaceae (III). 3 N trop. S Am. R: BJ 113(1991)178

Poecilanthe Benth. Leguminosae (III 4). Excl. *Amphiodon*, 9 trop. S Am. R: Rodriguesia 58(2007)256

Poecilocalyx Bremek. = *Pauridiantha*

Poecilochroma Miers = *Saracha*

Poecilolepis Grau (~ *Aster*). Compositae (Ast.-Hom.). 2 Cape. R: Strelitzia 9(2000)352

Poeciloneuron Beddome. Calophyllaceae (Guttiferae (I1). Excl. *Agasthiyumalaia*, 1 W Ghats, S Ind.: ***P. indicum*** Beddome – locellate anthers, locally monospecific forests, timber for electricity poles

Poecilostachys Hackel. Gramineae (XXIV 2). 19 Madag., trop. Afr. Forest understorey

Poellnitzia Uitew. = *Astroloba*

Poeppigia C. Presl. Leguminosae (I 3). 1 trop. Am.: ***P. procera*** (Spreng.) C. Presl – timber, local medic.

Poga Pierre. Anisophylleaceae. 1 W trop. Afr.: ***P. oleosa*** Pierre (afo, inoy or inoi nut, poga) – timber & oilseed. R: Phytotaxa 229(2015)176

Poggea Guerke. Achariaceae (Flacourtiaceae). 4 trop. W & C Afr. R: BJ 94(1974)296

Pogogyne Benth. Labiatae (VII 2b). 7 W N Am. R: PCAS IV, 20(1931)105. Aquatics with 'Hippuris syndrome'

Pogonachne Bor. Gramineae (XXII). 1 W India: ***P. racemosa*** Bor. R: KB 1949:176

Pogonanthera Blume = *Pachycentria*

Pogonanthera H.W. Li = *Sinopogonanthera*

Pogonarthria Stapf = *Eragrostis*

Pogonatherum P. Beauv. Gramineae (XXII 6). 3 trop. As. (China 3 – R: FOC 22(2006)591). A 1 or 2. ***P. paniceum*** (Lam.) Hackel (baby panda bamboo) – cult. orn.

Pogonia Juss. Orchidaceae (II 1). 5 temp. As., N Am. R: Gen. Orch. 3(2003)294. Cult. orn. terr. orchids (beard fl.) with bearded lip, esp. ***P. ophioglossoides*** (L.) Ker-Gawler (N Am.)

Pogoniopsis Reichb.f. Orchidaceae (V 4b). 2 Braz. R: Gen. Orch. 3(2003)297. Mycotrophic

Pogonochloa C. Hubb. Gramineae (Chlorid.). 1 S trop. Afr.: ***P. greenwayi*** C. Hubb. R: HIP 35(1940)t.3421

Pogonolepis Steetz (~ *Angianthus*). Compositae (Gnap.-Ang.). 2 Aus. R: Muelleria 6(1986)237

Pogonolobus F. Muell. = *Caelospermum*

Pogononeura Napper. Gramineae (XXIX). 1 trop. E Afr.: ***P. biflora*** Napper. R: Kirkia 3(1963)112

Pogonophora Miers ex Benth. Peraceae (Pog.; Euphorbiaceae Acal.-Pog.). 1 trop. S Am., 2 Gabon

Pogonopus Klotzsch. Rubiaceae (Cond.). 2–3 trop. Am. Alks. ***P. speciosus*** (Jacq.) K. Schum. (C Am.) – cult. orn. with brightly coloured enlarged K-lobe

Pogonorrhinum Betsche = *Nanorrhinum*

Pogonotium Dransf. = *Calamus* (but see KB 34(1980)761, Principes 26(1982)174)

Pogostemon Desf. Labiatae (VI). Incl. *Eusteralis*, 85 S & E As. to Jap. & Aus., S trop. Afr. (few). R: BBMNHB 27(1997)86. Alks. ***P. benghalensis*** (Burm.f.) Kuntze (*P. parviflorus*, E As.) – imp. honey pl. (panagol h.) also for patchouli; ***P. cablin*** (Blanco) Benth. (patchouli, patchouly, Indomal.) – oil (80% prod. Sumatra; Indonesian prod. 1300 t a yr by 2010) used in scenting cosmetics, shampoos & cigarettes, in insecticides (insect-proofing clothes), antidandruff, & leech-repellent & local medic. incl. to soothe menstrual pains, but in commerce oft. replaced by ***P. indicus*** (Roth) Kuntze (*P. heyneanus*, Ind. p., Indomal.); ***P. mutamba*** (Hiern) G. Taylor (S trop. Afr.) – starchy ed. tubers

Pohlidium Davidse & al. = *Zeugites* (but see SB 11(1986)131)

Pohliella Engl. = *Saxicolella*

pohutukawa *Metrosideros excelsa*

Poicilla Griseb. = *Jacaima*

Poicillopsis Schltr. ex Rendle = *Matelea*

Poidium Nees = *Chascolytrum*

Poikilacanthus Lindau. Acanthaceae (III 2c). 6 trop. Am.

Poikilogyne Bak. f. Melastomataceae. 20 Borneo, NG, Aus. (1). R: GBS 35(1982) 223

Poikilospermum Zipp. ex Miq. Urticaceae (I; Cecropiaceae). 20 E Himal. to Mal. R: GBS 20(1963)1. Myrmecophily, local fibres. ***P. suaveolens*** (Blume) Merr. (Indomal.) – medic., aerial roots coll. & smoked (Philippines)

Poilanedora Gagnepain. Family? (not Capparaceae). 1 SE As.: ***P. unijuga*** Gagnepain – K5, C5 etc. R: BSBF 95(1948)27

Poilaniella Gagnepain = *Trigonstemon*

Poilannammia C. Hansen. Melastomataceae. 4 Vietnam. R: BMNHN 4, 9(1987)263

Poinciana Tourn. ex L. = *Caesalpinia*

Poincianella Britton & Rose = *Erythrostemon*

poinsettia *Euphorbia pulcherrima*

Poinsettia J. Graham = *Euphorbia*

pointvetch *Oxytropis* spp.

Poiretia Vent. Leguminosae (III 11). 11 trop. Am. ***P. bahiana*** Cl. Müll. (Braz.) – medic. herbal tea

poison ivy or **oak** *Toxicodendron radicans*. Names used loosely for other *T.* spp. in N Am.; **crow p.** *Stenanthium densum*; **prickly p.** *Gastrolobium spinosum*; **p. walnut** *Cryptocarya pleurosperma*

Poissonia Baill. (~ *Coursetia*). Leguminosae (III 23). 5 Peru, Bolivia, Arg. R: SB 36(2011) 63

Poitea Vent. Leguminosae (III 23). (Incl. *Sabinea*) 12 WI. R: SBM 37(1993). ***P. carinalis*** (Griseb.) Lavin (Dominica) – cult. orn. tree with bright red fls

Poivrea Comm. ex Thouars = *Combretum*

Pojarkovia Askerova. Compositae (Sen.-Sen.). 1 Cauc.: ***P. stenocephala*** (Boiss.) Askerova

poka, banana *Passiflora tarminiana*

pokaka *Elaeocarpus hookerianus*

poke or **p.weed** *Phytolacca americana*

pokosola *Ochrosia elliptica*

Polakia Stapf = *Salvia*

Polakowskia Pittier = *Frantzia*

Polanisia Raf. = *Cleome*

Polaskia Backeb. (~ *Lemaireocereus*). Cactaceae (III 1). 2 S Mex. Links *Stenocereus* & *Myrtillocactus*. Fr. ed.

Polemannia Ecklon & Zeyher. Umbelliferae (III 8). 3 SE Afr. mts. R: EJB 48(1991) 242

Polemanniopsis B.L. Burtt (~ *Polemannia*). Umbelliferae (II). 3 W Cape R: NRBGE 45(1988)498. Woody; mericarps heteromorphic

Polemoniaceae Juss. Magnoliidae – Ericales. 22/375 Am. (esp. W N), Euras. Shrubs, lianes (*Cobaea*), small trees (*Cantua*) or, usu., herbs, usu. storing carbohydrate as inulin & oft. stinking; xylem usu. forming continuous ring. Lvs simple to pinnatisect, palmatisect or pinnate, in spirals (opp. in *Linanthus* & *Phlox*, whorled in *Gymnosteris*); stip. 0. Fls bisex., (4)5(6)-merous, (solit. or, usu.,) in head-like cymes; (K) (K in *Cobaea*), lobes s.t. unequal, (C) reg. or ± 2-labiate, the lobes convolute in bud, A attached to C-tube alt. with lobes, s.t. at diff. levels, anthers with longit. slits, annular nectary-disk usu. around $\underline{G}$ ((2)3(4)) with as many loc. & axile placentas & term. style with as many stigma-lobes, each loc. with 1–∞ anatropous to hemitropous unitegmic ovules in 2 rows. Fr. a capsule usu. loculicidally dehiscent (septicidally in *Acanthogilia*, *Cobaea*), sometimes explosive (*Collomia*, *Phlox*), rarely ± indehiscent; seeds 1–∞, oft. mucilaginous when wetted, with ± straight embryo in (scanty–) copious oily endosperm. n = 9 (sometimes reduced to 6)

Classification & principal genera:

I. **Acanthogilioideae** (spiny shrub; capsule septicidal): *Acanthogilia* (only)
II. **Cobaeoideae** (small trees & lianes or woody perennials; capsule loculicidal or septicidal): *Cantua, Cobaea* (sometimes segregated as monogeneric fam.)
III. **Polemonioideae** (usu. herbs, s.t. ann.; capsule loculicidal): *Aliciella, Collomia, Eriastrum, Gilia, Ipomopsis, Linanthus, Navarretia, Phlox, Polemonium*

Diverse pollen structure & sculpturing app. not correlated with diverse poll. systems (bees from which poll. by hummingbirds, flies & beetles seems to have arisen independently in several genera) incl. bats (*Cobaea*) & Lepidoptera. The trop. genera usu. woody pls with large to medium-sized C & winged seeds with little or 0 endosperm whereas temp. ones mostly herbaceous with medium to small C & wingless seeds with endosperm. Many cult. orn. esp. *Linanthus, Phlox*

Polemonium Tourn. ex L. Polemoniaceae (III). 28 N temp. (Eur. 3) to Mex., Chile (2). R: UCPB 23(1950)209. Cult. orn. herbs (moss pink, N Am.) esp. ***P. caeruleum*** L. (Jacob's ladder, Greek valerian, Euras.) – pinnate lvs form a continuum of leaf & shoot processes

polenta *Zea mays*

Polevansia De Winter. Gramineae (XXIX 1). 1 SW Afr.: ***P. rigida*** De Winter. R: Strelitzia 10(2000)712

Polhillia Stirton. Leguminosae (III 9). 7 SW Cape. R: SAJB 52(1986)167

Polianthes L. = *Agave*

Polianthion Thiele. Rhamnaceae (5). 4 Aus. (3 SW, 1 S Qld). R: AusSB 19(2006)170

policeman's buttons *Jasione montana*

Poliomintha A. Gray. Labiatae (VII 2b). 7 SW N Am. R: Sida 5(1972)8, Phytol. 74(1993)164. ***P. incana*** (Torrey) A. Gray – locally used potherb & fls used for flavouring

Poliophyton O. Schulz = *Halimolobos*

Poliothyrsis Oliv. Salicaceae (Flacourtiaceae). 1 China: ***P. sinensis*** Oliv. – cult. orn. decid. tree with winged seeds

Poljakanthema Kamelin = *Chrysanthemum*

Poljakovia Grubov & Filat. = *Tanacetum* (but see NSL 33(2001)226)

polka-dot plant *Hypoestes phyllostachya*

Pollalesta Kunth = *Piptocoma*

Pollia Thunb. Commelinaceae (II 2). 18 OW trop. (Afr. 3, Madag. 1, Aus. 2), Panamá (1). Blue fr. (colour from light reflecting off tightly coiled cellulose of cell-walls) attractive to pheasants in Mal. forest undergrowth

Pollichia Aiton. Caryophyllaceae (I 2). 1 trop. & S Afr., Arabia: ***P. campestris*** Aiton

Polliniopsis Hayata = *Leptatherum*

Polpoda C. Presl. Molluginaceae. 2 SW & W Cape. R: JSAB 21(1955)93

Polyachyrus Lag. Compositae (Mut.-Nass.). 7 Peru, Chile (W Andean slope). R: Gayana 26(1974)3. Herbs

Polyadoa Stapf = *Hunteria*

Polyalthia Blume. Annonaceae (IV 7). Incl. *Haplostichanthus*, excl. *Hubera, Maasia, Monoon*, c. 80 OW trop. (Madag. 15 (R: BMNHN 4,12(1990)113; Afr. spp. = *Greenwayodendron*), Aus. 4 (3 endemic)). Local medic. ***P. cerasoides*** (Roxb.) Bedd. (Ind. redwood, Ind., SE As.) – cordage, timber for boats & carpentry; ***P. micrantha*** (Hassk.) Boerl. – living fence in Java; ***P. suberosa*** (Roxb.) Thw. (Ind. to C Mal.) – extracts with anti-HIV activity. See also *Fenerivia*

Polyandra Leal = ? *Cleidion*

Polyandrococos Barb. Rodr. = *Allagoptera*

Polyanthina R. King & H. Robinson (~ *Eupatorium*). Compositae (Eup.-Ayap.). 1 trop. Am. along Andes to Bolivia: ***P. nemorosa*** (Klatt) R. King & H. Robinson. R: MSBMBG 22(1987)199

polyanthus *Primula × polyantha*

Polyanthus C.H. Hu = *Pleioblastus*

Polyarrhena Cass. (~ *Felicia*). Compositae (Ast.-Hom.). 4 SW Cape. R: MBSM 7(1970)347

Polyaster Hook.f. Rutaceae (I 5). 1 Mex.: ***P. boronioides*** Hook.f.

Polyaulax Backer = *Meiogyne*

Polybactrum Salisb. = *Pseudorchis*

Polybotrya Humb. & Bonpl. ex Willd. Dryopteridaceae (II). 35 trop. Am. esp. Andes. R: INHSB 34(1987)1. Large cli. ferns with dimorphic fronds

Polycalymma F. Muell. & Sonder (~ *Myriocephalus*). Compositae (Gnap.-Ang.). 1 Aus.: ***P. stuartii*** F. Muell. & Sonder. R: OB 104(1991)115

Polycardia Juss. Celastraceae (I). 5 Madag. R: FMad 116(1946)8. Epiphyllous infls in ***P. aquifolium*** Tul., ***P. phyllanthoides*** (Lam.) DC. etc.

Polycarena Benth. Scrophulariaceae (Man.). 17 SW Cape. R: O. Hilliard, *Manuleeae* (1994)395. See also *Phyllopodium*

Polycarpaea Lam. Caryophyllaceae (I 1). 50 trop. & warm esp. OW. Some locally medic. in China. ***P. spirostylis*** F. Muell. (Aus.) – copper indicator

Polycarpon L. Caryophyllaceae (I 1). 1 Eur. & Medit., widely natur.: ***P. tetraphyllum*** (L.) L. (four-leaved allseed) – v. variable

Polycephalium Engl. = *Pyrenacantha*

Polyceratocarpus Engl. & Diels. Annonaceae (IV 1). 9 trop. Afr., esp. W

Polychilos Breda = *Phalaenopsis*

Polychrysum (Tzvelev) Kovalevsk. Compositae (Anth.-Han.). 1 C As., Afghanistan: ***P. tadschikorum*** (Kudjr.) Kovalevsk. R: BBMNHB 23(1993)109

Polyclathra Bertol. Cucurbitaceae (XV). 1 (variable) C Am.: ***P. cucumerina*** Bertol.

Polyclita A.C. Sm. Ericaceae (VIII 5). 1 Bolivia: ***P. turbinata*** (Kuntze) A.C. Sm.

Polycnemum L. Amaranthaceae (Chenopodiaceae IV 1). 6 C & S Eur. (4), Medit. to C As. R: T 62(2013)108

Polycodium Raf. ex Greene = *Vaccinium*

Polycoryne Keay = *Pleiocoryne*

Polyctenium Greene. Cruciferae (11). 1 NW Am.: ***P. fremontii*** (S. Watson) Greene. R: FNA 7(2010)416

Polycycliska Ridl. = *Lerchea*

Polycniopsis Szlach. = *Polycycnis* (but see PJB 51(2006)35)

Polycycnis Reichb. f. Orchidaceae (V 12i). 17 trop. Am. Cult. orn.

Polydora Fenzl (~ *Vernonia*). Compositae (Vern.-Erl.). 8 trop. Afr.

Polygala Tourn. ex L. Polygalaceae (IV). Excl. *Asemeia, Caamembeca, Gymnosiphon, Hebecarpa, Heterosamara, Polygaloides, Rhinopteris*, c. 200 subcosmop. (not NZ). Milkwort (once thought to increase cow-milk yields). Trees, shrubs & herbs (Chamberlain's & Scarrone's Models) with irreg. fls (inner 2 K petaloid (wings); C 3–5, oft. united, the lowermost (keel) oft. crested), the A & pistil emerging when insect alights (cf. Leguminosae), & arillate (s.t. reduced to hairs etc.) seeds. Some cult. orn., local medic., oilseeds etc.: ***P. arillata*** D. Don (Himal.) – roots fermented for alcoholic drinks in Nepal; ***P. butyracea*** Heckel (Afr., not known wild) – source of fibre used in sacking & of Beni-seed from which malukang butter made; ***P.*** × ***dalmaisiana*** T. Baines (*P. fruticosa* Bergius (*P. oppositifolia*, S Afr.) – lvs opp. × *P. myrtifolia* L. (S Afr., invasive SE Aus.) lvs in spirals) – shrub with almost continuously prod. purplish or rose fls, lvs opp. &/or in spirals on same pl., cult. Medit. climates; ***P. paniculata*** L. (trop. Am.) – introd. (& natur.) OW as medic. (snakebite, etc.); ***P. senega*** L. (senega root, N Am.) – dried root medic. esp. for snakebite; ***P. sibirica*** L. (E Eur. to As.) – 'root' with liquorice used to treat depression, irritability & insomnia in TCM; ***P. tenuifolia*** Willd. (yuan zhi, E As.) – imp. in TCM esp. as expectorant; ***P. virgata*** Thunb. (S Afr.) – invasive in Aus.; ***P. vulgaris*** L. (gang fl., Eur. & Medit.) – wings red, white or blue (red crossed with white gives blue), pollinator unknown, *Lasius niger* ants take seeds (with elaiosomes) to 2 m; others locally medic. & in comm. cough mixtures etc. See also *Badiera*

Polygalaceae Hoffmsgg. & Link. Magnoliidae – Fabales. Incl. Xanthophyllaceae, excl. Emblingiaceae (K (5), C 2), 24/700 subcosmop. Trees, shrubs, lianes (e.g. *Barnhartia*) & herbs, oft. accum. aluminium, s.t. parasitic (*Salomonia*) & oft. with extrafl. nectaries. Xylem usu. in closed ring even in herbs. Lvs usu. in spirals, simple & entire; stip. usu. 0, s.t. a pair of glands or spines. Fls bisex., 2-bracteolate, usu. strongly irreg. & hypogynous, in spikes, racemes or panicles, rarely solit.: K 5, rarely basally connate or 2 lower ones so, the 2 inner oft. petaloid, C (3)5 oft. adnate to A to form tube, when 3 the 2 upper ones + lower median boat-shaped (oft. contrasting colour) & oft. apically fringed, A 4 + 4 or 10 or 3–8 usu. basally connate with anthers with apical pores to longit. slits, annular disk s.t. around G̲ (2–5(–8)) with axile placentas (s.t. pseudomonomerous & 1-loc. as in *Xanthophyllum*) with term. style, oft. curved & 2-lobed (1-lobe stigmatic, other ending in tuft of hairs), each loc. with 1(–40 in *X.*) pendulous, epitropous, anatropous to hemitropous bitegmic ovules. Fr. a loculicidal capsule, nut, samara, berry or drupe; seeds oft. arillate (caruncle) or hairy with 0–copious oily, proteinaceous endosperm. x = 5–11+

Classification & principal genera (4 tribes):

I. **Xanthophylloideae** (woody, berry or irreg. dehisc. caps., trop.): *Xanthophyllum* (only)

II. **Moutabeae** (woody, berry, trop.): *Moutabea*

III. **Carpolobieae** (woody, berry, trop. Afr.): *Carpolobia*

IV. **Polygaleae** (woody to ann. herbs; caps., drupe or samara): *Bredemeyera, Comesperma, Monnina, Muraltia, Polygala, Securidaca*

Some cult. orn. & medic. etc. (*Polygala, Polygaloides, Securidaca*), some dyes & oilseeds (*Xanthophyllum*)

Polygaloides Haller (~ *Polygala*). Polygalaceae (IV). 6–7 Eur., N Afr., N Am. ***P. chamaebuxus*** (L.) O. Schwarz (*Polygala c.*, bastard box, C Eur. to Italy) – cult. orn. everg. shrubby rock-pl. esp. '**Atropurpurea**' (*Polygala c.* var. *grandiflora*) with purple wings & C yellow

Polygonaceae Juss. Magnoliidae – Caryophyllales. 52/1200 ± cosmop., esp. (N) temp. Trees, shrubs, lianes or oft. herbs oft. with unusual vascular structure & app. swollen nodes, producing anthocyanin & not betalain. Shoots monopodial, branching from previous flush. Lvs simple, usu. entire & revolute, usu. in spirals; stip. usu. conspic. & oft. united as a scarious sheath (ocrea) around stem, or ± 0 (e.g. *Eriogonum*). Fls usu. bisex. (or pl. dioec.), reg., 2-, 3- (orig.?) or 5-merous, small, oft. with pseudopedicel above articulation with pedicel oft. in involucrate fascicles subtended by persistent ocreola in simple or branched infls; P (2–6) with minute tube, green to ± petaloid, oft. 3 + 3 but not recognizably K & C, or 5 but always arising spirally, persistent & oft. accrescent in fr., A (2,)3 + 3, 8(9–), arising in front of P (1 or 2 usu. per P), filaments s.t. basally connate, oft. of 2 lengths, anthers usu. versatile, introrse with longit. slits, pollen grains v. variable, annular nectary-disk around G or nectaries between A, $\underline{G}$ ((2)3(4)), 1-loc. with ± united styles & 1 basal, orthotropous (–anatropous) bi-(or ± uni-)tegmic ovule. Fr. oft. 3-gonous achene or nut, sometimes encl. in persistent P oft. forming a wing (wind-disp.) or fleshy hypanthium; seed with straight to curved usu. eccentric embryo in starchy & oily (s.t. ruminate) endosperm. n = 7–13

Classification & principal genera:

I. **Eriogonoideae** (oft. woody, ocrea 0; 6 tribes; R: Britt. 63(2011)516):
  1. **Eriogoneae** (infl. involucre tubular or series of 3 to many bracts; R: SB 37(2012)723 – 20/325): *Chorizanthe, Eriogonum*
  2. **Pterostegieae** (involucre a single bract encl. mature achene): *Harfordia*

II. **Polygonoideae** (ocrea present; 5 tribes; R: T 64(2015)1199):
  1. **Triplareae** (trees & shrubs, oft. dioec., P 3 + 3; – **I.**): *Ruprechtia, Triplaris; Symmeria* now seg. as **Symmeriodeae**
  2. **Coccolobeae** (trees, shrubs & lianes, P 5- **I.**): *Antigonon, Coccoloba*] *Muehlenbeckia*]
  3. **Rumiceae** (herbs, P (2)3 + (2)3): *Oxyria, Rheum, Rumex*
  4. **Polygoneae** (shrubs & herbs, P 5, outer oft. winged; R: T 60(2011)1253): *Atraphaxis, Calligonum, Fallopia, Polygonum* [*Oxygonum* now seg. as **Oxygoneae**]
  5. **Persicarieae** (herbs, P 5, outer rarely winged; R: BJLS 98(1988)321): *Fagopyrum* [*Pteroxygonum* now seg. as **Pteroxygoneae**], *Persicaria*

Pentamery app. primitive condition, even though 3-merous P with (frequently) 2 midveins rather than 1 in the 5-merous suggest recent amalgamation of 3 + 3

Some ed. products (*Coccoloba, Fagopyrum* (buckwheat), *Fallopia, Persicaria, Rheum, Rumex*), timber (*Triplaris*), charcoal & honey (*Gymnopodium*), tanning material (*Rumex*), dye (*Persicaria*), cult. orn. (*Antigonon, Bistorta, Eriogonum, Fallopia, Muehlenbeckia, Persicaria, Reynoutria, Rheum*) & some bad weeds (*Fallopia, Persicaria, Polygonum* – knotweeds, *Reynoutria, Rumex* – docks)

Polygonanthus Ducke. Anisophylleaceae. 2 Amazonia. R: Phytotaxa 229(2015)179

Polygonatum Mill. Asparagaceae (Convallariaceae). 57 N temp. (Eur. 5, N Am. 3) esp. SW China. Herbs with robust horiz. rhiz., locally used as food & medic. but widely cult. orn. (Solomon's seal i.e. mender (use in treatment of bruises)); usu. woodland pls (***P. hookeri*** Bak. also in high alt. Himal. grasslands, cult. orn. dwarf sp.) visited by bees esp. solit. *Anthophora* spp. Most commonly cult. is ***P. × hybridum*** Bruegger (*P. odoratum* (Mill.) Druce (2n = 20, Euras. – ingredient of 'natural viagra') × *P. multiflorum* (L.) All. (2n = 18, Euras., W & N N Am.)) – usu. sterile (2n = 19 with usu. 8 bivalents & 1 trivalent at meiosis, though it, like its parents, may have up to 2n = 30)

Polygonella Michaux = *Polygonum* (but see Britt. 15(1963)177, FNA 5(2005)534)

Polygonum Tourn. ex L. Polygonaceae (II 4). Incl. *Polygonella*, excl. *Bistorta* c. 75 N temp. (33 incl. natur. N Am.; ***P. maritimum*** L. also in S S Am.). Shrubs to ann. herbs (knotweeds) usu. insect-poll. Some bad weeds; for cult. orn. see *Bistorta, Fallopia & Persicaria*. ***P. aviculare*** L. (knotweed, knotgrass, Euras., widely natur., invasive in S Afr., SE US) – form. widely used medic. esp. as tea in treatment of asthma etc., now bad weed (app. most common weed of all – Al-Shehbaz) resistant to herbicide in many places; ***P. perfoliatum*** L. (Indomal.) – invasive SE US

Polylepis Ruíz & Pavón. Rosaceae (8). c. 20 Andes. R: SCB 43(1979)1. Shrubs & trees to 5000 m in mts, the highest alt. reached by an arborescent angiosperm genus. Firewood, bark local medic. (coughs, arthritis etc.), cult. orn. esp. ***P. australis*** Bitter – spectacular flaking bark

Polylophium Boiss. Umbelliferae (III 12). 2 W As.

Polylychnis Bremek. = *Ruellia*

Polymeria R. Br. Convolvulaceae (3). 7 Aus. (6 endemic) to E Mal. & New Caled.

Polymita N.E. Br. = *Schlechteranthus* (but see H. E. K. Hartmann, *Ill. Handb. Succ. Pls*, Aiz. F–Z(2002)223)

Polymnia Kalm. Compositae (Polym.). 4 E N Am. R: FNA 21(2006)39, SB 36(2011)485

Polyosma Blume. Escalloniaceae (Grossulariaceae s.l.). c. 80 E Himalaya to trop. Aus. & New Caled. (5). Many aluminium accumulators

Polyosmaceae Blume = Escalloniaceae

Polyotidium Garay. Orchidaceae (V 12h). 1 Colombia: ***P. huebneri*** (Mansf.) Garay. R: Gen. Orch. 5(2009)328

Polyphlebium Copel. (~ *Trichomanes*). Hymenophyllaceae (Hym.). 15 Mal. to NZ (***P. venosum*** (R. Br.) Copel.), Afr., Madag.

Polypleurella Engl. Podostemaceae. 1 SE As.: ***P. schmidtiana*** (Warm.) Engl.

Polypleurum (Tul.) Warm. Podostemaceae. Excl. *Saxicolella* 8 Ind., Sri Lanka to Thailand

Polypodiaceae J. Presl & C. Presl. Polypodiidae – Polypodiales. Incl. Grammitidaceae, 54/1625 trop. with few temp. R: T 5(2006)719. Usu. epiphytic (few terr.) with creeping dictyostelic rhizome covered with peltate scales. Fronds sometimes dimorphic, simple, lobed to pinnate (bipinnate, pedate etc.) with sori usu. round, spreading along adaxial veins (rarely marginal) or on v. diff. fronds (acrostichoid); indusium 0. n = (25) 35, 36, 37
Classification & principal genera (Phytotaxa 19(2011):18):

I. **Loxogrammoideae** (all but roots lacking internal sclerenchyma): *Loxogramme*
II. **Drynarioideae** (fronds dimorphic): *Aglaomorpha, Drynaria*
III. **Platycerioideae** (fronds with stellate hairs): *Platycerium, Pyrrosia*
IV. **Microsoroideae** (exospore thin): *Lecanopteris, Microsorum*
V. **Polypodioideae** (scales &/or hairs but never stellate): *Belvisia, Lepisorus, Pecluma, Polypodium, Selliguea*; incl. grammitids (without inclusion of Grammitidaceae fam. paraphyletic; veins usu. free, gametophytes ribbon-shaped; 18/550 trop. esp. cloud forest, oft. minute epiphytes with simple fronds): *Calymmodon, Ctenopteris, Grammitis, Lellingeria, Terpsichore*

Fossils from Triassic. Some ant-pls (*Lecanopteris, Microgramma*); some ed. & medic. esp. *Polypodium*, & some cult. orn. esp. *Aglaomorpha, Drynaria, Microsorum, Platycerium, Polypodium*

Polypodiastrum Ching = *Goniophlebium*

Polypodiidae Cronq. & al. (Filicopsida). 260/10 500 in 42 fams. Bulk of ferns (see also Ophioglossaceae, Psilotaceae, Marattiaceae). Sporophytes with roots, stems & lvs (megaphylls) in spirals, oft. markedly compound, protostelic, solenostelic or dictyostelic, sometimes polycyclic; some with restricted sec. thickening. Sporangia homosporous or heterosporous borne term. on axis or on fronds (marginal or superficial abaxially); antherozoids multiflagellate
From tree ferns (Cyatheaceae) to delicate filmy ferns (Hymenophyllaceae) & free-floating aquatics (Salviniaceae), mostly rhiz. perennials, e.g. Aspleniaceae, Dryopteridaceae, Polypodiaceae, Pteridaceae. Not as biochemically variable as angiosperms but with a range of toxic sec. compounds: condensed tannins & other phenolics, thiaminase, phytoecdysones, cyanogenic glycosides & sesquiterpenes

Polypodiodes Ching = *Gonophlebium*

Polypodiopteris C. Reed. Polypodiaceae (II). 3 Borneo. R: Blumea 39(1994)365. Montane forests

Polypodium Tourn. ex L. Polypodiaceae (V). Incl. *Phlebodium*, excl. *Serpocaulon*, c. 110 cosmop. (Eur. 1–4) esp. trop. Am. Polypody. Some cult. orn. (though many now referred to other genera e.g. *Goniophlebium, Microgramma* & *Pleopeltis*) incl. ***P. aureum*** L. (*Phlebodium a.*, trop. Am.) – many cvs & ***P. vulgare*** L. (Euras.) – hardy epiphyte form. used to flavour tobacco (liquorice taste) & containing, in small amounts, ostadin, a steroid saponin, 3000 times as sweet as sucrose – one of sweetest of all known compounds, rhiz. mild laxative known since classical times, fronds used in Ireland for coughs, colds & asthma; closely allied ***P. glycyrrhiza*** D. Eaton (N Am.) with liquorice-flavoured rhiz. eaten by Native Americans, while similar ***P. virginianum*** L. (E As., E N Am.) does not have sweet taste (both medic.), but it & *P. vulgare* etc. have phytoecdysones (25 mg per 2.5 g of rhiz. (same as 0.5 t silkworms!))

polypody *Polypodium* spp. esp. *P. vulgare*

Polypogon Desf. Gramineae (XVI 5). 26 warm temp. (Eur. 3), trop. mts. Hybrids with *Agrostis* spp. ***P. monspeliensis*** (L.) Desf. (beard-grass, Eur.) – cult. orn., invasive W US, Aus.

Polypompholyx Lehm. = *Utricularia*

Polyporandra Becc. = *Iodes*

Polypremum L. Tetrachondraceae (Loganiaceae s.l.). 1 warm Am.: ***P. procumbens*** L.

Polypsecadium O. Schulz. Cruciferae (46). 15 N & W S Am. R: Darw. 44(2006)353

Polypteris Nutt. = *Palafoxia*

Polyradicion Garay = *Dendrophylax*

Polyrhabda C. Towns. Amaranthaceae (I 2). 1 Somalia: ***P. atriplicifolia*** C. Towns. R: KB 39(1984)775

Polyschemone Schott, Nyman & Kotschy = *Silene*

Polyscias Forst. & Forst. f. Araliaceae. Incl. *Arthrophyllum, Cuphocarpus, Gastonia, Nothopanax, Tetraplasandra*, 159 OW trop. esp. Madag., Masc., NG, New Caled. R: PDE 128(2010)61. Unarmed shrubs & trees (all A. with pinnate lvs & articulated pedicels), oft. (Leeuwenberg's Model) with reg. pseudo-di(–5)-chotmous candelabriform branching (umbrella trees) or Chamberlain's Model, some with cryptic dioec. Herb. specimens to 100 yrs old smell of fenugreek. Many cult. orn. W Pac. esp. varieg. forms as hedges & locally as salad veg., esp. cvs of ***P. fruticosa*** (L.) Harms (Ming aralia, cultigen or series of cs, orig. ? W Pac.) – pot-pl. USA & ***P. guilfoylei*** (W. Bull) L. Bailey (cultigen orig. ? E Mal.) – toothed varieg. leaflets, rarely flowering in cult. ***P. gymnocarpa*** (Hillebr.) Lowry & G. Plunkett (*T. g.*, Hawaii) – only completely superior G in A.; ***P. murrayi*** (F. Muell.) Harms (pencil pine, NE Aus.) – cult. orn.; ***P. scutellaria*** (Burm.f.) Fosb. (? wild form = *P. pinnata* Forst. & Forst. f., SW Pacific) – source of scent & medic.; other spp. used to stupefy fish; ***P. spectabilis*** (Harms) Lowry & G. Plunkett (*G. s.*, NG) – at 40 m the tallest araliad, timber for light carpentry

Polysolen Rauschert = *Leptomischus*

Polysolenia Hook.f. = *Leptomischus*

Polyspatha Benth. Commelinaceae (II 2). 3 trop. Afr. R: FTEA Commel.(2012)57

Polysphaeria Hook.f. Rubiaceae (II 5). 21 trop. Afr., Madag., Comoro Is. R: KB 35(1980)97

Polystachya Hook. Orchidaceae (V 16b). c. 230 trop. (1 ext. to As.: ***P. concreta*** (Jacq.) Garay & H. Sweet) & warm (Florida 1). Epiphytes with swollen stems or pseudobulbs; some with false pollen rich in protein & or starch on lip as pollinator attractant. Some cult. orn.

Polystemma Decne. Apocynaceae (V c; Asclepiadaceae III 3). 3 C Am.

Polystemonanthus Harms. Leguminosae (I 2). 1 W Afr.: ***P. dinklagei*** Harms

Polystichopsis (J. Sm.) Holttum = *Arachniodes*

Polystichum Roth. Dryopteridaceae (I). Incl. *Cyrtomium, Phanerophlebia*, c. 400 cosmop. (Eur. 4; Afr. 16 – R: BBMNHB 30(2000)35). Terr. woodland ferns, some cult. orn. (holly f.) incl. ***P. falcatum*** (L.f.) Diels (*C. f.*, trop. As. to Hawaii, rare in trop. lowlands but natur. GB, Azores, Netherlands, S Afr., Sydney Harbour, NZ, US etc.) – tough greenhouse fern & ***P. munitum*** (Kaulf.) C. Presl (W N Am.) – peeled rhizome with banana taste. ***P. acrostichoides*** (Michaux) Schott (N Am.) – medic. esp antirheumatic; ***P. minutissimum*** L.B. Zhang & H. He (S C China) – in karst caves growing on active stalctites

Polytaenia DC. Umbelliferae (III 10). 3 N Am. R: Phytoneuron 2012–84:5

Polytaenium Desv. (~ *Antrophyum*). Pteridaceae (Vitt.). c. 10 trop. Am.

Polytaxis Bunge (~ *Jurinea*; = *Saussurea*). Compositae (Card.-Card.). 2 C As.

Polytepalum Süsseng. & Beyerle. Caryophyllaceae (I 1). 1 Angola: ***P. angolense*** Süsseng. & Beyerle

Polytoca R. Br. Gramineae (XXII). Incl. *Chionachne, Sclerachne*, 10 Ind. to E Aus. R: Blumea 47(2002)554, 571. Maize-like fr. imp. in diet of large Aus. birds. ***P. cyathopoda*** (F. Muell.) Bailey (*C. c.*, Aus.) – good fodder; ***P. gigantea*** (Koenig) Mabb. (*C. g.*, Ind. to Vietnam) – ♀ spikelets used as rosary beads (cf. *Coix*); ***P. macrophylla*** Benth. (*C. m.*, E Mal.) – ant-disp. (elaiosome), good fodder; ***P. punctata*** (R. Br.) Stapf (*C. p.*, *S. p.*, Thailand to W Mal.) – ant-disp. (elaiosome)

Polytrema C.B. Clarke = *Ptyssiglottis*

Polytrias Hackel (~ *Eulalia*). Gramineae (XXII). 1 SE As.: ***P. indica*** (Houtt.) Veldk. (*P. amaura*, Java grass, Mal.) – lawn grass escaping pantrop. R: FOC 22(2006)592

Polytrichum Hedw. Polytrichaceae (Musci). c. 70 cosmop. (N Am. 7). Usu acid-indicators. Pls used for mattress-stuffing. ***P. commune*** Hedw. (cosmop.) – local medic., plaited in Roman times, ropes in Bronze & Iron Ages, bedding (used by Linnaeus), cult. bowl pl. & as lawns Jap. (also Wordsworth's Dove Cottage garden, England); ***P. strictum*** Menzies ex Brid. (N temp.) – indicator of drying out of bog upper layers

Polyura Hook.f. Rubiaceae (IV 1). 1 Assam: ***P. geminata*** Hook.f.

Polyxena Kunth = *Lachenalia*

Polyzygus Dalz. Umbelliferae (III 11). 1 S Ind.: ***P. tuberosus*** Walp.

pomade (pomatum) scented ointment esp. for hair, orig. alleged to contain apple pulp

Pomaderris Labill. Rhamnaceae (5). c. 75 SE As. (1), Aus. (esp. SE), NZ (8). Some NZ spp. triploid. Cult. orn. shrubs & trees incl. ***P. apetala*** Labill. (Aus.) – ? introd. NZ as used for canoe skids & ***P. elliptica*** Labill. (Victoria, Tasmania) – locally medic.

pomander orig. a mixture of aromatic subs. moulded into balls & suspended as preservatives against infection, latterly an orange stuck with cloves

Pomaria Cav. (~ *Caesalpinia*). Leguminosae (I 1). 16 Am., trop. Afr. (4). R: PhytoKeys 71(2016)111

Pomatocalpa Breda. Orchidaceae (V 16c). 13 China to Mal., Aus. & Polynesia. R: HPB 11(2006)220. Cult. orn.

Pomatosace Maxim. = *Androsace*

Pomatostoma Stapf = *Anerincleistus*

pomatum = pomade

Pomax Sol. ex DC. Rubiaceae (IV 13). 1 Aus.: ***P. umbellata*** (Gaertn.) A. Rich.

Pomazota Ridl. = *Coptophyllum*

Pombalia Vand. (~ *Hybanthus*). Violaceae. 41 S Am. R: Phytotaxa 183(2014)1. Alks. ***P. calceolaria*** (L.) Paula-Souza (*H. c.*, white ipecanuanha, trop. S Am.) – used like ipecacuanha, ***P. parviflora*** (L.f.) Paula-Souza (*H. p.*, cuicunchulli, warm S Am.), ***P. prunifolia*** (Willd.) Paula-Souza (*H. p.*, trop. S Am.) – understorey shrub induced to mass-flowering by 12 mm rain after dry spell

pomegranate *Punica granatum*

Pomelina (Maire) Güemes & Raymond = *Fumana*

pomelo *Citrus maxima*

pomerac *Syzygium malaccense*

Pometia Forst. & Forst. f. Sapindaceae. 2 Indomal. R: Reinwardtia 6(1962)109. ***P. pinnata*** Forst. & Forst. f. (Mal.) – v. variable, fr. ed. (much valued in Bismarcks & Tonga), timber locally used, local medic.

Pommereschea Wittm. Zingiberaceae (II 4). 2 Myanmar, Thailand

Pommereulla L.f. Gramineae (XXIX). 1 S Ind., Sri Lanka: ***P. cornucopiae*** L.f.

pompano *Calathea lutea*

pompelmous *Citrus maxima*

pomfrets *Glycyrrhiza glabra*

pomion *Cucurbita pepo*

pom-pom weed *Campuloclinium macrocephalum*

pompon See *Dahlia*

Ponapea Becc. (~ *Ptychosperma*). Palmae (V 14i). 4 Caroline Is. R: JJB 12(1936)731

Poncirus Raf. = *Citrus*

pond apple *Annona glabra*; **p.weed** *Potamogeton* spp.; **Canadian p.w.** *Elodea canadensis*; **Cape p.w.** *Aponogeton distachyos*; **sago p.w.** *P. pectinatus*

Ponera Lindl. Orchidaceae (V 13d). Excl. *Nemaconia*, 2 trop. Am.

Ponerorchis Reichb.f. = *Hemipilia*

ponga *Cyathea medullaris*; **p. powder** *C. dealbata*

Pongamia Vent. = *Millettia*

Pongamiopsis R. Viguier. Leguminosae (III 16). 3 Madag.

pongoware *Cyathea medullaris*

Pontechium Böhle & Hilger (~ *Echium*). Boraginaceae (B.2.2). 1 S Eur., W As.: ***P. maculatum*** (L.) Böhle & Hilger. R: T 49(2000)743

Pontederia L. Pontederiaceae. Excl. *Reussia*, 6 Canada to Arg. R: Rhodora 75(1973)426. Tristyly. ***P. cordata*** L. (pickerel weed, wampee, E N Am. to Carib.) – pollen trimorphism but short-style morph rarely legit. poll. N Am., among pollinators a bee (*Dufourea novae-angliae* Robertson) which visits no other pl., its emergence coinciding with anthesis of *P. c.*; cult. orn. with ed. 1-seeded fr.

Pontederiaceae Kunth. Magnoliidae – Commelinales. 9/35 trop. & warm (esp. Am.) to N temp. (few). Hydrophytes, s.t. free-floating, occ. ann., with sympodial branching, vegetatively glabrous. Lvs usu. with sheath, distinct (s.t. inflated with aerenchyma) petiole & expanded lamina with parallel curved-convergent veins (filiform in *Hydrothrix*), in basal rosettes, distich. or in spirals along stem; stip. forming a ocrea, sheath, ligule or 0. Roots with vessel-elements (oft. in stem too). Fls bisex., reg. or not, oft. tristylous, solit. or in term. racemes, spikes or panicles subtended by a sheath, on determinate shoots, usu. insect-poll.; P 3 + 3 (4 in *Scholleropsis*), petaloid, ± basally connate, A (1 with 2 staminodes – *Hydrothrix*) 3 + 3 or 3 ± staminodes, adnate to P-tube, anthers with longit. slits or term. pores (*Monochoria*), G (3), 3-loc. with axile placentas or 1-loc. with intruded parietal placentas (pseudomonomerous in *Pontederia*), usu. with septal nectaries & 1 style & (1 pendulous (*Pontederia*)–)∞ anatropous, bitegmic ovules. Fr. a loculicidal capsule (nut in *Pontederia*), maturing underwater; seeds longit. ribbed with red-coloured tegmen, axile cylindrical embryo in copious starchy endosperm with outer aleurone layer, germ. underwater. n = (7) 8 (-13)

Genera: *Eichhornia, Eurystemon, Heteranthera, Hydrothrix, Monochoria, Pontederia, Reussia, Scholleropsis, Zosterella*

Some bad weeds esp. *Eichhornia*, ed. parts (*Monochoria, Pontederia*) & cult. orn. esp. *Eichhornia, Heteranthera & Pontederia*

Pontefract cakes *Glycyrrhiza glabra*

Ponthieva R. Br. Orchidaceae (IV 2b). c. 60 trop. & warm Am. (US 2). ***P. glandulosa*** (Sims) R. Br. (C Am.) – roots an ipecacuanha subs. in Costa Rica

pontianak, black *Shorea* spp.

ponytail palm *Beaucarnea recurvata*

poon(ga) (oil) *Millettia pinnata*

poor man's orchid *Schizanthus* spp.; **p. m. weather glass** *Anagallis arvensis*

popinac *Vachellia farnesiana*

poplar *Populus* spp.; (US), p. (**yellow**) used for *Liriodendron tulipifera*; **balsam p.** *P. balsamifera*; **black p.** *P. nigra*: **Canada p.** *P. balsamifera*; **grey p.** *P. × canescens*; **Italian** or **Lombardy p.** *P. nigra* 'Italica'; **Manchester p.** *P. nigra* subsp. *betulifolia*; **white p.** *P. alba*

Popoviocodonia Fed. = *Campanula*

Popoviolimon Lincz. Plumbaginaceae (II). 1 C As.: ***P. turcomanicum*** (Lincz.) Lincz. R: BZ 56(1971)1633

Popowia Endl. Annonaceae (IV 7). 30 trop. As. to Aus. Some locally used for scent & ed. fr.

poppy *Papaver* spp.; **alpine p.** *P. alpinum* agg.; **blue p.** *Meconopsis* spp.; **Californian p.** *Eschscholzia californica*; **common, corn, field** or **Flanders p.** *P. rhoeas*; **horned p.** *Glaucium flavum*; **Iceland p.** *P. nudicaule*; **p. mallow** *Callirhoe papaver*; **Matilija p.** *Romneya coulteri*; **Mexican p.** *Argemone mexicana*; **opium p.** *P. somniferum*; **oriental p.** *P. orientale, P. setiferum*; **plume p.** *Macleaya cordata*; **prickly p.** *Argemone mexicana*; **Shirley p.** *P. rhoeas* cvs; **water p.** *Hydrocleys nymphoides*; **Welsh p.** *P. cambricum*

Populina Baill. = *Ecbolium*

Populus Tourn. ex L. Salicaceae (1). c. 30 N temp. (Eur. 3) with ***P. ilicifolia*** (Engl.) Rouleau (*Tsavo i.*, allied to *P. euphratica*) endangered sp. at low alts in E Afr. R: R.F. Stettler & al., *Biology of P.* (1996)7. Dioec. trees – poplars (name in Horace from their growing around public squares & meeting-places), cottonwoods (in US for rough-barked spp., aspen for smooth); many hybrids. Wind-poll. fls in catkins before lvs expand. Fast-growing trees (first to have genome sequenced) for shelter-belts, pulp, comm. coppice-fuel etc., esp. hybrids, being fastest-growing trees in N temp., though suckers & roots oft. troublesome in drains, paving or house-walls (recommended to be planted at least 40m away); timber odourless & of low flammability, used for brake-blocks, allegedly oast-house floors, matches, icecream sticks, veneers, chips, form. for painting-boards (used by e.g. Caravaggio, Poussin), etc., esp. ***P. alba*** L. (abele, white p., Euras., N Afr.) – invasive in Aus., wood resistance to splitting making it suitable for matchboxes, punnets & trugs, folded bark a double reed for wind instruments (Turkey), ***P. balsamifera*** L. (balsam or Canada p., tacamahac, hackmatack, N Am., temp. As.) – plywood, boxes, pulp, excelsior, medic. (analgaesic, antirheumatic), cambium ed., bark-cloth, ***P. × canadensis*** Moench (*P. × euromericana, P. deltoides* (introd. France 1870s) × *P. nigra*) – fast-growing cvs imp. in match industry like '**Serotina**' widely planted, '**Regenerata**' tolerant of urban pollution & oft. pollarded, favoured host for comm. mistletoe Br., Fr.; ***P. × canescens*** (Aiton) Sm.

(grey p., *P. alba* × *P. tremula*), ***P. deltoides*** Bartram ex Marshall (cottonwood, SE US) – local medic., yellow dye, ***P. euphratica*** Olivier (Arabia to Himal.) – orig. 'weeping willow', ***P. × generosa*** A. Henry (*P. deltoides* × *P. trichocarpa*) – imp. plantation tree Eur. (but now less planted because of susceptibility to bacterial canker), NW Am., fastest-growing p. (by volume increment), ***P. grandidentata*** Michaux (Canadian aspen, E N Am.), ***P. × hastata*** Dode (*P. balsamifera* × *P. trichocarpa* esp. '**Balsam Spire**' (fastigiate fem.) form. much planted Br. esp. as windbreak (but susceptible to canker), ***P. heterophylla*** L. (swamp cottonwood, E US) – timber for interior joinery, ***P. × jackii*** Sarg. (*P. × gileadensis*, poss. *P. balsamifera* (*P. candicans*) × *P. deltoides*) esp. '**Gileadensis**' (balm of Gilead) – suckering & known only as female, ***P. maximowiczii*** A. Henry (doronoki, NE As.) – timber for matches & pulp, ***P. nigra*** L. (black p., Euras., rare in GB where 2000 (subsp. *betulifolia*) now left, though in C19 imp. for making wagons, clothes-pegs etc. & featured in Constable's *Haywain* (1821) etc.) – poss. hybrid, timber & pulp, apical buds a source of resin used medic. (some 'balm of Gilead' coll. Eur.) & sought by bees for propolis, '**Italica**' (var. *italica*, Lombardy or Italian p.) – fastigiate male mutant of As. form, planted for centuries (1758 introd. GB) but prone to collapse ('**Plantierensis**' better & now preferred), subsp. ***betulifolia*** (Pursh) Buttler & Hand (Manchester p.) – much planted in Am., ***P. tremula*** L. (aspen, Euras., N Afr.) – trembling lvs due to versatile petiolar anatomy, seeds viable a few days, germ. in 6, timber imp. in match industry & gunpowder charcoal, form. as pattens & arrows (1415 Eng. legislation to favour fletchers' use), catkin hairs used comm. in Germany as kapok subs., stake (cf. *Acer*, *Crataegus*) used to strike through hearts of vampires, ***P. trichocarpa*** Torr. & A. Gray ex Hook. (*P. balsamifera* subsp. *t.*, cottonwood, W N Am.) – most commonly planted 'balsam poplar', first genome sequenced (2006); ***P. tomentosa*** Carr. (China) – hybrid origin, *P. alba* × *P. tremula* (?); ***P. tremuloides*** Michaux (Canadian a., N Am.) – clones to 10 000 yrs old (Utah), root-suckers only mode of reprod. in northern prairies, inner bark ed., timber & medic., used in Blackfoot ceremonial e.g. as head-wreaths

Porana Burm.f. Convolvulaceae (6). 2 SE As. to Philippines (1), Mex. (1!). R: Blumea 51(2006)455. See also *Poranopsis*

Porandra Hong = *Amischotolype*

Poranopsis Roberty (~ *Porana*). Convolvulaceae (2). 3 S & SE As. R: Blumea 51(2006)459. ***P. paniculata*** (Roxb.) Roberty (*Porana p.*, bridal bouquet, N Ind. & Myanmar) – cult. liane with white fls

Poranthera Rudge. Phyllanthaceae (Por.; Euphorbiaceae s.l.). 16 Aus. (most), NZ. R: Austrobaileya 7(2005)3

Poraqueiba Aubl. Metteniusaceae (Icacinaceae). 3 trop. S Am. Bat-disp. in Amazon. R: CGH 142(1942)49. Locally eaten fr. ***P. sericea*** Tul. – a cultigen poss. derived from ***P. guianensis*** Aubl., lvs for treating dysentery, seed-oil for frying fish

Porcelia Ruíz & Pavón. Annonaceae (III 1). 7 trop. Am. R: SBM 40(1993)89. Scarab-poll. ***P. ponderosa*** (Rusby) Rusby (*P. saffordiana*, N Bolivia) – ed. fr. to 29 kg

porcupine wood *Cocos nucifera*; see also *Centrolobium robustum*

Porlieria Ruíz & Pavón. Zygophyllaceae (II). 6 Texas & Mex. (1), Andes (5). Some locally medic. & cult. orn. shrubs incl. ***P. hygrometra*** Ruíz & Pavón (Andes) – leaflets fold together at night (nyctinasty)

Porocystis Radlk. Sapindaceae. 3 trop. S Am.

Porodittia G. Don ex Kraenzlin = *Calceolaria*

Porolabium Tang & F.T. Wang (~ *Herminium*). Orchidaceae (IV 4d). 1 China: ***P. biporosum*** (Maxim.) Tang & F.T. Wang. R: Gen. Orch. 2(2001)355

Porophyllum Guett. (~ *Pectis*). Compositae (Tag.-Pect.). Excl. *Bajacalia*, c. 30 warm Am. R: UKSB 48(1969)225. Some v. smelly; some local medic. Arg. ***P. ruderale*** (Jacq.) Cass. (trop. Am.) – seasoning (Bolivian coriander)

Porospermum F. Muell. = *Delarbrea*

Porpax Lindl. Orchidaceae (V 15). 13 trop. As. R: BT 72(1977)1

Porphyra Agardh. Bangiaceae (Rhodophyta). Excl. *Pyropia*, 6 temp. R: J Phycol. 47(2011)1138. ***P. umbilicalis*** Kützing – fronds only 1 cell thick, rich in iodine, in Br. Is., esp. Wales, basis of laver (bread)

Porphyrocoma Scheidw. ex Hook. = *Justicia*

Porphyrodesme Schltr. = *Renanthera*

Porphyroglottis Ridl. Orchidaceae (V 12a). 1 W Mal.: ***P. maxwelliae*** Ridl.

Porphyroscias Miq. = *Angelica*

Porphyrospatha Engl. = *Syngonium*

Porphyrostachys Reichb.f. Orchidaceae (IV 2b). 2 Andes of Ecuador & Peru. ***P. pilifera*** (Kunth) Reichb.f. – poss. poll. hummingbirds

Porphyrostemma Benth. ex Oliv. Compositae (Inul.-Pluch.). 4 trop. Afr.

porridge *Avena sativa*

Porroglossum Schltr. Orchidaceae (V 13c). c. 40 Andes. R: MSB 24(1987)25, 26(1988)108, 31(1989)124, 39(1991)147. Cult. orn. incl. ***P. echidnum*** (Reichb.f.) Garay (*Masdevallia muscosa*, Colombia) – prob. poll. by Diptera, which settle on a ridge on distal triangular part of lip; ridge is sensitive & lifts, carrying insect up into funnel formed by united sepal-bases so that it can only escape through tube between lip & column where pollinia & stigma are, the 'trap' remaining closed for c. 30 mins

Porrorhachis Garay. Orchidaceae (V 16c). 2 Mal.

Port Jackson *Acacia saligna*; **P. J. fig** *Ficus rubiginosa*; **P. J. willow** *A. saligna*

Portea Brongn. ex K. Koch. Bromeliaceae (3). 9 E Braz. R: FN 14(1979)2038. Infls used in spectacular fl. arrangements

Portenschlagiella Tutin = *Athamanta* (but see FR 74(1967)32)

Porterandia Ridl. Rubiaceae (II 1). c. 22 W Mal. See also *Aoranthe* for Afr. spp.

Porteranthus Britton = *Gillenia*

Porterella Torrey (~ *Solenopsis*). Campanulaceae (III 2). 1 W N Am.: ***P. carnosula*** (Hook. & Arn.) Torrey

Porteresia Tateoka = *Oryza*

Portillia Koeniger = *Masdevallia*

Portlandia P. Browne. Rubiaceae (I 7). 6 C Am., WI. R: JAA 60(1979)99

Portuguese cabbage *Brassica oleracea* Tronchuda Group

Portulaca Tourn. ex L. Portulacaceae. 116 trop. & warm (Eur. 1). R: U. Eggli, *Ill. Handb. Succ. Pls*, Dicots (2002)400. Succ. trailing usu. ann. herbs with semi-inf. G & A 8–∞, s.t. sensitive to contact. Many selfing lines (jordanons) form. described as spp. now referred to ***P. oleracea*** L. (polyploid cosmop. weed – 1 of 10 worst); ***P. oleracea*** 'subsp. *sativa* (Haw.) Čelak' (purslane, postelein (Netherlands)) – salad or potherb (highest content of omega-43 fatty acids & antioxidants of any green veg.) & locally medic. Some cult. orn. esp. ***P. grandiflora*** Hook. (N S Am.) – tender ann. with short-lived pink, red, yellow or white fls incl. doubles (Lemoine at Nancy 1852)

Portulacaceae Juss. (~ Cactaceae). Magnoliidae – Caryophyllales. Excl. Anacampserotaceae, Hectorellaceae &/= Montiaceae, Talinaceae, 1/116 cosmop. esp. W Am. R: T 59(2010)235. Usu. succ. herbs; betalains Lvs usu. in spirals, simple & with Kranz anatomy & C_4 photosynthesis; stip.(?) tufts of hairs or 0. Fls usu. reg. & bisex., solit. in term. infls; K 2, C (45(–8) delicate & ephemeral, basally connate, A 4–∞ $\underline{G}$ semi-inf. ((4)5–8), 1-loc. ∞ bitegmic ovules. Fr. a circumscissile capsule (pyxidium), its lid (operculum) falling with marcescent P; seeds shiny with spongy aril & ± curved embryo around abundant starchy perisperm (endosperm 0)
Genus: *Portulaca*

Portulacaria Jacq. Didiereaceae (Portulacaceae s.l.). Incl. *Ceraria*, 7 S Afr. Trees & shrubs. R: T 63(2014)1060. ***P. afra*** Jacq. – cult. orn. succ. shrub useful as fodder in S Afr.

Posadaea Cogn. = *Cucumeropsis*

posh-te *Annona scleroderma*

Posidonia C. Koenig. Posidoniaceae. 1 cosmop., 8 Aus. Tapeweeds. R: AB 20(1984)267. Poll. hydrophilous. ***P. oceanica*** (L.) Del. – to depths of 40 m in Medit., stems used in glass-packing & fibre in coarse sacking etc., dead pl. fibres s.t. seen on seashores round & worn – posidonia balls of which tonnes bulldozed from Medit. beaches to 'clean' them (used to fertilize fields)

Posidoniaceae Vines (~ Juncaginaceae, Potamogetonaceae). Magnoliidae – Alismatales. 1/9 Medit. & Aus. coasts. Marine or estuarine perenn. glabrous herbs with flattened rhiz. & erect monopodial stems term. in lvs; vessels & stomata 0. Lvs distich. with open persistent sheath, ligule & linear blade; intravaginal scales at nodes; stomata 0. Fls small, reg., bisex., protandrous, water-poll., in term. spikes with 2–4 bracts, on long flattened peduncle; P 0, A 3 (4) with sessile anthers with ± laminar connectives & longit. slits, pollen grains filamentous (pollen tubes) with 0 exine, $\underline{G}$ 1 with irreg. sessile stigma & 1 orthotropous pendulous ovule. Fr. a drupe with spongy pericarp, free-floating but eventually dehiscent, seed with ventral wing remaining attached to young pl. for 1–2 yrs after germ., straight embryo & 0 endosperm
Genus: *Posidonia*
Marine forms of Potamogetonaceae to which fam. they should be moved

Poskea Vatke. Plantaginaceae (Glob.; Globulariaceae). 3 Somalia, Socotra. R: Webbia 24(1970)624. Two spp. coll. only once

Posoqueria Aubl. Rubiaceae (?II 1). c. 25 trop. Am. Cult. orn. shrubs & trees with fragrant fls incl. ***P. latifolia*** (Rudge) Roemer & Schultes (Brazilian oak, trop. Am.) – when anthers touched, pollen expelled to at least 50 cm towards pollinator moth & C-tube closes; in fls with pollen removed tube reopens allowing cross-poll.; fr. ± ed., timber for walking-sticks, powdered fls used to repel fleas, bark yields blood-clotting agent used after wound from poison arrows in Amazon & recently tried as AIDS vaccine. ***P. grandiflora*** Standl. (C Am.) – fls every 3 yrs, fr. taking 32 months to mature

possumwood *Quintinia sieberi*

post oak *Quercus stellata*

postelein *Portulaca oleracea*

Postelsia Rupr. Laminariaceae (Ochrophyta). 1 W N Am. coasts: ***P. palmiformis*** Rupr. (sea palm) – ann., most often not submerged, ed. Calif. but overharvested

Postia Boiss. & Blanche = *Rhanteriopsis*

Postiella Kljuykov. Umbelliferae (III 5). 1 Turkey: ***P. capillifolia*** (Post) Kljuykov

pot *Cannabis sativa* 'subsp. *indica*'

Potalia Aubl. Gentianaceae. 9 trop. Am. R: SB 29(2004)674. ***P. amara*** Aubl. – Chamberlain's Model, used against snakebite in Amazonia

Potaliaceae Mart. = Gentianaceae

Potameia Thouars (~ *Beilschmiedia*). Lauraceae (I). 20 Madag.

Potamoganos Sandw. = *Bignonia*

Potamogeton Tourn. ex L. Potamogetonaceae. Incl. *Stuckenia*, c. 60 cosmop. esp. N (Eur. 22; N Am. 33 – R: FNA 22(2000)48). Pondweeds; oft. hybridizing (esp. GB, Scandinavia, Jap. – 46 known). Some with floating as well as submerged lvs; some with overwintering rhiz., or tubers borne on branches or winter buds. Nitrogen-fixing bacteria in rhizosphere; useful manure (lime in lvs & animal protein also attached). Some (lvs) used in burn ointments; some cult. orn. in aquaria incl. ***P. natans*** L. (temp. & warm) – floating lvs in E Anglia form. believed to give rise to young pikes, hence local name pickerel weed, rhiz. possible starch-source. ***P. diversifolius*** Raf. (N Am.) – fibre for nets; ***P. pectinatus*** L. (*S. p.*, sago pondweed, Euras., Aus., Am.) – v. imp. food for waterfowl

Potamogetonaceae Bercht. & J. Presl (~ Juncaginaceae). Magnoliidae – Alismatales. Incl. Zannichelliaceae (R: T 25(1976)273), 6/70 cosmop. Freshwater to brackish or alkaline, perenn. glabrous herbs rooted in substrate (roots with vessel-elements) with creeping sympodial rhiz. & erect leafy shoots. Lvs usu. in spirals, with basal open sheath, ligule & parallel-veined (1+) lamina, linear or expanded atop a petiole in floating ones, s.t. attached some way down sheath which appears as a stipule. Fls unisexual or bisex., small, usu. wind-poll. (some water-), reg., 4-merous in rather fleshy bractless (bracts s.t. initiated but not developed) spikes held a little above water or complex axillary sympodial infls underwater); P 0, 3 or 4 valvate, fleshy, with short claws or scale-like, A 1- 4 (opp. & adnate to claws), anthers sessile with longit. slits, $\underline{G}$ (1-)3 or 4 (-9) alt. with A, each with term. style or sessile stigma & 1 orthotropous bitegmic ovule oft. becoming campylotropous or anatropous at maturity. Fr. a head of achenes or drupelets, usu. floating (pericarp aerenchymatous); endosperm 0. n = 6–8, 12, 14–18

Genera: *Althenia, Groenlandia, Lepilaena, Potamogeton, Pseudalthenia, Zannichellia; Ruppia* s.t. incl.; Cymodoceaceae, Posidoniaceae & Zosteraceae (? marine derivatives of this fam.) poss. to be incl.

Some cult. orn. etc. (*Potamogeton*)

Potamophila R. Br. Gramineae (III 3 ii). 1 N NSW: ***P. parviflora*** R. Br. – on gravel banks in streams, ? diplosporous apomict

Potaninia Maxim. Rosaceae (9). 1 Mongolia: ***P. mongolica*** Maxim. – epicalyx 0; K, C & A 3; G 1. R: FOC 9(2003)381

Potarophytum Sandw. Rapateaceae (II 2). 1 NE S Am.: ***P. riparium*** Sandw. R: KB 1939:21

potato *Solanum tuberosum*; **air p.** *Dioscorea bulbifera*; **p. bean** *Apios americana*; **Chinese p.** *D. polystachya*; **p. chips/crisps** *S. tuberosum* esp. 'Record'; **couch p.** *Homo sapiens* (sedentary form); **fairy p.** *Claytonia* spp.; **p. fern** *Microgramma bifrons*; **Hausa p.** *Coleus rotundifolius*; **Livingstone p.** *C. esculentus*; **p. onion** *Allium cepa* Aggregatum Group; **p. orchid** *Gastrodia sesamoides*; **sweet p.** *Ipomoea batatas*; **p. vine** *I. pandurata, S. laxum*; **p. yam** *D. esculenta*

poteen illicitly distilled whisky (Ireland)

Potentilla L. Rosaceae (9). Incl. *Duchesnea, Horkelia, Horkeliella, Ivesia,* excl. *Comarum, Dasiphora, Drymocallis, Fragaria,* c. 450 N temp. (N Am. 151) & boreal, few S temp. R: BB 71(1908). Herbs (s.t. with rooting runners) with bractlets (epicalyx) alt. with K; some with swollen receptacle (*Duchesnea*). Some facultative apomictic groups (aposopory, pseudogamy) in Eur. Some cult. orn. esp. herb. perenn. ***P. atrosanguinea*** Lodd. (*P. argyrophylla,* W & C Himal.) cvs, ***P. indica*** (Andr.) T. Wolf (*D. i.,* Ind. strawberry, Ind., natur. Eur., N Am.) – hanging basket pl., ***P. lineata*** Trev. (*Argentina l., P. fulgens,* Himal.) – cult. orn., ***P. nepalensis*** Hook. (Pakistan to C Nepal) – locally medic. & ed., **'Miss Willmott'** apomictic so 'true' from seed. ***P. anserina*** L. (*A. a.,* silverweed, N temp. to S Am., Aus, NZ) – roots ed., form. cosmetic to clear spots, sun tan etc., lvs form. placed in shoes to reduce effects of sweating; ***P. erecta*** (L.) Räusch. (tormentil, Euras.) – dried rhiz. medic. (astringent tannins), also used for diarrhoea in cattle & horses etc.; ***P. reptans*** L. (cinquefoil, Euras., natur. N Am.) – locally medic., some 'doubles' with petalody of A & G; ***P. rivalis*** Nutt. ex Torrey & A. Gray (N Am.) – cross-poll. by bees or self-poll. by thrips; ***P. sterilis*** (L.) Garcke (barren strawberry, Eur., Medit.); ***P. sundaica*** (Blume) Theob. (Himal, China & Jap., W Mal mts) – weedy

Poteranthera Bong. Melastomataceae. 3 trop. S Am. R: Britt. 64(2012)7, Phytotaxa 263(2016)226. Savanna ann., fls 5-merous

Poteridium Spach (~ *Sanguisorba*). Rosaceae (Agrim.). 2 N Am. R: FNA 9(2014)319

Poterium L. (~ *Sanguisorba*). Rosaceae (Agrim.). Incl. *Bencomia, Sarcopoterium,* 13 N temp. to Canary Is. Candelabriform (Canary Is.) shrubs, s.t. spiny, to herb. perennials; some cult. orn. ***P. sanguisorba*** L. var. ***polygamum*** (Waldst. & Kit.) Vis. (*Sanguisorba minor* subsp. *balearicum,* S Eur.) – form. fodder-pl. in GB; ***P. spinosum*** L. (*Sarcopoterium s.,* E Medit.) – oft. dominant in the phrygana, v. common nr Jerusalem & poss. the crown of thorns of Jesus Christ, fr. oft. water-disp.

Pothoidium Schott. Araceae (III 1). 1 C Mal.: ***P. lobbianum*** Schott – v. short or 0 spathe; fibre for fish-traps in Philippines. R: Telopea 9(2001)558

Pothomorphe Miq. = *Piper*

Pothos L. Araceae (III 1). c. 70 Madag. to NE Aus. (Mal. & W Pacific 41 – R: Telopea 9(2001)449). Usu. lianes with lvs in 2 rows, P 4–6, in bisex. fls. Some cult. orn., local medic. ***P. tener*** Wall. (*P. rumphii,* C Mal. to Vanuatu) – lvs form. shredded to make 'pubic aprons' in New Ireland

Pothuava Gaudich. ex L.B. Sm. & Kress = *Aechmea*

Potoxylon Kosterm. (~ *Eusideroxylon*). Lauraceae (I). 1 Borneo: ***P. melagangai*** (Sym.) Kosterm.

pottery tree of Pará *Licania octandra*

Pottingeria Prain. Celastraceae (inc. sed.; Pottingeriaceae). 1 Assam to NW Thailand: ***P. acuminata*** Prain – venation like *Cinnamomum,* septicidal caps.

Pottingeriaceae Takht. = Celastraceae

Pottsia Hook. & Arn. Apocynaceae (II e). 3 Ind. to Java. Local fibres

Pouchetia A. Rich. ex DC. Rubiaceae (II 5). 6 trop. Afr.

Poulsenia Eggers. Moraceae (II). 2 trop. Am. R: FN 83(2001)138. Prickles (unique in fam.). ***P. armata*** (Miq.) Standl. – young lvs imp. protein source for red spider-monkeys, bark fibre used for cloth, mats etc.

pounce *Tetraclinis articulata*

Pounguia Benoist = *Whitfieldia*

Poupartia Comm. ex Juss. Anacardiaceae (II). Excl. *Sclerocarya,* 7 Madag., Masc.

Poupartiopsis Capuron ex J.D. Mitchell & Daly. Anacardiaceae (II). 1 E Madag.: ***P. spondiocarpus*** Capuron ex J.D. Mitchell & Daly – water-disp. fr. in littoral forests. R: SB 31(2006)338

Pourouma Aubl. Urticaceae (Cecropiaceae). 27 trop. Am. R: FN 50(1990)116. Some ed. fr. esp. ***P. cecropiifolia*** Mart. (uvilla, Braz.) – excellent grape flavour

Pourthiaea Decne (~ *Photinia*). Rosaceae (Ros.-Mal.). c. 25 E As. Some cult. orn.

Pouteria Aubl. Sapotaceae (IV). Incl. *Aningeria, Calocarpum, Lucuma, Neoxythece,* excl. *Beccariella, Planchonella, Pleioluma,* c. 140 trop. Am. (120 Amaz. & Atlantic forest), 60 As. to Pac., 5 Afr. R: T.D. Pennington, S. (1991)184. Trees & shrubs incl. some rheophytes, some fish-disp. Amazon. Alks. Some timber (anegré, aniègre) & ed. fr. (abiu) esp. ***P. caimito*** (Ruíz & Pavón) Radlk. (*P. cainito,* trop. Am.); ***P. campechiana*** (Kunth) Baehni (canistel, mamey, C Am., WI) – fr. adapted to disp. by ?extinct megafauna, ed. pulp (e.g. in Florida smoothies) & ***P. sapota*** (Jacq.) H. Moore & Stearn (sapote, marmalade plum, mammee zapote, C Am.) – imp. Carib. fr., fresh or preserved. ***P. cuspidata*** (A. DC.) Baehni (*A. r.*, Braz.) – comm. timber (asanfona); ***P. lucuma*** (Ruíz & Pavón) Kuntze (*Lucuma obovata*, lucuma, Peruvian Andes) – in comm. skin creams & icecream flavouring, rich in carotene, a 'superfood'

Pouzolzia Gaudich. Urticaceae (III). 35 trop. (OW 24 – R: NJB 24(2006)10; Am. 14 – R: OB 129(1996)61). Some local medic.; fibres esp. for fishing-nets – ***P. mixta*** Solms (*P. hypoleuca*, tingo, trop. Afr.) & ***P. sanguinea*** (Blume) Merr. (*P. viminea*, Mal.). ***P. tuberosa*** Wight (Indomal.) – ed. tubers

Povedadaphne Burger (~ *Ocotea*). Lauraceae (I). 1 Costa Rica: ***P. quadriporata*** Burger. R: Britt. 40(1988)276

poverty bush *Eremophila* spp.

poyok *Afrolicania elaeosperma*

Pozoa Lag. Umbelliferae (Azor.). 2 Andes of Chile & Arg. R: UCPB 33(1962)131

pracaxi fat *Pentaclethra macroloba*

Pradosia Liais. Sapotaceae (IV). 23 trop. S Am. (R: FN 52(1990)639, 1 ext. to C Am.), 1 Congo

Praecereus F. Buxb. (~ *Cereus*). Cactaceae (III 3). 2 trop. Am.

Praecitrullus Pang. = *Benincasa*

Praecoxanthus Hopper & A.P. Brown. Orchidaceae (IV 3b) 1 SW W Aus.: ***P. aphyllus*** (Benth.) Hopper & A.P. Brown. R: Gen. Orch. 2(2001)113

Prainea King ex Hook.f. = *Artocarpus* (but see JAA 40(1959)30)

prairie flax *Linum lewisii*; **p. gentian** *Eustoma grandiflorum*; **p. senna** *Chamaecrista fasciculata*

Pranceacanthus Wasshausen (~ *Juruasia*). Acanthaceae (III 2c). 1 Amaz. Braz.: ***P. coccineus*** Wasshausen. R: Britt. 36(1984)1

Prangos Lindl. (~ *Cachrys*). Umbelliferae (III 5). c. 30 Euras., Medit. R: Boissiera 26(1977)

Praravinia Korth. (~ *Urophyllum*) Rubiaceae (I 9). 49 C Mal.

Prasium L. Labiatae (VI). 1 Medit. incl. Eur., Canary Is.: ***P. majus*** L. – drupes

Prasophyllum R. Br. Orchidaceae (IV 2g). c. 80 Aus. & NZ (2–3). Leek orchids; fls not resupinate, mainly wasp-poll. with rewards unlike most in tribe. Some cult. orn.

Pratia Gaudich. = *Lobelia*

Pravinaria Bremek. Rubiaceae (I 9). 2 Borneo. R: RTBN 37(1940)184

Praxeliopsis G. Barroso. Compositae (Eup.-Prax.). 1 Braz.: ***P. mattogrossensis*** G. Barroso. R: MSBMBG 22(1987)392

Praxelis Cass. (~ *Eupatorium*). Compositae (Eup.-Prax.). 16 trop. Am. R: MSBMBG 22(1987)380. ***P. clematidea*** (Griseb.) R. King & H. Robinson (*E. catarium*, S Am., Florida) – foetid weed spreading in Hong Kong & Queensland

prayer plant *Maranta leuconeura*

Premna L. Labiatae (Premn.). c. 25 trop. & warm OW Mal. 14 – R: KB 68(2013)58. Cryptic dioecy; alks. Pyrogenous forms, e.g. ***P. herbacea*** Roxb. (trop. As.), form. distinguished as *Pygmaeopremna*. Some locally medic. (green 'jelly' made from lvs in Mal.) & some timbers esp. ***P. serratifolia*** L. (*P. integrifolia*, headache tree, Indomal. to Pac.) – Champagnat's Model, timber beautifully marked & used for cutlery handles

Prenanthella Rydb. Compositae (Cich.-Cich. (Micr.)). 1 SW N Am. deserts: ***P. exigua*** (A. Gray) Rydb. R: FNA 19(2006)359

× Prenanthenia Svent. = *Prenanthes* × *Sventenia* (= *Sonchus*)

Prenanthes Vaill. ex L. Compositae (Cich.-Cich.). 30 N temp. (Eur. 1, N Am. 13) to Afr. mts. ***P. subpeltata*** Stebb. (Afr. mts) – cli. Other spp. locally medic.

Prenia N.E. Br. = *Mesembryanthemum* (but see R: BJ 118(1996)25,419)

Prepodesma N.E. Br. (~ *Aloinopsis*). Aizoaceae (V). 1 S Afr.: ***P. orpenii*** (N.E. Br.) N.E. Br. R: H. E. K. Hartmann, *Ill. Handb. Succ. Pls*, Aiz. F–Z(2002)228

Prepusa Mart. Gentianaceae. 6 Braz. mts. R: KB 63(2008)173, Phytotaxa 163(2014)292

Prescotia Lindl. = seq.

Prescottia Lindl. (*Prescotia*). Orchidaceae (IV 2b). 15 trop. Am. (N Am. 1). R: Phytotaxa 178(2014)233. Some autogamy & (?) apomixis. ***P. plantaginifolia*** Lindl. – shoots form from roots, first orchid to develop into seedlings in cult.

Preslia Opiz = *Mentha*

Preslianthus Iltis & Cornejo (~ *Capparis*). 4 trop. Am. R: HPB 16(2011)68

Presliophytum (Urb. & Gilg) Weigend. Loasaceae (I 1). 3 W Peru deserts

Prestelia Schultz-Bip. (~ *Eremanthus*). Compositae (Vern.-Vern.). 2 SE Braz. Pachycaul rosette-trees

Prestoea Hook.f. Palmae (V 10). 10 C Am. R: FN 72(1996)46. Palm-hearts

Prestonia R. Br. Apocynaceae (II e). c. 55 trop. Am. Lianes with alks, some hallucinogens

Preussiella Gilg. Melastomataceae. 2 trop. W Afr. R: Adansonia 16(1976)405

Preussiodora Keay. Rubiaceae (II 1). 1 trop. W Afr.: ***P. sulphurea*** (K. Schum.) Keay. R: BJBB 28(1958)31

Priamosia Urb.= *Xylosma*

prickly ash *Orites excelsa*; **p. pear** *Opuntia* spp.; **p. poison** *Gastrolobium spinosum*; **p. poppy** *Argemone mexicana*

Pridania Gagnepain = *Pycnarrhena*

pride of Burma *Amherstia nobilis*; **p. of India** *Lagerstroemia speciosa, Koelreuteria paniculata*

Priestleya DC. = *Liparia*

Prieurella Pierre = *Chrysophyllum*

primavera *Roseodendron donnell-smithii*

primrose *Primula vulgaris*; **bird's-eye p.** *P. farinosa*; **Cape p.** *Streptocarpus* spp.; **Chinese p.** *P. sinensis*; **evening p.** *Oenothera* spp.; **water p.** *Ludwigia grandiflora*

Primula L. Primulaceae. Incl. *Cortusa, Dodecatheon* (Am. cowslips, shooting-stars – C reflexed & self-poll. poss. (usu. buzz-poll.)), 430 N hemisph. (Eur. 34, E Sino-Himalaya 225, N Am. 20 – R: FNA 9(2009)286), incl. Ethiopia & trop. As. mts to Java & NG, S S Am. (***P. magellanica*** Lehm. (*P. farinosa* var. *m.*)). R: J. Richards (2003) *P.*, 2nd ed.: 71; arr. in 37 sects (26 cult. Eur.); *Dionysia* & *Kaufmannia* prob. referable here. Herbs with rhiz. & term. infls, s.t. a series of whorls as in *P. verticillata*, an umbel in *P. veris* where 'stalk' is peduncle & fls erect until anthesis when pendent but becoming erect again in fr., or reduced peduncle hidden in basal rosette of lvs as in *P. vulgaris* where 'stalk' is an elongated pedicel. Heterostyly (91% of spp.), with pl. either bearing fls with short styles & anthers at mouth of fl. ('thrum') or long styles with anthers down inside fl. ('pin'), the diff. morphs with diff. pollen grains & stigmatic surfaces, the whole system promoting outcrossing when pollinators (bees & Lepidoptera) receive pollen in a position appropriate to depositing it on stigmas in the fls of second morph; homostyles, where system has broken down, incl. ***P. scotica*** Hook. (Scotland, one of v. few Br. endemic pls, now in decline; 2n = 54, *P. farinosa* (Euras., 2n = 18, 36) × *P. halleri* Honck. (C & SE Eur.), back-crossed with *P. f.* giving ***P. scandinavica*** Brunn), also populations of ***P. prolifera*** Wall. in Java, the pl. in Himal. being heterostyled, & ***P. pumila*** (Ledeb.) Pax (*P. eximia*) a widespread colonist in Alaska & nearby Russia derived from heterostylous *P. tschuktschorum* Kjellm. (Bering Strait). Many spp. with 'farina', a wax from glands on lvs, abaxial surfaces of which frequently bearing glands with primin, a flavone allergenic to c. 6% humans (0.001 mg may suffice), leading to frequently delayed dermatitis & further complications. Many cult. orn. hardy (e.g. 300 cvs grown at Pavlovsk, St Petersburg 1780s–90s) & greenhouse pls, the latter esp. ***P. obconica*** Hance (SW China, Thailand) with umbels of fls with notched petals, ***P. sinensis*** Sabine ex Lindl. (*P. praenitens*, Chinese primrose, anc. cultigen in China) with whorls of large fls & ***P. malacoides*** Franch. (NE Myanmar, SW China, weed of cult.) with whorls of small fls & less allergenic than *P. obconica*, all treated as annuals for winter-flowering; most familiar hardy pls incl. ***P. vulgaris*** Hudson (primrose, Eur. [in Netherlands reduced to 3 pops] & Med. to Iran) – elaiosomes, many cvs incl. doubles (A & G C-like), 'hose in hose' with C inside C-like K, '**Jackanapes**' with K apically leafy, basally C-like, & '**Jack in the Green**' with large leafy K, & ***P. ×polyantha*** Mill. (polyanthus, known since 1660, hybrid group involving *P. vulgaris*, (in some, ***P. elatior*** (L.) Hill (oxlip, Eur. to Iran) – indicator of woods >100 yrs old in E England) & *P. veris*) – many cvs. Other cult. orn. incl. ***P. allionii*** Lois. (Alps, rare) – many cvs & hybrids; ***P. auricula*** L. (auricula, bear's ears, Alps, Carpathians, Apennines) – largely calcicole, fls yellow, form. much cult. as 'florist's fl. 'esp. in mining communities in GB, esp. forms with farina & most of hybrid origin (***P. × pubescens*** Jacq., other parent *P. hirsuta* All. (*P. rubra*, C Alps, Pyrenees, calcifuge, fls pink with white eye), form. medic., first cult. Vienna c. mid 16 cent., like polyanthus the subjects of florists' societies esp. in 17 cent. & 18 cent. with cvs costing up to £20 by 1682 (cf. *Tulipa*) & those with green-edged C selected by mid 18 cent., exhibited in 'auricula theatres', ***P. sieboldii*** C.J. Morren (NE As., Jap.) a similar cult in 18 cent. Jap.; ***P. denticulata*** Sm. (Himal.) – 'drumsticks' of pink or purple yellow-eyed fls; ***P. farinosa*** L. (bird's-eye primrose, sub-boreal & boreal OW with closely allied pls in N Am.); ***P. florindae*** Kingdon-Ward (SE Tibet) – 'cowslip' over 1m tall; '***P. kewensis***' (enigmatic greenhouse pl. alleged to be allopolyploid (2n = 36) derived from sterile hybrid *P. × kewensis* W. Watson (2n = 18), *P. verticillata* Forssk. (Ethiopia, S Arabia, 2n = 18) or poss. *P. simensis* Hochst. (Ethiopia) × *P. floribunda* Wall. (Himal., 2n = 18) but sterile & with 'thrum' fls) – the pl. now grown as '*P. kewensis*' allegedly arising as a cross between a shoot bearing a 'pin' fl. & the original (the chance of the double recessive 'pin' arising as a somatic mutation on double dominant 'thrum' plants is so slim that the story is prob. more complex), though the tetraploid arose at least 3 times; ***P. marginata*** Curtis (Alps) – many cvs & hybrids with 'auriculas'; ***P. media*** (L.) Mast & Reveal (*Dodecatheon m.*, E US) – cult. orn.; ***P. parryi*** A. Gray (SW US) – fls carrion-scented; ***P. soldanelloides*** Watt (E Himal.) – minute with nodding fls; ***P. veris*** L. (cowslip (*primula veris* = Lat. firstling of spring), paigle or palsywort (arch.), Eur. to Iran) – form. medic. incl. tea from lvs, fls (fragrant) form. used in cowslip wine; ***P. vialii*** Delavay ex Franch. (China) – dense spikes of small pinkish fls resembling an *Orchis* sp.; also many hybrids of complex ancestry incl. **Candelabra Group** ('C. Hybrids', sect. *Proliferae*, E As.), **Petiolares Group** ('P. Hybrids', Himalaya), **Pruhonicensis Group** ('P. Hybrids', *P. juliae* Kuzn. (E Cauc., decid.) × spp. in sect.'Vernales' (sect. Primula)) esp. '**Wanda**'

Primulaceae Batsch. Magnoliidae – Ericales. Incl. Myrsinaceae, Theophrastaceae, 57/2645 subcosmop. but esp. trop. to N temp. Everg. trees, s.t. strikingly pachycaul (*Clavija, Oncostemum, Tapeinosperma, Theophrasta*), or lianes, subshrubs or herbs, with schizogenous secretory canals. Lvs in spirals, opp. or whorled, usu. simple & ± toothed (pinnatisect in *Hottonia*), oft. basal in herbs; stip. 0. Fls bisex., reg. (irreg. in *Coris*), (3–) 5(–9)-merous, s.t. heterostylous, in umbels, panicles, heads or solit.; (K) oft. persistent, (C) with imbr. or convolute lobes (C 3 in some *Anagallis*; C 0 in '*Glaux*'), A = & opp. C & attached in tube, anthers introrse, with longit. slits or term. pores, staminodes alt. with A s.t. present, $\underline{G}$ ((3-)5), half-inf. in *Maesa* & *Samolus*, 1-loc. with rudimentary partitions at base, 1 hollow style & (5–)∞ hemitropous or anatropous, bitegmic ovules on free-central placenta (campylotropous & unitegmic in *Cyclamen*). Fr. a capsule (s.t. circumscissile), rarely indehiscent, or a drupe, with (1–) ± ∞ angled seeds; embryo linear or short, straight (1 cotyledon in *Cyclamen*), embedded in copious starchless endosperm with reserves of oil, protein & amylose. n = 5, 8–15, 17, 19, 22

Classification & principal genera:

Form. recog. tribes in P. s.s. prob. best united as follows with Myrsinaceae, (Maesaceae) & Theophrastaceae as subfams (cf. Malvaceae & Preface), otherwise *Anagallis, Coris, Cyclamen, Lysimachia* (incl. *Glaux*), *Trientalis* = Myrsinaceae (s.s. with berries, drupes – *Ardisia, Cybianthus, Discocalyx* (herbaceous), *Embelia, Geissanthus, Myrsine, Oncostemum, Parathesis, Stylogyne, Tapeinosperma*), *Samolus* goes to Theophrastaceae (*Bonellia, Clavija, Deherainia, Jacquinia*), or own fam., & *Maesa* (ex Myrsinaceae; G half-inf., fr. many-seeded) goes to own fam. Primulaceae *s.s.*:

Primuleae ($\underline{G}$, C imbr., capsule with valves): *Androsace, Dionysia, Hottonia, Primula, Soldanella*

Cyclamineae ($\underline{G}$, C convolute, capsule with valves, seeds unitegmic & monocot., tubers = Myrsinaceae): *Cyclamen* (only)

Lysimachieae ($\underline{G}$, C convolute, capsule with valves or circumscissile & bitegmic seeds = Myrsinaceae): *Anagallis, Lysimachia*

Samoleae (G half-inf. = Samolaceae, Theophrastaceae): *Samolus*

Corideae (fls irreg., subshrubby): *Coris* (only; s.t. segregated as sep. fam., = Myrsinaceae).

One of 10 fams with elaiophores attractive to bees (*Lysimachia*); *Aegiceras* in mangove

Timber (*Jacquinia, Myrsine*), some local medic. (*Labisia, Maesa*); fish-poisons (*Aegiceras, Jacquinia*)

Many cult. orn. esp. *Anagallis, Androsace, Ardisia, Cyclamen, Lysimachia, Myrsine, Primula*

Primularia Brenan = *Cincinnobotrys*

Primulina Hance. Gesneriaceae (III 2j). c. 131 W & S China, Vietnam, limestone. R: T 60(2011)780. ***P. tabacum*** Hance (S China) – leaf glands said to smell of tobacco

Prince Albert's yew *Saxegothaea conspicua*; **p. [of Wales]'s feather** *Amaranthus hybridus*

Princea Dubard & Dop = *Triainolepis*

princess vine *Cissus verticillata*

princewood, princes wood *Cordia gerascanthus, Dalbergia cearensis, Exostema caribaeum*

Principina Uittien. Cyperaceae (I 1). 1 Principé (W Afr.): ***P. grandis*** Uittien – coll. there once (recollected São Tomé 2007). R: RTBN 32(1935)282

Pringlea T. Anderson ex Hook.f. Cruciferae (46). 1 Kerguélen & Crozet Is. (S Ind. Ocean): ***P. antiscorbutica*** R. Br. ex Hook.f. (Kerguélen cabbage) – pachycaul giant cabbage with lat. infls of usu. wind-poll. C-less fls, though s.t. (allegedly in shaded sites) with C, app. s.t. also visited by thrips; no septum in G. Pollen evidence supports theory of its being a relic of Tertiary flora; antiscorbutic

Pringleochloa Scribner = *Bouteloua*

Prinsepia Royle. Rosaceae (14). 3–4 Himal. to N China & Taiwan. R: Taiwania 11(1965)101. Thorny decid. shrubs with lat. styles & differing from *Prunus* in having chambered pith in stems. Some oilseeds & cult. orn. esp. ***P. sinensis*** (Oliv.) Oliv. (E As.) with yellow fls

Printzia Cass. Compositae (Ast.). 6 Afr. R: MBSM 16(1980)108. Heterogeneous?

Priogymnanthus P. Green (~ *Chionanthus*). Oleaceae (4d). 2 S Am. R: KB 49(1994)280

Prionanthium Desv. = *Pentameris* (but see Bothalia 18(1988)143)

Prioniaceae S. Munro & Linder = Thurniaceae

Prionium E. Meyer. Thurniaceae (Juncaceae s.l.). 1 S Afr. mt. streams: ***P. serratum*** (L.f.) E. Meyer (palmiet) – pachycaul aloe-like pl. (Tomlinson's Model) to 2 m with stems clothed in marcescent leaf-bases & runners. R: FOW 5(2001)2

Prionophyllum K. Koch = *Dyckia*

Prionopsis Nutt. = *Grindelia*

Prionosciadium S. Watson. Umbelliferae (III 9). 8 Mex.

Prionostemma Miers (~ *Hippocratea*). Celastraceae (III). 5 trop. Am., Afr., Ind. R: BMNHN Adans. 3,1(1981)7

Prionotaceae Hutch. = Ericaceae

Prionotes R. Br. Ericaceae (VII 1; Epacridaceae). 1 Tasmania: ***P. cerinthoides*** (Labill.) R. Br. – scrambling epiphyte, cult. orn.

Prionotrichon Botsch. & Vved. = *Rhammatophyllum*

Prioria Griseb. Leguminosae (I 2). Excl. *Gossweilerodendron*, *Kingiodendron*, *Oxystigma*, 1 C Am.: ***P. copaifera*** Griseb. (cativo) – resins used in euglossine bees' nests, timber for coarse construction, veneer etc.

Prismatocarpus L'Hérit. Campanulaceae (I 2). 29 trop. (1) & S esp. SW Afr. R: JSAB 17(1952)93. Some poll. monkey beetles

Prismatomeris Thw. Rubiaceae (Prismat.). 15 Sri Lanka to S China & W Mal. R: OB 94(1987)5. Wood anatomy suggests exclusion from tribe. ***P. tetrandra*** (Roxb.) K. Schum. (Indomal.) – local medic.

Pristiglottis Cretz. & J.J. Sm. = *Odontochilus*

Pristimera Miers (~ *Hippocratea*). Celastraceae (III). 24 trop. R: BMNHNAdans. 3, 1(1981)8. Heterogeneous. ***P. celastroides*** (Kunth) A.C. Sm. (Mex.) – source of insecticide

Pritchardia Seemann & H. Wendl. Palmae (III 4b). c. 27 Fiji, Samoa, Tonga, Tuamotus, Hawaii (19). R: Principes 24(1980)65. Fan palms with bisex. fls; some cult. orn. & many endangered spp. known only from trees in cult. (though much-vaunted ***P. maideniana*** Becc. (Hawaii) in Royal Botanic Garden Sydney not, as in fact (older name for) *P. affinis* Becc.)

Pritchardiopsis Becc. = *Saribus* (but see Allertonia 3(1984)317)

Pritzelago Kuntze = *Hornungia*

Priva Adans. Verbenaceae (Priv.). 20 trop. & warm. R: FR 41(1936)1

privet *Ligustrum* spp., esp. *L. vulgare* (**common** or **European p.**) & *L. ovalifolium*; **Chinese p.** *L. sinense*; **glossy p.** *L. lucidum*; **golden p.** *L. vulgare* 'Aureum'; **Japanese p.** *L. japonicum*

Proboscidea Schmidel. Martyniaceae (Pedaliaceae s.l.). 8 warm Am. R: P.K. Bretting, *Syst. P.* (1981) 119. Orig. fr. distrib. on legs of megafauna? Cult. orn. & ed. fr. esp. ***P. louisianica*** (Mill.) Thell. (S N Am., natur. Eur., Aus.) – fr. interferes with shearing of sheep & can get in eyes or even clamp mouth of sheep, but young fr. pickled & pl. grown comm. for this; strips from endocarp used by Native Americans in black or dark designs in basketry; some oilseeds esp. ***P. parviflora*** (Wooton) Wooton & Standl. (S N Am.) & ***P. triloba*** (Schldl. & Cham.) Decne (Mex.) – poss. crops

Proatriplex (W. Weber) Stutz & G.L. Chu (~ *Atriplex*). Amaranthaceae (Atrip.). 1 SW US: ***P. pleiantha*** (W. Weber) Stutz & G.L. Chu. R: SB 35(2010)852

Prochnyanthes S. Watson = *Agave*

Prockia P. Br. ex L. Salicaceae (Flacourtiaceae). 2 trop. Am. (1 restricted to Venez.). R: FN 22(1980)66

Prockiopsis Baill. Achariaceae (Flacourtiaceae). 3 Madag. R: Adansonia III,25(2003)46

Procopiania Guşul. = *Symphytum*

Procris Comm. ex Juss. = *Elatostema* (but see FR 45(1938)179, 257)

Proferea C. Presl = *Sphaerostephanos*

Proiphys Herb. (*Eurycles*). Amaryllidaceae. 4 Mal. (***P. alba*** (R. Br.) Mabb.) to trop. & E Aus. R: Austrobaileya 6(2001)121. 'Seeds' actually bulbils. Some cult. orn. incl. ***P. amboinensis*** (L.) Herb. (Mal. to Aus.) & ***P. cunninghamii*** (Lindl.) Mabb. (Brisbane lily, SE Queensland, NSW)

Prolobus R. King & H. Robinson. Compositae (Eup.-Gyp.). 1 Bahia (Braz.): ***P. nitidulus*** (Bak.) R. King & H. Robinson. R: MSBMBG 22(1987)100

Prolongoa Boiss. (~ *Chrysanthemum*). Compositae (Anth.-Leucanthemops.). 1 Spain: ***P. hispanica*** G. López & Jarvis. R: BBMNHB 23(1993)141

Promenaea Lindl. Orchidaceae (V 12j). c. 20 Braz. Cult. orn. epiphytes

Prometheum (A. Berger) H. Ohba (~ *Sedum*). Crassulaceae. 8 Euras. R: U. Eggli, *Ill. Handb. Succ. Pls*, Crass. (2005)204

Pronaya Huegel = *Billardiera*

Pronephrium C. Presl = *Cyclosorus*

prophet flower *Arnebia pulchra*

propolis resin coll. from buds of poplars, horse-chestnut etc. by bees used to fix combs in hives (bee-glue)

Prosanerpis S.F. Blake = *Clidemia*

Prosaptia C. Presl (~ *Grammitis*). Polypodiaceae (V; Grammitidaceae). c. 60 trop. As.

Prosartes D. Don (~ *Disporum*). Liliaceae (Convallariaceae). 5 N Am. (5). R: FNA 26(2002)142. Fairy-bells. ***P. lanuginosa*** (Michaux) D. Don (*D. l.*, EC US) – myrmechory, cult. orn.

Proscephaleium Korth. = *Chassalia*

Proserpinaca L. Haloragaceae. 2 E N Am. to WI (common fossil in Euras.). Fls 3-merous. ***P. palustris*** L. (N Am. to WI) – sometimes cult. in aquaria

proso millet *Panicum miliaceum*

Prosopanche Bary. Aristolochiaceae (Hydnoraceae). 1 Costa Rica (rainforest), 2 Arg. (dry forest), 1 NE Braz. R: Kurtziana 2(1965)60

Prosopidastrum Burkart. Leguminosae (II 1). 1 Mex., 4 S S Am.

Prosopis L. Leguminosae (II 1). 44 warm Am. (most), SW As., Afr. R: JAA 57(1976)219, 450. Heterogeneous. Usu. spiny trees or shrubs of value in dry country for browse, agroforestry, ed. fr., comm. (mesquite) honey, etc., s.t. invasive (e.g. S Afr.), incl. hybrids (Aus.). Alks. ***P. chilensis*** (Molina) Stuntz (algarroba, Arg. & Chile) – fast-growing in dry conditions, locally cult. for cattle fodder & sweet-tasting legumes; ***P. cineraria*** (L.) Druce (*Mimosa cinerea*, *P. spicigera*, Iran to Ind.) – ed. fr., bark fibre & local timber; ***P. glandulosa*** Torrey (algaroba, mesquite, SW N Am., invasive Aus., S Afr.) – timber, fuel, gum, seeds ed. cooked; ***P. juliflora*** (Sw.) DC. (trop. Am., invasive Ethiopia, Masc.) – pods (screw-beans) ed.; ***P. pallida*** (Willd.) Kunth (S Am., natur. Hawaii (derived from 1 pl. 1828) & Aus.) – salt-tolerant, seeds disp. animals (endozoic), legume ed. & source of sweet syrup used in drinks; ***P. tamarugo*** F. Philippi (N Chile) – only tree to survive in arid salt flats of N C., where ann. rainfall 70 mm, water-table at 20 m & salt 0.5 m thick; ***P. velutina*** Wooton (SW N Am.) – a million fls per tree, attracting hundreds of insect spp. incl. 60 spp. of solitary bee, local medic. & ed. seeds

Prosopostelma Baill. = *Cynanchum*

Prospero Salisb. = *Scilla*

Prosphytochloa Schweick. Gramineae (III 3i). 1 S Afr.: ***P. prehensilis*** (Nees) Schweick. – cli. to 10 m by wind bringing leaf-tips with retrorse scabrid surface in contact with supports. R: Strelitzia 10(2000)712

Prostanthera Labill. Labiatae (IV 2). c. 100 Aus. (R (sect. *Klanderia*, 15): JABG 6(1984)207). Many natural hybrids. Some cult. orn. (mintbush)

Prosthechea Knowles & Westc. (~ *Encyclia*). Orchidaceae (V 13b). c. 110 trop. Am. (Florida 3). R: Phytol. 82(1977)375

Prosthecidiscus F.D. Sm. Apocynaceae (Vc; Asclepiadaceae III 3). 1 C Am.: ***P. guatemalensis*** F.D. Sm. R: MSBMBG 85(2001)269

Protarum Engl. Araceae (VII 25). 1 Seychelles: ***P. sechellarum*** Engl. R: Aroideana 29(2006)37

Protasparagus Oberm. = *Asparagus*

Protea L. Proteaceae (IV 3). c. 103 trop. (high alts, 21) & S Afr. (82 (R: J.P. Rourke (1980) *The Ps. of S Afr.*), 69 endemic in SW Cape, ***P. gaguedi*** J.F. Gmel. from Ethiopia to Namibia). Fls in large heads with much nectar (sugar bushes esp. ***P. repens*** (L.) L. (*P. mellifera*, Cape – visited by Cape Sugar Bird, good bee-pl., nectar form. boiled down to syrup), 35 fynbos spp. rodent-poll. (yeast-scented, nectar-rich infls nr ground-level); some cult. orn. & used as long-lasting cut-fls esp. ***P. cynaroides*** (L.) L. (Cape) with heads to 20 cm across & hybrids favoured by florists, esp. involving ***P. neriifolia*** R. Br. (Cape) – first Cape pl. specimen to reach Europe (1597), e.g. **'Pink Ice'**, 1 of world's most pop. cut-fls. ***P. nitida***

Mill. (waboom, S Afr.) – form. wood for charcoal, wagon parts, furniture, bark for tan, lvs for ink

Proteaceae Juss. Magnoliidae – Proteales. 75/1710 trop. & subtrop. esp. S hemisph. (most in Aus. & S Afr.). R: Telopea 11(2006)329. Everg. trees & shrubs usu. with 3-celled hairs, oft. accum. aluminium; roots usu. without mycorrhizae but oft. with short lat. ('proteoid') roots releasing organic acids making otherwise inaccessible phosphorus available for absorption. Lvs usu. in spirals, simple (usu.) to pinnate or bipinnate, oft. xeromorphic; stip. 0. Fls usu. bisex. (or pl. monoec. or dioec. as in *Leucadendron*), reg. or not, protandrous, poll. by insects, birds or marsupials, 4-merous, solit. or paired in axil of bract, in racemes, umbels or involucrate heads or primary 2-fld infls in sec. racemes; K valvate, oft. petaloid, usu. with basal tube s.t. cleft most on 1 side or 3 connate & 1 free, C app. an annular ± 4-lobed nectary-disk or (2–)4 scales or glands alt. with K, or 0, A opp. K with broad filaments usu. adnate to K & anthers with longit. slits & 1 theca oft. sterile & oft. elongated connective, $\underline{G}$ 1, s.t. not closed, with elongate style s.t. a pollen-presenter & 1 or 2 (– ± ∞) marginal (anatropous to) hemitropous or amphitropous (or orthotropous) bitegmic ovules. Fr. a follicle, nut, achene or drupe, oft. 1-seeded when indehiscent; seeds oft. winged, embryo straight & oily with 2(–8 in *Persoonia*) large cotyledons & usu. 0 endosperm. n = 5, 7, 10–13 (? orig. 7), the chromosomes s.t. v. large

Classification & principal genera:

I. **Bellendenoideae** (A free, ovules 2, fr. dry indehiscent, winged, n = 5, Aus.) – *Bellendena* (only)

II. **Persoonioideae** (A largely or completely adnate to K, fr. unwinged, n = 7(14); 2 tribes, SW Pacific): 1. Placospermeae – *Placospermum* (only); 2. Persoonieae – *Persoonia*

III. **Symphionematoideae** (A basally adnate to K, ovules 1 or 2, fr. indehiscent, dry, n = 10, 14, Aus.): *Symphionema*

IV. **Proteoideae** (A ± adnate to K, ovule 1, fr. indehiscent; 4 tribes, OW): 1. Conospermeae (Aus.) – *Conospermum*; 2. Petrophileae (Aus., Cape – *Aulax, Petrophile*; 3. Proteeae (Afr.) – *Faurea, Protea*; 4. Leucadendreae (S Aus., Cape) – *Adenanthos, Isopogon, Leucadendron, Leucospermum*

V. **Grevilleoideae** (incl. Carnarvonioideae, Sphalmioideae; A basally to usu. completely adnate to K, (1) 2 ovules, fr. not winged, n = 10–14; 4 tribes, mostly Aus., some Am., 1 Madag.): 1. Roupaleae – *Helicia, Knightia, Lambertia, Roupala*; 2. Banksieae – *Banksia* (incl. *Dryandra*); 3. Embothrieae – *Hakea, Grevillea, Stenocarpus, Telopea*; 4. Macadamieae – *Brabejum, Euplassa, Gevuina, Macadamia, Malagasia*

Palmate-leaved fossils from Aus. Eocene. App. primitively rainforest trees, the xeromorphic features (incl. serotiny & lignotubers) sec. & of parallel origin in Aus., S Afr., New Caled. & to lesser extent S Am. The geoflorous habit assoc. with mammalian poll. & burning also in parallel in *Banksia* (SW Aus.) & *Protea* (S Afr.).

Some cult. orn. esp. *Banksia, Embothrium, Gevuina, Grevillea, Persoonia, Protea* etc., some timber (esp. *Cardwellia, Grevillea, Knightia, Orites, Panopsis & Stenocarpus*), medic. (*Conospermum*) & ed. seeds, esp. *Macadamia* but also *Brabejum, Finschia, Floydia, Gevuina & Hicksbeachia*; some weedy e.g. *Hakea* in S Afr.

Proteopsis Mart. & Zucc. ex Schultz-Bip. Compositae (Vern.- Lych.). 1 S Braz. campos: ***P. argentea*** Mart. & Zucc. ex Schultz-Bip. – pachycaul rosette-tree

Protium Burm.f. Burseraceae (II). c. 180 trop. Am. (Peru 38), Madag., Mal. Source of resins (used by euglossine bees in their nest-making) esp. in trop. Am. esp. ***P. carana*** (Humb.) Marchand (Carana elemi) & ***P. heptaphyllum*** (Aubl.) Marchand (Brazilian e., Cayenne incense, i. tree); ***P. javanicum*** Burm.f. (Mal.) – durable timber

Protocyrtandra Hosok. = *Cyrtandra*

Protoedraianthus Lakusic = *Edraianthus*

Protogabunia Boit. = *Tabernaemontana*

Protolirion Ridl. = *Petrosavia*

Protomarattia Hayata = *Marattia*

Protomegabaria Hutch. Phyllanthaceae (Scep.; Euphorbiaceae s.l.). 3 W & C Afr. R: Adans. 36(2014)105. Rauh's Model; noisy slash

Protorhus Engl. Anacardiaceae (I). 1 E S Afr.: ***P. longifolia*** (Bernh.) Engl. See also *Abrahamia*

Protoschwenkia Soler. Solanaceae (II). 1 Andean Bolivia & adjacent Braz.: ***P. mandonii*** Soler.

Prototulbaghia Vosa = *Tulbaghia*

Protowoodsia Ching = *Woodsia*

Proustia Lag. Compositae (Mut.-Nass.). 3 Bolivia, Chile & Arg. R: RMLP 11(1968)23. Shrubs

Provancheria B. Boivin (*Provencheria*) = *Dichodon*

pru see *Gouania polygama*

Prumnopityaceae A. Bobrov & Melikian = Podocarpaceae

Prumnopitys Philippi (~ *Podocarpus*). Podocarpaceae. Incl. *Sundacarpus*, 9 Costa Rica to Venez., S Chile, Mal., NZ, New Caled. (Palaeocene of As., Eocene of S England, Cretaceous of N Hemisph.) R: Blumea 24(1978)189, Phytol.M 7(1984)68. Timber. ***P. andina*** (Endl.) Laubenf. (Chile) – seed ed. (no resins); ***P. ferruginea*** (D. Don) Laubenf. (miro, NZ) – timber for cabinet-work, gum medic.; ***P. taxifolia*** (D. Don) Laubenf. (*Podocarpus spicatus*, matai, NZ) – timber for flooring, railway sleepers etc.

prune *Prunus* × *domestica*

Prunella Tourn. ex L. Labiatae (VII 2b). c. 7 Euras. (Eur. 4), N Afr. Some cvs for ground-cover etc. esp. ***P. grandiflora*** (L.) Scholler (*P. webbiana*, Eur.). ***P. vulgaris*** L. (selfheal, Euras., natur. Aus., Afr., Am.) – form. medic. esp. for sore throat (Doctrine of Signatures – fl. shape), staunching bleeding (until WW II), heart disease etc.

Prunus Tourn. ex L. Rosaceae (14). Incl. *Maddenia*, 200+ temp. (esp. N; Eur. 17), trop. mts. Orig. OW. Subg. s.t. recog. as separate genera, though unity confirmed by molecular findings: *Prunus* s.s. (*P. spinosa* & allies), *Cerasus* Mill. (*P. avium* & allies), *Armenaica* Scop. (*P. armenaica*), *Persica* Mill. (*P. persica* & allies), *Amygdalus* L. (*P. dulcis* & allies), *Padus* Mill. (*P. padus* & allies) & *Laurocerasus* Duham. (*P. laurocerasus* & allies). Decid. & everg. trees & shrubs, s.t. spiny, s.t. with domatia or extrafl. nectaries (e.g. on leaves of *P. serotina* which attract *Formica obscuripes* ants repelling caterpillar attack, the nectar flow ceasing once caterpillars too big to be killed by ants); imp. fr. trees – plums, cherries, almonds, peaches, apricots etc., cult. orn. & gums; lvs & fls salt-pickled in Jap. or used in sweets & tea; many hybrids incl. most 'flowering cherries' (W. Kuitert (1999) *Japanese flowering cherries*), most common cvs (e.g. the ubiquitous '**Kwanzan**' ('Kanzan') & fastigiate '**Amano-gawa**' of suburbia) grouped as ***P.* Sato-zakura Group**. ***P. africana*** (Hook.f.) Kalkman (*Pygeum a.*, kanda (Cameroun), kotofihy (Madag.), red stinkwood, trop. & S Afr.) – trad. medic., now comm. treatment for prostitis (phytosterols, now largest vol. of any Afr. medic. pl. in world trade), timber form. used for wagons etc.; ***P. americana*** Marshall (N Am.) – local medic., fr. ed. (s.t. dried); ***P. angustifolia*** Marshall (Chickasaw plum, E & S US) – cult. ed. fr.; ***P. armeniaca*** L. (apricot, N China where cult. since 2000 BC) – fr. (4 M t a yr by 2011; rich in vitamin A) eaten fresh (10 monoterpenes active in flavour compared with peach) or tinned, juice, brandy & liqueurs (e.g. amaretto from endocarps), seed-oil used in cosmetics etc., wood for reed wind instruments for Ezerum market, old cvs incl. '**Moorpark**' mentioned by Jane Austen (*Mansfield Park*); ***P. avium*** (L.) L. (gean, mazzard, wild cherry, hagberry, hegberry. Euras.) – self-compatible, 2n = 16 (cf. *P. cerasus*), source of sweet cherries (1.1 M t a yr by 2009), timber (form. soaked in limewater to give mahogany subs.) for veneers, walking-sticks etc., said to lower urate levels (& therefore gout) in blood; ***P. bokhariensis*** Royle ex Schneider (Bokhara plum, Ind.); ***P. brigantina*** Villars (Briançon apricot, SE France) – seed-oil scented; ***P. cerasifera*** Ehrh. (myrobalan, mirabelle, cherry plum, C As. to Balkans (or cultigen?)) – small ed. fr., oft. candied, stock for plums, invasive Aus., '**Pissardii**' – ubiquitous red-leaved cv. of suburbia, hybrids with *P. salicina* = ***P.* x *rossica*** Erem. (Russian ps) – many cvs in Russia; ***P. cerasus*** L. (morello or amarelle cherry, origin unclear, widely natur.) – self-incompatible, 2n = 32, source of sour cooking cherries, '**Marasca**' (var. *marasca*, maraschino cherry or marasco) used in liqueurs & dried (melatonin source so poss. relief from jet-lag), & crossed with *P. avium* to give Duke cherries (***P.* × *gondouinii*** (Poit. & Turpin) Rehder), lvs a tea-source, gum form. used in cotton-printing (cherry gum), infusion of stalks still sold as diuretic in Paris markets (1997), wood-chips (like apple) used for smoking salmon; ***P.* × *domestica*** L. (plum, gage, anc. cultigen, poss. orig. SW As. but hexaploid & of complex background poss. involving *P. cerasifera* & *P. spinosa*) – the familiar plum (7 M t a yr by 2007), with many cvs used fresh (form. much dried), tinned, or as juice, in liqueurs etc., timber used for turnery, gum to flavour cider, but plums in US s.t. native Am. spp. or (or, as in UK, hybrids with) *P. salicina* Lindl. (China, Korea) – all diploids, dried fr. esp. of *P.* × *domestica* being prunes (poss. preventing osteoporosis), an inf. type being the quetsche(n) plum, '**Prune d'Agen**' (sugar ps exported from Aus. to Jap.), while the '**Victoria**' p. is a self-fertile form found in a garden in Sussex in 1840 (or orig. Glos., or Fr.), 'subsp. *institia*' (*P. institia*) – sweet forms = 'mirabelle', others being damson (bullace) – usu. preserved, 'subsp. *italica*' = greengage, hybrids with apricots = plumcots (pluots); ***P. dulcis*** (Mill.) D. Webb (*Amygdalus communis*, almond, W As.) – amygdalin in nectar poss. deterring some pollinators, attracting others, ed. seed for dessert & confectionery, 'var. *dulcis*' sweet (sweetness a result of a dominant gene) the most grown of all 'nuts' (Calif. biggest producer, drawing more than half US bee colonies in Feb.), seed syrup (orgeat – orig. *Hordeum vulgare*) a flavouring as in gin, ground for macaroons (amaretti biscuits) & cakes or coated in sugar ('sugared a.s'), blanched, pulverized & soaked in water giving almond milk for rich infants in Eur. until end C18 (now comm. as is a. butter, peanut subs. in US), immature fr. sold in Arab markets, 'var. *amara* (DC.) H. Moore' (bitter a.) – bitterer, source of almond oil used medic. (e.g. earache, allegedly keeping cholesterol levels low) & as hair-conditioner, both much cult. in Medit., Calif., Aus. etc., the 2nd s.t. as stock for first, endocarps ground for fibre used in clay-tilemaking in SW Eur.; ***P. emarginata*** (Douglas) Eaton (N Am.) – imp. local medic., ed. fr., bark for basketry; ***P. gracilis*** Engelm. & A. Gray (Oklahoma plum, prairie cherry, Arkansas to Texas) – fr. ed. & form. much dried by Native Americans for winter use; ***P. hortulana*** L. Bailey (wild goose plum, C & SE US) – several cvs with ed. fr.; ***P. incisa*** Thunb. (Fuji cherry, Japan) – shrubby, many cvs incl. dwarf sold as potpls; ***P. laurocerasus*** L. (cherry laurel, SE Eur., SW As., invasive C Eur., Aus.) – the laurel of suburbia (superseded *Laurus nobilis* in wreaths), usu. clipped as hedge but a tree to 14 m with many cvs, lvs with alks & the glycoside prulaurasine which is broken down to release HCN by an enzyme when lvs damaged, so used in 'killing' bottles by entomologists, glands at base of lamina extrafl. nectaries visited by ants poss. homologous with lat. leaflets of pinnate-leaved Rosaceae; ***P. lusitanica*** L. (Portugal laurel, SW Eur., Macaronesia) – everg. resembling bay; ***P. mahaleb*** L. (mahaleb (cherry), St Lucie c., Euras., introd. N Am.) – timber for turnery & tobacco pipes, lvs a flavouring incl. in tobacco; ***P. maritima*** Marshall (beach plum, E N Am.) – ed. fr. & sand-binder; ***P. mume*** (K. Koch) Sieb. & Zucc. (mume, Jap. apricot, ume, SW Jap.) – winter-flowering 'plum blossom' of Jap. (ginger jars, mah jong tiles, etc.) much grown & illus. there (one of most depicted of all fls in China, where used to flavour tea), used as bonsai & ed. fr. (acid, used like vinegar & basis of plum sauce), for which first planted, later around temples with citrus but now replaced by *P. serrulata*; ***P. nigra*** Aiton (Canada p., E N Am.) – ed. fr., some cvs, planted by Iroquois in New York State; ***P. padus*** L. (bird cherry, hagberry, hegberry, Eur. to Jap.) – timber for interior work, boats etc.; ***P. persica*** (L.) Batsch (peach, China, poss. cultigen derived from '*P. davidiana* (Carr.) N.E. Br.' (China) – used as stock for peaches, plums, cherries, apricots etc., endocarps used in Buddhist rosaries, early natur. N Am. incl. Manhattan, prob. via Florida in C16 but in Carolinas & Georgia before English arrived) – flavour due to undecalactone alone (cf. apricot), next to apple world's most widely grown tree-fr. (19 M t a yr by 2010), dwarf forms produced by budding on *P. tomentosa* etc., many cvs (Aus. alone 160; fashionable incl. flattened '**Saturne**' or do[ugh]nut ps, 'paragueiros' in Spain) for fr. eaten fresh ('free-stoned'), dried, tinned ('cling-stoned'), source of juice, brandy, ground seeds in certain Italian sweetmeats (e.g. 'Amaretto'), blossom used in Chinese New Year celebrations (UK) instead of *Enkianthus quinqueflorus*, 'var. *nucipersica* C. Schneider' (var. *nectarina*, nectarine) with smooth fr. skin spontaneus mutants on peach shoots, v. rarely reverting; smooth skin recessive, yellow flesh recessive), ? hybrids being 'peacharines'; ***P. pseudocerasus*** Lindl. (China) – cult. for cherries in China; ***P. pumila*** L. (sand cherry, NE US) – shrub sometimes prostrate (*P. depressa*) forming mats to 2m across; ***P. rivularis*** Scheele (*P. munsoniana*, wild goose plum, N Am.) – ed. fr., several cvs; ***P. salicina*** Lindl. (Jap. plum, Chinese p., China, Korea) – ed. fr., see *P.* × *domestica*; ***P. serotina*** Ehrh. (American or black or cabinet or rum cherry, capulin (c.), E N Am., introd. UK 1629, invasive C & E Eur.) – timber for cabinet-making, bark form. medic., flavouring for rum & brandy; ***P. serrulata*** Lindl. (Oshima cherry, NE As.) – many cvs, prob. contributing towards the bulk of the Jap. flowering cherries ('var. *speciosa*', '*P. lannesiana*', *P. speciosa*), spreading habit, incl 'doubles', the wild form ('var. *spontanea* (Maxim.) E. Wilson') with upright habit with some 'doubles' being the Jap. cherry of poetry & painting, planted as votive offerings C7 onwards); ***P. simonii*** Carr. (apricot plum, N China, not known wild) – bitter but ed. fr. (= plumcots?); ***P. spinosa*** L. (sloe, blackthorn, Eur., W As.) – spiny shrub or small tree with fr. eaten by birds, badgers & foxes in GB – disp. agents, used in liqueurs incl. sloe gin, etc., wood for turnery, hay-rake teeth, shillelaghs etc., lvs form. tea subs., medic. (warts S Eng., coughs etc.); ***P.* × *subhirtella*** Miq. (triploid, prob. *P. incisa* × *P. pendula* Maxim. f. *ascendens* (Makino) Ohwi, NE As.) – winter-flowering esp. '**Autumnalis**' & '**A. Rosea**' & hybrids with other spp.; ***P. tenella*** Batsch (dwarf Russian almond, C Eur. to Siberia) – dwarf suckering cult. orn.; ***P. tomentosa*** Thunb. (Nanking cherry, temp. E As.) – sweet ed.

fr.; ***P. umbellata*** Elliott (*P. alleghaniensis*, Alleghany plum, black sloe, flatwoods plum, E N Am.); ***P. virginiana*** L. (chokecherry, E N Am.) – fr. ed. cooked, bark medic. esp. dysentery & resp. problems; ***P. webbii*** (Spach) Fritsch (S & E Eur.) – seeds ed. Greece before *P. dulcis* introd.; ***P.*** × ***yedoensis*** Matsum. (Yoshino cherry, most frequently cult. flowering c. in Jap.) – Jap. national fl., poss. hybrid between *P. serrulata* (*P. speciosa*) & *P. pendula* f. *ascendens* (or *P. subhirtella*) with fls before lvs expand so v. spectacular (as c. 1000 trees at Ueno Park, Tokyo), **'Somei-yoshino'** = Yoshino c.

pry *Tilia cordata*

Przewalskia Maxim. (~ *Scopolia*). Solanaceae (IV 2). 1 W China: ***P. tangutica*** Maxim. R: BBMNHB 27(1997)27

Psacadopaepale Bremek. = *Strobilanthes*

Psacaliopsis H. Robinson & Brettell (~ *Senecio*). Compositae (Sen.-Tuss.). 5 Mex., Guatemala

Psacalium Cass. (~ *Senecio*). Compositae (Sen.-Tuss.). 42 SW US (1), C Am. R: Phytol. 27(1973)254. ***P. decompositum*** (A. Gray) H. Robinson & Brettell (matarique, Mex.) – roots medic.

Psammagrostis C. Gardner & C. Hubb. = *Eragrostis* (but see FA 44B(2005)411)

Psammetes Hepper = *Lindernia* (but see KB 52(1997)750)

Psammiosorus C. Chr. = *Arthropteris*

Psammisia Klotzsch. Ericaceae (VIII 5). c. 60 trop. Am.

Psammochloa A. Hitchc. Gramineae (X). 1 Gobi Desert: ***P. villosa*** (Trin.) Bor. R: KB 6(1951)196. Allied to *Stipa*

Psammogeton Edgew. Umbelliferae (III 8). 3–5 SW As. R: BDBG 69(1956)227

Psammomoya Diels & Loes. Celastraceae (I). 4 SW Aus. R: Nuytsia 14(2002)385

Psammophiliella Ikonn. (~ *Gypsophila*). Caryophyllaceae. 4 C As.

Psammosilene W.C. Wu & C.Y. Wu (~ *Silene*). Caryophyllaceae. 1 Yunnan: ***P. tunicoides*** W.C. Wu & C.Y. Wu

Psammophora Dinter & Schwantes. Aizoaceae (V). 4 S Afr. R: H. E. K. Hartmann, *Ill. Handb. Succ. Pls*, Aiz. F–Z(2002)228. Cult. orn. mat-forming succ. pls, the stems s.t. underground, leaf secretions capture sand-grain camouflage

Psammotropha Ecklon & Zeyher. Molluginaceae. 11 SE Afr. with 2 ext. to trop. Afr. R: JSAB 25(1959)51

Psathura Comm. ex Juss. Rubiaceae (IV 7). 8 Madag. & Masc.

Psathyranthus Ule = *Psittacanthus*

Psathyrostachys Nevski (~ *Hordeum*). Gramineae (XV). 9 E Med. (Eur. 1) to C As. R: NJB 11(1991)3. ***P. fragilis*** (Boiss.) Nevski crossed with *Hordeum vulgare* L. leads to chromosome elimination resulting in haploid (*Hordeum*) tillers; ***P. juncea*** (Fischer) Nevski (Russian wild rye) – cult. N Am.

Psathyrotes (Nutt.) A. Gray. Compositae (Hele.-Psat.). 3 SW N Am. R: Madroño 23(1975)24. ***P. ramosissima*** (Torrey) A. Gray – local medic.

Psathyrotopsis Rydb. (~ *Psathyrotes*). Compositae (Bah.). 3 SW US, Mex.

Psectra (Endl.) Tomšovic = *Echinops*

Psednotrichia Hiern (*Xyridopsis*). Compositae (Sen.-Sen.). 2 Angola. R: NJB 15(1995)378. Annuals

Psephellus Cass. (~ *Centaurea*). Compositae (Card.-Cent.). c. 100 E Medit. to Siberia

Pseudabutilon R. Fries. Malvaceae (Malv.-Malv.). 19 warm Am. R: CUMH 21(1997)176

Pseudacanthopale Benoist = *Strobilanthopsis*

Pseudacoridium Ames = *Dendrochilum*

Pseudactis S. Moore = *Emilia*

Pseudaechmanthera Bremek. = *Strobilanthes*

Pseudaechmea L.B. Sm. & Read = *Billbergia* (but see Phytol. 52(1982)53)

Pseudaegiphila Rusby = *Aegiphila*

Pseudagrostistachys Pax & K. Hoffm. Euphorbiaceae (Acal.-Agr.). 2 trop. Afr.

Pseudaidia Tirv. (~ *Aidia*). Rubiaceae (II 1). 1 Ind.: ***P. speciosa*** (Bedd.) Tirv. R: BMNHN 4,8(1986)286

Pseudais Decne = *Phaleria*

Pseudalthenia Nakai (*Vleisia*). Potamogetonaceae (Zannichelliaceae). 1 SW Cape: ***P. aschersoniana*** (Graebner) den Hartog – in vleis. R: Strelitzia 10(2000)738

Pseudammi H. Wolff = *Seseli*

Pseudanamomis Kausel (~ *Myrcianthes*). Myrtaceae (II 10). 1 C Am.: ***P. umbellifera*** (Kunth) Kausel (*M. u.*, monos plum) – ed. fr.

Pseudananas Hassler ex Harms. Bromeliaceae (3). 1 S trop. Am.: ***P. sagenarius*** (Arruda) Camargo, like *Ananas* but with stolons, textiles in Gran Chaco, cult. orn. terr. herb. R: FN 14,3(1979)2048

Pseudannona (Baill.) Saff. = *Xylopia*

Pseudanthistiria (Hackel) Hook.f. Gramineae (XXII 7). 4 Ind., S China, Thailand

Pseudanthus Sieber ex A. Spreng. Picrodendraceae (Cal.-Pseud.; Euphorbiaceae s.l.). Excl. *Stachystemon*, 9 Aus. R: Austrobaileya 6(2003)499

Pseudarabidella O. Schulz = *Andrzeiowskia*

Pseudarrhenatherum Rouy = *Arrhenatherum*

Pseudartabotrys Pellegrin. Annonaceae (III 3). 1 W trop. Afr.: ***P. letestui*** Pellegrin – liane

Pseudarthria Wight & Arn. Leguminosae (XXIV 2). 3–4 trop. OW

Pseudechinolaena Stapf. Gramineae (34b). 6 Madag. with 1 extending pantrop. R: Adansonia 15(1975)123

Pseudelephantopus Rohr (~ *Elephantopus*). Compositae (Vern.-Ele.). 2 trop. Am., ***P. spicatus*** (Aubl.) C. Baker natur. OW trop.

Pseudelleanthus Brieger = *Sertifera*

Pseudellipanthus Schellenb. = *Ellipanthus*

Pseudeminia Verdc. Leguminosae (III 18). 4 trop. Afr.

Pseudencyclia Chiron & V. Castro = *Prosthechea*

Pseudephedranthus Aristeg. Annonaceae (IV 2). 1 Braz., Venez.: ***P. fragrans*** (R.E. Fries) Aristeg. R: BMPEG 15(1999)161

Pseuderanthemum Radlk. Acanthaceae (III 2c). c. 50 trop. local medic.; cult. orn. greenhouse shrubs. ***P. racemosum*** (Roxb.) Radlk. (Mal.) – veg. in Moluccas. See also *Streblacanthus*

Pseuderemostachys Popov = *Phlomoides*

Pseuderia Schltr. Orchidaceae (V 15). c. 20 Mal., W Pac.

Pseuderucaria (Boiss.) O. Schulz. Cruciferae (2). 3 Morocco to Israel

Pseudeugenia Legrand & Mattos. Myrtaceae. 1 Braz.: ***P. stolonifera*** Legrand & Mattos

Pseudibatia Malme = *Matelea*

Pseudima Radlk. Sapindaceae. 1 trop. Am.: ***P. frutescens*** (Aubl.) Radlk. – Leeuwenberg's Model

Pseudiosma DC. = ?

Pseudoacanthocereus F. Ritter (~ *Acanthocereus*). Cactaceae (III 1). 2 NE S Am.

Pseudoarabidopsis Al-Shehbaz & al. (~ *Arabidopsis*). Cruciferae (15). 1 C As. to W China: ***P. toxophylla*** (Bieb.) Al-Shehbaz & al. R: Novon 9(1999)314

Pseudobaccharis Cabrera = *Baccharis*

Pseudobaeckea Niedenzu = *Brunia* + *Thamnea* spp. (but see Strelitzia 9(2000)385)

Pseudobahia (A. Gray) Rydb. Compositae (Mad.-Baer.). 3 Calif. R: Novon 1(1991)122

Pseudobambusa Nguyen (~ *Bambusa*). Gramineae (V 4). 1 Vietnam, S Myanmar, Andamans: ***P. schizostachyoides*** (Kurz) Nguyen

Pseudobartlettia Rydb. = *Psathyrotopsis*

Pseudobartsia Hong = *Parentucellia*

Pseudoberlinia Duvign. = *Julbernardia*

Pseudobersama Verdc. = *Trichilia*

Pseudobetckea (Hoeck) Lincz. = *Valerianella*

Pseudoblepharispermum Lebrun & Stork. Compositae (Inul.-Pluch.). 1 Ethiopia, 1 Somalia – known only from type specimens. R: KB 57(2002)214

Pseudoboivinella Aubrév. & Pellegrin = *Englerophytum*

Pseudobombax Dugand. Malvaceae (Bomb.; Bombacaceae). 20 trop. Am. R: BJBB 33(1963)28. Some cult. orn. spineless trees incl. ***P. ellipticum*** (Kunth) Dugand (*Bombax e.*, '*Pachira insignis*', shaving-brush, C Am.) – decoction used in treatment of toothache

Pseudobotrys Moeser. Cardiopteridaceae (Icacinaceae s.l.). 2 NG

Pseudobrachiaria Launert = *Brachiaria*

Pseudobrassaiopsis Banerjee = *Trevesia*

Pseudobravoa Rose = *Agave*

Pseudobrickellia R. King & H. Robinson (~ *Eupatorium*). Compositae (Eup.-Alom.). 2 Braz. R: MSBMBG 22(1987)256

Pseudobromus K. Schum. (~ *Festuca*). Gramineae (XVI 10). 6 trop. Afr., Madag.

Pseudobrownanthus Ihlenf. & Bittrich = *Mesembryanthemum*

Pseudocadiscus Lisowski = *Stenops*

Pseudocalymma A. Samp. & Kuhlm. = *Mansoa*

Pseudocalyx Radlk. Acanthaceae (II). 6 trop. Afr., Madag. R: Adans. 20(1998)271
Pseudocamelina (Boiss.) N. Busch. Cruciferae (47). 3 Iran. R: NRBGE 36(1978)26
Pseudocampanula Kolak. = *Campanula*
Pseudocannaboides van Wyk (~ *Heteromorpha*). Umbelliferae (III 8). 1 Madag.: ***P. andingitrensis*** (Humbert) van Wyk. R: T 48(1999)742
Pseudocarapa Hemsl. = *Dysoxylum*
Pseudocarpidium Millsp. Labiatae (I; Verbenaceae). 8 WI
Pseudocarum Norman. Umbelliferae (III 8). 2 trop. E Afr., Madag. R: T 48(1999)743. ***P. eminii*** (Engl.) H. Wolff (E Afr.) – prehensile petioles
Pseudocaryophyllus O. Berg = *Pimenta*
Pseudocaryopteris (Briq.) Cantino (~ *Caryopteris*). Labiatae (III). 3 Ind., SE As., S China. ***P. bicolor*** (Hardw.) Cantino (*C. b.*, Himal.) – dried bark used for scent
Pseudocatalpa A. Gentry = *Tanaecium* (but see Britt. 25(1973)241)
Pseudocedrela Harms. Meliaceae (I). 1 trop. Afr.: ***P. kotschyi*** (Schweinf.) Harms – gum, dyes & local medic.
Pseudocentrum Lindl. Orchidaceae (IV 2b). 7 trop. Am.
Pseudocerastium C.Y. Wu & al. Caryophyllaceae (II 1). 1 E China: ***P. stellarioides*** Guo & X.P. Zhang. R: ABY 20(1998)395
Pseudochaetochloa A. Hitchc. Gramineae (XXIV 4). 1 W Aus.: ***P. australiensis*** A. Hitchc.
Pseudochamaesphacos Parsa = ? *Chamaesphacos*
Pseudocherleria Dillenb. & Kadereit (~ *Minuartia*). Caryophyllaceae. 12 C & E As., W N Am. R: T 63(2014)84
Pseudochimarrhis Ducke = *Chimarrhis*
Pseudochirita W.T. Wang. Gesneriaceae (III 2j). 1 Vietnam, S China: ***P. guangxiensis*** (Huang) W.T. Wang. R: FOC 18(1998)293
Pseudocinchona A. Chev. = *Corynanthe*
Pseudoclappia Rydb. Compositae (Tag. – Pect.). 2 SW US & N Mex. R: FNA 21(2006)252
Pseudoclausena T. Clark (~ *Walsura*). Meliaceae (II). 1 SE As., Mal.: ***P. chrysogyne*** (Miq.) T. Clark. R: Blumea 38(1994)291
Pseudoclausia Popov = *Parrya*
Pseudocodon Hong & H. Sun (~ *Codonopsis*). Campanulaceae (I). 8 Sinohimal. R: D.Y. Hong, *Mongr. Codonopsis* (2015)190. C deeply cleft cf. *Codonopsis*. Cult. orn.
Pseudocoeloglossum (Szlach. & Olsz.) Szlach. = *Habenaria*
Pseudocoix A. Camus = *Hickelia*
Pseudocolysis Gómez = *Pleopeltis*
Pseudoconnarus Radlk. Connaraceae (IV). 5 trop. S Am.
Pseudoconyza Cuatrec. (~ *Blumea*). Compositae (Inul.-Pluch.). 1 trop.: ***P. viscosa*** (Mill.) D'Arcy (*P. lyrata*). R: Strelitzia 10(2000)157
Pseudocopaiva Britton & P. Wilson = *Guibourtia*
Pseudocorchorus Capuron (~ *Corchorus*). Malvaceae (Grew.-Apeib.; Tiliaceae). 6 Madag.
Pseudocranichis Garay. Orchidaceae (Orch.-Cran.). 1 Mex.: ***P. thysanochila*** (Robins. & Greenm.) Garay. R: HBML 28(1982)347
Pseudocroton Muell. Arg. = *Capparis*
Pseudocrupina Velen. = *Leysera*
Pseudoctomeria Kraenzlin = *Specklinia*
Pseudocunila Brade = *Hedeoma*
Pseudocyclanthera Mart. Crov. = *Cyclanthera*
Pseudocyclosorus Ching = *Cyclosorus*
Pseudocydonia (C. Schneider) C. Schneider (~ *Chaenomeles*). Rosaceae (Ros.-Mal.). 1 China: ***P. sinensis*** (Thouin) C. Schneider. R: FNA 9(2014)487
Pseudocylosorus Airy Shaw = *Thelypteris*
Pseudocymbidium Szlach. & Sitko = *Maxillaria*
Pseudocymopterus J. Coulter & Rose. Umbelliferae (III 9). 2 SW N Am.
Pseudodacryodes Pierlot. Burseraceae (V). 1 Congo: ***P. leonardiana*** Pierlot – only female fls known. R: BJBB 66(1997) 182
Pseudodanthonia Bor & Hubb. Gramineae (IX). Excl. *Sinochasea* 1 NW Himal.: ***P. himalaica*** (Hook.f.) Bor & Hubb. R: KBAS 13(1986)122
Pseudodelphinium Duman & al. (~ *Delphinium*). Ranunculaceae (I 3/4). 1 Turkey: ***P. turcicum*** Duman – halophyte, fls reg., K 0. R: TJB 36(2012)428
Pseudodichanthium Bor. Gramineae (XXII 7). 1 W Ind.: ***P. serrafalcoides*** (Cooke & Stapf) Bor

Pseudodicliptera Benoist. Acanthaceae (III 2c). 2 Madag.
Pseudodigera Chiov. = *Digera*
Pseudodiphasium Holub = *Lycopodium*
Pseudodiplospora Deb. Rubiaceae. 1 Andamans: ***P. andamanica*** (Balakr. & N.G. Nair) Deb
Pseudodiphryllum Nevski = *Platanthera*
Pseudodissochaeta Nayar. Melastomataceae. 5 N Ind. to Hainan
Pseudodraba Al-Shehbaz & al. Cruciferae (11). 1 Afghanistan, Pakistan: ***P. hystrix*** (Hook.f. & Thoms.) Al-Shehbaz & al. R: PDE 129(2011)73
Pseudodracontium N.E. Br. = *Amorphophallus*
Pseudodrimys Doweld = *Tasmannia*
Pseudodrynaria (C. Chr.) C. Chr. = *Aglaomorpha*
Pseudo-Elephantopus Rohr = *Pseudelephantopus*
Pseudoentada Britton & Rose = *Adenopodia*
Pseudoeriosema Hauman. Leguminosae (III 18). 4 trop. Afr. Some locally eaten roots
Pseudoernestia (Cogn.) Krasser. Melastomataceae. 1 Venez.: ***P. cordifolia*** (Triana) Kasser
Pseudoeugenia Scort. = *Syzygium*
Pseudoeurya Yamamoto = *Eurya*
Pseudoeurystyles Hoehne = *Eurystyles*
Pseudoeverardia Gilly = *Everardia*
Pseudofortuynia Hedge. Cruciferae. 1 Iran: ***P. esfandiarii*** Hedge
Pseudofumaria Medik. (~ *Corydalis*). Papaveraceae (Fumariaceae). 2 NW Balkans, S Alps, natur. elsewhere. R: OB 88(1986)32. Stems branched & perenn. unlike *Corydalis*. ***P. lutea*** (L.) Borkh. (*C. lutea*, Eur., widely natur. inc. GB) – cult. orn.
Pseudogaillonia Lincz. = *Plocama*
Pseudogaltonia (Kuntze) Engl. (~ *Ornithogalum*). Asparagaceae. 1 Namibia, Botswana: ***P. clavata*** (Mast.) E. Phillips. R: T 58(2009)98
Pseudogardenia Keay = *Adenorandia*
Pseudoglossanthis Poljak. = *Trichanthemis*
Pseudognaphalium Kirpiczn. (~ *Gnaphalium*). Compositae (Gnap.- Gnap.). c. 90 Am. (most, N Am. 21), warm OW (few), ***P. luteoalbum*** (L.) Hilliard & B.L. Burtt weedy. R: OB 104(1991)146. ***P. obtusifolium*** (L.) Hilliard & B.L. Burtt (*G. o.*, rabbit tobacco, N Am.) – local medic. for asthma (smoked!)
Pseudognidia E. Phillips = *Gnidia*
Pseudogomphrena R. Fries (~ *Gomphrena*). Amaranthaceae (II 2). 1 Braz.: ***P. scandens*** R. Fries
Pseudogonocalyx Bisse & Berazaín = *Schoepfia*
Pseudogoodyera Schltr. Orchidaceae (IV 2h). 1 Cuba, C Am.: ***P. wrightii*** (Reichb.f.) Schltr.
Pseudogynoxys (Greenman) Cabrera (~ *Jacobaea*). Compositae (Sen.-Sen.). Excl. *Garcibarrigoa*, *Talamancalia*, 14 trop. S Am. R: Phytol. 36(1977)177, 80(1996)253. ***P. chenopodioides*** (Kunth) Cabrera (*Senecio confusus*, *S. tagetes*, C Am.) – cult. orn. cli. with fragrant orange fl. heads
Pseudohamelia Wernham. Rubiaceae (inc. sed.). 1 Andes: ***P. hirsuta*** Wernham
Pseudohandelia Tzvelev. Compositae (Anth.-Han.). 1 C As., Iran, Afghanistan, China: ***P. umbellifera*** (Boiss.)Tzvelev
Pseudohemipilia Szlach. = *Habenaria*
Pseudohexadesmia Brieger = *Scaphyglottis*
Pseudohydrosme Engl. Araceae (VII 12). 2 trop. W Afr. Seasonally dormant
Pseudojacobaea (Hook.f.) Mathur = *Senecio*
Pseudokyrsteniopsis R. King & H. Robinson = *Kyrsteniopsis*
Pseudolabatia Aubrév. & Pellegrin = *Pouteria*
Pseudolachnostylis Pax. Phyllanthaceae (Bro.-Pseud.; Euphorbiaceae s.l.). 1 trop. Afr.: ***P. maprouneifolia*** Pax. R: A.R.-Sm., *Gen. Euph.* (2001)28
Pseudolaelia Porto & Brade (~ *Schomburgkia*). Orchidaceae (V 13b). 18 trop. Am.
Pseudolarix Gordon. Pinaceae. 1 C & NE China: ***P. amabilis*** (Nelson) Rehder (*P. kaempferi*, golden larch or pine) – like *Larix* but male cones clustered (not solit.) & females with decid. scales, etc. R: RV 121(1990)123
Pseudolasiacis (A. Camus) A. Camus = *Lasiacis*. But see Adansonia III,21(1999)232
Pseudoligandra Dillon & Sagást. = *Chionolaena*
Pseudolinosyris Novopokr. = *Crinitaria*
Pseudoliparis Finet = *Malaxis*
Pseudolitchi Danguy & Choux = *Stadmania*

Pseudolithos Bally. Apocynaceae (V b; Asclepiadaceae III 5). 6 NE trop. Afr. R: F. Albers & U. Meve, *Ill. Handb. Succ. Pls*, Asclep. (2002)212. Plants subspherical

Pseudolmedia Trécul. Moraceae (III). 9 trop. Am. R: FN 7(1972)20, 83(2001)246. Cyanogenesis reported

Pseudolobivia (Backeb.) Backeb. = *Echinopsis*

Pseudolopezia Rose = *Lopezia*

Pseudolophanthus Levin = *Marmoritis*

Pseudolotus Rech.f. (~ *Lotus*). Leguminosae (III 22). 1 Oman to Pakistan: ***P. villosus*** (Blatter & Hallb.) Ali & Sokoloff. R: KB 56(2001)762

Pseudoludovia Harling = *Sphaeradenia*

Pseudolycopodiella Holub (~ *Lycopodiella*). Lycopodiaceae. 12 cosmop.

Pseudolycopodium Holub = *Lycopodium*

Pseudolysimachion (Koch) Opiz = *Veronica*

Pseudomachaerium Hassler = *Nissolia*

Pseudomacrolobium Hauman. Leguminosae (I 2). 1 Congo basin: ***P. mengei*** (De Wild.) Hauman – timber, fish-poison

Pseudomaihuenopsis Guiggi = *Tephrocactus*

Pseudomalmea Chatrou (~ *Malmea*). Annonaceae (IV 2). 4 trop. Am. R: Blumea 51(2006)219

Pseudomantalania J. Leroy. Rubiaceae (II). 1 Madag.: ***P. macrophylla*** J. Leroy – Corner's Model

Pseudomariscus Rauschert = *Courtoisina*

Pseudomarrubium Popov. Labiatae (VI). 1 Kazakhstan: ***P. eremostachyoides*** Popov

Pseudomarsdenia Baill. = *Marsdenia*

Pseudomaxillaria Hoehne = *Maxillaria* (but see BJ 97(1977)551)

Pseudomelasma E. Fischer. Orobanchaceae (Esc.). 1 Madag.: ***P. pedunculariodes*** (Bak.) E. Fischer. R: Adansonia 18(1996)61

Pseudomertensia Riedl = *Decalepidanthus*

Pseudomiltemia Borhidi. Rubiaceae. 2 Mex. R: Britt. 63(2011)197

Pseudomisopates Güemes (~ *Misopates*). Plantaginaceae (Ant.-Ant.; Scrophulariaceae s.l.). 1 Iberia: ***P. rivas-martinezii*** (Sánchez Mata) Güemes. R: AJBM 55(1997)493

Pseudomitrocereus Bravo & Buxb. (~ *Cephalocereus*). Cactaceae. 1 Mex.: ***P. fulviceps*** (K. Schum.) Bravo & Buxb.

Pseudomonotes Londoño & al. Dipterocarpaceae (Monot.). 1 Amaz. Colombia: ***P. tropenbosii*** Londoño & al. R: Britt. 47(1995)230

Pseudomuscari Garb. & Greuter = *Muscari*

Pseudomussaenda Wernham (~ *Mussaenda*). Rubiaceae (I 8). 6 trop. Afr.

Pseudomyrcianthes Kausel = *Myrcianthes*

Pseudonemacladus McVaugh. Campanulaceae (II). 1 Mex.: ***P. oppositifolius*** (Robinson) McVaugh

Pseudonesohedyotis Tenn. Rubiaceae (IV 1). 1 Uluguru Mts, Tanzania: ***P. bremekampii*** Tenn.

Pseudonoseris H. Robinson & Brettell. Compositae (Liab). 2 Andean Peru & Bolivia. R: Phytoneuron 2012–113:3

Pseudopachystela Aubrév. & Pellegrin = *Synsepalum*

Pseudopaegma Urb. = *Anemopaegma*

Pseudopanax K. Koch. Araliaceae. Incl. *Neopanax* 12 Tasmania, NZ, Chile. Dioec. or monoec. cult. orn. shrubs & trees with lvs simple to compound at diff. ages of pl. in some spp. e.g. ***P. crassifolius*** (Cunn.) K. Koch (lancewood, NZ) where 4 distinct forms of foliage. See also *Raukaua*

Pseudopancovia Pellegrin. Sapindaceae. 1 trop. W Afr.: ***P. heteropetala*** Pellegrin. R: BSBF 102(1955)228

Pseudoparis Perrier. Commelinaceae (II 2). 2–3 Madag. R: SCB 76(1991)155

Pseudopavonia Hassler = *Hibiscus*

Pseudopectinaria Lavranos = *Echidnopsis*

Pseudopegolettia H. Robinson & al. (~ *Vernonia*). Compositae (Vern.). 2 S Afr. R: PhytoKeys 60(2016)107

Pseudopentameris Conert (~ *Danthonia*). Gramineae (Danth.). 4 S Afr. R: Bothalia 25(1995)144

Pseudopentatropis Costantin. = *Pentatropis*

Pseudopeponidium Homolle ex Arènes = *Pyrostria*

Pseudoperistylus (P. Hunt) Szlach. & Olsz. = *Habenaria*
Pseudophegopteris Ching. Thelypteridaceae. 20 OW trop. R: Blumea 17(1969)112. No indusia
Pseudophleum Doğan (~ *Phleum*). Gramineae. 2 Turkey. R: SB 40(2015)455
Pseudophoenix H. Wendl. ex Sarg. Palmae (IV 1). 4 Carib. R: GH 10(1968)169, Principes 13(1969)77, Palms 46(2002)19. Tapped for wine but ***P. ekmanii*** Burret (Hispaniola), poss. 'wine palm' of early explorers, prob. extinct since 1926
Pseudophyllanthus (Muell.Arg.) Voronts. & Petra Hoffm. (~ *Andrachne*). Phyllanthaceae (Por.). 1 S Afr.: ***P. ovalis*** (Sond.) Vronts. & Petra Hoffm. R: KB 63(2008)50
Pseudopilocereus F. Buxb. = *Pilosocereus*
Pseudopinanga Burret = *Pinanga*
Pseudopiptadenia Rauschert (*Monoschisma*). Leguminosae (II 1). 11 trop. Am. Some timbers
Pseudopiptocarpha H. Robinson (~ *Vernonia*). Compositae (Vern.-Lep.). 2 Colombia
Pseudopipturus Skottsb. Older name for *Nothocnide*
Pseudoplantago Süsseng. Amaranthaceae (II 1). 2 Arg., Venez. Pollen cuboid or prismatic (unique in fam.)
Pseudopodospermum (Lipsch. & H. Kraschen.) Kuth. = *Lasiospora*
Pseudopogonatherum A. Camus = *Eulalia*
Pseudoponera Brieger = *Ponera*
Pseudoprosopis Harms. Leguminosae (II 1). 7 trop. Afr. R: BJBB 53(1983)417. Fibre, fish-poisons
Pseudoprospero Speta. Asparagaceae (Hyacinthaceae). 1 E Cape: ***P. firmifolium*** (Bak.) Speta. R: Phyton 38(1998)116
Pseudoprotorhus Perrier = *Filicium*
Pseudopteris Baill. Sapindaceae. 3 Madag.
Pseudopyxis Miq. Rubiaceae (IV 12). 3 Jap. R: EJB 64(2007)304
Pseudoraphis Griff. ex Pilg. Gramineae (XXIV 4). 10 Ind. to Jap. (China 3) & Aus. Aquatic or semi-a.
Pseudorchis Ség. (*Polybactrum*, *Leucorchis*). Orchidaceae (IV 4d). 3 E N Am. to Eur. (2). ***P. albida*** (L.) Löve & D. Löve (Eur., W Siberia) – hybridizes with *Platanthera montana* in Scotland
Pseudorhipsalis Britton & Rose (~ *Disocactus*). Cactaceae (III 2). 6 trop. Am.
Pseudorlaya (Murb.) Murb. = *Daucus*
Pseudorobanche Rouy = *Alectra*
Pseudoroegneria (Nevski) Löve (~ *Elytrigia*). Gramineae (XV). c. 15 Euras., N Am. (1)
Pseudorontium (A. Gray) Rothm. (~ *Antirrhinum*). Plantaginaceae (Ant.-Ant.; Scrophulariaceae s.l.). 1 SW N Am.: ***P. cyathiferum*** (Benth.) Rothm.
Pseudoruellia Benoist = *Ruellia* (but see BSBF 109(1962)181)
Pseudosabicea Hallé = *Sabicea*
Pseudosagotia Secco = *Croizatia*
Pseudosalacia Codd. Celastraceae (I). 1 Natal: ***P. streyi*** Codd. R: Bothalia 11(1972)565
Pseudosamanea Harms (~ *Albizia*). Leguminosae (II 4). 2 trop. Am. R: MNYBG 74,1(1996)113. Timber
Pseudosaponaria (F.N. Williams) Ikonn. = *Gypsophila*
Pseudosarcolobus Costantin = *Marsdenia*
Pseudosarcopera Giraldo-Cañas = *Sarcopera* (but see Caldasia 29(2007)205)
Pseudosasa Makino ex Nakai. Gramineae (IV). 19 E As. (China 18, 17 endemic – R: FOC 22(2006)115). R: KBAS 13(1986)47. App. type sp. hybrid between *Sasamorpha* & *Pleioblastus* spp. Like *Sasa* (A 6) but A 3(4). ***P. amabilis*** (McClure) Keng f. (*Arundinaria a.*, Tongking cane or bamboo, cultigen from China) – valued for split-cane fishing-rods; ***P. japonica*** (Steud.) Nakai (E As.) – hybrid cultigen, ed. shoots, form. arrows in Jap., much-cult. bamboo to 5 m for screening & hedges, infls for brooms in Kerala
Pseudosassafras Lecomte = *Sassafras*
Pseudosbeckia Fernandes & R. Fernandes. Melastomataceae. 1 trop. E Afr.: ***P. swynnertonii*** (Bak.f.) Fernandes & R. Fernandes
Pseudoscabiosa Devesa (~ *Scabiosa*). Caprifoliaceae (Dipsacaceae). Excl. *Pterocephalidium*, 4 C & W Med. R: Lagascalia 12(1984)213
Pseudoschoenus (C.B. Clarke) Oteng-Yeboah (~ *Scirpus*). Cyperaceae (II 2). 1 S Afr.: ***P. inanis*** (Thunb.) Oteng-Yeboah. R: Strelitzia 10(2000)601
Pseudosciadium Baill. = *Delarbrea*

Pseudosclerochloa Tzvelev (~ *Puccinellia*). Gramineae (XVI 14). 3 W Eur., China. R: BZ 89(2004)840

Pseudoscleropodium (Limpr.) Broth. Brachytheciaceae (Musci). 1 Eur., Macar.: ***P. purum*** (Limpr.) Fleisch. (widely naturalized temp. regions, as in New York cemeteries) – form. used as braids in women's hats

Pseudoscolopia Gilg. Salicaceae (Flacourtiaceae). 1 S Afr.: ***P. polyantha*** Gilg. R: Strelitzia 9(2000)513

Pseudosedum (Boiss.) A. Berger. Crassulaceae (I 2). 12 C As. R: U. Eggli, *Ill. Handb. Succ. Pls*, Crass (2005)207

Pseudoselago Hilliard (~ *Selago*). Scrophulariaceae (Man.). 28 SW Cape. R: EJB 52(1995)245. ***P. spuria*** (L.) Hilliard (*S. s.*, blue haze) – cult. orn.

Pseudoselinum Norman. Umbelliferae (III 10). 1 Angola: ***P. angolense*** (Norman) Norman

Pseudosempervivum (Boiss.) Grossh. (~ *Cochlearia*). Cruciferae (19). 6 Turkey, Armenia

Pseudosenefeldera Esser = *Senefeldera* (but see A. R.-Sm., *Gen. Euphorb.* (2001)387)

Pseudosericocoma Cavaco. Amaranthaceae (I 2). 1 SW & S Afr.: ***P. pungens*** (Fenzl) Cavaco

Pseudosicydium Harms = *Pteropepon*

Pseudosindora Sym. = *Copaifera*

Pseudosmelia Sleumer. Salicaceae (Samydaceae, Flacourtiaceae). 1 Moluccas: ***P. moluccana*** Sleumer

Pseudosmilax Hayata = *Smilax*

Pseudosmodingium Engl. Anacardiaceae. 4 Mex. R: Rhodora 106(2004)350

Pseudosopubia Engl. Orobanchaceae (Buch.; Scrophulariaceae s.l.). 3–4 trop. Afr.

Pseudosorghum A. Camus. Gramineae (XXII 5). 2 Indomal. R: FOC 22(2006)602

Pseudospigelia Klett = *Spigelia*

Pseudospondias Engl. Anacardiaceae (II). 2 W & C trop. Afr. Bark medic. ***P. microcarpa*** (A. Rich.) Engl. – ed. pulp

Pseudostachyum Munro (~ *Schizostachyum*). Gramineae (V 5). 1 Himal. to S China: ***P. polymorpha*** Munro (*S. p.*, Himal. to Borneo) – paper pulp

Pseudostelis Schltr. = *Pleurothallis*

Pseudostellaria Pax. Caryophyllaceae (II 1). 21 Eur. (1) & C As., Afghanistan to Jap. & N Am. (3). R: JapJB 9(1937)95

Pseudostenomesson Velarde = *Ismene*

Pseudostenosiphonium Lindau = *Strobilanthes*

Pseudostifftia H. Robinson (~ *Moquinia*). Compositae (Moq.). 1 Bahia: ***P. kingii*** H. Robinson. R: T 43(1994)42

Pseudostreptogyne A. Camus = *Koordersiochloa*

Pseudostriga Bonati. Orobanchaceae (Buch.; Scrophulariaceae s.l.). 1 SE As.: ***P. cambodiana*** Bonati

Pseudotaenidia Mackenzie = *Taenidia*

Pseudotaxus W.C. Cheng. Taxaceae. 1 EC China: ***P. chienii*** (W.C. Cheng) W.C. Cheng. R: BR 64(1998)399

Pseudotectaria Tard. (~ *Tectaria*). Dryopteridaceae. 6 Madag. R: KB 45(1990)259

Pseudotenathera Majumdar = *Schizostachyum*

Pseudotrachydium (Kljuykov & al.) Pim. & Kljuykov = *Aulacospermum* (but see FR 111(2000)526)

Pseudotrillium S. Farmer (~ *Trillium*). Melanthiaceae. 1 W N Am.: ***P. rivale*** (S. Watson) S. Farmer. R: SB 27(2002)687

Pseudotrimezia R. Foster. Iridaceae (VII 4). c. 12 E Braz. (Minas Gerais)

Pseudotsuga Carr. Pinaceae. 4 E As. (China 1, S Jap. 1), W N Am. (2). R: A. Farjon, *World checklist ... conifers*, ed. 2(2001)239. Allied to *Picea* but with conspic. bracts between cone-scales. ***P. menziesii*** (Mirb.) Franco (*P. douglasii*, *P. taxifolia*, Douglas fir, Oregon (pine), W N Am.) – to 1300 yrs old, major timber tree (133 m specimen felled in Br. Columbia in 1895, 126.5 m in 1902; at 140m photosynthesis would cease as pressure too high for water to enter cells, though tree 138 m claimed – world's tallest) with bole to 4.4 m diam., introd. GB 1827 (by 2014 UK's tallest tree – 66.4 m nr Inverness, tallest conifer in Eur.; 67.36 m specimen erected as Christmas tree Seattle 1950), used for telegraph poles, railway sleepers, plywood & pulp, source of Oregon balsam, medic., resin a chewing gum, lvs a coffee subs., sugar exudates collected & stored by Native Americans

Pseudoturritis Al-Shehbaz (~ *Arabis*). Cruciferae (45). 1 S & SE Eur.: ***P. turrita*** (L.) Al-Shehbaz (*A. t.*). R: Novon 15(2005)522

Pseudourceolina Vargas = *Urceolina*

Pseudovanilla Garay (~ *Galeola*). Orchidaceae (II 2). 8 Mal. to Pac. (Aus. 1). ***P. foliata*** (F. Muell.) Garay (E As.) – liane to 15 m long

Pseudovesicaria (Boiss.) Rupr. Cruciferae (47). 1 Caucasus: ***P. digitata*** (C. Meyer) Rupr.

Pseudovigna (Harms) Verdc. Leguminosae (III 18). 3 trop. Afr. R: KB 66(2011)590

Pseudovossia A. Camus = *Phacelurus*

Pseudovouapa Britton & Killip = *Macrolobium*

Pseudoweinmannia Engl. Cunoniaceae (V). 2 NE Aus. R: Austrobaileya 8(2011)260. ***P. lachnocarpa*** (F. Muell.) Engl. (marara) – local timber

Pseudowillughbeia Markgraf = *Melodinus*

Pseudowintera Dandy. Winteraceae. 4 NZ. R: Blumea 18(1970)227, NZJB 44(2006)94. ***P. colorata*** (Raoul) Dandy – self-incompatible, visited by stigma exudate-eating chironomids & pollen-eating beetles

Pseudowolffia Hartog & van der Plas = *Wolffiella*

Pseudoxandra R. Fries. Annonaceae (IV 2). 22 trop. S Am. R: Blumea 48(2003)205. Some fish-disp. in Amazon

Pseudoxytenanthera Soderstrom & R. Ellis (~ *Schizostachyum*). Gramineae (V 4). 12 S Ind., Sri Lanka

Pseudoxythece Aubrév. = *Pouteria*

Pseudoyoungia Maity & Maiti (~ *Youngia*). Compositae (Cich.-Crep.) 9 Himal. R: CN 48(2010)29

Pseudoziziphus Hauenschild (~ *Ziziphus*). Rhamnaceae (6). 2 S N Am. R: T 65(2016)53

Pseudozoysia Chiov. Gramineae (Chlorid.). 1 Somalia: ***P. sessilis*** Chiov., on coastal dunes

Pseuduvaria Miq. Annonaceae (IV 7). Incl. *Craibella*, *Oreomitra*, 56 Ind. to Aus. R: SBM 79 (2006)41. ***P. megalopus*** (K. Schum.) Y. Su & Mols (*Petalolophus m.*, NE NG) – extended P wings prob. mimic carrion, fly-poll.

Psiadia Jacq. ex Willd. Compositae (Ast.). 60 OW trop. (Madag. 28, 27 endemic). Leeuwenberg's Model

Psiadiella Humbert. Compositae (Ast.). 1 Madag.: ***P. humilis*** Humbert

Psidiopsis O. Berg = *Calycolpus*

Psidium L. Myrtaceae (II 10). c. 70 trop. Am. Guavas, everg. trees & shrubs (Troll's Model) with ed. berries, esp. ***P. guajava*** L. (common g.) – cult. & natur. (oft. a pest, incl. in S Afr. ***P. × hasslerianum*** Barb. Rodr. (*P. × durbanense*, hybrid with *P. guineense*), Aus., NZ, Galápagos, Hawaii) throughout trop. & subtrop., fr. usu. tinned or made into jelly, jam & chutney, used to counter diarrhoea in Burundi; evergrowing shoots with capacity to fl. at base. Other spp. with ed. fr. include ***P. cattleyanum*** Sabine (*'P. cattleianum'*, *P. littorale*, purple g., strawberry g., Braz.) – sweet purplish red. fr., serious allelopathic weed in Hawaii, also Masc., Seychelles, Aus., W Pacific, SE US; ***P. friedrichsthalianum*** (O. Berg) Niedenzu (Costa Rican g., C Am.) – fr. smaller, ***P. guineense*** Sw. (guisaro, trop. Am.) – small tart fr. for jelly

Psiguria Necker ex Arn. Cucurbitaceae (XIII). 15 trop. Am. Monoec. but changing sex so as to appear dioec.; tough fls visited by birds but also butterflies that mix pollen with nectar, causing it to pre-germinate & release amino-acids taken up by butterflies, which, thereby, live up to 6 months (much more than allied spp.)

Psila Philippi = *Baccharis*

Psilactis A. Gray (~ *Machaeranthera*). Compositae (Ast.-Sym.). 6 SW N Am. R: SB 18(1993)290, Phytol. 77(1994)265

Psilantha (K. Koch) Tzvelev = *Eragrostis*

Psilanthele Lindau. Acanthaceae (III 2c). 1 Ecuador: ***P. eggersii*** Lindau

Psilanthopsis A. Chev. = *Coffea*

Psilanthus Hook.f. = *Coffea* (but see BBAS 33(1992)209)

Psilathera Link = *Sesleria*

Psilocarphus Nutt. Compositae (Gnap.-Gnap.). 8 trop., W US (5), S S Am. R: FNA 19(2006)456

Psilocarya Torrey = *Rhynchospora*

Psilocaulon N.E. Br. = *Mesembryanthemum* (but see BJ 120(1998)346)

Psilochilus Barb. Rodr. (~ *Pogonia*). Orchidaceae (V 4b). 7 trop. Am. R: SB 37(2012)354

Psilochloa Launert = *Panicum*

Psiloesthes Benoist = *Peristrophe*

Psilolaemus I.M. Johnston = *Lithospermum* (but see JAA 35(1954)33)

Psilolemma S. Phillips. Gramineae (XXVIII). 1 E Afr.: ***P. jaegeri*** (Pilg.) S. Phillips. R: KB 29(1974)267

Psilopeganum Hemsl. Rutaceae (I 10). 1 China: ***P. sinense*** Hemsl. – shrub used in Chinese med. for dropsy. R: FOC 11(2008)74

Psilosiphon Goldbl. & Manning = *Afrosolen*

Psilostrophe DC. Compositae (Hele.-Tet.). 7 SW N Am. R: Rhodora 79(1977)169. Local medic. & yellow dyes, esp. ***P. tagetina*** (Nutt.) Greene

Psilotaceae J. Griff. & Henfrey. Ophioglossidae – Psilotales. 2/17 trop. to warm. Sporophytes oft. epiphytic, rootless, with dichot. endophytic mycorrhizal rhiz. & aerial branches with protostele or solenostele. Leaf-like or scale-like lat. appendages in spirals or appearing distich. Sporangia homosporous, on short lat. branches; antherozoids multiflagellate. Gametophyte subterr. n = 52

Genera: *Psilotum, Tmesipteris* (sometimes placed in its own fam.)

Sister to Ophioglossaceae. Structure much resembling some of the earliest land pls

Psilothonna (DC.) E. Phillips = *Steirodiscus*

Psilotrichopsis C. Towns. Amaranthaceae (I 2). 3 Thailand, Malay Pen.

Psilotrichum Blume. Amaranthaceae (I 2). 18 trop. OW esp. Afr. Heterogeneous

Psilotum Sw. Psilotaceae. 2 trop., C Jap. & SW Spain. Whisk ferns. ***P. nudum*** (L.) P. Beauv. (trop. to SW Spain, Hawaii & NZ) – terr. or epiphytic herb with dichot. branches (due to loss of mother apical cell & appearance of 2 daughter apical cells) & scales without leaf-traces, cult. orn. Jap. for 400 yrs incl. cvs with term. sporangia & no leaf-like appendages thus much resembling fossil *Rhynia*!; ***P. complanatum*** Sw. (*P. flaccidum*, Mex., Jamaica, Pac.) – epiphyte with flattened branches & leaf-traces from stele but not reaching scales (cf. *Tmesipteris*)

Psiloxylaceae Croizat = Myrtaceae–Psiloxyloideae

Psiloxylon Thouars ex Tul. Myrtaceae (I 1; Psiloxylaceae). 1 Mauritius, Réunion: ***P. mauritianum*** (Hook.f.) Baill. – allied to *Heteropyxis*

Psilurus Trin. = *Festuca*

Psittacanthus Mart. Loranthaceae (4 4). 120+ trop. Am. R: SBM 86(2009)47. Hummingbird-poll. 'Woodroses' prod. at junction with host sold as curios

Psittacoglossum Lex. = *Maxillaria*

Psomiocarpa C. Presl. Tectariaceae (Dryopteridaceae s.l.). 1 Philippines: ***P. apiifolia*** (Kunze) C. Presl – hairs 'intestiniform'. R: FM II,2(1991)100

Psophocarpus Necker ex DC. Leguminosae (III 18). c. 10 trop. OW. R: KB 33(1978)191. ***P. tetragonolobus*** (L.) DC. (winged bean, Goa b., dambala, trop. As.) – cult. for ed. legumes high in protein & ed. roots, poss. derived from *P. grandiflorus* Wilczek in Congo

Psoralea Royen ex L. Leguminosae (III 2). Incl. *Hallea*, excl. *Bituminaria, Cullen, Otholobium*, 50 S Afr. esp. Cape. Some cult. orn. incl. ***P. pinnata*** L. (blue pea, Dally pine) – fls blue & white, aggressive weedy tree in SW Aus. (& NZ) but good bee-fodder

Psoralidium Rydb. = *Pediomelum* (but see MNYBG 61(1990)32; see also *Ladeania*)

Psorospermum Spach = *Vismia*

Psorothamnus Rydb. Leguminosae (III 10). 9 deserts SW N Am. ***P. polydenius*** (S. Watson) Rydb. – local medic.

Psychanthus (K. Schum.) Ridl. = *Pleuranthodium*

Psychilis Raf. = *Epidendrum* (but see C.L. Withner, *The Cattleyas & their relatives* IV(1996) 89)

Psychine Desf. Cruciferae (12). 1 N Afr.: ***P. stylosa*** Desf.

Psychopsiella Lückel & Braem = *Psychopsis*

Psychopsis Raf. (~ *Oncidium*). Orchidaceae (V 10). Incl. *Psychopsiella*, 6 trop. Am. Cult. orn. ***P. papilio*** (Lindl.) H. Jones – introd. from Trinidad 1826, start of C19 'orchidomania'

Psychotria L. Rubiaceae (IV 7). Excl. *Carapichea, Cephaelis, Margaritopsis, Notopleura, Ronabea*, 800–2000 trop. (Philippines 112 (27 extinct), Papuasia 115, New Caledonia 59 – R: Adans. 35(2013)235). Trees & shrubs (architectural branching a useful char. at subgeneric or species-group rank), incl. epiphytes (some with adventitious roots), rheophytes (some fish-disp. in Amazonia), & litter-gathering monopodial treelets, e.g. ***P. kupensis*** Cheek (Cameroun) in Afr., with alks & some heterostyly, some with bacterial leaf-nodules (see *Pavetta*). Some medic. & cult. orn. (oft. with coloured infl. axes) incl. ***P. emetica*** L.f. (trop. Am.) – roots source of inferior subs. for ipecacuanha. ***P. douarrei*** (Beauvisage) Däniker (New Caled.) – lvs accum. nickel to 4.7% dry wt., the highest recorded in any pl.; ***P. ulviformis*** Steyerm. (Fr. Guiana) – lvs appear brown, poss. avoiding predation as appearing dead; ***P. viridis*** Ruíz & Pavón (Amazonia) – mixed with other hallucinogens, inhibits aminase so allowing tryptamine in ayahuasca to function

Psychrogeton Boiss. Compositae (Ast.-Ast.). 20 SW & C As. R: NRBGE 27(1967)101

Psychrophila (DC.) Bercht. & J. Presl = *Caltha*

Psychrophyton Beauverd = *Raoulia* (but see OB 104(1991)61)

Psydrax Gaertn. (~ *Canthium*). Rubiaceae (III 2). c. 100 OW trop. (Aus. 23 – R: Austrobaileya 6(2004)826; Afr. 34 – R: KB 40(1985)687). Myrmecophily in Afr., New Caled. ***P. odorata*** (Forst.f.) A.C. Sm. & Darwin (Pacific) – black dye from lvs (Hawaii)

Psygmorchis Dodson & Dressler = *Erycina*

Psylliostachys (Jaub. & Spach) Nevski. Plumbaginaceae (II). 2–3 E Med. to C As. Some cult. orn. like *Limonium* esp. ***P. suworowii*** (Regel) Roshk. (Iran & C As.) – C pink

psyllium *Plantago afra*

Psyllocarpus Mart. & Zucc. Rubiaceae (IV 15). 9 Braz., white sands & savanna. R: SCB 41(1979)

Ptaeroxylaceae J. Leroy = Rutaceae

Ptaeroxylon Ecklon & Zeyher. Rutaceae (III; Ptaeroxylaceae). 1 S Afr. to NE Tanzania: ***P. obliquum*** (Thunb.) Radlk. (*P. utile*, sneezewood) – termite-resistant timber for external use (also outlasts metal as machine bearings), pepper-scented sawdust a medic. snuff. R: Strelitzia 9(2000)635

Ptelea L. Rutaceae (I). 3 N Am. R: Britt. 14(1962)3. Polygamous trees & shrubs with cordate reticulate-winged samaras & alks. Cult. orn. esp. ***P. trifoliata*** L. (hop tree, C & E US) – some antifungal alks, frs a hop subs. in brewing beer

Pteleocarpa Oliv. Gelsemiaceae. 1 W Mal.: ***P. lamponga*** (Miq.) Heyne – tree, form. in Boraginaceae but 2 ovules per loc. with upper ascending v. odd there; style double-forked

Pteleocarpaceae Brummitt = Gelsemiaceae

Pteleopsis Engl. Combretaceae (II 2a). 9 trop. Afr.

Ptelidium Thouars. Celastraceae (I). 4 Madag. R: FMad 116(1946)32

Pteracanthus (Nees) Bremek. = *Strobilanthes*

Pterachaenia (Benth.) Lipsch. Compositae (Lact.-Scor.). 1 Iran, Pakistan & Afghanistan: ***P. stewartii*** (Hook. f.) R. R. Stewart. R: BZ 56(1971)1150

Pteralyxia K. Schum. Apocynaceae (Ii). 2 Hawaii. R: W.L. Wagner & al., *Man. Fl. Pls Hawaii* (1990)219

Pterandra A. Juss. Malpighiaceae. 6 trop. Am. R: CUMH 21(1997)5

Pteranthus Forssk. Caryophyllaceae (I 2). 1 N Afr. & Cyprus to Iran: ***P. dichotomus*** Forssk.

Pterichis Lindl. Orchidaceae (IV 2b). 20 Andes, Costa Rica (1), Jamaica (1)

Pteridaceae Kirchn. (Parkeriaceae). Polypodiidae – Polypodiales. Incl. Adiantaceae, Platyzomataceae, Vittariaceae, 68/1375 cosmop. esp. trop. & arid. R: T 55(2006)715. Terr. or aquatic (*Ceratopteris*) ferns, usu. small. Stem with medullated protostele to dictyostele with trichomes &/or scales. Frond usu. pinnate (simple in vittarioids), indefinite apical growth in *Jamesonia*. Sporangia in soral lines along veins (indusium 0) or marginal when covered with marginal indusium. n = usu. 29, 30

Classification & chief genera:

I. **Cryptogrammoideae**: *Cryptogramma*

II. **Ceratopteridoideae** (Parkerioideae; Acrostichaceae, Parkeriaceae s.s.; aquatic): *Acrostichum*, *Ceratopteris*

III. **Pteridoideae** (incl. Platyzomatoideae, Taenitidoideae): *Eriosorus*, *Pityrogramma*, *Pteris*, *Taenitis*

IV. **Cheilanthoideae** (petiole with 1 vasc. bundle or if 2 then frond not farinose abaxially): *Cheilanthes*, *Notholaena*, *Pellaea*

V. **Vittarioideae** (Adiantoideae; sporangia on a circular to elongate strongly recurved margin; incl. minutest of all ferns (*Microgramma*, cf. grammitid Polypodiaceae): *Adiantum*, *Antrophyum*, *Doryopteris*, *Vittaria*)

Cult. orn. esp. *Adiantum* & *Pteris*; *Pityrogramma* petioles used in basketware

Pteridiaceae Ching = Dennstaedtiaceae

Pteridium Gled. ex Scop. Dennstaedtiaceae. 3 cosmop. (1 diploid – ***P. aquilinum*** (L.) Kuhn s.s. (bracken, eagle fern, (hog) brake ('Brache' = Old German for wasted land) N hemisphere to Afr., N Aus.; 1 diploid S hemisphere (***P. esculentum*** (Forst.f.) Cockayne subsp. *e.* in Australasia, subsp. ***arachnoideum*** (Kaulf.) J.A. Thomson in trop. Am.); 1 allotetraploid where they meet (***P. semihastatum*** (Agardh) S.B. Andrews, As.). R: R.T. Smith & J.A. Taylor (eds, 1986) *Bracken*; T 63(2014)509. Creeping rhizome (the vasc. strands in section alleged to resemble G O D, J C, or I H S) with usu. ann. fronds (the croziers poss. inspiration for the capitals of Ionic columns) toxic to stock through thiaminase causing acute symptoms of vitamin B_1 deficiency, also producing ecdysones & bearing nectaries (in N Yorks populations with attracted wood-ants had reduced insect predators esp. suckers); rhiz. with tracheid end-walls dissolving to give vessel-like elements, forming clones to 1400 yrs old; allelopathic, the toxins leaching from green fronds all year round in Costa Rica to affect surrounding pls, from litter in W N Am. preventing germination of seeds & from standing fronds in Calif. countering germination in wet season. Oft. serious weeds (by 1990 occupying 1.2 – 2.7% of total GB land-surface); rhiz. & young fronds tinned esp. in Jap. ('sawarabi', the ed. starch known as 'warabi-ko'), form. much eaten by Maoris & Aus. Aborigines (*P. esculentum*), N Am. Indians (*P. aquilinum*) etc. (ground with oats in Canary Is. – 'goflo') but shikimic acid in them known to promote stomach cancers being carcinogenic & mutagenic; form. locally medic., astringent rhiz. also used in preparation of kid & chamois leathers & form. in glass- & soap-making (potash), the fronds used for thatch & paper; hop subs. in brewing; pl. a source of a green dye but also dark yellow in some Scottish tartans & black when rhiz. boiled; boiled rhiz. also used in fibrework (brown when unboiled) of Native Americans

Pteridoblechnum Hennipman = *Blechnum* (but see PRSQ 87(1978)98)

Pteridocalyx Wernham. Rubiaceae (I 6). 2 Guyana

Pteridophyllaceae Nakai ex Reveal & Hoogl. = Papaveraceae

Pteridophyllum Sieb. & Zucc. Papaveraceae (Pteridophyllaceae). 1 Jap.: ***P. racemosum*** Sieb. & Zucc. – rare in coniferous woods, cult. orn. with alks

Pteridophyta Schimp. Old name to cover Lycopdiopsida & ferns (see Appendix). Vasc. pls with spores, having achieved a terr. level of organization without seeds. Many fossil groups & the pls which are the basis of coal deposits (Carboniferous spore-forests). The spores have 'pre-adapted' the group (esp. Polypodiopsida & clubmosses) to living in the rising canopy of rain forests where they are v. successful (cf. Orchidaceae)

Pteridrys C. Chr. & Ching. Tectariaceae (Dryopteridaceae). 8 trop. As. & Mal.

Pterigeron (DC.) Benth. = *Streptoglossa*

Pteris L. Pteridaceae (III). Incl. *Nevrocallis, Platyzoma*, 250 cosmop. (Eur. 3). Cult. orn. (R: BFG 10(1970)143) terr. ferns esp. ***P. cretica*** L. (trop. & warm OW) – common pot-pl. with many cvs incl. crested & varieg. ones; ***P. ensiformis*** Burm.f. (Indomal. to Jap. & W Pac.) – young fronds ed., locally medic.; ***P. multifida*** Poir. (E As., natur. Medit., E Afr., warm Am.) – locally medic. (vermifuge), many cvs; ***P. recurva*** (Desv.) Christenh. (*P. platyzomopsis, Platyzoma microphyllum*, N & NE Aus.) – terr. xerophyte, juveniles with filiform fronds also in zones on adults, heterospory unique in ferns with megaspores twice size of microspores, the latter germ. as filamentous prothalli bearing antheridia, the former giving spathulate prothalli with archegonia when young & antheridia later if fert. has not occurred (cf. *Equisetum*); ***P. vittata*** L. (trop. & warm OW, introd. Am.) – arsenic hyperaccumulator (to 13.8 g per kg in fronds), natur. hot slag-heaps of Forest of Dean, S England, poss. use in remediating contaminated soils

Pterisanthes Blume (~ *Vitis*). Vitaceae. 20 W & C Mal.

Pternandra Jack. Melastomataceae. 15 Mal. R: GBS 34(1981)1. Berry

Pternopetalum Franch. Umbelliferae (III 8). 15 China. R: JSE 50(2012)551

Pterobesleria Morton = *Besleria*

Pterocactus K. Schum. Cactaceae (II). 9 S & W Arg. R: CSJ 44(1982)51. Aril forming papery wing round seed – wind-disp.

Pterocarpus Jacq. Leguminosae (III 11). c. 35 trop. (esp. Afr.). R: PM 5(1972). Troll's Model; some water-disp. in Amazonia. Sources of timber (padauk or padouk, vermilion wood, bloodwood) & kino. ***P. angolensis*** DC. (ambila, bleedwood tree, kiaat, mukwa, muninga, trop. & S Afr.) – timber for furniture, drums, canoes etc., form. for railway sleepers in C Afr., recycled as furniture sold in London; ***P. dalbergioides*** Roxb. ex DC. (Andaman padauk, E I mahogany, Indomal.) – cabinet-work; ***P. erinaceus*** Poir. ((W) African rosewood, barwood, African kino, W Afr.) – cabinet-work; ***P. indicus*** Willd. (angsana, Burmese or (Papua) NG rosewood, Andaman redwood, Amboyna wood, narra, E & SE As. to Mal.) – fl. buds open 3 days after sudden drop in temp., rose-scented timber (poss. most econ. imp. legume timber) for furniture, cult. orn. shade-tree rapidly growing from v. long cuttings as in Singapore City; ***P. macrocarpus*** Kurz (Burma padauk, maidu, SE As.) – inlay etc.; ***P. marsupium*** Roxb. (gamalu, E Ind. or Malabar kino, Ind.) – timber for outdoor use esp. window- & door-frames, also wood-carving in Sri Lanka; ***P. santalinus*** L.f. (zitan, red sandalwood, sanderswood or saunderswood, confused with algum or almug wood of Bible, Ind.) – cabinet-work (esp. classic Chinese furniture), red dye used to mark castes in Hindu religion; ***P. soyauxii*** Taubert (W Afr. padauk, barwood, camwood, W & C Afr.) – wood for canoes, dye for body-paint

Pterocarya Kunth. Juglandaceae. 6 Cauc. to E & SE As. R: AMBG 65(1978)1073. Wingnut. Cult. orn. trees with winged fr. esp. ***P. fraxinifolia*** (Poir.) Spach (Cauc. to Iran), ***P. stenoptera*** C. DC. (China) & their hybrid, ***P. × rehderiana*** C. Schneider (raised New York 1908), hardier than either & v. fast-growing

Pterocaulon Elliott. Compositae (Inul.-Pluch.). Incl. *Neojeffreya*, 12 warm Am., 14 SE As. to Aus. & New Caled. (3), 1 Afr. R: Darwiniana 21(1978)185. ***P. serrulatum*** (Montr.) Guillaumin (Aus., New Caled.) – vapour from crushed lvs inhaled for resp. problems by Aborigines. ***P. virgatum*** (L.) DC. (*P. undulatum*, US) – medic.

Pterocelastrus Meissn. (*Asterocarpus*). Celastraceae (I). 3 SE Afr. Sneezewood (termite-resistant building timber)

Pteroceltis Maxim. Cannabaceae (Ulmaceae II). 1 N & C China: ***P. tatarinowii*** Maxim. – fibre in paper-making. R: FOC 5(2003)9

Pterocephalidium G. López (~ *Pterocephalus*). Caprifoliaceae (Dipsacaceae). 1 C Spain, Portugal: ***P. diandrum*** (Lag.) G. López. R: Fl. Ib. 15(2007)317

Pterocephalodes V. Mayer & Ehrend. = *Bassecoia*

Pterocephalus Vaill. ex Adans. Caprifoliaceae (Dipsacaceae). Excl. *Bassecoia*, 25 Macar., Medit. (Eur. 5) to C As., Himal. & W China, trop. Afr. K with 10 or more pappus-like awns aiding wind-disp. Some cult. orn. rock-pls

Pteroceras Hasselt ex Hassk. (~ *Sarcochilus*). Orchidaceae (V 16c). 25 Indomal. R: OB 117(1993). Fls last 1 day

Pterocereus MacDoug. & Miranda. Cactaceae. 1 Mex.: ***P. gaumeri*** (Britton & Rose) MacDoug. & Miranda

Pterochaeta Steetz (~ *Waitzia*). Compositae (Gnap.-Ang.). 1 SW Aus.: ***P. paniculata*** Steetz. R: Nuytsia 8(1992)422

Pterochloris (A. Camus) A. Camus = *Chloris*

Pterocissus Urb. & Ekman = *Cissus*

Pterocladon Hook.f. = *Miconia*

Pterococcus Hassk. = *Plukenetia*

Pterocyclus Klotzsch (~ *Pleurospermum*). Umbelliferae (III 5). 4 Himal., SW China. R: FR 111(2000)521

Pterocymbium R. Br. Malvaceae (Sterc.; Sterculiaceae). 15 SE As. to Fiji. Winged ovaries aiding wind-disp.

Pterocypsela Shih = *Lactuca* (but see APS 26(1988)385)

Pterodiscus Hook. Pedaliaceae (2). c. 13 trop. & S Afr. R: MIABH 28–9(2000)13. Cult. orn. esp. those with swollen caudex

Pterodon Vogel. Leguminosae (III 3). c. 3 Braz., Bolivia

Pterogaillonia Lincz. = *Plocama*

Pterogastra Naud. Melastomataceae. 2 N trop. S Am. R: NJB 14(1994)66

Pteroglossa Schltr. (~ *Stenorrhynchos*). Orchidaceae (IV 2h). 11 S Am. R: HBML 28(1982)349

Pteroglossaspis Reichb.f. = *Orthochilus*

Pterogonum Gross = *Eriogonum*

Pterogyne Tul. Leguminosae (I 4). 1 trop. S Am.: ***P. nitens*** Tul. – valuable timber for construction, furniture, barrels etc.

Pterolepis (DC.) Miq. (~ *Tibouchina*). Melastomataceae. 14 trop. Am. esp. Brazilian savanna. R: NJB 14(1994)80

Pterolobium R. Br. ex Wight & Arn. Leguminosae (I 4). 10 OW trop. (Afr. 1). R: BMNHN 3e Bot. 15(1974)

Pteroloma Desv. ex Benth. Older name for *Tadehagi*

Pteromanes Pichi-Serm. = *Crepidomanes*

Pteromonnina Eriksen. See *Monnina*

Pteronia L. Compositae (Ast.-Hin.). c. 80 S Afr. R: ASAM 9(1917)277. Imparting particular flavour to lamb grazed on it. ***P. incana*** DC. – oil poss. use in perfumery

Pteropepon (Cogn.) Cogn. Cucurbitaceae (II). 5 S Am.

Pteroptychia Bremek. = *Strobilanthes*

Pteropyrum Jaub. & Spach. Polygonaceae (II 4). 6 SW As. Shrubs

Pterorhachis Harms. Meliaceae (II-Trich.). 1–2 W trop. Afr. ***P. zenkeri*** Harms – bark (nut flavour) a local aphrodisiac

Pteroscleria Nees = *Diplacrum*

Pteroselinum Reichb. = *Peucedanum*

Pterosicyos Brandegee = *Sicyos*

Pterospartum (Spach) K. Koch = *Genista*

Pterospermum Schreb. Malvaceae (Dom.; Sterculiaceae). c. 30 trop. As. Troll's Model; fragrant fls opening for 1 night. Timber esp. ***P. acerifolium*** (L.) Willd. ([anc.] karnikar, Ind. to Java) – durable like teak, fl. fragrance diminishing on picking

Pterospora Nutt. Ericaceae (II 2; Monotropaceae). 1 W N Am.: ***P. andromeda*** Nutt. – root-parasite to 1 m tall, local medic. R: FN 66(1995)25

Pterostegia Fischer & C. Meyer. Polygonaceae (Erio.-Erio.). 1 W & SW N Am.: ***P. drymarioides*** Fischer & C. Meyer. R: FNA 5(2005)477

Pterostemma Kraenzlin. Orchidaceae (V 12h). 3 Colombia

Pterostemon Schauer. Iteaceae (Pterostemonaceae, Grossulariaceae s.l.). 3 Mex. R: ABM 41(1997)21

Pterostemonaceae Small = Iteaceae

Pterostylis R. Br. Orchidaceae (IV 2f). c. 210 E Mal. to New Caled. & NZ esp. Aus. Fls usu. green ('greenhoods') poll. small flies or fungus-gnats. ***P. acuminata*** R. Br. (E Aus.) – tubers ed. Aborigines nr Sydney

Pterostyrax Sieb. & Zucc. Styracaceae. 4 Myanmar to Jap. Cult. orn.

Pterotaberna Stapf = *Tabernaemontana*

Pterothamnus V. Mayer & Ehrend. (~ *Pterocephalus*). Caprifoliaceae. 1 Mozambique: ***P. centenii*** (M. Cannon) V. Mayer & Ehrend. R: T 62(2013)124

Pterothrix DC. = *Amphiglossa*

Pteroxygonum Dammer & Diels (~ *Fagopyrum*). Polygonaceae (II-Pterox.). 2 China

Pterozonium Fée. Pteridaceae (III). 14 Roraima sandstones (Venez.), some ext. to Surinam, Colombia & N Peru R: MNYBG 17(1967)2

Pterygiella Oliv. Orobanchaceae (Rhin.; Scrophulariaceae). 4 S China

Pterygiosperma O. Schulz = *Sibara*

Pterygocalyx Maxim. Gentianaceae. 1 E As.: ***P. volubilis*** Maxim. R: APS 36(1998)58. Poss. incl. *Gentianopsis*

Pterygodium Sw. Orchidaceae (IV 4b). 19 E trop. (1) & S Afr. Elaiophores attractive to bees. ***P. pentherianum*** Schltr. & ***P. schelpei*** Linder (S Afr.) – sympatric but pollen deposited on diff. parts of same pollinator

Pterygopappus Hook.f. Compositae (Gnap.-Lor.). 1 Tasmania: ***P. lawrencei*** Hook.f. R: OB 104(1991)62

Pterygopleurum Kitagawa. Umbelliferae (III 8). 1 Korea, S Jap.: ***P. neurophyllum*** (Maxim.) Kitagawa. R: FOC 14(2005)135

Pterygopodium Harms = *Prioria*

Pterygostemon V.V. Botsch. = *Fibigia*

Pterygota Schott & Endl. Malvaceae (Sterc.; Sterculiaceae). 17 trop. esp. OW. Like *Brachychiton* & *Sterculia* but seeds winged, those of ***P. alata*** (Roxb.) R. Br., opium subs. in N Ind., with wings to 6 cm long. Timber (koto) for furniture esp. ***P. bequaertii*** De Wild. (awari, W Afr.)

Pteryxia (Torrey & A. Gray) J. Coulter & Rose. Umbelliferae (III 9). 5 W N Am.

Ptilagrostis Griseb. (~ *Stipa*). Gramineae (X). 8 Russia to China, mts

Ptilanthelium Steud. = *Schoenus*

Ptilanthus Gleason = *Graffenrieda*

Ptilimnium Raf. Umbelliferae (III 8). 5 E N Am. R: T 61(2012)414

Ptilochaeta Turcz. Malpighiaceae. 5 warm S Am.

Ptilopteris Hance = *Monachosorum*

Ptilostemon Cass. = *Jurinea* (but see Boissiera 22(1973))

Ptilothrix K.A. Wilson (*'Ptilanthelium'*). Cyperaceae. (II 7) 1 E Aus.: ***P. deusta*** (R. Br.) K.A. Wilson. R: Telopea 5(1994)612

Ptilotrichum C. Meyer = *Stevenia*, but see also *Hormathophylla*

Ptilotus R. Br. Amaranthaceae (I 2). 90 arid & semi-arid Aus., 1 ext. to Mal. Mulla mulla. Some cult. orn. (pussy tail) esp. ***P. manglesii*** (Lindl.) F. Muell. – remarkable jointed hairs on P, & ***P. nobilis*** (Lindl.) F. Muell. (*P. exaltatus*, inland E Aus.) – potpl. in N hemisph.

Ptisana Murdock (~ *Marattia*). Marattiaceae. 20 OW trop. esp. NG. R: T 57(2008)744. ***P. oreades*** (Domin) Murdock (*'M. fraxinea'*, Aus.) & ***P. salicina*** (Sm.) Murdock (*M. s.*, NZ, SW Pacific) – cult. orn., the thick stem much eaten by Aus. Aborigines & Maori

Ptycanthera Decne = *Matelea*

Ptychandra R. Scheffer = *Heterospathe*

Ptychococcus Becc. Palmae (V 14i). 2 NG, Solomon Is. R: SB 20(2005)523

Ptychogyne Pfitzer = *Coelogyne*

Ptycholobium Harms (~ *Tephrosia*). Leguminosae (III 16). 3 dry trop. Afr. to S Arabia. R: KB 35(1980)461

Ptychomeria Benth. = *Gymnosiphon*

Ptychopetalum Benth. Olacaceae. 4 trop. Am. (2) & Afr. (2). R.-parasites. Locally medic. ***P. olacoides*** Benth. (muira puama, Braz.) – trad. aphrodisiac from bark & roots, since 1930s comm. (Eur.)

Ptychopyxis Miq. Euphorbiaceae (Acal.-Acal.-Blum.). 13 Thailand to W Mal., E NG

Ptychosema Benth. Leguminosae (III 14). 1 C & W Aus.: ***P. pusillum*** Benth. R: Muelleria 29(2011)183

Ptychosperma Labill. Palmae (V 14i). 29 E Mal., Papuasia & trop. Aus. R: Allertonia 1(1978)415. Cult. orn. unarmed monoec. palms, readily hybridizing in cult.

Ptychostomum Hornsch. (~ *Bryum*). Bryaceae (Musci). c. 40 cosmop. ***P. pendulum*** Hornsch. (*B. p*, cosmop.) – sand-dune colonist growing up through layers of deposited sand

Ptychotis Koch. Umbelliferae (III 8). 1 C & S Eur.: ***P. saxifraga*** (L.) Loret & Barrandon

Ptyssiglottis T. Anderson. Acanthaceae (III 2c). (Incl. *Hallieracantha*) 33 Indomal. R: OB 116(1992)23. Local medic.

pua *Typha orientalis*

Pubistylus Thoth. (~ *Diplospora*). Rubiaceae (?II 5). 1 Andamans: ***P. andamanensis*** Thoth. R: Reinw. 7(1966)283

Pucara Ravenna = *Stenomesson*

Puccinellia Parl. (~ *Phippsia*). Gramineae (XVI 14). c. 110 temp. & arctic (Eur. 13; N Am. 21 (18 native) – R: FNA 24(2007)459) esp. As. (China 50, 14 endemic – R: FOC 22(2006)246), S Afr., Aus. (1). Alkali grass (N Am.), planted to ameliorate salinity in Aus. Hybrids with *Phippsia* spp.

Puccionia Chiov. Cleomaceae (Capparaceae s.l.). 1 N Somalia: ***P. macradenia*** Chiov., on gypsum; = *Cleome*?

puccoon, red *Sanguinaria canadensis*; **yellow p.** *Lithospermum canescens*

puchurin nut *Licaria pucheri*

pudding pipe tree *Cassia fistula*

Puebloa Doweld = *Pediocactus*

Puelia Franch. Gramineae (III). 5 trop. Afr. R: HIP 37 t. 3642(1967)3. Ligules external. ***P. coriacea*** Pilg. (Cameroun) with root tubers, ***P. schumanniana*** W. Clayton (Zaire) with only 1 leaf at top of culm

Pueraria DC. Leguminosae (III 18). Excl. *Haymondia, Neustanthus, Toxicopueraria*, c. 8 trop. & E As. Twiners with extrafl. nectaries. R: AUWP 85–1(1985)9, Phytotaxa 218(2015)201. ***P. candollei*** Wall. ex Benth. var. ***mirifica*** (Airy Shaw & Suvatabandhu) Niyomdham (Thailand) – miroestrol like oestrone but stronger & used in abortion in Thailand; ***P. montana*** (Lour.) Merr. var. ***thomsonii*** (Benth.) D.B. Ward (var. *lobata, P. lobata, P. thunbergiana*, kudzu vine, Jap. arrowroot, E Ind., China & Jap.) – fodder & cover crop, erosion control, ed. roots, stems to 20 m long with fibres for textiles (anciently used in China) & cordage, poss. comm. source medic. compounds, invasive pest in parts of S Eur., S Afr., US (where planting originally encouraged by officialdom), Hawaii, produces nitrate (& isoprenes forming ozone in air)

Pugionium Gaertn. Cruciferae (36). 2 Russia, Mongolia, N China. ***P. cornutum*** (L.) Gaertn. – veg.

pukatea *Laurelia novae-zelandiae*

pukeweed *Lobelia inflata*

pulasan *Nephelium ramboutan-ake*

Pulchranthus Baum, Reveal & Nowicke. Acanthaceae (III 2c). 4 trop. S Am. R: SB 8(1983)211

Pulicaria Gaertn. Compositae (Inul.-Inul.). 85 temp. (Eur. 5) & warm Euras. R: PM 14(1981)64. Fleabane. Heterogeneous?

Pullea Schltr. (~ *Codia*). Cunoniaceae (VII). 3+ E Mal. & trop. Aus. to Fiji. R: Blumea 25(1979)490

Pulmonaria Tourn. ex L. Boraginaceae (B.2.1.1). Incl. *Paraskevia*, 17 Eur. R: BB 131(1975). Cult. orn bristly herbs incl. ***P. angustifolia*** L. (lvs green) & ***P. officinalis*** L. (lungwort; lvs white-spotted) with heterostyly & fls changing from red to blue as they mature; *P. officinalis* form. medic. in bronchial complaints (dried lvs, the speckled lvs suggesting diseased lungs in Doctrine of Signatures)

pulque See *Agave*

Pulsatilla Mill. = *Anemone*

Pulsatilloides Starodubtzev = *Anemone*

Pultenaea Sm. Leguminosae (III 13). (s.s.) 112 Aus. (91 E Aus.; s.l. = all III 13 – 19/470). R: AusSB 16(2002)81, 16(2003)229, 17(2004)273, 18(2004)149. Bush peas

pulu *Cibotium glaucum, Sadleria cyatheoides*

Pulvinaria Fourn. = *Lhotzkyella*

pulza *Jatropha curcas*

pumilio pine oil *Pinus mugo*

pummelo *Citrus maxima*

pumpkin *Cucurbita moschata, C. pepo, C. argyrosperma*

punah *Tetramerista glabra*

Punica Tourn. ex L. Lythraceae (Punicaceae). 2 SE Eur. to Himal. (*P. granatum*), Socotra (*P. protopunica*). G.M. Levin (2006) *Pomegranate*. ***P. granatum*** L. (pomegranate) – each leaf with single apical nectary exuding fructose, glucose & sucrose; cult anc. cultigen (? wild in S Caspian & NE Turkey) known from Bronze Age Jericho (figured in capitals of the temple in Jerusalem, when busting open indicative of hope & immortality, a motif used by Grinling Gibbons & impersonators), grown for refreshing fr. (ed. pulp higher in polyphenol antioxidants than citrus or blueberries poss. reducing heart disease, rich in vitamin C) around seeds (which contain 17 mg per kg oestrone (beta-sitosterol poss. use in promoting uterine contractions) & when dried (anardana) used to acidify curries in Ind.), easily fermented to grenadine, a cordial, extract effective against HIV & allegedly effective against prostate cancer, bark medic. (alks) & it & fr. pericarp used in tanning leather in Egypt, the fls medic. (balustine fls), persistent calyx the inspiration for King Solomon's crown (& hence other crowns in Eur.; name (from fr.) also cognate with 'grenade'!), allegedly the fr. given to Venus by Paris, on Ming porcelain meaning fertility; cult. orn. esp. '**Nana**' to 1.5 m. ***P. protopunica*** Balf.f. – poss. ancestor (cf. *Ceratonia*), now reduced to a few trees & endangered

Punicaceae Bercht. & J. Presl = Lythraceae

Punjuba Britton & Rose = *Abarema*

Punotia D. Hunt (~ *Austrocylindropuntia*). Cactaceae. 1 Peru: ***P. lagopus*** (K. Schum.) D. Hunt

puntarelle *Cichorium intybus* Witloof Group

Puntia Hedge = *Endostemon*

Pupalia Juss. Amaranthaceae (I 2). 4 trop. OW. R: KB 34(1979)131

Pupilla Rizz. = *Justicia*

Purdiaea Planch. Clethraceae (Cyrillaceae s.l.). 14 trop. Am. esp. Cuba (12). R: CGH 186(1960)47

Purdieanthus Gilg = *Lehmanniella*

purging buckthorn *Rhamnus cathartica*; **p. cassia** *Cassia fistula*; **p. flax** *Linum catharticum*; **p. nut** *Jatropha curcas*

Puria Nair = *Cissus*

puriri *Vitex lucens*

purple apple berry *Billardiera macrantha*; **p. beech** *Fagus sylvatica* Atropunicea Group; **p. guava** *Psidium cattleyanum*; **p. heart, p.-wood** *Peltogyne paniculata*; **p. sprouting** *Brassica oleracea* Italica Group

Purpureostemon Gugerli. Myrtaceae (II 1). 1 New Caled.: ***P. ciliatus*** (Forst. & Forst.f.) Gugerli

Purpusia Brandegee = *Potentilla*

Purshia DC. ex Poir. Rosaceae (Dryad.). Incl. *Cowania*, 6 W N Am. R: J.A. Young & C.D. Clements (2002) *P*. Decid. shrubs (antelope bush) with nitrogen-fixing actinobacterial root-symbionts, inner layer of outer integument with stomata. Cult. orn. esp. ***P. tridentata*** (Pursh) DC. – imp. local medic., bark used for moccasins

purslane, common *Portulaca oleracea* 'subsp. *sativa*'; **sea p.** *Honckenya peploides*; **water p.** *Lythrum portula*; **winter p.** *Claytonia perfoliata*

Puschkinia Adams. Asparagaceae (Hyacinthaceae, Liliaceae s.l.). 3–4 Syria, Lebanon, Turkey & Cauc. Cult. orn. esp. ***P. scilloides*** Adams (striped or Lebanon squill) – oily ant-disp. seeds

Pusillanthus Kuijt. Loranthaceae. 1 Venez.: ***P. pubescens*** (Rizz.) Caires (*P. trichodes*). R: Novon 18(2008)372, ABB 26(2012)668

pussy tail *Ptilotus* spp.; **p. willow** *Salix caprea*

pussy's-toes *Antennaria* spp.

Putoria Pers. = *Plocama*

Putranjiva Wall. (~ *Drypetes*). Putranjivaceae. 4 trop. As. ***P. roxburghii*** Wall. (*D. r.*, Ind., Myanmar) – seeds in rosaries

Putranjivaceae Meissn. (Euphorbiaceae s.l.). Magnoliidae – Malpighiales. 2/205 trop., to E As. & S Afr. (Usu.) dioec. evergreen trees with simple hairs & glucosinolates. Lvs simple with asymmetric base, in spirals becoming distich.; stip. decid. or not. Infls fasciculate (s.t. cauliflorous) or fls solit. K (3)4 or 5(6), imbr.; C 0; A (2)3–20(–50), usu. introrse with longit. slits; disk intrastaminal or 0; $\underline{G}$ 1–3(-6)-loc. with axile placentation & 2 anatropous bitegmic ovules; fr. a drupe, embryo large, straight. n = 7, 19(20)21
Genera: *Drypetes, Putranjiva*
Chemistry, embryology, seed structure & fr. distinguish P. from E.

Putterlickia Endl. = *Gymnosporia* (but see SAJB 64(1998)323)

puttyroot *Aplectrum hyemale*

putumuju *Centrolobium* spp.

Puya Molina. Bromeliaceae (3). c. 219 highland S Am. R: FN 14(1974)66. Terr. oft. pachycaul pls with spiny-margined lvs. Some cult. orn. incl. ***P. alpestris*** (Poepp.) Gay ('*P. berteroana*') & (& hybrids with) ***P. chilensis*** Molina (Chile) – lvs yield a fibre used for rot-resistant fishing-nets, but many grubbed up as hazardous to sheep which get entangled in them, as do birds, the nutrients from which, as well as those from their droppings may be absorbed (? foliar). ***P. raimondii*** Harms (Peru, Bolivia) – to 10.7 m, hapaxanthic (Holttum's Model, cf. *Lobelia*) after 80–150 yrs (world's slowest) in Andes, but after 28 yrs when grown at sea-level, with 15 000–20 000 fls & infl. branch extensions acting as perches for poll. birds (cf. *Babiana*), v. uniform (inbreeding), now endangered

Pyankovia Akhani & Roalson (~ *Salsola*). Amaranthaceae. 3+ SE Eur. to C As.

Pycnandra Benth. Sapotaceae (IV). Incl. *Leptostylis*, c. 65 New Caled. R (subgg. *P., Achradotypus*): AusSB 22(2009)441, (2010)188. Mostly pachycaul treelets

Pycnantha Ravenna = ? *Malaxis*

Pycnanthaceae Ravenna = Orchidaceae

Pycnanthemum Michaux. Labiatae (VII 2b). 19 N Am. (esp. SE). R: UCPB 20(1943)202, JBRIT 2(2008)194. Some local medic. & flavourings incl. ***P. incanum*** (L.) Michaux – poss. source of natural rubber

Pycnanthus Warb. Myristicaceae. 3–4 trop. Afr. Valuable fat from endosperm. ***P. angolensis*** (Welw.) Warb. (*P. kombo*, akomu, eteng, ilomba, kombo, false nutmeg) – seeds a source of oil, timber useful (plywood in Uganda)

Pycnarrhena Miers ex Hook.f. & Thomson. Menispermaceae (I). 9 Indomal. to Aus. R: KB 26(1972)405, 30(1975)97. ***P. manillensis*** S. Vidal (Philipp.) – powdered root used against snakebite

Pycnobotrya Benth. Apocynaceae (I e). 1 trop. Afr.: ***P. nitida*** Benth.

Pycnobregma Baill. = *Matelea*

Pycnocephalum (Less.) DC. = *Chresta* (but see SB 10 (1985)461)

Pycnocoma Benth. Euphorbiaceae (Acal.-Pyc.-Pyc.). 18 trop. Afr. (Congo 10 – R: BJBB 65(1996)38) to Masc. Some with Corner's Model, litter-collecting rosettes of lvs

Pycnocomon Hoffsgg. & Link (~ *Lomelosia*). Caprifoliaceae (Dipsacaceae). 2 W Medit. R: Fl. Ib. 15(2007)347

Pycnocycla Lindl. Umbelliferae (III 1). 12 trop. W Afr. to NW Ind.

Pycnoloma C. Chr. = *Seliguea*

Pycnoneurum Decne. = *Cynanchum*

Pycnonia L. Johnson & B. Briggs = *Persoonia*

Pycnophyllopsis Skottsb. Caryophyllaceae (II 1). 2 Andes

Pycnophyllum Remy. Caryophyllaceae (II 2). 17 Andes. R: FR 18(1922)167

Pycnoplinthopsis Jafri. Cruciferae (26). 1 Bhutan: ***P. bhutanica*** Jafri

Pycnoplinthus O. Schulz. Cruciferae (26). 1 Himal.: ***P. uniflorus*** (Hook.f. & Thomson) O. Schulz. R: FOC 8(2001)170

Pycnorhachis Benth. Apocynaceae (V; Asclepiadaceae). 1 Malay Peninsula: ***P. maingayi*** Hook.f.

Pycnosorus Benth. (~ *Craspedia*). Compositae (Gnaph.). c. 6 Aus. R: Telopea 5(1992)39

Pycnospatha Thorel ex Gagnepain. Araceae (V). 2 SE As. R: OBZ 122(1973)202

Pycnosphaera Gilg. Gentianaceae. 1 trop. Afr.: ***P. buchananii*** (Bak.) N.E. Br. R: Strelitzia 10(2000)311

Pycnospora R. Br. ex Wight & Arn. Leguminosae (III 19). 1 OW trop.: ***P. lutescens*** (Poir.) Schindler

Pycnostachys Hook. Labiatae (VII 3e). c. 35 trop. & S Afr., Madag. R: KB 1939:563. Cult orn. herbs esp. ***P. urticifolia*** Hook. (blue boys, trop. & S Afr.) – fls bright blue

Pycnostelma Bunge ex Decne. = *Vincetoxicum*

Pycnostylis Pierre = *Triclisia*

Pycreus P. Beauv. = *Cyperus*

Pygeum Gaertn. = *Prunus*

Pygmaea B.D. Jackson = *Veronica*

Pygmaeocereus H. Johnson & Backeb. = *Haageocereus*

Pygmaeopremna Merr. = *Premna*

Pygmaeorchis Brade. Orchidaceae (V 13b). 2 Braz.

Pygmaeothamnus F. Robyns. Rubiaceae (III 2). 2 trop. & S Afr. Geoxylic suffrutices

Pygmea Hook.f. = *Veronica*

pygmyweed, NZ *Crassula helmsii*

pyinkado *Xylia xylocarpa*

pyinma *Lagerstroemia speciosa*; **Andaman p.** *L. hypoleuca*

pyjama lily *Crinum kirkii*

Pynaertiodendron De Wild. = *Cryptosepalum*

Pyracantha M. Roemer. Rosaceae (Ros.-Mal.). 10 SE Eur. (1) & As. R: CJB 68(1990)2230. Cult. orn. usu. thorny shrubs esp. ***P. coccinea*** M. Roemer (*Mespilus pauciflora*, firethorn, firebush, SE Eur., native in GB in warmer interglacials) – many cvs, oft. grown espaliered for persistent colourful fr. in winter. Many named cvs hybrids, some unaffected by fireblight &/or with frs not taken by birds. Some invasive Aus., S Afr., Hawaii

Pyragra Bremek. (~ *Psychotria*). Rubiaceae (IV 7). 2 Madag. R: Candollea 16(1958)174

Pyramia Cham. = *Cambessedesia*

Pyramidanthe Miq. (~ *Fissistigma*). Annonaceae (III 7). 1 W Mal.: ***P. prismatica*** (Hook.f. & Thomson) Sinclair – local medic.

Pyramidium Boiss.= *Veselskya*

Pyramidoptera Boiss. Umbelliferae (III 7). 1 Afghanistan: ***P. cabulica*** Boiss.

Pyranthus Dupuy & Labat. Leguminosae (III 16). 6 Madag. R: KB 50(1995)73

Pyrenacantha Hook. Icacinaceae. Incl. *Chlamydocarya*, *Polycephalium*, 25 OW trop.

Pyrenaria Blume. Theaceae. 42 China, SE As., W Mal.

pyrene oil *Olea europaea*

Pyrenocarpa H.T. Chang & Miao = *Decaspermum*

Pyrenoglyphis Karsten = *Bactris*

pyrethrum *Tanacetum cinerariifolium* (insecticide), *T. coccineum* (cut-fl.)

Pyrethrum Zinn = *Tanacetum*

Pyrgophyllum (Gagnep.) T.L. Wu & Z.Y. Chen. Zingiberaceae (II 1). 1 China: ***P. yunnanense*** (Gagnepain) T.L. Wu & Z.Y. Chen. R: APS 27(1989)124

Pyriluma Baill. ex Aubrév. = *Planchonella*

Pyrola L. Ericaceae (II 1; Pyrolaceae). 35 N hemisph. (Eur. 7) to Sumatra, temp. S Am. Wintergreen. Seeds wind-disp. Some locally used to heal wounds & cult. orn. incl. ***P. asarifolia*** Michaux (N Am. & As.) – lvs a source of a tea for N Am. Indians in E Canada

Pyrolaceae Lindl. = Ericaceae

Pyrolirion Herb. (~ *Zephyranthes*). Amaryllidaceae. 5–10 Andes. R: BSB 2,53(1981)1197

Pyropia J. Agardh (~ *Porphyra*). Bangiaceae (Rhodophyta). 75+ cosmop. R: JPhycol. 47(2011)1142. ***P.*** spp. esp. ***P. tenera*** (Kjellm.) N. Kikuchi & al. (*Porphyra t.*, *P. yezoensis*, subcosmop.) – sold (typically toasted) as 'nori' or 'seaweed' for sushi, 800 km^2 off Jap. prod. 350 K t (worth $1 billion) a yr (3 × China prod.)

Pyrorchis D.L. Jones & M.A. Clem. (~ *Lyperanthus*). Orchidaceae (IV 3f). 2 SW & SE Aus. R: Gen. Orch. 2(2001)167

Pyrostegia C. Presl. Bignoniaceae (3). 2 trop. S Am. R: AMBG 99(2014)460. Lianes with 6–8-ribbed branchlets. ***P. venusta*** (Ker-Gawler) Miers (*P. ignea*, flame vine, golden shower, Braz. & Paraguay) – commonly planted in trop. for orange fls poll. by hummingbirds

Pyrostria Comm. ex Juss. Rubiaceae (III 2). Incl. *Scyphochlamys*, 70+ Afr., Madag., Mauritius, Rodrigues (? to Mal.). ***P. bibracteata*** (Baill.) Cavaco (E Afr. to Seychelles) – fr. ed.

Pyrrhanthera Zotov = *Rytidosperma* (but see NZJB 1(1963)125)

Pyrrhocactus (A. Berger) Backeb. & Knuth = *Eriosyce*

Pyrrhopappus DC. Compositae (Lact.-Mic.). 1–5 N Am. R: FNA 19(2006)376. Some roots eaten by N Am. Indians

Pyrrocoma Hook. (~ *Haplopappus*). Compositae (Ast.-Mac.). 14 W US & Canada. R: FNA 20(2006)413

Pyrrorhiza Maguire & Wurd. Haemodoraceae (I). 1 Venez.: ***P. nebliniae*** Maguire & Wurd. R: MNYBG 9(1957)318

Pyrrosia Mirb. Polypodiaceae (III). c. 50 trop. OW to NE As. & NZ. R: AFJ 73(1983)73, LBS 9(1986)139. Scandent epiphytes oft. with crassulacean acid metabolism & usu. simple leathery/fleshy fronds covered in stellate hairs. Some local medic. (diuretics etc.) esp. ***P. lingua*** (Thunb.) Farwell (Ind. to Korea) – antiviral, diuretic in anc. Chinese medic, cult. orn. as is ***P. nummulariifolia*** (Sw.) Ching (Indomal.) – pest of coffee in webbing branches together & poss. causing rotting

Pyrrothrix Bremek. = *Strobilanthes*

Pyrularia Michaux. Santalaceae (Cervantesiaceae). 2 SE US (1), China, Himal. R.-parasites. ***P. pubera*** Michaux (buffalo or elk nut, SE US) – ed. fr. but oil acid & poisonous

Pyrus Tourn. ex L. Rosaceae (Ros.-Mal.). c. 15 Euras. (esp. Armenia (>20), Eur. 13), Medit. R: CJB 68(1990)2238. Pears. Form. (not on unreasonable grounds) spp. referred here now placed in *Malus* & *Sorbus*, but *P.* now restricted to trees with sclerenchyma cells in pome (i.e. gritty texture in pear flesh); extrafl. nectaries. Ed. pears (1000 cvs (Theophrastus described 3, Pliny 39, C16 Italy 232, 622 in GB 1826, 677 by 1831), prop. by grafting (no cuttings), some parthenocarpic) found in Neolithic & Bronze Age Eur. deposits, derive from ***P. communis*** L. (Eur., W As., ? hybrid origin) – incl. Bartlett pears (i.e. '**Williams' Bon Chrétien**', 1760s, Aldermaston, England, introd. Aus. 1797/9) which are those most commonly tinned, 'corella' ps in Aus. artificially reddened by cold treatent, most requiring cross-poll. (self-incompatible) & some triploids; this has been crossed with ***P. pyrifolia*** (Burm.f.) Nakai (*P. serotina*, Chinese or Jap. 'Asian' p., nashi (= Jap. for pear), China, natur. Jap. & grown in trop. As.) to give ***P.* × *lecontei*** Rehder esp. '**Kieffer**' with enhanced disease resistance; fr. of *P. pyrifolia* (also hybridized with ***P. ussuriensis*** Maxim. (NE As.) to give Chinese 'Asian' ps), usu. ± apple-shaped & noted for excellent keeping qualities; 'fragrant ps' oft. of complex hybrid ancestry involving Eur. & As. spp. (cf. *Malus*!). Trees living to 300 yrs (US); fr. prod. 18 M t a yr by 2005; best-flavoured (flavour due to ethyl *trans*-2, *cis*-4-decadienoate) being '**Doyenne du Comice**' (1849) with buds less prone (higher concs of phenolic acids) to bullfinch attack than '**Conference**' (1894), the most widely planted of c. 20 in UK commerce; perry (cf. cider from apples) made from fr. with trees 200–300 yrs old surviving in Gloucestershire, UK, though in decline until 1940s when controlled fermentation used at Shepton Mallett, Somerset to produce comm. 'Babycham'; perry vinegar prod. in W Aus. etc.; bark with antibacterial action & timber used for turnery, cutlery handles, inlay & (stained) piano keys – form. much favoured for carving as by the incomparable Grinling Gibbons (1648–1721) & woodblocks for printing. Some cult. orn. esp. ***P. calleryana*** Decne (China) – more salt-tolerant in urban sites than *Platanus* × *hispanica*, many cvs tolerant of poorly drained land grown as street-trees in US esp. thornless '**Bradford**' (Bradford pear), & ***P. salicifolia*** Pallas (E Turkey, Cauc., NW Iran) – habit of willow with greyish lvs

Pytinicarpa Nesom. Compositae (Ast.-Lag.). 2 New Caled., 1 Fiji. R: Sida 19(2001)516

Pyxidanthera Michaux. Diapensiaceae. 1–2 E US incl. ***P. barbulata*** Michaux. R: Britt. 27(1975)115

Qaisera Omer = *Gentiana*

qat *Catha edulis*

qinghao, qinghaosu *Artemisia annua*

Qiongzhuea (T.H. Wen & Ohrnberger) Hsueh & Yi = *Chimonobambusa*

Quadrella (DC.) J. Presl (~ *Capparis*). Capparaceae. c. 25 warm Am.

Quadribractea (~ *Wedelia*). Compositae (Helia.). 1 E Mal.: ***Q. moluccana*** (Blume) Orchard. R: Nuytsia 23(2013) 433

Quadricasaea Woodson = *Bonafousia*

Quadripterygium Tard. = *Euonymus*

quake or **quaking grass** *Briza* spp.

Qualea Aubl. Vochysiaceae (I). 59 trop. Am. R: ABN 2(1953)150. Massart's Model; G

Qualup bell *Pimelea physodes*

quamash *Camassia* spp.

Quamoclidion Choisy = *Mirabilis*
Quamoclit Mill. = *Ipomoea*
quandong *Santalum acuminatum*
Quapoya Aubl. = *Clusia*
Quaqua N.E. Br. (~ *Caralluma*). Apocynaceae (Vb; Asclepiadaceae III 5). 29 SW Afr. R: F. Albers & U. Meve, *Ill. Handb. Succ. Pls*, Asclep. (2002)213. Cult. orn. succ.
Quararibea Aubl. Malvaceae (Bomb.; Bombacaceae). 20+ trop. Am. Fagerlind's Model. ***Q. funebris*** (Llave) Vischer (Mex.) & other spp. – fls dried & added as spice to flavour chocolate drinks in Mex.; ***Q. pumila*** Alverson (Costa Rica) – cauliflorous pachycaul. See also *Matisia*
quaruba *Vochysia hondurensis*
Quassia L. Simaroubaceae. Excl. *Hannoa, Odyendea, Pierrodendron, Samadera, Simaba, Simarouba*, 1 W Afr., 1 Braz.: ***Q. amara*** L. (marupa, simaruba, Surinam quassia wood, stave-wood, Brazil) – source of bitters, vermifuge & poison in fly-papers
quassia wood *Picrasma excelsa*; **Surinam q. w.** *Quassia amara*
Quaternella Pedersen (~ *Pfaffia*). Amaranthaceae (II 2). 3 S Am. R: BMNHN 4,12(1990)92
quebracha *Aspidosperma tomentosum*
quebracho *Schinopsis* spp.
Quechua Salazar & Jost (~ *Spiranthes*). Orchidaceae (IV 2h). 1 Peru, Ecuador: ***Q. glabrescens*** (T. Hashim.) Salazar & Jost. R: SB 37(2012)80
Quechualia H. Robinson (~ *Vernonia*). Compositae (Vern.-Vern.). 4 Peru to Arg.
Queen Anne's lace *Anthriscus sylvestris*; **q. of the night** *Epiphyllum oxypetalum, Peniocereus greggii*
queen's flower *Lagerstroemia speciosa*
Queensland arrowroot *Canna indica*; **Q. bean** *Entada* spp.; **Q. hemp** *Sida rhombifolia*; **Q. lignum-vitae** *Vitex lignum-vitae*; **Q. maple** *Flindersia brayleyana*; **Q. nut** *Macadamia* spp.; **Q. walnut** *Endiandra palmerstonii*
Queenslandiella Domin = *Cyperus*
queenwood *Daviesia arborea*
Quekettia Lindl. Orchidaceae (V 12h). Incl. *Stictophyllorchis*, excl. *Suarezia*, 4 trop. S Am. to Trinidad
Quelchia N.E. Br. Compositae (Stifft.). 4 Venez., Guyana. Shrubs & trees
quenette *Melicoccus bijugatus*
Quercifilix Copel. = *Tectaria*
quercitron yellow dye from inner bark of *Quercus* spp. esp. *Q. velutina*
Quercus Tourn. ex L. Fagaceae. c. 530 N temp. (Eur. 20, N Am. c. 90 with many hybrids.) S to Mal. & Colombia (***Q. humboldtii*** Bonpl.) at alt. Oaks. R: A. Camus (1934–54) *Les chênes*; journal – *International Oaks*. Everg. or decid. trees (Rauh's Model), few shrubs, some with domatia; trop. spp. with intermittent growth of apical buds developing fls & in next flush lvs photosynthesizing while acorns develop, with acorn maturity apical buds producing more fls, lvs, etc.; in spp. where winter stops growth, everg. oaks fl. in spring & fr. in autumn, decid. ones with wind-poll. fls before new lvs fully grown & shed lvs when acorns have fallen; in subg. *Erythrobalanus* (sect. *Lobatae*, red oaks, Am.) fert. in spring the year after poll. so fr. matures in 2nd autumn. Cult. orn. & timber trees, acorns used for swine-food (ground for human food N Am.), local medic. (preserved at Pompeii) & as tanning agents, bark for dye (quercitron; purple for cardinals replaced 1467 by red from cochineal), gargle for sore throats, remedy sweaty feet, tannins & cork, galls (oak apples; Oak Apple Day = 29 May, anniv. of return of Charles II (who had hidden in an o. tree, 1651) to London, 1660) for tanning (gall-wasp. deliberately introd. Devon from Middle E. c. 1730 as galls 17% tannin used in dyes & inks) & lvs for feeding silkworms; timber for construction (& orig. wainscot[ing]) but also whisky casks (bourbon in new ones, whisky in old, some 1M a yr in US (*Q. alba*) & 100 000 in Eur.), shingles (N Am.) & inlay (brown o. = timber stained with mycelium of bracket fungus (*Fistulina hepatica*) & green, as in Tunbridge ware, by *Chlorociboria* (*Chlorosplenium*) *aeruginascens* due to a quinone, xylindein); worshipped in S Eur. from earliest times & now national tree of Eire, Ger. & US. ***Q. acuta*** Thunb. (Jap.) – favoured timber of Walter Burley Griffin; ***Q. alba*** L. (white or Quebec oak, E N Am.) – imp. local medic., timber & fuel source; ***Q. canariensis*** Willd. (Algerian oak, Iberia, N Afr.) – not Canary Is.!; ***Q. cerris*** L. (Turkey or manna o., C & S Eur., W As.) – timber & source of sweetmeats etc.; ***Q. coccifera*** L. (Kermes o., Medit.) – to 20 m but tolerates grazing to be a few cm tall, form. used for feeding cochineal insects, 3 sprigs the crest of the Dyers' Company; ***Q. coccinea*** Muenchh.

(scarlet o., NC & E US) – timber, lvs bright red in autumn; ***Q. emoryi*** Torr. (black o., SW US) – sweet-tasting acorns much traded; ***Q. falcata*** Michaux ((Amer.) red o., C & E US); ***Q. gambelii*** Nutt. (shin o., S N Am.) – acorns ed.; ***Q. garryana*** Douglas ex Hook. (Oregon white o., W N Am.) – imp. timber & fuel; ***Q. ilex*** L. (incl. *Q. rotundifolia* (*Q. i.* subsp. *ballota* (Desf.) Samp., W Medit.), holm o., holly o., ilex, Medit.) – much-planted 'evergreen o.' with ed. acorns, galls (prob. Morea, Greek, Marmora or Italian g.) used in tanning (ilex an old Roman name now transferred to holly, i.e. *Ilex*, while holm, Old English name for holly now transferred to *Q. ilex*!), hybrids (***Q. × turneri*** Willd.) with *Q. robur* raised c. 1783; ***Q. ilicifolia*** Wangenh. (bear o., SE Ontario, E US) – usu. shrubby; ***Q. imbricaria*** Michaux (shingle o., C & E US) – imp. timber for clapboards, shingles etc.; ***Q. ithaburensis*** Decne subsp. ***macrolepis*** (Kotschy) Hedge & Yalt. (*Q. m.*, '*Q. aegilops*', S Eur., W As.) – dried cupules used in tanning (valonea, palamut) esp. for quality heavy leathers; ***Q. lobata*** Née (Californian o., Calif.) – acorns form. eaten by Native Amer.; ***Q. lusitanica*** Lam. (Portugal, Spain, Morocco) – shrub, oft. creeping; ***Q. macrocarpa*** Michaux (burr o., E N Am.) – timber for ship-building, furniture etc.; ***Q. marilandica*** (L.) Muenchh. (jack o., C & SE US) – charcoal; ***Q. minima*** (Sarg.) Small (SE US) – rhiz.; ***Q. mongolica*** Fischer ex Ledeb. (Japanese or Manchurian o., NE As.) – imp. local timber, exported for furniture-making; ***Q. montana*** Willd. ('*Q. prinus*', chestnut o., (Amer.) white o., E US) – imp. tanbark, timber, ed. acorns; ***Q. nigra*** L. (possum o., C & E US) – acorns form. Native Amer. food; ***Q. palustris*** Muenchh. (pin o., E N Am.) – imp. street-tree US; ***Q. pubescens*** Willd. (*Q. infectoria*, gall, E Med.) – galls (Aleppo, Levant, Mecca or Turkish galls) with 36–58% tannin coll. for high tannin content & used in ointments & suppositories esp. in treatment of piles; ***Q. robur*** L. (English, F., Polish or Slavonian o., Eur., Medit.) – principal hardwood of Br., selected as 'standards' in coppice & standards silviculture, imp. timber for construction of houses & furniture of past, also warships, flooring, wine-barrels ('French' oak ones [?*Q. petraea*] in US cost $900), charcoal etc., cupules yield a yellow dye (Turkey), acorns eaten by pigs (pannage) & ground as coffee subs., replaced in N & W GB by ***Q. petraea*** (Mattuschka) Liebl. (*Q. esculus* L., durmast or sessile o., Eur., W As.) with similar uses but many trees in GB at least appear to be of hybrid origin (***Q. × rosacea*** Bechst.), the 'spp.' being imperfectly isolated genetically, growth-rates a good indication of soil quality & depth, sometimes distrib. by jays which select largest acorns they can swallow; ***Q. rubra*** L. (*Q. borealis*, (Amer.) red o., E N Am.) – imp. timber for general construction; ***Q. stellata*** Wangenh. (*Q. obtusiloba*, iron or post o., C & E US) – timber, ed. acorns, lvs used as cigarette-papers; ***Q. suber*** L. (cork o., S Eur., N Afr.) – thick bark the cork of commerce, stripped off trees older than 20 yrs every 8–10 until c. 200 yrs old (35% suberin by weight in first cut, later 38% as used in stoppers), used for insulation, tiles & bottle-corks ('corking' = tainting of wine due to 2,4,6-trichloroanisole prod. by action of chlorine on cork), floats, form. cork-tipped cigarettes, when ground a constituent of linoleum & used in steam-cleaning buildings, 250 000 t per annum (54% in Portugal), hybrid with *Q. cerris* = ***Q. × crenata*** Lam. (*Q. × hispanica*, *Q. × pseudosuber*, Lucombe o.) hardy GB; ***Q. variabilis*** Blume (China) – corks from bark for Great Wall wine; ***Q. velutina*** Lam. (*Q. tinctoria*, quercitron, E N Am.) – tanbark, yellow dye used in printing calico; ***Q. virginiana*** Mill. (live o., SE US) – trad. shade-tree of S US, to 1200 yrs old, form. imp. ship-building timber (& therefore strategic), seed oil form. used, tuberous hypocotyls fried like potatoes

Quesnelia Gaudich. Bromeliaceae (3). 19 SE Brazil. R: FN 14(1979)1956, SB 34(2009)660. Some cult. orn.

quetsche(n) plum *Prunus × domestica*

Quetzalia Lundell. Celastraceae (I). 11 Mex., C Am.

Quezelia Scholz = *Quezeliantha*

Quezeliantha Scholz ex Rauschert. Cruciferae (12). 1 Sahara: ***Q. tibestica*** (Scholz) Rauschert. R: T 31(1982)558

Quiabentia Britton & Rose (~ *Pereskiopsis*). Cactaceae (II). 2 Arg. & Paraguay (Chaco)

quickbeam *Sorbus aucuparia*

quickset, quickthorn *Crataegus monogyna*

Quidproquo Greuter & Burdet (~ *Raphanus*). Cruciferae. 1 Syria, Iran: ***Q. confusum*** Greuter & Burdet

Quiducia Gagnepain = *Silvianthus*

quihuicha *Amaranthus caudatus*

Quiina Aubl. Ochnaceae (Quiinaceae). 34 trop. S Am. Chamberlain's Model; some fish-disp. in Amazon

Quiinaceae Choisy = Ochnaceae

quillai *Quillaja saponaria*

Quillaja Molina. Quillajaceae (Rosaceae s.l.). 2 temp. S Am. Everg. trees with bark used as soap & medic. esp. ***Q. saponaria*** Molina (bois de Panama, quillai, soapbark tree, Chile) – source of q. bark with 9% saponin used in fire-extinguishers & 'whipped cream' ed. Middle East (cf. *Saponaria officinalis*)

Quillajaceae D. Don (~ Rosaceae). Magnoliidae – Fabales. 1/2 temp. S Am. Small everg. glabrous trees with saponins. Lvs simple, in spirals, toothed (? hydathodes); stip. petiolar. Infls few-flowered botryoids with bisex. term. fl., lat. male. Fls 5-merous, tomentose. K valvate somewhat accrescent with thick nectary disk on lower half; C clawed; A 5 opp. K on outer edge of disk, 5 opp. C from near ovary base; G connate axially, opp. K, ovules bitegmic, numerous in 2 ranks. Fr. asymmetrically lobed follicles spreading star-like, dehiscing along 2 sutures. Seeds many, exotestal, winged; cotyledons convolute, endosperm thin. n = 14, 17
Genus: *Quillaja*

quillings See *Cinnamomum verum*

quince *Cydonia oblonga*; **Bengal q.** *Aegle marmelos*; **Japanese q.** *Chaenomeles speciosa*

Quinchamalium Molina. Santalaceae (Schoepfiaceae). 1(–20) Chile: ***Q. chilense*** Molina – sold as folk med. R: SB 40(2015)1045

Quincula Raf. (~ *Physalis*). Solanaceae (IV 5c). 1 N Am.: ***Q. lobata*** (Torrey) Raf. R: Kurtzeana 25(2000)74

Quinetia Cass. Compositae (Gnap.-Ang.). 1 S & SW Aus.: ***Q. urvillei*** Cass. R: OB 104(1991)124

quinine *Cinchona* spp., see also *Hintonia lateriflora*; **q. bush** or **tree** *Petalostigma* spp. esp. *P. pubescens*

quinoa, quinua *Chenopodium quinoa*

Quinqueremulus Paul G. Wilson. Compositae (Gnap.-Ang.). 1 W Aus.: ***Q. linearis*** Paul G. Wilson. R: Nuytsia 6(1987)1

quinsyberry *Ribes nigrum*

Quintinia A. DC. Paracryphiaceae (Grossulariaceae s.l.). 25 Philippines (1), New Guinea (12), Aus. (4), Vanuatu (1), New Caled. (6) & NZ (1). Trees & shrubs incl. ***Q. sieberi*** A. DC. (opossum wood, possumwood, Aus.)

Quintiniaceae Doweld = Paracryphiaceae

Quiotania Zarucchi. Apocynaceae (II d). 1 Colombia: ***Q. colombiana*** Zarucchi. R: Novon 1(1991)33

quipo *Cavanillesia platanifolia*

Quipuanthus Michelang. & Ulloa. Melastomataceae. 1 Andes of Ecuador, Peru: ***Q. epipetricus*** Michelang. & Ulloa. R: SB 39(2014)533

Quisqualis L. = *Combretum*

Quisqueya Dod (~ *Broughtonia*). Orchidaceae (V 13b). 4 WI. R: C.L. Withner, *The Cattleyas & their relatives* IV(1996)105

Quisumbingia Merr. = *Sarcolobus*

Quito orange *Solanum quitoense*

Quivisianthe Baill. Meliaceae (II). 2 Madag.

Quoya Gaudich. (~ *Pityrodia*). Labiatae (IV 1). 7 W Aus. R: AusSB 24(2011)7

quticha *Areca catechu*

R

rabbit tobacco *Pseudognaphalium obtusifolium*; **r.-ears** *Thelymitra antennifera*; **r.-eye blueberry** *Vaccinium ashei*; **r.'s foot** *Lagurus ovatus*

Rabdosia (Blume) Hassk. = *Isodon*

Rabdosiella Codd = *Plectranthus*

Rabiea N.E. Br. (~ *Nananthus*). Aizoaceae (V). 6 S Afr. R: H. E. K. Hartmann, *Ill. Handb. Succ. Pls*, Aiz. F–Z(2002)237

Racemobambos Holttum. Gramineae (V 7). Excl. *Neomicrocalamus*, 17 Indomal. Montane forests. R: KB 37(1982)661, 44(1989)364

Rachelia J. Ward & Breitw. Compositae (Gnap.). 1 NZ (S Is.): ***R. glaria*** J. Ward & Breitw. R: NZJB 35(1997)146

Rachicallis DC. = *Arcytophyllum*

Raciborskanthos Szlach. = *Cleisostoma*

Racinaea M. Spencer & L.B. Sm. (~ *Tillandsia*). Bromeliaceae (2). 56 trop. Am. R: Phytol. 74(1993)151, 429

Racosperma Mart. = *Acacia*

Radamaea Benth. Orobanchaceae ('Butt.'; Scrophulariaceae s.l.). 5 Madag. Shrubs

Radcliffea Petra Hoffm. & K. Wurd. Euphorbiaceae (Crot.). 1 Madag.: ***R. smithii*** Petra Hoffm. & K. Wurd. R: KB 61(2006)194

Raddia Bertol. Gramineae (VI 3). 9 E Braz. & Guyana. R: KBAS 13(1986)63. Oft. fern-like; disp. ballistic

Raddiella Swallen. Gramineae (VI 3). 8 trop. Am. R: CUSNH 39(2002)108. V. small bamboos, like dwarf grasses (at 2 cm, ***R. vanessae*** Judz. (Fr. Guiana) world's smallest b.), mostly in spray zone of waterfalls

Radermachera Zoll. & Moritzi. Bignoniaceae (1). 15 SE As., Mal. R: Blumea 23(1976)121. Some locally used timber & cult. orn. trees & shrubs esp. ***R. sinica*** (Hance) Hemsl. (China) – foliage pot-pl. in Eur. (bipinnate lvs)

radiata *Pinus radiata*

radicchio (rosso), radiccio *Cichorium intybus* Radicchio Group

Radinosiphon N.E. Br. Iridaceae (V 5). 2–3 S trop. & S Afr. R: Strel. 10(2000)634

Radiogrammitis Parris (~ *Grammitis*). Polypodiaceae (V). 28 Sri Lanka to Polynesia. R: GBS 58(2007)240

Radiola Hill (~ *Linum*). Linaceae (II). 1 Eur., Medit., temp. As., trop. Afr. mts: ***R. linoides*** Roth (*L. radiola*, allseed)

Radiovittaria (Benedict) E. Crane (~ *Vittaria*). Pteridaceae (V). 9 trop. Am.

radish *Raphanus sativus*; **Chinese** or **Japanese r.** *R. s.* 'Longipinnatus'; **wild r.** *R. raphanistrum*

radium weed (Aus.) *Euphorbia peplus*

Radlkofera Gilg. Sapindaceae. 1 trop. Afr.: ***R. calodendron*** Gilg – Corner's Model

Radlkoferotoma Kuntze. Compositae (Eup.-Ager.). 3 Urug., S Braz. R: MSBMBG 22(1987)142. Hairy shrubs with opp. lvs

Radyera Bullock (~ *Hibiscus*). Malvaceae (Malv.-Hib.). 1 S Afr., 1 Aus. R: BG 132(1971)57. Embryo straight, endosperm copious, cf. most M. (M.)

Raffenaldia Godron. Cruciferae (12). 2 Morocco, Algeria

raffia *Raphia farinifera*

Rafflesia R. Br. ex Thomson Rafflesiaceae. 27+ W Mal. (Philippines 10, all endemic – R: Blumea 54(2009)79). R: FM I,13(1997)13, Blumea 44(1999)343. Parasitic (only through wounds, or poss. germinating with host seeds to form a chimaerical dual organism?) on *Tetrastigma* spp. (Vitaceae) with at least some genes also transferred from *T.* to *R.*; stomata abnormal with 3 or more guard-cells; fls carrion-scented, ? poll. carrion-flies (*Lucilia* spp. with life-cycle c. 3–4.5 yrs), though claimed to be inefficient & poss. orig. poll. carrion- or dung-beetles now absent because dung of extinct megaherbivores (poss. disp. agents) absent; attempts at germ. seeds, transplanting pls, etc., unsuccessful. ***R. arnoldii*** R. Br. (*R. titan*, Sumatra) – fl. buds appear 19–21 months before anthesis when fl. to 80 cm diam. (largest fl. known) & 7 kg, now endangered; ***R. manillana*** Teschem. (Philippines) – fl. buds take 1 yr to mature on host (*T. harmandii* Planch.); ***R. patma*** Blume (? *R. horsfieldii* R. Br., Java) – fls 40–50 cm diam., poll. dipterans

Rafflesiaceae Dumort. Magnoliidae – Malpighiales. Excl. Apodanthaceae, Cytinaceae, Mitrastemonaceae, 3/34+ trop. As. to W Mal. Tanniniferous, chlorophyll-less parasites of *Tetrastigma* spp. (Vitaceae), usu. on r., the veg. body ± filamentous like a fungal mycelium through the host, giving rise to infls endogenously. Fls solit. & oft. v. large, fleshy & stinking (some thermogenic), unisexual & reg.; P 5 petaloid, basally connate, A 12–40 adnate to stylar column, extrorse, anthers with term. pores, pollen without pores or ornament, stylar column in large fls apically expanded above anthers to form large disk, G half-inf., 1-loc. with parietal placentas, ovules ∞. Fr. baccate; seeds ∞ with undifferentiated embryo in endosperm. n = 11, 12

Genera: *Rafflesia, Rhizanthes, Sapria*

Re-allied with Passifloraceae, as proposed by Robert Brown (1820). Perhaps the most remarkable of all angiosperms, parasitic on Vitaceae (from which some mitochondrial genes app. received through horizontal gene transfer). Other highly simplified parasites on Cistaceae, Fagaceae, Leguminosae, etc. form. incl. here now known to be unrelated & in diff. orders, showing that this habit has evolved independently several times

Rafinesquia Nutt. Compositae (Cich.-Cich. (Micr.)). 2 SW N Am. R: FNA 19(2006)348

Rafnia Thunb. Leguminosae (III 18). 19 S Afr. esp. SW Cape. R: SAJB 67(2001)109. ***R. acuminata*** (E. Mey.) G. Camb. & B. Wyk ('*R. perfoliata*') – diuretic; ***R. amplexicaulis*** Thunb. – form. liquorice subs.

Ragala Pierre = *Chrysophyllum*

ragged robin *Silene flos-cuculi*

ragi *Eleusine coracana*

ragweed *Ambrosia* & *Artemisia* spp.

ragwort *Senecio* & *Jacobaea* spp. esp. *J. vulgaris*; **Oxford r.** *S. squalidus*

Rahowardiana D'Arcy = *Juanulloa*

rai *Brassica juncea*

Raillardella (A. Gray) Benth. Compositae (Mad.-Mad.). 3 W US. R: FNA 21(2006)256

Railliardia Gaudich. = *Dubautia*

Raillardiopsis Rydb. = *Anisocarpus*

railway creeper *Ipomoea cairica*

Raimondia Saff. = *Annona*

Raimondianthus Harms = *Chaetocalyx*

Raimundochloa A. Molina = *Koeleria*

rain lily *Zephyranthes minima*; **r. tree** *Samanea saman*; **r.bow shower** *Cassia* × *nealiae*

Rainiera Greene (~ *Luina*). Compositae (Sen.-Tuss.). 1 NW US: ***R. stricta*** Greene. R: FNA 20(2006)628

raisin *Vitis vinifera*; **r. tree** *Hovenia dulcis*

rajah cane *Eugeissona minor*

Rajania L. = *Dioscorea*

raki *Pimpinella anisum* (also *Foeniculum vulgare*)

rakkyō *Allium chinense*

rakum *Salacca wallichiana*

Ramatuela Kunth = *Terminalia*

rambai *Baccaurea motleyana* & *B. ramiflora*

rambutan *Nephelium lappaceum*

Ramelia Baill. = *Bocquillonia*

Rameya Baill. = *Triclisia*

ramie *Boehmeria nivea*

ramin *Gonystylus bancanus*

Ramirezella Rose (~ *Oxyrhynchus*). Leguminosae (III 18). 7 Mex., C Am.

Ramisia Glaz. ex Baill. Nyctaginaceae (1). 1 SE Braz.: ***R. brasiliensis*** Oliv.

ramón *Brosimum alicastrum*

Ramonda Rich. Gesneriaceae (III 2f). 3 Pyrenees, Balkans. Fls almost reg., A 5. Cult. orn. rock rosette-pls esp. ***R. myconi*** (L.) Reichb. (Pyrenees) – can survive 2–3 yrs drought & 55°C, dehydration-tolerance due to sucrose & raffinose 'vitrification'

ramontchi *Flacourtia indica*

Ramorinoa Speg. Leguminosae (III 11). 1 W Arg.: ***R. girolae*** Speg. (chica) – wood for furniture & musical instruments, seeds a coffee-subs.

Ramosmania Tirv. & Verdc. (~ *Randia*). Rubiaceae (II 5). 2 Rodrigues: ***R. heterophylla*** (Balf.f.) Tirv. & Verdc. – marked heterophylly (? prot. from tortoise-grazing) last seen 1938 until 1 pl. found 1999, & ***R. rodriguesii*** Tirv. – thought extinct until 1 pl. found 1980, now prod. seed & cuttings prop. Kew & re-introd. R. R: Comptes Rendus Soc. Biogeogr. 65(1989)13

rampion *Campanula rapunculus*, *Phyteuma orbiculare*

ramps *Allium tricoccum*

ramsons *Allium ursinum*

ramtil *Guizotia abyssinica*

Ranalisma Stapf (~ *Echinodorus*). Alismataceae. 1 SE As. & Mal., 1 trop. W Afr. Fls often replaced by veg. buds, infls can take root

Randia Houston ex L. Rubiaceae (II 1). Incl. *Basanacantha*, c. 20 trop. & warm Am. to Florida & Bolivia. Usu. spiny. Some ed. fr. & cult. orn. incl. ***R. formosa*** (Jacq.) K. Schum. OW spp. form. here referred to other genera (see e.g. *Catunaregam*, *Kailarsenia* & *Rothmannia*)

Randonia Cosson (~ *Ochradenus*). Resedaceae. 1 Sahara to Egypt: ***R. africana*** Cosson – camel fodder. R: MLW 67–8(1967)93

Ranevea L. Bailey = *Ravenea*

Rangaeris (Schltr.) Summerh. Orchidaceae (V 16d). 6 trop. & S Afr.

Rangoon creeper *Combretum indicum*; **R. rubber** *Urceola brachysepala*

Ranopisoa J. Leroy (~ *Oftia*). Scrophulariaceae (Teed.). 1 Madag.: ***R. rakotosonii*** (Capuron) J. Leroy

Ranunculaceae Juss. Magnoliidae – Ranunculales. Incl. Glaucidiaceae, Hydrastidaceae & Kingdoniaceae, 56/2100 temp. (esp. N) & boreal. R: APG 41(1990)93, (Am.) Phytol. 70(1991)24. Lianes (*Clematis* spp.), small shrubs (*Xanthorhiza*), & (usu.) herbs, s.t. aquatic, usu. with benzyl-isoquinoline or aporphine alks or var. other toxins. Lvs simple to compound, in spirals (opp. in *Clematis*), oft. with broad bases; stip. minute or 0. Fls usu. bisex. & insect-poll. (wind-poll. in *Thalictrum* etc.), (solit. or) in basically cymose infls, with ± elongate receptacle & P in spirals; K (3–)5–8 or more oft. petaloid & caducous, 'C' (0) few–∞, app. staminodal & oft. with basal nectaries (nectar sucrose-rich), A usu. ∞ in spirals, centripetal (1 or 2 whorls of 5 in *Xanthorhiza*, 6–10 such in *Aquilegia*) with unprolonged connective & anthers with longit. slits, G (1–) ∞ (? 10 000 in *Laccopetalum*), rarely ± connate (e.g. *Nigella*) with axile placentation, each carpel with several – ∞ marginal, or 1 nearly basal pend., anatr. or hemitropous, bitegmic or unitegmic ovules. Fr. usu. follicles, achenes or berries (capsule in *Nigella*); seeds with minute undeveloped embryo or linear & with cotyledons oft. basally connate, s.t. 1 suppressed, in copious oily & proteinaceous (s.t. starchy) endosperm. Germ. epigeal. n = 6–10, 13

Classification & principal genera (modified from Kubitzki):

a. **Hydrastidoideae** (Hydrastidaceae; C 0, G several, 2–4 ovules per carpel, n = 13): *Hydrastis (only)*

b. **Glaucidioideae** (Glaucidiaceae; C 0, G 2, ∞ ovules per carpel, n = 10): *Glaucidium* (only)

[I. Helleboroideae (G with many ovules; 4 tribes) = **e.**:

1. **Helleboreae** (lvs simple to palmate or pedate, fls reg.): *Caltha, Eranthis, Helleborus*
2. **Actaeeae** (Cimicifugeae; lvs ternately compound, fls usu. reg.): *Actaea*
3. **Nigelleae** (lvs pinnately compound, fls reg.): *Nigella*
4. **Delphinieae** (fls zygomorphic, K spurred): *Aconitum, Consolida, Delphinium*]

II. **e. Ranunculoideae** (G with 1(2) ovules; 3 tribes):

1. **Adonideae** (K petaloid, C longer): *Adonis, Trollius*
2. **Anemoneae** (K petaloid, C 0 or small): *Anemone, Clematis, Kingdonia* (poss. to be moved to Circaeasteraceae)
3. **Ranunculeae** (K sepaloid or petaloid, C usu. present): *Ficaria, Ranunculus*

[III. Isopyroideae] (= **d.**; G with 2 to many ovules; 3 tribes):

1. **Isopyreae** (lvs ternately compound): *Aquilegia, Isopyrum*
2. **Dichocarpeae** (lvs usu. pedately compound): *Dichocarpum*
3. [Coptideae] (= **c. Coptoideae;** lvs variously compound, follicles without transverse veins): *Coptis, Xanthorhiza*

IV. **d. Thalictroideae** (C 0, 1 ovule per carpel): *Thalictrum*

Glaucidium back here (from Paeoniaceae)

Many cult. orn. & some med. (also ed. seeds) etc., though many are v. poisonous. Some bad weeds e.g. *Ficaria, Ranunculus* spp.

Ranunculus Tourn. ex L. Ranunculaceae (II 3). c. 600 temp. (excl. *Beckwithia, Coptidium, Ficaria*, Eur. 128), NZ 43) incl. trop. mts & boreal (*Myosurus* poss. here). Buttercups, crowfoot, paigle. Herbs with yellow, white or red fls, all poisonous (when lvs bruised ranunculin (glucoside) hydrolyzed enzymatically to protoanemonin, vesicant leading to lesions, see also *Anemone*) & usu. avoided by stock though harmless in hay. Cult. orn. & weeds, some apomicts (aposospory, pseudogamy, alpine & weedy groups), some aquatics (water crowfoot), esp. ***R. aquatilis*** L. (Eur. & bipolar) with submerged dissected lvs & lobed floating ones, while ***R. fluitans*** Lam. (Eur.) with only dissected ones & ***R. omiophyllus*** Ten. (Eur.) on mud with only lobed ones (***R. flammula*** L. (N temp.) – gene controlling leaf development with 2 alleles: 1 expressed when submerged (v. narrow lvs), the other when aerial (broader). ***R. aconitifolius*** L. (Fair maids of F. (or Kent), C Eur. mts) – cult. orn. with tubers & white fls; ***R. acris*** L. (Euras.) – double-fld forms ('**Flore Pleno**') cult. orn. ('bachelor's buttons'); ***R. asiaticus*** L. (garden ranunculus, SE Eur. & SW As.) – cult. orn. with claw-like tubers & variously coloured, oft. double fls, form. imp. florists' fls first dev. by Turks & 1100 cvs listed for GB by 1777, some 'broken' forms (cf. *Tulipa*) due to up to 10 viruses (incl. cucumber mosaic v.) in 1 pl.; ***R. auricomus*** L. agg. (goldilocks, Euras.) – many lines (prob. 4 sexual diploid spp. in Alps, Carpathians, SE Austria & Appennines, but many polyploids, facultative or obligate apomicts of hybrid orig. (facultative sexual reprod. & recombination imp.) in anthropogenic meadows – 600–800 aposporous or pseudogamous, some introd. N Am.) recog. as spp.

by some authorities; ***R. lingua*** L. (spearwort, Eur., Siberia) – cult. orn. with ± ovate lvs; ***R. lobatus*** Jacquem. ex Camb. (Himal.) – with *Desideria himalayensis* the fl. pl. reaching highest alts (7756 m); ***R. ophioglossifolius*** Villars (Euras., Med., natur. Aus., NZ) – ann. sp., subject of world's smallest nature reserve (0.04 ha) at Badgeworth, Gloucestershire, GB; ***R. polyphyllus*** Waldst. & Kit. ex Willd. (E Eur., W As.) – aquatic with 'Hippuris syndrome'; ***R. repens*** L. (creeping b., Euras., invasive Aus., natur. N Am.) – rain-poll. (rain washes pollen to base of glossy C & capillary action draws it up to stigmas), bad weed but non-acrid & used as 'kenning herb' in Cornwall for eye ulcers (kennings) & for jaundice in Ireland

Ranzania Itô. Berberidaceae (II 2). 1 Jap.: ***R. japonica*** (Maxim.) Itô – cult. orn. herb with nodding purple fls

Raoulia Hook.f. ex Raoul. Compositae (Gnap.-Cass.). Incl. *Psychrophyton*, 23 NZ (Aus. spp. = *Ewartia*). R: OB 104(1991)83. Tufted or creeping herbs & subshrubs; some cult. orn. incl. ***R. australis*** Hook.f. ex Raoul with silver-hairy lvs forming mat. ***R. mammillaris*** Hook.f. (*P. m.*) – forming dense rounded cushions (vegetable sheep), clones to 1000 yrs old

Raouliopsis S.F. Blake. Compositae (Gnap.-Lor.). 2 Colombian Andes. R: OB 104(1991)63

rapadura *Saccharum officinarum*

Rapanea Aubl. = *Myrsine*

Rapatea Aubl. Rapateaceae (II 1). 18 trop. S Am. R: MNYBG 12,3(1965)91

Rapateaceae Dumort. Magnoliidae – Poales. 16/90 trop. S Am. esp. Guayana Highland, with monospecific *Maschalocephalus* in W Afr. Perenn. oft. large rhizomatous herbs oft. accum. aluminium & with vessel-elements throughout stems & r. Lvs distich., basal, with folded open sheath & narrow parallel-veined lamina. Fls reg., 3-merous, bisex., insect-poll. in spikelets of single term. fl. with imbr. bracts in spirals beneath it, the spikelets forming a head or 1-sided raceme oft. with involucre of (1) 2 – several large bracts; K 3, ± basally connate, imbr., C 3 usu. basally connate forming tube with imbr. lobes, yellow (red), nectaries & nectar 0, A 3 + 3 with short filaments oft. connate & adnate to C-tube, anthers with 1, 2 or 4 apical pores or short slits, pollen grains usu. monosulcate, $\underline{G}$ (3), 3-loc. with 1 style & several–∞ (or 1–2 axile – basal) anatr. ovules on axile placentas in each loc. Fr. a septicidal capsule; seeds with small embryo near hilum alongside copious starchy (mostly simple grains) endosperm & s.t. with caruncle or elaiosome. n = 11, 26

Classification & principal genera:

I. **Saxifridericoideae**
  1. Saxofridericieae (several ovules per loc., C yellow; 5 genera): *Saxofridericia, Stegolepis*
  2. Schoenocephalieae (several ovules per loc., C reddish; 3 genera): *Schoenocephalium*

II. **Rapateoideae**
  1. Rapateeae (1 ovule per loc., seeds longit. striate; 4 genera): *Rapatea*
  2. Monotremeae (1 ovule per loc.; seeds not striate; 4 genera): *Maschalocephalus, Monotrema*

Perhaps closest to Xyridaceae

rape *Brassica* spp. esp. *B. napus*; **oilseed r.** *B. napus* (summer races)

Raphanorhyncha Rollins. Cruciferae. 1 Mex.: ***R. crassa*** Rollins. R: CGH 206(1976)10

Raphanus Tourn. ex L. (~ *Brassica*). Cruciferae (12). Excl. *Quidproquo*, 2(–15) W & C Eur. (1). Medit. to C As. Radish. R: JAA 66(1985)328, UBZ 65(2008)816, 66(2009)15. Prob. *B.* hybrids with *B. nigra* as 1 parent. ***R. sativus*** L. (cultigen grown since time of Assyrians & prob. selected from '*R. raphanistrum* L. subsp. *landra* (DC.) Bonnier & Layens' (*R. landra*)) – taproots cherry-sized to 1 m long & 60 cm diam., common garden radish with ed. taproot used in salad, oilseed in anc. Egypt; sprouted seedlings ed. (jaba); some cvs cooked, lvs used like spinach & young lomenta (fr.) pickled – '**Caudatus**' ('Rat's Tail') with lomenta to 30 cm long, pickled, '**Longipinnatus**' (daikon, mooli, mula, muli, Chinese or Jap. r.) with large, long-lasting r. to 50 kg much grown for human (e.g. 'turnip cake' eaten at Chinese New Year) & stock food in E

Raphia P. Beauv. Palmae (I 2b). 20 trop. Am. (1, also in Afr.), Afr. & Madag. [?natur.]. R: J.Nig.Inst.Oil Palm Res. 6(1982)145. Solit. or clustered hapaxanthic stems (Schoute's Model) with fr. covered with prominent shiny scales. Fibre (raffia), cooking-oil & palm wine, those with small fr. a source of beads, petioles a bamboo subs., some cult. orn. (bamboo palm). ***R. farinifera*** (Gaertn.) N. Hylander (*R. ruffia*, raphia palm, trop. Afr., Madag. [natur.?]) – cuticle & hypodermis of young lvs the source of raffia form. imp. hort. tying material also used in basketry etc., older lvs source of raffia wax, now afforded 'special protection' in Zimbabwe; ***R. hookeri*** G. Mann & H. Wendl. (W Afr.) – wine palm, source (with ***R. palma-pinus*** (Gaertn.) Hutch., W Afr.) of Afr. piassava; ***R. regalis*** Becc. (trop. Afr.) – at 25.11 m+, world's longest leaf

Raphidiocystis Hook.f. Cucurbitaceae (XIV). 5 trop. Afr. & Madag. R: BJBB 37(1967)319

Raphiocarpus Chun (~ *Didissandra*). Gesneriaceae (III 2j). 11 Vietnam, S China

Raphiolepis Lindl. = *Rhaphiolepis*

Raphionacme Harv. Apocynaceae (III; Asclepiadaceae I). 37 trop. & S Afr., Arabia (1: ***R. arabica*** A.G. Miller & Biagi – rootstock ed.). R: F. Albers & U. Meve, *Ill. Handb. Succ. Pls*, Asclep. (2002)220. Mostly herbaceous geophytes, some in desert. ***R. utilis*** N.E. Br. & Stapf (Bitinga rubber, S trop. Afr.) – rubber source

Raphistemma Wall. = *Cynanchum*

Rapicactus F. Buxb. & Oehme (~ *Turbinicarpus*). Cactaceae. 5 Mex.

rapini *Brassica rapa* Ruvo Group

Rapistrum Crantz (~ *Brassica*). Cruciferae (12). 2 C Eur. (2), Medit. & W As. R: JAA 66(1985)335

Rapona Baill. Convolvulaceae (?6). 1 Madag.: ***R. tiliifolia*** (Bak.) Verdc. – seeds with irritant hairs

Raputia Aubl. Rutaceae (I 8). Incl. *Achuaria*, excl. *Neoraputia, Raputiarana, Sigmatanthus*, 11 trop. S Am. R: Britt. 46(1994)281. Medic. bark

Raputiarana Emmerich (~ *Raputia*). Rutaceae (I 8). 2 trop. Am.

Raritebe Wernham. Rubiaceae (I Uroph.). 1 trop. Am.: ***R. palicoureoides*** Wernham. R: Britt. 31(1979)304

rasamala *Liquidambar excelsa*

raseita-so *Boehmeria splitgerbera*

rashmato *Coccinia grandis*

Raspalia Brongn. = *Brunia*

raspberry *Rubus idaeus*; **black r.** *R. occidentalis*; **r. jam (tree)** *Acacia acuminata*; **Mauritius r.** *R. rosifolius*; **Mysore r.** *R. niveus*; **purple r.** *R.* × *neglectus*

Rastrophyllum Wild & Pope. Compositae (Vern.-Erl.). 2 Zambia, Tanzania

rata *Metrosideros robusta*

Rathbunia Britton & Rose = *Stenocereus*

rati *Abrus precatorius*

Ratibida Raf. Compositae (Helia.-Rudb.). 7 N Am. R: Rhodora 70(1968)348. Cult. orn. coarse-hairy herbs (cone flowers) esp. ***R. columnifera*** (Nutt.) Wooton & Standl. (W N Am., natur. E N Am.)

rat-ki-rani *Cestrum nocturnum*

rat's tail (cactus) *Disocactus flagelliformis*

rattan Stripped stems of lianoid palms esp. spp. of *Calamus, Daemonorops, Eremospatha, Korthalsia, Laccosperma, Oncocalamus, Plectocomia*

rattle, yellow *Rhinanthus minor* **r.pod** *Crotalaria* spp. (Aus.)

rattlesnakemaster *Eryngium aquaticum*

Rattraya J. Phipps = *Danthoniopsis*

Ratzeburgia Kunth = *Mnesithea*

rau ram *Persicaria odorata*

Rauhia Traub (~ *Eucrosia*). Amaryllidaceae. 2–3 N Peru. R: U. Eggli, *Ill. Handb. Succ. Pls*, Monocots (2001)227. Lge fleshy lvs

Rauhiella Pabst & Braga (~ *Ornithocephalus*). Orchidaceae (V 12h). 3 Braz.

Rauhocereus Backeb. (~ *Weberbauerocereus*. Cactaceae (III 4). 1 Peru: ***R. riosaniensis*** Backeb.

Rauia Nees & Mart. (~ *Angostura*). Rutaceae (I 8). c. 10 Braz., Peru

Raukaua Seemann (~ *Pseudopanax*). Araliaceae. 6 Tasmania (1), NZ (3 – R: NZJB 35(1997)311), S S Am. (2). Heterogeneous. ***R. simplex*** (Forst.f.) A.D. Mitchell & al. (*P. s.*, NZ) – lvs on r.-suckers more juvenile in form, the further from trunk

raulí *Nothofagus alpina*

Raulinoa Cowan. Rutaceae (I 6). 1 Braz.: ***R. echinata*** Cowan. R: FN 33(1982)130

Raulinoreitzia R. King & H. Robinson (~ *Eupatorium*). Compositae (Eup.-Dis.). 3 E S Am. R: MSBMBG 22(1987)74

raupo *Typha orientalis*

Rautanenia Buchenau = *Burnatia*

Rauvolfia Plum. ex L. Apocynaceae (I b). c. 60 trop. (Mal. 9 – R: Blumea 44(1999)449, Afr. 7, Madag. 3 – R: BJBB 61(1991)21). Leeuwenberg's Model; lvs oft. in whorls of 3–5; v. many alks. Cult. orn.; tranquillizing drugs from r.; some inks & dyes from fr. ***R. serpentina*** (L.) Kurz (Indomal.) – source of medic. alks esp. reserpine; ***R. vomitoria*** Afzel. (trop. Afr.) – similar, though this defence chem. overcome by colobine monkeys; reserpine reduces high blood pressure & form. used in treatment of mental illness, but can cause severe depression

Rauwenhoffia R. Scheffer = *Uvaria*

Ravenala Adans. Strelitziaceae. 1 Madag.: ***R. madagascariensis*** Sonn. (traveller's palm, t. tree, invasive Mascarenes) – clump-forming tree (Tomlinson's Model) with palm-like trunks (pith ed., also stockfeed) usu. in sec. forest but widely planted trop. & warm; lvs in 2 ranks; fl. bracts & leaf-bases hold rainwater useful in emergency; fls explosively bird- & lemur-poll.; aril blue, laciniate (seeds ed.)

Ravenea H. Wendl ex Bouché. Palmae (IV 2). 20 Madag., Comoro Is. (2). R: KB 49(1994)628. Some for constr., hat-making, cult. orn. ***R. dransfieldii*** Beentje (Madag.) – lvs for hat-making; ***R. musicalis*** Beentje (Madag. – 1 river) – rheophyte, starts as submerged aquatic (seed germ. in fr.); ***R. rivularis*** Jum. & Perrier (majestic palm, Madag.) – local timber (termite-proof), lvs for hat-making, exploited for drugs, much cult. orn esp. Aus.; ***R. xerophila*** Jum. (Madag.) – water-storing tuberous rs unique in palms

Ravenia Vell. Conc. Rutaceae (I 8). 14 trop. Am. Alks.

Raveniopsis Gleason. Rutaceae (I 8). 19 Guayana Highland to Braz. R: BMPEG 7(1992)146

Ravensara Sonn. = *Cryptocarya*

Ravnia Oersted = *Hillia*

Rawsonia Harv. & Sonder. Achariaceae (Flacourtiaceae). 2 trop. Afr.

Raycadenco Dodson = *Fernandezia* (but see R: Gen.Orch. 5(2009)339)

Rayjacksonia R. Hartman & M. Lane. Compositae (Ast.-Mach.). 3 N Am. R: AJB 83(1996)368

Rayleya Cristóbal. Malvaceae (Bytt.-Bytt.; Sterculiaceae). 1 Bahia: ***R. bahiensis*** Cristóbal

Raynalia Soják = *Alinula*

rayon Usu wood-pulp regenerated cellulose in which substitutes have replaced no more than 15% of the hydroxyl groups

Razafimandimbisonia Kainul. & B. Bremer (~ *Alberta*). Rubiaceae (Alb.). 5 Madag. R: T 58(2009)765

Razisea Oersted. Acanthaceae (III 2c). Incl. *Kolobochilus* 5 C Am.

Rea Bertero ex Decne. = *Dendroseris*

Readea Gillespie = *Eumachia*

Reaumuria Hasselq. ex L. Tamaricaceae. 13 E Medit. (Eur. 1) to C As. & Pakistan. R: BZ 51(1966)1057. Halophytes

Reboudia Cosson & Durieu = *Erucaria*

Rebutia K. Schum. Cactaceae (III 4). excl. *Aylostera, Weingartia,* 12 S Bolivia, NW Arg. R: PJB 43(2011)2776. Cult. orn. small tubercled cacti

Recchia Moçiño & Sessé ex DC. Surianaceae (Simaroubaceae s.l.). 3 Mex. R: Britt. 37(1985)219. Stip.

Rechsteineria Regel = *Sinningia*

Recordia Mold. Verbenaceae (Dur.). 2 Bolivia, Braz. R: SB 38(2013)815

Recordoxylon Ducke. Leguminosae (I 4). 3 Amazonia. R: ABV 21(1984)3. ***R. amazonicum*** (Ducke) Ducke – timber v. hard used for durable constr. work

Rectanthera Degener = *Callisia*

red almond, r. ash (Aus.) *Alphitonia* spp.; **r. ash (Am.)** *Fraxinus pennsylvanica*; **r. bartsia** *Odontites vernus*; **r. bean** *Dysoxylum mollissimum, Phaseolus vulgaris*; **r. beech** *Nothofagus fusca*; **r. birch** *Betula nigra*; **r. boppel nut** *Hicksbeachia pinnatifida*; **r. box** *Lophostemon confertus*; **r. buckeye** *Aesculus pavia*; **r. bud** *Cercis canadensis*; **r. bugles** *Conostylis canescens*; **r. cedar** *Toona ciliata*, (US) *Juniperus virginiana*; **r.c.** (, **western**) *Thuja plicata*; **r. chokeberry** *Aronia arbutifolia*; **r. clover** *Trifolium pratense*; **r.currant** *Ribes rubrum*; **r. deal** *Pinus sylvestris*; **r. dhup** *Parishia insignis*; **r. fir** *Abies magnifica*; **r. gum, Cape York** *Eucalyptus brassiana*; **r. g., forest** *E. tereticornis*; **r. g., Murray** or **river** *E. camaldulensis*; **r. g., Sydney** *Angophora costata*; **r. hot poker** *Kniphofia* spp.; **r. lauan** *Shorea* spp. esp. *S. negrosensis*; **r. louro** *Sextonia rubra*; **r. mahogany** *E. resinifera*; **r. maple** *Acer rubrum*; **r. oak** *Quercus rubra*; **r. pepper(corns)** *Schinus terebinthifolius*; **r. pine** *Dacrydium cupressinum, Pinus* spp. esp. *P. resinosa*; **r. puccoon** *Sanguinaria canadensis*; **r. rice** *Oryza rufipogon*; **r. sandalwood** *Adenanthera pavonina, Pterocarpus santalinus*; **r. siris** *Falcataria toona*; **r. spruce** *Picea rubens*; **r. squill** *Drimia maritima*; **r. top** *Agrostis alba, A. gigantea*

Redfieldia Vasey = *Muhlenbergia*

redleg *Persicaria maculosa*

Redowskia Cham. & Schldl. = *Smelowskia*

redshank *Persicaria maculosa*

redwood *Sequoia sempervirens*, (St Helena) *Trochetiopsis erythroxylon*; **Andaman r.** *Pterocarpus indicus*; **Baltic r.** *Pinus sylvestris*; **Borneo r.** *Shorea* spp.; **Brazilian r.** *Brosimum rubescens*, *Paubrasilia echinata*; **Californian r.** *S. sempervirens*; **Honduras r.** *Erythroxylum affine*; **Indian r.** *Soymida febrifuga*, *Chukrasia tabularis*, *Polyalthia cerasoides*; **WI r.** *Guarea guidonia*; **Zambesi r.** *Baikiaea plurijuga*

reed *Phragmites australis*, (musical instruments) *Arundo donax*; **r. canarygrass** *Phalaris arundinacea*; **Devon** or **wheat r.** *Triticum aestivum*; **r. mace** *Typha* spp.; **Norfolk** or **water r.** *Phragmites australis*

Reederochloa Soderstrom & H. Decker = *Distichlis* (but see SCB 87(1997)38)

Reedia F. Muell. Cyperaceae (II 7). 1 SW Aus.: ***R. spathacea*** F. Muell.

Reediella Pichi-Serm. = *Crepidomanes*

Reedrollinsia J. Walker = *Stenanona*

Reesia Ewart = *Polycarpaea*

Reevesia Lindl. Malvaceae (Hel.; Sterculiaceae). 25 Indomal., C Am. R: NRBGE 15(1926)121

Regelia Schauer = *Melaleuca*

Registaniella Rech.f. Umbelliferae (III 3). 1 Afghanistan: ***R. hapaxlegomena*** Rech.f.

Regnellidium Lindman. Marsileaceae. 1 SE Braz. & adjacent Arg. (form. Euras. etc. so relict now): ***R. diphyllum*** Lindman – like *Marsilea* but latex present & pinnae 2, cult. orn. aquatic

Rehdera Mold. Verbenaceae (Cith.). 3 C Am. R: FR 39(1935)48

Rehderodendron Hu. Styracaceae. 5 Myanmar, Vietnam, S & W China

Rehderophoenix Burret = *Veitchia*

Rehia Fijten (~ *Olyra*). Gramineae (VI 3). 1 trop. NE S Am.: ***R. nervata*** Fijten. R: Blumea 22(1975)416

Rehmannia Libosch. ex Fischer & C. Meyer. Orobanchaceae. 9 E As. R: Taiwania 1(1948)71, Plantsman 8(1987)193. Some cult. orn. form. referred to Gesneriaceae, esp. ***R. elata*** N.E. Br. ('*R. angulata*', Chinese foxglove, China) to 2 m; ***R. glutinosa*** (Gaertn.) Libosch. ex Fischer & C. Meyer (shu di huang, China) – medic. in modern Chinese herbalism incl. 'natural Viagra'

Rehmanniaceae Reveal = Orobanchaceae

Rehsonia Stritch = *Wisteria*

Reichantha Luer = *Masdevallia*

Reichardia Roth. Compositae (Cich.-Cich.(Hyos.)). 8 Medit. (Eur. 4). R: Lagascalia 9(1980)159. ***R. picroides*** Roth – local veg.

Reichea Kausel = *Myrcianthes*

Reicheella Pax. Caryophyllaceae (II 1). 1 Chile: ***R. andicola*** (Philippi) Pax

Reicheia Kausel = *Myrcianthes*

Reichenbachanthus Barb. Rodr. = *Scaphyglottis*

Reichenbachia Spreng. Nyctaginaceae (1). 2 trop. S Am.

Reimaria Humb. & Bonpl. ex Fluegge = *Paspalum*

Reimarochloa A. Hitchc. = *Paspalum* (but see JAA suppl. 1(1991)290)

reindeer bush or **wattle** *Acacia aphylla*

Reineckea Kunth. Asparagaceae (Convallariaceae). 1 China & Jap.: ***R. carnea*** (Andrews) Kunth – cult. orn. like *Ophiopogon*

Reinhardtia Liebm. Palmae (V 7). 6 C Am. R: GH 8(1957)541, AJB 89(2002)1491. Small understorey palms suited to house cult. esp. ***R. gracilis*** (H. Wendl.) Burret (C Am.)

Reinwardtia Dumort. Linaceae (II). 1 mts N Ind. & China: ***R. indica*** Dumort. – glabrous subshrub with tristylous yellow fls., cult. orn.

Reinwardtiodendron Koord. Meliaceae (II -Trich.). 7 Indomal. R: Blumea 31(1985)144

Reissantia Hallé (~ *Hippocratea*). Celastraceae (III). 6 OW trop. ***R. indica*** (Willd.) N. Hallé (Indomal.) – extract for fever

Reissekia Endl. Rhamnaceae (4). 1 Braz.: ***R. smilacina*** (Sm.) Steud.

Reitzia Swallen (~ *Olyra*). Gramineae (VI 3). 1 S Braz.: ***R. smithii*** Swallen. R: Sellowia 7(1956)602

Rejoua Gaudich. = *Tabernaemontana*

Relbunium (Endl.) Hook.f. = *Galium*

Relchela Steud. (~ *Briza*). Gramineae (XVI 5). 1 Arg., Chile: ***R. panicoides*** Steud.

Reldia Wiehler. Gesneriaceae (II 3a). 7 Panamá to N Peru. R: NJB 8(1989)601, Britt. 67(2015)290

Relhania L'Hérit. Compositae (Gnap.-Rel.). 13 S Afr. R: OB 40(1976)

Remijia DC. Rubiaceae (I 1). Excl. *Ciliosemina*, c. 38 trop. S Am. Some with ant-infested hollow stem domatia. Some quinine subs in tonic water

Remirea Aubl. = *Cyperus* (but see KB 38(1983)479)

Remirema Kerr (~ *Operculina*). Convolvulaceae (?12). 1 SE As.: ***R. bracteata*** Kerr. R: HIP 35(1943)t.3435

Remusatia Schott (~ *Colocasia*). Araceae (VII 25). 4 trop. Afr., Himal. to Taiwan. R: ABY 13(1991)113. ***R. vivipara*** (Roxb.) Schott (Afr. to S Arabia, As.) – bulbils in special shoots, the bulbils with hooks which aid animal-disp., rarely fls in As. (fruiting not recorded in Ind. or Java, where pollen sterile), never in Afr. or Dhofar, tubers ed. (toxic untreated), medic.

Remya Hillebrand ex Benth. (~ *Olearia*). Compositae (Ast.-Hin.). 3 Hawaii (Kaui & Maui). R: SB 12(1987)601. ***R. kauiensis*** Hillebrand (Kaui) – long believed extinct but re-found 1983

ren shen *Panax ginseng*

× **Renantanda** Vacherot & Lecoufle. Orchidaceae. *Renanthera* × *Vanda* – 334 grexes

Renanthera Lour. Orchidaceae (V 16d). c. 20 Indomal. Cult. orn. epiphytes with many-fld infls, hybridized with spp. in other genera

Renantherella Ridl. = *Renanthera*

Renata Ruschi = *Pseudolaelia*

Rendlia Chiov. = *Microchloa* (but see Strel. 10(2000)713)

Renealmia L.f. Zingiberaceae (II 4). c. 75 trop. Am. (59; R: FN 18(1977), NRBGE 44(1987)237, 46(1990)315), trop. Afr. Some ant-disp.?; some cult. orn. & locally medic. (piles). ***R. alpinia*** (Rottb.) Maas (Mex.) – ed. (aril), sold in markets, dye from pericarp used on bark cloth

rengas *Gluta renghas*, but also used for many Anacardiaceae in Mal.

Renggeria Meissn. = *Clusia*

Rennellia Korth. (~ *Morinda*). Rubiaceae (Prismat.). Incl. *Didymoecium*, 5 W Mal. R: Blumea 34(1989)3. Wood anatomy suggests diff. tribal placement. Local medic.

Rennera Merxm. Compositae (Anth.-Pen.). 4 Namibia. R: BotJLS 96(1988)308, 129(1999)373

rennet, vegetable *Pinguicula* spp., *Withania coagulans*

Renschia Vatke. Labiatae (V). 1 N Somalia: ***R. heterotypica*** (S. Moore) Vatke. R: BT 70(1975)56

Rensonia S.F. Blake. Compositae (Helia.-Ecl.). 1 C Am.: ***R. salvadorica*** S.F. Blake

Renvoizea Zuloaga & Morrone (~ *Panicum*). Gramineae (XXIII 3). 10 Braz. R: SB 33(2008)294

Renzorchis Szlach. & Olsz. = *Habenaria*

Reptonia A. DC. = *Sideroxylon*

Requienia DC. (~ *Tephrosia*). Leguminosae (III 16). 3 drier trop. & S Afr. R: KB 35(1980)469

rescue grass *Bromus catharticus*

Reseda Tourn. ex L. Resedaceae. 55 Eur. (20), Medit. to C As. R: MLW 78–14(1978). Herbs with alks & source of yellow dyes (luteolin); cult. orn. esp. ***R. lutea*** L. (wild mignonette, Eur., Medit., natur. N Am.), ***R. luteola*** L. (dyer's rocket or weld, Medit., Macaronesia) – yellow dye used since Neolithic esp. for silk & wedding garments by Romans; ***R. odorata*** L. (mignonette, N Afr. [poss. hybrid origin]) – cult. orn. & for ess. oil used in scent-making, though fls scentless to some

Resedaceae Martinov. Magnoliidae – Brassicales. Incl. Borthwickiaceae, Stixidaceae, 11/85 N hemisph. esp. OW. R: Belmontia ns. 8, 26A(1967) B(1978). Herbs & (few) shrubs with 1-cellular hairs & scattered myrosin cells throughout & prod. mustard-oil glucosides. Lvs entire to deeply lobed, in spirals; stip. rep. by glands. Fls ± strongly zygomorphic, bisex., with short androgynophore or gynophore, in spikes or racemes. K (4–)6(–8), valvate or slightly imbr.; C (0, 2)4–8 (usu. 6), unequal, valvate, yellow to white, innermost (upper) large & usu. fringed, outer ones progressively smaller & with fewer appendages, androgynophore oft. with dilated nectary-disk below A 3–50+, anthers with longit. slits; $\underline{G}$ ((2)3–6(7)), 1-loc. but open apically & bearing small stigmas (G free in *Sesamoides* where only 1(2) ovule, united only basally in *Caylusea* where several ovules crowded on short axile placentas) but usu. ovules campylotropous, bitegmic, on parietal placentas. Fr. an open capsule, berry or (*Sesamoides*) distinct radiating carpels; seeds reniform, usu. arillate, with large curved or folded oily embryo & ± 0 endosperm. n = 6–15

Genera: *Caylusea, Ochradenus, Oligomeris, Randonia, Reseda, Sesamoides* (all exc. *S.* occur in Sahara), *Borthwickia, Forchhameria, Neothorelia, Stixis, Tirania* (all form. Capparaceae); Gyrostemonaceae poss. here

The open ovary of *Reseda* & allies unique. Some dyes & cult. orn. (*Reseda*)

reserpine *Rauvolfia serpentina*

Resia H. Moore. Gesneriaceae (II 3b). 2 Colombia, Venez. R: Biollania esp. ed. 6(1997)518

Resnova J. Merwe = *Ledebouria*

Restella Pobed. = *Wikstroemia*

restharrow *Ononis* spp.

Restio Rottb. Restionaceae. Incl. *Calopsis, Ischyrolepis*, c. 160 S (Cape 85 with 82 endemic – R: Strel. 9(2000)210) & trop. Afr., Madag. R: Bothalia 15(1985)437

Restionaceae R. Br. Magnoliidae – Poales. Incl. Anarthriaceae, Centrolepidaceae, excl. Ecdeiocoleaceae, 47/530 S hemisph. See Telopea 12(2009)333. OW esp. Aus. (19 endemic genera in SW) & S Afr. (11 endemic genera & 180 endemic spp. in Cape region) to Vietnam (1 sp.) & Hainan; Chile (1 sp.). Everg. perenn. (rarely ann.) usu. xeromorphic herbs with rhiz. & solid or hollow internodes, the simple or branched stems photosynthetic; vessel-elements generally throughout; silica-bodies. Lvs usu. open sheaths; ligule usu. 0. Fls small, reg., usu. unisexual & pl. dioec., wind-poll. in axils of chaffy bracts in (1–) several – ∞-fld spikelets or much-branched infls s.t. with leafy bracts & bracteoles, spikelets usu. with sheath-like basal spathe, the male & female infls usu. dissimilar, the fls oft. with rudiments of opp. sex; P (0–)3 + (0–)3 scale-like, A (1–)3(4) opp. inner P, the filaments usu. free with anthers usu. 2-sporangiate (rarely 4-) & monothecal (rarely 2-), usu. introrse, with longit. slits & monoporate ± graminoid pollen, $\underline{G}$ ((1–)3), (1–) 3-loc., the styles s.t. basally connate, with 1 pend., apical-axile, orthotropous, bitegmic ovule per loc. Fr. an achene, small nut or loculicidal capsule; seeds 1–3 with copious starchy endosperm. n = 6–13+

Principal genera: *Calopsis, Centrolepis, Chordifex, Elegia, Lepidobolus, Restio, Thamnochortus*

Form. distinguished Centrolepidaceae (tufted grass- or moss-like monoec. herbs, s.t. ann.) poss. neotenous (Stevens) cf. Lemnaceae in Araceae. *Ecdeiocolea* with 4-sporangiate, 2-thecal anthers split off as segregate family

Taxonomy difficult, the diff. sexes oft. v. diff. in aspect. Typical of nutrient-poor soils & imp. component of veg. of SW Cape, where germ. improved by smoke. Wind-disp. aided by persistent P in *Restio* etc., ant-disp. aided by elaiosomes on nuts of *Cannomois, Hypodiscus* & *Willdenowia* (S Afr.) or on persistent P in some *Restio* spp. Some locally used for thatching, broom-making etc.

Restrepia Kunth. Orchidaceae (V 13c). c. 50 trop. Am. Cult. orn. epiphytes

Restrepiella Garay & Dunsterv. Orchidaceae (V 13c). 2 C Am.: ***R. ophiocephala*** (Lindl.) Garay & Dunsterv. – cult. orn. tufted epiphyte. R: MSB 39(1991)83, Willd. 37(2007) 327

Restrepiopsis Luer. Orchidaceae (V 13). 15 trop. Am. R: MSB 39(1991)87

resurrection plant *Anastatica hierochuntica, Selaginella lepidophylla*

Retama Raf. Leguminosae (III 9). 4 Canary Is. & Medit. (Eur. 3, '*Lygos*') to Middle E. Alks. ***R. monosperma*** (L.) Boiss. (white broom, Medit.) – cult. orn. cutflower with white scented fls; ***R. raetam*** (Forssk.) Webb (Canary Is., Medit.) – retem, the juniper of the Bible, a desert shrub with wood used for charcoal, dangerous med.

retamo wax *Bulnesia retamo*

Retanilla (DC.) Brongn. Rhamnaceae (2). 4 Peru, Chile. R: Darw. 31(1992)232. *Frankia* symbionts. Locally medic. incl. ***R. trinervia*** (Hook.) Hook. & Arn. (*Trevoa t.*, Chile) – bark used on burns

retem *Retama raetam*

Retiniphyllum Bonpl. Rubiaceae (III 1). 20 trop. S Am. Locally medic.

Retinispora Sieb. & Zucc. = *Chamaecyparis*. Name form. used for many scale-leaved conifers in the juvenile needle-leaved stage

Retispatha Dransf. = *Calamus* (but see KB 34(1979)529)

Retrophyllum Page (*Decussocarpus*). Podocarpaceae. 2 New Caled., 1 Moluccas to Fiji, 2 S Am. R: NRBGE 45(1988)379

Retzia Thunb. Stilbaceae (Stilb.; Retziaceae). 1 W Cape mts: ***R. capensis*** Thunb. R: Strel. 10(2000)542

Retziaceae Choisy = Stilbaceae

Reussia Endl. (~ *Pontederia*). Pontederiaceae. 2 C Am., E Arg.

Reutealis Airy Shaw (~ *Aleurites*). Euphorbiaceae (Crot.-Al.). 1 Philippines: ***R. trisperma*** (Blanco) Airy Shaw – seed-oil form. soap source. R: Blumea 44(1999)85

Reutera Boiss. = *Pimpinella*

Revealia R. King & H. Robinson = *Carphochaete*

Reverchonia A. Gray = *Phyllanthus*

Revwattsia D.L. Jones = *Dryopteris* (but see FA 48(1998)401)

rewa-rewa *Knightia excelsa*

Reyemia Hilliard = *Zaluzianskya* (But see EJB 49(1993)291)

Reyesia C. Gay (~ *Salpiglossis*). Solanaceae (II 3). 4 N Chile. R: HBML 27(1980)24

Reynaudia Kunth. Gramineae (Panicoid.). 1 Cuba, Hispaniola: ***R. filiformis*** (Schultes) Kunth

Reynoldsia A. Gray. Araliaceae. 3 Samoa, 2 Marquesas, 2 Society Is., 1 Hawaii

Reynosia Griseb. Rhamnaceae (6). c. 15 Florida (1), C Am. (1), WI. Endosperm ruminate. Some timber & ed. fr.

Reynoutria Houtt. (~ *Fallopia*). Polygonaceae (II 4). 6 temp. OW. ***R. japonica*** Houtt. (*F. japonica*, *Polygonum cuspidatum*, Jap. knotweed, Jap., invasive in Eur., NZ, N Am.) – cult. orn. with bamboo-like stems to 3 m, locally medic., cattle fodder, young shoots a veg. or treated like rhubarb, pulp for moderate quality handmade paper, 'growpots' & hanging-basket liners, rhiz. to 20 m long source of yellow dye, oft. becoming aggressive weed in orn. plantations, introd. GB by 1850 (single male-sterile clone, now the biggest female on earth? By 2003 control estimated at £1.5 bn; now using *Aphalara itador*, a Japanese psyllid – first EU sanctioned biol. control) from von Siebold's Nursery (Leiden, Netherlands; once neglected, nursery over-run by it) & wild by 1886, but in Br. Is. all pls are octoploid & almost all seed set is of hybrid orig. involving *F. baldschuanica* giving **× *Reyllopia conollyana*** (J.P. Bailey) Galasso first found in Wales (1983) or ***Reynoutria sachalinensis*** (Maxim.) Nakai (E As.) introd. 1869 giving ***R.* × *bohemica*** Chřtek & Chřtova (? *P.* × *cookii*); ***R. multiflora*** (Thunb.) Moldenke (fo-ti, China) – rhiz. medic. (comm. Vietnam for weakness after childbirth) but after long dosage causing hair incl. beard to grow black

Rhabdadenia Muell-Arg. Apocynaceae (II e). 3 trop. Am. R: JBRIT 3(2009)547. Alks

Rhabdocaulon (Benth.) Epling. Labiatae (VII 2b). 7 trop. S Am. R: FRB 115(1939)9

Rhabdodendraceae Prance. Magnoliidae – Caryophyllales. 1/3 trop. S Am. Shrubs oft. with unusual sec. growth (successive cambia like some other C.), parenchyma with resin-filled scattered secretory cavities. Lvs simple, large, revolute, leathery, in spirals, with peltate hairs & glandular dots; stip. minute (? petiole base) or 0. Fls reg., usu. bisex., in racemes or racemoid cymes; K ± entire or with 5 imbr. lobes, C 5 sepal-like, A 25–50 in ± 3 whorls with short flat filaments & anthers with longit. slits, disk 0, $\underline{G}$ 1 with long almost basal style & 1 (s.t. another aborted) basal campylotropous (?) unitegmic ovule. Fr. a drupe with somewhat woody endocarp, surrounded by persistent K & swollen pedicel apex; seed with 2 fleshy cotyledons & 0 endosperm. n = 10

Genus: *Rhabdodendron*

Form. in Rosales (or Rutales) but gene sequences show similarities with Caryophyllales (though seed & sieve-tube plastid char. aberrant)

Rhabdodendron Gilg & Pilg. Rhabdodendraceae. 3 trop. S Am. R: FN 11(1972)3

Rhabdophyllum Tieghem (~ *Ouratea*). Ochnaceae. 8 trop. Afr. R: Adans. 30(2008)121

Rhabdosciadium Boiss. Umbelliferae (III 2). 5 Eur. to W As.

Rhabdothamnopsis Hemsl. Gesneriaceae (III 2i). 1 S China: ***R. sinensis*** Hemsl.

Rhabdothamnus Cunn. Gesneriaceae (II 4a). 1 NZ (N Is.): ***R. solandri*** Cunn.

Rhabdotosperma Hartl (~ *Verbascum*). Scrophulariaceae (Scoph.). 7 Arabia, trop. Afr. mts. R: BBP 53(1977)57

Rhachicallis Spach = *Arcytophyllum*

Rhachidosoraceae X.C. Zhang (~ Athyriaceae). 1/8 As. R: T 61(2012)520. Terr. with rhiz. rarely branched; rhiz. scales clathrate. Lvs monomorphic, in spirals, 2- or 3-pinnate-pinnatifid; petioles with 2 vasc. bundles, veins free, term. before margin. Sori dorsal along veins, elongate; indusia lateral. x = 41. Genus: *Rhachidisorus*

Rhachidosorus Ching (~ *Diplazium*). Rhachidosoraceae. 8 As., esp. limestone

Rhacodiscus Lindau = *Justicia*

Rhadamanthopsis (Oberm.) Speta = *Drimia*

Rhadamanthus Salisb. = *Drimia*

Rhadinopus S. Moore. Rubiaceae (II 5). 2 NG. R: Blumea 25(1979)297

Rhadinothamnus Paul G. Wilson (~ *Nematolepis*). Rutaceae (I 3). 3 SW Aus. R: Nuytsia 12(1998)286

Rhaesteria Summerh. Orchidaceae (V 16d). 1 Uganda: ***R. eggelingii*** Summerh. – coll. only once

Rhagadiolus Juss. (~ *Crepis*). Compositae (Cich.-Cich.(Crep.)). 2 Medit. (Eur. 1) to Iran. R: T 28(1979)133. Fr. linear encased in involucral bract, pappus 0

Rhagodia R. Br. = *Chenopodium* (but see FA 4(1984)164)

Rhammatophyllum O. Schulz (~ *Erysimum*). Cruciferae (26). 8 C As. R: Novon 12 (2002)2

Rhamnaceae Juss. Magnoliidae – Rosales. 63/1050 cosmop. esp. trop. & warm. R: KB 54(2005)628. Trees & shrubs, oft. armed, s.t. lianoid (hooks in *Ventilago*, tendrils in *Gouania*), rarely herbs (*Crumenaria*), oft. with anthraquinone glycosides & alks; some with nitrogen-fixing actinobacteria in r. (e.g. *Ceanothus, Colletia, Discaria, Trevoa*). Lvs simple (s.t. much reduced & stems photosynthetic), toothed, in spirals or opp., pinnately veined or with 3–5 main veins from base; stip. small or spiny or 0. Fls usu. small & unisexual, (4 or) 5-merous with hypanthium resembling K-tube to epigyny with disk adnate to G & hypanthium, which is oft. circumscissile & decid. above middle, the fls in cymes, thyrses or fascicles (–solit.); K lobes valvate, C oft. hooded & concave holding anthers or 0 (rare), A alt. with K, filaments adnate to C, anthers minute with longit. slits & pollen grains oft. ± triangular, disk within A, oft. adnate to hypanthium & s.t. G ((1)2–4), ± pluriloc. with term. deeply cleft style, each loc. with 1 (2 in *Karwinskia*) anatropous, bitegmic ovule. Fr. a drupe with sep. stones or 1 pluriloc. one, or dry dehiscent or separating into mericarps, etc.; seeds exotestal, s.t. with dorsal groove, embryo usu. straight, large & oily with little or 0 (ruminate in *Reynosia*) endosperm. n = (9–)12(13, 23)

Classification & principal genera (some incl. *Ceanothus, Colubrina* inc. sed.):

1. **Paliureae** (s.t. arm ed; lvs in spirals): *Hovenia, Paliurus, Ziziphus*
2. **Colletieae** (usu. armed; lvs opp.: R: NZJB 43(2005)868): *Colletia, Discaria*
3. **Phyliceae** (unarmed; lvs in spirals or opp.; fr. splitting into 3 1-seeded portions): *Phylica*
4. **Gouanieae** (climbers or herbs with tendrils; $\overline{G}$): *Gouania*
5. **Pomaderreae** (s.t. armed; indumentum stellate; fr. a schizocarp, seeds arillate; R: T 54(2005)628): *Cryptandra, Pomaderris, Spyridium*
6. **Rhamneae** (usu. unarmed; drupes): *Frangula, Rhamnus, Sageretia*
7. **Maesopsideae** (unarmed large trees; drupes): *Maesopsis* (only)
8. **Ventilagineae** (unarmed trees or tendril-less climbers; fr. a samara or capsule): *Smythea, Ventilago*
9. **Ampelozizypheae** (tendril-less climber; fr. an explosive capsule): *Ampelozizyphus* (only)
10. **Doerpfeldieae** (unarmed tree; drupe): *Doerpfeldia*
11. **Bathiorhamneae** (unarmed tree; fr. of 3 indehiscent mericarps): *Bathiorhamnus* (only)

Medic. (*Colubrina, Frangula* (cascara), *Hovenia, Karwinskia, Krugiodendron, Retanilla, Rhamnus*) etc., soap (*Gouania*), dyes (*Alphitonia, Rhamnus, Ventilago* – also oilseed), timber (*Alphitonia, Emmenosperma, Karwinskia, Maesopsis, Reynosia, Ziziphus*), ed. fr. (*Berchemia, Hovenia, Ziziphus*) & cult. orn. esp. *Ceanothus, Colletia, Pommaderris*

Rhamnella Miq. Rhamnaceae (6). 10 W Himal. to Jap., Fiji, trop. Aus. & New Caled. (1)

Rhamnidium Reissek. Rhamnaceae (6). 12 trop. Am.

Rhamnoneuron Gilg. Thymelaeaceae (Thym.-Daph.). 2 SE As.

Rhamnus Tourn. ex L. Rhamnaceae (6). Excl. *Atadinus, Endotropis, Frangula* c. 55 N hemisph. to Braz. & S Afr. Buckthorns; R: TBIANSSSR I, 8(1949)243. Usu. decid. trees (to 50 m; ***R. glandulosa*** Ait. part of laurel forest of Madeira) & shrubs with lvs in spirals or opp. & 4-merous polygamous & dioec. (usu.; cf. 5-merous bisex. usu. in *Frangula*) fls. Dyes (fr. gives blue-green, bark yellow) & medic. (purgatives); some timbers for turnery & cult. orn. ***R. cathartica*** L. (common, purging or European b., Euras., invasive C US) – drupes (Rhine berries) medic. & source of artist's pigment 'sap green'; ***R. davurica*** Pallas (Jap., Korea) & ***R. utilis*** Decne. (C & E China) – sources of Chinese green indigo used in dyeing silk; ***R. erythroxyloides*** Hoffsgg. (*R. pallasii*, Pallas's b., W As.) – cult. orn.; ***R. prinoides*** L'Hérit. (Ethiopia to S Afr.) – flavour in tej (Ethiopia) & in some beers instead of hops; ***R. saxatilis*** Jacq. (*R. infectoria*, F. or Avignon or Persian or yellow berry, S & SC Eur.) – form. imp. source of yellow dye

Rhamphicarpa Benth. Orobanchaceae (Buch.; Scrophulariaceae s.l.). 6 Russia, Turkey, trop. & S Afr., Ind., trop. Aus. R: BT 70(1975)103

Rhamphogyne S. Moore. Compositae (Ast.-Lag.). 2 Rodrigues & NG

Rhamphorhynchus Garay. Orchidaceae (IV 2d). 1 Braz.: ***R. mendoncae*** (Brade & Pabst) Garay. R: Gen. Orch. 3(2003)141

Rhanteriopsis Rauschert (*Postia*). Compositae (Inul.-Inul.). 2 Lebanon & Syria, 2 Iran & Iraq. R: BotJLS 95(1987)27

Rhanterium Desf. Compositae (Inul.-Inul.). 3 NW Afr. to Pakistan. R: BotJLS 93(1986)231

Rhaphidophora Hassk. Araceae (IV 4). c. 120 trop. Afr. to Pacific. Lvs s.t. perforate; fls bisex. ***R. beccarii*** Engl. (Borneo) – rheophyte. Some cult. orn. & locally medic. (e.g. ***R. decursiva*** (Roxb.) Schott (Himal. to Vietnam) antimalarial), long r. for cordage. ***R. pertusa*** (Roxb.) Schott (Ind., Sri Lanka) – veg. indistinguishable from juvenile or small cvs of *Monstera deliciosa*

Rhaphidophyton Iljin. Amaranthaceae (Chenopodiaceae III 3). 1 C As.: ***R. regelii*** (Bunge) Iljin

Rhaphidospora Nees = *Justicia*

Rhaphidura Bremek. Rubiaceae (I 9). 1 Borneo: ***R. lowii*** (Ridl.) Bremek.

Rhaphiodon Schauer (~ *Hyptis*). Labiatae (VII 3c). 1 NE Braz.: ***R. echinus*** (Nees & Mart.) Schauer. R: FRB 85(1936)186

Rhaphiolepis Lindl. Rosaceae (16). 9 E & SE As. R: CJB 68(1990)2240. Everg. shrubs incl. some rheophytes & cult. orn. with white or pink fls esp. ***R. indica*** (L.) Lindl. (Ind. hawthorn, S China), ***R. umbellata*** (Thunb.) Makino (Korea, Jap.) – bark a source of brown dye, & their hybrids = ***R. × delacourii*** André. Hybrids with *Eriobotrya* = ***× Rhaphiobotrya*** Coombes

Rhaphiostylis Planch. ex Benth. Metteniusaceae (Icacinaceae). 10 trop. Afr. Mangenot's Model

Rhaphispermum Benth. Orobanchaceae ('Butt.'; Scrophulariaceae s.l.). 1 C Madag.: ***R. gerardioides*** Benth.

Rhaphithamnus Miers. Verbenaceae (Cith.). 2 Chile, Arg., Juan Fernandez. Cult. orn. trees & shrubs, ***R. spinosus*** (Juss.) Moldenke (mainland) – herm.; ***R. venustus*** Robinson (Juan Fernandez) – gynodioec.

Rhapidophyllum H. Wendl. & Drude. Palmae (III 4a). 1 SE US: ***R. hystrix*** (Thouin) H. Wendl. & Drude – low polygamo-dioec. or dioec. (rarely monoec.), weevil-poll. palm with needle-like appendages on leaf-sheaths & smelly hairy fr. taken by black bears, v. hardy

Rhapis L.f. Palmae (III 4a). 11 S China to W Mal. R: Phytotaxa 258(2016)137. Cult. orn. tub-pls esp. ***R. excelsa*** (Thunb.) A. Henry (*R. flabelliformis*, Jap. peace palm, lady p., S China to N Vietnam) – culture a cult in Jap. on a par with tulipomania, esp. varieg. forms today, purchased as hedge against inflation (kan-non-chiku)

Rhaponticoides Vaill. (~ *Centaurea*). Compositae (Card.-Cent.). 17 Euras., Medit.

Rhaponticum Ludw. Compositae (Card.-Cent.). Incl. *Acroptilon*, *Leuzea*, *Stemmacantha* (R: Cand. 39(1984)45), c. 20 Medit., As., Aus. (1). ***R. carthamoides*** (Willd.) Iljin (*S. c.*, Siberia) – ecdysteroids; ***R. repens*** (L.) Hidalgo (*A. r.*, Russian knapweed, SW Eur. to C As.) – noxious allelopathic weed introd. N Am., Aus.

Rhaptonema Miers. Menispermaceae (V). 4 Madag.

Rhaptopetalum Oliv. Lecythidaceae (? V; Scytopetalaceae II). 11 trop. W Afr. Troll's Model

rhatany (root) *Krameria* spp.

Rhazya Decne. Apocynaceae (I g). 1 SE Eur. to Pakistan & Arabia: ***R. stricta*** Decne. (*R. orientalis*, *Amsonia o.*) – alks, form. most imp. desert medic. in Arabia, source of lacquer, wax from fr. used in candles, cult. orn. perenn. with blue to lilac fls

rhea *Boehmeria nivea* 'var. *tenacissima*'

Rheedia L. = *Garcinia*

Rhektophyllum N.E. Br. = *Cercestis*

Rheochloa Filgueiras & al. Gramineae (XXIX 5). 1 Braz.: ***R. scabriflora*** Filgueiras & al. R: SB 24(1999)123

Rheome Goldbl. = *Moraea*

Rheopteris Alston. Pteridaceae (V; Vittariaceae). 1 NG: ***R. cheesemaniae*** Alston

Rhetinantha M. Blanco = *Maxillaria* (but see Lankesteriana 7(2007)534)

Rhetinocarpha Paul G. Wilson & M.A. Wilson. Compositae (Gnap.). 1 W Aus.: ***R. suffruticosa*** (Benth.) Paul G. Wilson & M.A. Wilson. R: Nuytsia 16(2006)255

Rhetinodendron Meissn. = *Senecio* (*Robinsonia*)

Rhetinolepis Cosson. Compositae (Anth.-Sant.). 1 Algeria, Tunisia, Libya: ***R. lonadioides*** Cosson. R: BBMNHB 23(1993)130

Rhetinosperma Radlk. = *Chisocheton*

Rheum L. Polygonaceae (II 3). c. 60 Eur. (2), temp. & warm As. Rhubarb; R: TBIANSSSR I, 3(1936)67; C.M. Foust (1992) *Rhubarb: the wondrous drug*. Coarse herbs with large lvs (with anthroquinones & oxalates) & entomophilous fls (cf. *Rumex* but large stigmas like anemophilous (? ancestors)). Medic., ed. & cult. orn. ***R. alexandrae*** Batalin (3400–4500 m, W China) – chlorophyll-less bracts prot. infl. more efficient at absorbing ultraviolet radiation than are rosette-lvs; ***R. australe*** D. Don (*R. emodi*, Himal. or Ind. rhubarb, Himal.) – drug (r.); ***R. nobile*** Hook.f. & Thomson (Nepal to SE Tibet) – infls densely covered with bracts 15 cm across; ***R. officinale*** Baill. (W China & Tibet) – r. & rhiz. medic. (purgative); ***R. palmatum*** L. (Chinese or Turkish r., NW China) – medic. (e.g. dysentery in 1920s); ***R. rhabarbarum*** L. (*R.* × *hybridum*, ? orig.) – 2n = 44 & ***R. rhaponticum*** L. (? *R.* × *cultorum*, 'Eur.' r., ? orig.) – 2n = 22, petioles (garden rhubarb) with malic acid stewed as pudding, used in jam & wine; ***R. ribes*** L. (SW As.) – stalks taste like currants

Rhexia Gronov. Melastomataceae. 13 N Am. R: Sida 3(1969)397, Phytoneuron 2012–15:6. Cult. orn. herbs incl. ***R. virginica*** L. – lvs a source of a tea for N Am. Indians in E Canada

Rhigiocarya Miers. Menispermaceae (III). 3 trop. W Afr.

Rhigiophyllum Hochst. Campanulaceae (I 2). 1 W Cape: ***R. squarrosum*** Hochst. R: Strel. 9(2000)395

Rhigospira Miers (~ *Tabernaemontana*). Apocynaceae (I d). 1 N S Am. to Andes. R: AUWP 87–1(1987)

Rhigozum Burchell. Bignoniaceae (1). 7 NE (1) & S (5) Afr., Madag. (1). Much-branched spiny shrubs

Rhinacanthus Nees. Acanthaceae (III 2c). c. 25 trop. OW. Medic. (skin diseases)

Rhinactinidia Novopokr. = *Aster*

Rhinanthaceae Vent. = Orobanchaceae

Rhinanthus L. Orobanchaceae (Rhin.; Scrophulariaceae s.l.). 45 N hemisph. esp. Eur. (25 incl. many ecotypes). Hemiparasitic ann. herbs esp. ***R. minor*** L. (yellow or hay rattle, Eur., E N Am.)

Rhine berry *Rhamnus cathartica*

Rhinephyllum N.E. Br. Aizoaceae (V). 10 Cape. R: H. E. K. Hartmann, *Ill. Handb. Succ. Pls*, Aiz. F–Z(2002)239. Cult. orn.

Rhinerrhiza Rupp. Orchidaceae (V 16c). 1 Aus., NG, Solomon Is.: ***R. divitiflora*** (Benth.) Rupp

Rhinerrhizopsis Ormerod (~ *Thrixspermum*). Orchidaceae (V 16c). 3 Indomal.

Rhinocerotidium Szlach. = *Gomesa* (but see PJB 51(2006)40)

rhinoceros bush *Elytropappus rhinocerotis*

Rhinocidium Baptista = *Gomesa*

Rhinopetalum Fischer ex D. Don = *Fritillaria*

Rhinopterys Niedenzu = *Acridocarpus*

Rhinorchis Szlach. = *Habenaria*

Rhinotropis (S.F. Blake) J. Abbott (~ *Polygala*). Polygalaceae. 17 SW US to Guatemala (1). R: JBRIT 5(2011)134

Rhipidantha Bremek. Rubiaceae (I 10). 1 Uluguru Mts, Tanzania: ***R. chlorantha*** (K. Schum.) Bremek.

Rhipidia Markgraf = *Condylocarpon*

Rhipidocladum McClure. Gramineae (V 2). 15 trop. Am. R: CUSNH 39(2000)109. Bamboos. ***R. racemiflorum*** (Steud.) McClure (Mex.) – furniture & handicrafts

Rhipidoglossum Schltr. (~ *Diaphananthe*). Orchidaceae (V 16d). c. 35 trop. Afr.

Rhipogonum Spreng. = *Ripogonum*

Rhipsalidopsis Britton & Rose = *Hatiora*

Rhipsalis Gaertn. Cactaceae (III 6). 35 trop. Am. (esp. Braz.), 1 ext to Afr., Madag. & Sri Lanka. R: SB 37(2012)987. Epiphytic or saxicolous jointed cacti with cylindrical, angled or flattened joints & spherical translucent mucilaginous fr. Cult. orn., specific limits unclear (mistletoe cactus), incl. ***R. baccifera*** (Sol.) Stearn (*R. cassutha*, *R. cassytha*, trop. Am. to Sri Lanka); ***R. juengeri*** Barthlott & N.P. Taylor (E Braz.) – forming 3-m-long 'curtains', with bat-disp. (other spp. by birds), black-currant-scented fr.

Rhizanthella R. Rogers. Orchidaceae (IV 3h). 2 SW & E Aus. R: Gen. Orch. 2(2001)193. ***R. gardneri*** R. Rogers almost completely subterr., known from 5 localities always in stands of *Melaleuca uncinata* R. Br., infl. growing up to soil surface with bracts pushing away soil to expose fls poss. poll. by tiny flies or termites, infr. emerging & seeds disp. by wind

Rhizanthes Dumort. Rafflesiaceae. 4 Thailand, W Mal. R: NHBSS 48(2000)121. Parasites of *Tetrastigma* spp., mimicking carrion brood-sites & poll. blowflies (but see *Rafflesia*), fls purple, sour-smelling, emetic. ***R. infanticida*** Bänziger & Hansen (Thailand) – column smelling like 'stuffy room', caudate appendages like excrement/cheese

Rhizobotrya Tausch. Cruciferae (34). 1 W Dolomites: ***R. alpina*** Tausch – cult. orn. rock-pl.

Rhizocephalum Wedd. = *Lysipomia*

Rhizocephalus Boiss. Gramineae (XVI 15). 1 Medit. to Iran: ***R. orientalis*** Boiss.

Rhizoglossum C. Presl = *Ophioglossum*

Rhizomatopteris A.P. Khokhr. = *Cystopteris*

Rhizomonanthes Danser = *Amyema*

Rhizophora L. Rhizophoraceae (III). 6 trop. coasts (Aus. 4) + 2 nat. hybrids. Mangrove trees (Attims's Model) with arching aerial r. from stem & branches, their subaerial parts with conspic. lenticels; app. moving from animal- to wind-poll.; seeds germ. in fr. & prod. elongate radicle before being dropped, those of ***R. apiculata*** Blume (E Afr., Ind. to W Pacific, local) surviving up to 89 days in seawater, those of *R. mucronata* 150 days. ***R. mangle*** L. (SW Pac. & NW) – ? distrib. achieved in Eocene. Bark used in tanning, timber, esp. from ***R. mucronata*** Lam. (boriti poles, OW) for building but also for pulp & rayon in Mal.

Rhizophoraceae Pers. Magnoliidae – Malpighiales. 15/125 trop. esp. OW. R: AMBG 75(1988)1278. Trees & shrubs, oft. mangroves s.t. with aerial roots, at least s.t. with alks. Lvs simple, oft. serrate, opp. or verticillate; stip. interpetiolar, caducous, sheathing term. bud. Fls reg., 4- or 5-merous, s.t. unisexual, s.t. with elongated hypanthium, in axillary infls or solit. pedicels articulated; K (3)4, 5(–16), valvate, thick, C alt. with K, oft. fleshy & shorter than K, convolute or infolded in bud, A 2(–4) × K or ∞, oft. paired opp. C in single whorl usu. each petal enwrapping 1–5 A, filaments oft. basally connate, attached around or on to disk or disk 0 or anthers sessile, G ((2)3–5(-20)) sup. to inf., with as many loc. or partitions aborting to give 1-loc., with term. style & 2(-6) pend., anatropous, bitegmic ovules with zig-zag micropyles per loc. Fr. a berry with 1 seed (or 1 per loc.), rarely a capsule; seeds s.t. arillate or winged with straight, linear, oft. green (viviparous in mangroves) embryo in copious oily endosperm. n = (13) 14, 16, 18, 21

Principal genera: *Bruguiera, Carallia, Cassipourea, Ceriops, Crossostylis, Pellacalyx, Rhizophora*

Embryologically heterogeneous, this & many other features against trad. position in Myrtales; seed-structure suggests segregation of *Gynotroches* & *Pellacalyx* as Legnotidaceae

Timber, charcoal & tanbarks

Rhodalsine Gay (*Psammanthe*; ~ *Minuartia*). Caryophyllaceae. 1 S Medit. to Canary Is. & Somalia: ***R. geniculata*** (Poir.) F.N. Williams. R: T 63(2014)79

Rhodamnia Jack. Myrtaceae (II 10). c. 28 Indomal. to W Pac. (Aus. 13 endemic – R: SBM 82(2007)16). R: KB 33(1979)429. ***R. cinerea*** Jack (SE As., W Mal.) – timber & charcoal, wood for wayang golek puppets in Java

Rhodanthe Lindl. ('*Helipterum*'). Compositae (Gnap.-Ang.). 46 Aus. (36 W Aus., 24 endemic). R: Nuytsia 8(1992)383. Usu. annuals (poss. 6 sep. lineages, 1 assoc. with *Podotheca*). Cult. orn. 'everlastings' esp. ***R. manglesii*** Lindl. (*H. m.*, W Aus.) – heads pink

Rhodanthemum (Vogt) Wilcox & al. (~ *Chrysanthemum*). Compositae (Anth.-Leuc.). 14 Spain (1), Morocco, Algeria. R: BBMNHB 23(1993)141. Some cult. orn. incl. ***R. hosmariense*** (Ball) Wilcox & al. (*C. h.*, '*Pyrethropsis h.*', Moroccan daisy, Morocco)

Rhodax Spach = *Helianthemum*

Rhodiola L. See *Sedum*

Rhodobryum (Schimp.) Hampe. Bryaceae (Musci). 25 trop. to N temp. (2) & S Am. Local medic. incl. ***R. giganteum*** (Schwaegr.) Paris (huixincao, Madag. to Mal.) – trad. cure for angina (China), ether extacts increasing blood flow in mouse aorta by 38%, cult bowl-pl. in Jap.

Rhodocactus (A. Berger) F.M. Knuth = *Pereskia*

Rhodocalyx Muell. Arg. = *Prestonia*

Rhodochiton Zucc. ex Otto & Dietr. Plantaginaceae (Ant.-Ant.; Scrophulariaceae s.l.). 3 Mex., C Am. R: D.A. Sutton, *Rev. Antirrhineae* (1988)505. ***R. atrosanguineum*** (Zucc.) Rothm. (Mex.) – twiner with sensitive petioles & peduncles, K persistent

Rhodocodon Bak. = *Drimia* (but see Phytotaxa 195(2015)101)

Rhodocolea Baill. Bignoniaceae (6). 10–14 Madag. R: FMad. 178 (1938)54. ***R. perrieri*** Capuron with reg. fls & A5

Rhodocoma Nees (~ *Restio*). Restionaceae. 8 W Cape to Natal. R: PSE 175(1991)154, Bothalia 40(2010)10

Rhododendron L. Ericaceae (V 2). Incl. *Ledum* (marsh rosemary), *Menziesia*, *Tsusiophyllum* (A with slits, 3-carpellate G), c. 1000 temp. N hemisph. (Eur. 7, N Am. 20 – R:

FNA 9(2009)455, China 650, largest pl. genus there) esp. Himal., SE As. & Mal. mts (155 endemics in NG), Aus. (2). R: NRBGE 39(1980)1, (1982)209, 40(1982)225, 44(1986)1, EJB 47(1990)89, 50(1993)249, 52(1995)1, A.C. Leslie (2004) *Int. R. register & checklist*, ed. 2. Everg. to decid. trees & shrubs (pachycaul to leptocaul, Leeuwenberg's & Scarrone's Models) incl. epiphytes, poisonous (emetic cardiac toxins, cyclic diterpenes esp. andromedotoxin (acetylandromedol)), in ruminants causing staggers & death; nectar in some spp. containing antibiotic andromedotoxin as low sugar content otherwise does not prevent bacterial activity, so that excessive eating of rhododendron honey by humans can lower blood pressure, promoting dizziness ('Mad honey poisoning', US) with lvs in term. rosettes & var. coloured fls. Several (some monospecific) subg. recog. but most spp. in 3: subg. ***Rhododendron*** (lepidote r.) – some 500 everg. spp. incl. most Mal. spp., e.g. sect. ***Schistanthe*** Schltr (*Vireya*, cult. esp. Aus.), c. 300 spp. with glandular hairs (scales) & lvs adaxially rolled in bud, polyploidy common, incl. all epiphytes, the most northerly & southerly spp. & all 'blue'-fld spp.; subg. ***Hymenanthes*** (Blume) K. Koch (incl. subg. *Pentanthera* G. Don (subg. *Azalea*) – elepidote & usu. decid., azaleas of commerce), *H. s.s.* with abaxially rolled lvs, unique nodal anatomy & ± winged seeds, i.e. most spp. in cult. incl. ***R. ponticum*** L. (W & E Medit., in Ireland in last interglacial) – introd. from Gibraltar region to GB 1763 & now unintentionally hybridized to form aggressive weed, ***R. × superponticum*** Cullen (*R. p.* × *R. catawbiense* Michaux (E US), *R. maximum* L. (E N Am.) and/or *R. macrophyllum* G. Don (W N Am.)) in forestry plantations on light acid soils (by 1988 cost to clear Snowdonia alone was £30 M); but up to 120 spp. used in hybridizing with over 1000 named cvs in commerce; subg. ***Azaleastrum*** Planch. (incl. *Menziesia*) – like *Hymenanthes* but infls axillary. Rhododendrons (name from the Gk. for similar-leaved oleander) & azaleas of gardens & florists intimately assoc. with an almost unparalleled vainglory & exclusiveness in hort. of the C19 through the concentrated efforts to introd. & hybridize particularly the Asiatic spp. (though first hybrid (Waterer, England, 1811) between Am. spp., *R. catawbiense* Michaux & *R. maximum*) by the wealthy & privileged of W Eur., over 500 spp. having been cult. in GB alone. Decid. hybrids sold as pot-pls (azaleas) for a wider market (by 2005 US trade worth $36M+) incl. crosses between Am. (e.g. ***R. calendulaceum*** (Michaux) Torrey (US) & esp. ***R. occidentale*** A. Gray (W N Am., bringing scent to many Exbury & Knaphill azaleas), Eur. (esp. ***R. luteum*** Sweet (E Eur. to Cauc.) – oft. used as grafting stock) & As. spp., orig. the latter alone & long cult. China & Jap., esp. ***R. mucronatum*** G. Don (?orig.) incl. doubles, ***R. indicum*** (L.) Sweet (S Jap.) – long cult. Jap. where over 200 cvs by 1900, & their hybrid ***R. × pulchrum*** Sweet (raised Jap.), later ***R. × obtusum*** (Lindl.) Planch. (*R. kaempferi* Planch. × *R. kiusianum* Mak. (both Jap.) ? & × *R. indicum*) – kurume azaleas; ***R. simsii*** Planch. (NE Myanmar, China, Taiwan) – orig. of widely cult. 'Ind. azalea' (subg. Tsutsusi) usu. with ± double fls. Notable wild & useful spp. incl. ***R. anthopogon*** D. Don (E Himal.) – lvs & twigs burnt as incense in temples in Sikkim; ***R. arboreum*** Sm. (Himal., China to Thailand, S Ind., Sri Lanka (subsp. ***zeylanicum*** (Booth) Tagg – 1st pl. record for the is. (1343–4!)); ***R. falconeri*** Hook.f. (Bhutan to Nepal) – lvs to 30 cm long, rusty-tomentose abaxially; ***R. ferrugineum*** L. (alpenrose, alpine rose, C Eur.) – form. medic. but dangerous; ***R. groenlandicum*** (Oeder) Kron & Judd (*Ledum g.*, N Am., Greenland) – imp. local medic. for dermatitis etc.; ***R. hirsutum*** L. (Alps) – (cf. most) tolerant of mildly calcareous soil as is **'Inkarho'** rootstock; ***R. molle*** (Blume) G. Don (C & E China) – leaf-extracts effective insect-repellents; ***R. stenopetalum*** (Hogg) Mabb. (*R. macrosepalum*, Jap.) – cult. incl. '**Linearifolium**' mutant (spider azalea), with narrow lvs & distinct petals; ***R. tanakae*** (Maxim.) Ohwi (*Tsusiophyllum t.*, Jap.) – rock-garden shrub; ***R. tomentosum*** Harmaja (*Ledum palustre*, circumpolar) – lvs used as 'Labrador tea' by Native Americans, form. used like mothballs (fends off moths from birch trees nearby); ***R. viscosum*** (L.) Torrey (swamp honeysuckle, E US)

Rhododon Epling. Labiatae (VIII 2). 2 Texas. R: Phytol. 78(1995)448

Rhodogeron Griseb. (~ *Sachsia*). Compositae (Inul.-Pluch.). 1 Cuba: ***R. coronopifolius*** Griseb.

Rhodognaphalon (Ulbr.) Roberty = *Pachira*

Rhodognaphalopsis Robyns = *Pachira*

Rhodohypoxis Nel. Hypoxidaceae. 6 SE Afr. R: NRBGE 36(1978)43, Plantsman 6(1984)53. Cult. orn. rock-pls esp. ***R. baurii*** (Bak.) Nel & incl. hybrids with *Hypoxis* spp. = ***× Rhodoxis*** B. Mathew

Rhodolaena Thouars. Sarcolaenaceae. 7 Madag. R: Adansonia III, 22(2000)241

Rhodoleia Champ. ex Hook. Hamamelidaceae (III). 1–7 S China to Sumatra. Stip. on young shoots; bird-poll.

Rhodoleiaceae Nakai = Hamamelidaceae

Rhodomyrtus (DC.) Reichb. Myrtaceae (II 10.). 25 Indomal. to Aus. (7 endemic) & New Caled. (*Kanakomyrtus*). R: KB 33(1978)311, HPB 15(2010)69. Heterogeneous. Esp. susceptible to myrtle rust introd. from S Am. Some ed. fr. esp. ***R. tomentosa*** (Aiton) Hassk. (SE As., invasive SE US) but also ***R. macrocarpa*** Benth. (finger cherry, Queensland) though alleged to promote blindness

Rhodophiala C. Presl (~ *Hippeastrum*). Amaryllidaceae. c. 30 Chile (most) to Bolivia, Arg., Urug. & Braz.

Rhodopis Urb. Leguminosae (III 18). 2 Hispaniola

Rhodosciadium S. Watson. Umbelliferae (III 10). 14 Mex.

Rhodoscirpus Léveillé-Bourret & al. (~ *Scirpus*). Cyperaceae. 1 S Am.: ***R. asper*** (J. Presl & C. Presl) Léveillé-Bourret & al. R: T 64(2015)940

Rhodosepala Bak. = *Dissotis*

Rhodospatha Poeppig. Araceae (IV 4). c. 70 trop. Am. Cult. orn. climbers

Rhodosphaera Engl. Anacardiaceae (I). 1 NE Aus.: ***R. rhodanthema*** (F. Muell.) Engl. (yellow cedar) – allied to *Rhus*, timber beautifully marked & used in cabinet-making. R: FA 25(1985)183

Rhodostachys Philippi = *Ochagavia*

Rhodostegiella Li = *Cynanchum*

Rhodostemonodaphne Rohwer & Kubitzki (~ *Ocotea*). Lauraceae (I). c. 38 trop. S Am. R: MIABH 20(1986)82

Rhodothamnus Reichb. Ericaceae (V 3). 1 E Alps (***R. chamaecistus*** (L.) Reichb. – cult. orn.) & 1 NE Turkey (stomata on both sides of lvs). Hybrids with *Phyllodoce*

Rhodothyrsus Esser = *Senefeldera* (but see Britt. 51(1999)177)

Rhodotypos Sieb. & Zucc. Rosaceae (Ros.-Kerr.). 1 Korea, China, Jap.: ***R. scandens*** (Thunb.) Mak. – cult. orn. decid. shrub with *opp.* lvs (unknown otherwise in Rosaceae exc. *Lyonothamnus* & seedling *Prunus* spp.), epicalyx & solit. term. white fls with C 4, nectarless intrastaminal disk & (1–)4(–6) shining black achenes (modified drupelets). R: FNA 9(2014)386

Rhoeo Hance = *Tradescantia*

Rhoiacarpos A. DC. Santalaceae. 1 S Afr.: ***R. capensis*** (Harv.) A. DC. R: Strelitzia 10(2000)502

Rhoicissus Planch. (~ *Ampelopsis*) Vitaceae. 12 trop. & S Afr. Lianes & shrubs with tendrils, like *Cissus* but fls 5-merous, like *Vitis* but disk adnate to G. ***R. tomentosa*** (Lam.) Wild & R. Drumm. (*R. capensis*, S Afr.) – cult. orn. house-pl. with fr. ed. when cooked

Rhoiptelea Diels & Hand.-Mazz. Juglandaceae (Rhoipteleaceae). 1 SW China & N Vietnam: ***R. chiliantha*** Diels & Hand.-Mazz. – 'living fossil'

Rhoipteleaceae Hand.-Mazz. = Juglandaceae

Rhombochlamys Lindau. Acanthaceae (III 1). 2 Colombia

Rhomboda Lindl. (~ *Zeuxine*). Orchidaceae (IV 2d). 22 Himal., Indomal. to New Caled.

Rhombolytrum Link = *Chascolytrum* (but see KBAS 13(1986)100)

Rhombonema Schltr. = *Parapodium*

Rhombophyllum (Schwantes) Schwantes. Aizoaceae (V). 5 E Cape. R: H. E. K. Hartmann, *Ill. Handb. Succ. Pls*, Aiz. F–Z(2002)241. Cult. orn. succ. ± shrubby pl.

Rhoogeton Leeuwenb. Gesneriaceae (II 5c). 4 Guyana

Rhopalephora Hassk. (~ *Aneilema*). Commelinaceae (II 2). 3–4 Madag. & Ind. to Vanuatu. R: Phytologia 37(1977)479

Rhopaloblaste R. Scheffer. Palmae (V 14). 6 Nicobars & Malay Pen. to Solomon Is. R: KB 59(2004)51. Some cult. orn. monoec. unarmed palms

Rhopalobrachium Schltr. & K. Krause = *Cyclopyllum* (but see R: Candollea 46(1991)253)

Rhopalocarpaceae Hemsl. ex Takht. = Sphaerosepalaceae

Rhopalocarpus Bojer. Sphaerosepalaceae (Diegodendraceae, Rhopalocarpaceae). 17 Madag. R: Adansonia III,21(1999)112. Troll's Model

Rhopalocnemis Junghuhn. Balanophoraceae. 1 Indomal.: ***R. phalloides*** Jungh. R: FM I,7(1976)785

Rhopalocyclus Schwantes = *Leipoldtia*

Rhopalopilia Pierre. Opiliaceae. 3 C Afr. R.-parasites. R: BJ 108(1987)273

Rhopalopodium Ulbr. = *Ranunculus*

Rhopalosciadium Rech.f. = ? *Torilis*

Rhopalostylis H. Wendl. & Drude. Palmae (V 14j). 2 NZ (1: ***R. sapida*** (Forst. f.) H. Wendl. & Drude (nikau palm) – southernmost palm (Chatham Is., 44° 20'S), stemless for

40–50 yrs & mature at 150–250, lvs used for hut-building by Maoris), Norfolk Is. (1: ***R. baueri*** (Hook.f.) H. Wendl. & Drude – palm-hearts ed.), Raoul (Sunday) Is. (1)

Rhopalota N.E. Br = *Crassula*

Rhuacophila Blume (~ *Dianella*). Asphodelaceae (Hemerocallidaceae). 5+ Mal., New Caled., Fiji

rhubarb *Rheum* spp. esp. *R. rhaponticum*; **r. chard** *Beta vulgaris*; **Chinese r.** *R. palmatum*; **French r.** *Angelica archangelica*; **Himalayan** or **Indian r.** *R. australe*; **monk's r.** *Rumex pseudoalpinus*; **Turkish r.** *Rheum palmatum*

Rhus Tourn. ex L. Anacardiaceae (I). Excl. *Baronia, Malosma, Melanococca, Searsia, Terminthia, Toxicodendron*, c. 35 N temp. & warm (Medit. to As. 1, As. 6, Hawaii 1, N & C Am. 27). Usu. dioec. trees, shrubs or lianes (Leeuwenberg's & Bell's Models) with clinging r., simple or pinnate lvs & drupes. Dyes, lacquers & tanning materials, local medic., others cult. orn. (sumac(h)s). ***R. chinensis*** Mill. (*R. javanica* var. *c.*, *R. semialata*, Chinese sumac, E & SE As. to Sumatra, cult. Java) – Chinese or Jap. galls (gall nuts) promoted by aphis attack (aphids overwintering in moss, so more moss = more galls) ed. & medic., known in Eur. since 15 cent., used in tanning & dyeing blue silk; ***R. copallinum*** L. (E US) – local medic., red dye from fr., lvs used for tanning & dyeing; ***R. coriaria*** L. (sumac, S Eur. to As.) – Middle E spice, ground dried lvs form. v. imp. tanning material for sheepskin etc. in S Italy (tannin content c. 26%), dye for Cordoba & Morocco leather; ***R. glabra*** L. (E N Am.) – drink made from fr., local medic., dyes from fr. (red, black) & r. (yellow); ***R. trilobata*** Nutt. (Calif.) – stems used for 3500 yrs in basketry & for figures, bushes burnt to prod. the necessary long sucker shoots, imp. local medic., fr. ed., dyes from fr. & r., lvs smoked with tobacco; ***R. typhina*** L. (staghorn or velvet sumac, E N Am.) – high in oils & polyphenols, imp. tannin source, fr. used in drinks, local medic., cult. orn. (Leeuwenberg's Model) & serrate to laciniate ('**Dissecta**', f. *laciniata*, *R. hirta*) leaflets with brilliant autumn tints, young shoots used as pipes for tapping sugar maple

× **Rhyncatlaelia** J.M.H. Shaw. Orchidaceae. *Cattleya* × *Laelia* × *Rhyncholaelia* – 163 grexes

× **Rhyncattleanthe** J.M.H. Shaw. Orchidaceae. *Cattleya* × *Guarianthe* × *Rhyncholaelia* – 2364 grexes

Rhynchanthera DC. Melastomataceae. 15 Mex. to Bolivia & Paraguay

Rhynchanthus Hook.f. Zingiberaceae (II 4). 5–6 Myanmar, SW China, W Mal.

Rhyncharrhena F. Muell. = *Vincetoxicum* (but see FA 28(1996)229)

Rhynchelytrum Nees = *Melinis*

× **Rhynchobrassoleya** J.M.H. Shaw. Orchidaceae. *Brassavola* × *Cattleya* × *Rhyncholaelia* – 2364 grexes

Rhynchocalycaceae L. Johnson & Briggs = Crypteroniaceae

Rhynchocalyx Oliv. Crypteroniaceae (Rhynchocalycaceae). 1 Natal & Transkei: ***R. lawsonioides*** Oliv. – coll. 1884, rediscovered 1966. R: Strel. 10(2000)469

Rhynchocladium T. Koyama. Cyperaceae (II 7). 1 Venez.: ***R. steyermarkii*** (T. Koyama) T. Koyama. R: MNYBG 23(1972)86

Rhynchocorys Griseb. Orobanchaceae (Rhin.; Scrophulariaceae s.l.). 6 S Eur. (1) to Iran. R: NRBGE 30(1970)97. ***R. elephas*** (L.) Griseb. (Medit., W As.) – seedlings are underground parasites for at least 2 yrs

Rhynchodia Benth. = *Chonemorpha*

Rhynchoglossum Blume. Gesneriaceae (III 1a). c. 10 Indomal., 1 (?+) trop. Am. Marked anisophylly, fls in 1-sided racemes. Cult. orn. (*Klugia*)

Rhynchogyna Seidenf. & Garay (~ *Cleisostoma*). Orchidaceae (V 16c). 3 SE As., Mal. R: OB 95(1988)192

Rhyncholacis Tul. Podostemaceae (III). 25 N trop. S Am. ***R. nobilis*** Van Royen – powdered lvs used like salt

Rhyncholaelia Schltr. (~ *Brassavola*). Orchidaceae (V 13b). 2 Mex., C Am. R: C.L. Withner, *The Cattleyas & their relatives* V(1998)173. Cult. orn. esp. many intergeneric hybrids, esp. seq. ***R. digbyana*** (Lindl.) Schltr. (C Am.) – fragrant white moth-poll. fls

× **Rhyncholaeliocattleya** H.G. Jones. Orchidaceae. *Cattleya* × *Rhyncholaelia* – 13 190 grexes

Rhynchophora Arènes. Malpighiaceae. 2 Madag. R: CUMH 23(2001)53

Rhynchophreatia Schltr. (~ *Phreatia*). Orchidaceae (V 14). 5 NG, Micronesia

Rhynchopsidium DC. & A. DC. (~ *Relhania*). Compositae (Gnap.-Rel.). 2 S Afr. R: Strel. 9(2000)356

Rhynchoryza Baill. (~ *Oryza*). Gramineae (III 3 ii). 1 Paraguay to Arg.: ***R. subulata*** (Nees) Baill. – like *Oryza* but lemma with aerenchymatous beak, allowing flotation. R: CUSNH 39(2000)111

Rhynchosia Lour. Leguminosae (III 18). c. 230 trop. (Afr.-Madag. c. 140). R. used like *Eminia* spp. in brewing beer (amylase). ***R. pyramidalis*** (Lam.) Urb. (C Am.) – seeds consumed with hallucinogenic mushrooms

Rhynchosida Fryx. Malvaceae (Malv.-Malv.). 2 S Texas, N Mex., Bolivia, Arg. R: Britt. 30(1979)458

Rhynchosinapis Hayek = *Coincya*

Rhynchospermum Reinw. = *Aster*

Rhynchospora Vahl. Cyperaceae (II 7). Incl. *Dichromena* (= sect. *D.* – R: MNYBG 37(1984)23, with at least 7 spp. insect-poll.), c. 350 subcosmop. (Eur. 3) esp. trop. & warm S Am. ('Mesoamerica' 81, R: Britt. 44(1992)14). Beak rush. R: BJ 74(1949)375, 75(1950)90, (1951)273, (1952)451. Some ant-disp. frs

Rhynchostele Reichb.f. (~ *Leochilus*). Orchidaceae (V 12h). Incl. *Amparoa*, *Lemboglossum*, 17 trop. Am.

Rhynchostigma Benth. = *Secamone*

Rhynchostylis Blume (~ *Aerides*). Orchidaceae (V 16c). 3 trop. As. R: Selbyana 9(1986)169. Cult. orn. epiphytes

Rhynchotechum Blume (*Rhynchotoechum*). Gesneriaceae (III 2d). 16 Indomal. R: EJB 70(2013)128 – some local med. & leaf veg.

Rhynchotheca Ruíz & Pavón. Francoaceae (Ledocarpaceae). 1 Andes: ***R. spinosa*** Ruíz & Pavón: spiny, C 0

Rhynchothecaceae Endl. = Francoaceae

Rhynchotoechum Blume = *Rhynchotechum*

Rhynchotropis Harms. Leguminosae (III 15). 2 SC Afr.

Rhynea DC. = *Tenrhynea*

Rhysolepis S.F. Blake. Compositae (Helia.-Helia.). 3 Mex. R: PBSW 117(2004)323

Rhysopterus J. Coulter & Rose. Umbelliferae (III 9). 3 N Am.

Rhysotoechia Radlk. Sapindaceae. 14–15 C Mal. to trop. Aus. (5). R: Blumea 39(1994)41

Rhyssocarpus Endl. = *Melanopsidium*

Rhyssolobium E. Meyer. Apocynaceae (Va; Asclepiadaceae III 4). 1 S Afr.: ***R. dumosum*** E. Meyer

Rhyssopteris Blume ex A. Juss. = *Stigmaphyllon*

Rhyssostelma Decne = *Oxypetalum*

Rhytachne Desv. Gramineae (XXII 2). 12 trop. Am. & Afr. R: KB 32(1978)767

Rhyticalymma Bremek. = *Justicia*

Rhyticarpus Sonder = *Anginon*

Rhyticaryum Becc. Icacinaceae. 12 E Mal., W Pac. Mangenot's Model. Some local potherbs esp. ***R. oleraceum*** Becc. in Moluccas & W NG & ***R. longifolium*** K. Schum. & Lauterb. in Solomon Is.

Rhyticocos Becc. = *Syagrus*

Rhytidanthera (Planch.) Tieghem (~ *Godoya*). Ochnaceae (III). 5 Colombia. R: BJ 113(1991)173. Lvs pinnate

Rhytidea Lindl. = *Dichelostemma*

Rhytidiadelphus (Limpr.) Warnst. Hylocomiaceae (Musci). 6 N temp. ***R. squarrosus*** (Hedw.) Warnst. – dominant in heavily grazed pastures & mown fairways of golfcourses

Rhytidocaulon Bally (~ *Echidnopsis*). Apocynaceae (Vb; Asclepiadaceae III 5). 10 NE trop. Afr., Arabia. R: F. Albers & U. Meve, *Ill. Handb. Succ. Pls*, Asclep. (2002)229. Diff. cult. orn. succ.

Rhytidophyllum Mart. (~ *Gesneria*). Gesneriaceae (II 5a). 21 WI, S Am. (1). Trees & shrubs, usu. unbranched; pseudostipules; mucilage from lvs scented & (?) attractive to bat-pollinators

Rhtyidosporum F. Muell. (~ *Billardiera*). Pittosporaceae. 5 Aus.

Rhytionanthus Garay & al. = *Bulbophyllum*

ribbon grass *Phalaris arundinacea* 'Picta'

riberry *Syzygium luehmannii*

Ribes L. Grossulariaceae. c. 160 temp. N hemisph. (Eur. 9, N Am. 53 – R: FNA 8(2009)9), Andes to Tierra del Fuego (***R. magellanicum*** Poir.). Currants. Low shrubs (some everg.), ± prickles with hermaphr. fls or dioec. (evolved several times) & fr. crowned with persistent K. Many ed. fr. & some cult. orn. ***R. americanum*** Mill. (American blackc., N US) – fr. ed.; ***R. curvatum*** Small (granite gooseberry, S & SE US); ***R. cynosbati*** L. (American or prickly g., E N Am.) – fr. prickly, unarmed cvs cult. for fr. used in pies etc.; ***R. divaricatum*** Douglas (Worcesterberry, W N Am.) – nat. hybrid?, fr. form. dried; ***R. glandulosum***

Grauer (skunk c., US mts) – foetid; ***R. grossularioides*** Maxim. (catberry, Jap.) – cult. orn.; ***R. hirtellum*** Michaux (E N Am.) – ed. gooseberry used in hybridizing; ***R. lacustre*** (Pers.) Poir. (N Am.) – ed. fr., local medic.; ***R. laurifolium*** Jancz. (W China) – everg. cult. orn.; ***R. nigrum*** L. (blackcurrant, Euras.) – fr. stewed or made into jam or a liqueur (cassis) or soft drinks with reputation for antioxidants & higher levels of vitamin C than citrus, but active flavour principles unknown (s.t. adulterated with buchu which has similar 'catty' odour), since Mediaeval times used to colour wine, a trad. treatment for colds (quinsyberry), alt. host for pine blister rust (*Cronartium ribicola* J. Fischer) prob. originally restricted to *Pinus cembra* in W Siberia & poss. Swiss Alps, though modern blackcurrants involving *R. bracteosum* Douglas (NW N Am.) & *R. sanguineum* etc.; ***R. aureum*** Pursh (*R. odoratum*, golden, Missouri or buffalo c., US to Mex.) – fr. ed., form. used with buffalo meat as pemmican, cult. for large yellow fls; ***R. rubrum*** L. (*R. sativum*, *R. sylvestre*, redcurrant, garnetberry, Euras.) – cult. for fr. like blackc. in Eur., rarely in Am., esp. for jelly eaten with turkey, whitecurrant a form with greenish fr., many redcurrants also involving *R. petraeum* Wulf. (W & C Eur.), *R. spicatum* E. Robson (N Eur.), etc.; ***R. sanguineum*** Pursh (blood, flowering or American c., W N Am.) – in Canada fls in March as first hummingbirds arrive, cult. orn. shrub of suburbia; ***R. speciosum*** Pursh (Californian fuchsia, Calif.) – bright red drooping fls in winter, visited by Anna hummingbirds which breed before other spp. in Calif.; ***R. uva-crispa*** L. (*R. grossularia*, gooseberry, goosegog, (arch.) feaberry, Eur., natur. or poss. native GB) – spiny bushes reaching great age, lvs form. used for salads, fr. (fresh or preserved) much used in pies & jam, subject of 'Gooseberry Clubs' (first in Manchester area 1740s) in C & N England in late C18 & early C19 so that greatly increased fr. size (wild = c. 10 g; by 1852 70 g+) was selected for through competitions, over 171 g. (rumbullion, i.e. Fr. Rambouillet, large ones for bottling) & 720 cvs by 1850, by 1989 still 10 shows; ***R. viscosum*** Dum.-Cours. (*R. orientale*, SE Eur., temp. As.) – fr. purgative. Hybrid berries incl. ***R. × nidigrolaria*** Rud. Bauer & A. Bauer ('*R.* × *culverwellii*', jostaberry) – a gooseberry–blackcurrant cross × a blackcurrant–*R. divaricatum* cross

ribwort *Plantago* spp. esp. *P. lanceolata*

Riccia L. Ricciaceae (Hepaticae-Marchantiales). c. 175 cosmop. Thallose liverworts, some floating, incl. subcosmop. ***R. fluitans*** L. – in eutrophic ponds & ditches oft. with *Lemna* spp., veg. prop. by 'adventive branches' oft. from underside of midrib, comm. aquarium pl.

Ricciocarpos Corda. Ricciaceae (Hepaticae-Marchantiales). 2 subcosmop. ***R. natans*** (L.) Corda – forming dense scums upsetting fisheries, e.g. Kenya

rice *Oryza sativa*; **Canadian wild r.** *Zizania palustris*; **r. flower** *Pimelea* spp.; **r. grass** *Spartina* spp., *Paspalum scrobiculatum*; **hungry r.** *Digitaria exilis*; **Indian r.** *Z. palustris*; **jungle r.** *Echinochloa colona*, *E. frumentacea*; **Manchurian water r.** *Z. palustris*; **mountain r.** *Piptatherum* spp.; **r. paper** *Tetrapanax papyrifer*, though s.t. bamboo, *Broussonetia papyrifera* or *Scaevola sericea*, (ed.) *O. sativa*; **Malayan r. p.** *S. s.*; **paddy r. herb** *Limnophila aromatica*; **red r.** *Oryza glaberrima*, *O. rufipogon*; **Tuscarora** or **wild r.** *Z. palustris*

Richardella Pierre = *Pouteria*

Richardia Kunth = *Zantedeschia*

Richardia Houston ex L. Rubiaceae (IV 5). 15 trop. Am. R: Britt. 26(1974)271. Some cult. orn. incl. ***R. scabra*** L. – source of false ipecacuanha

Richardsiella Elffers & Kenn.-O'Byrne. Gramineae (XXVII 2). 1 Zambia: ***R. eruciformis*** Elffers & Kenn.-O'Byrne

Richea R. Br. (~ *Dracophyllum*). Ericaceae (VII 6; Epacridaceae). 11 SE Aus. R: AusSB 13(2000)775. Some wind-poll., some only poll. insects once snow-skinks (*Niveoscincus* spp., SE Aus.) have torn open fls to feed on nectar; lvs with parallel veins; some pls arborescent incl. ***R. pandanifolia*** Hook.f. (pandani, pandanni, pandanny, Tasmania) to 15 m.

Richella A. Gray = *Goniothalamus* (but see Blumea 12(1964)356)

Richeria Vahl. Phyllanthaceae (Scep.; Euphorbiaceae s.l.). 2 trop. S Am. R: BMPEG 6(1990)143. ***R. grandis*** Vahl (WI to Peru) – Aubréville's Model, timber & local medic.

Richeriella Pax & K. Hoffm. = *Flueggea*

Richterago Kuntze. Compositae (Gochn.). Incl. *Actinoseris*, 1 SE Braz., Urug., NE Arg. R: SB 39(2014)998

Richteria Karelin & Kir. (~ *Chrysanthemum*). Compositae (Anth.-Han.). 3 SW to E As. R: BBMNHB 34(1993)98

Ricinocarpodendron Amman ex Boehmer. Poss. oldest name for *Aphanamixis*

Ricinocarpos Desf. Euphorbiaceae (Crot.-Ricinoc.). 28 Aus. R: Austrobaileya 7(2007)392. ***R. pinifolius*** Desf. (wedding bush) – cult. orn.

Ricinodendron Heckel. Euphorbiaceae (Ricinod.). Incl. *Schinziophyton*, 2+ trop. & S Afr. ***R. heudelotii*** (Baill.) Heckel (erimado) – Rauh's Model, trade timber, ed. nuts & oilseeds; ***R. rautanenii*** Schinz (*S. r.*, mongongo, mugongo) – elephant-disp. seeds a staple of Sou bushmen (!Kung) & source of manketti nut oil used in food, varnishes etc., timber a balsa subs. & poss. use in paper-making

Ricinus Tourn. ex L. Euphorbiaceae (Acal.-Ric.). 1 E & NE Afr. to Middle E, now natur. throughout trop., invasive in Aus., Hawaii, S Afr., US, Mex., Galapagos: ***R. communis*** L. (castor oil, palma-Christi) – fast-growing (Jonah alleged to have sat under one, which then grew rapidly) colonist shrub (Leeuwenberg's Model) to 4 m, the term. infls usu. with male fls (open for 1 day) at apex & females (open for 14 days) at base, the males with much-branched A, protogynous, at least partially insect-poll., extrafl. nectaries, seed with caruncle (aril) that absorbs & temporarily retains water enabling germ. where those without cannot. Many alks & toxic principles incl. ricin, a toxalbumin binding to carbohydrates (10^{-7} body wt. fatal (vaccine against it now dev.); used with a ?sharpened umbrella-ferrule in a 1978 Bulgarian political assassination in London & suggested use in cancer treatment), 2–6 seeds being a fatal dose. Cult. for at least 6000 yrs (by 2010 1 M t a yr), the anc. Egyptians using the seed-oil as an illuminant, but most familiar as purgative; diff. fractions used in high speed aero-engines (viscosity changes little with temp.), soap, paint, varnish, candles, cosmetics, crayons, carbon paper, mole-smokes & in dyeing textiles, preserving leather & waterproofing fabrics, 'Rilson' being a nylon-type polyamide fibre prep. from oil; residue used for fert. & stems for fibre & board; silkworms (Eri) fed on lvs in Afr.; some forms cult. orn. esp. red-leaved '**Gibsonii**'

Ricotia L. Cruciferae. 9 E Med. (Eur. 2). R: KB 1951(1952)123

Riddellia Nutt. = *Psilostrophe*

Ridleyandra A. Weber & B.L. Burtt (~ *Didissandra*). Gesneriaceae (III 2j). 30 W Mal.

Ridleyella Schltr. Orchidaceae (V 15). 1 NG: ***R. paniculata*** Schltr.

Ridolfia Moris (~ *Carum*). Umbelliferae (III 8). 1 Medit. incl. Eur.: ***R. segetum*** (Guss.) Moris

Riedelia Oliv. Zingiberaceae (II 4). 60 E Mal.

Riedeliella Harms. Leguminosae (III 11). 3 Paraguay, Braz. R: Rodriguesia 58(1984)9

Riencourtia Cass. Compositae (Helia.-Ecl.). 6 E trop. S Am.

Riesenbachia C. Presl = *Lopezia*

Rigidella Lindl. = *Tigridia*

Rigiolepis Hook.f. = *Vaccinium*

Rigiopappus A. Gray. Compositae (Ast.- Pent.). 1 SW N Am.: ***R. leptocladus*** A. Gray. R: SBM 20(2000)78

Rikliella Raynal = *Lipocarpha*

Rimacactus Mottram = *Eriosyce*

Rimacola Rupp. Orchidaceae (IV 3f). 1 NSW: ***R. elliptica*** (R. Br.) Rupp. R: Gen. Orch. 2(2001)170

Rimaria N.E. Br. = *Gibbaeum*

rimu *Dacrydium cupressinum*; **mountain r.** *Lepidothamnus laxifolius*

Rindera Pallas = *Cynoglossum*

ringworm cassia or senna *Senna alata*

Rinorea Aubl. Violaceae (III 1a). c. 225 trop. (As. c. 30, Madag. c. 45, Am. 49 – R: FN 46(1988)35) esp. Afr. Heterogeneous. Fagerlind's (with sympodial trunk, Am.) & Troll's (As.) Models; some 'litterbox' pls. Some potherbs in S Am., local medic. As., Afr. ***R. bengalensis*** (Wall.) Kuntze (Mal.) – hyperaccumulator

Rinoreocarpus Ducke. Violaceae (III 1a). 1 Amazonia: ***R. ulei*** (Melch.) Ducke. FN 46(1988)183

Rinzia Schauer (~ *Baeckea*). Myrtaceae (II 15). 12 SW Aus. R: Nuytsia 5(1986)415. Heterogeneous. Most with arils

Riocreuxia Decne. (~ *Ceropegia*). Apocynaceae (Vb; Asclepiadaceae III 5). 8 trop. & S Afr. R: KB 60(2005)410

Riodocea Delprete. Rubiaceae (II Cord.). 1 Braz.: ***R. pulcherrima*** Delprete – dioec. R: Britt. 51(1999)17

ripgut *Bromus rigidus*

Ripidium Trin. = *Saccharum*

Ripogonaceae J. Conran & H. Clifford (Rhipogonaceae; ~ Smilacaceae). Magnoliidae – Liliales. 1/6 Aus., NZ. Prickly dextrorse-twining lianes with woody rhiz.; tendrils 0. Lvs simple, opp., with 3 parallel veins & reticulate sec. venation. Fls small. G with 2 ovules per carpel. Fr. a berry. Seeds with starchy endosperm. n = 15

Ripogonum Forst. & Forst. f. Ripogonaceae (Smilacaceae s.l.). 5 E Aus. (1 ext. to NG), 1 NZ. R: FA 46(1986)188. McClure's Model. ***R. scandens*** Forst. & Forst. f. (NZ) – cult. orn. liane locally medic. & used in basketry & lashings for canoes (Chatham Is.)

Riqueuria Ruíz & Pavón. Rubiaceae (inc. sed.). 1 Peru: ***R. avenia*** Ruíz & Pavón

Risleya King & Pantl. Orchidaceae (V 14). 1 Himal., W China: ***R. atropurpurea*** King & Pantling. R: Fl. Bhutan 3,3(2002)239

Ristantia Peter G. Wilson & Waterhouse (~ *Tristania*). Myrtaceae (II 5). 3 Queensland. R: Telopea 3(1988)265

Ritchiea R. Br. ex G. Don. Capparaceae. 30 trop. Afr. Shrubs & lianes; A 35–10 all fert. & equal. ***R. pygmaea*** (Gilg) De Wolf – geoxylic suffrutex, C 0

Ritonia Benoist. Acanthaceae (III 2c). 3 Madag.

Ritterocactus Doweld = *Parodia*

Rivasmartinezia Fern. Prieto & Cires. Umbelliferae. 2 Spain. R: PB 148(2014)982

Rivea Choisy (~ *Ipomoea*). Convolvulaceae. 3 S & SE As. R: EJB 64(2007)214. ***R. hypocrateriformis*** Choisy (Ind., Sri Lanka) – fls fragrant

river oak *Casuarina cunninghamiana*; **r. red gum** *Eucalyptus camaldulensis*; **r. wattle** *Acacia subporosa*

Rivina L. Petiveriaceae (Phytolaccaceae II). 1 S US to trop. Am.: ***R. humilis*** L. (bloodberry) – cult. orn. herb with red fr., source of red dye. R: JAA 66(1985)26

roast beef plant *Iris foetidissima*

Robbairea Boiss. = *Polycarpaea*

Robbrechtia De Block. Rubiaceae (II 2). 2 Madag. R: SB 28(2003)146

Robeschia Hochst. ex O. Schulz = *Descurainia*

robin, ragged *Silene flos-cuculi*

Robinia L. Leguminosae (III 23). 4 N Am. R: Castanea 49(1984)187. Locust. Decid. trees & shrubs with extrafl. nectaries & thorny stipules, cult. orn. incl. ***R. hispida*** L. (rose acacia, SE US) – fls rose or purple & ***R. pseudoacacia*** L. (false or bastard a., black locust, C & E US, widely planted (in Paris since 1635; promoted in GB by William Cobbett), invasive C, S & E Eur., Aus., S Afr.) – dominating early succession in Appalachians, shoots growing to 8 m in 3 yrs, timber (locustwood) for constr. (weatherproof house-fronts in US, fenceposts in E Fr.), furniture, veneer & fuel, fls good bee-forage & ed. fresh or dried, a manganese indicator in Arkansas, many orn. cvs incl. the bilious yellow-leaved '**Frisia**' of suburbia & the twisted-stemmed '**Tortuosa**', the 350 yr-old tree of the wild form from Robin's garden still at Jardin des Plantes, Paris, in 1980s (now sucker shoots), oldest Paris tree now nr Notre Dame

Robinsonecio T. Barkley & Janovec. Compositae (Sen.-Tuss.). 1 Mex., Guatemala: ***R. gerberifolius*** (Hemsl.) T. Barkey & Janovec. R: Phytoneuron 2012–38:5

Robinsonella Rose & Bak. f. Malvaceae (Malv.-Malv.). 15 C Am. R: GH 11(1973)1. Some cult. orn. shrubs & trees

Robinsonia DC. = *Senecio*

Robinsoniodendron Merr. = *Leucosyke*

Robiquetia Gaudich. Orchidaceae (V 16c). Incl. *Abdominea, India, Malleola, Megalotus*, c. 45 Indomal. to Tonga.

roble Furniture timber of var. spp. incl. *Nothofagus obliqua, Platymiscium pinnatum, Tabebuia rosea*; **coral r.** *Terminalia amazonica*

Roborowskia Batalin = *Corydalis*

Robsonodendron R. Archer (~ *Maytenus*). Celastraceae (I). 2 E S Afr. R: SAJB 63(1997) 116

robusta (coffee) *Coffea canephora*

Robynsia Hutch. Rubiaceae. 1 trop. W Afr.: ***R. glabrata*** Hutch.

Robynsiella Süsseng. = *Centemopsis*

Robynsiochloa Jacq.-Fél. = *Chasmopodium*

Robynsiophyton R. Wilczek. Leguminosae (III 8). 1 SC Afr.: ***R. vanderystii*** R. Wilczek – A9 (5 fert.)

rocambole *Allium sativum* var. *ophioscorodon*

Rochea DC. = *Crassula*

Rochefortia Sw. Boraginaceae (Ehretiaceae). 8 WI, 1 C Am. to Colombia. Dioec.

Rochelia Reichb. Boraginaceae (B.3.5.1). 15 Euras. (Eur. 1)

Rochonia DC. Compositae (Ast.-Hin.). 4 Madag.

rock cress *Arabis* spp.; **r. elm** *Milicia excelsa*; **r.foil** *Saxifraga* spp.; **r. jasmine** *Androsace* spp.; **r. lily** *Dendrobium speciosum*; **r. maple** *Acer saccharum*; **r. melon** (NSW) *Cucumis melo* Reticulatus Group; **r. orchid** *D. speciosum*; **r. rose** *Cistus* spp., *Helianthemum* spp.

rocket *Eruca vesicaria* (see also *Diplotaxis tenuifolia*); **r.** or **dame's r.** *Hesperis matronalis*; **dyer's r.** *Reseda luteola*; **garden r.** *E. vesicaria*; **London r.** *Sisymbrium irio*; **salad r.** *E. vesicaria*; **sea r.** *Cakile maritima*; **wall r.** *D. muralis*; **white r.** *D. erucoides*; **wild r.** *D. tenuifolia*; **yellow r.** *Barbarea vulgaris*

Rockia Heimerl = *Pisonia*

Rockinghamia Airy Shaw. Euphorbiaceae (Acal.-Epi.-Rock.). 2 NE Queensland

Rodgersia A. Gray. Saxifragaceae (I 4). 5 Nepal & E As. R: APS 32(1994)321. Pls with stout rhiz. & peltate, palmate or pinnate lvs & large term. infls of small white fls; most cult. orn. incl. ***R. pinnata*** Franch. (SW China) – cattle fodder in Tibet

Rodrigoa Braas = *Masdevallia*

Rodriguezia Ruíz & Pavón. Orchidaceae (V 12h). 48 trop. Am. esp. Braz. Cult. orn. epiphytes

Rodrigueziella Kuntze (~ *Theodora*). Orchidaceae (V 10). 5 Braz.

Rodrigueziopsis Schltr. Orchidaceae (V 10). 2 Braz.

× **Rodrumnia** J.M.H. Shaw. Orchidaceae. *Rodriquezia* × *Tolumnea* – 243 grexes

rodwood *Miconia* (*Calycogonium*) spp.

Roebuckia Short = *seq.*

Roebuckiella Short (~ *Brachycome*). Compositae. 9 Aus. R: JABG 28(2014)169, (2015)221

Roegneria K. Koch = *Elymus*

Roella L. Campanulaceae (I 2). 20 S Afr. R: JSAB 17(1952)93

Roemeria Medik. (~ *Papaver*). Papaveraceae (IV). 3 Medit. (Eur. 1) to Afghanistan (all in Turkey). R: Flora 179(1987)135. Annuals with solit. red or violet fls; ***R. hybrida*** (L.) DC. (*R. violacea*, Medit. to SW As.) – cult. orn.

Roentgenia Urb. = *Bignonia*

Roepera A. Juss. (~ *Zygophyllum*). Zygophyllaceae (I). c. 60 S Afr., Aus. (31 – R: FA 26(2013)544). R: PSE 240(2003)29

Roeperocharis Reichb.f. Orchidaceae (IV 4d). 5 trop. E Afr.

Roezliella Schltr. = *Oncidium*

Roger, stinking *Tagetes minuta*

Rogeria Gay ex Del. Pedaliaceae (2). 4 trop. & S Afr. R: Webbia 68(2013)105. Records from Braz. app. erroneous

Rogersonanthus B. Maguire & Boom (~ *Irlbachia*). Gentianaceae. 2 NE S Am., Trinidad. R: MNYBG 51(1989)3. Small trees

Roggeveldia Goldbl. = *Moraea*

Rogiera Planch. = *Rondeletia*

Rohdea Roth. Asparagaceae (Convallariaceae). Incl. *Campylandra*, 11 SE China & Jap. ***R. japonica*** (Thunb.) Roth ex Kunth (omoto) – poll. by slugs & snails feeding on fleshy P smelling of bad bread, national fl. of Manchus, cult. orn. esp. in Jap. (signifying fortitude & endurance) for 500 yrs with 1500 named cvs incl. forms with twisted, varieg. or curled lvs

Rorhrbachia (Riedl) Mavrodiev = *Typha*

Roifia Verdc. = *Hibiscus*

Roigella Borh. & Fernández (~ *Rondeletia*). Rubiaceae (I 5). 1 W Cuba: ***R. correifolia*** (Griseb.) Borh. & Fernández

Rojasia Malme (~ *Oxypetalum*). Apocynaceae. 1 S Am.: ***R. gracilis*** (Morong) Malme. R: KB 61(2006)32

Rojasianthe Standl. & Steyerm. Compositae (Helia.-Rojas.). 1 C Am.: ***R. superba*** Standl. & Steyerm.

Rojasimalva Fryxell = *Hibiscus* (but see Ernstia 28(1984)11)

rokshi *Canna indica*

Rolandra Rottb. Compositae (Vern.-Rol.). 1 trop. Am., introd. SE As., Jap.: ***R. fruticosa*** (L.) Kuntze. R: AMBG 62(1975)885

Roldana Llave (~ *Senecio*). Compositae (Sen.-Tuss.). 48 S US (1), Mex. & C Am. R: AMBG 95(2008)289. Trees to herbs; some cult. orn. esp. ***R. petasitis*** (Sims) H. Robinson & Brettell (*S. p.*)

Rolfeella Schltr. = *Benthamia*

Rollandia Gaudich. = *Cyanea*

Rollinia A. St-Hil. = *Annona* (but see AHB 12(1934)112,190)

Rolliniopsis Saff. = *Annona*

Rollinsia Al-Shehbaz = *Dryopetalon* (but see T 31(1982)422)

Roman hyacinth *Bellevalia romana*

Romanoa Trevis. (*Anabaena*). Euphorbiaceae (Acal.-Pluk.-Pluk.). 1 E & S Braz., Paraguay: ***R. tamnoides*** (A. Juss.) A.R. Sm. R: A.R. Sm., *Gen. Euphorb.* (2001)244

Romanschulzia O. Schulz. Cruciferae (46). 14 C Am. R: HPB 18(2013)2

Romanzoffia Cham. Boraginaceae (Hydrophyllaceae-Rom.). 5 W N Am. to Aleutians. R: Pittonia 5(1902)34. Some cult. orn. rock-pls

Romeroa Dugand. Bignoniaceae (1). 1 Colombia: ***R. verticillata*** Dugand. R: FN 25,2(1992)113

Romnalda P. Stevens. Asparagaceae (Lomandraceae). 5 Papuasia & Queensland. R: Telopea 12(2008)176

Romneya Harv. Papaveraceae (IV). 2 S Calif. (2) & N Baja Calif. R: FNA 3(1997)334. ***R. coulteri*** Harv. (Matilija poppy) – cult. orn. glaucous rabbit-proof perenn. capable of penetrating bricks & mortar, with large white fls & alks, poll. bees become 'fuddled' in fls & only slowly recover

Romulea Maratti. Iridaceae (VI 5). 93 Eur. (8), Medit., Socotra, E Afr. mts, S Afr. (83 – R: Adansonia III,23(2001)68). Cormous herbs like *Crocus* but without keeled lvs & with long peduncle above ground. Fls close as temp. falls & will open in dark if warmed; some gynodioecy, some S Afr. spp. poll. monkey beetles. Cult. orn., several natur. Aus. ('onion-grass') e.g. invasive ***R. rosea*** (L.) Ecklon (Guildford grass, SW Cape) also natur. Channel Is. & St Helena

Ronabea Aubl. (~ *Psychotria*). Rubiaceae (Lasianth.). 3 trop. Am. R: SGP 74(2004)39

Rondeletia Plum. ex L. Rubiaceae (I 5). c. 170 WI, 1 Panamá, 7 S Am. Incl. *Arachnothryx*, *Rogiera* (R: ABAH 28(1982)66, 33(1987)301, 35(1989)311). Cult. orn. trees & shrubs esp. ***R. odorata*** Jacq. (Cuba). ***R. anguillensis*** Howard & Kellogg (Anguilla) – long-styled fls with 3-& 4-colpate pollen, short-styled with only 3-colpate; ***R. strigosa*** Hemsl. (Guatemala) – fls with protein- & starch-rich false pollen

Rondonanthus Herzog (~ *Paepalanthus*). Eriocaulaceae. 5 Guyana Shield, NE S Am. R: AMBG 78(1991)441)

Ronnbergia C.J. Morren & André (~ *Aechmea*). Bromeliaceae (3). 12 C Am., NW S Am. R: FN 14(1979)1497. ***R. explodens*** L.B. Sm. (Panamá & Peru) – fr. explosive

Roodebergia R. Nordenstam (~ *Felicia*). Compositae (Ast.-Fel.). 1 Cape: ***R. kitamurae*** R. Nordenstam. R: APG 53(2002)101

Roodia N.E. Br. = *Argyroderma*

rooibos tea *Aspalathus linearis*

rooigras *Themeda triandra*

Rooseveltia Cook = *Euterpe*

root beer cold drink (esp. N Am.) either alcoholic (made from liquorice & sassafras r. bark etc.) or soft (made from var. syrups etc., flavour largely due to *Betula lenta*)

Roraimaea Struwe & al. Gentianaceae (Hel.). 2 N Braz., S Venez. R: HPB 13(2008)36

Roridula Burm.f. ex L. Roridulaceae (Byblidaceae s.l.). 2 W Cape. R: Strel. 9(2000)608. Viscid subshrubs trapping, but not absorbing contents of, insects (? defence) & dying insects preyed on by crab spiders which are not caught by pl., but also by hemipteran *Pameridea* spp. which eat trapped insects & whose droppings cont. nutrients absorbed by pl.; occ. bee-poll., but usu. self-poll. by *Pameridea* showered with pollen when they touch sensitive anthers which suddenly spring upright

Roridulaceae Martinov (~ Byblidaceae). Magnoliidae – Ericales. 1/2 Cape. Shrublets with glandular hairs secreting resin. Lvs linear, sessile, not circinnate in bud nor with sensitive hairs, in spirals; stip. 0. Infl. with term. 5-merous fls; C free, imbr.; A = & opp. K, anthers introrse, irritable (swinging up to become erect), dehiscing by 4 apical pores or slits; $\underline{G}$ (3) with undivided style, apical placentation & 1–4 anatropous, unitegmic ovules per carpel. Fr. a 3-valved loculicidal capsule. Seeds exotestal; embryo straight; endosperm copious, n = 6

Genus: *Roridula*

Remarkable parallelism with Byblidaceae (Lamiales), comparable with parallel evolution in parasitic former 'Rafflesiaceae'

Roripella (Maire) Greuter & Burdet = *Rorippa*

Rorippa Scop. Cruciferae (16). Excl. *Nasturtium*, *Sisymbrella*, c. 85 subcosmop. R: JAA 69(1988)144. Seeds mucilaginous when wet. ***R. montana*** (Wall.) Small (China) – effective in resp. complaints; ***R. palustris*** (L.) Besser (Eur.) – invasive Aus.

Rosa Tourn. ex L. Rosaceae (Ros.-Ros.). c. 140 N temp. (Eur. c. 45, N Am. 33 – R FNA 9(2014)75), esp. As., to trop. mts (Philippines, Ethiopia (***R. abyssinica*** R. Br. ex Lindl.), Mex.). Roses. Prickly shrubs (Champagnat's Model) s.t. climbing or trailing, usu. with decid. pinnate lvs (***R. persica*** Juss. (*Hulthemia p.*, As.) 1-foliolate, stip. o, hypanthium armed) & fls in panicles or solit., followed by hips – achenes encl. in fleshy receptacle; K quincuncial: 2 'bearded', 2 not, 1 half so ('the five brethren of the rose'). Sources of imp. ess. oils used in scent-making, also local medic. but esp. imp. as cult. orn. (assoc. with Rosicrucians & Virgin Mary, though early Christians considered rs decadent, & later Med. 'Mystery of the Rose'; 'sub rosa' [i.e. in private] from trad. of putting r. above a council table where secrecy expected, this poss. from legend that Cupid gave Harpocrates, god of silence, a r. to keep him from revealing Venus's indiscretions), 'double' ones (scent, insect-attracting terpenoids & benzenoids (sesquiterpenes from K), assoc. with C, so stronger smell per fl. in 'doubles', though in 'Synstylae' roses doubling due to modified A, so singles usu. better) discussed by Herodotus, 484 BC, ***R. × richardii*** Rehder (holy rose, *R. gallica* × *R. phoenicia* Boiss./*R. arvensis*) figured in Minoan murals at Knossos; so many thousands of cvs that formal class. is now inexact & several hort. schemes proposed. Many of the orig. 'spp.' usu. anc. hybrids of complex & disputed ancestry (see W.J. Bean (1980) *Trees & shrubs hardy in the British Isles* ed. 8, vol. 4: 131). Essentially hybridization has brought together the genomes of eastern diploid (2n = 14) 'tea' roses. & western tetraploid 'cabbage' roses (brought to Mesopotamia from beyond Taurus C24 BCE; fls in anc. Egyptian tombs). Generally these are the modern garden roses used for display & cutting (incl. Hybrid Teas derived from crosses between older Hybrid Perpetuals (see below) & the Teas & Chinas (**'Slater's Crimson'** c. 1790, oldest surviving cv.), the first cv. so classified being '**La France**' (1867) but many hundreds of new ones now raised each yr; Floribundas diff. in usu. clustered fls & derived from crosses between Hybrid Teas & Hybrid Polyanthas (these being crosses between Polyanthas, Hybrid Teas etc.); such cvs as the pop. '**Queen Elizabeth**' (c. 1950) & **Peace** (trade name for 'Madame A. Meilland') somewhat intermediate, while Miniature Roses derived from crosses between Hybrid Teas & ***R. chinensis*** Jacq. '**Minima**' (*R. roulettii*, fairy rose, China); Cli. Roses incl. mutants of the above groups as well as hybrids between wild cli. spp.). Old Garden Roses incl. Hybrid Perpetuals, *the* roses late C19 prob. derived from a no. of rs incl. ***R. × borboniana*** Desp. (Bourbon roses, *R. chinensis* (a Chinese cv., the wild type in W China) × ***R. × damascena*** Mill. (damask or Persian r. (incl. 'mossed' mutant **'Quatre Saison(s) Blanc Mousseux'**), poss. hybrid between ***R. gallica*** L. – Eur., W As. (red r. of Lancaster ('**Officinalis**') in War of the Roses, still used in potpourri, '**Versicolor**' (*R. mundi*) – fls striped) & ***R. × moschata*** Herrm. (musk r., poss. same parentage as & = *R. × centifolia*) – parent with *R. chinensis* of ***R. × noisettiana*** Thory (Noisette r.)) – source of rose-water in Iran, exported to Syria, used for washing hands before Koran handled, medic. (incl. hangovers), '**Celsiana**' proliferous with fl. arising from centre of a fl.); Tea Roses (? so-called because ***R. × odorata*** (Andrews) Sweet (also *R. rugosa*) petals used to scent tea) derived largely from *R. × odorata* (*R. chinensis* × *R. gigantea* Crépin (NE Myanmar to China) raised in China), Polyanthas from ***R. × rehderiana*** Blackburn (*R. chinensis* × *R. multiflora*). More recent developments include Hybrid Musks, 'English Roses' (David Austin, UK) combining old scented fls-forms with repeat-fls, Patio Roses bred from Floribunda-miniature crosses, & ground-cover roses; gene for delphinidin prod. introd. from pansies to give 'blue' roses c. 2004, **'Suntory Applause'** sold as cut-fl. in Jap. Modern roses with repeat-flowering (due to a recessive gene) basis of enormous cut-fl. & automated rose-bush industry (providing the quintessence of modern romanticism, such that at St Valentine's day in UK (& winter elsewhere), (often scentless) roses airfreighted from Colombia, Ethiopia) worth \$38M in US by 2005 with \$795M turnover in Netherlands by 2007; essences (otto or attar of roses, rose-oil c. 4 t a yr) still obtained by steam distillation of rose petals esp. of *R. × damascena* & others esp. ***R. × alba*** L. (?*R. gallica* × *R. canina*/*R. arvensis* Hudson (C & SW Eur. to Turkey)) – incl. white r. of York ('**Alba Semi Plena**') as in War of the Roses & the Jacobite rose of Bonnie Prince Charlie ('**Alba Maxima**'), & roses cult. by anc. Romans. Some local medic. (in Middle Ages stamens believed to be styptic); ***R. acicularis*** Lindl. (N As., N Am.) – local medic., ed. hips; ***R. banksiae*** Aiton f. (Banksian r., W & C China) – wild type cult. for 80 yrs without fls & in meantime double cv. intr. & named, fls yellow or white, bark of r. used in tanning in China, at Tombstone Arizona allegedly the biggest of all roses (covering 740 m^2, bole 3.7 m circumf., from a 1884 cutting); ***R. bracteata*** Wendl. (Macartney r., SE China) – everg.; ***R. canina*** L. (dog rose, Euras., Medit., natur. N Am., invasive Aus.) & other Br. spp. – hips coll. by children during (1941–5 – 2000 t, yielding some 10 M bottles National Rose-hip Syrup) & after World War II to boost supplies of vitamin C (syrup still used in S Afr., S Am.), hairs & achenes a vermifuge & 'itching powder', form. used as rootstock (largely replaced by ***R. corymbifera*** Borkh. (Euras., Medit.) '**Laxa**' that rarely suckers), bad weed in NSW where controlled by goats; ***R. carolina*** L. (pasture r., C & E N Am.); ***R.* × *centifolia*** L. (cabbage r.) – no seeds, introd. Eur. C17 from Iran via Netherlands, prob. derived from *R. gallica* & Damask roses etc., source of essence, cvs incl. '**Muscosa**' (moss r.) – K & pedicels with 'mossy' outgrowths (much-branched scent-glands); ***R. cinnamomea*** L. (*R. majalis*, cinnamon r., Euras.); ***R. davurica*** Pallas (shi mei, NE As.) – fls used as herbal tea; ***R. filipes*** Rehder & E. Wilson (W China) – rampant esp. '**Kiftsgate**' growing several m. per annum; ***R. glauca*** Pourret (*R. rubrifolia*, C & S Eur. mts) – reddish shoots; ***R.* × *harisonii*** Rivers ('Harrison's [sic] Yellow', yellow rose of Texas, *R.* × *foetida* Herrm. (brought yellow colour into Eur. r. breeding) × *R. spinosissima*, raised NY 1820s, planted by US pioneers travelling W; ***R. laevigata*** Michaux (Cherokee r., China) – med. & hair-colour restorer, introd. US c. 1792 via UK, now invasive SE US; ***R. lucieae*** Franch. & Rochebr. ex Crépin (*'R. luciae'*, *R. wichuraiana*, E As., natur. N Am.) – scrambler from which common 'ramblers' '**Dorothy Perkins**' & '**American Pillar**' (*R. l.* × *R. setigera* crossed with 'Red Letter Day') derived; ***R. moyesii*** A. Henry & E. Wilson (W China) – favoured for spectacular hips; ***R. multiflora*** Thunb. (Korea, Jap.) – imp. rootstock, achenes a purgative in Jap. med., invasive N Am. '**Watsoniana**' (*R. w.*) with sublinear leaflets (cf. *Rhododendron stenopetalum*); ***R. nutkana*** C. Presl (W N Am.) – hips ed. locally; ***R.* × *odorata*** '**Viridiflora**' (*R. chinensis* 'V.', green rose) – C green streaked reddish, A & G replaced by toothed leafy organs, cult. orn. curiosity; ***R. roxburghii*** Tratt. (bur r., W China) – prickly hips; ***R. rubiginosa*** L. (*R. eglanteria*, eglantine, sweetbriar, Euras., Medit., invasive Aus., NZ, S Afr., US) – foliage scented, fls medic. ***R. rugosa*** Thunb. (Jap. r., Korea, China, Jap., natur. GB, US (hybridizes with native *R. blanda* Aiton in Canada), invasive N Eur.) – v. hardy r. planted in Russia etc., used to flavour some Chinese wines, hips eaten by Ainu; ***R. setigera*** Michaux (prairie r., C & E N Am.) – dioec. (unique) but cryptic though more fls per infl. in males & greater petal expansion in females; ***R. spinosissima*** L. (*R. pimpinellifolia*, burnet or Scotch r., Euras.) – form. many cvs cult.; ***R. woodsii*** Lindl. (N Am.) – hips ed. locally

Rosaceae Juss. Magnoliidae – Rosales. Excl. Guamatelaceae, Quillajaceae, 83/3575 (but many of these apomictic lines) subcosmop. but esp. temp. & warm N. Trees, shrubs & herbs, v. rarely with alks. Lvs simple, compound or dissected, usu. in spirals (opp. in *Lyonothamnus*, *Rhodotypos*); stip. usu. present & oft. adnate to petiole (0 in *Spiraea* etc.). Fls usu. reg. & bisex., perigynous with a hypanthium to epigynous, solit. or in cymes; K (0–)4 or 5(–12) imbr., oft. appearing as lobes on hypanthium, C(0–)4 or 5(–10) free, imbr., oft. large, rarely 0, A (1,5)20 – 130(–220) usu. in sets of 5 oft. originating centripetally, filaments usu. free & attached to hypanthium, anthers with longit. slits, rarely term. pores, inner surface of hypanthium oft. nectariferous, $\underline{G}$ (usu. inf. in Maleae) 1 (mainly Pruneae) –250(–450) free or united with 2–5 sep. styles & axile placentas, 1 or 2–(5+), (on marginal placenta in Spiraeeae), anatr. to hemitropous or campylotropous bi-(uni-)tegmic ovules. Fr. a head of follicles or achenes (s.t. on enlarged fleshy receptacle) or encl. in swollen hypanthium (as in rose hips) or a drupe or head of drupelets (as in *Rubus*) or a pome (Maleae), rarely a capsule (*Lindleya*); seeds 1 or 2(–12+) mesotestal with embryo straight or bent usu. without endosperm. x = 7–9, 17+

Classification & principal genera (PSE 266(2007)36):

1. **Exochordeae** (woody, unarmed; stip.; epicalyx 0; G 5 free or basally connate; seeds winged): *Exochorda*, *Lindleya*
2. **Spiraeeae** (usu. woody; stip. 0; epicalyx 0; G2–5): *Aruncus*, *Spiraea*
3. **Neillieae** (unarmed shrubs; stip. free; epicalyx 0; follicles large): *Neillia*, *Physocarpus*
4. **Gillenieae** (unarmed; lvs compound; epicalyx 0; follicles): *Sorbaria*
5. **Kerrieae** (woody, unarmed; lvs simple; stip. free; achenes somewhat drupe-like): *Kerria*, *Rhodotypos*
6. **Dryadeae** (woody, unarmed, with actinorhizal symbiosis; lvs simple; epicalyx 0; achenes with persistent stylodia or beak): *Dryas*, *Purshia*
7. **Ulmarieae** (herbs, lvs pinnate; stip.; epicalyx 0; fr. indehiscent): *Filipendula* (only)
8. **Sanguisorbeae** (trees to herbs; lvs imparipinnate to 3-foliolate; stip. adnate to petiole; achenes): *Acaena*, *Agrimonia*, *Cliffortia*, *Hagenia*, *Podolepis*, *Sanguisorba*
9. **Potentilleae** (unarmed; lvs compound; stip. adnate to petiole; epicalyx usu. present; achenes): *Potentilla*, *Fragaria*

10. Geum group (herbs, unarmed; lvs imparipinnate; stip. usu. adnate to petiole; epicalyx present; C usu. yellow; A many; ovule 1, basal; achenes): *Geum*
11. **Rubeae** (stip. free; epicalyx 0; ovules 2, pend.; head of drupelets): *Rubus* (only)
12. **Roseae** (woody, lvs pinnate; stip. adnate to petiole; epicalyx 0; A many; achenes): *Rosa* (only)
13. Alchemilla group (lvs usu. palmatiform; stip. adnate to petiole; epicalyx almost always present; C 0; A & G few to 1): *Alchemilla* (only)
14. **Pruneae** (woody, lvs simple epicalyx 0; drupes): *Prunus*
15. Cydonia group (~ 16; woody; stip. at twig-petiole junction; epicalyx 0; pomes): *Chaenomeles, Cydonia*
16. **Maleae** (Pyreae, Maloideae, Pyroideae; ectomycorrhizae, epicalyx 0; $\overline{G}$ (2–5) ripening as pome; x = 17; R: CJB 68(1990)2209, SB 6(1991)376): *Amelanchier, Aronia, Eriobotrya, Malus, Pyrus, Sorbus*
17. **Crataegeae** (~ 16; fr. a fleshy hypanthium): *Cotoneaster, Crataegus, Mespilus, Pyracantha*

Taxonomy of the fam. confounded by apomixis (e.g. *Rubus, Sorbus, Cotoneaster, Amelanchier, Alchemilla*), non-mixing of parental chromosomes (*Rosa*) & at generic level by excessive splitting of temperate genera of economic imp. e.g. *Pyrus* & *Malus, Crataegus* & *Mespilus* (cf. *Rosmarinus* & *Salvia*, Labiatae), a hangover from pre-Linnaean folk taxonomies unfortunately revived in C18 by Philip Miller (as with *Citrus*, Rutaceae), now rectified in *Prunus* (as with *Citrus*) – but much still to be done

Many comm. imp. frs esp. *Malus* (apples), *Prunus* (almonds, apricots, cherries, peaches & plums), *Rubus* (blackberries, loganberries, raspberries etc.), *Pyrus* (pears), *Fragaria* (strawberries), *Cydonia* (quince), *Eriobotrya* (loquat), *Mespilus* (medlar), *Oemleria* etc.; scents (*Rosa*), local medic. (esp. *Filipendula*) & many cult. orn. shrubs & herbs esp. *Acaena, Alchemilla, Amelanchier, Aronia, Aruncus, Chaenomeles, Cotoneaster, Crataegus, Dasiphora, Dryas, Exochorda, Geum, Kerria, Malus, Neillia, Photinia, Potentilla, Prunus, Pyracantha, Rosa, Rubus, Sorbaria, Sorbus, Spiraea*. Temperate gardens would be v. diff. if R. did not exist; but many bird-disp. garden spp. invasive in e.g. Aus.

rosary pea *Abrus precatorius*

Roscheria H. Wendl. ex Balf.f. Palmae (V 14k). 1 Seychelles: ***R. melanochaetes*** (H. Wendl.) Balf.f.

Roscoea Sm. Zingiberaceae (II 1). 20 Himal. & W China. R: J. Cowley (2007) *The genus R.* Herbs with thick fleshy r., cult. orn.

rose *Rosa* spp. but in compound names oft. applied to pls with fls superficially like *R.*; **r. acacia** *Robinia hispida*; **alpine r.** *Rhododendron ferrugineum*; **Andes r.** *Bejaria* spp.; **r. apple** *Syzygium jambos, S. malaccense*; **Banksian r.** *Rosa banksiae*; **r. bay** *Nerium oleander*; **r.b. willowherb** *Chamaenerion angustifolium*; **bur r.** *R. roxburghii*; **burnet r.** *R. spinosissima*; **cabbage r.** *R.* × *centifolia*; **r. campion** *Silene* spp.; **Cherokee r.** *R. laevigata*; **Chinese r.** *Hibiscus rosa-sinensis*; **Christmas r.** *Helleborus niger*; **cinnamon r.** *R. cinnamomea*; **Confederate r.** *Hibiscus mutabilis*; **damask r.** *R.* × *damascena*; **desert r.** *Adenium* spp.; **Dowerin r.** *Eucalyptus pyriformis*; **eglantine r.** *R. rubiginosa*; **fairy r.** *R. chinensis* 'Minima'; **green r.** *R.* × *odorata* 'Viridiflora'; **guelder r.** *Viburnum opulus*; **r. gum** *E. grandis*; **holy r.** *R.* × *richardii*; **Honolulu r.** *Clerodendrum chinense*; **Japanese r.** *R. rugosa*; **r. of Jericho** *Anastatica hierochuntica, Selaginella lepidophylla*; **kiwi r.** *Telopea speciosissima*; **Lady Nugent's r.** *C. chinense*; **Lenten r.** *Helleborus orientalis*; **Macartney r.** *R. bracteata*; **r. mallow** *Hibiscus* spp. esp. *H. rosa-sinensis*; **marsh r.** *Mimetes zeyheri*; **moss r.** *R.* × *centifolia* 'Muscosa'; **musk r.** *R.* × *moschata*; **noisette r.** *R.* × *noisettiana*; **Persian r.** *R.* × *damascena*; **r. pink** *Sabatia angularis*; **r. plantain** *Plantago major* 'Rosularis'; **prairie r.** *R. setigera*; **rock r.** *Helianthemum* & *Cistus* spp.; **r. root** *Sedum rosea*; **Scotch r.** *R. spinosissima*; **r. of Sharon** *Hypericum calycinum*, (in Bible) *Gladiolus italicus, Lilium candidum, Pancratium maritimum* or *Tulipa agenensis*, (in N Am.) *Hibiscus syriacus*; **stock r.** *Sparrmannia africana*; **Sturt('s) desert r.** *Gossypium sturtianum*; **r. of Texas, yellow** *R.* × *harisonii*; **r. vervain** *Glandularia canadensis*; **wood(en) r.** *Argyreia nervosa, Dactylanthus taylorii, Merremia tuberosa, Pedistylis galpinii*

roseau, black *Bactris major*; **white r.** *Gynerium sagittatum*

rosecoco bean *Phaseolus vulgaris*

rosella *Hibiscus heterophyllus*

roselle *Hibiscus sabdariffa*

rosemary *Rosmarinus officinalis*; **bog r.** *Andromeda polifolia*; **marsh r.** *Limonium vulgare, Rhododendron* ssp.

Rosenbergiodendron Fagerl. (~ *Randia*). Rubiaceae (II 1). 3 trop. Am. R: Britt. 50(1998)455. Not spiny (cf. *Randia*)

Rosenia Thunb. Compositae (Gnap.-Rel.). 4 S Afr. Shrubs with opp. lvs. R: BN 129(1976)97

Rosenstockia Copel. = *Hymenophyllum*

Roseodendron Miranda (~ *Tabebuia*). Bignoniaceae (1). 2 trop. Am. R: SB 32(2007)666. ***R. donnell-smithii*** (Rose) Miranda (*T. d., Cybistax d.*, primavera, C Am.) -imp. furniture timber, form. railway carriage fittings in UK, US

roseroot *Sedum rosea*

Rosetta wood *Dalbergia latifolia*

rosewood (? corruption of Rhodes-wood, *lignum Rhodiium*, app. from *Convolvulus* (*Rhodorhiza*) spp.), orig. *Amyris* spp., then *Pterocarpus* spp. & then *Dalbergia* spp. esp. *D. nigra* (**Brazilian, Rio** or **Bahia r.**) but used for timbers with similar qualities; **African r.** *Guibourtia demeusei, P. erinaceus*; **Australian r.** *Dysoxylum fraserianum, Alectryon oleifolius*; **bastard r.** *Synoum glandulosum*; **Burmese r.** *P. indicus*; **Honduras r.** *Dalbergia stevensonii*; **Indian** or **Malabar r.** *D. latifolia*; (**Papua) New Guinea r.** *P. indicus*; **scentless r.** *S. glandulosum*; **W Afr. r.** *P. erinaceus*

rosha *Cymbopogon martini*

Rosifax C. Townsend. Amaranthaceae (I 2). 1 Somalia: ***R. sabuletorum*** C. Townsend. R: KB 46(1991)101

Rosilla Less. = *Dyssodia*

rosin *Pinus* spp.

Rosmarinus Tourn. ex L. (~ *Salvia*). Labiatae (VII 2a). 3 (hybridizing) Medit. incl. Eur. Rosemary. ***R. officinalis*** L. (*S. rosmarinus* Schleiden) – cult. orn. & herb traded from S to N Eur. since C13, with lvs dried for use in stews, sausages etc., the fl. shoots distilled for oil used in scent-making (with bergamot (citrus) & neroli oil the chief constituent of eau-de-Cologne) & medic. (incl. jet-lag aromatherapy), excellent honey-pl., form. burnt as fuel, worn at funerals & weddings, & thought to aid memory

Rosselia Forman. Burseraceae (4). 1 Louisade Arch. (SE NG): ***R. bracteata*** Forman. R: KB 49(1994)603

Rossioglossum (Schltr.) Garay & G. Kenn. (~ *Odontoglossum*). Orchidaceae (V 12h). Incl. *Chelyorchis, Ticoglossum*, 9 C Am. Cult. orn. showy epiphytes

Rostellularia Reichb. = *Justicia*

Rostkovia Desv. Juncaceae. 1 Tristan da Cunha, 1 temp. S Am., NZ. R: FOW 6(2002)2

Rostraria Trin. = *Trisetaria*

Rostrinucula Kudô = *Comanthosphace*

Rosularia (DC.) Stapf. Crassulaceae (I 5a). Excl. *Sempervivella* (= *Sedum*), 17 Euras., Medit. R: U. Eggli, *Ill. Handb. Succ. Pls*, Crass. (2002)227. Some cult. orn.

Rotala L. Lythraceae. 44 temp. to trop. (As. 24, Mal. 68, Aus. 1). Wet places, several with 'Hippuris syndrome' incl. ***R. hippuris*** Mak. (Jap.) & ***R. mexicana*** Cham. & Schldl. (pantrop.). R: Boissiera 29(1979). ***R. indica*** (Willd.) Koehne (Indomal. to Jap.) – s.t. grown in aquaria; ***R. repens*** (Hochst.) Koehne (E & NE Afr. highlands) – on contact with water, seeds put out hairs which become viscid & can be attached to animal-disp. agents

rotan(g) = rattan

Rotheca Raf. (*Cyclonema*; ~ *Clerodendrum*). Labiatae (III). c. 35 trop. OW (As. 3). Cult. orn. esp. ***R. myricoides*** (Hochst.) Steane & Mabb. '**Ugandensis**' (*C. m., C. u.*, Oxford & Cambridge bush, trop. E Afr.) – local medic. Tanzania; ***R. serrata*** (L.) Steane & Mabb. (*C. s.*, trop. As.) – r. local medic.

Rothia Pers. Leguminosae (III 8). 1 dry Afr., 1 Baluchistan to Aus.: ***R. indica*** (L.) Druce. R: AusSB 21(2008)424

Rothmaleria Font Quer. Compositae (Cich.-Cich.(Cich.)). 1 S Spain: ***R. granatensis*** (DC.) Font Quer. R: Willd. 10(1980)38

Rothmannia Thunb. Rubiaceae (II 1). c. 30 trop. & S Afr. to Seychelles & Malay Pen. (3) to NG (non-Afr. spp. to be excl.). Cult. orn. gardenia-like small trees (Fagerlind's Model), some Afr. spp. with ant-infested stems. ***R. capensis*** Thunb. (S Afr.) – C overlapping to right (others to left)

Rothrockia A. Gray = *Matelea*

Rottboellia L.f. Gramineae (XXII 3). Incl. *Coelorachis*, excl. *Manisuris*, 33 trop. ***R. cochinchinensis*** (Lour.) W. Clayton (Kelly grass) – spikelets sunk in short axis disintegrating into 1-fruited segments exposing elaiosome growing from diaphragm in shoot, minor cereal, fodder, serious trop. cane-like weed since 1970s

Rotula Lour. = *Ehretia*

rou cong rong *Cistanche deserticola*

rou gui *Cinnamomum aromaticum*

Roubieva Moq. = *Dysphania*

Roucheria Planch. Linaceae (I). Excl. *Hebepetalum*, c. 7 trop. S Am.

roucou *Bixa orellana*

Rouen lilac *Syringa* × *persica*

rough lemon *Citrus* × *taitensis*

Rouliniella Vail = *Cynanchum*

roundwood *Sorbus americana*

Roupala Aubl. Proteaceae (V 1). 33 trop. Am. Some locally used timber, cult. orn. & medic.

Roupellina (Baill.) Pichon = *Strophanthus*

Rourea Aubl. Connaraceae (IV). 55–60 trop. (Afr. 12 – R: AUWP 89–6(1989)310; Am. 42). Mostly lianes. Some with poisonous seeds, some local medic. incl. ***R. induta*** Planch. (Braz.) – abortifacient; ***R. mimosoides*** (Vahl) Planch. (Mal.) – sleep movements

Roureopsis Planch. = *Rourea*

Roussea Sm. Rousseaceae (Grossulariaceae s.l.). 1 Mauritius: ***R. simplex*** Sm. – 2 small pops, endangered, poll. & disp. by *Phelsuma cepediana* (gecko), but attacked by mealybug-farming *Technomyrex* ants

Rousseaceae DC. (~ Grossulariaceae). Magnoliidae – Asterales. 4/6 Mauritius, NG, Aus., NZ. Everg. trees to climbers. Lvs opp. or in spirals; stip. 0. Fls 4–6(7)-merous; K small free or basally united; P free, valvate; A opp. K, free s.t. with intrastaminal nectary-disk. G̲ (3–7), placentation axile. Fr. a berry or loculicidal capsule; seeds numerous with hard testa, embryo minute in copious endosperm. n = 14, 15

Classification & genera:

Carpodetoideae (Carpodetaceae; SW Pacific; R: AusSB 10(1997)859) – *Abrophyllum*, *Carpodetus*, *Cuttsia*

Rousseoideae (Mauritius) – *Roussea*

Form. assoc. with Celastraceae, Saxifragaceae or what are now Apiales, etc., poss. sister to Campanulaceae

Rousseauxia DC. Melastomataceae. 13 Madag. R: Adansonia 13(1973)180

Rousselia Gaudich. Urticaceae (IV). 3 C Am. to Colombia & WI

Rouya Coincy = *Daucus*

rowan *Sorbus aucuparia*

royal fern *Osmunda* spp. esp. *O. regalis*; **r. palm** *Roystonea* spp. esp. *R. regia*

Roycea C. Gardner. Amaranthaceae (Chenopodiaceae I 5). 3 temp. & subtrop. W Aus. R: FA 4(1984)216

Royena L. = *Diospyros* (but poss. distinct)

Roylea Wall. ex Benth. Labiatae (VI). 1 Himal.: ***R. cinerea*** (D. Don) Baill.

Roystonea Cook. Palmae (V 6). 10 Carib. & NE S Am. R: GH 8(1949)114, FN 71(1996). Unarmed monoec. palms with pinnate lvs. Leaf-bases for thatch; some ed. fr., sago, oilseed; cult. orn. esp. as avenues in trop., usu. ***R. regia*** (Kunth) Cook (*R. elata*, royal palm, Am. to Caribb., Florida) – source of palm-hearts ('row-crop' in Honduras) as is ***R. oleracea*** (Jacq.) Cook (palmiste, Lesser Antilles to NE Colombia)

Ruagea Karsten. Meliaceae (II-Trich.). c. 10 trop. Am. montane rainforest. R: FN 28(1981)242

rubber Coagulated latex of var. trees, usu. *Hevea brasiliensis*; **abbo r.** (W Afr.) *Ficus lutea*; **African r.** *Funtumia elastica*, *Landolphia* spp. etc.; **Assam r.** *Ficus elastica*; **Bitinga r.** *Raphionacme utilis*; **Bolivian r.** *Sapium glandulosum*; **Borneo r.** *Willughbeia coriacea*; **ceara r.** *Manihot* spp. esp. *M. glaziovii*; **C. American r.** *Castilla elastica*; **chilte r.** *Cnidoscolus elasticus*; **Congo r.** *F. lutea*; **couma r.** *Couma* spp.; **Dahomey r.** *F. lutea*; **E African r.** *L.* spp.; **Esmeralda r.** *S. jenmanii*; **guayule r.** *Parthenium argentatum*; **India(n) r.** *F. elastica*; **intisy r.** *Euphorbia intisy*; **Iré r.** *Funtumia elastica*; **kok-saghyz r.** *Taraxacum bicorne*; **krim-saghyz r.** *T. megalorhizon*; **Lagos r.** *F. elastica*; **Madagascar r.** *Cryptostegia*, *Landolphia*, *Marsdenia* & *Mascarenhasia* spp.; **mangabeira r.** *Hancornia speciosa*; **Manicoba r.** *Manihot glaziovii*; **milkweed r.** *Asclepias* spp.; **Panamá r.** *Castilla elastica*; **Pará r.** *Hevea brasiliensis*; **Pernambuco r.** *Hancornia speciosa*; **r. plant** *Ficus elastica*; **Rangoon r.** *Urceola brachysepala*; **silk r.** *Funtumia elastica*; **tau-saghyz r.** *Scorzonera tausaghyz*; **teke-saghyz r.** *S. acanthoclada*; **tirucalli r.** *Euphorbia tirucalli*; **Tongking r.** *Streblus tonkinensis*; **Ulé r.** *Castilla elastica*; **r. vine** *Cryptostegia grandiflora*

Rubellia (Luer) Luer = *Platystele*

Rubia Tourn. ex L. Rubiaceae (IV 16). c. 80 Medit. (Eur. 4), Afr., temp. As. (China 38), Am. Source of dyes & local medic. etc. esp. ***R. cordifolia*** L. (incl. *R. manjith* Roxb. ex Fleming), Ind. madder, munjeet, Afr. to As.) – red dye from r., medic. in China (menstrual disorders); ***R. peregrina*** L. (Levant m., wild m., Eur., Medit.) – medic.; ***R. tinctorum*** L. (madder, S Eur., W As.) – form. cult. for alizarin dye (Turkish red, synth. 1868) from r. fixed by mordants (alum to give dark red, iron alum for brown-red, chrome alum for red-violet), known to Persians & anc. Egyptians, declining in modern times until mass intr. of cotton, but now supplanted by aniline dyes (a derivative of anthracene in coal tar 1868), though used until recently for 'rose madder' & 'm. brown' tints in paint-boxes; ***R. yunnanensis*** Diels – subpachycaul scrambler with long lvs

Rubiaceae Juss. Magnoliidae – Gentianales. Incl. Dialypetalanthaceae, 576/12 000 cosmop. but esp. trop. & warm. Trees (many with Roux's Model), shrubs, lianes & few herbs, tanniniferous, s.t. ant-inhabited (as *Cuviera, Duroia, Hydnophytum, Myrmecodia, Myrmeconauclea* etc., a third of all vasc. pl. myrmecophytes), epiphytic or rarely aquatic (1 *Limnosipanea* sp.; *Scyphiphora* in mangrove), s.t. with unusual sec. growth & with wide range of chemical repellents incl. isoquinoline alks, oft. drying black. Lvs simple & usu. entire, usu. decussate & with usu. connate stip. or less oft. app. whorled with lvs in place of interpetiolar stip. (e.g. *Galium*), rarely (e.g. *Didymochlamys*) lvs in spirals through suppression of 1 member of each pr; stip. oft. bearing colleters with slime prot. growing buds, petiole s.t. with medullary bundles. Fls usu. bisex. & epigynous & insect-poll., oft. heterostylous, in cymes or rarely solit.; K (4 or 5), the lobes open & oft. small or 0 (enlarged & brightly coloured in *Mussaenda* etc.), C ((3)4 or 5(8–10)) reg. to 2-labiate, lobes valvate, imbr. or convolute, A attached to C-tube & alt. with lobes (in *Dialypetalanthus*), anthers with longit. slits, nectary-disk oft. present at top of $\overline{G}$ (2(3–5+)) (but secondarily $\underline{G}$ in *Gaertnera* & *Pagamea*) with axile placentas, rarely 1- loc. with parietal placentas (in *Gardenia* etc.) with term. style (free styles in *Galium* etc.), each loc. with 1–∞ anatr.to hemitropous unitegmic ovules oft. with a funicular obturator. Fr. a capsule, berry, drupe or schizocarp etc.; seeds with usu. straight embryo embedded in usu. copious oily endosperm s.t. with reserves of starch & hemicellulose, 0 in Guettardeae etc. n = (6–)11(–17)

Classification & principal genera (still debated & being modified – see below. Here based on OBB 1(1988)178, though some genera (e.g. *Manettia*) not clearly placed & some other tribes e.g. IV. – **Danaideae & Putorieae** recently recog., while only subfams I. & IV. recog. by some – see SGP 76(2006)131):

I. **Cinchonoideae** (usu. woody, oft. large trees; stip. usu. entire; raphides usu. 0; fr. usu. dry, endosperm present; 10 tribes incl. **Strumpfieae** (*Strumpfia* only)):
  1. **Cinchoneae** (seeds winged; excl. **Hymenodictyeae** (*Hymenodictyon, Paracorynanthe;* R: BJLS 152(2006)337, SB 39(2014)XX); **Dunnieae, Foonchewieae**): *Cinchona, Ladenbergia, Uncaria*
  2. **Naucleeae** (Naucleaceae; fls in heads; 3 subtribes, R: Blumea 24(1977)315; SB 39(2014)304): *Nauclea, Neolamarckia, Neonauclea*
  3. **Hillieae** (seeds plumose): *Hillia*
  [4. **Henriquezieae** (Henriqueziaceae; now thought to be 'basal' II.): *Platycarpum*]
  [5. **Rondeletieae** (not clearly distinct from 7.; some to go to II.): *Arachnothryx, Rondeletia, Simira, Wendlandia*]
  [6. **Sipaneeae** (herbaceous forms of 5., to be referred to II.): *Sipanea*]
  [7. **Dialypetalantheae** (Condamineeae; capsules with many horizontal seeds; R: AJB 97(2010)1976; to **II**): *Portlandia, Dialypetalanthus*]
  8. **Isertieae** (fr. fleshy with many small angular seeds): *Isertia, Kerianthera* (only, = 1.; rest = II. 6 & 7)
  [9. **Urophylleae** (like 8. but raphides present; = **IV**.; R: T 57(2008)24): *Urophyllum*]
  [10. **Pauridiantheae** (= 9; prob. = IV.): *Pauridiantha*]

II. **Dialypetalanthoideae** (Ixoroideae; woody; stip. usu. entire; raphides usu. 0; fr. usu. fleshy; 7 tribes; R: BJLS 173(2013)394):
  1. **Gardenieae** (C lobes usu. contorted to left; [excl. **Sherbournieae** (*Oxyanthus*), **Cordiereae** (*Alibertia, Duroia*) – R: T 63(2014)814]; R: T 63(2014)814): *Aidia, Gardenia, Randia, Rothmannia*
  2. **Pavetteae** (like 1. but fr. fleshy; **Ixoreae** to be dist.; R: T 64(2015)79): *Ixora, Pavetta, Tarenna*
  3. **Coffeeae** (like 2. but 2 seeds per fr.; R: AJB 94(2007)321): *Coffea*
  [4. **Aulacocalyceae** (like 3. but seeds without seed-coat = **1.**): *Heinsenia*]
  5. Hypobathreae (= **Octotropideae**; like 4. but toughened exotesta): *Hypobathrum*

6. **Mussaendeae** (stip. bifid): *Mussaenda*
7. **Sabiceeae** (like 6. but no C-like K; R: T 57(2008)16): *Hekistocarpa, Sabicea, Tamiridaea, Virectaria* (only)

III. **Antirheoideae** (like II. but exotestal cells parenchyma-like; 8 tribes, but nested in I. & II., some to IV.):

[1. **Retiniphylleae** (infls term.; = **II.**): *Retiniphyllum*]
[2. **Vanguerieae** (infls axillary; = **II.**): *Canthium, Cuviera, Fadogia, Keetia, Psydrax, Pyrostria*]
3. **Guettardeae** (Guettardiodeae; like 1. & 2. but seeds oft. elongated; = I.): *Antirhea, Bobea, Guettarda, Timonius*
4. **Chiococceae** (like 3. but endosperm oily; = I.): *Chiococca*
[5. **Alberteae** (fls ±irreg.; = **II.**): *Nematostylis*]
6. **Cephalantheae** (like 3. but seeds with apical stony aril; prob. = IV.): *Cephalanthus*
[7. **Craterispermeae** (trees or shrubs, C valvate; = IV. 10): *Craterispermum, Rudgea*]
[8. **Knoxieae** (like 7. but herbs; = IV.): *Knoxia*]

IV. **Rubioideae** (raphides usu. present, heterostyly common; R: PSE 223(2000)55; 16 tribes):

[1. **Hedyotideae** (usu. herbs; exotestal cells usu. parenchyma-like; heterogeneous, most to **15.** some to **II 7**): *Hedyotis, Kohautia, Oldenlandia, Pentas*]
2. **Ophiorrhizeae** (woody to herbaceous; exotestal cells (?)thick-walled): *Ophiorrhiza*
3. **Coccocypseleae** (like 2. but oft. creeping herbs; prob. = **11.**): *Coccocypselum*
4. **Argostemmateae** (like 3. but unbranched plantlets; 4 genera; T 58(2009)804, 64(2015)293): *Argostemma*
5. **Hamelieae** (shrubs or small trees, fls monomorphic; = **I.**, incl. I. 3): *Hoffmania*
6. **Schradereae** (scramblers with adhesive r.; R: Blumea 43(1998)283): *Schradera*
7. **Psychotrieae** (stip. inter- to intra-petiolar, C valvate): *Amaracarpus, Carapichea, Chassalia, Gaertnera* (with *Pagamea* distinct at tribal level (**Gaertnereae**) acc. to wood anatomy & G), *Hydnophytum, Myrmecodia, Palicourea, Psychotria*
8. **Triainolepideae** (like 7. but 2 or 3 ovules per loc.; = 1'III.8'): *Triainolepis*
9. **Lathraeocarpeae** (fls solit.): *Lathraeocarpa*
10. **Morindeae** (fr. oft. connate in syncarps; heterogeneous – some to **Prismatomerideae**): *Gynochthodes, Lasianthus* (to **Lasiantheae**), *Morinda*
11. **Coussareeae** (fr. 1-loc., 1-seeded): *Coussarea, Faramea*
12. **Paederieae** (fr. dry dehiscent with mericarps or operculum or fleshy indehiscent; R: T 58(2009)804): *Paederia; Plocama* now **Putorieae** (R: T 56(2007)322)
13. **Anthospermeae** (fls unisexual, wind-poll.; 3 subtribes not supported by molecular findings): *Anthospermum, Coprosma*
14. **Theligoneae** (Theligonaceae; like 13. but marked anisophylly): *Theligonum*
15. **Spermacoceae** (herbs, fr. dry with mericarps; prob. incl. most of 1., 8.; excl. *Cyanoneuron* (**Cyanoneuroneae**; R: T 64(2015)294)): *Spermacoce*
16. **Rubieae** (herbs with lvs & leaflike stip. appearing whorled): *Asperula, Crucianella, Galium, Relbunium, Rubia*

Although so widespread & common, particularly as trees & shrubs (freq. poll. Lepidoptera) in the lower reaches of trop. forests, & varying from pachycaul treelets (*Ixora* (*Captaincookia*)) to creeping herbs, this huge fam. contributes little to human welfare (cf. Compositae) exc. as drugs & stimulants (*Carapichea* – ipecacuanha, *Cinchona* – quinine, *Coffea* – coffee, *Pausinystalia* – yohimbine, *Pinckneya*), some (usu. light) timbers (*Burchellia, Calycophyllum, Canthium, Hymenodictyon, Mitragyna, Mussaendopsis, Nauclea, Neolamarckia*), ed. fr. (*Guettarda, Sarcocephalus, Vangueria*), dyes (*Asperula, Genipa, Morinda, Oldenlandia, Rubia*) & tan (*Uncaria*), as well as cult. orn. – *Alberta, Asperula, Bouvardia, Coprosma, Galium, Gardenia, Hillia, Hoffmannia, Ixora, Manettia, Mussaenda, Pavetta, Pentas, Rothmannia, Wendlandia* etc.

Rubimons Sun = *Miscanthus*

Rubiteucris Kudô (~ *Teucrium*). Labiatae (III). 2 Ind., Myanmar, China

Rubovietnamia Tirveng. Rubiaceae (II 1). 2 S China, N Vietnam, Philippines. R: GBS 65(2013)108

Rubrivena Král = *Koenigia*

Rubus Tourn. ex L. Rosaceae (Ros.-Rub.). c. 250 (+ ∞ apomictic lines – apospory, pseudogamy) cosmop. esp. N (N Am. 37 – FNA 9(2014)28). R: BBot 72(1910–11)1, 83(1914)1. Shrubs (Champagnat's Model), oft. stoloniferous with subpachycaul shoots (brambles, many serious weeds), thorns, conspic. insect- poll. fls & bird- or other animal-disp. fr. of aggregates of drupelets, some with nitrogen-fixing actinobacterial r. symbionts, NZ spp. dioec.; x = 7. Divisible into 12 subgg. of which subg. ***R.*** (brambles, blackberries – drupelets adhere to abscising receptacle) has sexual diploids & apomicts (S(C) Eur., S(W) Am.) & higher ploidies apomicts (C & N Eur., N Am.), subg. ***Idaeobatus*** Focke (raspberries – drupelets falling free of marcescent receptacle) has sexual diploids (trop. to temp.), triploid apomicts & sexual tetraploids, & subg. ***Malachobatus*** Focke (trop. & subtrop.) has sexual 4–14-ploids. Many cult. fr. & orn., though many complex hybrids, some of obscure ancestry. ***R. allegheniensis*** Porter (Alleghany blackberry, E N Am.) – ed. usu. long or oblong fr., local medic.; ***R. arcticus*** L. (nagoonberry, N Euras., N Am.) – subherbaceous, crossed with *R. a.* subsp. *stellatus* (Sm.) B. Boivin (*R. s.*, Alaska) in 1950s to give comm. fr. crop cvs; ***R. armeniacus*** Focke (Himal. blackberry, Caucasus) – natur. Eur., S. Afr., NZ, NSW US, Chile; ***R. australis*** Forst. f. (bush lawyer, NZ) – dioec. liane; ***R. bifrons*** Vest (Eur.) – bad weed N Am.; ***R. caesius*** L. (dewberry, Eur., SW As.) – ed.; ***R. chamaemorus*** L. (cloudberry, circumboreal) – prob. anc. polyploid (2n = 56), dioec. (rarely fr. in GB as mostly male pls), fr. ed., coll. from wild, eaten as sauce with reindeer (Sweden), fermented drink comm. Finland & NE N Am.; ***R. cuneifolius*** Pursh (E N Am.) – invasive Aus., S Afr.; ***R. ellipticus*** Sm. (cheeseberry, Indomal. to Aus.) – N-fixing *Frankia* filamentous bacteria in r., ed. yellow fr., invasive trop. Afr., Hawaii; ***R. flagellaris*** Willd. (dewberry (Am.), E N Am.) – black round ed. fr. with many cvs; ***R. franchetianus*** Léveillé (*R. fragarioides* Bertol., Himal.) – subherbaceous; ***R. fruticosus*** L. (bramble, blackberry, Eur., Med., invasive Aus., NZ, S Afr.) – aggregate of pseudogamous facultative apomicts & other forms incl. tetraploids & hexaploids as well as triploids & pentaploids (precisely applied *R. fruticosus* s.s. is corr. name for ***R. plicatus*** Weihe & Nees (C & NW Eur.)), also sexual ***R. ulmifolius*** Schott ('*R. inermis*', SW & C Eur., invasive Aus., W Pacific, W US) – most toothsome fr. ('**Bellidiflorus**' cult. orn. with double fls) but in autumn in Spain fl. buds prod. lvs instead of P & G, the rest in Eur. 'not profitable to treat fully & by conventional means' (*Flora Europaea*, where the whole array of agamospecies grouped about 66 'circle-spp.' used as 'nodes' of variation), aggressive weeds in Aus. (where usu. ***R. anglocandicans*** A. Newton (C & E England) controlled by goats) & NZ (controlled by blackberry rust *Phragmidium violaceum* introd. US) – the thornless cvs of blackberry derived from crosses between *R. ulmifolius* & *R. hastiformis* W.C.R. Watson (GB), local medic. (colds, shingles, acne, diarrhoea), 5-sided stolons form. split into 5 strips for skep-making (Cotswolds, England); ***R. glaucus*** Benth. (Andes berry, trop. Am.) – raspberry cult. Ecuador & Colombia (comm. fr. juice) but retains 'plug' so poss. hybrid with blackberry; ***R. idaeus*** L. (raspberry, Euras., N Am. – subsp. ***strigosus*** (Michaux) Focke) – ed. fr. (rich in ellagic acid, binding with DNA prot. it from carcinogens) red or s.t. (no anthocyanin) yellow ('golden' or 'amber') or white, orig. coll. for flavouring wine, the leaves for making tea, local medic. (N Am. & Eur. – trad. infusion taken in pregnancy to alleviate labour pains & by 1940s used thus in maternity hospitals in England, as fragarine relaxes womb contractions), now comm. imp. as fresh fr. (450 K t by 2010; best flavour – due to 1-(*p*-hydroxyphenyl)-3-butanone – in '**Lloyd George**' found in a wood at Corfe Castle, Dorset, England, but most seen in UK now is '**Glen Moy**' (1982), the leading Eur. early-fruiting thornless cv.), in jams, pies, sweets, drinks, liqueurs (in Fr. = Chambord) & vinegar, subsp. *strigosus* also imp. (cross between them prob. ancestry of '**Cuthbert**' standard comm. cv. for nearly 100 yrs in US), purple raspberries being crosses between subsp. *strigosus* & ***R. occidentalis*** L. (black r., thimbleberry, C & E N Am., local medic.) = ***R.* × *neglectus*** Peck, selection of 2nd generation cross between 'Cuthbert' & *R. ursinus* being 'phenomenal berry' & this crossed with *R. caesius* 'Austin Mayes' = youngberry much grown in S Afr. as 'S Afr. loganberry'; *R. idaeus* 'November Abundance' **×** *R. rusticanus* E. Merc. (*R. ulmifolius* s.l.) = ***R.* 'Veitchberry'** (veitchberry, a parent of '**Bedford Giant**' blackberry), 'Malling Jewel' (first virus-resistant cv) **×** *R. ursinus* = sunberry, tetraploid form with an Am. blackberry ('Aurora') = tayberry ('**Buckingham**' thornless) & this crossed with a sister seedling = tummelberry; ***R. laciniatus*** Willd. (? Eur., natur. N Am.) – leaflets deeply lobed, parent of some garden blackberries; ***R. loganobaccus*** L. Bailey (*R. ursinus* var. *loganobaccus*, loganberry, cross (6n, Calif. 1883) between *R. u.* 'Aughinbaugh' (8n) & *R. idaeus* 'Red Antwerp') – large red ed. fr. high in vitamin C, some thornless cvs, nectarberry being a form (? mutant), '**Boysenberry**' (8n, 1920s, fr. to 2 cm) backcross with raspberry, **'Karaka Black**' (karaka berry, backcross with blackberry); ***R.* 'Marion**' (1956, marionberry) similar but a blackberry selection, most widely grown b. in US incl. for syrup (Oregon), crossed with 'Boysenberry' in Aus. = ***R.* 'Silvan'** (silvanberry), while **R. 'Hildaberry'** (hildaberry) = tayberry × boysenberry, & laxtonberry is backcrossed with *R. idaeus* 'Superlative'; ***R. niveus*** Thunb. (*R. albescens*, Mysore r., Ind., W China) – fr. red, ed. '**Mysore**' comm. cult. Florida; ***R. pedatus*** Sm. (NE As., NW N Am., strawberry bramble) – runners like a s.; ***R. pensilvanicus*** Poir. (*R. argutus*, N Am.) – invasive Hawaii; ***R. phoenicolasius*** Maxim. ((Jap.) wineberry, Korea, China, Jap.) – fr. small, ed., invasive N Am.; ***R. rosifolius*** Sm. (Mauritius r., E As.) – ed. fr. cult. trop., r. med.; ***R. spectabilis*** Pursh (salmonberry, W N Am.) – fls coincide with arrival of hummingbirds from S winter quarters, fr. ingredient of pemmican; ***R. ursinus*** Cham. & Schldl. (*R. vitifolius*, dewberry (Am.), W N Am.) – ed. fr., parent of loganberry etc. Cult. orn. incl. ***R. cissoides*** A. Cunn. (NZ) – liane to 10 m, juvenile form (poss. *R. australis*) with lvs reduced to prickly midribs forming a bolster-like bush without fls., like other 4 endemic NZ spp. heteroblastic in habit (in raspberries juvenile features may be maintained through infection with crown gall) cult. in cool house, ***R. cockburnianus*** Hemsl. (N & C China) – cult. orn. (canes white), ***R. tricolor*** Focke (China) – ground-cover, **R. 'Benenden'** ('*R. tridel*', *R. trilobus* Ser. (S Mex.) × *R. deliciosus* Torrey (W US) poss. not specifically distinct) – form. vogue hardy shrub

Ruckeria DC. = *Euryops*

Rudbeckia L. Compositae (Helia.-Rudb.). (Incl. *Dracopsis*) 23 N Am. R: FNA 21(2006)44. Coneflowers, black-eyed Susan, Mex. hat. Cult. orn. herbs esp. 'ann.' ***R. hirta*** L. & perenn. ***R. laciniata*** L. (esp. 'double' '**Hortensia**') – toxic, medic. E Canadian Indians, also ***R. fulgida*** Aiton (E N Am.), '**Goldsturm**' – apomictic so true from seed, ***R. occidentalis*** Nutt. (W US) – ray florets 0, phyllaries to 3 cm, green

Ruddia Yakovlev. See *Ormosia*

ruddles *Calendula officinalis*

Rudgea Salisb. Rubiaceae (IV 7). c. 70 trop. Am. Some heterostyly, some domatia

Rudolfiella Hoehne (~ *Bifrenaria*). Orchidaceae (V 12g). 6 trop. Am.

Rudua F. Maek. = *Phaseolus*

rue *Ruta graveolens*

Ruehssia Karsten ex Schldl. = *Marsdenia*

Ruellia Plum. ex L. Acanthaceae (III 2b). Incl. *Aporuellia, Blechum, Dipteracanthus, Eusiphon, Polylychnis*, c. 350 trop., temp. N Am. R: T 22(1973)543. Heterogeneous? ***R. eurycodon*** Lindau (Braz.) & other spp. prob. buzz-poll. Seeds mucilaginous-pubescent when wet though appearing glabrous when dry. Local medic. (some diuretic); cult. orn. shrubs & herbs incl. ***R. baikiei*** (Hook.) N.E. Br. (Braz.), ***R. makoyana*** Hort. Makoy ex Closon (monkey pl., Braz.) & ***R. pearcei*** (Hook.f.) N.E. Br. (Bolivia). ***R. prostrata*** Poir. (*D. p.*, OW trop.) – C falls before midday; ***R. radicans*** (Nees) Lindau (NE S Am.) – poll. hummingbirds attracted by nectar but after C falls nectar higher in sugars & attractive to (? prot.) ants; ***R. tuberosa*** L. (trop. Am.) – natur. Indomal.

Ruelliola Baill. = *Brillantaisia*

Ruelliopsis C.B. Clarke. Acanthaceae (III 2f). 2–3 trop. & S Afr.

Rufodorsia Wiehler. Gesneriaceae (II 5c). 4 C Am. R: Selbyana 1(1975)138

Rugelia Shuttlew. ex Chapman (~ *Senecio*). Compositae (Sen.-Tuss.). 1 SE US: ***R. nudicaulis*** Shuttlew. ex Chapman. R: FNA 20(2006)625

Rugoloa Zuloaga. Gramineae (XXIII 2). 3 trop. Am.

Ruilopezia Cuatrec. See *Espeletia*

Ruizia Cav. (~ *Dombeya*). Malvaceae (Malv.-Dom.; Sterculiaceae). 1 Réunion: ***R. cordata*** Cav. – almost extinct

Ruizodendron R. Fries. Annonaceae (IV 2). 1 Colombia, Ecuador, Bolivia, Peru, Amaz. Braz.: ***R. ovale*** (Ruíz & Pavón) R. Fries

Ruizterania Marcano-Berti = *Qualea*

rukam *Flacourtia rukam*

Rulingia R. Br. = *Commersonia*

rum spirit distilled from sugar-cane; **r.berry** *Myrciaria floribunda*; **r. cherry** *Prunus serotina*

rumbullion *Ribes uva-crispa*

Rumex L. Polygonaceae (II 3). Incl. *Emex*, c. 200 temp. esp. N (Eur. 44, N Am. 63 – R: FNA 5(2005)488). Docks, sorrel. Sex chromosomes (pops with female-biased sex-ratios). Weeds (e.g. natur. Aus. from contents of camel-saddles) esp. subg. *R.* but this subg. parasitized by leaf-fungus (*Ramularia rubella* (Bon.) Nannf.) a poss. mycoherbicide, culinary herbs & tannin, lvs trad. (esp. ***R. obtusifolius*** L. (Eur.; app. allelopathic to germ. of competitors)) applied to nettle-stings. ***R. acetosa*** L. (sorrel, Euras.) – ed. lvs cooked or in salads, sauces etc. ('*R. rugosus* Campd.', cultigen derived from it), juice form. used to remove rust stains on linen, also to clear acne; ***R. acetosella*** L. (sheep('s) s., N temp., invasive Aus., W US) – complex of diploids, tetra- & hexa-ploids, commonest form natur. cosmop. acid soils a hexaploid poss. from S Eur.; ***R. conglomeratus*** Murr. (Euras.) – invasive Aus.; ***R. crispus*** L. (yellow dock, Euras., invasive Aus.) – weed with long-lived seeds, these lasting 80 but not 90 yrs in Dr Beal's experiment set up in 1879, contains hydroxyanthraquinone (like *Senna*) used as laxative & in arthritis treatment; ***R. hymenosepalus*** Torrey (canaigre, ganagra, tanner's dock, SW N Am.) – tubers rich in tannin (30–35%) form. used on leather, also yellow dye, lvs ed., petioles like rhubarb; ***R. lunaria*** L. (Canary Is.) – dioec. everg. shrub planted as hedge in S Eur.; ***R. nervosus*** Vahl (Ethiopia) – shrub; ***R. patientia*** L. (patience (dock), Euras., natur. N Am.) – lvs like sorrel; ***R. pseudoalpinus*** Hoefft ('*R. alpinus*', monk's rhubarb, Eur. mts) – lvs ed. cooked or salad; ***R. pulcher*** L. (fiddle d., Eur.) – natur. N Am.; ***R. sagittatus*** Thunb. (S Afr.) – invasive Aus.; ***R. scutatus*** L. ((F.) s., Euras.) – used like *R. acetosa*; ***R. spinosus*** L. (*E. s.*, *E. australis*, Cape spinach, warm OW) – medic. in S Afr.

Rumfordia DC. Compositae (Mill.-Mill.). 12 Mex. & C Am. mts. R: SB 2(1977)302

Rumia Hoffm. = *Trinia*

Rumicastrum Ulbr. Type = Amaranthacea, but most spp. = *Calandrinia*

Rumicicarpus Chiov. = *Triumfetta*

Rumohra Raddi. Dryopteridaceae (II). 7: 1 circumaustral to Bermuda, Zimbabwe, Madag., NG: ***R. adiantiformis*** (Forst. f.) Ching – cult. orn. (most prominent 'fern' with cut fls since 1960s, replacing asparagus f.; by 1999 worth over $62 M to Florida alone, Costa Rica an imp. exporter), 5 Madag. (1 obligate epiphyte on *Pandanus*), 1 Juan Fernandez

Rumphia L. Older name for *Canarium*

runch *Sinapis arvensis*

Rungia Nees = *Justicia*

running buffalo clover *Trifolium stoloniferum*; **r. pop** *Passiflora foetida*

Runyonia Rose = *Agave*

Rupertia Grimes. Leguminosae (III 20). 3 W N Am. R: MNYBG 61(1990)52

Rupestrea R. Goldenb. & al. (~ *Miconia*). Melastomataceae. 2 Bahia. R: SB 40(2015)562. Hydrochory?

Rupicapnos Pomel. Papaveraceae (Fumariaceae I 2). 7 S Spain (1), NW Afr., on cliffs. R: OB 88(1986)92

Rupichloa Salariato & Morrone (~ *Urochloa*). Gramineae (XXIV 5). 2 E Braz. R: T 58(2009)388

Rupicola Maiden & Betche = *Epacris*

Rupiphila Pimenov & Lavrova = *Ligusticum*

Ruppia L. Ruppiaceae. 9 temp. (Eur. 2) & subtrop. salt, alkaline or brackish water. R: Strel. 9(2000)217. Pollen in rafts to stigma, cf. *Halodule*, *Halophila*, *Lepilaena*. Duckfood. ***R. maritima*** L. – in Calif. brackish lakes loose pls rolled into balls by wave-action

Ruppiaceae Horan. (~ Potamogetonaceae, Juncaginaceae). Magnoliidae – Alismatales. 1/9 temp. & subtrop. coastal waters, S Am. & NZ. R: PCAS IV, 25(1946)469. Submerged, fibrous-r., glabrous, usu. ann., monopodial herbs without vessel-elements; stem terete, a sympodium formed by development of lateral buds beneath term. infls, usu. with 4 lvs to each segment. Lvs opp. or in spirals, ± linear (serrulate) with midvein & basal open sheath & 2 intravaginal scales at each axil. Fls small, reg., bisex., in usu. 2-fld axillary or term. spikes, subtended by small sheathing prophyll s.t. later assoc. with upper fl. only, the peduncle usu. eventually elongated bringing fls to water surface, later twisting to bring developing fr. underwater; P 0 (or a tiny appendage near tip of anther connective, A 2 opp. with subsessile anthers with expanded connective & longit. slits, extrorse, $\underline{G}$ (2–)4(–16), each stipitate in fr. & with 1 pend., campylotropous, bitegmic ovule. Fr. of oft. asymmetric drupelets; seeds with 0 endosperm. n = 8, 10–12

Genus: *Ruppia*

Some in ± fresh water in NZ & Andes, where found at up to 4000 m. Poll. underwater or pollen floating in some races to floating stigmas. Specific distinctions dubious

Ruprechtia C. Meyer. Polygonaceae (Erio.-Trip.). Excl. *Magoniella*, *Salta*, 34 trop. Am. R: SBM 67(2004)17. Dioec. trees, shrubs & lianes

Ruptiliocarpon Hammel & Zamora (~ *Lepidobotrys*). Lepidobotryaceae. 2 Costa Rica, Peru. R: Novon 3(1993)408

rupturewort *Herniaria glabra*

Rusbya Britton. Ericaceae (VIII 5). 1 N Bolivia: ***R. taxifolia*** Britton

Rusbyanthus Gilg = *Macrocarpaea*

Rusbyella Rolfe ex Rusby = *Cyrtochilum*

Ruscaceae M. Roemer = Asparagaceae

Ruschia Schwantes. Aizoaceae (V). Incl. spiny '*Eberlanzia*' (q. v.) spp., 206 S Afr. (Cape 88, 79 endemic – R: Strel. 9(2000)258). R: H. E. K. Hartmann, *Ill. Handb. Succ. Pls*, Aiz. F–Z(2002)244. Succ. shrubs, older stems oft. bearing dry leaf-sheaths of old lvs; some cult. orn. & natur. S Eur.

Ruschianthemum Friedrich = *Stoeberia*

Ruschianthus L. Bolus. Aizoaceae (V). 1 S Namibia: ***R. falcatus*** L. Bolus. R: H. E. K. Hartmann, *Ill. Handb. Succ. Pls*, Aiz. F–Z(2002)285

Ruschiella Klak (~ *Mesembryanthemum*). Aizoaceae. 4 S Afr.

Ruscus Tourn. ex L. Asparagaceae (Ruscaceae). 6 Macaronesia & W Eur. (3), Medit. to Iran. R: NRBGE 28(1968)237. Dioec. (some herm. clones in cult.) everg. shrubs with shoots from rhiz. behaving rather like perenn. infls. of asparagus, the term. leaf-like 'phylloclade' a flattened apex, the laterals similar in axils of scale lvs; each 'phylloclade' bearing a scale-leaf with axillary fl. on midrib. ***R. aculeatus*** L. (butcher's broom (form. used to decorate meat), pettigree, pettigrue, range of genus) – form. young shoots ed. like asparagus (Greeks & Romans), form. Plantagenet emblem, dead shoots sold with wax fls by gypsies, cult. orn. ('**Hermaphrodite**' breeds true) as is ***R. hypoglossum*** L. (S Eur. to Turkey) – the 'laurel' of Caesar

rush *Juncus* spp. esp. *J. effusus*; **beak r.** *Rhynchospora* spp.; **chair-maker's r.** *Schoenoplectus pungens*; **corkscrew r.** *J. effusus* 'Spiralis'; **Dutch r.** *Equisetum hyemale, S. lacustris*; **flowering r.** *Butomus umbellatus*; **r. grass** *Sporobolus* spp.; **scouring r.** *Equisetum* spp.; **spike r.** *Eleocharis* spp.; **toad r.** *J. bufonius*; **wood r.** *Luzula* spp.

rusha (grass) *Cymbopogon martini*

Ruspolia Lindau. Acanthaceae (III 2c). 5 trop. Afr., Madag. Cult. orn. shrubs; moth-poll. but hybrids formed with bird-poll. & butterfly-poll. spp. of *Ruttya*

Russelia Jacq. Plantaginaceae (Russ.; Scrophulariaceae s.l.). 52 Cuba & Mex. to Colombia. R: FMBot. 29(1957)231. Stems s.t. pendent & bearing scale lvs as in ***R. equisetiformis*** Schldl. & Cham. (*R. juncea*, firecracker pl., coral pl., fountain pl., Mex., widely natur.) – cult. orn. with red fls, shoots s.t. bearing broad lvs

Russian garlic *Allium ampeloprasum*; **R. knapweed** *Rhaponticum repens*; **R. olive** *Elaeagnus angustifolia*; **R. plum** *Prunus* × *rossica*; **R. thistle** *Salsola tragus*; **R. vine** *Fallopia baldschuanica*; **R. wild rye** *Thinopyrum junceum*

Russowia Winkler. Compositae (Card.-Cent.) 1 Turkestan: ***R. sogdiana*** (Bunge) B. Fedtsch.

Rustia Klotzsch. Rubiaceae (Cond.). c. 15 trop. Am.

rusty-back fern *Asplenium officinarum*

Ruta Tourn. ex L. Rutaceae (I 10). 7 Macaronesia, Medit. (Eur. 5) to SW As. Rue. Alks. Cult. orn. & (esp. form.) for flavouring & form. medic. esp. ***R. chalepensis*** L. (Medit.) – rue of Bible, but usu. ***R. graveolens*** L. (S Eur.) – rue of gardens ('herb of grace'), strong-smelling (ethereal oils in lvs) with infls where term. fl. 5-merous & laterals 4-merous, stamens bend over stigma & dehisce, fall back while stigma matures & return effecting self-poll., cult. since time of Pliny for strong flavour (form. much used in sausages) & medic. but causing dermatitis in some; ***R. montana*** (L.) L. (Medit.) – in US causes blisters & ulcers on skin when absorbed & exposed to sun (furcocoumarins) when thiamine & pyrimidine bond covalently causing cross-linking in DNA strands leading to antimitotic effect

rutabaga *Brassica napus* Napobrassica Group

Rutaceae Juss. Magnoliidae – Sapindales. Incl. Cneoraceae, Ptaeroxylaceae, 153/1975 cosmop. esp. trop., most diverse in Australasia. Aromatic trees, sometimes pachycaul, & shrubs, s.t. lianoid, rarely herbs, s.t. spiny (shoots?) & oft. with bitter terpenoids, alks & schizogenous secretory cavities with aromatic ethereal oils scattered through parenchyma & pericarp. Lvs usu. in spirals, pinnate or 3-foliolate, rarely simple or pinnatisect, with pellucid gland-dots; stip. 0. Fls usu. bisex. & reg., hypogynous, in thyrses or racemes, umbels etc., rarely solit. or epiphyllous (*Erythrochiton*); K (2–)5, s.t. basally connate, oft. imbr., C alt. with & 2 × K (3 or 4 in *Cneorum* & some *Citrus* etc.), s.t. basally connate, imbr. or valvate, rarely 0, A usu. diplostemonous (1 whorl s.t. staminodes, or those alt. C absent, or 3–4 × C to 60, rarely only 2 or 3 fert.), filaments ± basally connate (s.t. completely; c.f. Meliaceae), anthers with longit. slits, nectary-disk annular, s.t. 1-sided or a gynophore, around $\underline{G}$ (1–)4 or 5(–∞), ± united & pluriloc. or sep. exc. for coherent style, rarely 1-loc. or G 1, each loc. with usu. parietal placentation & (1)2(–∞) usu. superposed, anatr. or hemitropous, usu. bitegmic (unitegmic in *Glycosmis*) ovules. Fr. schizocarps, berries, drupes etc.; seed with large straight or curved embryo ± endosperm (oily). n = 7–11+, anciently x = (?) 9

Classification & principal genera (AMBG 99(2014)637)):

I. **Rutoideae** ([incl. Toddalioideae (G (5–)2 or 1, each with 2–1 ovules; fr. a drupe or dry & winged) – Toddalieae: *Ptelea, Skimmia*] G (1–)4 or 5(+), oft. united only by styles; fr. with loculicidal dehiscence, rarely fleshy drupes) – 10 groups with some e.g. *Casimiroa, Dictamnus, Flindersia* (Flindersiaceae), *Pilocarpus, Ptelea, Skimmia* unplaced:

1. 'Bosistoa tribe' (rainforest trees; loc. usu. with 8–4 ovules, explosive fr. disp.; 5 genera, Mal., Aus.): *Bosistoa*
2. **Zanthoxyleae** (woody pls; fls usu. reg., oft. unisexual; loc. usu. with 2 ovules; 32 genera esp. rainforest SW Pacif.): *Euodia, Melicope, Tetradium, Vepris, Zanthoxylum*
3. **Boronieae** (shrubs, usu. with reg. bisex. fls., 18 genera Aus.): *Boronia, Correa, Eriostemon, Phebalium, Zieria*
4. **Diosmeae** (usu. shrubby; lvs simple; endosperm usu. 0; 11 genera esp. S Afr. with 10/150 endemics): *Agathosma, Calodendrum, Diosma*
5. 'Polyaster alliance' (~ 2; lvs simple, styles usu. joined; 5 genera S Am.): ? *Choisya*
6. 'Esenbeckia alliance' (capsules; 3 genera esp. S Am.): *Esenbeckia*
7. 'Balfourodendron alliance' (samaras; 2 genera trop. Am.): *Helietta*
8. 'Angostura alliance' ('Cusparieae'; some zygomorphy, oft. 2 fert. A; trop. Am.): *Angostura, Conchocarpus, Galipea*
9. 'Amyris alliance' (G1, drupe, endosperm 0; 2 genera trop. & warm Am.): *Amyris*
10. **Ruteae** (usu. ± herbs, fls bisex.; loc. usu. with 4–8 ovules; 7 genera esp. N hemisph. oft. mesic to arid): *Haplophyllum, Ruta, ?Chloroxylon*

II. **Aurantioideae** (fr. a berry with pulp derived from hairs within ovary (hesperidium); endosperm 0; trop. & warm OW)

1. **Clauseneae** (unarmed; lvs pinnate with unwinged petioles; hesperidium 0; 4 genera Indomal.): *Bergera, Clausena, Glycosmis*
2. **Aurantieae** (Citreae; pinnate to unifoliolate lvs, oft. axillary spines, hesperidium; 20 genera OW): *Aegle, Citrus, Limonia, Murraya, Triphasia*

III. **Cneoroideae** (Dictyolomatoideae, Spathelioideae, Cneoraceae, Ptaeroxylaceae; some hapaxanthic pachycauls, oil-glands s.t. only at leaf margins, ovules usu. 1 or 2 per loc.; 7 genera trop. to warm temp.): *Cedrelopsis, Cneorum, Harrisonia, Ptaeroxylon, Spathelia*

Close to Meliaceae & Simaroubaceae (from which *Harrisonia* acquired). Fam. outstanding in no. of sec. pl. constituents esp. quinolones & acridones (& other alks), coumarins, but esp. limonoids & volatile oils

Imp. fr. trees (esp. *Citrus*, but also *Aegle, Casimiroa, Clausena, Glycosmis, Limonia*) & flavourings (*Agathosma, Angostura, Bergera, Casimiroa, Esenbeckia, Ruta, Zanthoxylum*); some timbers esp. *Amyris, Balfourodendron, Bouchardatia, Chloroxylon, Euxylophora* & *Flindersia* (s.t. these 2 put in sep. Flindersiaceae), *Euxylophora, Geijera, Halfordia, Limonia, Ptaeroxylon* & *Zanthoxylum*, & medic. incl. *Agathosma, Pilocarpus* (jaborandi) & *Raputia* with some cult. orn. esp. *Boronia* (also for scent), *Choisya, Citrus* (also scent), *Correa, Dictamnus, Euodia, Melicope, Murraya, Orixa, Ptelea, Skimmia, Tetradium, Triphasia*

Rutaneblina Steyerm. & Luteyn. Rutaceae (I ?8). 1 Venez.: ***R. pusilla*** Steyerm. & Luteyn. R: AMBG 71(1984)314

Ruthalicia C. Jeffrey. Cucurbitaceae (XIV). 2 trop. W Afr. KB 15(1962)360

Ruthea Bolle = *Rutheopsis*

Rutheopsis Hansen & Kunkel (*Ruthea, Gliopsis*). Umbelliferae (III 8). 1 Canary Is.: ***R. herbanica*** (Bolle) Hansen & Kunkel

Ruthiella Steenis (*Phyllocharis*). Campanulaceae (III 1). 4 NG. R: FM I,6(1960)137. Like *Lobelia* but fls epiphyllous

Rutidea DC. Rubiaceae (II 2). 22 trop. Afr. & Madag. R: KB 33(1978)243

Rutidosis DC. Compositae (Gnap.-Ang.). 9 Aus. R: Austrobaileya 5(1999)565

rutin material used in haemorrhage treatment, derived from several pls incl. *Eucalyptus macrorhyncha, Fagopyrum esculentum* & *Stypholobium japonicum*

Ruttya Harv. Acanthaceae (III 2c). 3 trop. & S Afr. to Yemen, 3 Madag. Cult. orn. shrubs esp. ***R. fruticosa*** Lindau (E Afr.) – fls yellow to scarlet with darker markings

Ruyschia Jacq. Marcgraviaceae. 7 trop. Am. (C 3; N Andes 3, WI 1)

Ryania Vahl. Salicaceae (Samydaceae, Flacourtiaceae (9)). 8 trop. Am. R: FN 22(1980)258. Cook's & Roux's Models. Toxic: ***R. angustifolia*** (Turcz.) Monach. used to poison alligators; ryanodine also an efficient insecticide

Rydingia Scheen & V. Albert (~ *Otostegia*). Labiatae (VI). 4 SW As., NE Afr. R: SGP 77(2007)234

rye *Secale cereale*; **giant r.** *Triticum turgidum* Polonicum Group; **Russia wild r.** *Thinopyrum junceum*

ryegrass *Lolium* spp.; **Italian r.** *L. multiflorum*; **perennial r.** *L. perenne*

Rylstonea R. Bak. = *Homoranthus*

Ryncholeucaena Britton & Rose = *Leucaena*

Ryparosa Blume. Achariaceae (Flacourtiaceae). 27 Mal. Some myrmecophytes

Ryssopterys Blume ex A. Juss. = *Stigmaphyllon*

Ryticaryum Becc. = *Rhyticaryum*

Rytidocarpus Cosson. Cruciferae. 1 Morocco: ***R. moricandiodes*** Cosson

Rytidosperma Steud. (~ *Danthonia*). Gramineae (Danth.). Incl. *Austrodanthonia, Erythranthera*, 35 Aus., NZ, S Am. Some forage (wallaby grass)

Rytidostylis Hook. & Arn. = *Cyclanthera*

Rytigynia Blume (~ *Vangueria*). Rubiaceae (III 2). 50–60 trop. Afr. Links to *Canthium, Vangueria* etc. Some ed. fr. & vermifuge Uganda

Rytilix Raf. ex A. Hitchc. = *Hackelochloa*

Rzedowskia Gonz. Medr. Celastraceae (I). 1 Mex.: ***R. tolantonguensis*** Gonz. Medr. R: BSBM 41(1981)41

S

saba nut *Pachira aquatica*

Saba (Pichon) Pichon. Apocynaceae (I c). 3 trop. Afr. to Madag. R: BJBB 59(1989)189. ***S. comorensis*** (Bojer) Pichon (*S. florida*, trop. Afr.) – handsome liane with strong-smelling fls & ed. fr. (markets), source of rubber in World War II E Afr., 'sponges' from stem-fibres

sabadilla *Schoenocaulon officinale*

Sabal Adans. Palmae (III 1). 16 SE US to S Am. (also Eocene of Br.). R: Aliso 12(1990)583. Dwarf or stout unarmed palms with bisex. fls. Cult. orn. with comm. imp. fibres & thatch esp. ***S. causiarum*** (Cook) Becc. (Hispaniola, Virgin Is., Puerto Rico) – lvs split for matting, basketry & hat-making; ***S. palmetto*** (Walter) Schultes & Schultes f. (palmetto, cabbage palmetto palm, SE US to Bahamas & Cuba) – palm cabbage, lvs for thatch, mats etc., stems for furniture, wharf-piles

Sabatia Adans. Gentianaceae. c. 20 N Am., WI. R: Rhodora 73(1971)309. Cult. orn. (rose pinks) esp. ***S. angularis*** (L.) Pursh (E US) – cut-fl. in Netherlands, bitter principle medic.

Sabaudia Buscal. & Muschler = *Lavandula*

Sabaudiella Chiov. = *Hildebrandtia*

Sabazia Cass. (~ *Galinsoga*). Compositae (Mill.-Gal.). 17 C Am. mts. R: PMSUBS 4(1970)321

sabi grass *Urochloa mosambicensis*

Sabia Colebr. Sabiaceae. 19 SE As., Mal. R: Blumea 26(1980)1

Sabiaceae Blume. Magnoliidae – Sabiales. 3/c. 50 SE As., Mal., trop. Am. Trees, shrubs & lianes. Lvs simple or pinnate, in spirals, s.t. becoming 2-ranked; stip. 0. Fls small, bisex. (or pl. polygamo-dioec.) in panicles or thyrses; K (3–)5, imbr., unequal (2 outer small ones in *Meliosma* considered bracteoles by some), s.t. basally connate, C (4 or) 5 opp. or alt. with K, the 2 inner oft. smaller, A opp. C, all fert. (*Sabia*) or only 2 inner so, nectary-disk small, annular, around $\underline{G}$ (2(3)) with ± connate styles & 2(3) loc. each with 1 or 2 axile, hemitropous, unitegmic ovules. Fr. 1-loc. or ± dicoccous, fleshy or not, indehiscent; seed 1 with large oily embryo with curved radicle & 2 folded or coiled cotyledons in little or 0 endosperm. n = 12, 16

Genera: *Meliosma, Ophiocaryon, Sabia*

Fossils from Cretaceous of Europe. *Ophiocaryon* & *Meliosma* s.t. sep. as Meliosmaceae, *Sabia* diff. in its 5 fert. stamens

Sabicea Aubl. Rubiaceae (II 7). Incl. *Ecpoma, Pseudosabicea, Schizostigma*, c. 170 trop. Sri Lanka (1), Afr. & Madag. (115), Am. (54). R: H.F. Wernham (1914) *Monog. genus S.* Some spp. with fls on prostrate shoots. ***S. amazonensis*** Wernham (Braz.) – frs used to symbolize hearts in manhood initiations in Amazon

sabicú *Lysiloma* spp. esp. *L. sabicu*

Sabina Mill. = *Juniperus*

Sabinaria R. Bernal & Galeano. Palmae. 1 NW Colombia: ***S. magnifica*** Galeano & R. Bernal. R: Phytotaxa 144–2(2013)27

Sabinea DC. = *Poitea*

sabre bean *Canavalia ensiformis*

Sabulina Reichb. (~ *Minuartia*). Caryophyllaceae. 65 N hemisp., Chile (2). R: T 63(2014)85. ***S. stricta*** (Sw.) Reichb. (*M. s.*, Teesdale sandwort, Arctic & N Eur. mts) – protected in GB

Saccardophytum Speg. = *Benthamiella*

Saccellium Bonpl. = *Cordia*

Saccharum L. Gramineae (XXII 5). Incl. *Erianthus*, sect. *Ripidium* (= *Tripidium*), 37 trop. & warm (Eur. 3, China 12), s.s. 5 (3 wild, 2 cultigens; R: Baileya 23(1991)109). Sugar-cane the source of half of world's sugar (by 2010 1.68 billion t a yr), in last cent. almost all cvs grown derived from ***S. officinarum*** L., a complex agg. of hybrids derived from '*S. robustum* Brandes & Jeswiet ex Grassl' domesticated in NG ('noble' canes) where used for weaving house-walls, one selection being ***S. edule*** Hassk. ((lowland) pit pit) – immature infls a veg. in NG, but now hybrids with other spp. (e.g. ***S. spontaneum*** L. (Mal.) leading to most modern cvs) or cultigens also grown esp. in WI, Hawaii etc.; hybrids also formed with *Sorghum* spp. Crusaders took sugar from Israel to Medit. is. & Iberia; later taken by colonists to Madeira, Canary Is. & then Am. & is. of Ind. Ocean where plantations have done much to destroy the orig. veg.; cult. responsible for transmigration of peoples from Afr. to Carib., China to C Am., Polynesia to Aus. etc. C. 1660 discovered that fr. could be preserved with sugar; jam made before 1730. Dutch gave up New York for sugar lands of Surinam (1670s) & Fr. swapped Canada for sugar is. of Guadeloupe, WI (1763). By 1980s ann. consumption in US 57 kg per person (UK 36 kg): major contributor to obesity, caries etc. In refining sugar (raw sugar = rapadura, diet fad; until end C19 refined sold as conical sugar-loaves, sugar-cubes invented 1843 Moravia), uncrystallizable sugars a byproduct– molasses or treacle from unrefined (muscovado); fermented juice gives betsa-betsa (Madag.) & when distilled is rum (distilled in Am. colonies to pay for slaves from W Afr. to pay for sugar-cane; restrictions on molasses trade (Molasses Act 1730) a factor in push for US independence) depth of its colour dependent on level of caramel, & aguadent (cachaça) in Braz.; crushed fibre after extraction (bagasse) used as fuel in mills, garden mulch (Aus.), fibreboard & some kinds of paper, mixed with molasses used as cattle-feed; wax from lvs; at times of high oil-prices (e.g. 1970s–80s, 2000s), much grown in S Am. used as alcohol source for automobiles (10% Colombian fuel), also promoted as replacement for paraffin in greenhouse heaters; medic. (diarrhoea as water & salt absorption improved in presence of sugar); uba cane an old cv. now used only for fodder. Cult. orn. incl. ***S. bengalense*** Retz. (ekar, ekra, Iran to N Ind.); all spp. can be used as hosts for orn. parasite *Aeginetia indica*

Saccifoliaceae Maguire & Pires = Gentianaceae

Saccifolium Maguire & Pires. Gentianaceae (Saccifoliaceae). 1 Guayana Highland: ***S. bandeirae*** Maguire & Pires – shrub coll. 3 times, lvs saccate

Sacciolepis Nash. Gramineae (XXIV). 27 trop. & warm esp. Afr. R: JAA suppl. 1(1991)252

Saccocalyx Cosson & Durieu (~ *Thymus*). Labiatae (VII 2b). 1 Algeria: ***S. saturejoides*** Cosson & Durieu

Saccoglossum Schltr. Orchidaceae (V 16). 5 NG

Saccoglottis Walp. = *Sacoglottis*

Saccolabiopsis J.J. Sm. Orchidaceae (V 16c). 14 Indomal. to Fiji

Saccolabium Blume (~ *Gastrolobium*). Orchidaceae (V 16c). 5 Mal. R: KB 41(1986)833

Saccolena Gleason = *Salpinga*

Saccoloma Kaulf. Saccolomataceae (Dennstaedtiaceae s.l.). 12 trop. exc. Afr. (3 Am., 1 Madag.). ? Heterogeneous

Saccolomataceae Doweld (~ Dennstaedtiaceae). Polypodiidae – Polypodiales. 1/12 trop. R: T 55(2006)714.Terr. ferns with short or trunk-like dictyostelic rhiz. Petiole with omega-shaped vasc. strand (t.s.); blade pinnate to decompound without articulated hairs; veins free. Sori term. on veins; indusia cup- or pouch-shaped. n = c. 63

Genus: *Saccoloma*

Saccopetalum Benn. = *Miliusa*

Saccularia Kellogg = *Gambelia*

sacha inchi *Plukenetia* spp.

Sachokiella Kolak. (~ *Campanula*). Campanulaceae (I 7). 1 Caucasus: ***S. macrochlamys*** (Boiss. & Huet) Kolak. – capsule splitting into 5 long segments

Sachsia Griseb. Compositae (Inul.-Pluch.). Excl. *Rhodogeron*, 3 Florida, Bahamas, Cuba

Sacleuxia Baill. Apocynaceae (III; Asclepiadaceae I). 2 trop. E Afr. R: FTEA Apoc.,2(2012)135

Sacoglottis Mart. Humiriaceae. 8 trop. Am., 1 W Afr. (*S. **gabonensis*** (Baill.) Urb. (Liberian cherry)). R: CUSNH 35(1961)161. Noisy slash; bat-disp. in Amazon

Sacoila Raf. (~ *Spiranthes*). Orchidaceae (IV 2h). 7 warm Am. (US 2). R: HBML 28(1982)351. Hummingbird-poll., but some apomicts

Sacosperma G. Taylor. Rubiaceae (IV 1). 2 trop. Afr.

sacred bamboo, Chinese *Dracaena sanderiana*

sadao *Azadirachta indica*

Sadiria Mez. Primulaceae (Myrsinaceae). 4 E Himal. & Assam

Sadleria Kaulf. (~ *Blechnum*). Blechnaceae (Blechnoideae). 5–6 Hawaii. R: Pac. Sci. 51(1997)295. Low tree-ferns with 2-pinnate fronds, oft. on lava; only endemic fern genus in Hawaii. S.t. cult. esp. ***S. cyatheoides*** Kaulf. – form. source of red dye; scales of this & other spp. the 'pulu' used as packing material

safflower *Carthamus tinctorius*

Saffordia Maxon = *Trachypteris*

Saffordiella Merr. = *Myrtella*

saffron *Crocus sativus*; **s. thistle** *Carthamus tinctorius*

safu *Dacryodes edulis*

sagapenum *Ferula* spp.

sage *Salvia officinalis*; **blue s.** *S. azurea*; **s. brush** *Artemisia tridentata*; **Jerusalem s.** *Phlomis fruticosa*; **pineapple (-scented) s.** *S. elegans*; **red** or **scarlet s.** *S. coccinea, S. splendens*; **Spanish s.** *Coleus amboinicus, S. lavandulifolia*; **wood s.** *Teucrium scorodonia*

Sageraea Dalz. Annonaceae (IV 7). 9 Ind. & Sri Lanka to Philippines. R: NJB 17(1997)41

Sageretia Brongn. Rhamnaceae (6). 35 Somalia & SW As. to Taiwan, trop. & warm Am. Some ed. fr. & bonsai subjects imported to Eur. ***S. thea*** (Osbeck) M. Johnston (*S. theezans*, C & E Ind. & China to Jap.) – lvs used as tea in Vietnam

Sagina L. Caryophyllaceae (II 1). c. 30 N temp. (Eur. 12, N Am. 10), trop. mts. Herbs, usu. tufted, s.t. C 0. Pearlwort (some cult. orn. rock-pls) esp. ***S. procumbens*** L. (N temp.) – bad weed of paving etc., but '**Boydii**' (*S. b.*, Scotland) cult.

Sagittanthera Mart.-Azorin & al. (~ *Drimia*). Asparagaceae. 3 E Cape. R: Phytotaxa 98(2013)46. Buzz-poll.

Sagittaria Rupp. ex L. Alismataceae. c. 30 cosmop. (Eur. 4) esp. Am. (N Am. 24 – R: FNA 22(2000)11). R: MNYBG 9(1955)179, AZBB 76(1972)1, 78(1972)1. Monoec. stoloniferous, oft. tuberiferous herbs of aquatic habitats; lvs of var. types, submerged being ribbon-shaped, floating ones with ovate lamina, the projecting ones sagittate. Some with ed. tubers incl. ***S. cuneata*** E. Sheldon (also horse medic.) & ***S. latifolia*** Willd. (wapato, N Am.) – dioec. in ever-wet habitats, monoec. in ephemeral wet. Some cult. orn. esp. ***S. sagittifolia*** L. (*S. trifolia*, arrowhead, Euras.) – ed. tubers (kyor, Kashmir) usu. fed to pigs in Mal., where grown in paddy-fields

Saglorithys Rizz. = *Justicia*

sago palm *Metroxylon* spp. (see also *Arenga, Caryota, Cycas, Encephalartos, Eugeissona, Mauritia* & *Syagrus*); **s. pondweed** *Potamogeton pectinatus*

Sagotia Baill. Euphorbiaceae (Crot.). 2 NE S Am. R: AA 15, supp. (1985)81

Sagraea DC. = *Miconia*

saguaragy bark *Colubrina glandulosa*

saguaro *Carnegiea gigantea*

Sahagunia Liebm. = *Clarisia*

Saharanthus Crespo & Lledó (*Lerrouxia*; ~ *Limoniastrum*). Plumbaginaceae (II). 1 S Morocco, N Sahara: ***S. ifniensis*** (Caball.) Crespo & Lledó. R: BotJLS 132(2000)169

sailor, blue *Cichorium intybus*

sainfoin *Onobrychis viciifolia*

Saint Augustine grass *Stenotaphrum secundatum*; **S. Barnaby's thistle** *Centaurea solstitialis*; **S. Bernard's lily** *Anthericum liliago*; **S. Bruno's l.** *Paradisea liliastrum*; **S. Dabeoc's heath** *Daboecia cantabrica*; **S. Ignatius's bean** *Strychnos ignatii*; **S. John's wort** *Hypericum* spp.; **S. Lucia fir** *Abies bracteata*; **S. Lucie cherry** *Prunus mahaleb*; **S. Patrick's cabbage** *Saxifraga spathularis*; **S. Vincent arrowroot** *Maranta arundinacea*

Saintpaulia H. Wendl. = *Streptocarpus* (but see FTEA Gesn. (2006)50)

Saintpauliopsis Staner (~ *Staurogyne*). Acanthaceae (I). 1 trop. Afr.: ***S. lebrunii*** Staner. R: Aliso 32(2014)28

Saionia Hatusima (~ *Oxygyne*). Burmanniaceae. 3 Jap. R: JJB 90(2015)116

Sairocarpus D.A. Sutton (~ *Antirrhinum*). Plantaginaceae (Ant.-Ant.; Scrophulariaceae s.l.). 13 SW N Am. R: D.A. Sutton, *Rev. Antirrh.* (1988)461

Sajanella Soják. Umbelliferae (III 8). 1 S Siberia: ***S. monstrosa*** (Willd.) Soják

Sajania Pim. = *Sajanella*

sajna *Moringa oleifera*

sakaki *Cleyera japonica*

sake *Oryza sativa*

Sakersia Hook.f. = *Dichaetanthera*

Sakoanala R. Viguier. Leguminosae (III 2). 2 NW & E Madag. R: NSL 17(1980)164

sal *Shorea robusta*

salab-misri = salep

Salacca Reinw. Palmae (I 3b). 21 Indomal. (Mal. 17). Subsessile spiny dioec. palms (salak). Infls r. to form new pls in ***S. flabellata*** Furtado (Malay Pen.) & ***S. wallichiana*** Mart. (rakum) – several cvs, sarcotesta ed., petioles used for walls & hen-coops in S Thailand. ***S. zalacca*** (Gaertn.) Voss (*S. edulis*, snake fr., Mal.) – poll. weevils, cult. for ed. fr. (esp. salads), s.t. pickled

Salacia L. Celastraceae (IV). c. 200 trop. Roux's Model; oft. lianes with dimorphic branching. Some locally ed. fr.; some known to reduce glucose levels in (rat) blood

Salacicratea Loes. = *Salacia*

Salacighia Loes. Celastraceae (IV). 2 W trop. Afr. R: KB 11(1956)246

Salaciopsis Bak. f. Celastraceae (I). 6 New Caled.

salad, millionaire's = palm hearts

saladini mixed salad lvs of various spp.

salago *Wikstroemia* spp.

salak *Salacca* spp.

salal *Gaultheria shallon*

Salaxis Salisb. = *Erica*

Salazaria Torrey = *Scutellaria*

Salcedoa Jiménez & Katinas. Compositae (Stifft.). 1 Hispaniola: ***S. mirabiliarum*** Jiménez & Katinas. R: SB 29(2004)991

Saldanhaea Bur. = *Cuspidaria*

Saldinia A. Rich. ex DC. Rubiaceae (IV 7). 22 Madag. R: Candollea 16(1957)95

salep Dried starchy tubers of var. orchids form. used in cooking & medic., esp. from spp. of *Eulophia* (As.), *Dactylorhiza* (Iran) & *Orchis* (Eur.)

Salicaceae Mirb. Magnoliidae – Malpighiales. Incl. Flacourtiaceae p.p., Samydaceae, Scyphostegiaceae, 54/1250 trop. to temp. (esp. N, few in Aus., NZ 0) to Arctic. R: KB 57(2002)165. Trees & shrubs, s.t. creeping, s.t. decid., oft. dioec., s.t. with phenolic heterosides like salicin, rarely with alks; ectomycorrhizae recorded (1.). Lvs simple, serrate, in spirals but s.t. forming planar sprays in maturity (*Salix*) to opp. (7.); stip. usu. present. Fls insect- or wind-poll., in term. or axillary infls, oft. small. K (0-) 3–8 (-15), valvate or imbr., free to basally connate, s.t. K obconical or turbinate; C 0 or = K, free; disk usu. cupular & adnate to adaxial surface of K, or annular & extrastaminal (or – gynoecial in female fls), s.t. forming sep. glands; A (1)2 (most *Salix*) – ∞, s.t. ± connate in groups on receptacle or even forming a column, usu. developing centrifugally; $\underline{G}$ (2–5(-13)), 1-loc. s.t. incompletely septate, with usu. parietal placentation with usu. orthotropous ovules & 1–8 distinct styles, s.t. branched. Fr. a berry or (2)3(–6)-valved capsule, less oft. a drupe or samara; seeds oft. arillate, when wind-disp. (1.) minute, not long-lived, with tuft of hairs derived from funicle or placenta, embryo straight in little or 0 oily endosperm. n = 9–12, 19

Classification & principal genera (*Oncoba* unplaced):

1. **Saliceae** (Salicaceae s.s.; dioec.; infl. a spike or catkin, P 0, capsule 2–4-valved): *Populus*, *Salix* (only)
2. **Flacourtieae** (Flacourtiaceae s.s.; usu. dioec., K 3–8(-15), C 0, fr. a berry less oft. drupe or capsule): *Azara, Flacourtia, Idesia, Ludia, Xylosma*
3. **Samydeae** (Samydaceae; rarely dioec.; infls usu. axillary, K 3–7, C 0, fr. a capsule rarely berry): *Casearia, Lunania, Samyda*
4. **Homalieae** (rarely dioec., K 4–8, C = K or rarely 0, fr. usu. capsule): *Calantica, Homalium*
5. **Scolopieae** (s.t. dioec., fls s.t. solit., K 3–5(6), C = K or more, fr. a berry or capsule): *Scolopia*

6. **Prockieae** (rarely dioec.; K 3–5, C = K or 0, fr. a berry or capsule): *Banara*
7. **Abatieae** (hermaph., lvs opp., infl. term., K 4 or 5, C 0, fr. a capsule, trop. Am.): *Abatia*
8. **Bembicieae** (hermaph., infl. axillary, K 5, C 5, capsule 3-valved, Madag.): *Bembicia* (only)
9. **Scyphostegieae** (Scyphostegiaceae; dioec.; infl. term. with telescoped accrescent bracts, P 3 + 3, fr. a capsule, Borneo): *Scyphostegia* (only)

Salicin of S. s.s. long known from former Flacourtiaceae (*Idesia*), which also are hosts to similar rust-fungi & fed on by similar Lepidoptera. The simple fls of S. s.s. were once considered primitive but now seen as simplified & advanced (cf. similar trends assoc. with wind-poll. in Plantaginaceae s.l.); fossil *Pseudosalix* of Eocene N Am. morph. intermediate between 1. & rest of S. Some workers maintain 3. & 9. as sep. fams (see also Malvaceae, Primulaceae & Preface). For disposition of rest of form. Flacourtiaceae see there. Timber pulp, osiers, medic. (*Casearia, Homalium*), but esp. *Populus* (aspen, poplars) & *Salix* (sallows, willows; wood separable by Dutch clog-makers spitting on heel, as tyloses of *S.* prevent air passing though, *P.*, without, permits) with many hybrids, also cult. orn. as are spp. of *Azara* & *Idesia*; ed. fr. (*Dovyalis, Flacourtia*) & dyes (*Abatia, Pineda*)

Salicornia Tourn. ex L. Amaranthaceae (Chenopodiaceae II 3). S.s. 10; (incl. *Sarcocornia*) 25 cosmop. (Eur. 12, Aus. 3 (*Sarcocornia*), sea-coasts & other salt habitats (glasswort, marsh samphire). Succ. herbs with scale-like lvs & jointed nodes; fls v. small, in 3s in axils, sunken, K 4 fleshy, C 0, A 1 or 2, in ***S. europaea*** L. (chicken claws) the median fls largest prod. 1 large seed (0.78 ± 0.1 mg), the smaller laterals each 1 small seed (0.24 ± 0.4 mg), the larger ones more salt-tolerant (floating seeds viable for up to 4 months). Diploids usu. homozygous inbreeding lines. Some eaten as spinach or pickled; oil-rich seeds, tried comm. in SW US (esp. ***S. bigelovii*** Torrey)

Salicorniaceae Martinov = Amaranthaceae

saligna gum *Eucalyptus saligna*

saligot *Trapa natans*

Salix Tourn. ex L. Salicaceae (1). c. 450 cold (***S. arctica*** Pallas (circumpolar) with *Papaver radicatum* most northerly vasc. pl. (83° N)) & temp. N (Eur. 64, China 257, N Am. 113) few S (Aus. 0). Willows, osiers, sallows; v. many hybrids (59 combinations (10 involving 3 spp.) in GB alone). R: C. Newsholme (1992) *Willows*. Trees 30 m high to creeping shrublets of 1–2 cm, some rheophytes; usu. insect-poll. (attractants incl. benzoids & isoprenoids), creeping ***S. herbacea*** L. (circumboreal) poll. by 'biting midges', catkins of Arctic spp. with temp. several degrees higher within than outside; foliage oft. disfigured by sawfly galls in summer; male seedlings of *S. myrsinifolia* Salisb.–*S. phylicifolia* L. complex in N Sweden consumed by voles 3 times as much as females, latitudinal variation in maleness assoc. with vole herbivory (also more males on vole-less is.); some timbers, coppice for comm. fuel, pliable branches (withies or withes) esp. of *S.* ×*fragilis* & *S. viminalis* used in basketry (earliest weave for chairs & still the best for hot-air balloon baskets, trugs – since 1825 though now with chestnut frame), where coppice shoots from trees grown in osier-beds oft. used (in GB usu. *S. viminalis*, *S. triandra* (*S. amygdalina*) & *S. purpurea* with locally favoured clones; medic. since Hippocrates, but Rev. Edward Stone promoted *S. alba* bark as effective against 'agues' (c. 1717), from 1878 used for rheumatic fever having effects similar to quinine, bark the source of analgesic aspirin (see *Filipendula*; purified 1897, now synth., though action still not understood; by 1999 some 100 billion tablets a yr. consumed), active salicin (converted to salicylic acid in body) suppressing cyclooxygenase & therefore thromboxane which causes blood-clots, hence used to keep blood 'thin'; r. yield blue-red dyes; folded willow-bark used as double reeds for wind instruments in Turkey; many cult. orn. readily prop. by cuttings (r. in water); long considered a symbol of chastity ('spontaneous generation' & (males) no seeds. ***S. alaxensis*** (Anderss.) Coville (N Am.) – young shoots ed. Alaska (high in vitamin C); ***S. alba*** L. (white willow, Euras., Medit.) – twigs for basketry, timber form. for Dutch clogs, lvs form. used as tea, cvs incl. '**Caerulea**' (cricket-bat w.) – timber for c.bs (handles of cane), imp. industry in Kashmir valley (10 000 employed in making 1–1.2 M bats a yr.), '**Britzensis**', '**Chermesina**' & '**Vitellina**' (golden w.) – grown for coloured bark (orange & yellow) consp. in winter; ***S. babylonica*** L. (*S. matsudana*, N As., invasive Aus.) – inspiration for willow-pattern plates ('Napoleon's w.' at N.'s grave on St Helena (though poss. orig. *Acacia vestita*) prop. by cuttings & grown in Scotland, Aus., NZ, S Afr., USA), parent of common weeping willows (*S.* × *sepulcralis*) & said to be the tree under which children of Israel mourned & wept (prob. *Populus euphratica*) though 'weeping' to describe trees with trailing growth may stem from this, while all willows are assoc. with sadness, '**Tortuosa**' (*S. m.* 'T.', corkscrew w.) – branchlets spirally twisted, cult. 'orn.'; ***S. caprea*** L. (pussy or goat w., Eur.) – charcoal, poss. source of polyphenols, the usu. 'palm' of Palm Sunday in rural GB, the week before known as w. week in Russia, '**Pendula**' (Kilmarnock w.) – crooked drooping branches; ***S. canariensis*** Buch (Macaronesia) – part of Madeiran laurel forest; ***S. caroliniana*** Michaux (N Am.) – imp. local medic.; ***S. cinerea*** L. (Euras.) – invasive Aus., NZ, though hybrid with *S. alba* = ***S.* × *reichardtii*** A. Kern., sterile male triploid not); ***S.* ×*fragilis*** L. (*S. alba* × *S. euxina* Belyaeva, crack or brittle w., Euras., invasive Aus., NZ, S Afr.) – basketry & charcoal, '*S.* × *rubens* Schrank' a back cross with *S. alba*, basketry in Braz.; ***S. humboldtiana*** Willd. ('*S. chilensis*', Humboldt w., Mex. & trop. Am.) – fastigiate forms like Lombardy poplars; ***S. humilis*** Marshall (prairie w., E N Am.); ***S. lasiolepis*** Benth. (W N Am.) – shrubby growth-form in Arizona maintained by galling sawflies; ***S. mucronata*** Thunb. (*S. capensis*, Cape w., Afr.) – locally medic., subsp. ***subserrata*** (Willd.) R. Archer & Jordaan (*S. subserrata*, N Afr.) – lvs used in garlands in Tutankhamun's tomb; ***S. nigra*** Marshall (black w., N Am.) – catkins appear with lvs, pulp, timber for artificial limbs etc.; ***S. purpurea*** L. (basket w., Eur. & Medit. to Jap.) – osiers; ***S. repens*** L. (creeping w., Euras.); ***S.* × *sepulcralis*** Simonkai (*S. alba* × *S. babylonica*, weeping w.) – hardier than *S. babylonica*, esp. '**Chrysocoma**' (*S. alba* 'Tristis', *S.* 'Chrysocoma'), largest weeping tree in cult.; ***S. tetrasperma*** Roxb. (? subtrop. As.) – cult. Mal. where all male; ***S. triandra*** L. (Eur.) – osier; ***S. viminalis*** L. (basket w., Euras.) – sp. most used for basketry, many cvs for biomass prod.

sallow *Salix* spp. esp. *S. caprea* etc.

Sally-my-handsome *Carpobrotus edulis*; **sally wattle** *Acacia longifolia*

Salmalia Schott & Endl. = *Bombax*

Salmea DC. Compositae (Helia.- Spil.). 10 trop. Am. R: SB 16(1991)462

Salmeopsis Benth. = *Salmea*

Salmiopuntia Frič ex Guiggi (~ *Opuntia*). Cactaceae. 1 Braz.: ***S. salmiana*** (Pfeiff.) Guiggi

salmon gum *Eucalyptus salmonophloia*

salmonberry *Rubus spectabilis*

Salmonopuntia P. Heath = *Austrocylindropuntia*

salmwood *Cordia alliodora*

Salomonia Lour. Polygalaceae (IV). Excl. *Epirixanthes*, 6 Indomal. to Aus. R: BNSMTokyo B21(1995)1

salpani *Desmodium gangeticum*

Salpianthus Bonpl. Nyctaginaceae (2). 3 Mex. & C Am.

Salpichlaena Hook. Blechnaceae (Blechnoideae). 3 trop. Am. R: AFJ 98(2008)53. ***S. volubilis*** (Kaulf.) Hook. – rachis scandent, twining

Salpichroa Miers. Solanaceae (IV). c. 15 SW US to S Am. ***S. origanifolia*** (Lam.) Baill. (*S. rhomboidea*, SE Bolivia, Paraguay, Urug., Arg., invasive Aus., natur. UK, Medit., US) – ed. fr. of poor flavour but cult. S US as ground-cover & bee-forage, antifeedant (fly larvae) due to withanolides

Salpiglossis Ruíz & Pavón. Solanaceae (II 3). 2 S Andes. R: HBML 27(1980)4. ***S. sinuata*** Ruíz & Pavón – pollen in tetrads, cult. orn. with yellow or dark purple to scarlet or almost blue fls with markings, many cvs

Salpinctes Woodson = *Mandevilla*

Salpinctium T. Edwards = *Asystasia* (but see SAJB 55(1989)6)

Salpinga Mart. ex DC. Melastomataceae. 8 trop. S Am.

Salpingostylis Small (~ *Calydorea*). Iridaceae (VII 5). 1 Florida: ***S. coelestina*** (Bartram) Small. R: Britt. 27(1975)374

Salpistele Dressler (V 13). Orchidaceae. 6 C Am. to Ecuador. R: MSB 39(1991)123

Salpixantha Hook. Acanthaceae (III 1). 1 Jamaica: ***S. coccinea*** Hook.

salsify *Tragopogon porrifolius*; **Spanish s.** *Scorzonera hispanica*

salsilla *Bomarea edulis*

Salsola L. Amaranthaceae (Chenopodiaceae III 3). Excl. *Caroxylon*, *Climacoptera*, *Kali* (q.v.), *Oreosalsola*, c. 100 cosmop. sea-coasts or other saline habitats. Prob. still heterogeneous. Alks; phytoecdysones in some Egyptian spp.; glassworts – ash form. used in soap- & glass-making (barilla); some locally eaten or cattle fodder. ***S. kali*** L. (= *Kali*; blackbush, coastal Euras., N Afr., introd. Aus., N Am., etc.) – weedy, can grow away from saline habitats; ***S. nitraria*** Pallas (Russia, SW & C As.) – used as boron indicator in Russia; ***S. soda*** L. (agretti, Medit.) – salad veg.; ***S. tragus*** L. (*S. australis*, *S. kali* subsp. *ruthenica*, *S. pestifer*, Russian thistle, Euras., natur. Aus., S Afr., Am.) – a tumbleweed poll. by bees & wasps, since c. 1900 an agricultural pest in N Am., where introd. 1871 (S Dakota) in flaxseed

salt fern *Asplenium acrobryum*

Salta A. Sanchez (~ *Ruprechtia*). Polygonaceae. 1 S S Am.: ***S. triflora*** (Griseb.) A. Sanchez. R: SB 36(2011)708

saltbush *Atriplex* spp. esp. *A. vesicaria* (Aus.); **old man s.** *A. nummularia*

Saltera Bullock. Crypteroniaceae (Penaeaceae). 1 SW Cape: ***S. sarcocolla*** (L.) Bullock. R: Strelitzia 10(2000)441

Saltia R. Br. ex Moq. Amaranthaceae (I 2). 1 S Arabia: ***S. papposa*** (Forssk.) Moq.

Saltugilia (V. Grant) L.A. Johnson. Polemoniaceae (III). 4 Calif. R: Aliso 19(2000)69, Madroño 48(2001)198

saltwort *Batis maritima*, *Lysimachia maritima*

Salvador henequen *Agave vivipara* var. *letonae*

Salvadora Garcin ex L. Salvadoraceae. 4 warm Afr. to trop. As. ***S. persica*** L. (trop. Afr. to As.) – 'tooth-brush tree' (Champagnat's Model), the branches with bundles of vessels in thick-walled fibres, sep. by thin-walled parenchyma with included phloem which two are destroyed when twigs beaten such that fibres & vessels form 'bristles' of chewing-stick much used by Bedouin & others as toothbrushes, antiseptic reducing tooth decay, healing & hardening gums (chlorine, trimethylamine, resin, silica, sulphur & vitamin C; 'peelu extract' used in comm. toothpastes in UK); shoots ed. & camel fodder; ash yields salt, seeds a wax used on skin & for candles, fr. ed. (sweet but peppery, poss. the mustard seed of Bible), wood useful (poss. fuelwood source on saline soils)

Salvadoraceae Lindl. Magnoliidae – Brassicales. 3/10 warm OW esp. dry. Small trees or shrubs, s.t. scrambling, s.t. with mustard oils (but no myrosin cells) or alks; intraxylary phloem oft. present. Lvs simple, opp., oft. leathery; stip. small or 0, axillary thorns s.t. present. Fls small, bisex. or not (pls then dioec. or polygamous), reg., in varied infls; K (2–4(5)) with oft. imbr. lobes, C 4(5), imbr., free (basally connate in *Salvadora*), A alt. with C, free (*Azima*), basally united (*Dobera*) or adnate to C base (*Salvadora*), anthers with longit. slits, disk 0 or glands alt. with A, $\underline{G}$ (2) with term. short style, 1-loc. (2- in *Azima*), each loc. with 1 or 2 basal erect anatr. bitegmic ovules. Fr. usu. drupe or 1-seeded berry; embryo with thick cordate oily cotyledons in 0 endosperm. n = 12

Genera: *Azima, Dobera, Salvadora*

Affinities unclear. Some locally useful pls

Salvadoropsis Perrier. Celastraceae (I). 3 Madag. R: BSBF 91(1944)96

Salvertia A. St-Hil. Vochysiaceae (I). 1 S Braz., campos: ***S. convallariodora*** A. St- Hil. – hawkmoth-poll. R: RTBN 41(1948)397

Salvia Tourn. ex L. Labiatae (VII 2a). c. 950 trop. (Afr. 59) to temp. (Eur. 36; Aus. 1) esp. Am. (591, Mex. c. 250), Sino-Himal. & SW As. (Turkey 86, 50% endemics). ?Heterogeneous, prob. incl. *Perovskia, Rosmarinus*.; 8 subg. – R. (subg. *Jungia* (*Calosphace*) prob. to be excl.): FRB 110(1938)1. Shrubs & herbs with A 2, each with elongate connective & single fert. anther cell brought into contact with back of poll. bees etc. (c. 185 Am., 3 OW, bird-poll. – usu. red) by other end of versatile anther being pushed upwards by pollinator probing for nectar; protandry, position of A being taken up by style & stigma later; ***S. brandegeei*** Munz (Calif.) – heterostylous (only zygomorphic example known). Cult. orn. & medic. (Lat.: *salveo* = I heal), some culinary (chia, sage) etc., though ***S. reflexa*** Hornem. (N to C Am.) & other spp. toxic to stock in Aus.; ***S. sclarea*** L. & ***S. verbenaca*** L. (Eur., SW As., invasive Aus.) & other spp. with seeds swelling in water so used to remove foreign bodies from eyes (*sclarea* = *clarus*, clear). ***S. apiana*** Jepson (? = *S. camphorata* Hort. Huber, Calif.) – local medic., lvs & seeds ed., bee-forage; ***S. azurea*** Michaux ex Lam. (blue sage, SE US) – cult. orn.; ***S. coccinea*** Etl. (scarlet s., trop. S Am.) – cult. orn., many cvs; ***S. columbariae*** Benth. (SW US), ***S. hispanica*** L. (C Mex.), etc. – chia, seeds high in omega-3 fatty acids, antioxidants, protein & fibre, form. staple in pre-Columbian C Am., made into a cake, soups & a drink, now esp. for cholesterol control etc., cult. 'superfood' esp. W Aus. for bread, breakfast cereals etc., oil in painting etc.; ***S. divinorum*** Epling & Játiva (Mex., cultigen allied to ***S. venulosa*** Epling (Colombia)) – hallucinogen (salvinorin A) used only in ceremony by Mazatecs, recreational drug in Eur. & US (banned in Aus.), also contains loliolide (potent ant-repellent); ***S. dombeyi*** Epling (cultigen Bolivia, Peru) – cli. to 6 m, red fls 9–13 cm long, fr. unknown; ***S. elegans*** Vahl ('*S. rutilans*', pineapple (-scented) s., Mex., Guatemala) – cult. orn.; ***S. lavandulifolia*** Vahl (Spanish sage, S Eur.) – oil alleged to retard Alzheimer's disease; ***S. microphylla*** Kunth (*S. grahamii*, Mex.) – cult. orn. hardy shrub with red fls; ***S. miltiorhiza*** Bunge (dan shen, den shan, China) – TCM used in heart conditions; ***S. officinalis*** L. (sage, S Eur. & Medit.) – lvs used, esp. dried, in cooking (that in US oft. adulterated with 50–95% *S. fruticosa* Mill. (*S. cypria*, *S. triloba*, Sicily to Syria) –

anti-oxidant, folk-med. for heart disease in Crete), incl. in cheese (e.g. sage Derby in UK), form. cleaning teeth, local drinks (chahomilia, Cyprus), flavour due to terpineol & thujone, sage oil (c. 35 t a yr by 1990s) capsules allegedly aiding memory; ***S. sclarea*** (clary, S Eur. to C As.) – aromatic oil used in soaps & scent (esp. lavender & bergamot), eau-de-cologne, wine (nutmeg flavour), vermouth & liqueurs; ***S. splendens*** Sellow ex Schultes (scarlet s., S Braz.) – the common red bedding salvia of gardens, several cvs; ***S. viridis*** L. (*S. horminum*, bluebeard, S Eur.) – cult. orn.

Salviastrum Scheele (*non* Heister ex Fabr.) = *Salvia*

Salvinia Séguier. Salviniaceae. 10 trop. & warm temp. (Eur. 1 with 2 escapes; not native SE As. to Aus.). R: Hedwigia 74(1935)257. Water ferns with water-repellent hairs on adaxial surface – those of ***S. auriculata*** Aubl. (trop. Am.) with the tips of 4 arms united like a lantern. Sporocarp on submerged lvs rep. a single sorus with indusium forming sporocarp wall, the first-formed with up to 25 megasporangia, the later with microsporangia in large numbers; all but 1 megaspore abort & this is surrounded by thick perispore resembling integument of gymnosperm seed; 64(32) microspores lie on surface of a frothy massula, the male prothalli projecting all round after germ. & each 1 with 2 antheridia giving 8 sperms. Megaspore remains in sporangium, the prothallus protruding after dispersal of sporocarp & bearing several archegonia. Some cult. orn. incl. ***S. natans*** (L.) All. (Euras., N Afr.) – in GB in earlier interglacials; ***S. molesta*** D. Mitch. ('*S. auriculata*', Kariba weed) – sterile pentaploid, poss. of hybrid orig. in SE Braz., a pestilential weed of waterways in Aus., Am. (1985), W Pacific, trop. (doubling time of 4.5 days in Lake Naivasha in 1980s) & S Afr., where prob. introd. as cult. orn. but reduced by salvinia weevil (*Cyrtobagous salviniae*) introd. Afr. from Braz. c. 1930. High in tannin, lignin etc. so restricted use as fodder

Salviniaceae Martinov. Polypodiidae – Salviniales. Incl. Azollaceae, 2/15 trop. & warm. R: T 55(2006)711. Heterosporous free-floating water ferns, simple decid. chlorophyllous r. in *Azolla*. Stems dichot. with rudimentary protostele & sessile lvs spiral (becoming 2-ranked in *Azolla*, in whorls of 3, of which 2 floating & 1 submerged & finely divided so as to resemble r. in *Salvinia*). n = 9 (*Salvinia*; lowest known in ferns; 22 (*Azolla*)
Genera: *Azolla, Salvinia*
Known from Cretaceous (5 genera incl. *Salvinia* & *Azolla*)

Salweenia Bak. f. Leguminosae (III 2). 2 SW China. R: T 60(2011)1370

Salzmannia DC. Rubiaceae (III 4). 2 E Braz.

sama *Panicum sumatrense*

Samadera Gaertn. (~ *Quassia*). Simaroubaceae. 5–6 Madag. to Aus. ***S. indica*** Gaertn. (*Q. i.*) – medic. oil & insecticide, wood for knife-handles (Sarawak)

Samaipaticereus Cárdenas. Cactaceae (III 4). 1 Bolivia: ***S. corroanus*** Cárdenas

saman *Samanea saman*

Samanea (Benth.) Merr. (~ *Albizia*). Leguminosae (II 4). 3 trop. Am. R: MNYBG 74, 1(1996)117. ***S. saman*** (Jacq.) Merr. (*A. s.*, rain tree, N S Am.) – decid. annually but lvs last only 6 months in Mal., distinctive leaf-movements (circadian oscillations & light-change-induced through massive fluxes of K^+, Cl^-, H^+ ions through pulvinar motor-cell membranes), poss. form. disp. by extinct Pleistocene mammals, good bee-fodder, timber (guango) esp. Jamaica, much-planted trop. street-tree with shallow grooved bark colonized by epiphytes & 'rain' excreted by Homoptera, assoc. with *Azteca* ants in Costa Rica

Samarorchis Ormerod = *Robiquetia* (but see Taiwania 53(2008)160)

Sambirania Tard. = *Lindsaea*

Sambucaceae Batsch ex Borkh. = Viburnaceae

Sambucus Tourn. ex L. Viburnaceae (Caprifoliaceae s.l.). 9 temp. (Eur. 3) & subtrop. R: DB 223(1994). Elder. Shrubs & small trees (Champagnat's Model) with pithy stems & alks; extrafl. nectaries. Cult. orn., some ed. fr. (some toxic), some locally med. (sambucol said to shorten duration of influenza symptoms by 56% & reduce muscle pain). ***S. caerulea*** Raf. (blue e., W N Am.) – local medic., cult. orn., ed. fr.; ***S. canadensis*** L. (*S. nigra* subsp./var. *c.*, American e., E N Am.) – imp. local medic., cult. for fr. (the elderberry of US) used in jellies, pies, sauces & wines, several cvs, also cult. orn. incl. '**Maxima**' with fl. heads to 40 cm diam.; ***S. ebulus*** L. (danewort, dwarf e., Euras., Medit. with distinct subsp. on trop. Afr. mts, a subpachycaul shrubby pl. with marcescent lvs) – ± herbaceous, not setting good seed in GB, local medic. (dropsy), drupes source of blue dye used in colouring leather etc.; ***S. nigra*** L. (elder, e.berry, Euras., Medit., invasive NZ) – foetid pithy shrub, weed assoc. with superstition, drupes used for making wine, form. (Romans) used as hair-dye (Pan dyed himself with it), fls form. medic. (colds, burns, insect-bites, gout; e.flower water a skin cleanser – 'Eau de Sureau' & eye-lotion) & ed. as elderflower pancakes, as a cordial (e.f. pressé) & fermented to e.f. 'champagne' & liqueur (St Germain Liqueur), wood for mouth-pieces of Dutch wind-instrument (midwinterhoorn) blown around Christmastide, soft pith (hardwood very hard) used for holding specimens when sectioning botanical material, fly-control since C2 in W Eur.; ***S. pubens*** Michaux (*S. racemosa* subsp. *p.*, stinking e., E N Am.) – red fr. ined.; ***S. racemosa*** L. (Euras.) – cult. orn. with several cvs incl. '**Tenuifolia**', dwarf suckering shrub with lvs like a Jap. maple

Sameraria Desv. = *Isatis*

Samolaceae Raf. = Primulaceae

Samolus Tourn. ex L. Primulaceae (Samolaceae). 10–12 cosmop. esp. salt-marshes (Eur. 1: ***S. valerandi*** L.)

Sampaiella J.C. Gomes = *Adenocalymma* (but see RBB 27(2004)782)

Sampantaea Airy Shaw. Euphorbiaceae (Acal.-Acal.-Cleid.). 1 SE As.: ***S. amentiflora*** (Airy Shaw) Airy Shaw. R: KB 26(1972)328

Sampera Funk & H. Robinson (~ *Liabum*). Compositae (Liab.). 8 Colombia, Ecuador, Peru. R: PBSW 122(2009)156

samphire *Crithmum maritimum*; **golden s.** *Limbarda crithmoides*; **marsh s.** *Salicornia* spp.

Samuela Trel. = *Yucca*

Samuelssonia Urb. & Ekman. Acanthaceae (III 2c). 1 Hispaniola: ***S. verrucosa*** Urb. & Ekman

Samyda Jacq. = *Casearia* (but see FN 22(1980)225)

Samydaceae Vent. = Salicaceae

San Domingo apricot *Mammea americana*; **S. D. boxwood** *Phyllostylon brasiliensis*; **S. Pedro cactus** *Echinopsis pachanoi*

Sanango Bunting & Duke. Gesneriaceae (I). 1 Ecuador, Peru: ***S. racemosum*** (Ruíz & Pavón) Barringer – tree to 15 m form. referred to Loganiaceae etc.

Sanblasia L. Andersson = *Calathea* (but see NJB 4(1984)21)

Sanchezia Ruíz & Pavón. Acanthaceae (III 2c). 55 trop. Am., esp. Andes. R: JWAS 16(1926)484, Novon 24(2015)214. Some cult. orn. ***S. thinophyllum*** Leonard – decoction used to bathe heads of girls depilated (by pulling) during initiation ceremonies in NW Amaz.

sanchi *Panax pseudoginseng*

Sanctambrosia Skottsb. Caryophyllaceae (I 1). 1 S Ambrosio Is. (Chile): ***S. manicata*** (Skottsb.) Skottsb. – small tree or shrub, K 4, C 4, A 4 + 4, G (3)

sand box tree *Hura crepitans*; **s. cherry** *Prunus pumila*; **s. leek** *Allium scorodoprasum*; **s. myrtle** *Kalmia buxifolia*; **s. spurrey** *Spergularia* spp.; **s.wort** *Arenaria* spp.; **Teesdale s.w.** *Minuartia stricta*

sandalwood or Indian s. *Santalum album*; **Afr. s.** *Excoecaria africana*; **Australian s.** *S.* spp.; **E Afr. s.** *Osyris lanceolata*; **s. fan** *Chrysopogon zizanioides*; **red s.** *Adenanthera pavonina*, *Pterocarpus santalinus*

sandarac *Tetraclinis articulata*; **Australian s.** *Callitris endlicheri*

Sandbergia Greene (~ *Halimolobos*). Cruciferae (11). 2 NW US. R: HPB 12(2007)425

Sandemania Gleason. Melastomataceae. 1 Amaz. savannas of Braz., Venez., Peru: ***S. hoehnei*** (Cogn.) Wurdack. R: Britt. 39(1987)441

Sanderella Kuntze. Orchidaceae (V 12h). 2 Braz.

Sandersonia Hook. Colchicaceae (Liliaceae s.l.). 1 Natal: ***S. aurantiaca*** Hook. (Christmas bells) – bird-poll., cult. orn. R: Strel. 10(2000)590

sanderswood *Santalum album*; **red s.** *Pterocarpus santalinus*

Sandoricum Cav. Meliaceae (II). 5 Mal. R: Blumea 31(1985)146. ***S. borneense*** Miq. (Borneo) – riparian (? fish-disp.); ***S. koetjape*** (Burm.f.) Merr. (kechapi, santol, sentul, Mal.) – cvs with ed. fr. prob. orig. W Mal. (flowering form. cue for rice-sowing in Mal.), now widely planted in Ind. Ocean, Florida etc., some timber (katon; used in wayang golek puppets in Java), antifeedant limonoids in seeds, cytotoxic triterpenes in stems

Sandwithia Lanj. (~ *Sagotia*). Euphorbiaceae (Crot.). 2 Amaz.

Sandwithiodoxa Aubrév. & Pellegrin = *Pouteria*

sang *Panax ginseng*

Sanguilluma Plowes = *Monolluma*

Sanguinaria Dill. ex L. Papaveraceae (I). 1 E N Am.: ***S. canadensis*** L. (bloodroot, red puccoon) – woodland herb with alks & rhiz. prod. 1 leaf & 1 term. fl., K 2 ephemeral, C 8–16,

A ∞, seeds arillate, ant-disp.; medic. (abortifacient, anthelminthic, emetic etc.), red dye, cult. orn. esp. '**Multiplex**' with A & G replaced by C. R: FNA 3(1997)305

Sanguisorba L. Rosaceae (Agrim.). Excl. *Dendriopoterium, Poteridium, Poterium*, c. 15 N temp. (N Am. 4). Wind- to insect-poll. Some supposed to have styptic qualities, some cult. orn. & ed. lvs esp. ***S. officinalis*** L. (burnet, Euras., natur. N Am.) – infls used for wine-making in NW England until at least 1950s

Sanhilaria Baill. = *Tanaecium*

sanicle, wood *Sanicula europaea*

Sanicula Tourn. ex L. Umbelliferae (II 1). 39 subcosmop. (Eur. 2; not Papuasia, Australasia). R: UCPB 25(1951)1. Herbs with fr. covered with hooked bristles or tubercles (animal-disp.) & alks. Some form. medic. esp. ***S. europaea*** L. (wood sanicle, Euras.)

Saniculaceae Bercht. & J. Presl = Umbelliferae (II)

Saniculiphyllum C.Y. Wu & Ku. Saxifragaceae (Saniculiphyllaceae). 1 China: ***S. guangxiense*** C.Y. Wu & Ku. R: APS 30(1992)194

Saniella Hilliard & B.L. Burtt = *Pauridia*

sanjeevani *Selaginella bryopteris*

Sankowskia P.I. Forst. Picrodendraceae (Cal.-Diss.). 1 NE Queensland: ***S. stipularis*** P.I. Forst. R: Austrobaileya 4(1995)329

Sannantha Peter G. Wilson (~ *Babingtonia*). Myrtaceae (II 15). 15 E Aus., New Caled. R: AusSB 20(2007)313

Sanrafaelia Verdc. (~ ? *Hexalobium*). Annonaceae (III 6). 1 NE Tanzania: ***S. ruffonammari*** Verdc. R: GOB 13(1996)43

Sansevieria Thunb. = *Dracaena*

sansho *Zanthoxylum piperitum*

Sansonia Chiron (~ *Pleurothallis*). Orchidaceae (V 13c). 2 trop. Am. R: Richardiana 12(2012)80

sant pods *Vachellia nilotica*

Santa Maria *Calophyllum brasiliense*

Santalaceae R. Br. Magnoliidae – Santalales. Incl. Amphorogynaceae, Cervantesiaceae, Comandraceae, Eremolepidaceae, Nanodeaceae, Thesiaceae, Viscaceae, 48/1025 subcosmop. esp. trop. & warm dry. Hemiparasitic trees, shrubs & herbs on r. or branches, s.t. thorny or xeromorphic. Lvs simple, entire, s.t. scale-like, in spirals to opp.; stip. 0. Fls small, reg., bisex. or not, hypogynous to epigynous, oft. greenish (some wind-poll.), in var. types of infl. (oft. small dichasium in axil of each bract); P (?) (3)4 or 5(–8) free or valvate lobes of a fleshy tube, A opp. & oft. adnate to base of C, anthers with longit. slits or 1 apical pore, disk oft. present on ovary etc., $\underline{G}$–$\overline{G}$ ((2)3(–5)) with term. style, 1-loc. or with basal partitions, the free-C placenta with 1–4 pend., orthotropous to anatropous, unitegmic (st. integument not differentiated) ovules. Fr. a nut or drupe; seed 1 without testa, embryo straight, embedded in copious fleshy, oily or starchy (*Thesium*) endosperm. n = 5–7, 12, 13+

Principal genera: *Arceuthobium, Dendrophthora, Phoradendron, Santalum, Thesium, Viscum; Okoubaka* – a large tree with largest seeds of any parasite

This classification is not satisfactory (see BJLS 181(2016)18) and the relationship with Balanophoraceae is unclear; fam. likely to be split up. See also Schoepfiaceae

Timber (*Exocarpos, Osyris, Santalum* – sandalwood, also oil), ed. fr. (*Acanthosyris, Leptomeria*), ed. tubers (*Arjona*), tanning material (*Iodina*), local medic.; *Phoradendron* & *Viscum* used as (superstitious) winter decorations

Santaloides Schellenb. = *Rourea*

Santalum L. Santalaceae. Incl. *Eucarya*, 19 Indomal. to Aus. (6, 5 endemic) & Hawaii (***S. freycinetianum*** Gaudich. – basis of Hawaiian sandalwood industry from 1791, peaking 1810s, exhausted 1840, now almost extinct). R.-parasites. Fragrant (scent close to that of testosterone) timber, esp. ***S. album*** L. ((Ind.) sandalwood (form. sanderswood), Ind., widely cult., natur. Florida), used for making chests, the distilled oil used in scent-making & medic., ground wood used in cosmetics & one of the pigments used in making caste-marks, the timber burnt at Buddhist funerals; ***S. acuminatum*** (R. Br.) A. DC. (*E. acuminata*, Aus.) – ed. fr. & seeds (dong or quandong nuts), used in C19 jewellery; ***S. lanceolatum*** R. Br. (plumwood, trop. Aus.) – local timber; ***S. spicatum*** (R. Br.) A. DC. (*E. s.*, S & E Aus.) – now much reduced through exploitation, & other spp. = Aus. s., distilled for incense & medic.; ***S. yasi*** Seem. (W Pacific) – similar uses

Santanderella P. Ortiz = *Notyliopsis*

Santapaua Balakr. & Subram. = *Hygrophila*

Santiria Blume. Burseraceae (V). 22 OW trop. (Afr. 1: ***S. trimera*** (Oliv.) Aubrév.). Some locally used timbers & ed. fr., which also yields oil

Santiriopsis Engl. = *Scutinanthe*

Santisukia Brummitt (*Barnettia* Santisuk). Bignoniaceae (1). 2 Thailand. R: NHBSS 33(1986)82

santo serrated *Brassica rapa* Pekinensis Group

santol *Sandoricum koetjape*

Santolina Tourn. ex L. Compositae (Anth.-Sant.). c. 10 Medit. (Eur. 5). R: BBMNHB 23(1993)128. Holy flax, lavender cotton esp. ***S. chamaecyparissus*** L. (Medit.) – cult. orn. silvery grey shrub (poss. only 1 clone in cult.), app. insecticidal, form. used as vermifuge, in scent & brushed over clothes by Victorians; ***S. rosmarinifolia*** L. (S Eur.) – sold in Algarve markets for flavouring

Santomasia N. Robson = *Hypericum* (but see BBMNHB 8(1981)61)

santonica *Artemisia cina*

Santosia R. King & H. Robinson. Compositae (Eup.-Crit.). 1 E Braz.: ***S. talmonii*** R. King & H. Robinson – liane. R: Phytol. 45(1980)463

Santotomasia Ormerod. Orchidaceae (V 16c). 1 Luzon: ***S. wardiana*** Ormerod. R: Taiwania 53(2008)162

Sanvitalia Lam. Compositae (Helia.-Zinn.). 7 SW US to Arg. R: Britt. 16(1964)417. ***S. procumbens*** Lam. – cult. orn. ann. with dark purple disk-florets & yellow to orange ray-florets

Sanwa millet *Echinochloa esculenta*

sapele *Entandrophragma* spp.

sapgreen *Rhamnus cathartica*

Saphesia N.E. Br. (= ?). Aizoaceae (V). 1 SW Cape: ***S. flaccida*** (Jacq.) N.E. Br. R: H. E. K. Hartmann, *Ill. Handb. Succ. Pls*, Aiz. F–Z(2002)286

Sapindaceae Juss. Magnoliidae – Sapindales. Incl. Aceraceae, Hippocastanaceae, 142/1850 trop. & warm, few temp. Trees, shrubs, lianes & herbaceous climbers, these oft. with tendrils in place of infls & with unusual vasc. structure; toxic saponins usu. present. Lvs (bi-)pinnate (oft. with apical tip), palmate (*Aesculus*), 3-foliolate or simple (s.t. palmately lobed – *Acer*), usu. in spirals; petiolules oft. swollen basally, stip. 0 (exc. climbers). Fls small, usu. reg. & unisexual in cymes or thyrses, rarely solit.-axillary; K 4 or 5, s.t. basally connate, usu. imbr., C (3)4 or 5(+) or 0, oft. with basal scale-like appendage concealing a nectary, annular or oft. 1-sided disk around A or bearing A, rarely (*Dodonaea*) intrastaminal, A (3–)5–8(–30), filaments oft. hairy, anthers with longit. slits, pollen-exine v. variable, $\underline{G}$ ((2)3(–6)), usu. pluriloc. with distinct or 1 style(s) & 1(2; 7 or 8 in *Xanthoceras*, *Magonia*) ascending (pend.) or several spreading, bitegmic, anatr. to hemitropous or campylotropous ovules, oft. without a clear funicle but attached to placental protuberance. Fr. fleshy or dry, dehiscent or not; seeds with arils or sarcotestas, embryo oily & starchy, curved, oft. with plicate or twisted cotyledons, endosperm 0. n = 10–12 (climbers), 14–16 (others)

Classification & principal genera:

I. **Xanthoceroideae** (lvs in spirals, ovules 7 or 8 per loc., disk with orange horn-like apps; R: SB 30(2005)366): *Xanthoceras* (only), 'sister' to rest of fam.

II. **Hippocastanoideae** (Aceraceae, Hippocastanaceae; lvs opp., usu. palmate, ovules 2 per loc.) – 2 tribes:
   1. **Acereae** (fls reg.): *Acer* (only)
   2. **Hippocastaneae** (fls zygomorphic): *Aesculus*

III. **Dodonaeoideae** (lvs in spirals) – 2 tribes:
   1. **Dodonaeaeae** (disk (semi)annular, ovules (1)2(3 or 8) per loc., fr. dehiscent): *Dodonaea*, *Harpullia*
   2. **Doratoxyleae** (disk annular, ovules (1)2(3) per loc., fr. indehiscent): *Filicium*

IV. **Sapindoideae** (lvs in spirals; ovule 1 per loc., erect) – tribes
   1. **Paullinieae** (herb. vines, trees & shrubs, lvs imparipinnate, fls zygomorphic): *Allophylus*, *Paullinia*, *Serjania*
   2. **Melicocceae** (fl. reg., fr. indehiscent, 1-seeded): *Melicoccus*, *Talisia*
   3. [remaining genera]: *Alectryon*, *Arytera*, *Blighia*, *Cupania*, *Cupaniopsis*, *Dimocarpus*, *Guioa*, *Koelreuteria*, *Lepisanthes*, *Litchi*, *Matayba*, *Nephelium*, *Pappea*, *Sapindus*

Over half of genera with 3 or fewer (great majority of these 1) spp.

Toxicity in some due to non-protein amino-acids (cf. unrelated Leguminosae); many seem to have male phase then female & finally male ((duo)dichogamy)

Many ed. fr. esp. *Blighia* (akee), *Cubilia, Dimocarpus* (longan), *Litchi, Melicoccus, Pancovia, Nephelium* (rambutan), *Pappea*; timber (*Acer, Cupania, Harpullia, Hypelate, Matayba, Melicoccus, Pappea, Schleichera*), sugar (*Acer*), oilseeds (*Dilodendron, Pappea, Paranephelium* (medic.), *Schleichera*), beads (*Cardiospermum, Sapindus*), soap subs. (*Sapindus*), fish-poisons (*Jagera*), stimulating drinks (*Paullinia*) & cult. orn. esp. *Acer, Aesculus* (medic.), *Koelreuteria* & *Xanthoceras*

Sapindopsis How & Ho = *Lepisanthes*

Sapindus Tourn. ex L. Sapindaceae (IV). 10 trop. & warm OW (6 As., 1 Hawaii), 1 trop. Am., natur. OW trop. Berries rich in saponins & used as soap subs. (*sap-indus* = Ind. soap), soap nuts, soapberry, & local medic. (esp. skin diseases, lice etc.), esp. ***S. marginatus*** Willd. (S N Am.) – split timber used for cotton baskets, ***S. rarak*** DC. (Assam to C Mal.) with antifungal & molluscicidal saponins, ***S. saponaria*** L. (trop. Am., natur. OW (*S. mukorossi*)) – fr. also used as beads (lei & necklaces in Hawaii), & ***S. trifoliatus*** L. (Ind., Sri Lanka), all cult. orn.

Sapium Jacq. Euphorbiaceae (Euph.-Hipp.). Excl. *Balakata, Falconeria, Neoshirakia, Sclerocroton, Shirakiopsis, Triadica* (q.v.), c. 25 trop. Am. Some timber, some latex good as rubber & bird-lime esp. ***S. glandulosum*** (L.) Morong (*S. aucuparium*, Bolivian rubber, trop. Am.) & ***S. jenmanii*** Hemsl. (Esmeralda r., NE S Am.)

Sapium P. Browne = *Gymnanthes*

sapodilla (plum) *Manilkara zapota*

Saponaria L. Caryophyllaceae (III 1). 40 temp. Euras. (Eur. 10). R: DAWW 85(1910)433. Soapwort, form. used as soap subs. & still in bodyscrubs esp. ***S. officinalis*** L. (bouncing bet, fuller's herb, Euras., natur. N Am.) – cult. orn., form. medic., dried rootstock used like bois de Panama (*Quillaja saponaria*) in Middle E.; ***S. ocymoides*** L. (S Eur. mts) – common rock-pl.

Saposhnikovia Schischkin (~ *Ledebouriella*). Umbelliferae (III 10). 1 NE As.: ***S. divaricata*** (Turcz.) Schischkin

Sapotaceae Juss. Magnoliidae – Ericales. 58/1100 trop., few temp. R: T.D. Pennington (1991) *S.*; R. Govaerts & al. (2001) *World checklist & bibliography of S.* Trees (many with Aubréville's Model) & shrubs (rarely lianes or geoxylic suffrutices) with well-dev. latex-system (latex usu. white, rarely yellow or blue) & 2-armed hairs (1 arm oft. reduced or 0). Lvs simple, conduplicate, usu. in spirals, s.t. becoming distich.; stip. rare. Fls usu. reg., usu. bisex., in cymes (occ. solit.; large panicles in *Sarcosperma*); K (4-)6(–12) ± free & imbr., s.t. in a tight spiral, or in 2 whorls of 2–4, C (4–18) with imbr. lobes, entire to variously lobed, A 4–35(-43) in 1–3 whorls, epipet., s.t. incl. petaloid staminodes, anthers with longit. slits, $\underline{G}$ (1–15(–30)), usu. hairy, with axile or axile-basal placentation (1-loc. in *Diploon*), with 1 style & 1 anatr. to hemitropous, unitegmic ovule per loc. Fr. a berry, rarely a drupe or tardily dehiscent; seeds 1-∞, globose to oblong, oft. with large hollow hilum & shiny integument, embryo large with flat cotyledons in oily endosperm or thick ones without endosperm. n = (10-)13(14)

Classification & principal genera:

I. **Mimusopeae** (K usu 2 whorls of 3 or 4 with A same no.; 17 genera in 3 subtribes): 1. Mimusopinae (*Baillonella, Mimusops, Tieghemella, Vitellaria*); 2. Manilkarinae (*Manilkara, Northia*); 3. Glueminae (*Lecomtedoxa, Neolemonniera*)

II. **Isonandreae** (K usu. 2 whorls of 2 or 3 with A 2–3 times no.; 7 genera): *Madhuca, Palaquium*

III. **Sarcospermateae** (Sideroxyleae; K 5, C rotate or cyathiform with lobes oft. 3-fid; 6 genera): *Argania, Sideroxylon* [now held that *Sarcosperma* sister (= **Sarcospermatoideae**) to rest of I.–III. comprising **Sapotoideae**, IV. & V. **Chrysophylloideae**]

IV. **Chrysophylleae** (K 4 or 5(-11), C tubular to rotate with entire lobes; 19 genera): *Chrysophyllum, Micropholis, Pouteria, Synsepalum; Capurodendron* (with *Bemangidia, Tsebona*), excl. as **Tseboneae** (R: T 62(2013)979)

V. **Omphalocarpeae** (like 4. but A in groups of 2–6 opp. each C lobe); 4 genera): *Omphalocarpum*

Many ed. fr. (esp. *Chrysophyllum, Manilkara* (sapodilla, chewing-gum), *Pouteria* (sapote)), oils (*Argania, Diploknema, Madhuca, Palaquium, Vitellaria*), timber (*Argania, Baillonella, Madhuca, Manilkara, Micropholis, Palaquium, Sideroxylon, Tieghemella*), gutta-percha (*Palaquium, Payena*) & sweetener (*Synsepalum*)

sapote *Pouteria sapota* (see also *Matisia cordata*); **black s.** *Diospyros nigra*; **white s.** *Casimiroa edulis*

sappan wood *Biancaea sappan*

Sapphoa Urb. Acanthaceae (III 2c). 2 Cuba

Sapranthus Seemann. Annonaceae (IV 7). 8 C Am.

sapree wood *Widdringtonia nodiflora*

Sapria Griff. Rafflesiaceae. 3 trop. SE As. R: NHBSS 45(1997)164

Saprosma Blume. Rubiaceae (IV 12). c. 30 Indomal. Foetid smell

sapu *Magnolia champaca*

sapucaia nut *Lecythis zabucajo*

Sapucaya F. Knuth = *Lecythis*

Saraca L. Leguminosae (I 2). 11 Indomal., char. of particular streams. R: Blumea 15(1967)413, 27(1981)235. Troll's Model. Young lvs limp (cf. *Amherstia*, *Brownea*); fls with brightly coloured bracts & K; C 0 with some A derived from C primordia. Fls used in temple offerings, esp. ***S. asoca*** (Roxb.) Wilde ([anc.] asoka, Indomal.) – form. medic. & the tree under which Buddha was born (but see *Shorea*)

Saracha Ruíz & Pavón. Solanaceae (IV 5b). 2 S Am. R: A.T. Hunziker, *Gen. Solan.* (2001)226

Saranthe (Regel & Koern.) Eichler. Marantaceae. 10 Braz. R: Britt. 13(1961)212

Sararanga Hemsl. Pandanaceae. 2 Philippines, Solomon Is. R: Britt. 13(1961)214. Trees (Leeuwenberg's Model) to 20 m

Sarathrostachys Klotzsch = *Gymnanthes*

Sarawakodendron Ding Hou. Celastraceae (I). 1 Borneo: ***S. filamentosum*** Ding Hou – links the form. recog. fam. Hippocrateaceae to C., seed with filamentous aril. R: FM I,6(1972)930

Sarcandra Gardner. Chloranthaceae. 2 E As., Indomal. Vessels only in primary xylem of stems & sec. of r.

Sarcanthemum Cass. (~ *Psiadia*). Compositae (Ast.). 1 Rodrigues: ***S. coronopus*** (Lam.) Cass.

Sarcanthidion Baill. = *Citronella*

Sarcanthopsis Garay. Orchidaceae (V 16c). 5 Indomal. to New Caled. (1)

Sarcanthus Lindl. = *Cleisostoma*

Sarcaulus Radlk. Sapotaceae (IV). 5 trop. S Am. R: FN 52(1990)232

Sarcinanthus Oersted = *Asplundia*

Sarcinula Luer = *Specklinia*

Sarcobataceae Behnke (Chenopodiaceae III 1). Magnoliidae – Caryophyllales. 1/1–2 W N Am. R: FNA 4(2003)387. Monoec. decid., thorny, divaricate shrubs. Lvs linear. Male infl. a catkin, female fls solit.; fls with peltate scales; P tubular, 2-lipped (? bracteoles), accrescent, A (1)2–4(5) with long anthers, G (2). Fr. an achene; seeds flat. n = 9

Genus: *Sarcobatus*

Poss. incl. *Agdestis* (Phytolaccaceae)

Sarcobatus Nees. Sarcobataceae (Chenopodiaceae s.l.). 1–2 W N Am. ***S. vermiculatus*** (Hook.) Torrey (greasewood) – hard yellow fuelwood (also arrow-shafts), poss. shellac source, fodder but toxic in excess, local medic.

Sarcoca Raf. = *Phytolacca* (but see Preslia 57(1985)372)

Sarcocadetia (Schltr.) M.L. Clem. & D.L. Jones = *Cadetia*

Sarcocapnos DC. Papaveraceae (Fumariaceae I 2). 4 W Medit. incl. Eur. R: OB 88(1986)33

Sarcocaulon (DC.) Sweet = *Monsonia*

Sarcocephalus Afzel. ex Sabine = *Nauclea*

Sarcochilus R. Br. Orchidaceae (V 16c). c. 25 Aus. to New Caled. (1). Some cult. orn. diminutive epiphytes

Sarcochlamys Gaudich. Urticaceae (III). 1 Indomal.: ***S. pulcherrima*** Gaudich. – ramie subs.

Sarcococca Lindl. Buxaceae. 11 Afghanistan to China & Philippines. R: BotJLS 92(1986)117. Everg. monoec. shrubs (Mangenot's Model), male fls with nectar, females without; like *Buxus* but lvs in spirals, fr. a drupe etc. Cult. orn. esp. ground-cover; alks

sarcocolla or gum s. *Astragalus gummifer*

Sarcocornia A.J. Scott = *Salicornia* (but see BotJLS 75(1978)366)

Sarcodes Torrey. Ericaceae (II 3; Monotropaceae). 1 W US, Mex.: ***S. sanguinea*** Torrey (snow pl.). R: FNA 9(2009)390

Sarcodraba Gilg & Muschler. Cruciferae (46). 5 Andes. R: Darw. 44(2006)360

Sarcodum Lour. (~ *Clianthus*). Leguminosae (III 16). 3 SE As. to Solomon Is. R: KB 63(2008)155

Sarcoglottis C. Presl. Orchidaceae (IV 2h). c. 45 trop. Am. R: HBML 28(1982)352. Poll. euglossine bees

Sarcoglyphis Garay. Orchidaceae (V 16c). 12 Indomal.

Sarcolaena Thouars. Sarcolaenaceae. 8 Madag. R: AMBG 86(1999)711

Sarcolaenaceae Caruel (Chlaenaceae; ~ Dipterocarpaceae). Magnoliidae – Malvales. 10/71 Madag. Trees & shrubs usu. with stellate indumentum & mucilage-cells in pith, cortex etc. Lvs simple, in spirals becoming distich., petioles oft. with medullary bundles; stip. usu. caducous. Fls bisex., ± reg., usu. with involucel of bracteoles around 1 or 2 fls, these in thyrsoid infls, rarely solit., the involucel oft. cupulate (? pedicel tip); K 3 (–5, when outer ones smaller), C 5(6), convolute in bud, s.t. basally connate, A (10–)∞ usu. inside a ± cupulate disk (disk rarely 0), filaments s.t. basally connate in 5–10 bundles, anthers with longit. slits, pollen in tetrads, $\underline{G}$ ((1)2–4(5)) with term. style & (1)2–∞ erect to pend. anatr. ovules per loc. on (basal, apical or) axile placentas. Fr. a many-seeded loculicidal capsule or indehiscent with 1 or few oft. ruminate seeds (only 1 carpel maturing); embryo straight, endosperm starchy, copious to (rarely) 0. n = 11
Principal genera: *Leptolaena, Sarcolaena, Schizolaena*
Fossils from Tertiary of the Cape. Beautiful trees of the E rain forests of Madag. & form. dominating the western slopes of the high plateaux until destroyed by burning & replaced by *Heteropogon* grassland

Sarcolobus R. Br. Apocynaceae (Va; Asclepiadaceae III 4). 14 Indomal., W Pac. R: Blumea 26(1980)65. Some pericarps eaten in Malay Pen., also candied (toxic seeds removed); some poisons (glycosides), alks for killing dogs etc., said to have exterminated Javanese tigers. ***S. cambogensis*** McHone & Livsh. (Cambodia) – rheophyte

Sarcolophium Troupin. Menispermaceae (III). 1 trop. Afr.: ***S. tuberosum*** (Diels) Troupin. R: BJBB 30(1960)30

Sarcomelicope Engl. Rutaceae (I 2). 9 New Caled., E Aus. (1) to Fiji. R: Aus. JB 30(1982)359, BMNHN 4,8(1986)183

Sarcomphalus P. Browne (~ *Ziziphus*). Rhamnaceae. 32 trop. & warm Am. R: T 65(2016)54. ***S. chloroxylon*** (L.) Hauenschild (*Z. c.*, cogwood, Jamaica) – tough wood; ***S. joazeiro*** (Mart.) Hauenschild (*Z. j.*, Braz.) – fodder

Sarcopera Bedell (~ *Norantea*). Marcgraviaceae. c. 10 Honduras, Guyana highlands, N Andes

Sarcopetalum F. Muell. Menispermaceae (V). 1 S NG, E Aus.: ***S. harveyanum*** F. Muell. R: FA 2(2007)378

Sarcophagophilus Dinter = *Quaqua*

Sarcopharyngia (Stapf) Boit. = *Tabernaemontana*

Sarcophrynium K. Schum. Marantaceae. 3 trop. Afr.

Sarcophytaceae Kerner = Balanophoraceae

Sarcophyte Sparrmann. Balanophoraceae (Sarcophytaceae). 1 NE & E trop. Afr.: ***S. sanguinea*** Sparrmann. R: BJ 106(1986)364

Sarcophyton Garay. Orchidaceae (V 16c). 3 SE As.

Sarcopilea Urb. = *Pilea* (but see U. Eggli, *Ill. Handb. Succ. Pls*, Dicots (2002)449)

Sarcopoterium Spach = *Poterium*

Sarcopteryx Radlk. Sapindaceae. 12 E Mal., Aus. Some myrmecophytes. R: Austrobaileya 2(1984)53, Blumea 36(1991)87. ***S. stipata*** (F. Muell.) Radlk. (NE Aus.) – timber (corduroy)

Sarcopygme Setch. & Christoph. = *Morinda*

Sarcopyramis Wall. Melastomataceae. 1–3 Indomal. Disp. by raindrops forcing seeds from capsule

Sarcorhachis Trel. = *Manekia*

Sarcorhynchus Schltr. = *Diaphananthe*

Sarcorrhiza Bullock. Apocynaceae (III; Asclepiadaceae 1). 1 trop. Afr.: ***S. epiphytica*** Bullock. R: FTEA Apoc. (2012)164

Sarcosperma Hook.f. Sapotaceae (III; Sarcospermataceae). 8 Indomal. R: T.D. Pennington, *S.* (1991)179

Sarcospermataceae H.J. Lam = Sapotaceae

Sarcostemma R. Br. = *Cynanchum*

Sarcostigma Wight & Arn. Icacinaceae. 2 Indomal. ***S. kleinii*** Wight & Arn. (Indomal.) – seed-oil used in rheumatism

Sarcostoma Blume. Orchidaceae (V 15). 5 W Mal.

Sarcotheca Blume. Oxalidaceae. 11 W Mal. Acid fr. eaten with curry. ***S. celebica*** Veldk. (Sulawesi) – app. evolving dioecy from distyly with short-styled pls contributing genes largely through pollen, long-styled through ovules; ***S. ochracea*** Hallier f. (Sarawak) – flagelliflory

Sarcotoechia Radlk. Sapindaceae. 10–11 NG (5, R: Blumea 33(1988)198), NE Aus. (R: Austrobaileya 33(1988)198)

Sarcotoxicum Cornejo & Iltis (~ *Capparis*). Capparaceae. 1 Bolivia, Parag., N Arg.: ***S. salicifolium*** (Griseb.) Cornejo & Iltis. R: HPB 13(2008)104

Sarcoyucca (Trel.) Lindinger = *Yucca*

Sarcozona J. Black (~ *Carpobrotus*). Aizoaceae. 2 Aus.

Sarcozygium Bunge = *Zygophyllum*

Sarga Ewart (~ *Sorghum*). Gramineae (XXII 5). 9 E Mal., Aus., Mex., C Am. (1). R: AusSB 16(2003)287. Some fodders (Aus.)

Sargentia S. Watson = *Casimiroa*

Sargentodoxa Rehder & E. Wilson. Lardizabalaceae (Sargentodoxaceae). 1 China, Laos, Vietnam (mid-Eocene N Am.): ***S. cuneata*** (Oliv.) Rehder & E. Wilson – lvs simple to trifoliolate, monoec. or dioec., cult. orn. R: CBM n.s. 29(2012)243

Sargentodoxaceae Stapf = Lardizabalaceae

sari kuning, s. tijina *Styphnolobium japonicum*

Saribus Blume (~ *Livistona*). Palmae (III 4b). Incl. *Pritchardiopsis*, 8 C Mal. to New Caled. R: Palms 55(2011)112. ***S. jeanneneyi*** (Becc.) Bacon & W. Baker (*P. j.*, New Caled.) – almost extinct (over-exploitation of apical bud as palm cabbage) save 3 small pops on serpentine; ***S. rotundifolius*** (Lam.) Blume (*L. r.*, Mal.) – seeds sold in Amsterdam as 'Buddha palm'

Sarinia Cook = *Attalea*

Saritaea Dugand = *Bignonia*

Sarmenticola Senghas & Garay = *Macroclinium* (but see Willd. 25(1996)657)

Sarmienta Ruíz & Pavón. Gesneriaceae (II 4b). 1 S Chile: ***S. scandens*** (Molina) Pers. – cult. orn. liane with reddish fls; only Am. G. with A2

Sarocalamus Stapleton (~ *Arundinaria*). Gramineae (IV). 3 Himal. R: Novon 14(2004)346

Sarojusticia Bremek. = *Justicia*

Saropsis Briggs & L. Johnson = *Chordifex*

Sarothamnus Wimmer = *Cytisus*

Sarothrostachys Klotzsch. ? Older name for *Anomostachys*

Sarracenella Luer = *Pleurothallis*

Sarracenia Tourn. ex L. Sarraceniaceae. 8 E N Am. Pitcher pls. of swamps; many natural hybrids. R: FNA 9(2009)350. Pitchers (with blue fluorescence attractive to arthropods) in place of lvs (cf. *Nepenthes*) & absorbing nutrients from insects, in ***S. flava*** L. (SE US) at least, imp. for nitrogen & phosphorus but not calcium, magnesium & potassium, trapping flying insects (***S. psittacina*** Michaux (SE US) submerged in winter & trapping creeping ones); pitcher lips with nectariferous veins inside, pitcher inner surfaces with 5 zones – just inside a zone of downward-pointing hairs, then one of cells with elongated downward-pointing processes, then smooth zone, then a long zone of sinuous hairs & a 2nd smooth zone. Cult. orn. esp. ***S. purpurea*** L. (huntsman's cup or horn, E N Am., natur. Ireland (introd. late C19 & planted in Roscommon 1906) & Switzerland) – fls purple poll. newly-emerged *Bombus* queen bees, carnivory understood by 1791, proteolytic enzyme = leucine aminopeptidase, prey-capture rate generally low (not clear that prod. of nectar attracting ant prey compensated by ant-capture), obligate commensals incl. midge larvae feeding on mixed detritus in bottom of pitchers of this & other spp., while a maggot living at the surface of the water inside eats up to half of the captured prey

Sarraceniaceae Dumort. Magnoliidae – Ericales. 3/32 W & E N Am., NE S Am. R: S. McPherson & al. (2011), *S. of S Am.*, *S. of N. Am.* Perenn., usu. stemless, carnivorous herbs with rhiz. with rather irreg. anatomy; alks s.t. present. Lvs rep. by rosetted (or on scrambling stem in *Heliamphora*) pitcher-like traps with digestive (not proved for *H.*) liquid, short petioles, a well-dev. wing or ridge adaxially & flattened but rather small hood-like prolongation abaxially, the traps with nectar-glands without & within, where retrorse stiff hairs near top & smooth lower down; stip. 0. Fls large, reg., bisex., nodding, solit. on scape (few-fld racemes in *H.*); K (3–)5(6), imbr., persistent & ± petaloid, C (0 – *H.*) 5 imbr., decid., A 10–20(–∞), anthers basifixed (versatile in *Sarracenia*), $\underline{G}$ ((3–*H.*) 5), basally placentation axile, apically intruded-parietal, with term. style, with 5 short branches with term. stigmas in *Darlingtonia* or branches peltate with stigmas underneath in *Sarracenia*, each loc. with ∞ anatr. unitegmic or bitegmic ovules. Fr. a loculicidal capsule with ∞ small, oft. winged seeds with linear embryos in copious endosperm rich in oil & protein. n = 13 (*Sarracenia*), 15 (*Darlingtonia*), 21 (*Heliamphora*)

Genera: *Darlingtonia, Heliamphora, Sarracenia*
Wood of *Heliamphora* is like that in other Ericales (T 24(1975)297). Alleged early Cretaceous fossils (*Archaeamphora*) from NE China

sarsaparilla *Smilax* spp.; **bristly s.** *Aralia hispida*; **false s.** *Hardenbergia violacea*; **wild s.** *S. glauca*, (US) *A. nudicaulis*

sarson *Brassica rapa*

Sartidia De Winter. Gramineae (Aristid.). 6 trop. & S Afr. (***S. dewinteri*** Munday & Fish on serpentine), Madag.

Sartoria Boiss. & Heldr. = *Hedysarum*

Sartorina R. King & H. Robinson (~ *Eupatorium*). Compositae (Eup.-Flei.). 1 Mex. (?): ***S. schultzii*** R. King & H. Robinson. R: MSBMBG 22(1987)289

Sartwellia A. Gray. Compositae (Tag.-Flav.). 43 S N Am. R: Sida 4(1971)265

Saruma Oliv. Aristolochiaceae (I). 1 NW to SW China: ***S. henryi*** Oliv. – apocarpous, pollen sulcate. R: CBM 22(2005)203

Sarx H. St. John = *Sicyos*

Sasa Mak. & Shib. Gramineae (IV). Incl. *Sasaella* (*Sasa* × *Pleioblastus* hybrids?), excl. *Sasamorpha*, c. 40 temp. E As. esp. Jap. Small long-lived hapaxanthic bamboos; some cult. orn. incl. ***S. palmata*** (Burb.) Camus (Jap., Sakhalin) – rapid spreader to 2.5 m tall, s.t. used in cardboard-making & ***S. ramosa*** (Mak.) Mak. & Shib. (*Arundinaria vagans, Sasaella ramosa*, Jap.) – first flowered outside Jap. 1981, 89 yrs after introd.

Sasaella Mak. = *Sasa*

sasah *Aporosa frutescens*

Sasamorpha Nakai (~ *Sasa*). Gramineae (IV). 5 E As.

saskatoon *Amelanchier* spp.

sassafras *Sassafras albidum*; **Brazilian s.** *Aniba* spp.; **Chinese s.** *S. tzumu*; **Orinoco s.** *Ocotea cymbarum*

Sassafras Nees & Eberm. Lauraceae (II). 3 E As. (2, 1 endemic Taiwan), E N Am. (1); allied *Sassafrasoxylon* in late Cretaceous Antarctic. ***S. albidum*** (Nutt.) Nees (*S. officinale*, sassafras, E N Am.) – Aubréville's Model, the first export of N Am. colonies (2 shiploads saturated the market), antiscorbutic but abortifacient & now linked to liver cancers so banned from pharmacy & cosmetics throughout EU (toxin = safrole) though elsewhere powdered lvs (filé) used to thicken sauces, in r. beer, etc., oil used medic. incl. killing lice & for insect-bites, & in scent-making, bark & twigs made into a tea by E Canadian Indians, light timber, cult. orn.; ***S. tzumu*** (Hemsl.) Hemsl. (Chinese s., C China) – imp. local timber

Sassia Molina = *Oxalis*

sasswood, sassy bark *Erythrophleum suaveolens*

Satakentia H. Moore (~ *Clinostigma*). Palmae (V 14d). 1 Ryukyu Is.: ***S. liukiuensis*** (Hatusima) H. Moore – cult. orn. monoec. palm with swollen trunk-base. R: Principes 13(1969)5

Satanocrater Schweinf. Acanthaceae (III 2b). 4 trop. Afr. R: NJB 24(2007)385

satin *Gossypium* spp.; **s. oak** *Alloxylon* spp.; **s. walnut** *Liquidambar styraciflua*; **s.(wood)** *Chloroxylon swietenia*; **African s.w.** *Zanthoxylum gillettii*; **Cairns s.w.** *Dysoxylum pettigrewianum*; **concha s.w.** *Z. caribaeum*; **Jamaican** or **WI s.w.** *Z. flavum*; **Nigerian s.w.** *Distemonanthus benthamianus*

satiné *Brosimum rubescens*

Satorkis Thouars. Older name for *Dactylorhiza* (*Coeloglossum*)

Satranala Dransf. & Beentje. Palmae (III 8a). 1 NE Madag. (1 site with 30 trees, 40 juveniles & seedlings): ***S. decussilvae*** Dransf. & Beentje – strongly winged endocarp poss. protection for seed in gizzard of (extinct) elephant-bird (? disp. agent). R: KB 50(1995) 87

satsuma *Citrus reticulata*

Sattadia Fourn. = *Tassadia*

Saturday-night pepper *Euphorbia helioscopia*

Satureja Tourn. ex L. Labiatae (VII 2b). c. 38 Medit. to Iran. Gynodioec. herbs cult. for condiments (see also *Clinopodium, Micromeria*) esp. ***S. hortensis*** L. (summer savory, Medit.) – flavouring (dried lvs) like sage oft. in 'mixed herbs', trad. used with legumes esp. broad beans; ***S. montana*** L. (winter s., S Eur.) – less used, flavour poorer

Saturn peach *Prunus persica* cv.

Satyria Klotzsch. Ericaceae (VIII 5). 25 trop. Am. R: OB 92(1987)121

Satyridium Lindl. = seq.

Satyrium Sw. Orchidaceae (IV 4d). 92 trop. OW to Arabia & S Afr., Madag. (5), As. (4). Some cult. terr. orchids, some ed. tubers Zambia (chikanda). ***S. bicallosum*** Thunb. (S Afr.) – poll. sciarid flies (fungus gnats) & ***S. pumilum*** Thunb. – smell of rotting flesh, poll. sarcophagid (flesh) flies while 3 other S Afr. spp. poll. sunbirds

sau *Falcataria moluccana*

sauerkraut *Brassica oleracea*

Saugetia A. Hitchc. & Chase = *Tetrapogon*

Saundersia Reichb.f. Orchidaceae (V 12h). 2 Braz.

Saurauia Willd. Actinidiaceae (Saurauiaceae). 300 trop. Am. (32, R: FMBot. n.s. 2(1980)), trop. As. to Aus. (1). Cryptic dioecy. Some rheophytes. Some locally eaten fr. ***S. callithrix*** Miq. (Celebes) – habit like geocarpic figs; ***S. myrmecoidea*** Merr. (Borneo) – myrmecophyte; ***S. purgans*** B.L. Burtt (Solomon Is.) – fls make a crying noise when opening

Saurauiaceae Griseb. = Actinidiaceae

Sauria Bajtenov = *Eritrichium*

Sauroglossum Lindl. Orchidaceae (IV 2h). 11 S Am. R: HBML 28(1982)355

Saurolophorkis Marg. & Szlach. = *Malaxis* (But see PBJ 46(2001)7)

Saurolluma Plowes = *Caralluma*

Sauromatum Schott (~ *Typhonium*). Araceae (VII 23). 9 Indomal. R: T 59(2010)445. ***S. venosum*** (Aiton) Kunth (*Arum cornutum, S. guttatum, S. nubicum, T. v.*, voodoo lily, monarch-of-the-east) – weedy, lvs pedate, solit., emerging after disgustingly foetid infls, sold in Eur. as dried tubers which prod. infl. without water, roasted tubers ed.

Sauropus Blume = *Breynia*

Saururaceae F. Voigt. Magnoliidae – Piperales. 4/6 E As., N Am. R: FOW 11(2005). Perenn. aromatic herbs with articulated stems, rhiz. or stolons & vasc. bundles in 1 or 2 concentric rings (cambium in *Anemopsis*); alks 0. Lvs simple, in spirals with stip. adnate to petioles, venation palmate. Fls bisex., small, in dense bracteate spikes or racemes resembling a single fl. when basal bracts large & petaloid; P 0, A3 or 3 + 3 or 4 + 4, s.t. adnate to G, anthers with longit. slits, G (3 or 4) or G 3 or 4, 1-loc. with parietal placentation (or distinct above connate bases in *Saururus*; sunk in infl. axis in *Anemopsis*) with free styles & 6–10 ((1)2–4 in *Saururus*) hemitropous to orthotropous, bitegmic ovules on each placenta. Fr. an apically dehiscent capsule (head of 1-seeded carpels in *S.*); seeds endotegmic with minute embryo in little endosperm & copious perisperm. n = 9, 11,12
Genera: *Anemopsis, Gymnotheca, Houttuynia, Saururus*
Some cult. orn.

Saururus L. Saururaceae. 1 E N Am., 1 E As. Bog pls, locally medic. incl. ***S. cernuus*** L. (E N Am.) – cult. orn. with fragrant white fls. R: FOW 11(2005)1

sausage tree *Kigelia africana*

Saussurea DC. Compositae (Card.-Card.). Excl. *Hemisteptia,. Himalaiella, Shangwua*, c. 300 Euras. (China 250, Eur. 9). R: S. Lipschitz, *Rod Saussurea* (1979). Some prob. referable to *Lipschitzella*. Some sources of TCM drugs, some cult. orn. esp. ***S. alpina*** (L.) DC. (Euras.) – sweetly-scented purple fls; ***S. costus*** (Falc.) Lipsch. (*S. lappa*, costus r., kuth, mu xiang, E Himal.) – medic. but esp. used in scents because of its strong lingering smell; ***S. gossypiphora*** D. Don (Himal.) – dwarf pachycaul with solit. hollow hairy stem in alpine belt (cf. *Lobelia*); ***S. laniceps*** Hand.-Mazz. (Himal.) – pops shorter than 100 yrs ago (big ones coll.; genetic erosion)

Sautiera Decne. = *Dyschoriste*

Sauvagesia L. Ochnaceae (III). 39 trop. (SE As. 1, Mal. 2, Afr. 1, Am. c. 35). R: BJ 113(1991)184. Some with Corner's Model; buzz-poll. by euglossine bees in Am. ***S. erecta*** L. (pantrop. ruderal) – creole tea, abortifacient (Braz.)

Sauvagesiaceae Dumort. = Ochnaceae

Sauvallea C. Wright (~ *Commelina*). Commelinaceae (II 1). 1 Cuba: ***S. blainii*** C. Wright

Sauvallella Rydb. = *Poitea*

Sauvetrea Szlach. = *Maxillaria* (but see Lankesteriana 7(2007)535)

sauwi *Panicum sonorum*

Savannosiphon Goldbl. & Marais. Iridaceae (VI 2). 1 S trop. Afr.: ***S. euryphylla*** (Harms) Goldbl. & Marais. R: AMBG 66(1979)849

Savia Willd. Phyllanthaceae (Brid.-Sav.; Euphorbiaceae s.l.). 2–3 warm Am.

Savignya DC. Cruciferae (12). 1 N Afr. to Middle E deserts: ***S. parviflora*** (Del.) Webb

savin *Juniperus sabina*

savonette *Lonchocarpus sericeus*

savory, summer *Satureja hortensis*; **winter s.** *S. montana*
Savoy cabbage *Brassica oleracea* Capitata Group
saw *Millettia pinnata*
sawai *Eulaliopsis binata*
saw-wort *Serratula tinctoria*
saxaul *Haloxylon persicum, H. aphyllum*
Saxegothaea Lindl. Podocarpaceae (Saxegothaeaceae). 1 S Chile & Arg.: ***S. conspicua*** Lindl. (Prince Albert's yew) – cult. orn. monoec. everg. with some features reminiscent of Araucariaceae; timber used locally
Saxegothaeaceae Gaussen ex Doweld & Reveal = Podocarpaceae
Saxicolella Engl. Podostemaceae (III). Incl. *Butumia, Pohliella*, c. 6 trop. W Afr.
Saxifraga Tourn. ex L. Saxifragaceae (II). Excl. *Micranthes* (seeds usu. unitegmic), c. 370 N temp., esp. Eur. (123, R: D.A. Webb & R.J. Gornall (1989) *Saxifrages of Eur.*), Himal. & E As. (China 200) & W N Am. (25 – R: FNA 9(2009)132), to Arctic, few S to Thailand, Ethiopia & Andes to Tierra del Fuego. Saxifrages, rockfoil. F. Kohlein (1984) *Saxifrages & related genera*; M. McGregor & W. Harding (1998) *Saxifrages, the complete list of species*. M. McGregor (2008) *Saxifrages*. 2n = 10–200+. Usu. perenn. herbs, some rather pachycaul & hapaxanthic with large rosettes, few annuals; insect-poll. with $\underline{G}$ to $\overline{G}$. Cult. orn., some alpine spp. used in Chartreuse, some local medic. (e.g. earache in Vietnam, China; form. with other saxicolous pls considered to break up kidney-stones – Doctrine of Signatures). 13 sects (R: BJLS 95(1987)273) recog. incl. *Irregulares* (*Diptera*, IRR) – fls irreg., *Gymnopera* (*Robertsonia*, GYM) – lvs not pitted nor lime-secreting, obovate or orbicular, *Cymbalaria* (CYM) – annuals, fls yellow or white, *Saxifraga* (SAX) – usu. bienn. with bulbils & white fls (incl. *Dactyloides* – cushion-forming without bulbils (mossy s.), *Ligulatae* (*Euaizoonia*, LIG) – lvs lime-secreting, offsets separating, *Porphyrion* (*Kabschia*, POR) incl. *Engleria* – offsets remaining attached. Cult. orn. spp. incl.: ***S. burseriana*** L. (POR, E Alps) – fls large, white; ***S. callosa*** Sm. (LIG, *S. lingulata*, S Eur.) – fls in large much-branched panicles; ***S. cernua*** L. (SAX, Arctic to N Am. & Eur.) – in GB a protected sp. with no seed or fr., reprod. by bulbils; ***S. cespitosa*** L. (SAX, N N temp.) – prot. sp. in GB; ***S. cymbalaria*** L. (CYM, SE Eur., SW As., N Afr.) – self-sowing ann. with yellow fls; ***S. granulata*** L. (SAX, meadow s., Eur.) – bulbils, fls white, gynodioec. & in N England females veget. more vigorous but prod. only 57% seeds of hermaphr. though seeds 1.28 times 'fitter'; ***S. hypnoides*** L. (SAX, NW Eur.) – fls white, one of orig. mossy s. spp.; ***S. longifolia*** Lapeyr. (LIG, Pyrenees, E Spain, Morocco) – hapaxanthic unbranched calcicole, some cvs; ***S. paniculata*** Mill. (LIG, *S. aizoon*, N N temp.) – many cvs of encrusted s.; ***S. stolonifera*** Curtis (IRR, *S. sarmentosa*, mother-of-thousands, wandering Jew, China, Jap.) – hanging baskets; ***S. × superba*** Rouy & E.G. Camus (LIG, *S. longifolia* × *S. cotyledon* L. (Eur. mts)) – cult. hybrids esp. '**Tumbling Waters**' with rosettes to 30 cm diam.; ***S. × urbium*** D. Webb (GYM, *S. spathularis* Brot. (W Eur.) × *S. umbrosa* L. (Pyrenees)) – the common London pride (after London & Wise, C18 nurserymen) or none-so-pretty, confused with *S. umbrosa* but diff. in larger more deeply crenated lvs & larger fls etc.; besides the last the most freq. seen are the mossy s. (SAX) – many hybrids & the encrusted s. (LIG, POR) – choice rock-pls (usu. hybrids in cult.) with hydathodes secreting chalky water on to surface where chalk deposited
Saxifragaceae Juss. Magnoliidae – Saxifragales. Excl. Penthoraceae, 34/640 subcosmop. esp. N temp. & cold. Usu. perenn. herbs, s.t. rather succ., rarely suffrutescent; cortical &/or medullary bundles s.t. present. Lvs usu. in spirals, oft. basal rosettes, simple to pinnate or palmate, oft. with hydathodes; stip. usu. 0. Fls usu. reg. & bisex., in cymes, racemes or solit., ± hypo- to epigynous; K (3–)5(–10) imbr. or valvate, oft. lobes on hypanthium, C (0) same as & alt. with K, imbr. or convolute, s.t. lobed or early decid. or small, A in 2 whorls with 1 whorl s.t. staminodes or 0, anthers with longit. slits, nectary-disk or annulus oft. around $\underline{G}$ (2(3)), rarely ± free, each lobe term. with a stigma & with marginal, axile or parietal placentas etc. with several–∞ anatr. (uni- or) bi-tegmic ovules s.t. with zig-zag micropyle. Fr. dry dehiscent usu. septicidally; seeds small with straight embryo in copious (rarely 0) oily endosperm. n = (5-)7(+)
Classification & principal genera:
I. '**Heucheroids**' – 8 groups: 1. Cascadia (*C., Saxifragodes* (only); 2. Micranthes (*M.* only); 3. Peltoboykinia (*Chrysosplenium*); 4. Darmera (*Bergenia, Darmera, Rodgersia*); 5. Heuchera (*Heuchera, Mitella*); 6. Boykinia (*Boykinia*); 7. Astilbe (*Astilbe*); 8. Leptarrhena (*Tanakaea*)
II. '**Saxifragoids**': *Saxifraga* (the bulk of the fam.)

Generic diversity greatest in W N Am.; over half of genera monospecific. *Astilbe* (*q.v.*), superf. like Rosaceae, oft. confused with *Filipendula*, an interesting convergence Many cult. orn. in above genera & *Elmera, Lithophragma, Tellima, Tiarella, Tolmiea* etc.

saxifrage *Saxifraga* spp.; **burnet s.** *Pimpinella major*; **golden s.** *Chrysosplenium oppositifolium*; **meadow s.** *S. granulata*; **mossy s.** *S.* sect. *Saxifraga*

Saxifragella Engl. = *Saxifraga*

Saxifragodes D.M. Moore. Saxifragaceae (I 1). 1 Tierra del Fuego to 51° S in Chile: ***S. albowiana*** (Kurtz) D.M. Moore – 'sister' to *Cascadia nuttallii* (NW N Am.!). R: BN 122(1969)323

Saxifragopsis Small (~ *Saxifraga*). Saxifragaceae (I 7). 1 N Calif., S Oregon: ***S. fragarioides*** (Greene) Small. R: FNA 9(2009)130

Saxiglossum Ching = *Pyrrosia*

Saxipoa Soreng & al. Gramineae (XVI 15). 1 Aus.: ***S. saxicola*** (R. Br.) Soreng & al. R: AusSB 22(2009)407

Saxofridericia Schomb. Rapateaceae (I 1). 9 trop. S Am.

Sayeria Kraenzl. = *Dendrobium*

Scabiosa Tourn. ex L. Caprifoliaceae (Dipsacaceae). Excl. *Lomelosia, Sixalix*, 30+ temp. Euras., Medit. R [Iberia]: Lagascalia 12(1984)143. Scabious; herbs with epicalyx extension acting as an umbrella in wind-disp. of fr. Some cult. orn. incl. ***S. columbaria*** L. (Euras., Medit.) – powdered roots for baby-powder, local medic. S Afr.

Scabiosella Tieghem = *Scabiosa*

Scabiosiopsis Rech.f. = *Lomelosia*

scabious *Scabiosa* spp.; **devil's bit s.** *Succisa pratensis*; **field s.** *Knautia arvensis*; **sweet s.** *Sixalix atropurpurea*

Scabrethia W. Weber = *Wyethia* (but see FNA 21(2006)99)

scabrin *Heliopsis* spp.

Scadoxus Raf. (~ *Haemanthus*). Amaryllidaceae. 9 trop. Afr. to Yemen. Like *Haemanthus* but 2n = 18 (16 in *H.*) & lvs with distinct midvein & not distich.; alks toxic to stock; large chromosomes suited to cytological study. Cult. orn. with red (at least some bird-poll.) fls esp. ***S. multiflorus*** (T. Martyn) Raf. (blood fls, trop. Afr.) & ***S. puniceus*** (L.) I. Friis & Nordal (Ethiopian highlands, Tanz., S Afr.)

Scaevola L. Goodeniaceae. Incl. *Diaspasis, Selliera*, 102 trop. Indo-Pac. esp. Aus. (R: Telopea 3(1990)489, FA 35(1992)84) with 2 widespread trop. beach spp. (Aubréville's Model): ***S. plumieri*** (L.) Vahl (Indo-Atlantic) & ***S. taccada*** (Gaertn.) Roxb. (*S. sericea*, Indo-Pac.), their seeds viable for long periods in seawater but germinate only in freshwater, i.e. when washed up on a rainy beach; extrafl. nectaries. Pith of *S. taccada* (taccada) used for making Malayan rice-paper, artificial fls etc., bark contraceptive (NG), invasive SE US (introd. for erosion control); ***S. aemula*** R. Br. (SE Aus.) – cult. orn. for hanging baskets etc. US. etc

scag *Papaver somniferum*

Scagea McPherson. Picrodendraceae (Cal.-Pseud.; Euphorbiaceae s.l.). 2 New Caled. R: BMNHN 4,7(1986)247

scald *Cuscuta* spp.

Scalesia Arn. Compositae (Helia.-Helia.). 11 Galápagos. R: OB 36(1974). Name an error for '*Stablesia*', commemorating W.A. Stables. All allopatric, allied (4 tree spp. forming forests in highlands) to *Helianthus*, showing adaptive radiation comparable with Darwin's finches. ***S. pedunculata*** Hook. f. – shade-intolerant even-aged monocultures dying ± all at once & followed by same

Scaligeria DC. Umbelliferae (III 5). 4 Medit. (Eur. 1) to SW As.

scallion Orig. = shallot, later = spring onion, (US) leek

scallopini Variously used for small forms of courgette

Scambopus O. Schulz. Cruciferae (37). 1 S & E Aus.: ***S. curvipes*** (F. Muell.) O. Schulz. R: TRSSA 89(1965)219

scammony *Ipomoea orizabensis*, **(Levant s.)** *Convolvulus scammonia*

Scandentia Cabral & Bacog. = *Denscantia*

Scandia Dawson. Older name for *Gingidia*

Scandicium (K. Koch) Thell. = *Scandix*

Scandivepres Loes. = *Acanthothamnus*

Scandix Tourn. ex L. Umbelliferae (III 2). 15–20 Eur. (3), Medit. R: EJB 58(2001)340. ***S. pecten-veneris*** L. (shepherd's needle, Venus's or Lady's comb, Euras.) – v. long mericarps, which sep. violently

Scaphiophora Schltr. = *Thismia*

Scaphispatha Brongn. ex Schott. Araceae (VII 11). 2 Bolivia, Braz. R: Rodriguesia 56(2005)55

Scaphium Schott & Endl. Malvaceae (Sterc.; Sterculiaceae). 8 trop. As. R: EJB 66(2009)291. Ovaries winged (wind-disp.). Monkeys eat jelly formed around germ. seeds; in Malay Pen. seeds soaked overnight swell in prod. mucilage which is drunk as febrifuge etc.; Laos exports 1 K t seeds – imp. medic. (sore throats) in China etc.

Scaphocalyx Ridl. Achariaceae (Flacourtiaceae). 2 Sumatra, Malay Pen.

Scaphochlamys Bak. Zingiberaceae (II 1). 31 S Thailand to Sumatra, Borneo. R: EJB 67(2010)80

Scaphopetalum Masters. Malvaceae (Bytt.-Bytt.; Sterculiaceae). 20 trop. Afr. ***S. amoenum*** A. Chev. (W Afr.) – Champagnat's Model, axis with vertical growth then leaning over, the tip rooting & thickest of epicormics growing up like orig. trunk; ***S. mannii*** Mast. (W Afr.) – 'litterbox' pl.

Scaphosepalum Pfitzer. Orchidaceae (V 13c). c. 45 trop. Am. R: MSB 26(1988)21, 39(1991)147, 44(1992)42. Cult. orn.

Scaphospermum Korovin = *Parasilaus*

Scaphyglottis Poeppig & Endl. Orchidaceae (V 13b). Incl. *Hexisea*, c. 70 trop. Am. Cult. orn. epiphytes

Scapiarabis M. Koch & al. (~ *Arabis*). Cruciferae (7). 4 C As. to China. R: T 61(2012) 965

Scapicephalus Ovcz. & Chukavina = *Decalepidanthus*

Scarborough lily *Cyrtanthus elatus*

scaredy cat plant *Coleus caninus*

Scariola F.W. Schmidt = *Lactuca*

scarlet globe mallow *Sphaeralcea coccinea*, *S. angustifolia* subsp. *cuspidata*; **s. runner bean** *Phaseolus coccineus*

Scassellatia Chiov. = *Lannea*

Sceletium N.E. Br. = *Mesembryanthemum* (but see BJ 118(1996)9)

Scelochiloides Dodson & M. Chase = *Comparettia*

Scelochilopsis Dodson & M. Chase = *Comparettia* (but see Orquideologia 21(1998)61)

Scelochilus Klotzsch = *Comparettia*

scented myrhh *Commiphora guidotii*; **s. orchid** *Gymnadenia conopsea*

scentless rosewood *Synoum glandulosum*

Schachtia Karsten = *Duroia*

Schaefferia Jacq. Celastraceae (I). 23 trop. & warm Am. Some box subs. esp. ***S. frutescens*** Jacq. (Florida boxwood, Florida to WI)

Schaenomorphus Thorel ex Gagnepain = *Tropidia*

Schaetzellia Schultz-Bip. = *Macvaughiella*

Schaffnerella Nash = *Muhlenbergia*

Schaffneria Fée ex T. Moore = *Asplenium*

Schaueria Nees. Acanthaceae (III 2c). 8 Braz. ***S. flavicoma*** (Lindl.) N.E. Br. – cult. orn. with yellow fls

Schaueriopsis Champluvier & I. Darbysh. Acanthaceae (IIId). 1 Congo: ***S. variabilis*** Champluvier & I. Darbysh. R: PEE 145(2012)283

Schedonnardus Steud. = *Muhlenbergia*

Schedonorus P. Beauv. = *Lolium*

Scheelea Karsten = *Attalea*

Schefferomitra Diels. Annonaceae (III 7). 1 NG: ***S. subaequalis*** (Scheffer) Diels

Schefflera Forst. & Forst. f. Araliaceae (1). Incl. *Brassaia*, *Crepinella*, *Didymopanax*, *Tupidanthus* (A & G to 100 allegedly due to fasciation), excl. *Plerandra*, c. 550 trop. & warm (s.s. 8 NZ, SW Pacific). Heterogenous. Trees (Corner's & Leeuwenberg's Models), shrubs, lianes & epiphytes (***S. gemma*** Frodin (NG) a herb). Many cult. orn. trees (trop.) & house-pls esp. ***S. actinophylla*** (Endl.) Harms (*Brassaia a.*, umbrella tree, NG, trop. Aus., invasive SE US, Hawaii – 1 of worst there) – s.t. epiphytic, street-tree, ***S. arboricola*** (Hayata) Merr. (Hainan, Taiwan) – freq. pot-pl., ***S. morototoni*** (Aubl.) Maguire & al. (*Didymopanax m.*, jereton, maramara, trop. Am.) – timber used for pulp, drums, matches etc., seeds in maracas & made into women's aprons; ***S. octophylla*** (Lour.) Harms (SE As.) – comm. med. in Vietnam; ***S. pueckleri*** (K. Koch) Frodin (*Tupidanthus calyptratus*, trop. As.) – cult. orn. liane or tree

Schefflerodendron Harms. Leguminosae (III 16). 4 trop. Afr.

Scheffleropsis Ridl. = *Schefflera*

Schelhammera R. Br. Colchicaceae. 2 E Aus., 1 ext. to NG. R: FA 45(1987)412. Form. referred to Asparagaceae (Convallariaceae). Alks

Schellenbergia C.E. Parkinson = *Vismianthus*

Schellolepis J. Sm. = *Goniophlebium*

Schenckia K. Schum. = *Deppea*

Schenckochloa Ortíz. Gramineae (XXIX 5). 1 NE Braz.: ***S. barbata*** (Hackel) Ortíz. R: Candollea 46(1991)241

Schenkia Griseb. (~ *Centaurium*). Gentianaceae. 6 Euras. & Medit. (1), Aus. & Pacific (Hawaii 1). R: T 53(2004)724

Scherya R. King & H. Robinson. Compositae (Eup.-Ager.). 1 E Braz.: ***S. bahiensis*** R. King & H. Robinson – known only from 1 coll. (1944) until 2007. R: MSBMBG 22(1987) 150

Scheuchzeria L. Scheuchzeriaceae. 1 N temp. & Arctic: ***S. palustris*** L.– bogs

Scheuchzeriaceae Rudolphi (~ Juncaginaceae). Magnoliidae – Alismatales. 1/1 N temp. & Arctic, in *Sphagnum* bogs. Sympodial perenn. herb with rhiz.; vessels confined to r. & rhiz. Stem with long intravaginal hairs at nodes. Lvs with long semi-terete lamina (pore at apex), open sheath & ligule at their junction, in spirals, becoming distich. Fls wind-poll., bisex., in term. bracteate racemes; P 3 + 3, yellow-green, A 6, with elongate anthers with longit. slits, pollen in diads, $\underline{G}$ 3(–6), basally weakly connate, each with sessile stigma & 1(2) erect, anatropous, bitegmic ovules. Fr. a head of follicles, each with 1(2) seeds with 0 endosperm. n = 11
Genus: *Scheuchzeria*

Schickendantzia Pax = *Alstroemeria*

Schickendantziella Speg. Alliaceae (Gill.). 1 Arg.: ***S. trichosepala*** (Speg.) Speg.

Schidorhynchus Szlach. = *Sauroglossum*

Schiedea Cham. & Schldl. Caryophyllaceae (II 1). Incl. *Alsinidendron*, 34 Hawaii. R: SBM 72(2005)33. Shrubs & lianes with coloured nectar; moth-poll. herm. spp. in forest, wind-poll. dioec. spp. in dry habitats. ***S. salicaria*** Hillebrand (Maui) long believed extinct but lately rediscovered, the similar ***S. adamantis*** St John (Oahu) restricted to 60–65 individuals, cryptically dioec.

Schiedeella Schltr. Orchidaceae (IV 2h). Incl. *Physogyne*, 27 montane trop. Am. to Arizona (1). R: FFG 37(1992)165

Schiekia Meissn. Haemodoraceae (I). 1 trop. S. Am.: ***S. orinocensis*** (Kunth) Meissn. – soap-subs., local medic. R: FN 61(1993)17

Schima Reinw. ex Blume. Theaceae. 1 (v. variable or 10–15 recog.) trop. & warm As.: ***S. wallichii*** (DC.) Korth.– plantation timber (needlewood) for constr. & toys, imp. tree for reafforestation in Java; contact with bark (source of a fish-poison) causes intense itching, used (gatal-gatal) to enlarge penis in NG. R: Reinw. 2(1952)133. Deliberately crossed hybrids (1999–2000) with *Franklinia* = **× *Schimlinia*** Ranney & Fantz

Schimpera Hochst. & Steud. ex Endl. Cruciferae (33). 1 E Medit. to S Iran, deserts: ***S. arabica*** Hochst. & Steud.

Schimperella H. Wolff = *Oreoschimperella*

Schindleria H. Walter. Petiveriaceae (Phytolaccaceae II). 2 Peru, Bolivia. R: AMBG 55(1968)337

Schinopsis Engl. Anacardiaceae (I). 7 S Am. R: Lilloa 33(1973)205. ***S. quebrachocolorado*** (Schldl.) F. Barkley & T. Meyer (*Aspidosperma q.*) & other spp. with bark & timber (quebracho) high in tannin used in tanning

Schinus L. Anacardiaceae (I). 33 trop. Am. R: Britt. 5(1944)160, Lilloa 28(1957)7. Usu. dioec. trees, some planted in trop. & warm as shade etc. esp. ***S. molle*** L. (*S. m.* var. *areira*, *S. a.*, pepper tree, Calif. p., Peruvian Andes, natur. Aus.) – lvs 7–20-jugate, cult. shade-tree in Medit. etc., locally medic., fr. used to adulterate pepper, exudates chewed (American mastic), fert. control in Urug., harbours black scale of citrus, *molle* a rendering of native name 'mulli'; ***S. terebinthifolius*** Raddi (Brazilian pepper tree, Christmas berry (Hawaii), Braz.) – lvs 2– or 3(–6)-jugate with domatia, cult. orn. natur. Florida, ? allelopathic, where a cause of dermatitis & resp. problems, intoxicating birds but has good nectar for bees, cult. for fr. (red pepper, r. p.corns, pink p.c.) esp. Réunion, invasive in Aus., S Afr., Masc., SE US, Hawaii (first coll. 1911)

Schinziella Gilg. Gentianaceae. 1 trop. Afr.: ***S. tetragona*** (Schinz) Gilg. R: Blumea 48(2003)39

Schinziophyton Hutch. ex A.R.-Sm. = *Ricinodendron* (but see KB 45(1990)157)

Schippia Burret. Palmae (IIII 2). 1 Belize, Guatemala: ***S. concolor*** Burret – cult. orn. R: J. Dransfield & al., *Gen. Palm.* (2008)219

Schisandra Michaux. Schisandraceae. 23 E As. & E N Am. (1). R: FOW 4(2001)2. Poll. female gall-midges (*Megommata* sp.). Cult. orn. dioec. or monoec. (***S. chinensis*** (Turcz.) Baill. (China, Jap.) with labile sexuality) lianes with attractive fr.; some effective tonics (ingredient in 'natural Viagra') & sedatives

Schisandraceae Blume. Magnoliidae – Austrobaileyales. Incl. Illiciaceae, 3/73 Indomal., E As. & E N Am. to Carib. R (p.p.): FOW 4(2001). Dioec. or monoec. aromatic, glabrous everg. trees, shrubs & lianes with scattered ethereal oil-cells. Lvs simple, s.t. toothed & oft. with pellucid dots, in spirals (s.t. app. whorled); stip. 0. Fls small, bisex., reg. with ± elongate receptacle, solit. or in few-fld infls. in axils (or supra-axillary); P 5–24 (–33) in spirals in 2–several series, s.t. outer ± K-like & inner C-like & s.t. transitional to A 4–80 ± in spirals, ± connate basally, anthers basifixed with longit. slits & ± expanded connectives, $\underline{G}$ (5-)12–100(–300) conduplicate, unsealed, in spiral with decurrent stigmas & each with 1–5(–11) marginal, anatr. to campylotropous, bitegmic ovules. Fr. a head (*Illicium*, *Kadsura*) or elongate axis of berries or follicles (*Illicium*) with 1 or 2 seeds; embryo minute in copious endosperm rich in oil & starch. n = 13, 14

Genera: *Illicium, Kadsura, Schisandra*

E As. – E N Am. disjunctions in *Illicium* & *Schisandra*. Fr. analogous to that in *Sargentodoxa*

Comm. oils & flavourings (*Illicium*); cult. orn

Schischkinia Iljin. Compositae (Card.-Cent.). 1 SW & C As.: ***S. albispina*** (Bunge) Iljin

Schischkiniella Steenis = *Silene*

Schismatoclada Bak. (~ *Payera*). Rubiaceae (IV-Dan.). 20 Madag.

Schismatoglottis Zoll. & Moritzi. Araceae (VII 8). Excl. *Apoballis*, c. 250 S China to Vanuatu (Mal. 89 – R: Telopea 9(2000)1) esp. Borneo (Am. sp. = *Philonotion*). Herbs, some rheophytic, with spathe-tube persistent around female fls, the blade dropped early exposing males (organization into male fls obscure) & term. sterile appendix of spadix; some cult. orn. incl varieg. cvs. ***S. corneri*** A. Hay (Anambas, Borneo) – massive arborescent pachycaul; ***S. prietoi*** Boyce & al. (Philippines) – aquatic; ***S. roseospatha*** Bogner (Sarawak) – rheophytic aquarium pl.

Schismocarpus S.F. Blake. Loasaceae (II). 1 Mex.: ***S. pachypus*** S.F. Blake. R: U. Eggli, *Ill. Handb. Succ. Pls*, Dicots (2002)307

Schismus P. Beauv. Gramineae (Danth.). 5 Afr., Medit. (Eur. 2) to China. R: Abh. Senck. Nat. Ges. 532(1974)1. ***S. arabicus*** Nees (Arabia) – invasive W US

Schistocarpaea F. Muell. Rhamnaceae (inc. sed.). 1 Queensland: ***S. johnsonii*** F. Muell.

Schistocarpha Less. Compositae (Mill.-Gal.). 16 trop. Am. R: Phytol. 59 (1986)272

Schistocaryum Franch. = *Microula*

Schistogyne Hook. & Arn. = *Oxypetalum*

Schistolobos W.T. Wang = *Opithandra*

Schistonema Schltr. Apocynaceae (Vc; Asclepiadaceae III 1). 1 Peru: ***S. weberbaueri*** Schltr.

Schistophragma Benth. ex Endl. (~ *Leucocarpus*). Plantaginaceae (Grat.-Stem.; Scrophulariaceae s.l.). 2 W N to C Am.

Schistostemon (Urb.) Cuatrec. Humiriaceae. 9 trop. S Am. R: CUSNH 35(1961)146

Schistostephium Less. Compositae (Anth.-Col.). 8 trop. & S Afr. R: KB 68(2013)108

Schistotylus Dockr. Orchidaceae (V 16c). 1 NSW: ***S. purpuratus*** (Rupp) Dockr.

Schivereckia Andrz. ex DC. = *Draba*

Schizachne Hackel. Gramineae (XI). 1 Arctic Eur., NE As., temp. N Am., SW US (mts): ***S. purpurascens*** (Torr.) Swallen (*S. callosa*). R: CUSNH 48(2003)607; FNA 24(2007)103

Schizachyrium Nees. Gramineae (XXII 7). 64 trop. savannas. ***S. condensatum*** (Kunth) Nees (trop. Am.) – invasive Hawaii; ***S. scoparium*** (Michaux) Nash (blue-stem, bunchgrass, N Am.) – imp. erosion control & grazing in Great Plains (form. major constituent of orig. grasslands)

Schizacme Dunlop (~ *Mitrasacme*). Loganiaceae. 3–4 Aus., NZ. R: FA 28(1996)58

Schizaea Sm. Schizaeaceae. Excl. *Actinostachys* 28 trop., S temp., N Am. Usu. on nutrient-poor soils, s.t. on decaying wood. Fronds simple or dichot. lobed with sporangia in rows along segments. Prothalli merely uniseriate filaments, in ***S. dichotoma*** (L.) Sm. (Tanzania, Madag. to Pac. – local medic.) multiseriate & subterr. or cylindrical, subterr. & becoming tuberous with age

Schizaeaceae Kaulf. Polypodiidae – Schizaeales. Excl. Anemiaceae, Lygodiaceae, 2/44 trop., N Am., S temp. R: T 55(2006)711. Terr. ferns with simple or fan-shaped fronds with dichot. veins. Sporangia borne singly on modified segments of whole fronds & not in sori, the annulus merely a group of thick-walled cells, dehiscence longit. Gametophytes green, filamentous (*Schizaea*) or subterr., echlorophyllous & tuberous (*Actinostachys*). n = 77, 94, 103 [!]

Genera: *Actinostachys, Schizaea*

Cretaceous fossils (*Schizaeopsis*), others from Jurassic, referred here

Schizanthus Ruíz & Pavón. Solanaceae (inc. sed.). 12 Chile. R: MBSM 20(1984)121. Sister to rest of fam. Fls irreg., resupinate, with upper 2 petals forming 3–4 lobed lip, laterals 4-lobed, lower one forming simple or weakly bilobed upper lip, A 4 (2 sterile); fl. mechanism like Leguminosae (III) with explosion (cf. *Genista*). Cult. orn. (poor man's orchid, butterfly fls) esp. ***S. pinnatus*** Ruíz & Pavón & its hybrids

Schizeilema (Hook.f.) Domin. Umbelliferae (Azor.). 1 S E Aus. alpine: ***S. fragoseum*** (F. Muell.) Domin – allied to *Diplaspis* & *Huanaca*

Schizenterospermum Homolle ex Arènes (~ *Tarenna*). Rubiaceae (II 2). 4 Madag. R: NS 16(1960)9

Schizobasis Bak. = *Drimia*

Schizoboea (Fritsch) B.L. Burtt = *Streptocarpus* (but see NRBGE 33(1974)266)

Schizocaena J. Sm. ex Hook. = *Cyathea*

Schizocalomyrtus Kausel = *Calycorectes*

Schizocalyx Wedd. (~ *Bathysa*). Rubiaceae (Cond.). 9 trop. Am. R: Novon 21(2011)499

Schizocapsa Hance = *Tacca*

Schizocarphus J. Merwe (~ *Scilla*). Asparagaceae. 1 trop. E & S Afr.: ***S. nervosus*** (Burchell) J. Merwe. R: EJB 66(2004)559

Schizocarpum Schrad. Cucurbitaceae (XV). 11 Mex., Guatemala

Schizochilus Sonder. Orchidaceae (IV 4d). 11 trop. & S Afr. R: JSAB 46(1980)379

Schizococcus Eastw. = *Arctostaphylos*

Schizocodon Sieb. & Zucc. = *Shortia*

Schizocolea Bremek. Rubiaceae (IV 11). 2 trop. W Afr.

Schizodium Lindl. = *Disa* (but see JSAB 47(1981)339)

Schizoglossum E. Meyer. Apocynaceae (Vc; Asclepiadaceae III 1). 15 S Afr. R: F. Albers & U. Meve, *Ill. Handb. Succ. Pls*, Asclep. (2002)236

Schizogyne Cass. (~ *Inula*). Compositae (Inul.-Inul.). 2 Canary Is.

Schizolaena Thouars. Sarcolaenaceae. 20 Madag. R: Adansonia III,21(1999)185

Schizolepton Fée = *Taenitis*

Schizolobium Vogel. Leguminosae (I 4). 1 trop. Am.: ***S. parahyba*** (Vell. Conc.) S.F. Blake (guapiruvu) – spectacular cult. orn. pachycaul tree (Rauh's Model, to 60 cm diam. bole in 3 yrs), unbranched when young & bearing bipinnate lvs to 150 cm long, fls almost reg., 5-merous. R: Britt. 48(1996)178

Schizoloma Gaudich. = *Lindsaea*

Schizomeria D. Don. Cunoniaceae (III). 10 Moluccas, Papuasia, E Aus. ***S. ovata*** D. Don (E Aus.) – timber for coffins, veneers etc.

Schizomeryta R. Viguier = *Meryta*

Schizomussaenda Li = *Mussaenda*

Schizonepeta (Benth.) Briq. Labiatae (VII 2b). 3 steppes of S Siberia, Mongolia & China. R: BZ 78,2(1993)112. ***S. tenuifolia*** (L.) Briq. (China) – painkiller used with *Stenocoelium divaricatum* in treatment of arthritis & toothache in modern Chinese herbalism

Schizopepon Maxim. Cucurbitaceae (XI). 6–8 Himal. to E As. R: APS 23(1985)110. Fls s.t. bisex.

Schizopetalon Sims. Cruciferae (41). 10 NC Chile & nearby Arg. R: HPB 1(1989)31. Pinnatsect C unique in fam. ***S. walkeri*** Sims (Chile) – cult. orn. ann. with bifid twisted cotyledons as have 2 other spp.

Schizophragma Sieb. & Zucc. = *Hydrangea* (but see BJLS 165(2011)295)

Schizopsera Turcz. Compositae (Helia.-Ecl.). 1 Ecuador: ***S. peduncularis*** (Benth.) S.F. Blake

Schizoptera Benth. = *Schizopsera*

Schizorhiza Goldbl. & Manning (~ *Lapeirousia*). Iridaceae (III 1). 1 SW Cape: ***S. neglecta*** (Golbl.) Goldbl. R: Strelitzia 35(2015)104

Schizoscyphus K. Schum. ex Taubert = *Maniltoa*

Schizosepala G. Barroso. Plantaginaceae (Grat.-Stem.; Scrophulariaceae). 1 Braz.: ***S. glandulosa*** G. Barroso

Schizosiphon K. Schum. = *Maniltoa*

Schizospatha Furt. = *Calamus*

Schizostachyum Nees. Gramineae (1c). Incl. *Neohouzeaua, Teinostachyum,* excl. *Cephalostachyum, Davidsea, Neohouzeaua, Pseudostachyum, Sirochloa, Teinostachyum,* c. 50 SE As. R: KBAS 13(1986)56. Bamboos; some with 3-yr cycle of fls; some extrafloral nectaries on auricles. Some used for plaited mats in Bali. ***S. brachycladum*** Kurz (Mal.) – stems used for flutes in Sumatra; ***S. funghomii*** McClure (China) – imp. comm. bamboo shoots in Vietnam; ***S. jaculans*** Holttum (Malay Pen.) – blowpipes; ***S. latifolium*** Gamble (Indomal.) – up to 10 lodicules grading into staminodes

Schizostegopsis Copel. = *Pteris*

Schizostephanus Hochst. ex K. Schum. = *Cynanchum*

Schizostigma Arn. ex Meissn. Rubiaceae (II 7). 1 Sri Lanka: ***S. hirsutum*** Arn. ex Meissn.- G 5–7-loc.

Schizostylis Backh. & Harv. = *Hesperantha*

Schizotorenia Yamaz. (~ *Lindernia*). Linderniaceae (Plantaginaceae (Grat.-Lind.); Scrophulariaceae s.l.). 2 SE As. R: JJB 53(1978)101

Schizotrichia Benth. Compositae (Tag.-Pect.). 2(–5) Peru

Schizozygia Baill. Apocynaceae (I d). 1 trop. E Afr., Comoro Is.: ***S. coffeoides*** Baill. R: MLW 83,7(1983)47

Schkuhria Roth. Compositae (Bah.). 6 warm Am., ***S. pinnata*** (Lam.) Thell. weedy elsewhere, invasive S Afr. R: FNA 21(2006)381

Schlagintweitiella Ulbr. = *Thalictrum*

Schlechtendalia Less. Compositae (Barn.). 1 (variable) Braz., Urug., Arg.: ***S. luzulifolia*** Less. – 'basal' in C., habit *Eryngium*-like with opp. or whorled linear lvs

Schlechteranthus Schwantes. Aizoaceae (V). 14 Cape. R: T 65(2016)258

Schlechterella K. Schum. Apocynaceae (III; Asclepiadaceae I). 2 E Afr. R: SAJB 64(1998)350

Schlechteria Bolus ex Schltr. = *Heliophila*

Schlechterina Harms (~ *Crossostemma*). Passifloraceae. 1 trop. E Afr.: ***S. mitostemmatoides*** Harms. R: Strel. 10(2000)436

Schlechterorchis Szlach. = *Habenaria*

Schlechterosciadium H. Wolff = *Chamarea*

Schlegelia Miq. Schlegeliaceae (Scrophulariaceae s.l.). 12 trop. Am. Form. incl. in Bignoniaceae

Schlegeliaceae Reveal (~ Bignoniaceae; Scrophulariaceae s.l.). Magnoliidae – Lamiales. 4/25 trop. Am. Trees (bark white), shrubs, lianes & epiphytes. Lvs entire to serrate, opp. Fls large in clustered leafy or bracteate axillary cymes with nectaries outside K. C with 2-lipped limb; A with 2 sep. divergent thecae ± staminode. Fr. a berry with persistent K; seeds ∞, compressed, angular, s.t. winged, embryo straight. Germ. epigeal. n = 20

Genera: *Exarata, Gibsoniothamnus, Schlegelia, Synapsis*

App. close to Thomandersiaceae

Schleichera Willd. Sapindaceae. 1 Indomal.: ***S. oleosa*** (Lour.) Oken (*S. trijuga,* kussum, kosumba) – timber (Ceylon oak) hard & used for mortars, etc., bark for tanning, lvs ed. as veg. with rice, unripe fr. pickled, seeds the source of orig. Macassar oil, used for candles, hairdressing, batik-work, soap & illumination; tree a host of lac insects. R: FM I,11(1994)727

Schleidenia Endl. = *Heliotropium*

Schleinitzia Warb. ex Guinet (~ *Prosopis*). Leguminosae (II 1). 4 Pac. (Mal. 2; Guam 1, Vanuatu to Tahiti 1). R: Adansonia 18(1978)345

Schliebenia Mildbr. = *Isoglossa*

Schlimia Planch. & Linden. Orchidaceae (V 12i). 7 N Andes

Schlimmia Planch. & Linden = *praec.*

Schlumbergera Lem. (*Zygocactus*). Cactaceae (III 6). 6 SE Braz. mts (all near Rio). R: A.J.S. McMillan & J.F. Horobin *Christmas cacti* (1995)18. Some cult. orn. hummingbird-poll. epiphytic cacti with flat-jointed stems (millions a yr sold in Denmark & Holland – some 200 cvs incl. some with fls yellow under heat but pink when cool) esp. ***S. x buckleyi*** (T. Moore) Tjaden (*S. bridgesii, S. truncata* × *S. russelliana* (Hook.) Britton & Rose, Christmas cactus [Organ Mts]) – stem-joints crenate, fls almost reg., ovary 4–5-angled, fls in winter & ***S. truncata*** (Haw.) Moran (crab cactus) – stem-joints 2–4-serrate, fls irreg., ovary cylindrical, fls in autumn

Schlumbergera C.J. Morren = *Guzmania*

Schmalhausenia Winkler = *Arctium*

Schmaltzia Desv. ex Small = *Rhus*

Schmardaea Karsten. Meliaceae (I). 1 Andes: ***S. microphylla*** (Hook.) C. Mueller. R: FN 28(1981)387

Schmidtia Steud. ex J.A. Schmidt (~ *Enneapogon*). Gramineae (XXVII 1). 2 Afr., Cape Verde Is., Pakistan. R: BSB II,39(1965)303

Schmidtottia Urb. Rubiaceae (inc. sed.). 14 E Cuba (serpentine)

Schnabelia Hand.-Mazz. Labiatae (III; Verbenaceae). 5 S & SW China. Cleistogamy

schnap(p)s type of gin oft. flavoured with caraway or cumin

Schnarfia Speta = *Scilla*

Schnella Raddi (~ *Bauhinia*). Leguminosae (I 1). 47 trop. Am. R: Phytotaxa 204(2015)237. Lianes with tendrils

Schoenefeldia Kunth. Gramineae (XXIX 5). 2 Afr. to Ind. Awns from adjacent lemmas 'braided'

Schoenia Steetz (~ *Helichrysum*). Compositae (Gnap.-Ang.). 5 temp. Aus. R: Nuytsia 8(1992)371. ***S. cassiniana*** (Gaudich.) Steetz – cult. orn. ann. 'everlasting'

Schoenobiblus Mart. Thymelaeaceae (Thym.-Daph.). 8 trop. Am. Dioec. ***S. peruvianus*** Standl. (Amaz.) – fish- & arrow-poison

Schoenocaulon A. Gray. Melanthiaceae. 26 Florida (N Am. 3) to Peru, esp. Mex. R: AMBG 29(1942)292. Alks incl. veratrin. ***S. officinale*** (Schldl. & Cham.) A. Gray (sabadilla, cevadilla, Mex. To Peru) – seeds insecticidal, used in veterinary medicine

Schoenocephalium Seub. Rapateaceae (I 2). 5 NE S Am.

Schoenocrambe Greene = *Sisymbrium*

Schoenoides Seberg = *Oreobolus*

Schoenolaena Bunge (~ *Xanthosia*). Umbelliferae (? Mack.). 2 W Aus.

Schoenolirion Torrey. Asparagaceae (Hyacinthaceae). 3 S US. R: Madroño 38(1991)132

Schoenomorphus Thorel ex Gagnepain = *Tropidia*

Schoenoplectus (Reichb.) Palla (~ *Scirpus*). Cyperaceae (II 2). c. 77 cosmop. (Aus. 11). ***S. acutus*** (Bigelow) Löve & D. Löve (*Scirpus a.*, N Am.) – stembases & lvs used for teepee matting; ***S. californicus*** (C. Meyer) Soják (*Scirpus c.*, totora, S US to S Am., Easter Is., Hawaii, introd. NZ) – used for mats, houses, floats, boats etc. & minor food-source at Lake Titicaca; ***S. corymbosus*** (Roemer & Schultes) Raynal (OW warm & trop.) – used by anc. Egyptians in funeral wreaths; ***S. lacustris*** (L.) Palla (*S. tabernaemontani*, *Scirpus l.*, clubrush, bulrush, Dutch rush, N temp.) – shoots regenerate in total absence of oxygen for up to 90 days allowing spread into anaerobic muds denied to other spp., chair-seats, mats, hassocks, baskets, rhiz. eaten by Native Americans, used in water purification in Netherlands & Ger.; ***S. mucronatus*** (L.) Palla (*Scirpus m.*, warm OW) – cult. Sumatra for bag prod.; ***S. pungens*** (Vahl) Palla (*Scirpus americanus*, sword grass, chairmaker's rush, N Am.) – seating rush; ***S. supinus*** (L.) Palla subsp. ***lateriflorus*** (J.F. Gmelin) Soják (*S. l.*, Ind. to Aus.) – hat-making in Taiwan; ***S. triqueter*** (L.) Palla (*Scirpus t.*, OW) – matting

Schoenorchis Reinw. ex Blume. Orchidaceae (V 16c). c. 25 Indomal. & China to Samoa. R: OB 95(1988)66. Some cult. orn. with narrow lvs

Schoenoselinum Jim. Mejías & P. Vargas (~ *Anethum*). Umbelliferae. 1 Morocco: ***S. foeniculoides*** (Maire & Wilczek) Jim. Mejías & P. Vargas. R: Phytotaxa 212(2015)75

Schoenoxiphium Nees = *Carex*

Schoenus L. Cyperaceae (II 7). 120+ subcosmop. (not N Am.; Eur. 2) esp. Mal. & Aus. (90, 85 endemic). In a single spikelet fls ± P. ***S. asperocarpus*** F. Muell. (Aus.) – one of few Aus. sedges reputed to be toxic (cattle); ***S. clandestinus*** S.T. Blake (SW Aus.) – mats of lvs & styles above ground; ***S. nigricans*** L. (black bog rush, subcosmop.) – still used for thatch in Ireland

Schoepfia Schreb. Schoepfiaceae (Olacaceae s.l.). c. 25 Indomal., trop. Am. (20 – R: FN 38(1984)19). Oft. r.-parasites

Schoepfiaceae Blume (~ Olacaceae). Magnoliidae – Santalales. 1/c. 25 Indomal., trop. Am. Semi-parasitic trees & shrubs. Lvs simple, in spirals. Fls bisexual, pedicellate in short axillary raceme with persistent bracts. K obscure; C (4)5(6)-lobed tube; A connate with C; G inf. topped with fleshy disk, 3-loc., each loc. with 1 pend. unitegmic ovule. Fr. a drupe; seeds with 2 or 3 cotyledons. n = 12

Genus: *Schoepfia*

Arjona, *Quinchamalium* returned to Santalaceae

Scholleropsis Perrier. Pontederiaceae. 1 N Cameroun (Lake Chad), Madag.: ***S. lutea*** Perrier

Scholtzia Schauer. Myrtaceae (II 15). 13 SW Aus.

Schomburgkia Lindl. = *Laelia*

Schotia Jacq. Leguminosae (I 2). c. 4 S Afr., open woodland & scrub. Cult. orn. trees. Some ed. seeds (Boer beans) esp. ***S. afra*** (L.) Thunb. (*S. speciosa*)

Scottariella Boyce & S. Wong. Araceae. 1 Sarawak: ***S. mirifica*** Boyce & S. Wong

Schoutenia Korth. Malvaceae (Dom.; Tiliaceae). 9 Thailand to C Mal. & N Aus. Some local timbers

Schouwia DC. Cruciferae (12). 1 N Afr., Arabia: ***S. purpurea*** (Forssk.) Schweinf. – cult. orn. ann., lvs ed. Sahara

Schradera Vahl. Rubiaceae (IV 6). Incl. *Lucinaea*, c. 55 Mal. (16 – R: Blumea 43(1998)287), Am. (R: MNYBG 10(1963)259). Some epiphytes

Schraderanthus Averett (~ *Chamaesaracha*). Solanaceae (IV 5d). 1 Mex., C Am.: ***S. viscosus*** (Schrad.) Averett. R: Phytologia 91(2009)54

Schrameckia Danguy = *Tambourissa*

Schranckiastrum Hassler = *Mimosa*

Schrankia Willd. = *Mimosa*

Schrebera Roxb. Oleaceae (4b). 8 trop. (SE As. 2, Afr. 3, 1 ext. to Madag. (+2 endemic), Peru 1)

Schreiteria Carolin (~ *Calandrinia*). Montiaceae (Portulacaceae s.l.). 1 Arg.: ***S. macrocarpa*** (Speg.) Carolin. R: Parodiana 3(1985)330

Schrenkia Fischer & C. Meyer. Umbelliferae (III 4). Incl. *Kosopoljanskia*, 15 C As.

Schrenkiella German & Al-Shehbaz. Cruciferae. 1 NW China: ***S. parvula*** (Schrenk) German & Al-Shehbaz. R: NJB 28(2010)648

Schtschurowskia Regel & Schmalh. Umbelliferae (III 4). 2 C As.

Schubertia Mart. Apocynaceae (Vc; Asclepiadaceae III 3). 6 S Am.

Schuitemania Ormerod (~ *Platylepis*). Orchidaceae (IV 2d). 1 Philippines: ***S. merrillii*** (Ames) Ormerod. R: Lindleyana 17(2002)228

Schultesia Mart. Gentianaceae. 20 trop. Am.

Schultesianthus Hunz. Solanaceae (IV). 8 trop. Am.

Schultesiophytum Harling. Cyclanthaceae. 1 NW trop. S Am.: ***S. chorianthum*** Harling

Schulzia Spreng. Umbelliferae (III 8). 2 C As., ? 2 NW Ind.

Schumacheria Vahl. Dilleniaceae (IV). 3 Sri Lanka. ***S. castaneifolia*** Vahl – 'throw-away' branches, stilt-r. in wet sites

Schumannia Kuntze = *Ferula*

Schumannianthus Gagnepain (~ *Donax*). Marantaceae. 2 Indomal. (Borneo 2)

Schumanniophyton Harms. Rubiaceae (II 1?). 3 trop. Afr. Petit's & Cook's Models. Stimulants & fish-poison esp. ***S. magnificum*** (K. Schum.) Harms (W Afr.)

Schumeria Iljin = *Klasea*

Schunkea Senghas. Orchidaceae (V 12h). 1 Peru: ***S. vierlingii*** Senghas. R: Gen. Orch. 5(2009)354

Schuurmansia Blume. Ochnaceae (III). 3 C Mal. to Papuasia. Leeuwenberg's Model

Schuurmansiella Hallier. Ochnaceae (III). 1 NW Borneo: ***S. angustifolia*** (Hook.f.) Hallier. R: BJ 113(1991)181

Schwabea Endl. & Fenzl = *Monechma*

Schwackaea Cogn. Melastomataceae. 1 Mex., C Am.: ***S. cupheoides*** (Benth.) Cogn. R: NJB 14(1994)69

Schwalbea Gronov. ex L. Orobanchaceae (Cymb.; Scrophulariaceae s.l.). 1 E US: ***S. americana*** L. – hemiparasitic perenn. of savannas, not host-specific

Schwannia Endl. = *Janusia*

Schwantesia Dinter (~ *Lithops*). Aizoaceae (V). 11 SW Afr. R: H. E. K. Hartmann, *Ill. Handb. Succ. Pls*, Aiz. F–Z(2002)291. Dwarf succ. with marcescent lvs around internodes, cult. orn.

Schwartzia Vell. (~ *Norantea*). Marcgraviaceae. 14 trop. Am. R: JBRIT 3(2009)694

Schwartzkopffia Kraenzlin = *Brachycorythis*

Schweiggeria Spreng. Violaceae (III 2b). 2 Mex., Braz.

Schweinfurthia A. Braun. Plantaginaceae (Ant.-Ant.; Scrophulariaceae s.l.). 6 NE Afr. to Ind., desert & semi-d. R: NRBGE 40(1982)23

Schwenckia L. Solanaceae (Schw.). 25 trop. Am., 1 (***S. americana*** L. long natur. trop. Afr.)

Schwenckiopsis Dammer = *Protoschwenckia*

Schwendenera K. Schum. Rubiaceae (IV 15). 1 SE Braz.: ***S. tetrapyxis*** K. Schum.

Schwenkia L. = *Schwenckia*

Schwenkiopsis Dammer = *Protoschwenckia*

Sciadocephala Mattf. Compositae (Eup.-Aden.). 6 Panamá & N S Am. to Braz. R: KB 68(2013)311

Sciadodendron Griseb. = *Aralia*

Sciadopanax Seemann = *Polyscias*

Sciadophyllum P. Browne = *Schefflera*

Sciadopityaceae Luerss. (~ Taxodiaceae). Pinidae – Pinales. 1/1 C & S Jap. R: A. Farjon (2005) *Monogr. Cupressaceae & Sciadopitys*: 547. Resinous monoec. tree with short shoots borne on principal ones. Photosynthetic organs appear to be united needles which occ. branch – pigeonholed as 'cladodes' or 'phylloclades' they are best considered organs intermediate between lvs & stems, flattened, in spirals in seedlings, adults with scale-like spurs subtending 'phylloclades' forming false whorls. Male cones subglobose in term. raceme-like clusters; females solit. with ∞ scales each with 7–9 anatr. ovules. Seeds ovate-elliptic with v. narrow wing & 2 cotyledons. Gametophytes rather like Pinaceae & Podocarpaceae, while n = 10 (11, 33 etc. in Taxodiaceae = Cupressaceae)
Genus: *Sciadopitys*
Not like other Cupressaceae (Taxodiaceae) on rbcl data & app. distinct since Upper Triassic (?) when more widespread

Sciadopitys Sieb. & Zucc. Sciadopityaceae. 1 C & S Jap.: ***S. verticillata*** (Thunb.) Sieb. & Zucc. (parasol pine, umbrella fir) – now widely planted around temples & perhaps native only in 2 small areas of C Honshu, though in Tertiary common in Eur. & a char. fossil of some Brown Coal, cones mature over 2 yrs, timber used for ship-building etc., oil for varnish, cult. orn. for acid soil

Sciadotenia Miers. Menispermaceae (I). 19 trop. Am. R: MNYBG 22, 2(1971)15. Curare sources

Sciaphila Blume. Triuridaceae. Incl. *Andruris*, *Hyalisma*, c. 40 trop. (esp. NG) & warm. R: PR IV, 18(1938)30, BB 140(1991)9. Endotrophic mycorrhiza

Sciaphyllum Bremek. = *Streblacanthus*

Sciaplea Rauschert = *Dialium*

Scilla L. Asparagaceae (Hyacinthaceae). (Incl. *Chionodoxa*, excl. *Barnardia*, *Ledebouria*) 46 Euras. (Eur. 18), temp. Afr. S.t. split up into many small genera, e.g. *Pfosseria*, *Prospero*, etc. etc. In fr. peduncle of many spp. extends & flops on ground when ant-disp. seeds released. Cult. orn. bulbous pls esp. ***S. autumnalis*** L. (autumn squill, Eur. to N Afr. & Cauc.), ***S. forbesii*** (Bak.) Speta (*S. siehei*, *C. s.*, '*C. luciliae*' [true *C. l.* (W Turkey) with 1–3 fls per infl.], glory of the snow, W Turkey) – fls 4–12 blue with white centre (wild type), white or pink, ***S. mischtschenkoana*** Grossh. (*S. tubergeniana*, NW Iran & Transcaucasia) – v. early white fls with blue stripes, the pls in cult. prob. a clone prod. v. few seeds, suggesting meiotic irregularities, ***S. peruviana*** L. (Cuban lily, W Medit.(!)), ***S. sardensis*** (Barr & Sugden) Speta (*C. s.*, W Turkey) – blue fls; ***S. siberica*** Haw. (Iran–Russia) esp. '**Spring Beauty**' (sterile so fls last longer), ***S. verna*** Huds. (spring squill, W Eur., Morocco)

Scindapsus Schott. Araceae (IV 4). 35 Indomal., Pacific. Some cult. orn. lianes esp. ***S. pictus*** Hassk. '**Argyraeus**' (W Mal.) & local medic.

Sciodaphyllum P. Browne = *Schefflera*

Sciothamnus Endl. (~ *Peucedanum*). Umbelliferae (III 10). 4 S Afr.

Scirpodendron Zipp. ex Kurz. Cyperaceae (I 1). 2 Sri Lanka to Polynesia, tidal swamps etc. Fls truly term. ***S. ghaeri*** (Gaertn.) Merr. (*S. costatum*) – hats & mats locally, seeds ed. (Samoa)

Scirpoides Séguier (*Holoschoenus*; ~ *Scirpus*). Cyperaceae (II 5). c. 5 warm Euras., S Afr., ? Mex.

Scirpus Micheli ex L. Cyperaceae (II 1). Excl. *Bolboschoenus*, *Dracoscirpoides*, *Isolepis*, *Rhodoscirpus*, *Schoenoplectus*, *Scirpoides*, *Trichophorum*, 64 subcosmop. (N Am. 18). R (s.l.): JFSUTB 3, 7(1958)271. Heterogeneous? ***S. atrovirens*** Willd. (N Am., bakana) – alks, cult. said to lead to insanity (Mex.); ***S. paludosus*** Nelson (bayonet grass, N Am.) – rhiz. form. eaten

Sclerachne R. Br. = *Polytoca*

Sclerandrium Stapf & C. Hubb. = *Germainia*

Scleranthera Pichon = *Wrightia*

Scleranthopsis Rech.f. = *Acanthophyllum*

Scleranthus L. Caryophyllaceae (II 5). 12 Eur. (3), As., Afr., Aus. C 0, autogamous, weedy esp. ***S. annuus*** L. (knawel, temp. Euras., Medit., introd. N Am.) – local medic. (urinary probs) in Ireland

Scleria P. Bergius. Cyperaceae (III 3). Excl. *Diplacrum*, c. 220 trop. & warm (Aus. 23, S Afr. 23 – R: Bothalia 15(1985)505, N Am. 14). R: T 65(2016)458. Local medic., lvs for polishing wood & mats. Scrambling spp. oft. a fire-risk & objectionable, e.g. ***S. boivinii*** Steud. (W Afr.) with cutting leaf-edges; ***S. biflora*** Roxb. (Indomal.) – veg. in Java

Sclerobassia Ulbr. = *Bassia*

Scleroblitum Ulbr. (~ *Chenopodium*). Amaranthaceae (Chenopodiaceae I 2). 1 SE Aus.: ***S. atriplicinum*** (F. Muell.) Ulbr. R: FA 4(1984)175

Sclerocactus Britton & Rose (~ *Pediocactus*). Cactaceae (III 9). (Incl. *Ancistrocactus*) 17 SW N Am. R: CSJ 38(1966)50, 100. Small cult. orn. undulate-ribbed cacti s.t. with areoles with extrafl. nectaries

Sclerocarpus Jacq. Compositae (Helia.-Helia.). 8 trop. & warm Am., Afr. (1)

Sclerocarya Hochst. (~ *Poupartia*). Anacardiaceae (II). 2–3 trop. & S Afr. ***S. birrea*** (A. Rich.) Hochst. subsp. ***caffra*** (Sond.) Kokwaro (*P. b.*, maroola plum, marula) – elephants attracted to trees with fermenting fruit (but reports of drunkenness denied) and also disp. seed, fr. brewed to a potent beer esp. Swaziland where a serious social problem at fruiting time, flavour in comm. liqueur in Eur.

Sclerocaryopsis Brand = *Lappula*

Sclerocephalus Boiss. = *Gymnocarpos*

Sclerochiton Harv. Acanthaceae (III 1). 19 trop. & S Afr. R: KB 46(1991)7

Sclerochlamys F. Muell. = *Sclerolaena*

Sclerochlamys Morrone & Zuloaga = *Keratochlaena* (but see T 58(2009)373)

Sclerochloa P. Beauv. Gramineae (XVI 14). 2 S Eur. to China, ***S. dura*** (L.) P. Beauv. now a cosmop. weed

Sclerochorton Boiss. Umbelliferae (III 8). 1 SE Eur., 1 SW Aus.

Sclerocroton Hochst. (~ *Sapium*). Euphorbiaceae (Euph.-Hipp.). 6 trop. Afr. (5), Madag. (1)

Sclerodactylon Stapf. Gramineae (XXIX 5). 1 E Afr. coast, is. of Ind. Ocean on coral rocks near shore: ***S. macrostachyum*** (Benth.) A. Camus

Sclerodeyeuxia Pilg. = *Calamagrostis*

Scleroglossum Alderw. (~ *Grammitis*). Polypodiaceae (V; Grammitidaceae). c. 7 Indomal. to Aus. & W Pac. Superficially resembling small *Vittaria* spp. but spores chlorophyllous etc.

Sclerolaena R. Br. Amaranthaceae (Chenopodiaceae I 5). Incl. *Osteocarpum*, *Stelligera*, *Threlkeldia*, c. 70 Aus. (not Tasmania). Copper-burr. R: FA 4(1984)236, Telopea 3(1988)142

Sclerolepis Cass. Compositae (Eup.-Tric.). 1 E US: ***S. uniflora*** (Walter) Britton & al. R: MSBMBG 22(1987)192

Sclerolinon C. Rogers = *Linum*

Sclerolobium Vogel = *Tachigali*

Scleromitrion (Wight & Arn.) Meisn. (~ *Oldenlandia*). Rubiaceae (IV 15). Spp.? ***S. diffusum*** (Willd.) R. Wang (*O. d.*, Indomal.) – constituent of a Chinese herbal tea

Scleronema Benth. Malvaceae (Bomb.; Bombacaceae). 5 trop. S Am.

Sclerophylacaceae Miers = Solanaceae

Sclerophylax Miers. Solanaceae (IV; Sclerophyllacaceae). 14 Arg. R: Kurtziana 1(1961)9. Succ. halophytes; lvs opp.

Scleropoa Griseb. = *Desmazeria*

Scleropodium Bruch & Schimper. Brachytheciaceae (Musci). 6 Eur., Medit., W N Am. R: SB 39(2014)682

Scleropogon Philippi. Gramineae (XXIX 1). 1 SW US, Mex., Arg., Chile: ***S. brevifolius*** Philippi (burrograss) – fr. buried by awn of floret. R: Phytol. 62(1987)267

Scleropyrum Arn. Santalaceae (Cervantesiaceae). 6 Indomal. R.-parasites. ***S. pentandrum*** (Dennst.) Mabb. (*S. wallichianum*, Ind.) – local medic.

Sclerorhachis (Rech.f.) Rech.f. Compositae (Anth.-Han.). 4 Afghanistan. R: BBMNHB 23(1993)110

Sclerosciadium Koch ex DC = *Capnophyllum*

Sclerosiphon Nevski = *Iris* (but see BZ 91(2006)1897)

Sclerosperma G. Mann & H. Wendl. Palmae (V 5). 3 trop. W Afr. R: KB 63(2008)77. Some 'litterbox' pls

Sclerostachya (Hackel) A. Camus = *Miscanthus*

Sclerostegia Paul G. Wilson = *Tecticornia* (but see Nuytsia 3(1980)1)

Sclerostephane Chiov. (~ *Pulicaria*). Compositae (Inul.). 5 Somalia. R: BJ 104(1983)91

Sclerotheca A. DC. Campanulaceae (III 4). 6 Cook Is., Society Is. Woody

Sclerothrix C. Presl = *Klaprothia*

Sclerotiaria Korovin. Umbelliferae (III 4). 1 C As.: ***S. pentaceros*** (Korovin) Korovin

Scobinaria Seib. = *Arrabidaea*

Scoliaxon Payson. Cruciferae (42). 1 NE Mex.: ***S. mexicanus*** (S. Watson) Payson

Scoliopus Torrey. Liliaceae. 2 W N Am. R: AMBG 79(1992)137. Form. in Trilliaceae (= Melanthiaceae). Cult. orn. with foetid fls

Scoliosorus T. Moore (~ *Antrophyum*). Pteridaceae (Vitt.) 1 trop. Am.: ***S. ensiformis*** (Hook.) T. Moore. R: T 65(2016)718

Scoliotheca Baill. = *Monopyle*

Scolochloa Link (~ *Festuca*). Gramineae (XVI 6). 1 N temp. inc. Eur.: ***S. festucacea*** (Willd.) Link (sprangletop) – fodder & hay

Scolopendrogyne Szlach. & Mytnik = *Quekettia*

Scolophyllum Yamaz. (~ *Lindernia*). Linderniaceae (Plantaginaceae (Grat.-Lind.); Scrophulariaceae s.l.). 3 SE As. R: Willd. 43(2013)229

Scolopia Schreb. Salicaceae (Flacourtiaceae). 37 OW trop. R: Blumea 20(1972)25. Some S Afr. spp. with hard timber for axles etc. esp. ***S. braunii*** (Klotzsch.) Sleumer (*S. brownii*, flintwood, E Aus.) – poss. walnut subs., ***S. ecklonii*** (Nees) Harv. (*S. zeyheri*, thorn pear). ***S. nitida*** C. White (NG) – fls open all at once so trees white for 2–3 days

Scolosanthus Vahl. Rubiaceae (III 4). 21 WI

Scolymus Tourn. ex L. Compositae (Lact.-?). 3 Medit. incl. Eur. ***S. hispanicus*** L. (cardillo, Spanish oyster, golden thistle, S Eur. to NW F.) – r. ed. (like salsify), coffee subs., fls saffron subs. R: AJBM 58(2000)84

Scoparia L. Plantaginaceae (Grat. –Grat.; Scrophulariaceae s.l.). 20 trop. Am., ***S. dulcis*** L. a pantrop. weed used to sweeten well-water & for snakebite (C Am.). Some cult. orn.

Scopella W. de Wilde & Duyfjes = seq.

Scopellaria W. de Wilde & Duyfjes (*Scopella*, ~ *Zehneria*). Cucurbitaceae (XIV). 2 S China, W Mal. R: Blumea 51(2006)34,297

Scopelogena L. Bolus (~ *Ruschia*). Aizoaceae (V). 2 N & W Cape. R: H. E. K. Hartmann, *Ill. Handb. Succ. Pls*, Aiz. F–Z(2002)294

Scopolia Jacq. Solanaceae (IV 2). Excl. *Anisodus*, 1 Alps, Carpathians, Cauc., 1 Korea & Jap. R: BBMNB 27(1972)27. Alks, some sources of atropine

Scopulophila M.E. Jones (~ *Achyronychia*). Caryophyllaceae (I 2). 2 SW US, Mex.

Scorodocarpus Becc. Erythropalaceae (Olacaceae). 1 Mal.: ***S. borneensis*** (Baill.) Becc. – hard onion-scented timber for constr., ed. fr. R: FPM 2(2012)317

Scorodophloeus Harms. Leguminosae (I 2). 1 Guinea coast, 1 E Afr. coast. Timber

scorpion grass *Myosotis* spp.; **s. senna** *Hippocrepis emerus*

Scorpiothyrsus Li. Melastomataceae. 6 Hainan, SE As.

Scorpiurus L. Leguminosae (III 22). 2 Macaronesia & Medit. (Eur. 2) to Iran. R: Lagascalia 4(1974)259. Legume twisted, indehiscent. Some cult. fodder pls (Aus.)

Scorzonera Tourn. ex L. Compositae (Lact.-Scor.). Excl. *Lasiopora*, *Podospermum*, c. 175 Medit. (Eur. 28) to C As. ***S. acanthoclada*** Franch. (Turkestan) – teke-saghyz rubber; ***S. hispanica*** L. (scorzonera, Spanish salsify, viper's grass, Eur.) – vanilla-scented (a.m.) fls, ed. r. like salsify, coffee subs.; ***S. tausaghyz*** Lipsch. & Bosse (former USSR) – poss. rubber (tau-saghyz. r.) source tried in World War II

Scorzoneroides Moench (~ *Leontodon*). Compositae (Cich.-Hypo). c. 25 temp. Euras. ***S. autumnalis*** (L.) Moench (*L. a.*) – lawn weed

Scotch, Scots or **Scottish asphodel** *Tofieldia pusilla*; **S. broom** *Cytisus scoparius*; **S. fir** or **pine** *Pinus sylvestris*; **S. (spear)mint** *Mentha* × *gracilis*; **S. thistle** *Cirsium vulgare* or *Carduus nutans* now tending to be applied to *Onopordum acanthium*

Scottellia Oliv. Achariaceae (Flacourtiaceae). 3 trop. Afr. ***S. coriacea*** A. Chev. ex Hutch. & Dalz. (W Afr.) – comm. timber (odoko)

screwbean *Prosopis juliflora*

screwpines *Pandanus* spp.

Scribneria Hackel. Gramineae (XVI 7). 1 W US: ***S. bolanderi*** (Thurber) Hackel. R: FNA 24(2007)689

Scrithacola Alava. Umbelliferae (III 11). 1 Afghanistan, Pakistan: ***S. kurramensis*** (Kitam.) Alava. R: NRBGE 38(1980)260

Scrobicaria Cass. (~ *Gynoxys*). Compositae (Sen.-Tuss.). 2 NE S Am.

Scrobicularia Mansf. = *Poikilogyne*

Scrofella Maxim. Plantaginaceae (Dig.-Ver.; Scrophulariaceae s.l.). 1 NW China: ***S. chinensis*** Maxim.

Scrophucephalus Khokhr. = *Scrophularia*

Scrophularia Tourn. ex L. Scrophulariaceae (Scroph.). 200 N temp. (Eur. 30), esp. Iran & Afghanistan, to trop. Am. Coarse foetid herbs & shrubs (figwort); style & stamens lying along lower lip of C, posterior stamen rep. by staminode; some poll. wasps. Few cult. orn. e.g. ***S. auriculata*** L. ('*S. aquatica*', water f., Eur., Med.). Local medic. e.g. ***S. ningpoensis*** Hemsl. (hexuanshen, China) – TCM, ingredient of comm. 'natural Viagra'; ***S. nodosa*** L. (N temp.) – venerated in anc. Ireland, powdered 'roots' used for piles, coughs etc.

Scrophulariaceae Juss. Magnoliidae – Lamiales. Incl. Buddlejaceae, Myoporaceae, excl. Calceolariaceae, Linderniaceae, Mazaceae, Paulowniaceae, Schlegeliaceae (& many genera to Orobanchaceae, Phrymaceae, Plantaginaceae & Stilbaceae), 66/1800 trop. & warm (esp. S Afr.), few temp. Small trees, shrubs (*Dermabotrys* epiphytic) & herbs incl. annuals (& aquatics – *Limosella*). Lvs simple, in spirals or opp.; stip. 0. Fls usu. irreg. & bisex., solit. or in spikes, racemes or thyrses; K (2)4- or 5-lobed with valvate or imbr. segments, C ((0,4)5(–8)), oft. bilabiate, s.t. basally spurred, lobes valvate or imbr., A 5 or 4 ± staminode (uppermost) or 2(3) with lower pr. reduced or 0, anthers with longit. slits, disk oft. around $\underline{G}$ (2(3)), 2(3)-loc. with term. style, each loc. with (1,2–)±∞ anatr. or hemitropous (rarely amphitropous or campylotropous) unitegmic ovules on axile placentas. Fr. oft. a septicidal capsule (s.t. loculicidal or with pores), rarely a drupe or berry; seeds winged or angled with straight or weakly curved embryo in oily endosperm (s.t. 0). x = 6 – 10, 12, 13, 17–19

Classification & principal genera:

[Verbascoideae (2 posterior C-lobes outside laterals in bud; lvs in spirals, A oft. 5; s.s. now incl. in Scrophularieae)]: *Verbascum* etc.; Scrophularioideae (C-lobes similar; at least lower lvs opp., A usu. 4 ± staminode; 7 tribes – *Calceolaria* to Calceolariaceae, others to Linderniaceae, Phrymaceae, etc.]

Now 10 tribes recog. (some genera unassigned), incl. [Buddlejaceae]: *Buddleja*; ***Hemimerideae***: *Diascia, Nemesia*; ***Selagineae*** (Limoselleae, Manuleae [R: O.M. Hilliard (1994) *The M.*, (1999) *The Tribe S.* [in part (& Antirrhineae etc.) referred to Plantaginaceae]): *Hebenstretia, Jamesbrittenia, Limosella, Manulea, Selago, Sutera, Zaluzianskya*; ***Myoporeae*** (Myoporaceae): *Eremophila, Myoporum*; ***Scrophularieae***: *Scrophularia, Verbascum*

[Rhinanthoideae (2 posterior C-lobes covered by 1 or both laterals in bud; many hemiparasites; 3 tribes – most referred to Orobanchaceae, others to Plantaginaceae)]

Molecular work has greatly clarified fam. limits, confirming inclusion of Buddlejaceae (form. in Loganiaceae) & Myoporaceae (some genera from which had already been included), & exclusion of hemiparasitic genera to form. wholly parasitic Orobanchaceae, other genera to Stilbaceae etc. (see above)

Many with oil-secreting hairs attractive to pollinators

Some timber (*Buddleja, Eremophila*), cult. orn. (*Alonsoa, Buddleja* (some invasive), *Diascia, Freylinia, Hebenstretia, Manulea, Myoporum, Nemesia, Phygelius, Sutera, Verbascum, Zaluzianskya*)

Scrotochloa Judz. (~ *Leptaspis*). Gramineae (II). 2 Indomal. R: Phytol. 56(1984)299

scrub bloodwood *Baloghia inophylla*; **horizontal s.** *Anodopetalum biglandulosum*

scullcap *Scutellaria* spp.

scuppernong *Vitis rotundifolia*

Scurrula L. Loranthaceae (5 4). 50 S As. to China & Moluccas (Mal. 8; R: Blumea 36(1991)65). Local medic.

scurvy grass *Cochlearia officinalis*

Scutachne A. Hitchc. & Chase = *Urochloa*

Scutellaria L. Labiatae (V). c. 360 cosmop. (Eur. 13) exc. S Afr. R: NRBGE 46(1990)345. Some locally medic. & cult. orn. (scullcap, erroneously skull cap, helmet fls) incl. ***S. baicalensis*** Georgi (Siberia to Jap.) – medic. (flavonoids) in modern Chinese herbalism & ***S. mexicana*** (Torrey) A. Paton (*Salazaria* m., SW N Am.) – shrub for arid areas

Scutia (DC.) Brongn. Rhamnaceae (6). 4 trop. S Am., 1 OW trop. to S Afr. R: BTBC 101(1974)64. Alks

Scuticaria Lindl. Orchidaceae (V 12g). 11 trop. S Am. Cult. orn. epiphytes

Scutinanthe Thw. Burseraceae (V). 2 Sri Lanka to Sulawesi

Scybalium Schott & Endl. Balanophoraceae. 4 trop. Am. R: FN 23(1980)25

Scyphanthus D. Don (~ *Cajophora*). Loasaceae (I 1). 1–2 Chile

Scyphellandra Thw. = *Rinorea*

Scyphiphora Gaertn. f. Rubiaceae (Vang.). 1 coasts of Indomal. to Aus. & New Caled., Madag.: ***S. hydrophyllacea*** Gaertn.f. – only R. mangrove. R: FOC 19(2011)323

Scyphocephalium Warb. Myristicaceae. 4 trop. W Afr.

Scyphochlamys Balf.f. = *Pyrostria* (but see KM 6(1989)102)

Scyphocoronis A. Gray (~ *Millotia*). Compositae (Gnap.). 2 Aus. R: ASB 8(1995)10

Scyphogyne Decne. = *Erica*

Scyphonychium Radlk. Sapindaceae. 1 NE Braz.: ***S. multiflorum*** (Mart.) Radlk. – monoec. R: Bonpl. 6(1989)117

Scyphopappus R. Nordenstam = *Argyranthemum*

Scyphostachys Thw. Rubiaceae (II 5). 2 Sri Lanka

Scyphostegia Stapf. Salicaceae (9; Scyphostegiaceae). 1 Borneo: ***S. borneensis*** Stapf. R: FM I,5(1957)297

Scyphostegiaceae Hutch. = Salicaceae

Scyphostelma Baill. (~ *Cynanchum*). Apocynaceae. 27(–50) trop. Am. R: AMBG 99(2013)67

Scyphostrychnos S. Moore = *Strychnos*

Scyphosyce Baill. Moraceae (IV). 2 W trop. Afr. R: BJBB 47(1977)283

Scyphularia Fée = *Davallia*

Scythian lamb *Cibotium barometz*

Scytopetalaceae Engl. = Lecythidaceae

Scytopetalum Pierre ex Engl. Lecythidaceae (IV; Scytopetalaceae). 3 trop. W Afr. R: KB 70,6(2015)43. Troll's Model. Some locally used timber

sea aster *Tripolium pannonicum*; **s. bean** *Entada gigas*; **s. beet** *Beta vulgaris* 'subsp. *maritima*'; **s. blite** *Suaeda vera, S. maritima*; **s. buckthorn** *Hippophae rhamnoides*; **s. campion** *Silene uniflora*; **s. celery** *Apium prostratum*; **s. daffodil** *Pancratium maritimum*; **s. heath** *Frankenia* spp. esp. *F. laevis*; **s. holly** *Eryngium maritimum*; **S. Island cotton** *Gossypium barbadense*; **s. kale** *Crambe maritima*; **s. lavender** *Limonium* spp.; **s. milkwort** *Lysimachia maritima*; **s. onion** *Drimia maritima*; **s. parsley** *A. prostratum*; **s. palm** *Postellsia palmiformis*; **s. pink** *Armeria maritima*; **s. poppy** *Glaucium flavum*; **s. purslane** *Honckenya peploides*; **s. rocket** *Cakile maritima*; **s. samphire** *Crithmum maritimum*; **s. weed** (sushi) *Pyropia* spp. esp. *P. tenera*; **s. weed, land** *Kali komarovii*; **s. wormwood** *Artemisia maritima*

sea-grass *Zostera marina*; **s. matting** (Aus.) *Oryza sativa*

sealing-wax palm *Cyrtostachys renda*

Searsia F. Barkley (~ *Rhus*). Anacardiaceae (I). 120+ Medit. to S Afr. & China. R: Bothalia 37(2007)166. ***S. natalensis*** (Krauss) Moffett (*R. n.*, trop. & S Afr.) – fr. ed. Uganda; ***S. tripartita*** (Ucria) F. Barkley (*R. t.*, '*R. oxyacantha*', Sahara) – fr. ed.

Sebaea Sol. ex R. Br. Gentianaceae. Excl. *Exochaetium, Lagenias*, 60 trop. & warm OW esp. Afr. Some cult. orn., usu. annuals e.g. ***S. albens*** (L.f.) Sm. & ***S. aurea*** (L.f.) Sm. (S Afr.)

Sebastiania Spreng. Euphorbiaceae (Hipp.). c. 25 trop. Am. For OW spp. see *Gymnanthes, Microstachys*. Some, esp. ***S. pavoniana*** Muell. Arg. (Mex.), with fr. when occupied by larvae of a small moth, *Cydia* (*Carpocapsa*) *deshaisiana*, give 'jumping beans'; ***S. bilocularis*** S. Watson (yerba de la flecha, SW N Am.) – source of arrow-poison

Sebastiano-schaueria Nees. Acanthaceae (III 2c). 1 Braz.: ***S. oblongata*** Nees

sebastião-de-arruda *Dalbergia decipularis*

Sebertia Pierre ex Engl. = *Pycnandra*

sebesten plum *Cordia myxa*

Secale L. (~ *Triticum*). Gramineae (XV). 8 Euras (Eur. 2 + ***S. cereale*** L., rye), S Afr. R: NJB 18(1998)405. Rye derived from '*S. strictum* (C. Presl) C. Presl' (*S. montanum*), a weed of wheat & barley, giving rise to non-shattering-eared forms, from which large-fruited forms selected in E Turkey, where rye orig. used as crop in places diff. for other cereals – grain (14 M t a yr by 2008) rich in gluten & used in Schwarzbröt & crispbreads, in (grain) whisky (US), gin (Holland), vodka (orig. from wine, finally potatoes) & beer (Russia); attacked by ergot but good stock fodder & stems used for the best straw-matting & archery targets; r.-system in 0.051 m^3 found to be 622.8 km long, increasing by 5 km a day, incl. r.-hairs, 10 620 km, daily increasing 90 km!

Secamone R. Br. Apocynaceae (IV; Asclepiadaceae II). Incl. *Genianthus, Toxocarpus*, c. 90 Afr. (21 – R: KB 47(1992)439) Madag. (esp.), As. to Aus. (c. 12 incl. ***S. emetica*** (Retz.) Sm., R: KB 47(1992)597)

Secamonopsis Jum. Apocynaceae (IV; Asclepiadaceae II). 2 Madag. R: Novon 6(1996)144. Some rubber

Sechiopsis Naudin = *Sicyos* (but see SB 17(1992)395)

Sechium P. Browne = *Sicyos*

Secondatia A. DC. Apocynaceae (II d). 4 trop. Am. R: Candollea 58(2003)306

Securidaca L. Polygalaceae (IV). 80 trop. (Afr. 2; R: SAJB 59(1987)5) exc. Aus. Trees (Troll's Model) & scramblers; fr. a samara with dorsal wing. Some cult. orn. for showy fragrant fls incl. ***S. longepedunculata*** Fres. (trop. Afr.) – small tree with allegedly up to 100 medical uses, twigs source of a fibre (buaze) used in fishing-nets

Securigera DC. = *Coronilla* (but see Willd. 19(1989)59)

Securinega Comm. ex Juss. Phyllanthaceae (Phyll.-Brid.-Sec.; Euphorbiaceae s.l.). 5 Madag., Masc. (1). Alks. ***S. durissima*** J. Gmelin (Otaheite myrtle, Masc.). See also *Flueggea*

Sedastrum Rose = *Sedum*

Seddera Hochst. & Steud. Convolvulaceae (4). 31 trop. & warm Afr., Madag. & Arabia. R: KB 64(2009)198

Sedella Fourr. (~ *Sedum*). Crassulaceae (I 5a). 3 C Calif. to Oregon. Ephemerals, oft. in rockpools

sedge *Carex* spp. & other Cyperaceae

Sedirea Garay & H. Sweet = *Phalaenopsis*

Sedobassia Freitag & G. Kadereit (~ *Bassia*). Amaranthaceae. 1 E Eur. to Siberia: ***S. sedoides*** (Pall.) Freitag & G. Kadereit. R: T 60(2011)72

Sedopsis (Legrand) Exell & Mendonça = *Portulaca*

sedra *Ziziphus jujuba*

Sedum Tourn. ex L. Crassulaceae (I 5b). Incl. *Cremnophila*, excl. *Petrosedum*, *Phedimus*, *Sedella*, c. 450 N temp. (s.s., Eur. 36 – R: H. 't Hart (2003) *Ss of Europe*), trop. mts, Madag., Mex. Stonecrops, live-long. R: U. Eggli, *Ill. Handb. Succ. Pls*, Crass. (2005)135, 196, 210, 234, 235; R. Stephenson (1994) *S. Cultivated stonecrops*. Generic limits still disputed (*Diamorpha*, *Mucizonia*, *Ohbaea*, incl. here, though *Rhodiola* (c. 80 N temp., see JFSUTB III, 12(1978)182) & *Hylotelephium* (26, but partly = *Orostachys*), & *Telmissa* prob. also to be excl.); hybrids formed with spp. of *Echeveria* & *Villadia*. Many cult. orn. succ. herbs & shrublets, some locally eaten in salads or medic. ***S. acre*** L. (wall pepper, stonecrop, Euras., Medit.) – invasive NZ, natur. N Am.; ***S. album*** L. (Eur.) – cult. orn., natur. US since 1934; ***S. morganianum*** Walther (Mex.) – trailing tail-like shoots, hanging-basket pl.; ***S. populifolium*** Pall. (*H. p.*, W Siberia) – decid. subshrub to 30 cm; ***S. praealtum*** DC. (*S. dendroideum* subsp. *p.*, Mex.) – shrub to 2 m or trailing to 6 m with trunk to 10 cm diam. & branches with term. bunches of lvs, natur. Medit. & S GB; ***S. rosea*** (L.) Scop. (*Rhodiola r.*, '*S. roseum*', roseroot, N hemisph.) – lvs eaten, ant-inflammatory extracts in comm. med. preps; ***S. rupestre*** L. (*S. reflexum*, W & C Eur.) – lvs (tripmadam) eaten in salads; ***S. smallii*** (Britton) H. Ahles (*D. s.*, SE US) – on granite outcrops, easily manipulated experimental pl., ant-poll.; ***S. spectabile*** Boreau (*H. s.*, E As.) – border pl., a parent with *S. telephium* of pop. '**Herbstfreude**' ('Autumn Joy') favoured by late butterflies, A 0; ***S. suaveolens*** Kimnach (Mex.) – 2n = 640! [highest no. in angiosperms]; ***S. telephium*** L. (*H. t.*, orpine, live-forever, Eur. to Siberia)

seedcake *Carum carvi*

Seegeriella Senghas. Orchidaceae (V 12h). 2 Ecuador, Bolivia. R: Selbyana 30(2009)75

Seemannaralia R. Viguier. Araliaceae. 1 S Afr.: ***S. gerrardii*** (Seem.) R. Viguier. R: Strel. 10(2000)100

Seemannia Regel (~ *Gloxinia*). Gesneriaceae (II 5b). 4 S Am.

seersucker plant *Geogenanthus poeppigii*

Seetzenia R. Br. Zygophyllaceae. 1 N & S Afr. to Afghanistan: ***S. lanata*** (Willd.) Bullock (*S. africana*)

Segetella Desv. = *Spergularia*

sego lily *Calochortus* spp.

Seguieria Loefl. Petiveriaceae (Phytolaccaceae II). 6 trop. S Am. R: MBSM 18(1982)244. Trees, shrubs, lianes; stip. thorny, G1, fr. a samara; strong garlic odour

Sehima Forssk. Gramineae (XXII). 5 trop. OW

Seidelia Baill. Euphorbiaceae (Acal.-Acal.-Merc.). 2 S Afr.

Seidenfadenia Garay. Orchidaceae (V 16c). 1 Myanmar & Thailand: ***S. mitrata*** (Reichb.f.) Garay. R: OB 95(1988)212

Seidenfadeniella Sathish (~ *Cleisostomopsis*). Orchidaceae (V 16c). 2 Ind., Sri Lanka

Seidenfia Szlach. = *Malaxis*

Seidenforchis Marg. = *Crepidium* (but see ActaSoc.Bot.Pol. 75(2006)302)

Seidlitzia Bunge ex Boiss. = *Salsola*

seje *Oenocarpus bataua*

Sekanama Speta = *Drimia*

Selaginaceae Choisy = Scrophulariaceae

Selaginella P. Beauv. Selaginellaceae. Incl. *Bryodesma*, c. 800 trop. & warm, few temp. (Eur. 4). Some dendroid woody spp. Some local medic. esp. ***S. bryopteris*** (L.) Bak. (sanjeevani), a 'resurrection' pl., in India & ***S. doederleinii*** Hieron. (E As.) in China. Cult. orn. esp. ***S. kraussiana*** (Kunze) A. Braun (S Afr.) – creeping & rooting sp. much grown in greenhouses, natur. Cornwall, Ireland etc.; ***S. lepidophylla*** (Hook. & Grev.) Spring (rose of Jericho, resurrection pl., S US to Peru) – tufted pl. with branches curling up into ball when dry (chloroplasts, in crushed form, retain some chlorophyll content), but re-opening when wet (also in some other spp, e.g. ***S. serpens*** (Poir.) Spring (Carib.) – turns silver-grey as chloroplasts (1 per cell) contract to small spheres making cells largely transparent), sold as curiosity, locally medic.; ***S. tamariscina*** (P. Beauv.) Spring (Ind. to E As.) – veg. in NG, local medic. esp. China & Ger. in a tea against brittle nails; ***S. tuberosa*** McAlpin & Lellinger (Costa Rica) – ann. reprod. from aestivating tubers arising on aerial stems; ***S. willdenowii*** (Poir.) Bak. (SE As. to C Mal.) – cult. orn. scandent to 6 m or more, lvs appearing blue due to reflections of cuticle & outer epidermal walls differentially reflecting blue, a phenomenon not fully dev. in full light

Selaginellaceae Milde. Lycopodiidae – Selaginellales. 1/800 cosmop. Erect, prostrate, tufted, creeping & rooting or cli. moss-like pls with dichot. branching, s.t. frond-like, & dichot. r. Lvs scale-like, all similar & in spirals or dimorphic & 4-ranked. Sporophylls in term. strobili bearing microsporangia with many microspores & megasporangia usu. with 4 megaspores; embryology v. varied. n = usu. 9
Genus: *Selaginella* (& fossil *Selaginellites* (Carboniferous))

Selago L. Scrophulariaceae (Man.). Excl. *Pseudselago*, incl. *Walafrida*, 190 trop. (few) & S Afr., Madag. (1). R: O. Hilliard, *Selagineae* (1999)24

selangan *Shorea* spp. esp. *S. guiso*

Selbyana Archila = *Lycaste*

Selenia Nutt. Cruciferae (16). 5 S N Am. R: JAA 69(1988)127. ***S. aurea*** Nutt. (Montana, Kansas to Texas) – cult. orn. ann. with sweetly-scented fls

Selenicereus (A. Berger) Britton & Rose. Cactaceae (III 2). 15 S US to trop. Am. R: Bradleya 7(1989)89. Prob. incl. *Hylocereus*. Climbers with cardiac glycosides, aerial r. & nocturnal fls (see also *Epiphyllum oxypetalum*); cult. orn. esp. ***S. grandiflorus*** (L.) Britton & Rose (night-flowering cactus, Cuba & Jamaica, natur. trop. Am.) – cult. in Mex. for drug used in rheumatism, used as heart-stimulant in Costa Rica & now cult. comm. for this in Ger. etc. (in China then dried for soup), crossed early C19 with *Disocactus speciosus* to give **× *Disoselenicereus maynardiae*** (Paxton) E. Meier

Selenipedium Reichb.f. Orchidaceae (III). 5 trop. Am. R: BM 26 (2009)7. ***S. chica*** Reichhb.f. (Panamá) to 5 m tall (some form. vanilla subs.)

Selenodesmium (Prant) Copel. = *Abrodictyum*

Selenothamnus Melville = *Lawrencia*

Selera Ulbr. = *Gossypium*

self-heal *Prunella vulgaris*

Selinocarpus A. Gray = *Acleisanthes*

Selinopsis Coss. & Durieu ex Batt. & Trab. (~ *Carum*). Umbelliferae (III 8). 2 Medit.

Selinum L. Umbelliferae (III 8). Incl. *Cnidiocarpa*, *Kadenia*, excl. *Cnidium*, c. 8 Euras.

Selkirkia Hemsl. (~ *Omphalodes*). Boraginaceae (B.3.4.). 3 S Am., 1 Juan Fernandez

Selleola Urb. = *Mononeuria*

Selleophytum Urb. (~ *Coreopsis*). Compositae (Cor.). 1 Hispaniola: ***S. buchii*** Urb. R: NJB 24(2006)163

Selliera Cav. = *Goodenia* (but see FA 35(1992)281)

Selliguea Bory (~ *Polypodium*). Polypodiaceae (II). c. 60 Ind. to Jap. & NG (most), Aus. (1) to Fiji, ? Madag.(1). R (Mal. to Fiji – 53): Blumea 43(1998)1. ***S. feei*** Bory (Mal.) – consp. terr. sp. near fumaroles on volcanoes in Java, local medic., cult. orn.

Selloa Kunth. Compositae (Mill.-Gal.). 1 Mex.: ***S. plantaginea*** Kunth. R: PMMSUBS 4(1970)371

Sellocharis Taubert. Leguminosae (III 9). 1 SE Braz.: ***S. paradoxa*** Taubert – 5–7 leaf-like structures in whorls at nodes (unique in L.)

Sellulocalamus W.T. Lin = *Dendrocalamus*

Selysia Cogn. = *Cayaponia* (but see Novon 4(1994)37)

semaphore plant *Codariocalyx motorius*

Semaphyllanthe L. Andersson (~ *Calycophyllum*). Rubiaceae (Cond.). 6 trop. Am.

Semecarpus L.f. Anacardiaceae (I). 75 Indomal. to New Caled. (c. 5) & Fiji. Trees, some pachycaul treelets (Corner's Model) with detritus-coll. heads of large lvs oft. occupied by ants; some rheophytes. Some with fleshy fr. ed. after roasting. ***S. anacardium*** L.f. (trop. As. to Aus.) – unripe fr. (marking nut) has sap drying as a black resin used as an ink or dye for linen when mixed with lime, green fr. used as bird-lime & in tanning etc.; *S. australiensis* Engl. (tar tree, trop. Aus.) – sap dermatological, swollen peduncle ed. but pericarp toxic

Semeiandra Hook. & Arn. = *Lopezia*

Semeiocardium Zoll. = *Impatiens*

Semele Kunth. Asparagaceae (Convallariaceae). 1 Macaronesia: ***S. androgyna*** (L.) Kunth – cult. orn. dioec. liane to 20 m with shoots arising from underground & bearing flattened leaf-like organs in axils of scale lvs, fls borne in small infls around edge of the leaf-like organs

Semenovia Regel & Herder (*Platytaenia*; ~ *Heracleum*). Umbelliferae (III 11). 24 As.

Semialarium Hallé ('*Hemiangium*'). Celastraceae (III). 2 trop. Am. R: BMNHN 4,5(1983)24

Semiaquilegia Mak. (~ *Aquilegia*). Ranunculaceae (III 1). 6 E As. R: KB 1920:165. Like *A.* but fls ± spurless; some cult. orn. esp. ***S. adoxoides*** (DC.) Mak. (E China, Korea, Jap.) with cream & chocolate coloured fls

Semiarundinaria Mak. ex Nakai. Gramineae (1b). c. 5 China & Jap. R: KBAS 13(1986)28. Cult. orn. bamboos (? hybrids between *Phyllostachys* & *Pleioblastus* spp.)

Semibegoniella C. DC. = *Begonia*

Semiliquidambar H.T. Chang = *Liquidambar*

semilla de jícaro *Crescentia cujete*

Seminole tea *Asimina reticulata*

Semiramisia Klotzsch. Ericaceae (VIII 5). 4 N Andes. R: SB 9(1984)359

Semiria Hind. Compositae (Eup.-Gyp.). 1 Braz. (Bahia): ***S. viscosa*** Hind. R: KB 54(1999)425

Semnanthe N.E. Br. = *Erepsia*

Semnostachya Bremek. = *Strobilanthes*

Semnothyrsus Bremek. = *Strobilanthes*

semolina *Triticum turgidum* Durum Group

Semonvillea Gay = *Limeum*

Sempervivella Stapf = *Sedum*

Sempervivum L. Crassulaceae (I 3). Incl. *Jovibarba*, 63 (Eur. 19), Morocco, W As. R: U. Eggli, *Ill. Handb Succ. Pls*, Crass. (2005)332. Houseleeks. Alks incl. nicotine. Cult. orn. (over 1000 named cvs.) rosetted succ. usu. with offsets, grown in rock-gardens etc. esp. ***S. tectorum*** L. (Eur.) – cultigen ? or hybrid from C Eur., form. much planted on roofs to keep slates in place & held to ward off thunder, form. crushed in lard for treatment of sore eyes, burns, chilblains & piles & used since time of Pliny in pl.-pest control, variable & forms hybrids readily, as in Pyrenees with ***S. arachnoideum*** L. (S Eur. mts) – commonly cult., its smaller rosettes with cobwebby strands connecting leaf-tips, so that some 'spp.' may rep. anc. hybrids; ***S. heuffelii*** Schott (*J. h.*, E Eur.) – new rosettes arising by division of old ones, not as offsets. Journal: S. Society J.

sempilor *Dacrydium elatum*

sen *Eleutherococcus septemlobus*

Senaea Taubert. Gentianaceae. 2 Braz. mts. R: KB 63(2008)184

senat seed *Cucumis melo* subsp. *agrestis*

Senecio Tourn. ex L. Compositae (Sen.-Sen.). Incl. *Aetheolaena*, *Cadiscus*, *Culcitium*, *Hasteola*, *Iocenes*, *Robinsonia*, c. 1000 cosmop. (Eur. c. 60, Afr. c. 350 [s.l., Cape 110, 58 endemic: R: Strel. 9(2000)356], N Am. 55, S Am. c. 500) exc. WI, Antarctica. Trees (e.g. woody, pachycaul, s.t. epiphytic on tree-ferns, on Juan Fernandez where dioec. (*R.* spp.)), shrubs, lianes & herbs incl. aquatics, e.g. ***S. cadiscus*** R. Nordenstam & Pelser (*Cadiscus aquaticus*, S Afr.) but often weedy; although recently shorn (or reshorn) of a no. of 'satellite' genera (see *Bethencourtia*, *Brachyglottis*, *Brachyrhynchos*, *Cineraria*, *Crassocephalum*, *Curio*, *Delairea*, *Dendrosenecio*, *Gynura*, *Jacobaea*, *Kleinia*, *Lachanodes*, *Ligularia*, *Madaractis*, *Othonna*, *Parasenecio*, *Pericallis*, *Pladaroxylon*, *Roldana*, *Sinacalia*, *Sinosenecio*, *Solanecio*, *Tephroseris* etc.; certain Afr. succ. prob. to be excl.), still one of the largest genera of seed-pls & limits not fully clear. ***S. angulatus*** L.f. (S Afr.) – invasive Aus., NZ; ***S. crassissimus*** Humbert (? *Curio*, Madag.) – succ. sp. with lvs held vertically; ***S. glastifolius*** L.f. (S Afr.) – invasive NSW; ***S. inaequidens*** DC. (S Afr.) – 2n & 4n, latter natur. along railway lines in Ger. etc.; ***S. madagascariensis*** Poir. (fireweed, S Afr.) – pest in Aus. introd. early 20 cent.; ***S. mohavensis*** A. Gray – subsp. ***m.*** in SW N Am., subsp. ***breviflorus*** (Kadereit) M. Coleman (Middle E to Pakistan)!; ***S. smithii*** DC. (S Patagonia) – introd. to Orkney & Shetland by whalers; ***S. squalidus*** L. (Oxford ragwort, S Eur.) – natur. GB, having escaped from Oxford Botanic Garden (prob. hybrid from Mt Etna (*S. aethnensis* Jan ex DC. (*S. s.* subsp. *a.* (DC.) Greuter) × *S. chrysanthemifolius* Poir. (*S. s.* subsp. *c.* (Poit.) Greuter) introd.

Badminton, Glos. 1700–02 & in Oxford by 1719) along railway lines to London etc., hybridizing with *S. vulgaris* (tetraploid) in N Wales & Edinburgh to form sexual hexaploid sp., ***S. cambrensis*** Rosser, a sp. which has thereby arisen polytopically (cf. *Sporobolus anglicus*, *Tragopogon minus*), & tetraploid ***S. eboracensis*** R. Abbott & A Lowe (discovered 1979, extinct by 2010); ***S inaequidens*** DC. (S Afr.) – introd. Ger. in sheep's wool natur. since 1889, Netherlands since 1942, since 1985 disp. by trains (cf. *S. squalidus* in UK); ***S. viscosus*** L. (Eur., Medit.) – weedy self-incompatible derived from self-incompatible *S. nebrodensis* L. (mts of Spain); ***S. vulgaris*** L. (groundsel (= 'pus-absorber'), Euras., now pantemp. weed) – 2n = 40, rayless, short-lived, seeds with no dormancy, derived from subsp. ***denticulatus*** (O.F. Muell.) Sell (W Eur. coasts, mts of Spain & Italy) or poss. allopolypoid or autotetraploid of *S. vernalis* Waldst. & Kit. (*S. leucanthemifolius* subsp. *v.*, self-incompat. sp., 2n = 20) in E Medit., form. medic. incl. purgative

Senefeldera Mart. Euphorbiaceae (Euph. – Hipp.). 6 trop. Am. ***S. inclinata*** Muell. Arg. – bark for toothache in Amazon

Senefelderopsis Steyerm. Euphorbiaceae (Euph. – Hipp.). 5 N S Am. R: MIABH 25(1995)127

senega root *Polygala senega*

Senegal tea *Gymnocoronis spilanthoides*

Senegalia Raf. (~ *Acacia*). Leguminosae (II 4). 183 trop. (As. 43 (7 also in Afr.), Aus. 2, Afr. 69, 97 Am. – R: Phytol. 88(2006)41, 91(2009)27). Form. *A.* subg. *Aculeiferum* – lvs bipinnate, unarmed or with prickles, peduncle without involucre. Imp. honey sources in Afr. ***S. berlandieri*** (Benth.) Britton & Rose (*A. b.*, Mex.) has toxic amines (though imp. honey source in Texas), while tannins of ***S. nigrescens*** (Oliv.) P. Hurter (*A. n.*, SE Afr.) imp. killer of browsing kudu, dying of indigestion, levels of tannins allegedly rising when pls attacked, while in ***S. caffra*** (Thunb.) P. Hurter & Mabb. (*A. c.*, S. Afr.) damaged lvs release ethylene promoting tannin prod. in trees downwind (so kudu graze upwind!). Dyestuffs, gums, fuelwood. ***S. catechu*** (L.f.) P. Hurter & Mabb. (*A. c.*, catechu, cutch, Ind. to China) – heartwood for tanning, dyeing (true khaki cloth dyed & shrunk with it), for treating fishing-nets & sails, med. & as masticatory with betel, fuelwood (khayer); ***S. greggii*** (A. Gray) Britton & Rose (*A. g.*, cat's-claw, N Am.) – common desert pl., imp. Honey source in SW US, fr. & seeds ed., building timber for Native Americans; ***S. modesta*** (Wall.) P. Hurter (*A. m.*, N Ind., Afghanistan, Pakistan) – Amritsar gum; ***S. pennata*** (L.) Maslin (*A. p.*, cha-oon, trop. As.) – shoots in curry, soups, Thai lamb dishes; ***S. senegal*** (L.) Britton (*A. s.*, arid trop. Afr.) – tapped trees source of true gum arabic (kolhol g., 60 K t a yr by 2008) for lozenges, gum sweets, adhesives, inks, watercolours & med. (mediaeval stauncher of blood), exports to US (for Coca-Cola) not barred during sanctions against Sudan

Senghasia Szlach. = *Kefersteinia*

Senghasiella Szlach. = *Habenaria*

Senisetum Honda = *Agrostis*

senna *Senna italica*; **Alexandrian s.** *S. alexandrina*; **American s.** *S. marilandica*; **bastard s.** *Coronilla valentina*; **bladder s.** *Colutea arborescens*; **coffee s.** *S. occidentalis*; **dog** or **Italian s.** *S. italica*; **prairie s.** *Chamaecrista fasciculata*; **ringworm s.** *S. alata*; **scorpion s.** *Hippocrepis emerus*; **Spanish s.** *S. italica*; **Tinnevelly s.** *S. alexandrina*; **wild s.** *S. marilandica*

Senna Mill. (~ *Cassia*). Leguminosae (I 3). c. 300 trop. esp. Am. (Afr. 24, R: KB 43(1988)338) & warm temp. Form. *Cassia* subg. *Senna* (Mill.) Benth. but diff. from *Cassia* s.s. in filaments of all A straight & bracteoles 0, pods flattened. Trees (Scarrone's Model), shrubs & herbs, buzz-poll., many imp. medic. (anthraquinones) & cult. orn. (senna; colonizing spp. in Aus. called 'acacia'). ***S. alata*** (L.) Roxb. (*C. a.*, ringworm cassia or senna, candlebush, craw-craw, trop. Am., natur. trop. OW – Java by C17, invasive Aus.) – for skin disease & vermifuge, cult.; ***S. alexandrina*** Mill. (*C. senna*, Alexandrian or Tinnevelly s., NE Afr. Middle E) – lvs & pods some of the s. of commerce (7–10 K t a yr); ***S. artemisioides*** (DC.) Randell (*C. sturtii*, Aus.) – part of hybrid complex due to apomixis, polyploidy & hybridization, drought-tolerant fodder shrub esp. in Israel; ***S. auriculata*** (L.) Roxb. (*C. a.*, avaram, Matara tea, turwad bark, tanner's c., Tanz., Ind., Sri Lanka); ***S. bicapsularis*** (L.) Roxb. (trop. Am.) – invasive S Afr.; ***S. corymbosa*** (Lam.) Irwin & Barneby (*C. c.*, N S Am., natur. N Am.) – much cult.; ***S. didymobotrya*** (Fres.) Irwin & Barneby (*C. d.*, trop. Afr., natur. As. & Am., invasive Aus., S Afr., SE US) – cult. as green manure in As., cult. orn. shrub, lvs & r. medic., used in curses in Kenya; ***S. italica*** Mill. (*C. i.*, Italian or Spanish or dog s., OW trop.) – lvs & pods some of the s. of commerce; ***S. lactea*** (Vatke) Du Puy (Madag.) – coffee-shade, yellow wood for furniture, fuel; ***S. marilandica*** (L.) Link (*C. m.*, Am. or wild s., E US); ***S. obtusifolia*** (L.) Irwin & Barneby (*C. o.*, trop. & warm, invasive

Aus., Galápagos) – green lvs fermented to give protein source (20% by wt., kawal) by bacteria & *Rhizopus* fungi in Sudan; ***S. occidentalis*** (L.) Link (*C. o.*, coffeeweed, trop. OW) – medic., seeds a coffee subs.; ***S. pendula*** (Willd.) Irwin & Barneby (trop. Am.) – invasive Aus., SE US; ***S. siamea*** (Lam.) Irwin & Barneby (*C. s.*, minjiri, Myanmar to Mal.) – cult. orn. & firewood crop, timber hard & strong for bridges, mine props, telegraph poles etc.; ***S. spectabilis*** (DC.) Irwin & Barneby (*C. s.*, trop. Am.) – cult. orn., source of spectaline, analgesic & anti-inflammatory, derivatives inhibiting acetylcholine esterase; ***S. surattensis*** (Burm.f.) Irwin & Barneby (? Aus.) – cult. orn., widely natur.; ***S. tora*** (L.) Roxb. (*C. t.*, sicklepod, Indomal.) – medic. (cult. Ind.), seeds used as mordant in dyeing blue cloth, seeds a coffee subs. & in a Vietnam tea with liquorice & fls of *Styphnolobium japonicum*

Sennia Chiov. = *Dialium*

Senniella Aellen = *Atriplex*

sennit Polynesian cordage of plaited grass, coir etc.; **s. hat** = boater, 'straw' hat made of straw, etc.

senposai *Brassica juncea* 'Crispifolia'

Senra Cav. = *Hibiscus*

sensitive plant *Mimosa pudica*

sentul *Sandoricum koetjape*

Senyumia Kiew & al. Gesneriaceae (III 2i). 1 Malay Pen.: ***S. minutiflora*** (Ridl.) Kiew & al. – inside limestone caves, seriously endangered. R: BBP 76(1998)400

Seorsus Rye & Trudgen (~ *Baeckea*). Myrtaceae. 2 Borneo, 2 Aus. R: Nuytsia 18(2008)248

Sepalosaccus Schltr. = *Maxillaria*

Sepalosiphon Schltr. = *Glomera*

Separotheca Waterf. = *Tradescantia*

sepetir *Sindora* spp.

Sepikea Schltr. (~ *Cyrtandra*). Gesneriaceae (III 2j). 1 NG: ***S. cylindrocarpa*** Schltr.

Septas L. = *Crassula*

Septimia P.V. Heath = *Crassula*

Septogarcinia Kosterm. = *Garcinia*

Septotheca Ulbr. Bombacaceae. 1 Peru: ***S. tessmannii*** Ulbr. – wood for furniture & general constr.

Septulina Tieghem (~ *Taxillus*). Loranthaceae (5 7). 2 W Cape, S Namibia. R: R. Polhill & D. Wiens, *Mistletoes of Afr.* (1998)223

Sequencia Givnish = *Brocchinia*

Sequoia Endl. Cupressaceae (Taxodiaceae). 1 Oregon to Calif. in foothills within 30 km of coast (Palo Alto [= lone tree] named after tree still standing 1950): ***S. sempervirens*** (D. Don) Endl. (Calif. redwood) – massive tree (Massart's Model) to 2000 yrs old & 116.6 m tall (tallest of all surviving trees?), 1760 m^3 with bole to 8.5 m diam. & r.-suckers, cotyledons 2–4, hexaploid (2n = 66), mitochondrial DNA paternally inherited (? unique), valuable timber, form. much used for shingles, fences & general building etc., over-exploited & now restricted to a few reserves etc., cult. orn. incl. dwarf cvs ('**Prostrata**' a witch's broom c. 1927 Cambridge Botanic Garden, UK, from which erect shoots gave '**Cantab**'). R: A. Farjon, *Monogr. Cupressaceae & Sciadopitys* (2005)108. Closely allied *Austrosequoia* fossils from Aus. Cretaceous

Sequoiadendron Buchholz. Cupressaceae (Taxodiaceae). 1 W slopes of Sierra Nevada of Calif. (common in W Nevada Pliocene): ***S. giganteum*** (Lindl.) Buchholz (wellingtonia, mammoth or big tree) – discovered 1850, introd. Eur. 1853, reaching 105 m (Rauh's Model) with bole to 12 m diam. (the bulkiest of all trees) & 3206 (oldest recorded; 2890 oldest living, 2005) yrs ('General Sherman' 84 m tall, 25 m girth with trunk vol. c. 1400 m^3 & wt. c. 2500 t), reduced to 72 groves (some v. small), sets seed only when dry season shorter than average, cotyledons usu. 4, timber only good for shingles (& fencing), cult. orn. (the biggest tree of every county in GB, never blown down & indifferent to frost & drought) incl. dwarf cvs & '**Pendulum**' with sinuous trunk s.t. 'looping the loop'. R: A. Farjon, *Monogr. Cupressaceae & Sciadopitys* (2005)104

serai *Cymbopogon citratus*

Seraphyta Fischer & C. Meyer = *Epidendrum*

Serapias L. Orchidaceae (IV 4d). 13 Azores to Medit. (Eur. 5). R: QBAGS 43(1975)188. Terr. ***S. vomeracea*** (Burm.f.) Briq. (Medit.) – in Israel bees sleep inside fls & are warmed 3°C more than outside temperature in morning

seraya, red *Shorea* spp.; **white s.** *Parashorea* spp.; **yellow s.** *S.* spp.

sereh *Cymbopogon citratus*

Serenoa Hook.f. Palmae (II 4b). 1 SE US: ***S. repens*** (Bartram) Small (*S. serrulata*, saw palmetto (palm)) – colonial ± stemless palm with bisex. fls & toothed petiole margins oft. forming pestilential thickets, fossils in Eocene of London Clay, fibre, wax, thatch, fr. ed. (incl. cattle, bears, humans – form. staple) & a tea, allegedly aphrodisiac (male; actually with a steroidal drug inhibiting conversion of testosterone to dihydrotestosterone which binds to receptors in hair-follicles & prostate so used in treating baldness & prostate problems), also treating candida (caprylic acid antifungal). R: J. Dransfield & al., *Gen. Palm.* (2008)274

Seretoberlinia Duvign. = *Julbernardia*

Sergia Fed. (~ *Campanula*). Campanulaceae (I 9). 2 C As.

Serialbizzia Kosterm. = *Albizia*

Serianthes Benth. Leguminosae (II 4). 18 Thailand to W Polynesia esp. New Caled. (Mal. 4). R: BMNHN4,6(1984)84. Some timber

Sericanthe Robbrecht (~ *Tricalysia*). Rubiaceae (II 3). 17 trop. Afr. R: BJBB 48(1978)3, 51(1981)171. Some bacterial leaf nodules (see *Pavetta*)

Serichonus Thiele. Rhamnaceae. 1 SW Aus.: ***S. gracilipes*** (Diels) Thiele. R: Nuytsia 16(2007)310

Sericocalyx Bremek. = *Strobilanthes*

Sericocarpus Nees (~ *Aster*). Compositae (Ast.-Sym.). 5 N Am. R: Phytol. 75(1993)45, 77(1994)266

Sericocoma Fenzl. Amaranthaceae (I 2). 6 trop. & A Afr.

Sericocomopsis Schinz. Amaranthaceae (I 2). 2 trop. E Afr.

Sericodes A. Gray. Zygophyllaceae (V). 1 N Mex.: ***S. greggii*** A. Gray

Sericographis Nees = *Justicia*

Sericolea Schltr. Elaeocarpaceae. 16 E Mal. R: Blumea 28(1982)103. ***S. micans*** Schltr (New Guinea) – cli.

Sericorema (Hook.f.) Lopr. Amaranthaceae (I 2). 2 S Afr.

Sericospora Nees. Acanthaceae. 1 WI: ***S. crinita*** Nees (= ?)

Sericostachys Gilg & Lopr. Amaranthaceae (I 2). 1 trop. Afr.: ***S. scandens*** Gilg & Lopr. – mass-flowering every 10–15 yrs (triggered by dry spells?)

Sericostoma Stocks = *Echiochilon*

Seringia Gay. Malvaceae (Bytt.-Lasio.; Sterculiaceae s.l.). 1 NG, E Aus.: ***S. arborescens*** (Aiton) Druce

Seriphidium (Hook.) Fourr. = *Artemisia*

Serissa Comm. ex Juss. Rubiaceae (IV 12). 1–2 S China (? introd. Jap.). R: FOC 19(2011)324. ***S. japonica*** (Thunb.) Thunb. (*S. foetida*) – early introd. or native to Jap., cult. orn. shrub with foetid lvs (when bruised), esp. cvs with double fls & varieg. lvs, also sold as bonsai in Eur.

Serjania Plum. ex Mill. Sapindaceae (IV 1). c. 230 trop. & warm Am. 6 sects. R (sect. *Platycoccus* – 13): MNYBG 67(1993)49). Polygamous lianes with 'watch-spring' tendrils & 3-winged schizocarps; some cordage & fish-poisons, ***S. ferruginea*** (Lindl.) Mabb. (*S. cuspidata*, Braz.) grown on trop. pergolas & fences

serpentaria or **serpentary** = snakeroot

Serpenticaulis M.A. Clem. & D.L. Jones = *Bulbophyllum*

Serpocaulon A.R. Sm. (~ *Polypodium*). Polypodiaceae (V). 40–45 trop. Am. R: T 55(2006)924

serpolet oil *Thymus polytrichus*

Serpyllopsis Bosch = *Hymenophyllum*

serradella *Ornithopus sativus*

serrated tussock *Nassella trichotoma*

Serratula Dill. ex L. Compositae (Card.-Cent.). Excl. *Klasea*, *Klaseopsis*, c. 50 temp. Euras. ***S. tinctoria*** L. (*S. seoanei*, saw-wort, Euras., N Afr.) – lvs with alum give green or yellow dye for wool, form. medic. (wounds & piles), cult. orn. cvs esp. dwarf '**Shawii**' ('*S. shawii*')

Serruria Burm. ex Salisb. Proteaceae (IV 4). 50 SW Cape. R: Strel. 9(2000)592. Some cult. orn. shrubs

Sersalisia R. Br. (~ *Pouteria*). Sapotaceae (IV). Excl. *Pichonia*, c. 5 C Mal. to Aus.

Sertifera Lindl. & Reichb.f. Orchidaceae (V 2). 7 Andes. Bird-poll.

service berry *Amelanchier* spp.; **s. tree** *Sorbus torminalis*

sesame or sesamum *Sesamum indicum*; **black s.** *Hyptis spicigera*

Sesamoides All. Resedaceae. 1 (?5) W Medit. (incl. Eur.): ***S. canescens*** (L.) Kuntze

Sesamothamnus Welw. Pedaliaceae (1). 6 trop. Afr. (disjunct: 2 NE Afr., 2 SW & 1 SE). Some with water-storing trunks, cult. orn.

Sesamum L. Pedaliaceae (3). 20+ OW trop. & S Afr. Heterogeneous. ***S. indicum*** L. (*S. orientale*, sesame, sesamum, simsim, gingelly, halvah, trop. OW, domesticated Ind. from 'subsp. *malabaricum*') – widely natur. (taken from Afr. to Am. where magic pl., long cult. (oldest grown oilseed since c. 3500–3050 BC in Indus & anc. Mesopotamia, by 2010 3.84 M t a yr), polyunsaturated omega-6 oil (poss. help prevent stomach cancer) used like olive o., the seeds sprinkled on bread & cakes like poppy s. or made into sticky sweetmeats (paste of crushed seeds used in Lebanese cooking = tahini; toasted seeds + salt a condiment (gomas(h)io) in Jap.) & used in cosmetics. R: JAA suppl. 1(1991)328

sesban *Sesbania sesban*

Sesbania Scop. Leguminosae (III 21). Incl. *Glottidium* c. 60 warm & usu. wet. Some cult. orn. herbs, shrubs & shade-trees (Rauh's Model); some poss. firewood crops incl. ***S. aculeata*** (Schreb.) Poir. (*S. bispinosa*, dhaincha, OW trop.) – poss. source of guar, fibres used for sails & fishing-nets, ***S. cannabina*** (Retz.) Poir. ('*S. aculeata*', '*S. bispinosa*', orig. Aus.) – widely cult., ***S. eremurus*** (Aubl.) Urb. (trop. Am.) – ann. growing to 6 m, ***S. grandiflora*** (L.) Poir. (bakphul, trop. As., natur. Am.) – cult. orn. with large (to 10 cm) red or white fls, ed. as salad (fr. ined.), lvs & bark medic., ***S. herbacea*** (Mill.) McVaugh (*S. exaltata*, N Am.) – form. imp. Native American fibre, ***S. punicea*** (Cav.) Benth. (S Am.) – natur. SE US & pestilential in S Afr. fynbos where controlled by weevils, ***S. sesban*** (L.) Merr. (sesban, *S. aegyptiaca*, OW trop.) – molluscicidal saponins, imp. fodder & fibre, cult. for fls in anc. Egypt, pl. green manure, wood form. used for gunpowder charcoal. ***S. rostrata*** Bremek. & Oberm. (S Afr.) – nitrogen-fixing nodules on stems; ***S. vesicaria*** (Jacq.) Elliott (*G. v.*, bag-pod, SE US) – poisonous to stock

Seseli Boerh. ex L. Umbelliferae (III 8). Incl. *Eriocycla* 100–120 Eur. (34) to C As. & N trop. Afr. ***S. buchtormense*** (Hornem.) Koch (Siberia) – ess. oil

Seselopsis Schischkin. Umbelliferae (III 8). 2 C As.

Seshagiria Ansari & Hemadri = *Cynanchum*

Sesleria Scop. Gramineae (XVI 13). 26 Eur. (26) esp. Balkans, W As. R: OBC 3(1946)1. Spp. closely related & introgressing e.g. ***S. caerulea*** (L.) Ard. (*S. albicans*, blue moorgrass, W & C Eur.) – cult. N Am.

Sesleriella Deyl = *Sesleria*

Sessea Ruíz & Pavón (~ *Cestrum*). Solanaceae (II 1). 15 Andes

Sesseopsis Hassler = *Sessea*

Sessilanthera Molseed & Cruden = *Tigridia*

Sessilibulbum Brieger = *Scaphyglottis*

Sessilistigma Goldbl. = *Moraea*

Sestochilos Breda = *Bulbophyllum*

Sesuvium L. Aizoaceae (II). c. 15 warm, esp. coastal. R: H. E. K. Hartmann, *Ill. Handb. Succ. Pls*, Aiz. F–Z(2002)296. Halophytes – 1 pantrop.: ***S. portulacastrum*** (L.) L. (mboga (Afr.), veg. sold in markets in As.) – pollution-tolerant (like *Iresine vermicularis*) replacing mangroves nr Rio, 1 Galápagos, 4 Angola area

Setaria P. Beauv. Gramineae (XXIV 4). Incl. *Paspalidium*, 115 trop. & warm (Eur. 3 + *S. italica*; N Am. 15 native). R: Sida 15(1993)447. Grains (e.g. ground to paste & roasted by Aus. Aborigines), imp. fodders, some local medic. Some hosts for maize & sugar-cane mosaic viruses. ***S. italica*** (L.) P. Beauv. (*S. viridis* subsp. *italica*, foxtail, Ger., Hungarian or Italian millet, Bronze Age cereal cultigen prob. derived from '*S. viridis* (L.) P. Beauv.' (temp. & warm, like similar *S. verticillata* (L.) P. Beauv. oft. a noxious weed) in China or poss. several centres in Eur. & As. c. 5000 BC) – hay, pasture & cereal (27 Mt a yr by 2005), oft. seen as bird-seed, the millet-spray of petshops; ***S. parviflora*** (Poir.) Kerguélen (*S. geniculata*, US to Chile) – domesticated C Am., ? before maize; ***S. pumila*** (Poir.) Roemer & Schultes ('*S. glauca*', yellow foxtail, pigeon grass, cat-tail millet, warm) – cattle fodder, bird-seed; ***S. palmifolia*** (Koenig) Stapf (Indomal.) – young shoots eaten with rice in Java & NG, where domesticated (highland pitpit), cult. orn. in Eur.; ***S. sphacelata*** (Schum.) Stapf & C. Hubb. (S Afr.) – diploids to decaploids with hybridization between levels, imp. silage crop

Setariopsis Scribner. Gramineae (XXIV 4). 2 Arizona, Mex. to Colombia & Venez.

Setchellanthaceae Iltis (~ Capparaceae). Magnoliidae – Brassicales. 1/1 Mex. R: T 48(1999)257. Pungent shrub with long & short shoots, T-shaped hairs. Lvs simple, small, in spirals; stip. 0. Fls reg., large, (5)6(7)-merous, solit, axillary on long shoots. K (5–7), connate, irreg. splitting into 1 or 2 flaps; C 5–7 blue to lilac, clawed, imbr.; A ∞, centrifugal in 5–7 groups on elongated axis, anthers dehiscing introrsely by longit. slits; disk 0; gynophore short; $\underline{G}$ (3), placentation axile with 10–14 ovules in 2 ranks per loc. Fr. a

septifragal capsule with persistent C columella. Seeds 3–10 per carpel, with soft-spongy wing, endosperm scanty
Genus: *Setchellanthus*

Setchellanthus Brandegee. Setchellanthaceae (Capparaceae s.l.). 1 N & S-C Mex.: ***S. caeruleus*** Brandegee

Setcreasea K. Schum. & Sydow = *Tradescantia*

Setiacis S.L. Chen & Y.X. Jin = *Acroceras* (but see FOC 22(2006)514)

Seticleistocactus Backeb. = *Cleistocactus*

Setiechinopsis Backeb. = *Echinopsis*

Setilobus Baill. = *Cuspidaria*

Setiscapella Barnhart = *Utricularia*

setterwort *Helleborus foetidus*

Setulocarya R. Mill & Long = *Lasiocaryum* (but see EJB 53(1996)113)

Seutera Reichb. (~ *Vincetoxicum*). Apocynaceae. 2 N Am. R: Novon 15(2005)592

Sevada Moq. (~ *Suaeda*). Amaranthaceae (Chenopodiaceae III). 1 Sudan, Ethiopia, Somalia, Arabia: ***S. schimperi*** Moq. R: NJB 11(1991)315

sevadilla *Schoenocaulon officinale*

sevendara (grass) *Chrysopogon zizanioides*

Severinia Ten. ex Endl. = *Atalantia*

Sextonia van der Werff (~ *Ocotea*). Lauraceae. 2 trop. S Am. R: Novon 7(1997)437. ***S. rubra*** (Mez) van der Werff (*O. r.*, determa, red louro, NE S Am.) – timber

Seychellaria Hemsl. Triuridaceae. 1 Tanzania, 2 Madag., 1 Seychelles

Seymeria Pursh. Orobanchaceae (Gerard.; Scrophulariaceae s.l.). 25 S N Am. R: PANSP 77(1926)349. ***S. cassioides*** (J. Gmelin) S.F. Blake (SE US) – hemiparasite on pine r., serious for p. seedlings in SE US

Seymeriopsis Tzvelev (~ *Seymeria*). Orobanchaceae (Gerard.). 1 Cuba: ***S. bissei*** Tzvelev

Seyrigia Keraudren. Cucurbitaceae (XIII). 6 Madag. R: U. Eggli, *Ill. Handb. Succ. Pls*, Dicots (2002)90. Dioec. succ. lianes

sha shen *Adenophora tetraphylla*

shad or s. bush *Amelanchier* spp.

shaddock *Citrus maxima*

Shafera Greenm. Compositae (Sen.-Sen.). 1 E Cuba (serpentine): ***S. platyphylla*** Greenm. – multiseriate involucral bracts odd in tribe

Shaferocharis Urb. Rubiaceae (III 4). 3 E Cuba (serpentines)

Shaferodendron Gilly = *Manilkara*

shagbark hickory *Carya ovata*

shaggy pea *Oxylobium* spp.

shaking grass *Briza* spp.

shallon bush *Gaultheria shallon*

shallot *Allium cepa* Aggregatum Group

Shama millet *Echinochloa colona*

shamrock usu. *Trifolium dubium* but s.t. *T. repens* or *Medicago lupulina* while in US *Oxalis* spp. oft. used; **Cape s.** *Oxalis* spp.

shan yao *Dioscorea polystachya*; **s. zhu yu** *Cornus officinalis*

Shangrilaia Al-Shehbaz & al. Cruciferae (26). 1 Yunnan: ***S. nana*** Al-Shehbaz & al. R: Novon 14(2004)271

Shangwua Y.J. Wang & al. (~ *Saussurea*). Compositae (Card.-Card.). 3 As. mts. R: T 62(2013)992

shanshi *Coriaria ruscifolia* subsp. *microphylla*

Shaniodendron M.B. Deng, H.T. Wei & X.K. Wang = *Parrotia*

Shantung cabbage *Brassica rapa* Chinensis Group

sharifa *Annon cherimola*

shark wood *Dysoxylum bijugum*

Sharon fruit early-fruiting seedless persimmons; **rose of S.** poss. orig. *Lilium candidum, Gladiolus italicus, Pancratium maritimum*, or *Tulipa agenensis*, lately *Hypericum androsaemum, H. calycinum* & in Am. *Hibiscus syriacus*

shaving brush *Pseudobombax ellipticum*

shawl, Spanish *Heterocentron elegans*

she balsam *Abies fraseri*; **s. oak** *Casuarina* & *Allocasuarina* spp.

shea butter *Vitellaria paradoxa*

Sheareria S. Moore. Compositae (Ast.-Ast.). 1 China: ***S. nana*** S. Moore – monoec. rheophyte. R: T 58(2009)777

shedua *Guibourtia ehie*

sheep's bit scabious *Jasione montana;* **s. bush** *Geijera parviflora;* **s.'s fescue** *Festuca ovina;* **s. laurel** *Kalmia angustifolia;* **s.('s) sorrel** *Rumex acetosella;* **vegetable s.** *Raoulia* spp. esp. *R. mammillaris*

sheepberry *Viburnum lentago*

sheesham *Dalbergia sissoo*

Sheilanthera I.J. Williams. Rutaceae (I 4). 1 SW Cape mts: ***S. pubens*** I.J. Williams. R: Strel. 9(2000)636

shell flower *Moluccella laevis*

shellac or lac a resin derived from secretion of lac insect (*Laccifer lacca*) feeding on trees such as *Butea monosperma, Schleichera oleosa* & *Ziziphus jujuba* etc. in trop. As., now of little imp. because of synthetics

sheng jiang *Zingiber officinale*

Shepherdia Nutt. Elaeagnaceae. 3 N Am. Dioec. shrubs & small trees with a drupe-like swollen receptacle in fr. ***S. argentea*** (Pursh) Nutt. (buffaloberry, beef-suet tree, silverberry, E N Am.) – cult. orn. v. hardy hedge-pl. with ed. fr. used in jelly or dried by Native Americans & eaten in winter with buffalo meat, touted as 'superfood' rich in anti-oxidant lycopene; ***S. canadensis*** (L.) Nutt. (N Am.) – imp. local medic., fr. ed. (& whipped with sugar to give 'Ind. icecream')

shepherd's needle *Scandix pecten-veneris;* **s.'s cress** *Teesdalia nudicaulis;* **s.'s purse** *Capsella bursa-pastoris;* **s.'s rod** *Dipsacus pilosus;* **s.'s weather-glass** *Anagallis arvensis*

Sherardia Dill. ex L. Rubiaceae (IV 16). 1 E Medit.: ***S. arvensis*** L. (field madder) – now N temp. weed natur. S Afr., Am.

Sherbournia G. Don. Rubiaceae (II Sherb.). 13 trop. Afr. R: SGP 75(2005)66. K convolute

shi mei *Rosa davurica*

Shibataea Mak. ex Nakai. Gramineae (IV). 7 SE China & SW Jap. (1). R: APS 26(1988)130, FOC 22(2006)161. ***S. kumasaca*** (Steud.) Nakai (Jap., where anc. common garden pl.) – small bamboo for ground-cover

shield fern *Dryopteris* spp.; **Persian s.** *Strobilanthes auriculata* var. *dyeriana*

shingle oak *Quercus imbricaria*

Shinnersia R. King & H. Robinson (~ *Trichocoronis*). Compositae (Eup.-Tric.). 1 SW N Am.: ***S. rivularis*** (A. Gray) R. King & H. Robinson

Shinnersoseris Tomb (~ *Lygodesmia*). Compositae (Cich.-Cich. (Micr.)). 1 N Am.: ***S. rostrata*** (A. Gray) Tomb. R: FNA 19(2006)368

shipova × *Sorbopyrus auricularis*

Shirakiopsis Esser = *Gymnanthes*

shisham wood *Dalbergia sissoo*

shiso *Perilla frutescens*

shittim wood *Vachellia seyal*

Shiuyinghua Paclt (~ *Paulownia*). Paulowniaceae (Scrophulariaceae s.l.). 1 C China: ***S. silvestrii*** (Pampan. & Bonati) Paclt. R: JAA 43(1962)217

shoe-flower *Hibiscus rosa-sinensis*

shola pith *Aeschynomene aspera, A. indica*

Shonia R. Henderson & Halford (~ *Beyeria*). Euphorbiaceae (Ricinoc.-Be.). 4 N Aus. R: Austrobaileya 7(2005)2018

shoo-fly *Nicandra physalodes*

shooting-star *Primula* (*Dodecatheon*) spp.

shore weed *Plantago uniflora*

Shorea Roxb. ex Gaertn. f. Dipterocarpaceae (Dipt.-Shor.). 196 Sri Lanka to S China, Moluccas & Lesser Sunda Is. (Mal. 163; R: FM I,9(1982)436; Borneo 138, 91 endemic). Heterogeneous. Usu. emergent trees (Massart's & Roux's Models) of rainforest – the most imp. timber genus in trop. As. (***S. acuminata*** Dyer (W Mal.) – 1 tree can give timber for a house & all its furniture in Sumatra): Borneo redwood, doon, red meranti, r. seraya, r. lauan (luan; esp. ***S. negrosensis*** Foxw., chan, Philipp.), being medium or lightweight reddish timbers (sects *Brachypterae, Mutica, Ovalis, Pachycarpae, Rubella*) for light constr. & veneers, white m. (sect. *Anthoshorea*) imp. veneers; imp. trade timbers incl. ***S. albida*** Sym. (alan, NW Borneo), ***S. almon*** Foxw. (almon, C Mal.), ***S. contorta*** S. Vidal (white lauan, Philippines), ***S. glauca*** King (balan, W Mal.), ***S. guiso*** (Blanco) Blume (selangan, red balau, SE As. to Philippines), ***S. kunstleri*** King (red balau, W Mal.); ***S. polysperma***

(Blanco) Merr. (Bataan or Philippine mahogany, bataan, tangile, Philipp.), ***S. robusta*** Roxb. ex Gaertn. f. (sal, N Ind.) – coppice fire-resistant, general constr., bark a source of a black dye, a dammar (resin, 10 K t a yr by 1994) used in typewriter ribbons, carbon paper, incense, Buddha born in a grove of it (but see *Saraca*) & buried between 2 trees of it. Dammars of other spp. form. imp. in varnish manufacture, e.g. ***S. javanica*** Koord. & Val. in S Sumatra where retained when forest cleared for cult. & also planted out, oft. in agroforestry with coffee & pepper. Fr. boiled as veg. but also exploited (as illipe nuts, when soaked known as black pontianak in trade) for oil (up to 70%) esp. ***S. macrophylla*** (Vriese) Ashton & other Bornean spp. used as subs. for cocoa butter in chocolate-making & also in cosmetics, form. imp. for soap, candles etc. e.g. Borneo tallow (***S. palembanica*** Miq., W Mal.); ***S. ovalis*** (Korth.) Blume (W Mal.) – tetraploid apomict of hybrid orig. & ***S. resinosa*** Foxw. (W Mal.) triploid; ***S. trapezifolia*** (Thw.) Ashton (Sri Lanka) – fls annually unlike rest of genus in aseasonal trop. SE As. 1 month after increase in night temp. (23+°C), poss. use as plywood sp. in enrichment planting. See footnote on p. 67

shoreweed *Plantago uniflora*

Shortia Torrey & A. Gray. Diapensiaceae. 1 N Am., 5 E As. ('*Schizocodon*'). R: T 32(1983)420, Plantsman 12(1990)23. Cult. orn. everg. herbs

Shoshonea Evert & Constance. Umbelliferae (III 8). 1 Wyoming: ***S. pulvinata*** Evert & Constance. R: SB 7(1982)471

shot weed *Cardamine oligosperma*

shrimp plant *Justicia brandegeeana*

shu di huang *Rehmannia glutinosa*

Shuaria D. Neill & J.L. Clark. Gesneriaceae (II 3b). 1 SE Ecuador: ***S. ecuadorica*** D. Neill & J.L. Clark – arborescent. R: SB 35(2010)670

shungiku *Glebionis coronaria*

Shuteria Wight & Arn. Leguminosae (III 18). 4 –5 Indomal. R: Adansonia 12(1972)291. ***S. involucrata*** (Wall.) Benth. (Indomal.) – cover crop under *Cinchona*

Siagonanthus Poepp. & Endl. = *Maxillaria*

Siam weed *Chromolaena odorata*

Siamanthus Larsen & Mood. Zingiberaceae (II 4). 1 Thailand: ***S. siliquosus*** Larsen & Mood. R: NJB 18(1998)393

Siamese balsa *Alstonia ? spatulata*

Siamosia Larsen & T.M. Pedersen. Amaranthaceae (I 2). 1 Thailand: ***S. thailandica*** Larsen & T.M. Pedersen. R: NJB 7(1987)271

Siapaea Pruski. Compositae (Eup.-Ayap.). 1 Venez.: ***S. liesneri*** Pruski

Sibangea Oliv. = *Drypetes*

Sibara Greene. Cruciferae (46). 13 E & S N Am. R: HBP 15(2010)140

Sibaropsis S. Boyd & T. Ross = *Streptanthus* (but see Madroño 44(1997)30)

Sibbaldia L. (~ *Potentilla*). Rosaceae (Pot.). Incl. *Sibbaldiopsis*, 6 N temp. R: PakJB 29(1997)9

Sibbaldianthe Juz. = *Potentilla*

Sibbaldiopsis Rydb. = *Sibbaldia*

Siberian elm *Ulmus pumila*; **S. saxifrage** *Bergenia* spp.; **S. wallflower** *Erysimum* × *marshallii*

Sibiraea Maxim. (= *Eleiosina*). Rosaceae (2). c. 4 SE Eur. (1), C & E As. ***S. laevigata*** (L.) Maxim. (= *E. i.* (L.) Raf.; *S. altaiensis*, Balkans, Siberia, China) – cult. orn. decid. shrub

Sibthorpia L. Plantaginaceae (Dig.-Sib.; Scrophulariaceae s.l.). 5 trop. & S Am., Macaronesia, Eur. (2), Afr. mts. R: BN 108(1955)161. Creeping pls like *Hydrocotyle* spp. esp. ***S. europaea*** L. (Cornish moneywort, pennywort, W Eur.)

Sicana Naud. Cucurbitaceae (XV). c. 4 trop. Am. R: Brenesia 35(1991)37. ***S. odorifera*** (Vell. Conc.) Naud. (casabanana, cassabanana, orig. Braz.?) – monoec. perenn. liane with ed. fragrant fr. used for preserves & scenting linen

Siccobaccatus P. Braun & Esteves = *Micranthocereus*

Sichuan pepper *Zanthoxylum simulans*

Sichuania M. Gilbert & P.T. Li = *Cynanchum* (but see Novon 5(1995)12)

Sickingia Willd. = *Simira*

sicklepod *Senna tora*

Sicrea (Baill.) Hallier f. (~ *Schoutenia*). Malvaceae (Dom.; Tiliaceae). 1 Cambodia: ***S. godefroyana*** (Baill.) Hall. f.

Siculosciadium C. Brullo & al. (~ *Peucedanum*). Umbelliferae. 1 Sicily: ***S. nebrodense*** (Guss.) C. Brullo & al.

Sicydium Schldl. Cucurbitaceae (II). c. 7 trop. Am.

Sicyocarya (A. Gray) H. St John = *Sicyos*

Sicyocaulis Wiggins = *Sicyos*

Sicyos L. Cucurbitaceae (XII). Incl. *Microsechium, Parasicyos, Sechiopsis, Sechium, Sicyosperma*, c. 75 Aus., NZ, Pac. (incl. Hawaii 24), trop. & warm Am. ***S. angulatus*** L. (bur cucumber, E N Am.) – grown as screen; ***S. edulis*** Jacq. (*Sechium e.*, chayote, choco, cho-cho, choko, chow chow, christophine, Madeira marrow, veg. pear) – fr. & tubers cooked, each fr. with 1 v. large seed, in Java grown as cover for fish-ponds

Sicyosperma A. Gray = *Sicyos*

Sida L. Malvaceae (Malv.-Malv.). c. 250 trop. & warm (Aus. c. 90, allied to *Meximalva, Sidastrum*) esp. Am. R: CGH 180(1957)1. Shrubs & herbs with schizocarps with 1-seeded mericarps; alks. Some fibre-pls esp. ***S. rhombifolia*** L. (Queensland hemp, Paddy's lucerne, trop.) – used e.g. for art basketry as by Margaret Preston (Aus.), stimulant (ephedrine), medic. (piles) in Bali. Some weedy, controlled by ladybird beetles in NG. ***S. acuta*** Burm.f. (C Am.) – cannabis subs. in Mex.; ***S. cordifolia*** L. (Ind., natur. warm countries) – in slimming tablets; ***S. fallax*** Walp. ('ilima, Pacific) – leis (form. for nobility only) Hawaii; ***S. jatrophoides*** L'Hérit. (Peru & Galápagos) & ***S. palmata*** Cav. (Peru & Ecuador) with glandular trichomes prod. secretions toxic to ants & cockroaches

Sidalcea A. Gray. Malvaceae (Malv.-Malv.). 20 W N Am. R: AMBG 18(1931)117. Cult. orn. herbs esp. ***S. malviflora*** (DC.) Benth. with pink fls, many cvs & hybrids

Sidasodes Fryx. & Fuertes (~ *Sida*). Malvaceae (Malv.-Malv.). 2 Colombia to Peru. R: Britt. 44(1992)438

Sidastrum Bak. f. (~ *Sida*). Malvaceae (Malv.-Malv.). 8+ trop. Am. (+ Oceania, Aus.). R: Britt. 30(1978)449)

sidder *Ziziphus spina-christi*

Siderasis Raf. Commelinaceae (II 1e). 2–3 Braz. ***S. fuscata*** (Lodd.) H. Moore (E Braz.) – cult. orn. rosette-herb with violet fls

Sideria Ewart & Petrie = *Melhania*

Sideritis Tourn. ex L. (~ *Stachys*). Labiatae (VI). c. 140 N temp. OW, Macaronesia. R (sect. *S.* – 69 spp.): PM 20 & 21 (1992, 1994). Heterogeneous. Some cult. orn. herbs & shrubs; lvs of spp. in sect. *Empedoclea* used as tea in Turkey

Siderobombyx Bremek. = *Xanthophytum*

Sideropogon Pichon = *Cuspidaria*

Sideroxylon Dill. ex L. Sapotaceae (III). Incl. *Bumelia, Dipholis* & *Mastichodendron*, 75 trop. (Am. 49) to E C US (N Am. 11). R: T.D. Pennington, *S.* (1991)169. Hard wood; sources of a kind of chicle e.g. ***S. foetidissimum*** Jacq. (*Mastichodendron sloaneanum*, Barbados mastic, WI) & some ed. fr., e.g. ***S. obtusifolium*** (Roemer & Schultes) Penn. (trop. Am.) – most delicious fr. of Paraguayan chaco. ***S. inerme*** L. (S Afr.) – at Cape Agulhas, W Cape, 'post office tree' for C17 mariners; ***S. mastichodendron*** Jacq. (Am.) – ship- & boat-building as timber not attacked by *Teredo*; ***S. mirmulano*** R. Br. (Macaronesia, Cape Verde Is.) – part of Madeiran laurel forest; ***S. sessiliflorum*** (Poir.) Aubrév. (Mauritius) – v. rarely regenerating & seeds rot if pulp not removed, so supposed to have been disp. by dodos extinct for some 300 yrs (or other extinct animals), so turkeys force-fed with fr. & germ. enhanced (but introd. monkeys also eat unripe fr.)

Sidotheca Reveal (~ *Oxytheca*). Polygonaceae (Erio.-Erio.). 3 Calif., NW Mex. R: FNA 5(2005)439

sidra *Ziziphus spina-christi*

Siebera Gay. Compositae (Card.-Card.(Card.)). 2 Middle E to Afghanistan

Siederella Szlach. & al. = *Cyrtochilum*

Siegfriedia C. Gardner. Rhamnaceae (5). 1 SW Aus. (restricted): ***S. darwinioides*** C.A. Gardner. R: Nuytsia 11(1996)116

Sieglingia Bernh. = *Danthonia*

Siemensia Urb. Rubiaceae (? V 1). 1 W Cuba: ***S. pendula*** (Griseb.) Urb.

Sierra Leone butter *Pentadesma butyracea*; **S. L. copal** *Guibourtia copallifera*

Sievekingia Reichb.f. Orchidaceae (V 12i). 16 trop. Am. R: Schlechteriana 3(1992)55, 105,4(1993)30. Cult. orn. incl. ***S. suavis*** Reichb.f. (Costa Rica, Panamá) – floral fragrance largely cineole attractive to male euglossine bees

Sieversandreas Eb. Fischer. Orobanchaceae (Buch.). 1 S Madag.: ***S. madagascariensis*** Eb. Fischer

Sieversia Willd. (~ *Geum*). Rosaceae (Colur.). 1 NE As. to Jap.: ***S. pentapetala*** (L.) Greene. R: FNA 9(2014)57

Sigesbeckia L. Compositae (Mill.-Mill.). 8 trop. (orig. OW), Macaronesia (Eur. 2 natur.). Stone's Model; capitula disp. as units, the 5 bracts forming involucre covered with sticky glands. ***S. orientalis*** L. (OW) – locally medic. in Ind., poss. oilseed, used in comm. skin prods

Sigmatanthus Huber ex Emmerich (~ *Raputia*). Rubiaceae (I 8). 1 NE Braz.: ***S. trifoliatus*** Huber ex Emmerich

Sigmatochilus Rolfe = *Chelonistele*

Sigmatogyne Pfitzer = *Panisea*

Sigmatostalix Reichb.f. = *Oncdium*

Sigmoidotropis (Piper) Delgado (~ *Vigna*). Leguminosae (III 18). 9+ trop. Am. R: AJB 98(2011)1710

silage green fodder preserved by pressure in silo or stack

Silaum Mill. Umbelliferae (III 8). 1 temp. Euras.: ***S. silaus*** (L.) Schinz & Thell. (pepper saxifrage)

Silene L. Caryophyllaceae (III 3). Incl. *Cucubalus, Lychnis, Melandrium, Petrocoptis, Uebelinia, Viscaria*, c. 700 N hemisph. (Eur. 199; N Am. 70 incl. introd.) esp. S Balkan Pen. (Greece 119) & SW As., trop. & S Afr. (8 – R: Bothalia 42(2012)148, incl. ***S. burchellii*** Otth (*S. pilosellifolia*, throughout Afr.). R: T 44(1995)543. Campion, catchfly. Herbs (some ± aquatic), rarely shrubby (e.g. Hawaii), with extrafl. nectaries, some with ecdysteroids. Sex chromosomes in dioec. spp., pops with female-biased sex ratios. Cult. orn. annuals & perennials: ***S. acaulis*** (L.) Jacq. (moss campion, Arctic Euras., N N Am., C Eur.) – rock-pl.; ***S. armeria*** L. (none-so-pretty, C & S Eur., natur. N Am.) – ann. or bienn. with flat-topped cymes; ***S. baccifera*** (L.) Roth (*Cucubalus b.*, Euras., N Afr.) – form. medic., fr. a berry (dry when mature); ***S. banksia*** (Meerb.) Mabb. (*S. fulgens, L. f.*, E As.) – ecdysteroids, cult. as '*L.* × *haageana*', '*L. senno*' an autotriploid cv. with white fls form. in cult.; ***S. chalcedonica*** (L.) E. Krause (*L. c.*, Maltese cross, c. of Jerusalem, N Russia 49–56 °N, natur. N Am.) – fls vivid scarlet, hybrids with ? *S. banksia* = 'L. × *arkwrightii* Heydt' esp. '**Vesuvius**'; ***S. coronaria*** (L.) Clairv. (*L. c.*, rose campion, mullein pink, S Eur. to N Iran but widely natur.) – woolly herb with purplish fls; ***S. dioica*** (L.) Clairv. (red campion, Eur., natur. N Am.) – corrosive juice used on warts & corns W England, pl. of woods & hedges but hybridizing (= ***S.* × *hampeana*** Meusel & K. Werner) with ***S. latifolia*** Poir. (*S. alba, S. pratensis*, white campion or cockle, Eur., Medit., natur. subcosmop.) – weed of open ground & fields with dimorphic pollen & sex chromosomes XX, XY with large dominant Y (when half Y removed pl. hermaphr. so 1 region involved in female suppression; silver thiosulphate enhances A development in female, rust fungus activates male genes in female but pollen grains replaced by rust spores!); ***S. flos-cuculi*** (L.) Clairv. (*L. f.*, ragged robin, Eur. to C As., natur. N Am.) – fls rose-red; ***S. nivalis*** (Kit.) Rohrb. (Roumania) – rare cushion-pl. rarely cult.; ***S. otites*** (L.) Wibel (Eur. to Siberia) – fls poll. by nocturnal Lepidoptera & mosquitoes in Dutch sand-dunes; ***S. schafta*** C. Gmelin ex Hohen. (Cauc.) – commonly cult. rock-pl. with pink fls; ***S. stenophylla*** Ledeb. (As.) – seeds 32 000 yrs old frozen in NE Siberia successfully germ.; ***S. suecica*** (Lodd.) Greuter & Burdet (*L. alpina*, alpine c., Eur., NE N Am.) – used as copper indicator in Norway & nickel indicator in Finland; ***S. uniflora*** Roth (*S. maritima*, sea campion, Eur. mts & coasts); ***S. vulgaris*** (Moench) Garcke (bladder campion, Euras., Medit., natur. N Am.) – night-scented fls, shoots a veg. (silene, stridolo, strigoli etc., Italy)

Silentvalleya V.J. Nair & al. Gramineae (Chlorid). 2 Ind. R: KB 67(2012)547

Siler Mill. (~ *Laserpitium*). Umbelliferae (III 2). 1 Eur.: ***S. montanum*** Crantz (*L. siler*). R: T 65(2016)585

Silicularia Compton = *Heliophila*

Siliquamomum Baill. Zingiberaceae (II 1) 3 SW China, Vietnam. R: GBS 66(2014)45

silk cotton tree *Bombax ceiba, Cochlospermum religiosum*; **Indian s. c. t.** *Bombax ceiba*; **s. tree** *Albizia julibrissin*; **s. vine** *Periploca graeca*; **s. weed** *Asclepias syriaca*; **silky bent grass** *Apera* spp.; **s. oak** *Grevillea robusta, Cardwellia sublimis*

silkwood *Flindersia pimenteliana*

Siloxerus Labill. (~ *Angianthus*). Compositae (Gnap.-Ang.). 3 SW Aus., 1 ext. to Tasmania. R: ASBSN 78(1994)6

Silphiodiscus (Koso-Polj.) Spalik & al. (~ *Daucus*). Umbelliferae (III 2). 2 Euras. R: T 65(2016)578

silphion, silphium ? *Ferula* sp.

Silphium L. Compositae (Helia.-Eng.). 12 E N Am. R: FNA 21(2006)77. ***S. laciniatum*** L. (compass pl., prairies) – young pl. with lvs tipped vertically, avoiding full incidence of midday sun, the surfaces facing N & S (cf. *Lactuca serriola*), locally medic.

Silvaea Philippi (*Philippiamra*) = *Cistanthe*

silvanberry see *Rubus*

silver beech *Nothofagus menziesii*; **s. beet** = spinach b.; **s. bell** *Halesia carolina*; **s. birch** *Betula pendula*; **s. fir** *Abies* spp.; **s. greywood** *Terminalia bialata*; **s. maple** *Acer saccharinum*; **s. pine** *Lagarostrobos colensoi*; **s. top ash** *Eucalyptus sieberi*; **s. wattle** *Acacia dealbata*; **s.weed** *Potentilla anserina*

silverballi *Nectandra* spp.

silverberry *Shepherdia argentea*

silverseed gourd *Cucurbita argyrosperma*

silversword *Argyroxiphium* spp.

Silvianthus Hook.f. Carlemanniaceae (Caprifoliaceae s.l.). 2 E Ind. to SE As. R: FOC 19(2011)478

Silviella Pennell. Orobanchaceae (Gerard.; Scrophulariaceae s.l.). 2 Mex. R: PANSP 86(1928)434

Silvorchis J.J. Sm. Orchidaceae (IV 4d). 3 Mal. ***S. colorata*** J.J. Sm. (Java, ? extinct) – mycotrophic

Silybum Vaill. ex Adans. Compositae (Card.-Card.). 2 Medit. inc. Eur. R: JAA 71(1990)426. ***S. marianum*** (L.) Gaertn. (holy or milk thistle) – ann. or bienn. with white-blotched lvs found in nutrient-rich sites, the cypselas with oil-bodies attractive to harvester ants which deposit them in rich organic material from their nests, cult. orn. now widely natur. N & S Am. incl. pampas & bad weed in NSW, lvs & stems ed. as salad, fr. (kenguel seed) form. a coffee subs. & locally medic. considered to prot. liver from hepatitis since time of Dioscorides, now known to cont. flavonoids effective as antidotes to *Amanita* poisoning, their being able to displace phalloidin from membrane receptors, also used for premenstrual tension (affects oestrogen levels), hangovers etc.

Simaba Aubl. (~ *Quassia*). Simaroubaceae. 25 trop. Am. ***S. cedron*** Planch. (*Q. c.*, cedron. Amaz.) – bitter seeds vermifuge; ***S. guianensis*** Aubl. (*Q. g.*, trop. Am.) – fish-disp. in Amaz.

Simarouba Aubl. (~ *Quassia*). Simaroubaceae. 6 trop. Am. ***S. amara*** Aubl. (*Q. simarouba, S. glauca*, aceituna, C Am. WI) – oilseed

Simaroubaceae DC. Magnoliidae – Sapindales. Incl. Leitneriaceae, 21/145 trop. (*Ailanthus altissima* & *Picrasma* extending to temp. As., *Leitneria* to SE US). R: T 44(1995)177. Trees & shrubs usu. with bitter bark (quassinoids – triterpenoid derivatives), wood & seeds (triterpenoid lactones – simaroubilides – present) & pithy shoots. Lvs pinnate, usu. in spirals, without gland dots; leaflets opp., not articulated, stip. 0 exc. *Picrasma*. Fls reg., usu. unisexual & oft. with rudiments of opposite sex (monoec. or dioec.), 3–5(–)8)-merous, small, in racemes, cymes or thyrses; K usu. 5 (0 in *Leitneria*) & basally connate, lobes imbr. or valvate, C usu. 4 or 5 (0 in *Leitneria*) free, imbr. or valvate, A (as many as & alt. with) twice C (or –18), filaments oft. with basal appendages, anthers with longit. slits, disk usu. present, s.t. a gynophore or androgynophore, $\underline{G}$ (1)2–5 rarely free (*Picrolemma*), usu. ± united (s.t. only styles connate), s.t. forming pluriloc. ovary with axile placentas, each (rarely only 1) loc. with 1(2) apical or basal, anatr. to hemitropous, bitegmic ovule. Fr. with 1–3(–5) samaroid or drupaceous mericarps; embryo straight or curved, oily, endosperm ± 0. x = 8, 13+

Principal genera: *Ailanthus, Castela, Eurycoma, Quassia, Simaba, Soulamea*

Fam. restricted to form. subfam. Simarouboideae; *Harrisonia* referred to Rutaceae, *Irvingia* & allies separated as Irvingiaceae, *Kirkia* & allies as Kirkiaceae; *Suriana* (without simaroubalides) & other genera with simple lvs to Surianaceae, while *Alvaradoa* & *Picramnia* put in own fam. Picramniaceae. In many respects close to Rutaceae, but with similarities to Meliaceae (both with limonoids biosnthetically related to quassinoids) & perhaps more similar to the ancestral group from which these families have been derived. Some medic. (*Brucea, Picrasma, Quassia*), oilseeds & timber (*Quassia*) & cult. orn. esp. *Ailanthus* (s.t. invasive)

simaruba *Quassia amara*

Simenia Z. Szabó = *Dipsacus*

Simethis Kunth. Asphodelaceae. 1 W Eur. & Medit.: ***S. mattiazzii*** (Vand.) Sacc. (*S. planifolia*, Kerry lily)

Simicratea Hallé (~ *Hippocratea*). Celastraceae (III). 1 Angola: ***S. welwitschii*** (Oliv.) Hallé. R: BMNHN 4,5(1983)20

Similisinocarum Cawet-Marc & Farille = *Pimpinella*

Simira Aubl. Rubiaceae (Cond.). 44 trop. Am. R: MNYBG 23(1972)299. Some local timbers, dyes (esp. ***S. cordifolia*** (Hook.f.) Steyerm. for bark cloth) & febrifuges ('*Sickingia*')

Simirestis Hallé (~ *Hippocratea*). Celastraceae (III). 8 trop. Afr. R: BMNHN 4,6(1984)4

Simmondsia Nutt. Simmondsiaceae. 1 SW N Am.: ***S. chinensis*** (Link) C. Schneider (*S. californica*, jojoba, goat nut, pig n.) – wind-poll. shrub or small tree tolerant of arid sites & source of jojoba oil, a subs. for sperm oil used in cosmetics etc., the seeds avoided by desert rodents (cyanoglucosides) but form. ground as coffee subs. (poss. 'slimming food' as wax indigestible), the foliage imp. browse

Simmondsiaceae Tiegh. (~ Buxaceae). Magnoliidae – Caryophyllales. 1/1 SW N Am. Everg. dioec. shrub with unusual sec. growth. Lvs small, leathery, simple but with jointed base, opp.; stip. 0. Fls small, reg., C 0, the males in axillary clusters, the females usu. solit.; K (4) 5 (6), imbr., accrescent in females, disk 0, A (8)10(-16) free, anthers elongate, with longit. slits, $\underline{G}$ (3(4)), 3-loc. with 3 long feathery stigmas, each loc. with 1 apical-axile, pend., anatr. ovule. Fr. a loculicidal capsule with usu. 2 empty locules & 1 seed; embryo straight, cotyledons fleshy & with much liquid wax, endosperm ± 0. n = 13

Genus: *Simmondsia*

Form. thought allied to Buxaceae but diff. in in 5-mery, A 10 & structure, pollen, solit. ovule & unusual sec. thickening (concentric rings of vasc. bundles), sieve-tube plastids etc. Anthocyanins found

Simocheilus Klotzsch = *Erica*

Simplicia Kirk. Gramineae (XVI 15). 2 NZ. R: NZJB 9(1971)539

Simpliglottis Szlach. = *Chiloglottis*

Simsia Pers. Compositae (Helia.-Helia.). 22 trop. Am. to S US. R: SBM 30(1990). Shrubs to annuals

simsim *Sesamum indicum*

simul *Bombax ceiba*

Sinacalia H. Robinson & Brettell (~ *Ligularia*). Compositae (Sen.-Tuss.). 4 China. R: KB 39(1984)215. ***S. tangutica*** (Maxim.) R. Nordenstam (*Senecio t.*) – cult. orn. with panicles of capitula

Sinadoxa C.Y. Wu, Z.L. Wu & R.F. Huang. Viburnaceae (Adoxaceae). 1 China: ***S. corydalifolia*** C.Y. Wu, Z.L. Wu & R.F. Huang (to 4000 m). R: BBR 7,4(1987)100

Sinalliaria X.F. Jin & al. (~ *Orychophragmus*). Cruciferae. 1 E China: ***S. limprichtiana*** (Pax) X.F. Jin & al. R: Phytotaxa 186(2014)192

Sinapidendron Lowe (~ *Brassica*). Cruciferae (12). 5 Macaronesia

Sinapis L. (~ *Brassica*). Cruciferae (12). 5 Eur. (4), Medit. R: JAA 66(1985)312; G. Baillargeon, *Tax. Rev. Gattung S.* (1986)96. All but *S. aucheri* (Boiss.) O. Schulze close to *B. nigra*. **S. *alba*** L. (*B. hirta* Moench, black, yellow or white mustard, Medit.) – mustard of mustard & cress, when grown to cotyledon stage only; ***S. arvensis*** L. (*B. kaber* (DC.) Wheeler, charlock, runch) – farmweed, form. famine food (Ireland), form. v. common in cornfields, the seeds remaining viable for at least 10 years even after ingestion by animals

Sinarundinaria Nakai = *Fargesia*; see also *Chimonocalamus*, *Otatea*, *Thamnocalamus*, *Yushania*

Sinclairia Hook. & Arn. Compositae (Liab.). Excl. *Liabellum*, 20 Mex. to Colombia. R: Phytol. 67(1989)168

Sinclairiopsis Rydb. (~ *Sinclairia*). Compositae (Liab.). 2 Mex.

Sincoraea Ule = *Orthophytum*

Sindechites Oliv. Apocynaceae (II c). 1 subtrop. China: ***S. henryi*** Oliv. R: AUWP 88–6(1988)29

Sindora Miq. Leguminosae (I 2). 18–20 SE As., Mal. See also seq. Some timbers (sepetir) & medic. oils (skin disease). ***S. supa*** Merr. (Luzon) – imp. comm. timber, wood-oil for illumination, paint & varnish

Sindoropsis Léonard. Leguminosae (I 2). 1 Gabon: ***S. letestui*** (Pellegr.) Léonard (gheombi) – timber for flooring, furniture, veneers

Sindroa Jum. = *Orania*

Sineoperculum Jaarsveld = *Cleretum*

Sinephropteris Mickel = *Asplenium*

Singana Aubl. = ? Leguminosae

Singapore daisy *Sphagneticola trilobata*; **S. holly** *Malpighia coccigera*

Singchia Z.J. Liu & L.J. Chen (~ *Pteroceras*). Orchidaceae (V 16c). 1 SE Yunnan: ***S. malipoensis*** Z.J. Liu & L.J. Chen. R: JSE 47(2009)600

Singhara nut *Trapa natans*

sing-kwa *Luffa acutangula*

Singularybas Molloy & al. = *Corybas*
Sinia Diels = *Sauvagesia*
sinicuichi *Heimia salicifolia*
Sinningia Nees. Gesneriaceae (II 5e). 60 trop. Am., esp. E & S Braz. Herbs & shrubs, usu. tuberous (tubers derived from perenn. stems), s.t. with scented fls (rare in fam.), some bat- ('*Lietzia*') & bird-poll. ('*Rechsteineria*') or both ('*Paliavana*'), others (e.g. '*Vanhouttea*') visited by bees (bee- & hummingbird-poll. evolving several times), the glossy texture of C due to fine hairs. Many cult. orn. (prop. by leaf-cuttings prod. tubers rather than plantlets at cut surface) esp. 'gloxinia', ***S. speciosa*** (Lodd.) Hiern (Braz.), of which florist's g. differs from wild pl. in having large erect campanulate (i.e. peloric; arisen once, 1844 Scotland) fls (**Fyfiana Group**), the wild ones having nodding fls; other cult. spp. incl. ***S. canescens*** (Mart.) Wiehler (*S. leucotricha, Rechsteineria l.*, Braz.) – tubers may flower without planting; ***S. cardinalis*** (Lehm.) H. Moore (*R. c.*, Braz.) – scarlet 2-lipped bird-poll. fls, though many selected lines referred to this sp. may be hybrids
Sinoadina Ridsd. = *Adina* (but see Blumea 24(1978)352)
Sinoarabis Karl & al. (~ *Arabis*). Cruciferae (7). 1 Tibet: ***S. setosifolia*** (Al-Shehbaz) Karl & al.
Sinobacopa Hong = *Bacopa*
Sinobaijiania C. Jeffrey & de Wilde = *Baijiania* (but see BZ 91(2006)769)
Sinobambusa Mak. ex Nakai. Gramineae (IV). 10 N Vietnam, SW & S China (11, 9 endemic – R: FOC 22(2006)147). R: JBR 1(1982)140. Forest bamboos; cult. orn. incl. ***S. tootsik*** (Mak.) Nakai (Vietnam, China) introd. Jap.
Sinocalamus McClure = *Dendrocalamus*
Sinocalycanthus (W.C. Cheng & S.Y. Chang) W.C. Cheng & S.Y. Chang = *Calycanthus*
Sinocarum H. Wolff ex Shan & Pu. Umbelliferae (III 8). Incl. *Dactylaea* c. 24 China
Sinochasea Keng (~ *Pseudodanthonia*). Gramineae (IX). 1 W China: ***S. trigyna*** Keng. R: FOC 22(2006)191
Sinocrassula A. Berger (~ *Orostachys*). Crassulaceae (I 1). 7 Himal. to SW China. R: U. Eggli, *Ill. Handb. Succ. Pls*, Crass. (2005)350. Some cult. orn. esp. ***S. yunnanensis*** (Franch.) A. Berger (Yunnan)
Sinodielsia H. Wolff = *Meeboldia*
Sinodolichos Verdc. (~ *Glycine*). Leguminosae (III 10). 2 Myanmar, S China
Sinofranchetia (Diels) Hemsl. Lardizabalaceae. 1 C & W China: ***S. chinensis*** (Franch.) Hemsl. – cult. orn. liane, monoec. to dioec, with purple berries & black seeds. R: CBM n.s. 29(2012)249
Sinoga S.T. Blake = *Asteromyrtus*
Sinogentiana Favre & Y.M. Yuan (~ *Metagentiana*). Gentianaceae. 2 SW China. R: T 63(2014)351
Sinojackia Hu. Styracaceae. 5 S China. Cult. orn. shrubs
Sinojohnstonia Hu (~ *Omphalodes*). Boraginaceae (B.3.8.2?). 4 W China. R: Novon 23(2014)254
Sinoleontopodium Y.L. Chen (~ *Leontopodium*). Compositae. 1 China: ***S. lingianum*** Y.L. Chen. R: APS 23(1985)457
Sinolimprichtia H. Wolff. Umbelliferae (III 5). 1 E Tibet: ***S. alpina*** H. Wolff
Sinomanglietia Z.X. Yu = *Magnolia*
Sinomarsdenia P.T. Li & Chen = *Marsdenia*
Sinomenium Diels. Menispermaceae (V). 1 C China, Jap.: ***S. acutum*** (Thunb.) Rehder & E. Wilson – alks
Sinomerrillia Hu = *Neuropeltis*
Sinopanax Li. Araliaceae. 1 Taiwan: ***S. formosanus*** (Hayata) Li
Sinopimelodendron Tsiang = *Cleidiocarpon*
Sinoplagiospermum Rauschert = *Prinsepia*
Sinopodophyllum Ying = *Podophyllum*
Sinopogonanthera Li = *Paraphlomis*
Sinopora J. Li & al. (~ *Syndiclis*). Lauraceae. 1 Hong Kong: ***S. hongkongensis*** (N. Xia & al.) J. Li & al. R: Novon 18(2008)199
Sinopteridaceae Koidz. = Pteridaceae
Sinopteris C. Chr. & Ching = *Cheilanthes*
Sinoradlkofera F. Meyer = *Boniodendron*
Sinorchis S.C. Chen = *Cephalanthera*
Sinosassafras H.W. Li = *Parasassafras*

Sinosenecio R. Nordenstam (~ *Senecio*). Compositae (Sen.-Tuss.). 36 China to SE As. R: KB 39(1984)222

Sinosideroxylon (Engl.) Aubrév. = *Sideroxylon*

Sinosophiopsis Al-Shehbaz = *Smelowskia*

Sinoswertia T.N. Ho & al. Gentianaceae. 1 China: ***S. tetraptera*** (Maxim.) T.N. Ho & al. R: T.N. Ho & S.W. Liu, *Worldwide monog. Swertia* (2015)303

Sinowilsonia Hemsl. Hamamelidaceae (I 3). 1 C & W China: ***S. henryi*** Hemsl. – cult. orn. tree

sinqua (melon) *Luffa acutangula*

Sinthroblastes Bremek. = *Strobilanthes*

Siolmatra Baill. Cucurbitaceae (III). 2 S Am. R: Sida 21(2005)1961. Wind-disp. (cf. allied *Fevillea*)

Sipanea Aubl. Rubiaceae (I 6). 19 trop. S Am.

Sipaneopsis Steyerm. Rubiaceae (I 6). 6 NW trop. S Am.

Sipapoa Maguire = *Diacidia*

Sipapoantha Maguire & Boom. Gentianaceae. 1 Venez.: ***S. ostrina*** Maguire & Boom. R: MNYBG 51(1989)23

Siparuna Aubl. Siparunaceae (Monimiaceae s.l.). 53 trop. S Am. Poss. heterogeneous. Champagnat's, Mangenot's, Massart's & Nozeran's Models. Almost closed hypanthium (cf. *Ficus*) poll. gall-midges. ***S. cujabana*** (Mart.) A. DC. (Braz.) – contraceptive

Siparunaceae Schodde (~ Monimiaceae). Magnoliopsida – Laurales. 2/57 trop. Am., W Afr. R: FN 95(2005)36. Dioec. or monoec. trees & lianes, accum. aluminium. Lvs opp., serrate; stip. 0. P 4–6(7) or obscure, with calyptra; A(1)2-∞ (in pairs); $\underline{G}$3-∞ with unitegmic ovules. Hypanthium fleshy, splitting. n = 22

Genera: *Glossocalyx, Siparuna*

Form. 2 tribes in Monimiaceae s.l.

Siphanthera Pohl ex DC. Melastomataceae. 16 NE S Am. R: SBM 93(2011)25

Siphantheropsis Brade = *Macairea*

Siphocampylus Pohl. Campanulaceae (III 3). 231 trop. Am. R: Pfl. IV, 276b(1953)264, 276c(1968)845. Prob. incl. *Burmeistera, Centropogon* (baccate fr. prob. evolved several times). Shrubs; many bird-, some bat-poll. but ***S. sulfureus*** E. Wimmer (SE Braz.) both – smelly yellow fls attractive to bats, but open in day with nectar high in sucrose attractive to hummingbirds. Some poss. rubber sources

Siphocodon Turcz. Campanulaceae (I 2). 2 SW Cape. R: Strel. 9(2000)396

Siphocranion Kudô (~ *Hanceola*). Labiatae (VII 3b). 2 China, SE As.

Siphokentia Burret = *Hydriastele*

Siphonandra Klotzsch. Ericaceae (VIII 5). 5 Bolivia, Peru. R: JBRIT 2(2008)250

Siphonandrium K. Schum. Rubiaceae (IV 10). 1 NG: ***S. intricatum*** K. Schum.

Siphonella Small = *Valerianella*

Siphoneugena O. Berg (~ *Eugenia*). Myrtaceae (II 10). 9 trop. Am. R: EJB 47(1990)239

Siphonochilus J.M. Wood & Franks. Zingiberaceae (II 1). 20 trop. & S Afr. R: NRBGE 40(1982)372. Like *Kaempferia* but infls lat. etc.; dry forests & savannas. Some cult. orn. ***S. aethiopicus*** (Schweinf.) B.L. Burtt (indungulu) – imp. Zulu medic., over-exploited

Siphonodon Griff. Celastraceae (I; Siphonodontaceae). 7 Indomal., Aus. ***S. australis*** Benth. (ivorywood, Aus.) – timber for rulers, turnery, engraving, etc.

Siphonodontaceae Gagnepain & Tard. = Celastraceae

Siphonoglossa Oersted = *Justicia*

Siphonosmanthus Stapf = *Osmanthus*

Siphonostegia Benth. Orobanchaceae (Rhin.; Scrophulariaceae s.l.). 1 E Med. incl. Eur., 2 E As.

Siphonostelma Schltr. = *Brachystelma*

Siphonostylis W. Schulze = *Iris*

Siphonychia Torrey & A. Gray = *Paronychia*

Sipolisia Glaz. ex Oliv. = *Heterocoma*

Siraitia Merr. Cucurbitaceae (VII). 3–4 SE As. R: APS 31(1993)54, Blumea 51(2006)409. ***S. africana*** (C. Jeffrey) A.M. Lu & Z.Y. Zhang (SE Nigeria & S Tanz.) – not re-collected since 1966, local sweetener (glycosides); ***S. grosvenorii*** (Swingle) A.M. Liu & Z.Y. Zhang (China) – sweet, hearty tea from fr.

Sirdavidia Couvreur & Sauquet. Annonaceae (IV 1). 1 Gabon: ***S. solannona*** Couvreur & Sauquet – buzz.-poll., 'sister' to *Mwasumbia* 3000 km away. R: PhytoKeys 46(2015)4

Sirhookera Kuntze. Orchidaceae (V 16a). 2 S Ind., Sri Lanka

Sirindhornia H. Pedersen & Suksathan. Orchidaceae (IV 4d). 3 Myanmar, Yunnan, Thailand. R: NJB 22(2002)393

siris *Albizia lebbeck*; **red s.** *Falcataria toona*

Sirochloa S. Dransf. (~ *Schizostachyum*). Gramineae (V6). 1 Madag.: ***S. parvifolia*** (Munro) S. Dransf. R: KB 57(2002)965

sisal *Agave sisalana*

Sison L. Umbelliferae (III 8). 2 Eur. (1), Medit. ***S. amomum*** L. (Eur., Medit.) – local medic. & food-flavouring

sissoo *Dalbergia sissoo*

Sisymbrella Spach (~ *Rorippa*). Cruciferae (16). 1 ***S. aspera*** (L.) Spach

Sisymbriopsis Botsch. & Tzvelev. Cruciferae (26). 5 Tajikstan, China. R: Novon 9(1999)309

Sisymbrium Tourn. ex L. Cruciferae (43). 41 Euras. (Eur. 20, incl. *Lycocarpus*), Medit., S Afr., N Am. (1 native). R: Pfr. IV,105(1924)46. ***S. altissimum*** L. (Eur.) – tumbleweed in US; ***S. irio*** L. (London rocket, Medit.) – appeared conspic. in London after Great Fire of 1666 as did ***S. orientale*** L. (Medit.) after the blitz of World War II; ***S. officinale*** (L.) Scop. (hedge mustard, Eur., Medit.) – form. medic. & antiscorbutic

Sisyndite E. Meyer ex Sonder. Zygophyllaceae (IV). 1 S Afr.: ***S. spartea*** E. Meyer ex Sonder. R: Strel. 10(2000)564

Sisyranthus E. Meyer ex Sonder. Apocynaceae (Vb; Asclepiadaceae III 5). 12 trop. & S Afr.

Sisyrinchium L. Iridaceae (VII 3). c. 140 Am. (esp. C & S; N Am. 37) with 1 ext. to Ireland (poss. natur.). See also *Olsynium*; Australasian spp. referred to allied *Libertia* (P not ± similar). 40 spp. with elaiophores attractive to bees; some with black seeds (unusual in I.). Some used in tattooing, some cult. orn. (blue-eyed grass) incl. ***S. bermudiana*** L. (? actually *S. angustifolium* Mill. [natur. subcosmop.], E N Am. to Ireland) poss. a pre-glacial relic in Eur. (Godwin), but known in BI only since 1845, oft. confused with ***S. montanum*** Greene (E N Am.) with larger fls & erect (not nodding) pedicels in fr., natur. Eur.; ***S. striatum*** Sm. (Chile & Arg.) – sturdy garden pl. with greenish-yellow fls, natur. Scilly Is.; ***S. vaginatum*** Spreng. (trop. Am.) – fert. control in Urug.

Sisyrolepis Radlk. (*Delpya*). Sapindaceae. 1 Thailand, Cambodia: ***S. muricata*** (Pierre) Leenh.

sita phal *Annona squarrosa*

Sitanion Raf. = *Elymus* (but see Britt. 15(1963)303)

Sitella L. Bailey = *Waltheria*

Sitka cypress *Xanthocyparis nootkatensis*; **S. spruce** *Picea sitchensis*

Sitopsis (Jaub. & Spach) Löve = *Triticum*

Sium Tourn. ex L. (~ *Berula*). Umbelliferae (III 8). 4 N hemisph. (Eur. 2). Some cult. orn. (water parsnip), lvs locally eaten, esp. ***S. sisarum*** L. (crummock, skirret, C Eur. to E As.) – derived from '*S. sisaroideum* DC.', grown for ed. r. eaten like salsify or coffee subs. ***S. suave*** Walter (NE As., N Am.) – tubers a major food for some Native Americans, local medic.

Sivadasania Mohanna & Pim. Umbelliferae (III 5). 1 S Ind.: ***S. josephiania*** (Wadhwa & Chowdhery)

Sixalix Raf. (~ *Scabiosa*). Caprifoliaceae (Dipsacaceae). 8 Medit. esp. N Afr., SW As. Cult. orn. esp. ann. ***S. atropurpurea*** (L.) Greuter & Burdet (*Scabiosa a.*, sweet scabious, Medit., SW As.) – fls dark purple to white

Skapanthus C.Y. Wu & Li = *Isodon*

skeleton weed *Chondrilla juncea*

Skeptrostachys Garay (~ *Stenorrhynchos*). Orchidaceae (IV 2h). 13 trop. S Am. R: HBML 28(1982)358

skewerwood *Cornus sanguinea*

Skiatophytum L. Bolus. Aizoaceae (V). 3 SW Cape: R: T 64(2015)519

Skimmia Thunb. Rutaceae (I). 4 Himal., China & Jap. to Philippines. R: KM 4(1987)168. Everg. shrubs with alks. Cult. orn. incl. hybrids, with fragrant fls esp. ***S. japonica*** Thunb. (Jap., Sakhalin, Philipp.) – dioec., the males with more strongly scented fls, the females with bright red drupes in winter (occ. 'males' prod. fr.), many cvs. ***S. laureola*** (DC.) Decne. (Sino-Himal.) – fr. an abortifacient in Ind.

skirret *Sium sisarum*

Skoliopterys Cuatrec. = *Clonodia*

Skottsbergianthus Boelcke = *Xerodraba*

Skottsbergiella Boelcke = *Xerodraba*

Skottsbergiliana H. St John = *Sicyos*

skull cap [(sic, error for scullcap] *Scutellaria* spp.

skunk *Cannabis sativa*; **s. cabbage** *Lysichiton americanus*, *Symplocarpus foetidus*; **s. currant** *Ribes glandulosum*; **s.weed** *Croton texensis*, *Navarretia squarrosa*

Skytanthus Meyen. Apocynaceae (I g). 2 Braz., Chile. Alks

Sladenia Kurz. Sladeniaceae (Theaceae s.l.). 2 SE As. (*Sladenoxylon* wood from Cretaceous of Sudan)

Sladeniaceae Airy Shaw (~ Pentaphylacaceae; Theaceae). Magnoliidae – Ericales. 2/3 SE As., E Afr. mts. Everg. trees. Lvs usu. serrate in spirals s.t. becoming distich.; stip. 0. Fls small, 5(6)-merous, in axillary cymes. K & C imbr., usu. free; A (8-)10–15 free or adnate to C, anthers basifixed, dehiscing by apical pores or slits; disk 0; $\underline{G}$ (3, 5), each carpel with 2 apical pend. epitropous ovules or placentation axile with many ovules per carpel. Fr. a schizocarp or loculicidal capsule with persistent K. Seeds s.t. winged (*Sladenia*). n = 24

Genera: *Ficalhoa, Sladenia*

Poss. best joined with Pentaphylacaceae

slash pine *Pinus elliottii*

Sleumeria Utteridge & al. Icacinaceae. 1 N Borneo: ***S. auriculata*** Utteridge & al. R: SB 30(2005)638

Sleumerodendron Virot. Proteaceae (V 4). 1 New Caled.: ***S. austrocaledonicum*** (Brongn. & Gris) Virot

slipper flower *Calceolaria* spp.; **s. orchid** *Cypripedium* & *Paphiopedilum* spp.

slippery elm *Ulmus rubra*

Sloanea L. Elaeocarpaceae. c. 150 trop. (OW c. 50 – R: KB 38(1983)347, incl. Madag. (3), Aus. (4), New Caled. (18); Am. c. 100 – R: CGH 175(1954)1). Aubréville's Model. Some locally used timbers esp. ***S. australis*** (Benth.) F. Muell. (maiden's blush, E Aus.) – turnery

sloe *Prunus spinosa*

Sloetia Kurz = *Streblus*

Sloetiopsis Engl. = *Streblus*

smack *Papaver somniferum*

Smallanthus Mackenzie ex Small (~ *Polymnia*). Compositae (Mill.-Mill.). 23 trop. & warm Am. R: Phytotaxa 214(2015)1. Disk-florets with pedicels as in *P.* ***S. sonchifolius*** (Poeppig & Endl.) H. Robinson (earth apple, *P. s., P. edulis*, S Am. esp. Peru) – source of yacon (yakon), ed. tubers (inulin so poss. 'slimming' food) used as alcohol source etc.; ***S. uvedalia*** (L.) Mackenzie (E N Am.) – local stimulant & medic.

smartweed *Persicaria hydropiper*

Smeathmannia Sol. ex R. Br. Passifloraceae. 2 trop. W Afr. R: EJB 66(2009)40. Cook's Model

Smelophyllum Radlk. (~ *Stadmania*). Sapindaceae. 1 E Cape: ***S. capense*** Radlk. R: Strel. 9(2000)643

Smelowskia C. Meyer. Cruciferae (44). Incl. *Gorodkovia, Redowskia, Sophiopsis*, etc. 25 Arctic-alpine E As., W N Am. R: HPB 11(2006)92. Cult. orn. rock-pls

Smicrostigma N.E. Br. (~ *Ruschia*). Aizoaceae (V). 1 W Cape: ***S. viride*** (Haw.) N.E. Br. R: H. E. K. Hartmann, *Ill. Handb. Succ. Pls*, Aiz. F–Z(2002)302

Smilacaceae Vent. Magnoliopsida – Liliales. 1/260 trop. & warm esp. S. (Usu.) prickly lianes or branching shrubs with starchy woody rhiz. & usu. vessel-elements throughout veg. body. R. mycorrhizal, without r.-hairs. Lvs in spirals becoming distich., oft. with pr. of tendrils from base of petiole near sheath, lamina usu. with 3–7 curved-convergent veins, cross-connected. Fls reg., usu. unisexual (dioecy iat least in *Smilax*), 3-merous, solit. or in axillary or term. infls, s.t. foetid.; P 3 + 3, usu. petaloid, s.t. basally connate (a 3- or 6-toothed P-tube in '*Heterosmilax*'), A (3)6(–12) s.t. adnate to P-tube or forming a column, anthers with longit. slits, $\underline{G}$ usu. (3), 1- or 3-loc. with parietal or axile placentation & 1 or 3 styles, each loc. with 1–∞ anatr. to campylotropous or orthotropous bitegmic ovules. Fr. a berry with 1–3(–∞) seeds; embryo usu. small in hard endosperm rich in protein, oil & hemicellulose. n = 14–16

Genus: *Smilax*

Genera form. here now in Philesiaceae, *Ripogonum* & *Petermannia* now both in other monogeneric fams, but *Eustrephus* unplaced

Smilacina Desf. = *Maianthemum*

smilax Of florists, *Asparagus asparagoides*

Smilax Tourn. ex L. Smilacaceae. Inc. *Heterosmilax*, c. 260 trop. & temp. (Eur. 3), Afr. (2), Aus. (2), Am. + Caribbean 29 – R: Willd. 40(2010)228. Dioec. lianes (green or cat briar, bamboo vines) or herb. climbers with paired stipular tendrils & oft. recurved hooks on stems, though ***S. biflora*** Sieb. ex Miq. (S Jap.) a bush 10 cm tall. N Am. spp. – young shoots ed. like asparagus, seeds for beads ('Ind. coral'), brown dye from rhiz.; dried rhiz.

of some trop. Am. spp. the sarsaparilla of commerce, used in some r. beer, medic. (esp. rheumatism) form. as tonic e.g. ***S. china*** L. (Chinese r., E As.) & for gym enthusiasts in US, ***S. regelii*** Killip & Morton ('*S. officinalis*', N C Am.) – Jamaica s., while rhiz. of other spp. a carbohydrate source e.g. ***S. bona-nox*** L. (China briar, SE US to Texas), while ***S. glyciphylla*** Sm. (E Aus.) with ed lvs & fr. (21 mg vitamin C per 100 g equivalent to tomatoes), sold as 'sweet tea' in Sydney markets until 1930s. Some cult. orn. esp. ***S. anceps*** Willd. (*S. kraussiana*, Afr.), ***S. glauca*** Walter (wild sarsaparilla, SE US) & ***S. rotundifolia*** L. (bull briar, b. brier, E US)

Smirnowia Bunge. Leguminosae (III 24). 1 Turkestan: ***S. turkestana*** Bunge – allied to *Sphaerophysa*; alks

Smithanthe Szlach. & Marg. = *Habenaria*

Smithatris Kress & Larsen = *Curcuma* (But see SB 26(2001)226, Novon 13(2003)68)

Smithia Aiton. Leguminosae (III 11). c. 20 OW trop. esp. As. & Madag. ***S. sensitiva*** Aiton (Indomal.) – lvs a potherb., medic. (urinary problems)

Smithiantha Kuntze. Gesneriaceae (II 5b). 7 Mex., Guatemala. R: Gloxinian 20(1970)11. Cult. orn. with creeping scaly rhiz.

Smithiella Dunn = *Pilea*

Smithorchis Tang & Wang = *Platanthera* (but see Gen. Orch. 2(2001)367)

Smithsonia J. Saldanha (*Loxoma*). Orchidaceae (V 16c). 3 Ind. R: JBNHS 71(1974)73

Smitinandia Holttum. Orchidaceae (V 16c). 3 Indomal. R: GBS 25(1969)105

Smodingium E. Meyer ex Sonder. Anacardiaceae (I). 1 S Afr.: ***S. argutum*** E. Meyer ex Sonder. R: Strel. 10(2000)59

smoke bush *Conospermum* spp.; **s. b.** or **tree** *Cotinus coggygria*

Smyrniopsis Boiss. Umbelliferae (III 5). 4 E Medit. to Iran

Smyrnium Tourn. ex L. Umbelliferae (III 5). 8 Eur. (5), Medit. ***S. olusatrum*** L. (alexanders, W Eur., Medit., natur. GB, Netherlands & Bermuda) – aggressive pl. in SW GB, form. used like celery (fl. buds also ed.) & r. like parsnips

Smythea Seemann ex A. Gray. Rhamnaceae (8). 10 SE As. to Polynesia. Endosperm 0

snail flower *Cochliasanthus caracalla*

snake bark *Colubrina arborescens*; **s. bean** *Vigna unguiculata* subsp. *sesquipedalis*; **s. climber** *Bauhinia scandens* var. *anguina*; **s. fruit** *Salacca zalacca*; **s. gourd** *Trichosanthes cucumerina*; **s.'s head** *Fritillaria* spp. esp. *F. meleagris*, *Iris tuberosa*; **s. melon** *Cucumis melo* Flexuosus Group; **s. tongue** *Ophioglossum* spp.; **s. vine** *Hibbertia scandens*; **s. wood** *Brosimum guianense*, *Strychnos minor* etc.

snakeroot, serpentaria or **serpentary** Name used for many pls in US esp. *Actaea racemosa* (black s.) & *Aristolochia serpentaria* (Virginian s.)

snapdragon *Antirrhinum majus*

sneezeweed *Achillea ptarmica*, *Centipeda* spp., *Helenium autumnale*, *Hymenoxys hoopesii*

sneezewood *Ptaeroxylon obliquum*, *Pterocelastrus* spp.

sneezewort *Achillea ptarmica*

snodgollion, snot-goblin (!) or **snottygobble** (!!) *Persoonia* spp.

snow, glory of the *Scilla siehei*; **s. gum** *Eucalyptus niphophila*; **s. on the mountain** *Euphorbia marginata*; **s. pea** = mangetout; **s. plant** *Sarcodea sanguinea*; **s. in summer** *Cerastium tomentosum*

snowball tree *Viburnum opulus* 'Roseum'

snowberry *Symphoricarpos alba*

snowbush *Breynia disticha* 'Roseopicta'

Snowdenia C. Hubb. Gramineae (XXIV 4). 4 trop. E Afr.

Snowdon lily *Gagea serotina*

snowdrop *Galanthus* spp. esp. *G. nivalis*; **Barbados s.** *Habranthus tubispathus*; **s. tree** *Halesia carolina*

snowflake *Leucojum* & *Acis* spp.; **autumn s.** *A. autumnalis*; **spring s.** *L. vernum*; **summer s.** *L. aestivum*

snowy mespilus *Amelanchier* spp. esp. *A. ovalis*

soap tree *Alphitonia excelsa*

soapbark tree *Quillaja saponaria*

soapberry *Sapindus* spp.

soapwood *Caryocar glabrum*

soapwort *Saponaria officinalis*

Soaresia Schultz-Bip. Compositae (Vern.-Chr.). 1 Braz.: ***S. velutina*** Schultz-Bip.

soba (**noodles**) *Fagopyrum esculentum*

Sobennikoffia Schltr. Orchidaceae (V 16d). 4 Mascarenes

Sobolewskia M. Bieb. Cruciferae (47). 4 E Med. (Eur. 1) to Cauc. Some cult. rock-pls

Sobralia Ruíz & Pavón. Orchidaceae (V 2). c. 150 trop. Am. Cult. orn. terr. & epiphytic, usu. bee-poll. orchids (***S. altissima*** D.E. Bennett & Christenson (Peru) to 13.4 m height) with large fls esp. ***S. macrantha*** Lindl. (Mex. to Costa Rica) – fls to 25 cm across lasting 1 day

society garlic *Tulbaghia* spp.

Socotora Balf.f. = *Periploca*

Socotranthus Kuntze = *Cryptolepis*

Socotrella Bruyns & A. Miller (~ *Caralluma*). Apocynaceae (Asclepiadaceae). 1 Socotra: ***S. dolichocnema*** Bruyns & A. Miller. R: Novon 12(2002)330

Socotria Levin = *Punica*

Socratea Karsten. Palmae (V 1). 5 N S Am. R: Britt. 38(1986)55. Beetle-poll. cf. allied *Iriartea*. ***S. exorrhiza*** (Mart.) H. Wendl. (paxiuba palm) – base of trunk with stilt-r., the trunk eventually free of ground such that pl. can 'walk' from under obstacles as it grows, wood used for constr.; ***S. salazarii*** H.E. Moore (Peru) – flagelliform infls rooting at tips

Socratina Balle. Loranthaceae (5 7). 3 Madag. R: Candollea 69(2014)66

Soderstromia C. Morton = *Bouteloua*

Sodiroa André = *Guzmania*

Sodiroella Schltr. = *Telipogon*

Sodom apple *Solanum* spp. esp. *S. aculeatissimum* & *S. linnaeanum*, *Calotropis procera*

Soehrensia Backeb. = *Echinopsis*

Soejatmia Wong. Gramineae (V 4). 1 Malay Pen.: ***S. ridleyi*** (Gamble) Wong – bamboo. R: KB 48(1993)530

Soemmeringia Mart. Leguminosae (III 11). 1 trop. S Am.: ***S. semperflorens*** Mart.

softgrass, creeping *Holcus mollis*

soga *Peltophorum pterocarpum*

Sogerianthe Danser. Loranthaceae (5 3). 45 E NG to Solomon Is. R: AusJB 22(1974)599

Sohnsia Airy Shaw. Gramineae (XXIX 1). 1 Mex.: ***S. filifolia*** (E. Fourn.) Airy Shaw – dioec. R: SCB 87(1997)39

soh-phlong *Flemingia vestita*

soja *Glycine max*

sola pith *Aeschynomene aspera*

Solanaceae Juss. Magnoliidae – Solanales. 88/2650 (well over half in *Solanum*) subcosmop. esp. trop. Am. R: J.G. Hawkes & al. (1979) *The biology & taxonomy of the S.*; A.T. Hunziker (2001) *Genera Solanacearum*. FZ 8(2005)2–3. Shrubs, trees (e.g. *Acnistus*), lianes & herbs (oft. foetid) oft. with branched hairs &/or prickles & alks (esp. nicotine & steroid groups); internal phloem around pith; some dioec. Lvs simple, or lobed to pinnate or 3-foliolate, usu. in spirals; stip. 0. Fls solit. or in app. basically cymose infls; K (5), persistent, C ((4)5(6)), rotate to tubular with lobes usu. plicate (& s.t. convolute) in bud, convolute, imbr. or valvate, s.t. irreg. or even bilabiate, A usu. alt. with C & attached to tube, s.t. 4 (6 in *Goetzea*) or (*Schizanthus*) 2 fert. with staminodes, anthers (with calcium deposits transferred by pollinators – function unclear) oft. connivent, with longit. slits, term. pores or slits, usu. a disk around $\underline{G}$ (2) orientated diagonally in fl. (s.t. 4-loc. or irreg. 3–5-loc. (*Nicandra* etc. to 30- in *Nolana*) or pseudomonomerous with 1 ovule (*Henoonia*)) with (1–) $\pm\infty$ anatr. to hemitropous or amphitropous unitegmic ovules on axile placentas per loc. Fr. a berry (or drupe) or dehiscent (oft. a septicidal capsule); seeds 1–c. 5000 with usu. linear straight to $\pm$ curved embryo in (0 or) usu. oily & proteinaceous (rarely starchy) endosperm. n = 7–12+. Journal: *Solanaceae Newsletter*

Classification & principal genera (but some unplaced incl. *Brunfelsia, Duckeodendron, Nierembergia, Petunia, Schizanthus*):

I. **Goetzeoideae** (trees & shrubs, few-seeded capsules or berries): *Goetzea, Espadaea, Coeloneurum, Henoonia, Metternichia*, trop. Am.; *Tsoala* Madag.!)

II. **Cestroideae** (Browallioideae; K not accrescent, C oft. irreg. capsular fr., embryo straight, n = rarely 12; Am.) – 3 tribes & some unplaced: 1. **Cestreae** (*Cestrum, Sessea*); 2. **Browallieae** (*Browallia*); 3. **Salpiglossideae** (*Salpiglossis*)

III. **Nicotianoideae** (usu. shrubs & herbs; A 4 + staminode) – 2 tribes: 1. **Anthocercideae** (Aus.; *Anthocercis, Duboisia*); 2. **Nicotianieae** (*Nicotiana* only)

IV. **Solanoideae** (incl. Nolanaceae; small trees to herbs, K accrescent, C reg., fr. usu. baccate, seeds flat, n = 12) – 7 tribes but some unplaced e.g. *Jaborosa, Mandragora, Nicandra, Nolana, Solandra*: 1. **Lycieae** (*Lycium* only); 2. **Hyoscyameae** (*Atropa,*

Hyoscyamus, Scopolia); 3. **Juanulloeae** (*Juanulloa, Markea*); 4. **Datureae** (*Brugmansia, Datura*); 5. **Physalideae** (some unplaced; a. Withaniinae – *Withania*, b. Iochrominae – *Iochroma*, c. Physalidinae – *Physalis, Witheringia*; 6. **Capsiceae** (*Capsicum, Lycianthes*); 7. **Solaneae** (*Jaltomata, Solanum*)

1 of only 10 fams with elaiophores attractive to bees (*Nierembergia*).

Subfam. IV. provides many ed. fr. (*Capsicum* (peppers), *Physalis, Solanum* (tomato, tree-tomato etc., also ed. tubers = potatoes)). Many yield medic. or hallucinogenic alks, s.t. fish-poisons or insecticidal (*Atropa, Brugmansia, Cestrum, Brunfelsia, Datura, Duboisia, Latua, Mandragora, Nicotiana* (tobacco), *Solandra, Withania*, atropine being the standard antidote to some systemic herbicides) while many of these genera, *Browallia, Cestrum, Fabiana, Juanulloa, Lycianthes, Lycium, Nicandra, Nierembergia, Petunia, Salpiglossis, Schizanthus, Solandra* incl. cult. orn. spp. Some noxious weeds esp. spp. of *Physalis, Solanum*

Solandra Sw. Solanaceae. (IV). 10 trop. Am. R: NJB 7(1987)639. Alks – source of sacred hallucinogens in Mex. Cult. orn. shrubs & lianes with showy bat-poll. fls fragrant at night esp. ***S. maxima*** (Sessé & Moçiño) P. Green (*S. hartwegii*, '*S. guttata*', chalice vine, Mex. to Colombia, Venez.) – glabrous liane with yellow fls to 24 cm long, hybrid with *S. grandiflora* Sw. (Carib.) = '*S.* × *nairobiensis*'

Solanecio (Schultz-Bip). Walp. (~ *Kleinia*). Compositae (Sen.-Sen.). 16 Yemen, trop. Afr. Madag. R: KB 41(1986)920. Climbers, herbs & pachycaul trees incl. ***S. gigas*** (Vatke) C. Jeffrey (*Senecio g.*, Ethiopia) – hedging in Ethiopia, & ***S. mannii*** (Hook.f.) C. Jeffrey (*Senecio m.*, trop. Afr.) – fast-growing, candelabriform (Leeuwenberg's Model), used for hedging in Kenya, local medic. Some potherbs (Afr.)

Solanoa Greene = *Asclepias*

Solanopteris Copel. = *Microgramma*

Solanum Tourn. ex L. Solanaceae (IV 7). Incl. *Cyphomandra* (32 trop. Am. – R: FN 63(1994)42), *Lycopersicon* (= sect. *L.*, W S Am.: 13 – R: SB 30(2005)425), *Normania*, c. 1400 subcosmop. (Eur. 3 + many natur.; Afr. + Madag. 106, Am. c. 850 – R: M. Nee & al., Solanaceae IV (1999)292)) esp. warm (orig. Am.). R: R.G. van den Berg & al., Solanaceae V(2001)40; AMBG 59(1972)274, (subg. *Leptostemonum*) GH 12(1984)179, (sect. *S.*) J.M. Edmonds & J.A. Chweya (1997) *Black nightshades*; (sect. *Petota*; 107 + 4 cultigens warm Am.) BR 80(2014)286. See also *Lycianthes*. Trees (Leeuwenberg's Model), shrubs & herbs, s.t. lianoid, oft. prickly (incl. ***S. monachophyllum*** Dunal (*S. sacupanense*, NE S Am.) – only known thorny rheophyte) with prickliness assoc. with marsupial (esp. wallabies) grazing in Aus., from coasts to upland forest & semi-desert. Many alks so oft. toxic though fr. of many cultigens ed.; some with extrafl. nectaries visited by ants; 'buzz'-poll. fls, fr. disp. by animals, e.g. ***S. lycocarpum*** St.-Hil. (Braz.) – principal food of maned wolf in cerrado, but usu. birds. V. imp. food-pls (many cultigens), cult. orn. & weeds, presence oft. indicating overgrazing (e.g. ***S. incanum*** L. (carcinogenic *N*-nitrosodimethylamine in fr.) in S Arabia). ***S. aculeatissimum*** Jacq. (Sodom apple (name used for other spp. incl. ***S. linnaeanum*** Hepper & Jaeger ('*S. ciliatum*', *S. hermannii*, S Afr.)), warm Am., invasive Aus., SE US, Hawaii) – sliced fr. consumed by cockroaches (fatal) in Puerto Rico, medic.; ***S. aethiopicum*** L. (*S. gilo*, '*S. indicum*', bitterberry, trop. Afr., prickly 'wild' forms called *S. anguivi* Lam.) – bitter pea-sized imp. ed. fr. & lvs Afr., ed. curried in S As. once toxic seeds (alks) removed, medic.; ***S. americanum*** Mill. (trop. & warm) – locally cult. for ed. lvs & fr.; ***S. berthaultii*** J. Hawkes (Bolivia) – traps small insects & mites in hairs & emits pheromone mimicking scent of aphids in distress; ***S. betaceum*** Cav. (*Cyphomandra b., C. crassifolia*, tree-tomato, tamarillo, Bolivia or NW Arg.) – long cult. Peru, & now widespread, for egg-shaped red fr. used for jelly etc., also ed. raw; ***S. cajanumense*** Kunth (*C. c.*, '*C. hartwegii*', casana, chambala, N Andes) – peach- & passionfr.-flavoured fr. crop being tried in NZ; ***S. campylacanthum*** Hochst. ex A. Rich. (Afr.) – imp. medic., common in disturbed habitats to 2000m; ***S. capsicoides*** All. (*S. ciliatum*, cockroach berry, E Braz.) – pantrop. weed; ***S. centrale*** J. Black (bush tomato, kutjera, arid Aus.) – fr. fresh or dried an Aboriginal staple (also other spp.), 'planted' along A. walking tracks, now comm., used in chutneys; ***S. cinereum*** R. Br. (Narrawa burr, SE Aus.) – noxious weed; ***S. dulcamara*** L. (bittersweet [chewed stems firstly bitter, then sweet], woody nightshade, Euras., natur. N Am.) – mediaeval symbol of fidelity (also of Jesus Christ), locally medic. (e.g. chillblain ointment from preserved berries in Cotswolds, UK), twigs form. chewed by Cumberland (UK) schoolboys as tobacco subs.; ***S. erianthum*** D. Don. ('*S. verbascifolium*', trop. Am., widely natur.) – lvs used for washing clothes in Amaz. but causes skin rash; ***S. jamesii*** Torrey (W N Am.) – tubers ed. Native Americans; ***S. laciniatum*** Aiton (~ *S. aviculare* Forst. f. (distinct sp.), kangaroo apple, NG, Aus., NZ) – cult. in E Eur. & Hawaii for lvs with 1–2% solasodine, steroidal alk. like diosgenin used in contraceptive pills, ripe fr. ed. Aboriginal people, poisonous when green; ***S. laxum*** Spreng. (*S. jasminoides*, potato vine, Braz., invasive NZ) – cult. orn. hardy liane; ***S. lycopersicum*** L. (*Lycopersicon l., L. esculentum*, tomato, love-apple [orig. Ital. *pomi d'oro* = golden apple 1544 (early ones in Eur. poss. orange-yellow, red colour due to lycopene, allied to carotene), then Span. *pome dei Moro* = Moor's a., then Fr. *pomme d'amour*], *S. pimpinellifolium* L. (*L. p.*, currant tomato, tomberry, coastal Peru) being 'wild' forms, fr. 1 cm diam.) – solanine in lvs interfering with acetylcholinesterase activity (nerve function) so insecticidal, taken from Peru (where no depictions of or word for it) to Mex. ('tomatl'), orig. weed of Aztec maize & bean fields, shorter stigmas selected so selfing easier, with increase in locule no. (*fas* gene) selected, fr. first cooked Eur., esp. pasta sauce, now comm. t. (26% world prod. China, 6% Italy) eaten as salad veg. (multilocular 'beefsteaks' 120 – 450 g, UK record – 2.537 kg), cooked & tinned (esp. irreg. 'Italian' ts), puree (anti-oxidant lycopene in paste allegedly reducing by 40% incidence of prostate & other cancers, though questioned by some research) or ketchup, soup (first tinned by Joseph Campbell, New Jersey, 1869) or juice, or green in chutney & pickle, seeds yielding a cooking oil also used in soap with resultant cake as cattle-food, some 7500 (500 comm.) cvs incl. fashionable small 'cherry' ts ('var. *cerasiforme*') & kumato ('**Olmeca**'), seeds passing through herbivores incl. humans & germ. in dung, sewage farms etc., naturally hybridizing with ***S. galapagense*** S. Darwin & Peralta & ***S. cheesemaniae*** (L. Riley) Fosb. (Galápagos) – seeds germ. after passage through giant tortoises, these used to introd. genes promoting ease of mechanized picking and salt-tolerance, '**Flavr Savr**' (1987) first GM food with ripening gene turned off, so remaining firm after harvesting; ***S. macrocarpon*** L. (W Afr.) – cult. Afr. & As. for orange-yellow fr., lvs potherb; ***S. mammosum*** L. (macaw bush, nipple fr., trop. Am.) – molluscicidal glycoalks, shrub cult. for orn. but poison fr. (used against cockroaches & rats), dispersers unknown; ***S. mauritianum*** Scop. (S Am.) – invasive Azores, Aus., trop. & S Afr.; ***S. melongena*** L. (aubergine, egg pl., Jew's apple, brinjal, badinjan, nasubi, trop. OW– first domesticated SE As., but orig. Afr.) – fr. an imp. veg. (31 M t a yr; c. 1500 cvs, round pale forms = 'Thai e.-p.'), ess. ingredient of moussaka, but in drought conditions increased alks leading to nightmares in eaters; ***S. muricatum*** Aiton (pepino, melon pear, tree m., Andes, cultigen, hybrid or poss. wild forms = *S. basendopogon* Bitter, *S. caripense* Dunal or *S. tabanoense* Correll, C Am.) – fr. eaten fresh; ***S. nigrum*** L. (black nightshade, Euras., widely natur. (neophyte in Eur.?)), – poss. trigenomic hexaploid, some forms used as potherbs, some with ed. fr. used in pies; ***S. phlomoides*** Cunn. ex Benth. (NW Aus.) – pericarp imp. part of Aboriginal diet in W Aus.; ***S. pseudocapsicum*** L. (*S. capsicastrum, S. diflorum*, Jerusalem cherry, winter c., SE temp. S Am., natur. Madeira since 1650, now widely natur., invasive Aus.) – cult. pot-pl. for orn. poison fr. in winter ('winter cherry') widely natur. & self-sown on London's pavements, several cvs incl. varieg.; ***S. quitoense*** Lam. (naranjilla, lulo, Quito orange, Andes) – cult. orn. & for fr. juice, hybrid with *S. sessiliforum* cult. for fr. Ecuador; ***S. rostratum*** Dunal (Kansas thistle, buffalo bur, SW N Am.) – pestilential weed in N Am.; ***S. scabrum*** Mill. (*S. burbankii, S. intrusum, S. melanocerasum*, huckleberry, sun berry, wonderberry, Afr. hexaploid cultigen poss. involving tetraploid *S. retroflexum* Dunal) – ann. cult. (esp. W Afr.) for ed. fr. resembling blackcurrants (also source of ink) & lvs; ***S. sessiliflorum*** Dunal (*S. hyporhodium, S. topiro*, cocoña, peach tomato, Upper Amaz.) – cult. orn. & comm. for fr. juice; ***S. sisymbriifolium*** Lam. (Arg.) – 'trap crop' for potato cyst nematode (eelworm); ***S. tampicense*** Dunal (Mex.) – invasive SE US; ***S. tarderemotum*** Bitter (trop. & S Afr.) – tetraploid ally of *S. nigrum*, potherb; ***S. terminale*** Forssk. (Afr.) – local medic.; ***S. torvum*** Sw. (pea egg pl., susumber, trop. As., noxious weed US) – unripe fr. used with curry, comm. source of solasodine; ***S. trilobatum*** L. (trop. As.) – imp. medic. pl. Ind.; ***S. trisectum*** Dunal (*Normania triphyllum*, Madeira) – thought extinct since 1855, 1 pl. found 1991, now prop. by seed; ***S. tuberosum*** L. (potato, anc. cultigen (known to Spanish by 1537, established Canary Is. 1562), derived from pls domesticated c. 8000 BP (tuber alks bred out), prob. arising at c. 3500–4000 m from diploid *S. brevicaule* Bitter complex (esp. *S. candolleanum* Berthault (*S. bukasovii, S. canasense*) in S Peru, giving diploid to pentaploid forms incl. autotetraploid *S. tuberosum* Andigenum Group (W Venez. to N Arg. landraces, *S. stenotomum*; frost resistance from high alt. pops), prob. also with introgression from *S. berthaultii* J. Hawkes (Bolivia & Arg.) etc., giving *S. t.* Chilotanum Group (Chilean landraces) introd. from Chile 1840s; L. Zuckerman (1999) *The potato*) – world's 4th (poss. 3rd) crop (c. 4000 cvs; 376 M t. a yr by 2013 – biggest prod. As., worth $100 billion a yr by 2000, though long assoc. with so-called 'lower classes'; *Phytophthora infestans* (potato blight (an oomycete),

Mex., prob. arr. Eur. via Antwerp 1840s) causing Great Famine of 1845–1852 when 1–1.5 M in Ireland died, 1 M migrated) after wheat, rice & maize, many cvs (for history see NP 94(1983)479) prop. vegetatively from tubers (UK tuber record 3.2 kg) developing from underground runners from leaf-axils (hence value of ridging potatoes, where more axils covered) or from their 'eyes', i.e. buds in axils of rudimentary lvs, first grown in Eur. in 16 cent. (Ireland in 1566), though orig. short-day tuber-forming introductions now scarce & replaced by day-neutral ones, & providing all nutrients needed by humans exc. Ca, vitamins A & D (so with milk a complete diet), now providing up to 5% of protein, 7–10% of iron & riboflavin & 25% of vitamin C in Europeans' diet (Britons eat over 100 kg each per annum (1995 – Irish more than Br., e.g. p. bread incl. boxty) though overconsumption (of solanine) by pregnant women alleged to promote spina bifida in babies), eaten boiled (**'International Kidney'**, (Jersey Royal) c. 1879, Jersey, Channel Is., fashionable, though virus-ridden, 'new p.' – 30–40 K t a yr) or fried (chips, a Fr. innovation introd. c. 1870 (now e.g. '**Russet Burbank**' ('Idaho Baker') raised by Luther Burbank & used for McDonald's, Kentucky Fried Chicken & Burger King F. fries (UK chips) worldwide, though '**Maris Piper**' favoured in Br. for nat. dish, 'fish & chips', record chip 24.5 cm long from '**Amandine**'), crisps (US chips invented C19) esp. '**Record**'), now GM so as not to turn brown when cut or bruised, so lowering costs of making crisps & chips, powdered when dried for 'instant' potato, freeze-dried chuño stored for up to 10 yrs, also comm. source of starch (as used in biodegradable crockery & cutlery) & alcohol (e.g. Polish [and UK] vodka) & GM '**Amflora**' (Germany; 2013 first of all GM pls comm. in Eur.) for amylopectin prod. (for glossy paper etc.), now GM also for oral vaccine for hepatitis B; ***S. viarum*** Dunal (Braz., SE S Am.) – comm. source of solasodine, intr. Ind., invasive SE US; ***S. villosum*** Mill. (~ *S. nigrum*, orig. S Eur.) – tetraploid now widespread (tobacco weed in Aus.), local salad & ed. fr., medic. etc.; ***S. viride*** Sol. ex Forst.f. (*S. uporo*, Melanesia to Polynesia) '**Anthropophagorum**' (*S. a.*, cannibal's tomato) – tomato-like ed. fr. used in cerem., basis of 'Cannibal Chutney' (Fijian gimmick)

Solaria Philippi. Amaryllidaceae (Alliaceae-Gill.). 5 S Chile, S Arg.

Soldanella Tourn. ex L. Primulaceae. 16 C & S Eur. mts. R: NJB 22(2002)136. Rock-pls oft. growing up through & fls in snow, cult. orn. incl. hybrids

Soleirolia Gaudich. Urticaceae (IV). 1 W Medit. Is., Italy: ***S. soleirolii*** (Req.) Dandy (*Helxine s.*, helxine, mind-your-own-business, mother-of-thousands) – cult. orn. creeping pl. with pale green tiny lvs used in greenhouses etc.

Solena Lour. Cucurbitaceae (XIV). 3 trop. As. to Mal. R: Blumea 49(2004)70. ***S. amplexicaulis*** (Lam.) Gandhi (S Ind.) – lvs, r. & fr. ed, r. & seeds sources of purgatives & stimulants

Solenachne Steud. = *Sporobolus*

Solenangis Schltr. Orchidaceae (V 16d). 8 trop. Afr. to Masc. ***S. aphylla*** Schltr. leafless, others not & cult. orn.

Solenanthus Ledeb. = *Cynoglossum*

Solenidiopsis Senghas = *Oncidium* (but see Selbyana 20(1999)7, 23(2002)197)

Solenidium Lindl. Orchidaceae (V 12h). 3 N S Am.

Solenixora Baill. = *Coffea*

Solenocarpus Wight & Arn. (~ *Spondias*). Anacardiaceae (II). 1 S Ind., 1 Mal.

Solenocentrum Schltr. Orchidaceae (IV 2b). 4 trop. Am.

Solenogyne Cass. Compositae (Ast.-Lag.). 3 Aus. R: Brunonia 2(1979)43

Solenomelus Miers (~ *Sisyrinchium*). Iridaceae (VII 3). 2 Chile, Arg. Cult. orn. like *Sisyrinchium* but with P-tube; seeds half-winged (cf. *Gladiolus*)

Solenophora Benth. Gesneriaceae (II 5b). 16 C Am. R: HPB 7(2002)41. Some trees to 12 m tall with boles to 30 cm diam.

Solenopsis C. Presl (*Laurentia*). Campanulaceae (III 1). 6 Medit., Canary Is. R: PSE 200(1998)216 [Aus. spp. = *Isotoma*, S Afr. spp. = *Wimmerella* (though 1 long-presumed S Afr. sp. truly from Balearics = ***S. minima*** (Sims) Crespo & al. (*S. balearica*)), Am. spp. = *Palmerella, Porterella*]

Solenoruellia Baill. = *Henrya*

Solenospermum Zoll. = *Lophopetalum*

Solenostemma Hayne. Apocynaceae (V c; Asclepiadaceae III 1). 1 Egypt, Arabia: ***S. argel*** (Del.) Hayne

Solenostemon Thonn. = *Coleus*

Solfia Rech. = *Balaka*

Solidago Vaill. ex L. Compositae (Ast.-Sol.). Incl. *Oligoneuron* c. 100 N Am. (77 – R: FNA 20(2006)107), S Am. (1), Macaronesia & Euras. (Eur. 1: *S. **virgaurea*** L. (Euras.) – form. medic. (incl. for wounds), diuretic action due to flavonol glycosides & saponins). R: Phytol. 75(1993)3. Cult. orn. (goldenrod) & locally medic. esp. ***S. canadensis*** L. (N Am.) – commonest garden sp. in Eur. & hybrids between it & *S. virgaurea* but oft. weedy & becoming natur. (invasive Euras.); ***S. gigantea*** Aiton (N Am.) – invasive C & E Eur.; ***S. × lutea*** (Dress) Brouillet & Semple (× *Solidaster luteus*, solidaster, ? *Solidago canadensis* × *S. ptarmicoides* (Torrey & A. Gray) Boivin (*Oligoneuron album*, N Am.) – spontaneous hybrid found 1910 in a Lyon nursery, now pop. cut-fl. esp. Netherlands; ***S. leavenworthii*** Torrey & A. Gray (N Am.) – in 1920s Thomas Edison selected forms to 4 m with 12% rubber; ***S. odorata*** Aiton (N Am.) – State herb Delaware, local medic., exported as tea 19 cent. See also *Euthamia*

solidaster *Solidago × lutea*

× **Solidaster** Wehrh. Compositae. *Solidago × Aster* spp. (in cult. actually *Solidago × lutea*)

Solitaria (McNeill) Saleghian & Zarre (~ *Minuartia*). Caryophyllaceae. 7 Himal.

Soliva Ruíz & Pavón. Compositae (Anth.-Cot.). (Incl. *Gymnostyles*) 8 S Am. R: NMLP 14 Bot 70(1949)123. ***S. anthemifolia*** (Juss.) Sweet & others esp. ***S. sessilis*** Ruíz & Pavón (*S. pterosperma*, bindyi, bindy-eye, jo-jo weed) – troublesome weeds natur. in Aus., Calif. etc. in lawns etc.

Sollya Lindl. = *Billardiera*

Solmsia Baill. Thymelaeaceae (Oct.). 2 New Caled.

Solms-laubachia Muschler. Cruciferae (26). Incl. *Desideria*, 33 C As. & Himal. to China. R: AMBG 95(2008)530. ***S. himalayensis*** (Camb.) J.P. Yue & al. (*Christolea h.*, *D. h.*, Himal.) – with *Ranunculus lobatus* the pl. reaching highest alt. (7756 m)

Solomon nut oil *Canarium indicum*; **S.'s seal** *Polygonatum* spp.

Solonia Urb. Primulaceae (Myrsinaceae). 1 E Cuba: ***S. reflexa*** Urb.

Solori Adans. (*Brachypterum*; ~ *Derris*). Leguminosae (III 16). 12 Indomal. to Aus. R: T 63(2014)532

soma *Ephedra* spp. (?), *Cynanchum sarcomedium*; name used by Aldous Huxley in *Brave New World* (1932)

Somalluma Plowes = *Caralluma*

Sommera Schldl. Rubiaceae (Cond.). 10 trop. Am.

Sommerfeltia Less. Compositae (Ast.-Pod.). 2 S Am.: R: Phytol. 76(1994)104

Sommieria Becc. Palmae (V 13). 1 NG: ***S. leucophylla*** Becc. R: KB 57(2002)606

Somphoxylon Eichler = *Odontocarya*

sompong *Tetrameles nudiflora*

Somrania Middleton (~ *Loxocarpus*, *Damrongia*). Gesneriaceae (III 2i). 3 Thailand (limestone). R: GBS 65(2013)181, T 65(2016)288

somura *Trachyspermum roxburghianum*

Sonchella Sennikov. Compositae (Cich.-Cich.(-Crep.)). 2 C As. to China

Sonchus Tourn. ex L. Compositae (Cich.-Cich.(-Hyos.)). Incl. *Actites*, *Aetheorhiza*, *Dendroseris*, *Embergeria*, *Kirkianella*, *Thamnoseris*, c. 80 Euras. (Eur. 9) to Australasia, trop. Afr. (30 Macaronesia, incl. subg. *Dendrosonchus*, pachycaul shrubs & treelets) & Am. (Chile, 11 Juan Fernàndez, palm-like pachycauls (Holttum's & Leeuwenberg's Models, incl. ***S. sinuatus*** S.C. Kim & Mejías (*D. macrantha*), extinct 1980, & ***S. brassicifolius*** S.C. Kim & Mejías (*D. litoralis*), reduced to 3 wild pls in 1980s but now street-tree in Chile & cult. elsewhere). R: BN 125(1972)287, 126(1973)155, 127(1974)7, 407. Diploids to octoploids. Herbs = milk thistles; young lvs ed as salad (Pliny had Theseus eating them before going to kill Minotaur); many weedy spp. now subcosmop. esp. ***S. oleraceus*** L. (sowt., Euras., Medit., invasive Aus.). ***S. crassifolius*** Pourr. ex Willd. (C Spain) – young shoots local veg.; ***S. grandifolius*** Kirk (NZ) – lvs to 1 m × 20 cm

soncoya *Annona purpurea*

Sonderina H. Wolff (~ *Stoibrax*). Umbelliferae (III 8). 4 S Afr. R: EJB 48(1991)248

Sonderothamnus R. Dahlgren. Crypteroniaceae (Penaeaceae). 2 W Cape. R: Strel. 9(2000)561; OB 18(1968)41

Sondottia Short. Compositae (Gnap.-Ang.). 2 Aus. R: Muelleria 7(1989)113

Sonerila Roxb. Melastomataceae. c. 175 trop. As. Herbs & shrubs with 3-merous fls (only M. thus). Local medic.; some cult. orn. esp. ***S. margaritacea*** Lindl. (Myanmar to Java) – herb with lvs bearing rows of puckered pearly spots between veins, & hybrids with *Bertolonia* spp.

Sonnea Greene = *Plagiobothrys*

Sonneratia L.f. Lythraceae (Sonneratiaceae). 7 mangroves of Ind. & Pac. Oceans. R: APS 23(1985)313. Trees (Attims's & Rauh's Models) with aerial 'breathing' r. arising from ordinary ones, their aerenchyma used as cork subs. for fishermen's floats etc. when boiled; fls ephemeral, foetid, nocturnal, those of ***S. caseolaris*** (L.) Engl. (Indomal., Attims's Model) visited by birds, bats & moths, lvs & fr. ed.

Sonneratiaceae Engl. = Lythraceae

Sooia Pócs = *Epiclastopelma*

Sophiopsis O. Schulz = *Smelowskia*

Sophora L. Leguminosae (III 2). Excl. *Ammothamnus, Calia, Styphnolobium,* c. 50 trop. & mostly N temp. (Eur. 2; NZ 5 – R: NZJB 39(2001)35). R: APS 19(1981)1, 143. Alks. ***S. nuttalliana*** B.L. Turner (*S. sericea,* N Am.) – r. chewed for sweet taste; ***S. tetraptera*** J.S. Muell. (kowhai, NZ (N Is.)) – useful timber; ***S. toromiro*** Skottsb. (Easter Is.) – only pl. on Easter Is. suitable for building, carving etc., reduced to 1 tree by 1917 & exterminated by grazing by 1962 but seeds grown from herbarium spec. coll. Thor Heyerdahl so that sp. survives in at least 18 botanic gardens & being re-introd. to E. Is. via Kew & Melbourne

Sophronanthe Benth. (~ *Gratiola*). Plantaginaceae (Grat.-Grat.). 1 SE US: ***S. hispida*** Benth.

Sophronitella Schltr. = *Isabelia*

Sophronitis Lindl. = *Cattleya* (but see C. Withner (1993) *The Cattleyas* III: 57)

Sopubia Buch.-Ham. ex D. Don. Orobanchaceae (Buch.; Scrophulariaceae s.l.). c. 30 OW trop. esp. Afr. (Indomal. 1, Madag. 4)

Soranthus Ledeb. = *Ferula*

Sorbaria (Seringe) A. Braun. Rosaceae (Ros.-Sorb.). 4 E As. R: NJB 8(1989)557. Cult. orn. decid. shrubs, esp. ***S. sorbifolia*** (L.) A. Braun – stout to 2 m

× **Sorbopyrus** C. Schneider. Rosaceae (Ros.-Mal.). *Sorbus* × *Pyrus* hybrids. × ***S. auricularis*** (Knoop) C. Schneider (× *S. pollveria, Sorbus aria* × *P. communis,* Bollwiller pear) – mostly clonal, orig. Alsace end C16, fr. ed.

Sorbus Tourn. ex L. Rosaceae (Ros.-Mal.). 135 N temp. 4 subg. (treated as genera by some): subg. *Aria* (*A.* (Pers.) Host; incl. *Chamaemespilus* & *Micromeles;* R: SBM 69(2004)42) – lvs simple (39, Euras., N Afr.), subg. *Cormus* (*C.* Spach) – lvs pinnate (*S. domestica*), subg. *Torminaria* (*Torminalis* Medik.; R: SBM 69(2004)123) – lvs simple-lobed (3 incl. *S. torminalis*) & subg. *S.* (*Sorbus* s.s.) – lvs pinnate (92 N temp.). Trees & shrubs to shrublets a few cm tall with simple or pinnate lvs; fls usu. foetid (trimethylamine like privet); many apomicts (41 in W Eur. alone incl. ***S. latifolia*** (Lam.) Pers. (2n = 51) poss. polytopically derived from *S. torminalis* & *S. aria* group (2n = 34)). Cult. orn. (esp. for brightly coloured frs e.g. '**Joseph Rock**' (form of an undescribed sp. from Yunnan)) & some ed. fr. etc. ***S. americana*** Marshall (roundwood, C & E US) – fr. locally medic.; ***S. aria*** (L.) Crantz (whitebeam, Eur., N Afr.) – fr. used in brandy & vinegar etc.; ***S. aucuparia*** L. (rowan, mt. ash, quickbeam, Eur., SW As.) – fr. used in jellies, brandy etc., against scurvy in C17 Wales, ed cvs selected Russia, though allegedly carcinogenic, wood for turnery etc., assoc. with superstitions Scotland (& NZ, introd. 1904), hybridizes with *Aronia* spp. in N. Am.; ***S. domestica*** L. (Whitty pear, S Wales (?), S Eur., N Afr., SW As.) – timber useful, bark for tanning leather, fr. ed. when bletted like a medlar, the tree grown for this in F. & near Genoa, flavour in Ger. apple wine, nr Frankfurt fermented as additive in a kind of cider & distilled as a spirit; ***S. intermedia*** (Ehrh.) Pers. (Swedish whitebeam, Eur.) – cult. orn.; ***S. torminalis*** (L.) Crantz ((wild) service tree, Eur., Medit. to Iran) – fr. used like that of *S. domestica,* Neolithic staple, in drink called checkers or chequers (so pubs displaying a chessboard prob. solecism)

Soredium Miers ex Henfrey = *Peltophyllum*

Sorengia Zuloaga & Morrone = *Coleataenia*

Sorghastrum Nash. Gramineae (XXII 5). 21 Afr. (2), trop. & warm Am. (15; N Am. 4 native). ***S. nutans*** (L.) Nash (*S. avenaceum,* Canada to Mex.) – imp. forage grass, 1 of only 4 principal grass spp. of former tall-grass prairie

Sorghum Moench. Gramineae (XXII 5). Excl. *Sarga,* 8 OW trop. R (sect. S.): AJB 65(1978)477. Guinea corn, kaolang (China), sorghum esp. cvs of ***S. bicolor*** (L.) Moench (great millet, imphee) domesticated in Sudan c. 3000 BP (ground seeds found in Stone Age [100 000 yrs ago] Mozambique cave, poss. foliage used for bedding or tinder), the wild progenitor prob. '*S. arundinaceum* (Desv.) Stapf' & now (65 M t by 2010) world's 4th most imp. cereal after wheat, rice & maize, a staple in Afr., Ind. & China, thriving under drier conditions than does maize, game-cover plots in S Br., poll. coll. by bees; 8 major groups of cvs grown for grain (non-saccharine sorghums), sweet juice or forage (saccharine s.) & brush manufacture (broom corns), incl. **Caudatum Group** (feterita) – large white, yellow or red grains, **Durra Group** (durra, dari, jowar) – principal grain sorghum of Afr. & Ind., **Saccharatum Group** – culms to 4 m with juicy sweet pith, cult. for syrup, forage & silage in US, sorghum wine (kaoliang) in Taiwan, **Subglabrescens Group** (milo) – imp. grain in C N Am., **Technicum Group** (Italian, Florence or Venetian whisk, broom corn) – principal source of domestic brooms & brushes; backcrosses with '*S. arundinaceum*' gave '*S.* × *drummondii* (Steud.) Millsp. & Chase' incl. Sudan grass (*S. sudanense* (Piper) Stapf), cult. for forage. ***S.* × *almum*** Parodi (raised Arg.) – introd. Aus. as forage but weedy; ***S. halepense*** (L.) Pers. (Johnson grass, Medit.) – widely natur. fodder pl., oft. weedy, oft. cyanogenic, hybridizes with *S. bicolor*; ***S. macrospermum*** Garber (*Vacoparis m.*, N Aus.) – Aboriginal food-pl.

Soridium Miers. Triuridaceae. 1 N S Am.: ***S. spruceanum*** Miers. R: BB 140(1991)10

Sorindeia Thouars. Anacardiaceae (I). 9 trop. Afr., Madag., Masc. (Am. spp. = *Mauria*). R: Adansonia III, 25(2003)94. Locally used timber & ed. fr. pulp incl. ***S. grandifolia*** Engl. (*S. warneckei*, W Afr.) – twigs used as (sweet-tasting) chewing-sticks in Nigeria

Sorocea A. St-Hil. Moraceae (II?). c. 24 trop. Am. R: PKNAW, C 88(1985)381, FN 83(2001)92

Sorocephalus R. Br. = *Spatalla*

Sorolepidium Christ = *Polystichum*

Soromanes Fée = *Polybotrya*

Soroseris Stebb. Compositae (Lact.-Crep.). 9 Himal. to W China. R: APS 31(1993)444

Sorostachys Steud. = *Cyperus*

Soroveta Linder & C. Hardy (~ *Restio*). Restionaceae. 1 Cape: ***S. ambigua*** (Masters) Linder & C. Hardy. R: Bothalia 40(2010)6. 'Sister' to rest of fam.

sorrel *Rumex acetosa*; **French s.** *R. scutatus*; **red s.** *Hibiscus sabdariffa*; **sheep's s.** *R. acetosella*; **tree s.** *Oxydendrum arboreum*; **wood s.** *Oxalis acetosella*

sorva *Couma macrocarpa*

Sosnovskya Takht. = *Centaurea*

Soterosanthus Lehm. ex Jenny. Orchidaceae (V 12i). 1 trop. Am.: ***S. shepheardii*** (Rolfe) Jenny. R: Gen. Orch. 5(2009)442

Sotoa Salazar (~ *Deiregyne*). Orchidaceae (IV 2h). 1 S US & Mex.: ***S. confusa*** (Garay) Salazar. R: Lankesteriana 9(2010)501

sotol *Dasylirion* spp.

souchong *Camellia sinensis*

soufrière plant *Spachea elegans*

Soulamea Lam. Simaroubaceae. Excl. *Amaroria*, 13 Mahé (Seychelles, 1), Mal. (1), New Caled. (11). ***S. amara*** Lam. (Mal.) – locally medic. (fevers)

Souliea Franch. = *Actaea*

souphlong *Flemingia vestita*

sour cherry *Prunus cerasus*; **s.-figs** *Carpobrotus* spp.; **s. orange** *Citrus* × *aurantium*; **s.sob** *Oxalis pes-caprae*; **s.sop** *Annona muricata*; **s. wood** *Oxydendrum arboreum*

Souroubea Aubl. Marcgraviaceae. 19 Mex. to Bolivia (not WI). G 5-loc. Local medic.

southern pine *Pinus echinata, P. elliottii, P. palustris, P. taeda*

southernwood *Artemisia abrotanum*

Southland beech *Nothofagus menziesii*

sow bread *Cyclamen hederifolium*; **s. thistle** *Sonchus oleraceus*; **s.t., alpine** *Cicerbita alpina*

Sowerbaea Sm. Asparagaceae (Lomandraceae). 5 Aus. R: FA 45(1987)264. Form. referred to Anthericaceae, Liliaceae

soy or soya bean *Glycine max*

Soyauxia Oliv. Peridiscaceae. 9 trop. W Afr.

Soymida A. Juss. Meliaceae (I). 1 Ind., (? natur. Sri Lanka): ***S. febrifuga*** (Roxb.) A. Juss. (Ind. redwood, bastard cedar) – locally medic., bark a fibre-source & tanning, timber for building, frs used in pot-pourri as 'lily flowers'

Spachea A. Juss. Malpighiaceae. 10 trop. Am. At least 1 sp. functionally dioec. ***S. elegans*** (G.Meyer) A. Juss. (soufrière pl., of St Vincent) – long thought exterminated by volcano but actually native in Guyana

spaghetti *Triticum turgidum* Durum Group; **vegetable s.** *Cucurbita pepo*

Spananthe Jacq. Umbelliferae (Azor.). 1 Andes: ***S. paniculata*** Jacq. – ann. to 2 m tall

Spaniopappus Robinson. Compositae (Eup.-Oxy.). 5 Cuba. R: MSBMBG 22(1987)444

Spanish bayonet *Yucca aloifolia*; **S. bluebell** *Hyacinthoides hispanica*; **S. broom** *Spartium junceum*; **S. cane** *Arundo donax*; **S. chestnut** *Castanea sativa*; **S. elm** *Cordia* spp. esp. *C. gerascanthus*; **S. iris** *Iris xiphium*; **S. jasmine** *Jasminum grandiflorum*; **S. moss** *Tillandsia* *usneoides*; **S. nut** *Moraea sisyrinchium*; **S. reed** *Arundo donax*; **S. sage** *Coleus amboinicus, Salvia lavandulifolia*; **S. senna** *Senna italica*; **S. shawl** *Heterocentron elegans*

Sparattanthelium Mart. Hernandiaceae (Gyrocarpaceae). 13 trop. S Am. R: BJ 89(1969)193. Liane with recurved stem-hooks

Sparattosperma Mart. ex Meissn. Bignoniaceae (1). 2 trop. S Am. R: FN 25,2(1992)115. Cult. orn. trees. ***S. leucanthum*** (Vell.) K. Schum. – source of brown dye to stain cotton

Sparattosyce Bur. Moraceae (II). 1–2 New Caled.

Sparaxis Ker-Gawler. Iridaceae (VI 5). Incl. *Synnotia*, 16 SW Cape & W Karoo. R: Strelitzia 32(2013)18. 3 overlapping poll. systems: bees ('gullet' fls), long-proboscis flies & generalists; some poll. monkey beetles. Cult. orn. cormous pls esp. ***S. tricolor*** (Schneev.) Ker-Gawler (& hybrids with ***S. elegans*** (Sweet) Goldbl. (SW Cape) by 1820s, and with ***S. bulbifera*** (L.) Ker-Gawler (S & SW Cape), esp. **'Queen Victoria'**), though *S. bulbifera* with bulbils in leaf-axils leading to formation of dense stands & resistant weeds in S Aus., poss. serious in native veg.

Sparganiaceae Hanin = Typhaceae

Sparganium Tourn. ex L. Typhaceae (Sparganiaceae). 14 N temp. (Eur. 7, N Am. 10), Mal. to Aus. & NZ. R: BH 96(1986)213, 97(1987)1. Bur reed. Fls wind-poll. & s.t. by syrphids. Fr. an imp. part of wildfowl diet in autumn & early winter

Sparganophoros Vaill. = seq.

Sparganophorus Boehmer (*Struchium*). Compositae (Vern. – Vern.). 1 trop. Am., natur. W Afr.: ***S. sparganophora*** (L.) C. Jeffrey (*Struchium s.*) – local medic. Guyana

Sparmannia auctt. = seq.

Sparrmannia L.f. Malvaceae (Grew.-Apeib.; Tiliaceae). 7 Afr. & Madag. Cult. orn. esp. ***S. africana*** L.f. (Afr. hemp, house lime, stock rose, S Afr.) – house-pl. tolerant of rough treatment, fls inverted with anthers app. exposed but C reflexed forming a cup behind A – this fills with water during rain & gently overflows a drop at a time, the anthers remaining dry, an insect visiting the fls touches the stamens, which are irritable, springing apart & depositing pollen on insect, bark a fibre-source, orn. cvs incl. 'doubles'

sparrow grass = asparagus

Spartidium Pomel (~ *Genista*). Leguminosae (III 8). 1 N Afr.: ***S. saharae*** (Cosson & Dur.) Pomel – like *Lebeckia*

Spartina Schreb. = *Sporobolus* (but see R: ISCJS 30(1956)471)

Spartium L. Leguminosae (III 9). 1 Medit. (incl. Eur.): ***S. junceum*** L. (Spanish broom) – cult. orn. with alks, stems form. used in basketry & as fibre-source, fragrant fls a poss. constituent in scent-making & source of yellow dye, invasive S Afr., S Am. (now occupying large area in Andean highlands)

Spartochloa C. Hubb. Gramineae (XXI). 1 SW Aus.: ***S. scirpoidea*** (Steud.) C. Hubb. – laminaless sheaths

Spartocytisus Webb. & Berth. = *Cytisus*

Spartothamnella Briq. = *Teucrium* (but see JABG 1(1976)1, Nuytsia 24(2014)184)

Spatalla Salisb. Proteaceae (IV 4). Incl. *Sorocephalus* (R: JSAB 7 suppl.(1969)21), 31 SW & S Cape

Spatallopsis E. Phillips = *Spatalla*

Spathacanthus Baill. (~ *Justicia*). Acanthaceae (III 2c). 3 C Am. R: CUMH 22(1999)37

Spathandra Guillemin & Perrottet (~ *Memecylon*). Melastomataceae. 1 trop. Afr.: ***S. blakeoides*** (G. Don) Jacq.-Fél. R: Adansonia 18(1978)245

Spathantheum Schott. Araceae (VII 4). 2 Bolivia. R: RMLPB 11(1971)266. Infls epiphyllous; G 6–8-loc. Cult orn.

Spathanthus Desv. Rapateaceae (II 1). 2 N S Am. Lvs to 1.5 m long

Spathelia L. Rutaceae (III). c. 18 WI to N S Am. Unbranched or sparsely branched hapaxanthic pachycaul (Holttum's Model) with pinnate lvs & term. infls. ***S. sorbifolia*** (L.) Fawcett & Rendle (*S. simplex*, mt. pride, Jamaica) – fls after c. 8–10 yrs

Spathia Ewart. Gramineae (XXII 5). 1 trop. Aus.: ***S. neurosa*** Ewart & M. Archer

Spathicalyx J. Gómes = *Tanaecium*

Spathicarpa Hook. Araceae (VII 4). 7 trop. S Am. Infls adnate to spathe. Cult. orn. esp. ***S. hastifolia*** Hook. (*S. sagittifolia*, caterpillar pl.)

Spathichlamys R. Parker. Rubiaceae (I 5). 1 Myanmar: ***S. oblonga*** R. Parker

Spathidolepis Schltr. = *Dischidia*

Spathionema Taubert. Leguminosae (III 18). 1 trop. Afr.: ***S. kilimandscharicum*** Taubert

Spathiostemon Blume. Euphorbiaceae (Acal.-Acal.-Lasioc.). Incl. *Clonostylis* 2 S Thailand, W Mal. R: Blumea 43(1998)145

Spathipappus Tzvelev = *Tanacetum*

Spathiphyllum Schott. Araceae (IV 1). c. 60 trop. Am., C Mal. to Solomon Is. (1). R: MNYBG 10,3(1960)1. Peace lilies. Spadix partly adnate to spathe; scent & starchless pollen-coll. euglossine bees (cannot digest starch); some with elaiosomes. Cult. orn. with usu. white spathes esp. '**Clevelandii**' used in cut-fl. trade (poss. a form of *S. wallisii* Regel, C Am.) & hybrids esp. '**Mauna Loa**', also ***S. cannifolium*** (Dryander) Schott (N S Am., Trinidad) – ash gives a superior alkaline powder for coating pellets of hallucinogenic *Virola elongata*

Spathodea P. Beauv. Bignoniaceae (1). 1 trop. Afr.: ***S. campanulata*** P. Beauv. (*S. nilotica*, flame tree, Afr. tulip t., fountain t., Nandi flame) – cult. orn. street-tree in trop., invasive in lowland forests in Hawaii, NG, N Aus.; extrafl. nectaries & 3 or 4 buds in each leaf axil, K inflated & full of secreted (?) water (so buds used as water-pistols), elaiophores attracting bees

Spathodeopsis Dop = *Fernandoa*

Spathoglottis Blume. Orchidaceae (V 14). 48 trop. As. to Aus., Samoa, Niue. Cult. orn. terr. orchids with pseudobulbs; some locally medic. ***S. plicata*** (As.) – competes & ousts *Bletia patula* Hook. in Puerto Rico

Spatholirion Ridl. Commelinaceae (II 1b). 3–4 SE As. 1 erect, 2 cli.

Spatholobus Hassk. Leguminosae (III 18). 29 SE As. to C Mal. R: Reinw. 10(1985)139. Rough cordage, local medic. (esp. ***S. suberectus*** Dunn in Vietnam) assoc. with red sap

Spathulata (Boriss.) Löve & D. Löve = *Phedimus*

Spathulopetalum Chiov. = *Caralluma*

speargrass *Aciphylla squarrosa*

spearmint *Mentha spicata*; **Scotch s.** *M.* × *gracilis*

spearwort *Ranunculus lingua*

Specklinia Lindl. (~ *Pleurothallis*). Orchidaceae (V 13). Incl. *Acostaea*, c. 130 trop. Am. R: Phytotaxa 272(2016)

Speculantha D. L. Jones & M. Clements = *Pterostylis*

Specularia Heist. ex A. DC. = *Legousia*

Speea Loes. Amaryllidaceae (Alliaceae-Gill.). 2 Chile

speedwell *Veronica* spp.; **germander s.** *V. chamaedrys*

Spegazziniophytum Esser (~ *Colliguaja*). Euphorbiaceae (Euph.-Hipp.). 1 Arg. (Patagonia): ***S. patagonicum*** (Speg.) Esser. R: A.R.-Sm., *Gen. Euphorb.* (2001)371

Speirantha Bak. Asparagaceae (Convallariaceae). 1 E China: ***S. convallarioides*** Bak. – cult. orn.

Spelaeanthus Kiew & al. Gesneriaceae (III 2i). 1 Malay Pen. (limestone caves): ***S. chinii*** Kiew & al. R: BBP 70(1998)401

spelt *Triticum aestivum* Spelta Group

Spenceria Trimen. Rosaceae (8). 1 W China: ***S. ramalana*** Trimen – cult. orn. herb like *Agrimonia*. R: JJB 81(2006)159

Speranskia Baill. Euphorbiaceae (Acal.-Chro.-Sper.). 2 Myanmar, China. R: FOC 11(2008)223

Spergella Reichb. = *Sagina*

Spergula L. Caryophyllaceae (I 1). 5 temp. (Eur. 4; 1 endemic N Patagonia). ***S. arvensis*** L. ((corn) spurrey, Eur., widely natur.) – some forms cult. for forage & green manure as in Holland

Spergularia (Pers.) J. Presl & C. Presl (~ *Spergula*). Caryophyllaceae (I 1). c. 60 cosmop. esp. halophytes (Eur. 17, N Am. 11). Sand spurrey esp. ***S. rubra*** (L.) J. Presl & C. Presl (Eur., natur. N Am.) – seeds a famine food. ***S. marina*** (L.) Besser (*S. salina*, Euras., S S Am.) – disjunct pops hybridize but hybrids almost sterile, form. medic. in Hawaii

Sperihedium Dorofeev = *Hesperis*

Spermacoce Dill. ex L. Rubiaceae (IV 15). Incl. *Borreria*, c. 250 warm (Aus. 57 – R: AusSB 18(2005)297). Buttonweeds – some troublesome weeds in S Afr., local medic.

Spermadictyon Roxb. (*Hamiltonia*). Rubiaceae (IV 12). 1 Ind., Himal.: ***S. suaveolens*** Roxb. – cult. orn. shrub with fragrant blue, pink or white fls. R: FOC 19(2011)324

Spermatophyta Britton & A. Brown. Seed-pls (dividing megaspore receiving nutrition from parental sporophyte; embryo surrounded by parental tissues & disp. as such), classically divided into Angiospermae (= Magnoliidae) & gymnosperms (q.v.). Extant taxa arr. as 13 484/286 500 in 425 fams

Spermolepis Raf. Umbelliferae (III 8). 11 N Am. (9), Arg. (1), Hawaii (1: ***S. hawaiiensis*** Wolff – believed extinct but lately re-found). R: Phytoneuron 2012–87:14

Spetaea Wetschnig & Pfosser (~ *Scilla*). Asparagaceae (Hyacinthaceae). 1 W Cape: ***S. lachenaliiflora*** Wetschnig & Pfosser. R: T 52(2003)87

Sphacanthus Benoist. Acanthaceae (III). 2 Madag.

Sphacele Benth. = *Lepechinia*
Sphaenolobium Pim. Umbelliferae (III 8). 3 C As.
Sphaeradenia Harling. Cyclanthaceae. 50 trop. Am. R: OB 126(1995)25
Sphaeralcea A. St-Hil. Malvaceae (Malv.-Malv.). Excl. *Iliamna*, c. 40 arid Am. R: UCPB 19 (1935)1. Local medic. (N Am.); cult. orn. herbs & subshrubs incl. ***S. coccinea*** (Pursh) Rydb. & ***S. angustifolia*** (Cav.) G. Don subsp. ***cuspidata*** (A. Gray) Kearney (scarlet globe mallow, yerba de la negrita, N Am.) – negrita extract used for hair conditioners in commerce
Sphaeranthus Vaill. ex L. Compositae (Pluch). 41 OW trop. to Iran & Egypt (all but 4 in Afr.). R: HIP 36(1955)tt.3501–25. Local medic.
Sphaerantia Peter G. Wilson & Hyland. Myrtaceae (II 5). 2 N Queensland. R: Telopea 3(1988)260
Sphaereupatorium (O. Hoffm.) Robinson. Compositae (Eup.-Crit.). 1 S Am.: ***S. scandens*** (Gardner) R. King & H. Robinson. R: MSBMBG 22(1987)321
Sphaerobambos S. Dransf. (~ *Bambusa*). Gramineae (V 4). 3 Mal. R: KB 44(1989)428
Sphaerocardamum Schauer. Cruciferae (29). 4 Mex. R: SB 32(2007)153
Sphaerocaryum Nees ex Hook.f. Gramineae (Micr.). 1 Ind. to Taiwan & W Mal.: ***S. malaccense*** (Trin.) Pilger. R: FOC 22(2006)560
Sphaerocionium C. Presl = *Hymenophyllum* (but see JFSUTB III,13(1982)207)
Sphaeroclinium (DC.) Schultz-Bip. = *Cotula*
Sphaerocodon Benth. = *Vincetoxicum*
Sphaerocoma T. Anderson. Caryophyllaceae (I 2). 1 Middle E to Pakistan: ***S. hookeri*** T. Anderson. R: T 61(2012)72
Sphaerocoryne (Boerl.) Ridl. (~ *Melodorum*). Annonaceae. 3 SE As. to Mal.
Sphaerocyperus Lye = *Cyperus* (but see NJB 13(1995)507)
Sphaerolobium Sm. Leguminosae (III 13). 22 Aus. esp. SW
Sphaeromariscus Camus = *Cyperus*
Sphaeromeria Nutt. = *Artemisia*
Sphaeromorphaea DC. (~ *Epaltes*). Compositae (Pluch.). 6 Indopacific (Aus. 6). R: Austrobaileya 9(2013)35
Sphaerophora Blume = *Gynochthodes*
Sphaerophysa DC. Leguminosae (III 24). 2 E Medit. to C As. Halophytes
Sphaeropteris Bernh. = *Cyathea*
Sphaerorrhiza Roalson & Boggan (~ *Achimenes*). Gesneriaceae (II 5e). 2 Braz.
Sphaerosacme Wall. ex Royle (~ *Lansium*). Meliaceae (II-Trich.). 1 Himal.: ***S. decandra*** (Wall.) Penn. R: Blumea 22(1975)489
Sphaerosciadium Pim. & Kljuykov. Umbelliferae (III 5). 1 C As.: ***S. denaense*** (Schischk.) Pim. & Kljuykov
Sphaerosepalaceae Tiegh. ex Bullock (~ Ochnaceae). Magnoliidae – Malvales. 2/18 Madag. Decid. trees with simple hairs. Lvs in spirals, s.t. becoming distich.; stip. broad, ± encircling stem. Fls usu. 4-merous in infls of subumbelliform cymules; K usu. 2 + 2, imbr., inner larger; C (3)4(-9), clawed, imbr., when 4 opp. K, A ∞ with broad connective, pollen usu. spiny, $\underline{G}$ (2(-5)) with basal placentation & 2–9 epitropous ovules per carpel. Fr. ± baccate with 1 (2) seeds per carpel, s.t. arillate, exotesta mucilaginous, endosperm copious, s.t. ruminate
Genera: *Dialyceras, Rhopalocarpus*
Sphaerostephanos J. Sm. = *Cyclosorus*
Sphaerostylis Baill. Euphorbiaceae (Acal.-Pluk.-Trag.). 2 Madag. Stinging
Sphaerothylax Bisch. ex Krauss. Podostemaceae (III). 2+ trop. & S Afr., Madag.
Sphaerotorrhiza (O. Schulz) Khokr. = *Cardamine*
Sphaerotylos C.J. Chen = *Sarcochlamys*
Sphagneticola O. Hoffm. Compositae (Helia.-Ecl.). 5 trop. As., Am. R: Novon 6(1996)411. ***S. trilobata*** (L.) Pruski (*Wedelia t.*, Singapore daisy, Florida, trop. Am.) – medic. but causing dermatitis, invasive Sumatra, W Pacific (pestilential in Queensland where introd. for bank stabilization), S Afr., Hawaii
Sphagnum Dill. ex L. Sphagnaceae (Musci). 285 subcosmop. (N Am. 89 – R: FNA 27(2007)45); form. many ecotypes recog. as spp. R: BB 160(2011). Unique branching (fascicles); caps. explosive-dehisc. Char. of acid bogs, assoc. with N-fixing cyanobacteria, pls with up to 30 times wt. in water, insulator (used by crocodiles for nests in Palawan), medic. (China), form. used in wool for cheap clothes (Ger.), as tampons (comm. in Canada) & for babies' napkins (N Am. & Scotland, best absorption in ***S. papillosum*** Lindb, dioec., major peat-former), form. for 'dry' toilets, wound dressings in WW I (1 M a

month; sup. to cotton because somewhat antibiotic [*Penicillium* spp. inside?] & absorbing 3+ times as much liquid 3 times faster, besides releasing c. for gunpowder-making), still in some hiking-boots for cushioning & absorption. Decomposing pls basis of much peat, giving flavour to Scotch whisky & filtering & absorption of effluents incl. acids, heavy metals, oils, detergents & dyes, & used both for fuel (70 M t burnt in USSR alone in 1975) & in hort. where undecomposed pls used in air-layering & as potting medium

Sphallerocarpus Besser ex DC. Umbelliferae (III 2). 1 S As.: ***S. gracilis*** (Trev.) Koso-Polj.

Sphalmanthus N.E. Br. = *Mesembryanthemum*

Sphalmium Briggs & al. Proteaceae (V). 1 NE Queensland: ***S. racemosum*** (C. White) Briggs & al. R: FA 16(1995)342

Sphedamnocarpus Planch. ex Benth. Malpighiaceae. 12 trop. & S Afr., Madag.

Sphenandra Benth. = *Sutera*

Spheneria Kuhlm. = *Paspalum* (but see CUSNH 46(2003)606)

Sphenocarpus Korovin = *Seseli*

Sphenocentrum Pierre. Menispermaceae (I). 1 trop. W Afr.: ***S. jollyanum*** Pierre – unbranched pachycaul treelet, r. used as chewing-sticks rendering food eaten afterwards sweet, locally medic. R: NJBB 53(1983)59

Sphenoclea Gaertn. Sphenocleaceae. 1 OW pantrop. (? orig. Afr.), natur. Am.: ***S. zeylanica*** Gaertn. – Stone's & Leeuwenberg's Models, a weed of rice (r. exudates control rice r. nematode) in 17 countries, shoots locally eaten (e.g. Java) with rice. R: JAA 67(1986)1

Sphenocleaceae Baskerv. (~ Campanulaceae). Magnoliidae – Asterales. 1/1 trop. Rather fleshy glabrous ann. herb of wet places, with cortical air-passages. Lvs simple, entire, in spirals; stip. 0. Fls reg., bisex., bibracteolate, in axils of small bracts on term. spikes; K (5) with imbr. lobes, C (5) with urceolate-campanulate tube & imbr. lobes, A alt. with C & attached to tube, anthers with longit. slits, nectary 0, $\overline{G}(2)$ or semi-inf., 2-loc., stigma without pollen-coll. hairs, axile placentas with ∞ anatr. unitegmic ovules. Fr. a circumscissile capsule encl. by accrescent K; seeds ∞ small with straight embryo in ± 0 endosperm. n = 12, 16, 20 etc.

Genus: *Sphenoclea* – habit like *Phytolacca*

Sphenodesme Jack. Labiatae (I; Symphoremataceae). (6–)14 SE As., W Mal. R: GBS 21(1966)315

Sphenomeris Maxon (~ *Odontosoria*). Lindsaeaceae. 1 trop. Am.: ***S. clavata*** (L.) Maxon. R: BJLS 163(2010)333

Sphenopholis Scribner. Gramineae (XVI 2). 8 Canada to Mex. incl. ***S. pensylvanica*** (L.) Hitchc. (*Trisetum p.*, swamp oat, N Am.). R: ISJS 39(1965)289, FNA 24(2007)620

Sphenopsida Engl. = Equisetopsida

Sphenopus Trin. Gramineae (XVI 12). 2 Medit. (Eur. 1). Halophytes

Sphenosciadium A. Gray = *Angelica*

Sphenostemon Baill. Paracryphiaceae (Sphenostemonaceae, Aquifoliaceae s.l.). 9 C Mal. to trop. Aus. & New Caled.

Sphenostemonaceae P. Royen & Airy Shaw = Paracryphiaceae

Sphenostigma Bak. = *Gelasine*

Sphenostylis E. Meyer. Leguminosae (III 18). 7 Afr., Ind. (1). Some ed. seeds & tubers esp. ***S. stenocarpa*** (A. Rich.) Harms (girigiri, trop. Afr.) – tubers twice as rich as seeds in protein

Sphenotoma (R. Br.) Sweet. Ericaceae (VII 6; Epacridaceae). 6 SW Aus.

Sphinctacanthus Benth. Acanthaceae (III 2c). 1 NE Ind. to Myanmar: ***S. griffithii*** (T. Anderson) Benth. R: NJB 5(1985)226

Sphinctanthus Benth. Rubiaceae (II 1). 8 S Am.

Sphinctospermum Rose (~ *Tephrosia*). Leguminosae (III 23). 1 SW N Am.: ***S. constrictum*** (S. Watson) Rose. R: SB 15(1990)544, 16(1991)162

Sphinga Barneby & Grimes (~ *Pithecellobium*). Leguminosae (II 4). 3 trop. Am. R: MNYBG 74,1(1996)160

Sphingiphila A. Gentry = *Tanaecium* (but see SB 15(1990)277)

Sphyranthera Hook.f. Euphorbiaceae (Acal.-Sph.). 2 Andaman & Nicobar Is.

Sphyrarhynchus Mansf. Orchidaceae (V 16d). 1 Tanzania: ***S. schliebenii*** Mansf.

Sphyrastylis Schltr. = *Ornithocephalus*

Sphyrospermum Poeppig & Endl. Ericaceae (VIII 5). 21 trop. Am. mts. R: Phytotaxa 79(2013)79

spice bush *Lindera benzoin*

Spiculaea Lindl. Orchidaceae (IV 3e). 1 W Aus.: ***S. ciliata*** Lindl. – poll. by sexual deceit (thynnine wasps). R: Gen. Orchid. 2(2001)152

spider azalea *Rhododendron stenopetalum*; **s. flower** *Cleome houtteana*; **s.f., grey** *Grevillea buxifolia*; **s. lily** *Ismene × festalis*; **s. orchid** *Ophrys sphegodes*; **s. plant** *Chlorophytum comosum*; **s.wort** *Tradescantia virginiana*

Spigelia L. Loganiaceae (Strychnaceae). 50 trop. & warm Am., 1 natur. OW: ***S. anthelmia*** L. – vermifuge & criminal poison. ***S. longiflora*** Martens & Galeotti (Mex.) – insect antifeedant. Other spp. medic. esp. ***S. marilandica*** (L.) L. (Maryland pinkroot, SE US) – vermifuge (dried rhiz. etc.), Leeuwenberg's Model, cult. orn.

Spigeliaceae Mart. = Loganiaceae

spignel *Meum athamanticum*

spiked millet *Cenchrus spicatus*

spikenard *Nardostachys jatamansi*; **American, false** or **wild s.** *Maianthemum racemosum*; **ploughman's s.** *Inula conyzae*

Spilanthes Jacq. Compositae (Helia.-Spil.). 6 trop. R: SB 6(1981)231. See also *Acmella*

Spiloxene Salisb. = *Pauridia*

spinach *Spinacia oleracea*; **s. beet** *Beta vulgaris*; **Cape s.** *Rumex spinosus*; **Ceylon** or **Indian** or **Malabar s.** *Basella alba*; **Chinese s.** *Amaranthus tricolor*; **Cuban s.** *Claytonia perfoliata*; **mustard s.** *Brassica campestris*; **NZ s.** *Tetragonia tetragonoides*; **Okinawa s.** *Gynura bicolor*; **tree s.** *Hibiscus manihot*; **water s.** *Ipomoea aquatica*

Spinacia Tourn. ex L. Amaranthaceae (Chenopodiaceae I 3). 3 SW As., N Afr. R: BSBG 48(1938)485. ***S. oleracea*** L. (spinach, orig. SW As.?) – dioec. (exc. 'var. *americana*') potherb introd. Eur. c. AD 1000, rich in vitamins A, B, C, E & K (also oxalic acid & alks), carotenoids, calcium & iron (cannot be assimilated by body), alleged to reduce incidence of cataract (lutein filters sun's blue light & reduces ultraviolet), also used in soup & tinned, but in excess oxalates can inhibit calcium absorption

spindle tree *Euonymus europaeus*

spinifex *Triodia* spp.; **buck s.** *T. longiceps*

Spinifex L. Gramineae (XXIV 4). 5 Indomal. to E As. & Pac. beaches. R: ASBSN 56(1988)13. Dioec. grasses, female spikelets 1-fld with spiny bracts, in heads, which break off at maturity & are blown about like tumbleweeds, dispersing fr.

Spiniluma Baill. ex Aubrév. = *Sideroxylon*

spinks *Cardamine pratensis*

Spiracantha Kunth. Compositae (Vern.-Rol.). 1 C Am. to Venez.: ***S. cornifolia*** Kunth

Spiradiclis Blume. Rubiaceae (IV 2). 40+ India & SW China (35 - R: FOC 19(2011)330) to SE As. & Java. R: ABAS 1(1983)32, Cand. 44(1989)225

Spiraea Tourn. ex L. Rosaceae (2). Incl. *Pentactinia* c. 110 N temp. (Eur. 8, N Am. 14) to Mex. & Himal. (c. 30). Cult. orn. decid. shrubs esp. ***S. × arguta*** Zabel (*S. × multiflora* Zabel (i.e. *S. crenata* L. (SE Eur. to C As.) × *S. hypericifolia* L. (SE Eur. to Siberia)) × *S. thunbergii* Sieb. ex Blume (China, Jap.)) – shrub with arching branches of white fls, ***S. cantoniensis*** Lour. (China) – 'may' in E Aus., ***S. japonica*** L.f. (incl. '*S. × bumalda*', Himal. to Jap.), ***S. tomentosa*** L. (hardhack, steeplebush, E N Am.), etc.

Spiraeanthemum A. Gray. Cunoniaceae (VI). (Incl. *Acsmithia*) 19 Mal. (4) to Samoa (Aus. 1, New Caled. 7, Fiji 4). R: Blumea 25(1979)492, 501. Some locally used timbers

Spiraeanthus (Fischer & C. Meyer) Maxim. Rosaceae (4). 1 Kazakhstan: ***S. schrenkianus*** (Fischer & C. Meyer) Maxim.

Spiraeopsis Miq. (~ *Caldcluvia*). Cunoniaceae (VI). 6 Philippines to Samoa. R: FM I,16(2002)128

Spiralluma Plowes = *Caralluma*

Spiranthera A. St-Hil. Rutaceae (I 8). 4 N S Am. R: BJBB 54(1984)485

Spiranthes Rich. Orchidaceae (IV 2h). 34 N temp. (Eur. 3–4), few in trop. Am., Mal., Aus. (1) & Pac. R: HBML 28(1982)360. Exclusively sexual to selfed or apomictic spp. Cult. orn. terr. orchids (lady's tresses) incl. ***S. romanzoffiana*** Cham. & Schldl. (N Am., Ireland & coasts of W Br. but absent from continental Eur.; cf. *Eriocaulon aquaticum*)

Spirella Costantin. Apocynaceae (V a; Asclepiadaceae III 4). 2 SE As.

Spiroceratium H. Wolff = *Pimpinella*

Spirodela Schleiden. Araceae (Lemnaceae). 3 cosmop. (Eur. 1, 1 restricted to S Am.). R: IBM 34(1965)8. Least reduced of duckweeds (cf. *Pistia*). In US, grown on dairy farm wastewater & subs. for alfalfa in cattle & pig feed. See also *Landoltia*

Spirogardnera Stauffer (~ *Choretrum*). Santalaceae (Amphorogynaceae). 1 SW Aus.: ***S. rubescens*** Stauffer – r.-parasite. R: FA 22(1984)49

Spirolobium Baill. Apocynaceae (II b). 1 SE As. to Borneo: ***S. cambodianum*** Baill.

Spiropetalum Gilg = *Rourea*

Spirorhynchus Karelin & Kir. = *Goldbachia*

Spiroseris Rech.f. Compositae (Lact.-Crep.). 1 Pakistan: ***S. phyllocephala*** Rech.f.

Spirospermum Thouars. Menispermaceae (V). 1 Madag.: ***S. penduliflorum*** Thouars

Spirostachys Sonder = *Excoecaria*

Spirostegia Ivanina = *Triaenophora*

Spirostigma Nees. = *Ruellia*

Spirotecoma Baill. ex Dalla Torre & Harms. Bignoniaceae (1). 5 Cuba, Hispaniola

Spirotheca Ulbr. (~ *Ceiba*). Malvaceae (Bomb.). 6 trop. Am. R: SB 37(2012)982. Some stranglers

Spirotropis Tul. Leguminosae (III 2). 2–3 NE S Am.

spleenwort *Asplenium* spp.; **maidenhair s.** *A. trichomanes*; **mountain s.** *A. montanum*; **sea s.** *A. marinum*; **wall s.** *A. ruta-muraria*

Spodiopogon Trin. Gramineae (XXII). 18 Turkey to Jap. (China 9, 6 endemic), ***S. sibiricus*** Trin. ext. to Irkutsk, Russia

Spondianthus Engl. Phyllanthaceae (Spond.; Euphorbiaceae s.l.). 1 trop. Afr.: ***S. preussii*** Engl. – resin exudate toxic (fluoroacetic acid) cf. *Uapaca*. R: BJBB 59(1989)133

Spondias L. Anacardiaceae (II). Excl. *Solenocarpus*, 19 Madag. (1), Indomal. to SE As., trop. Am. (10). R: PhytoKeys 55(2015)511. Trees (Scarrone's Model, some stranglers, cf. *Ficus*) with resins coll. euglossine bees for nests & drupes, some ed. (hog or Jamaica plum, Otaheite apple, imbu (Braz.)) esp. ***S. dulcis*** Park. (*S. cytherea*, ambarella, Jew's plum, golden or Otatheite apple, li tree, Pac.) – fr. for jellies, pickles etc.; ***S. mombin*** L. (*S. lutea*, caja fr., yellow mombin, jobo, trop. Am.) – bat-disp. fr. eaten fresh incl. juice, in icecream & liqueurs; ***S. purpurea*** L. (ciruela, jocote, ovo, Spanish plum, red mombin, trop. Am.) – parthenocarpic in SE As., app. domesticated several times, eaten fresh, boiled or dried, fuelwood; ***S. tuberosa*** Arruda (umbu, Braz.) -imp. local fr.

Spondogona Raf. = *Sideroxylon*

Spongiola J.J. Wood & Lamb (~ *Aerides*). Orchidaceae (V 16c). 1 Borneo: ***S. lohokii*** J.J. Wood & Lamb

Spongiocarpella Yakolev & Ulzjkhumag = *Chesneya*

Spongiosperma Zarucchi (~ *Ambalania*). Apocynaceae (I d). 6 N S Am. R: AUWP 87–1(1987)

Spongiosyndesmus Gilli = *Ladyginia*

Sporadanthus F. Muell. (~ *Lepeyrodia*). Restionaceae. 9 Aus. (E 5, SW 2), 1 NZ, 1 Chatham Is. salt-marshes & peatswamps

Sporobolus R. Br. Gramineae (XXVIII). Incl. *Calamovilfa*, *Crypsis*, *Spartina*, 230 trop. & warm (Aus. 26 – R: AusSB 12(1999)382) to temp (few). R [p.p.] T 63(2014)1232. Sect. *Crypsis* – pericarp swelling when wet, extruding fr.; sect. *Spartina* (marsh or cord grass) 16 coastal Am. (N Am. 9 + 2 natur.), Eur. (3 + 2 natur.), N Afr. – halophytes with salt-excreting hydathodes in epidermis, some mooted as biofuel crops for poor soils. Some resurrection pls Aus., Afr.: lvs of ***S. festivus*** Hochst., ***S. lampranthus*** Pilg. & ***S. stapfianus*** Gand. (Afr.) revive after reduction of water-content to 5–13%. Some used in revegetation (rush grass), as fodder & ed. grains (ground to paste & baked Aus.): ***S. cryptandrus*** (Torrey) A. Gray (N Am.) – grains consumed by Native Americans. ***S. anglicus*** (C. Hubb.) Peterson & Saarela (*Spartina a.*, 2n = 120–124) – derived c. 1890 from *S.* × *townsendii* (Groves & J. Groves) Peterson & Saarela (*Spartina* × *t.*, 2n = 62), hybrid arising c. 1870 at Hythe in Southampton Water, England (& Bay of Biscay ('*Spartina* × *neyrautii*' in 1892)) between *S. alterniflorus* (Loisel.) Peterson & Saarela (*Spartina a.*, 2n = 62, N Am., introd. c. 1820s; invasive NZ, W US, introd. San Francisco Bay 1970s & now hybridizing with *S. foliosus* (Trin.) Peterson Saarela (*Spartina f.*) there) & *S. maritimus* (Curtis) Peterson & Saarela (*Spartina m.*, 2n = 60, Eur., Afr. – introd.?, invasive Aus.) – classic case of allopolyploid sp. orig. but polytopic (cf. *Tragopogon minus*, *Senecio cambrensis*), its parents almost extinct in GB, now invasive Tasmania, NZ (spreading up to 5.3 m a yr), W US; ***S. fimbriatus*** Nees (dropseed, S Afr.) – potential pasture, soil stabilizer; ***S. indicus*** (L.) R. Br. (incl. *S. africanus*, Parramatta grass, trop. & S Afr.) – invasive Azores, NZ, Hawaii, agric. weed in Aus.; ***S. michauxianus*** (Hitchc.) Peterson & Saarela (*Spartina pectinata*, N Am.) – thatch, despite leaf saw-tooth margins

Sporoxeia W.W. Sm. Melastomataceae. 6 Myanmar to SE As.

spotted gum *Corymbia maculata*

Spraguea Torrey = *Calyptridium*

Spragueanella Balle (~ *Oncocalyx*). Loranthaceae (5 6). 2 E & SC Afr. R: R. Polhill & D. Wiens, *Mistletoes of Afr.* (1998)112

sprangletop *Scolochloa festucacea*; **green s.** *Disakisperma dubium*

Sprekelia Heister (~ *Hippeastrum*). Amaryllidaceae. 1 Mex., Guatemala: ***S. formosissima*** (L.) Herb. (Jacobean lily) – cult. orn. with crimson fls & alks

Sprengelia Sm. Ericaceae (VII 3; Epacridaceae). 8 E Aus. (1 NZ – introd.?). R: Telopea 15(2013)60. ***S. incarnata*** Sm. – prot. sp.

spring beauty *Claytonia virginica*

sprouting, purple or s. broccoli *Brassica oleracea* Italica Group

sprouts (**, Brussels**) *Brassica oleracea* Gemmifera Group

spruce (Spruce an old name for Prussia) *Picea* spp.; **s. beer** *P. abies*; (**Can.**) **black** or **bog s.** *P. mariana*; **s. gum** *P. rubens*; **hemlock s.** *Tsuga* spp.; **Norway s.** *P. abies*; **red s.** *P. rubens*; **Sitka s.** *P. sitchensis*; **white s.** *P. glauca*

Spryginia Popov (~ *Orychophragmus*). Cruciferae (26). 7 C As.

spurge *Euphorbia* spp.; **caper s.** *E. lathyrus*; **cypress s.** *E. cyparissias*; **ipecacuanha s.** *E. ipecacuanhae*; **s. laurel** *Daphne laureola*; **s. olive** *Cneorum tricoccon*; **wood s.** *E. amygdaloides*

Spuriacianthus Szlach. & Marg. = *Acianthus*

Spuriodaucus Norman. Umbelliferae (inc. sed.). 3 trop. Afr.

Spuriopimpinella Kitag. = *Pimpinella*

spurrey *Spergula arvensis*; **sand s.** *Spergularia* spp.

Spyridium Fenzl. Rhamnaceae (5). 45 temp. Aus. R: Nuytsia 11(1996)116

Squamellaria Becc. Rubiaceae (IV 7). 4 Fiji. R: Blumea 36(1991)53. ? Ant-pls. 'Male' fls force up & snap off stigma, females with reduced anthers

Squamopappus R. Jansen & al. (~ *Podachaenium*). Compositae (Helia.-Verb.). 1 SE Mex., Guatemala: ***S. skutchii*** (S.F. Blake) R. Jansen & al. R: SB 7(1982)480

squash *Cucurbita* spp.; **butternut** or **coquina s.** *C. moschata*

squaw root *Caulophyllum thalictroides*, *Conopholis americana*

squill or red s. *Drimia maritima*; **autumn s.** *Scilla autumnalis*; **Lebanon** or **striped s.** *Puschkinia scilloides*; **spring s.** *S. verna*

squinancywort *Asperula cynanchica*

squirting cucumber *Ecballium elaterium*

Sredinskya (Stein) Fed. = *Primula*

Sreemadhavana Rauschert = *Aphelandra*

St. See Saint

Staavia Dahl. Bruniaceae (3). 11 W Cape. R: Bothalia 15(1985)396, Strel. 9(2000)386

Staberoha Kunth. Restionaceae. 9 W Cape. R: Strel. 9(2000)215

Stachyacanthus Nees = ? *Eranthemum*

Stachyandra Leroy ex R.-Sm. = *Androstachys* (but see KB 45(1990)562)

Stachyanthus DC. = *Chesta*

Stachyanthus Engl. Icacinaceae. 6 trop. Afr.

Stachyarrhena Hook.f. Rubiaceae (II Cord.). 13 trop. Am. R: RBB 6(1983)109

Stachycephalum Schultz-Bip. ex Benth. Compositae (Mill.-Mill.). 3 Mex., Ecuador, Arg.

Stachydeoma Small (~ *Hedeoma*). Labiatae (VII 2b). 1 Florida: ***S. graveolens*** (A. Gray) Small

Stachyococcus Standl. = *Carapichea*

Stachyophorbe (Mart.) Liebm. = *Chamaedorea*

Stachyopsis Popov & Vved. = *Eriophyton*

Stachyothyrsus Harms. Leguminosae (I 4). 2 W trop. Afr.

Stachyphrynium K. Schum. (*Phyllodes*). Marantaceae. 9 Indomal. R: GBS 13(1951)273

Stachypitys A. Bobrov & Melikian = *Prumnopitys*

Stachys Tourn. ex L. Labiatae (VI). Excl. *Betonica*, c. 300 temp. (Eur. 58; Turkey 72) & warm exc. Australasia, trop. mts. R: NRBGE 38(1980)71. Prob. heterogeneous (some allied to *Prasium*, *Sideritis* prob. to be incl. etc.). N Am. spp. prob. gave rise to polyploids, thence Hawaiian *Haplostachys*, *Phyllostegia* & *Stenogyne*. Cult. orn. herbs & shrubs with alks, some with ed. tubers (hedge nettle or woundwort). ***S. affinis*** Bunge (*S. sieboldii*, Chinese or Jap. artichoke, crosnes, China) – ed. white tubers salted, pickled or boiled; ***S. × ambigua*** Sm. (*S. palustris* × *S. sylvatica* L. (Euras.)) – usu. sterile but now spread beyond parents, poss. grown as veg. or for medic. (woundwort); ***S. arvensis*** (L.) L. (Eur., Medit.) – natur. N Am. & Aus. where toxic to sheep (stagger-weed); ***S. byzantina*** K. Koch (*S. lanata*, '*S. olympica*', lamb's ears, SW As.) – much cult. rabbit-proof perenn. with silvery densely tomentose lvs; ***S. palustris*** L. (marsh betony, N temp.) – locally medic. & eaten tubers

Stachystemon Planch. (~ *Pseudanthus*). Picrodendraceae (Cal.-Pseud.; Euphorbiaceae s.l.). 9 SW Aus. R: Austrobaileya 6(2003)515

Stachytarpheta Vahl. Verbenaceae (Dur.). c. 130 trop. & warm Am., weedy trop. OW. Fls last 1 day but if spike picked C shed in few mins (traumatochory). Some cult. orn. but usu. widespread weeds, ***S. jamaicensis*** (L.) Vahl medic. & tea subs. ('devil's coachwhip'), weedy; ***S. urticifolia*** Sims – Leeuwenberg's Model, hedge-pl. in Afr., natur. OW

Stachyuraceae J. Agardh. Magnoliidae – Crossosomatales. 1/16 Himal. to Jap. Evergreen to decid. small trees & shrubs. Lvs simple, toothed, in spirals; stip. small, decid. Fls usu. bisex. (pls s.t. dioec.), small, 4-merous, bibracteolate, in axillary spikes or racemes; K 2 + 2 (outer 2 smaller), imbr., C 4, imbr., A 4 + 4 with deeply sagittate anthers with longit. slits, nectary at base of G, $\underline{G}$ (4), 4-loc. basally but partitions not reaching apex where placentation parietal distally, axile proximally, with ∞ anatropous, bitegmic ovules. Fr. a 4-loc. dry berry; seeds ∞ arillate, with straight embryo in hard oily proteinaceous endosperm. n = 12

Genus: *Stachyurus*

Form. considered near Flacourtiaceae (= Salicaceae), Malpighiales

Stachyurus Sieb. & Zucc. Stachyuraceae. 16 Himal. to Jap. R: ABY 2(1981)125. Cult. orn. shrubs fls before lvs expand

Stackhousia Sm. Celastraceae (II; Stackhousiaceae). 16 Australasia. R: FA 22(1984)186. ***S. monogyna*** Labill. (candles, E Aus.) – cult. orn.; ***S. tryonii*** Bailey (Queensland) – endemic of serpentine soils of Port Curtis Dist., nickel hyperaccum.

Stackhousiaceae R. Br. = Celastraceae (II)

Stacyella Szlach. = *Erycina*

Stadiochilus R.M. Sm. (~ *Hedychium*). Zingiberaceae (II – tribe?). 1 Myanmar: ***S. burmanicus*** R.M. Sm. R: NRBGE 38(1980)15

Stadmania Lam. Sapindaceae. 6 Madag., ***S. oppositifolia*** Poir. ext. to E Afr., Mauritius. R: MMNHNP 19(1969)151. See also *Smelophyllum*

Stadmannia Radlk. = *Stadmania*

Staehelina L. Compositae (Card.-Card.). Incl. *Hirtellina*, 8 Medit.

Staelia Cham. & Schldl. Rubiaceae (IV 15). 14 N S Am.

stagger-weed *Stachys arvensis*

stag(s)horn fern *Platycerium* spp.; **s. sumach** *Rhus hirta*

Stahelia Jonker = *Tapeinostemon*

Stahlia Bello = *Libidibia*

Stahlianthus Kuntze = *Curcuma*

Staintoniella H. Hara = *Aphragmus*

Stalagmitis Murray = *Garcinia*. Genus based on specimen comprising 2 diff. spp. stuck together with sealing-wax (cf. *Actinotinus* & *Papilionopsis*)

Stalkya Garay. Orchidaceae (IV 2h). 1 Venez. Andes: ***S. muscicola*** (Garay & Dunst.) Garay. R: Gen. Orchid. 3(2003)266

Staminodianthus D. Cardoso & al. (~ *Diplotropis*). Leguminsae (III 2). 3 Braz. R: Phytotaxa 110(2013)7

Standleya Brade. Rubiaceae (I 5). 4 Braz.

Standleyacanthus Leonard = *Herpetacanthus*

Standleyanthus R. King & H. Robinson (~ *Eupatorium*). Compositae (Eup.-Oxy.). 1 Costa Rica: ***S. triptychus*** (Robinson) R. King & H. Robinson. R: MSBMBG 22(1987)446

Stanfieldiella Brenan. Commelinaceae (II 2). 4 trop. Afr.

Stanfordia S. Watson = *Streptanthus*

Stangea Graebner = *Valeriana*

Stangeria T. Moore. Zamiaceae (Stangeriaceae). 1 S Afr.: ***S. eriopus*** (Kunze) Baill. (*S. paradoxa*) – first described as a fern (*Lomaria e.*), prob. close to *Ceratozamia*; buds develop from r., leaflets with dichotomously branching veins. R: FOW 2(1999)2

Stangeriaceae Schimp. & Schenk = Zamiaceae

Stanhopea Frost ex Hook. Orchidaceae (V 12i). c. 60 trop. Am. Cult. orn. epiphytes with large v. fragrant fls to 20 cm diam. (***S. tigrina*** Bateman ex Lindl., E Mex.– fls form. used in tortillas) incl. ***S. candida*** Barb. Rodr. (NE S Am.) – poll. euglossine bee, *Eulaema mocsaryi*, which scrapes for scent & brushes off pollinium which is deposited in stigmatic cavity of next fl. visited but *Euglossa ignita* also collects scent but does not brush off pollinia so is a robber. R: R. Jenny (2010) *The S. book*

Stanleya Nutt. Cruciferae (46). 7 W N Am. R: FNA 7(2010)695. Some potherbs used by Native Americans

Stanleyella Rydb. = *Thelypodium*

Stanmarkia Almeda. Melastomataceae. 2 W Guatemala & Mex. R: Britt. 45(1993)187

Stapelia L. Apocynaceae (V b; Asclepiadaceae III 5). c. 40 trop. & S Afr. R: F. Albers & U. Meve, *Ill. Handb. Succ. Pls*, Asclep. (2002)242. Cult. orn. succ. with photosynthetic stems & small or 0 lvs. Fls to 46 cm diam. in ***S. gigantea*** N.E. Br., oft. luridly marked & fleshy, foetid, attractive to flies which lay eggs there & carry out poll. (carrion fls). The anthers cont. the pollen united into masses, 'pollinia' (cf. Orchidaceae), & are embedded below the C disk which is the style, bearing the stigmatic surfaces on the underside; poll. effected by legs or proboscides of the flies falling or being pushed through cracks overlying anthers, pollinia being clipped on & finally deposited on stigmas through other cracks. See also *Orbea, Stapelianthus, Tridentea, Tromotriche* (somewhat 'one-legged' segregates ('genuslets'))

Stapelianthus Choux ex A. White & B. Sloane. Apocynaceae (V b; Asclepiadaceae III 5). 7 Madag. R: AMBG 91(2004)425

Stapeliopsis Pill. (~ *Orbea*). Apocynaceae (Vb). 8 SW Afr. R: BJLS 148(2005)142

Stapfiella Gilg. Passifloraceae (Turneraceae). 6 trop. Afr.

Stapfiola Kuntze = *Desmostachya*

Stapfiophyton Li = *Fordiophyton*

Stapfochloa H. Scholz (~ *Chloris*). Gramineae (XXIX 5). 6 Afr. (1), S Am. R: T 64(2015)459

Staphisagria Hill (~ *Delphinium*). Ranunculaceae (I 4). 3 Medit. R: PhytoKeys 7(2011)21. ***S. macrosperma*** Spach (*D. staphisagria*, stavesacre, S Eur., SW As.) – from 1788 seeds form. used to control ectoparasites, rats & ants in W Eur.

Staphylea L. Staphyleaceae. Incl. *Euscaphis, Turpinia*, 23 N temp. (Eur. 1) to Hainan & S Am. R [s. s.]: Arnoldia 40(1980)76. Some timbers; some cult. orn. trees & shrubs (bladder nuts) esp. ***S. pinnata*** L. (Eur., SW As.) with white fls, ***S. japonica*** (Thunb.) Mabb. (*E. j.*, temp. E As.)

Staphyleaceae Martinov. Magnoliidae – Crossosomatales. 2/48 Am., Euras. to Mal. Everg. to decid. trees & shrubs. Lvs pinnate (to unifoliolate), opp., leaflets usu. toothed; stip. decid. or 0. Fls bisex. or not (pls s.t. dioec.), reg., 5-merous, in racemes or panicles; K imbr., oft. petaloid, unequal, C imbr., A alt. with C on or outside annular nectary-disk, anthers with longit. slits, $\underline{G}$ (2 or 3(4)), pluriloc., styles free or ± united, each loc. with (1 or 2–)6–12 anatropous, bitegmic ovules in 2 rows on axile or basal-axile placentas. Fr. a head of follicles, a drupe or berry or an inflated capsule, each loc. oft. with only 1 or 2 seeds; embryo straight in copious oily endosperm. n = (11-)13(14)
Genera: *Dalrymplea, Staphylea* (Tapiscioideae = Tapisciaceae, Huerteales)
Cult. orn. & timbers

Stapletonia P. Singh & al. (~ *Schizostachyum*). Gramineae (V 5). 2 Ind.

star anise *Illicium verum*; **s. apple** *Chrysophyllum cainito*; **s. of Bethlehem** *Ornithogalum angustifolium* & other spp.; **yellow s. of B.** *Gagea lutea*; **s. fruit** *Averrhoa carambola*; **s. gooseberry** *Breynia androgyna*; **s. grass** *Cynodon* spp. esp. *C. dactylon*; **s. jasmine** *Trachelospermum jasminoides*; **Texas s.** *Lindheimera texana*; **s. thistle, yellow** *Centaurea solstitialis*; **s.-vine** *Hydrilla verticillata*; **s.wort** *Callitriche* spp., *Stellaria* spp., *Symphyotrichum* spp.

starch comm. prep. from potatoes, cassava, maize, wheat & other cereals

starflower oil *Borago officinalis*

starry stonewort *Nitellopsis obtusa*

Stathmostelma K. Schum. (~ *Asclepias*). Apocynaceae (V c; Asclepiadaceae III 1). 13 trop. Afr. R: F. Albers & U. Meve, *Ill. Handb. Succ. Pls*, Asclep. (2002)260. ***S. spectabile*** (N.E. Br.) Schltr (E Afr.) – cut-fl. in Netherlands, esp. '**Beatrix**'

statice *Limonium* spp.

Staudtia Warb. Myristicaceae. 1 W Afr. to W Uganda: ***S. kamarunensis*** Warb. (*S. stipitata*, niové) – timber, little-known oilseeds, locally medic.; 1 São Tomé. R: FTEA Myrist. (1997)7

Staufferia Z. Rogers & al. Santalaceae (Cervantesiaceae). 1 Madag.: ***S. capuronii*** Z. Rogers & al. R: AMBG 95(2008)394. Poss. incl. *Pilgerina*

Stauntonia DC. Lardizabalaceae. Incl. *Holboellia, Parvatia*, 29 E As. R: CBM n.s. 29 (2012)260. Everg. monoec. lianes. ***S. hexaphylla*** (Thunb.) Decne. (mube, Jap.) – cult. orn. with fragrant white fls & ed. purple fr. favoured in Jap.

Stauracanthus Link. Leguminosae (III 9). 3 SW Eur. (3), NW Afr. R: Fl. Iberica 7,1(1999)240. Gorse-like prickly shrubs with 3-foliolate seedling lvs like *Ulex*

Stauranthera Benth. Gesneriaceae (III 1c). 8 SE As. to NG. R: JAA 65(1984)129

Stauranthus Liebm. Rutaceae (I 9). 1 S Mex. to Panamá: ***S. perforatus*** Liebm.

Staurochilus Ridl. ex Pfitzer = *Trichoglottis*

Staurochlamys Bak. Compositae (Helia.-Mel.). 1 N Braz.: ***S. burchellii*** Bak.

Staurogyne Wall. Acanthaceae (I). 145 trop. esp. Mal. & Am. (Afr. 5; R: BJBB 61(1991)98). Excl. those with epiphyllous infls (= *Saintpauliopsis*). Some ed. lvs chewed with betel (sweet flavour), local medic.

Staurophragma Fischer & C. Meyer = *Verbascum*

stavesacre *Staphisagria macrosperma*

stave-wood *Quassia amara*

Stawellia F. Muell. Asphodelaceae (Johnsoniaceae). 2 SW Aus. R: FA 45(1987)252. Form. referred to Anthericaceae & Liliaceae

Stayneria L. Bolus (~ *Ruschia*). Aizoaceae (V). 1 W Cape: ***S. neilii*** (L. Bolus) L. Bolus. R: H. E. K. Hartmann, *Ill. Handb. Succ. Pls*, Aiz. F–Z(2002)304

Stebbinsia Lipsch. = *Soroseris*

Stebbinsoseris K. Chambers = *Microseris* (but see FNA 19(2006)346)

steel grass *Xanthorrhoea johnsonii*

Steenisia Bakh.f. Rubiaceae (I 5). 5 Mal. Pen., Borneo, Natuna Is. R: NJB 4(1984)333

Steenisioblechnum Hennipman = *Blechnum*

steeplebush *Spiraea tomentosa*

Stefanoffia H. Wolff. Umbelliferae (III 8). 2 E Medit. (Eur. 1)

Steganotaenia Hochst. Umbelliferae (II). 3 Ethiopia to S Afr. ***S. araliacea*** Hochst. – tree (no evidence for 'secondary woodiness') with fls prod. before lvs; ***S. hockii*** (Norman) Norman – perenn. herb

Steganthera Perkins (~ *Matthaea*). Monimiaceae (V 2). Incl. *Anthobembix*, 18 E Mal. to Solomon Is. & trop. Aus. R: Blumea 29(1984)481. Some myrmecophytes

Stegia DC. = *Lavatera*

Stegnogramma Blume = *Cyclosorus*

Stegnosperma Benth. Stegnospermataceae (Phytolaccaceae s.l.). 3 Baja Calif. to WI & C Am. R: BSBM 46(1984)37. Some r. a soap subs

Stegnospermataceae Nakai (~ Phytolaccaceae). Magnoliidae – Caryophyllales. 1/3 SW N & C Am. Glabrous small trees & shrubs. Lvs simple, fleshy, in spirals; stip. 0. Fls in thyrses, racemes or cymules, reg., bisex. K 5, C (2-) 5 (? staminodes), A 5(–10) united at base, anthers with longit. slits, $\underline{G}$ 2–5, becoming 1-loc. with C column, each with 1 basal amphitropous bitegmic ovule. Fr. a capsule with 3–5 valves; seeds (1–)3–5 covered with red or white aril, perisperm copious, embryo curved
Genus: *Stegnosperma*

Stegocedrus Doweld = *Libocedrus*

Stegolepis Klotzsch ex Koern. Rapateaceae (I 1). 30+ N S Am.

Stegostyla D.L. Jones & M.A. Clem. = *Caladenia*

Steinbachiella Harms (~ *Diphysa*). Leguminosae (III 11). 1 Bolivia: ***S. leptoclada*** Harms. R: KB 67(2012)793

Steinchisma Raf. = *Otachyrium* (but see AJB 90(2003)817)

Steinheilia Decne. = *Cynanchum*

Steirachne Ekman. Gramineae (XXVII 2). 2 NE S Am.

Steiractinia S.F. Blake. Compositae (Helia.-Ecl.). 12 Venez., Colombia, Ecuador

Steirodiscus Less. (*Gamolepis*, *Psilothonna*). Compositae (Sen.-Sen.). 5 S Afr. R: Bothalia 43(2013)110. Some cult. orn.

Steiropteris (C. Chr.) Pichi-Serm. = *Cyclosorus*

Steirosanchezia Lindau = *Sanchezia*

Steirotis Raf. = *Struthanthus*

Stelbophyllum M.A. Clem. & D.L. Jones = *Bulbophyllum*

Stelechantha Bremek. = *Pauridantha* (but see KB 57(2002)401)

Stelechocarpus Hook.f. & Thomson. Annonaceae (IV 7). Incl. *Winitia*, 3 SE As., Mal. R: Blumea 40(1995)429. Roux's Model. Dioec., females cauliflorous. Ed. fr. esp. ***S. burahol*** (Blume) Hook.f. & Thomson (keppel fr., Mal.) – cult., gives all bodily secretions violet scent temporarily so a favourite of Javanese sultans' harems

Steleocodon Gilli = *Phalacraea*

Steleostemma Schltr. = *Philibertia*

Steliopsis Brieger = *Stelis*

Stelestylis Drude. Cyclanthaceae. 4 N S Am.

Stelis Sw. Orchidaceae (V 13c). c. 850 trop. Am. R: HBML 26(1980)167. Few cult. orn. epiphytes with small dingy fls

Stellaria L. Caryophyllaceae (II 1). c. 120 cosmop. (Eur. 18; N Am. 29 incl. introd.). Some cult. orn. herbs & weeds (stitchwort, chickweed, starwort) esp. ***S. holostea*** L. (greater stitchwort, adder's meat, Eur., Medit.) – chewed for 'stitch' & other muscle pains (until at least 1930s Somerset), ***S. media*** (L.) Vill. (common c., (?) orig. S Eur.) – cosmop. weed usu. autogamous, poss. also s.t. thrips-poll. in Netherlands, s.t. cleistogamous, ed. Greeks &

Romans & anc. Jap. as winter greenstuff, local cure-all & poultice for reducing swellings, ingredient of wt.-loss herbal teas, saponins basis of efficacy in treatment of skin disorders, & ***S. neglecta*** Weihe (greater c., Eur.) – since 1990s spreading weed US

Stellarioides Medik. = *Ornithogalum*

Stellariopsis (Baill.) Rydb. = *Potentilla*

Stellera Gmelin. Thymelaeaceae (Thym.-Daph.). 1 Iran to China: ***S. chamaejasme*** L. (C As., Himal.) – cult. shrub. R: NRBGE 40(1982)216

Stelleropsis Pobed. = *Diarthron*

Stelligera A.J. Scott = *Sclerolaena*

Stellilabium Schltr. = *Telipogon*

Stellularia Benth. = *Buchnera*

Stelmacrypton Baill. Apocynaceae (III; Asclepiadaceae I). 1 E Ind., S China: ***S. khasianum*** (Kurz) Baill.

Stelmagonum Baill. Apocynaceae (V c; Asclepiadaceae III 3). 2 trop. Am.

Stelmanis Raf. = *Heterotheca*

Stelmation Fourn. = *Metastelma*

Stelmatocodon Schltr. = *Philibertia*

Stelmatocrypton Baill. = *Pentanura*

Stemmacantha Cass. = *Rhaponticum*

Stemmadenia Benth. = *Tabernaemontana* (but see Candollea 60(2005)348)

Stemmatella Wedd. ex Benth. = *Galinsoga*

Stemmatodaphne Gamble = *Alseodaphne*

Stemodia L. Plantaginaceae (Grat.-Stem.; Scrophulariaceae s.l.). Excl. *Darcya, Leucospora* 56 trop. (Am. 32, R: Phytol. 74(1993)61, 75(1993)281)

Stemodiopsis Engl. Linderniaceae (Plantaginaceae (Grat.-Stem.); Scrophulariaceae s.l.). 6 trop. Afr. R: BJ 119(1997)310

Stemona Lour. Stemonaceae. c. 20 Indomal. to E As. & trop. Aus. Alks (some effective insecticides). ***S. tuberosa*** Lour. (Indomal.) – looks like hop pl., fls carrion-scented & visited by flies, tubers medic.

Stemonaceae Caruel. Magnoliidae – Pandanales. Incl. Pentastemonaceae, 4/32 Indomal. to E As. & trop. Aus., SE US. R: Blumea 36(1991)239. Usu. glabrous herbs, cli. or erect, or shrublets with rhiz. or tubers; vasc. bundles in 1 or 2 rings (vessels in at least r.); alks. Lvs with non-sheathing petioles & entire cordate laminas with 5–15 arching, convergent main veins with cross-veins, in spirals, distich., opp. or verticillate. Fls usu. bisex., axillary, solit. or in few-flowered axillary infls, 2-merous (5-merous in *Pentastemona*); P 2 + 2, s.t. basally united, A 2 + 2, filaments s.t. united or adnate to P, connective s.t. elongate, G (2), semi-inf. in *Stichoneuron*, 1-loc. with usu. few basal (*Stemona*) or apical pend., anatr. or orthotropous (*Stemona*) bitegmic ovules. Fr. a 2-valved capsule or 10-ribbed berry (*Pentastemona*); seeds 1–few, longit. ribbed bearing arillate structures (elaiosomes, caruncles etc.), embryo small in copious oily proteinaceous endosperm s.t. with hemicellulose. n = 7, 9, 12

Genera: *Croomia, Pentastemona* (ever-wet forests cf. other genera in seasonal habitats), *Stemona, Stichoneuron*

Pentastemona, form. excl. to own fam., is unique in monocots in reg. 5-merous 3-whorled fls

Some insecticides (*Stemona*)

Stemonocoleus Harms. Leguminosae (I 2). 1 Guineo-Congolian forests: ***S. micranthus*** Harms – timber, resins, medic.

Stemonoporus Thw. Dipterocarpaceae (Dipt.-Dipt.). c. 20 Sri Lanka. R: Blumea 20(1972) 363

Stemonuraceae Kårehed (~ Icacinaceae). Magnoliidae – Aquifoliales. 10/90 trop. OW. Trees & shrubs; term. buds abruptly contracted. Lvs simple, in spirals s.t. appearing distichous; stip. Fls bisexual, s.t. functionally unisex., reg, usu. small, in cymose infls to solit. K 4 or 5, connate, artic. with pedicel; C 4 or 5(7), usu. free, valvate; A opp. K, anthers with longit. slits; disk cupular etc. or 0; G 1 (? pseudomonomerous), 1-loc with 2 apical pend. anatr. unitegmic ovules. Fr. a drupe, oft. laterally compressed; seed 1 with small embryo. Recent split from Icacinaceae (q.v.), cf. Cardiopteridaceae

Principal genera: *Cantleya* (timber), *Gomphandra, Medusanthera, Stemonurus*

Stemonurus Blume. Stemonuraceae (Icacinaceae s.l). Incl. *Urandra*, c. 14 Indomal. R: Blumea 17(1969)255. Some timber

Stemotria Wettst. & Harms ex Engl. = *Calceolaria*

Stenachaenium Benth. Compositae (Inul.-Pluch.). 5 S Braz., Arg., Urug., Paraguay
Stenactis Cass. = *Erigeron*
Stenadenium Pax = *Euphorbia*
Stenandriopsis S. Moore = *Stenandrium*
Stenandrium Nees. Acanthaceae (III). 8 Afr., 10 Madag. (R: KB 47(1992)176), 25 warm Am. Some cult. orn.
Stenanona Standl. Annonaceae (IV 7). 15 Costa Rica, Panamá. R: Blumea 55(2010)206
Stenanthella Rydb. = *Anticlea*
Stenanthemum Reissek (~ *Cryptandra*). Rhamnaceae (5). 30–35 temp. & arid Aus., esp. SW. R: Nuytsia 13(2001)496, 16(2007)370
Stenanthium (A. Gray) Kunth (~ *Zigadenus*). Melanthiaceae. 3 SE US. R: Novon 12(2002)303. Some cult. orn., esp. ***S. densum*** (Desv.) Zomm. & Judd (*Z. d.*, crow poison)
Stenaria (Raf.) Terrell (~ *Hedyotis*). Rubiaceae (IV 1). 5 N Am. R: Sida 19(2001)592
Stenia Lindl. Orchidaceae (V 12j). Incl. *Dodsonia*, 22 N S Am. to Trinidad. Some cult. orn.
Stenocactus (K. Schum.) A. Berger = *Ferocactus*
Stenocarpha S.F. Blake = *Galinsoga*
Stenocarpus R. Br. Proteaceae (V 3). 21 W Pac. (Aus. 9, 7 endemic; poss. incl. *Strangea*). Some timbers esp. ***S. salignus*** R. Br. (beefwood, E Aus.) – used for furniture; some cult. orn. incl. ***S. sinuatus*** Endl. (firewheel tree, E Aus.) with red fls, foliage a 'cut-fl.' crop in NE Aus.
Stenocephalum Schultz-Bip. (~ *Vernonia*). Compositae (Vern.-Lep.). 5 trop. Am. R: PBSW 100(1987)582
Stenocereus (A. Berger) Riccob. (*Rathbunia*; ~ *Lemaireocereus*). Cactaceae (III 1). 24 S US (1), Mex. to WI. Ed. fr. (pitaya) incl. cult. ***S. queretaroensis*** (Mathsson) Buxbaum (Mex.) – fr. glochid-free & ***S. stellatus*** (Pfeiffer) Riccob. (C Mex.); ***S. thurberi*** (Engelm.) Buxbaum (*L. t.*) – fr. in 'cactus jam'
Stenochasma Miq. = *Broussonetia*
Stenochilus R. Br. = *Eremophila*
Stenochlaena J. Sm. Blechnaceae (Stenochlaenoideae). 6–7 trop. OW (Afr. 2 ext. to Madag. & S Afr.). R: Telopea 15(2013)15. Young fronds of some ed. & cult. orn. cli. epiphytes
Stenochlaenaceae Ching = Blechnaceae
Stenochlamys Griff. = ? *Davallia*
Stenocline DC. Compositae (Gnap.-Gnap.). 3 Madag., ? Mauritius. R: OB 104(1991)138. Some locally medic.
Stenocoelium Ledeb. Umbelliferae (III 8). 3 C As. ***S. divaricatum*** Turcz. – used with *Schizonepeta tenuifolia* in treatment of toothache & colds in modern Chinese herbalism
Stenocoryne Lindl. = *Bifrenaria*
Stenodon Naud. Melastomataceae. 1 S Braz.: ***S. suberosus*** Naud.
Stenodraba O. Schulz = *Weberbauera*
Stenodrepanum Harms. Leguminosae (I 4). 1 Arg.: ***S. bergii*** Harms. R: PhytoKeys 71(2016)105
Stenofestuca (Honda) Nakai = *Bromus*
Stenoglottis Lindl. Orchidaceae (IV 4d). 7 E & S Afr. R: KM 6(1989)9. Cult. orn. terr. orchids
Stenogonum Nutt. (~ *Eriogonum*). Polygonaceae (Erio.-Erio.). 2 W N Am. R: FNA 5(2005)431
Stenogrammitis Labiak (~ *Lellingeria*). Polypodiaceae (V.). 24 trop. R: Britt. 63(2011)141
Stenogyne Benth. Labiatae (VI). 21 Hawaii. W. Wagner & al., *Man. Fl. Pl. Hawaii* (1990)831, PS 45(1991)30. Lianes & herbs, app. derived from N Am. *Stachys*
Stenolepia Alderw. Dryopteridaceae. 1 C & E Mal.: ***S. tristis*** (Blume) Alderw.
Stenolirion Bak. = *Ammocharis*
Stenoloma Fée = *Odontosoria*
Stenomeria Turcz. = *Tassadia* (but see ABV 16(1990)82)
Stenomeris Planch. Dioscoreaceae. 2 W & C Mal. R: T 51(2002)111. Rhiz. to 2 m long
Stenomesson Herb. Amaryllidaceae. Excl. *Clinanthus* c. 13 Andes. R: PL 27(1971)73
Stenopadus S.F. Blake. Compositae (Wund./Mut.-Wund.). 14 Guayana Highland. R: AMBG 76(1989)1002. Pachycaul shrubs & trees
Stenopetalum R. Br. ex DC. Cruciferae (37). 10 Aus. R: JAA 53(1972)52. 2 spp. with 2n reduced to 4. ***S. velutinum*** F. Muell. – Aboriginal veg.
Stenophalium A. Anderb. = *Achyrocline* (but see OB 104(1991)141)
Stenops R. Nordenstam (~ *Senecio*). Compositae (Sen.-Sen.). 2 trop. Afr. ***S. zairensis*** (Lisiowski) R. Nordenstam (Congo) – floating aquatic

Stenoptera C. Presl. Orchidaceae (IV 2b). 7 Andes

Stenorrhynchos Rich. ex Spreng. Orchidaceae (IV 2h). 5 trop. Am. R: HBML 28(1982)372. ***S. lanceolatum*** (Aubl.) Rich. (Braz.) – hummingbird-poll.

Stenoschista Bremek. = *Ruellia*

Stenosemia C. Presl = *Tectaria*

Stenosemis E. Meyer ex Harv. & Sonder (~ *Annesorhiza*). Umbelliferae (III 8). 2 S Afr.

Stenosepala Persson. Rubiaceae (II Cord.). 1 Panamá, Colombia: ***S. hirsuta*** Persson. R: Novon 10(2000)403

Stenoseris Shih = *Notoseris* (but see APS 29(1991)411)

Stenosiphon Spach = *Oenothera*

Stenosiphonium Nees = *Strobilanthes* but see BotJLS 133(2000)101

Stenosolen (Muell. Arg.) Markgraf = *Tabernaemontana*

Stenosolenium Turcz. (~ *Arnebia*). Boraginaceae (B.2.2). 1 C As.: ***S. saxatile*** (Pall.)Turcz. R: JAA 35(1954)46

Stenospermation Schott. Araceae (IV 4). 36 (–c. 250) trop. Am.

Stenostachys Turcz. (~ *Elymus*). Gramineae (XV). 4 NZ. R: Telopea 13(2011)43

Stenostegia Bean. Myrtaceae (II 15). 1 N Aus.: ***S. congesta*** Bean. R: Muelleria 11(1998) 127

Stenostelma Schltr. (~ *Schizoglossum*). Apocynaceae (V c; Asclepiadaceae III 1). 4 S Afr. R: F. Albers & U. Meve, *Ill. Handb. Succ. Pls*, Asclep. (2002)263

Stenostomum Gaertn. f. NW spp. of *Antirhea*

Stenostephanus Nees. Acanthaceae (III 2c). Incl. *Habracanthus*, *Kalbreyeracanthus*, 75 trop. S Am.

Stenotaenia Boiss. (~ *Heracleum*). Umbelliferae (III 11). 5–6 SW As.

Stenotalis Briggs & L. Johnson = *Leptocarpus* (but see Telopea 7(1998)368)

Stenotaphrum Trin. Gramineae (XXIV 4). 7 trop. & warm. R: Britt. 24(1972)202. Some fodders; ***S. secundatum*** (Walter) Kuntze (St Augustine grass, buffalo grass (Aus.), warm Am., invasive S Eur., Azores, Aus.) – lawn-grass & for binding sand, varieg. '**Variegatum**' a hanging-basket pl.

Stenothyrsus C.B. Clarke. Acanthaceae (III 2). 1 Malay Pen.: ***S. ridleyi*** C.B. Clarke

Stenotis Terrell = *Hedyotis* (but see Sida 19(2001)901)

Stenotus Nutt. (~ *Haplopappus*). Compositae (Ast.-Sol.). 4 W N Am. R: FNA 20(2002)174

Stenotyla Dressler = *Chondrorhyncha*

Stephanachne Keng. Gramineae (IX). 3 C As., W China (3 – R: FOC 22(2006)189)

Stephanandra Sieb. & Zucc. = *Neillia*

Stephanbeckia H. Robinson & Funk. Compositae (Liab.). 1 S Bolivia: ***S. plumosa*** H. Robinson & Funk – coll. once. R: Britt. 63(2011)78

Stephania Lour. Menispermaceae (V). 35 trop. OW (Afr. 5). R: KB 11(1956)43, 22(1960)352. Many alks; some locally medic. incl. ***S. japonica*** (Thunb.) Miers (*S. hernandiifolia*, Indomal. to E Aus.) – fish-poison

Stephanocaryum Popov = *Trigonotis*

Stephanocereus A. Berger (~ *Cephalocereus*). Cactaceae (III 3). 2 Bahia. R: Bradleya 9(1991)91. Cult. orn. tree or unbranched

Stephanochilus Cosson & Durieu ex Maire = *Centaurea*

Stephanococcus Bremek. Rubiaceae (IV 1). 1 trop. Afr.: ***S. crepinianus*** (K. Schum.) Bremek.

Stephanodaphne Baill. Thymelaeaceae (Thym.-Daph.). 9 Madag., Comoro Is. R: Adansonia 26(2004)12

Stephanodoria Greene (~ *Xanthocephalum*). Compositae (Ast.-Mac.). 1 Mex.: ***S. tomentella*** (Robinson) Greene. R: SBM 20(2000)82

Stephanolepis S. Moore = *Erlangea*

Stephanomeria Nutt. Compositae (Cich.-Cich.(-Micr.)). 22 W N Am. (Canada & US 14 – R: FNA 19(2006)350). Local medic. ***S. diegensis*** Gottlieb (*S. exigua* Nutt. × *S. virgata* Benth.) – 1 of only 8 confirmed homoploid hybrid pl. spp

Stephanopholis S.F. Blake = *Chromolepis*

Stephanophysum Pohl = *Ruellia*

Stephanopodium Poeppig. Dichapetalaceae. 13 trop. S Am. R: FN 10(1972)36, KB 50(1995)295. Infls on petioles

Stephanorossia Chiov. = *Oenanthe*

Stephanostachys Zuloaga & Morrone (~ *Panicum*). Gramineae (XXIII 1). 1 trop. Am.: ***S. mertensii*** (Roth) Zuloaga & Morrone

Stephanostegia Baill. Apocynaceae (I e). 2 Madag. R: WAUP 97–2(1997)95

Stephanostema K. Schum. Apocynaceae (II a). 1 Tanz.: *S.* ***stenocarpum*** K. Schum. (?) extinct in wild (in cult. Kew, Wageningen), aestivation to left!

Stephanotella Fourn. = *Marsdenia*

Stephanothelys Garay. Orchidaceae (IV 2d). 5 Andes

Stephanotis Thouars = *Marsdenia*

Steptorhamphus Bunge = *Lactuca*

Sterculia L. Malvaceae (Sterc.; Sterculiaceae). c. 150 trop. Monoec. or polygamous trees (Aubréville's Model; for New Caled. remarkable unbranched pachycauls see *Acropogon*). Many spp. with sugar-secreting hairs within K, attracting lots of bugs & cicadas which lead to enlarging & sterilizing of some floral parts. Some timbers, gums (used in comm. bowel regulators), ed. seeds & cult. orn. ***S. africana*** (Lour.) Fiori (S Arabia) – resin coll. to make a lathering agent against lice; ***S. apetala*** (Jacq.) Karsten (trop. Am.) – seeds ed. roasted, national tree Panamá; ***S. balanghas*** L. (SE As. to trop. Aus.) – seeds ed.; ***S. chicha*** A. St-Hil. (~ *S. a.*, NE S Am.) – ed. seeds, oil used in lubrication etc.; ***S. cinerea*** A. Rich. (NE Afr.) – Tartar gum; ***S. fanaiho*** Setch. (Tonga) – fibre for textiles & mats; ***S. foetida*** L. (Java olive, trop. OW) – ed. seeds, oil, fls smell of decaying meat, bole to 2 m girth; ***S. monosperma*** Vent. (Chinese chestnut, China) – ed. seeds; ***S. oblonga*** Mart. (trop. Afr.) – timber (eyong, yellow sterculia) dense, for tables, boats, sleepers etc.; ***S. rhinopetala*** K. Schum. (W Afr.) – timber (aye, brown sterculia); ***S. urens*** Roxb. (Indomal.) – source of karaya (or kutira) gum or Ind. tragacanth used as tragacanth subs. in cosmetics & ice cream; ***S. villosa*** Roxb. (Ind., Myanmar) – timber for tea-chests, bark-fibre for elephant harness, coarse canvas, bags etc. & form. for paper; ***S. vitiensis*** Seemann (Vanuatu, Fiji) – seed ed.

Sterculiaceae Vent. = Malvaceae

Stereocaryum Burret (~ *Eugenia*). Myrtaceae (II 10). 3 New Caled.

Stereochilus Lindl. (~ *Sarcochilus*). Orchidaceae (V 16c). 7 Ind. to Thailand & Vietnam

Stereochlaena Hackel. Gramineae (XXIV 4). 4 trop. to S Afr. R: KB 33(1978)295

Stereosandra Blume. Orchidaceae (V 8b). 1 SE As., W Mal.: ***S. javanica*** Blume – mycotrophic

Stereospermum Cham. Bignoniaceae (1). 20 trop. OW (Indomal 4, Afr. 4, Madag, 12). Cult. orn. trees with good timber (cf. most B.) esp. ***S. colais*** (Dillwyn) Mabb. (*S. personatum*, padri, Ind.) – tea-chests, furniture, constr., imp. local medic.

Sterigmapetalum Kuhlm. Rhizophoraceae (I). 9 trop. S Am. R: AMBG 70(1983)179

Sterigmostemum M. Bieb. Cruciferae (5). Incl. *Oreoloma*, 11 Eur. (1), SW As. to China. R: EJB 62(2006)115

Steriphoma Spreng. Capparaceae. 8 Trinidad & Guatemala to Peru

Steris Adans. = *Silene* (*Viscaria*)

Sternbergia Waldst. & Kit. Amaryllidaceae. 7 SE Eur. (2) to SW As. & Kashmir. R: Plantsman 5(1983)1. Some spp. with ant-disp. seeds with fleshy arils remaining soft & sticky on herbarium specimens to 35 yrs old! Cult. orn. with alks, a yellow-fld 'version' of *Colchicum*, esp. ***S. lutea*** (L.) Spreng. (SE Eur. to Iran) – contender for 'lilies of the field', also ***S. vernalis*** (Mill.) Gorer & J. Harvey (*S. fischeriana*, Cauc. to Kashmir) with larger fls

Sterropetalum N.E. Br. = *Nelia*

Stethoma Raf. = *Justicia*

Stetsonia Britton & Rose. Cactaceae (III 1). 1 Bolivia, Paraguay, Arg.: ***S. coryne*** (Salm-Dyck) Britton & Rose – fr. ed., hedge-pl., cult. orn. tree-like

Steudnera K. Koch. Araceae (VII 25). 9 Himal. to SE As. & Malay Pen.

Stevenia Adams ex Fischer. Cruciferae (45). Incl. *Ptilotrichum*, 8 Euras.

Steveniella Schltr. Orchidaceae (IV 4d). 1 Crimea to W As.: ***S. satyroides*** (Spreng.) Schltr. (*S. caucasica*)

Stevensia Poit. Rubiaceae (I 5). 11 WI. R: Phytol. 70(1991)151

stevia *Piqueria trinervia*

Stevia Cav. Compositae (Eup.-Piq.). c. 200 trop. & warm Am. R: MSBMBG 22(1987) 170. ***S. rebaudiana*** (Bertoni) Bertoni (caa-ehe, Paraguay) – long used by indigenous people for sweetening drinks, a diterpene glycoside (stevioside) up to 300 times as sweet as sucrose (i.e. up to three fifths as effective as saccharine) now much used in Jap., also 'Coca-Cola Life' (2013), the shrub being prop. by r.-cuttings (seeds diff.)

Steviopsis R. King & H. Robinson (~ *Eupatorium*). Compositae (Eup.-Alom.). 10 SW US, Mex. R: MSBMBG 22(1987)247, Phytol. 76(1994)389

Stewartia Lawson. Theaceae. c. 20 China, Jap. (3), SE US (2). R: JAA 55(1974)182. Cult. orn. decid. trees & shrubs

Stewartiella Nasir. Umbelliferae (III 5). 2 Pakistan

Steyerbromelia L.B. Sm. Bromeliaceae (1). 9 C Amazonas, Venez. R: AMBG 73(1986)699, SB 40(2015)743

Steyermarkia Standl. Rubiaceae (I 6). 1 C Am.: ***S. guatemalensis*** Standl.

Steyermarkina R. King & H. Robinson (~ *Eupatorium*). Compositae (Eup.-Crit.). 4 Venez. & Braz. R: MSBMBG 22(1987)368

Steyermarkochloa Davidse & R. Ellis. Gramineae (XVII). 1 Venez. & Colombia: ***S. angustifolia*** (Spreng.) Judz. – forming monospecific tribe. R: AMBG 77(1990)204

Stiburus Stapf (~ *Eragrostis*). Gramineae (XXVII 3). 2 S Afr.

Sticherus C. Presl. Gleicheniaceae (Gleich.). c. 80 trop. (Afr. 3) & S temp. Some cult. orn.

Stichianthus Valeton. Rubiaceae (I 9). 1 Borneo: ***S. minutiflorus*** Valeton

Stichoneuron Hook.f. Stemonaceae (Croomiaceae). 5 Assam to Malay Pen. R: EJB 66(2009)217

Stichorkis Thouars (~ *Liparis*). Orchidaceae (V 11b). 8 OW trop.

sticktight *Bidens* spp.

sticky wattle *Acacia howittii*; **s. weed** *Galium aparine*

Stictocardia Hallier f. (~ *Ipomoea*). Convolvulaceae (9). 10 trop. OW, some natur. Am. R: KB 52(1997)166, Willd. 31(2001)80. ***S. beraviensis*** (Vatke) Hallier f. (Afr.) – liane with crimson fls, cult. orn.

Stictophyllorchis Carnevali & Dodson = *Quekettia* (but see Lindleyana 8(1993)101)

Stictophyllum Dodson & M. Chase = *Quekettia*

Stifftia Mikan. Compositae (Stiff.-Stiff.). 6 NE Braz., Fr. Guiana. R: SB 16(1991)685. Trees, shrubs & lianes

Stigmaphyllon A. Juss. Malpighiaceae. Incl. *Ryssopterys*, c. 120 W Pacific (21), trop. Am. R: SBM 51(1997)18, [subg. *R.*]: Blumea 56(2011)76. Lianes, some cult. orn. esp. ***S. ciliatum*** (Lam.) A. Juss. (Belize to Urug.)

Stigmatella Eig = *Eigia*

Stigmatodactylus Maxim. ex Makino (*Pantlingia*). Orchidaceae (IV 3a). 10 Himal. to Jap. & New Caled. Some autogamy

Stigmatopteris C. Chr. Dryopteridaceae (II). c. 28 trop. Am. R: AMBG 78(1991)868

Stigmatorhynchus Schltr. Apocynaceae (V a; Asclepiadaceae III 4). 2–3 E & S Afr.

Stigmatorthos M. Chase & D. Bennett = *Comparettia* (but see Lindleyana 8(1993)3)

Stigmatosema Garay = *Cyclopogon*

Stilaginaceae Agardh = Phyllanthaceae

Stilbaceae Kunth (~ Verbenaceae). Magnoliidae – Lamiales. Incl. Retziaceae, 12/41 Cape, *Nuxia* ext. to S Arabia & Masc. Small trees to ericoid shrubs. Lvs simple, opp. ternate or in dense pseudowhorls of 3–7, rarely in spirals; stip. 0 or a ring. Fls (4)5(-7)-merous, bisex., reg. to zygomorphic, in usu. term. thyrses; bracteoles as long as K. K (4 or 5), s.t. bilobed; C (4 or 5) ± equal lobes & cylindrical to funnel-shaped tube; A = & opp. K; $\underline{G}$ 2-loc., apically 1-loc. with 1–3 (or many, *Nuxia*) ovules per carpel. Fr. a usu. loculicidal capsule, s.t. septicidal or a berry (*Halleria*); seeds with endosperm. n = 10, 12, 19

Principal genera: *Bowkeria, Euthystachys, Halleria, Nuxia, Retzia, Stilbe*

Split from Verbenaceae s.l. & united with *Nuxia* & *Retzia* excl. from Loganiaceae as lvs usu. opp. acicular, & Bowkerieae from Scrophulariaceae s.l. Some with elaiophores attractive to bees (cf. Scrophulariaceae s.s.)

Cult. orn. (*Bowkeria, Halleria, Nuxia*)

Stilbanthus Hook.f. Amaranthaceae (I 2). 1 Himal.: ***S. scandens*** Hook.

Stilbe P. Bergius. Stilbaceae. Incl. *Eurylobium, Xeroplana*, 7 W Cape. R: Bothalia 30(2000)11

Stilbocarpa (Hook.f.) Decne. & Planch. (~ *Huanaca*). Umbelliferae (Azor.). 1 islands off S NZ: ***S. polaris*** (Hombr. & Jacquinot) A. Gray – form referred to Araliaceae, stout herb form. antiscorbutic for sailors

Stilbophyllum M.A. Clem. & D.L. Jones = *Bulbophyllum*

Stilifolium Königer & Pongratz = *Oncidium*

Stillingia Garden ex L. Euphorbiaceae (Euph.-Hipp.). c. 30 trop. & warm Am. (27) incl. US, Madag. (1–2), Masc., E Mal. & Fiji (1). R: AMBG 38(1951)207, 75(1988)1666. Alks, those of ***S. sylvatica*** Garden ex L. (SE US) rhiz. medic.

Stilpnogyne DC. Compositae (Sen.-Sen.). 1 S Afr.: ***S. bellidioides*** DC. R: Strel. 10(2000)161

Stilpnolepis H. Kraschen. (~ *Artemisia*). Compositae (Anth.-Art.). 1 China, Mongolia: ***S. centiflora*** (Maxim.) Krasch. R: APS 23(1985)470

Stilpnopappus Mart. ex DC. Compositae (Vern.-Lep.). 20 trop. S Am.

Stilpnophleum Nevski = *Calamagrostis*

Stilpnophyllum Hook.f. Rubiaceae (I 1). 4 Andes of Ecuador & Peru

Stilpnophyton Less. = *Athanasia*

Stimpsonia C. Wright ex A. Gray (~ *Androsace*). Primulaceae (Myrsinaceae). 1 E As.: ***S. chamaedryoides*** C. Wright ex A. Gray

stinger *Urtica dioica, Dendrocnide* spp.

stinking cedar or yew *Torreya taxifolia*; **s. gladwin** *Iris foetidissima*; **s. Roger** *Tagetes minuta*; **s. Willie** *Jacobaea vulgaris*

stinkweed *Thlaspi arvense*

stinkwood *Coprosma foetidissima, Gustavia augusta, Jacksonia scoparia, Ocotea bullata* (**black s.**), *Zieria arborescens* etc.; **red s.** *Prunus africana*

stinkwort *Helleborus foetidus*

Stipa L. Gramineae (X). Excl. *Achnatherum, Amelichloa, Anemanthele, Austrostipa, Celtica, Hesperostipa, Jarava, Macrochloa, Nassella, Orthoraphium, Patis, Piptochaetium, Ptilagrostis, Stipellula*, q.v., c. 110 warm & temp. Euras. (SW & S As. 42, R: NRBGE 42(1985)355), N Afr., oft. dry. Feather grass, corkscrew grass (lvs oft. inrolling in dry conditions covering stomata & green tissues (on adaxial surface only)); awns long (to 50 mm in ***S. pulcherrima*** K. Koch (C & S Eur.) – cult. orn.) & feathered, the caryopsis with backward-pointing hairs, such that fr. is driven into skin, eyes & mouths of stock (cf. *Heteropogon*) & contaminates wool. Some cult. orn. e.g. ***S. pennata*** L. (Euras.), fibres etc.

Stipagrostis Nees. Gramineae (Arist.). 56 desert & semi-desert OW (Eur. 2, China 2). Like *Aristida* but with plumose awns. ***S. plumosa*** (L.) T. Anders. (warm OW) – much favoured by horses in Middle E

Stipecoma Muell. Arg. Apocynaceae (II e). 1 Braz.: ***S. peltigera*** (Stadelm.) Muell. Arg. R: Candollea 60(2005)309

Stipella (Tzvelev) Röser & Hamasha = *Stipellula*

Stipellula Röser & Hamasha (~ *Stipa*). Gramineae (X). 2 Medit. R: PSE 298(2012)365

Stiptanthus (Benth.) Briq. = *Anisochilus*

Stipularia P. Beauv. = *Sabicea*

Stipulicida Michaux. Caryophyllaceae (I 1). 2 SE US, Cuba. R: FNA 5(2005)27

Stirlingia Endl. Proteaceae (IV 1). 7 Aus. R: FA 16(1995)136

Stironeurum Radlk. = *Synsepalum*

Stirtonanthus Wyk & Schutte (~ *Podalyria*). Leguminosae (III 7.). 3 W Cape. R: NJB 15(1995)67

Stirtonia Wyk & Schutte = praec.

stitchwort *Stellaria* spp.; **greater s.** *S. holostea*

Stixis Lour. Resedaceae. 7 E Himal. to Hainan & Lesser Sunda Is. R: Blumea 12(1963)5. Excluded from Capparaceae (G usu. 3-loc., placentation axile etc.), once put in own fam. ('Stixaceae')

Stixidaceae Doweld = Resedaceae

Stizolobium P. Browne = *Mucuna*

Stizolophus Cass. (~ *Centaurea*). Compositae (Card.-Cent.). 2 E Medit. to C As.

Stizophyllum Miers. Bignoniaceae (3). 3 trop. Am. R: AMBG 99(2014)462. Twigs hollow

stock, Brompton, garden or hoary *Matthiola incana*; **night-scented s.** *M. longipetala* subsp. *bicornis*; **s. rose** *Sparrmannia africana*; **Virginia s.** *Malcolmia maritima*

Stocksia Benth. Sapindaceae. 1 E Iran, Afghanistan: ***S. brahuica*** Benth.

Stockwellia D.J. Carr & al. Myrtaceae (II 11). 1 Atherton Tableland, Queensland: ***S. quadrifida*** D.J. Carr & al. R: BotJLS 139(2002)416

Stoebe L. Compositae (Gnap.-Rel.). 34 trop. & S Afr. to Masc. R: JSAB 3(1937)1

Stoeberia Dinter & Schwantes (~ *Ruschia*). Aizoaceae (V). 5 SW Afr. R: H. E. K. Hartmann, *Ill. Handb. Succ. Pls*, Aiz. F–Z(2002)305

Stoibrax Raf. (*Brachyapium*). Umbelliferae (III 8). 4 Iberia (1), W Cape. R: EJB 48(1991)250

Stokesia L'Hérit. Compositae (Vern.-Sto.). 1 SE US: ***S. laevis*** (Hill) Greene (*S. cyanea*) – cult. orn. herb with blue fls, poss. comm. oilseed. R: EB 28(1974)130, (cvs) Plantsman 5(2006) 80

Stokoeanthus E. Oliv. = *Erica*

Stolzia Schltr. Orchidaceae (V 15). 15 trop. Afr. R: KB 33(1978)79

Stomandra Standl. = *Rustia*

Stomatanthes R. King & H. Robinson (~ *Eupatorium*). Compositae (Eup.-Gyp.). Excl. Am. spp. (see MSBMBG 22(1987)69), 4 trop. Afr. R: SB 38(2013)837

Stomatium Schwantes. Aizoaceae (V). 39 SW Cape to Namaqualand & Karoo. R: H. E. K. Hartmann, *Ill. Handb. Succ. Pls*, Aiz. F–Z(2002)308. Cult. orn. tufted succ.

Stomatochaeta (S.F. Blake) Maguire & Wurd. Compositae (Wund./Mut.-Wund.). 6 Venez., Guyana, Braz. R: Britt. 41(1989)37. Shrubs

Stomatostemma N.E. Br. = *Cryptolepis*

stone pine *Pinus pinea*; **s. root** *Collinsonia canadensis*

stonecrop *Sedum acre* & other *S.* spp.

Stonesia G. Taylor. Podostemaceae (III). 4 trop. W Afr. R: Adansonia 13(1973)307

Stonesiella Crisp & P. Weston (~ *Pultenaea*). Leguminosae (III 13). 1 Tasmania: ***S. selaginoides*** (Hook.f.) Crisp & P. Weston. R: T 48(1999)711

stonewort Spp. of Characeae; **starry s.** *Nitellopsis obtusa*

storax *Styrax officinalis*, see also *Liquidambar orientalis*; **American s.** *L. styraciflua*

Storckiella Seemann. Leguminosae (I 3). 4 New Caled. (2), Queensland (1), Fiji (1). R: Adans. 27(2005)218. Allied to *Koompassia* (Mal.), *Baudouinia* & *Mendoravia* (Madag.), app. an arch. group. Winged fr. effectively samara

storksbill *Erodium* spp.

Storthocalyx Radlk. Sapindaceae. 5 New Caled. R: SB 41(2016)388

Stracheya Benth. = *Hedysarum*

Strailia T. Durand = *Lecythis*

Stramentopappus H. Robinson & Funk (~ *Vernonia*). Compositae (Vern.-Leib.). 1 Mex.: ***S. pooleae*** (B. Turner) H. Robinson & Funk. R: BJ 108(1987)227

stramonium *Datura stramonium*

Strangea Meissn. (~ *Stenocarpus*). Proteaceae (V 3). 3 Aus. R: FA 16(1995)360

Strangweja Bertol. = *Bellevalia*

Stranvaesia Lindl. = *Photinia* (but see CJB 68(1990)2248)

strapwort *Corrigiola litoralis*

Strasburg turpentine *Abies alba*

Strasburgeria Baill. Strasburgeriaceae (Ochnaceae s.l.). 1 New Caled. (ultramafic soils): ***S. robusta*** (Pancher & Sebert) Guillaumin – sister to *Ixerba*, subpachycaul, Attims's Model, 2n = c. 500

Strasburgeriaceae Tiegh. (~ Ochnaceae). Magnoliidae – Crossosomatales. Incl. Ixerbaceae, 2/2 New Caled., NZ. Everg. trees. Lvs simple, toothed, in spirals, opp. or veticillate (*Ixerba*); stip. intrapetiolar, basally connate, or 0 (*Ixerba*). Fls solit., axillary or in term. corymbose infls (*Ixerba*); K 8–10 imbr., in spiral, C 5(6) imbr., A 5 + 5, with long filaments & sagittate anthers with longit. slits, $\underline{G}$ (4–7) with usu. 1 anatr. bitegmic ovule per loc. Fr. indehiscent or loculicidal caps. (*Ixerba*), K persistent; seed arillate & scant endosperm. n (*Strasburgeria*) = 250(!)

Genera: *Ixerba*, *Strasburgeria*

Pollen referred here found in Tertiary of Aus. & NZ

Strateuma Raf. = *Zeuxine*

Stratiotes L. Hydrocharitaceae. 1 Euras.: ***S. aloides*** L. (water soldier) – dioec. stoloniferous rosetted perenn. aquatic, floating to surface & fls in summer, sinking in autumn, this rise & fall alleged to be due to levels of lime in lvs; in GB usu. female, occ. bisex. but poss. just 1 clone; cult. orn. & where abundant used as manure. R: AB 16(1983)213

Straussiella Hausskn. Cruciferae (2). 1 Iran: ***S. purpurea*** (Bunge) Hausskn.

strawberry *Fragaria ananassa*; **alpine s.** *F. vesca*; **barren s.** *P. sterilis*; **s. bramble** *Rubus pedatus*; **s. guava** *Psidium cattleyanum*; **hautbois s.** *F. moschata*; **Indian s.** *P. indica*; **s. myrtle** *Ugni molinae*; **Plymouth s.** *F. vesca* 'Muricata'; **s. tomato** *Physalis pruinosa*; **s. tree** *Arbutus unedo*; **wild s.** *F. vesca*

strawflower *Xerochrysum bracteatum*, (on porcelain) pattern derived from Chinese 'Wheel of the Year'

Streblacanthus Kuntze (~ *Schaueria*). Acanthaceae (III 2c). 7 C Am. ***S. roseus*** (Radlk.) B.L. Burtt (*Pseuderanthemum r.*, *Sciaphyllum amoenum*, Peru) – cult. orn.

Streblochaete Hochst. ex Pilg. (~ *Koordersiochloa*). Gramineae (XI). 1 trop. Afr., Réunion, & Mal. mts: ***S. longiarista*** (A. Rich.) Pilg. – coiling of awn draws out floret exposing a callus which adheres to clothes & pelts so that tangles of florets are disp. R: Strel. 10(2000)718

Streblorrhiza Endl. Leguminosae (III 24). 1 Philip Is. (nr. Norfolk Is.): ***S. speciosa*** Endl. – extinct 19 cent. through grazing by feral animals. R: FA 49(1994)186

Streblosa Korth. (~ *Psychotria*). Rubiaceae (IV 7). 25 W Mal. R: JAA 28(1947)145

Streblosiopsis Valeton. Rubiaceae (I 8?). 1 Borneo: ***S. cupulata*** Valeton

Streblus Lour. Moraceae (I). Excl. *Bleekrodea*, 14 Afr. (1), Madag., Indomal. to Solomon Is. & Norfolk Is. (1, endemic). R: PKNAW C91(1988)356. Troll's Model. ***S. asper*** Lour. (paper-bark, trop. As.) – locally medic., used for paper-making in Thailand, where also a topiary subject in temples; S. ***brunonianus*** F. Muell. (whalebone tree, NE & E Aus.) – wind-poll.?, timber

Strelitzia Sowerby. Strelitziaceae (Musaceae s.l.). 4–5 S Afr. Woody pls of riverbanks & forest glades with alks, & 2-ranked banana-like lvs; infl. encl. in spathe from which fls emerge one at a time, inner tepals forming arrow-like structure traversed by a longit. groove in which the 5 stamens & style lie, pollen united by filaments into viscous mass; arils orange, hairy, containing bilirubin (an 'animal' pigment, as in jaundice). Cult. orn. greenhouse pls esp. ***S. reginae*** Sowerby (bird-of-paradise, crane fl., Cape) – dichotomizing trunk to 1 m (Schoute's Model), infl. with a conspic. droplet & bright orange & purple fls attractive to *Nectarinia afra* (sunbird), which alights on perch-like spathe &, in getting to nectar, presses against swollen stigma, depositing pollen from another fl.; at same time, pollen lying in the 'arrow' deposited on bird's feathers to be taken to another fl. (poll. poss. also effected by birds' feet); '**Prolifera**' – cv with 1 infl. emerging from another; var. ***juncea*** (Ker Gawl.) H.E. Moore (*S. juncea*, *S.* × *kewensis* 'Juncea') – wild pops with lvs rush-like as lamina suppressed, cult. orn.;. ***S. nicolai*** Regel & Koern. (S Natal, NE Cape) – trunk to 10 m (Tomlinson's Model)

Strelitziaceae Hutch. (~ Musaceae). Magnoliidae – Zingiberales. 3/6–7 S Afr., Madag. E S Am. Trees with suckers, or long-lived rhiz. perennials. Lvs entire (tearing in wind (as Musaceae), distich. with long petioles or sheaths & midrib with many close-set lat. veins. Fls bisex. in thyrses subtended by leathery boat-shaped bracts. C basally connate, A 5 or 6, staminodes 0, $\overline{G}$ 3-loc; ovules ∞, anatropous, on axile placentas. Fr. a woody capsule; seeds with fimbriate arils & copious starchy endosperm. n = (7) 11

Genera: *Phenakospermum, Ravenala, Strelitzia*

Cult. orn.

Strempelia A. Rich. ex DC. = *Psychotria*

Strempeliopsis Benth. Apocynaceae (Ia). 2 Cuba, Jamaica

Strephium Schrad. ex Nees = *Raddia*

Strephonema Hook.f. Combretaceae (I). 3 trop. W Afr. R: AMBG 82(1995)535. Troll's Model

Streptachne R. Br. = *Aristida*

Streptanthella Rydb. = *Streptanthus* (but see FNA 7(2010)699)

Streptanthera Sweet = *Sparaxis*

Streptanthus Nutt. Cruciferae (46). Incl. *Caulanthus*, 54 W & S N Am. R: FNA 7(2010)700. A of 3 unequal lengths. Some nickel accumulators on serpentine; some cult. orn. annuals. ***S. flavescens*** Hook. (*C. procerus*) – ed. by Native Americans (Nevada)

Streptocalyx Beer (~ *Aechmea*). Bromeliaceae (3). 20 Braz. R: FN 14(1979)1513

Streptocarpus Lindl. Gesneriaceae (III 2g). Incl. *Linnaeopsis, Saintpaulia*, c. 140 trop. & S Afr., Madag. R: O. Hilliard & B.L. Burtt (1971) *S.*, T 64(2015)1259,1272. Subg. *Streptocarpus* (x = 16), Cape primrose etc.; subg. *Strepocarpella* (x = 15) incl. sect. *Saintpaulia*, Afr. violets (journal: *Afr. Violet Mag.*; enantiostyly like *Exacum* (q.v.). cult. orn. with over 2000 cvs (spreading, trailing or dwarf, 'boy-type' plain, 'girl-type' with basal white spot on leaf; fls single to double, petals entire to ruffled, white or pink to violet or deep blue) derived esp. from ***S. ionanthus*** (H. Wendl.) Christenh. (*Saintpaulia i.*, coastal, Kenya, Tanzania) – now rare in wild but many cvs & followers; usu. pls of wet places in forest but ***S. i.*** subsp. ***rupicola*** (B.L. Burtt) Christenh. (*Saintpaulia r.*) on limestone outcrops in more open country in SE Kenya; some smaller spp. now pop. as novelties & involved in hybridization). Ann. or perenn., s.t. hapaxanthic, herbs & subshrubs. Cotyledons grow unequally after germ. & in ***S. grandis*** N.E. Br. (SE & S Afr.) & other spp. (as in *Monophyllaea*) 1 cotyledon enlarges by basal growth (to 76 cm in some spp.) while plumule development is suppressed, the hypocotyl increasing in thickness to form 'stem' of pl. (evolution of these 'unifoliates' from rosette-pls occurred several times); fls borne at junction of hypocotyl & midrib with adventitious r. from lower part of hypocotyl. Other spp. develop new 'leaf' & 'petiole' units (phyllomorphs) on the first & these may behave as hapaxanthic units so that a pl., while appearing perenn., may be considered a colony of unifoliates; there are other elaborations of these morphological novelties & the morphological flexibility of these pls is evident in the way that they may be prop. from small fragments of the 'leaf'; phyllomorph development can be suppressed & caulescent growth typical of other spp. induced by supplying gibberellic acid or inhibiting auxin transport. Cult. orn. esp. phyllomorphic ***S. x hybridus*** Voss, the florist's streptocarpus, a complex hybrid group with ***S. rexii*** (Hook.) Lindl. (Natal) predominating, early hybrids with ***S. polyanthus*** Hook. (S Afr.) = ***S.*** × ***bruantii*** Carr. & André, modern ones incl. '**Constant Nymph**' (*S.* × *hybridus*

'Merton Blue' x *S. johannis* L. Britten (S Afr.), 1947); caulescent cult. orn. incl. ***S. caulescens*** Vatke & ***S. stomandrus*** B.L. Burtt (E Afr.)

Streptocaulon Wight & Arn. Apocynaceae (III; Asclepiadaceae I). 9 Indomal. Local medic. (fever, dysentery) incl. rs of ***S. juventas*** (Lour.) Merr. (*S. griffithii*, SE As.) for stomache-ache in Vietnam. ***S. baumii*** Decne (Philippines) – used for baskets & handicrafts

Streptochaeta Schrad. ex Nees. Gramineae (I). 3 trop. Am. R: SCB 68(1989)29

Streptochaetaceae Nakai = Gramineae

Streptoechites Middleton & Livsh. (~ *Sindechites*). Apocynaceae. 1 SE As.: ***S. chinensis*** (Merr.) Middleton & Livsh. R: Adans. 34(2012)370

Streptoglossa Steetz. (~ *Oliganthemum*). Compositae (Pluch.). 8 Aus. R: JABG 3(1981)167. ***S. bubakii*** (Domin) Dunlop (NW Aus.) – fly-repellent high in caryophyllene & gamma-elemine

Streptogyna P. Beauv. Gramineae (III). 1 OW trop., 1 Am. R: AMBG 74(1987)871. At maturity disarticulating florets dangle from infl. by long tangled stigmas

Streptolirion Edgew. Commelinaceae (II 1b). 1 E Himal. to Korea & SE As.: ***S. volubile*** Edgew. – cli. with infl. penetrating sheath

Streptoloma Bunge. Cruciferae (26). 2 C As. to Afghanistan

Streptolophus Hughes. Gramineae (XXIV 4). 1 Angola: ***S. sagittifolius*** Hughes

Streptomanes K. Schum. = *Cryptolepis*

Streptopetalum Hochst. Passifloraceae (Turneraceae). 6 trop. & S Afr.

Streptopus Michaux. Liliaceae. 7 N temp. (Eur. 1, N Am. 3) to Himal. & S US. R: Rhodora 37(1935)88. Local medic., potherbs; some cult. orn.

Streptosiphon Mildbr. Acanthaceae (III 1). 1 Tanzania: ***S. hirsutus*** Mildbr. R: KB 49(1994)405

Streptosolen Miers = *Browallia*

Streptostachys Desv. (~ *Panicum*). Gramineae (XXIII 3). Excl. *Oncorachis*, *Sclerochlamys*, 2 trop. Am. R: T 58(2009)370

Streptothamnus F. Muell. (~ *Berberidopsis*). Berberidopsidaceae (Flacourtiaceae s.l.). 1 E Aus.: ***S. beckleri*** F. Muell.

Streptotrachelus Greenman = *Laubertia*

stretchberry *Forestiera pubescens*

Stricklandia Bak. = *Phaedranassa*

stridolo, strigoli *Silene vulgaris*

Striga Lour. Orobanchaceae (Buch.; Scrophulariaceae s.l.). 33 OW trop. (Afr. 28 – R: AMBG 88(2001)64) to S Afr. Hemiparasites with minute seeds viable for up to 20 yrs germ. in response to host r. exudates (some spp. with strains restricted to a particular host genus), oft. pestilential (esp. in C_4 cereals where loss can be total) in crops (witchweed) esp. ***S. asiatica*** (L.) Kuntze ('*S. lutea*', trop. OW – local medic.) on sugar-cane, maize, sorghum etc. in Afr., some introd. Am. populations cleistogamous; application of nitrogen to infested crop improves yield because parasites then photosynthesize more & take less from hosts. ***S. hermonthica*** (Del.) Benth. (Afr.) – germ. improved when seed passed through rumen, local medic., poss. stock feed in semi-arid regions

Strigina Engl. = *Lindernia*

Strigosella Boiss. (~ *Malcolmia*). Cruciferae (26). 23 Euras., N Afr.

stringybark *Eucalyptus* spp. with fibrous bark; **brown s.** *E. capitellata*; **red. s.** *E. macrorhyncha*; **yellow s.** *E. acmenoides*, *E. muelleriana*

Striolaria Ducke = *Pentagonia*

striped squill *Puschkinia scilloides*

Strobilacanthus Griseb. Acanthaceae (III 1). 1 Panamá: ***S. lepidospermus*** Griseb.

Strobilanthes Blume. Acanthaceae (III 2g). Incl. *Aechmanthera*, *Hemigraphis* & *Stenosiphonium*, 400+ China (128, 57 endemic), Indomal. (Java 25 – R: KB 58(2003)17) to Melanesia. Poss. heterogeneous & s.t. split into many unsatisfactory segregate genera. Many spp. with gregarious fls in forest undergrowth: ***S. cernua*** Blume has a cycle of 5–12 yrs in Java, ***S. gossypina*** (Nees) T. Anderson (*A. g.*) every 12 yrs in Mussoorie Hills, Ind., ***S. kunthiana*** (Nees) T. Anderson (neelakorinji) a 12-yr one in Ind. but with sporadic fls in between, ***S. pulcherrima*** T. Anderson (*Leptacanthus walkeri*) every 12 yrs, as does ***S. sexennis*** Nees – to 8 m; when stigma touched, moves downwards becoming pressed against lower lip of fl. Cult. orn. (esp. '*Hemigraphis*' for ground-cover), local medic. & dye-sources: **S. sp.** (*H. alternata* (Burm.f.) T. Anderson., trop. As.) – local medic., esp. diuretic (high potassium content); ***S. auriculata*** Nees var. ***dyeriana*** (Mast.) J.R.I. Wood (Persian [!] shield, Myanmar, Thailand) – cult. orn. foliage pl.; ***S. crispa*** Blume (W Mal.) – anticancer properties, form. used for snakebite; ***S. cusia*** (Nees) Kuntze (*S. flaccidifolia*, Ind. to S China & SE As.) – cult. for blue dye (maigyee, Assam indigo) & Chinese medic. (malanye) for leukaemia (though oft. adulerated with other spp.) ***S. persicifolia*** (Lindl.) J.R.I. Wood (*S. anisophylla*, goldfussia, Assam) – shrub with lvs in unequal-sized prs;

Strobilanthopsis S. Moore. Acanthaceae (III 2f). 1 trop. Afr.: ***S. linifolia*** (Clarke) Milne-Redh.

Strobilocarpus Klotzsch = *Grubbia*

Strobilopanax R. Viguier = *Meryta*

Strobilopsis Hilliard & B.L. Burtt. Scrophulariaceae (Man.). 1 Lesotho & Natal: ***S. wrightii*** Hilliard & B.L. Burtt. R: O.M. Hilliard (1994) *Manuleeae*: 532

Strobocalyx (DC.) Spach. Compositae (Vern.). c. 10 S & SE As.

Strobopetalum N.E. Br. = *Pentatropis*

Stroganowia Karelin & Kir. = *Lepidium*

Stromanthe Sonder. Marantaceae. 13 trop. S Am. Cult. orn. herbs with distich. lvs & racemes or panicles with decid. coloured bracts & zig-zag rachises

Stromatopteridaceae Bierh. = Gleicheniaceae (Stromatopteridoideae)

Stromatopteris Mett. Gleicheniaceae (Stromatopteridaceae). 1 New Caled.: ***S. moniliformis*** Mett. – frond continuous with 'stem' axis, not a lat. appendage, so that there is no clear distinction between stem & leaf. R: Phytomorphology 18(1968)232

Strombocactus Britton & Rose. Cactaceae (III 9). 2 Mex. Cult. orn. with arillate seeds. R: SB 9(1984)42

Strombocarpa (Benth.) A. Gray = *Prosopis*

Strombosia Blume. Erythropalaceae (Olacaceae s.l.). c. 16 OW trop. (As. 7). Roux's Model. ***S. javanica*** Blume (W Mal.) – young lvs ed., useful timber

Strombosiaceae Tiegh. = Erythropylaceae

Strombosiopsis Engl. Olacaceae (Strombosiaceae). 2 trop. Afr. Useful timber

Strongylocaryum Burret = *Ptychosperma*

Strongylodon Vogel. Leguminosae (III 18). 12 Madag. to Polynesia, esp. Philipp. R: AUWP 90–8(1990)1. ***S. craveniae*** Baron & Bak. (Madag.) – lemur-poll.; ***S. lucidus*** (Forst. f.) Seemann (range of genus) – distrib. sea currents; ***S. macrobotrys*** A. Gray (jade vine, Philipp.) – liane to 18 m with bluish green fls 8 cm long & large indehiscent 3–10-seeded fr., cult. orn.

Strophacanthus Lindau = *Isoglossa*

Strophanthus DC. Apocynaceae (II a). 38 trop. OW. R: MLW 82–4(1982). Usu. lianoid shrubs (some with Prévost's model) with long, oft. twisted C-lobes & spreading follicles; some cult. orn. but seeds imp. sources of arrow-poisons & strophanthin, a cardiac drug, esp. ***S. gratus*** (Wall. & Hook.) Baill. (W Afr.) & ***S. kombe*** Oliv. (trop. & S Afr.), ***S. sarmentosus*** DC. (trop. Afr.) being a comm. source of cortisone

Strophioblachia Boerl. Euphorbiaceae (Cod.-Cod.). 2 SE As., Hainan to C Mal. R: Blumea 43(1998)479. ***S. fimbricalyx*** Boerl. – seeds used in fermented drinks in Philippines

Strophiodiscus Choux = *Plagioscyphus*

Strophocactus Britton & Rose (~ *Selenicereus*). Cactaceae (III 1). 3 trop. Am.

Stropholirion Torrey = *Dichelostemma*

Strophostyles Elliott (~ *Phaseolus*). Leguminosae (III 18). 3 N Am. R: SB 29(2004)637

Strotheria B. Turner (~ *Thymophylla*). Compositae (Hele.-Pect.). 1 N & C Mex.: ***S. gypsophila*** B. Turner – dwarf succ. R: AJB 59(1975)180

Struchium P. Browne (~ *Sparganophorus*). Compositae (Vern.-Lep.). 2 trop.

Strumaria Jacq. Amaryllidaceae. Incl. *Carpolyza*, 27 S Afr., semi-arid. R: CBH 16(1994)82. Some cult. orn.

Strumpfia Jacq. Rubiaceae (Cinch.-Strum.). 1 WI: ***S. maritima*** Jacq.

Struthanthus Mart. Loranthaceae (4 4). Excl. *Peristethium*, 45 trop. Am. Heterogeneous? ***S. densiflorus*** Mart. – 'wood roses' prod. at junction with host sold as curios

Struthiola L. Thymelaeaceae (Thym.-Daph.). 35 trop. & S Afr. (Cape 21 – R: Strel. 9(2000)684). Cult. orn. everg. shrubs s.t. with scented fls

Struthiolopsis E. Phillips = *Gnidia*

Strychnaceae DC. ex Perleb = Loganiaceae

Strychnopsis Baill. Menispermaceae (V). 1 Madag.: ***S. thouarsii*** Baill.

Strychnos L. Loganiaceae (Strychnaceae). 190 trop. & warm (Afr. c. 75). Trees, shrubs & lianes (Mangenot's, Massart's & Troll's Models) with v. many alks & axillary thorns or (lianes) axillary hooks which twine around support & lignify; fr. berry-like (monkey apples), suggested as the inspiration for the 'apple' of the Garden of Eden, the flesh allegedly harmless (large-fr. spp. cult. or coll. e.g. Madag.: ***S. madagascariensis*** Poir., *S. spinosa*) but seeds extremely toxic because of strychnine in integuments, causing tetanus-like convulsions & form. used to poison rats etc., also in S Am. arrow-poisons & as ordeal-poisons in Afr. (properties discovered independently (?) in OW & Am.). Some timbers & cult. orn., locally medic. ***S. ignatii*** P. Bergius (St Ignatius's bean, Ignatius b., Vietnam, Mal.) – fr. eaten by monkeys & civets, seeds a source of strychnine; ***S. minor*** Dennst. (*S. colubrina*, snakewood, trop. As.) – bark & wood used medic. (malaria etc.); ***S. nux-vomica*** L. (nux-vomica, S As.) – comm. source of strychnine for rodent control in W Eur. since 1802; ***S. potatorum*** L.f. (clearing nut, kataka, E Ind., Myanmar) – rubbed inside water-vessels, it causes precipitation of impurities in cloudy water, locally medic., timber useful; ***S. spinosa*** Lam. (trop. & S Afr., Madag.) – fr. ed., to 11 cm diam., r. used against chiggers on Mafia Is.; ***S. toxifera*** R. Schomb. (trop. Am.) – a source of curare obtained by scraping & macerating bark, used medic. (see *Chondrodendron*) but form. imp. arrow-poison

Stryphnodendron Mart. Leguminosae (II 1). c. 30 trop. Am. (Amazonia 14, R: Leandra 10/11(1981)3). Some locally used timbers, tanbarks (esp. ***S. adstringens*** (Mart.) Cov. (*S. barbatimao*, barbatimão, Braz.) – 20–35% tannin) & medic.

Stuartia L'Hérit. = *Stewartia*

Stuartina Sonder. Compositae (Gnap.-Gnap.). 2 S & E Aus. R: Muelleria 6(1986)255

Stubendorffia Schrenk ex Fischer & al. = *Lepidium*

Stuckenia Boerner = *Potamogeton* (but see Novon 8(1998)241)

Stuckertia Kuntze = *Araujia*

Stuckertiella Beauverd. Compositae (Gnap.-Gnap.). 2 Arg. R: OB 104(1991)157

Stuessya B. Turner & Davies. Compositae (Helia.-Helia.). 3 Mex. R: Britt. 32(1980)209

stuffing gourd *Cyclanthera pedata*

Stuhlmannia Taubert (~ *Caesalpinia*). Leguminosae (I 4). 1 coastal forest of Tanz.: ***S. moavi*** Taubert. R: KB 51(1996)377

Stultitia E. Phillips = *Orbea*

Sturt('s) desert pea *Swainsona formosa*; **S('s). d. rose** *Gossypium sturtianum*

Sturtia R. Br. = *Gossypium*

Stussenia C. Hansen (~ *Neodriessenia*). Melastomataceae. 1 Vietnam: ***S. membranifolia*** (Li) C. Hansen. R: Willd. 15(1985)175

Stutzia E. Zacharias (*Endolepis*). Amaranthaceae (Atrip.). 2 US to N Mex. R: SB 35(2010)851

Styasasia S. Moore = *Asystasia*

Stylapterus A. Juss. (~ *Penaea*). Crypteroniaceae (Penaeaceae). 8 SW Cape. R: Strel. 9(2000)561, OB 15(1967)3

Stylidiaceae R. Br. Magnoliidae – Asterales. Incl. Donatiaceae, 4/240 S & SE As., Australasia, S S Am. R: R. Erickson (1958) *Triggerpls*. Small herbs, rarely shrublets, usu. with basal rosette of linear lvs, storing inulin; glandular hairs prod. proteases, so poss. carnivorous; laticifers 0. Lvs simple, in spirals; stip. 0. Fls usu. bisex., in term. bracteate cymes or racemes or solit. in upper axils or term. (*Donatia*); K ((3–)5(–7)), C (5(-10)) (free in *Donatia*), irreg. with imbr. lobes & resupinate or half so, or C reg., A 2 or 3 (*Donatia*) free from C but adnate to style forming a column, anthers extrorse with longit. slits, nectary-disk (or pr. of glands) atop $\overline{G}$ (2 or 3), ± 2- or 3-loc. with free styles & axile to free-C placentation or posterior loc. reduced or 0 (G pseudomonomerous), the stylar column oft. irritable moving rapidly from bent position on 1 side of fl. to opp. (trigger pl. due to phloem cells in small groups separating from xylem while other cells prob. change shape & size on bending), ovules ∞ anatropous, unitegmic. Fr. usu. a capsule; seeds (few–)∞ with small (freq. monocot.) embryo in oily endosperm. n = 15, 18

Genera: *Donatia, Forstera, Levenhookia, Stylidium*

Some cult. orn. (*Forstera, Stylidium*)

Stylidium Sw. Stylidiaceae. Excl. *Phyllachne*, c. 220 Sri Lanka (1), SE As., Aus. (almost all) esp. SW, NZ. Trigger pls (see fam.); some with explosive poll. mechanism in that column showers insect with pollen, then after A shrivel again stigma getting pollen from insects. n = 5–16, 26, 28, 30. Few cult. orn.

Stylisma Raf. Convolvulaceae (4). 6 S & E US. R: Britt. 18(1966)97

Stylites Amstutz = *Isoetes*

stylo *Stylosanthes* spp.; **Townsville s.** *S. humilis*

Stylobasiaceae J. Agardh = Surianaceae

Stylobasium Desf. Surianaceae (Stylobasiaceae). 2 N & W Aus.

Styloceras Kunth ex A. Juss. Buxaceae (Stylocerataceae). 5 Venez., N Andes. R: Novon 3(1993)142. Some locally eaten fr., timber for joinery

Stylocerataceae Takht. ex Rev. & Hoogl. = Buxaceae

Stylochaeton Lepr. Araceae (VII 2). 18 trop. & warm Afr. esp. Tanz. R. swollen; infl. below ground with only tip emergent. Some local medic.

Stylochiton Schott = *Stylochaeton*

Stylocline Nutt. Compositae (Gnap.-Gnap.). 7 SW N Am. R: Madroño 39(1992)114. Tiny annuals

Styloconus Baill. = *Blancoa*

Stylodon Raf. = *Verbena*

Stylogyne A. DC. Primulaceae (Myrsinaceae). 35 trop. Am. (Braz. 18 – R: SB 37(2012)478). R: Novon 20(2010)438. Rauh's Model; dioec.

Stylolepis Lehm. = *Podolepis*

Stylomecon G. Taylor = *Papaver* (but see FNA 3(1997)336)

Stylophorum Nutt. Papaveraceae (I). 2 E As., 1 E US: ***S. diphyllum*** (Michaux) Nutt. (E US) – cult. orn. with yellow fls. Alks, seeds arillate

Stylophyllum Britton & Rose = *Dudleya*

Stylosanthes Sw. Leguminosae (III 11). 25 trop. & warm. R: AMBG 44(1958)299. Cult. fodder pls (stylo); mycorrhizal, phosphate accumulators, esp. ***S. erecta*** P. Beauv. (W Afr.) – only 6x, genome as 4x *S. scabra* Vogel (Braz.) + 2x *S. angustifolia* Vogel (Braz.), & ***S. guianensis*** (Aubl.) Sw. (trop. Am.); ***S. humilis*** Kunth (*S. sundaica*, Townsville stylo, Braz., natur. E Mal. to trop. Aus.) – poss. early introd. by Portuguese, form. imp. fodder in N Aus., where cattle ticks immobilized & killed on fls stems

Stylosiphonia Brandegee. Rubiaceae. 2 C Am. = ?

Stylotrichium Mattf. Compositae (Eup.-Gyp.). 5 E Braz. R: MSBMBG 22(1987)122

Stylurus Salisb. ex J. Knight = *Grevillea*

Stypandra R. Br. Asphodelaceae (Phormiaceae). 2 temp. Aus. (? & New Caled.). ***S. glauca*** R. Br. (blind grass, Aus.) – causing blindness in grazing goats & sheep through degeneration of optic nerve etc.

Styphelia Sm. Ericaceae (VII 7; Epacridaceae). Excl. *Cyathodes* etc., 15 S Aus. Some ed. fr. (***S. acerosa*** Sol. ex Gaertn. imp. emu food)

Styphnolobium Schott (~ *Sophora*). Leguminosae (III 2). 9 C Am. & China (1) R: AMBG 80(1993)273. ***S. japonicum*** (L.) Schott (*Sophora j.*, pagoda tree, China) – fls source of a yellow dye, dried fls (sold in Java as sari kuning or s. tijina) medic. (rutin used in haemorrhagic problems) used in a Vietnam tea (with *Senna tora* & liquorice seeds), cult. orn. decid. tree (several cvs incl. '**Tortuosa**' with twisted branches)

Styppeiochloa De Winter (~ *Zenkeria*). Gramineae (Arund.). 3 S Afr. & Madag. R: Bothalia 9(1966)134

Styracaceae DC. & Spreng. Magnoliidae – Ericales. 11/170 warm temp. & trop. Am., Medit., SE As., W Mal. R: IDSY 2014:160. Trees & shrubs with resinous bark & usu. stellate or peltate trichomes. Lvs in spirals, simple, usu. serrate; stip. 0. Fls usu. bisex., reg., ebracteolate in racemose or cymose infls, rarely solit.; K (4,5(–9)) with open or valvate lobes or 0, C (4, 5(–9)) tubular proximally, the lobes imbr. or valvate, shorter than tube in *Halesia*, C free in *Bruinsmia*, A 5 (*Styrax* sect. *Pamphilia*) or 2(–4) × C, filaments usu. adnate to C-tube & basally connate with a tube, anthers with longit. slits & connective s.t. prolonged, $\underline{G}$ to $\overline{G}$(2–4 (5), basally pluriloc., but apically oft. 1-loc. with slender style & axile placentation, each placenta with (1–)4–9(–c. 30) erect or pend., anatr. to hemitropous uni- or bitegmic ovules (not more than 2 becoming seeds). Fr. usu. a capsule, s.t. samaroid or a drupe (*Parastyrax*) with persistent K; seeds with large straight to weakly curved embryo in copious oily endosperm. n = 8, 12

Principal genera: *Huodendron, Pterostyrax, Rehderodendron, Sinojackia, Styrax*

Resins (*Styrax*) & cult. orn. (esp. *Halesia, Sinojackia, Styrax*)

Styrax Tourn. ex L. Styracaceae. Incl. *Pamphilia*, c. 140 Medit. (incl. Eur.; 1), SE As., Mal., trop. Am. R: Pflanzenr. IV 241(1907)17; Sida 5(1974)191. Trees & shrubs (Am. spp. with erect twigs; As spp. with plagiotropic) with resins used medic. (benzoin, gum Benjamin (a corruption) used in friar's balsam; benzoic acid esters haemolytic) & in incense, obtained by wounding bark of trop. As. spp. esp. ***S. benzoin*** Dryander (Sumatra), the resin being principally 2 alcohols combined with cinnamic acid & free cinnamonic & benzoic acids, & used in treatment of coughs, as antiseptic, in flavouring cigarettes & in cerem. Some with remarkable aphid-induced galls; some cult. orn. ***S. officinalis*** L.

(storax, styrax, Medit. with close ally, ***S. redivivus*** (Torrey) L. Wheeler in Calif.) – orig. balsam in med., seeds used as beads; ***S. tessmannii*** Perkins (trop. Am.) – crushed lvs used against fungal infections of feet in Colombia; ***S. tonkinensis*** (Pierre) Hartwich (SE As.) – pulpwood, benzoin for perfumery

Styrophyton S.Y. Hu = *Allomorphia*

Suaeda Forssk. ex J.F. Gmelin. Amaranthaceae (Chenopodiaceae III 2). c. 110 cosmop. coasts & salt steppe (Eur. 15, Aus. 5), in GB ***S. vera*** Forssk. ex J. F. Gmelin ('*S. fruticosa*', Euras.) – camel fodder & ***S. maritima*** (L.) Dumort. (Am., Euras. etc.) – seablite. Some with C4 physiology but no Krantz anatomy; alks. Ash (barilla) high in sodium carbonate form. used in glass-making; some ed. seeds & local medic. ***S. monoica*** Forssk. ex J.F. Gmelin (Medit. to As. & dry trop. Afr.) – imp. camel fodder in dry season in N Kenya; ***S. suffrutescens*** S. Watson (SW US) – poll. by bees, butterflies & thrips, source of a black dye used by Native Americans

suan zau (zao) ren *Ziziphus spinosa*

Suarezia Dodson (~ *Quekettia*). Orchidaceae (V 12h). 1 Ecuador: ***S. ecuadorana*** Dodson. R: Gen. Orch. 5(2009)358

suari *Caryocar amygdaliferum*

sub clover *Trifolium subterraneum*

Suberanthus Borh. & Fernández (~ *Rondeletia*). Rubiaceae (I 1). 7 Cuba & Hispaniola. R: ABH 29(1983)29

Subularia L. Cruciferae. 1 N temp. incl. Eur.: ***S. aquatica*** L. (awlwort) – one of few aquatic annuals; 1 mts trop. E Afr. R: Rhodora 66(1964)127

Succisa Haller. Caprifoliaceae (Dipsacaceae). 1 Eur. & W Siberia, N Afr.: ***S. pratensis*** L. (devil's bit (scabious), blue buttons) – locally medic. (antiseptic); 1 NW Spain; 1 Cameroun Mt. R: AMH n.s. 2(1952)237

Succisella G. Beck. Caprifoliaceae (Dipsacaceae). 5 Eur. R (Iberia): BotJLS 144(2004)351

succory *Cichorium intybus*; **lamb's** or **swine s.** *Arnoseris minima*

Succowia Medik. Cruciferae (12). 1 Canary Is., W Medit.: ***S. balearica*** (L.) Medik.

Suchtelenia Karelin ex Meissn. (~ *Cynoglossum*). Boraginaceae (B.3.5.2). 1 Cauc. to C As.: ***S. calycina*** (C. Meyer) A. DC.

Suckleya A. Gray. Amaranthaceae (Chenopodiaceae I 4). 1 Rocky Mts: ***S. suckleyana*** (Torrey) Rydb. R: FNA 4(2003)305

Sucrea Soderstrom. Gramineae (VI 3). 3 Braz. R: KBAS 13(1986)63. Intermediate between *Olyra* & *Raddia*. Ballistic disp. ***S. monophylla*** Soderstrom (Braz.) – sterile culms each with only 1 leaf; ***S. sampaiana*** (Hitchc.) Soderstrom (Braz.) – r. tubers

sucupira timber from *Bowdichia*, *Diplotropis* & *Sweetia* spp.

sudachi *Citrus* ?x *junos*

Sudamerlycaste Archila (*Ida*). Orchidaceae (V 12g). c. 40 trop. Am. R: H.F. Oakeley (2008) *Lycaste, I. & Anguloa*

Sudan grass *Sorghum bicolor*

Suddia Renvoize. Gramineae (IV). 1 Sudan, Uganda: ***S. sagittifolia*** Renvoize – aquatic in sudd, blades sagittate, false petioles to 1 m. R: KB 39(1984)455

Suessenguthia Merxm. (~ *Sanchezia*). Acanthaceae (III 2c). 6 E Andes of Peru (4) & Bolivia. R: Candollea 58(2003)108

Suessenguthiella Friedrich (~ *Pharnaceum*). Molluginaceae. 1 NW Cape & Namibia: ***S. scleranthoides*** (Sonder) Friedrich. R: MBSM 2(1955)60

Suffrenia Bellardi = *Rotala*

sugar apple *Annona* spp.; **s. beet** *Beta vulgaris* cv.; **s. berry** *Celtis* spp.; **s. bush** *Protea* spp. esp. *P. repens*; **s. cane** *Saccharum officinarum*; **s. grass** *Eulalia area*; **s. loaves** *S. officinarum*; **s. maple** *Acer saccharum*; **s. pine** *Pinus lambertiana*; **s. plum** *Prunus* × *domestica* cvs esp. 'Prune d'Agen', *Uapaca guineensis*

sugi *Cryptomeria japonica*

sui mui *Wrightia religiosa*

Suksdorfia A. Gray. Saxifragaceae (I 6). Excl. *Hieronymusia*, 2 NW N Am. & Andes of N Arg. & S Bolivia. R: BotJLS 90(1985)60. Small cult. orn. rock-pls

Sukunia A.C. Sm. = *Atractocarpus*

Sulaimania Hedge & Rech.f. = *Moluccella*

Sulcolluma Plowes = *Caralluma*

Sulcorebutia Backeb. = *Weingartia*

Sulitia Merr. = *Atractocarpus*

sulla *Sulla coronaria*

Sulla Medik. (~ *Hedysarum*). Leguminosae (III 25). 7 Medit. R: T 52(2003)574. ***S. coronaria*** (L.) Medik. (*H. c.*, F. honeysuckle, W Medit., natur. rest of S Eur.) – s.t. cult. as fodder (espercet, sulla)

Sullivantia Torrey & A. Gray. Saxifragaceae (I 6). 3 C US. R: FNA 9(2009)121. Most pops restricted to unglaciated area nr Pleistocene glacial margins. Cult. orn.

sultan, sweet *Amberboa moschata*

sultana *Vitis vinifera* 'Sultana'

sultan's flower *Impatiens walleriana*

sumac(h) *Rhus coriaria*; **Chinese s.** *R. chinensis*; **Hungarian, Indian, Turkish, Tyrolean** or **Venetian s.** *Cotinus coggygria*; **staghorn** or **velvet s.** *R. typhina*

Sumatra camphor *Dryobalanops aromatica*

Sumatroscirpus Oteng-Yeboah (~ *Scirpus*). Cyperaceae (II 6). 1 N Sumatra: ***S. junghuhnii*** (Miq.) Oteng-Yeboah. R: NRBGE 33(1974)307

Sumbaviopsis J.J. Sm. Euphorbiaceae (Acal.-Chro.-Chro.). 1 Assam to W Mal.: ***S. albicans*** (Blume) J.J. Sm. – fr. ed., medic. R: Blumea 44(1999)426

sumbul *Ferula sumbul*

summer cypress *Bassia scoparia*; **s. grass** *Digitaria sanguinalis*; **s. hyacinth** *Ornithogalum candicans*

Summerhayesia Cribb. Orchidaceae (V 16d). 2 trop. Afr. Cult. orn. epiphytes

sumpweed *Iva annua*

sun berry *Physalis minima, Solanum scabrum*; see also *Rubus*; **s.dew** *Drosera* spp.; **s.flower** *Helianthus* spp. esp. *H. annuus*; **Mexican s.flower** *Tithonia diversifolia*; **s.(n) hemp** *Crotalaria juncea*

Sundacarpus (Buchholz & N.E. Gray) Page = *Prumnopitys*

sungkai *Peronema canescens*

Sunipia Buch.-Ham. ex Sm. Orchidaceae (V 15). Incl. *Ione*, 22 S & SE As

sunrise lime *Citrus* × *oliveri*

suapari *Areca catechu*

Supushpa Suryan. = *Strobilanthes*

sura *Amorphophallus paeoniifolius*

Suregada Roxb. ex Rottler. Euphorbiaceae (Crot.-Gel.). c. 30 OW trop. (As. 9, Afr. 8, Madag. 14). local medic.

surette *Byrsonima* spp.

Surfacea Moldenke = *Premna*

surfboard Orig. made from *Acacia koa*

Suriana Plum. ex L. Surianaceae. 1 pantrop., littoral: ***S. maritima*** L. – Koriba's Model

Surianaceae Arn. Magnoliidae – Fabales. 5/8 trop. & warm esp. Aus. Trees & shrubs. Lvs usu. simple, in spirals s.t. becoming distich.; stip. small & decid. or 0. Fls reg., usu. bisex., 5-merous, in thyrses or solit. in axils; K 5(–7) ± free, imbr., C 5 free & imbr. or 0 (*Stylobasium* – wind-poll.), A in 2 whorls but some or all of whorl opp. C in *Suriana* staminodal or 0, anthers with longit. slits, disk 0, G̲ 1 (*Stylobasium*), 2 (*Guilfoylia*), 5 (*Cadellia, Suriana*) with 2 collateral, campylotropous unitegmic ovules per carpel. Fr. a berry, drupe, or nut-like; seeds with curved or folded embryo with starchy or oily cotyledons in little or 0 endosperm; germ. epigeal

Genera: *Cadellia, Guilfoylia, Recchia, Stylobasium, Suriana*

Some genera form. in Simaroubaceae

Surreya R. Masson & Kadereit (~ *Hemichroa*). Amaranthaceae (Chenopodiaceae IV 1). 2 Aus. R: T 62(2013)109

Susanna E. Phillips = *Amellus*

sushi see *Pyropia*

Susilkumara Bennet = *Eriophytum*

Sussenia C. Hansen. Melastomataceae. 1 As.: ***S. membranifolia*** (Li) C. Hansen. R: Willd. 15(1985)175

susumber *Solanum torvum*

Sutera Roth. Scrophulariaceae (Man.). Excl. *Chaenostoma*, 49 trop. & S Afr. R: O. Hilliard (1994) *Manuleeae*: 220. See also *Camptoloma* & *Jamesbrittenia*

Sutherlandia R. Br. = *Lessertia*

Sutrina Lindl. Orchidaceae (V 12h). 2 Peru, Bolivia

Suttonia A. Rich. = *Myrsine*

Suzukia Kudô. Labiatae (VI). 2 Ryukyu Is., Taiwan

Svenkoeltzia Burns-Bal. (~ *Funkiella*). Orchidaceae (IV 2h). 3 Mex.

Svensonia Moldenke = *Chascanum*

Sventenia Font Quer = *Sonchus*

Svitramia Cham. Melastomataceae. 6 S Braz. R: KB 58(2003)403

Swainsona Salisb. Leguminosae (III 24). Incl. *Montigena*, 84 dry Aus. R: CNSWNH 1(1948)131; Telopea 5(1993)427. Toxic because of mannose-based oligosaccharides which animals cannot break down. Some cult. orn. esp. ***S. formosa*** (G. Don) J. Thompson (*Clianthus f.*, (Sturt's) desert pea, Aus.) – prot. sp., favoured camel (introd.) food: ***S. galegifolia*** (Andrews) R. Br. (Darling pea, E Aus.) – perenn. herb with red fls., cont. swainsonine, potential chemotherapy drug. See also *Montigena*

Swallenia Soderstrom & H. Decker. Gramineae (XXIX 1). 1 Calif. Sand-dunes (4 sites): ***S. alexandrae*** (Swallen) Soderstrom & H. Decker – caryopses (without lemma & palea) abscized from pl. R: SCB 87(1997)40

Swallenochloa McClure = *Chusquea* (but see Britt. 30(1978)303)

swallow wort *Asclepias curassavica, Chelidonium majus*

swamp cottonwood *Populus heterophylla*; **s. cypress** *Taxodium distichum*; **s. mahogany** *Eucalyptus botryoides* & *E. robusta*; **s. oak** *Casuarina* spp.; **s. oat** *Sphenopholis pensylvanica*

swan orchis *Cycnoches* spp.; **S. River daisy** *Brachycome iberidifolia*

swarri nut *Caryocar amygdaliferum*

Swartzia Schreb. Leguminosae (III 1). c. 180 trop. Am. (R: FN 1(1968); Afr. spp. = *Bobgunnia*). Imp. lowland rainforest trees (Troll's Model); apogeotropic r. grow up stems of other trees to 13–14 m (Venez.) poss. attracted by calcium ions. Some timber (wamara) & cult. orn.; many local medic. NW Amaz. ***S. polyphylla*** DC. (Amazon) – air-filled cavity between cotyledons allowing hydrochory; ***S. simplex*** (Sw.) Spreng. (trop. Am.) – molluscicidal saponins (triterpenes)

swede *Brassica napus* Napobrassica Group

Swedish ivy *Plectranthus verticillatus*; **S. whitebeam** *Sorbus intermedia*

sweet acacia *Vachellia farnesiana*; **s. alyssum** *Lobularia maritima*; **s. balm** *Melissa officinalis*; **s. basil** *Ocimum* × *africanum*; **s. bay** *Laurus nobilis*; **s.briar** *Rosa rubiginosa*; **s. chestnut** *Castanea sativa*; **s. cicely** *Myrrhis odorata*; **s.corn** *Zea mays*; **s. flag** *Acorus calamus*; **s. gale** *Myrica gale*; **s. galingale** *Cyperus longus*; **s. grass** *Glyceria* spp. esp. *G. fluitans, Muhlenbergia sericea*; **s. vernal g.** *Anthoxanthum odoratum*; **s. gum** *Liquidambar styraciflua*; **s. hearts** *Galium aparine*; **s. John** *Dianthus barbatus*; **s. marjoram** *Origanum majorana*; **s. pea** *Lathyrus odoratus*; **s. p. shoots** *Pisum sativum* + *Triticum aestivum*; **s. potato** *Ipomoea batatas*; **s. rocket** *Hesperis matronalis*; **s. scabious** *Sixalix atropurpurea*; **s.shrub** *Calycanthus* spp.; **s.sop** *Annona squamosa*; **s. sultan** *Amberboa moschata*; **s. tea** *Smilax glyciphylla*; **s. william** *D. barbatus*

Sweetia Spreng. Leguminosae (III 11). 1 trop. S Am.: ***S. fruticosa*** Spreng. – timber (sucupira). R: NRBGE 29(1969)348

sweetie *Citrus limetta*

Swertia L. Gentianaceae. Incl. *Frasera*, 168 N temp. (Eur. 1), Afr. & Mal. mts. R: T.N. Ho & S.W. Liu, *Worldwide monog. S.* (2015)87. Herbs s.t. with lvs in spirals, fls with 1 or 2 nectar-prod. pits at base of each C lobe; alks. – some medic. (chirata, chiretta) as tonics; few cult. orn. ***S. radiata*** (Kellogg) Kuntze (*F. speciosa*, Rocky Mts) – sporadic synchronous flowering every 2–4 yrs (fls preformed at least 3 yrs before anthesis) suggesting that massive seedcrops necessary to overcome pressure of seed-predators

Swida Opiz = *Cornus*

Swietenia Jacq. Meliaceae (I). 3 trop. Am. R: FN 28(1981)389. CITES-listed emergent rainforest trees (Rauh's Model): mahogany (name poss. derived from 'm'oganwo' (i.e. *Khaya ivorensis* in Yoruba, SW Nigeria) modified by Portuguese & English in C17 Jamaica where Yoruba were transported as slaves), esp. ***S. mahagoni*** (L.) Jacq. (Cuban m.) – much exploited for fine cabinet timber, first used extensively by Chippendale & Hepplewhite, & form. much used for ship-building etc., but trade reduced to zero 1908–60 & wild populations now of poor form through selection of best pls for felling (genetic erosion); ***S. macrophylla*** King (Honduras m., baywood, invasive trop. As.) & ***S. humilis*** Zucc. (Mexican m.) – similar uses, the first in plantation & described from cult. material in As.

swine('s) cress *Coronopus didymus*; **s. succory** *Arnoseris minima*

Swinglea Merr. Rutaceae (II 2). 1 Philippines: ***S. glutinosa*** (Blanco) Merr. – locally medic. (skin disease), live fence in Colombia, cult. orn.

Swintonia Griff. Anacardiaceae (I). 12 Indomal. C accrescent & persistent, forming wings on fr. Some timbers, esp. ***S. floribunda*** Griff. (SE As., W Mal.) – match-boxes & -sticks in Myanmar

Swiss chard *Beta vulgaris* 'subsp. *vulgaris*'; **S. cheese plant** *Monstera deliciosa;* **S. pine** (timber) *Pinus cembra*, see also *Abies alba*

switchgrass *Panicum virgatum*

sword bean *Canavalia ensiformis;* **s. fern** *Nephrolepis* spp.; **s. grass** *Schoenoplectus pungens;* **s. lily** *Gladiolus murielae;* **s. plant** *Echinodorus* spp.; **s. sedge** *Lepidosperma gladiatum*

Swynnertonia S. Moore = *Neoschumannia*

Syagrus Mart. Palmae (V 8a). Incl. *Lytocaryum*, 36 trop. S Am. R: IBM 56(1987)16. ? Heterogeneous. Monoec. palms with lvs arr. as to appear in 3 ranks; many cult. orn. incl. hybrids, even with *Butia capitata* (Mart.) Becc. Sources of a palm kernel oil esp. ***S. coronata*** (Mart.) Becc. (ouricuri, nicuri (palm nut), arid Braz.) – oil (urucury wax) used in soap & subs. for carnauba wax, there being perhaps 5 billion trees (many more than carnauba), lvs for weaving into sleeping mats, found in archaeol. sites 8000–11 000 yrs old; ***S. cocoides*** Burret (Guyana to Braz.) – 1 pop. in NE Braz. with dichot. trunks; ***S. romanzoffiana*** (Cham.) Glassman (cocos palm, Braz. to NE Arg.) – fr. disp. tapirs, source of sago, cult. orn. (overplanted) esp. Aus.; ***S. weddellianum*** (H. Wendl.) Becc. (*Cocos w., L. w., Microcoelum w.*, SE Braz.) – much cult. house-pl.

sycamore *Acer pseudoplatanus*, (Am.) *Platanus* spp. esp. *P. occidentalis*

sycomore *Ficus sycomorus*

Sycopsis Oliv. Hamamelidaceae (I 4). 2–3 Assam to Taiwan. ***S. sinensis*** Oliv. (C China) – cult. orn.

Sydney blue gum *Eucalyptus saligna;* **S. peppermint** *E. piperita;* **S. red gum** *Angophora costata*

Sylphia Luer = *Specklinia*

Sylvichadsia Du Puy & Labat. Leguminosae (III 16). 4 Madag. Fish-poisons

Sylvipoa Soreng & al. Gramineae (XVI 15). 1 NE Aus.: ***S. queenslandica*** (C. Hubb.) Soreng & al. R: AusSB 22(2009)404

Symbegonia Warb. = *Begonia* (But see SCB 60(1986)252)

Symbolanthus G. Don. Gentianaceae. c. 30 trop. Am. ***S. latifolius*** Gilg – bat-poll.

Symingtonia Steenis = *Exbucklandia*

Symmeria Benth. Polygonaceae (Polyg.-Symm.). 1 N S Am., W Afr.: ***S. paniculata*** Benth. – small tree

Symonanthus Haegi. Solanaceae (III 1). 2 SW S Aus. R: A.T. Hunziker, *Gen. Solanacearum* (2001)394. Dioec.

Sympa Ravenna = *Herbertia*

Sympagis (Nees) Bremek. = *Strobilanthes*

Sympegma Bunge. Amaranthaceae (Chenopodiaceae III 3). 1 C As.: ***S. regelii*** Bunge

Sympetalandra Stapf. Leguminosae (I 4). 5 W Mal. R: Blumea 22(1975)159

Sympetaleia A. Gray = *Eucnide*

Symphionema R. Br. Proteaceae (III). 2 NSW. R: FA 16(1995)133

Symphonia L.f. Guttiferae (3.). 17 – 25 Afr. (1–2), Madag. (15–21), trop. Am. (1–2 incl. ***S. globulifera*** L.f. (hog plum) – Roux's Model, knee-like pneumatophores, hummingbird-poll., timber (boarwood), resin for caulking boats, locally medic.).

Symphorema Roxb. Labiatae (I; Symphoremataceae). 3 Indomal.

Symphoremataceae Moldenke ex Rev. & Hoogl. = Labiatae (I)

Symphoricarpos Duhamel. Caprifoliaceae. 16 N Am. to Mex. (15), China (1). R: JAA 21(1940)201. Local medic.; cult. orn. decid. suckering shrubs esp. ***S. albus*** (L.) S.F. Blake (snowberry, E N Am.) – spreading by suckers in GB, rarely by seed, fr. a mushy white berry; ***S. orbiculatus*** Moench (coral berry, N Am.) – lvs locally medic., fr. red

Symphostemon Hiern = *Plectranthus*

Symphyandra A. DC. = *Campanula* (but see QBAGS 45(1977)246)

Symphyglossum Schltr. = *Oncidium*

Symphyllarion Gagnepain = *Hedyotis*

Symphyllia Baill. (~ *Epiprinus*). Euphorbiaceae (Acal.-Epi.-Epi.). 3 trop. As.

Symphyllocarpus Maxim. Compositae (? Athro.). 1 Manchuria, E Siberia: ***S. exilis*** Maxim.

Symphyllophyton Gilg. Gentianaceae. 2 Braz.

Symphyobasis K. Krause = *Goodenia*

Symphyochaeta (DC.) Skottsb. = *Senecio* (*Robinsonia*)

Symphyochlamys Guerke. Malvaceae (Malv.-Hib.). 1 NE trop. Afr.: ***S. erlangeri*** Guerke

Symphyoloma C. Meyer = *Heracleum*

Symphyonema Spreng. = *Symphionema*

Symphyopappus Turcz. Compositae (Eup.-Dis.). 12 Braz., campos. R: MSBMBG 22(1987)81

Symphyosepalum Hand.-Mazz. Orchidaceae (Orch.-Orch.). 1 China: ***S. gymnadenioides*** Hand.-Mazz.

Symphyotrichum Nees (~ *Aster*). Compositae Ast.-Sym.). Incl. *Brachyactis, Conyzanthus,* 92 E As., N Am. (76 – R: FNA 20(2006)465) to S S Am. (few). R: Phytol. 77(1994)267. Starworts, frost fls. Local medic.; cult. orn. herb. perennials (Michaemas daisies – see *Aster*; 5 major groups [incl. 1 derived from *A. amellus*]) esp. from ***S. novae-angliae*** (L.) Nesom (*A. n.*) – lvs dull, slightly hairy, c. 30 cvs, few hybrids, & ***S. novi-belgii*** (L.) Nesom (*A. n.*, E N Am. esp. coasts) – lvs shiny, glabrous, c. 300 cvs (form. over 1000, though many being hybrids, incl. with *S. lanceolatum* (Willd.) Nesom (N Am.) = ***S. × salignum*** (Willd.) Nesom, & with *S. laeve* (L.) Löve & D. Löve (N Am.) = ***S. × versicolor*** (Willd.) Nesom – tall, late-flowering) 20–150 cm, oft. natur. (actually oft. hybrids); ***S. oblongifolium*** (Nutt.) Nesom (*A. o.*, E US) – cult. orn., lvs scented

Symphysia C. Presl (~ *Vaccinium*). Ericaceae (VIII 5). 15 C Am., WI. R: T 53(2004)96

Symphytonema Schltr. = *Camptocarpus*

Symphytum Tourn. ex L. Boraginaceae (B.2.1.1). 35 Eur. (11), Medit. to Cauc. Comfrey; L.D. Hills (1976) *C.* Rich in allantoin promoting healing in connective tissue through proliferation of new cells; mucilage in lvs form. used like plaster of Paris. Liver failure from tea or tablets prep. from c. due to pyrrolizidine alks which break down in an hour giving toxins blocking veins in liver. ***S. grandiflorum*** DC. (Caucasus) – ground-cover, hybrid cvs with *S. × uplandicum* = ***S. × hidcotense*** Sell; ***S. officinale*** L. (Abraham, Isaac & Jacob, Euras., natur. N Am.) – form. used as styptic, young shoots eaten like asparagus; hybrid with *S. asperum* Lepechin (Russia to Iran) – cult. as forage, = ***S. × uplandicum*** Nyman (Russian or blue c.), a common pl. of roadsides in GB (introd. early 19 cent.), resynthesized through artificial hybridization of parents; also hybridizing with *S. tuberosum* L. (Eur., W As.)

Sympieza Lichtenst. ex Roemer & Schultes = *Erica*

Symplectochilus Lindau = *Anisotes*

Symplectrodia Lazarides (~ *Triodia*). Gramineae (XXIX 4). 2 N Territory, Aus. R: Nuytsia 5(1984)273. Sandy soils

Symplocaceae Desf. Magnoliidae – Ericales. 2/320 trop. & warm Am. & E OW. R: T 57(2008)841. Usu. everg. trees oft. accumu. aluminium. Lvs in spirals (s.t. becoming distich.), oft. sweet-tasting & drying yellowish; stip. 0. Fls usu. bisex., reg., fragrant, usu. bibracteolate, in axils of bracts in racemes (less oft. thyrses, solit. etc.); K (3–)5, basally connate, the lobes imbr., C (3–)5(–15) with short tube & imbr. lobes s.t. in ± 2 rows, A (4–15)40–100 attached to C-tube, usu. in 2 or more whorls or in bundles alt. with C, s.t. forming a tube with anthers inside, anthers with longit. slits, $\overline{G}$, rarely half-inf. (2)3(–5), usu. pluriloc., the style oft. surrounded by nectary-disk, each loc. with 2–4 pend. anatr. unitegmic ovules on axile or deeply intruded placentas. Fr. a (usu. blue) drupe with apical persistent K, endocarp with 1 apical germ. pore per loc.; seeds with large straight or curved embryo in copious oily & proteinaceous endosperm (s.t. also with starch). n = 11–14, c. 45

Genera: *Cordyloblaste, Symplocos*

S. subg. *Hopea* common as fossils in Eocene Eur.

Symplocarpus Salisb. ex W. Barton. Araceae (II). 4 NE As., NE N Am. ***S. foetidus*** (L.) W. Barton (skunk cabbage, N Am.) – cult. orn. herb (one of earliest spring pls., spathe emerging from snow) with alks, spadix reaching 35°C above ambient, local medic.

Symplococarpon Airy Shaw. Pentaphylacaceae (3; Theaceae s.l.). 9 trop. Am. R: JAA 22(1941)188. G inf.

Symplocos Jacq. Symplocaceae. 318 trop. & warm Am. & E OW (108 excl. New Caled.). R: T 57(2008)842. Some polyembryony. All aluminium accumulators have yellow-green lvs & blue fr.; spp. with sweet foliage relished by stock. Local medic., chewing-sticks; some sources of dye e.g. bark for dyeing batik red or brown in Java, mordant (aluminium) in turkey red etc.; ***S. racemosa*** Roxb. (lodh bark, S & SE As.) – lvs & bark for dye, bark medic.; ***S. theiformis*** (L.f.) Oken (Bogota tea, C Am.) – tea subs. in Colombia; ***S. tinctoria*** (L.) L'Hérit. (E N Am.) – lvs & fr. sources of yellow dyes, wood for turnery

Synadenium Boiss. = *Euphorbia*

Synammia C. Presl (~ *Polypodium*). Polypodiaceae (V). 3 S Am.

Synandra Nutt. Labiatae (VI). 1 E US: ***S. hispidula*** (Michaux) Baill.

Synandrina Standl. & L.O. Williams = *Casearia*
Synandrodaphne Gilg. Thymelaeaceae (Thym.-Syn.). 1 W trop. Afr.: ***S. paradoxa*** Gilg
Synandrogyne Buchet = *Arophyton*
Synandropus A.C. Sm. = *Odontocarya*
Synandrospadix Engl. Araceae (VII 4). 1 NE Andes (N Arg. & Bolivia): ***S. vermitoxicus*** (Griseb.) Engl. – v. poisonous
Synanthes Burns-Bal., H. Robinson & S. Foster = *Eurystyles*
Synaphea R. Br. Proteaceae (IV 1). 10 SW Aus. R: FA 16(1995)271. Lignotubers
Synapsis Griseb. Schlegeliaceae (Bignoniaceae s.l.). 1 E Cuba: ***S. ilicifolia*** Griseb.
Synaptantha Hook.f. (~ *Hedyotis*). Rubiaceae (IV 1). 2 warm Aus.
Synaptolepis Oliv. Thymelaeaceae (Thym.-Daph.). 4–5 trop. Afr., 1 Madag.
Synaptophyllum N.E. Br. = *Mesembryanthemum*
Synardisia (Mez) Lundell = *Ardisia*
Synarmosepalum Garay & al. = *Bulbophyllum* (But see NJB 14(1994)639)
Synassa Lindl. = *Sauroglossum*
Syncalathium Lipsch. Compositae (Cich.-Cich.(-Crep.)). 7 Tibet, China. R: APS 10(1965) 283
Syncarpha DC. ('*Helipterum*' *p.p.*). Compositae (Gnap.-Gnap.). 28 S Afr. R: OB 104(1991) 150. Everlastings
Syncarpia Ten. Myrtaceae (II 12). 3 NE Aus. R: Austrobaileya 4(1994)338. Specialized fire-resistant epicormic systems. ***S. glomulifera*** (Sm.) Niedenzu (*S. laurifolia*, turpentine wood, NE Aus.) – timber resistant to white ants & marine borers, used for ship-building, pole-houses (Queensland), etc., cult. shade-tree in S US
Syncephalantha Bartling = *Dyssodia*
Syncephalum DC. Compositae (Gnap.-Gnap.). 5 Madag. R: OB 104(1991)137
Synchaeta Kirpiczn. = *Gnaphalium*
Synchoriste Baill. = *Lasiocladus*
Synclisia Benth. Menispermaceae (I). 1 C Afr.: ***S. scabrida*** Miers
Syncolostemon E. Meyer ex Benth. Labiatae (VII 3d). Incl. *Hemizygia*, 35 SE Afr., R: Bothalia 12(1976)21
Syncretocarpus S.F. Blake. Compositae (Helia.-Helia.). 2 Peru. Xeromorphic
Syndesmanthus Klotzsch = *Erica*
Syndiclis Hook.f. = *Potameia*
Syndyophyllum K. Schum. & Lauterb. Euphorbiaceae (Acal.-Eris.). 2 Mal. R: Blumea 40(1995)388
Synechanthus H. Wendl. Palmae (V 2). 2 trop. Am. R: Principes 15(1971)10. Unarmed monoec. palms, cult. orn.
Synedrella Gaertn. Compositae (Helia.-Ecl.). 1 trop. Am., natur. OW: ***S. nodiflora*** (L.) Gaertn. – Leeuwenberg's Model, fr. disp. on clothes, weedy but young lvs eaten as potherb in Indonesia, local medic. R: FA 37(2015)493
Synedrellopsis Hieron. & Kuntze. Compositae (Helia.-Ecl.). 1 Bolivia, Parag, Arg.: ***S. grisebachii*** Hieron. & Kuntze – bad glyphosate-tolerant weed in Aus. R: FA 37(2015) 495
Syneilesis Maxim. (= ? *Ligularia*). Compositae (Sen.-Tuss.). 7 E As. R: Phytol. 27(1973)269. Cotyledons app. 0. ***S. palmata*** Maxim. (Jap.) – cult. orn.
Synelcosciadium Boiss. = *Tordylium*
Synepilaena Baill. = *Kohleria*
Syngonanthus Ruhl. Eriocaulaceae. Excl. *Comanthera*, c. 160 trop. Am., Afr. (14) & Madag. ***S. nitens*** (Bong.) Ruhl. (Braz.) – v. imp. in comm. handicrafts, the golden scapes combined with *Mauritia flexuosa*
Syngonium Schott. Araceae (VII 11). 36 trop. Am. R: AMBG 68(1981)565. Cult. orn. lianes with milky latex (used to counter *Paraponera* ant bites) esp. ***S. podophyllum*** Schott with several cvs
Syngramma J. Sm. Pteridaceae (III). 15 Mal. (Borneo 7) to Carolines
Syngrammatopsis Alston = *Pterozonium*
Synima Radlk. Sapindaceae. 2 SE NG, NE Aus. R: FA 25(1985)81,201
Synisoon Baill. = *Retiniphyllum*
Synnema Benth. = *Hygrophila*
Synnotia Sweet = *Sparaxis*
Synosma Raf. ex Britton & Brown = *Senecio*
Synostemon F. Muell. = *Breynia* (but prob. to be excluded)

Synotis (C.B. Clarke) C. Jeffrey & Y.L. Chen. Compositae (Sen.-Sen.). c. 55 Sino-Himal. & China (1)

Synotoma (G. Don) R. Schulz = *Physoplexis*

Synoum A. Juss. Meliaceae (II-Trich.). 1 NE Aus.: ***S. glandulosum*** (Sm.) A. Juss. (bastard or scentless rosewood) – dark red rose-scented timber for cabinet-work; seeds united by common 'aril'

Synsepalum (A. DC.) Daniell. Sapotaceae (IV). 20 trop. Afr. R: T.D. Pennington, *S.* (1991)242. Aubréville's Model. ***S. dulcificum*** (Schum. & Thonn.) Daniell (miraculous berry, W Afr.) – berries with miraculin causing sour & salt things to taste sweet by affecting taste-buds, a glycoprotein also depressing appetite, poss. a sugar-mimic disp. adaptation

Synstemon Botsch. Cruciferae (5). 2 China. R: Novon 19(2000)100

Synstylis Cusset = *Hydrobryum*

Synthlipsis A. Gray. Cruciferae (40). 2 S N Am. R: Rhodora 61(1959)253

Synthyris Benth. = *Veronica* (But see SB 29(2004)734)

Syntriandrium Engl. Menispermaceae (III). 1 trop. W Afr.: ***S. preussii*** Engl.

Syntrichopappus A. Gray. Compositae (Mad.-Bae.). 2 SW US. R: Novon 1(1991)123

Syntrinema H. Pfeiffer = *Rhynchospora*

Synurus Iljin. Compositae (Card.-Card.(Card.)). 5 E As.

Sypharissa Salisb. = *Tenicroa*

Syreitschikovia Pavlov. Compositae (Card.-Card.(Card.)). 2 C As.

Syrenia Andrz. ex Besser = *Erysimum*

Syrenopsis Jaub. & Spach = *Noccaea*

syringa *Philadelphus* spp., *Melia azedarach*

Syringa L. Oleaceae (4c). 21 SE Eur. (2) to E As. (esp. China). R: J.L. Fiala (2008) *Lilacs*, rev. ed., Cathaya 17–18(2008)40. Prob. incl. *Ligustrum* (berries, as ser. *Ligustrae*), linked by *S.* sect. *Ligustrina*. Decid. shrubs & small trees with showy thyrses or panicles of usu. strongly fragrant fls (synthetic lilac fragrance = anisyl acetate) & extrafl. nectaries; cult. orn. (lilac; nilak = bluish in Persian) esp. ***S. vulgaris*** L. (common l., SE Eur.) – some 1500 cvs incl. double-fld (since 1843) esp. due to Lemoine & fils (1850–1955) at Nancy, Fr., & US cvs with C 5 or 6 in 1 whorl, depithed stems form used for pipes (p. tree); other cult. hybrids & spp. incl. ***S. × chinensis*** Willd. (Rouen lilac, *S. vulgaris* × *S. persica*) – first synth. Rouen (Fr.) c. 1777 but also since, **S. 'Correlata'** – graft hybrid (Berlin Botanic Garden 1873), perclinal chimaera with outer tissue = *S. vulgaris*, inner ? *S.* × *chinensis*, ***S. emodi*** Wall. ex Royle (Himal. l., Afghanistan, Himal.), ***S. × hyacinthiflora*** (Lemoine) Rehder (*S. oblata* × *S. vulgaris*), many cvs, ***S. oblata*** Lindl. (China) – most commonly seen lilac in China; ***S. pubescens*** Turcz. (*S. meyeri*, N China) – esp. '**Palibin**' commonly seen dwarf l., ***S. persica*** L. (*S. afghanica*, *S. buxifolia*, *S. protolaciniata*, Ind., Afghanistan, China) – incl. cvs ('*S. laciniata*') with pinnately lobed juvenile lvs, introd. to Eur. via Iran 17C, ***S. pinnatifolia*** Hemsl. (W China) – only sp. with pinnate lvs, & ***S. reticulata*** (Blume) H. Hara (*S. amurensis*, Jap. l., N Jap.)

Syringantha Standl. Rubiaceae (I 1). 1 Mex.: ***S. coulteri*** (Hook.f.) McDowell. R: Novon 6(1996)277

Syringidium Lindau = *Stenostephanus*

Syringodea Hook.f. Iridaceae (VI 5). 7 S Afr. R: JSAB 40(1974)201. Capsules open in *wet* weather, close in dry (unique in I.)

Syringodium Kütz. Cymodoceaceae. 1 Ind. & W Pac. Oceans, 1 Carib. R: JB 77(1939)114. R: VKNAW II,59(1970)176. ***S. filiforme*** Kütz (Carib.) – dioec. submarine poll., pollen clumps released over some hours & collide with filiform stigmas, but also floats so surface-poll. also poss.

Syrmatium Vogel = *Acmispon*

Syrrheonema Miers. Menispermaceae (I). 3 trop. W Afr.

Systellantha Stone. Primulaceae (Myrsinaceae). 5 Borneo. R: MNJ 46(1992)14

Systeloglossum Schltr. Orchidaceae (V 12h). 5 Costa Rica to S Am.

Systemonodaphne Mez. Lauraceae (I). 1 Guianas, Amazon: ***S. geminiflora*** (Meissn.) Mez. R: BJ 110(1988)161

Systenotheca Rev. & Hardham (~ *Centrostegia*). Polygonaceae (I 1). 1 Calif.: ***S. vortriedei*** (Brandegee) Rev. & Hardham. R: FNA 5(2005)472

Syzygiopsis Ducke = *Pouteria*

Syzygium Gaertn. Myrtaceae (II 9). *S.l.*, incl. *Acmena*, *Acmenospora*, *Anetholea*, *Cleistocalyx*, *Cupheanthus*, *Piliocalyx*, *Waterhousea*, c. 1120 SE Afr. (30, Madag. 20; SE As. 56 – Adans.

37(2015)179) to Pacific (s.s. c. 550). R: Blumea 55(2010)96. Everg. canopy & emergent trees (Mangenot's Model), treelets (e.g. ***S. acre*** (Guillaumin) J.W. Dawson (New Caled.) – Corner's Model) & shrubs, incl. some rheophytes, from littoral & swamps, lowland to montane forests, diff. from superficially similar but entirely unrelated *Eugenia* in axile vasc. supply to ovules (not transseptal) & a set of differential morphological char. Some with ant-infested hollow stems. Cult. orn. & fr. trees (jambu, esp. *S. jambos*) incl. ***S. aqueum*** (Burm.f.) Alston (water rose apple, W Mal.), ***S. cumini*** (L.) Skeels (jambolan, jamun, Java plum, Indomal., invasive Hawaii, S Afr., SE US) – ed. fr. much used for juice, tanbark, fuelwood, ***S. jambos*** (L.) Alston (rose apple, SE As., invasive S Afr., Masc., Seychelles, Hawaii, Galápagos, Caribbean) – rheophytic when young, fr. best preserved, ***S. malaccense*** (L.) Merr. & Perry (rose or Malay apple, pomerac, Otaheite a., ohia, Malay Pen., natur. Sri Lanka to Hawaii, Galápagos) – oft. polyembryonic, introd. WI with breadfruit (Capt. Bligh), fr. ed. fresh or cooked, wood carved Hawaii for idols, ***S. samarangense*** (Blume) Merr. & Perry (wax jambu, Mal.) – ed. fresh, ***S. suborbiculare*** (Benth.) T. Hartley & L. Perry (lady apple, trop. Aus.) – ed. fr. ***S. aromaticum*** (L.) Merr. & Perry (*Eugenia caryophyllus*, cloves [from Lat. *clavus*, nail – shape of buds], Moluccas) – sundried fl. buds [cf. *Capparis*!] the cloves of comm. (140 K t a yr), much exported from Zanzibar (Z. red heads) etc. (all in Réunion & Madag. thought to derive from 1 tree), known to Chinese (& in Alexandria) by 200 BC when courtiers had to have them in their mouths when addressing the Emperor, now used in flavouring pickles, cakes & apple pies etc., in making pomanders, the oil (eugenol) having antibacterial, antiviral (herpes), & analgesic properties much used in relief of toothache, in toothpaste, as well as principal source of vanillin, fragments added to kretek cigarettes in Indonesia (some 30K t per annum); ***S. anisatum*** (Vickery) Craven & Biffin (*Anetholea a., Backhousia a.*, NSW) – aniseed odour (anethole) ***S. australe*** (Link) Hyland (E Aus.) – cult. orn. esp. '**Tiny Trev**' – clipped pot-pl.; ***S. cordatum*** Hochst. ex Krauss (E Afr.) – used for heart palpitations & stomach problems in Tanzania; ***S. guineense*** (Willd.) DC. (trop. & S Afr.) – 30 m tree to pyrophytic subshrub, poss. most variable Afr. pl.; ***S. luehmannii*** (F. Muell.) L. Johnson (riberry, Aus.) – fr. used in comm. skin moisturizers; ***S. polyanthum*** (Wight) Walp. (daun salam, Indomal.) – used like bay in e.g. nasi goreng; ***S. smithii*** (Poir.) Niedenzu (*Acmena s.*, lillypilly, lilli-pilli, E Aus.) – street-tree, fr. ed., extract a comm. facial cleanser

Szovitsia Fischer & C. Meyer. Umbelliferae (III 8). 1 Cauc., Armenia, Iran: ***S. callicarpa*** Fischer & C. Meyer

T

tabardillo *Piqueria trinervia*

Tabaroa Queiroz & al. Leguminosae (III 4). 1 Bahia: ***T. caatingicola*** Queiroz & al. R: KB 65(2010)193

Tabascina Baill. = *Justicia*

tabasco *Capsicum frutescens*

tabasheer or **tabashir** Siliceous concretions found in hollow stems of *Bambusa bambos* & other As. bamboos – alleged medic. value

Tabebuia Gomes ex DC. Bignoniaceae (1). Excl. *Cybistax, Handroanthus, Roseodendron* 67 trop. Am. R: SB 32(2007)667. ***T. aquatilis*** (E. Meyer) Sprague & Sandwith (Amazon) – air-filled seed-coat promoting water-disp.; ***T. pallida*** (Lindl.) Miers (WI, invasive Seychelles) – cult. orn. with 4 or 5 buds per axil; ***T. rosea*** (Bertol.) DC. ('*T. pentaphylla*', roble, Mex. to Venez.) – cult. orn. with large pink fls (nat. fl. El Salvador), timber valued

Taberna Miers = *Tabernaemontana*

Tabernaemontana Plum. ex L. Apocynaceae (I d). Incl. *Stemmadenia*, 110 trop. (OW 55, incl. *Conopharyngia, Ervatamia*). R: A.J.M. Leeuwenberg, *Rev. of T. I, II* (1991, 1994). Trees & shrubs with alks; some rubber sources & local medic., some masticatories. Leeuwenberg's Model: false dichotomy in branching due in at least ***T. crassa*** Benth. (W & C Afr.) to slowing of apical meristem & growing out of laterals. ***T. divaricata*** (L.) Roemer & Schultes (*T. coronaria, Ervatamia c.*, crepe jasmine, moon beam, Ind.) – cult. orn.; ***T. pachysiphon*** Stapf (trop. E Afr.) – fr. ed. chimpanzees, bark source of dodo cloth, lvs of a black dye for hair

Tabernanthe Baill. Apocynaceae (I d). 2 C Afr. R: AUWP 89–4(1989)3. Alks: ***T. iboga*** Baill. (W Afr.) – alks incl. ibogaine, hallucinogen (iboga), used in treatment of opiate addiction

tacamahac *Populus balsamifera*

Tacarcuna Huft. Phyllanthaceae (Brid.-Sav.; Euphorbiaceae s.l.). 3 trop. Am. R: AMBG 76(1989)1080

tacay *Caryodendron orinocense*

Tacazzea Decne. Apocynacee (III; Asclepiadaceae I 1). 4 trop. & S Afr. R: SAJB 56(1990)93. Riparian lianes & shrubs

Tacca Forst. & Forst. f. Taccaceae (Dioscoreaceae s.l.). 13 OW trop. (Mal. to Papuasia 9). R: Blumea 20(1972)374. Fls poll. by flies & may act as traps, thermogenesis to 40°C with scent in filiform bracts (whiskers); fr. disp. birds & mammals. Tubers a source of starch used for bread etc., once bitter taccalin removed, esp. ***T. integrifolia*** Ker-Gawler (bat fl., b. lily, b. pl., Ind. to W Mal.) – rhiz. pounded for treating amoebic dysentery (Sumatra) & ***T. leontopetaloides*** (L.) Kuntze (*T. pinnatifida*, E Ind. or Tahiti arrowroot, trop. OW) – ? orig. a beach pl. (fr. float for many months), lvs used in hat-making; ***T. palmata*** Blume (Mal.) – locally medic.

Taccaceae Dumort. (~ Dioscoreaceae). Magnoliidae – Dioscoreales. 1/13 trop. R: Blumea 20(1972)366. Perenn. herbs with starchy rhiz.; vessels restricted to r.; indumentum of minute pluricellular hairs. Lvs basal, with long petoles & elliptic to dissected laminas; venation parallel or palmate. Fls bisex., reg., 3-merous in cymose umbels with involucre of (2-)4(-12) bracts at apex of scape. Floral bracts ∞, filiform, drooping, s.t. inconspic.; P 3 + 3, ± petaloid, s.t. basally connate, brown-purple to greenish, A 3 + 3 attached to base of P with short, flat ± petaloid filaments forming hoods over anthers which have long slits, $\overline{G}(3)$, 1-loc., 6-ribbed with ± intruded parietal placentas & ∞ anatr. to campylotropous bitegmic ovules, style 1 with 3 stigmas, oft. petaloid. Fr. usu. a berry (or loculicidal capsule) with 10-∞ ribbed seeds, s.t. with thin fleshy aril; embryo small, in copious starchless endosperm rich in fat & protein. n = 15
Genus: *Tacca*

taccada pith *Scaevola taccada*

Taccarum Brongn. ex Schott. Araceae (VII 4). 4 N S Am. R: Willd. 19(1989)191

Tachia Aubl. Gentianaceae. 13 trop. S Am. R: SB 38(2013)1145. Some tree-like, often pachycaul, myrmecophytic; some locally medic. (antimalarials)

Tachiadenus Griseb. Gentianaceae. 11 Madag. R: BMNHN 4,9(1987)43. Bitters used in beer- making

Tachigali Aubl. Leguminosae (I 4). Incl. *Sclerolobium* (R: Lloydia 20(1957)67, 266) 60 trop. Am. esp. Amaz. Trees (some with Petit's Model) with extrafloral nectaries (many myrmecophytes); some hapaxanthic but much-branched (cf. *Cerberiopsis*); some with lvs with continuous growth (cf. *Chisocheton*). Some medic., dyes etc.

Tachigalia Juss. = *Tachigali*

Tacinga Britton & Rose. Cactaceae (II). 8 E Braz., Venez.(1). ***T. funalis*** Britton & Rose – cult. orn. clambering to 13 m like *Opuntia* but A long-exerted

Tacitus Moran = *Graptopetalum*

Tacoanthus Baill. = *Ruellia*

tade *Persicaria hydropiper*

Tadeastrum Szlach. = *Stanhopea*

Tadehagi Ohashi (*Pteroloma*). Leguminosae (III 19). 6 Indomal. R: JJB 78(2003)276. Intergrades with *Droogmansia*. ***T. triquetrum*** (L.) Ohashi – medic. (piles)

Taeckholmia Boulos = *Sonchus*

Taeniandra Bremek. = *Strobilanthes*

Taenianthera Burret = *Geonoma*

Taeniatherum Nevski (~ *Triticum*). Gramineae (XV). 1 Spain to C As. & Pakistan: ***T. caput-medusae*** (L.) Nevski (Medusa head) – noxious weed W US. R: NJB 6(1986)389

Taenidia (Torrey & Gray) Drude. Umbelliferae (III 5). 2 E US. R: Britt. 34(1982)365

Taeniopetalum Vis. = *Peucedanum.*

Taeniophyllum Blume. Orchidaceae (V 16c). c. 170 trop. Afr. (1) to Jap., Aus. (8 – R: Austrobaileya 9(2015)383) & Tahiti. Photosynthetic r.

Taeniopleurum J. Coulter & Rose = *Perideridia*

Taeniorhachis Cope (~ *Digitaria*). Gramineae (XXIV 1). 1 Somalia: ***T. repens*** Cope. R: KB 48(1993)403

Taeniorrhiza Summerh. Orchidaceae (V 16d). 1 Gabon: ***T. gabonensis*** Summerh. – leafless

Taenitidaceae Pichi-Serm. = Pteridaceae

Taenitis Willd. ex Schkuhr. Pteridaceae (III; Taenitidaceae). 15 Indomal. to trop. Aus. & Fiji. R: Blumea 16(1968)87

tag (**hemp**) *Crotalaria juncea*

tagasaste *Cytisus proliferus*

Tagetes Tourn. ex L. Compositae (Tag.-Pect.). c. 50 trop. & warm Am. (Mex. 24). Foetid herbs, some medic. etc. (Aztec use incl. for tick-removal & as incense); alpha-terphenyl considered poss. effective against AIDS & as effective as DDT in controlling mosquito larvae; some cult. orn. ***T. erecta*** L. (incl. *T. patula*, Afr. & F. marigolds, Mex. & C Am., widely natur.) – cvs (wild form = '*T. elongata* Willd.') to 30 cm with yellow, orange & brownish fls ('French', tetraploid) to 1 m ('Afr.'; triploid hybrids = 'Afro-French'), cult. orn. in pre-Conquest Mex., still used in cerem., fls source of a poor quality yellow dye, locally medic., many cvs for formal bedding, petals eaten in salads; ***T. lucida*** Cav. (Mexicon tarragon, Mex., Guatemala) – tea a tonic in Mex., hallucinogenic (agent unidentified but not alk.), flavour for Aztecs' chocolate, tarragon subs. (Aus.); ***T. minuta*** L. (stinking Roger, S Am., invasive Hawaii, trop. & S Afr.) – locally medic. & insecticidal, alleged to prot. potatoes from eelworm & to be a weedkiller (allelopathic) & effective at controlling 'damping off' (*Pythium* spp.), ess. oil crop promoted in Ind.

taggar *Cinnamosma fragrans*

tagua *Phytelephas* spp. esp. *P. macrocarpa*

tah(onal) *Viguiera dentata*

taheebo (tea) *Handroanthus impetiginosus*

Tahina Dransf. & Rakot. Palmae (III 5). 1 NW Madag.: ***T. spectabilis*** Dransf. & Rakot. – hapaxanthic, only 91 trees + seedlings (2008). R: BJLS 156(2008)81

tahini *Sesamum indicum*

Tahiti arrowroot *Tacca leontopetaloides*; **T. chestnut** *Inocarpus fagifer*; **T. vanilla** *V.* × *tahitensis*

Tahitia Burret = *Christiana*

Tahitian bridal veil *Gibasis pellucida*; **T. lime** *Citrus* × *latifolia*

Taihangia T.T. Yu & C.L. Li = *Geum*

Tainia Blume. Orchidaceae (V 14). Excl. *Ania*, *Ascidieria*, 24 Indomal. to China & Aus. R: OM 6(1992)73

Tainionema Schltr. = ? *Secamone*

Tainiopsis Schltr. = *Eriodes*

Taitonia Yamamoto = *Gomphostemma*

Taiwania Hayata. Cupressaceae (Taxodiaceae). 2–3 NE Myanmar, NE Vietnam, Yunnan, Taiwan. R: IDSY 2010(2011)24. ***T. cryptomerioides*** Hayata – cult. orn. to 70 m allied to *Sequoiadendron* but rare as much felled (e.g. for coffin planks)

Takeikadzuchia Kitagawa & Kitam. = *Olgaea*

Takhtajania M. Baranova & J. Leroy. Winteraceae (Takhtajaniaceae). 1 Madag.: ***T. perrieri*** (Capuron) M. Baranova & J. Leroy – only W. in Afr. or Madag., coll. once until 50 trees (Scarrone's Model) found 1990s; whorls 2-, 4-, 5-merous respectively. R: AMBG 87(2000)297

Takhtajaniaceae J. Leroy = Winteraceae

Takhtajaniantha Nazarova = *Scorzonera* but polyploid & poss. distinct

Takhtajanianthus De = *Rhanteriopsis*

Takhtajaniella Avet. = *Alyssum*

tak-out galls tamarisk g.

Takulumena Szlach. = *Epidendrum*

Talamancalia H. Robinson & Cuatrec. (~ *Pseudogynoxis*). Compositae (Sen.-Sen.). 4 C & N S Am. R: Novon 4(1994)50, CN 27(1995)34)

talas *Colocasia esculenta*

Talauma Juss. = *Magnolia*

Talbotia S. Moore = *Afrofittonia*

Talbotia Balf. = *Xerophyta*

Talbotiella Bak. f. (~ *Hymenostegia*). Leguminosae (I 2). 8 trop. W Afr. R: KB 65(2010)401. ***T. gentii*** Hutch. & Greenway – forms monospecific stands in drier forests of Ghana (canopy & understorey) but endangered (charcoal)

Talbotiopsis L.B. Sm. = *Xerophyta*

Talguenea Miers ex Endl. = *Trevoa*

talh gum *Vachellia seyal*

Talinaceae Doweld (~ Portulaceae). Magnoliidae – Caryophyllales. 2/28 S Afr, Am. R: T 59(2010)236. Scrambling shrubs, to herbs oft. with tubers. Lvs simple, vernation revolute, in spirals. Fls reg., bisex., in term. panicles or solit. in axils. P quincuncial of 2 'sepaloids' & usu. 5 'petaloids' s.t. fewer & not clearly diff. from sepaloids; A c. 15–35; $\underline{G\ (3(-5))}$, septate at least initially. Fr. loculicidal caps. or mucilaginous berry; seeds usu. black, (?)arillate with curved embryo. n = 8.

Genera: *Amphipetalum, Talinum*

Talinaria Brandegee = *Anacampseros*

Talinella Baill. = *Talinum* (but see Adansonia 27(2005)50)

Talinopsis A. Gray. Anacampserotaceae (Portulacaceae s.l.). 1 arid S US & Mex.: ***T. frutescens*** A. Gray – subshrub. R: FNA 4(2003)502

Talinum Adans. Talinaceae (Portulacaceae s.l.). Incl. *Talinella*, excl. *Phemeranthus*, c. 27 S Afr., Am. Herbs, ± succ.; R: FR 35(1934)1. Some with molluscicidal saponins. Some cult. orn. & potherbs esp. ***T. fruticosum*** (L.) Juss. (*T. triangulare*, ? orig. trop. Am.) – tastes like purslane

Talipariti Fryxell = *Hibiscus*

talipot palm *Corypha umbraculifera*

Talisia Aubl. Sapindaceae (IV 2). 52 trop. Am. R: FN 87(2003)50. Chamberlain's Model. Some fish-poisons (saponins), some cult. for ed. fr. sold in markets esp. ***T. oliviformis*** (Kunth) Radlk. (Mex. to Colombia, Trinidad)

tallicona *Carapa guianensis*

tallow, Mafura *Trichilia emetica*; **Malabar t.** *Vateria indica*; **t. nut** *Ximenia americana*; **t. shrub** *Morella cerifera*; **t. tree** *Detarium senegalense, Pentadesma butyracea*; **Chinese** or **vegetable t.t.** *Triadica sebifera*; **Japanese t.t.** *Toxiocodendron succedaneum*; **t. wood** *Eucalyptus microcorys*

tally See *Ulmus*

Taltalia E. Bayer = *Alstroemeria* (but see Sendtnera 5(1998)7)

tamal *Garcinia xanthochymus*

Tamamschjanella Pim. & Kljuykov. Umbelliferae (III 8). 3 Balkans, Cauc.

Tamamschjania Pim. & Kljuykov = *Eleutherospermum* (but see FR 105(1994)434)

Tamananthus Badillo. Compositae (Mill.-Esp.). 1 Venez.: ***T. crinitus*** Badillo

Tamania Cuatrec. See *Espeletia*

tamanu *Calophyllum inophyllum*

tamarack larch *Larix laricina*

Tamaricaceae Link (~ Frankeniaceae). Magnoliidae– Caryophyllales. 5/88 Euras. & Afr. esp. Medit. to C As. Shrubs & trees, usu. halophytes, xerophytes or rheophytes with slender branches, occ. dioec. Lvs small, oft. scale-like & centric & with salt-excreting glands, in spirals; stip. 0. Fls small, ebracteolate, solit. or usu. in bracteate spikes to panicles; K 4 or 5(6) rarely basally connate, imbr., persistent with C alt. & s.t. persistent but oft. seated on nectary-disk with A, or disk intrapetiolar or 0, A usu. 2 × C or more & connate in 5 bundles, anthers with longit. slits, $\underline{G}$ ((2)3 or 4(5)), 1-loc. with parietal or (*Tamarix*) basal or parietal basal placentation s.t. almost pluriloc. with 2–∞ anatropous, bitegmic ovules per placenta. Fr. a loculicidal capsule; seeds hairy or the hairs forming a tuft at 1 end, embryo straight in 0 or scanty starchy endosperm oft. with thin perisperm. n = 12

Genera: *Hololachna, Myricaria, Myrtama, Reaumuria, Tamarix*

Dyes, medic., manna & cult. orn. (*Tamarix*)

Tamaricaria Qaiser & Ali = *Myrtama*

tamarilla, tamarillo *Solanum betaceum*

tamarind *Tamarindus indica*; **Manila t.** *Pithecellobium dulce*; **native t.** *Diploglottis australis*; **velvet t.** *Dialium guineense, D. ovoideum*

Tamarindus Tourn. ex L. Leguminosae (I 2). 1 cultigen (? orig. trop. Afr.): ***T. indica*** L. (tamarind, Ind. date) – nyctitropic movements described by Theophrastus, fr. disp. by ruminants like gazelles (fr. high in vitamin C, protein, tartaric, malic & citric acids), cult. orn. shade tree (presence in N Aus. poss. indicator of Macassar trepang-collector sites) with useful timber & ed. pulp (comm. 0.5 M t by 1999) around seeds (laxative), sharp-tasting & used in drinks, chutney etc. (an ess. ingredient of Worcestershire sauce), when overripe used to clean brass & copper, clipped into gnarled 'bonsai' in Thailand, walking-sticks made from r.

tamarira (oil) *Eruca vesicaria*

tamarisk *Tamarix* spp.; **t. galls** *T. aphylla & T. gallica*; **manna t.** *T. mannifera*

Tamarix L. Tamaricaceae. c. 60 Euras. (Eur. 14), Afr. R: B.R. Baum (1978) *The genus T.* Tamarisk. Spp. diff. to delimit; many rheophytes, cult. orn. & galls (teggaout or tak-out g.) for tanning; some manna sources esp. ***T. mannifera*** Ehrenb. (manna t., Iran to Arabia) – manna of the Bedouin produced as result of punctures in stems made by scale-insects. Most commonly cult. orn. is ***T. chinensis*** Lour. (temp. E As., natur. N Am.) – stems used for making lobster-pots in Cornwall, oft. confused with ***T. gallica*** L. (*T. anglica*, Medit.) diff. in decid. C etc., with active principle combating liver damage, parasitized fr. behaving like 'jumping beans' (cf. *Sebastiana*). Morocco leather tanned with galls of ***T. aphylla*** (L.) Karsten (athel pine, a. tree, S & E Med. to Ind., invasive Aus.) & *T. gallica* (40–45% tannin); ***T. ramosissima*** Ledeb. (E Eur.) – invasive Aus., S Afr., W US

Tamaulipa R. King & H. Robinson (~ *Eupatorium*). Compositae (Eup.-Gyp.). 1 Texas, N Mex.: ***T. azurea*** (DC.) R. King & H. Robinson. R: MSBMBG 22(1987)132

Tamayoa Badillo = *Lepidesmia*

Tamayorkis Szlach. = *Malaxis* (but see ABF 35(1998)21)

Tambourissa Sonn. Monimiaceae (V 1). 43 Madag., Masc. R: AMBG 72(1985)90. Mangenot's & Nozeran's Models. G to 2000!

Tamia Rav. = *Calydorea*

Tamijia Sakai & Nagam. Zingiberaceae (Tam.). 1 Sarawak: ***T. flagellaris*** Sakai & Nagam. R: EJB 57(2000)245

Tamilnadia Tirv. & Sastre (~ *Gardenia*). Rubiaceae (II 1). 1 S & SE As.: ***T. uliginosa*** (Retz.) Tirv. & Sastre

tämjölk milk fermented with *Pinguicula* lvs

Tammsia Karsten. Rubiaceae (Cond.). 1 Venez., Colombia: ***T. anomala*** Karsten

Tamonea Aubl. (*Ghinia*). Verbenaceae (Cass.). 6 trop. Am. R: BJLS 157(2008)358

tampala *Amaranthus tricolor*

tampico fibre *Agave* spp.

Tamridaea Thulin & B. Bremer. Rubiaceae (II 7). 1 Socotra: ***T. capsulifera*** (Balf.f.) Thulin & B. Bremer. R: PSE 211(1998)85

Tamuria Starod. = *Anemone*

Tamus L. = *Dioscorea*

Tana B.-E. van Wyk. (~ *Heteromorpha*). Umbelliferae. 1 Madag.: ***T. bojerianum*** (Bak.) B.-E. van Wyk. R: T: 48(1999)743

Tanacetopsis (Tzvelev) Kovalevsk. (~ *Cancrinia*). Compositae (Anth.-Han.). 21 Iran to C As. R: BBMNHB 23(1993)104

Tanacetum Tourn. ex L. (~ *Chrysanthemum*). Compositae (Anth.-Anth.). (Incl. *Balsamita*) 160 N temp. esp. OW (Eur. 14). R: BBMNHNB 23(1993)100. Cult. orn. aromatic herbs esp. ***T. balsamita*** L. (*Balsamita major*, *C. b.*, alecost, camphor pl., costmary, Eur. to C As.) – aromatic lvs form. flavour in ale, ***T. coccineum*** (Willd.) Grierson (pyrethrum of gardens, Cauc., Iran), ***T. parthenium*** (L.) Schultz-Bip. (feverfew, SE Eur. & N Afr. to C As.) – dried capitula used medic. (febrifuge & efficacious in treatment of migraine, period pains) & as tea etc., '**Aureum**' (golden feather) – commonly cult. orn. ***T. cinerariifolium*** (Trev.) Schultz-Bip. (SE Eur. to China) – self-incompatible diploid clones in cult., fl. heads (cypselas) the source (10 K t a yr by 2011) of pyrethrum insecticide (monoterpenes called pyrethrins), cult. esp. highlands of E Afr.; ***T. corymbosum*** (L.) Schultz-Bip. (Medit. to C As.) – oil antibacterial; ***T. ptarmiciflorum*** Schultz-Bip. ('Silver Lace', Canary Is.) – confused with *Jacobaea maritima*; ***T. vulgare*** L. (tansy, Euras., introd. Br. Is., taken to US by early settlers for remedies & 'Easter puddings') – tea form. a tonic, abortifacient, febrifuge (redwater fever in Ireland & Hebrides) & vermifuge, now used against herpes, crystallized for treatment of gout, lvs form. placed in winding-sheets as deterrent to worms, rubbed in to meat to keep off flies, mouse-deterrent, ingredient of drisheen, an Irish sausage made of sheep's blood & in modern cosmetics, 16 cent. cakes (tansies) eaten in remembrance of 'bitter herbs' of Passover

Tanaecium Sw. (~ *Bignonia*). Bignoniaceae (3). Incl. *Paragonia*, 17 trop. Am. R: AMBG 99(2014)463. ***T. pyramidata*** (Rich.) L. Lohmann (*P. p.*) – local med. in NW Amaz.

Tanakaea Franch. & Sav. Saxifragaceae (I 8). 1 China, Jap.: ***T. radicans*** Franch. & Sav. – cult. orn. dioec. everg. perenn. herb

Tanaosolen N.E. Br. = *Tritoniopsis*

tanekaha *Phyllocladus trichomanoides*

tangelo *Citrus* × *aurantium* Tangelo Group

tangerine *Citrus reticulata*

tanget *Cordyline fruticosa*

Tanghinia Thouars = *Cerbera*

tangile *Shorea polysperma*

tanglehead *Heteropogon contortus*

tangor *Citrus* × *aurantium* Tangor Group

tang-shen *Codonopsis tangshen*

Tangtsinia S.C. Chen = *Cephalanthera*
Tanner grass *Urochloa arrecta*
tanner's cassia *Senna auriculata*; **t. dock** *Rumex hymenosepalus*
tannia *Xanthosoma sagittifolium*
tannins astringent (as in wine) phenolics from bark etc. (seeds, epicarp & infl. axes in wine), used to tan leather
Tannodia Baill. Euphorbiaceae (Al.-Gross.). 9 trop. Afr. (3), Madag. (6). R: KB 53(1998)174
Tanquana H. Hartmann & Liede (~ *Pleiospilos*). Aizoaceae (V). 3 Cape. R: BJ 106(1986)479
Tansaniochloa Rauschert = *Setaria*
tansy *Tanacetum vulgare*
Tanulepis Balf.f. = *Camptocarpus*
tapa cloth *Broussonetia papyrifera*
táparos *Attalea cuatrecasana*
Tapeinanthus Herb. = *Narcissus*
Tapeinia Comm. ex Juss. Iridaceae (VII 3). 1 S Chile & Arg.: ***T. pumila*** (Forst.f.) Baill.
Tapeinidium (C. Presl) C. Chr. Lindsaeaceae (Dennstaedtiaceae s.l.). 18 SE As. to Samoa
Tapeinochilos Miq. Costaceae (Zingiberaceae s.l.). 16 Moluccas to Aus. (esp. NG). McClure's Model. ***T. ananassae*** (Hassk.) K. Schum. (Moluccas to Queensland) – local medic., form. veget., cult. orn.
Tapeinoglossum Schltr. = *Bulbophyllum*
Tapeinosperma Hook.f. Primulaceae (Myrsinaceae). 5 E Mal. & trop. Aus., 39 New Caled., 11 Fiji. Pachycaul treelets (Corner's, Leeuwenberg's & Rauh's Models)
Tapeinostemon Benth. (~ *Curtia*). Gentianaceae. 6 Guyana Highlands, 2 Andes. R: MNYBG 32(1981)356
tapeweed *Posidonia* spp.
Tapheocarpa Conrau (~ *Aneilema*). Commelinaceae (II). 1 N Queensland: ***T. calandrinioides*** (F. Muell.) Conrau – fr. hypogynous. R: Aus SB 7(1994)585
Taphrospermum C. Meyer = *Eutrema* (but see HPB 5(2000)101)
Tapinanthus (Blume) Reichb. Loranthaceae (5 7). 30 trop. & S Afr., 1 ext. to N Yemen. R: R. Polhill & D. Wiens, *Mistletoes of Africa* (1998)183. Some with explosive fls like *Globimetula* (q.v.), C-bud darkening on maturity & secreting nectar, signalling birds to peck & release targeted spray of pollen; 2 S Afr. spp. decid. See also *Agelanthus*, *Phragmanthera*
Tapinopentas Bremek. = *Otomeria*
Tapinostemma (Benth.) Tieghem = *Plicosepalus*
tapioca *Manihot esculenta*
Tapiphyllum F. Robyns = *Vangueria*
Tapirira Aubl. Anacardiaceae (II). 8+ trop. Am. ***T. guianensis*** Aubl. – 96 spp. of insect-visitors to fls
Tapirocarpus Sagot = *Talisia*
Tapiscia Oliv. Tapisciaceae. 3 S & SE China, N Vietnam
Tapisciaceae Takht. (~ Staphyleaceae). Magnoliidae – Huerteales. 2/7 S China to Vietnam, WI & N S Am. Trees. Lvs in spirals, imparipinnate to trifoliolate with glands or stipels at articulations. Fls 5(6)-merous, small, in s.t. polygamous panicles; K s.t. connate; C imbr.; intrastaminal disk s.t. present; G (2) with 1 style & 1(2) basal erect anatr. bitegmic ovule per carpel. Fr. a drupe or berry; endosperm copious. n = 13
Genera: *Huertea*, *Tapiscia*
Form. in Staphyleaceae but lvs in spirals etc., the 2 fams actually in diff. orders
Taplinia Lander. Compositae (Gnap.-Ang.). 1 W Aus.: ***T. saxatilis*** Lander. R: Nuytsia 7(1989)37
Tapoides Airy Shaw. Euphorbiaceae (Al.-Gross.). 1 Borneo: ***T. villamilii*** (Merr.) Airy Shaw. R: KB 14(1960)473
Taprobanea E. Christ. (~ *Vanda*). Orchidaceae (V 16c). 1 S Ind., Sri Lanka: ***T. spathulata*** (L.) E. Christ. R: Lindleyana 7(1992)90
Tapura Aubl. Dichapetalaceae. 28 trop. Am. (20), trop. Afr. (8). R: Britt. 35(1983)49, AUWP 86–3(1986)43. Cook's Model; infls axillary, on petioles or midribs. Some timber
tar tree *Semecarpus australiensis*
tara *Tara spinosa*; **t. gum** *Cassia fistula*
Tara Molina (~ *Caesalpinia*). Leguminosae (I 4). 3 trop. Am. R: PhytoKeys 71(2016)48. ***T. spinosa*** (Molina) Britton & Rose (*C. s.*, tara, deserts W S Am.) – imp. tannin source in Peru
taraire *Beilschmiedia tarairi*

Taraktogenos Hassk. = *Hydnocarpus*

Taralea Aubl. (~ *Dipteryx*). Leguminosae (III 3). 7 trop. Am. Wood v. hard. ***T. oppositifolia*** Aubl. – industrial oil

Tarasa Philippi (~ *Nototriche*). Malvaceae (Malv.-Malv.). 30 Andes to Mex. (2). Heterogeneous? Unlike in most genera polyploids are annuals, are smaller & with smaller pl. features incl. pollen

tarata *Pittosporum eugenioides*

Taravalia Greene = *Ptelea*

Taraxacum Weber ex Wigg. Compositae (Lact.-Crep.). 60 temp. inc. S Am. (2), some cosmop. weeds; c. 1600–2500 apomictic lines named ('microspp.' in c. 52 sects; 30 'groups' in Eur.; hybrids between intr. agamospermous triploids (3n = 24) & native sexual diploids common in urban Jap.), *T.* poss. orig. Himal. R: Biblioth.Konink.Ned.Natuurhist.Veren. 42(1987), T 60(2011)216. Some cult. orn., medic., salad & rubber pls. ***T. bicorne*** Dahlst. (*T. koksaghyz*, kok-saghyz, Russian dandelion, Turkestan) – obligate outbreeding diploid cult. for rubber in Russia & N China; ***T. megalorhizon*** (Forssk.) Hand.-Mazz. (krim-saghyz, W As.) – similar; ***T. officinale*** Weber ex Wigg. s.l. (*T. vulgare* (Lam.) Schrank, dandelion, Euras., cosmop. triploid weed where endosperm does not necessarily develop before embryo (all identical!), *T. o.* s.s. (***T. campylodes*** Hagerup) restricted to Lapland; the debate over the corr. name for this most common pl. should be resolved by conservation with a new type) – cult. (blanched) for salad lvs (several cvs), taproot ground as coffee subs., somewhat diuretic (Fr.: *pisse-en-lit*) though agent unknown, allegedly a liver & kidney remedy, used for warts, colds & as blood 'cleanser', unopened fl. heads used like capers, open heads dried for dandelion wine-making & comm. dandelion & burdock tea, good bee-fodder (in orchards can distract bees from poll. fr. trees), comm. jelly, form with pitcher-shaped lvs found in England 1980s

Taraxia (Torrey & A. Gray) Raim. (~ *Camissonia*). Onagraceae. c. 10 N Am. R: CUSNH 37(1969)161. Stemless perennials

Taraxis Briggs & L. Johnson. Restionaceae. 1 SW Aus.: ***T. grossa*** Briggs & L. Johnson – to 2 m tall. R: Telopea 7(1998)363

Tarchonanthus L. Compositae (Card.-Tarch.). 2 Afr., Arabia. R: KB 54(1999)82. Poss. incl. *Brachylaena*. Dioec. ***T. camphoratus*** L. (camphor wood, E & S Afr.) – lvs chewed medic., wood for musical instruments etc.

tare *Vicia* spp. esp. *V. sativa*

Tarenaya Raf. = *Cleome*

Tarenna Gaertn. Rubiaceae (II 2). Incl. *Chomelia, Cladoceras, Diplospora*, excl. *Coptosperma* c. 370 trop. As. Some timber; alks

Tarennoidea Tirv. & Sastre (~ *Randia*). Rubiaceae (II 1). 2 Indomal.

tares *Lolium temulentum*

tarhui *Lupinus mutabilis*

Tarigidia Stent. Gramineae (XXIV 1). 1 S Afr., 1 Puerto Rico. ? *Anthephora* × *Digitaria*

Tarlmounia H. Robinson & al. Compositae (Vern.). 1 S & SE As. (natur. Pacific): ***T. elliptica*** (DC.) H. Robinson & al. (*Vernonia e.*) – liane planted e.g. Singapore (sterile there) to obscure unsightly views. R: FA 37(2015)187

taro *Colocasia esculenta*; **giant t.** *Alocasia macrorrhizos*; **swamp t.** *Cyrtosperma merkusii*

Tarphochlamys Bremek. = *Strobilanthes*

tarragon *Artemisia dracunculus*; **Mexican t.** *Tagetes lucida*

Tarrietia Blume = *Heritiera*

Tartagalia (Robyns) T. Meyer = *Eriotheca*

Tartarian lamb *Cibotium barometz*

tarweed *Madia* spp.

tarwi *Lupinus mutabilis*

Tashiroea Matsum. = *Bredia*

Tasmanian beech *Nothofagus cunninghamii*; **T. blackwood** *Acacia melanoxylon*; **T. cedar** *Athrotaxis* spp.; **T. myrtle** *N. cunninghamii*; **T. oak** *Eucalyptus delegatensis*

Tasmannia R. Br. ex DC. (~ *Drimys*). Winteraceae. 15 C Mal. to Pacific. Usu. dioec. ***T. lanceolata*** (Poir.) A.C. Sm. (*D. l., D. aromatica*, mountain pepper, Tasmania) – dried fr. a pepper subs. (also a liqueur in Tasm.), planted as hedges in Ireland

Tassadia Decne. (~ *Cynanchum*). Apocynaceae (V c; Asclepiadaceae III 1). Incl. *Sattadia*, 23 trop. Am. R: AJBRJ 21(1977)235. Mostly riparian

tasua *Aglaia cucullata, Aphanamixis polystachya*

tatami *Juncus effusus*
Tateanthus Gleason. Melastomataceae. 1 Venez.: ***T. duidae*** Gleason
Tatianyx Zuloaga & Soderstrom. Gramineae (XXIII 1). 1 Braz.: ***T. arnacites*** (Trin.) Zuloaga & Soderstrom – savanna. R: SCB 59(1985)56
tatsoi *Brassica rapa* Chinensis Group
tau foo *Glycine max*
tau miu 'sweet pea sprouts' = pea + wheat sprouts
Taubertia K. Schum. = *Disciphania*
tau-saghyz *Scorzonera tausaghyz*
Tauscheria Fischer ex DC. = *Isatis*
Tauschia Schldl. Umbelliferae (III 5). 31 W US to W trop. S Am.
Tavaresia Welw. (~ *Decabelone*). Apocynaceae (V b; Asclepiadaceae III 5). 3 trop. & S Afr. R: F. Albers & U. Meve, *Ill. Handb. Succ. Pls*, Asclep. (2002)265. Cult. orn. succ.
Taverniera DC. Leguminosae (III 25). 15 SW As., NE Afr. R: SBU 25,1(1985)45
Taveunia Burret = *Cyphosperma*
tawa *Beilschmiedia tawa*
tawara-shibo *Chamaecyparis obtusa*
Taxaceae Gray. Pinidae – Pinales. Incl. Cephalotaxaceae, 6/47 N temp. to Mal. & New Caled. R: BR 64(1998)291. Dioec. everg. trees & shrubs s.t. without resin-canals. Lvs linear, needle-like, in spirals but oft. appearing 2-ranked to decussate. Microsporangiophores 6–14 in small cones, scale-like, with 3–6 pollen-sacs; ovules solit., arillate, usu. term. dwarf shoots. Embryo with 2 cotyledons
Genera: *Amentotaxus, Austrotaxus, Cephalotaxus, Pseudotaxus, Taxus, Torreya*
Other genera form. here now in Podocarpaceae. Timber & cult. orn.
Taxandria (Benth.) J.R. Wheeler & N. Marchant (~ *Agonis*). Myrtaceae (II 14). 11 SW Aus. R: Nuytsia 16(2007)406
Taxillus Tieghem. Loranthaceae (5 4). c. 30 trop. As. to C Mal., Kenya coast (1: ***T. wiensii*** Polhill)
Taxiphyllum Fleisch. Musci (Hypnaceae). c. 10 cosmop. ***T. barbieri*** (Cardot & Copp.) Z. Iwats. (Java moss, Indomal.) – common aquarium pl. worldwide, oft. confused with *Vesicularia dubyana* Fleisch.
Taxodiaceae Saporta = Cupressaceae
× **Taxodiomeria** Z.J. Ye & al. Cupressaceae. *Taxodium* × *Cryptomeria*. R: Sida 20(2003)1001. **× *T. peizhongii*** Z.J. Ye & al. (*Taxodium mucronatum* Ten. (Mex.) × *Cryptomeria fortunei* Hooibrenk ex Otto & Dietr. (China)) – orig. Nanjing 1963, veg. prop., poss. good urban tree
Taxodium Rich. Cupressaceae (Taxodiaceae). 2 E N Am., highland Mex. R: A. Farjon, *Monogr. Cupressaceae & Sciadopitys* (2005)123. Cult. orn. timber trees esp. ***T. distichum*** (L.) Rich. (swamp, S or bald cypress, E N Am.) – to 45 m with conspic. aerating 'knees', seeds water-disp., timber for sleepers etc.; ***T. mucronatum*** Ten. (Texas, C Plateau, Mex., Guatemala) – to 2000 ys old, 'el Arbol del Tule' in Oaxaca 45.7 m circumference at breast height (largest tree-girth known), timber & medic. resins, sacred in Oaxaca
taxol *Taxus* spp. esp. *T. brevifolia*
Taxopsida Lotsy = Pinidae
Taxus Tourn. ex L. Taxaceae. 24 N temp. (Eur. 1, China 5) to C Mal. & Mex. R: JBRIT 1(2007)208. Dioec. (after 300 yrs 1 *T. baccata* changed sex at Oxford Botanic Garden) evergreens (Massart's Model) with simple male strobilus derived from compound as in *Cephalotaxus*, fleshy scarlet aril; foliage & seeds toxic (alks; *toxon* = Greek for bow & poss. cognate with toxin), though bird-disp., the aril being harmless. Local medic. esp. N Am. Yew esp. ***T. baccata*** L. (English y., Eur., Medit.) – part of Madeiran laurel forest, cult. orn. sombre slow-growing tree (to 2000 yrs, allegedly 1500–3000-yr-old tree nr Aberfeldy, Scotland, 4000–5000-yr-old tree in Clwyd, Wales) assoc. with churchyards (though some poss. pre-Christian & older than churches nearby), but excellent hedge, preserved in anc. Egyptian tombs, form. 'palm' in N Eur., wood form. much used for bows in med. Ger. (over 500 000 over 60 yrs from a single Nuremberg 'sales firm') & GB (though English y. too brittle & mostly imported fom Spain & Italy), bagpipes & knife-handles (250 000-yr-old spear from Clacton, Essex, UK oldest known wooden artefact), lvs rich in beta-ecdysones, body putrefaction delayed in early 19 cent. by spray of leaf infusion, many cvs commonly planted incl. '**Aurea**' (golden y.) & '**Fastigiata**' ('Hibernica', 'Stricta', Irish y.) – female, all trees from cuttings from one at Florence Court, County Fermanagh (where

2 trees found on moors in 1778), Ireland, columnar habit; other cult. orn. etc. incl. ***T. brevifolia*** Nutt. (Calif. y., W N Am.) – bark much exploited for taxol (less conc. in lvs of other *T.* spp.) effective in control of ovarian cancer due to phenylisoserine side-chain (stabilizes microtubules in cells) but 6 trees needed for 1 dose (10 000 for 1 kg) though baccatin-3 from *T. baccata* lvs now used in taxol synthesis, ***T. canadensis*** Marshall (American y., E N Am.), ***T. chinensis*** (Pilg.) Rehder (Chinese y., S China), ***T. cuspidata*** Sieb. & Zucc. (Jap. y., E As.) – cult. orn., timber for furniture, marquetry, pencils etc., brown dye from heartwood, & ***T. x media*** Rehder (*T. baccata* × *T. cuspidata*) – many cvs (esp. '**Hicksii**' often used in hedging instead of *T. b.*), c. 4 times as much taxol as in *T. b.*; ***T. wallichiana*** Zucc. – needles & bark source of anticancer drug paclitaxel

tayberry see *Rubus*

Tayloriophyton Nayar. Melastomataceae. 2 W Mal. R: BSBI 10(1968)92

Tchihatchewia Boiss. (*Neotchihatchewia*). Cruciferae (31). 1 C Turkey: ***T. isatidea*** Boiss. (*N. i.*) – cult. orn. subpachycaul rosette-pl. with tall term. infl.

tchirisch *Asphodelus aestivus*

tchitola *Oxystigma oxyphyllum*

tea *Camellia sinensis*; **t. balm** *Melissa officinalis*; **Bogotá t.** *Symplocos theiformis*; **Cape Barren t.** *Correa alba*; **chamomile t.** *Chamaemelum nobile*; **Fukien t.** *Ehretia microphylla*; **Hopi t.** *Thelesperma megapoticum*; **Hottentot t.** *Helichrysum serpyllifolium*; **Jesuits' t.** *Otholobium glandulosum*; **Labrador t.** *Rhododendron tomentosum*; **Matara t.** *Senna auriculata*; **New Jersey t.** *Ceanothus americanus*; **Oswego t.** *Monarda didyma*; **Paraguay t.** *Ilex paraguariensis*; **Rooibos t.** *Aspalathus linearis*; **t. seed-oil** *Camellia sasanqua*; **Senegal t.** *Gymnocoronis spilanthoides*; **sweet t.** *Smilax glyciphylla*; **t. tree, t.-tree** *Leptospermum* spp., *Melaleuca alternifolia*; **Duke of Argyll's t. t.** *Lycium barbarum*; **yaupon t.** *Ilex vomitoria*; **yerba maté t.** *I. paraguariensis*

Teagueia (Luer) Luer (~ *Platystele*). Orchidaceae (V 13c). 28 Andes. R: MSB 39(1991)139. In 2000 4 new spp. found in 1 m^2 in Ecuador

teak *Tectona grandis*; **African t.** *Milicia excelsa*, *Oldfieldia africana*; **Australian t.** *Flindersia australis*; **bastard t.** *Butea monosperma*; **Borneo t.** *Intsia bijuga*; **Brunei t.** *Dryobalanops* spp.; **grey t.** *Gmelina* spp.; **Malacca t.** *Intsia palembanica*; **New Guinea t.** *Vitex cofassus*; **Sudan t.** *Cordia myxa*

teasel (teazel or **teazle)** or **fuller's t.** *Dipsacus sativus*

Teclea Del. = *Vepris*

Tecleopsis Hoyle & Leakey = *Vepris*

Tecoma Juss. Bignoniaceae (1). Incl. *Tecomaria*, 14 trop. & S Afr. (2), trop. Am. esp. Andes, to Arizona. Cult. orn. trees & shrubs with alks & extrafl. nectaries esp. ***T. capensis*** (Thunb.) Lindl. (*Tecomaria c.*, Cape honeysuckle, E & S Afr.) – fls scarlet, powdered bark medic. in S Afr. & ***T. stans*** (L.) Kunth (trop. Am., invasive S Afr., Masc., Bahamas, Virgin Is., Chile, Arg.) – fls bright yellow, locally used diuretic

Tecomanthe Baill. Bignoniaceae (I). 5 Mal., Aus., NZ (Three Kings Is.: ***T. speciosa*** W. Oliv. now reduced to a single tree from which being propagated). R: New Plantsman 3(1996)183

Tecomaria Spach = *Tecoma*

Tecomella Seemann. Bignoniaceae (1). 1 Arabia to W Ind.: ***T. undulata*** (Sm.) Seemann – cult. orn. with orange fls

Tecophilaea Bertero ex Colla. Tecophilaeaceae (Liliaceae s.l.). 2 Chilean Andes. Cult. orn. esp. ***T. cyanocrocus*** Leyb. (Chilean crocus), recently refound in wild

Tecophilaeaceae Leyb. Magnoliidae – Asparagales. Excl. Ixioliriaceae, 9/27 trop. & S Afr., Calif., Chile. Herb. pls with usu. tunicated corms to rhiz. & fibrous r. with vessels. Lvs lanceolate to linear, parallel-veined, glabrous, s.t. petiolate. Fls bisex., solit. or in racemes or thyrses, s.t. umbel-like, ± reg.; P 3 + 3 imbr., shortly connate at base, A 3 + 3 at mouth of P tube, some (esp. upper) sterile, anthers with apical slits or pores. G usu. semi-inf., 3-loc. each with 2 –∞ ovules in 2 rows. Fr. a loculicidal capsule with small black to yellow seeds

Genera: *Conanthera, Cyanastrum, Cyanella, Eremiolirion, Kabuyea, Odontostomum, Tecophilaea, Walleria, Zephyra*; Doryanthaceae (& Ixioliriaceae) poss. here too

Tectaria Cav. Tectariaceae (Dryopteridaceae s.l.). Incl. *Amphiblestra, Dictyoxiphium, Fadyenia, Stenosemia* etc., c. 230 trop. (Afr. 12, Am. 40). Terrestrial, some acrostichoid, some with simple fronds, some with apical buds on fronds. Some cult. orn., ***T. incisa*** Cav. (trop. Am.) – invasive SE US

Tectariaceae Panigrahi (~ Dryopteridaceae). Polypodiidae – Polypodiales. 11/390 trop. R: T 55(2006)718, 65(2016)732. Terr. ferns with dictyostelic rhiz. bearing scales. Petioles marcescent; fronds simple to bipinnate, s.t. decompound; indumentum of jointed, usu. short hairs. Indusia reniform or pelate to 0. n = (39)40(41)

Genera: *Arthropteris, Draconopteris, Hypoderris, Malaifilix, Pteridrys, Tectaria, Triplophyllum*

Pleocnemia removed to Dryopteridaceae

Tectaridium Copel. = *Tectaria*

Tecticornia Hook.f. Amaranthaceae (Chenopodiaceae II 2). Incl. *Holosarcia, Pachycornia, Sclerostegia, Tegicornia,* 33 Aus., 1 ext. to NG, 1 ext. to Afr. R: AusSB 20(2007)324

Tectiphiala H. Moore. Palmae (V 14h). 1 Mauritius: ***T. ferox*** H. Moore. R: GH 11(1978)285

Tectona L.f. Labiatae (inc. sed.; Verbenaceae). 3 SE As. to Mal. ***T. grandis*** L.f. (teak, Ind. to Laos, introd. to Indonesia 400–600 yrs ago & natur.) – Champagnat's Model, imp. timber much used for ship-building, bridges, flooring, furniture etc. (incl. cone-shaped vasthu wood, with conch shells buried at main entrance, both repelling negative energy); sinks in water unless dried so in Ind. tree 'girdled' i.e. a ring of bark & living tissue removed near base, so that it dies & is left standing for 2 yrs before extraction; sawdust in water an antidote to *Gluta* lesions; lvs source of brown-red dye in C Mal. D.N. Tewari (1992) *A monograph on teak*

Tecunumania Standl. & Steyerm. Cucurbitaceae (XV). 1 Guatemala: ***T. quetzalteca*** Standl. & Steyerm.

tedera *Bituminaria bituminosa* var. *albomarginata*

Tedingea D. & U. Mueller-Doblies = *Strumaria*

Teedia Rudolphi. Scrophulariaceae (Teed.). 4 S Afr. Berries & drupes. Cult. orn. foetid shrubs

Teesdalia R. Br. Cruciferae (32). 3 Eur. (3), Medit. incl. ***T. nudicaulis*** (L.) R. Br. (shepherd's cress). R: Novon 8(1998)218

Teesdaliopsis (Willk.) Rothm. = *Teesdalia*

teeta *Coptis teeta*

tef, t'ef, teff *Eragrostis tef*

teg(g)aout galls *Tamarix* spp.

Tegicornia Paul G. Wilson = *Tecticornia* (but see FA 4(1984)300)

Tehuana Panero & Villaseñor. Compositae (Helia.-Zinn.). 1 Mex.: ***T. calzadae*** Panero & Villaseñor. R: SB 21(1997)555

Teijsmanniodendron Koord. (~ *Vitex*). Labiatae (II). 23 SE As., Mal. (exc. S; Borneo 23). R: KB 64(2010)589. Champagnat's Model

Teinosolen Hook.f. = *Heterophyllaea*

Teinostachyum Munro (~ *Schizostachyum*). Gramineae (V 5). 2 Myanmar, Ind.

Teixeiranthus R. King & H. Robinson (~ *Eupatorium*). Compositae (Eup.-Ager.). 2 Braz.

Telanthophora H. Robinson & Brettell (~ *Senecio*). Compositae (Sen.-Tuss.). 14 C Am. R: Phytologia 27(1974)424. Trees & shrubs. ***T. grandifolia*** (Less.) H. Robinson & Brettell (Mex.) – cult. orn. shrub

Telectadium Baill. Apocynaceae (I 3; Asclepiadaceae I). 3 SE As. Rheophytes. ***T. edule*** Baill. (Laos) – ed.

telegraph plant *Codariocalyx motorius*

Telekia Baumg. (~ *Inula*). Compositae (Inul.-Inul.). 1 C Eur. to Cauc.: ***T. speciosa*** (Schreb.) Baumg. (SE Eur. to S Russia) – cult. orn. coarse herb

Telemachia Urb. = *Elaeodendron*

Telephium L. Caryophyllaceae (I 3). 5 Medit. (Eur. 1), Madag. R: JB 44(1906)289. ***T. imperati*** L. (W Med.) – cult. orn. rock-pl.

Telesonix Raf. (~ *Boykinia*). Saxifragaceae (I 6). 2 W US. R: FNA 9(2009)116

Telfairia Hook. Cucurbitaceae (IX). 3 trop. Afr. Ed. oil seeds esp. ***T. occidentalis*** Hook.f. (krobonko) & ***T. pedata*** (Sm.) Hook. (oyster nut, kweme) – oil used in soap- & candle-making

Teline Medik. = *Genista*

Teliostachya Nees = *Lepidagathis*

Telipogon Mutis ex Kunth. Orchidaceae (V 12h). Incl. *Stellilabium,* c. 200 C & trop. S Am. Some v. small pls with minute fls, some leafless at maturity; pseudocopulation by flies. Some cult. orn.

Telitoxicum Mold. Menispermaceae (II). 6 trop. S Am. R: MNYBG 22,2(1971)76. Form. curare sources

Tellichery bark *Holarrhena pubescens*

Tellima R. Br. Saxifragaceae. 1 W N Am.: ***T. grandiflora*** (Pursh) Douglas ex Lindl. – cult. orn. ground-cover

Telmatoblechnum Perrie & al. (~ *Blechnum*). Blechnaceae. 2 trop. R: T 63(2014)755. ***T. indicum*** (Burmf.) Perrie & al. (*B. i.*, OW trop.) – rhiz. form. trad. foodstuff (bungwall) N & NE Aus.

Telmatophila Mart. ex Bak. Compositae (Vern.-Mes.). 1 NE Braz.: ***T. scolymastrum*** Mart. ex Bak.

Telminostelma Fourn. = *Cynanchum*

Telmissa Fenzl = *Sedum*

Telopea R. Br. Proteaceae (V 3). 5 E Aus. R: FA 16(1995)386. Nectar drunk by Aborigines, lvs used as protection against fire. Cult. orn with red fls in dense heads surrounded by coloured involucres esp. ***T. speciosissima*** (Sm.) R. Br. (waratah, NSW) – prot. (floral emblem NSW, favourite motif in nationalistic architecture, pottery etc., but cut-fl. exported from NZ as kiwi rose!), & hybrids like '**Gembrook**' (*T. s.* × ***T. oreades*** F. Muell. (E Victoria) – veneers), used in Olympic medalists' bouquets (Sydney 2000)

Telosiphonia (Woodson) Henrickson = *Mandevilla*

Telosma Cov. Apocynaceae (V a; Asclepiadaceae III 4). 5 trop. OW (Afr. 1). ***T. cordata*** (Burm.f.) Merr. (Ind. to SE As.) – liane with r., lvs & night-scented fls (as in Thai omelettes) locally eaten, cult. orn. as for *lei* in Hawaii

Teloxys Moq. (~ *Dysphania*). Amaranthaceae. 1 E As.: ***T. aristata*** (L.) Moq.

Temburongia S. Dransf. & Wong. Gramineae (V 4). 1 Temburong R., Brunei: ***T. simplex*** S. Dransf. & Wong. R: Sandakania 7(1996)53

temiche cap *Manicaria saccifera*

Temmodaphne Kosterm. = *Cinnamomum*

Temnadenia Miers. Apocynaceae (II e). 4 trop. S Am. R: Candollea 60(2005)211

Temnocalyx F. Robyns. Rubiaceae (III 2). 1 Tanzania: ***T. nodulosa*** Robyns

Temnopteryx Hook.f. Rubiaceae (Uroph.). 1 trop. W Afr.: ***T. sericea*** Hook.f.

Temochloa S. Dransf. Gramineae (V 4). 1 S Thailand: ***T. liliana*** S. Dransf. R: TFB 28(2000)179

temple grass *Zoysia tenuifolia*; **t. tree** *Plumeria rubra*

Templetonia R. Br. Leguminosae (III 4). 10 Aus. R: Muelleria 5(1982)1. Cult. orn. shrubs with alks esp. ***T. retusa*** (Vent.) R. Br. (coral bush, S & W Aus.)

Temu O. Berg = *Blepharocalyx*

Tenagocharis Hochst. = *Butomopsis*

Tenaris E. Meyer = *Brachystelma*

Tenaxia N. Barker & Linder (~ *Danthonia*). Gramineae (Danth.). 8 Afr. to Himal.

tendu *Diospyros melanoxylon*

Tengia Chun = *Petrocodon*

tengkawang (oil) *Diploknema sebifera*

Tenicroa Raf. = *Ornithogalum*

Tennantia Verdc. Rubiaceae (II 2). 1 E Afr.: ***T. sennii*** (Chiov.) Verdc. & Bridson. R: KB 36(1981)511

Tenrhynea Hilliard & B.L. Burtt (*Rhynea*). Compositae (Gnap.-Gnap.). 1 S Afr.: ***T. phylicifolia*** (DC.) Hilliard & B.L. Burtt. R: OB 104(1991)140

teosinte *Zea mays* 'subsp. *mexicana'*

tepa *Laureliopsis philippiana*

tepary bean *Phaseolus acutifolius*

tepescohuite *Mimosa tenuiflora*

Tephrocactus Lem. (~ *Opuntia*). Cactaceae (II). 7 S Am. R: Schumannia 2(1998)95. Cult. orn. esp. ***T. articulatus*** (Pfeiffer) Backeb. (*O. a.*, *O. diademata*, *O. papyracantha*, W Arg.) – some cvs with long papery spines

Tephroseris (Reichb.) Reichb. (~ *Senecio*). Compositae (Sen.-Tuss.). 50 temp. & Arctic Euras., N Am. (6). Some cult. orn. incl. ***T. integrifolia*** (L.) Holub (*Senecio i.*, fleawort, Euras.)

Tephrosia Pers. Leguminosae (III 16). c. 350 seasonal trop. esp. Afr. (incl. *Lupinophyllum* – geocarpic fr.). Cult. cover crops & green manures, fish-poisons & cult. orn. ***T. candida*** DC. (boga medalo(a), Ind.) – green manure; ***T. macropoda*** Harv. (lozane, S Afr.) – r. a fish-poison & insecticide; ***T. purpurea*** (L.) Pers. (OW trop.) – seeds ed. Mafia Is., Tanz.; ***T. sinapou*** (Buc'hoz) A. Chev. (*T. toxicaria*, yarroconalli, trop. Am.) – r. a fish-poison, green manure; ***T. virginiana*** (L.) Pers. (goat's rue, cat gut, E N Am.) – r. medic. etc.; ***T. vogelii*** Hook. (trop. Afr.) – cult. for lvs used as fish-poison in small-scale coastal fishing in Comoro Is.

Tephrothamnus Schultz-Bip. (~ *Piptocarpha*). Compositae (Vern.). 1 Venez.: ***T. paradoxus*** Schultz-Bip.

Tepualia Griseb. = *Metrosideros*

Tepuia Camp. Ericaceae (VIII 4). 7 Guayana Highland. R: MNYBG 29(1978)153

Tepuianthaceae Maguire & Steyerm. = Thymelaeaceae

Tepuianthus Maguire & Steyerm. Thymelaeaceae (Tepuianthaceae). 7 Guayana Highland. R: MNYBG 32(1981)9

tequila *Agave* spp. esp. *A. tequilana*

Teramnus P. Browne. Leguminosae (III 18). 9 trop. esp. OW

terap *Artocarpus odoratissimus*

Teratophyllum Mett. ex Kuhn. Dryopteridaceae (II; Lomariopsidaceae s.l.). 13 S Myanmar to Queensland & E Polynesia

Terauchia Nakai = *Anemarrhena* (smut-infested)

terebinth *Pistacia terebinthus*

Terebraria Kuntze = *Neolaugeria*

Terminalia L. Combretaceae (II 2a). Incl. *Bucida*, c. 190 trop. (Am. c. 34). Timber, dyes, tannin, gums (Ashanti g.) & cult. orn. with char. planar foliage (pagoda tree shape; Aubréville's Model, also Roux's), some ('*Bucida*') used for bonsai. ***T. alata*** Roth (asna, Ind. laurel, Ind.) – comm. timber, food for silkworms; ***T. amazonia*** (J.F. Gmel.) Exell (trop. Am.) – imp. timber (nargusta or roble coral in int. trade); ***T. arjuna*** (DC.) Wight & Arn. (arjun, kumbuk, kahua, Ind., Sri Lanka) – timber for constr., bark with heart stimulants, lime extractable for chewing with betel; ***T. arostrata*** Ewart & O. Davies (nutwood, NW Aus.) – ed. seeds; ***T. australis*** Camb. (Braz., Arg., Paraguay) – rheophyte used for erosion control; ***T. bellirica*** (Gaertn.) Roxb. (Indomal.) – fr. a tannin source & black dyes & inks, timber good; ***T. bialata*** (Roxb.) Steud. (silver greywood, Myanmar, Andamans) – grey heartwood for fancy work like picture-framing (chuglam) & constr. (white chuglam wood); ***T. buceras*** (L.) Wright (*Bucida b.*, C Am., WI) – pop. bonsai pl. in Hawaii etc., sap stains masonry; ***T. catappa*** L. (Ind., Barbados or wild almond, badam, Indomal.) – widely planted & natur., salt-tolerant street-tree, sea- & bat-disp. (rat-disp. on Krakatoa), timber red & good, lvs (with domatia) for silkworms, r. & bark for tanning, seeds ed. like almonds with similar oil; ***T. chebula*** Retz. (myrobalan (name also used for other spp.), Indomal.) – dried fr. used in tanning, giving soft mellow leather (tannin c. 32%), medic.; ***T. elliptica*** Willd. (Ind. laurel, Ind.) – timber; ***T. ferdinandiana*** Exell (*T. latipes* Benth. subsp. *psilocarpa* Pedley, Kakadu plum, Aus.) – small ed. fr. ('wild plum') with 31 mg vitamin C per g (oranges with 0.5 mg), used in comm. skin products; ***T. glaucescens*** Planch. ex Benth. & ***T. laxiflora*** Engl. (W Afr.) – r. chewing-sticks in Nigeria; ***T. ivorensis*** A. Chev. (black afara, emeri, idigbo, W Afr.) – weather-resistant timber for shingles etc. & plywood; ***T. kaernbachii*** Warb. (*T. okari*, okari, S NG, (?) natur. Polynesia) – ed. fr., the kernel a good flavoured 'nut'; ***T. lucida*** Hoffsgg. ex Mart. & Zucc. (trop. Am.) – assoc. with *Azteca* ants in C Am.; ***T. mollis*** Teijsm. & Binn. (E Mal.) – antibacterial, tried for STD; ***T. muelleri*** Benth. (Aus.) – Aboriginal symbols drawn on back of lvs; ***T. oblonga*** (Ruíz & Pavón) Steud. (trop. Am.) – timber for flooring & panelling (up to 13% calcium by wt.); ***T. procera*** Roxb. (white bombway, Andamans) – timber; ***T. superba*** Engl. & Diels (afara, akom, limba, ofram, white afara, W Afr.) – general constr. timber

Terminaliopsis Danguy = *Terminalia*

Terminthia Bernh. (~ *Rhus*). Anacardiaceae. 1 Himal. to Yunnan: ***T. paniculata*** (G. Don) V.Y. Wu & T.L. Ming. R: FOC 11(2008)348

Terminthodia Ridl. = *Tetractomia*

Terniola Tul. = *Dalzellia*

Terniopsis H.C. Chao (~ *Dalzellia*). Podostemaceae. c. 7 Indomal.

Ternstroemia Mutis ex L.f. Pentaphylacaceae (2; Ternstroemiaceae, Theaceae s.l.). c. 100 trop. (Afr. 2). ***T. gymnanthera*** (Wight & Arn.) Beddome ('*T. japonica*', S & E As.) – hedge-pl.

Ternstroemiaceae Mirb. ex DC. = Pentaphylacaceae

Ternstroemiopsis Urb. = *Eurya*

Terpsichore A.R. Sm. (~ *Grammitis*). Polypodiaceae (V; Grammitidaceae). Excl. *Alansmia*, *Ascogrammitis*, *Galactadenia*, *Mycopteris*, Spp? (but see Novon 3(1993)478)

Terrlluia Lunell = *Elymus*

terrycloth *Gossypium* spp.

Tersonia Moq. Gyrostemonaceae. 1 SW Aus.: ***T. cyathiflora*** (Fenzl) A.S. George

Terua Standl. & F.J. Herm. = *Lonchocarpus*

tesota *Olneya tesota*

Tessarandra Miers = *Chionanthus*

Tessaria Ruíz & Pavón (~ *Pluchea*). Compositae (Inul.-Pluch.). 1 warm & trop. Am.: ***T. integrifolia*** Ruíz & Pavón (arrow-weed) – shrub or small tree with r.-suckers, forming almost pure stands along rivers

Tessmannia Harms. Leguminosae (I 2). c. 12 trop. Afr. Furniture timbers, resins

Tessmanniacanthus Mildbr. Acanthaceae (III 2c). 1 E Peru: ***T. chlamydocardiodes*** Mildbr.

Tessmannianthus Markgraf. Melastomataceae. c. 6 trop. Am. Trees to 45 m

Tessmanniodoxa Burret = *Chelyocarpus*

Testudinaria Salisb. ex Burch. = *Dioscorea*

Testudipes Markgraf = *Tabernaemontana*

Testulea Pellegrin. Ochnaceae (III). 1 Gabon: ***T. gabonensis*** Pellegrin. R: BJ 113(1991)175

Tetilla DC. Francoaceae (Saxifragaceae s.l.). 1 Chile: ***T. hydrocotylifolia*** DC. R: U. Eggli, *Ill. Handb. Succ. Pls*, Dicots (2002)443

Tetrabaculum M.A. Clem. & D.L. Jones = *Dendrobium*

Tetraberlinia (Harms) Hauman. Leguminosae (I 2). 7 Guineo-Congolian forest. R: WAUP 99–3(1999)255. Timbers. ***T. moreliana*** Aubrév. (Gabon) – exploding fr. expelling seeds to 60 m at 70 km/hr

Tetracanthus A. Rich. = *Pectis*

Tetracarpaea Hook. Haloragaceae (Tetracarpaeaceae). 1 Tasmania: ***T. tasmannica*** Hook.

Tetracarpaeaceae Nakai = Haloragaceae

Tetracarpidium Pax = *Plukenetia*

Tetracentraceae A.C. Sm. = Trochodendraceae

Tetracentron Oliv. Trochodendraceae (Tetracentraceae). 1 Nepal, SW & C China, N Myanmar: ***T. sinense*** Oliv. – vessels, with tracheids to 45 mm long

Tetracera L. Dilleniaceae (I). c. 44 trop. (Am. 14). R: MBSM 8(1970)29. Some cryptic dioecy. Rough lvs of some spp. used like glasspaper. ***T. alnifolia*** Willd. (trop. Afr.) – aerenchymatous r. floating limply in small pools; ***T. scandens*** (L.) Merr. (Madag.) – local medic.

Tetrachaete Chiov. Gramineae (XXVII). 1 Tanzania, Ethiopia, Arabia: ***T. elionuroides*** Chiov.

Tetrachne Nees (~ *Uniola*). Gramineae (XXVII 3). 1 S Afr. & Pakistan: ***T. dregei*** Nees – introd. Am. for forage. R: SCB 87(1997)41

Tetrachondra Petrie ex Oliv. Tetrachondraceae (Labiatae s.l.). 2 NZ (1), temp. S Am. (1). R: BSAB 13(1970)2

Tetrachondraceae Wettst. (~ Labiatae). Magnoliidae– Lamiales. 2/3 NZ, temp. S Am., C Am. Perenn. herbs. Lvs opp. with bases connate or connected by membaneous stip. Fls reg., small, 4-merous, usu. solit. or in leafy cymes. K valvate, persistent; C with short tube, lobes imbr.; A opp. K, anthers sep., introrse, longit. dehiscent, with pollen in tetrads; nectary 0; G 4, partly inferior or sup., each carpel with 2 – ∞ anatropous, unitegmic ovules. Fr. 4 nutlets or 2-valved capsule with persistent green K; seeds with straight embryo & copious endosperm. n = 10, 11

Genera: *Polypremum, Tetrachondra*, form. Loganiaceae & Labiatae resp.

Tetrachyron Schldl. (~ *Calea*). Compositae (Helia.-Verb.). c. 7 E Mex., Guatemala. R: SB 4(1979)297

Tetraclea A. Gray (~ *Clerodendrum*). Labiatae (II). 2 S N Am. R: BSAB 13(1970)2

Tetraclinis Masters. Cupressaceae. 1 S Spain, Malta, N Afr.: ***T. articulata*** (Vahl) Masters (arar, alerce, thyine, thuya) – timber used for building since antiquity (& burnt on Calypso's fire in the Odyssey), the citrus-wood of antiquity particularly valued for tables (Pliny) made from sections of coppicing stumps, resin (sandarac(h)) used in preparing mummies, picture varnish, for glazing paper, for pounce & incense. R: A. Farjon, *Monogr. Cupressaceae & Sciadopitys* (2005)410

Tetraclis Hiern = *Diospyros*

Tetracme Bunge. Cruciferae (26). 9 E Medit. (Eur. 1) to C As. & Baluchistan

Tetracoccus Engelm. ex C. Parry. Picrodendraceae (Euphorbiaceae s.l.). 5 SW N Am. R: Rhodora 56(1954)49. Dioec. shrubs

Tetractomia Hook.f. (~ *Melicope*). Rutaceae (I 2). 6 Mal. R: JAA 60(1979)127

Tetracustelma Baill. = *Matelea*

Tetradema Schltr. = *Agalmyla*

Tetradenia Benth. Labiatae (VII 3e). c. 25 trop. Afr. (10 – R: Adans. 30(2008)179), Madag. R: Bothalia 14(1983)177, 15(1984)1. Dioec. (s.t. hermaphr.) cult. orn. (incl. *Iboza*)

Tetradiclidaceae Takht. = Nitrariaceae

Tetradiclis Steven ex M. Bieb. Nitrariaceae (Zygophyllaceae s.l.). 2 SE Russia & E Med. to C As.

Tetradium Lour. Rutaceae (I 2). 9 Himal. to Jap., Philippines, Sumatra & Java. R: GBS 34(1981)91. ***T. daniellii*** (Bennett) Hartley (*Euodia d., Evodia d.*, SW China to Korea) – cult. orn. tree; ***T. fraxinifolium*** (Hook.) Hartley (Nepal to SE As.) – fr. for chutney & medic., cult. orn.; ***T. ruticarpum*** (A. Juss.) Hartley (China, Taiwan) – drug (wu-chu-yu) used for 2000 yrs, stimulant to anthelminthic, alks with uterotonic activity

Tetradoa Pichon = *Hunteria*

Tetradoxa C.Y. Wu = *Adoxa*

Tetradyas Danser = *Cyne*

Tetradymia DC. Compositae (Sen.-Tuss.). 10 N Am. R: Britt. 26(1974)177. Some cold remedies etc. but toxic to sheep – if eaten in sunlight leads to 'bighead' swelling, liver damage & photosensitization

Tetraedrocarpus O. Schwarz = *Echiochilon*

Tetraena Maxim. Zygophyllaceae (I). Incl. *Petrusia*, 40 Canary Is. & S Afr. to China. R: PSE 240(2003)34

Tetragamestus Reichb.f. = *Scaphyglottis*

Tetragastris Gaertn. Burseraceae (II). 9 trop. Am. R: KB 45(1990)179. Rauh's Model; some fish-disp. in Amazon. ***T. balsamifera*** (Sw.) Oken (pig balsam, WI) – timber for cabinet-work, panelling, flooring etc.

Tetraglochidium Bremek. = *Strobilanthes*

Tetraglochin Poeppig = *Margyricarpus*

Tetragoga Bremek. = *Strobilanthes*

Tetragompha Bremek. = *Strobilanthes*

Tetragonia L. Aizoaceae (I; Tetragoniaceae). 57 trop. & warm S esp. S Afr., N Am. R: H. E. K. Hartmann, *Ill. Handb. Succ. Pls*, Aiz. F–Z(2002)316. Fr. with thorny projections s.t. prod. fls. ***T. tetragonoides*** (Pallas) Kuntze (ice-pl., NZ spinach, warrigal (cabbage), Jap. & Pac. to NZ & temp. S Am.) – adv. buds s.t. develop on persistent P giving small s.t. fert. fls, grown as spinach (though high in oxalates) in sites too hot for true s., fed to his crew by Captain Cook & cult. in Eur. before Aus. settlement by whites

Tetragoniaceae Lindl. = Aizoaceae

Tetragonocalamus Nakai = *Bambusa*

Tetragonolobus Scop. = *Lotus*

Tetragonotheca Dill. ex L. Compositae (Mill.-Dysc.). 4 SE US. R: Sida 8(1980)296

Tetralix Griseb. Malvaceae (Grew.-Grew.; Tiliaceae). 5 Cuba (serpentine). Dioec. shrubs

Tetralocularia O'Don. Convolvulaceae (I). 1 Colombia: ***T. pennellii*** O'Don. R: Lilloa 30(1960)66

Tetralopha Hook.f. = *Gynochthodes*

Tetramelaceae Airy Shaw = Datiscaceae

Tetrameles R. Br. Datiscaceae (Tetramelaceae). 1 Indomal. to Aus.: ***T. nudiflora*** R. Br. (kapong, sompong) – buttresses large, allegedly the trees with trunks 'dripping' over ruins (but see *Cochlospermum*), wood for matches, canoes (NG) etc.; seeds winged

Tetrameranthus R. Fries. Annonaceae (II). 7 trop. S Am. R: PKNAW C88(1985)449. Lvs in spirals

Tetramerista Miq. Tetrameristaceae. 1 Mal. ***T. glabra*** Miq. (punah) – timber for beams, ed. fr.

Tetrameristaceae Hutch. Magnoliidae – Ericales. Incl. Pellicieraceae, 3/3 Mal., C Am. shores, Guayana Highland. Everg. trees & shrubs. Lvs simple, bases decurrent, in spirals; stip. 0. Fls bisex., bibracteolate in axillary racemes or solit. (*Pelliciera*); K 4 or 5, imbr., persistent or not, with 2 red decid. petaloid bracts larger than C, C 4 or 5 imbr., A 4 or 5 alt. with C, filaments s.t. basally connate, anthers with longit. slits, G ((2,) 4 or 5), 2 (*Pelliciera*), 4- or 5-loc. with term. style, each loc. with 1 (or 0) axile-basal anatr. to campylotropous, bitegmic ovule. Fr. a 4- or 5-seeded berry or dry (*Pelliciera*); seeds with straight embryo in copious or 0 (*Pelliciera*) endosperm

Genera: *Pelliciera, Pentamerista, Tetramerista*

Pelliciera a mangrove with reduced distrib., *Pentamerista* discovered in 1972, adding to the remarkable known distrib. & demonstrating the relic nature of the Guayana Highland flora

Tetramerium Nees. Acanthaceae (III 2c). 29 C Am. R: SBM 12(1986)

Tetramicra Lindl. Orchidaceae (V 13b). 14 WI. R: C.L. Withner, *The Cattleyas & their relatives* IV(1996)113. Cult. orn. terr. orchids

Tetramolopium Nees = *Vittadinia*

Tetranema Benth. Plantaginaceae (Ant.-Chel.; Scrophulariaceae s.l.). 4 C Am. R: ABM 32(1995)59. Cult. orn. esp. ***T. roseum*** (M. Martens & Galeotti) Standl. & Steyerm. (*T. mexicanum*, Mex. foxglove)

Tetraneuris Greene (~ *Hymenoxys*). Compositae (Hele-Tet.). 9 N Am. R: FNA 21(2006)447

Tetranthera Jacq. = *Litsea*

Tetranthus Sw. Compositae (Helia.-Spil.). 2–4 WI

Tetrapanax (K. Koch) K. Koch. Araliaceae (1). 1 S China (? native), Taiwan: ***T. papyrifer*** (Hook.) K. Koch – cult. orn. unarmed clump-forming shrub, pith an imp. source of fine rice-paper in China, sliced pith used for artificial fls, for lens paper in Taiwan

Tetrapathaea (DC.) Reichb. = *Passiflora*

Tetraperone Urb. Compositae (Helia.-Pin.). 1 Cuba: ***T. bellioides*** Urb.

Tetrapetalum Miq. = *Cyathostemma*

Tetraphyle Ecklon & Zeyher = *Crassula*

Tetraphyllaster Gilg. Melastomataceae. 1 trop. W Afr.: ***T. rosaceum*** Gilg

Tetraphyllum Griff. ex C.B. Clarke. Gesneriaceae (III 2c). c. 6 NE Ind. to Thailand

Tetraphysa Schltr. = *Pentacyphus* (but see ABM 52(1994)38)

Tetrapilus Lour. = *Olea*

Tetraplandra Baill. = *Algernonia*

Tetraplasandra A. Gray = *Polyscias* (but see W L. Wagner & al., *Man. Fl. Pls Hawaii* (1990) 232)

Tetraplasia Rehder = *Damnacanthus*

Tetrapleura Benth. Leguminosae (II 1). 2 trop. Afr. Timber. ***T. tetraptera*** (Schum.) Taubert (aridan) – smelly fr. elephant-disp., saponins (triterpenes) antifungal & molluscicidal in field trials

Tetrapodenia Gleason = *Burdachia*

Tetrapogon Desf. Gramineae (XXIX 5). 10 warm Afr., Middle E, Ind., Mex. (4), Cuba (1)

Tetrapollinia Maguire & Boom (~ *Lisianthus*). Gentianaceae. 1 NE trop. Am.: ***T. caerulescens*** (Aubl.) Maguire & Boom. R: MNYBG 51(1989)31

Tetrapteris Cav. = *Tetrapterys*

Tetrapterocarpon Humbert. Leguminosae (I 4). 2 Madag.

Tetrapteron (Munz) W.L. Wagner & Hoch (~ *Camissonia*). 2 W N Am. R: CUSNH 37(1969)161. Annuals

Tetrapterys Cav. (*Tetrapteris*). Malpighiaceae. 90 trop. Am. Hallucinogenic bark (cf. *Banisteriopsis*); some medic.

Tetrardisia Mez = *Ardisia*

Tetraria P. Beauv. Cyperaceae (II 7). c. 52 S Afr. (45), C Afr., SW Aus. (6), NZ (1)

Tetrariopsis C.B. Clarke = *Tetraria*

Tetrarrhena R. Br. = *Ehrharta*

Tetraselago Jun. Scrophulariaceae (Manul.). 4 S Afr. R: NRBGE 35(1977)175

Tetrasida Ulbr. Malvaceae (Malv.-Malv.). 2 Peru. R: Britt. 44(1992)444

Tetrasiphon Urb. Celastraceae (I). 1 Jamaica: ***T. jamaicensis*** Urb.

Tetraspidium Bak. Orobanchaceae (Buch.; Scrophulariaceae s.l.). 1 Madag.: ***T. laxiflorum*** Bak.

Tetrastigma (Miq.) Planch. (~ *Cayratia*). Vitaceae. 95 Indomal. to trop. Aus. (5). Decid. or everg. dioec. lianes ± tendrils, hosts for *Rafflesia*, *Rhizanthes* & *Sapria* spp. Cult. orn. housepls esp. ***T. harmandii*** Planch. (SE As. to Philippines) – ed. fr. for jellies etc. & ***T. voinierianum*** (Mottet) Gagnepain (chestnut vine, Laos) – screening pl. in Calif. ***T. lauterbachianum*** Gilg (Papuasia) – liane to 40 m with aerial r. 10–40 m long used for weaving in Bismarck Is.; ***T. obtectum*** (M. Lawson) Franch. (Sinohimal.) – promoted as ivy subs.

Tetrastylidium Engl. Erythropalaceae (Olacaceae s.l.). 2 S Braz. R: FN 38(1984)102

Tetrastylis Barb. Rodr. = *Passiflora*

Tetrasynandra Perkins. Monimiaceae (V 2). 3 NE Aus.

Tetrataenium (DC.) Manden. Umbelliferae (III 11). 7–8 SW to C As.

Tetrataxis Hook.f. Lythraceae. 1 Mauritius: ***T. salicifolia*** (Tul.) Bak. – only 7 individuals left of this tree rediscovered in 1970s

Tetratelia Sonder = *Cleome*

Tetrathalamus Lauterb. = *Zygogynum*

Tetratheca Sm. Elaeocarpaceae (Tremandraceae). c. 50 Aus. R: Telopea 1(1976)139. Cult. orn. small shrubs (pink-eyes)

Tetrathylacium Poeppig. Salicaceae (Samydaceae, Flacourtiaceae). 2 trop. Am. R: FN 22(1980)219. ***T. costaricense*** Standl. (Costa Rica) – ant-infested twigs in 68% of pls

Tetrathyrium Benth. = *Loropetalum*

Tetraulacium Turcz. Plantaginaceae (Grat.-Stem.; Scrophulariaceae s.l.). 1 Braz.: ***T. veroniciforme*** Turcz.

Tetrazygia Rich. ex DC. = *Miconia*

Tetrazygiopsis Borh. = *Miconia*

Tetrodon (Kraenzl.) M.A. Clem. & D.L. Jones = *Eria*

Tetroncium Willd. Juncaginaceae. 1 S S Am north to Falklands & 40°S in Andes, Gough Is. (3300 km away in C S Atlantic!): ***T. magellanicum*** Willd. – dioec., fls 2-merous. R: Willd. 43(2013)14

Tetrorchidiopsis Rauschert = seq.

Tetrorchidium Poeppig. Euphorbiaceae (Aden.-Aden.). 23 trop. Am. (19) & Afr. (4 – R: Adansonia III,21(1999)98). Petit's Model

tetter-berry *Bryonia dioica*

Teucridium Hook.f. = *Teucrium*

Teucrium Tourn. ex L. Labiatae (II). Incl. *Spartothamnella, Teucridium*, c. 250 cosmop. esp. Medit. (Eur. 49). C with 1 5-lobed lip; some switch pls with succ. drupes (*Spartothamnella*). Germander, esp. ***T. chamaedrys*** L. (Eur. to Cauc.) – locally medic. tea used since antiquity for digestive problems (though actually causing liver disease). Cult. orn. herbs & shrubs incl. ***T. canadense*** L. (N Am.) – poss. source of natural rubber; ***T. × lucidrys*** Boom (*T. chamaedrys* × *T. lucidum* L. (Alps)) – evergreen subs. for diseased box-edging UK; ***T. marum*** L. (cat thyme, W Medit.; is.) – attractive to cats; ***T. polium*** L. (Medit., SW As.) – imp. medic. for anc. Greeks, an infusion now used for stomach ulcers; ***T. scordium*** L. (Euras., Medit.) – source of yellow-green dye for cloth in Danube area; ***T. scorodonia*** L. (wood g. or w. sage, Eur., natur. N Am.) – form. medic.

Teuscheria Garay. Orchidaceae (V 12g). 7 N S Am.

Texas blue grass *Poa arachnifera*; **T. millet** *Urochloa texana*; **T. star** *Lindheimera texana*; **yellow rose of T.** *Rosa × harisonii*

Teyleria Backer. Leguminosae (III 18). 3 SE As., Sumatra, Java. R: AUWP 85–1(1985)119

Thai basil *Ocimum tenuiflorum*; **T. coriander** *Eryngium foetidum*; **T. eggplant** *Solanum melongena*

Thaia Seidenf. Orchidaceae (V 9). 1 Thailand: ***T. saprophytica*** Seidenf. – plicate lvs, corms, doubtfully mycotrophic. R: T 61(2012)52

Thailentadopsis Kosterm. (~ *Cathormion*). Leguminosae (II 4). 3 Sri Lanka, Thailand, Vietnam (1 each). R: KB 58(2003)492

Thalassia Banks ex C. Koenig. Hydrocharitaceae (Thalassiaceae). 1 Carib., 1 trop. Ind. & Pac. Oceans. R: VKNAW II,59(1970)222. Submerged marine aquatics (Bell's Model) to 30m depth; nitrogen-fixing bacteria in rhizosphere. ***T. hemprichii*** (Ehrenb.) Asch. (OW) – mulch for paddy-fields & coconuts; ***T. testudinum*** C. Koenig (Carib.) – dioec., some frs detaching & floating for 10 days & poss. up to 100 km

Thalassiaceae Nakai = Hydrocharitaceae

Thalassodendron Hartog. Cymodoceaceae. 1 Red Sea, W Ind. Ocean, E Mal., NE Aus.: ***T. ciliatum*** (Forssk.) Hartog – submarine poll. & disp. by free-floating detached fruiting branchlets; 1 SW Aus. Bell's Model. R: Novon 22(2012)21

Thaleropia Peter G. Wilson (~ *Metrosideros*). Myrtaceae (II 8). 3 NG, Queensland. R: AusSB 6(1993)257

Thalestris Rizz. = *Justicia*

Thalia L. Marantaceae. 6 trop. Am., ***T. geniculata*** L. ext. to trop. Afr. – pollen deposited on style in bud & insect proboscis touching style promotes explosive S-shaped movement, in 0.03 secs style becomes erect, scrapes pollen from proboscis into stigmatic hollow & deposits home pollen on proboscis; cult. orn.

Thalictrum Tourn. ex L. Ranunculaceae (IV). 120–200 N temp. (most; Eur. 15, N Am. 22), NG, trop. Am., trop. & S Afr. R: BSRBB 24(1885)78. Meadow rue. Herbs with alks; C 0, A oft. coloured & conspic., insect-poll., others wind-poll. Some locally medic. (powerful laxatives in r.); cult. orn. esp. ***T. delavayi*** Franch. (W China) – K red or lilac, A yellow

Thaminophyllum Harv. Compositae (Anth.-Cot.). 3 SW Cape. R: JSAB 46(1980)157

Thamnea Sol. ex Brongn. Bruniaceae (2). 9 SW Cape mts. R: T 60(2011)1145

Thamnobryum Nieuw. Neckeraceae (Musci). 42 cosmop. ***T. alopecurum*** (Hedw.) Gangulee – dendroid shoots borne on creeping ones, can form moss-balls

Thamnocalamus Munro (~ *Arundinaria*). Gramineae (IV). Excl. *Bergbambos, Fargesia*, 2 Himal. R: KBAS 13(1986)43. Forest bamboos. Cult. orn. while ***T. spathiflorus*** (Trin.) Munro (NW Himal.) fl. every 16 or 17

Thamnocharis W.T. Wang = *Oreocharis*

Thamnochortus P. Bergius. Restionaceae. 32 SW Afr. to E Cape & Namaqualand. R: Bothalia 15(1985)471. ***T. insignis*** Masters – imp. for thatching (coll. from wild), S Cape

Thamnojusticia Mildbr. = *Justicia*

Thamnoldenlandia Groeninckx. Rubiaceae (IV 15). 1 Madag.: ***T. ambovombensis*** Groeninckx. R: BJLS 163(2010)458

Thamnosciadium Hartvig (~ *Seseli*). Umbelliferae (III 8). 1 Greece: ***T. junceum*** (Sm.) Hartvig. R: Willd. 14(1984)321

Thamnoseris F. Philippi = *Sonchus*

Thamnosma Torrey & Frémont. Rutaceae (I 10). 8 SW N Am. (2), Afr. (6 – R: NJB 19(1999)6: S W Afr. deserts 3, Horn of Afr. 3)

Thamnus Klotzsch = *Erica*

thanaka *Murraya paniculata*

Thapsia Tourn. ex L. Umbelliferae (III 12). Incl. *Ammodaucus, Distichoselinum, Guillonea, Margotia*, 19 Medit. R: T 65(2016)585. ***T. garganica*** L. (Spanish turpeth r., W & S Medit.) – resin used in plasters, being trialled for prostate cancer treatment; ***T. leucotricha*** (Cosson & Durieu) Simonsen & al. (trop. to N Afr., Macaronesia) – oft. sold in markets, cult. as condiment and med.

Thaspium Nutt. Umbelliferae (III 8). 3 N Am.

Thaumasianthes Danser. Loranthaceae (3). 1 Philippines: ***T. amplifolia*** (Merr.) Danser. R: FM I,13(1997)394

Thaumastochloa C. Hubb. Gramineae (XXII 3). 7 Aus. to NG. R: GBS 36(1983)137

Thaumatocaryon Baill. Boraginaceae (B.2.1.2). 3 SW S Am.

Thaumatococcus Benth. Marantaceae. 1 trop. W Afr.: ***T. daniellii*** (Bennett) Benth. (miraculous fr.) – aril contains thaumatin a protein with a taste 1600 times as sweet as sucrose; thought a promising sweetener but breaks down on heating food, grown in tissue culture & genetic engineering using bacteria attempted but now dipeptides synth. from aspartate used in preference to natural prod.

Thaumatophyllum Schott = *Philodendron*

Thawatchaia Kato & al. Podostemaceae (III). 1 N Thailand: ***T. trilobata*** Kato & al. R: APG 55(2004)66

Thayeria Copel. = *Aglaomorpha*

Theaceae Mirb. ex Ker-Gawl. Magnoliidae – Ericales. Excl. Sladeniaceae, 7/c. 225 trop. with few warm temp. Usu. everg. trees & shrubs, accum. aluminium, hairs unicellular or 0. Lvs simple, entire to toothed, usu. in spirals & coriaceous, oft. withering red; stip. 0. Fls usu. large & bisex., hypogynous to (rarer) epigynous, solit. & axillary, bibracteolate, bracteoles to K to C intergrading spiral or K 5 (+) + C 5(- ∞) imbr., K usu. basally connate, s.t. persistent, C s.t. basally connate, A usu. ±∞ developing centrifugally, filaments long, free or basally connate in a ring or in 5 bundles opp. & adnate to C, oft. basally nectariferous, anthers short, dorsifixed with longit. slits, $\underline{G}$ ((3–)5(–10)) with as many loc. (united only basally in some *Camellia*) & axile placentation, styles s.t. basally united, each loc. with 2–few anatr. or weakly campylotropous bitegmic ovules. Fr. a loculicidal capsule with persistent columella or drupe; 2–few mesotestal seeds (s.t. winged) per loc., embryo straight in little or 0 endosperm. n = 15, 18

Classification & genera (after Cronquist, but Bonnetioideae (C convolute, capsule septicidal, seed structure diff. etc.) here = Bonnetiaceae):

[1. Asteropeioideae (fr. dry-indehiscent; K accrescent & wing-like in fr., Madag.) = Asteropeiaceae]

[2. **Theoideae**:] *Apterosperma, Camellia, Franklinia, Gordonia, Pyrenaria, Schima, Stewartia*

[3. Ternstroemioideae (fr. baccate or dry-indehiscent; K not accrescent; anthers usu. long & basifixed; embryo usu. ± curved) = Pentaphylacaceae]

Tea is *Camellia sinensis*, other spp. of *C.* oil-sources & cult. orn. as are spp. of *Franklinia, Gordonia, Stewartia*, etc. (some intergeneric hybrids); some timbers (*Schima*)

Theana Aver. = *Grossourdya*

Thecacoris A. Juss. Phyllanthaceae (Ant.-Ant.; Euphorbiaceae s.l.). c. 25 trop. Afr., Madag. (2)

Thecagonum Babu = *Leptopetalum*

Thecanthes Wikström = *Pimelea*

Thecocarpus Boiss. Umbelliferae (III 1). 2 Turkey, Iran. Mericarps not separating

Thecophyllum André = *Guzmania*

Thecopus Seidenf. (~ *Thecostele*). Orchidaceae (V 12a). 2 Indomal. R: OB 72(1983)101

Thecorchus Bremek. Rubiaceae (IV 1). 1 trop. Afr.: ***T. wauensis*** (Hiern) Bremek.

Thecostele Reichb.f. Orchidaceae (V 12a). 1 Myanmar to W Mal.: ***T. alata*** (Roxb.) Parish & Reichb.f. R: OB 72(1983)98

Thedachloa S. Jacobs. Gramineae (XXIV). 1 NW Aus.: ***T. annua*** S. Jacobs. R: Telopea 10(2004)635

theetsee *Gluta usitata*

Theileamea Baill. = *Phaulopsis*

Theilera E. Phillips (~ *Wahlenbergia*). Campanulaceae (I 2). 2 Cape. R: T 51(2002)732

Thelasis Blume. Orchidaceae (V 15). c. 25 Indomal. to S China & Solomon Is.

Thelechitonia Cuatrec. = *Sphagneticola* (but see Phytologia 72(1992)142)

Theleophyton (Hook.f.) Moq. = *Atriplex*

Thelepaepale Bremek. = *Strobilanthes*

Thelepogon Roth ex Roemer & Schultes. Gramineae (XXII). 2 OW trop. ***T. elegans*** Roth ex Roemer & Schultes (Ethiopia to Ind.) – alks, local veterinary medic. R: Strel. 12(2002)719

Thelesperma Less. Compositae (Cor.). 15 W N Am. (9), S S Am. Involucral bracts in 2 rows, the inner united in lower 1/3 or more; pappus of 2 recurved barbed awns like *Bidens*. Cult. orn. (*Cosmidium*) esp. ***T. burridgeanum*** (Regel, Koern. & Rach) S.F. Blake (~ *T. filifolium* (Hook.) A. Gray, Texas). ***T. megapoticum*** (Spreng.) Kuntze (Hopi tea, W N Am.) – tea, local medic., dyes

Thelethylax C. Cusset. Podostemaceae (III). 2 Madag. R: Adans. 12(1973)564

Theligonaceae Dumort. = Rubiaceae

Theligonum L. Rubiaceae (IV 14; Theligonaceae). 4 Macaronesia & Medit. (Eur. 1), SW China & Jap. Form. assoc. with Haloragaceae, Hippuridaceae, Portulacaceae etc. Seeds disp. by ants which feed on an elaiosome attached to seed & derived from pericarp. ***T. cynocrambe*** L. (Medit.) – ann., switches from decussate to spiral phyllotaxis, fls oft. 2- or 3-merous, wind-poll., K 0, young shoots a veg., form. laxative

Thelionema R. Henderson (~ *Stypandra*). Asphodelaceae (Phormiaceae). 3 E Aus. R: FA 45(1987)228

Thellungia Stapf = *Sporobolus*

Thellungiella O. Schulz = *Eutrema*

Thelocactus (K. Schum.) Britton & Rose = *Ferocactus* (but see R: Bradleya 5(1987)49)

Thelychiton Endl. = *Dendrobium*

Thelycrania (Dumort.) Fourr. = *Cornus*

Thelymitra Forst. & Forst. f. Orchidaceae (IV 3i). c. 50–100 Mal., Aus., New Caled., NZ (12). Range from obligate insect-poll. (app. food mimics) to facultative selfing to cleistogamy, most with fls closing at night, many blue. Some cult. orn. terr. orchids. ***T. antennifera*** Hook.f. (rabbit-ears, S Aus.) – general 'mimic' of yellow- or cream-fld spp. of *Hibbertia*, *Goodenia* etc., brown 'ears' on top of column; ***T. crinita*** Lindl. & ***T. macrophylla*** Lindl. (W Aus.) – Batesian mimics (scent & colour) of *Orthrosanthus* laxus (Endl.) Benth. as poll. by same female bees with no reward; ***T. pauciflora*** R. Br. (E Aus.) – tubers eaten by Sydney Aborigines

Thelypodiopsis Rydb. (~ *Thelypodium*). Cruciferae (46). 7 N Am. to Guatemala. R: CGH 212(1982)74, 214(1984)26

Thelypodium Endl. Cruciferae (46). Excl. *Thelypodiopsis*, 16 W & C Am. R: CGH 204(1973)79

Thelypteridaceae Ching ex Pichi-Serm. Polypodiidae – Polypodiales. 5/c. 950 subcosmop. (Eur. 5 spp., Mal. c. 440, NW c. 300, Afr. 55). R: T 61(2012)524. Terr. ferns without clathrate scales, s.t. with upright trunks, with simple to pinnate (to 3-pinnate-pinnatifid) fronds (rarely dimorphic), the pinnae almost equal in length; petiole usu. with 2 vasc. bundles, veins simple, less oft. forked; unicellular acicular hairs on adaxial surfaces. Sori round to oblong (rarely elongate), on abaxial surface of veins, ± indusium. n = 27, 29–36

Genera: *Cyclosorus* (most form. recog. genera now incl. here), *Macrothelypteris*, *Phegopteris*, *Pseudophegopteris*, *Thelypteris*. Some ed. (*Cyclosorus*)

Thelypteris Schmidel. Thelypteridaceae. (S.l.) c. 300 trop. & temp. (incl. *Amauropelta*, *'Lastrea'*, *Metathelypteris*, *Parathelypteris*) in 6 subgg. (s.s. 2 subpantemp.)

Thelyschista Garay. Orchidaceae (IV 2h). 1 E Braz.: ***T. ghillanyi*** (Pabst) Garay. R: Gen. Orch. 3(2003)272

Themeda Forssk. Gramineae (XXII 7). 29 trop. OW (China 13). S.t. forming the principal cover in tropical fire-climax 'steppe' as ***T. triandra*** Forssk. (rooigras) in trop. & S Afr. – largely apomictic, ***T. australis*** (R. Br.) Stapf (? *Avena novae-valliae*; ~ *T. triandra*) – 75–95% of diet of red & grey kangaroos (with other spp. in Aus. = kangaroo grass). ***T. arguens*** (L.) Hackel (pianograss, SE As.) – introd. to Jamaica (where now natur.) in a piano packing-case; ***T. gigantea*** (Cav.) Hackel (ulla, trop. As.) – poss. use in paper-making; ***T. villosa*** (Poir.) A. Camus (Indomal.) – huge cult. orn., assoc. with burials in E Indonesia

Themelium (T. Moore) Parris. Polypodiaceae (V; Grammitidaceae). c. 20 trop. As. R: KB 52(1997)737

Themidaceae Salisb. = Asparagaceae

Themistoclesia Klotzsch. Ericaceae (VIII 5). 25 N Andes

Thenardia Kunth. Apocynaceae (II e). 3 C Am. R: Lundellia 1(1998)84

Theobroma L. Malvaceae (Bytt.-Theob.; Sterculiaceae). 20 trop. Am. R: CUSNH 35(1964) 379. Trees (Corner's, Massart's & Nozeran's Models) with cauliflorous infl. & usu. large woody fr. of monkey-disp. seeds; some cult. for cocoa esp. ***T. cacao*** L. (cacao, Andean foothills) – genome sequenced 2010, poll. by biting midges bred in the decaying fr., cult. since antiquity (at least by Mayas, who added chillies, honey & water to make a drink, by 600 BC, Aztecs mixing it with *Enterolobium cyclocarpum* & *Piper amalago*) in Am. for pulp fermented to a beer since 1100 BCE but esp. seeds, the source of (addictive, esp. in women) chocolate cont. saturated fats & stimulating alks (up to 4 'squares' of chocolate improve mind action [flavanols boost blood-flow to brain] & strengthen immune responses but more reverses this, & excess leads to arrhythmia) incl. theobromine (so toxic to dogs), caffeine & theophylline (muscle stimulant) & over 700 compounds (incl. phenolics (as in red wine) efficacious in slowing fat build-up in arteries but also tyramine & phenylethylamine responsible for migraine in some by causing platelets to clump, releasing serotonin which constricts blood vessels reducing blood to brain) analyzed but still unclear what is responsible for flavour, the butter also used medic. (flavanols usu. removed in prod. but now attempts to retain them as they allow blood vessels to dilate, cf. Viagra) & in cosmetics though causing urticaria in some, the seeds currency in Yucatán until 1850 (in 1540s 10 would pay for a prostitute's services) & still valued in 1923; now much cult. in W Afr., Mal., Braz. etc.: usu. '**Criollo**' (now rare) & '**Forasteros**' (c. 80%, esp. grown Braz., W Afr. – Ivory Coast (2002–3) providing 43% world's prod. (oft. using poorly-paid child-workers)) & hybrid = '**Trinitario**' (Sri Lanka, Indonesia, Venez.) with the good taste of 'C.' & disease-resistance of 'F.', by 2010 world market 4 M t (40% Ivory Coast), worth £3.6 billion, 400 000 t. a yr traded through Amsterdam alone. Consumption in Switzerland 9.6 kg (UK 9 kg) per person per annum: industry based on confectionery & drinks worth £3 billion per annum (1993) in UK alone also utilizing chocolate derived from cvs (s.t. polyembryonous) of *T. cacao* s.t. crossed with, or the prod. mixed with that of, ***T. angustifolium*** DC. (monkey c.), ***T. bicolor*** Bonpl. (tiger c.) & ***T. grandiflorum*** (Spreng.) K. Schum. (cupuaçu) – used as pudding flavouring in Braz. Rotting (fermenting) fr. leads to fruity esters, then dried & roasted (caramelization). Tasted by Spanish 1502, shipments going to Spain by 1585; chocolate houses were fashionable in GB before coffee h. but the way of separating the fat from the drink had not been perfected before introduction; milk chocolate is 19 cent. invention (less effective med. than dark as absorption of polyphenols (in 1 'square' a day more effective than beta-blockers in lowering blood pressure) prevented by milk proteins, deriving from a glut of milk & thus milk-powder used in its manufacture (first c. bar 1870), now supplemented with palm or soybean oil to prevent 'whitening'; diff. sorts due to balance between chocolate liquor & cocoa butter (e.g. white c. has no liquor only c. butter, oft. with other veg. oil), while bitter c. has less sugar than milk c.; many favourite 'bars' etc. incl. 'Flake' (1920; oft. in icecream = '99'), 'Crunchie' (1930), 'Milky Way' (1932), 'Mars Bar' (1932, now 3M a day made), 'Black Magic' (1933), 'Aero' (1935), 'Maltesers' & 'Quality Street' (1936), 'Kit-Kat', 'Poppets', 'Rolo' & 'Smarties' (1937) released pre-WW II; now 'single-orig.' c. (cf. wine) fashionable, e.g. from Dominica, Grenada, Jamaica, St Lucia, St Vincent & Grenadines, Samoa, Surinam, Trinidad & Tobago. N. Bailleux & al. (1996) *The book of chocolate*; S.D. & M.D. Coe (1996) *The true hist. of c.*

Theodorovia Kolak. (~ *Campanula*). Campanulaceae (I 7). 1 Caucasus: ***T. karakuschensis*** (Gossh.) Kolak.

Theophrasta L. Primulaceae (Theophrastaceae). 2 Hispaniola. R: NJB 7(1987)529. Pachycaul trees with serial buds in axils & thorny scales in upper parts of stems; sapromyophilous fls in axils of scale lvs of buds

Theophrastaceae G. Don = Primulaceae

Thepparatia Phuph. (~ *Thespesia*). Malvaceae (Malv.-Godd.). 1 Thailand: ***T. thailandica*** Phuph. – liane to 20 m. R: TFB 34(2006)195

Thereianthus G. Lewis. Iridaceae (VI 2). 11 W Cape. R: Bothalia 41(2011)245

Theriophonum Blume. Araceae (VII 23). 5 C & S Ind., Sri Lanka. R: KB 37(1982)277

Thermopsis R. Br. Leguminosae (III 6). 13 E As., 10 N Am. (R: AMBG 81(1994)718), usu. montane. Cult. orn. with alks

Therocistus Holub = *Tuberaria*

Theropogon Maxim. Asparagaceae (Convallariaceae). 1 Himal.: ***T. pallidus*** (Kunth) Maxim. – cult. orn. ground-cover

Therorhodion Small (~ *Rhododendron*). Ericaceae (V 2). 3 NE As., W Alaska

Thesiaceae Vest. See Santalaceae

Thesidium Sonder (~ *Thesium*). Santalaceae. 7 SW to E Cape

Thesium L. Santalaceae (Thesiaceae). 240+ OW esp. trop. & S Afr. (Eur. 25; Cape 81, 35 endemic – R: Strel. 9(2000)637; SW Aus. 1). Herbaceous r.-parasites with epiphyllous infls & alks, incl. ***T. humifusum*** DC. (bastard toadflax, W Eur.). ***T. humile*** Vahl (Medit.) – s.t. a bad weed of barley in Iraq

Thesmophora Rourke. Stilbaceae. 1 SW Cape: ***T. scopulosa*** Rourke. R: EJB 50(1993)89

Thespesia Sol. ex Corr. Serr. Malvaceae (Malv.-Goss.). 17 (incl. *Azanza*, poss. distinct) trop. R: P.A. Fryxell (n.d.) *Nat. hist. Cotton tribe*: 84. Some cult. orn. incl. ***T. garckeana*** F. Hoffm. (E & S Afr.) – fr. ed. & ***T. grandiflora*** DC. (*Montezuma speciosissima*, Puerto Rico – cult. orn., street-tree in Hawaii & S Florida, timber for furniture; ***T. lampas*** (Cav.) Dalz. & A. Gibson (E Afr. to Philipp.) – fibre like *Crotalaria juncea*; ***T. populnea*** (L.) Corr. Serr. (mahoe, milo, pantrop., littoral, invasive SE US) – hard wood for bowls, gunstocks, wheel-frames (Sri Lanka where used also for 'hopper' bowls), base-ball bats (Cuba) etc., street-tree, fls turning yellow to purple in 24 hrs, eradicated from cotton-growing areas as host to cotton-stainer, an insect discolouring young cotton fibres

Thespesiopsis Exell & Hillc. = *Thespesia*

Thespidium F. Muell. ex Benth. Compositae (Inul.-Pluch.). 1 trop. Aus.: ***T. basiflorum*** (F. Muell.) Benth. R: FA 27(2015)418

Thespis DC. Compositae (Ast.-Lag.). 3 SE As.

Thevenotia DC. (~ *Carlina*). Compositae (Card.-Carl.). 2 SW As.

Thevetia L. Apocynaceae (I 9). 8 trop. Am. Scarrone's Model; lvs in spirals. Cult. orn. trees (all but type (***T. ahouai*** (L.) DC.) s.t. considered a sep. genus, *Cascabela*) esp. ***T. peruviana*** (Pers.) K. Schum. (*T. neriifolia*, yellow oleander, lucky bean or nut, trop. Am.) – rheophyte with thevetin (glucoside), heart-depressant, taken by suicides in S Ind.

Thibaudia Ruíz & Pavón. Ericaceae (VIII 5). 60 trop. Am.

thickhead *Crassocephalum crepidioides*

Thieleodoxa Cham. = *Alibertia*

Thilachium Lour. Capparaceae. 13 Somalia to SE Afr., Madag., Masc.

Thiloa Eichler = *Combretum*

thimbleberry *Rubus occidentalis*

thingadu *Parashorea stellata*

thingan *Hopea odorata*

Thinicola J. Ross (~ *Templetonia*). Leguminosae (III 4). 1 W Aus.: ***T. incana*** (J. Ross) J. Ross. R: Muelleria 15(2001)11

Thinopyrum Löve (~ *Triticum*). Gramineae (XV). 6 Euras. some polyploids with genomes from *Agropyron*, *Pseudoroegneria*. ***T. bessarabicum*** (Săvul. & Rayss.) Löve (W As.) – poss. salt-tolerant cereal, 1 diploid parent of tetraploid ***T. junceiforme*** (Löve & D. Löve) Löve & of *T. sartorii* (Boiss. & Heldr.) Löve, while hexaploid ***T. junceum*** (L.) Löve (*Elytrigia j.*, Russian wild rye, Medit. Eur. to Black Sea; invasive Aus.) has all 3 genomes – cf. *Triticum aestivum*!; ***T. ponticum*** (Podw.) Barkw. & D. Dewey (*Triticum p.* Podw., Medit.) – planted Aus. to ameliorate dry-land salinity & in US along roadsides as tolerant of salt used to de-ice roads

Thinouia Triana & Planch. Sapindaceae (IV 1). 12 warm S Am.

thinwin *Millettia pinnata*

Thiollierea Montr. (~ *Bikkia*). 12 New Caled. R: Adans. 33(2011)120,137

Thiseltonia Hemsl. (~ *Hyalosperma*). Compositae (Gnap.-Ang.). 1 W Aus.: ***T. gracillima*** (F. Muell. & Tate) Paul G. Wilson. R: Nuytsia 8(1992)481

Thismia Griff. Thismiaceae (Burmanniaceae). c. 55 Jap. (1), trop. As. (20) incl. Hong Kong, Taiwan, Aus. (4 – R: Telopea 16(2014)173) & NZ (1), trop. Am. (12) & nr Chicago (*T.*

americana N. Pfeiffer – disc. in prairie 1912 but not seen since 1917, prob. closest ally *T. rodwayi* F. Muell. in Aus., NZ)

Thismiaceae J. Agardh (~ Dioscoreaceae). Magnoliidae – Dioscoreales. Excl. *Saionia*, 5/77 trop. ext. to NZ & US. Aclorophyllous mycotrophs diff. from Burmanniaceae in circumscissile P, A 6 (3 in *Oxygyne*). n = 6–8, 11–13

Genera: *Afrothismia, Haplothismia, Oxygyne, Thismia, Tiputinia*

Remarkable disjunct distribs as in B.

thistle usu. *Cirsium* or *Carlina* spp., (Bible) *Centaurea iberica*; **blessed t.** *Centaurea benedicta*; **bull t.** *Cirsium vulgare*; **Canadian t.** *C. arvense*; **carline t.** *Carlina vulgaris*; **cotton t.** *Onopordum acanthium*; **creeping t.** *Cirsium arvense*; **globe t.** *Echinops* spp.; **golden t.** *Scolymus hispanicus*; **holy t.** *Silybum marianum*; **marsh t.** *C. palustre*; **milk t.** *Sonchus* spp., *Silybum marianum*; **musk t.** *Carduus nutans*; **Russian t.** *Salsola tragus*; **saffron t.** *Carthamus lanatus, C. tinctorius*; **St Barnaby's t.** *Centaurea solstitialis*; **Scotch t.** *Cirsium vulgare, Carduus nutans* but now tending to be applied to *Onopordum acanthium*; **sow t.** *Sonchus* spp. esp. *S. oleraceus*; **spear t.** *Cirsium vulgare*; **star t.** *Centaurea calcitrapa*; **s.t., yellow** *C. solstitialis*

thitka *Pentace burmannica*

thitsi *Gluta usitata*

Thladiantha Bunge. Cucurbitaceae (VI). 23 Afr. (1), E As. (China 23, 19 endemic – R: FOC 19(2011)21) to Mal. Dioec. tendril-climbers with r.-tubers; elaiophores attractive to bees. Some cult. orn.

Thlaspeocarpa C.A. Sm. = *Heliophila*

Thlaspi Tourn. ex L. Cruciferae (47). s.s. 6 N temp. (most N Am. etc. spp. = *Noccaea*). *T. arvense* L. (penny cress, Mithridate mustard, stinkweed, Eur., natur. N Am.) – form medic., seeds with 30–40% oil suitable for illumination

Thlaspidium (Lipsky) Rassulova = *Astragalus*

Thodaya Compton = *Euryops*

Thogsennia Aiello. Rubiaceae (I 7). 1 E Cuba, Hispaniola: ***T. lindeniana*** (A. Rich.) Aiello. R: JAA 60(1979)117

Thomandersia Baill. Thomandersiaceae (Acanthaceae s.l.). 6 trop. Afr. R: BJBB 36(1966)207. Trad. medic. & magic, wood for local constr.

Thomandersiaceae Sreemad. (Acanthaceae s.l.). Magnoliidae – Lamiales. 1/6 trop. Afr. Small trees (Prévost's Model) s.t. with stilt-r., shrubs or lianes. Stems terete. Lvs simple, anisophyllous opp.; stip. & cystoliths 0. Fls opp. or in triads in term. or axillary racemes; K 5-lobed, campanulate with nectaries without, C 5, 2-lipped, white or yellow to red or purple, A 4 didynamous + 1 staminode, arising from C tube, $\underline{G}$ 2-loc. with axile placentation & 2 or 3 ovules per loc. Fr. a capsule (not explosive) subtended by accrescent K; seeds with 'jaculators' & covered with triangular scales or warts in spirals, endosperm 0

Genus: *Thomandersia*

Form. in Acanthaceae but cystoliths 0 etc.; prob. allied with Schlegeliaceae

Thomasia Gay. Malvaceae (Bytt.-Lasio.; Sterculiaceae). 32 Aus. (all but 1 endemic SW)

Thompsonella Britton & Rose (~ *Echeveria*). Crassulaceae (I 5b). 8 Mex., usu. limestone. R: NSL (2013)98. Cult. orn.

Thomsonia Wall. = *Amorphophallus*

Thonandia Linder & Verboom = *Notodanthonia*

Thonningia Vahl. Balanophoraceae. 1 trop. Afr.: ***T. sanguinea*** Vahl – tubers sold in W Afr. markets for medic., fr. with alks. R: BJ 106(1986)367

Thoracocarpus Harling. Cyclanthaceae. 1 trop. S Am.: ***T. bissectus*** (Vell.) Harling. R: ANB 18(1958)254

Thoracosperma Klotzsch = *Erica*

Thoracostachyum Kurz = *Mapania*

Thoreauea J.K. Williams. Apocynaceae (II e). 2 Mex. R: Britt. 57(2005)259

Thoreldora Pierre = *Glycosmis*

Thorella Briq. = *Caropsis*

thorn *Crataegus* spp.; **t. apple** *Datura stramonium*; **blackt.** *Prunus spinosa*; **buffalo** or **Cape t.** *Ziziphus mucronata*; **Christ's t.** *Z. spina-christi, Paliurus s.*; **cockspur t.** *C. crus-galli, Vachellia eburnea*; **crown of t.s** *Euphorbia milii*, (Bible) *Poterium spinosum* or *Ziziphus spina-christi*; **devil t.** *Tribulus terrestris*; **holy t.** *C. monogyna* 'Biflora'; **kangaroo t.** or **t. tree** *Acacia paradoxa*; **lemon t.** *Cassinopsis* spp.; **t. pear** *Scolopia ecklonii*; **whistling t.** *V. drepanolobium*

Thorncroftia N.E. Br. Labiatae (VIII 3e). 4 Transvaal. R: Bothalia 7(1961)429

Thornea Breedlove & McClint. Hypericaceae (II; Guttiferae I). 2 Mex., Guatemala

Thorntonia Reichb. = *Hibiscus*
thorow-wax *Bupleurum rotundifolium*
Thottea Rottb. Aristolochiaceae (II 1). Incl. *Apama*, *Asiphonia*, c. 45 Indomal. Troll's Model; fly-poll. Some medic. (alpam r.) incl. ***T. piperiformis*** (Griff.) Mabb. (*T. corymbosa*, W Mal.) – lvs pounded for toothache relief, ***T. tomentosa*** (Blume) Ding Hou (Indomal.) – also used as clothes soap
Thouarsiora Homolle ex Arènes = *Ixora*
Thouinia Poit. Sapindaceae (V 1). 30 Mex. to WI. Lianes
Thouinidium Radlk. Sapindaceae. 6 Mex. to WI
Thozetia F. Muell. ex Benth. = *Marsdenia*
Thrasya Kunth = *Paspalum* (but see ABV 14,4(1987)7)
Thrasyopsis L. Parodi = *Pasalum* (but see Phyton 23(1983)101)
Thraulococcus Radlk. = *Lepisanthes*
Threlkeldia R. Br. = *Sclerolaena* (but see FA 4(1984)230)
thrift *Armeria maritima*
Thrinax L. f. ex Sw. Palmae (III 2). Excl. *Hemithrinax*, *Leucothrinax*, 3 Carib. R: SCB 19(1975), ABH 3(1985)225. Some wind-poll. (rare in palms). Lvs for thatch; fibres. Cult. orn. esp. ***T. parviflora*** Sw. (thatch palm, Jamaica) – dried lvs used for decoration in temp. regions; ***T. radiata*** Lodd. ex Schultes & Schultes f. (*T. wendlandiana*, Florida to N S Am.) – fr. ed., fibre for mattress-stuffing
Thrincia Roth = *Leontodon*
Thrixspermum Lour. Orchidaceae (V 16c). c. 150 Indomal. to Taiwan & W Pac. Some cult. orn.
thrumwort *Damasonium alisma*
Thryallis Mart. Malpighiaceae. 5 Braz., Paraguay, Bolivia. R: CUMH 20(1995)6
Thryothamnus Philippi = *Junellia*
Thryptomene Endl. Myrtaceae (II 15). c. 45 S, C & NE Aus. See also *Aluta*
Thuarea Pers. Gramineae (XXIV 5). 2 Madag. to Polynesia. ***T. involuta*** (Forst.f.) Sm. – leafy structure enveloping ripening fr., sandbinder
Thuja Tourn. ex L. Cupressaceae. 5 E As., W & E N Am. R: A. Farjon, *Monogr. Cupressaceae & Sciadopitys* (2005)141. Arbor-vitae; monoec. medic. (esp. N Am., analgesic) cult. orn. & timber trees (Attims's Model) esp. ***T. occidentalis*** L. (white cedar, American a.-v., E N Am.) – soft fragrant wood used for fencing, railway sleepers, etc., oil medic., planted as screens & windbreaks, many orn. cvs incl. varieg. & dwarf ones; ***T. plicata*** Donn ex D. Don (western red cedar, w. or giant a.-v., W N Am.) – ? hybrid origin, to 1400 (? 2000) yrs old, 71 m tall with bole to 6.31 m diam., pl. sp. with greatest no. of uses by Native Americans, timber weather-resistant & used for shingles, 'cedar' greenhouses & frames, bee-hives, boats, bungalows, ladders, totem poles, though sawdust can cause asthma (plicatic acid), bark form. used for clothing, foliage used in floristry, introd. GB 1853 & now much planted for hedging, '**Gracilis**' haploid (n = 11); ***T. standishii*** (Gordon) Carrière (Jap. a.-v., C Jap.). See also *Platycladus*
Thujopsis Sieb. & Zucc. ex Endl. Cupressaceae. 1 N Jap.: ***T. dolabrata*** (L.f.) Sieb. & Zucc. (hiba) – cult. orn., durable timber for constr., cabinet-work, railway sleepers etc. R: A. Farjon, *Monogr. Cupressaceae & Sciadopitys* (2005)137
Thulinia Cribb. Orchidaceae (IV 4d). 1 Nguru Mts, Tanzania: ***T. albolutea*** Cribb. R: Gen. Orch. 2(2001)375
Thunbergia Retz. Acanthaceae (II; Thunbergiaceae). 100–150 OW trop. Cli. or erect herbs & shrubs (Champagnat's Model) with bracteoles encl. K. Cult. orn. incl. ***T. alata*** Bojer ex Sims (black-eyed Susan, trop. Afr.) – precise orig. unclear as long cult. in Afr., common greenhouse cli. with cream, white or orange fls with darker middle, natur. As., Mal., S Am.; ***T. grandiflora*** (Rottler) Roxb. (Bengal clock vine, N Ind.) – liane with blue 2-lipped fls, extrafl. nectaries attract ants keeping off infestation by pyralid caterpillars, grown as arbour vine, invasive Aus., Hawaii; ***T. mysorensis*** (Wight) T. Anderson (Nilgiris, Ind.) – similar but with red & yellow fls
Thunbergiaceae Lilja = Acanthaceae (II)
Thunbergianthus Engl. Orobanchaceae ('Butt.'; Scrophulariaceae s.l.). 1 São Tomé (W Afr.), 1 trop. E Afr.
Thunbergiella H. Wolff = *Itasina*
thunderbolt See Gramineae
Thunia Reichb.f. Orchidaceae (V 10b). 5 Himal. to Myanmar. Cult. orn. esp. ***T. alba*** (Lindl.) Reichb.f. (incl. *T. marshalliae* B.S. Williams)

Thuniopsis L. Li & al. Orchidaceae. 1 Yunnan: ***T. cleistogama*** L. Li & al.

Thuranthos C.H. Wright = *Drimia*

Thurberia A. Gray = *Gossypium*

Thurnia Hook.f. Thurniaceae. 3 Amazon basin & Guyana. R: BSVCN 8(55)(1943)241

Thurniaceae Engl. (incl. Prioniaceae; R: FOW 5(2001)). Magnoliidae – Poales. 2/4 S Afr., N S Am. Tough perenn. herbs with upright rhiz. & no sec. growth; vessels in all veg. organs. Lvs with sheathing base & long leathery flat or canaliculate lamina s.t. with marginal prickles, basal, in spirals. Fls bisex., small, in dense term. racemose heads subtended by leafy bracts, wind-poll.; P 3 + 3, chaffy, A 3 + 3, filament ± adnate to P-base, anthers basifixed with longit. slits, pollen (monoporate) in tetrads, G̲ (3), 3-loc., each loc. with 1 or more erect anatr. ovules. Fr. a loculicidal capsule with 1 seed per loc.; seeds hispid with processes at both ends, embryo small, ± cylindric, embedded in copious mealy starchy endosperm

Genera: *Prionium, Thurnia*; kept sep. from Juncaceae because leaf vasc. bundles are paired 1 above another with phloem strands adjacent (unique in monocots), presence of silica-bodies in lvs (*Thurnia*) as well as seed-char. above

Thurovia Rose (~ *Gutierrezia*). Compositae (Ast.-Sol.). 1 SE Texas: ***T. triflora*** Rose. R: SBM 20(2000)87

Thurya Boiss. & Bal. Caryophyllaceae (II 1). 1 SW As.: ***T. capitata*** Boiss. & Bal.

Thuspeinanta T. Durand. Labiatae (VI). 2 C As. to Iran, Afghanistan & Pakistan

thuya *Tetraclinis articulata*; see also *Thuja*

thyine *Tetraclinis articulata*

Thylacanthus Tul. = *Julbernardia*

Thylacodraba (Náb.) O. Schulz = *Draba*

Thylacophora Ridl. = *Riedelia*

Thylacopteris Kunze ex J. Sm. Polypodiaceae (IV). 2 Mal. R: Blumea 39(1994)351

Thylacospermum Fenzl. Caryophyllaceae (II 1). 2 C As. to Himal. & W China. R: JJB 90(2015)352

Thymbra L. Labiatae (VII 2b). Incl. *Coridothymus* 4 Medit. (Eur. 3) to SW As. R: AJBM 44(1987)348. Cult. orn. shrubs. ***T. capitata*** (L.) Cav. (*C. c.*, Medit.) – anti-oxidants, heart-disease folk. med. (Crete)

thyme *Thymus vulgaris*; **basil t.** *Clinopodium acinos*; **caraway t.** *T. herba-barona*; **cat t.** *Teucrium marum*; **lemon t.** *T.* × *carolipaui* 'Culinary Lemon' (also a form of *T. pulegioides*); **Pennsylvanian Dutch t.** *T. pulegioides*; **wild t.** *T. praecox* subsp. *britannicus*

Thymelaea Mill. Thymelaeaceae (Thym.-Daph.). 30 N temp. OW (Eur. 17) esp. Medit. R: NRBGE 38(1980)89. ***T. hirsuta*** (L.) Endl. (*Passerina h.*, mituan, Medit.) – dioec. or monoec. with male fls first, bark fibre for rope & (since 1979) paper in Israel; ***T. tartonraira*** (L.) All. (Medit.) – ship cordage in anc. Greece

Thymelaeaceae Juss. Magnoliidae – Malvales. Incl. Tepuianthaceae, 46/860 cosmop. esp. Aus. & trop. Afr. Toxic trees & shrubs (oft. smelly) with fibrous bark, rarely lianes or herbs, prod. glycosides & accum. daphnin (a coumarin) or related substances; usu. internal phloem next to pith; hairs simple (oft. silky-adpressed). Lvs simple, entire, s.t. ericoid or even sheath-like (*Struthiola*), in spirals, opp. or ± whorled; stip. ± 0. Fls usu. bisex., reg. or not, oft. ± perigynous with hypanthium s.t. coloured, in racemes or heads or solit., (3)4 or 5(6)-merous; K imbr. or valvate lobes on hypanthium or ± free (fl. hypogynous), C small & oft. scale-like in throat of hypanthium, alt. with K or prs opp. K (∞ in Octolepidoideae) or oft. 0, A opp. K or in 2 whorls or ∞ (Octolepidoideae) or even 2 (*Pimelea*), filaments short or 0, anthers with longit. slits, disk oft. present around G̲ (2–5(–12)) with as many loc. or (G 2) pseudomonomerous & 1-loc., style oft. eccentric, each loc. with 1 pend. anatr. to hemitropous bitegmic ovule. Fr. dry-indehiscent, baccate or drupaceous, less oft. a loculicidal capsule; seeds oft. carunculate, embryo oily, straight in (0 or copious) endosperm. n = 9 (oft.)

Classification & principal genera (after Cronquist):

1. **Gonystyloideae** (Octolepidoideae; capsule; C usu. ∞, A 8–∞; disk usu. 0; hypanthium ± 0; lvs oft. dotted; internal phloem 0): *Gonystylus, Lethedon*

[2. Aquilarioideae = 4. – Aqu.]

[3. Synandrodaphnoideae = 4. – Syn.]

4. **Thymelaeoideae** (fr. capsule or indehiscent, disk usu. present): Synandrodaphneae (*Synandrodaphne*, only); Aquilarieae (*Aquilaria, Gyrinops*, only); Daphneae (*Daphne, Daphnopsis, Dicranolepis, Gnidia, Lachnaea, Passerina, Pimelea, Struthiola, Thymelaea, Wikstroemia*)

App. affintities with Malvales (as opposed to former disposition in Myrtales or Malpighiales) confirmed by DNA work. *Tepuianthus* (trees & shrubs with resin cells, infls androdioec., disk of 5–10 suborbicular fleshy glands, A in 1–3 whorls, anthers versatile, fr. a 3-loc. loculicidal capsule; seeds with consp. ridged raphe; R: MNYBG 32(1981)8) app. sister to rest of fam.

Timber (*Gonystylus* – ramin, *Gyrinops*), incense (*Aquilaria, Wikstroemia*), bark fibre for paper (*Daphne, Edgeworthia, Thymelaea, Wikstroemia*), cordage (*Dais, Daphne, Daphnopsis, Dirca, Gyrinops, Thymelaea*) or ornament (*Lagetta*) & cult. orn. esp. *Daphne, Dirca, Pimelea* etc.

Thymocarpus Nicolson & al. = *Goeppertia*

Thymophylla Lag. (~ *Dyssodia*). Compositae (Tag.-Pect.). c. 18 SW N Am. (US 8). R: Sida 11(1986)371. Some cult. orn. esp. ***T. tenuiloba*** (DC.) Small (*D. t.*, Dahlberg daisy) – strongly scented bedding pl.

Thymopsis Benth. Compositae (Bah.). 2 WI

Thymus Tourn. ex L. Labiatae (VII 2b). c. 220 temp. Euras. (Eur. 66). Herbs & shrubs, gynodioec., protandrous. Cult. orn., sedative teas ('nerve'-calming) & flavourings esp. ***T. vulgaris*** L. (thyme, W Medit. to SE Italy, with females prod. more seeds than hermaphr. & on burnt sites percentage of females high at first but then declining) – dried lvs used to flavour meat dishes, sausages etc., oil medic. (suppressing lung infection, easing bronchial spasms & helping sufferers to expel phlegm from lungs), anc. Romans considered t. honey the best. Also cult.: ***T. × carolipaui*** Mateo & Crispo **'Culinary Lemon'** ('*T. × citriodorus*', lemon thyme = *T. pulegioides × T. vulgaris*), ***T. herba-barona*** Lois. (caraway t., Majorca, Sardinia & Corsica), ***T. praecox*** Opiz (Eur.) esp. subsp. ***britannicus*** (Ronn.) Kerguélen (subsp. *arcticus*, *T. drucei*, *T. polytrichus* subsp. *b.*, wild t., W & N Eur.) – fl. heads fed on by larvae of Large Blue (*Phengaris* (*Maculinea*) *arion*) before red ants carry them to nests where they feed on grubs, oil medic. (serpolet oil), ***T. pulegioides*** L. (Pennsylvanian Dutch t., Eur.), one form a 'lemon thyme'; ***T. zygis*** Loefl. ex L. (Iberia) – used to flavour olives in C Spain

Thyrasperma N.E. Br. = *Hymenogyne*

Thyridachne C. Hubb. Gramineae (XXIV). 1 trop. Afr.: ***T. tisserantii*** C. Hubb.

Thyridia W.R. Barker & Beardsley (~ *Mimulus*). Phrymaceae. 1 Aus.-NZ: ***T. repens*** (R. Br.) W.R. Barker & Beardsley. R: Phytoneuron 2012–39: 20

Thyridocalyx Bremek. (~ *Trianolepis*). Rubiaceae (IV 8). 1 Madag.: ***T. ampandrandavae***

Thyridolepis S.T. Blake. Gramineae (XXIV 3). 3 arid Aus. Fodder

Thyrocarpus Hance. Boraginaceae (B.3.8.2). 3 China

Thyrsacanthus Moric. (~ *Anisacanthus*). Acanthaceae (III 2c). 5 S Am. R: T 59(2010)967

Thyrsanthella Pichon = *Trachelospermum*

Thyrsanthemum Pichon. Commelinaceae (II 1f). 3 Mex.

Thyrsanthera Pierre ex Gagnep. Euphorbiaceae (Acal.-Chro.-Chro.). 1 SE As.: ***T. suborbicularis*** Pierre ex Gagnep. R: Blumea 44(1999)431

Thyrsia Stapf = *Rottboellia*

Thyrsodium Salzm. ex Benth. Anacardiaceae (I). 7 trop. Am. R: Britt. 45(1993)115

Thyrsopteridaceae C. Presl (~ Dicksoniaceae). Polypodiidae – Cyatheales. 1/1 Juan Fernandez. R: T 55(2006)712. Ascending to erect solenostelic rhiz. bearing runners & clad with stiff pluricellular hairs. Fronds 3–5-pinnate, large, subdimorphic; veins free with terminal sori. Outer & inner indusia united to form asymmetric cuplike unit, each sorus with a columnar, clavate receptacle; spores globose-tetrahedral. n = c. 78

Genus: *Thyrsopteris*

Thyrsopteris Kunze. Thyrsopteridaceae (Dicksoniaceae). 1 Juan Fernandez (400–700 m): ***T. elegans*** Kunze – long-lived in cult. R: M.F. Large & J.E. Braggins, *Tree ferns*(2004)305. Fossils referred to the genus widespread, *T. elegans* a relic

Thyrsosalacia Loes. Celastraceae (IV). 4 W trop. Afr.

Thyrsostachys Gamble. Gramineae (V 4). 2 Myanmar to China (2) & Thailand: ***T. siamensis*** Gamble (*T. regia*) – stems used for umbrella handles; ***T. oliveri*** Gamble – fls cycle of 48 yrs

Thysanella A. Gray = *Polygonella*

Thysanocarpus Hook. Cruciferae (46). 7 N Am. R: SB 35(2010)566

Thysanoglossa Porto & Brade. Orchidaceae (V 12h). 3 Braz.

Thysanolaena Nees. Gramineae (XXI). 1 SE As.: ***T. latifolia*** (Hornem.) Honda (*T. maxima*, tiger grass) – panicles used as brooms. R: FOC 22(2006)446

Thysanosoria Gepp (~ *Lomariopsis*). Lomariopsidaceae (Dryopteridaceae s.l.). 1 NW NG: ***T. pteridiformis*** (Cesati) C. Chr. (? atavistic form of *Lomariopsis*). R: LBS 2(1977)1

Thysanostemon Maguire. Guttiferae (3.). 2 Venez. R: MNYBG 10(1964)132

Thysanostigma Imlay. Acanthaceae (III 2c). 2 S Thailand, Malay Pen. R: NJB 8(1988)227

Thysanotus R. Br. Asparagaceae (Anthericaceae). Incl. *Murchisonia*, 51 Aus., esp. SW, 2 ext. to NG, 1 to China. R: FA 45(1987)308. Cult. orn. (fringed lilies). Some climbers incl. Aus. ***T. patersonii*** R. Br. & ***T. volublis*** R. Br. (*T. manglesianus*)

Thysanurus O. Hoffm. = *Geigeria*

Thyselium Raf. = *Peucedanum*

Thysselinum Adans. = *Peucedanum*

ti *Cordyline fruticosa*

Tianschaniella B. Fedtsch. ex Popov = *Eritrichium*

Tiarella L. Saxifragaceae (I 5). 2 N Am. (R: FNA 9(2009)114), 1 E As. Cult. orn. rhizomatous herbs esp. ***T. cordifolia*** L. (coolwort, foamflower, E N Am.) – alleged diuretic. See also × *Heucherella*

Tiarocarpus Rech.f. = *Cousinia*

Tibestina Maire = *Dicoma*

Tibetia (Ali) Tsui (~ *Gueldenstaedtia*). Leguminosae (III 24). 5 Sino-himalaya. R: BJLS 148(2005)476

Tibetoseris Sennikov (~ *Youngia*). Compositae (Cich.-Cich.(Crep.). Excl. *Pseudoyoungia*, 1 Himal.: ***T. depressa*** (Hook.f. & Thomson) Sennikov. R: CN 48(2010)28

tibirisi *Mauritia flexuosa*

Tibouchina Aubl. Melastomataceae. Excl. *Pleroma*, c. 240 trop. Am. Diff. from *Melastoma* (fr. fleshy) only in dry fr. Cult. orn. shrubs esp. ***T. urvilleana*** (DC.) Cogn. ('*T. semidecandra*', glory bush, Braz., weedy in e.g. Hawaii) – fls purple to violet with purple anthers

Tibouchinopsis Markgraf. Melastomataceae. 2 NE Braz.

tic bean *Vicia faba*

Ticanto Adans. (~ *Caesalpinia*). Leguminosae (I 14). 15 As. R: PhytoKeys 71(2016)127

tick bush *Kunzea ambigua*; **t. seed** *Bidens* spp.

Ticodendraceae Gómez-Laurito & L.D. Gómez. Magnoliidae – Fagales. 1/1 C Am. Dioec. (polygamodioec.) everg. Lvs serrate; stip. encircling stem, decid. Male fls in spike-like thyrses of 1–3-fld cymules in trimerous whorls subtended by a bract, P 0, A 8–10 with longit. dehiscence. Female fls solit. subtended by pr of bracteoles, P an inconsp. rim, $\overline{G}$ (2), 4-loc., each loc. with 1 hemitropous ovule, styles 2(3) stigmatic throughout. Fr. a drupe; seed with thin endosperm & massive oily embryo

Genus: *Ticodendron*

Most primitive wood in Fagales. Fr. in Eocene of Oregon & England

Ticodendron Gómez-Laurito & L.D. Gómez. Ticodendraceae. 1 S Mex. to C Panamá: ***T. incognitum*** Gómez-Laurito & L.D. Gómez – overlooked until 1980s

Ticoglossum Lucas Rodr. ex Halb. = *Rossioglossum*

Ticorea Aubl. Rutaceae (I 8). 5 Costa Rica to NE S Am. R: Britt. 50(1998)504

Tidestromia Standl. Amaranthaceae (II 2). 6 SW N Am. deserts

tidy-tips *Layia platyglossa*

Tiedemannia DC. (~ *Oxypolis*). Umbelliferae (III 8). 2 SE US. R: T 61(2012)413

Tieghemella Pierre. Sapotaceae (I 1). 2 trop. W Afr. ***T. heckelii*** (A. Chev.) Heine (makoré, Afr. cherry, c. mahogany, bacu, baku) – Aubréville's Model, fr. eaten by elephant, timber a mahogany subs., oilseed for soap ('Dumori butter')

Tieghemia Balle = *Oncocalyx*

Tieghemopanax R. Viguier = *Polyscias*

Tienmuia Hu = *Phacellanthus*

Tietkensia Short. Compositae (Gnap.-Ang.). 1 C & WC Aus.: ***T. corrickiae*** Short. R: Muelleria 7(1990)248

tigasco oil *Campnosperma* spp.

tiger cocoa *Theobroma bicolor*; **t. flower** *Tigridia pavonia*; **t. grass** *Thysanolaena latifolia*; **t.'s jaws** *Faucaria tigrina*; **t. lily** *Lilium lancifolium*; **t. maple** *Acer* spp. (US) esp. *A. macrophyllum* (W), *A. rubrum* (E); **t. nut** *Cyperus esculentus*; **t. orchid** *Grammatophyllum speciosum*; **t. pear** *Opuntia aurantiaca*; **t. wood** *Astronium fraxinifolium*, *Lovoa trichilioides*

Tigridia Juss. Iridaceae (VII 5). Incl. *Ainea*, *Sessilanthera* (buzz-poll. with vestigial filaments, porose anthers & 0 nectaries), *Fosteria*, *Rigidella* (hummingbird-poll. with red fls, exserted style-branches, nectar (oil 0), c. 55 C & S Am. R: UCPB 54(1970)1. Elaiophores attractive to bees in some. Cult. orn. bulbous herbs with plicate lvs esp. ***T. pavonia*** (L.f.) Ker-Gawler ((peacock) tiger fl., Mex., natur. Guatemala etc.) – lvs plicate, wild pl. with red fls to 15 cm across (yellow & purple spots in tube), starchy bulbs eaten since Aztec times, selected cvs with white, yellow, pink & orange fls, each lasting only 8–12 hrs

Tigridiopalma C. Chen = *Phyllagathis*

Tikalia Lundell = *Blomia*

Tilesia G. Meyer (*Wulffia*). Compositae (Helia.-Ecl.). 3 trop. Am. R: Novon 6(1996)413. ***T. baccata*** (L.) Pruski (*W. b., W. stenoglossa*) – weedy, infr. a head of bird-disp. fleshy cypselas resembling a blackberry

Tilia Tourn. ex L. Malvaceae (Til.; Tiliaceae). 23 N temp. (Eur. + W As. 4; E As. 17, Am. 2). Lime, linden. R: Plantsman 5(1984)206; D. Piggott (2012) *Lime-trees & basswoods*. Decid. trees (Troll's Model) with 2-ranked lvs (some with domatia) & fragrant fls on infl. emergent from a large bract; imp. timbers, cult. orn., bee-fodder (though some sugars (mannose) toxic in excess, so dead bees (oft. bumblebees) oft. found below fl. trees), phloem bast used for cordage, ed. oil extractable from fr. (endosperm) used in comm. cough mixtures. Timber pale & used for piano-keys & decorative carving as by the incomparable Grinling Gibbons (1648–1721). ***T. americana*** L. (American lime or basswood, whitewood, C & E N Am.) – wood for cheap furniture & excelsior, inner bark for mats etc., local medic.; ***T. cordata*** Mill. (*T. officinarum*, small-leaved l., pry, Eur.) – form. imp. tree in lowland England (a circle of c. 60 trees at Silk Wood, Glos., part of a clone 2000–6000 yrs old) characterizing the orig. forest as the last major tree to enter the island after the last Ice Age & not reaching Ireland & Scotland (seeds sterile in NW England where too cold for pollen-tube growth for successful fert.), timber for tables, plates, spoons, musical instruments, excelsior, charcoal, fls used in a medic. tea also used as mouthwash, bark for 'Archangel' mats; ***T. × europaea*** L. (*T. × vulgaris*, common l., *T. cordata × T. platyphyllos* Scop. (C & S Eur.), natural hybrid) – much cult. street-tree but disagreeable because aphids secrete large amounts of honeydew on pavements & parked motors at a rate of 1 kg of sugars per m^2 per annum, poss. stimulating growth of nitrogen-fixing bacteria in ground around tree & thus enhancing its nitrogen availability at the cost of some carbohydrate taken from the phloem by the aphids, '**Pallida**' with larger lvs the linden of Unter den Linden, Berlin; ***T. japonica*** (Miq.) Simonkai (Jap. l., Jap.) – much like Eur. but smaller; ***T. miqueliana*** Maxim. (China) – oft. planted nr Jap. temples; ***T. mongolica*** Maxim. (China, Mongolia) – ground fls & fr. said to be good chocolate subs.; ***T. tomentosa*** Moench (SE Eur., As.) esp. '**Petiolaris**' (silver or weeping l., poss. hybrid) with pendent branches & fls opening later than *T. cordata*

Tiliaceae Juss. = Malvaceae. See also Muntingiaceae

Tiliacora Colebr. Menispermaceae (II). 22 OW trop. (Afr. 19, SE As., 2, Aus. 1). R: KB 30(1975)89, 37(1972)369. Alks. ***T. triandra*** (Colebr.) Diels (SE As.) – cordage, local medic., flavouring (Thailand)

Tilingia Regel & Tiling = *Ligusticum*

Tillaea Micheli ex L. = *Crassula*

Tillandsia L. Bromeliaceae (2). Excl. *Racinaea*, c. 610 trop. Am. (s.s. 240 esp. Mex.). R: FN 14(1977)665, 1392. ***T. capillaris*** Ruíz & Pavón – cleistogamy. Cult. orn. epiphytes with rosetted lvs & poor r.-systems (nutrients being absorbed all over pls) but some with pendent stems esp. ***T. usneoides*** (L.) L. (Spanish or Florida moss, old man's beard, S Virginia to Arg., over 8000 km latitude, an almost unique distrib.) – hanging in festoons from trees & overhead wires like a lichen such as *Usnea*, the base attached to support but dying as apex grows downwards leaving an axile strand of sclerenchyma, the whole pl. absorbing water & nutrients (esp. calcium) dripping over it; rarely fls but distrib. by wind as fragments & by birds using it as nesting material; dried pl. used as packing material & like horsehair in upholstery (e.g. car seats in Model T Ford), by 1930s some 10 K t used a yr; ***T. australis*** Mez (*T. maxima*, Bolivia, Arg.) – cult. orn., rosettes to 1 m diam.; ***T. crocata*** (C.J. Morren) N.E. Br. (Braz.) – cult. orn. with strongly scented fls; ***T. latifolia*** Meyen (Peru & N Chile desert) – rootless & blown about; ***T. recurvata*** (L.) L. (trop. Am.) – in Baja Calif. epiphyte on columnar cacti (& cables) with N-fixing bacterium, *Pseudomonas stutzeri*

Tillospermum Salisb. = *Kunzea*

timbo *Lonchocarpus* spp. esp. *L. nicou*

Timonius DC. Rubiaceae (III 3). c. 170 Seychelles & Mauritius (2), Sri Lanka (1), Andamans (1), Mal. (esp. NG) to Pac. R (subg. *T.* – 8): Allertonia 7(1993)15. Some stranglers (cf. *Ficus*). ***T. timon*** (Spreng.) Merr. (*T. rumphii*, Mal. to Aus.) – ed. fr.

timothy *Phleum pratense*

Timouria Rosch. (~ *Stipa*). Gramineae (X). 5 As.

Tina Schultes. Sapindaceae. Incl. *Neotina, Tinopsis,* 20 Madag. R: Cand. 66(2011)125. Leeuwenberg's Model

Tinadendron Achille (~ *Guettarda*). Rubiaceae ((III 3). 2 New Caled. & Vanuatu. R: Adansonia III,28(2006)169

Tinantia Scheidw. Commelinaceae (II 1f). c. 14 Texas to trop. Am. Cult. orn. esp. ***T. erecta*** (Jacq.) Fenzl – fls pink to blue

tinda *Benincasa fistulosa*

tindora *Coccinia grandis*

tingo fibre *Pouzolzia mixta*

Tinguarra Parl. = *Athamanta*

tinker's weed *Triosteum perfoliatum*

Tinnea Kotschy ex Hook.f. Labiatae (V). 19 trop. Afr. R: BT 70(1975)1. Some fish-poisons

Tinnevelly senna *Senna alexandrina*

Tinomiscium Miers ex Hook.f. & Thomson. Menispermaceae (IV). 1 Indomal. to SE As.: ***T. petiolare*** Miers ex Hook.f. & Thomson – alks (fr. a fish-poison but seeds ed.), medic. e.g. dental caries. R: KB 40(1985)542

Tinopsis Radlk. = *Tina*

Tinospora Miers. Menispermaceae (III). 32 trop. OW. R (not Afr.): KB 36(1981)379. Locally medic., ***T. crispa*** (L.) Hook.f. & Thomson (*T. rumphii,* Indomal.) – diabetes, hypertension, lumbago in Sabah, used to flavour cocktails & cordials; ***T. sinensis*** (Lour.) Merr. (Ind. to SE As.) & other spp. act as host-pls for noctuid moths attacking *Dimocarpus longan* & citrus & are typical of sec. forest so clearance of primary forest leads to their increase & more damage to fr. crops

Tintinabulum Rydb. = *Linanthus*

Tintinnabularia Woodson. Apocynaceae (II d). 3 Mex., Honduras

tipo (oil of) *Minthostachys mollis*

tipu *Tipuana tipu*

Tipuana (Benth.) Benth. Leguminosae (III 11). 1 Braz., Bolivia & Arg.: ***T. tipu*** (Benth.) Kuntze (tipu) – widely planted street-tree in trop. (invasive S Afr.) with samara-like 1–3-seeded winged legumes & useful furniture timber

Tipularia Nutt. Orchidaceae (V 13e). 7 Himal., Jap., E N Am. (1). Cult. orn. terr. orchids with corm & 1 leaf

Tiputinia P. Berry & C. Woodward. Thismiaceae. 1 Ecuador: ***T. foetida*** P. Berry & C. Woodward – known from 1 carrion-scented pl. R: T 56(2007)158

Tiquilia Pers. (~ *Coldenia*). Boraginaceae (Ehr.; Ehretiaceae). 28 Am. deserts. R: Rhodora 79(1977)467

Tirania Pierre. Resedaceae.1 S Vietnam: ***T. purpurea*** Pierre – form. placed in Capparaceae but K6, C6

tirite *Ischnosiphon arouma*

Tirpitzia Hallier f. Linaceae (II). 3 SW China, SE As. R: TFB 34(2006)202

tirucalli *Euphorbia tirucalli*

Tischleria Schwantes = *Carruanthus*

Tisonia Baill. Salicaceae (Flacourtiaceae). 14 Madag. R: Adansonia 10(1970)339

Tisserantia Humbert = *Sphaeranthus*

Tisserantiodoxa Aubrév. & Pellegrin = *Englerophytum*

Tisserantodendron Sillans = *Fernandoa*

tisso flowers *Butea monosperma*

Titanopsis Schwantes (~ *Aloinopsis*). Aizoaceae (V). 3 S Afr. R: H. E. K. Hartmann, *Ill. Handb. Succ. Pls,* Aiz. F–Z(2002)328. Cult. orn. stemless succ. with coloured lvs

Titanotrichum Soler. Gesneriaceae (II). 1 China, Taiwan, S Jap.: ***T. oldhamii*** (Hemsl.) Soler. – s.t. referred to Scophulariaceae s.l., cult. orn. terr. herb with fls s.t. replaced by green scale-like veg. propagules

Tithonia Desf. ex Juss. Compositae (Helia.-Helia.). 11 SW US to C Am. R: Rhodora 84(1982) 453. Cult. orn. herbs & shrubs esp. ***T. diversifolia*** (Hemsl.) A. Gray (Mex. sunflower) – natur. OW where much cult. as hedge-pl. esp. in Kenya, green manure in Sri Lanka, fodder in Colombia

Tithymalus Gaertn. = *Euphorbia*

titoki *Alectryon excelsus*

Tittmannia Brongn. = *Audouinia* (but see Strel. 9(2000)387)

tjintjau *Cyclea barbata*

tlanochtle *Lycianthes moziniana*

Tmesipteridaceae Nakai = Psilotaceae

Tmesipteris Bernh. Psilotaceae (Tmesipteridaceae). c. 15 SE As. to (esp.) Aus. & NZ (both with tetraploids & octoploids) to Tahiti. Terr. or oft. epiphytic & then oft. on tree ferns with dichotomizing rhiz. with rhizoids & mycorrhiza; aerial stem usu. unbranched, basally with scale lvs like *Psilotum* but elsewhere with bilaterally symmetrical (not dorsiventral) broadly lanceolate lvs to 2 cm long; prothalli like sporophytic rhiz.; n = 102–105, 204–210. All Aus. spp. prot.

toad lily *Tricyrtis* spp.; **t. rush** *Juncus bufonius*

toadflax *Linaria* spp.; **bastard t.** *Thesium humifusum*; **common t.** *L. vulgaris*; **ivy-leaved t.** *Cymbalaria muralis*

tobacco *Nicotiana* spp. esp. *N. tabacum*; **Aztec t.** *N. rustica*; **Indian t.** *Lobelia inflata*; **mountain t.** *Arnica montana*; **t. plant** or **flowering t.** *N. alata*, but usu. *N.* × *sanderae*; **rabbit t.** *Pseudognaphalium obtusifolium*; **wild t.** *N. rustica*

Tobagoa Urb. Rubiaceae (IV 15). 1 Panamá, Tobago, Venez.: ***T. maleolens*** Urb.

tobira *Pittosporum tobira*

tobosa grass *Hilaria mutica*

Tocantinia Ravenna. Amaryllidaceae. 1 Braz.: ***T. mira*** Ravenna. R: Onira 5(2000)9

Tococa Aubl. Older name for *Miconia* (but see FN 98(2005)23)

Tocoyena Aubl. Rubiaceae (II 1). 12 trop. Am. R: PSE 181(1992)158. Alks – r. an ipecacuanha subs. ***T. formosa*** (Cham. & Schldl) K. Schum. (Braz.) – poss. hybrid orig.

Todaroa Parl. Umbelliferae (III 8). 1 Canary Is.: ***T. aurea*** (Sol.) Parl.

Toddalia Juss. Rutaceae (I 2). 1 OW trop.: ***T. asiatica*** (L.) Lam. – prickly scrambler, source of Lopez r., medic. (alks) & yellow dye. R: SBU 30,1(1992)68

Toddaliopsis Engl. = *Vepris*

toddy palm wine esp. from *Arenga, Borassus, Caryota, Cocos, Elaeis, Hyphaene, Nypa, Phoenix* & *Raphia* spp. etc.; **t. palm** *Caryota urens*

Todea Willd. ex Bernh. (~ *Osmunda*). Osmundaceae. 1 NG, 1 Aus., NZ, S Afr. (Lower Cretaceous of SW Canada): ***T. barbara*** (L.) T. Moore (crape fern) – cult. orn. with massive erect rhiz. to 3 m tall & fronds to 2.5 m long, pls weighing up to 1.5 t (specimens over 1 t exported from Victoria to Eur. C19, now prot. sp. in Aus.)

Toechima Radlk. Sapindaceae. 8 Aus. (6, 5 endemic; R: FA 25(1985)77), NG, Flores

Toelkenia P.V. Heath = *Crassula*

toffee sweetmeat made from boiling sugar (or molasses/treacle) with butter & milk

Tofieldia Hudson. Tofieldiaceae (Melanthiaceae s.l.). Incl. *Triantha*, 7 or 8 N temp. (Eur. 2, N Am. 3) to subArctic. R: JLSB 53(1947)194. Tufted rhizomatous pls with colchicine; some cult. orn. incl. ***T. pusilla*** (Michaux) Pers. (Scotch asphodel, Alps, Carpathians, Andorra) – 3-lobed involucre beneath K

Tofieldiaceae Takht. (~ Nartheciaceae, q.v.). Magnoliidae – Alismatales. 3/c. 20 N temp. to NW S Am. Small tufted herbs with isobifacial lvs in 2 ranks. Infl. a bracteate raceme (fl. solit. in *Harperocallis*) oft. with 1–3-merous calyculus below fl. P 3 + 3, free, G with sep. styles. Fr. usu. a septicidal capsule

Genera: *Harperocallis, Pleea, Tofieldia*

tofu *Glycine max*

Toiyabea R. Roberts & al. (~ *Haplopappus*). Compositae (Ast.-Sol.). 1 S Nevada: ***T. alpina*** (L. Anderson & Goodrich) R. Roberts & al. R: FNA 20(2006)172

tola wood *Gossweilerodendron balsamiferum*

Tolbonia Kuntze = *Calotis*

Toliara Judz. = *Perotis* (but see Adans. 31(2009)274)

tollon *Heteromeles arbutifolia*

Tolmiea Torrey & A. Gray. Saxifragaceae (I 5). 1(or 2) W N Am.: ***T. menziesii*** (Pursh) Torrey & A. Gray (pickaback or piggyback pl.) – poll. fungus-gnats, cult. orn. herb with plantlets developing at junction of petiole & lamina. R: FNA 9(2009)107

Tolpis Adans. Compositae (Cich.-Cich.(Cich.)). 15 Macaronesia (most), Medit. (Eur. 5) to Ethiopia, Somalia & S Afr. with ***T. barbata*** (L.) Gaertn. (Medit.) – widely natur. Primitively woody, herbaceous taxa re-colonizing Afr. from Macaronesia

tolu balsam *Myroxylon balsamum*

Tolumnia Raf. (~ *Oncidium*). Orchidaceae (V 12h). 27 trop. Am., 1 ext. to Florida. Fan-shaped

Tolypanthus (Blume) Reichb. Loranthaceae. 1 SW Sri Lanka, c. 30 SE As.

Tomanthea DC. = *Centaurea*

Tomanthera Raf. (~ *Agalinis*). Orobanchaceae (Gerard.). 2 C N Am. R: PANSP 80(1928)439

tomatillo *Physalis philadelphica*

tomato *Solanum lycopersicum*; **bush t.** *S. centrale*; **cannibal's t.** *S. viride* 'Anthropophagorum'; **cherry t.** *S. l.* 'var. *cerasiforme*'; **currant t.** *S. pimpinellifolium*; **husk** or **strawberry t.** *Physalis pruinosa*; **peach t.** *S. topira*

tomberry *Solanum pimpinellifolium*

Tomentaurum Nesom (~ *Heterotheca*). Compositae (Ast.-Chr.). 1 NW Mex.: ***T. niveum*** (S. Watson) Nesom (*T. vandevenderorum*). R: SBM 20(2000)87

Tommasinia Bertol. = *Peucedanum*

Tomophyllum (E. Fourn.) Parris (~ *Ctenopteris*). Polypodiaceae (V). 22 Ind. to Melanesia. R: GBS 58(2007)245

Tomostima Raf. (~ *Draba*). Cruciferae (7). 6 N Am.

Tomzanonia Nir (~ *Dilomilis*). Orchidaceae (V 13c). 1 Haiti: ***T. filicina*** (Dod) Nir. R: Lindleyana 12(1997)186

Tonalanthus Brandegee = *Calea*

Tonduzia Pittier (~ *Alstonia*). Apocynaceae (Ia). 2 Mex., C Am.

Tonella Nutt. ex A. Gray. Plantaginaceae (Ant.-Chel.; Scrophulariaceae s.l.). 2 W N Am. Cult. orn. annuals like *Collinsia* but C not strongly 2-lipped

Tonestus A. Nelson (~ *Haplopappus*). Compositae (Ast.). 4 W N Am. R: FNA 20 (2006)181. Some cult. orn.

tong ho *Glebionis coronaria*

Tongan oil *Calophyllum inophyllum*

Tongking bamboo, T. cane *Pseudosasa amabilis*

tongkut ali *Eurycoma longifolia*

Tongoloa H. Wolff. Umbelliferae (III 8). 10 – 15 C As. to Ind.

tongue, mother-in-law's *Dracaena trifasciata*

Tonina Aubl. = *Paepalanthus*

Tonka bean *Dipteryx odorata*

Tonningia Necker ex A. Juss. = *Cyanotis*

Tontelea Aubl. = *Elachyptera*

Tontelea Miers. Celastraceae (IV). 31 trop. Am. R: Britt. 3(1940)463

toon *Toona ciliata*

Toona (Endl.) M. Roemer. Meliaceae (I). 4–5 Indomal. to N Aus. (Mal. 3 – R: FM I,12(1995)358). Decid. timber trees esp. ***T. ciliata*** M. Roemer (*T. australis*, *Cedrela toona*, toon, Aus. (or) red cedar, Burma or Moulmein c., Indomal. to Aus., invasive S Afr., Hawaii) – furniture & building, form. principal timber in Aus., by 1859 2557 t. a yr exported to UK alone; ***T. sinensis*** (A. Juss.) M. Roemer (*Cedrela s.*, China to W Mal. mts) – similar, coffee-shade, street-tree in Eur., shoots ed. China; ***T. sureni*** (Blume) Merr. (Mal.) – similar timber, bark medic.

toothache grass *Ctenium aromaticum*

toothbrush tree *Salvadora persica*

toothwort *Lathraea squamaria*

Toowoomba canary grass *Phalaris aquatica*

topee-tampo, topinambour, topi-tamboo *Goeppertia allouia*

Topobea Aubl. = *Blakea*

toquilla *Carludovica palmata*

tor grass *Brachypodium pinnatum*

toraji *Platycodon grandiflorus*

torch ginger *Etlingera elatior*; **t. lily** *Kniphofia* spp.; **t. wood** *Amyris elemifera*

Tordyliopsis Wall. ex DC. Umbelliferae (III 11). 1 Himal., Tibet: ***T. brunonis*** Wall. ex DC. R: Willd. 30(2000)364

Tordylium Tourn. ex L. Umbelliferae (III 11). Incl. *Ainsworthia*, 18 Eur. (5), Medit., SW As. R: BJLS 97(1988)357. ***T. apulum*** L. (Medit.) – lvs veg. in Greece

Torenia L. Linderniaceae (Plantaginaceae (Grat.-Lind.); Scrophulariaceae s.l.). Excl. *Legazpia*, 51 trop. OW (Afr. 3), ***T. thouarsii*** (Cham. & Schldl.) Kuntze (OW trop.) natur. C & S Am. R (Indochina – 19): JFSUTB III,13(1985)603; Willd. 43(2013)230. Cult. orn. esp. ***T. fournieri*** Linden ex Fourn. (wishbone fl., SE As.) – pot-pl. with pale blue, purple & yellow fls, anthers shed pollen by lever action of flange-like outgrowths of lat. pollen-sac wall – when pressed causes buckling of wall nearby & pollen forced out (1–1.5 g force against 4 levers of anther pr. releases <3000 grains)

Toricellia DC. = *Torricellia*

Torilis Adans. Umbelliferae (III 3). 15 Canary Is., Medit. (Eur. 6) to E As. trop. & S Afr. ***T. japonica*** (Houtt.) DC. (hedge parsley, Euras.)

tormentil *Potentilla erecta*

Torminalis Medik. = *Sorbus*

Tornabenea Parl. = *Daucus*

Toronia L. Johnson & B. Briggs (~ *Persoonia*). Proteaceae (II 2). 1 NZ: ***T. toru*** (A. Cunn.) L. Johnson & B. Briggs – some 'males' s.t. prod. fr. R: BotJLS 70(1975)174

Torralbasia Krug & Urb. (~ *Euonymus*). Celastraceae (I). 1 WI: ***T. cuneifolia*** (Wright) Krug & Urb.

Torrenticola Domin = *Cladopus*

Torresea Allemão = *Amburana*

Torreya Arn. Taxaceae. 8 E As., S US (2 – stinking yew). R: BR 64(1998)310. Dioec. & monoec. trees with 'simple' strobili derived from compound as in *Cephalotaxus* & drupe-like seeds covered in fleshy arils; fossils from mid-Jurassic referred here. ***T. californica*** Torrey (Calif. nutmeg, Calif.) – local medic. (tuberculosis etc.), cult. orn.; ***T. nucifera*** (L.) Sieb. & Zucc. (kaya nut, Jap.) – seeds ed., oil used for cooking in Jap.; ***T. taxifolia*** Arn. (stinking cedar or yew, 3 counties of Florida, 1 site in Georgia) – bruised foliage foetid, timber for fencing, endangered sp. due to fungal attack in 1950s, so by 1993 no sexually mature trees left

Torreyaceae Nakai = Taxaceae

Torreycactus Doweld = *Thelocactus*

Torreyochloa Church (~ *Puccinellia*). Gramineae (XVI 1). 4 N Am., NE As. R: KBAS 13(1986)99

Torricellia DC. Torricelliaceae (Cornaceae s.l.). 3 E Himal. to W China

Torricelliaceae Hu (~ Cornaceae). Magnoliidae – Apiales. 3/ 11 Madag., Himal. to China, W Mal. Trees. Lvs toothed or lobed. K imbr. or with tube; G (2–4). n = 12, 20+
Genera: *Aralidium, Melanophylla, Torricellia*

Torrubia Vell. = *Guapira*

Tortuella Urb. Rubiaceae (IV 15). 1 Isle Tortue (nr Hispaniola): ***T. abietifolia*** Urb. & Ekman

Torularia O. Schulz = *Neotorularia*

Torulinium Desv. = *Cyperus*

Tostimontia S. Díaz = *Jungia*

totara *Podocarpus totara*

totora *Schoenoplectus californicus*

totorilla *Juncus balticus* subsp. *andicola*

Toubaouate Aubrév. & Pellegrin = *Didelotia*

Toubasuate Airy Shaw = *Didelotia*

Touchardia Gaudich. Urticaceae (III). 1 Hawaii: ***T. latifolia*** Gaudich. (olona) – fibre used for fishing-lines & nets

touch-me-not *Impatiens* spp.

tou-fou *Glycine max*

Toulicia Aubl. Sapindaceae. 12 N S Am. Leeuwenberg's Model

Tournefortia L. = *Heliotropium*

Tournefortiopsis Rusby = *Guettarda*

tournesol *Chrozophora tinctoria*

Tourneuxia Cosson. Compositae (Lact.-Scor.). 1 Algeria: ***T. variifolia*** Cosson

Tournonia Moq. Basellaceae. 1 Colombia: ***T. hookeriana*** Moq. R: KB 62(2007)307

Touroulia Aubl. Ochnaceae (Quiinaceae). 2 trop. S Am.

Tourrettia Foug. Bignoniaceae (5). 1 Andes to Mex.: ***T. lappacea*** (L'Hérit.) Willd. – ann. cli. with some char. like Pedaliaceae. R: FN 25(1980)110

tous-les-mois *Canna indica*

Toussaintia Boutique. Annonaceae (III 7). 3 trop. Afr.

Tovaria Ruíz & Pavón. Tovariaceae. 1–2 trop. Am.

Tovariaceae Pax. Magnoliidae – Brassicales. 1/2 trop. Am. Foetid herbs & shrubs with mustard-oil glucosides. Lvs 3-foliolate, in spirals; stip. minute. Fls bisex., reg., (6–)8(9)-merous in long term. racemes; K imbr., decid., C shortly clawed, A usu. 8, within lobed nectary-disk, anthers sagittate with longit. slits, $\underline{G}$ ((5)6(–8)), pluriloc., ± gynophore, style short, with ∞ ± campylotropous bitegmic ovules with zig-zag micropyles on expanded placenta. Fr. a berry with C placental mass separating from pericarp & ∞

small seeds with embryo curved around periphery in thin layer of oily endosperm. n = 14

Genus: *Tovaria*

Close to Capparaceae but endosperm, pluriloc. ovary & polymery different

Tovarochloa T. Macfarl. & But = *Poa* (but see Britt. 34(1982)478)

Tovomita Aubl. Guttiferae (1.). c. 25 trop. Am. Fagerlind's Model. Fl. tea for treatment of diarrhoea

Tovomitidium Ducke = *Tovomita*

Tovomitopsis Planch. & Triana (~ *Chrysochlamys*). Guttiferae (1.). c. 3 S Am.

tow waste fibre after preparation by scutching of flax, hemp, jute etc. for spinning

towai bark *Weinmannia racemosa*

towel gourd *Luffa aegyptiaca*

tower mustard *Turritis glabra*

Townsendia Hook. Compositae (Ast.-Astr.). 26 W N Am. (US & Canada 26 – R: FNA 20(2006)193), 2 ext to Mex. R: CGH 183(1957)1. Some cult. orn. herbs

Townsonia Cheeseman (~ *Acianthus*). Orchidaceae (IV 3a). 2 Tasmania, NZ

Townsville stylo *Stylosanthes humilis*

Toxanthes Turcz. (~ *Millotia*). Compositae (Gnap.). 3 Aus. R: AusSB 8(1995)10

Toxicodendron Mill. (~ *Rhus*). Anacardiaceae (I). 22 Ind. to E As. & NG, Canada to Bolivia. ***T. radicans*** (L.) Kuntze (*R. r.*, *R. toxicodendron*, poison ivy, p. oak, C China to Jap., S Canada to Guatemala) – liane, shrub or tree causing dermatitis due to 3-*n* pentadecylcatechol in over 350 K cases a yr in US alone; ***T. succedaneum*** (L.) Kuntze (*R. s.*, Jap. tallow or wax tree, E As.) – cult. for wax from fr. used as subs. for beeswax in polishes etc., stem exudes a natural lacquer; ***T. vernicifluum*** (Stokes) F.A. Barkley (*R. v.*, *R. vernicifera*, (Chinese or Jap.) lacquer tree, temp. E As.) – used for lacquerware in Japan for 7000 yrs, cult. SW Jap., where a major source (lactone causing allergy), obtained by cutting bark, also wax source for candles

Toxicopueraria Egan & B. Pan (~ *Pueraria*). Leguminosae (III 18). 2 SE As. R: Phytotaxa 218(2015)214

Toxicoscordion Rydb. (~ *Zigadenus*). Melanthiaceae. 8 W & C N Am. R: Novon 12(2002)304. Many poisonous (alks) to stock (esp. W US); few cult. orn. incl. ***T. nuttallii*** (A. Gray) Rydb. (death c.)

Toxocarpus Wight & Arn. = *Secamone*

Toxosiphon Baill. (~ *Erythrochiton*). Rutaceae (I 8). 4 trop. Am. R: Britt. 44(1992)117

toyon *Heteromeles arbutifolia*

Tozzia Micheli ex L. Orobanchaceae (Rhin.; Scrophulariaceae s.l.). 1 Alps, Carpathians, Pyrenees: ***T. alpina*** L. – hemiparasite, fly-poll.

trac *Dalbergia cochinchinensis*

Trachelanthus Kunze = *Cynoglossum*

Tracheliopsis Buser = *Campanula*

Trachelium Tourn. ex L. Campanulaceae (I 3). 2 Macar., W Medit. R: BHB 2(1894)511. Cult. orn. esp. cut-fls

Trachelospermum Lem. Apocynaceae (II c). 10 Ind. to Jap. (R: Sunyatsenia 3(1936)65), SE US (1). Cult. orn. with fragrant fls esp. scrambling ***T. jasminoides*** (Lindl.) Lem. (star jasmine, China)

Trachoma Garay (~ *Tuberolabium*). Orchidaceae (V 16c). 14 Indomal.

Trachomitum Woodson = *Apocynum* (but see AMBG 17(1930)156)

Trachyandra Kunth (~ *Anthericum*). Asphodelaceae. c. 50 trop. & S Afr. (49 esp. W Cape), Madag. (1). R: Bothalia 7(1962)669. Some tumbleweeds. ***T. adamsonii*** (Compton) Oberm. (S Afr.) – woody stem to 180 cm tall; ***T. divaricata*** (Jacq.) Kunth (S Afr.) – camouflaged by sand-grains adhering to leaf secretions

Trachycalymma (K. Schum.) Bullock = *Asclepias* (but see KB 56(2001)132)

Trachycarpus H. Wendl. Palmae (III 4a). 9 Himal. to China, N Thailand. R: Principes 21(1977)155. Usu. dioec. palms, the trunks oft. covered with marcescent lvs. Cult. orn. esp. ***T. fortunei*** (Hook.) H. Wendl. ('*T. excelsus*', Chusan or windmill palm, N Myanmar, C & E China) – hardy in GB, fibre (Chinese coir) of leaf-bases used for cordage, brooms & capes (NW Yunnan), lvs used for hats, fls for food, drugs from seeds, wax from fr.

Trachydium Lindl. Umbelliferae (III 5). Incl. *Pseudotrachydium*, 6 SW As., Himal. R: FR 111(2000)522,526

Trachylobium Hayne = *Hymenaea*

Trachymene Rudge (*Didiscus*). Araliaceae (Hydroc.; Umbelliferae s.l.). 56 SE As. to Aus. (39, 38 endemic – R: SB 19(2006)18), New Caled. & Fiji. Cult. orn. esp. ***T. coerulea*** Graham (blue lace fls, W Aus.)

Trachynia Link = *Brachypodium*

Trachyphrynium Benth. Older name for *Hypselodelphys*

Trachypogon Nees. Gramineae (XXII 7). 4 trop. Am., Afr. ***T. plumosus*** (Humb. & Bonpl.) Nees – used for thatch in pre-Columbian San Salvador

Trachypteris André ex Christ. Pteridaceae (IV). 3 S Am., Galápagos, Madag. (***T. drakeana*** Christ – sterile lvs with apical buds (veg. reproduction))

Trachyrhizum (Schltr.) Brieger = *Dendrobium*

Trachys Pers. Gramineae (XXIV 1). 2 Tanzania, S Ind. & Sri Lanka to Myanmar, coastal

Trachyspermum Link. Umbelliferae (III 8). 20 trop. & NE Afr. to C As., Ind. & W China. ***T. ammi*** (L.) Sprague (*T. copticum*, ajowan, ? Egypt, Ethiopia) – fr. medic. & spice (carom seeds) cont. thymol, the principal flavour of 'Bombay Mix' of nuts, pulses & crisp sticks; ***T. roxburghianum*** (DC.) H. Wolff (ajmud, gandini, somura Indomal.) – fr. used in pickles

Trachystemon D. Don. Boraginaceae (B.2.1.1). 2 Medit. (Eur. 1). ***T. orientalis*** (L.) G. Don (E Bulgaria & As. Minor) – cult. orn. perenn. with bold foliage

Trachystigma C.B. Clarke = *Streptocarpus*

Trachystoma O. Schulz (~ *Sinapis*). Cruciferae (12). 3 Morocco

Trachystylis S.T. Blake. Cyperaceae (II 7). 1 E Aus.: ***T. stadbrokensis*** (Domin) Kükenthal. R: BJ 75(1952)493

Tractema Raf. = *Scilla*

Tractocopevodia Raiz. & Naray = *Melicope*

Tracyina S.F. Blake. Compositae (Ast.-Pent.). 1 Calif.: ***T. rostrata*** S.F. Blake. R: SBM 20(2000)89

Tradescantia L. Commelinaceae (II 1g). 70 Am. R: KB 35(1980)437 (incl. *Cymbispatha*, *Rhoeo*, *Setcreasea* & *Zebrina*); erect or trailing, rather pachycaul to v. slender cult. orn. esp. hardy ***T. virginiana*** L. (spiderwort, E N Am.) – erect with violet-purple fls though most garden material is referable to '*T.* × *andersoniana* W. Ludwig & Rohw.' (invalid name, *T. v.* × *T. ohiensis* Raf. (blue jacket, E N Am.) × *T. subaspera* Ker-Gawler (E N Am.)) – many cvs; long staminal hairs, in which Robert Brown first observed & described protoplasmic streaming (1828), eaten by poll. insects. Many greenhouse or house pls ('wandering Jews'; see also *Gibasis*) of trailing habit incl. ***T. fluminensis*** Vell. Conc. (*T. albiflora*, S Am., aggressive weed S Eur., Aus. (no seed), NZ, SE US) – fls white, & ***T. zebrina*** Heynh. ex Bosse (*Zebrina pendula*, *T. p.*, Mex.) – fls red-purple, these some of the commonest of all housepls. Other cult. spp. incl. ***T. cerinthoides*** Kunth (*T. blossfeldiana*, SE Braz.) – trailing & ascending purplish stems, densely white-villous; ***T. pallida*** (Rose) D. Hunt (E Mex.) – erect or sprawling esp. '**Purple Heart**' ('Purpurea', *Setcreasea purpurea*) with intense violet-purple lvs for bedding; ***T. sillamontana*** Matuda (Mex.) – house-pl., bedding; ***T. spathacea*** Sw. (*T. discolor*, *Rhoeo d.*, *R. s.*, boat lily, C Am., WI, invasive SE US) – pachycaul usu. unbranched with lvs to 30 cm. See also *Callisia*

tragacanth *Astragalus* spp. esp. *A. gummifer*

Traganopsis Maire & Wilczek. Amaranthaceae (Chenopodiaceae III 3). 1 Morocco: ***T. glomerata*** Maire & Wilczek

Traganum Del. Amaranthaceae (Chenopodiaceae III 3). 2 N Afr., E Medit.

tragasol *Ceratonia siliqua*

Tragia Plum. ex L. Euphorbiaceae (Acal.-Pluk.-Trag.). Excl. *Bia*, *Ctenomeria*, *Zuckertia*, c. 170 trop. & warm. Usu. stinging

Tragiella Pax & K. Hoffm. (~ *Sphaerostylis*). Euphorbiaceae (Acal.-Pluk.-Trag.). 5 NE trop. to S Afr.

Tragiola Small & Pennell = *Gratiola*

Tragiopsis Pomel = *Stoibrax*

Tragoceros Kunth = *Zinnia*

Tragopogon Vaill. ex L. Compositae (Lact.-Scor.). c. 110 temp. Euras. (Eur. 20), Medit. Taprooted herbs with monocot-like lvs & solit. capitula opening only in morning, so weedy ***T. pratensis*** L. (goat's-beard, Eur., natur. N Am.) known as Johnny-go-to-bed (-at-noon), r. form. ed.; ***T. porrifolius*** L. (salsify, veg. oyster, S Eur.) – cult. for ed. r. & for young fl. shoots ('chards'), parent with *T. dubius* Scop. (Eur.) of ***T. mirus*** Ownbey, an allopolyploid (like 6 OW spp.) with multiple origins (cf. *Senecio cambrensis*, *Spartina anglica*) in US, as is ***T. miscellus*** Ownbey (*T. d.* × *T. pratensis*)

Tragus Haller. Gramineae (XXIX 1). 8 OW trop., esp. Afr., 1 ext. to Eur. R: KB 36(1981)55. ***T. racemosus*** (L.) All. (S Eur., widely natur.) – colonized much of Aus. (fr. carried on sheep fleece)

Trailliaedoxa W.W. Sm. & Forrest. Rubiaceae (Vang.). 1 SW China: ***T. gracilis*** W.W. Sm. & Forrest. R: FOC 19(2011)347

Transberingia Al-Shehbaz & O'Kane (*Beringia*). Cruciferae (22). 1 E Russia, W N Am., Greenland: ***T. bursifolia*** (DC.) Al-Shehbaz & O'Kane. R: Novon 13(2003)396

Transcaucasia Hiroe = *Astrantia*

Transvaal daisy *Gerbera jamesonii*

Trapa L. Lythraceae (Trapaceae). 1 polymorphic warm. OW: ***T. natans*** L. (incl. *T. bicornis*, *T. bispinosa*, water chestnut, w. caltrops, saligot, horn or Jesuit's or ling or singhara nut, weedy as in Caspian where a threat to sturgeon feeding-grounds, invasive E N Am.) – ann. usu. floating aquatic, vasc. bundles with internal phloem, submerged stem with ± opp. elongate dissected green organs ('lvs', 'stipules', 'photosynthetic roots') & adventitious r., lvs with elongate petiole with aerenchymatous float & rhombic blade on short aerial stem. Indehiscent fr. falls to bottom of pond (seeds viable for up to 12 yrs), seed rich in starch & fat, staple in As., flour mixed with honey & sugar a pastry in Indochina, used in GB for food in Neolithic, trad. medic. incl. sunstroke in China

Trapaceae Dumort. = Lythraceae

Trapella Oliv. Plantaginaceae (Grat.-Grat.; Trapellaceae). 1–2 E As. R: BJ 71(1940)267. Form. referred to Pedaliaceae. Aquatics with oblong submerged lvs, deltoid-round floating ones, & submerged cleistogamous fls (aerial ones normal); inf. G, fr. with 3–5 long curved appendages & 1 seed

Trapellaceae Honda & Sakis. = Plantaginaceae

Trattinnickia Willd. Burseraceae (V). 14 N S Am. R: KB 54(1999)131. Resins coll. euglossine bees for their nests

Traubia Mold. Amaryllidaceae. 1 Chile: ***T. modesta*** Mold.

Traunia K. Schum. = *Marsdenia*

Traunsteinera Reichb. Orchidaceae (IV 4d). 2 Eur., Medit. R: NSL 44(2013)61. ***T. globosa*** (L.) Reichb. – poss. poll. mimic of *Scabiosa columbaria*

Trautvetteria Fischer & C. Meyer. Ranunculaceae (II 3). 1 E As., W & S N Am.: ***T. caroliniensis*** (Walter) Vail. R: FNA 3(1997)138

traveller's joy *Clematis vitalba*; **t. palm** or **tree** *Ravenala madagascariensis*

Traversia Hook.f. = *Brachyglottis*

trazel *Corylus* × *colurnoides*

treacle = molasses; **t. mustard** *Erysimum cheiranthoides*

Trechonaetes Miers = *Jaborosa*

Treculia Decne. ex Trécul. Moraceae (II). 3 trop. Afr. & Madag. R: BJBB 47(1977)378. Troll's Model. ***T. africana*** Decne. ex Trécul (Afr. breadfruit) – biggest Afr. fr. (to 25 cm diam.), elephant-disp. seeds ed., ground into flour

tree cotton *Gossypium arboreum*; **t. dahlia** *Dahlia excelsa*; **t. daisy** *Montanoa* spp., *Olearia* spp.; **t.-fern** Cyatheaceae, Dicksoniaceae etc.; **t. lucerne** *Cytisus proliferus*; **t. mallow** *Malva dendromorpha*; **t. medick** *Medicago arborea*; **t. melon** *Solanum muricatum*; **t. of heaven** *Ailanthus altissima*; **t. onion** *Allium* × *proliferum*; **t. peony** *Paeonia suffruticosa*; **t. spinach** *Hibiscus manihot*; **t. tomato** *Solanum betaceum*

trefoil, bird's-foot *Lotus corniculatus*; **hop t.** *Trifolium campestre*; **yellow t.** *Medicago lupulina*

Treichelia Vatke. Campanulaceae (I 2). 2 SW Cape. R: KB 66(2011)618

Trema Lour. Cannabaceae (Ulmaceae s.l.). Incl. *Paraponia*, 15–20 trop. & warm. Fast-growing pioneer trees (Roux's Model) with alks, incl. ***T. orientalis*** (L.) Blume (*T. guineensis*, OW) – lvs (mpesi) 8% dry wt. tannin so avoided by colobus monkeys but used for tanning fish-nets in W Afr. & elsewhere, wood for charcoal & fireworks, & ***T. micrantha*** (L.) Blume (guacimilla, Florida trema, trop. Am.) – growing in forest gaps at least 376 m^2 & reaching 13.5 m in 2 yrs, pre-Hispanic barkcloth now for tourism & export (San Pablito, Mex.), with soft timbers for tea-chests & matches, tea & coffee-shade & poss. in soil conservation (as in S Afr.)

Tremacanthus S. Moore. Acanthaceae. 1 Braz.: ***T. roberti*** S. Moore

Tremacron Craib = *Oreocharis* (but see FOC 18(1998)261)

Tremandra R. Br. ex DC. Elaeocarpaceae (Tremandraceae). 2 SW Aus.

Tremandraceae R. Br. ex DC. = Elaeocarpaceae

Tremastelma Raf. = *Lomelosia*
Trematocarpus A. Zahlbr. = *Trematolobelia*
Trematolobelia A. Zahlbr. ex Rock (~ *Lobelia*). Campanulaceae (III 4). 8 Hawaii. R: Britt. 61(2009)130. Prob. derived from *L.* subg. *Tupa* of As.
Trembleya DC. Melastomataceae. 23 S Braz.
Tremulina Briggs & L. Johnson (~ *Restio*). Restionaceae. 2 SW W Aus. R: Telopea 7(1998)361, 9(2001)256
Trepadonia H. Robinson (~ *Vernonia*). Compositae (Vern.-Vern.). 2 Peru. R: Sida 19(2000)112. Scandent
Trepocarpus Nutt. ex DC. Umbelliferae (III 8). 1 S US: ***T. aethusae*** Nutt. ex DC.
Tresanthera Karsten = *Rustia*
Treutlera Hook.f. Apocynaceae (V; Asclepiadaceae III 4). 1 E Himal.: ***T. insignis*** Hook.f.
Trevesia Vis. Araliaceae. Incl. *Brassaiopsis* c. 40 Indomal. R (*T.* s.s. – 7): Glasra 3(1998)90. Local medic., young infls a potherb; cult. orn. shrubs with palmate leaflets attached to a lamina
Trevia L. = *Mallotus*
Trevoa Miers. Rhamnaceae (2). 1 Andes: ***T. quinquenervia*** Gillies & Hook. – spiny tree with actinobacterial r.-symbionts. R: Darw. 31(1992)230. See also *Retanilla*
Trevoria F. Lehm. Orchidaceae (V 12i). 5 Colombia, Ecuador. R: Orquideologia 5(1970)3
Trewia L. = *Mallotus*
Triadenum Raf. Hypericaceae (II; Guttiferae s.l.). 6 E As., E N Am.
Triadica Lour. (~ *Sapium*). Euphorbiaceae (Hipp.). 3 Indomal. R: HPB 7(2002)17. ***T. sebifera*** (L.) Small (*S. s.*, *Stillingia s.*, Chinese or veg. tallow-tree, China, Jap.) – frost- & shade-tolerant 'pioneer' of poplar habit invasive US forests Texas to Florida, black dye from lvs, fatty seed-covering (? caruncle, 54.5% tallow) used for candlewax & soap
Triadodaphne Kosterm. (~ *Endiandra*). Lauraceae (I). 3 Mal.
Triaenacanthus Nees = *Strobilanthes*
Triaenanthus Nees = *Strobilanthes*
Triaenophora (Hook.f.) Soler. Orobanchaceae. 3 NE As.
Triainolepis Hook.f. Rubiaceae (IV 8). 2 trop. E Afr., Madag.
Trianaea Planch. & Linden. Solanaceae (IV). c. 6 trop. Am. ***T. speciosa*** (Drake) Soler. (Ecuador) – bat-poll.
Trianaeopiper Trel. = *Piper*
triangle palm *Dypsis decaryi*
Trianoptiles Fenzl. Cyperaceae (II 7). 3 SW Cape. R: BN 130(1977)235
Triantha (Nutt.) Bak. = *Tofieldia*
Trianthema Sauvag. ex L. Aizoaceae (II). 28 warm esp. Aus. (12, 10 endemic). R: H. E. K. Hartmann, *Ill. Handb. Succ. Pls*, Aiz. F–Z(2002)330, (subg. *Papularia*; 17 – PEE 144(2011)194). Alks – some medic., soap, lvs used like spinach
Triaristella Brieger = *Trisetella*
Triaristellina Rauschert = *Trisetella*
Trias Lindl. (~ *Bulbophyllum*). Orchidaceae (V 15). 14 Indomal. R: BT 71(1976)19
Triaspis Burchell. Malpighiaceae. 12 trop. & S (2) Afr.
Tribelaceae Airy Shaw = Escalloniaceae
Tribeles Philippi. Escalloniaceae (Grossulariaceae s.l.). 1 temp. S Am.: ***T. australis*** Philippi – prostrate shrublet with contorted C
Triblemma (J. Sm.) Ching = *Diplazium*
Tribolium Desv. Gramineae (Danth.). Incl. *Karroochloa*, *Lasiochloa*, 16 S Afr. bushland. R: BJ 119(1997)468
Tribonanthes Endl. Haemodoraceae (II). 5 SW Aus. R: FA 45(1987)132
Tribounia Middleton (~ *Didymocarpus*). Gesneriaceae (III 2h). 2 Thailand, limestone. R: T 61(2012)1287
Tribroma Cook = *Theobroma*
Tribulago Luer = *Specklinia*
Tribulocarpus S. Moore. Aizoaceae (II; Tetragoniaceae). 2 SW & NE Afr. R: T 61(2012)61
Tribulopis R. Br. (~ *Tribulus*). Zygophyllaceae (IV). 10 trop. Aus. R: FA 26(2013)516. Geocarpy (unique in fam.)
Tribulus Tourn. ex L. Zygophyllaceae (IV). 25 trop. & warm (Eur. 1), esp. dry Afr. R: Taeckholmia 9(1978)59. C4 photosynthesis. Fr. with G 3–5 with sharp spines separating when mature & disp. on animals & feet (burnut). Weeds esp. ***T. terrestris*** L. (caltrops, devil's thorn, OW (orig. Medit.?), widely natur.) – whole life-cycle in 6 wks, r. reaching water

at 2.5 m, bad weed in Calif., where, after 50 yrs of herbicides, Calif. Dept. of Agriculture intr. 2 weevil spp. from Ind. (1961) & these fed selectively on *T. t.*, controlling it; ingestion in stock leads to hepatogenic photosensitization perhaps involving nitrate & selenium poisoning manifest as 'bighead', though imp. local medic. (steroidal saponins acting on testosterone levels affecting libido & athletes' stamina), cult. Bulgaria & US & sold as 'anabolic nutrients' for body-building

Tricalistra Ridl. = *Tupistra*

Tricalysia A. Rich. ex DC. Rubiaceae (II 3). Excl. *Empogona*, 73 trop. Afr. (most), Madag. R: BJBB 49(1979)239, 52(1982)311, 53(1983)299, 57(1987)39 (As. spp. = *Diplospora*, *Discospermum*). Roux's Model

Tricardia Torrey. Boraginaceae (Hydrophyllaceae-Rom.). 1 SW US: ***T. watsonii*** Torrey

Tricarpelema J.K. Morton. Commelinaceae (II 2). 8 As., Afr. R: Novon 17(2007)166

Tricarpha Longpre = *Sabazia*

Triceratella Brenan. Commelinaceae (II 1). 1 Zimbabwe: ***T. drummondii*** Brenan. R: Bothalia 31(2001)37

Triceratorhynchus Summerh. Orchidaceae (V 16d). 1 trop. E Afr.: ***T. viridiflorus*** Summerh.

Triceratostris (Szlach.) Szlach. & Tamayo = *Deiregyne*

Tricerma Liebm. = *Maytenus* (but see Wrightia 4(1971)158)

Trichacanthus Zoll. = *Blepharis*

Trichachne Nees = *Digitaria*

Trichadenia Thw. Achariaceae (Flacourtiaceae). 1 Sri Lanka, 1 E Mal.

Trichantha Hook. = *Columnea*

Trichanthecium Zuloaga & Morrone (~ *Panicum*). Gramineae (XXIV). 38 trop. Afr., Am. R: SBM 94(2011)

Trichanthemis Regel & Schmalh. Compositae (Anth.-Han.). 9 C As. R: BBMNHB 23(1993)97

Trichanthera Kunth. Acanthaceae (III 2c). 2 N S Am. ***T. gigantea*** (Humb. & Bonpl.) Nees – fodder-tree in Colombia

Trichanthodium Sonder & F. Muell. (~ *Gnephosis*). Compositae (Gnap.-Ang.). 4 Aus. R: Muelleria 7(1990)213

Trichapium Gilli = *Clibadium*

Trichaulax Vollesen. Acanthaceae (III 2c). 1 E Afr. coast: ***T. mwasumbii*** Vollesen. R: KB 47(1992)613

Trichilia P. Browne. Meliaceae (II-Trich.). 107 trop. (Am. 81 (R: Phytotaxa 259(2016)18), Afr. 20, Madag. 6). For As. spp. see *Heynea*. Timber, poss. comm. wood-oil, & oilseeds esp. ***T. emetica*** Vahl (Afr.) – timber (Cape or Natal mahogany) & seed-oil (mafura or mafoureira tallow) used for candle- & soap-making; ***T. moschata*** Sw. (trop. Am.) – source of pameroon bark

Trichipteris C. Presl = *Cyathea*

Trichlora Bak. Amaryllidaceae (Alliaceae-Gill.). 2 Peru

Trichloris Fourn. ex Benth. (~ *Chloris*). Gramineae (XXIX 5). 2 both disjunct S N Am. & Arg. ***T. crinita*** (Lag.) L. Parodi – cult. orn.

Trichocalyx Balf.f. Acanthaceae (III 2c). 2 Socotra

Trichocaulon N.E. Br. = *Hoodia*. See also *Lavrania*

Trichocentrum Poepping & Endl. (~ *Oncidium*). Orchidaceae (V 12h). c. 70 trop. Am. Some poll. oil-bees

Trichocephalus Brongn. (~ *Phylica*). Rhamnaceae (3). 1 W Cape: ***T. stipularis*** (L.) Brongn. R: Strel. 9(2000)607

Trichocereus (A. Berger) Riccob. = *Echinopsis*

Trichoceros Kunth. Orchidaceae (V 12h). 9 N S Am. ***T. antennifer*** Kunth (Ecuador) – pseudocopulation by *Paragymnomma* flies

Trichochiton Komarov = *Cryptospora*

Trichocladus Pers. Hamamelidaceae (I 1). 5–6 trop. & S Afr.

Trichocline Cass. Compositae (Mut.-Mut.). 21 S Am., SW Aus. (1). R: Darwiniana 19(1975)618. Herbs; some used like tobacco, esp. ***T. reptans*** (Wedd.) Robinson (coro)

Trichocoronis A. Gray. Compositae (Eup.-Tric.). Excl. *Shinnersia*, 2 SW N Am. R: MSBMBG 22(1987)188, 190. Wholly aquatic

Trichocoryne S.F. Blake. Compositae (Helia.-Zinn.). 1 NW Mex.: ***T. connata*** S.F. Blake

Trichocyamos Yakovlev. See *Ormosia*

Trichodesma R. Br. Boraginaceae (B.3.1). c. 45 trop. & warm OW. Alks. ***T. africanum*** (L.) Sm. (trop. Afr.) – local medic; ***T. indicum*** (L.) Sm. (Iran to Ind.) – Ayurvedic medic;

T. zeylanicum (Burm.f.) R. Br. (camel bush, trop. As. to Aus.) – reputedly favoured by introd. grazing camels in Aus., poss. comm. oilseed

Trichodiadema Schwantes. Aizoaceae (V). 32 Ethiopia, S Afr. R: H. E. K. Hartmann, *Ill. Handb. Succ. Pls*, Aiz. F–Z(2002)338. Some cult. orn. oft. with turnip-like r.

Trichodrymonia Oerst. (~ *Episcia*). Gesneriaceae (II g). 40 trop. Am. R: SB 41(2016)96

Trichodypsis Baill. = *Dypsis*

Trichoglottis Blume. Orchidaceae (V 16c). c. 60 Indomal. (Philippines 21) to Taiwan & Polynesia. Cult. orn. epiphytes

Trichogonia (DC.) Gardner (~ *Eupatorium*). Compositae (Eup.-Trich.). 20 N S Am. (Braz. 17). R: SB 37(2012)526

Trichogoniopsis R. King & H. Robinson (~ *Eupatorium*). Compositae (Eup.-Trich.). 4 Braz. R: MSBMBG 22(1987)104

Trichogyne Less. = *Ifloga* (but see OB 104(1991)155)

Tricholaena Schrad. Gramineae (XXIV 5). 4 Medit. (Eur. 1) & Macaronesia to Afr. R: BB 138(1988)36. ***T. vestita*** (Balf.f.) Stapf & C.E. Hubb. (Socotra) – only known from type. See also *Melinis*

Tricholaser Gilli. Umbelliferae (III 11). 2 S & SW As.

Tricholemma (Röser) Röser = *Helictotrichon*

Tricholepidium Ching = *Microsorum* (but see APG 29(1978)41)

Tricholepis DC. Compositae (Card.-Cent.). 18 C As. to Myanmar. Some medic. (skin disease, 'seminal debility')

Trichomanaceae Burmeister = Hymenophyllaceae

Trichomanes L. Hymenophyllaceae (Hym.). Incl. *Didymoglossum*, *Microgonium*, c. 300 trop. & warm Am., few OW; (s.s.) 65 trop. & warm Am. R: PJSB 51(1933)119. Bristle or kidney ferns. Some with ± peltate fronds (*Microgonium*). Some cult. orn. See also *Crepidomanes*

Trichomeriopsis auctt. = *Trochomeriopsis*

Trichoneura Andersson. Gramineae (XXIX 3). 8 Arabia, trop. Afr., S US (1), Peru (1), Galápagos (1). R: AB 11,9(1912)8

Trichoneuron Ching (~ *Asplenium*). Aspleniaceae (Tectarioideae). 1 SW China: ***T. microlepioides*** Ching

Trichopetalum Lindl. (*Bottionea*). Asparagaceae (Anthericaceae, Lomandraceae). 2 Chile. ***T. plumosum*** (Ruíz & Pavón) Macbr. – cult. orn.

Trichophorum Pers. (~ *Scirpus*.). Cyperaceae (II 1). 9 N temp., boreal, Andes & trop. SE As. mts

Trichopilia Lindl. Orchidaceae (V 12h). c. 40 trop. Am. Cult. orn. epiphytes

Trichopodaceae Hutch. = Dioscoreaceae

Trichopteryx Nees. Gramineae (XVIII). 5 trop. & S Afr., Madag.

Trichoptilium A. Gray. Compositae (Hele.-Psat.). 1 SW US, NW Mex.: ***T. incisum*** (A. Gray) A. Gray. R: FNA 21(2006)418

Trichopuntia Guiggi = *Austrocylindropuntia*

Trichopus Gaertn. Dioscoreaceae (Trichopodaceae). Incl *Avetra*, 1 E Madag., 1 Indomal.: ***T. zeylanicus*** Gaertn. (arogyapaccha) – stem with 1 app. term. leaf, comm. medic. R: T 51(2002)110

Trichosacme Zucc. Apocynaceae (V c; Asclepiadaceae III 3). 1 Mex.: ***T. lanata*** Zucc.

Trichosalpinx Luer (~ *Pleurothallis*). Orchidaceae (V 13c). c. 110 trop. Am. R: MSBMBG 64(1997)1. Some cult. orn.

Trichosanchezia Mildbr. Acanthaceae (III 2d). 1 E Peru: ***T. chrysothrix*** Mildbr.

Trichosandra Decne. Apocynaceae (IV; Asclepiadaceae II). 1 Réunion: ***T. borbonica*** Decne. R: BMNHN 4,12(1990)131

Trichosanthes L. Cucurbitaceae (XII). Incl. *Gymnopetalum*, 91 Indomal. to Pac. R: PhytoKeys 12(2012)23. ***T. cucumerina*** L. (*T. anguina*, snake gourd, Ind.) – monoec. ann. cli. cult. for ed. slender fr. 30–200 cm long, oft. coiled, peptides used as abortifacient in China; ***T. edulis*** Rugayah (E Mal.) – cult. for ed. fr.; ***T. kirilowii*** Maxim. (*T. japonica*, E As.) – r. a starch source; ***T. pilosa*** Lour. (*T. cucumeroides*, *T. ovigera*, Indomal.) – dried fr. a soap subs.; others locally medic.

Trichoschoenus Raynal. Cyperaceae (II 7). 1 Madag.: ***T. bosseri*** Raynal. R: Adans. II, 8(1968)223

Trichoscypha Hook.f. Anacardiaceae (I). 32 trop. Afr. R (part): Adansonia III, 23(2001)248, 26(2004)100. Some with Corner's Model

Trichospermum Blume. Malvaceae (Grew.-Grew.; Tiliaceae). (Incl. *Belotia*) 36 Mal., W Pac., 3 trop. Am. R: TBSE 41(1972)401. Bark for cord & rope

Trichospira Kunth. Compositae (Vern.-Trich.). 1 trop. Am.: ***T. verticillata*** (L.) S.F. Blake

Trichostachys Hook.f. Rubiaceae (IV 10). 14 trop. Afr.

Trichostelma Baill. = *Gonolobus*

Trichostema L. Labiatae (III). 18 N Am. R: Britt. 5(1945)276. ***T. lanceolatum*** Benth. – local medic., tea, insecticide. Cult. orn. herbs & shrubs esp. ***T. lanatum*** Benth. (blue-curls, Calif.) – char. of chaparral, toxic volatiles inhibiting other pl. growth, fls in woolly infls

Trichostephania Tard. = *Ellipanthus*

Trichostephanus Gilg. Achariaceae (inc. sed.; Flacourtiaceae). 2 W trop. Afr. R: BJBB 60(1990)143. Corona!

Trichostigma A. Rich. (*Villamilla*, ~ *Rivina*). Petiveriaceae (Phytolaccaceae II). 3 trop. Am. Glabrous shrubs, some locally medic., cult. orn. ***T. octandrum*** (L.) H. Walter – potherb, bark strips woven in basketry (WI)

Trichostomanthemum Domin = *Melodinus*

Trichotaenia Yamaz. = *Vandellia*

Trichothalamus Spreng. = *Potentilla*

Trichotolinum O. Schulz (~ *Descurainia*). Cruciferae (9). 1 Patagonia: ***T. deserticola*** (Speg.) O. Schulz

Trichotosia Blume (~ *Eria*). Orchidaceae (V 15). c. 75 Ind. to Vanuatu. Cult. orn. epiphytes

Trichovaselia Tieghem = *Elvasia*

Trichuriella Bennet (*Trichurus* C. Towns., ~ *Aerva*). Amaranthaceae (I 2). 1 S & SE As.: ***T. monsoniae*** (L.f.) Bennet

Trichurus C. Towns. = praec.

Tricliceras Thonn. ex DC. (*Wormskioldia*). Passifloraceae (Turneraceae). 16 warm Afr.

Triclisia Benth. Menispermaceae (I). 10 trop. Afr. & Madag. Alks – some arrow-poisons & medic. (allegedly antimalarial)

Tricomaria Gillies ex Hook. & Arn. Malpighiaceae. 1 Arg.: ***T. usillo*** Hook. & Arn.

Tricoryne R. Br. Asphodelaceae (Johnsoniaceae). 7 Aus., 1 ext. to NG. R: FA 45(1987)292. Form. referred to Anthericaceae, Liliaceae. Aerial parts app. all inflorescence, P 6 subequal, persistent & spirally twisted after flowering

Tricostularia Nees. Cyperaceae (II 7). 6 Aus., 1 ext. to S As., ***T. undulata*** (Thwaites) Kern. R: FR 53(1944)212

Tricuspidaria Ruíz & Pavón = *Crinodendron*

Tricycla Cav. = *Bougainvillea*

Tricyclandra Keraudren. Cucurbitaceae (IX). Incl. *Odosicyos*, 2 Madag R: BSBF 112(1965)327. Dioec. with large tuberous rootstocks to 1 m or more

Tricyrtidaceae Takht. = Liliaceae

Tricyrtis Wall. Liliaceae (Tricyrtidaceae). 18 Himal. to Taiwan & Jap. R: Plantsman 6(1985)193. App. links L. to Colchicaceae. Cult. orn. (toad lilies) esp. ***T. hirta*** (Thunb.) Hook. (Jap. t. l., Jap.) with purple-spotted P, several cvs

Tridactyle Schltr. Orchidaceae (V 16d). 47 trop. & S Afr. R: KB 3(1948)282

Tridactylina (DC.) Schultz-Bip. (~ *Chrysanthemum*). Compositae (Anth.-Art.). 1 E Siberia: ***T. kirilowii*** Schultz-Bip. R: BBMNHB 23(1993)115

Tridax L. Compositae (Mill.-Dysc.). Excl. *Cymophora*, 30 Am. esp. Mex. R: Britt. 17(1967)47. ***T. procumbens*** L. (Mex.) – now pantrop. weed, mooted as biopesticide

Tridens Roemer & Schultes. Gramineae (XXIX 7). Excl. *Tridentopsis*, 15 E US to Arg., Angola (1). R: AJB 48(1961)565

Tridelta Luer = *Specklinia*

Tridentea Haw. (~ *Stapelia*). Apocynaceae (Vb; Asclepiadaceae III 5). 8 S Afr. R: SAJB 61(1995)192. Cult. orn. succ.

Tridentopsis Peterson (~ *Tridens*). Gramineae (XXIX 3). 2 SW N Am. to Carib. R: T 63(2014)284

Tridesmostemon Engl. Sapotaceae (V). 2–3 C Afr. R: T.D. Pennington, *S.* (1991)261

Tridianisia Baill. = *Cassinopsis*

Tridimeris Baill. Annonaceae (IV 7). 1 Mex.: ***T. hahniana*** Baill. R: Phytoneuron 2013–15:2

Tridynamia Gagnepain (~ *Porana*). Convolvulaceae (2). 4 Ind. to Hainan & W Mal. R: Blumea 51(2006)467

Trieenea Hilliard (~ *Phyllopodium*). Scrophulariaceae (Manul.). 9 S Afr. R: NRBGE 45(1988)489. Commemorates E.E. Esterhuysen!

Trientalis L. (~ *Lysimachia*). Primulaceae (Myrsinaceae). 3 N temp. (Eur. 1). R: FNA 9(2009)303. Perenn. rhizomatous herbs with term. tufts of 4–7 lvs, K (5), A 5; cult. orn. esp. ***T. europaea*** L. (chickweed wintergreen, Euras.)

triffid weed *Chromolaena odorata*

Trifidacanthus Merr. (~ *Desmodium*). Leguminosae (III 19). 1 Vietnam, Hainan & C Mal.: *T. **unifoliolatus*** Merr. R: JJB 71(1996)63

Triflorensia S. Reynolds (~ *Tarenna*). Rubiaceae (II 2). 3 N Aus. R: Austrobaileya 7(2005)43

trifoliata *Citrus trifoliata*

Trifolium Tourn. ex L. Leguminosae (III 27). 238 temp. (Eur. 99) & subtrop. exc. Aus. R: M. Zohary & D. Heller (1984) *The genus T.*, Acta Univ. Carol. 33(1989)257. Clover. Bee-poll. herbs with A & style emerging when keel depressed by insect, returning when insect leaves; imp. fodder pls, 6 cyanogenic esp. ***T. nigrescens*** Viv. (W Medit. to Iran, Russia), ann., rich in isoflavones affecting mammal reproductive capacity if taken in excess (oestrogenic). ***T. amabile*** Kunth (Aztec c., Mex. to S Am.) – eaten with maize etc.; ***T. arvense*** L. (Euras.) – invasive Aus.; ***T. burchellianum*** Ser. subsp. ***johnstonii*** (Oliv.) J.B. Gillett (Uganda c., E Afr.); ***T. campestre*** Schreb. (hop trefoil, Eur., Medit., natur. N Am.) – forage; ***T. dubium*** Sibth. (Eur. to Cauc.) – the shamrock (*seamróg*, though many mimics & pretenders); ***T. fragiferum*** L. (strawberry c., Medit.) – salt-tolerant fodder, bladdery wing formed by persistent K (cf. persistent C in ***T. badium*** Schreb., C & S Eur. mts); ***T. hybridum*** L. (alsike c., Eur., natur. US) – fodder; ***T. incarnatum*** L. (Italian c., S & W Eur.) – fodder; ***T. medium*** L. (zig-zag c., Eur. to Iran) – fodder; ***T. pannonicum*** Jacq. (Hungarian c., E Eur.) – drought-resistant forage; ***T. pratense*** L. (red or purple c., Eur. to Afghanistan, natur. N Am.) – bumblebee-poll. fodder, nectar sucked out by country children, 'sprouts' sold in Aus., antispasmodic cigarettes, isoflavones active in Promensil promoted for menopause alleviation, but leading to infertility in grazing cattle; ***T. reflexum*** L. (buffalo c., US) – fodder; ***T. repens*** L. (white or Dutch c., Eur., Medit., natur. N Am., invasive Aus.) – amphidiploid (*T. nigrescens* × *T. occidentale* Coombe (NW Eur. coasts)), most imp. fodder & rotational (10% cover gives 50 kg nitrogen per ha per annum, enough to make a sward self-sustaining) widely introd. (early in Peru) with a number of strains adapted to co-existence with particular other spp. such that a field will hold a no. of discrete 'clones', resembling a polymorphism with diff. leaf-markings, cyanogenesis capacity, fl. colour etc., imp. bee-forage (clover honey), lvs s.t. used as shamrock (cf. *T. dubium*), '**Purpurascens Quadrifolium**' – cult. orn. with lvs with 4 bronzy-red green-margined leaflets (up to 14 leaflets recorded in wild pops; also in *T. pratense*), f. ***lodigense*** Gams (Ladino c.) a 'giant' form used in pastures, some forms with plantlets in axils of petals; ***T. resupinatum*** L. (Persian c., Medit. to Iran) – fodder; ***T. stoloniferum*** Muhl. ex Eaton (running buffalo clover, N Am.) – almost extinct with 2 pops in W Virginia & 1 in Kentucky poss. due to disappearance of bison; ***T. subterraneum*** L. (subterr. or sub c., N Eur. to Med., invasive Aus.) – aerial infls & geocarpic with few fls the rest with hooked K acting as grapnels, fodder e.g. worth up to £200M per annum in 1930s though orig. intr. as packing around wine-bottles; ***T. wormskioldii*** Lehm. (NW Am.) – rhiz. eaten

Trifurcia Herb. = *Herbertia*

trigger plant *Stylidium* spp.

Triglochin L. Juncaginaceae. Excl. *Cycnogeton*, 25 cosmop. (Eur. 4) esp. S temp. (Aus. 16), freshwater & salt-marshes. Arrowgrass. Plantain-like herbs, wind-poll. pseudanthial/euanthial infls, spiny prob. animal-disp. fr. Lvs allegedly toxic (hydrogen cyanide) in W US, though young lvs of ***T. maritima*** L., N temp.) cooked & fr. sold as birdseed in Paris markets while some spp. tubers ed. N Am.

Trigonachras Radlk. Sapindaceae. 9 Mal. R: Blumea 33(1988)204

Trigonanthe (Schltr.) Brieger = *Dryadella*

Trigonastrotheca F. Muell. (~ *Mollugo*). Molluginaceae. 3 As. to Aus. R: T 65(2016)704

Trigonella L. Leguminosae (III 27). c. 55 Medit. (spp. with explosive fls referred to *Medicago*), Macaronesia, S Afr., Aus. (1). R: TBIANSSR I, 10(1953)124. ***T. foenum-graecum*** L. (fenugreek, methi, S Eur., W As.) – cult. since time of Assyrians for yellow seeds (mathai, mathe) used like lentils, colouring curry yellow, & medic. (allegedly contraceptive – steroids (diosgenin), seed extract Testofen^{TM} (with furostanol saponins incl. fenuside known s.t. to raise testosterone levels) used in body-building), dyes etc., preserved in Tutankhamun's tomb (1325 BC)

Trigonia Aubl. Trigoniaceae. 27 trop. Am.

Trigoniaceae A. Juss. Magnoliidae – Malpighiales. 5/31 trop. Everg. trees, shrubs & lianes without internal phloem. Lvs simple, opp. or in spirals (*Trigoniastrum*); stip. usu. decid., oft. united. Fls bisex., irreg., 2- or 3-bracteolate, in racemes, panicles, cymes or thyrses; K (5) with unequal imbr. lobes, C papilionoid convolute in bud with 2 lower ones forming a keel, the upper a spurred or saccate standard, the laterals spathulate, A 5–13 on anterior (lower) side of fl. (up to 5 staminodal) ± united as a tube, anthers with longit. slits, (0-)2 nectary glands in front of standard, G ((3)4), pluriloc. with axile placentas or 1-loc. with ± deeply intruded parietal placentas, style simple, each loc. with 1 or 2 to ∞ anatr. bitegmic ovules. Fr. of 3 1-winged samaras or (*Trigonia*) a septicidal capsule; seeds with straight embryo & 0 endosperm. n = c. 10

Genera: *Humbertiodendron* (Madag.), *Isidodendron*, *Trigonia* & *Trigoniodendron* (trop. Am.), *Trigoniastrum* (W Mal.) – a distribution diff. to explain

Ovule & seed structure suggested this position; DNA shows v. close affinity with Chrysobalanaceae

Trigoniastrum Miq. Trigoniaceae. 1 W Mal.: ***T. hypoleucum*** Miq. – timber for furniture-making. R: FPM II,1(2010)277

Trigonidium Lindl. = *Maxillaria*

Trigoniodendron E. Guim. & Miguel. Trigoniaceae. 1 E Braz.: ***T. spiritusanctense*** E. Guim. & Miguel. R: RBB 47(1987)559

Trigonobalanus Forman. Fagaceae. 3 W Mal. (1), Thailand, Laos & China (1), Colombia (1), fossils in Eocene of Eur. Attims's Model. Suckering. coppicing. Fr. like *Fagus* & eaten by rhinoceros

Trigonocapnos Schltr. Papaveraceae (Fumariaceae I 2). 1 S Afr.: ***T. lichtensteinii*** (Cham. & Schldl.) Lidén. R: Strel. 9(2000)514

Trigonocaryum Trautv. = *Myosotis*

Trigonochilum Königer & Schildh. = *Cyrtochilum*

Trigonochloa Peterson & N. Snow (~ *Leptochloa*). Gramineae (XXIX 3). 2 OW trop. R: PhytoKeys 13(2012)25. ***T. uniflora*** (A. Rich.) Peterson & N. Snow (*L. u.*, NE trop. Afr.) – minor cereal

Trigonophyllum (Prantl) Pichi-Serm. = *Trichomanes*

Trigonopleura Hook.f. (~ *Chaetocarpus*). Peraceae (Chaet.; Euphorbiaceae Acal.-Chaet.). 3 W & C Mal. R: Blumea 40(1995)363

Trigonopterum Steetz ex Anderss. (*Macraea*). Compositae (Helia.-Ecl.). 1 Galápagos: ***T. laricifolium*** (Hook.f.) W.L. Wagner & H. Robinson. R: Britt. 53(2001)559

Trigonopyren Bremek. = *Psychotria*

Trigonosciadium Boiss. Umbelliferae (III 11). 3 W As.

Trigonospermum Less. Compositae (Mill.-Mill.). 4 Mex., C Am. R: CUMH 9, 6(1972)495

Trigonospora Holttum = *Cyclosorus*

Trigonostemon Blume. Euphorbiaceae (Cod.-Trig.). c. 60 Indomal. (Mal. 37), Pacific. Some unbranched pachycauls. Local medic.

Trigonotis Steven. Boraginaceae (B.3.7). 60 E Eur. (1), C As. to NG

Triguera Cav. = *Solanum*

Trigynaea Schldl. (~ *Hornschuchia*). Annonaceae (III 1). 8 N S Am. R: Britt. 47(1995)273. Fls supra-axillary

Trigynia Jacq.-Fél. = *Leandra*

Trihaloragis Moody & Les (~ *Gonocarpus*). Haloragaceae. 1 Aus.: ***T. hexandra*** (F. Muell) Moody & Les

Trihesperus Herb. = *Echeandia*

Trikeraia Bor. Gramineae (X). 3 Pakistan to China (3 – R: FOC 22(2006)190)

Trilepidea Tieghem. Loranthaceae (3). 1 NZ (N Is.): ***T. adamsii*** (Cheeseman) Tieghem – extinct. R: Conserv. Biol. 5(1991)52

Trilepis Nees. Cyperaceae (III 2). c. 3 NE S Am. R: MNYBG 12,3(1965)14. Erect pls with woody stems & 3-ranked persistent ligulate lvs

Trilepisium Thouars (*Bosqueia*). Moraceae (IV). 1 trop. & S Afr. to Masc.: ***T. madagascariense*** DC. – Troll's Model

Trilisa (Cass.) Cass. (~ *Carpephorus*). Compositae (Eup.-Gyp.). 2 SE US. R: MSBMBG 22(1987)275

Trillesanthus Pierre = *Marquesia* (but see Blumea 55(2010)90)

Trilliaceae Chevall. = Melanthiaceae

Trillidium Kunth = *Paris*

Trillium L. Melanthiaceae (Trilliaceae). Excl. *Pseudotrillium*, 43 N Am. (38, esp. S Appalachians), Himalaya, E As. R: Plantsman 10(1989)221, 11(1989)67, 133, 12(1990)44, 13(1992)219; F.W. & R.B. Case (1997) *Ts*. Poss. excl. *Delostylis*. Overground parts = infl., so 'lvs' = bracts; some ant-disp. (elaiosomes). Sapogenins used medic. e.g. uterine stimulants; cult. orn. (wake robin, esp. ***T. grandiflorum*** (Michaux) Salisb. (E N Am.) with freq.

wild abnormal variants with bizarre fls etc.). ***T. apetalon*** Mak. (~ *T. smallii* Maxim., Jap.) – P replaced by outer A, outer A by inner, & inner by G; ***T. chloropetalum*** (Torrey) Howell (Calif.) – much used in chromosome studies where, after cold treatment, segments of heterochromatin stick together at anaphase, the adhesion persisting to telophase forming a bridge between daughter nuclei, a phenomenon prob. caused by subchromatic breakage followed by partial inverted reunion; ***T. erectum*** L. (birthroot, stinking Benjamin, E N Am.) – locally medic.

Trilobachne Schenck ex Henrard = *Chionachne* (but see Blumea 47(2002)573)

Trimenia Seemann. Trimeniaceae. (Incl. *Piptocalyx*) 8 C Mal. to SE Aus., Marquesas & Samoa. R: Blumea 19(1971)3

Trimeniaceae Gibbs. Magnoliidae – Austrobaileyales. 1/8 C Mal. to SE Aus., Marquesas & Samoa. R: JAA 64(1983)447 Polygamous trees (Stone's Model), shrubs & lianes oft. accum. aluminium. Lvs simple, ± opp., with translucent dots; stip. 0. Fls small, in cymes or panicles, wind-poll., with ± flat receptacle continuous with pedicel bearing 2–38 opp. bracteoles (?) passing imperceptibly to P in spiral, all decid. at anthesis, A 6–25 in (1) 2 or 3 rows inserted spirally, anthers elongate, basifixed, with longit. slits, weakly extended connective, $\underline{G}$ 1(2) with 1 pend., anatr. ovule per carpel. Fr. a berry; embryo small in copious endosperm. n = 9
Genus: *Trimenia*

Trimeria Harv. Salicaceae (Flacourtiaceae). 2 trop. & S Afr. R: BJ 94(1974)302

Trimeris C. Presl (~ *Lobelia*). Campanulaceae (III 1). 1 St Helena: ***T. scaevolifolia*** (Roxb.) Mabb. – shrub (Attims's Model), affinities with Pac. & S Am. taxa, so poss. incl. *L.* sect. *Colensoa*. R: KB 29(1974) 579

Trimerocalyx (Murb.) Murb. = *Linaria*

Trimezia Salisb. ex Herb. Iridaceae (VIi 4). Excl. dubiously distinct *Neomarica* & *Pseudotrimezia*, 20 trop. Am. S.t. viviparous with plantlets in old infrs (cf. *Dietes*, *N.*). Elaiophores attractive to bees. Local laxatives; cult. orn. with fugacious fls, incl. ***T. martinicensis*** (Jacq.) Herb. – widely natur. trop.

Trimorpha Cass. = *Erigeron* (But see Phytol. 67(1989)63)

Trimorphopetalum Bak. = *Impatiens*

Trinacte Gaertn. = *Jungia*

Trincomali wood *Berrya cordifolia*

Trinia Hoffm. Umbelliferae (III 8). c. 10 Eur. (9), Medit. to C As. ***T. glauca*** (L.) Dumort. (honewort, C & S Eur.) – ant-poll.

Triniochloa A. Hitchc. Gramineae (XI). 6 trop. Am. R: Novon 5(1995)36, 8(1998)146

Triniteurybia Brouillet & al. (~ *Haplopappus*). Compositae (Ast.-?Mac.). 1 NW US: ***T. aberrans*** (A. Nelson) Brouillet & al. R: FNA 20(2006)382

Triocles Salisb. = *Kniphofia*

Triodanis Raf. Campanulaceae (I 8). 6 N & S Am., ***T. perfoliata*** (L.) Nieuw. introd. Medit., China. R: JAA 67(1986)33

Triodia R. Br. Gramineae (XXIX 4). Incl. *Plectrachne*, 66 Aus. R: FA 44B(2005)203. Hummock-grasses, flammable & fuel-source for wildfires; the pungent lvs dominate much of Aus. veg. – 'spinifex', e.g. ***T. irritans*** R. Br. (S Aus.) – needling like *Heteropogon* & ***T. longiceps*** J. Black (black s., S Aus.). Extracts used medic. Aborigines, smoke a mosquito-repellent, gums used to affix spearheads

Triodoglossum Bullock = *Schlechterella*

Triolena Naud. Melastomataceae. 22 trop. Am. Some cult. orn. like *Bertolonia* but with 3 thread-like appendages on connectives of larger A & 3-winged capsule

Triomma Hook.f. Burseraceae (V). 1 W Mal.: ***T. malaccensis*** Hook.f. – soft timber. R: FM I,5(1956)218

Trioncinia (F. Muell.) Veldk. (~ *Glossocardia*). Compositae (Cor.). 2 C Queensland. R: FA 37(2015)447

Triopteris L. = seq.

Triopterys L. = *Mascagnia*

Triosteum L. Caprifoliaceae. 3 E As., 3 E N Am. Perenn. woody herbs incl. ***T. perfoliatum*** L. (fever r., tinker's weed, E N Am.) – local medic., form. a coffee subs.

Tripetaleia Sieb. & Zucc. = *Elliottia*

Tripetalum K. Schum. = *Garcinia*

Triphasia Lour. Rutaceae (II 2). 3 SE As., Philippines. R: W.T. Swingle (1943) *Bot. Citrus*: 236. ***T. trifolia*** (Burm.f.) P. Wilson (limeberry, Chinese or myrtle lime, orig. unclear (? Malay Pen.)) – widely cult. hedge-pl. in trop., fr. ed. esp. preserved (China)

Triphora Nutt. Orchidaceae (V 4b). 18 Am. (US & Canada 5). ***T. gentianodes*** (Sw.) Ames & Schltr. (Jamaica) – almost leafless

Triphylleion Süsseng. = *Niphogeton*

Triphyophyllum Airy Shaw. Dioncophyllaceae. 1 Sierra Leone, Liberia, Ivory Coast: ***T. peltatum*** (Hutch. & Dalz.) Airy Shaw – carnivorous liane to 40 m long on soil of pH 4.2, lvs on sterile branches ephemeral, elongate, circinnate & with stalked & sessile glands; other lvs with 2 small hooks (long shoots) or larger & hookless (short shoots) hence name; glands with hydrolytic enzymes the most elaborate of all known angiosperm glands, mucilage blocks insects' spiracles; alk. dioncophylline larvicidal, fungicidal, molluscicidal. R: KB 1951(1952)341

Triphysaria Fischer & C. Meyer (~ *Orthocarpus*). Orobanchaceae ('Cast.'; Scrophulariaceae s.l.). 5 Calif., with 1 to Br. Columbia: ***T. pusilla*** (Benth.) Chuang & Heckard – ant-poll.?

Tripidium H. Scholz = *Saccharum*

Triplachne Link. Gramineae (XVI 5). 1 Medit.: ***T. nitens*** (Guss.) Link

Tripladenia D. Don (~ *Schelhammera*). Colchicaceae. 1 E Aus.: ***T. cunninghamii*** D. Don. R: FA 45(1987)416. Form. referred to Convallariaceae

Triplarina Raf. (~ *Baeckea*). Myrtaceae (II 15). 7 E Aus. R: Austrobaileya 4(1995)356

Triplaris Loefl. Polygonaceae (Erio.-Trip.). c. 17 trop. Am. R: NJB 6(1986)545. Dioec. trees with hollow stems inhabited by ants; outer P accrescent as wings on fr. – wind-disp. Cult. orn. (long jack) with fls in v. long panicles incl. ***T. weigeltiana*** (Reichb.f.) Kuntze (*T. surinamensis*, NE S Am.) – timber for interior work

Triplasis P. Beauv. Gramineae (XXIX 3). 2 C & E US to Costa Rica

Triplateia Bartl. (~ *Minuartia*). Caryophyllaceae. 1 C Mex.: ***T. moehringioides*** (DC.) Kuntze. R: T 63(2014)79

Tripleurospermum Schultz-Bip. (~ *Matricaria*). Compositae (Anth.-Anth.). 38 N temp. (Eur. 8 incl. ***T. maritimum*** (L.) Koch (*T. inodorum*, *T. perforatum*). R: BBMNHB 23(1993)155

Triplisomeris Aubrév. & Pellegrin = *Isomacrolobium*

Triplocephalum O. Hoffm. Compositae (Inul.-Pluch.). 1 trop. E Afr.: ***T. holstii*** O. Hoffm. – capitula in sec. heads

Triplochiton K. Schum. Malvaceae (Hel.; Sterculiaceae). 3 trop. Afr. ***T. scleroxylon*** K. Schum. (obeche, arere, ayous, wawa, whitewood, W & C Afr.) – plantation timber tree (Rauh's Model), wood used for plywood & veneers (when stained, superficially like mahogany), induced to fl. when a seedling by hormonal treatment in breeding programmes, young lvs used as a sauce

Triplochlamys Ulbr. = ? *Hibiscus*

Triplophyllum Holttum (~ *Ctenitis*). Tectariaceae (Dryopteridaceae s.l.). 9 trop. S Am. R: Britt. 60(2008)105

Triplopogon Bor. Gramineae (XXII). 1 W Ind.: ***T. ramosissima*** (Hack.) Bor

Triplostegia Wall. ex DC. Caprifoliaceae (Triplostegiaceae). 2 E As., Mal. R: T 46(1997)23

Triplostegiaceae Bobrov ex Airy Shaw = Caprifoliaceae

Triplotaxis Hutch. = *Cyanthillium*

tripmadam *Sedum rupestre*

Tripodandra Baill. = *Rhaptonema*

Tripodanthus (Eichler) Tieghem (~ *Loranthus*). Loranthaceae (4 4). 3 S Am. R: Novon 15(2005)207. ***T. acutifolius*** (Ruíz & Pavón) Tieghem – lvs yield a blackish dye

Tripodion Medik. (~ *Anthyllis*). Leguminosae (III 22). 1 Medit.: ***T. tetraphyllum*** (L.) Fourr.

Tripodium auctt. = *praec.*

Tripogandra Raf. Commelinaceae (II 1g). 22 trop. Am. R: Rhodora 77(1975)213. Cult. orn. incl. ***T. grandiflora*** (J.D. Sm.) Woodson (C Am.) – cli. to 3 m

Tripogon Roemer & Schultes. Gramineae (XXIX 6). (? incl. *Eragrostiella*) 44 OW trop. (Ind. 5, 3 endemic, China 11) to Aus (1). R (Afr. – 9): KB 25(1971)301. Wet flushes; some resurrection pls incl. Ind. spp. & ***T. minimus*** (A. Rich.) Steud. (Afr.) – lvs revive after reduction of water-content to 7%

Tripolium Nees (~ *Aster*). Compositae (Ast.-Bell.). 1 Euras., N Afr., N Am.: ***T. pannonicum*** (Jacq.) Dobrocz. (*T. vulgare*, *A. t.*, sea aster) – fleshy halophyte, orig. 'Michaelmas daisy' (see *A.*), 'var. *discoideus*' increasing in UK. R: SBM 20(2000)90.

Tripora Cantino (~ *Caryopteris*). Labiatae (III). 1 E As.: ***T. divaricata*** (Maxim.) Cantino

Tripsacum L. Gramineae (XXII 2). 16 S US (3) to Paraguay esp. C Am. R: Phytologia 33(1976)203. Some pseudogamy. ***T. dactyloides*** (L.) L. (gama grass, C Am.) – form. considered to have contributed to evolution of maize, though only ***T. andersonii*** J.R. Gray (cult. in Guatemala) may have some *Zea* chromosomes, some hybrids having them only in endosperm; *T. fasciculatum* Trin. ex Asch. (*T. laxum*, Guatemala grass, C Am.) – fodder

Tripteris Less. (~ *Osteospermum*). Compositae (Calend.). 20 S Afr. to Jordan

Tripterocalyx Hook. ex Standl. = *Abronia*

Tripterococcus Endl. (~ *Stackhousia*). Celastraceae (II; Stackhousiaceae). 2 SW Aus.

Tripterodendron Radlk. (~ *Dilodendron*). Sapindaceae. 1 Braz.: ***T. filicifolium*** (Linden) Radlk.

Tripterospermum Blume (~ *Gentiana*). Gentianaceae. 34 E As. R: SB 38(2013)231. Twiners

Tripterygium Hook.f. Celastraceae (I). 1 E China to Taiwan: ***T. wilfordii*** Hook.f. – scandent shrub, r. with insecticidal & antitumour uses, sesquiterpene alks with significant anti-HIV activity, cult. orn. R: EJB 56(1999)37

Triptilion Ruíz & Pavón = *Nassauvia* (but see BSBC 63(1992)101)

Triptilodiscus Turcz. (~ '*Helipterum*'). Compositae (Gnap.-Ang.). 1 Aus.: ***T. pygmaeus*** Turcz. R: Nuytsia 8(1993)420

Triraphis R. Br. Gramineae (XXVI). 8 trop. Afr., Arabia & Aus. (1), 1 C Braz. (1!)

Trirostellum C.P. Wang & Xie = *Gynostemma*

Triscenia Griseb. (~ *Panicum*). Gramineae (XXIII 1). 1 Cuba: ***T. ovina*** Griseb.

Trischidium Tul. (~ *Bocoa*). Leguminosae (III 1). 5 S Am. R: KB 62(2007)334

Triscyphus Taubert ex Warm. = *Thismia*

Trisepalum C.B. Clarke = *Paraboea* (but see NRBGE 41(1984)441)

Trisetaria Forssk. Gramineae (XVI 2). Incl. *Avellinia, Gaudinia, Koeleria, Parvotrisetum, Rostraria, Trisetum*, c. 180 temp. Hairgrass esp. ***T. flavescens*** (L.) Baumg. (*Trisetum f.*, Eur., Med.) – forage grass, prod. vitamin D in ultraviolet as animals do; ***T. sp.*** (*Koeleria macrantha*, N temp.) – form. staple bread grain for some Native Americans; ***T. sp.*** (*K. phleoides*, cat-tail grass, Medit.) – natur. GB, US. See also *Sphenopholis*

Trisetella Luer. Orchidaceae (V 13c). c. 20 trop. Am. R: MSB 31(1989)69. Cult. orn. small epiphytes

Trisetobromus Nevski = *Bromus*

Trisetum Pers. Gramineae (21b). c. 75 temp. (Eur. 24; N Am. 8) excl. Afr. R: Webbia 17(1963)569, BN 118(1965)210. Like *Helictotrichon* but G glabrous, lemma thin & palea free.

Trismeria Fée = *Pityogramma*

Tristachya Nees. Gramineae (XVIII). Incl. *Loudetiopsis*, c. 40 trop. & S Afr., Madag., trop. Am. Some fodders esp. ***T. leiostachya*** Nees (trop. Am.)

Tristagma Poeppig. Amaryllidaceae (Alliaceae). Excl. *Ipheion*, 12 temp. S Am. R: Phytotaxa 297(2016)21. Some cult. orn.

Tristania R. Br. Myrtaceae (II 8). (s.s.) 1 NSW: ***T. neriifolia*** (Sims) R. Br. (water gum). See *Lophostemon, Ristantia, Tristaniopsis, Welchiodendron* for other spp.

Tristaniopsis Brongn. & Gris (~ *Tristania*). Myrtaceae (II 5). 40 SE As., Mal. to E Aus. & New Caled. (13). Specialized fire-resistant epicormic systems; some rheophytes

Tristellateia Thouars. Malpighiaceae. c. 20 Madag., Afr. (1), Indomal. to Aus. & New Caled. (1). R: MMNHN n.s. 21(1947)275

Tristemma Juss. Melastomataceae. 15 trop. Afr. to Masc. R: BMNHN 3° Bot 28(1977)137

Tristemonanthus Loes. (~ *Campylostemon*). Celastraceae (III). 2 trop. W Afr.

Tristerix Mart. (~ *Macrosolen*). Loranthaceae (4 3). 12 Andes. R: SBM 19(1988)12. ***T. aphyllus*** (DC.) Barlow & Wiens (Chile) – disp. only by Chilean mocking-birds, germ. seed with radicle to 91 mm in 43 days (unique in L.) poss. mechanism that overcomes length of host spines (see *Echinopsis*) as barrier to infection; in adult only infls of bright red fls appear above host surface; ***T. corymbosus*** (L.) Kuijt (S Arg.) – disp. (endozoochory) by nocturnal marsupial *Dromiciops gliroides (D. australis)*, poss. ancestral condition for L.

Tristicha Thouars. Podostemaceae (II; Tristichaceae). Incl. *Malaccotristicha*, 1–6 trop. Am., Afr. to Ind., N Aus. (***T. trifaria*** (Willd.) Spreng. pantrop. W of Wallace's Line). Some described *Frullania* spp. (mosses) & *Crassula* truly *T.*

Tristichaceae Willis = Podostemaceae

Tristira Radlk. Sapindaceae. 1 C Mal.: ***T. triptera*** (Blanco) Radlk. R: FM I,11(1994)740

Tristiropsis Radlk. Sapindaceae. 3 C & E Mal., W Pac. R: FM I,11(1994)742

Triteleia Douglas ex Lindl. (~ *Brodiaea*). Asparagaceae (Alliaceae s.l.). 15 W N Am. (14 – N Mex. R: Four Seasons 6,1(1980)17, FNA 26(2003)? Cult. orn. cormous pls esp. ***T. laxa*** Benth. with blue fls; corms form. ed Native Americans. See also *Ipheion*

Triteleiopsis Hoover (~ *Brodiaea*). Asparagaceae (Alliaceae s.l.). 1 Baja Calif., Arizona: ***T. palmeri*** (S. Watson) Hoover. R: FNA 26(2002)332

Trithecanthera Tieghem. Loranthaceae. 5 W Mal. R: FM I,13(1997)396

Trithrinax Mart. Palmae (III 2). 3 Braz., Bolivia, Paraguay, Urug., Arg. R: Phytotaxa 136(2013)1. Many allegedly primitive features, fls bisex. Ed. fr. & oilseeds; some cult. orn. ***T. schizophylla*** Drude (Braz. to Uruguay) – ash a salt-source in chaco of Paraguay

Trithuria Hook.f. Hydatellaceae. 12 Ind. (1), Aus. (10), NZ (1). R: T 57(2008)192

triticale × *Triticosecale*

× **Triticosecale** Wittm. ex A. Camus. Hybrids between spp. of *Triticum* & *Secale* (triticale). Hexaploid forms approaching wheat yields with consistently higher contents of proteins & ess. amino-acids

Triticum Tourn. ex L. Gramineae (XV). Incl. *Aegilops*, 41 Macar., Medit. (Eur. 11) to China. R: WAUP 94–7(1994)88,111,139. Wheat (*Secale* prob. referable here). Most imp. temp. cereals (17 000 cvs) of complex ancestry involving spp. s.t. separated as *Aegilops*, the most. cult. of all pls (713 M t a yr by 2013, 20% of all food calories consumed by humans; 8–14% protein with gluten allowing 'rising' of dough; gluten fermented for prod. monosodium glutamate food-'flavouring', but g. intolerance in 1 in 1000 people (coeliacs, 1 in 300 in UK)); dried grains of diff. spp. sold as farro. ***T. monococcum*** L. (einkorn, 2n = 14, E Medit.) – diploid fodder wheat of Palaeolithic derived from forms ('subsp. *aegilopoides* (*baeoticum*)', Near E esp. Turkey) with brittle ear, hulled grain, prob. coll. before grown, little cult. now; ***T. turgidum*** L. (2n = 28, a genome from *T. urartu* Thumanjan ex Gandilyan (2n = 14, Caucasus), the other from *T. speltoides* (Tausch) Gren. (*A. s.*, 2n = 14, W As.)) – the allotetraploid wheats selected from forms ('subsp. *dicoccoides*') with brittle ears & hulled grains (? orig. SE Turkey & poss. Upper Jordan Valley) arr. in 8 groups incl. **Dicoccon Group** (*T. dicoccon*, emmer) – cult. by Babylonians & Swiss lake-dwellers, now usu. for livestock or breakfast food, **Durum Group** (*T. durum*, durum w., flint, hard or macaroni w.) – gluten-rich, cult. for cracked wheat (boughal, bulgur, burghul, esp. N Afr.), couscous (s.t. with barley; N Afr.), udon noodles (Jap.), macaroni (being traded by Arabs by 12 cent., when known in Italy), semolina [US = cream of wheat], spaghetti, vermicelli, orzo (rice-shaped) & other pasta (27 kg per head eaten annually in Italy, 2 kg in GB), atta flour used in most S As. flatbreads (chapati, naan, puri, roti), **Polonicum Group** (*T. polonicum*, Polish w., giant or Jerusalem rye), **Turgidum Group** (Poulard w.) – tall winter or spring wheats, little cult.; ***T. aestivum*** L. (common bread wheat, 2n = 42, as *T. turgidum* with addition of genome (cf. *Thinopyrum junceum*) from *T. aegilops* P. Beauv. ex Roem. & Schultes (*A. tauschii*, '*A. squarrosa*', W As., bringing dependence on mycorrhiza & increased gluten) once tetraploids taken into its range & probably arose several times (cf. *Sporobolus anglicus*)) – genome sequenced 2010, allohexaploid wheats arr. in 6 groups incl. **Aestivum Group** – spring & winter grain wheats for bread & fermenting to 'white' beer (witte b. flavoured with *Galium odoratum*), distilling to vodka (Russia; 180 bottles per person there drunk per year) & grain whisky, the frumenty (wheat porridge) of 19 cent. England, **Compactum Group** – protein-poor cvs used for pastry-flour, **Spelta Group** (*T. spelta*, spelt) – anc. cvs with hulled grains tolerant of poor soils; wheatgerm a health food & used in cosmetics, seedlings blended as fashionable health drinks; Devon or wheat reed for thatch lasting up to 40 yrs

Tritonia Ker-Gawler. Iridaceae (VI 5). c. 30 S trop. Afr. to Cape. R: JSAB 48(1982)105, 49(1983)347. Some poll. monkey-beetles; cult. orn. but see *Crocosmia*

Tritoniopsis L. Bolus. Iridaceae (VI 1). 23 S Afr. R: SAJB 56(1990)580. Cult. orn.; some ('*Anapalina*') bird-poll. with long tube & exserted A, others pink short-tubed bee-poll., others by flies, moths or bees (***T. parviflora*** (Jacq.) G. Lewis with floral oils)

Triumfetta Plum. ex L. Malvaceae (Grew.-Apeib.; Tiliaceae). 150 trop. (Aus. 61 (3 introd.) – R: Austrobaileya **5** (1997)500). Some locally imp. fibres, but many weedy. ***T. bartramia*** L. (*T. rhomboidea*, OW trop.) – now pantrop. weed, fodder, famine food, local medic., fibre; ***T. cordifolia*** Guillemin & al. (Afr.) – Leeuwenberg's Model, fibre for ropes & sacks (Angola); ***T. rotundifolia*** Lam. (? *T. malabarica* Rottb., trop. As.) – weedy; ***T. tomentosa*** Bojer (pantrop.) – cult. Mauritius for rope

Triumfettoides Rauschert = *Triumfetta*

Triunia L. Johnson & B. Briggs (~ *Helicia*). Proteaceae IV 1). 4 warm E Aus. R: Muelleria 6(1986)195, (1987)302. seeds highly toxic, though those of ***T. erythrocarpa*** Foreman (Queensland) eaten by musky rat-kangaroos

Triuranthera Backer = *Driessenia*

Triuridaceae Gardner. Magnoliidae – Pandanales. Incl. Lacandoniaceae, 8/56 trop. & warm. R: BB 140(1991), KB 58(2003)940. Glabrous usu. monoec. or dioec. achlorophyllous mycotrophic herbs to 1 m tall (usu. much smaller), white, yellow or purplish; slender rhiz. with scales; vessel-elements only in r.; stomata 0. Fls reg., usu. in term. bracteate racemes; P (3–)6(–10) valvate, s.t. basally connate & with apical appendages, A (2)3 or 6(8), some staminodes, anthers ± sessile & connate with longit. or tranverse slits, the connective oft. exended into long appendage, G 10–50, each with 1 (2 – *Kupea* (Kupeaeae)) basal erect anatr. bitegmic ovule. Fr. a head of achenes; seeds with small undifferentiated embryo in copious oily & proteinaceous endosperm. n = 9, 11, 12, 14 (–16)

Genera: *Kihansia, Kupea, Peltophyllum, Sciaphila, Seychellaria, Soridium, Triuridopsis, Triuris*

So reduced in structure as to obscure their relationships, the high no. of free carpels suggesting they retain primitive features typical of early monocotyledons (fossils from Upper Cretaceous of New Jersey). Lacandoniaceae with G outside A now seen as homoeotic mutation in *Triuris*

Triuridopsis H. Maas & Maas (~ *Triuris*). Triuridaceae. 2 Bolivia, Peru incl. ***T. peruviana*** H. Maas & Maas. R: PSE 192(1994)257

Triuris Miers. Triuridaceae. 4 Guatemala, Guyana, Braz. R: BB 140(1991)10

Triurocodon Schltr. = *Thismia*

Trivalvaria (Miq.) Miq. Annonaceae (IV 7). 4 Assam to W Mal. R: NJB 17(1997)172

Trixis P. Browne. Compositae (Mut.-Nass.). 38 SW N Am. (MNYB 22(1972)11) to Chile (S Am. 21 – R: Darw. 34(1996)44). Some fert. control in Urug.

Trizeuxis Lindl. Orchidaceae (V 12h). 1 trop. Am.: ***T. falcata*** Lindl. – small cult. orn. epiphyte. R: Gen. Orch. 5(2009)383

Trocdaris Raf. (~ *Carum*). Umbelliferae (III 8). 1 W Eur., Medit.: ***T. verticillatum*** (L.) Raf. R: Willd. 42(2012)159

Trochetia DC. Malvaceae (Bomb.; Sterculiaceae). 6 Mauritius, Réunion. Nectar coloured (lizard poll. syndrome), ***T. blackburniana*** Bojer ex Bak. (Mauritius) poll. endemic gecko (*Phelsuma cepediana*). Some cult. orn. small trees, ***T. boutoniana*** Friedmann (Mauritius) – lost for 100 yrs but now nat. fl.

Trochetiopsis Marais (~ *Trochetia*). Malvaceae (Bomb.; Sterculiaceae). 3 St Helena: ***T. erythroxylon*** (Forst. f.) Marais (redwood) – form. building timber, extinct in wild & true ***T. melanoxylon*** (Sims) Marais extinct since end 18 cent., pls rediscovered 1970 (form. common 200–500 m, 2 trees left 1993) referred to ***T. ebenus*** Cronk – 2000+ now re-introd. to wild, sec. pollen presentation on P, hybridized with *T. erythroxylon* in cult. R: EJB 52(1995)205. Affinities with Madag. & Masc. pls. Dead trunks used to burn lime for mortar for fortifications in Napoleonic era & wood from old r. still used in inlay work

Trochiscanthes Koch. Umbelliferae (III 8). 1 S Eur.: ***T. nodiflora*** (Vill.) Koch

Trochiscus O. Schulz = *Rorippa*

Trochocarpa R. Br. Ericaceae (VII 7; Epacridaceae). 12 Mal., E Aus. (7–8)

Trochocodon Candargy = ? *Campanula*

Trochodendraceae Eichler. Magnoliidae – Trochodendrales. Incl. Tetracentraceae, 2/2 Himal. to Taiwan & Jap. Everg. androdioec. glabrous trees without vessels. Lvs simple, serrate, in term. spirals; stip. 0 or small & adnate to petiole. Fls in term. racemoid cymes, bisex., with small scales around swollen pedicel; P 0 (rudimentary in *Trochodendron*), A 4 or 40–70 developing centripetally in a spiral, anthers basifixed with longit. slits, connective not prolonged, G 4–11(–17) laterally connate but adaxial surfaces nectariferous, each with 5–30 anatropous, bitegmic ovules in 2 lat. series. Fr. a head of laterally cohering follicles, styles becoming basal; seeds dust-like, testa thin, embryo straight & minute in copious oily & proteinaceous endosperm. 2n = 40

Genera: *Tetracenton, Trochodendron*

Trochodendron Sieb. & Zucc. Trochodendraceae. 1 Korea & Jap. to Taiwan: ***T. aralioides*** Sieb. & Zucc. – cult. orn. tree with trunk to 20 m, s.t. an epiphyte on *Cryptomeria japonica* at first & growing at alts to 3000 m, cryptic dioecy (heterodichogamy recorded), bark locally used as bird-lime. R: IDSY 2009:28

Trochomeria Hook.f. Cucurbitaceae (XIV). 8 Afr. Some ed. r.

Trochomeriopsis Cogn. Cucurbitaceae (XIII). 1 Madag.: ***T. diversifolia*** Cogn. R: U. Eggli, *Ill. Handb. Succ. Pls*, Dicots (2002)93

Troglophyton Hilliard & B.L. Burtt (~ *Gnaphalium*). Compositae (Gnap.-Gnap.). 6 S Afr. R: OB 104(1991)168

Trogostolon Copel. = *Davallia*

Trollius L. Ranunculaceae (II 1). 31 N temp. (Eur. 2). R: MB 41(1974)1. Herbs with alks & ranunculin (see *Anemone*); cult. orn. (globe fls) with term. usu. solit. fls with petaloid K 5–15 & C 5+ rep. by nectaries, esp. ***T. ranunculinus*** (Sm.) Stearn (*T. caucasicus*, Turkey, Cauc. & NW Iran) – K 5–10 yellow, ***T. asiaticus*** L. (Siberia to NW China) – K 10–15 orange but usu. its hybrids (***T.* × *cultorum*** Bergmans) with ***T. europaeus*** L. (N temp.), ***T. chinensis*** Bunge (NE As.) – to 1 m tall

Trommsdorffia Bernh. = *Hypochaeris*

Trommsdorffia Mart. = *Pedersenia*

Tromotriche Haw. (~ *Stapelia*). Apocynaceae (Vb; Asclepiadaceae III 5). 11 S Afr. R: SAJB 61(1995)198

troolie *Manicaria saccifera*

Tropaeastrum auctt. = *Tropaeolum*

Tropaeolaceae Juss. ex DC. Magnoliidae – Brassicales. 1/92 C & S Am. R: OB 108(1991)1. Herbs, ± succ. & usu. cli. (petioles twining), s.t. with tuberous r.; mustard-oils. Lvs peltate or palmately lobed or divided, long-petiolate, in spirals, ± stip. esp. in seedlings. Fls large, bisex., ± irreg., solit. & usu. axillary; K 5 imbr., the adaxial 1 (or 3) extended into nectariferous spur (rarely almost 0), C 5 imbr., clawed, the 3 abaxial usu. diff. (or 0) oft. with hairy claw, A 4 + 4 with small anthers with longit. slits, $\underline{G}$ (3), 3-loc. with 3-fid style, each loc. with 1 (a second aborted early) pend., apical-axile, anatropous, bitegmic ovule. Fr. separating into 1-seeded mericarps, fleshy or dry (rarely samaroid with only 1 maturing); seed with straight embryo in 0 endosperm. n = 12–15
Genus: *Tropaeolum*

Tropaeolum L. Tropaeolaceae. Incl. *Magallana, Trophaeastrum*, 92 S Mex. to Braz. & Patagonia. R: OB 108(1991)18,19,125, T 49(2000)733. 'Superhydrophobic' analogue of water-repelling surface-ridges on lvs, most efficient synthetic yet (cf. *Nelumbo*). ***T. majus*** L. (nasturtium (because of mustard-oil like Cruciferae), Ind. cress, cultigen (poss. spontaneous hybrid between Peruvian spp., *T. minus* L. (cult. Eur. C16) & *T. ferreyae* Sparre, in Lima area) introd. Eur. from Peru 1684 (prob. the orig. of all modern Eur. stock), though after 1845 crossed with *T. peltophorum* & backcrossed with *T. minus* (non-cli.), fast-growing cult. orn. cli. ann., high in vitamin C, with fibre capable of being made into delicate lace & signifying patriotism in 'Language of Fls'; ***T. peltophorum*** Benth. (Andes) – cult. orn., antibiotic action, fl.-buds & young fr. used as caper subs.; ***T. peregrinum*** L. (*T. canariense*, canary creeper, 'canariense', C & S Peru) – cult. orn. with adaxial C fimbriate, yellow; ***T. speciosum*** Poeppig & Endl. (Chile) – cult. orn. perenn. with red fls & blue fr.; ***T. tuberosum*** Ruíz & Pavón (añu, mashua, osañu, ysaño, Andes) – tubers ed. when boiled, also cont. anti-aphrodisiacs (reducing testosterone levels, allowing Inca soldiers to concentrate on job in hand), poss. *T. tricolor* Sweet wild form

Trophaeastrum Sparre = *Tropaeolum*

Trophis P. Browne. Moraceae (I). 9 trop. Am., Madag., Mal. to New Caled. R: PKNAW C91(1988)352. ***T. involucrata*** Burger (Costa Rica) – dioec., wind-poll.; ***T. scandens*** (Lour.) Hook. & Arn. (*Malaisia* s., Indopacific) – wind-poll., sand-papery stems can cause lasting burning sensation

Tropidia Lindl. Orchidaceae (V 3). c. 30 Indomal. to Taiwan & W Pac., Florida (1) & C Am. to WI. ***T. curculigoides*** Lindl. (Indomal.) – local medic. (diarrhoea, malaria)

Tropidocarpum Hook. Cruciferae (23). Incl. *Agallis* & *Twisselmannia*, 4 Calif., Mex., Chile (1). R: Novon 13(2003)393

Tropilis Raf. = *Dendrobium* (but see Willdenowia 12(1982)249)

Trouettia Pierre ex Baill. = *Pycnandra*

Trudelia Garay = *Vanda*

Trukia Kanehira = *Atractocarpus*

trumpet climber, creeper or **vine** *Campsis radicans*; **Chinese t. flower** *C. grandiflora*; **t. tree** *Cecropia peltata*

Trungboa Rauschert (*Cyphocalyx*). ? Labiatae. 1 SE As.: ***T. poilanei*** (Gagnep.) Rauschert – coll. once, form. referred to Scrophulariaceae

Trybliocalyx Lindau = *Chileranthemum*

Trychinolepis Robinson = *Ophryosporus*

Tryginia Jacq.-Fél. = *Leandra*

Trymalium Fenzl. Rhamnaceae (5). 13 S Aus., esp. SW (12, endemic). R: Nuytsia 11(1996)124, 13(2000)339

Trymatococcus Poeppig & Endl. Moraceae (IV). 2 trop. Am. R: FN 7(1972)208, 83(2001)241. Troll's Model

Tryonella Pichi-Serm. (~ *Doryopteris*). Pteridaceae (IV). 2 Braz.

Tryonia Schuettp. & al. (~ *Eriosorus*). Pteridaceae. 4 SE Braz., Urug, R: PhytoKeys 35(2014)35

Tryphostemma Harv. = *Basananthe*

Tryssophyton Wurd. Melastomataceae. 1 Guyana: ***T. merumense*** Wurd.

Tsaiorchis Tang & Wang = *Platanthera*

tsampa *Hordeum vulgare*

Tsavo Jarmol. = *Populus*

Tsebona Capuron. Sapotaceae (Tseb.). 1 Madag.: ***T. macrantha*** Capuron. R: T.D. Pennington, *S.* (1991)257

Tsiangia But, Hsue & P.T. Li = *Ixora* (but see Blumea 31(1986)311)

Tsimatimia Jum. & Perrier = *Garcinia*

Tsingya Capuron. Sapindaceae. 1 Madag.: ***T. bemarana*** Capuron. R: Cand. 69(2014)196

Tsoala Bosser & D'Arcy. Solanaceae (I; Goetzeaceae). 1 Madag.: ***T. tubiflora*** Bosser & D'Arcy – not seen since 1959

Tsoongia Merr. = *Vitex* (but see FOC 17(1994)27)

Tsoongiodendron Chun = *Magnolia*

tsubaki oil *Camellia japonica*

Tsuga (Endl.) Carr. Pinaceae. 9 temp. N Am. (4) & E As. S to Vietnam. R: A. Farjon, *World Checkl.... Confers* ed. 2(2001)244. Hemlock (spruce). Everg. monoec. timber trees ('palm' subs. for P. Sunday), imp. local medic. N Am., esp. ***T. canadensis*** (L.) Carr. (Canada or E or white h., E N Am.) – to 350 yrs old in Connecticut, coarse timber, tanbark (form. most imp. in N Am. – 'hemlock bark'), bark & twigs made into refreshing medic. tea by N Am. Ind. in E Canada, many orn. cvs; ***T. caroliniana*** Engelm. (Carolina h., SE US mts) – cult. orn.; ***T. heterophylla*** (Raf.) Sarg. (western h., Alaskan pine, W N Am.) – to 1238 yrs old, 75 m tall with bole to 2.75 m diam., imp. timber & pulp tree introd. GB 1851; ***T. sieboldii*** Carr. (Jap. h., S Jap.) – cult. orn. See also *Hesperopeuce*

tsukemono *Pteridium aquilinum*

Tsusiophyllum Maxim. = *Rhododendron*

tu si zi *Cuscuta chinensis*

tualang *Koompassia excelsa*

tuart *Eucalypus gomphocephala*

tuba root *Paraderris elliptica*

Tuberaria (Dunal) Spach (*Xolantha*). Cistaceae. 12 W & C Eur. (10), Medit. Herbs like *Helianthemum* but lvs in basal rosette & stigma sessile. A sensitive. Some cult. orn.

Tuberculocarpus Pruski. Compositae (Heli.-Ecl.). 1 S Venez.: ***T. ruber*** (Aristeg.) Pruski. R: Novon 6(1996)415

Tuberolabium Yamamoto. Orchidaceae (V 16c). Incl. *Parapteroceras*, 11 Indomal. R: Orchids 75(2006)922

tuberose *Agave polianthes* (*Polianthes tuberosa* – corruption of specific name, cf. japonica)

Tuberostylis Steetz. Compositae (Eup.-Crit.). 2 Panamá, Colombia. R: MSBMBG 22(1987)376. Epiphytes with fleshy lvs in mangrove

Tubilabium J.J. Sm. = *Myrmechis*

Tubocapsicum (Wettst.) Mak. = *Capsicum* (but see A.T. Hunziker, *Gen. Solan.* (2001)230)

tuckeroo *Cupaniopsis anarcardioides*

Tucma Ravenna = *Ennealophus*

Tuctoria Reeder. Gramineae (XXIX 2). 3 Calif., Mex. in spring-pools. R: AJB 69(1982)1090

tucuma palm *Astrocaryum aculeatum*

tucupi *Manihot esculenta*

Tuerckheimocharis Urb. = *Scrophularia*

Tugarinovia Iljin. Compositae (Card.-Carl.). 1 Mongolia: ***T. mongolica*** Iljin – dioec. R: BJ 108(1987)167

tula ixtle *Agave lechugilla*

Tulasnea Naud. = *Siphanthera*

Tulasneantha P. Royen = *Mourera* (but see ABN 2(1953)17)

Tulbaghia L. Alliaceae (Tulb.). 22 trop. & S Afr. R: NRBGE 36(1978)77. Some cult. orn. (society garlic) with delicate, s.t. sweetly scented, fls

tule *Cyperus canus*

Tulestea Aubrév. & Pellegrin = *Synsepalum*

tulip *Tulipa* spp., (NG) *Gnetum* spp.; **t. oak** *Argyrodendron* spp.; **t. poplar** or **t. tree** *Liriodendron tulipifera*; **African t. t.** *Spathodea campanulata*; **Chinese t. t.** *L. chinense*; **waterlily t.** *Tulipa kaufmanniana*

Tulipa Tourn. ex L. Liliaceae. Excl. *Amana*, 77 SW Eur. & N Afr. to C As. (Turkey 8) & W China (14). Tulips. Close to *Erythronium*. Cult. since 13 cent. in Iran when there were multicoloured forms; nothing of them in Eur. art before mid-16 cent. suggesting that all Eur. tulips (exc. ***T. sylvestris*** L. (tetraploid; incl. subsp. ***australis*** (Link) Pamp. (*T. australis*, diploid)), Medit. to W China) are escapes from cult. & weeds of crops etc. Z.P. Botschantzeva (1982) *Tulips*; A. Pavord (1998) *The tulip*; BJLS 172(2013)296,301, D. Everett (2013) *The genus T.* 4 subgg. – *Ornythia* (4), *Tulipa* (53 – T), *Clusianae* (4), *Eriostemones* (19 – E). Cult. forms introd. via Turkey (not grown there before 1500 (though allegedly discussed as Andalucian orn. crop in C11) but prominent in design etc.; *tulbend* = Turkish for turban in which ts were worn) to Eur. (Portugal 1530, Vienna 1554, in Holland (promoted by Clusius 1571) leading to 'tulipomania' (actually first in Turkey, then France; by 1635 1 '**Semper Augustus**' bulb costing equivalent of c. £1 M), as late as 1836 a bulb of '**Citadel of Antwerp**' cost £650) esp. those (now called Rembrandt ts) with 'broken' fls (i.e. infested with aphid-transmitted tulip-breaking virus poss. from *Lilium candidum*, though modern broken ts uninfected mutants); these were already complex garden hybrids (early ones = ***T.*** × ***gesneriana*** L.) & have given rise to the common tulip of modern gardens, but by 1950s Dutch growers had selected forms of wild spp. collected in C As. (esp. ***T. fosteriana*** Hoog ex W. Irv. (T; scarlet black-blotched fls; & its crosses with garden tulips), ***T. kaufmanniana*** Regel (T; waterlily tulip, fls opening flat), ***T. greigii*** Regel (T; lvs streaked purple-brown etc.)) since World War I & hybridized them with one another & old cvs to give e.g. 'Darwin Hybrids'. Besides these groups, garden cvs arr. according to flowering-time, the 'late or May-fls' ts incl. Single Late ts (incl. Darwin ts, single fls usu. rectangular in outline at base, bred at Lille), Rembrandt ts (fls striped or marked), parrot ts (P laciniate) & Double Late or Peony-fld ts with many P, used in pots. With pinks the first pls to have cv. names – now some 6500 named & c. 300 grown on comm. scale, though c. 10 dominate; imp. cut-fl. industry (worth $US42M a yr by 2005 in US alone) esp. in Netherlands (also 2 M bulbs exported per annum – worth £1.7 M by 2000, but also Spalding area, E England since 1890 & NW N Am.), the bulbs eaten during the Nazi occupation (disturbing menstrual cycles). Form. many Engl. t. societies (cf. auriculas, gooseberries) esp. C19 (only 1 by 1998), the most widely grown cvs now being '**Monte Carlo**' (Double Early, yellow, 1955) & '**Apeldoorn**' (Darwin Hybrid (Mid-season) scarlet, 1951). Many 'wild' spp. cult. in rock gardens etc. incl. ***T. agenensis*** Redouté (T; W & S Turkey, NW Iran etc., natur. W Medit., perhaps the 'Rose of Sharon'), ***T. persica*** (Lindl.) Sweet (T; '*T. eichleri*', '*T. undulatifolia*', E Medit. to C As.), ***T. praestans*** H.B. May (T; C As.), etc.

tulipwood *Dalbergia cearensis*; **Australian t.w.** *Harpullia pendula*; **Brazilian t.w.** *D. decipularis*

tulsi *Ocimum tenuiflorum*

Tulista Raf. (~ *Haworthia*). Asphodelaceae. 4 S Afr. R: SB 39(2014)69

Tuloclinia Raf. = *Metalasia*

Tulotis Raf. = *Platanthera*

Tumamoca Rose = *Ibervillea* (but see Madroño 41(1994)25)

tumble grass *Muhlenbergia paniculata*; **t. weed** *Amaranthus caudatus*, *Salsola kali*, (of Bible, poss.) *Gundelia tournefortii*

Tumidinodus Li = *Anna*

tummelberry *Rubus* 'Tummelberry'

tuna *Opuntia* spp.

Tunaria Kuntze = *Cantua*

tung oil *Vernicia* spp.

Tunilla D. Hunt & J. Iliff. (~ *Opuntia*). Cactaceae (II). 12 S Am.

Tupa G. Don = *Lobelia* L. (but prob. to be re-separated)

Tupacamaria Archila = *Galeandra*

Tupeia Cham. & Schldl. Loranthaceae (4 1). 1 NZ: ***T. antarctica*** (Forst.f.) Cham. & Schldl.

tupelo *Nyssa* spp. esp. *N. sylvatica*

Tupidanthus Hook.f. & Thomson = *Schefflera*

Tupistra Ker-Gawler. Asparagaceae (Convallariaceae). Excl. *Campylandra* (= *Rohdea*) 13 E Himal. & China to W Mal. ***T. nutans*** Wall. (Himal.) – fl. spikes (cf. *Ornithogalum*) veg. in Sikkim, petioles smoked in hookahs in Bhutan

Turanecio Hamzaoglu (*Senecio*). 10 SW As. R: TJB 35(2011)484

Turanga (Bunge) Kimura = *Populus*

Turania Akhani & Roalson (~ *Salsola*). Amaranthaceae. 4 C As.

Turaniphytum Polj. = *Artemisia*

Turbina Raf. (~ *Ipomoea*). Convolvulaceae (9). 15 trop. Am. (5), trop. & S Afr. (9), New Caled. (1!). R: BTBC 118(1991)265. Heterogeneous. ***T. corymbosa*** (L.) Raf. (trop. Am., invasive NE Aus.) – cult. orn. cli., seeds used by Indians (source not identified until 1940s)

as hallucinogen (smoked by Aztecs), active principles being ergoline alks & lysergic acid derivatives previously known in ergot (leading to 'St Anthony's fire' in rye-flour thus tainted)

Turbinicarpus (Backeb.) F. Buxb. & Backeb. Cactaceae. (~ *Neolloydia*). Excl. *Rapicactus*, 11 Mex.

Turczaninowia DC. = *Aster*

Turczaninowiella Kozo-Polj. Umbelliferae. Spp.?

Turgenia Hoffm. = *Daucus* (but see Fl. Iraq 5,2(2013)272)

Turgeniopsis Boiss. = *Glochidotheca*

Turkey oak *Quercus cerris*; **T. red** *Peganum harmala, Morinda citrifolia*

Turkish beech *Fagus sylvatica* subsp. *orientalis*; **T. delight** form. salep (*Ophrys* & *Orchis* spp.); **T. galls** *Quercus pubescens*; **T. hazel** *Corylus colurna*

Turk's cap (cactus) *Melocactus intortus*; **T. c. gourd** *Cucurbita maxima* 'Turbaniformis'

turmeric *Curcuma longa*; **t. root** *Hydrastis canadensis*

Turnera Plum. ex L. Passifloraceae (Turneraceae). 130 trop. & warm Am., E (1) & SW (1) Afr. Herbs & shrubs with alks; some with epiphyllous fls. ***T. diffusa*** Willd. (damiana, trop. Am.) – dried lvs laxative & stimulant (sold as aphrodisiac in NL), tea, liqueur flavour; ***T. ulmifolia*** L. (trop. Am.) – variable weed (Attims's Model; several spp. recog. by some authors) s.t. cult. orn., in Puerto Rico with distyly & incompatibility mechanisms but in Jamaica with variable anther heights though no 'thrum' pl.

Turneraceae Kunth ex DC. = Passifloraceae

turnip *Brassica rapa* Rapifera Group; **t. wood** *Akania bidwillii*

turnsole *Chrozophora tinctoria*

turpentine (cognate with terebinth), *Pinus* spp.; **t. bush** *Beyeria* spp.; **Chian t.** *Pistacia terebinthus*; **Jura t.** *Picea abies*; **Strasburg t.** *Abies alba*; **Venice t.** *Larix decidua*; **t. wood** *Syncarpia glomulifera*

turpeth root, turpethum *Operculina turpethum*; **Spanish t. r.** *Thapsia garganica*

Turpinia Vent. = *Staphylea*

Turraea L. Meliaceae (II-Trich.). 60 trop. & S Afr. (24), Madag., Masc., Socotra (1), trop. As. to Aus. (1: *T. pubescens* Hell.). Arillate seeds disp. by birds, those of grassland spp. (***T. pulchella*** (Harms) Penn. etc.) in S Afr. small & ant-disp.

Turraeanthus Baill. Meliaceae (II-Trich.). 2–3 trop. W Afr. ***T. africanus*** (C. DC.) Pellegrin (avodiré, C & W Afr.) – furniture timber

Turricula Macbr. (~ *Eriodictyon*). Boraginaceae (Nam.; Namaceae) 1 SW N Am.: ***T. parryi*** (A. Gray) Macbr.

Turrigera Decne = *Tweedia*

Turrillia A.C. Sm. = *Bleasdalea*

Turritis Tourn. ex L. (~ *Arabis*). Cruciferae (48). 2 Euras. R: Novon 15(2005)523. ***T. glabra*** L. (*A.g.*, tower mustard) – natur. N Am.

turtle bean *Phaseolus vulgaris*; **t. head** *Chelone* spp.

Tuscan cabbage, T. kale *Brassica oleracea* Acephala Group

Tuscarora rice *Zizania palustris*

Tussacia Benth. = *Chrysothemis*

turwad bark *Senna auriculata*

Tussilago Tourn. ex L. Compositae (Sen.-Tuss.). 1 Euras., N Afr.: ***T. farfara*** L. (coltsfoot, Euras., Medit., natur. E N Am.) – infls prod. in spring before lvs, each head with 40 nectariferous male fls with pollen-presenting styles but no stigmas surrounded by c. 300 nectarless females, highly protogynous; lvs smoked, form. in treatment of asthma, coughs; tincture an ingredient of comm. hairsprays

tussock grass *Deschampsia cespitosa*; **serrated t.** *Nassella trichotoma*

Tutcheria Dunn = *Pyrenaria*

tutsan *Hypericum androsaemum*

Tuxtla Villaseñor & Strother. Compositae (Helia.-Ecl.). 1 Mex., Costa Rica: ***T. pittieri*** (Greenm.) Villaseñor & Strother. R: SB 14(1989)529

Tuyamaea Yamaz. = *Lindernia*

twayblade *Neottia ovata*

Tweedia Hook. & Arn. (~ *Oxypetalum*). Apocynaceae. 6 Chile, Bolivia, Arg. R: T 63(2014)1272

twill *Gossypium* spp.

twinberry *Myrcianthes fragrans*

twinflower *Linnaea borealis*

twinleaf *Zygophyllum* spp. (Aus.)

Twisselmannia Al-Shehbaz = *Tropidocarpum*

twistwood *Viburnum lantana*

twitch *Elymus repens*

Tylanthera C. Hanson = *Phyllagathis*

Tylecodon Toelken (~ *Cotyledon*). Crassulaceae (II). 46 S Afr. R: U. Eggli, *Ill. Handb. Succ. Pls*, Crass. (2005)354. ***T. paniculatus*** (L.f.) Toelken – to 2 m. Some cult. orn. succ. esp. ***T. papillaris*** (L.) Rowley (*T. cacalioides, C. c.*, Cape) – toxic

Tyleria Gleason. Ochnaceae (III). 13 Guayana Higland, Venez. R: BJ 113(1991)182

Tyleropappus Greenman = *Calea*

Tylocarya Nelmes = *Fimbristylis*

Tylodontia Griseb. = *Cynanchum*

Tylomium C. Presl = *Lobelia* (but prob. to be resegregated)

Tylopetalum Barneby & Krukoff = *Sciadotenia*

Tylophora R. Br. = *Vincetoxicum*

Tylophoropsis N.E. Br. = *Vincetoxicum* (but see KB 48(1994)749)

Tylopsacas Leeuwenb. Gesneriaceae (II 3b). 1 Guayana Highland: ***T. cuneata*** (Gleason) Leeuwenb. R: T 9(1960)220

Tylosema (Schweinf.) Torre & Hillc. (~ *Bauhinia*). Leguminosae (I 1). 4 Afr., esp. NE. ***T. esculentum*** (Burch.) Schreib. (*B. e.*, marama bean, morama b., S Afr.) – oil-rich seed with as much protein as soya

Tylosperma Botsch. = *Potentilla*

Tylostigma Schltr. Orchidaceae (IV 4d). 8 Madag.

Tynanthus Miers. Bignoniaceae (3). 15 trop. Am. R: AMBG 99(2014)467. Clove-smelling lianes used as condiments

Typha Tourn. ex L. Typhaceae. 10–12 cosmop. (Eur. 6). Reed mace, cat-tail, bulrush (q.v.), cumbungi (reeds); starch-rich rhiz. emergency food (starch from stems ed. 30K yrs ago), plush of female fls form. used as kapok subs., pollen (unusual carbohydrates) eaten locally, fibrous lvs for mats, basketry & chair-seating (also for boat-building – berdi, clothing & paper), esp. ***T. angustifolia*** L. (N temp.) – bulk of natural 'rush' available in US, ***T. domingensis*** Pers. (trop.) – pollen imp. in Chaco; ***T. elephantina*** Roxb. (OW trop.) etc.; some used for paper-making & dried infl. orn. ***T. latifolia*** L. (N temp., invasive Aus., Canada) – imp. local medic. N Am., pollen (golden) in honey sold as a sweetmeat in anc. China; ***T. orientalis*** C. Presl (W Pacific) – used as sails (sewn with *Phormium tenax*) & pollen as food (pua) by Maori (NZ), raupo thatch for trad. Maori houses

Typhaceae Juss. Magnoliidae – Poales. Incl. Sparganiaceae, 2/26–28 cosmop. Perenn., monoec., wind-poll., glabrous, marsh-herbs, with scaly, sympodial, starchy rhiz.; stems erect, term. by infl., interior pithy; vessel-elements throughout veg. body. Lvs with sheathing base & narrow parallel-veined spongy lamina, distich., mostly basal. Infl. of ∞ fls forming dense cylindrical spike, the upper part distinct & male, the lower female (*Typha*) or globular unisexual heads with males uppermost (*Sparganium*); males with P 0–6(–8) slender bristles (some being bracts?), A 1–8 with short basally to largely connate filaments, pollen in monads or tetrads; females with P (2)3 or 4(5) to ± ∞ (*Typha*) slender bristles or scales in 1–4 irreg. whorls (some bracts?), s.t. ± connate in groups & adnate to gynophore bearing G 1 (pseudomonomerous) with 1 pend., anatropous, bitegmic ovule (some with abortive ovaries). Fr. with slender stipe formed by accrescent gynophore & accrescent style, wind-disp., 1-seeded, eventually dehiscent (*Typha*) or a spongy drupe (*Sparganium*); embryo straight in copious starchy endosperm (with protein & oil) & thin perisperm. n = 15

Genera: *Sparganium, Typha*

The disp. unit is the whole fl., the P acting as 'pappus'

Typhoides Moench = *Phalaris*

Typhonium Schott. Araceae (VII 23). Excl. *Lazarum, Sauromatum* c. 50 SE As., Indomal., NE Aus. (Australasia 14, R: Blumea 37(1993)345). R: JFSUTB 15(1994)288. Some weedy in trop., some local medic. ***T. venosum*** (Aiton) Hett. & Boyce (*Arum cornutum, S. guttatum, S. nubicum*, voodoo lily, monarch-of-the-east, OW, weedy in wild) – lvs pedate, solit., emerging after infls, which are disgustingly foetid, sold in Eur. as dried tubers which prod. infl. without water, roasted tubers eaten locally

Typhonodorum Schott. Araceae (VII 20). 1 E Afr. coast, Madag., Masc.: ***T. lindleyanum*** Schott – viviparous water-pl. of banana-like aspect to 4 m tall, with 'trunk' (pseudostem) to 30 cm diam. & huge lvs, allied to *Peltandra*; seeds & tuber ed.

Tyrbastes Briggs & L. Johnson. Restionaceae. 1 SW Aus.: ***T. glaucescens*** Briggs & L. Johnson. R: Telopea 7(1998)365

Tyrimnus (Cass.) Bosc. Compositae (Card.-Card.). 1 Medit. incl. S Eur.: ***T. leucographus*** (L.) Cass.

Tysonia Bolus = *Afrotysonia*

Tytonia G. Don = *Hydrocera*

Tytthostemma Nevski = *Stellaria*

Tzelevia Alexeev = *Poa*

Tzellemtinia Chiov. = *Bridelia*

Tzeltalia Estrada & Martínez = *Physalis* (but see Britt. 50(1998)289)

Tzvelevopyrethrum Kamelin = ? *Tanacetum*

Tzvelevia Alexeev = *Festuca*

U

Uapaca Baill. Phyllanthaceae (Ant.; Euphorbiaceae s.l.). c. 50 trop. Afr. & Madag. (8 endemic – R: Adansonia 33(2011)222). Rauh's Model; pachycaul shoots; pseudanthia, dioec. ***U. guineensis*** Muell. Arg. (sugar plum, trop. Afr.) – medlar-flavoured fr., timber valuable; ***U. kirkiana*** Muell. Arg. (wild loquat, C Afr.) – ed. fr., timber good, charcoal

Uapacaceae Airy Shaw = Phyllanthaceae

uba cane *Saccharum officinarum*

Ubochea Baill. = *Stachytarpheta*

ucahuba or **ucuhuba** *Virola surinamensis*

udo *Aralia cordata*

udon (**noodles**) *Triticum turgidum*

Uebelinia Hochst. = *Silene* (but see BJ 120(1998)153)

Uebelmannia Buin. Cactaceae (III 5). 3 E Braz. mts. Cult. orn.

Uechtritzia Freyn. Compositae (Mut.-Mut.). 3 Armenia & C As. to W Himal. & China. R: NJB 8(1988)72. Herbs

Ugamia Pavlov. Compositae (Anth.-?Han.). 1 C As.: ***U. angrenica*** (H. Kraschen.) Tzvelev. R: BBMNHB 23(1993)98

Uganda grass *Cynodon transvaalensis*

Ugli proprietary name of a hybrid citrus fr., widely misapplied to *Citrus* × *aurantium* Tangelo Group

Ugni Turcz. Myrtaceae (II 10). 4 trop. & warm Am. ***U. molinae*** Turcz. (*Myrtus ugni*, Chilean guava, NZ cranberry [!], strawberry myrtle) – cult. shrub with blue-black berries c. 6 mm diam. used for jam, weedy Juan Fernandez replacing native ***U. selkirkii*** (Hook. & Arn.) O. Berg

Uittienia Steenis (~ *Dialium*). Leguminosae (I 3). 1 W Mal.: ***U. modesta*** Steenis. R: FM I,12(1996)714

Uladendron Marc.-Berti. Malvaceae (Brown./Malv.). 1 Venez.: ***U. codesuri*** Marc.-Berti

Ulantha Hook. = *Chloraea*

Ulbrichia Urb. = *Thespesia*

Uldinia J. Black = *Trachymene*

Uleanthus Harms. Leguminosae (III 2). 1 Amaz. Braz.: ***U. erythrinoides*** Harms

Ulearum Engl. Araceae (VII 10). 2 Amazonia

Uleiorchis Hoehne. Orchidaceae (V 7). 2 trop. Am. R: BJ 121(1999)67. Mycotrophs

Uleodendron Rauschert = *Naucleopsis*

Uleophytum Hieron. Compositae (Eup.-Crit.). 1 Peru: ***U. scandens*** Hieron. – liane. R: MSBMBG 22(1987)373

Ulex L. Leguminosae (III 9). 10–20 W Eur. (7), N Afr. R: BJ 72(1941)69. Seedlings with 3-foliolate lvs, later stages with them v. reduced with branches forming spines. Poll. mechanism explosive – in newly opened fls keel C adhere by upper edges & keel held straight by A tube & style, no nectar but foraging bees cause keel to break apart releasing A & G; fr. dehisces explosively with valves curving back & becoming twisted, expelling seeds bearing elaiosomes carried off by ants. Shrubs (furze) with alks, dominating much of the fire-climax heathland of Atlantic Eur., a formation replacing the orig. decid. forest long-cleared, esp. ***U. europaeus*** L. (gorse, whin) – seeds persist 30+ ys in seed-bank, form. imp. fuel & crushed for animal fodder, now weedy elsewhere (invasive Azores, temp. &

trop. As., Aus., NZ, Hawaii, Masc., W US; cult. forms with 'double' fls), but also smaller ***U. minor*** Roth & ***U. gallii*** Planch. – dwarf gorse (W Eur.)

uli *Castilla ulei*

ulla grass *Themeda gigantea*

Ulleria Bremek. = *Ruellia*

ullucu *Ullucus tuberosus*

Ullucus Caldas. Basellaceae. 1 Andes: ***U. tuberosus*** Caldas (chioca, ullucu, oca quina) – prostrate pl. with delicious ed. potato-like tubers from rhiz. (cult. NZ), anthers with apical pores, fr. a berry. R: KB 62(2007)308

Ulmaceae Mirb. Magnoliidae – Rosales. 7/c. 50 N temp. to trop. Am. Trees without laticifers but oft. with mucilage cells or canals. Lvs simple, serrate, oft. basally oblique, usu. in spirals, oft. becoming distich., lat. veins running into teeth (exc. *Ampelocera*); decid. stip. extrapetiolar, prot. bud. Fls small, wind-poll., unisexual (monoec.) or bisex. (e.g. *Ulmus*), ± reg., solit. (females) & axillary or in cymes to panicles; P (2–)5(–9) in spiral, s.t. connate, anthers dorsifixed & somewhat versatile, G (2(3)), a pistillode in males, 1- (or 2-) loc. with 3-bundled sep. styles, each loc. with 1 pend., anatr. to amphitropous bitegmic ovule. Fr. a samara; seeds with straight or curved embryo in little or 0 endosperm. x = 14

Genera: *Ampelocera, Hemiptelea, Holoptelea, Phyllostylon, Planera, Ulmus, Zelkova* (subfam. **Celtidoideae** (styles 1-bundled, fr. a drupe, x = 10, 11) = Cannabaceae)

Imp. timber trees, esp. *Ulmus*, but also *Holoptelea, Phyllostylon, Zelkova* spp. etc., fibres & cult. orn.

ulmer pipes *Acer campestre*

ulmo *Eucryphia cordifolia*

Ulmus Tourn. ex L. Ulmaceae. Incl. *Chaetoptelea*, 25–30 N temp. (Eur. 6) to N Mex. Elms. R.H. Richens (1983) *Elm*. Troll's Model; most fl. before lvs expand, though ***U. parvifolia*** Jacq. (Jap., China) etc. autumn-flowering while some spp. everg. (cf. *Quercus*). Form. imp. timber (wood stained to resemble mahogany) & street-trees but many susceptible to beetle-borne Dutch Elm Disease (D.E.D. below, *Ophiostoma* (*Ceratocystis*) *ulmi* etc., *O. novo-ulmi* most virulent) spread by bark-beetles, orig. As., introd. US from Netherlands c. 1928; rapid disappearance of Eur. elm in pollen record (The Elm Decline) c. 3000 yrs BC poss. also partly attributable to D.E.D., but also to selective pollarding & other management by Neolithic humans, ***U. wallichiana*** Planch. (NW Himal.) still imp. cattle fodder tree, cf. *Ficus semicordata*. Since c. 1965 most in N Am. & Eur. killed by virulent strain introd. from US (11M in GB 1970–8 alone; 20 M in 1905 reduced to a few hundred by 2005, though still surviving in Brighton, on Isle of Wight and in Channel Is.; 90% of Fr. elms killed; best surviving N Eur. pops in Amsterdam and The Hague; disease in NZ but not Aus. yet). Timber rot-resistant underwater, that of Eur. spp. for beams where oak too dear, furniture, bellows, coffins, chair-seats, wheel-hubs, mallet-heads, village pumps & orig. water-pipes as in London before metal & now plastic ones; bark form. used for Native American houses in Kansas, imp. local medic. N Am. Elm sticks, with transverse notches indicating amount of debt, split lengthwise for lender & borrower to be 'tallied' on payment; e. tallies used by Br. government ordered to be burnt 1834, overheating flues leading to destruction of Houses of Parliament! ***U. alata*** Michaux (red elm, E N Am.) – timber, bast used to tie cotton-bales; ***U. americana*** L. (American or white elm, E N Am.) – polyploid complex, timber (orham [i.e. Fr. *orme*] wood) but largely killed by D.E.D., seedlings from 1920s surviving to 2004 as huge underground stems with shoots continually dying back, tree (Washington e.) at Cambridge Mass. under which George Washington allegedly assumed control of rebel army (1775) d. 1923 but cutting had been taken (1900) to Univ. of W., Seattle, from which cutting taken back to Mass., many orn. cvs, '**Princeton**' app. resistant to D.E.D.; ***U. crassifolia*** Nutt. (cedar e., S US) – allegedly resistant to D.E.D.; ***U. davidiana*** Planch. (N China, Jap.) incl. Jap. e. – timber; ***U. glabra*** Hudson (wych [i.e. pliant] or Scotch e., N Eur. to SW As.) – mucilage (cambium layers) form. for scalds & burns (Ireland), timber durable in water so used in boat-planking (esp. imp. in WW II), form. used for desks ('wyches'), form. common woodland elm in GB, not suckering, many orn. cvs incl. '**Camperdownii**' with drooping branches usu. grafted on an erect stock, & '**Clusius**' (hybrid with *U. wallichiana*), selected Netherlands) – app. resistant to D.E.D.; ***U. laevis*** Pallas (C & E Eur.) – unattractive to bark-beetles because of alnulin (triterpenes) in bark, so less suceptible to D.E.D.; ***U. mexicana*** (Liebm.) Planch. (Mex.) – timber used for railway-sleepers, cart-wheels etc.; ***U. minor*** Mill. (W to S & C Eur.) – ? imported timber used for chariots in Tutankhamun's tomb, some clones prop.

veget., e.g. '*U. plotii*' (Eng.) single clone planted in Eng., '**Stricta**' (Cornish elm) form. common in UK, but esp. '**Atinia**' (var. *vulgaris*, *U. sativa*, *U. procera*, English e., ? orig., allegedly introd. N Spain by Celts from S Fr. but otherwise known only from England & S Wales), form. hour-glass-shaped tree char. of English landscape as depicted by Constable, etc. but almost all lost to D.E.D. in 1970s–80s, clonal & largely reprod. by suckers (fr. usu. sterile); ***U. pumila*** L. (Siberian e., E As.) – resistant to D.E.D., invasive US; ***U. rubra*** Muhlenberg (*U. fulva*, slippery e., E N Am.) – medic.; ***U. thomasii*** Sarg. (cork or hickory or rock e., NE N Am.) – timber for chairs, tool-handles, etc.; many hybrids incl. ***U.* × *hollandica*** Mill. (*U. minor* × *U. glabra*, Dutch e.) – many clones form. in cult., backcross with *U. m.* hybridized with *U. chenmoui* W.C. Cheng (China) = '**Morfeo**' (Firenze, 2000), also resistant to D.E.D., ***U.* × *vegeta*** (Loudon) Ley (*U.* 'Vegetata', ? = *U.* × *hollandica*, Huntingdon e.) – cult. orn. with ascending branches, a magnificent avenue tree susceptible to D.E.D., & '**Sapporo Autumn Gold**' (*U. davidiana* var. *japonica* (Rehder) Nakai × *U. pumila*) raised Japan 1958 app. resistant

Ultragossypium Roberty = *Gossypium*

Ulugbekia Zak. = *Lithospermum*

ulva marina *Zostera marina*

Umbelliferae Juss. (Apiaceae). Magnoliidae – Apiales. 431/3700 cosmop. esp. N temp. & trop. mts. R: V.H. Heywood (ed. 1971) *The biology & chemistry of the U.*; M.G. Pimenov & M.V. Leonov (1993) *The genera of the U.* Herbs, oft. with pithy scapes, or less oft. shrubs or even trees (*Steganotaenia*), some pachycaul (*Dahliaphyllum*), oft. with unusual sec. thickening, aromatic & s.t. poisonous (17-carbon-skeletoned poly-acetylenes, rarely alks, e.g. *Conium*). Lvs pinnately or ternately compound or dissected (rarely palmately so or simple & even phyllodic), s.t. spiny, usu. in spirals, usu. with broad sheathing base s.t. with stipular flanges. Fls usu. small, bisex. (*Acronema* dioec.), 5-merous exc. G, reg. in compound umbels (Apioideae) when oft. subtended by involucre of free or united bracts & the marginal fls s.t. sterile with C expanded marginally, heads or simple umbels s.t. reduced to single fls, or dichasia; K oft. small teeth around ovary apex to 0, C oft. white, yellow, purple etc., valvate (rarely 0), A alt. with C on nectary-disk, anthers with longit. slits, $\overline{G}$ (2), 2-loc. (rarely 1-loc & pseudomonomerous) with distinct styles, each loc. with 1 (+ 1 abortive) apical-axile, pend., anatr. unitegmic ovule. Fr. usu. a schizocarp of 2 mericarps facially united; integument oft. adherent to pericarp, embryo usu. small, in copious oily endosperm. n = (4)8(–)11(12)

Classification & principal genera (modified after Drude – in need of modern review; some genera not placed in this scheme):

Mackinlayoideae (Actinotaceae; Hydocotyleae – Hydocotylinae p.p. + Araliaceae – Mack., some woody, most S Pacific rim): *Actinotus*, *Apiopetalum*, *Centella*, *Mackinlaya*, *Micropleura*, *Xanthosia* (only)

Azorelloideae ([incl. most of 'I. Hydrocotyloideae' (stip. present; fr. with woody endocarp; mainly S temp., 60% in Am., 90% of those in S; x = ?8) – 2 tribes: 1. 'Hydocotyleae' (true H. = Araliaceae – Hydrocotyloideae; *Centella* & b, Xanthosiinae = Mack.); 2. Mulineae (3 subtribes)] temp. S Pacific to China, *Drusa* Afr.): *Azorella*, *Bolax*, *Bowlesia*, *Hermas*, *Mulinum*, *Stilbocarpa*

[II.] **Saniculoideae** (lvs usu. undivided with teeth with prickles or hairs, stip. 0; infl. usu. a simple umbel or head, fr. prickly or scaly with soft endocarp, style surrounded by ring-like disk; some basal taxa woody – *Polemanniopsis*, *Steganotaenia*

1. **Saniculeae** (Eryngieae; R: T 57(2008)365); n = 8 (9, 11, 12)): *Alepidea*, *Astrantia*, *Eryngium*, *Sanicula*

[2. Lagoecieae = III 8.]

[III.] **Apioideae** (lvs usu. finely divided, stip. 0, infls usu. compound umbels, style on apex of disk, fr. with soft endocarp; n= 11; some woody (*Bupleurum*, *Myrrhidendron*); orig. ? S Afr., 80% in OW) – 12 tribes with details of principal ones:

1. **Echinophoreae**: *Thecocarpus*
2. **Scandiceae** (parenchyma around carpophore with crystal layer): *Anthriscus*, *Chaerophyllum*, *Myrrhis*, *Osmorhiza*, *Scandix*
3. **Caucalideae** (Dauceae; mericarps with spines on ridges): *Caucalis*, *Cuminum*, *Daucus* [with *Cuminum*, *Thapsia* to 2. as Daucinae – R: T 65(2016)585], *Torilis*
4. **Coriandreae** (parenchyma without crystal layer; heterogeneous): *Coriandrum*
5. **Smyrnieae** (mericarps rounded outwards): *Arracacia*, *Cachrys*, *Conium*, *Oreomyrrhis*, *Prangos*, *Scaligeria*, *Smyrnium*, *Tauschia*
6. **Hohenackerieae**: *Hohenackeria* (only)

7. **Pyramidoptereae**: *Cyclosperma, Pyramidoptera*
8. **Apieae** (incl. II. 2; primary ridges of mericarps all similar; seeds semi-circular in cross-sect.; poss. splittable into subtribes): *Aciphylla, Acronema, Aegopodium, Anisotome, Apium, Bunium, Bupleurum, Elaeosticta, Foeniculum, Heteromorpha, Ligusticum, Oenanthe, Pimpinella, Seseli*
9. **Angeliceae**: *Angelica, Cymopterus*
10. **Peucedaneae**: *Anethum, Ferula, Ferulago, Lomatium, Peucedanum, Steganotaenia*
11. **Tordylieae**: *Heracleum, Pastinaca*
12. **Laserpitieae** (Thapsieae): *Laserpitium, Thapsia*

App. an 'old Southern family'. Division between U. & Araliaceae resolved though dismemberment of heterogeneous Hydrocotyloideae (parallel reductions to habit of *Centella, Hydocotyle* = Araliaceae), import of Araliaceae-Mack., exclusion of Myodocarpaceae from A., etc. using molecular systematics (amalgamation of all in a greater U. would prob. draw in Pittosporaceae; papilionid caterpillars eat U. s.s. but not A. (incl. *Hydrocotyle*) nor P.). U. have been long recog. as an assemblage based on morphological (& chem.) char. & poss. first designated a fam. in 16 cent.; Morison (1672) wrote the first (Western) systematic study (of any group) on them, utilizing fr. char. which have been v. imp. in their class., though perhaps overstressed. Only c. 20% of genera have more than 9 spp. (many tribes with 1 large genus & many 'satellites') & many are unispecific & some of these with odd affinities, their closest allies being geographically v. distant now. The narrow generic concept (cf. Cruciferae) prob. result of pre-Linnaean folk taxonomy in Apioideae because of many temp. genera recog. for distinctive flavours: future investigation of the phylogeny of the group could well reduce the no. of genera recog. in temp. Floras. Poll. by insects, the enlarged marginal fls act as attractants (e.g. *Daucus*; cf. *Hydrangea, Viburnum*) or involucral bracts resembling a C such that infl. looks like single fl. (e.g. *Astrantia, Bupleurum, Eryngium, Hacquetia*), in *Mathiasella* the true fls unisexual with females without C. Hybrids v. rare (but see *Aciphylla*)

Many imp. foods (V.E. Rubtatzky & al. (1999) *Carrots & related vegetable Umbelliferae*), herbs, spices & flavourings incl. *Anethum* (dill), *Angelica, Anthriscus* (chervil), *Apium* (celery), *Arracacia, Carum* (caraway), *Chaerophyllum, Conopodium, Coriandrum* (coriander), *Crithmum, Cryptotaenia* (mitsuba), *Cuminum* (cumin), *Daucus* (carrot), *Foeniculum* (fennel), *Laserpitium, Levisticum* (lovage), *Lomatium, Myrrhis, Oenanthe, Opopanax, Pastinaca* (parsnip), *Petroselinum* (parsley), *Pimpinella* (aniseed) etc., though many are toxic (e.g. *Aethusa, Cicuta, Conium* (hemlock), *Oenanthe*) while *Deverra* & *Heracleum* can cause serious dermatitis; medic. & other gums are derived from *Ferula*, toothpicks from *Ammi*, scent from *Dorema*; few cult. orn. incl. *Aciphylla, Ammi, Astrantia, Eryngium, Hacquetia*, etc. *Aegopodium, Aethusa* & *Cyclospermum* weedy

Umbellularia (Nees) Nutt. Lauraceae (II). 1 W N Am.: ***U. californica*** (Hook. & Arn.) Nutt. (Calif. laurel or olive, Calif.) – fine timber (myrtlewood, pepperwood) used even in jewelery & C19–20 Oregon coins, local medic., incl. disinfectant, sniffing crushed lvs can lead to sneezing & headaches

Umbilicus DC. Crassulaceae (I 2). Incl. *Chiastophyllum*, 14 Eur. (6) & Medit. to Iran & Afr. mts. R: U. Eggli, *Ill. Handb. Succ. Pls*, Crass. (2005)364. Some cult. orn. esp. ***U. rupestris*** (Salisb.) Dandy (? = *U. umbilicatus*, navelwort, penny pies, (wall) pennywort, GB to Macaronesia & SW As.) – form. ed., medic. (skin problems) & used in expulsion of afterbirth in cows, sore udders etc., also ***U. oppositifolius*** Ledeb. (*C. o.*, lamb's-tail, Cauc.) – yellow fls in drooping infls

umbrella bamboo *Thamnocalamus spathaceus*; **u. fir** *Sciadopitys verticillata*; **u. pine** *Pinus pinea*; **u. palm** *Hedyscepe canterburyana*; **u. plant** *Darmera peltata, Cyperus alternifolius*; **u. tree** *Cussonia* spp., *Musanga cecropioides, Polyscias* spp., *Schefflera actinophylla*

umbu *Spondias tuberosa*

umburana *Amburana* spp.

ume *Prunus mume*

umiry balsam *Humiria* spp.

Umtiza Sim. Leguminosae (I 4). 1 E Cape: ***U. listeriana*** Sim – local medic., wood (oily) form. used as housings of boat propellor-shafts. R: Strelitzia 10(2000)298

umzimbeet *Millettia grandis*

Unanuea Ruíz & Pavón ex Pennell = *Stemodia*

Uncaria Schreb. Rubiaceae (II 2; Naucleaceae). c. 34 trop. (Afr. & Med. 3, Am. 2). R: Blumea 24(1978)68. Lianes (Roux's Model) with alks & accrescent clasping hooks in place of infl. axes, 1 Afr. sp. with hollow ant-infested stems. Some locally medic. & dye-pls; ***U. gambir***

(Hunter) Roxb. (gambier, gambir, kath(a), cultigen in trop. As.) – form. imp. tan-source grown in plantation, now scarcely cult., in 17 cent. lvs & young shoots used to adulterate tea ('catechu'), now boiled lvs a mulch for pepper (W Mal.)

Uncarina (Baill.) Stapf. Pedaliaceae (2). 11 Madag. R: F Mad. 179(1979)7. Fleshy r. medic.; frs make effective mouse-traps

Uncifera Lindl. Orchidaceae (V 16c). 6 Ind. to W Mal.

Uncinia Pers. = *Carex*

Undaria Suringar. Alariaceae (Brown Algae). 3–4 E As. ***U. pinnatifida*** (Harv.) Suringar (wakame) – highly invasive, now in UK, Aus., NZ, US etc., long cult. Jap., since 1983 also Fr., as veg. & in soups (macrobiotic movement), fucoxanthin helping 'burn fat'

Ungeria Schott & Endl. Malvaceae (Hel.; Sterculiaceae). 1 Norfolk Is.: ***U. floribunda*** Schott & Endl. R: FA 49(1994)108

Ungernia Bunge. Amaryllidaceae. 6 C As. (steppes) to Jap. Alks

Ungnadia Endl. Sapindaceae. 1 S N Am.: ***U. speciosa*** Endl. (Mex. buck-eye) – cult. orn. decid. tree

Ungula Barlow = *Amyema*

Ungulipetalum Mold. Menispermaceae (I). 1 Braz.: ***U. filipendulum*** (Mart.) Mold.

unicorn root *Aletris farinosa*, *Chamaelirium luteum*

Unigenes F. Wimmer (~ *Lobelia*). Campanulaceae (III 1). 1 S Afr.: ***U. humifusa*** (A. DC.) F. Wimmer

Uniola L. Gramineae (XXVII 3). 8 trop. & warm Am. Sand-binders (esp. ***U. paniculata*** L.), poss. material for paper

union nut *Bouchardatia neurococca*

Univiscidiatus (Kores) Szlach. = *Acianthus*

Unona L.f. = *Xylopia*

Unonopsis R. Fries. Annonaceae (IV 2). 36 trop. Am. (Amazon 15 (Peru 12) – R: BMPEG 5(1989)207). ***U. guatterioides*** (A. DC.) R. Fries (Amaz.) – poll. male euglossine bees; ***U. veneficiorum*** (Mart.) R. Fries (Amazonia) – arrow-poison & contraceptive

Unxia L.f. (~ *Villanova*). Compositae (Mill.-Mill.). 3 Panamá & N S Am. R: Britt. 21(1969)314

upas tree *Antiaris toxicaria*

upland cress *Barbarea verna*

Upudalia Raf. = *Eranthemum*

Upuna Sym. Dipterocarpaceae (Dipt.-Dipt.). 1 Borneo: ***U. borneensis*** Sym. – many app. primitive features incl. arillate seeds; heavy constr. timber. R: TFSS 5(2004)348

Uralepis Nutt. = *Triplasis*

Urandra Thw. = *Stemonurus*

Uranodactylus Gilli = *Lepidium*

Uranthoecium Stapf. Gramineae (~ *Setaria*; XXIV 4). 1 trop. Aus.: ***U. truncatum*** (Maiden & Betche) Stapf

Uraria Desv. Leguminosae (III 19). 20 trop. OW. Some locally medic.

Urariopsis Schindler = *Uraria*

Urbananthus R. King & H. Robinson (~ *Eupatorium*). Compositae (Eup.-Crit.). 2 Cuba, Jamaica. R: MSBMBG 22(1987)300

Urbania Philippi = *Junellia* (but see BSAB 25(1988)478)

Urbanodendron Mez. Lauraceae (I). 1 E trop. S Am. R: BJ 110(1988)165

Urbanodoxa Muschler = *Cremolobus*

Urbanoguarea Harms = *Guarea*

Urbanolophium Melchior = *Amphilophium*

Urbanosciadium H. Wolff = *Niphogeton*

Urbinella Greenman. Compositae (Helia.-Pect.). 1 Mex.: ***U. palmeri*** Greenman

Urceodiscus W. de Wilde & Duyfjes = *Papuasicyos* (but see Blumea 51(2006)38)

Urceola Roxb. Apocynaceae (II c). 16 Indomal. R: Blumea 41(1996)82. Rubber sources esp. ***U. brachysepala*** Hook. f. (*U. maingayi*, Rangoon rubber, Indomal.) & ***U. javanica*** (Blume) Boerl. (*Hymenolophus romburghii*, Mal.). ***U. lucida*** (D. Don) Kurz (*U. esculenta*, kyetpaung, Myanmar) – fr. ed., source of a blue dye

Urceolina Reichb. Amaryllidaceae. 5–7 S C Peruvian Andes. Cult. orn. rain-forest bulbous pls. See also *Stenomesson*

urd *Vigna mungo*

Urechites Muell. Arg. = *Pentalinon*

Urelytrum Hackel. Gramineae (XXII 2). 7 trop. Afr.

Urena Dill. ex L. = *Hibiscus*

Urera Gaudich. Urticaceae (I). 35 trop. & S Afr., Madag., Hawaii, trop. Am. Heterogeneous. Trees, shrubs & lianes with powerful stinging hairs with fr. encl. in persistent fleshy P; alks; oft. used for hedges, local medic. ***U. baccifera*** (L.) Wedd. (S Am.) – bark used for amate paintings sold to tourists in Mex.; ***U. glabra*** (Hook. & Arn.) Wedd. (Hawaii) – fibre for fishing-nets & networks for feather cloaks & *Cordyline* raincoats; ***U. hypselodendron*** Hochst. (E & NE Afr.) – liane to 25 m, fibre for bags in Ruwenzori; ***U. laciniata*** (Goudot) Wedd. (Panamá) – large spines

Urginavia Speta = *Drimia*

Urginea Steinh. = *Drimia*

Urgineopsis Compton = *Drimia*

Uribea Dugand & Romero. Leguminosae (III 2). 1 C Am.: ***U. tamarindoides*** Dugand & Romero

Urmenetea Philippi. Compositae (Mut.-Onis.). 1 N Chile & NW Arg.: ***U. atacamensis*** Philippi – Andean ann. herb, coca subs. used in a tea for altitude sickness

urn plant *Aechmea fasciata*

Urnularia Stapf = *Willughbeia*

Urobotrya Stapf. Opiliaceae. 7 trop. Afr. (2), SE As. to Flores (5). R: BJ 107(1985)137. R.-parasites

Urocarpidium Ulbr. (~ *Tarasa*). Malvaceae (Malv.-Malv.). 1 Peru: ***U. albiflorum*** Ulbr. R: T 60(2011)1335

Urocarpus J.L. Drumm. ex Harv. = *Asterolasia*

Urochilus D.L. Jones & M. Clements = *Pterostylis*

Urochlaena Nees = *Tribolium*

Urochloa P. Beauv. (~ *Panicum*). Gramineae (XXIV 5). Incl. *Brachiaria, Eriochloa p.p., Megathyrsus*, excl. *Moorochloa, Rupichloa*, 135 trop. & warm (N Am. 6 native). Some fodder esp. ***U. arrecta*** (T. Durand & Schinz) Morron & Zuloaga (*B. a.*, Tanner grass, Afr.), ***U. brizantha*** (A. Rich.) R. Webster (*B. b.*, palisade g., Afr.), ***U. maxima*** (Jacq.) R. Webster (*Megathyrsus m., P. m.*, Guinea grass, Afr., natur. Aus., Am.) – most imp. lowland trop. Am. cult. forage, facultative apomict, tetraploid races commonest; ***U. mosambicensis*** (Hackel) Dandy (sabi grass, trop. & S Afr.) – mooted for land reclamation, ***U. mutica*** (Forssk.) T.Q. Nguyen (*B. m.*, Pará or para g., trop.) – pasture improvement but invasive N Aus. floodplains, SE US, trop. S Am., Hawaii, Galapagos, ***U. polystachya*** (Kunth) Mabb. (*Eriochloa p.*, Carib grass, WI) – esp. in US, & ***U. texana*** (Buckley) R. Webster (Colorado g., Texas millet, S N Am.)

Urochondra C. Hubb. Gramineae (XXVIII). 1 Sudan & Somalia to Pakistan: ***U. setulosa*** (Trin.) C. Hubb. – grain with beak (accrescent connate style-bases)

Urodon Turcz. (~ *Pultenaea*). Leguminosae (III 13). 4 SW Aus.

Urogentias Gilg & C. Benedict. Gentianaceae. 1 Uluguru & Nguru Mts, Tanz.: ***U. ulugurensis*** Gilg & C. Benedict – C with v. long fimbriate processes

Urolepis (DC.) R. King & H. Robinson (~ *Eupatorium*). Compositae (Eup.-Gyp.) 1 S Am.: ***U. hecatantha*** (DC.) R. King & H. Robinson

Uromyrtus Burret. Myrtaceae (II 10). c. 20 Borneo to NG, New Caled., trop. Aus. (4 – SB 26(2001)733)

Uropappus Nutt. (~ *Microseris*). Compositae (Lact.-Mic.). 1 SW N Am.: ***U. lindleyi*** (DC.) Nutt. R: FNA 19(2006)322

Urophyllum Jack ex Wall. Rubiaceae (I 9). Incl. *Pleiocarpidia, Pravinaria*, 150 OW trop. to Jap.

Urophysa Ulbr. Ranunculaceae (III 1). 2 China

Uroskinnera Lindl. Plantaginaceae (Ant.- Chel.; Scrophulariaceae s.l.). 4 Mex. & C Am. R: Madroño 39(1992)131

Urospatha Schott. Araceae (V). 10 trop. Am. Seeds corky, water-disp. ***U. antisylleptica*** R. Schultes (Colombia) – ground spadix an oral contraceptive; ***U. caudata*** (Poepp.) Schott (Peru, Braz.) – spongy rhiz. ed.

Urospathella Bunting = praec.

Urospermum Scop. Compositae (Lact.-Hypo.). 2 Medit. incl. Eur. to Pakistan, ***U. picroides*** (L.) F.W. Schmidt weedy Am.

Urostachya (Lindl.) Brieger = *Eria*

Urostemon R. Nordenstam = *Brachyglottis*

Urostephanus Robinson & Greenman. Apocynaceae (Asclepiadaceae III 3). 1 Mex.: ***U. gonoloboides*** Robinson & Greenman

Urotheca Gilg. = *Gravesia*

Ursia Vassilcz. = *Trifolium*

Ursifolium Doweld = *Trifolium*

Ursinia Gaertn. Compositae (Anth.-Urs.). 43 S Afr. & Ethiopia (1). R: MBSM 6(1967)363, 531. Shrubs to herbs; some ray-florets absent exposing iridescent involucral bracts poss. mimicking beetles (cf. *Gorteria*) & some poll. monkey beetles. Some ann. spp. cult. for orange or yellow fls

Ursiniopsis E. Phillips = *Ursinia*

Ursopuntia Heath = *Tephrocactus*

Ursulaea Read & Baensch = *Aechmea*

Urtica Tourn. ex L. Urticaceae (I). Incl. *Hesperocnide*, 80 subcosmop. esp. N temp. (Eur. 11). ? Heterogeneous. Nettles, Eur. spp. larval food-pls for red admiral (*Vanessa atalanta*), peacock (*Aglais* [*Inachis*] *io*) & small tortoiseshell (*A. urticae*) butterflies. Usu. herbs, incl. annuals, with opp. lvs, monoec. or dioec., usu. with alks & stinging hairs (bulb-like projections which break off along a weak line leaving sharp bevelled edge which pierces skin & injects (in ***U. urens*** L. (small n., N temp.) at least), histamine causing itching, a neurotoxin (prob. a sodium channel toxin), 5-hydroxytryptamine & acetylcholine a burning sensation effective even in dried pl. In *U. urens*, panicle of male & female fls at each node, in ***U. pilulifera*** L. (Roman n., S Eur.) male catkin-like infl. & female pseudohead infl. at each node, in ***U. dioica*** L. (stinging n., stinger, Euras., widely natur., tetraploid weed poss. derived from stingless diploid (subsp. ***galeopsifolia*** (Opiz) Chrtek, 'var. *inermis*') of forests) usu. dioec.; A bent down in bud & when ripe spring upwards violently, the anthers turning inside out & ejecting a cloud of smooth poll. borne away by wind. Some young shoots eaten like spinach (nettle pudding) or in certain Dutch cheeses, some fibre-sources esp. for fishermen's nets: *U. dioica* used for fixing Stone Age arrowheads & for cloth until 18 cent. & in Silesia until 20 cent., but also WW I Ger. military uniforms (40 kg nettles needed for 1 shirt) when cotton unavailable, green dye (as in WW II for wool & camouflage nets), imp. fibre & local medic. N Am. (& Nepal), form. an infusion used as a massage to cure dandruff, also spring tonic to clear boils, eczema & pimples, & used for rheumatism, coughs, lung disease etc. in UK, rhiz. extracts effective treatment for urinary symptoms assoc. with benign prostate enlargement, seeds steeped in wine alleged to promote lechery (Galen); ***U. massaica*** Mildbr. (E Afr.) – used for diabetes & other kidney problems in Kenya; ***U. parviflora*** Roxb. (Himal.) – shoots sold as vegetable in Sikkim; ***U. pilulifera*** 'var. *dodartii*' passed off as 'Spanish marjoram' in 18 cent. as a 'joke'

Urticaceae Juss. Magnoliidae – Rosales. Incl. Cecropiaceae, 48/1600 trop. to temp. (few); over half spp. in *Elatostema* & *Pilea*. Usu. dioec. or monoec. wind-poll. herbs & shrubs, lianes or trees (few; some pachycaul – *Dendrocnide*, *Obetia*, '*Sarcopilea*'), oft. with stinging hairs but usu. without milky latex. Lvs simple (deeply lobed to palmate in *Cecropia* & allies), in spirals or opp., usu. with 3 subequal veins from base & with stip. Fls small, usu. reg., in axillary cymes s.t. reduced to 1 fl., s.t. unisexual; males with P(1–)4 or 5(6) valvate, s.t. united by tube & A (1 or) as many as & opp. them, filaments usu. violently reflexed at pollen maturity, anthers with longit. slits, G vestigial; females with P 4(5), oft. unequal, free to ± united or 0, $\underline{G}$ pseudomonomerous (1 carpel v. reduced), 1-loc. with 1 style & 1 basal orthotropous (to hemitropous) bitegmic ovule. Fr. an achene, nut or drupe oft. encl. in accrescent P (s.t. fleshy), rudimentary A in females acting to eject achene; seeds with reduced testa & straight embryo in thin oily or starchy endosperm (or 0). n = 7–14

Classification & principal genera:

Cecropieae (Cecropiaceae; trees with deeply lobed or palmate lvs, non-explosive A; R: T 27(1978)39) – *Cecropia, Musanga*

I. **Urticeae** (trees to herbs, stinging hairs): *Dendrocnide, Laportea, Obetia, Urera, Urtica*

II. **Elatostemateae** (Lecantheae; Procrideae; no stinging hairs; P of females 3-lobed, stigma like a paint-brush): *Elatostema, Pilea*

III. **Boehmerieae** (no stinging hairs; males usu. with A 4 or 5; involucre 0): *Boehmeria, Leucosyke* (females with P 0)

IV. **Parietarieae** (no stinging hairs; P present; bracts oft. forming involucre): *Parietaria* (fls bisex., stip. 0)

V. **Forsskaoleeae** (no stinging hairs; A 1; R: NJB 8(1988)34): *Forsskaolea*

Close to Moraceae which should prob. be reunited with U.

Fibres (*Boehmeria* – ramie, *Debregeasia, Forsskaolea, Girardinia, Laportea, Oreocnide, Pipturus, Poikilospermum, Pouzolzia, Sarcochlamys, Touchardia, Urera, Urtica* etc.), some ed. as spinach (*Laportea, Urtica* etc.) or tubers (*Pouzolzia*), or fr. (*Pourouma*), dyes (*Oreocnide*), some light timber for pulp etc. (*Musanga, Myrianthus*) & some cult. orn. esp. spp. of *Pilea & Soleirolia*. Stinging hairs in *Dendrocnide* potentially v. dangerous; some bad weeds – *Parietaria, Urtica*

urucu *Bixa orellana*

urunday *Astronium urundeuva*

Urvillea Kunth. Sapindaceae (IV 1). 15 trop. Am. Lianes

Usteria Willd. Loganiaceae (Strychnaceae). 1 W & C Afr.: ***U. guineensis*** Willd.

Utania G. Don (~ *Fagraea*). Gentianaceae. c. 15 SE As. to NG. Roux's Model

utile *Entandrophragma utile*

Utleria Beddome ex Benth. = *Decalepis*

Utleya Wilbur & Luteyn. Ericaceae (VIII 5). 1 Costa Rica: ***U. costaricensis*** Wilbur & Luteyn. R: Britt. 29(1977)267

Utricularia L. Lentibulariaceae. c. 220 cosmop. (Eur. 6; Ind. 33; W Aus. 55; Braz., Venez. & Guyana 59) esp. trop. R: KBAS 14(1989)75. Aquatics, epiphytes or even twiners; carnivores, most terr. spp. preying on Protozoa (primitive condition?). Many spp. with cleistogamous fls in spring & resting buds (turions) tolerant of dry conditions. ***U. vulgaris*** L. (bladderwort, N temp.) – submerged aquatic with no r. (not even in seedling), the stem bearing photosynthetic appendages carrying bladders with trap-doors, glands pump out water from traps, generating negative pressure allowing elastic energy to be stored in walls, trapdoor buckles reversing curvature when triggered by small animals, opening & closing in less than 1 millisecond, relaxation of walls leading to sucking in of water & prey, the pitchers secreting digestive enzymes & absorbing the nutrients from the corpse; other spp. (trop.) with runners bearing bladders or (epiphytes) tuberous branches holding water; in aquatics fls aerial on axes with small lvs. No clear distinction between r., stems & lvs; all lvs, bladders, runners, tubers, erect shoots etc.'homologous', developing from similar primordia under diff. conditions, though not all primordia are totipotent; stolons of ***U. longifolia*** Gardner (Braz.) & terr. spp. strongly resemble adventitious r. like some rootcapless ones in *Pinguicula*. Cult. orn. for enthusiasts. ***U. humboldtii*** Schomb. (trop. S Am.) – grows in *Brocchinia* pitchers at high alts; ***U. purpurea*** Walter (N Am.) – few prey but many algae, zooplankton etc. & debris (? symbiosis)

Utsetela Pellegrin. Moraceae (IV). 2 Gabon, Congo. R: BJBB 64(1995)179

uva grass *Gynerium sagittatum*

uvalha *Eugenia pyriformis*

Uvaria L. Annonaceae (III 7). Incl. *Anomianthus, Cyathostemma, Ellipeia, Ellipeiopsis*, 100+ OW trop. (Afr. 69). Usu. lianes (Roux's & Troll's Models) with recurved infl.-axis hooks. ***U. concava*** Teijsm. & Binn. (Mal. to Queensland) – poll. pollen-coll. meliponid bees; ***U. dulcis*** Dunal (*A. d.*, SE As., Java) – sweet fr.; ***U. elmeri*** Merr. (Borneo) – cauliflorous liane, poll. cockroaches. Acetogenins with antitumour action; some local medic. esp. ***U. chamae*** P. Beauv. (finger r., W Afr.) – eyewash etc. & ***U. grandiflora*** Roxb. ex Hornem. (trop. As.). Some ed. fr.

Uvariastrum Engl. Annonaceae (III 6). 5 trop. Afr. R: PhytoKeys 33(2014)9

Uvariodendron (Engl. & Diels) R. Fries. Annonaceae (III 6). 12 trop. Afr. Troll's Model

Uvariopsis Engl. Annonaceae (III 6). 17 trop. Afr. R: Novon 13(2003)447. Roux's Model

uva-ursi *Arctostaphylos uva-ursi*

Uvedalia R. Br. (~ *Mimulus*). Phrymaceae. 2 Aus., Timor, ?NG. R: Phytoneuron 2012–39:21

uvilla *Pourouma cecropiifolia*

Uvularia L. Colchicaceae (Uvulariaceae). 5 E N Am. R: FNA 26(2002)147. Bellwort. local medic.; cult. orn. rhizomatous herbs incl. ***U. sessilifolia*** L. – young shoots ed. like asparagus (cf. *Ornithogalum*)

Uvulariaceae A. Gray ex Kunth = Colchicaceae

V

Vaccaria Wolf. Caryophyllaceae (III 1). 1 Euras., Medit.: ***V. hispanica*** (Mill.) Rauschert (*V. pyramidata, V. segetalis*, cow cockle) – cult. orn., appearing from sown birdseed, natur. Aus., S Afr., N & S Am.

Vacciniaceae DC. ex Perleb = Ericaceae
Vacciniopsis Rusby = *Disterigma*
Vaccinium L. Ericaceae (VIII 5). c. 500 circumpolar (4) & N temp. (Eur. 8, Jap. 22, N Am. 65 – S.P. Vander Kloet (1988) *The genus V. in N Am.*), Indomal. (= ? *Agapetes*), C & SE Afr. mts & Madag. (6), trop. Am. (30). Heterogeneous? Decid. or everg. shrubs, small trees & lianes (Mangenot's Model), many with ed. fr. (blueberries, buckberries, huckleberries, bluets, N Am.) used in orig. pemmican, some cult. orn., some clones (Pennsylvania) allegedly to 13 000 yrs old. ***V. angustifolium*** Aiton (NE N Am.) – fr. much-coll. from wild pls managed for prod.; ***V. arboreum*** Marshall (farkleberry, S & SE US) – fr. ined.; ***V. arctostaphylos*** L. (Broussa tea, Cauc.); ***V. corymbosum*** L. (E US) – cult. blueberry in N Am. (s.t. hybrids with *V. angustifolium* or ***V. darrowii*** Camp (SE US), though 'rabbit-eye' bs derived from ***V. ashei*** Reade, SE US)) & the blueberry grown comm. in W & C Eur. (even hydroponically in Aus.) for pies, syrups, tinning etc., antioxidant anthocyanins alleged to increase communication between brain cells so may stem age-related memory-loss; ***V. floribundum*** Kunth (*V. mortinia*, mortiña, Andes) – fr. in local commerce, also inhibiting formation of fat-cells, so poss. useful in treating obesity; ***V. macrocarpon*** Aiton (cranberry, N As., E N Am.) – fr. float (water-disp.?), acidic, harvested by flooding, eaten with poultry, grown comm. by Joseph Banks in GB 1808, before US (1840s), dried (& flavoured with elderberry juice) as snacks, US, vogue medic. (long used against cystitis etc., poss. useful for gout & arthritis, febrifuge, etc.); ***V. myrtillus*** L. (bilberry, whinberry, whortleberry, blaeberry, Euras., N Am.) – ed. fr. for pies etc., wine-making, fr. skin alleged to improve night vision, juice for dissolving kidney-stones & diarrhoea (Scottish Highlands); ***V. ovalifolium*** Sm. (mathers, N N Am.) – cult. orn., fls opening before lvs; ***V. oxycoccos*** L. (cranberry, N temp.) – lvs revolute, anti-bacterial properties, fr. an inferior subs. for *V. macrocarpon*; ***V. reticulatum*** Sm. ('ohelo berry', Hawaii) – disp. Hawaiian geese (nénés), fr. in comm. preserves; ***V. stramineum*** L. (deerberry, E US) – buzz-poll., cult. orn., fr. ined.; ***V. vitis-idaea*** L. (cowberry, ling(on)berry, foxberry, Euras., N Am.) – fr. a cranberry subs. See also *Agapetes*, *Dimorphanthera*
Vachellia Wight & Arn. (~ *Acacia*). Leguminosae (II 3). c. 160 trop. (As. 36, Aus. 7, Afr. 73, Am. 60). Lvs bipinnate, paired stipular spines, s.t. swollen & inhabited by ants; peduncle with involucre. Trees & shrubs of dry regions, where they can dominate e.g. ***V. xanthophloea*** (Benth.) Banfi & Galasso (fever trees of Kenya Rift Valley), V. *nilotica* (L.) P. Hurter & Mabb. (*A. n.*, Ind. plains where almost only tree) & ***V. gerrardii*** (Benth.) P. Hurter (*A. g.*, trop. Afr. to Iraq) – only native tree sp. in Kuwait. Myrmecophilous spp. like ***V. cornigera*** (L.) Siegler & Ebinger (*A. c.*, bull-horn thorn, C Am., natur. US) & ***V. sphaerocephala*** (Cham. & Schldl.) Siegler & Ebinger (C Am.) with extrafl. nectaries on petioles & yellow sausage-shaped food-bodies on leaflet tips, attractive to ants which discourage cli. plants & attack would-be grazers but are not exclusive to any particular sp. Many imp. prods – timber, fuel, forage, tanbark (some tannins molluscicidal), gums, scents, honey-sources (Afr.) & cult. orn.: ***V. drepanolobium*** (B. Y. Sjöstedt) P. Hurter (*A. d.*, whistling thorn, myrmecophyte (whistling due to wind blowing over ant-holes), E Afr.) – source of gum arabic subs.; ***V. eburnea*** (L.f.) P. Hurter & Mabb. (*A. e.*, cockspur thorn, Arabia to Ind.) – cult. orn., long spines; ***V. erioloba*** (E. Mey.) P. Hurter (*A. e.*, SW Afr.) – only tree in 0–50 mm rain per annum zone, seeds passing through stock & germ. after floods; ***V. farnesiana*** (L.) Wight & Arn. (cassie, opopanax, popinac, sweet acacia, warm Am. now pantrop., introd. Aus. before Eur. settlement) – source of ess. oil (cassie ancienne) for perfumery (used in 'violet' scent), esp. cult. in S France (Farnese Palace, Rome 1611); ***V. horrida*** (L.) Kyal. & Boatwr. (*A. latronum*, S Afr.) – spines form. used as gramophone needles; ***V. karroo*** (Hayne) Banfi & Galasso (*A. karroo*, ? *Mimosa capensis*, S Afr.) – source of Cape gum; ***V. nilotica*** (L.) P. Hurter & Mabb. (*A. n.*, babul, trop. Afr., natur. Ind., weed of nat. significance in Aus., E Java) – imp. tanbark & gum arabic subs. (Amrad gum), timber & forage, used by Anc. Egyptians, fr. (Gambia or sant pods) ed. & tanning (30% tannin, molluscicidal (used on field scale) & algicidal), local medic.; ***V. seyal*** (Del.) P. Hurter (*A. s.*, shittim, trop. Afr. to Egypt) – ed. gum (talha, talh g.), contender for timber used for Ark of the Covenant & coffins of pharaohs (also ***V. tortilis*** (Forssk.) Galasso & Banfi (*A. raddiana*))
Vacoparis Spangler = *Sorghum* (but see AusSB 16(2003)297)
Vagaria Herb. Amaryllidaceae. 1–2 Syria & Israel (1: ***V. parviflora*** Herb. – cult. orn.)
Vaginularia Fée (~ *Monogramma*). Pteridaceae (Vitt.). 4 trop. As., Pacific
Vahadenia Stapf. Apocynaceae (I c). 2 trop. W Afr. R: BJBB 63(1994)320
Vahlia Thunb. Vahliaceae (Saxifragaceae s.l.). 5 trop. & S Afr. R: KB 30(1975)163

Vahliaceae Dandy (~ Saxifragaceae). Magnoliidae – Vahliales. 1/5 trop. & S Afr. R: FTEA Vahliaceae (1975). Densely glandular-hairy to glabrous small shrubs & herbs. Lvs entire, opp. Fls reg. small, protogynous, in axillary cymose infls; K valvate, C free, valvate, A5 free, inserted on disk, $\overline{G}(2(3))$ with disk, 1-loc., placentation apical with ∞ anatr. bitegmic ovules per placenta. Fr. a septicidal caps. with ∞ seeds & persistent K. n = 6, 9

Genus: *Vahlia*; fossil fls from Upper Cretaceous of Sweden (*Scandianthus*)

Vahlodea Fries (~ *Deschampsia*). Gramineae (XVI 7). 1 circumboreal (disjunct), S S Am.: ***V. atropurpurea*** (Wahlenb.) Hartm. (*D. a.*). R: FNA 24(2007)691

vahy *Landolphia madagascariensis*

Vailia Rusby. Apocynaceae (V c; Asclepiadaceae III 4). 2 S Am.

vakul *Mimusops elengi*

Valantia Tourn. ex L. Rubiaceae (IV 16). 8 Macaronesia to Iran (Eur. 3)

Valdivia C. Gay ex J. Rémy (~ *Escallonia*). Escalloniaceae. 1 S & C Chile: ***V. gayana*** J. Rémy – rare & endangered

Valdiviesoa Szlach. & Kolan. = *Fernandezia*

Valentiana Raf. = *Thunbergia*

Valentiniella Speg. = *Heliotropium*

Valenzuelia Bertero ex Cambess. = *Guindilia*

Valeria Minod = *Stemodia*

valerian *Valeriana officinalis*; **Greek v.** *Polemonium caeruleum*; **red v.** *Centranthus ruber*

Valeriana Tourn. ex L. Caprifoliaceae (Valerianaceae). c. 270 N temp. (Eur. 20), S Afr., Andes (incl. *Phyllactis*, R: NJB 6(1986)435). Herbs incl. pachycaul rosette-pls (*Phyllactis*) & shrubs with alks; K forms pappus on fr. Valerian esp. ***V. officinalis*** L. (Euras.) – dried rhiz. medic. (nerve relaxant since anc. times, sedative (form. used in psychoneuroses, now in proprietary herbal tranquillisers), hangover tonic & for staunching bleeding in folk medic., until 20 cent. an infusion in wine used for epilepsy), form. used in scent etc., rancid smell (valeric acid, also in human sweat), attractive to cats, Canidae & rats (? Pied Piper of Hamelin), poss. due to actinidine; other spp. locally medic. esp. ***V. jatamansi*** Jones (Himal.) used like *V. o.* & *Nardostachys j.*, or cult. orn. e.g. ***V. pratensis*** (Benth.) Steud. in Mex.

Valerianaceae Batsch = Caprifoliaceae

Valerianella Mill. Caprifoliaceae (Valerianaceae). Incl. *Fedia*, *Pseudobetckea*, 64 N temp. to N Afr. (Eur. 24). Infl. appearing dichot. due to acrotonic development. Fr. disp. mech. various – inflated K, or sterile loc., K parachute or hooks (diff. spp.). Some potherbs etc. esp. ***V. locusta*** (L.) Laterr. (cornsalad, lamb's-lettuce, mache, Eur. & Medit.) – salad, as is ***V. cornucopiae*** (L.) Loisel. (*F. c.*, horn of plenty, Medit.) – also cult. orn.

Valerioa Standl. & Steyerm. = *Peltanthera*

Valerioanthus Lundell = *Ardisia* (But see Wrightia 7(1982)50)

Valiha S. Dransf. (~ *Ochlandra*). Gramineae (V 6). 2 Madag., R: KB 53(1998)380

Vallariopsis Woodson. Apocynaceae (II c). 1 W Mal.: ***V. lancifolia*** (Hook.f.) Woodson. R: AUWP 86–5(1986)89

Vallaris Burm.f. Apocynaceae (II c). 3 trop. As. R: MLW 82–11(1982)

Vallea Mutis ex L.f. Elaeocarpaceae. 1–2 Colombia to Bolivia. R: NJB 8(1988)19. In Andean closed forest

Vallesia Ruíz & Pavón. Apocynaceae (Ia). 5 trop. Am. Shrubs & small trees with alks

vallis See *Vallisneria*

Vallisneria Micheli ex L. Hydrocharitaceae. Incl. *Maidenia*, 14–15 trop. & warm temp. (Eur. 1). R: AB 13(1992)283, Telopea 7(1997)111, SB 33(2008)49. Submerged dioec. grass-like herbs; male fls in head subtended by spathe, breaking off & floating up to open on surface; female fls solit. & sessile in tubular spathe on long peduncle, tightly spiralling in fr. & maturing fr. at bottom; good oxygenators ('vallis') for aquaria

Vallota Salisb. ex Herb. = *Cyrtanthus*

valonea *Quercus ithaburensis* subsp. *macrolepis*

Valvanthera C. White = *Hernandia*

Vanasushava P.K. Mukherjee & Constance. Umbelliferae (III 11). 1 S Ind.: ***V. pedata*** (Wight) P.K. Mukherjee & Constance. R: KB 29(1974)595

Vanclevea Greene = *Chrysothamnus*

Vancouveria Morren & Decne. (~ *Epimedium*). Berberidaceae (II 2). 3 W N Am. R: JLSBot. 51(1938)445. Cult. orn. herbs with ant-disp. seeds (elaiosomes) incl. ***V. hexandra*** (Hook.) Morren & Decne (inside-out fls, W US) – follicles open before seeds mature, green seeds continuing to grow thereafter, seeds also distrib. *Vespula vulgaris* (L.), common yellow jacket

Vanda Jones ex R. Br. Orchidaceae (V 16). Incl. *Ascocentrum, Christensonia, Eparmatostigma, Neofinetia, Trudelia*, 73 Himal. to Aus. (Cape York). R: BJLS 173 (2013)568. Cult. orn. epiphytes (some sweet-scented) incl. many hybrids (many intergeneric), with extrafl. nectaries; see also *Papilionanthe* & *Taprobanea*

× **Vandachostylis** Guillaumin. Orchidaceae. *Rhynchostylis* × *Vanda* – 461 grexes

× **Vandaenopsis** Guillaumin. Orchidaceae. *Phalaenopsis* × *Vanda* – 234 grexes

Vandasia Domin = *Vandasina*

Vandasina Rauschert (*Vandasia*). Leguminosae (III 18). 1 NG, Queensland: ***V. retusa*** (Benth.) Rauschert – allied to *Kennedia*

Vandellia P. Browne ex L. (~ *Lindernia*). Linderniaceae. 52 OW trop. (***V. diffusa*** L. (Afr., Madag.) natur. trop. Am.). R: Willd. 43(2013)232. ***V. micrantha*** (D. Don) Eb. Fisch. & al. (*L. m.*, Himal. to Jap., NG) – sold as condiment in Laos

Vandenboschia Copel. (~ *Crepidomanes*). Hymenophyllaceae. 15 trop. & N temp.

Vandopsis Pfitzer. Orchidaceae (V 16c). 4 SE As. to Mal. Some cult. orn.

Vangueria Comm. ex Juss. Rubiaceae (III 2). 58 trop. Afr. & Madag. R: PSE 253(2005)179. Trees & shrubs, some with ed. fr. esp. ***V. madagascariensis*** J. Gmelin (*V. edulis*, Madag.)

Vangueriella Verdc. (~ *Vangueriopsis*). Rubiaceae (III 2). 18 W trop. Afr. R: KB 42(1987)189

Vangueriopsis F. Robyns. Rubiaceae (III 2). 4 trop. Afr. R: KB 42(1987)187

Vanheerdea L. Bolus ex Hartmann. Aizoaceae (V). 2 Cape. R: Bradleya 10(1992)15

Vanhouttea Lem. (~ *Sinningia*). Gesneriaceae (II 5e). 9 Braz.

vanilla *Vanilla planifolia* & other spp.; **v. lily** *Arthropodium* spp.

Vanilla Plum. ex Mill. Orchidaceae ((II 2). Incl. *Dictyophylloria* (non-climbing), 106 pantrop. exc. Aus. (Indomal. 28, Afr.-Madag. 24, Am. 60). R: K. Cameron (2011) V. *orchids*. Epiphytic usu. scandent & s.t. leafless lianes to 30+ m, poll. euglossine bees; usu. endozoochorous; vanillin flavour (smell close to milky breast) due to anti-oxidants. By preventing haemoglobin S molecules sticking together, vanillin stops blood cells morphing into sickle-cells, but ingested v. is broken down before reaching bloodstream, so synthetic MX-1520, which breaks down to v. when ingested, more effective. ***V. grandiflora*** Lindl. (Amaz.) – fleshy fr. attractive to bees (? disp. agents, suggesting disp. may have antedated poll.); ***V. planifolia*** Andrews ('*V. mexicana*', trop. Am.) – first Am. orchid figured in Eur. (*Codex Badianus*, 1552) used by Aztecs to flavour cocoa, cult. for orn. & for long fr. (berry; 'pod', without vanillin when unripe (6% when mature), vanilloside breaking down to v. & glucose on ripening over 9 months), when cured source of v. extract (esp. in Madag. where anther & stigma have to be pressed together as poll. euglossine bee absent (world's only hand-poll. crop) – pioneered 1841 by 12-yr-old slave from Réunion; all W I Ocean stock allegedly from a single cutting in Jardin des Plantes, Paris, coll. Veracruz)), although vanillin synth. from eugenol from cloves in 1891, pods still used in cooking by the discriminating & worth $60–80 M a yr, (Madag. alone 3 K t a yr by 2005), but most flavouring today ('vanilla essence') is from wood pulp as a byproduct of paper-making & from coal-tar (toluene); ***V.* × *tahitensis*** J.W. Moore (*V. p.* × *V. odorata* C. Presl (trop. Am.), Tahiti vanilla,) also imp., esp. Pacific is.

Vanillosmopsis Schultz-Bip. = *Eremanthus*

Vanoverberghia Merr. Zingiberaceae (II 4). 1 Philippines: ***V. sepulchrei*** Merr. – ? extinct, fr. & seeds said to be ed. R: Taiwania 45(2000)270

Vanroyena Aubrév. (~ *Pouteria*). Sapotaceae (IV). 1 NE Aus.: ***V. castanosperma*** (C.T. White) Aubrév. R: T 62(2013)766

Vanroyenella Novelo & Philbrick = *Marathrum* (but see SB 18(1993)64)

Vantanea Aubl. Humiriaceae. 16 trop. Am. R: CUSNH 35(1961)49. Bat-disp. in Amazon. ***V. barbourii*** Standl. (Costa Rica) – to 65 m tall & 2 m diam.; ***V. obovata*** (Nees & Mart.) Benth. (Braz.) – fr. used as bobbins in lace-making

Vantieghemia Bobrov & Melikian = *Prumnopitys*

Vanwykia Wiens. Loranthaceae (5 7). 1–2 SE Afr. R: R. Polhill & D. Wiens, *Mistletoes of Afr.* (1998)221. Fls non-explosive

Vanzijlia L. Bolus. Aizoaceae (V). 1 SE Cape to Namaqualand: ***V. annulata*** (A. Berger) L. Bolus. R: H. E. K. Hartmann, *Ill. Handb. Succ. Pls*, Aiz. F–Z(2002)348

Vappodes M.A. Clem. & D.L. Jones = *Dendrobium*

Vargasiella C. Schweinf. Orchidaceae (V 12j). 2 Venez., Peru

Varilla A. Gray. Compositae (Tag.-Var.). 2 S N Am. R: Phytol. 69(1990)4

Varronia P. Browne (~ *Cordia*). Boraginaceae (Cord.; Cordiaceae). c. 100 trop. Am.

Varthemia DC. Compositae (Inul.-Inul.). 1 Iran, Afghanistan, Pakistan: *V. persica* DC.

Vasconcellea A. St.-Hil. (~ *Carica*). Caricaceae. 20 trop. Am., *V. chilensis* Planch. in temp. coastal Chile. R: Ernstia 10(2000)75, 11(2001)75. ***V. pubescens*** A. DC. (*C. candamarcensis, C. pubescens, V. cundinamarcensis*, mt. papaw or pawpaw, Andes) – small papaya fr., usu. candied or preserved, hybrid with *V. stipulata* (Badillo) Badillo (*C. s.*, Ecuador) = ***V. × pentagona*** (Heilborn) Mabb. (*C. × p., V. × heilbornii*), the babacó of fr. stalls, cult. Aus., & NZ, all fls female so parthenocarpic & fr. seedless; ***V. monoica*** (Desf.) A. DC. (*C. m.*) – fr. pulp dry but single-gene mutants with fleshy

Vaseyanthus Cogn. = *Echinopepon*

Vaseyochloa A. Hitchc. Gramineae (XXIX 3). 1 Texas: ***V. multinervosa*** (Vasey) A. Hitchc. R: SCB 87(1997)44

Vasivaea Baill. Malvaceae (Grew.-Grew.; Tiliaceae). 2 Braz., Peru. ***V. alchornioides*** Baill. – hydrochory

Vasquezia Philippi = *Villanova*

Vasqueziella Dodson. Orchidaceae (V 12i). 1 Bolivia: ***V. boliviana*** Dodson

Vassilczenkoa Lincz. Plumbaginaceae (II). 1 C As., Afghanistan: ***V. sogdiana*** (Lincz.) Lincz.

Vassobia Rusby (~ *Witheringia*). Solanaceae (IV 5b). 2 S Am. R: A.T. Hunziker, *Gen. Solan.* (2001)254

Vatairea Aubl. Leguminosae (III 11). 8 trop. Am. R: AJBRJ 26(1982)188. Wood for flooring etc.

Vataireopsis Ducke (~ *Vatairea*). Leguminosae (III 11). 4 N S Am. Medic. (psoriasis, esp. ***V. araraoba*** (Aguiar) Ducke – yellow granular inclusions in wood source of araroba or Goa powder; dermatitis & leishmaniasis)

Vateria L. Dipterocarpaceae (Dipt.-Dipt.). 2 S Ind. (1), Sri Lanka (1). ***V. indica*** L. (S Ind.) – resin (white dammar, piney varnish) form. imp., seed-fat (Malabar tallow, dhupa fat) used for candles etc.

Vateriopsis (Dyer) Heim. Dipterocarpaceae (Dipt.-Dipt.). 1 Seychelles: ***V. seychellarum*** (Dyer) Heim reduced to 50 trees through felling for timber

Vatica L. Dipterocarpaceae (Dipt.-Dipt.). 67 Indomal. Beetle-poll. (Borneo). Some timbers & dammars

Vatovaea Chiov. (~ *Spathionema*). Leguminosae (III 18). 1 E trop. Afr. to Oman: ***V. pseudolablab*** (Harms) Gillett – pods & lvs ed. R: KB 20(1966)104

Vatricania Backeb. (~ *Espostoa*). Cactaceae (III 4). 1 Bolivia: ***V. guentheri*** (Kupper) Backeb.

Vauanthes Haw. = *Crassula* (but see NSL 42(2010)168)

Vaughania S. Moore = *Indigofera*

Vaupelia Brand = *Cystostemon*

Vaupesia R. Schultes. Euphorbiaceae (Crot.-Jat.). 1 Colombia, W Braz.: ***V. cataractarum*** R.E. Schultes – seeds ed. once boiled. R: A. R.-Sm., *Gen. Euphorb.* (2001)290

Vauquelinia Corr. Serr. ex Bonpl. Rosaceae (Ros.-Mal.). 3 SW N Am. R: Sida 12(1987)101

Vausagesia Baill. = *Sauvagesia*

Vavaea Benth. Meliaceae (II). 4 Philippines to W Pac. R: Blumea 17(1969)351. Pagoda trees (Aubréville's Model), some bird-poll.?

Vavara Benoist. Acanthaceae (III). 1 Madag.: ***V. breviflora*** Benoist

Vavilovia Fed. (~ *Pisum*). Leguminosae (III 28). 1 Turkey to Cauc.: ***V. formosa*** (Steven) Fed.

Vazquezella Szlach. & Sitko = *Maxillaria*

Veconcibea (Muell. Arg.) Pax & K. Hoffm. = *Conceveiba*

Veeresia Monach. & Mold. = *Reevesia*

Vegaea Urb. Primulaceae (Myrsinaceae). 1 Hispaniola; ***V. pungens*** Urb. – diageotropic-rooted epiphyte

vegetable gold *Coptis trifolia*; **v. hair** *Chamaerops humilis, Carex brizoides*; **v. ivory** *Ammandra decasperma, Phytelephas aequatorialis, P. macrocarpa, Palandra aequatorialis*; **v. i. substitute** *Hyphaene thebaica*; **v. lamb of Tartary** *Cibotium barometz*; **v. marrow** *Cucurbita pepo*; **v. oyster** *Tragopogon porrifolius*; **v. pear** *Sicyos edulis*; **v. sheep** *Raoulia* spp. esp. *R. mammillaris*; **v. spaghetti** *C. pepo*; **v. sponge** *Luffa aegyptiaca*; **v. sulphur** *Lycopodium clavatum*; **v. tallow-tree** *Triadica sebifera*

Veillonia H. Moore = *Cyphophoenix* (but see Allertonia 3(1984)390)

veitchberry *Rubus* 'Veitchberry'

Veitchia H. Wendl. Palmae (V 14i). Excl. *Adonidia*, 11 Solomon Is., Vanuatu, Fiji, Tonga. R: HPB 4(1999)545. Some for house-constr., some cult. orn.

Velascoa Caldéron & Rzed. Crossosomataceae. 1 Mex.: ***V. recondita*** Caldéron & Rzed. R: ABM 39(1997)54

Veldkampia Ibarangi & Kobayashi. Gramineae. 1 Myanmar: *V. **sagaingensis*** Ibaragi & Kobayashi. R: JJB 83(2008)108

Velezia L. Caryophyllaceae (III 1). 6 Medit. (Eur. 2) to Afghanistan. R: FT 2(1967)135

Vella L. Cruciferae (12). Incl. *Boleum, Euzomodendron* 7 W Medit. (Eur. 5). R: Novon 8(1998)323. Some spiny

Velleia Sm. Goodeniaceae. 21 Aus. (20 endemic), NG. R: PLSNSW 92(1967)27. G ± sup.

Vellereophyton Hilliard & B.L. Burtt (~ *Helichrysum*). Compositae (Gnap.-Gnap.). 7 S Afr., *V. **dealbatum*** (Thunb.) Hilliard & B.L. Burtt natur. Aus. & NZ. R: OB 104(1991)168

Vellosiella Baill. Orobanchaceae (Esc.; Scrophulariaceae s.l.). 2 Braz.

Vellozia Vand. Velloziaceae. 124 trop. Am., esp. campos. Some hummingbird poll. in SE Braz. Dead stems sold as kindling in Bahia markets

Velloziaceae J. Agardh. Magnoliidae – Pandanales. Incl. Acanthochlamydaceae, 7/330 S Am., Afr., Madag. & S Arabia, SW China. R: SCB 30(1976)3. Oft. ± pachycaul xeromorphic shrubs (app. v. long-lived) usu. with marcescent leaf-sheaths & adventitious r. passing down through them making 'trunk', rarely herbaceous; vessel-elements in r. & s.t. shoots; sec. growth 0. Lvs narrow, parallel-veined, oft. dentate-spinulose, tristichous to spiro-tristichous, clustered at branch-tips, blade eventually falling from sheath. Fls usu. bisex., reg., 3-merous, usu. consp., usu. solit. term. becoming axillary; P 3 + 3 petaloid, ± basally united with tube oft. bearing 6 united appendages forming a corona outside & s.t. adnate to A 3 + 3 (6 bundles of 2 – ∞ in *Vellozia*) free of or adnate to P, anthers with longit. slits with pollen oft. in tetrads, $\overline{G}$ (3), 3-loc. with axile placentation & slender style with 3 stigmas, the placentae lamellar bearing ∞ ovules. Fr. a loculicidal capsule with compressed hard bitegmic seeds; embryo small in copious hard endosperm with hemicellulose, protein & oil or s.t. starchy. n = 7, 8, 17, 24 (x = 12?)

Genera: *Acanthochlamys* (rhiz. herb, SW China, form. Anthericaceae, prob. in own subfam.), *Barbacenia, Barbaceniopsis, Burlemarxia, Pleurostima, Vellozia, Xerophyta*

Form. allied with Liliaceae but molecular systematics place fam. here. Leaf-sheaths absorb moisture from fogs & prot. stem from fires

Veltheimia Gled. Asparagaceae (Hyacinthaceae). 2 S Afr. R: JRHS 97(1952)483. Cult. orn. greenhouse-pls

velvet bean *Mucuna pruriens* var. *utilis*; **v. leaf** *Abutilon theophrasti*; **v. sumac(h)** *Rhus typhina*; **v. tamarind** *Dialium guineense, D. ovoideum*

Velvitsia Hiern = *Melasma*

Venegasia DC. Compositae (Mad.-Ven.). 1 SW N Am.: *V. **carpesioides*** DC. R: Sida 15(1992)223

Venice turpentine *Larix decidua*

Venidium Less. = *Arctotis*

Ventenata Koeler. Gramineae (XVI 15). 13 S Eur. (2), Medit. to Caspian

Ventia Hauenschild = *Endotropis*

Ventilago Gaertn. Rhamnaceae (8). c. 40 OW trop. (Aus. 3, Afr. 1, Madag. 1). Roux's Model; some hook-climbers; fr. winged (accrescent style). Local medic. *V. **calyculata*** Tul. (Ind.) – cooking oil from seeds; *V. **madraspatana*** Gaertn. (trop. As.) – bark a source of a red dye for textiles

Ventricularia Garay = *Trichoglottis* (but see OB 95(1988)56)

Venus's comb *Scandix pecten-veneris*; **V. flytrap** *Dionaea muscipula*; **V. looking-glass** *Legousia hybrida*

Veprecella Naud. = *Gravesia*

Vepris Comm. ex A. Juss. Rutaceae (I 2). Incl. *Diphasia, Teclea*, c. 80 Arabia (1), Afr., Madag. (c. 30), SW Ind. (1). R: SBU 30,1(1992)69. Alks; wood hard, used for tool-handles

Verapazia Archila = *Specklinia*

Veratrilla Baill. ex Franch. Gentianaceae. 2 Himal. R: T.N. Ho & S.W. Liu, *Worldwide monog. Swertia* (2015)315

veratrin *Schoenocaulon officinale*

Veratrum Tourn. ex L. Melanthiaceae. Excl. *Melanthium*, 45–50 N temp. (Eur. 2, N Am. 5). R: Plantsman 11(1989)35; FR 24(1927)61, 25(1928)1. Coarse rhizomatous herbs (to 100 yrs old) of wet places. Cult. orn. with plicate lvs & term. panicles of white, green, brown or purplish fls, the lowermost bisex., the upper male, occ. with all-male pls. V. many alks: locally medic. esp. *V. **album*** L. (white hellebore, Euras.) – poisonous to stock in Scandinavia where Sámi used r. as snuff (before tobacco), lambs born of ewes which have fed on it have a single central eye, suggesting an inspiration for the Polyphemus of

Homer, used in control of rodents & pl. pests in Eur. since 1 cent.; *V.* ***californicum*** Durand (W N Am.) – medic. incl. contraceptive ('sterility for life'); *V.* ***viride*** Aiton (N Am.) – form. used for hypertension, esp. in pregnancy

verawood *Bulnesia arborea*

Verbascum Tourn. ex L. Scrophulariaceae (Scroph.). Incl. *Rhabdotosperma*, c. 320 Euras. (Eur. 87, Turkey 228), Ethiopian & E Afr. highlands. R: LUA n.s. 29,2(1933–4), 32,1(1936). Incl. *Celsia* (oft. with long pedicels & deeply lobed lvs, e.g. *V.* ***daenzeri*** (Fauché & Chaub.) Fenzl (*C. d.*, SE Eur.)). Mulleins; usu. bienn. herbs, rarely ann. or shrubby, with large rosettes of lvs & pachycaul infls (Holttum's Model) typical of much dry Medit. veg. as in Greece & Turkey. Many hybrids though usu. sterile. Some fr. & seeds used to kill fish in Greece & form. in Spain (verbasco, barbasco); cult. orn. *V.* ***blattaria*** L. (moth m., Euras., natur. N Am.) – cult. orn., seeds long-lived (germ. after 90 yrs in Dr W.J. Beal's experiment set up US in 1879 with 20% success & some still viable after 120 yrs); *V.* ***lychnitis*** L. (white m., W & C Eur., W As.) – cult. orn.; *V.* ***nigrum*** L. (black or dark m., Euras.); *V.* ***thapsus*** L. (Aaron's rod, common m., flannel pl., hag-taper, Euras., invasive Aus., Hawaii (where 31% pls fasciated), W US) – seeds germ. after 100 yrs in Dr Beal's experiment, lvs form. used medic. (anc. use in lung & chest complaints) incl. in cigarettes for the asthmatic, diarrhoea & bovine TB (pulmonary TB in Ireland – mucilage drunk); *V.* ***virgatum*** Stokes (W Eur.) – invasive Aus.

verbena, garden v. *Glandularia × hybrida*; **lemon v.** *Aloysia citrodora*

Verbena Tourn. ex L. Verbenaceae (Verb.). 200 trop. & temp. Am., 2–3 OW (Eur. 2 incl. *V.* ***officinalis*** L. (vervain, Juno's tears, Euras., Afr., Am.) – form. medic., bright-eyed fls suggesting cure for eye disease ('Doctrine of Signatures'), a divinatory in pre-Christian Eur. & magic charm against 'devils', verbenaline acting like quinine (fevers, tonic); *V.* ***bonariensis*** L. (*V. b.* var. *conglomerata*, now pantrop.) & *V.* ***incompta*** P.W. Michael (with smaller fls), temp. and warm S. Am., invasive E Aus., S Afr.; *V.* ***litoralis*** Kunth – used in fert. control in Urug.; *V.* ***rigida*** Spreng. (S Am.) – invasive colourful weed esp. in NSW and Qld. See also *Aloysia*, *Glandularia*

Verbenaceae J. St-Hil. Magnoliidae – Lamiales. Excl. Symphoremataceae, 32/1000 trop. esp. S Am. with few temp. R: AJB 97(2010)1650. Trees, lianes, shrubs & herbs, oft. aromatic, s.t. dioec. (*Citharexylum*, *Lippia*), s.t.thorny, young twigs oft. square in t.s. Lvs usu. simple & serrate, usu. opp.; stip. 0. Fls usu. bisex. in racemes, cymes or heads, oft. with involucre of coloured bracts; K ((4)5) with teeth or lobes s.t. irreg., C ((4)5) with imbr. lobes, ± irreg. to 2- lipped, oft. with slender tube, A 4(5) s.t. with staminodes, filaments arising from C-tube alt. with lobes, anthers with longit. slits, $\underline{G}$ usu. (2), initially 2-loc. but s.t. later subdivided by intrusive partitions into 4 uni-ovulate, style 1 with unpointed stigmatic lobes (cf. Labiatae); ovules 1 or 2 per loc., usu. anatropous, unitegmic. Fr. a head of 1-seeded separating mericarps, a drupe with 1, 2 or 4 pyrenes; seeds with straight oily embryo usu. in 0 endosperm (occ. oily). n = 5–12

Classification & principal genera:

1. **Verbeneae** (schizocarp with 4 1-seeded mericarps): *Glandularia*, *Junellia*, *Verbena*
2. **Lantaneae** (drupe or schizocarp of 2 1-seeded mericarps): *Aloysia*, *Lantana*, *Lippia*, *Stachytarpheta*
3. **Priveae** (schizocarp with 2 2-loc. 2-seeded mericarps): *Priva*
4. **Petreeae** (drupes): *Petrea*
5. **Casselieae** (fr. fleshy): *Casselia*
6. **Citharexyleae** (drupe or schizocarp with 2-loc. 2-seeded mericarps): *Citharexylum*, *Duranta*

The limits of the fam. have been much debated but recent evaluations of Lamiales remove many genera to Labiatae s.l., Acanthaceae & Oleaceae & split off smaller groups as fams – Phrymaceae & Stilbaceae (q.v.)

Timber (*Citharexylum*), flavourings & medic. teas etc. (*Aloysia*, *Nashia*, *Stachytarpheta* etc.), cult. orn. (*Aloysia*, *Duranta*, *Glandularia* (garden verbena), *Lantana*, *Petrea*, *Verbena*) but some widespread weeds of warm countries esp. *Lantana*, *Lippia*, *Stachytarpheta* & *Verbena* spp.

Verbenoxylum Tronc. = *Recordia* (but see Darw. 16(1971)622)

Verbesina L. Compositae (Helia.-Verb.). c. 300 warm Am. (esp. Mex. & trop. Andes). R: PAAAS 34(1899)536. Trees (to 25 m), shrubs & herbs. Some local medic., cult. orn. (crown-beard) incl. *V.* ***helianthoides*** Michaux (*Actinomeris h.*, S & SC US) – caruncles

Verdcourtia R. Wilczek = *Dipogon*

Verdesmum Ohashi & K. Ohashi (~ *Desmodium*). Leguminosae (III 19). 1 Yunnan, 1 Borneo, NG. R: JJB 87(2012)301, 88(2013)156

Verdickia De Wild. = *Chlorophytum*

Verena Minod = *Stemodia*

Verhuellia Miq. (~ *Peperomia*). Piperaceae (1). 3 Cuba, Hispaniola. R: T 57(2008)584

verjuice *Vitis vinifera*

Verlotia Fourn. = *Marsdenia*

Vermeulenia Löve & D. Löve = *Anacamptis*

vermicelli *Triticum turgidum* Durum Group & *Oryza sativa*

Vermifrux J.B. Gillett = *Dorycnopsis*

vermilion wood *Pterocarpus* spp.

vernal grass, sweet *Anthoxanthum odoratum*

Vernicia Lour. (~ *Aleurites*). Euphorbiaceae (Al.-Al.). 4 Myanmar & SE As. to S China & Jap. R: Blumea 44(1999)89. *V. **fordii*** (Hemsl.) Airy Shaw (*A. f.*) & *V.* ***montana*** Lour. (*A. m.*) – poorer quality – sources of tung or wood oil used in paints & quick-drying varnishes, widely cult.

Vernonanthura H. Robinson (~ *Vernonia*). Compositae (Vern.-Vern.). c. 70 trop. Am. R: Phytol. 73(1992)66. Some weedy

Vernonia Schreb. Compositae (Vern.-Vern.). Excl. *Aedesia, Baccharoides, Cyanthillium, Distephanus, Gymnanthemum, Lepidaploa, Lessingianthus, Vernonanthura, Vernoniastrum*, etc., 22 SE US to C Mex., 2 S Am. Iron weed (US)

Vernoniastrum H. Robinson (~ *Vernonia*). Compositae (Vern.-Vern.). 13 trop. Afr.

Vernoniopsis Humbert. Compositae (Ast.). 2 Madag. R: Cand. 66(2011)409. Trees

Veronica Tourn. ex L. Plantaginaceae (Dig.-Veron.; Scrophulariaceae s.l.). Incl. *Besseya, Chionohebe, Derwentia, Detzneria, Hebe* (90 Australasia, prob. orig. Aus. with NZ spp. derived from 1 intr. & long-distance disp. to S Am., 1 to NG; incl. hybrid swarms), *Heliohebe, Odicardis, Paederota, Parahebe, Pseudolysimachion* (19 temp. & subtemp. – R: OB 75(1984)57), *Synthyris* (19 W N Am. mts – R: SB 29(2004)734), 450 N temp., few trop. mts & S with 9 subcosmop. weeds. R: BMOIPB 82,1(1977)151, OB 75(1984)57, T 53(2004)437, 56(2007) 574 – 13 subgenera. Speedwell (i.e. 'goodbye' because C falls as soon as gathered), bird's-eye. Trees & shrubs (hebes, some NZ spp. with small adpressed lvs resembling conifers, e.g. *V.* ***cupressoides*** Hook.f. (*H. c.*) & *V.* ***lycopodioides*** Hook.f. (*H. l.*)) to herbs; lvs spiral to opp. (hebes); C rotate & A 2 held laterally & grasped by visiting insects such that they come together dusting poll. agents with pollen; hebes with dehiscent fr. Some spp. turn black after c. 100 yrs as herbarium specimens. Some with lvs used for diarrhoea & dysentery; cult. orn. esp. shrubby hebes (several cvs derived from *V.* ***speciosa*** Cunn. (*H. s.*, NZ), incl. *V.* **× *franciscana*** Eastw. (*H.* ×*f.*; *V. s.* × *V. elliptica* Forst.f. (*H. e., V. decussata*, NZ to S Am.)), shrub used as hedging in SW GB), shrublets incl. *V.* ***catarractae*** Forst.f. (*Parahebe c.*, NZ) – rock-pl., & herbs incl. *V.* ***beccabunga*** L. (brooklime, Euras., Medit.) – helophyte, locally medic. (diuretic, etc.; 18 cent. cure for scrofula in Co. Cork, Eire), shoots ed.; *V.* ***bonarota*** L. (*Paederota b.*, E Alps) – cult. orn.; *V.* ***catenata*** Pennell (water speedwell, N Temp. to E Afr.); *V.* ***chamaedrys*** L. (germander speedwell, Euras., natur. N Am.); *V.* ***filiformis*** Sm. (As. Minor & Cauc.) – escaped in GB in 1927 (first record 1838), a persistent weed with rooting stems in lawns in GB & US, esp. common in churchyards; *V.* ***longifolia*** L. (Euras., natur. N Am.) – common garden perenn.; *V.* ***officinalis*** L. (fluellen (arch.), N temp.) – form. medic. tea etc.; *V.* ***persica*** Poir. (subcosmop.) – allotetraploid weed (2n = 28) derived from *V. polita* E. Fries (2n = 14, N Iran, a Neolithic weed spreading to Europe) × *V. ceratocarpa* C. Meyer (2n = 14, Cauc. & Iran); *V.* ***prostrata*** L. (Eur.) – common rock-pl. with deep blue fls; *V.* ***salicifolia*** Forst.f. (*H. s.*, S NZ to S Am.) – hardiest hebe, lvs used for diarrhoea; *V.* ***spicata*** L. (*Pseudolysimachion s.*, N Euras.) – prot. in GB

Veronicaceae Cassel = Plantaginaceae

Veronicastrum Heister ex Fabr. Plantaginaceae (Dig.-Veron.; Scrophulariaceae s.l.). 1 temp. NE As.; 1 temp. NE Am.: *V.* ***virginicum*** (L.) Farw. (blackroot, Culver's r.) – cult. orn., violent purgative, emetic & cathartic (!). R: OB 75(1984)56

Verreauxia Benth. Goodeniaceae. 3 SW Aus. R: FA 35(1992)298. Nut

Verrucifera N.E. Br. = *Titanopsis*

Verrucularia A. Juss. Malpighiaceae. 2 Braz. R: MNYBG 32(1981)45

Verrucularina Rauschert = *Verrucularia*

Verschaffeltia H. Wendl. Palmae (V 14k). 1 Seychelles: *V.* ***splendida*** H. Wendl. – monoec. with spiny trunk & stilt-r.

Versteegia Valeton = *Ixora*

Verticordia DC. Myrtaceae (II 15). c. 97 Aus. esp. SW. R: Nuytsia 7(1991)291; E. George (2002) *V.: the turner of hearts*. Heterogeneous. Cult. orn. heath-like shrubs (feather fls, morrison). ***V. aurea*** A.S. George & ***V. nitens*** (Lindl.) Endl. – sole sources of pollen & nectar for the 2 sole pollinator solitary bees (*Euryglossa* spp.)

vervain *Verbena officinalis;* **rose v.** *Glandularia canadensis*

Vesalia Martens & Galeotti (~ *Abelia*). Caprifoliaceae. 2–3 Mex.

Vescisepalum (J.J. Sm.) Garay, Hamer & Siegerist = *Bulbophyllum* (But see NJB 14(1994)641)

Veselskya Opiz (*Pyramidium*). Cruciferae (21?). 1 Afghanistan: ***V. griffithiana*** (Boiss.) Opiz

Veseyochloa J. Phipps = *Tristachya*

Vesicarex Steyerm. = *Carex*

Vesicaria Adans. = *Alyssoides*

Vesper R. Hartman & Nesom (~ *Cymopterus*). Umbelliferae (III 9). 6 W N Am. R: Phytoneuron 2012–94:2

Vesselowskya Pampan. Cunoniaceae (VIII). 2 NSW. R: AusSB 14(2001)181

Vestia Willd. (~ *Sessea*). Solanaceae (II 1). 1 Chile: ***V. foetida*** (Ruíz & Pavón) Hoffsgg. (*V. lycioides*) – cult. orn. shrub

vetch *Vicia* spp. esp. *V. sativa;* **bitter v.** *V. ervilea;* **bush v.** *V. sepium;* **chickling v.** *Lathyrus sativus;* **horseshoe v.** *Hippocrepis comosa;* **kidney v.** *Anthyllis vulneraria;* **milk v.** *Astragalus glycophyllos;* **point-v.** *Oxytropis* spp.; **Russian** or **Siberian v.** *V. villosa;* **tufted v.** *V. cracca*

vetchling, common or **meadow** *Lathyrus pratensis;* **yellow v.** *L. aphaca*

vetiver *Chrysopogon zizanioides*

Vetiveria Bory = *Chrysopogon*

Vexatorella Rourke = *Paranomus*

Vexillabium Maekawa = *Kuhlhasseltia*

Veyretella Szlach. & Olsz. (~ *Habenaria*). Orchidaceae (IV 4d). 2 Gabon. R: Gen. Orch. 2(2001)381

Veyretia Szlach. (~ *Pelexia*). Orchidaceae (IV 2h). 11 trop. Am.

Viburnaceae Raf. (Adoxaceae, Sambucaceae). Magnoliidae – Dipsacales. 4/195 temp. to warm, trop. mts. Shrubs & small trees to perenn. herbs. Lvs simple to 3+-foliolate, toothed, opp. (rarely 3-verticillate). Infl. usu. corymbose. Fls mostly 5-merous, herm.; K (2–5), open during development, C (3–5(6)) lobes imbr. to valvate, each with a nectary at base. A attached to C-tube, s.t. divided almost to base giving twice as many unithecal, $\overline{G}$ ((2)3–5), 1 pend. anatrop. bitegmic ovule per loc. Fr. a drupe with 1 (*Viburnum*) or few pyrenes; seeds with copious oily endosperm. x = 9 (large). R: BBR 7,4(1987)93

Genera: *Adoxa, Sinadoxa* (Adoxoideae-Adoxeae), *Sambucus* (Adoxoideae-Sambuceae), *Viburnum* (Viburnoideae)

Some cult. orn. esp. *Viburnum*

Viburnum Tourn. ex L. Viburnaceae (Adoxaceae). c. 180 temp. (Eur. 3) & warm esp. As. (China 73) & N Am. L. Kenyon (2003) *V.* Small trees & shrubs, s.t. with domatia, with panicles or umbel-like cymes, these s.t. with pollinator-attractant marginal sterile fls with enlarged C (cf. K in *Hydrangea* with which s.t. confused), in some garden forms all fls thus, extrafl. nectaries, ed. or poisonous drupes; v. imp. cult. orn. garden shrubs with consp. oft. scented fls & orn. fr., locally medic. etc. ***V. acerifolium*** L. (arrowwood, E N Am.) – bark medic., wood for arrows; ***V. × bodnantense*** Aberc. ex Stearn (*V. farreri* × *V. grandiflorum* Wall. ex DC., Himal.) – fragrant winter fls.; ***V. carlesii*** Hemsl. (Korea, China, Jap.) – white v. fragrant fls; ***V. costaricense*** Hemsl. (C Am.) – pH of nectar c. 10 (most nectar acidic); ***V. davidii*** Franch. (W Sichuan) – much (? over-) planted everg. shrub; ***V. dentatum*** L. (arrowwood, E N Am.) – wood for arrows; ***V. edule*** (Michaux) Raf. (N Am.) – local medic. incl. colds, ed. fr.; ***V. ellipticum*** Hook. (W N Am.) – with stip.! (cf. × *Fatshedera*); ***V. farreri*** Stearn (*V. fragrans*, N China) – fragrant winter fls; ***V. lantana*** L. (wayfaring tree, twistwood, Euras., Medit.) – arrowshafts found with Neolithic 'Iceman' preserved in Alps ice; ***V. lantanoides*** Michaux (*V. alnifolium*, hobblebush, mooseberry, moosewood, E N Am.) – marginal sterile fls; ***V. lentago*** L. (nannyberry, sheepberry, E N Am.) – fr. cooked by Native Americans; ***V. macrocephalum*** Fortune (China) –f. ***m.***, a cult. orn. 'snowball tree' (see *V. opulus*) of China, f. ***keteleeri*** (Carr.) Rehder the wild pl. (jade fl.) – floral emblem of Yangzhou; ***V. opulus*** L. (guelder rose, crampbark, Euras., Medit.) – wood form. for skewers, fr. ed. & subs. for cranberries, bark medic., '**Compactum**' & '**Nanum**' – grotesque dwarf cvs rarely flowering, '**Roseum**' (Whitsuntide boss, snowball tree) – all fls (white) sterile making a globose infl. the orig. Guelder rose cult.

near Guelders, inner fls of wild form with nectar cont. indole & a sweet-fishy scent attractive to Diptera; ***V. plicatum*** Thunb. (China & Jap.) – long cult. in sterile form & first intr. thus, resembling *V. o.* 'Roseum' but lvs unlobed, the wild pl. (f. ***tomentosum*** (Thunb.) Rehder) oft. taken for a hydrangea; ***V. prunifolium*** L. ((black) haw, E & EC N Am.) – fr. ed. esp. after frost; ***V. setigerum*** Hance (*V. theiferum*, C & W China) – local 'tea'; ***V. tinus*** L. (laurustinus, Medit.) – imp. constituent of orig. Medit. veg. (the 'plane' of Isaiah) – winter-flowering everg. with poisonous fr.; ***V. trilobum*** Marsh. (N Am.) – fr. for preserves etc.

Vicatia DC. Umbelliferae (III 8). 4 Himalaya to W China. R: FR 102(1991)376,383

Vicia Tourn. ex L. Leguminosae (III 28). 160 N temp. (Eur. 54) with extensions to S Am. (27), Hawaii (***V. menziesii*** Spreng. – endangered) & trop. E Afr. but paraphyletic. Leaflet vernation conduplicate (in *Lathyrus* supervolute), A tube oblique (*L.* – truncate), styles pubescent around apex (*L.* – on 1 side only). Usu. scrambling herbs (vetch, tare) with alks. Forage, green manure, ed. seeds etc. ***V. cracca*** L. (tufted v., N temp.) – cult. orn. scrambler, forage; ***V. ervilia*** (L.) Willd. (bitter v., Macaronesia & Medit. to Afghanistan, ? orig. SW As.) – fodder; ***V. faba*** L. (broad bean, field b., foulia b., horse b., lupini b. (2n = 12), usu. but erroneously considered cultigen derived from *V. narbonensis* L. (Medit. to C As., 2n = 14) & poss. domesticated in C As.) – 4 M t a yr by 2011, primitively black-seeded, the bean of antiquity & that of 'peas & beans' in mediaeval crop-rotation in GB (Celts had beanos & beanfeasts to honour the fairies), ed. seeds but when uncooked can lead to hepatitis in Italians & some Jewish people due to biochemical deficiency in red blood cells (anaemia = favism), fl. smell allegedly aphrodisiac, white fluff lining pods form. used to treat warts, 'Faba' orig. of Lat. first name Fabius, 'var. *equina* Pers.' (pigeon bean, tic b.) – small-seeded form fed to racing pigeons; ***V. sativa*** L. (vetch, tare, Euras., natur. N Am.) – allegedly weed of lentils becoming a crop or spread from Middle East, green manure & fodder; ***V. sepium*** L. (bush v., Eur. to Himal.); ***V. villosa*** Roth (Russian or Siberian v., Euras.) – manure, hay & silage

Vicoa Cass. = *Pentanema*

Victoria Lindl. (~ *Nymphaea*). Nymphaeaceae (Euryalaceae). 2 trop. S Am. R: T. Anisko (2013) *V*. Rooted aquatics prickly (? defence against grazing manatees) except on adaxial leaf surfaces, lvs peltate, floating with upturned margin & remarkably architectural leaf-venation, perforated with holes (stomatodes) 0.2–0.3 mm diam. draining rainfall on leaf & air beneath to escape, the fragrant fls opening white late in afternoon & staying open but turning pink on 2nd day. In temp. regions cult. as annuals under glass, a patriotic vogue ('*V. regia*') in 19 cent. leading to constr. of Victoria Houses as at Chatsworth (Derbyshire) & Oxford Botanic Garden, their designs being forerunners of Paxton's Crystal Palace of the Great Exhibition (1851). ***V. amazonica*** (Poeppig) Klotzsch (*V. regia*, giant waterlily, Guyana & Amazonia, found by Haenke in 1801) – germ. at 30°C, lvs to 2.93 m diam., if not punctured, supporting 345 kg bricks (Ghent in 1885), 4 spp. of dynastid beetle attracted by scent & temp. (11°C above ambient) are trapped for 24 hrs & feed on starchy carpellary appendages before leaving with pollen (similar pollinators in *Nymphaea* subg. *Hydrocallis*, *Lotos*), seeds ed. if roasted; ***V. cruziana*** Orb. (N Arg., Paraguay, Bolivia) – germ. at 20°C, oft. cult. in place of *V. amazonica* (upturned leaf margin to 20 cm, outer P prickly only at base) as is their hybrid, '**Longwood Hybrid**'

Victorinia Léon = *Cnidoscolus*

Vidalasia Tirveng. (~ *Randia*). Rubiaceae (II 1). 5 SE As., Philippines

Vieraea Schultz-Bip. (*Vieria*). Compositae (Inul.-Inul.). 1 Canary Is.: ***V. laevigata*** Webb & Berth.

Viereckia R. King & H. Robinson (~ *Eupatorium*). Compositae (Eup.-Crit.). 1 Mex. (hybrid?, coll. once): ***V. tamaulipasensis*** R. King & H. Robinson. R: MSBMBG 22(1987)312

Vieria Webb & Berth. = *Vieraea*

Vierlingia Königer = *Cyrtochilum*

Vietnam(ese) balm *Elsholtzia ciliata*; **V. coriander** or **mint** *Persicaria odorata*

Vietnamia P.T. Li. Apocynaceae (Asclepiadaceae). 1 Vietnam: ***V. inflexa*** P.T. Li

Vietnamocalamus Nguyen. Gramineae. 1 Vietnam: ***V. catbaensis*** Nguyen

Vietnamochloa Veldk. & Nowack. Gramineae. 1 Vietnam: ***V. aurea*** Veldk. & Nowack

Vietnamosasa Nguyen (~ *Racemobambos*). Gramineae (V 4). 3 Vietnam

Vietsenia C. Hansen. Melastomataceae. 4 Vietnam. R: BMNHN 4,9(1987)259

Vigethia W. Weber. Compositae (Helia.-Eng.). 1 Mex.: ***V. mexicana*** (S. Watson) W. Weber

Vigia Vell. = *Plukenetia*

Vigna Savi. Leguminosae (III 18). Excl. *Ancistrotropis, Cochliasanthus, Condylostylis, Leptospron*, 90+ trop. esp. OW. R: ARES 20(1989)199. Erect or twining herbs with extrafl. nectaries, like *Phaseolus* but stip. oft. with appendages, thickened part of style less strongly twisted, etc.; incl. *Voandzeia* (spp. with subterr. poll. by ants & geocarpy). Imp. pulses, green manure etc. ***V. aconitifolia*** (Jacq.) Maréchal (*Phaseolus a.*, moth bean, S As.) – beans ed., forage; ***V. angularis*** (Willd.) Ohwi & Ohashi (*P. a.*, adzuki or azuki b., As.) – long cult. for beans boiled or made into curd, constituent of comm. 'washing grains'; ***V. lanceolata*** Benth. (maloga bean, Aus.) – taproot to 40 cm ed.; ***V. mungo*** (L.) Hepper (*P. m.*, urd, black gram, trop. As.) – imp. pulse in As.; ***V. radiata*** (L.) R. Wilczek (*P. r., P. aureus*, mung bean, green or golden gram, ? Indonesia) – 'var. *setulosa* (Dalz.) Ohwi & Ohashi' (var. *sublobata*) the poss. ancestor, seedlings the 'bean sprouts' of commerce, ed. seeds & pods, source of cellophane, bean thread, or glass noodles; ***V. subterranea*** (L.) Verdc. (*Voandzeia s.*, Bambara groundnut, W Afr.) – geocarpic fr. with imp. ed. seeds; ***V. umbellata*** (Thunb.) Ohwi & Ohashi (*P. calcaratus*, rice bean, S As.) – beans ed.; ***V. unguiculata*** (L.) Walp. (black-eye(d) b. or pea, cowpea, yawa, OW) – imp. pulse (chowlee – Ind., gubgub – WI), forage & fibre poss. domesticated Ethiopia, 'subsp. *sesquipedalis* (L.) Verdc.' (asparagus b., bora, long. b., snake b., yard-long b.) – legume to 90 cm eaten; ***V. vexillata*** (L.) A. Rich. (trop. OW) – r. ed.

Viguiera Kunth. Compositae (Helia.-Helia.). Excl. *Bahiopsis, Calanticaria, Hymenostephium*, c. 140 warm & trop. Am. R: CGH 54(1918)1. Few cult. orn. with yellow fls. ***V. dentata*** (Cav.) Spreng. (tah(onal), Mex.) – used in honey exported to UK

Viguieranthus Villiers (~ *Calliandra*). Leguminosae (II 4). 23 Madag. (18), trop. As. (5)

Viguierella A. Camus. Gramineae (Chlorid). 1 Madag.: ***V. madagascariensis*** A. Camus

Villadia Rose (~ *Sedum*). Crassulaceae (I 5b). 21 Texas to Peru. R: U. Eggli, *Ill. Handb. Succ. Pls*, Crass. (2005)367. Cult. orn. succ.

Villamillia Ruíz & Pavón ex Moq. = *Trichostigma*

Villanova Lag. Compositae (Per.-Gal.). Excl. *Unxia*, 10 Mex. to Chile

Villaresia Ruíz & Pavón (1802) = *Citronella*

Villaresiopsis Sleumer = *Citronella*

Villarealia Nesom. Umbelliferae. 1 N Mex.: ***V. calcicola*** (Mathias & Constance) Nesom. R: Phytoneuron 2012–85:2

Villaria Rolfe. Rubiaceae (II 5). 6 Philippines. R: Acta Manilana 39(1991)54

Villarsia Vent. Menyanthaceae. Excl. *Liparophyllum, Ornduffia*, 3 S Afr. R: PSE 296(2011)13

Villasenoria B.L. Clark (~ *Senecio*). Compositae (Sen.-Sen.). 1 Mex.: ***W. orcuttii*** (Greenm.) B.L. Clark. R: Sida 18(1999)632

Villebrunea Gaudich. ex Wedd. = *Oreocnide*

Villocuspis (A.DC.) Aubrév. & Pellegrin = *Chrysophyllum*

Vilobia Strother = *Tagetes*

Viminaria Sm. Leguminosae (III 13). 1 Aus.: ***V. juncea*** (Schrad.) Hoffsgg. (*V. denudata*) – cult. orn. shrub with broom-like branches of lvs mostly reduced to thread-like petioles

Vinca L. Apocynaceae (Ib). 5 Eur. (5) to N Afr. & C As. R: BMNHN II, 23(1951)439, W.I. Taylor & N. Farnsworth (1973) *The V. alkaloids*. Periwinkles. Local medic. (lvs for cuts etc., sedative & toothache); cult. orn. esp. ***V. major*** L. (greater p., invasive Aus., NZ, W US) – lvs ciliate, & ***V. minor*** L. (lesser p., Eur., natur. GB) – lvs glabrous, both with v. many alks., locally medic., vincamine used in cerebral vascular disorders though now comm. synth. from tabersonine from *Voacanga* spp. See also *Catharanthus*

Vincentella Pierre = *Synsepalum*

Vincentia Gaudich. = *Machaerina*

Vincetoxicopsis Costantin. Apocynaceae (V c; Asclepiadaceae III 4). 1 SE As.: ***V. harmandii*** Costantin

Vincetoxicum Wolf (~ *Cynanchum*). Apocynaceae (Vc; Asclepiadaceae III 1). Incl. *Biondia, Tylophora*, c. 180 temp. Euras. (Eur. 11) esp. Caucasus, China, OW trop. (Mal. c. 20), S Afr. Seeds with wing allowing water-disp.; alks. ***V. indicum*** (Burm.f.) Mabb. (*T. i.*, Ind.) – medic. r. with effect of ipecacuanha

Vindasia Benoist. Acanthaceae (III). 1 Madag.: ***V. virgata*** Benoist

vine *Vitis vinifera*, (US) any liane; **balloon v.** *Cardiospermum halicacabum*; **kangaroo v.** *Cissus antarctica*; **kudzu v.** *Pueraria montana* var. *thomsonii*; **Madeira** or **mignonette v.** *Anredera cordifolia*; **potato v.** *Ipomoea pandurata, Solanum laxum*; **Russian v.** *Fallopia baldschuanica*; **silk v.** *Periploca graeca*; **wonga-wonga v.** *Pandorea pandorana*

vinhatico *Plathymenia reticulata*

Vinicia Dematteis. Compositae (Vern.-Lynch.). 1 Braz.: *V. **tomentosa*** Dematteis. R: Bonplandia 16(2007)260

Vinkia Meijden = *Myriophyllum* (but see Blumea 22(1975)251)

Vinkiella R. Johns = *Gnetum*

Vinticena Steud. = *Grewia*

Viola Tourn. ex L. Violaceae (III 2b). c. 400 temp. (Eur. 91) esp. N & Andes. Violets. R: JRHS 55(1930)223, 57(1932)212; R. Fuller (1990) *Pansies, violets and violettas: the complete guide*: 188. Herbs, rarely subshrubs, sect. *Andinum* 'rosulates' (100 spp.), oft. with cleistogamous fls at end of season. Consp. fls with insect landing-place formed by anterior petal oft. spurred & cont. nectar; in some spp. pollen shed on to that petal & lower edge of stigma prot. from it by a flap which closes as insect leaves fl. thus preventing self-poll. (in weedy spp. like *V. **arvensis*** Murray (field pansy, Eur.) no flap & self-poll. freq.); some with ant-disp. seeds with caruncles (arils). Cult. orn. (violets, violas (rayless = violettas, '**Violetta**' (white 1887 leading to vs with vanilla scent now raised by crossing *V. cornuta* (treated with colchicine) with violas) incl. winter-flowering 'Universal Hybrids', pansies) comm. since 400 BC in Attica (worth $112M in US alone by 2005) & local medic., scent incl. ionine which deadens scent receptors temporarily (see *Vachellia farnesiana*), hence 'fleeting' sense. ***V. alba*** Besser (C & E Eur.) – cvs of subsp. ***dehnhardtii*** (Ten.) W. Becker (Medit.) incl. Parma vs with stolons & fragrant abnormal fls; ***V. arborescens*** L. (W Medit.) – stems woody; ***V. canina*** L. (dog violet, Euras.) – like *V. lactea* Sm. (W Eur.) & *V. pumila* Chaix (C & E Eur.), alloploid with 1 genome from *V. stagnina* Kit. (Eur.; of which *V. elatior* Fries (Euras.) an autotetraploid derivative), cult.; ***V. cornuta*** L. (viola, Spain & Pyrenees) – tufted or bedding pansies & violas largely from this, bred into pansies to give familiar winter-flowering cvs; ***V. lilliputana*** Iltis & H. Ballard (SE Peruvian Andes) – only up to 1.1 cm tall; ***V. oahuensis*** C. Forbes (Oahu, Hawaii) – unbranched shrub to 40 cm in cloud forest; ***V. odorata*** L. (sweet violet, Euras. to Afr.) – essential oil used in flavouring & in scent-making (e.g. '**Victoria**' near Grasse, Fr.), 100 kg of fls giving 31 g of oil, fls crystallized as food decorations etc. (still comm. Toulouse, F. – 'violettes de Toulouse'), incl. Devon v. & the common v. of florists (prob. hybrids), signifying modesty in 'Language of Fls', Br. folk medic. for cancerous tumours, form. scattered on graves; ***V. palustris*** L. (marsh v., N temp.); ***V. riviniana*** Reichb. (Eur.) – common scentless violet 2–3 wks later than ***V. reichenbachiana*** Boreau (W, C & S Eur.), '**Purpurea** ('*V. labradorica*') – cult. patio weed; ***V. tricolor*** L. (heart's-ease, love-in-idleness, Johnny-jump-up, Eur. & natur. in N Am.) – locally medic. (risky) & a parent of garden pansies (***V. × wittrockiana*** Gams ex Nauenb. & Buttler) raised c. 1830 & involving *V. lutea* Hudson subsp. *sudetica* (Willd.) W. Becker (C Eur. mts) & *V. altaica* Ker-Gawler (As. Minor), backcrossing with *V. tricolor* in wild

Violaceae Batsch. Magnoliidae – Malpighiales. 29/815 cosmop. but only *Viola* temp. SB 39(2014)240. Trees, shrubs, s.t. lianes or herbs, oft. with alks, oft. accum. aluminium. Lvs simple, entire to (rarely) dissected, usu. in spirals with stip. Fls reg. to irreg. (when oft. resupinate so saccate C lowermost), usu. bisex. (s.t. cleistogamous), bibracteolate, solit. & axillary or in racemes, thyrses or heads; K 5, imbr., ± free, oft. persistent, C 5 imbr. or convolute, the lowermost oft. with spur in irreg. fls, A (3)5 with filaments free or ± connate & anthers oft. connivent around G & with nectaries on back, $\overline{G}$ ((2)3(–5)), 1-loc. with parietal placentation & 1 style, each placenta with 1 – ∞ anatropous, bitegmic ovules. Fr. a berry or loculicidal capsule, rarely a nut (*Leonia*); seeds oft. arillate (s.t. a wing) with straight embryo embedded in oily endosperm. n = 6–13, 17, 21, 23

Classification & principal genera:

I. **Fusispermoideae** (trees, A 3, caps.; trop. Am.): *Fusispermum* (only)
II. **Leonioideae** (trees, A 3 or 5; nut or berry: trop. Am.): *Leonia* (only)
III. **Violoideae** (trees, shrubs, herbs, usu. 3-valved caps.):
   1. **Rinoreeae** (C usu. reg.): *Rinorea* (a. Rinoreinae), *Melicytus* (b. Hymenantherinae), *Isodendrion* (c. Isodendriinae), *Paypayrola* (d. Paypayrolinae)
   2. **Violeae** (C usu. zygom.): *Hybanthus*, *Pombalia* (a. Hybanthinae), *Viola* (b. Violinae)
   Phyllanoa form. in Euphorbiaceae

Some medic. esp. *Corynostylis* & *Pombalia*; cult. orn. esp. *Melicytus* (also timber) & *Viola*, which also yields ess. oils

violet *Viola* spp.; **African v.** *Streptocarpus ionanthus*; **Arabian v.** *Exacum affine*; **bog v.** *V. palustris*, *Pinguicula vulgaris*; **v. cress** *Ionopsidium acaule*; **dame's v.** *Hesperis matronalis*; **Devon**

v. see *V. odorata;* **dog v.** *V. canina;* **d. tooth v.** *Erythronium dens-canis;* **essence of v.** *Iris* 'Florentina'; **Parma v.** see *V. alba;* **Persian v.** *Exacum affine;* **Philippine v.** *Barleria cristata;* **sweet v.** *V. odorata;* **Usambara v.** *S. ionantha;* **water v.** *Hottonia palustris;* **v. wood** *Acacia omalophylla*

violetta See *Viola*

viper's bugloss *Echium vulgare;* **v. grass** *Scorzonera hispanica*

Viposia Lundell = *Plenckia*

Viracocha Szlach. & Sitko = *Maxillaria*

Virecta Sm. = *Virectaria*

Virectaria Bremek. Rubiaceae (II 7). 8 trop. Afr. R: BotJLS 137(2001)16

Virga Hill = *Dipsacus*

Virgilia Poir. Leguminosae (III 7). 2 S & SW Cape. R: SAJB 52(1986)347. ***V. oroboides*** (P. Bergius) Salter (*V. capensis*) – cult. orn., timber for rafters etc.; alks

virgin oil *Olea europaea*

Virginia(n) bluebell or **cowslip** *Mertensia virginica;* **V. creeper** *Parthenocissus quinquefolia;* **V. snakeroot** *Aristolochia serpentaria;* **V. stock** *Malcolmia maritima*

Virgulaster Semple = *Symphyotrichum*

Virgulus Raf. = *Symphyotrichum*

Viridantha Espejo (~ *Tillandsia*). Bromeliaceae (2). 6 Mex. R: ABM 60(2002)27

Viridivia J.H. Hemsl. & Verdc. Passifloraceae. 1 Zambia, SW Tanz.: ***V. suberosa*** J.H. Hemsl. & Verdc.

Virola Aubl. Myristicaceae. c. 40 trop. Am. (Peru c. 20). R (Braz.): AA 10,1 suppl.(1980)30. Timber, hallucinogenic snuff (active principles in cambium) blown up nostrils through *Ischnosiphon* tubes (though oft. taken as pills covered in Lecythidaceous (e.g. *Eschweilera*), Cyclanthaceous or (best) *Spathiphyllum cannifolium* ash) & comm. fats (see also *Bicuiba*). Alks. ***V. elongata*** (Benth.) Warb. (*V. theiodora*, NE S Am.) – resin boiled, dried & powdered to give hallucinogenic snuff rich in tryptamines, also arrow-poison; ***V. koschnyi*** Warb. (banak, C & N S Am.) – timber, promising tree for reafforestation; ***V. sebifera*** Aubl. (N S Am.) – seed-oil used for candles & soap; ***V. surinamensis*** (Rottb.) Warb. (NE S Am.) – bird- & water-disp. (seeds 15.4% dry-wt. soluble tannins – highest % defensive compounds known), timber (baboen, dalli) for plywood etc., seed-oil (caihuba, ucahuba, ucuhuba) like cocoa butter

Virotia L. Johnson & B. Briggs (~ *Macadamia*). Proteaceae (V 4). 6 New Caled. Pachycauls

Viscaceae Batsch = Santalaceae

Viscainoa Greene. Zygophyllaceae (V). 1 Baja California: ***V. geniculata*** (Kellogg) Greene

Viscaria Bernh. = *Silene*

viscose rayon filaments made of regimented cellulose coagulated from cellulose xanthate solution

Viscum Tourn. ex L. Santalaceae (Viscaceae). c. 150 OW trop. to temp. (few; Eur. 2, Aus. 4, Afr. 45 – R: R. Polhill & D. Wiens (1998) *Mistletoes of Afr.*: 279). Shrubby, rarely minute herbs. ***V. album*** L. (mistletoe, Eur. (introd. to Ireland), W As.) – parasitic (3 subspp., 2 restricted to conifers in C & S Eur. & N Turkey) esp. on lime, hawthorn, poplars, apple & other aliens, v. rarely on oak (etc. (even hyperparasitizing *Loranthus europaeus* app. reducing it in oakwoods when artificially increased using black-caps)), its distrib. poss. increased by humans, dioec. shrub (Leeuwenberg's Model) with fly-poll. fls (pollen & ? nectar for honeybees in Feb.), fr. taken by birds but viscid tissue on seed prevents seed being swallowed & bird scrapes it off into bark-crevice etc. where it germinates, berries used as bird-lime & locally medic. (cytotoxic & immunostimulatory, & relaxing nervous system as known to Druids & Pliny the Elder etc.) but with cardiotoxic polypeptide (viscotoxin) affecting mammals but not birds, v. large chromosomes (record for eudicot genome size), since mid 19 cent. (Herefordshire) with everg. holly & ivy anc. wonder at evergreen-ness in N winter accommodated in Christianized winter festival of Christmas, poss. 'Golden Bough' of Virgil's Aeneid also imp. in Viking sagas but modern significance from assoc. with Druids since time of Pliny (the kissing habit app. English & poss. assoc. with general views on fertility symbol presented by paired frs); ***V. articulatum*** Burm.f. (Indomal.) – on other Santalaceae & Loranthaceae in Aus., lvs sold as paste for fractures & bruises in Sikkim; ***V. minimum*** Harv. (E Cape) – shoots c. 3 mm long

Vismia Vand. Hypericaceae (I; Guttiferae I). c. 52 trop. Am. (Peru 19). Mangenot's, Roux's & Troll's Models. Some resins locally medic. See also *Harungana*

Vismianthus Mildbr. Connaraceae (I). 1 SE Tanz., 1 SW Myanmar. R: AUWP 89–6(1989)369

visnaga *Visnaga daucoides*

Visnaga Mill. (~ *Ammi*). Umbelliferae. 1 S Eur., SW As.: ***V. daucoides*** Gaertn. (*A. visnaga*, khella, natur. N Am.) – cult. by Assyrians & ever since for skin & kidney complaints, & esp. lung (khellin effective against angina discovered 1945 leading to synth. drug), e.g. asthma (visnaga, basis of 'Intal'), pedicels sold in Egypt for tooth-picks

Visnea L.f. Pentaphylacaceae (3; Ternstroemiaceae, Theaceae s.l.). 1 Canary Is., Madeira: ***V. mocanera*** L.f. – cult. orn.

Vitaceae Juss. (Vitidaceae). Magnoliidae – Vitales. Incl. Leeaceae (R: Blumea 22(1974)57), 16/900 trop. & warm. Lianes usu. with tendrils (simple to 12-branched) opp. lvs, rarely succ. treelets or herbs without (*Leea*); alks 0. Nodes oft. swollen. Lvs simple, less oft. palmate or pinnate (to 4-pinnate in *Leea*) but oft. palmately lobed or veined; stip. oft. decid. Fls reg., small, bisex. or not, (3)4 or 5(–7)-merous, in cymes (oft. corymbs) or panicles, term., opp. lvs or rarely axillary; K small & ± reduced to lobes or a collar, C valvate, usu. free (apically a decid. calyptra in *Vitis* etc.), A opp. C, nectary-disk of 5 glands or annular to cupulate around $\underline{G}$ (2; 3 or 4 with false partitons in *Leea*)), 2-loc. with simple style & s.t. ± sunk in disk, each loc. with 2 anatr. bitegmic ovules ascending from carpellary margins. Fr. a 1- or 2(-4)-loc. berry, each loc. with 1 or 2 seeds; seeds endotestal with a deep groove either side of raphe, the embryo small, straight embedded in ruminate oily & proteinaceous endosperm. n = 11–20

Principal genera: *Ampelocissus, Ampelopsis, Cayratia, Cissus, Cyphostemma, Leea* (own subfam. – usu. treelets), *Tetrastigma* (sole hosts for *Rafflesia*), *Vitis*

Seed structure suggested move from Rhamnales. Grapes & currants (*Vitis*) & cult. orn. esp. *Ampelopsis, Cissus, Cyphostemma, Leea, Parthenocissus* & *Rhoicissus*

Vitaliana Sesler = *Androsace*

Vitekorchis Romowicz & Szlach. (~ *Oncidium*). Orchidaceae (V 12h). 4 N Andes. R: PJB 51(2006)45

Vitellaria Gaertn. f. (*Butyrospermum*). Sapotaceae (I1). 1(–2) W trop. Afr. & Cameroun. Aubréville's Model. ***V. paradoxa*** Gaertn. f. (*B. parkii*, karité) – pulp ed. (sweet), kernel source of shea butter rich in vitamin E, used in food (form. in chocolate as has high melting-point) & illumination, cosmetics & aromatherapy, r. a chewing-stick in Nigeria

Vitellariopsis Baill. ex Dubard. Sapotaceae (I 1). 6 E Afr.

Vitex Tourn. ex L. Labiatae (II; Verbenaceae s.l.). Incl. *Paravitex, Petitia* (R: FR 42(1937)230), *Tsoongia, Viticipremna*, 250 trop. (Mal. 16) to temp. (few; Eur. 1). Trees & shrubs (e.g. Champagnat's Model) incl. monocaul pachycauls (Madag., but pollen distinctive), some with domatia, some with ecdysteroids (e.g. ***V. glabrata*** R. Br. (Indomal.) – medic. in Thailand). Some timber (fiddle-wood, molave [Philippines]) & cult. orn. ***V. agnus-castus*** L. (chaste tree, S Eur., widely natur.) – rheophyte, twigs s.t. used in basketwork (Odysseus's companions tied under sheep with it to escape Polyphemus), fr. a pepper subs. & with low levels of progesterone etc. so used to alleviate PMT, regulate menstrual cycles etc., '**Alba**' with white fls long considered a symbol of chastity; ***V. altissima*** L.f. (Pakistan to Sri Lanka) – timber for constr. esp. window-frames & cabinet-work; ***V. cofassus*** Reinw. ex Blume (NG teak, vitex, Mal. to W Pac.) – wood for carving, drums, pestles; ***V. cymosa*** Bertero ex Spreng. (trop. Am.) – water-disp. in Amazon (fr. with aerenchyma); ***V. divaricata*** Sw. (Venez., WI) – timber for shingles, lvs for tanning; ***V. doniana*** Sweet (trop. Afr.) – common savanna tree, local timber, fr. ed.; ***V. lignum-vitae*** Schauer (*Premna l.*, Queensland lignum-vitae, NE Aus.) – timber incl. gunstocks; ***V. lucens*** Kirk (puriri, NZ) – timber for general construction, cult. orn. tree; ***V. negundo*** L. (lagundi, Indomal.) – shrub, largely sterile, trad. remedy coughs & colds (Philippines, but often adulterated), soil-stabilizer in Nepal; ***V. pinnata*** L.f. (*V. pubescens*, milla, trop. As.) – timber for constr., r.-suckers, living tree carrying many epiphytic orchids in Malay Pen. & so much searched by collectors; ***V. rotundifolia*** L.f. (*V. trifolia* subsp. *litoralis*, Indomal.) – Bell's Model, used for 'flu, colds & sore eyes in modern Chinese herbalism; ***V. triflora*** Vahl (trop. S Am.) – sweet-tasting fr. poss. for conserves; ***V. sp.*** (*P. domingensis* Jacq., fiddlewood (corruption of *bois fidèle*)) – good timber for poles, furniture etc.

Viticipremna H.J. Lam = *Vitex* (but see JABG 7(1985)181)

Vitidaceae Juss. = Vitaceae

Vitiphoenix Becc. = *Veitchia*

Vitis Tourn. ex L. Vitaceae. c. 60 N hemisph. H. Wilson (2003) *Wine & words.*; G. Kerridge & A. Gackle (2005) *Vines for wines*; J. Robinson & al. (2012) *Wine grapes*, H. Johnson &

J. Robinson (2013) *World atlas of wine*, ed. 7, I. Tattersall & R. DeSalle (2015) *A nat. hist. wine*. Lianes cult. for fr. eaten fresh, dried or drunk fermented (wine) etc. (distrib. birds, foxes, bears, box turtles etc. incl. primates attracted by alcohol; wine known to be beneficial in reasonable quantities – esp. allegedly due to resveratrol (mimics tyrosine & binds to TyprRS allegedly protecting DNA from damage) in red wine in cardiovascular & brain health, poss. minimizing damage from stroke, in mice at least reducing weight-gain & diabetes), local medic. (N Am.) & orn. with domatia (oft.) & tendrils in place of infls, esp. ***V. vinifera*** L. (grape vine, poss. derived from 'subsp. *sylvestris*' (E Medit., though that name strictly refers to escaped pls in Rhine Valley) believed brought to Mesopotamia from beyond Taurus in C24 BCE (Caucasus, but also Italy, refugia during glaciation) & now cult. to 52° N in Poland, 70 M t a yr [50 M t for wine, 26 M l., from 7.4 M ha], $600 billion industry, by 2009 Italy 8 billion bottles a yr (Spain bigger area – 1 M ha+), France 7 billion, Aus. 166 M, England 2–3 M; max. consumption Luxemburg 53.5l per person a yr), Eur. prod. gravely affected (60–90% of vineyards lost) in C19 by introduction (via UK) of phylloxera root louse (*Daktulosphaira vitifoliae*, E N Am.) deforming roots leading to sec. infection by fungi cutting off xylem-flow, avoided by grafting on stocks of Am. spp. exuding sticky sap gumming up insect mouthparts – longevity of wines due to intrinsic preservative sugar, alcohol & tannin, longest-lived being sweet ones (e.g. Malmsey), cult. clones (long-lived e.g. '**Schiva Grossa**' ('Black Hamburg(h)') at Hampton Court, UK planted 1769 still alive), prob. arose in SW As. (all with bisex. fls & not dioec.) grown from 4th millenium BCE in Syria & Egypt (though 9000-yr-old evidence (but poss. hawthorn brew) known from China), from 2500 BCE in Aegean, but after outbreak of attack by phylloxera rootlouse (1867) Am. spp. intr. & used as resistant stocks (today notably **'Chambourcin'** (poss. involving 8 US spp. in parentage) good in wet autumns e.g. Aus., as fungus- & louse-resistant); these esp. *V. labrusca* also involved in hybrids giving American cultivar grapes much grown in US & backcrossed with *V. vinifera* to give French hybrids, largely wine grapes, but also v. hardy dessert grape '**Brant**' (= *V.* × *novae-angliae* Fernald 'Clinton' (*V. labrusca* × *V. riparia* Michaux (C N Am.) – cult. now banned in Eur.) × *V. vinifera* 'Black St Peters' (= **'Zinfandel'** ('Primitivo', most imp. in Calif.)); for early ripening necessary in e.g. UK hybrids with *V. amurensis* Rupr. (NE As.) grown. Classical cvs of *V. vinifera* grown for wine incl. '**Cabernet Sauvignon**' ('Cabarnet Franc' × 'Sauvignon Blanc', prop. veget. for over 800 yrs) & '**Pinot Noir**' (red, 'Pinot Blanc' derived by deletion of VvmybA1 allele), '**Chardonnay**' & '**Riesling**' (white) etc., claret (or red Bordeaux) being made from 'Cabernet Sauvignon', '**Merlot**' etc., white Bordeaux & sancerre from '**Sauvignon Blanc**', sauternes & barsac from '**Sémillon**' etc. with 'noble rot', burgundy from 'Pinot Noir', beaujolais from '**Gamay**', chablis from 'Chardonnay', champagne (fermenting in bottle; first 'sparkling' 1662 (C. Merret), before Dom Perignan) from 'Chardonnay' & 'Pinot Noir' (white & pink), fitou from '**Carignan**', vin jaune (Jura) from **'Savagnin'**, Alsace from '**Gewürztraminer**', '**Pinot Blanc**', '**Sylvaner**' etc.; elsewhere 'Riesling' for Mosel & 'hock' (Eiswein from grapes frozen on vines, Ger.), '**Sangiovese**' for chianti & '**Corvina**' (& others)) for valpolicella, '**Gorganega**' for soave, '**Malvasia**' & '**Trebbiano Toscano**' (prob. source of more red wine than is any other cv.) for frascati (Italy, where leading grape is '**Barbera**'), '**Harslevelu**' & '**Furmint**' blended to give tokay (Hungary), '**Müller-Thurgau**' – leading white grape in Ger., '**Malbec**' – imp. black g. in Arg.; world's most planted red cv. is '**Grenache**' (esp. Spain, as 'Garnacha', being a component of rioja, in France for Banyuls & Châteauneuf-du-Pape) with '**Pinotage**' widely grown in S Afr., NZ etc. (monoterpenes responsible for aroma & flavour of Muscat grapes, also 'Riesling', '**Traminer**' & '**Müller-Thurgau**'; 2-methoxy-3-isobutylpyrazine in 'Cabernet Sauvignon', 'Sauvignon Blanc' & '**Shiraz**' gives 'vegetative' flavour (also in broad beans, spinach & parsnips), poss. connected with anti-herbivory & disp. attraction etc.), the concentration of organic acids (largely tartaric acid converted from vitamin C) esp. malic acid, anthocyanin & phenolics etc. affected by rain régime so that latter 2 higher in water deficit; with added brandy or other alcohol such wines are 'fortified' giving sherry (90% '**Palomieno**' in Spain), madeira, port ('**Touriga Nacional**' most imp. in Douro, white port from '**Verdelho**' also used in Madeira though used for table-wine in France & Aus.) etc.; with added flavourings they give vermouths & martinis; with quinine, iron etc., they are medicinal; distilled to brandy (e.g. '**Ugni Blanc**' ('Trebbiano' in Italy) for armagnac ('**Ondenc**') & cognac (France); distillate of infl. axes, seeds etc. = grappa (Italy), marc (France). Records of grape-harvests from Med. times used to reconstruct weather-patterns (e.g. 1520s & 1630–80 allegedly as warm as 'human-induced', warmed late 20 cent.). Grape wine has been assoc. with religion &

other ceremonial since antiquity (known 3000 BC Mesopotamia & Egypt, though drunk diluted by anc. Greeks) & despite its dangers (in excess) is an intrinsic feature of W. society (by 1880 80% of Italian pop. ± reliant on wine for a living; by 1998 Br. third biggest market – 980 M bottles per annum = 9.5% of world prod.). Fresh dessert grapes imp. export of S Afr., Israel etc., '**Chasselas**' being imp. white g. in France. Excess wine used for prod. industrial alcohol; unfermented grape-juice imp. industry, juice boiled down to thick sweet syrup = dibs, & basis of balsamic vinegar & extract used for coughs & in ear-drops (Oman); dried fr. = raisins used in cooking (form. used as subs. for plums, so constituents of 'plum duff', 'plum pudding'; Aus. biggest producer esp. **'Muscat Gordo'** & **'Waltham Cross'**), sultanas being seedless forms ('**Sultana**' ('Kishmish,' 'Thompson Seedless') – esp. Turkey) & the most widely grown (also 80% of fresh table-grapes), currants being smaller ones much used in cakes etc. (currant [cf. *Ribes*] a corruption of Corinth(ian grape), passum being raisin wine; muscatel grapes are those left to dry on vine in Med.; seed-oil used in cosmetics & cooking (grapeseed oil); verjuice (vert jus) from crushed unripe grapes early in season, esp. for salad dressing; lvs used in e.g. Greek cuisine (dolmades). Although early S Am. plantings of *V. vinifera* survived, those in N Am. died out (r.-louse, etc.) though spontaneous hybrids, ***V.* × *alexanderi*** Prince ex Jacques (*V. prolifera*, first 'American Hybrids') with *V. labrusca* survived, '**Alexander**' basis of first comm. wine-growing in US (Indiana 1806), '**Concord**' imp. in US jam ('jelly')-making, '**Isabella**' successful in trop. e.g. Hawaii but banned France (foxy genes) 1934; ***V.* × *bourquiniana*** Munson (*V. vinifera* × *V. aestivalis* Michaux) incl. '**Jacquez**' ('Le Noir', 'Troya') orig. Georgia or Carolinas. Other cult. spp., several used in modern hybrids, incl. ***V. acerifolia*** Raf. (bush g., SC US) & ***V. arizonica*** Engelm. (canyon g., SW N Am.) – fr. sweet; ***V. coignetiae*** Pull. ex Planch. (Jap., Korea) – rapid-growing screening liane with red autumn colours; ***V. labrusca*** L. (fox or skunk g., NE US) – fr. with musky taste, basis of early 11 cent. Norse name, 'Vinland', for N Am.; ***V. monticola*** Buckley (mt. g., SW Texas); ***V. palmata*** Vahl (cat g., C & S US); ***V. riparia*** Michaux (N Am.) – hybridized with 'French hybrids' to give cold-resistant grapes for Mid-West US, e.g. '**Frontenac**' tolerant of -33°F; ***V. rotundifolia*** Michaux (bullace or fox g., muscadine, scuppernong, SE US) – fr. thick-skinned, ed., used in a sweet US wine since C16 Florida, v. high in resveratrol (40 mg/l cf. 0.2–5.8 mg in *V. vinifera*); ***V. rupestris*** Scheele (sand g., SC US) – fr. sweet; ***V. vulpina*** L. (chicken or frost g., C & E US) – fr. v. acid but becoming sweet & ed. after frost

Vittadinia A. Rich. Compositae (Ast.-Pod.). Incl. *Peripleura, Tetramolopium*, 60 NG (most), Aus., New Caled. (1), NZ (1), Hawaii (11, some extinct). R: Brunonia 5(1982)1. Some cult. orn.

Vittaria Sm. Pteridaceae (V; Vittariaceae). 7 trop., temp. N Am. Epiphytes with linear pendent fronds with linear sori in parallel rows; some cult. orn.

Vittariaceae Ching = Pteridaceae

Vittetia R. King & H. Robinson (~ *Eupatorium*). Compositae (Eup.-Gyp.). 2 Braz. R: MSBMBG 22(1987)110

Viviania Cav. Francoaceae (Vivianiaceae; Geraniaceae s.l.). 6 S Braz., Chile. R: UCOP 2(1975)231

Vivianiaceae Klotzsch = Francoaceae

Vladimiria Iljin = *Dolomiaea*

Vleisia Toml. & Posluszny = *Pseudalthenia*

Vlokia S. Hammer. Aizoaceae. 1 W Cape: ***V. ater*** S. Hammer. R: H. E. K. Hartmann, *Ill. Handb. Succ. Pls*, Aiz. F–Z(2002)350

Voacanga Thouars. Apocynaceae (I d). 12 trop. OW (Afr. 7, As. 5). R: AUWP 85–3(1985)5. Leeuwenberg's Model. Seeds of ***V. africana*** Stapf & ***V. thouarsii*** Roemer & Schultes (trop. Afr.). coll. for industrial prod. of tabersonine (alk.) used as depressor of C nervous system activity in geriatric patients & source of precursor for vincamine (see *Vinca*) for cerebral vascular disorders

Voandzeia Thouars = *Vigna*

Voanioala Dransf. Palmae (V 8a). 1 Madag.: ***V. gerardii*** Dransf. – 2n = 596+ (twice that in any other monocot; polyploidy rare in palms), endangered (large sclerified endocarps accum. beneath trees so ? no effective disperser). R: KB 44(1989)191.

Voatamalo Capuron ex Bosser. Picrodendraceae (Pic.-Misch.; Euphorbiaceae s.l.). 2 Madag. R: Adansonia 15(1976)333

Vochysia Aubl. Vochysiaceae (I). c. 110 trop. Am. R: RTBN 41(1948)423. Some timbers esp. ***V. hondurensis*** Sprague (quaruba, C Am.); many local medic.

Vochysiaceae A. St-Hil. Magnoliidae – Myrtales. Excl. Euphroniaceae, 7/200 trop. Am. & W Afr. (2/3). Trees, shrubs, rarely herbs, with resin, oft. with internal phloem & accum. aluminium. Lvs simple usu. opp. or whorled; stip. small to 0. Fls bisex., obliquely irreg., bibracteolate, in racemes or thyrses; K 5 basally connate, imbr. with 1 oft. large & spurred basally, C (0) 1, 3 or 5, ± unequal, convolute or imbr., A 1 –5(–7) usu. only 1 fert. with filaments free or connate in 2 groups, anthers with longit. slits, $\underline{G}$ to $\overline{G}$ (3(4)), 3-loc. with axile placentas or pseudomonomerous with 1 loc., style 1, each loc. with (1)2 – ∞ anatr. or hemitropous bitegmic ovules. Fr. a loculicidal capsule, nut-like or samaroid, winged with accrescent K; seeds 1–several oft. winged or hairy with straight embryo ± endosperm. n = 11, 12

Classification & genera:

I. **Vochysieae** (simple hairs; G 3-loc., seeds 3-several, winged): *Callisthene, Qualea, Salvertia, Vochysia*

II. **Erismeae** (usu. stellate hairs, G 1-loc., fr. samaroid or nut-like, seed 1, wingless): *Erisma, Erismadelphus, Korupodendron*

Some timbers

vodka *Solanum tuberosum* (Poland), *Triticum aestivum* (Russia)

Voharanga Costantin & Bois = *Cynanchum*

Vohemaria Buchenau = *Cynanchum*

Voladeria Benoist = *Oreobolus*

Volkameria L. (~ *Clerodendrum*). Labiatae (III). c. 25–30 trop. (As. 1: *V. **inermis*** L. (*C. i.*, Indomal. to W Pac., coastal) – widely introd. trop. as sand-binder, fire-sticks in N Aus., topiary in Ind., canework in Tonga). R: T 59(2010)215

Volkensia O. Hoffm. = *Bothriocline*

Volkensiella H. Wolff = *Oenanthe*

Volkensinia Schinz. Amaranthaceae (I 2). 1 E Afr.: *V. **prostrata*** (Gilg) Schinz. R: KB 33(1979)417

Volkeranthus Gerbaulet = *Mesembryanthemum*

Volkiella Merxm. & Czech. = *Cyperus* (*Lipocarpha*; but see KB 41(1986)945)

Volutaria Cass. (~ *Centaurea*). Compositae (Card.-Cent.). 18 Medit. (Eur. 1), SW As., trop. E Afr.

Vonitra Becc. = *Dypsis*

voodoo lily *Sauromatum venosum*

Vossia Wall. & Griff. Gramineae (XXII 2). 1 trop. & SW Afr., E Ind. to Myanmar: *V. **cuspidata*** (Roxb.) Griff. – floating grass, part of the Nile sudd & poss. the orig. bulrush (q.v.) of the Bible. R: Strel. 19(2000)423

Votomita Aubl. Melastomataceae. c. 6 trop. Am. R: FN 16(1976)256. Elaiophores attractive to bees

Votschia Ståhl (~ *Jacquinia*). Primulaceae (Theophrastaceae). 1 NE Panamá: *V. **nemophila*** (Pittier) Ståhl. R: Britt. 45(1993)204

Vouacapoua Aubl. (~*Andira*). Leguminosae (I 4). 3 trop. S Am. Troll's Model. Good timber (brownheart, wacapou) esp. *V. **americana*** Aubl. (*A. excelsa*)

Vouarana Aubl. Sapindaceae. 2 NE S Am.

Voyria Aubl. Gentianaceae. 19 trop. Am., W Afr. (1: *V. **primuloides*** Bak.). R: Britt. 49(1997)473. Mycotrophic (some epiphytic Amaz.). Seeds unitegmic & anatr. to ategmic & orthotropous

Voyriella Miq. Gentianaceae. 2 NE S Am. R: MNYBG 51(1989)47. Mycotrophs

Vriesea Lindl. Bromeliaceae (2). c. 265 trop. Am. but few in Amazonia. R: FN 14(1977)1068, 1395. Heterogeneous. Usu. large epiphytes with conspic. bracts much cult. incl. many hybrids, some of unknown parentage. *V. **fosteriana*** L.B. Sm. (Braz.) – centrepiece in US 'summer bedding'; *V. **imperialis*** Carr. (Rio State, Braz.) – 3–5 m in montane grassland taking 20 yrs to flower; *V. **incurva*** (Griseb.) Read (trop. Am.) – leaf-bases secrete a mucilage said to be proteolytic; *V. **ranifera*** L.B. Sm. (Costa Rica cloud forest) – spp. of frog restricted to the pl. where they breed

Vrydagzynea Blume. Orchidaceae (IV 2d). c. 40 Indomal. to Taiwan & W Pac.

Vulpia C. Gmelin = *Festuca* (but see KBAS 13(1986)97)

Vulpiella (Battand. & Trabut) Burollet. Gramineae (XVI 12). 1 W Medit. incl. Eur.: *V. **stipoides*** (L.) Maire (*V. tenuis*). R: BSBF 124(1977)347

Vvedenskya Korovin (~ *Phaeonychium*). Umbelliferae. 1 C As.: *V. **pinnatifolia*** Korovin

Vvedenskyella Botsch. = *Solms-laubachia*

waboom *Protea nitida*

wacapou *Vouacapoua* spp.

Wachendorfia Burm. Haemodoraceae (I). 4 SW & S Cape. R: Strel. 9(2000)93. Rhiz. red; lvs plicate; app. 'sup.' ovary actually inferior with elongated receptacle

Wagatea Dalz. = *Moullava*

Wagenitzia Dostál = *Centaurea*

Wagneriopteris Löve & D. Löve (~ *Thelypteris*). Dryopteridaceae (II). 7 trop. & warm

Wahlenbergia Schrad. ex Roth. Campanulaceae (I 2). c. 260 subcosmop. esp. S temp. (Eur. 2, trop. & S Afr. (39 – R: T 44(1995)334) & Madag. 51, St Helena 4 (2 extinct since 1870s – ***W. burchellii*** A. DC. & ***W. roxburghii*** A.DC., with ***W. linifolia*** (Roxb.) A. DC. reduced to 25 pls), Aus. 26 (21 endemic – R: Telopea 5(1992)91), NZ 10 (9 endemic – R: NZJB 35(1997)16). Heterogeneous. Like *Campanula* & *Edraianthus* but capsule with apical valves; on St Helena pachycaul at high alts; some S Afr. spp. poll. monkey-beetles. Some cult. orn. ('bluebell', harebell, NZ, S Afr.) incl. ***W. hederacea*** (L.) Reichb. (ivy-leaved bell fl., W Eur.). ***W. marginata*** (Thunb.) A. DC. (Indomal. to NZ) – med. (skin problems)

wainscot orig., oak boards

Waireia D.L. Jones & al. (~ *Thelymitra*). Orchidaceae (IV 3f). 1 NZ: ***W. stenopetala*** (Hook.f.) D.L. Jones & al. R: Gen. Orch. 2(2001)172

Waitzia Wendl. Compositae (Gnap.-Ang.). 5 temp. S & W Aus. R: Nuytsia 8(1992)461. Annuals with 'everlasting' capitula & ectomycorrhizae

Wajira Thulin. Leguminosae (III 18). 5 Afr., Ind., Sri Lanka. R: SB 29(2004)906

wakame *Undaria pinnatifida*

wake-robin *Trillium grandiflorum*, *Arum maculatum*

Wakilia Gilli = *Solms-laubachia*

Walafrida E. Meyer = *Selago*

Waldheimia Karelin & Kir. = *Allardia*

Waldsteinia Willd. = *Geum* (but see FRB 72(1933)93)

Walidda (A. DC.) Pichon = *Wrightia*

walking fern *Asplenium rhizophyllum*

wall cress *Arabis* spp.; **w. lettuce** *Cicerbita muralis*; **w. pennywort** *Umbilicus rupestris*; **w. pepper** *Sedum acre*; **w. rocket** *Diplotaxis muralis*; **w. rue** or **spleenwort** *Adiantum ruta-muraria*

wallaba *Eperua* spp.

wallaby grass *Rytidosperma* spp.

Wallacea Spruce ex Hook.f. Ochnaceae (III). 2 N S Am. R: BJ 113(1991)179

Wallaceodendron Koord. Leguminosae (II 4). 1 Philipp., Sulawesi: ***W. celebicum*** Koord. – high-quality furniture. R: BMNHN 4,5(1983)347

Wallaceodoxa Heatubun & W. Baker. Palmae (V 14i). 1 W NG: ***W. raja-ampat*** Heatubun & W. Baker. R: KB 69[9525](2014)13

wallapata *Gyrinops walla*

Wallenia Sw. (~ *Cybianthus*). Primulaceae (Myrsinaceae). c. 20 WI. Rauh's Model

Walleniella P. Wilson = *Solonia*

Walleria J. Kirk. Tecophilaeaceae. 3 trop. & S Afr. R: Bothalia 42(2012)22

Walleriaceae H. Huber ex Takht. = Tecophilaeaceae

wallflower *Erysimum cheiri*; **alpine** or **fairy w.** *E.* spp.; **Siberian w.** *E.* × *marshallii*; **western w.** *E. asperum*

Wallichia Roxb. Palmae (III 6). 9 E Himal. to S China. R: Taiwania 52(2007)1. Monoec., hapaxanthic (Holttum's Model)

Wallnoeferia Szlach. = *Helonoma*

wallum *Banksia aemula*

walnut *Juglans regia*; **African or Benin w.** *Lovoa trichilioides*; **American** or **black w.** *J. nigra*; **w. bean** *Endiandra palmerstonii*; **Black Sea, Carpathian, European** or **Persian w.** *J. regia*; **claro w.** *J. hindsii*; **E Indian w.** *Albizia lebbeck*; **false w.** *Neoguillauminia cleopatra*; **Japanese w.** *J. ailanthifolia*; **Manchurian w.** *J. mandshurica*; **New Guinea w.** *Dracontomelon* spp.; **Nigerian golden w.** *L. trichilioides*; **Otaheite w.** *Aleurites moluccanus*; **Pacific** or **Papuan w.** *D.* spp.; **poison w.** *Cryptocarya pleurosperma*; **Queensland** or **Australian w.** *Endiandra palmerstonii*; **satin w.** *Liquidambar styraciflua*; **yellow w.** *Beilschmiedia bancroftii*

Walsura Roxb. Meliaceae (II). 16 Indomal. (to Sulawesi). R: Blumea 38(1994)257. ***W. monophylla*** Elmer ex Merr. (Palawan) – on ultramafics accum. nickel to 7000 μg/g & tested for use in AIDS & cancer treatment; *W.* ***trifoliolata*** (A. Juss.) Harms ('*W. trifoliata*', Ind., Sri Lanka) – local medic.

Walteranthus Keighery. Gyrostemonaceae. 1 W Aus.: ***W. erectus*** Keighery. R: BJ 106(1985)108

Waltheria L. Malvaceae (Bytt.-Herm.; Sterculiaceae). c. 50–60 trop. Am., Afr. (1), Madag. (1), Malay Pen. (1), Taiwan (1). Cult. orn. with alks esp. ***W. indica*** L. (*W. americana*, pantrop.) – Petit's Model, form. medic. (Hawaii), fibre like jute

Walwhalleya Bruhl = *Homopholis* (but see AusSB 19(2006)315, Telopea 13(2011)88)

Wamalchitamia Strother. Compositae (Helia.-Ecl.). 7 C Am. R: SBM 33(1991)30, Lundellia 16(2003)1

wamara *Swartzia* spp.

wampee *Pontederia cordata*

wampi *Clausena lansium*

wand flower *Dierama* spp.

wandering Jew *Tradescantia* spp., *Saxifraga stolonifera*; **w. sailor** *Cymbalaria muralis*

Wandersong D.W. Taylor (*Colleteria*; ~ *Chione*). Rubiaceae. 2 Carib. R: JBRIT 8(2014)530

wandoo *Eucalyptus redunca*

wangee *Phyllostachys nigra*

Wangenheimia Moench = *Festuca* (but see KBAS 13(1986)97)

Wangerinia Franz = *Microphyes*

wanzee *Cordia africana*

wapato *Sagittaria cuneata, S. latifolia*

wara *Calotropis gigantea*

warabi *Pteridium aquilinum*

waragi *Musa* cvs.

waras *Flemingia* spp.

waratah *Telopea speciosissima* & hybrids

Warburgia Engl. Canellaceae. 3 E Afr. ***W. salutaris*** (Bertol.f.) Chiov. (*W. ugandensis*) – resin for fixing tool-handles, bark a purgative & used as antimalarial, & with antifeedant sesquiterpenes (e.g. warburganal) active against army worm, lvs eaten in curries

Warburgina Eig = *Callipeltis*

Warczewiczella Reichb.f. Orchidaceae (V 12j). 10–12 trop. am,

Wardaster Small = *Aster*

Wardenia King = *Trevesia*

Ward's weed *Carrichtera annua*

Warea Nutt. Cruciferae (46). 4 SE US. R: FNA 7(2010)742

Warionia Benth. & Cosson. Compositae (Cich.-Cich.). 1 NW Sahara: ***W. saharae*** Benth. & Cosson – shrublet. R: AJBM 65(2008)369

Warmingia Reichb.f. Orchidaceae (V 12h). 4 trop. Am. R: Lindleyana 7(1992)196. Cult. orn.

Warneckea Gilg (~ *Memecylon*). Melastomataceae. c. 50 trop. Afr., Madag. R: Adansonia 18(1978)228. Elaiophores attractive to bees

Warnockia M. Turner. Labiatae (VI). 1 S US, NW Mex.: ***W. scutellarioides*** (A. Gray) M. Turner. R: PSE 203(1996)78

Warpuria Stapf = *Podorungia*

Warrea Lindl. Orchidaceae (V 12j). 4 trop. Am. Cult. orn. terr. orchids with large fls

Warreella Schltr. Orchidaceae (V 12j). 2 Colombia. ***W. cyanea*** (Lindl.). Schltr. – cult. orn.

Warreopsis Garay. Orchidaceae (V 12j). 4 trop. Am.

warrigal (cabbage) *Tetragonia tetragonoides*

warrus *Flemingia* spp.

Warscaea Szlach. = *Cyclopogon*

Warszewiczia Klotzsch. Rubiaceae (Cond.). 8 trop. Am. ***W. coccinea*** (Vahl) Klotzsch – cult. orn. greenhouse shrub with 1 sepal in a group of fls enlarged to 6 cm long, red, a mutant known with all fls with such a lobe

wartcress *Lepidium coronopus*

wasabi *Eutrema japonicum*

Wasabia Matsum. = *Eutrema*

washiba (wood) *Handroanthus serratifolius*

Washington elm *Ulmus americana*; **W. thorn** *Crataegus phaenopyrum*

Washingtonia H. Wendl. Palmae (III 4b). 2 arid SW N Am. R: GH 4(1936)53. Massive palms with bisex. fls & palmate lvs with spiny petioles (spineless above 14m, over height of present (or past?) herbivores). Used as street trees. ***W. filifera*** (André) Bary (SW US) –

the palm of Palm Springs, fr. & seeds ed., leaf-fibre for basketry; *W.* ***robusta*** H. Wendl. (*W. sonorae*, Mex.) – fr. ed.; their hybrid = *W.* ×***filibusta*** Hodel – common in nursery trade

water arum *Calla palustris*; **w. ash** *Fraxinus caroliniana*; **w. avens** *Geum rivale*; **w. beech** *Carpinus caroliniana*; **w. blinks** *Montia fontana*; **w. caltrop** *Trapa natans*; **w. chestnut** *T.* spp.; **Chinese w. c.** *Eleocharis dulcis*; **chinquapin w. c.** *Nelumbo lutea*; **w.cress** *Nasturtium* spp. esp. *N. officinale*; **w. crowfoot** *Ranunculus aquatilis*; **w. dropwort** *Oenanthe* spp.; **w. elm** *Planera aquatica*; **w. feather** *Myriophyllum brasiliense*; **w. fern** *Azolla* spp., *Salvinia* spp.; **w. figwort** *Scrophularia auriculata*; **w. forget-me-not** *Myosotis scorpioides*; **w. fringe** *Nymphoides peltata*; **w. grass** *Panicum molle*; **w. gum** *Tristania neriifolia*; **w. hair** *Catabrosa aquatica*; **w. hawthorn** *Aponogeton distachyos*; **w. hemlock** *Oenanthe crocata*; **w. hyacinth** *Eichhornia crassipes*; **w. lemon** *Passiflora laurifolia*; **w. lettuce** *Pistia stratiotes*; **w.lily** *Nymphaea* spp.; **w.l., giant** *Victoria amazonica*; **w.l., yellow** *Nuphar lutea*; **w. lobelia** *Lobelia dortmanna*; **w. meal** *Wolffia* spp.; **w. melon** *Citrullus lanatus*; **w. milfoil** *Myriophyllum* spp.; **w. mimosa** *Neptunia oleracea*; **w. mint** *Mentha aquatica*; **w. oat** *Zizania* spp.; **w. parsnip** *Sium* spp.; **w. pepper** *Persicaria hydropiper*; **w. phlox** *Wurmbea* (*Onixotis*) spp.; **w. plantain** *Alisma* spp.; **w. primrose** *Ludwigia grandiflora*; **w. rice** *Z. palustris*; **w. rose apple** *Syzygium aqueum*; **w. soldier** *Stratiotes aloides*; **w. speedwell** *Veronica catenata*; **w. spinach** *Ipomoea aquatica*; **w. starwort** *Callitriche stagnalis*; **w. tupelo** *Nyssa aquatica*; **w. violet** *Hottonia palustris*; **w.weed, Canadian** or **American** *Elodea canadensis*; **w.wheel** *Aldrovanda vesiculosa*; **w. wisteria** *Hygrophila difformis*; **w.wort** *Elatine* spp.; **w. yam** *Dioscorea alata*

Waterhousea B. Hyland = *Syzygium*

Watsonia Mill. Iridaceae (III 1). 51 S Afr. R: P. Goldblatt (1989) *The genus W.* Cult. orn. like *Gladiolus* but style-branches 2-fid, etc. esp. *W.* ***borbonica*** (Pourr.) Goldbl. (*W. ardernei*, *W. pyramidata*, SW Cape). Some noxious weeds in temp. Aus. esp. *W.* ***meriana*** (L.) Mill. '**Bulbillifera**' – large-fld sterile triploid with axillary cormlets, but occ. fert.

Wattakaka Hassk. = ?*Marsdenia*

wattle *Acacia* spp.; **black w.** *A. decurrens*, *A. mearnsii*, *Callicoma serratifolia*; **blue w.** *A. dealbata*; **w.cino** *A. victoriae* etc.; **Cootamundra w.** *A. baileyana*; **downy w.** *A. pubescens*; **golden w.** *A. pycnantha*; **golden rain w.** *A. prominens*; **green w.** *A. decurrens*; **hairy w.** *A. pubescens*; **Mudgee w.** *A. spectabilis*; **reindeer w.** *A. aphylla*; **river w.** *A. subporosa*; **sally w.** *A. longifolia*; **salt w.** *A. ampliceps*; **silver w.** *A. dealbata*; **sticky w.** *A. howittii*; **sunshine w.** *A. terminalis*; **Sydney golden w.** *A. longifolia*

wawa *Triplochiton scleroxylon*

wax bean *Phaseolus vulgaris*; **Carnauba w.** *Copernicia prunifera*; **w. flower** *Hoya carnosa*, *Chamelaucium*, *Crowea*, *Eriostemon* & *Philotheca* spp.; **Geraldton w.(f.)** *Chamelaucium uncinatum*; **w. gourd** *Benincasa hispida*; **w. jambu** *Syzygium samarangense*; **Japanese w.** *Toxicodendron succedaneum*; **w. myrtle** *Morella* spp., **w. palm** *Ceroxylon* spp.; **w. plant** *Hoya carnosa*, (Aus.) *Eriostemon* spp.; **w. tree, Japanese** *T. succedaneum*; **urucury w.** *Syagrus coronata*

wax-lip (orchid) *Glossodia* spp.

wayfaring tree *Viburnum lantana*

weasel's snout *Misopates orontium*

Weberaster Löve & D. Löve = *Eurybia*

Weberbauera Gilg & Muschler. Cruciferae (46). Incl. *Stenodraba*, 24 W & S S Am. R: Novon 14(2004)258. Some cult. orn. alpines

Weberbauerella Ulbr. Leguminosae (III 11). 3 coastal Peru

Weberbauriella Ferreyra = *Chucoa*

Weberbauerocereus Backeb. (~ *Haageocereus*). Cactaceae (III 4). 5 Peru

Weberocereus Britton & Rose. Cactaceae (III 2). 8 C Am. esp. Costa Rica. Cult. orn. climbers with aerial r., night-fls intergrading with *Selenicereus*. *W.* ***tunilla*** (Weber) Britton & Rose (C Am.) – flagelliflory, infl. to 2 m, with foul-smelling bat-poll. fls

Websteria S. Wright = *Eleocharis* (but see KB 26(1972)582)

Weddellina Tul. Podostemaceae (Tristichaceae). 1 N S Am.: *W.* ***squamulosa*** Tul. – shoots to 75cm & branched with short unbranched flowering ones. R: BMNHN 4,10(1988)169

wedding bush *Ricinocarpos* spp. esp. *R. pinifolius*; **w. flower** *Francoa sonchifolia*

Wedelia Jacq. Compositae (Helia.-Ecl.). Incl. *Aspilia*, poss. to be re-excl., 110 trop. Afr. & Am. (c. 75; R (N Am.): SBM 33(1991)38). Some spp. with elaiosomes, *W.* ***acapulcensis*** Kunth var. ***hispida*** (Kunth) Strother (*W. h.*, Mex.) at least, myrmechorous. Many weeds, e.g. *W.* ***glauca*** (Ortega) Hicken (pascalia weed, Chile) – bad weed in Aus. & Arg., toxic to some stock; some locally medic., stopping bleeding, while some ed. chimpanzees (e.g.

some forms of ***W. mossambicensis*** Oliv. (*A. m.*, trop. Afr.)) cont. thiarubrine A effective against bacteria & intestinal parasites, some with molluscicidal saponins; extracts of *W. mossambicensis* in vitro effective against AIDS & herpes viruses, kaurenoic & grandiflorenic acids effective uterostimulators. See also *Sphagneticola*

wedge pea *Gompholobium* spp.

weed *Cannabis sativa* 'subsp. *indica*'

weeooka *Eremophila oppositifolia*

weeping ash *Fraxinus excelsior* 'Pendula'; **w. willow** *Salix* × *sepulcralis* esp. 'Chrysocoma', (willow-pattern) *S. babylonica*, (Bible) *Populus euphratica*

weevil lily *Molineria capitulata*

Wehlia F. Muell. = *Homalocalyx*

Weidmannia G. Romero & Carnevali = *Zygosepalum* (but see HPB 15(2010)191)

Weigela Thunb. Caprifoliaceae (Diervillaceae). 10 E As. R: Gingkoana 5(1983)137. Poss. heterogeneous. Cult. orn. shrubs with some hybrids, esp. ***W. florida*** (Bunge) A. DC. (*W. rosea*, N China, Korea, Jap.) – pink fls, many cvs

Weigeltia A. DC. = *Cybianthus*

Weihea Spreng. = *Cassipourea*

Weingartia Werderm. (~ *Rebutia*). Incl. *Cintia*, *Sulcorebutia*, 8 Bolivia, Peru

Weinmannia L. Cunoniaceae (VIII). c. 155 Andes (c. 75), Madag. & Masc. (c. 40), Mal. & Pac. incl. NZ (c. 40) but not Afr., Aus. or most As. Timbers, local medic., bee-forage for honey, & tanbarks esp. ***W. racemosa*** L.f. (towai, NZ). ***W. fraxinea*** Sm. ex D. Don (E Mal.) – bark to colour & flavour sago

Welchiodendron Peter G. Wilson & Waterhouse (~ *Tristania*). Myrtaceae (II 2). 1 NG & Queensland: ***W. longivalve*** (F. Muell.) Peter G. Wilson & Waterhouse. R: AusJB 30(1982)440

weld *Reseda luteola*

Weldenia Schultes f. Commelinaceae (II 1f). 1 Mex., Guatemala: ***W. candida*** Schultes f. – cult. orn. with tuberous r., underground stem & rosette of linear-lanceolate lvs

Welfia H. Wendl. Palmae (V 11). 2 C Am. R: Phytotaxa 119,1(2013)33. ***W. regia*** H. Wendl. (*W. georgii*) – fronds 6–8 m long

wellingtonia *Sequoiadendron giganteum*

Wellstedia Balf.f. Boraginaceae (Wellstediaceae). 6 Somalia, Socotra, Ethiopia, S Afr. (1)

Wellstediaceae Novák = Boraginaceae

Welsh onion *Allium fistulosum*; **W. poppy** *Papaver cambricum*

Welwitschia Hook.f. Welwitschiaceae. 1 S Angola & Namibia, deserts extending into mopane woodland: ***W. mirabilis*** Hook.f. (*W. bainesii*). R: C.H. Bornman (1978) *W.*, FOW 3(1999)2. Long-lived (at least 1500 yrs with stem to 1 m across), deriving moisture from sea-fog dew. Winged seeds germ. in wet years, the cotyledons photosynthesizing for 1.5 yrs; foliage lvs decussate with respect to cotyledons, of indefinite growth (up to 13.8 cm a year), wearing away at tips, the apical meristem being lost; 2 scale-like lvs develop & engulf shoot apex which then becomes meristematically inactive. Successive cambia prod. sec phloem with rays & sec xylem with rays. Fert. buds prod. on the 'crowns' between leaf-bases

Welwitschiaceae Caruel. Gnetidae – Gnetales. 1/1 SW Afr. (Lower Cretaceous of Braz.). R: FOW 3(1999). Dioec. perenn. (Corner's Model) with short stem & taproot. Calcium oxalate crystals in intercellular spaces (as in *Ephedra*). Lvs 2 opp., parallel-veined. Infls dichasial; insect-poll. Females subtended by cone-scales making up a red cone & consisting of single nucellus encl. in an integument & another layer derived from 2 confluent primordia ('perianth') with 2 'bracts'; megaspore mother-cell grows into prothallus without archegonia. Male fls (prob. reduced compound cones) subtended by cone-scales making up a red cone & consisting of 2 lat. 'bracts' & a 'perianth' formed by union of 2 bract-like organs, 6 microsporangiophores & a sterile unitegmic ovule; male gametophyte merely a tube-nucleus, an abortive sterile cell & a cell giving rise to 2 sperm nuclei; pollen-wall development like that in angiosperms. As pollen tubes penetrate prothallus, prothallial tubes grow up to meet them in the nucellus, fusion of females & males allegedly occurring in the tubes; many zygotes but only 1 matures & is disp. with 'perianth' as wing. n = 24

Genus: *Welwitschia*

No double fertilization, *cf. Ephedra*, *Gnetum*

Welwitschiella O. Hoffm. Compositae (Ast.-Grang.). 1 Angola, Zambia: ***W. neriifolia*** O. Hoffm. R: KB 64(2010)655

Wenchengia C.Y. Wu & S. Chow. Labiatae (V). 1 Hainan (SE China): *W. **alternifolia*** C.Y. Wu & S. Chow – thought extinct, but re-found 2010. R: APS 10(1965)250

Wendelboa Soest = *Taraxacum*

Wendlandia Bartling ex DC. Rubiaceae (?I 5). 90+ warm As. (Iraq 1), Indomal. R: NRBGE 16(1932)233. Lvs opp. or whorled (stip. leafy); some cult. orn. shrubs & trees

Wendlandiella Dammer. Palmae (V 2). 1 Peru, Braz.: *W. **gracilis*** Wendl. R: J. Dransfield & al., *Gen. Palm.* (2008)372

Wendtia Meyen = *Balbisia*

wenge *Millettia laurentii*

Wentsaiboea Fang & D.H. Qin = *Primulina* (but see APS 42(2004)533)

Wenzelia Merr. Rutaceae (II 2). 9 Philippines to Solomons, Fiji. R: W.T. Swingle, *Bot. Citrus* (1943)214. Prob. incl. *Monanthocitrus*

Werauhia J.R. Grant (~ *Vriesea*). Bromeliaceae (2). 66 trop. S Am. esp. montane & cloud forests. R: TSP 91(1995)28

Wercklea Pittier & Standl. Malvaceae (Malv.-Hib.). 12 trop. Am. R: JAA 62(1981)457. Shrubs & trees, *W. **insignis*** Pittier & Standl. (Costa Rica) – cult. orn.

Werckleocereus Britton & Rose = *Weberocereus*

Werdermannia O. Schulz = *Sibara*

Werneria Kunth. Compositae (Sen.-Sen.). Excl. *Xenophyllum*, c. 25 Andes

Wernhamia S. Moore = *Simira*

West Indian arrowroot *Maranta arundinacea*; **WI birch** *Bursera simaruba*; **WI boxwood** *Casearia praecox*; **WI cherry** *Malpighia glabra*; **WI ebony** *Brya ebenus*; **WI elemi** *Bursera simaruba*; **WI gherkin** *Cucumis anguria*; **WI gooseberry** *Pereskia aculeata*; **WI locust-tree** *Hymenaea courbaril*; **WI mahogany** *Swietenia mahagoni*; **WI redwood** *Guarea guidonia*; **WI sandalwood** *Amyris balsamifera*

western arbor-vitae or **red cedar** *Thuja plicata*; **w. hemlock** *Tsuga heterophylla*

Westoniella Cuatrec. Compositae (Ast.-Hin.). 6 Panamá, Costa Rica, paramos. R: Phytol. 35(1977)471

Westphalina Robyns & Bamps = *Mortoniodendron*

Westringia Sm. Labiatae (IV 2). 25 Aus., Lord Howe Is. R: PRSQ 60(1949)99, Nuytsia 6(1988)335. Shrubs

Wetria Baill. Euphorbiaceae (Acal.-Acal.-Cleid.). 2 S Myanmar & Thailand to W Mal., NG & Queensland. R: Blumea 43(1998)156. Double-forked style

Wettinia Poeppig. Palmae (V 1). Incl. *Catoblastus*, 21 trop. S Am. esp. Colombia. R: NRBGE 36(1978)259, Caldasia 17(1995)367

Wettiniicarpus Burret = *Wettinia*

Wettsteiniola Süsseng. Podostemaceae (III). 3 S Braz., Arg.

Weymouth pine *Pinus strobus*

whalebone tree *Streblus brunonianus*

Whalleya K.Wills & Bruhl = *Homopholis* (but see AusSB 13(2000)462)

whangee *Phyllostachys nigra*

wheat (bread w.) *Triticum aestivum*; **crack** or **durum w.** *T. turgidum* Durum Group; **emmer w.** *T. t.* Dicoccon Group; **flint, hard** or **macaroni w.** *T. t.* Durum Group; **Polish w.** *T. t.* Polonicum Group; **Poulard w.** *T. t.* Turgidum Group

whin *Ulex europaeus*

Whipplea Torrey. Hydrangeaceae (II 1; Philadelphaceae). 1 W N Am.: *W. **modesta*** Torrey – decid. trailing shrub for rock garden

whipple-tree (Chaucer, *Knight's Tale*) poss. *Cornus sanguinea*

whisk fern *Psilotum* spp.; **Florence, Italian** or **Venetian w.** *Sorghum bicolor* Technicum Group; **French w.** *Chrysopogon gryllus*

whisk(e)y *Hordeum vulgare* (see also *Triticum aestivum, Zea mays*)

whistling jacks *Gladiolus* spp.; **w. pine** *Casuarina equisetifolia*; **w. thorn** *Vachellia drepanolobium*

white ash *Fraxinus americana*; **w.beam** *Sorbus aria* & other *S.* spp.; **w.b., Swedish** *S. intermedia*; **w. bean** *Phaseolus vulgaris*; **w. beech** (Aus.) *Gmelina* spp. esp. *G. leichhardtii*, *Elaeocarpus kirktonii*; **w. bombway** *Terminalia procera*; **w. broom** *Retama monosperma*; **w. bryony** *Bryonia dioica*; **w. camas(h)** *Anticlea elegans*; **w. cedar** *Calocedrus decurrens*, *Chamaecyparis thyoides*, *Melia azedarach*; **w. chuglam (wood)** *T. bialata*; **w. cinnamon** *Alyxia* spp.; **w. clover** *Trifolium repens*; **w.currant** *Ribes rubrum* cv.; **w. cypress pine** *Callitris glauca*; **w. dammar** *Vateria indica*; **w. dead nettle** *Lamium album*; **w. deal** *Picea abies*; **w. elm** *Ulmus americana*; **w. fir** *Abies grandis*; **w. gum** *Eucalyptus nobilis*, *E. viminalis*; **w. hellebore** *Veratrum album*;

w. horehound *Marrubium vulgare*; **w. ipecacuanha** *Pombalia calceolaria*; **w. mahogany** *E. acmenoides, E. robusta*; **w. maple** *Acer saccharinum*; **w. melilot** *Melilotus albus*; **w. mulberry** *Morus alba*; **w. mullein** *Verbascum lychnitis*; **w. mustard** *Sinapis alba*; **w. oak** *Quercus alba, Q. montana*; **w. pear** *Apodytes dimidiata*; **w. pepper** *Piper nigrum*; **w. pine** *Pinus strobus* (**Am. w. p.**), *Podocarpus* spp.; **w.p., Japanese** *P. parviflora* var. *pentaphylla*; **w. poplar** *Populus alba*; **w. Sally** *Eucalyptus pauciflora*; **w. spruce** *Picea glauca*; **w. thorn** *Crataegus monogyna*; **w. willow** *Salix alba*; **w.wood** *Abies alba, Atalaya hemiglauca, Liriodendron tulipifera, Triplochiton scleroxylon, Tilia americana*; **w.w., African** *Annickia chlorantha*; **w.w., American** *L. tulipifera*; **w.w., Baltic** *Picea abies*; **w.w., Canary** *Magnolia* spp.; **w. yam** *Dioscorea alata*

Whiteheadia Harv. = *Massonia*

Whiteochloa C. Hubb. Gramineae (XXIV 4). 5 trop. Aus., Aru Is. R: Brunonia 1(1978)69

Whiteodendron Steenis. Myrtaceae (II 2). 1 Borneo: ***W. moultonianum*** (W.W. Sm.) Steenis. R: TFSS 7(2011)325

Whitesloanea Chiov. Apocynaceae (V b; Asclepiadaceae III 5). 1 Somalia: ***W. crassa*** (N.E. Br.) Chiov. R: F. Albers & U. Meve, *Ill. Handb. Succ. Pls*, Asclep. (2002) 273

Whitfieldia Hook. Acanthaceae (III). 12 trop. Afr. Some cult. orn. shrubs & local dyes

Whitfordiodendron Elmer = *Callerya*

whitlow-grass *Draba verna*

Whitmorea Sleumer. Stemonuraceae (Icacinaceae s.l.). 1 Solomon Is.: ***W. grandiflora*** Sleumer. R: Blumea 17(1969)263

Whitneya A. Gray = *Arnica*

Whittonia Sandw. Peridiscaceae. 1 Guyana: ***W. guianensis*** Sandw. R: KB 15(1962)468

Whitsuntide boss *Viburnum opulus* 'Roseum'

whortleberry *Vaccinium myrtillus*

Whyanbeelia Airy Shaw & B. Hyland. Picrodendraceae (Cal.-Diss.; Euphorbiaceae s.l.). 1 Queensland: ***W. terrae-reginae*** Airy Shaw & B. Hyland. R: KB 31(1976)375

Whytockia W.W. Sm. Gesneriaceae (III 1b). 8 S China & Taiwan. R: NRBGE 40(1982)359. Form like that from which *Monophyllaea* (q.v.) derived

Wiasemskya Klotzsch = *Tammsia*

Wiborgia Thunb. Leguminosae (III 8). 10 Cape. R: OB 28(1975). Close to *Lebeckia*

wickup *Chamaenerion angustifolium*

Widdringtonia Endl. Cupressaceae. 4 trop. & S Afr. R: A. Farjon (2005) *Monogr. Cupressaceae & Sciadopitys*: 469. Afr. cypress; some timber & cult. orn. esp. ***W. cedarbergensis*** J. Marsh (*W. juniperoides*, Clanwilliam cedar, SW Cape), ***W. nodiflora*** (L.) Powrie (*W. cupressoides*, sapree wood, S & S trop. Afr.) – coppicing sp., germ. enhanced by smoke, & ***W. whytei*** Rendle (Mlanje c., Mt. Mulanje [S Malawi])

widdy *Potentilla fruticosa*

Widgrenia Malme = *Oxypetalum* (but see Darw. 30(1990)279)

widow, black or **mourning** *Geranium phaeum*; **grass w.** *Olsynium douglasii*

Wiedemannia Fischer & C. Meyer = *Lamium*

Wielandia Baill. Phyllanthaceae (Wiel.-Wiel.; Euphorbiaceae s.l.). Incl. *Blotia, Petalodiscus*, 13 SE Kenya, Madag. (10 endemic), Comores, Seychelles. R: AMBG 94(2007)538

Wiesneria M. Micheli. Alismataceae. 3 trop. Afr. to Ind. Like *Aponogeton*

wig tree *Cotinus coggygria*

Wigandia Kunth. Boraginaceae (Nam.; Hydrophyllaceae-Nam.). 6 trop. Am. R: Biollania esp. ed. 6(1997)326. Shrubs & trees grown for large lvs (covered in stinging hairs) in sapling stages & used in 'subtrop.' gardening, ***W. caracasana*** Kunth ('*W. urens*') now weedy on the Riviera

Wigginsia D. Porter = *Parodia*

Wightia Wall. Paulowniaceae (Scrophulariaceae s.l.). 2 Indomal. Epiphytic shrubs, eventually becoming independent trees, like strangling figs

Wikstroemia Endl. (~ *Daphne*). Thymelaeaceae (Thym.-Daph.). c. 70 SE As. to Pac. Some cryptic dioec. (dioecy evolved twice in Hawaii), ***W. indica*** (L.) C. Meyer (Indomal. to Aus.) 3n apomictic. Bark (fibre = salago) used for rope, bank-notes & strong paper, Jap. sliding-doors (shoji) & kimonos, wood a source of incense. ***W. ovata*** C. Meyer (C Mal.) – strong purge; ***W. sikokiana*** Franch. & Sav. (*Diplomorpha s.*, gampi, Korea) – bark used for paper in Jap.

Wilbrandia J. Silva Manso. Cucurbitaceae (XIII). 5 N S Am.

Wilcoxia Britton & Rose = *Echinocereus*

Wilczekra M.P. Simmons (~ *Euonymus*). Celastraceae. 1 trop. Afr.: ***W. congolensis*** (Wilczek) M.P. Simmons. R: SB 38(2013)150

wild goose plum *Prunus rivularis;* **w. oat** *Avena fatua;* **w. rice** *Zizania aquatica;* **w. thyme** *Thymus praecox* subsp. *britannicus*

Wildemaniodoxa Aubrév. & Pellegrin = *Englerophytum*

Wildpretia U. Riefenb. & A. Reifenb. = *Sonchus*

wilga *Geijera parviflora*

Wilhelminia Hochr. = *Hibiscus*

Wilhelmsia Reichb. (~ *Arenaria*). Caryophyllaceae (II 1). 1 NE As., NW Am.: *W.* ***physodes*** (Ser.) McNeill – close to *Honckenya*. R: FNA 5(2005)137

wiliwili *Erythrina sandwichensis*

Wilkesia A. Gray. Compositae (Mad.-Mad.). 1–2 Kauai (Hawaii). R: Allertonia 4,1(1985)59. *W.* ***gymnoxiphium*** A. Gray (iliau), not or weakly branched pachycaul tree, s.t. hapaxanthic (Holttum's Model), hybridizes with *Argyroxiphium* & *Dubautia* spp.

Wilkiea F. Muell. (~ *Kibara*) Monimiaceae (V 2). 12 + SE NG (1), E Aus.

Willardia Rose = *Lonchocarpus*

Willbleibia Herter = *Willkommia*

Willdampia A.S. George = *Swainsona*

Willdenowia Thunb. Restionaceae. c. 6 S & SW Cape, Namaqualand. R: Bothalia 15(1985)493

Willemetia Necker (*Calycocorsus* ~ *Chondrilla*). Compositae (Cich.-Cich. (Chon.)). 2 C & S Eur. to Iran. R: T 45(1996)628

Williamodendron Kubitzki & H. Richter (~ *Mezilaurus*). Lauraceae (I). 3 trop. Am. R: BJ 109(1987)49

Williamsia Merr. = *Praravinia*

Willie, stinking *Jacobaea vulgaris*

Willisia Warm. Podostemaceae (III). 1–2 S Ind.

Willkommia Hackel. Gramineae (XXIX 1). 3 trop. & S Afr., 1 Texas, introd. Arg. R: KBAS 13(1986)239

willow *Salix* spp.; **basket w.** *S. viminalis, S. purpurea;* **black w.** *S. nigra;* **brittle w.** *S. fragilis;* **Cape w.** *S. mucronata;* **corkscrew w.** *S. matsudana* 'Tortuosa'; **crack w.** *S. fragilis;* **creeping w.** *S. repens;* **cricket-bat w.** *S. alba* 'Caerulea'; **flowering w.** *Chilopsis linearis;* **goat w.** *S. caprea;* **golden w.** *S. alba* 'Vitellina'; **w.herb** *Epilobium* spp.; **w.h., rose-bay** *Chamaenerion angustifolium;* **Humboldt w.** *S. humboldtiana;* **Indian w.** *Monoon longifolium* cv.; **Kilmarnock w.** *S. caprea* 'Pendula'; **w. myrtle** *Agonis* spp.; **Napoleon's w.** *S. babylonica;* **Port Jackson w.** *Acacia saligna;* **prairie w.** *S. humilis;* **pussy w.** *S. caprea;* **weeping w.** *S. × sepulcralis,* (willow-pattern) *S. babylonica,* (Bible) *Populus euphratica;* **white w.** *S. alba*

Willughbeia Roxb. Apocynaceae (Ic). 16 Indomal. R: Blumea 38(1993)1, 40(1996)123. Some rubber sources esp. *W.* ***coriacea*** Wall. (*W. firma,* Borneo rubber, W Mal.) – cult. before *Hevea brasiliensis* introd.

Willughbeiopsis Rauschert = *Willughbeia*

Wilsonia R. Br. Convolvulaceae (4). 3 Aus. R: Fl. Vict. 4(1999)366

Wimmerella Serra & al. (~ *Laurentia*). Campanulaceae (III 1). 10 S Afr. R: Novon 9(1999)415

Wimmeria Schldl. & Cham. Celastraceae (I). 12 C Am. R: CUMH 3(1939)6. Lvs torn across hang together because of pulled-out xylem wall-thickenings (cf. *Cornus, Maytenus*)

Winchia A. DC. = *Alstonia*

wind flower [confused name; see *Anemone*]

windmill grass *Chloris* spp. esp. *C. truncata;* **w. palm** *Trachycarpus fortunei*

Windsorina Gleason. Rapateaceae (II 2). 1 Guyana: *W.* ***guianensis*** Gleason

wine see *Vitis vinifera*

wineberry *Rubus phoenicolasius*

wingnut *Pterocarya* spp.

Winifredia L. Johnson & B. Briggs. Restionaceae. 1 Tasmania: *W.* ***sola*** L. Johnson & B. Briggs. R: Telopea 2(1986)737

Winika M.A. Clem. & al. = *Dendrobium*

Winitia Chaowasku = *Stelechocarpus*

Winklera Regel = *Lepidium*

Winklerella Engl. Podostemaceae (III). 1 Cameroun: *W.* ***dichotoma*** Engl.

winter aconite *Eranthis hyemalis;* **W.'s bark** *Drimys winteri;* **w. cherry** *Alkekengi officinarum, Solanum pseudocapsicum;* **w. cress** *Barbarea* spp.; **w.fat** *Krascheninnikovia lanata;* **w. grass** (Aus.) *Poa annua;* **w.green** *Gaultheria* spp. but now usu. *Betula lenta, Pyrola* spp.; **w. heliotrope** *Petasites fragrans;* **w. marjoram** *Origanum dictamus* subsp. *viride;* **w. purslane** *Claytonia perfoliata;* **w. sweet** *Chimonanthus praecox*

Winteraceae R. Br. ex Lindl. Magnoliidae – Winterales. 5/75 montane S Am., Madag. (1/1), Aus., NG, SW Pac. (fossil wood from Cretaceous Antarctica). Aromatic everg. trees & shrubs, s.t. with alks (?); hairs (exc. on G) & vessels 0. Lvs simple, entire, in spirals, oft. glaucous abaxially; stip. 0. Fls usu. bisex., reg. to irreg., poll. by insects or wind, solit. & term. or in cymes, receptacle short; K 2–4(–6) valvate, s.t. basally connate or completely so & forming calyptra, C (2–)5–∞, oft. in 2 or more whorls, A 3 – ∞, in spiral, initiated centripetally but maturing centrifugally, oft. ± laminar though s.t. with distinct filament, oft. with term. bisporangiate pollen-sacs, pollen grains usu. in tetrads, $\underline{G}$ (1–) usually ∞ in 1 whorl, s.t. weakly connate, conduplicate & oft. not completely closed with stigma along margins to closed with term. stigma, 1 – several anatropous, bitegmic ovules laminar or near margin. Fr. berry-like or follicles in heads, s.t. ± connate into multiloc. capsule or syncarp; seeds exotestal with small embryo in copious oily endosperm. n = 13, 18 (*Takhtajania*), 43

Genera: *Drimys, Pseudowintera, Takhtajania* (prob. best in own subfam. **Takhtajanioideae**), *Tasmannia, Zygogonum*

Fossils in Cretaceous of Antarctica, Tertiary of Cape; wood from Maastrichtian of Calif. Oft. recognizable in the field by abaxial dotting of lvs due to plugging of stomata with wax. Hydrolysis of polysaccharide granules in cells of petals of several *Zygogynum* spp. leads to rapid uptake of water & subsequent opening or closing of fls; may also act as pollen rewards for beetles (also some thermogenesis)

Some cult. orn. & medic. (*Drimys*)

Winterocereus Backeb. = *Cleistocactus* (but see T 56(2007)226)

wirilda *Acacia retinodes*

wishbone flower *Torenia fournieri*

Wislizenia Engelm. = *Cleome* (but see Britt. 31(1979)333)

Wissadula Medik. Malvaceae (Malv.-Malv.). 26 trop. esp. Am. R: KSVAH n.s. 43,4(1908)1. Some locally used bark-fibres esp. ***W. spicata*** C. Presl (paco-paco, trop. Am.) & ***W. amplissima*** (L.) R. Fries (*W. rostrata*, maholtine, trop.) – fibre like jute

Wissmannia Burret = *Livistona*

wisteria *Wisteria* spp.; **Chinese w.** *W. sinensis*; **Japanese w.** *W. floribunda*; **water w.** *Hygrophila difformis*

Wisteria Nutt. Leguminosae (III 16). 5 E As., N Am. (***W. frutescens*** (L.) Poir). R: P. Valder (1995) *Ws.*, CBM 22(2015)189. Poss. incl. *Callerya*. Lianes with poisonous lvs, fr. & seeds, & pendent racemes of showy oft. scented fls (incl. 'doubles'), much planted against houses, esp. ***W. floribunda*** (Willd.) DC. (Jap. wisteria, Jap., invasive SE US) cli. clockwise with scented racemes to 45 (–120 in f. ***multijuga*** (van Houtte) J. Compton & Thijsse) cm long & lvs with 13–19 leaflets, many cvs (some actually hybrids (***W. × valderi*** J. Compton) with *W. brachybotrys* Sieb. & Zucc. (Jap.)), & ***W. sinensis*** (Sims) Sweet (Chinese w., China, invasive SE US) cli. anticlockwise with scentless racemes to 30 cm & lvs with usu. 11 leaflets, a specimen at Sierra Madré, Calif. the largest of all blossoming pls in cult. (planted 1892–4, 152 m long with 1.5 M infls, covering 4000 m^2 & weighing some 250 t – 30 000 people a yr pay to see it on the 1 day poss.), tortured into *penjing* & medic. China (but many in cult. in As. & US (esp. naturalized ones) actually hybrids with *W. floribunda*), fls ed. fried & in cakes. Hybrids all climb clockwise (dominant char.)

witch grass *Elymus repens*; **w. hazel** *Hamamelis virginiana*; **w.weed** *Striga* spp. esp. *S. asiatica*

Withania Pauquy. Solanaceae (IC 5a). Incl. *Mellissia* (***W. begoniifolia*** (Roxb.) Hunz. & Barbosa (St Helena) – believed extinct since 1875 but re-found 1998) c. 20 OW (Eur. 2). Dioec.; alks. ***W. coagulans*** (Stocks) Dunal (Iran, Afghanistan, Nepal) – fr. (panirband, veg. rennet) used to coagulate milk in cheese-making; ***W. somnifera*** (L.) Dunal (Afr., Medit. to Ind.) – source of ashwagandha, a drug used in Ayurvedic med., narcotic & diuretic, fr. in floral collar of innermost coffin of Tutankhamum, always threaded on strips of date-palm leaf & used thus until time of Pliny

Witheringia L'Hérit. Solanaceae (IV 5c). 11 SE Mex. to Bolivia (Antilles 1). R: Kurtziana 5(1969)110

witloof *Cichorium intybus* Witloof Group

Witsenia Thunb. Iridaceae (V). 1 Cape: ***W. maura*** (L.) Thunb. – rare shrubby, bird-poll. marsh pl. to 3 m, seed water-disp., cult. orn. R: P. Goldblatt, *Woody I.* (1993)117

Wittia K. Schum. = *Disocactus*

Wittiocactus Rauschert = *Pseudorhipsalis*

Wittmackanthus Kuntze (*Pallasia*). Rubiaceae (Cond.). 1 trop. Am.: ***W. stanleyanus*** (Schomb.) Kuntze

Wittmackia Mez = *Aechmea*

Wittrockia Lindman. Bromeliaceae (3). 11 E Braz. R: FN 14(1979)1725. Some cult. orn. like *Canistrum* but C united

Wittsteinia F. Muell. Alseuosmiaceae. Excl. *Periomphale*, 2 New Caled., NG, SE Aus. Leeuwenberg's Model

woad *Isatis tinctoria*

Wodyetia Irvine. Palmae (V 14i). 1 Queensland (Melville Range): ***W. bifurcata*** Irvine (foxtail palm) – fr. eaten by rats & feral pigs. R: FA 39(2011)200

Woehleria Griseb. (~ *Iresine*). Amaranthaceae (II 2). 1 Cuba: ***W. serpyllifolia*** Griseb.

Woikoia Baehni = *Pouteria*

Wokoia Baehni = *Pichonia*

wolfberry *Lycium barbarum*

Wolffia Horkel ex Schleiden. Araceae (Lemnaceae). 11 trop. & warm to temp. R: IBM 34(1965)41. Water meal. Minute thalloid aquatics rarely fls, ***W. arrhiza*** (L.) Wimmer (Euras., Afr., Aus.) with 1 orthotropous bitegmic ovule per fl., pl. c. 1.5 mm long & the smallest angiosperm by repute but ***W. brasiliensis*** Wedd. (Braz.), growing with *Victoria amazonica* (!), & ***W. microscopica*** (Griff.) Kurz (Ind.) even smaller & poss. ***W. angusta*** Landolt (SE Aus.) described 1980 the minutest (0.6 mm × 0.33 mm). *W. arrhiza* has stomata in parallel rows, a deep reproductive pocket & a meristem for new thalloid growths which are 'budded' off, the pl. entirely veg. in Eur. & SE As. where (?) introd., imp. cheap food (khai-nam) SE As. (20+% protein), used as a quantitative test for herbicide pollution in Poland; *W. brasiliensis* distrib. on birds' feathers

Wolffiella (Hegelm.) Hegelm. Araceae (Lemnaceae). Incl. *Wolffiopsis*, 10 trop. & S Afr. (5), trop. & warm Am. (6), 1 in both. R: IBM 34(1965)34

Wolffiopsis Hartog & van der Plas = praec.

wolfsbane *Aconitum lycoctonum*

Wollastonia DC. ex Decne. (~ *Melanthera*). Compositae (Helia.-Ecl.). Incl. *Lipochaeta* p.p., c. 20 E Afr. to Jap., Fiji & Hawaii (16). R: Nuytsia 23(2013)386. ***W. biflora*** (L.) DC. (*M. b.*, *Wedelia b.*) – Indopacific strand-pl.

Wollemi pine *Wollemia nobilis*

Wollemia W.G. Jones & al. (~ *Agathis*). Araucariaceae. 1 NSW (v. rare in wild – c. 45 genetically v. uniform individuals in 3 stands to ?1000 yrs old): ***W. nobilis*** W.G. Jones & al. (Wollemi pine) – tree to c. 40 m undetected until 1994; lvs, pollen & seeds similar to fossils from Lower Cretaceous to Miocene of Aus. & NZ; Massart's Model, char. spongy nodular bark, branches rather than individual lvs shed (zone weakened by stranded xylem with profusion of ray parenchyma & bordered pits), lvs trimorphic (cf. *Glyptostrobos*); prod. taxol, (unruly cf. *Araucaraia araucana*) cult. orn. R: Telopea 6(1995)173

wollum wollum *Hymenosporum flavum*

wombat berry *Eustrephus latifolius*

wonderberry *Solanum scabrum*

wonga-wonga vine *Pandorea pandorana*

wong-bok *Brassica rapa* Pekinensis Group

wood apple *Limonia acidissima*; **w. avens** *Geum urbanum*; **w.bine** *Lonicera periclymenum*; **w. calamint** *Clinopodium menthifolium*; **w. flowers** Loranthaceae esp. *Phoradendron* spp.; **w. melick** *Melica uniflora*; **w. oil** *Vernicia fordii*; **w.o., China** *Aleurites moluccana*; **w. rose** *Argyreia nervosa*, *Merremia tuberosa* (see also *Dactylanthus*, *Erianthemum*, *Myzodendron*, *Pedistylis*, *Psittacanthus*, *Struthanthus*); **w. rush** *Luzula* spp.; **w. sorrel** *Oxalis acetosella*; **w. spurge** *Euphorbia amygdaloides*

Woodburnia Prain. Araliaceae. 1 Myanmar: ***W. penduliflora*** Prain – coll. once (poss. not A.)

Woodfordia Salisb. Lythraceae. 1 NE Afr., S Arabia: ***W. uniflora*** (A. Rich.) Koehne – used in tanning, 1 Madag. to China & Timor: ***W. fruticosa*** (L.) Kurz (*W. floribunda*) – source of tragacanth-like gum & dye (from fls), imp. local medic. K tubular, red (C v. small)

Woodia R. Br. = *Woodsia*

Woodia Schltr. Apocynaceae (IV c; Asclepiadaceae III 1). 3 S Afr.

Woodiella Merr. = *Monoon*

Woodiellantha Rauschert = *Monoon* (but see Sandakania 11(1998)49)

Woodrowia Stapf = *Dimeria*

woodruff *Galium odoratum*

Woodsia R. Br. Woodsiaceae. c. 35 temp. (not Aus.) & cool-temp. (Eur. 3). R: BNH 16(1964). Tufted saxicolous ferns, some cult. orn. incl. ***W. alpina*** (Bolton) Gray (Euras., N Am.) & ***W. ilvensis*** (L.) R. Br. (Euras., N Am.), both prot. spp. in GB

Woodsiaceae Herter. Polypodiidae – Polypodiales. Excl. Athyriaceae, Cystopteridaceae, Diplaziopsidaceae, Hemidictyaceae, Rhachidosoraceae, 1/35 temp. R: T 61(2012)525. Usu. saxic., s.t. terr. with short-creeping rhiz. with non-clathrate scales. Lvs pinnate to bipinnate-pinnatifid, monomorphic; petioles with 2 vasc. bundles; veins usu. ending in hydathodes. Sori abaxial along veins, round, indusiate (basal indusium of scale-like filamentous segments). n = 33, 38, 39, 41

Genus, *Woodsia* (indusium unique)

Woodsonia L. Bailey = *Neonicholsonia*

Woodwardia Sm. (~ *Blechnum*). Blechnaceae (Blechnoideae). Incl. *Lorinseria*, 14 N hemisph. (S Eur. 1) esp. E As. (some gaps) S to Costa Rica & Mal. ***W. orientalis*** Sw. (China, Jap.) – buds on upper surface of frond, above sori, which develop only if buds do not; ***W. radicans*** (L.) Sm. (SW Eur., Macaronesia) – frond-tips r., leading to 'walking' fern. Cult. orn. incl. ***W. fimbriata*** Sm. (W N Am.) stems dyed red & those of ***W. spinulosa*** M. Martens & Galeotti (Mex., Guatemala) white before use in Native American weaving

woody pear *Xylomelum* spp.

Wooleya L. Bolus. Aizoaceae (V). 1 Namaqualand: ***W. farinosa*** (L. Bolus) L. Bolus. R: H. E. K. Hartmann, *Ill. Handb. Succ. Pls*, Aiz. F–Z(2002)351

Woollsia F. Muell. (~ *Lysinema*). Ericaceae (VII 4; Epacridaceae). 1 NE Aus.: ***W. pungens*** (Cav.) F. Muell.

woolly bush *Adenanthos* spp.; **w.butt** *Eucalyptus longifolia*

Woonyoungia Law = *Magnolia*

Wootonella Standl. = *Verbesina*

Wootonia Greene = *Dicranocarpus*

Worcesterberry *Ribes divaricatum*

Wormia Rottb. = *Dillenia*

wormseed *Dysphania ambrosioides*; **Levant w.** *Artemisia cina*

Wormskioldia Thonn. = *Tricliceras*

wormwood *Artemisia absinthium*, (Bible) *A. herba-alba*; **Roman w.** *A. pontica*; **sea w.** *A. maritima*

Woronowia Juz. = *Geum*

Worsleya (Traub) Traub (~ *Griffinia*). Amaryllidaceae. 1 Braz. (Organ Mts): ***W. procera*** (Lem.) Traub (*W. rayneri*, blue amaryllis) – rare in wild but cult. orn. with falcate lvs, lilac fls & bulb-neck to 1.5 m

woundwort *Stachys* spp.

Woytkowskia Woodson. Apocynaceae (Id). 2 trop. S Am. R: MMNHN 30(1985)120

Wrightia R. Br. Apocynaceae (II a). 29 trop. OW (Afr. (S & E) 2, R: AUWP 87–5(1987)35). R: AMBG 52(1965)114. Troll's Model. Aestivation to left! Cult. orn. trees & shrubs, some timber. ***W. arborea*** (Dennst.) Mabb. (*W. tomentosa*, Ind.) – fls smell of decaying fr. (? poll. flies), white wood for carving etc., bark medic., yellow dye used as styptic by Nepalese; ***W. coccinea*** (Hornem.) Sims (Pakistan to Yunnan) – cult. orn. with dark red fls; ***W. religiosa*** (Teijsm. & Binn.) Kurz (sui mui, SE As.) – grown in Chinese temple-gardens in Mal. for scented nodding fls (oft. topiary); ***W. tinctoria*** (Roxb.) R. Br. (India) – building timber, indigo-like dye from lvs

Wrixonia F. Muell. = *Prostanthera* (but see JABG 1(1976)27)

wu-chu-yu *Tetradium ruticarpum*

Wulfenia Jacq. Plantaginaceae (Dig.-Ver.; Scrophulariaceae s.l.). 4 SE Eur., Turkey. R: Wulfenia 1(1992)27, T 63(2014)843. Cult. orn. rock-pls

Wulfeniopsis Hong (~ *Wulfenia*). Plantaginaceae (Dig.-Ver.; Scrophulariaceae s.l.). 2 E Afghanistan, Pakistan, N Ind., Nepal. R: OB 75(1984)56

Wulffia Necker ex Cass. = *Tilesia*

Wullschlaegelia Reichb.f. Orchidaceae (V 6). 2 trop. Am. R: BJ 121(1999)57. Mycotrophs

Wunderlichia Riedel ex Benth. Compositae (Wund./Mut.-Wund.). 5 Braz. R: RBB 33 (1973)379. Pachycaul shrubs & trees

Wunschmannia Urb. = *Amphilophium*

Wurdackanthus Maguire = *Symbolanthus* (but see MNYBG 51(1989)8)

Wurdackia Mold. = *Rondonanthus*

Wurdastom Wallnöfer (~ *Alloneuron*). Melastomataceae. 8 S Am. R: IJPS 172 (2011)1175

Wurmbea Thunb. Colchicaceae (Liliaceae s.l.). Incl. *Anguillaria*, *Neodregea*, *Onixotis* (poss. to be excl.), 37 Afr. (24), Aus. (R: FA 45(1987)387). Hermaphrodite to dioec. spp. (via gynodioecy); male fls conspic., females not; alks. Cult. orn. incl. water phlox (*O.* spp.) & ***W. dioica*** (R. Br.) F. Muell. (*A. d.*, early Nancy, Aus.) – dioec. in stressed conditions

wutong *Firmiana simplex*

wych elm *Ulmus glabra*

Wyethia Nutt. Compositae (Helia.-Eng.). Incl. *Agnorhiza, Balsamorhiza, Scabrethia*, c. 28 W N Am. R: FNA 21(2006)100. Coarse herbs; cypselas & taproots (esp. ***W. sagittata*** (Pursh) Mabb. (*B. s.*, balsam r.) – imp. local medic., oil & incense source) form. eaten by Native Americans. Some cult. orn.

Xantheranthemum Lindau. Acanthaceae (III 1). 1 Peruvian Andes: ***X. igneum*** (Regel) Lindau – cult. orn. foliage pl.

Xanthisma DC. (~ *Machaeranthera*). Compositae (Ast.-Mac.). 17 W N Am. (9), Mex. R: Sida 20(2003) 1397, 1585. ***X. texana*** DC. (Texas) – cult. orn. with capitula closing at night

Xanthium Tourn. ex L. Compositae (Helia.-Amb.). c. 3 now cosmop. Monoec., the female fls in prs in a prickly involucre with only styles projecting; fr. enclosed in accrescent involucre covered with hooks aiding disp. on animal fur (burweed) but lowering value of fleeces etc. e.g. ***X. occidentale*** Bertol. (*X. strumarium* s.l., Noogoora burr, orig. WI?) in Aus. Some locally medic.: ***X. spinosum*** L. (clotbur, bastard or Bathurst burr, orig. S Am., invasive Aus., S Afr.) & ***X. strumarium*** L. (*X. canadense*, cocklebur, Am.) – much used in genetic analysis (cf. *Datura*), lactone acting as growth inhibitor in *Drosophila melanogaster* larvae

Xanthocephalum Willd. (~ *Gutierrezia*). Compositae (Ast.-Mac.). 6 Mex. to SW US (1: ***X. sarothrae*** (Pursh) Shinners (*G. s.*) – local medic. (protein with antitumour activity), insecticidal, cerem. etc.

Xanthoceras Bunge. Sapindaceae (I; Xanthocerataceae). 1 N China: ***X. sorbifolium*** Bunge – small cult. orn. decid. tree with white fls & ed. seeds (oil mooted as biodiesel)

Xanthocerataceae Bierki & al. = Sapindaceae (I)

Xanthocercis Baill. Leguminosae (III 2). 3 Gabon (1), S Afr. (1: ***X. zambesiaca*** (Bak.) Dumaz (nyala tree) – typical tree of low veld), N Madag. (1: ***X. madagascariensis*** Baill. – hard wood for constr., seed-pulp ed.). R: BZ 95(2010)1153

Xanthocyparis Farjon & Hiep (*Callitropsis*; ~ *Cupressus*). Cupressaceae. 2 Vietnam, China (Guangxii; ***X. vietnamensis*** Farjon & Nguyen (*Callitropsis v.*) – discovered 1999, a few 100s in wild), W N Am. (***X. nootkatensis*** (D. Don) Farjon & Harder (*Callitropsis n.*, Nootka cypress, (Alaska) yellow cedar or cypress, Sitka cypress) – to 1824 (?2000) yrs old, 62 m tall with bole to 3.65 m diam., lvs without white markings (cf. *Chamaecyparis lawsoniana*), timber similar to *C. l.*, a parent of × *Cuprocyparis leylandii*, local medic. & fibre, several cvs incl. drooping '**Pendula**'). R: A. Farjon, *Monogr. Cupressaceae & Sciadopitys* (2005) 424, Novon 22(2012)12)

Xanthogalum Avé-Lall. = *Angelica*

Xanthomyrtus Diels. Myrtaceae (II 8). 24 Borneo to New Caled. R: KB 33(1979)461

Xanthopappus Winkler = *Alfredia*

Xanthophthalmum Schultz-Bip. = *Glebionis*

Xanthophyllaceae Gagnepain ex Reveal & Hoogland = Polygalaceae

Xanthophyllum Roxb. Polygalaceae (I; Xanthophyllaceae). 94 Indomal. R: LBS 7(1982). Genus originating in Aus. (van der Meijden). Some dyes & oilseeds

Xanthophytopsis Pitard = *Xanthophytum*

Xanthophytum Reinw. ex Blume. Rubiaceae (IV 2). 32 Java & Borneo to Fiji. R: Blumea 34(1990)425, NJB 15(1995)575. Some monocaul treelets

Xanthorhiza Marshall. Ranunculaceae (III 3). 1 E N Am.: ***X. simplicissima*** Marshall. Decid. *shrub* with alks, 1–2-pinnate lvs, brownish-purple fls & bitter yellow r., source of a yellow dye. R: FNA 3(1997)245

Xanthorrhoea Sm. Asphodelaceae (Xanthorrhoeaceae). 28 Aus. Yacca, (Aus.) grass tree, black boy (blackened by fire). R: FA 46(1986)148. Slow-growing, long-lived fire-tolerant pachycaul pls (Chamberlain's Model) with narrowly oblong to linear, xeromorphic. spine-tipped lvs, contractile rs, sec. thickening & stout stems prod. acaroid resins at bases of old lvs – resin form. used to fix spear-heads to shafts, now used to varnish or lacquer metals & leather; wood used for making bowls etc. ***X. australis*** R. Br. (SE Aus.) – resistant to fire when young as bud 12 cm below ground, later bud prot. by old lvs; ***X. johnsonii*** A.T. Lee (Queensland) – poisons stock, cut foliage crop (steel grass); ***X. preissii*** Endl. (SW Aus.) – lives to at least 350 yrs & fls may be delayed until 200 yrs old, though stimulated by fire, other spp. at least fls most freq. near habitation as fires are more freq. there; ***X. pumilio*** R. Br. (Queensland) – lvs only 3 cm long; ***X. resinosa*** Pers. (SE Aus.) – infl. axis used as harpoon-shaft, yellow resin ('yellow gum', also from other spp.) medic. & in glue, used to fix ornaments in Aborigines' hair; ***X. semiplana*** F. Muell. subsp. ***tateana*** (F. Muell.) Bedford (*X. tateana*, yakka, S Aus.) – yacca gum industry of 1000 t per annum on Kangaroo Is., S Aus. in decline (1990s)

Xanthorrhoeaceae Dumort. = Asphodelaceae

Xanthoselinum Schur = *Peucedanum*

Xanthosia Rudge. Umbelliferae (Mack.). 25 Aus. esp. SW. Umbels in some spp. 1-fld

Xanthosoma Schott. Araceae (VII 11). c. 75 trop. Am. Stemless or caulescent herbs with thick rhiz. or tubers & milky sap; lvs sagittate or hastate (cf. *Colocasia*); cult. orn. & food-pls (malanga, yautia) esp. ***X. violaceum*** Schott (*X. nigrum*, ~ *X. sagittifolium*) & ***X. sagittifolium*** (L.) Schott (tannia, tania, orig. range unclear) – grown for ed. tubers (350 M t a yr by 2007) & young lvs ed. like spinach

Xanthostachya Bremek. = *Strobilanthes*

Xanthostemon F. Muell. Myrtaceae (II 1). c. 45 C Mal. to NE Aus. (13 – R: Telopea 3(1990)451) & New Caled. ***X. oppositifolius*** Bailey (penda, SE Queensland) – timber tree

Xanthoxerampellia Szlach. & Sitko = *Maxillaria*

Xanthoxylum Mill. = *Zanthoxylum*

Xantolis Raf. Sapotaceae (IV). 14 S Ind., SE As., Philippines. R: Blumea 8(1957)207

Xantonnea Pierre ex Pitard. Rubiaceae (II 3). 2 SE As.

Xantonneopsis Pitard. Rubiaceae (II 1). 1 SE As.: ***X. robinsonii*** Pitard

Xatardia Meissn. & Zeyher. Umbelliferae (III 9). 1 Pyrenees: ***X. scabra*** (Lapeyr.) Meissn.

Xenacanthus Bremek. = *Strobilanthes*

Xenia Gerbaulet = *Anacampseros* (but see BJ 113(1992)552)

Xenikophyton Garay = *Schoenorchis* (but see KB 57(2002)228)

Xenophya Schott = *Alocasia*

Xenophyllum Funk (~ *Werneria*). Compositae (Sen.-Sen.). 21 Andes. R: Novon 7(1997)235. Mat- & hummock-formers

Xenopoma Willd. = *Micromeria*

Xenoscapa (Goldbl.) Goldbl. & Manning (~ *Anomatheca*). Iridaceae (VI 4). 3 SW Afr. R: SB 20(1995)172, Bothalia 41(2011)284

Xenostegia D. Austin & Staples (~ *Merremia*). Convolvulaceae (12). 2 trop. OW

Xeranthemum Tourn. ex L. Compositae (Card.-Card. (Card.)). 5 Medit. (Eur. 3) to SW As. Non-spiny ann. herbs, cult. orn. 'everlastings' esp. ***X. annuum*** L. (SW Eur. (natur.) to Iran)

Xeroaloysia Tronc. = *Aloysia* (but see Darw. 12(1960)48)

Xerocarpa H.J. Lam = *Teijsmanniodendron*

Xerochlamys Bak. (~ *Leptolaena*). Sarcolaenaceae. 8 Madag. R: Adans. 31(2009)319

Xerochloa R. Br. Gramineae (XXIV 4). 3 Aus. R: BJ 35(1904)64

Xerochrysum Tzvelev (*Bracteantha*). Compositae (Gnap.-Ang.). 6 Aus., widely natur. Cult. orn. 'everlastings' esp. ***X. bracteatum*** (Vent.) Tzvelev (*B. b.*, *Helichrysum b.*, strawflower) – natur. St Helena, allegedly escaped from Napoleon's garden

Xerocladia Harv. Leguminosae (II 1). 1 SW Afr.: ***X. viridiramis*** (Burch.) Taubert. R: Strel. 10(2000)299

Xerococcus Oersted = *Hoffmannia*

Xerodanthia J. Phipps = *Danthoniopsis*

Xeroderris Roberty = *Aganope* (but see BSB II,43(1968)273)

Xerodraba Skottsb. Cruciferae (27). 8 S Arg. R: Phytotaxa 207(2015)39. Some like veg. sheep (see *Raoulia*)

Xerolekia A. Anderb. = *Buphthalmum* (but see PSE 176(1991)93)

Xerolirion A.S. George. Asparagaceae (Lomandraceae). 1 SW Aus.: ***X. divaricata*** A.S. George. R: FA 46(1986)98

Xeromphis Raf. = *Catunaregam*

Xeronema Brongn. & Gris. Xeronemataceae. 2 New Caled. & Poor Knights Is. (N NZ – ***X. callistemon*** W. Oliver, P.K. lily). Form. referred to Phormiaceae (= Asphodelaceae)

Xeronemataceae M. Chase & al. Magnoliidae – Asparagales. 1/2 SW Pacific. R: KB 55(2000)869. Large herbs with rhiz. Lvs isobifacial, in 2 ranks. Infl. a dense spike of large reg. upward-facing fls; P6; A 6 strongly exserted, with versatile anthers dehiscing longitudinally. n = 17, 18

Genus: *Xeronema*

Xerophyllum Michaux. Melanthiaceae. 2 N Am. R: FNA 26(2000)71. Cult. orn. tall herbs incl. ***X. tenax*** (Pursh) Nutt. (bear grass, W N Am.) – resistant to light fires, lvs form. used by Native Americans to make water-tight baskets, inner lvs used in floristry internationally

Xerophyta Juss. Velloziaceae. Incl. *Talbotia*, 72 Arabia (1), trop. Afr. (46: BJLS 172(2013)32) & Madag. (25). R: KB 29(1974)184. Some cult. orn. esp. ***X. elegans*** (Hook.f.) Bak. (*T. e.*, S Afr.)

Xeroplana Briq. = *Stilbe*

Xerorchis Schltr. Orchidaceae (V 5). 2 N S Am.

Xerosicyos Humbert. Cucurbitaceae (III). Incl. *Zygosicyos*, 5 Madag. R: Ashingtonia 2(1977)177; Adans. 18(1996)161

Xerosiphon Turcz. (~ *Gomphrena*). Amaranthaceae (II 2). 2 S Am. R: BMNHN 4,12(1990)94

Xerospermum Blume. Sapindaceae. 2 Indomal. R: Blumea 28(1983)389. Some cryptic dioecy. Some tough timbers

Xerospiraea Henrickson (~ *Spiraea*). Rosaceae (2). 1 Mex.: ***X. hartwegiana*** (Rydb.) Henrickson – on limestone. R: Aliso 11(1986)206

Xerotecoma J. Gómes = *Godmania*

Xerothamnella C. White. Acanthaceae (III 2c). 2 E Aus. R: JABG 9(1986)166

Xerotia Oliv. (~ *Polycarpaea*). Caryophyllaceae (I 1). 1 Arabia: ***X. arabica*** Oliv. – *Ephedra*-like subshrub

Xerxes J.R. Grant = *Heterocoma* (but see NJB 14(1994)287)

xhoba *Hoodia* spp.

xi shu *Camptotheca acuminata*

xianmu *Burretiodendron hsienmu*

Ximenia Plum. ex L. Olacaceae (Ximeniaceae). 10+ trop. R.-parasites. Champagnat's Model. In Afr. fr. (high in vitamin C) with 65% oil used for softening hides. ***X. americana*** L. (hog, monkey or spiny plum, tallow nut, trop.) – haustoria on a range of host-trees, timber a sandalwood subs., fr. ed. though kernel contains a strong purgative, oilseed a subs. for ghee, first Aus. pl. record (1606, Torres Strait is.); ***X. caffra*** Sond. (monkey plum, trop. & S Afr., s.t. weedy) – fr. ed. (just about), seed-oil useful, local medic.

Ximeniaceae Horan. See Olacaceae

Ximeniopsis Alain = *Ximenia*

Xiphidium Aubl. Haemodoraceae (I). 2 trop. Am. ***X. coeruleum*** Aubl. – buzz-fls visited by male euglossine bees collecting oils

Xiphion Mill. = *Iris*

Xiphochaeta Poeppig (~ *Stilpnopappus*). Compositae (Vern.-Vern.). 1 NE S Am.: ***X. aquatica*** Poeppig, ± aquatic

Xiphopterella Parris (~ *Xiphopteris*). Polypodiaceae (V). 6 Mal. R: GBS 58(2007)249

Xiphopteris Kaulf. = *Cochlidium*

Xiphotheca Ecklon & Zeyher (~ *Priestleya*). Leguminosae (III 7). 10 Cape fynbos. R: AMBG 84(1997)93

Xizangia Hong = *Pterygiella* (But see APS 24(1986)139)

Xolantha Raf. = *Tuberaria*

Xolocotzia Miranda (~ *Petrea*). Verbenaceae (Verb.-Petr.). 1 Mex.: ***X. asperifolia*** Miranda. R: BSBM 29(1965)40

xuduan *Dipsacus* spp., esp. *D. asper*

Xylanche G. Beck (~ *Boschniakia*). Orobanchaceae (Orob.). 2 Himalaya to Taiwan. Parasitic on *Rhodoodendron* spp.

Xylanthemum Tzvelev (~ *Chrysanthemum*). Compositae (Anth.-Han.). 8 S & C As. R: BMNHB 23(1993)105

Xylia Benth. (*Esclerona*). Leguminosae (II 1). 9 trop. OW esp. Afr. ***X. xylocarpa*** (Roxb.) Theob. (*X. dolabriformis*, pyinkado, ironwood, Myanmar to Mal.) – hard reddish timber second only to teak in Myanmar, used for ship-building, bridges etc.

Xylinabaria Pierre = *Urceola*

Xylinabariopsis Pitard = *Urceola*

Xylobium Lindl. Orchidaceae (V 12g). c. 30 trop. Am. Cult. orn. epiphytic orchids

Xylocalyx Balf.f. Orobanchaceae ('Xylo.'; Scrophulariaceae s.l.). 5 Somalia (3), Socotra (2). R: KB 16(1962)147, NJB 7(1987)267

Xylocarpus Koenig. Meliaceae (I-Xyl.). 3 coasts trop. OW, E Afr. eastwards to Pac. R: MF 45(1982)448, FM I,12(1995)371. Rauh's Model; fr. with corky seeds fitted together like a

Chinese puzzle; these irreg. shaped & sized seeds float just beneath surface in seawater; timber hard; tannin for nets etc. ***X. granatum*** Koenig (range of genus) sympatric in mangroves with ***X. moluccensis*** (Lam.) M. Roemer (Indomal. to trop. Aus.) the first with snakelike surface r., the second with pneumatophores; ***X. rumphii*** (Kostel.) Mabb. restricted to rocky coasts through range of genus

Xylococcus Nutt. (~ *Arctostaphylos*). Ericaceae (III). 1 S Calif., NW Mex.: ***X. bicolor*** Nutt. R: FNA 9(2009)404

Xylomelum Sm. Proteaceae (V 1). 6 SW & SE Aus. R: FA 16(1995)399. Woody pears, prot. spp.; fr. woody, pear-shaped, splitting to release winged seeds

Xylonagra J.D. Sm. & Rose. Onagraceae. 1 Baja Calif.: ***X. arborea*** J.D. Sm. & Rose

Xylonymus Kalkman ex Ding Hou. Celastraceae (I). 1 W NG: ***X. versteeghii*** Kalkman ex Ding Hou. R: FM I,6(1962)243

Xyloolaena Baill. Sarcolaenaceae. 5 Madag. R: Adansonia III,24(2002)9

Xylophragma Sprague. Bignoniaceae (3). 7 trop. Am. R: AMBG 99(2014)470

Xylopia L. Annonaceae (III 2). c. 100 trop. (Afr. 61, Am. c. 55). Only pantrop. genus in fam.; Roux's Model; perigyny unique. Timbers, fr. used as condiments, local med.; alks, soporifics in Amazonia. ***X. aethiopica*** (Dunal) A. Rich. (Afr., Guinea or negro pepper, trop. W Afr.) – form. much used in Eur., now shown to have antimalarial activity; ***X. amazonica*** R.E. Fries (Amaz.) – thrips-poll.; ***X. brasiliensis*** Spreng. (trop. S Am.) – pepper source (piperine)

Xylopodia Weigend. Loasaceae (I 2). 1 Peru: ***X. klaprothioides*** Weigend. R: T 55(2006)467

Xylorhiza Nutt. (~ *Machaeranthera*). Compositae (Ast.-Mac.). 10 W US, Mex. R: Britt. 29(1977)199. ***X. venusta*** (M.E. Jones) A. Heller (*M. v.*, Colorado, Utah) – selenium & uranium indicator

Xylosalsola Tzvelev (~ *Salsola*). Amaranthaceae (Chenopodiaceae). 1 C As.: ***X. arbuscula*** (Pall.) Tzvelev

Xylosma Forst.f. Salicaceae (2; Flacourtiaceae 8). 85 trop. (Am. 49, R: FN 22(1980)128; SE As. 5, Mal. 4, Aus. 4, New Caled. 15, Vanuatu 1, Polynesia 7, Guam 1) to E As. Usu. dioec. trees; nickel accumulators in New Caled. ***X. congesta*** (Lour.) Merr. (*X. japonica*, *X. senticosa*, E As.) – hedge-pl. in e.g. Aus.; ***X. intermedia*** (Seem.) Griseb. (trop. Am.) – dioec., wind-poll.

Xylosterculia Kosterm. = *Sterculia*

Xylothamia Nesom & al. = *Gundlachia* (but see Sida 14(1990)101). See also *Chihuahuana*, *Medranoa*, *Neonesomia*, *Xylovirgata*

Xylotheca Hochst. (~ *Oncoba*). Achariaceae (Flacourtiaceae). c. 10 E & S Afr. ***X. tettensis*** (Klotzsch) Gilg (E Afr.) – supposed aphrodisiac on Mafia Is.

Xylovirgata Urbatsch & R. Roberts. Compositae (Ast.-Sol.). 1 Mex.: ***X. pseudobaccharis*** (S.F. Blake) Urbatsch & R. Roberts

Xymalos Baill. Monimiaceae (V 1). 1 trop. & S Afr.: ***X. monospora*** (Harv.) Baill. (lemonwood) – Mangenot's Model, timber useful

Xyridaceae Agardh. Magnoliidae – Poales. 5/275 trop. & warm, few temp. R: AMBG 71(1984)300. Herbs, usu. in wet places, s.t. rhizomatous; vessel-elements throughout veg. body. Lvs with open sheath & narrow flat to terete, usu. distich. & basal. Fls bisex., 3-merous, poll. by pollen-coll. bees (no nectar or scent), usu. sessile in axils of tough imbr. bracts in cylindrical heads or dense spikes; K (2)3, outer 1 thin, membranous & reflexed at anthesis, to 0, other 2 chaffy, boat-shaped, C 3 clawed or basally connate, usu. yellow & ephemeral, A 3 (6, or + 3 staminodes), anthers with longit. slits, $\underline{G}$ (3), 1(3)-loc. with parietal (to axile) placentation, simple or 3-fid style, each placenta with (1–)∞ orthotropous to anatr. or weakly campylotropous bitegmic ovules. Fr. a loculicidal to irreg. dehiscent capsule, s.t. encl. within persistent C-tube; seeds (1-)many, small with scarcely differentiated lenticular to scutelliform embryo lying alongside starchy, proteinaceous endosperm s.t. with oil. n = 8–10, 13, 17

Genera: *Abolboda*, *Achlyphila*, *Aratitiyopea*, *Orectanthe*, *Xyris*

Abolboda & *Orectanthe* with lvs usu. in spirals, etc. have been seg. in own fam. *Orectanthe* with poss. largest angiosperm pollen

Xyridion (Tausch) Fourr. = *Iris* (but see BZ 90(2005)57)

Xyridopsis Welw. ex R. Nordenstam = *Psednotrichia*

Xyris Gronov. ex L. Xyridaceae. 250+ trop. (esp. Guayana Highlands & Amaz.) & warm (Aus. 20, 17 endemic; Afr. 25). Few cult. orn. (yellow-eyed grass) & locally medic. ***X. capensis*** Thunb. (OW trop.) – culms used to make figures for Hindu temples in Java

Xyropteris Kramer (~ *Tapeinidium*). Lindsaeaceae (Dennstaedtiaceae-Lindsaeoideae). 1 Sumatra, Borneo: ***X. stortii*** (Alderw.) Kramer

Xyroerix Airy Shaw = *Xyropteris*

Xysmalobium R. Br. (~ *Asclepias*). Apocynaceae (V c; Asclepiadaceae III 1). c. 30 trop. & S Afr. Some ed. r.

Y

Yabea Kozo-Polj. (~ *Caucalis*). Umbelliferae (III 3). 1 W N Am.: ***Y. microcarpa*** (Hook. & Arn.) Kozo-Polj.

yacca *Xanthorrhoea* spp.

yacon *Smallanthus sonchifolius*

Yadakeya Makino = *Pseudosasa*

yajé *Banisteriopsis* spp. esp. *B. caapi*

Yakirra Lazarides & R. Webster = *Panicum* (but see Brunonia 7(1984)289)

yakka *Xanthorrhoea* spp.

yakon *Smallanthus sonchifolius*

yala *Ipomoea costata*

yalka *Cyperus bulbosus*

yam *Dioscorea* spp., (US) *Ipomoea batatas*; **acom** or **aerial y.** *D. bulbifera*; **y. bean** *Pachyrhizus* spp.; **Chinese y.** *D. polystachya*; **cush-cush y.** *D. trifida*; **y. daisy** *Microseris lanceolata*; **Guinea y.** *D.* × *cayenensis*; **long y.** *D. transversa*; **Otaheite y.** *D. bulbifera*; **potato-y.** *D. esculenta*; **water** or **white y.** *D. alata*; **yampi y.** *D. trifida*; **yellow y.** *D.* × *cayenensis*

yampah *Perideridia gairdneri*

yampee *Dioscorea trifida*

yanagi *Debregeasia longifolia*

yang *Dipterocarpus* spp.

yang(g)ona *Piper methystichum*

yang-mei *Morella rubra*

yangtao *Actinidia deliciosa*

Yanomamua J.R. Grant & al. (~ *Chelonanthus*). Gentianaceae. 1 Amaz. Braz.: ***Y. araca*** J.R. Grant & al. – fls unknown! R: HPB 11(2006)31

yar *Casuarina equisetifolia*

yard grass *Eleusine indica*; **y.-long bean** *Vigna unguiculata* subsp. *sesquipedalis*

yareta *Azorella compacta*

Yarina Cook = *Phytelephas*

yarran *Acacia homalophylla*

yarroconalli *Tephrosia sinapou*

yarrow *Achillea millefolium*

Yasunia van der Werff = *Beilschmiedia* (but see Novon 20(2010)494)

yate *Eucalyptus cornuta*

yaupon *Ilex vomitoria*

yautia *Xanthosoma* spp., *Caladium lindenii*

Yavia Kiesling & Piltz. Cactaceae. 1 Arg.: ***Y. cryptocarpa*** Kiesling & Piltz

yawa *Vigna unguiculata*

Yeatesia Small. Acanthaceae (III 2c). 3 SE US to NE Mex. R: SB 14(1989)427

ye'eb nut *Cordeauxia edulis*

yeenga *Geodorum densiflorum*

yegoma oil *Perilla frutescens*

yellow archangel *Lamium galeobdolon*; **y. ash** *Cladrastis kentukea, Eucalyptus luehmanniana*; **y. avens** *Geum urbanum*; **y. bark** *Cinchona calisaya*; **y. bartsia** *Bellardia viscosa*; **y. berries** *Rhamnus saxatilis*; **y. birch** *Betula alleghaniensis*; **y. bird's-nest** *Hypopitys monotropa*; **y. buckeye** *Aesculus flava*; **y. cedar** *Xanthocyparis nootkatensis, Rhodosphaera rhodanthema*; **y. chamomile** *Cota tinctoria*; **y. cypress** *X. n.*; **y. deal** or **Baltic y. d.** *Pinus sylvestris*; **y.-eyed grass** *Xyris* spp.; **y. flag** *Iris pseudacorus*; **y. foxtail** *Setaria pumila*; **y. gentian** *Gentiana lutea*; **y. gum** *Xanthorrhoea resinosa*; **y. loosestrife** *Lysimachia vulgaris*; **y. monkey flower** *Mimulus luteus*; **y. nutsedge** *Cyperus esculentus*; **y. pereira** *Aristolochia glaucescens*; **y. pimpernel** *Anagallis nemorum*; **y. pine** *Pinus echinata*; **y. p., Am.** *P. strobus*; **y. poplar** *Liriodendron tulipifera*; **y. rocket** *Barbarea verna*; **y. rose of Texas** *Rosa* × *harisonii*; **y. star**

thistle *Centaurea solstitialis*; **y. stringybark** see **stringybark**; **y. walnut** *Beilschmiedia bancroftii*; **y. waterlily** *Nuphar lutea*; **y.wood** spp. of *Cladrastis, Flindersia, Milicia, Ochrosia, Podocarpus, Zanthoxylum* etc.; **y.w., Kentucky** *Cladrastis kentukea*; **y.wort** *Blackstonia perfoliata*; **y. yam** *Dioscorea × cayenensis*

yemani *Gmelina arborea*

yerba buena *Clinopodium douglasii*; **y. de la feche** *Sebastiania bilocularis*; **y. de hicotea** *Hygrophila costata*; **y. mate** *Ilex paraguariensis*; **y. de la negrita** *Sphaeralcea coccinea, S. angustifolia* subsp. *cuspidata*; **y. reuma** *Frankenia salina*; **y. santa** *Eriodictyon californicum*; **y. del zorillo** *Dysphania graveolens*

yercum *Calotropis gigantea*

Yermo Dorn. Compositae (Sen.-Tuss.). 1 Wyoming: ***Y. xanthocephalus*** Dorn. R: Madroño 38(1991)198

Yersinochloa H.N. Nguyen & Tran. Gramineae (Bamb.). 1 S Vietnam: ***Y. dalatensis*** H.N. Nguyen & Tran. R: NJB 34(2016)400

yesterday, today and tomorrow *Brunfelsia australis, B. pauciflora*

yeung chi ging *Bauhinia × blakeana*

yew *Taxus baccata*; **American y.** *T. canadensis*; **Californian y.** *T. brevifolia*; **Chinese y.** *T. mairei*; **English y.** *T. baccata*; **golden y.** *T. b.* 'Aurea'; **Irish y.** *T. b.* 'Fastigiata'; **Japanese y.** *T. cuspidata*; **plum y.** *Cephalotaxus* spp.; **Prince Albert's y.** *Saxegothaea conspicua*; **stinking y.** *Torreya californica*

yggdrasill Scandinavian mythological 'world-tree' (based on *Fraxinus excelsior*) with branches spread over the world & reaching above heaven

yin yang huo *Epimedium sagittatum*

yinma *Chukrasia tabularis*

Yinquania Z.Y. Zhu = *Cornus*

Yinshania Ma & Zhao (~ *Sophiopsis*). Cruciferae (49). 13 Vietnam, China. R: HPB 3(1998)81

ylang-ylang *Cananga odorata*

Ynesa Cook = *Attalea*

Yoania Maxim. Orchidaceae (V 13c). 4 Himal., Jap., NZ (1). Mycotrophic

yoco *Paullinia yoco*

yohimbe bark *Pausinystalia johimbe*

Yolanda Hoehne = *Brachionidium*

yomhin *Chukrasia tabularis*

yomogi soba (noodles) *Fagopyrum esculentum* with *Artemisia vulgaris*

yon *Anogeissus acuminata*

yopo *Anadenanthera peregrina*

York road poison *Gastrolobium calycinum*

Yorkshire fog *Holcus lanatus*

Yosemitea P. Alexander & Windham (~ *Boechera*). Cruciferae. 1 Sierra Nevada: ***Y. repanda*** (S. Watson) P. Alexander & Windham. R: SB 38 (2013)203

Yoshino cherry *Prunus × yedoensis*

young fustic *Cotinus coggygria*; **y.berry** See *Rubus*

Youngia Cass. Compositae (Lact.-Crep.). c. 30 As. R: CIWP 484(1939). ***Y. japonica*** (L.) DC. – lawn weed S US

Ypsilandra Franch. (~ *Helonias*). Melanthiaceae. 3 Himal. to W China. R: FOC 24(1998) 86

Ypsilopus Summerh. (~ *Tridactyle*). Orchidaceae (V 16d). 5 E to S (1) Afr. R: KB 40(1985)417. Cult. orn.

Ypsilorchis Z.J. Liu & al. Orchidaceae. 1 NE Chongqing: ***Y. fissipetala*** (Finet) Z.J. Liu & al. R: JSE 46(2008)623

ysaño *Tropaeolum tuberosum*

Ystia Compère = *Schizachyrium*

Yua C.L. Li (~ *Parthenocissus*). Vitaceae. 3 Himal.

yuan zhi *Polygala tenuifolia*

yuca *Manihot esculenta*

Yucaratonia Burkart = *Gliricidia*

Yucca Dill. ex L. Asparagaceae (Agavaceae). Incl. *Samuela*, excl. *Hesperoyucca* (loculicidal capsule), 47 warm N Am. (US 28), C Am., WI. R: U. Eggli, *Ill Handb. Succ. Pls*, Monocots (2001) 87. Woody pachycauls ± trunks with stiff sword-like lvs (? protection against extinct megafauna, poss. form. fr.-dispersers) & panicles of usu. white fls & septicidal capsules (beargrass, Adam's needle; *Yucca* being taken from yuca which Gerard thought was these pls). Most spp. app. dependent on yucca moths (allied to moths visiting

Dasylirion & *Nolina* spp.) for poll. (though seed set in cult. in other countries), visited by sibling spp. of & incl. *Tegeticula* (*Pronuba*) *yuccasella*, which is active by night, resting in fls (similar colour) by day; fls most strongly scented at night & nectar s.t. secreted at base of ovary though moths do not take it (poss. serves to attract other insects away from stigma); female y. moth climbs up a stamen & bends her head over the anther steadied by her uncoiling tongue, the pollen then being scraped together in a ball under the head by the maxillary palps; as many as 4 stamens may be processed thus before moth flies to another fl., where the ovary is inspected & if it is of the right age & does not already have eggs in it, the moth usu. lays 1 egg in each loc. & after laying each 1 deposits some pollen into the tube formed by the stigmas. As unpoll. fls are soon dropped, deposition of pollen ensures that there will be a continuing food supply for larvae provided by abnormal growth of ovules near larvae; other ovules develop normally & larvae emerge to pupate in ground when seeds are ripe. Adults emerge over a period up to 3 yrs with effect that even if the *Y.* does not fl. every yr, some moths survive to reproduce. Bogus y. moths (*Prodoxus*) also breed in ovaries but do not pollinate, relying on the true y. moth for survival. Some fibres, consituents of comm. shampoos, ed. parts etc. (see Excelsa 7(1977)45) & cult. orn. incl. hybrids esp. ***Y. aloifolia*** L. (Spanish bayonet, SE US, C Am., WI) – Leeuwenberg's Model, in absence of moths effecively bee-poll., rope-fibre & soap from lvs, molluscicidal glycosides; ***Y. baccata*** Torrey (SW US, N Mex.) – tough fibre, fr. ed., fl. buds roasted; ***Y. brevifolia*** Engelm. (Joshua tree, SW US) – Schoute's Model, grows c. 10 cm per annum, lvs function for c. 12–20 yrs, fibre can be used as newsprint, form. seeds spread by giant sloths (extinct 13 000 yrs BP, poss. due to human hunting); ***Y. carnerosana*** (Trel.) McKelvey (*Samuela c.*, SW N Am.) – fls & fr. ed., lvs source of palma fibre used for twine, brushes, etc.; ***Y. elata*** Engelm. (SW N Am.) – fibre, 'root' extract mooted as foaming agent for drinks; ***Y. elephantipes*** Regel (*Y. guatemalensis*, ? C Am., Mex). – fuelwood; ***Y. filamentosa*** L. (lvs to 2.5 cm wide) & ***Y. gloriosa*** L. (lvs to 7 cm wide, homoploid hybrid, *Y. aloifolia* × *V. filamentosa*) of E N Am. – commonly seen cult. spp. in Eur.; ***Y. glauca*** Nutt. (WC US) – imp. local medic. & shampoo, fibre used in kraft paper; ***Y. schidigera*** Roezl ex Ortgies (palmilla, SW US) – juice a source of preservative, deodorant etc

yuchán *Ceiba speciosa*

yulan *Magnolia denudata*

yumberry *Morella rubra*

Yunckeria Lundell = *Ctenardisia*

Yungasocereus Ritter (~*Samaipaticereus*). Cactaceae (III 4). 1 Bolivia: ***Y. inquisivensis*** (Cárdenas) Ritter

Yunnanopilia C.Y. Wu & D.Z. Li = *Champereia* (but see ABY 22(2000)248)

Yunquea Skottsb. = *Centaurodendron*

Yunorchis X.J. Liu & al. = *Yoania*

Yushania Keng f. (*Sinarundinaria* auctt.). Gramineae (IV). Excl. *Oldeania*, 77 OW trop. R: FOC 22(2006)57. S.t. with r.-spines. ***Y. anceps*** (Mitford) W.C. Lin (*A. a.*, *S. a.*, ? China) – ed. shoots; ***Y. microphylla*** (Munro) R.B. Majumdar (*A. m.*, *S. m.*, Himal.) – brooms in Bhutan

Yutajea Steyerm. = *Isertia*

yuthika *Jasminum auriculatum*

Yuyba (Barb.-Rodr.) L. Bailey = *Bactris*

yuzu *Citrus* × *junos*

Yvesia A. Camus (~ *Brachiaria*). Gramineae (XXIV 5). 1 Madag.: ***Y. madagascariensis*** A. Camus

Z

za'atar *Origanum syriacum*

Zabelia (Rehder) Makino (~ *Abelia*). Caprifoliaceae. 6 temp. Afghanistan to Japan. 'Sister' to former Dipsacaceae + Morinaceae + Valerianaceae. Cult. orn. shrubs esp. ***Z. triflora*** (R. Br.) Makino (*A. t.*, NW Himal.)

zacate *Nolina longifolia*

Zacateza Bullock. Apocynaceae (III; Asclepiadaceae I). 1 trop. Afr.: ***Z. pedicellata*** (K. Schum.) Bullock. R: KB 9(1954)361

Zaczatea Baill. = *Raphionacme*

Zagrosia Speta = *Scilla*

Zahlbrucknera Reichb. = *Saxifraga*

zakaton *Muhlenbergia macroura*

zalacca Blume = *Salacca*

Zalaccella Becc. = *Calamus*

Zaleya Burm.f. (~ *Trianthema*). Aizoaceae (II). 7 NE & E Afr., Ind., Aus. R: H. E. K. Hartmann, *Ill. Handb. Succ. Pls*, Aiz. F–Z(2002)351

zalil *Delphinium semibarbatum*

zallouh *Ferula harmonis*

Zaluzania Pers. Compositae (Helia.-Zal.). 10 SW US, Mex. R: Rhodora 81(1979)449. See also *Kingianthus*

Zaluzianskya F.W. Schmidt. Scrophulariaceae (Manul.). Incl. *Reyemia* (R: EJB 49(1993)291) 57 W S Afr., E Afr. (Mt Elgon, 1). R: O. Hilliard (1994) *Manuleeae*: 460. Some cult. orn. herbs with evening-fragrant fls superfic. resembling Caryophyllaceae

zambac *Jasminum sambac*

Zambesi redwood *Baikiaea plurijuga*

Zameioscirpus Dhooge & Goetghebeur. Cyperaceae (II 1). 3 Andes. R: PSE 243(2003) 75

Zamia L. Zamiaceae (III). Incl. *Chigua*, 62 trop. & warm Am. R: JAA suppl. 1(1991)371, BR 70(2004)291. Stomata at nucellus apex. Cult. orn. & starch sources (oft. toxic until boiled) – Florida arrowroot (esp. ***Z. pumila*** L. (*Z. integrifolia*, Florida, WI) – survives repeated scrub-fires in Florida, host-specific beetle-poll. (see next); ***Z. furfuracea*** L.f. (E Mex.) – poll. host-specific *Rhopalotria mollis* Sharp (snout weevil) mating, feeding & ovipositing in male cones rich in starch, larvae pupating & adults chewing through pollen-sacs before merely visiting starch-poor females & thereby effecting poll.; ***Z. pseudoparasitica*** Yates (Panama) – oft. epiphytic (rare in gymnosperms); ***Z. roezlii*** Linden (*Z. chigma*, Mex.) – stem rots at base, adventitious r. forming above so pl. potentially immortal, egg to 3 mm long, spermatozoids to 400 μm diam. & with 40 000 flagella

Zamiaceae Horan. Cycadidae – Cycadales. Incl. Boweniaceae, Stangeriaceae (R: FOW 2(1999)), 9/185 trop. & warm Afr., Aus. & Am. Dioec. pachycaul trees to 18 m or with underground stems. Lvs pinnate (pinnae rarely dichot. divided as in *Stangeria*), leaflets midrib-less (exc. *Stangeria*; cf. Cycadaceae) but with ± parallel longit. veins. Sporophylls in determinate cones, scales on females ± peltate & with 2(3+ – see *Microcycas*) ovules on adaxial margins

Classification & genera:

1. **Encephalarteae** (sporophylls imbr., pinnae narrowing to attachment point): *Encephalartos, Lepidozamia, Macrozamia*
2. **Dioeae** ((like 1. but pinnae not narrrowing): *Dioon*
3. **Zamieae** (? Stangeriaceae; sporophylls app. valvate): ? *Bowenia, Ceratozamia, Microcycas*, ? *Stangeria, Zamia*

Stangeria dioec. fern-like with underground stem diff. from Zamiaceae s.s. in convolute (not imbr.) leaf vernation, pinnae with definite midrib & many dichot. branched costae Cult. orn., oft. poisonous but starch & seeds in many ed. after treatment

Zamioculcas Schott. Araceae (VII 1). 2 SE trop. Afr. R: FTEA Arac.(1985)15. ***Z. zamiifolia*** (Lodd.) Engl. – lvs pinnate, vogue pot-pl. of airports & offices etc. (poss. the aspidistra of 21 cent.)

Zandera D.L. Schulz. Compositae (Helia.-Mel.). 3 Mex. R: Haussknechtia 4(1988)32

Zanha Hiern. Sapindaceae (III 2). 3 trop. Afr. (2), Madag. (1). ***Z. africana*** (Radlk.) Exell (Afr.) – local medic. in E Afr. for all ailments, fr. for soap

Zannichellia Micheli ex L. Potamogetonaceae (Zannichelliaceae). 1 cosmop.: ***Z. palustris*** L. (horned pondweed) – monoec. herb of fresh or brackish water with male & female fls in same axil, poll. underwater, when selfed pollen dropped in gelatinous mass directly on stigma. R: Lagascalia 14(1986)241, JAA 68(1987)264

Zannichelliaceae Chevall. = Potamogetonaceae

Zanonia L. Cucurbitaceae (III). 1 Indomal.: ***Z. indica*** L. R: Blumea 52(2007)282

Zantedeschia Spreng. (*Richardia*). Araceae (VII 16). 8 trop. & S Afr. R: Bothalia 11(1973)5. Cult. orn. rhizomatous pls with showy spathes (calla of florists) esp. ***Z. aethiopica*** (L.) Spreng. (arum lily, S Afr., invasive Aus., natur. Portugal etc.) – rhiz. favoured by porcupines, spathes white, much used at & assoc. with funerals in GB, the 'Easter lilies' of Irish Nationalists; also ***Z. elliottiana*** (W. Watson) Engl. (yellow a. l., cultigen, poss. hybrid) –

spathe yellow, lvs spotted white; ***Z. odorata*** P. Perry (S Afr.) – freesia-scented, crossed with *Z. a.* to give scented 'arum lilies'

Zanthorhiza L'Hérit. = *Xanthorhiza*

Zanthoxylum Colden ex L. (*Xanthoxylum*). Rutaceae (I 2). Incl. *Fagara*, c. 225 trop. (incl. Mal., R: JAA 47(1986)221), E As., Aus. Decid. or everg. prickly shrubs & trees (Scarrone's Model) with aromatic bark, some dioec. wind-poll. (Mex.), s.t. with knobs tipped with spines (knobthorn); some myrmecophytes (Mal.). Local medic. (seeds used as diuretic in China (jiaomu) esp. ***Z. bungeanum*** Maxim. – alks incl. some with antitumour activity); some timbers (yellowwood), spices etc. ***Z. acanthopodium*** DC. (Himal.) – fr. sold as spice in Sikkim; ***Z. americanum*** Mill. (E N Am.) – hardy shrub with locally medic. bark; ***Z. armatum*** DC. (*Z. alatum*, Indomal.) – spice (Pakistan) & medic. incl. for toothache, insect-repellent; ***Z. beecheyanum*** K. Koch (Okinawa & Bonin Is.) – used like *Z. piperitum*; ***Z. capense*** (Thunb.) Harv. (knobthorn, knobwood, S Afr.) – locally medic. ('fever tree'); ***Z. caribaeum*** Lam. (trop. Am.) – timber (concha satinwood); ***Z. clava-herculis*** L. (Hercules' club, C & S US) – toothache cure & other local medic.; ***Z. flavum*** Vahl (trop. Am.) – timber (Jamaican or WI satinwood) used for cabinet-work etc.; ***Z. gillettii*** (De Wild.) Waterm. (*Z. macrophyllum*, trop. Afr., Afr. satinwood) – fine timber; ***Z. piperitum*** (L.) DC. (Jap. pepper, sansho, E As.) – fr. used as condiment in Jap., one of the few in Jap. cuisine; ***Z. simulans*** Hance (*Z. bungei*, Sichuan or Chinese pepper, China) – cult. as condiment; ***Z. xanthoxyloides*** (Lam.) Zepernick & Timler (W Afr.) – chewing-stick from r. with marked antimicrobial activity against oral flora

Zanzibar copal *Hymenaea verrucosa*; **Z. redheads** *Syzygium aromaticum*

zapallo *Cucurbita moschata*

zapatero *Casearia praecox*

Zapoteca H. Hernández (~ *Calliandra*). Leguminosae (II 4). 20 SW N Am. to N Arg. R: AMBG 76(1989)806. Moth-poll. but moths preyed on by crab spiders

Zataria Boiss. Labiatae (VII 2b). 1 Iran, Afghanistan, Pakistan: ***Z. multiflora*** Boiss.

Zauschneria C. Presl = *Epilobium*

zawa *Lophira* spp.

Zdravetz oil *Geranium macrorrhizum*

Zea L. Gramineae (XXII 2). 7 Mex. & C Am., Nicaragua (1). R: AJB 67(1980)1000. ***Z. mays*** L. (maize) – tetraploid (2n [4x] = 20; genome sequenced 2009 – 2.3 billion base-pairs) cultigen with controversial history in that geneticists form. attempted to explain its orig. through the then fashionable phenomena of introgression (some, limited, transfer still considered poss.) etc. involving *Tripsacum* grasses ('jumping gene' repressing branching, changing pl. architecture), though app. largely derived (poss. more than once) merely by selection of homoeotic sexually transformed cvs of teosinte ('subsp. *mexicana* (Schrad.) Iltis' orig. cult. for sugary pith) called 'subsp. *parviglumis* Iltis & Doebley' with distich. lat. ears, once considered to belong in a sep. genus (*Euchlaena*) so that orig. of the cultigen form. thought to have given rise to a new genus. When attacked by caterpillars (GM 'MON 810' resistant) releases terpenes attractive to female *Cotesia marginiventris* wasps that lay eggs in caterpillars; cult. pl. with male fls in term. plumes (tassels), females in lower axils & bearing silks (styles) which pollen-grains penetrate, fert. leading to development of cob from which seeds cannot escape exc. through human activities. Earliest known forms (c. 5600 yrs old) with small cobs but each grain subtended by long glumes as in modern popcorn; planted NY State 1075–1285 AD; sculptures of 12 & 13 cent. in Ind. app. erroneously claimed to rep. it; Mayan maize god personifying cycle of life. Planted in Afr. by Portuguese to provide stores for slaves being shipped to Am. Next to wheat & rice, the most imp. cereal (1016 M t. [35% GM] a yr from c. 160 M ha by 2013) cult. in trop. & warm ('corn' (US), mealies, i.e. Port. 'milho [i.e. millet] (maiz)' (Afr.), Ind. corn) with many cvs used as human (incl. cornflakes, polenta, tortilla chips, cornflour (= corn starch in US), corn on the cob, babycorn (immature cobs for stir-fry oft. from Taiwan)) & esp. animal feed, cooking oil (poly-unsaturated, omega-6 (not 3) fatty acids, in margarine etc.), comm. starch (as, with sugar, in custard powder – Alfred Bird of Birmingham, UK (1837), in baking-powder (Harvard, 1850s), in biodegradable plastics for cups, glasses, carrier-bags, & in Goodyear 'Biotred' car tyres, but esp. a thickener in many processed foods, e.g. milk-shakes), sweetener (HFCS = high-fructose corn syrup) ubiquitous in US diet & blamed for much obesity there, beer (incl. to 'extend' barley) & spirits (esp. bourbon, a 'grain whisky', national spirit of US first made (not then aged in oak barrels) in B. County (none there today), Kentucky; by 1830s average US adult consumption almost 22 l per annum – now 24 times less, but basis of 'mint julep' in S US), industrial alcohol (e.g. motor fuel in US incl. Ford's orginal Model T; 42% US crop for ethanol by 2012) etc.; GM cvs used for prod. gastric fibrase in treatment of cystic fibrosis; locally the staple carbohydrate esp. in Afr. where eaten like dough made from m. flour, crushed grains (US) being hominy (grits), though excess leads to pellagra (no vitamin C, nicotinic acid). Cvs with hard endosperm exploding when heated because thin sheets of cellulose crystallize so that water cannot escape = popcorn ('var. *praecox*') – 17 billion 'quarts' consumed annually USA, those with ± translucent horny endosperm = sweetcorn ('var. *rugosa*'), usu. harvested when immature & tinned; gall prod. *Ustilago maydis* (corn smut, a rust) a veg.; some orn. cvs with grains of diff. colours in same cob, all red ('strawberry c.') etc., blue-black ones ('blue c.' used in blue tortilla chips) held sacred by Pueblos & now extracts used in US cosmetics

zebrano *Microberlinia* spp.

zebrawood *Astronium fraxinifolium, Centrolobium robustum, Diospyros* spp., *Microberlinia* spp.

Zebrina Schnizl. = *Tradescantia*

Zederbauera H.P. Fuchs = *Erysimum*

zedoary *Curcuma zedoaria*

Zehnderia Cusset. Podostemaceae (III). 1 Cameroun: ***Z. microgyna*** Cusset

Zehneria Endl. Cucurbitaceae (XIV). Incl. *Neoachmandra, Pilogyne*, c. 55 OW trop. (Aus. 1). Allied to *Melothria* (Am.). ***Z. anomala*** C. Jeffrey (NE Afr. to S Arabia) – fr. sweet

Zehntnerella Britton & Rose = *Facheiroa*

Zelenkoa M. Chase & N. Williams (~ *Oncidium*). Orchidaceae (V 12h). 1 Panama: ***Z. venusta*** (Lindl.) M. Chase & N. Williams. R: Lindleyana 16(2001)139

Zelkova Spach. Ulmaceae. 4 Crete (1), W & E As. R: Plantsman 11(1989)80. Cult. orn. like elms (some now attacked by Dutch Elm disease) with some timber esp. ***Z. serrata*** (Thunb.) Makino (Jap. elm, keaki, keyaki, E China, Taiwan, Jap.) – imp. timber tree for building, also a bonsai subject

Zeltnera Mansion (~ *Centaurium*). Gentianaceae. 26 SW N Am. R: T 53(2004)727

Zemisia R. Nordenstam (~ *Senecio*). Compositae (Sen.-Sen.). 1 Jamaica: ***Z. discolor*** (Sw.) R. Nordenstam. R: CN 44(2006)71

Zemisne Degener & Sherff = *Scalesia*

Zenia Chun. Leguminosae (I 3). 1 S China, Vietnam, Thailand: ***Z. insignis*** Chun

Zenkerella Taubert. Leguminosae (I 3). 5 trop. Afr. with 1 on Guinea coast, rest in E Afr. coastal mts

Zenkeria Trin. Gramineae (Arund.). 4 Ind., Sri Lanka. R: HIP (1962) t. 3597

Zenobia D. Don. Ericaceae (VIII 3). 1 SE US: ***Z. pulverulenta*** (Willd.) Pollard (*Z. speciosa*) – cult. orn. shrub. R: Castanea 61(1996)115

zephyr lily *Zephyranthes atamasca*

Zephyra D. Don. Tecophilaeaceae. 2 Chile

Zephyranthella (Pax) Pax = *Habranthus*

Zephyranthes Herb. Amaryllidaceae. Incl. *Haylockia* 70 SE US (16) to Arg. R: JRHS 62(1937)195, Plantsman 2(1980)8. Cult. orn. bulbous herbs with *Colchicum*-like white, yellow, pink or red fls esp. ***Z. atamasca*** (L.) Herb. (atamasco or zephyr lily, SE US) – fls white, bulbs eaten in emergency, ***Z. minima***(Kunth) D. Dietr. (*Z. carinata, Z. grandiflora, 'Z. rosea'*, rain lily, S Mex. to Guatemala) – most commonly cult. sp., fls pink after stimulus of rain & associated cooling. See also *Habranthus*

Zeravschania Korovin. Umbelliferae (III 10). 11 SW to C As. R: Willd. 37(2007)494

Zerdana Boiss. Cruciferae (5). 1 mts of Iran: ***Z. anchonioides*** Boiss. R: Cand. 40(1985)363

Zerna Panzer = *Festuca* but oft. applied to *Bromus* (*Bromopsis*)

Zetagyne Ridl. = *Panisea*

Zeugandra P. Davis. Campanulaceae (I 6). 2 Iran

Zeugites P. Browne. Gramineae (XIX). 12 trop. Am. Ovate blades & false petioles much resembling dicot lvs

Zeuktophyllum N.E. Br. Aizoaceae (V). 2 W Cape. R: H. E. K. Hartmann, *Ill. Handb. Succ. Pls*, Aiz. A–Z(2002)354

Zeuxanthe Ridl. = *Prismatomeris*

Zeuxine Lindl. Orchidaceae (IV 2d). c. 70 trop. & warm OW (Aus. 1). Some cult. orn. incl. ***Z. strateumatica*** (L.) Schltr. (Indomal.) – apomictic, one of few 'ann.' orchids, weedy Florida (since 1930s), Braz., Hawaii, wash used by Seminoles to overcome impotence

Zexmenia Llave (~ *Lasianthaea, Wedelia*). Compositae (Helia.-Ecl.). 2 C Am. R: SBM 3(1991)86

Zeyherella (Engl.) Aubrév. & Pellegrin = *Chrysophyllum*

Zeyheria Mart. Bignoniaceae (1). 2 Braz.: ***Z. montana*** Mart. – hummingbird-poll.; ***Z. tuberculosa*** Bur. – constr. timber

Zeylanidium (Tul.) Engl. (~ *Hydrobryum*). Podostemaceae (III). 5–6 Sri Lanka to Myanmar

Zhengyia T. Deng & al. Urticaceae. 1 C China: ***Z. shennongensis*** T. Deng & al. – bulbils. R: T 62(2013)94

zhexuanshen *Scrophularia ningpoensis*

zhi fu zi *Aconitum* spp.

Zhukowskia Szlach. & al. = *Sarcoglottis*

Zhumeria Rech.f. & Wendelbo. Labiatae (VII 2a). 1 S Iran: ***Z. majdae*** Rech.f. & Wendelbo. R: IJB 1(1976)2

Zieria Sm. Rutaceae (I 3). 61 Aus. (60 endemic), New Caled. (1). R: AusSB 15(2002)295. ***Z. arborescens*** Sims (stinkwood, SE Aus.) – lvs unpleasantly scented when crushed

Zieridium Baill. = *Picrella*

Zigadenus Michaux. Melanthiaceae. Excl. *Amianthum, Anticlea, Stenanthium, Toxicoscordion* (q.v.), 1 SE US: ***Z. glaberrimus*** Michaux. R: Novon 12(2002)305

Zilla Forssk. Cruciferae. 2 N Afr. to Arabia

Zimmermannia Pax = *Meineckia*

Zimmermanniopsis Radcl.-Sm. = *Meineckia*

zingana *Microberlinia brazzavillensis*

Zingeria Smirnov. Gramineae (XVI 14). 5 Eur. (2) to Iran. x = 2, 2n = 4

Zingeriopsis Probat. = *Zingeria*

Zingiber Mill. Zingiberaceae (II 3). c. 100 Indomal. (Borneo 19; R: NRBGE 45(1988)409) to E As. & trop. Aus. Herbs with aromatic rhiz. esp. ***Z. officinale*** Roscoe (ginger [= Tamil *inji ver*], cultigen of Ind. orig. (? SE As.) – 50 cvs in Ind.; cult. E Afr. by time of Dioscorides) – sterile, the rhiz. used fresh (green g.) or preserved in syrup or crystallized (esp. in Hong Kong) or dried. & powdered, used in biscuits (g. nuts), cakes, sweets, g. beer, g. wine & brandy, also shampoos & cosmetics, app. alleviating pain & inflammation in arthritis by inhibiting cycloxygenase responsible for it (TCM as sheng jiang), also much prescribed for travel sickness & much cult. in Jamaica (first exported 1547), now fastest-growing Hawaiian prod. industry; other spp. used incl. ***Z. mioga*** (Thunb.) Roscoe (Jap. or mioga g., Jap.) – bergamot flavour, anc. medic. in China (vermifuge (earliest recorded), antimalarial, insect-bites etc.), ***Z. montanum*** (Koenig) A. Dietr. (*Z. cassumunar, Z. purpureum*, cassumar g., Indomal.) & ***Z. zerumbet*** (L.) Sm. (Indomal.) – ethereal oils etc. shown to have antineoplastic principles & to inhibit prostaglandin synthesis, soapy fluid in infl. stalks trad. shampoo & now comm. ***Z. squarrosum*** Roxb. (Indomal.) – used to tranquillize *Apis dorsata* bees in Andaman Is. so honey can be coll.

Zingiberaceae Martinov. Magnoliidae – Zingiberales. Excl. Costoideae (= Costaceae), 51/1400 trop. esp. Indomal. R: NRBGE 31(1972)171. Aromatic usu. caulescent herbs (Tomlinson's Model) with thickened rhiz. & secretory cells with ethereal oils; vessel-elements in r. & s.t. also in stems etc.; aerial stem short. Lvs in spirals or distich. with usu. open sheath, the sheaths forming a pseudostem, ± petiole & with simple lamina rolled from 1 side to another, usu. with prominent midrib & costae pinnate-parallel; adaxial ligule usu. at junction of sheath & lamina. Fls bisex., zygomorphic, insect- & bird-poll., usu. ephemeral (1 day), solit. or in thyrses etc. in axils of bracts, s.t. on a short sep. sheath-covered stem from rhiz.; K 3 not petaloid, basally united with a tube or spathe-like & split to base on 1 side only, C (3) with short lobes the median adaxial oft. larger than rest, A 1 (median adaxial inner of a presumed 3 + 3 now rep. by consp. labellum (other 2 inner; cf. Orchidaceae where lip is P, or Costaceae all other 5) & staminodes), ± petaloid with elongate thecae & longit. slits, $\bar{G}$(3), (1–) 3- loc. with (parietal or) axile placentation & usu. apical nectaries & ± ∞ anatropous, bitegmic ovules. Fr. a loculicidal capsule or indehiscent (fleshy or dry); seeds with arils, embryo straight in copious starchy endosperm & perisperm. n = (7–) 9–26

Classification & principal genera:

[I. Costoideae = Costaceae

II. Zingiberoideae s.l.]

1. **Hedychieae** (G 3-loc., lvs parallel to rhiz., style not far exserted beyond anther; *Siphonochilus* prob. distinct as **Siphonochiloideae** with *Tamijia* as **Tamijioideae** (basal groups), rest to go to **Zingiberoideae** s.s.): 19 genera incl. *Boesenbergia, Curcuma, Hedychium, Kaempferia, Roscoea* & *Scaphochlamys*

2. Globbeae (G 1-loc.; anther usu. long-exserted on arched ascending filament): 4 genera incl. *Globba* – to Zingiberoideae s.s.
3. Zingibereae (**Zingiberoideae** s.s.; G 3-loc., lvs parallel to rhiz., style exerted beyond anther & enveloped by elongate process (anther crest): *Zingiber* (+ **2.** & most of **1.**)
4. Alpinieae (**Alpinioideae**; G usu. 3-loc., lvs at right angles to rhiz.): *Aframomum, Alpinia, Amomum, Elettaria, Hornstedtia, Renealmia, Riedelia*

Many imp. spice-pls incl. spp. of *Aframomum, Amomum, Curcuma* (turmeric), *Elettaria* (cardamom), *Kaempferia, Zingiber* (ginger); scent (*Hedychium, Kaempferia*), dyes (*Curcuma*) & cult. orn. in these genera & *Cautleya, Etlingera, Roscoea* etc.

Zinnia L. Compositae (Helia.-Zinn.). c. 25 US to Arg. esp. Mex. R: Britt. 15(1963)4, 293. Cult. orn. herbs (& shrubs) with opp. or whorled lvs & alks, esp. ***Z. elegans*** Jacq. (*Z. violacea*, Mex.) – allopolyploid with *Z. angustifolia* Kunth bringing disease resistance, cult. by Aztecs & now with v. many cvs (some with black-tipped receptacle scales derived from ***Z. haageana*** Regel (Mex.)) with fls of all colours save blue. C of female fls of some spp. (*Tragoceros*, R: Britt. 15 (1963)290) persistent & hooked aiding disp.

Zinowiewia Turcz. Celastraceae (I). 17 Mex. to Venez. R: CUMH 3(1939)35

Zippelia Blume (~ *Piper*). Piperaceae (3). 1 SE As., W Mal.: ***Z. begoniifolia*** Blume – n = 19 (13 in *P.*). R: FOC 4(1999)110

ziricote *Cordia dodecandra*

zitan *Pterocarpus santalinus*

Zizania Gronov. ex L. Gramineae (III 3ii). 5 E Ind. to E As. (1), N Am. R: CUSNH 39(2000)116. Water oats; aquatic grasses. ***Z. palustris*** L. (~ *Z. aquatica* L., (Canadian) wild rice, (Manchurian) water rice, Tuscarora r., N Am.) – trad. Native American food, still grown comm. in Minnesota & Far E, s.t. 'popped' (cf. popcorn); ***Z. latifolia*** (Griseb.) Stapf (As., invasive Aus.) – young shoots eaten in Chinese food to give texture, esp. forms with culms swollen by infection by a smut (*Ustilago esculenta*, 'gaausun'), spores from infecting smut used as eyebrow & hair black, lvs made into mats

Zizaniopsis Doell & Asch. Gramineae (III 3ii). 5 warm & trop. Am. R: Hickenia 1(1976)39

Zizia Koch. Umbelliferae (III 8). 4 N Am. ***Z. aurea*** (L.) Koch – cult. orn.

Ziziphora L. Labiatae (VII 2b). 8 Medit. (Eur. 6, dry places) to C As., Afghanistan & Himal. Some local medic. (e.g. ***Z. tenuior*** L. in Turkey – oil mainly pulegone) & flavouring for yoghurt etc.

Ziziphus Mill. (*Zizyphus*). Rhamnaceae (I). Excl. *Pseudoziziphus, Sarcomphalus*, c. 65 trop. (esp. Mal.) & warm (Eur. 1, Aus. 4). Decid. & everg. shrubs & trees (Roux's Model), with alks & usu. with stipular thorns, 1 straight, the other recurved; heterodichogamy recorded. Some timber & ed. fr. (jujubes), oft. dried like dates. ***Z. jujuba*** Mill. (*Z. zizyphus*, sedra, F. jujube, Chinese date, SE Eur. to China) – tree to 1000 yrs old, fr. eaten fresh or cooked, form. imported to GB for cough-cures & extracts for comm. hair-dyes ('neutral henna'), source of shellac, many pomological cvs esp. in E; ***Z. lotus*** (L.) Lam. ((Afr.) lotus, Medit.) – lotus fr. of the Ancients; ***Z. mauritiana*** Lam. (ber, Ind. jujube, OW trop.) – ed. fr., source of lac, lvs for tanning, invasive Aus. forming continuous shrub-layer in previously 'open' *Eucalyptus* woodland; ***Z. mistol*** Griseb. (Andes) – ed. fr. (mistol) used in alcoholic drink; ***Z. mucronata*** Willd. (buffalo or Cape thorn, trop. Afr.) – medic., seeds used in rosaries; ***Z. spina-christi*** (L.) Willd. (ilb, Christ's thorn, crown of thorns, Medit. to Arabia) – poss. Christ's crown of thorns (allegedly thorns removed from 'Christ's' chaplet now in Notre Dame, Paris), fr. in Tutankhamun's tomb, ed. & medic., wood used in anc. Egypt for coffins etc., form. on Socotra for keys & locks & as building timber; ***Z. spinosa*** Hu (China) – seeds (suan zao/u ren) imp. trad. Chinese med. for insomnia etc.

Zizyphus Adans. = *Ziziphus*

Zoegea L. Compositae (Card.-Cent.). 3 SW to C As., Egypt

Zoellnerallium Crosa = *Nothoscordum*

Zollernia Wied-Neuw. & Nees. Leguminosae (III 1). 10 trop. Am. R: KB 59(2004)499. Some timber for cutlery-handles, brush-backs etc.

Zollikoferiastrum Kamelin = *Cicerbita*

Zollingeria Kurz. Sapindaceae. 2–3 SE As., 1 Borneo

Zombia L. Bailey. Palmae (III 2). 1 Hispaniola: ***Z. antillarum*** (Desc.) L. Bailey – suckering palm with persistent fibrous sheaths with spine-like marginal fibres & bisex. fls

Zombitsia Keraudren = *Ctenolepis* (but see R: Adans. 3(1963)167)

Zomicarpa Schott. Araceae (VII 10). 2 NE Braz. R: KB 67(2012)444

Zomicarpella N.E. Br. Araceae (VII 10). 2 Colombia, Braz. R: Willd. 37(2007)526. Seasonally dormant geophytes

Zonanthus Griseb. Gentianaceae. 1 Cuba: ***Z. cubensis*** Griseb.

Zonotriche (C. Hubb.) J. Phipps = *Tristachya*

Zootrophion Luer (~ *Pleurothallis*). Orchidaceae (V 13c) 22 trop. Am. R: Selbyana 7(1982)80

Zornia J. Gmelin. Leguminosae (III 11). c. 80 warm (Aus. 19 endemic). R: Webbia 16(1961)1

Zosima Hoffm. Umbelliferae (III 11). 4 W As. R: ANWM 103B(2001)566

Zostera L. Zosteraceae. Incl. *Heterozostera*, 9 warm to cool (Eur. 3, Aus. 3) to NG. R: SB 27(2002)483. Nitrogen-fixing bacteria in rhizosphere. ***Z. marina*** L. (eelgrass, grass-wrack, alva or ulva marina, N temp.) – pop. crash in N Am 1930s due to slime-mould (*Labyrinthula zosterae*), dried lvs used for matting etc. (sea-grass m.), packing (as for glass at Venice), pillows etc., comm. (Nova Scotia) insulating material from early 1800s incl. in early US skyscrapers, promising grain (imp. bird food in winter); ***Z. tasmanica*** Martens ex Asch. (*H. t.*, coasts of temp. Aus. & Chile) to 31 m deep

Zosteraceae Dumort. (~ Potamogetonaceae, Juncaginaceae). Magnoliidae – Alismatales. 3/22 warm to cool coasts. Monoec. or dioec. (*Phyllospadix*) perenn. submerged halophytes to depths of 50 m or exposed at high tide, glabrous & with creeping rhiz.; vessel-elements, stomata & lignin 0. Adventitious r. unbranched. Lvs linear or filiform with open or closed sheath oft. with stipuloid flanges, ligule at junction with parallel-veined lamina (s.t. with midrib); intravaginal scales at nodes. Fls small, water-poll., sessile in 2 rows (in each alt. male & female in *Zostera*) on 1 side of spadix encl. in spathe, the axis with bract-like lobes (retinacula) that fold over fl(s); P 0, A 1, G(2), 1-loc. with basally united styles & 1 pend. orthotropous bitegmic ovule. Fr. a small drupe or irreg. dehiscing; endosperm 0. n = 6, 9, 10

Genera: *Phyllospadix, Zostera*

Poll. with grains same density as seawater, captured by feathery stigmas. Marine derivatives of Potamogetonaceae with which Z. should be united

Zosterella Small (~ *Heteranthera*). Pontederiaceae. 2 temp. & subtrop. N Am.

Zosterophyllanthos Szlach. & Marg. = *Pleurothallis*

Zotovia Edgar & Connor = *Ehrharta* (but see NZJB 36(1998)566)

Zoutpansbergia Hutch. = *Callilepis*

Zoysia Willd. Gramineae (XXVIII). 11 Mauritius to Polynesia (China 5). Some lawn-grasses esp. ***Z. matrella*** (L.) Merr. (Manila grass, trop. As.) & its hybrids in S US & ***Z. tenuifolia*** Willd. ex Trin. (Korean grass, temple grass, cv of *Z. m.*?) in S As.

Zschokkea Muell. Arg. = *Lacmellea*

Zuccagnia Cav. Leguminosae (I 4). 1 Chile, Arg.: ***Z. punctata*** Cav. – yellow dye from lvs. R: PhytoKeys 71(2016)101

Zuccarinia Blume. Rubiaceae (II 5). 1 W Mal. Cook's Model: ***Z. macrophylla*** Blume

zucchini *Cucurbita pepo* 'Zucchini'

Zuckertia Baill. (~ *Bia*). Euphorbiaceae (Acal.-Pluk.-Trag.). 1 Mex.: ***Z. caudata*** Baill. R: NJB 31(2013)595

Zuckia Standl. = *Grayia* (but see FNA 4(2003)303)

Zuelania A. Rich. (~ *Casearia*). Salicaceae (Flacourtiaceae). 1 trop. Am.: ***Z. guidonia*** (Sw.) Britton & Millsp.

Zuloagaea Bess (~ *Panicum*). Gramineae (XXIV 4). 1 Ariz. to N S Am.: ***Z. bulbosa*** (Kunth) Bess. R: SB 31(2006)666

Zuloagocardamum Solariato & Al-Shehbaz. Cruciferae (Thel.). 1 Arg.: ***Z. jujuyensis*** Solariato & Al-Shehbaz. R: SB 39(2014)571

Zunilia Lundell = *Ardisia*

Zuvanda (Dvořák) Askerova (~ *Malcolmia*). Cruciferae (26). 3 SW As. R: HPB 19(2014)53

Zycona Kuntze = *Schistocarpha*

Zygella S. Moore = *Larentia*

Zygia P. Browne. Leguminosae (II 4). c. 45 trop. Am. R: MNYBG 74,2(1996)60. ***Z. inaequalis*** (Willd.) Pittier (*Pithecellobium i.*) – water-disp. due to air-filled cavity between cotyledons; ***Z. lehmannii*** (Harms) Britton & Rose (Colombia) – almost exterminated as used for tool-handles

Zygocactus K. Schum. = *Schlumbergera*

Zygocarpon Thulin & Lavin (~ *Ormocarpum*). Leguminosae (III 11). 6 Horn of Afr. (Somalia 3, Socotra 1, Yemen & Oman 2). R: SB 26(2001)308

Zygochloa S.T. Blake. Gramineae (XXIV 4). 1 C Aus.: ***Z. paradoxa*** (R. Br.) S.T. Blake – shrubby, dioec., on desert sand-dunes

Zygodia Benth. = *Baissea*

Zygogynum Baill. Winteraceae. Incl. *Belliolum, Bubbia* & *Exospermum*, c. 50 C Mal. to New Caled. & Aus. R: Blumea 31(1985)39, 48(2003)183. Koriba's, Leeuwenberg's & Scarrone's Models.; some rheophytes; poll. by thrips, Coleoptera & primitive moths etc. Bark used in tobacco in Papua NG to induce dream-like state

Zygonerion Baill. = *Strophanthus*

Zygoon Hiern = *Tarenna*

Zygopetalon auctt. = seq.

Zygopetalum Hook. Orchidaceae (V 12j). 14 trop. Am. Cult. orn. epiphytes (esp. ***Z. intermedium*** Lindl. ('*Z. mackaii*')), some with scented fls

Zygophlebia L.E. Bishop. Polypodiaceae (V; Grammitidaceae). 22 trop. Am. (7), Afr. (15) R: AFJ 79(1989)103

Zygophyllaceae R. Br. Magnoliidae – Zygophyllales. Incl. Balanitaceae, excl. Nitrariaceae, 27/300 trop. & warm esp. arid, s.t. saline. Small trees, shrubs (usu.) to herbs, s.t. with alks or mustard-oils; stem oft. swollen at nodes & s.t. with unusual sec. thickening. Lvs usu. paripinnate, oft. 2-foliolate, less oft. simple, usu. opp. (or in spirals), oft. resinous & usu. with stip., oft. persistent & s.t. spiny. Fls usu. bisex. & reg., (4)5(6)-merous solit., paired or in small cymes; K s.t. basally connate, imbr. or valvate (*Seetzenia*), C usu. free, imbr. or convolute, rarely valvate or 0, A in (1)2(3) whorls, filaments s.t. with basal glands, anthers with longit. slits, disk usu. intrastaminal & s.t. a gynophore, $\underline{G}$ ((2–)4 or 5(–12)), pluriloc. with axile placentas & 1 style, each loc. with 1–10 usu. pend., anatr. to s.t. hemitropous, campylotropous or orthotropous, bitegmic ovules. Fr. oft. a capsule or schizocarp, rarely a drupe (*Balanites*); embryo straight to weakly curved, in hard oily (or 0) endosperm. x = 6, 8–13+

Classification & principal genera (subfams as yet diff. to characterize morphologically):

I. **Zygophylloideae** (R: PSE 240(2003)29): *Augea, Fagonia, Tetraena, Zygophyllum*

II. **Larreoideae** (warm Am.): *Bulnesia, Guaiacum, Larrea, Porlieria*

III. **Seetzenioideae** (prostrate herb): *Seetzenia* (only)

IV. **Tribuloideae**: *Balanites* (stip. minute, spines axillary, drupes – Balanitaceae), *Kallstroemia, Sisyndite, Tribulus*

V. **Morkillioideae** (SW N Am.): *Morkillia*

Krameriaceae poss. to come here. Some (*Augea, Sisyndite*) at the moisture limits of plant growth. Major generic disjunctions esp. in N & S Am. – *Fagonia, Larrea, Porlieria*, but also *Seetzenia* (Afr. etc.)

Timber (*Guaiacum* – lignum-vitae), wax (*Bulnesia*), ed. fr. (*Balanites*), weeds (*Tribulus*)

Zygophyllum L. Zygophyllaceae (I). Excl. *Melocarpum, Roepera*, c. 50 Middle E to Iran, C As. & China, oft. in deserts or other arid areas. R: PSE 240(2003)33. C_3 & C_4 pl. with alks & succ. shoots. Some ed. fl. buds a caper subs.

Zygoruellia Baill. Acanthaceae (III [Whitf.]). 1 Madag.: ***Z. benthamii*** Baill. R: FMad. 182(1967)111

Zygosepalum Reichb.f. Orchidaceae (V 12j). 8 trop. S Am. Cult. orn. epiphytes

Zygosicyos Humbert = *Xerosicyos*

Zygostates Lindl. Orchidaceae (V 12h). Incl. *Dipteranthus*, c. 20 trop. S Am. (Braz. 14). Cult. orn. dwarf epiphytes

Zygostelma Benth. Apocynaceae (III; Asclepiadaceae I). 1 Thailand: ***Z. benthamii*** Baill.

Zygostigma Griseb. Gentianaceae. 2 Braz., Arg.

Zygotritonia Mildbr. Iridaceae (VI 2). 4 trop. Afr. R: BMNHN 4,11(1989)199. Some corms ± ed.

Zyrphelis Cass. (~ *Mairia*). Compositae (Ast.-Hom.). 10 S Afr. R: Phytol. 76(1994)91

Zyzyura H. Robinson & Pruski (~ *Fleischmannia*). Compositae (Eup.-Flei.). 1 Belize: ***Z. mayana*** (Pruski) H. Robinson & Pruski. T: PhytoKeys 20(2013)3

Zyzyxia Strother. Compositae (Helia.-Ecl.). 1 Mex., Belize: ***Z. lundellii*** (H. Robinson) Strother. R: SBM 33(1991)91

Appendix: System for Arrangement of Extant Vascular Plants

This book follows, in general, K. Kubitzki's *The Families and Genera of Vascular Plants* (1990–), modified to take account of recent findings, especially from molecular systematics, the overall framework following M.W. Chase & J.L. Reveal (2009), A phylogenetic classification of the land plants to accompany APG III, *Botanical Journal of the Linnean Society* 161: 122–127, with some modifications for lycophytes and ferns from M.J.M. Christenhusz & al. (2011), A linear sequence of extant families and genera of lycophytes and ferns, *Phytotaxa* 19: 7–54. The vascular plants, as set out below, and bryophytes, including Subclass Bryidae Engl. (mosses), make up Class Equisetopsida C. Agardh (the land plants). Class Equisetopsida is now known to be sister (L.A. Lewis & R.M. McCourt (2004), Green algae and the origin of land plants, *American Journal of Botany* 91: 1535–1556) to Class Charopsida, comprising the charophytes (stoneworts), an affinity recognized in nineteenth-century Floras.

The concordance below shows moves of families from their position in *Mabberley's Plant-book* ('PB III', i.e. ed. 3, 2008). When there is doubt over DNA findings or no new information from DNA work [or anything else], no change has been made. Where there is continuing debate, a broad view of families is taken (following van Steenis's lead – see Introduction): this means that where separating sister 'families' would lead to a change from PB III, sisters are kept together. There are 446 recognized families included in 100 orders.

N.B. Where families (or orders) described or resurrected since PB III are not recognized here, they do not appear below, even in synonymy (though the family names will appear as synonyms in the text above). For a key to the classic literature, see: M.T. Davis (1957) A guide [to] & an analysis of Engler 's "Das Pflanzenreich" in *Taxon* 6: 161–182

Subclass **LYCOPODIIDAE** Bek.
Clubmosses and quillworts
According to M.J.M. Christenhusz & M.W. Chase, Trends and concepts in fern classification, *Annals of Botany* 113(2014)571
[Three families in 3 orders]

Order 1. Lycopodiales DC. ex Bercht. & J. Presl
Lycopodiaceae
Order 2. Selaginellales Prantl
Selaginellaceae
Order 3. Isoetales Prantl
Isoetaceae

[*FERNS*]
After A.R. Smith & al., A classification for extant ferns, *Taxon* 55(2006)705–731 with some modifications according to M.J.M. Christenhusz & al., A linear sequence of extant families and genera of lycophytes and ferns, *Phytotaxa* 19(2011)7 and M.J.M. Christenhusz & M.W. Chase, Trends and concepts in fern classification, *Annals of Botany* 113(2014)571
[46 families in 11 orders]

Subclass **EQUISETIDAE** Warm.

Order Equisetales DC.
Equisetaceae

Subclass **OPHIOGLOSSIDAE** Klinge (incl. Psilotidae)

Order 1. Ophioglossales Link
Ophioglossaceae

Order 2. Psilotales Prantl
Psilotaceae

Subclass **MARATTIIDAE** Klinge

Order Marattiales Link
Marattiaceae

Sublass **POLYPODIIDAE** Cronq. & al.
[42 families in 7 orders]

Order 1. Osmundales Link
Osmundaceae
Order 2. Hymenophyllales Frank
Hymenophyllaceae
Order 3. Gleicheniales Schimper
1. **Gleicheniaceae**
2. **Dipteridaceae**
3. **Matoniaceae**

Order 4. Schizaeales Schimper
1. **Lygodiaceae** [*sometimes incl. in* 2.]
2. **Schizaeaceae**
3. **Anemiaceae** [*sometimes incl. in* 2.]

Order 5. Salviniales Bartl.
1. **Marsileaceae**
2. **Salviniaceae**

Order 6. Cyatheales Frank [*all sometimes incl. in* 6.]
1. **Thyrsopteridaceae**
2. **Loxsomataceae**
3. **Culcitaceae**
4. **Plagiogyriaceae**
5. **Cibotiaceae**
6. **Cyatheaceae**
7. **Dicksoniaceae**
8. **Metaxyaceae**

Order 7. Polypodiales Link
1. **Cystodiaceae** [*form. incl. in* 3.]
2. **Lonchitidaceae** [*form. incl. in* 3.]
3. **Lindsaeaceae**
4. **Saccolomataceae**
5. **Dennstaedtiaceae**
6. **Pteridaceae**
7. **Cystopteridaceae** [*form. incl. in* 14.]
8. **Rhachidosoraceae** [*form. incl. in* 14.]
9. **Diplaziopsidaceae** [*form. incl. in* 14.]
10. **Aspleniaceae** [*poss. incl.* 7.–9., 11.–17. *sometimes incl. in* 10. *as subfams*]
11. **Thelypteridaceae**
12. **Desmophlebiaceae** [*form. incl. in* 14.]
13. **Hemidictyaceae** [*form. incl. in* 14.]
14. **Woodsiaceae**
15. **Athyriaceae**
16. **Onocleaceae** [*poss. to be incl. in* 17.]
17. **Blechnaceae**
18. **Hypodematiaceae** [*poss. excl.* Didymochlaenaceae; 18.–24. *sometimes incl. in* 25. *as subfams*]
19. **Dryopteridaceae**
20. **Lomariopsidaceae**
21. **Tectariaceae**
22. **Oleandraceae**
23. **Davalliaceae**
24. **Polypodiaceae**

[*SEED-PLANTS; SPERMATOPHYTA; 425 families in 68 orders in 5 subclasses*]

GYMNOSPERMS

From molecular data, extant gymnosperms apparently 'sister' to angiosperms, though many argue the following classes are independent lines from ancestral pteridosperms (seed-ferns); 12 families in 4 orders

Subclass **GINKGOIDAE** Engl.

Order Ginkgoales Gorozh.

Ginkgoaceae

Subclass **CYCADIDAE** Pax (cycads)

Order Cycadales Pers. ex Bercht. & J. Presl

1. **Cycadaceae**
2. **Zamiaceae**

Subclass **PINIDAE** Cronq. & al. (conifers)

Order Pinales Gorozh.

1. **Pinaceae**
2. **Araucariaceae**
3. **Podocarpaceae**
4. **Sciadopityaceae**
5. **Taxaceae**
6. **Cupressaceae**

i

Subclass **GNETIDAE** Pax [form. incl. in Pinidae]

Order Gnetales Pax [*form. incl. in* Pinales]

1. **Gnetaceae**
2. **Ephedraceae**
3. **Welwitschiaceae**

Subclass **MAGNOLIIDAE** Novák ex Takht. [angiosperms]

Following Angiosperm Phylogeny Group (2016), An update of the Angiosperm Phylogeny Group classification for the orders and families of flowering plants: APG IV. *Botanical Journal of the Linnean Society* 181: 1–20 (2016; though some PB III families split there not followed here). It is instructive to see that by comparison with the differences between PB II and PB III, the current system has very few substantial changes in ordinal position of families, showing that the findings from molecular systematics are leading to stability, though the arrangement of orders is somewhat modified. However, there has been considerable splitting in Caryophyllales and there remain a small number of other problematic orders, notably Santalales, where future changes are likely to be made. It is worth noting that many of the recent departures from traditional 'consensus' groupings of the last century were presaged by seed-anatomy work published long ago (Corner 1976). There are 413 families in 64 orders.

Order 1. Amborellales Melikian & al.

Amborellaceae

Order 2. Nymphaeales Salisb. ex Bercht. & J. Presl

1. **Hydatellaceae**
2. **Cabombaceae**
3. **Nymphaeaceae** [*the position was part of Corner's underlying philosophy – Durian Theory*]

Order 3. Austrobaileyales Takht. ex Reveal

1. **Austrobaileyaceae**
2. **Trimeniaceae**
3. **Schisandraceae**

[Magnoliids]

Order 1. Canellales Cronq.

1. **Canellaceae**
2. **Winteraceae**

Order 2. Piperales Bercht. & J. Presl

1. **Saururaceae**

2. **Piperaceae**
3. **Aristolochiaceae** (incl. Hydnoraceae, Lactoridaceae)

Order 3. Magnoliales Juss. ex Bercht. & J. Presl

1. **Myristicaceae**
2. **Magnoliaceae**
3. **Degeneriaceae**
4. **Himantandraceae**
5. **Eupomatiaceae**
6. **Annonaceae**

Order 4. Laurales Juss. ex Bercht. & J. Presl

1. **Calycanthaceae**
2. **Siparunaceae**
3. **Gomortegaceae**
4. **Atherospermataceae**
5. **Hernandiaceae**
6. **Monimiaceae**
7. **Lauraceae**

Order Chloranthales Mart.

Chloranthaceae

MONOCOTYLEDONS

79 families, arranged in 11 orders.

Order 1. Acorales Martinov [*formerly in Alismatales*]

Acoraceae

Order 2. Alismatales R. Br. ex Bercht. & J. Presl

1. **Araceae**
2. **Tofieldiaceae**
3. **Alismataceae**
4. **Butomaceae**
5. **Hydrocharitaceae**
6. **Scheuchzeriaceae**
7. **Aponogetonaceae**
8. **Juncaginaceae** [*poss. incl.* 9.–14.]
9. **Maundiaceae** [*formerly in* 8.]
10. **Zosteraceae**
11. **Potamogetonaceae**
12. **Posidoniaceae**
13. **Ruppiaceae**
14. **Cymodoceaceae**

Order 3. Petrosaviales Takht.

Petrosaviaceae

Order 4. Dioscoreales R. Br. ex Mart.

1. **Nartheciaceae**
2. **Taccaceae**
3. **Thismiaceae**
4. **Dioscoreaceae** [*poss. incl.* 2.,3.,5.]
5. **Burmanniaceae**

Order 5. Pandanales R. Br. ex Bercht. & J. Presl

1. **Triuridaceae**
2. **Velloziaceae**
3. **Stemonaceae**
4. **Cyclanthaceae**
5. **Pandanaceae**

Order 6. Liliales Perleb (= Liliaceae *s.l.*)

1. **Campynemataceae**
2. **Corsiaceae**
3. **Melanthiaceae**
4. **Petermanniaceae**
5. **Alstroemeriaceae** (incl. Luzuriagaceae)
6. **Colchicaceae**

7. **Philesiaceae**
8. **Ripogonaceae**
9. **Smilacaceae**
10. **Liliaceae**

Order 7. Asparagales Link

1. **Orchidaceae**
2. **Boryaceae**
3. **Blandfordiaceae**
4. **Asteliaceae**
5. **Lanariaceae** [*poss. to be incl. in* 6.]
6. **Hypoxidaceae**
7. **Doryanthaceae** [*poss. to be incl. in* 9.]
8. **Ixioliriaceae** [*poss. to be incl. in* 9.]
9. **Tecophilaeaceae**
10. **Iridaceae**
11. **Xeronemataceae**
12. **Asphodelaceae** (incl. Hemerocallidaceae, Xanthorrhoeaceae)
13. **Amaryllidaceae** (incl. Agapanthaceae, Alliaceae)
14. **Asparagaceae**

Order 8. Arecales Bromhead (incl. Dasypogonales)

1. **Dasypogonaceae**
2. **Palmae** (Arecaceae)

Order 9. Commelinales Mirb. ex Bercht. & J. Presl

1. **Hanguanaceae**
2. **Commelinaceae**
3. **Philydraceae**
4. **Pontederiaceae**
5. **Haemodoraceae**

Order 10. Zingiberales Griseb.

1. **Strelitziaceae**
2. **Lowiaceae**
3. **Heliconiaceae**
4. **Musaceae**
5. **Cannaceae**
6. **Marantaceae**
7. **Costaceae**
8. **Zingiberaceae**

Order 11. Poales Small

1. **Typhaceae** (incl. Sparganiaceae)
2. **Bromeliaceae**
3. **Rapateaceae**
4. **Xyridaceae**
5. **Eriocaulaceae**
6. **Mayacaceae**
7. **Thurniaceae**
8. **Juncaceae**
9. **Cyperaceae**
10. **Restionaceae** (incl. Anarthriaceae, Centrolepidaceae)
11. **Flagellariaceae**
12. **Joinvilleaceae**
13. **Ecdeiocoleaceae**
14. **Gramineae** (Poaceae)

Order Ceratophyllales Link

Ceratophyllaceae [*Corner: 95*]

'EUDICOTS' [*Dicotyledons s.s.*]

307 families arranged in 44 orders

Order 1. Ranunculales Juss. ex Bercht. & J. Presl

1. **Eupteleaceae**
2. **Papaveraceae**

3. **Circaeasteraceae**
4. **Lardizabalaceae**
5. **Menispermaceae**
6. **Berberidaceae**
7. **Ranunculaceae**

Order 2. Proteales Juss. ex Bercht. & J. Presl (incl. Sabiales)

1. **Sabiaceae**
2. **Nelumbonaceae**
3. **Platanaceae**
4. **Proteaceae**

Order 3. Trochodendrales Takht. ex Cronq.

Trochodendraceae

Order 4. Buxales Takht. ex Reveal

Buxaceae (incl. Haptanthaceae)

['Core Eudicots']

Order 5. Gunnerales Takht. ex Reveal

1. **Myrothamnaceae**
2. **Gunneraceae**

Order 6. Dilleniales DC. ex Bercht. & J. Presl

Dilleniaceae

Order 7. Saxifragales Bercht. & J. Presl

1. **Peridiscaceae** (incl. Medusandraceae)
2. **Paeoniaceae**
3. **Hamamelidaceae**
4. **Cercidiphyllaceae**
5. **Daphniphyllaceae**
6. **Iteaceae**
7. **Grossulariaceae**
8. **Saxifragaceae**
9. **Crassulaceae**
10. **Aphanopetalaceae**
11. **Penthoraceae** [*form. in* 12.]
12. **Haloragaceae**
13. **Cynomoriaceae**

[*Rosids*]

Order 8. Vitales Juss. ex Bercht. & J. Presl

Vitaceae

Order 9. Zygophyllales Link

1. **Krameriaceae** [*perhaps to be incl. in* 2.]
2. **Zygophyllaceae**

Order 10. Fabales Bromhead

1. **Quillajaceae**
2. **Leguminosae** (Fabaceae)
3. **Surianaceae**
4. **Polygalaceae**

Order 11. Rosales Bercht. & J. Presl

1. **Rosaceae**
2. **Barbeyaceae**
3. **Dirachmaceae**
4. **Elaeagnaceae**
5. **Rhamnaceae**
6. **Ulmaceae**
7. **Cannabaceae**
8. **Moraceae** [*poss. to be incl. in* 9.]
9. **Urticaceae**

Order 12. Fagales Engl.

1. **Nothofagaceae**
2. **Fagaceae**
3. **Myricaceae**
4. **Juglandaceae** (incl. Rhoipteleaceae)
5. **Casuarinaceae**

6. **Ticodendraceae**
7. **Betulaceae**

Order 13. Cucurbitales Juss. ex Bercht. & J. Presl

1. **Apodanthaceae** [*formerly in Malvales*]
2. **Anisophylleaceae**
3. **Corynocarpaceae**
4. **Coriariaceae**
5. **Cucurbitaceae**
6. **Datiscaceae**
7. **Begoniaceae**

Order 14. Celastrales Link

1. **Lepidobotryaceae**
2. **Celastraceae**

Order 15. Oxalidales Bercht. & J. Presl (incl. Huales)

1. **Huaceae**
2. **Connaraceae**
3. **Oxalidaceae**
4. **Cunoniaceae**
5. **Elaeocarpaceae**
6. **Cephalotaceae**
7. **Brunelliaceae**

Order 16. Malpighiales Juss. ex Bercht. & J. Presl

1. **Pandaceae**
2. **Irvingiaceae**
3. **Ctenolophonaceae**
4. **Rhizophoraceae**
5. **Erythroxylaceae**
6. **Ochnaceae**
7. **Bonnetiaceae**
8. **Guttiferae** (Clusiaceae)
9. **Calophyllaceae** [*formerly in* 8.]
10. **Podostemaceae**
11. **Hypericaceae**
12. **Caryocaraceae**
13. **Lophopyxidaceae**
14. **Putranjivaceae**
15. **Centroplacaceae**
16. **Elatinaceae**
17. **Malpighiaceae**
18. **Balanopaceae**
19. **Trigoniaceae**
20. **Dichapetalaceae**
21. **Euphroniaceae**
22. **Chrysobalanaceae**
23. **Humiriaceae**
24. **Achariaceae**
25. **Violaceae**
26. **Goupiaceae**
27. **Passifloraceae**
28. **Lacistemataceae**
29. **Salicaceae**
30. **Peraceae** [*formerly in* 32.]
31. **Rafflesiaceae**
32. **Euphorbiaceae**
33. **Linaceae**
34. **Ixonanthaceae**
35. **Picrodendraceae**
36. **Phyllanthaceae**

Order 17. Geraniales Juss. ex Bercht. & J. Presl

1. **Geraniaceae**
2. **Francoaceae** [incl Ledocarpaceae, Melianthaceae, Vivianiaceae]

Order 18. Myrtales Juss. ex Bercht. & J. Presl

1. **Combretaceae**
2. **Lythraceae**
3. **Onagraceae**
4. **Vochysiaceae**
5. **Myrtaceae**
6. **Melastomataceae**
7. **Crypteroniaceae**

Order 19. Crossosomatales Takht. ex Reveal

1. **Aphloiaceae**
2. **Geissolomataceae**
3. **Strasburgeriaceae** (incl. Ixerbaceae)
4. **Staphyleaceae**
5. **Guamatelaceae**
6. **Stachyuraceae**
7. **Crossosomataceae**

Order 20. Picramniales Doweld

Picramniaceae

Order 21. Huerteales Doweld

1. **Gerrardinaceae**
2. **Petenaeaceae** [*form. in Malvales*]
3. **Tapisciaceae**
4. **Dipentodontaceae**

Order 22. Sapindales Juss. ex Bercht. & J. Presl

1. **Biebersteiniaceae**
2. **Nitrariaceae**
3. **Kirkiaceae**
4. **Burseraceae**
5. **Anacardiaceae**
6. **Sapindaceae**
7. **Rutaceae**
8. **Simaroubaceae**
9. **Meliaceae**

Order 23. Malvales Juss. ex Bercht. & J. Presl

1. **Cytinaceae**
2. **Muntingiaceae**
3. **Neuradaceae**
4. **Malvaceae**
5. **Sphaerosepalaceae**
6. **Thymelaeaceae**
7. **Bixaceae**
8. **Cistaceae** [*perhaps incl.* 9., 10.]
9. **Sarcolaenaceae**
10. **Dipterocarpaceae**

Order 24. Brassicales Bromhead

1. **Akaniaceae**
2. **Tropaeolaceae**
3. **Moringaceae**
4. **Caricaceae**
5. **Limnanthaceae**
6. **Setchellanthaceae**
7. **Koeberliniaceae**
8. **Bataceae**
9. **Salvadoraceae**
10. **Emblingiaceae**
11. **Tovariaceae**
12. **Pentadiplandraceae**
13. **Gyrostemonaceae** [*perhaps to be incl. in* 14.]
14. **Resedaceae**
15. **Capparaceae**

16. **Cleomaceae** [*perhaps to be incl. in* 17.]
17. **Cruciferae** (Brassicaceae)

[*'SuperAsterids'*]

Order 25. Berberidopsidales Doweld [*formerly in 'Core Eudicots'*)

1. **Aextoxicaceae**
2. **Berberidopsidaceae**

Order 26. Santalales R. Br. ex Bercht. & J. Presl [*formerly in Rosids*]

1. **Erythropalaceae**
2. **Olacaceae** [*prob. to be split up*]
3. **Opiliaceae**
4. **Balanophoraceae**
5. **Santalaceae** [*to be split up*]
6. **Misodendraceae**
7. **Schoepfiaceae**
8. **Loranthaceae**

Order 27. Caryophyllales Juss. ex Bercht. & J. Presl [*formerly in 'Core Eudicots'*] – R: Willd. 45(2015)295

1. **Frankeniaceae**
2. **Tamaricaceae** [*perhaps to be incl. in 1.*]
3. **Plumbaginaceae**
4. **Polygonaceae**
5. **Droseraceae**
6. **Nepenthaceae**
7. **Drosophyllaceae**
8. **Dioncophyllaceae**
9. **Ancistrocladaceae**
10. **Rhabdodendraceae**
11. **Simmondsiaceae**
12. **Physenaceae**
13. **Asteropeiaceae**
14. **Macarthuriaceae** [*formerly in* 20.]
15. **Microteaceae** [*formerly in* 26.]
16. **Caryophyllaceae**
17. **Achatocarpaceae**
18. **Amaranthaceae**
19. **Stegnospermataceae**
20. **Limeaceae**
21. **Lophiocarpaceae** [*formerly in* 26.]
22. **Kewaceae** [*formerly in* 30.]
23. **Barbeuiaceae**
24. **Gisekiaceae** [*formerly in* 28.]
25. **Aizoaceae**
26. **Phytolaccaceae**
27. **Petiveriaceae** [*formerly in 26.*]
28. **Sarcobataceae**
29. **Nyctaginaceae**
30. **Molluginaceae**
31. **Montiaceae** [*formerly in 36.*]
32. **Didiereaceae**
33. **Basellaceae**
34. **Halophytaceae**
35. **Talinaceae** [*formerly incl. in* 36.; *perhaps best incl. in* 38.]
36. **Portulacaceae** [*perhaps best incl. in* 38.]
37. **Anacampserotaceae** [*formerly in* 36.; *perhaps best incl. in* 38.]
38. **Cactaceae**

Order 28. Cornales Link

1. **Nyssaceae**
2. **Hydrostachyaceae**
3. **Hydrangeaceae**
4. **Loasaceae**

5. **Curtisiaceae**
6. **Grubbiaceae**
7. **Cornaceae**

Order 29. Ericales Bercht. & J. Presl

1. **Balsaminaceae**
2. **Marcgraviaceae**
3. **Tetrameristaceae**
4. **Fouquieriaceae** [*perhaps to be incl. in* 5.]
5. **Polemoniaceae**
6. **Lecythidaceae**
7. **Sladeniaceae** [*perhaps to be incl. in* 8.]
8. **Pentaphylacaceae**
9. **Sapotaceae**
10. **Ebenaceae**
11. **Primulaceae**
12. **Theaceae**
13. **Symplocaceae**
14. **Diapensiaceae**
15. **Styracaceae**
16. **Sarraceniaceae**
17. **Roridulaceae**
18. **Actinidiaceae**
19. **Clethraceae**
20. **Cyrillaceae**
21. **Ericaceae**
22. **Mitrastemonaceae**

Order 30. Icacinales Tiegh.

Icacinaceae

Order 31. Metteniusales Takht.

Metteniusaceae (incl. Oncothecaceae)

Order 32. Garryales Mart.

1. **Eucommiaceae**
2. **Garryaceae**

Order 33. Gentianales Juss. ex Bercht. & J. Presl

1. **Rubiaceae**
2. **Gentianaceae**
3. **Loganiaceae**
4. **Gelsemiaceae**
5. **Apocynaceae**

Order 34. Boraginales Juss. ex Bercht. & J. Presl

Boraginaceae (incl. Hoplestigmataceae)

Order 35. Vahliales Doweld

Vahliaceae

Order 36. Solanales Juss. ex Bercht. & J. Presl

1. **Convolvulaceae**
2. **Solanaceae**
3. **Montiniaceae**
4. **Sphenocleaceae**
5. **Hydroleaceae**

Order 37. Lamiales Bromhead

1. **Plocospermataceae**
2. **Carlemanniaceae**
3. **Oleaceae**
4. **Tetrachondraceae**
5. **Calceolariaceae** [*perhaps to be incl. in* 6.]
6. **Gesneriaceae**
7. **Plantaginaceae**
8. **Scrophulariaceae**
9. **Stilbaceae**
10. **Linderniaceae** [*perhaps to be incl. in* 11.]
11. **Byblidaceae**

12. **Martyniaceae**
13. **Pedaliaceae**
14. **Acanthaceae**
15. **Bignoniaceae**
16. **Lentibulariaceae**
17. **Schlegeliaceae**
18. **Thomandersiaceae**
19. **Verbenaceae**
20. **Labiatae** (Lamiaceae)
21. **Mazaceae** [*form. in* 22. *perhaps to be incl. in* 24.]
22. **Phrymaceae** [*perhaps to be incl. in* 24.]
23. **Paulowniaceae** [*perhaps to be incl. in* 24.]
24. **Orobanchaceae**

Order 38. Aquifoliales Senft
1. **Stemonuraceae**
2. **Cardiopteridaceae**
3. **Phyllonomaceae** [*perhaps to be incl. in* 5.]
4. **Helwingiaceae** [*perhaps to be incl. in* 5.]
5. **Aquifoliaceae**

Order 39. Asterales Link
1. **Rousseaceae**
2. **Campanulaceae**
3. **Pentaphragmataceae**
4. **Stylidiaceae**
5. **Alseuosmiaceae**
6. **Phellinaceae**
7. **Argophyllaceae**
8. **Menyanthaceae**
9. **Goodeniaceae**
10. **Calyceraceae**
11. **Compositae** (Asteraceae)

Order 40. Escalloniales Link
Escalloniaceae

Order 41. Bruniales Dumort.
1. **Columelliaceae** (incl. Desfontainiaceae)
2. **Bruniaceae**

Order 42. Paracryphiales Takht. ex Reveal
Paracryphiaceae

Order 43. Dipsacales Juss. ex Bercht. & J. Presl
1. **Viburnaceae** (Adoxaceae)
2. **Caprifoliaceae**

Order 44. Apiales Nakai
1. **Pennantiaceae**
2. **Torricelliaceae**
3. **Griseliniaceae**
4. **Pittosporaceae**
5. **Araliaceae** [*perhaps to be incl. in* 7.]
6. **Myodocarpaceae** [*perhaps to be incl. in* 7.]
7. **Umbelliferae** (Apiaceae)

Acknowledgement of Sources

1. Floras and Handbooks; Websites

Ackerley, J.R. (1940). *Hindoo holiday*. Penguin, Harmondsworth

Adams, C. D. (1972). *Flowering plants of Jamaica*. University of the West Indies, Mona

Agnew, A. D. Q. (2003). *Upland Kenya wild flowers*, ed. 3. Nature Kenya, Nairobi

Alexander, S. (1996). *The cook's companion*. Viking, Ringwood, Victoria

Allan, H. H. (1961). *Flora of New Zealand*. Volume I. Government Printer, Wellington

Allen, D. E. & G. Hatfield (2004). *Medicinal plants in folk tradition: an ethnobotany of Britain and Ireland*. Timber Press, Portland & Cambridge

Amherst, A. (1895). *A history of gardening in England*. Quaritch, London

Andrews, S. *et al.* (eds, 1999). *Taxonomy of cultivated plants*. Royal Botanic Gardens, Kew

Aubréville, A. *et al.* (eds, 1961–). *Flore du Gabon*. Muséum national d'Histoire naturelle, Paris

Aubréville, A. *et al.* (eds, 1963–). *Flore du Cameroun*. Muséum national d'Histoire naturelle, Paris

Aubréville, A. *et al.* (eds, 1967–). *Flore de la Nouvelle-Calédonie et Dépendances*. Muséum national d'Histoire naturelle, Paris

Backer, C. & R.C. Bakhuizen Van denBrink (1963–1968). *Flora of Java*. 3 vols. Noordhoff, Groningen, etc.

Baker, J. G. (1887). *Handbook of the fern-allies*. Bell, London

Baker, R.T. (1919). *The hardwoods of Australia and their economics*. Technological Museum, Sydney [Technological Education Series 23]

Balgooy, M.M.J. van *& al.* (2015). *Spot-characters for the identification of Malesian seed plants. A guide*. Natural History Publications (Borneo), Kota Kinabalu

Barlow, C. (2000). *The ghosts of evolution*. Basic Books, New York

Baumann, H. (1982). *Die griechische Pflanzenwelt in Mythos, Kunst und Literatur*. Hirmer, Munich [trans. W.T. & E.R. Stearn (1993) as *The Greek plant-world in myth, art and literature*. Timber Press, Portland, Oregon]

Beadle, N. C. W. *et al.* (1982). *Flora of the Sydney Region*. Reed, Balgowlah, NSW

Bean, W. J. (1970–1988). *Trees and shrubs hardy in the British Isles*, ed. 8. 4 vols + suppl. Murray, London

Benedix, E.H. *et al.* (1986). *Rudolf Mansfeld's Verzeichnis Landwirtschaftlicher und Gärtnerischer Kulturpflanzen (ohne Zierpflanzen)*. 4 vols. Akademie-Verlag, Berlin

Bentham, G. (1863–1878). *Flora australiensis*. 7 vols. Reeve, London

Bews, J. W. (1925). *Plant forms and their evolution in South Africa*. Longman, London, etc.

Blakeless, J. (1950). *The eyes of discovery*. Lippincott, New York

Blamey, M. & C. Grey-Wilson (1989). *The illustrated flora of Britain and northern Europe*. Hodder, London

Borg, J. (1959). *Cacti*, ed. 3. Blandford, London

Bosser, J. *et al.* (eds, 1976–). *Flore des Mascareignes*. Royal Botanic Gardens, Kew; ORSTOM, Paris, etc.

Bowett, A. (2012). *Woods in British furniture-making 1400–1900*. Oblong Creative & Royal Botanic Gardens Kew, London

Bramwell, D. & Z. I. Bramwell (1974). *Wild flowers of the Canary Islands*. Thornes, London & Burford

Briggs, D. & S. M. Walters (1984). *Plant variation and evolution*, ed. 2. Cambridge University Press

Bureau of Flora and Fauna, etc. (1981–). *Flora of Australia*. Canberra

Burbidge, N. (1963). *Dictionary of Australian plant genera (gymnosperms and angiosperms)*. Angus & Robertson, Sydney

Burbidge, N. T. & M. Gray (1979). *Flora of the Australian Capital Territory*. Australian National University, Canberra

Burkill, I. H. (1935). *A dictionary of the economic products of the Malay Peninsula*. 2 vols. Crown Agents, London

Buurman, P. (1988). *Wayang Golek*. Oxford University Press, Singapore

Castrovejo, S. *et al.* (1986–). *Flora iberica*. CSIC, Madrid

Chih Min, B., O.-H. Kartini & C.L. Ou-Yang (2006). *Garden plants in Singapore*, ed. 2. National Parks Board, Singapore

Chittenden, F. J. (ed., 1951). *Dictionary of gardening*. 4 vols. Clarendon Press, Oxford

Christensen, C. *et al.* (1906–1965). *Index filicum*, with four supplements. Hagerup, Copenhagen, etc.

Church, A. H. (1908). *Types of floral mechanism*. Part I. Clarendon Press, Oxford

Clapham, A. R., T. G. Tutin & E. F. Warburg (1962). *Flora of the British Isles*. Cambridge University Press

Clapham, A. R., T. G. Tutin & E. F. Warburg (1981). *Excursion flora of the British Isles*, ed. 3. Cambridge University Press
Clarke, P.A. (2008). *Aboriginal plant collectors*. Rosenberg, Kenthurst, NSW
Clay, S. (1937). *The present-day rock garden*. Jack, London
Cook, C. D. C. *et al.* (1974). *Water plants of the world*. Junk, The Hague
Cooke, M. C. (1882). *Freaks and marvels of plant life; or, curiosities of vegetation*. SPCK, London
Cooper W. (illust. W.T. Cooper, 2004). *Fruits of the Australian tropical rainforest*. Nokomis, Melbourne
Corner, E.J.H. (1952). *Wayside trees of Malaya*, ed. 2. 2 vols. Government Printer, Singapore; (1988) *idem*, ed. 3. 2 vols. Malayan Nature Society, Kuala Lumpur
Corner, E. J. H. (1964). *The life of plants*. Weidenfeld & Nicholson, London
Corner, E. J. H. (1966). *The natural history of palms*. Weidenfeld & Nicolson, London
Corner, E. J. H. (1976). *The seeds of dicotyledons*. 2 vols. Cambridge University Press
Cowie, I. D. *et al.* (2000). *Floodplain flora: Flora of the coastal plains of the Northern Territory, Australia*. ABRS, Canberra
Cranbrook, Earl of (1988). *Malaysia*. Pergamon, Oxford
Craw, R. C. *et al.* (1999). *Panbiogeography. Tracking the history of life*. Oxford University Press, Oxford, New York, etc.
Crawford, R. M. M. (1989). *Studies in plant survival*. Blackwell, Oxford
Cronk, Q. C. B. (2000). *The endemic flora of St. Helena*. Nelson, Oswestry
Cronquist, A. (1981). *An integrated system of classification of flowering plants*. Columbia University Press, New York
Cubey, J. (ed. in chief, 2016). *RHS plant finder 2016*. Royal Horticultural Society, London
Dalby, A. (2000). *Dangerous tastes. The story of spices*. British Museum Press, London
Dale, I. R. & P. J. Greenway (1961). *Kenya trees and shrubs*. Buchanan's Kenya Estates Ltd
Dallimore, W. & A. B. Jackson (1966). *A handbook of Coniferae and Ginkgoaceae*, ed. 4. Arnold, London
Dassanayake, M. D. (ed., 1980–2006). *A revised handbook to the flora of Ceylon*. 15 vols. Amerind Publishing, New Delhi, etc.
Davey, G. B. & G. Seal (1993). *The Oxford companion to Australian folklore*. Oxford University Press, Melbourne
Davidson, A. (1995). *The Oxford companion to food*. Oxford University Press, Oxford etc.
Davies, R. & S. Ollier (1989). *Allergy. The facts*. Oxford University Press, Oxford etc.
Davis, P.H., etc. (1965–2000). *Flora of Turkey and the East Aegean Islands*. 11 vols, Edinburgh University Press
Desmomd, R. (1995). *Kew*. Harvill, London
Drummond, J. C. & A. Wilbraham (1958). *The Englishman's food*. Revised by D. Hollingsworth. London
Dudareva, N. & E. Pichersky (2006). *Biology of floral scent*. Taylor & Francis, London etc.
Duke, J.A. & al. (2002). *Handbook of medicinal herbs*, ed. 2. CRC Press, Boca Raton, Florida, etc.
Duthie, R. (1988). *Florists' flowers and societies*. Shire, Princes Risborough
Dyer, R. A. (1975–1976). *The genera of southern African flowering plants*. 2 vols. Department of Agricultural Technical Services, Pretoria
Dyer, R. A. *& al.* (1963–). *Flora of southern Africa*. Government Printer, Pretoria
Edgar, E. & H. Connor (2000). *Flora of New Zealand*. Vol. V. Manaaki Whenua Press, Lincoln, NZ
Eggli, U. & H. E. K. Hartmann (eds, 2001–2005). *Illustrated handbook of succulent plants*. 6 vols. Springer, Berlin, etc.
Engler, A. *& al.* (eds, 1900–). *Das Pflanzenreich*. Engelmann, Leipzig. See M.T. Davis (1957) A guide and an analysis of Engler's 'Das Pflanzenreich'. *Taxon* 6: 161–182
Engler, A. *& al.* (eds, 1924). *Die Natürlichen Pflanzenfamilien*, ed. 2. Engelmann, Leipzig, etc.
Exell, A.W. *& al.* (eds, 1960–). *Flora Zambesiaca*. Crown Agents, London
Epple, A. E. (1995). *A field guide to the plants of Arizona*. Globe Pequot, Guilford, Connecticut
Ewan, J. (1969). *A short history of botany in the United States*. Hafner, New York & London
Fahn, A. (1974). *Plant anatomy*, ed. 2. Pergamon Press, Oxford, etc.
Farr, E., J. A. Leussink & F. A. Stafleu (eds, 1979). *Index nominum genericorum (plantarum)*. 3 vols. Bohn, Scheltema & Holkema, Utrecht, etc.
Farr, E., J.A. Leussink & G. Zijlstra (eds, 1986). *Idem*, Supplementum I (see http://rathbun.si.edu/botany/ing)
Farrar, L. (1998). *Ancient Roman gardens*. Sutton, Stroud, Glos, UK
Farrer, R. (1928). *The English rock garden*, 4th impr. 2 vols. Jack, London
Favretti, R.F. & G.F. DeWolf (1972). *Colonial gardens*. Barre Publishers, Barre, Mass.
Fernald, M. L. (1950). *Gray's manual of botany*. ed. 8 (centennial). American Book Co., New York, etc.
Fernando, E. S. *& al.* (2004). *Flowering plants and ferns of Mt. Makiling*. AKECU, Seoul
Fischer, E. & D. Killmann (2008). *Illustrated field guide to the plants of Ngungwe National Park, Rwanda*. University of Kobelenz-Landau
Fisher, C. (2011). *Flowers of the Renaissance*. Frances Lincoln, London
Flora of North America Editorial Committee (1993–). *Flora of North America north of Mexico*. Oxford University Press, New York and Oxford
Florence, J. (1997–). *Flore de la Polynésie française*. ORSTOM, Paris
Folkard, R. (1892). *Plant lore, legends, and lyric*, ed. 2. Sampson Low, London
Forey, P. L. (1981). *The evolving biosphere*. British Museum (Natural History), London
Francis, S.A. (2009). *British field crops*, ed. 2. Francis, Bury St. Edmunds

Fraser, M. & L. Fraser (2011). *The smallest kingdom*. Kew Publishing, Richmond, Surrey, UK
Frodin, D.G. (1984). *Guide to the standard floras of the world*. Cambridge University Press
Funch, Ligia Silveira & al. (2004). *Plantas úteis – Chapada Diamantina*. Rima, São Carlos, Brazil
Gardner, C.M. & B. G. Gardner (2014). *Flora of the Silk Road*. Tauris, London & New York
Gates, W. (1939). *An Aztec herbal*. Maga Society, Baltimore
Glare, P. G. W. (ed., 1982). *Oxford Latin dictionary*. Clarendon Press, Oxford
Glime, J. M. & D. Saxena (1991). *Uses of bryophytes*. Today & Tomorrow's Printers & Publishers, New Delhi
Godwin, H. (1975). *The history of the British flora*, ed. 2. Cambridge University Press
Good, R. (1974). *The geography of the flowering plants*, ed. 4. Longman, London
Goody, J. (1993). *The culture of flowers*. Cambridge University Press, Cambridge
Grey-Wilson, C. & P. Cribb (2011). *Guide to the flowers of western China*. Kew Publishing, Richmond, Surrey, UK
Grierson, A. J. C. *& al*. (1983–2000). *Flora of Bhutan*. Royal Botanic Garden Edinburgh
Grigson, G. (1958). *The Englishman's flora*. Phoenix House, London
Grimshaw, K. & R. Bayton (2009). *New trees*. Royal Botanic Gardens, Kew, Richmond, Surrey, UK
Grove, A.T. & O. Rackham (2001). *The nature of Mediterranean Europe: an ecological history*. Yale University Press, New Haven & London
Hallé, F. (2002). *In praise of plants*. Timber Press, Portland, Oregon
Hallé, F. (2004). *Architectures de plantes*. The author, Montpellier
Hallé, F., R. A. A. Oldeman & P. B. Tomlinson (1978). *Tropical trees and forests: an architectural analysis*. Springer, Berlin, etc.
Hammond, N. G. L. & H. H. Scullard (1970). *The Oxford classical dictionary*. Ed. 2, Clarendon Press, Oxford
Hara, H., W. T. Stearn, L. H. J. Williams & A. O. Chater (1978–1982). *Enumeration of the flowering plants of Nepal*. 3 vols. British Museum (Natural History), London
Harborne, J. B. (1988). *Introduction to ecological biochemistry*, ed. 3. Academic Press, London
Harborne, J. B. & F. A. Tomas-Barberan (1991). *Ecological chemistry and biochemistry of plant terpenoids*. Clarendon Press, Oxford
Harden, G. J. (ed., 1990–1993). *Flora of New South Wales*. 4 vols [ed. 2, vols 1 & 2, 2000–2002]. NSW University Press
Harden, G., B. McDonald & J. Williams (2006). *Rainforest trees and shrubs*. Gwen Harden Publishing, Nambucca Heads, New South Wales
Harling, G. & B. Sparre (eds, 1973–). *Flora of Ecuador. Opera botanica* B. Lund
Harrison, S.G., G.B. Masefield & M. Wallis (1969). *The Oxford book of food plants*. Oxford University Press
Harvey, J. (1981). *Mediaeval gardens*. Batsford, London
Haslam, S. M., P. D. Sell & P. A. Wolseley (1977). *A flora of the Maltese islands*. Malta University Press
Hawthorne, W. (1990). *Field guide to the forest trees of Ghana*. ODA, London
Heath, A. & R. Heath (2009). *Field Guide to the plants of Northern Botswana including the Okavango Delta*. Kew Publishing, Richmond, Surrey, UK
Henley, A. J. & E. Edgar (1980). *Flora of New Zealand*. Vol. III. Govt. Printer, Wellington, NZ
Hepper, F. N. (1990). *Pharaoh's flowers. The botanical treasures of Tutankhamun*. HMSO, London
Heywood, V. H. (ed., 1978). *Flowering plants of the world*. Oxford University Press
Heywood, V. H. & S. R. Chant (1982). *Popular encyclopedia of plants*. Cambridge University Press
Hirono, I. (ed., 1987). *Naturally occurring carcinogens of plant origin*. Elsevier, Tokyo etc.
Holman, A. (1996). *Plants in medicine*. Chelsea Physic Garden, London
Holm, L. G. *& al*. (1979). *The world's worst weeds*. Honolulu
Hooker, J. D. (ed., 1872–1897). *The flora of British India*. 7 vols. Reeve, London
[Hough, R.B. & al.] (2007). *The woodbook*. Taschen, Köln
Hubbard, C. E. (1968). *Grasses*, ed. 2. Penguin, Harmondsworth
Humbert, H. *& al*. (eds, 1936–). *Flore de Madagascar et des Comores (Plantes vasculaires)*. Muséum national d'Histoire naturelle, Paris
Hunt, T. (1989). *Plant names of mediaeval England*. Brewer, Cambridge
Hutchinson, J. (1964–1967). *The genera of flowering plants*. 2 vols. Clarendon Press, Oxford
Hutchinson, J. & J. M. Dalziel (1954–1972). *Flora of west tropical Africa*, ed. 2, revised by R.W.J. Keay & F.N. Hepper. 3 vols with *The ferns and fern allies* (1959, by A.H.G. Alston) as supplement. Crown Agents, London
Huxley, A. (1967). *Mountain flowers*. Blandford, London
Index Nominum Genericorum. http://botany.si.edu/ing/
Index Nominum Supragenericorum Plantarum Vascularium. http://www.plantsystematics.org/reveal/pbio/fam/allspgnames.html
International Plant Name Index. www.ipni.org
Jackson, B. D. *& al*. (1895–). *Index kewensis plantarum phanerogamarum*. Clarendon Press, Oxford
Jacobs, R. (2009). *Trees of Belgium. Revisited*. BAI, Belgium
Jacobsen, H. (1960). *A handbook of succulent plants*. 3 vols. Blandford, London
Janzen, D. H. (ed., 1983). *Costa Rican natural history*. University of Chicago Press
Jarvis, C. (2007). *Order out of chaos*. Linnean Society, London
Jex-Blake, A. J. (1950). *Gardening in East Africa*, ed. 3. Longman, London
Johnson, O. (2011). *Champion trees of Britain and Ireland*. Kew Publishing, Richmond, Surrey, UK
Juniper, B. E., R. J. Robins & D. M. Joel (1989). *The carnivorous plants*. Academic Press, London

Acknowledgement of Sources

Kennedy, D.O. (2014). *Plants and the human brain*. Oxford University Press, Oxford

Kennett, D.J. & B. Winterhalder (eds, 2006). *Behavioural ecology and the transition to agriculture*. University of California Press, Berkeley, etc.

Kerner von Marilaun, A. (1904). *The natural history of plants*, trans. F. W. Oliver. 2 vols. Gresham, London

Kesseler, R. & W. Stuppy (2006). *Seeds. Time capsules of life*. Papadakis, London

Kiew, R. & al. (2010–). *Flora of Peninsular Malaysia*. Forest Research Institute Malaysia, Kepong, Malaysia

Kingdon-Ward, F. (1960). *Pilgrimage for plants*. Harrap, London etc.

Komarov, V. L. (ed., 1968). *Flora of the U.S.S.R.*, translated from the Russian. Israel Program for Scientific Translations, Jerusalem

Kruckeberg, A.R. (1982). *Gardening with native plants of the Pacific Northwest*. University of Washington Press, Seattle

Kubitzki, K. (ed., 1990–). *The families and genera of vascular plants*. Springer, Heidelberg etc.

Large, M. F. & J. E. Braggins (2004). *Tree ferns*. Timber Press, Portland, Oregon & Cambridge, UK

Lee, D. (2007). *Nature's palette. The science of plant color*. Chicago University Press, Chicago & London

Liberty Hyde Bailey Hortorium (1976). *Hortus third. A concise dictionary of plants cultivated in the United States and Canada*. Macmillan, New York, etc.

Lock, S. *& al.* (2001). *The Oxford illustrated companion to medicine*. Oxford University Press, Oxford

Lord, T. (chief consultant, 2003). *Flora*. 2 vols. Cassell, London (also in Aus. as *ABC Gardening Australia Flora*, 1 vol.)

Lucas, G. & H. Synge (1978). *The IUCN Red Data Book*. IUCN, Morges

Mabberley, D. J. (ed., 1981). *Revolutionary botany: Thalassiophyta and other essays of A.H. Church*. Clarendon, Oxford

Mabberley, D. J. (1991, '1992'). *Tropical rain forest ecology*, ed. 2. Blackie, Glasgow & Chapman & Hall, New York

Mabberley, D. J. (1995). *Plants and prejudice*. Rijksuniversiteit Leiden, Leiden

Mabberley, D. J. (1999, '1998'). *Paradisus. Hawaiian plant illustrations by Geraldine King Tam*. Honolulu Academy of Arts, Honolulu

Mabberley, D. J. (2000). *Arthur Harry Church: the anatomy of flowers*. Merrell, London

Mabberley, D. J. & P. J. Placito (1993). *Algarve plants and landscape*. Oxford University Press, Oxford

Mabey, R. (1996). *Flora britannica*. Sinclair-Stevenson, London

McFarlane, D. (ed., 1988). *The Guiness Book of Records*. Guiness, Enfield

MacGregor, N. (2010). *A history of the world in 100 objects*. Allen Lane, London

Macmillan, H. F. (1935). *Tropical planting and gardening*, ed. 4. Macmillan, London

Macmillan, H.F. (1991). *Idem*, ed. 6 (revised by H.S. Barlow, I. Enoch & R.A. Russell). Malayan Nature Society, Kuala Lumpur

McNeill, J. & al. (eds, 2012). *International code of nomenclature for algae, fungi and plants* (Melbourne Code) [*Regnum vegetabile* 154]. Koeltz, Oberreifenberg

Mahady, G.B., H.F.S. Fong & N. Farnsworth (2001). *Botanical dietary supplements: quality, safety and efficiency*. Swets & Zeitlinger, Lisse, etc., The Netherlands

Manning, J. (2007). *Field guide to the fynbos*. Struik, Cape Town

Marticorena, C. & R. Rodriguez (eds, 1995–). *Flora de Chile*. Universidad de Concepción, Concepción

Matthews, J. R. (1955). *Origin and distribution of the British flora*. Hutchinson, London

Mayer, H.E. (1988). *The Crusades*, ed. 2. Oxford University Press, Oxford

Meiggs, R. (1982). *Trees and timbers in the ancient Mediterranean world*. Clarendon Press, Oxford

Van Der Meijden, R. (2005). *Heukels' Flora van Nederland*, 23 ed. Wolters-Noordhoff, Groningen/Houten, Netherlands

Meikle, R. D. (1977–1985). *Flora of Cyprus*. 2 vols. Bentham-Moxon Trust, Kew

Menninger, E. A. (1967). *Fantastic trees*. Viking, New York

Menninger, E. A. (1977). *Edible nuts of the world*. Horticultural Books, Stuart, Florida

Metcalfe, C. R. & L. Chalk (1950). *Anatomy of the Dicotyledons*. 2 vols. Clarendon Press, Oxford; ed. 2 (1979–), Clarendon Press, Oxford

Miller, A. & T.A. Cope (1996–). *Flora of the Arabian Peninsula and Socotra*. Edinburgh University Press, Edinburgh

Miller, A. G. & M. Morris (1988). *Plants of Dhofar*. Office of the Advisor for Conservation of the Environment, Oman

Miller, A. G. & M. Morris (2004). *Ethnoflora of the Soqotra archipelago*. Royal Botanic Garden, Edinburgh

Miodownik, M. (2014). *Stuff matters*. Penguin, London

Mitchell, A. (1974). *A field guide to the trees of Britain and northern Europe*. Collins, London

Moerman, D. E. (1998). *Native American ethnobotany*. Timber Press, Portland, Oregon

Moore, D. M. (1983). *Flora of Tierra del Fuego*. Nelson, Oswestry

Moore, L. B. & E. Edgar (1970). *Flora of New Zealand*. Volume II. Government Printer, Wellington

Morley, B. D. & H. R. Toelken (eds, 1983). *Flowering plants in Australia*. Rigby, Adelaide, etc.

Morran, R. C. (2004). *A natural history of ferns*. Timber Press, Portland & Cambridge

Munz, P. A. (1959). *A California flora*. University of California Press

Musselman, L.J. (2007). *Figs, dates, laurel and myrrh. Plants of the Bible and the Quran*. Timber Press, Portland, Oregon

Needham, J. (1986). *Science and civilization in China* 6(1) Botany. Cambridge University Press

Ng, F.S.P. (2006). *Tropical horticulture and gardening*. Clearwater Publications, Kuala Lumpur

Noltie, H. J. (2002). *The Dapuri drawings*. Antique Collectors' Club, London & Royal Botanic Garden, Edinburgh
Oakeley, H. (2009). *A year in the medicinal garden of the Royal College of Physicians*. Royal College of Physicians, London
Ohwi, J. (1965). *Flora of Japan*, ed. by F.G. Meyer & E.H. Walker. Smithsonian Institution, Washington, D.C.
Orchard, A. E. (Exec. ed., 1999–). *Species plantarum: Flora of the world*. ABRS, Canberra
Organisation for Flora Neotropica (1968–). *Flora neotropica*. Hafner, New York, etc.
Ozenda, P. (1977). *Flore du Sahara*, ed. 2. CNRS, Paris
Page, M. & W. T. Stearn (1974). *Culinary herbs*. Royal Horticultural Society, London
Page, R. (1983). *The education of a gardener*, new ed. Collins, London etc.
Pearce, N. R. & P. J. Cribb (2002). *The orchids of Bhutan*. Royal Botanic Garden Edinburgh & Royal Govt of Bhutan
Peekel, P. G. (1984). *Flora of the Bismarck Archipelago for naturalists*. Trans. E.E. Henty. Office of Forests, Lae
Pennington, T. D. *& al.* (2004). *Illustrated guide to the trees of Peru*. Hunt, Sherborne, UK
Pericin, C. (2001). *Fiori e piante dell'Istria distribuiti per ambiente* [Collana degli Atti Centro di Ricerche Storiche – Rovigno extra ser. n. 3]. Unione Italiana etc., Trieste
Perry, F. (1972). *Collins' guide to border plants*, ed. 3. Collins, London
Pickering, H. & E. Roe (2009). *Wild flowers of the Victoria Falls area*. Privately published
Plant Resources of South-East Asia [PROSEA] (1989–). Pudoc, Wageningen; etc.
Pojar, J. & A. MacKinnon (compilers & eds, 2004). *Plants of Pacific Northwest Coast*. Revised ed. Lone Pine, Vancouver, Canada
Polunin, N. (1959). *Circumpolar Arctic flora*. Clarendon Press, Oxford
Polunin, O. & A. Stainton (1984). *Flowers of the Himalaya*. Oxford University Press
Prance, G. T. & T. E. Lovejoy (eds, 1984). *Amazonia*. Pergamon, Oxford
Pratt, A. (1899–1905). *Flowering plants, grasses, sedges and ferns of Great Britain*. New ed. by E. Step. 4 vols. Warne, London
Preston, C. D. & J. M. Croft (1997). *Aquatic plants in Britain and Ireland*. Harley, Colchester
Proctor, M. & P. Yeo (1973). *The pollination of flowers*. Collins, London
Purseglove, J. (1968–1972). *Tropical crops*. 4 vols. Longman, London
Putz, F. E. & H. A. Mooney (1991). *The biology of vines*. Cambridge University Press
Quezel, P. & S. Santa (1962–1963). *Nouvelle flore de l'Algérie et des régions désertiques meridionales*. CNRS, Paris
Rackham, O. (1986). *History of the British countryside*. Dent, London
Rätsch, C. (2004). *The encyclopedia of psychoactive plants*. Park St. Press, Rochester, Vermont
Ramson, W. S. (ed., 1988). *The Australian national dictionary*. Oxford University Press, Melbourne
Ranson, F. (1949). *British herbs*. Penguin, Harmondsworth
Rattauf, R. F. (1970). *A handbook of alkaloids and alkaloid-containing plants*. Wiley, New York, etc.
Rechinger, K. H. (ed., 1963–). *Flora Iranica*. Akademische Druck-and Verlagsanstalt, Graz
Reynolds, J. & J. Tampion (1983). *Double flowers*. Pembridge, London, & Van Nostrand Reinhold, New York
Richards, P. W. (1996). *The tropical rain forest. An ecological study*. ed. 2. Cambridge University Press, Cambridge
Ridley, H. N. (1922–1925). *The flora of the Malay Peninsula*. 5 vols. Reeve, London
Ridley, H. N. (1930). *The dispersal of plants throughout the world*. Reeve, Ashford
Roach, F.A. (1985). *Cultivated fruits of Britain*. Blackwell, Oxford & New York
Robins, J. (1996). *Wild lime*. Allen & Unwin, St. Leonards, NSW
[Ruskin, F.R.] (ed., 1975). *Underexploited tropical plants with promising economic value*. National Academy of Sciences, Washington, D.C.
Sainty, G.R. & S.W.L. Jacobs (2003). *Waterplants in Australia*, expanded ed. 4. Sainty, Potts Point, NSW
Schatz, G. E. (2001). *Generic tree Flora of Madagascar*. Royal Botanic Gardens, Kew, and Missouri Botanical Garden, St Louis
Schultes, R.E. & R.F. Rattauf (1990). *The healing forest*. Dioscorides, Portland
Sculthorpe, C.D. (1967). *The biology of aquatic vascular plants*. Arnold, London
Sharma, B.D. & al. (eds, 1993–), *Flora of India*. Botanical Survey of India, Calcutta
Shulkina, T. (2004). *Ornamental plants from Russia and adjacent states of the former Soviet Union*. St. Petersburg
Simmonds, N. W. (ed., 1976). *Evolution of crop plants*. Longman, Harlow
Simpson, J. A. & E. S. C. Weiner (preps, 1989). *The Oxford English Dictionary*, ed. 2. Clarendon Press, Oxford
Smith, A. C. (1979–1997). *Flora vitiensis nova*. National Tropical Botanical Garden, Lawai
Smith, D. C. & A. E. Douglas (1987). *The biology of symbiosis*. Arnold, London
Soepadmo, E. & K. M. Wong (eds, 1995–). *Tree Flora of Sabah and Sarawak*. FRIM etc., Kuala Lumpur
Spencer, R. (ed., 1995–2005). *Horticultural Flora of south-eastern Australia*. 5 vols. UNSW Press, Sydney
Sporne, K. R. (1974). *The morphology of gymnosperms*, ed. 2. Hutchinson, London
Sporne, K. R. (1975). *The morphology of pteridophytes*, ed. 4. Hutchinson, London
Stace, C. (2010). *New flora of the British Isles*, ed. 3. Cambridge University Press
Stainton, A. (1988). *Flowers of the Himalaya. A supplement*. Oxford University Press
Staples, G. & D. Herbst (2005). *A tropical garden Flora*. Bishop Museum, Honolulu
Staples, G. W. & M. S. Kristiansen (1999). *Ethnic culinary herbs*. University of Hawaii Press, Honolulu

van Steenis, C.G.G.J. (etc., eds, 1948–). *Flora malesiana*. Noordhoff-Kolff, Jakarta, etc.
van Steenis, C.G.G.J. (1972). *The mountain flora of Java*. Brill, Leiden
van Steenis, C.G.G.J. (1981). *Rheophytes of the world*. Sijthoff & Noordhoff, Alphen aan den Rijn
Stevens, P.F. (2001 onwards). *Angiosperm phylogeny website*. www.mobot.org/MOBOT/research/APweb/
Stones, M. & L. Urbatsch (1991). *Flora of Louisiana*. Louisiana State University Press, Baton Rouge
Strid, A. (ed., 1986–1991). *Mountain Flora of Greece*. Vol. 1, Cambridge University Press; vol. 2 (ed. with Kit Tan), Edinburgh University Press
Sudworth, G. B. (1908). *Forest trees of the Pacific slope*. US Department of Agriculture, Washington, D.C.
Synge, P. M. (1969). *Supplement to the dictionary of gardening*. Clarendon Press, Oxford
Tang, S. & M. Palmer (1986). *Chinese herbal prescriptions*. Rider, London
Tannahill, R. (1988). *Food in history*. New ed. Penguin Books, London
Tansley, A. G. (1939). *The British Islands and their vegetation*. Cambridge University Press
The Plant List http://www.theplantlist.org/
Thulin, M. (ed., 1993–2006). *Flora of Somalia*. Royal Botanic Gardens, Kew
Titley, N. & F. Wood (1991). *Oriental gardens*. British Library, London
Tomlinson, P. B. (1986). *The botany of mangroves*. Cambridge University Press
Tomlinson, P. B. & M. H. Zimmermann (eds, 1978). *Tropical trees as living systems*. Cambridge University Press
Turrill, W. B. *& al.* (eds, 1952–). *Flora of tropical East Africa*. Crown Agents, London
Tutin, T. G., V. H. Heywood *& al.* (1964–1980). *Flora europaea*. 5 vols. Ed. 2 (1993–). Cambridge University Press
Uphof, J. C. T. (1968). *Dictionary of economic plants*, ed. 2. Cramer, Lehre
Valder, P. (1999). *The garden plants of China*. Florilegium, Balmain, NSW
Valder, P. (2002). *Gardens in China*. Florilegium, Glebe, New South Wales
Valdés, B. *& al.* (1987). *Flora vascular de Andalucia Occidental*. 3 vols. Barcelona
Verdcourt, B. & E. C. Trump (1969). *Common poisonous plants of East Africa*. Collins, London
Vickery, R. (1995). *A dictionary of plant lore*. Oxford University Press, Oxford
Vilayleck, B. & B. Strobel (2012). *Les fleurs de la dévotion*. Pha Tad Ke Botanical Garden, Luang Prabang, Laos
Wagner, W. L. *& al.* (1990). *Manual of the flowering plants of Hawaii*. 2 vols. University of Hawaii Press
Walter, A. & C. Sam (1999). *Fruits d'Océanie*. Institut de Recherche pour le Développement, Paris
Walters, S. M. *& al.* (eds, 1984–2000). *The European garden flora*. 6 vols. Cambridge University Press
Watson, E.V. (1967). *The structure and life of bryophytes*, ed. 2. Hutchinson, London
Watson, E.V. (1981). *British mosses and liverworts*, ed. 3. Cambridge University Press
Watt, J. M. & M. G. Breyer-Brandwijk (1962). *The medicinal and poisonous plants of Southern and Eastern Africa*, ed. 2. Livingstone, Edinburgh and London
Webb, C. J. *& al.* (1988). *Flora of New Zealand*. Vol. IV. DSIR, Christchurch
Weber, E. (2003). *Invasive plant species of the world. A reference guide to environmental weeds*. CABI, Wallingford, UK
White, F. (1983). *The vegetation of Africa*. UNESCO, Paris
Whitmore, T. C. (1966). *Guide to the forests of the British Solomon Islands*. Oxford University Press, Oxford
Whitmore, T. C. & F. S. P. Ng (eds, 1972–1989). *Tree flora of Malaya*. 4 vols. Longman Malaysia, Petaling Jaya
Wiggins, I. R. & D. L. Porter (1971). *Flora of the Galápagos Islands*. Stanford University Press
Williams, R. O. (1949). *Useful and ornamental plants of Zanzibar and Pemba*. Zanzibar Protectorate
Willis, J. C. (1931). *A dictionary of the flowering plants and ferns*, ed. 6. Cambridge University Press
Willis, J. C. (1973). *A dictionary of the flowering plants and ferns*, ed. 8, revised by H.K. Airy Shaw. Cambridge University Press
Willmer, P. (2011). *Pollination and floral ecology*. Princeton University Press
Wrigley, J.W. & M. Fagg (2003). *Australian native plants*, ed. 5. New Holland, Frenchs Forest, NSW
Wu, Z. Y. & P. H. Raven (1994–2013). *Flora of China*. Science Press, Beijing, & Missouri Botanical Garden, St Louis
Wunderlin, R. P. (1998). *Guide to the vascular plants of Florida*. University Press of Florida, Gainesville
Zohary, D. & M. Hopf (1988, 1993). *Domestication of plants in the Old World*, eds 1 & 2. Clarendon Press, Oxford
Zohary, M. (1982). *Plants of the Bible*. Cambridge University Press

2. Periodicals

Note: Increasingly, the contents of these periodicals are available on the Internet and in many cases, recent journals and recent issues of established ones are only available thus.

Abhandlungen der Akademie der Wissenschaften und der Literatur Mainz:
Mathematisch-Naturwissenschaftliche Klasse: tropische und subtropische Pflanzenwelt
Abhandlungen der zoologisch-botanischen Gesellschaft in Österreich
Acanthus
ACIAR technical Reports
Acta agrobotanica
Acta amazonica

Acta biologica Colombiana
Acta biologica Cracoviensia Series Botanica
Acta biologica Paranaense, formerly Boletim Universidade do Paraná Botânica
Acta biologica Venezuelica
Acta botanica Academiae Scientiarum Hungaricae, later Acta botanica Hungarica
Acta botanica Austro Sinica
Acta botanica Barcinonensia, formerly Acta phytotaxonomica Barcinonensia
Acta botanica Brasilica
Acta botanica Croatica
Acta botanica Cubana
Acta botanica Fennica
Acta botanica Gallica, formerly Bulletin de la Société botanique de France, continued as Botany Letters
Acta botanica Hungarica, formerly Acta botanica Academiae Scientarum Hungaricae
Acta botanica Islandica
Acta botanica Malacitana
Acta botanica Mexicana
Acta botanica Neerlandica, later merged in Plant Biology
Acta botanica Sinica
Acta botanica Slovaca Academiae Scientiarum Slovacae, Ser A. Taxonomica Geobotanica
Acta botanica Venezuelica
Acta botanica Yunnanica, later Plant Diversity and Resources
Acta cientifica Potosina
Acta Facultatis rerum naturalium Universitatis Comenianae (Bratislava) Series Botanica; Series Physiologia Plantarum
Acta geobotanica Barcinonensia, later Acta botanica Barcinonensia
Acta Horti Bergiani
Acta Manilana
Acta Musei Reginaehradecensis, A
Acta phytogeographica Suecica
Acta phytotaxonomica Barcinonensia, continued as Acta botanica Barcinonensia
Acta phytotaxonomica et geobotanica
Acta phytotaxonomica Sinica, later Journal of Systematics and Evolution
Acta Reginae Societatis Scientarum et Litterarum Gothoburgensis, Botanica
Acta Societatis Botanicorum Poloniae
Acta Universitatis Carolinae Biologica Prague
Adansonia, formerly Notulae systematicae, later Bulletin du Muséum national d'Histoire naturelle. Section B, Adansonia
Advances in Botanical Research
Advances in Ecological Research
Advances in Economic Botany
Advances in Genetics
Agricultural University, Wageningen. Papers, formerly Mededelingen van de Landbouwhogeschool te Wageningen, later Wageningen Agricultural University Papers
Agrobotanika
Agronomia lusitana
Albertoa
Aliso
Allertonia
Allionia
Amazoniana
Ambio
American Fern Journal
American Journal of Botany
American Midland Naturalist
American Naturalist
Anales de la Escuela Nacional de Ciencias biológicas
Anales del Instituto de Biologia, ser. Botánica. Universidad Nacional autónoma de México
Anales del Instituto botánico A.J. Cavanilles
Anales del Jardín botánico de Madrid
Anales del Museo de Historia natural de Valparaiso
Anales de la Sociedad cientifica Argentina
Anales de la Sociedad Mexicana de Historia de la Ciencia de la Teconologia
Annalen des (K.K.) naturhistorischen Museum (Hofmuseums) in Wien
Annales Bogorienses
Annales botanici Fennici, formerly Annales botanici Societatis zoologicae-botanicae fennicae Vanamo
Annales botanici Societatis zoologicae-botanicae fennicae Vanamo, continued as Annales botanici Fennici
Annales de la Faculté des Sciences, Université de Dakar

Acknowledgement of Sources

Annales Musei Goulandris
Annales de Physiologie Végétale, Bruxelles
Annales des Sciences naturelles (Paris)
Annales scientifiques de L'Université de Besançon
Annali di Botanica (Roma)
Annals of Applied Biology
Annals of Botany
Annals Kirstenbosch Botanical Garden, formerly Journal of South African Botany, supplement
Annals of the Missouri Botanical Garden
Annals of the Tsukuba Botanical Garden
Annonaceae Newsletter
Annual Report Huntingdonshire Fauna and Flora Society
Annual Review of Biochemistry
Annual Review of Ecology and Systematics
Annual Review of Genetics
Annual Review of Plant Physiology
Aqua Planta
Aquarium
Aquatic Botany
Aquilo, Seria botanica
Arab Gulf Journal of Scientific Research
Arboretum Kórnickie
Arboretum Leaves
Arboricultural Journal
Archives of Natural History, formerly Journal of the Society for the Bibliography of Natural History
Archivio botânico e biogeograpfico Italiano
Arena
Arkiv för Botanik
Arnaldoa
Aroideana
Arquivos do Jardim botânico do Rio de Janeiro
Asclepiadaceae, continued as Asklepios
Ashingtonia
Asklepios, formerly Asclepiadaceae
ASPT Newsletter
Astarte
Atas de Sociedade Botânica do Brasil, Secção Rio de Janeiro
Atoll Research Bulletin
Atti della Società Italiana di Scienze naturali e del Museo civico di Storia naturale di/in Milano
Atti dell'Istituto botânico della Università e Laboratorio crittogamico di Pavia
Austral Ecology, formerly Australian Journal of Ecology
Australian Acacias
Australian Journal of Agricultural Research
Australian Journal of Botany
Australian Journal of Ecology, later Austral Ecology
Australian Orchid Research
Australian Systematic Botany, formerly Brunonia
Australian [later Australasian] Systematic Botany Society, Newsletter
Austrobaileya, formerly Contributions from the Queensland Herbarium
Baileya
Balduinia
Bamboo Science and Culture
Bangladesh Journal of Botany
Bangladesh Journal of Plant Taxonomy
Bartonia
Bauhinia
Bean Bag
Beiträge zur Biologie der Pflanzen
Belgian Journal of Botany, formerly Bulletin de la Société Royale de Botanique de Belgique, later Plant Ecology and Evolution
Belmontia
Berichte der Deutschen botanischen Gesellschaft, continued as Botanica Acta
Berichte der Schweizerischen botanischen Gesellschaft, continued as Botanica Helvetica
Bibliotheca botanica
Biblioteca José Jerónimo Triana
Biochemical Systematics and Ecology
BioLlania
Biologia

Biologia Plantarum
Biological Conservation
Biological Invasions
Biological Journal of the Linnean Society of London, formerly Proceedings of the Linnean Society
Biological Reviews
Biological Sciences Review
Biologiske Schrifter
Biology Letters
Biota
Biotica
Biotropia
Biotropica
Bishop Museum Bulletins in Botany
Bishop Museum Occasional Papers, formerly Occasional Papers of the Bernice P. Bishop Museum
Blancoana
Blumea
Blyttia
Bocconea
Boissiera
Boletim de Botânica (da Universidade de São Paulo)
Boletim do Instituto de Botânica
Boletim Museu botânico Municipal. Curitiba
Boletim do Museu Botânico Kuhlmann
Boletim do Museu municipal do Funchal
Boletim do Museu nacional de Rio de Janerio. Botânica
Boletim do Museu Paraense 'Emílio Goedi' New Ser.
Boletim da Sociedade Broteriana
Boletim técnico Instituto agronomico do Norte
Boletim Universidade do Paraná Botânica, continued as Acta biológica Paranaense
Boletín de los Jardines Botánicos de América Latina
Boletín Museo nacional de Historia natural, Chile
Boletín de la Sociedad Argentina de Botánica
Boletín de la Sociedad botánica de México, later Botanical Sciences
Bollettino, Museo Regionale di Scienze naturali
Bonplandia
Botanica Acta, formerly Berichte der Deutschen botanischen Gesellschaft, later merged in Plant Biology
Botanica Complutensis, formerly Trabajos del Departamento de Botanica y Fisiologia Vegetal, Universidad de Madrid
Botanica Helvetica, formerly Berichte der Schweizerischen botanischen Gesellschaft
Botanica Lithuanica
Botanica Macaronesica
Botanica Serbica
Botanical Bulletin of Academia Sinica
Botanical Gazette, later International Journal of Plant Sciences
Botanical Journal of the Linnean Society, formerly Journal of the Linnean Society, Botany
Botanical Journal of Scotland, formerly Transactions of the Botanical Society of Edinburgh
Botanical Journal of South China
Botanical Magazine
Botanical Magazine (Tokyo), continued as Journal of Plant Research
Botanical Museum Leaflets. Harvard University
Botanical Review
Botanical Sciences, formerly Boletín de la Sociedad botánica de México
Botanical Society of the British Isles News
Botaničeskie Materialȳ Gerbarija Instituta, botaniki Akademii nauk Uzbekskoj SSR
Botanicheskii Zhurnal
Botanika Chronika
Botanische Jahrbücher für Systematik, Pflanzengeschichte und Pflanzengeographie, continued as Plant Diversity and Evolution
Botaniska Notiser
Botanisk Tidsskrift
Botany Bulletin, Department of Forests (Papua New Guinea)
Botany Letters, formerly Acta botanica Gallica
Bothalia
Bradea
Bradleya
Brazilian Journal of Botany, formerly Revista Brasileira de Botânica
Breeding Science

Acknowledgement of Sources

Brenesia
British Archaeology
British Cactus and Succulent Journal, formerly National Cactus and Succulent Journal
British Fern Gazette, continued as Fern Gazette
British Medical Journal
British Museum technical Bulletin
Brittonia
Bromélia
Brunonia, formerly Contributions from Herbarium Australiense, later Australian Systematic Botany
Bryologist
Buletin Kebun Raya
Bulletin Auckland Institute and Museum
Bulletin of Botanical Laboratory [Research] of North-Eastern Forestry Institute (China), continued as Bulletin of Botanical Research
Bulletin of Botanical Research, formerly Bulletin of Botanical Laboratory [Research] of North-Eastern Forestry Institute (China)
Bulletin of the Botanical Society of Bengal, later Journal of the National Botanical Society
Bulletin of the Botanical Survey of India, later Nelumbo
Bulletin of the British Museum (Natural History) (Series E). Botany, later Bulletin of the Natural History Museum. Botany Series
Bulletin. Fairchild Tropical Garden
Bulletin of the Forest Research Institute Chittagong, Plant Taxonomy Series
Bulletin de l'Institut fondamental d'Afrique noire. Sér. A
Bulletin of the International Group for the Study of Mimosoideae
Bulletin du Jardin botanique national de Belgique, later Systematics and Geography of Plants
Bulletin mensuel de la Société Linnéenne de Lyon
Bulletin of [the] Moscow Society of Naturalists
Bulletin du Muséum national d'Histoire naturelle
Bulletin du Muséum national d'Histoire naturelle. Section B, Adansonia. Formerly Adansonia
Bulletin of the Nanjing Botanical Garden
Bulletin National Tropical Botanical Garden, formerly Bulletin Pacific Tropical Garden
Bulletin of the Natural History Museum. Botany Series, formerly Bulletin of the British Museum (Natural History) (Series E). Botany
Bulletin. New Zealand Department of Scientific and Industrial Research
Bulletin de la Société botanique de France, later Acta botanica Gallica, incl. Actualités botaniques & Lettres botaniques
Bulletin de la Société Royale de Botanique de Belgique, later Belgian Journal of Botany
Bulletin of the Sugadaira Biological Laboratory
Bulletin of the Torrey Botanical Club, later Journal of the Torrey Botanical Society
Bulletin of the Wellington Botanical Society
Byulleten Gosudarstvennogo Nikitskogo Opytnogo Botanicheskogo Sada
Byulleten Moskovskogo Obshchestva Ispȳtateleĭ Prirodȳ. Biol.
Byulleten Vsesoyuznogo Ordena Lenina Instituta Rastenievodstva Imeni N.I. Vavilova, continued as Nauchno-tekhnicheskii Byulleten Vsesoyuznogo Ordena Lenina i Ordena Drozhby Narodov Instituta Rastenievodstva Imeni N.I. Vavilova
Cactaceas y Suculentas Mexicanas
Cactician, The
Cactus and Succulent Journal of Great Britain
Cactus and Succulent Journal (US)
Caldasia
California Agriculture
Calyx
Camas Quarterly
Canadian Field-Naturalist
Canadian Journal of Botany
Canadian Journal of Plant Science
Candollea
Carnegie Institution of Washington Publications
Castanea
Castanea. Occasional Papers
Cathaya
Ceiba
Ceylon Forester
Ceylon Journal of Science, Biological Sciences
Chemical Plant Taxonomy Newsletter
Chinese Bulletin of Botany
Chinese Journal of Botany
Chronica Horticulturae

Ciencias. Ser. 4. Ciencias biológicas, Universidad de la Habana
Ciencias. Ser. 10. Botánica, Universidad de la Habana
Ciencias biológicas, Academia de Ciencias de Cuba
Cladistics
Codon
Collectanea Botanica
Commentationes biologicae
Commonwealth Forestry Review
Communicaciones Botanicas del Museo de Historia natural de Montevideo, later Communicaciones Botanicas Museos Nacionales de Historia Natural y Antropologia
Communicaciones del Museo Argentino de Ciencias Naturales e Instituto nacional de Investigación de las Ciencias naturales
Compositae Newsletter
Compte rendu des Séances mensuelles. Société des Sciences naturelles du Maroc
Compte rendu des Séances. Société de Biogéographie
Conservation Biology
Contributions from the Bolus Herbarium
Contributions from the Gray Herbarium of Harvard University
Contributions from Herbarium Australiense, continued as Brunonia
Contributions from the National Botanic Garden Glasnevin, continued as Glasra
Contributions from the New South Wales National Herbarium, continued as Telopea
Contributions from the New York Botanical Garden
Contributions from the Queensland Herbarium, continued as Austrobaileya
Contributions in Science
Contributions from the United States National Herbarium
Contributions from the University of Michigan Herbarium
Cuadernos de Botanica Canaria
Cunninghamia
Current Biology
Cuttings
Cuscatlania
Cyperaceae Newsletter
Dansk Botanisk Arkiv
Darwiniana
Davidsonia
Delpinoa
Desert Plants
Deserta
Development
Dinteria
Diversity and Distributions
DNA
Dominguezia
Dumortiera
eBioNews
Ecological Monographs
Ecological Review
Ecology
Economic Botany
Ecotropica
Edinburgh Journal of Botany, formerly Notes from the Royal Botanic Garden Edinburgh
Egyptian Journal of Botany, formerly United Arab Republic Journal of Botany
Elliottia
Endeavour
Englera
Environmental Conservation
Environmental Monitoring and Assessment
Ernstia
Essex Naturalist
ETRFN News
Evolution
Evolutionary Biology
Evolutionary Ecology
Evolutionary Trends in Plants
Excelsa
Excelsa, Taxonomic Series
Families and Genera of Vascular Plants Dilleniid Newsletter
Fauna och Flora

Acknowledgement of Sources

Feddes Repertorium Zeitschrift für botanische Taxonomie und Geobotanik
Fern Gazette, formerly British Fern Gazette
Field Studies
Fieldiana (Botany)
Financial Times
Fitoterapia
Flavour and Fragrance Journal
Flora
Flora of China Newsletter
Flora og Fauna
Flora Malesiana Bulletin
Flora Mediterranea
Floribunda
Folia botanica Extramadurensis
Folia botanica et geobotanica Correntesiana
Folia botanica miscellanea
Folia geobotanica et phyto-taxonomica Bohemoslovaca, later Folia geobotanica
Folia Malaysiana
Folia Musei rerum Naturalium Bohemiae Occidentalis
Fontqueria
Forestry
Four Seasons
Fragmenta floristica et geobotanica, later Polish Botanical Journal
Fritschiana
Frontiers in Ecology and the Environment
Functional Ecology
Future Farm
Garcia de Orta. Serie de Botanica
Garden, The
Gardens' Bulletin Singapore
Gardenwise (Singapore)
Gaussenia, formerly Travaux de Laboratoire forestier de Toulouse
Gayana. Botánica
Genetic Resources Plant Evolution
Genetica
Gentes Herbarum
Ginkgoana
Giornale botanico Italiano, later Plant Biosystems
Glasgow Naturalist
Glasra, formerly Contributions from the National Botanic Garden Glasnevin
Gleditschia
Global Ecology and Biogeography [formerly] Letters
Göttinger floristische Rundbriefe
Gorteria
Grana
Great Basin Naturalist, later Western North American Naturalist
Great Basin Naturalist. Memoirs
[The] Guardian
Guihaia
Gunneria
Gymnocalycium
Hacquetia
Hanburyana
Harvard Papers in Botany
Haussknechtia
Hebe News
Hemispheres
Herbertia
Herbs, Spices and medicinal Plants
Hercynia
Heringeriana
Hickenia. Boletín del Darwinion
Hikobia
Hoehnea
Home Fort Lauderdale
Hooker's Icones Plantarum
Hortax News
Horticultural Science

Human Ecology
Huntia
IAWA Bulletin, later IAWA Journal
Iheringia, série Botânica
Illinois Biological Monographs
Independent, The
Independent on Sunday, The
Indian Forest Records, (new Series) Botany
Indian Forester
Indian Journal of Botany
Indian Journal of Forestry
Insula
International Dendrology Society Yearbook
International Journal of Plant Sciences, formerly Botanical Gazette
International Tree Crops Journal
Iowa State Journal of Research
Iranian Journal of Botany
Irish Naturalists' Journal
Iselya
Israel Journal of Botany, later Israel Journal of Plant Sciences
Istanbul Üniversitesi Eczacilik Fakültesi Mecmuasi
Istanbul Üniversitesi Fen Fakültesi Mecmuasi, Seri B: Tabii Ilimber, later Biyoloji Dergisi
Itinera geobotanica
Izvestiya Akademii Nauk Kazakhskoi SSR, Seriya biologicheskaya
Izvestiya Akademii Nauk Turkmenskoi SSR, Seriya biologicheskikh Nauk
Izvestiya na Botanicheskiya Insitut (Sofiya), continued as Phytology
Jahrbuch des Bochumer Botanischen Vereins
Jakarta Post
Japanese Journal of Botany
Japanese Journal of Historical Botany
JARE, continued as Memoirs of National Institute of Polar Research, Series E
Journal of the Adelaide Botanic Gardens
Journal d'Agriculture tropicale et de Botanique appliquée, later Journal d'Agriculture traditionelle et de Botanique appliquée
Journal of Animal Ecology
Journal of applied Ecology
Journal of archaeological Research
Journal of Arid Environments
Journal of the Arnold Arboretum
Journal of Bamboo Research
Journal of Biogeography
Journal of Biological Sciences Research (Baghdad)
Journal of Biology (Vietnam)
Journal of Biosciences
Journal of the Bombay Natural History Society
Journal of the Botanical Research Institute of Texas, formerly Sida
Journal of the Botanical Society of South Africa
[le] Journal de Botanique de la Societé botanique de France
Journal of Classification
Journal of clinical Pharmacology
Journal of the College of Arts and Sciences, Chiba University. Natural Sciences Series
Journal of the East Africa Natural History Society and National Museum
Journal of Ecology
Journal of economic and taxonomic Botany
Journal of the Elisha Mitchell Scientific Society
Journal of evolutionary Biology
Journal of experimental Botany
Journal of the Faculty of Science, University of Tokyo. Botany
Journal of Forestry
Journal of Garden History
Journal of Geobotany, continued as Journal of Phytogeography and Taxonomy
Journal of Herbs, Spices and Medicinal Plants
Journal of the Indian Botanical Society
Journal of Japanese Botany
Journal of the Kew Guild
Journal of the Korean Forestry Society
Journal of Korean Plant Taxonomy, later Korean Journal of Plant Taxonomy
Journal of Life Sciences, Royal Dublin Society

Journal of the Linnean Society. Botany, continued as Botanical Journal of the Linnean Society
Journal of Nanjing Technological College of Forest Products
Journal of the National Botanical Society, formerly Bulletin of the Botanical Society of Bengal
Journal of the National Taiwan Museum, formerly Journal of the Taiwan Museum
Journal of Natural Products, formerly Lloydia
Journal of the Orissa Botanical Society
Journal of Phycology
Journal of Phytogeography and Taxonomy, formerly Journal of Geobotany
Journal of Plant Anatomy and Morphology
Journal of Plant Research, formerly Botanical Magazine (Tokyo)
Journal of Plant Resources and Environment
Journal and Proceedings of the Royal Society of New South Wales
Journal of the Royal Society of Western Australia
Journal of Science of the Hiroshima University, Series B, Division 2 Botany
Journal of the Society for the Bibliography of Natural History, continued as Archives of Natural History
Journal of the South African Biological Society
Journal of South African Botany
Journal of Systematics and Evolution, formerly Acta phytotaxonomica Sinica
Journal of the Taiwan Museum, formerly Quarterly Journal of the Taiwan Museum, later Journal of the National Taiwan Museum
Journal of the Torrey Botanical Society, formerly Bulletin of the Torrey Botanical Club
Journal of Tropical Ecology
Journal of tropical Forest Products
Journal of Wuhan botanical Research
Journal of Zhejiang Forestry College
Kakteen und andere Sukkulenten
Kalikasan
Kalmia
Kew Bulletin
Kew Bulletin, Additional Series
Kew Magazine, later (Curtis's) Botanical Magazine
Kew Magazine, Monograph Series
Kew Record
Kew Scientist
Kingia
Kings Park Research Notes
Kirkia
Kochia
Komarovia
Kongelige Danske Videnskabernes Selskabs Skrifter, Biological Series
Korean Journal of Botany
Korean Journal of Plant Taxonomy, formerly Journal of Korean Plant Taxonomy
Kulturpflanze
Kurtziana
Kurzmitteilungen der Deutschen dendrologischen Gesellschaft
Lagascalia
Laitsch
Lamiales Newsletter, later Vitex
Landbouwhogeschool te Wageningen, Miscellaneous Papers
Landscope
Lankesteriana
Lasca Leaves
Lasianthera
Laurales Newsletter
Lavori dell' Istituto botanico dell' Università di Milano
Lavori dell' Istituto e Orto botanico dell' Università di Palermo
La-Ya'aran
Lazaroa
Leandra
Leiden Botanical Series
Lejeunia
Lesovedenie
Lidia
Lilloa
Lindleyana, later continued in Orchids
Linnean, The
Linnean Society Symposium Series

Lloydia, continued as Journal of Natural Products
Loefgrenia
London Evening Standard
London Naturalist
Lorentzia
Ludoviciana
Lundellia
Lutukka
Lyonia
Madroño
Makinoa
Malayan Forest Records
Malayan Nature Journal
Malaysian Forester
Mededelingen van de Landbouwhogeschool te Wageningen continued as Agricultural University, Wageningen. Papers
Medical Daily
Mediterranéa
Mémoires de l'Academie Royale des Sciences d'Outre-Mer, Classe des Sciences naturelles et médicales
Mémoires du Muséum nationale d'Histoire naturelle
Mémoires de la Société botanique de France
Mémoires de la Société botanique de Genève
Mémoires de la Société royale botanique de Belgique
Memoirs of the Botanical Survey of South Africa
Memoirs of the Ehime University Section II, Natural Science, Series B (Biology)
Memoirs of the Faculty of Agriculture, Kagoshima University
Memoirs of the Hong Kong Natural History Society
Memoirs of National Institute of Polar Research, Series E, formerly JARE
Memoirs of the National Science Museum, Tokyo
Memoirs of the New York Botanical Garden
Memoirs and Proceedings of the Manchester Literary and Philosophical Society
Memoirs of the Torrey Botanical Club
Memórias da Sociedade Broteriana
Memórias de la Sociedad de Ciencias Naturales la Salle
Mentzelia
Michigan Botanist
Micronesica
Mitteilungen aus der botanischen Staatssammlung München, later Sendtnera
Mitteilungen der Deutschen dendrologischen Gesellschaft
Mitteilungen aus dem Institut für Allgemeine Botanik Hamburg
Mitteilungen aus dem Staatsinstitut für allgemeine Botanik in Hamburg
Miyabea
Molecular Biology and Evolution
Molecular Phylogenetics and Evolution
Monografias del Jardín botánico de Córdoba
Monographiae biologicae Canarienses
Monographiae botanicae
Monographs in systematic Botany from the Missouri Botanical Garden
Morris Arboretum Bulletin
Morton Arboretum Quarterly
Moscosoa
Muelleria
Musées de Genève
Mutisia
Napaea
National Cactus and Succulent Journal, continued as British Cactus and Succulent Journal
Natural History Bulletin of the Siam Society
Natural History Research
Naturalia Monspeliensia, Série botanique
Naturalist
Naturaliste Canadien
Nature in Wales
Nauchno-tekhnicheskii Byulleten Vsesoyuznogo Ordena Lenina i Ordena Drozhby Narodov Instituta Rastenievodstva Imeni N.I. Vavilova, formerly Byulleten Vsesoyuznogo Ordena Lenina Instituta Rastenievodstva Imeni N.I. Vavilova
Neilreichia
Nelumbo, formerly Bulletin of the Botanical Survey of India
New Botanist

New Journal of Botany, formerly Watsonia
New Phytologist
New Plantsman, formerly and later Plantsman
New Scientist
New York Times, The
New Yorker, The
New Zealand Journal of Botany
New Zealand Journal of Ecology
New Zealand Journal of Science
New Zealand Natural Sciences
Newsletter. Botanical Society of Otago
Newsletter. Friends of the Royal Botanic Gardens Sydney
Newsletter. Hawaiian Botanical Society
Newsletter of Himalayan Botany
Nigerian Field
Nordic Journal of Botany
Norrlinia
Northern Territory Botanical Bulletin
Norwegian Journal of Botany
Notas del Museo de la Plata. Botánica
Notes from the Royal Botanic Garden, Edinburgh, continued as Edinburgh Journal of Botany
Notulae botanicae Horti agrobotanici Cluj-Napoca
Notulae Naturae of the Academy of Natural Sciences of Philadelphia
Notulae systematicae, continued as Adansonia
Nova Acta Regiae Societatis Scientiarum Upsaliensis
Novitates botanicae Universitatis Carolinae
Novon
Novosti Sistematiki Vȳsshikh Rastenii
Nuytsia
Occasional Papers of the Bernice P. Bishop Museum, later Bishop Museum Occasional Papers
Occasional Papers of the Californian Academy of Sciences
Oecologia Plantarum
Österreichische botanische Zeitschrift, continued as Plant Systematics and Evolution
Ohio Journal of Science
Oikos
Onira
Opera botanica
Opera botanica Belgica
OPTIMA Leaflets OPTIMA Newsletter
Opuscula botanica Pharmaciae complutensis, later Rivasgodaya
Orchid Monographs
Orchids, later including Lindleyana
Oréades
Organisms, Diversity and Evolution
Orquidea
Orquideología
Oxford Plant Systematics
Oxford Today
Pabstia
Pacific Conservation Biology
Pacific Science
Pakistan Journal of Botany
Pakistan Systematics
Paleobiology
Palmengarten
Palms, formerly Principes
Parodiana
Perspectives in Plant Ecology, Evolution and Systematics
Pesquisas
Phanerogamarum Monographiae
Philippine Agriculturist
Philippine Flora Newsletter
Philosophical Transactions of the Royal Society
Phyta
Phytochemical Society of Europe Symposium Series
PhytoKeys
Phytologia
Phytologia Balcanica

Phytologia Memoirs
Phytology, formerly Izvestiya na Botanischeskiya Institut (Sofiya)
Phytomorphology
Phyton (Argentina)
Phyton (Austria)
Phytoneuron
Phytotaxa
Pittieria
Pittonia
Plant Biology, formerly Acta botanica Neerlandica and Botanica Acta
Plant Biosystems, formerly Giornale botanica Italiano
Plant, Cell and Environment
Plant Diversity and Evolution, formerly Botanische Jahrbücher für Systematik, Pflanzengeschichte und Pflanzengeographie
Plant Diversity and Resources, formerly Acta botanica Yunnanica
Plant Ecology and Diversity
Plant Ecology and Evolution, formerly Belgian Journal of Botany & incorporating Systematics and Geography of Plants
Plant Physiology
Plant Press
Plant Species Biology
Plant Systematics and Evolution, formerly Österreiche botanische Zeitschrift
Planta
Planta medica
Plants and Gardens
Plantsman, continued as New Plantsman
Plantula
Pleione
Polish Botanical Journal, formerly Fragmenta floristica et geobotanica
Polish Botanical Studies
Portugaliae Acta biologia
Prace Botaniczne
Preslia
Principes, later Palms
Private Eye
Proceedings of the Academy of Natural Sciences of Philadelphia
Proceedings of the Biological Society of Washington
Proceedings of the California Academy of Sciences
Proceedings of the Ecological Society of Australia
Proceedings of the Indian Academy of Sciences, Plant Sciences
Proceedings of the Iowa Academy of Science
Proceedings Koninklijke Nederlandse Akademie van Wetenschappen Series C, formerly Proceedings of the Section of Sciences, Koninklijke Nederlandse Akademie van Wetenschappen
Proceedings of the Linnean Society, continued as Biological Journal of the Linnean Society of London
Proceedings of the Linnean Society of New South Wales
Proceedings of the Nova Scotian Institute of Science
Proceedings of the Royal Microscopical Society
Proceedings of the Royal Society, Series B
Proceedings of the Royal Society of New Zealand
Proceedings of the Royal Society of Queensland
Proceedings of the Royal Society of Victoria
Proceedings of the Section of Sciences, Koninklijke Nederlandse Akademie van Wetenschappen, continued as Proceedings Koninklijke Nederlandse Akademie van Wetenschappen Series C
PROTA Newsletter
Protecção de Natureza
Publications in Botany, National Museum of Natural Sciences (Canada)
Publications of the Cairo University Herbarium, continued as Taeckholmia
Publications from the Department of Botany, University of Helsinki
Quarterly Bulletin of the Alpine Garden Society
Quarterly Journal of Forestry
Quarterly Journal of the Taiwan Museum, continued as Journal of the Taiwan Museum
Queensland Botany Bulletin
Queensland Naturalist
Raymondiana
Rea
Records of the Auckland Museum
Regnum vegetabile
Reinwardtia

Acknowledgement of Sources

Revista de Biologia
Revista Brasileira de Botânica, later Brazilian Journal of Botany
Revista del Jardín Botánico nacional, Universidad de la Habana
Revista del Museo Argentino de Ciencias Naturalis 'Bernardino Rivadavia'
Revista del Museo de la Plata (Botanica)
Revista de la Sociedad Boliviana de Botanica
Revista de la Sociedad Mexicana de Historia Natural
Revue de Cytologie et de Biologie Végétales, continued as Revue de Cytologie et de Biologie Végétales, le Botaniste
Revue de Cytologie et de Biologie Végétales, le Botaniste, formerly Revue de Cytologie et de Biologie Végétales and Botaniste
Revue général de Botanique
Revue Roumaine de Biologie, Serie de Botanique
Rheedea
Rhodora
Richardiana
Rivasgodaya, formerly Opuscula botanica Pharmaciae complutensis
Rock Garden Quarterly
Rodriguésia
Ruizia
Sabah Parks Nature Journal
Sabonet News
San Francisco Chronicle, The
Sandakania
Sarawak Museum Journal
Sarsia
Saussurea
Schlechtendahlia
Schlechteriana
Schumannia
Science Monthly
Science in New Guinea [later] Journal
Science and Public Affairs
Science Reports of the Tôhoku University, Fourth Series, Biology
Scientia Horticulturae
Scientific Proceedings of the Royal Dublin Society, Series A
Seed Science Research
Selbyana
Sellowia
Sendtnera, formerly Mitteilungen aus der botanischen Staatsammlungen München
Sepilok Bulletin
Sibbaldia
Sida, later Journal of the Botanical Research Institute of Texas
Sida Botanical Miscellany
Silva Fennica
Silvae genetica
Sind University Research Journal (Science Series)
Sistematik Botanik Dergesi
Smithsonian Contributions to Botany
Solanaceae Newsletter
Sommerfeltia
South African Journal of Botany
South Australian Naturalist
Southwestern Naturalist
Stapfia
Star (Kuala Lumpur), The
Strelitzia
Studia Botanica (Salamanca)
Studies from the Herbarium, California State University Chico
Stuttgarter Beiträge zur Naturkunde ser. A (Biologie)
Sultania
Surinaamse Landbouw
Svensk botanisk Tidskrift
Sydney Morning Herald, The
Syesis
Symbolae botanicae Upsalienses
Systematic Botany
Systematic Botany Monographs

Systematics and Biodiversity
Systematics and Geography of Plants, formerly Bulletin du Jardin botanique national de Belgique, incorporated in Plant Ecology and Evolution
Taeckholmia, formerly Publications of the Cairo University Herbarium
Taiwania
Taxon
Telopea, formerly Contributions from the New South Wales National Herbarium
Thai Forest Bulletin (Botany)
Thai Journal of Botany
Thaiszia
Theoretical and Applied Genetics
Threatened Plants Committee Newsletter
Trabajos del Departamento de Botanica y Fisiologia vegetal, Universidad de Madrid, later Botanica Complutensis
Trabajos del Departamento de Botanica (Salamanca)
Transactions (and Proceedings) of the Botanical Society of Edinburgh, continued as Botanical Journal of Scotland
Transactions of the Royal Society of Canada
Transactions of the Royal Society of South Africa
Transactions of the Royal Society of South Australia
Transactions of the Wisconsin Academy of Sciences, Arts and Letters
Travaux de l'Institut scientifique, Série botanique (Rabat)
Travaux du Laboratoire forestier de Toulouse, continued as Gaussenia
Travaux de la Section scientifique et Technique, Institut Français de Pondichéry
Treballs de l'Institut Botànic de Barcelona
Trees in South Africa
Trends in Ecology and Evolution
Trianea
Tropical Agriculture
Tropical Ecology
Tropische und subtropische Pflanzenwelt
Tuatara
Tulane Studies in Zoology and Botany
Turkish Journal of Botany
Turrialba
Ukraïnskii Botanichnïi Zhurnal, later Ukrainian Botanical Journal
Unasylva
United Arab Republic Journal of Botany, continued as Egyptian Journal of Botany
United States Department of Agriculture Publications
University of California Publications in Botany
Utafiti
Växtodling
Vegetatio
Verslagen en Mededelingen van de Koninklijke Nederlandse botanische Vereniging, Jaarboek
Verslagen en Mededelingen Plantenziektekundige Dienst, Wageningen
Vieraea
Vitex, formerly Lamiales Newsletter
Wageningen Agricultural University Papers, formerly Agricultural University, Wageningen. Papers
Wahlenbergia
Watsonia, later New Journal of Botany
Webbia
Wentia
Western Australian Herbarium Research Notes
Western Australian Naturalist
Western North American Naturalist, formerly Great Basin Naturalist
Willdenowia
Wood and Fibre
Wrightia
Wulfenia

Abbreviations and Symbols Used in This Book

1. General

NB A number of abbreviations refer to journals not published since 1970 and therefore not appearing under 'Acknowledgement of sources' – a list of materials which has been thoroughly scanned. Acknowledgement to the editors and proprietors of the additional journals listed is made here.

A	androecium, stamens
AA	Acta Amazonica
AB	Aquatic Botany
ABAS	Acta botanica Austro Sinica
ABA(S)H	Acta botanica Academiae Scientarum Hungaricae
ABF	Acta botanica Fennica
ABG	Acta botanica Gallica
ABH	Ata botanica Hungarica
ABiV	Acta biologica Venezuelica
ABM	Acta botanica Mexicana
ABN	Acta botanica Neerlandica
ABV	Acta botanica Venezuelica
ABY	Acta botanica Yunnanica
accum.	accumulating
AD	Anno Domini
AFJ	American Fern Journal
Afr.	Africa(n)
AHB	Acta Horti Bergiani
AHP	Acta Horti Petropolitani
AIBC	Anales del Instituto botánico A.J. Cavanilles
AIBU	Anales del Instituto de biologia, Universidad nacional autónoma de México
AJB	American Journal of Botany
AJBM	Anales del Jardin botanico de Madrid
AJBRJ	Arquivos do Jardim botânico do Rio de Janeiro
AK	Arboretum Kórnickie
AKBG	Annals of Kirstenbosch Botanic Garden
AL	Annonaceae Newsletter
alks	alkaloids
alt.	alternative, alternate, altitude
Am(er).	America(n)
Amaz.	Amazon(ia(n))
AMBG	Annals of the Missouri Botanical Garden
AMH	Annales historico naturales-musei nationalis Hungarici
AMHNV	Anales del Museo de Historia natural de Valparaiso
AMN	American Midland Naturalist
anatr.	anatropous
anc.	ancient
ANMW	Annalen des K. K. naturhistorischen (Museums) Hofmuseums. Wien
ann.	annual, anniversary
AnnBF	Annales Botanici Fennici
AOSB	American Orchid Society Bulletin
AP	Australian Plants
APG	Acta phytotaxonomica et geobotanica
app.	apparent(ly)
APS	Acta phytotaxonomica Sinica

Abbreviations

ARBGC	Annals of the Royal Botanic Garden Calcutta
arch.	archaic, archipelago
ARES	Annual Review of Ecology and Systematics
Arg(ent).	Argentina
Ark.B.	Arkiv für Botanik
Aroid.	Aroideana
arr.	arranged, arrangement
ARSSL	Acta Reginae Societatis Scientarum et Litterarum Gothoburgensis Botanica
As.	Asia(tic)
ASAM	Annals of the South African Museum
ASBP	Acta Societas Botanicorum Poloniae
ASBSN	Australian Systematic Botany Society Newsletter
ASNB	Annales des Sciences naturelles, Botanique
assoc.	associated
attrib.	attributed, attribution
auctt.	of authors
Aus.	Australia(n)
AusJB	Australian Journal of Botany
AusSB	Australian Systematic Botany
Austr.Pl.	Australian Plants
AUWP	Agricultural University, Wageningen. Papers
AZBB	Annotationes zoologicae et botanicae. Bratislava
BAIM	Bulletin Auckland Institute and Museum
BANCC	Boletín de Academia nacional de Ciencias en Córdoba
BB	Bibliotheca botanica
BBAS	Botanical Bulletin of Academia Sinica
BBC	Beihefte zum Botanischen Centralblatt
BBGB	Bulletin of the Botanic Gardens Buitenzorg
BBLNEFI	Bulletin of the Botanical Laboratory of the North-Eastern Forestry Institute
BBMNH(B)	Bulletin of the British Museum (Natural History), series E – Botany
BBP	Beiträge zur Biologie der Pflanzen
BBPBM	Bulletin of the Bernice P. Bishop Museum
BBR	Bulletin of Botanical Research
BBSI	Bulletin of the Botanical Survey of India
BC	Biological Conservation, before Christ
BDBG	Berichte der Deutschen Botanischen Gesellschaft
BdeB	Boletim de Botânica
BFG	British Fern Gazette
BG	Botanical Gazette
BH	Botanica Helvetica
BHB	Bulletin de l'Herbier Boissier
BI	British Isles
bienn.	biennial
bisex.	bisexual
BJ	Botanische Jahrbücher
BJB	Bangladesh Journal of Botany
BJBB	Bulletin du Jardin botanique à Bruxelles (*later* national de Belgique)
BJBBuit.	Bulletin du Jardin botanique de Buitenzorg
B(ot.)JLS	Botanical Journal of the Linnean Society
BM	Botanical Magazine
BMNHNAdans.	Bulletin du Muséum national d'Histoire naturelle, Adansonia
BMNR(J)B	Boletim do Museu nacional de Rio de Janeiro. Botânica
BMOIPB	Byulleten Moskovskogo Obshchestva Ispȳtateleĭ Priorodȳ. Biol.
BMPEG	Boletim do Museu Paraense 'Emílio Goedi', Ser. Botânica
BM(T)	Botanical Magazine (Tokyo)
BMac.	Botanica Macaronesica
BMSN	Bullein of [the] Moscow Society of Naturalists
BN	Botaniska Notiser
BNH	Beihefte zur Nova Hedwigia
BolB	Boletim de Botânica (da Universidade de São Paulo)
BP	before present
Br.	Britain, British
BR	Botanical Review
Braz.	Brazil(ian)
Britt.	Brittonia
BRCI	Bulletin of the Reseach Council of Israel

BS	Botanical Sciences
BSAB	Boletin de la Sociedad Argentina de Botánica
BSB	Boletim da Sociedade Broteriana
BSBC	Boletin de la Sociedad de Biología de Concepción
BSBF	Bulletin de la Société botanique de France
BSBG	Berichte der Schweizerischen botanischen Gesellschaft
BSBM	Boletín de la Sociedad botánica de México
BSRBB	Bulletin de la Société Royale de Botanique de Belgique
BSVCN	Boletín de la Sociedad Venezolana de Ciencias Naturales
BT	Botanisk Tidsskrift
BTBC	Bulletin of the Torrey Botanical Club
BYUSBB	Brigham Young University Science Bulletin, Biology
BZ	Botanicheskii Zhurnal
c.	about
C	corolla (members), central, century
Calif.	California(n)
campylot.	Campylotropous
caps.	capsule
Carib.	Caribbean
Cauc.	Caucasus, Caucasian
CB	Chronica botanica
CBH	Contributions from the Bolus Herbarium
CBLUP	Contributions from the Botanical Laboratory of the University of Pennsylvania
CDH	Contributions from the Dudley Herbarium (of Stanford University)
cent.	century
cerem.	ceremony, -ies, -ial
cf.	compare
CGH	Contributions from the Gray Herbarium of Harvard University
CHA	Contributions from Herbarium Australiense
char(ac).	character(s), characteristic(ally)
chem.	chemical(ly), chemistry
CITES	Convention on International Trade in Endangered Species of Wild Fauna and Flora
CIWP	Carnegie Institution of Washington Publications
CJB	Canadian Journal of Botany
CJPS	Canadian Journal of Plant Science
Clad.	Cladistics
class.	classified, classification
cli.	climber, climbing
CN	Compositae Newsletter
CNS	central nervous system
CNSWNH	Contributions from the New South Wales National Herbarium
coll.	collected, collecting
Coll. B	Collectanea Botanica
comm.	commercial(ly)
Cons. Biol.	Conservation Biology
consp(ic).	conspicuous(ly)
constr.	construction
cont.	contain(ing)
Cont.	The Continent of Europe, excluding British Isles
corr.	correct(ly)
cosmop.	cosmopolitan
CQH	Contributions from the Queensland Herbarium
CRSS(S)B	Compte rendu Sommaire des Séances. Société de Biogéographie
CSJ	Cactus and Succulent Journal of Great Britain
CSJAm.	Cactus and Succulent Journal of the Cactus and Succulent Society of America
CSM	Cactaceas y Succulentas Mexicanas
CTRFBS	Contributions from the Texas Research Foundation, Botanical Studies
cult.	cultivated, cultivation
CUMH	Contributions from the University of Michigan Herbarium
CUSNH	Contributions from the United States National Herbarium
cv.	cultivar
Darw.	Darwiniana
DAWW	Denkschriften der (kaiserlichen) Akademie der Wissenschaften. Wien
DB	Dissertationes botanicae
DBA	Dansk botanisk Arkiv

decid.	deciduous
dev.	developed
diam.	diameter
dichot.	dichotomous(ly)
dicots	dicotyledons
diff.	difference(s), different, differing, difficult
dioec.	dioecious
disp.	dispersal, dispersed (by)
distich.	distichous
distr(ib).	distribution(s), distributed
do	as above
DP	Desert Plants
DTYB	Daffodil and Tulip Yearbook
E	eastern, East
EB	Economic Botany
ed.	editor, edited, edible (or eaten by)
EFN	extrafloral nectary
EI	East Indies, East Indian
EJB	Edinburgh Journal of Botany
encl.	enclosed, enclosing
epipet.	epipetalous
esp.	especially
ess.	essential(ly)
EU	European Union
Eur.	Europe(an)
Euras.	Eurasia(n)
everg.	evergreen
ex	validly published by (appears after the name of an authority)
exc.	except
excl.	excluding
exstip.	exstipulate
ext.	extending
extrafl.	extrafloral
f.	forma, filius
F.	France, French
FA	Flora of Australia
fam(s).	family(-ies)
fav.	favourite
FBA	Folia biologica Andina
FE	Flora Europaea
fert.	fertilized (by), fertilization, fertile
FFG	Fragmenta floristica et geobotanica
FG	Fern Gazette
FGeo/FGP	Folia geobotanica phytotaxonomica
fl.	flower, flowering, active
fls	flowers
Fl. Cap.	Flora Capensis
-fld	-flowered
Fl. Ib.	Flora Iberica
Fl. Iraq	Flora of Iraq
Fl. Masc.	Flore des Mascareignes
FM	Flora Malesiana
FMB	Flora Malesiana Bulletin
FMBot.	Field Museum of Natural History. Botanical Series.
F Mad(ag).	Flore de Madagascar
FMNHB	Field Museum of Natural History. Botanical Series.
FN	Flora neotropica
FNA	Flora of North America
FNC	Flore de la Nouvelle-Calédonie et Dépendances
FNSW	Flora of New South Wales
FOC	Flora of China
form.	formerly
FOW	Species Plantarum: Flora of the World
FPM	Flora of Peninsular Malaysia
fr.	fruit
Fr.	French
FR	Feddes Repertorium. Zeitschrift für botanische Taxonomie und Geobotanik

FRB(eih.)	Feddes Repertorium. Zeitschrift für botanische Taxonomie und Geobotanik. Beiheft
freq.	frequent(ly), frequency
frs	fruits
FSA	Flora of Southern Africa
FT	Flora of Turkey
FTEA	Flora of Tropical East Africa
FZ	Flora Zambesiaca
G	gynoecium ($\underline{G}$ superior, $\overline{G}$ inferior)
GB	Great Britain
GBIUS	Godišnjak biološkog Institute u Sarajevu
GBN	Great Basin Naturalist
GBS	Gardens' Bulletin Singapore
Gen. Orch.	Genera Orchidacearum
Ger(m).	German(y)
germ.	germination, germinating
GH	Gentes Herbarum
Gk.	Greek
GM	genetically modified (crops)
GOB	Garcia de Orta. Serie de Botanica
h, hr	hour
half-inf.	half-inferior
hapax.	hapaxanthic
HBML	Harvard Botanical Museum Leaflets
hemisph.	hemisphere
herm(aphr).	hermaphrodite
Himal.	Himalaya(n)
HIP	Hooker's Icones Plantarum
horiz.	horizontal(ly)
hort.	horticulture, horticulturally, Hortus
HPB	Harvard Papers in Botany
hypog.	hypogynous
IBM	Illinois Biological Monographs
IDSY	International Dendrology Society Yearbook
IJPS	International Journal of Plant Sciences
IF	Indian Forester
illus./ill.	Illustrated, illustration
imbr.	imbricate
imp.	important, importance
inc(l).	including, included
inconsp.	inconspicuous
inc. sed.	of uncertain affinity (incerta(e) sedis)
Ind.	India(n), Indies
indehisc.	indehiscent
Indom(al).	Indomalesia(n), i.e. India to New Guinea
ined.	unpublished, inedible
infl(s).	inflorescence(s)
infr(s).	infructescence(s)
INHSB	Illinois Natural History Survey Bulletin
intr(od).	introduced, introduction
Ir.JB	Iranian Journal of Botany
irreg.	irregular(ly)
is.	island(s)
ISCJS	Iowa State College Journal of Science
IS(U)JS	Iowa State (University) Journal of Science
ITCJ	International Tree Crops Journal
JAA	Journal of the Arnold Arboretum
JABG	Journal of the Adelaide Botanic Gardens
Jap.	Japan(ese)
JapJB	Japanese Journal of Botany
JB	Journal of Botany
JBNHS	Journal of the Bombay Natural History Society
JBR	Journal of Bamboo Research
JBRIT	Journal of the Botanical Research Institute of Texas
JEMSS	Journal of the Elisha Mitchell Scientific Society
JETB	Journal of Economic and Taxonomic Botany
JFSUTB	Journal of the Faculty of Science, University of Tokyo. Botany
JG	Journal of Geobotany

JJB	Journal of Japanese Botany
JLSBot.	Journal of the Linnean Society. Botany
JNTCFP	Journal of Nanjing Technological College of Forest Products
JPhycol	Journal of Phycology
JRHS	Journal of the Royal Horticultural Society
JRSNZ	Journal of the Royal Society of New Zealand
JRSWA	Journal of the Royal Society of Western Australia
JSAB	Journal of South African Botany
JSE	Journal of Systematics and Evolution
JTBS	Journal of the Torrey Botanical Society
JWAS	Journal of the Washington Academy of Sciences
K	calyx (members)
KB	Kew Bulletin
KBAS	Kew Bulletin, Additional Series
KM	Kew Magazine
KMMS	Kew Magazine, Monograph Series
KNAWC	Koninklijke Nederlandse Akademie van Wetenschappen, Series C
KSV(A)H	Kungliga Svenska Vetenskapsacademiens Handlingar
KUSB	Kansas University Science Bulletin
l	litre
lat.	lateral
Lat.	Latin
LBS	Leiden Botanical Series
linr.	linear
loc.	locule, loculate
longit.	longitudinal
LUA	Lund Universitets Årsskrift
lvs	leaves
LWB	Leaflets of Western Botany
m	metre
M	million
Madag.	Madagascar
MAISP	Mémoires de l'Académie impériale des Sciences de St.- Pétersbourg
Mal.	Malesia(n)
MAM	Morris Arboretum Monographs
Masc.	Mascarenes
MB	Monographiae botanicae
MBMHR	Mededeelingen van het botanisch Laboratorium der Rijksuniversiteit te Utrecht
MBMUZ	Mitteilungen aus dem botanischen Museum der Universität Zürich
MBSM	Mitteilungen aus der botanischen Staatssammlung München
MCSUK	Memoirs of the College of Science. University of Kyoto
MedBotUtrecht	Mededelingen van het Botanisch Museum en Herbarium van de Rijksuniversiteit te Utrecht
med(ic).	medicine, medicinal(ly), medical(ly)
Med(it).	Mediterranean
Mex.	Mexico, Mexican
MF	Malaysian Forester
MHB	Mémoires de l'Herbier Boissier
MIABH	Mitteilungen aus dem Institut für allgemeine Botanik in Hamburg
microscop.	microscopic
MLW	Mededelingen van de Landbouwhogeschool te Wageningen
MMNHN(P)	Mémoires du Muséum nationale d'Histoire naturelle (Paris)
MNJ	Malayan Nature Journal
MNYBG	Memoirs of the New York Botanical Garden
monocot.	monocotyledonous
monocots	monocotyledons (Monocotyledonae)
monoec.	monoecious
morph.	morphological(ly), morphology
MPE	Molecular Phylogenetics and Evolution
MRL	Mededelingen van's Rijksherbarium Leiden
MS(S)	manuscript(s)
MSABH	Mitteilungen aus dem Staatsinstitut für allgemeine Botanik in Hamburg
MSB	Memórias da Sociedade Broteriana
MSBMBG	Monographs in systematic Botany, Missouri Botanical Garden
MSPS	Minnesota Studies in Plant Science
mt(s).	mountain(s)
MTBC	Memoirs of the Torrey Botanical Club

n	haploid chromosome number
N	northern, North
natur.	naturalized
NB	note that
NBGB	Notizblatt des botanischen Gartens und Museums zu Berlin
NBP	Novitationes botanicae. Prague
New Caled.	New Caledonia(n)
NCSJ	National Cactus and Succulent Journal
n.d.	no date
NDSNG	Neue Denkschriften der Schweizerischen Naturforschenden Gesellschaft
NG	New Guinea
NHBSS	Natural History Bulletin of the Siam Society
NJB	Nordic Journal of Botany
nm.	nothomorph
NM	Naturalia monspeliensia
NMBA	Notas del Museo (de la Plata) Buenos Aires
no.	number
nom. illeg.	name not in accordance with the rules of nomenclature (nomen illegitimum)
NP	New Phytologist
nr	near
NRBGE	Notes from the Royal Botanic Garden, Edinburgh
n.s.	new series
NS, NS Paris	Notulae systematicae (Paris)
NSL, NSPV	Notulae systematicae (Leningrad)
NSW	New South Wales
NW	New World
NZ	New Zealand
NZJB	New Zealand Journal of Botany
OB	Opera botanica
OBB	Opera botanica Belgica
OBC	Opera botanica Cechica
OBPC	Opuscula botanica Pharmaciae Complutensis
obs.	obscure(ly), obsolete
OBZ	Österreichische botanische Zeitschrift
occ.	occasional(ly)
OD	Orchid Digest
oft.	often
OJS	Ohio Journal of Science
OM	Orchid Monographs
O(p).L(ill).	Opera Lilloana
opp.	opposite
orig.	origin(s), original(ly)
orn.	ornamental(ly)
orthotr.	orthotropous
OW	Old World
p(p).	page(s)
P	perianth (members)
p.a.	per annum
PAAAS	Proceedings of the American Academy of Arts and Sciences
PAB	Portugaliae Acta Biologica
Pac(if).	Pacific
Pac. Sci.	Pacific Science
PakJB	Pakistan Journal of Botany
PANSP	Proceedings of the Academy of Natural Sciences of Philadelphia
palaeotrop.	palaeotropics, palaeotropical
pantrop.	pantropical
PAPS	Proceedings of the American Philosophical Society
Papuas.	Papuasia(n)
Par.	Parodiana
PB	Plant Biosystems
PBJ	Polish Botanical Journal
PBS	Polish Botanical Studies
PBSW	Proceedings of the Biological Society of Washington
PCAS	Proceedings of the California Academy of Sciences
PCIW	Publications. Carnegie Institute, Washington
PCUH	Publications of the Cairo University Herbarium
PDE	Plant Diversity and Evolution

PDR	Plant Diversity and Resources
PEE	Plant Ecology and Evolution
pen(in).	peninsula
pend.	pendulous
perenn.	perennial
Pfl.	Die natürlichen Pflanzenfamilien
Pflr.	Das Pflanzenreich
PFMB	Publications of the Field Museum of Natural History, Botanical Series
PFSUM	Publications de la Faculté des Sciences de l'Université Masaryk
Philipp.	Philippines
Phytol.	Phytologia
Phytol.M	Phytologia Memoirs
PIAS	Proceedings of the Indiana Academy of Science
PJB(J(S))	Palestine Journal of Botany. Jerusalem Series
PJS(ci).	Philippine Journal of Science
PJSB	Philippine Journal of Science, Section C. Botany
PKNAW	Proceedings Koninklijke Nederlandse Akademie van Wetenschappen, Series C
pl(s).	plant(s)
PL	Plant Life
PLPH	Pengumuman istimewa. Lembaga Pusat Penjelidikan Kehutanan
PLSNSW	Proceedings of the Linnean Society of New South Wales
PM	Phanerogamarum Monographiae
PMSUB(S)	Publications. Michigan State University Museum (Biological Series)
poll.	pollination, pollinating, pollinated (by)
pop.	population, popular
poss.	possible, possibly
p.p.	in part
p.p.m.	parts per million
PPEES	Perspectives in Plant Ecology, Evolution & Systematics
pr(s)	pair(s)
PR	Das Pflanzenreich
praec.	preceding
prep.	preparing, preparation, prepared
prob.	probable, probably
prod(s).	product(s), producing, production
prop.	propagated, propagation
prot.	protected, protection, protecting, protect
PRSQ	Proceedings of the Royal Society of Queensland
PSB	Plant Species Biology
PSE	Plant Systematics and Evolution
publ.	publication(s), published
Publ. Univ. Penn.	Contributions from the Botanical Laboratory, University of Pennsylvania
QBAGS	Quarterly Bulletin of the Alpine Garden Society
QJTM	Quarterly Journal of the Taiwan Museum
q.v.	see
r.	root(s)
R	review, revision, synopsis, key, monograph
RAA	Revista Argentina de Agronomía
RBB	Revista Brasileira de Botânica
rbcl	gene coding for ribulose 1, 5 bisphosphate carboxylase (Calvin Cycle)
recog.	recognize(d)
reg.	regular(ly)
rep.	represented, representing
repr.	reprinted
reprod.	reproduced, reproducing, reproduction
resp.	respective(ly), respiratory
rhiz.	rhizome(s), rhizomatous
RJBN	Revista del Jardín Botánico Nacional (Cuba)
RMBG	Report Missouri Botanical Garden
RM(L)P	Revista del Museo de la Plata (Botanica)
RTBN	Recueil des Travaux botaniques Néerlandais
RV	Regnum Vegetabile
S	southern, South
SA[&, and]B	Systematics and Biodiversity
SAJB	South African Journal of Botany
saxic.	saxicole, saxicolous
SB	Systematic Botany

SBD	Sistematik Botanik Dergisi
SBi	Senckenbergiana biologica
SBM	Systematic Botany Monographs
SBNat.	Stuttgarter Beiträge zur Naturkunde aus dem Staatlichen Museum für Naturkunde in Stuttgart
SBU	Symbolae botanicae Upsalienses
SCB	Smithsonian Contributions to Botany
SE As.	mainland Asia centred on Indochina & Thailand
sec.	secondary
sect.	section
seg.	segregated
Selb.	Selbyana
semi-inf.	semi-inferior
semi-sup.	semi-superior
sep.	separate
seq.	following
S.G.	specific gravity
SGP	Systematics and Geography of Plants
SiBM	Sida Botanical Miscellany
s.l.	in the broad sense
solit.	solitary
sp.	species
spp.	species (plural)
s.s.	in the narrow sense
s.t.	sometimes
stip.	stipule(s), stipular
Strel.	Strelitzia
subg.	subgenus, subgenera
subs.	substitute(s)
subsp.	subspecies
subseq.	subsequent(ly)
subtemp.	subtemperate
subterr.	subterranean
subtrop.	subtropical
succ.	succulent(s)
sup.	superior
superf.	superficial(ly)
suppl.	supplement(ary)
SWNat.	Southwestern Naturalist
symp.	sympodial(ly)
synth.	synthesized
Syst.B	Systematic Botany
T	Taxon
Tasm.	Tasmania(n)
TBIANSSR	Trud ȳ Botanicheskogo Instituta. Akademiy nauk SSR
TBSE	Transactions of the Botanical Society of Edinburgh
TCM	Traditional Chinese Medicine
temp.	temperate, temperature
term.	terminal, terminating
terr.	terrestrial
TFB	Thai Forest Bulletin (Botany)
TFSS	Tree Flora of Sabah and Sarawak
TJB	Turkish Journal of Botany
TLS	Transactions of the Linnean Society
trad.	traditional(ly)
trop.	tropical, tropics
TRSNZ	Transactions of the Royal Society of New Zealand
TRSSA	Transactions of the Royal Society of South Australia
t.s.	transverse section
TSDSNH	Transactions of the San Diego Society for Natural History
TSP	Abhandlungen der Akademie . . . Mainz . . . tropische und subtropische Pflanzenwelt
TSSTIF	Travaux de la Section Scientifique et Technique, Institut Français de Pondichéry
UCOP	University of Connecticut Occasional Papers, Biological Science Series
UBZ	Ukraïnskii Botanichnïi Zhurnal
UCPB	University of California Publications in Botany
UKSB	University of Kansas Science Bulletin

Urug.	Uruguay
US	United States
USDA Agric. Handbk	United States Department of Agriculture, Agriculture Handbooks
USDA Agric. Mon.	United States Department of Agriculture, Agriculture Monographs
USDATB	United States Department of Agriculture, Technical Bulletin
USK	University Studies, University of Karachi
usu.	usually
UWPSB	University of Wyoming Publications in Science. Botany
v.	very
var.	variety, various
varieg.	variegated
vasc.	vascular
veg.	vegetable(s), vegetative, vegetation
Venez.	Venezuela(n)
vern.	vernacular
VGIETHZ	Veröffentlichungen des Geobotanischen Instituts, Eidgenössische technische Hochschule Rübel in Zürich
viz.	namely
VKNAW	Verhandelingen der Koninklijke Nederlandsche Akademie van Wetenschappen
VKZGW	Verhandlungen der kaiserlich-königlichen zoologisch-botanischen Gesellschaft in Wien
vol.	volume
W	western, West
WAUP	Wageningen agricultural University Papers
WI	West Indies
WJB	Wasmann Journal of Biology
wt.	weight
x	basic chromosome number
xeroph.	xerophyte, xerophytic
yr(s)	year(s)
zygom.	zygomorphic
2n	diploid chromosome number
±	more-or-less, approximately, with or without
×	hybrid
♀♂	bisexual
∞	numerous
+	graft hybrid

2. Authors' Names

As far as could be ascertained with respect to full names and dates, they are largely as follows. (*Draft index of author abbreviations compiled at the Herbarium, Royal Botanic Gardens Kew* (second imprint 1984 with further corrections found in R.K. Brummitt & C.E. Powell (eds, 1992) *Authors of plant names*, Royal Botanic Gardens, Kew & the IPNI on-line version).

Abbott, J.	Abbott, J. Richard (1968–)
Abbott, R.	Abbott, Richard J. (fl. 2003)
Abel	Abel, Gottlieb Friedrich (1763–?)
Abel, C.	Abel, Clarke (1789–1826)
Aberc.	McLaren, Henry Duncan, Lord Aberconway (1879–1953)
Abrams	Abrams, Le Roy (1874–1956)
Acev.-Rodr.	Acevedo-Rodrígues, Pedro (1954-)
Achille	Achille, Frédéric (1971–)
Ackerfield	Ackerfield, Jennifer (fl. 2002)
Ackerman	Ackerman, James D. (1950–)
Acuña	Acuña Galé, Julián Baldomero (1900–1973)
Adams	Adams, Michael Friedrich (1780–1833)
Adams, C.	Adams, Charles Dennis (1920–2005)
Adams, L.G.	Adams, Laurence George (1929–)
Adamson	Adamson, Robert Stephen (1885–1965)
Adans.	Adanson, Michel (1727–1806)
Adelb.	Adelbert, Aalbert George Ludwig (1914–1972)
Adema	Adema, Fredericus Arnoldus Constantin Basil (1939–)
Adlam	Adlam, Richard Wills (1853–1903)
Aellen	Aellen, Paul (1896–1973)

Afzel.	Afzelius, Adam (1750–1837)
Afzel., K.	Afzelius, Karl Rudolf (1887–1971)
Agardh	Agardh, Carl Adolf (1785–1859)
Agardh, J.	Agardh, Jakob Georg (1813–1901)
Aguiar	do Aguiar, Joaquim Macedo (1854–1882)
Ahles, H.	Ahles, Harry E. (1924–1981)
Ai	Ai, Tie Min (1946–)
Aiello	Aiello, Annette (1941–)
Airy Shaw	Airy Shaw, Herbert Kenneth (1902–1985)
Aitch.	Aitchison, James Edward Tierney (1836–1898)
Aiton	Aiton, William (1733–1793)
Aiton f.	Aiton, William Townsend (1766–1849)
Akers	Akers, John F. (1906–)
Akhani	Akhani, Hossain (1965–)
Alain	Alain, Brother (né Liogier de Sereys Allut, Henri Eugene) (1916–2009)
Alava	Alava, Reino Olavi (1915–)
Alavi	Alavi, S. A. (1934–)
Albers	Albers, Focko (1940–)
Albert, V.	Albert, Victor Anthony (1958–)
Albov	Albov (Alboff), Nikolai Michaïlovich (1866–1897)
Aldasoro	Aldasoro, Juan José (fl. 1999)
Alderw.	van Alderwerelt van Rosenburgh, Cornelis Rogier Willem Karel (1863–1936)
Alef.	Alefeld, Friedrich Georg Christoph (1820–1872)
Alejandro	Alejandro, Grecebio Jonathan D. (1973–)
Alexander	Alexander, Edward Johnston (1901–1985)
Alexander, P.	Alexander, Patrick J. (1981–)
Alexeev	Alexeev, E.B. (1946–1976)
Alford	Alford, Mac Haverson (1975–)
Ali	Ali, Syed Irtifaq (1930–)
All.	Allioni, Carlo (1728–1804)
Allam.	Allamand, Jean Nicholas Sébastien (1731–1793)
Allam., F.	Allamand, Frédérique (Frédéric) Louis (1735–1803)
Allan	Allan, Harry Howard Barton (1882–1957)
Allemão	Allemão e Cysneiro, Francisco Friera (1797–1874)
Allemão, M.	Allemão, Manoel (?–1863)
Allen, C.	Allen, Caroline Kathryn (1904–1975)
Allen, J.	Allen, James (c. 1830–1906)
Allen, P.	Allen, Paul Hamilton (1911–1963)
Alm	Alm, Carl Gustav (1888–)
Almeda	Almeda, Frank (1946–)
Almeida, M.R.L.	Almeida, Marselin Rusario (1939–)
Almeida, S.M.	Almeida, Sarah M. (1940–)
Al-Shehbaz	Al-Shehbaz, Ihsan Ali (1939–)
Alston	Alston, Arthur Hugh Garfit (1902–1958)
Alverson	Alverson, William Surprison (1953–)
Ames	Ames, Oakes (1874–1950)
Amirahm.	Amirahmadi, Atefe (fl. 2013)
Amman	Amman, Johann (1707–1741)
Amo	Amo, Mariano de Amo y Mora (1820–1896)
Amstutz	Amstutz, Erika (fl. 1957)
Anderb.	Anderberg, Arne A. (1954–)
Anderson, L.	Anderson, Loran Crittendon (1936–)
Anderson, R.	Anderson, Robert Henry (1899–1969)
Anderson, T.	Anderson, Thomas (1832–1870)
Anderson, W.	Anderson, William (1750–1778)
Anderson, W.R.	Anderson, William Russel (1942–2013)
Andersson	Andersson, Nils Johan (1821–1880)
Andersson, L.	Andersson, Bengt Lennart (1948–2005)
André	André, Edouard-François (1840–1911)
Andréanszky	Andréanszky, Gábor (1895–1967)
Andres	Andres, Heinrich (1883–1970)
Andrews	Andrews, Henry Charles (fl. 1794–1830)
Andrews, S.B.	Andrews, Spencer Bruce (fl. 1977)
Andrz.	Andrzejowski, Antoni Lukianowicz (1785–1868)
Ångstr.	Ångstrom, Johan (1813–1879)
Añon	Añon, Delia C. Suarez de Cullen (1917–)
Ansari	Ansari, M.Y. (1929–)

Antoine	Antoine, Franz (1815–1886)
Antonelli	Antonelli, Alexandre (fl. 2005)
Appan	Appan, Subramanian G. (1937–)
Araujo	Onofre de Araujo, Andréa (fl. 2007)
Arbo	Arbo, Maria Mercedes (1945–)
Archer, R.	Archer, Robert H. (1965-)
Archer, W. [bis]	Archer, William (1820–1874)
Archila	Archila, Fredy (1973–)
Ard.	Arduino, Pietro (1728–1805)
Arech(av).	Arechavaleta, José (1838–1912)
Arends	Arends, Georg (1862–1952)
Arènes	Arènes, Jean (1898–1960)
Aristeg.	Aristeguieta, Leandro (1923–)
Ariza	Ariza Espinar, Luis (1933–)
Arm., J.D.	Armitage, James D. (fl. 2011)
Armstr.	Armstrong, Joseph Beattie (1850–1926)
Armstr., J.A.	Armstrong, James Andrew (1950–)
Arn.	Arnott, George Arnott Walker (1799–1868)
Arn., S.	Arnott, Samuel (1852–1930)
Arnold, J.F.	Arnold, Johann Franz Xaver (1730–1801)
Arriaga	Arriaga, Mirta (fl. 1995)
Arrill.	Arri(l)laga de Maffei, Blanca Renée (1917–)
Arruda	Arruda de Cámara Manoel (1752–1810)
Asch.	Ascherson, Paul Friedrich August (1834–1913)
Ashburner	Ashburner, Kenneth (1927–2010)
Ashton	Ashton, Peter Shaw (1934–)
Askerova	Askerova, Rosa K. (1929–)
Aspl.	Asplund, Erik (1888–1974)
Asso	de Asso y del Rio Ignacio Jordán (1742–1814)
Athiê -Souza	Athiê-Souza, Sarah Maria (fl. 2015)
Attigala	Attigala, Lakshmi (fl. 2014)
Aubl.	Aublet, Jean Baptiste Christophore Fusée (1720–1778)
Aubrév.	Aubréville, André (1897–1982)
Audubon	Audubon, John James (1785–1851)
Auq.	Auquier, Paul Henri (1939–1980)
Austin, D.	Austin, Daniel Frank (1943–2015)
Autran	Autran, Eugène John Benjamin (1855–1912)
Avé-Lall.	Avé-Lallemant, Julius Léopold Edouard (1803–1867)
Aver., Averyanov	Averyanov, Leonid V. (1955–)
Averett	Averett, John Earl (1943–)
Avet.	Avetisyan, Vandika Ervandovna (1928–)
Aymard	Aymard Corredor, Gerardo Antonio (1959–)
Ayensu	Ayensu, Edward Solomon (1935–)
Azevedo	Goulart de Azevedo, Ana Maria (fl. 1982)
Baas	Baas, Pieter (1944–)
Bab.	Babington, Charles Cardale (1808–1895)
Babu	Babu, Cherukuri Raghavendra (1940–)
Bacig.	Bacigalupo, Nélida María (1924–)
Backeb.	Backeberg, Curt (1894–1966)
Backer	Backer, Cornelis Andries (1874–1963)
Backh.	Backhouse, James (1794–1869)
Backlund	Backlund, Anders (fl. 1993)
Bacon	Bacon, Christine D. (fl. 2011)
Baden	Baden, Claus (1952–)
Badillo	Badillo Franceri, Victor Manuel (1920–2008)
Baehni	Baehni, Charles (1906–1964)
Baensch	Baensch, H. Ulrich (fl. 1994)
Bänziger	Bänziger, Hans (fl. 1997)
Baijnath	Baijnath, Himansu (1943–)
Bailey	Bailey, Frederick Manson (1827–1915)
Bailey, C.	Bailey, C. Donovan (fl. 2001)
Bailey, D.	Bailey, Dana K. (1916–)
Bailey, I.	Bailey, Irving Widmer (1884–1967)
Bailey, J.	Bailey, John Frederick (1866–1938)
Bailey, J.P.	Bailey, John Paul (1951–)
Bailey, L.	Bailey, Liberty Hyde (1858–1954)
Baill.	Baillon, Henri Ernest (1827–1895)
Baird	Baird, Gary Innes (1955-)

Bajtenov	Bajtenov, M.S. (1927–)
Bak.	Baker, John Gilbert (1834–1920)
Baker, A.	Baker, Alan John Martin (1948–)
Bak(er), C.	Baker, Charles Fuller (1872–1927)
Bak. f.	Baker, Edmund Gilbert (1864–1949)
Baker, R.	Baker, Richard Thomas (1854–1941)
Baker, W.	Baker, William John (1972–)
Bakh.f.	Bakhuizen van den Brink, Reinier Cornelis (1911–1987)
Bakker, den	den Bakker, Hendrik Cornelis (1972–)
Bal.	Balansa, Benedict (1825–1892)
Balakr.	Balakrishnan, Nambiyath Puthansurayil (1935–)
Baldwin	Baldwin, Bruce Gregg (1957–)
Balf.	Balfour, John Hutton (1808–1884)
Balf.f.	Balfour, Isaac Bayley (1853–1922)
Ball	Ball, John (1818–1889)
Ball, P.	Ball, Peter William (1932–)
Ballard	Ballard, Francis (1896–1976)
Ballard, H.	Ballard, Harvey Eugene (1958–)
Balle	Balle, Simone (1906–)
Bally	Bally, Peter René Oscar (1895–1980)
Balslev	Balslev, Henrik (1951–)
Baltet	Baltet, Charles (1830–1908)
Bamps	Bamps, Paul Rodolphe Joseph (1932–)
Bancr.	Bancroft, Edward Nathaniel (1772–1842)
Banerjee	Banerjee, Rabindra Nath (1935–)
Banfi	Banfi, Enrico Augusto (1948–)
Bange	Bange, A.J. (1896–1950)
Banks	Banks, Joseph (1743–1820)
Baptista	Baptista, Dalton Holland (1962–)
Baranova, M.	Baranova, M.V. (1932–)
Barbey	Barbey, William (1842–1914)
Barbier	Barbier, E. (fl. 1973)
Barb. Rodr.	Barbosa Rodrigues, João (1842–1909)
Barfod	Barfod, Anders S. (1957–)
Barker, N.	Barker, Nigel P. (fl. 1993)
Barker, W.	Barker, Winsome Fanny (1907–1994)
Barker, W.R.	Barker, William Robert (1948–)
Barkley, F.	Barkley, Fred Alexander (1908–1989)
Barkley, T.	Barkley, Theodore Mitchell (1934–2004)
Barkworth	Barkworth, Mary Elizabeth (1941–)
Barlow	Barlow, Bryan Alwyn (1933–)
Barnades	Barnades (Barnadez), Miguel (1717–1771)
Barneby	Barneby, Rupert Charles (1911–2000)
Barnéoud	Barnéoud, François Marius (1821–)
Barney	Barney, E.E. (fl. 1877–9)
Barney, V.	Barney, Victoria E. (fl. 2001)
Barnhart	Barnhart, John Hendley (1871–1949)
Baron	Baron, Richard (1847–1907)
Barra	Barra, Lazáro, Alfredo (1946–)
Barrabé	Barrabé, Laure (fl. 2011)
Barrandon	Barrandon, Auguste (1814–1897)
Barratte	Barratte, Jean François Gustave (1857–1920)
Barringer	Barringer, Kerry A. (1954–)
Barros	de Barros, Fábio (1956–)
Barroso	Barroso, Liberato Joaquim (1900–1949)
Barroso, G.	Barroso, Graziela Maciel (1912–2003)
Barthlott	Barthlott, Wilhelm A. (1946–)
Bartlett	Bartlett, Harley Harris (1886–1960)
Bartling	Bartling, Friedrich, Gottlieb (1798–1875)
Barton	Barton, Benjamin Smith (1766–1815)
Barton, W.	Barton, William Paul Crillon (1786–1856)
Bartram	Bartram, William (1739–1823)
Bary	de Bary, Heinrich Anton (1831–1888)
Basiner	Basiner, Theodor, Friedrich Julius (1817–1862)
Baskerville	Baskerville, Thomas (1812–1840)
Bassi	Bassi, Ferdinando (1710–1774)
Batalin	Batalin, Alexander Theodorowicz (1847–1896)
Bateman	Bateman, James (1811–1897)

Bateman, R.	Bateman, Richard M. (fl. 1983)
Bates (, D.)	Bates, David Martin (1935–)
Batista	Batista, Augusto Chaves (1916–1967)
Batsch	Batsch, August Johann Georg Karl (1761–1802)
Battand.	Battandier, Jules Aimé (1848–1922)
Baudet	Baudet, Jean C. (1944–)
Bauer	Bauer, Franz Andreas (1758–1840)
Bauer, A.	Bauer, Annelise (fl. 1989)
Bauer, Rud.	Bauer, Rudolf (1910–1982)
Baum	Baum, Vicki M. (fl. 1982)
Baum, D.	Baum, David Alastair (1964–)
Baum.-Bodenh.	Baumann-Bodenheim, Marcel Gustav (1920–1996)
Baumg.	Baumgarten, Johann Christian Gottlob (1765–1843)
Bausch	Bausch, Jan (1917–)
Baxter	Baxter, William Hart (1816?–1890)
Bayer, E.	Bayer, Ehrenhaut (1953–)
Baylis	Baylis, Geoffrey Thomas Sandford (1913–2003)
Beadle	Beadle, Chauncey Delos (1866–1950)
Beal, E.O.	Beal, Ernest Oscar (1928–1980)
Beaman	Beaman, John Homer (1929–)
Bean	Bean, William Jackson (1863–1947)
Bean, A.	Bean, Anthony R. (1957–)
Bean, P.	Bean, Patricia Anne (1930–)
Beardsley	Beardsley, Paul M. (fl. 2012)
Beauv., P.	Palisot de Beauvois, Ambroise Marie François Joseph (1752–1820)
Beauverd	Beauverd, Gustave (1867–1942)
Beauvis.	Beauvisage, Georges Eugène Charles (1852–1925)
Becc.	Beccari, Odoardo (1843–1920)
Beck, G.	Beck, Günther Ritter (Beck) von Mannagetta und Lerchenau (1856–1931)
Beck, H.	Beck, Hans T. (fl. 1992)
Becker, W.	Becker, Wilhelm (1874–1928)
Beddome	Beddome, Richard Henry (1830–1911)
Bedell	Bedell, Hollis G. (fl. 1989)
Bedford	Bedford, David John (1952–)
Beentje	Beentje, Henk Jaap (1951–)
Beer	Beer, Johann Georg (1803–1873)
Beer, E.	Beer, Eva (?1892–?1977)
Bég.	Béguinot, Augusto (1875–1940)
Behnke	Behnke, H.-Dietmar (fl. 1997)
Behr	Behr, Hans Hermann (1818–1904)
Beier	Beier, Björn-Axel (1965–)
Beille	Beille, Lucien (1862–1946)
Bek.	Beketow, Andrej Nikolaevich (1825–1902)
Bél.	Bélanger, Charles Paulus (1805–1881)
Bell, H.	Bell, Hester L. (fl. 2008)
Bellair	Bellair, Georges Adolphe (1860–1939)
Bellardi	Bellardi, Carlo Antonio Lodovico (1741–1826)
Bello	Bello y Espinosa, Domingo (1817–1884)
Beltrán, H.	Beltrán, Hamilton (1963–)
Belyaeva	Belyaeva, Irina Veniaminovna (1957–)
Benedict	Benedict, Ralph Curtiss (1883–1965)
Benedict, C.	Gilg-Benedict, Charlotte (1872–1936)
Benham	Benham, Dale Maurice (1957–)
Benj.	Benjamin, Ludwig (1825–1848)
Benl	Benl, Gerhard (1910–2001)
Benn., A.	Bennett, Alfred William (1833–1902)
Bennet	Bennet, S.S.R. (1940–2009)
Bennett	Bennett, John Joseph (1801–1876)
Benn(ett), D.	Bennett, David Edward (1923–2009)
Bennett, E.	Bennett, Eleanor Marion (1942–)
Benn(ett), G.	Bennett, George (1804–1893)
Benoist	Benoist, Raymond (1881–1970)
Benson, L.	Benson, Lyman David (1909–1993)
Benth.	Bentham, George (1800–1884)
Berazaín	Berazaín, Rosalina (1947–)
Bercht.	von Berchtold, Friedrich (1781–1876)
Berg	von Berg, Ernst (1782–1855)
Berg, C.	Berg, Cornelis Christiaan (1934–2012)

Berg, O.	Berg, Otto Carl (1815–1866)
Berger, A.	Berger, Alwin (1871–1931)
Berggren, S.	Berggren, Sven (1837–1917)
Bergius	Bergius, Benedictus (1723–1784)
Bergius, P.	Bergius, Peter Jonas (1730–1790)
Bergmans	Bergmans, John (Johannes Baptista?) (1892–1980)
Berk.	Berkeley, Miles Joseph (1803–1889)
Berkut.	Berkutenko, A.N. (1950–)
Berland.	Berlandier, Jean Louis (1805–1851)
Bernal, R.	Bernal, Rodrigo (1959–)
Bernh.	Bernhardi, Johann Jakob (1774–1850)
Berry	Berry, Andrew (1764–1833)
Berry, L.	Berry, L.A. (fl. 1991)
Berry, P.	Berry, Paul Edward (1952–)
Bertero	Bertero, Carlo Luigi Giuseppe (1789–1831)
Berth.	Berthelot, Sabin (1794–1880)
Bertol.	Bertoloni, Antonio (1775–1869)
Bertol. f.	Bertoloni, Giuseppe (1804–1879)
Bertoni	Bertoni, Moisés de Santiago (1857–1929)
Bertrand	Bertrand, Marcel C.
Besler	Besler, Basilius (1561–1629)
Bess	Bess, Emilie C. (fl. 2006)
Besser	von Besser, Gilibald Swibert Joseph Gottlieb (1784–1842)
Betche	Betche, Ernst (1851–1913)
Betcke	Betcke, Ernst Friedrich (1815–1865)
Betsche	Betsche, I. (fl. 1984)
Beyerle	Beyerle, C. Richard (fl. 1938)
Bezerra	Bezerra, José Luiz (fl. 1970)
Bhand.	Bhandari, Madan Mal (1929–2011)
Bhargavan	Bhargavan, P. (1939–)
Bhide	Bhide, R.K. (fl. 1911)
Bieb., M.	Marschall von Bieberstein, Friedrich August (1768–1826)
Bien.	Bienert (Binert), Theophil (?–1873)
Bierh.	Bierhorst, David William (1924–1997)
Bierner	Bierner, Mark William (1946–)
Biffin	Biffin, Edward Sturt (1967–)
Bigazzi	Bigazzi, Massimo (1953–2006)
Bigelow	Bigelow, Jacob (1787–1879)
Binnend.	Binnendijk, Simon (1821–1883)
Birdsey	Birdsey, Monroe Roberts (1922–2000)
Birdw.	Birdwood, George Christopher Molesworth (1832–1917)
Bisch.	Bischoff, Gottlieb T.G. (1797–1854)
Bishop, L.E.	Bishop, Luther Earl (1943–1991)
Bisse	Bisse, Johannes (1935–1984)
Bitter	Bitter, Friedrich August Georg (1873–1927)
Bittrich	Bittrich, Volker (1954–)
Bizzarri	Bizzarri, Maria Paola (1937–)
Bjurzon	Bjurzon, Jonas (1810–1882)
Black, G.A.	Black, George Alexander (1916–1957)
Black, J.	Black, John McConnell (1855–1951)
Blackburn	Blackburn, Benjamin Coleman (1908–)
Blake, S.F.	Blake, Sidney Fay (1892–1959)
Blake, S.T.	Blake, Stanley Thatcher (1910–1973)
Blakeley	Blakeley, William Faris (1875–1941)
Blanca	Blanca, Gabriel (1954–)
Blanche	Blanche, Emanuel (1824–1908)
Blanco	Blanco, Francisco Manuel (1778–1845)
Blanco, M.	Blanco, Mario Albero (1972–)
Blatter	Blatter, Ethelbert (1877–1934)
Blaxell	Blaxell, Donald Frederick (1934–)
Bleck	Bleck, M.B. (fl. 1984–)
Bluff	Bluff, Matthias Joseph (1805–1837)
Blume	von Blume, Carl Ludwig (1796–1862)
Boatwr.	Boatwright, James S. (fl. 2007)
Bobrov	Bobrov, Evgenij Grigorievićz (1920–1983)
Bobrov, A.	Bobrov, Alexey Vladimir F. Ch. (1969–)
Bock, B.	Bock, Benoît (1972–)
Bockem.	Bockemühl, Leonore (1927–)

Bocquillon	Bocquillon, Henri Théophile (1834–1883)
Bod., M.	Bodard, Marcel (1927–1988)
Bodin	Bodin, Nicolaus Gustavus (fl. 1798)
Boeck	Boeck, Christian Peter Bianco (1798–1877)
Boeckeler	Boeckeler, Johann Otto (1803–1899)
Bödecker	Bödecker, Friedrich (1867–1937)
Böhle	Böhle, Uta-Regina (fl. 2000)
Boehmer	Boehmer, Georg Rudolf (1723–1803)
Boelcke	Boelcke, Osvaldo (1920–1990)
Boender	Boender, Ronald (fl. 2001)
Boenn.	von Boenninghausen (Bönninghausen), Clemens Maria Franz (1785–1864)
Boerh.	Boerhaave, Herman (1668–1739)
Boerl.	Boerlage, Jacob Gijsbert (1849–1900)
Boerner	Boerner (Börner), Karl Julius Bernhard (1880–1953)
Boggan	Boggan, John Kendall (1962–)
Bogner	Bogner, Josef (1939–)
Bois	Bois, Désiré Georges Jean Marie (1856–1946)
Boiss.	Boissier, Pierre Edmond (1810–1885)
Boissev.	Boissevain, Charles Hercules (1893–1946)
Boissieu	Boissieu (de la Martinière), Claude Victor (1784–1868)
Boissieu, H.	de Boissieu, Henri (1871–1912)
Boit.	Boiteau, Pierre L. (1911–1980)
Boivin	Boivin, Louis Hyacinthe (1808–1852)
Boivin, B.	Boivin, Joseph Robert Bernard (1916–1985)
Bojer	Bojer, Wenceslas or Wenzel (1797–1856)
Boker	Boker, Gili (f. 2007)
Bol.	Bolander, Henry Nicholas (1832–1897)
Bolle	Bolle, Karl August (1821–1909)
Bolle, F.	Bolle, Friedrich Franz August Albrecht (1905–1999)
Bolli	Bolli, Richard (fl. 1994)
Bolliger	Bolliger, Markus (1951–)
Bolton	Bolton, James (c.1758–1799)
Bolus	Bolus, Harry (1834–1911)
Bolus, L.	Bolus, Harriet Margaret Louisa (née Kensit) (1877–1970)
Bommer	Bommer, Joseph Edouard (1829–1895)
Bonap.	Bonaparte, Roland Napoléon (1858–1924)
Bonati	Bonati, Gustave Henri (1873–1927)
Bondar	Bondar, Gregório Gregorievich (1881–1959)
Bong.	von Bongard, August (Gustav) Heinrich (1786–1839)
Bonifacino	Bonifacino, José Maurico (fl. 2004)
Bonnet	Bonnet, Edmond (1848–1922)
Bonnier	Bonnier, Gaston Eugène Marie (1851–1922)
Bonpl.	Bonpland, Aimé Jacques Alexandre (né Goujaud) (1773–1858)
Boom	Boom, Boudewijn Karel (1903–1980)
Boom, B.	Boom, Brian Morley (1954–)
Booth	Booth, William Beattie (c. 1804–1874)
Bor	Bor, Norman Loftus (1893–1972)
Borbás	von Borbás, Vinczé (1844–1905)
Borchs.	Borchsenius, Finn (1959–)
Bordère	Bordère, Henri (1825–1889)
Boreau	Boreau, Alexandre (1803–1875)
Borgen	Borgen, Liv (1943–)
Borh.	Borhidi, Attila L. (1932–)
Boriss.	Borissova-Bekrjasheva, Antonina Georgievna (1903–1970)
Borkh.	Borkhausen, Mortiz (Moriz) Balthasar (1760–1806)
Bornm.	Bornmüller, Joseph Friedrich Nicolaus (1862–1948)
Borsch	Borsch, Thomas (1969–)
Bort	Bort, Katherine Stephens (1870–?)
Bory	Bory de Saint-Vincent, Jean Baptiste Georges (Geneviève) Marcellin (1778–1846)
Borzi	Borzi, Antonino (1852–1921)
Bosc	Bosc, Louis Augustin Guillaume (1759–1828)
Bosch	van den Bosch, Roelof Benjamin (1810–1862)
Bosse	Bosse, Julius Friedrich Wilhelm (1788–1864)
Bosser	Bosser, Jean M. (1922–2013)
Botsch.	Botschantzev, Victor Petrovič (1910–1990)
Botsch., V.V.	Botschantzeva, Vera Viktorovna (1946–)
Bouché	Bouché, Peter Carl (1783–1856)

Boulger	Boulger, George Edward Simmonds (1853–1922)
Boulos	Boulos, Loutfy (1932–2015)
Boutelje	Boutelje, Julius B. (fl. 1954)
Boutique	Boutique, Raymond (1906–1985)
Bove	Bove, Claudia Petean (fl. 2001)
Bowd., S.	Bowdich, Sarah (1791–1856)
Bowdich	Bowdich, Thomas Edward (1791–1824)
Bower	Bower, Frederick Orpen (1855–1948)
Bowles	Bowles, Edward Augustus (1865–1954)
Box	Box, Harold Edmund (1898–)
Boyce	Boyce, Peter Charles (1964–)
Boyd, S.	Boyd, Steve (fl. 1997)
Boynton	Boynton, Kenneth Rowland (1891–)
Br., E.	Brown, Elizabeth Dorothy (Wuist) (1880–1972)
Br., F.	Brown, Forest Buffen Harkness (1873–1954)
Br., N.E.	Brown, Nicholas Edward (1849–1934)
Br., R.	Brown, Robert (1773–1858)
Br., R.W.	Brown, Roland Wilbur (1893–1961)
Braas	Braas, Lothar Alfred (1942–1995)
Brackenr.	Brackenridge, William Dunlop (1810–1893)
Brade	Brade, Alexander Curt (1881–1971)
Bräuchler	Bräuchler, Christian (1975–)
Braem	Braem, Guido Jozef (1944–)
Braga	Braga, Ruby (fl. 1964)
Braggins	Braggins, John E. (1944–)
Bramw.	Bramwell, David (1942–)
Brand	Brand, August (1863–1930)
Brandegee	Brandegee, Townshend Stith (1843–1925)
Brandegee, M.	Brandegee, Mary Katharine (formerly Curran: née Layne) (1844–1920)
Brandes	Brandes (alias Schöpfer), Edvard (fl. 1845)
Brandis	Brandis, Dietrich (1824–1907)
Braun, A.	Braun, Alexander Karl Heinrich (1805–1877)
Braun, P.	Braun, Pierre Josef (1959–)
Braun-Blanquet	Braun-Blanquet, Josias (1884–1980)
Bravo	Bravo Hollis, Helia (1901–2001)
Brayshaw	Brayshaw, Thomas Christopher (1919–2014)
Breda	van Breda, Jacob Gijsbert Samuel (1788–1867)
Breedlove	Breedlove, Dennis Eugene (1939–2012)
Breitw.	Breitwieser, Ilse (fl. 1986)
Bremek.	Bremekamp, Cornelis Elisa Bertus (1888–1984)
Bremer	Bremer, Kåre (1948–)
Bremer, B.	Bremer, Birgitta (1950–)
Brenan	Brenan, John Patrick Micklethwait (1917–1985)
Breteler	Breteler, Franciscus Joseph (1932–)
Brettell	Brettell, Robert D. (1934–)
Brickell	Brickell, John (1748–1809)
Brickell, C.	Brickell, Christopher David (1932–)
Brid., Bridel	von Bridel, Samuel Élisée (1761–1828)
Bridson	Bridson, Diane Mary (1942–)
Brieger	Brieger, Friedrich Gustav (1900–1985)
Briggs, B.	Briggs, Barbara Gillian (1934–)
Brign.	de Brignoli di Brunnhoff, Giovanni (1774–1857)
Bringel	Bringel, João Bernardo de Azevedo (fl. 2011)
Briq.	Briquet, John Isaac (1870–1931)
Brittan	Brittan, Norman Henry (1920–)
Britten, L.	Britten, Lillian Louisa (1886–1952)
Britton	Britton, Nathaniel Lord (1859–1934)
Briz.	Brizicky, George Konstantine (1901–1968)
Bromf.	Bromfield, William Arnold (1801–1851)
Bromhead	Bromhead, Edward Ffrench (1789–1855)
Brongn.	Brongniart, Adolphe Théodore (1801–1851)
Brooker	Brooker, Murray Ian Hill (1934–2016)
Brot.	Brotero, Felix de (Silva) Avellar (1744–1828)
Broth.	Brotherus, Viktor Ferdinand (1849–1929)
Brouss.	Broussonet, Pierre Auguste Marie (1761–1807)
Browicz	Browicz, Kasimierz (1925–2009)
Brown, A.	Brown, Addison (1830–1913)
Brown, A.P.	Brown, Andrew Phillip (1951–)

Browne, P.	Browne, Patrick (1720–1790)
Browning	Browning, J. (fl. 1994)
Bruce	Bruce, James (1730–1794)
Bruce, E.A.	Bruce, Eileen Adelaide (1905–1955)
Bruegger	Bruegger (von Churwalden), Christian Georg (1833–1899)
Brühl	Brühl, Paul Johannes (1855–1935)
Bruggen	van Bruggen, Heinrich Wilhelm Eduard (1927–2010)
Bruhl	Bruhl, Jeremy James (1956–)
Brullo	Brullo, Salvatore (1947–)
Brullo, C.	Brullo, Cristian (1979–)
Brummitt	Brummitt, Richard Kenneth (1937–2013)
Brunel, J.	Brunel, Jean F. (fl. 1996)
Bruyns	Bruyns, Peter Vincent (1957–)
Bubani	Bubani, Pietro (1806–1888)
Buch., J., Buchanan	Buchanan, John (1855–1896)
Buchenau	Buchenau, Franz Georg Philipp (1831–1906)
Buchet	Buchet, Samuel (1875–1956)
Buch.-Ham.	Buchanan-Hamilton, Francis (1762–1829)
Buchheim	Buchheim, Arno Fritz Günther (1924–2007)
Buchholz	Buchholz, Fedor Vladimirovic (1872–1924)
Buc'hoz	Buc'hoz, Pierre Joseph (1731–1807)
Buckley	Buckley, Samuel Botsford (1809–1884)
Buek	Buek, Heinrich Wilhelm (1796–1878)
Buen	de Buen y del Cos, O. (1863–1945)
Buerki	Buerki, Sven (fl. 2006)
Büscher	Büscher, D. (fl. 2010)
Buhse	Buhse, Fedor Aleksandrovich (1821–1898)
Buijsen	Buijsen, J.R.M. (fl. 1988)
Buin.	Buining, Albert Frederick Hendrik (1901–1980)
Buist	Buist, George (1805–1860)
Bukasov	Bukasov, Sergej (Sergei) Mikhailovich (1891–1983)
Bull, W.	Bull, William (1828–1902)
Bullock	Bullock, Arthur Allman (1906–1980)
Bunge	von Bunge, Alexander Andrejewitsch (Andreevic, Aleksandrovic) (1803–1890)
Bunting	Bunting, George Sydney (1927–2015)
Burb.	Burbidge, Frederick William Thomas (1847–1905)
Burb., N.	Burbidge, Nancy Tyson (1912–1977)
Burchell	Burchell, William John (1781–1863)
Burck	Burck, William (1848–1910)
Burdet	Burdet, Hervé Maurice (1939–)
Bureau (or Bur.)	Bureau, Louis Edouard (1830–1918)
Burger	Burger, William Carl (1932–)
Burgoyne	Burgoyne, Priscilla M. (fl. 1998)
Burkart	Burkart, Arturo Erhado (1906–1975)
Burkill	Burkill, Isaac Henry (1870–1965)
Burm(an)	Burman, Johannes (1707–1779)
Burm.f.	Burman, Nicolaas Laurens (1734–1793)
Burnett	Burnett, Gilbert Thomas (1800–1835)
Burns-Balogh	Burns-Balogh, Pamela (1949–)
Burollet	Burollet, Pierre-Andre (1889–1961)
Burret	Burret, (Maximilian) Karl Ewald (1883–1964)
Burtt, B.L.	Burtt, Brian Lawrence (1913–2008)
Buscal.	Buscalioni, Luigi (1863–1954)
Busch	Busch, Anton (1823–1895)
Busch, N.	Busch, Nicolai Adolfowitsch (1869–1941)
Buse	Buse, Lodewijk Hendrik (1819–1888)
Buser	Buser, Robert (1857–1931)
Bush	Bush, Benjamin Franklin (1858–1937)
But	But, Paul Pui-Hay (fl. 1982)
Buttler	Buttler, Karl Peter (1942–)
Butzin	Butzin, Friedhelm Reinhold (1936–)
Buxb., F.	Buxbaum, Franz (1900–1979)
Buxton	Buxton, Bertram Henry (1852–1934)
Byalt	Byalt, V.V. (fl. 1995)
Byles	Byles, Ronald Stewart (fl. 1957)
Byng	Byng, James W. (fl. 2014)
Byrnes	Byrnes, Norman Bryce (1922–1998)

Caball.	Caballero, Arturo (1877–1950)
Cabezudo	Cabezudo, Baltasar (1946–)
Cabral	Cabral, Elsa Leonor (1951–)
Cabrera	Cabrera, Angel Lulio (1908–1998)
Caddick	Caddick, Lizabeth R. (fl. 2002)
Caldas	de Caldas y Tenorio, Francisco José (1771–1816)
Calderón	Calderón, Graciela de Rzedowski (1931–)
Calderón, C.	Calderón, Cleofé Elsa (1929–2007)
Calest.	Calestani, Vittorio (1882–1949)
Callm.	Callmander, Martin W. (1975–)
Camargo	de Camargo, Felisberto Cardoso (1896–1943)
Cambess.	Cambessèdes, Jacques (1799–1863)
Camp	Camp, Wendell Holmes (1904–1963)
Campacci	Campacci, Marcos Antonio (1948–)
Campb., G.	Campbell, Gael Jean (1973–)
Campderá	Campderá i Camins, Francisco (Francesc) (1793–1862)
Camus	Camus, Edmond Gustav(e) (1852–1915)
Camus, A.	Camus, Aimée Antoinette (1879–1965)
Candargy	Candargy, Paléologos C. (1870–?)
Cannon	Cannon, John Francis Michael (1930–2008)
Cannon, M.	Cannon, Margaret Joy (1928–2002)
Cao, T.R.	Cao, Te-Ru (1940–)
Capanema	Capanema, Guilherme Schüch (1824–1908)
Capurro	Capurro, Robert Horacio (1910–)
Capuron	Capuron, René (1921–1971)
Cárdenas	Cárdenas, Hermosa Martin (1899–1973)
Cardot	Cardot, Jules (1860–1934)
Carlq.	Carlquist, Sherwin (1930–)
Carnevali	Carnevali, Fernández-Concha, Germán (1955–)
Caro	Caro, José Aristida (1919–1985)
Carolin	Carolin, Roger Charles (1929–)
Carr	Carr, Cedric Errol (1892–1936)
Carr, D.J.	Carr, Denis John (1915–2008)
Carr, S.	Carr, Stella Grace Maisie (1912–1988)
Carrick	Carrick, John (1914–1978)
Carrière	Carrière, Élie Abel (1818–1896)
Carse	Carse, Harry (1857–1930)
Carter, A.	Carter, Annetta Mary (1907–1991)
Carter, C.	Carter, Charles R. (fl. 1980s)
Caruel	Caruel, Théodore (1830–1898)
Casar.	Casaretto, Giovanni (1812–1879)
Caspary	Caspary, Johann Xaver (Robert) (1818–1887)
Cass.	de Cassini, Alexandre Henri Gabriel (1781–1832)
Cassel	Cassel, Franz Peter (1784–1821)
Castro, V.	Castro, Vitorino Paiva (1942–)
Catharino	Catharino, Eduardo Luis Martins (1960–)
Cauwet-Marc	Cauwet-Marc, Ann Marie
Cav.	Cavanilles Palop, Antonio José (Antoni Josep) (1745–1804)
Cavaco	Cavaco, Alberto Judice Leote (1916–)
Cavalc.	Cavalcanti, Wlandemir de Albuquerque (fl. 1967)
Cavalerie	Cavalerie, Pierre Julien (1869–1927)
Cavara	Cavara, Fridiano (1857–1929)
Cavill.	Cavillier, François Georges (1868–1953)
Cayzer	Cayzer, Lindy W. (fl. 1999)
Čelak.	Čelakovský, Ladislav Josef (1834–1902)
Cellin.	Cellinese, Nicoletta (fl. 1997)
Cerv.	de Cervantes, Vicente (Vincente) (1755–1829)
Cesati	Cesati, Vicenzo de (1806–1883)
Chaix	Chaix, Dominique (1730–1799)
Chakrab.	Chakrabarty, Tapas (1954-)
Chakrav.	Chakravarty, Hira Lal (1907– ?)
Chalk	Chalk, Laurence (1895–1979)
Cham.	von Chamisso, Ludolf Karl Adelbert (1781–1838)
Chambers, K.	Chambers, Kenton Lee (1929–)
Champ.	Champion, John George (1815–1854)
Champluvier	Champluvier, Dominique (1953–)
Chandler, B.	Chandler, Bertha (?1885–1961)
Chandra	Chandra, Vinod (1953–)

Chaney	Chaney, Ralph Works (1890–1971)
Chang, C.C.	Chang, Chao-Chien (1900–)
Chang, H.T.	Chang, Hung-Ta (1914–2016)
Chang, M.C.	Chang, Mei-Chen (1933–)
Chang, S.Y.	Chang, Shao-Yao (fl. 1983)
Chao, A.C.	Chao, Ai-Cheng (fl. 1958)
Chao, C.S.	Chao, Chi-Son (1936–)
Chao, H.C.	Chao, Hsin-Chien (1918–)
Chaowasku	Chaowasku, Tanawat (fl. 2006)
Chapman	Chapman, Alvin Wentworth (1809–1899)
Charadze	Charadze, Anna Lukianovna (1905–1977)
Charif	Charif (fl. 1952)
Chase	Chase, Mary Agnes (née Merrill) (1869–1963)
Chase, M.	Chase, Mark Wayne (1951–)
Chassot	Chassot, Philippe (1972–)
Châtel.	Châtelain, Jean Jacques (1736–1822)
Chatin	Chatin, Gaspard Adolphe (1813–1901)
Chatrou	Chatrou, Laurentius Willem (1966–)
Chatterjee	Chatterjee, Debabarta (1911–1960)
Chaub.	Chaubard, Louis Athanase (1785–1854)
Chav.	Chavannes, Edouard Louis (1805–1861)
Cheek	Cheek, Martin Roy (1960–)
Cheeseman	Cheeseman, Thomas Frederick (1846–1923)
Chen, C.	Chen, Cheih (1928–)
Chen, C.J.	Chen, Chia-Jui (1935–)
Chen, C.T.	Chen, Ching-Tao (fl. 1973)
Chen, L.J.	Chen, Li-Jun (1980–)
Chen, S.C.	Chen, Sing-Chi (1931–)
Chen, S.H.	Chen, Su-Hwa (1948–)
Chen, S.J.	Chen, Sen-Jen (1933–)
Chen, S.L.	Chen, Shou-Liang (1921–)
Chen, W.H.	Chen, Wen-Hong (fl. 2002)
Chen, Y.L.	Chen, Yi-Ling (1930–)
Chen, Z.D.	Chen, Z.D. (fl. 2013)
Chen, Z.Y.	Chen, Zhong-Yi (fl. 1989)
Cheng, W.C.	Cheng, Wan-Chun (1904–1983)
Cheng f.	Cheng, Sze-Hsu (1931–1967)
Chermezon	Chermezon, Henri (1885–1939)
Chesselet	Chesselet, Pascale (1959–)
Chev., A.	Chevalier, Auguste Jean Baptiste (1873–1956)
Chevall.	Chevallier, François Fulgis (1796–1840)
Chew	Chew, Wee-Lek (1932–)
Chia	Chia, Liang-Chi (1921–)
Chien	Chien, Jian-Ju (fl. 1957–1984)
Ching	Ching, Ren-Chang (1898–1986)
Chinnock	Chinnock, Robert James (1943–)
Chiov.	Chiovenda, Emilio (1871–1941)
Chippendale	Chippendale, George McCartney (1921–2010)
Chiron	Chiron, Guy Robert (1944–)
Chitt.	Chittenden, Frederick James (1873–1950)
Chodat	Chodat, Robert Hippolyte (1865–1934)
Choi, H.G.	Choi, Han-Gu (fl. 2000)
Choisy	Choisy, Jacques Denys (Denis) (1799–1859)
Chou, Y.H.	Chou, Y.H. (fl. 1980)
Chouard	Chouard, Pierre (1903–1983)
Choux	Choux, Pierre (1890–1983)
Chow, S.	Chow, Shuan (fl. 1962)
Chowdhery	Chowdhery, Harsh Jeet (1949–)
Chr., C.	Christensen, Carl Frederick Albert (1872–1942)
Christ	Christ, Konrad Hermann Heinrich (1833–1933)
Christenh.	Christenhusz, Maarten Joost Maria (1976–)
Christenson	Christenson, Eric Alston (1956–2011)
Christm.	Christmann, Gottlieb Friedrich (1752–1836)
Christoph.	Christophersen, Erling (1898–1994)
Chrshan.	Chrshanovski (Khrzhanowski, Chrzanovskij), Vladimir Gennadievich (1912–1985)
Chrtek	Chrtek, Jindřich (1930–2008)
Chrtková	Chrtková, Anna (1930–2010)

Chu, G.L.	Chu, Ge-Lin (1934–)
Chu, W.M.	Chu, Wei-Ming (1930–)
Chuang	Chuang, Tsan-Iang (1933–1994)
Chukavina	Chukavina, Anna Prokofevna (1929–1985)
Chun	Chun, Woon-Young (1890–1971)
Church	Church, George Lyle (1903–)
Churchill, D.	Churchill, David Maughan (1933–)
Cif.	Ciferri, Raffaele (1897–1964)
Cires	Cires, Eduardo (fl. 2012)
Cirillo	Cirillo, Domenico Maria Leone (1739–1799)
Clairv.	de Clairville, Joseph Philippe (1742–1830)
Clapham	Clapham, Arthur Roy (1904–1990)
Clarion	Clarion, Jacques (1776–1844)
Clark, B.L.	Clark, Bonnie Lynne (1966–)
Clark, J.L.	Clark, John Littner (1969–)
Clark, L.G.	Clark, Lynn G. (1956–)
Clarke, C.B.	Clarke, Charles Baron (1832–1906)
Clarkson	Clarkson, John Richard (1950–)
Clausen, R.T.	Clausen, Robert Theodore (1911–1981)
Clayton	Clayton, John (1686–1733)
Clayton, W.	Clayton, William Derek (1926–)
Clemants	Clemants, Steven Earl (1954–2008)
Clement, E.J.	Clement, E.J. (fl. 2004)
Clements	Clements, Frederic Edward (1874–1945)
Clifford	Clifford, Harold Trevor (1927–)
Clos	Clos, Dominique (1821–1908)
Closon	Closon, Jules (fl. 1897)
Coaz	Coaz, Johann Wilhelm Fortunat (1822–1918)
Cochet	Cochet, Pierre Charles Marie (known as 'Cochet Cochet') (1866–1936)
Cochrane	Cochrane, Theodore Stuart (1942–)
Cockayne	Cockayne, Leonard C. (1855–1934)
Cockerell	Cockerell, Theodore Dru Alison (1866–1948)
Codd	Codd, Leslie Edward Wostall (1908–1999)
Coem.	Coemans, Henri Eugène Lucien Gaëtan (1825–1871)
Cogn.	Cogniaux, Célestin Alfred (1841–1916)
Cohen, J.	Cohen, James I. (fl. 2009)
Coiffard	Coiffard, Clément (fl. 2014)
Coincy	de Coincy, Auguste Henri Cornut de la Fontaine (1837–1903)
Colden	Colden, Jane (1724–1766)
Colebr.	Colebrooke, Henry Thomas (1765–1837)
Coleman	Coleman, M. (fl. 2001)
Colenso	Colenso, John William (1811–1899)
Colg., Colgan	Colgan, Nathaniel (1851–1919)
Colla	Colla, Luigi Aloysius (1766–1848)
Collett	Collett, Henry (1836–1901)
Columbus	Columbus, James Travis (1962–)
Comm.	Commerçon, Philibert (1727–1773)
Compère	Compère, Pierre (1934–)
Compton	Compton, Robert Harold (1886–1979)
Compton, J.	Compton, James A. (1953–)
Conant	Conant, David Stoughton (1949–)
Conert	Conert, Hans Joachim (1929–)
Conran	Conran, John Godfrey (1960–)
Console	Console, Michelangelo (1812–1897)
Constance	Constance, Lincoln (1909–2001)
Contandr.	Contandriopoulos, Juliette (1922–)
Contreras	Contreras Jiménez, José Luis Regino (1952–)
Cook	Cook, Orator Fuller (1867–1949)
Cook, C.	Cook, Christopher David Kentish (1933–)
Cook, V.J.	Cook, Varner James (1904–)
Cooke	Cooke, David Alan (1949–)
Coombes	Coombes, Allen J. (fl. 2003)
Cooper	Cooper, Daniel (?1817–1842)
Cope	Cope, Thomas Arthur (1949–)
Copel.	Copeland, Edwin Bingham (1873–1964)
Copel., H.	Copeland, Herbert Faulkner (1902–1968)
Copp.	Coppey, Amedee (1874–1913)
Coppens	Coppens d'Eeckenbrugge, Geo (fl. 2001)

Corb.	Corbière, François Marie Louis (1850–1941)
Corda	Corda, August Karl Joseph (1809–1849)
Cordeiro	Cordeiro, Inês (1958–)
Cordemoy	de Cordemoy, Philippe Eugène Jacob (1837–1911)
Core	Core, Earl Lemley (1902–1984)
Cornejo	Cornejo, Xavier (fl. 2002)
Corner	Corner, Edred John Henry (1906–1996)
Correa	Correa, Maevia Noemi (1914–2005)
Corr. Serr.	Corrêa de Serra, José Francisco (1751–1823)
Correll	Correll, Donovan Stewart (1908–1983)
Cortés	Cortés, Santiago (1854–1924)
Cory	Cory, Victor Louis (1880–1964)
Cosson	Cosson, Ernest Saint-Charles (1819–1889)
Costa	Costa i Cuxart, Antonio Cipriano (Antoni Cebrià) (1817–1886)
Costantin	Costantin, Julien Noël (1857–1936)
Costich	Costich, Denise E. (fl. 1989)
Coulter	Coulter, Thomas (1793–1843)
Coulter, J.	Coulter, John Merle (1851–1928)
Courchet	Courchet, Lucien Désiré Joseph (1851–1924)
Courtois	Courtois, Richard Joseph (1806–1835)
Cout.	Coutinho, António Xavier Pereira (1851–1939)
Coutts	Coutts, John (1872–1952)
Couvreur	Couvreur, Thomas Louis Peter (1979–)
Cov.	Coville, Frederick Vernon (1867–1937)
Covas	Covas, Guillermo (1915–1995)
Cowan	Cowan, John Macqueen (1891–1960)
Cowan, C.	Cowan, Clark C. (fl. 1993)
Cowan, R.	Cowan, Richard Sumner (1921–1997)
Craib	Craib, William Grant (1882–1933)
Crane, E.	Crane, Edmund H. (fl. 1997)
Crantz	von Crantz, Heinrich Johann Nepomuk (1722–1797)
Craven	Craven, Lyndley Alan (1945–2014)
Crawford	Crawford, Daniel J. (1942–)
Crepet	Crepet, W.L. (fl. 1990)
Crépin	Crépin, François (1830–1903)
Crespo	Crespo, Manuel Benito (1962–)
Cretz.	Cretzoiu, Paul (1909–1946)
Cribb	Cribb, Phillip James (1946–)
Crisci	Crisci, Jorge Victor (1945–)
Crisp	Crisp, Michael Douglas (1950–)
Cristóbal	Cristóbal, Carmen Lelia (1932–)
Critchf.	Critchfield, William Burke (1923–1989)
Croat	Croat, Thomas Bernard (1938–)
Croizat	Croizat, Léon Camille Marius (1894–1982)
Cron	Cron, Glynis V. (fl. 1994)
Cronk	Cronk, Quentin C.B. (1959–)
Cronq.	Cronquist, Arthur John (1919–1992)
Crosa	Crosa, Orfeo (fl. 1975)
Crosswh.	Crosswhite, Frank Samuel (1940–)
Cruden	Cruden, Robert William (1936–)
Crüger	Crüger, Hermann (1818–1864)
Cruz	Cruz Durán, Ramiro (fl. 2011)
Cuatrec.	Cuatrecasas i Arumí, José (Josep) (1903–1996)
Cuervo	Cuervo Márquez, Carlos (1858–1930)
Cuf.	Cufodontis, Georg (1896–1974)
Cullen	Cullen, James (1936–2013)
Cunn. (, A.)	Cunningham, Allan (1791–1839)
Cunn., R.	Cunningham, Richard (1793–1835)
Curran	Curran, Mary Katherine (later Brandegee) (1844–1920)
Curtis	Curtis, William (1746–1799)
Curtis, C.	Curtis, Charles Henry (1869–1958)
Curtis, M.	Curtis, Moses Ashley (1808–1872)
Curtis, W.	Curtis, Winifred Mary (1905–2005)
Cusset, C.	Cusset, Colette (1944–)
Cusset, G.	Cusset, Gérard Henri Jean (1936–)
Cusson	Cusson, Pierre (1727–1783)
Cust.	Custer, Jakob Laurenz (1755–1828)
Cutler	Cutler, Hugh Carson (1912–1998)

Cutler, D.	Cutler, David Frederick (1939–)
Czech	Czech, Gerald (1930–2013)
Czerep.	Czerepanov, Sergei Kirillovich (1921–1995)
Czerniak.	Czerniakowska, Ekaterina Georgiewna (née Reineke) (1892–1942)
Czuk.	Czukavina, Anna Prokofevna (1929–1985)
Dahl	Dahl, Andreas or Anders (1751–1789)
Dahlgren, R.	Dahlgren, Rolf Martin Theodor (1932–1987)
Dahlst.	Dahlstedt, Hugo Gustav Adolf (or Gustav Adolf Hugo) (1856–1934)
Dai	Dai, Chan-Din (fl. 1985)
Dai, Q.H.	Dai, Qi-Hui (fl. 1986)
Dale	Dale, Ivan Robert (1904–1963)
Dalla Torre	von Dalla Torre, Karl Wilhelm (1850–1928)
Dallimore	Dallimore, William (1871–1959)
Dalström	Dalström, Stig (fl. 1983)
Daly	Daly, Douglas C. (1953–)
Dalz.	Dalzell, Nicolas Alexander (1817–1878)
Dalziel	Dalziel, John McEwan (1872–1948)
Dammer	Dammer, Karl Lebrecht Udo (1860–1920)
Dandy	Dandy, James Edgar (1903–1976)
Danert	Danert, Siegfried (1926–1973)
Danguy	Danguy, Paul Auguste (1862–1942)
Daniel, T.	Daniel, Thomas Franklin (1954–)
Daniell	Daniell, William Freeman (1818–1865)
Danser	Danser, Benedictus Hubertus (1891–1943)
Darbysh.	Darbyshire, S.J. (1953–)
Darbysh., I.	Darbyshire, Iain (fl. 2004)
D'Arcy	D'Arcy, William Gerald (1931–1999)
Darl. C.	Darlington, Cyril Dean (1903–1981)
Darwin, S.	Darwin, Steven P. (1949–)
Dasuki	Dasuki, Undang A. (1943–)
Daubeny	Daubeny, Charles Giles Bridle (1795–1867)
Davidse	Davidse, Gerrit (1942–)
Davidson	Davidson, Anstruther (1860–1932)
Davidson, C.	Davidson, Carol (1944–)
Davies	Davies, Frances G. (1944–)
Davies, O.	Davies, Olive Blanche (1884–1976)
Davis, C.	Davis, Charles Carroll (1911–)
Davis, P.	Davis, Peter Hadland (1918–1992)
Davy	Davy, Joseph Burtt (1870–1940)
Dawe	Dawe, Morley Thomas (1880–1943)
Dawson	Dawson, John Wyndham (1928–)
Day, A.	Day, Alva George (later Grant, Alva Day) (1920–)
Dayton	Dayton, William Adams (1885–1958)
DC.	de Candolle, Augustin-Pyramus (1778–1841)
DC., A.	de Candolle, Alphonse Louis Pierre Pyramus (1806–1893)
DC., C.	de Candolle, Anne Casimir Pyramus (1836–1918)
De	De, A.B. (fl. 1983)
De Block	De Block, Petra (fl. 1998)
De Smet, Y.	De Smet, Yannick (fl. 2015)
De Wild.	De Wildeman, Emile August(e) Joseph (1866–1947)
De Winter	De Winter, Bernard (1924–)
De Wolf	De Wolf, Gordon Barker (1927–)
Dean	Dean, Richard (1830–1905)
Deane	Deane, Henry (1847–1924)
Deb	Deb, Debandra Bijoy (1924–2013)
Deble	Deble, Leonardo Paz (fl. 2004)
Decker	Decker, Paul (1867–?)
Decker, D.	Decker, Deena S. (fl. 1988)
Decker, H.	Decker, Henry Fleming (1930–)
Decne	Decaisne, Joseph (1807–1882)
DeFilipps	DeFilipps, Robert Anthony (1939–2004)
Deflers	Deflers, Albert (1841–1921)
Degener	Degener, Otto (1899–1988)
Degener, I.	Degener, Isa (1924–)
Dehnh.	Dehnhardt, Friedrich (1787–1870)
Del.	Delile, Alire Raffeneau (1778–1850)
Delafosse	Delafosse, Gabriel (1796–1878)
Delarbre	Delarbre, Antoine (1724–1813)

Delaroche	Delaroche, Daniel (1743–1813)
Delavay	Delavay, Pierre Jean Marie (1834–1895)
Delchevalerie	Delchevalerie, Gustav (fl. 1867–1899)
Delforge, P.	Delforge, Pierre (1945–)
Delgado	Delgado Salinas, Alfonso (1950–)
Delprete	Delprete, Piero G. (fl. 1995)
Dematteis	Dematteis, Massimiliano (1970–)
Deng, M.B.	Deng, Mao-Bin (fl. 1987)
Deng, T.	Deng, Tao (fl. 2013)
Deng, Y.F.	Deng, Yun-Fei (fl. 2001)
Dennst.	Dennstaedt, August Wilhelm (1776–1826)
Der	Der, Joshua P. (fl. 2010)
Derx	Derx, H.G. (fl. 1930)
Des Moul.	Des Moulins, Charles Robert Alexandre (1798–1875)
Desc.	Descourtilz, Michel Etienne (1775–1836)
Descoings	Descoings, Bernard M. (1931–)
Desf.	Desfontaines, Réné Louiche (1750–1833)
Desp.	Desportes, Jean Baptiste Réné Pouppé (1704–1748)
Desr.	Desrousseaux, Louis Auguste Joseph (1753–1838)
Dessein	Dessein, Steven (1976–)
Desv.	Desvaux, Auguste Nicaise (Augustin) (1784–1856)
Desv., E.	Desvaux, Etienne-Emile (1830–1854)
Determann	Determann, Ronald Oskar (1951–)
Deuter	Deuter, M. (fl. 1993)
Devesa	Devesa, Juan Antonio (1955–)
Dewar	Dewar, Daniel (c. 1860–1905)
Dewey, D.	Dewey, Douglas R. (1929–)
Dewey, L.	Dewey, Lyster Hoxie (1865–1944)
Deyl	Deyl, Miloš (1906–1985)
Dhooge	Dhooge, Sandra (1976–)
Díaz, S.	Díaz, S.
Díaz de la Guardia	Díaz de la Guardia Guerrero, Consuelo (1952–)
Díaz-Miranda	Díaz-Miranda, David (1946–)
Dickinson	Dickinson, Timothy Adam (1946-)
Dicks.	Dickson, James (1738–1822)
Didr.	Didrichsen, Didrik Ferdinand (1814–1887)
Diego	Diego Perez, Nelly (fl. 2004)
Diels	Diels, Friedrich Ludwig Emil (1874–1945)
Dierb.	Dierbach, Johann Heinrich (1788–1845)
Dieterle	Dieterle, Jennie van Akkeren (1909–1999)
Dietr.	Dietrich, Friedrich Gottlieb (1768–1850)
Dietr., A.	Dietrich, Albert Gottfried (1795–1856)
Dietr., D.	Dietrich, David Nathaniel Friedrich (1799–1888)
Dill., Dillenius	Dillenius, Johann Jacob (1684–1747)
Dillon	Dillon, Michael O. (1947–)
Dillwyn	Dillwyn, Lewis Weston (1778–1855)
Ding Hou	Ding Hou (1920–2008)
Diniz, A.	Diniz, Manuel (de) Assunção (fl. 1957)
Dinsm.	Dinsmore, John Edward (1862–1951)
Dinter	Dinter, Moritz Kurt (1868–1945)
Dippel	Dippel, Leopold (1827–1914)
Dittr.	Dittrich, Manfred (1934–)
Dixit	Dixit, Ram Das (1942–2006)
Dobrocz.	Dobroczajaeva, Dariya Nikitichna (1916–1995)
Docha Neto	Docha Neto, Americo (1946–)
Dockr.	Dockrill, Alick William (1915–2011)
Dod	Dod, Donald Dungan (1912–2008)
Dode	Dode, Louis-Albert (1875–1943)
Dodson	Dodson, Calaway H. (1928–)
Doebley	Doebley, John F. (fl. 1980)
Doell	Döll, Johann(es) Christoph (Christian) (also Doell) (1808–1885)
Doerfler	Doerfler, Ignaz (also Dörfler) (1866–1950)
Dogan	Dogan, Musa (fl. 1982)
Dombey	Dombey, Joseph (1742–1796)
Domin	Domin, Karel (1882–1953)
Domke	Domke, Friedrich Walter (1899–1988)
Don, D.	Don, David (1799/1800–1841)
Don, G.	Don, George (1798–1856)

Donn	Donn, James (1758–1813)
Donnell	Donnell, Aliya A. (fl. 2012)
Dop	Dop, Paul Louis Amans (1876–1954)
Dorn	Dorn, Robert Donald (1943–)
Dorofeev	Dorofeev, V.F. (1919–1988)
Dorr	Dorr, Laurence Joseph (1953–)
Dostál	Dostál, Josef (1903–1999)
Doty	Doty, Maxwell Stanford (1911–1996)
Douglas	Douglas, David (1798–1834)
Douglas, A.	Douglas, A.W. (fl. 1996)
Doweld	Doweld, Alexander Borissovitch (1973–)
Downie	Downie, Stephen R. (fl. 1986)
Downs	Downs, Robert Jack (1923–)
Doyle	Doyle, Conrad Bartling (1884–1973)
Drake	Drake del Castillo, Emmanuel (1855–1904)
Dransf.	Dransfield, John (1945–)
Dransf., S.	Dransfield, Soejatmi (1939–)
Drapiez	Drapiez, Pierre Auguste Joseph (1778–1856)
Drejer	Drejer, Solomon (Salomon) Thomas Nicolai (1813–1842)
Dress	Dress, William John (1918–)
Dressler	Dressler, Robert Louis (1927–)
Dressler, S.	Dressler, Stefan (1964–)
Drobov	Drobov, Vasilii Petrovich (1885–1956)
Druce	Druce, George Claridge (1850–1932)
Drude	Drude, Carl Georg Oscar (1852–1933)
Drumm.	Drummond, Thomas (c. 1780–1835)
Drumm. J.L.	Drummond, James Lawson (1783–1853)
Drumm. J.R.	Drummond, James Ramsey (1851–1921)
Dryander	Dryander, Jonas Carlsson (1748–1810)
Du Puy	Du Puy, David J. (1958–)
Du Roi	Du Roi, Johann Philipp (1741–1785)
Dubard	Dubard, Marcel Marie Maurice (1873–1914)
Duby	Duby, Jean Etienne (1798–1885)
Duchartre	Duchartre, Pierre Etienne Simon (1811–1894)
Duchesne	Duchesne, Antoine Nicolas (1747–1827)
Ducke	Ducke, Walter Adolpho (1876–1959)
Dudley, T.	Dudley, Theodore Robert (1936–1994)
Dufr.	Dufresne, Pierre (1786–1836)
Dugand	Dugand, Armando (1906–1971)
Duhamel	Duhamel du Monceau, Henri Louis (1700–1782)
Duke	Duke, James A. (1929–)
Duman	Duman, Hayri (fl. 1990)
Dumaz	Dumaz-le-Grand, Noëlle (fl. 1953)
Dum.-Cours.	Dumont de Courset, George(s) Louis Marie (1746–1824)
Dummer	Dummer, Richard Arnold (1887–1922)
Dumort.	Dumortier, Barthélemy Charles Joseph (1797–1878)
Dunal	Dunal, Michel Félix (1789–1856)
Dunkley	Dunkley, Harvey Lawrence (1910–1999)
Dunlop	Dunlop, Clyde Robert (1946–)
Dunn	Dunn, Stephen Troyte (1868–1938)
Dunsterv.	Dunsterville, Galfrid Clemens Keyworth (1905–1988)
Durand	Durand, Elias Magloire (1794–1873)
Durand, T.	Durand, Théophile Alexis (1855–1912)
Durande	Durande, Jean François (1732–1794)
Durazz.	Durazzini, Antonio (1740–1810)
Durieu	Durieu de Maisonneuve, Michel Charles (1796/97–1878)
Dusén	Dusén, Per Karl Hjalmar (1855–1926)
Duthie	Duthie, John Firminger (1845–1922)
Dutta	Dutta, S. (fl. 1951)
Duval	Duval, Charles Jeunet (1751–1828)
Duvign.	Duvigneaud, Paul Auguste (1913–1991)
Duyfjes	Duyfjes, Brigitta Emma Elisabeth (1936–)
Dvorák	Dvořák, František (1921–)
Dwyer	Dwyer, John Duncan (1915–2005)
Dyer	Dyer, William Turner Thiselton (or Thistleton-) (1843–1928)
Dyer, R.A.	Dyer, Robert Allen (1900–1987)
Eastw.	Eastwood, Alice (1859–1953)
Eaton	Eaton, Amos (1776–1842)

Eaton, D.	Eaton, Daniel Cady (1834–1895)
Eberh.	Eberhardt, Philippe Albert (1874–1942)
Eberm.	Ebermaier, Johann Erdwin Christopher (1769–1825)
Eberwein	Eberwein, Roland K. (fl. 2004)
Ebinger	Ebinger, John Edwin (1933–)
Ecklon	Ecklon, Christian Frederich (1795–1868)
Edgew.	Edgeworth, Michael Pakenham (1812–1881)
Edmondson	Edmondson, John Richard (1948–)
Edwall	Edwall, Gustavo (Gustaf) (1862–1946)
Edwards	Edwards, Trevor J. (1960–)
Egan, A.	Egan, Ashley Noel (1977–)
Eggers	von Eggers, Henrik Franz Alexander (1844–1903)
Eggli	Eggli, Urs (1959–)
Egli	Egli, Bernhard (fl. 1990)
Egorova	Egorova, Tatiana Vladimirovna (1930–2007)
Ehrenb.	Ehrenberg, Christian Gottfried (1795–1876)
Ehrenb., C.	Ehrenberg, Carl August (1801–1845)
Ehrend.	Ehrendorfer, Friedrich (1927–)
Ehrh.	Ehrhart, (Jacob) Friedrich (1742–1795)
Eichler	Eichler, August Wilhelm (1839–1887)
Eifert	Eifert, Imre János (1934–)
Eig	Eig, Alexander (1894–1938)
Eiten	Eiten, George (1923–2012)
Ekman	Ekman, Hedda Maria Emerence Elisabeth (née Akerhielm) (1862–1936)
Ekman, E.	Ekman, Erik Leonard (1883–1931)
El Gazzar	El Gazzar, Adel Ibrahim Hamed (1942–)
Elffers	Elffers, Joan (1928–)
Elias, T.S.	Elias, Thomas Sam (1942–)
Elliott	Elliott, Stephen (1771–1830)
Ellis	Ellis, John (1711–1776)
Ellis, R.	Ellis, Roger P. (fl. 1984)
Elmer	Elmer, Adolph Daniel Edward (1870–1942)
Emb.	Emberger, Marie Louis (1897–1969)
Emeric	Emeric (fl. c. 1828)
Emmerich	Emmerich, Margarete (1933–)
Endl.	Endlicher, Stephan Friedrich Ladislaus (1804–1849)
Endress	Endress, Peter Karl (1942–)
Engelhorn	Engelhorn, Tamra (1945–)
Engelm.	Engelmann, Georg (1809–1884)
Engl.	Engler, Heinrich Gustav Adolf (1844–1930)
Epling	Epling, Carl Clawson (1894–1968)
Erdtman	Erdtman, Otto Gunnar Elias (1897–1973)
Erem.	Eremin, G.V. (fl. 1979–)
Eriksen	Eriksen, Bente (1960–)
Eriksson	Eriksson, Roger (1958–)
Ernst	Ernst, Alfons (1875–1968)
Ertter	Ertter, Barbara Jean (1953–)
Escal.	Escalante, Manuel Gaspar (1920–1993)
Eschsch.	von Eschscholtz, Johann Friedrich Gustav (1793–1831)
Escobar	Escobar, Rodrigo (1935–2009)
Eselt.	van Eseltine, Glen Parker (1888–1938)
Eskuche	Eskuche, Ulrich Georg (1926–2008)
Espejo	Espejo-Serna, Mario Adolfo (1951–)
Esser	Esser, Hans-Joachim (1960–)
Estrada	Estrada, Enrique (fl. 1998)
Ettingsh.	von Ettingshausen, Constantin (1826–1897)
Evans, R.	Evans, R. E.
Evans, W.E.	Evans, William Edgar (1882–1963)
Everett (J.)	Everett, Joy (1953–)
Evert	Evert, Erwin Frank Charles (1940–2010)
Ewan	Ewan, Joseph Andorfer (1909–1999)
Ewart	Ewart, Alfred James (1872–1937)
Exell	Exell, Arthur Wallis (1901–1993)
Eyma	Eyma, Pierre Joseph (1903–1945)
Fabr.	Fabricius, Philipp Conrad (1714–1774)
Fabris	Fabris, Humberto Antonio (1924–1976)
Faden	Faden, Robert B. (1942–)
Fagerl.	Fagerlind, Folke (1907–1996)

Falc.	Falconer, Hugh (1808–1865)
Fang, D.	Fang, Ding (1920–)
Fantz	Fantz, Paul R. (1941–)
Farille	Farille, Michel A. (1945–)
Farjon	Farjon, Aljos (1946–)
Farmer, S.	Farmer, Susan B. (fl. 2002)
Farron	Farron, Claude (1935–)
Farw.	Farwell, Oliver Atkins (1867–1944)
Fasano	Fasano, Angelo (fl. 1787)
Fassett	Fassett, Norman Carter (1900–1954)
Fauché	Fauché, Jean-Baptiste (fl. 1832)
Favre	Favre, Adrien (1980–)
Fawcett	Fawcett, William (1851–1926)
Fayed	Fayed, A.-A. (fl. 1979)
Fed.	Fedorov, Andrej Aleksandrovich (1909–1987)
Fedde	Fedde, Friedrich Karl Georg (1873–1942)
Fedtsch., B.	Fedtschenko, Boris Alexeevich (or Alexjewitsch) (1872–1947)
Fée	Fée, Antoine Laurent Apollinaire (1789–1874)
Feer	Feer, Heinrich (1857–1892)
Feinbrun	Feinbrun, Naomi (1900–1995)
Felger	Felger, Richard Stephen (fl. 1968)
Fenzi	Fenzi, Emanuele Orazio (1843–1924)
Fenzl	Fenzl, Eduard (1808–1879)
Fern.	Fernald, Merritt Lyndon (1873–1950)
Fern. (A.)	Fernandes, Abílio (1906–1994)
Fern., R.	Fernandes, Rosette Mercedes Saraiva Batarda (1916–2005)
Fern. Casas.	Fernández Casas, Francisco Javier (1945–)
Fern. Prieto	Fernández Prieto, J. A. (fl. 2014)
Fernández	Fernández, Z. Mayra (1948–)
Fernández, A.	Fernández- Pérez, Alvaro (1928–1994)
Fernández-Alonso	Fernández-Alonso, José Luis (1959–)
Fernández-Villar	Fernández-Villar, Celestino (1838–1907)
Fernando	Fernando, Edwino S. (1953–)
Ferreyra	Ferreyra, Ramón Alejandro (1912–2005)
Fessel	Fessel, Hans (1929-)
Feuillé	Feuillé, Louis Éconches (1660–1732)
Feuillet	Feuillet, Christian Patrice Georges-André (1948–)
Fiaschi	Fiaschi, Pedro (fl. 2003)
Ficalho	de Ficalho, Francisco Manoel Carlos de Mello (1837–1903)
Field	Field, Barron (1786–1846)
Field, D.V.	Field, David Vincent (1937–)
Figlar	Figlar, Richard B. (fl. 2000)
Fijten	Fijten, Femmy (fl. 1975)
Filat.	Filatenko, Anna A. (1937–)
Filg.(ueiras)	Filgueiras, Tarciso S. (1950–)
Finet	Finet, Achille Eugène (1863–1913)
Fingerh.	Fingerhuth, Carl Anton (1802–1876)
Fiori	Fiori, Adriano (1865–1950)
Fischer	von Fischer, Friedrich Ernst Ludwig (1782–1854)
Fischer, C.	Fischer, Cecil Ernest Claude (1874–1950)
Fischer, E(b).	Fischer, Eberhard (1961–)
Fischer, J.	Fischer, Johann Carl (1804–1885)
Fish	Fish, Lyn (fl. 2006)
Fitzg.	Fitzgerald, Robert Desmond (David) (1830–1892)
Fitzg., W.	Fitzgerald, William Vincent (1867–1929)
FitzMaurice	Fitz Maurice, Walter A. (fl. 1995)
Fleisch.	Fleischer, Max (1861–1930)
Fleming	Fleming, John (1785–1857)
Fletcher	Fletcher, Harold Roy (1907–1978)
Florence	Florence, E. Jacques M. (1951–)
Floret	Floret, Jean-Jacques (1939–)
Florin	Florin, Carl Rudolf (1894–1965)
Flueckiger	Flueckiger, Friedrich August (1828–1894)
Fluegge	Fluegge, Johannes (also Flügge) (1775–1816)
Focke	Focke, Wilhelm Olbers (1834–1922)
Focke, H.	Focke, Hendrik Charles (1802–1856)
Förther	Förther, Harald (1963–)
Fomin	Fomin, Aleksander Vasiljevich (1869–1935)

Fonnegra	Fonnegra G., Ramiro (fl. 1985)
Font Quer	Font y Quer, Pio (Pius) (1888–1964)
Forbes	Forbes, James (1773–1861)
Forbes, C.	Forbes, Charles Noyes (1883–1920)
Forbes, F.B.	Forbes, Francis Blackwell (1839–1908)
Forbes, H.M.	Forbes, Helena M.L. (1900–1959)
Ford	Ford, Neridah Clifton (1926–2006)
Ford-Lloyd	Ford-Lloyd, Brian V. (fl. 1976–7)
Foreman	Foreman, Donald Bruce (1945–2004)
Forest, F.	Forest, Felix (1972–)
Forman	Forman, Lewis Leonard (1929–1998)
Forrest	Forrest, George (1873–1932)
Forssk.	Forsskål, Pehr (1732–1763)
Forst.	Forster, Johann Reinhold (1729–1798)
Forst. f. (or G. Forster)	Forster, Johann Georg Adam (1754–1794)
Forsyth f.	Forsyth, William (1772–1835)
Fortunato	Fortunato, Renée Hersilia (1957–)
Fosb.	Fosberg, Francis Raymond (1908–1993)
Foster, M.	Foster, Mercedes S. (fl. 1984)
Foster, R.	Foster, Robert Crichton (1904–1986)
Foug.	Fougeroux de Bondaroy, Auguste Denis (1732–1789)
Fourc.	Fourcade, Henry Georges (1866–1948)
Fourn.	Fournier, Eugène Pierre Nicolas (1834–1884)
Fourn., P.	Fournier, Paul-Victor (1877–1964)
Fourr.	Fourreau, Jules Pierre (1844–1871)
Foxw.	Foxworthy, Frederick William (1877–1950)
Fraga	Fraga, Naomi S. (fl. 2012)
Franceschi	Franceschi, Francesco (1843–1924)
Franchet	Franchet, Adrien René (1834–1900)
Francis	Francis, William Douglas (1889–1959)
Franco	do Amaral Franco, João Manuel Antonio Paes (1921–2009)
Frank	Frank, Joseph C. (1782–1835)
Frank, A.	Frank, Albert Bernhard (1839–1900)
Franks	Franks, M. (1886–1961)
Franquet	Franquet, Robert Fernand (1897–1930)
Franz	Franz, C.
Frappier	Frappier, Charles (fl. 1853–95)
Fraser	Fraser, John (1750–1811)
Fraser, J.	Fraser, James (1854–1935)
Frederiksen	Frederiksen, Signe (1942–)
Freeman, G.	Freeman, George Fouché (1876–1930)
Freire	Freire, Susana Edith (1954–)
Freitag	Freitag, Helmut E. (1932–)
Frémont	Frémont, John Charles (1813–1890)
Fres.	Fresenius, Johann Baptist Georg Wolfgang (1808–1866)
Freyn	Freyn, Josef Franz (1845–1903)
Frič	Frič, Alberto Vojtech (1882–1944)
Friedmann	Friedmann, Francis (1941–)
Friedrich	Friedrich, Hans-Christian (1925–1992)
Friedrichsthal	von Friedrichsthal, Emanuel (1809–1842)
Fries	Fries, Elias Magnus (1794–1878)
Fries, R.(E.)	Fries, Klas Robert Elias (1876–1966)
Fries, T.C.E.	Fries, Thore Christian Elias (1886–1930)
Friis, I.	Friis, Ib (1945–)
Fritsch	Fritsch, Karl (1864–1934)
Friv.	Frivaldsky von Frivald, Emerich (1799–1870)
Frodin	Frodin, David Gamman (1940–)
Froebel	Froebel Karl Otto (1844–1906)
Fröhner	Fröhner, Eugene (1858–1940)
Frost	Frost, John (1803–1840)
Fryer	Fryer, Jeanette (fl. 1993)
Fryx.	Fryxell, Paul Arnold (1927–2011)
Fu, D.Z.	Fu, De-Zhi (1952–)
Fu, K.T.	Fu, Kun-Tsun (1912–)
Fuchs, H.P.	Fuchs, Hans-Peter (1928–1999)
Fuentes, S.	Fuentes-Bazàn, Susy (1977–)
Fuertes	Fuertes, Javier (1960–)
Fukuhara	Fukuhara, Tatsundo (fl. 1997)

Fung	Fung Hok-Lam (fl. 1988)
Funk	Funk, Victoria Ann (1947–)
Furnari	Furnari, Francesco (1933–)
Fursa	Fursa, T.B (fl. 1972)
Furt.	Furtado, Caetano Xavier, Dos Remedios (1897–1980)
Fuss	Fuss, Johann Mihály (1814–1883)
Gaertn.	Gaertner, Joseph (1732–1791)
Gaertn., P.	Gaertner, Philipp Gottfried (1754–1825)
Gaertn. f.	von Gaertner, Carl Friedrich (1772–1850)
Gagnebin	Gagnebin, Abraham (1707–1800)
Gagnepain	Gagnepain, François (1866–1952)
Gagnon	Gagnon, Edeline (fl. 2015)
Galasso	Galasso, Gabriele (1967–)
Galbany	Galbany-Casals, Mercè (fl. 2004)
Galeotti	Galeotti, Henri Guillaume (1814–1858)
Gallaud	Gallaud, Ernest-Isidore (fl. 1907)
Galushko	Galushko, Anatol I. (1926–1993)
Gamble	Gamble, James Sykes (1847–1925)
Gams	Gams, Helmut (1893–1976)
Gand.	Gandoger, Michel (1850–1926)
Gandhi	Gandhi, Kanchi Natarajan (1948–)
Gandilyan	Gandilyan, P.A. (1929–)
Gangulee	Gangulee, Hirendra Chandra (1914–1992)
Gao	Gao, Cheng-Zhi (1939–)
Garay	Garay, Leslie Andrew (1924–)
Garb.	Garbari, Fabio (1937–)
García-Madrid	García-Madrid, Ana S. (fl. 2015)
García-Martín	García-Martín, Felipe (1954–)
Garcke	Garcke, Christian August Friedrich (1819–1904)
Garden	Garden, Alexander (1730–1792)
Garden, J.	Garden, Joy, later J. Thompson (1923–)
Gardner	Gardner, George (1812–1849)
Gardner, C.	Gardner, Charles Austin (1896–1970)
Gardner, H.	Gardner, H.M.
Garnock-Jones	Garnock-Jones, Philip John (1950–)
Garsault	de Garsault, François Alexandre Pierre (1691–1778)
Gâteblé	Gâteblé, Gildas (fl. 2014)
Gates	Gates, David M. (1921–)
Gauba	Gauba, Erwin (1891–1964)
Gaudich.	Gaudichaud-Beaupré, Charles (1789–1854)
Gaudin	Gaudin, Jean François Aimé Gottlieb Philippe (1766–1833)
Gaussen	Gaussen, Henri Marcel (1891–1981)
Gautier	Gautier, Laurent (1960–)
Gay	Gay, Jacques Etienne (1786–1864)
Gay, C.	Gay, Claude (1800–1873)
Geel	van Geel, Petrus Cornelius (or Pierre Corneille) (1796–1836)
Geer.	Geerinck, Daniel (1945–)
Geesink	Geesink, Robert (1945–1992)
Geiseler	Geiseler, Eduard Ferdinand (1781–1827)
Geld.	Geldikhanov, A.M. (1953–)
Gemeinholzer	Gemeinholzer, Birgit (fl. 2006)
Gentry, A.	Gentry, Alwyn Howard (1945–1993)
George, A.S.	George, Alexander Segger (1939–)
Georgi	Georgi, Johann Gottlieb (1729–1802)
Gepp	Gepp, Anthony (1862–1955)
Gérardin	Gérardin, Sébastien (1751–1816)
Gerbaulet	Gerbaulet, Maike (1962–)
Gereau	Gereau, Roy E. (1947–)
Gerlach	Gerlach, Günther (1953–)
German, D.	German, Dmitry Aleksandrovich (1972–)
Gerrard	Gerrard, William Tyrer (?–1866)
Gerstb.	Gerstberger, Pedro (1951–)
Giac.	Giacomini, Valerio (1914–1981)
Gibbs	Gibbs, Lilian Suzette (1870–1925)
Gibbs, P.	Gibbs, Peter Edward (1938–)
Gibson, A.	Gibson, Alexander (1800–1867)
Giesenh.	Giesenhagen, Karl Friedrich Georg (1860–1928)
Gilbert, M.	Gilbert, Michael George (1943–)

Gift	Gift, N. (fl. 2002)
Gilg	Gilg, Ernst Friedrich (1867–1933)
Gilib.	Gilibert, Jean-Emmanuel (1741–1814)
Gillespie	Gillespie, John Wynn (?–1932)
Gillespie, L.	Gillespie, Lynn J. (1955–)
Gillett, G.	Gillett, George Wilson (1917–1976)
Gillett, J.B.	Gillett, Jan Bevington (1911–1995)
Gilli	Gilli, Alexander (1904–2007)
Gillies	Gillies, John (1792–1834)
Gillis	Gillis, William Thomas (1933–1979)
Gilly	Gilly, Charles Louis (1911–1970)
Gilmour, C.	Gilmour, C.N. (fl. 2013)
Ging.	Gingins de la Sarraz, Frédéric Charles Jean (1790–1863)
Girard	de Girard, Frédéric (1810–1851)
Giri	Giri, Girija Sankar (1950–)
Giroux	Giroux, Mathilde (fl. 1933)
Giseke	Giseke, Paul Dietrich (1741–1796)
Giul.	Giulietti, Ana Maria (1945–)
Given	Given, David Roger (1943–2005)
Givnish	Givnish, Thomas Joseph (1951–)
Glass	Glass, Charles Edward (1934–1998)
Glassman	Glassman, Sidney Frederick (1919–2008)
Glaz.	Glaziou, Auguste François Marie (1828–1906)
Gleason	Gleason, Henry Allan (1882–1975)
Gled.	Gleditsch, Johann Gottlieb (1714–1786)
Glen	Glen, Hugh Francis (1950–)
Glenny	Glenny, David Steven (1955–)
Gmelin	Gmelin, Johann Georg (1709–1755)
Gmelin, C.	Gmelin, Carl Christian (1762–1837)
Gmelin, J.	Gmelin, Johann Friedrich (1748–1804)
Gmelin, S.	Gmelin, Samuel Gottlieb (1743/5–1774)
God.-Leb.	Godefroy-Lebeuf, Alexandre (1852–1903)
Godron	Godron, Dominique Alexandre (1807–1880)
Goebel	von Goebel, Karl Immanuel Eberhard (1855–1932)
Goeppert	Göppert, Johann Heinrich Robert (1800–1884)
Goetgh.	Goetghebeur, Paul (1952–)
Goldbl(att)	Goldblatt, Peter (1943–)
Goldenb, R., Goldenberg	Goldenberg, Renato (1968–)
Goldie	Goldie, John (1793–1886)
Golosk.	Goloskokov, Vitaliĭ Petrovich (1913–1999)
Gomes	Gomes, Bernardino António (1769–1823)
Gomes, B.A.	Gomes, Bernardino António (1806–1877)
Gómes, J.	Gómes, José Corrêa (1919–1965)
Gómez (P.), L.D.	Gómez (Pignataro), Luis Diego (1944–)
Gómez-Campo	Gómez-Campo, César (1933–2009)
Gonç. (E.G.)	Gonçalves, Eduardo G. (fl. 1997)
Gontsch.	Gontscharow, Nikolai Fedorovich (1900–1942)
González	González, Francisco (fl. 1877)
González, R.	González Tamayo, Roberto (1945–)
González, S.	del Socorro González Elizondo, Maria (1953–)
Gonz. Medr.	González Medrano, Francisco (1939–)
Good, R.	Good, Ronald d'Oyley (1896–1992)
Goodman	Goodman, George Jones (1904–1999)
Goodrich	Goodrich, Sherel (fl. 1980)
Goodspeed	Goodspeed, Thomas Harper (1887–1966)
Goossens	Goossens, Antonie Petrus Gerhardy (1896–1972)
Gopalan	Gopalan, Rangasamy (1947–)
Gordon	Gordon, George (1806–1879)
Gorer	Gorer, Richard (1913–)
Górniak	Górniak, Marcin (1960–)
Gorozh.	Gorozhankin, Ivan Nikolaevich (1848–1904)
Gottl.-Tann.	von Gottlieb-Tannenhain, Paul (1879–1945)
Gottschling	Gottschling, Marc (1971–)
Gouan	Gouan, Antoine (1733–1821)
Goudot	Goudot, Justin (1822–1845)
Gould	Gould, Frank Walton (1913–1981)
Govaerts	Govaerts, Rafaël Herman Anna (1968–)
Goy	Goy, Doris Alma (1912–1999)

Goyder	Goyder, David John (1959–)
Goyena	Goyena, Miguel Ramírez (1857–1927)
Graebner	Graebner, Karl Otto Robert Peter Paul (1871–1933)
Graham	Graham, Robert C. (1786–1845)
Graham, J.	Graham, John (1805–1839)
Graham, S.	Graham, Shirley Ann Tousch (1935–)
Granados	Granados Mendoza, Carolina (fl. 2015)
Grande	Grande, Loreto (1878–1965)
Grande, J.	Grande Allende, José Ramón (1983–)
Grant, A.D.	Grant, Alva Day (formerly Day, Alva George) (1920–)
Grant, A.L.	Grant, Adele Lewis (1881–1969)
Grant, J.R.	Grant, Jason Randall (1969–)
Grant, M.	Grant, Michael Livingstone (1962–)
Grant, V.	Grant, Verne Edwin (1917–2007)
Grashoff	Grashoff, J.L.
Grassl	Grassl, Carl Otto (1908–)
Grau	Grau, Hans Rudolph Jürke (1937–)
Grauer	Grauer, Sebastian (1758–1820)
Gravely	Gravely, Frederic Henry (1885–1965)
Gray	Gray, Samuel Frederick (1766–1828)
Gray, A.	Gray, Asa (1810–1888)
Gray, J.R.	Gray, J.R. (fl. 1976)
Gray, N.	Gray, Netta Elizabeth (1913–1970)
Grayum	Grayum, Michael Howard (1949–)
Greb.	Grebenshchikov, Igor Sergeevich (1912–1986)
Greef	Greef, J.M. (fl. 1993)
Green	Green, Mary Letitia (1886–1978)
Green, J.	Green, John William (1930–)
Green, P.	Green, Peter Shaw (1920–2009)
Greene	Greene, Edward Lee (1843–1915)
Greenman	Greenman, Jesse More (1867–1951)
Gregory, W.	Gregory, Walton Carlyle (1910–1998)
Gremli	Gremli, August(e) (1833–1899)
Gren.	Grenier, Jean Charles Marie (1808–1875)
Gress	Gress, Anisa (fl. 2014)
Greuter	Greuter, Werner Rodolfo (1938–)
Grev.	Greville, Robert Kaye (1794–1866)
Greves	Greves, S. (fl. 1923–7)
Grey-Wilson	Grey-Wilson, Christopher (1944–)
Grierson	Grierson, Andrew John Charles (1929–1990)
Griess.	Griesselich, Ludwig (1804–1848)
Griff.	Griffith, William (1810–1845)
Griff., J.	Griffith, John Edward (1843–1933)
Griffiths	Griffiths, David (1867–1935)
Grimes	Grimes, James Walter (1953–)
Gris	Gris, Jean Antoine Arthur (1829–1872)
Griseb.	Grisebach, August Heinrich Rudolf (1814–1879)
Groeninckx	Groeninckx, Inge (fl. 2010)
Groenl.	Groenland, Johannes (1824–1891)
Grondona	Grondona, Eduardo M. (1911–)
Gronov.	Gronovius, Jan Fredrik (1686–1762)
Groppo	Groppo, Milton (fl. 2002)
Grosourdy	de Grosourdy, René (1807–1864)
Grose, S.	Grose, Susan Oviat (1974–)
Gross	Gross, Hugo (1888–1968)
Grossh.	Grossheim, Alexander Alfonsovich (1888–1948)
Groves	Groves, Henry (1855–1891)
Groves, J.	Groves, James (1858–1933)
Grubov	Grubov, Valery Ivanovich (1917–2009)
Grüning	Grüning, G.R. (1862–1926)
Grulich	Grulich, Vít (1956–)
Guag.	Guaglianone, Encarnación Rosa (1932–2014)
Guala	Guala, Gerald F. (fl. 1995)
Guédès	Guédès, Michel (1942–1985)
Guého	Guého, E.L. Joseph (1937–2008)
Gueldenst.	(von) Gueldenstaedt, Anton Johann (also Güldenstädt) (1745–1781)
Güemes	Güemes, Jaime
Guerke	Gürke, Robert Louis August Maximilian (1854–1911)

Guerra	Guerra, P. (fl. 1938)
Guersent	Guersent, Louis Ben (1776–1848)
Guett.	Guettard, Jean Etienne (1715–1786)
Gugerli	Gugerli, Karl (fl. 1939)
Guiard	Guiard, Josiane (1949–)
Guid.	Guidotti, Rolando (fl. 1930)
Guiggi	Guiggi, Alessandro (1973–)
Guill.	Guillarmod, Amy Frances May Gordon Jacot (1911–1992)
Guillaumin	Guillaumin, André (1885–1974)
Guillem.	Guillemin, Jean Baptiste Antoine (1796–1842)
Guinea	Guinea, Emilio (1907–1985)
Guinet	Guinet, Philippe (1925–)
Guinier	Guinier, Philibert (1876–1962)
Gumbleton	Gumbleton, William Edward (1840–1911)
Gunnerus	Gunnerus, Johan Ernst (1718–1773)
Guo	Guo, Xin Hu (fl. 1988)
Guo, Y.P.	Guo, Yan-Ping (fl. 2005)
Guss.	Gussone, Giovanni (1787–1866)
Guşul.	Guşuleac, Michail (1887–1960)
Ha	Ha, Thi Dung (fl. 1970)
Haager	Haager, Jïrí (1943–)
Hackel	Hackel, Eduard (1850–1926)
Hadač	Hadač, Emil (1914–2003)
Haegens	Haegens, Raoul Martin Anne Peter (1969–)
Haegi	Haegi, Laurence Arnold Robert (1952–)
Hágsater	Hágsater, Eric (1945–)
Haines	Haines, Henry Haselfoot I. (1867–1945)
Haines, R.	Haines, Richard Wheeler (1906–1982)
Halácsy	von Halácsy, Eugen (1842–1913)
Halb.	Halbinger, Federico (1925–2007)
Halford	Halford, David A. (fl. 1992)
Hall, H.M.	Hall, Harvey Monroe (1874–1932)
Hall, J.B.	Hall, John Bartholomew (1932–1984)
Hallberg	Hallberg, F. (1924–)
Hallé	Hallé, Nicolas (1927–)
Haller	von Haller, Victor Albrecht (1708–1777)
Hallier, H. or Hallier f.	Hallier, Johannes Gottfried ('Hans') (1868–1932)
Hamann	Hamann, Ole Jorgen (1944–)
Hamasha	Hamasha Hassan R. (fl. 2012)
Hamer	Hamer, Fritz (1912–2004)
Hamet	Raymond-Hamet (1890–1972)
Hammer	Hammer, K. (1944–)
Hammer, S.	Hammer, Steven Allen (1951–)
Hampe	Hampe, Georg Ernest Ludwig (1795–1880)
Hamzaoğlu	Hamzaoğlu, Ergin (1963–)
Hance	Hance, Henry Fletcher (1827–1886)
Hancock	Hancock, Thomas (1783–1849)
Hand.-Mazz.	Handel-Mazzetti, Heinrich (1882–1940)
Handro	Handro, Osvaldo (1908–1986)
Hanford	Hanford, William H. (fl. 1854)
Hanin	Hanin, L. (fl. 1800)
Hansen (B.)	Hansen, Bertel (1932–2005)
Hansen, A.	Hansen, Alfred (1925–2008)
Hansen, C.	Hansen, Carlo (1932–1991)
Hansen, H.	Hansen, Hans Vilhelm (1951–)
Hanst.	Hanstein, Johannes Ludwig Emil Robert von (1822–1880)
Hara	Hara, Kanesuke (1885–1962)
Hara, H.	Hara, Hiroshi (1911–1986)
Haraldson	Haraldson, Kerstin (fl. 1978)
Harder	Harder, Daniel Kenneth (1960–)
Hardham	Hardham, Clare Butterworth (1918–2010)
Hardw.	Hardwicke, Thomas (1757–1835)
Hardy, C.	Hardy, Christopher Ross (1971–)
Harley	Harley, Raymond Mervyn (1936–)
Harling	Harling, Gunnar Wilhelm (1920–2010)
Harmaja	Harmaja, Harri (1944–)
Harms	Harms, Hermann August Theodor (1870–1942)
Harper	Harper, Roland McMillan (1878–1966)

Harriman	Harriman, Neil A. (1938–)
Harris, D.J.	Harris, David J. (fl. 1999)
Hart	Hart, J.M. (fl. 1998)
Hartl	Hartl, Dimitri (1926–)
Hartley	Hartley, Thomas Gordon (1931–2016)
Hartman	Hartman, Carl Johann (1790–1849)
Hartman, R.	Hartman, Ronald Lee (1945–)
Hartmann, H.	Hartmann, Heidrun Elsbeth Klara Osterwald (1942–2016)
Hartog	Hartog, Cornelis den (1931–)
Hartvig	Hartvig, Per (1941–)
Hartweg	Hartweg, Karl Theodor (1812–1871)
Hartwiss	von Hartwiss, Nicolai Anders (1791–1860)
Harv.	Harvey, William Henry (1811–1866)
Harvey, J.	Harvey, John H.
Harz	Harz, Carl Otto (1842–1906)
Hashim.(oto)	Hashimoto, Goro (1913–)
Hashim, T.	Hashimoto, Tamotsu (1933–)
Hasselq.	Hasselquist, Fredric (1722–1752)
Hasselt	van Hasselt, Johan Coenraad (1797–1823)
Hassk.	Hasskarl, Justus Carl (1811–1894)
Hassler	Hassler, Emile (1861–1937)
Hatch	Hatch, Edwin Daniel (1919–2008)
Hatch, S.	Hatch, Stephan LaVor (1945–)
Hatusima	Hatusima, Sumihiko (1906–2008)
Hauman	Hauman, Lucien Leon (Hauman-Merck) (1880–1965)
Hausskn.	Haussknecht, Heinrich Carl (1838–1903)
Havil.	Haviland, George Darby (1857–1901)
Haw.	Haworth, Adrian Hardy (1768–1833)
Hawkes, A.	Hawkes, Alex Drum (1927–1977)
Hawkes, J.	Hawkes, John Gregory (1915–2007)
Hay, A.	Hay, Alistair James Montagu (1955–)
Hayata	Hayata, Bunzô (1874–1934)
Hayek	von Hayek, August Edler (1871–1928)
Hayne	Hayne, Friedrich Gottlob (1763–1832)
Haynes, R.	Haynes, Robert Ralph (1945–)
He, H.	He, Hai (fl. 1996)
Heads	Heads, Michael J. (1957–)
Heatubun	Heatubun, Charlie Danny (fl. 2000)
Heath	Heath, Paul V. (fl. 1983–94)
Hebenstr.	Hebenstreit, Johann Christian (1720–1795)
Heckard	Heckard, Lawrence Ray (1923–1991)
Heckel	Heckel, Edouard Marie (1843–1916)
Hedb.	Hedberg, Karl Olov (1923–2007)
Hedge	Hedge, Ian Charleson (1928–)
Hedr.	Hedrick, Ulysses Prentiss (1870–1951)
Hedwig f. (or R. Hedwig)	Hedwig, Romanus Adolf (1772–1806)
Heenan	Heenan, Peter B. (fl. 1993)
Heer	von Heer, Oswald (1809–1883)
Heering	Heering, Wilhelm Christian August (1876–1916)
Heese	Heese, Emil (1862–1914)
Hegelm.	Hegelmaier, Christoph Friedrich (1833–1906)
Heil	Heil, Hans Albrecht (1899–?)
Heilborn	Heilborn, Otto (1892–1943)
Heim	Heim, Frédéric Louis (1869–?)
Heimerl	Heimerl, Anton (1857–1942)
Heine	Heine, Heino Hermann (1923–1996)
Heintze	Heintze, Sven August (1881–1941)
Heiser	Heiser, Charles Bixler (1920–2010)
Heister	Heister, Lorenz (1683–1758)
Heklau	Heklau, Heike (fl. 2008)
Heldr.	von Heldreich, Theodor Heinrich Hermann (1822–1902)
Heller	Heller, Franz Xaver (1775–1840)
Heller, A.A.	Heller, Amos Arthur (1867–1944)
Hellwig	Hellwig, Frank H. (1958–)
Helwig	Helwig, Burghard (fl.1927)
Hemadri	Hemadri, Koppula (1938–)
Hemsl.	Hemsley, William Botting (1843–1924)
Hemsl., J.H.	Hemsley, James Hatton (1923–)

Henckel	Henckel von Donnersmarck, Leo Victor Felix (1785–1861)
Hend.	Henderson, Edward George (1782–1876)
Hend., A., Henderson, A.	Henderson, Archibald (1879–1921)
Hend., A.J.	Henderson, Andrew James (1950–)
Hend., Andr.	Henderson, Andrew (fl. 1857)
Henderson, L.	Henderson, Louis Forniquet (1853–1942)
Henderson, M.D.	Henderson, Mayda Doris (1928–)
Henderson, N.C.	Henderson, Norlan C. (1915–2016)
Henderson, R.	Henderson, Rodney John Francis (1938–)
Hendrych	Hendrych, Radovan (1926–2004)
Henfrey	Hentrey, Arthur (1819–1859)
Henkel	Henkel, Heinrich (fl. 1897–1914)
Hennipman	Hennipman, Elbert (1937–)
Henrard	Henrard, Jan Theodoor (1881–1974)
Henrickson	Henrickson, James Solberg (1940–)
Henry, A.	Henry, Augustine (1857–1930)
Henry, A.N.	Henry, Ambrose Nathaniel (1936–)
Henry, L.	Henry, Louis (1853–1903)
Henschel	Henschel, August Wilhelm Eduard Theodor (1790–1856)
Henwood	Henwood, Murray J. (fl. 1998)
Hepper	Hepper, Frank Nigel (1929–2013)
Herb.	Herbert, William (1778–1847)
Herbich	Herbich, Franz (1791–1865)
Herbst	Herbst, Derral Raymon (1934–)
Herder	von Herder, Ferdinand Gottfried Maximilian Theobold (1828–1896)
Herm., F.J.	Hermann, Frederick Joseph (1906–1987)
Hermans	Hermans, Johan (1956–)
Hern., Lizb.	Hernández Hernández, Lizbeth (fl. 2008)
Hernández, H.	Hernández, Héctor Manuel (1954–)
Herre	Herre, Adolar Gottlieb Julius (Hans) (1895–1979)
Herrera	Herrera, Alarcón de Loja, Berta (1930–)
Herrm.	Herrmann, Johann (or Jean) (1738–1800)
Hershk.	Hershkovitz, Mark A. (fl. 1990)
Herter	Herter, Wilhelm Gustav Franz (Guillermo Gustavo Francisco) (1884–1958)
Herzog	Herzog, Theodor Karl Julius (1880–1961)
Hett.	Hetterscheid, Wilbert L.A. (1957–)
Heward	Heward, Robert (1791–1877)
Hewson	Hewson, Helen Joan (1938–2007)
Heydt	Heydt, Adam (fl. 1932)
Heyne	Heyne, Benjamin (1770–1819)
Heyne, K.	Heyne, Karel (1877–1947)
Heynhold	Heynhold, Gustav (1800–1860)
Heyw.	Heywood, Vernon Hilton (1927–)
Hibb.	Hibberd, James Shirley (1825–1890)
Hicken	Hicken, Cristóbal Mariá (1875–1933)
Hiep	Nguyễn, Tiến Hiệp (1947–)
Hiern	Hiern, William Philip (1839–1925)
Hieron.	Hieronymus, Georg Hans Emmo Wolfgang (1846–1921)
Higgins	Higgins, Wesley Ervin (1949–)
Hildebr.	Hildebrand, Friedrich Hermann Gustav (1835–1915)
Hildm.	Hildmann, H. (?–1895)
Hilger	Hilger, Hartmut H. (1948–)
Hill	Hill, John (1716–1775)
Hill, A.W.	Hill, Arthur William (1875–1941)
Hill, K.	Hill, Kenneth David (1948–2010)
Hill, W.	Hill, Walter (1820–1904)
Hillc.	Hillcoat, Jean Olive Dorothy (1904–1990)
Hillebrand	Hillebrand, Wilhelm B. (1821–1886)
Hilliard	Hilliard, Olive Mary (1926–)
Hilu	Hilu, Khidir W. (fl. 1981)
Himmelb.	Himmelbauer, Wolfgang (1886–1937)
Hind	Hind, David John Nicholas (1957–)
Hinsley	Hinsley, Stewart Robert (1957–)
Hirai	Hirai, Regina Yoshie (fl. 2011)
Hiriart	Hiriart, Patricia (fl. 1981)
Hiroe	Hiroe, Minosuke (1914–2000)
Hitchc., A.	Hitchcock, Albert Spear (né Jennings) (1865–1935)
Hitchc., C.	Hitchcock, Charles Leo (1902–1986)

Hjelmq.	Hjelmquist, Karl Jesper Hakon (1905–1999)
Hjert.	Hjerting, Jens Peter Knudsen (1917–2012)
Ho	Ho, Chun-Nien (fl. 1955)
Ho, T.N.	Ho, Ting-Nung (1938–2011)
Hoch	Hoch, Peter C. (fl. 1992)
Hochr.	Hochreutiner, Bénédict Pierre Georges (1873–1959)
Hochst.	Hochstetter, Christian Ferdinand Friedrich (1787–1860)
Hochst., W.	Hochstetter, Wilhelm Christian (1825–1881)
Hodel	Hodel, Donald R. (fl. 1985)
Hodkinson	Hodkinson, Trevor R. (fl. 2001)
Hoehne	Hoehne, Frederico Carlos (1882–1959)
Hoeck	Hoeck, Fernando (1858–1915)
Hoefft	Höfft, Franz M.S.V. (fl. 1826)
Hoffm.	Hoffmann, George Franz (1761–1826)
Hoffm., F.	Hoffmann, Ferdinand (1860–1914)
Hoffm., J.J.	Hoffmann, Johann Joseph (1805–1878)
Hoffm., K.	Hoffmann, Käthe (1883–c. 1931)
Hoffm., O.	Hoffmann, Karl August Otto (1853–1909)
Hoffm., P.	Hoffmann, Philipp (fl. 1868)
Hoffm., Petra	Hoffmann, Petra (fl. 1997)
Hoffsgg.	von Hoffmannsegg, Johann Centurius (1766–1849)
Hofm., H-P.	Hofmann, Hans-Peter (fl. 1998)
Hogg	Hogg, Robert (1818–1897)
Hohen.	Hohenacker, Rudolph Friedrich (1798–1874)
Holl	Holl, Friedrich (fl. 1820–50)
Holland	Holland, John Henry (1869–1950)
Holmboe	Holmboe, Jens (1880–1943)
Holttum	Holttum, Richard Eric (1895–1990)
Holub	Holub, Josef Ludvík (1930–1999)
Hombr.	Hombron, Jacques Bernard (1800–1852)
Homolle	Homolle, Anne-Marie (1905–1988)
Honck.	Honckeny, Gerhard August (1724–1805)
Honda	Honda, Masaji (1897–1984)
Hong	Hong, De-Yuan (1936–)
Hong, S.P.	Hong, Suk-Pyo (fl. 1989)
Hong, T.	Hong, Tao (fl. 1963–1992)
Hoog	Hoog, Johannes Marius Cornelis (1865–1950)
Hoogl.	Hoogland, Ruurd Dirk (1922–1994)
Hooibrenk	Hooibrenk, Daniel (fl. 1848–1861)
Hook.	Hooker, William Jackson (1785–1865)
Hook.f.	Hooker, Joseph Dalton (1817–1911)
Hooper, S.	Hooper, Sheila Spenser (1925–)
Hoover	Hoover, Robert Francis (1913–1970)
Hoppe	Hoppe, David Heinrich (1760–1846)
Hopper	Hopper, Stephen Donald (1951–)
Horan.	Horaninow, Paul Fedorowitsch (1796–1865)
Horkel	Horkel, Johann (1769–1846)
Hornem.	Hornemann, Jens Wilken (1770–1841)
Hornsch.	Hornschuch, Christian Friedrich (1793–1850)
Hort. Allw.	Allwood Bros, Haywards Heath, England
Hort. Burkw. & Skipw.	Burkwood & Skipwith Ltd, Kingston-on-Thames, England (mid-C20)
Hort. Gand.	Hortus Gandaviensis
Hort. Huber	C. Huber & Cie, Hyères, France (C19)
Hort. Lemoine	V. Lemoine & fils, Nancy, France (1850–C20)
Hort. Makoy	Jacob-Makoy, Liège, Belgium (C19)
Hort. Sander	F. Sander & Co., St Albans, England
Hort. Tubergen	C.G. van Tubergen Ltd, Haarlem, Netherlands (1869–)
Hort. Veitch	James Veitch & Sons, Chelsea, England (early C19–1914)
Hort. Vilm.	Vilmorin-Andrieux & Cie, Paris, France (c. 1745–)
Horvat	Horvat, Ivo (1897–1963)
Hosok.	Hosokawa, Takahide (1909–1981)
Hossain	Hossain, Mosharraf (1928–)
Host	Host, Nicolaus Thomas (1761–1834)
Hotta	Hotta, Teikichi (1899–1976)
Houllet	Houllet, Jean-Baptiste (1815–1890)
Houlston	Houlston, John (fl. 1848–52)
Houston	Houston, Byron Robinson (1914–)
Houtt.	Houttuyn, Martin (1720–1798)

van Houtte	van Houtte, Louis Benoît (1810–1876)
Houz.	Houzeau de Lehaie, Jean (1867–1959)
Hovenkamp	Hovenkamp, Peter Hans (1953–)
How	How, Foon-Chew (1908–1959)
Howard, R.	Howard, Richard Alden (1917–2003)
Howell	Howell, Thomas Jefferson (1842–1912)
Howell, J.	Howell, John Thomas (1903–1994)
Howitt	Howitt, Alfred William (1830–1908)
Hoyle	Hoyle, Arthur Clagne (1905–1986)
Hsiao	Hsiao, Pei-Ken (1931–)
Hsue	Hsue, Hsiang-Hao (fl. 1963)
Hsueh	Hsueh, Chi-Ju (1921–)
Hu	Hu, Hsen-Hsu (1894–1968)
Hu, S.Y.	Hu, Shiu-Ying (1910–2012)
Hu, T.	Hu, Ta-Wei (fl. 1975)
Hua	Hua, Henri (1861–1919)
Huang	Huang, Rong-Fu (1940–)
Huang, J.Y.	Huang, Jing-Yi (1963–)
Huang, S.H.	Huang, Shu-Hua (1941–)
Hubb.	Hubbard, Frederic Tracy (1875–1962)
Hubb., C.(E.)	Hubbard, Charles Edward (1900–1980)
Huber	Huber, Jakob (Jacques) E. (1867–1914)
Huber, H.	Huber, Herbert Franz Josef (1931–2005)
Hub.-Mor.	Huber-Morath, Arthur (1901–1990)
Hudson	Hudson, William (1730–1793)
Huegel	von Huegel, Carl Alexander Anselm (Hügel) (1794–1870)
Hürl.	Hürlimann, Hans (1921–2004)
Huet, A.	Huet du Pavillon, Alfred (1829–1907)
Hughes	Hughes, Dorothy Kate (1899–1932)
Hull	Hull, John G. (1761–1843)
Hultén	Hultén, Oskar Eric Gunnar (1894–1981)
Humb.	von Humboldt, Friedrich Wilhelm Heinrich Alexander (1769–1859)
Humbert	Humbert, Jean-Henri (1887–1967)
Humbl.	Humblot, Léon (1852–1914)
Hummelinck	Hummelinck, Pieter Wagenaar (1907–)
Humphries	Humphries, Christopher John (1947–2009)
Hunt	Hunt, Trevor Edgar (1913–1970)
Hunt, D.	Hunt, David Richard (1938–)
Hunt, P.	Hunt, Peter Francis (1936–2013)
Hunter	Hunter, William (1755–1812)
Hunz.	Hunziker, Armando Theodoro (1919–2001)
Hurcombe	Hurcombe, Ruth
Hurst	Hurst, Charles Chamberlain (1870–1947)
Hurter, P.	Hurter, Petrus Johannes Hendrik (1963–)
Hurus.	Hurusawa, Isao (1916–2001)
Hutch.	Hutchinson, John (1884–1972)
Hutch., J.B.	Hutchinson, Joseph Burtt (1902–1988)
Huth	Huth, Ernst (1845–1897)
Huxley	Huxley, Camilla Rose (1952–)
Hwang, M.S.	Hwang, M.S.
Hy	Hy, Félix Charles (1853–1918)
Hyland, B.	Hyland, Bernard Patrick Matthew (1937–)
Hylander	Hylander, Hjalmer (1877–1965)
Hylander, N.	Hylander, Nils (1904–1970)
Hylmö	Hylmö, Bertil (1915–2001)
Ibaragi	Ibaragi, Yasushi (fl. 2008)
Ignatov	Ignatov, Mikhail S. (1956–)
Ihlenf.	Ihlenfeldt, Hans-Dieter (1932–)
Iinuma	Iinuma, Yokusai (1782–1865)
Ikonn.	Ikonnikov, Sergei Sergeevich (1931–2005)
Iljin	Iljin, Modest Mikhailovich (Ilyin) (1889–1967)
Iljinsk.	Iljinskaja, Irina Alekseevna (1921–2011)
Iltis	Iltis, Hugh Hellmut (1925–)
Imlay	Imlay, Joan B. (fl. 1939)
Ingram, J.	Ingram, John William (1924–)
Irmscher	Irmscher, Edgar (1887–1968)
Irv., W.	Irving, Walter (1867–1934)
Irvine	Irvine, Alexander (1793–1873)

Irvine, A.	Irvine, Anthony Kyle (1937–)
Irwin	Irwin, Howard Samuel (1928–)
Ising	Ising, Ernest Horace (1884–1973)
Itô	Itô, Tokutarô (1868–1941)
Itô, H.	Itô, Hiroshi (1909–2006)
Itô, Y.	Itô, Yoshi (1907–1992)
Ivanina	Ivanina, Lyudmila Ivanovna (1917–)
Iwarsson	Iwarsson, M. (1948–)
Iwats., Z.	Iwatsuki, Zennoske (1929–)
Iwatsuki	Iwatsuki, Kunio (1934–)
Jaarsveld	van Jaarsveld, Ernst Jacobus (1953–)
Jabl.	Jablonszky, Eugene, also Jablonski, E. (1892–1975)
Jack	Jack, William (1795–1822)
Jackson	Jackson, George (1779/1780–1811)
Jackson, A.B.	Jackson, Albert Bruce (1876–1947)
Jackson, B.D.	Jackson, Benjamin Daydon (1846–1927)
Jackson, R.	Jackson, Raymond Carl (1928–2008)
Jacobs	Jacobs, Marius (1929–1983)
Jacobs, S.	Jacobs, Surrey Wilfrid Laurance (1946–2009)
Jacobsen, H.J.	Jacobsen, Hermann Johannes Heinrich (1898–1978)
Jacq.	von Jacquin, Nikolaus (or Nicolaas) Joseph (1727–1817)
Jacq.f.	von Jacquin, Joseph Franz (1766–1839)
Jacques	Jacques, Henri Antoine (1782–1866)
Jacq.-Fél.	Jacques-Félix, Henri (1907–)
Jacquinot	Jacquinot, Honoré (1814–1887)
Jafri	Jafri, Saiyad Masudal Hasan (1927–1986)
Jain	Jain, Sudhanshu Kumar (1926–)
James	James, Lois Elsie (1914–)
Jan	Jan, Georg (1791–1866)
Janarth.	Janarthanam, Malapati Kuppuswamy (1960–)
Janchen	Janchen, Erwin Emil Alfred (1882–1970)
Jankalski	Jankalski, Stephen (fl. 1993)
Jannink	Jannink, T.A. (fl. 2002)
Janovec	Janovec, John Paul (1970–)
Jansen	Jansen, Pieter (1882–1955)
Jansen, M.	Jansen, M.E. (fl. 1979)
Jansen, R.	Jansen, Robert K. (1954–)
Járai-Komlódi	Járai-Komlódi, Magda (1931–)
Jaramillo	Jaramillo, Víctor (fl. 1984)
Jarmol.	Jarmolenko, A.V. (1905–1944)
Jarrett	Jarrett, Frances Mary (1931–2014)
Jarvis	Jarvis, Charles Edward (1954–)
Játiva	Játiva, Carlos D. (fl. 1963)
Jaub.	Jaubert, Hippolyte François (1798–1874)
Jayaw.	Jayaweera, Don Martin Arthur (1912–1982)
Jebb	Jebb, Matthew (1958–)
Jeffrey, C.	Jeffrey, Charles (1934–)
Jenjittikul	Jenjittikul, Thaya (fl. 2001)
Jenny	Jenny, Rudolph (1953–)
Jepson	Jepson, Willis Linn (1867–1946)
Jérémie	Jérémie, Joël (1944–)
Jessen	Jessen, Karl Friedrich Wilhelm (1821–1889)
Jessop	Jessop, John Peter (1939–)
Jestrow	Jestrow, Brett (fl. 2008)
Jeswiet	Jeswiet, Jacob (1879–1966)
Jiménez	Jiménez Rodríguez, Francisco
Jiménez, J.	Jiménez Ramírez, Jaime (fl. 2011)
Jin	Jin, Yue-Xing (1934–)
Jin, X.H.	Jin, Xiao-Hua (1975–)
Jir., V.	Jirásek, Václav (1906–1991)
Johansson	Johansson, Jan Thomas (fl. 1988)
Johns, R.	Johns, Robert James (1944–)
Johnson, A.M.	Johnson, Arthur Monrad (1878–1943)
Johnson, D.M.	Johnson, David Mark (1955–)
Johnson, J.(H.)	Johnson, Joseph Harry (1894–1987)
Johnson, L.	Johnson, Lawrence Alexander Sidney (1925–1997)
Johnson, L.A.	Johnson, Leigh Alma (1966–)
Johnson, P.	Johnson, Peter Neville (1946–)

Johnson, R.W.	Johnson, Robert William (1930–2012)
Johnston, I.M.	Johnston, Ivan Murray (1898–1960)
Johnston, J.R.	Johnston, John Robert (1880–1953)
Johnston, M.	Johnston, Marshall Conring (1930–)
Johow	Johow, Friedrich (or Federico) Richard Adelbert (1859–1933)
Jonch.	de Joncheere, Gerardus J. (1909–1989)
Jones	Jones, William (1746–1794)
Jones, B.M.	Jones, Brian Michael Glyn (1933–)
Jones, D.L.	Jones, David Lloyd (1944–)
Jones, H.	Jones, Henry Gordon (1939–1987)
Jones, M.E.	Jones, Marcus Eugene (1852–1934)
Jones, S.B.	Jones, Samuel Boscom (1933–)
Jones, W.G.	Jones, W.G. (fl. 1995)
Jongkind	Jongkind, Carel Christiaan Hugo (1954–)
Jonker	Jonker, Fredrik Pieter (1912–1995)
Jonker, A.	Jonker, (Verhoef) Anni Margriet Emma (1920–)
Jonss., L.	Jonsson, Lars (1946–)
Jordaan	Jordaan, Marie (1948–)
Jordan	Jordan, Claude Thomas Alexis (1814–1897)
Joseph	Joseph, J. (1928–2000)
Jost	Jost, Lou (fl. 2012)
Joubert, L.	Joubert, Lize (fl. 2008)
Jovet	Jovet, Paul Albert (1896–1991)
Jowitt	Jowitt, John F.
Judz.	Judziewicz, Emmet J. (1953–)
Juel	Juel, Hans Oscar (1863–1931)
Jum.	Jumelle, Henri Lucien (1866–1935)
Jun.	Junell, Sven Albert Brynolt (1901–)
Junghuhn	Junghuhn, Franz Wilhelm (1809–1864)
Jung-Mendaçolli	Jung-Mendaçolli, Sigrid Luiza (1952–)
Junussov	Junussov, Sabir Junussovicz (1934–)
Jusl.	Juslenius, Abrahamus Danielis (1732–1803)
Juss.	de Jussieu, Antoine Laurent (1748–1836)
Juss., A.	de Jussieu, Adrien Henri Laurent (1797–1853)
Juz.	Juzepczuk, Sergei Vasilievich (1893–1959)
Kadereit	Kadereit, Joachim Walter (1956–)
Kadereit, G.	Kadereit, Gudrun (fl. 2006)
Källersjö	Källersjö, Mari (1954–)
Kainul.	Kainulainen, Kent (fl. 2009)
Kalkman	Kalkman, Cornelis (1928–1998)
Kallunki	Kallunki, Jacquelyn A. (1948–)
Kam	Kam, Yee-Kiew (?–1981)
Kamari	Kamari, Georgia (1943–)
Kamelin	Kamelin, R.V. (1938–)
Kamm.	Kammathy, R.V. (1932–)
Kaneh.	Kanehira, Ryözö (1882–1948)
Kanis	Kanis, Andrew (1934–1986)
Kanitz	Kanitz, August (1843–1896)
Kao, P.C.	Kao, Pao-Chun (1935–)
Kappert	Kappert, Hans (1890–1976)
Karav.	Karavaev, Mikhail Nikolaevich (1903–1992)
Kårehed	Kårehed, Jesper (fl. 2001)
Karelin	Karelin, Grigorij Silyč (1801–1872)
Karl	Karl, Robert (fl. 2012)
Karis	Karis, Per Ola (1955–)
Karney	Karney, Alexander P. (fl. 2012)
Karsten	Karsten, Gustav Karl Wilhelm Hermann (1817–1908)
Kartesz	Kartesz, John T. (fl. 1990)
Karw.	von Karwinsky von Karwin, Wilhelm Friedrich (1780–1855)
Kasapl.	Kasapligil, Baki (1918–1992)
Katimas	Katimas, Liliana (fl. 1994)
Kato	Kato, Masahiro (1946–)
Kaulf.	Kaulfuss, Georg Friedrich (1786–1830)
Kausel	Kausel, Eberhard Max Leopold (1910–1972)
Kaz. Osaloo	Kazempour Osaloo, Shahrokh (1966–)
Kazmi	Kazmi, Syed Muhammad Anwar (1926–)
Kearney	Kearney, Thomas Henry (1874–1956)
Keay	Keay, Ronald William John (1920–1998)

Keck	Keck, David Daniels (1903–1995)
Keeley	Keeley, Sterling C. (1948–)
Keigh.	Keighery, Gregory John (1950–)
Kellerman	Kellerman, Maude (1888–)
Kellogg	Kellogg, Albert (1813–1887)
Kem.-Nat.	Kemularia-Natadze, Liubov Manucharovna (1891–1985)
Keng	Keng, Yi-Li (1897–1975)
Keng, H.	Keng, Hsuan (1923–2009)
Keng f.	Keng, Pai-Chieh (previously Keng, Kwan-Hou) (1917–)
Kenn., G.	Kennedy, George Clayton (1919–1980)
Kenn., H.	Kennedy, Helen Alberta (1944–)
Kenn., J.	Kennedy, John (1759–1842)
Kenn.-O'Byrne	Kennedy-O'Byrne, John Kevin Patrick (1927–)
Kensit	Kensit, Harriet Margaret Louisa (later Bolus) (1877–1970)
Kent	Kent, Douglas Henry (1920–1998)
Keraudren	Keraudren, Monique (1928–1981)
Ker(-Gawler)	Ker-Gawler, John Bellenden (also J. Gawler) (1764–1842)
Kern	Kern, Johannes Hendrikus (1903–1974)
Kern., A.	Kerner, Anton Joseph (1831–1898)
Kerner	von Kerner, Johann Simon (1755–1830)
Kerr	Kerr, Arthur Francis George (1877–1942)
Kessler, M.	Kessler, Michael (1967–)
Keyserl.	von Keyserling, (Andreëvich) Alexander Friedrich Michael Leberecht Arthur (1815–1891)
Khassanov	Khassanov, Furkat Oruncaevich (fl. 1992)
Khokhr.	Khokhrjakov, Michael Kuzmich (1905–)
Khokhr., A.P.	Khokhrjakov, Andrej Pavlovich (1933–1998)
Kiaerskov	Kiaerskov, Hjalmar Frederik Christian (1835–1900)
Kiesling	Kiesling, Roberto (1941–)
Kiew	Kiew, Ruth (1946–)
Kikuchi, N.	Kikuchi, N.
Kilian	Kilisan, Norbert (1957–)
Killeen	Killeen, Timothy John (1952–)
Killip	Killip, Ellsworth Paine (1890–1968)
Kim	Kim, Muyeol (fl. 1996)
Kim, S.C.	Kim, Seung-Chul (fl. 2012)
Kimnach	Kimnach, Myron William (1922–)
Kimura	Kimura, Arika (1900–1996)
King	King, George (1840–1909)
King, R.	King, Robert Merrill (1930–2007)
Kingdon-Ward	Kingdon-Ward, Francis (or Ward, Frank Kingdon) (1885–1958)
Kinzik.	Kinzikaëva, G.K. (1931–)
Kipp.	Kippist, Richard (1812–1882)
Kir.	Kirilov, Ivan Petrovich (1821–1842)
Kiran Raj	Kiran Raj, M.S. (fl. 1998)
Kirchner	Kirchner, Georg (1837–1885)
Kirk	Kirk, Thomas (1828–1898)
Kirk, J.	Kirk, John (1832–1922)
Kirkbr.	Kirkbride, Joseph Harold (1943–)
Kirpiczn.	Kirpicznikov, Moisey Elevich (1913–1995)
Kirschner	Kirschner, Jan (1955–)
Kissling	Kissling, Jonathan (fl. 2009)
Kit.	Kitaibel, Paul (1757–1817)
Kit Tan	Kit Tan (1953–)
Kitagawa	Kitagawa, Masao (1910–1995)
Kitam.	Kitamura, Siro (1906–2002)
Kite	Kite, Charles (1768–1811)
Kjellm.	Kjellman, Frans Reinhold (1846–1907)
Klack.	Klackenberg, Jens (1951–)
Klad.	Kladiwa, Leo (1920–1987)
Klak	Klak, Cornelia (1968–)
Klatt	Klatt, Friedrich Wilhelm (1825–1897)
Klein, E.	Klein, Erich (1931–)
Klett	Klett, Gustav Theodor (?–1827)
Kljuykov	Kljuykov, Evgeniy Vasilyevich (1950–)
Klokov	Klokov, Michail Vasiljevich (1896–1981)
Klopper	Klopper, Ronell Renett (1974–)
Klotzsch	Klotzsch, Johann Friedrich (1805–1860)

Knight, J.	Knight, Joseph (1777?–1855)
Kníže	Kníže, Karel (fl. 1969)
Knobloch	Knobloch, Irving William (1907–1999)
Knoche	Knoche, Edward Louis Herman (1870–1945)
Knoop	Knoop, Johann Herman(n) (1700–1769)
Knorr.	Knorring, O.E. (later Knorring-Neustrvjeva) (1896–1979)
Knowles	Knowles, George Beauchamp (1790–1862)
Knuth, (F.)	Knuth, Frederik Marcus (1904–1970)
Kobayashi	Kobayashi, Shiro (fl. 1997)
Kobuski	Kobuski, Clarence Emmeren (1900–1963)
Koch	Koch, Wilhelm Daniel Joseph (1771–1849)
Koch, K.	Koch, Karl Heinrich Emil Ludwig (1809–1879)
Koch, M.	Koch, Marcus A. (1967–)
Kocyan	Kocyan, Alexander (1965–)
Koechl.	Koechlin, Jean (1926–)
Koehne	Koehne, Bernhard Adalbert Emil (1848–1918)
Koeler	Koeler, Georg Ludwig (1765–1807)
Koelle	Koelle, Johann Ludwig Christian (1763–1797)
Koenen	Koenen, Erik J.M. (fl. 2012)
Koenig, (J.G.)	König (Koenig), Johann Gerhard (1728–1785)
Koenig, C.	Koenig, Karl Dietrich Eberhard (1774–1851)
Königer	Königer, Willibald (1934–)
Koern.	Koernicke (Körnicke), Friedrich August (1828–1908)
Koerte	Koerte, Franz Friedrich Ernst (1782–1845)
Kofoid	Kofoid, Charles Atwood (1865–1947)
Koi	Koi, Satoshi (1977–)
Koidz.	Koidzumi, Gen'ichi (1883–1953)
Kokwaro	Kokwaro, John Ongayo (1940–)
Kolak.	Kolakovsky, Alfred Alekseevich (1906–1997)
Komarov	Komarov, Nikolai Fedrovič (1901–1942)
Konno	Konno, Tatiana Ungaretti Paleo (fl. 2002)
Konstantinova	Konstantinova, A.I. (fl. 2005)
Koord.	Koorders, Sijfert Hendrik (1863–1919)
Korall	Korall, Petra (fl. 2006)
Kores	Kores, Paul J. (1950–)
Korovin	Korovin, Eugenii (or Yevgeni) Petrovich (1891–1963)
Korsh.	Korshinsky (Korzinskij), Sergei Ivanovitsch (1861–1900)
Korth.	Korthals, Pieter Willem (1807–1892)
Kostel.	Kosteletzky, Vincenz Franz (1801–1887)
Koster, J.	Koster, Joséphine Thérèse (1902–1986)
Kosterm.	Kostermans, André Joseph Guillaume Henri (1907–1994)
Kostr.	Kostrikin, Dmitry S. (fl. 1999)
Kotschy	Kotschy, Karl Georg Theodor (1813–1866)
Kovalevsk.	Kovalevskaja, S.S. (1929–)
Koyama, H.	Koyama, Hiroshige (1937–)
Koyama, T.	Koyama, Tetsuo Michael (1935–)
Kozo-Polj.	Kozo-Poljansky, Boris Mikhailovic (1890–1957)
Kraenzlin	Kränzlin, Friedrich Wilhelm Ludwig (1847–1934)
Krajina	Krajina, Vladimír Josef (1905–1993)
Král	Král, Milos (1932–)
Kramer	Kramer, Karl Ulrich (1928–1994)
Kramina	Kramina, Tatjana E. (fl. 1997)
Krapov.	Krapovickas, Antonio (1921–2015)
Kraschen. H.	Krascheninnikov, Ippolit Mikhailovich (1884–1947)
Krasser	Krasser, Fridolin (1863–1923)
Krause, E.	Krause, Ernst (?–1858)
Krause, K.	Krause, Kurt (1883–1963)
Krauss	Krauss, Otto (?–1935)
Krauss, C.	von Krauss, Christian Ferdinand Friedrich (1812–1890)
Krecz.	Kreczetovicz, Lev Melkhisedekovich (1878–1956)
Kress	Kress, Walter John (1951–)
Krestovsk.	Krestovskaja, Tatyana Valerievna (1953–)
Kreuz.	Kreuzinger, Kurt G. (1905–1989)
Krüger	Krüger, Åsa (fl. 2014)
Krug	Krug, Carl Wilhelm Leopold (1833–1898)
Kruijt	Kruijt, Robert C. (fl. 1996)
Krukoff	Krukoff, Boris Alexander (1898–1983)
Ku	Ku, Tsue-Chih (1931–)

Kuang	Kuang, Ko-Rjên (1914–1977)
Kuber	Kuber, G. (fl. 1964)
Kubitzki	Kubitzki, Klaus (1933–)
Kudô	Kudô, Yshun (1887–1932)
Kudr.	Kudrjaschev, S.N. (1907–1943)
Kük., Kükenthal	Kükenthal, Georg (1864–1955)
Kündig	Küundig, Jakob (1863–1933)
Küpfer, P.	Küpfer, Philippe (1942–)
Kütz., Kützing	Kützing, Friedrich Traugott (1807–1893)
Kuhl	Kuhl, Heinrich (1796–1821)
Kuhlm.	Kuhlmann, Joâo Geraldo (1882–1958)
Kuhn	Kuhn, Maximilian Friedrich Adalbert (1842–1894)
Kuijt	Kuijt, Job (1930–)
Kulju	Kulju, Kristo K.M. (fl. 2006)
Kumar	Kumar, Venkatachalam Sampath (1966–)
Kumar, M.	Kumar, Muktesh (1951–)
Kung, H.W.	Kung, Hsien-Wu (1897–)
Kunkel	Kunkel, Günther W.H. (1928–2007)
Kunth	Kunth, Karl Sigismund (1788–1850)
Kuntze	Kuntze, Carl Ernst (sometimes Eduard) Otto (1843–1907)
Kunze	Kunze, Gustav (1793–1851)
Kupicha	Kupicha, Frances Kristina (1947–)
Kupper	Kupper, Walter (1874–1953)
Kuprian.	Kuprianova, Lyndmila Andreyevna (1914–1987)
Kurz	Kurz, Wilhelm Sulpiz (1834–1878)
Kurzweil	Kurzweil, Hubert (1958–)
Kuschel	Kuschel, G. (fl. 1963)
Kuth.	Kuthatheladze, Schushana Ilyinichna (1905–?)
Kuzn.	Kuznetsov, Nikolai Ivanovich (1864–1932)
Kvist	Kvist, Lars Peter (1955–)
L.	Linnaeus (von Linné), Carl (1707–1778)
L.f.	Linnaeus (von Linné), Carl (1741–1783)
Laan, van der	van der Laan, F.M. (fl. 1986)
Lab(ill).	de Labillardière, Jacques Julien Houtou (1775–1834)
Labat	Labat, Jean-Noël (1959–2011)
Labiak	Labiak, Paulo Henrique (fl. 2000)
Labouret	Labouret, J. (fl. 1853–8)
Lacerda	de Lacerda, Kleber Garcia (1950–)
Lack	Lack, Hans Walter (1949–)
Lackey	Lackey, James A. (fl. 1978)
Ladiz.	Ladizinsky, Gideon (1936–)
Lakušić	Lakušić, Radomir (fl. 1988)
Laferr.	Laferrière, Joseph E. (fl. 1990)
LaFrankie	LaFrankie, James V. (fl. 1986)
Lagasca	Lagasca y Segura, Mariano (1776–1839)
Lagerh.	Lagerheim, Nils Gustaf (1860–1926)
Lagr.-Fossat	Lagrèze-Fossat, Adrian Rose Arnaud (1818–1874)
Lam.	de Lamarck, Jean Baptiste Antoine Pierre de Monnet (1744–1829)
Lam, H.J.	Lam, Herman Johannes (1892–1977)
Lamb	Lamb, Anthony L. (1942–)
Lamb.	Lambert, Aylmer Bourke (1761–1842)
Lambinon	Lambinon, Jacques (1936–)
Lamond	Lamond, Jenifer M. (1936–)
Lamont, E.	Lamont, Eric E. (fl. 1990)
Lander	Lander, Nicholas Sean (1948–)
Landon	Landon, John Waddell (1949–)
Landrein	Landrein, Sven (1977–)
Landrum	Landrum, Leslie Roger (1946–)
Lane, M.	Lane, Meredith A. (1951–)
Lang, K.Y.	Lang, Kai-Yung (1936–)
Lange	Lange, Johan Martin Christian (1818–1898)
Lanj.	Lanjouw, Joseph (1902–1984)
Lapeyr.	de Lapeyrouse, Philippe Picot (1744–1818)
Large	Large, Mark Frederick (1959–)
Larréat.	Larréategui, José Dionisio (fl. 1795–c. 1805)
Larsen	Larsen, Kai (1926–2012)
Lassen	Lassen, Per (1942–)
Laubenf.	de Laubenfels, David John (1925–)

Launert	Launert, Georg Oskar Edmund (1926–)
Lauterb.	Lauterbach, Carl Adolf Georg (1864–1937)
Lavallée	Lavallée, Pierre Alphonse Martin (1836–1884)
Lavin	Lavin, Matt (1956–)
Lavranos	Lavranos, John Jacob (1926–2011)
Lavrova	Lavrova, T.V. (1949–)
Law	Law, Yuh-Wu (1917–2004)
Lawalrée	Lawalrée, André Gilles Célestin (1921–2005)
Lawson	Lawson, Charles (1794–1873)
Lawson, C.	Lawson, Cheryl A. (1947–)
Lawson, P.	Lawson, Peter (fl. 1770s–1820)
Laxm.	Laxmann, Erik G. (1737–1796)
Layens	Layens, Georges (1834–1897)
Lazarides	Lazarides, Michael (1928–2011)
Le Jolis	Le Jolis, Auguste François (1823–1904)
Leach	Leach, Leslie Charles (1909–1996)
Leakey	Leakey, D.G.B. (fl. 1932)
Leal	Leal, Carlos G. (fl. 1951)
Leandri	Leandri, Jacques Désiré (1903–1982)
Leandro	Leandro do Sacromento, P. (1778–1829)
Lebrun	Lebrun, Jean-Paul Antoine (1906–1985)
Lecomte	Lecomte, Paul Henri (1856–1934)
Ledeb.	von Ledebour, Carl Friedrich (1785–1851)
Ledoux	Ledoux, E.P. (1898–)
Lee, A.T.	Lee, Alma Theodora (1912–1990)
Lee, S.K.	Lee, Shu-Kang (1915–)
Lee, W.	Lee, Wootchul (fl. 1995)
Leeke	Leeke, Georg Gustav Paul (1883–1933)
Leenh.	Leenhouts, Pieter Willem (1926–2004)
Leeuwenb.	Leeuwenberg, Anthonius Josephus Maria (1930–2010)
Lefor	Lefor, Michael William (fl. 1975)
Legrand, Legrand, D.	Legrand, Carlos Maria Diego Enrique (1901–1986)
Lehm.	Lehmann, Johann Georg Christian (1792–1860)
Lehm. F.	Lehmann, Friedrich Carl (1850–1903)
Lehtonen	Lehtonen, Samuli (fl. 2008)
Leichtlin	Leichtlin, Maximilian (1831–1910)
Leighton, F.M.	Leighton, Frances Margaret (later Mrs William Edwin Isaac) (1909–2006)
Leistner	Leistner, Otto Albrecht (1931–)
Lejoly	Lejoly, Jean (1945–)
Lellinger	Lellinger, David Bruce (1937–)
Lem.	Lemaire, Antoine Charles (1801–1871)
Leme	Leme, Elton Martinez Carvalho (1960–)
Lemmon	Lemmon, John Gill (1832–1908)
Lemoine	Lemoine, Pierre Louis Victor (1823–1911)
Lenné	Lenné, Peter Joseph (1789–1866)
Lenz, L.	Lenz, Wayne Lee (1915–)
Léon	Léon, Jorge (fl. 1949)
Leonard	Leonard, Emery Clarence (1892–1968)
Léonard	Léonard, Jean Joseph Gustave (1920–2013)
Lepechin	Lepechin, Ivan Ivanovich (1737–1802)
Lepr.	Leprieur, M.F.R. (1799–1869)
Leroy, J.	Leroy, Jean-François (1915–1999)
Les	Les, Donald H. (1954–)
Lesc.	Lescuyer, O.H. (fl. 1855–1872)
Leschen.	Leschenault de la Tour, Jean Baptiste Louis Claude Théodore (1773–1826)
Less.	Lessing, Christian Friedrich (1809–1862)
Lestib.(oudois)	Lestiboudois, Thémistocle Gaspard (1797–1876)
Letouzey	Letouzey, Réné (1918–1989)
Leute	Leute, Gerfried Horand (1941–)
Léveillé	Léveillé, Joseph-Henri (1796–1870)
Léveillé, A.	Léveillé, Augustin(e) Abel Hector (1863–1918)
Levier	Levier, Emilio (1839–1911)
Levin	Levin, Ernst Ivar (1868–)
Levyns	Levyns, Margaret Rutherford Bryan (née Mitchell) (1890–1975)
Lewis (, G.)	Lewis, Gwendoline Joyce (1909–1967)
Lewis, C.	Lewis, Carl E. (fl. 2006)
Lewis, G.P.	Lewis, Gwilym Peter (1952–)
Lewis, J.	Lewis, John (1921–)

Lewis, W.	Lewis, Walter Hepworth (1930–)
Lewton	Lewton, Frederick Lewis (1874–1959)
Lex.	de Lexarza, Juan José Martinez (1785–1824)
Leyb.	Leybold, Friedrich (1827–1879)
Leysser	von Leysser (or Leyser), Friedrich Wilhelm (1731–1815)
L'Hér(it).	L'Héritier (de Brutelle), Charles Louis (1746–1800)
Li	Li, Xi-Wen (1902–)
Li, A.J.	Li, An-Jen (fl. 1981)
Li, C.L.	Li, Chao-Luan(g) (1938–)
Li, D.Z.	Li, De-Zhu (1963–)
Li, H.L.	Li, Hui-Lin (1911–2002)
Li, J.	Li, Jie (fl. 2008)
Li, J.Q.	Li, Jian-Qiang (fl. 1968)
Li, L.	Li, Lin (fl. 2008)
Li, P.T.	Li, Ping-Tao (1936–)
Li, W.Z.	Li, Wen-Zheng (fl. 1987)
Li, Z.Y.	Li, Zheng-Yu (fl. 1987)
Liais	Liais, Emmanuel (1826–1900)
Liben	Liben, Louis (1926–2006)
Libosch.	Liboschitz, Joseph (1783–1824)
Lichtenst.	von Lichtenstein, Martin Heinrich Karl (1780–1857)
Lidén	Lidén, Magnus (1951–)
Liebl.	Lieblein, Franz Caspar (1744–1810)
Liebm.	Liebmann, Frederick Michael (1813–1856)
Liede	Liede, Sigrid (1957–)
Lilja	Lilja, Nils (1808–1870)
Lillo	Lillo, Miguel (1862–1931)
Lim, C.K.	Lim, Chong Keat (1930–)
Lim, C.L.	Lim, C.L. (fl. 2011)
Lima	de Lima, Haroldo Cavalcante (1955–)
Limpr.	Limpricht, Kurt Gustav (1834–1902)
Lin, L.	Lin, Lang-Yin
Lin, W.T.	Lin, Wan-Tao (1927–)
Lincz.	Linczevski, Igor Alexandrovich (1908–1997)
Lindau	Lindau, Gustav (1866–1923)
Lindb.	Lindberg, Sextus Otto (1835–1889)
Lindb.f.	Lindberg, Harald (1871–1963)
Linde-Laursen	Linde-Laursen, Ib (fl. 1996)
Linden	Linden, Jean Jules (1817–1898)
Linden, L.	Linden, Lucien (1851–1940)
Linden, van der	van der Linden, B.L. (fl. 1959)
Linder	Linder, Hans Peter (1954–)
Lindinger	Lindinger, Karl Hermann Leonhard (1879–1956)
Lindl.	Lindley, John (1799–1865)
Lindman	Lindman, Carl Axel Magnus (1856–1928)
Lindsay, G.	Lindsay, George Edmund (1916–2002)
Ling	Ling, Yong-Yuan (1903–1981)
Ling, Y.R.	Ling, Yeou-Ruenn (1937–)
Lingels.	von Lingelsheim, Alexander (1874–1937)
Link	Link, Johann Heinrich Friedrich (1767–1851)
Liou	Liou, L. (fl. 1997)
Lipsch.	Lipschitz, Sergej Julievitsch (1905–1983)
Lipsky	Lipsky, Vladimir Ippolitovich (also Lipskij) (1863–1937)
Lisowski	Lisowski, Stanisław (1924–2002)
Litard.	de Litardière, René Verriet (1888–1957)
Litt	Litt, Amy (fl. 2002)
Little	Little, Elbert Luther (Jr.) (1907–2004)
Little, D.P.	Little, Damon P. (fl. 2004)
Litv.	Litvinov, Dmitrij Ivanovitsch (1854–1929)
Liu	Liu, Tang-Shui (1911–)
Liu, H.	Liu, Hou (fl. 1932)
Liu, H.M,	Liu, Hong-Mei (fl. 2013)
Liu, S.W.	Liu, Sang-Wu (1934–)
Liu, Z.J.	Liu, Zhong-Jian (1958–)
Liv.	Livera, E.J.
Livsh.	Livshultz, Tatyana (1973–)
Llave	de la Llave, Pablo (1773–1833)
Lledó	Lledó, Maria Dolores (1967–)

Lo, E.	Lo, Eugenia Y. Y. (fl. 2007)
Lo, H.S.	Lo, Hsien-Shui (1927–)
Lo Presti	Lo Presti, Rosa Maria (fl. 2010)
Lodd.	Loddiges, Conrad (1738–1826)
Lodé	Lodé, Joël (1952–)
Loefgren	Loefgren (Löfgren), Johan Alberto Constantin (1854–1918)
Loefl.	Loefling, Pehr (1729–1756)
Löfstrand	Löfstrand, Stefan D. (fl. 2014)
Loes.	Loesener, Ludwig Eduard Theodor (1865–1941)
Löve	Löve, Askell (1916–1994)
Löve, D.	Löve, Doris Benta Maria (née Wahlen) (1918–2000)
Lomelí	Lomelí-Seneción, José Aquileo (fl. 1993)
Lorea-Hernández	Lorea-Hernández, Francisco G. (1956–)
Loher	Loher, August (1874–1930)
Lois. (Loisel.)	Loiseleur–Deslongchamps, Jean Louis August(e) (1774–1849)
Lojac.	Lojacono-Pojero, Michele (1853–1919)
Londoño	Londoño, Ximena (fl. 1987)
Long	Long, Bayard Henry (1885–1969)
Long, G.R.	Long, Guang-Ri (fl. 1999)
Longpre	Longpre, Edwin Keith (fl. 1970)
Loos	Loos, Götz Heinrich (1970–)
Looser	Looser, Gualterio (1898–1982)
López	López González, Ginés Alejandro (1950–)
López, A.	López, Alicia (fl. 2011)
Lopr.	Lopriore, Guiseppe (1865–1928)
Lorence	Lorence, David H. (1946–)
Lorentz	Lorentz, Paul Günther (1835–1881)
Loret	Loret, Henri (1811–1888)
Loscos	Loscos y Bernál, Francisco (1823–1886)
Losink.	Losinkaja, A.S. Losina (1903–1958)
Lotsy	Lotsy, Johannes Paulus (1867–1931)
Lott	Lott, Emily Jane (1947–)
Loud.(on)	Loudon, John Claudius (1783–1843)
Loudon, J.W.	Loudon, Jane Wells (1807–1858)
Louis	Louis, Jean Laurent Prosper (1903–1947)
Lour.	de Loureiro, Joâo (1717–1791)
Lourenço, A.R.	Lourenço, Ana Raquel Lima (fl. 2013)
Lourteig	Lourteig, Alicia (1913–2003)
Louzada	Louzada, Rafael Batista (fl. 2008)
Lowe	Lowe, Richard Thomas (1802–1874)
Lowe, A.	Lowe, A.J. (fl. 2003)
Lowry	Lowry, Porter Peter (1956–)
Lu, A.M.	Lu, An-Min(g) (1939–)
Lu, P.L.	Lu, Pei-Luen (fl. 2014)
Lu, S.G.	Lu, Shu-Gang (1957–)
Lubbers	Lubbers, Louis (1832–1905)
Lucas Rodr.	Lucas Rodríguez, Rafael (1915–1981)
Ludwig	Ludwig, Christian Gottlieb (1709–1773)
Ludwig, W.	Ludwig, Wolfgang (1923–1913)
Lückel	Lückel, Emil (1927-)
Lückh.	Lückhoff, Carl August (1914–1960)
Luer	Luer, Carlyle A. (1922–)
Luerssen	Luerssen, Christian (1843–1916)
Lundell	Lundell, Cyrus Longworth (1903–1994)
Lundin	Lundin, Roger (1955–2005)
Lunell	Lunell, Joël (1851–1920)
Luo	Luo, Yi-Bo (1964–)
Lush.	Lushington, Alfred Wyndham (c. 1860–1920)
Luteyn	Luteyn, James Leonard (1948–)
Lý	Lý, Ngoc Sâm (fl. 2008)
Lye	Lye, Kåre Arnstein (1940–)
Ma	Ma, Yu-Chuan (1916–)
Ma, X.T.	Ma, Xin-Tang (fl. 1994)
Maack	Maack, Richard Karlovich (1825–1886)
Maas	Maas, Paulus Johannes Maria (1939–)
Maas, H.	Maas-van de Kamer, Hillegonda (1941–)
Maassoumi	Maassoumi, Ali Asghan (1948–)
Mabb.	Mabberley, David John (1948–)

McAlpin	McAlpin, Bruce (fl. 1986)
McAll., McAllister	McAllister, Hugh A. (fl. 1993)
Macarthur	Macarthur, William (1800–1882)
Macbr.	Macbride, James (1784–1817)
Macbr., J.F.	Macbride, James Francis (1892–1976)
McClint.	McClintock, Elizabeth May (1912–2004)
McClint., D.	McClintock, David Charles (1913–2001)
McClure	McClure, Floyd Alonzo (1897–1970)
McDaniel	McDaniel, Sidney T. (1940–)
McDonald	McDonald, J. Andrew (fl. 1989)
MacDoug.	MacDougal, John (1954–)
McDowell	McDowell, Tim (fl. 1996)
Macfad.	Macfadyen, James (1798–1850)
Macfarl., R.	Macfarlane, Roger M. (1938–)
Macfarl., T.	Macfarlane, Terry D. (1953–)
McGillivray	McGillivray, Donald John (1935–2012)
McKelvey	McKelvey, Susan Adams (1888–1964)
McKen	McKen, Mark Johnston (1823–1872)
Mackenzie	Mackenzie, Kenneth Kent (1877–1934)
Mackinder	Mackinder, Barbara Ann (1958–)
MacLeish	MacLeish, Nanda F. Fleming (1953–)
McNeill	McNeill, John (1933–)
MacOwan	MacOwan, Peter (1830–1909)
McPherson	McPherson, Gordon (1947–)
McVaugh	McVaugh, Rogers (1909–2009)
Madison	Madison, Michael T. (1948–)
Maek., F.	Maekawa, Fumio (1908–1984)
Maekawa	Maekawa, Tokujirô (1886–1977)
Maesen	van der Maesen, Laurentius Josephus Gerardus (Jos) (1944–)
Magee	Magee, Anthony R. (fl. 2008)
Maguire	Maguire, Bassett (1904–1991)
Maiden	Maiden, Joseph Henry (1859–1925)
Maingay	Maingay, Alexander Carroll (1836–1869)
Maire	Maire, René Charles Joseph Ernest (1878–1949)
Maiti	Maiti, Gour Gopal (1947–)
Maity	Maity, Debabrata (1976–)
Majumdar	Majumdar, Radha Binod (1928–)
Makino	Makino, Tomitarô (1862–1957)
Maleki	Maleki, Zeynol-Abedin (1913–)
Malinv.	Malinvaud, Louis Jules Ernst (1836–1913)
Malme	Malme, Gustaf Oskar Andersson (1864–1937)
Malzev	Malzev, I.I. (1948–)
Manden.	Mandenova, Ida P. (1907–1995)
Mani	Mani, K.J. (fl. 1985)
Mann, G.	Mann, Gustav (1836–1916)
Mann, H.	Mann, Horace (1844–1868)
Mannheimer	Mannheimer, C.A. (fl. 2005)
Manning	Manning, John C. (fl. 1985)
Manns	Manns, Ulrike (fl. 2009)
Mansf.	Mansfeld, Rudolf (1901–1960)
Mansion	Mansion, Guilhem (1968–)
Marais	Marais, Wessel (1929–)
Maratti	Maratti, Giovanni Francesco (1723–1777)
Marcano-Berti	Marcano-Berti, Luis (fl. 1967)
Marchal	Marchal, Elie (1839–1923)
Marchand	Marchand, Nestor Léon (1833–1911)
Marchant	Marchant, William James (1886–1952)
Marchant, N.	Marchant, Neville Graeme (1939–)
Marchesi	Marchesi, Eduardo H. (1943–)
Maréchal	Maréchal, Robert Joseph Jean-Marie (1926–)
Marg.	Margońska, Hanna Bogna (1968–)
Margot	Margot, Henri (1807–1894)
Mariz	de Mariz, Joaquim (1847–1916)
Markgraf	Markgraf, Friedrich (1897–1987)
Marloth	Marloth, Hermann Wilhelm Rudolf (1855–1931)
Marquand	Marquand, Cecil Victor Boley (1897–1943)
Marquis	Marquis, Alexandre Louis (1777–1828)
Marsh, J.	Marsh, Judith Anne (1951–)

Marshall	Marshall, Humphry (1722–1801)
Marsili	Marsili, Giovanni M. (1727–1794)
Mart.	von Martius, Carl Friedrich Phillip (1794–1868)
Mart. Crov.	Martinez Crovetto, Raul (1921–1988)
Martelli	Martelli, Ugolino (1860–1934)
Martens	von Martens, Georg Matthias (1788–1872)
Martens, M.	Martens, Martin (1797–1863)
Martic.	Marticorena, Clodomiro Fidel Segundo (1929–2013)
Martínez	Martínez, Maximino (1888–1964)
Martínez G.	Martínez Garcia, Julieta (fl. 1989)
Martínez, M.	Martínez, Mahinda (fl. 1992)
Martinov	Martinov, Ivan Ivanovič (1771–1833)
Martins, L.	Martins, Ludvig (fl. 2005)
Martyn, J.	Martyn, John (1699–1768)
Martyn, T.	Martyn, Thomas M. (1736–1825)
Masam.	Masamune, Genkei (1899–1993)
Masf.	Masferrer i Arquimbau, Ramon (1850–1884)
Maslin	Maslin, Bruce Roger (1946–)
Mason, C.	Mason, Charles Thomas (1918–2012)
Masson	Masson, Francis (1741–1805)
Masson, R.	Masson, Rüdiger (fl. 2013)
Mast	Mast, Austin R. (1972–)
Masters	Masters, Maxwell Tylden (1833–1907)
Masters, J.	Masters, John William (c. 1792–1873)
Mateo	Mateo, Gonzalo (1953–)
Mathew, B.	Mathew, Brian Frederick (1936–)
Mathez	Mathez, Joël (fl. 1969)
Mathias	Mathias, Mildred Esther (1906–1995)
Mathiesen	Mathiesen, Cecilie (fl. 2010)
Mathsson	Mathsson, Albert (?–1898)
Mathur	Mathur, R. (1948–)
Maton	Maton, William George (1774–1835)
Matsum.	Matsumura, Jinzô (1856–1928)
Mattei	Mattei, Giovanni Ettore (1865–1943)
Mattf.	Mattfeld, Johannes (1895–1951)
Mattos	de Mattos, Joâo Rodrigues (1926–)
Mattos, A.	de Mattos, Armando (fl. 1968)
Mattuschka	von Mattuschka, Heinrich Gottfried (1734–1779)
Matuda	Matuda, Eizi (1894–1978)
Mavrodiev	Mavrodiev, Evgenij Vladimirovich (fl. 1999)
Maxim.	Maximowicz, Carl Johann (1827–1891)
Maxon	Maxon, William Ralph (1877–1948)
May, H.B.	May, Henry Benjamin (c. 1845–1936)
Mayer, V.	Mayer, Veronika (1959–)
Mayr	Mayr, Heinrich (1850–1911)
Mayur.	Mayuranathan, Pallassana Vaithi Pattar (1893–1939)
Mazel	Mazel, Eugène (fl. 1981)
Mazzucc.	Mazzuccato, Giovanni (1787–1814)
Medik.	Medikus, Friedrich Casimir (1736–1808)
Medrono	Medrono, Francisco González (1939–)
Medw.	Medwedew, Jakob Sergejevitsch (1847–1923)
Meerb.	Meerburgh, Nicolaas (1734–1814)
Meeuse (A.)	Meeuse, Adrianus Dirk Jacob (1914–)
Meeuwen, van	van Meeuwen, M.S. Knaap (1936–)
Meijden	van der Meijden, Ruud (1945–2007)
Meijer	Meijer, Willem (1923–2003)
Meikle	Meikle, Robert Desmond (1923–)
Meir, E.	Meir, R. (fl. 2006)
Meissn.	Meissner, Carl Daniel Friedrich (né Meisner) (1800–1874)
Mejías	Mejías, J.A. (fl. 1990)
Melchior	Melchior, Hans (1894–1984)
Melikian	Melikian, Alexander Pavlovich (1935–2008)
Méllo	Correia de Méllo, Joaquim (1816–1877)
Melville	Melville, Ronald (1903–1985)
Mendonça	Mendonça, Francisco de Ascençâo (1889–1982)
Menezes, N.	de Menezes, Nanuza Luiza (1934–)
Menzies	Menzies, Archibald (1754–1842)
Merc.	Mercier, Philippe (1781–1831)

Merc., E.	Mercier (de Coppet), Elysée (1802–1863)
Merr.	Merrill, Elmer Drew (1876–1956)
Merwe, J.	van der Merwe, Jacoba Johanna Maria (1946–)
Merxm.	Merxmüller, Hermann (1920–1988)
Messel	Messel, L.R. (fl. 1933)
Mett.	Mettenius, Georg Heinrich (1823–1866)
Meve	Meve, Ulrich (1958–)
Meyen	Meyen, Franz Julius Ferdinand (1804–1840)
Meyer	Meyer, Bernhard (1767–1836)
Meyer, C.	von Meyer, Carol Anton Andreevič (1795–1855)
Meyer, E.	Meyer, Ernst Heinrich Friedrich (1791–1858)
Meyer, F.	Meyer, Frederick Gustav (1917–2006)
Meyer, F.K.	Meyer, Friedrich Karl (1926–2012)
Meyer, G.	Meyer, Georg Friedrich Wilhelm (1782–1856)
Meyer, G.L.	Meyer, G.L. (fl. 1881)
Meyer, T.	Meyer, Teodore (1910–1974)
Mez	Mez, Carl Christian (1866–1944)
Miau	Miau, Ru-Huai (1943–)
Michael, P.W.	Michael, Peter William (fl. 1978)
Michaux	Michaux, André (1746–1802)
Michaux f.	Michaux, François André (1770–1855)
Michelang.	Michelangeli, Fabián Armando (1970–)
Micheli, M.	Micheli, Marc (1844–1902)
Micheli, P.	Micheli, Pier (Pietro) Antonio (1679–1737)
Mickel	Mickel, John Thomas (1934–)
Middleton	Middleton, David John (1963–)
Miégev.	Miégeville, Joseph (1819–1901)
Mielcarek	Mielcarek, R. (fl. 1982)
Miellez	Miellez, Auguste (?–1860)
Miers	Miers, John (1789–1879)
Miguel	Miguel, J.R. (fl. 1987)
Migula	Migula, Emil Friedrich August Walther (1863–1938)
Mikan	Mikan, Josef Gottfried (1743–1814)
Miki	Miki, Shigeru (1901–1974)
Milano	Milano, V.A. (1921–)
Mildbr.	Mildbraed, Gottfried Wilhelm Johannes (1879–1954)
Milde	Milde, Carl August Julius (1824–1871)
Millán	Millán, Aníbal Roberto (1892–)
Mill.	Miller, Philip (1691–1771)
Miller, A.	Miller, A.G. (1951–)
Miller, J.F.	Miller, John Frederick (1715–1794)
Miller, J.S.	Miller, James Spencer (1953–)
Millsp.	Millspaugh, Charles Frederick (1854–1923)
Milne	Milne, Colin (c. 1743–1815)
Milne-Redh.	Milne-Redhead, Edgar Wolston Bertram Handsley (1906–1996)
Mimeur	Mimeur, Geneviève
Ming, T.L.	Ming, Tien-Lu (1937–)
Minkw.	Minkwitz, Zenaida Alexandrovna (1878–1918/19)
Minod	Minod, Marcel Maurice (1887–1939)
Miq.	Miquel, Friedrich Anton Wilhelm (1811–1871)
Miranda	Miranda González, Faustino (1905–1964)
Mirb.	de Mirbel, Charles François Brisseau (1776–1854)
Mitch.	Mitchell, John (1711–1768)
Mitch., D.	Mitchell, David Searle (1935–)
Mitchell, A.D.	Mitchell, A.D. (fl. 1997)
Mitchell, J.D.	Mitchell, John D. (fl. 1993)
Mitchell, T.	Mitchell, T.L. (fl. 1927)
Mitford	Mitford, Algernon Bertram Freeman (1837–1916)
Mittler	Mittler, Ludwig (fl. 1844)
Miyabe	Miyabe, Kingo (1860–1951)
Mizg.	Mizgireva, Olga Fominichna (1908–199X)
Mob.	Mobayen, Sadegh (1919–2015)
Moçiño	Moçiño, José Mariano (1757–1820)
Möller, H.	Möller, Hjalmar August (1866–1941)
Möller, M.	Möller, Michael (fl. 2009)
Moench	Moench, Conrad (1744–1805)
Moeser	Moeser, Walter
Moffett	Moffett, Rodney Oliver (1937–)

Mohanan	Mohanan, Narayanan Nair (fl. 1989)
Mohr	Mohr, Daniel Matthias Heinrich (1780–1808)
Mohr, B.	Mohr, Barbara Adelheid Rosina (1953–)
Mohr, C.	Mohr, Charles Theodore (Karl Theodor) (1824–1901)
Moir	Moir, William Whitmore Goodale (1896–1985)
Molau	Molau, Ulf (1949–)
Mold.	Moldenke, Harold Norman (1909–1996)
Molero	Molero, Julián (Julià) (1946–)
Molina	Molina, Giovanni Ignazio (1737–1829)
Molina, A.	Molina, Ana María (1947–)
Moline	Moline, Philip M. (fl. 2005)
Molloy	Molloy, Brian Peter John (1930–)
Mols	Mols, Johan B. (fl. 2000)
Molseed	Molseed, Elwood Wendell (1938–1967)
Momose	Momose, Sizuo (1906–1968)
Monach.	Monachino, Joseph Vincent (1911–1962)
Moncada	Moncada Ferrera, Milagros (1937–)
Monro, A.K.	Monro, Alexandre Kenneth (1968–)
Monteiro, H.	da Costa Monteiro, Honória (1923–)
Monti	Monti, Gaetano Lorenzo (1712–1797)
Montr.	Montrouzier, Jean Xavier Hyacinthe (1820–1897)
Monts., J.	Montserrat-Marti, Josep Maria (1955–)
Monv.	de Monville, Hyppolite Boisel, Baron (1794–1863)
Mood	Mood, John Donald (1945–)
Moody	Moody, Michael L. (fl. 2005)
Moon	Moon, Alexander (?–1825)
Moore, C.	Moore, Charles (1820–1905)
Moore, D.M.	Moore, David Moresby (1933–2013)
Moore, H.	Moore, Harold Emery (1917–1980)
Moore, J.	Moore, John William (1901–1990)
Moore, L.	Moore, Lucy Beatrice (1906–1987)
Moore, S.	Moore, Spencer le Marchant (1850–1931)
Moore, T.	Moore, Thomas (1821–1887)
Moq.	Moquin-Tandon, Christian Horace Bénédict Alfred (1804–1863)
Morales	Morales, Sebastiàn Alfredo de (1823–1900)
Morales, J.	Morales, Juan Francisco (1970–)
Morales Torres	Morales Torres, Concepción (1944–)
Moran	Moran, Reid Venable (1916–2010)
Moran, R.C.	Moran, Robbin C. (fl. 1986)
Morandi	Morandi, Giambattista (fl. 1744)
Morat	Morat, Philippe (1937–)
Morden	Morden, Clifford W. (1955–)
Morelet	Morelet, Pierre Marie Arthur (1809–1892)
Morgan	Morgan, David R. (fl. 1990)
Moric.	Moricand, Stefano (1779–1854)
Morillo	Morillo, Gilberto N. (1944–)
Moris	Moris, Giuseppe Giacinto (1796–1869)
Moritz	Moritz, Otto (1904–)
Moritzi	Moritzi, Alexander (1807–1850)
Morong	Morong, Thomas (1827–1894)
Morren	Morren, Charles François Antoine (1807–1858)
Morren, C.J.	Morren, Charles Jacques Edouard (1833–1886)
Morris	Morris, Daniel (1844–1933)
Morrone	Morrone, Osvaldo (1957–2011)
Morton	Morton, Julius Sterling (1832–1902)
Morton, C.	Morton, Conrad Vernon (1905–1972)
Morton, J.K.	Morton, John Kenneth (1928–2011)
Mosco	Mosco, Alessandro (fl. 1997)
Moss	Moss, Charles Edward (1870–1930)
Mosyakin	Mosyakin, Sergei L. (1961–)
Mottet	Mottet, Séraphin Joseph (1861–1930)
Mottram	Mottram, Roy (1940–)
Mozaff.	Mozaffarian, Valiolah (1953–)
Mudie	Mudie, Robert (1777–1842)
Muell., C.	Müller, Carl Alfred (1855–)
Müll. Cl.	Müller, Claudio (fl. 1986)
Muell., F.	von Mueller, Ferdinand Jacob Heinrich (1825–1896)
Muell., J.S.	Mueller, Johann Sebastian (1715–c.1790)

Muell., O.F.	Müller, Otto Friedrich (Friderich, Fridrich, Frederik) (1730–1784)
Muell. Arg. (or J. Mueller)	Mueller, Jean (Müller) of Aargau (1828–1896)
Müll.-Doblies, D.	Müller-Doblies, Dietrich (1938–)
Müll.-Doblies, U.	Müller-Doblies, Ute (1938–)
Münchh.	von Münchhausen, Otto (1716–1774)
Muhlenb.	Muhlenberg, Gotthilf Heinrich Ernest (1753–1815)
Mukherjee, P.K.	Mukherjee, Pronob Kumar (1934–)
Muldashev	Muldashev, A.A. (1954–)
Munday	Munday, J. (1928–)
Muñoz Garm.	Muñoz Garmendia, José Felix (1949–)
Muñoz - Schick	Muñoz-Schick, Mélica (1941–)
Munro	Munro, William (1818–1889)
Munro, S.	Munro, Sioban L. (fl. 1998)
Munster	Munster, R. (fl. 1990)
Munz	Munz, Philip Alexander (1892–1974)
Munzinger	Munzinger, Jérôme (fl. 2000)
Murb.	Murbeck, Svante Samuel (1859–1946)
Murdock	Murdock, Andrew G. (fl. 2008)
Murillo	Murillo, Adolfo (1840–1899)
Murillo, J.	Murillo, José (1964–)
Murillo, M.	Murillo, María Teresz (1929–)
Muroi	Muroi, Hiroshi (1914–)
Murr., A.	Murray, Andrew (1812–1878)
Murray	Murray, Johan Andreas (1740–1791)
Murray, A. bis	Murray, Alexander (1798–1838)
Muschler	Muschler, Reinhold (Reno) Conrad (1882–1957)
Mutis	Mutis, José Celestino (1732–1808)
Mutis, S.	Mutis, Sinforoso (Sinforoso Mutis y Conswegra) (1773–1822)
Mytnik	Mytnik-Ejsmont, Joanna (1975–)
Náb.	Nábelek, František (1884–1965)
Nabiev	Nabiev, M.M. (1926–)
Nad.	Nadeaud, Jean (1834–1898)
Nagam.(asu)	Nagamasu, Hidetoshi (fl. 1986)
Nair	Nair, Velukutty Jayachandran (1940–)
Nair, N.G.	Nair, N.G. (1948–)
Naithani	Naithani, H.B. (1944–)
Nakai	Nakai, Takenoshin (1882–1952)
Nakaike	Nakaike, Toshiyuki (1943–)
Nannenga-Bremek.	Nannenga-Bremekamp, Neeltje Elizabeth (1916–1996)
Napper	Napper, Diana Margaret (1930–1972)
Naray.	Narayanaswami, V. (fl. 1949)
Nard.	Nardina, N.S. (fl. 1965)
Nash	Nash, George Valentine (1864–1921)
Nasir	Nasir, Eugene (1908–1991)
Naud.	Naudin, Charles Victor (1815–1899)
Nauenb.	Nauenburg, Johannes Dietrich (1951–2010)
Navarro	Navarro, Gonzalo (1955–)
Navarro, C.	Navarro Aranda, Carmen (1949–)
Nayar	Nayar, Madhavan Parameswaran (1932–)
Naz.(arova)	Nazarova, Estella A. (1936–)
Necker	de Necker, Noël Martin Joseph (1730–1793)
Née	Née, Louis (1734–1807)
Nees	Nees von Esenbeck, Christian Gottfried Daniel (1776–1858)
Neger	Neger, Franz Wilhelm (1868–1923)
Neill, D.	Neill, David A. (1953–)
Neilr.	Neilreich, August (1803–1871)
Nel	Nel, Gert Cornelius (1885–1950)
Nelmes	Nelmes, Ernest (1895–1959)
Nelson	Nelson, Aven (1859–1952)
Nelson, C.	Nelson, Cirilo H. (1938–)
Nelson, E.	Nelson, Ernest Charles (1951–)
Nestler	Nestler, Christian Gottfried (1778–1832)
Neto	Guarim Neto, Germano (1950–)
Neubo	Neuba, Danho Fursy Rodelec (1976–)
Neupane	Neupane, Suman (fl. 2013)
Nevski	Nevski, Sergei Arsenjevic (1908–1938)
Newberry	Newberry, John Strong (1822–1892)
Newman	Newman, Edward (1801–1876)

Newman, M.	Newman, Mark Fleming (1959–)
Newton, A.	Newton, Alan (1927–)
Nguyen	Nguyen, To Quyen (fl. 1965)
Nguyen, H.N.	Nguyen, Hoang Nghia (fl. 2010)
Nguyen, T.H.	Nguyen, Tien Hiep (fl. 1980)
Nicholas	Nicholas, Ashley (1954–)
Nicholls	Nicholls, William Henry (1885–1951)
Nicholson	Nicholson, George (1847–1908)
Nickrent	Nickrent, Daniel L. (1956–)
NicLugh.	Nic Lughadha, Eimear M. (1965–)
Nicolson	Nicolson, Dan Henry (1933–2016)
Nicora	Nicora de Panza, Elisa G. (1912–2001)
Niedenzu	Niedenzu, Franz Josef (1857–1937)
Nielsen	Nielsen, Etlar Lester (1905–2000)
Nielsen, I.	Nielsen, Ivan Christian (1946–2007)
Nieuw.	Nieuwland, Julius Arthur (1878–1936)
Niezgoda	Niezgoda, Christine J. (1950–)
Nilsson, O.	Nilsson, Örjan Eric Gustaf (1933–)
Nimmo	Nimmo, Joseph (?–1854)
Nir	Nir, Mark A. (1954–)
Nissole	Nissole, Guillaume (1647–1735)
Nixon	Nixon, Kevin C. (1953–)
Noltie	Noltie, Henry John (1957–)
Noot(eb).	Nooteboom, Hans Peter (1934–)
Nordal	Nordal, Inger (1944–)
Nordenstam, R.	Nordenstam, Rune Bertil (1936–)
Norlindh	Norlindh, Nils Tycho (1906–1995)
Norman	Norman, Cecil (1872–1947)
Normand	Normand, Didier (1908–2002)
Noronha	Noronha, Francisco (c. 1748–1787)
Norton	Norton, John Bitting Smith (1872–1966)
Not(aris), de	de Notaris, Giuseppe (1805–1877)
Novák	Novák, František Antonín (1892–1964)
Novelo	Novelo Retana, Alejandro Luis (1951–2006)
Novopokr.	Novopokrovsky, Ivan Vassiljevich (1880–1951)
Nowicke	Nowicke, Joan W. (1938–)
Noyes	Noyes, Richard David (1962–)
Nutt.	Nuttall, Thomas (1786–1859)
Nyár.	Nyárády, Erasmus Julius (1881–1966)
Nyman	Nyman, Carl Fredrik (1820–1893)
Oakeley	Oakeley, Henry Francis (fl. 2007)
Oberm.	Obermejer-Mauve, Anna Amelia (Obermeyer) (1907–2001)
Oberprieler	Oberprieler, Christoph (1964–)
O'Brien	O'Brien, James (1842–1930)
O'Byrne	O'Byrne, Peter (1955–)
Occh.	Occhioni, Paul (1915–2000)
O'Don.	O'Donnell, Carlos Alberto (1912–1954)
Odyuo	Odyuo, Nripemo (1968–)
Oefelein	Oefelein, Hans (1905–1970)
Oehme	Oehme, Hans (fl. 1940)
Oersted	Ørsted, Anders Sandøe (1816–1872)
Oh	Oh, Sang-Hun (fl. 2006)
Ohashi	Ohashi, Hiroyoshi (1936–)
Ohashi, K.	Ohashi, Kazuaki (fl. 2007)
Ohba, H.	Ohba, Hideaki (1943–)
Ohrnberger	Ohrnberger, Dieter (fl. 1993)
Ohwi	Ohwi, Jisaburo (1905–1977)
Ojeda	Ojeda, Isidro (1960–)
Okada	Okada, Hiroshi (1963–)
Okamoto	Okamoto, Motoharu (1947–)
O'Kane	O'Kane, Steve Lawrence (1956–)
Oken	Oken, Lorenz (1779–1851)
O'Leary	O'Leary, Nataly (fl. 2007)
Oliv.	Oliver, Daniel (1830–1916)
Oliveira, R.P.	de Oliveira, Reyjane Patricia (1976–)
Oliver, E.	Oliver, Edward George Hudson (1938–)
Oliver, F.	Oliver, Francis Wall (1864–1951)
Oliver, W.	Oliver, Walter Reginald Brook (1883–1957)

Olivier	Olivier, Guillaume Antoine (1756–1814)
Olmstead	Olmstead, Richard Glenn (1951–)
Olsz.	Olszewski, Tomasz Sebastian (1972–)
Omer	Omer, Saood (1957–)
Onno	Onno, Max (1903–)
Oostr.	van Oostroom, Simon Jan (1906–1982)
Opiz	Opiz, Philipp Maximilian (1787–1858)
Orb.	d'Orbigny, Alcide Dessalines (1802–1857)
Orch.	Orchard, Anthony Edward (1946–)
Orcutt	Orcutt, Charles Russell (1864–1929)
Ornduff	Ornduff, Robert (1932–2000)
O'Rorke	O'Rorke (fl. 1857)
Orph.	Orphanides, Theodoros Georgios (1817–1886)
Ortega	de Ortega, Casimiro Gómez (1740–1818)
Ortgies	Ortgies, Karl Eduard (1829–1916)
Ortíz	Ortíz, J. Javier (1957–)
Ortíz, P.	Ortíz Valdivieso, Pedro (1926–2012)
Ortiz, S.	Ortiz Núñez, Santiago (1957–)
Ortíz-Catedral	Ortíz-Catedral, Luis (1977–)
Osbeck	Osbeck, Pehr (1723–1805)
Osp.	Ospina, Hernandez Mariano (1934–)
Ossa	de la Ossa, José Antonio (?–1829)
Osten	Osten, Cornelius (1863–1936)
Ostenf.	Ostenfeld, Carl Emil Hansen (1873–1931)
Oteng-Yeboah	Oteng-Yeboah, A.A. (fl. 1970)
Otth	Otth, Carl (Karl) Adolph (1803–1839)
Otto	Otto, Christoph Friedrich (1783–1856)
Ovcz.	Ovczinnikov, Pavel Nikolaevich (1903–1979)
Overk.	Overkott, Ortrud (1914–)
Ownbey	Ownbey, Francis Marion (1910–1974)
Oxley	Oxley, John (1781–1828)
Pabst	Pabst, Guido Frederico João (1914–1980)
Packer	Packer, John G. (1929–)
Paclt	Paclt, Jiří (1925–)
Page	Page, Christopher Nigel (1942–)
Painter	Painter, Joseph Hannum (1879–1908)
Paiva	Paiva, Jorge Américo Rodrigues (1933–)
Pak (J.H.)	Pak, Jae-Hong (fl. 1992)
Pakhomova	Pakhomova, M.G. (1925–)
Palau	Palau i Verdera, Antoni (1734–1793)
Palib.	Palibin, Ivan Vladimirovich (1872–1949)
Palla	Palla, Eduard (1864–1922)
Pallas	von Pallas, Peter Simon (1741–1811)
Pampan.	Pampanini, Renato (1875–1949)
Pan, J.T.	Pan, Jin-Tang (1935–)
Pan, K.Y.	Pan, Kai-Yu (1937–)
Pancher	Pancher, Jean Armand Isidore (1814–1877)
Panero	Panero, José L. (1959–)
Pang.	Pangalo, Konstantin Ivanovič (1883–1965)
Panigr.	Panigrahi, Gopinath (1924–2004)
Pantl.	Pantling, Robert (1856–1910)
Pansarin	Pansarin, Emerson R. (fl. 2004)
Panzer	Panzer, Georg Wolfgang Franz (1755–1829)
Paol.	Paoletti, Giulio (1865–1941)
Pappe	Pappe, Carl Wilhelm Ludwig (1803–1862)
Pardo	Pardo de Tavera, Trinidad Herménégilde José (1857–1925)
Paris	Paris, Jean Édouard Gabriel Narcisse (1827–1911)
Parish	Parish, Samuel Bonsall (1838–1928)
Parish, C.	Parish, Charles Samuel Pollock (1822–1897)
Parker, R.	Parker, Richard Neville (1884–1958)
Parkinson	Parkinson, Sydney C. (1745–1771)
Parkinson, C.E.	Parkinson, Charles Edward (1890–1945)
Parl.	Parlatore, Filippo (1816–1877)
Parnell, J.	Parnell, John Adrian Naicker (1954–)
Parodi	Parodi, Domingo (1823–1890)
Parodi, L.	Parodi, Lorenzo Raimundo (1895–1966)
Parolly	Parolly, Gerald (1964–)
Parra	Parra, Lara Regina (fl. 2010)

Parry	Parry, William Edward (1790–1855)
Parry, C.	Parry, Charles Christopher (1823–1890)
Parsa	Parsa, Ahmed (Ahmad) (1907–1997)
Pascher	Pascher, Adolf A. (1881–1945)
Pastore	Pastore, José Floriano Barêa (fl. 2006)
Paton, A.	Paton, Alan James (1963–)
Pau	Pau y Español, Carlos (1857–1937)
Paul	Paul, Tapas Kumar (1956–)
Paula-Souza	de Paula-Souza, Juliana (1975–)
Pauquy	Pauquy, Charles Louis Constant (1800–1854)
Pavlov	Pavlov, Nikolai Vasilievich (1893–1971)
Pavón	Pavón, José Antonio (1754–1844)
Pavone	Pavone, Petro (1948–)
Pax	Pax, Ferdinand Albin (1858–1942)
Paxton	Paxton, Joseph (1803–1865)
Payson	Payson, Edwin Blake (1893–1927)
Pearson	Pearson, William Henry (1849–1923)
Pease	Pease, Arthur Stanley (1881–1964)
Peck	Peck, Charles Horton (1833–1917)
Pedersen	Pedersen, Troels Myndel (1916–2000)
Pedersen, H.	Pedersen, Henrik Aerenlund (1966–)
Pedro	Pedro, José Gomes (1915–2010)
Peixoto	Peixoto, Ariane Luna (1947–)
Pellegrin	Pellegrin, François (1881–1965)
Pellet., J.	Pelletier, Pierre Joseph (1788–1842)
Pelser	Pelser, Pieter B. (1976–)
Pelt.(ier), M.	Peltier, M. (fl. 1965)
Peñailillo	Peñailillo Brito, Patricio (fl. 1996)
Penn.	Pennington, Terence Dale (1938–)
Pennell	Pennell, Francis Whittier (1886–1952)
Penny	Penny, George (? –1838)
Pépin	Pépin, Pierre Denis (c. 1802–1876)
Per. Moura	Pereira-Moura, Maria Verônica Leite (1960–)
Peralta	Peralta, Paola (fl. 2008)
Perkins	Perkins, Janet Russell (1853–1933)
Perleb	Perleb, Karl Julius (1794–1845)
Perr.	Perrottet, Georges Samuel (1793–1870)
Perrie	Perrie, Leon R. (fl. 2003)
Perrier	Perrier de la Bâthie, Eugène Pierre (1825–1916)
Perrot	Perrot, Émile Constant (1867–1951)
Perry	Perry, Lily May (1895–1992)
Perry, P.	Perry, Pauline Lesley (1927–)
Pers.	Persoon, Christiaan Hendrik (1761–1836)
Persson	Persson, Nathan Petter Herman (1893–1978)
Persson, C.	Persson, Claes Håkan (1960–)
Petagna	Petagna, Vincenzo (1734–1810)
Peter	Peter, Gustav Albert (1853–1937)
Petermann	Petermann, Wilhelm Ludwig (1806–1855)
Petersen	Petersen, Otto George (1847–1937)
Peterson	Peterson, Paul M. (1954–)
Petit	Petit, Felix (fl. 1824)
Petit, E.	Petit, Ernest Marie Antoine (1927–2007)
Petrie	Petrie, Donald (1846–1925)
Petrovsky	Petrovsky, V.V. (1930–)
Pettigrew	Pettigrew, Jack D. (fl. 2012)
Peyr.	Peyritsch, Johann Joseph (1835–1889)
Pfeiffer	Pfeiffer, Louis (Ludwig) Karl Georg (1805–1877)
Pfeiffer, H.	Pfeiffer, Hans Heinrich (1890–1970)
Pfeiffer, N.	Pfeiffer, Norma Etta (1889–1989)
Pfitzer	Pfitzer, Ernst Hugo Heinrich (1846–1906)
Pfosser	Pfosser, Martin (fl. 2003)
Phan	Phan Kê Lôc (1935–)
Philbrick	Philbrick, C. Thomas (fl. 1993)
Philcox	Philcox, David (1926–2003)
Philippi	Philippi, Rulolf Amandus (Rodolfo Amando) (1808–1904)
Philippi, F.	Philippi, Federico (Friedrich Heinrich Eunom) (1838–1910)
Philipson	Philipson, William Raymond (1911–1997)
Phillips, B.	Phillips, Barry W. (fl. 2011)

Phillips, E.	Phillips, Edwin Percy (1884–1967)
Phillips, S.	Phillips, Sylvia Mabel (1945–)
Phipps, C.	Phipps, Carlie J. (fl. 1996)
Phipps, J.	Phipps, James Bird (1934–)
Phukan	Phukan, Sandhyajyoti (1950–)
Phuph.	Phuphathanaphong, Leena (1936–)
Picheans.	Picheansoonthon, Chayan (fl. 2004)
Pichi-Serm.	Pichi-Sermolli, Rodolfo Emilio Giuseppe (1912–2005)
Pichon	Pichon, Marcel (1921–1954)
Pickersgill	Pickersgill, Barbara (1940–)
Pieper	Pieper, Gustav Robert (fl. 1908)
Pierce	Pierce, John Hwett (1912–)
Pierlot	Pierlot, R. (fl. 1996)
Pierre	Pierre, Jean Baptiste Louis (1833–1905)
Pilg.	Pilger, Robert Knud Friedrich (1876–1953)
Pill.	Pillans, Neville Stuart (1884–1964)
Piltz	Piltz, J. (fl. 1980)
Pim.	Pimenov, Michael Georgievich (1937–)
Pinkley	Pinkley, Homer Virgil (1938–)
Pintaud	Pintaud, Jean-Christophe (fl. 1998)
Pinter	Pinter, Michael (fl. 2013)
Piper	Piper, Charles Vancouver (1867–1926)
Pippen	Pippen, Richard W. (1935–)
Pires	Pires, Joâo Murça (1917–1994)
Pitard	Pitard, Charles-Joseph Marie (1873–1937)
Pittier	Pittier (de Fábrega), Henri François (1857–1950)
Planch.	Planchon, Jules Emile (1823–1888)
Plancke	Plancke, Jacqueline (1937–)
van der Plas	van der Plas, F. (fl. 1970)
Plitm.	Plitmann, Uzi (1936–)
Plowes	Plowes, Darrel Charles Herbert (1925–2016)
Plum.	Plumier, Charles (1646–1704)
Plunkett, G.	Plunkett, Gregory M. (1965–)
Pobed.	Pobedimova, Eugenia Georgievna (1898–1973)
Pócs	Pócs, Tamás (1933–)
Podlech	Podlech, Dieter (1931–)
Poelln.	von Poellnitz, Karl (1896–1945)
Poeppig	Poeppig, Eduard Friedrich (1798–1868)
Pogg.	Poggenburg, Justus Ferdinand (1840–1893)
Pohl	Pohl, Johann Baptist Emanuel (1782–1834)
Poir.	Poiret, Jean Louis Marie (1755–1834)
Poisson	Poisson, Jules (1833–1919)
Poit.	Poiteau, Pierre Antoine (1766–1854)
Poitr.	Poitrasson, R.P. (fl. 1873–8)
Pojark.	Pojarkova, Antonina Ivanovna (1897–1980)
Polatschek	Polatschek, Adolf (1932–)
Polj.	Poljakov, Petr Petrovich (1902–1974)
Pollard	Pollard, Charles Louis (1872–1945)
Pollard, B.	Pollard, Benedict John (1972–)
Pollard, G.	Pollard, Glenn E. (1901–1976)
Pomel	Pomel, Auguste Nicolas (1821–1898)
Pongratz	Pongratz, D. (fl. 1997)
Pope	Pope, Willis Thomas (1873–1961)
Popov	Popov, Mikhail Grigorévich (1893–1955)
Popovkin	Popovkin, Alex V. (fl. 2011)
Porta	Porta, Pietro (1832–1923)
Porter	Porter, Thomas Conrad (1822–1901)
Porter, C.L.	Porter, Charles Lyman (1889–?)
Porter, D.	Porter, Duncan MacNair (1937–)
Porter, R.	Porter, Robert Ker (1779–1842)
Porto	Porto (Porte), Paulo Campos (1889–1968)
Poselger	Poselger, Heinrich (1818–1883)
Posluszny	Posluszny, Usher (fl. 1976)
Post	Post, George Edward (1838–1909)
Potter, D.	Potter, Daniel (fl. 1994)
Potztal	Potztal, Eva Hedwig Ingeborg (1924–2000)
Pourret	Pourret de Figeac, Pierre André (1754–1818)
Powell	Powell, Richard (1767–1834)

Powell, A.M.	Powell, Albert Michael (1937–)
Powell, J.	Powell, Jocelyn Marie (1939–)
Powell, M.	Powell, Martyn (1975–)
Powrie	Powrie, Elizabeth (1925–1977)
Pradhan	Pradhan, Udai Chandra (1949–)
Prado	Prado, Jefferson (1964–)
Praeger	Praeger, Robert Lloyd (1865–1953)
Prain	Prain, David (1857–1944)
Praminik	Praminik, B.B. (1933–)
Prance	Prance, Ghillean ('Iain') Tolmie (1937–)
Prantl	Prantl, Karl Anton Eugen (1849–1893)
Presl, C.	Presl, Carel Bořivoj (1794–1852)
Presl, J.	Presl, Jan Swatopluk (1791–1849)
Preuss	Preuss, Paul Rudolf (1861–)
Price, M.	Price, Michael Greene (1941–)
Price, R.	Price, Robert A. (fl. 1988)
Pridgeon	Pridgeon, Alec Melton (1949–)
Prina	Prina, Anibal Oscar (1957–)
Prince, W.	Prince, William Robert (1795–1869)
Pringle, J.S.	Pringle, James Scott (1937–)
Pritzel, E.	Pritzel, Ernst Georg (1875–1946)
Probat.	Probatova, N.S. (1939–)
Proença	Proença, Carolyn Elinore Barnes (1956–)
Progel	Progel, August (1829–1889)
Pruesapan	Pruesapan, Kanchana (1973–)
Pruski	Pruski, John Francis (1955–)
Pryer	Pryer, Kathleen M. (fl. 1993)
Pu, F.T.	Pu, Fa-Ting (1936–)
Puff	Puff, Christian (1949–2013)
Pugsley	Pugsley, Herbert William (1868–1947)
Pull.	Pulliatt, Victor (1827–1866)
Purdy	Purdy, Carlton Elmer (1861–1945)
Pursh	Pursh, Frederick Traugott (Pursch, Friedrich Traugott) (1774–1820)
Pusalkar	Pusalkar, Prashant Keshav (1978–)
Putterl.	Putterlick, Alois (1810–1845)
Puttock	Puttock, Christopher Francis (1954–)
Putzeys	Putzeys, Jules (or Julius) Antoine Adolphe Henri (1809–1882)
Pyck	Pyck, Nancy (fl. 1998)
Qaiser	Qaiser, Mohammad (1946–)
Qi	Qi, Cheng-Jing (1932–)
Qin	Qin, Hai-Ning (1960–)
Qin, D.H.	Qin, De-Hai (fl. 1988)
Queiroz	de Queiroz, Luciano Paganucci (1958–)
Quesnay, du	du Quesnay, M.C. (fl. 1971)
Quezada	Quezada, Max (1936–)
Quinn	Quinn, Christopher John (1936–)
Quis.	Quisumbing y Argüelles, Eduardo (1895–1986)
Raab-Straube	von Raab-Straube, Eckhard (1970–)
Rach	Rach, Louis Theodor (1821–1859)
Raddi	Raddi, Giuseppe (1770–1829)
Radl	Radl, Florian (1862–1911)
Radlk.	Radlkofer, Ludwig Adolph Timotheus (1829–1927)
Räusch.	Räuschel, Ernst Adolf (fl. 1772–1797)
Raf.	Rafinesque-Schmaltz, Constantine Samuel (1783–1840)
Raim.	Raimann, Rudolf (1863–1896)
Raiz.	Raizada, Mukat Behari (1907–2007)
Rajkumar	Rajkumar, S. (fl. 2000)
Rakot.	Rakotoarinivo, Mijoro (1980–)
Ralph	Ralph, Thomas Shearman (1813–1891)
Ramach.	Ramachandran, Veerambakkam Srinivasan (1953–)
Ramam.	Ramamurthy, Kandasamy (1933–)
Ramamoorthy	Ramamoorthy, Thennilapuram Parasuraman (1945–)
Ramat.	d'Audibert de Ramatuelle, Thomas Albin Joseph (1750–1794)
Ramírez	Ramírez, José (1852–1904)
Ramírez, I.	Ramírez Morillo, Ivón Mercedes (fl. 1987)
Ramond	Ramond de Caronbonnière, Louis François Elisabeth (1753–1827)
Ramsay, H.	Ramsay, Helen Patricia (1928–)
Randrianasolo	Randrianasolo, Armand (1958–)

Ranney	Ranney, Thomas G. (fl. 2003)
Rao, A.	Rao, Aragula Sathyanarayana (1924–1983)
Rao, A.N.	Rao, Abbareddy Nageswara (1954–)
Rao, R.	Rao, Rolla Seshagiri (1921–2015)
Raoul	Raoul, Edouard Fiacre Louis (1815–1852)
Rapaics	Rapaics, Raymund (1885–1953)
Rapini	Rapini, Alessandro (fl. 2001)
Rasm.	Rasmussen, Finn Nygaard (1948–)
Rassulova	Rassulova, M.R. (1926–)
Rattray	Rattray, James McFarlane (1907–1974)
Raup	Raup, Hugh Miller (1901–1995)
Raus	Raus, Thomas (1949–)
Rauschert	Rauschert, Stephan (1931–1986)
Raven	Raven, Peter Hamilton (1936–)
Ravenna	Ravenna, Pedro Felix (1938–)
Ray, M.	Ray, Martin Forbes (fl. 1998)
Raynal	Raynal, Jean (1933–1979)
Raynal, A.	Raynal, Aline Marie Roques (1937–)
Raynaud	Raynaud, Christian (1939–)
Rayss	Rayss, Tscharna (1890–1965)
Razafim.	Razafimandimbison, Sylvain Georges (1964–)
Read	Read, Robert William (1931–2003)
Rec.	Record, Samuel James (1881–1945)
Rech.f.	Rechinger, Karl Heinz (1906–1998)
Redouté	Redouté, Pierre Joseph (1759–1840)
Reduron	Reduron, Jean-Pierre (1950–)
Reed, C.	Reed, Clyde Franklin (1918–1999)
Reeder	Reeder, John Raymond (1914–2009)
Reeder, C.	Reeder, Charlotte Olive (née Goodding) (1916–2009)
Rees, B.	Rees, Bertha (fl. 1912)
Reese	Reese, Heinrich (fl. 1931–1939)
Regel	von Regel, Eduard August (1815–1892)
Rehder	Rehder, Alfred (1863–1949)
R(ei)chb.	Reichenbach, Heinrich Gottlieb Ludwig (1793–1879)
R(ei)chb.f.	Reichenbach, Heinrich Gustav (1824–1889)
Reiche	Reiche, Carlos Frederico (1860–1929)
Reichst.	Reichstein, Tadeus (1897–1996)
Reifenb., A.	Reifenberger, Adam (fl. 1992)
Reifenb., U.	Reifenberger, Ursula (fl. 1992)
Reim.	Reimers, Hermann Johann O. (1893–1961)
Reinw.	Reinwardt, Caspar George Carl (1773–1854)
Reissek	Reissek, Siegfried (1819–1871)
Remesh	Remesh, M. (fl. 2001)
Remy	Remy, Esprit Alexandre (1826–1893)
Rendle	Rendle, Alfred Barton (1865–1938)
Renvoize	Renvoize, Stephen Andrew (1944–)
Renz	Renz, Jany (1907–1999)
Req.	Requien, Esprit (1788–1851)
Retz.	Retzius, Anders Johan (1742–1821)
Reuter	Reuter, Georges François (1805–1872)
Rev.	Reveal, James Lauritz (1941–2011)
Reynaud	Reynaud, A.A. (1804–?)
Reynolds, S.	Reynolds, Sally T. (1932–)
Rezn.	Reznicek, Anton Albert (1950–)
Rheede	van Reede tot Drakenstein, Hendrik Adriaan (1637–1691)
Ribeiro	Ribeiro, Pedro (fl. 2007)
Riccob.	Riccobono, Vincenzo (1861–1943)
Rice	Rice, Rod (1963–)
Rich.	Richard, Louis Claude Marie (1754–1821)
Rich., A.	Richard, Achille (1794–1852)
Richardson	Richardson, Ian Bertram Kay (1940–)
Richter	Richter, August Gottlieb (1742–1812)
Richter, H.	Richter, Hans Georg (fl. 1987)
Rickett	Rickett, Harold William (1896–1989)
Rico	Rico Arce, Maria de Lourdes (1955–)
Riddell	Riddell, John Leonard (1807–1865)
Ridl.	Ridley, Henry Nicholas (1855–1956)
Ridsd.	Ridsdale, Colin Ernest (1944–2017)

Riedel	Riedel, Ludwig (1790–1861)
Riedl	von Riedl, Harald Udo (1936–)
Rigo	Rigo, Gregorio (1841–1922)
Říha	Říha, Jan (1947–)
Riley	Riley, John (c. 1796–1846)
Riley, L.	Riley, Lawrence Athelstan Molesworth (1889–1928)
Risso	Risso, Joseph Antoine (1777–1845)
Ritter, F.	Ritter, Friedrich (1898–1989)
Riv.	Rivière, Charles Marie (1845– ?)
Riv., M.	Rivière, Marie Auguste (1821–1877)
Rivas-Martínez	Rivas-Martínez, Salvador (1935–)
Rivers	Rivers, Thomas (1798–1877)
Rizz.	Rizzini, Carlos Toledo (1921–1992)
Roalson	Roalson, Eric Howard (1969–)
Robbrecht	Robbrecht, Elmar (1946–)
Roberts, R.	Roberts, Roland P. (fl. 2004)
Robertson, K.	Robertson, Kenneth R. (1941–)
Roberty	Roberty, Guy Edouard (1907–1971)
Robin	Robin, Claude Cesar (1750– ?)
Robinson	Robinson, Benjamin Lincoln (1864–1935)
Robinson, C.	Robinson, Charles Budd (1871–1913)
Robinson, H.	Robinson, Harold Ernest (1932–)
Robson, N.	Robson, Norman Keith Bonner (1928–)
Robyns	Robyns, André Georges Marie Walter Albert (1935–2003)
Robyns, F.	Robyns, Frans Herbert Edouard Arthur Walter (1901–1986)
Rochel	Rochel, Anton (1770–1847)
Rock	Rock, Joseph Francis Charles (1884–1962)
Rodigas	Rodigas, Emile (1831–1902)
Rodionenko	Rodiónenko, Gueorgui Ivánovich (1913–2014)
Rodr.	Rodrigues, William Antonio (1928–)
Rodr., Aarón	Rodríguez, Aarón (1966–)
Rodrigues, L.	Rodrigues, L.
Rodríguez	Rodríguez, José Demetrio (1780–1846)
Rodríguez Femenías	Rodríguez y Femenías, Juan Joaquin (1839–1905)
Rodway	Rodway, Leonard (1853–1936)
Roehl.	Röhling, Johann Christoph (1757–1813)
Roem.	Roemer, Johann Jacob (1763–1819)
Roem., M.	Roemer, Max Joseph (1791–1849)
Röser	Röser, Martin (fl. 1989)
Roessler	Roessler, Helmut (1926–)
Roezl	Roezl, Benedikt (Benito) (1824–1885)
Rogers, C.	Rogers, Claude Marvin (1919–2015)
Rogers, D.	Rogers, David James (1918–2007)
Rogers, R.	Rogers, Richard Sanders (1862–1942)
Rogers, Z.	Rogers, Zachary Scott (1976–)
Rohr	von Rohr, Julius Bernard (1686–1742)
Rohr, J.P.	von Rohr, Julius Philip Benjamin (1737–1793)
Rohw.	Rohweder, Otto (1919–)
Rohwer	Rohwer, Jens Gunter (1958–)
Roiv.	Roivainen, Heikki (1900–1983)
Rojas	Rojas, Teodoro (1877–1954)
Rojas, N.	Rojas Acosta, Nicolás (1873–1947)
Rolander	Rolander, Daniel (1725–1793)
Rolfe	Rolfe, Robert Allen (1855–1921)
Rol.-Goss.	Roland-Gosselin, Robert (1854–1925)
Rollins	Rollins, Reed Clark (1911–1998)
Romasch.	Romaschenko, Konstantyn (fl. 2008)
Romero	Romero, Rafael Castaneda (1910–1973)
Romero, G.	Romero, Gustavo A. (1955–)
Romero García	Romero García, Ana Teresa (1957–)
Romero-Zarco	Romero-Zarco, Carlos (1954–)
Romowicz	Romowicz, Agnieszka (1960–)
Ronn.	Ronniger, Karl (1871–1954)
Roon	de Roon, Adrianus Cornelis (1928–)
Roos	Roos, Marco C. (1955–)
Roque	Roque, Nadia (fl. 1997)
Ros.	Rosenstock, Eduard (1856–1938)
Rosch.	Roschevicz, Roman Julievich (or Roshevitz, Rozevic) (1882–1949)

Roscoe	Roscoe, William (1753–1831)
Rose	Rose, Joseph Nelson (1862–1928)
Roseng.	Rosengurtt, Bernardo (1916–1985)
Roshk.	Roshkova, Olga Ivanovna (1909–1989)
Ross, J.	Ross, James Henderson (1941–)
Ross, R.	Ross, Robert (1912–2005)
Ross, T.	Ross, Timothy Samuel (1962–)
Rosser	Rosser, Effie Moira (1923–1987)
Rossow	Rossow, Ricardo A. (1956–)
Roth	Roth, Albrecht Wilhelm (1757–1834)
Rothm.	Rothmaler, Werner Hugo Paul (1908–1962)
Rothr.	Rothrock, Joseph Trimble (1839–1922)
Rottb.	Rottbøll (Rottboell), Christen Friis (1727–1797)
Rottler	Rottler, Johan Peter (1749–1836)
Roul.	Rouleau, Joseph Albert Ernest (1916–1991)
Rourke	Rourke, John Patrick (1942–)
Roussel	Roussel, Henri François Anne de (1747–1812)
Roux, J.	Roux, Jacobus Petrus (1954–2013)
Rouy	Rouy, Georges C. Chr. (1851–1924)
Rowley, G.	Rowley, Gordon Douglas (1921–)
Roxb.	Roxburgh, William (1751–1815)
Roy	Roy, G.P. (1939–)
Royen	van Royen, David (1727–1799)
Royen, P.	van Royen, Pieter (1923–2003)
Royle	Royle, John Forbes (1798–1858)
Rozier	Rozier, François (Jean-François) (1734–1793)
R.-Sm.	Radcliffe-Smith, Alan (1938–2007)
Rubtzov	Rubtzov, G.A. (1887–1942)
Rudge	Rudge, Edward (1763–1846)
Rudolphi	Rudolphi, Karl Asmund (1771–1832)
Rudolphi, F.	Rudolphi, Friedrich Karl Ludwig (1801–1849)
Ruempler	Ruempler (or Rümpler), Karl Theodor (1817–1891)
Rüssmann	Rüssmann, Martin (1940–)
Rugayah	Rugayah (1946–)
Rúgolo	Rúgolo de Agrasar, Sulma E. (1940–)
Ruhl.	Ruhland, Wilhelm Otto Eugen (1878–1960)
Ruíz	Ruíz López, Hipólito (1754–1815)
Rumphius	Rumpf, Georg Eberhard (1628–1702)
Rupp	Rupp, Herman Montague Rucker (1872–1956)
Ruppius	Ruppius, Heinrich Bernard (1688–1719)
Rupr.	Ruprecht, Franz Josef Ivanovich (1814–1870)
Rusby	Rusby, Henry Hurd (1855–1940)
Ruschi	Ruschi, Augusto (1915–1986)
Ryan, A.	Ryan, Angela (1955–)
Rydb.	Rydberg, Per Axel (1860–1931)
Rye	Rye, Barbara Lynette (1952–)
Rzazade	Rzazade, Rza Jakhja Ogly (1909–)
Rzed.	Rzedowski, Jerzy (1926–)
Saarela	Saarela, Jeffery Michael (1978–)
Sabine	Sabine, Joseph (1770–1837)
Sacc.	Saccardo, Pier Andrea (1845–1920)
Sachet	Sachet, Marie-Hélène (1922–1986)
Saff.	Safford, William Edwin (1859–1926)
Safina	Saphina, L.K. (1961–)
Sagást.	Sagástegui Alva, Abundio (1932–2012)
Sagot	Sagot, Paul Antoine (1821–1888)
Sahagúa	Sahagúa-Godínez, Eduardo (fl. 1997)
Sakai	Sakai, Shoko (fl. 1996)
Sakis.	Sakisaka, Michiji (1895–?)
Salariato	Salariato, Diego L. (fl. 2009)
Salazar	Salazar, Gerardo A. (1961–)
Saldanha, J.	Saldanha da Gama, José de (1839–1905)
Salisb.	Salisbury, Richard Anthony (né Markham) (1761–1829)
Salm-Dyck	Salm-Reifferscheid-Dyck, Joseph Franz Maria Anton Hubert Ignatz Fürst zu (1773–1861)
Salter	Salter, Terence Macleane (1882–1969)
Salywon	Salywon, Andrew M. (fl. 2008)
Salzm.	Salzmann, Philipp (1781–1851)

Samain	Samain, Marie-Stéphanie (fl. 2006)
Samp.	Sampaio, Gonçalo António da Silva Ferreira (1865–1937)
Samp., A.	de Sampaio, Alberto José (1881–1946)
Samuelsson	Samuelsson, Gunnar (1885–1944)
Sanchez, A.	Sanchez, Adriana (fl. 2011)
Sancho	Sancho, Gisela (fl. 1999)
Sand., S.	Sanderson, Stewart C. (fl. 1987)
Sander	Sander, Henry Frederick Conrad (1847–1920)
Sanders, R.	Sanders, Roger William (1950–)
Sandw.	Sandwith, Noel Yvri (1901–1965)
Sands, M.	Sands, Martin Jonathan Southgate (1938–)
Sano	Sano, Paulo Takeo (1966–)
Santi	Santi, Giorgio (1746–1822)
Santisuk	Santisuk, Thawatchai (1944–)
Sarg.	Sargent, Charles Sprague (1841–1927)
Sastry	Sastry, A.R.K. Ramakrishna (1938–)
Sath.(ish) Kumar	Sathish Kumar, C. (1957–)
Sauer, G.	Sauer, G. (fl. 1980)
Sauer, W.	Sauer, Wilhelm (1935–)
Saunders, R.	Saunders, Richard M.K. (1964–)
Sauquet	Sauquet, Hervé Jacques Xavier (1977–)
Sauvag.	de Sauvages, Pierre-Augustin Boissier (1710–1795)
Savat.	Savatier, Paul Alexandre (1824–1886)
Savat., P.A.L.	Savatier, Paul Amedée Ludovic (1830–1891)
Savi	Savi, C. Gaëtano (1769–1844)
Savigny	de Savigny, Marie Jules César Lélorgne (1777–1851)
Săvul.	Săvulescu, Traian (1889–1963)
Scataglini	Scataglini, M. Amalia (fl. 2012)
Sch. Rodr.	Schütz Rodrigues, Rodrigo (fl. 2006)
Schaef. (er), H.	Schaefer, Hanno (1975–)
Schäferhoff	Schäferhoff, Bastian (fl. 2009)
Schaeff.	Schäffer (Schaeffer), Jacob Christian (H. von) (1718–1790)
Schauer	Schauer, Johannes Conrad (1813–1848)
Scheele	Scheele, George Heinrich Adolf (1808–1864)
Scheen	Scheen, Anne-Cathrine (fl. 2007)
Scheffer, R.	Scheffer, Rudolph Herman Christiaan Carel (1844–1880)
Scheidw.	Scheidweiler, Michael Joseph François (1799–1861)
Schellenb.	Schellenberg, Gustav August Ludwig David (1882–1963)
Schenck, M.	Schenck, Martin (fl. 1907)
Scheng.	Schengelia, E.M. (fl. 1953)
Schenk	Schenk, Joseph August (1815–1891)
Scherb.	Scherbius, Johannes (1769–1813)
Schery	Schery, Robert Walker (1917–1987)
Scheuchz.	Scheuchzer, Johannes Gaspar (1684–1738)
Schiede	Schiede, Christian Julius Wilhelm (1798–1836)
Schildh.(auer)	Schildhauer, Herbert (1963–)
Schilling, E.	Schilling, Edward E. (1953–)
Schimp.	Schimper, Wilhelm Philipp (1808–1880)
Schindler	Schindler, Anton Karl (1879–1964)
Schinz	Schinz, Hans (1858–1941)
Schipcz.	Schipczinski, Nikolaj Valerianovich (1886–1955)
Schischkin	Schischkin, Boris Konstantinovich (1886–1963)
Schkuhr	Schkuhr, Christian (1741–1811)
Schldl.	von Schlechtendal, Diederich Franz Leonhard (1794–1866)
Schltr.	Schlechter, Friedrich Richard Rudolf (1872–1925)
Schleiden	Schleiden, Matthias Jacob (1804–1881)
Schmalh.	Schmalhausen, Johannes Theodor (1849–1894)
Schmarse	Schmarse, Helmut (fl. 1933)
Schmeiss	Schmeiss, Oskar (fl. 1906)
Schmid	Schmid, Maurice (1922–)
Schmidel	Schmidel, Casimir Christoph (1718–1792)
Schmidt	Schmidt, Franz (1751–1834)
Schmidt, F.W.	Schmidt, Franz Wilibald (1764–1796)
Schmidt, J.A.	Schmidt, Johann Anton (1823–1905)
Schnack	Schnack, Benno Julio Christian (1910–1981)
Schneev.	Schneevoogt, George Voorhelm (Schneevoight) (1775–1850)
Schneider, C.	Schneider, Camillo Karl (1876–1951)
Schneid.(er), H.	Schneider, Harald (1962–)

Schnitzl.	Schnitzlein, Adalbert Carl Friedrich Hellwig Conrad (1814–1868)
Schnizlein	Schnizlein, (Karl Friedrich Christoph) Wilhelm (1780–1856)
Schodde	Schodde, Richard (1936–)
Schönl.	Schönland, Selmar (1860–1940)
Scholler	Scholler, Markus (fl. 1996)
Scholz	Scholz, Joseph B. (1858–1915)
Scholz, H.	Scholz, Hildemar Wolfgang (1928–2012)
Schomb.	Schomburgk, Robert Hermann (1804–1865)
Schomb., M.	Schomburgk, Moritz Richard (1811–1891)
Schot, A.M.	Schot, Anne M. (fl. 1994)
Schott	Schott, Heinrich Wilhelm (1794–1865)
Schottky	Schottky, Ernst Max (1888–1915)
Schousboe	Schousboe, Peder Kofod Anker (1766–1832)
Schouten	Schouten, R.T.A. (fl. 1986)
Schrader	Schrader, Heinrich Adolph (1767–1836)
Schrank	von Paula von Schrank, Franz (1747–1835)
Schreb.	von Schreber, Johann Christian Daniel (1739–1810)
Schreib.	Schreiber, Annelis (1927–2010)
Schrenk	von Schrenk, Alexander Gustav (1816–1876)
Schrire	Schrire, Brian David (1953–)
Schröder	Schröder, Richard Iwanowitch (1822–1903)
Schroeder, F.	Schroeder, Fred-Günter (1930–)
Schroedinger	Schrödinger, Rudolf (1857–1919)
Schubert, B.G.	Schubert, Bernice Giduz (1913–2000)
Schuebler	Schübler (Schuebler), Gustav (1787–1834)
Schultes	Schultes, Josef August (1773–1831)
Schultes, J.H. bis	Schultes, Julius Hermann (1820–1887)
Schultes, R.	Schultes, Richard Evans (1915–2001)
Schultes f.	Schultes, Julius Hermann (1804–1840)
Schultz, C.H.	Schultz, Carl Heinrich 'Schultzenstein' (1798–1871)
Schultz, F.W.	Schultz, Friedrich Wilhelm (1804–1876)
Schultz-Bip.	Schultz, Carl Heinrich 'Bipontinus' (1805–1867)
Schulz, B.	Schulz, Bernd (1963–)
Schulz, D.L.	Schulz, Dorothea L. (1931–)
Schulz, O.	Schulz, Otto Eugen (1874–1936)
Schulz, R.	Schulz, Roman (1873–1926)
Schulze, G.	Schulze, Georg Martin (1906–1985)
Schulze, W.	Schulze, Werner (1930–)
Schum.	Schumacher, Heinrich Christian Friedrich (1757–1830)
Schum., K.	Schumann, Karl Moritz (1851–1904)
Schur	Schur, Philip Johann Ferdinand (1799–1878)
Schutte	Schutte, Anne-Lise (1962–)
Schwacke	Schwacke, Carl August Wilhelm (1848–1904)
Schwaegr.	Schwägrichen, Christian Friedrich (1775–1853)
Schwantes	Schwantes, Martin Heinrich Gustav Georg (1891–1960)
Schwartz	Schwartz, Oskar (1901–1945)
Schwarz	Schwarz, August Friedrich (1852–1915)
Schwarz, O.	Schwarz, Otto Karl Anton (1900–1983)
Schweick.	Schweickerdt, Herold Georg Wilhelm Johannes (1903–1977)
Schweigger	Schweigger, August Friedrich (1783–1821)
Schwein.	von Schweinitz, Ludwig David (1780–1834)
Schweinf.	Schweinfurth, Georg August (1836–1925)
Schweinf., C.	Schweinfurth, Charles (1890–1970)
Scop.	Scopoli, Giovanni Antonio (1723–1788)
Scort.	Scortechini, Benedetto (1845–1886)
Scotland	Scotland, Robert W. (fl. 1998)
Scott, A.J.	Scott, Andrew John (1950–)
Scott, D.H.	Scott, Dunkinfield Henry (1854–1934)
Scott, M.	Scott, Munro Briggs (1887–1917)
Scott-Elliot	Scott-Elliot, George Francis (1862–1934)
Scribner	Scribner, Frank Lamson (1851–1938)
Sealy	Sealy, Joseph Robert (1907–2000)
Seberg	Seberg, Ole (1952–)
Sébert	Sébert, Hippolyte (1839–1930)
Secco	de Sousa Secco, Ricardo (1950–)
Seemann	Seemann, Berthold Carl (1825–1871)
Séguier	Séguier, Jean François (1703–1784)
Seib.	Seibert, Russell Jacob (1914–)

Seidel	Seidel, Johann Heinrich (1774–1815)
Seidenf.	Seidenfaden, Gunnar (1908–2001)
Seidenschnur	Seidenschnur, Christiane Eva (1944–)
Seidl	Seidl, Wenzel Benno (1773–1842)
Seigler	Seigler, David Stanley (1940–)
Sell	Sell, Peter Derek (1929–2013)
Sello	Sello, Hermann Ludwig (1800–1876)
Sellow	Sellow, Friedrich (1789–1831)
Selvi	Selvi, Federico (1966–)
Selvi, B.	Selvi, Bedrettin (fl. 2002)
Semir	Semir, S. João (1937–)
Semple	Semple, John Cameron (1947–)
Sendtner	Sendtner, Otto (1813–1859)
Senft	Senft, Christian Carl Friedrich Ferdinand (1810–1893)
Senghas	Senghas, Karlheinz (1928–2004)
Sennikov	Sennikov, Alexander Nikolaevitsch (1972–)
Ser.	Seringe, Nicolas Charles (1776–1858)
Serdyuk.	Serdyukova, L.B. (fl. 1973)
Serra	Serra, Luis (1966–)
Sesler	Sesler, Leonard (?–1785)
Sessé	de Sessé y Lacasta, Martín (1751–1808)
Setch.	Setchell, William Albert (1864–1943)
Setten	van Setten, A.K. (fl. 1985)
Seub.	Seubert, Moritz August (1818–1878)
Seward	Seward, Albert Charles (1863–1941)
Shan	Shan, Ren Hwa (1909–1986)
Shang	Shang, Chih-Bei (1935–)
Shao, W.	Shao, Wen (1979–)
Sharman	Sharman, Percy J. (fl. 1916)
Sharp	Sharp, Aaron John (1904–1997)
Sharp, W.	Sharp, Ward McClintic (1904–1985)
Shaw, E.	Shaw, Elizabeth Anne (1938–)
Shaw, J.M.H.	Shaw, Julian Mark Hugh (1955–)
Sheh	Sheh, Men(g)-Lan (fl. 1986)
Sheldon, E.	Sheldon, Edmund Perry (1869–1947)
Shen	Shen, Lian-Dai (Tai) (fl. 1970)
Sherff	Sherff, Earl Edward (1886–1966)
Shevock	Shevock, James R. (1950–)
Shib.	Shibata, Keita (1877–1949)
Shih (C.)	Shih, Chu (1932–)
Shim, P.S.	Shim, Phyau-Soon (1942–)
Shing	Shing, Kung-Hsieh (1929–)
Shinn.	Shinners, Lloyd Herbert (1918–1971)
Shipunov	Shipunov, Aleksey Borisovich (fl. 2000)
Shmakov	Shmakov, Alexander I. (fl. 1999)
Short	Short, Philip Sydney (1955–)
Shrestha	Shrestha, T.B.
Shulkina	Shulkina, Tatyana V. (fl. 2015)
Shuttlew.	Shuttleworth, Robert James (1810–1874)
Sibth.	Sibthorp, John (1758–1796)
Sieb.	von Siebold, Philipp Franz (1796–1866)
Siebenl.	Siebenlist
Sieber	Sieber, Franz Wilhelm (1789–1844)
Siebert	Siebert, August (1854–1923)
Siegerist	Siegerist, E.S. (1925–)
Siegler	Siegler, Eugene Alfred (1891– ?)
Sillans	Sillans, Roger (fl. 1952)
Silva, C.	Silva, Christian (fl. 2013)
Silva Manso, A.	da Silva Manso, António-Luiz Patricio (1788–1818)
Silva Manso, J.	da Silva Manso, José
Silva Tarouca	Silva Tarouca, Ernst Emmanuel (1860–1936)
Silverside	Silverside, Alan James (1947–)
Silvestre	Silvestre Domingo, Santiago (1944–)
Sim	Sim, Robert (1791–1878)
Simon	Simon, Bryan Keith (1943–)
Simonkai	Simonkai, Lájos von (1851–1910)
Simon-Louis	Simon-Louis, Léon L. (1834–1913)

Simonsen	Simonsen, Henrik Toft
Simmons, M.P.	Simmons, Mark P. (fl. 1997)
Simpson, B.	Simpson, Beryl Britnall (1942–)
Simpson, D.	Simpson, David Alan (1955–)
Simpson, D. R.	Simpson, Donald Ray (1932–)
Sims	Sims, John (1749–1831)
Sincl., James	Sinclair, James (1913–1968)
Singer, R.	Singer, Rodrigo Bustos (1970–)
Singh	Singh, D.N.
Singh, P.	Singh, Paramjit (1958–)
Sitko	Sitko, Magdalena (fl. 2012)
Sivadasan	Sivadasan, Mayandy (1948–)
Sjöstedt, B. Y.	Sjöstedt, Bror Yngva (1866–1948)
Skeels	Skeels, Homer Collar (1873–1934)
Skema	Skema, Cynthia (fl. 2010)
Skip.	Skipworth, John Peyton (1934–)
Škoda	Škoda, Bohdan (fl. 1996)
Skog	Skog, Laurence Edgar (1943–)
Škornick.	Leong-Škorničková, Jana (fl. 2003)
Skottsb.	Skottsberg, Carl Johan Fredrik (1880–1963)
Skovsted	Skovsted, Åge Thorsen (1903–1983)
Skvarla	Skvarla, John Jerome (1935–2014)
Skvortsova	Skvortsova, Nina Timofeevna (1925–)
Skvortzov	Skvortzov, Boris Vassilievich (1890–1980)
Slavíková	Slavíková, Zdeňka (1935–)
Sleumer	Sleumer, Hermann Otto (1906–1993)
Sloane, B.	Sloane, Boyd Lincoln (1885–1955)
Slooten	van Slooten, Dirk Fox (1891–1953)
Sm.	Smith, James Edward (1759–1828)
Sm., A.C.	Smith, Albert Charles (1906–1999)
Sm., A.R.	Smith, Alan Reid (1943–)
Sm., C.	Smith, Christen (1785–1816)
Sm., C.A.	Smith, Christo Albertyn (1898–1956)
Sm., E.W.	Smith, Elmer William (1920–1981)
Sm., F.D.	Donnell Smith, F.
Sm., G.F.	Smith, George F. (fl. 1995)
Sm., H.	Smith, Karl August Harald ('Harry') (1889–1971)
Sm., J.	Smith, John (1798–1888)
Sm., J.D.	Donnell Smith, John (1829–1928)
Sm., J.F.	Smith, James Foley (fl. 1993)
Sm., J.G.	Smith, Jared Gage (1866–1957)
Sm., J.J.	Smith, Johannes Jacobus (1867–1947)
Sm., L.B.	Smith, Lyman Bradford (1904–1997)
Sm., L.S.	Smith, Lindsay Stewart (1917–1970)
Sm., R.E.	Smith, Ralph Elliott (Eliot) (1874–1953)
Sm., R.M.	Smith, Rosemary Margaret (1933–)
Sm., W.(W.)	Smith, William Wright (1875–1956)
Sm., W.G.	Smith, Worthington George (1835–1917)
Small	Small, John Kunkel (1869–1938)
Small, E.	Small, Ernest (1940–)
Smedmark	Smedmark, Jenny E.E. (fl. 2006)
Smirnov	Smirnov, Pavel Aleksandrovich (1896–1980)
Smit, A.	Smit, A. (fl. 1971)
Smoljan.	Smoljaninova, Liudmila A. (1904–1990)
Snijman	Snijman, Deidré A. (1949–)
Snogerup	Snogerup, Sven E. (1929–)
Snow, N.	Snow, Neil (1960–)
Sobral	Sobral, Marcos (1960–)
Soderstrom	Soderstrom, Thomas Robert (1936–1987)
Soeg.-Reks.	Soegeng-Reksodiharjo, Wertit (1935–)
Soest	van Soest, Johannes Leendert (1898–1983)
Sohmer	Sohmer, Seymour Hans (1941–)
Soják	Soják, Jiří (1936–)
Sokoloff	Sokoloff, Dmitry Dmitrievich (1973–)
Sokolova	Sokolova, I.V. (1963–)
Sol.	Solander, Daniel Carl (1733–1782)
Sole	Sole, William (1741–1802)

Soler.	Solereder, Hans (1860–1920)
Solms-Laub.	zu Solms-Laubach, Hermann Maximilian Carl Ludwig Friedrich (1842–1915)
Somers	Somers, Carl (1963–)
Sommier	Sommier, Carlo Pietro Stefano (1848–1922)
Sonder	Sonder, Otto Wilhelm (1812–1881)
Sonké	Sonké, Bonaventure (fl. 1990)
Sonn.	Sonnerat, Pierre (1748–1814)
Soó	Soó von Bere, Károly Rezső (1903–1980)
Soreng	Soreng, Robert John (1952–)
Soriano	Soriano, Alberto (1920–1998)
Sosef	Sosef, Marc Simon Maria (1960–)
Sosn.	Sosnowsky, Dimitrii Ivanovich (1885–1952)
Sota	de la Sota, Elías Ramón (1932–)
Soto	Soto Arenas, Miguel Ángel (1963–2009)
Sotuyo, S.	Sotuyo Vázquez, Jeny Solange (1975–)
Soul.-Bod.	Soulange-Bodin, Etienne (1774–1846)
Sousa	Sousa S., Mario (1940–)
Souster	Souster, John Eustace Sirett (1912–2000)
Southworth	Southworth, Effie Almira (1860–1947)
Souza	Souza, Vinicius Castro (1954–)
Souza, E.	Rodrigues de Souza, Élvia (fl. 2004)
Sowerby	Sowerby, James (1757–1822)
Spach	Spach, Edouard (1801–1879)
Spalik	Spalik, Krzysztof (fl. 1997)
Span.	Spanoghe, Johan Baptist (1798–1838)
Sparre	Sparre, Benkt (1918–1986)
Sparrman	Sparrman, Anders (1748–1820)
Specht, C.	Specht, Chelsea D. (fl. 2006)
Speg.	Spegazzini, Carlo Luigi (1858–1926)
Spence, J.	Spence, John R. (1956–)
Spencer, M.	Spencer, Michael A. (fl. 1992)
Spenn.	Spenner, Fridolin Carl Leopold (1798–1841)
Speta	Speta, Franz (1941–)
Spin	Spin (Marquis de) (fl. 1809)
Spire, A.	Spire, André (fl. 1903)
Spire, C.	Spire, Camille Joseph (fl. 1903)
Splitg.	Splitgerber, Frederik Louis (1801–1845)
Sprague	Sprague, Thomas Archibald (1877–1958)
Spreng.	Sprengel, Curt Polycarp Joachim (1766–1833)
Spreng., A.	Sprengel, Anton (1803–1851)
Spring	Spring, Anton Friedrich (1814–1872)
Spruce	Spruce, Richard (1817–1893)
Sreek.	Sreekumar, Puthenpurayil Viswanathan (1954–)
Sreemad.	Sreemadhaven, C.P.
St John, (H.)	St John, Harold (1892–1991)
Stace	Stace, Clive Anthony (1938–)
Stadelm.	Stadelman, F. (fl. 1975)
Stadman	Stadman (fl. 1810)
Stadtm.(ann)	Stadtmann, Jean Frédéric (1762–1807)
Stafford	Stafford, Peter J. (fl. 1998)
Stafleu	Stafleu, Frans Antonie (1921–1997)
Ståhl	Ståhl, Bertil (1957–)
St-Amans	de Saint-Amans, Jean Florimond Boudon (1748–1831)
Standl.	Standley, Paul Carpenter (1884–1963)
Staner	Staner, Pierre (1901–1984)
Stapf	Stapf, Otto (1857–1933)
Staples	Staples, George William (1953–)
Stapleton	Stapleton, Christopher Mark Adrian (1957–)
Starod.	Starodubtzev, V.N. (1948–)
Stauffer	Stauffer, Hans Ulrich (1929–1965)
Steane	Steane, Dorothy A. (fl. 1998)
Stearn	Stearn, William Thomas (1911–2001)
Stebb.	Stebbins, George Ledyard (1906–2000)
Steck	Steck, Abraham (fl. 1757)
Steenis	van Steenis, Cornelis Gijsbert Gerrit Jan (1901–1986)
Steetz	Steetz, Joachim (1804–1862)
Stefanoff	Stefanoff, Boris (1894–1979)

Stehlé	Stehlé, Henri (1909–1983)
Stein	Stein, Berthold (1847–1899)
Steinh.	Steinheil, Adolphe (1810–1839)
Stent	Stent, Sydney Margaret (1875–1942)
Stephan	Stephan, Christian Friedrich (1757–1814)
Stephens	Stephens, Edith Layard (1884–1966)
Stern	Stern, William Louis (1926–)
Stern, F.	Stern, Frederick Claude (1884–1967)
Sternb.	von Sternberg, Caspar Maria (1761–1838)
Sterns	Sterns, Emerson Ellick (1846–1926)
Steud.	von Steudel, Ernst Gottlieb (1783–1856)
Steven	von Steven, Christian (1781–1863)
Stevens, P.	Stevens, Peter F. (1944–)
Stevenson	Stevenson, Dennis William (1942–)
Stewart, R. R.	Stewart, Ralph Randles (1890–1993)
Steyerm.	Steyermark, Julian Alfred (1909–1988)
St-Hil., A.	de Saint-Hilaire, Auguste François César Prouvençal (1779–1853)
St-Hil., J.	Saint-Hilaire, Jean Henri Jaume (1772–1845)
Stirton	Stirton, Charles Howard (1946–)
St-Lager	Saint-Lager, Jean Baptiste (1825–1912)
Stocks	Stocks, John Ellerton (1822–1854)
Stoffelen	Stoffelen, Piet (fl. 1996)
Stokes	Stokes, Jonathan S. (1755–1831)
Stone	Stone, Benjamin Clemens Masterman (1933–1994)
Stopp	Stopp, Klaus Dieter (1926–)
Stork	Stork, Adélaïde Louise (1937–)
Stoy.	Stoyanoff, Nikolai Andreev (1883–1968)
Strack	Strack, Dieter (1945–)
Strasb.	Strasburger, Eduard Adolf (1844–1912)
Straw	Straw, Richard Myron (1926–2012)
Stritch	Stritch, Lawrence R. (fl. 1982)
Strong	Strong, Mark T. (fl. 1993)
Strother	Strother, John Lance (1941–)
Struwe	Struwe, Lena (1967–)
Stschegl.	Stschegleev, Sergei Sergeevich (1820–1859)
Stuntz	Stuntz, Stephen Conrad (1875–1918)
Stutz	Stutz, Howard Coombs (1918–2010)
Su, Y.	Su, Yvonne C.F. (fl. 2001)
Su, Z.Y.	Su, Zhi Yun (1936–)
Suárez, S.	Suárez Suárez, Luz Stella (fl. 2000)
Subils	Subils, Rosa (1929–)
Subram.	Subramanyam, Krishnaier (1915–1980)
Suckow	Suckow, Georg Adolph (1751–1813)
Suddee	Suddee, Somran (fl. 2004)
Sudw.	Sudworth, George Bishop (1864–1927)
Süsseng.	Süssenguth (Suessenguth), Karl (1893–1955)
Sugumaran	Sugumaran, M. (fl. 2012)
Suh	Suh, Young-Bae (1956–)
Suksathan	Suksathan, Piyakaset (fl. 2001)
Suksd.	Suksdorf, Wilhelm Nikolaus (1850–1932)
Summerh.	Summerhayes, Victor Samuel (1897–1974)
Sun	Sun, Bi-Sin (1921–2002)
Sun, B.Y.	Sun, Byung-Yun (fl. 1993)
Sun, H.	Sun, Hang (1963–)
Sun, Y.Z.	Sun, Yon-Zai (1898–1964)
Sundberg	Sundberg, Scott D. (1954–2004)
Sundue	Sundue, Michael A. (1975–)
Sungkaew	Sungkaew, Sarawood (fl. 2003)
Suresh	Suresh, C.R. (fl. 1988)
Suringar	Suringar, Willem Frederik Reinier (1832–1898)
Suryan.	Suryanarayana, M.C.
Susanna	Susanna, Alfonso (1956–)
Sutô	Sutô, Tiharu (1910–1968)
Sutorý	Sutorý, Karel (1947–)
Suvatabandhu	Suvatabandhu, Kasin (1916–)
Svent.	Sventenius, Eric R. Svensson (1910–1973)
Sw.	Swartz, Olof Peter (1760–1818)
Swallen	Swallen, Jason Richard (1903–1991)

Swart	Swart, Jan Johannes (1901–1974)
Sweet	Sweet, Robert (1783–1835)
Sweet, H.	Sweet, Herman Royden (1911–1991)
Swingle	Swingle, Walter Tennyson (1871–1952)
Sydow	Sydow, Hans (1879–1946)
Sym.	Symington, Colin Fraser (1905–1943)
Syme	Syme, John Thomas Irvine Boswell (1822–1888)
Szabó, Z.	Szabó, Zoltán von (1882–1944)
Szlach.	Szlachetko, Dariusz L. (1961–)
Szyszyl.	von Szyszylowicz, Ignaz (1857–1910)
Tabern.	Dietrich, Jakob (1525–1590), called Tabernaemontanus
Tag.	Tagawa, Motozi (1908–1977)
Tagg	Tagg, Harry Frank (1874–1933)
Takah.	Takahashi, Yoshinao (1872–1914)
Takano	Takano, Atsuko (1971–)
Takeda	Takeda, Hisayoshi (1883–1972)
Takht.	Takhtadjan, Armen Leonovich (1910–2009)
Tamamschjan	Tamamschjan, Sophia G. (1901–1981)
Tamura	Tamura, Michio (1927–2007)
Tanaka (T.)	Tanaka, Tyôzaburô (1885–1976)
Tanaka, N.	Tanaka, Noriyuki (fl. 1998)
Tanaka, Yu.	Tanaka, Yuichiro (1901–1983)
Tanf.	Tanfani, Enrico (1848–1892)
Tang	Tang, T. (1897–1984)
Tange	Tange, Christian (fl. 1994)
Tao, D.D.	Tao, De-Ding (1937–)
Tao, X.L.	Tao, Xiu-Liu (fl. 1982)
Tard.	Tardieu, Marie Laure (also Tardieu-Blot) (1902–1998)
Targ.-Tozz.	Targioni-Tozzetti, Ottaviano (1755–1829)
Tate	Tate, Ralph (1840–1901)
Tate, J.	Tate, Jennifer A. (fl. 2003)
Tateoka	Tateoka, Tuguo (1931–1994)
Tatew.	Tatewaki, Misao (1899–1976)
Taubert	Taubert, Paul Hermann Wilhelm (1862–1897)
Tausch	Tausch, Ignaz Friedrich (1793–1848)
Taylor	Taylor, Richard (1781–1858)
Taylor, D.	Taylor, Dean W. (fl. 1992)
Taylor, D.W.	Taylor, David Winship (fl. 1999)
Taylor, G.	Taylor, George (1904–1993)
Taylor, N.	Taylor, Norman (1883–1967)
Taylor, N.P.	Taylor, Nigel Paul (1956–)
Taylor, P.	Taylor, Peter Geoffrey (1926–2011)
Teerawat.	Teerawatananon, Atchara (fl. 2007)
Teijsm.	Teijsmann, Johannes Elias (1809–1882)
Ten.	Tenore, Michele (1780–1861)
Tenn.	Tennant, James Robert (1928–)
Terán	de Mier y Terán, Manuel (1789–1832)
Terrell	Terrell, Edward Everett (1923–2011)
Teschem.	Teschemacher, James Engelbert (1790–1853)
Tharp	Tharp, Benjamin Carroll (1885–1964)
Thell.	Thellung, Albert (1881–1928)
Thénint	Thénint, André (fl. 1936)
Theob.	Theobald, William (1829–1908)
Theob., W.L.	Theobold, William Louis (1936–)
Thiede	Thiede, Joachim (1963–)
Thiele	Thiele, Kevin R. (fl. 1988)
Thieret	Thieret, John William (1926–2005)
Thijsse	Thijsse, Gerard (fl. 2015)
Thiv	Thiv, Mike (1970–)
Thomas, D.W.	Thomas, Duncan W. (fl. 1986)
Thomas, H.	Thomas, Hugh Hamshaw (1885–1962)
Thomas, Le	Le Thomas-Hommay, Annick (1936–)
Thomas, W.	Thomas, William Wayt (1951–)
Thompson, C.	Thompson, Charles Henry (1870–1931)
Thompson, D.	Thompson, David M. (1957–)
Thompson, J.	Thompson, Joy (née Garden) (1923–)
Thomson	Thomson, Thomas (1817–1878)
Thonn.	Thonning, Peter (1775–1848)

Thore	Thore, Jean (1762–1823)
Thorel	Thorel, Clovis (1833–1911)
Thory	Thory, Claude Antoine (1759–1827)
Thoth.	Thothathri, Krishnamurthy (1929–)
Thouars	du Petit Thouars, Louis Marie Aubert (1758–1831)
Thuill.	Thuiller, Jean Louis (1757–1822)
Thulin	Thulin, Mats (1948–)
Thumanjan	Thumanjan (fl. 1930s)
Thunb.	Thunberg, Carl Peter (1743–1828)
Thurb.(er)	Thurber, George (1821–1890)
Thurn	im Thurn, Everard Ferdinand (1852–1932)
Thwaites	Thwaites, George Henry Kendrick (1812–1882)
Tichom.	Tichomirov, Vadim Nikolaevich (1932–1998)
Tieghem	van Tieghem, Phillippe Edouard Léon (1839–1914)
Tiên Bân	Bân, Nguyên Tiên (fl. 1973)
Tilloch	Tilloch, Alexander (1759–1825)
Timbrook	Timbrook, Steven (fl. 1986)
Timler	Timler, Friedrich Karl (1914–1995)
Tindale	Tindale, Mary Douglas (1920–2011)
Tineo	Tineo, Vincenzo (1791–1856)
Tippery	Tippery, Nicholas P. (fl. 2009)
Tirv.	Tirvengadum, Deva D. (fl. 1986)
Tixier	Tixier, Pierre (1918–1997)
Tjaden	Tjaden, William Louis (1913–2008)
Tobe	Tobe, Hiroshi (1948–)
Tobler	Tobler, Friedrich (1879–1957)
Tod.	Todaro, Agostino (1818–1892)
Toelken	Toelken, Helmut R. (1939–)
Tol.	de Toledo, Joaquim Franco (1905–1952)
Tomb	Tomb, Andrew Spencer (1943–)
Toml.	Tomlinson, Philip Barry (1932–)
Tong	Tong, Shao-Quan (1935–)
Torre	da Torre, António Rocha (1904–1995)
Torres	Torres, Maria Amelia (1931–)
Torrey	Torrey, John (1796–1873)
Tortosa	Tortosa, Roberto D. (1946–)
Tourn.	Pitton de Tournefort, Joseph (1656–1708)
Townrow	Townrow, John A. (1927–)
Towns., C.	Townsend, Clifford Charles (1926–)
Toyok.	Toyokuni, Hideo (1932–1992)
Trabut	Trabut, Louis (Charles) (1853–1929)
Tracey	Tracey, John Geoffrey (1930–2004)
Tracey, R.	Tracey, R. (1951–)
Trail	Trail, James William Helenus (1851–1919)
Tralau	Tralau, Hans (1932–1977)
Tran	Tran, Van Tien (fl. 2010)
Tratt.	Trattinick, Leopold (1764–1849)
Traub	Traub, Hamilton Paul (1890–1983)
Trautv.	von Trautvetter, Ernst Rudolf (1809–1889)
Trécul	Trécul, Auguste Adolphe Lucien (1818–1896)
Trel.	Trelease, William (1857–1945)
Trev.	Treviranus, Ludolf Christian (1799–1864)
Trevis.	Trevisan de Saint-Léon, Vittore Benedetto Antonio (1818–1897)
Trew	Trew, Christoph Jakob (1695–1769)
Triana	Triana, José Jéronimo (1834–1890)
Triest	Triest, Ludwig J. (1957–)
Trimen	Trimen, Henry (1843–1896)
Trin.	von Trinius, Carl Bernhard (1778–1844)
Trinajstić	Trinajstić, Ivo (1933–)
Trofimov	Trofimov, Dimitrij (fl. 2016)
Tronc.	Troncoso, Nélida Sara (1914–1988)
Trotter	Trotter, Alessandro (1874–1967)
Trotzky	Trotzky, Petrus Kornuch (1803–1877)
Troupin	Troupin, Georges M.D.J. (1923–)
Trudgen	Trudgen, Malcolm Eric (1951–)
Tryon	Tryon, Rolla Milton (1916–2001)
Tscherneva	Tscherneva, O.V. (1929–)
Tsi	Tsi, Zhan-Huo (1937–2001)

Tsiang	Tsiang, Ying (1898–1982)
Tsoong	Tsoong, Pu-Chiu (1906–1981)
Tsui	Tsui Yu-Wen (1907–1980)
Tsukaya	Tsukaya, Hirokazu (1964–)
Tul.	Tulasne, Louis René (1815–1885)
Türpe	Türpe, Anna Maria (1946–)
Turcz.	Turczaninow, Porphir Kiril Nicolai Stepanowitsch (1796–1863)
Turner, B.	Turner, Billie Lee (1925–)
Turner, H.	Turner, Hubert (1955–)
Turner, I.	Turner, Ian Mark (1963–)
Turner, M.	Turner, Melvin D. (fl. 1988)
Turner, M.W.	Turner, Matt W. (fl. 1996)
Turpin	Turpin, Pierre Jean François (1775–1840)
Turra	Turra, Antonio (1730–1796)
Turrill	Turrill, William Bertram (1890–1961)
Tussac	de Tussac, François Richard (1751–1837)
Tutcher	Tutcher, William James (1867–1920)
Tutin	Tutin, Thomas Gaskell (1908–1987)
Tuyama	Tuyama, Takasi (1910–2000)
Tyrell	Tyrell, C.D. (fl. 2012)
Tzvelev	Tzvelev, Nikolai Nikolaievich (1925–2015)
Ucria	da Ucria, Bernardino (1739–1796)
Ueda	Ueda, Saburo (1898–?)
Uhl	Uhl, Charles Harrison (1918–2010)
Uitew.	Uitewaal, Antonius Josephus Adrianus (1899–1963)
Uittien	Uittien, Hendrik (1898–1944)
Ulbr.	Ulbrich, Oskar Eberhard (1879–1952)
Ule	Ule, Ernst Heinrich Georg (1854–1915)
Ulloa	Ulloa, María del Carmen Ulloa (1963–)
Ulvinen	Ulvinen, Tauno (1930–)
Underw.	Underwood, John (?–1834)
Underw., L.	Underwood, Lucien Marcus (1853–1907)
Unger	Unger, Franz (Joseph Andreas Nicolaus) (1800–1870)
Ung.-Sternb.	Ungern-Sternberg, Franz (1808–1885)
Urb.	Urban, Ignatz (1848–1931)
Urbatsch	Urbatsch, Lowell Edward (1942–)
Urmi-König	Urmi-König, Katherina (fl. 1975)
Ursch	Ursch, Eugène (1882–1962)
Urv.	Dumont d' Urville, Jules Sébastian César (1790–1842)
Utteridge	Utteridge, Timothy Michael Arthur (1970–)
Vacherot	Vacherot, Michel (1922–2002)
Vahl	Vahl, Martin (1749–1804)
Vail	Vail, Anna Murray (1863–1955)
Vaill.	Vaillant, Sébastien (1669–1722)
Vajravelu	Vajravelu, E. (1936–)
Valcken.	Valckenier, Suringar, Jan (1865–1932)
Valdebenito	Valdebenito, Hugo A. (fl. 1986)
Valdés	Valdés, Benito (1942–)
Valdés R.	Valdés-Reyna, Jésus (1948–)
Valeton	Valeton, Theodoric (1855–1929)
Vall.-Marín, Vallejo-Marín	Vallejo-Marín, Mario (fl. 2012)
Van den Berg	van den Berg, Ronald G. (fl. 2004)
Vand.	Vandelli, Domingo (1735–1816)
Vanderplank	Vanderplank, R. John R. (fl. 1996)
Vanderyst	Vanderyst, Hyacinthe Julien Robert (1860–1934)
Vaniot	Vaniot, Eugène (1846–1913)
Vargas	Vargas Calderón, Julio César (1907–1960)
Vargas, P.	Vargas, Pablo (1965–)
Vasc.	de Carvalho e Vasconcellos, João (1897–1972)
Vasey	Vasey, George (1822–1893)
Vasinger	Vasinger, Antonina Vasilievna (1892–1940)
Vázquez	Vázquez, V.M. (fl. 1982)
Vassal	Vassal J. (1932–)
Vassilcz.	Vassilczenko, Ivan Tikhonovich (1903–1995)
Vassiljeva, A.	Vassiljeva, A.N. (fl. 1969)
Vasudeva	Vasudeva Rao, M.K. (fl. 1979)
Vatke	Vatke, Georg Carl Wilhelm (1849–1889)
Vatt.	de Vattimo-Gil, Ida (1928–1993)

Vaucher	Vaucher, Jean Pierre Étienne (1763–1841)
Vega	Vega Flores, Karla (fl. 2007)
Veitch	Veitch, John Gould (1839–1870)
Veken	van der Veken, Paul A.J.B. (1928–)
Velarde	Velarde, Octavio (Octavio Velarde Nuñez) (1918–1963)
Veldk.	Veldkamp, Jan Frederik (1941–)
Velen.	Velenovský, Josef (1858–1949)
Vell.	Velloso de Miranda, Joaquim (1733–1815)
Vell. Conc.	Vellozo, José Mariano da Conceição (1742–1811)
Vent, W.	Vent, Walter (1920–2008)
Vent.	Ventenat, Etienne Pierre (1757–1808)
Venter	Venter, Hendrik Johannes Tjaart (1938–)
Verdc.	Verdcourt, Bernard (1925–2011)
Verh.(oeven), R.L.	Verhoeven, Rudolf L. (1945–)
Verloove	Verloove, Filip (fl. 2004)
Verlot	Verlot, Jean-Baptiste (1825–1891)
Verm., J.J.	Vermeulen, Jaap J. (1955–)
Verma	Verma, Dinesh Mohan (1937–)
Vermeulen	Vermeulen, Pieter (1899–1981)
Vermoesen	Vermoesen, François Marie Camille (1882–1922)
Vernet	Vernet (fl. 1904)
Versieux	Versieux, Leonardo M. (fl. 2007)
Versteegh	Versteegh, Corstiaen P.C. (fl. 2007)
Vest	von Vest, Lorenz Chrysanth (1776–1840)
Viana	Viana, Pedro Lage (fl. 2003)
Vick.	Vickery, Joyce Winifred (1908–1979)
Vidal, J.E.	Vidal, Jules Eugène (1914–)
Vidal, S.	Vidal i Soler, Sebastian (Sebastià) (1842–1889)
Vieill.	Vieillard, Eugène (1819–1896)
Viguier	Viguier, L.G. Alexandre (1790–1867)
Viguier, R.	Viguier, René (1880–1931)
Villar	Huguet del Villar i Serratacó, Emile (Emili) (1871–1951)
Villars	Villars, Domínique (1745–1814)
Villaseñor	Villaseñor, José Luis (1954–)
Villiers	Villiers, Jean François (1943–2001)
Vilm.	de Vilmorin, Pierre Philippe André Lévêque (1776–1862)
Vilm., P.L.	de Vilmorin, Pierre Louis François Lévêque (1816–1860)
Vines	Vines, Sydney Howard (1849–1934)
Vink	Vink, Willem (1931–)
Vinnersten	Vinnersten, Annika (fl. 2007)
Virot	Virot, Robert (1915–2002)
Vis.	de Visiani, Roberto (1800–1878)
Vischer	Vischer, Wilhelm (1890–1960)
Viv.	Viviani, Domenico (1772–1840)
Vogel	Vogel, Benedict Christian (1745–1825)
Vogel, de	de Vogel, Eduard Ferdinand (1942–)
Vogel, J.	Vogel, Julius Rudolph Theodor (1812–1841)
Vogt	Vogt, Robert M. (1957–)
Voigt	Voigt, Friedrich Sigismund (1781–1850)
Voigt, J.	Voigt, Joachim Otto (1798–1843)
Volkens	Volkens, Georg Ludwig August (1855–1917)
Vollesen	Vollesen, Kaj Børge (1946–)
Vorontsova	Vorontsova, Maria Sergeevna (1979–)
Vos, de	de Vos, Miriam Phoebe (1912–2005)
Voss	Voss, Andreas (1857–1924)
Votsch	Votsch, Oskar Hermann Wilhelm (1879–1927)
Vural	Vural, Mecit (fl. 1983)
Vriese	de Vriese, Willem Hendrik (1806–1862)
Vved.	Vvedensky, Aleksej Ivanovič (1898–1972)
Wadhwa	Wadhwa, Brij Mohan (1933–2009)
Wagenitz	Wagenitz, Gerhard Werner Friedrich (1927–)
Wager	Wager, Vincent Athelstan (1904–1989)
Wagner	Wagner, Warren Herbert (1920–2000)
Wagner, W.L.	Wagner, Warren Lambert (1950–)
Wahlenb.	Wahlenberg, Georg (Göran) (1780–1851)
Waldst.	von Waldstein (-Wartemburg), Franz de la Paula Adam (1759–1823)
Walker, E.	Walker, Egbert Hamilton (1899–1991)
Walker, J.	Walker, James Willard (1943–)

Walker, T.	Walter, Trevor George (1927–2006)
Wall	Wall, Arnold (1869–1966)
Wall.	Wallich, Nathaniel (or Nathan Wolf) (1786–1854)
Wallace	Wallace, Alfred Russel (1823–1913)
Wallace, A.	Wallace, Alexander (1829–1899)
Wallnöfer	Wallnöfer, Bruno (1960–)
Wallr(oth)	Wallroth, Carl Friedrich Wilhelm (1792–1857)
Walp.	Walpers, Wilhelm Gerhard (1816–1853)
Walter	Walter, Thomas (1740–1789)
Walter, H.	Walter, Hans Paul Heinrich (1882–?)
Walther	Walther, Edward Eric (1892–1959)
Walton	Walton, Frederick Arthur (1853–1922)
Wang	Wang, Chen-Hwa (1908–)
Wang, C.P.	Wang, Cheng-Ping (fl. 1982)
Wang, F.T.	Wang, Fa-Tsuan (1899–1985)
Wang, H.C.	Wang, Huan-Chong (fl. 2013)
Wang, Q.	Wang, Qian (fl. 2014)
Wang, R.J.	Wang, Rui-Jiang (fl. 1999)
Wang, W.T.	Wang, Wen-Tsai (1926–)
Wang, Wei	Wang, Wei (fl. 2009)
Wang, X.Q.	Wang, Xi-Qu (fl. 1992)
Wang, Y.J.	Wang, Yu-Jin (fl. 2003)
Wang, Y.Z.	Wang, You-Zhi (fl. 2000)
Wang, Z.R.	Wang, Zhong-Ren (1939–)
Wang, Zhang R.	Wang, Zhang-Ron (fl. 2012)
Wangenh.	von Wangenheim, Friedrich Adam Julius (1749–1800)
Wangerin	Wangerin, Walther Leonhard (1884–1938)
Warb.	Warburg, Otto (1859–1938)
Warb., E.	Warburg, Edmund Frederic (1908–1966)
Ward, D.B.	Ward, Daniel Bertram (1928–)
Ward, J.M.	Ward, Josephine M. (fl. 1997)
Warder	Warder, John Aston (1812–1883)
Wardleworth	Wardleworth, Thomas Hatton (fl. 1893)
Warm.	Warming, Johannes Eugen Bülow (1841–1924)
Warnst.	Warnstorf, Carl (Friedrich E.) (1837–1921)
Warsc.	von Warscewicz, Josef (1812–1866)
Warwick	Warwick, Suzanne I. (fl. 1997)
Wassh.	Wasshausen, Dieter Carl (1938–)
Waterf.	Waterfall, Umaldy Theodore (1910–1971)
Waterhouse	Waterhouse, John Teast (1924–1983)
Watson, H.	Watson, Hewett Cottrell (1804–1881)
Watson, J.	Watson, John Forbes (1827–1892)
Watson, L.	Watson, Leslie (1938–)
Watson, S.	Watson, Sereno (1826–1892)
Watson, W.	Watson, William (1832–1912)
Watson, W.C.R.	Watson, William Charles Richard (1885–1954)
Watson, Will.	Watson, William (1832–1912)
Watt	Watt, David Allan Poe (1830–1917)
Watts	Watts, William Walter (1856–1920)
Wawra	Wawra von Fernsee, Heinrich (1831–1887)
Webb	Webb, Philip Barker (1793–1854)
Webb, D.	Webb, David Allardice (1912–1994)
Weber, A.	Weber, Anton (1947–)
Weber, C.	Weber, Jean-Germaine Claude (1922–)
Weber, F.	Weber, Frédéric Albert Constantin (1830–1903)
Weber, G.	Weber, Georg Heinrich (1752–1828)
Weber, O.	Weber, Odile (fl. 2005)
Weber, W.A.	Weber, William Alfred (1918–)
Weberb.	Weberbauer, Otto (1846–1881)
Webster	Webster, Grady Linder (1927–2005)
Webster, R.	Webster, Robert D. (1950–)
Wedd.	Weddell, Hugh Algernon (1819–1877)
Wehrh.	Wehrhahn, Heinrich Rudolf (1887–1940)
Wei, F.N.	Wei, Fa-Nan (1941–)
Wei, H.T.	Wei, Hong-Tu (fl. 1982)
Wei, Y.G.	Wei, Yi-Gang (fl. 1995)
Weigend	Weigend, Maximilian (1969–)
Weihe	Weihe, Carl Ernst August (1779–1834)

Weiller	Weiller, Marc (1880–1945)
Weiller, C.	Weiller, Carolyn M. (fl. 1995)
Welsh	Welsh, Stanley Larson (1928–)
Welw.	Welwitsch, Friedrich Martin Josef (1806–1872)
Wen, H.Z.	Wen, Hai-Zhen (fl. 2012)
Wen, J.	Wen, Jun (1963–)
Wen, T.H.	Wen, Tai-Hui (1924–)
Wendelbo	Wendelbo, Per Erland Berg (1927–1981)
Wender.	Wenderoth, George Wilhelm Franz (1774–1861)
Wendl.	Wendland, Johann Christoph (1755–1828)
Wendl., H.(A.)	Wendland, Hermann A. (1825–1903)
Wendl., H.L.	Wendland, Heinrich Ludolph (1792–1869)
Wendt	Wendt, Thomas Leighton (1950–)
Wenzig	Wenzig, (Johann) Theodor (1824–1892)
Werderm.	Werdermann, Erich (1892–1959)
van der Werff	van der Werff, Henk (1946–)
Werner, K.	Werner, Klaus (1928–)
Wernham	Wernham, Herbert Fuller (1879–1941)
Wernisch.	Wernischeck, Johann Jacob (1743–1804)
Wesm.	Wesmael, Alfred (1832–1905)
Westc.	Westcott, Frederic (?–1861)
Wester	Wester, Peter Jansen (1877–1931)
Weston	Weston, Richard (1733–1806)
Weston, P.	Weston, Peter Henry (1956–)
Westra	Westra, Lübbert Ybele Theodoor (1932–)
de Wet	de Wet, Johannes Martenis Jacob (1927–)
Wetschnig	Wetschnig, Wolfgang (1958–)
Wettst.	von Wettstein von Westersheim, Richard (1863–1931)
Wheeler, J.R.	Wheeler, Judith Roderick (1944–)
Wheeler, L.C.	Wheeler, Louis Cotter (1910–1980)
Whiffin	Whiffin, Trevor Paul (1947–)
White, A.	White, Alain Campbell (1880–1951)
White, C.	White, Cyril Tenison (1890–1950)
White, F.	White, Frank (1927–1994)
White, R.	White, Richard Alan (1935–)
Whitm.	Whitmore, Timothy Charles (1935–2002)
Whitson	Whitson, Mary Kathryn (1972–)
Whitten	Whitten, Mark (1954–)
Wibel	Wibel, August Wilhelm Eberhard Christoph (1775–1814)
Widjaja	Widjaja, Elizabeth A. (1951–)
Wied-Neuw.	zu Wied-Neuwied, Maximilian Alexander Philipp (1782–1867)
Wiehler	Wiehler, Hans Joachim (1931–2003)
Wiens	Wiens, Delbert (1932–)
Wieringa	Wieringa, Jan Johannes (1967–)
Wiersema	Wiersema, John H. (1950–)
Wierzb.	Wierzbicki, Piotr Pawlus (1794–1847)
Wigg.	Wiggers, Friedrich Heinrich (1746–1811)
Wiggins	Wiggins, Ira Loren (1899–1987)
Wight	Wight, Robert (1796–1872)
Wight, W.	Wight, William Franklin (1874–1954)
Wijnands	Wijnands, D. Onno (1945–1993)
Wiklund	Wiklund, Annette (1953–)
Wikström	Wikström, Johan Emanuel (1789–1856)
Wikström, N.	Wikström, Niklas (fl. 2013)
Wilbur	Wilbur, Robert Lynch (1925–)
Wilczek	Wilczek, Ernst (1867–1948)
Wilczek, R.	Wilczek, Rudolf (1903–1984)
Wild	Wild, Hiram (1917–1982)
Wilde	Wilde, Earle Irving (1888–1949)
De Wilde, J.	de Wilde, Jan Jacobus Friedrich Egmond (1932–)
De Wilde, W.	de Wilde, Willem Jan Jacobus Oswald (1936–)
Wildpret	Wildpret, Hermann (1834–1908)
Wilhelm	Wilhelm, Karl Adolf (1848–1933)
Wilkin	Wilkin, Paul (fl. 1995)
Wilkins, C.F.	Wilkins, Carolyn F. (fl. 1999)
Willd.	von Willdenow, Carl Ludwig (1765–1812)
Willemet	Willemet, Pierre Remi (1735–1807)
Billiams, B.S.	Williams, Benjamin Samuel (1824–1890)

Williams, F.N.	Williams, Frederic Newton (1862–1923)
Williams, I.(J.)	Williams, Ion James Muirhead (1912–2001)
Williams, J.	Williams, John Beaumont (1932–2005)
Williams, J.K.	Williams, Justin Kirk (fl. 1995)
Williams, L.O.	Williams, Louis Otho (1908–1991)
Williams, N.	Williams, Norris H. (1943–)
Williams, S.	Williams, S.A. (fl. 2003)
Williamson	Williamson, Phyllis Alison (1925–)
Willis	Willis, John Christopher (1868–1958)
Willis, J.H.	Willis, James Hamlyn (1910–1995)
Willk.	Willkomm, Heinrich Moritz (1821–1895)
Wills, K.E.	Wills, Karen Elizabeth (fl. 2000)
Wilson, E.	Wilson, Ernest Henry (1876–1930)
Wilson, G.	Wilson, George Fox (1896–1951)
Wilson, J.S.	Wilson, James Stewart (1932–)
Wilson, K.	Wilson, Karen Louise (1950–)
Wilson, M.A.	Wilson, Margaret Anne (1942–)
Wilson, P.	Wilson, Percy (1879–1944)
Wilson, Paul G.	Wilson, Paul Graham (1928–)
Wilson, Peter G.	Wilson, Peter Gordon (1950–)
Wimmer	Wimmer, Christian Friedrich Heinrich (1803–1868)
Wimmer, F.	Wimmer, Franz Elfried (1881–1961)
Windham	Windham, Michael D. (1954–)
Winkler	Winkler, Constantin Georg Alexander (1848–1900)
Winkler, H.	Winkler, H. (fl. 1868)
Winter, P.	Winter, Pieter J.D. (1964–)
Wisura	Wisura, W. (fl. 1991)
de Wit	de Wit, Hendrik Cornelis Dirk (1909–1999)
With.	Withering, William (1741–1799)
Withner	Withner, Carl Leslie (1918–2012)
Wittm.	Wittmack, Marx Carl Ludwig (1839–1929)
Wodehouse	Wodehouse, Roger Philip (1889–1978)
Wolf	von Wolf, Nathanael Matthaeus (1724–1784)
Wolff, H.	Wolff, Karl Friedrich August Hermann (1866–1929)
Wol.	Woloszczak, Eustach (1835–1918)
Wollenw.	Wollenweber, Hans Wilhelm (1879–1949)
Woltz	Woltz, Philippe (fl. 1970)
Wong	Wong Khoon Meng (1954–)
Wong, S.Y.	Wong, Sin-Yeng (1975–)
Wood, Alph.	Wood, Alphonso W. (1810–1881)
Wood, J.J.	Wood, Jeffrey James (1952–)
Wood, J.M.	Wood, John Medley (1827–1915)
Wood, J.R.I.	Wood, John Richard Ironside (1944–)
Wood, K.	Wood, Kenneth Richard (1953–)
Woodson	Woodson, Robert Everard (1904–1963)
Woodward, C.	Woodward, Catherine L. (fl. 2007)
Woolls	Woolls, William (1814–1893)
Wooton	Wooton, Elmer Ottis (1865–1945)
Wormsk.	Wormskjöld, Martin (1783–1845)
Woronow	Woronow, Georg Jurij Nikolaewitch (1874–1931)
Worsley	Worsley, Arthrington (1861–1944)
Wright	Wright, William (1735–1819)
Wright, C.	Wright, Charles (Carlos) (1811–1885)
Wright, C.H.	Wright, Charles Henry (1864–1941)
Wright, J. bis	Wright, John (1836–1916)
Wright, S.	Wright, Samuel Hart (1825–1905)
Wu, C.Y.	Wu, Cheng-Yih (1916–2013)
Wu, H.	Wu, Han (fl. 1986)
Wu, S.K.	Wu, Su-Kung (1935–2013)
Wu, T.L.	Wu, Te-Lin(g) (1934–)
Wu, W.C.	Wu, Wen-Chen(g)
Wu, Z.L.	Wu, Zhen-Lan (1939–)
Wuerttemb.	von Württemberg, Friedrich Paul Wilhelm (1797–1860)
Wulfen	von Wulfen, Franz Xavier (1728–1805)
Wulff	Wulff, Eugen Vladimirowitsch (1885–1941)
Wunderlin	Wunderlin, Richard P. (1939–)
Wurd.	Wurdack, John Julius (1921–1998)
Wurd., K.	Wurdack, Kenneth J. (fl. 2005)

Wurmb	von Wurmb, Friedrich (1742–1781)
Wydler	Wydler, Heinrich (1800–1883)
Wyk	van Wyk, Abraham Erasmus (1952–)
Wyk, B.	van Wyk, Ben-Erik (1956–)
Xia, N.	Xia, Nian-He (1963–)
Xia, Q.	Xia, Qun (Quan) (1957–)
Xu, C.C.	Xu, C.C.
Xie	Xie, Quan-Zheng (fl. 1981)
Xue, B.	Xue, Bine (fl. 2011)
Yakovlev	Yakovlev, G.P. (1938–)
Yalt.	Yaltink, Faik (1930–)
Yamamoto	Yamamoto, Yoshimatsu (1893–1947)
Yamaz.	Yamazaki, Takasi (1921–2007)
Yang, B.M.	Yang, Bao-Min (1928–)
Yang, J.L.	Yang, Jun-Liang (fl. 1988)
Yang, Y.C.	Yang, Yen-Chin (1913–1984)
Yang, Y.L.	Yang, Ya-Ling (1933–)
Yano	Yano, Olga (1946–)
Yatabe	Yatabe, Ryôkichi (1851–1899)
Yates	Yates, Harris Oliver (1934–)
Yatsk., Yatskievych	Yatskievych, George A. (1957–)
Yen	Yen, Chi (fl. 1983)
Yesilyurt	Yesilyurt, Jovita Cislinski (fl. 2008)
Yi	Yi, Tong-Pei (fl. 1980)
Yieh	Yieh, Pei-Chong (fl. 2012)
Yild.	Yildirimli, Şinasi (1949–)
Ying, S.S.	Ying, Shao-Shun (fl. 1970)
Yu, T.T.	Yu, Tse-Tsun (1908–1986)
Yu, Z.X.	Yu, Zhi-Xiong (fl. 1988)
Yuan	Yuan, Yong-Ming (fl. 1992)
Yue	Yue, Ji-Pei (fl. 2004)
Yukawa	Yukawa, Tomohisa (fl. 1992)
Yuncker	Yuncker, Truman George (1891–1964)
Zabel	Zabel, Hermann (1832–1912)
Zacharias, E.	Zacharias, Elizabeth H. (fl. 2010)
Zahlbr., A.	Zahlbruckner, Alexander (1860–1938)
Zak.	Zakirov, K.Z. (1906–)
Zamora	Zamora, Nelso A. (fl. 1988)
Zanov.	Zanovello, Carlo (1947–)
Zardini	Zardini, Else Matilde (1949–)
Zareh	Zareh, M. (fl. 1989)
Zarre	Zarre, Shahin (1966–)
Zauschner	Zauschner, Johann Baptista Josef (1737–1799)
Záveská Drábková	Záveská Drábková, Lenka (fl. 2013)
Zepernick	Zepernick, Bernhard (1926–)
Zeyher	Zeyher, Carl Ludwig Philip(p) (1799–1858)
Zeyher, J.	Zeyher, K. Johann Michael (1770–1843)
Zhai	Zhai, Jian-Wen (1985–)
Zhang	Zhang, Yuffna (fl. 1986)
Zhang, D.X.	Zhang, Dian-Xiang (1963–)
Zhang, J.X.	Zhang, Jia-Xun (fl. 1992)
Zhang, L.	Zhang, Liang (fl. 2013)
Zhang, L.B.	Zhang, Li-Bin (fl. 1989)
Zhang, X.C.	Zhang, Xian-Chun (1964–)
Zhang, X.P.	Zhang, Xiao-Ping (fl. 1987)
Zhang, Z.Y.	Zhang, Zhi-Yun (1950–)
Zhao	Zhao, Zheng-Yu (1928–)
Zhou, L.H.	Zhou, Li-Hua (1934–)
Zhu, G.	Zhu, Guanghua (1964–2005)
Zhu, Z.Y.	Zhu, Zheng-Yin (fl. 1982)
Zimm., A.	Zimmeter, Albert (1848–1897)
Zinn	Zinn, Johann Gottfried (1727–1759)
Zipp.	Zippelius, Alexander (1797–1828)
Zizka	Zizka, Georg (1955–)
Zoellner	Zoellner, Otto (1909–2007)
Zoll.	Zollinger, Heinrich (1818–1859)
Zona	Zona, Scott (1959–)
Zotov	Zotov, Victor Dmitrievich (1908–1977)

Zou, S.Q.	Zou, Shou-Qing (fl. 1984)
Zoz	Zoz, I.G. (1903–)
Zucc.	Zuccarini, Joseph Gerhard (1797–1848)
Zuccagni	Zuccagni, Attilio (1754–1807)
Zuloaga	Zuloaga, Fernando Omar (1951–)

New Names Used in This Book

The authorities for these new names are those as set out below. I am very grateful to Frank Almeda, Christine Bartram, David Harris, Walter Judd, Gudrun Kadereit, Elizabeth (Toby) Kellogg, Gwilym Lewis, Eve Lucas, Serena Marner, Fabián Michelangeli, Gonzalo Nieto Feliner, Henry Noltie, Thomas Pink, Julian Shaw, Jun Wen and Fernando Zuloaga for their help with several of these names or agreeing to publish them here.

Causonis trifolia (L.) Mabb. & J. Wen, **comb. nova**
Vitis trifolia L. Sp. Pl. 1(1753)203. Type: India, Chennai, 'Pearmedoor, about 16 or 17 miles from Fort St. George', 27/28 Apr. 1696, *Browne* '2.67' (BM-SL [165 f. 84], neo, cf. FNC 25(2004)14)

Cleome refracta (Engelm.) Mabb., **comb. nova**
Wislizenia refracta Engelm. in Wislizenus, Mem. Tour N. Mex. (1848)99. Type: USA, Texas, 'On the upper crossing of the Rio Grande near el Paso', 8 Aug. 1846, *Wislizenus s.n.* (MO [MO-277880]) holo)

Deyeuxia epigejos (L.) Mabb., **comb. nova**
Arundo epigejos L., Sp. Pl. 1(1753)81. Type: Herb. Linn. 97.11 (LINN lecto)

Dissochaeta muscosa (Blume) G. Kadereit, **comb. nova**
Melastoma muscosum Blume, Bijdr. Fl. Ned. Ind. 17(1826/1827)1070. Type: Indonesia, Java, *Blume s.n.* (L [908.129–975] holo; BR, K iso)

Dracaea hyacinthoides (L.) Mabb., **comb. nova**
Aloe hyacinthoides L., Sp. Pl. 1(1753)321.Type [icon]: 'Aloe Guineensis radice geniculata fol. variegato' in Commelin, Praeludio Bot.(1703)84, t. 33(lecto)

Dracaena trifasciata (Prain) Mabb., **comb. nova**
Sansevieria trifasciata hort. ex Prain, Bengal Pl. 2(1903)1054. Type: not seen

Dracaena zeylanica (L.) Mabb., **comb. nova**
Aloe hyacinthoides L. var. *zeylanica* L., Sp. Pl. 1(1753)321.Type [icon]: 'Aloe Zeylanica pumila foliis variegatis' in Commelin, Hort. Med. Amstelod. Pl. Rar. 2(1701)41, t. 21 (lecto)

Eleutherococcus septemlobus (Thunb.) Mabb., **comb. nova**
Acer septemlobum Thunb. in Murray, Syst. Nat. (1794)912. Type: Japan, *Thunberg s.n.* in Herb. Thunberg 24095 (UPS lecto)

Erythranthe × maculosa (T. Moore) Mabb., **comb. nova**
Mimulus × maculosus T. Moore in Florist & Pomologist 1863: 73 + t. Type: not selected but one of the images on the plate a good candidate for lectotype

× **Gastonialoe** J.M.H. Shaw, **nothogen. novum**
= *Gasteria* Duval × *Gonialoe* (Bak.) Boatwr. & Manning

Heliotropium arboreum (Blanco) Mabb., **comb. nova**
Tournefortia arborea Blanco, Fl. Filip. (1837)129. Type: Philippines, Luzon, Aurora, Dingalan Bay, 24 Aug. 1916 (US00623760), *Species Blancoanae* 1007 (US neo designated here; BM [BM001209145], K isoneo)

Hibiscus americanus (L.f.) Mabb., **comb. nova**
Urena americana L.f., Suppl. Pl. (1782)308. Type: [Surinam,] *Dahlberg* 113 (LINN [LINN-HL873–7] holo; S [S09–24465] iso?)

Hibiscus capitatus (L.) Mabb., **comb. nova**
Sida capitata L., Sp. Pl. 2(1753)685. Type: *Anon. s.n.* in Herb. Linn. 867.1 (LINN [LINN-HL867–1] lecto)

Hibiscus roxburghianus (Wight) Mabb., **comb. nova**
Kydia roxburghiana Wight, Icon. Pl. In. Orient. 3,2(1844–45)7 + t. 881. Type: India, Tamil Nadu, 'Neilgherries', *Wight s.n.* (K [K000659557] lecto selected here)

Hibiscus rudis (Benth.) Mabb., **comb. nova**
Malachra rudis Benth., Pl. Hartweg. (1845)164. Type: Colombia, Bogota, 'Inter Villeta ad Guaduas', *Hartweg* 915 (K [K000328842] holo; BM [BM000645339], LD [LD1221075] iso)

Justicia klossii (S. Moore) Mabb., **comb. nova**
Rungia klossii S. Moore in Trans. Linn. Soc. Lond. Bot. 9(1916)135. Type: Indonesia, Papua, Mt Carstenz [= Jayawijaya], 'Camps VIa, [3050', W bank of R. Tsingarong [= Utakwa]], VII [3585'] and VIII [4970' above

gorge of the Tsingarong]', *Boden Kloss* 5113, s.n. (BM [BM00095009, BM000950080] syn; K [K000884273] isosyn)

Miconia crenata (Vahl) Michelang. **comb. nova**
Melastoma crenatum Vahl, Eclog. Amer. 1(1797)41. Type: Brazil, *Ryan s.n.* (C [C10014542], holo)

Miconia hammelii (Almeda) Almeda, **comb. nova**
Clidemia hammelii Almeda in Proc. Calif. Acad. Sci. IV,46(1989)140. Type: Costa Rica, Heredia: Finca La Selva, OTS Field Station, *Hammel* 8692 (CAS holo; DUKE iso)

Miconia mayeta (D. Don) Michelang., **comb. nova**
Tococa mayeta D. Don in Mem. Wern. Nat. Hist. Soc. 4(1823)305. Type: French Guiana, *Aublet s.n.* (BM [BM001008829] holo)

Miconia rubra (Aubl.) Mabb., **comb. nova**
Melastoma rubra Aubl., Hist. Pl. Guiane 1(1775)416 [& 4(1775)t. 161]. Type: French Guiana, 'prope praedium Du Chassis', *Aublet s.n.* (BM [BM001008275] holo)

Miconia salicina (DC.) Mabb., **comb. nova**
Melastoma salicinum Ser. ex DC., Prodr. 3(1828)199. Type: Brazil, *Anon. s.n.* (G-DC [G00311242] holo)

Murraya lucida (G. Forst.) Mabb., **comb. nova**
Limonia lucida G. Forst., Fl. Ins. Austr. (1786)33. Type: Vanuatu, Malakula, 22–24 July 1774, *J.R. & G. Forster s.n.* (BM [BM000798433] lecto selected here; BM [BM00079834, BM00101580] syn)

Myrcia luquillensis (Alain) E. Lucas & A.R. Lourenço, **comb. nova**
Calyptranthes luquillensis Alain in Bull. Torrey Bot. Club 90(1963)189. Type: USA, Puerto Rico, Pizá, Luquillo Mts, *Holridge* 61 (NY [NY00084418] holo)

Persea barbujana (Cav.) Mabb. & Nieto Fel., **comb. nova**
Laurus barbujana Cav. in Anales Ci. Nat. 3(1801)52. Type: Spain, Canary Is., Tenerife, *Broussonet s.n* (MA [MA 230778, 230779] syn; SEV [SEV-H13] syn)

Polytoca gigantea (J. Koenig) Mabb., **comb. nova**
Coix gigantea J. Koenig in Naturforscher (Halle) 23(1788)211. Type: India, Andhra Pradesh, Circars, Sama[r]l[a]kot[a], *Koenig s.n.* (BM [BM000047232] holo)
N.B. The type species of *Chionachne* R. Br. (*C. barbata* (Roxb.) Aitch. = *C. gigantea* (J. Koenig) Veldk.) was transferred to the contemporaneous *Polytoca* R. Br. by Stapf in Hook.f., Fl. Brit. India 7(1896)102; the type sp. of *Polytoca* (*P. bracteata* R. Br.) has never been combined in *Chionachne*. The only other congeneric spp., yet so far without names in *Polytoca*, are merely: **Polytoca biaurita** (Hack.) Mabb., **comb. nova** (*Chionachne biaurita* Hack. in Philipp. J. Sci. 1, suppl. 4(1906)263); **Polytoca hubbardiana** (Henrard) Mabb., **comb. nova** (*Chionachne hubbardiana* Henrard in Blumea 3(1938)162)

Serjania ferruginea (Lindl.) Mabb., **comb. nova**
Urvillea ferruginea Lindl. in Edwards, Bot. Reg. 13(1827)t.1077. Type [icon]: Bot. Reg. 13(1827) t. 1077 [i.e. image of cult. material from Braz.] UK, London, Chiswick, Horticultural Society Garden, May 1826 (lecto designated here, as no specimen extant at CGE [fide Christine Bartram] or been found elsewhere cf. ICN Art. 9.12)

Staphylea japonica (Thunb.) Mabb., **comb. nova**
Sambucus japonica Thunb. in Murray, Syst. Veg. (1784)295. Type: Japan, *Thunberg s.n.* (UPS-THUNB [7444] holo)

Urochloa polystachya (Kunth) Mabb., **comb. nova**
Eriochloa polystachya Kunth, Nov. Gen. Sp. 1(1816)95 + t. 31. Type: Ecuador, nr Guayaquil, *Bonpland s.n.* (P [P00128931] holo)

Vasconcellea × pentagona (Heilborn) Mabb., **comb. nova**
Carica × pentagona Heilborn in Ark. Bot. 17,12(1922)2, pro sp. Type: Ecuador, Pichincha, Quito, in horto in "La Merced", *Heilborn s.n.* (S [S04-696] holo)

Vincetoxicum indicum (Burm.f.) Mabb., **comb. nova**
Cynanchum indicum Burm.f., Fl. Ind. (1768)70. Type: not found

Wyethia sagittata (Pursh) Mabb., **comb. nova**
Buphthalmum sagittatum Pursh, Fl. Amer. Sept. 2(1813)564. Type: USA, Montana, 7 July 1806, *Lewis s.n.* (PH [PH00043200] lecto)

Printed in the USA
CPSIA information can be obtained
at www.ICGtesting.com
CBHW070914100424
6620CB00025B/65

9 781107 115026